ADVANCED
ENGINEERING
MATHEMATICS

5th Edition

ADVANCED ENGINEERING MATHEMATICS

5th Edition

PETER V. O'NEIL

*University of Alabama
at Birmingham*

THOMSON
™
BROOKS/COLE

Australia • Canada • Mexico • Singapore • Spain • United Kingdom • United States

THOMSON

BROOKS/COLE

Publisher: *Bill Stenquist*
Editorial Coordinator: *Valerie Boyajian*
Marketing Team: *Darcie Pool & Tom Ziolkowski*
Advertising Project Manager: *Laura Hubrich*
Production Coordinator: *Mary Vezilich*
Production Service: *Martha Emry Production Services*
Permissions Editor: *Stephanie Keough-Hedges*
Copy Editor: *Pamela Rockwell*
Photo Researcher: *Quest Creative*

Text Designer: *Terri Wright*
Illustrator: *Techsetters, Inc.*
Cover Designer: *Cloyce Wall*
Cover Image: *Michael Pasdzior/Getty Images*
Compositor: *Techsetters, Inc.*
Print Buyer: *Vena Dyer*
Cover Printer: *Lehigh Press, Inc.*
Printing & Binding: *Quebecor World–Versailles*

Printed in the United States of America

1 2 3 4 5 6 7 06 05 04 03 02

For more information about our products, contact us at:
Thomson Learning Academic Resource Center
1-800-423-0563

For permission to use material from this text, contact us by:
Phone: 1-800-730-2214   **Fax:** 1-800-730-2215
Web: http://www.thomsonrights.com

Brooks/Cole–Thomson Learning
511 Forest Lodge Rd
Pacific Grove, CA 93950–5098
USA

Asia
Thomson Learning
5 Shenton Way #01-01
UIC Building
Singapore 068808

Australia
Nelson Thomson Learning
102 Dodds Street
South Melbourne, Victoria 3205
Australia

Canada
Nelson Thomson Learning
1120 Birchmount Road
Toronto, Ontario M1K 5G4
Canada

Europe/Middle East/Africa
Thomson Learning
High Holborn House
50/51 Bedford Row
London WC1R 4LR
United Kingdom

Latin America
Thomson Learning
Seneca, 53
Colonia Polanco
11560 Mexico D.F.
Mexico

Spain
Paraninfo Thomson Learning
Calle/Magallanes, 25
28015 Madrid, Spain

Library of Congress Cataloging-in-Publication Data

O'Neil, Peter V.
   Advanced engineering mathematics / Peter V. O'Neil.—5th ed.
    p. cm.
   ISBN 0-534-40077-9 (case)
   1. Engineering mathematics. I. Title

TA330 .O53 2003
515'.1--dc21

2002019429

Contents

PART 1   Ordinary Differential Equations   1

Chapter 1   First-Order Differential Equations   3

1.1   Preliminary Concepts   3
1.2   Separable Equations   11
1.3   Linear Differential Equations   23
1.4   Exact Differential Equations   28
1.5   Integrating Factors   34
1.6   Homogeneous, Bernoulli, and Riccati Equations   40
1.7   Applications to Mechanics, Electrical Circuits, and Orthogonal Trajectories   48
1.8   Existence and Uniqueness for Solutions of Initial Value Problems   61

Chapter 2   Second-Order Differential Equations   65

2.1   Preliminary Concepts   65
2.2   Theory of Solutions $y'' + p(x)y' + q(x)y = f(x)$ 66
2.3   Reduction of Order   74
2.4   The Constant Coefficient Homogeneous Linear Equation   77
2.5   Euler's Equation   82
2.6   The Nonhomogeneous Equation $y'' + p(x)y' + q(x)y = f(x)$ 86
2.7   Application of Second-Order Differential Equations to a Mechanical System   98

Chapter 3   The Laplace Transform   113

3.1   Definition and Basic Properties   113
3.2   Solution of Initial Value Problems Using the Laplace Transform   122
3.3   Shifting Theorems and the Heaviside Function   127
3.4   Convolution   142
3.5   Unit Impulses and the Dirac Delta Function   147
3.6   Laplace Transform Solution of Systems   152
3.7   Differential Equations with Polynomial Coefficients   157

Chapter 4   Series Solutions   163

4.1   Power Series Solutions of Initial Value Problems   164
4.2   Power Series Solutions Using Recurrence Relations   169

4.3 Singular Points and the Method of Frobenius 174
4.4 Second Solutions and Logarithm Factors 181
4.5 Appendix on Power Series 189

PART 2 Vectors and Linear Algebra 199

Chapter 5 Vectors and Vector Spaces 201
5.1 The Algebra and Geometry of Vectors 201
5.2 The Dot Product 209
5.3 The Cross Product 216
5.4 The Vector Space R^n 222
5.5 Linear Independence, Spanning Sets, and Dimension in R^n 228
5.6 Abstract Vector Spaces 235

Chapter 6 Matrices and Systems of Linear Equations 241
6.1 Matrices 242
6.2 Elementary Row Operations and Elementary Matrices 256
6.3 The Row Echelon Form of a Matrix 263
6.4 The Row and Column Spaces of a Matrix and Rank of a Matrix 271
6.5 Solution of Homogeneous Systems of Linear Equations 278
6.6 The Solution Space of $AX = O$ 287
6.7 Nonhomogeneous Systems of Linear Equations 290
6.8 Summary for Linear Systems 301
6.9 Matrix Inverses 304

Chapter 7 Determinants 311
7.1 Permutations 311
7.2 Definition of the Determinant 313
7.3 Properties of Determinants 315
7.4 Evaluation of Determinants by Elementary Row and Column Operations 319
7.5 Cofactor Expansions 324
7.6 Determinants of Triangular Matrices 328
7.7 A Determinant Formula for a Matrix Inverse 329
7.8 Cramer's Rule 332
7.9 The Matrix Tree Theorem 334

Chapter 8 Eigenvalues, Diagonalization, and Special Matrices 337
8.1 Eigenvalues and Eigenvectors 337
8.2 Diagonalization of Matrices 345
8.3 Orthogonal and Symmetric Matrices 354
8.4 Quadratic Forms 363
8.5 Unitary, Hermitian, and Skew-Hermitian Matrices 368

PART 3   **Systems of Differential Equations and Qualitative Methods**   **375**

Chapter 9   Systems of Linear Differential Equations   377
9.1   Theory of Systems of Linear First-Order Differential Equations   377
9.2   Solution of $\mathbf{X}' = \mathbf{AX}$ When $\mathbf{A}$ Is Constant   389
9.3   Solution of $\mathbf{X}' = \mathbf{AX} + \mathbf{G}$ 410

Chapter 10   Qualitative Methods and Systems of Nonlinear Differential Equations   425
10.1   Nonlinear Systems and Existence of Solutions   425
10.2   The Phase Plane, Phase Portraits, and Direction Fields   428
10.3   Phase Portraits of Linear Systems   435
10.4   Critical Points and Stability   446
10.5   Almost Linear Systems   453
10.6   Predator/Prey Population Models   474
10.7   Competing Species Models   480
10.8   Lyapunov's Stability Criteria   489
10.9   Limit Cycles and Periodic Solutions   498

PART 4   **Vector Analysis**   **509**

Chapter 11   Vector Differential Calculus   511
11.1   Vector Functions of One Variable   511
11.2   Velocity, Acceleration, Curvature, and Torsion   517
11.3   Vector Fields and Streamlines   528
11.4   The Gradient Field and Directional Derivatives   535
11.5   Divergence and Curl   547

Chapter 12   Vector Integral Calculus   553
12.1   Line Integrals   553
12.2   Green's Theorem   565
12.3   Independence of Path and Potential Theory in the Plane   572
12.4   Surfaces in 3-Space and Surface Integrals   583
12.5   Applications of Surface Integrals   596
12.6   Preparation for the Integral Theorems of Gauss and Stokes   602
12.7   The Divergence Theorem of Gauss   604
12.8   The Integral Theorem of Stokes   613

PART 5   Fourier Analysis, Orthogonal Expansions, and Wavelets   623

Chapter 13   Fourier Series   625
13.1   Why Fourier Series?   625
13.2   The Fourier Series of a Function   628
13.3   Convergence of Fourier Series   635
13.4   Fourier Cosine and Sine Series   651
13.5   Integration and Differentiation of Fourier Series   657
13.6   The Phase Angle Form of a Fourier Series   667
13.7   Complex Fourier Series and the Frequency Spectrum   673

Chapter 14   The Fourier Integral and Fourier Transforms   681
14.1   The Fourier Integral   681
14.2   Fourier Cosine and Sine Integrals   685
14.3   The Complex Fourier Integral and the Fourier Transform   687
14.4   Additional Properties and Applications of the Fourier Transform   698
14.5   The Fourier Cosine and Sine Transforms   717
14.6   The Finite Fourier Cosine and Sine Transforms   719
14.7   The Discrete Fourier Transform   726
14.8   Sampled Fourier Series   733
14.9   The Fast Fourier Transform   745

Chapter 15   Special Functions, Orthogonal Expansions, and Wavelets   765
15.1   Legendre Polynomials   765
15.2   Bessel Functions   783
15.3   Sturm–Liouville Theory and Eigenfunction Expansions   815
15.4   Orthogonal Polynomials   836
15.5   Wavelets   841

PART 6   Partial Differential Equations   855

Chapter 16   The Wave Equation   857
16.1   The Wave Equation and Initial and Boundary Conditions   857
16.2   Fourier Series Solutions of the Wave Equation   862
16.3   Wave Motion Along Infinite and Semi-infinite Strings   881
16.4   Characteristics and d'Alembert's Solution   895
16.5   Normal Modes of Vibration of a Circular Elastic Membrane   904
16.6   Vibrations of a Circular Elastic Membrane, Revisited   907
16.7   Vibrations of a Rectangular Membrane   910

Chapter 17   The Heat Equation   915
    17.1   The Heat Equation and Initial and Boundary Conditions   915
    17.2   Fourier Series Solutions of the Heat Equation   918
    17.3   Heat Conduction in Infinite Media   940
    17.4   Heat Conduction in an Infinite Cylinder   949
    17.5   Heat Conduction in a Rectangular Plate   953

Chapter 18   The Potential Equation   955
    18.1   Harmonic Functions and the Dirichlet Problem   955
    18.2   Dirichlet Problem for a Rectangle   957
    18.3   Dirichlet Problem for a Disk   959
    18.4   Poisson's Integral Formula for the Disk   962
    18.5   Dirichlet Problems in Unbounded Regions   964
    18.6   A Dirichlet Problem for a Cube   972
    18.7   The Steady-State Heat Equation for a Solid Sphere   974
    18.8   The Neumann Problem   978

Chapter 19   Canonical Forms, Existence and Uniqueness of Solutions, and Well-Posed Problems   987
    19.1   Canonical Forms   987
    19.2   Existence and Uniqueness of Solutions   996
    19.3   Well-Posed Problems   998

PART 7   Complex Analysis   1001

Chapter 20   Geometry and Arithmetic of Complex Numbers   1003
    20.1   Complex Numbers   1003
    20.2   Loci and Sets of Points in the Complex Plane   1012

Chapter 21   Complex Functions   1027
    21.1   Limits, Continuity, and Derivatives   1027
    21.2   Power Series   1040
    21.3   The Exponential and Trigonometric Functions   1047
    21.4   The Complex Logarithm   1056
    21.5   Powers   1059

Chapter 22   Complex Integration   1065
    22.1   Curves in the Plane   1065
    22.2   The Integral of a Complex Function   1070
    22.3   Cauchy's Theorem   1081
    22.4   Consequences of Cauchy's Theorem   1088

Chapter 23   Series Representations of Functions   1101
    23.1   Power Series Representations   1101
    23.2   The Laurent Expansion   1113

Chapter 24   Singularities and the Residue Theorem   1121
    24.1   Singularities   1121
    24.2   The Residue Theorem   1128
    24.3   Some Applications of the Residue Theorem   1136

Chapter 25   Conformal Mappings   1163
    25.1   Functions as Mappings   1163
    25.2   Conformal Mappings   1171
    25.3   Construction of Conformal Mappings Between Domains   1182
    25.4   Harmonic Functions and the Dirichlet Problem   1193
    25.5   Complex Function Models of Plane Fluid Flow   1200

PART 8   Historical Notes   1211

Chapter 26   Development of Areas of Mathematics   1213
    26.1   Ordinary Differential Equations   1213
    26.2   Matrices and Determinants   1217
    26.3   Vector Analysis   1218
    26.4   Fourier Analysis   1220
    26.5   Partial Differential Equations   1223
    26.6   Complex Function Theory   1223

Chapter 27   Biographical Sketches   1225
    27.1   Galileo Galilei (1564–1642)   1225
    27.2   Isaac Newton (1642–1727)   1227
    27.3   Gottfried Wilhelm Leibniz (1646–1716)   1228
    27.4   The Bernoulli Family   1229
    27.5   Leonhard Euler (1707–1783)   1230
    27.6   Carl Friedrich Gauss (1777–1855)   1230
    27.7   Joseph-Louis Lagrange (1736–1813)   1231
    27.8   Pierre-Simon de Laplace (1749–1827)   1232
    27.9   Augustin-Louis Cauchy (1789–1857)   1233
    27.10   Joseph Fourier (1768–1830)   1234
    27.11   Henri Poincaré (1854–1912)   1235

Answers and Solutions to Selected Odd-Numbered Problems   A1

Index   I1

Preface

This Fifth Edition of *Advanced Engineering Mathematics* has two primary objectives.

The first is to make available much of the post-calculus mathematics needed and used by today's scientists, engineers, and applied mathematicians, in a setting that is helpful to both students and faculty. Throughout, it is recognized that mathematics provides powerful ways of modeling physical processes, but that these can lead to false conclusions if misunderstood or misapplied. This warrants careful attention to details in making correct statements of theorems and methods and in analyzing results.

The second objective is to engage theory with computational facility. Scientists, engineers, mathematicians, economists, ecologists, and other professionals often need to calculate things, moving from theory to practice. This involves acquiring skills in manipulating series, integrals, transforms, conformal mappings, and other standard objects used in mathematical modeling, as well as in carrying out calculations to reach reliable conclusions. Successful applications of such skills can be seen in major projects all over the world—the space shuttle, the Golden Gate Bridge, Kuala Lumpur's Petronas Towers, the Odeillo Solar Furnace in southern France, the English Channel tunnel joining the United Kingdom and France, the Ganter Bridge in Switzerland, and many others.

To meet these objectives, the following changes have been made in this edition.

- The wide availability of powerful and convenient computer software is an invitation to probe the relationship between the mathematics and real-world conclusions, connecting theory with models and the phenomena they describe. Making use of this capability should be an important part of a student's experience. Throughout the text, the student is asked to experiment with computations. These include generating direction fields and phase portraits, seeing the convergence of Fourier series and eigenfunction expansions through graphs of partial sums, observing the effects of filters on signals, seeing how various parameters and forcing terms influence solutions of wave and heat equations, constructing waves as sums of forward and backward waves, and using the discrete Fourier transform to approximate Fourier transforms and to sample Fourier series.

- Partial differential equations are given a reorganized and more detailed treatment. Chapter 16 covers the wave equation, first developing Fourier series solutions on a bounded interval, then Fourier integral and transform techniques for problems on the line and half-line. Characteristics are used to solve the wave equations with a forcing term, and transformations are used to deal with nonhomogeneous boundary and initial conditions. Chapter 17 follows a similar program for the heat equation, considering first solutions on a bounded interval, then on the line and half-line. New material on the nonhomogeneous heat equation is included. Finally, a chapter on the Dirichlet and Neumann problems has been added.

- The importance of the Fourier transform in modern science and engineering has been acknowledged by extending its discussion to include windowing, filtering, and use of the N-point discrete Fourier transform to sample Fourier series and approximate Fourier transforms.

Organizational Overview

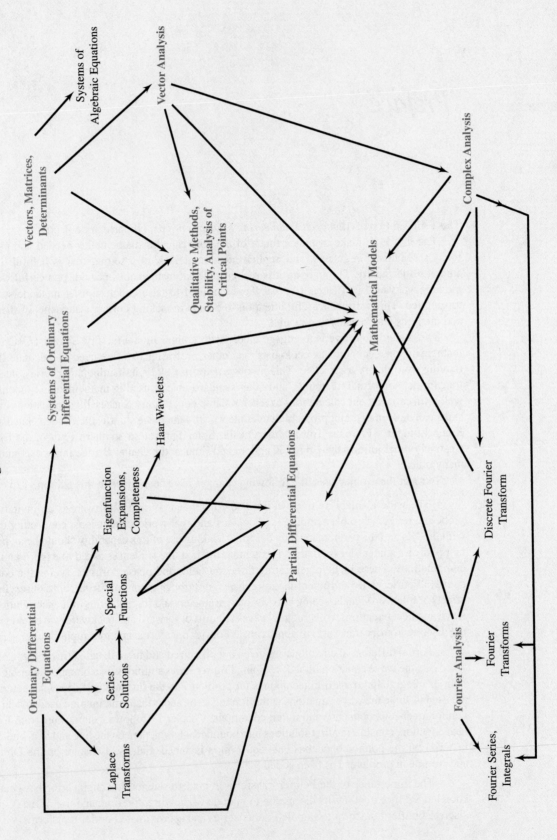

- Special functions are given a more thorough, unified treatment. Legendre polynomials, Bessel functions, and then other orthogonal polynomials are placed in the context of Sturm–Liouville theory and general eigenfunction expansions, with specific examples. These are followed by mean square convergence, the significance of completeness of the eigenfunctions, and the relationship between completeness and Parseval's theorem. Haar wavelets are discussed in the context of completeness and orthogonal expansions.

- The treatment of systems of linear, ordinary differential equations has additional material on the case of repeated eigenvalues of the coefficient matrix.

- The discussion of the qualitative behavior of nonlinear systems has more details on phase portraits and the classification of critical points and stability, including Lyapunov's criteria.

- The sections on determinants have been reorganized to provide greater clarity in understanding properties of determinants.

- The section containing answers to odd-numbered problems has been expanded, including details for some of the more difficult problems, as well as more illustrations.

- Behind the mathematics we usually find interesting people and events. Some of their stories are told in Chapters 26 and 27, but historical perspectives are also included throughout the text. For example, Section 14.9.1, on the fast Fourier transform, begins with a review of the personal interactions that led to the publication of the famous Cooley–Tukey paper, the first detailed description of the algorithm.

The chart opposite offers a complete organizational overview.

Acknowledgments

The production of a book of this size and scope requires much more than an author. Among those to whom I owe a debt of appreciation are Bill Stenquist, Pat Call, and Mary Vezilich of Brooks/Cole, Martha Emry of Martha Emry Production Services, designer Terri Wright, and the professionals at Techsetters, Inc., and Laurel Technical Services.

Dr. Thomas O'Neil of the California Polytechnic State University contributed material to previous editions, and much of this is continued in this edition. Dr. Fred Martens of the University of Alabama has helped with error checking in both text and problems and is the author of the solutions manual accompanying this edition. Finally, I want to acknowledge Rich Jones, who had the vision for the first edition of this text many years ago.

I would also like to acknowledge my debt to the reviewers, whose suggestions for improvements and clarifications are much appreciated:

Peter M. Bainum
Howard University

Robert C. Rogers
Virginia Tech

Harvey J. Charlton
North Carolina State University

Stephen Shipman
Duke University

Roland Mallier
University of Western Ohio

Peter V. O'Neil

PART 1

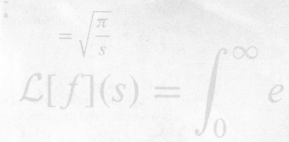

CHAPTER 1
First-Order Differential Equations

CHAPTER 2
Second-Order Differential Equations

CHAPTER 3
The Laplace Transform

CHAPTER 4
Series Solutions

Ordinary Differential Equations

A *differential equation* is an equation that contains one or more derivatives. For example,

$$y''(x) + y(x) = 4 \sin(3x)$$

and

$$\frac{d^4 w}{dt^4} - (w(t))^2 = e^{-t}$$

are differential equations. These are *ordinary* differential equations because they involve only total derivatives, rather than partial derivatives.

Differential equations are interesting and important because they express relationships involving rates of change. Such relationships form the basis for developing ideas and studying phenomena in the sciences, engineering, economics, and, increasingly, in other areas, such as the business world and the stock market. We will see examples of applications as we learn more about differential equations.

The *order* of a differential equation is the order of its highest derivative. The first example given above is of second order, while the second is of fourth order. The equation

$$xy' - y^2 = e^x$$

is of first order.

A *solution* of a differential equation is any function that satisfies it. A solution may be defined on the entire real line, or on only part of it, often an interval. For example,

$$y = \sin(2x)$$

is a solution of

$$y'' + 4y = 0,$$

because, by direct differentiation,

$$y'' + 4y = -4\sin(2x) + 4\sin(2x) = 0.$$

This solution is defined for all x (that is, on the whole real line).

By contrast,

$$y = x\ln(x) - x$$

is a solution of

$$y' = \frac{y}{x} + 1,$$

but this solution is defined only for $x > 0$. Indeed, the coefficient $1/x$ of y in this equation means that $x = 0$ is disallowed from the start.

We now begin a systematic development of ordinary differential equations, starting with the first-order case.

CHAPTER 1

First-Order Differential Equations

1.1 Preliminary Concepts

Before developing techniques for solving various kinds of differential equations, we will develop some terminology and geometric insight.

1.1.1 General and Particular Solutions

A first-order differential equation is any equation involving a first derivative, but no higher derivative. In its most general form, it has the appearance

$$F(x, y, y') = 0, \tag{1.1}$$

in which $y(x)$ is the function of interest and x is the independent variable. Examples are

$$y' - y^2 - e^y = 0,$$
$$y' - 2 = 0,$$

and

$$y' - \cos(x) = 0.$$

Note that y' must be present for an equation to qualify as a first-order differential equation, but x and/or y need not occur explicitly.

A *solution* of equation (1.1) on an interval I is a function φ that satisfies the equation for all x in I. That is,

$$F(x, \varphi(x), \varphi'(x)) = 0 \quad \text{for all } x \text{ in } I.$$

For example,

$$\varphi(x) = 2 + ke^{-x}$$

3

is a solution of

$$y' + y = 2$$

for all real x and for any number k. Here I can be chosen as the entire real line. And

$$\varphi(x) = x\ln(x) + cx$$

is a solution of

$$y' = \frac{y}{x} + 1$$

for all $x > 0$ and for any number c.

In both of these examples, the solution contained an arbitrary constant. This is a symbol independent of x and y that can be assigned any numerical value. Such a solution is called the *general solution* of the differential equation. Thus,

$$\varphi(x) = 2 + ke^{-x}$$

is the general solution of $y' + y = 2$.

Each choice of the constant in the general solution yields a *particular solution*. For example,

$$f(x) = 2 + e^{-x}, \quad g(x) = 2 - e^{-x},$$

and

$$h(x) = 2 - \sqrt{53}e^{-x}$$

are all particular solutions of $y' + y = 2$, obtained by choosing, respectively, $k = 1, -1$, and $-\sqrt{53}$ in the general solution.

1.1.2   Implicitly Defined Solutions

Sometimes we can write a solution explicitly giving y as a function of x. For example,

$$y = ke^{-x}$$

is the general solution of

$$y' = -y,$$

as can be verified by substitution. This general solution is explicit, with y isolated on one side of an equation and a function of x on the other.

By contrast, consider

$$y' = -\frac{2xy^3 + 2}{3x^2y^2 + 8e^{4y}}.$$

We claim that the general solution is the function $y(x)$ implicitly defined by the equation

$$x^2y^3 + 2x + 2e^{4y} = k, \tag{1.2}$$

in which k can be any number. To verify this, implicitly differentiate equation (1.2) with respect to x, remembering that y is a function of x. We obtain

$$2xy^3 + 3x^2y^2y' + 2 + 8e^{4y}y' = 0,$$

and solving for y' yields the differential equation.

In this example we are unable to solve equation (1.2) explicitly for y as a function of x, isolating y on one side. Equation (1.2), implicitly defining the general solution, was obtained by a technique we will develop shortly, but this technique cannot guarantee an explicit solution.

1.1.3 Integral Curves

A graph of a solution of a first-order differential equation is called an *integral curve* of the equation. If we know the general solution, we obtain an infinite family of integral curves, one for each choice of the arbitrary constant.

EXAMPLE 1.1

We have seen that the general solution of

$$y' + y = 2$$

is

$$y = 2 + ke^{-x}$$

for all x. The integral curves of $y' + y = 2$ are graphs of $y = 2 + ke^{-x}$ for different choices of k. Some of these are shown in Figure 1.1. ▧

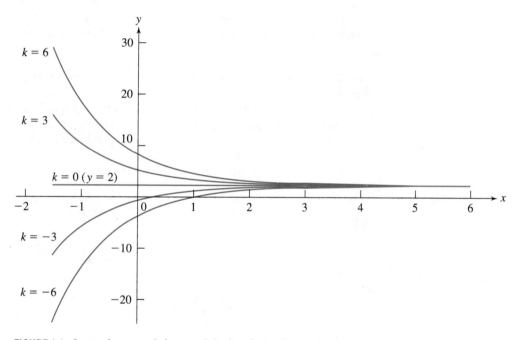

FIGURE 1.1 *Integral curves of $y' + y = 2$ for $k = 0, 3, -3, 6,$ and -6.*

EXAMPLE 1.2

It is routine to verify that the general solution of

$$y' + \frac{y}{x} = e^x$$

is

$$y = \frac{1}{x}\left(xe^x - e^x + c\right)$$

for $x \neq 0$. Graphs of some of these integral curves, obtained by making choices for c, are shown in Figure 1.2. ▪

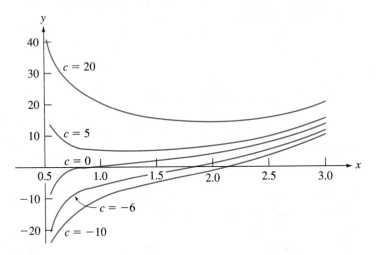

FIGURE 1.2 *Integral curves of $y' + \frac{1}{x}y = e^x$ for $c = 0, 5, 20, -6,$ and -10.*

We will see shortly how these general solutions are obtained. For the moment, we simply want to illustrate integral curves.

Although in simple cases integral curves can be sketched by hand, generally we need computer assistance. Computer packages such as MAPLE, MATHEMATICA, and MATLAB are widely available. Here is an example in which the need for computing assistance is clear.

EXAMPLE 1.3

The differential equation

$$y' + xy = 2$$

has general solution

$$y(x) = e^{-x^2/2} \int_0^x 2e^{\xi^2/2}\, d\xi + ke^{-x^2/2}.$$

Figure 1.3 shows computer-generated integral curves corresponding to $k = 0, 4, 13, -7, -15,$ and -11. ▪

1.1.4   The Initial Value Problem

The general solution of a first-order differential equation $F(x, y, y') = 0$ contains an arbitrary constant; hence there is an infinite family of integral curves, one for each choice of the constant. If we specify that a solution is to pass through a particular point (x_0, y_0), then we must find that particular integral curve (or curves) passing through this point. This is called an *initial value problem*. Thus, a first-order initial value problem has the form

$$F(x, y, y') = 0; \qquad y(x_0) = y_0,$$

in which x_0 and y_0 are given numbers. The condition $y(x_0) = y_0$ is called an *initial condition*.

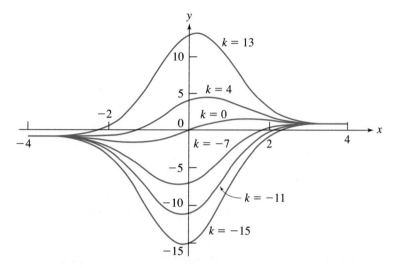

FIGURE 1.3 *Integral curves of $y' + xy = 2$ for $k = 0, 4, 13, -7, -15,$ and -11.*

EXAMPLE 1.4

Consider the initial value problem

$$y' + y = 2; \qquad y(1) = -5.$$

From Example 1.1, the general solution of $y' + y = 2$ is

$$y = 2 + ke^{-x}.$$

Graphs of this equation are the integral curves. We want the one passing through $(1, -5)$. Solve for k so that

$$y(1) = 2 + ke^{-1} = -5,$$

obtaining

$$k = -7e.$$

The solution of this initial value problem is

$$y = 2 - 7ee^{-x} = 2 - 7e^{-(x-1)}.$$

As a check, $y(1) = 2 - 7 = -5.$ ∎

The effect of the initial condition in this example was to pick out one special integral curve as the solution sought. This suggests that an initial value problem may be expected to have a unique solution. We will see later that this is the case, under mild conditions on the coefficients in the differential equation.

1.1.5 Direction Fields

Imagine a curve, as in Figure 1.4. If we choose some points on the curve and, at each point, draw a segment of the tangent to the curve there, then these segments give a rough outline of the shape of the curve. This simple observation is the key to a powerful device for envisioning integral curves of a differential equation.

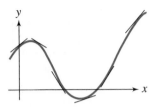

FIGURE 1.4 *Short tangent segments suggest the shape of the curve.*

The general first-order differential equation has the form

$$F(x, y, y') = 0.$$

Suppose we can solve for y' and write the differential equation as

$$y' = f(x, y).$$

Here f is a known function. Suppose $f(x, y)$ is defined for all points (x, y) in some region R of the plane. The slope of the integral curve through a given point (x_0, y_0) of R is $y'(x_0)$, which equals $f(x_0, y_0)$. If we compute $f(x, y)$ at selected points in R, and draw a small line segment having slope $f(x, y)$ at each (x, y), we obtain a collection of segments that trace out the shapes of the integral curves. This enables us to obtain important insight into the behavior of the solutions (such as where solutions are increasing or decreasing, limits they might have at various points, or behavior as x increases).

A drawing of the plane, with short line segments of slope $f(x, y)$ drawn at selected points (x, y), is called a *direction field* of the differential equation $y' = f(x, y)$. The name derives from the fact that at each point the line segment gives the direction of the integral curve through that point. The line segments are called *lineal elements*.

EXAMPLE 1.5

Consider the equation

$$y' = y^2.$$

Here $f(x, y) = y^2$, so the slope of the integral curve through (x, y) is y^2. Select some points and, through each, draw a short line segment having slope y^2. A computer-generated direction field is shown in Figure 1.5(a). The lineal elements form a profile of some integral curves and give us some insight into the behavior of solutions, at least in this part of the plane. Figure 1.5(b) reproduces this direction field, with graphs of the integral curves through $(0, 1)$, $(0, 2)$, $(0, 3)$, $(0, -1)$, $(0, -2)$, and $(0, -3)$.

By a method we will develop, the general solution of $y' = y^2$ is

$$y = -\frac{1}{x + k},$$

so the integral curves form a family of hyperbolas, as suggested by the curves sketched in Figure 1.5(b). ∎

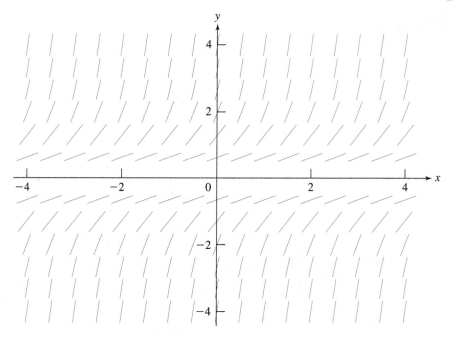

FIGURE 1.5(a) *A direction field for* $y' = y^2$.

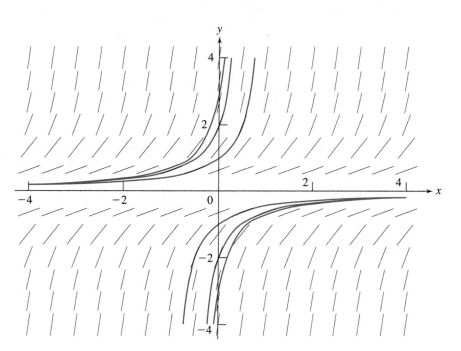

FIGURE 1.5(b) *Direction field for* $y' = y^2$ *and integral curves through* $(0, 1)$, $(0, 2)$, $(0, 3)(0, -1)$, $(0, -2)$, *and* $(0, -3)$.

EXAMPLE 1.6

Figure 1.6 shows a direction field for

$$y' = \sin(xy),$$

together with the integral curves through $(0, 1)$, $(0, 2)$, $(0, 3)$, $(0, -1)$, $(0, -2)$, and $(0, -3)$. In this case, we cannot write a simple expression for the general solution, and the direction field provides information about the behavior of solutions that is not otherwise readily apparent. ■

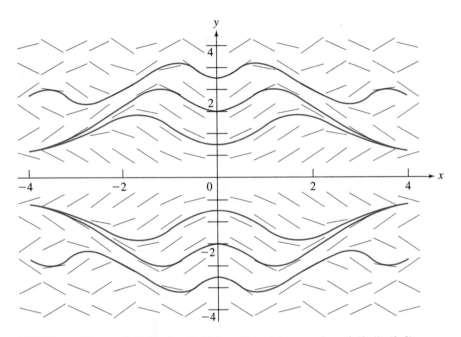

FIGURE 1.6 *Direction field for* $y' = \sin(xy)$ *and integral curves through* $(0, 1)$, $(0, 2)$, $(0, 3)$, $(0, -1)$, $(0, -2)$, *and* $(0, -3)$.

With this as background, we will begin a program of identifying special classes of first-order differential equations for which there are techniques for writing the general solution. This will occupy the next five sections.

SECTION 1.1 PROBLEMS

In each of Problems 1 through 10, determine whether the given function is a solution of the differential equation.

1. $2yy' = 1$; $\varphi(x) = \sqrt{x - 1}$ for $x > 1$

2. $y' + y = 0$; $\varphi(x) = Ce^{-x}$

3. $y' = -\dfrac{2y + e^x}{2x}$ for $x > 0$; $\varphi(x) = \dfrac{C - e^x}{2x}$

4. $y' = \dfrac{2xy}{2 - x^2}$ for $x \neq \pm\sqrt{2}$; $\varphi(x) = \dfrac{C}{x^2 - 2}$

5. $xy' = x - y$; $\varphi(x) = \dfrac{x^2 - 3}{2x}$ for $x \neq 0$

6. $y' + y = 1$; $\varphi(x) = 1 + Ce^{-x}$

7. $x^2yy' = -1 - xy^2$; $\varphi(x) = \dfrac{4 - x^2}{2x}$ for $x \neq 0$

8. $y' + 2y = 0$; $\varphi(x) = \sin(3x) - 4$

9. $\sinh(x)y' + \cosh(x)y = 0$; $\varphi(x) = -1/\sinh(x)$ for $x \neq 0$

10. $y' - 3y = x$; $\varphi(x) = -\frac{1}{3}x + \frac{1}{3} + Ce^{3x}$

In each of Problems 11 through 15, verify by implicit differentiation that the given equation implicitly defines a solution of the differential equation.

11. $y^2 + xy - 2x^2 - 3x - 2y = C$; $y - 4x - 2 + (x + 2y - 2)y' = 0$

12. $xy^3 - y = C$; $y^3 + (3xy^2 - 1)y' = 0$

13. $y^2 - 4x^2 + e^{xy} = C$; $8x - ye^{xy} - (2y + xe^{xy})y' = 0$

14. $8\ln|x - 2y + 4| - 2x + 6y = C$;

$$y' = \frac{x - 2y}{3x - 6y + 4}$$

15. $\tan^{-1}(y/x) + x^2 = C$;

$$\frac{2x^3 + 2xy^2 - y}{x^2 + y^2} + \frac{x}{x^2 + y^2}y' = 0$$

In each of Problems 16 through 20, solve the initial value problem and graph the solution. *Hint:* Each of these differential equations can be solved by direct integration. Use the initial condition to solve for the constant of integration.

16. $y' = 2x$; $y(2) = 1$

17. $y' = e^{-x}$; $y(0) = 2$

18. $y' = 2x + 2$; $y(-1) = 1$

19. $y' = 4\cos(x)\sin(x)$; $y(\pi/2) = 0$

20. $y' = 8x + \cos(2x)$; $y(0) = -3$

In each of Problems 21 through 26 draw some lineal elements of the differential equation for $-4 \leq x \leq 4$, $-4 \leq y \leq 4$. Use the resulting direction field to sketch a graph of the solution of the initial value problem. (These problems can be done by hand.)

21. $y' = x + y$; $y(2) = 2$

22. $y' = x - xy$; $y(0) = -1$

23. $y' = xy$; $y(0) = 2$

24. $y' = x - y + 1$; $y(0) = 1$

25. $y' = \sin(y)$; $y(1) = \pi/2$

26. $y' = y^2 - 2x^2$; $y(1) = 1$

In each of Problems 27 through 32, generate a direction field and some integral curves for the differential equation. Also draw the integral curve representing the solution of the initial value problem. These problems should be done by a software package.

27. $y' = e^{-x} + 2xy$; $y(0) = 1$

28. $y' = x\cos(2x) - y$; $y(1) = 0$

29. $y' = y\sin(x) - 3x^2$; $y(0) = 1$

30. $y' = e^x - y$; $y(-2) = 1$

31. $y' - y\cos(x) = 1 - x^2$; $y(2) = 2$

32. $y' = 2y + 3$; $y(0) = 1$

33. Show that for the differential equation $y' + p(x)y = q(x)$, the lineal elements on any vertical line $x = x_0$, with $p(x_0) \neq 0$, all pass through the single point (ξ, η), where

$$\xi = x_0 + \frac{1}{p(x_0)} \quad \text{and} \quad \eta = \frac{q(x_0)}{p(x_0)}.$$

1.2   Separable Equations

DEFINITION 1.1   *Separable Differential Equation*

A differential equation is called separable if it can be written

$$y' = A(x)B(y).$$

In this event, we can separate the variables and write, in differential form,

$$\frac{1}{B(y)}\,dy = A(x)\,dx$$

wherever $B(y) \neq 0$. We attempt to integrate this equation, writing

$$\int \frac{1}{B(y)}\, dy = \int A(x)\, dx.$$

This yields an equation in x, y, and a constant of integration. This equation implicitly defines the general solution $y(x)$. It may or may not be possible to solve explicitly for $y(x)$.

EXAMPLE 1.7

$y' = y^2 e^{-x}$ is separable. Write

$$\frac{dy}{dx} = y^2 e^{-x}$$

as

$$\frac{1}{y^2}\, dy = e^{-x}\, dx$$

for $y \neq 0$. Integrate this equation to obtain

$$-\frac{1}{y} = -e^{-x} + k,$$

an equation that implicitly defines the general solution. In this example, we can explicitly solve for y, obtaining the general solution

$$y = \frac{1}{e^{-x} - k}.$$

Now recall that we required that $y \neq 0$ in order to separate the variables by dividing by y^2. In fact, the zero function $y(x) = 0$ is a solution of $y' = y^2 e^x$, although it cannot be obtained from the general solution by any choice of k. For this reason, $y(x) = 0$ is called a singular solution of this equation.

Figure 1.7 shows graphs of particular solutions obtained by choosing k as 0, 3, −3, 6, and −6. ■

Whenever we use separation of variables, we must be alert to solutions potentially lost through conditions imposed by the algebra used to make the separation.

EXAMPLE 1.8

$x^2 y' = 1 + y$ is separable, and we can write

$$\frac{1}{1 + y}\, dy = \frac{1}{x^2}\, dx.$$

The algebra of separation has required that $x \neq 0$ and $y \neq -1$, even though we can put $x = 0$ and $y = -1$ into the differential equation to obtain the correct equation, $0 = 0$.

Now integrate the separated equation to obtain

$$\ln|1 + y| = -\frac{1}{x} + k.$$

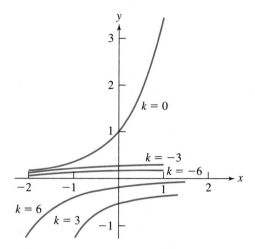

FIGURE 1.7 *Integral curves of $y' = y^2 e^{-x}$ for $k = 0, 3, -3, 6, and -6$.*

This implicitly defines the general solution. In this case, we can solve for $y(x)$ explicitly. Begin by taking the exponential of both sides to obtain

$$|1 + y| = e^k e^{-1/x} = Ae^{-1/x},$$

in which we have written $A = e^k$. Since k could be any number, A can be any positive number. Then

$$1 + y = \pm Ae^{-1/x} = Be^{-1/x},$$

in which $B = \pm A$ can be any nonzero number. The general solution is

$$y = -1 + Be^{-1/x},$$

in which B is any nonzero number.

Now revisit the assumption that $x \neq 0$ and $y \neq -1$. In the general solution, we actually obtain $y = -1$ if we allow $B = 0$. Further, the constant function $y(x) = -1$ does satisfy $x^2 y' = 1 + y$. Thus, by allowing B to be any number, including 0, the general solution

$$y(x) = -1 + Be^{-1/x}$$

contains all the solutions we have found. In this example, $y = -1$ is a solution, but not a singular solution, since it occurs as a special case of the general solution.

Figure 1.8 shows graphs of solutions corresponding to $B = -8, -5, 0, 4,$ and 7. ■

We often solve an initial value problem by finding the general solution of the differential equation, then solving for the appropriate choice of the constant.

EXAMPLE 1.9

Solve the initial value problem

$$y' = y^2 e^{-x}; \qquad y(1) = 4.$$

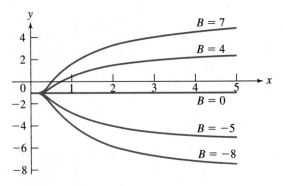

FIGURE 1.8 *Integral curves of $x^2 y' = 1 + y$ for $B = 0, 4, 7, -5,$ and -8.*

We know from Example 1.7 that the general solution of $y' = y^2 e^{-x}$ is

$$y(x) = \frac{1}{e^{-x} - k}.$$

Now we need to choose k so that

$$y(1) = \frac{1}{e^{-1} - k} = 4,$$

from which we get

$$k = e^{-1} - \frac{1}{4}.$$

The solution of the initial value problem is

$$y(x) = \frac{1}{e^{-x} + \frac{1}{4} - e^{-1}}. \ \blacksquare$$

EXAMPLE 1.10

The general solution of

$$y' = y \frac{(x-1)^2}{y+3}$$

is implicitly defined by

$$y + 3\ln|y| = \tfrac{1}{3}(x-1)^3 + k. \tag{1.3}$$

To obtain the solution satisfying $y(3) = -1$, put $x = 3$ and $y = -1$ into equation (1.3) to obtain

$$-1 = \tfrac{1}{3}(2)^3 + k,$$

hence

$$k = -\tfrac{11}{3}.$$

The solution of this initial value problem is implicitly defined by

$$y + 3\ln|y| = \tfrac{1}{3}(x-1)^3 - \tfrac{11}{3}. \ \blacksquare$$

1.2.1   Some Applications of Separable Differential Equations

Separable equations arise in many contexts, of which we will discuss three.

EXAMPLE 1.11   The Mathematical Policewoman

A murder victim is discovered, and a lieutenant from the forensic science laboratory is summoned to estimate the time of death.

The body is located in a room that is kept at a constant 68 degrees Fahrenheit. For some time after the death, the body will radiate heat into the cooler room, causing the body's temperature to decrease. Assuming (for want of better information) that the victim's temperature was a "normal" 98.6 at the time of death, the lieutenant will try to estimate this time by observing the body's current temperature and calculating how long it would have had to lose heat to reach this point.

According to Newton's law of cooling, the body will radiate heat energy into the room at a rate proportional to the difference in temperature between the body and the room. If $T(t)$ is the body temperature at time t, then for some constant of proportionality k,

$$T'(t) = k\,[T(t) - 68].$$

The lieutenant recognizes this as a separable differential equation and writes

$$\frac{1}{T - 68}\,dT = k\,dt.$$

Upon integrating, she gets

$$\ln|T - 68| = kt + C.$$

Taking exponentials, she gets

$$|T - 68| = e^{kt+C} = Ae^{kt},$$

in which $A = e^C$. Then

$$T - 68 = \pm Ae^{kt} = Be^{kt}.$$

Then

$$T(t) = 68 + Be^{kt}.$$

Now the constants k and B must be determined, and this requires information. The lieutenant arrived at 9:40 P.M. and immediately measured the body temperature, obtaining 94.4 degrees. Letting 9:40 be time zero for convenience, this means that

$$T(0) = 94.4 = 68 + B,$$

and so $B = 26.4$. Thus far,

$$T(t) = 68 + 26.4e^{kt}.$$

To determine k, the lieutenant makes another measurement. At 11:00 she finds that the body temperature is 89.2 degrees. Since 11:00 is 80 minutes past 9:40, this means that

$$T(80) = 89.2 = 68 + 26.4e^{80k}.$$

Then

$$e^{80k} = \frac{21.2}{26.4},$$

so

$$80k = \ln\left(\frac{21.2}{26.4}\right)$$

and

$$k = \frac{1}{80}\ln\left(\frac{21.2}{26.4}\right).$$

The lieutenant now has the temperature function:

$$T(t) = 68 + 26.4e^{\ln(21.2/26.4)t/80}.$$

In order to find when the last time the body was 98.6 (presumably the time of death), solve for the time in

$$T(t) = 98.6 = 68 + 26.4e^{\ln(21.2/26.4)t/80}.$$

To do this, the lieutenant writes

$$\frac{30.6}{26.4} = e^{\ln(21.2/26.4)t/80}$$

and takes the logarithm of both sides to obtain

$$\ln\left(\frac{30.6}{26.4}\right) = \frac{t}{80}\ln\left(\frac{21.2}{26.4}\right).$$

Therefore, the time of death, according to this mathematical model, was

$$t = \frac{80\ln(30.6/26.4)}{\ln(21.2/26.4)},$$

which is approximately -53.8 minutes. Death occurred approximately 53.8 minutes before (because of the negative sign) the first measurement at 9:40, which was chosen as time zero. This puts the murder at about 8:46 P.M. ■

EXAMPLE 1.12   Radioactive Decay and Carbon Dating

In radioactive decay, mass is converted to energy by radiation. It has been observed that the rate of change of the mass of a radioactive substance is proportional to the mass itself. This means that if $m(t)$ is the mass at time t, then for some constant of proportionality k that depends on the substance,

$$\frac{dm}{dt} = km.$$

This is a separable differential equation. Write it as

$$\frac{1}{m}dm = k\,dt$$

and integrate to obtain

$$\ln|m| = kt + c.$$

Since mass is positive, $|m| = m$ and

$$\ln(m) = kt + c.$$

Then

$$m(t) = e^{kt+c} = Ae^{kt},$$

in which A can be any positive number.

Determination of A and k for a given element requires two measurements. Suppose at some time, designated as time zero, there are M grams present. This is called the initial mass. Then

$$m(0) = A = M,$$

so

$$m(t) = Me^{kt}.$$

If at some later time T we find that there are M_T grams, then

$$m(T) = M_T = Me^{kT}.$$

Then

$$\ln\left(\frac{M_T}{M}\right) = kT,$$

hence

$$k = \frac{1}{T}\ln\left(\frac{M_T}{M}\right).$$

This gives us k and determines the mass at any time:

$$m(t) = Me^{\ln(M_T/M)t/T}.$$

We obtain a more convenient formula for the mass if we choose the time of the second measurement more carefully. Suppose we make the second measurement at that time $T = H$ at which exactly half of the mass has radiated away. At this time, half of the mass remains, so $M_T = M/2$ and $M_T/M = 1/2$. Now the expression for the mass becomes

$$m(t) = Me^{\ln(1/2)t/H},$$

or

$$m(t) = Me^{-\ln(2)t/H}. \tag{1.4}$$

This number H is called the half-life of the element. Although we took it to be the time needed for half of the original amount M to decay, in fact between any times t_1 and $t_1 + H$, exactly half of the mass of the element present at t_1 will radiate away. To see this, write

$$m(t_1 + H) = Me^{-\ln(2)(t_1+H)/H}$$

$$= Me^{-\ln(2)t_1/H}e^{-\ln(2)H/H} = e^{-\ln(2)}m(t_1)$$

$$= \tfrac{1}{2}m(t_1).$$

Equation (1.4) is the basis for an important technique used to estimate the ages of certain ancient artifacts. The earth's upper atmosphere is constantly bombarded by high-energy cosmic rays, producing large numbers of neutrons, which collide with nitrogen in the air, changing some of it into radioactive carbon-14, or ^{14}C. This element has a half-life of about 5730 years. Over the relatively recent period of the history of this planet in which life has evolved, the fraction of ^{14}C in the atmosphere, compared to regular carbon, has been essentially constant. This means that living matter (plant or animal) has ingested ^{14}C at about the same rate over a long historical period, and objects living, say, 2 million years ago would have had the same ratio of carbon-14 to carbon in their bodies as objects alive today. When an organism dies, it ceases its intake of

^{14}C, which then begins to decay. By measuring the ratio of ^{14}C to carbon in an artifact, we can estimate the amount of the decay, and hence the time it took, giving an estimate of the time the organism was alive. This process of estimating the age of an artifact is called *carbon dating*. Of course, in reality the ratio of ^{14}C in the atmosphere has only been approximately constant, and in addition a sample may have been contaminated by exposure to other living organisms, or even to the air, so carbon dating is a sensitive process that can lead to controversial results. Nevertheless, when applied rigorously and combined with other tests and information, it has proved a valuable tool in historical and archeological studies.

To apply equation (1.4) to carbon dating, use $H = 5730$ and compute

$$\frac{\ln(2)}{H} = \frac{\ln(2)}{5730} \approx 0.000120968$$

in which $\approx$ means "approximately equal" (not all decimal places are listed). Equation (1.4) becomes

$$m(t) = Me^{-0.000120968t}.$$

Now suppose we have an artifact, say a piece of fossilized wood, and measurements show that the ratio of ^{14}C to carbon in the sample is 37% of the current ratio. If we say that the wood died at time 0, then we want to compute the time T it would take for one gram of the radioactive carbon to decay this amount. Thus, solve for T in

$$0.37 = e^{-0.000120968T}.$$

We find that

$$T = -\frac{\ln(0.37)}{0.000120968} \approx 8219$$

years. This is a little less than one and one-half half-lives, a reasonable estimate if nearly $\frac{2}{3}$ of the ^{14}C has decayed. ∎

EXAMPLE 1.13   Torricelli's Law

Suppose we want to estimate how long it will take for a container to empty by discharging fluid through a drain hole. This is a simple enough problem for, say, a soda can, but not quite so easy for a large oil storage tank or chemical facility.

We need two principles from physics. The first is that the rate of discharge of a fluid flowing through an opening at the bottom of a container is given by

$$\frac{dV}{dt} = -kAv,$$

in which $V(t)$ is the volume of fluid in the container at time t, $v(t)$ is the discharge velocity of fluid through the opening, A is the cross-sectional area of the opening (assumed constant), and k is a constant determined by the viscosity of the fluid, the shape of the opening, and the fact that the cross-sectional area of fluid pouring out of the opening is slightly less than that of the opening itself. In practice, k must be determined for the particular fluid, container, and opening and is a number between 0 and 1.

We also need Torricelli's law, which states that $v(t)$ is equal to the velocity of a free-falling particle released from a height equal to the depth of the fluid at time t. (Free-falling means that the particle is influenced by gravity only). Now the work done by gravity in moving the particle

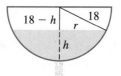

FIGURE 1.9

from its initial point by a distance $h(t)$ is $mgh(t)$, and this must equal the change in the kinetic energy, $(\frac{1}{2})mv^2$. Therefore,

$$v(t) = \sqrt{2gh(t)}.$$

Putting these two equations together yields

$$\frac{dV}{dt} = -kA\sqrt{2gh(t)}. \tag{1.5}$$

We will apply equation (1.5) to a specific case to illustrate its use. Suppose we have a hemispherical tank of water, as in Figure 1.9. The tank has radius 18 feet, and water drains through a circular hole of radius 3 inches at the bottom. How long will it take the tank to empty?

Equation 1.5 contains two unknown functions, $V(t)$ and $h(t)$, so one must be eliminated. Let $r(t)$ be the radius of the surface of the fluid at time t and consider an interval of time from t_0 to $t_1 = t_0 + \Delta t$. The volume ΔV of water draining from the tank in this time equals the volume of a disk of thickness Δh (the change in depth) and radius $r(t^*)$, for some t^* between t_0 and t_1. Therefore,

$$\Delta V = \pi \left[r(t^*) \right]^2 \Delta h$$

so

$$\frac{\Delta V}{\Delta t} = \pi \left[r(t^*) \right]^2 \frac{\Delta h}{\Delta t}.$$

In the limit as $\Delta t \to 0$,

$$\frac{dV}{dt} = \pi r^2 \frac{dh}{dt}.$$

Putting this into equation (1.5) yields

$$\pi r^2 \frac{dh}{dt} = -kA\sqrt{2gh}.$$

Now V has been eliminated, but at the cost of introducing $r(t)$. However, from Figure 1.9,

$$r^2 = 18^2 - (18 - h)^2 = 36h - h^2$$

so

$$\pi \left(36h - h^2 \right) \frac{dh}{dt} = -kA\sqrt{2gh}.$$

This is a separable differential equation, which we write as

$$\pi \frac{36h - h^2}{h^{1/2}} \, dh = -kA\sqrt{2g} \, dt.$$

Take g to be 32 feet per second per second. The radius of the circular opening is 3 inches, or $\frac{1}{4}$ feet, so its area is $A = \pi/16$ square feet. For water, and an opening of this shape and size, the experiment gives $k = 0.8$. The last equation becomes

$$\left(36h^{1/2} - h^{3/2}\right) dh = -(0.8)\left(\frac{1}{16}\right)\sqrt{64}\, dt,$$

or

$$\left(36h^{1/2} - h^{3/2}\right) dh = -0.4\, dt.$$

A routine integration yields

$$24h^{3/2} - \tfrac{2}{5}h^{5/2} = -\tfrac{2}{5}t + c,$$

or

$$60h^{3/2} - h^{5/2} = -t + k.$$

Now $h(0) = 18$, so

$$60(18)^{3/2} = (18)^{5/2} = k.$$

Thus, $k = 2268\sqrt{2}$ and $h(t)$ is implicitly determined by the equation

$$60h^{3/2} - h^{5/2} = 2268\sqrt{2} - t.$$

The tank is empty when $h = 0$, and this occurs when $t = 2268\sqrt{2}$ seconds, or about 53 minutes, 28 seconds. ■

The last three examples contain an important message. Differential equations can be used to solve a variety of problems, but a problem usually does not present itself as a differential equation. Normally, we have some event or process, and we must use whatever information we have about it to derive a differential equation and initial conditions. This process is called *mathematical modeling*. The model consists of the differential equation and other relevant information, such as initial conditions. We look for a function satisfying the differential equation and the other information, in the hope of being able to predict future behavior, or perhaps better understand the process being considered.

SECTION 1.2 PROBLEMS

In each of Problems 1 through 10, determine if the differential equation is separable. If it is, find the general solution (perhaps implicitly defined). If it is not separable, do not attempt a solution at this time.

1. $3y' = 4x/y^2$

2. $y + xy' = 0$

3. $\cos(y)y' = \sin(x + y)$

4. $e^{x+y}y' = 3x$

5. $xy' + y = y^2$

6. $y' = \dfrac{(x+1)^2 - 2y}{2y}$

7. $x\sin(y)y' = \cos(y)$

8. $\dfrac{x}{y}y' = \dfrac{2y^2 + 1}{x + 1}$

9. $y + y' = e^x - \sin(y)$

10. $[\cos(x + y) + \sin(x - y)]y' = \cos(2x)$

In each of Problems 11 through 15, solve the initial value problem.

11. $xy^2y' = y + 1$; $y(3e^2) = 2$

12. $y' = 3x^2(y + 2)$; $y(2) = 8$

13. $\ln(y^x)y' = 3x^2y$; $y(2) = e^3$

14. $2yy' = e^{x-y^2}$; $y(4) = -2$

15. $yy' = 2x\sec(3y)$; $y(2/3) = \pi/3$

16. An object having a temperature of 90 degrees Fahrenheit is placed into an environment kept at 60 degrees. Ten minutes later, the object has cooled to 88 degrees. What will be the temperature of the object after it has been in this environment for 20 minutes? How long will it take for the object to cool to 65 degrees?

17. A thermometer is carried outside a house whose ambient temperature is 70 degrees Fahrenheit. After five minutes, the thermometer reads 60 degrees, and fifteen minutes after this, 50.4 degrees. What is the outside temperature (which is assumed to be constant)?

18. Assume that the population of bacteria in a petri dish changes at a rate proportional to the population at that time. This means that, if $P(t)$ is the population at time t, then

$$\frac{dP}{dt} = kP$$

for some constant k. A particular culture has a population density of 100,000 bacteria per square inch. A culture that covered an area of 1 square inch at 10:00 A.M. on Tuesday was found to have grown to cover 3 square inches by noon the following Thursday. How many bacteria will be present at 3:00 P.M. the following Sunday? How many will be present on Monday at 4:00 P.M.? When will the world be overrun by these bacteria, assuming that they can live anywhere on the earth's surface? (Here you need to look up the land area of the earth.)

19. Assume that a sphere of ice melts at a rate proportional to its surface area, retaining a spherical shape. Interpret melting as a reduction of volume with respect to time. Determine an expression for the volume of the ice at any time t.

20. A radioactive element has a half-life of $\ln(2)$ weeks. If e^3 tons are present at a given time, how much will be left 3 weeks later?

21. The half-life of uranium-238 is approximately $4.5 \cdot 10^9$ years. How much of a 10-kilogram block of U-238 will be present 1 billion years from now?

22. Given that 12 grams of a radioactive element decays to 9.1 grams in 4 minutes, what is the half-life of this element?

23. Evaluate

$$\int_0^\infty e^{-t^2 - 9/t^2} \, dt.$$

Hint: Let

$$I(x) = \int_0^\infty e^{-t^2 - (x/t)^2} \, dt.$$

Calculate $I'(x)$ by differentiating under the integral sign, then let $u = x/t$. Show that $I'(x) = -2I(x)$ and solve for $I(x)$. Evaluate the constant by using the standard result that $\int_0^\infty e^{-t^2} dt = \sqrt{\pi}/2$. Finally, evaluate $I(3)$.

24. Find all real-valued functions $g(x)$ that are continuous and positive on $[0, \infty)$ and have the property that, for all $x > 0$, the average value of $g(x)$ on $[0, x]$ equals the y coordinate of the centroid of the region consisting of points (s, t), with $0 \le s \le x$ and $0 \le t \le g(s)$. This problem appeared on the 1984 William Lowell Putnam Mathematical Competition. *Hint:* Express the centroid and average as integrals and establish an equation involving these integrals, with fractions cleared. Use the fundamental theorem of calculus, then multiply by x and make a substitution from the original equation to obtain a quadratic equation for $\int_0^x g(t) \, dt$. Solve this and use the fundamental theorem again to obtain a separable differential equation for $g(t)$.

25. (*Logistic Model of Population Growth*) In 1837 the Dutch biologist Verhulst developed a differential equation to model the growth of populations, such as people on an island, bacteria in a petri dish, and others. (He was actually looking at fish populations in the Adriatic Sea). Verhulst reasoned that the rate of change of a population $P(t)$ with respect to time t should be influenced by growth factors such as the population itself, and also factors tending to retard the population, such as limitations on food and space. He formed a model by supposing that growth factors could be incorporated into a term $aP(t)$, and retarding factors into a term $-bP(t)^2$, with a and b positive constants whose values depend on the particular population. This led to his *logistic equation:*

$$P'(t) = aP(t) - bP(t)^2.$$

The initial population (at a time designated as time zero) is $P(0) = p_0$. Solve this initial value problem to get

$$P(t) = \frac{ap_0}{a - bp_0 + bp_0 e^{at}} e^{at}.$$

This is the *logistic model of population growth*. Its graph has the general appearance of Figure 1.10. Show that $\lim_{t \to \infty} P(t) = a/b$. This is the limiting value of the population, beyond which it cannot grow.

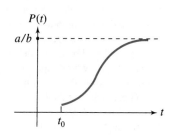

FIGURE 1.10 *Logistic model of population growth.*

		Logistic Model Predicted Figure	
Year	Population	$P(t)$	Percentage Error
1790	3,929,214		
1800	5,308,483		
1810	7,239,881		
1820	9,638,453		
1830	12,866,020		
1840	17,069,453		
1850	23,191,876		
1860	31,443,321		
1870	38,558,371		
1880	50,189,209		
1890	62,979,766		
1900	76,212,168		
1910	92,228,496		
1920	106,021,537		
1930	123,202,624		
1940	132,164,569		
1950	151,325,798		
1960	179,323,175		
1970	203,302,031		
1980	226,547,042		

TABLE 1.1 *Logistic Model of Population Growth for the United States*

In 1920 a study by Pearl and Reed in the *Proceedings of the National Academy of Sciences* suggested the values $a = 0.03134$ and $b = (1.5887)(10^{-10})$ for the population of the United States. Table 1.1 gives the census data for the United States in ten-year increments from 1790. Taking 1790 as year zero to determine p_0, use the Pearl/Reed numbers to show that the logistic model for the population of this country at time t is

$$P(t) = \frac{123{,}141.5668}{0.03072 + 0.00062e^{0.03134t}} e^{0.03134t}.$$

Calculate $P(t)$ in ten-year intervals from 1790 to find what the model gives for the U.S. population through 1980. Remember that 1790 is the base year, so put $t = 0$ for the population in 1790, $t = 10$ for the population in 1800, and so on. Plot these points together with the actual census data and draw smooth curves through the two sets of data points to compare actual population growth with that predicted by the model. You can see that the model is fairly accurate for a long period of time, then it diverges from the actual numbers. Show that the limiting population of the United States in this model is about 197,300,000 people, a number exceeded in 1970. This means that this model is not an accurate predictor over long periods of time. It may also be that the Pearl and Reed coefficients need to be adjusted, or perhaps other factors incorporated into the model.

Sometimes an exponential model is used for population growth. This assumes that the population changes at a rate proportional to the population, so that $P'(t) = kP(t)$ for some constant k. Using the 1790 and 1800 data, with 1790 again as time zero, solve this differential equation to get an exponential formula for the population growth. Use it to calculate predicted populations for the United States through 1980. It should be clear that these numbers soon diverge from the actual numbers and that the logistic model is a much better predictor of the actual population than the exponential model. Nevertheless, exponential models can be useful for certain populations over relatively short periods of time.

26. Derive the fact used in Example 1.13 that $v(t) = \sqrt{2gh(t)}$. *Hint:* Consider a free-falling particle having height $h(t)$ at time t. The work done by gravity in moving the particle from its starting point to a given point is $mgh(t)$, and this must equal the change in the kinetic energy, which is $(\frac{1}{2})mv^2$.

27. Calculate the time required to empty the hemispherical tank of Example 1.13 if the tank is positioned with its flat side down.

28. (*Draining a Hot Tub*) Consider a cylindrical hot tub with a 5-foot radius and height of 4 feet, placed on one of its circular ends. Water is draining from the tub

through a circular hole $\frac{5}{8}$ inches in diameter located in the base of the tub.

(a) Assume a value $k = 0.6$ to determine the rate at which the depth of the water is changing. Here it is useful to write

$$\frac{dh}{dt} = \frac{dh}{dV}\frac{dV}{dt} = \frac{dV/dt}{dV/dh}.$$

(b) Calculate the time T required to drain the hot tub if it is initially full. *Hint:* One way to do this is to write

$$T = \int_H^0 \frac{dt}{dh}\,dh.$$

(c) Determine how much longer it takes to drain the lower half than the upper half of the tub. *Hint:* Use the integral suggested in (b), with different limits for the two halves.

29. (*Draining a Cone*) A tank shaped like a right circular cone, with its vertex down, is 9 feet high and has a diameter of 8 feet. It is initially full of water.

(a) Determine the time required to drain the tank through a circular hole of diameter 2 inches at the vertex. Take $k = 0.6$.

(b) Determine the time it takes to drain the tank if it is inverted and the drain hole is of the same size and shape as in (a), but now located in the new base.

30. (*Drain Hole at Unknown Depth*) Determine the rate of change of the depth of water in the tank of Problem 29 (vertex at the bottom) if the drain hole is located in the side of the cone 2 feet above the bottom of the tank. What is the rate of change in the depth of the water when the drain hole is located in the bottom of the tank? Is it possible to determine the location of the drain hole if we are told the rate of change of the depth and the depth of the water in the tank? Can this be done without knowing the size of the drain opening?

31. Suppose the conical tank of Problem 29, vertex at the bottom, is initially empty and water is added at the constant rate of $\pi/10$ cubic feet per second. Does the tank ever overflow?

32. (*Draining a Sphere*) Determine the time it takes to completely drain a spherical tank of radius 18 feet if it is initially full of water and the water drains through a circular hole of radius 3 inches located in the bottom of the tank. Use $k = 0.8$.

33. (*Draining a Cylinder*) A cylindrical tank of radius 2 feet and length 6 feet lies on its side. The tank has a 1 inch by 1 inch square drain hole in its bottom (on the lateral side of the tank) and is initially full of water.

(a) Using $k = 0.6$, calculate the time required to drain the tank. *Hint:* The volume of fluid left in the tank can be expressed as the product of the length of the tank and an integral whose limits are a function of the depth of the fluid left in the tank. Use the fundamental theorem of calculus to find dV/dh from this integral.

(b) How much longer does it take to drain the bottom half than the top half of the tank?

34. Determine the shape of a container that is a surface of revolution and has the property that the rate at which the fluid level decreases is constant if the fluid is drained from the bottom of the container.

35. (*Plowing Snow*) Snow is falling at a constant rate throughout the day. A snowplow starts at noon to clear a road and manages to go 2 miles by 1:00 P.M. and one more mile by 2:00 P.M. Assuming that the plow removes a constant volume of snow per unit time, determine when it began to snow. *Hint:* Let L be the length of time it had been snowing when the plow started. Determine the depth of the snow at time t and assume that the product of this depth and the plow's rate of travel is constant.

1.3   Linear Differential Equations

DEFINITION 1.2   *Linear Differential Equation*

A first-order differential equation is linear if it has the form

$$y'(x) + p(x)y = q(x).$$

Assume that p and q are continuous on an interval I (possibly the whole real line). Because of the special form of the linear equation, we can obtain the general solution on I by a clever observation. Multiply the differential equation by $e^{\int p(x)\,dx}$ to get

$$e^{\int p(x)\,dx}y'(x) + p(x)e^{\int p(x)\,dx}y = q(x)e^{\int p(x)\,dx}.$$

The left side of this equation is the derivative of the product $y(x)e^{\int p(x)\,dx}$, enabling us to write

$$\frac{d}{dx}\left(y(x)e^{\int p(x)\,dx}\right) = q(x)e^{\int p(x)\,dx}.$$

Now integrate to obtain

$$y(x)e^{\int p(x)\,dx} = \int\left(q(x)e^{\int p(x)\,dx}\right)dx + C.$$

Finally, solve for $y(x)$:

$$y(x) = e^{-\int p(x)\,dx}\int\left(q(x)e^{\int p(x)\,dx}\right)dx + Ce^{-\int p(x)\,dx}. \tag{1.6}$$

The function $e^{\int p(x)\,dx}$ is called an *integrating factor* for the differential equation, because multiplication of the differential equation by this factor results in an equation that can be integrated to obtain the general solution. We do not recommend memorizing equation (1.6). Instead, recognize the form of the linear equation and understand the technique of solving it by multiplying by $e^{\int p(x)\,dx}$.

EXAMPLE 1.14

The equation $y' + y = \sin(x)$ is linear. Here $p(x) = 1$ and $q(x) = \sin(x)$, both continuous for all x. An integrating factor is

$$e^{\int dx},$$

or e^x. Multiply the differential equation by e^x to get

$$y'e^x + ye^x = e^x\sin(x),$$

or

$$\left(ye^x\right)' = e^x\sin(x).$$

Integrate to get

$$ye^x = \int e^x\sin(x)\,dx = \tfrac{1}{2}e^x\left[\sin(x) - \cos(x)\right] + C.$$

The general solution is

$$y(x) = \tfrac{1}{2}\left[\sin(x) - \cos(x)\right] + Ce^{-x}. \ \blacksquare$$

EXAMPLE 1.15

Solve the initial value problem

$$y' = 3x^2 - \frac{y}{x}; \qquad y(1) = 5.$$

First recognize that the differential equation can be written in linear form:

$$y' + \frac{1}{x}y = 3x^2.$$

An integrating factor is $e^{\int (1/x)\,dx} = e^{\ln(x)} = x$, for $x > 0$. Multiply the differential equation by x to get

$$xy' + y = 3x^3,$$

or

$$(xy)' = 3x^3.$$

Integrate to get

$$xy = \frac{3}{4}x^4 + C.$$

Then

$$y(x) = \frac{3}{4}x^3 + \frac{C}{x}$$

for $x > 0$. For the initial condition, we need

$$y(1) = 5 = \frac{3}{4} + C$$

so $C = \frac{17}{4}$ and the solution of the initial value problem is

$$y(x) = \frac{3}{4}x^3 + \frac{17}{4x}$$

for $x > 0$. ∎

Depending on p and q, it may not be possible to evaluate all of the integrals in the general solution 1.6 in closed form (as a finite algebraic combination of elementary functions). This occurs with

$$y' + xy = 2.$$

whose general solution is

$$y(x) = 2e^{-x^2/2} \int e^{x^2/2}\,dx + Ce^{-x^2/2}.$$

We cannot write $\int e^{x^2/2}\,dx$ in elementary terms. However, we could still use a software package to generate a direction field and integral curves, as is done in Figure 1.11. This provides some idea of the behavior of solutions, at least within the range of the diagram.

Linear differential equations arise in many contexts. Example 1.11, involving estimation of time of death, involved a separable differential equation that is also linear and could have been solved using an integrating factor.

EXAMPLE 1.16 A Mixing Problem

Sometimes we want to know how much of a given substance is present in a container in which various substances are being added, mixed, and removed. Such problems are called *mixing problems*, and they are frequently encountered in the chemical industry and in manufacturing processes.

As an example, suppose a tank contains 200 gallons of brine (salt mixed with water), in which 100 pounds of salt are dissolved. A mixture consisting of $\frac{1}{8}$ pound of salt per gallon is flowing into the tank at a rate of 3 gallons per minute, and the mixture is continuously stirred. Meanwhile,

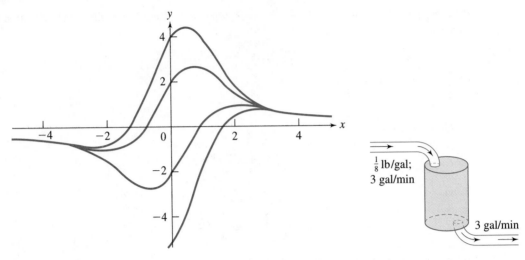

FIGURE 1.11 *Integral curves of $y' + xy = 2$ passing through $(0, 2)$, $(0, 4)$, $(0, -2)$, and $(0, -5)$.*   **FIGURE 1.12**

brine is allowed to empty out of the tank at the same rate of 3 gallons per minute (Figure 1.12). How much salt is in the tank at any time?

Before constructing a mathematical model, notice that the initial ratio of salt to brine in the tank is 100 pounds per 200 gallons, or $\frac{1}{2}$ pound per gallon. Since the mixture pumped in has a constant ratio of $\frac{1}{8}$ pound per gallon, we expect the brine mixture to dilute toward the incoming ratio, with a "terminal" amount of salt in the tank of $\frac{1}{8}$ pound per gallon times 200 gallons. This leads to the expectation that in the long term (as $t \to \infty$) the amount of salt in the tank should approach 25 pounds.

Now let $Q(t)$ be the amount of salt in the tank at time t. The rate of change of $Q(t)$ with time must equal the rate at which salt is pumped in, minus the rate at which it is pumped out. Thus,

$$\frac{dQ}{dt} = (\text{rate in}) - (\text{rate out})$$

$$= \left(\frac{1}{8}\frac{\text{pounds}}{\text{gallon}}\right)\left(3\frac{\text{gallons}}{\text{minute}}\right) - \left(\frac{Q(t)}{200}\frac{\text{pounds}}{\text{gallon}}\right)\left(3\frac{\text{gallons}}{\text{minute}}\right)$$

$$= \frac{3}{8} - \frac{3}{200}Q(t).$$

This is the linear equation

$$Q'(t) + \frac{3}{200}Q = \frac{3}{8}.$$

An integrating factor is $e^{\int (3/200)\, dt} = e^{3t/200}$. Multiply the differential equation by this factor to obtain

$$Q'e^{3t/200} + \frac{3}{200}e^{3t/200}Q = \frac{3}{8}e^{3t/200},$$

or

$$\left(Qe^{3t/200}\right)' = \frac{3}{8}e^{3t/200}.$$

Then

$$Qe^{3t/200} = \frac{3}{8}\frac{200}{3}e^{3t/200} + C,$$

so

$$Q(t) = 25 + Ce^{-3t/200}.$$

Now

$$Q(0) = 100 = 25 + C$$

so $C = 75$ and

$$Q(t) = 25 + 75e^{-3t/200}.$$

As we expected, as t increases, the amount of salt approaches the limiting value of 25 pounds. From the derivation of the differential equation for $Q(t)$, it is apparent that this limiting value depends on the rate at which salt is poured into the tank, but not on the initial amount of salt in the tank. The term 25 in the solution is called the steady-state part of the solution because it is independent of time, and the term $75e^{-3t/200}$ is the transient part. As t increases, the transient part exerts less influence on the amount of salt in the tank, and in the limit the solution approaches its steady-state part. ■

SECTION 1.3   PROBLEMS

In each of Problems 1 through 12, find the general solution. Not all integrals can be done in closed form.

1. $y' - \dfrac{3}{x}y = 2x^2$

2. $y' - y = \sinh(x)$

3. $y' + 2y = x$

4. $\sin(2x)y' + 2y\sin^2(x) = 2\sin(x)$

5. $y' - 2y = -8x^2$

6. $(x^2 - x - 2)y' + 3xy = x^2 - 4x + 4$

7. $y' + y = \dfrac{x-1}{x^2}$

8. $y' + \sec(x)y = \cos(x)$

9. $y' + \dfrac{1}{x}y = x^2 + 2$

10. $2y' + 3y = e^{2x}$

11. $y' + \dfrac{1}{x}y = 3e^{-x^2}$

12. $xy' + \dfrac{2}{x}y = 4$

13. Find the general solution of $(1 - 2xe^{2y})y' - e^{2y} = 0$. *Hint:* Reverse the traditional roles of x and y as independent and dependent variables and use the fact that $dy/dx = 1/(dx/dy)$, if $dx/dy \neq 0$.

14. Find the general solution of $(4y^3 - x)y' = y$.

In each of Problems 15 through 22, solve the initial value problem.

15. $y' + \dfrac{1}{x-2}y = 3x$; $y(3) = 4$

16. $y' + 3y = 5e^{2x} - 6$; $y(0) = 2$

17. $y' + \dfrac{2}{x+1}y = 3$; $y(0) = 5$

18. $(x^2 - 2x)y' + (x^2 - 5x + 4)y = (x^4 - 2x^3)e^{-x}$; $y(3) = 18e^{-3}$

19. $y' - y = 2e^{4x}$; $y(0) = -3$

20. $y' + \dfrac{5y}{9x} = 3x^3 + x$; $y(-1) = 4$

21. $y' + \dfrac{4}{x}y = 2$; $y(1) = -4$

22. $y' - 4y = x + \sin(x)$; $y(0) = 3$

23. Find all functions with the property that the y intercept of the tangent to the graph at (x, y) is $2x^2$.

24. A 500-gallon tank initially contains 50 gallons of brine solution in which 28 pounds of salt have been dissolved. Beginning at time zero, brine containing 2 pounds of salt per gallon is added at the rate of 3 gallons per minute, and the mixture is poured out of the tank at the rate of 2 gallons per minute. How much salt is in the tank when it contains 100 gallons of brine? *Hint:* The amount of brine in the tank at time t is $50 + t$.

25. A 500-gallon tank initially contains 100 gallons of brine in which 5 pounds of salt have been dissolved. Brine containing 2 pounds per gallon is added at the rate of 5 gallons per minute and mixture flows out at 3 gallons per minute. Determine how much salt is in the tank at the moment it overflows.

26. A 400-gallon tank is filled with brine containing 45 pounds of salt. At a certain time the brine begins draining out from an open valve at the bottom at a rate of 5 gallons per minute. Simultaneously, a brine mixture of $\frac{1}{8}$ pound per gallon is added to the tank at the rate of 3 gallons per minute. Three hours later, a freshwater valve is turned on, supplying 2 gallons per minute to the tank in addition to the brine mixture already being added. Calculate the amount of salt in the tank for any $t \geq 0$. What is the steady-state amount of salt in the tank? *Hint:* Treat the problem as a standard mixing problem in the first 3 hours. This determines the status of the mixture in the tank at the 3-hour mark, at which time a new problem emerges because new conditions are imposed.

27. Two tanks are cascaded as in Figure 1.13. Tank 1 initially contains 20 pounds of salt dissolved in 100 gal-

lons of brine, while tank 2 contains 150 gallons of brine in which 90 pounds of salt are dissolved. At time zero, a brine solution containing $\frac{1}{2}$ pound of salt per gallon is added to tank 1 at the rate of 5 gallons per minute. Tank 1 has an output that discharges brine into tank 2 at the rate of 5 gallons per minute, and tank 2 also has an output of 5 gallons per minute. Determine the amount of salt in each tank at any time t. Also determine when the concentration of salt in tank 2 is a minimum and how much salt is in the tank at that time. *Hint:* Solve for the amount of salt in tank 1 at time t first and then use this solution to determine the amount in tank 2.

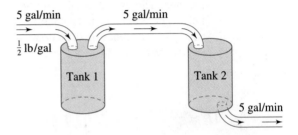

FIGURE 1.13 *Mixing between tanks in Problem 27.*

1.4 Exact Differential Equations

We continue the theme of identifying certain kinds of first-order differential equations for which there is a method leading to a solution.

We can write any first-order equation $y' = f(x, y)$ in the form

$$M(x, y) + N(x, y)y' = 0.$$

For example, put $M(x, y) = -f(x, y)$ and $N(x, y) = 1$. An interesting thing happens if there is a function φ such that

$$\frac{\partial \varphi}{\partial x} = M(x, y) \quad \text{and} \quad \frac{\partial \varphi}{\partial y} = N(x, y). \tag{1.7}$$

In this event, the differential equation becomes

$$\frac{\partial \varphi}{\partial x} + \frac{\partial \varphi}{\partial y} \frac{dy}{dx} = 0,$$

which, by the chain rule, is the same as

$$\frac{d}{dx} \varphi(x, y(x)) = 0.$$

But this means that

$$\varphi(x, y(x)) = C,$$

with C constant. If we now read this argument from the last line back to the first, the conclusion is that the equation

$$\varphi(x, y) = C$$

implicitly defines a function $y(x)$ that is the general solution of the differential equation. Thus, finding a function that satisfies equation (1.7) is equivalent to solving the differential equation.

Before taking this further, consider an example.

EXAMPLE 1.17

The differential equation

$$y' = -\frac{2xy^3 + 2}{3x^2y^2 + 8e^{4y}}$$

is neither separable nor linear. Write it in the form

$$M + Ny' = 2xy^3 + 2 + \left(3x^2y^2 + 8e^{4y}\right)y' = 0, \tag{1.8}$$

with

$$M(x, y) = 2xy^3 + 2 \quad \text{and} \quad N(x, y) = 3x^2y^2 + 8e^{4y}.$$

Equation (1.8) can in turn be written

$$M\,dx + N\,dy = (2xy^3 + 2)\,dx + (3x^2y^2 + 8e^{4y})\,dy = 0. \tag{1.9}$$

Now let

$$\varphi(x, y) = x^2y^3 + 2x + 2e^{4y}.$$

Soon we will see where this came from, but for now, observe that

$$\frac{\partial \varphi}{\partial x} = 2xy^3 + 2 = M \quad \text{and} \quad \frac{\partial \varphi}{\partial y} = 3x^2y^2 + 8e^{4y} = N.$$

With this choice of $\varphi(x, y)$, equation (1.9) becomes

$$\frac{\partial \varphi}{\partial x}\,dx + \frac{\partial \varphi}{\partial y}\,dy = 0,$$

or

$$d\varphi\,(x, y) = 0.$$

The general solution of this equation is

$$\varphi(x, y) = C,$$

or, in this example,

$$x^2y^3 + 2x + 2e^{4y} = C.$$

This implicitly defines the general solution of the differential equation (1.8).

To verify this, differentiate the last equation implicitly with respect to x:

$$2xy^3 + 3x^2y^2y' + 2 + 8e^{4y}y' = 0,$$

or

$$2xy^3 + 2 + (3x^2y^2 + 8e^{4y})y' = 0.$$

This is equivalent to the original differential equation

$$y' = -\frac{2xy^3 + 2}{3x^2y^2 + 8e^{4y}}. \quad \blacksquare$$

With this as background, we will make the following definitions.

DEFINITION 1.3   *Potential Function*

A function φ is a potential function for the differential equation $M(x, y) + N(x, y)y' = 0$ on a region R of the plane if, for each (x, y) in R,

$$\frac{\partial \varphi}{\partial x} = M(x, y) \quad \text{and} \quad \frac{\partial \varphi}{\partial y} = N(x, y).$$

DEFINITION 1.4   *Exact Differential Equation*

When a potential function exists on a region R for the differential equation $M + Ny' = 0$, then this equation is said to be exact on R.

The differential equation of Example 1.17 is exact (over the entire plane) because we exhibited a potential function for it, defined for all (x, y). Once a potential function is found, we can write an equation implicitly defining the general solution. Sometimes we can explicitly solve for the general solution, and sometimes we cannot.

Now go back to Example 1.17. We want to explore how the potential function that materialized there was found. Recall we required that

$$\frac{\partial \varphi}{\partial x} = 2xy^3 + 2 = M \quad \text{and} \quad \frac{\partial \varphi}{\partial y} = 3x^2y^2 + 8e^{4y} = N.$$

Pick either of these equations to begin and integrate it. Say we begin with the first. Then integrate with respect to x:

$$\varphi(x, y) = \int \frac{\partial \varphi}{\partial x} \, dx = \int \left(2xy^3 + 2\right) dx$$
$$= x^2y^3 + 2x + g(y).$$

In this integration with respect to x, we held y fixed, hence we must allow that y appears in the "constant" of integration. If we calculate $\partial \varphi / \partial x$, we get $2xy^2 + 2$ for any function $g(y)$.

Now we know φ to within this function g. Use the fact that we know $\partial \varphi / \partial y$ to write

$$\frac{\partial \varphi}{\partial y} = 3x^2y^2 + 8e^{4y}$$

$$= \frac{\partial}{\partial y}(x^2y^3 + 2x + g(y)) = 3x^2y^2 + g'(y).$$

This equation holds if $g'(y) = 8e^{4y}$; hence we may choose $g(y) = 2e^{4y}$. This gives the potential function

$$\varphi(x, y) = x^2y^3 + 2x + 2e^{4y}.$$

If we had chosen to integrate $\partial\varphi/\partial y$ first, we would have gotten

$$\varphi(x, y) = \int \left(3x^2y^2 + 8e^{4y}\right) dy$$

$$= x^2y^3 + 2e^{4y} + h(x).$$

Here h can be any function of one variable, because no matter how $h(x)$ is chosen,

$$\frac{\partial}{\partial y}\left(x^2y^3 + 2e^{4y} + h(x)\right) = 3x^2y^2 + 8e^{4y},$$

as required. Now we have two expressions for $\partial\varphi/\partial x$:

$$\frac{\partial\varphi}{\partial x} = 2xy^3 + 2$$

$$= \frac{\partial}{\partial x}\left(x^2y^3 + 2e^{4y} + h(x)\right) = 2xy^3 + h'(x).$$

This equation forces us to choose h so that $h'(x) = 2$, and we may therefore set $h(x) = 2x$. This gives

$$\varphi(x, y) = x^2y^3 + 2e^{4y} + 2x,$$

as we got before.

Not every first-order differential equation is exact. For example, consider

$$y + y' = 0.$$

If there were a potential function φ, then we would have

$$\frac{\partial\varphi}{\partial x} = y, \qquad \frac{\partial\varphi}{\partial y} = 1.$$

Integrate $\partial\varphi/\partial x = y$ with respect to x to get $\varphi(x, y) = xy + g(y)$. Substitute this into $\partial\varphi/\partial y = 1$ to get

$$\frac{\partial}{\partial y}(xy + g(y)) = x + g'(y) = 1.$$

But this can hold only if $g'(y) = 1 - x$, an impossibility if g is to be independent of x. Therefore, $y + y' = 0$ has no potential function. This differential equation is not exact (even though it is easily solved either as a separable or as a linear equation).

This example suggests the need for a convenient test for exactness. This is provided by the following theorem, in which a "rectangle in the plane" refers to the set of points on or inside any rectangle having sides parallel to the axes.

THEOREM 1.1 *Test for Exactness*

Suppose $M(x, y)$, $N(x, y)$, $\partial M/\partial y$, and $\partial N/\partial x$ are continuous for all (x, y) within a rectangle R in the plane. Then,

$$M(x, y) + N(x, y)y' = 0$$

is exact on R if and only if, for each (x, y) in R,

$$\frac{\partial M}{\partial y} = \frac{\partial N}{\partial x}.$$

Proof   If $M + Ny' = 0$ is exact, then there is a potential function φ and

$$\frac{\partial \varphi}{\partial x} = M(x, y) \quad \text{and} \quad \frac{\partial \varphi}{\partial y} = N(x, y).$$

Then, for (x, y) in R,

$$\frac{\partial M}{\partial y} = \frac{\partial}{\partial y}\left(\frac{\partial \varphi}{\partial x}\right) = \frac{\partial^2 \varphi}{\partial y \partial x} = \frac{\partial^2 \varphi}{\partial x \partial y} = \frac{\partial}{\partial x}\left(\frac{\partial \varphi}{\partial y}\right) = \frac{\partial N}{\partial x}.$$

Conversely, suppose $\partial M/\partial y$ and $\partial N/\partial x$ are continuous on R. Choose any (x_0, y_0) in R and define, for (x, y) in R,

$$\varphi(x, y) = \int_{x_0}^{x} M(\xi, y_0)\, d\xi + \int_{y_0}^{y} N(x, \eta)\, d\eta. \tag{1.10}$$

Immediately we have, from the fundamental theorem of calculus,

$$\frac{\partial \varphi}{\partial y} = N(x, y),$$

since the first integral in equation (1.10) is independent of y. Next, compute

$$\frac{\partial \varphi}{\partial x} = \frac{\partial}{\partial x}\int_{x_0}^{x} M(\xi, y_0)\, d\xi + \frac{\partial}{\partial x}\int_{y_0}^{y} N(x, \eta)\, d\eta$$

$$= M(x, y_0) + \int_{y_0}^{y} \frac{\partial N}{\partial x}(x, \eta)\, d\eta$$

$$= M(x, y_0) + \int_{y_0}^{y} \frac{\partial M}{\partial y}(x, \eta)\, d\eta$$

$$= M(x, y_0) + M(x, y) - M(x, y_0) = M(x, y),$$

and the proof is complete. ■

For example, consider again $y + y' = 0$. Here $M(x, y) = y$ and $N(x, y) = 1$, so

$$\frac{\partial N}{\partial x} = 0 \quad \text{and} \quad \frac{\partial M}{\partial y} = 1$$

throughout the entire plane. Thus, $y + y' = 0$ cannot be exact on any rectangle in the plane. We saw this previously by showing that this differential equation can have no potential function.

EXAMPLE 1.18

Consider

$$x^2 + 3xy + (4xy + 2x)y' = 0.$$

Here $M(x, y) = x^2 + 3xy$ and $N(x, y) = 4xy + 2x$. Now

$$\frac{\partial N}{\partial x} = 4y + 2 \quad \text{and} \quad \frac{\partial M}{\partial y} = 3x,$$

and

$$3x = 4y + 2$$

is satisfied by all (x, y) on a straight line. However, $\partial N/\partial x = \partial M/\partial y$ cannot hold for all (x, y) in an entire rectangle in the plane. Hence this differential equation is not exact on any rectangle. ∎

EXAMPLE 1.19

Consider

$$e^x \sin(y) - 2x + \left(e^x \cos(y) + 1\right) y' = 0.$$

With $M(x, y) = e^x \sin(y) - 2x$ and $N(x, y) = e^x \cos(y) + 1$, we have

$$\frac{\partial N}{\partial x} = e^x \cos(y) = \frac{\partial M}{\partial y}$$

for all (x, y). Therefore, this differential equation is exact. To find a potential function, set

$$\frac{\partial \varphi}{\partial x} = e^x \sin(y) - 2x \quad \text{and} \quad \frac{\partial \varphi}{\partial y} = e^x \cos(y) + 1.$$

Choose one of these equations and integrate it. Integrate the second equation with respect to y:

$$\varphi(x, y) = \int \left(e^x \cos(y) + 1\right) dy$$

$$= e^x \sin(y) + y + h(x).$$

Then we must have

$$\frac{\partial \varphi}{\partial x} = e^x \sin(y) - 2x$$

$$= \frac{\partial}{\partial x} (e^x \sin(y) + y + h(x)) = e^x \sin(y) + h'(x).$$

Then $h'(x) = -2x$ and we may choose $h(x) = -x^2$. A potential function is

$$\varphi(x, y) = e^x \sin(y) + y - x^2.$$

The general solution of the differential equation is defined implicitly by

$$e^x \sin(y) + y - x^2 = C. \quad ∎$$

Note of Caution: If φ is a potential function for $M + Ny' = 0$, φ itself is not the solution. The general solution is defined implicitly by the equation $\varphi(x, y) = C$.

SECTION 1.4   PROBLEMS

In each of Problems 1 through 12, determine where (if anywhere) in the plane the differential equation is exact. If it is exact, find a potential function and the general solution, perhaps implicitly defined. If the equation is not exact, do not attempt a solution at this time.

1. $2y^2 + ye^{xy} + (4xy + xe^{xy} + 2y)y' = 0$

2. $4xy + 2x + (2x^2 + 3y^2)y' = 0$

3. $4xy + 2x^2y + (2x^2 + 3y^2)y' = 0$

4. $2\cos(x + y) - 2x \sin(x + y) - 2x \sin(x + y)y' = 0$

5. $\dfrac{1}{x} + y + (3y^2 + x)y' = 0$

6. $e^x \sin(y^2) + xe^x \sin(y^2) + (2xye^x \sin(y^2) + e^y)y' = 0$

7. $\sinh(x)\sinh(y) + \cosh(x)\cosh(y)y' = 0$

8. $4y^4 + 3\cos(x) + (16y^3x - 3\cos(y))y' = 0$

9. $yx^{y-1} + x^y \ln(x)y' = 0$

10. $\dfrac{2x^3 + 2xy^2 - y}{x^2 + y^2} + \dfrac{x}{x^2 + y^2}y' = 0$

11. $y + e^x + xy' = 0$

12. $6x - ye^{xy} + (e^y - xe^{xy})y' = 0$

In each of Problems 13 through 20, determine if the differential equation is exact in some rectangle containing in its interior the point where the initial condition is given. If so, solve the initial value problem. This solution may be implicitly defined. If the differential equation is not exact, do not attempt a solution.

13. $3y^4 - 1 + 12xy^3y' = 0$; $y(1) = 2$

14. $2y - y^2 \sec^2(xy^2) + (2x - 2xy \sec^2(xy^2))y' = 0$; $y(1) = 2$

15. $x\cos(2y - x) - \sin(2y - x) - 2x\cos(2y - x)y' = 0$; $y(\pi/12) = \pi/8$

16. $1 + e^{y/x} - \dfrac{y}{x}e^{y/x} + e^{y/x}y' = 0$; $y(1) = -5$

17. $y\sinh(y - x) - \cosh(y - x) + y\sinh(y - x)y' = 0$; $y(4) = 4$

18. $e^y + (xe^y - 1)y' = 0$; $y(5) = 0$

19. $2x - y\sin(xy) + (3y^2 - x\sin(xy))y' = 0$; $y(0) = 2$

20. $\cosh(x - y) + x\sinh(x - y) - x\sinh(x - y)y' = 0$; $y(4) = 4$

In Problems 21 and 22, choose a constant α so that the differential equation is exact, then produce a potential function and obtain the general solution.

21. $2xy^3 - 3y - (3x + \alpha x^2y^2 - 2\alpha y)y' = 0$

22. $3x^2 + xy^\alpha - x^2y^{\alpha-1}y' = 0$

23. Let φ be a potential function for $M + Ny' = 0$ in some region R of the plane. Show that for any constant c, $\varphi + c$ is also a potential function. How does the general solution of $M + Ny' = 0$ obtained by using φ differ from that obtained using $\varphi + c$?

1.5 Integrating Factors

"Most" differential equations are not exact on any rectangle. But sometimes we can multiply the differential equation by a nonzero function $\mu(x, y)$ to obtain an exact equation. Here is an example that suggests why this might be useful.

EXAMPLE 1.20

The equation

$$y^2 - 6xy + (3xy - 6x^2)y' = 0 \tag{1.11}$$

is not exact on any rectangle. Multiply it by $\mu(x, y) = y$ to get

$$y^3 - 6xy^2 + (3xy^2 - 6x^2y)y' = 0. \tag{1.12}$$

Wherever $y \neq 0$, equations (1.11) and (1.12) have the same solution. The reason for this is that equation (1.12) is just

$$y\left[y^2 - 6xy + (3xy - 6x^2)y'\right] = 0,$$

and if $y \neq 0$, then necessarily $y^2 - 6xy + (3xy - 6x^2)y' = 0$.

Now notice that equation (1.12) is exact (over the entire plane), having potential function

$$\varphi(x, y) = xy^3 - 3x^2y^2.$$

Thus, the general solution of equation (1.12) is defined implicitly by

$$xy^3 - 3x^2y^2 = C,$$

and, wherever $y \neq 0$, this defines the general solution of equation (1.11) as well. ∎

To review what has just occurred, we began with a nonexact differential equation. We multiplied it by a function μ chosen so that the new equation was exact. We solved this exact equation, then found that this solution also worked for the original, nonexact equation. The function μ therefore enabled us to solve a nonexact equation by solving an exact one. This idea is worth pursuing, and we begin by giving a name to μ.

DEFINITION 1.5

Let $M(x, y)$ and $N(x, y)$ be defined on a region R of the plane. Then $\mu(x, y)$ is an integrating factor for $M + Ny' = 0$ if $\mu(x, y) \neq 0$ for all (x, y) in R, and $\mu M + \mu Ny' = 0$ is exact on R.

How do we find an integrating factor for $M + Ny' = 0$? For μ to be an integrating factor, $\mu M + \mu Ny' = 0$ must be exact (in some region of the plane), hence

$$\frac{\partial}{\partial x}(\mu N) = \frac{\partial}{\partial y}(\mu M) \tag{1.13}$$

in this region. This is a starting point. Depending on M and N, we may be able to determine μ from this equation.

Sometimes equation (1.13) becomes simple enough to solve if we try μ as a function of just x or just y.

EXAMPLE 1.21

The differential equation $x - xy - y' = 0$ is not exact. Here $M = x - xy$ and $N = -1$ and equation (1.13) is

$$\frac{\partial}{\partial x}(-\mu) = \frac{\partial}{\partial y}(\mu(x - xy)).$$

Write this as

$$-\frac{\partial \mu}{\partial x} = (x - xy)\frac{\partial \mu}{\partial y} - x\mu.$$

Now observe that this equation is simplified if we try to find μ as just a function of x, because in this event $\partial \mu / \partial y = 0$ and we are left with just

$$\frac{\partial \mu}{\partial x} = x\mu.$$

This is separable. Write

$$\frac{1}{\mu}\,d\mu = x\,dx$$

and integrate to obtain

$$\ln|\mu| = \tfrac{1}{2}x^2.$$

Here we let the constant of integration be zero because we need only one integrating factor. From the last equation, choose

$$\mu(x) = e^{x^2/2},$$

a nonzero function. Multiply the original differential equation by $e^{x^2/2}$ to obtain

$$(x - xy)e^{x^2/2} - e^{x^2/2}y' = 0.$$

This equation is exact over the entire plane, and we find the potential function $\varphi(x, y) = (1 - y)e^{x^2/2}$. The general solution of this exact equation is implicitly defined by

$$(1 - y)e^{x^2/2} = C.$$

In this case, we can explicitly solve for y to get

$$y(x) = 1 - Ce^{-x^2/2},$$

and this is also the general solution of the original equation $x - xy - y' = 0$. ∎

If we cannot find an integrating factor that is a function of just x or just y, then we must try something else. There is no template to follow, and often we must start with equation (1.13) and be observant.

EXAMPLE 1.22

Consider $2y^2 - 9xy + (3xy - 6x^2)y' = 0$. This is not exact. With $M = 2y^2 - 9xy$ and $N = 3xy - 6x^2$, begin looking for an integrating factor by writing equation (1.13):

$$\frac{\partial}{\partial x}\left[\mu(3xy - 6x^2)\right] = \frac{\partial}{\partial y}\left[\mu(2y^2 - 9xy)\right].$$

This is

$$(3xy - 6x^2)\frac{\partial\mu}{\partial x} + \mu(3y - 12x) = (2y^2 - 9xy)\frac{\partial\mu}{\partial y} + \mu(4y - 9x). \tag{1.14}$$

If we attempt $\mu = \mu(x)$, then $\partial\mu/\partial y = 0$ and we obtain

$$(3xy - 6x^2)\frac{\partial\mu}{\partial x} + \mu(3y - 12x) = \mu(4y - 9x)$$

which cannot be solved for μ as just a function of x. Similarly, if we try $\mu = \mu(y)$, so $\partial\mu/\partial x = 0$, we obtain an equation we cannot solve. We must try something else. Notice that equation (1.14) involves only integer powers of x and y. This suggests that we try $\mu(x, y) = x^a y^b$. Substitute this into equation (1.14) and attempt to choose a and b. The substitution gives us

$$3ax^a y^{b+1} - 6ax^{a+1}y^b + 3x^a y^{b+1} - 12x^{a+1}y^b = 2bx^a y^{b+1} - 9bx^{a+1}y^b + 4x^a y^{b+1} - 9x^{a+1}y^b.$$

Assume that $x \neq 0$ and $y \neq 0$. Then we can divide by $x^a y^b$ to get

$$3ay - 6ax + 3y - 12x = 2by - 9bx + 4y - 9x.$$

Rearrange terms to write

$$(1 + 2b - 3a)y = (-3 + 9b - 6a)x.$$

Since x and y are independent, this equation can hold for all x and y only if

$$1 + 2b - 3a = 0 \quad \text{and} \quad -3 + 9b - 6a = 0.$$

Solve these equations to obtain $a = b = 1$. An integrating factor is $\mu(x, y) = xy$. Multiply the differential equation by xy to get

$$2xy^3 - 9x^2 y^2 + (3x^2 y^2 - 6x^3 y)y' = 0.$$

This is exact with potential function $\varphi(x, y) = x^2y^3 - 3x^3y^2$. For $x \neq 0$ and $y \neq 0$, the solution of the original differential equation is given implicitly by

$$x^2y^3 - 3x^3y^2 = C. \quad \blacksquare$$

The manipulations used to find an integrating factor may fail to find some solutions, as we saw with singular solutions of separable equations. Here are two examples in which this occurs.

EXAMPLE 1.23

Consider

$$\frac{2xy}{y-1} - y' = 0. \tag{1.15}$$

We can solve this as a separable equation, but here we want to make a point about integrating factors. Equation (1.15) is not exact, but $\mu(x, y) = (y - 1)/y$ is an integrating factor for $y \neq 0$, a condition not required by the differential equation itself. Multiplying the differential equation by $\mu(x, y)$ yields the exact equation

$$2x - \frac{y-1}{y}y' = 0,$$

with potential function $\varphi(x, y) = x^2 - y + \ln|y|$ and general solution defined by

$$x^2 - y + \ln|y| = C \quad \text{for } y \neq 0.$$

This is also the general solution of equation (1.15), but the method used has required that $y \neq 0$. However, we see immediately that $y = 0$ is also a solution of equation (1.15). This singular solution is not contained in the expression for the general solution for any choice of C. $\blacksquare$

EXAMPLE 1.24

The equation

$$y - 3 - xy' = 0 \tag{1.16}$$

is not exact, but $\mu(x, y) = 1/x(y - 3)$ is an integrating factor for $x \neq 0$ and $y \neq 3$, conditions not required by the differential equation itself. Multiplying equation (1.16) by $\mu(x, y)$ yields the exact equation

$$\frac{1}{x} - \frac{1}{y-3}y' = 0,$$

with general solution defined by

$$\ln|x| + C = \ln|y - 3|.$$

This is also the general solution of equation (1.16) in any region of the plane not containing the lines $x = 0$ or $y = 3$.

This general solution can be solved for y explicitly in terms of x. First, any real number is the natural logarithm of some positive number, so write the arbitrary constant as $C = \ln(k)$, in which k can be any positive number. The equation for the general solution becomes

$$\ln|x| + \ln(k) = \ln|y - 3|,$$

or

$$\ln|kx| = \ln|y - 3|.$$

But then $y - 3 = \pm kx$. Replacing $\pm k$ with K, which can now be any nonzero real number, we obtain

$$y = 3 + Kx$$

as the general solution of equation (1.16). Now observe that $y = 3$ is a solution of equation (1.16). This solution was "lost," or at least not found, in using the integrating factor as a method of solution. However, $y = 3$ is not a singular solution because we can include it in the expression $y = 3 + Kx$ by allowing $K = 0$. Thus the general solution of equation (1.16) is $y = 3 + Kx$, with K any real number. ∎

1.5.1 Separable Equations and Integrating Factors

We will point out a connection between separable equations and integrating factors.

The separable equation $y' = A(x)B(y)$ is in general not exact. To see this, write it as

$$A(x)B(y) - y' = 0,$$

so in the present context we have $M(x, y) = A(x)B(y)$ and $N(x, y) = -1$. Now

$$\frac{\partial}{\partial x}(-1) = 0 \quad \text{and} \quad \frac{\partial}{\partial y}[A(x)B(y)] = A(x)B'(y),$$

and in general $A(x)B'(y) \neq 0$.

However, $\mu(y) = 1/B(y)$ is an integrating factor for the separable equation. If we multiply the differential equation by $1/B(y)$, we get

$$A(x) - \frac{1}{B(y)}y' = 0,$$

an exact equation because

$$\frac{\partial}{\partial x}\left[-\frac{1}{B(y)}\right] = \frac{\partial}{\partial y}[A(x)] = 0.$$

The act of separating the variables is the same as multiplying by the integrating factor $1/B(y)$.

1.5.2 Linear Equations and Integrating Factors

Consider the linear equation $y' + p(x)y = q(x)$. We can write this as $[p(x)y - q(x)] + y' = 0$, so in the present context, $M(x, y) = p(x)y - q(x)$ and $N(x, y) = 1$. Now

$$\frac{\partial}{\partial x}[1] = 0 \quad \text{and} \quad \frac{\partial}{\partial y}[p(x)y - q(x)] = p(x),$$

so the linear equation is not exact unless $p(x)$ is identically zero. However, $\mu(x, y) = e^{\int p(x)\,dx}$ is an integrating factor. Upon multiplying the linear equation by μ, we get

$$[p(x)y - q(x)]e^{\int p(x)\,dx} + e^{\int p(x)\,dx}y' = 0,$$

and this is exact because

$$\frac{\partial}{\partial x}e^{\int p(x)\,dx} = p(x)e^{\int p(x)\,dx} = \frac{\partial}{\partial y}\left[[p(x)y - q(x)]e^{\int p(x)\,dx}\right].$$

SECTION 1.5   PROBLEMS

1. Determine a test involving M and N to tell when $M + Ny' = 0$ has an integrating factor that is a function of y only.

2. Determine a test to determine when $M + Ny' = 0$ has an integrating factor of the form $\mu(x, y) = x^a y^b$ for some constants a and b.

3. Consider $y - xy' = 0$.

   (a) Show that this equation is not exact on any rectangle.

   (b) Find an integrating factor $\mu(x)$ that is a function of x alone.

   (c) Find an integrating factor $\nu(y)$ that is a function of y alone.

   (d) Show that there is also an integrating factor $\eta(x, y) = x^a y^b$ for some constants a and b. Find all such integrating factors.

In each of Problems 4 through 14, (a) show that the differential equation is not exact, (b) find an integrating factor, (c) find the general solution (perhaps implicitly defined), and (d) determine any singular solutions the differential equation might have.

4. $xy' - 3y = 2x^3$

5. $1 + (3x - e^{-2y})y' = 0$

6. $6x^2 y + 12xy + y^2 + (6x^2 + 2y)y' = 0$

7. $4xy + 6y^2 + (2x^2 + 6xy)y' = 0$

8. $y^2 + y - xy' = 0$

9. $2xy^2 + 2xy + (x^2 y + x^2)y' = 0$

10. $6xy + 2y + 8 + xy' = 0$

11. $2x - 2y - x^2 + 2xy + (2x^2 - 4xy - 2x)y' = 0$

    (*Hint:* Try $\mu(x, y) = e^{ax} e^{by}$.)

12. $2y^2 - 9xy + (3xy - 6x^2)y' = 0$

    (*Hint:* Try $\mu(x, y) = x^a y^b$.)

13. $y' + y = y^4$

    (*Hint:* Try $\mu(x, y) = e^{ax} y^b$.)

14. $x^2 y' + xy = -y^{-3/2}$

    (*Hint:* Try $\mu(x, y) = x^a y^b$.)

In each of Problems 15 through 22, find an integrating factor, use it to find the general solution of the differential equation, and then obtain the solution of the initial value problem.

15. $1 + xy' = 0;\ y(e^4) = 0$

16. $3y + 4xy' = 0;\ y(1) = 6$

17. $2(y^3 - 2) + 3xy^2 y' = 0;\ y(3) = 1$

18. $y(1 + x) + 2xy' = 0;\ y(4) = 6$

19. $2xy + 3y' = 0;\ y(0) = 4$

    (*Hint:* Try $\mu = y^a e^{bx^2}$.)

20. $2y(1 + x^2) + xy' = 0;\ y(2) = 3$

    (*Hint:* Try $\mu = x^a e^{bx^2}$.)

21. $\sin(x - y) + \cos(x - y) - \cos(x - y)y' = 0;$

    $y(0) = 7\pi/6$

22. $3x^2 y + y^3 + 2xy^2 y' = 0;\ y(2) = 1$

23. Show that any nonzero constant multiple of an integrating factor for $M + Ny' = 0$ is also an integrating factor.

24. Let $\mu(x, y)$ be an integrating factor for $M + Ny' = 0$ and suppose that the general solution is defined by $\varphi(x, y) = C$. Show that $\mu(x, y)G(\varphi(x, y))$ is also an integrating factor, for any differentiable function G of one variable.

25. Show that if $M + Ny' = 0$ has a solution, then it has an integrating factor. *Hint:* Suppose the general solution is implicitly defined by $\varphi(x, y) = C$. Show that

$$\frac{\partial \varphi / \partial x}{\partial \varphi / \partial y} = \frac{M(x, y)}{N(x, y)}$$

for all (x, y) such that $N(x, y) \neq 0$. Let $\mu(x, y) = (\partial \varphi / \partial x)/M(x, y)$.

26. Suppose that $\mu(x, y)$ and $\nu(x, y)$ are integrating factors for $M + Ny' = 0$ and neither is a constant multiple of the other. Show that the equation

$$\frac{\mu(x, y)}{\nu(x, y)} = C,$$

with C an arbitrary constant, implicitly defines the general solution of $M + Ny' = 0$. Explain why this fails to be true if one of μ or ν is a constant multiple of the other.

27. For $y - xy' = 0$, produce two integrating factors μ and ν that are not constant multiples of each other. Verify that $\mu(x, y)/\nu(x, y) = C$ defines the general solution.

1.6 Homogeneous, Bernoulli, and Riccati Equations

In this section we will consider three additional kinds of first-order differential equations for which techniques for finding solutions are available.

1.6.1 Homogeneous Differential Equations

DEFINITION 1.6 *Homogeneous Equation*

A first-order differential equation is homogeneous if it has the form

$$y' = f\left(\frac{y}{x}\right).$$

In a homogeneous equation, y' is isolated on one side, and the other side is some expression in which y and x must always appear in the combination y/x. For example,

$$y' = \frac{x}{y}\,\sin\left(\frac{y}{x}\right)$$

is homogeneous, while $y' = x^2 y$ is not.

Sometimes algebraic manipulation will put a first-order equation into the form of the homogeneous equation. For example,

$$y' = \frac{y}{x+y} \tag{1.17}$$

is not homogeneous. However, if $x \neq 0$, we can write this as

$$y' = \frac{y/x}{1+y/x}, \tag{1.18}$$

a homogeneous equation. Any technique we develop for homogeneous equations can therefore be used on equation (1.18). However, this solution assumes that $x \neq 0$, which is not required in equation (1.17). Thus, as we have seen before, when we perform manipulations on a differential equation, we must be careful that solutions have not been overlooked. A solution of equation (1.18) will also satisfy (1.17), but equation (1.17) may have other solutions as well.

Now to the point. A homogeneous equation is always transformed into a separable one by the transformation

$$y = ux.$$

To see this, compute $y' = u'x + x'u = u'x + u$ and write $u = y/x$. Then $y' = f(y/x)$ becomes

$$u'x + u = f(u).$$

We can write this as

$$\frac{1}{f(u) - u}\frac{du}{dx} = \frac{1}{x},$$

or, in differential form,

$$\frac{1}{f(u) - u} \, du = \frac{1}{x} \, dx,$$

and the variables (now x and u) have been separated. Upon integrating this equation, we obtain the general solution of the transformed equation. Substituting $u = y/x$ then gives the general solution of the original homogeneous equation.

EXAMPLE 1.25

Consider

$$xy' = \frac{y^2}{x} + y.$$

Write this as

$$y' = \left(\frac{y}{x}\right)^2 + \frac{y}{x}.$$

Let $y = ux$. Then

$$u'x + u = u^2 + u,$$

or

$$u'x = u^2.$$

Write this as

$$\frac{1}{u^2} \, du = \frac{1}{x} \, dx$$

and integrate to obtain

$$-\frac{1}{u} = \ln|x| + C.$$

Then

$$u(x) = \frac{-1}{\ln|x| + C},$$

the general solution of the transformed equation. The general solution of the original equation is

$$y = \frac{-x}{\ln|x| + C}. \quad \blacksquare$$

EXAMPLE 1.26   A Pursuit Problem

A pursuit problem is one of determining a trajectory so that one object intercepts another. Examples involving pursuit problems are missiles fired at airplanes and a rendezvous of a shuttle with a space station. These are complex problems that require numerical approximation techniques. We will consider a simple pursuit problem that can be solved explicitly.

Suppose a person jumps into a canal of constant width w and swims toward a fixed point directly opposite the point of entry into the canal. The person's speed is v and the water current's

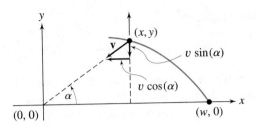

FIGURE 1.14 *The swimmer's path.*

speed is s. Assume that on the way across, the swimmer always orients to point toward the target. We want to determine the swimmer's trajectory.

Figure 1.14 shows a coordinate system drawn so that the swimmer's destination is the origin and the point of entry into the water is $(w, 0)$. At time t the swimmer is at the point $(x(t), y(t))$. The horizontal and vertical components of the swimmer's velocity are, respectively,

$$x'(t) = -v \cos(\alpha) \quad \text{and} \quad y'(t) = s - v \sin(\alpha),$$

with α the angle between the positive x axis and $(x(t), y(t))$ at time t. From these equations,

$$\frac{dy}{dx} = \frac{y'(t)}{x'(t)} = \frac{s - v \sin(\alpha)}{-v \cos(\alpha)} = \tan(\alpha) - \frac{s}{v} \sec(\alpha).$$

From Figure 1.14,

$$\tan(\alpha) = \frac{y}{x} \quad \text{and} \quad \sec(\alpha) = \frac{1}{x} \sqrt{x^2 + y^2}.$$

Therefore,

$$\frac{dy}{dx} = \frac{y}{x} - \frac{s}{v} \frac{1}{x} \sqrt{x^2 + y^2}.$$

Write this as the homogeneous equation

$$\frac{dy}{dx} = \frac{y}{x} - \frac{s}{v} \sqrt{1 + \left(\frac{y}{x}\right)^2}$$

and put $y = ux$ to obtain

$$\frac{1}{\sqrt{1 + u^2}} \, du = -\frac{s}{v} \frac{1}{x} \, dx.$$

Integrate to get

$$\ln \left| u + \sqrt{1 + u^2} \right| = -\frac{s}{v} \ln|x| + C.$$

Take the exponential of both sides of this equation:

$$\left| u + \sqrt{1 + u^2} \right| = e^C e^{-(s \ln|x|)/v}.$$

We can write this as

$$u + \sqrt{1 + u^2} = K x^{-s/v}.$$

This equation can be solved for u. First write

$$\sqrt{1 + u^2} = K x^{-s/v} - u$$

and square both sides to get

$$1 + u^2 = K^2 x^{-2s/v} - 2Kux^{-s/v} + u^2.$$

Now u^2 cancels and we can solve for u:

$$u(x) = \frac{1}{2}Kx^{-s/v} - \frac{1}{2}\frac{1}{K}x^{s/v}.$$

Finally, put $u = y/x$ to get

$$y(x) = \frac{1}{2}Kx^{1-s/v} - \frac{1}{2}\frac{1}{K}x^{1+s/v}.$$

To determine K, notice that $y(w) = 0$, since we put the origin at the point of destination. Thus,

$$\frac{1}{2}Kw^{1-s/v} - \frac{1}{2}\frac{1}{K}w^{1+s/v} = 0$$

and we obtain

$$K = w^{s/v}.$$

Therefore,

$$y(x) = \frac{w}{2}\left[\left(\frac{x}{w}\right)^{1-s/v} - \left(\frac{x}{w}\right)^{1+s/v}\right].$$

As might be expected, the path the swimmer takes depends on the width of the canal, the speed of the swimmer, and the speed of the current. Figure 1.15 shows trajectories corresponding to s/v equal to $\frac{1}{5}, \frac{1}{3}, \frac{1}{2}$ and $\frac{3}{4}$, with $w = 1$. ■

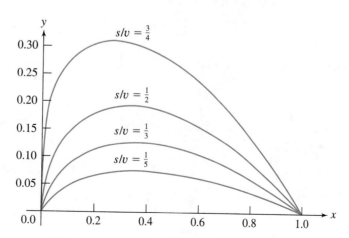

FIGURE 1.15 *Graphs of*

$$y = \frac{w}{2}\left[\left(\frac{x}{w}\right)^{1-s/v} - \left(\frac{x}{w}\right)^{1+s/v}\right]$$

for s/v equal to $\frac{1}{5}, \frac{1}{3}, \frac{1}{2}$ and $\frac{3}{4}$, and w chosen as 1.

1.6.2 The Bernoulli Equation

DEFINITION 1.7

A *Bernoulli equation* is a first-order equation,

$$y' + P(x)y = R(x)y^\alpha,$$

in which α is a real number.

A Bernoulli equation is separable if $\alpha = 0$ and linear if $\alpha = 1$. About 1696, Leibniz showed that a Bernoulli equation with $\alpha \neq 1$ transforms to a linear equation under the change of variables:

$$v = y^{1-\alpha}.$$

This is routine to verify. Here is an example.

EXAMPLE 1.27

Consider the equation

$$y' + \frac{1}{x}y = 3x^2y^3,$$

which is Bernoulli with $P(x) = 1/x$, $R(x) = 3x^2$, and $\alpha = 3$. Make the change of variables

$$v = y^{-2}.$$

Then $y = v^{-1/2}$ and

$$y'(x) = -\frac{1}{2}v^{-3/2}v'(x),$$

so the differential equation becomes

$$-\frac{1}{2}v^{-3/2}v'(x) + \frac{1}{x}v^{-1/2} = 3x^2v^{-3/2},$$

or, upon multiplying by $-2v^{3/2}$,

$$v' - \frac{2}{x}v = -6x^2,$$

a linear equation. An integrating factor is $e^{-\int (2/x)\,dx} = x^{-2}$. Multiply the last equation by this factor to get

$$x^{-2}v' - 2x^{-3}v = -6,$$

which is

$$(x^{-2}v)' = -6.$$

Integrate to get

$$x^{-2}v = -6x + C,$$

so

$$v = -6x^3 + Cx^2.$$

The general solution of the Bernoulli equation is

$$y(x) = \frac{1}{\sqrt{v(x)}} = \frac{1}{\sqrt{Cx^2 - 6x^3}}. \quad \blacksquare$$

1.6.3   The Riccati Equation

DEFINITION 1.8
_____

A differential equation of the form

$$y' = P(x)y^2 + Q(x)y + R(x)$$

is called a *Riccati equation*.

A Riccati equation is linear exactly when $P(x)$ is identically zero. If we can somehow obtain one solution, $S(x)$, of a Riccati equation, then the change of variables

$$y = S(x) + \frac{1}{z}$$

transforms the Riccati equation to a linear equation. The strategy is to find the general solution of this linear equation and from it produce the general solution of the original Riccati equation.

EXAMPLE 1.28
_____

Consider the Riccati equation

$$y' = \frac{1}{x}y^2 + \frac{1}{x}y - \frac{2}{x}.$$

By inspection, $y = S(x) = 1$ is one solution. Define a new variable z by putting

$$y = 1 + \frac{1}{z}.$$

Then

$$y' = -\frac{1}{z^2}z'.$$

Substitute these into the Riccati equation to get

$$-\frac{1}{z^2}z' = \frac{1}{x}\left(1 + \frac{1}{z}\right)^2 + \frac{1}{x}\left(1 + \frac{1}{z}\right) - \frac{2}{x},$$

or

$$z' + \frac{3}{x}z = -\frac{1}{x}.$$

This is linear. An integrating factor is $e^{\int (3/x)\,dx} = x^3$. Multiply by x^3 to get

$$x^3 z' + 3x^2 z = (x^3 z)' = -x^2.$$

Integrate to get

$$x^3 z = -\frac{1}{3}x^3 + C,$$

so

$$z(x) = -\frac{1}{3} + \frac{C}{x^3}.$$

The general solution of the Riccati equation is

$$y(x) = 1 + \frac{1}{z(x)} = 1 + \frac{1}{-1/3 + C/x^3}.$$

This solution can also be written

$$y(x) = \frac{K + 2x^3}{K - x^3},$$

in which $K = 3C$ is an arbitrary constant. ∎

SECTION 1.6   PROBLEMS

In each of Problems 1 through 14, find the general solution. These problems include all types considered in this section.

1. $y' = \frac{1}{x^2}y^2 - \frac{1}{x}y + 1$

2. $y' + \frac{1}{x}y = \frac{2}{x^3}y^{-4/3}$

3. $y' + xy = xy^2$

4. $y' = \frac{x}{y} + \frac{y}{x}$

5. $y' = \frac{y}{x+y}$

6. $y' = \frac{1}{2x}y^2 - \frac{1}{x}y - \frac{4}{x}$

7. $(x - 2y)y' = 2x - y$

8. $xy' = x\cos(y/x) + y$

9. $y' + \frac{1}{x}y = \frac{1}{x^4}y^{-3/4}$

10. $x^2 y' = x^2 + y^2$

11. $y' = -\frac{1}{x}y^2 + \frac{2}{x}y$

12. $x^3 y' = x^2 y - y^3$

13. $y' = -e^{-x}y^2 + y + e^x$

14. $y' + \frac{2}{x}y = \frac{3}{x}y^2$

15. Consider the differential equation

$$y' = F\left(\frac{ax + by + c}{dx + ey + r}\right),$$

in which $a, b, c, d, e,$ and r are constants and F is a differentiable function of one variable.

(a) Show that this equation is homogeneous if and only if $c = r = 0$.

(b) If c and/or r is not zero, this equation is called *nearly homogeneous*. Assuming that $ae - bd \neq 0$, show that it is possible to choose constants h and k so that the transformation $X = x + h, Y = y + k$ converts this nearly homogeneous equation into a homogeneous one. *Hint:* Put $x = X - h, y = Y - k$ into the differential equation and obtain a differential equation in X and Y. Use the conclusion of (a) to choose h and k so that this equation is homogeneous.

In each of Problems 16 through 19, use the idea of Problem 15 to find the general solution.

16. $y' = \frac{y - 3}{x + y - 1}$

17. $y' = \frac{3x - y - 9}{x + y + 1}$

18. $y' = \dfrac{x + 2y + 7}{-2x + y - 9}$

19. $y' = \dfrac{2x - 5y - 9}{-4x + y + 9}$

20. Continuing from Problem 15, consider the case that $ae - bd = 0$. Now let $u = (ax + by)/a$, assuming that $a \neq 0$. Show that this transforms the differential equation of Problem 15 into the separable equation

$$\frac{du}{dx} = 1 + \frac{b}{a} F\left(\frac{au + c}{du + r}\right).$$

In each of Problems 21 through 24, use the method of Problem 20 to find the general solution.

21. $y' = \dfrac{x - y + 2}{x - y + 3}$

22. $y' = \dfrac{3x + y - 1}{6x + 2y - 3}$

23. $y' = \dfrac{x - 2y}{3x - 6y + 4}$

24. $y' = \dfrac{x - y + 6}{3x - 3y + 4}$

25. (*The Pursuing Dog*) A man stands at the junction of two perpendicular roads and his dog is watching him from one of the roads at a distance A feet away. At a given instant the man starts to walk with constant speed v along the other road, and at the same time the dog begins to run toward the man with speed $2v$. Determine the path the dog will take, assuming that it always moves so that it is facing the man. Also determine when the dog will eventually catch the man. (This is *American Mathematical Monthly* problem 3942, 1941).

26. (*Pursuing Bugs*) One bug is located at each corner of a square table of side length a. At a given time they begin moving at constant speed v, each pursuing its neighbor to the right.

(a) Determine the curve of pursuit of each bug. *Hint:* Use polar coordinates with the origin at the center of the table and the polar axis containing one of the corners. When a bug is at $(f(\theta), \theta)$, its target is at $(f(\theta), \theta + \pi/2)$. Use the chain rule to write

$$\frac{dy}{dx} = \frac{dy/d\theta}{dx/d\theta},$$

where $y(\theta) = f(\theta) \sin(\theta)$ and $x(\theta) = f(\theta) \cos(\theta)$.

(b) Determine the distance traveled by each bug.

(c) Does any bug actually catch its quarry?

27. (*The Spinning Bug*) A bug steps onto the edge of a disk of radius a that is spinning at a constant angular

speed ω. The bug moves toward the center of the disk at constant speed v.

(a) Derive a differential equation for the path of the bug, using polar coordinates.

(b) How many revolutions will the disk make before the bug reaches the center? (The solution will be in terms of the angular speed and radius of the disk.)

(c) Referring to (b), what is the total distance the bug will travel, taking into account the motion of the disk?

28. (*The Tractrix*) Determine the path of a boat being pulled by a rope of length L by a man walking along a straight shore, beginning at a point on shore directly opposite the boat. The boat is initially located L units offshore. This curve is called a *tractrix*.

29. (*Hunting Submarines*) A destroyer is hunting a submarine. The sub surfaces at a distance 9 kilometers from the destroyer and is seen by a sailor. The sub immediately submerges and travels at a constant speed v km/hour in a straight line whose direction is unknown to the hunters. Suppose the destroyer captain knows that this is standard evasive action for submarines. Plot a course for the destroyer to guarantee that it will pass directly over the sub at some time, assuming that the destroyer moves at a constant speed $2v$. Will the solution of this problem help the destroyer captain sink the sub?

30. Suppose $M(x, y) + N(x, y)y' = 0$ is homogeneous. Show that

$$\mu(x, y) = \frac{1}{x M(x, y) + y N(x, y)}$$

is an integrating factor over any region in which the denominator does not vanish.

31. Suppose $M(x, y) + N(x, y)y' = 0$ is both homogeneous and exact and that $x M(x, y) + y N(x, y)$ is not constant. Show that the general solution of $M + Ny' = 0$ is defined implicitly by the equation $x M(x, y) + y N(x, y) = C$.

32. Derive an integrating factor for the Bernoulli equation. *Hint:* Multiply the Bernoulli equation by $\mu(x, y) = f(x)y^b$ and apply the test for exactness. Notice that this results in an equation that simplifies if b is chosen in a certain way. Following this choice, a separable differential equation for $f(x)$ results.

33. Show that the Riccati equation has an integrating factor

$$\mu(x, y) = \frac{1}{[y - S(x)]^2} e^{\int [2P(x)S(x) + Q(x)]\, dx}.$$

34. Let $S_1(x)$ and $S_2(x)$ be solutions of the Riccati equation and suppose neither solution is a constant multiple of the other. Let

$$\mu_1(x, y) = \frac{1}{[y - S_1(x)]^2} e^{\int [2P(x)S_1(x) + Q(x)]\, dx}$$

and

$$\mu_2(x, y) = \frac{1}{[y - S_2(x)]^2} e^{\int [2P(x)S_2(x) + Q(x)]\, dx}.$$

Show that the general solution of the Riccati equation is implicitly defined by

$$\frac{\mu_1(x, y)}{\mu_2(x, y)} = C.$$

35. Use the result of Problem 34 to show that the general solution of a Riccati equation always has the form

$$y(x) = \frac{F(x) + CG(x)}{H(x) + CJ(x)},$$

for some functions F, G, H, and J and arbitrary constant C.

36. Consider the special Riccati equation $y' = ay^2 + bx^\alpha$, in which a and b are constants. Show that the general solution can be obtained in closed form when α is either $-4n/(2n + 1)$ or $-4n/(2n - 1)$ for some nonnegative integer n. (Closed form means the solution is a finite sum of algebraic combinations of elementary functions). It can be shown that if α is not of this form, then the general solution cannot be written in closed form.

37. Find all functions f defined on intervals having 0 as the left endpoint and the average value of which on $[0, x]$ is the geometric mean of $f(0)$ and $f(x)$. This problem appeared in the William Lowell Putnam Mathematics Competition for 1962. (Recall that the geometric mean of a and b is $\sqrt{ab}$.)

1.7 Applications to Mechanics, Electrical Circuits, and Orthogonal Trajectories

1.7.1 Mechanics

Before applying first-order differential equations to problems in mechanics, we will review some background.

Newton's second law of motion states that the rate of change of momentum (mass times velocity) of a body is proportional to the resultant force acting on the body. This is a vector equation, but we will for now consider only motion along a straight line. In this case Newton's law is

$$F = k \frac{d}{dt}(mv).$$

We will take $k = 1$, consistent with certain units of measurement, such as the English, MKS, or gcs systems.

The mass of a moving object need not be constant. For example, an airplane consumes fuel as it moves. If m is constant, then Newton's law is

$$F = m \frac{dv}{dt} = ma,$$

in which a is the acceleration of the object along the line of motion. If m is not constant, then

$$F = m \frac{dv}{dt} + v \frac{dm}{dt}.$$

Newton's law of gravitational attraction states that if two objects have masses m_1 and m_2, and they (or their center of masses) are at distance r from each other, then each attracts the other with a gravitational force of magnitude

$$F = G \frac{m_1 m_2}{r^2}.$$

This force is directed along the line between the centers of mass. G is the universal gravitational constant.

If one of the objects is the earth, then

$$F = G \frac{mM}{(R+x)^2},$$

where M is the mass of the earth, R is its radius (about 3960 miles), m is the mass of the second object, and x is its distance from the surface of the earth. This assumes that the earth is spherical and that its center of mass is at the center of this sphere, a good-enough approximation for some purposes. If x is small compared to R, then $R+x$ is approximately R and the force on the object is approximately

$$\frac{GM}{R^2} m,$$

which is often written as mg. Here $g = GM/R^2$ is approximately 32 feet per second per second, or 9.8 meters per second per second.

We are now ready to analyze some problems in mechanics.

Terminal Velocity

Consider an object that is falling under the influence of gravity in a medium such as water, air, or oil. This medium retards the downward motion of the object. Think, for example, of a brick dropped in a swimming pool or a ball bearing dropped in a tank of oil. We want to analyze the object's motion.

Let $v(t)$ be the velocity at time t. The force of gravity pulls the object down and has magnitude mg. The medium retards the motion. The magnitude of this retarding force is not obvious, but experiment has shown that its magnitude is proportional to the square of the velocity. If we choose downward as the positive direction and upward as negative, then Newton's law tells us that, for some constant α,

$$F = mg - \alpha v^2 = m \frac{dv}{dt}.$$

If we assume that the object begins its motion from rest (dropped, not thrown) and if we start the clock at this instant, then $v(0) = 0$. We now have an initial value problem for the velocity:

$$v' = g - \frac{\alpha}{m} v^2; \qquad v(0) = 0.$$

This differential equation is separable. In differential form,

$$\frac{1}{g - (\alpha/m)v^2} \, dv = dt.$$

Integrate to get

$$\sqrt{\frac{m}{\alpha g}} \, \tanh^{-1}\left(\sqrt{\frac{\alpha}{mg}} v \right) = t + C.$$

Solve for the velocity, obtaining

$$v(t) = \sqrt{\frac{mg}{\alpha}} \, \tanh\left(\sqrt{\frac{\alpha g}{m}} (t + C) \right).$$

Now use the initial condition to solve for the integration constant:

$$v(0) = \sqrt{\frac{mg}{\alpha}} \, \tanh\left(C \sqrt{\frac{\alpha g}{m}} \right) = 0.$$

Since $\tanh(\xi) = 0$ only if $\xi = 0$, this requires that $C = 0$, and the solution for the velocity is

$$v(t) = \sqrt{\frac{mg}{\alpha}} \, \tanh\left(\sqrt{\frac{\alpha g}{m}} t\right).$$

Even in this generality, we can draw an important conclusion about the motion. As t increases, $\tanh(\sqrt{\alpha g / mt})$ approaches 1. This means that

$$\lim_{t \to \infty} v(t) = \sqrt{\frac{mg}{\alpha}}.$$

This means that an object falling under the influence of gravity, through a retarding medium (with force proportional to the square of the velocity), will not increase in velocity indefinitely. Instead, the object's velocity approaches the limiting value $\sqrt{mg/\alpha}$. If the medium is deep enough, the object will settle into a descent of approximately constant velocity. This number, $\sqrt{mg/\alpha}$, is called the *terminal velocity* of the object. Skydivers experience this phenomenon.

Motion of a Chain on a Pulley

A 16-foot-long chain weighing ρ pounds per foot hangs over a small pulley, which is 20 feet above the floor. Initially, the chain is held at rest, with 7 feet on one side and 9 on the other, as in Figure 1.16. How long after the chain is released, and with what velocity, will it leave the pulley?

When 8 feet of chain hang on each side of the pulley, the chain is in equilibrium. Call this position $x = 0$ and let $x(t)$ be the distance the chain has fallen below this point at time t. The net force acting on the chain is $2x\rho$ and the mass of the chain is $16\rho/32$, or $\rho/2$ slugs. The ends of the chain have the same speed as its center of mass, so the acceleration of the chain at its center of mass is the same as it is at its ends. The equation of motion is

$$\frac{\rho}{2} \frac{dv}{dt} = 2x\rho,$$

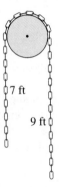

FIGURE 1.16
Chain on a pulley.

from which ρ cancels to yield

$$\frac{dv}{dt} = 4x.$$

A chain rule differentiation enables us to write this equation in terms of v as a function of x. Write

$$\frac{dv}{dt} = \frac{dv}{dx} \frac{dx}{dt} = v \frac{dv}{dx}.$$

Then

$$v \frac{dv}{dx} = 4x.$$

This is a separable equation, which we solve to get

$$v^2 = 4x^2 + K.$$

Now, $x = 1$ when $v = 0$, so $K = -4$ and

$$v^2 = 4x^2 - 4.$$

The chain leaves the pulley when $x = 8$. Whenever this occurs, $v^2 = 4(63) = 252$, so $v = \sqrt{252} = 6\sqrt{7}$ feet per second (about 15.87 feet per second).

To calculate the time t_f required for the chain to leave the pulley, compute

$$t_f = \int_0^{t_f} dt = \int_0^{6\sqrt{7}} \frac{dt}{dv} \, dv$$

$$= \int_1^8 \frac{dt}{dx} \, dx = \int_1^8 \frac{1}{v} \, dx.$$

Since $v(x) = 2\sqrt{x^2 - 1}$,

$$t_f = \frac{1}{2} \int_1^8 \frac{1}{\sqrt{x^2 - 1}} \, dx = \left[\frac{1}{2} \ln \left| x + \sqrt{x^2 - 1} \right| \right]_1^8$$

$$= \frac{1}{2} \ln(8 + \sqrt{63}),$$

about 1.38 seconds.

In this example the mass was constant, so $dm/dt = 0$ in Newton's law of motion. Next is an example in which the mass varies with time.

Chain Piling on the Floor

Suppose a 40-foot-long chain weighing ρ pounds per foot is supported in a pile several feet above the floor and begins to unwind when released from rest with 10 feet already played out. Determine the velocity with which the chain leaves the support.

The amount of chain that is actually in motion changes with time. Let $x(t)$ denote the length of that part of the chain that has left the support by time t and is currently in motion. The equation of motion is

$$m \frac{dv}{dt} + v \frac{dm}{dt} = F, \tag{1.19}$$

where F is the total external force acting on the chain. Now $F = x\rho = mg$, so $m = x\rho/g = x\rho/32$. Then

$$\frac{dm}{dt} = \frac{\rho}{32} \frac{dx}{dt} = \frac{\rho}{32} v.$$

Further,

$$\frac{dv}{dt} = v \frac{dv}{dx},$$

as in the preceding example. Put this information into equation (1.19) to get

$$\frac{x\rho}{32} v \frac{dv}{dx} + \frac{\rho}{32} v^2 = x\rho.$$

If we multiply this equation by $32/x\rho v$, we get

$$\frac{dv}{dx} + \frac{1}{x} v = \frac{32}{v}, \tag{1.20}$$

which we recognize as a Bernoulli equation with $\alpha = -1$. Make the transformation $w = v^{1-\alpha} = v^2$. Then $v = w^{1/2}$ and

$$\frac{dv}{dx} = \frac{1}{2} w^{-1/2} \frac{dw}{dx}.$$

Substitute these into equation (1.20) to get

$$\frac{1}{2} w^{-1/2} \frac{dw}{dx} + \frac{1}{x} w^{1/2} = 32 w^{-1/2}.$$

Upon multiplying this equation by $2w^{1/2}$, we get

$$w' + \frac{2}{x} w = 64,$$

a linear equation for $w(x)$. Solve this to get

$$w(x) = v(x)^2 = \frac{64}{3} x + \frac{C}{x^2}.$$

Since $v = 0$ when $x = 10, 0 = (64/3)(10) + C/100$, so $C = -64,000/3$. Therefore,

$$v(x)^2 = \frac{64}{3} \left[x - \frac{1000}{x^2} \right].$$

The chain leaves the support when $x = 40$. At this time,

$$v^2 = \frac{64}{3} \left[40 - \frac{1000}{1600} \right] = 4(210)$$

so at this time, the velocity is $v = 2\sqrt{210}$, or about 29 feet per second.

In these models involving chains, air resistance was neglected as having no significant impact on the outcome. This was quite different from the analysis of terminal velocity, in which air resistance is a key factor. Without it, skydivers dive only once!

Motion of a Block Sliding on an Inclined Plane

A block weighing 96 pounds is released from rest at the top of an inclined plane of slope length 50 feet and making an angle $\pi/6$ radians with the horizontal. Assume a coefficient of friction of $\mu = \sqrt{3}/4$. Assume also that air resistance acts to retard the block's descent down the ramp, with a force of magnitude equal to one-half the block's velocity. We want to determine the velocity $v(t)$ of the block at any time t.

Figure 1.17 shows the forces acting on the block. Gravity acts downward with magnitude $mg \sin(\theta)$, which is $96 \sin(\pi/6)$, or 48 pounds. Here $mg = 96$ is the weight of the block. The drag due to friction acts in the reverse direction and is, in pounds,

$$-\mu N = -\mu mg \cos(\theta) = -\frac{\sqrt{3}}{4}(96) \cos\left(\frac{\pi}{6}\right) = -36$$

The drag force due to air resistance is $-v/2$, the negative sign indicating that this is a retarding force. The total external force on the block is

$$F = 48 - 36 - \tfrac{1}{2}v = 12 - \tfrac{1}{2}v.$$

Since the block weighs 96 pounds, it has a mass of 96/32 slugs, or 3 slugs. From Newton's second law,

$$3 \frac{dv}{dt} = 12 - \frac{1}{2}v.$$

This is a linear equation, which we write as

$$v' + \tfrac{1}{6}v = 4.$$

An integrating factor is $e^{\int (1/6)\,dt} = e^{t/6}$. Multiply the differential equation by this factor to obtain

$$v'e^{t/6} + \tfrac{1}{6}e^{t/6}v = \left(ve^{t/6}\right)' = 4e^{t/6}$$

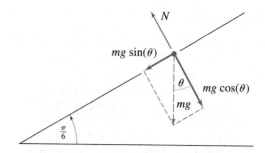

FIGURE 1.17 *Forces acting on a block on an inclined plane.*

and integrate to get

$$ve^{t/6} = 24e^{t/6} + C.$$

The velocity is

$$v(t) = 24 + Ce^{-t/6}.$$

Since the block starts from rest at time zero, $v(0) = 0 = 24 + C$, so $C = -24$ and

$$v(t) = 24\left(1 - e^{-t/6}\right).$$

Let $x(t)$ be the position of the block at any time, measured from the top of the plane. Since $v(t) = x'(t)$, we get

$$x(t) = \int v(t)\, dt = 24t + 144e^{-t/6} + K.$$

If we let the top of the block be the origin along the inclined plane, then $x(0) = 0 = 144 + K$, so

$$K = -144.$$

The position function is

$$x(t) = 24t + 144\left(e^{-t/6} - 1\right).$$

We can now determine the block's position and velocity at any time.

Suppose, for example, we want to know when the block reaches the bottom of the ramp. This happens when the block has gone 50 feet. If this occurs at time T, then

$$x(T) = 50 = 24T + 144\left(e^{-T/6} - 1\right).$$

This transcendental equation cannot be solved algebraically for T, but a computer approximation yields $T \approx 5.8$ seconds.

Notice that

$$\lim_{t \to \infty} v(t) = 24,$$

which means that the block sliding down the ramp has a terminal velocity. If the ramp is long enough, the block will eventually settle into a slide of approximately constant velocity.

The mathematical model we have constructed for the sliding block can be used to analyze the motion of the block under a variety of conditions. For example, we can solve the equations leaving θ arbitrary and determine the influence of the slope angle of the ramp on position and velocity. Or we could leave μ unspecified and study the influence of friction on the motion.

1.7.2   Electrical Circuits

Electrical engineers often use differential equations to model circuits. The mathematical model is used to analyze the behavior of circuits under various conditions and aids in the design of circuits having specific characteristics.

We will look at simple circuits having only resistors, inductors, and capacitors. A capacitor is a storage device consisting of two plates of conducting material isolated from one another by an insulating material, or dielectric. Electrons can be transferred from one plate to another via external circuitry by applying an electromotive force to the circuit. The charge on a capacitor is essentially a count of the difference between the numbers of electrons on the two plates. This charge is proportional to the applied electromotive force, and the constant of proportionality is the capacitance. Capacitance is usually a very small number, given in micro (10^{-6}) or pico (10^{-12}) farads. To simplify examples and problems, some of the capacitors in this book are assigned numerical values that would actually make them occupy large buildings.

An inductor is made by winding a conductor such as wire around a core of magnetic material. When a current is passed through the wire, a magnetic field is created in the core and around the inductor. The voltage drop across an inductor is proportional to the change in the current flow, and this constant of proportionality is the inductance of the inductor, measured in henrys.

Current is measured in amperes, with one amp equivalent to a rate of electron flow of one coulomb per second. Charge $q(t)$ and current $i(t)$ are related by

$$i(t) = q'(t).$$

The voltage drop across a resistor having resistance R is iR. The drop across a capacitor having capacitance C is q/C. And the voltage drop across an inductor having inductance L is $Li'(t)$.

We construct equations for a circuit by using Kirchhoff's current and voltage laws. Kirchhoff's current law states that the algebraic sum of the currents at any juncture of a circuit is zero. This means that the total current entering the junction must balance the current leaving (conservation of energy). Kirchhoff's voltage law states that the algebraic sum of the potential rises and drops around any closed loop in a circuit is zero.

As an example of modeling a circuit mathematically, consider the circuit of Figure 1.18. Starting at point A, move clockwise around the circuit, first crossing the battery, where there is an increase in potential of E volts. Next there is a decrease in potential of iR volts across the resistor. Finally, there is a decrease of $Li'(t)$ across the inductor, after which we return to point A. By Kirchhoff's voltage law,

$$E - iR - Li' = 0,$$

which is the linear equation

$$i' + \frac{E}{R}i = \frac{E}{L}.$$

Solve this to obtain

$$i(t) = \frac{E}{R} + Ke^{-Rt/L}.$$

To determine the constant K, we need to be given the current at some time. Even without this, we can tell from this equation that as $t \to \infty$, the current approaches the limiting value E/R. This is the steady-state value of the current in the circuit.

Another way to derive the differential equation of this circuit is to designate one of the components as a source, then set the voltage drop across that component equal to the sum of the voltage drops across the other components. To see this approach, consider the circuit of Figure 1.19. Suppose the switch is initially open so that no current flows and that the charge on the

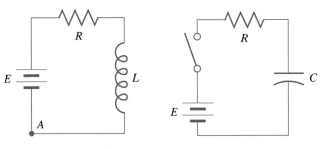

FIGURE 1.18 *RL Circuit.*   **FIGURE 1.19** *RC circuit.*

capacitor is zero. At time zero, close the switch. We want the charge on the capacitor. Notice that we have to close the switch before there is a loop. Using the battery as a source, write

$$iR + \frac{1}{C}q = E,$$

or

$$Rq' + \frac{1}{C}q = E.$$

This leads to the linear equation

$$q' + \frac{1}{RC}q = \frac{E}{R},$$

with solution

$$q(t) = EC\left(1 - e^{-t/RC}\right)$$

satisfying $q(0) = 0$. This equation provides a good deal of information about the circuit. Since the voltage on the capacitor at time t is $q(t)/C$, or $E(1 - e^{-t/RC})$, we can see that the voltage approaches E as $t \to \infty$. Since E is the battery potential, the difference between battery and capacitor voltages becomes negligible as time increases, indicating a very small voltage drop across the resistor.

The current in this circuit can be computed as

$$i(t) = q'(t) = \frac{E}{R}e^{-t/RC}$$

after the circuit is switched on. Thus, $i(t) \to E/R$ as $t \to 0$.

Often we encounter discontinuous currents and potential functions in dealing with circuits. These can be treated using Laplace transform techniques, which we will discuss in Chapter 3.

1.7.3 Orthogonal Trajectories

Two curves intersecting at a point P are said to be *orthogonal* if their tangents are perpendicular (orthogonal) at P. Two families of curves, or trajectories, are orthogonal if each curve of the first family is orthogonal to each curve of the second family wherever an intersection occurs. Orthogonal families occur in many contexts. Parallels and meridians on a globe are orthogonal, as are equipotential and electric lines of force.

A problem that occupied Newton and other early developers of the calculus was the determination of the family of orthogonal trajectories of a given family of curves. Suppose we are given a family $\mathfrak{F}$ of curves in the plane. We want to construct a second family $\mathfrak{G}$ of curves so that every curve in $\mathfrak{F}$ is orthogonal to every curve in $\mathfrak{G}$ wherever an intersection occurs. As a simple example, suppose $\mathfrak{F}$ consists of all circles about the origin. Then $\mathfrak{G}$ consists of all straight

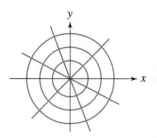

FIGURE 1.20 *Orthogonal families: circles and lines.*

lines through the origin (Figure 1.20). It is clear that each straight line is orthogonal to each circle wherever the two intersect.

In general, suppose we are given a family $\mathfrak{F}$ of curves. These must be described in some way, say by an equation

$$F(x, y, k) = 0,$$

giving a different curve for each choice of the constant k. Think of these curves as integral curves of a differential equation

$$y' = f(x, y),$$

which we determine from the equation $F(x, y, k) = 0$ by differentiation. At a point (x_0, y_0), the slope of the curve C in $\mathfrak{F}$ through this point is $f(x_0, y_0)$. Assuming that this is nonzero, any curve through (x_0, y_0) and orthogonal to C at this point must have slope $-1/f(x_0, y_0)$. (Here we use the fact that two lines are orthogonal if and only if their slopes are negative reciprocals.) The family $\mathfrak{G}$ of orthogonal trajectories of $\mathfrak{F}$ therefore consists of the integral curves of the differential equation

$$y' = -\frac{1}{f(x, y)}.$$

Solve this differential equation for the curves in $\mathfrak{G}$.

EXAMPLE 1.29

Consider the family $\mathfrak{F}$ of curves that are graphs of

$$F(x, y, k) = y - kx^2 = 0.$$

This is a family of parabolas. We want the family of orthogonal trajectories.

First obtain the differential equation of $\mathfrak{F}$. Differentiate $y - kx^2 = 0$ to get

$$y' - 2kx = 0.$$

To eliminate k, use the equation $y - kx^2 = 0$ to write

$$k = \frac{y}{x^2}.$$

Then

$$y' - 2\left(\frac{y}{x^2}\right)x = 0,$$

or

$$y' = 2\frac{y}{x} = f(x, y).$$

This is the differential equation of the family $\mathfrak{F}$. Curves in $\mathfrak{F}$ are integral curves of this differential equation, which is of the form $y' = f(x, y)$, with $f(x, y) = 2y/x$. The family $\mathfrak{G}$ of orthogonal trajectories therefore has differential equation

$$y' = -\frac{1}{f(x, y)} = -\frac{x}{2y}.$$

This equation is separable, since

$$2y \, dy = -x \, dx.$$

Integrate to get

$$y^2 = -\tfrac{1}{2}x^2 + C.$$

This is a family of ellipses

$$\tfrac{1}{2}x^2 + y^2 = C.$$

Some of the parabolas and ellipses from $\mathfrak{F}$ and $\mathfrak{G}$ are shown in Figure 1.21. Each parabola in $\mathfrak{F}$ is orthogonal to each ellipse in $\mathfrak{G}$ wherever these curves intersect. ■

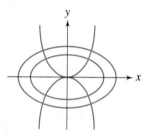

FIGURE 1.21 *Orthogonal families: parabolas and ellipses.*

SECTION 1.7   PROBLEMS

Mechanical Systems

1. Suppose that the pulley described in this section is only 9 feet above the floor. Assuming the same initial conditions as in the discussion, determine the velocity with which the chain leaves the pulley. *Hint:* The mass of the part of the chain that is in motion is $(16 - x)\rho/32$.

2. Determine the time it takes for the chain in Problem 1 to leave the pulley.

3. Suppose the support is only 10 feet above the floor in the discussion of the chain piling on the floor. Calcu-

late the velocity of the moving part of the chain as it leaves the support. (Note the hint to Problem 1.)

4. (*Chain and Weight on a Pulley*) An 8ρ-pound weight is attached to one end of a 40-foot chain that weighs ρ pounds per foot. The chain is supported by a small frictionless pulley located more than 40 feet above the floor. Initially, the chain is held at rest with 23 feet hanging on one side of the pulley with the remainder of the chain, along with the weight, on the other side. How long after the chain is released, and with what velocity, will it leave the pulley?

5. (*Chain on a Table*) A 24-foot chain weighing ρ pounds per foot is stretched out on a very tall, frictionless table with 6 feet hanging off the edge. If the chain is released from rest, determine the time it takes for the end of the chain to fall off the table and also the velocity of the chain at this instant.

6. (*Variable Mass Chain on a Low Table*) Suppose the chain in Problem 5 is placed on a table that is only 4 feet high, so that the chain accumulates on the floor as it slides off the table. Two feet of chain are already piled up on the floor at the time that the rest of the chain is released. Determine the velocity of the moving end of the chain at the instant it leaves the table top. *Hint:* The mass of that part of the chain that is moving changes with time. Newton's law applies to the center of mass of the moving system.

7. Determine the time it takes for the chain to leave the support in the discussion of the chain piling on the floor.

8. Use the conservation of energy principle (potential energy plus kinetic energy of a conservative system is a constant of the motion) to obtain the velocity of the chain in the discussion involving the chain on the pulley.

9. Use the conservation of energy principle to give an alternate derivation of the conclusion of the discussion of the chain piling on the floor.

10. (*Paraboloid of Revolution*) Determine the shape assumed by the surface of a liquid being spun in a circular bowl at constant angular velocity ω. *Hint:* Consider a particle of liquid located at (x, y) on the surface of the liquid, as in Figure 1.22. The forces acting on the particle are the horizontal force having magnitude $m\omega^2 x$ and a vertical force of magnitude mg. Since the particle is in radial equilibrium, the resultant vector is normal to the curve.

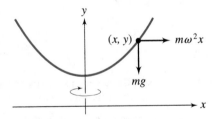

FIGURE 1.22 *Particle on the surface of a spinning liquid.*

Properties of spinning liquids have found application in astronomy. A Canadian astronomer has constructed a telescope by spinning a bowl of mercury, creating a reflective surface free of the defects obtained by the usual grinding of a solid lens. He claims that the idea was probably known to Newton but that he is the first to carry it out in practice. Roger Angel, a University of Arizona astronomer, has developed this idea into a technique for producing telescope mirrors called *spin casting*. As reported in *Time* (April 27, 1992), "... a complex ceramic mold is assembled inside the furnace and filled with glittering chunks of Pyrex-type glass. Once the furnace lid is sealed, the temperature will slowly ratchet up over a period of several days, at times rising no more than 2 degrees Centigrade in an hour. At 750 degrees C (1382 degrees Fahrenheit), when the glass is a smooth, shiny lake, the furnace starts to whirl like a merry-go-round, an innovation that automatically spins the glass into the parabolic shape traditionally achieved by grinding." The result is a parabolic surface requiring little or no grinding before a reflective coat is applied. Professor Angel believes that the method will allow the construction of much larger mirrors than are possible by conventional techniques. Supporting this claim is his recent production of one of the world's largest telescope mirrors, a 6.5 meter (about 21 feet) mirror to be placed in an observatory atop Mount Hopkins in Arizona.

11. A 10-pound ballast bag is dropped from a hot air balloon that is at an altitude of 342 feet and ascending at a rate of 4 feet per second. Assuming that air resistance is not a factor, determine the maximum height attained by the bag, how long it remains aloft, and the speed with which it strikes the ground.

12. A 6-ounce softball is thrown upward from a height of 7 feet with an initial velocity of 84 feet per second. If the ball is subjected to air resistance equal (in pounds) to $1/128$ times the speed of the ball (in feet per second), how high will it rise before falling back to the ground?

13. A ship weighing 86,400 tons starts from rest under a constant propeller thrust of 66,000 pounds. If the resistance due to the water is $4500v$ pounds, where v is the velocity in feet per second, find the velocity of the ship as a function of time, its terminal velocity in miles per hour, and how far it will have traveled by the time it has achieved 85% of its terminal velocity.

14. A girl and her sled weigh a total of 96 pounds. They begin at rest at the top of a long slope making an angle of 30 degrees with the horizontal. The coefficient of friction between the sled runners and the snow is $\sqrt{3}/24$, and there is a drag due to air resistance that is equal to the velocity. Nine seconds after she starts at the top of the slope, she glides out onto level ground. How far from the base of the hill will she travel before coming to rest?

15. A 96-pound box is placed on a long inclined plane making an angle of 30 degrees with the horizontal and then given an initial push of 2 feet per second down the

plane. If the coefficient of friction between the box and the plane is $\sqrt{3}/12$ and the drag due to air resistance is twice the velocity, what is the velocity of the box and how far has it traveled for any time t?

16. A 48-pound box is given an initial push of 16 feet per second down an inclined plane that has a gradient of $\frac{7}{24}$. If there is a coefficient of friction of $\frac{1}{3}$ between the box and the plane, and an air resistance equal to $\frac{3}{2}$ the velocity of the box, determine how far the box will travel before coming to rest.

17. A skydiver and her equipment together weigh 192 pounds. Before the parachute is opened, there is an air drag equal to six times her velocity. Four seconds after stepping from the plane, the skydiver opens the parachute, producing a drag equal to three times the square of the velocity. Determine the velocity and how far the skydiver has fallen at time t. What is the terminal velocity?

18. Archimedes' principle of buoyancy states that an object submerged in a fluid is buoyed up by a force equal to the weight of the fluid that is displaced by the object. A rectangular box 1 by 2 by 3 feet, and weighing 384 pounds, is dropped into a 100-foot-deep freshwater lake. The box begins to sink with a drag due to the water having magnitude equal to $\frac{1}{2}$ the velocity. Calculate the terminal velocity of the box. Will the box have achieved a velocity of 10 feet per second by the time it reaches bottom? Assume that the density of the water is 62.5 pounds per cubic foot.

19. Suppose the box in Problem 18 cracks open upon hitting the bottom of the lake and 32 pounds of its contents fall out. Approximate the velocity with which the box surfaces.

20. A horizontal spring of negligible mass has one of its ends attached to a wall and a weight of mass m attached to its free end. The spring exerts a restoring force of kx pounds when stretched by an amount x (Hooke's law). Assume that the friction between the weight and the surface upon which it glides is negligible. Find the relationship between the velocity and position of the mass at time $t > 0$ if the spring is stretched x_0 feet initially and then released from rest. *Hint:* Observe that t does not appear explicitly in the equation of motion, and write

$$\frac{dv}{dt} = \frac{dv}{dx}\frac{dx}{dt} = v\frac{dv}{dx}$$

in Newton's law.

21. How much would an object weighing 100 pounds at sea level weigh at an altitude of 990 miles? How much would it weigh at 3690 miles?

22. The acceleration due to gravity inside the earth is proportional to the distance from the center of the earth. An object is dropped from the surface of the earth into a hole extending through the earth's center. Calculate the speed the object achieves by the time it reaches the center.

23. A particle starts from rest at the highest point of a vertical circle and slides under only the influence of gravity along a chord to another point on the circle. Show that the time taken is independent of the choice of the terminal point. What is this common time?

24. A ball of mass m is thrown upward from the surface of the earth. The initial velocity is v_0, and the forces acting on the ball are gravity and air resistance. The latter force is proportional to the square of the velocity.

(a) Derive and solve a differential equation for the height of the ball at time t.

(b) Find the maximum height achieved by the ball and the time it takes to reach this height.

(c) Is it true that the time it takes the ball to reach its maximum height is the same as the time it takes for the ball to fall back to earth?

25. An oil tanker of mass M is sailing in a straight line. At time zero it shuts off its engines and coasts. Assume that the water tends to slow the tanker with a force proportional to $v(t)^\alpha$, in which $v(t)$ is the velocity and α is constant.

(a) Derive a differential equation for $v(t)$.

(b) Show that the tanker moves in a straight line and eventually comes to a full stop if $0 < \alpha < 1$. What happens if $\alpha \geq 1$?

Circuits

26. Determine each of the currents in the circuit of Figure 1.23.

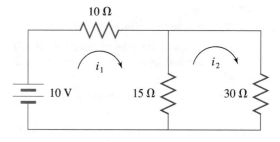

FIGURE 1.23

27. In the circuit of Figure 1.24, the capacitor is initially discharged. How long after the switch is closed will the

capacitor voltage be 76 volts? Determine the current in the resistor at that time. (Here $k\Omega$ denotes 1000 ohms and μF denotes 10^{-6} farads.)

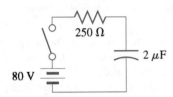

FIGURE 1.24

28. Suppose in Problem 27 the capacitor had a potential of 50 volts when the switch was closed. How long would it take for the capacitor voltage to reach 76 volts?

29. For the circuit of Figure 1.25, find all currents immediately after the switch is closed, assuming that all of these currents and the charges on the capacitors are zero just prior to closing the switch.

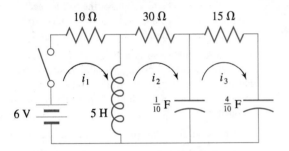

FIGURE 1.25

30. In a constant electromotive force RL circuit, we find that the current is given by

$$i(t) = \frac{E}{R}\left(1 - e^{-Rt/L}\right) + i(0)e^{-Rt/L}.$$

Let $i(0) = 0$.

(a) Show that the current increases with time.

(b) Find a time t_0 at which the current is 63% of E/R. This time is called the *inductive time constant* of the circuit.

(c) Does the inductive time constant depend on $i(0)$? If so, in what way?

31. Recall that the charge $q(t)$ in an RC circuit satisfies the linear differential equation

$$q' + \frac{1}{RC}q = \frac{1}{R}E(t).$$

(a) Solve for the charge in the case that $E(t) = E$, constant. Evaluate the constant of integration by using the condition $q(0) = q_0$.

(b) Determine $\lim_{t\to\infty} q(t)$ and show that this limit is independent of q_0.

(c) Graph $q(t)$. Determine when the charge has its maximum and minimum values.

(d) Determine at what time $q(t)$ is within 1% of its steady-state value (the limiting value requested in (b)).

32. Using the differential equation of Problem 31, determine the charge in an RC circuit having electromotive force $E(t) = A\cos(\omega t)$, with A and ω positive constants. Evaluate the constant in the general solution by using the condition $q(0) = q_0$.

33. Solve for the current $i(t)$ in an RL circuit if $R = 2$ ohms, $L = 25$ henrys, and $E(t) = Ae^{-t}$, with A a positive constant. Use the initial condition $i(0) = 0$. Graph the current as a function of time.

34. Solve for the current in an RL circuit in which $E(t) = A\cos(\omega t) + Be^{-t}$, with A, B, and ω positive constants and $i(0) = 0$. Choose values for A and B and graph the solution for different values of ω to gauge the influence this frequency term has on the solution.

35. Solve for the current in an RL circuit having electromotive force $E(t) = A\sin(\omega_1 t) + B\cos(\omega_2 t)$, with A, B, ω_1, and ω_2 positive constants. Use the initial condition $i(0) = 0$. Let $A = B = 1$ in the solution and graph solutions corresponding to different values of ω_1 and ω_2 to visualize the relative influences of these two frequency terms on the behavior of the solution.

36. Solve for the current in an RC circuit having a resistance of R ohms, capacitance of C farads, and an electromotive force $E(t) = 1 - \cos(2t)$. Assume that $q(0) = 0$.

37. Find the current in an RL circuit if $i(0) = 0$ and $E(t) = 1 + e^{-t}$.

Orthogonal Trajectories

In each of Problems 38 through 47, find the family of orthogonal trajectories of the given family of curves. If software is available, graph some curves in the given family and some curves in the family of orthogonal trajectories.

38. $x + 2y = K$

39. $2x^2 - 3y = K$

40. $x^2 + 2y^2 = K$

41. $y = Kx^2 + 1$

42. $x^2 - Ky^2 = 1$

43. $y = e^{kx}$

44. $y = ke^x$

45. $y = (x - k)^2$

46. $y^2 = Kx^3$

47. $x^2 - Ky = 1$

1.8   Existence and Uniqueness for Solutions of Initial Value Problems

We have solved several initial value problems

$$y' = f(x, y); \qquad y(x_0) = y_0,$$

and have always found that there is just one solution. That is, the solution existed, and it was unique. Can either existence or uniqueness fail to occur? The answer is yes, as the following examples show.

EXAMPLE 1.30

Consider the initial value problem

$$y' = 2y^{1/2}; \qquad y(0) = -1.$$

The differential equation is separable and has general solution

$$y(x) = (x + C)^2.$$

To satisfy the initial condition, we must choose C so that

$$y(0) = C^2 = -1,$$

and this is impossible if C is to be a real number. This initial value problem has no real-valued solution. ■

EXAMPLE 1.31

Consider the problem

$$y' = 2y^{1/2}; \qquad y(2) = 0.$$

One solution is the trivial function

$$y = \varphi(x) = 0 \quad \text{for all } x.$$

But there is another solution. Define

$$\psi(x) = \begin{cases} 0 & \text{for } x \leq 2 \\ (x - 2)^2 & \text{for } x \geq 2 \end{cases}.$$

Graphs of both solutions are shown in Figure 1.26. Uniqueness fails in this example. ■

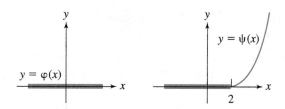

FIGURE 1.26 *Graphs of solutions of* $y' = 2\sqrt{y}$; $y(2) = 0.$

Because of examples such as these, we look for conditions that ensure that an initial value problem has a unique solution. The following theorem provides a convenient set of conditions.

THEOREM 1.2 Existence and Uniqueness

Let f and $\partial f/\partial y$ be continuous for all (x, y) in a closed rectangle R centered at (x_0, y_0). Then there exists a positive number h such that the initial value problem

$$y' = f(x, y); \qquad y(x_0) = y_0$$

has a unique solution defined over the interval $(x_0 - h, x_0 + h)$. ∎

As with the test for exactness (Theorem 1.1), by a closed rectangle we mean all points on or inside a rectangle in the plane, having sides parallel to the axes. Geometrically, existence of a solution of the initial value problem means that there is an integral curve of the differential equation passing through (x_0, y_0). Uniqueness means that there is only one such curve.

This is an example of a *local theorem*, in the following sense. The theorem guarantees existence of a unique solution that is defined on some interval of width $2h$, but it says nothing about how large h is. Depending on f and x_0, h may be small, giving us existence and uniqueness "near" x_0. This is dramatically demonstrated by the initial value problem

$$y' = y^2; \qquad y(0) = n,$$

in which n is any positive integer. Here $f(x, y) = y^2$ and $\partial f/\partial y = 2y$, both continuous over the entire plane, hence on any closed rectangle about $(0, n)$. The theorem tells us that there is a unique solution of this initial value problem in *some* interval $(-h, h)$.

In this case, we can solve the initial value problem explicitly, obtaining

$$y(x) = -\frac{1}{x - 1/n}.$$

This solution is valid for $-1/n < x < 1/n$, so we can take $h = 1/n$ in this example. This means that the size of n in the initial value controls the size of the interval for the solution. The larger n is, the smaller this interval must be. This fact is certainly not apparent from the initial value problem itself!

In the special case that the differential equation is linear, we can improve considerably on the existence/uniqueness theorem.

THEOREM 1.3

Let p and q be continuous on an open interval I and let x_0 be in I. Let y_0 be any number. Then the initial value problem

$$y' + p(x)y = q(x); \qquad y(x_0) = y_0$$

has a unique solution defined for all x in I. ∎

In particular, if p and q are continuous for all x, then there is a unique solution defined over the entire real line.

Proof Equation (1.6) of Section 1.3 gives the general solution of the linear equation. Using this, we can write the solution of the initial value problem:

$$y(x) = e^{-\int_{x_0}^{x} p(\xi)\,d\xi} \left[\int_{x_0}^{x} q(\xi) e^{\int_{x_0}^{x} p(\xi)\,d\xi}\,d\xi + y_0 \right].$$

Because p and q are continuous on I, this solution is defined for all x in I. ∎

Therefore, in the case that the differential equation is linear, the initial value problem has a unique solution in the largest open interval containing x_0, in which both p and q are continuous.

SECTION 1.8   PROBLEMS

In each of Problems 1 through 5, show that the conditions of Theorem 1.2 are satisfied by the initial value problem. Assume familiar facts from the calculus about continuity of real functions of one and two variables.

1. $y' = 2y^2 + 3xe^y \sin(xy); \ y(2) = 4$

2. $y' = 4xy + \cosh(x); \ y(1) = -1$

3. $y' = (xy)^3 - \sin(y); \ y(2) = 2$

4. $y' = x^5 - y^5 + 2xe^y; \ y(3) = \pi$

5. $y' = x^2ye^{-2x} + y^2; \ y(3) = 8$

6. Consider the initial value problem $|y'| = 2y;$ $y(x_0) = y_0$.

    (a) Find two solutions, assuming that $y_0 > 0$.

    (b) Explain why part (a) does not violate Theorem 1.2.

Theorem 1.2 can be proved using Picard iterates, which we will discuss briefly. Suppose f and $\partial f / \partial y$ are continuous in a closed rectangle R having (x_0, y_0) in its interior and sides parallel to the axes. Consider the initial value problem $y' = f(x, y); \ y(x_0) = y_0$. For each positive integer n, define

$$y_n(x) = y_0 + \int_{x_0}^x f(t, y_{n-1}(t)) \, dt.$$

This is a recursive definition, giving $y_1(x)$ in terms of y_0, then $y_2(x)$ in terms of $y_1(x)$, and so on. The functions $y_n(x)$ for $n = 1, 2, \ldots$ are called *Picard iterates* for the initial value problem. Under the assumptions made on f, the sequence $\{y_n(x)\}$ converges for all x in some interval about x_0, and the limit of this sequence is the solution of the initial value problem on this interval.

In each of Problems 7 through 10, (a) use Theorem 1.2 to show that the problem has a solution in some interval about x_0, (b) find this solution, (c) compute Picard iterates $y_1(x)$ through $y_6(x)$, and from these guess $y_n(x)$ in general, and (d) find the Taylor series of the solution from (b) about x_0. You should find that the iterates computed in (c) are partial sums of the series of (d). Conclude that in these examples the Picard iterates converge to the solution.

7. $y' = 2 - y; \ y(0) = 1$

8. $y' = 4 + y; \ y(0) = 3$

9. $y' = 2x^2; \ y(1) = 3$

10. $y' = \cos(x); \ y(\pi) = 1$

CHAPTER 1   ADDITIONAL PROBLEMS

In each of Problems 1 through 30, find the general solution. These differential equations include all types discussed in this chapter.

1. $4x^3y - 6e^y + (x^4 - 6xe^y)y' = 0$

2. $y' = 8x^3 - 3y$

3. $2y - 7x - 2(y - x)y' = 0$

4. $y^2 + 2xy - x^2y' = 0$

5. $xy' - y = \dfrac{y}{\ln(y) - \ln(x)}$

6. $(x^2 - 4)y' = y + 3$

7. $5x - y + 4 + (x - 5y - 4)y' = 0$

8. $xy' + y = 2y^{3/2}$

9. $6x - 2yy' = 0$

10. $y' + y^2 = x^{-2} - \dfrac{y}{x}$

11. $y = (3y^5 + 2x)y'$

12. $y' = xy^2 - 2x^2y + x^3 + 1$

13. $y' = \dfrac{4y}{4x - y}$

14. $y' = xy^2 + (1 - 2x)y + x - 1$

15. $y = xy' + \dfrac{y}{1 + y}$

16. $x^2y' = xy + e^x y^3$

17. $y' = \dfrac{1 - x}{y}$

18. $xy' + 4y = \dfrac{\cos(x)}{x^2}$

19. $2xyy' = x + 8y^2$

20. $xyy' + x^2 + y^2 = 0$

21. $y = (y^4 + 3x)y'$

22. $y' = \dfrac{e^{x-y}}{e^{x-y} - 1}$

23. $xy' + 3y = x^2 \sin(x)$

24. $y' = \dfrac{2x + y}{x - y}$

25. $y' = \dfrac{2y + y\cos(x)}{2x + \sin(x)}$

26. $(x - 2)y' = x - y$

27. $y' = (x + y)^2$

28. $\cos(x)y' - y = 5$

29. $y' = 2 + (y/x)^2$

30. $y' - 5y = e^{5x}\sin(x)$

31. Determine values of C and K so that the following equation is exact:

$$Cx^2 ye^y + 2\cos(y) + \left(x^3 e^y y + x^3 e^y + Kx\sin(y)\right)y' = 0.$$

32. The differential equation $y = xy' + f(y')$ is called *Clairaut's equation.*

(a) Show that, for any constant k such that $f(k)$ is defined, $y = kx + f(k)$ is a solution.

(b) Use (a) to find the general solution of $y = xy' + (y')^2$. What is the function f for this example?

(c) Graph some integral curves of the equation in (b).

(d) Show that $y = -x^2/4$ is a solution of the equation in (b) but cannot be obtained as a solution of the form $y = kx + f(k)$ for this equation by any choice of the constant k. How does the graph of this solution relate to the integral curves drawn in (c)? The graph of $y = -x^2/4$ is called the *envelope* of the integral curves of $y = xy' + (y')^2$.

CHAPTER 2

Second-Order Differential Equations

2.1   Preliminary Concepts

A *second-order differential equation* is an equation that contains a second derivative, but no higher derivative. Most generally, it has the form

$$F(x, y, y', y'') = 0,$$

although only a term involving y'' need appear explicitly. For example,

$$y'' = x^3,$$

$$xy'' - \cos(y) = e^x$$

and

$$y'' - 4xy' + y = 2$$

are second-order differential equations.

A *solution* of $F(x, y, y', y'') = 0$ on an interval I (perhaps the whole real line) is a function φ that satisfies the differential equation at each point of I:

$$F(x, \varphi(x), \varphi'(x), \varphi''(x)) = 0 \quad \text{for } x \text{ in } I.$$

For example, $\varphi(x) = 6\cos(4x) - 17\sin(4x)$ is a solution of

$$y'' + 16y = 0$$

for all real x. And $\varphi(x) = x^3 \cos(\ln(x))$ is a solution of

$$x^2 y'' - 5xy' + 10y = 0$$

for $x > 0$. These can be checked by substitution into the differential equation.

The *linear second-order differential equation* has the form

$$R(x)y'' + P(x)y' + Q(x)y = S(x),$$

65

in which R, P, Q, and S are continuous in some interval. On any interval where $R(x) \neq 0$, we can divide this equation by $R(x)$ and obtain the special linear equation

$$y'' + p(x)y' + q(x)y = f(x). \tag{2.1}$$

For the remainder of this chapter, we will concentrate on this equation. We want to know:

1. What can we expect in the way of existence and uniqueness of solutions of equation 2.1?
2. How can we produce all solutions of equation (2.1), at least in some cases that occur frequently and have important applications?

We begin with the underlying theory that will guide us in developing techniques for explicitly producing solutions of equation (2.1).

2.2   Theory of Solutions of $y'' + p(x)y' + q(x)y = f(x)$

To get some feeling for what we are dealing with, and what we should be looking for, consider the simple, linear second-order equation

$$y'' - 12x = 0.$$

We can write this as

$$y'' = 12x$$

and integrate to obtain

$$y' = \int y''(x)\, dx = \int 12x\, dx = 6x^2 + C.$$

Integrate again:

$$y(x) = \int y'(x)\, dx = \int (6x^2 + C)\, dx = 2x^3 + Cx + K.$$

This solution is defined for all x, and contains two arbitrary constants. If we recall that the general solution of a first-order equation contained one arbitrary constant, it seems natural that the solution of a second-order equation, involving two integrations, should contain two arbitrary constants.

For any choices of C and K, we can graph the integral curves $y = 2x^3 + Cx + K$ as curves in the plane. Figure 2.1 shows some of these curves for different choices of these constants.

Unlike the first-order case, there are many integral curves through each point in the plane. For example, suppose we want a solution satisfying the initial condition

$$y(0) = 3.$$

Then we need

$$y(0) = K = 3,$$

but are still free to choose C as any number. All solutions

$$y(x) = 2x^3 + Cx + 3$$

pass through $(0, 3)$. Some of these curves are shown in Figure 2.2.

We single out exactly one of these curves if we specify its slope at $(0, 3)$. Suppose, for example, we also specify the initial condition

$$y'(0) = -1.$$

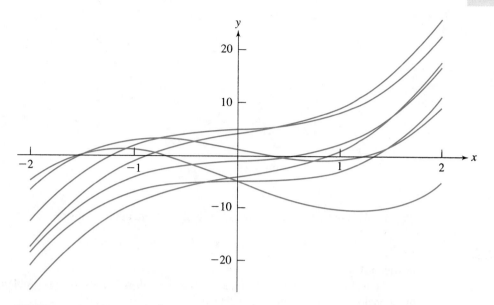

FIGURE 2.1 *Graphs of* $y = 2x^3 + Cx + K$ *for various values of* C *and* K.

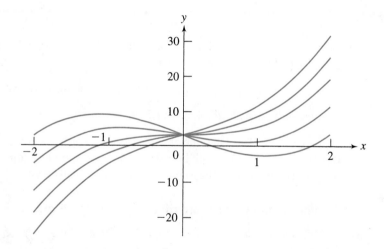

FIGURE 2.2 *Graphs of* $y = 2x^3 + Cx + 3$ *for various values of* C.

Since $y'(x) = 6x^2 + C$, this requires that $C = -1$. There is exactly one solution satisfying both initial conditions (going through a given point with given slope), and it is

$$y(x) = 6x^2 - x + 3.$$

A graph of this solution is given in Figure 2.3.

To sum up, at least in this example, the general solution of the differential equation involved two arbitrary constants. An initial condition, $y(0) = 3$, specifying that the solution curve must pass through $(0, 3)$, determined one of these constants. However, that left infinitely many solution curves passing through $(0, 3)$. The other initial condition, $y'(0) = -1$, picked out that solution curve through $(0, 3)$ having slope -1 and gave a unique solution of this problem.

This suggests that we define the initial value problem for equation (2.1) to be the differential equation, defined on some interval, together with two initial conditions, one specifying a point

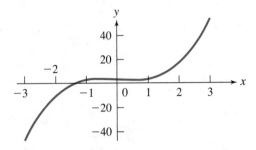

FIGURE 2.3   *Graph of* $y = 2x^3 - x + 3$.

lying on the solution curve and the other its slope at that point. This problem has the form

$$y'' + p(x)y' + q(x)y = f(x); \qquad y(x_0) = A, \, y'(x_0) = B,$$

in which A and B are given real numbers.

The main theorem on existence and uniqueness of solutions for this problem is the second-order analogue of Theorem 1.3 in Chapter 1.

THEOREM 2.1

Let p, q, and f be continuous on an open interval I. Let x_0 be in I and let A and B be any real numbers. Then the initial value problem

$$y'' + p(x)y' + q(x)y = f(x); \qquad y(x_0) = A, \, y'(x_0) = B$$

has a unique solution defined for all x in I. ∎

This gives us an idea of the kind of information needed to specify a unique solution of equation (2.1). Now we need a framework in which to proceed in finding solutions. We will provide this in two steps, beginning with the case that $f(x)$ is identically zero.

2.2.1   The Homogeneous Equation $y'' + p(x)y' + q(x) = 0$

When $f(x)$ is identically zero in equation (2.1), the resulting equation

$$y'' + p(x)y' + q(x) = 0 \tag{2.2}$$

is called *homogeneous*. This term was used in a different context with first-order equations, and its use here is unrelated to that. Here it simply means that the right side of equation (2.1) is zero.

A *linear combination* of solutions $y_1(x)$ and $y_2(x)$ of equation (2.2) is a sum of constant multiples of these functions:

$$c_1 y_1(x) + c_2 y_2(x)$$

with c_1 and c_2 real numbers. It is an important property of the homogeneous linear equation that linear combinations of solutions are again solutions.

THEOREM 2.2

Let y_1 and y_2 be solutions of $y'' + p(x)y' + q(x)y = 0$ on an interval I. Then any linear combination of these solutions is also a solution.

Proof   Let c_1 and c_2 be real numbers. Substituting $y(x) = c_1 y_1(x) + c_2 y_2(x)$ into the differential equation, we obtain

$$(c_1 y_1 + c_2 y_2)'' + p(x)(c_1 y_1 + c_2 y_2)' + q(x)(c_1 y_1 + c_2 y_2)$$
$$= c_1 y_1'' + c_2 y_2'' + c_1 p(x)y_1' + c_2 p(x)y_2' + c_1 q(x)y_1 + c_2 q(x)y_2$$
$$= c_1[y_1'' + p(x)y_1' + q(x)y_1] + c_2[y_2'' + p(x)y_2' + q(x)y_2]$$
$$= 0 + 0 = 0,$$

because of the assumption that y_1 and y_2 are both solutions. ∎

Of course, as a special case ($c_2 = 0$), this theorem tells us also that, for the homogeneous equation, a constant multiple of a solution is a solution. Even this special case of the theorem fails for a nonhomogeneous equation. For example, $y_1(x) = 4e^{2x}/5$ is a solution of

$$y'' + 2y' - 3y = 4e^{2x},$$

but $5y_1(x) = 4e^{2x}$ is not.

The point to taking linear combinations $c_1 y_1 + c_2 y_2$ is to obtain more solutions from just two solutions of equation (2.2). However, if y_2 is already a constant multiple of y_1, then

$$c_1 y_1 + c_2 y_2 = c_1 y_1 + c_2 k y_1 = (c_1 + k c_2)y_1,$$

just another constant multiple of y_1. In this event, y_2 is superfluous, providing us nothing we did not know from just y_1. This leads us to distinguish the case in which one solution is a constant multiple of another from the case in which the two solutions are not multiples of each other.

DEFINITION 2.1   *Linear Dependence, Independence*

Two functions f and g are linearly dependent on an open interval I if, for some constant c, either $f(x) = cg(x)$ for all x in I, or $g(x) = cf(x)$ for all x in I.

If f and g are not linearly dependent on I, then they are said to be linearly independent on the interval.

EXAMPLE 2.1

$y_1(x) = \cos(x)$ and $y_2(x) = \sin(x)$ are solutions of $y'' + y = 0$, over the real line. Neither of these functions is a constant multiple of the other. Indeed, if $\cos(x) = k \sin(x)$ for all x, then in particular

$$\cos\left(\frac{\pi}{4}\right) = \frac{\sqrt{2}}{2} = k \sin\left(\frac{\pi}{4}\right) = k\frac{\sqrt{2}}{2},$$

so k must be 1. But then $\cos(x) = \sin(x)$ for all x, a clear absurdity (for example, let $x = 0$). These solutions are linearly independent. Now we know from Theorem 2.2 that

$$a \cos(x) + b \sin(x)$$

is a solution for any numbers a and b. Because $\cos(x)$ and $\sin(x)$ are linearly independent, this linear combination provides an infinity of new solutions, instead of just constant multiples of one we already know. ∎

There is a simple test to tell whether two solutions of equation (2.2) are linearly independent on an interval. Define the *Wronskian* of solutions y_1 and y_2 to be

$$W(x) = y_1(x)y_2'(x) - y_1'(x)y_2(x).$$

This is the 2×2 determinant

$$W(x) = \begin{vmatrix} y_1(x) & y_2(x) \\ y_1'(x) & y_2'(x) \end{vmatrix}.$$

THEOREM 2.3 *Wronskian Test*

Let y_1 and y_2 be solutions of $y'' + p(x)y' + q(x)y = 0$ on an open interval I. Then,

1. Either $W(x) = 0$ for all x in I, or $W(x) \neq 0$ for all x in I.
2. y_1 and y_2 are linearly independent on I if and only if $W(x) \neq 0$ on I. ■

A proof of this theorem is outlined in the exercises. Conclusion 1 means that the Wronskian of two solutions cannot be nonzero at some points of I and zero at others. Either the Wronskian vanishes over the entire interval, or it is nonzero at every point of the interval. Conclusion 2 states that nonvanishing of the Wronskian is equivalent to linear independence of the solutions. Putting both conclusions together, it is therefore enough to test $W(x)$ at just one point of I to determine linear dependence or independence of these solutions. This gives us great latitude to choose a point at which the Wronskian is easy to evaluate.

EXAMPLE 2.2

In Example 2.1, we considered the solutions $y_1(x) = \cos(x)$ and $y_2(x) = \sin(x)$ of $y'' + y = 0$, for all x. In this case, linear independence was obvious. The Wronskian of these solutions is

$$W(x) = \begin{vmatrix} \cos(x) & \sin(x) \\ -\sin(x) & \cos(x) \end{vmatrix}$$

$$= \cos^2(x) + \sin^2(x) = 1 \neq 0. \quad ■$$

EXAMPLE 2.3

It is not always obvious whether two solutions are linearly independent or dependent on an interval. Consider the equation $y'' + xy = 0$. This equation appears simple but is not easy to solve. By a power series method we will develop later, we can write two solutions

$$y_1(x) = 1 - \frac{1}{6}x^3 + \frac{1}{180}x^6 - \frac{1}{12,960}x^9 + \cdots$$

and

$$y_2(x) = x - \frac{1}{12}x^4 + \frac{1}{504}x^7 - \frac{1}{45,360}x^{10} + \cdots,$$

with both series converging for all x. Here I is the entire real line. The Wronskian of these solutions at any nonzero x would be difficult to evaluate, but at $x = 0$ we easily obtain

$$W(0) = y_1(0)y_2'(0) - y_1'(0)y_2(0) = (1)(1) - (0)(0) = 1.$$

Nonvanishing of the Wronskian at this one point is enough to conclude linear independence of these solutions. ■

We are now ready to use the machinery we have built up to determine what is needed to find all solutions of $y'' + p(x)y' + q(x) = 0$.

THEOREM 2.4

Let y_1 and y_2 be linearly independent solutions of $y'' + p(x)y' + q(x)y = 0$ on an open interval I. Then, every solution of this differential equation on I is a linear combination of y_1 and y_2. ■

This fundamental theorem provides a strategy for finding all solutions of $y'' + p(x)y' + q(x)y = 0$ on I. Find two linearly independent solutions. Depending on p and q, this may be difficult, but at least we have a specific goal. If necessary, use the Wronskian to test for independence. The general linear combination $c_1 y_1 + c_2 y_2$, with c_1 and c_2 arbitrary constants, then contains all possible solutions.

We will prove the theorem following introduction of some standard terminology.

DEFINITION 2.2

Let y_1 and y_2 be solutions of $y'' + p(x)y' + q(x)y = 0$ on an open interval I.

1. y_1 and y_2 form a fundamental set of solutions on I if y_1 and y_2 are linearly independent on I.

2. When y_1 and y_2 form a fundamental set of solutions, we call $c_1 y_1 + c_2 y_2$, with c_1 and c_2 arbitrary constants, the general solution of the differential equation on I.

In these terms, we find the general solution by finding a fundamental set of solutions. Here is a proof of Theorem 2.4.

Proof   Let φ be any solution of $y'' + p(x)y' + q(x)y = 0$ on I. We want to show that there must be numbers c_1 and c_2 such that

$$\varphi(x) = c_1 y_1(x) + c_2 y_2(x).$$

Choose any x_0 in I. Let $\varphi(x_0) = A$ and $\varphi'(x_0) = B$. By Theorem 2.1, φ is the unique solution on I of the initial value problem

$$y'' + p(x)y' + q(x)y = 0; \qquad y(x_0) = A, \, y'(x_0) = B.$$

Now consider the system of two algebraic equations in two unknowns:

$$y_1(x_0)c_1 + y_2(x_0)c_2 = A$$
$$y_1'(x_0)c_1 + y_2'(x_0)c_2 = B.$$

It is routine to solve these algebraic equations. Assuming that $W(x_0) \neq 0$, we find that

$$c_1 = \frac{Ay_2'(x_0) - By_2(x_0)}{W(x_0)}, \qquad c_2 = \frac{By_1(x_0) - Ay_1'(x_0)}{W(x_0)}.$$

With this choice of c_1 and c_2, the function $c_1 y_1 + c_2 y_2$ is a solution of the initial value problem. By uniqueness of the solution of this problem, $\varphi(x) = c_1 y_1(x) + c_2 y_2(x)$ on I, and the proof is complete. ■

The proof reinforces the importance of having a fundamental set of solutions, since the nonvanishing of the Wronskian plays a vital role in showing that an arbitrary solution must be a linear combination of the fundamental solutions.

2.2.2   The Nonhomogeneous Equation $y'' + p(x)y' + q(x)y = f(x)$

The ideas just developed for the homogeneous equation (2.2) also provide the key to solving the nonhomogeneous equation

$$y'' + p(x)y' + q(x)y = f(x). \tag{2.3}$$

THEOREM 2.5

Let y_1 and y_2 be a fundamental set of solutions of $y'' + p(x)y' + q(x)y = 0$ on an open interval I. Let y_p be any solution of equation (2.3) on I. Then, for any solution φ of equation (2.3), there exist numbers c_1 and c_2 such that

$$\varphi = c_1 y_1 + c_2 y_2 + y_p. \blacksquare$$

This conclusion leads us to call $c_1 y_1 + c_2 y_2 + y_p$ the *general solution* of equation (2.3) and suggests the following strategy.

To solve $y'' + p(x)y' + q(x)y = f(x)$:

1. find the general solution $c_1 y_1 + c_2 y_2$ of the associated homogeneous equation $y'' + p(x)y' + q(x)y = 0$,

2. find *any* solution y_p of $y'' + p(x)y' + q(x)y = f(x)$, and

3. write the general solution $c_1 y_1 + c_2 y_2 + y_p$. This expressions contains all possible solutions of equation (2.3) on the interval.

Again, depending on p, q, and f, the first two steps may be formidable. Nevertheless, the theorem tells us what to look for and provides a clear way to proceed. Here is a proof of the theorem.

Proof   Since φ and y_p are both solutions of equation (2.3), then

$$(\varphi - y_p)'' + p(\varphi - y_p)' + q(\varphi - y_p) = \varphi'' + p\varphi' + q\varphi - (y_p'' + py_p' + qy_p)$$
$$= f - f = 0.$$

Therefore, $\varphi - y_p$ is a solution of $y'' + py' + qy = 0$. Since y_1 and y_2 form a fundamental set of solutions for this homogeneous equation, there are constants c_1 and c_2 such that

$$\varphi - y_p = c_1 y_1 + c_2 y_2,$$

and this is what we wanted to show. $\blacksquare$

The remainder of this chapter is devoted to techniques for carrying out the strategies just developed. For the general solution of the homogeneous equation (2.2) we must produce a fundamental set of solutions. And for the nonhomogeneous equation (2.3) we need to find one particular solution, together with a fundamental set of solutions of the associated homogeneous equation (2.2).

SECTION 2.2   PROBLEMS

In each of Problems 1 through 6, (a) verify that y_1 and y_2 are solutions of the differential equation, (b) show that their Wronskian is not zero, (c) write the general solution of the differential equation, and (d) find the solution of the initial value problem.

1. $y'' - 4y = 0$; $y(0) = 1$, $y'(0) = 0$

    $y_1(x) = \cosh(2x)$, $y_2(x) = \sinh(2x)$

2. $y'' + 9y = 0$; $y(\pi/3) = 0$, $y'(\pi/3) = 1$

    $y_1(x) = \cos(3x)$, $y_2(x) = \sin(3x)$

3. $y'' + 11y' + 24y = 0$; $y(0) = 1$, $y'(0) = 4$

    $y_1(x) = e^{-3x}$, $y_2(x) = e^{-8x}$

4. $y'' + 2y' + 8y = 0$; $y(0) = 2$, $y'(0) = 0$

    $y_1(x) = e^{-x}\cos(\sqrt{7}x)$, $y_2(x) = e^{-x}\sin(\sqrt{7}x)$

5. $y'' - \dfrac{7}{x}y' + \dfrac{16}{x^2}y = 0$; $y(1) = 2$, $y'(1) = 4$

    $y_1(x) = x^4$, $y_2(x) = x^4 \ln(x)$

6. $y'' + \dfrac{1}{x}y' + \left(1 - \dfrac{1}{4x^2}\right)y = 0$;

    $y(\pi) = -5$, $y'(\pi) = 8$

    $y_1(x) = \sqrt{\dfrac{2}{\pi x}}\cos(x)$, $y_2(x) = \sqrt{\dfrac{2}{\pi x}}\sin(x)$

7. Let $y_1(x) = x^2$ and $y_2(x) = x^3$. Show that $W(x) = x^4$ for all real x. Then $W(0) = 0$, but $W(x)$ is not identically zero. Why does this not contradict Theorem 2.3.1, with the interval I chosen as the entire real line?

8. Show that $y_1(x) = x$ and $y_2(x) = x^2$ are linearly independent solutions of $x^2 y'' - 2xy' + 2y = 0$ on $[-1, 1]$, but that $W(0) = 0$. Why does this not contradict Theorem 2.3.1 on this interval?

9. Give an example to show that the product of two solutions of $y'' + p(x)y' + q(x)y = 0$ need not be a solution.

10. Show that $y_1(x) = 3e^{2x} - 1$ and $y_2(x) = e^{-x} + 2$ are solutions of $yy'' + 2y' - (y')^2 = 0$, but that neither $2y_1$ nor $y_1 + y_2$ is a solution. Why does this not contradict Theorem 2.2?

11. Suppose y_1 and y_2 are solutions of $y'' + p(x)y' + q(x)y = 0$ on $[a, b]$, and that p and q are continuous on this interval. Suppose y_1 and y_2 both have a relative extremum at x_0 in (a, b). Prove that y_1 and y_2 are linearly dependent on $[a, b]$.

12. Let φ be a solution of $y'' + p(x)y' + q(x)y = 0$ on an open interval I and suppose $\varphi(x_0) = 0$ for some x_0 in I. Suppose $\varphi(x)$ is not identically zero. Prove that $\varphi'(x_0) \neq 0$.

13. Let y_1 and y_2 be distinct solutions of $y'' + p(x)y' + q(x)y = 0$ on an open interval I. Let x_0 be in I and suppose $y_1(x_0) = y_2(x_0) = 0$. Prove that y_1 and y_2 are linearly dependent on I. Thus, linearly independent solutions cannot share a common zero.

14. Let φ be a solution of the initial value problem

    $$y'' + p(x)y' + q(x)y = 0; \qquad y(x_0) = y'(x_0) = 0$$

    on an open interval I on which p and q are continuous. Prove that $\varphi(x) = 0$ for all x in I.

15. Prove Theorem 2.3.1. *Hint:* Begin with $y_1'' + p(x)y_1' + q(x)y_1 = 0$ and $y_2'' + p(x)y_2' + q(x)y_2 = 0$. Multiply the first equation by $-y_2$ and the second by y_1 and add the resulting equations to show that $W' + p(x)W = 0$. Solve this first-order equation for $W(x)$.

16. Prove Theorem 2.3.2. *Hint:* Suppose y_1 and y_2 are linearly dependent solutions on I. It is routine to check by direct calculation that $W(x) = 0$ for all x in I. Conversely, suppose $W(x) = 0$ for all x in I. Assuming that $y_2(x_0) \neq 0$ for some x_0 in I, there is a subinterval J of I on which $y_2(x) \neq 0$. Show that $[1/(y_2(x)^2)]W(x) = (d/dx)[y_1(x)/y_2(x)] = 0$, so $y_1(x)/y_2(x) = C$ for x in J, for some constant C. Then $y_1(x) = Cy_2(x)$ for all x in J. Now show that this holds for all x in I by showing that y_1 and Cy_2 are both solutions of the same initial value problem on I.

17. Suppose φ is a solution of

    $$y'' + p(x)y' + q(x)y = 0; \qquad y(x_0) = A, \, y'(x_0) = 0$$

    on an open interval I. Let ψ be a solution of

    $$y'' + p(x)y' + q(x)y = 0; \qquad y(x_0) = 0, \, y'(x_0) = B$$

    on I. Show that $\varphi + \psi$ is a solution of

    $$y'' + p(x)y' + q(x)y = 0; \quad y(x_0) = A, \, y'(x_0) = B.$$

18. Prove the Sturm separation theorem: Suppose φ and ψ are linearly independent solutions of $y'' + p(x)y' + q(x)y = 0$ on an open interval I. Then φ and ψ can have no common zero on I. Further, between any pair of consecutive zeros of φ on I, there must be exactly one zero of ψ. (A zero of f is a number w such that $f(w) = 0$). *Hint:* First observe that φ and ψ cannot have a common zero on I (note Problem 13). Now let x_1 and x_2 be consecutive zeros of φ on I, with $x_1 < x_2$. Since $W(x)$ cannot change sign on I, show that $\psi(x_1)$ and $\psi(x_2)$ must have opposite signs, hence (intermediate value theorem) ψ has a zero between x_1 and x_2. Finally, show that ψ cannot have more than one zero on $[x_1, x_2]$ by interchanging the roles of φ and ψ in this argument.

2.3   Reduction of Order

Given $y'' + p(x)y' + q(x)y = 0$, we want two independent solutions. Reduction of order is a technique for finding a second solution, if we can somehow produce a first solution.

Suppose we know a solution y_1, which is not identically zero. We will look for a second solution of the form $y_2(x) = u(x)y_1(x)$. Compute

$$y_2' = u'y_1 + uy_1', \quad y_2'' = u''y_1 + 2u'y_1' + uy_1''.$$

In order for y_2 to be a solution, we need

$$u''y_1 + 2u'y_1' + uy_1'' + p[u'y_1 + uy_1'] + quy_1 = 0.$$

Rearrange terms to write this equation as

$$u''y_1 + u'[2y_1' + py_1] + u[y_1'' + py_1' + qy_1] = 0.$$

The coefficient of u is zero because y_1 is a solution. Thus, we need to choose u so that

$$u''y_1 + u'[2y_1' + py_1] = 0.$$

On any interval in which $y_1(x) \neq 0$, we can write

$$u'' + \frac{2y_1' + py_1}{y_1}u' = 0.$$

To help focus on the problem of determining u, denote

$$g(x) = \frac{2y_1'(x) + p(x)y_1(x)}{y_1(x)},$$

a known function because $y_1(x)$ and $p(x)$ are known. Then

$$u'' + g(x)u' = 0.$$

Let $v = u'$ to get

$$v' + g(x)v = 0.$$

This is a linear first-order differential equation for v, with general solution

$$v(x) = Ce^{-\int g(x)\,dx}.$$

Since we need only one second solution y_2, we will take $C = 1$, so

$$v(x) = e^{-\int g(x)\,dx}.$$

Finally, since $v = u'$,

$$u(x) = \int e^{-\int g(x)\,dx}\,dx.$$

If we can perform these integrations and obtain $u(x)$, then $y_2 = uy_1$ is a second solution of $y'' + py' + qy = 0$. Further,

$$W(x) = y_1y_2' - y_1'y_2 = y_1(uy_1' + u'y_1) - y_1'uy_1 = u'y_1^2 = vy_1^2.$$

Since $v(x)$ is an exponential function, $v(x) \neq 0$. And the preceding derivation was carried out on an interval in which $y_1(x) \neq 0$. Thus, $W(x) \neq 0$ and y_1 and y_2 form a fundamental set of solutions on this interval. The general solution of $y'' + py' + qy = 0$ is $c_1y_1 + c_2y_2$.

We do not recommend memorizing formulas for g, v, and then u. Given one solution y_1, substitute $y_2 = uy_1$ into the differential equation and, after the cancellations that occur because y_1 is one solution, solve the resulting equation for $u(x)$.

EXAMPLE 2.4

Suppose we are given that $y_1(x) = e^{-2x}$ is one solution of $y'' + 4y' + 4y = 0$. To find a second solution, let $y_2(x) = u(x)e^{-2x}$. Then

$$y_2' = u'e^{-2x} - 2e^{-2x}u \quad \text{and} \quad y_2'' = u''e^{-2x} + 4e^{-2x}u - 4u'e^{-2x}.$$

Substitute these into the differential equation to get

$$u''e^{-2x} + 4e^{-2x}u - 4u'e^{-2x} + 4(u'e^{-2x} - 2e^{-2x}u) + 4ue^{-2x} = 0.$$

Some cancellations occur because e^{-2x} is one solution, leaving

$$u''e^{-2x} = 0,$$

or

$$u'' = 0.$$

Two integrations yield $u(x) = cx + d$. Since we need only one second solution y_2, we need only one u, so we will choose $c = 1$ and $d = 0$. This gives $u(x) = x$ and

$$y_2(x) = xe^{-2x}.$$

Now

$$W(x) = \begin{vmatrix} e^{-2x} & xe^{-2x} \\ -2e^{-2x} & e^{-2x} - 2xe^{-2x} \end{vmatrix} = e^{-4x} \neq 0$$

for all x. Therefore, y_1 and y_2 form a fundamental set of solutions for all x, and the general solution of $y'' + 4y' + 4y = 0$ is

$$y(x) = c_1 e^{-2x} + c_2 xe^{-2x}. \quad \blacksquare$$

EXAMPLE 2.5

Suppose we want the general solution of $y'' - (3/x)y' + (4/x^2)y = 0$ for $x > 0$, and somehow we find one solution $y_1(x) = x^2$. Put $y_2(x) = x^2 u(x)$ and compute

$$y_2' = 2xu + x^2 u' \quad \text{and} \quad y_2'' = 2u + 4xu' + x^2 u''.$$

Substitute into the differential equation to get

$$2u + 4xu' + x^2 u'' - \frac{3}{x}(2xu + x^2 u') + \frac{4}{x^2}(x^2 u) = 0.$$

Then

$$x^2 u'' + xu' = 0.$$

Since the interval of interest is $x > 0$, we can write this as

$$xu'' + u' = 0.$$

With $v = u'$, this is

$$xv' + v = (xv)' = 0,$$

so $xv = c$. We will choose $c = 1$. Then

$$v = u' = \frac{1}{x}$$

so

$$u = \ln(x) + d,$$

and we choose $d = 0$ because we need only one suitable u. Then $y_2(x) = x^2 \ln(x)$ is a second solution. Further, for $x > 0$,

$$W(x) = \begin{vmatrix} x^2 & x^2 \ln(x) \\ 2x & 2x \ln(x) + x \end{vmatrix} = x^3 \neq 0.$$

Then x^2 and $x^2 \ln(x)$ form a fundamental set of solutions for $x > 0$. The general solution for $x > 0$ is

$$y(x) = c_1 x^2 + c_2 x^2 \ln(x). \quad \blacksquare$$

SECTION 2.3   PROBLEMS

In each of Problems 1 through 10, verify that the given function is a solution of the differential equation, find a second solution by reduction of order, and finally write the general solution.

1. $y'' + 4y = 0$; $y_1(x) = \cos(2x)$

2. $y'' - 9y = 0$; $y_1(x) = e^{3x}$

3. $y'' - 10y' + 25y = 0$; $y_1(x) = e^{5x}$

4. $x^2 y'' - 7xy' + 16y = 0$; $y_1(x) = x^4$ for $x > 0$

5. $x^2 y'' - 3xy' + 4y = 0$; $y_1(x) = x^2$ for $x > 0$

6. $(2x^2 + 1)y'' - 4xy' + 4y = 0$; $y_1(x) = x$ for $x > 0$

7. $y'' - \dfrac{1}{x} y' - \dfrac{8}{x^2} y = 0$; $y_1(x) = x^4$ for $x > 0$

8. $y'' - \dfrac{2x}{1+x^2} y' + \dfrac{2}{1+x^2} y = 0$; $y_1(x) = x$

9. $y'' + \dfrac{1}{x} y' + \left(1 - \dfrac{1}{4x^2}\right) y = 0$; $y_1(x) = \dfrac{1}{\sqrt{x}} \cos(x)$ for $x > 0$

10. $(2x^2 + 3x + 1)y'' + 2xy' - 2y = 0$; $y_1(x) = x$ on any interval not containing -1 or $-\frac{1}{2}$

11. Verify that, for any nonzero constant a, $y_1(x) = e^{-ax}$ is a solution of $y'' + 2ay' + a^2 y = 0$. Write the general solution.

12. Find an equation for a curve passing through $(0, 2)$ with slope zero and whose curvature at any point (x, y) is $\cos(x)$.

13. Determine a two-parameter family of curves having constant curvature. That is, determine an equation involving x, y, and two arbitrary constants whose graphs all have constant curvature.

14. A second-order equation $F(x, y, y', y'') = 0$ in which y is not explicitly present can sometimes be solved by

putting $u = y'$. This results in a first-order equation $G(x, u, u') = 0$. If this can be solved for $u(x)$, then $y_1(x) = \int u(x)\, dx$ is a solution of the given second-order equation. Use this method to find one solution, then find a second solution, and finally the general solution of the following.

(a) $xy'' = 2 + y'$

(b) $xy'' + 2y' = x$

(c) $1 - y' = 4y''$

(d) $y'' + (y')^2 = 0$

(e) $y'' = 1 + (y')^2$

15. A second-order equation in which x does not explicitly appear can sometimes be solved by putting $u = y'$ and thinking of y as the independent variable and u as a function of y. Write

$$y'' = \frac{d}{dx}\left[\frac{dy}{dx}\right] = \frac{du}{dx} = \frac{du}{dy}\frac{dy}{dx} = u\frac{du}{dy}$$

to convert $F(y, y', y'') = 0$ into the first-order equation $F(y, u, u(du/dy)) = 0$. Solve this equation for $u(y)$ and then set $u = y'$ to solve for y as a function of x. Use this method to find a solution (perhaps implicitly defined) of each of the following.

(a) $yy'' + 3(y')^2 = 0$

(b) $yy'' + (y + 1)(y')^2 = 0$

(c) $yy'' = y^2 y' + (y')^2$

(d) $y'' = 1 + (y')^2$

(e) $y'' + (y')^2 = 0$

16. Consider $y'' + Ay' + By = 0$, in which A and B are constants and $A^2 - 4B = 0$. Show that $y_1(x) = e^{-Ax/2}$ is one solution and use reduction of order to find the second solution $y_2(x) = xe^{-Ax/2}$.

17. Consider $y'' + (A/x)y' + (B/x^2)y = 0$ for $x > 0$, with A and B constants such that $(A - 1)^2 - 4B = 0$. Verify that $y_1(x) = x^{(1-A)/2}$ is one solution and use reduction of order to derive the second solution $y_2(x) = x^{(1-A)/2} \ln(x)$.

18. The second-order differential equation $A(x)y'' + B(x)y' + C(x)y = 0$ is said to be *exact* if it can be written in the form

$$[A(x)y']' + [F(x)y]' = 0$$

for some $F(x)$. This is a different use of the term *exact* than that of Section 1.4 for first-order differential equations.

(a) Show that if the differential equation is exact, then $A''(x) - B'(x) + C(x) = 0$.

(b) Write a formula for the general solution of the differential equation in the case that it is exact. *Hint:* Integrate $[A(x)y']' + [F(x)y]' = 0$ once to obtain a first-order differential equation for y and then solve this equation.

19. In each of Problems (a) through (e), show that the differential equation is exact (Problem 18) by producing an appropriate $F(x)$. Use the procedure of Problem 18(b) to find the general solution.

(a) $2x^2y'' + 5xy' + y = 0$

(b) $x^2y'' + 3xy' + y = 0$

(c) $2x^2y'' + 8xy' + 4y = 0$

(d) $xy'' + (3x + 1)y' + 3y = 0$

(e) $4xy'' + x^2y' + 2xy = 0$

2.4   The Constant Coefficient Homogeneous Linear Equation

The linear homogeneous equation

$$y'' + Ay' + By = 0, \tag{2.4}$$

in which A and B are numbers, occurs frequently in important applications. There is a standard approach to solving this equation.

The form of equation (2.4) requires that constant multiples of derivatives of $y(x)$ must sum to zero. Since the derivative of an exponential function $e^{\lambda x}$ is a constant multiple of $e^{\lambda x}$, we will look for solutions $y(x) = e^{\lambda x}$. To see how to choose λ, substitute $e^{\lambda x}$ into equation (2.4) to get

$$\lambda^2 e^{\lambda x} + A\lambda e^{\lambda x} + Be^{\lambda x} = 0.$$

This can only be true if

$$\lambda^2 + A\lambda + B = 0.$$

This is called the *characteristic equation* of equation (2.4). Its roots are

$$\lambda = \frac{-A \pm \sqrt{A^2 - 4B}}{2},$$

leading to three cases.

2.4.1   Case 1: $A^2 - 4B > 0$

In this case the characteristic equation has two real, distinct roots,

$$a = \frac{-A + \sqrt{A^2 - 4B}}{2} \quad \text{and} \quad b = \frac{-A - \sqrt{A^2 - 4B}}{2}$$

yielding solutions $y_1(x) = e^{ax}$ and $y_2(x) = e^{bx}$ for equation (2.4). These form a fundamental set of solutions on the real line, since

$$W(x) = e^{ax}be^{bx} - e^{bx}ae^{bx} = (b - a)e^{(a+b)x}$$

and this is nonzero because $a \neq b$. The general solution in this case is

$$y(x) = c_1 e^{ax} + c_2 e^{bx}.$$

EXAMPLE 2.6

The characteristic equation of $y'' - y' - 6y = 0$ is

$$\lambda^2 - \lambda - 6 = 0,$$

with roots $a = -2$ and $b = 3$. The general solution is

$$y = c_1 e^{-2x} + c_2 e^{3x}. \quad \blacksquare$$

2.4.2 Case 2: $A^2 - 4B = 0$

Now the characteristic equation has the repeated root $\lambda = -A/2$, so $y_1(x) = e^{-Ax/2}$ is one solution. This method does not provide a second solution, but we have reduction of order for just such a circumstance. Try $y_2(x) = u(x)e^{-Ax/2}$ and substitute into the differential equation to get

$$\frac{A^2}{4} u e^{-Ax/2} - A u' e^{-Ax/2} + u'' e^{-Ax/2} + A \left(-\frac{A}{2} u e^{-Ax/2} + u' e^{-Ax/2} \right) + B u e^{-Ax/2} = 0.$$

Divide by $e^{-Ax/2}$ and rearrange terms to get

$$u'' + \left(B - \frac{A^2}{4} \right) u = 0.$$

Because in the current case $A^2 - 4B = 0$, this differential equation reduces to just $u'' = 0$, and we can choose $u(x) = x$. A second solution in this case is $y_2(x) = xe^{-Ax/2}$. Since y_1 and y_2 are linearly independent, they form a fundamental set and the general solution is

$$y(x) = c_1 e^{-Ax/2} + c_2 x e^{-Ax/2} = e^{-Ax/2}(c_1 + c_2 x).$$

EXAMPLE 2.7

The characteristic equation of $y'' - 6y' + 9y = 0$ is $\lambda^2 - 6\lambda + 9 = 0$, with repeated root $\lambda = 3$. The general solution is

$$y(x) = e^{3x}(c_1 + c_2 x). \quad \blacksquare$$

2.4.3 Case 3: $A^2 - 4B < 0$

Now the characteristic equation has complex roots:

$$\frac{-A \pm \sqrt{4B - A^2} i}{2}.$$

For convenience, write

$$p = -\frac{A}{2}, \quad q = \frac{1}{2}\sqrt{4B - A^2},$$

so the roots of the characteristic equation are $p \pm iq$. This yields two solutions

$$y_1(x) = e^{(p+iq)x} \quad \text{and} \quad y_2(x) = e^{(p-iq)x}.$$

These are linearly independent because their Wronskian is

$$W(x) = \begin{vmatrix} e^{(p+iq)x} & e^{(p-iq)x} \\ (p+iq)e^{(p+iq)x} & (p-iq)e^{(p-iq)x} \end{vmatrix}$$

$$= (p-iq)e^{2px} - (p+iq)e^{2px} = -2iqe^{2pq},$$

and this is nonzero in the current case in which $q \neq 0$. Therefore, the general solution is

$$y(x) = c_1 e^{(p+iq)x} + c_2 e^{(p-iq)x}. \tag{2.5}$$

EXAMPLE 2.8

The characteristic equation of $y'' + 2y' + 6y = 0$ is $\lambda^2 + 2\lambda + 6 = 0$, with roots $-1 \pm \sqrt{5}i$. The general solution is

$$y(x) = c_1 e^{(-1+\sqrt{5}i)x} + c_2 e^{(-1-\sqrt{5}i)x}. \ \blacksquare$$

2.4.4 An Alternative General Solution in the Complex Root Case

When the characteristic equation has complex roots, we can write a general solution in terms of complex exponential functions. This is sometimes inconvenient, for example, in graphing the solutions. But recall that *any* two linearly independent solutions form a fundamental set. We will therefore show how to use the general solution (2.5) to find a fundamental set of real-valued solutions. Begin by recalling the Maclaurin expansions of e^x, $\cos(x)$, and $\sin(x)$:

$$e^x = \sum_{n=0}^{\infty} \frac{1}{n!} x^n = 1 + x + \frac{1}{2!}x^2 + \frac{1}{3!}x^3 + \frac{1}{4!}x^4 + \frac{1}{5!}x^5 + \cdots,$$

$$\cos(x) = \sum_{n=0}^{\infty} \frac{(-1)^n}{(2n)!} x^{2n} = 1 - \frac{1}{2!}x^2 + \frac{1}{4!}x^4 - \frac{1}{6!}x^6 + \cdots,$$

and

$$\sin(x) = \sum_{n=0}^{\infty} \frac{(-1)^n}{(2n+1)!} x^{2n+1} = x - \frac{1}{3!}x^3 + \frac{1}{5!}x^5 - \frac{1}{7!}x^7 + \cdots,$$

with each series convergent for all real x. The eighteenth-century Swiss mathematician Leonhard Euler experimented with replacing x with ix in the exponential series and noticed an interesting relationship among the series for e^x, $\cos(x)$, and $\sin(x)$. First,

$$e^{ix} = \sum_{n=0}^{\infty} \frac{1}{n!} (ix)^n = 1 + ix + \frac{1}{2!}(ix)^2 + \frac{1}{3!}(ix)^3 + \frac{1}{4!}(ix)^4 + \frac{1}{5!}(ix)^5 + \frac{1}{6!}(ix)^6 + \cdots.$$

Now, integer powers of i repeat the values i, -1, $-i$, 1 with a period of four:

$$i^2 = -1, \quad i^3 = -i, \quad i^4 = 1, \quad i^5 = i^4 i = i, \quad i^6 = i^4 i^2 = -1, \quad i^7 = i^4 i^3 = -i,$$

and so on, continuing in cyclic fashion. Using this fact in the Maclaurin series for e^{ix}, we obtain

$$e^{ix} = 1 + ix - \frac{1}{2!}x^2 - \frac{i}{3!}x^3 + \frac{1}{4!}x^4 + \frac{i}{5!}x^5 - \frac{1}{6!}x^6 - \cdots$$

$$= \left(1 - \frac{1}{2!}x^2 + \frac{1}{4!}x^4 - \frac{1}{6!}x^6 + \cdots\right) + i \left(x - \frac{1}{3!}x^3 + \frac{1}{5!}x^5 - \cdots\right) \tag{2.6}$$

$$= \cos(x) + i\sin(x).$$

This is *Euler's formula*. In a different form, it was discovered a few years earlier by Newton's contemporary Roger Cotes (1682–1716). Cotes is not of the stature of Euler, but Newton's high opinion of him is reflected in Newton's remark, "If Cotes had lived, we would have known something."

Since $\cos(-x) = \cos(x)$ and $\sin(-x) = -\sin(x)$, replacing x by $-x$ in Euler's formula yields

$$e^{-ix} = \cos(x) - i\sin(x).$$

Now return to the problem of solving $y'' + Ay' + By = 0$ when the characteristic equation has complex roots $p \pm iq$. Since p and q are real numbers, we have

$$e^{(p+iq)x} = e^{px}e^{iqx} = e^{px}(\cos(qx) + i\sin(qx))$$

$$= e^{px}\cos(qx) + ie^{px}\sin(qx)$$

and

$$e^{(p-iq)x} = e^{px}e^{-iqx} = e^{px}(\cos(qx) - i\sin(qx))$$

$$= e^{px}\cos(qx) - ie^{px}\sin(qx).$$

The general solution (2.5) can therefore be written

$$y(x) = c_1 e^{px}\cos(qx) + ic_1 e^{px}\sin(qx) + c_2 e^{px}\cos(qx) - ic_2 e^{px}\sin(qx)$$

$$= (c_1 + c_2)e^{px}\cos(qx) + (c_1 - c_2)ie^{px}\sin(qx).$$

We obtain solutions for any numerical choices of c_1 and c_2. In particular, if we choose $c_1 = c_2 = \frac{1}{2}$, we obtain the solution

$$y_3(x) = e^{px}\cos(qx).$$

And if we put $c_1 = \frac{1}{2i}$ and $c_2 = -\frac{1}{2i}$, we obtain still another solution:

$$y_4(x) = e^{px}\sin(qx).$$

Further, these last two solutions are linearly independent on the real line, since

$$W(x) = \begin{vmatrix} e^{px}\cos(qx) & e^{px}\sin(qx) \\ pe^{px}\cos(qx) - qe^{px}\sin(qx) & pe^{px}\sin(qx) + qe^{px}\cos(qx) \end{vmatrix}$$

$$= e^{2px}(\sin(qx)\cos(qx) + \cos^2(qx) - \sin(qx)\cos(qx) + \sin^2(qx))$$

$$= e^{2px} \neq 0 \quad \text{for all real } x.$$

We can therefore, if we prefer, form a fundamental set of solutions using y_3 and y_4, writing the general solution of $y'' + Ay' + By = 0$ in this case as

$$y(x) = e^{px}(c_1\cos(qx) + c_2\sin(qx)).$$

This is simply another way of writing the general solution of equation (2.4) in the complex root case.

EXAMPLE 2.9

Revisiting the equation $y'' + 2y' + 6y = 0$ of Example 2.8, we can also write the general solution

$$y(x) = e^{-x}\left(c_1\cos\left(\sqrt{5}x\right) + c_2\sin\left(\sqrt{5}x\right)\right). \quad \blacksquare$$

We now have the general solution of the constant coefficient linear homogeneous equation $y'' + Ay' + By = 0$ in all cases. As usual, we can solve an initial value problem by first finding the general solution of the differential equation, then solving for the constants to satisfy the initial conditions.

EXAMPLE 2.10

Solve the initial value problem

$$y'' - 4y' + 53y = 0; \qquad y(\pi) = -3, y'(\pi) = 2.$$

First solve the differential equation. The characteristic equation is

$$\lambda^2 - 4\lambda + 53 = 0,$$

with complex roots $2 \pm 7i$. The general solution is

$$y(x) = c_1 e^{2x} \cos(7x) + c_2 e^{2x} \sin(7x).$$

Now

$$y(\pi) = c_1 e^{2\pi} \cos(7\pi) + c_2 e^{2\pi} \sin(7\pi) = -c_1 e^{2\pi} = -3,$$

so

$$c_1 = 3e^{-2\pi}.$$

Thus far

$$y(x) = 3e^{-2\pi} e^{2x} \cos(7x) + c_2 e^{2x} \sin(7x).$$

Compute

$$y'(x) = 3e^{-2\pi}[2e^{2x} \cos(7x) - 7e^{2x} \sin(7x)] + 2c_2 e^{2x} \sin(7x) + 7c_2 e^{2x} \cos(7x).$$

Then

$$y'(\pi) = 3e^{-2\pi} 2e^{2\pi}(-1) + 7c_2 e^{2\pi}(-1) = 2,$$

so

$$c_2 = -\tfrac{8}{7} e^{-2\pi}.$$

The solution of the initial value problem is

$$y(x) = 3e^{-2\pi} e^{2x} \cos(7x) - \tfrac{8}{7} e^{-2\pi} e^{2x} \sin(7x)$$

$$= e^{2(x-\pi)} \left[3 \cos(7x) - \tfrac{8}{7} \sin(7x) \right]. \quad \blacksquare$$

SECTION 2.4 PROBLEMS

In each of Problems 1 through 15, find the general solution.

1. $y'' - y' - 6y = 0$

2. $y'' - 2y' + 10y = 0$

3. $y'' + 6y' + 9y = 0$

4. $y'' - 3y' = 0$

5. $y'' + 10y' + 26y = 0$

6. $y'' + 6y' - 40y = 0$

7. $y'' + 3y' + 18y = 0$

8. $y'' + 16y' + 64y = 0$

9. $y'' - 14y' + 49y = 0$

10. $y'' - 6y' + 7y = 0$

11. $y'' + 4y' + 9y = 0$

12. $y'' + 5y' = 0$

13. $y'' - y' - y = 0$

14. $y'' + 12y' + 6y = 0$

15. $y'' - 9y' - 9y = 0$

In each of Problems 16 through 25, solve the initial value problem.

16. $y'' + 2y' - 3y = 0;\ y(0) = 6,\ y'(0) = -2$

17. $y'' + 3y' = 0;\ y(0) = 3,\ y'(0) = 6$

18. $y'' + 2y' - 3y = 0;\ y(0) = 6,\ y'(0) = -2$

19. $y'' - 2y' + y = 0;\ y(1) = y'(1) = 0$

20. $y'' - 4y' + 4y = 0;\ y(0) = 3,\ y'(0) = 5$

21. $y'' + y' - 12y = 0;\ y(2) = 2,\ y'(2) = 1$

22. $y'' - 2y' - 5y = 0;\ y(0) = 0,\ y'(0) = 3$

23. $y'' - 2y' + y = 0;\ y(1) = 12,\ y'(1) = -5$

24. $y'' - 5y' + 12y = 0;\ y(2) = 0,\ y'(2) = -4$

25. $y'' - y' + 4y = 0;\ y(-2) = 1,\ y'(-2) = 3$

26. Let $k > 0$. Write the general solution of $y'' - k^2 y = 0$ in terms of hyperbolic functions.

27. Show that, if $A^2 - 4B > 0$, then the general solution of $y'' + Ay' + By = 0$ can be written

$$y(x) = e^{\alpha x}[c_1 \cosh(\beta x) + c_2 \sinh(\beta x)]$$

for choices of α and β that depend on A and B.

28. This problem illustrates how small changes in the coefficients of a differential equation may cause dramatic changes in the solutions.

(a) Find the general solution $\varphi(x)$ of $y'' - 2ay' + a^2 y = 0$, with a a nonzero constant.

(b) Find the general solution $\varphi_\epsilon(x)$ of $y'' - 2ay' + (a^2 - \epsilon^2)y = 0$, in which ϵ is a positive constant.

(c) Show that as $\epsilon \to 0$, the differential equation in (b) approaches in a limit sense the differential equation in (a), but the solution $\varphi_\epsilon(x)$ for (b) does not in general approach the solution $\varphi(x)$ for (a).

29. (a) Find the solution ψ of the initial value problem

$$y'' - 2ay' + a^2 y = 0; \qquad y(0) = c,\ y'(0) = d,$$

with a, c, and d constants and $a \neq 0$.

(b) Find the solution ψ_ϵ of the initial value problem

$$y'' - 2ay' + (a^2 - \epsilon^2)y = 0; \qquad y(0) = c,\ y'(0) = d.$$

Here ϵ is any positive number.

(c) Is it true that $\lim_{\epsilon \to 0} \psi_\epsilon(x) = \psi(x)$? How does this answer differ, if at all, from the conclusion in Problem 28(c)?

30. Suppose φ is a solution of

$$y'' + Ay' + By = 0; \qquad y(x_0) = a,\ y'(x_0) = b.$$

Here A, B, a, and b are constants. Suppose A and B are positive. Prove that $\lim_{x \to \infty} \varphi(x) = 0$.

31. Show that $a\cos(\omega x) + b\sin(\omega x)$ can always be written as $C\cos(\omega x + \delta)$ for some C and δ, with $0 \leq \delta \leq 2\pi$. The latter expression is called the *phase angle*, or *harmonic*, form of the former. In each of (a) through (d), find the phase angle form of the function.

(a) $-4\sqrt{3}\cos(6x) + 12\sin(6x)$

(b) $-3\cos(\pi x) - \sqrt{3}\sin(\pi x)$

(c) $8\cos(2x) - 8\sin(2x)$

(d) $5\sin(4x)$

32. In each of (a) through (d), find a second-order differential equation having the function as general solution.

(a) $c_1 e^{-2x} + c_2 e^{3x}$

(b) $c_1 e^{-3x}\cos(2x) + c_2 e^{-3x}\sin(2x)$

(c) $c_1 e^{-4x} + c_2 x e^{-4x}$

(d) $c_1 + c_2 e^{-8x}$

2.5   Euler's Equation

In this section we will define another class of second-order differential equations for which there is an elementary technique for finding the general solution.

The second-order homogeneous equation

$$y'' + \frac{1}{x}Ay' + \frac{1}{x^2}By = 0, \tag{2.7}$$

with A and B constant, is called *Euler's equation*. It is defined on the half-lines $x > 0$ and $x < 0$. We will assume for this section that $x > 0$.

We will solve Euler's equation by transforming it to a constant coefficient linear equation, which we can solve easily. Recall that any positive number x can be written as e^t for some t (namely, for $t = \ln(x)$). Make the change of variables

$$x = e^t \quad \text{or, equivalently, } t = \ln(x)$$

and let

$$Y(t) = y(e^t).$$

That is, in the function $y(x)$, replace x by e^t, obtaining a new function of t. For example, if $y(x) = x^3$, then $Y(t) = (e^t)^3 = e^{3t}$. Now compute chain-rule derivatives. First,

$$y'(x) = \frac{dY}{dt}\frac{dt}{dx} = Y'(t)\frac{1}{x}$$

so

$$Y'(t) = xy'(x).$$

Next,

$$y''(x) = \frac{d}{dx}\, y'(x) = \frac{d}{dx}\left(\frac{1}{x}Y'(t)\right)$$

$$= -\frac{1}{x^2}Y'(t) + \frac{1}{x}\frac{d}{dx}\, Y'(t)$$

$$= -\frac{1}{x^2}Y'(t) + \frac{1}{x}\frac{dY'}{dt}\frac{dt}{dx}$$

$$= -\frac{1}{x^2}Y'(t) + \frac{1}{x}Y''(t)\frac{1}{x}$$

$$= \frac{1}{x^2}(Y''(t) - Y'(t)).$$

Therefore,

$$x^2y''(x) = Y''(t) - Y'(t).$$

If we write Euler's equation as

$$x^2y''(x) + Axy'(x) + By(x) = 0,$$

then these substitutions yield

$$Y''(t) - Y'(t) + AY'(t) + BY(t) = 0,$$

or

$$Y'' + (A - 1)Y' + BY = 0. \tag{2.8}$$

This is a constant coefficient homogeneous linear differential equation for $Y(t)$. Solve this equation, then let $t = \ln(x)$ in the solution $Y(t)$ to obtain $y(x)$ satisfying the Euler equation.

We need not repeat this derivation each time we want to solve an Euler equation, since the coefficients $A - 1$ and B for the transformed equation (2.8) are easily read from the Euler equation (2.7).

In carrying out this strategy, it is useful to recall that for $x > 0$,

$$x^r = e^{r \ln(x)}.$$

EXAMPLE 2.11

Find the general solution of $x^2 y'' + 2xy' - 6y = 0$.

Upon letting $x = e^t$, this differential equation transforms to

$$Y'' + Y' - 6Y = 0.$$

The coefficient of Y' is $A - 1$, with $A = 2$ in Euler's equation. The general solution of this linear homogeneous differential equation is

$$Y(t) = c_1 e^{-3t} + c_2 e^{2t}$$

for all real t. Putting $t = \ln(x)$ with $x > 0$, we obtain

$$y(x) = c_1 e^{-3 \ln(x)} + c_2 e^{2 \ln(x)} = c_1 x^{-3} + c_2 x^2,$$

and this is the general solution of the Euler equation. ∎

EXAMPLE 2.12

Consider the Euler equation $x^2 y'' - 5xy' + 9y = 0$. The transformed equation is

$$Y'' - 6Y' + 9Y = 0,$$

with general solution

$$Y(t) = c_1 e^{3t} + c_2 t e^{3t}.$$

Let $t = \ln(x)$ to obtain

$$y(x) = c_1 x^3 + c_2 x^3 \ln(x)$$

for $x > 0$. This is the general solution of the Euler equation. ∎

EXAMPLE 2.13

Solve $x^2 y'' + 3xy' + 10y = 0$

This transforms to

$$Y'' + 2Y' + 10Y = 0,$$

with general solution

$$Y(t) = c_1 e^{-t} \cos(3t) + c_2 e^{-t} \sin(3t).$$

Then

$$y(x) = c_1 x^{-1} \cos(3 \ln(x)) + c_2 x^{-1} \sin(3 \ln(x))$$

$$= \frac{1}{x} (c_1 \cos(3 \ln(x)) + c_2 \sin(3 \ln(x)))$$

for $x > 0$. ∎

As usual, we can solve an initial value problem by finding the general solution of the differential equation, then solving for the constants to satisfy the initial conditions.

EXAMPLE 2.14

Solve the initial value problem

$$x^2 y'' - 5xy' + 10y = 0; \qquad y(1) = 4, y'(1) = -6.$$

We will first find the general solution of the Euler equation, then determine the constants to satisfy the initial conditions. With $t = \ln(x)$, we obtain

$$Y'' - 6Y' + 10Y = 0,$$

having general solution

$$Y(t) = c_1 e^{3t} \cos(t) + c_2 e^{3t} \sin(t).$$

The general solution of the Euler equation is

$$y(x) = c_1 x^3 \cos(\ln(x)) + c_2 x^3 \sin(\ln(x)).$$

For the first initial condition, we need

$$y(1) = 4 = c_1.$$

Thus far,

$$y(x) = 4x^3 \cos(\ln(x)) + c_2 x^3 \sin(\ln(x)).$$

Then

$$y'(x) = 12x^2 \cos(\ln(x)) - 4x^2 \sin(\ln(x)) + 3c_2 x^2 \sin(\ln(x)) + c_2 x^2 \cos(\ln(x)),$$

so

$$y'(1) = 12 + c_2 = -6.$$

Then $c_2 = -18$ and the solution of the initial value problem is

$$y(x) = 4x^3 \cos(\ln(x)) - 18x^3 \sin(\ln(x)). \quad \blacksquare$$

Observe the structure of the solutions of different kinds of differential equations. Solutions of the constant coefficient linear equation $y'' + Ay' + By = 0$ must have the forms $e^{\alpha x}$, $xe^{\alpha x}$, $e^{\alpha x} \cos(\beta x)$, or $e^{\alpha x} \sin(\beta x)$, depending on the coefficients. And solutions of an Euler equation $x^2 y'' + Axy' + By = 0$ must have the forms x^r, $x^r \ln(x)$, $x^p \cos(q \ln(x))$, or $x^p \sin(q \ln(x))$. For example, x^3 could never be a solution of the constant coefficient linear equation and e^{-6x} could never be the solution of an Euler equation.

SECTION 2.5   PROBLEMS

In each of Problems 1 through 15, find the general solution.

1. $x^2 y'' + 2xy' - 6y = 0$

2. $x^2 y'' + 3xy' + y = 0$

3. $x^2 y'' + xy' + 4y = 0$

4. $x^2 y'' + xy' - 4y = 0$

5. $x^2 y'' + xy' - 16y = 0$

6. $x^2y'' + 3xy' + 10y = 0$

7. $x^2y'' + 6xy' + 6y = 0$

8. $x^2y'' - 5xy' + 58y = 0$

9. $x^2y'' + 25xy' + 144y = 0$

10. $x^2y'' - 11xy' + 35y = 0$

11. $x^2y'' - 2xy' + 12y = 0$

12. $x^2y'' + 4y = 0$

13. $x^2y'' + 7xy' - 2y = 0$

14. $x^2y'' + xy' = 0$

15. $x^2y'' - 8xy' - 8y = 0$

In each of Problems 16 through 25, solve the initial value problem.

16. $x^2y'' + 3xy' + 2y = 0; \ y(1) = 3, \ y'(1) = 3$

17. $x^2y'' + 5xy' + 20y = 0; \ y(-1) = 3, \ y'(-1) = 2$
(Here the solution of Euler's equation for $x < 0$ is needed.)

18. $x^2y'' + 5xy' - 21y = 0; \ y(2) = 1, \ y'(2) = 0$

19. $x^2y'' - xy' = 0; \ y(2) = 5, \ y'(2) = 8$

20. $x^2y'' - 3xy' + 4y = 0; \ y(1) = 4, \ y'(1) = 5$

21. $x^2y'' + 7xy' + 13y = 0; \ y(-1) = 1, \ y'(-1) = 3$

22. $x^2y'' + xy' - y = 0; \ y(2) = 1, \ y'(2) = -3$

23. $x^2y'' + 25xy' + 144y = 0; \ y(1) = -4, \ y'(1) = 0$

24. $x^2y'' - 9xy' + 24y = 0; \ y(1) = 1, \ y'(1) = 10$

25. $x^2y'' + xy' - 4y = 0; \ y(1) = 7, \ y'(1) = -3$

26. Here is another approach to solving an Euler equation. For $x > 0$, substitute $y = x^r$ and obtain values of r to make this a solution. Show how this leads in all cases to the same general solution as obtained by the transformation method.

27. The transformation method used to solve the Euler equation is a special case of a more general technique. Determine conditions on $p(x)$ and $q(x)$ so that there exists a transformation converting $y'' + p(x)y' +$ $q(x)y = 0$ into a constant coefficient second-order linear differential equation. *Hint:* Suppose $t = f(x)$ is such a transformation, with f twice differentiable and having an inverse, $x = f^{-1}(t)$. Let $Y(t) = y(f^{-1}(x))$. Show that $y'(x) = Y'(t)f'(x)$ and $y''(x) = Y''(t)[f'(x)]^2 + Y'(t)f''(x)$. Thus obtain

$$Y''(t) + \frac{f''(x) + p(x)f'(x)}{[f'(x)]^2}Y'(t) + \frac{q(x)}{[f'(x)]^2}Y(t) = 0.$$

This will be a constant coefficient equation exactly when, for some constants a and b,

$$\frac{f''(x) + p(x)f'(x)}{[f'(x)]^2} = a \quad \text{and} \quad \frac{q(x)}{[f'(x)]^2} = b.$$

Differentiate the second equation and solve for $f''(x)$, then obtain $f'(x)$ in terms of known functions. Use these expressions for $f'(x)$ and $f''(x)$ in the first equation to obtain

$$\frac{q'(x) + 2p(x)q(x)}{[q(x)]^{3/2}} = \frac{2a}{\sqrt{b}} = \text{constant}.$$

Verify that this condition is satisfied in the case of the Euler equation. Finally, use $q(x) = b[f'(x)]^2$ to obtain the transformation function $f(x)$.

In each of Problems 28 through 31, use the idea of Problem 27 to transform the given equation into a constant coefficient differential equation. Find the general solution of the transformed equation and use it to write the general solution of the given equation.

28. $x(1 - x^2)^2y'' - (1 - x^2)^2y' + x^3y = 0$

29. $y'' + (e^x - 1)y' + e^{2x}y = 0$

30. $4y'' + 4(e^x - 1)y' + e^{2x}y = 0$

31. $xy'' + (x^2 - 1)y' + x^3y = 0$

In each of Problems 32 through 34, find a differential equation having the given function as general solution.

32. $c_1x^2 + c_2x^{-3}$

33. $x^{-2}[c_1\cos(3\ln(x)) + c_2\sin(3\ln(x))]$

34. $c_1x^4 + c_2x^4\ln(x)$

2.6   The Nonhomogeneous Equation $y'' + p(x)y' + q(x)y = f(x)$

In view of Theorem 2.5, if we are able to find the general solution y_h of the linear homogeneous equation $y'' + p(x)y' + q(x)y = 0$, then the general solution of the linear nonhomogeneous equation

$$y'' + p(x)y' + q(x)y = f(x) \tag{2.9}$$

is $y = y_h + y_p$, in which y_p is *any* solution of equation (2.9). This section is devoted to two methods for finding such a particular solution y_p.

2.6.1 The Method of Variation of Parameters

Suppose we can find a fundamental set of solutions y_1 and y_2 for the homogeneous equation. The general solution of this homogeneous equation has the form $y_h(x) = c_1 y_1(x) + c_2 y_2(x)$. The method of variation of parameters consists of attempting a particular solution of the nonhomogeneous equation by replacing the constants c_1 and c_2 with functions of x. Thus, attempt to find $u(x)$ and $v(x)$ so that

$$y_p(x) = u(x)y_1(x) + v(x)y_2(x)$$

is a solution of equation (2.9). How should we choose u and v?

First compute

$$y_p' = uy_1' + vy_2' + u'y_1 + v'y_2.$$

In order to simplify this expression, the first condition we will impose on u and v is that

$$u'y_1 + v'y_2 = 0. \tag{2.10}$$

Now

$$y_p' = uy_1' + vy_2'.$$

Next compute

$$y_p'' = u'y_1' + v'y_2' + uy_1'' + vy_2''.$$

Substitute these expressions for y_p' and y_p'' into equation (2.9):

$$u'y_1' + v'y_2' + uy_1'' + vy_2'' + p(x)(uy_1' + vy_2') + q(x)(uy_1 + vy_2) = f(x).$$

Rearrange terms in this equation to get

$$u[y_1'' + p(x)y_1' + q(x)y_1] + v[y_2'' + p(x)y_2' + q(x)y_2] + u'y_1' + v'y_2' = f(x).$$

The two terms in square brackets vanish because y_1 and y_2 are solutions of the homogeneous equation. This leaves

$$u'y_1' + v'y_2' = f(x). \tag{2.11}$$

Now solve equations (2.10) and (2.11) for u' and v' to get

$$u'(x) = -\frac{y_2(x)f(x)}{W(x)} \quad \text{and} \quad v'(x) = \frac{y_1(x)f(x)}{W(x)} \tag{2.12}$$

in which W is the Wronskian of y_1 and y_2. If we can integrate these equations to determine u and v, then we have y_p.

EXAMPLE 2.15

We will find the general solution of $y'' + 4y = \sec(x)$ for $-\pi/4 < x < \pi/4$.

The characteristic equation of $y'' + 4y = 0$ is $\lambda^2 + 4 = 0$, with roots $\pm 2i$. We may therefore choose $y_1(x) = \cos(2x)$ and $y_2(x) = \sin(2x)$. The Wronskian of these solutions of the homogeneous equation is

$$W(x) = \begin{vmatrix} \cos(2x) & \sin(2x) \\ -2\sin(2x) & 2\cos(2x) \end{vmatrix} = 2.$$

With $f(x) = \sec(x)$, equations (2.12) give us

$$u'(x) = -\frac{1}{2}\sin(2x)\sec(x)$$

$$= -\frac{1}{2}2\sin(x)\cos(x)\frac{1}{\cos(x)} = -\sin(x)$$

and

$$v'(x) = \frac{1}{2}\cos(2x)\sec(x) = \frac{1}{2}[2\cos^2(x) - 1]\frac{1}{\cos(x)}$$

$$= \cos(x) - \frac{1}{2}\sec(x).$$

Then

$$u(x) = \int -\sin(x)\,dx = \cos(x)$$

and

$$v(x) = \int \cos(x)\,dx - \frac{1}{2}\int \sec(x)\,dx$$

$$= \sin(x) - \frac{1}{2}\ln|\sec(x) + \tan(x)|.$$

Here we have let the constants of integration be zero because we need only one u and one v. Now we have the particular solution

$$y_p(x) = u(x)y_1(x) + v(x)y_2(x)$$

$$= \cos(x)\cos(2x) + \left(\sin(x) - \frac{1}{2}\ln|\sec(x) + \tan(x)|\right)\sin(2x).$$

The general solution of $y'' + 4y = \sec(x)$ is

$$y(x) = y_h(x) + y_p(x) = c_1\cos(2x) + c_2\sin(2x) + \cos(x)\cos(2x)$$

$$+ \left(\sin(x) - \frac{1}{2}\ln|\sec(x) + \tan(x)|\right)\sin(2x). \blacksquare$$

EXAMPLE 2.16

Suppose we want the general solution of

$$y'' - \frac{4}{x}y' + \frac{4}{x^2}y = x^2 + 1$$

for $x > 0$. The associated homogeneous equation is

$$y'' - \frac{4}{x}y' + \frac{4}{x^2}y = 0,$$

which we recognize as an Euler equation, with fundamental solutions $y_1(x) = x$ and $y_2(x) = x^4$ for $x > 0$.

The Wronskian of these solutions is

$$W(x) = \begin{vmatrix} x & x^4 \\ 1 & 4x^3 \end{vmatrix} = 3x^4$$

and this is nonzero for $x > 0$. From equations (2.12),

$$u'(x) = -\frac{x^4(x^2+1)}{3x^4} = -\frac{1}{3}(x^2+1)$$

and

$$v'(x) = \frac{x(x^2+1)}{3x^4} = \frac{1}{3}\left(\frac{1}{x} + \frac{1}{x^3}\right).$$

Integrate to get

$$u(x) = -\frac{1}{9}x^3 - \frac{1}{3}x$$

and

$$v(x) = \frac{1}{3}\ln(x) - \frac{1}{6x^2}.$$

A particular solution is

$$y_p(x) = \left(-\frac{1}{9}x^3 - \frac{1}{3}x\right)x + \left(\frac{1}{3}\ln(x) - \frac{1}{6x^2}\right)x^4.$$

The general solution is

$$y(x) = y_h(x) + y_p(x)$$

$$= c_1x + c_2x^4 - \frac{1}{9}x^4 - \frac{1}{3}x^2 + \frac{1}{3}x^4\ln(x) - \frac{1}{6}x^2$$

$$= c_1x + c_2x^4 - \frac{1}{9}x^4 - \frac{1}{2}x^2 + \frac{1}{3}x^4\ln(x).$$

for $x > 0$. ∎

2.6.2 The Method of Undetermined Coefficients

Here is a second method for finding a particular solution y_p, but it only applies if $p(x)$ and $q(x)$ are constant. Thus consider

$$y'' + Ay' + By = f(x).$$

Sometimes we can guess the general form of a solution y_p from the form of $f(x)$. For example, suppose $f(x)$ is a polynomial. Since derivatives of polynomials are polynomials, we might try a polynomial for $y_p(x)$. Substitute a polynomial with unknown coefficients into the differential equation and then choose the coefficients to match $y'' + Ay' + By$ with $f(x)$. Or suppose $f(x)$ is an exponential function, say $f(x) = e^{-2x}$. Since derivatives of e^{-2x} are just constant multiples of e^{-2x}, we would attempt a solution of the form $y_p = Ce^{-2x}$, substitute into the differential equation and solve for C to match the left and right sides of the differential equation.

Here are some examples of this method.

EXAMPLE 2.17

Solve $y'' - 4y = 8x^2 - 2x$.

Since $f(x) = 8x^2 - 2x$ is a polynomial of degree 2, we will attempt a solution

$$y_p(x) = ax^2 + bx + c.$$

We do not need to try a higher-degree polynomial, since the degree of $y'' - 4y$ must be 2. If, for example, we included an x^3 term in y_p, then $y_p'' - 4y_p$ would have an x^3 term, and we know that it does not.

Compute

$$y_p' = 2ax + b \quad \text{and} \quad y_p'' = 2a$$

and substitute into the differential equation to get

$$2a - 4(ax^2 + bx + c) = 8x^2 - 2x.$$

Collect coefficients of like powers of x to write

$$(-4a - 8)x^2 + (-4b + 2)x + (2a - 4c) = 0.$$

For y_p to be a solution for all x, the polynomial on the left must be zero for all x. But a second-degree polynomial can have only two roots, unless it is the zero polynomial. Thus all the coefficients must vanish, and we have the equations

$$-4a - 8 = 0,$$
$$-4b + 2 = 0,$$
$$2a - 4c = 0.$$

Solve these to obtain

$$a = -2, \quad b = \tfrac{1}{2}, \quad c = -1.$$

Thus a solution is

$$y_p(x) = -2x^2 + \tfrac{1}{2}x - 1,$$

as can be verified by substitution into the differential equation.

If we want the general solution of the differential equation, we need the general solution y_h of $y'' - 4y = 0$. This is

$$y_h(x) = c_1 e^{2x} + c_2 e^{-2x}.$$

The general solution of $y'' - 4y = 8x^2 - 2x$ is

$$y(x) = c_1 e^{2x} + c_2 e^{-2x} - 2x^2 + \tfrac{1}{2}x - 1. \quad \blacksquare$$

The method we have just illustrated is called the *method of undetermined coefficients*, because the idea is to guess a general form for y_p and then solve for the coefficients to make a solution. Here are two more examples, after which we will point out a circumstance in which we must supplement the method.

EXAMPLE 2.18

Solve $y'' + 2y' - 3y = 4e^{2x}$.

Because $f(x)$ is a constant times an exponential, and the derivative of such a function is always a constant times the same function, we attempt $y_p = ae^{2x}$. Then $y_p' = 2ae^{2x}$ and

$y_p'' = 4ae^{2x}$. Substitute into the differential equation to get

$$4ae^{2x} + 4ae^{2x} - 3ae^{2x} = 4e^{2x}.$$

Then $5ae^{2x} = 4e^{2x}$, so choose $a = \frac{4}{5}$ to get the solution

$$y_p(x) = \tfrac{4}{5}e^{2x}.$$

Again, if we wish we can write the general solution

$$y(x) = c_1 e^{-3x} + c_2 e^x + \tfrac{4}{5}e^{2x}. \quad \blacksquare$$

EXAMPLE 2.19

Solve $y'' - 5y' + 6y = -3\sin(2x)$.

Here $f(x) = -3\sin(2x)$. Now we must be careful, because derivatives of $\sin(2x)$ can be multiples $\sin(2x)$ or $\cos(2x)$, depending on how many times we differentiate. This leads us to include both possibilities in a proposed solution:

$$y_p(x) = c\cos(2x) + d\sin(2x).$$

Compute

$$y_p' = -2c\sin(2x) + 2d\cos(2x), \quad y_p'' = -4c\cos(2x) - 4d\sin(2x).$$

Substitute into the differential equation to get

$$-4c\cos(2x) - 4d\sin(2x) - 5[-2c\sin(2x) + 2d\cos(2x)]$$
$$+ 6[c\cos(2x) + d\sin(2x)] = -3\sin(2x).$$

Collecting the cosine terms on one side and the sine terms on the other:

$$[2d + 10c + 3]\sin(2x) = [-2c + 10d]\cos(2x).$$

For y_p to be a solution for all real x, this equation must hold for all x. But $\sin(2x)$ and $\cos(2x)$ are linearly independent (they are solutions of $y'' + 4y = 0$, and their Wronskian is nonzero). Therefore, neither can be a constant multiple of the other. The only way the last equation can hold for all x is for the coefficient to be zero on both sides:

$$2d + 10c = -3$$

and

$$10d - 2c = 0.$$

Then

$$d = -\frac{3}{52} \quad \text{and} \quad c = -\frac{15}{52}$$

and we have found a solution:

$$y_p(x) = -\frac{15}{52}\cos(2x) - \frac{3}{52}\sin(2x).$$

The general solution of this differential equation is

$$y(x) = c_1 e^{3x} + c_2 e^{2x} - \frac{15}{52}\cos(2x) - \frac{3}{52}\sin(2x). \quad \blacksquare$$

As effective as this method is, there are several circumstances in it that may fail to produce a solution or may require an additional step.

1. The differential equation must be of the form $y'' + Ay' + By = f(x)$, with A and B constant. If either of these coefficients is a function of x, this method does not apply. We could not solve $y'' - xy' + y = \cos(2x)$ this way.

2. $f(x)$ must suggest a general form for y_p. For example, it is not clear what we would try for y_p to solve $y'' + 2y' + 6y = \tan(3x)$.

3. Items 1 and 2 are limitations on the differential equation and on f. But there is also a difficulty that is intrinsic to the method. It can be successfuly met, but one must be aware of it and know how to proceed.

Consider the following example.

EXAMPLE 2.20

Solve $y'' + 2y' - 3y = 8e^x$. The coefficients on the left side are constant, and $f(x) = 8e^x$ seems simple enough, so we proceed with

$$y_p(x) = ce^x.$$

Substitute into the differential equation to get

$$ce^x + 2ce^x - 3ce^x = 8e^x,$$

or

$$0 = 8e^x.$$

Something is wrong. What happened? ∎

The problem in this example is that e^x is a solution of $y'' + 2y' - 3y = 0$, so if we substitute ce^x into $y'' + 2y' - 3y = 8e^x$, the left side will equal zero, not $8e^x$. This difficulty will occur whenever the proposed y_p contains a term that is a solution of the homogeneous equation $y'' + Ay' + By = 0$, because then this term (which may be all of the proposed y_p) will vanish when substituted into $y'' + Ay' + By$.

There is a way out of this difficulty. If a term of the proposed y_p is a solution of $y'' + Ay' + By = 0$, multiply the proposed solution by x and try the modified function as y_p. If this also contains a term, or by itself satisfies $y'' + Ay' + By = 0$, then multiply by x again to try x^2 times the original proposed solution. This is as far as we will have to go in the case of second-order differential equations.

Now continue Example 2.20 with this strategy.

EXAMPLE 2.21

Consider again $y'' + 2y' - 3y = 8e^x$. We saw that $y_p = ce^x$ does not work, because e^x, and hence also ce^x, satisfies $y'' + 2y' - 3y = 0$. Try $y_p = cxe^x$. Compute

$$y'_p = ce^x + cxe^x, \quad y''_p = 2ce^x + cxe^x$$

and substitute into the differential equation to get

$$2ce^x + cxe^x + 2[ce^x + cxe^x] - 3cxe^x = 8e^x.$$

Some terms cancel and we are left with

$$4ce^x = 8e^x.$$

Choose $c = 2$ to obtain the particular solution $y_p(x) = 2xe^x$. ∎

EXAMPLE 2.22

Solve $y'' - 6y' + 9y = 5e^{3x}$.

Our first impulse is to try $y_p = ce^{3x}$. But this is a solution of $y'' - 6y' + 9y = 0$. If we try $y_p = cxe^{3x}$, we also obtain an equation that cannot be solved for c. The reason is that the characteristic equation of $y'' - 6y' + 9y = 0$ is $(\lambda - 3)^2 = 0$, with repeated root 3. This means that e^{3x} and xe^{3x} are both solutions of the homogeneous equation $y'' - 6y' + 9y = 0$. Thus try $y_p(x) = cx^2 e^{3x}$. Compute

$$y_p' = 2cxe^{3x} + 3cx^2 e^{3x}, \quad y_p'' = 2ce^{3x} + 12cxe^{3x} + 9cx^2 e^{3x}.$$

Substitute into the differential equation to get

$$2ce^{3x} + 12cxe^{3x} + 9cx^2 e^{3x} - 6[2cxe^{3x} + 3cx^2 e^{3x}] + 9cx^2 e^{3x} = 5e^{3x}.$$

After cancellations we have

$$2ce^{3x} = 5e^{3x}$$

so $c = 5/2$. We have found a particular solution $y_p(x) = 5x^2 e^{3x}/2$. ∎

The last two examples suggest that in applying undetermined coefficients to $y'' + Ay' + By = f(x)$, we should first obtain the general solution of $y'' + Ay' + By = 0$. We need this anyway for a general solution of the nonhomogeneous equation, but it also tells us whether to multiply our first choice for y_p by x or x^2 before proceeding.

Here is a summary of the method of undetermined coefficients.

1. From $f(x)$, make a first conjecture for the form of y_p.
2. Solve $y'' + Ay' + By = 0$. If a solution of this equation appears in any term of the conjectured form for y_p, modify this form by multiplying it by x. If this modified function still occurs in a solution of $y'' + Ay' + By = 0$, multiply by x again (so the original y_p is multiplied by x^2 in this case).
3. Substitute the final proposed y_p into $y'' + Ay' + By = f(x)$ and solve for its coefficients.

Here is a list of functions to try in the initial stage (1) of formulating y_p. In this list $P(x)$ indicates a given polynomial of degree n, and $Q(x)$ and $R(x)$ polynomials with undetermined coefficients, of degree n.

$f(x)$	Initial Guess for y_p
$P(x)$	$Q(x)$
ce^{ax}	de^{ax}
$\alpha \cos(bx)$ or $\beta \sin(bx)$	$c \cos(bx) + d \sin(bx)$
$P(x)e^{ax}$	$Q(x)e^{ax}$
$P(x) \cos(bx)$ or $P(x) \sin(bx)$	$Q(x) \cos(bx) + R(x) \sin(bx)$
$P(x)e^{ax} \cos(bx)$ or $P(x)e^{ax} \sin(bx)$	$Q(x)e^{ax} \cos(bx) + R(x)e^{ax} \sin(bx)$

EXAMPLE 2.23

Solve $y'' + 9y = -4x \sin(3x)$.

With $f(x) = -4x \sin(3x)$, the preceding list suggests that we attempt a particular solution of the form

$$y_p(x) = (ax + b) \cos(3x) + (cx + d) \sin(3x).$$

Now solve $y'' + 9y = 0$ to obtain the fundamental set of solutions $\cos(3x)$ and $\sin(3x)$. The proposed y_p includes terms $b \cos(3x)$ and $d \sin(3x)$, which are also solutions of $y'' + 9y = 0$. Therefore, modify the proposed y_p by multiplying it by x, trying instead

$$y_p(x) = (ax^2 + bx) \cos(3x) + (cx^2 + dx) \sin(3x).$$

Compute

$$y_p' = (2ax + b) \cos(3x) - (3ax^2 + 3bx) \sin(3x) + (2cx + d) \sin(3x) + (3cx^2 + 3dx) \cos(3x)$$

and

$$\begin{aligned} y_p'' = {}& 2a \cos(3x) - (6ax + 3b) \sin(3x) - (6ax + 3b) \sin(3x) \\ & - (9ax^2 + 9bx) \cos(3x) + 2c \sin(3x) + (6cx + 3d) \cos(3x) \\ & + (6cx + 3d) \cos(3x) - (9cx^2 + 9dx) \sin(3x). \end{aligned}$$

Substitute these into the differential equation to obtain

$$\begin{aligned} & 2a \cos(3x) - (6ax + 3b) \sin(3x) - (6ax + 3b) \sin(3x) \\ & - (9ax^2 + 9bx) \cos(3x) + 2c \sin(3x) + (6cx + 3d) \cos(3x) \\ & + (6cx + 3d) \cos(3x) - (9cx^2 + 9dx) \sin(3x) \\ & + (9ax^2 + 9bx) \cos(3x) + (9cx^2 + 9dx) \sin(3x) = -4x \sin(3x). \end{aligned}$$

Now collect coefficients of "like" terms ($\sin(3x)$, $x \sin(3x)$, $x^2 \sin(3x)$ and so on). We get

$$(2a + 6d) \cos(3x) + (-6b + 2c) \sin(3x) + 12cx \cos(3x) + (-12a + 4)x \sin(3x) = 0,$$

with all other terms canceling. For this linear combination of $\cos(3x)$, $\sin(3x)$, $x \cos(3x)$, and $x \sin(3x)$ to be zero for all x, each coefficient must be zero. Therefore,

$$2a + 6d = 0,$$
$$-6b + 2c = 0,$$
$$12c = 0$$

and

$$-12a + 4 = 0.$$

Then $a = \frac{1}{3}$, $c = 0$, $b = 0$ and $d = -\frac{1}{9}$. We have found the particular solution

$$y_p(x) = \tfrac{1}{3}x^2 \cos(3x) - \tfrac{1}{9}x \sin(3x).$$

The general solution is

$$y(x) = c_1 \cos(3x) + c_2 \sin(3x) + \tfrac{1}{3}x^2 \cos(3x) - \tfrac{1}{9}x \sin(3x). \quad \blacksquare$$

Sometimes a differential equation has nonconstant coefficients but transforms to a constant coefficient equation. We may then be able to use the method of undetermined coefficients on the transformed equation and then use the results to obtain solutions of the original equation.

EXAMPLE 2.24

Solve $x^2 y'' - 5xy' + 8y = 2\ln(x)$.

The method of undetermined coefficients does not apply here, since the differential equation has nonconstant coefficients. However, from our experience with the Euler equation, apply the transformation $t = \ln(x)$ and let $Y(t) = y(e^t)$. Using results from Section 2.5, the differential equation transforms to

$$Y''(t) - 6Y'(t) + 8Y(t) = 2t,$$

which has constant coefficients on the left side.

The homogeneous equation $Y'' - 6Y' + 8Y = 0$ has general solution

$$Y_h(t) = c_1 e^{2t} + c_2 e^{4t}$$

and, by the method of undetermined coefficients, we find one solution of $Y'' - 6Y' + 8Y = 2t$ to be

$$Y_p(t) = \tfrac{1}{4}t + \tfrac{3}{16}.$$

The general solution for Y is

$$Y(t) = c_1 e^{2t} + c_2 e^{4t} + \tfrac{1}{4}t + \tfrac{3}{16}.$$

Since $t = \ln(x)$, the original differential equation for y has general solution

$$y(x) = c_1 e^{2\ln(x)} + c_2 e^{4\ln(x)} + \tfrac{1}{4}\ln(x) + \tfrac{3}{16}$$

$$= c_1 x^2 + c_2 x^4 + \tfrac{1}{4}\ln(x) + \tfrac{3}{16}. \quad \blacksquare$$

2.6.3 The Principle of Superposition

Consider the equation

$$y'' + p(x)y' + q(x)y = f_1(x) + f_2(x) + \cdots + f_N(x). \tag{2.13}$$

Suppose y_{pj} is a solution of

$$y'' + p(x)y' + q(x)y = f_j(x).$$

We claim that

$$y_{p1} + y_{p2} + \cdots + y_{pN}$$

is a solution of equation (2.13). This is easy to check by direct substitution into the differential equation:

$$(y_{p1} + y_{p2} + \cdots + y_{pN})'' + p(x)(y_{p1} + y_{p2} + \cdots + y_{pN})'$$

$$+ q(x)(y_{p1} + y_{p2} + \cdots + y_{pN})$$

$$= (y_{p1}'' + p(x)y_{p1}' + q(x)y_{p1}) + \cdots + (y_{pN}'' + p(x)y_{pN}' + q(x)y_{pN})$$

$$= f_1(x) + f_2(x) + \cdots + f_N(x).$$

This means that we can solve each equation $y'' + p(x)y' + q(x)y = f_j(x)$ individually, and the sum of these solutions is a solution of equation (2.13). This is called the *principle of superposition*, and it sometimes enables us to solve a problem by breaking it into a sum of "smaller" problems that are easier to handle individually.

EXAMPLE 2.25

Solve $y'' + 4y = x + 2e^{-2x}$.

Consider two problems:

Problem 1: $y'' + 4y = x$, and

Problem 2: $y'' + 4y = 2e^{-2x}$.

Using undetermined coefficients, we find that a solution of Problem 1 is $y_{p1}(x) = x/4$, and that a solution of Problem 2 is $y_{p2}(x) = e^{-2x}/4$. Therefore,

$$y_p(x) = \tfrac{1}{4}(x + e^{-2x})$$

is a solution of $y'' + 4y = x + 2e^{-2x}$. The general solution of this differential equation is

$$y(x) = c_1 \cos(2x) + c_2 \sin(2x) + \tfrac{1}{4}(x + e^{-2x}). \;\blacksquare$$

2.6.4   Higher-Order Differential Equations

The methods we now have for solving $y'' + p(x)y' + q(x)y = f(x)$ under certain conditions can also be applied to higher-order differential equations, at least in theory. However, there are practical difficulties to this approach. Consider the following example.

EXAMPLE 2.26

Solve

$$\frac{d^6 y}{dx^6} - 4\frac{d^4 y}{dx^4} + 2\frac{dy}{dx} + 15y = 0.$$

If we take a cue from the second-order case, we attempt solutions $y = e^{\lambda x}$. Upon substituting this into the differential equation, we obtain an equation for λ:

$$\lambda^6 - 4\lambda^4 + 2\lambda + 15 = 0.$$

In the second-order case, the characteristic polynomial is always of degree 2 and easily solved. Here we encounter a sixth-degree polynomial whose roots are not obvious. They are, approximately,

$$-1.685798616 \pm 0.2107428331i,$$

$$-0.04747911354 \pm 1.279046854i,$$

and

$$1.733277730 \pm 0.4099384482i. \;\blacksquare$$

When the order of the differential equation is $n > 2$, having to find the roots of an nth degree polynomial is enough of a barrier to make this approach impractical except in special cases. A better approach is to convert this sixth-order equation to a system of first-order equations as follows. Define new variables

$$z_1 = y, \quad z_2 = y', \quad z_3 = y'', \quad z_4 = \frac{d^3 y}{dx^3}, \quad z_5 = \frac{d^4 y}{dx^4}, \quad z_6 = \frac{d^5 y}{dx^5}.$$

Now we have a system of six first-order differential equations:

$$z_1' = z_2$$
$$z_2' = z_3$$
$$z_3' = z_4$$
$$z_4' = z_5$$
$$z_5' = z_6$$
$$z_6' = 4z_5 - 2z_2 - 15z_1.$$

The last equation in this system is exactly the original differential equation, stated in terms of the new quantities z_j.

The point to reformulating the problem in this way is that powerful matrix techniques can be invoked to find solutions. We therefore put off discussion of differential equations of order higher than 2 until we have developed the matrix machinery needed to exploit this approach.

SECTION 2.6   PROBLEMS

In each of Problems 1 through 6, find the general solution using the method of variation of parameters.

1. $y'' + y = \tan(x)$

2. $y'' - 4y' + 3y = 2\cos(x + 3)$

3. $y'' + 9y = 12\sec(3x)$

4. $y'' - 2y' - 3y = 2\sin^2(x)$

5. $y'' - 3y' + 2y = \cos(e^{-x})$

6. $y'' - 5y' + 6y = 8\sin^2(4x)$

In each of Problems 7 through 26, find the general solution using the method of undetermined coefficients.

7. $y'' - y' - 2y = 2x^2 + 5$

8. $y'' - y' - 6y = 8e^{2x}$

9. $y'' - 2y' + 10y = 20x^2 + 2x - 8$

10. $y'' - 4y' + 5y = 21e^{2x}$

11. $y'' - 6y' + 8y = 3e^x$

12. $y'' + 6y' + 9y = 9\cos(3x)$

13. $y'' - 3y' + 2y = 10\sin(x)$

14. $y'' - 4y' = 8x^2 + 2e^{3x}$

15. $y'' - 4y' + 13y = 3e^{2x} - 5e^{3x}$

16. $y'' - 2y' + y = 3x + 25\sin(3x)$

17. $y'' - 4y' = 36x^2 - 2x + 24$

18. $y'' - y' - 6y = 12xe^x$

19. $y'' + 12y' = -3\cos(3x) + \sin(2x)$

20. $y'' + 7y' - 12y = x + \cos(3x)$

21. $y'' + 2y' + y = -3e^{-x} + 8xe^{-x} + 1$

22. $y'' - 3y' + 2y = 60e^{2x}\cos(3x)$

23. $y'' + 4y = 5\sinh(2x)$

24. $y'' + y' - 6y = 50xe^{2x}$

25. $y'' + 4y' + 4y = 7x - 3\cos(2x) + 5xe^{-2x}$

26. $y'' + 5y' = xe^{-x}\sin(3x)$

In each of Problems 27 through 44, find the general solution of the differential equation, using any method.

27. $y'' - y' - 2y = e^{2x}$

28. $x^2 y'' + 5xy' - 12y = \ln(x)$

29. $y'' + y' - 6y = x$

30. $y'' - y - 12y = 2\sinh^2(x)$

31. $x^2 y'' - 5xy' + 8y = 3x$

32. $x^2 y'' + 3xy' + y = \dfrac{4}{x}$

33. $x^2 y'' + xy' + 4y = \sin(2\ln(x))$

34. $x^2 y'' + 2xy' - 6y = x^2 - 2$

35. $y'' - 4y' + 4y = e^{3x} - 1$

36. $y'' - y' - 2y = x$

37. $x^2 y'' - xy' + y = 6x$

38. $y'' + 2y' + y = \dfrac{1}{x}e^{-x}$

39. $x^2 y'' + 3xy' + y = 9x^2 + 8x + 5$

40. $x^2 y'' - 3xy' + 3y = 6x^4 e^{-3x}$

41. $x^2 y'' - 12y = 4x$

42. $x^2 y'' - 2xy' + 2y = \ln(x) + 1$

43. $x^2 y'' - 4xy' + 6y = x^2 - x$

44. $x^2 y'' - xy' - 8y = x^4 - 3\ln(x)$

In each of Problems 41 through 61, solve the initial value problem.

45. $y'' - 4y = -7e^{2x} + x$; $y(0) = 1$, $y'(0) = 3$

46. $y'' + 4y' = 8 + 34\cos(x)$; $y(0) = 3$, $y'(0) = 2$

47. $y'' + 8y' + 12y = e^{-x} + 7$; $y(0) = 1$, $y'(0) = 0$

48. $y'' - 3y' = 2e^{2x}\sin(x)$; $y(0) = 1$, $y'(0) = 2$

49. $y'' - 2y' - 8y = 10e^{-x} + 8e^{2x}$; $y(0) = 1$, $y'(0) = 4$

50. $y'' - 6y' + 9y = 4e^{3x}$; $y(0) = 1$, $y'(0) = 2$

51. $y'' - 5y' + 6y = \cos(2x)$; $y(0) = 0$, $y'(0) = 4$

52. $y'' - y' + y = 1$; $y(1) = 4$, $y'(1) = -2$

53. $y'' - 8y' + 2y = e^{-x}$; $y(-1) = 5$, $y'(-1) = 2$

54. $y'' + 6y' + 9y = -\cos(x)$; $y(0) = 1$, $y'(0) = -6$

55. $y'' - y = 5\sin^2(x)$; $y(0) = 2$, $y'(0) = -4$

56. $y'' + y = \tan(x)$; $y(0) = 4$, $y'(0) = 3$

57. $x^2 y'' - 6y = 8x^2$; $y(1) = 1$, $y'(1) = 0$

58. $x^2 y'' + 7xy' + 9y = 27\ln(x)$; $y(1) = 1$, $y'(1) = -4$

59. $x^2 y'' - 2xy' + 2y = 10\sin(\ln(x))$;
$y(1) = 3$, $y'(1) = 0$

60. $x^2 y'' - 4xy' + 6y = x^4 e^x$; $y(2) = 2$, $y'(2) = 7$

61. Assuming that the principle of superposition extends to convergent infinite series, find a particular solution of

$$y'' - 4y = \sum_{n=1}^{\infty} \frac{1}{n}\sin(nx).$$

62. Show that $y_1(x) = x$ and $y_2(x) = x^2 - 1$ are solutions of

$$(x^2 + 1)y'' - 2xy' + 2y = 0.$$

Use this fact to find the general solution of

$$(x^2 + 1)y'' - 2xy' + 2y = 6(x^2 + 1)^2.$$

63. Show that $y_1(x) = x^2$ and $y_2(x) = x - 1$ are solutions of

$$(x^2 - 2x)y'' + 2(1 - x)y' + 2y = 0.$$

Use this to find the general solution of

$$(x^2 - 2x)y'' + 2(1 - x)y' + 2y = 6(x^2 - 2x)^2.$$

64. Evaluate

$$\int_1^x \int_1^t \frac{\ln(z)}{z^2}\, dz\, dt.$$

Hint: Think of this integral as a function $y(x)$ of x. Use the fundamental theorem of calculus to compute $y'(x)$ and $y''(x)$ and show that y satisfies a nonhomogeneous Euler differential equation. Evaluate $y(1)$ and $y'(1)$ and solve the initial value problem for y.

2.7 Application of Second-Order Differential Equations to a Mechanical System

Envision a spring of natural (unstretched) length L and spring constant k. This constant quantifies the "stiffness" of the spring. The spring is suspended vertically. An object of mass m is attached at the lower end, stretching the spring d units past its rest length. The object comes to rest in its equilibrium position. It is then displaced vertically a distance y_0 units (up or down), and released, possibly with an initial velocity (Figure 2.4). We want to construct a mathematical model allowing us to analyze the motion of the object.

Let $y(t)$ be the displacement of the object from the equilibrium position at time t. As a convenience, take this equilibrium position to be $y = 0$. Choose down as the positive direction. Both of these choices are arbitrary.

Now consider the forces acting on the object. Gravity pulls it downward with a force of magnitude mg. By Hooke's law, the force the spring exerts on the object has magnitude ky. At the equilibrium position, the force of the spring is $-kd$, negative because it acts upward. If the object is pulled downward a distance y from this position, an additional force $-ky$ is exerted on it. Thus, the total force on the object due to the spring is

$$-kd - ky.$$

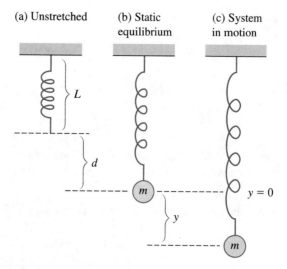

FIGURE 2.4 *Mass/spring system.*

The total force due to gravity and the spring is

$$mg - kd - ky.$$

Since at the equilibrium point ($y = 0$) this force is zero, then $mg = kd$. The net force acting on the object due to gravity and the spring is therefore just $-ky$.

Finally, there are forces tending to retard or damp out the motion. These include air resistance or viscosity of the medium if the object is suspended in some fluid such as oil. A standard assumption, arising from experiment, is that the retarding forces have magnitude proportional to the velocity y'. Thus, for some constant c called the *damping constant*, the retarding forces have magnitude cy'. The total force acting on the object due to gravity, damping, and the spring itself therefore have magnitude

$$-ky - cy'.$$

Finally, there may be a driving force of magnitude $f(t)$ on the object. Now the total external force acting on the object has magnitude

$$F = -ky - cy' + f(t).$$

Assuming that the mass is constant, Newton's second law of motion enables us to write

$$my'' = -ky - cy' + f(t),$$

or

$$y'' + \frac{c}{m}y' + \frac{k}{m}y = f(t). \tag{2.14}$$

This is the *spring equation*. We will analyze the motion described by solutions of this equation, under various conditions.

2.7.1 Unforced Motion

Suppose first that $f(t) = 0$, so there is no driving force. Now the spring equation is

$$y'' + \frac{c}{m}y' + \frac{k}{m}y = 0$$

with characteristic equation

$$\lambda^2 + \frac{c}{m}\lambda + \frac{k}{m} = 0.$$

This has roots

$$\lambda = -\frac{c}{2m} \pm \frac{1}{2m}\sqrt{c^2 - 4km}.$$

As we might expect, the general solution, hence the motion of the object, will depend on its mass, the amount of damping, and the stiffness of the spring. Consider the following cases.

Case 1 $c^2 - 4km > 0$
In this event, the characteristic equation equation has two real, distinct roots:

$$\lambda_1 = -\frac{c}{2m} + \frac{1}{2m}\sqrt{c^2 - 4km} \quad \text{and} \quad \lambda_2 = -\frac{c}{2m} - \frac{1}{2m}\sqrt{c^2 - 4km}.$$

The general solution of equation (2.14) in this case is

$$y(t) = c_1 e^{\lambda_1 t} + c_2 e^{\lambda_2 t}.$$

Clearly, $\lambda_2 < 0$. Since m and k are positive, $c^2 - 4km < c^2$, so $\sqrt{c^2 - 4km} < c$ and λ_1 is negative also. Therefore,

$$\lim_{t \to \infty} y(t) = 0,$$

regardless of initial conditions. In the case $c^2 - 4km > 0$, the motion of the object decays to zero as time increases. This case is called *overdamping*, and it occurs when the square of the damping constant exceeds four times the product of the mass and spring constant.

EXAMPLE 2.27   Overdamping

Suppose $c = 6$, $k = 5$, and $m = 1$. Now the general solution is

$$y(t) = c_1 e^{-t} + c_2 e^{-5t}.$$

Suppose, to be specific, the object was initially (at $t = 0$) drawn upward 4 units from the equilibrium position and released downward with a speed of 2 units per second. Then $y(0) = -4$ and $y'(0) = 2$, and we obtain

$$y(t) = \tfrac{1}{2}e^{-t}(-9 + e^{-4t}).$$

A graph of this solution is shown in Figure 2.5.
What does the solution tell us about the motion? Since $-9 + e^{-4t} < 0$ for $t > 0$, then $y(t) < 0$ and the object always remains above the equilibrium point. Its velocity $y'(t) = e^{-t}(9 - 5e^{-4t})/2$ decreases to zero as t increases, and $y(t) \to 0$ as t increases, so the object moves downward toward equilibrium with ever-decreasing velocity, approaching closer to but never reaching the equilibrium point, and never coming to rest. ∎

Case 2 $c^2 - 4km = 0$
Now the general solution of the spring equation (2.14) is

$$y(t) = (c_1 + c_2 t)e^{-ct/2m}.$$

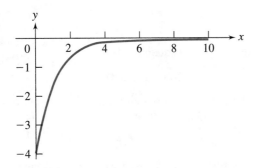

FIGURE 2.5 *An example of overdamped motion, no driving force.*

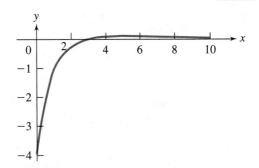

FIGURE 2.6 *An example of critical damped motion, no driving force.*

This case is called *critical damping*. While $y(t) \to 0$ as $t \to \infty$, as in the overdamping case, we will see an important difference between critical and overdamping.

EXAMPLE 2.28

Let $c = 2$ and $k = m = 1$. Now $y(t) = (c_1 + c_2 t)e^{-t}$. Suppose the object is initially pulled up four units above the equilibrium position and then pushed downward with a speed of 5 units per second. Then $y(0) = -4$ and $y'(0) = 5$, so

$$y(t) = (-4 + t)e^{-t}.$$

Observe that $y(4) = 0$, so, unlike what we saw with overdamping, the object actually reaches the equilibrium position, four seconds after it was released, and then passes through it. In fact, $y(t)$ reaches its maximum when $t = 5$ seconds, and this maximum value is $y(5) = e^{-5}$, about 0.007 unit below the equilibrium point. The velocity $y'(t) = (5 - t)e^{-t}$ is negative for $t > 5$, so the object's velocity decreases after this 5-second point. Since $y(t) \to 0$ as $t \to \infty$, the object moves with decreasing velocity back toward the equilibrium point as time increases. Figure 2.6 shows a graph of the displacement function in this case. ■

In general, when critical damping occurs, the object either passes through the equilibrium point exactly once, as just seen, or never reaches it at all, depending on the initial conditions.

Case 3 $c^2 - 4km < 0$

Now the spring constant and mass together are sufficiently large that $c^2 < 4km$, and the damping is less dominant. This case is called *underdamping*. The general solution now is

$$y(t) = e^{-ct/2m}[c_1 \cos(\beta t) + c_2 \sin(\beta t)],$$

in which

$$\beta = \frac{1}{2m}\sqrt{4km - c^2}.$$

Because c and m are positive, $y(t) \to 0$ as $t \to \infty$. However, now the motion is oscillatory because of the sine and cosine terms in the solution. The motion is not, however, periodic, because of the exponential factor, which causes the amplitude of the oscillations to decay to zero as time increases.

EXAMPLE 2.29

Suppose $c = k = 2$ and $m = 1$. Now the general solution is

$$y(t) = e^{-t}[c_1 \cos(t) + c_2 \sin(t)].$$

Suppose the object is driven downward from a point 3 units above equilibrium, with an initial speed of 2 units per second. Then $y(0) = -3$ and $y'(0) = 2$ and the solution is

$$y(t) = -e^{-t}(3 \cos(t) + \sin(t)).$$

The behavior of this solution is more easily visualized if we write it in phase angle form. We want to choose C and δ so that

$$3 \cos(t) + \sin(t) = C \cos(t + \delta).$$

For this, we need

$$3 \cos(t) + \sin(t) = C \cos(t) \cos(\delta) - C \sin(t) \sin(\delta),$$

so

$$C \cos(\delta) = 3 \quad \text{and} \quad C \sin(\delta) = -1.$$

Then

$$\frac{C \sin(\delta)}{C \cos(\delta)} = \tan(\delta) = -\frac{1}{3},$$

so

$$\delta = \tan^{-1}\left(-\frac{1}{3}\right) = -\tan^{-1}\left(\frac{1}{3}\right).$$

To solve for C, write

$$C^2 \cos^2(\delta) + C^2 \sin^2(\delta) = C^2 = 3^2 + 1^2 = 10$$

so $C = \sqrt{10}$. Now we can write the solution as

$$y(t) = \sqrt{10}e^{-t} \cos(t - \tan^{-1}(1/3)).$$

The graph is therefore a cosine curve with decaying amplitude, squashed between graphs of $y = \sqrt{10}e^{-t}$ and $y = -\sqrt{10}e^{-t}$. The solution is shown in Figure 2.7, with these two exponential functions shown as reference curves. Because of the oscillatory cosine term, the object passes

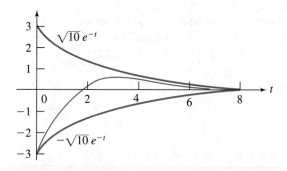

FIGURE 2.7 *An example of underdamped motion, no driving force.*

back and forth through the equilibrium point. In fact, it passes through equilibrium exactly when $y(t) = 0$, or

$$t = \tan^{-1}\left(\frac{1}{3}\right) + \frac{2n+1}{2}\pi$$

for $n = 0, 1, 2, 3, \ldots$. In theory, the object oscillates through the equilibrium infinitely often in this underdamping case, although the amplitudes of the oscillations decrease to zero as time increases. ∎

2.7.2 Forced Motion

Now suppose an external driving force of magnitude $f(t)$ acts on the object. Of course, different forces will cause different kinds of motion. As an illustration, we will analyze the motion under the influence of a periodic driving force $f(t) = A\cos(\omega t)$, with A and ω positive constants. Now the spring equation is

$$y'' + \frac{c}{m}y' + \frac{k}{m}y = \frac{A}{m}\cos(\omega t). \tag{2.15}$$

We know how to solve this nonhomogeneous linear equation. Begin by finding a particular solution, using the method of undetermined coefficients. Attempt a solution

$$y_p(x) = a\cos(\omega t) + b\sin(\omega t).$$

Substitution of this into equation (2.15) and rearrangement of terms yields

$$\left[-a\omega^2 + \frac{b\omega c}{m} + a\frac{k}{m} - \frac{A}{m}\right]\cos(\omega t) = \left[b\omega^2 + \frac{a\omega c}{m} - b\frac{k}{m}\right]\sin(\omega t).$$

Since $\sin(\omega t)$ and $\cos(\omega t)$ are not constant multiples of each other, the only way this can be true for all $t \geq 0$ is for the coefficient on each side of the equation to be zero. Therefore,

$$-a\omega^2 + \frac{b\omega c}{m} + a\frac{k}{m} - \frac{A}{m} = 0$$

and

$$b\omega^2 + \frac{a\omega c}{m} - b\frac{k}{m} = 0.$$

Solve these for a and b, keeping in mind that A, c, k, and m are given. We get

$$a = \frac{A(k - m\omega^2)}{(k - m\omega^2)^2 + \omega^2 c^2} \quad \text{and} \quad b = \frac{A\omega c}{(k - m\omega^2)^2 + \omega^2 c^2}.$$

Let $\omega_0 = \sqrt{k/m}$. Then a particular solution of equation (2.15), for this forcing function, is given by

$$y_p(x) = \frac{mA(\omega_0^2 - \omega^2)}{m^2(\omega_0^2 - \omega^2)^2 + \omega^2 c^2}\cos(\omega t)$$

$$+ \frac{A\omega c}{m^2(\omega_0^2 - \omega^2)^2 + \omega^2 c^2}\sin(\omega t), \tag{2.16}$$

assuming that $c \neq 0$ or $\omega \neq \omega_0$.

We will now examine some specific cases to get some insight into the motion with this forcing function.

Overdamped Forced Motion

Suppose $c = 6$, $k = 5$, and $m = 1$, as we had previously in the overdamping case. Suppose also that $A = 6\sqrt{5}$ and $\omega = \sqrt{5}$. If the object is released from rest from the equilibrium position, then the displacement function satisfies the initial value problem

$$y'' + 6y' + 5y = 6\sqrt{5}\cos(\sqrt{5}t); \qquad y(0) = y'(0) = 0.$$

This problem has the unique solution

$$y(t) = \frac{\sqrt{5}}{4}(-e^{-t} + e^{-5t}) + \sin(\sqrt{5}t),$$

a graph of which is shown in Figure 2.8. As time increases, the exponential terms decrease to zero, exerting less influence on the motion, while the sine term oscillates. Thus, as t increases, the solution tends to behave more like $\sin(\sqrt{5}t)$ and the object moves up and down through the equilibrium point, with approximate period $2\pi/\sqrt{5}$. Contrast this with the overdamped motion with no forcing function, in which the object began above the equilibrium point and moved with decreasing velocity down toward it, but never reached it.

Critically Damped Forced Motion

Let $c = 2$ and $m = k = 1$. Suppose $\omega = 1$ and $A = 2$. Assume that the object is released from rest from the equilibrium position. Now the initial value problem for the position function is

$$y'' + 2y' + y = 2\cos(t); \qquad y(0) = y'(0) = 0$$

with solution

$$y(t) = -te^{-t} + \sin(t).$$

A graph of this solution is shown in Figure 2.9. The exponential term exerts a significant influence at first, but decreases to zero as time increases. The term $-te^{-t}$ decreases to zero as t increases, but not as quickly as the corresponding term $(\sqrt{5}/4)(-e^{-t} + e^{-5t})$ in the overdamping case. Nevertheless, after a while the motion settles into nearly (but not exactly, because $-te^{-t}$ is never actually zero for positive t) a sinusoidal motion back and forth through the equilibrium point. This is an example of critically damped forced motion.

Underdamped Forced Motion

Suppose now that $c = k = 2$, $m = 1$, $\omega = \sqrt{2}$, and $A = 2\sqrt{2}$. Now $c^2 - 4km < 0$, and we have underdamped motion, but this time with a forcing function. If the object is released from

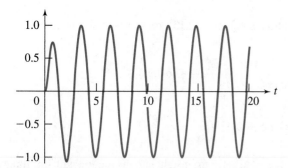

FIGURE 2.8 *An example of overdamped motion driven by $6\sqrt{5}\,\cos(\sqrt{5}t)$.*

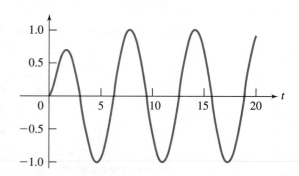

FIGURE 2.9 *An example of critical damped motion driven by $2\cos(t)$.*

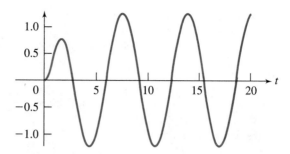

FIGURE 2.10 *An example of underdamped motion driven by $2\sqrt{2}\cos(\sqrt{2}t)$.*

rest from the equilibrium position, then the initial value problem for the displacement function is

$$y'' + 2y' + 2y = 2\sqrt{2}\cos(\sqrt{2}t); \qquad y(0) = y'(0) = 0,$$

with solution

$$y(t) = -\sqrt{2}e^{-t}\sin(t) + \sin(\sqrt{2}t).$$

Unlike the other two cases, the exponential factor in this solution has a $\sin(t)$ factor. Figure 2.10 shows a graph of this function. As time increases, the term $-\sqrt{2}e^{-t}\sin(t)$ becomes less influential and the motion settles nearly into an oscillation back and forth through the equilibrium point, with period nearly $2\pi/\sqrt{2}$.

2.7.3 Resonance

In the absence of damping, an interesting phenomenon called resonance can occur. Suppose $c = 0$ but that there is still a periodic driving force $f(t) = A\cos(\omega t)$. Now the spring equation is

$$y'' + \frac{k}{m}y = \frac{A}{m}\cos(\omega t).$$

From equation (2.16) with $c = 0$, this equation has general solution

$$y(t) = c_1\cos(\omega_0 t) + c_2\sin(\omega_0 t) + \frac{A}{m(\omega_0^2 - \omega^2)}\cos(\omega t), \qquad (2.17)$$

in which $\omega_0 = \sqrt{k/m}$. This number is called the *natural frequency* of the spring system and is a function of the stiffness of the spring and mass of the object, while ω is the *input frequency* and is contained in the driving force. This general solution assumes that the natural and input frequencies are different. Of course, the closer we choose the natural and input frequencies, the larger the amplitude of the $\cos(\omega t)$ term in the solution.

Consider the case that the natural and input frequencies are the same. Now the differential equation is

$$y'' + \frac{k}{m}y = \frac{A}{m}\cos(\omega_0 t) \qquad (2.18)$$

and the function given by equation (2.17) is not a solution. To solve equation (2.18), first write the general solution y_h of $y'' + (k/m)y = 0$:

$$y_h(t) = c_1\cos(\omega_0 t) + c_2\sin(\omega_0 t).$$

For a particular solution of equation (2.18), we will proceed by the method of undetermined coefficients. Since the forcing function contains a term found in y_h, we will attempt a particular solution of the form

$$y_p(t) = at\cos(\omega_0 t) + bt\sin(\omega_0 t).$$

Substitute this into equation (2.18) to obtain

$$-2a\omega_0 \sin(\omega_0 t) + 2b\omega_0 \cos(\omega_0 t) = \frac{A}{m} \cos(\omega_0 t).$$

Thus choose

$$a = 0 \quad \text{and} \quad 2b\omega_0 = \frac{A}{m},$$

leading to the particular solution

$$y_p(t) = \frac{A}{2m\omega_0} t \sin(\omega_0 t).$$

The general solution of equation (2.18) is therefore

$$y(t) = c_1 \cos(\omega_0 t) + c_2 \sin(\omega_0 t) + \frac{A}{2m\omega_0} t \sin(\omega_0 t).$$

This solution differs from that in the case $\omega \neq \omega_0$ in the factor of t in $y_p(t)$. Because of this, solutions increase in amplitude as t increases. This phenomenon is called *resonance*.

As a specific example, let $c_1 = c_2 = \omega_0 = 1$ and $A/2m = 1$ to write the solution as

$$y(t) = \cos(t) + \sin(t) + t \sin(t).$$

A graph of this function is shown in Figure 2.11, clearly revealing the increasing magnitude of the oscillations with time.

While there is always some damping in the real world, if the damping constant is close to zero compared to other factors, such as the mass, and if the natural and input frequencies are (nearly) equal, then oscillations can build up to a sufficiently large amplitude to cause resonance-like behavior and damage a system. This can occur with soldiers marching in step across a bridge. If the cadence of the march (input frequency) is near enough to the natural frequency of the material of the bridge, vibrations can build up to dangerous levels. This occurred near Manchester, England, in 1831 when a column of soldiers marching across the Broughton Bridge caused it to collapse. More recently, the Tacoma Narrows Bridge in Washington experienced increasing oscillations driven by energy from the wind, causing it to whip about in sensational fashion before its collapse into the river. Videos of the wild thrashing about of the bridge are available in some libraries and engineering and science departments.

2.7.4   Beats

In the absence of damping, an oscillatory driving force can also cause a phenomenon called beats. Suppose $\omega \neq \omega_0$ and consider

$$y'' + \omega_0^2 y = \frac{A}{m} \cos(\omega_0 t).$$

Assuming that the object is released from rest at the equilibrium position, then $y(0) = y'(0) = 0$ and from equation (2.17) we have the solution

$$y(t) = \frac{A}{m(\omega_0^2 - \omega^2)} [\cos(\omega t) - \cos(\omega_0 t)].$$

The behavior of this solution reveals itself more clearly if we write it as

$$y(t) = \frac{2A}{m(\omega_0^2 - \omega^2)} \sin\left(\frac{1}{2}(\omega_0 + \omega)t\right) \sin\left(\frac{1}{2}(\omega_0 - \omega)t\right).$$

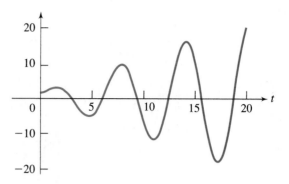

FIGURE 2.11 *Resonance.*

The Tacoma Narrows Bridge was completed in 1940 and stood as a new standard of combined artistry and functionality. The bridge soon became known for its tendency to sway in high winds, but no one suspected what was about to occur. On November 7, 1940, energy provided by unusually strong winds, coupled with a resonating effect in the bridge's material and design, caused the oscillations in the bridge to be reinfored and build to dangerous levels. Soon, the twisting caused one side of the sidewalk to rise 28 feet above that of the other side. Concrete dropped out of the roadway, and a section of the suspension span completely rotated and fell away. Shortly thereafter, the entire center span collapsed into Puget Sound. This sensational construction failure motivated new mathematical treatments of vibration and wave phenomena in the design of bridges and other large structures. The forces that brought down this bridge are a more complicated version of the resonance phenomenon discussed in Section 2.7.3.

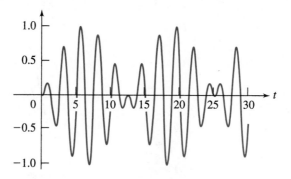

FIGURE 2.12 *Beats.*

This formulation reveals a periodic variation of amplitude in the solution, depending on the relative sizes of $\omega_0 + \omega$ and $\omega_0 - \omega$. It is this periodic variation of amplitude that is called a *beat*. As a specific example, suppose $\omega_0 + \omega = 5$ and $\omega_0 - \omega = \frac{1}{2}$, and the constants are chosen so that $2A/[m(\omega_0^2 - \omega^2)] = 1$. In this case, the displacement function is

$$y(t) = \sin\left(\frac{5t}{2}\right)\sin\left(\frac{t}{4}\right).$$

The beats are apparent in the graph of this solution in Figure 2.12.

2.7.5 Analogy with an Electrical Circuit

If a circuit contains a resistance R, inductance L, and capacitance C, and the electromotive force is $E(t)$, then the impressed voltage is obtained as a sum of the voltage drops in the circuit:

$$E(t) = Li'(t) + Ri(t) + \frac{1}{C}q(t).$$

Here, $i(t)$ is the current at time t, and $q(t)$ is the charge. Since $i = q'$, we can write the second-order linear differential equation

$$q'' + \frac{R}{L}q' + \frac{1}{LC}q = \frac{1}{L}E.$$

If R, L, and C are constant, this is a linear equation of the type we have solved for various choices of $E(t)$. It is interesting to observe that this equation is of exactly the same form as the equation for the displacement of an object attached to a spring, which is

$$y'' + \frac{c}{m}y' + \frac{k}{m}y = \frac{1}{m}f(t).$$

This means that solutions of one equation readily translate into solutions of the other and suggests the following equivalences between electrical and mechanical quantities:

displacement function $y(t) \Longleftrightarrow$ charge $q(t)$

velocity $y'(t) \Longleftrightarrow$ current $i(t)$

driving force $f(t) \Longleftrightarrow$ electromotive force $E(t)$

mass $m \Longleftrightarrow$ inductance L

damping constant $c \Longleftrightarrow$ resistance R

spring modulus $k \Longleftrightarrow$ reciprocal $1/C$ of the capacitance

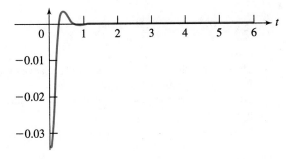

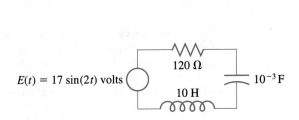

FIGURE 2.13

FIGURE 2.14 *Transient part of the current for the circuit of Figure 2.13.*

EXAMPLE 2.30

Consider the circuit of Figure 2.13, driven by a potential of $E(t) = 17\sin(2t)$ volts. At time zero the current is zero and the charge on the capacitor is $1/2000$ coulomb. The charge $q(t)$ on the capacitor for $t > 0$ is obtained by solving the initial value problem

$$10q'' + 120q' + 1000q = 17\sin(2t); \qquad q(0) = \tfrac{1}{2000}, q'(0) = 0.$$

The solution is

$$q(t) = \tfrac{1}{1500}e^{-6t}[7\cos(8t) - \sin(8t)] + \tfrac{1}{240}[-\cos(2t) + 4\sin(2t)].$$

The current can be calculated as

$$i(t) = q'(t) = -\tfrac{1}{30}e^{-6t}[\cos(8t) + \sin(8t)] + \tfrac{1}{120}[4\cos(2t) + \sin(2t)].$$

The current is a sum of a transient part

$$-\tfrac{1}{30}e^{-6t}[\cos(8t) + \sin(8t)],$$

named for the fact that it decays to zero as t increases, and a steady-state part

$$\tfrac{1}{120}[4\cos(2t) + \sin(2t)].$$

The transient and steady-state parts are shown in Figures 2.14 and 2.15, and their sum, the current, is shown in Figure 2.16. ■

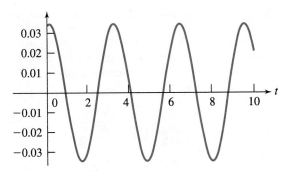

FIGURE 2.15 *Steady-state part of the current for the circuit of Figure 2.13.*

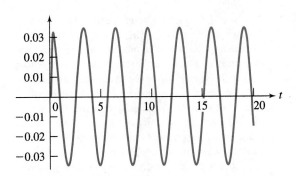

FIGURE 2.16 *Current function for the circuit of Figure 2.13.*

SECTION 2.7 PROBLEMS

1. The object of this problem is to gauge the relative effects of initial position and velocity on the motion in the unforced overdamped case. Solve the initial value problems

$$y'' + 4y' + 2y = 0; \qquad y(0) = 5, \, y'(0) = 0$$

and

$$y'' + 4y' + 2y = 0; \qquad y(0) = 0, \, y'(0) = 5.$$

Graph the solutions on the same set of axes. What conclusions can be drawn from these solutions about the influence of initial position and velocity?

2. Repeat the experiment of Problem 1, except now use the critically damped unforced equation $y'' + 4y' + 4y = 0$.

3. Repeat the experiment of Problem 1 for the underdamped unforced case $y'' + 2y' + 5y = 0$.

Problems 4 through 9 explore the effects of changing the initial position or initial velocity on the motion of the bob. In each, use the same set of axes to graph the solution of the initial value problem for the given values of A and observe the effect that these changes cause in the solution.

4. $y'' + 4y' + 2y = 0$; $y(0) = A$, $y'(0) = 0$; A has values 1, 3, 6, 10, −4, and −7.

5. $y'' + 4y' + 2y = 0$; $y(0) = 0$, $y'(0) = A$; A has values 1, 3, 6, 10, −4, and −7.

6. $y'' + 4y' + 4y = 0$; $y(0) = A$, $y'(0) = 0$; A has values 1, 3, 6, 10, −4, and −7.

7. $y'' + 4y' + 4y = 0$; $y(0) = 0$, $y'(0) = A$; A has values 1, 3, 6, 10, −4, and −7.

8. $y'' + 2y' + 5y = 0$; $y(0) = A$, $y'(0) = 0$; A has values 1, 3, 6, 10, −4, and −7.

9. $y'' + 2y' + 5y = 0$; $y(0) = 0$, $y'(0) = A$; A has values 1, 3, 6, 10, −4, and −7.

10. An object having mass 1 gram is attached to the lower end of a spring having spring modulus 29 dynes per centimeter. The bob is, in turn, adhered to a dashpot that imposes a damping force of $10v$ dynes, where $v(t)$ is the velocity at time t in centimeters per second. Determine the motion of the bob if it is pulled down 3 centimeters from equilibrium and then struck upward with a blow sufficient to impart a velocity of 1 centimeter per second. Graph the solution. Solve the problem when the initial velocity is, in turn, 2, 4, 7, and 12 centimeters per second. Graph these solutions on the same set of axes to visualize the influence of the initial velocity on the motion.

11. An object having mass 1 kilogram is suspended from a spring having a spring constant of 24 newtons per meter. Attached to the object is a shock absorber, which induces a drag of $11v$ newtons (velocity is in meters per second). The system is set in motion by lowering the bob $\frac{25}{3}$ centimeters and then striking it hard enough to impart an upward velocity of 5 meters per second. Solve for and graph the displacement function. Obtain the solution for the cases that the bob is lowered, in turn, 12, 20, 30, and 45 centimeters and graph the displacement functions for the five cases on the same set of axes to see the effect of the distance lowered.

12. When an 8-pound weight is suspended from a spring, it stretches the spring 2 inches. Determine the equation of motion when an object with a mass of 7 kilograms is suspended from this spring and the system is set in motion by striking the object an upward blow, imparting a velocity of 4 meters per second.

13. How many times can the bob pass through the equilibrium point in the case of overdamped motion? What condition can be placed on the initial displacement $y(0)$ to guarantee that the bob never passes through equilibrium?

14. How many times can the bob pass through the equilibrium point in the case of critical damping? What condition can be placed on $y(0)$ to ensure that the bob never passes through this position? How does the initial velocity influence whether the bob passes through the equilibrium position?

15. In underdamped motion, what effect does the damping constant c have on the frequency of the oscillations of motion?

16. Suppose $y(0) = y'(0) \neq 0$. Determine the maximum displacement of the bob in the critically damped case and show that the time at which this maximum occurs is independent of the initial displacement.

17. As the mass increases in the underdamped case, the solution appears to die out more quickly as time increases. Intuitively, however, it would seem that a heavier bob would stretch the spring more and cause greater oscillations. Which view is correct?

18. In the case of underdamped motion, the general solution is

$$y(t) = e^{-ct/2m} \left[c_1 \cos\left(\frac{\sqrt{4km - c^2}\, t}{2m}\right) \right.$$
$$\left. + c_2 \sin\left(\frac{\sqrt{4km - c^2}\, t}{2m}\right) \right].$$

Show how ω^* and δ can be chosen to write this solution in phase angle form

$$y(t) = de^{-ct/2m} \cos(\omega^* t + \delta).$$

19. Consider underdamped motion with the general solution written in the phase angle form of Problem 18.

(a) The *natural period* of the undamped system is defined to be $T = 2\pi \sqrt{m/k}$. An underdamped system does not exhibit this natural period. However, define the *quasi period* T_q as twice the time between successive zeros of $y(t)$. Show that

$$T_q = \frac{4m\pi}{\sqrt{4km - c^2}}.$$

(b) Show that

$$T_q = \frac{T}{\sqrt{1 - c^2/4mk}}.$$

Hence conclude that when c^2/mk is sufficiently small, effects of damping are negligible in computing the quasi period, and then $T \approx T_q$.

20. Prove the *principle of conservation of energy* for the undamped harmonic oscillator. This states that the total energy of the system remains constant through the motion. *Hint:* The energy is a sum of the kinetic energy $mv^2/2$ and the potential energy $ky^2/2$. Write the differential equation $my'' + ky = 0$ as $mv(dv/dy) = -ky$ and integrate.

21. Suppose the acceleration of the bob on the spring at distance d from the equilibrium position is a. Prove that the period of the motion is $2\pi \sqrt{d/a}$ in the case of undamped motion.

22. A mass m_1 is attached to a spring and allowed to vibrate with undamped motion having period p. At some later time, a second mass m_2 is instantaneously fused with m_1. Prove that the new object, having mass $m_1 + m_2$, exhibits simple harmonic motion with period $p/\sqrt{1 + m_2/m_1}$.

23. Let $y(t)$ be the solution of $y'' + \omega_0^2 y = (A/m)\cos(\omega t)$, with $y(0) = y'(0) = 0$. Assuming that $\omega \neq \omega_0$, find $\lim_{\omega \to \omega_0} y(t)$. How does this limit compare with the solution of $y'' + \omega_0^2 y = (A/m)\cos(\omega_0 t)$, with $y(0) = y'(0) = 0$?

24. A 16-pound weight is suspended from a spring, stretching it $\frac{8}{11}$ feet. Then the weight is submerged in a fluid that imposes a drag of $2v$ pounds. The entire system is subjected to an external force $4\cos(\omega t)$. Determine the value of ω that maximizes the amplitude of the steady-state oscillation. What is this maximum amplitude?

25. Consider overdamped forced motion governed by $y'' + 6y' + 2y = 4\cos(3t)$.

(a) Find the solution satisfying $y(0) = 6$, $y'(0) = 0$.

(b) Find the solution satisfying $y(0) = 0$, $y'(0) = 6$.

(c) Graph these solutions on the same set of axes to compare the effect of initial displacement with that of initial velocity.

26. Carry out the program of Problem 25 for the critically damped forced system governed by $y'' + 4y' + 4y = 4\cos(3t)$.

27. Carry out the program of Problem 25 for the underdamped forced system governed by $y'' + y' + 3y = 4\cos(3t)$.

28. Analyze the motion of a body of mass m suspended from a spring with spring constant k and damping constant c, if the external driving force is $F(t) = e^{-t}$. Consider all the cases, depending on relative sizes of m, c, and k, and discuss what the solution for the displacement function reveals about the motion of the object.

29. Carry out the instructions of Problem 28 if $F(t) = t$.

30. Carry out the instructions of Problem 28 if $F(t) = Ae^{-t}\sin(\omega t)$.

31. Show that the distance between successive maxima of $y(t)$ in the case of beats is $4\pi/(\omega + \omega_0)$.

32. Show that the damped forced motion of a mass on a spring, with forcing function $A\cos(\omega t)$, is always bounded in magnitude.

33. Consider the damped forced motion governed by $my'' + cy' + ky = A\cos(\omega t)$. Show that the maximum amplitude of the steady-state solution is achieved if ω is chosen so that

$$\omega^2 = \frac{k}{m} - \frac{c^2}{2m^2}.$$

In building a seismic detector, we would try to choose k, c, and m so that ω^2 is as near to this value as possible in order to maximize the response of the instrument.

34. A 64-pound block is attached to a spring with spring modulus 26 pounds per foot. The block and spring are then placed on an inclined plane having a 30-degree slope. The spring is attached to the top of the plane and the block is allowed to come to rest. The coefficient of friction between the block and the plane is $\sqrt{3}/8$ pound-second per foot. The block is pulled down the plane 6 inches beyond the equilibrium position and released from rest at time zero. Assuming that the drag due to air resistance is negligible, determine the position of the block relative to the equilibrium position for $t \geq 0$.

35. A cylindrical buoy of radius $\pi/5$ feet and height 6 feet weighs $10\pi^3$ pounds. The buoy is put into a freshwater lake and pushed down until it is just submerged and then released from rest. Calculate the resultant amplitude and frequency of the buoy's oscillations. Take the density of water to be 62.5 pounds per cubic foot.

In each of Problems 36 through 39, use the information to find the current in the RLC circuit of Figure 2.17. Assume zero initial current and capacitor charge.

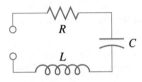

FIGURE 2.17 *RLC circuit.*

36. $R = 200\ \Omega,\ L = 0.1\ \text{H},\ C = 0.006\ \text{F},\ E(t) = te^{-t}\ \text{V}$

37. $R = 400\ \Omega,\ L = 0.12\ \text{H},\ C = 0.04\ \text{F},$
$E(t) = 120\sin(20t)\ \text{V}$

38. $R = 150\ \Omega,\ L = 0.2\ \text{H},\ C = 0.05\ \text{F},\ E(t) = 1 - e^{-t}\ \text{V}$

39. $R = 450\ \Omega,\ L = 0.95\ \text{H},\ C = 0.007\ \text{F},$
$E(t) = e^{-t}\sin^2(3t)\ \text{V}$

CHAPTER 2   ADDITIONAL PROBLEMS

In each of Problems 1 through 20, find the general solution of the differential equation or the solution of the initial value problem.

1. $x^2 y'' - 9xy' + 25y = 0$

2. $yy'' = (y')^2$ (Note Problem 15 of Section 2.3.)

3. $y'' - 14y' + 49y = 0$

4. $yy'' + 2y' = (y')^2$

5. $y'' - y' - 2y = 5e^{4x} + 6x$

6. $y'' + 22y' + 121y = 28;\ y(0) = -5,\ y'(0) = 2$

7. $x^2 y'' - 7xy' + 16y = 0$

8. $x^2 y'' + 13xy' + 45y = 0;\ y(1) = 2,\ y'(1) = 2$

9. $yy'' - 2(y')^2 = 0$

10. $y'' + 4y' - 96y = 3e^{8x} - \cos(3x);\ y(0) = 0,\ y'(0) = 0$

11. $y'' + y = \tan(x)$

12. $y'' + 4y = 4\sec^2(x)$

13. $y'' + 2y' - 3y = 13\cos(2x)$

14. $y'' - y' - 20y = x^2 - 1;\ y(0) = 0,\ y'(0) = 0$

15. $y'' - 2y' + y = 2\sin(3x);\ y(0) = 2,\ y'(0) = 1$

16. $y'' - 4y' + 3y = -3\sin(x + 2);\ y(-2) = 2,$
$y'(-2) = 2$

17. $y'' + 8y' + 16y = x\cos(2x)$

18. $y'' - 5y' + 9y = x^2 - \sin(x)$

19. $x^2 y'' - xy' - 2y = x^3 + 4\ln(x);\ y(1) = 9,\ y'(1) = 7$

20. $y'' + y = \sec^3(x);\ y(0) = 4,\ y'(0) = 2$

21. Assuming that the earth is a sphere, a particle of mass m inside the earth at a distance r from the center experiences a gravitational force $F = -mgr/R$, where

R is the radius of the earth. Show that a particle in an evacuated tube through the earth's center will execute simple harmonic motion and determine the period of the motion.

22. A ball of mass m is thrown vertically downward from a stationary dirigible hovering h feet above the ground. The initial velocity of the ball is v_0. Neglecting air resistance, show that the ball will impact the ground $(1/g)(\sqrt{v_0^2 + 2gh} - v_0)$ seconds after it is released.

23. A 4-pound brick is dropped from the top of a building h feet high. After it has fallen k feet ($k < h$), a 6-pound brick is dropped from the same release point. Neglecting air resistance, show that, when the first brick strikes the ground, the second still has $2\sqrt{kh} - k$ feet to fall. Of what importance is the weight of each brick in drawing this conclusion?

24. For a simple pendulum, the differential equation for the angle of displacement $\theta(t)$ of the arm from the vertical is $\theta''(t) + (g/L)\sin(\theta(t)) = 0$. Assume that the bob is released from rest at time $t = 0$, from the position $\theta = -\alpha$, where $0 < \alpha < \pi/2$. (That is, the bob is drawn back α radians to the left and released from rest.) Show that on the first half-swing,

$$t = \sqrt{\frac{L}{2g}} \int_{-\alpha}^{\theta} \frac{1}{\sqrt{\cos(\varphi) - \cos(\alpha)}}\, d\varphi v.$$

25. Prove that the motion of the simple pendulum obeys the conservation of energy law (the sum of the kinetic and potential energies is a constant of the motion). *Hint:* The kinetic energy is $(m/2)(ds/dt)^2$, where $s = L\theta$ is the length of arc through which the pendulum swings and the potential energy is $mgL[1 - \cos(\theta)]$.

CHAPTER 3

The Laplace Transform

3.1 Definition and Basic Properties

In mathematics, a transform is usually a device that converts one type of problem into another type, presumably easier to solve. The strategy is to solve the transformed problem, then transform back the other way to obtain the solution of the original problem. In the case of the Laplace transform, initial value problems are often converted to algebra problems, a process we can diagram as follows:

initial value problem

$\Downarrow$

algebra problem

$\Downarrow$

solution of the algebra problem

$\Downarrow$

solution of the initial value problem.

DEFINITION 3.1 *Laplace Transform*

The Laplace transform $\mathcal{L}[f]$ of f is a function defined by

$$\mathcal{L}[f](s) = \int_0^\infty e^{-st} f(t)\, dt,$$

for all s such that this integral converges.

The Laplace transform converts a function f to a new function called $\mathcal{L}[f]$. Often we use t as the independent variable for f and s for the independent variable of $\mathcal{L}[f]$. Thus, $f(t)$ is the function f evaluated at t, and $\mathcal{L}[f](s)$ is the function $\mathcal{L}[f]$ evaluated at s.

It is often convenient to agree to use lowercase letters for a function put into the Laplace transform, and its uppercase for the function that comes out. In this notation,

$$F = \mathcal{L}[f], \quad G = \mathcal{L}[g], \quad H = \mathcal{L}[h],$$

and so on.

EXAMPLE 3.1

Let $f(t) = e^{at}$, with a any real number. Then

$$\mathcal{L}[f](s) = F(s) = \int_0^\infty e^{-st} e^{at}\, dt = \int_0^\infty e^{(a-s)t}\, dt$$

$$= \lim_{k \to \infty} \int_0^k e^{(a-s)t}\, dt = \lim_{k \to \infty} \left[\frac{1}{a-s} e^{(a-s)t} \right]_0^k$$

$$= \lim_{k \to \infty} \left[\frac{1}{a-s} e^{(a-s)k} - \frac{1}{a-s} \right]$$

$$= -\frac{1}{a-s} = \frac{1}{s-a}$$

provided that $a - s < 0$, or $s > a$. The Laplace transform of $f(t) = e^{at}$ is $F(s) = 1/(s - a)$, defined for $s > a$. ▪

EXAMPLE 3.2

Let $g(t) = \sin(t)$. Then

$$\mathcal{L}[g](s) = G(s) = \int_0^\infty e^{-st} \sin(t)\, dt$$

$$= \lim_{k \to \infty} \int_0^k e^{-st} \sin(t)\, dt$$

$$= \lim_{k \to \infty} \left[-\frac{e^{-ks} \cos(k) + s e^{-ks} \sin(k) - 1}{s^2 + 1} \right] = \frac{1}{s^2 + 1}.$$

$G(s)$ is defined for all real s. ▪

A Laplace transform is rarely computed by referring directly to the definition and integrating. Instead, we use tables of Laplace transforms of commonly used functions (such as Table 3.1) or computer software. We will also develop methods that are used to find the Laplace transform of a shifted or translated function, step functions, pulses, and various other functions that arise frequently in applications.

The Laplace transform is linear, which means that constants factor through the transform, and the transform of a sum of functions is the sum of the transform of these functions.

TABLE 3.1   *Table of Laplace Transforms of Functions*

	$f(t)$	$F(s) = \mathcal{L}[f(t)](s)$
1.	1	$\dfrac{1}{s}$
2.	t	$\dfrac{1}{s^2}$
3.	$t^n \ (n = 1, 2, 3, \cdots)$	$\dfrac{n!}{s^{n+1}}$
4.	$\dfrac{1}{\sqrt{t}}$	$\sqrt{\dfrac{\pi}{s}}$
5.	e^{at}	$\dfrac{1}{s-a}$
6.	te^{at}	$\dfrac{1}{(s-a)^2}$
7.	$t^n e^{at}$	$\dfrac{n!}{(s-a)^{n+1}}$
8.	$\dfrac{1}{a-b}(e^{at} - e^{bt})$	$\dfrac{1}{(s-a)(s-b)}$
9.	$\dfrac{1}{a-b}(ae^{at} - be^{bt})$	$\dfrac{s}{(s-a)(s-b)}$
10.	$\dfrac{(c-b)e^{at} + (a-c)e^{bt} + (b-a)e^{ct}}{(a-b)(b-c)(c-a)}$	$\dfrac{1}{(s-a)(s-b)(s-c)}$
11.	$\sin(at)$	$\dfrac{a}{s^2 + a^2}$
12.	$\cos(at)$	$\dfrac{s}{s^2 + a^2}$
13.	$1 - \cos(at)$	$\dfrac{a^2}{s(s^2 + a^2)}$
14.	$at - \sin(at)$	$\dfrac{a^3}{s^2(s^2 + a^2)}$
15.	$\sin(at) - at\cos(at)$	$\dfrac{2a^3}{(s^2 + a^2)^2}$
16	$\sin(at) + at\cos(at)$	$\dfrac{2as^2}{(s^2 + a^2)^2}$
17.	$t\sin(at)$	$\dfrac{2as}{(s^2 + a^2)^2}$
18.	$t\cos(at)$	$\dfrac{(s^2 - a^2)}{(s^2 + a^2)^2}$
19.	$\dfrac{\cos(at) - \cos(bt)}{(b-a)(b+a)}$	$\dfrac{s}{(s^2 + a^2)(s^2 + b^2)}$
20.	$e^{at}\sin(bt)$	$\dfrac{b}{(s-a)^2 + b^2}$
21.	$e^{at}\cos(bt)$	$\dfrac{s-a}{(s-a)^2 + b^2}$
22.	$\sinh(at)$	$\dfrac{a}{s^2 - a^2}$
23.	$\cosh(at)$	$\dfrac{s}{s^2 - a^2}$
24.	$\sin(at)\cosh(at) - \cos(at)\sinh(at)$	$\dfrac{4a^3}{s^4 + 4a^4}$
25.	$\sin(at)\sinh(at)$	$\dfrac{2a^2 s}{s^4 + 4a^4}$

TABLE 3.1   *(continued)*

$f(t)$	$F(s) = \mathcal{L}[f(t)](s)$
26. $\sinh(at) - \sin(at)$	$\dfrac{2a^3}{s^4 - a^4}$
27. $\cosh(at) - \cos(at)$	$\dfrac{2a^2 s}{s^4 - a^4}$
28. $\dfrac{1}{\sqrt{\pi t}} e^{at}(1 + 2at)$	$\dfrac{s}{(s-a)^{3/2}}$
29. $J_0(at)$	$\dfrac{1}{\sqrt{s^2 + a^2}}$
30. $J_n(at)$	$\dfrac{1}{a^n} \dfrac{\left(\sqrt{s^2 + a^2} - s\right)^n}{\sqrt{s^2 + a^2}}$
31. $J_0(2\sqrt{at})$	$\dfrac{1}{s} e^{-a/s}$
32. $\dfrac{1}{t} \sin(at)$	$\tan^{-1}\left(\dfrac{a}{s}\right)$
33. $\dfrac{2}{t}[1 - \cos(at)]$	$\ln\left(\dfrac{s^2 + a^2}{s^2}\right)$
34. $\dfrac{2}{t}[1 - \cosh(at)]$	$\ln\left(\dfrac{s^2 - a^2}{s^2}\right)$
35. $\dfrac{1}{\sqrt{\pi t}} - a e^{a^2 t}\, \mathrm{erfc}\left(\dfrac{a}{\sqrt{t}}\right)$	$\dfrac{1}{\sqrt{s} + a}$
36. $\dfrac{1}{\sqrt{\pi t}} + a e^{a^2 t}\, \mathrm{erf}\left(\dfrac{a}{\sqrt{t}}\right)$	$\dfrac{\sqrt{s}}{s - a^2}$
37. $e^{a^2 t}\, \mathrm{erf}(a\sqrt{t})$	$\dfrac{a}{\sqrt{s}(s - a^2)}$
38. $e^{a^2 t}\, \mathrm{erfc}(a\sqrt{t})$	$\dfrac{1}{\sqrt{s}\,(\sqrt{s} + a)}$
39. $\mathrm{erfc}\left(\dfrac{a}{2\sqrt{t}}\right)$	$\dfrac{1}{s} e^{-a\sqrt{s}}$
40. $\dfrac{1}{\sqrt{\pi t}} e^{-a^2/4t}$	$\dfrac{1}{\sqrt{s}} e^{-a\sqrt{s}}$
41. $\dfrac{1}{\sqrt{\pi(t + a)}}$	$\dfrac{1}{\sqrt{s}} e^{as}\, \mathrm{erfc}(\sqrt{as})$
42. $\dfrac{1}{\pi t} \sin(2a\sqrt{t})$	$\mathrm{erf}\left(\dfrac{a}{\sqrt{s}}\right)$
43. $f\left(\dfrac{t}{a}\right)$	$a F(as)$
44. $e^{bt/a} f\left(\dfrac{t}{a}\right)$	$a F(as - b)$
45. $\delta_\epsilon(t)$	$\dfrac{e^{-\epsilon s}(1 - e^{-\epsilon s})}{\epsilon s}$
46. $\delta(t - a)$	e^{-as}
47. $L_n(t)$ (Laguerre polynomial)	$\dfrac{1}{s}\left(\dfrac{s - 1}{s}\right)^n$

TABLE 3.1   *(continued)*

$f(t)$	$F(s) = \mathcal{L}[f(t)](s)$
48. $\dfrac{n!}{(2n)!\sqrt{\pi t}} H_{2n}(t)$ (Hermite polynomial)	$\dfrac{(1-s)^n}{s^{n+1/2}}$
49. $\dfrac{-n!}{\sqrt{\pi}(2n+1)!} H_{2n+1}(t)$ (Hermite polynomial)	$\dfrac{(1-s)^n}{s^{n+3/2}}$
50. triangular wave 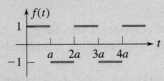	$\dfrac{1}{as^2}\left[\dfrac{1-e^{-as}}{1+e^{-as}}\right]\left(=\dfrac{1}{as^2}\tanh\left(\dfrac{as}{2}\right)\right)$
51. square wave 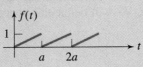	$\dfrac{1}{s}\tanh\left(\dfrac{as}{2}\right)$
52. sawtooth wave 	$\dfrac{1}{as^2}-\dfrac{e^{-as}}{s(1-e^{-as})}$

Operational Formulas

$f(t)$	$F(s)$
$af(t)+bg(t)$	$aF(s)+bG(s)$
$f'(t)$	$sF(s)-f(0+)$
$f^{(n)}(t)$	$s^nF(s)-s^{n-1}f(0)-\cdots-f^{(n-1)}(0)$
$\displaystyle\int_0^t f(\tau)\,d\tau$	$\dfrac{1}{s}F(s)$
$tf(t)$	$-F'(s)$
$t^nf(t)$	$(-1)^nF^{(n)}(s)$
$\dfrac{1}{t}f(t)$	$\displaystyle\int_s^\infty F(\sigma)\,d\sigma$
$e^{at}f(t)$	$F(s-a)$
$f(t-a)H(t-a)$	$e^{-as}F(s)$
$f(t+\tau)=f(t)$ (periodic)	$\dfrac{1}{1-e^{-\tau s}}\displaystyle\int_0^\tau e^{-st}f(t)\,dt$

THEOREM 3.1   *Linearity of the Laplace Transform*

Suppose $\mathcal{L}[f](s)$ and $\mathcal{L}[g](s)$ are defined for $s > a$, and α and β are real numbers. Then

$$\mathcal{L}[\alpha f + \beta g](s) = \alpha F(s) + \beta G(s)$$

for $s > a$.

Proof   By assumption, $\int_0^\infty e^{-st} f(t)\, dt$ and $\int_0^\infty e^{-st} g(t)\, dt$ converge for $s > a$. Then

$$\mathcal{L}[\alpha f + \beta g](s) = \int_0^\infty e^{-st}(\alpha f(t) + \beta g(t))\, dt$$

$$= \alpha \int_0^\infty e^{-st} f(t)\, dt + \beta \int_0^\infty e^{-st} g(t)\, dt = \alpha F(s) + \beta G(s)$$

for $s > a$.   ∎

This conclusion extends to any finite sum:

$$\mathcal{L}[\alpha_1 f_1 + \cdots + \alpha_n f_n](s) = \alpha_1 F_1(s) + \cdots + \alpha_n F_n(s),$$

for all s such that each $F_j(s)$ is defined.

Not every function has a Laplace transform, because $\int_0^\infty e^{-st} f(t)\, dt$ may not converge for any real values of s. We will consider conditions that can be placed on f to ensure that f has a Laplace transform.

An obvious necessary condition is that $\int_0^k e^{-st} f(t)\, dt$ must be defined for every $k > 0$, because $\mathcal{L}[f](s) = \int_0^\infty e^{-st} f(t)\, dt$. For this to occur, it is enough that f be piecewise continuous on $[0, k]$ for every positive number k. We will define this concept in general terms because it occurs in other contexts as well.

DEFINITION 3.2   *Piecewise Continuity*

f is piecewise continuous on $[a, b]$ if there are points

$$a < t_1 < t_2 < \cdots < t_n < b$$

such that f is continuous on each open interval (a, t_1), (t_{j-1}, t_j), and (t_n, b), and all of the following one-sided limits are finite:

$$\lim_{t \to a+} f(t),\ \lim_{t \to t_j-} f(t),\ \lim_{t \to t_j+} f(t),\quad \text{and}\quad \lim_{t \to b-} f(t).$$

This means that f is continuous on $[a, b]$ except perhaps at finitely many points, at each of which f has finite one-sided limits from within the interval. The only discontinuities a piecewise continuous function f can experience on $[a, b]$ are finitely many jump discontinuities (gaps of finite width in the graph). Figure 3.1 shows typical jump discontinuities in a graph.

For example, let

$$f(t) = \begin{cases} t^2 & \text{for } 0 \leq t < 2 \\ 2 & \text{at } t = 2 \\ 1 & \text{for } 2 < t \leq 3 \\ -1 & \text{for } 3 < t \leq 4 \end{cases}.$$

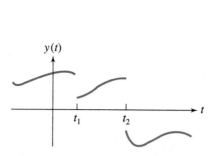

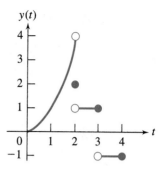

FIGURE 3.1 *A function having jump discontinuities at t_1 and t_2.*

FIGURE 3.2

$$f(t) = \begin{cases} t^2 & \text{if } 0 \le t < 2 \\ 2 & \text{if } t = 2 \\ 1 & \text{if } 2 < t \le 3 \\ -1 & \text{if } 3 < t \le 4 \end{cases}.$$

Then f is continuous $[0, 4]$ except at 2 and 3, where f has jump discontinuities. A graph of this function is shown in Figure 3.2.

If f is piecewise continuous on $[0, k]$, then so is $e^{-st} f(t)$ and $\int_0^k e^{-st} f(t)\, dt$ exists.

Existence of $\int_0^k e^{-st} f(t)\, dt$ for every positive k does not ensure existence of $\lim_{k \to \infty} \int_0^k e^{-st} f(t)\, dt$. For example, $f(t) = e^{t^2}$ is continuous on every interval $[0, k]$, but $\int_0^\infty e^{-st} e^{t^2} dt$ diverges for every real value of s. Thus, for convergence of $\int_0^\infty e^{-st} f(t)\, dt$, we need another condition on f. The form of this integral suggests one condition that is sufficient. If, for some numbers M and b, we have $|f(t)| \le M e^{bt}$, then

$$e^{-st} |f(t)| \le M e^{(b-s)t} \quad \text{for } s \ge b.$$

But

$$\int_0^\infty M e^{(b-s)t} \, dt$$

converges (to $M/(s - b)$) if $b - s < 0$ or $s > b$. Then, by comparison, $\int_0^\infty e^{-st} |f(t)|\, dt$ also converges if $s > b$, hence $\int_0^\infty e^{-st} f(t)\, dt$ converges if $s > b$.

This line of reasoning suggests a set of conditions that are sufficient for a function to have a Laplace transform.

THEOREM 3.2   *Existence of $\mathfrak{L}[f]$*

Suppose f is piecewise continuous on $[0, k]$ for every positive k. Suppose also that there are numbers M and b such that $|f(t)| \le M e^{bt}$ for $t \ge 0$. Then $\int_0^\infty e^{-st} f(t)\, dt$ converges for $s > b$, hence $\mathfrak{L}[f](s)$ is defined for $s > b$. ∎

Many functions satisfy these conditions, including polynomials, $\sin(at)$, $\cos(at)$, e^{at}, and others.

The conditions of the theorem are sufficient, but not necessary, for a function to have a Laplace transform. Consider, for example, $f(t) = t^{-1/2}$ for $t > 0$. This function is not piecewise continuous on any $[0, k]$ because $\lim_{t \to 0+} t^{-1/2} = \infty$. Nevertheless, $\int_0^k e^{-st} t^{-1/2} dt$ exists for every positive k and $s > 0$. Further,

$$\mathcal{L}[f](s) = \int_0^\infty e^{-st} t^{-1/2} \, dt = 2 \int_0^\infty e^{-sx^2} \, dx \qquad (\text{let } x = t^{1/2})$$

$$= \frac{2}{\sqrt{s}} \int_0^\infty e^{-z^2} \, dz \qquad (\text{let } z = x\sqrt{s})$$

$$= \sqrt{\frac{\pi}{s}},$$

in which we have used the fact (found in some standard integral tables) that $\int_0^\infty e^{-z^2} dz = \sqrt{\pi}/2$.

Now revisit the flow chart at the start of this chapter. Taking the Laplace transform of a function is the first step in solving certain kinds of problems. The bottom of the flow chart suggests that at some point we must be able to go back the other way. After we find some function $G(s)$, we will need to produce a function g whose Laplace transform is G. This is the process of taking an inverse Laplace transform.

DEFINITION 3.3   *Inverse Laplace Transform*

Given a function G, a function g such that $\mathcal{L}[g] = G$ is called an inverse Laplace transform of G.

In this event, we write

$$g = \mathcal{L}^{-1}[G].$$

For example,

$$\mathcal{L}^{-1}\left[\frac{1}{s-a}\right](t) = e^{at}$$

and

$$\mathcal{L}^{-1}\left[\frac{1}{s^2+1}\right](t) = \sin(t).$$

This inverse process is ambiguous because, given G, there will be be many functions whose Laplace transform is G. For example, we know that the Laplace transform of e^{-t} is $1/(s+1)$ for $s > -1$. However, if we change $f(t)$ at just one point, letting

$$h(t) = \begin{cases} e^{-t} & \text{for } t \neq 3 \\ 0 & \text{for } t = 3 \end{cases},$$

then $\int_0^\infty e^{-st} f(t) \, dt = \int_0^\infty e^{-st} h(t) \, dt$ and h has the same Laplace transform as f. In such a case, which one do we call the inverse Laplace transform of $1/(s+1)$?

One answer is provided by Lerch's theorem, which states that two *continuous* functions having the same Laplace transform must be equal.

THEOREM 3.3   Lerch

Let f and g be continuous on $[0, \infty)$ and suppose that $\mathcal{L}[f] = \mathcal{L}[g]$. Then $f = g$. ∎

In view of this, we will partially resolve the ambiguity in taking the inverse Laplace transform by agreeing that, given $F(s)$, we seek a continuous f whose Laplace transform is F. If there is no continuous inverse transform function, then we simply have to make some agreement as to which of several possible candidates we will call $\mathcal{L}^{-1}[F]$. In applications, context will often make this choice obvious.

Because of the linearity of the Laplace transform, its inverse is also linear.

THEOREM 3.4

If $\mathcal{L}^{-1}[F] = f$ and $\mathcal{L}^{-1}[G] = g$ and α and β are real numbers, then

$$\mathcal{L}^{-1}[\alpha F + \beta G] = \alpha f + \beta g. \quad \blacksquare$$

If Table 3.1 is used to find $\mathcal{L}[f]$, look up f in the left column and read $\mathcal{L}[f]$ from the right column. For $\mathcal{L}^{-1}[F]$, look up F in the right column and match it with f in the left.

SECTION 3.1 PROBLEMS

In each of Problems 1 through 10, use the linearity of the Laplace transform and Table 3.1 to find the Laplace transform of the function.

1. $2\sinh(t) - 4$

2. $\cos(t) - \sin(t)$

3. $4t\sin(2t)$

4. $t^2 - 3t + 5$

5. $t - \cos(5t)$

6. $2t^2 e^{-3t} - 4t + 1$

7. $(t + 4)^2$

8. $3e^{-t} + \sin(6t)$

9. $t^3 - 3t + \cos(4t)$

10. $-3\cos(2t) + 5\sin(4t)$

In each of Problems 11 through 18, use the linearity of the inverse Laplace transform and Table 3.1 to find the (continuous) inverse Laplace transform of the function.

11. $\dfrac{-2}{s + 16}$

12. $\dfrac{4s}{s^2 - 14}$

13. $\dfrac{2s - 5}{s^2 + 16}$

14. $\dfrac{3s + 17}{s^2 - 7}$

15. $\dfrac{3}{s - 7} + \dfrac{1}{s^2}$

16. $\dfrac{5}{(s + 7)^2}$

17. $\dfrac{1}{s - 4} - \dfrac{6}{(s - 4)^2}$

18. $\dfrac{2}{s^4}\left[\dfrac{1}{s} - \dfrac{3}{s^2} + \dfrac{4}{s^6}\right]$

19. Let $\mathcal{L}[f](s) = F(s)$. Prove that for any positive number c, $\mathcal{L}[f(ct)](s) = (1/c)F(s/c)$.

20. Suppose f is piecewise continuous on $[0, k]$ for every $k > 0$. Suppose there are numbers M and b such that $|f(t)| \leq Me^{bt}$ for $t \geq 0$. Prove that $\lim_{s \to \infty} F(s) = 0$, where $F = \mathcal{L}[f]$. *Hint:* Write

$$|\mathcal{L}[f](s)| = \left|\int_0^\infty e^{-st}f(t)\,dt\right|$$

$$= \left|\int_0^{t_0} e^{-st}f(t)\,dt + \int_{t_0}^\infty e^{-st}f(t)\,dt\right|$$

$$\leq \left|\int_0^{t_0} e^{-st}f(t)\,dt\right| + \left|\int_{t_0}^\infty e^{-st}f(t)\,dt\right|.$$

Show that for some K, $|f(t)| \leq K$ for $0 \leq t \leq t_0$ and take the limit in the last two integrals as $s \to \infty$.

Suppose that $f(t)$ is defined for all $t \geq 0$. Then f is *periodic* with period T if $f(t + T) = f(t)$ for all $t \geq 0$. For example, $\sin(t)$ has period 2π. In Problems 21, 22, and 23, assume that f has period T.

21. Show that

$$\mathcal{L}[f](s) = \sum_{n=0}^\infty \int_{nT}^{(n+1)T} e^{-st}f(t)\,dt.$$

22. Show that

$$\int_{nT}^{(n+1)T} e^{-st} f(t)\, dt = e^{-nsT} \int_0^T e^{-st} f(t)\, dt.$$

23. From Problems 21 and 22, show that

$$\mathcal{L}[f](s) = \left[\sum_{n=0}^{\infty} e^{-nsT} \right] \int_0^T e^{-st} f(t)\, dt.$$

24. Use the geometric series $\sum_{n=0}^{\infty} r^n = 1/(1 - r)$ for $|r| < 1$, together with the result of Problem 23, to show that

$$\mathcal{L}[f](s) = \frac{1}{1 - e^{-sT}} \int_0^T e^{-st} f(t)\, dt.$$

In each of Problems 25 through 32, a periodic function is given, sometimes by a graph. Find $\mathcal{L}[f]$, using the result of Problem 24.

25. f has period 6 and

$$f(t) = \begin{cases} 5 & \text{for } 0 < t \le 3 \\ 0 & \text{for } 3 < t \le 6 \end{cases}$$

26. $f(t) = |E \sin(\omega t)|$, with E and ω positive constants. (Here f has period π/ω).

27. f has the graph of Figure 3.3.

FIGURE 3.3

28. f has the graph of Figure 3.4.

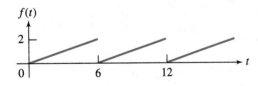

FIGURE 3.4

29. f has the graph of Figure 3.5.

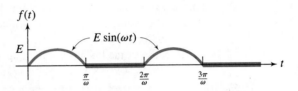

FIGURE 3.5

30. f has the graph of Figure 3.6.

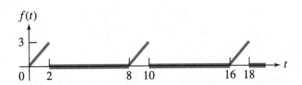

FIGURE 3.6

31. f has the graph of Figure 3.7.

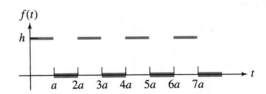

FIGURE 3.7

32. f has the graph of Figure 3.8.

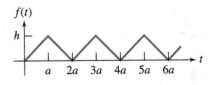

FIGURE 3.8

3.2   Solution of Initial Value Problems Using the Laplace Transform

The Laplace transform is a powerful tool for solving some kinds of initial value problems. The technique depends on the following fact about the Laplace transform of a derivative.

THEOREM 3.5 Laplace Transform of a Derivative

Let f be continuous on $[0, \infty)$ and suppose f' is piecewise continuous on $[0, k]$ for every positive k. Suppose also that $\lim_{k \to \infty} e^{-sk} f(k) = 0$ if $s > 0$. Then

$$\mathcal{L}[f'](s) = sF(s) - f(0). \quad \blacksquare \tag{3.1}$$

That is, the Laplace transform of the derivative of f is s times the Laplace transform of f at s, minus f at zero.

Proof Begin with an integration by parts, with $u = e^{-st}$ and $dv = f'(t) \, dt$. For $k > 0$,

$$\int_0^k e^{-st} f'(t) \, dt = [e^{-st} f(t)]_0^k - \int_0^k -se^{-st} f(t) \, dt$$

$$= e^{-sk} f(k) - f(0) + s \int_0^k e^{-st} f(t) \, dt.$$

Take the limit as $k \to \infty$ and use the assumption that $e^{-sk} f(k) \to 0$ to obtain

$$\mathcal{L}[f'](s) = \lim_{k \to \infty} \left[e^{-sk} f(k) - f(0) + s \int_0^k e^{-st} f(t) \, dt \right]$$

$$= -f(0) + s \int_0^\infty e^{-st} f(t) \, dt = -f(0) + sF(s). \quad \blacksquare$$

If f has a jump discontinuity at 0 (as occurs, for example, if f is an electromotive force that is switched on at time zero), then this conclusion can be amended to read

$$\mathcal{L}[f'](s) = sF(s) - f(0+),$$

where

$$f(0+) = \lim_{t \to 0+} f(t)$$

is the right limit of $f(t)$ at 0.

For problems involving differential equations of order 2 or higher, we need a higher derivative version of the theorem. Let $f^{(j)}$ denote the jth derivative of f. As a notational convenience, we let $f^{(0)} = f$.

THEOREM 3.6 Laplace Transform of a Higher Derivative

Suppose $f, f', \ldots, f^{n-1}$ are continuous on $[0, \infty)$ and $f^{(n)}$ is piecewise continuous on $[0, k]$ for every positive k. Suppose also that $\lim_{k \to \infty} e^{-sk} f^{(j)}(k) = 0$ for $s > 0$ and for $j = 1, 2, \ldots, n-1$. Then

$$\mathcal{L}[f^{(n)}](s) = s^n F(s) - s^{n-1} f(0) - s^{n-2} f'(0) - \cdots - sf^{(n-2)}(0) - f^{(n-1)}(0). \quad \blacksquare \tag{3.2}$$

The second derivative case ($n = 2$) occurs sufficiently often that we will record it separately. Under the conditions of the theorem,

$$\mathcal{L}[f''](s) = s^2 F(s) - sf(0) - f'(0). \tag{3.3}$$

We are now ready to use the Laplace transform to solve certain initial value problems.

EXAMPLE 3.3

Solve $y' - 4y = 1$; $y(0) = 1$.

We know how to solve this problem, but we will use the Laplace transform to illustrate the technique. Write $\mathcal{L}[y](s) = Y(s)$. Take the Laplace transform of the differential equation, using the linearity of $\mathcal{L}$ and equation (3.1), with $y(t)$ in place of $f(t)$:

$$\mathcal{L}[y' - 4y](s) = \mathcal{L}[y'](s) - 4\mathcal{L}[y](s)$$

$$= (sY(s) - y(0)) - 4Y(s) = \mathcal{L}[1](s) = \frac{1}{s}.$$

Here we used the fact (from Table 3.1) that $\mathcal{L}[1](s) = 1/s$ for $s > 0$. Since $y(0) = 1$, we now have

$$(s - 4)Y(s) = y(0) + \frac{1}{s} = 1 + \frac{1}{s}.$$

At this point we have an algebra problem to solve for $Y(s)$, obtaining

$$Y(s) = \frac{1}{(s - 4)} + \frac{1}{s(s - 4)}$$

(note the flow chart at the beginning of this chapter). The solution of the initial value problem is

$$y = \mathcal{L}^{-1}[Y] = \mathcal{L}^{-1}\left[\frac{1}{s - 4}\right] + \mathcal{L}^{-1}\left[\frac{1}{s(s - 4)}\right].$$

From entry 5 of Table 3.1, with $a = 4$,

$$\mathcal{L}^{-1}\left[\frac{1}{s - 4}\right] = e^{4t}.$$

And from entry 8, with $a = 0$ and $b = 4$,

$$\mathcal{L}^{-1}\left[\frac{1}{s(s - 4)}\right] = \frac{1}{-4}(e^{0t} - e^{4t}) = \frac{1}{4}(e^{4t} - 1).$$

The solution of the initial value problem is

$$y(t) = e^{4t} + \frac{1}{4}(e^{4t} - 1)$$

$$= \frac{5}{4}e^{4t} - \frac{1}{4}. \quad \blacksquare$$

One feature of this Laplace transform technique is that the initial value given in the problem is naturally incorporated into the solution process through equation (3.1). We need not find the general solution first, then solve for the constant to satisfy the initial condition.

EXAMPLE 3.4

Solve

$$y'' + 4y' + 3y = e^t; \qquad y(0) = 0, y'(0) = 2.$$

Apply $\mathcal{L}$ to the differential equation to get $\mathcal{L}[y''] + 4\mathcal{L}[y'] + 3\mathcal{L}[y] = \mathcal{L}[e^t].$

Now

$$\mathcal{L}[y''] = s^2 Y - sy(0) - y'(0) = s^2 Y - 2$$

and

$$\mathcal{L}[y'] = sY - y(0) = sY.$$

Therefore,

$$s^2 Y - 2 + 4sY + 3Y = \frac{1}{s-1}.$$

Solve for Y to obtain

$$Y(s) = \frac{2s-1}{(s-1)(s^2 + 4s + 3)}.$$

The solution is the inverse Laplace transform of this function. Some software will produce this inverse. If we want to use Table 3.1, we must use a partial fractions decomposition to write $Y(s)$ as a sum of simpler functions. Write

$$Y(s) = \frac{2s-1}{(s-1)(s^2 + 4s + 3)}$$

$$= \frac{2s-1}{(s-1)(s+1)(s+3)} = \frac{A}{s-1} + \frac{B}{s+1} + \frac{C}{s+3}.$$

This equation can hold only if, for all s,

$$A(s+1)(s+3) + B(s-1)(s+3) + C(s-1)(s+1) = 2s - 1.$$

Now choose values of s to simplify the task of determining A, B, and C. Let $s = 1$ to get $8A = 1$, so $A = \frac{1}{8}$. Let $s = -1$ to get $-4B = -3$, so $B = 3/4$. Choose $s = -3$ to get $8C = -7$, so $C = -\frac{7}{8}$. Then

$$Y(s) = \frac{1}{8} \frac{1}{s-1} + \frac{3}{4} \frac{1}{s+1} - \frac{7}{8} \frac{1}{s+3}.$$

Now read from Table 3.1 that

$$y(t) = \frac{1}{8} e^t + \frac{3}{4} e^{-t} - \frac{7}{8} e^{-3t}. \quad \blacksquare$$

Again, the Laplace transform has converted an initial value problem to an algebra problem, incorporating the initial conditions into the algebraic manipulations. Once we obtain $Y(s)$, the problem becomes one of inverting the transformed function to obtain $y(t)$.

EXAMPLE 3.5

Solve

$$y'' + y = t; \qquad y(0) = 1, y'(0) = 0.$$

Take the Laplace transform of the differential equation to obtain

$$s^2 Y - sy(0) - y'(0) + Y = \mathcal{L}[t](s) = \frac{1}{s^2}.$$

Inserting the initial values, we obtain

$$s^2Y - s + Y = \frac{1}{s^2},$$

or

$$Y(s) = \frac{1}{s^2(s^2+1)} + \frac{s}{s^2+1}.$$

We can read inverses of the functions on the right from entries 14 and 12 of Table 3.1 (reading right to left), obtaining the solution

$$y(t) = t - \sin(t) + \cos(t). \ \blacksquare$$

Equation (3.1) has an interesting consequence that will be useful later. Under the conditions of the theorem, we know that

$$\mathcal{L}[f'] = s\mathcal{L}[f] - f(0).$$

Suppose $f(t)$ is defined by an integral, say,

$$f(t) = \int_0^t g(\tau)\,d\tau.$$

Now $f(0) = 0$ and, assuming continuity of g, $f'(t) = g(t)$. Then

$$\mathcal{L}[f'] = \mathcal{L}[g] = s\mathcal{L}\left[\int_0^t g(\tau)\,d\tau\right].$$

This means that

$$\mathcal{L}\left[\int_0^t g(\tau)\,d\tau\right] = \frac{1}{s}\mathcal{L}[g], \tag{3.4}$$

enabling us to take the Laplace transform of a function defined by an integral. We will use this equation later in dealing with circuits having discontinuous electromotive forces.

Thus far we have illustrated a Laplace transform technique for solving initial value problems with constant coefficients. However, we could have solved the problems in these examples by other means. In the next three sections, we will develop the machinery needed to apply the Laplace transform to problems that defy previous methods.

SECTION 3.2   PROBLEMS

In each of Problems 1 through 10, use the Laplace transform to solve the initial value problem.

1. $y' + 4y = 1$; $y(0) = -3$

2. $y' - 9y = t$; $y(0) = 5$

3. $y' + 4y = \cos(t)$; $y(0) = 0$

4. $y' + 2y = e^{-t}$; $y(0) = 1$

5. $y' - 2y = 1 - t$; $y(0) = 4$

6. $y'' + y = 1$; $y(0) = 6$, $y'(0) = 0$

7. $y'' - 4y' + 4y = \cos(t)$; $y(0) = 1$, $y'(0) = -1$

8. $y'' + 9y = t^2$; $y(0) = y'(0) = 0$

9. $y'' + 16y = 1 + t$; $y(0) = -2$, $y'(0) = 1$

10. $y'' - 5y' + 6y = e^{-t}$; $y(0) = 0$, $y'(0) = 2$

11. Suppose f satisfies the hypotheses of Theorem 3.5, except for a jump discontinuity at 0. Show that

$\mathcal{L}[f'](s) = sF(s) - f(0+)$, where $f(0+) = \lim_{t \to 0+} f(t)$.

12. Suppose f satisfies the hypotheses of Theorem 3.5, except for a jump discontinuity at a positive number c. Prove that

$$\mathcal{L}[f'](s) = sF(s) - f(0) - e^{-cs}[f(c+) - f(c-)],$$

where $f(c-) = \lim_{t \to c-} f(t)$.

13. Suppose g is piecewise continuous on $[0, k]$ for every $k > 0$ and that there are numbers M, b, and a such that $|g(t)| \le Me^{bt}$ for $t \ge a$. Let $\mathcal{L}[g] = G$. Show that

$$\mathcal{L}\left[\int_0^t g(w)\, dw\right](s) = \frac{1}{s}G(s) - \frac{1}{s}\int_0^a g(w)\, dw.$$

3.3 Shifting Theorems and the Heaviside Function

One point to developing the Laplace transform is to broaden the class of problems we are able to solve. Methods of Chapters 1 and 2 are primarily aimed at problems involving continuous functions. But many mathematical models deal with discontinuous processes (for example, switches thrown on and off in a circuit). For these, the Laplace transform is often effective, but we must learn more about representing discontinuous functions and applying both the transform and its inverse to them.

3.3.1 The First Shifting Theorem

We will show that the Laplace transform of $e^{at} f(t)$ is nothing more than the Laplace transform of $f(t)$ shifted a units to the right. This is achieved by replacing s by $s - a$ in $F(s)$ to obtain $F(s - a)$.

THEOREM 3.7 *First Shifting Theorem, or Shifting in the s Variable*

Let $\mathcal{L}[f](s) = F(s)$ for $s > b \ge 0$. Let a be any number. Then

$$\mathcal{L}[e^{at} f(t)](s) = F(s - a) \quad \text{for } s > a + b.$$

Proof Compute

$$\mathcal{L}[e^{at} f(t)](s) = \int_0^\infty e^{at} e^{-st} f(s)\, ds$$

$$= \int_0^\infty e^{-(s-a)t} f(t)\, dt = F(s - a)$$

for $s - a > b$, or $s > a + b$. ∎

EXAMPLE 3.6

We know from Table 3.1 that $\mathcal{L}[\cos(bt)] = s/(s^2 + b^2)$. For the Laplace transform of $e^{at} \cos(bt)$, replace s with $s - a$ to get

$$\mathcal{L}[e^{at} \cos(bt)](s) = \frac{s - a}{(s - a)^2 + b^2}. \quad ∎$$

EXAMPLE 3.7

Since $\mathcal{L}[t^3] = 6/s^4$, then

$$\mathcal{L}[t^3 e^{7t}](s) = \frac{6}{(s-7)^4}. \quad \blacksquare$$

The first shifting theorem suggests a corresponding formula for the inverse Laplace transform: If $\mathcal{L}[f] = F$, then

$$\mathcal{L}^{-1}[F(s-a)] = e^{at} f(t).$$

Sometimes it is convenient to write this result as

$$\mathcal{L}^{-1}[F(s-a)] = e^{at} \mathcal{L}^{-1}[F(s)]. \tag{3.5}$$

EXAMPLE 3.8

Suppose we want to compute

$$\mathcal{L}^{-1}\left[\frac{4}{s^2 + 4s + 20}\right].$$

We will manipulate the quotient into a form to which we can apply the shifting theorem. Complete the square in the denominator to write

$$\frac{4}{s^2 + 4s + 20} = \frac{4}{(s+2)^2 + 16}.$$

Think of the quotient on the right as a function of $s + 2$:

$$F(s+2) = \frac{4}{(s+2)^2 + 16}.$$

This means we should choose

$$F(s) = \frac{4}{s^2 + 16}.$$

Now the shifting theorem tells us that

$$\mathcal{L}[e^{-2t} \sin(4t)] = F(s - (-2)) = F(s+2) = \frac{4}{(s+2)^2 + 16}$$

and therefore

$$\mathcal{L}^{-1}\left[\frac{4}{(s+2)^2 + 16}\right] = e^{-2t} \sin(4t). \quad \blacksquare$$

EXAMPLE 3.9

Compute

$$\mathcal{L}^{-1}\left[\frac{3s - 1}{s^2 - 6s + 2}\right].$$

Again, begin with some manipulation into the form of a function of $s - a$ for some a:

$$\frac{3s - 1}{s^2 - 6s + 2} = \frac{3s - 1}{(s - 3)^2 - 7}$$

$$= \frac{3(s - 3)}{(s - 3)^2 - 7} + \frac{8}{(s - 3)^2 - 7} = G(s - 3) + K(s - 3)$$

if we choose

$$G(s) = \frac{3s}{s^2 - 7} \quad \text{and} \quad K(s) = \frac{8}{s^2 - 7}.$$

Now apply equation (3.5) (in the second line) to write

$$\mathcal{L}^{-1}\left[\frac{3s - 1}{s^2 - 6s + 2}\right] = \mathcal{L}^{-1}[G(s - 3)] + \mathcal{L}^{-1}[K(s - 3)]$$

$$= e^{3t}\mathcal{L}^{-1}[G(s)] + e^{3t}\mathcal{L}^{-1}[K(s)]$$

$$= e^{3t}\mathcal{L}^{-1}\left[\frac{3s}{s^2 - 7}\right] + e^{3t}\mathcal{L}^{-1}\left[\frac{8}{s^2 - 7}\right]$$

$$= 3e^{3t}\mathcal{L}^{-1}\left[\frac{s}{s^2 - 7}\right] + 8e^{3t}\mathcal{L}^{-1}\left[\frac{1}{s^2 - 7}\right]$$

$$= 3e^{3t}\cosh(\sqrt{7}t) + \frac{8}{\sqrt{7}}e^{3t}\sinh(\sqrt{7}t). \quad \blacksquare$$

3.3.2 The Heaviside Function and Pulses

We will now lay the foundations for solving certain initial value problems having discontinuous forcing functions. To do this, we will use the Heaviside function.

Recall that f has a jump discontinuity at a if $\lim_{t \to a-} f(t)$ and $\lim_{t \to a+} f(t)$ both exist and are finite, but unequal. Figure 3.9 shows a typical jump discontinuity. The magnitude of the jump discontinuity is the "width of the gap" in the graph at a. This width is

$$\left| \lim_{t \to a+} f(t) - \lim_{t \to a-} f(t) \right|.$$

Functions with jump discontinuities can be treated very efficiently using the unit step function, or Heaviside function.

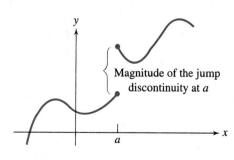

FIGURE 3.9

DEFINITION 3.4   *Heaviside Function*

The Heaviside function H is defined by

$$H(t) = \begin{cases} 0 & \text{if } t < 0 \\ 1 & \text{if } t \geq 0 \end{cases}.$$

Oliver Heaviside (1850–1925) was an English electrical engineer who did much to introduce Laplace transform methods into engineering practice. A graph of H is shown in Figure 3.10. It has a jump discontinuity of magnitude 1 at 0.

The Heaviside function may be thought of as a flat switching function, "on" when $t \geq 0$, where $H(t) = 1$, and "off" when $t < 0$, where $H(t) = 0$. We will use it to achieve a variety of effects, including switching functions on and off at different times, shifting functions along the axis, and combining functions with pulses.

To begin this program, if a is any number, then $H(t - a)$ is the Heaviside function shifted a units to the right, as shown in Figure 3.11, since

$$H(t - a) = \begin{cases} 0 & \text{if } t < a \\ 1 & \text{if } t \geq a \end{cases}.$$

$H(t - a)$ models a flat signal of magnitude 1, turned off until time $t = a$ and then switched on.

We can use $H(t - a)$ to achieve the effect of turning a given function g off until time $t = a$, at which time it is switched on. In particular,

$$H(t - a)g(t) = \begin{cases} 0 & \text{if } t < a \\ g(t) & \text{if } t \geq a \end{cases}$$

is zero until time $t = a$, at which time it switches on $g(t)$. To see this in a specific case, let $g(t) = \cos(t)$ for all t. Then

$$H(t - \pi)g(t) = H(t - \pi)\cos(t) = \begin{cases} 0 & \text{if } t < \pi \\ \cos(t) & \text{if } t \geq \pi \end{cases}.$$

Graphs of $\cos(t)$ and $H(t - \pi)\cos(t)$ are shown in Figure 3.12 for comparison.

We can also use the Heaviside function to describe a pulse.

FIGURE 3.10   *The Heaviside function $H(t)$.*

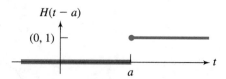

FIGURE 3.11   *A shifted Heaviside function.*

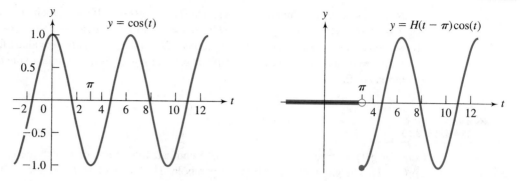

FIGURE 3.12 *Comparison of $y = \cos(t)$ and $y = H(t - \pi)\cos(t)$.*

DEFINITION 3.5 Pulse

A pulse is a function of the form

$$H(t - a) - H(t - b),$$

in which $a < b$.

This pulse function is graphed in Figure 3.13. It has value 0 if $t < a$ (where $H(t - a) = H(t - b) = 0$), value 1 if $a \le t < b$ (where $H(t - a) = 1$ and $H(t - b) = 0$), and value 0 if $t \ge b$ (where $H(t - a) = H(t - b) = 1$).

Multiplying a function g by this pulse has the effect of leaving $g(t)$ switched off until time a. The function is then turned on until time b, when it is switched off again. For example, let $g(t) = e^t$. Then

$$[H(t - 1) - H(t - 2)]e^t = \begin{cases} 0 & \text{if } t < 1 \\ e^t & \text{if } 1 \le t < 2 \\ 0 & \text{if } t \ge 2 \end{cases}.$$

Figure 3.14 shows a graph of this function.

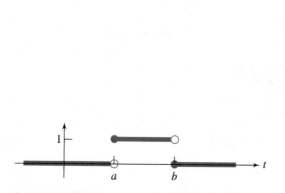

FIGURE 3.13 *Pulse function $H(t - a) - H(t - b)$.*

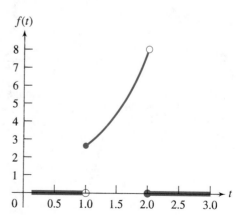

FIGURE 3.14 *Graph of*
$f(t) = [H(t - 1) - H(t - 2)]e^t$.

Next consider shifted functions of the form $H(t-a)g(t-a)$. If $t < a$, then $g(t-a)H(t-a) = 0$ because $H(t-a) = 0$. If $t \geq a$, then $H(t-a) = 1$ and $H(t-a)g(t-a) = g(t-a)$, which is $g(t)$ shifted a units to the right. Thus the graph of $H(t-a)g(t-a)$ is zero along the horizontal axis until $t = a$, and for $t \geq a$, is the graph of $g(t)$ for $t \geq 0$, shifted a units to the right to begin at a instead of 0.

EXAMPLE 3.10

Consider $g(t) = t^2$ and $a = 2$. Figure 3.15 compares the graph of g with the graph of $H(t - 2)g(t - 2)$. The graph of g is a familiar parabola. The graph of $H(t - 2)g(t - 2)$ is zero until time 2, then has the shape of the graph of t^2 for $t \geq 0$, but shifted 2 units to the right to start at $t = 2$. ▪

It is important to understand the difference between $g(t)$, $H(t-a)g(t)$, and $H(t-a)g(t-a)$. Figure 3.16 shows graphs of these three functions for $g(t) = t^2$ and $a = 3$.

3.3.3   The Second Shifting Theorem

Sometimes $H(t - a)g(t - a)$ is referred to as a *shifted function*, although it is more than that because this graph is also zero for $t < a$. The second shifting theorem deals with the Laplace transform of such a function.

THEOREM 3.8   *Second Shifting Theorem, or Shifting in the t Variable*

Let $\mathcal{L}[f](s) = F(s)$ for $s > b$. Then

$$\mathcal{L}[H(t - a)f(t - a)](s) = e^{-as}F(s)$$

for $s > b$. ▪

That is, we obtain the Laplace transform of $H(t - a)f(t - a)$ by multiplying the Laplace transform of $f(t)$ by e^{-as}.

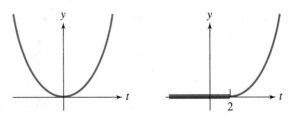

FIGURE 3.15   *Comparison of $y = t^2$ and $y = (t - 2)^2 H(t - 2)$.*

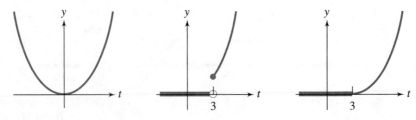

FIGURE 3.16   *Comparison of $y = t^2$, $y = t^2 H(t - 3)$, and $y = (t - 3)^2 H(t - 3)$.*

Proof Proceeding from the definition,

$$\mathcal{L}[H(t-a)f(t-a)](s) = \int_0^\infty e^{-st} H(t-a)f(t-a)\,dt$$

$$= \int_a^\infty e^{-st} f(t-a)\,dt$$

because $H(t-a) = 0$ for $t < a$ and $H(t-a) = 1$ for $t \geq a$. Now let $w = t - a$ in the last integral to obtain

$$\mathcal{L}[H(t-a)f(t-a)](s) = \int_0^\infty e^{-s(a+w)} f(w)\,dw$$

$$= e^{-as} \int_0^\infty e^{-sw} f(w)\,dw = e^{-as} F(s).$$

EXAMPLE 3.11

Suppose we want the Laplace transform of $H(t-a)$. Write this as $H(t-a)f(t-a)$, with $f(t) = 1$ for all t. Since $F(s) = 1/s$ (from Table 3.1 or by direct computation from the definition), then

$$\mathcal{L}[H(t-a)](s) = e^{-as}\mathcal{L}[1](s) = \frac{1}{s}e^{-as}. \quad \blacksquare$$

EXAMPLE 3.12

Compute $\mathcal{L}[g]$, where $g(t) = 0$ for $0 \leq t < 2$ and $g(t) = t^2 + 1$ for $t \geq 2$.

Since $g(t)$ is zero until time $t = 2$, and is then $t^2 + 1$, we may write $g(t) = H(t-2)(t^2+1)$.

To apply the second shifting theorem, we must write $g(t)$ as a function, or perhaps sum of functions, of the form $f(t-2)H(t-2)$. This necessitates writing $t^2 + 1$ as a sum of functions of $t - 2$. One way to do this is to expand $t^2 + 1$ in a Taylor series about 2. In this simple case we can achieve the same result by algebraic manipulation:

$$t^2 + 1 = (t-2+2)^2 + 1 = (t-2)^2 + 4(t-2) + 5.$$

Then

$$g(t) = (t^2+1)H(t-2)$$

$$= (t-2)^2 H(t-2) + 4(t-2)H(t-2) + 5H(t-2).$$

Now we can apply the second shifting theorem:

$$\mathcal{L}[g] = \mathcal{L}[(t-2)^2 H(t-2)] + 4\mathcal{L}[(t-2)H(t-2)] + 5\mathcal{L}[H(t-2)]$$

$$= e^{-2s}\mathcal{L}[t^2] + 4e^{-2s}\mathcal{L}[t] + 5e^{-2s}\mathcal{L}[1]$$

$$= e^{-2s}\left[\frac{2}{s^3} + \frac{4}{s^2} + \frac{5}{s}\right]. \quad \blacksquare$$

As usual, any formula for the Laplace transform of a class of functions can also be read as a formula for an inverse Laplace transform. The inverse version of the second shifting theorem is

$$\mathcal{L}^{-1}[e^{-as}F(s)](t) = H(t-a)f(t-a). \tag{3.6}$$

This enables us to compute the inverse Laplace transform of a known transformed function multipled by an exponential e^{-as}.

EXAMPLE 3.13

Compute

$$\mathcal{L}^{-1}\left[\frac{se^{-3s}}{s^2+4}\right].$$

The presence of the exponential factor suggests the use of equation (3.6). Concentrate on finding

$$\mathcal{L}^{-1}\left[\frac{s}{s^2+4}\right].$$

This inverse can be read directly from Table 3.1 and is $f(t) = \cos(2t)$. Therefore,

$$\mathcal{L}^{-1}\left[\frac{se^{-3s}}{s^2+4}\right](t) = H(t-3)\cos(2(t-3)). \blacksquare$$

We are now prepared to solve certain initial value problems involving discontinuous forcing functions.

EXAMPLE 3.14

Solve the initial value problem

$$y'' + 4y = f(t); \qquad y(0) = y'(0) = 0,$$

in which

$$f(t) = \begin{cases} 0 & \text{for } t < 3 \\ t & \text{for } t \geq 3 \end{cases}.$$

Because of the discontinuity in f, methods developed in Chapter 2 do not apply. First recognize that

$$f(t) = H(t-3)t.$$

Apply the Laplace transform to the differential equation to get

$$\mathcal{L}[y''] + 4\mathcal{L}[y] = s^2Y(s) - sy(0) - y'(0) + 4Y(s)$$

$$= (s^2+4)Y(s) = \mathcal{L}[H(t-3)t],$$

in which we have inserted the initial conditions $y(0) = y'(0) = 0$. In order to use the second shifting theorem to compute $\mathcal{L}[H(t-3)t]$, write

$$\mathcal{L}[H(t-3)t] = \mathcal{L}[H(t-3)(t-3+3)]$$

$$= \mathcal{L}[H(t-3)(t-3)] + 3\mathcal{L}[H(t-3)]$$

$$= e^{-3s}\mathcal{L}[t] + 3e^{-3s}\mathcal{L}[1] = \frac{1}{s^2}e^{-3s} + \frac{3}{s}e^{-3s}.$$

We now have

$$(s^2 + 4)Y = \frac{1}{s^2}e^{-3s} + \frac{3}{s}e^{-3s}.$$

The transform of the solution is

$$Y(s) = \frac{3s + 1}{s^2(s^2 + 4)}e^{-3s}.$$

The solution is within reach. We must take the inverse Laplace transform of $Y(s)$. To do this, first use a partial fractions decomposition to write

$$\frac{3s + 1}{s^2(s^2 + 4)}e^{-3s} = \frac{3}{4}\frac{1}{s}e^{-3s} - \frac{3}{4}\frac{s}{s^2 + 4}e^{-3s} + \frac{1}{4}\frac{1}{s^2}e^{-3s} - \frac{1}{4}\frac{1}{s^2 + 4}e^{-3s}.$$

Each term is an exponential times a function whose Laplace transform we know, and we can apply equation (3.6) to write

$$y(t) = \frac{3}{4}H(t - 3) - \frac{3}{4}H(t - 3)\cos(2(t - 3))$$

$$+ \frac{1}{4}H(t - 3)(t - 3) - \frac{1}{4}H(t - 3)\frac{1}{2}\sin(2(t - 3)).$$

Because of the $H(t - 3)$ factor in each term, this solution is zero until time $t = 3$, and we may write

$$y(t) = \begin{cases} 0 & \text{for } t < 3 \\ \frac{3}{4} - \frac{3}{4}\cos(2(t - 3)) + \frac{1}{4}(t - 3) - \frac{1}{8}\sin(2(t - 3)) & \text{for } t \geq 3 \end{cases},$$

or, upon combining terms,

$$y(t) = \begin{cases} 0 & \text{for } t < 3 \\ \frac{1}{8}[2t - 6\cos(2(t - 3)) - \sin(2(t - 3))] & \text{for } t \geq 3 \end{cases}.$$

A graph of this solution is shown in Figure 3.17. ■

In this example, it is interesting to observe that the solution is differentiable everywhere, even though the function f occurring in the differential equation had a jump discontinuity at 3.

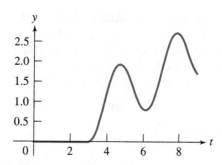

FIGURE 3.17 *Solution of*

$$y'' + 4y = \begin{cases} 0 & \text{if } 0 \leq t < 3 \\ t & \text{if } t \leq 3 \end{cases}; \quad (y(0) = y'(0) = 0).$$

This behavior is typical of initial value problems having a discontinuous forcing function. If the differential equation has order n and φ is a solution, then φ and its first $n - 1$ derivatives will be continuous, while the nth derivative will have a jump discontinuity wherever f does, and these jump discontinuities will agree in magnitude with the corresponding jump discontinuities of f.

Often we need to write a function having several jump discontinuities in terms of Heaviside functions in order to use the shifting theorems. Here is an example.

EXAMPLE 3.15

Let

$$
f(t) = \begin{cases} 0 & \text{if } t < 2 \\ t - 1 & \text{if } 2 \le t < 3 \\ -4 & \text{if } t \ge 3 \end{cases}.
$$

A graph of f is shown in Figure 3.18. There are jump discontinuities of magnitude 1 at $t = 2$ and magnitude 6 at $t = 3$.

Think of $f(t)$ as consisting of two nonzero parts, the part that is $t - 1$ on $[2, 3)$ and the part that is -4 on $[3, \infty)$. We want to turn on $t - 1$ at time 2 and turn it off at time 3, then turn on -4 at time 3 and leave it on.

The first effect is achieved by multiplying the pulse function $H(t - 2) - H(t - 3)$ by $t - 1$. The second is achieved by multiplying $H(t - 3)$ by 4. Therefore,

$$
f(t) = [H(t - 2) - H(t - 3)](t - 1) - 4H(t - 3).
$$

As a check, this gives $f(t) = 0$ if $t < 2$ because all of the shifted Heaviside functions are zero for $t < 2$. For $2 \le t < 3$, $H(t - 2) = 1$ but $H(t - 3) = 0$, so $f(t) = t - 1$. And for $t \ge 3$, $H(t - 2) = H(t - 3) = 1$, so $f(t) = -4$. ∎

3.3.4 Analysis of Electrical Circuits

The Heaviside function is important in many kinds of problems, including the analysis of electrical circuits, where we anticipate turning switches on and off. Here are two examples.

EXAMPLE 3.16

Suppose the capacitor in the circuit of Figure 3.19 initially has zero charge and that there is no initial current. At time $t = 2$ seconds, the switch is thrown from position B to A, held there for 1 second, then switched back to B. We want the output voltage E_{out} on the capacitor.

From the circuit diagram, the forcing function is zero until $t = 2$, then has value 10 volts until $t = 3$, and then is zero again. Thus E is the pulse function

$$
E(t) = 10[H(t - 2) - H(t - 3)].
$$

By Kirchhoff's voltage law,

$$
Ri(t) + \frac{1}{C}q(t) = E(t),
$$

or

$$
250{,}000q'(t) + 10^6 q(t) = E(t).
$$

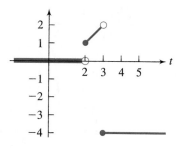

FIGURE 3.18 *Graph of*

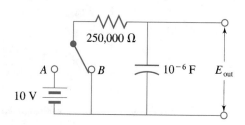

FIGURE 3.19

$$f(t) = \begin{cases} 0 & \text{if } t < 2 \\ t - 1 & \text{if } 2 \le t < 3 \\ -4 & \text{if } t \ge 3 \end{cases} .$$

We want to solve for q subject to the initial condition $q(0) = 0$. Apply the Laplace transform to the differential equation, incorporating the initial condition, to write

$$250{,}000[s\,Q(t) - q(0)] + 10^6 Q(t) = 250{,}000s\,Q + 10^6 Q = \mathcal{L}[E(t)].$$

Now

$$\mathcal{L}[E(t)](s) = 10\mathcal{L}[H(t-2)](s) - 10\mathcal{L}[H(t-3)](s)$$

$$= \frac{10}{s}e^{-2s} - \frac{10}{s}e^{-3s}.$$

We now have the following equation for Q:

$$2.5(10^5)s\,Q(s) + 10^6 Q(s) = \frac{10}{s}e^{-2s} - \frac{10}{s}e^{-3s}$$

or

$$Q(s) = 4(10^{-5})\frac{1}{s(s+4)}e^{-2s} - 4(10)^{-5}\frac{1}{s(s+4)}e^{-3s}.$$

Use a partial fractions decomposition to write

$$Q(s) = 10^{-5}\left[\frac{1}{s}e^{-2s} - \frac{1}{s+4}e^{-2s}\right] - 10^{-5}\left[\frac{1}{s}e^{-3s} - \frac{1}{s+4}e^{-3s}\right].$$

By the second shifting theorem,

$$\mathcal{L}^{-1}\left[\frac{1}{s}e^{-2s}\right](t) = H(t-2)$$

and

$$\mathcal{L}^{-1}\left[\frac{1}{s+4}e^{-2s}\right] = H(t-2)f(t-2),$$

where $f(t) = \mathcal{L}^{-1}[1/(s+4)] = e^{-4t}$. Thus,

$$\mathcal{L}^{-1}\left[\frac{1}{s+4}e^{-2s}\right] = H(t-2)e^{-4(t-2)}.$$

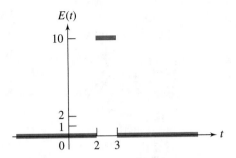

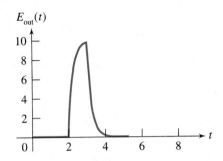

FIGURE 3.20 *Input voltage for the circuit of Figure 3.19.*

FIGURE 3.21 *Output voltage for the circuit of Figure 3.19.*

The other two terms in $Q(s)$ are treated similarly, and we obtain

$$q(t) = 10^{-5}[H(t-2) - H(t-2)e^{-4(t-2)}] - 10^{-5}[H(t-3) - H(t-3)e^{-4(t-3)}]$$

$$= 10^{-5}H(t-2)[1 - e^{-4(t-2)}] - 10^{-5}H(t-3)[1 - e^{-4(t-3)}].$$

Finally, since the output voltage is $E_{out}(t) = 10^6 q(t)$,

$$E_{out}(t) = 10H(t-2)[1 - e^{-4(t-2)}] - 10H(t-3)[1 - e^{-4(t-3)}].$$

The input and output voltages are graphed in Figures 3.20 and 3.21. ■

EXAMPLE 3.17

The circuit of Figure 3.22 has the roles of resistor and capacitor interchanged from the circuit of the preceding example. We want to know the output voltage $i(t)R$ at any time.

The differential equation of the preceding example applies to this circuit, but now we are interested in the current. Since $i = q'$, then

$$(2.5)(10^5)i(t) + 10^6 q(t) = E(t); \qquad i(0) = q(0) = 0.$$

The strategy of eliminating q by differentiating and using $i = q'$ does not apply here, because $E(t)$ is not differentiable. To eliminate $q(t)$ in the present case, write

$$q(t) = \int_0^t i(\tau)\,d\tau + q(0) = \int_0^t i(\tau)\,d\tau.$$

We now have the following problem to solve for the current:

$$(2.5)(10^5)i(t) + 10^6 \int_0^t i(\tau)\,d\tau = E(t); \qquad i(0) = 0.$$

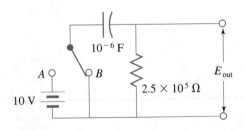

FIGURE 3.22

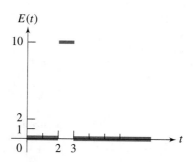

FIGURE 3.23 *Input voltage for the circuit of Figure 3.22.*

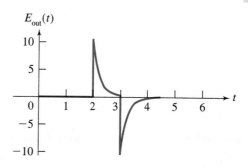

FIGURE 3.24 *Output voltage for the circuit of Figure 3.22.*

This is not a differential equation. Nevertheless, we have the means to solve it. Take the Laplace transform of the equation, using equation (3.4), to obtain

$$(2.5)(10^5)I(s) + 10^6\frac{1}{s}I(s) = \mathcal{L}[E](s)$$

$$= 10\frac{1}{s}e^{-2s} - 10\frac{1}{s}e^{-3s}.$$

Here $I = \mathcal{L}[i]$. Solve for $I(s)$ to get

$$I(s) = 4(10^{-5})\frac{1}{s+4}e^{-2s} - 4(10^{-5})\frac{1}{s+4}e^{-3s}.$$

Take the inverse Laplace transform to obtain

$$i(t) = 4(10^{-5})H(t-2)e^{-4(t-2)} - 4(10^{-5})H(t-3)e^{-4(t-3)}.$$

The input and output voltages are graphed in Figures 3.23 and 3.24. ■

SECTION 3.3   PROBLEMS

In each of Problems 1 through 15, find the Laplace transform of the function.

1. $(t^3 - 3t + 2)e^{-2t}$

2. $e^{-3t}(t - 2)$

3. $f(t) = \begin{cases} 1 & \text{for } 0 \le t < 7 \\ \cos(t) & \text{for } t \ge 7 \end{cases}$

4. $e^{4t}[t - \cos(t)]$

5. $f(t) = \begin{cases} t & \text{for } 0 \le t < 3 \\ 1 - 3t & \text{for } t \ge 3 \end{cases}$

6. $f(t) = \begin{cases} 2t - \sin(t) & \text{for } 0 \le t < \pi \\ 0 & \text{for } t \ge \pi \end{cases}$

7. $e^{-t}[1 - t^2 + \sin(t)]$

8. $f(t) = \begin{cases} t^2 & \text{for } 0 \le t < 2 \\ 1 - t - 3t^2 & \text{for } t \ge 2 \end{cases}$

9. $f(t) = \begin{cases} \cos(t) & \text{for } 0 \le t < 2\pi \\ 2 - \sin(t) & \text{for } t \ge 2\pi \end{cases}$

10. $f(t) = \begin{cases} -4 & \text{for } 0 \le t < 1 \\ 0 & \text{for } 1 \le t < 3 \\ e^{-t} & \text{for } t \ge 3 \end{cases}$

11. $te^{-2t}\cos(3t)$

12. $e^{t}[1 - \cosh(t)]$

13. $f(t) = \begin{cases} t - 2 & \text{for } 0 \le t < 16 \\ -1 & \text{for } t \ge 16 \end{cases}$

14. $f(t) = \begin{cases} 1 - \cos(2t) & \text{for } 0 \le t < 3\pi \\ 0 & \text{for } t \ge 3\pi \end{cases}$

15. $e^{-5t}(t^4 + 2t^2 + t)$

In each of Problems 16 through 29, find the inverse Laplace transform of the function.

16. $\dfrac{1}{s^2 + 4s + 12}$

17. $\dfrac{1}{s^2 - 4s + 5}$

18. $\dfrac{1}{s^3}e^{-5s}$

19. $\dfrac{se^{-2s}}{s^2 + 9}$

20. $\dfrac{3}{s + 2}e^{-4s}$

21. $\dfrac{1}{s^2 + 6s + 7}$

22. $\dfrac{s - 4}{s^2 - 8s + 10}$

23. $\dfrac{s + 2}{s^2 + 6s + 1}$

24. $\dfrac{1}{(s - 5)^3}e^{-s}$

25. $\dfrac{1}{s(s^2 + 16)}e^{-21s}$

26. $\dfrac{s - 3}{s^2 + 10s + 9}$

27. $\dfrac{2s + 4}{s^2 - 4s + 4}$

28. $\dfrac{se^{-10s}}{(s^2 + 4)^2}$

29. $\dfrac{s}{s^2 - 14s + 1}$

30. Determine $\mathcal{L}[e^{-2t}\int_0^t e^{2w}\cos(3w)\,dw]$. *Hint:* Use the first shifting theorem.

In each of Problems 31 through 40, solve the initial value problem by using the Laplace transform.

31. $y'' + 4y = f(t); y(0) = 1, y'(0) = 0$, with
$$f(t) = \begin{cases} 0 & \text{for } 0 \le t < 4 \\ 3 & \text{for } t \ge 4 \end{cases}$$

32. $y'' - 2y' - 3y = f(t); y(0) = 1, y'(0) = 0$, with
$$f(t) = \begin{cases} 0 & \text{for } 0 \le t < 4 \\ 12 & \text{for } t \ge 4 \end{cases}$$

33. $y^{(3)} - 8y = g(t); y(0) = y'(0) = y''(0) = 0$, with
$$g(t) = \begin{cases} 0 & \text{for } 0 \le t < 6 \\ 2 & \text{for } t \ge 6 \end{cases}$$

34. $y'' + 5y' + 6y = f(t); y(0) = y'(0) = 0$, with
$$f(t) = \begin{cases} -2 & \text{for } 0 \le t < 3 \\ 0 & \text{for } t \ge 3 \end{cases}$$

35. $y^{(3)} - y'' + 4y' - 4y = f(t); y(0) = y'(0) = 0,$
$y''(0) = 1$, with $f(t) = \begin{cases} 1 & \text{for } 0 \le t < 5 \\ 2 & \text{for } t \ge 5 \end{cases}$

36. $y'' - 4y' + 4y = f(t); y(0) = -2, y'(0) = 1$, with
$$f(t) = \begin{cases} t & \text{for } 0 \le t < 3 \\ t + 2 & \text{for } t \ge 3 \end{cases}$$

37. $y'' + 2y' - 7y = f(t); y(0) = -2, y'(0) = 0$, with
$$f(t) = \begin{cases} 0 & \text{for } 0 \le t < 5 \\ 2 & \text{for } t \ge 5 \end{cases}$$

38. $y'' + 9y = f(t); y(0) = y'(0) = 1$, with
$$f(t) = \begin{cases} 0 & \text{for } 0 \le t < \pi \\ \cos(t) & \text{for } t \ge \pi \end{cases}$$

39. $y'' + 4y' + 4y = f(t); y(0) = 1, y'(0) = 2$, with
$$f(t) = \begin{cases} 1 & \text{for } 0 \le t < 2 \\ 0 & \text{for } t \ge 2 \end{cases}$$

40. $y'' + 5y' + 6y = f(t); y(0) = 0, y'(0) = -4$, with
$$f(t) = \begin{cases} t^2 & \text{for } 0 \le t < 3 \\ 0 & \text{for } t \ge 3 \end{cases}$$

41. Find the solution φ of the initial value problem
$$y'' - 3y' + 2y = g(t) = \begin{cases} 0 & \text{for } 0 \le t < 3 \\ 2 & \text{for } t \ge 3 \end{cases};$$
$$y(0) = y'(0) = 0.$$

Show that φ and φ' are continuous for $t \ge 0$ but that $\varphi''(t)$ does not exist at $t = 3$. Use l'Hôpital's rule to show that

$$\lim_{t \to 3+} \frac{\varphi'(t) - \varphi'(3)}{t - 3} = 2$$

while

$$\lim_{t \to 3-} \frac{\varphi'(t) - \varphi'(3)}{t - 3} = 0.$$

Thus, the second derivative has a jump discontinuity at 3 equal in direction and magnitude to the jump discontinuity of g at 3.

42. Consider the initial value problem

$$y' - 3y = g(t); \qquad y(0) = 2,$$

where

$$g(t) = \begin{cases} 0 & \text{for } 0 \le t < 4 \\ 3 & \text{for } t \ge 4 \end{cases}.$$

Solve this problem using the Laplace transform. Next, solve the problem as follows. First find the solution φ of $y' - 3y = 0$; $y(0) = 2$, and then find a solution ψ of $y' - 3y = 4$. The latter will involve an arbitrary constant. Choose this constant so that $\lim_{t \to 4+} \psi(t) = \lim_{t \to 4-} \varphi(t)$. Then define $\xi(t)$ to equal $\varphi(t)$ if $0 \le t \le 4$ and to equal $\psi(t)$ if $t > 4$. Show that ξ is a solution of the initial value problem and compare it with the solution obtained using the Laplace transform.

43. Calculate and graph the output voltage in the circuit of Figure 3.19, assuming that at time zero the capacitor is charged to a potential of 5 volts and the switch is opened at 0 and closed 5 seconds later.

44. Calculate and graph the output voltage in the RL circuit of Figure 3.25 if the current is initially zero and

$$E(t) = \begin{cases} 0 & \text{for } 0 \le t < 5 \\ 2 & \text{for } t \ge 5 \end{cases}.$$

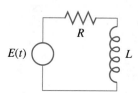

FIGURE 3.25

45. Solve for the current in the RL circuit of Problem 44 if the current is initially zero and

$$E(t) = \begin{cases} k & \text{for } 0 \le t < 5 \\ 0 & \text{for } t \ge 5 \end{cases}.$$

46. Solve for the current in the RL circuit of Problem 44 if the initial current is zero and

$$E(t) = \begin{cases} 0 & \text{for } 0 \le t < 4 \\ Ae^{-t} & \text{for } t \ge 4 \end{cases}.$$

47. Write the function graphed in Figure 3.26 in terms of the Heaviside function and find its Laplace transform.

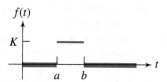

FIGURE 3.26

48. Write the function graphed in Figure 3.27 in terms of the Heaviside function and find its Laplace transform.

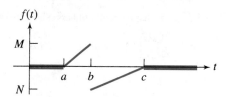

FIGURE 3.27

49. Write the function graphed in Figure 3.28 in terms of the Heaviside function and find its Laplace transform.

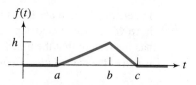

FIGURE 3.28

50. Solve for the current in the RL circuit of Figure 3.29 if the initial current is zero, $E(t)$ has period 4, and

$$E(t) = \begin{cases} 10 & \text{for } 0 \le t < 2 \\ 0 & \text{for } 2 \le t < 4 \end{cases}.$$

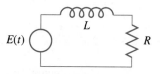

FIGURE 3.29

Hint: See Problem 24 of Section 3.1 for the Laplace transform of a periodic function. You should find that $I(s) = F(s)/(1 + e^{-2s})$ for some $F(s)$. Use a geometric series to write

$$\frac{1}{1 + e^{-2s}} = \sum_{n=0}^{\infty} (-1)^n e^{-2ns}$$

to write $I(s)$ as an infinite series, then take the inverse transform term by term by using a shifting theorem. Graph the current for $0 \le t < 8$.

51. (*Heaviside's Formulas*) Oliver Heaviside developed formulas for computing inverse Laplace transforms of quotients $P(s)/Q(s)$ of polynomials having no common factors. The idea is to construct terms of the inverse transform corresponding to factors of $Q(s)$. We will illustrate the method for the case that $Q(s)$ can be completely factored into linear factors. Thus suppose that

$$Q(s) = k(s - a_1)(s - a_2) \cdots (s - a_n),$$

with $a_1, \ldots, a_n$ distinct real numbers. Show that

$$\mathcal{L}^{-1} \left[\frac{P(s)}{Q(s)} \right](t) = Z(a_1)e^{a_1 t} + Z(a_2)e^{a_2 t}$$

$$+ \cdots + Z(a_n)e^{a_n t},$$

where $Z(s) = P(s)/Q'(s)$. *Hint:* Use a partial fractions decomposition to write

$$\frac{P(s)}{Q(s)} = \frac{A_1}{s - a_1} + \frac{A_2}{s - a_2} + \cdots + \frac{A_n}{s - a_n}.$$

To determine A_j, write

$$\frac{P(s)(s - a_j)}{Q(s) - Q(a_j)} = A_j + (s - a_j)G(s),$$

in which $G(a_j) \ne 0$, and take the limit of both sides of this equation as $s \to a_j$. This method is similar in form to a method of calculating residues of complex functions at simple poles, as we will do later.

In each of the following, use the Heaviside method to determine the inverse Laplace transform of the function.

(a) $\dfrac{s}{s^2 - 3s + 2}$

(b) $\dfrac{s + 4}{s^2 - 5s - 6}$

(c) $\dfrac{s + 8}{s^3 + 6s^2 - s - 30}$

(d) $\dfrac{s^2 + 4s + 16}{s^3 + 8s^2 - 9s - 72}$

3.4   Convolution

In general, the Laplace transform of two functions is not the product of their transforms. There is, however, a special kind of product, denoted $f * g$, called the *convolution* of f with g. Convolution has the feature that the transform of $f * g$ is the product of the transforms of f and g. This fact is called the *convolution theorem*.

DEFINITION 3.6   *Convolution*

If f and g are defined on $[0, \infty)$, then the convolution $f * g$ of f with g is the function defined by

$$(f * g)(t) = \int_0^t f(t - \tau)g(\tau) \, d\tau$$

for $t \ge 0$.

THEOREM 3.9   *Convolution Theorem*

If $f * g$ is defined, then

$$\mathcal{L}[f * g] = \mathcal{L}[f]\mathcal{L}[g].$$

Proof Let $F = \mathcal{L}[f]$ and $G = \mathcal{L}[g]$. Then

$$F(s)G(s) = F(s) \int_0^\infty e^{-st} g(t)\, dt = \int_0^\infty F(s) e^{-s\tau} g(\tau)\, d\tau,$$

in which we changed the variable of integration to τ and brought $F(s)$ within the integral. Now recall that

$$e^{-s\tau} F(s) = \mathcal{L}[H(t - \tau) f(t - \tau)](s).$$

Substitute this into the integral for $F(s)G(s)$ to get

$$F(s)G(s) = \int_0^\infty \mathcal{L}[H(t - \tau) f(t - \tau)](s) g(\tau)\, d\tau. \tag{3.7}$$

But, from the definition of the Laplace transform,

$$\mathcal{L}[H(t - \tau) f(t - \tau)] = \int_0^\infty e^{-st} H(t - \tau) f(t - \tau)\, dt.$$

Substitute this into equation (3.7) to get

$$F(s)G(s) = \int_0^\infty \left[\int_0^\infty e^{-st} H(t - \tau) f(t - \tau)\, dt \right] g(\tau)\, d\tau$$

$$= \int_0^\infty \int_0^\infty e^{-st} g(\tau) H(t - \tau) f(t - \tau)\, dt\, d\tau.$$

Now recall that $H(t - \tau) = 0$ if $0 \le t < \tau$, while $H(t - \tau) = 1$ if $t \ge \tau$. Therefore,

$$F(s)G(s) = \int_0^\infty \int_\tau^\infty e^{-st} g(\tau) f(t - \tau)\, dt\, d\tau.$$

Figure 3.30 shows the $t\tau$ plane. The last integration is over the shaded region, consisting of points (t, τ) satisfying $0 \le \tau \le t < \infty$. Reverse the order of integration to write

$$F(s)G(s) = \int_0^\infty \int_0^t e^{-st} g(\tau) f(t - \tau)\, d\tau$$

$$= \int_0^\infty e^{-st} \left[\int_0^t g(\tau) f(t - \tau)\, d\tau \right] dt$$

$$= \int_0^\infty e^{-st} (f * g)(t)\, dt = \mathcal{L}[f * g](s).$$

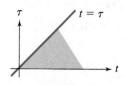

FIGURE 3.30

Therefore,

$$F(s)G(s) = \mathcal{L}[f * g](s),$$

as we wanted to show. ∎

The inverse version of the convolution theorem is useful when we want to find the inverse transform of a function that is a product and we know the inverse transform of each factor.

THEOREM 3.10

Let $\mathcal{L}^{-1}[F] = f$ and $\mathcal{L}^{-1}[G] = g$. Then

$$\mathcal{L}^{-1}[FG] = f * g. \quad \blacksquare$$

EXAMPLE 3.18

Compute

$$\mathcal{L}^{-1}\left[\frac{1}{s(s-4)^2}\right].$$

We can do this several ways (a table, a program, a partial fractions decomposition). But we can also write

$$\mathcal{L}^{-1}\left[\frac{1}{s(s-4)^2}\right] = \mathcal{L}^{-1}\left[\frac{1}{s}\frac{1}{(s-4)^2}\right] = \mathcal{L}^{-1}[F(s)G(s)].$$

Now

$$\mathcal{L}^{-1}\left[\frac{1}{s}\right] = 1 = f(t) \quad \text{and} \quad \mathcal{L}^{-1}\left[\frac{1}{(s-4)^2}\right] = te^{4t} = g(t).$$

Therefore,

$$\mathcal{L}^{-1}\left[\frac{1}{s(s-4)^2}\right] = f(t) * g(t) = 1 * te^{4t}$$

$$= \int_0^t \tau e^{4\tau}\,d\tau = \frac{1}{4}te^{4t} - \frac{1}{16}e^{4t} + \frac{1}{16}. \quad \blacksquare$$

The convolution operation is commutative.

THEOREM 3.11

If $f * g$ is defined, so is $g * f$, and $f * g = g * f$.

Proof   Let $z = t - \tau$ in the integral defining the convolution to get

$$(f * g)(t) = \int_0^t f(t - \tau)g(\tau)\,d\tau$$

$$= \int_t^0 f(z)g(t - z)(-1)\,dz = \int_0^t f(z)g(t - z)\,dz = (g * f)(t). \quad \blacksquare$$

Commutativity can have practical importance, since the integral defining $g * f$ may be easier to evaluate than the integral defining $f * g$ in specific cases.

Convolution can sometimes enable us to write solutions of problems that are stated in very general terms.

EXAMPLE 3.19

We will solve the problem

$$y'' - 2y' - 8y = f(t); \qquad y(0) = 1, y'(0) = 0.$$

Apply the Laplace transform, inserting the initial values, to obtain

$$\mathcal{L}[y'' - 2y' - 8y](s) = (s^2 Y(s) - s) - 2(sY(s) - 1) - 8Y(s) = \mathcal{L}[f](s) = F(s).$$

Then

$$(s^2 - 2s - 8)Y(s) - s + 2 = F(s),$$

so

$$Y(s) = \frac{1}{s^2 - 2s - 8} F(s) + \frac{s - 2}{s^2 - 2s - 8}.$$

Use a partial fractions decomposition to write

$$Y(s) = \frac{1}{6} \frac{1}{s - 4} F(s) - \frac{1}{6} \frac{1}{s + 2} F(s) + \frac{1}{3} \frac{1}{s - 4} + \frac{2}{3} \frac{1}{s + 2}.$$

Then

$$y(t) = \frac{1}{6} e^{4t} * f(t) - \frac{1}{6} e^{-2t} * f(t) + \frac{1}{3} e^{4t} + \frac{2}{3} e^{-2t}.$$

This is the solution, for any function f having a convolution with e^{4t} and e^{-2t}. ∎

Convolution is also used to solve certain kinds of integral equations, in which the function to be determined occurs in an integral. We saw an example of this in solving for the current in Example 3.17.

EXAMPLE 3.20

Determine f such that

$$f(t) = 2t^2 + \int_0^t f(t - \tau) e^{-\tau} \, d\tau.$$

Recognize the integral on the right as the convolution of f with e^{-t}. Thus the equation has the form

$$f(t) = 2t^2 + f(t) * e^{-t}.$$

Taking the Laplace transform of this equation yields

$$F(s) = \frac{4}{s^3} + F(s) \frac{1}{s + 1}.$$

Then

$$F(s) = \frac{4}{s^3} + \frac{4}{s^4},$$

and from this we easily invert to obtain

$$f(t) = 2t^2 + \frac{2}{3}t^3. \ \blacksquare$$

SECTION 3.4 PROBLEMS

In each of Problems 1 through 8, use the convolution theorem to compute the inverse Laplace transform of the function (even if another method would work). Wherever they occur, *a* and *b* are positive constants.

1. $\dfrac{1}{(s^2+4)(s^2-4)}$

2. $\dfrac{1}{s^2+16}e^{-2s}$

3. $\dfrac{s}{(s^2+a^2)(s^2+b^2)}$

4. $\dfrac{s^2}{(s-3)(s^2+5)}$

5. $\dfrac{1}{s(s^2+a^2)^2}$

6. $\dfrac{1}{s^4(s-5)}$

7. $\dfrac{1}{s(s+2)}e^{-4s}$

8. $\dfrac{2}{s^3(s^2+5)}$

In each of Problems 9 through 16, use the convolution theorem to write a formula for the solution of the initial value problem in terms of $f(t)$.

9. $y'' - 5y' + 6y = f(t); y(0) = y'(0) = 0$

10. $y'' + 10y' + 24y = f(t); y(0) = 1, y'(0) = 0$

11. $y'' - 8y' + 12y = f(t); y(0) = -3, y'(0) = 2$

12. $y'' - 4y' - 5y = f(t); y(0) = 2, y'(0) = 1$

13. $y'' + 9y = f(t); y(0) = -1, y'(0) = 1$

14. $y'' - k^2y = f(t); y(0) = 2, y'(0) = -4$

15. $y^{(3)} - y'' - 4y' + 4y = f(t); y(0) = y'(0) = 1,$
$y''(0) = 0$

16. $y^{(4)} - 11y'' + 18y = f(t);$
$y(0) = y'(0) = y''(0) = y^{(3)}(0) = 0$

In each of Problems 17 through 23, solve the integral equation.

17. $f(t) = -1 + \int_0^t f(t-\alpha)e^{-3\alpha}\,d\alpha$

18. $f(t) = -t + \int_0^t f(t-\alpha)\sin(\alpha)\,d\alpha$

19. $f(t) = e^{-t} + \int_0^t f(t-\alpha)\,d\alpha$

20. $f(t) = -1 + t - 2\int_0^t f(t-\alpha)\sin(\alpha)\,d\alpha$

21. $f(t) = 3 + \int_0^t f(\alpha)\cos[2(t-\alpha)]\,d\alpha$

22. $f(t) = \cos(t) + e^{-2t}\int_0^t f(\alpha)e^{2\alpha}\,d\alpha$

23. $f(t) = e^{-3t}\left[e^t - 3\int_0^t f(\alpha)e^{3\alpha}\,d\alpha\right]$

24. Use the convolution theorem to derive the formula $\mathcal{L}[\int_0^t f(w)\,dw](s) = (1/s)F(s)$. What assumptions are needed about $f(t)$?

25. Prove that convolution is distributive: $f * (g + h) = f * g + f * h$.

26. Prove that convolution is associative: $f * (g * h) = (f * g) * h$.

27. Show that $0 * f = f * 0 = 0$, where 0 denotes the function that is identically zero for all t.

28. Show by example that in general $f * 1 \neq f$, where 1 denotes the function that is identically 1 for all t. *Hint:* Consider $f(t) = \cos(t)$.

29. Show by example that $f * f$ need not be nonnegative. *Hint:* Let

$$f(t) = \begin{cases} \sin(t) & \text{for } \pi \le t \le 3\pi/2 \\ 0 & \text{for all other real } t. \end{cases}$$

30. Use the convolution theorem to determine the Laplace transform of $e^{-2t}\int_0^t e^{2w}\cos(3w)\,dw$.

31. Use the convolution theorem to show that

$$\mathcal{L}^{-1}\left[\frac{1}{s^2}F(s)\right](t) = \int_0^t \int_0^w f(\alpha)\,d\alpha\,dw.$$

3.5 Unit Impulses and the Dirac Delta Function

Sometimes we encounter the concept of an impulse, which may be intuitively understood as a force of large magnitude applied over an instant of time. We can model an impulse as follows. For any positive number ϵ, consider the pulse δ_ϵ defined by

$$\delta_\epsilon(t) = \frac{1}{\epsilon}[H(t) - H(t - \epsilon)].$$

As shown in Figure 3.31, this is a pulse of magnitude $1/\epsilon$ and duration ϵ. By letting ϵ approach zero, we obtain pulses of increasing magnitude over shorter time intervals.

Dirac's delta function is thought of as a pulse of "infinite magnitude" over an "infinitely short" duration and is defined to be

$$\delta(t) = \lim_{\epsilon \to 0+} \delta_\epsilon(t).$$

This is not really a function in the conventional sense, but is a more general object called a *distribution*. Nevertheless, for historical reasons it continues to be referred to as the delta function. It is also named for the Nobel laureate physicist P. A. M. Dirac. The shifted delta function $\delta(t - a)$ is zero except for $t = a$, where it has its infinite spike.

We can define the Laplace transform of the delta function as follows. Begin with

$$\delta_\epsilon(t - a) = \frac{1}{\epsilon}[H(t - a) - H(t - a - \epsilon)].$$

Then

$$\mathcal{L}[\delta_\epsilon(t - a)] = \frac{1}{\epsilon}\left[\frac{1}{s}e^{-as} - \frac{1}{s}e^{-(a+\epsilon)s}\right] = \frac{e^{-as}(1 - e^{-\epsilon s})}{\epsilon s}.$$

This suggests that we define

$$\mathcal{L}[\delta(t - a)] = \lim_{\epsilon \to 0+} \frac{e^{-as}(1 - e^{-\epsilon s})}{\epsilon s} = e^{-as}.$$

In particular, upon choosing $a = 0$ we have

$$\mathcal{L}[\delta(t)] = 1.$$

Thus we think of the delta function as having constant Laplace transform equal to 1.

The following result is called the *filtering* property of the delta function. If at time a, a signal (function) is hit with an impulse, by multiplying it by $\delta(t - a)$, and the resulting signal is summed over all positive time by integrating from zero to infinity, then we obtain exactly the signal value $f(a)$.

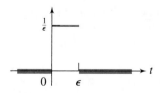

FIGURE 3.31 *Graph of* $\delta_\epsilon(t - a).$

THEOREM 3.12 **Filtering Property**

Let $a > 0$ and let f be integrable on $[0, \infty)$ and continuous at a. Then

$$\int_0^\infty f(t)\delta(t-a)\,dt = f(a).$$

Proof First calculate

$$\int_0^\infty f(t)\delta_\epsilon(t-a)\,dt = \int_0^\infty \frac{1}{\epsilon}[H(t-a) - H(t-a-\epsilon)]f(t)\,dt$$

$$= \frac{1}{\epsilon}\int_a^{a+\epsilon} f(t)\,dt.$$

By the mean value theorem for integrals, there is some t_ϵ between a and $a + \epsilon$ such that

$$\int_a^{a+\epsilon} f(t)\,dt = \frac{1}{\epsilon}f(t_\epsilon).$$

Then

$$\int_0^\infty f(t)\delta_\epsilon(t-a)\,dt = f(t_\epsilon).$$

As $\epsilon \to 0+$, $a + \epsilon \to a$, so $t_\epsilon \to a$ and, by continuity, $f(t_\epsilon) \to f(a)$. Then

$$\lim_{\epsilon \to 0+}\int_0^\infty f(t)\delta_\epsilon(t-a)\,dt = \int_0^\infty f(t)\lim_{\epsilon \to 0+}\delta_\epsilon(t-a)\,dt$$

$$= \int_0^\infty f(t)\delta(t-a)\,dt = \lim_{\epsilon \to 0+} f(t_\epsilon) = f(a),$$

as we wanted to show. ∎

If we apply the filtering property to $f(t) = e^{-st}$, we get

$$\int_0^\infty e^{-st}\delta(t-a)\,dt = e^{-as},$$

consistent with the definition of the Laplace transform of the delta function. Further, if we change notation in the filtering property and write it as

$$\int_0^\infty f(\tau)\delta(\tau - t)\,d\tau = f(t),$$

then we can recognize the convolution of f with δ and read the last equation as

$$f * \delta = f.$$

The delta function therefore acts as an identity for the "product" defined by the convolution of two functions.

Here is an example of a boundary value problem involving the delta function.

EXAMPLE 3.21

Solve

$$y'' + 2y' + 2y = \delta(t-3); \qquad y(0) = y'(0) = 0.$$

Apply the Laplace transform to the differential equation to get

$$s^2 Y(s) + 2s Y(s) + 2Y(s) = e^{-3s},$$

hence

$$Y(s) = \frac{e^{-3s}}{s^2 + 2s + 2}.$$

To find the inverse transform of the function on the right, first write

$$Y(s) = \frac{1}{(s+1)^2 + 1} e^{-3s}.$$

Now use both shifting theorems. Because $\mathcal{L}^{-1}[1/(s^2 + 1)] = \sin(t)$, a shift in the s variable gives us

$$\mathcal{L}^{-1}\left[\frac{1}{(s+1)^2 + 1}\right] = e^{-t}\sin(t).$$

Now shift the t variable to obtain

$$y(t) = H(t-3)e^{-(t-3)}\sin(t-3).$$

A graph of this solution is shown in Figure 3.32. The solution is differentiable for $t > 0$, but $y'(t)$ has a jump discontinuity of magnitude 1 at $t = 3$. The magnitude of the jump is the coefficient of $\delta(t-3)$ in the differential equation. ■

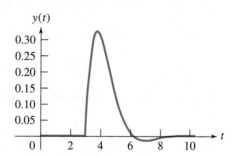

FIGURE 3.32 *Graph of*

$$y(t) = \begin{cases} 0 & \text{if } 0 \le t < 3 \\ e^{-(t-3)}\sin(t-3) & \text{if } t \ge 3 \end{cases}.$$

The delta function may be used to study the behavior of a circuit that has been subjected to transients. These are generated during switching, and the high input voltages associated with them can create excessive current in the components, damaging the circuit. Transients can also be harmful because they contain a broad spectrum of frequencies. Introducing a transient into a circuit can therefore have the effect of forcing the circuit with a range of frequencies. If one of these is near the natural frequency of the system, resonance may occur, resulting in oscillations large enough to damage the system.

For this reason, before a circuit is built, engineers sometimes use a delta function to model a transient and study its effect on the circuit.

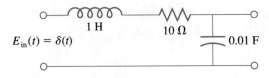

FIGURE 3.33

EXAMPLE 3.22

Suppose, in the circuit of Figure 3.33, the current and charge on the capacitor are zero at time zero. We want to determine the output voltage response to a transient modeled by $\delta(t)$.

The output voltage is $q(t)/C$, so we will determine $q(t)$. By Kirchhoff's voltage law,

$$Li' + Ri + \frac{1}{C}q = i' + 10i + 100q = \delta(t).$$

Since $i = q'$,

$$q'' + 10q' + 100q = \delta(t).$$

We assume initial conditions $q(0) = q'(0) = 0$.

Apply the Laplace transform to the differential equation and use the initial conditions to obtain

$$s^2 Q(s) + 10s\,Q(s) + 100Q(s) = 1.$$

Then

$$Q(s) = \frac{1}{s^2 + 10s + 100}.$$

In order to invert this by using a shifting theorem, complete the square to write

$$Q(s) = \frac{1}{(s+5)^2 + 75}.$$

Since

$$\mathcal{L}^{-1}\left[\frac{1}{s^2 + 75}\right] = \frac{1}{5\sqrt{3}}\sin(5\sqrt{3}t),$$

then

$$q(t) = \mathcal{L}^{-1}\left[\frac{1}{(s+5)^2 + 75}\right] = \frac{1}{5\sqrt{3}}e^{-5t}\sin(5\sqrt{3}t).$$

The output voltage is

$$\frac{1}{C}q(t) = 100q(t) = \frac{20}{\sqrt{3}}e^{-5t}\sin(5\sqrt{3}t).$$

A graph of this output is shown in Figure 3.34. The circuit output displays damped oscillations at its natural frequency, even though it was not explicitly forced by oscillations of this frequency. If we wish, we can obtain the current by $i(t) = q'(t)$. ∎

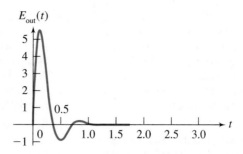

FIGURE 3.34 *Output of the circuit of Figure 3.32.*

SECTION 3.5   PROBLEMS

In each of Problems 1 through 5, solve the initial value problem and graph the solution.

1. $y'' + 5y' + 6y = 3\delta(t - 2) - 4\delta(t - 5)$;
   $y(0) = y'(0) = 0$

2. $y'' - 4y' + 13y = 4\delta(t - 3)$; $y(0) = y'(0) = 0$

3. $y^{(3)} + 4y'' + 5y' + 2y = 6\delta(t)$;
   $y(0) = y'(0) = y''(0) = 0$

4. $y'' + 16y' = 12\delta(t - 5\pi/8)$; $y(0) = 3$, $y'(0) = 0$

5. $y'' + 5y' + 6y = B\delta(t)$; $y(0) = 3$, $y'(0) = 0$. Call the solution φ. What are $\varphi(0)$ and $\varphi'(0)$? Using this information, what physical phenomenon does the Dirac delta function model?

6. Suppose f is not continuous at a, but $\lim_{t \to a+} f(t) = f(a+)$ is finite. Prove that $\int_0^\infty f(t)\delta(t - a)\,dt = f(a+)$.

7. Evaluate $\int_0^\infty (\sin(t)/t)\delta(t - \pi/6)\,dt$.

8. Evaluate $\int_0^2 t^2\delta(t - 3)\,dt$.

9. Evaluate $\int_0^\infty f(t)\delta(t - 2)\,dt$, where

$$f(t) = \begin{cases} t & \text{for } 0 \le t < 2 \\ t^2 & \text{for } t > 2 \\ 5 & \text{for } t = 2 \end{cases}.$$

10. It is sometimes convenient to consider $\delta(t)$ as the derivative of the Heaviside function $H(t)$. Use the definitions of the derivative, the Heaviside function, and the delta function (as a limit of δ_ϵ) to give a heuristic justification for this.

11. Use the idea that $H'(t) = \delta(t)$ from Problem 10 to determine the output voltage of the circuit of Example 3.17 by differentiating the relevant equation to obtain an equation in i rather than writing the charge as an integral.

12. If $H'(t) = \delta(t)$, then $\mathcal{L}[H'(t)](s) = 1$. Show that not all of the operational rules for the Laplace transform are compatible with this expression. *Hint:* Check to see whether $\mathcal{L}[H'(t)](s) = s\mathcal{L}[H(t)](s) - H(0+)$.

13. Evaluate $\delta(t - a) * f(t)$.

14. Solve

$$y^{(4)}(x) = \frac{M}{EI}\delta(x - a);$$

$$y(0) = 0, y'(0) = B, y^{(3)}(0) = F_0, y(L) = 0.$$

This problem (which is not a standard initial value problem) models the deflection of a beam of length L, horizontally restrained at both ends, with a load M at $x = a$ and with the weight of the beam neglected. E is Young's modulus and I is the moment of inertia of the cross section at x with respect to a horizontal line through the centroid of the beam.

15. Solve

$$y^{(4)}(x) = \frac{M}{EI}\delta(x - a);$$

$$y(0) = y''(0) = y(L) = 0, y^{(3)}(0) = F_0.$$

Here F_0 measures the shearing force of the beam at the left end ($x = 0$).

16. An object of mass m is attached to the lower end of a spring of modulus k. Assume that there is no damping. Derive and solve an equation of motion for the position of the object at time $t > 0$, assuming that, at time zero, the object is pushed down from the equilibrium position with an initial velocity v_0. With what momentum does the object leave the equilibrium position?

17. Suppose an object of mass m is attached to the lower end of a spring having modulus k. Assume that there is no damping. Solve the equation of motion for the position of the object for any time $t \geq 0$ if, at time zero, the weight is struck a downward blow of magnitude mv_0. How does the position of the object in Problem 16 compare with that of the object in this problem for any positive time?

18. A 2-pound weight is attached to the lower end of a spring, stretching it $\frac{8}{3}$ inches. The weight is allowed to come to rest in the equilibrium position. At some later time, which is called time zero, the weight is struck a downward blow of magnitude $\frac{1}{4}$ pound (an impulse). Assume that there is no damping in the system. Determine the velocity with which the weight leaves the equilibrium position as well as the frequency and magnitude of the resulting oscillations.

19. An object weighing 8 pounds is suspended from a spring, stretching it 6 inches. The weight, which is at rest in the equilibrium position, is struck an upward blow of magnitude 1 pound. Assuming no damping, determine the velocity with which the weight leaves the equilibrium position, as well as the frequency and magnitude of the oscillations.

20. An object of mass 1 kilogram is suspended from a spring that has spring constant 24 newtons per meter. Attached to the object is a shock absorber that induces a drag of $11v$ newtons, where v is the velocity of the object. The system is set in motion from the equilibrium position by striking the weight a downward blow of magnitude 4 newtons. Determine the velocity with which the weight leaves the equilibrium position, the time at which it obtains its maximum displacement from the equilibrium position, and the displacement at this time.

3.6   Laplace Transform Solution of Systems

The Laplace transform can be of use in solving systems of equations involving derivatives and integrals.

EXAMPLE 3.23

Consider the system of differential equations and initial conditions for the functions x and y:

$$x'' - 2x' + 3y' + 2y = 4,$$
$$2y' - x' + 3y = 0,$$
$$x(0) = x'(0) = y(0) = 0.$$

Begin by applying the Laplace transform to the differential equations, incorporating the initial conditions. We get

$$s^2 X - 2sX + 3sY + 2Y = \frac{4}{s},$$

$$2sY - sX + 3Y = 0.$$

Solve these equations for $X(s)$ and $Y(s)$ to get

$$X(s) = \frac{4s + 6}{s^2(s + 2)(s - 1)} \quad \text{and} \quad Y(s) = \frac{2}{s(s + 2)(s - 1)}.$$

A partial fractions decomposition yields

$$X(s) = -\frac{7}{2}\frac{1}{s} - 3\frac{1}{s^2} + \frac{1}{6}\frac{1}{s + 2} + \frac{10}{3}\frac{1}{s - 1}$$

and

$$Y(s) = -\frac{1}{s} + \frac{1}{3}\frac{1}{s+2} + \frac{2}{3}\frac{1}{s-1}.$$

Upon applying the inverse Laplace transform, we obtain the solution

$$x(t) = -\frac{7}{2} - 3t + \frac{1}{6}e^{-2t} + \frac{10}{3}e^t$$

and

$$y(t) = -1 + \frac{1}{3}e^{-2t} + \frac{2}{3}e^t. \quad \blacksquare$$

The analysis of mechanical and electrical systems having several components can lead to systems of differential equations that can be solved using the Laplace transform.

EXAMPLE 3.24

Consider the spring/mass system of Figure 3.35. Let $x_1 = x_2 = 0$ at the equilibrium position, where the weights are at rest. Choose the direction to the right as positive and suppose the weights are at positions $x_1(t)$ and $x_2(t)$ at time t.

By two applications of Hooke's law, the restoring force on m_1 is

$$-k_1x_1 + k_2(x_2 - x_1)$$

and that on m_2 is

$$-k_2(x_2 - x_1) - k_3x_2.$$

By Newton's second law of motion,

$$m_1x_1'' = -(k_1 + k_2)x_1 + k_2x_2 + f_1(t)$$

and

$$m_2x_2'' = k_2x_1 - (k_2 + k_3)x_2 + f_2(t).$$

These equations assume that damping is negligible but allow for forcing functions acting on each mass.

As a specific example, suppose $m_1 = m_2 = 1$ and $k_1 = k_3 = 4$ while $k_2 = \frac{5}{2}$. Suppose $f_2(t) = 0$, so no external driving force acts on the second mass, while a force of magnitude $f_1(t) = 2[1 - H(t - 3)]$ acts on the second. This hits the first mass with a force of constant

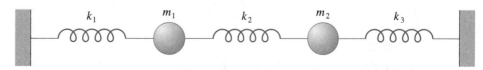

FIGURE 3.35

magnitude 2 for the first 3 seconds, then turns off. Now the system of equations for the displacement functions is

$$x_1'' = -\frac{13}{2}x_1 + \frac{5}{2}x_2 + 2[1 - H(t - 3)],$$

$$x_2'' = \frac{5}{2}x_1 - \frac{13}{2}x_2.$$

If the masses are initially at rest at the equilibrium position, then

$$x_1(0) = x_2(0) = x_1'(0) = x_2'(0) = 0.$$

Apply the Laplace transform to each equation of the system to get

$$s^2 X_1 = -\frac{13}{2}X_1 + \frac{5}{2}X_2 + \frac{2(1 - e^{-3s})}{s},$$

$$s^2 X_2 = \frac{5}{2}X_1 - \frac{13}{2}X_2.$$

Solve these to obtain

$$X_1(s) = \frac{2}{(s^2 + 9)(s^2 + 4)}\left(s^2 + \frac{13}{2}\right)\frac{1}{s}(1 - e^{-3s})$$

and

$$X_2(s) = \frac{5}{(s^2 + 9)(s^2 + 4)}\frac{1}{s}(1 - e^{-3s}).$$

In preparation for applying the inverse Laplace transform, use a partial fractions decomposition to write

$$X_1(s) = \frac{13}{36}\frac{1}{s} - \frac{1}{4}\frac{s}{s^2 + 4} - \frac{1}{9}\frac{s}{s^2 + 9} - \frac{13}{36}\frac{1}{s}e^{-3s} + \frac{1}{4}\frac{s}{s^2 + 4}e^{-3s} + \frac{1}{9}\frac{s}{s^2 + 9}e^{-3s}$$

and

$$X_2(s) = \frac{5}{36}\frac{1}{s} - \frac{1}{4}\frac{s}{s^2 + 4} + \frac{1}{9}\frac{s}{s^2 + 9} - \frac{5}{36}\frac{1}{s}e^{-3s} + \frac{1}{4}\frac{s}{s^2 + 4}e^{-3s} - \frac{1}{9}\frac{s}{s^2 + 9}e^{-3s}$$

Now it is routine to apply the inverse Laplace transform to obtain the solution:

$$x_1(t) = \frac{13}{36} - \frac{1}{4}\cos(2t) - \frac{1}{9}\cos(3t)$$

$$+ \left[-\frac{13}{36} + \frac{1}{4}\cos(2(t - 3)) - \frac{1}{9}\cos(3(t - 3))\right]H(t - 3),$$

$$x_2(t) = \frac{5}{36} - \frac{1}{4}\cos(2t) + \frac{1}{9}\cos(3t)$$

$$+ \left[-\frac{5}{36} + \frac{1}{4}\cos(2(t - 3)) - \frac{1}{9}\cos(3(t - 3))\right]H(t - 3). \quad\blacksquare$$

EXAMPLE 3.25

In the circuit of Figure 3.36, suppose the switch is closed at time zero. We want to know the current in each loop. Assume that both loop currents and the charges on the capacitors are initially zero. Apply Kirchhoff's laws to each loop to get

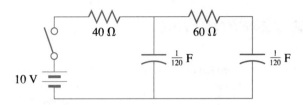

FIGURE 3.36

$$40i_1 + 120(q_1 - q_2) = 10$$

$$60i_2 + 120q_2 = 120(q_1 - q_2).$$

Since $i = q'$, we can write $q(t) = \int_0^t i(\tau)\, d\tau + q(0)$. Put into the two circuit equations, we get

$$40i_1 + 120 \int_0^t [i_1(\tau) - i_2(\tau)]\, d\tau + 120[q_1(0) - q_2(0)] = 10$$

$$60i_2 + 120 \int_0^t i_2(\tau) d\tau + 120q_2(0) = 120 \int_0^t [i_1(\tau) - i_2(\tau)]\, d\tau + 120[q_1(0) - q_2(0)].$$

Put $q_1(0) = q_2(0) = 0$ in this system to get

$$40i_1 + 120 \int_0^t [i_1(\tau) - i_2(\tau)]\, d\tau = 10$$

$$60i_2 + 120 \int_0^t i_2(\tau)\, d\tau = 120 \int_0^t [i_1(\tau) - i_2(\tau)]\, d\tau.$$

Apply the Laplace transform to each equation to get

$$40I_1 + \frac{120}{s} I_1 - \frac{120}{s} I_2 = \frac{10}{s}$$

$$60I_2 + \frac{120}{s} I_2 = \frac{120}{s} I_1 - \frac{120}{s} I_2.$$

After some rearrangement, we have

$$(s + 3)I_1 - 3I_2 = \tfrac{1}{4}$$

$$2I_1 - (s + 4)I_2 = 0.$$

Solve these to get

$$I_1(s) = \frac{s + 4}{4(s + 1)(s + 6)} = \frac{3}{20} \frac{1}{s + 1} + \frac{1}{10} \frac{1}{s + 6}$$

and

$$I_2(s) = \frac{1}{2(s + 1)(s + 6)} = \frac{1}{10} \frac{1}{s + 1} - \frac{1}{10} \frac{1}{s + 6}.$$

Now use the inverse Laplace transform to find the solution

$$i_1(t) = \frac{3}{20} e^{-t} + \frac{1}{10} e^{-6t}, \quad i_2(t) = \frac{1}{10} e^{-t} - \frac{1}{10} e^{-6t}. \quad \blacksquare$$

SECTION 3.6 PROBLEMS

In each of Problems 1 through 10, use the Laplace transform to solve the initial value problem for the system.

1. $x' - 2y' = 1, x' + y - x = 0; x(0) = y(0) = 0$

2. $2x' - 3y + y' = 0, x' + y' = t; x(0) = y(0) = 0$

3. $x' + 2y' - y = 1, 2x' + y = 0; x(0) = y(0) = 0$

4. $x' + y' - x = \cos(2t), x' + 2y' = 0; x(0) = y(0) = 0$

5. $3x' - y = 2t, x' + y' - y = 0; x(0) = y(0) = 0$

6. $x' + 4y' - y = 0, x' + 2y = e^{-t}; x(0) = y(0) = 0$

7. $x' + 2x - y' = 0, x' + y + x = t^2; x(0) = y(0) = 0$

8. $x' + 4x - y = 0, x' + y' = t; x(0) = y(0) = 0$

9. $x' + y' + x - y = 0, x' + 2y' + x = 1; x(0) = y(0) = 0$

10. $x' + 2y' - x = 0, 4x' + 3y' + y = -6; x(0) = y(0) = 0$

11. Use the Laplace transform to solve the system

$$y_1' - 2y_2' + 3y_3 = 0,$$
$$y_1 - 4y_2' + 3y_3' = t,$$
$$y_1 - 2y_2' + 3y_3' = -1,$$
$$y_1(0) = y_2(0) = y_3(0) = 0.$$

12. Solve for the currents in the circuit of Figure 3.37, assuming that the currents and charges are initially zero and that $E(t) = 2H(t-4) - H(t-5)$.

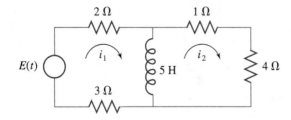

FIGURE 3.37

13. Solve for the currents in the circuit of Figure 3.37 if the currents are initially zero and $E(t) = 1 - H(t - 4)\sin(2(t-4))$.

14. Solve for the displacement functions of the masses in the system of Figure 3.38. Neglect damping and assume zero initial displacements and velocities and external forces $f_1(t) = 2$ and $f_2(t) = 0$.

15. Solve for the displacement functions in the system of Figure 3.38 if $f_1(t) = 1 - H(t - 2)$ and $f_2(t) = 0$. Assume zero initial displacements and velocities.

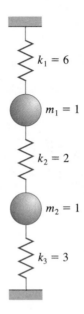

FIGURE 3.38

16. Consider the system of Figure 3.39. Let M be subjected to a periodic driving force $f(t) = A\sin(\omega t)$. The masses are initially at rest in the equilibrium position.

(a) Derive and solve the initial value problem for the displacement functions.

(b) Show that if m and k_2 are chosen so that $\omega = \sqrt{k_2/m}$, then the mass m cancels the forced vibrations of M. In this case we call m a *vibration absorber*.

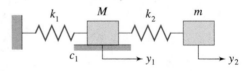

FIGURE 3.39

17. Two objects of masses m_1 and m_2 are attached to opposite ends of a spring having spring constant k (Figure 3.40). The entire apparatus is placed on a highly varnished table. Show that, if stretched and released from rest, the masses oscillate with respect to each other with period

$$2\pi\sqrt{\frac{m_1 m_2}{k(m_1 + m_2)}}.$$

FIGURE 3.40

18. Solve for the currents in the circuit of Figure 3.41 if $E(t) = 5H(t - 2)$ and the initial currents are zero.

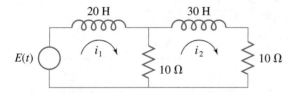

FIGURE 3.41

19. Solve for the currents in the circuit of Figure 3.41 if $E(t) = 5\delta(t - 1)$.

20. Two tanks are connected by a series of pipes as shown in Figure 3.42. Tank 1 initially contains 60 gallons of brine in which 11 pounds of salt are dissolved. Tank 2 initially contains 7 pounds of salt dissolved in 18 gallons of brine. Beginning at time zero a mixture containing $\frac{1}{6}$ pound of salt for each gallon of water is pumped into tank 1 at the rate of 2 gallons per minute, while saltwater solutions are interchanged between the two tanks and also flow out of tank 2 at the rates shown in the diagram. Four minutes after time zero, salt is poured into tank 2 at the rate of 11 pounds per minute for a period of 2 minutes. Determine the amount of salt in each tank for any time $t \geq 0$.

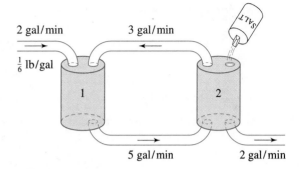

FIGURE 3.42

21. Two tanks are connected by a series of pipes as shown in Figure 3.43. Tank 1 initially contains 200 gallons of brine in which 10 pounds of salt are dissolved. Tank 2 initially contains 5 pounds of salt dissolved in 100 gallons of water. Beginning at time zero, pure water is pumped into tank 1 at the rate of 3 gallons per minute, while brine solutions are interchanged between the tanks at the rates shown in the diagram. Three minutes after time zero, 5 pounds of salt are dumped into tank 2. Determine the amount of salt in each tank for any time $t \geq 0$.

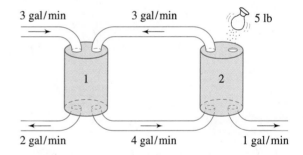

FIGURE 3.43

3.7   **Differential Equations with Polynomial Coefficients**

The Laplace transform can sometimes be used to solve linear differential equations having polynomials as coefficients. For this we need the fact that the Laplace transform of $tf(t)$ is the negative of the derivative of the Laplace transform of $f(t)$.

THEOREM 3.13

Let $\mathcal{L}[f](s) = F(s)$ for $s > b$ and suppose that F is differentiable. Then

$$\mathcal{L}[tf(t)](s) = -F'(s)$$

for $s > b$.

Proof   Differentiate under the integral sign to calculate

$$F'(s) = \frac{d}{ds}\int_0^\infty e^{-st}f(t)\,dt = \int_0^\infty \frac{d}{ds}(e^{-st}f(t))\,dt$$

$$= \int_0^\infty -te^{-st}f(t)\,dt = \int_0^\infty e^{-st}[-tf(t)]\,dt$$

$$= \mathcal{L}[-tf(t)](s),$$

and this is equivalent to the conclusion of the theorem. ∎

By applying this result n times, we reach the following.

COROLLARY 3.1

Let $\mathcal{L}[f](s) = F(s)$ for $s > b$ and let n be a positive integer. Suppose F is n times differentiable. Then, for $s > b$,

$$\mathcal{L}[t^n f(t)](s) = (-1)^n \frac{d^n}{ds^n}F(s). \quad ∎$$

EXAMPLE 3.26

Consider the problem

$$ty'' + (4t - 2)y' - 4y = 0; \qquad y(0) = 1.$$

If we write this differential equation in the form $y'' + p(t)y' + q(t)y = 0$, then we must choose $p(t) = (4t-2)/t$, and this is not defined at $t = 0$, where the initial condition is given. This problem is not of the type for which we proved an existence/uniqueness theorem in Chapter 2. Further, we have only one initial condition. Nevertheless, we will look for functions satisfying the problem as stated.

Apply the Laplace transform to the differential equation to get

$$\mathcal{L}[ty''] + 4\mathcal{L}[ty'] - 2\mathcal{L}[y'] - 4\mathcal{L}[y] = 0.$$

Calculate the first three terms as follows. First,

$$\mathcal{L}[ty''] = -\frac{d}{ds}\mathcal{L}[y''] = -\frac{d}{ds}[s^2Y - sy(0) - y'(0)]$$

$$= -2sY - s^2Y' + 1$$

because $y(0) = 1$ and $y'(0)$, though unknown, is constant and has zero derivative. Next,

$$\mathcal{L}[ty'] = -\frac{d}{ds}[y']$$

$$= -\frac{d}{ds}[sY - y(0)] = -Y - sY'.$$

Finally,

$$\mathcal{L}[y'] = sY - y(0) = sY - 1.$$

The transform of the differential equation is therefore

$$-2sY - s^2Y' + 1 - 4Y - 4sY' - 2sY + 2 - 4Y = 0.$$

Then

$$Y' + \frac{4s + 8}{s(s + 4)}Y = \frac{3}{s(s + 4)}.$$

This is a linear first-order differential equation, and we will find an integrating factor. First compute

$$\int \frac{4s + 8}{s(s + 4)}\, ds = \ln[s^2(s + 4)^2].$$

Then

$$e^{\ln[(s^2(s+4)^2)]} = s^2(s + 4)^2$$

is an integrating factor. Multiply the differential equation by this factor to obtain

$$s^2(s + 4)^2Y' + (4s + 8)s(s + 4)Y = 3s(s + 4),$$

or

$$[s^2(s + 4)^2Y]' = 3s(s + 4).$$

Integrate to get

$$s^2(s + 4)^2Y = s^3 + 6s^2 + C.$$

Then

$$Y(s) = \frac{s}{(s + 4)^2} + \frac{6}{(s + 4)^2} + \frac{C}{s^2(s + 4)^2}.$$

Upon applying the inverse Laplace transform, we obtain

$$y(t) = e^{-4t} + 2te^{-4t} + \frac{C}{32}[-1 + 2t + e^{-4t} + 2te^{-4t}].$$

This function satisfies the differential equation and the condition $y(0) = 1$ for any real number C. This problem does not have a unique solution. ■

When we applied the Laplace transform to a constant coefficient differential equation $y'' + Ay' + By = f(t)$, we obtained an algebraic expression for Y. In this example, with polynomials occurring as coefficients, we obtained a differential equation for Y because the process of computing the transform of $t^k y(t)$ involves differentiating $Y(s)$.

In the next example, we will need the following fact.

THEOREM 3.14

Let f be piecewise continuous on $[0, k]$ for every positive number k and suppose there are numbers M and b such that $|f(t)| \leq Me^{bt}$ for $t \geq 0$. Let $\mathcal{L}[f] = F$. Then

$$\lim_{s \to \infty} F(s) = 0.$$

Proof   Write

$$|F(s)| = \left| \int_0^\infty e^{-st} f(t) \, dt \right| \leq \int_0^\infty e^{-st} Me^{bt} \, dt$$

$$= \frac{M}{b-s} e^{-(s-b)t} \bigg]_0^\infty = \frac{M}{s-b} \to 0$$

as $s \to \infty$. ∎

This result will enable us to solve the following initial value problem.

EXAMPLE 3.27

Suppose we want to solve

$$y'' + 2ty' - 4y = 1; \qquad y(0) = y'(0) = 0.$$

Unlike the preceding example, this problem satisfies the hypotheses of the existence/uniqueness theorem in Chapter 2.

Apply the Laplace transform to the differential equation to get

$$s^2 Y(s) - sy(0) - y'(0) + 2\mathcal{L}[ty'](s) - 4Y(s) = \frac{1}{s}.$$

Now $y(0) = y'(0) = 0$ and

$$\mathcal{L}[ty'](s) = -\frac{d}{ds} [\mathcal{L}[y'](s)]$$

$$= -\frac{d}{ds} [sY(s) - y(0)] = -Y(s) - sY'(s).$$

We therefore have

$$s^2 Y(s) - 2Y(s) - 2sY'(s) - 4Y(s) = \frac{1}{s},$$

or

$$Y' + \left(\frac{3}{s} - \frac{s}{2} \right) Y = -\frac{1}{2s^2}.$$

This is a linear first-order differential equation for Y. To find an integrating factor, first compute

$$\int \left(\frac{3}{s} - \frac{s}{2} \right) ds = 3\ln(s) - \frac{1}{4}s^2.$$

The exponential of this function, or

$$s^3 e^{-s^2/4},$$

is an integrating factor. Multiply the differential equation by this function to obtain

$$(s^3 e^{-s^2/4} Y)' = -\frac{1}{2} s e^{-s^2/4}.$$

Then

$$s^3 e^{-s^2/4} Y = e^{-s^2/4} + C$$

so

$$Y(s) = \frac{1}{s^3} + \frac{C}{s^3} e^{s^2/4}.$$

We do not have any further initial conditions to determine C. However, in order to have $\lim_{s \to \infty} Y(s) = 0$, we must choose $C = 0$. Then $Y(s) = 1/s^3$ so

$$y(t) = \frac{1}{2} t^2. \ \blacksquare$$

SECTION 3.7   PROBLEMS

Use the Laplace transform to solve each of Problems 1 through 10.

1. $t^2 y' - 2y = 2$

2. $y'' + 4ty' - 4y = 0; \ y(0) = 0, \ y'(0) = -7$

3. $y'' - 16ty' + 32y = 14; \ y(0) = y'(0) = 0$

4. $y'' + 8ty' - 8y = 0; \ y(0) = 0, \ y'(0) = -4$

5. $ty'' + (t-1)y' + y = 0; \ y(0) = 0$

6. $y'' + 2ty' - 4y = 6; \ y(0) = 0, \ y'(0) = 0$

7. $y'' + 8ty' = 0; \ y(0) = 4, \ y'(0) = 0$

8. $y'' - 4ty' + 4y = 0; \ y(0) = 0, \ y'(0) = 10$

9. $y'' - 8ty' + 16y = 3; \ y(0) = 0, \ y'(0) = 0$

10. $(1-t)y'' + ty' - y = 0; \ y(0) = 3, \ y'(0) = -1$

11. Use the Laplace transform to solve the Euler equation $t^2 y'' + Bty' + Cy = 0$.

12. Use the Laplace transform to solve *Laguerre's equation* $ty'' + (1-t)y' + ny = 0$, in which n is any positive integer.

13. Use the Laplace transform to solve *Bessel's equation of order zero*: $ty'' + y' + ty = 0$, subject to the condition $y(0) = 1$. *Hint:* Obtain a differential equation for $Y(s)$ having general solution $Y(s) = C/\sqrt{1 + s^2}$. Write this

as $Y(s) = C/s\sqrt{1 + (1/s)^2}$. Assume that $s > 1$ and use the binomial expansion to obtain an infinite series for $Y(s)$. Take the inverse Laplace transform term by term to obtain a series for $y(t)$.

14. Show that the Laplace transform, applied to $t^2 y''(t) + 2ty'(t) + ky(t) = 0$, with k constant, simply returns this differential equation. *Hint:* Show that under appropriate conditions on y, y', and y'',

$$\mathcal{L}[t^2 y''(t)](s) = s^2 Y''(s) + 4s Y'(s) + 2Y(s).$$

15. Assume that f is piecewise continuous on $[0, k]$ for every positive k and $|f(t)| \le Me^{bt}$ for some numbers M and b. Assume also that $\lim_{t \to 0+}(1/t)f(t)$ exists and is finite. Prove that

$$\mathcal{L}\left[\frac{1}{t} f(t)\right](s) = \int_s^\infty F(w)\, dw.$$

Hint: Let $g(t) = f(t)/t$ and show that g has a Laplace transform. Show that

$$F(s) = \mathcal{L}[f(t)](s) = \mathcal{L}[tg(t)](s) = -\frac{d}{ds} \mathcal{L}[g(t)](s).$$

Conclude that $\mathcal{L}[g](s) = -\int_0^s F(w)\, dw + C$. Evaluate C by using the condition $\lim_{s \to \infty} G(s) = 0$.

In each of Problems 16 through 18, use the result of Problem 15 to find the Laplace transform of the function.

16. $\sin(t)/t$

17. $\sinh(t)/t$

18. $(e^{-at} - e^{-bt})/t$

19. Use the Laplace transform to solve the boundary value problem

$$t(1-t)y'' + 2y' + 2y = 6t; \qquad y(0) = 0, \, y(2) = 0.$$

This is called a *boundary value problem* because conditions are given on the function at the endpoints of an interval.

CHAPTER 4

Series Solutions

Sometimes we can find an explicit, closed form solution of a differential equation or initial value problem. This occurs with

$$y' + 2y = 1; \qquad y(0) = 3,$$

which has the unique solution

$$y(x) = \tfrac{1}{2}(1 + 5e^{-2x}).$$

This solution is explicit, giving $y(x)$ as a function of x, and is in closed form because it is a finite algebraic combination of elementary functions (which are functions such as polynomials, trigonometric functions, and exponential functions).

Sometimes standard methods do not yield a solution in closed form. For example, the problem

$$y' + e^x y = x^2; \qquad y(0) = 4$$

has the unique solution

$$y(x) = e^{-e^x} \int_0^x \xi^2 e^{e^\xi} \, d\xi + 4ee^{-e^x}.$$

This solution is explicit, but it is not in closed form because of the integral. It is difficult to analyze this solution, or even to evaluate it at specific points.

Sometimes a series solution is a good strategy for solving an initial value problem. Such a solution is explicit, giving $y(x)$ as an infinite series involving constants times powers of x. It may also reveal important information about the behavior of the solution—for example, whether it passes through the origin, whether it is an even or odd function, or whether the function is increasing or decreasing on a given interval. It may also be possible to make good approximations to function values from a series representation.

We will begin with power series solutions for differential equations admitting such solutions. Following this, we will develop another kind of series for problems whose solutions do not have power series expansions about a particular point.

This chapter assumes familiarity with basic facts about power series. These are reviewed at the end of the chapter.

4.1 Power Series Solutions of Initial Value Problems

Consider the linear first-order initial value problem

$$y' + p(x)y = q(x); \qquad y(x_0) = y_0.$$

If p and q are continuous on an open interval I about x_0, we are guaranteed by Theorem 1.3 that this problem has a unique solution defined for all x in I.

With a stronger condition on these coefficients, we can infer that the solution will have a stronger property, which we now define.

DEFINITION 4.1 *Analytic Function*

A function f is analytic at x_0 if $f(x)$ has a power series representation in some open interval about x_0:

$$f(x) = \sum_{n=0}^{\infty} a_n(x - x_0)^n$$

in some interval $(x_0 - h, x_0 + h)$.

For example, $\sin(x)$ is analytic at 0, having the power series representation

$$\sin(x) = \sum_{n=0}^{\infty} \frac{(-1)^n}{(2n + 1)!} x^{2n+1}.$$

This series converges for all real x.

Analyticity requires at least that f be infinitely differentiable at x_0, although this by itself is not sufficient for f to be analytic at x_0.

We claim that when the coefficients of an initial value problem are analytic, then the solution is as well.

THEOREM 4.1

Let p and q be analytic at x_0. Then the initial value problem

$$y' + p(x)y = q(x); \qquad y(x_0) = y_0$$

has a solution that is analytic at x_0. ∎

This means that an initial value problem whose coefficients are analytic at x_0 has an analytic solution at x_0. This justifies attempting to expand the solution in a power series about x_0, where the initial condition is specified. This expansion has the form

$$y(x) = \sum_{n=0}^{\infty} a_n(x - x_0)^n, \tag{4.1}$$

in which

$$a_n = \frac{1}{n!} y^{(n)}(x_0).$$

One strategy to solve the initial value problem of the theorem is to use the differential equation and the initial condition to calculate these derivatives, hence obtain coefficients in the expansion (4.1) of the solution.

EXAMPLE 4.1

Consider again the problem

$$y' + e^x y = x^2; \qquad y(0) = 4.$$

The theorem guarantees an analytic solution at 0:

$$y(x) = \sum_{n=0}^{\infty} \frac{1}{n!} y^{(n)}(0) x^n$$

$$= y(0) + y'(0)x + \frac{1}{2!} y''(0)x^2 + \frac{1}{3!} y^{(3)}(0)x^3 + \cdots$$

We will know this series if we can determine the terms $y(0)$, $y'(0)$, $y''(0)$, ...

The initial condition gives us $y(0) = 4$. Put $x = 0$ into the differential equation to get

$$y'(0) + y(0) = 0,$$

or

$$y'(0) + 4 = 0.$$

Then

$$y'(0) = -4.$$

Next determine $y''(0)$. Differentiate the differential equation to get

$$y'' + e^x y' + e^x y = 2x \tag{4.2}$$

and put $x = 0$ to get

$$y''(0) + y'(0) + y(0) = 0.$$

Then

$$y''(0) = -y'(0) - y(0) = -(-4) - 4 = 0.$$

Next we will find $y^{(3)}(x)$. Differentiate equation (4.2) to get

$$y^{(3)} + 2e^x y' + e^x y'' + e^x y = 2. \tag{4.3}$$

Then

$$y^{(3)}(0) + 2y'(0) + y''(0) + y(0) = 2,$$

or

$$y^{(3)}(0) + 2(-4) + 4 = 2.$$

Then

$$y^{(3)}(0) = 6.$$

Next differentiate equation (4.3):

$$y^{(4)} + 3e^x y' + 3e^x y'' + e^x y^{(3)} + e^x y = 0.$$

Evaluate this at 0 to get

$$y^{(4)}(0) + 3(-4) + 3(0) + 6 + 4 = 0,$$

so

$$y^{(4)}(0) = 2.$$

At this point we have the first five terms of the Maclaurin expansion of the solution:

$$y(x) = y(0) + y'(0)x + \tfrac{1}{2}y''(0)x^2 + \tfrac{1}{6}y^{(3)}(0)x^3 + \tfrac{1}{24}y^{(4)}(0)x^4 + \cdots$$

$$= 4 - 4x + x^3 + \tfrac{1}{12}x^4 + \cdots .$$

By differentiating more times, we can write as many terms of this series as we want. ∎

EXAMPLE 4.2

Consider the initial value problem

$$y' + \sin(x)y = 1 - x; \qquad y(\pi) = -3.$$

Since the initial condition is given at $x = \pi$, we will seek terms in the Taylor expansion of the solution about π. This series has the form

$$y(x) = y(\pi) + y'(\pi)(x - \pi) + \tfrac{1}{2}y''(\pi)(x - \pi)^2$$

$$+ \tfrac{1}{6}y^{(3)}(\pi)(x - \pi)^3 + \tfrac{1}{24}y^{(4)}(\pi)(x - \pi)^4 + \cdots .$$

We know the first term, $y(\pi) = -3$. From the differential equation,

$$y'(\pi) = 1 - \pi + 3\sin(\pi) = 1 - \pi.$$

Now differentiate the differential equation:

$$y''(x) + \cos(x)y + \sin(x)y' = -1. \tag{4.4}$$

Substitute $x = \pi$ to get

$$y''(\pi) - (-3) = -1,$$

so

$$y''(\pi) = -4.$$

Next differentiate equation (4.4):

$$y^{(3)}(x) - \sin(x)y + 2\cos(x)y' + \sin(x)y'' = 0.$$

Substitute $x = \pi$ to get

$$y^{(3)}(\pi) - 2(1 - \pi) = 0,$$

so

$$y^{(3)}(\pi) = 2(1 - \pi).$$

Up to this point we have four terms of the expansion of the solution about π:

$$y(x) = -3 + (1 - \pi)(x - \pi) - \frac{4}{2!}(x - \pi)^2 + \frac{2(1 - \pi)}{3!}(x - \pi)^3 + \cdots$$

$$= -3 + (1 - \pi)(x - \pi) - 2(x - \pi)^2 + \tfrac{1}{3}(1 - \pi)(x - \pi)^3 + \cdots.$$

Again, with more work we can compute more terms. ■

This method for generating a series solution of a first-order linear initial value problem extends readily to second-order problems, justified by the following theorem.

THEOREM 4.2

Let p, q and f be analytic at x_0. Then the initial value problem

$$y'' + p(x)y' + q(x)y = f(x); \qquad y(x_0) = A, \, y'(x_0) = B$$

has a unique solution that is also analytic at x_0. ■

EXAMPLE 4.3

Solve

$$y'' - xy' + e^x y = 4; \qquad y(0) = 1, \, y'(0) = 4.$$

Methods from preceding chapters do not apply to this problem. Since $-x$, e^x, and 4 are analytic at 0, the problem has a series solution expanded about 0. The solution has the form

$$y(x) = y(0) + y'(0)x + \tfrac{1}{2!}y''(0)x^2 + \tfrac{1}{3!}y^{(3)}(0)x^3 + \cdots.$$

We already know the first two coefficients from the initial conditions. From the differential equation,

$$y''(0) = 4 - y(0) = 3.$$

Now differentiate the differential equation to get

$$y^{(3)} - y' - xy'' + e^x y + e^x y' = 0.$$

Then

$$y^{(3)}(0) = y'(0) - y(0) - y'(0) = -1.$$

Thus far we have four terms of the series solution about 0:

$$y(x) = 1 + 4x + \tfrac{3}{2}x^2 - \tfrac{1}{6}x^3 + \cdots. \quad ■$$

Although we have illustrated the series method for initial value problems, we can also use it to find general solutions.

EXAMPLE 4.4

We will find the general solution of

$$y'' + \cos(x)y' + 4y = 2x - 1.$$

The idea is to think of this as an initial value problem,

$$y'' + \cos(x)y' + 4y = 2x - 1; \qquad y(0) = a, \ y'(0) = b,$$

with a and b arbitrary (these will be the two arbitrary constants in the general solution). Now proceed as we have been doing. We will determine terms of a solution expanded about 0. The first two coefficients are a and b. For the coefficient of x^2, we find, from the differential equation

$$y''(0) = -y'(0) - 4y(0) - 1 = -b - 4a - 1.$$

Next, differentiate the differential equation:

$$y^{(3)} - \sin(x)y' + \cos(x)y'' + 4y' = 2,$$

so

$$y^{(3)}(0) = -y''(0) - 4y'(0) + 2$$

$$= b + 4a + 1 - 4b + 2 = 4a - 3b + 3.$$

Continuing in this way, we obtain (with details omitted)

$$y(x) = a + bx + \frac{-1 - 4a - b}{2}x^2 + \frac{3 + 4a - 3b}{6}x^3$$

$$+ \frac{1 + 12a + 8b}{24}x^4 + \frac{-16 - 40a + b}{120}x^5 + \cdots. \ \blacksquare$$

In the next section, we will revisit power series solutions, but from a different perspective.

SECTION 4.1   PROBLEMS

In each of Problems 1 through 14, find the first five nonzero terms of the power series solution of the initial value problem, about the point where the initial conditions are given.

1. $y'' + y' - xy = 0$; $y(0) = -2, \ y'(0) = 0$

2. $y'' + 2xy' + (x - 1)y = 0$; $y(0) = 1, \ y'(0) = 2$

3. $y'' - xy = 2x$; $y(1) = 3, \ y'(1) = 0$

4. $y'' + xy' = -1 + x$; $y(2) = 1, \ y'(2) = -4$

5. $y'' - \frac{1}{x^2}y' + \frac{1}{x}y = 0$; $y(1) = 7, \ y'(1) = 3$

6. $y'' + x^2 y = e^x$; $y(0) = -2, \ y'(0) = 7$

7. $y'' - e^x y' + 2y = 1$; $y(0) = -3, \ y'(0) = 1$

8. $y'' + y' - x^4 y = \sin(2x)$; $y(0) = 0, \ y'(0) = -2$

9. $y'' + \frac{1}{x+2}y' - xy = 0$; $y(0) = y'(0) = 1$

10. $y'' - y' + \frac{1}{x}y = 1$; $y(4), \ y'(4) = 2$

11. $y'' - \ln(x)y' = -1 + x$; $y(1) = 1, \ y'(1) = \pi$

12. $y'' + x^2 y' - xy = e^x$; $y(-1) = 14, \ y'(-1) = 0$

13. $y'' + \frac{1}{x-1}y' + \frac{1}{x+2}y = 2$; $y(0) = y'(0) = 3$

14. $(x - 2)y'' + xy' - y = 0$; $y(0) = -3, \ y'(0) = 5$

In each of Problems 15 through 30, find the first five nonzero terms of the Maclaurin expansion of the general solution.

15. $y' + \sin(x)y = -x$

16. $y' - x^2 y = 1$

17. $y' + xy = 1 - x + x^2$

18. $y' - y = \ln(x + 1)$

19. $y'' + xy = 0$

20. $y'' - 2y' + xy = 0$

21. $y'' - x^3 y = 1$

22. $y'' + (1 - x)y' + 2xy = 0$

23. $y'' + y' - x^2 y = 0$

24. $y'' - 8xy = 1 + 2x^9$

25. $y'' - 2xy' + 4x^2 y = 0$

26. $y'' + 12y' + e^x y = 0$

27. $y'' + 2\cos(x)y' = x$

28. $y'' - \tan(x)y' + y = 0$

29. $y'' - e^{-3x} y = 2x^2$

30. $y'' - x^2 y' + xy = e^x \sin(x)$

31. Find the first five terms of the Maclaurin series solution of Airy's equation $y'' + xy = 0$, satisfying $y(0) = a$, $y'(0) = b$.

In each of Problems 32 through 35, the initial value problem can be solved in closed form using methods from Chapters 1 and 2. Find this solution and expand it in a Maclaurin series. Then find the Maclaurin series solution using methods of Section 4.1. The two series should agree.

32. $y'' + y = 1$; $y(0) = 0$, $y'(0) = 0$

33. $y' + y = 2$; $y(0) = -1$

34. $y'' + 3y' + 2y = x$; $y(0) = 0$, $y'(0) = 1$

35. $y'' - 4y' + 5y = 1$; $y(0) = -1$, $y'(0) = 4$

4.2   Power Series Solutions Using Recurrence Relations

We have just seen one way to utilize the differential equation and initial conditions to generate terms of a series solution, expanded about the point where the initial conditions are specified. Another way to generate coefficients is to develop a recurrence relation, which allows us to produce coefficients once certain preceding ones are known. We will consider three examples of this method. In these examples we will use manipulations, such as shifting of indices, which are discussed in the appendix to this chapter.

EXAMPLE 4.5

Consider $y'' + x^2 y = 0$. Suppose we want a solution expanded about 0.

Instead of computing successive derivatives at 0, as we did before, now begin by substituting $y(x) = \sum_{n=0}^{\infty} a_n x^n$ into the differential equation. To do this, we need

$$y' = \sum_{n=1}^{\infty} na^n x^{n-1} \quad \text{and} \quad y'' = \sum_{n=2}^{\infty} n(n-1)a_n x^{n-2}.$$

Notice that the series for y' begins at $n = 1$, and that for y'' at $n = 2$. Put these series into the differential equation to get

$$y'' + x^2 y = \sum_{n=2}^{\infty} n(n-1)a_n x^{n-2} + \sum_{n=0}^{\infty} a_n x^{n+2} = 0. \tag{4.5}$$

Shift indices in both summations so that the power of x occurring in each series is the same. One way to do this is to write

$$\sum_{n=2}^{\infty} n(n-1)a_n x^{n-2} = \sum_{n=0}^{\infty} (n+2)(n+1)a_{n+2} x^n$$

and

$$\sum_{n=0}^{\infty} a_n x^{n+2} = \sum_{n=2}^{\infty} a_{n-2} x^n.$$

Using these series, we can write equation (4.5) as

$$\sum_{n=0}^{\infty}(n+2)(n+1)a_{n+2}x^n + \sum_{n=2}^{\infty} a_{n-2} x^n = 0.$$

We can combine the terms for $n \geq 2$ under one summation and factor out the common x^n (this was the reason for rewriting the series). When we do this, we must list the $n = 0$ and $n = 1$ terms of the first summation separately, or else we lose terms. We get

$$2(1)a_2 x^0 + 3(2)a_3 x + \sum_{n=2}^{\infty}[(n+2)(n+1)a_{n+2} + a_{n-2}]x^n = 0.$$

The only way for this series to be zero for all x in some open interval about 0 is for the coefficient of each power of x to be zero. Therefore,

$$a_2 = a_3 = 0$$

and, for $n = 2, 3, \ldots,$

$$(n+2)(n+1)a_{n+2} + a_{n-2} = 0.$$

This implies that

$$a_{n+2} = -\frac{1}{(n+2)(n+1)}a_{n-2} \quad \text{for} \ \ n = 2, 3, \ldots. \tag{4.6}$$

This is a recurrence relation for this differential equation. In this example, it gives a_{n+2} in terms of a_{n-2} for $n = 2, 3, \ldots$. Thus, we know a_4 in terms of a_0, a_5 in terms of a_1, a_6 in terms of a_2, and so on. The form of the recurrence relation will vary with the differential equation, but it always gives coefficients in terms of one or more previously indexed ones. Using equation (4.6), we proceed:

$$a_4 = -\tfrac{1}{4(3)}a_0 = -\tfrac{1}{12}a_0$$

(by putting $n = 2$);

$$a_5 = -\tfrac{1}{5(4)}a_1 = -\tfrac{1}{20}a_1$$

(by putting $n = 3$);

$$a_6 = -\tfrac{1}{6(5)}a_2 = 0 \quad \text{(because } a_2 = 0)$$

$$a_7 = -\tfrac{1}{7(6)}a_3 = 0 \quad \text{(because } a_3 = 0)$$

$$a_8 = -\tfrac{1}{8(7)}a_4 = \tfrac{1}{(56)(12)}a_0$$

$$a_9 = -\tfrac{1}{9(8)}a_5 = \tfrac{1}{(72)(20)}a_1$$

and so on. The first few terms of the series solution expanded about 0 are

$$y(x) = a_0 + a_1 x + 0x^2 + 0x^3 - \tfrac{1}{12} a_0 x^4$$

$$- \tfrac{1}{20} a_1 x^5 + 0x^6 + 0x^7 + \tfrac{1}{672} a_0 x^8 + \tfrac{1}{1440} a_1 x^9 + \cdots$$

$$= a_0 \left(1 - \tfrac{1}{12} x^4 + \tfrac{1}{672} x^6 + \cdots \right) + a_1 \left(x - \tfrac{1}{20} x^5 + \tfrac{1}{1440} x^9 + \cdots \right).$$

This is actually the general solution, since a_0 and a_1 are arbitrary constants. Note that $a_0 = y(0)$ and $a_1 = y'(0)$, so a solution is completely specified by giving $y(0)$ and $y'(0)$. ■

EXAMPLE 4.6

Consider the nonhomogeneous differential equation

$$y'' + x^2 y' + 4y = 1 - x^2.$$

Attempt a solution $y(x) = \sum_{n=0}^{\infty} a_n x^n$. Substitute this series into the differential equation to get

$$\sum_{n=2}^{\infty} n(n-1)a_n x^{n-2} + x^2 \sum_{n=1}^{\infty} n a_n x^{n-1} + 4 \sum_{n=0}^{\infty} a_n x^n = 1 - x^2.$$

Then

$$\sum_{n=2}^{\infty} n(n-1)a_n x^{n-2} + \sum_{n=1}^{\infty} n a_n x^{n+1} + \sum_{n=0}^{\infty} 4 a_n x^n = 1 - x^2. \qquad (4.7)$$

Shift indices in the first and second summation so that the power of x occurring in each is x^n:

$$\sum_{n=2}^{\infty} n(n-1)a_n x^{n-2} = \sum_{n=0}^{\infty} (n+2)(n+1)a_{n+2} x^n$$

and

$$\sum_{n=1}^{\infty} n a_n x^{n+1} = \sum_{n=2}^{\infty} (n-1)a_{n-1} x^n.$$

Equation (4.7) becomes

$$\sum_{n=0}^{\infty} (n+2)(n+1)a_{n+2} x^n + \sum_{n=2}^{\infty} (n-1)a_{n-1} x^n + \sum_{n=0}^{\infty} 4 a_n x^n = 1 - x^2.$$

We can combine summations from $n = 2$ on, writing the $n = 0$ and $n = 1$ terms from the first and third summations separately. Then

$$2a_2 x^0 + 6a_3 x + 4a_0 x^0 + 4a_1 x + \sum_{n=2}^{\infty} [(n+2)(n+1)a_{n+2} + (n-1)a_{n-1} + 4a_n] x^n = 1 - x^2.$$

For this to hold for all x in some interval about 0, the coefficient of x^n on the left must match the coefficient of x^n on the right. By matching these coefficients, we get:

$$2a_2 + 4a_0 = 1$$

(from x^0),

$$6a_3 + 4a_1 = 0$$

(from x),

$$4(3)a_4 + a_1 + 4a_2 = -1$$

(from x^2), and, for $n \geq 3$,

$$(n+2)(n+1)a_{n+2} + (n-1)a_{n-1} + 4a_n = 0.$$

From these equations we get, in turn,

$$a_2 = \tfrac{1}{2} - 2a_0,$$

$$a_3 = -\tfrac{2}{3}a_1,$$

$$a_4 = \tfrac{1}{12}(-1 - a_1 - 4a_2)$$

$$= -\tfrac{1}{12} - \tfrac{1}{12}a_1 - \tfrac{1}{3}\left(\tfrac{1}{2} - 2a_0\right)$$

$$= -\tfrac{1}{4} + \tfrac{2}{3}a_0 - \tfrac{1}{12}a_1,$$

and, for $n = 3, 4, \ldots$,

$$a_{n+2} = -\frac{4a_n + (n-1)a_{n-1}}{(n+2)(n+1)}.$$

This is the recurrence relation for this differential equation, and it enables us to determine a_{n+2} if we know the two previous coefficients a_n and a_{n-1}. With $n = 3$ we get

$$a_5 = -\frac{4a_3 + 2a_2}{20} = -\frac{1}{20}\left(-\frac{8}{3}a_1 + 1 - 4a_0\right)$$

$$= -\frac{1}{20} + \frac{1}{5}a_0 + \frac{2}{15}a_1.$$

With $n = 4$ the recurrence relation gives us

$$a_6 = -\frac{1}{30}(4a_4 + 3a_3) = -\frac{1}{30}\left(-1 + \frac{8}{3}a_0 - \frac{1}{3}a_1 - 2a_1\right)$$

$$= \frac{1}{30} - \frac{4}{45}a_0 + \frac{7}{90}a_1.$$

Thus far we have six terms of the solution:

$$y(x) = a_0 + a_1 x + \left(\frac{1}{2} - 2a_0\right)x^2 - \frac{2}{3}a_1 x^3 + \left(-\frac{1}{4} + \frac{2}{3}a_0 - \frac{1}{12}a_1\right)x^4$$

$$+ \left(-\frac{1}{20} + \frac{1}{5}a_0 + \frac{2}{15}a_1\right)x^5 + \left(\frac{1}{30} - \frac{4}{45}a_0 + \frac{7}{90}a_1\right)x^6 + \cdots.$$

Using the recurrence relation, we can produce as many terms of this series as we wish. A recurrence relation is particularly suited to computer generation of coefficients. Because this recurrence relation specifies each a_n (for $n \geq 3$) in terms of two preceding coefficients, it is called a *two-term recurrence relation*. It will give each a_n for $n \geq 3$ in terms of a_0 and a_1, which are arbitrary constants. Indeed, $y(0) = a_1$ and $y'(0) = a_1$, so assigning values to these constants uniquely determines the solution. ■

Sometimes we must represent one or more coefficients as power series to apply the current method. This does not alter the basic idea of collecting coefficients of like powers of x and solving for the coefficients.

EXAMPLE 4.7

Solve

$$y'' + xy' - y = e^{3x}.$$

Each coefficient is analytic at 0, so we will look for a power series solution expanded about 0. Substitute $y = \sum_{n=0}^{\infty} a_n x^n$ and also $e^{3x} = \sum_{n=0}^{\infty} (3^n/n!) x^n$ into the differential equation to get:

$$\sum_{n=2}^{\infty} n(n-1)a_n x^{n-2} + \sum_{n=1}^{\infty} na_n x^n - \sum_{n=0}^{\infty} a_n x^n = \sum_{n=0}^{\infty} \frac{3^n}{n!} x^n.$$

Shift indices in the first summation to write this equation as

$$\sum_{n=0}^{\infty} (n+2)(n+1)a_{n+2} x^n + \sum_{n=1}^{\infty} na_n x^n - \sum_{n=0}^{\infty} a_n x^n = \sum_{n=0}^{\infty} \frac{3^n}{n!} x^n.$$

We can collect terms from $n = 1$ on under one summation, obtaining

$$\sum_{n=1}^{\infty} [(n+2)(n+1)a_{n+2} + (n-1)a_n] x^n + 2a_2 - a_0 = 1 + \sum_{n=1}^{\infty} \frac{3^n}{n!} x^n.$$

Equate coefficients of like powers of x on both sides of the equation to obtain

$$2a_2 - a_0 = 1$$

and, for $n = 1, 2, \ldots,$

$$(n+2)(n+1)a_{n+2} + (n-1)a_n = \frac{3^n}{n!}.$$

This gives

$$a_2 = \frac{1}{2}(1 + a_0)$$

and, for $n = 1, 2, \ldots,$ we have the one-term recurrence relation (in terms of one preceding coefficient)

$$a_{n+2} = \frac{(3^n/n!) + (1-n)a_n}{(n+2)(n+1)}.$$

Using this relationship, we can generate as many coefficients as we want in the solution series, in terms of the arbitrary constants a_0 and a_1. The first few terms are

$$y(x) = a_0 + a_1 x + \frac{1 + a_0}{2} x^2 + \frac{1}{2} x^3$$

$$+ \left(\frac{1}{3} - \frac{a_0}{24} \right) x^4 + \frac{7}{40} x^5 + \left(\frac{19}{240} + \frac{1}{240} a_0 \right) x^6 + \cdots. \quad ■$$

SECTION 4.2   PROBLEMS

In each of Problems 1 through 15, find the recurrence relation and use it to generate the first five terms of the Maclaurin series of the general solution.

1. $y' - xy = 1 - x$

2. $y' - x^3 y = 4$

3. $y' + (1 - x^2)y = x$

4. $y'' + 2y' + xy = 0$

5. $y'' - xy' + y = 3$

6. $y'' + xy' + xy = 0$

7. $y'' - x^2 y' + 2y = x$

8. $y'' + x^2 y' + 2y = 0$

9. $y'' + (1 - x)y' + 2y = 1 - x^2$

10. $y'' + y' - (1 - x + x^2)y = -5$

11. $y' + xy = \cos(x)$

12. $y'' + xy' = 1 - e^x$

13. $y'' + y' - y = 1 - x^2 + x^4$

14. $y'' + x^2 y = \sin(x)$

15. $y'' + y = -x \cos(x)$

4.3   Singular Points and the Method of Frobenius

In this section we will consider the second-order linear differential equation

$$P(x)y'' + Q(x)y' + R(x)y = F(x). \tag{4.8}$$

If we can divide this equation by $P(x)$ and obtain an equation of the form

$$y'' + p(x)y' + q(x)y = f(x), \tag{4.9}$$

with analytic coefficients in some open interval about x_0, then we can proceed to a power series solution of equation (4.9) by methods already developed and thereby solve equation (4.8). In this case we call x_0 an ordinary point of the differential equation. If, however, x_0 is not an ordinary point, then this strategy fails and we must develop some new machinery.

DEFINITION 4.2   *Ordinary and Singular Points*

x_0 is an ordinary point of equation (4.8) if $P(x_0) \neq 0$ and $Q(x)/P(x)$, $R(x)/P(x)$, and $F(x)/P(x)$ are analytic at x_0.

x_0 is a singular point of equation (4.8) if x_0 is not an ordinary point.

Thus, x_0 is a singular point if $P(x_0) = 0$, or if any one of $Q(x)/P(x)$, $R(x)/P(x)$, or $F(x)/P(x)$ fails to be analytic at x_0.

EXAMPLE 4.8

The differential equation

$$x^3(x - 2)^2 y'' + 5(x + 2)(x - 2)y' + 3x^2 y = 0$$

has singular points at 0 and 2, because $P(x) = x^3(x - 2)^2$ and $P(0) = P(2) = 0$. Every other real number is a regular point of this equation. ∎

In an interval about a singular point, solutions can exhibit behavior that is quite different from what we have seen in an interval about an ordinary point. In particular, the general solution of equation (4.8) may contain a logarithm term, which will tend toward ∞ in magnitude as x approaches x_0.

In order to seek some understanding of the behavior of solutions near a singular point, we will concentrate on the homogeneous equation

$$P(x)y'' + Q(x)y' + R(x)y = 0. \tag{4.10}$$

Once this case is understood, it does not add substantial further difficulty to consider the non-homogeneous equation (4.8). Experience and research have shown that some singular points are "worse" than others, in the sense that the subtleties they bring to attempts at solution are deepened. We therefore distinguish two kinds of singular points.

DEFINITION 4.3 *Regular and Irregular Singular Points*

x_0 is a regular singular point of equation (4.10) if x_0 is a singular point, and the functions

$$(x - x_0)\frac{Q(x)}{P(x)} \quad \text{and} \quad (x - x_0)^2 \frac{R(x)}{P(x)}$$

are analytic at x_0.

A singular point that is not regular is said to be an irregular singular point.

EXAMPLE 4.9

We have already noted that

$$x^3(x - 2)^2 y'' + 5(x + 2)(x - 2)y' + 3x^2 y = 0$$

has singular points at 0 and 2. We will classify these singular points.

In this example, $P(x) = x^3(x - 2)^2$, $Q(x) = 5(x + 2)(x - 2)$, and $R(x) = 3x^2$. First consider $x_0 = 0$. Now

$$(x - x_0)\frac{Q(x)}{P(x)} = \frac{5x(x + 2)(x - 2)}{x^3(x - 2)^2} = \frac{5}{x^2}\frac{x + 2}{x - 2}$$

is not defined at 0, hence is not analytic there. This is enough to conclude that 0 is an irregular singular point of this differential equation.

Next let $x_0 = 2$ and consider

$$(x - 2)\frac{Q(x)}{P(x)} = 5\frac{x + 2}{x^3}$$

and

$$(x - 2)^2\frac{R(x)}{P(x)} = \frac{3}{x}.$$

Both of these functions are analytic at 2. Therefore, 2 is a regular singular point of the differential equation. ∎

Suppose now that equation (4.10) has a regular singular point at x_0. Then there may be no solution as a power series about x_0. In this case we attempt to choose numbers c_n and a number r so that

$$y(x) = \sum_{n=0}^{\infty} c_n(x - x_0)^{n+r} \tag{4.11}$$

is a solution. This series is called a *Frobenius series*, and the strategy of attempting a solution of this form is called the *method of Frobenius*. A Frobenius series need not be a power series, since r may be negative or may be a noninteger.

A Frobenius series "begins" with $c_0 x^r$, which is constant only if $r = 0$. Thus, in computing the derivative of the Frobenius series (4.11), we get

$$y'(x) = \sum_{n=0}^{\infty} (n + r)c_n(x - x_0)^{n+r-1},$$

and this summation begins at zero again because the derivative of the $n = 0$ term need not be zero. Similarly,

$$y''(x) = \sum_{n=0}^{\infty} (n + r)(n + r - 1)c_n(x - x_0)^{n+r-2}.$$

We will now illustrate the method of Frobenius.

EXAMPLE 4.10

We want to solve

$$x^2 y'' + x \left(\tfrac{1}{2} + 2x \right) y' + \left(x - \tfrac{1}{2} \right) y = 0.$$

It is routine to show that 0 is a regular singular point. Substitute a Frobenius series $y(x) = \sum_{n=0}^{\infty} c_n x^{n+r}$ into the differential equation to get

$$\sum_{n=0}^{\infty} (n + r)(n + r - 1)c_n x^{n+r} + \sum_{n=0}^{\infty} \tfrac{1}{2}(n + r)c_n x^{n+r}$$

$$+ \sum_{n=0}^{\infty} 2(n + r)c_n x^{n+r+1} + \sum_{n=0}^{\infty} c_n x^{n+r+1} - \sum_{n=0}^{\infty} \tfrac{1}{2} c_n x^{n+r} = 0.$$

Shift indices in the third and fourth summations to write this equation as

$$\left[r(r - 1)c_0 + \tfrac{1}{2}c_0 r - \tfrac{1}{2}c_0 \right] x^r + \sum_{n=1}^{\infty} \left[(n + r)(n + r - 1)c_n + \tfrac{1}{2}(n + r)c_n \right.$$

$$\left. + 2(n + r - 1)c_{n-1} + c_{n-1} - \tfrac{1}{2}c_n \right] x^{n+r} = 0.$$

This equation will hold if the coefficient of each x^{n+r} is zero. This gives us the equations

$$r(r - 1)c_0 + \tfrac{1}{2}c_0 r - \tfrac{1}{2}c_0 = 0 \tag{4.12}$$

and

$$(n+r)(n+r-1)c_n + \tfrac{1}{2}(n+r)c_n + 2(n+r-1)c_{n-1} + c_{n-1} - \tfrac{1}{2}c_n = 0 \qquad (4.13)$$

for $n = 1, 2, \ldots$. Assuming that $c_0 \neq 0$, an essential requirement in the method, equation (4.12) implies that

$$r(r-1) + \tfrac{1}{2}r - \tfrac{1}{2} = 0. \qquad (4.14)$$

This is the indicial equation for this differential equation, and it determines the values of r we can use. Solve it to obtain $r_1 = 1$ and $r_2 = -\tfrac{1}{2}$. Equation (4.13) enables us to solve for c_n in terms of c_{n-1} to get the recurrence relation

$$c_n = -\frac{1 + 2(n+r-1)}{(n+r)(n+r-1) + \tfrac{1}{2}(n+r) - \tfrac{1}{2}}c_{n-1}$$

for $n = 1, 2, \ldots$.

First put $r = r_1 = 1$ into the recurrence relation to obtain

$$c_n = -\frac{1+2n}{n\left(n + \tfrac{3}{2}\right)}c_{n-1} \quad \text{for } n = 1, 2, \ldots.$$

Some of these coefficients are

$$c_1 = -\frac{3}{\frac{5}{2}}c_0 = -\tfrac{6}{5}c_0,$$

$$c_2 = -\tfrac{5}{7}c_1 = -\tfrac{5}{7}\left(-\tfrac{6}{5}c_0\right) = \tfrac{6}{7}c_0,$$

$$c_3 = -\frac{7}{\frac{27}{2}}c_2 = -\tfrac{14}{27}\left(\tfrac{6}{7}c_0\right) = -\tfrac{4}{9}c_0,$$

and so on. One Frobenius solution is

$$y_1(x) = c_0\left(x - \tfrac{6}{5}x^2 + \tfrac{6}{7}x^3 - \tfrac{4}{9}x^4 + \cdots\right).$$

Because r_1 is a nonnegative integer, this first Frobenius solution is actually a power series about 0.

For a second Frobenius solution, substitute $r = r_2 = -\tfrac{1}{2}$ into the recurrence relation. To avoid confusion we will replace c_n with c_n^* in this relation. We get

$$c_n^* = -\frac{1 + 2\left(n - \tfrac{3}{2}\right)}{\left(n - \tfrac{1}{2}\right)\left(n - \tfrac{3}{2}\right) + \tfrac{1}{2}\left(n - \tfrac{1}{2}\right) - \tfrac{1}{2}}c_{n-1}^*$$

for $n = 1, 2, \ldots$. This simplifies to

$$c_n^* = -\frac{2n-2}{n\left(n - \tfrac{3}{2}\right)}c_{n-1}^*.$$

It happens in this example that $c_1^* = 0$, so each $c_n^* = 0$ for $n = 1, 2, \ldots$ and the second Frobenius solution is

$$y_2(x) = \sum_{n=0}^{\infty} c_n^* x^{n-1/2} = c_0^* x^{-1/2} \quad \text{for } x > 0. \ \blacksquare$$

The method of Frobenius is justified by the following theorem.

THEOREM 4.3   Method of Frobenius

Suppose x_0 is a regular singular point of $P(x)y'' + Q(x)y' + R(x)y = 0$. Then there exists at least one Frobenius solution

$$y(x) = \sum_{n=0}^{\infty} c_n(x - x_0)^r$$

with $c_0 \neq 0$. Further, if the Taylor expansions of $(x - x_0)Q(x)/R(x)$ and $(x - x_0)^2 R(x)/P(x)$ about x_0 converge in an open interval $(x_0 - h, x_0 + h)$, then this Frobenius series also converges in this interval, except perhaps at x_0 itself. ■

It is significant that the theorem only guarantees the existence of one Frobenius solution. Although we obtained two such solutions in the preceding example, the next example shows that there may be only one.

EXAMPLE 4.11

Suppose we want to solve

$$x^2 y'' + 5xy' + (x + 4)y = 0.$$

Zero is a regular singular point, so attempt a Frobenius solution $y(x) = \sum_{n=0}^{\infty} c_n x^{n+r}$. Substitute into the differential equation to get

$$\sum_{n=0}^{\infty} (n + r)(n + r - 1)c_n x^{n+r} + \sum_{n=0}^{\infty} 5(n + r)c_n x^{n+r} + \sum_{n=0}^{\infty} c_n x^{n+r+1} + \sum_{n=0}^{\infty} 4c_n x^{n+r} = 0.$$

Shift indices in the third summation to write this equation as

$$\sum_{n=0}^{\infty} (n + r)(n + r - 1)c_n x^{n+r} + \sum_{n=0}^{\infty} 5(n + r)c_n x^{n+r} + \sum_{n=1}^{\infty} c_{n-1} x^{n+r} + \sum_{n=0}^{\infty} 4c_n x^{n+r} = 0.$$

Now combine terms to write

$$[r(r - 1) + 5r + 4]c_0 x^r + \sum_{n=1}^{\infty} [(n + r)(n + r - 1)c_n + 5(n + r)c_n + c_{n-1} + 4c_n]x^{n+r} = 0.$$

Setting the coefficient of x^r equal to zero (since $c_0 \neq 0$ as part of the method), we get the indicial equation

$$r(r - 1) + 5r + 4 = 0$$

with the repeated root $r = -2$. The coefficient of x^{n+r} in the series, with $r = -2$ inserted, gives us the recurrence relation

$$(n - 2)(n - 3)c_n + 5(n - 2)c_n + c_{n-1} + 4c_n = 0$$

or

$$c_n = -\frac{1}{(n - 2)(n - 3) + 5(n - 2) + 4}c_{n-1}$$

for $n = 1, 2, \ldots$. This simplifies to

$$c_n = -\frac{1}{n^2}c_{n-1} \quad \text{for } n = 1, 2, 3, \ldots.$$

Some of the coefficients are

$$c_1 = -c_0$$

$$c_2 = -\frac{1}{4}c_1 = \frac{1}{4}c_0 = \frac{1}{(2)^2}c_0$$

$$c_3 = -\frac{1}{9}c_2 = -\frac{1}{4 \cdot 9}c_0 = -\frac{1}{(2 \cdot 3)^2}c_0$$

$$c_4 = -\frac{1}{16}c_3 = \frac{1}{4 \cdot 9 \cdot 16}c_0 = \frac{1}{(2 \cdot 3 \cdot 4)^2}c_0$$

and so on. In general,

$$c_n = (-1)^n \frac{1}{(n!)^2}c_0$$

for $n = 1, 2, 3, \ldots$. The Frobenius solution we have found is

$$y(x) = c_0[x^{-2} - x^{-1} + \frac{1}{4} - \frac{1}{36}x + \frac{1}{576}x^2 + \cdots]$$

$$= c_0 \sum_{n=0}^{\infty} (-1)^n \frac{1}{(n!)^2}x^{n-2}.$$

In this example, $xQ(x)/P(x) = x(5x/x^2) = 5$ and $x^2R(x)/P(x) = x^2(x+4)/x^2 = x + 4$. These polynomials are their own Maclaurin series about 0, and these series, being finite, converge for all x. By Theorem 4.3, the Frobenius series solution converges for all x, except $x = 0$.

In this example the method of Frobenius produces only one solution. ∎

In the last example the recurrence relation produced a simple formula for c_n in terms of c_0. Depending on the coefficients in the differential equation, a formula for c_n in terms of c_0 may be quite complicated, or it may even not be possible to write a formula in terms of elementary algebraic expressions. We will give another example, having some importance for later work, in which the Frobenius method may produce only one solution.

EXAMPLE 4.12　Bessel Functions of the First Kind

The differential equation

$$x^2y'' + xy' + (x^2 - \nu^2)y = 0$$

is called Bessel's equation of order ν, for $\nu \geq 0$. Although it is a second-order differential equation, this description of it as being of order ν refers to the parameter ν appearing in it, and is traditional. Solutions of Bessel's equation are called *Bessel functions*, and we will encounter them in Chapter 16 when we treat special functions, and again in Chapter 18 when we analyze heat conduction in an infinite cylinder.

Zero is a regular singular point of Bessel's equation, so attempt a solution:

$$y(x) = \sum_{n=0}^{\infty} c_n x^{n+r}.$$

Upon substituting this series into Bessel's equation, we obtain

$$[r(r-1)+r-v^2]c_0 x^r + [r(r+1)+(r+1)-v^2]c_1 x^{r+1}$$

$$+ \sum_{n=2}^{\infty}[[(n+r)(n+r-1)+(n+r)-v^2]c_n + c_{n-2}]x^{n+r} = 0. \tag{4.15}$$

Set the coefficient of each power of x equal to zero. Assuming that $c_0 \neq 0$, we obtain the indicial equation

$$r^2 - v^2 = 0,$$

with roots $\pm v$. Let $r = v$ in the coefficient of x^{r+1} in equation (4.16) to get

$$(2v+1)c_1 = 0.$$

Since $2v + 1 \neq 0$, we conclude that $c_1 = 0$.

From the coefficient of x^{n+r} in equation (4.16), we get

$$[(n+r)(n+r-1)+(n+r)-v^2]c_n + c_{n-2} = 0$$

for $n = 2, 3, \dots$. Set $r = v$ in this equation and solve for c_n to get

$$c_n = -\frac{1}{n(n+2v)}c_{n-2}$$

for $n = 2, 3, \dots$. Since $c_1 = 0$, this equation yields

$$c_3 = c_5 = \cdots = c_{\text{odd}} = 0.$$

For the even-indexed coefficients, write

$$c_{2n} = -\frac{1}{2n(2n+2v)}c_{2n-2} = -\frac{1}{2^2 n(n+v)}c_{2n-2}$$

$$= -\frac{1}{2^2 n(n+v)}\frac{-1}{2(n-1)[(2(n-1)+2v]}c_{2n-4}$$

$$= \frac{1}{2^4 n(n-1)(n+v)(n+v-1)}c_{2n-4}$$

$$= \cdots = \frac{(-1)^n}{2^{2n}n(n-1)\cdots(2)(1)(n+v)(n-1+v)\cdots(1+v)}c_0$$

$$= \frac{(-1)^n}{2^{2n}n!(1+v)(2+v)\cdots(n+v)}c_0.$$

One Frobenius solution of Bessel's equation of order v is therefore

$$y_1(x) = c_0 \sum_{n=0}^{\infty}\frac{(-1)^n}{2^{2n}n!(1+v)(2+v)\cdots(n+v)}x^{2n+v}. \tag{4.16}$$

These functions are called *Bessel functions of the first kind of order v*. ∎

The roots of the indicial equation for Bessel's equation are $\pm v$. Depending on v, we may or may not obtain two linearly independent solutions by using v and $-v$ in the series solution (4.16). We will discuss this in more detail when we treat Bessel functions in Chapter 16, where we will see that when v is a positive integer, the functions obtained by using v and then $-v$ in the recurrence relation are linearly dependent.

SECTION 4.3   PROBLEMS

In each of Problems 1 through 6, find all of the singular points and classify each singular point as regular or singular.

1. $x^2(x-3)^2y'' + 4x(x^2-x-6)y' + (x^2-x-2)y = 0$

2. $(x^3-2x^2-7x-4)y'' - 2(x^2+1)y' + (5x^2-2x)y = 0$

3. $x^2(x-2)y'' + (5x-7)y' + 2(3+5x^2)y = 0$

4. $[(9-x^2)y']' + (2+x^2)y = 0$

5. $[(x-2)^{-1}y']' + x^{-5/2}y = 0$

6. $x^2\sin^2(x-\pi)y'' + \tan(x-\pi)\tan(x)y' + (7x-2)\cos(x)y = 0$

In each of Problems 7 through 20, (a) show that zero is a regular singular point of the differential equation, (b) find and solve the indicial equation, (c) determine the recurrence relation, and (d) use the results of (b) and (c) to find the first five nonzero terms of two linearly independent Frobenius solutions.

7. $4x^2y'' + 2xy' - xy = 0$

8. $16x^2y'' - 4x^2y' + 3y = 0$

9. $9x^2y'' + 2(2x+1)y = 0$

10. $12x^2y'' + 5xy' + (1-2x^2)y = 0$

11. $2xy'' + (2x+1)y' + 2y = 0$

12. $2x^2y'' - xy' + (1-x^2)y = 0$

13. $2x^2y'' + x(2x+1)y' - (2x^2+1)y = 0$

14. $3x^2y'' + 4xy' - (3x+2)y = 0$

15. $9x^2y'' + 9xy' + (9x^2-4)y = 0$

16. $3x^2y'' + x(1-2x^2)y' - 4x^2y = 0$

17. $2xy'' + y' + 6xy = 0$

18. $12x^2y'' + x(12x^3-5)y' + 6y = 0$

19. $2x^2y'' - 3xy' - (2x^2+3)y = 0$

20. $6x^2y'' + x(5-3x)y' + x(3-2x)y = 0$

21. The differential equation
$$x(1-x)y'' + [c-(1+a+b)x]y' - aby = 0$$
is called the *hypergeometric equation*. Here a, b, and c are constants.

(a) Show that 0 is a regular singular point.

(b) Assuming that c is not an integer, find the first five nonzero terms of two linearly independent Frobenius solutions.

22. Show that the Euler differential equation $x^2y'' + axy' + by = 0$ has a regular singular point at 0. Attempt a solution using the method of Frobenius.

4.4   Second Solutions and Logarithm Factors

In the preceding section we saw that under certain conditions we can always produce a Frobenius series solution of equation (4.10), but possibly not a second, linearly independent solution. This may occur if the indicial equation has a repeated root, or even if it has distinct roots that differ by a positive integer.

In the case that the method of Frobenius only produces one solution, there is a method for finding a second, linearly independent solution. The key is to know what form to expect this solution to have, so that this template can be substituted into the differential equation to determine the coefficients. This template is provided by the following theorem. We will state the theorem with $x_0 = 0$ to simplify the notation. To apply it to a differential equation having a singular point $x_0 \neq 0$, use the change of variables $z = x - x_0$.

THEOREM 4.4   *A Second Solution in the Method of Frobenius*

Suppose 0 is a regular singular point of
$$P(x)y'' + Q(x)y' + R(x)y = 0.$$

Let r_1 and r_2 be roots of the indicial equation. If these are real, suppose $r_1 \geq r_2$. Then

1. If $r_1 - r_2$ is not an integer, there are two linearly independent Frobenius solutions:

$$y_1(x) = \sum_{n=0}^{\infty} c_n x^{n+r_1} \quad \text{and} \quad y_2(x) = \sum_{n=0}^{\infty} c_n^* x^{n+r_2},$$

   with $c_0 \neq 0$ and $c_0^* \neq 0$. These solutions are valid in some interval $(0, h)$ or $(-h, 0)$.

2. If $r_1 - r_2 = 0$, there is a Frobenius solution $y_1(x) = \sum_{n=0}^{\infty} c_n x^{n+r_1}$ with $c_0 \neq 0$ as well as a second solution:

$$y_2(x) = y_1(x) \ln(x) + \sum_{n=1}^{\infty} c_n^* x^{n+r_1}.$$

   Further, y_1 and y_2 form a fundamental set of solutions on some interval $(0, h)$.

3. If $r_1 - r_2$ is a positive integer, then there is a Frobenius series solution:

$$y_1(x) = \sum_{n=0}^{\infty} c_n x^{n+r_1}.$$

   In this case there is a second solution of the form

$$y_2(x) = k y_1(x) \ln(x) + \sum_{n=0}^{\infty} c_n^* x^{n+r_2}.$$

   If $k = 0$ this is a second Frobenius series solution; if not, the solution contains a logarithm term. In either event, y_1 and y_2 form a fundamental set on some interval $(0, h)$. ∎

We may now summarize the method of Frobenius as follows, for the equation $P(x)y'' + Q(x)y' + R(x)y = 0$. Suppose 0 is a regular singular point.

Substitute $y(x) = \sum_{n=0}^{\infty} c_n x^{n+r}$ into the differential equation. From the indicial equation, determine the values of r. If these are distinct and do not differ by an integer, we are guaranteed two linearly independent Frobenius solutions.

If the indicial equation has repeated roots, then there is just one Frobenius solution, y_1. But there is a second solution:

$$y_2(x) = y_1(x) \ln(x) + \sum_{n=1}^{\infty} c_n^* x^{n+r_1}.$$

The series on the right starts its summation at $n = 1$, not $n = 0$. Substitute $y_2(x)$ into the differential equation and obtain a recurrence relation for the coefficients c_n^*. Because this solution has a logarithm term, y_1 and y_2 are linearly independent.

If $r_1 - r_2$ is a positive integer, there may or may not be a second Frobenius solution. In this case there is a second solution of the form

$$y_2(x) = k y_1(x) \ln(x) + \sum_{n=0}^{\infty} c_n^* x^{n+r_2}.$$

Substitute y_2 into the differential equation and obtain an equation for k and a recurrence relation for the coefficients c_n^*. If $k = 0$, we obtain a second Frobenius solution; if not, then this second solution has a logarithm term. In either case y_1 and y_2 are linearly independent.

In the preceding section we saw in Example 4.10 a differential equation in which $r_1 - r_2$ was not an integer. There we found two linearly independent Frobenius solutions. We will illustrate cases (2) and (3) of the theorem.

EXAMPLE 4.13   Conclusion (2), Equal Roots

Consider again $x^2 y'' + 5xy' + (x+4)y = 0$. In Example 4.11 we found one Frobenius solution:

$$y_1(x) = c_0 \sum_{n=0}^{\infty} (-1)^n \frac{1}{(n!)^2} x^{n-2}.$$

The indicial equation is $(r+2)^2 = 0$ with the repeated root $r = -2$. Conclusion (2) of the theorem suggests that we attempt a second solution of the form

$$y_2(x) = y_1(x) \ln(x) + \sum_{n=1}^{\infty} c_n^* x^{n-2}.$$

Note that the series on the right begins at $n = 1$, not $n = 0$. Substitute this series into the differential equation to get, after some rearrangement of terms,

$$4y_1 + 2xy_1' + \sum_{n=1}^{\infty} (n-2)(n-3)c_n^* x^{n-2} + \sum_{n=1}^{\infty} 5(n-2)c_n^* x^{n-2}$$

$$+ \sum_{n=1}^{\infty} c_n^* x^{n-1} + \sum_{n=1}^{\infty} 4c_n^* x^{n-2} + \ln(x)[x^2 y_1'' + 5xy_1' + (x+4)y_1] = 0.$$

The bracketed coefficient of $\ln(x)$ is zero because y_1 is a solution of the differential equation. In the last equation, choose $c_0 = 1$ (we need only one second solution), shift indices to write $\sum_{n=1}^{\infty} c_n^* x^{n-1} = \sum_{n=2}^{\infty} c_{n-1}^* x^{n-2}$, and substitute the series obtained for $y_1(x)$ to get

$$-2x^{-1} + c_1^* x^{-1} + \sum_{n=2}^{\infty} \left[\frac{4(-1)^n}{(n!)^2} + \frac{2(-1)^n}{(n!)^2}(n-2) \right.$$

$$+ (n-2)(n-3)c_n^* + 5(n-2)c_n^* + c_{n-1}^* + 4c_n^* \Big] x^{n-2} = 0.$$

Set the coefficient of each power of x equal to zero. From the coefficient of x^{-1} we get

$$c_1^* = 2.$$

From the coefficient of x^{n-2} in the summation we get, after some routine algebra,

$$\frac{2(-1)^n}{(n!)^2} n + n^2 c_n^* + c_{n-1}^* = 0,$$

or

$$c_n^* = -\frac{1}{n^2} c_{n-1}^* - \frac{2(-1)^n}{n(n!)^2}$$

for $n = 2, 3, 4, \ldots$. This enables us to calculate as many coefficients as we wish. Some of the terms of the resulting solution are

$$y_2(x) = y_1(x) \ln(x) + \frac{2}{x} - \frac{3}{4} + \frac{11}{108} x - \frac{25}{3456} x^2 + \frac{137}{432,000} x^3 + \cdots.$$

Because of the logarithm term, it is obvious that this solution is not a constant multiple of y_1, so y_1 and y_2 form a fundamental set of solutions (on some interval $(0, h)$). The general solution is

$$y(x) = [C_1 + C_2 \ln(x)] \sum_{n=0}^{\infty} \frac{(-1)^n}{(n!)^2} x^{n-2}$$

$$+ C_2 \left[\frac{2}{x} - \frac{3}{4} + \frac{11}{108}x - \frac{25}{3456}x^2 + \frac{137}{432,000}x^3 + \cdots \right]. \quad \blacksquare$$

EXAMPLE 4.14 Conclusion (3), with $k = 0$

The equation $x^2 y'' + x^2 y' - 2y = 0$ has a regular singular point at 0. Substitute $y(x) = \sum_{n=0}^{\infty} c_n x^{n+r}$ and shift indices to obtain

$$[r(r-1) - 2]c_0 x^r + \sum_{n=1}^{\infty} [(n+r)(n+r-1)c_n + (n+r-1)c_{n-1} - 2c_n]x^{n+r} = 0.$$

Assume that $c_0 \neq 0$. The indicial equation is $r^2 - r - 2 = 0$, with roots $r_1 = 2$ and $r_2 = -1$. Now $r_1 - r_2 = 3$ and case (3) of the theorem applies.

For a first solution, set the coefficient of x^{n+r} equal to zero to get

$$(n+r)(n+r-1)c_n + (n+r-1)c_{n-1} - 2c_n = 0. \tag{4.17}$$

Let $r = 2$ to get

$$(n+2)(n+1)c_n + (n+1)c_{n-1} - 2c_n = 0,$$

or

$$c_n = -\frac{n+1}{n(n+3)}c_{n-1} \quad \text{for } n = 1, 2, \ldots.$$

Using this recurrence relation to generate terms of the series, we obtain

$$y_1(x) = c_0 x^2 \left[1 - \frac{1}{2}x + \frac{3}{20}x^2 - \frac{1}{30}x^3 + \frac{1}{168}x^4 - \frac{1}{1120}x^5 + \frac{1}{8640}x^6 + \cdots \right].$$

Now try the second root $r = -1$ in the recurrence relation (4.17). We get

$$(n-1)(n-2)c_n^* + (n-2)c_{n-1}^* - 2c_n^* = 0$$

for $n = 1, 2, \ldots$. When $n = 3$, this gives $c_2^* = 0$, which forces $c_n^* = 0$ for $n \geq 2$. But then

$$y_2(x) = c_0^* \frac{1}{x} + c_1^*.$$

Substitute this into the differential equation to get

$$x^2(2c_0^* x^{-3}) + x^2(-c_0^* x^{-2}) - 2\left(c_1^* + c_0^* \frac{1}{x}\right) = -c_0^* - 2c_1^* = 0.$$

Then $c_1^* = -\frac{1}{2}c_0^*$ and we obtain the second solution

$$y_2(x) = c_0^* \left(\frac{1}{x} - \frac{1}{2} \right),$$

with c_0^* nonzero but otherwise arbitrary. The functions y_1 and y_2 form a fundamental set of solutions. ∎

EXAMPLE 4.15   Conclusion (3), $k \neq 0$

Consider the differential equation $xy'' - y = 0$, which has a regular singular point at 0. Substitute $y(x) = \sum_{n=0}^{\infty} c_n x^{n+r}$ to obtain

$$\sum_{n=0}^{\infty} (n+r)(n+r-1)c_n x^{n+r-1} - \sum_{n=0}^{\infty} c_n x^{n+r} = 0.$$

Shift indices in the second summation to write this equation as

$$(r^2 - r)c_0 x^{r-1} + \sum_{n=1}^{\infty} [(n+r)(n+r-1)c_n - c_{n-1}]x^{n+r-1} = 0.$$

The indicial equation is $r^2 - r = 0$, with roots $r_1 = 1$, $r_2 = 0$. Here $r_1 - r_2 = 1$, a positive integer, so we are in case (3) of the theorem. The recurrence relation is

$$(n+r)(n+r-1)c_n - c_{n-1} = 0$$

for $n = 1, 2, \ldots$. Let $r = 1$ and solve for c_n:

$$c_n = \frac{1}{n(n+1)}c_{n-1} \quad \text{for } n = 1, 2, 3, \ldots.$$

Some of the coefficients are

$$c_1 = \frac{1}{1(2)}c_0,$$

$$c_2 = \frac{1}{2(3)}c_1 = \frac{1}{2(2)(3)}c_0,$$

$$c_3 = \frac{1}{3(4)}c_2 = \frac{1}{2(3)(2)(3)(4)}c_0.$$

In general, we find that

$$c_n = \frac{1}{n!(n+1)!}c_0$$

for $n = 1, 2, \ldots$. This gives us a Frobenius series solution

$$y_1(x) = c_0 \sum_{n=0}^{\infty} \frac{1}{n!(n+1)!}x^{n+1}$$

$$= c_0 \left[x + \frac{1}{2}x^2 + \frac{1}{12}x^3 + \frac{1}{144}x^4 + \cdots \right].$$

In this example, if we put $r = 0$ into the recurrence relation, we get

$$n(n-1)c_n - c_{n-1} = 0$$

for $n = 1, 2, \ldots$. But if we put $n = 1$ into this equation, we get $c_0 = 0$, contrary to the assumption that $c_0 \neq 0$. Unlike the preceding example, we cannot find a second Frobenius solution by simply putting r_2 into the recurrence relation.

Try a second solution:

$$y_2(x) = ky_1(x)\ln(x) + \sum_{n=0}^{\infty} c_n^* x^n$$

(here $x^{n+r_2} = x^n$ because $r_2 = 0$). Substitute this into the differential equation to get

$$x\left[ky_1''\ln(x) + 2ky_1'\frac{1}{x} - ky_1\frac{1}{x^2} + \sum_{n=2}^{\infty} n(n-1)c_n^* x^{n-2}\right]$$

$$- ky_1\ln(x) - \sum_{n=0}^{\infty} c_n^* x^n = 0.$$
(4.18)

Now

$$k\ln(x)[xy_1'' - y_1] = 0$$

because y_1 is a solution of the differential equation. For the remaining terms in equation (4.19), insert the series for $y_1(x)$ (with $c_0 = 1$ for convenience) to get

$$2k\sum_{n=0}^{\infty} \frac{1}{(n!)^2} x^n - k\sum_{n=0}^{\infty} \frac{1}{n!(n+1)!} x^n + \sum_{n=2}^{\infty} c_n^* n(n-1)x^{n-1} - \sum_{n=0}^{\infty} c_n^* x^n = 0.$$

Shift indices in the third summation to write this equation as

$$2k\sum_{n=0}^{\infty} \frac{1}{(n!)^2} x^n - k\sum_{n=0}^{\infty} \frac{1}{n!(n+1)!} x^n + \sum_{n=1}^{\infty} c_{n+1}^* (n+1)nx^n - \sum_{n=0}^{\infty} c_n^* x^n = 0.$$

Then

$$(2k - k - c_0^*)x^0 + \sum_{n=1}^{\infty} \left[\frac{2k}{(n!)^2} - \frac{k}{n!(n+1)!} + n(n+1)c_{n+1}^* - c_n^*\right]x^n = 0.$$

Then

$$k - c_0^* = 0$$

and, for $n = 1, 2, \ldots$,

$$\frac{2k}{(n!)^2} - \frac{k}{n!(n+1)!} + n(n+1)c_{n+1}^* - c_n^* = 0.$$

This gives us $k = c_0^*$ and the recurrence relation

$$c_{n+1}^* = \frac{1}{n(n+1)}\left[c_n^* - \frac{(2n+1)k}{n!(n+1)!}\right]$$

for $n = 1, 2, 3, \ldots$. Since c_0^* can be any nonzero real number, we may choose $c_0^* = 1$. Then $k = 1$. For a particular second solution, let $c_1^* = 0$, obtaining:

$$y_2(x) = y_1(x)\ln(x) + 1 - \frac{3}{4}x^2 - \frac{7}{36}x^3 - \frac{35}{1728}x^4 - \cdots. \quad \blacksquare$$

To conclude this section, we will produce a second solution for Bessel's equation, in a case where the Frobenius method yields only one solution. This will be of use later when we study Bessel functions.

EXAMPLE 4.16 Bessel Function of the Second Kind

Consider Bessel's equation of zero order ($\nu = 0$). From Example 4.12, this is

$$x^2 y'' + xy' + x^2 y = 0.$$

We know from that example that the indicial equation has only one root, $r = 0$. From equation (4.16), with $c_0 = 1$, one Frobenius solution is

$$y_1(x) = \sum_{k=0}^{\infty} (-1)^k \frac{1}{2^{2k}(k!)^2} x^{2k}.$$

Attempt a second, linearly independent solution of the form

$$y_2(x) = y_1(x) \ln(x) + \sum_{k=1}^{\infty} c_k^* x^k.$$

Substitute $y_2(x)$ into the differential equation to get

$$xy_1'' \ln(x) + 2y_1' - \frac{1}{x} y_1 + \sum_{k=2}^{\infty} k(k-1)c_k^* x^{k-1}$$

$$+ y_1' \ln(x) + \frac{1}{x} y_1 + \sum_{k=1}^{\infty} kc_k^* x^{k-1} + xy_1 \ln(x) + \sum_{k=1}^{\infty} c_k^* x^{k+1} = 0.$$

Terms involving $\ln(x)$ and $y_1(x)$ cancel, and we are left with

$$2y_1' + \sum_{k=2}^{\infty} k(k-1)c_k^* x^{k-1} + \sum_{k=1}^{\infty} kc_k^* x^{k-1} + \sum_{k=1}^{\infty} c_k^* x^{k+1} = 0.$$

Since $k(k-1) = k^2 - k$, part of the first summation cancels all terms except the $k = 1$ term in the second summation, and we have

$$2y_1' + \sum_{k=2}^{\infty} k^2 c_k^* x^{k-1} + c_1^* + \sum_{k=1}^{\infty} c_k^* x^{k+1} = 0.$$

Substitute the series for y_1' into this equation to get

$$2 \sum_{k=1}^{\infty} \frac{(-1)^k}{2^{2k-1} k!(k-1)!} x^{2k-1} + \sum_{k=2}^{\infty} k^2 c_k^* x^{k-1} + c_1^* + \sum_{k=1}^{\infty} c_k^* x^{k+1} = 0.$$

Shift indices in the last series to write this equation as

$$\sum_{k=1}^{\infty} \frac{(-1)^k}{2^{2k-2} k!(k-1)!} x^{2k-1} + c_1^* + 4c_2^* x + \sum_{k=3}^{\infty} (k^2 c_k^* + c_{k-2}^*) x^{k-1} = 0. \qquad (4.19)$$

The only constant term on the left side of this equation is c_1^*, which must therefore be zero. The only even powers of x appearing in equation (4.19) are in the rightmost series when k is odd. The coefficients of these powers of x must be zero, hence

$$k^2 c_k^* + c_{k-2}^* = 0 \quad \text{for} \ k = 3, 5, 7, \ldots.$$

But then all odd-indexed coefficients are multiples of c_1^*, which is zero, so

$$c_{2k+1}^* = 0 \quad \text{for } k = 0, 1, 2, \ldots.$$

To determine the even-indexed coefficients, replace k by $2j$ in the second summation of equation (4.19) and k with j in the first summation to get

$$\sum_{j=1}^{\infty} \frac{(-1)^j}{2^{2j-2} j!(j-1)!} x^{2j-1} + 4c_2^* x + \sum_{j=2}^{\infty} (4j^2 c_{2j}^* + c_{2j-2}^*) x^{2j-1} = 0.$$

Now combine terms and write this equation as

$$(4c_2^* - 1)x + \sum_{j=2}^{\infty} \left[\frac{(-1)^j}{2^{2j-2} j!(j-1)!} + 4j^2 c_{2j}^* + c_{2j-2}^* \right] x^{2j-1} = 0.$$

Equate the coefficient of each power of x to zero. We get

$$c_2^* = \frac{1}{4}$$

and the recurrence relation

$$c_{2j}^* = \frac{(-1)^{j+1}}{2^{2j}[j!]^2 j} - \frac{1}{4j^2} c_{2j-2}^* \quad \text{for } j = 2, 3, 4, \ldots.$$

If we write some of these coefficients, a pattern emerges:

$$c_4^* = \frac{-1}{2^2 4^2} \left[1 + \frac{1}{2} \right],$$

$$c_6^* = \frac{1}{2^2 4^2 6^2} \left[1 + \frac{1}{2} + \frac{1}{3} \right],$$

and, in general,

$$c_{2j}^* = \frac{(-1)^{j+1}}{2^2 4^2 \cdots (2j)^2} \left[1 + \frac{1}{2} + \cdots + \frac{1}{j} \right] = \frac{(-1)^{j+1}}{2^{2j}(j!)^2} \emptyset(j),$$

where

$$\emptyset(j) = 1 + \frac{1}{2} + \cdots + \frac{1}{j} \quad \text{for } j = 1, 2, \ldots.$$

We therefore have a second solution of Bessel's equation of order zero:

$$y_2(x) = y_1(x) \ln(x) + \sum_{k=1}^{\infty} \frac{(-1)^{k+1}}{2^{2k}(k!)^2} \emptyset(k) x^{2k}$$

for $x > 0$. This solution is linearly independent from $y_1(x)$ for $x > 0$. ■

When a differential equation with a regular singular point has only one Frobenius series solution expanded about that point, it is tempting to try reduction of order to find a second solution. This is a workable strategy if we can write $y_1(x)$ in closed form. But if $y_1(x)$ is an infinite series, it may be better to substitute the appropriate form of the second solution from Theorem 4.4 and solve for the coefficients.

This concludes our introduction to differential equations. We will return to systems of differential equations and quantitative methods for obtaining information about solutions when we have developed the necessary machinery from linear algebra.

SECTION 4.4   PROBLEMS

In each of Problems 1 through 14, (a) find the indicial equation, (b) determine the appropriate form of each of two linearly independent solutions, and (c) find the first five terms of each of two linearly independent solutions. In Problems 15 through 20, find only the form that two linearly independent solutions should take.

1. $xy'' + (1 - x)y' + y = 0$

2. $xy'' - 2xy' + 2y = 0$

3. $x(x - 1)y'' + 3y' - 2y = 0$

4. $4x^2y'' + 4xy' + (4x^2 - 9)y = 0$

5. $4xy'' + 2y' + y = 0$

6. $4x^2y'' + 4xy' - y = 0$

7. $x^2y'' - 2xy' - (x^2 - 2)y = 0$

8. $xy'' - y' + 2y = 0$

9. $x(2 - x)y'' - 2(x - 1)y' + 2y = 0$

10. $x^2y'' + x(x^3 + 1)y' - y = 0$

11. $x^2y'' + x(x - 2)y' + (x^2 + 2)y = 0$

12. $3x^2y'' + (6x^2 - 7x)y' + 3(1 + x^3)y = 0$

13. $x^2y'' + (x^2 - 3x)y' + (x - 4)y = 0$

14. $x^2y'' - 3xy' + 4(1 + x)y = 0$

15. $25x(1 - x^2)y'' - 20(5x - 2)y' + \left(25x - \dfrac{4}{x}\right)y = 0$

16. $6(3x - 4)(5x + 8)y'' + (2x - 21)\left(3 + \dfrac{16}{x}\right)y' + \left(4 - \dfrac{27}{x^2}\right)y = 0$

17. $12x(4 + 3x)y'' - 2(5x + 7)(7x - 2)y' + 24\left(5 - \dfrac{1}{3x}\right)y = 0$

18. $3x(2x + 3)y'' + 2(6 - 5x)y' + 7\left(2x - \dfrac{8}{x}\right)y = 0$

19. $x(x + 4)y'' - 3(x - 2)y' + 2y = 0$

20. $(3x^3 + x^2)y'' - x(10x + 1)y' + (x^2 + 2)y = 0$

4.5   Appendix on Power Series

4.5.1   Convergence of Power Series

DEFINITION 4.4   *Power Series*
_____

A power series about x_0 is a series of the form

$$\sum_{n=0}^{\infty} a_n(x - x_0)^n.$$

The number x_0 is the center of the series, and the numbers a_n are the coefficients.

Certainly a power series $\sum_{n=0}^{\infty} a_n(x - x_0)^n$ converges for $x = x_0$, because then every term is zero except possibly the constant term a_0. This may be the only point at which it converges. This occurs, for example, with

$$\sum_{n=0}^{\infty} n!x^n,$$

which converges for $x = 0$ but diverges for any $x \neq 0$. If, however, the series $\sum_{n=0}^{\infty} a_n(x - x_0)^n$ converges for any x_1 different from x_0, then it must converge for all x closer to x_0 than x_1. This conclusion is stated as the next theorem.

THEOREM 4.5

Suppose $\sum_{n=0}^{\infty} a_n(x - x_0)^n$ converges for $x = x_1 \neq x_0$. Then the series converges absolutely for all x such that

$$|x - x_0| < |x_1 - x_0|. \quad \blacksquare$$

As a consequence of this result, there are three possibilities for convergence of $\sum_{n=0}^{\infty} a_n(x - x_0)^n$:

1. The series may converge only for $x = x_0$. (This is not a very interesting power series.)
2. The series may converge for all real numbers x. This occurs, for example, with

$$\sum_{n=0}^{\infty} \frac{1}{n!} x^n.$$

3. The third possibility is that there is a positive number r, called the *radius of convergence* of the series, such that $\sum_{n=0}^{\infty} a_n(x - x_0)^n$ converges if $|x - x_0| < r$ and diverges if $|x - x_0| > r$. The series converges for x closer to x_0 than r and diverges for x at distance greater than r from x_0. In this case, r is called the *radius of convergence* of the power series, and $(x_0 - r, x_0 + r)$ is called the *open interval of convergence*. The series may or may not converge at $x_0 + r$ and $x_0 - r$, the endpoints of this interval, depending on the particular series.

Often we combine cases (2) and (3) by saying that $r = \infty$ in case (2). In this event, the interval of convergence is the entire real line. We can include case (1) as well if we say in that case $r = 0$. However, in this case, in which the series converges only for $x = x_0$, we do not speak of an interval of convergence.

Sometimes we can find the radius of convergence of a power series by applying the ratio test for series of constants, which we recall.

THEOREM 4.6   *Ratio Test*

Suppose $b_n \neq 0$ for $n = 0, 1, 2, \ldots$ and that

$$\lim_{n \to \infty} \left| \frac{b_{n+1}}{b_n} \right| = L,$$

in which L may be infinity. Then $\sum_{n=0}^{\infty} b_n$ converges absolutely if $L < 1$ and diverges if $L > 1$. If $L = 1$, then this test allows no conclusion. $\blacksquare$

We can use this test with a power series as follows.

EXAMPLE 4.17

We will find the open interval of convergence of

$$\sum_{n=0}^{\infty} \frac{(-1)^n}{(n + 1)9^n} (x - 2)^{2n}.$$

Put $b_n = (-1)^n (x-2)^{2n} / (n+1)9^n$ and compute the ratio

$$\left| \frac{\frac{(-1)^{n+1}}{(n+2)9^{n+1}} (x-2)^{2n+2}}{\frac{(-1)^n}{(n+1)9^n} (x-2)^{2n}} \right| = \frac{n+1}{9(n+2)} |x-2|^2.$$

Then

$$\lim_{n \to \infty} \left| \frac{b_{n+1}}{b_n} \right| = \lim_{n \to \infty} \frac{n+1}{9(n+2)} |x-2|^2$$

$$= |x-2|^2 \lim_{n \to \infty} \frac{n+1}{9(n+2)} = \frac{1}{9} |x-2|^2.$$

The series converges absolutely if this limit is less than 1, which occurs if $|x-2|^2 < 9$, or $|x-2| < 3$. The power series therefore converges for $-1 < x < 5$. If $x < -1$ or $x > 5$ (that is, $|x-2| > 3$), then the power series diverges.

We can determine convergence or divergence at the endpoints of this interval if we wish. At $x = -1$ the series is

$$\sum_{n=0}^{\infty} \frac{(-1)^n}{(n+1)9^n} (-3)^{2n},$$

which is the same as

$$\sum_{n=0}^{\infty} \frac{(-1)^n}{n+1},$$

a convergent alternating series. At $x = 5$ the series is

$$\sum_{n=0}^{\infty} \frac{(-1)^n}{(n+1)9^n} (3)^{2n},$$

which is the same alternating series because $3^{2n} = 9^n$. Therefore, this power series converges on $[-1, 5]$ and diverges for $x < -1$ and $x > 5$. ■

4.5.2 Algebra and Calculus of Power Series

Assuming that a power series has a nonzero radius of convergence, then within its open interval of convergence the series defines a function. Suppose

$$f(x) = \sum_{n=0}^{\infty} a_n (x-x_0)^n \quad \text{and} \quad g(x) = \sum_{n=0}^{\infty} b_n (x-x_0)^n,$$

with both series convergent at least on some interval $(x_0 - r, x_0 + r)$. We can add and subtract power series as follows:

$$(f+g)(x) = f(x) + g(x) = \sum_{n=0}^{\infty} (a_n + b_n)(x-x_0)^n$$

and

$$(f-g)(x) = f(x) - g(x) = \sum_{n=0}^{\infty} (a_n - b_n)(x-x_0)^n$$

with both of these series on the right converging at least on $(x_0 - r, x_0 + r)$. We can also multiply a power series by a constant:

$$kf(x) = \sum_{n=0}^{\infty} ka_n(x - x_0)^n,$$

again convergent at least on $(x_0 - r, x_0 + r)$.

Power series can also be multiplied according to the rule

$$(fg)(x) = f(x)g(x) = \sum_{n=0}^{\infty} c_n(x - x_0)^k$$

with

$$c_n = \sum_{j=0}^{n} a_n b_{n-j}.$$

A few terms of this power series are

$$f(x)g(x) = a_0 b_0 + (a_0 b_1 + a_1 b_0)(x - x_0) + (a_0 b_2 + a_1 b_1 + a_2 b_0)(x - x_0)^2$$
$$+ (a_0 b_3 + a_1 b_2 + a_2 b_1 + a_3 b_0)(x - x_0)^3 + \cdots .$$

These are exactly the coefficients we get if we multiply

$$[a_0 + a_1(x - x_0) + a_2(x - x_0)^2 + \cdots][b_0 + b_1(x - x_0) + b_2(x - x_0)^2 + \cdots]$$

term-by-term and then collect the coefficient of each power of $x - x_0$.

In any open interval in which the power series converges, we can differentiate the series term-by-term to write

$$f'(x) = \sum_{n=1}^{\infty} na_n(x - x_0)^{n-1} = a_1 + 2a_2(x - x_0) + 3a_3(x - x_0)^2 + \cdots . \tag{4.20}$$

This summation begins at $n = 1$ because the derivative of the constant term a_0 in $\sum_{n=0}^{\infty} a_n(x - x_0)^n$ is zero. This new power series for $f'(x)$ has the same radius of convergence as the original series for $f(x)$. We can then continue to differentiate as many times as we like:

$$f''(x) = \sum_{n=2}^{\infty} n(n - 1)a_n(x - x_0)^{n-2}, \tag{4.21}$$

$$f^{(3)}(x) = \sum_{n=3}^{\infty} n(n - 1)(n - 2)a_n(x - x_0)^{n-3}, \tag{4.22}$$

and so on. The kth derivative is

$$f^{(k)}(x) = \sum_{n=k}^{\infty} n(n - 1)(n - 2) \cdots (n - k + 1)a_n(x - x_0)^{n-k}. \tag{4.23}$$

Each of these series has the same radius of convergence as the series for $f(x)$.

Within the open interval of convergence, we can also integrate a power series term-by-term:

$$\int f(x) \, dx = \sum_{n=0}^{\infty} a_n \int (x - x_0)^n \, dx = \sum_{n=0}^{\infty} \frac{a_n}{n + 1}(x - x_0)^{n+1} + c.$$

4.5.3   Taylor and Maclaurin Expansions

Instead of defining a function by a convergent power series, suppose we begin with a function and seek a power series representation about some point x_0. Suppose there is such an expansion, say

$$f(x) = \sum_{n=0}^{\infty} a_n (x - x_0)^n.$$

in some interval $(x_0 - r, x_0 + r)$. We can compute the coefficients in terms of f and its derivatives at x_0 as follows. First,

$$f(x_0) = a_0$$

because all terms except the constant term vanish if $x = x_0$. Next, from equation (4.20),

$$f'(x_0) = a_1.$$

From equation (4.21),

$$f''(x_0) = 2a_2;$$

from equation (4.22),

$$f^{(3)}(x_0) = 3 \cdot 2a_3;$$

and from equation (4.23),

$$f^{(k)}(x_0) = k(k-1) \cdots (2)(1)a_k.$$

In general,

$$f^{(k)}(x_0) = k!a_k,$$

hence

$$a_k = \frac{1}{k!} f^{(k)}(x_0).$$

This number is called the *k*th *Taylor coefficient* of f at x_0, and the series

$$\sum_{n=0}^{\infty} \frac{1}{n!} f^{(n)}(x_0)(x - x_0)^n$$

is the *Taylor series* for $f(x)$ about x_0. Our argument thus far has shown that if $f(x)$ has a power series expansion about x_0, then this series must be the Taylor series, having the Taylor coefficients. As a corollary to this, if two power series are equal in some open interval, then their coefficients must be equal. Suppose, in some $(x_0 - r, x_0 + r)$,

$$\sum_{n=0}^{\infty} a_n (x - x_0)^n = \sum_{n=0}^{\infty} b_n (x - x_0)^n.$$

If we let $f(x)$ be the function defined by these series, then the series must be the Taylor series of f, so

$$a_n = \frac{1}{n!} f^{(n)}(x_0) = b_n.$$

Not every function has a Taylor series expansion about a point. If such an expansion exists, we call the function *analytic* there. Clearly, f must be infinitely differentiable at x_0 to be analytic at x_0. However, this necessary condition is not sufficient. There exist functions that are infinitely differentiable, for example, at 0, but for which the Taylor series does not converge to the function. One such function is $f(x) = e^{-x^2}$. We will not pursue the details of this here.

Many familiar functions are analytic about specific points and have Taylor series expansions (for example, e^x, $\sin(x)$, and $\cos(x)$). If $x_0 = 0$, the Taylor expansion is about 0 and is often called a *Maclaurin expansion*. For example,

$$e^x = \sum_{n=0}^{\infty} \frac{1}{n!} x^n$$

and

$$\sin(x) = \sum_{n=0}^{\infty} \frac{(-1)^n}{(2n+1)!} x^{2n+1}$$

are Maclaurin expansions.

EXAMPLE 4.18

Suppose we want to expand $g(x) = \ln(1 + x)$ in a power series about 1. Here are two ways we can proceed.

First, we can compute the Taylor coefficients directly. Some derivatives of g are

$$g'(x) = \frac{1}{1+x} = (1+x)^{-1}$$

$$g''(x) = -(1+x)^{-2},$$

$$g^{(3)}(x) = 2(1+x)^{-3},$$

$$g^{(4)}(x) = -2 \cdot 3(1+x)^{-4},$$

and so on. The pattern is apparent:

$$g^{(n)}(x) = (-1)^{n+1}(n-1)!(1+x)^{-n} \quad \text{for } n = 1, 2, \ldots.$$

Then

$$g^{(n)}(1) = (-1)^{n+1}(n-1)!2^{-n} \quad \text{for } n = 1, 2, 3, \ldots.$$

For $n = 1, 2, \ldots$, the nth Taylor coefficient of $g(x)$ about 1 is

$$\frac{1}{n!} g^{(n)}(1) = \frac{1}{n!} (-1)^{n-1} \frac{(n-1)!}{2^n} = \frac{(-1)^{n-1}}{n 2^n}.$$

Since $g(1) = \ln(2)$, the Taylor series for $\ln(1 + x)$ about 1 is

$$\ln(2) + \sum_{n=1}^{\infty} \frac{(-1)^{n-1}}{n 2^n} (x-1)^n.$$

It can be shown that this series converges to $\ln(1 + x)$ for $-1 < x < 3$.

Another way to proceed is to begin with the familiar geometric series

$$\frac{1}{1+x} = \sum_{n=0}^{\infty} (-1)^n x^n$$

for $|x| < 1$. Rearrange this to form an expansion about 1:

$$\frac{1}{1+x} = \frac{1}{2 + (x-1)} = \frac{1}{2} \frac{1}{1 + \frac{x-1}{2}}$$

$$= \frac{1}{2} \sum_{n=0}^{\infty} (-1)^n \left(\frac{x-1}{2}\right)^n = \sum_{n=0}^{\infty} \frac{(-1)^n}{2^{n+1}} (x-1)^n$$

for $|x - 1| < 2$. Now integrate this series term-by-term:

$$\int \frac{1}{1+x} dx = \ln(1+x) = \sum_{n=0}^{\infty} \frac{(-1)^n}{2^{n+1}} \int (x-1)^n + c$$

$$= \sum_{n=0}^{\infty} \frac{(-1)^n}{(n+1)2^{n+1}} (x-1)^{n+1} + c.$$

To evaluate the constant of integration, put $x = 1$ into this expansion to get $\ln(2) = c$, so

$$\ln(1+x) = \ln(2) + \sum_{n=0}^{\infty} \frac{(-1)^n}{(n+1)2^{n+1}} (x-1)^{n+1}.$$

Finally, recognize that

$$\sum_{n=1}^{\infty} \frac{(-1)^{n-1}}{n2^n} (x-1)^n = \sum_{n=0}^{\infty} \frac{(-1)^n}{(n+1)2^{n+1}} (x-1)^{n+1}$$

since both series equal

$$\frac{1}{2}(x-1) - \frac{1}{8}(x-1)^2 + \frac{1}{24}(x-1)^3 - \cdots . \quad \blacksquare$$

4.5.4 Shifting Indices

At the end of the last example we had two series that looked different, $\sum_{n=1}^{\infty}[(-1)^{n-1}/n2^n](x-1)^n$ and $\sum_{n=0}^{\infty}[(-1)^n/(n+1)2^{n+1}](x-1)^{n+1}$, but were in fact identical, having the same terms. If we put $n = k$ into the series on the left and $n = k - 1$ into the series on the right, we get exactly the same terms. This seems reasonable, since the first series starts at $n = 1$ and the second at $n = 0$.

Another way of looking at this is that the summation index is a dummy variable. Any convenient letter can be used for it, and the index can be changed in a way analogous to a change of variables in an integral. To illustrate, let $m = n - 1$ in the series $\sum_{n=1}^{\infty}[(-1)^{n-1}/n2^n](x-1)^n$. When $n = 1$, $m = 0$. Thus

$$\sum_{n=1}^{\infty} \frac{(-1)^{n-1}}{n2^n} (x-1)^n = \sum_{m=0}^{\infty} \frac{(-1)^m}{(m+1)2^{m+1}} (x-1)^{m+1}.$$

Further, since any letter can be used for the index of summation on the right, we can write

$$\sum_{n=1}^{\infty} \frac{(-1)^{n-1}}{n2^n} (x-1)^n = \sum_{n=0}^{\infty} \frac{(-1)^n}{(n+1)2^{n+1}} (x-1)^{n+1},$$

again reaffirming the previous observation.

What we have just done is shift indices in a summation. This is often of use in using series to solve differential equations. Substitution of a power series into the differential equation and carrying out various differentiations will in general yield series having different powers of $x - x_0$. We would like to write all of the series so that common powers of $x - x_0$ can be factored out and various terms in the equation combined. Here is an example of shifting indices.

EXAMPLE 4.19

Consider the following sum of two series:

$$\sum_{n=0}^{\infty} a_n x^{n+2} + \sum_{n=0}^{\infty} b_n x^n. \tag{4.24}$$

We would like to combine these series under one summation in such a way that we can factor out coefficients of like powers of x. To do this, we will rewrite the first series so that powers x^n, rather than x^{n+2}, occur.

One way to do this is to simply recognize that

$$\sum_{n=0}^{\infty} a_n x^{n+2} = a_0 x^2 + a_1 x^3 + a_2 x^4 + \cdots = \sum_{n=2}^{\infty} a_{n-2} x^n.$$

The key is that the index on the coefficient lags behind the exponent on x by 2. When written out term-by-term, these series are the same, except the form on the right is more suited to the original objective of combining two series. Now write equation (4.24) as

$$\sum_{n=0}^{\infty} a_n x^{n+2} + \sum_{n=0}^{\infty} b_n x^n = \sum_{n=2}^{\infty} a_{n-2} x^n + \sum_{n=0}^{\infty} b_n x^n$$

$$= \sum_{n=2}^{\infty} a_{n-2} x^n + \sum_{n=2}^{\infty} b_n x^n + b_0 + b_1 x$$

$$= \sum_{n=2}^{\infty} (a_{n-2} + b_n) x^n + b_0 + b_1 x. \ \blacksquare$$

In more complicated instances, we may need a more formal process to rewrite a series. This can be done by a change of the summation index, analogous to a change of variables in an integral. For example, consider again $\sum_{n=0}^{\infty} a_n x^{n+2}$. Let $m = n + 2$. Then $m = 2$ when $n = 0$. Further, $n = m - 2$, so $a_n = a_{m-2}$ and $x^{n+2} = x^m$. Then

$$\sum_{n=0}^{\infty} a_n x^{n+2} = \sum_{m=2}^{\infty} a_{m-2} x^m,$$

as we got previously. Finally, if we wish to use the same index of summation in all series, we can recognize that any reasonable letter can be used in place of m in the last summation. In particular,

$$\sum_{m=2}^{\infty} a_{m-2} x^m = \sum_{n=2}^{\infty} a_{n-2} x^n.$$

SECTION 4.5 PROBLEMS

In each of Problems 1 through 14, find the radius of convergence and interval of convergence of the power series.

1. $\displaystyle\sum_{n=0}^{\infty} \frac{(-1)^n}{n+1}(x-4)^n$

2. $\displaystyle\sum_{n=0}^{\infty} \frac{2^n}{n!} x^n$

3. $\displaystyle\sum_{n=0}^{\infty} \frac{1}{n+2}(x+1)^n$

4. $\displaystyle\sum_{n=0}^{\infty} n^2 x^n$

5. $\displaystyle\sum_{n=0}^{\infty} \frac{2n+1}{2n-1} x^n$

6. $\displaystyle\sum_{n=0}^{\infty} n^n x^n$

7. $\displaystyle\sum_{n=0}^{\infty} \left(-\frac{3}{2}\right)^n \left(x - \frac{5}{2}\right)^n$

8. $\displaystyle\sum_{n=0}^{\infty} \left(\frac{n^2 - 3n}{n^2 + 4}\right) x^n$

9. $\displaystyle\sum_{n=0}^{\infty} \left(\frac{n+1}{n}\right)^n x^n$

10. $\displaystyle\sum_{n=0}^{\infty} \frac{3^n}{(2n)!} x^{2n}$

11. $\displaystyle\sum_{n=0}^{\infty} \frac{(-1)^n}{n^2 3^n}(x-2)^n$

12. $\displaystyle\sum_{n=1}^{\infty} \frac{n!}{n^n} x^n$

13. $\displaystyle\sum_{n=2}^{\infty} \frac{\ln(n)}{n} x^n$

14. $\displaystyle\sum_{n=0}^{\infty} \frac{e^n}{n!} x^{n+2}$

In each of Problems 15 through 18, shift indices so that all powers of x appearing in a summation are x^n.

15. $\displaystyle\sum_{n=0}^{\infty} \frac{(-1)^{n+1}}{2n+4} x^{n+1}$

16. $\displaystyle\sum_{n=0}^{\infty} \frac{(n+1)^n}{2^n} x^{n+1}$

17. $\displaystyle\sum_{n=1}^{\infty} \frac{2n+3}{n} x^{n-1}$

18. $\displaystyle\sum_{n=4}^{\infty} \frac{(-1)^{n+1}}{2+n^2} x^{n-3}$

In each of Problems 19 through 25, combine as many terms as possible of the two series under one summation by shifting indices. There are various ways of doing this, depending on the index shift chosen.

19. $\displaystyle\sum_{n=1}^{\infty} 2^n x^{n+1} + \sum_{n=0}^{\infty} (n+1) x^n$

20. $\displaystyle\sum_{n=0}^{\infty} \frac{n!}{2^n} x^{n+3} + \sum_{n=1}^{\infty} \frac{1}{n+1} x^{n-1}$

21. $\displaystyle\sum_{n=1}^{\infty} \frac{n!}{n^2} x^{n-1} + \sum_{n=2}^{\infty} 2^n x^n$

22. $\displaystyle\sum_{n=2}^{\infty} \frac{1}{n} x^n + \sum_{n=1}^{\infty} (-1)^{n+1} x^{n+2}$

23. $\displaystyle\sum_{n=1}^{\infty} \frac{1}{2n} x^n - \sum_{n=3}^{\infty} n^n x^{n-3}$

24. $\displaystyle\sum_{n=1}^{\infty} x^{n+3} + \sum_{n=3}^{\infty} n! x^{n+2}$

25. $\displaystyle\sum_{n=1}^{\infty} (2n-1) x^{n+3} + \sum_{n=0}^{\infty} \frac{1}{n+1} x^n$

26. Recall that $e^x = \sum_{n=0}^{\infty} (1/n!) x^n$ for all real x. Multiply this series with the series for e^y and use the formula for the coefficients in a product series to prove that $e^{x+y} = e^x e^y$. *Hint:* Recall the binomial expansion of $(x+y)^n$.

27. Recall that $\cos(x) = \sum_{n=0}^{\infty}[(-1)^n/(2n)!]x^{2n}$ and $\sin(x) = \sum_{n=0}^{\infty}[(-1)^n/(2n+1)!]x^{2n+1}$. Multiply these series to show that

$$\sin(x)\cos(x) = \tfrac{1}{2}\sin(2x).$$

28. Determine the first four nonzero terms in the Maclaurin series for $\tan(x)$ by using long division to divide the Maclaurin series for $\cos(x)$ into the Maclaurin series for $\sin(x)$.

29. Write the Taylor series for $1/(1 + x)$ about 2. *Hint:* Write

$$\frac{1}{1+x} = \frac{1}{3}\frac{1}{1 + \dfrac{x-2}{3}}$$

and use a geometric series.

PART 2

Vectors and Linear Algebra

CHAPTER 5
Vectors and Vector Spaces

CHAPTER 6
Matrices and Systems of Linear Equations

CHAPTER 7
Determinants

CHAPTER 8
Eigenvalues, Diagonalization, and Special Matrices

Some quantities are completely determined by their magnitude, or "size." This is true of temperature and mass, which are numbers referred to some scale or measurement system. Such quantities are called *scalars*. Length, volume, and distance are other scalars.

By contrast, a vector carries with it a sense of both magnitude and direction. The effect of a push against an object will depend not only on the magnitude or strength of the push but also on the direction in which it is exerted.

This part is concerned with the notation and algebra of vectors and objects called *matrices*. This algebra will be used to solve systems of linear algebraic equations and systems of linear differential equations. It will also give us the machinery needed for the quantitative study of systems of differential equations (Part 3), in which we attempt to determine the behavior and properties of solutions when we cannot write these solutions explicitly or in closed form. In Part 4, vector algebra will be used to develop vector calculus, which extends derivatives and integrals to higher dimensions, with applications to models of physical systems, partial differential equations, and the analysis of complex-valued functions.

CHAPTER 5

Vectors and Vector Spaces

5.1 ## The Algebra and Geometry of Vectors

When dealing with vectors, a real number is often called a *scalar*. The temperature of an object and the grade of a motor oil are scalars.

We want to define the concept of a vector in such a way that the package contains information about both direction and magnitude. One way to do this is to define a vector (in 3-dimensional space) as an ordered triple of real numbers.

DEFINITION 5.1 *Vector*

A vector is an ordered triple (a, b, c), in which a, b, and c are real numbers.

We represent the vector (a, b, c) as an arrow from the origin $(0, 0, 0)$ to the point (a, b, c) in 3-space, as in Figure 5.1. In this way, the direction indicated by the arrow, as viewed from the origin, gives the direction of the vector. The length of the arrow is the magnitude (or norm) of the vector—a longer arrow represents a vector of greater strength. Since the distance from the origin to the point (a, b, c) is $\sqrt{a^2 + b^2 + c^2}$, we will define this number to be the magnitude of the vector (a, b, c).

DEFINITION 5.2 *Norm of a Vector*

The norm, or magnitude, of a vector (a, b, c) is the number $\|(a, b, c)\|$ defined by

$$\|(a, b, c)\| = \sqrt{a^2 + b^2 + c^2}.$$

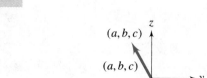

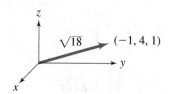

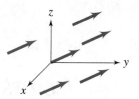

FIGURE 5.1 *The vector (a, b, c) is represented by the arrow from $(0, 0, 0)$ to the point (a, b, c).*

FIGURE 5.2 $\|(-1, 4, 1)\| = \sqrt{18}$.

FIGURE 5.3 *Parallel representations of the same vector.*

For example, the norm of $(-1, 4, 1)$ is $\|(-1, 4, 1)\| = \sqrt{1 + 16 + 1} = \sqrt{18}$. This is the length of the arrow from the origin to the point $(-1, 4, 1)$ (Figure 5.2).

The only vector that is not represented by an arrow from the origin is the zero vector $(0, 0, 0)$, which has zero magnitude and no direction. It is, however, useful to have a zero vector, because various forces in a physical process may cancel each other, resulting in a zero force or vector.

The number a is the *first component* of (a, b, c), b is the *second component*, and c the *third component*. Two vectors are *equal* if and only if each of their respective components is equal:

$$(a, b, c) = (u, v, w)$$

if and only if

$$a = u, \quad b = v, \quad c = w.$$

We will usually denote scalars (real numbers) by letters in regular typeface (a, b, c, A, B, ...) and vectors by letters in boldface ($\mathbf{a}$, $\mathbf{b}$, $\mathbf{c}$, $\mathbf{A}$, $\mathbf{B}$, ...). The zero vector is denoted $\mathbf{O}$.

Although there is a difference between a vector (ordered triple) and an arrow (visual representation of a vector), we often speak of vectors and arrows interchangeably. This is useful in giving geometric interpretations to vector operations. However, any two arrows having the same length and same direction are said to represent the same vector. In Figure 5.3, all the arrows represent the same vector.

We will now develop algebraic operations with vectors and relate them to the norm.

DEFINITION 5.3   *Product of a Scalar and Vector*

The product of a real number α with a vector $\mathbf{F} = (a, b, c)$ is denoted $\alpha\mathbf{F}$ and is defined by

$$\alpha\mathbf{F} = (\alpha a, \alpha b, \alpha c).$$

Thus, a vector is multiplied by a scalar by multiplying each component by the scalar. For example,

$$3(2, -5, 1) = (6, -15, 3) \quad \text{and} \quad -5(-4, 2, 10) = (20, -10, -50).$$

The following relationship between the norm and the product of a scalar with a vector leads to a simple geometric interpretation of this operation.

THEOREM 5.1

Let $\mathbf{F}$ be a vector and α a scalar. Then

1. $\|\alpha\mathbf{F}\| = |\alpha|\|\mathbf{F}\|$.
2. $\|\mathbf{F}\| = 0$ if and only if $\mathbf{F} = \mathbf{O}$.

Proof If $\mathbf{F} = (a, b, c)$, then $\alpha\mathbf{F} = (\alpha a, \alpha b, \alpha c)$, so

$$\|\alpha\mathbf{F}\| = \sqrt{\alpha^2 a^2 + \alpha^2 b^2 + \alpha^2 c^2}$$
$$= |\alpha|\sqrt{a^2 + b^2 + c^2} = |\alpha|\|\mathbf{F}\|.$$

This proves conclusion (1). For (2), first recall that $\mathbf{O} = (0, 0, 0)$, so

$$\|\mathbf{O}\| = \sqrt{0^2 + 0^2 + 0^2} = 0.$$

Conversely, if $\|\mathbf{F}\| = 0$, then $a^2 + b^2 + c^2 = 0$, hence $a = b = c = 0$ and $\mathbf{F} = \mathbf{O}$. ■

Consider this product of a scalar with a vector from a geometric point of view. By (1) of the theorem, the length of $\alpha\mathbf{F}$ is $|\alpha|$ times the length of $\mathbf{F}$. Multiplying by α lengthens the arrow representing $\mathbf{F}$ if $|\alpha| > 1$ and shrinks it to a shorter arrow if $0 < |\alpha| < 1$. Of course, if $\alpha = 0$ then $\alpha\mathbf{F}$ is the zero vector, with zero length. But the algebraic sign of α has an effect as well. If α is positive, then $\alpha\mathbf{F}$ has the same direction as $\mathbf{F}$, while if α is negative, $\alpha\mathbf{F}$ has the opposite direction.

EXAMPLE 5.1

Let $\mathbf{F} = (2, 4, 1)$, as shown in Figure 5.4. $3\mathbf{F} = (6, 12, 3)$ is along the same direction as $\mathbf{F}$, but is represented as an arrow three times longer. But $-3\mathbf{F} = (-6, -12, -3)$, while being three times longer than $\mathbf{F}$, is in the direction opposite that of $\mathbf{F}$ through the origin. And $\frac{1}{2}\mathbf{F} = (1, 2, \frac{1}{2})$ is in the same direction as $\mathbf{F}$, but half as long. ■

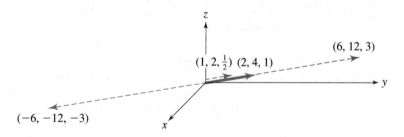

FIGURE 5.4 *Scalar multiples of a vector.*

In particular, the scalar product of -1 with $\mathbf{F} = (a, b, c)$ is the vector $(-a, -b, -c)$, having the same length as $\mathbf{F}$, but the opposite direction. This vector is called "*minus* $\mathbf{F}$," or the *negative of* $\mathbf{F}$, and is denoted $-\mathbf{F}$.

Consistent with the interpretation of multiplication of a vector by a scalar, we define two vectors $\mathbf{F}$ and $\mathbf{G}$ to be *parallel* if each is a nonzero scalar multiple of the other. Of course, if $\mathbf{F} = \alpha\mathbf{G}$ and $\alpha \neq 0$, then $\mathbf{G} = (1/\alpha)\mathbf{F}$. Parallel vectors may differ in length, and even be in opposite directions, but the straight lines through arrows representing these vectors are parallel lines in 3-space.

The algebraic sum of two vectors is defined as follows.

DEFINITION 5.4 Vector Sum

The sum of $\mathbf{F} = (a_1, b_1, c_1)$ and $\mathbf{G} = (a_2, b_2, c_2)$ is the vector

$$\mathbf{F} + \mathbf{G} = (a_1 + a_2, b_1 + b_2, c_1 + c_2).$$

That is, we add vectors by adding respective components. For example,

$$(-4, \pi, 2) + (16, 1, -5) = (12, \pi + 1, -3).$$

If $\mathbf{F} = (a_1, b_1, c_1)$ and $\mathbf{G} = (a_2, b_2, c_2)$, then the sum of $\mathbf{F}$ with $-\mathbf{G}$ is $(a_1 - a_2, b_1 - b_2, c_1 - c_2)$. It is natural to denote this vector as $\mathbf{F} - \mathbf{G}$ and refer to it as "$\mathbf{F}$ minus $\mathbf{G}$." For example, $(-4, \pi, 2)$ minus $(16, 1, -5)$ is

$$(-4, \pi, 2) - (16, 1, -5) = (-20, \pi - 1, 7).$$

We therefore subtract two vectors by subtracting their respective components.

Vector addition and multiplication of a vector by a scalar have the following computational properties.

THEOREM 5.2 Algebra of Vectors

Let $\mathbf{F}$, $\mathbf{G}$, and $\mathbf{H}$ be vectors and let α and β be scalars. Then

1. $\mathbf{F} + \mathbf{G} = \mathbf{G} + \mathbf{F}$.
2. $(\mathbf{F} + \mathbf{G}) + \mathbf{H} = \mathbf{F} + (\mathbf{G} + \mathbf{H})$.
3. $\mathbf{F} + \mathbf{O} = \mathbf{F}$.
4. $\alpha(\mathbf{F} + \mathbf{G}) = \alpha\mathbf{F} + \alpha\mathbf{G}$.
5. $(\alpha\beta)\mathbf{F} = \alpha(\beta\mathbf{F})$.
6. $(\alpha + \beta)\mathbf{F} = \alpha\mathbf{F} + \beta\mathbf{F}$. ∎

Conclusion (1) is the commutative law for vector addition, and (2) is the associative law. Conclusion (3) states that the zero vector behaves with vectors like the number zero does with real numbers, as far as addition is concerned. The theorem is proved by routine calculations, using the properties of real-number arithmetic. For example, to prove (4), write $\mathbf{F} = (a_1, b_1, c_1)$ and $\mathbf{G} = (a_2, b_2, c_2)$. Then

$$
\begin{aligned}
\alpha(\mathbf{F} + \mathbf{G}) &= \alpha(a_1 + a_2, b_1 + b_2, c_1 + c_2) \\
&= (\alpha(a_1 + a_2), \alpha(b_1 + b_2), \alpha(c_1 + c_2)) \\
&= (\alpha a_1 + \alpha a_2, \alpha b_1 + \alpha b_2, \alpha c_1 + \alpha c_2) \\
&= (\alpha a_1, \alpha b_1, \alpha c_1) + (\alpha a_2, \alpha b_2, \alpha c_2) \\
&= \alpha(a_1, b_1, c_1) + \alpha(a_2, b_2, c_2) = \alpha\mathbf{F} + \alpha\mathbf{G}.
\end{aligned}
$$

Vector addition has a simple geometric interpretation. If $\mathbf{F}$ and $\mathbf{G}$ are represented as arrows from the same point P, as in Figure 5.5, then $\mathbf{F} + \mathbf{G}$ is represented as the arrow from P to the opposite vertex of the parallelogram having $\mathbf{F}$ and $\mathbf{G}$ as two incident sides. This is called the *parallelogram law* for vector addition.

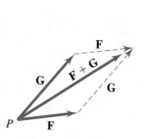

FIGURE 5.5 *Parallelogram law for vector addition.*

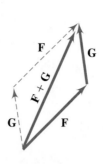

FIGURE 5.6 *Another way of visualizing the parallelogram law.*

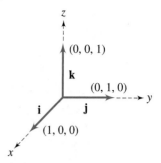

FIGURE 5.7 *Unit vectors along the axes.*

The parallelogram law suggests a strategy for visualizing addition that is sometimes useful. Since two arrows having the same direction and length represent the same vector, we could apply the parallelogram law to form $\mathbf{F} + \mathbf{G}$ as in Figure 5.6, in which the arrow representing $\mathbf{G}$ is drawn from the tip of $\mathbf{F}$, rather than from a common initial point with $\mathbf{F}$. We often do this in visualizing computations with vectors.

Any vector can be written as a sum of scalar multiples of "standard" vectors as follows. Define

$$\mathbf{i} = (1, 0, 0), \quad \mathbf{j} = (0, 1, 0), \quad \mathbf{k} = (0, 0, 1).$$

These are *unit vectors* (length 1) aligned along the three coordinate axes in the positive direction (Figure 5.7). In terms of these vectors,

$$\mathbf{F} = (a, b, c) = a(1, 0, 0) + b(0, 1, 0) + c(0, 0, 1) = a\mathbf{i} + b\mathbf{j} + c\mathbf{k}.$$

This is called the *standard representation* of $\mathbf{F}$. When a component of $\mathbf{F}$ is zero, we usually just omit it in the standard representation. For example,

$$(-3, 0, 1) = -3\mathbf{i} + \mathbf{k}.$$

Figure 5.8 shows two points: P_1 (a_1, b_1, c_1) and P_2 (a_2, b_2, c_2). It will be useful to know the vector represented by the arrow from P_1 to P_2. Let $\mathbf{H}$ be this vector. Denote

$$\mathbf{G} = a_1\mathbf{i} + b_1\mathbf{j} + c_1\mathbf{k} \quad \text{and} \quad \mathbf{F} = a_2\mathbf{i} + b_2\mathbf{j} + c_2\mathbf{k}.$$

By the parallelogram law (Figure 5.9),

$$\mathbf{G} + \mathbf{H} = \mathbf{F}.$$

Hence,

$$\mathbf{H} = \mathbf{F} - \mathbf{G} = (a_2 - a_1)\mathbf{i} + (b_2 - b_1)\mathbf{j} + (c_2 - c_1)\mathbf{k}.$$

For example, the vector represented by the arrow from $(-2, 4, 1)$ to $(14, 5, -7)$ is $16\mathbf{i} + \mathbf{j} - 8\mathbf{k}$. The vector from $(14, 5, -7)$ to $(-2, 4, 1)$ is the negative of this, or $-16\mathbf{i} - \mathbf{j} + 8\mathbf{k}$.

Vector notation and algebra are often useful in solving problems in geometry. This is not our goal here, but the reasoning involved is often useful in solving problems in the sciences and engineering. We will give three examples to demonstrate the efficiency of thinking in terms of vectors.

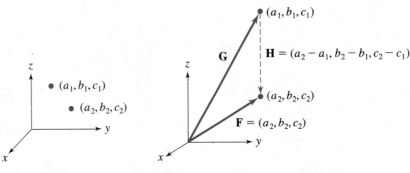

FIGURE 5.8

FIGURE 5.9 *The arrow from (a_1, b_1, c_1) to (a_2, b_2, c_2) is $(a_2 - a_1)\mathbf{i} + (b_2 - b_1)\mathbf{j} + (c_2 - c_1)\mathbf{k}$.*

EXAMPLE 5.2

Suppose we want the equation of the line L through the points $(1, -2, 4)$ and $(6, 2, -3)$.

This problem is more subtle in 3-space than in the plane because in three dimensions there is no point–slope formula. Reason as follows. Let (x, y, z) be any point on L. Then (Figure 5.10), the vector represented by the arrow from $(1, -2, 4)$ to (x, y, z) must be parallel to the vector from $(1, -2, 4)$ to $(6, 2, -3)$, because arrows representing these vectors are both along L. This means that $(x - 1)\mathbf{i} + (y + 2)\mathbf{j} + (z - 4)\mathbf{k}$ is parallel to $5\mathbf{i} + 4\mathbf{j} - 7\mathbf{k}$. Then, for some scalar t,

$$(x - 1)\mathbf{i} + (y + 2)\mathbf{j} + (z - 4)\mathbf{k} = t[5\mathbf{i} + 4\mathbf{j} - 7\mathbf{k}].$$

But then the respective components of these vectors must be equal:

$$x - 1 = 5t, \quad y + 2 = 4t, \quad z - 4 = -7t.$$

Then,

$$x = 1 + 5t, \quad y = -2 + 4t, \quad z = 4 - 7t. \tag{5.1}$$

A point is on L if and only if its coordinates are $(1 + 5t, -2 + 4t, 4 - 7t)$ for some real number t (Figure 5.11). Equations 5.1 are parametric equations of the line, with t, which can be assigned any real value, as parameter. When $t = 0$, we get $(1, -2, 4)$, and when $t = 1$ we get $(6, 2, -3)$.

We can also write the equation of this line in what is called *normal form*. By eliminating t, this form is

$$\frac{x - 1}{5} = \frac{y + 2}{4} = \frac{z - 4}{-7}.$$

We may also envision the line as swept out by the arrow pivoted at the origin and extending to the point $(1 + 5t, -2 + 4t, 4 - 7t)$ as t varies over the real numbers. ■

Some care must be taken in writing the normal form of a straight line. For example, the line through $(2, -1, 6)$ and $(-4, -1, 2)$ has parametric equations

$$x = 2 - 6t, \quad y = -1, \quad z = 6 - 4t.$$

If we eliminate t, we get

$$\frac{x - 2}{-6} = \frac{z - 6}{-4}, \quad y = -1.$$

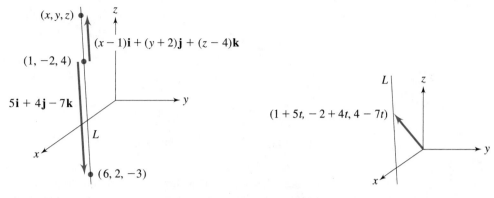

FIGURE 5.10   FIGURE 5.11

Every point on the line has the second coordinate -1, and this is independent of t. This information must not be omitted from the equations of the line. If we omit $y = -1$ and write just

$$\frac{x-2}{-6} = \frac{z-6}{-4},$$

then we have the equation of a plane, not a line.

EXAMPLE 5.3

Suppose we want a vector $\mathbf{F}$ in the x, y plane, making an angle of $\pi/7$ with the positive x axis and having magnitude 19.

By "find a vector" we mean determine its components. Let $\mathbf{F} = a\mathbf{i} + b\mathbf{j}$. From the right triangle in Figure 5.12,

$$\cos\left(\frac{\pi}{7}\right) = \frac{a}{19} \quad \text{and} \quad \sin\left(\frac{\pi}{7}\right) = \frac{b}{19}.$$

Then,

$$\mathbf{F} = 10\cos\left(\frac{\pi}{7}\right)\mathbf{i} + 19\sin\left(\frac{\pi}{7}\right)\mathbf{j}. \quad \blacksquare$$

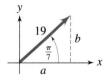

FIGURE 5.12

EXAMPLE 5.4

We will prove that the line segments formed by connecting successive midpoints of the sides of a quadrilateral form a parallelogram. Again, our overall objective is not to prove theorems of geometry, but this argument is good practice in the use of vectors.

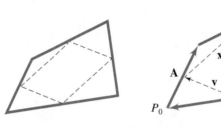

FIGURE 5.13   FIGURE 5.14

Figure 5.13 illustrates what we want to show. Draw the quadrilateral again, with arrows (vectors) **A**, **B**, **C**, and **D** as sides. The vectors **x**, **y**, **u**, and **v** drawn with dashed lines connect the midpoints of successive sides (Figure 5.14). We want to show that **x** and **u** are parallel and of the same length and that **y** and **v** are also parallel and of the same length.

From the parallelogram law for vector addition, and the definitions of **x** and **u**,

$$\mathbf{x} = \tfrac{1}{2}\mathbf{A} + \tfrac{1}{2}\mathbf{B}$$

and

$$\mathbf{u} = \tfrac{1}{2}\mathbf{C} + \tfrac{1}{2}\mathbf{D}.$$

But also by the parallelogram law, $\mathbf{A} + \mathbf{B}$ is the arrow from P_0 to P_1, while $\mathbf{C} + \mathbf{D}$ is the arrow from P_1 to P_0. These arrows have the same length, and opposite directions. This means that

$$\mathbf{A} + \mathbf{B} = -(\mathbf{C} + \mathbf{D}).$$

But then $\mathbf{x} = -\mathbf{u}$, so these vectors are parallel and of the same length (just opposite in direction).

A similar argument shows that **y** and **v** are also parallel and of the same length, completing the proof. ■

SECTION 5.1   PROBLEMS

In each of Problems 1 through 5, compute $\mathbf{F} + \mathbf{G}$, $\mathbf{F} - \mathbf{G}$, $\|\mathbf{F}\|$, $\|\mathbf{G}\|$, $2\mathbf{F}$, and $3\mathbf{G}$.

1. $\mathbf{F} = 2\mathbf{i} - 3\mathbf{j} + 5\mathbf{k}, \mathbf{G} = \sqrt{2}\mathbf{i} + 6\mathbf{j} - 5\mathbf{k}$

2. $\mathbf{F} = \mathbf{i} - 3\mathbf{k}, \mathbf{G} = 4\mathbf{j}$

3. $\mathbf{F} = 2\mathbf{i} - 5\mathbf{j}, \mathbf{G} = \mathbf{i} + 5\mathbf{j} - \mathbf{k}$

4. $\mathbf{F} = \sqrt{2}\mathbf{i} + \mathbf{j} - 6\mathbf{k}, \mathbf{G} = 8\mathbf{i} + 2\mathbf{k}$

5. $\mathbf{F} = \mathbf{i} + \mathbf{j} + \mathbf{k}, \mathbf{G} = 2\mathbf{i} - 2\mathbf{j} + 2\mathbf{k}$

In each of Problems 6 through 10, calculate $\mathbf{F} + \mathbf{G}$ and $\mathbf{F} - \mathbf{G}$ by representing the vectors as arrows and using the parallelogram law.

6. $\mathbf{F} = \mathbf{i}, \mathbf{G} = 6\mathbf{j}$

7. $\mathbf{F} = 2\mathbf{i} - \mathbf{j}, \mathbf{G} = \mathbf{i} - \mathbf{j}$

8. $\mathbf{F} = -3\mathbf{i} + \mathbf{j}, \mathbf{G} = 4\mathbf{j}$

9. $\mathbf{F} = \mathbf{i} - 2\mathbf{j}, \mathbf{G} = \mathbf{i} - 3\mathbf{j}$

10. $\mathbf{F} = -\mathbf{i} + 4\mathbf{j}, \mathbf{G} = -2\mathbf{i} - 2\mathbf{j}$

In each of Problems 11 through 15, determine $\alpha \mathbf{F}$ and represent $\mathbf{F}$ and $\alpha \mathbf{F}$ as arrows.

11. $\mathbf{F} = \mathbf{i} + \mathbf{j}, \alpha = -\tfrac{1}{2}$

12. $\mathbf{F} = 6\mathbf{i} - 2\mathbf{j}, \alpha = 2$

13. $\mathbf{F} = -3\mathbf{j}, \alpha = -4$

14. $\mathbf{F} = 6\mathbf{i} - 6\mathbf{j}, \alpha = \tfrac{1}{2}$

15. $\mathbf{F} = -3\mathbf{i} + 2\mathbf{j}, \alpha = 3$

In each of Problems 16 through 25, find parametric equations of the straight line containing the given points. Also find the normal form of this line.

16. $(1, 0, 4), (2, 1, 1)$

17. $(3, 0, 0), (-3, 1, 0)$

18. $(2, 1, 1), (2, 1, -2)$

19. $(0, 1, 3), (0, 0, 1)$

20. $(1, 0, -4), (-2, -2, 5)$

21. $(2, -3, 6), (-1, 6, 4)$

22. $(-4, -2, 5), (1, 1, -5)$

23. $(3, 3, -5), (2, -6, 1)$

24. $(0, -3, 0), (1, -1, 5)$

25. $(4, -8, 1), (-1, 0, 0)$

In each of Problems 26 through 30, find a vector $\mathbf{F}$ in the x, y plane having the given length and making the angle (given in radians) with the positive x axis. Represent the vector as an arrow in the plane.

26. $\sqrt{5}, \pi/4$

27. $6, \pi/3$

28. $5, 3\pi/5$

29. $15, 7\pi/4$

30. $25, 3\pi/2$

31. Let $P_1, P_2, \ldots, P_n$ be distinct points in 3-space, with $n \geq 3$. Let $\mathbf{F}_i$ be the vector represented by the arrow from P_i to P_{i+1} for $i = 1, 2, \ldots, n-1$ and let $\mathbf{F}_n$ be the vector represented by the arrow from P_n to P_1. Prove that $\mathbf{F}_1 + \mathbf{F}_2 + \cdots + \mathbf{F}_n = \mathbf{O}$.

32. Let $\mathbf{F}$ be any nonzero vector. Determine a scalar t such that $\|t\mathbf{F}\| = 1$.

33. Use vectors to prove that the altitudes of any triangle intersect in a single point. (Recall that an altitude is a line from a vertex, perpendicular to the opposite side of the triangle).

34. Let P, Q, R be any three points, not all on a line. Let $\mathbf{A}$ be the vector from P to Q and $\mathbf{B}$ the vector from P to R. Determine the vector from P to the midpoint of the segment $\overline{RQ}$. The answer should be a sum of scalar multiples of $\mathbf{A}$ and $\mathbf{B}$.

35. Use vectors to prove that the lines drawn from a selected vertex of a parallelogram to the midpoints of the two opposite sides trisect a diagonal of the parallelogram.

36. Let $\mathbf{F}$ and $\mathbf{G}$ be nonzero vectors that are not parallel. Suppose $\alpha\mathbf{F} + \beta\mathbf{G} = \mathbf{O}$ for some scalars α and β. Prove that $\alpha = \beta = 0$.

5.2 The Dot Product

Throughout this section, let $\mathbf{F} = a_1\mathbf{i} + b_1\mathbf{j} + c_1\mathbf{k}$ and $\mathbf{G} = a_2\mathbf{i} + b_2\mathbf{j} + c_2\mathbf{k}$.

DEFINITION 5.5   *Dot Product*

The dot product of $\mathbf{F}$ and $\mathbf{G}$ is the number $\mathbf{F} \cdot \mathbf{G}$ defined by

$$\mathbf{F} \cdot \mathbf{G} = a_1a_2 + b_1b_2 + c_1c_2.$$

For example,

$$(\sqrt{3}\mathbf{i} + 4\mathbf{j} - \pi\mathbf{k}) \cdot (-2\mathbf{i} + 6\mathbf{j} + 3\mathbf{k}) = -2\sqrt{3} + 24 - 3\pi.$$

Sometimes the dot product is referred to as a scalar product, since the dot product of two vectors is a scalar (real number). This must not be confused with the product of a vector with a scalar. Here are some rules for operating with the dot product.

THEOREM 5.3   *Properties of the Dot Product*

Let $\mathbf{F}$, $\mathbf{G}$, and $\mathbf{H}$ be vectors, and α a scalar. Then

1. $\mathbf{F} \cdot \mathbf{G} = \mathbf{G} \cdot \mathbf{F}$.
2. $(\mathbf{F} + \mathbf{G}) \cdot \mathbf{H} = \mathbf{F} \cdot \mathbf{H} + \mathbf{G} \cdot \mathbf{H}$.
3. $\alpha(\mathbf{F} \cdot \mathbf{G}) = (\alpha \mathbf{F}) \cdot \mathbf{G} = \mathbf{F} \cdot (\alpha \mathbf{G})$.
4. $\mathbf{F} \cdot \mathbf{F} = \|\mathbf{F}\|^2$.
5. $\mathbf{F} \cdot \mathbf{F} = 0$ if and only if $\mathbf{F} = \mathbf{O}$. ∎

Conclusion (1) is the commutativity of the dot product (we can perform the operation in either order), and (2) is a distributive law. Conclusion (3) states that a constant factors through a dot product. Conclusion (4) is very useful in some kinds of calculations, as we will see shortly. A proof of the theorem involves routine calculations, two of which we will illustrate.

Proof   For (3), write

$$\alpha(\mathbf{F} \cdot \mathbf{G}) = \alpha(a_1 a_2 + b_1 b_2 + c_1 c_2) = (\alpha a_1)a_2 + (\alpha b_1)b_2 + (\alpha c_1)c_2$$

$$= (\alpha \mathbf{F}) \cdot \mathbf{G} = a_1(\alpha a_2) + b_1(\alpha b_2) + c_1(\alpha c_2) = \mathbf{F} \cdot (\alpha \mathbf{G}).$$

For (4), we have

$$\mathbf{F} \cdot \mathbf{F} = (a_1 \mathbf{i} + b_1 \mathbf{j} + c_1 \mathbf{k}) \cdot (a_1 \mathbf{i} + b_1 \mathbf{j} + c_1 \mathbf{k})$$

$$= a_1^2 + b_1^2 + c_1^2 = \|\mathbf{F}\|^2. \ ∎$$

Using conclusion (4) of the theorem, we can derive a relationship we will use frequently.

LEMMA 5.1

Let $\mathbf{F}$ and $\mathbf{G}$ be vectors, and let α and β be scalars. Then

$$\|\alpha \mathbf{F} + \beta \mathbf{G}\|^2 = \alpha^2 \|\mathbf{F}\|^2 + 2\alpha\beta \mathbf{F} \cdot \mathbf{G} + \beta^2 \|\mathbf{G}\|.$$

Proof   By using Theorem 5.3, we have

$$\|\alpha \mathbf{F} + \beta \mathbf{G}\|^2 = (\alpha \mathbf{F} + \beta \mathbf{G}) \cdot (\alpha \mathbf{F} + \beta \mathbf{G})$$

$$= \alpha^2 \mathbf{F} \cdot \mathbf{F} + \alpha\beta \mathbf{F} \cdot \mathbf{G} + \alpha\beta \mathbf{G} \cdot \mathbf{F} + \beta^2 \mathbf{G} \cdot \mathbf{G}$$

$$= \alpha^2 \mathbf{F} \cdot \mathbf{F} + 2\alpha\beta \mathbf{F} \cdot \mathbf{G} + \beta^2 \mathbf{G} \cdot \mathbf{G}$$

$$= \alpha^2 \|\mathbf{F}\|^2 + 2\alpha\beta \mathbf{F} \cdot \mathbf{G} + \beta^2 \|\mathbf{G}\|^2. \ ∎$$

The dot product can be used to determine the angle between vectors. Represent $\mathbf{F}$ and $\mathbf{G}$ as arrows from a common point, as in Figure 5.15. Let θ be the angle between $\mathbf{F}$ and $\mathbf{G}$. The arrow from the tip of $\mathbf{F}$ to the tip of $\mathbf{G}$ represents $\mathbf{G} - \mathbf{F}$, and these three vectors form the sides of a triangle. Now recall the law of cosines, which states, for the triangle of Figure 5.16, that

$$a^2 + b^2 - 2ab\cos(\theta) = c^2. \tag{5.2}$$

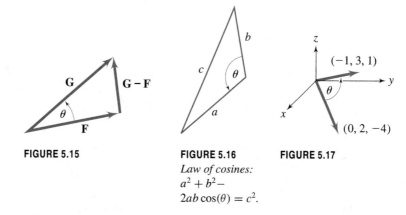

FIGURE 5.15

FIGURE 5.16
Law of cosines:
$a^2 + b^2 - 2ab\cos(\theta) = c^2.$

FIGURE 5.17

Apply this to the vector triangle of Figure 5.15, with sides of length $a = \|\mathbf{G}\|$, $b = \|\mathbf{F}\|$, and $c = \|\mathbf{G} - \mathbf{F}\|$. By using Lemma 5.1 with $\alpha = -1$ and $\beta = 1$, equation (5.2) becomes

$$\|\mathbf{G}\|^2 + \|\mathbf{F}\|^2 - 2\|\mathbf{G}\|\|\mathbf{F}\|\cos(\theta) = \|\mathbf{G} - \mathbf{F}\|^2$$

$$= \|\mathbf{G}\|^2 + \|\mathbf{F}\|^2 - 2\mathbf{G}\cdot\mathbf{F}.$$

Then,

$$\mathbf{F}\cdot\mathbf{G} = \|\mathbf{F}\|\|\mathbf{G}\|\cos(\theta).$$

Assuming that neither $\mathbf{F}$ nor $\mathbf{G}$ is the zero vector, then

$$\cos(\theta) = \frac{\mathbf{F}\cdot\mathbf{G}}{\|\mathbf{F}\|\|\mathbf{G}\|}. \tag{5.3}$$

This provides a simple way of computing the cosine of the angle between two arrows representing vectors. Since vectors can be drawn along straight lines, this also lets us calculate the angle between two intersecting lines.

EXAMPLE 5.5

Let $\mathbf{F} = -\mathbf{i} + 3\mathbf{j} + \mathbf{k}$ and $\mathbf{G} = 2\mathbf{j} - 4\mathbf{k}$. The cosine of the angle between these vectors (Figure 5.17) is

$$\cos(\theta) = \frac{(-\mathbf{i} + 3\mathbf{j} + \mathbf{k})\cdot(2\mathbf{j} - 4\mathbf{k})}{\|-\mathbf{i} + 3\mathbf{j} + \mathbf{k}\|\|2\mathbf{j} - 4\mathbf{k}\|}$$

$$= \frac{(-1)(0) + (3)(2) + (1)(-4)}{\sqrt{1^2 + 3^2 + 1^2}\sqrt{2^2 + 4^2}} = \frac{2}{\sqrt{220}}.$$

Then,

$$\theta = \arccos\left(\frac{2}{\sqrt{220}}\right),$$

which is that unique number in $[0, \pi]$ whose cosine is $2/\sqrt{220}$. θ is approximately 1.436 radians. ∎

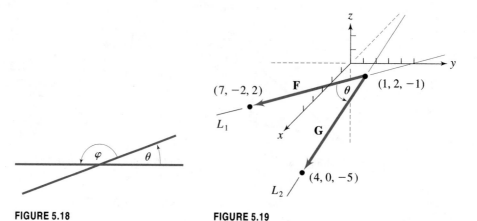

FIGURE 5.18   FIGURE 5.19

EXAMPLE 5.6

Lines L_1 and L_2 are given, respectively, by the parametric equations

$$x = 1 + 6t, \quad y = 2 - 4t, \quad z = -1 + 3t$$

and

$$x = 4 - 3p, \quad y = 2p, \quad z = -5 + 4p$$

in which the parameters t and p take on all real values. We want the angle between these lines at their point of intersection, which is $(1, 2, -1)$ (on L_1 for $t = 0$ and on L_2 for $p = 1$).

Of course, two intersecting lines have two angles between them (Figure 5.18). However, the sum of these angles is π, so either angle determines the other.

The strategy for solving this problem is to find a vector along each line, then find the angle between these vectors. For a vector $\mathbf{F}$ along L_1, choose any two points on this line, say $(1, 2, -1)$ and, with $t = 1$, $(7, -2, 2)$. The vector from the first to the second point is

$$\mathbf{F} = (7 - 1)\mathbf{i} + (-2 - 2)\mathbf{j} + (2 - (-1))\mathbf{k} = 6\mathbf{i} - 4\mathbf{j} + 3\mathbf{k}.$$

Two points on L_2 are $(1, 2, -1)$ and, with $p = 0$, $(4, 0, -5)$. The vector $\mathbf{G}$ from the first to the second of these points is

$$\mathbf{G} = (4 - 1)\mathbf{i} + (0 - 2)\mathbf{j} + (-5 - (-1))\mathbf{k} = 3\mathbf{i} - 2\mathbf{j} - 4\mathbf{k}.$$

These vectors are shown in Figure 5.19. The cosine of the angle between $\mathbf{F}$ and $\mathbf{G}$ is

$$\cos(\theta) = \frac{\mathbf{F} \cdot \mathbf{G}}{\|\mathbf{F}\|\|\mathbf{G}\|} = \frac{6(3) - 4(-2) + 3(-4)}{\sqrt{36 + 16 + 9}\sqrt{9 + 4 + 16}} = \frac{14}{\sqrt{1769}}.$$

One angle between the lines is $\theta = \arccos(14/\sqrt{1769})$, approximately 1.23 radians. ∎

If we had used $-\mathbf{G}$ in place of $\mathbf{G}$ in this calculation, we would have gotten $\theta = \arccos(-14/\sqrt{1769})$, or about 1.91 radians. This is the supplement of the angle found in the example.

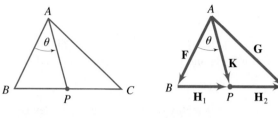

FIGURE 5.20   FIGURE 5.21

EXAMPLE 5.7

The points A $(1, -2, 1)$, B $(0, 1, 6)$, and C $(-3, 4, -2)$ form the vertices of a triangle. Suppose we want the angle between the line $\overline{AB}$ and the line from A to the midpoint of $\overline{BC}$. This line is a median of the triangle and is shown in Figure 5.20

Visualize the sides of the triangle as vectors, as in Figure 5.21. If P is the midpoint of $\overline{BC}$, then $\mathbf{H}_1 = \mathbf{H}_2$ because both vectors have the same direction and length. From the coordinates of the vertices, calculate

$$\mathbf{F} = -\mathbf{i} + 3\mathbf{j} + 5\mathbf{k} \quad \text{and} \quad \mathbf{G} = -4\mathbf{i} + 6\mathbf{j} - 3\mathbf{k}.$$

We want the angle between $\mathbf{F}$ and $\mathbf{K}$, so we need $\mathbf{K}$. By the parallelogram law,

$$\mathbf{F} + \mathbf{H}_1 = \mathbf{K} \quad \text{and} \quad \mathbf{K} + \mathbf{H}_2 = \mathbf{G}.$$

Since $\mathbf{H}_1 = \mathbf{H}_2$, these equations imply that

$$\mathbf{K} = \mathbf{F} + \mathbf{H}_1 = \mathbf{F} + (\mathbf{G} - \mathbf{K}).$$

Therefore,

$$\mathbf{K} = \tfrac{1}{2}(\mathbf{F} + \mathbf{G}) = -\tfrac{5}{2}\mathbf{i} + \tfrac{9}{2}\mathbf{j} + \mathbf{k}.$$

Now the cosine of the angle we want is

$$\cos(\theta) = \frac{\mathbf{F} \cdot \mathbf{K}}{\|\mathbf{F}\|\|\mathbf{K}\|} = \frac{42}{\sqrt{35}\sqrt{110}} = \frac{42}{\sqrt{3850}}.$$

θ is approximately 0.83 radians. ∎

The arrows representing two nonzero vectors $\mathbf{F}$ and $\mathbf{G}$ are perpendicular exactly when the cosine of the angle between them is zero, and by equation (5.3) this occurs when $\mathbf{F} \cdot \mathbf{G} = 0$. This suggests we use this condition to define orthogonality (perpendicularity) of vectors. If we agree to the convention that the zero vector is orthogonal to every vector, then this dot product condition allows a general definition without requiring that the vectors be nonzero.

DEFINITION 5.6 *Orthogonal Vectors*

Vectors $\mathbf{F}$ and $\mathbf{G}$ are orthogonal if and only if $\mathbf{F} \cdot \mathbf{G} = 0$.

EXAMPLE 5.8

Let $\mathbf{F} = -4\mathbf{i} + \mathbf{j} + 2\mathbf{k}$, $\mathbf{G} = 2\mathbf{i} + 4\mathbf{k}$, and $\mathbf{H} = 6\mathbf{i} - \mathbf{j} - 2\mathbf{k}$. Then $\mathbf{F} \cdot \mathbf{G} = 0$, so $\mathbf{F}$ and $\mathbf{G}$ are orthogonal. But $\mathbf{F} \cdot \mathbf{H}$ and $\mathbf{G} \cdot \mathbf{H}$ are nonzero, so $\mathbf{F}$ and $\mathbf{H}$ are not orthogonal, and $\mathbf{G}$ and $\mathbf{H}$ are not orthogonal. ■

Sometimes orthogonality of vectors is a useful device for dealing with lines and planes in 3-dimensional space.

EXAMPLE 5.9

Two lines are given parametrically by

$$L_1: x = 2 - 4t, \quad y = 6 + t, \quad z = 3t$$

and

$$L_2: x = -2 + p, \quad y = 7 + 2p, \quad z = 3 - 4p.$$

We want to know whether these lines are perpendicular. (It does not matter whether the lines intersect).

The idea is to form a vector along each line and test these vectors for orthogonality. For a vector along L_1, choose two points on this line, say $(2, 6, 0)$ when $t = 0$ and $(-2, 7, 3)$ when $t = 1$. Then $\mathbf{F} = -4\mathbf{i} + \mathbf{j} + 3\mathbf{k}$ is along L_1. Two points on L_2 are $(-2, 7, 3)$ when $p = 0$ and $(-1, 9, -1)$ when $p = 1$. Then $\mathbf{G} = \mathbf{i} + 2\mathbf{j} - 4\mathbf{k}$ is along L_2. Since $\mathbf{F} \cdot \mathbf{G} = -14 \neq 0$, these vectors, hence these lines, are not orthogonal. ■

EXAMPLE 5.10

Suppose we want the equation of a plane Π containing the point $(-6, 1, 1)$ and perpendicular to the vector $\mathbf{N} = -2\mathbf{i} + 4\mathbf{j} + \mathbf{k}$.

A strategy to find such an equation is suggested by Figure 5.22. A point (x, y, z) is on Π if and only if the vector from $(-6, 1, 1)$ to (x, y, z) is in Π and therefore is orthogonal to $\mathbf{N}$. This means that

$$((x + 6)\mathbf{i} + (y - 1)\mathbf{j} + (z - 1)\mathbf{k}) \cdot \mathbf{N} = 0.$$

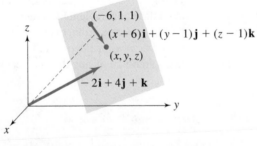

FIGURE 5.22

Carrying out this dot product, we get the equation

$$-2(x+6) + 4(y-1) + (z-1) = 0,$$

or

$$-2x + 4y + z = 17.$$

This is the equation of the plane. Of course the given point $(-6, 1, 1)$ satisfies this equation. ∎

We will conclude this section with the important Cauchy–Schwarz inequality, which states that the dot product of two vectors cannot be greater in absolute value than the product of the lengths of the vectors.

THEOREM 5.4 *Cauchy–Schwarz Inequality*

Let **F** and **G** be vectors. Then

$$|\mathbf{F} \cdot \mathbf{G}| \leq \|\mathbf{F}\|\|\mathbf{G}\|.$$

Proof If either vector is the zero vector, then both sides of the proposed inequality are zero. Thus, suppose neither vector is the zero vector. In this event,

$$\cos(\theta) = \frac{\mathbf{F} \cdot \mathbf{G}}{\|\mathbf{F}\|\|\mathbf{G}\|},$$

where θ is the angle between **F** and **G**. But then

$$-1 \leq \frac{\mathbf{F} \cdot \mathbf{G}}{\|\mathbf{F}\|\|\mathbf{G}\|} \leq 1,$$

so

$$-\|\mathbf{F}\|\|\mathbf{G}\| \leq \mathbf{F} \cdot \mathbf{G} \leq \|\mathbf{F}\|\|\mathbf{G}\|,$$

which is equivalent to the Cauchy–Schwarz inequality. ∎

SECTION 5.2   PROBLEMS

In each of Problems 1 through 6, compute the dot product of the vectors and the cosine of the angle between them. Also determine if they are orthogonal and verify the Cauchy–Schwarz inequality for these vectors.

1. $\mathbf{i}, 2\mathbf{i} - 3\mathbf{j} + \mathbf{k}$

2. $2\mathbf{i} - 6\mathbf{j} + \mathbf{k}, \mathbf{i} - \mathbf{j}$

3. $-4\mathbf{i} - 2\mathbf{j} + 3\mathbf{k}, 6\mathbf{i} - 2\mathbf{j} - \mathbf{k}$

4. $8\mathbf{i} - 3\mathbf{j} + 2\mathbf{k}, -8\mathbf{i} - 3\mathbf{j} + \mathbf{k}$

5. $\mathbf{i} - 3\mathbf{k}, 2\mathbf{j} + 6\mathbf{k}$

6. $\mathbf{i} + \mathbf{j} + 2\mathbf{k}, \mathbf{i} - \mathbf{j} + 2\mathbf{k}$

In each of Problems 7 through 12, find the equation of the plane containing the given point and having the given vector as normal vector.

7. $(-1, 1, 2), 3\mathbf{i} - \mathbf{j} + 4\mathbf{k}$

8. $(-1, 0, 0), \mathbf{i} - 2\mathbf{j}$

9. $(2, -3, 4), 8\mathbf{i} - 6\mathbf{j} + 4\mathbf{k}$

10. $(-1, -1, -5), -3\mathbf{i} + 2\mathbf{j}$

11. $(0, -1, 4), 7\mathbf{i} + 6\mathbf{j} - 5\mathbf{k}$

12. $(-2, 1, -1), 4\mathbf{i} + 3\mathbf{j} + \mathbf{k}$

In each of Problems 13 through 16, find the cosine of the angle between $\overline{AB}$ and the line from A to the midpoint of $\overline{BC}$.

13. $A = (1, -2, 6), B = (3, 0, 1), C = (4, 2, -7)$

14. $A = (3, -2, -3), B = (-2, 0, 1), C = (1, 1, 7)$

15. $A = (1, -2, 6), B = (0, 4, -3), C = (-3, -2, 7)$

16. $A = (0, 5, -1), B = (1, -2, 5), C = (7, 0, -1)$

17. Suppose $\mathbf{F} \cdot \mathbf{X} = 0$ for every vector $\mathbf{X}$. What can be concluded about $\mathbf{F}$?

18. Suppose $\mathbf{F} \cdot \mathbf{i} = \mathbf{F} \cdot \mathbf{j} = \mathbf{F} \cdot \mathbf{k} = 0$. What can be concluded about $\mathbf{F}$?

19. Suppose $\mathbf{F} \neq \mathbf{O}$. Prove that the unit vector $\mathbf{u}$ for which $|\mathbf{F} \cdot \mathbf{u}|$ is a maximum must be parallel to $\mathbf{F}$.

20. Prove that for any vector $\mathbf{F}$,

$$\mathbf{F} = (\mathbf{F} \cdot \mathbf{i})\mathbf{i} + (\mathbf{F} \cdot \mathbf{j})\mathbf{j} + (\mathbf{F} \cdot \mathbf{k})\mathbf{k}.$$

21. Let $\mathbf{A}$ and $\mathbf{B}$ be vectors, with at least $\mathbf{B}$ not the zero vector. Prove that there are vectors $\mathbf{F}$ and $\mathbf{G}$ such that $\mathbf{F}$ is orthogonal to $\mathbf{B}$, $\mathbf{G}$ is orthogonal to $\mathbf{B}$, and $\mathbf{A} = \mathbf{F} + \mathbf{G}$.

22. Let $\mathbf{F}$, $\mathbf{G}$, and $\mathbf{H}$ be nonzero vectors, each orthogonal to the other two. Let $\mathbf{A}$ be any vector. Show that there are unique scalars α, β, and γ such that $\mathbf{A} = \alpha\mathbf{F} + \beta\mathbf{G} + \gamma\mathbf{H}$. *Hint:* Think about the idea behind Problem 20.

23. Suppose the dot product had been defined by

$$(a_1\mathbf{i} + b_1\mathbf{j} + c_1\mathbf{k}) \cdot (a_2\mathbf{i} + b_2\mathbf{j} + c_2\mathbf{k})$$
$$= 2a_1a_2 + b_1b_2 + c_1c_2.$$

What properties of the dot product would remain valid, and which would now fail to be true?

24. Suppose the dot product had been defined by

$$(a_1\mathbf{i} + b_1\mathbf{j} + c_1\mathbf{k}) \cdot (a_2\mathbf{i} + b_2\mathbf{j} + c_2\mathbf{k})$$
$$= a_1a_2 - b_1b_2 + c_1c_2.$$

Which properties of the dot product would remain valid, and which would now fail to hold?

25. Prove that the diagonals of a rhombus are perpendicular. (A rhombus is a parallelogram whose sides have equal length.)

26. Let $\mathbf{F}$ and $\mathbf{G}$ be nonzero vectors. Prove that the vector $\|\mathbf{G}\|\mathbf{F} + \|\mathbf{F}\|\mathbf{G}$ bisects the angle between $\mathbf{F}$ and $\mathbf{G}$. *Hint:* Compute the angle this vector makes with $\mathbf{F}$ and then the angle it makes with $\mathbf{G}$.

27. Prove Theorem 5.3(1).

28. Prove Theorem 5.3(2).

29. Prove Theorem 5.3(5).

5.3   The Cross Product

The dot product produces a scalar from two vectors. We will now define the cross product, which produces a vector from two vectors.

For this section, let $\mathbf{F} = a_1\mathbf{i} + b_1\mathbf{j} + c_1\mathbf{k}$ and $\mathbf{G} = a_2\mathbf{i} + b_2\mathbf{j} + c_2\mathbf{k}$.

DEFINITION 5.7   *Cross Product*

The cross product of $\mathbf{F}$ with $\mathbf{G}$ is the vector $\mathbf{F} \times \mathbf{G}$ defined by

$$\mathbf{F} \times \mathbf{G} = (b_1c_2 - b_2c_1)\mathbf{i} + (a_2c_1 - a_1c_2)\mathbf{j} + (a_1b_2 - a_2b_1)\mathbf{k}.$$

This vector is read "$\mathbf{F}$ cross $\mathbf{G}$." For example,

$$(\mathbf{i} + 2\mathbf{j} - 3\mathbf{k}) \times (-2\mathbf{i} + \mathbf{j} + 4\mathbf{k}) = (8 + 3)\mathbf{i} + (6 - 4)\mathbf{j} + (1 + 4)\mathbf{k} = 11\mathbf{i} + 2\mathbf{j} + 5\mathbf{k}.$$

A cross product is often computed as a 3×3 (read "3 by 3") "determinant," with the unit vectors in the first row, components of $\mathbf{F}$ in the second row, and components of $\mathbf{G}$ in the third. If expanded by the first row, this determinant gives $\mathbf{F} \times \mathbf{G}$. For example,

$$\begin{vmatrix} \mathbf{i} & \mathbf{j} & \mathbf{k} \\ 1 & 2 & -3 \\ -2 & 1 & 4 \end{vmatrix} = \begin{vmatrix} 2 & -3 \\ 1 & 4 \end{vmatrix} \mathbf{i} - \begin{vmatrix} 1 & -3 \\ -2 & 4 \end{vmatrix} \mathbf{j} + \begin{vmatrix} 1 & 2 \\ -2 & 1 \end{vmatrix} \mathbf{k}$$

$$= 11\mathbf{i} + 2\mathbf{j} + 5\mathbf{k} = \mathbf{F} \times \mathbf{G}.$$

The interchange of two rows in a determinant results in a change of sign. This means that interchanging $\mathbf{F}$ and $\mathbf{G}$ in the cross product results in a change of sign:

$$\mathbf{F} \times \mathbf{G} = -\mathbf{G} \times \mathbf{F}.$$

This is also apparent from the definition. Unlike addition and multiplication of real numbers, and the dot product operation, the cross product is not commutative, and the order in which it is performed makes a difference. This is true of many physical processes, for example, the order in which chemicals are combined may make a significant difference.

Some of the rules we need to compute with cross products are given in the next theorem.

THEOREM 5.5   *Properties of the Cross Product*

Let $\mathbf{F}$, $\mathbf{G}$, and $\mathbf{H}$ be vectors and let α be a scalar.

1. $\mathbf{F} \times \mathbf{G} = -\mathbf{G} \times \mathbf{F}$.
2. $\mathbf{F} \times \mathbf{G}$ is orthogonal to both $\mathbf{F}$ and $\mathbf{G}$.
3. $\|\mathbf{F} \times \mathbf{G}\| = \|\mathbf{F}\|\|\mathbf{G}\| \sin(\theta)$, in which θ is the angle between $\mathbf{F}$ and $\mathbf{G}$.
4. If $\mathbf{F}$ and $\mathbf{G}$ are not zero vectors, then $\mathbf{F} \times \mathbf{G} = \mathbf{O}$ if and only $\mathbf{F}$ and $\mathbf{G}$ are parallel.
5. $\mathbf{F} \times (\mathbf{G} + \mathbf{H}) = \mathbf{F} \times \mathbf{G} + \mathbf{F} \times \mathbf{H}$.
6. $\alpha(\mathbf{F} \times \mathbf{G}) = (\alpha\mathbf{F}) \times \mathbf{G} = \mathbf{F} \times (\alpha\mathbf{G})$. ∎

Proofs of these statements are for the most part routine calculations. We will prove (2) and (3).

Proof   For (2), compute

$$\mathbf{F} \cdot (\mathbf{F} \times \mathbf{G}) = a_1(b_1c_2 - b_2c_1) + b_1(a_2c_1 - a_1c_2) + c_1(a_1b_2 - a_2b_1) = 0.$$

Therefore, $\mathbf{F}$ and $\mathbf{F} \times \mathbf{G}$ are orthogonal. A similar argument holds for $\mathbf{G}$.

For (3), compute

$$\begin{aligned} \|\mathbf{F} \times \mathbf{G}\|^2 &= (b_1c_2 - b_2c_1)^2 + (a_2c_1 - a_1c_2)^2 + (a_1b_2 - a_2b_1)^2 \\ &= (a_1^2 + b_1^2 + c_1^2)(a_2^2 + b_2^2 + c_2^2) - (a_1a_2 + b_1b_2 + c_1c_2)^2 \\ &= \|\mathbf{F}\|^2\|\mathbf{G}\|^2 - (\mathbf{F} \cdot \mathbf{G})^2 \\ &= \|\mathbf{F}\|^2\|\mathbf{G}\|^2 - \|\mathbf{F}\|^2\|\mathbf{G}\|^2 \cos^2(\theta) \\ &= \|\mathbf{F}\|^2\|\mathbf{G}\|^2(1 - \cos^2(\theta)) \\ &= \|\mathbf{F}\|^2\|\mathbf{G}\|^2 \sin^2(\theta). \end{aligned}$$

Because $0 \le \theta \le \pi$, all of the factors whose squares appear in this equation are nonnegative, and upon taking square roots we obtain conclusion (3). ∎

If $\mathbf{F}$ and $\mathbf{G}$ are nonzero and not parallel, then arrows representing these vectors determine a plane in 3-dimensional space (Figure 5.23). $\mathbf{F} \times \mathbf{G}$ is orthogonal to this plane and oriented as in

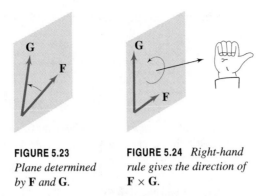

FIGURE 5.23
*Plane determined by **F** and **G**.*

FIGURE 5.24 *Right-hand rule gives the direction of **F** × **G**.*

Figure 5.24. If a person's right hand is placed so that the fingers curl from **F** to **G**, then the thumb points up along **F** × **G**. This is referred to as the *right-hand rule*. **G** × **F** = −**F** × **G** points in the opposite direction. As a simple example, **i** × **j** = **k**, and these three vectors define a standard right-handed coordinate system in 3-space.

The fact that **F** × **G** is orthogonal to both **F** and **G** is often useful. If they are not parallel, then vectors **F** and **G** determine a plane Π (Figure 5.23). This is consistent with the fact that three points, not on the same straight line, determine a plane. One point forms a base point for drawing the arrows representing **F** and **G**, and the other two points are the terminal points of these arrows. If we know a vector orthogonal to both **F** and **G**, then this vector is orthogonal to every vector in Π. Such a vector is said to be *normal* to Π. In Example 5.10 we showed how to find the equation of a plane given a point in the plane and a normal vector. Now we can find the equation of a plane given three points in it (not all on a line), because we can use the cross product to produce a normal vector.

EXAMPLE 5.11

Suppose we want the equation of the plane Π containing the points $(1, 2, 1)$, $(−1, 1, 3)$, and $(−2, −2, −2)$.

Begin by finding a vector normal to Π. We will do this by finding two vectors in Π and taking their cross product. The vectors from $(1, 2, 1)$ to the other two given points are in Π (Figure 5.25). These vectors are

$$\mathbf{F} = -2\mathbf{i} - \mathbf{j} + 2\mathbf{k} \quad \text{and} \quad \mathbf{G} = -3\mathbf{i} - 4\mathbf{j} - 3\mathbf{k}.$$

Form

$$\mathbf{N} = \mathbf{F} \times \mathbf{G} = \begin{vmatrix} \mathbf{i} & \mathbf{j} & \mathbf{k} \\ -2 & -1 & 2 \\ -3 & -4 & -3 \end{vmatrix} = 11\mathbf{i} - 12\mathbf{j} + 5\mathbf{k}.$$

This vector is normal to Π (orthogonal to every vector lying in Π). Now proceed as in Example 5.10. If (x, y, z) is any point in Π, then $(x − 1)\mathbf{i} + (y − 2)\mathbf{j} + (z − 1)\mathbf{k}$ is in Π and so is orthogonal to **N**. Therefore,

$$[(x - 1)\mathbf{i} + (y - 2)\mathbf{j} + (z - 1)\mathbf{k}] \cdot \mathbf{N} = 11(x - 1) - 12(y - 2) + 5(z - 1) = 0.$$

This gives

$$11x - 12y + 5z = -8.$$

This is the equation of the plane in the sense that a point (x, y, z) is in the plane if and only if its coordinates satisfy this equation. ∎

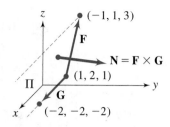

FIGURE 5.25

If we had specified three points lying on a line (collinear) in this example, then we would have found that **F** and **G** are parallel, hence $\mathbf{F} \times \mathbf{G} = \mathbf{O}$. When we calculated this cross product and got a nonzero vector, we knew that the points were not collinear.

The cross product also has geometric interpretations as an area or volume.

THEOREM 5.6

Let **F** and **G** be represented by arrows lying along incident sides of a parallelogram (Figure 5.26). Then the area of this parallelogram is $\|\mathbf{F} \times \mathbf{G}\|$.

Proof The area of a parallelogram is the product of the lengths of two incident sides and the sine of the angle between them. Draw vectors **F** and **G** along two incident sides. Then these sides have length $\|\mathbf{F}\|$ and $\|\mathbf{G}\|$. If θ is the angle between them, then the area of the parallelogram is $\|\mathbf{F}\|\|\mathbf{G}\| \sin(\theta)$. But this is exactly $\|\mathbf{F} \times \mathbf{G}\|$. ∎

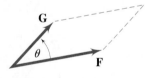

FIGURE 5.26 *Area*
$= \|\mathbf{F} \times \mathbf{G}\|$.

EXAMPLE 5.12

A parallelogram has two sides extending from $(0, 1, -2)$ to $(1, 2, 2)$ and from $(0, 1, -2)$ to $(1, 4, 1)$. We want to find the area of this parallelogram.

Form vectors along these sides:

$$\mathbf{F} = \mathbf{i} + \mathbf{j} + 4\mathbf{k}, \quad \mathbf{G} = \mathbf{i} + 3\mathbf{j} + 3\mathbf{k}.$$

Calculate

$$\mathbf{F} \times \mathbf{G} = \begin{vmatrix} \mathbf{i} & \mathbf{j} & \mathbf{k} \\ 1 & 1 & 4 \\ 1 & 3 & 3 \end{vmatrix} = -9\mathbf{i} + \mathbf{j} + 2\mathbf{k}$$

and the area of the parallelogram is $\|\mathbf{F} \times \mathbf{G}\| = \sqrt{86}$ square units. ∎

If a rectangular box is skewed as in Figure 5.27, the resulting solid is called a *rectangular parallelepiped*. All of its faces are parallelograms. We can find the volume of such a solid by combining dot and cross products, as follows.

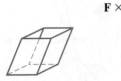

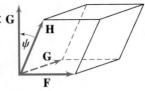

FIGURE 5.27
Parallelo-
piped.

FIGURE 5.28   *Volume*
$= |\mathbf{H} \cdot (\mathbf{F} \times \mathbf{G})|.$

THEOREM 5.7

Let $\mathbf{F}$, $\mathbf{G}$, and $\mathbf{H}$ be vectors along incident sides of a rectangular parallelepiped. Then the volume of the parallelepiped is $|\mathbf{H} \cdot (\mathbf{F} \times \mathbf{G})|$. ∎

This is the absolute value of the real number formed by taking the dot product of $\mathbf{H}$ with $\mathbf{F} \times \mathbf{G}$.

Proof   Figure 5.28 shows the parallelepiped. $\mathbf{F} \times \mathbf{G}$ is normal to the plane of $\mathbf{F}$ and $\mathbf{G}$ and oriented as shown in the diagram, according to the right-hand rule. If $\mathbf{H}$ is along the third side of the parallelepiped, and ψ is the angle between $\mathbf{H}$ and $\mathbf{F} \times \mathbf{G}$, then $\|\mathbf{H}\| \cos(\psi)$ is the altitude of the parallelepiped. The area of the base parallelogram is $\|\mathbf{F} \times \mathbf{G}\|$ by Theorem 5.6. Thus the volume of the parallelepiped is

$$\|\mathbf{H}\|\|\mathbf{F} \times \mathbf{G}\| \cos(\psi).$$

But this is $|\mathbf{H} \cdot (\mathbf{F} \times \mathbf{G})|$. ∎

EXAMPLE 5.13

One corner of a rectangular parallelepiped is at $(-1, 2, 2)$, and three incident sides extend from this point to $(0, 1, 1)$, $(-4, 6, 8)$, and $(-3, -2, 4)$. To find the volume of this solid, form the vectors

$$\mathbf{F} = (0 - (-1))\mathbf{i} + (1 - 2)\mathbf{j} + (1 - 2)\mathbf{k} = \mathbf{i} - \mathbf{j} - \mathbf{k},$$

$$\mathbf{G} = (-4 - (-1))\mathbf{i} + (6 - 2)\mathbf{j} + (8 - 2)\mathbf{k} = -3\mathbf{i} + 4\mathbf{j} + 6\mathbf{k},$$

and

$$\mathbf{H} = (-3 - (-1))\mathbf{i} + (-2 - 2)\mathbf{j} + (4 - 2)\mathbf{k} = -2\mathbf{i} - 4\mathbf{j} + 2\mathbf{k}.$$

Calculate

$$\mathbf{F} \times \mathbf{G} = \begin{vmatrix} \mathbf{i} & \mathbf{j} & \mathbf{k} \\ 1 & -1 & -1 \\ -3 & 4 & 6 \end{vmatrix} = -2\mathbf{i} - 3\mathbf{j} + \mathbf{k}.$$

Then,

$$\mathbf{H} \cdot (\mathbf{F} \times \mathbf{G}) = (-2)(-2) + (-4)(-3) + (2)(1) = 18,$$

and the volume is 18 cubic units. ∎

The quantity $\mathbf{H} \cdot (\mathbf{F} \times \mathbf{G})$ is called a *scalar triple product*. We will outline some of its properties in the problems.

SECTION 5.3   PROBLEMS

In each of Problems 1 through 10, compute $\mathbf{F} \times \mathbf{G}$ and, independently, $\mathbf{G} \times \mathbf{F}$, verifying that one is the negative of the other. Use the dot product to compute the cosine of the angle θ between $\mathbf{F}$ and $\mathbf{G}$ and use this to determine $\sin(\theta)$. Then calculate $\|\mathbf{F}\|\|\mathbf{G}\| \sin(\theta)$ and verify that this gives $\|\mathbf{F} \times \mathbf{G}\|$.

1. $\mathbf{F} = -3\mathbf{i} + 6\mathbf{j} + \mathbf{k}, \mathbf{G} = -\mathbf{i} - 2\mathbf{j} + \mathbf{k}$

2. $\mathbf{F} = 6\mathbf{i} - \mathbf{k}, \mathbf{G} = \mathbf{j} + 2\mathbf{k}$

3. $\mathbf{F} = 2\mathbf{i} - 3\mathbf{j} + 4\mathbf{k}, \mathbf{G} = -3\mathbf{i} + 2\mathbf{j}$

4. $\mathbf{F} = 8\mathbf{i} + 6\mathbf{j}, \mathbf{G} = 14\mathbf{j}$

5. $\mathbf{F} = 5\mathbf{i} + 3\mathbf{j} + 4\mathbf{k}, \mathbf{G} = 20\mathbf{i} + 6\mathbf{k}$

6. $\mathbf{F} = 2\mathbf{k}, \mathbf{G} = 8\mathbf{i} - \mathbf{j}$

7. $\mathbf{F} = 18\mathbf{i} - 3\mathbf{j} + 4\mathbf{k}, \mathbf{G} = 22\mathbf{j} - \mathbf{k}$

8. $\mathbf{F} = \mathbf{i} - 3\mathbf{j} - \mathbf{k}, \mathbf{G} = 18\mathbf{i} - 21\mathbf{j}$

9. $\mathbf{F} = -4\mathbf{i} + 6\mathbf{k}, \mathbf{G} = \mathbf{i} - 2\mathbf{j} + 7\mathbf{k}$

10. $\mathbf{F} = -3\mathbf{i} + 2\mathbf{j} + \mathbf{k}, \mathbf{G} = 8\mathbf{i} - 6\mathbf{j} + 2\mathbf{k}$

In each of Problems 11 through 15, determine whether the points are collinear. If they are not, find an equation of the plane containing all three points.

11. $(-1, 1, 6), (2, 0, 1), (3, 0, 0)$

12. $(4, 1, 1), (-2, -2, 3), (6, 0, 1)$

13. $(1, 0, -2), (0, 0, 0), (5, 1, 1)$

14. $(0, 0, 2), (-4, 1, 0), (2, -1, 1)$

15. $(-4, 2, -6), (1, 1, 3), (-2, 4, 5)$

In each of Problems 16 through 20, find the area of the parallelogram having incident sides extending from the first point to each of the other two.

16. $(1, -3, 7), (2, 1, 1), (6, -1, 2)$

17. $(6, 1, 1), (7, -2, 4), (8, -4, 3)$

18. $(-2, 1, 6), (2, 1, -7), (4, 1, 1)$

19. $(4, 2, -3), (6, 2, -1), (2, -6, 4)$

20. $(1, 1, -8), (9, -3, 0), (-2, 5, 2)$

In each of Problems 21 through 25, find the volume of the parallepiped whose incident sides extend from the first point to each of the other three.

21. $(1, 1, 1), (-4, 2, 7), (3, 5, 7), (0, 1, 6)$

22. $(0, 1, -6), (-3, 1, 4), (1, 7, 2), (-3, 0, 4)$

23. $(1, 6, 1), (-2, 4, 2), (3, 0, 0), (2, 2, -4)$

24. $(0, 1, 7), (9, 1, 3), (-2, 4, 1), (3, 0, -3)$

25. $(1, 1, 1), (2, 2, 2), (6, 1, 3), (-2, 4, 6)$

In each of Problems 26 through 30, find a vector normal to the given plane. There are infinitely many such vectors.

26. $8x - y + z = 12$

27. $x - y + 2z = 0$

28. $x - 3y + 2z = 9$

29. $7x + y - 7z = 7$

30. $4x + 6y + 4z = -5$

31. Prove that $\mathbf{F} \times (\mathbf{G} + \mathbf{H}) = \mathbf{F} \times \mathbf{G} + \mathbf{F} \times \mathbf{H}$.

32. Prove that $(\alpha\mathbf{F}) \times \mathbf{G} = \mathbf{F} \times (\alpha\mathbf{G}) = \alpha(\mathbf{F} \times \mathbf{G})$.

33. Prove that $\mathbf{F} \times (\mathbf{G} \times \mathbf{H}) = (\mathbf{F} \cdot \mathbf{H})\mathbf{G} - (\mathbf{F} \cdot \mathbf{G})\mathbf{H}$.

34. Give a geometrical reason why $\mathbf{F} \times (\mathbf{G} \times \mathbf{H})$ is represented by an arrow in the plane determined by $\mathbf{G} \times \mathbf{H}$, assuming that these vectors are not parallel.

35. Prove that
$$\mathbf{F} \cdot (\mathbf{G} \times \mathbf{H}) = \mathbf{G} \cdot (\mathbf{H} \times \mathbf{F}) = \mathbf{H} \cdot (\mathbf{F} \times \mathbf{G}).$$

36. Suppose $\mathbf{F}, \mathbf{G}, \mathbf{H}$, and $\mathbf{K}$ lie in the same plane. What can be said about $(\mathbf{F} \times \mathbf{G}) \times (\mathbf{H} \times \mathbf{K})$?

37. Prove *Lagrange's vector identity*:
$$(\mathbf{F} \times \mathbf{G}) \cdot (\mathbf{H} \times \mathbf{K}) = (\mathbf{F} \cdot \mathbf{H})(\mathbf{G} \cdot \mathbf{K})$$
$$- (\mathbf{F} \cdot \mathbf{K})(\mathbf{G} \cdot \mathbf{H}).$$

38. Use vector operations to find a formula for the area of the triangle having vertices (a_i, b_i, c_i) for $i = 1, 2, 3$. What conditions must be placed on these coordinates to ensure that the points are not collinear (all on a line)?

The *scalar triple product* of $\mathbf{F}, \mathbf{G}$, and $\mathbf{H}$ is defined to be $[\mathbf{F}, \mathbf{G}, \mathbf{H}] = \mathbf{F} \cdot (\mathbf{G} \times \mathbf{H})$.

39. Let $\mathbf{F} = a_1\mathbf{i} + b_1\mathbf{j} + c_1\mathbf{k}, \mathbf{G} = a_2\mathbf{i} + b_2\mathbf{j} + c_2\mathbf{k}$, and $\mathbf{H} = a_3\mathbf{i} + b_3\mathbf{j} + c_3\mathbf{k}$. Prove that
$$[\mathbf{F}, \mathbf{G}, \mathbf{H}] = \begin{vmatrix} a_1 & b_1 & c_1 \\ a_2 & b_2 & c_2 \\ a_3 & b_3 & c_3 \end{vmatrix}.$$

40. Prove that $[\mathbf{F}, \mathbf{G}, \mathbf{H}] = [\mathbf{G}, \mathbf{H}, \mathbf{F}] = [\mathbf{H}, \mathbf{F}, \mathbf{G}]$, but that $[\mathbf{F}, \mathbf{G}, \mathbf{H}] = -[\mathbf{F}, \mathbf{H}, \mathbf{G}]$.

The pattern is in the diagram

As long as the terms in the scalar triple product retain the order shown by following arrows in the diagram (regardless of starting point), the sign of the scalar triple product remains the same. If, however, an arrow is reversed, the sign of the scalar triple product is changed.

41. Prove that, for any scalars α and β,

$$[\alpha \mathbf{F} + \beta \mathbf{K}, \mathbf{G}, \mathbf{H}] = \alpha[\mathbf{F}, \mathbf{G}, \mathbf{H}] + \beta[\mathbf{K}, \mathbf{G}, \mathbf{H}].$$

42. Prove that, if any one of $\mathbf{F}, \mathbf{G},$ and $\mathbf{H}$ is a sum of scalar multiples of the other two, then $[\mathbf{F}, \mathbf{G}, \mathbf{H}] = 0$.

5.4 The Vector Space R^n

The world of everyday experience has three space dimensions. But often we encounter settings in which more dimensions occur. If we want to specify not only the location of a particle but the time in which it occupies a particular point, we need four coordinates (x, y, z, t). And specifying the location of each particle in a system of particles may require any number of coordinates. The natural setting for such problems is R^n, the space of points having n coordinates.

DEFINITION 5.8 n-vector

If n is a positive integer, an n-vector is an n-tuple $(x_1, x_2, \ldots, x_n)$, with each coordinate x_j a real number. The set of all n-vectors is denoted R^n.

R^1 is the real line, consisting of all real numbers. We can think of real numbers as 1-vectors, but there is is no advantage to doing this. R^2 consists of ordered pairs (x, y) of real numbers, and each such ordered pair (or 2-vector) can be identified with a point in the plane. R^3 consists of all 3-vectors, or points in 3-space. If $n \geq 4$, we can no longer draw a set of mutually independent coordinate axes, one for each coordinate, but we can still work with vectors in R^n according to rules we will now describe.

DEFINITION 5.9 Algebra of R^n

1. Two n-vectors are added by adding their respective components:

$$(x_1, x_2, \ldots, x_n) + (y_1, y_2, \ldots, y_n) = (x_1 + y_1, x_2 + y_2, \ldots, x_n + y_n).$$

2. An n-vector is multiplied by a scalar by multiplying each component by the scalar:

$$\alpha(x_1, x_2, \ldots, x_n) = (\alpha x_1, \alpha x_2, \ldots, \alpha x_n).$$

The zero vector in R^n is the n-vector $\mathbf{O} = (0, 0, \ldots, 0)$ having each coordinate equal to zero.

The negative of $\mathbf{F} = (x_1, x_2, \ldots, x_n)$ is $-\mathbf{F} = (-x_1, -x_2, \ldots, -x_n)$. As we did with $n = 3$, we denote $\mathbf{G} + (-\mathbf{F})$ as $\mathbf{G} - \mathbf{F}$.

The algebraic rules in R^n mirror those we saw for R^3.

THEOREM 5.8

Let **F**, **G**, and **H** be in R^n and let α and β be real numbers. Then

1. $\mathbf{F} + \mathbf{G} = \mathbf{G} + \mathbf{F}$.
2. $\mathbf{F} + (\mathbf{G} + \mathbf{H}) = (\mathbf{F} + \mathbf{G}) + \mathbf{H}$.
3. $\mathbf{F} + \mathbf{O} = \mathbf{F}$.
4. $(\alpha + \beta)\mathbf{F} = \alpha\mathbf{F} + \beta\mathbf{F}$.
5. $(\alpha\beta)\mathbf{F} = \alpha(\beta\mathbf{F})$.
6. $\alpha(\mathbf{F} + \mathbf{G}) = \alpha\mathbf{F} + \beta\mathbf{G}$.
7. $\alpha\mathbf{O} = \mathbf{O}$. ∎

Because of these properties of the operations of addition of n-vectors, and multiplication of an n-vector by a scalar, we call R^n a *vector space*. In the next section we will clarify the sense in which R^n can be said to have dimension n.

The length (norm, magnitude) of $\mathbf{F} = (x_1, x_2, \ldots, x_n)$ is defined by a direct generalization from the plane and 3-space:

$$\|\mathbf{F}\| = \sqrt{x_1^2 + x_2^2 + \cdots + x_n^2}.$$

There is no analogue of the cross product for vectors in R^n when $n > 3$. However, the dot product readily extends to n-vectors.

DEFINITION 5.10   *Dot Product of n-Vectors*

The dot product of $(x_1, x_2, \ldots, x_n)$ and $(y_1, y_2, \ldots, y_n)$ is defined by

$$(x_1, x_2, \ldots, x_n) \cdot (y_1, y_2, \ldots, y_n) = x_1 y_1 + x_2 y_2 + \cdots + x_n y_n.$$

All of the conclusions of Theorem 5.3 remain true for n-vectors, as does Lemma 5.1. We will record these results for completeness.

THEOREM 5.9

Let **F**, **G**, and **H** be n-vectors, and let α and β be real numbers. Then

1. $\mathbf{F} \cdot \mathbf{G} = \mathbf{G} \cdot \mathbf{F}$.
2. $(\mathbf{F} + \mathbf{G}) \cdot \mathbf{H} = \mathbf{F} \cdot \mathbf{H} + \mathbf{G} \cdot \mathbf{H}$.
3. $\alpha(\mathbf{F} \cdot \mathbf{G}) = (\alpha\mathbf{F}) \cdot \mathbf{G} = \mathbf{F} \cdot (\alpha\mathbf{G})$.
4. $\mathbf{F} \cdot \mathbf{F} = \|\mathbf{F}\|^2$.
5. $\mathbf{F} \cdot \mathbf{F} = 0$ if and only if $\mathbf{F} = \mathbf{O}$.
6. $\|\alpha\mathbf{F} + \beta\mathbf{G}\|^2 = \alpha^2\|\mathbf{F}\|^2 + 2\alpha\beta\mathbf{F} \cdot \mathbf{G} + \beta^2\|\mathbf{G}\|^2$. ∎

The Cauchy–Schwarz inequality holds for n-vectors, but the proof given previously for 3-vectors does not generalize to R^n. We will therefore give a proof that is valid for any n.

THEOREM 5.10   *Cauchy–Schwarz Inequality in R^n*

Let $\mathbf{F}$ and $\mathbf{G}$ be in R^n. Then

$$|\mathbf{F} \cdot \mathbf{G}| \leq \|\mathbf{F}\|\|\mathbf{G}\|.$$

Proof   The inequality reduces to $0 \leq 0$ if either vector is the zero vector. Thus, suppose $\mathbf{F} \neq \mathbf{O}$ and $\mathbf{G} \neq \mathbf{O}$.

Choose $\alpha = \|\mathbf{G}\|$ and $\beta = -\|\mathbf{F}\|$ in Theorem 5.9(6). We get

$$0 \leq \|\alpha\mathbf{F} + \beta\mathbf{G}\|^2 = \|\mathbf{G}\|^2\|\mathbf{F}\|^2 - 2\|\mathbf{G}\|\|\mathbf{F}\|\mathbf{F} \cdot \mathbf{G} + \|\mathbf{F}\|^2\|\mathbf{G}\|^2.$$

Upon dividing this inequality by $2\|\mathbf{F}\|\|\mathbf{G}\|$, we obtain

$$\mathbf{F} \cdot \mathbf{G} \leq \|\mathbf{F}\|\|\mathbf{G}\|.$$

Now go back to Theorem 5.9(6), but this time choose $\alpha = \|\mathbf{G}\|$ and $\beta = \|\mathbf{F}\|$ to get

$$0 \leq \|\mathbf{G}\|^2\|\mathbf{F}\|^2 + 2\|\mathbf{G}\|\|\mathbf{F}\|\mathbf{F} \cdot \mathbf{G} + \|\mathbf{G}\|^2\|\mathbf{F}\|^2,$$

and upon dividing by $2\|\mathbf{F}\|\|\mathbf{G}\|$ we get

$$-\|\mathbf{F}\|\|\mathbf{G}\| \leq \mathbf{F} \cdot \mathbf{G}.$$

We have now shown that

$$-\|\mathbf{F}\|\|\mathbf{G}\| \leq \mathbf{F} \cdot \mathbf{G} \leq \|\mathbf{F}\|\|\mathbf{G}\|,$$

and this is equivalent to the Cauchy–Schwarz inequality.   ∎

In view of the Cauchy–Schwarz inequality, we can define the cosine of the angle between vectors $\mathbf{F}$ and $\mathbf{G}$ in R^n by

$$\cos(\theta) = \begin{cases} 0 & \text{if } \mathbf{F} \text{ or } \mathbf{G} \text{ equals the zero vector} \\ (\mathbf{F} \cdot \mathbf{G})/(\|\mathbf{F}\|\|\mathbf{G}\|) & \text{if } \mathbf{F} \neq \mathbf{O} \text{ and } \mathbf{G} \neq \mathbf{O} \end{cases}$$

This is sometimes useful in bringing some geometric intuition to R^n. For example, it is natural to define $\mathbf{F}$ and $\mathbf{G}$ to be orthogonal if the angle between them is $\pi/2$, and by this definition of $\cos(\theta)$, this is equivalent to requiring that $\mathbf{F} \cdot \mathbf{G} = 0$, consistent with orthogonality in R^2 and R^3.

We can define a *standard representation* of vectors in R^n by defining unit vectors along the n directions:

$$\mathbf{e}_1 = (1, 0, 0, \ldots, 0)$$

$$\mathbf{e}_2 = (0, 1, 0, \ldots, 0)$$

$$\vdots$$

$$\mathbf{e}_n = (0, 0, \ldots, 0, 1).$$

Now any n-vector can be written

$$(x_1, x_2, \ldots, x_n) = x_1\mathbf{e}_1 + x_2\mathbf{e}_2 + \cdots + x_n\mathbf{e}_n$$

$$= \sum_{j=1}^{n} x_j\mathbf{e}_j.$$

A set of n-vectors containing the zero vector, as well as sums of vectors in the set, and scalar multiples of vectors in the set, is called a *subspace* of R^n.

DEFINITION 5.11 *Subspace*

A set S of n-vectors is a subspace of R^n if:

1. $\mathbf{O}$ is in S.
2. The sum of any vectors in S is in S.
3. The product of any vector in S with any real number is also in S.

Conditions (2) and (3) of the definition can be combined by requiring that $\alpha\mathbf{F} + \beta\mathbf{G}$ be in S for any vectors $\mathbf{F}$ and $\mathbf{G}$ in S and any real numbers α and β.

EXAMPLE 5.14

Let S consist of all vectors in R^n having norm 1. In R^2 (the plane), this is the set of points on the unit circle about the origin; in R^3 this is the set of points on the unit sphere about the origin.

S is not a subspace of R^n for several reasons. First, $\mathbf{O}$ is not in S because $\mathbf{O}$ does not have norm 1. Further, a sum of vectors in S is not in S (a sum of vectors having norm 1 does not have norm 1). And, if $\alpha \neq 1$, and $\mathbf{F}$ has norm 1, then $\alpha\mathbf{F}$ does not have norm 1, so $\alpha\mathbf{F}$ is not in S. ■

In this example S failed all three criteria for being a subspace. It is enough to fail one to disqualify a set of vectors from being a subspace.

EXAMPLE 5.15

Let K consist of all scalar multiples of $(-1, 4, 2, 0)$ in R^4. We want to know if K is a subspace of R^4.

First, $\mathbf{O}$ is in K, because $\mathbf{O} = 0(-1, 4, 2, 0) = (0, 0, 0, 0)$.

Next, if $\mathbf{F}$ and $\mathbf{G}$ are in K, then $\mathbf{F} = \alpha(-1, 4, 2, 0)$ for some α and $\mathbf{G} = \beta(-1, 4, 2, 0)$ for some β, so

$$\mathbf{F} + \mathbf{G} = (\alpha + \beta)(-1, 4, 2, 0)$$

is a scalar multiple of $(-1, 4, 2, 0)$ and therefore is in K.

Finally, if $\mathbf{F} = \alpha(-1, 4, 2, 0)$ is any vector in K, and β is any scalar, then

$$\beta\mathbf{F} = (\beta\alpha)(-1, 4, 2, 0)$$

is a scalar multiple of $(-1, 4, 2, 0)$ and hence is in K. Thus K is a subspace of R^4. ■

EXAMPLE 5.16

Let S consist of just the zero vector $\mathbf{O}$ in R^n. Then S is a subspace of R^n. This is called the *trivial subspace*. At the other extreme, R^n is also a subspace of R^n. ■

In R^2 and R^3 there are simple geometric characterizations of all possible subspaces.

Begin with R^2. We claim that only subspaces are the trivial subspace consisting of just the zero vector, or R^2 itself, or all vectors lying along a single line through the origin. To demonstrate this, we need the following fact.

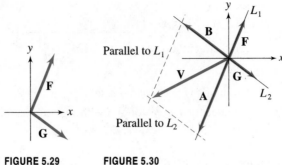

FIGURE 5.29   FIGURE 5.30

LEMMA 5.2

Let $\mathbf{F}$ and $\mathbf{G}$ be nonzero vectors in R^2 that are not parallel. Then every vector in R^2 can be written in the form $\alpha\mathbf{F} + \beta\mathbf{G}$ for some scalars α and β.

Proof   Represent $\mathbf{F}$ and $\mathbf{G}$ as arrows from the origin (Figure 5.29). These determine nonparallel lines L_1 and L_2, respectively, through the origin, because $\mathbf{F}$ and $\mathbf{G}$ are assumed to be nonparallel. Let $\mathbf{V}$ be any 2-vector. If $\mathbf{V} = \mathbf{O}$, then $\mathbf{V} = 0\mathbf{F} + 0\mathbf{G}$. We therefore consider the case that $\mathbf{V} \neq \mathbf{O}$ and represent it as an arrow from the origin as well. We want to show that $\mathbf{V}$ must be the sum of scalar multiples of $\mathbf{F}$ and $\mathbf{G}$.

If $\mathbf{V}$ is along L_1, then $\mathbf{V} = \alpha\mathbf{F}$ for some real number α, and then $\mathbf{V} = \alpha\mathbf{F} + 0\mathbf{G}$. Similarly, if $\mathbf{V}$ is along L_2, then $\mathbf{V} = \beta\mathbf{G} = 0\mathbf{F} + \beta\mathbf{G}$.

Thus, suppose that $\mathbf{V}$ is not a scalar multiple of either $\mathbf{F}$ or $\mathbf{G}$. Then the arrow representing $\mathbf{V}$ is not along L_1 or L_2. Now carry out the construction shown in Figure 5.30. Draw lines parallel to L_2 and L_2 from the tip of $\mathbf{V}$. Arrows from the origin to where these parallels intersect L_2 and L_1 determine, respectively, vectors $\mathbf{A}$ and $\mathbf{B}$. By the parallelogram law,

$$\mathbf{V} = \mathbf{A} + \mathbf{B}.$$

But $\mathbf{A}$ is along L_1, so $\mathbf{A} = \alpha\mathbf{F}$ for some scalar α. And $\mathbf{B}$ is along L_2 and so $\mathbf{B} = \beta\mathbf{G}$ for some scalar β. Thus, $\mathbf{V} = \alpha\mathbf{F} + \beta\mathbf{G}$, completing the proof. ■

We can now completely characterize the subspaces of R^2.

THEOREM 5.11   *The Subspaces of R^2*

Let S be a subspace of R^2. Then one of the following three possibilities must hold:

1. $S = R^2$, or
2. S consists of just the zero vector, or
3. S consists of all vectors along to some straight line through the origin.

Proof   Suppose cases (1) and (2) do not hold. Because S is not the trivial subspace, S must contain at least one nonzero vector $\mathbf{F}$. We will show that every vector in S is a scalar multiple of $\mathbf{F}$.

Suppose instead that there is a nonzero vector $\mathbf{G}$ in S that is not a scalar multiple of $\mathbf{F}$. If $\mathbf{V}$ is any vector in R^2, then by Lemma 5.2, $\mathbf{V} = \alpha\mathbf{F} + \beta\mathbf{G}$ for some scalars α and β. But S is a subspace of R^2, so $\alpha\mathbf{F} + \beta\mathbf{G}$ is in S. This would imply that every vector in R^2 is in S, hence that $S = R^2$, a contradiction. Therefore, there is no vector in S that is not a scalar multiple of $\mathbf{F}$.

We conclude that every vector in S is a scalar multiple of $\mathbf{F}$, and the proof is complete. ■

By a similar argument involving more cases, we can prove that any subspace of R^3 must be either the trivial subspace, R^3 itself, all vectors along some line through the origin, or all vectors on some plane through the origin.

SECTION 5.4   PROBLEMS

In each of Problems 1 through 6, find the sum of the vectors and express this sum in standard form. Calculate the dot product of the vectors and the angle between them. The latter may be expressed as an inverse cosine of a number.

1. $(-1, 6, 2, 4, 0), (6, -1, 4, 1, 1)$

2. $(0, 1, 4, -3), (2, 8, 6, -4)$

3. $(1, -4, 3, 2), (16, 0, 0, 4)$

4. $(6, 1, 1, -1, 2), (4, 3, 5, 1, -2)$

5. $(0, 1, 6, -4, 1, 2, 9, -3), (6, 6, -12, 4, -3, -3, 2, 7)$

6. $(-5, 2, 2, -7, -8), (1, 1, 1, -8, 7)$

In each of Problems 7 through 16, determine whether the set of vectors is a subspace of R^n for the appropriate n.

7. S consists of all vectors (x, y, z, x, x) in R^5.

8. S consists of all vectors $(x, 2x, 3x, y)$ in R^4.

9. S consists of all vectors $(x, 0, 0, 1, 0, y)$ in R^6.

10. S consists of all vectors $(0, x, y)$ in R^3.

11. S consists of all vectors $(x, y, x + y, x - y)$ in R^4.

12. S consists of all vectors in R^7 having zero third and fifth components.

13. S consists of all vectors in R^4 whose first and second components are equal.

14. S consists of all vectors in R^{93} having third component equal to 2.

15. S consists of all vectors in R^7 whose seventh component is the sum of the first six components.

16. S consists of all vectors in R^8 with first, second, and fourth components equal to zero, and the third component equal to the sixth component.

17. Let S consist of all vectors in R^3 on or parallel to the plane $ax + by + cz = 0$, with $a, b,$ and c real numbers, at least one of which is nonzero. Is S a subspace of R^3?

18. Let S consist of all vectors in R^3 on or parallel to the plane $ax + by + cz = k$, with $a, b,$ and c real numbers, at least one of which is nonzero and $k \neq 0$. Is S a subspace of R^3?

19. Let $\mathbf{F}$ and $\mathbf{G}$ be in R^n. Prove that

$$\|\mathbf{F} + \mathbf{G}\|^2 + \|\mathbf{F} - \mathbf{G}\|^2 = 2(\|\mathbf{F}\|^2 + \|\mathbf{G}\|^2).$$

Hint: Use the fact that the square of the norm of a vector is the dot product of the vector with itself.

20. Let $\mathbf{F}$ and $\mathbf{G}$ be orthogonal vectors in R^n. Prove that $\|\mathbf{F} + \mathbf{G}\|^2 = \|\mathbf{F}\|^2 + \|\mathbf{G}\|^2$. This is called *Pythagoras's theorem*.

21. Suppose $\mathbf{F}$ and $\mathbf{G}$ are vectors in R^n satisfying the relationship of Pythagoras's theorem (Problem 20). Does it follow that $\mathbf{F}$ and $\mathbf{G}$ are orthogonal?

22. How does the famous Pythagorean theorem of geometry relate to Pythagoras's theorem of Problem 20? In view of Problem 21, is the converse of the Pythagorean theorem true? That is, if a triangle has sides of length $a, b,$ and c, and $a^2 + b^2 = c^2$, does this triangle have to be a right triangle?

23. Prove the triangle inequality in R^n. This states that if $\mathbf{F}$ and $\mathbf{G}$ are in R^n, then $\|\mathbf{F} + \mathbf{G}\| \leq \|\mathbf{F}\| + \|\mathbf{G}\|$.

24. What conditions on $\mathbf{F}$ and $\mathbf{G}$ are sufficient for there to be equality in the triangle inequality (Problem 23)?

25. Use the Cauchy–Schwarz inequality to prove that if $x_1, \ldots, x_n$ and $y_1, \ldots, y_n$ are real numbers, then

$$\left| \sum_{j=1}^{n} x_j y_j \right|^2 \leq \left(\sum_{j=1}^{n} x_j^2 \right) \left(\sum_{j=1}^{n} y_j^2 \right).$$

26. Give an example of a set K of vectors in R^n such that $\mathbf{O}$ is in K, and any scalar multiple of a vector in K is in K, but K is not a subspace of R^n.

5.5 Linear Independence, Spanning Sets, and Dimension in R^n

In solving systems of linear algebraic equations and systems of linear differential equations, as well as for later work in Fourier analysis, we will use terminology and ideas from linear algebra. We will define these terms first in R^n, where we have some geometric intuition, and then extend them to a more abstract setting in the next section.

DEFINITION 5.12   *Linear Combinations in R^n*

A linear combination of k vectors $\mathbf{F}_1, \ldots, \mathbf{F}_k$ in R^n is a sum

$$\alpha_1 \mathbf{F}_1 + \cdots + \alpha_k \mathbf{F}_k$$

in which each α_j is a real number.

For example,

$$-8(-2, 4, 1, 0) + 6(1, 1, -1, 7) - \pi(8, 0, 0, 0)$$

is a linear combination of $(-2, 4, 1, 0)$, $(1, 1, -1, 7)$, and $(8, 0, 0, 0)$ in R^4. This linear combination is equal to the 4-vector

$$(22 - 8\pi, -26, -14, 42).$$

The set of all linear combinations of any given (finite) number of vectors in R^n is always a subspace of R^n.

THEOREM 5.12

Let $\mathbf{F}_1, \ldots, \mathbf{F}_k$ be in R^n, and let V consist of all vectors $\alpha_1 \mathbf{F}_1 + \cdots + \alpha_k \mathbf{F}_k$, in which each α_j can be any real number. Then V is a subspace of R^n.

Proof   First, $\mathbf{O}$ is in V (choose $\alpha_1 = \alpha_2 = \cdots = \alpha_k = 0$).

Next, suppose $\mathbf{G}$ and $\mathbf{H}$ are in V. Then

$$\mathbf{G} = \alpha_1 \mathbf{F}_1 + \cdots + \alpha_k \mathbf{F}_k \quad \text{and} \quad \mathbf{H} = \beta_1 \mathbf{F}_1 + \cdots + \beta_k \mathbf{F}_k$$

for some real numbers $\alpha_1, \ldots, \alpha_k, \beta_1, \ldots, \beta_k$. Then

$$\mathbf{G} + \mathbf{H} = (\alpha_1 + \beta_1)\mathbf{F}_1 + \cdots + (\alpha_k + \beta_k)\mathbf{F}_k$$

is again a linear combination of $\mathbf{F}_1, \ldots, \mathbf{F}_k$, and so is in V.

Finally, let $\mathbf{G} = \alpha_1 \mathbf{F}_1 + \cdots + \alpha_k \mathbf{F}_k$ be in V. If c is any real number, then

$$c\mathbf{G} = (c\alpha_1)\mathbf{F}_1 + \cdots + (c\alpha_k)\mathbf{F}_k$$

is also a linear combination of $\mathbf{F}_1, \ldots, \mathbf{F}_k$, and is therefore in V. Therefore, V is a subspace of R^n. ∎

Whenever we form a subspace by taking a linear combination of given vectors, we say that these vectors *span* the subspace.

DEFINITION 5.13   *Spanning Set*

Let $\mathbf{F}_1, \ldots, \mathbf{F}_k$ be vectors in a subspace S of R^n. Then $\mathbf{F}_1, \ldots, \mathbf{F}_k$ form a spanning set for S if every vector in S is a linear combination of $\mathbf{F}_1, \ldots, \mathbf{F}_k$.

In this case we say that S is spanned by $\mathbf{F}_1, \ldots, \mathbf{F}_k$, or that $\mathbf{F}_1, \ldots, \mathbf{F}_k$ span S.

For example, $\mathbf{i}$, $\mathbf{j}$, and $\mathbf{k}$ span R^3, because every vector in R^3 can be written $a\mathbf{i} + b\mathbf{j} + c\mathbf{k}$. The vector $\mathbf{i} + \mathbf{j}$ in R^2 spans the subspace consisting of all vectors $\alpha(\mathbf{i} + \mathbf{j})$, with α any scalar. These vectors all lie along the straight line $y = x$ through the origin in the plane.

Different sets of vectors may span the same subspace of R^n. Consider the following example.

EXAMPLE 5.17

Let V be the subspace of R^4 consisting of all vectors $(\alpha, \beta, 0, 0)$. Every vector in S can be written

$$(\alpha, \beta, 0, 0) = \alpha(1, 0, 0, 0) + \beta(0, 1, 0, 0),$$

so $(1, 0, 0, 0)$ and $(0, 1, 0, 0)$ span V. But we can also write any vector in V as

$$(\alpha, \beta, 0, 0) = \frac{\alpha}{2}(2, 0, 0, 0) + \frac{\beta}{\pi}(0, \pi, 0, 0),$$

so the vectors $(2, 0, 0, 0)$ and $(0, \pi, 0, 0)$ also span V.

Different numbers of vectors may also span the same subspace. The vectors

$$(4, 0, 0, 0), (0, 3, 0, 0), (1, 2, 0, 0)$$

also span V. To see this, write an arbitrary vector in V as

$$(\alpha, \beta, 0, 0) = \frac{\alpha - 2}{4}(4, 0, 0, 0) + \frac{\beta - 4}{3}(0, 3, 0, 0) + 2(1, 2, 0, 0). \quad \blacksquare$$

The last example suggests that some spanning sets are more efficient than others. If two vectors will span a subspace V, why should we choose a spanning set with three vectors in it? Indeed, in the last example, the last vector in the spanning set $(4, 0, 0, 0), (0, 3, 0, 0), (1, 2, 0, 0)$ is a linear combination of the first two:

$$(1, 2, 0, 0) = \tfrac{1}{4}(4, 0, 0, 0) + \tfrac{2}{3}(0, 3, 0, 0).$$

Thus any linear combination of these three vectors can always be written as a linear combination of just $(4, 0, 0, 0)$ and $(0, 3, 0, 0)$. The third vector, being a linear combination of the first two, is "extraneous information."

These ideas suggest the following definition.

DEFINITION 5.14   *Linear Dependence and Independence*

Let $\mathbf{F}_1, \ldots, \mathbf{F}_k$ be vectors in R^n.

1. $\mathbf{F}_1, \ldots, \mathbf{F}_k$ are linearly dependent if and only if one of these vectors is a linear combination of the others.

2. $\mathbf{F}_1, \ldots, \mathbf{F}_k$ are linearly independent if and only if they are not linearly dependent.

Linear dependence of $\mathbf{F}_1, \ldots, \mathbf{F}_k$ means that, whatever information these vectors carry, not all of them are needed, because at least one of them can be written in terms of the others. For example, if

$$\mathbf{F}_k = \alpha_1 \mathbf{F}_1 + \cdots + \alpha_{k-1} \mathbf{F}_{k-1},$$

then knowing just $\mathbf{F}_1, \ldots, \mathbf{F}_{k-1}$ gives us $\mathbf{F}_k$ as well. In this sense, linearly dependent vectors are redundant. We can remove at least one $\mathbf{F}_j$, and the remaining $k - 1$ vectors will span the same subspace of R^n that $\mathbf{F}_1, \ldots, \mathbf{F}_k$ do.

Linear independence means that no one of the vectors $\mathbf{F}_1, \ldots, \mathbf{F}_k$ is a linear combination of the others. Whatever these vectors are telling us (for example, specifying a subspace), we need all of them or we lose information. If we omit $\mathbf{F}_k$, we cannot retrieve it from $\mathbf{F}_1, \ldots, \mathbf{F}_{k-1}$.

EXAMPLE 5.18

The vectors $(1, 1, 0)$ and $(-2, 0, 3)$ are linearly independent in R^3. To prove this, suppose instead that these vectors are linearly dependent. Then one is a linear combination of the other, say

$$(-2, 0, 3) = \alpha(1, 1, 0).$$

But then, from the first components, $\alpha = -2$, while from the second components, $\alpha = 0$, an impossibility.

These vectors span the subspace V of R^3 consisting of all vectors

$$\alpha(-2, 0, 3) + \beta(1, 1, 0).$$

Both of the vectors $(1, 1, 0)$ and $(-2, 0, 3)$ are needed to describe V. If we omit one, say $(1, 1, 0)$, then the subspace of R^3 spanned by the remaining vector, $(-2, 0, 3)$, is different from V. For example, it does not have $(1, 1, 0)$ in it. ∎

The following is a useful characterization of linear dependence and independence.

THEOREM 5.13

Let $\mathbf{F}_1, \ldots, \mathbf{F}_k$ be vectors in R^n. Then

1. $\mathbf{F}_1, \ldots, \mathbf{F}_k$ are linearly dependent if and only if there are real numbers $\alpha_1, \ldots, \alpha_k$, not all zero, such that

$$\alpha_1 \mathbf{F}_1 + \alpha_2 \mathbf{F}_2 + \cdots + \alpha_k \mathbf{F}_k = \mathbf{O}.$$

2. $\mathbf{F}_1, \ldots, \mathbf{F}_k$ are linearly independent if and only if an equation

$$\alpha_1 \mathbf{F}_1 + \alpha_2 \mathbf{F}_2 + \cdots + \alpha_k \mathbf{F}_k = \mathbf{O}$$

can hold only if $\alpha_1 = \alpha_2 = \cdots = \alpha_k = 0$.

Proof   To prove (1), suppose first that $\mathbf{F}_1, \ldots, \mathbf{F}_k$ are linearly dependent. Then at least one of these vectors is a linear combination of the others. Say, to be specific, that

$$\mathbf{F}_1 = \alpha_2 \mathbf{F}_2 + \cdots + \alpha_k \mathbf{F}_k.$$

Then

$$\mathbf{F}_1 - \alpha_2 \mathbf{F}_2 + \cdots + \alpha_k \mathbf{F}_k = \mathbf{O}.$$

But this is a linear combination of $\mathbf{F}_1, \ldots, \mathbf{F}_k$ adding up to the zero vector and having a nonzero coefficient (the coefficient of $\mathbf{F}_1$ is 1).

Conversely, suppose

$$\alpha_1 \mathbf{F}_1 + \alpha_2 \mathbf{F}_2 + \cdots + \alpha_k \mathbf{F}_k = \mathbf{O}$$

with at least some $\alpha_j \neq 0$. We want to show that $\mathbf{F}_1, \ldots, \mathbf{F}_k$ are linearly dependent. By renaming the vectors if necessary, we may suppose for convenience that $\alpha_1 \neq 0$. But then

$$\mathbf{F}_1 = -\frac{\alpha_2}{\alpha_1}\mathbf{F}_2 - \cdots - \frac{\alpha_k}{\alpha_1}\mathbf{F}_k,$$

so $\mathbf{F}_1$ is a linear combination of $\mathbf{F}_2, \ldots, \mathbf{F}_k$ and hence $\mathbf{F}_1, \ldots, \mathbf{F}_k$ are linearly dependent. This completes the proof of (1).

Conclusion (2) follows from (1) and the fact that $\mathbf{F}_1, \ldots, \mathbf{F}_k$ are linearly independent exactly when these vectors are not linearly dependent. ∎

This theorem suggests a strategy for determining whether a given set of vectors is linearly dependent or independent. Given $\mathbf{F}_1, \ldots, \mathbf{F}_k$, set

$$\alpha_1 \mathbf{F}_1 + \alpha_2 \mathbf{F}_2 + \cdots + \alpha_k \mathbf{F}_k = \mathbf{O} \qquad (5.4)$$

and attempt to solve for the coefficients $\alpha_1, \ldots, \alpha_k$. If equation (5.4) forces $\alpha_1 = \cdots = \alpha_k = 0$, then $\mathbf{F}_1, \ldots, \mathbf{F}_k$ are linearly independent. If we can find at least one nonzero α_j so that equation (5.4) is true, then $\mathbf{F}_1, \ldots, \mathbf{F}_k$ are linearly dependent.

EXAMPLE 5.19

Consider $(1, 0, 3, 1)$, $(0, 1, -6, -1)$, and $(0, 2, 1, 0)$ in R^4. We want to know whether these vectors are linearly dependent or independent. Look at a linear combination

$$c_1(1, 0, 3, 1) + c_2(0, 1, -6, -1) + c_3(0, 2, 1, 0) = (0, 0, 0, 0).$$

If this is to hold, then each component of the vector $(c_1, c_2 + 2c_3, 3c_1 - 6c_2 + c_3, c_1 - c_2)$ must be zero:

$$c_1 = 0$$
$$c_2 + 2c_3 = 0$$
$$3c_1 - 6c_2 + c_3 = 0$$
$$c_1 - c_2 = 0.$$

The first equation gives $c_1 = 0$, so the fourth equation tells us that $c_2 = 0$. But then the second equation requires that $c_3 = 0$. Therefore, the only linear combination of these three vectors that equals the zero vector is the trivial linear combination (all coefficients zero), and by (2) of Theorem 5.13, the vectors are linearly independent. ∎

In the plane R^2, two vectors are linearly dependent if and only if they are parallel. In R^3, two vectors are linearly dependent if and only if they are parallel, and three vectors are linearly dependent if and only if they are in the same plane.

Any set of vectors that includes the zero vector must be linearly dependent. Consider, for example, the vectors $\mathbf{O}, \mathbf{F}_2, \ldots, \mathbf{F}_k$. Then the linear combination

$$1\mathbf{O} + 0\mathbf{F}_2 + \cdots + 0\mathbf{F}_k = \mathbf{O}$$

is a linear combination of these vectors that adds up to the zero vector but has a nonzero coefficient (the coefficient of $\mathbf{O}$ is 1). By Theorem 5.13(1), these vectors are linearly dependent.

There is a special circumstance in which it is particularly easy to tell that a set of vectors is linearly independent. This is given in the following lemma, which we will use later.

LEMMA 5.3

Let $\mathbf{F}_1, \ldots, \mathbf{F}_k$ be vectors in R^n. Suppose each $\mathbf{F}_j$ has a nonzero element in some component where each of the other $\mathbf{F}_i$s has a zero component. Then $\mathbf{F}_1, \ldots, \mathbf{F}_k$ are linearly independent. ■

An example will clarify why this is true.

EXAMPLE 5.20

Consider the vectors

$$\mathbf{F}_1 = (0, 4, 0, 0, 2), \quad \mathbf{F}_2 = (0, 0, 6, 0, -5), \quad \mathbf{F}_3 = (0, 0, 0, -4, 12).$$

To see why these are linearly independent, suppose

$$\alpha \mathbf{F}_1 + \beta \mathbf{F}_2 + \gamma \mathbf{F}_3 = (0, 0, 0, 0, 0).$$

Then

$$(0, 4\alpha, 6\beta, -4\gamma, 2\alpha - 5\beta + 12\gamma) = (0, 0, 0, 0, 0).$$

From the second components, $4\alpha = 0$ so $\alpha = 0$. From the third components, $6\beta = 0$ so $\beta = 0$. And from the fourth components, $-4\gamma = 0$ so $\gamma = 0$. Then the vectors are linearly independent by Theorem 5.13(2). The fact that each of the vectors has a nonzero element where all the others have only zero components makes it particularly easy to conclude that $\alpha = \beta = \gamma = 0$, and that is what is needed to apply Theorem 5.13. ■

There is another important setting in which it is easy to tell that vectors are linearly independent. Nonzero vectors $\mathbf{F}_1, \ldots, \mathbf{F}_k$ in R^n are said to be *mutually orthogonal* if each is orthogonal to each of the other vectors in the set. That is, $\mathbf{F}_i \cdot \mathbf{F}_j = 0$ if $i \neq j$. Mutually orthogonal nonzero vectors are necessarily linearly independent.

THEOREM 5.14

Let $\mathbf{F}_1, \ldots, \mathbf{F}_k$ be mutually orthogonal nonzero vectors in R^n. Then $\mathbf{F}_1, \ldots, \mathbf{F}_k$ are linearly independent.

Proof Suppose

$$\alpha_1 \mathbf{F}_1 + \alpha_2 \mathbf{F}_2 + \cdots + \alpha_k \mathbf{F}_k = \mathbf{O}.$$

For any $j = 1, \ldots, k$,

$(\alpha_1 \mathbf{F}_1 + \alpha_2 \mathbf{F}_2 + \cdots + \alpha_k \mathbf{F}_k) \cdot \mathbf{F}_j = 0$

$$= \alpha_1 \mathbf{F}_1 \cdot \mathbf{F}_j + \alpha_2 \mathbf{F}_2 \cdot \mathbf{F}_j + \cdots + \alpha_j \mathbf{F}_j \cdot \mathbf{F}_j + \cdots + \alpha_k \mathbf{F}_k \cdot \mathbf{F}_j$$

$$= c_j \mathbf{F}_j \cdot \mathbf{F}_j = c_j \|\mathbf{F}_j\|^2.$$

because $\mathbf{F}_i \cdot \mathbf{F}_j = 0$ if $i \neq j$. But $\mathbf{F}_j$ is not the zero vector, so $\|\mathbf{F}_j\|^2 \neq 0$, hence $c_j = 0$. Therefore, each coefficient is zero and $\mathbf{F}_1, \ldots, \mathbf{F}_k$ are linearly independent by Theorem 5.13(2). ∎

EXAMPLE 5.21

The vectors $(-4, 0, 0)$, $(0, -2, 1)$, $(0, 1, 2)$ are linearly independent in R^3, because each is orthogonal to the other two. ∎

A "smallest" spanning set for a subspace of R^n is called a *basis* for that subspace.

DEFINITION 5.15 Basis

Let V be a subspace of R^n. A set of vectors $\mathbf{F}_1, \ldots, \mathbf{F}_k$ in V form a basis for V if $\mathbf{F}_1, \ldots, \mathbf{F}_k$ are linearly independent and also span V.

Thus, for $\mathbf{F}_1, \ldots, \mathbf{F}_k$ to be a basis for V, every vector in V must be a linear combination of $\mathbf{F}_1, \ldots, \mathbf{F}_k$, and if any $\mathbf{F}_j$ is omitted from the list $\mathbf{F}_1, \ldots, \mathbf{F}_k$, the remaining vectors do not span V. In particular, if $\mathbf{F}_j$ is omitted, then the subspace spanned by $\mathbf{F}_1, \ldots, \mathbf{F}_{j-1}, \mathbf{F}_{j+1}, \ldots, \mathbf{F}_k$ cannot contain $\mathbf{F}_j$, because by linear independence, $\mathbf{F}_j$ is not a linear combination of $\mathbf{F}_1, \ldots, \mathbf{F}_{j-1}, \mathbf{F}_{j+1}, \ldots, \mathbf{F}_k$.

EXAMPLE 5.22

$\mathbf{i}, \mathbf{j}$, and $\mathbf{k}$ form a basis for R^3, and $\mathbf{e}_1, \mathbf{e}_2, \ldots, \mathbf{e}_n$ form a basis for R^n. ∎

EXAMPLE 5.23

Let V be the subspace of R^n consisting of all n-vectors with zero first component. Then $\mathbf{e}_2, \ldots, \mathbf{e}_n$ form a basis for V. ∎

EXAMPLE 5.24

In R^2, let V consist of all vectors parallel to the line $y = 4x$. Every vector in V is a multiple of $(1, 4)$. This vector by itself forms a basis for V. In fact, any vector $(\alpha, 4\alpha)$ with $\alpha \neq 0$ forms a basis for V. ∎

EXAMPLE 5.25

In R^3, let M be the subspace of all vectors on or parallel to the plane $x + y + z = 0$. A vector (x, y, z) in R^3 is in M exactly when $z = -x - y$, so such a vector can be written

$$(x, y, z) = (x, y, -x - y) = x(1, 0, -1) + z(0, 1, -1).$$

The vectors $(1, 0, -1)$ and $(0, 1, -1)$ therefore span M. Since these two vectors are linearly independent, they form a basis for M. ∎

EXAMPLE 5.26

Let W consist of all vectors $(0, x, y, 0, y)$ in R^5. Then W is a subspace of R^5, and every vector in W has the form

$$(0, x, y, 0, y) = x(0, 1, 0, 0, 0) + y(0, 0, 1, 0, 1).$$

The vectors $(0, 1, 0, 0, 0)$ and $(0, 0, 1, 0, 1)$ span W. Since these two vectors are linearly independent (neither is a scalar multiple of the other), they form a basis for W. ■

EXAMPLE 5.27

Let K be the subspace of R^6 consisting of all vectors $(x, y, 0, x - y, x + y, z)$. Then every vector in K has the form

$$(x, y, 0, x - y, x + y, z) = x(1, 0, 0, 1, 1, 0) + y(0, 1, 0, -1, 1, 0) + z(0, 0, 0, 0, 0, 1).$$

Therefore, the three vectors $(1, 0, 0, 1, 1, 0)$, $(0, 1, 0, -1, 1, 0)$, and $(0, 0, 0, 0, 0, 1)$ span K. Since these vectors are linearly independent, they form a basis for K. ■

We may think of a basis of V as a minimal linearly independent spanning set $\mathbf{F}_1, \ldots, \mathbf{F}_k$ for V. If we omit any of these vectors, the remaining vectors will not be enough to span V. And if we use additional vectors, say the set $\mathbf{F}_1, \ldots, \mathbf{F}_k, \mathbf{H}$, then this set also spans V but is not linearly independent (because $\mathbf{H}$ is a linear combination of $\mathbf{F}_1, \ldots, \mathbf{F}_k$).

There is nothing unique about a basis for a subspace of R^n. Any nontrivial subspace of R^n has infinitely many different bases. However, it is a theorem of linear algebra, which we will not prove, that for a given subspace V of R^n, every basis has the same *number* of vectors in it. This number is the dimension of the subspace.

> **DEFINITION 5.16   *Dimension***
>
> The dimension of a subspace of R^n is the number of vectors in any basis for the subspace.

In particular, R^n (which is a subspace of itself) has dimension n, a basis consisting of the n-vectors $\mathbf{e}_1, \ldots, \mathbf{e}_n$. The subspace K of R^6 in Example 5.27 has dimension 3. The subspaces in Examples 5.25 and 5.26 each has dimension 2.

SECTION 5.5   PROBLEMS

In each of Problems 1 through 10, determine whether the given vectors are linearly independent or dependent in R^n for appropriate n.

1. $3\mathbf{i} + 2\mathbf{j}, \mathbf{i} - \mathbf{j}$ in R^3

2. $2\mathbf{i}, 3\mathbf{j}, 5\mathbf{i} - 12\mathbf{k}, \mathbf{i} + \mathbf{j} + \mathbf{k}$ in R^3

3. $(8, 0, 2, 0, 0, 0, 0), (0, 0, 0, 0, 1, -1, 0)$ in R^7

4. $(1, 0, 0, 0), (0, 1, 1, 0), (-4, 6, 6, 0)$ in R^4

5. $(1, 2, -3, 1), (4, 0, 0, 2), (6, 4, -6, 4)$ in R^4

6. $(0, 1, 1, 1), (-3, 2, 4, 4), (-2, 2, 34, 2), (1, 1, -6-2)$ in R^4

7. $(1, -2), (4, 1), (6, 6)$ in R^2

8. $(-1, 1, 0, 0, 0), (0, -1, 1, 0, 0), (0, 1, 1, 1, 0)$ in R^5

9. $(-2, 0, 0, 1, 1)$, $(1, 0, 0, 0, 0)$, $(0, 0, 0, 0, 2)$, $(1, -1, 3, 3, 1)$ in R^5

10. $(3, 0, 0, 4)$, $(2, 0, 0, 8)$ in R^4

11. Prove that three vectors in R^3 are linearly dependent if and only if their scalar triple product is zero. (See Problems 39–42 in Section 5.3).

In each of Problems 12 through 16, use the result of Problem 11 to determine whether the three vectors in R^3 are linearly dependent or independent.

12. $3\mathbf{i} + 6\mathbf{j} - \mathbf{k}$, $8\mathbf{i} + 2\mathbf{j} - 4\mathbf{k}$, $\mathbf{i} - \mathbf{j} + \mathbf{k}$

13. $\mathbf{i} + 6\mathbf{j} - 2\mathbf{k}$, $-\mathbf{i} + 4\mathbf{j} - 3\mathbf{k}$, $\mathbf{i} + 16\mathbf{j} - 7\mathbf{k}$

14. $4\mathbf{i} - 3\mathbf{j} + \mathbf{k}$, $10\mathbf{i} - 3\mathbf{j}$, $2\mathbf{i} - 6\mathbf{j} + 3\mathbf{k}$

15. $8\mathbf{i} + 6\mathbf{j}$, $2\mathbf{i} - 4\mathbf{j}$, $\mathbf{i} + \mathbf{k}$

16. $12\mathbf{i} - 3\mathbf{k}$, $\mathbf{i} + 2\mathbf{j} - \mathbf{k}$, $-3\mathbf{i} + 4\mathbf{j}$

In each of Problems 17 through 24, determine a basis for the subspace S of R^n and determine the dimension of the subspace.

17. S consists of all vectors $(x, y, -y, -x)$ in R^4.

18. S consists of all vectors $(x, y, 2x, 3y)$ in R^4.

19. S consists of all vectors in the plane $2x - y + z = 0$.

20. S consists of all vectors $(x, y, -y, x - y, z)$ in R^5.

21. S consists of all vectors in R^4 with zero second component.

22. S consists of all vectors $(-x, x, y, 2y)$ in R^4.

23. S consists of all vectors parallel to the line $y = 4x$ in R^2.

24. S consists of all vectors parallel to the plane $4x + 2y - z = 0$ in R^3.

25. Determine the dimension of the subspace of R^2 consisting of all vectors parallel to a given line through the origin. Is the dimension different if the line is thought of as being in R^3 rather than R^2?

26. Determine the dimension of the subspace of R^3 consisting of all vectors parallel to a given plane through the origin.

27. Prove that any three vectors in R^2 are linearly dependent.

28. Prove that any four vectors in R^3 are linearly dependent.

29. Suppose we have a set of k linearly independent vectors in R^n. Prove that any $k - 1$ of these vectors are also linearly independent.

30. Suppose we are given a set of k linearly dependent vectors in R^n. Prove that any finite set of vectors in R^n containing these k vectors is also linearly dependent.

31. Let $\mathbf{F}$, $\mathbf{G}$, and $\mathbf{H}$ be linearly independent vectors in R^3. Let $\mathbf{V}$ be any vector in R^3. Prove that

$$\mathbf{V} = \frac{[\mathbf{V}, \mathbf{G}, \mathbf{H}]}{[\mathbf{F}, \mathbf{G}, \mathbf{H}]}\,\mathbf{F} + \frac{[\mathbf{V}, \mathbf{H}, \mathbf{F}]}{[\mathbf{F}, \mathbf{G}, \mathbf{H}]}\,\mathbf{G} + \frac{[\mathbf{V}, \mathbf{F}, \mathbf{G}]}{[\mathbf{F}, \mathbf{G}, \mathbf{H}]}\,\mathbf{H}.$$

(Recall scalar triple product from Problems 39–42 in Section 5.3).

5.6   Abstract Vector Spaces

We will now place the concepts of the preceding two sections in a more general setting. This will enable us to consider vector spaces of functions, with important ramifications for studying systems of linear differential equations, Fourier series, and other important topics. Keeping in mind that R^n forms the prototype for the definitions we are about to make should be helpful in understanding the ideas involved.

Throughout this section, the term *set* refers to any collection of objects. We can speak, for example, of a set of vectors in R^n, or a set of functions that are differentiable on an interval, or a set of functions that are continuous at the origin.

DEFINITION 5.17   *Vector Space*

Let V be a set. Suppose there are two operations, one that combines pairs of objects in V and is called *addition*, and one that combines real numbers with objects in V and is called *multiplication by scalars*. The addition operation is denoted $\oplus$ and obeys the following rules. For x, y, and z in V:

1. $x \oplus y$ is in V.
2. $x \oplus y = y \oplus x$.
3. $x \oplus (y \oplus z) = (x \oplus y) \oplus z$.
4. There is some object, called θ, in V such that $x \oplus \theta = x$ for every x in V.
5. For each x in V, there is some $\hat{x}$ in V such that $x \oplus \hat{x} = \theta$.

The operation of multiplication by scalars is denoted $\otimes$ and obeys the following rules: For any real numbers α and β:

6. $\alpha \otimes x$ is in V.
7. $(\alpha + \beta) \otimes x = (\alpha \otimes x) \oplus (\beta \otimes x)$.
8. $(\alpha\beta) \otimes x = \alpha \otimes (\beta \otimes x)$.
9. $\alpha \otimes (x \oplus y) = (\alpha \otimes x) \oplus (\alpha \otimes y)$.
10. $1 \otimes x = x$.

Then V is called a *vector space*, with respect to the operations of addition $\oplus$ and scalar multiplication $\otimes$.

The notion of a vector space therefore involves three things—the set of objects and the two operations, one for adding objects in the set, the other for multiplying objects in the set by real numbers. If either the set or one or both of these operations is changed, but together they still satisfy the conditions of the definition, then we have a different vector space.

It is customary to refer to objects in a vector space as vectors, whether or not the vector space under consideration consists of vectors in the classical sense of objects in R^n. We will not, however, use boldface print for members of an abstract vector space. Often such spaces have functions as their elements, and functions are generally not written that way.

The special vector θ whose addition to x results in x again, is called the *zero vector* of the vector space. If x is in V, the object $\hat{x}$ such that $x \oplus \hat{x} = \theta$, whose existence is guaranteed by (5) of the definition, is called the *negative* of x and is denoted $-x$. Thus we can write $x \oplus (-x) = \theta$. A sum $x \oplus (-y)$ is often denoted $x \ominus y$.

We will look at some examples of vector spaces.

EXAMPLE 5.28

R^n is a vector space. Here the set consists of all n-tuples of real numbers, $\oplus$ is the operation of adding n-vectors (by adding respective components), and $\otimes$ is the operation of multiplying an n-vector by a real number. In this vector space, the zero vector is $\theta = (0, 0, \ldots, 0)$, and $-(x_1, \ldots, x_n) = (-x_1, \ldots, -x_n)$. ∎

EXAMPLE 5.29

Let V consist of all real-valued functions that are defined and continuous on $[0, 1]$. Functions are added in the usual way. This means that $f \oplus g$ is the function defined by

$$(f \oplus g)(x) = f(x) + g(x) \quad \text{for } 0 \leq x \leq 1.$$

Since a sum of continuous functions is continuous, $f \oplus g$ is in V whenever f and g are. The zero vector in this space is the zero function defined by

$$\theta(x) = 0 \quad \text{for } 0 \le x \le 1.$$

The negative of f is the function $-f$ defined by

$$(-f)(x) = -f(x) \quad \text{for } 0 \le x \le 1.$$

The product $\alpha \otimes f$ of a real number with a function in this vector space is the function defined by

$$(\alpha \otimes f)(x) = \alpha f(x) \quad \text{for } 0 \le x \le 1.$$

It is routine to check that the conditions of the definition are fulfilled. ■

EXAMPLE 5.30

Let D consist of all real-valued functions defined and differentiable on the real line. With the usual addition of functions and multiplication by real numbers (as in the preceding example), D is a vector space. Here we are using the fact that a sum of differentiable functions is differentiable, and the product of a real number with a differentiable function yields a differentiable function. ■

EXAMPLE 5.31

Let S consist of all solutions of $y'' + 4xy' + x^2 y = 0$, with $\oplus$ the usual addition of functions and $\otimes$ the usual product of a function by a real number. A sum of solutions of this differential equation is a solution, and a product of a solution with a real number is also a solution. The zero function is a solution and is the zero vector θ of S. It is routine to verify that the conditions of the definition are satisfied. S is called the solution space of this differential equation. ■

We can take all of the definitions and concepts developed throughout this chapter for the vector space R^n and bring them over verbatim to the general vector space notion we have just defined. In particular, linear combinations of vectors, linear dependence and independence, spanning sets, basis, and dimension carry over directly.

EXAMPLE 5.32

Let S consist of all solutions of $y'' - 5y' + 6y = 0$. S forms a vector space under the usual addition of functions and multiplication by constants. We find that $y_1(x) = e^{2x}$ and $y_2(x) = e^{3x}$ form a fundamental set of solutions, in the sense defined in Section 2.2. This means that every solution is a linear combination of y_1 and y_2, and these solutions, being linearly independent, form a basis for the solution space. This space has dimension 2, the order of the differential equation. ■

EXAMPLE 5.33

Let M consist of all symbols

$$\begin{pmatrix} a & b \\ c & d \end{pmatrix},$$

with a, b, c, and d real numbers. Define

$$\begin{pmatrix} a_1 & b_1 \\ c_1 & d_1 \end{pmatrix} \oplus \begin{pmatrix} a_2 & b_2 \\ c_2 & d_2 \end{pmatrix} = \begin{pmatrix} a_1 + a_2 & b_1 + b_2 \\ c_1 + c_2 & d_1 + d_2 \end{pmatrix}$$

and

$$\alpha \otimes \begin{pmatrix} a & b \\ c & d \end{pmatrix} = \begin{pmatrix} \alpha a & \alpha b \\ \alpha c & \alpha d \end{pmatrix}.$$

With

$$\theta = \begin{pmatrix} 0 & 0 \\ 0 & 0 \end{pmatrix},$$

M is a vector space with these operations. This is called the *space of* 2×2 *matrices*. This vector space has dimension 4, because the vectors

$$\begin{pmatrix} 1 & 0 \\ 0 & 0 \end{pmatrix}, \begin{pmatrix} 0 & 1 \\ 0 & 0 \end{pmatrix}, \begin{pmatrix} 0 & 0 \\ 1 & 0 \end{pmatrix}, \quad \text{and} \quad \begin{pmatrix} 0 & 0 \\ 0 & 1 \end{pmatrix}$$

form a basis. ■

The vector space of the last example is identical in all of its properties to R^4. Simply take a 4-vector (a, b, c, d) and write it as

$$\begin{pmatrix} a & b \\ c & d \end{pmatrix}.$$

In other words, M and R^4 differ only notationally. Such spaces are said to be *isomorphic*.

In some vector spaces, it is possible to mimic the notions of norm and inner product seen in R^n. This will be important when we deal with Fourier series and eigenfunction expansions in solving partial differential equations.

EXAMPLE 5.34

Let V be the vector space of functions that are continuous on $[-\pi, \pi]$, with piecewise continuous derivatives. (We encountered piecewise continuous functions in treating the Laplace transform.) If f and g are in V, define

$$f \cdot g = \int_{-\pi}^{\pi} f(x) g(x) \, dx.$$

This operation satisfies all of the conditions we saw for the dot product of n-vectors in Theorem 5.9(1), (2), (3), and (5) . We therefore refer to $f \cdot g$ as the *dot product of vectors (functions) in V*. Once we have a dot product, we can define a norm by recalling Theorem 5.3(4) or Theorem 5.9(4). Define

$$\| f \| = (f \cdot f)^{1/2} = \left(\int_{-\pi}^{\pi} f(x)^2 \, dx \right)^{1/2}.$$

This defines a concept of length for functions, the length of f being $\|f\|$. We can also speak of functions as being orthogonal if $f \cdot g = 0$.

In terms of this norm, the triangle inequality takes the form

$$\|f + g\| = \left(\int_{-\pi}^{\pi} [f(x) + g(x)]^2 \, dx \right)^{1/2}$$

$$\leq \|f\| + \|g\| = \left(\int_{-\pi}^{\pi} [f(x)]^2 \, dx \right)^{1/2} + \left(\int_{-\pi}^{\pi} [g(x)]^2 \, dx \right)^{1/2}.$$

This integral inequality may not be obvious outside of this vector space context, but in terms of the triangle inequality for norms it follows immediately. The Cauchy–Schwarz inequality takes the form

$$|f \cdot g| = \left| \int_{-\pi}^{\pi} f(x)g(x) \, dx \right| \leq \|f\| \|g\| \leq \left(\int_{-\pi}^{\pi} f(x)^2 \, dx \right)^{1/2} \left(\int_{-\pi}^{\pi} g(x)^2 \, dx \right)^{1/2}.$$

When we do Fourier analysis, it will be important to observe that the functions 1, $\cos(nx)$, and $\sin(nx)$ (for $n = 1, 2, 3, \ldots$) form an orthogonal set in V. This means that each one of these functions is orthogonal to the others, as we see by a straightforward integration. For positive integers n and m:

$$1 \cdot \cos(nx) = \int_{-\pi}^{\pi} \cos(nx) \, dx = 0,$$

$$1 \cdot \sin(nx) = \int_{-\pi}^{\pi} \sin(nx) \, dx = 0,$$

$$\cos(nx) \cdot \cos(mx) = \int_{-\pi}^{\pi} \cos(nx)\cos(mx) \, dx = 0 \quad \text{if } n \neq m,$$

$$\sin(nx) \cdot \sin(mx) = \int_{-\pi}^{\pi} \sin(nx)\sin(mx) \, dx = 0 \quad \text{if } n \neq m,$$

and

$$\cos(nx) \cdot \sin(mx) = \int_{-\pi}^{\pi} \cos(nx)\sin(mx) \, dx = 0.$$

These functions are therefore linearly independent (same reasoning as in Theorem 5.14). We will see later that these functions (for all positive integers n) form a basis for V, which is an infinite dimensional vector space. This means that any function in V can be expanded in infinite series of these functions, having the form

$$f(x) = A_0 + \sum_{n=1}^{\infty} A_n \cos(nx) + B_n \sin(nx).$$

This expansion is the Fourier series of f and will be taken up in Chapter 13. ∎

In the next chapter, we will develop the matrix machinery needed to treat systems of linear algebraic equations and linear differential equations.

SECTION 5.6   PROBLEMS

1. Let P_m be the set of all polynomials with real coefficients and degree not exceeding m. Show that P_m is a vector space, with the usual addition of polynomials and multiplication of polynomials by real numbers as operations. Determine a basis for P_m, hence its dimension.

2. Let V consist of all real-valued functions defined and continuous on $[a, b]$ such that $f(a) = f(b) = 0$. Show that V is a vector space, with the usual addition of functions and multiplication of functions by real numbers. Does this space have a finite dimension (that is, are there finitely many functions $f_1, \ldots, f_n$ such that every function f in V is a linear combination of $f_1, \ldots, f_n$)? To answer this question, construct infinitely many functions $f_1, f_2, \ldots$ such that each f_j is in V but no f_m is a linear combination of $f_1, \ldots, f_{m-1}$.

3. Let V consist of all real-valued functions defined and continuous on $[a, b]$ such that $f(b) = 1$, with the usual addition of functions and multiplication by real numbers. Show that V is not a vector space.

4. Let P consist of all polynomials with real coefficients, with the usual addition of polynomials and multiplication by constants. Show that P is a vector space. Show that P does not have finite dimension. Show that P_m of Problem 1 is a subspace of P.

5. Let D_n consist of all polynomials with real coefficients and having degree n. With the usual addition of polynomials and multiplication of polynomials by constants, show that D_n is not a vector space.

6. Let S consist of all solutions of $y'' - 8xy = 0$. With the usual addition of functions and multiplication of functions by constants, show that S is a vector space. It is not necessary to solve this differential equation to do this.

7. Let T consist of all solutions of $y'' - 8xy = x$. With the usual addition of functions and multiplication of functions by constants, show that T is not a vector space.

8. Let K consist of all symbols $\langle x, y \rangle$ with x and y real numbers. Define addition in K by

$$\langle x, y \rangle \oplus \langle u, v \rangle = \langle 2x + 2u, y + v \rangle.$$

Define multiplication of objects in K by scalars by

$$\alpha \otimes \langle x, y \rangle = \left\langle \frac{\alpha x}{2}, \frac{\alpha y}{2} \right\rangle.$$

Is K a vector space, using these operations?

9. Let W consist of all $\langle x, y, z \rangle$ such that x, y, and z are real numbers. Define

$$\langle x, y, z \rangle \oplus \langle u, v, w \rangle = \langle 3x + 2u, y + v, z - w \rangle.$$

Define

$$\alpha \otimes \langle x, y, z \rangle = \left\langle \alpha x, \alpha y, \frac{\alpha z}{2} \right\rangle.$$

Is W a vector space using these operations?

10. Let Q consist of all real-valued functions continuous on $[0, 1]$ such that $\int_0^1 f(x)\, dx = 0$. Using the usual addition of functions and multiplication of functions by constants, is Q a vector space?

11. Let Y consist of all real polynomials having degree ≤ 5 and 1 as a root. With the usual addition of polynomials and multiplication of polynomials by constants, is Y a vector space?

12. Let U consist of all real-valued functions f continuous on $[0, 1]$ such that $f(0) = f'(\frac{1}{2}) = f(1) = 0$. With the usual addition of functions and multiplication of functions by constants, is U vector space?

In each of Problems 13 through 18, list all the conditions that are violated in the definition of a vector space.

13. Q is the set of rational numbers, with the usual addition and multiplication of real numbers.

14. S is the set of solutions of $y'' - 4y = 8$, with the usual addition of functions and multiplication of functions by constants.

15. P is the set of real polynomials having degree 4 or 8, with the usual addition of polynomials and multiplication of polynomials by constants.

16. V consists of all symbols $\langle x, y \rangle$ with x and y real numbers, with the operations defined by

$$\langle x, y \rangle \oplus \langle u, v \rangle = \langle x + u, y - v \rangle$$

and

$$\alpha \otimes \langle x, y \rangle = \langle \alpha x, y \rangle.$$

17. P consists of all polynomials with real coefficients. Addition is the usual addition of polynomials, but α times any polynomial is defined to be x.

18. V consists of all real-valued functions f continuous on $[0, 1]$ and such that $f(\frac{1}{3}) = 1$, with the usual addition of functions and multiplication of functions by constants.

CHAPTER 6

Matrices and Systems of Linear Equations

This chapter is devoted to the notation and algebra of matrices, as well as their use in solving systems of linear algebraic equations.

To illustrate the idea of a matrix, consider a system of linear equations:

$$x_1 + 2x_2 - x_3 + 4x_4 = 0$$

$$3x_1 - 4x_2 + 2x_3 - 6x_4 = 0$$

$$x_1 - 3x_2 - 2x_3 + x_4 = 0.$$

All of the information needed to solve this system lies in its coefficients. Whether the first unknown is called x_1, or y_1, or some other name, is unimportant. It is important, however, that the coefficient of the first unknown in the second equation is 3. If we change this number we may change the solutions of the system.

We can therefore work with such a system by storing its coefficients in an array called a matrix:

$$\begin{pmatrix} 1 & 2 & -1 & 4 \\ 3 & -4 & 2 & -6 \\ 1 & -3 & -2 & 1 \end{pmatrix}.$$

This matrix displays the coefficients in the pattern in which they appear in the system of equations. The coefficients of the ith equation are in row i, and the coefficients of the jth unknown x_j are in column j. The number in row i, column j is the coefficient of x_j in equation i.

But matrices provide more than a visual aid or storage device. The algebra and calculus of matrices will form the basis for methods of solving systems of linear algebraic equations, and later for solving systems of linear differential equations and analyzing solutions of systems of nonlinear differential equations.

241

6.1 Matrices

DEFINITION 6.1 *Matrix*

An n by m matrix is an array of objects arranged in n rows and m columns.

We will denote matrices by boldface type, as was done with vectors. When $\mathbf{A}$ is an n by m matrix, we often write that $\mathbf{A}$ is $n \times m$ (read "n by m"). The first integer is the number of rows in the matrix, and the second integer is the number of columns. The objects in the matrix may be numbers, functions, or other quantities. For example,

$$\begin{pmatrix} 2 & 1 & \pi \\ 1 & \sqrt{2} & -5 \end{pmatrix}$$

is a 2×3 matrix,

$$\begin{pmatrix} e^{2x} & e^{-4x} \\ \cos(x) & x^2 \end{pmatrix}$$

is a 2×2 matrix, and

$$\begin{pmatrix} 0 \\ -4 \\ x^3 \\ 2 \end{pmatrix}$$

is a 4×1 matrix.

A matrix having the same number of rows as columns is called a *square matrix*. The 2×2 matrix shown above is square.

The object in row i and column j of a matrix is called the i, j *element*, or i, j *entry* of the matrix. If a matrix is denoted by an uppercase letter, say $\mathbf{A}$, then its i, j element is often denoted a_{ij} and we write $\mathbf{A} = [a_{ij}]$. For example, if

$$\mathbf{H} = [h_{ij}] = \begin{pmatrix} 0 & x \\ 1 - \sin(x) & 1 - 2i \\ x^2 & i \end{pmatrix}$$

then $\mathbf{H}$ is a 3×2 matrix, and $h_{11} = 0$, $h_{12} = x$, $h_{21} = 1 - \sin(x)$, $h_{22} = 1 - 2i$, $h_{31} = x^2$, and $h_{32} = i$. We will be dealing with matrices whose elements are real or complex numbers, or functions.

Sometimes it is also convenient to denote the i, j element of $\mathbf{A}$ by $(\mathbf{A})_{ij}$. In the matrix $\mathbf{H}$,

$$(\mathbf{H})_{22} = 1 - 2i \quad \text{and} \quad (\mathbf{H})_{31} = x^2.$$

DEFINITION 6.2 *Equality*

Matrices $\mathbf{A} = [a_{ij}]$ and $\mathbf{B} = [b_{ij}]$ are equal if and only if they have the same number of rows, the same number of columns, and for each i and j, $a_{ij} = b_{ij}$.

If two matrices have different numbers of rows or columns, or if the objects in a particular location in the matrices are different, then the matrices are unequal.

6.1.1 Matrix Algebra

We will develop the operations of addition and multiplication of matrices and multiplication of a matrix by a number.

DEFINITION 6.3 *Matrix Addition*

If $\mathbf{A} = [a_{ij}]$ and $\mathbf{B} = [b_{ij}]$ are $n \times m$ matrices, then their sum is the $n \times m$ matrix

$$\mathbf{A} + \mathbf{B} = [a_{ij} + b_{ij}].$$

We therefore add matrices by adding corresponding elements. For example,

$$\begin{pmatrix} 1 & 2 & -3 \\ 4 & 0 & 2 \end{pmatrix} + \begin{pmatrix} -1 & 6 & 3 \\ 8 & 12 & 14 \end{pmatrix} = \begin{pmatrix} 0 & 8 & 0 \\ 12 & 12 & 16 \end{pmatrix}.$$

If two matrices are of different dimensions (different numbers of rows or columns), then they cannot be added, just as we do not add 4-vectors and 7-vectors.

DEFINITION 6.4 *Product of a Matrix and a Scalar*

If $\mathbf{A} = [a_{ij}]$ and α is a scalar, then $\alpha\mathbf{A}$ is the matrix defined by

$$\alpha\mathbf{A} = [\alpha a_{ij}].$$

This means that we multiply a matrix by α by multiplying each element of the matrix by α. For example,

$$3 \begin{pmatrix} 2 & 0 \\ 0 & 0 \\ 1 & 4 \\ 2 & 6 \end{pmatrix} = \begin{pmatrix} 6 & 0 \\ 0 & 0 \\ 3 & 12 \\ 6 & 18 \end{pmatrix}$$

and

$$x \begin{pmatrix} 1 & x \\ -x & \cos(x) \end{pmatrix} = \begin{pmatrix} x & x^2 \\ -x^2 & x\cos(x) \end{pmatrix}.$$

Some, but not all, pairs of matrices can be multiplied.

DEFINITION 6.5 *Multiplication of Matrices*

Let $\mathbf{A} = [a_{ij}]$ be an $n \times r$ matrix and $\mathbf{B} = [b_{ij}]$ an $r \times m$ matrix. Then the matrix product $\mathbf{AB}$ is the $n \times m$ matrix whose i, j element is

$$a_{i1}b_{1j} + a_{i2}b_{2j} + \cdots + a_{ir}b_{rj}.$$

That is,

$$\mathbf{AB} = \left[\sum_{k=1}^{r} a_{ik}b_{kj} \right].$$

If we think of each row of **A** as an r-vector, and each column of **B** as an r-vector, then the i, j element of **AB** is the dot product of row i of **A** with column j of **B**:

$$i, j \text{ element of } \mathbf{AB} = (\text{row } i \text{ of } \mathbf{A}) \cdot (\text{column } j \text{ of } \mathbf{B}).$$

This is why the number of columns of **A** must equal the number of rows of **B** for **AB** to be defined. These rows of **A** and columns of **B** must be vectors of the same length in order to take this dot product. Thus, not every pair of matrices can be multiplied. Further, even when **AB** is defined, **BA** need not be.

We will give one rationale for defining matrix multiplication in this way shortly. First we will look at some examples of matrix products and then develop the rules of matrix algebra.

EXAMPLE 6.1

Let

$$\mathbf{A} = \begin{pmatrix} 1 & 3 \\ 2 & 5 \end{pmatrix} \quad \text{and} \quad \mathbf{B} = \begin{pmatrix} 1 & 1 & 3 \\ 2 & 1 & 4 \end{pmatrix}.$$

Then **A** is 2×2 and **B** is 2×3, so **AB** is defined (number of columns of **A** equals the number of rows of **B**). Further, **AB** is 2×3 (number of rows of **A**, number of columns of **B**).

Now compute

$$\mathbf{AB} = \begin{pmatrix} 1 & 3 \\ 2 & 5 \end{pmatrix} \begin{pmatrix} 1 & 1 & 3 \\ 2 & 1 & 4 \end{pmatrix}$$

$$= \begin{pmatrix} (1, 3) \cdot (1, 2) & (1, 3) \cdot (1, 1) & (1, 3) \cdot (3, 4) \\ (2, 5) \cdot (1, 2) & (2, 5) \cdot (1, 1) & (2, 5) \cdot 3, 4) \end{pmatrix}$$

$$= \begin{pmatrix} 7 & 4 & 15 \\ 12 & 7 & 26 \end{pmatrix}.$$

In this example, **BA** is not defined, because the number of columns of **B**, which is 3, does not equal the number of rows of **A**, which is 2. ∎

EXAMPLE 6.2

Let

$$\mathbf{A} = \begin{pmatrix} 1 & 1 & 2 & 1 \\ 4 & 1 & 6 & 2 \end{pmatrix} \quad \text{and} \quad \mathbf{B} = \begin{pmatrix} -1 & 8 \\ 2 & 1 \\ 1 & 1 \\ 12 & 6 \end{pmatrix}.$$

Since **A** is 2×4 and **B** is 4×2, **AB** is defined and is a 2×2 matrix:

$$\mathbf{AB} = \begin{pmatrix} (1, 1, 2, 1) \cdot (-1, 2, 1, 12) & (1, 1, 2, 1) \cdot (8, 1, 1, 6) \\ (4, 1, 6, 2) \cdot (-1, 2, 1, 12) & (4, 1, 6, 2) \cdot (8, 1, 1, 6) \end{pmatrix}$$

$$= \begin{pmatrix} 15 & 17 \\ 28 & 51 \end{pmatrix}.$$

In this example, **BA** is also defined and is 4×4:

$$\mathbf{BA} = \begin{pmatrix} -1 & 8 \\ 2 & 1 \\ 1 & 1 \\ 12 & 6 \end{pmatrix} \begin{pmatrix} 1 & 1 & 2 & 1 \\ 4 & 1 & 6 & 2 \end{pmatrix}$$

$$= \begin{pmatrix} 31 & 7 & 46 & 15 \\ 6 & 3 & 10 & 4 \\ 5 & 2 & 8 & 3 \\ 36 & 18 & 60 & 24 \end{pmatrix}. \quad \blacksquare$$

As the last example shows, even when both **AB** and **BA** are defined, these may be matrices of different dimensions. Matrix multiplication is noncommutative, and it is the exception rather than the rule to have **AB** equal **BA**.

If **A** is a square matrix, then **AA** is defined and is also square. Denote **AA** as $\mathbf{A}^2$. Similarly, $\mathbf{A}(\mathbf{A}^2) = \mathbf{A}^3$ and, for any positive integer k, $\mathbf{A}^k = \mathbf{AA} \cdots \mathbf{A}$, a product with k factors.

Some of the rules for manipulating matrices are like those for real numbers.

THEOREM 6.1

Let **A**, **B**, and **C** be matrices. Then, whenever the indicated operations are defined, we have:

1. $\mathbf{A} + \mathbf{B} = \mathbf{B} + \mathbf{A}$.
2. $\mathbf{A}(\mathbf{B} + \mathbf{C}) = \mathbf{AB} + \mathbf{AC}$.
3. $(\mathbf{A} + \mathbf{B})\mathbf{C} = \mathbf{AC} + \mathbf{BC}$.
4. $\mathbf{A}(\mathbf{BC}) = (\mathbf{AB})\mathbf{C}$. $\blacksquare$

For (1), both matrices must have the same dimensions, say $n \times m$. For (2), **B** and **C** must have the same dimensions, and the number of columns in **A** must equal the number of rows in **B** and in **C**. For (4), **A** must be $n \times r$, **B** must be $r \times k$, and **C** must be $k \times m$. Then $\mathbf{A}(\mathbf{BC})$ and $(\mathbf{AB})\mathbf{C}$ are $n \times m$.

Proof   The theorem is proved by direct appeal to the definitions. We will provide the details for (1) and (2). To prove (1), let $\mathbf{A} = [a_{ij}]$ and $\mathbf{B} = [b_{ij}]$. Then

$$\mathbf{A} + \mathbf{B} = [a_{ij} + b_{ij}] = [b_{ij} + a_{ij}] = \mathbf{B} + \mathbf{A},$$

because each a_{ij} and b_{ij} is a number or function and the addition of these objects is commutative.

For (2), let $\mathbf{A} = [a_{ij}]$, $\mathbf{B} = [b_{ij}]$, and $\mathbf{C} = [c_{ij}]$. Suppose **A** is $n \times k$ and **B** and **C** are $k \times m$. Then $\mathbf{B} + \mathbf{C}$ is $k \times m$, so $\mathbf{A}(\mathbf{B} + \mathbf{C})$ is defined and is $n \times m$. And **AB** and **BC** are both defined and $n \times m$. There remains to show that the i, j element of $\mathbf{AB} + \mathbf{AC}$ is the same as the i, j element of $\mathbf{A}(\mathbf{B} + \mathbf{C})$.

Row i of **A**, and columns j of **B** and **C**, are k-vectors, and from properties of the dot product,

i, j element of $\mathbf{A}(\mathbf{B} + \mathbf{C})$ = (row i of **A**) · (column j of $\mathbf{B} + \mathbf{C}$)

$\qquad$ = (row i of **A**) · (column j of **B**) + (row i of **A**) · (column j of **C**)

$\qquad$ = (i, j element of **AB**) + (i, j element of **AC**)

$\qquad$ = i, j element of $\mathbf{AB} + \mathbf{AC}$. $\blacksquare$

We have already noted that matrix multiplication does not behave in some ways like multiplication of numbers. Here is a summary of three significant differences.

Difference 1 For matrices, even when **AB** and **BA** are both defined, possibly **AB** ≠ **BA**.

EXAMPLE 6.3

$$\begin{pmatrix} 1 & 0 \\ -2 & 4 \end{pmatrix} \begin{pmatrix} -2 & 6 \\ 1 & 3 \end{pmatrix} = \begin{pmatrix} -2 & 6 \\ 8 & 0 \end{pmatrix}$$

but

$$\begin{pmatrix} -2 & 6 \\ 1 & 3 \end{pmatrix} \begin{pmatrix} 1 & 0 \\ -2 & 4 \end{pmatrix} = \begin{pmatrix} -14 & 24 \\ -5 & 12 \end{pmatrix}. \blacksquare$$

Difference 2 There is no cancellation in products. If **AB** = **AC**, we cannot infer that **B** = **C**.

EXAMPLE 6.4

$$\begin{pmatrix} 1 & 1 \\ 3 & 3 \end{pmatrix} \begin{pmatrix} 4 & 2 \\ 3 & 16 \end{pmatrix} = \begin{pmatrix} 1 & 1 \\ 3 & 3 \end{pmatrix} \begin{pmatrix} 2 & 7 \\ 5 & 11 \end{pmatrix}$$

$$= \begin{pmatrix} 7 & 18 \\ 21 & 54 \end{pmatrix}.$$

But

$$\begin{pmatrix} 4 & 2 \\ 3 & 16 \end{pmatrix} \neq \begin{pmatrix} 2 & 7 \\ 5 & 11 \end{pmatrix}. \blacksquare$$

Difference 3 The product of two nonzero matrices may be zero.

EXAMPLE 6.5

$$\begin{pmatrix} 1 & 2 \\ 0 & 0 \end{pmatrix} \begin{pmatrix} 6 & 4 \\ -3 & -2 \end{pmatrix} = \begin{pmatrix} 0 & 0 \\ 0 & 0 \end{pmatrix}. \blacksquare$$

6.1.2 Matrix Notation for Systems of Linear Equations

Matrix notation is very efficient for writing systems of linear algebraic equations. Consider, for example, the system

$$2x_1 - x_2 + 3x_3 + x_4 = 1$$
$$x_1 + 3x_2 - 2x_4 = 0$$
$$-4x_1 - x_2 + 2x_3 - 9x_4 = -3.$$

The matrix of coefficients of this system is the 3 × 4 matrix

$$\mathbf{A} = \begin{pmatrix} 2 & -1 & 3 & 1 \\ 1 & 3 & 0 & -2 \\ -4 & -1 & 2 & -9 \end{pmatrix}.$$

Row i contains the coefficients of the ith equation, and column j contains the coefficients of x_j. Define

$$\mathbf{X} = \begin{pmatrix} x_1 \\ x_2 \\ x_3 \\ x_4 \end{pmatrix} \quad \text{and} \quad \mathbf{B} = \begin{pmatrix} 1 \\ 0 \\ -3 \end{pmatrix}.$$

Then

$$\mathbf{AX} = \begin{pmatrix} 2 & -1 & 3 & 1 \\ 1 & 3 & 0 & -2 \\ -4 & -1 & 2 & -9 \end{pmatrix} \begin{pmatrix} x_1 \\ x_2 \\ x_3 \\ x_4 \end{pmatrix}$$

$$= \begin{pmatrix} 2x_1 - x_2 + 3x_3 + x_4 \\ x_1 + 3x_2 - 2x_4 \\ -4x_1 - x_2 + 2x_3 - 9x_4 \end{pmatrix} = \begin{pmatrix} 1 \\ 0 \\ -3 \end{pmatrix}.$$

We can therefore write the system of equations in matrix form as

$$\mathbf{AX} = \mathbf{B}.$$

This is more than just notation. Soon this matrix formulation will enable us to use matrix operations to solve the system.

A similar approach can be taken toward systems of linear differential equations. Consider the system

$$x_1' + tx_2' - x_3' = 2t - 1$$
$$t^2 x_1' - \cos(t)x_2' - x_3' = e^t.$$

Let

$$\mathbf{A} = \begin{pmatrix} 1 & t & -1 \\ t^2 & -\cos(t) & -1 \end{pmatrix}, \quad \mathbf{X} = \begin{pmatrix} x_1 \\ x_2 \\ x_3 \end{pmatrix}, \quad \text{and} \quad \mathbf{F} = \begin{pmatrix} 2t - 1 \\ e^t \end{pmatrix}.$$

Then the system can be written

$$\mathbf{AX'} = \mathbf{F},$$

in which $\mathbf{X'}$ is formed by differentiating each matrix element of $\mathbf{X}$. As with systems of linear algebraic equations, this formulation will enable us to bring matrix methods to bear on solving the system of differential equations.

In both of these formulations, the definition of matrix product played a key role. Matrix multiplication may seem unmotivated at first, but it is just right for converting a system of linear algebraic or differential equations to a matrix equation.

6.1.3 Some Special Matrices

Some matrices occur often enough to warrant special names and notation.

DEFINITION 6.6   *Zero Matrix*

$\mathbf{O}_{n,m}$ denotes the $n \times m$ zero matrix, having each element equal to zero.

For example,

$$\mathbf{O}_{2,3} = \begin{pmatrix} 0 & 0 & 0 \\ 0 & 0 & 0 \end{pmatrix}.$$

If $\mathbf{A}$ is $n \times m$, then

$$\mathbf{A} + \mathbf{O}_{n,m} = \mathbf{O}_{n,m} + \mathbf{A} = \mathbf{A}.$$

The negative of $\mathbf{A}$ is the matrix obtained by replacing each element of $\mathbf{A}$ with its negative. This matrix is denoted $-\mathbf{A}$. If $\mathbf{A} = [a_{ij}]$, then $-\mathbf{A} = [-a_{ij}]$. If $\mathbf{A}$ is $n \times m$, then

$$\mathbf{A} + (-\mathbf{A}) = \mathbf{O}_{n,m}.$$

Usually, we write $\mathbf{A} + (-\mathbf{B})$ as $\mathbf{A} - \mathbf{B}$.

DEFINITION 6.7 *Identity Matrix*

The $n \times n$ identity matrix is the matrix $\mathbf{I}_n$ having each i, j element equal to zero if $i \neq j$ and each i, i element equal to 1.

For example,

$$\mathbf{I}_2 = \begin{pmatrix} 1 & 0 \\ 0 & 1 \end{pmatrix} \quad \text{and} \quad \mathbf{I}_3 = \begin{pmatrix} 1 & 0 & 0 \\ 0 & 1 & 0 \\ 0 & 0 & 1 \end{pmatrix}.$$

THEOREM 6.2

If $\mathbf{A}$ is $n \times m$, then

$$\mathbf{A}\mathbf{I}_m = \mathbf{I}_n\mathbf{A} = \mathbf{A}. \quad \blacksquare$$

We leave a proof of this to the student.

EXAMPLE 6.6

Let

$$\mathbf{A} = \begin{pmatrix} 1 & 0 \\ 2 & 1 \\ -1 & 8 \end{pmatrix}.$$

Then

$$\mathbf{I}_3\mathbf{A} = \begin{pmatrix} 1 & 0 & 0 \\ 0 & 1 & 0 \\ 0 & 0 & 1 \end{pmatrix} \begin{pmatrix} 1 & 0 \\ 2 & 1 \\ -1 & 8 \end{pmatrix} = \begin{pmatrix} 1 & 0 \\ 2 & 1 \\ -1 & 8 \end{pmatrix} = \mathbf{A}$$

and

$$\mathbf{A}\mathbf{I}_2 = \begin{pmatrix} 1 & 0 \\ 2 & 1 \\ -1 & 8 \end{pmatrix} \begin{pmatrix} 1 & 0 \\ 0 & 1 \end{pmatrix} = \begin{pmatrix} 1 & 0 \\ 2 & 1 \\ -1 & 8 \end{pmatrix} = \mathbf{A}. \ \blacksquare$$

DEFINITION 6.8 *Transpose*

If $\mathbf{A} = [a_{ij}]$ is an $n \times m$ matrix, then the transpose of $\mathbf{A}$ is the $m \times n$ matrix $\mathbf{A}^t = [a_{ji}]$.

The transpose of $\mathbf{A}$ is formed by making row k of $\mathbf{A}$, column k of $\mathbf{A}^t$.

EXAMPLE 6.7

Let

$$\mathbf{A} = \begin{pmatrix} -1 & 6 & 3 & 3 \\ 0 & \pi & 12 & -5 \end{pmatrix}.$$

This is a 2×4 matrix. The transpose is the 4×2 matrix

$$\mathbf{A}^t = \begin{pmatrix} -1 & 0 \\ 6 & \pi \\ 3 & 12 \\ 3 & -5 \end{pmatrix}. \ \blacksquare$$

THEOREM 6.3

1. $(\mathbf{I}_n)^t = \mathbf{I}_n$.
2. For any matrix $\mathbf{A}$, $(\mathbf{A}^t)^t = \mathbf{A}$.
3. If $\mathbf{AB}$ is defined, then $(\mathbf{AB})^t = \mathbf{B}^t\mathbf{A}^t$. $\blacksquare$

Conclusion (1) should not be surprising, since row i of $\mathbf{I}_n$ is the same as column i, so interchanging rows and columns has no effect.

Similarly, (2) is intuitively clear. If we interchange rows and columns of $\mathbf{A}$ to form $\mathbf{A}^t$, and then interchange the rows and columns of $\mathbf{A}^t$, we should put everything back where it was, resulting in $\mathbf{A}$ again.

We will prove conclusion (3).

Proof Let $\mathbf{A} = [a_{ij}]$ be $n \times k$ and let $\mathbf{B} = [b_{ij}]$ be $k \times m$. Then $\mathbf{AB}$ is defined and is $n \times m$. Since $\mathbf{B}^t$ is $m \times k$ and $\mathbf{A}^t$ is $k \times n$, then $\mathbf{B}^t\mathbf{A}^t$ is defined and is $m \times n$. Thus $(\mathbf{AB})^t$ and $\mathbf{B}^t\mathbf{A}^t$ have the same dimensions. Now we must show that the i, j element of $(\mathbf{AB})^t$ equals the i, j element of $\mathbf{B}^t\mathbf{A}^t$. Falling back on the definition of matrix product, we have

$$i, j \text{ element of } \mathbf{B}^t \mathbf{A}^t = \sum_{s=1}^{k} (\mathbf{B}^t)_{is} (\mathbf{A}^t)_{sj} = \sum_{s=1}^{k} b_{si} a_{js}$$

$$= \sum_{s=1}^{k} a_{js} b_{si} = j, i \text{ element of } \mathbf{AB} = i, j \text{ element of } (\mathbf{AB})^t.$$

This completes the proof of (3). ∎

In some calculations it is convenient to write the dot product of two n-vectors as a matrix product, using the transpose. Write the n-vector $(x_1, x_2, \ldots, x_n)$ as an $n \times 1$ column matrix:

$$\mathbf{X} = \begin{pmatrix} x_1 \\ x_2 \\ \vdots \\ x_n \end{pmatrix}.$$

Then

$$\mathbf{X}^t = \begin{pmatrix} x_1 & x_2 & \cdots & x_n \end{pmatrix},$$

a $1 \times n$ matrix. Let $(y_1, y_2, \ldots, y_n)$ also be an n-vector, which we write as an $n \times 1$ column matrix:

$$\mathbf{Y} = \begin{pmatrix} y_1 \\ y_2 \\ \vdots \\ y_n \end{pmatrix}.$$

Then $\mathbf{X}^t \mathbf{Y}$ is the 1×1 matrix

$$\mathbf{X}^t \mathbf{Y} = \begin{pmatrix} x_1 & x_2 & \cdots & x_n \end{pmatrix} \begin{pmatrix} y_1 \\ y_2 \\ \vdots \\ y_n \end{pmatrix}$$

$$= (x_1 y_1 + x_2 y_2 + \cdots + x_n y_n) = \mathbf{X} \cdot \mathbf{Y}.$$

Here we have written the resulting 1×1 matrix as just its single element, without the matrix brackets. This is common practice for 1×1 matrices. We now have the dot product of two n-vectors, written as $n \times 1$ column vectors, as the matrix product

$$\mathbf{X}^t \mathbf{Y}.$$

This will prove particularly useful when we treat eigenvalues of matrices.

6.1.4   Another Rationale for the Definition of Matrix Multiplication

We have seen that matrix multiplication allows us to write linear systems of algebraic and differential equations in compact matrix form as $\mathbf{AX} = \mathbf{B}$ or $\mathbf{AX}' = \mathbf{F}$.

Matrix products are also tailored to other purposes, such as changes of variables in linear equations. To illustrate, consider a 2×2 linear system

$$a_{11} x_1 + a_{12} x_2 = c_1$$

$$a_{21} x_1 + a_{22} x_2 = c_2.$$

(6.1)

Change variables by putting

$$x_1 = h_{11}y_1 + h_{12}y_2$$
$$x_2 = h_{21}y_1 + h_{22}y_2.$$

(6.2)

Then

$$a_{11}(h_{11}y_1 + h_{12}y_2) + a_{12}(h_{21}y_1 + h_{22}y_2) = c_1$$

and

$$a_{21}(h_{11}y_1 + h_{12}y_2) + a_{22}(h_{21}y_1 + h_{22}y_2) = c_2.$$

After rearranging terms, the transformed system is

$$(a_{11}h_{11} + a_{12}h_{21})y_1 + (a_{11}h_{12} + a_{12}h_{22})y_2 = c_1$$
$$(a_{21}h_{11} + a_{22}h_{21})y_1 + (a_{21}h_{12} + a_{22}h_{22})y_2 = c_2.$$

(6.3)

Now carry out the same transformation using matrices. Write the original system (6.1) as **AX = C**, where

$$\mathbf{A} = \begin{pmatrix} a_{11} & a_{12} \\ a_{21} & a_{22} \end{pmatrix}, \quad \mathbf{X} = \begin{pmatrix} x_1 \\ x_2 \end{pmatrix} \quad \text{and} \quad \mathbf{C} = \begin{pmatrix} c_1 \\ c_2 \end{pmatrix}$$

and the equations of the transformation (6.2) as **X = HY**, where

$$\mathbf{H} = \begin{pmatrix} h_{11} & h_{12} \\ h_{21} & h_{22} \end{pmatrix} \quad \text{and} \quad \mathbf{Y} = \begin{pmatrix} y_1 \\ y_2 \end{pmatrix}.$$

Then

$$\mathbf{AX} = \mathbf{A(HY)} = \mathbf{(AH)Y} = \mathbf{C}.$$

Now observe that

$$\mathbf{AH} = \begin{pmatrix} a_{11} & a_{12} \\ a_{21} & a_{22} \end{pmatrix} \begin{pmatrix} h_{11} & h_{12} \\ h_{21} & h_{22} \end{pmatrix}$$
$$= \begin{pmatrix} a_{11}h_{11} + a_{12}h_{21} & a_{11}h_{12} + a_{12}h_{22} \\ a_{21}h_{11} + a_{22}h_{21} & a_{21}h_{12} + a_{22}h_{22} \end{pmatrix},$$

exactly as we found in the system (6.3) by term-by-term substitution. The definition of matrix product is just what is needed to carry out a linear change of variables. This idea also applies to linear transformations in systems of differential equations.

6.1.5   Random Walks in Crystals

We will conclude this section with another application of matrix multiplication, this time to the problem of enumerating the paths atoms can take through a crystal lattice.

Crystals have sites arranged in a lattice pattern. An atom may jump from a site it occupies to any adjacent, vacant site. If more than one adjacent site is vacant, the atom "selects" its target site at random. The path such an atom makes through the crystal is called a *random walk*.

We can represent the lattice of locations and adjacencies by drawing a point for each location, with a line between two points only if an atom can move directly from one to the other in the crystal. Such a diagram is called a *graph*. Figure 6.1 shows a typical graph. In this graph an atom could move from point v_1 to v_2 or v_3, to which it is connected by lines, but not directly to v_6 because there is no line between v_1 and v_6.

Two points are called *adjacent* in G if there is a line between them in the graph. A point is not considered adjacent to itself—there are no lines starting and ending at the same point.

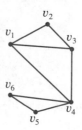

FIGURE 6.1 *A typical graph.*

A *walk* of length n in such a graph is a sequence $t_1, \ldots, t_{n+1}$ of points (not necessarily different) with each t_j adjacent to t_{j+1} in the graph. Such a walk represents a possible path an atom might take through various sites in the crystal. Points may repeat in a walk because an atom may return to the same site any number of times. A v_i–v_j walk is a walk that begins at v_i and ends at v_j.

Physicists and materials engineers who study crystals are interested in the following question: Given a crystal with n sites labeled $v_1, \ldots, v_n$, how many different walks of length k are there between any two sites (or from a site back to itself)?

Matrices enter into the solution of this problem as follows. Define the *adjacency matrix* $\mathbf{A}$ of the graph to be the $n \times n$ matrix having each i, i element zero, and for $i \neq j$, the i, j element equal to 1 if there is a line in the graph between v_i and v_j, and 0 if there is no such line. For example, the graph of Figure 6.1 has adjacency matrix

$$\mathbf{A} = \begin{pmatrix} 0 & 1 & 1 & 1 & 0 & 0 \\ 1 & 0 & 1 & 0 & 0 & 0 \\ 1 & 1 & 0 & 1 & 0 & 0 \\ 1 & 0 & 1 & 0 & 1 & 1 \\ 0 & 0 & 0 & 1 & 0 & 1 \\ 0 & 0 & 0 & 1 & 1 & 0 \end{pmatrix}.$$

The 1, 2 element of $\mathbf{A}$ is 1 because there is a line between v_1 and v_2, while the 1, 5 element is zero because there is no line between v_1 and v_5.

The following remarkable theorem uses the adjacency matrix to solve the walk-enumeration problem.

THEOREM 6.4

Let $\mathbf{A} = [a_{ij}]$ be the adjacency matrix of a graph G having points $v_1, \ldots, v_n$. Let k be any positive integer. Then the number of distinct v_i–v_j walks of length k in G is equal to the i, j element of $\mathbf{A}^k$. ∎

We can therefore calculate the number of random walks of length k between any two points (or from any point back to itself) by reading the elements of the kth power of the adjacency matrix.

Proof Proceed by mathematical induction on k. First consider the case $k = 1$. If $i \neq j$, there is a v_i–v_j walk of length 1 in G exactly when there is a line between v_i and v_j, and in this case $a_{ij} = 1$. There is no v_i–v_j walk of length 1 if v_i and v_j have no line between them, and in this case $a_{ij} = 0$. If $i = j$, there is no v_i–v_i walk of length 1, and $a_{ii} = 0$. Thus, in the case $k = 1$, the i, j element of $\mathbf{A}$ gives the number of walks of length 1 from v_i to v_j, and the conclusion of the theorem is true.

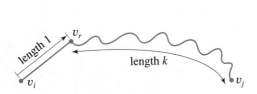

FIGURE 6.2

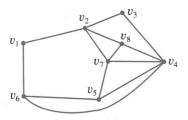

FIGURE 6.3

Now assume that the conclusion of the theorem is true for walks of length k. We will prove that the conclusion holds for walks of length $k + 1$. Thus, we are assuming that the i, j element of $\mathbf{A}^k$ is the number of distinct v_i–v_j walks of length k in G, and we want to prove that the i, j element of $\mathbf{A}^{k+1}$ is the number of distinct v_i–v_j walks of length $k + 1$.

Consider how a v_i–v_j walk of length $k + 1$ is formed. First, there must be a v_i–v_r walk of length 1 from v_i to some point v_r adjacent to v_i, followed by a v_r–v_j walk of length k (Figure 6.2). Therefore,

$$\text{Number of distinct } v_i\text{–}v_j \text{ walks of length } k + 1$$

$$= \text{Sum of the number of distinct } v_r\text{–}v_j \text{ walks of length } k,$$

with the sum taken over all points v_r adjacent to v_i. Now $a_{ir} = 1$ if v_r is adjacent to v_i, and $a_{ir} = 0$ otherwise. Further, by the inductive hypothesis, the number of distinct v_r–v_j walks of length k is the r, j element of $\mathbf{A}^k$. Denote $\mathbf{A}^k = \mathbf{B} = [b_{ij}]$. Then, for $r = 1, \ldots, n$,

$$a_{ir}b_{rj} = 0 \quad \text{if } v_r \text{ is not adjacent to } v_i$$

and

$$a_{ir}b_{rj} = \text{the number of distinct } v_i\text{–}v_j \text{ walks of length } k + 1$$

$$\text{passing through } v_r, \text{ if } v_r \text{ is adjacent to } v_i.$$

Therefore, the number of v_i–v_j walks of length $k + 1$ in G is

$$a_{i1}b_{1j} + a_{i2}b_{2j} + \cdots + a_{in}b_{nj}$$

because this counts the number of walks of length k from v_r to v_j for each point v_r adjacent to v_i. But this sum is exactly the i, j element of $\mathbf{AB}$, which is $\mathbf{A}^{k+1}$. This completes the proof by induction. ∎

For example, the adjacency matrix of the graph of Figure 6.3 is

$$\mathbf{A} = \begin{pmatrix} 0 & 1 & 0 & 0 & 0 & 1 & 0 & 0 \\ 1 & 0 & 1 & 0 & 0 & 0 & 1 & 1 \\ 0 & 1 & 0 & 1 & 0 & 0 & 0 & 0 \\ 0 & 0 & 1 & 0 & 1 & 1 & 1 & 1 \\ 0 & 0 & 0 & 1 & 0 & 1 & 1 & 0 \\ 1 & 0 & 0 & 1 & 1 & 0 & 0 & 0 \\ 0 & 1 & 0 & 1 & 1 & 0 & 0 & 1 \\ 0 & 1 & 0 & 1 & 0 & 0 & 1 & 0 \end{pmatrix}.$$

Suppose we want the number of v_4–v_7 walks of length 3 in G. Calculate

$$\mathbf{A}^3 = \begin{pmatrix} 0 & 5 & 1 & 4 & 2 & 4 & 3 & 2 \\ 6 & 2 & 7 & 4 & 5 & 4 & 9 & 8 \\ 1 & 7 & 0 & 8 & 3 & 2 & 3 & 2 \\ 4 & 4 & 8 & 6 & 8 & 8 & 11 & 10 \\ 2 & 5 & 3 & 8 & 4 & 6 & 8 & 4 \\ 4 & 4 & 2 & 8 & 6 & 2 & 4 & 4 \\ 3 & 9 & 3 & 11 & 8 & 4 & 6 & 7 \\ 2 & 8 & 2 & 10 & 4 & 4 & 7 & 4 \end{pmatrix}.$$

We read from the 4, 7 element of $\mathbf{A}^3$ that there are 11 walks of length 3 from v_4 to v_7. For this relatively simple graph, we can actually list all these walks:

$$v_4v_7v_4v_7; \ v_4v_3v_4v_7; \ v_4v_8v_4v_7; \ v_4v_5v_4v_7; \ v_4v_6v_4v_7;$$

$$v_4v_7v_8v_7; \ v_4v_7v_5v_7; \ v_4v_7v_2v_7; \ v_4v_3v_2v_7; \ v_4v_8v_2v_7; \ v_4v_6v_5v_7.$$

Obviously, it would not be practical to determine the number of v_i–v_j walks of length k by explicitly listing them if k or n is large. Software routines for matrix calculations make this theorem a practical solution to the random walk counting problem.

SECTION 6.1   PROBLEMS

In each of Problems 1 through 6, carry out the requested computation with the given matrices $\mathbf{A}$ and $\mathbf{B}$.

1. $\mathbf{A} = \begin{pmatrix} 1 & -1 & 3 \\ 2 & -4 & 6 \\ -1 & 1 & 2 \end{pmatrix}, \mathbf{B} = \begin{pmatrix} -4 & 0 & 0 \\ -2 & -1 & 6 \\ 8 & 15 & 4 \end{pmatrix};$

$2\mathbf{A} - 3\mathbf{B}$

2. $\mathbf{A} = \begin{pmatrix} -2 & 2 \\ 0 & 1 \\ 14 & 2 \\ 6 & 8 \end{pmatrix}, \mathbf{B} = \begin{pmatrix} 3 & 4 \\ 2 & 1 \\ 14 & 16 \\ 1 & 25 \end{pmatrix};$

$-5\mathbf{A} + 3\mathbf{B}$

3. $\mathbf{A} = \begin{pmatrix} x & 1-x \\ 2 & e^x \end{pmatrix}, \mathbf{B} = \begin{pmatrix} 1 & -6 \\ x & \cos(x) \end{pmatrix};$

$\mathbf{A}^2 + 2\mathbf{AB}$

4. $\mathbf{A} = (14), \mathbf{B} = (-12); -3\mathbf{A} - 5\mathbf{B}$

5. $\mathbf{A} = \begin{pmatrix} 1 & -2 & 1 & 7 & -9 \\ 8 & 2 & -5 & 0 & 0 \end{pmatrix},$

$\mathbf{B} = \begin{pmatrix} -5 & 1 & 8 & 21 & 7 \\ 12 & -6 & -2 & -1 & 9 \end{pmatrix};$

$4\mathbf{A} + 8\mathbf{B}$

6. $\mathbf{A} = \begin{pmatrix} -2 & 3 \\ 1 & 1 \end{pmatrix}, \mathbf{B} = \begin{pmatrix} 0 & 8 \\ -5 & 1 \end{pmatrix}; \mathbf{A}^3 - \mathbf{B}^2$

In each of Problems 7 through 20, determine which of $\mathbf{AB}$ and $\mathbf{BA}$ are defined. Carry out the products that are defined.

7. $\mathbf{A} = \begin{pmatrix} -4 & 6 & 2 \\ -2 & -2 & 3 \\ 1 & 1 & 8 \end{pmatrix},$

$\mathbf{B} = \begin{pmatrix} -2 & 4 & 6 & 12 & 5 \\ -3 & -3 & 1 & 1 & 4 \\ 0 & 0 & 1 & 6 & -9 \end{pmatrix}$

8. $\mathbf{A} = \begin{pmatrix} -2 & -4 \\ 3 & -1 \end{pmatrix}, \mathbf{B} = \begin{pmatrix} 6 & 8 \\ 1 & -4 \end{pmatrix}$

9. $\mathbf{A} = (-1 \ \ 6 \ \ 2 \ \ 14 \ \ -22), \mathbf{B} = \begin{pmatrix} -3 \\ 2 \\ 6 \\ 0 \\ -4 \end{pmatrix}$

10. $\mathbf{A} = \begin{pmatrix} -3 & 1 \\ 6 & 2 \\ 18 & -22 \\ 1 & 6 \end{pmatrix}, \mathbf{B} = \begin{pmatrix} -16 & 0 & 0 & 28 \\ 0 & 1 & 1 & 26 \end{pmatrix}$

11. $\mathbf{A} = \begin{pmatrix} -21 & 4 & 8 & -3 \\ 12 & 1 & 0 & 14 \\ 1 & 16 & 0 & -8 \\ 13 & 4 & 8 & 0 \end{pmatrix}$,

$\mathbf{B} = \begin{pmatrix} -9 & 16 & 3 & 2 \\ 5 & 9 & 14 & 0 \end{pmatrix}$

12. $\mathbf{A} = \begin{pmatrix} -2 & 4 \\ 3 & 9 \end{pmatrix}$, $\mathbf{B} = \begin{pmatrix} 1 & -3 & 7 & 2 \\ -5 & 6 & 1 & 0 \end{pmatrix}$

13. $\mathbf{A} = \begin{pmatrix} -4 & -2 & 0 \\ 0 & 5 & 3 \\ -3 & 1 & 1 \end{pmatrix}$, $\mathbf{B} = \begin{pmatrix} 1 & -3 & 4 \end{pmatrix}$

14. $\mathbf{A} = \begin{pmatrix} 3 \\ 0 \\ -1 \\ 4 \end{pmatrix}$, $\mathbf{B} = \begin{pmatrix} 3 & -2 & 7 \end{pmatrix}$

15. $\mathbf{A} = \begin{pmatrix} 7 & -8 \\ 1 & 6 \end{pmatrix}$, $\mathbf{B} = \begin{pmatrix} 1 & -4 & 3 \\ -4 & 7 & 0 \end{pmatrix}$

16. $\mathbf{A} = \begin{pmatrix} -3 & 2 \\ 0 & -2 \\ 1 & 8 \\ 3 & -3 \end{pmatrix}$, $\mathbf{B} = \begin{pmatrix} -5 & 5 & 7 & 2 \end{pmatrix}$

17. $\mathbf{A} = \begin{pmatrix} 4 & -2 & 8 \\ 1 & 6 & -4 \\ 2 & 2 & 0 \end{pmatrix}$, $\mathbf{B} = \begin{pmatrix} 2 & -4 & 6 \\ 1 & 9 & -5 \\ 1 & 1 & 1 \end{pmatrix}$

18. $\mathbf{A} = \begin{pmatrix} 1 \\ 0 \\ -4 \end{pmatrix}$, $\mathbf{B} = \begin{pmatrix} 2 & -6 \end{pmatrix}$

19. $\mathbf{A} = \begin{pmatrix} 1 & -2 \\ 2 & 4 \end{pmatrix}$,

$\mathbf{B} = \begin{pmatrix} -1 & 3 & 2 & 9 & -4 & 6 \\ 0 & -1 & 6 & 0 & 9 & -4 \end{pmatrix}$

20. $\mathbf{A} = \begin{pmatrix} -2 & 1 \\ 2 & 0 \\ 0 & 9 \\ 6 & -5 \end{pmatrix}$, $\mathbf{B} = \begin{pmatrix} 1 & 1 & -5 \\ 0 & 4 & 2 \end{pmatrix}$

In each of Problems 21 through 25, determine if **AB** is defined and if **BA** is defined. For those products that are defined, give the dimensions of the product matrix.

21. **A** is 14×21, **B** is 21×14.

22. **A** is 18×4, **B** is 18×4.

23. **A** is 6×2, **B** is 4×6.

24. **A** is 1×3, **B** is 3×3.

25. **A** is 7×6, **B** is 7×7.

26. Find nonzero 2×2 matrices **A**, **B**, and **C** such that **BA** = **CA** but **B** $\neq$ **C**.

27. A matrix is called *symmetric* if it equals its transpose. Prove that any symmetric matrix must be square.

28. Let **A** be any square matrix. Prove that $\mathbf{A} + \mathbf{A}^t$ is symmetric.

29. Let **A** and **B** be $n \times n$ symmetric matrices.

(a) Give an example to show that **AB** need not be symmetric.

(b) Prove that **AB** is symmetric if and only if **AB** = **BA**.

30. Let **A** be an $n \times m$ matrix. Prove that $\mathbf{AI}_m = \mathbf{A}$ and $\mathbf{I}_n\mathbf{A} = \mathbf{A}$.

31. Let G be the graph of Figure 6.4. Determine the number of v_1–v_4 walks of lengths 3 and 4, the number of v_2–v_3 walks of length 3, and the number of v_2–v_4 walks of length 4 in G.

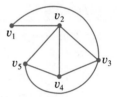

FIGURE 6.4

32. Let G be the graph of Figure 6.5. Determine the number of v_1–v_4 walks of length 4 in G. Determine the number of v_2–v_3 walks of length 2.

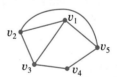

FIGURE 6.5

33. Let G be the graph of Figure 6.6. Determine the number of v_4–v_5 walks of length 2, the number of v_2–v_3 walks of length 3, and the number of v_1–v_2 and v_4–v_5 walks of length 4 in G.

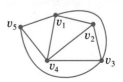

FIGURE 6.6

34. Let **A** be the adjacency matrix of a graph G.

(a) Prove that the i, j element of $\mathbf{A}^2$ equals the number of points of G that are neighbors of v_i in G. This number is called the *degree* of v_i.

(b) Prove that the i, j element of $\mathbf{A}^3$ equals twice the number of triangles in G containing v_i as a vertex. A triangle in G consists of three points, each a neighbor of the other two.

6.2 Elementary Row Operations and Elementary Matrices

When we solve a system of linear algebraic equations by elimination of unknowns, we routinely perform three kinds of operations: interchange of equations, multiplication of an equation by a nonzero constant, and addition of a constant multiple of one equation to another equation.

When we write a homogeneous system in matrix form $\mathbf{AX} = \mathbf{O}$, row k of **A** lists the coefficients in equation k of the system. The three operations on equations correspond, respectively, to the interchange of two rows of **A**, multiplication of a row **A** by a constant, and addition of a scalar multiple of one row of **A** to another row of **A**. We will focus on these row operations in anticipation of using them to solve the system.

DEFINITION 6.9

Let **A** be an $n \times m$ matrix. The three elementary row operations that can be performed on **A** are

1. Type I operation: Interchange two rows of **A**.
2. Type II operation: Multiply a row of **A** by a nonzero constant.
3. Type III operation: Add a scalar multiple of one row to another row.

The rows of **A** are m-vectors. In a Type II operation, multiply a row by a nonzero constant by multiplying this row vector by the number. That is, multiply each element of the row by that number. Similarly, in a Type III operation, we add a scalar multiple of one row vector to another row vector.

EXAMPLE 6.8

Let

$$\mathbf{A} = \begin{pmatrix} -2 & 1 & 6 \\ 1 & 1 & 2 \\ 0 & 1 & 3 \\ 2 & -3 & 4 \end{pmatrix}.$$

Type I operation: If we interchange rows 2 and 4 of **A**, we obtain the new matrix

$$\begin{pmatrix} -2 & 1 & 6 \\ 2 & -3 & 4 \\ 0 & 1 & 3 \\ 1 & 1 & 2 \end{pmatrix}.$$

Type II operation: Multiply row 2 of **A** by 7 to get

$$\begin{pmatrix} -2 & 1 & 6 \\ 7 & 7 & 14 \\ 0 & 1 & 3 \\ 2 & -3 & 4 \end{pmatrix}.$$

Type III operation: Add 2 times row 1 to row 3 of **A**, obtaining

$$\begin{pmatrix} -2 & 1 & 6 \\ 1 & 1 & 2 \\ -4 & 3 & 15 \\ 2 & -3 & 4 \end{pmatrix}. \ \blacksquare$$

Elementary row operations can be performed on any matrix. When performed on an identity matrix, we obtain special matrices that will be particularly useful. We therefore give matrices formed in this way a name.

DEFINITION 6.10 *Elementary Matrix*

An elementary matrix is a matrix formed by performing an elementary row operation on $\mathbf{I}_n$.

For example,

$$\begin{pmatrix} 0 & 1 & 0 \\ 1 & 0 & 0 \\ 0 & 0 & 1 \end{pmatrix}$$

is an elementary matrix, obtained from $\mathbf{I}_3$ by interchanging rows 1 and 2. And

$$\begin{pmatrix} 1 & 0 & 0 \\ 0 & 1 & 0 \\ -4 & 0 & 1 \end{pmatrix}$$

is the elementary matrix formed by adding -4 times row 1 of $\mathbf{I}_3$ to row 3.

The following theorem is the reason why elementary matrices are interesting. It says that each elementary row operation on **A** can be performed by multiplying **A** on the left by an elementary matrix.

THEOREM 6.5
_____

Let **A** be an $n \times m$ matrix. Let **B** be formed from **A** by an elementary row operation. Let **E** be the elementary matrix formed by performing this elementary row operation on $\mathbf{I}_n$. Then

$$\mathbf{B} = \mathbf{EA}. \ \blacksquare$$

We leave a proof to the exercises. It is instructive to see the theorem in practice.

EXAMPLE 6.9

Let

$$\mathbf{A} = \begin{pmatrix} 1 & -5 \\ 9 & 4 \\ -3 & 2 \end{pmatrix}.$$

Suppose we form $\mathbf{B}$ from $\mathbf{A}$ by interchanging rows 2 and 3 of $\mathbf{A}$. We can do this directly. But we can also form an elementary matrix by performing this operation on $\mathbf{I}_3$ to form

$$\mathbf{E} = \begin{pmatrix} 1 & 0 & 0 \\ 0 & 0 & 1 \\ 0 & 1 & 0 \end{pmatrix}.$$

Then

$$\mathbf{EA} = \begin{pmatrix} 1 & 0 & 0 \\ 0 & 0 & 1 \\ 0 & 1 & 0 \end{pmatrix} \begin{pmatrix} 1 & -5 \\ 9 & 4 \\ -3 & 2 \end{pmatrix} = \begin{pmatrix} 1 & -5 \\ -3 & 2 \\ 9 & 4 \end{pmatrix} = \mathbf{B}. \ \blacksquare$$

EXAMPLE 6.10

Let

$$\mathbf{A} = \begin{pmatrix} 0 & -7 & 3 & 6 \\ 5 & 1 & -11 & 3 \end{pmatrix}.$$

Form $\mathbf{C}$ from $\mathbf{A}$ by multiplying row 2 by -8. Again, we can do this directly. However, if we form $\mathbf{E}$ by performing this operation on $\mathbf{I}_2$, then

$$\mathbf{E} = \begin{pmatrix} 1 & 0 \\ 0 & -8 \end{pmatrix}$$

and

$$\mathbf{EA} = \begin{pmatrix} 1 & 0 \\ 0 & -8 \end{pmatrix} \begin{pmatrix} 0 & -7 & 3 & 6 \\ 5 & 1 & -11 & 3 \end{pmatrix}$$

$$= \begin{pmatrix} 0 & -7 & 3 & 6 \\ -40 & -8 & 88 & -24 \end{pmatrix} = \mathbf{C}. \ \blacksquare$$

EXAMPLE 6.11

Let

$$\mathbf{A} = \begin{pmatrix} -6 & 14 & 2 \\ 4 & 4 & -9 \\ -3 & 2 & 13 \end{pmatrix}.$$

Form $\mathbf{D}$ from $\mathbf{A}$ by adding 6 times row 1 to row 2. If we perform this operation on $\mathbf{I}_3$ to form

$$\mathbf{E} = \begin{pmatrix} 1 & 0 & 0 \\ 6 & 1 & 0 \\ 0 & 0 & 1 \end{pmatrix},$$

then

$$\mathbf{EA} = \begin{pmatrix} 1 & 0 & 0 \\ 6 & 1 & 0 \\ 0 & 0 & 1 \end{pmatrix} \begin{pmatrix} -6 & 14 & 2 \\ 4 & 4 & -9 \\ -3 & 2 & 13 \end{pmatrix}$$

$$= \begin{pmatrix} -6 & 14 & 2 \\ -32 & 88 & 3 \\ -3 & 2 & 13 \end{pmatrix} = \mathbf{B}. \; \blacksquare$$

Later we will want to perform not just one elementary row operation, but a sequence of such operations. Suppose we perform operation $\mathcal{O}_1$ on $\mathbf{A}$ to form $\mathbf{A}_1$, then operation $\mathcal{O}_2$ on $\mathbf{A}_1$ to form $\mathbf{A}_2$, and so on until finally we perform $\mathcal{O}_r$ on $\mathbf{A}_{r-1}$ to get $\mathbf{A}_r$. This process can be diagrammed:

$$\mathbf{A} \underset{\mathcal{O}_1}{\rightarrow} \mathbf{A}_1 \underset{\mathcal{O}_2}{\rightarrow} \mathbf{A}_2 \underset{\mathcal{O}_3}{\rightarrow} \cdots \underset{\mathcal{O}_{r-1}}{\rightarrow} \mathbf{A}_{r-1} \underset{\mathcal{O}_r}{\rightarrow} \mathbf{A}_r.$$

Let $\mathbf{E}_j$ be the elementary matrix obtained by performing operation $\mathcal{O}_j$ on $\mathbf{I}_n$. Then

$$\mathbf{A}_1 = \mathbf{E}_1\mathbf{A},$$

$$\mathbf{A}_2 = \mathbf{E}_2\mathbf{A}_1 = (\mathbf{E}_2\mathbf{E}_1)\mathbf{A},$$

$$\mathbf{A}_3 = \mathbf{E}_3\mathbf{A}_2 = (\mathbf{E}_3\mathbf{E}_2\mathbf{E}_1)\mathbf{A},$$

$$\vdots$$

$$\mathbf{A}_r = \mathbf{E}_r\mathbf{A}_{r-1} = (\mathbf{E}_r\mathbf{E}_{r-1} \cdots \mathbf{E}_3\mathbf{E}_2\mathbf{E}_1)\mathbf{A}.$$

This forms a matrix $\mathbf{\Omega} = \mathbf{E}_r\mathbf{E}_{r-1} \cdots \mathbf{E}_2\mathbf{E}_1$ such that

$$\mathbf{A}_r = \mathbf{\Omega}\mathbf{A}.$$

The significance of this equation is that we have produced a matrix $\mathbf{\Omega}$ such that multiplying $\mathbf{A}$ on the left by $\mathbf{\Omega}$ performs a given sequence of elementary row operations. $\mathbf{\Omega}$ is formed as a product $\mathbf{E}_r\mathbf{E}_{r-1} \cdots \mathbf{E}_2\mathbf{E}_1$ of elementary matrices, *in the correct order*, with each elementary matrix performing one of the prescribed elementary row operations in the sequence ($\mathbf{E}_1$ performs the first operation, $\mathbf{E}_2$ the second, and so on until $\mathbf{E}_r$ performs the last). We will record this result as a theorem.

THEOREM 6.6

Let $\mathbf{A}$ be an $n \times m$ matrix. If $\mathbf{B}$ is produced from $\mathbf{A}$ by any finite sequence of elementary row operations, then there is an $n \times n$ matrix $\mathbf{\Omega}$ such that

$$\mathbf{B} = \mathbf{\Omega}\mathbf{A}. \; \blacksquare$$

The proof of the theorem is contained in the line of reasoning outlined just prior to its statement.

EXAMPLE 6.12

Let

$$\mathbf{A} = \begin{pmatrix} 2 & 1 & 0 \\ 0 & 1 & 2 \\ -1 & 3 & 2 \end{pmatrix}.$$

We will form a new matrix **B** from **A** by performing, in order, the following operations:

$\mathcal{O}_1$: Interchange rows 1 and 2 of **A** to form $\mathbf{A}_1$.

$\mathcal{O}_2$: Multiply row 3 of $\mathbf{A}_1$ by 2 to form $\mathbf{A}_2$.

$\mathcal{O}_3$: Add 2 times row 1 to row 3 of $\mathbf{A}_2$ to get $\mathbf{A}_3 = \mathbf{B}$.

If we perform this sequence in order, starting with **A**, we get

$$\mathbf{A} \underset{\mathcal{O}_1}{\to} \mathbf{A}_1 = \begin{pmatrix} 0 & 1 & 2 \\ 2 & 1 & 0 \\ -1 & 3 & 2 \end{pmatrix} \underset{\mathcal{O}_2}{\to} \mathbf{A}_2 = \begin{pmatrix} 0 & 1 & 2 \\ 2 & 1 & 0 \\ -2 & 6 & 4 \end{pmatrix}$$

$$\underset{\mathcal{O}_3}{\to} \mathbf{A}_3 = \begin{pmatrix} 0 & 1 & 2 \\ 2 & 1 & 0 \\ -2 & 8 & 8 \end{pmatrix} = \mathbf{B}.$$

To produce $\mathbf{\Omega}$ such that $\mathbf{B} = \mathbf{\Omega} \mathbf{A}$, perform this sequence of operations in turn, beginning with $\mathbf{I}_3$:

$$\mathbf{I}_3 \underset{\mathcal{O}_1}{\to} \begin{pmatrix} 0 & 1 & 0 \\ 1 & 0 & 0 \\ 0 & 0 & 1 \end{pmatrix} \underset{\mathcal{O}_2}{\to} \begin{pmatrix} 0 & 1 & 0 \\ 1 & 0 & 0 \\ 0 & 0 & 2 \end{pmatrix} \underset{\mathcal{O}_3}{\to} \begin{pmatrix} 0 & 1 & 0 \\ 1 & 0 & 0 \\ 0 & 2 & 2 \end{pmatrix} = \mathbf{\Omega}.$$

Now check that

$$\mathbf{\Omega}\mathbf{A} = \begin{pmatrix} 0 & 1 & 0 \\ 1 & 0 & 0 \\ 0 & 2 & 2 \end{pmatrix} \begin{pmatrix} 2 & 1 & 0 \\ 0 & 1 & 2 \\ -1 & 3 & 2 \end{pmatrix}$$

$$= \begin{pmatrix} 0 & 1 & 2 \\ 2 & 1 & 0 \\ -2 & 8 & 8 \end{pmatrix} = \mathbf{B}.$$

It is also easy to check that $\mathbf{\Omega} = \mathbf{E}_3 \mathbf{E}_2 \mathbf{E}_1$, where $\mathbf{E}_j$ is the elementary matrix obtained by performing operation $\mathcal{O}_j$ on $\mathbf{I}_3$. ∎

EXAMPLE 6.13

Let

$$\mathbf{A} = \begin{pmatrix} 6 & -1 & 1 & 4 \\ 9 & 3 & 7 & -7 \\ 0 & 2 & 1 & 5 \end{pmatrix}.$$

We want to perform, in succession and in the given order, the following operations:

$\mathcal{O}_1$: Add (-3)(row 2) to row 3.

$\mathcal{O}_2$: Add 2(row 1) to row 2.

$\mathcal{O}_3$: Interchange rows 1 and 3.

$\mathcal{O}_4$: Multiply row 2 by -4.

Suppose the end result of these operations is the matrix $\mathbf{B}$. We will produce a 3×3 matrix $\mathbf{\Omega}$ such that $\mathbf{B} = \mathbf{\Omega A}$. Perform the sequence of operations, starting with $\mathbf{I}_3$:

$$\mathbf{I}_3 \underset{\mathcal{O}_1}{\to} \begin{pmatrix} 1 & 0 & 0 \\ 0 & 1 & 0 \\ 0 & -3 & 1 \end{pmatrix} \underset{\mathcal{O}_2}{\to} \begin{pmatrix} 1 & 0 & 0 \\ 2 & 1 & 0 \\ 0 & -3 & 1 \end{pmatrix} \underset{\mathcal{O}_3}{\to} \begin{pmatrix} 0 & -3 & 1 \\ 2 & 1 & 0 \\ 1 & 0 & 0 \end{pmatrix} \underset{\mathcal{O}_4}{\to} \begin{pmatrix} 0 & -3 & 1 \\ -8 & -4 & 0 \\ 1 & 0 & 0 \end{pmatrix} = \mathbf{\Omega}.$$

Then

$$\mathbf{\Omega A} = \begin{pmatrix} 0 & -3 & 1 \\ -8 & -4 & 0 \\ 1 & 0 & 0 \end{pmatrix} \begin{pmatrix} 6 & -1 & 1 & 4 \\ 9 & 3 & 7 & -7 \\ 0 & 2 & 1 & 5 \end{pmatrix} = \begin{pmatrix} -27 & -7 & -20 & 26 \\ -84 & -4 & -36 & -4 \\ 6 & -1 & 1 & 4 \end{pmatrix} = \mathbf{B}.$$

It is straightforward to check that $\mathbf{\Omega} = \mathbf{E}_4\mathbf{E}_3\mathbf{E}_2\mathbf{E}_1$, where $\mathbf{E}_j$ is the elementary matrix obtained from $\mathbf{I}_3$ by applying operation $\mathcal{O}_j$. If the operations $\mathcal{O}_j$ are performed in succession, starting with $\mathbf{A}$, then $\mathbf{B}$ results. ∎

DEFINITION 6.11   *Row Equivalence*

Two matrices are row equivalent if and only if one can be obtained from the other by a sequence of elementary row operations.

In each of the last two examples, $\mathbf{B}$ is row equivalent to $\mathbf{A}$. The relationship of row equivalence has the following properties:

THEOREM 6.7

1. Every matrix is row equivalent to itself. (This is the reflexive property.)
2. If $\mathbf{A}$ is row equivalent to $\mathbf{B}$, then $\mathbf{B}$ is row equivalent to $\mathbf{A}$. (This is the symmetry property.)
3. If $\mathbf{A}$ is row equivalent to $\mathbf{B}$, and $\mathbf{B}$ to $\mathbf{C}$, then $\mathbf{A}$ is row equivalent to $\mathbf{C}$. (This is transitivity.) ∎

A proof of the theorem is outlined in the exercises.

It is sometimes of interest to undo the effect of an elementary row operation. This can always be done by the same kind of elementary row operation. Consider each kind of operation in turn.

If we interchange rows i and j of $\mathbf{A}$ to form $\mathbf{B}$, then interchanging rows i and j of $\mathbf{B}$ yields $\mathbf{A}$ again. Thus a Type I operation can reverse a Type I operation.

If we form $\mathbf{C}$ from $\mathbf{A}$ by multiplying row i by nonzero α, then multiplying row i of $\mathbf{C}$ by $1/\alpha$ brings us back to $\mathbf{A}$. A Type II operation can reverse a Type II operation.

Finally, suppose we form $\mathbf{D}$ from $\mathbf{A}$ by adding $\alpha(\text{row } i)$ to row j. Then

$$\mathbf{A} = \begin{pmatrix} a_{11} & a_{12} & \cdots & a_{1m} \\ \cdots & \cdots & \cdots & \cdots \\ a_{i1} & a_{i2} & \cdots & a_{im} \\ \cdots & \cdots & \cdots & \cdots \\ a_{j1} & a_{j2} & \cdots & a_{jm} \\ \cdots & \cdots & \cdots & \cdots \\ a_{n1} & a_{n2} & \cdots & a_{nm} \end{pmatrix}$$

and

$$
\mathbf{D} =
\begin{pmatrix}
a_{11} & a_{12} & \cdots & a_{1m} \\
\cdots & \cdots & \cdots & \cdots \\
a_{i1} & a_{i2} & \cdots & a_{im} \\
\cdots & \cdots & \cdots & \cdots \\
\alpha a_{i1} + a_{j1} & \alpha a_{i2} + a_{j2} & \cdots & \alpha a_{im} + a_{jm} \\
\cdots & \cdots & \cdots & \cdots \\
a_{n1} & a_{n2} & \cdots & a_{nm}
\end{pmatrix}.
$$

Now we can get from $\mathbf{D}$ back to $\mathbf{A}$ by adding $-\alpha(\text{row } i)$ to row j of $\mathbf{D}$. Thus a Type III operation can be used to reverse a Type III operation.

This ability to reverse the effects of elementary row operations will be useful later, and we will record it as a theorem.

THEOREM 6.8

Let $\mathbf{E}_1$ be an elementary matrix that performs an elementary row operation on a matrix $\mathbf{A}$. Then there is an elementary matrix $\mathbf{E}_2$ such that $\mathbf{E}_2(\mathbf{E}_1\mathbf{A}) = \mathbf{A}$. ∎

In fact, $\mathbf{E}_2\mathbf{E}_1 = \mathbf{I}_n$.

SECTION 6.2   PROBLEMS

In each of Problems 1 through 12, perform the row operation, or sequence of row operations, directly on $\mathbf{A}$ and then find a matrix $\boldsymbol{\Omega}$ such that the final result is $\boldsymbol{\Omega}\mathbf{A}$.

1. $\mathbf{A} = \begin{pmatrix} -2 & 1 & 4 & 2 \\ 0 & 1 & 16 & 3 \\ 1 & -2 & 4 & 8 \end{pmatrix}$; multiply row 2 by $\sqrt{3}$.

2. $\mathbf{A} = \begin{pmatrix} 3 & -6 \\ 1 & 1 \\ 8 & -2 \\ 0 & 5 \end{pmatrix}$; add 6 times row 2 to row 3.

3. $\mathbf{A} = \begin{pmatrix} -2 & 14 & 6 \\ 8 & 1 & -3 \\ 2 & 9 & 5 \end{pmatrix}$; add $\sqrt{13}$ times row 3 to row 1, then interchange rows 2 and 1, then multiply row 1 by 5.

4. $\mathbf{A} = \begin{pmatrix} -4 & 6 & -3 \\ 12 & 4 & -4 \\ 1 & 3 & 0 \end{pmatrix}$; interchange rows 2 and 3, then add negative row 1 to row 2.

5. $\mathbf{A} = \begin{pmatrix} -3 & 15 \\ 2 & 8 \end{pmatrix}$; add $\sqrt{3}$ times row 2 to row 1, then multiply row 2 by 15, then interchange rows 1 and 2.

6. $\mathbf{A} = \begin{pmatrix} 3 & -4 & 5 & 9 \\ 2 & 1 & 3 & -6 \\ 1 & 13 & 2 & 6 \end{pmatrix}$; add row 1 to row 3, then add $\sqrt{3}$ times row 1 to row 2, then multiply row 3 by row 4, then add row 2 to row 3.

7. $\mathbf{A} = \begin{pmatrix} -1 & 0 & 3 & 0 \\ 1 & 3 & 2 & 9 \\ -9 & 7 & -5 & 7 \end{pmatrix}$; multiply row 3 by 4, then add 14 times row 1 to row 2, then interchange rows 3 and 2.

8. $\mathbf{A} = \begin{pmatrix} 0 & -9 & 14 \\ 1 & 5 & 2 \\ 9 & 15 & 0 \end{pmatrix}$; interchange rows 2 and 3, then add 3 times row 2 to row 3, then interchange rows 1 and 3, then multiply row 3 by 4.

9. $\mathbf{A} = \begin{pmatrix} -3 & 7 & 1 & 1 \\ 0 & 3 & 3 & -5 \\ 2 & 1 & -5 & 3 \end{pmatrix}$; add 2 times row 1 to

row 2, then multiply row 3 by -5, then interchange rows 2 and 3.

10. $\mathbf{A} = \begin{pmatrix} 2 & -6 & 5 & 8 \\ 0 & 1 & -3 & 5 \\ 0 & -4 & 2 & -6 \\ 1 & 7 & 3 & -3 \end{pmatrix}$; multiply row 4 by -5,

then add $\sqrt{2}$ times row 4 to row 1, then interchange rows 1 and 3, then multiply row 3 by -1.

11. $\mathbf{A} = \begin{pmatrix} 2 & -3 & 1 \\ 0 & 0 & 0 \\ 1 & -5 & 0 \end{pmatrix}$; interchange rows 1 and 2, then

multiply row 2 by 5, then add -3 times row 3 to row 1.

12. $\mathbf{A} = \begin{pmatrix} -5 & 1 & -4 \\ 0 & 3 & -2 \\ 1 & 2 & 2 \end{pmatrix}$; add 4 times row 2 to row 3,

then interchange rows 1 and 3, then multiply row 1 by -2.

In Problems 13, 14, and 15, $\mathbf{A}$ is an $n \times m$ matrix.

13. Let $\mathbf{B}$ be formed from $\mathbf{A}$ by interchanging rows s and t. Let $\mathbf{E}$ be formed from $\mathbf{I}_n$ by interchanging rows s and t. Prove that $\mathbf{B} = \mathbf{EA}$.

14. Let $\mathbf{B}$ be formed from $\mathbf{A}$ by multiplying row s by α and let $\mathbf{E}$ be formed from $\mathbf{I}_n$ by multiplying row s by α. Prove that $\mathbf{B} = \mathbf{EA}$.

15. Let $\mathbf{B}$ be formed from $\mathbf{A}$ by adding α times row s to row t. Let $\mathbf{E}$ be formed from $\mathbf{I}_n$ by adding α times row s to row t. Prove that $\mathbf{B} = \mathbf{EA}$.

16. Prove that any matrix is row equivalent to itself. *Hint:* $\mathbf{A}$ is obtained from $\mathbf{A}$ by multiplying row 1 by 1.

17. Suppose $\mathbf{A}$ is row equivalent to $\mathbf{B}$ and $\mathbf{B}$ is row equivalent to $\mathbf{C}$. Prove that $\mathbf{A}$ is row equivalent to $\mathbf{C}$. *Hint:* For some elementary matrices $\mathbf{E}_1, \ldots, \mathbf{E}_s$, $\mathbf{B} = \mathbf{E}_s \mathbf{E}_{s-1} \cdots \mathbf{E}_1 \mathbf{A}$. For some elementary matrices $\mathbf{M}_1, \ldots, \mathbf{M}_r$, $\mathbf{C} = \mathbf{M}_r \cdots \mathbf{M}_1 \mathbf{B}$. Now produce a product $\mathbf{K}$ of elementary matrices such that $\mathbf{C} = \mathbf{KA}$.

18. Let $\mathbf{E}$ be an elementary $n \times n$ matrix. Prove that there exists a matrix $\mathbf{K}$ such that $\mathbf{EK} = \mathbf{KE} = \mathbf{I}_n$. This means in effect that any elementary row operation can be undone by multiplying on the left by a matrix. *Hint:* Consider each of the three kinds of elementary row operations in turn. If $\mathbf{E}$ is obtained from $\mathbf{I}_n$ by interchanging rows s and t, prove that $\mathbf{E}^2 = \mathbf{I}_n$. Next, suppose $\mathbf{E}$ is obtained from $\mathbf{I}_n$ by multiplying row s by a nonzero number α. Form $\mathbf{K}$ from $\mathbf{I}_n$ by multiplying row s by $1/\alpha$ and show that $\mathbf{KE} = \mathbf{EK} = \mathbf{I}_n$. Finally, suppose $\mathbf{E}$ is formed from $\mathbf{I}_n$ by adding α times row s to row t. Form $\mathbf{K}$ from $\mathbf{I}_n$ by adding $-\alpha$ row s to row t. Prove that $\mathbf{EK} = \mathbf{KE} = \mathbf{I}_n$.

19. Suppose $\mathbf{A}$ is row equivalent to $\mathbf{B}$. Prove that $\mathbf{B}$ is row equivalent to $\mathbf{A}$. *Hint:* First argue that there are elementary matrices $\mathbf{E}_1, \cdots, \mathbf{E}_r$ such that $\mathbf{B} = \mathbf{E}_r \cdots \mathbf{E}_1 \mathbf{A}$. Produce $\mathbf{K}_j$ such that $\mathbf{K}_j \mathbf{E}_j = \mathbf{I}_n$ (note Problem 18). Show that $\mathbf{A} = \mathbf{K}_1 \cdots \mathbf{K}_r \mathbf{B}$.

6.3 The Row Echelon Form of a Matrix

Sometimes a matrix has a special form that makes it convenient to work with in solving certain problems. For solving systems of linear algebraic equations, we want the *reduced row echelon form*, or *reduced form*, of a matrix.

Let $\mathbf{A}$ be an $n \times m$ matrix. A *zero row* of $\mathbf{A}$ is a row having each element equal to zero. If at least one element of a row is nonzero, that row is a *nonzero row*. The *leading entry* of a nonzero row is its first nonzero element, reading from left to right. For example, if

$$\mathbf{A} = \begin{pmatrix} 0 & 2 & 7 \\ 1 & -2 & 0 \\ 0 & 0 & 0 \\ 0 & 0 & 9 \end{pmatrix},$$

then row three is a zero row and rows one, two, and four are nonzero rows. The leading entry of row 1 is 2, the leading entry of row 2 is 1, and the leading entry of row 4 is 9. We do not speak of a leading entry of a zero row.

We can now define a reduced row echelon matrix.

DEFINITION 6.12 *Reduced Row Echelon Matrix*

A matrix is in reduced row echelon form if it satisfies the following conditions:

1. The leading entry of any nonzero row is 1.
2. If any row has its leading entry in column j, then all other elements of column j are zero.
3. If row i is a nonzero row and row k is a zero row, then $i < k$.
4. If the leading entry of row r_1 is in column c_1, and the leading entry of row r_2 is in column c_2, and if $r_1 < r_2$, then $c_1 < c_2$.

A matrix in reduced row echelon form is said to be in reduced form, or to be a reduced matrix.

A reduced matrix has a very special structure. By condition (1), if we move from left to right along a nonzero row, the first nonzero number we see is 1. Condition (2) means that if we stand at the leading entry 1 of any row and look straight up or down, we see only zeros in the rest of this column. A reduced matrix need not have any zero rows. But if there is a zero row, it must be below any nonzero row. That is, all the zero rows are at the bottom of the matrix. Condition (4) means that the leading entries move downward to the right as we look at the matrix.

EXAMPLE 6.14

The following four matrices are all reduced:

$$\begin{pmatrix} 1 & -4 & 1 & 0 \\ 0 & 0 & 0 & 1 \end{pmatrix}, \begin{pmatrix} 0 & 1 & 3 & 2 \\ 0 & 0 & 0 & 1 \\ 0 & 0 & 0 & 0 \end{pmatrix},$$

$$\begin{pmatrix} 0 & 1 & 2 & 0 & 0 \\ 0 & 0 & 0 & 1 & 0 \\ 0 & 0 & 0 & 0 & 0 \\ 0 & 0 & 0 & 0 & 0 \end{pmatrix}, \quad \text{and} \quad \begin{pmatrix} 1 & 0 & 0 & 3 & 1 \\ 0 & 1 & 0 & -2 & 4 \\ 0 & 0 & 1 & 0 & 1 \\ 0 & 0 & 0 & 0 & 0 \end{pmatrix}. \quad \blacksquare$$

EXAMPLE 6.15

To see one context in which reduced matrices are interesting, consider the last matrix of the preceding example and suppose it is the matrix of coefficients of a system of homogeneous linear equations. This system is $\mathbf{AX} = \mathbf{O}$, and the equations are

$$x_1 + 3x_4 + x_5 = 0$$
$$x_2 - 2x_4 + 4x_5 = 0$$
$$x_3 + x_5 = 0.$$

The fourth row represents the equation $0x_1 + 0x_2 + 0x_3 + 0x_4 + 0x_5 = 0$, which we do not write out (it is satisfied by any numbers x_1 through x_5, and so provides no information). Because the matrix of coefficients is in reduced form, this system is particularly easy to solve. From the third equation,

$$x_3 = -x_5.$$

From the second equation,

$$x_2 = 2x_4 - 4x_5.$$

And from the first equation,

$$x_1 = -3x_4 - x_5.$$

We can therefore choose $x_4 = \alpha$, any number, and $x_5 = \beta$, any number, and obtain a solution by choosing the other unknowns as

$$x_1 = -3\alpha - \beta, \quad x_2 = 2\alpha - 4\beta, \quad x_3 = -\beta.$$

The form of the reduced matrix is selected just so that as a matrix of coefficients of a system of linear equations, the solution of these equations can be read by inspection. ∎

EXAMPLE 6.16

The matrix

$$A = \begin{pmatrix} 0 & 1 & 5 & 0 & 0 \\ 0 & 0 & 1 & 0 & 0 \\ 0 & 0 & 0 & 1 & 0 \\ 0 & 0 & 0 & 0 & 1 \end{pmatrix}$$

is not reduced. The leading entry of row 2 is 1, as it must be, but there is a nonzero element in the column containing this leading entry. However, **A** is row equivalent to a reduced matrix. If we add -5(row 2) to row 1, we obtain

$$B = \begin{pmatrix} 0 & 1 & 0 & 0 & 0 \\ 0 & 0 & 1 & 0 & 0 \\ 0 & 0 & 0 & 1 & 0 \\ 0 & 0 & 0 & 0 & 1 \end{pmatrix},$$

and this is a reduced matrix. ∎

EXAMPLE 6.17

The matrix

$$C = \begin{pmatrix} 2 & 0 & 0 \\ 0 & 1 & 0 \\ 1 & 0 & 1 \end{pmatrix}$$

is not reduced. The leading entry of the first row is not 1, and the first column, containing this leading entry of row 1, has another nonzero element. In addition, the leading entry of row 3 is to the left of the leading entry of row 2, and this violates condition (4). However, $\mathbf{C}$ is row equivalent to a reduced matrix. First form $\mathbf{D}$ by multiplying row 1 by $\frac{1}{2}$:

$$\mathbf{D} = \begin{pmatrix} 1 & 0 & 0 \\ 0 & 1 & 0 \\ 1 & 0 & 1 \end{pmatrix}.$$

Now form $\mathbf{F}$ from $\mathbf{D}$ by adding $-(\text{row } 1)$ to row 3:

$$\mathbf{F} = \begin{pmatrix} 1 & 0 & 0 \\ 0 & 1 & 0 \\ 0 & 0 & 1 \end{pmatrix}.$$

Then $\mathbf{F}$ is a reduced matrix that is row equivalent to $\mathbf{C}$, since it was formed by a sequence of elementary row operations, starting with $\mathbf{C}$. ■

In the last two examples we had matrices that were not in reduced form, but could in both cases proceed to a reduced matrix by elementary row operations. We claim that this is always possible (although in general more operations may be needed than in these two examples).

THEOREM 6.9

Every matrix is row equivalent to a reduced matrix.

Proof   The proof consists of exhibiting a sequence of elementary row operations that will produce a reduced matrix. Let $\mathbf{A}$ be any matrix.

If $\mathbf{A}$ is a zero matrix, we are done. Thus suppose that $\mathbf{A}$ has at least one nonzero row.

Reading from left to right across the matrix, find the first column having a nonzero element. Suppose this is in column c_1. Reading from top to bottom in this column, suppose α is the top nonzero element. Say α is in row r_1. Multiply this row by $1/\alpha$ to obtain a matrix $\mathbf{B}$ in which column c_1 has its top nonzero element equal to 1, and this is in row r_1. If any row below r_1 in $\mathbf{B}$ has a nonzero element β in column c_1, add $-\beta$ times row r_1 to this row. In this way, we obtain a matrix $\mathbf{C}$ that is row equivalent to $\mathbf{A}$, having 1 in the r_1, c_1 position and all other elements of column c_1 equal to zero.

Now interchange, if necessary, rows 1 and r_1 of $\mathbf{C}$ to obtain a matrix $\mathbf{D}$ having leading entry 1 in row 1 and column c_1 and all other elements of this column equal to zero. Further, by choice of c_1, any column of $\mathbf{D}$ to the left of column c_1 has all zero elements (if there is such a column). $\mathbf{D}$ is row equivalent to $\mathbf{A}$.

If $\mathbf{D}$ is reduced, we are done. If not, repeat this procedure, but now look for the first column, say column c_2, to the right of column c_1 having a nonzero element below row 1. Let γ be the top nonzero element of this column lying below row 1. Say this element occurs in row r_2. Multiply row r_2 by $1/\gamma$ to obtain a new matrix $\mathbf{E}$ having 1 in the r_2, c_2 position. If this column has a nonzero element δ above or below row r_2, add $-\delta(\text{row } r_2)$ to this row. In this way, we obtain a matrix $\mathbf{F}$ that is row equivalent to $\mathbf{A}$ and has leading entry 1 in row r_2 and all other elements of column c_2 equal to zero. Finally, form $\mathbf{G}$ from $\mathbf{F}$ by interchanging rows r_2 and 2, if necessary.

If $\mathbf{G}$ is reduced, we are done. If not, locate the first column to the right of column c_2 having a nonzero element and repeat the procedure done to form the first two rows of $\mathbf{G}$.

Since $\mathbf{A}$ has only finitely many columns, eventually this process terminates in a reduced matrix $\mathbf{R}$. Since $\mathbf{R}$ was obtained from $\mathbf{A}$ by elementary row operations, $\mathbf{R}$ is row equivalent to $\mathbf{A}$ and the proof is complete. ■

The process of obtaining a reduced matrix row equivalent to a given matrix $\mathbf{A}$ is referred to as *reducing* $\mathbf{A}$. It is possible to reduce a matrix in many different ways (that is, by different sequences of elementary row operations). We claim that this does not matter and that for a given **A** any reduction process will result in the same reduced matrix.

THEOREM 6.10

Let **A** be a matrix. Then there is exactly one reduced matrix $\mathbf{A}_R$ that is row equivalent to **A**. ■

We leave a proof of this result to the student. In view of this theorem, we can speak of *the* reduced form of a given matrix **A**. We will denote this matrix $\mathbf{A}_R$.

EXAMPLE 6.18

Let

$$\mathbf{A} = \begin{pmatrix} -2 & 1 & 3 \\ 0 & 1 & 1 \\ 2 & 0 & 1 \end{pmatrix}.$$

We want to find $\mathbf{A}_R$. Column 1 has a nonzero element in row 1. Begin with

$$\mathbf{A} = \begin{pmatrix} -2 & 1 & 3 \\ 0 & 1 & 1 \\ 2 & 0 & 1 \end{pmatrix}.$$

Perform the operations:

$$\text{multiply row 1 by } -\frac{1}{2} \rightarrow \begin{pmatrix} 1 & -\frac{1}{2} & -\frac{3}{2} \\ 0 & 1 & 1 \\ 2 & 0 & 1 \end{pmatrix}$$

$$\rightarrow \text{add } (-2)(\text{row 1}) \text{ to row 3} \rightarrow \begin{pmatrix} 1 & -\frac{1}{2} & -\frac{3}{2} \\ 0 & 1 & 1 \\ 0 & 1 & 4 \end{pmatrix}.$$

In the last matrix, column 2 has a nonzero element below row 1, the highest being 1 in the 2, 2 position. Since we want a 1 here, we do not have to multiply this row by anything. However, we want zeros above and below this 1, in the 1, 2 and 3, 2 positions. Thus add $\frac{1}{2}$ times row 2 to row 1, and $-$row 2 to row 3 in the last matrix to obtain

$$\begin{pmatrix} 1 & 0 & -1 \\ 0 & 1 & 1 \\ 0 & 0 & 3 \end{pmatrix}.$$

In this matrix column 3 has a nonzero element below row 2, in the 3, 3 location. Multiply row 3 by $\frac{1}{3}$ to obtain

$$\begin{pmatrix} 1 & 0 & -1 \\ 0 & 1 & 1 \\ 0 & 0 & 1 \end{pmatrix}.$$

Finally, we want zeros above the 3, 3 position in column 3. Add row 3 to row 1 and −row 3 to row 2 to get

$$\mathbf{A}_R = \begin{pmatrix} 1 & 0 & 0 \\ 0 & 1 & 0 \\ 0 & 0 & 1 \end{pmatrix}.$$

This is $\mathbf{A}_R$ because it is a reduced matrix and it is row equivalent to $\mathbf{A}$. ∎

To illustrate the last theorem, we will use a different sequence of elementary row operations to reduce $\mathbf{A}$, arriving at the same final result. Proceed in this way:

$$\mathbf{A} \to (\text{add row 3 to row 1}) \to \begin{pmatrix} 0 & 1 & 4 \\ 0 & 1 & 1 \\ 2 & 0 & 1 \end{pmatrix}$$

$$\text{add } (-1)(\text{row 2}) \text{ to row 1} \to \begin{pmatrix} 0 & 0 & 3 \\ 0 & 1 & 1 \\ 2 & 0 & 1 \end{pmatrix}$$

$$\frac{1}{3}(\text{row 1}) \to \begin{pmatrix} 0 & 0 & 1 \\ 0 & 1 & 1 \\ 2 & 0 & 1 \end{pmatrix}$$

$$\text{add } (-1)(\text{row 1}) \text{ to rows 2 and 3} \to \begin{pmatrix} 0 & 0 & 1 \\ 0 & 1 & 0 \\ 2 & 0 & 0 \end{pmatrix}$$

$$\text{interchange rows 1 and 3} \to \begin{pmatrix} 2 & 0 & 0 \\ 0 & 1 & 0 \\ 0 & 0 & 1 \end{pmatrix}$$

$$\frac{1}{2}(\text{row 1}) \to \begin{pmatrix} 1 & 0 & 0 \\ 0 & 1 & 0 \\ 0 & 0 & 1 \end{pmatrix} = \mathbf{A}_R.$$

EXAMPLE 6.19

Let

$$\mathbf{B} = \begin{pmatrix} 0 & 0 & 0 & 0 & 0 \\ 0 & 0 & 2 & 0 & 0 \\ 0 & 1 & 0 & 1 & 1 \\ 0 & 4 & 3 & 4 & 0 \end{pmatrix}.$$

Reduce **B** as follows:

$$\text{add } -4(\text{row 3}) \text{ to row 4} \rightarrow \begin{pmatrix} 0 & 0 & 0 & 0 & 0 \\ 0 & 0 & 2 & 0 & 0 \\ 0 & 1 & 0 & 1 & 1 \\ 0 & 0 & 3 & 0 & -4 \end{pmatrix}$$

$$\text{interchange rows 3 and 1} \rightarrow \begin{pmatrix} 0 & 1 & 0 & 1 & 1 \\ 0 & 0 & 2 & 0 & 0 \\ 0 & 0 & 0 & 0 & 0 \\ 0 & 0 & 3 & 0 & -4 \end{pmatrix}$$

$$\frac{1}{2}(\text{row 2}) \rightarrow \begin{pmatrix} 0 & 1 & 0 & 1 & 1 \\ 0 & 0 & 1 & 0 & 0 \\ 0 & 0 & 0 & 0 & 0 \\ 0 & 0 & 3 & 0 & -4 \end{pmatrix}$$

$$\text{add } (-3)(\text{row 2}) \text{ to row 4} \rightarrow \begin{pmatrix} 0 & 1 & 0 & 1 & 1 \\ 0 & 0 & 1 & 0 & 0 \\ 0 & 0 & 0 & 0 & 0 \\ 0 & 0 & 0 & 0 & -4 \end{pmatrix}$$

$$-\frac{1}{4}(\text{row 4}) \rightarrow \begin{pmatrix} 0 & 1 & 0 & 1 & 1 \\ 0 & 0 & 1 & 0 & 0 \\ 0 & 0 & 0 & 0 & 0 \\ 0 & 0 & 0 & 0 & 1 \end{pmatrix}$$

$$\text{add } (-1)(\text{row 4}) \text{ to row 1} \rightarrow \begin{pmatrix} 0 & 1 & 0 & 1 & 0 \\ 0 & 0 & 1 & 0 & 0 \\ 0 & 0 & 0 & 0 & 0 \\ 0 & 0 & 0 & 0 & 1 \end{pmatrix}$$

$$\text{interchange rows 3 and 4} \rightarrow \begin{pmatrix} 0 & 1 & 0 & 1 & 0 \\ 0 & 0 & 1 & 0 & 0 \\ 0 & 0 & 0 & 0 & 1 \\ 0 & 0 & 0 & 0 & 0 \end{pmatrix}.$$

This is a reduced matrix; hence it is the reduced matrix $\mathbf{B}_R$ of **B**. ∎

In view of Theorem 6.6 of the preceding section, we immediately have the following.

THEOREM 6.11

Let **A** be an $n \times m$ matrix. Then there is an $n \times n$ matrix $\mathbf{\Omega}$ such that $\mathbf{\Omega A} = \mathbf{A}_R$. ∎

There is a convenient notational device that enables us to find both $\mathbf{\Omega}$ and $\mathbf{A}_R$ together. We know what $\mathbf{\Omega}$ is. If **A** is $n = n \times m$, then $\mathbf{\Omega}$ is an $n \times n$ matrix formed by starting with $\mathbf{I}_n$ and carrying out, in order, the same sequence of elementary row operations used to reduce **A**. A simple way to form $\mathbf{\Omega}$ while reducing **A** is to form an $n \times (n + m)$ matrix $[\mathbf{I}_n \vdots \mathbf{A}]$ by placing $\mathbf{I}_n$ alongside **A** on its left. The first n columns of this matrix $[\mathbf{I}_n \vdots \mathbf{A}]$ are just $\mathbf{I}_n$, and the last m columns are

A. Now reduce **A** by elementary row operations, performing the same operations on the first n columns ($\mathbf{I}_n$) as well. When **A** is reduced, the resulting $n \times (n + m)$ matrix will have the form $[\mathbf{\Omega} \vdots \mathbf{A}_R]$, and we read $\mathbf{\Omega}$ as the first n columns.

EXAMPLE 6.20

Let

$$\mathbf{A} = \begin{pmatrix} -3 & 1 & 0 \\ 4 & -2 & 1 \end{pmatrix}.$$

We want to find a 2×2 matrix $\mathbf{\Omega}$ such that $\mathbf{\Omega A} = \mathbf{A}_R$. Since **A** is 2×3, form the matrix

$$[\mathbf{I}_2 \vdots \mathbf{A}] = \begin{pmatrix} 1 & 0 & | & -3 & 1 & 0 \\ 0 & 1 & | & 4 & -2 & 1 \end{pmatrix}.$$

Now reduce the last three columns, performing the same operations on the first two. The column of dashes is just a bookkeeping device to separate **A** from $\mathbf{I}_2$. Proceed:

$$[\mathbf{I}_2 \vdots \mathbf{A}] \rightarrow -\frac{1}{3}(\text{row } 1) \rightarrow \begin{pmatrix} -\frac{1}{3} & 0 & | & 1 & -\frac{1}{3} & 0 \\ 0 & 1 & | & 4 & -2 & 1 \end{pmatrix}$$

$$\rightarrow \frac{1}{4}(\text{row } 2) \rightarrow \begin{pmatrix} -\frac{1}{3} & 0 & | & 1 & -\frac{1}{3} & 0 \\ 0 & \frac{1}{4} & | & 1 & -\frac{1}{2} & \frac{1}{4} \end{pmatrix}$$

$$\rightarrow (\text{row } 2 - \text{row } 1) \rightarrow \begin{pmatrix} -\frac{1}{3} & 0 & | & 1 & -\frac{1}{3} & 0 \\ \frac{1}{3} & \frac{1}{4} & | & 0 & -\frac{1}{6} & \frac{1}{4} \end{pmatrix}$$

$$\rightarrow (-6)(\text{row } 2) \rightarrow \begin{pmatrix} -\frac{1}{3} & 0 & | & 1 & -\frac{1}{3} & 0 \\ -2 & -\frac{3}{2} & | & 0 & 1 & -\frac{3}{2} \end{pmatrix}$$

$$\rightarrow \frac{1}{3}(\text{row } 2) + (\text{row } 1) \rightarrow \begin{pmatrix} -1 & -\frac{1}{2} & | & 1 & 0 & -\frac{1}{2} \\ -2 & -\frac{3}{2} & | & 0 & 1 & -\frac{3}{2} \end{pmatrix}.$$

The last three columns are in reduced form, so they form $\mathbf{A}_R$. The first two columns form $\mathbf{\Omega}$:

$$\mathbf{\Omega} = \begin{pmatrix} -1 & -\frac{1}{2} \\ -2 & -\frac{3}{2} \end{pmatrix}.$$

As a check on this, form the product

$$\mathbf{\Omega A} = \begin{pmatrix} -1 & -\frac{1}{2} \\ -2 & -\frac{3}{2} \end{pmatrix} \begin{pmatrix} -3 & 1 & 0 \\ 4 & -2 & 1 \end{pmatrix} = \begin{pmatrix} 1 & 0 & -\frac{1}{2} \\ 0 & 1 & -\frac{3}{2} \end{pmatrix} = \mathbf{A}_R. \ \blacksquare$$

SECTION 6.3   PROBLEMS

In each of the following, find the reduced form of **A** and produce a matrix $\mathbf{\Omega}$ such that $\mathbf{\Omega A} = \mathbf{A}_R$.

1. $\mathbf{A} = \begin{pmatrix} 1 & -1 & 3 \\ 0 & 1 & 2 \\ 0 & 0 & 0 \end{pmatrix}$

2. $A = \begin{pmatrix} 3 & 1 & 1 & 4 \\ 0 & 1 & 0 & 0 \end{pmatrix}$

3. $A = \begin{pmatrix} -1 & 4 & 1 & 1 \\ 0 & 0 & 0 & 0 \\ 0 & 0 & 0 & 0 \\ 0 & 0 & 0 & 1 \end{pmatrix}$

4. $A = \begin{pmatrix} 1 & 0 & 1 & 1 & -1 \\ 0 & 1 & 0 & 0 & 2 \end{pmatrix}$

5. $A = \begin{pmatrix} 6 & 1 \\ 0 & 0 \\ 1 & 3 \\ 0 & 1 \end{pmatrix}$

6. $A = \begin{pmatrix} 2 & 2 \\ 1 & 1 \end{pmatrix}$

7. $A = \begin{pmatrix} -1 & 4 & 6 \\ 2 & 3 & -5 \\ 7 & 1 & 1 \end{pmatrix}$

8. $A = \begin{pmatrix} -3 & 4 & 4 \\ 0 & 0 & 0 \end{pmatrix}$

9. $A = \begin{pmatrix} -1 & 2 & 3 & 1 \\ 1 & 0 & 0 & 0 \end{pmatrix}$

10. $A = \begin{pmatrix} 8 & 2 & 1 & 0 \\ 0 & 1 & 1 & 3 \\ 4 & 0 & 0 & -3 \end{pmatrix}$

11. $A = \begin{pmatrix} 4 & 1 & -7 \\ 2 & 2 & 0 \\ 0 & 1 & 0 \end{pmatrix}$

12. $A = \begin{pmatrix} 6 \\ -3 \\ 1 \\ 1 \end{pmatrix}$

13. $A = \begin{pmatrix} 1 & 0 & -4 & 0 & 6 \\ 5 & 1 & -3 & -3 & 9 \\ 6 & 3 & 7 & -3 & 1 \end{pmatrix}$

14. $A = \begin{pmatrix} 0 & -1 & 5 & 2 \\ 1 & 1 & -5 & 2 \\ -5 & 3 & 7 & 3 \\ 0 & 2 & 7 & 0 \end{pmatrix}$

15. $A = \begin{pmatrix} -4 & 0 & 1 & 1 & -5 \end{pmatrix}$

16. $A = \begin{pmatrix} 5 & -2 & 1 & 5 \\ 0 & 3 & 3 & -7 \\ 7 & -4 & 1 & 5 \\ 9 & 5 & 3 & -8 \end{pmatrix}$

17. $A = \begin{pmatrix} -3 & 6 & 1 \\ 0 & -6 & 4 \\ 1 & -1 & 7 \\ 9 & -6 & 4 \end{pmatrix}$

18. $A = \begin{pmatrix} -2 & 3 & 8 & 5 \\ 1 & -5 & 3 & 3 \end{pmatrix}$

19. $A = \begin{pmatrix} 5 & -2 & 3 \\ 0 & 1 & -6 \\ -3 & 5 & 11 \end{pmatrix}$

20. $A = \begin{pmatrix} 0 & 1 & 2 & 0 & 0 & -3 \\ 0 & 0 & 0 & 0 & 1 & 7 \end{pmatrix}$

6.4 The Row and Column Spaces of a Matrix and Rank of a Matrix

In this section we will develop three numbers associated with matrices that play a significant role in the solution of systems of linear equations.

Suppose A is an $n \times m$ matrix with real number elements. Each row of A has m elements and can be thought of as a vector in R^m. There are n such vectors. The set of all linear combinations of these row vectors is a subspace of R^m called the *row space* of A. This space is spanned by the row vectors. If these row vectors are linearly independent, they form a basis for this row space, and this space has dimension n. If they are not linearly independent, then some subset of them forms a basis for the row space, and this space has dimension $<n$.

If we look down instead of across, we can think of each column of A as a vector in R^n. We often write these vectors as columns simply to keep in mind their origin, although they can be written in standard vector notation. The set of all linear combinations of these columns forms a subspace of R^n. This is the *column space* of **A**. If these columns are linearly independent, they form a basis for this column space, which then has dimension m; otherwise, this dimension is less than m.

EXAMPLE 6.21

Let

$$
\mathbf{B} = \begin{pmatrix} -2 & 6 & 1 \\ 2 & 2 & -4 \\ 10 & -8 & 12 \\ 3 & 1 & -2 \\ 5 & -5 & 7 \end{pmatrix}.
$$

The row space is the subspace of R^3 spanned by the row vectors of **B**. This row space consists of all vectors

$$\alpha(-2, 6, 1) + \beta(2, 2, -4) + \gamma(10, -8, 12) + \delta(3, 1, -2) + \epsilon(5, -5, 7).$$

The first three row vectors are linearly independent. The last two are linear combinations of the first three. Specifically,

$$(3, 1, -2) = \tfrac{4}{101}(-2, 6, 1) + \tfrac{181}{202}(2, 2, -4) + \tfrac{13}{101}(10, -8, 12)$$

and

$$(5, -5, 7) = -\tfrac{7}{101}(-2, 6, 1) - \tfrac{39}{202}(2, 2, -4) + \tfrac{53}{101}(10, -8, 12).$$

The first three row vectors form a basis for the row space, which therefore has dimension 3. The row space of **B** is all of R^3.

The column space of **B** is the subspace of R^5 consisting of all vectors

$$
\alpha \begin{pmatrix} -2 \\ 2 \\ 10 \\ 3 \\ 5 \end{pmatrix} + \beta \begin{pmatrix} 6 \\ 2 \\ -8 \\ 1 \\ -5 \end{pmatrix} + \gamma \begin{pmatrix} 1 \\ -4 \\ 12 \\ -2 \\ 7 \end{pmatrix}.
$$

These three column vectors are linearly independent in R^5. Neither is a linear combination of the other two, or, equivalently, the only way this linear combination can be the zero vector is for $\alpha = \beta = \gamma = 0$. Therefore, the column space of **B** has dimension 3 and is a subspace of the dimension 5 space R^5. ∎

EXAMPLE 6.22

Let

$$
\mathbf{A} = \begin{pmatrix} -4 & 2 & 2 & 1 & -5 & 4 \\ 0 & 4 & -8 & 7 & 15 & 16 \\ 2 & 1 & -5 & 3 & 10 & 6 \end{pmatrix}.
$$

The row space is the subspace of R^6 spanned by $(-4, 2, 2, 1, -5, 4)$, $(0, 4, -8, 7, 15, 16)$ and $(2, 1, -5, 3, 10, 6)$. This space consists of all linear combinations of these vectors, that is, all vectors

$$\alpha(-4, 2, 2, 1, -5, 4) + \beta(0, 4, -8, 7, 15, 16) + \gamma(2, 1, -5, 3, 10, 6).$$

However, the three row vectors are not linearly independent. The second is twice the third added to the first:

$$(0, 4, -8, 7, 15, 16) = 2(2, 1, -5, 3, 10, 6) + (-4, 2, 2, 1, -5, 4).$$

Therefore, just the two vectors $(-4, 2, 2, 1, -5, 4)$ and $(2, 1, -5, 3, 10, 6)$ are enough to span the row space. Further, these two vectors are linearly independent, since neither is a multiple of the other. They form a basis for the row space, which has dimension 2.

The column space of $\mathbf{A}$ is spanned by the column vectors and consists of all linear combinations

$$\alpha \begin{pmatrix} -4 \\ 0 \\ 2 \end{pmatrix} + \beta \begin{pmatrix} 2 \\ 4 \\ 1 \end{pmatrix} + \gamma \begin{pmatrix} 2 \\ -8 \\ -5 \end{pmatrix} + \delta \begin{pmatrix} 1 \\ 7 \\ 3 \end{pmatrix} + \epsilon \begin{pmatrix} -5 \\ 15 \\ 10 \end{pmatrix} + \eta \begin{pmatrix} 4 \\ 16 \\ 6 \end{pmatrix}.$$

This column space of $\mathbf{A}$ is a subspace of R^3. The first two column vectors, $\begin{pmatrix} -4 \\ 0 \\ 2 \end{pmatrix}$ and $\begin{pmatrix} 2 \\ 4 \\ 1 \end{pmatrix}$, are

linearly independent, since neither is a constant multiple of the other. It is routine to check that each of the other column vectors is a linear combination of these two vectors. For example, the third column vector is

$$\begin{pmatrix} 2 \\ -8 \\ -5 \end{pmatrix} = -\frac{3}{2} \begin{pmatrix} -4 \\ 0 \\ 2 \end{pmatrix} - 2 \begin{pmatrix} 2 \\ 4 \\ 1 \end{pmatrix}.$$

Therefore, $\begin{pmatrix} -4 \\ 0 \\ 2 \end{pmatrix}$ and $\begin{pmatrix} 2 \\ 4 \\ 1 \end{pmatrix}$ form a basis for the column space of $\mathbf{A}$, which also has dimension 2. ■

In these two examples, the row space of the matrix had the same dimension as the column space, even though the row vectors were in R^m and the column vectors in R^n, with $n \neq m$. This is not a coincidence.

THEOREM 6.12

For any matrix $\mathbf{A}$ having real numbers as elements, the row and column spaces have the same dimension.

Proof   Suppose $\mathbf{A}$ is $n \times m$:

$$\mathbf{A} = \begin{pmatrix} a_{11} & a_{12} & \cdots & a_{1r} & a_{1,r+1} & \cdots & a_{1m} \\ a_{21} & a_{22} & \cdots & a_{2r} & a_{2,r+1} & \cdots & a_{2m} \\ \vdots & \vdots & \vdots & \vdots & \vdots & \vdots & \vdots \\ a_{r1} & a_{r2} & \cdots & a_{rr} & a_{r,r+1} & \cdots & a_{rm} \\ a_{r+1,1} & a_{r+1,2} & \cdots & a_{r+1,r} & a_{r+1,r+1} & \cdots & a_{r+1,m} \\ \vdots & \vdots & \vdots & \vdots & \vdots & \vdots & \vdots \\ a_{n1} & a_{n2} & \cdots & a_{nr} & a_{n,r+1} & \cdots & a_{nm} \end{pmatrix}.$$

Denote the row vectors $\mathbf{R}_1, \ldots, \mathbf{R}_n$, so

$$\mathbf{R}_i = (a_{i1}, a_{i2}, \ldots, a_{im}) \text{ in } R^m.$$

Now suppose that the dimension of the row space of $\mathbf{A}$ is r. Then exactly r of these row vectors are linearly independent. As a notational convenience, suppose the first r rows $\mathbf{R}_1, \ldots, \mathbf{R}_r$ are linearly independent. Then each of $\mathbf{R}_{r+1}, \ldots, \mathbf{R}_n$ is a linear combination of these r vectors. Write

$$\mathbf{R}_{r+1} = \beta_{r+1,1}\mathbf{R}_1 + \cdots + \beta_{r+1,r}\mathbf{R}_r,$$

$$\mathbf{R}_{r+2} = \beta_{r+2,1}\mathbf{R}_1 + \cdots + \beta_{r+2,r}\mathbf{R}_r,$$

$$\vdots$$

$$\mathbf{R}_n = \beta_{n,1}\mathbf{R}_1 + \cdots + \beta_{n,r}\mathbf{R}_r.$$

Now observe that column j of $\mathbf{A}$ can be written

$$\begin{pmatrix} a_{1j} \\ a_{2j} \\ \vdots \\ a_{rj} \\ a_{r+1,j} \\ \vdots \\ a_{nj} \end{pmatrix} = a_{1j}\begin{pmatrix} 1 \\ 0 \\ \vdots \\ 0 \\ \beta_{r+1,1} \\ \vdots \\ \beta_{n1} \end{pmatrix} + a_{2j}\begin{pmatrix} 0 \\ 1 \\ \vdots \\ 0 \\ \beta_{r+1,2} \\ \vdots \\ \beta_{n,2} \end{pmatrix} + \cdots + a_{rj}\begin{pmatrix} 0 \\ 0 \\ \vdots \\ 1 \\ \beta_{r+1,r} \\ \vdots \\ \beta_{n,r} \end{pmatrix}.$$

This means that each column vector of $\mathbf{A}$ is a linear combination of the r n-vectors on the right side of this equation. These r vectors therefore span the column space of $\mathbf{A}$. If these vectors are linearly independent, then the dimension of the column space is r. If not, then remove from this list of vectors any that are linear combinations of the others and thus determine a basis for the column space having fewer than r vectors. In any event,

Dimension of the column space of $\mathbf{A}$ $\leq$ Dimension of the row space of $\mathbf{A}$.

By essentially repeating this argument, with row and column vectors interchanged, we obtain

Dimension of the row space of $\mathbf{A}$ $\leq$ Dimension of the column space of $\mathbf{A}$,

and these two inequalities together prove the theorem. ∎

It is interesting to ask what effect elementary row operations have on the row space of a matrix. The answer is—none! We will need this fact shortly.

THEOREM 6.13

Let $\mathbf{A}$ be an $n \times m$ matrix and let $\mathbf{B}$ be formed from $\mathbf{A}$ by an elementary row operation. Then the row space of $\mathbf{A}$ and the row space of $\mathbf{B}$ are the same.

Proof   If $\mathbf{B}$ is obtained by a Type I operation, we simply interchange two rows. Then $\mathbf{A}$ and $\mathbf{B}$ still have the same row vectors, just listed in a different order, so these row vectors span the same row space.

Suppose $\mathbf{B}$ is obtained by a Type II operation, multiplying row i by a nonzero constant c. Linear combinations of the rows of $\mathbf{A}$ have the form

$$\alpha_1\mathbf{R}_1 + \cdots + \alpha_i\mathbf{R}_i + \cdots + \alpha_n\mathbf{R}_n$$

while linear combinations of the rows of $\mathbf{B}$ are

$$\alpha_1\mathbf{R}_1 + \cdots + c\alpha_i\mathbf{R}_i + \cdots + \alpha_n\mathbf{R}_n$$

Since α_i can be any number, so can $c\alpha_i$, so these linear combinations yield the same vectors when the coefficients are chosen arbitrarily. Thus the row space of **A** and **B** are again the same.

Finally, suppose **B** is obtained from **A** by adding $c(\text{row } i)$ to row j. The column vectors of **B** are now

$$\mathbf{R}_1, \ldots, \mathbf{R}_{j-1}, c\mathbf{R}_i + \mathbf{R}_j, \mathbf{R}_{j+1}, \ldots, \mathbf{R}_n.$$

But we can write an arbitrary linear combination of these rows of **B** as

$$\alpha_1 \mathbf{R}_1 + \cdots + \alpha_{j-1}\mathbf{R}_{j-1} + \alpha_j \left(c\mathbf{R}_i + \mathbf{R}_j\right) + \alpha_{j+1}\mathbf{R}_{j+1} + \cdots + \alpha_n \mathbf{R}_n,$$

and this is

$$\alpha_1 \mathbf{R}_1 + \cdots + (\alpha_i + c\alpha_j)\mathbf{R}_i + \cdots + \alpha_j \mathbf{R}_j + \cdots + \alpha_n \mathbf{R}_n,$$

which is again just a linear combination of the row vectors of **A**. Thus again the row spaces of **A** and **B** are the same, and the theorem is proved. ∎

COROLLARY 6.1

For any matrix **A**, the row spaces of **A** and $\mathbf{A}_R$ are the same. ∎

This follows immediately from Theorem 6.13. Each time we perform an elementary row operation on a matrix, we leave the row space unchanged. Since we obtain $\mathbf{A}_R$ from **A** by elementary row operations, then **A** and $\mathbf{A}_R$ must have the same row spaces.

The dimensions of the row and column spaces will be important when we consider solutions of systems of linear equations. There is another number that will play a significant role in this, the rank of a matrix.

DEFINITION 6.13   *Rank*

The rank of a matrix **A** is the number of nonzero rows in $\mathbf{A}_R$.

We denote the rank of **A** as $rank(\mathbf{A})$. If **B** is a reduced matrix, then $\mathbf{B} = \mathbf{B}_R$, so the rank of **B** is just the number of nonzero rows of **B** itself. Further, for any matrix **A**

$$rank(\mathbf{A}) = \text{number of nonzero rows of } \mathbf{A}_R = rank(\mathbf{A}_R).$$

We claim that the rank of a matrix is equal to the dimension of its row space (or column space). First we will show this for reduced matrices.

LEMMA 6.1

Let **B** be a reduced matrix. Then the rank of **B** equals the dimension of the row space of **B**.

Proof   Let $\mathbf{R}_1, \ldots, \mathbf{R}_r$ be the nonzero row vectors of **B**. The row space consists of all linear combinations

$$c_1 \mathbf{R}_1 + \cdots + c_r \mathbf{R}_r.$$

If nonzero row j has its leading entry in column k, then the kth component of $\mathbf{R}_j$ is 1. Because **B** is reduced, all the other elements of column k are zero, hence each other $\mathbf{R}_i$ has kth component zero. By Lemma 5.3, $\mathbf{R}_1, \ldots, \mathbf{R}_r$ are linearly independent. Therefore these vectors form a basis for the row space of $\mathbf{B}$, and the dimension of this space is r. But

$$r = \text{number of nonzero rows of } \mathbf{B} = \text{number of nonzero rows of } \mathbf{B}_R = rank(\mathbf{B}). \quad \blacksquare$$

EXAMPLE 6.23

Let

$$\mathbf{B} = \begin{pmatrix} 0 & 1 & 0 & 0 & 3 & 0 & 6 \\ 0 & 0 & 1 & 0 & -2 & 1 & 5 \\ 0 & 0 & 0 & 1 & 2 & 0 & -4 \\ 0 & 0 & 0 & 0 & 0 & 0 & 0 \end{pmatrix}.$$

Then $\mathbf{B}$ is in reduced form, so $\mathbf{B} = \mathbf{B}_R$. The rank of $\mathbf{B}$ is its number of nonzero rows, which is 3. Further, the nonzero row vectors are

$$(0, 1, 0, 0, 3, 0, 6), (0, 0, 1, 0, -2, 1, 5), (0, 0, 0, 1, 2, 0, -4)$$

and these are linearly independent. Indeed, if a linear combination of these vectors yielded the zero vector, we would have

$$\alpha(0, 1, 0, 0, 3, 0, 6) + \beta(0, 0, 1, 0, -2, 1, 5) + \gamma(0, 0, 0, 1, 2, 0, -4) = (0, 0, 0, 0, 0, 0, 0).$$

But then

$$(0, \alpha, \beta, \gamma, 3\alpha - 2\beta + 2\gamma, \beta, 6\alpha + 5\beta - 4\gamma) = (0, 0, 0, 0, 0, 0, 0),$$

and from the second, third, and fourth components we read that $\alpha = \beta = \gamma = 0$. By Theorem 5.13(2), these three row vectors are linearly independent and form a basis for the row space, which therefore has dimension 3. $\blacksquare$

Using this as a stepping-stone, we can prove the result for arbitrary matrices.

THEOREM 6.14

For any matrix $\mathbf{A}$, the rank of $\mathbf{A}$ equals the dimension of the row space of $\mathbf{A}$.

Proof   From the lemma, we know that

$$rank(\mathbf{A}) = rank(\mathbf{A}_R) = \text{ dimension of the row space of } \mathbf{A}_R$$
$$= \text{ dimension of the row space of } \mathbf{A},$$

since $\mathbf{A}$ and $\mathbf{A}_R$ have the same row space. $\blacksquare$

Of course, we can also assert that

$$rank(\mathbf{A}) = \text{ dimension of the column space of } \mathbf{A}.$$

If $\mathbf{A}$ is $n \times m$, then so is $\mathbf{A}_R$. Now $\mathbf{A}_R$ cannot have more than n nonzero rows (because it has only n rows). This means that

$$rank(\mathbf{A}) \leq \text{ number of rows of } \mathbf{A}.$$

There is a special circumstance in which the rank of a square matrix actually equals its number of rows.

THEOREM 6.15

Let $\mathbf{A}$ be an $n \times n$ matrix. Then $rank(\mathbf{A}) = n$ if and only if $\mathbf{A}_R = \mathbf{I}_n$.

Proof If $\mathbf{A}_R = \mathbf{I}_n$, then the number of nonzero rows in $\mathbf{A}_R$ is n, since $\mathbf{I}_n$ has no zero rows. Hence in this case $rank(\mathbf{A}) = n$.

Conversely, suppose that $rank(\mathbf{A}) = n$. Then $\mathbf{A}_R$ has n nonzero rows, hence no zero rows. By definition of a reduced matrix, each row of $\mathbf{A}_R$ has leading entry 1. Since each row, being a nonzero row, has a leading entry, then the i, i elements of $\mathbf{A}_R$ are all equal to 1. But it is also required that if column j contains a leading entry, then all other elements of that column are zero. Thus $\mathbf{A}_R$ must have each i, j element equal to zero if $i \neq j$, so $\mathbf{A}_R = \mathbf{I}_n$. ∎

EXAMPLE 6.24

Let

$$\mathbf{A} = \begin{pmatrix} 1 & -1 & 4 & 2 \\ 0 & 1 & 3 & 2 \\ 3 & -2 & 15 & 8 \end{pmatrix}.$$

We find that

$$\mathbf{A}_R = \begin{pmatrix} 1 & 0 & 7 & 0 \\ 0 & 1 & 3 & 2 \\ 0 & 0 & 0 & 0 \end{pmatrix}.$$

Therefore, $rank(\mathbf{A}) = 2$. This is also the dimension of the row space of $\mathbf{A}$ and of the column space of $\mathbf{A}$. ∎

In the next section we will use the reduced form of a matrix to solve homogeneous systems of linear algebraic equations.

SECTION 6.4   PROBLEMS

In each of Problems 1 through 20, (a) find the reduced form of the matrix, and from this the rank, (b) find a basis for the row space of the matrix, and the dimension of this space, and (c) find a basis for the column space, and the dimension of this space.

1. $\begin{pmatrix} -4 & 1 & 3 \\ 2 & 2 & 0 \end{pmatrix}$

2. $\begin{pmatrix} 1 & -1 & 4 \\ 0 & 1 & 3 \\ 2 & -1 & 11 \end{pmatrix}$

3. $\begin{pmatrix} -3 & 1 \\ 2 & 2 \\ 4 & -3 \end{pmatrix}$

4. $\begin{pmatrix} 6 & 0 & 0 & 1 & 1 \\ 12 & 0 & 0 & 2 & 2 \\ 1 & -1 & 0 & 0 & 0 \end{pmatrix}$

5. $\begin{pmatrix} 8 & -4 & 3 & 2 \\ 1 & -1 & 1 & 0 \end{pmatrix}$

6. $\begin{pmatrix} 1 & 3 & 0 \\ 0 & 0 & 1 \end{pmatrix}$

7. $\begin{pmatrix} 2 & 2 & 1 \\ 1 & -1 & 3 \\ 0 & 0 & 1 \\ 4 & 0 & 7 \end{pmatrix}$

8. $\begin{pmatrix} 0 & -1 & 0 \\ 0 & 0 & -1 \\ 0 & 0 & 2 \end{pmatrix}$

9. $\begin{pmatrix} 0 & 4 & 3 \\ 6 & 1 & 0 \\ 2 & 2 & 2 \end{pmatrix}$

10. $\begin{pmatrix} 1 & 0 & 0 \\ 2 & 0 & 0 \\ 1 & 0 & -1 \\ 3 & 0 & 0 \end{pmatrix}$

11. $\begin{pmatrix} -3 & 2 & 2 \\ 1 & 0 & 5 \\ 0 & 0 & 2 \end{pmatrix}$

12. $\begin{pmatrix} -4 & -2 & 1 & 6 \\ 0 & 4 & -4 & 2 \\ 1 & 0 & 0 & 0 \end{pmatrix}$

13. $\begin{pmatrix} -2 & 5 & 7 \\ 0 & 1 & -3 \\ -4 & 11 & 11 \end{pmatrix}$

14. $\begin{pmatrix} -3 & 2 & 1 & 1 & 0 \\ 6 & -4 & -2 & -2 & 0 \end{pmatrix}$

15. $\begin{pmatrix} 7 & -2 & 1 & -2 \\ 0 & 2 & 6 & 3 \\ 7 & 2 & 13 & 4 \\ 7 & 0 & 7 & 1 \end{pmatrix}$

16. $\begin{pmatrix} -4 & 2 & 5 \\ 0 & 0 & 0 \\ 0 & 0 & 0 \end{pmatrix}$

17. $\begin{pmatrix} 4 & 1 & -3 & 5 \\ 2 & 0 & 0 & -2 \\ 13 & 2 & 0 & -1 \end{pmatrix}$

18. $\begin{pmatrix} -4 & 2 & 6 & 1 \\ 0 & 0 & 4 & 1 \\ 4 & -2 & -2 & 0 \end{pmatrix}$

19. $\begin{pmatrix} 5 & -2 & 5 & 6 & 1 \\ -2 & 0 & 1 & -1 & 3 \\ -1 & -2 & 8 & 3 & 10 \end{pmatrix}$

20. $\begin{pmatrix} 3 & -3 & 5 & 1 \\ 0 & 2 & 1 & -5 \\ 0 & 0 & 0 & 1 \end{pmatrix}$

21. Show that for any matrix $\mathbf{A}$, $rank(\mathbf{A}) = rank(\mathbf{A}^t)$.

22. Let V be the set of all $n \times m$ real matrices. Show that V is a vector space, with the usual addition of matrices and multiplication of matrices by real numbers. Show that the dimension of this vector space is nm. *Hint:* Look at simple cases, such as 1×2, 2×2, or 2×3 matrices, to see how to form a basis for V.

23. Let C be the set of all $n \times m$ matrices with complex numbers as elements. Show that C is a vector space, with the usual addition of matrices and multiplication of a matrix by a scalar. Show that the dimension of this vector space is $2nm$.

6.5   Solution of Homogeneous Systems of Linear Equations

We will apply the matrix machinery we have developed to the solution of systems of n linear homogeneous equations in m unknowns:

$$a_{11}x_1 + a_{12}x_2 + \cdots + a_{1m}x_m = 0$$

$$a_2x_1 + a_{22}x_2 + \cdots + a_{2m}x_m = 0$$

$$\vdots$$

$$a_{n1}x_1 + a_{n2}x_2 + \cdots + a_{nm}x_m = 0.$$

This term *homogeneous* applies here because the right side of each equation is zero.

As a prelude to a matrix approach to solving this system, consider the simple system

$$x_1 - 3x_2 + 2x_3 = 0$$

$$-2x_1 + x_2 - 3x_3 = 0.$$

We can solve this easily by "eliminating unknowns." Add 2(equation 1) to equation 2 to get

$$-5x_2 + x_3 = 0,$$

hence

$$x_2 = \tfrac{1}{5}x_3.$$

Now put this into the first equation of the system to get

$$x_1 - \tfrac{3}{5}x_3 + 2x_3 = 0,$$

or

$$x_1 + \tfrac{7}{5}x_3 = 0.$$

Then

$$x_1 = -\tfrac{7}{5}x_3.$$

We now have the solution:

$$x_1 = -\tfrac{7}{5}\alpha, \quad x_2 = \tfrac{1}{5}\alpha, \quad x_3 = \alpha,$$

in which α can be any number. For this system, two of the unknowns can be written as constant multiples of the third, which can be assigned any value. The system therefore has infinitely many solutions.

For this simple system we do not need matrices. However, it is instructive to see how matrices could be used here. First, write this system in matrix form as $\mathbf{AX} = \mathbf{O}$, where

$$\mathbf{A} = \begin{pmatrix} 1 & -3 & 2 \\ -2 & 1 & -3 \end{pmatrix} \quad \text{and} \quad \mathbf{X} = \begin{pmatrix} x_1 \\ x_2 \\ x_3 \end{pmatrix}.$$

Now reduce $\mathbf{A}$. We find that

$$\mathbf{A}_R = \begin{pmatrix} 1 & 0 & \tfrac{7}{5} \\ 0 & 1 & -\tfrac{1}{5} \end{pmatrix}.$$

The system $\mathbf{A}_R\mathbf{X} = \mathbf{0}$ is just

$$x_1 + \tfrac{7}{5}x_3 = 0$$

$$x_2 - \tfrac{1}{5}x_3 = 0.$$

This reduced system has the advantage of simplicity—we can solve it on sight, obtaining the same solutions that we got for the original system.

This is not a coincidence. $\mathbf{A}_R$ is formed from $\mathbf{A}$ by elementary row operations. Since each row of $\mathbf{A}$ contains the coefficients of an equation of the system, these row operations correspond in the system to interchanging equations, multiplying an equation by a nonzero constant, and adding a constant multiple of one equation to another equation of the system. This is why these elementary row operations were selected. But these operations always result in new systems having the same solutions as the original system (a proof of this will be given shortly). The reduced system

$A_R X = O$ therefore has the same solutions as $AX = O$. But A_R is defined in just such a way that we can just read the solutions, giving some unknowns in terms of others, as we saw in this simple case.

We will look at two more examples and then say more about the method in general.

EXAMPLE 6.25

Solve the system

$$x_1 - 3x_2 + x_3 - 7x_4 + 4x_5 = 0$$

$$x_1 + 2x_2 - 3x_3 = 0$$

$$x_2 - 4x_3 + x_5 = 0.$$

This is the system $AX = O$, with

$$A = \begin{pmatrix} 1 & -3 & 1 & -7 & 4 \\ 1 & 2 & -3 & 0 & 0 \\ 0 & 1 & -4 & 0 & 1 \end{pmatrix}.$$

We find that

$$A_R = \begin{pmatrix} 1 & 0 & 0 & -\frac{35}{16} & \frac{13}{16} \\ 0 & 1 & 0 & \frac{28}{16} & -\frac{20}{16} \\ 0 & 0 & 1 & \frac{7}{16} & -\frac{9}{16} \end{pmatrix}.$$

The systems $AX = O$ and $A_R X = O$ have the same solutions. But the equations of the reduced system $A_R X = O$ are

$$x_1 - \frac{35}{16}x_4 + \frac{13}{16}x_5 = 0$$

$$x_2 + \frac{28}{16}x_4 - \frac{20}{16}x_5 = 0$$

$$x_3 + \frac{7}{16}x_4 - \frac{9}{16}x_5 = 0.$$

From these we immediately read the solution. We can let $x_4 = \alpha$ and $x_5 = \beta$ (any numbers), and then

$$x_1 = \frac{35}{16}\alpha - \frac{13}{16}\beta, \quad x_2 = -\frac{28}{16}\alpha + \frac{20}{16}\beta, \quad x_3 = -\frac{7}{16}\alpha + \frac{9}{16}\beta. \ \blacksquare$$

Not only did we essentially have the solution once we obtained A_R but we also knew the number of arbitrary constants that appear in the solution. In the last example, this number was 2. This was the number of columns minus the number of rows having leading entries (or $m - rank(A)$).

It is convenient to write solutions of $AX = O$ as column vectors. In the last example, we could write

$$X = \begin{pmatrix} \frac{35}{16}\alpha - \frac{13}{16}\beta \\ -\frac{28}{16}\alpha + \frac{20}{16}\beta \\ -\frac{7}{16}\alpha + \frac{9}{16}\beta \\ \alpha \\ \beta \end{pmatrix}.$$

This formulation also makes it easy to display other information about solutions. In this example, we can also write

$$\mathbf{X} = \gamma \begin{pmatrix} 35 \\ -28 \\ -7 \\ 16 \\ 0 \end{pmatrix} + \delta \begin{pmatrix} -13 \\ 20 \\ 9 \\ 0 \\ 16 \end{pmatrix}$$

in which $\gamma = \alpha/16$ can be any number (since α can be any number), and $\delta = \beta/16$ is also any number. This displays the solution as a linear combination of two linearly independent vectors. We will say more about the significance of this in the next section.

EXAMPLE 6.26

Solve the system

$$-x_2 + 2x_3 + 4x_4 = 0$$
$$-x_3 + 3x_4 = 0$$
$$2x_1 + x_2 + 3x_3 + 7x_4 = 0$$
$$6x_1 + 2x_2 + 10x_3 + 28x_4 = 0.$$

Let

$$\mathbf{A} = \begin{pmatrix} 0 & -1 & 2 & 4 \\ 0 & 0 & -1 & 3 \\ 2 & 1 & 3 & 7 \\ 6 & 2 & 10 & 28 \end{pmatrix}.$$

We find that

$$\mathbf{A}_R = \begin{pmatrix} 1 & 0 & 0 & 13 \\ 0 & 1 & 0 & -10 \\ 0 & 0 & 1 & -3 \\ 0 & 0 & 0 & 0 \end{pmatrix}.$$

From the first three rows of $\mathbf{A}_R$, read that

$$x_1 + 13x_4 = 0$$
$$x_2 - 10x_4 = 0$$
$$x_3 - 3x_4 = 0.$$

Thus the solution is given by

$$x_1 = -13\alpha, \quad x_2 = 10\alpha, \quad x_3 = 3\alpha, \quad x_4 = \alpha,$$

in which α can be any number. We can write the solution as

$$\mathbf{X} = \alpha \begin{pmatrix} -13 \\ 10 \\ 3 \\ 1 \end{pmatrix}$$

with α any number. In this example, every solution is a constant multiple of one 4-vector. Note also that $m - rank(\mathbf{A}) = 4 - 3 = 1$. ∎

We will now firm up some of the ideas we have discussed informally and then look at additional examples.

First, everything we have done in this section has been based on the assertion that $\mathbf{AX} = \mathbf{O}$ and $\mathbf{A}_R\mathbf{X} = \mathbf{O}$ have the same solutions. We will prove this.

THEOREM 6.16

Let $\mathbf{A}$ be an $n \times m$ matrix. Then the linear homogeneous systems $\mathbf{AX} = \mathbf{O}$ and $\mathbf{A}_R\mathbf{X} = \mathbf{O}$ have the same solutions.

Proof   We know that there is an $n \times n$ matrix $\boldsymbol{\Omega}$ such that $\boldsymbol{\Omega}\mathbf{A} = \mathbf{A}_R$. Further, $\boldsymbol{\Omega}$ can be written as a product of elementary matrices $\mathbf{E}_1 \cdots \mathbf{E}_r$.

Suppose first that $\mathbf{X} = \mathbf{C}$ is a solution of $\mathbf{AX} = \mathbf{O}$. Then $\mathbf{AC} = \mathbf{O}$, so

$$\mathbf{A}_R\mathbf{C} = (\boldsymbol{\Omega}\mathbf{A})\mathbf{C} = \boldsymbol{\Omega}(\mathbf{AC}) = \boldsymbol{\Omega}\mathbf{O} = \mathbf{O}.$$

Then $\mathbf{C}$ is also a solution of $\mathbf{A}_R\mathbf{X} = \mathbf{O}$.

Conversely, suppose $\mathbf{K}$ is a solution of $\mathbf{A}_R\mathbf{X} = \mathbf{O}$. Then $\mathbf{A}_R\mathbf{K} = \mathbf{O}$. We want to show that $\mathbf{AK} = \mathbf{O}$ also. Because $\mathbf{A}_R\mathbf{K} = \mathbf{O}$, we have $(\boldsymbol{\Omega}\mathbf{A})\mathbf{K} = \mathbf{O}$, or

$$(\mathbf{E}_r \cdots \mathbf{E}_1\mathbf{A})\mathbf{K} = \mathbf{0}.$$

By Theorem 6.8, for each $\mathbf{E}_j$ there is an elementary matrix $\mathbf{E}_j^*$ that reverses the effect of $\mathbf{E}_j$. Then, from the last equation we have

$$\mathbf{E}_1^*\mathbf{E}_2^* \cdots \mathbf{E}_{r-1}^*\mathbf{E}_r^*(\mathbf{E}_r\mathbf{E}_{r-1} \cdots \mathbf{E}_2\mathbf{E}_1\mathbf{A})\mathbf{K} = \mathbf{0}.$$

But $\mathbf{E}_r^*\mathbf{E}_r = \mathbf{I}_n$, because $\mathbf{E}_r^*$ reverses the effect of $\mathbf{E}_r$. Similarly, $\mathbf{E}_{r-1}^*\mathbf{E}_{r-1} = \mathbf{I}_n$, until finally the last equation becomes

$$\mathbf{E}_1^*(\mathbf{E}_1\mathbf{A})\mathbf{K} = \mathbf{AK} = \mathbf{O}.$$

Thus $\mathbf{K}$ is a solution of $\mathbf{AX} = \mathbf{O}$ and the proof is complete. ∎

The method for solving $\mathbf{AX} = \mathbf{O}$, which we illustrated above, is called the *Gauss–Jordan method*, or *complete pivoting*. Here is an outline of the method. Keep in mind that in a system $\mathbf{AX} = \mathbf{O}$, row k gives the coefficients of equation k and column j contains the coefficients of x_j as we look down the set of equations.

Gauss–Jordan Method for Solving $\mathbf{AX} = \mathbf{O}$

1. Find $\mathbf{A}_R$.

2. Look down the columns of $\mathbf{A}_R$. If column j contains the leading entry of some row (so all other elements of this column are zero), then x_j is said to be *dependent*. Determine all the dependent unknowns. The remaining unknowns (if any) are said to be *independent*.

3. Each nonzero row of $\mathbf{A}_R$ represents an equation in the reduced system, having one dependent unknown (in the column having the leading entry 1) and all other unknowns in this equation (if any) independent. This enables us to write this dependent unknown in terms of the independent ones.

4. After step (3) is carried out for each nonzero row, we have each dependent unknown in terms of the independent ones. The independent unknowns can then be assigned any values, and these determine the dependent unknowns, solving the system. We can write the resulting solution as a linear combination of column solutions, one for each independent unknown. The resulting expression, containing an arbitrary constant for each independent unknown, is called the *general solution* of the system.

EXAMPLE 6.27

Solve the system

$$-x_1 + x_3 + x_4 + 2x_5 = 0$$
$$x_2 + 3x_3 + 4x_5 = 0$$
$$x_1 + 2x_2 + x_3 + x_4 + x_5 = 0$$
$$-3x_1 + x_2 + 4x_5 = 0.$$

The matrix of coefficients is

$$\mathbf{A} = \begin{pmatrix} -1 & 0 & 1 & 1 & 2 \\ 0 & 1 & 3 & 0 & 4 \\ 1 & 2 & 1 & 1 & 1 \\ -3 & 1 & 0 & 0 & 4 \end{pmatrix}.$$

We find that

$$\mathbf{A}_R = \begin{pmatrix} 1 & 0 & 0 & 0 & -\frac{9}{8} \\ 0 & 1 & 0 & 0 & \frac{5}{8} \\ 0 & 0 & 1 & 0 & \frac{9}{8} \\ 0 & 0 & 0 & 1 & -\frac{1}{4} \end{pmatrix}.$$

Because columns 1 through 4 of $\mathbf{A}_R$ contain leading entries of rows, x_1, x_2, x_3, and x_4 are dependent, while the remaining unknown, x_5, is independent. The equations of the reduced system (which has the same solutions as the original system) are

$$x_1 - \tfrac{9}{8}x_5 = 0$$

$$x_2 + \tfrac{5}{8}x_5 = 0$$

$$x_3 + \tfrac{9}{8}x_5 = 0$$

$$x_4 - \tfrac{1}{4}x_5 = 0.$$

We wrote these out for illustration, but in fact the solution can be read immediately from $\mathbf{A}_R$. We can choose $x_5 = \alpha$, any number, and then

$$x_1 = \tfrac{9}{8}\alpha, \quad x_2 = -\tfrac{5}{8}\alpha, \quad x_3 = -\tfrac{9}{8}\alpha, \quad x_4 = \tfrac{1}{4}\alpha.$$

The dependent unknowns are given by $\mathbf{A}_R$ in terms of the independent unknowns (only one in this case).

We can write this solution more neatly as

$$\mathbf{X} = \gamma \begin{pmatrix} 9 \\ -5 \\ -9 \\ 2 \\ 8 \end{pmatrix}$$

in which $\gamma = \alpha/8$ can be any number. This is the general solution of $\mathbf{AX} = \mathbf{O}$. In this example, $m - rank(\mathbf{A}) = 5 - 4 = 1$. ∎

EXAMPLE 6.28

Consider the system

$$3x_1 - 11x_2 + 5x_3 = 0$$
$$4x_1 + x_2 - 10x_3 = 0$$
$$4x_1 + 9x_2 - 6x_3 = 0.$$

The matrix of coefficients is

$$\mathbf{A} = \begin{pmatrix} 3 & -11 & 5 \\ 4 & 1 & -10 \\ 4 & 9 & -6 \end{pmatrix}.$$

The reduced matrix is

$$\mathbf{A}_R = \begin{pmatrix} 1 & 0 & 0 \\ 0 & 1 & 0 \\ 0 & 0 & 1 \end{pmatrix} = \mathbf{I}_3.$$

The reduced system is just

$$x_1 = 0, \quad x_2 = 0, \quad x_3 = 0.$$

This system has only the trivial solution, with each $x_j = 0$. Notice that in this example there are no independent unknowns. If there were, we could assign them any values and have infinitely many solutions. ∎

EXAMPLE 6.29

Consider the system

$$2x_1 - 4x_2 + x_3 + x_4 + 6x_5 + 12x_6 - 5x_7 = 0$$
$$-4x_1 + x_2 + 6x_3 + 3x_4 + 10x_5 - 9x_6 + 8x_7 = 0$$
$$7x_1 + 2x_2 + 4x_3 - 8x_4 + 6x_5 - 5x_6 + 15x_7 = 0$$
$$2x_1 + x_2 + 6x_3 + 3x_4 - 4x_5 - 2x_6 - 21x_7 = 0.$$

The coefficient matrix is

$$\mathbf{A} = \begin{pmatrix} 2 & -4 & 1 & 1 & 6 & 12 & -5 \\ -4 & 1 & 6 & 3 & 10 & -9 & 8 \\ 7 & 2 & 4 & -8 & 6 & -5 & 15 \\ 2 & 1 & 6 & 3 & -4 & -2 & -21 \end{pmatrix}.$$

We find that

$$\mathbf{A}_R = \begin{pmatrix} 1 & 0 & 0 & 0 & -\frac{7}{3} & \frac{7}{6} & -\frac{29}{6} \\ 0 & 1 & 0 & 0 & -\frac{233}{82} & -\frac{395}{164} & -\frac{375}{164} \\ 0 & 0 & 1 & 0 & \frac{1379}{738} & -\frac{995}{1476} & \frac{2161}{1476} \\ 0 & 0 & 0 & 1 & -\frac{1895}{738} & \frac{1043}{1476} & -\frac{8773}{1476} \end{pmatrix}.$$

From this matrix we see that x_1, x_2, x_3, x_4 are dependent, and x_5, x_6, x_7 are independent. We read immediately from $\mathbf{A}_R$ that

$$x_1 = \tfrac{7}{3}x_5 - \tfrac{7}{6}x_6 + \tfrac{29}{6}x_7$$

$$x_2 = \tfrac{233}{82}x_5 + \tfrac{395}{164}x_6 + \tfrac{375}{164}x_7$$

$$x_3 = -\tfrac{1379}{738}x_5 + \tfrac{995}{1476}x_6 - \tfrac{2161}{1476}x_7$$

$$x_4 = \tfrac{1895}{738}x_5 - \tfrac{1043}{1476}x_6 + \tfrac{8773}{1476}x_7,$$

while x_5, x_6, and x_7 can (independent of each other) be assigned any numerical values. To make the solution look neater, write $x_5 = 1476\alpha$, $x_6 = 1476\beta$, and $x_7 = 1476\gamma$, where α, β, and γ are any numbers. Now the solution can be written

$$x_1 = 3444\alpha - 1722\beta + 7134\gamma, \quad x_2 = 4194\alpha + 3555\beta + 3375\gamma$$

$$x_3 = -2758\alpha + 995\beta - 2161\gamma, \quad x_4 = 3790\alpha - 1043\beta + 8773\gamma$$

$$x_5 = 1476\alpha, \quad x_6 = 1476\beta, \quad x_7 = 1476\gamma,$$

with α, β, and γ any numbers. In column notation,

$$\mathbf{X} = \alpha \begin{pmatrix} 3444 \\ 4194 \\ -2758 \\ 3790 \\ 1476 \\ 0 \\ 0 \end{pmatrix} + \beta \begin{pmatrix} -1722 \\ 3555 \\ 995 \\ -1043 \\ 0 \\ 1476 \\ 0 \end{pmatrix} + \gamma \begin{pmatrix} 7134 \\ 3375 \\ -2161 \\ 8773 \\ 0 \\ 0 \\ 1476 \end{pmatrix}.$$

This is the general solution, being a linear combination of three linearly independent 7-vectors. In this example, $m - rank(\mathbf{A}) = 7 - 4 = 3$. ■

In each of these examples, after we found the general solution, we noted that the number $m - rank(\mathbf{A})$, the number of columns minus the rank of $\mathbf{A}$, coincided with the number of linearly independent column solutions in the general solution (the number of arbitrary unknowns in the general solution). We will see shortly that this is always true.

In the next section we will put the Gauss–Jordan method into a vector space context. This will result in an understanding of the algebraic structure of the solutions of a system $\mathbf{AX} = \mathbf{O}$, as well as practical criteria for determining when such a system has a nonzero solution.

SECTION 6.5 PROBLEMS

In each of Problems 1 through 20, find the general solution of the system and write it as a column matrix or sum of column matrices.

1. $x_1 + 2x_2 - x_3 + x_4 = 0$
$x_2 - x_3 + x_4 = 0$

2. $-3x_1 + x_2 - x_3 + x_4 + x_5 = 0$
$x_2 + x_3 + 4x_5 = 0$
$-3x_3 + 2x_4 + x_5 = 0$

3. $-2x_1 + x_2 + 2x_3 = 0$
$x_1 - x_2 = 0$
$x_1 + x_2 = 0$

4. $4x_1 + x_2 - 3x_3 + x_4 = 0$
$2x_1 - x_3 = 0$

5. $x_1 - x_2 + 3x_3 - x_4 + 4x_5 = 0$
$2x_1 - 2x_2 + x_3 + x_4 = 0$
$x_1 - 2x_3 + x_5 = 0$
$x_3 + x_4 - x_5 = 0$

6. $6x_1 - x_2 + x_3 = 0$
$x_1 - x_4 + 2x_5 = 0$
$x_1 - 2x_5 = 0$

7. $-10x_1 - x_2 + 4x_3 - x_4 + x_5 - x_6 = 0$
$x_2 - x_3 + 3x_4 = 0$
$2x_1 - x_2 + x_5 = 0$
$x_2 - x_4 + x_6 = 0$

8. $8x_1 - 2x_3 + x_6 = 0$
$2x_1 - x_2 + 3x_4 - x_6 = 0$
$x_2 + x_3 - 2x_5 - x_6 = 0$
$x_4 - 3x_5 + 2x_6 = 0$

9. $x_2 - 3x_4 + x_5 = 0$
$2x_1 - x_2 + x_4 = 0$
$2x_1 - 3x_2 + 4x_5 = 0$

10. $4x_1 - 3x_2 + x_4 + x_5 - 3x_6 = 0$
$2x_2 + 4x_4 - x_5 - 6x_6 = 0$
$3x_1 - 2x_2 + 4x_5 - x_6 = 0$
$2x_1 + x_2 - 3x_3 + 4x_4 = 0$

11. $x_1 - 2x_2 + x_5 - x_6 + x_7 = 0$
$x_3 - x_4 + x_5 - 2x_6 + 3x_7 = 0$
$x_1 - x_5 + 2x_6 = 0$
$2x_1 - 3x_4 + x_5 = 0$

12. $2x_1 - 4x_5 + x_7 + x_8 = 0$
$2x_2 - x_6 + x_7 - x_8 = 0$
$x_3 - 4x_4 + x_8 = 0$
$x_2 - x_3 + x_4 = 0$
$x_2 - x_5 + x_6 - x_7 = 0$

13. $x_1 - 4x_2 + x_5 = 0$
$2x_3 - x_4 = 0$
$x_2 - 5x_4 + 6x_5 = 0$

14. $12x_1 + 4x_2 - x_3 + x_4 = 0$
$-x_1 + 2x_2 + 5x_3 - 5x_4 = 0$

15. $-5x_1 + x_2 - 3x_3 + 4x_5 = 0$
$x_2 - 5x_3 + 7x_5 - x_4 = 0$

16. $-3x_1 - 4x_2 + x_3 + x_4 = 0$
$x_2 - 4x_3 + 2x_4 = 0$
$-2x_1 + 4x_2 - 5x_3 + 2x_4 = 0$

17. $9x_2 + x_3 - 5x_4 = 0$
$x_1 + x_2 - 4x_4 = 0$
$x_3 + 8x_4 = 0$

18. $-3x_1 + 3x_2 - 8x_3 + x_5 = 0$
$x_3 + 6x_4 - 2x_5 = 0$
$x_2 + x_4 + 5x_5 = 0$
$x_1 + x_2 + x_4 + 7x_5 = 0$

19. $5x_1 + x_2 - 4x_3 - x_6 = 0$

   $x_2 + x_4 + 6x_5 - x_6 = 0$

   $2x_1 + x_3 - 5x_4 + 11x_6 = 0$

20. $x_3 + 3x_4 - x_5 = 0$

   $2x_1 + 5x_2 + 3x_3 - x_5 = 0$

6.6   The Solution Space of AX = O

Suppose **A** is an $n \times m$ matrix. We have been writing solutions of $\mathbf{AX} = \mathbf{O}$ as column m-vectors. Now observe that the set of all solutions has the algebraic structure of a subspace of R^m.

THEOREM 6.17

Let **A** be an $n \times m$ matrix. Then the set of solutions of the system $\mathbf{AX} = \mathbf{O}$ is a subspace of R^m.

Proof   Let S be the set of all solutions of $\mathbf{AX} = \mathbf{O}$. Then S is a set of vectors in R^m.

Since $\mathbf{AO} = \mathbf{O}, \mathbf{O}$ is in S. Now suppose $\mathbf{X}_1$ and $\mathbf{X}_2$ are solutions and α and β are real numbers. Then

$$\mathbf{A}(\alpha\mathbf{X}_1 + \beta\mathbf{X}_2) = \alpha\mathbf{AX}_1 + \beta\mathbf{AX}_2 = \alpha\mathbf{O} + \beta\mathbf{O} = \mathbf{O},$$

so $\alpha\mathbf{X}_1 + \beta\mathbf{X}_2$ is also a solution, hence is in S. Therefore, S is a subspace of R^m. ■

We would like to know a basis for this solution space, because then every solution is a linear combination of the basis vectors. This is similar to finding a fundamental set of solutions for a linear homogeneous differential equation, because then every solution is a linear combination of these fundamental solutions.

In examples in the preceding section, we were always able to write the general solution as a linear combination of $m - rank(\mathbf{A})$ linearly independent solutions (vectors). This suggests that this number is the dimension of the solution space.

To see why this is true in general, notice that we obtain a dependent x_j corresponding to each row having a leading entry. Since only nonzero rows have leading entries, the number of dependent unknowns is the number of nonzero rows of $\mathbf{A}_R$. But then the number of independent unknowns is the total number of unknowns, m, minus the number of dependent unknowns (the number of nonzero rows of $\mathbf{A}_R$). Since the number of nonzero rows of $\mathbf{A}_R$ is the rank of $\mathbf{A}$, then the general solution can always be written as a linear combination of $m - rank(A)$ independent solutions.

Further, as a practical matter, solving the system $\mathbf{AX} = \mathbf{O}$ by solving the system $\mathbf{A}_R\mathbf{X} = \mathbf{O}$ automatically displays the general solution as a linear combination of this number of basis vectors for the solution space.

We will summarize this discussion as a theorem.

THEOREM 6.18

Let **A** be $n \times m$. Then the solution space of the system $\mathbf{AX} = \mathbf{O}$ has dimension

$$m - rank(\mathbf{A})$$

or, equivalently,

$$m - (\text{number of nonzero rows in } \mathbf{A}_R). ■$$

EXAMPLE 6.30

Consider the system

$$-4x_1 + x_2 + 3x_3 - 10x_4 + x_5 = 0$$
$$2x_1 + 8x_2 - x_3 - x_4 + 3x_5 = 0$$
$$-6x_1 + x_2 + x_3 - 5x_4 - 2x_5 = 0.$$

The matrix of coefficients is

$$\mathbf{A} = \begin{pmatrix} -4 & 1 & 3 & -10 & 1 \\ 2 & 8 & -1 & -1 & 3 \\ -6 & 1 & 1 & -5 & -2 \end{pmatrix},$$

and we find that

$$\mathbf{A}_R = \begin{pmatrix} 1 & 0 & 0 & \frac{33}{118} & \frac{65}{118} \\ 0 & 1 & 0 & -\frac{32}{59} & \frac{21}{59} \\ 0 & 0 & 1 & -\frac{164}{59} & \frac{56}{59} \end{pmatrix}.$$

Now $\mathbf{A}$ has $m = 5$ columns, and $\mathbf{A}_R$ has 3 nonzero rows, so $rank(\mathbf{A}) = 3$ and the solution space of $\mathbf{AX} = \mathbf{0}$ has dimension

$$m - rank(\mathbf{A}) = 5 - 3 = 2.$$

From the reduced system we read the solutions

$$x_1 = -\tfrac{33}{118}\alpha - \tfrac{65}{118}\beta, \quad x_2 = \tfrac{32}{59}\alpha - \tfrac{21}{59}\beta, \quad x_3 = \tfrac{164}{59}\alpha - \tfrac{56}{59}\beta, \quad x_4 = \alpha, \quad x_5 = \beta,$$

in which α and β are any numbers. It is neater to replace α with 118γ and β with 118δ (which still can be any numbers) and write the general solution as

$$\mathbf{X} = \begin{pmatrix} -33\gamma - 65\delta \\ 64\gamma - 42\delta \\ 328\gamma - 112\delta \\ 118\gamma \\ 118\delta \end{pmatrix} = \gamma \begin{pmatrix} -33 \\ 64 \\ 328 \\ 118 \\ 0 \end{pmatrix} + \delta \begin{pmatrix} -65 \\ -42 \\ -112 \\ 0 \\ 118 \end{pmatrix}.$$

This displays the general solution (arbitrary element of the solution space) as a linear combination of two linearly independent vectors that form a basis for the dimension 2 solution space. ■

We know that a system $\mathbf{AX} = \mathbf{O}$ always has at least the zero (trivial) solution. This may be the only solution it has. Rank provides a useful criterion for determining when a system $\mathbf{AX} = \mathbf{O}$ has a nontrivial solution.

THEOREM 6.19

Let $\mathbf{A}$ be $n \times m$. Then the system $\mathbf{AX} = \mathbf{O}$ has a nontrivial solution if and only if

$$m > rank(\mathbf{A}). \quad ■$$

This means that the system of homogeneous equations has a nontrivial solution exactly when the number of unknowns exceeds the rank of the coefficient matrix (the number of nonzero rows in the reduced matrix).

Proof   We have seen that the dimension of the solution space is $m - rank(\mathbf{A})$. There is a nontrivial solution if and only if this solution space has something in it besides the zero solution, and this occurs exactly when the dimension of this solution space is positive. But $m - rank(\mathbf{A}) > 0$ is equivalent to $m > rank(\mathbf{A})$. ■

This theorem has important consequences. First, suppose the number of unknowns exceeds the number of equations. Then $n < m$. But $rank(\mathbf{A}) \leq n$ is always true, so in this case

$$m - rank(\mathbf{A}) \geq m - n > 0$$

and by Theorem 6.19, the system has nontrivial solutions.

COROLLARY 6.2

A homogeneous system $\mathbf{AX} = \mathbf{O}$ with more unknowns than equations always has a nontrivial solution. ■

For another consequence of Theorem 6.19, suppose that $\mathbf{A}$ is square, so $n = m$. Now the dimension of the solution space of $\mathbf{AX} = \mathbf{O}$ is $n - rank(\mathbf{A})$. If this number is positive, the system has nontrivial solutions. If $n - rank(\mathbf{A})$ is not positive, then it must be zero, because $rank(\mathbf{A}) \leq n$ is always true. But $n - rank(\mathbf{A}) = 0$ corresponds to a solution space with only the zero solution. And it also corresponds, by Theorem 6.15, to $\mathbf{A}$ having the identity matrix as its reduced matrix. This means that a square system $\mathbf{AX} = \mathbf{O}$, having the same number of unknowns as equations, has only the trivial solution exactly when the reduced form of $\mathbf{A}$ is the identity matrix.

COROLLARY 6.3

Let $\mathbf{A}$ be an $n \times n$ matrix of real numbers. Then the system $\mathbf{AX} = \mathbf{O}$ has only the trivial solution exactly when $\mathbf{A}_R = \mathbf{I}_n$. ■

EXAMPLE 6.31

Consider the system

$$-4x_1 + x_2 - 7x_3 = 0$$
$$2x_1 + 9x_2 - 13x_3 = 0$$
$$x_1 + x_2 + 10x_3 = 0.$$

The matrix of coefficients is 3×3:

$$\mathbf{A} = \begin{pmatrix} -4 & 1 & -7 \\ 2 & 9 & -13 \\ 1 & 1 & 10 \end{pmatrix}.$$

We find that

$$\mathbf{A}_R = \begin{pmatrix} 1 & 0 & 0 \\ 0 & 1 & 0 \\ 0 & 0 & 1 \end{pmatrix} = \mathbf{I}_3.$$

This means that the system $\mathbf{AX} = \mathbf{O}$ has only the solution $x_1 = x_2 = x_3 = 0$. This makes sense in view of the fact that the system $\mathbf{AX} = \mathbf{O}$ has the same solutions as the reduced system $\mathbf{A}_R \mathbf{X} = \mathbf{O}$, and when $\mathbf{A}_R = \mathbf{I}_3$ this reduced system is just $\mathbf{X} = \mathbf{O}$. ■

SECTION 6.6   **PROBLEMS**

1–20. For $n = 1, \ldots, 20$, use the solution of Problem n, Section 6.5, to determine the dimension of the solution space of the system of homogeneous equations.

21. Can a system $\mathbf{AX} = \mathbf{O}$, in which there are at least as many equations as unknowns, have a nontrivial solution?

22. Prove Corollary 6.2.

23. Prove Corollary 6.3.

6.7   Nonhomogeneous Systems of Linear Equations

We will now consider nonhomogeneous linear systems of equations:

$$a_{11}x_1 + a_{12}x_2 + \cdots + a_{1m}x_m = b_1$$
$$a_{21}x_1 + a_{22}x_2 + \cdots + a_{2m}x_m = b_2$$
$$\vdots$$
$$a_{n1}x_1 + a_{n2}x_2 + \cdots + a_{nm}x_m = b_n.$$

We can write this system in matrix form as $\mathbf{AX} = \mathbf{B}$, in which $\mathbf{A} = [a_{ij}]$ is the $n \times m$ matrix of coefficients,

$$\mathbf{X} = \begin{pmatrix} x_1 \\ x_2 \\ \vdots \\ x_m \end{pmatrix} \quad \text{and} \quad \mathbf{B} = \begin{pmatrix} b_1 \\ b_2 \\ \vdots \\ b_n \end{pmatrix}.$$

This system has n equations in m unknowns. Of course, if each $b_j = 0$, then this is a homogeneous system $\mathbf{AX} = \mathbf{O}$.

A homogeneous system always has at least one solution, the zero solution. A nonhomogeneous system need not have any solution at all.

EXAMPLE 6.32

Consider the system

$$2x_1 - 3x_2 = 6$$
$$4x_1 - 6x_2 = 18.$$

If there were a solution $x_1 = \alpha$, $x_2 = \beta$, then from the first equation we would have $2\alpha - 3\beta = 6$. But then the second equation would give us $4\alpha - 6\beta = 18 = 2(2\alpha - 3\beta) = 12$, a contradiction. ∎

We therefore have an existence question to worry about with the nonhomogeneous system. Before treating this issue, we will ask, What must solutions of $\mathbf{AX} = \mathbf{B}$ look like?

6.7.1   The Structure of Solutions of AX = B

We can take a cue from linear second-order differential equations. There we saw that every solution of $y'' + py' + qy = f(x)$ is a sum of a solution of the homogeneous equation $y'' + py' + qy = 0$

and a particular solution of $y'' + py' + qy = f(x)$. We will show that the same idea holds true for linear algebraic systems of equations as well.

THEOREM 6.20

Let $\mathbf{U}_p$ be any solution of $\mathbf{AX} = \mathbf{B}$. Then every solution of $\mathbf{AX} = \mathbf{B}$ is of the form $\mathbf{U}_p + \mathbf{H}$, in which $\mathbf{H}$ is a solution of $\mathbf{AX} = \mathbf{O}$.

Proof   Let $\mathbf{W}$ be any solution of $\mathbf{AX} = \mathbf{B}$. Since $\mathbf{U}_p$ is also a solution of this system, then

$$\mathbf{A}(\mathbf{W} - \mathbf{U}_p) = \mathbf{AW} - \mathbf{AU}_p = \mathbf{B} - \mathbf{B} = \mathbf{O}.$$

Then $\mathbf{W} - \mathbf{U}_p$ is a solution of $\mathbf{AX} = \mathbf{O}$. Letting $\mathbf{H} = \mathbf{W} - \mathbf{U}_p$, then $\mathbf{W} = \mathbf{U}_p + \mathbf{H}$.
Conversely, if $\mathbf{W} = \mathbf{U}_p + \mathbf{H}$, where $\mathbf{H}$ is a solution of $\mathbf{AX} = \mathbf{O}$, then

$$\mathbf{AW} = \mathbf{A}(\mathbf{U}_p + \mathbf{H}) = \mathbf{AU}_p + \mathbf{AH} = \mathbf{B} + \mathbf{O} = \mathbf{B},$$

so $\mathbf{W}$ is a solution of $\mathbf{AX} = \mathbf{B}$. ∎

This means that if $\mathbf{U}_p$ is any solution of $\mathbf{AX} = \mathbf{B}$, and $\mathbf{H}$ is the general solution of $\mathbf{AX} = \mathbf{O}$, then the expression $\mathbf{U}_p + \mathbf{H}$ contains all possible solutions of $\mathbf{AX} = \mathbf{B}$. For this reason, we call such $\mathbf{U}_p + \mathbf{H}$ the *general solution* of $\mathbf{AX} = \mathbf{B}$, for any particular solution $\mathbf{U}_p$ of $\mathbf{AX} = \mathbf{B}$.

EXAMPLE 6.33

Consider the system

$$-x_1 + x_2 + 3x_3 = -2$$
$$x_2 + 2x_3 = 4.$$

Here

$$\mathbf{A} = \begin{pmatrix} -1 & 1 & 3 \\ 0 & 1 & 2 \end{pmatrix} \quad \text{and} \quad \mathbf{B} = \begin{pmatrix} -2 \\ 4 \end{pmatrix}.$$

We find from methods of the preceding sections that the general solution of $\mathbf{AX} = \mathbf{O}$ is

$$\alpha \begin{pmatrix} 1 \\ -2 \\ 1 \end{pmatrix}.$$

By a method we will describe shortly,

$$\mathbf{U}_p = \begin{pmatrix} 6 \\ 4 \\ 0 \end{pmatrix}$$

is a particular solution of $\mathbf{AX} = \mathbf{B}$. Therefore, every solution of $\mathbf{AX} = \mathbf{B}$ is contained in the expression

$$\alpha \begin{pmatrix} 1 \\ -2 \\ 1 \end{pmatrix} + \begin{pmatrix} 6 \\ 4 \\ 0 \end{pmatrix},$$

in which α is any number. This is the general solution of the system $\mathbf{AX} = \mathbf{B}$. ∎

6.7.2   Existence and Uniqueness of Solutions of $AX = B$

Now we know what to look for in solving $AX = B$. In this section, we will develop criteria to determine when a solution U_p exists, as well as a method that automatically produces the general solution in the form $X = H + U_p$, where H is the general solution of $AX = O$.

DEFINITION 6.14   *Consistent System of Equations*

A nonhomogeneous system $AX = B$ is said to be consistent if there exists a solution. If there is no solution, the system is *inconsistent*.

The difference between a system $AX = O$ and $AX = B$ is B. For the homogeneous system, it is enough to specify the coefficient matrix A when working with the system. But for $AX = B$, we must incorporate B into our computations. For this reason, we introduce the *augmented matrix* $[A \vdots B]$. If A is $n \times m$, $[A \vdots B]$ is the $n \times (m + 1)$ matrix formed by adjoining B to A as a new last column. For example, if

$$A = \begin{pmatrix} -3 & 2 & 6 & 1 \\ 0 & 3 & 3 & -5 \\ 2 & 4 & 4 & -6 \end{pmatrix} \quad \text{and} \quad B = \begin{pmatrix} 5 \\ 2 \\ -8 \end{pmatrix}$$

then

$$\left[A \vdots B \right] = \begin{pmatrix} -3 & 2 & 6 & 1 & \vdots & 5 \\ 0 & 3 & 3 & -5 & \vdots & 2 \\ 2 & 4 & 4 & -6 & \vdots & -8 \end{pmatrix}.$$

The column of dashes does not count in the dimension of the matrix and is simply a visual device to clarify that we are dealing with an augmented matrix giving both A and B for a system $AX = B$. If we just attached B as a last column without such an indicator, we might be dealing with a homogeneous system having three equations in five unknowns.

Continuing with these matrices for the moment, reduce A to find A_R:

$$A_R = \begin{pmatrix} 1 & 0 & 0 & \frac{1}{3} \\ 0 & 1 & 0 & -3 \\ 0 & 0 & 1 & \frac{4}{3} \end{pmatrix}.$$

Next, reduce $[A \vdots B]$ (ignore the dotted column in the row operations) to get

$$[A \vdots B]_R = \begin{pmatrix} 1 & 0 & 0 & \frac{1}{3} & \vdots & -\frac{16}{3} \\ 0 & 1 & 0 & -3 & \vdots & \frac{15}{4} \\ 0 & 0 & 1 & \frac{4}{3} & \vdots & -\frac{37}{12} \end{pmatrix}.$$

Notice that

$$[A \vdots B]_R = [A_R \vdots C].$$

If we reduce the augmented matrix $[A \vdots B]$, we obtain in the first m columns the reduced form of A, together with some new last column. The reason for this can be seen by reviewing how we reduce a matrix. Perform elementary row operations, beginning with the left-most column containing a leading entry, and work from left to right through the columns of the matrix. In finding the reduced form of the augmented matrix $[\mathbf{A} \vdots \mathbf{B}]$, we deal with columns $1, \ldots, m$, which constitute $\mathbf{A}$. The row operations used to reduce $[\mathbf{A} \vdots \mathbf{B}]$ will, of course, operate on the elements of the last column as well, eventually resulting in what is called $\mathbf{C}$ in the last equation. We will state this result as a theorem.

THEOREM 6.21

Let $\mathbf{A}$ be $n \times m$ and let $\mathbf{B}$ be $m \times 1$. Then for some $m \times 1$ matrix $\mathbf{C}$,

$$[\mathbf{A} \vdots \mathbf{B}]_R = [\mathbf{A}_R \vdots \mathbf{C}]. \quad \blacksquare$$

The reason this result is important is that the original system $\mathbf{AX} = \mathbf{B}$ and the reduced system $\mathbf{A}_R \mathbf{X} = \mathbf{C}$ have the same solutions (as in the homogeneous case, because the elementary row operations do not change the solutions of the system). But because of the special form of $\mathbf{A}_R$, the system $\mathbf{A}_R \mathbf{X} = \mathbf{C}$ is easy either to solve by inspection or to see that there is no solution.

EXAMPLE 6.34

Consider the system

$$\begin{pmatrix} -3 & 2 & 2 \\ 1 & 4 & -6 \\ 0 & -2 & 2 \end{pmatrix} \mathbf{X} = \begin{pmatrix} 8 \\ 1 \\ -2 \end{pmatrix}.$$

We will reduce the augmented matrix

$$[\mathbf{A} \vdots \mathbf{B}] = \begin{pmatrix} -3 & 2 & 2 & | & 8 \\ 1 & 4 & -6 & | & 1 \\ 0 & -2 & 2 & | & -2 \end{pmatrix}.$$

One way to proceed is

$$[\mathbf{A} \vdots \mathbf{B}] \to \text{interchange rows 1 and 2} \to \begin{pmatrix} 1 & 4 & -6 & | & 1 \\ -3 & 2 & 2 & | & 8 \\ 0 & -2 & 2 & | & -2 \end{pmatrix}$$

$$\to \text{add 3(row 1) to row 2} \to \begin{pmatrix} 1 & 4 & -6 & | & 1 \\ 0 & 14 & -16 & | & 11 \\ 0 & -2 & 2 & | & -2 \end{pmatrix}$$

$$\to \frac{1}{14}(\text{row 2}) \to \begin{pmatrix} 1 & 4 & -6 & | & 1 \\ 0 & 1 & -\frac{8}{7} & | & \frac{11}{14} \\ 0 & -2 & 2 & | & -2 \end{pmatrix}$$

$$\to -4(\text{row 2}) \text{ to row 1, } 2(\text{row 2}) \text{ to row 3} \to \begin{pmatrix} 1 & 0 & -\frac{10}{7} & | & -\frac{15}{7} \\ 0 & 1 & -\frac{8}{7} & | & \frac{11}{14} \\ 0 & 0 & -\frac{2}{7} & | & -\frac{3}{7} \end{pmatrix}.$$

$$\rightarrow -\frac{7}{2}(\text{row } 3) \rightarrow \begin{pmatrix} 1 & 0 & -\frac{10}{7} & | & -\frac{15}{7} \\ 0 & 1 & -\frac{8}{7} & | & \frac{11}{14} \\ 0 & 0 & 1 & | & \frac{3}{2} \end{pmatrix}$$

$$\rightarrow \frac{10}{7}(\text{row } 3) \text{ to row } 1, \frac{8}{7}(\text{row } 3) \text{ to row } 2 \rightarrow \begin{pmatrix} 1 & 0 & 0 & | & 0 \\ 0 & 1 & 0 & | & \frac{5}{2} \\ 0 & 0 & 1 & | & \frac{3}{2} \end{pmatrix}.$$

As can be seen in this process, we actually arrived at $\mathbf{A}_R$ in the first three rows and columns, and whatever ends up in the last column is what we call $\mathbf{C}$:

$$[\mathbf{A}\vdots\mathbf{B}]_R = [\mathbf{A}_R\vdots\mathbf{C}].$$

Notice that the reduced augmented matrix is $[\mathbf{I}_3\vdots\mathbf{C}]$ and represents the reduced system $\mathbf{I}_3\mathbf{X} = \mathbf{C}$. This is the system

$$\begin{pmatrix} 1 & 0 & 0 \\ 0 & 1 & 0 \\ 0 & 0 & 1 \end{pmatrix} \mathbf{X} = \begin{pmatrix} 0 \\ \frac{5}{2} \\ \frac{3}{2} \end{pmatrix},$$

which we solve by inspection to get $x_1 = 0$, $x_2 = \frac{5}{2}$, $x_3 = \frac{3}{2}$. Thus reducing $[\mathbf{A}\vdots\mathbf{B}]$ immediately yields the solution

$$\mathbf{U}_p = \begin{pmatrix} 0 \\ \frac{5}{2} \\ \frac{3}{2} \end{pmatrix}$$

of the original system $\mathbf{A}\mathbf{X} = \mathbf{B}$. Because $\mathbf{A}_R = \mathbf{I}_3$, Corollary 6.3 tells us that the homogeneous system $\mathbf{A}\mathbf{X} = \mathbf{O}$ has only the trivial solution, and therefore $\mathbf{H} = \mathbf{O}$ in Theorem 6.20 and $\mathbf{U}_p$ is the unique solution of $\mathbf{A}\mathbf{X} = \mathbf{B}$. ■

EXAMPLE 6.35

The system

$$2x_1 - 3x_2 = 6$$
$$4x_1 - 6x_2 = 18$$

is inconsistent, as we saw in Example 6.30. We will put the fact that this system has no solution into the context of the current discussion. Write the augmented matrix

$$[\mathbf{A}\vdots\mathbf{B}] = \begin{pmatrix} 2 & -3 & | & 6 \\ 4 & -6 & | & 18 \end{pmatrix}.$$

Reduce this matrix. We find that

$$[\mathbf{A}\vdots\mathbf{B}]_R = \begin{pmatrix} 1 & -\frac{3}{2} & | & 0 \\ 0 & 0 & | & 1 \end{pmatrix}.$$

From this we immediately read the reduced system $\mathbf{A}_R\mathbf{X} = \mathbf{C}$:

$$\mathbf{A}_R\mathbf{X} = \begin{pmatrix} 1 & -\frac{3}{2} \\ 0 & 0 \end{pmatrix} \mathbf{X} = \begin{pmatrix} 0 \\ 1 \end{pmatrix}.$$

This system has the same solutions as the original system. But the second equation of the reduced system is

$$0x_1 + 0x_2 = 1,$$

which has no solution. Therefore, $\mathbf{AX} = \mathbf{B}$ has no solution either. ■

In this example, the reduced system has an impossible equation because $\mathbf{A}_R$ has a zero second row, while the second row of $[\mathbf{A}\vdots\mathbf{B}]_R$ has a nonzero element in the augmented column. Whenever this happens, we obtain an equation having all zero coefficients of the unknowns but equal to a nonzero number. In such a case, the reduced system $\mathbf{A}_R\mathbf{X} = \mathbf{C}$, hence the original system $\mathbf{AX} = \mathbf{B}$ can have no solution. The key to recognizing when this will occur is that it happens when the rank of $\mathbf{A}_R$ (its number of nonzero rows) is less than the rank of $[\mathbf{A}\vdots\mathbf{B}]$.

THEOREM 6.22

The nonhomogeneous system $\mathbf{AX} = \mathbf{B}$ has a solution if and only if $\mathbf{A}$ and $[\mathbf{A}\vdots\mathbf{B}]$ have the same rank.

Proof Let $\mathbf{A}$ be $n \times m$. Suppose first that $rank(\mathbf{A}) = rank([\mathbf{A}\vdots\mathbf{B}]) = r$. By Theorems 6.12 and 6.14, the column space of $[\mathbf{A}\vdots\mathbf{B}]$ has dimension r. Certainly r cannot exceed the number of columns of $\mathbf{A}$, so $\mathbf{B}$, which is column $m + 1$ of $[\mathbf{A}\vdots\mathbf{B}]$, must be a linear combination of the first m columns of $[\mathbf{A}\vdots\mathbf{B}]$, which form $\mathbf{A}$. This means that, for some numbers $\alpha_1, \ldots, \alpha_m$,

$$\mathbf{B} = \alpha_1 \begin{pmatrix} a_{11} \\ a_{21} \\ \vdots \\ a_{n1} \end{pmatrix} + \alpha_2 \begin{pmatrix} a_{12} \\ a_{22} \\ \vdots \\ a_{n2} \end{pmatrix} + \cdots + \alpha_m \begin{pmatrix} a_{1m} \\ a_{2m} \\ \vdots \\ a_{nm} \end{pmatrix}$$

$$= \begin{pmatrix} \alpha_1 a_{11} + \alpha_2 a_{12} + \cdots + a_m a_{1m} \\ \alpha_1 a_{21} + \alpha_2 a_{22} + \cdots + a_m a_{2m} \\ \vdots \\ \alpha_1 a_{n1} + \alpha_2 a_{n2} + \cdots + a_m a_{nm} \end{pmatrix} = \mathbf{A} \begin{pmatrix} \alpha_1 \\ \alpha_2 \\ \vdots \\ \alpha_m \end{pmatrix}.$$

But then $\begin{pmatrix} \alpha_1 \\ \alpha_2 \\ \vdots \\ \alpha_m \end{pmatrix}$ is a solution of $\mathbf{AX} = \mathbf{B}$.

Conversely, suppose $\mathbf{AX} = \mathbf{B}$ has a solution $\begin{pmatrix} \alpha_1 \\ \alpha_2 \\ \vdots \\ \alpha_m \end{pmatrix}$. Then

$$\mathbf{B} = \mathbf{A} \begin{pmatrix} \alpha_1 \\ \alpha_2 \\ \vdots \\ \alpha_m \end{pmatrix} = \begin{pmatrix} \alpha_1 a_{11} + \alpha_2 a_{12} + \cdots + a_m a_{1m} \\ \alpha_1 a_{21} + \alpha_2 a_{22} + \cdots + a_m a_{2m} \\ \vdots \\ \alpha_1 a_{n1} + \alpha_2 a_{n2} + \cdots + a_m a_{nm} \end{pmatrix}$$

$$= \alpha_1 \begin{pmatrix} a_{11} \\ a_{21} \\ \vdots \\ a_{n1} \end{pmatrix} + \alpha_2 \begin{pmatrix} a_{12} \\ a_{22} \\ \vdots \\ a_{n2} \end{pmatrix} + \cdots + \alpha_m \begin{pmatrix} a_{1m} \\ a_{2m} \\ \vdots \\ a_{nm} \end{pmatrix}.$$

Then **B** is a linear combination of the columns of **A**, thought of as vectors in R^n. But then the column space of **A** is the same as the column space of [A:B]. Then

$$rank(\mathbf{A}) = \text{ dimension of the column space of } \mathbf{A}$$

$$= \text{ dimension of the column space of } [\mathbf{A} \vdots \mathbf{B}] = rank[\mathbf{A} \vdots \mathbf{B}],$$

and the proof is complete. ■

EXAMPLE 6.36

We will find the general solution of the system

$$-x_1 + x_2 + 3x_3 = -2$$

$$x_2 + 2x_3 = 4.$$

This is the system $\mathbf{AX} = \mathbf{B}$, with

$$\mathbf{A} = \begin{pmatrix} -1 & 1 & 3 \\ 0 & 1 & 2 \end{pmatrix} \quad \text{and} \quad [\mathbf{A} \vdots \mathbf{B}] = \begin{pmatrix} -1 & 1 & 3 & | & -2 \\ 0 & 1 & 2 & | & 4 \end{pmatrix}.$$

We find that

$$[\mathbf{A} \vdots \mathbf{B}]_R = \begin{pmatrix} 1 & 0 & -1 & | & 6 \\ 0 & 1 & 2 & | & 4 \end{pmatrix}.$$

From the first three columns we immediately have

$$\mathbf{A}_R = \begin{pmatrix} 1 & 0 & -1 \\ 0 & 1 & 2 \end{pmatrix},$$

so $rank(\mathbf{A}) = 2 = rank([\mathbf{A} \vdots \mathbf{B}])$. This system has a solution, which we will now find.

From $\mathbf{A}_R$ we see that x_1 and x_2 are dependent and x_3 is independent. From the rows of $[\mathbf{A} \vdots \mathbf{B}]_R$ we have the equations

$$x_1 - x_3 = 6$$

$$x_2 + 2x_3 = 2.$$

Then

$$x_1 = x_3 + 6$$

and

$$x_2 = -2x_3 + 4.$$

If we write $x_3 = \alpha$, any number, then the general solution of $\mathbf{AX} = \mathbf{B}$ is

$$\mathbf{X} = \begin{pmatrix} \alpha + 6 \\ -2\alpha + 4 \\ \alpha \end{pmatrix}. \quad ■$$

It is instructive to write this general solution as

$$\mathbf{X} = \begin{pmatrix} \alpha + 6 \\ -2\alpha + 4 \\ \alpha \end{pmatrix} = \alpha \begin{pmatrix} 1 \\ -2 \\ 1 \end{pmatrix} + \begin{pmatrix} 6 \\ 4 \\ 0 \end{pmatrix}.$$

The solution space of $\mathbf{AX} = \mathbf{O}$ has dimension $m - rank(\mathbf{A}) = 3 - 2 = 1$, and $\begin{pmatrix} 1 \\ -2 \\ 1 \end{pmatrix}$ is a

solution of $\mathbf{AX} = \mathbf{O}$. This means that $\mathbf{H} = \alpha \begin{pmatrix} 1 \\ 2 \\ 1 \end{pmatrix}$ is the general solution of $\mathbf{AX} = \mathbf{O}$. Since

$\mathbf{U}_p = \begin{pmatrix} 6 \\ 4 \\ 0 \end{pmatrix}$ is a particular solution of $\mathbf{AX} = \mathbf{B}$, the method of solution has actually displayed the

general solution in the form $\mathbf{H} + \mathbf{U}_p$, the general solution of the homogeneous system plus any
particular solution of the nonhomogeneous system.

EXAMPLE 6.37

Solve the system

$$x_1 - x_2 + 2x_3 = 3$$
$$-4x_1 + x_2 + 7x_3 = -5$$
$$-2x_1 - x_2 + 11x_3 = 14.$$

The augmented matrix is

$$[\mathbf{A}\!:\!\mathbf{B}] = \begin{pmatrix} 1 & -1 & 2 & | & 3 \\ -4 & 1 & 7 & | & -5 \\ -2 & -1 & 11 & \cdots & 14 \end{pmatrix}.$$

When we reduce this matrix we obtain

$$[\mathbf{A}\!:\!\mathbf{B}]_R = [\mathbf{A}_R\!:\!\mathbf{C}] = \begin{pmatrix} 1 & 0 & -3 & | & 0 \\ 0 & 1 & -5 & | & 0 \\ 0 & 0 & 0 & | & 1 \end{pmatrix}.$$

The first three columns of this reduced matrix make up $\mathbf{A}_R$. But

$$rank(\mathbf{A}) = 2 \quad \text{and} \quad rank([\mathbf{A}\!:\!\mathbf{B}]_R) = 3,$$

so this system has no solution. The last equation of the reduced system is

$$0x_1 + 0x_2 + 0x_3 = 1,$$

which can have no solution. ■

EXAMPLE 6.38

Solve

$$x_1 - x_3 + 2x_4 + x_5 + 6x_6 = -3$$
$$x_2 + x_3 + 3x_4 + 2x_5 + 4x_6 = 1$$
$$x_1 - 4x_2 + 3x_3 + x_4 + 2x_6 = 0.$$

The augmented matrix is

$$[\mathbf{A} \vdots \mathbf{B}] = \begin{pmatrix} 1 & 0 & -1 & 2 & 1 & 6 & \vdots & -3 \\ 0 & 1 & 1 & 3 & 2 & 4 & \vdots & 1 \\ 1 & -4 & 3 & 1 & 0 & 2 & \vdots & 0 \end{pmatrix}.$$

Reduce this to get

$$[\mathbf{A} \vdots \mathbf{B}]_R = \begin{pmatrix} 1 & 0 & 0 & \frac{27}{8} & \frac{15}{8} & \frac{60}{8} & \vdots & -\frac{17}{8} \\ 0 & 1 & 0 & \frac{13}{8} & \frac{9}{8} & \frac{20}{8} & \vdots & \frac{1}{8} \\ 0 & 0 & 1 & \frac{11}{8} & \frac{7}{8} & \frac{12}{8} & \vdots & \frac{7}{8} \end{pmatrix}.$$

The first six columns of this matrix form $\mathbf{A}_R$, and we read that $rank(\mathbf{A}) = 3 = rank([\mathbf{A} \vdots \mathbf{B}]_R)$. From $[\mathbf{A} \vdots \mathbf{B}]_R$, identify x_1, x_2, x_3 as dependent and x_4, x_5, x_6 as independent. The number of independent unknowns is $m - rank(\mathbf{A}) = 6 - 3 = 3$, and this is the dimension of the solution space of $\mathbf{AX} = \mathbf{O}$. From the reduced augmented matrix, the first equation of the reduced system is

$$x_1 + \tfrac{27}{8}x_4 + \tfrac{15}{8}x_5 + \tfrac{60}{8}x_6 = -\tfrac{17}{8},$$

so

$$x_1 = -\tfrac{27}{8}x_4 - \tfrac{15}{8}x_5 - \tfrac{60}{8}x_6 - \tfrac{17}{8}.$$

We will not write out all of the equations of the reduced system. The point is that we can read directly from $[\mathbf{A} \vdots \mathbf{B}]_R$ that

$$x_2 = -\tfrac{13}{8}x_4 - \tfrac{9}{8}x_5 - \tfrac{20}{8}x_6 + \tfrac{1}{8}$$

and

$$x_3 = -\tfrac{11}{8}x_4 - \tfrac{7}{8}x_5 - \tfrac{12}{8}x_6 + \tfrac{7}{8},$$

while x_4, x_5, x_6 can be assigned any numerical values. We can write this solution as

$$\mathbf{X} = \begin{pmatrix} -\frac{27}{8}x_4 - \frac{15}{8}x_5 - \frac{60}{8}x_6 - \frac{17}{8} \\ -\frac{13}{8}x_4 - \frac{9}{8}x_5 - \frac{20}{8}x_6 + \frac{1}{8} \\ -\frac{11}{8}x_4 - \frac{7}{8}x_5 - \frac{12}{8}x_6 + \frac{7}{8} \\ x_4 \\ x_5 \\ x_6 \end{pmatrix}.$$

If we let $x_4 = 8\alpha$, $x_5 = 8\beta$, and $x_6 = 8\gamma$, with $\alpha, \beta,$ and γ any numbers, then the general solution is

$$\mathbf{X} = \alpha \begin{pmatrix} -27 \\ -13 \\ -11 \\ 8 \\ 0 \\ 0 \end{pmatrix} + \beta \begin{pmatrix} -15 \\ -9 \\ -7 \\ 0 \\ 8 \\ 0 \end{pmatrix} + \gamma \begin{pmatrix} -60 \\ -20 \\ -12 \\ 0 \\ 0 \\ 8 \end{pmatrix} + \begin{pmatrix} -\frac{17}{8} \\ \frac{1}{8} \\ \frac{7}{8} \\ 0 \\ 0 \\ 0 \end{pmatrix}.$$

This is in the form $\mathbf{H} + \mathbf{U}_p$, with $\mathbf{H}$ the general solution of $\mathbf{AX} = \mathbf{O}$ and $\mathbf{U}_p$ a particular solution of $\mathbf{AX} = \mathbf{B}$. ■

Since the general solution is of the form $\mathbf{X} = \mathbf{H} + \mathbf{U}_p$, with $\mathbf{H}$ the general solution of $\mathbf{AX} = \mathbf{O}$, the only way $\mathbf{AX} = \mathbf{B}$ can have a unique solution is if $\mathbf{H} = \mathbf{O}$, that is, the homogeneous system must have only the trivial solution. But for a system with the same number of unknowns as equations, this can occur only if $\mathbf{A}_R$ is the identity matrix.

THEOREM 6.23

Let $\mathbf{A}$ be $n \times n$. Then the nonhomogeneous system $\mathbf{AX} = \mathbf{B}$ has a unique solution if and only if $\mathbf{A}_R = \mathbf{I}_n$. ■

This, in turn, occurs exactly when $rank(\mathbf{A}) = n$.

EXAMPLE 6.39

Consider the system

$$\begin{pmatrix} 2 & 1 & -11 \\ -5 & 1 & 9 \\ 1 & 1 & 14 \end{pmatrix} \mathbf{X} = \begin{pmatrix} -6 \\ 12 \\ -5 \end{pmatrix}.$$

The augmented matrix is

$$[\mathbf{A} \vdots \mathbf{B}] = \begin{pmatrix} 2 & 1 & -11 & | & -6 \\ -5 & 1 & 9 & | & 12 \\ 1 & 1 & 14 & | & -5 \end{pmatrix}$$

and we find that

$$[\mathbf{A} \vdots \mathbf{B}]_R = \begin{pmatrix} 1 & 0 & 0 & | & -\frac{86}{31} \\ 0 & 1 & 0 & | & -\frac{191}{155} \\ 0 & 0 & 1 & | & -\frac{11}{155} \end{pmatrix}.$$

The first three columns tell us that $\mathbf{A}_R = \mathbf{I}_3$. The homogeneous system $\mathbf{AX} = \mathbf{O}$ has only the trivial solution. Then $\mathbf{AX} = \mathbf{B}$ has a unique solution, which we read from $[\mathbf{A} \vdots \mathbf{B}]_R$:

$$\mathbf{X} = \begin{pmatrix} -\frac{86}{31} \\ -\frac{191}{155} \\ -\frac{11}{155} \end{pmatrix}.$$

Note that $rank(\mathbf{A}) = 3$ and the dimension of the solution space $\mathbf{AX} = \mathbf{O}$ is $n - rank(\mathbf{A}) = 3 - 3 = 0$, consistent with this solution space having no elements except the zero vector. ■

SECTION 6.7 PROBLEMS

In each of Problems 1 through 20, find the general solution of the system or show that the system has no solution.

1.
$$3x_1 - 2x_2 + x_3 = 6$$
$$x_1 + 10x_2 - x_3 = 2$$
$$-3x_1 - 2x_2 + x_3 = 0$$

2. $4x_1 - 2x_2 + 3x_3 + 10x_4 = 1$
$$x_1 - 3x_4 = 8$$
$$2x_1 - 3x_2 + x_4 = 16$$

3. $2x_1 - 3x_2 + x_4 - x_6 = 0$
$$3x_1 - 2x_3 + x_5 = 1$$
$$x_2 - x_4 + 6x_6 = 3$$

4. $2x_1 - 3x_2 = 1$
$$-x_1 + 3x_2 = 0$$
$$x_1 - 4x_2 = 3$$

5.
$$3x_2 - 4x_4 = 10$$
$$x_1 - 3x_2 + 4x_5 - x_6 = 8$$
$$x_2 + x_3 - 6x_4 + x_6 = -9$$
$$x_1 - x_2 + x_6 = 0$$

6. $2x_1 - 3x_2 + x_4 = 1$
$$3x_2 + x_3 - x_4 = 0$$
$$2x_1 - 3x_2 + 10x_3 = 0$$

7. $8x_2 - 4x_3 + 10x_6 = 1$
$$x_3 + x_5 - x_6 = 2$$
$$x_4 - 3x_5 + 2x_6 = 0$$

8. $2x_1 - 3x_3 = 1$
$$x_1 - x_2 + x_3 = 1$$
$$2x_1 - 4x_2 + x_3 = 2$$

9.
$$14x_3 - 3x_5 + x_7 = 2$$
$$x_1 + x_2 + x_3 - x_4 + x_6 = -4$$

10. $3x_1 - 2x_2 = -1$
$$4x_1 + 3x_2 = 4$$

11.
$$7x_1 - 3x_2 + 4x_3 = -7$$
$$2x_1 + x_2 - x_3 + 4x_4 = 6$$
$$x_2 - 3x_4 = -5$$

12.
$$-4x_1 + 5x_2 - 6x_3 = 2$$
$$2x_1 - 6x_2 + x_3 = -5$$
$$-6x_1 + 16x_2 - 11x_3 = 1$$

13.
$$4x_1 - x_2 + 4x_3 = 1$$
$$x_1 + x_2 - 5x_3 = 0$$
$$-2x_1 + x_2 + 7x_3 = 4$$

14. $-6x_1 + 2x_2 - x_3 + x_4 = 0$
$$x_1 + 4x_2 - x_4 = -5$$
$$x_1 + x_2 + x_3 - 7x_4 = 0$$

15.
$$4x_1 - 3x_2 + x_3 = -1$$
$$-3x_1 + x_2 - 5x_3 = 0$$
$$-5x_1 - 14x_3 = 10$$

16.
$$9x_1 + x_2 - 4x_5 = -1$$
$$x_2 + 4x_3 - 4x_5 = 2$$
$$x_2 + x_4 - x_5 = 0$$

17.
$$-5x_1 + 3x_2 - x_3 = 0$$
$$x_1 - 7x_2 + x_4 = -4$$
$$x_1 - x_2 + 3x_3 - 6x_4 = -11$$

18. $-6x_1 + x_2 - 4x_3 = 1$
$$2x_1 - x_2 - x_3 = 8$$
$$x_1 + 6x_2 - x_3 = -3$$

19. $-5x_1 + 3x_2 + x_3 - x_4 = -8$
$$4x_1 + 3x_2 - x_4 = 9$$
$$2x_1 + 3x_2 - 3x_3 + x_4 = -7$$

20.
$$3x_2 + x_4 - x_5 = 15$$
$$x_1 + 3x_2 + x_4 - 7x_5 = 10$$
$$-5x_1 + x_2 - 4x_3 + x_4 + 6x_5 = 1$$
$$2x_1 + 4x_2 - x_3 + 10x_4 + 8x_5 = 7$$

21. Let $\mathbf{A}$ be an $n \times m$ matrix with rank r. Prove that the reduced system $\mathbf{A}_R\mathbf{X} = \mathbf{B}$ has a solution if and only if $b_{r+1} = \cdots = b_n = 0$.

6.8 Summary for Linear Systems

We will summarize the terminology and methodology we have for solving systems of linear algebraic equations. Let $\mathbf{A}$ be an $n \times m$ matrix.

For the homogeneous system $\mathbf{AX} = \mathbf{O}$:

1. Compute $\mathbf{A}_R$.

2. Count the number of nonzero rows of $\mathbf{A}_R$. This is $rank(\mathbf{A})$. Let $r = rank(\mathbf{A})$. If $m - r = 0$, then $\mathbf{AX} = \mathbf{O}$ has only the trivial solution. If $m - r > 0$, then $\mathbf{AX} = \mathbf{O}$ has this number of linearly independent solutions, and these form a basis for the solution space.

3. From the reduced system $\mathbf{A}_R\mathbf{X} = \mathbf{O}$, solve for each dependent unknown $x_1, \ldots, x_r$ in terms of the independent unknowns $x_{r+1}, \ldots, x_m$.

4. Write the general solution as a sum of $m - r$ linearly independent solution column vectors, each having one of the independent unknowns as a coefficient. These independent unknowns can be assigned any numerical values.

EXAMPLE 6.40

Solve the system

$$-4x_1 + x_2 + x_3 - 5x_4 + x_5 - 3x_6 - 2x_7 = 0$$
$$3x_1 - x_2 - 2x_3 - 4x_4 + 2x_5 - 6x_6 + x_7 = 0$$
$$4x_1 + 4x_2 - x_3 - x_4 + 2x_5 - x_6 + 2x_7 = 0.$$

The matrix of coefficients is

$$\mathbf{A} = \begin{pmatrix} -4 & 1 & 1 & -5 & 1 & -3 & -2 \\ 3 & -1 & -2 & -4 & 2 & -6 & 1 \\ 4 & 4 & -1 & -1 & 2 & -1 & 2 \end{pmatrix}.$$

We find that

$$\mathbf{A}_R = \begin{pmatrix} 1 & 0 & 0 & \frac{64}{25} & -\frac{17}{25} & \frac{56}{25} & \frac{3}{5} \\ 0 & 1 & 0 & -\frac{6}{5} & \frac{3}{5} & -\frac{4}{5} & 0 \\ 0 & 0 & 1 & \frac{161}{25} & -\frac{58}{25} & \frac{169}{25} & \frac{2}{5} \end{pmatrix}.$$

Then $rank(\mathbf{A}) = r = 3$ and $m - r = 7 - 3 = 4$. The solution space has dimension 4, and there are three dependent unknowns x_1, x_2, x_3 and four independent unknowns x_4, x_5, x_6, x_7. From the reduced system $\mathbf{A}_R\mathbf{X} = \mathbf{O}$, we read the solutions for the dependent unknowns in terms of the independent unknowns:

$$x_1 = -\frac{64}{25}x_4 + \frac{17}{25}x_5 - \frac{56}{25}x_6 - \frac{3}{5}x_7$$
$$x_2 = \frac{6}{5}x_4 - \frac{3}{5}x_5 + \frac{4}{5}x_6$$
$$x_3 = -\frac{161}{25}x_4 + \frac{58}{25}x_5 - \frac{169}{25}x_6 - \frac{2}{5}x_7.$$

Now write the general solution:

$$
\mathbf{X} = \begin{pmatrix}
-\frac{64}{25}x_4 + \frac{17}{25}x_5 - \frac{56}{25}x_6 - \frac{3}{5}x_7 \\[4pt]
\frac{6}{5}x_4 - \frac{3}{5}x_5 + \frac{4}{5}x_6 \\[4pt]
-\frac{161}{25}x_4 + \frac{58}{25}x_5 - \frac{169}{25}x_6 - \frac{2}{5}x_7 \\[4pt]
x_4 \\[4pt]
x_5 \\[4pt]
x_6 \\[4pt]
x_7
\end{pmatrix}
$$

$$
= x_4 \begin{pmatrix} -\frac{64}{25} \\[4pt] \frac{6}{5} \\[4pt] -\frac{161}{25} \\[4pt] 1 \\[2pt] 0 \\[2pt] 0 \\[2pt] 0 \end{pmatrix}
+ x_5 \begin{pmatrix} \frac{17}{25} \\[4pt] -\frac{3}{5} \\[4pt] \frac{58}{25} \\[4pt] 0 \\[2pt] 1 \\[2pt] 0 \\[2pt] 0 \end{pmatrix}
+ x_6 \begin{pmatrix} -\frac{56}{25} \\[4pt] \frac{4}{5} \\[4pt] -\frac{169}{25} \\[4pt] 0 \\[2pt] 0 \\[2pt] 1 \\[2pt] 0 \end{pmatrix}
+ x_7 \begin{pmatrix} -\frac{3}{5} \\[4pt] 0 \\[4pt] -\frac{2}{5} \\[4pt] 0 \\[2pt] 0 \\[2pt] 0 \\[2pt] 1 \end{pmatrix}.
$$

If we wish, we can now write $x_4 = 25\alpha$, $x_5 = 25\beta$, $x_6 = 25\gamma$, and $x_7 = 5\delta$ to write the general solution

$$
\mathbf{X} = \alpha \begin{pmatrix} -64 \\ 30 \\ -161 \\ 25 \\ 0 \\ 0 \\ 0 \end{pmatrix}
+ \beta \begin{pmatrix} 17 \\ -15 \\ 58 \\ 0 \\ 25 \\ 0 \\ 0 \end{pmatrix}
+ \gamma \begin{pmatrix} -56 \\ 20 \\ -169 \\ 0 \\ 0 \\ 25 \\ 0 \end{pmatrix}
+ \delta \begin{pmatrix} -3 \\ 0 \\ -2 \\ 0 \\ 0 \\ 0 \\ 5 \end{pmatrix},
$$

with α, β, γ, and δ arbitrary numbers. This expresses the solution as a linear combination of four linearly independent solutions, consistent with the dimension of the solution space being 4. ∎

For the nonhomogeneous linear system $\mathbf{AX} = \mathbf{B}$,

1. Write the augmented matrix $[\mathbf{A} \vdots \mathbf{B}]$ and reduce it to $[\mathbf{A} \vdots \mathbf{B}]_R$, which has the form $[\mathbf{A}_R \vdots \mathbf{C}]$ for some $n \times 1$ matrix $\mathbf{C}$.

2. From $[\mathbf{A} \vdots \mathbf{B}]_R$, we can read both $rank(\mathbf{A})$ (from the first m columns) and $rank([\mathbf{A} \vdots \mathbf{B}])$. If these numbers are not equal, the system has no solution. If they are equal, there exists a solution.

3. Let $rank(\mathbf{A}) = r$. Then $x_1, \ldots, x_r$ are determined by the independent unknowns $x_{r+1}, \ldots, x_m$ and by the elements of $\mathbf{C} = \begin{pmatrix} c_1 \\ c_2 \\ \vdots \\ c_n \end{pmatrix}$. From the reduced matrix $[\mathbf{A}_R \vdots \mathbf{C}]$, write each dependent unknown x_j in terms of $x_{r+1}, \ldots, x_m$ and c_j.

4. The general solution can now be written in the form $\mathbf{X} = \mathbf{H} + \mathbf{U}_p$, where $\mathbf{H}$ is the general solution of $\mathbf{AX} = \mathbf{O}$ and $\mathbf{U}_p$ is the particular solution found (in terms of the c_j's) for $\mathbf{AX} = \mathbf{B}$.

EXAMPLE 6.41

Solve the system

$$3x_1 - x_2 - 2x_3 - 5x_4 + 6x_5 + x_6 + 3x_7 = 4$$
$$-x_1 + 4x_2 - x_3 + x_4 + 3x_5 + 3x_6 - x_7 = -2$$
$$5x_1 - x_2 + 4x_3 + x_4 - 2x_5 - x_6 + 3x_7 = 2.$$

The augmented matrix is

$$[\mathbf{A} \vdots \mathbf{B}] = \begin{pmatrix} 3 & -1 & -2 & -5 & 6 & 1 & 3 & | & 4 \\ -1 & 4 & -1 & 1 & 3 & 3 & -1 & | & -2 \\ 5 & -1 & 4 & 1 & -2 & -1 & 3 & | & 2 \end{pmatrix}.$$

This reduces to

$$[\mathbf{A} \vdots \mathbf{B}]_R = [\mathbf{A}_R \vdots \mathbf{C}]$$

$$= \begin{pmatrix} 1 & 0 & 0 & -\frac{5}{7} & \frac{15}{14} & \frac{2}{7} & \frac{11}{14} & | & \frac{11}{14} \\ 0 & 1 & 0 & \frac{8}{21} & \frac{25}{42} & \frac{5}{7} & -\frac{5}{42} & | & -\frac{19}{42} \\ 0 & 0 & 1 & \frac{26}{21} & -\frac{71}{42} & -\frac{3}{7} & -\frac{11}{42} & | & -\frac{25}{42} \end{pmatrix}.$$

Now x_1, x_2, and x_3 are dependent, while x_4, x_5, x_6, and x_7 are independent. The c_j's appear in the last column. From this reduced augmented matrix we immediately write the solution:

$$\mathbf{X} = \begin{pmatrix} \frac{5}{7}x_4 - \frac{15}{14}x_5 - \frac{2}{7}x_6 - \frac{11}{14}x_7 + \frac{11}{14} \\ -\frac{8}{21}x_4 - \frac{25}{42}x_5 - \frac{5}{7}x_6 + \frac{5}{42}x_7 - \frac{19}{42} \\ -\frac{26}{21}x_4 + \frac{71}{42}x_5 + \frac{3}{7}x_6 + \frac{11}{42}x_7 - \frac{25}{42} \\ x_4 \\ x_5 \\ x_6 \\ x_7 \end{pmatrix},$$

in which each of x_4, x_5, x_6, x_7 can be assigned any numerical values. We can write this solution as

$$X = \alpha \begin{pmatrix} 15 \\ -8 \\ -26 \\ 21 \\ 0 \\ 0 \\ 0 \end{pmatrix} + \beta \begin{pmatrix} -45 \\ -25 \\ 71 \\ 0 \\ 42 \\ 0 \\ 0 \end{pmatrix} + \gamma \begin{pmatrix} -2 \\ -5 \\ 3 \\ 0 \\ 0 \\ 7 \\ 0 \end{pmatrix} + \delta \begin{pmatrix} -33 \\ 5 \\ 11 \\ 0 \\ 0 \\ 0 \\ 42 \end{pmatrix} + \begin{pmatrix} \frac{11}{14} \\ -\frac{19}{42} \\ -\frac{25}{42} \\ 0 \\ 0 \\ 0 \\ 0 \end{pmatrix}.$$

in which $x_4 = 21\alpha$, $x_5 = 42\beta$, $x_6 = 7\gamma$, and $x_7 = 42\delta$, with α, β, γ, and δ any numbers. This gives the solution as $\mathbf{H} + \mathbf{U}_p$, the sum of the general solution of the homogeneous system $\mathbf{AX} = \mathbf{O}$ and a particular solution of $\mathbf{AX} = \mathbf{B}$. ∎

In each of Problems 1 through 15, if the system is homogeneous, write the general solution and the dimension of the solution space. If the system is nonhomogeneous, write the general solution or show that the system has no solution.

1. $3x_1 - x_2 + 5x_3 = 0$
   $2x_1 + x_2 - x_3 = 0$

2. $-x_1 + 4x_2 - 2x_3 = -2$
   $3x_1 + x_2 + x_3 = 5$
   $6x_1 - 11x_2 + 7x_3 = 8$

3. $5x_1 - x_2 + 2x_3 = 1$
   $x_2 + x_3 = 4$

4. $x_2 - 8x_3 + 3x_4 = 0$
   $x_1 - 5x_3 + x_4 = 0$

5. $2x_1 + 2x_2 - x_4 = 1$
   $x_1 - x_3 - x_4 = -3$
   $3x_1 + 4x_2 - x_3 + 2x_4 = 5$

6. $x_1 - x_2 + 4x_3 = 0$
   $x_1 + 5x_2 + x_3 = 0$
   $-7x_1 + 2x_2 + x_3 = 0$

7. $-x_1 - x_2 + 2x_3 = 5$
   $3x_1 + x_2 + 6x_3 = 1$
   $x_1 - x_2 + 10x_3 = -1$

8. $x_1 + 2x_2 + 6x_3 - x_4 = 0$
   $-3x_1 - x_2 + 8x_3 - x_4 = 0$

9. $4x_1 + 4x_2 + 3x_3 = 10$
   $x_1 + x_2 - 5x_3 = 5$

10. $7x_1 - 4x_2 + x_3 - x_4 = 0$
   $6x_1 - x_2 + x_3 + 2x_4 = 0$
   $x_1 + 6x_2 + 5x_3 - 2x_4 = 0$

11. $-x_1 + 4x_2 - x_3 = 1$
   $x_1 + 2x_2 + 6x_3 = 3$
   $-3x_1 - x_2 + 3x_3 = -2$

12. $x_1 + 5x_2 - x_3 - x_4 - x_5 = 1$
   $-2x_1 + x_3 - x_4 - 3x_5 = 6$

13. $8x_1 + 5x_2 - 2x_4 = 5$
   $x_1 - x_3 + 6x_4 = 1$
   $-x_1 + x_2 - 7x_3 + x_4 = 2$

14. $-4x_1 - x_2 + x_4 - 2x_5 = 0$
   $x_1 + x_2 - 4x_3 - x_5 = 0$

15. $-5x_1 + x_2 - 3x_3 = -6$
   $x_1 + 2x_2 = 4$
   $-2x_1 + x_2 - 5x_3 = -2$

6.9 Matrix Inverses

DEFINITION 6.15 *Matrix Inverse*

Let $\mathbf{A}$ be an $n \times n$ matrix. Then $\mathbf{B}$ is an inverse of $\mathbf{A}$ if

$$\mathbf{AB} = \mathbf{BA} = \mathbf{I}_n.$$

In this definition, $\mathbf{B}$ must also be $n \times n$ because both $\mathbf{AB}$ and $\mathbf{BA}$ must be defined. Further, if $\mathbf{B}$ is an inverse of $\mathbf{A}$, then $\mathbf{A}$ is also an inverse of $\mathbf{B}$.

It is easy to find nonzero square matrices that have no inverse. For example, let

$$\mathbf{A} = \begin{pmatrix} 1 & 0 \\ 2 & 0 \end{pmatrix}.$$

If **B** is an inverse of **A**, say $\mathbf{B} = \begin{pmatrix} a & b \\ c & d \end{pmatrix}$, then we must have

$$\mathbf{AB} = \begin{pmatrix} 1 & 0 \\ 2 & 0 \end{pmatrix} \begin{pmatrix} a & b \\ c & d \end{pmatrix}$$

$$= \begin{pmatrix} a & b \\ 2a & 2b \end{pmatrix} = \begin{pmatrix} 1 & 0 \\ 0 & 1 \end{pmatrix}.$$

But then

$$a = 1, b = 0, 2a = 0 \quad \text{and} \quad b = 1$$

which are impossible conditions. On the other hand, some matrices do have inverses. For example,

$$\begin{pmatrix} 2 & 1 \\ 1 & 4 \end{pmatrix} \begin{pmatrix} \frac{4}{7} & -\frac{1}{7} \\ -\frac{1}{7} & \frac{2}{7} \end{pmatrix} = \begin{pmatrix} \frac{4}{7} & -\frac{1}{7} \\ -\frac{1}{7} & \frac{2}{7} \end{pmatrix} \begin{pmatrix} 2 & 1 \\ 1 & 4 \end{pmatrix} = \begin{pmatrix} 1 & 0 \\ 0 & 1 \end{pmatrix}.$$

DEFINITION 6.16 *Nonsingular and Singular Matrices*
_____

A square matrix is said to be nonsingular if it has an inverse. If it has no inverse, the matrix is called singular.

If a matrix has an inverse, then it can have only one.

THEOREM 6.24 *Uniqueness of Inverses* _____

Let **B** and **C** be inverses of **A**. Then **B** = **C**.

Proof Write

$$\mathbf{B} = \mathbf{BI}_n = \mathbf{B(AC)} = \mathbf{(BA)C} = \mathbf{I}_n\mathbf{C} = \mathbf{C}. \quad \blacksquare$$

In view of this, we will denote the inverse of **A** as $\mathbf{A}^{-1}$. Here are properties of inverse matrices. In proving parts of the theorem, we repeatedly employ the strategy that if $\mathbf{AB} = \mathbf{BA} = \mathbf{I}_n$, then **B** must be the inverse of **A**.

THEOREM 6.25 _____

1. $\mathbf{I}_n$ is nonsingular and $\mathbf{I}_n^{-1} = \mathbf{I}_n$.
2. If **A** and **B** are nonsingular $n \times n$ matrices, then **AB** is nonsingular and

$$(\mathbf{AB})^{-1} = \mathbf{B}^{-1}\mathbf{A}^{-1}.$$

3. If **A** is nonsingular, so is $\mathbf{A}^{-1}$, and

$$(\mathbf{A}^{-1})^{-1} = \mathbf{A}.$$

4. If $\mathbf{A}$ is nonsingular, so is $\mathbf{A}^t$, and

$$(\mathbf{A}^t)^{-1} = (\mathbf{A}^{-1})^t.$$

5. If $\mathbf{A}$ and $\mathbf{B}$ are $n \times n$ and either is singular, then $\mathbf{AB}$ and $\mathbf{BA}$ are both singular.

Proof For (2), compute

$$(\mathbf{AB})(\mathbf{B}^{-1}\mathbf{A}^{-1}) = \mathbf{A}(\mathbf{BB}^{-1})\mathbf{A}^{-1} = \mathbf{AA}^{-1} = \mathbf{I}_n.$$

Similarly, $(\mathbf{B}^{-1}\mathbf{A}^{-1})(\mathbf{AB}) = \mathbf{I}_n$. Therefore, $(\mathbf{AB})^{-1} = \mathbf{B}^{-1}\mathbf{A}^{-1}$.

For (4), use Theorem 6.3(3) to write

$$(\mathbf{A}^t)(\mathbf{A}^{-1})^t = (\mathbf{A}^{-1}\mathbf{A})^t = (\mathbf{I}_n)^t = \mathbf{I}_n.$$

Similarly,

$$(\mathbf{A}^{-1})^t(\mathbf{A}^t) = (\mathbf{AA}^{-1})^t = \mathbf{I}_n.$$

Therefore, $(\mathbf{A}^t)^{-1} = (\mathbf{A}^{-1})^t$. ∎

We will be able to give a very short proof of (5) when we have developed determinants.

We saw before that not every matrix has an inverse. How can we tell whether a matrix is singular or nonsingular? The following theorem gives a reasonable test.

THEOREM 6.26

An $n \times n$ matrix $\mathbf{A}$ is nonsingular if and only if $\mathbf{A}_R = \mathbf{I}_n$. ∎

Alternatively, an $n \times n$ matrix is nonsingular if and only if its rank is n. The proof consists of understanding a relationship between a matrix having an inverse and its reduced form, being the identity matrix. The key lies in noticing that we can form the columns of a matrix product $\mathbf{AB}$ by multiplying, in turn, $\mathbf{A}$ by each column of $\mathbf{B}$:

$$\text{column } j \text{ of } \mathbf{AB} = \mathbf{A}(\text{column } j \text{ of } \mathbf{B}) = \mathbf{A}\begin{pmatrix} b_{1j} \\ b_{2j} \\ \vdots \\ b_{nj} \end{pmatrix}.$$

Proof We will build an inverse for $\mathbf{A}$ a column at a time. To have $\mathbf{AB} = \mathbf{I}_n$, we must be able to choose the columns of $\mathbf{B}$ so that

$$\text{column } j \text{ of } \mathbf{AB} = \mathbf{A}\begin{pmatrix} b_{1j} \\ b_{2j} \\ \vdots \\ b_{nj} \end{pmatrix} = \text{column } j \text{ of } \mathbf{I}_n = \begin{pmatrix} 0 \\ 0 \\ \vdots \\ 1 \\ \vdots \\ 0 \end{pmatrix}, \tag{6.4}$$

with a 1 in the jth place and zeros elsewhere.

Suppose now that $\mathbf{A}_R = \mathbf{I}_n$. Then, by Theorem 6.23, the system (6.4) has a unique solution for each $j = 1, \dots, n$. These solutions form the columns of a matrix $\mathbf{B}$ such that $\mathbf{AB} = \mathbf{I}_n$, and then $\mathbf{B} = \mathbf{A}^{-1}$. (Actually we must show that $\mathbf{BA} = \mathbf{I}_n$ also, but we leave this as an exercise.)

Conversely, suppose $\mathbf{A}$ is nonsingular. Then system (6.4) has a unique solution for $j = 1, \ldots, n$, because these solutions are the columns of $\mathbf{A}^{-1}$. Then, by Theorem 6.23, $\mathbf{A}_R = \mathbf{I}_n$. ∎

6.9.1 A Method for Finding $\mathbf{A}^{-1}$

We know some computational rules for working with matrix inverses, as well as a criterion for a matrix to have an inverse. Now we want an efficient way of computing $\mathbf{A}^{-1}$ from $\mathbf{A}$.

Theorem 6.26 suggests a strategy. We know that, in any event, there is an $n \times n$ matrix $\mathbf{\Omega}$ such that $\mathbf{\Omega A} = \mathbf{A}_R$. $\mathbf{\Omega}$ is a product of elementary matrices representing the elementary row operations used to reduce $\mathbf{A}$. Previously we found $\mathbf{\Omega}$ by adjoining $\mathbf{I}_n$ to the left of $\mathbf{A}$ to form an $n \times 2n$ matrix $[\mathbf{I}_n \vdots \mathbf{A}]$. Reduce $\mathbf{A}$, performing the elementary row operations on all of $[\mathbf{I}_n \vdots \mathbf{A}]$, to eventually arrive at $[\mathbf{\Omega} \vdots \mathbf{I}_n]$. This produces $\mathbf{\Omega}$ such that

$$\mathbf{\Omega A} = \mathbf{A}_R.$$

If $\mathbf{A}_R = \mathbf{I}_n$, then $\mathbf{\Omega} = \mathbf{A}^{-1}$. If $\mathbf{A}_R \neq \mathbf{I}_n$, then $\mathbf{A}$ has no inverse.

EXAMPLE 6.42

Let

$$\mathbf{A} = \begin{pmatrix} 5 & -1 \\ 6 & 8 \end{pmatrix}.$$

We want to know if $\mathbf{A}$ is nonsingular and if it is, produce its inverse.

Form

$$[\mathbf{I}_2 \vdots \mathbf{A}] = \begin{pmatrix} 1 & 0 & | & 5 & -1 \\ 0 & 1 & | & 6 & 8 \end{pmatrix}.$$

Reduce $\mathbf{A}$ (the last two columns), carrying out the same operations on the first two columns:

$$[\mathbf{I}_2 \vdots \mathbf{A}] \to \frac{1}{5}(\text{row } 1) \to \begin{pmatrix} \frac{1}{5} & 0 & | & 1 & -\frac{1}{5} \\ 0 & 1 & | & 6 & 8 \end{pmatrix}$$

$$\to -6(\text{row } 1) + (\text{row } 2) \to \begin{pmatrix} \frac{1}{5} & 0 & | & 1 & -\frac{1}{5} \\ -\frac{6}{5} & 1 & | & 0 & \frac{46}{5} \end{pmatrix}$$

$$\to \frac{5}{46}(\text{row } 2) \to \begin{pmatrix} \frac{1}{5} & 0 & | & 1 & -\frac{1}{5} \\ -\frac{6}{46} & \frac{5}{46} & | & 0 & 1 \end{pmatrix}$$

$$\to \frac{1}{5}(\text{row } 2) + (\text{row } 1) \to \begin{pmatrix} \frac{8}{46} & \frac{1}{46} & | & 1 & 0 \\ -\frac{6}{46} & \frac{5}{46} & | & 0 & 1 \end{pmatrix}.$$

In the last two columns we read $\mathbf{A}_R = \mathbf{I}_2$. This means that $\mathbf{A}$ is nonsingular. From the first two columns,

$$\mathbf{A}^{-1} = \frac{1}{46} \begin{pmatrix} 8 & 1 \\ -6 & 5 \end{pmatrix}. \quad ∎$$

EXAMPLE 6.43

Let

$$A = \begin{pmatrix} -3 & 21 \\ 4 & -28 \end{pmatrix}.$$

Perform a reduction:

$$[I_2 \vdots A] = \begin{pmatrix} 1 & 0 & | & -3 & 21 \\ 0 & 1 & | & 4 & -28 \end{pmatrix}$$

$$\rightarrow -\frac{1}{3}(\text{row } 1) \rightarrow \begin{pmatrix} -\frac{1}{3} & 0 & | & 1 & -7 \\ 0 & 1 & | & 4 & -28 \end{pmatrix}$$

$$\rightarrow -4(\text{row } 1) + (\text{row } 2) \rightarrow \begin{pmatrix} -\frac{1}{3} & 0 & | & 1 & -7 \\ \frac{4}{3} & 1 & | & 0 & 0 \end{pmatrix}.$$

We read A_R from the last two columns, which form a 2×2 reduced matrix. Since this is not I_2, A is singular and has no inverse. ∎

Here is how inverses relate to the solution of systems of linear equations in which the number of unknowns equals the number of equations.

THEOREM 6.27

Let A be an $n \times n$ matrix.

1. A homogeneous system $AX = O$ has a nontrivial solution if and only if A is singular.
2. A nonhomogeneous system $AX = B$ has a unique solution if and only if A is nonsingular. In this case the solution is $X = A^{-1}B$. ∎

For a homogeneous system $AX = O$, if A were nonsingular, then we could multiply the equation on the left by A^{-1} to get

$$X = A^{-1}O = O.$$

Thus, in the nonsingular case, a homogeneous system can have only a trivial solution. In the singular case, we know that $rank(A) < n$, so the solution space has positive dimension $n - rank(A)$ and therefore has nontrivial solutions in it.

In the nonsingular case, we can multiply a nonhomogeneous equation $AX = B$ on the left by A^{-1} to get the unique solution

$$X = A^{-1}B.$$

EXAMPLE 6.44

Consider the nonhomogeneous system

$$2x_1 - x_2 + 3x_3 = 4$$

$$x_1 + 9x_2 - 2x_3 = -8$$

$$4x_1 - 8x_2 + 11x_3 = 15.$$

The matrix of coefficients is

$$\mathbf{A} = \begin{pmatrix} 2 & -1 & 3 \\ 1 & 9 & -2 \\ 4 & -8 & 11 \end{pmatrix}$$

and we find that

$$\mathbf{A}^{-1} = \frac{1}{53} \begin{pmatrix} 83 & -13 & -25 \\ -19 & 10 & 7 \\ -44 & 12 & 19 \end{pmatrix}.$$

The unique solution of this system is

$$\mathbf{X} = \mathbf{A}^{-1}\mathbf{B} = \frac{1}{53} \begin{pmatrix} 83 & -13 & -25 \\ -19 & 10 & 7 \\ -44 & 12 & 19 \end{pmatrix} \begin{pmatrix} 4 \\ -8 \\ 15 \end{pmatrix}$$

$$= \begin{pmatrix} \frac{61}{53} \\ -\frac{51}{53} \\ \frac{13}{53} \end{pmatrix}. \quad \blacksquare$$

SECTION 6.9   PROBLEMS

In each of Problems 1 through 15, find the inverse of the matrix or show that the matrix is singular.

1. $\begin{pmatrix} -1 & 2 \\ 2 & 1 \end{pmatrix}$

2. $\begin{pmatrix} 12 & 3 \\ 4 & 1 \end{pmatrix}$

3. $\begin{pmatrix} -5 & 2 \\ 1 & 2 \end{pmatrix}$

4. $\begin{pmatrix} -1 & 0 \\ 4 & 4 \end{pmatrix}$

5. $\begin{pmatrix} 6 & 2 \\ 3 & 3 \end{pmatrix}$

6. $\begin{pmatrix} 1 & 1 & -3 \\ 2 & 16 & 1 \\ 0 & 0 & 4 \end{pmatrix}$

7. $\begin{pmatrix} -3 & 4 & 1 \\ 1 & 2 & 0 \\ 1 & 1 & 3 \end{pmatrix}$

8. $\begin{pmatrix} -2 & 1 & -5 \\ 1 & 1 & 4 \\ 0 & 3 & 3 \end{pmatrix}$

9. $\begin{pmatrix} -2 & 1 & 1 \\ 0 & 1 & 1 \\ -3 & 0 & 6 \end{pmatrix}$

10. $\begin{pmatrix} 12 & 1 & 14 \\ -3 & 2 & 0 \\ 0 & 9 & 14 \end{pmatrix}$

11. $\begin{pmatrix} 0 & 0 & -1 \\ 1 & 12 & 0 \\ 1 & -2 & 4 \end{pmatrix}$

12. $\begin{pmatrix} -1 & 1 & 1 & 0 \\ 1 & 0 & 2 & 0 \\ 1 & 1 & 1 & 1 \\ 3 & 0 & 0 & 1 \end{pmatrix}$

13. $\begin{pmatrix} -1 & 1 & 16 & 2 \\ 0 & 0 & 1 & 4 \\ 0 & 0 & 1 & 6 \\ 0 & 1 & 1 & -3 \end{pmatrix}$

14. $\begin{pmatrix} 4 & -3 & 2 & 1 \\ 0 & -2 & 4 & 3 \\ 8 & 1 & -3 & 1 \\ 2 & 3 & 3 & -5 \end{pmatrix}$

15. $\begin{pmatrix} -2 & 6 & 0 & 0 \\ 1 & 4 & 4 & 11 \\ 4 & -4 & -5 & 3 \\ -3 & 1 & 2 & -6 \end{pmatrix}$

In each of Problems 16 through 21, find the unique solution of the system, using Theorem 6.27(2).

16. $3x_1 - 4x_2 - 6x_3 = 0$
$x_1 + x_2 - 3x_3 = 4$
$2x_1 - x_2 + 6x_3 = -1$

17. $x_1 - x_2 + 3x_3 - x_4 = 1$
$x_2 - 3x_3 + 5x_4 = 2$
$x_1 - x_3 + x_4 = 0$
$x_1 + 2x_3 - x_4 = -5$

18. $8x_1 - x_2 - x_3 = 4$
$x_1 + 2x_2 - 3x_3 = 0$
$2x_1 - x_2 + 4x_3 = 5$

19. $2x_1 - 6x_2 + 3x_3 = -4$
$-x_1 + x_2 + x_3 = 5$
$2x_1 + 6x_2 - 5x_3 = 8$

20. $12x_1 + x_2 - 3x_3 = 4$
$x_1 - x_2 + 3x_3 = -5$
$-2x_1 + x_2 + x_3 = 0$

21. $4x_1 + 6x_2 - 3x_3 = 0$
$2x_1 + 3x_2 - 4x_3 = 0$
$x_1 - x_2 + 3x_3 = -7$

22. Let $\mathbf{A}$ be an $n \times n$ nonsingular real matrix. Prove that every vector in R^n can be written as a linear combination of the column vectors of $\mathbf{A}$. *Hint:* Use the fact that $rank(\mathbf{A}) = n$.

23. Let $\mathbf{A}$ be an $n \times n$ matrix of real numbers. Prove that $\mathbf{A}$ is nonsingular if and only if the row vectors of $\mathbf{A}$ form a basis for R^n.

24. Let $\mathbf{A}$ be nonsingular. Prove that for any positive integer k, $\mathbf{A}^k$ is nonsingular and $(\mathbf{A}^k)^{-1} = (\mathbf{A}^{-1})^k$.

25. Let $\mathbf{A}$, $\mathbf{B}$, and $\mathbf{C}$ be $n \times n$ real matrices. Suppose $\mathbf{BA} = \mathbf{AC} = \mathbf{I}_n$. Prove that $\mathbf{B} = \mathbf{C}$.

CHAPTER 7

Determinants

If **A** is a square matrix, the determinant of **A** is a sum of products of elements of **A**, formed according to a procedure we will now describe. First we need some information about permutations.

7.1 Permutations

If n is a positive integer, a *permutation* of order n is an arrangement of the integers $1, \ldots, n$ in any order. For example, suppose p is a permutation that reorders the integers $1, \ldots, 6$ as

$$3, 1, 4, 5, 2, 6.$$

Then

$$p(1) = 3, \quad p(2) = 1, \quad p(3) = 4, \quad p(4) = 5, \quad p(5) = 2, \quad p(6) = 6,$$

with $p(j)$ the number the permutation has put in place j.

For small n it is possible to list all permutations on $1, \ldots, n$. Here is a short list:

For $n = 2$ there are two permutations on the integers $1, 2$, one leaving them in place and the second interchanging them:

$$1, 2$$

$$2, 1.$$

For $n = 3$ there are six permutations on $1, 2, 3$, and they are

$$1, 2, 3$$

$$1, 3, 2$$

$$2, 1, 3$$

$$2, 3, 1$$

$$3, 1, 2$$

$$3, 2, 1.$$

For $n = 4$ there are 24 permutations on $1, 2, 3, 4$:

$$1, 2, 3, 4; \ 1, 2, 4, 3; \ 1, 3, 2, 4; \ 1, 3, 4, 2; \ 1, 4, 2, 3; \ 1, 4, 3, 2;$$

$$2, 1, 3, 4; \ 2, 1, 4, 3; \ 2, 3, 1, 4; \ 2, 3, 4, 1; \ 2, 4, 1, 3; \ 2, 4, 3, 1;$$

$$3, 1, 2, 4; \ 3, 1, 4, 2; \ 3, 2, 1, 4; \ 3, 2, 4, 1; \ 3, 4, 1, 2; \ 3, 4, 2, 1;$$

$$4, 1, 2, 3; \ 4, 1, 3, 2; \ 4, 2, 1, 3; \ 4, 2, 3, 1; \ 4, 3, 1, 2; \ 4, 3, 2, 1.$$

An examination of this list of permutations suggests a systematic approach by which they were all listed, and such an approach will work in theory for higher n. However, we can also observe that the number of permutations on $1, \ldots, n$ increases rapidly with n.

There are $n! = 1 \cdot 2 \cdots \cdot n$ permutations on $1, \ldots, n$. This fact is not difficult to derive. Imagine a row of n boxes and start putting the integers from 1 to n into the boxes, one to each box. There are n choices for a number to put into the first box, $n - 1$ choices for the second, $n - 2$ for the third, and so on until there is only one left to put in the last box. There is a total of $n(n - 1)(n - 2) \cdots 1 = n!$ ways to do this, hence $n!$ permutations on n objects.

A permutation is characterized as even or odd, according to a rule we will now illustrate. Consider the permutation

$$2, 5, 1, 4, 3$$

on the integers $1, \ldots, 5$. For each number k in the list, count the number of integers to its right that are smaller than k. In this way form a list

k	number of integers smaller than k to the right of k
2	1
5	3
1	0
4	1
3	0

Sum the integers in the right column to get 5, which is odd. We therefore call this permutation *odd*. As an example of an even permutation, consider

$$2, 1, 5, 4, 3.$$

Now the list is

k	number of integers smaller than k to the right of k
2	1
1	0
5	2
4	1
3	0

and the integers in the right column sum to 4, an even number. This permutation is even.

If p is a permutation, let

$$sgn(p) = \begin{cases} 0 & \text{if } p \text{ is even} \\ 1 & \text{if } p \text{ is odd} \end{cases}.$$

SECTION 7.1 PROBLEMS

1. The six permutations of 1, 2, 3 are given in the discussion. Which of these permutations are even and which are odd?

2. The 24 permutations of 1, 2, 3, 4 are given in the discussion. Which of these are even and which are odd?

3. Show that half of the permutations on $1, 2, \ldots, n$ are even and the other half are odd.

7.2   Definition of the Determinant

Let $\mathbf{A} = [a_{ij}]$ be an $n \times n$ matrix, with numbers or functions as elements.

DEFINITION 7.1 _____

The *determinant of* $\mathbf{A}$, denoted $\det(\mathbf{A})$, is the sum of all products

$$(-1)^{sgn(p)} a_{1p(1)} a_{2p(2)} \cdots a_{np(n)},$$

taken over all permutations p on $1, \ldots, n$. This sum is denoted

$$\sum_{p} (-1)^{sgn(p)} a_{1p(1)} a_{2p(2)} \cdots a_{np(n)}. \tag{7.1}$$

Each term in the defining sum (7.1) contains exactly one element from each row and from each column, chosen according to the indices j, $p(j)$ determined by the permutation. Each product in the sum is multiplied by 1 if the permutation is even and by -1 if p is odd.

Since there are $n!$ permutations on $1, \ldots, n$, this sum involves $n!$ terms and is therefore quite daunting for, say, $n \geq 4$. We will examine the small cases $n = 2$ and $n = 3$ and then look for ways of evaluating $\det(A)$ for larger n.

In the case $n = 2$,

$$\mathbf{A} = \begin{pmatrix} a_{11} & a_{12} \\ a_{21} & a_{22} \end{pmatrix}.$$

We have seen that there are two permutations on 1, 2, namely,

$$p : 1, 2,$$

which is an even permutation, and

$$q : 2, 1,$$

which is odd. Then

$$\det(\mathbf{A}) = (-1)^{sgn(p)} a_{1p(1)} a_{2p(2)} + (-1)^{sgn(q)} a_{1q(1)} a_{2q(2)}$$

$$= (-1)^0 a_{11} a_{22} + (-1)^1 a_{12} a_{21}$$

$$= a_{11} a_{22} - a_{12} a_{21}.$$

This rule for evaluating $\det(\mathbf{A})$ holds for any 2×2 matrix.

In the case $n = 3$,

$$\mathbf{A} = \begin{pmatrix} a_{11} & a_{12} & a_{13} \\ a_{21} & a_{22} & a_{23} \\ a_{31} & a_{32} & a_{33} \end{pmatrix}.$$

The permutations of $1, 2, 3$ are

$$p_1 : 1, 2, 3$$
$$p_2 : 1, 3, 2$$
$$p_3 : 2, 1, 3$$
$$p_4 : 2, 3, 1$$
$$p_5 : 3, 1, 2$$
$$p_6 : 3, 2, 1.$$

It is routine to check that p_1, p_5, and p_4 are even and p_2, p_3, and p_6 are odd. Then

$$\det(\mathbf{A}) = (-1)^{sgn(p_1)} a_{1p_1(1)} a_{2p_1(2)} a_{3p_1(3)} + (-1)^{sgn(p_2)} a_{1p_2(1)} a_{2p_2(2)} a_{3p_2(3)}$$

$$+ (-1)^{sgn(p_3)} a_{1p_3(1)} a_{2p_3(2)} a_{3p_3(3)} + (-1)^{sgn(p_4)} a_{1p_4(1)} a_{2p_4(2)} a_{3p_4(3)}$$

$$+ (-1)^{sgn(p_5)} a_{1p_5(1)} a_{2p_5(2)} a_{3p_5(3)} + (-1)^{sgn(p_6)} a_{1p_6(1)} a_{2p_6(2)} a_{3p_6(3)}$$

$$= a_{11} a_{22} a_{33} - a_{11} a_{23} a_{32} - a_{12} a_{21} a_{33} + a_{12} a_{23} a_{31} + a_{13} a_{21} a_{32} - a_{13} a_{22} a_{31}.$$

If $\mathbf{A}$ is 4×4, then evaluation of $\det(\mathbf{A})$ by direct recourse to the definition will involve 24 terms as well as explicitly listing all 24 permutations on $1, 2, 3, 4$. This is not practical. We will therefore develop some properties of determinants that will make their evaluation more efficient.

SECTION 7.2   PROBLEMS

In Problems 1 through 4, use the formula for $\det(\mathbf{A})$ in the 3×3 case to evaluate the determinant of the given matrix.

1. $\mathbf{A} = \begin{pmatrix} 1 & 6 & 0 \\ 1 & 2 & -1 \\ 0 & 1 & 1 \end{pmatrix}$

2. $\mathbf{A} = \begin{pmatrix} -1 & 3 & 1 \\ 2 & 2 & 0 \\ 1 & 1 & 4 \end{pmatrix}$

3. $\mathbf{A} = \begin{pmatrix} 6 & -3 & 5 \\ 2 & 1 & 4 \\ 0 & 1 & -4 \end{pmatrix}$

4. $\mathbf{A} = \begin{pmatrix} -4 & 0 & 1 \\ 0 & 1 & 1 \\ 0 & 0 & 0 \end{pmatrix}$

5. The permutations on $1, 2, 3, 4$ were listed in Section 7.1. Use this list to write a formula for $\det(\mathbf{A})$ when $\mathbf{A}$ is 4×4.

7.3   Properties of Determinants

We will develop some of the properties of determinants that are used in evaluating them and applying them to problems. There are effective computer routines for evaluating quite large determinants, but these are also based on the properties we will display.

First, it is standard to use vertical lines to denote determinants, so we will often write

$$\det(\mathbf{A}) = |\mathbf{A}|.$$

This should not be confused with absolute value. If $\mathbf{A}$ has numerical elements, then $|\mathbf{A}|$ is a number and can be positive, negative, or zero.

Throughout the rest of this chapter let $\mathbf{A}$ and $\mathbf{B}$ be $n \times n$ matrices. Our first result says that a matrix having a zero row has a zero determinant.

THEOREM 7.1

If $\mathbf{A}$ has a zero row, then $|\mathbf{A}| = 0$. ∎

This is easy to see from the defining sum (7.1). Suppose, for some i, each $a_{ij} = 0$. Each term of the sum (7.1) contains a factor $a_{ip_j(i)}$ from row i, hence each term in the sum is zero.

Next, we claim that multiplying a row of a matrix by a scalar α has the effect of multiplying the determinant of the matrix by α.

THEOREM 7.2

Let $\mathbf{B}$ be formed from $\mathbf{A}$ by multiplying row k by a scalar α. Then

$$|\mathbf{B}| = \alpha|\mathbf{A}|. \quad ∎$$

The effect of multiplying row k of $\mathbf{A}$ by α is to replace each a_{kj} by αa_{kj}. Then $b_{ij} = a_{ij}$ for $i \neq k$, and $b_{kj} = \alpha a_{kj}$, so

$$|\mathbf{A}| = \sum_p (-1)^{sgn(p)} a_{1p(1)} a_{2p(2)} \cdots a_{kp(k)} \cdots a_{np(n)}$$

and

$$|\mathbf{B}| = \sum_p (-1)^{sgn(p)} b_{1p(1)} b_{2p(2)} \cdots b_{kp(k)} \cdots b_{np(n)}$$

$$= \sum_p (-1)^{sgn(p)} a_{1p(1)} a_{2p(2)} \cdots (\alpha a_{kp(k)}) \cdots a_{np(n)}$$

$$= \alpha \sum_p (-1)^{sgn(p)} a_{1p(1)} a_{2p(2)} \cdots a_{kp(k)} \cdots a_{np(n)} = \alpha|\mathbf{A}|.$$

The next result states that the interchange of two rows in a matrix causes a sign change in the determinant.

THEOREM 7.3

Let $\mathbf{B}$ be formed from $\mathbf{A}$ by interchanging two rows. Then

$$|\mathbf{A}| = -|\mathbf{B}|. \quad ∎$$

A proof of this involves a close examination of the effect of a row interchange on the terms of the sum (7.1), and we will not go through these details. The result is easy to see in the case of 2×2 determinants. Let

$$\mathbf{A} = \begin{pmatrix} a_{11} & a_{12} \\ a_{21} & a_{22} \end{pmatrix} \quad \text{and} \quad \mathbf{B} = \begin{pmatrix} a_{21} & a_{22} \\ a_{11} & a_{12} \end{pmatrix}.$$

Then

$$|\mathbf{A}| = a_{11}a_{22} - a_{12}a_{21}$$

and

$$|\mathbf{B}| = a_{21}a_{12} - a_{22}a_{11} = -|\mathbf{A}|.$$

This result has two important consequences. The first is that the determinant of a matrix with two identical rows must be zero.

COROLLARY 7.1

If two rows of $\mathbf{A}$ are the same, then $|\mathbf{A}| = 0$. ∎

The reason for this is that if we form $\mathbf{B}$ from $\mathbf{A}$ by interchanging the identical rows, then $\mathbf{B} = \mathbf{A}$, so $|\mathbf{B}| = |\mathbf{A}|$. But by Theorem 7.3, $|\mathbf{B}| = -|\mathbf{A}|$, so $|\mathbf{A}| = 0$.

COROLLARY 7.2

If for some scalar α, row k of $\mathbf{A}$ is α times row i, then $|\mathbf{A}| = 0$. ∎

To see this, consider two cases. First, if $\alpha = 0$, then row k of A is a zero row, so $|\mathbf{A}| = 0$. If $\alpha \neq 0$, then we can multiply row k of $\mathbf{A}$ by $1/\alpha$ to obtain a matrix $\mathbf{B}$ having rows i and k the same. Then $|\mathbf{B}| = 0$. But $|\mathbf{B}| = (1/\alpha)|\mathbf{A}|$ by Theorem 7.2, so $|\mathbf{A}| = 0$.

Next, we claim that the determinant of a product is the product of the determinants.

THEOREM 7.4

Let $\mathbf{A}$ and $\mathbf{B}$ be $n \times n$ matrices. Then $|\mathbf{AB}| = |\mathbf{A}||\mathbf{B}|$. ∎

Obviously this extends to a product involving any finite number of $n \times n$ matrices. The theorem enables us to evaluate the determinant of such a product without carrying out the matrix multiplications of all the factors. We will illustrate the theorem when we have efficient ways of evaluating determinants.

The following theorem gives the determinant of a matrix that is written as a sum of matrices in a special way.

THEOREM 7.5

Suppose each element of row k of $\mathbf{A}$ is written as a sum $\alpha_{kj} + \beta_{kj}$. Form two matrices from $\mathbf{A}$. The first, $\mathbf{A}_1$, is identical to $\mathbf{A}$ except the elements of row k are α_{kj}. The second, $\mathbf{A}_2$, is identical to $\mathbf{A}$ except the elements of row k are β_{kj}. Then

$$|\mathbf{A}| = |\mathbf{A}_1| + |\mathbf{A}_2|. \quad ∎$$

If we display the elements of these matrices, the conclusion states that

$$
\begin{vmatrix}
a_{11} & \cdots & a_{1j} & \cdots & a_{1n} \\
\vdots & \vdots & \vdots & \vdots & \vdots \\
\alpha_{k1} + \beta_{k1} & \cdots & \alpha_{kj} + \beta_{kj} & \cdots & \alpha_{kn} + \beta_{kn} \\
\vdots & \vdots & \vdots & \vdots & \vdots \\
a_{n1} & \cdots & a_{nj} & \cdots & a_{nn}
\end{vmatrix}
$$

$$
=
\begin{vmatrix}
a_{11} & \cdots & a_{1j} & \cdots & a_{1n} \\
\vdots & \vdots & \vdots & \vdots & \vdots \\
\alpha_{k1} & \cdots & \alpha_{kj} & \cdots & \alpha_{kn} \\
\vdots & \vdots & \vdots & \vdots & \vdots \\
a_{n1} & \cdots & a_{nj} & \cdots & a_{nn}
\end{vmatrix}
+
\begin{vmatrix}
a_{11} & \cdots & a_{1j} & \cdots & a_{1n} \\
\vdots & \vdots & \vdots & \vdots & \vdots \\
\beta_{k1} & \cdots & \beta_{kj} & \cdots & \beta_{kn} \\
\vdots & \vdots & \vdots & \vdots & \vdots \\
a_{n1} & \cdots & a_{nj} & \cdots & a_{nn}
\end{vmatrix}.
$$

This result can be seen by examining the terms of (7.1) for each of these determinants:

$$
|\mathbf{A}| = \sum_p (-1)^{sgn(p)} a_{1p(1)} a_{2p(2)} \cdots (\alpha_{kp(k)} + \beta_{kp(k)}) \cdots a_{np(n)}
$$

$$
= \sum_p (-1)^{sgn(p)} a_{1p(1)} a_{2p(2)} \cdots \alpha_{kp(k)} \cdots a_{np(n)}
$$

$$
+ \sum_p (-1)^{sgn(p)} a_{1p(1)} a_{2p(2)} \cdots + \beta_{kp(k)} \cdots a_{np(n)} = |\mathbf{A}_1| + |\mathbf{A}_2|.
$$

As a corollary to this, adding a scalar multiple of one row to another of a matrix does not change the value of the determinant.

COROLLARY 7.3

Let **B** be formed from **A** by adding γ times row i to row k. Then $|\mathbf{B}| = |\mathbf{A}|$. ∎

This result follows immediately from the preceding theorem by noting that row k of **B** is $\gamma a_{ij} + a_{kj}$. Then

$$
|\mathbf{B}| =
\begin{vmatrix}
a_{11} & \cdots & a_{1j} & \cdots & a_{1n} \\
\vdots & \vdots & \vdots & \vdots & \vdots \\
a_{i1} & \cdots & a_{ij} & \cdots & a_{in} \\
\vdots & \vdots & \vdots & \vdots & \vdots \\
\gamma a_{i1} + a_{k1} & \cdots & \gamma a_{ij} + a_{kj} & \cdots & \gamma a_{kn} + a_{kn} \\
\vdots & \vdots & \vdots & \vdots & \vdots \\
a_{n1} & \cdots & a_{nj} & \cdots & a_{nn}
\end{vmatrix}
$$

$$
= \gamma
\begin{vmatrix}
a_{11} & \cdots & a_{1j} & \cdots & a_{1n} \\
\vdots & \vdots & \vdots & \vdots & \vdots \\
a_{i1} & \cdots & a_{ij} & \cdots & a_{in} \\
\vdots & \vdots & \vdots & \vdots & \vdots \\
a_{i1} & \cdots & a_{ij} & \cdots & a_{kn} \\
\vdots & \vdots & \vdots & \vdots & \vdots \\
a_{n1} & \cdots & a_{nj} & \cdots & a_{nn}
\end{vmatrix}
+
\begin{vmatrix}
a_{11} & \cdots & a_{1j} & \cdots & a_{1n} \\
\vdots & \vdots & \vdots & \vdots & \vdots \\
a_{i1} & \cdots & a_{ij} & \cdots & a_{in} \\
\vdots & \vdots & \vdots & \vdots & \vdots \\
a_{k1} & \cdots & a_{kj} & \cdots & a_{kn} \\
\vdots & \vdots & \vdots & \vdots & \vdots \\
a_{n1} & \cdots & a_{nj} & \cdots & a_{nn}
\end{vmatrix}
= |\mathbf{A}|.
$$

In the last line, the first term is γ times a determinant with rows i and k identical, hence is zero. The second term is just $|\mathbf{A}|$.

We now know the effect of elementary row operations on a determinant. In summary:

Type I operation—interchange of two rows. This changes the sign of the determinant.

Type II operation—multiplication of a row by a scalar α. This multiplies the determinant by α.

Type III operation—addition of a scalar multiple of one row to another row. This does not change the determinant.

Recall that the transpose $\mathbf{A}^t$ of a matrix $\mathbf{A}$ is obtained by writing the rows of $\mathbf{A}$ as the columns of $\mathbf{A}^t$. We claim that a matrix and its transpose have the same determinant.

THEOREM 7.6

$|\mathbf{A}| = |\mathbf{A}^t|$. ■

For example, consider the 2×2 case:

$$\mathbf{A} = \begin{pmatrix} a_{11} & a_{12} \\ a_{21} & a_{22} \end{pmatrix}, \quad \mathbf{A}^t = \begin{pmatrix} a_{11} & a_{21} \\ a_{12} & a_{22} \end{pmatrix}.$$

Then

$$|\mathbf{A}| = a_{11}a_{22} - a_{12}a_{21} \quad \text{and} \quad |\mathbf{A}^t| = a_{11}a_{22} - a_{21}a_{12} = |\mathbf{A}|.$$

A proof of this theorem consists of comparing terms of the determinants. If $\mathbf{A} = [a_{ij}]$ then $\mathbf{A}^t = [a_{ji}]$. Now, from the defining sum (7.1),

$$|\mathbf{A}| = \sum_p (-1)^{sgn(p)} a_{1p(1)} a_{2p(2)} \cdots a_{np(n)}$$

and

$$|\mathbf{A}^t| = \sum_p (-1)^{sgn(p)} (\mathbf{A}^t)_{ip(1)} (\mathbf{A}^t)_{2p(2)} \cdots (\mathbf{A}^t)_{np(n)}$$

$$= \sum_q (-1)^{sgn(q)} a_{q(1)1} a_{q(2)2} \cdots a_{q(n)n}.$$

One can show that each term $(-1)^{sgn(p)} a_{1p(1)} a_{2p(2)} \cdots a_{np(n)}$ in the sum for $|\mathbf{A}|$ is equal to a corresponding term $(-1)^{sgn(q)} a_{q(1)1} a_{q(2)2} \cdots a_{q(n)n}$ in the sum for $|\mathbf{A}^t|$. The key is to realize that because q is a permutation of $1, \ldots, n$, we can rearrange the terms in the latter product to write them in increasing order of the first (row) index. This induces a permutation on the second (column) index, and we can match this term up with a corresponding term in the sum for $|\mathbf{A}|$. We will not elaborate on the details of this argument.

One consequence of this result is that we can perform not only elementary row operations on a matrix but also the corresponding elementary column operations, and we know the effect of each operation on the determinant. In particular, from the column perspective:

If two columns of $\mathbf{A}$ are identical, or if one column is a zero column, then $|\mathbf{A}| = 0$.

Interchange of two columns of $\mathbf{A}$ changes the sign of the determinant.

Multiplication of a column by a scalar α multiplies the determinant by α.

And addition of a scalar multiple of one column to another column does not change the determinant.

These operations on rows and columns of a matrix, and their effect on the determinant of the newly formed matrix, form the basis for strategies to evaluate determinants.

1. Let $A = [a_{ij}]$ be an $n \times n$ matrix and let α be any scalar. Let $B = [\alpha a_{ij}]$. Thus B is formed by multiplying each element of A by α. Prove that $|B| = \alpha^n |A|$.

2. Let $A = [a_{ij}]$ be an $n \times n$ matrix. Let α be a nonzero number. Form a new matrix $B = [\alpha^{i-j} a_{ij}]$. How are $|A|$ and $|B|$ related? *Hint:* It is useful to examine the

2×2 and 3×3 cases to get some idea of what B looks like.

3. An $n \times n$ matrix is skew symmetric if $A = -A^t$. Prove that the determinant of a skew-symmetric matrix of odd order is zero.

7.4 Evaluation of Determinants by Elementary Row and Column Operations

The use of elementary row and column operations to evaluate a determinant is predicated on the following observation. If a row or column of an $n \times n$ matrix A has all zero elements except possibly for a_{ij} in row i and column j, then the determinant of A is $(-1)^{i+j} a_{ij}$ times the determinant of the $(n-1) \times (n-1)$ matrix obtained by deleting row i and column j from A. This reduces the problem of evaluating an $n \times n$ determinant to one of evaluating a smaller determinant, having one less row and one less column.

Here is a statement of this result, with (1) the row version and (2) the column version.

THEOREM 7.7

1. *Row Version*

$$
\begin{vmatrix}
a_{11} & \cdots & a_{1,j-1} & a_{1j} & a_{1,j+1} & \cdots & a_{1n} \\
\vdots & \vdots & \vdots & \vdots & \vdots & \vdots & \vdots \\
a_{i-1,1} & \cdots & a_{i-1,j-1} & a_{i-1,j} & a_{i-1,j+1} & \cdots & a_{i-1,n} \\
0 & \cdots & 0 & a_{ij} & 0 & \cdots & 0 \\
a_{i+1,1} & \cdots & a_{i+1,j-1} & a_{i+1,j} & a_{i+1,j+1} & \cdots & a_{i+1,n} \\
\vdots & \vdots & \vdots & \vdots & \vdots & \vdots & \vdots \\
a_{n1} & \cdots & a_{n,j-1} & a_{nj} & a_{n,j+1} & \cdots & a_{nn}
\end{vmatrix}
$$

$$
= (-1)^{i+j} a_{ij}
\begin{vmatrix}
a_{11} & \cdots & a_{1,j-1} & a_{1,j+1} & \cdots & a_{1n} \\
\vdots & \vdots & \vdots & \vdots & \vdots & \vdots \\
a_{i-1,1} & \cdots & a_{i-1,j-1} & a_{i-1,j+1} & \cdots & a_{i-1,n} \\
a_{i+1,1} & \cdots & a_{i+1,j-1} & a_{i+1,j+1} & \cdots & a_{i+1,n} \\
\vdots & \vdots & \vdots & \vdots & \vdots & \vdots \\
a_{n1} & \cdots & a_{n,j-1} & a_{n,j+1} & \cdots & a_{nn}
\end{vmatrix}.
$$

2. *Column Version*

$$
\begin{vmatrix}
a_{11} & \cdots & a_{1,j-1} & 0 & a_{1,j+1} & \cdots & a_{1n} \\
\vdots & \vdots & \vdots & \vdots & \vdots & \vdots & \vdots \\
a_{i-1,1} & \cdots & a_{i-1,j-1} & 0 & a_{i-1,j+1} & \cdots & a_{i-1,n} \\
a_{i1} & \cdots & a_{i,j-1} & a_{ij} & a_{i,j+1} & \cdots & a_{i,n} \\
a_{i+1,1} & \cdots & a_{i+1,j-1} & 0 & a_{i+1,j+1} & \cdots & a_{i+1,n} \\
\vdots & \vdots & \vdots & \vdots & \vdots & \vdots & \vdots \\
a_{n1} & \cdots & a_{n,j-1} & 0 & a_{n,j+1} & \cdots & a_{nn}
\end{vmatrix}
$$

$$
= (-1)^{i+j} a_{ij}
\begin{vmatrix}
a_{11} & \cdots & a_{1,j-1} & a_{1,j+1} & \cdots & a_{1n} \\
\vdots & \vdots & \vdots & \vdots & \vdots & \vdots \\
a_{i-1,1} & \cdots & a_{i-1,j-1} & a_{i-1,j+1} & \cdots & a_{i-1,n} \\
a_{i+1,1} & \cdots & a_{i+1,j-1} & a_{i+1,j+1} & \cdots & a_{i+1,n} \\
\vdots & \vdots & \vdots & \vdots & \vdots & \vdots \\
a_{n1} & \cdots & a_{n,j-1} & a_{n,j+1} & \cdots & a_{nn}
\end{vmatrix}.
$$

This result suggests one strategy for evaluating a determinant. Given an $n \times n$ matrix $\mathbf{A}$, use the row and/or column operations to obtain a new matrix $\mathbf{B}$ having at most one nonzero element in some row or column. Then $|\mathbf{A}|$ is a scalar multiple of $|\mathbf{B}|$, and $|\mathbf{B}|$ is a scalar multiple of the $(n-1) \times (n-1)$ determinant formed by deleting from $\mathbf{B}$ the row and column containing this nonzero element. We can then repeat this strategy on this $(n-1) \times (n-1)$ matrix, eventually reducing the problem to one of evaluating a "small" determinant.

Here are two illustrations of this process.

EXAMPLE 7.1

Let

$$
\mathbf{A} =
\begin{pmatrix}
-6 & 0 & 1 & 3 & 2 \\
-1 & 5 & 0 & 1 & 7 \\
8 & 3 & 2 & 1 & 7 \\
0 & 1 & 5 & -3 & 2 \\
1 & 15 & -3 & 9 & 4
\end{pmatrix}.
$$

We want to evaluate $|\mathbf{A}|$. There are many ways to proceed with the strategy we are illustrating. To begin, we can exploit the fact that $a_{13} = 1$ and use elementary row operations to get zeros in the rest of column 3. Of course, $a_{23} = 0$ to begin with, so we need only worry about column 3 entries in rows 3, 4, 5. Add $(-2)(\text{row } 1)$ to row 3, $-5(\text{row } 1)$ to row 4 and $3(\text{row } 1)$ to row 5 to get

$$
\mathbf{B} =
\begin{pmatrix}
-6 & 0 & 1 & 3 & 2 \\
-1 & 5 & 0 & 1 & 7 \\
20 & 3 & 0 & -5 & 3 \\
30 & 1 & 0 & -18 & -8 \\
-17 & 15 & 0 & 18 & 10
\end{pmatrix}.
$$

Because we have used Type III row operations,

$$
|\mathbf{A}| = |\mathbf{B}|.
$$

Further, by Theorem 7.7,

$$|\mathbf{B}| = (-1)^{1+3}b_{13}|\mathbf{C}| = (1)|\mathbf{C}| = |\mathbf{C}|,$$

where **C** is the 4×4 matrix obtained by deleting row 1 and column 3 of **B**:

$$\mathbf{C} = \begin{pmatrix} -1 & 5 & 1 & 7 \\ 20 & 3 & -5 & 3 \\ 30 & 1 & -18 & -8 \\ -17 & 15 & 18 & 10 \end{pmatrix}.$$

This is a 4×4 matrix, "smaller" than **A**. We will now apply the strategy to $|\mathbf{C}|$. We can, for example, exploit the -1 entry in the 1, 1 position of **C**, this time using column operations to get zeros in row 1, columns 2, 3, 4 of the new matrix. Specifically, add 5(column 1) to column 2, add column 1 to column 3, and add 7(column 1) to column 4 of **C** to get

$$\mathbf{D} = \begin{pmatrix} -1 & 0 & 0 & 0 \\ 20 & 103 & 15 & 143 \\ 30 & 151 & 12 & 202 \\ -17 & -70 & 1 & -109 \end{pmatrix}.$$

Again, because we used Type III operations (this time on columns) of **C**, then

$$|\mathbf{C}| = |\mathbf{D}|.$$

But by the theorem, because we are using the element $d_{11} = -1$ as the single nonzero element of row 1, we have

$$|\mathbf{D}| = (-1)^{1+1}d_{11}|\mathbf{E}| = -|\mathbf{E}|,$$

in which **E** is the 3×3 matrix obtained by deleting row 1 and column 1 from **D**:

$$\mathbf{E} = \begin{pmatrix} 103 & 15 & 143 \\ 151 & 12 & 202 \\ -70 & 1 & -109 \end{pmatrix}.$$

To evaluate $|\mathbf{E}|$, we can exploit the 3, 2 entry $e_{32} = 1$. Add -15(row 3 to row 1 and -12(row 3) to row 2 to get

$$\mathbf{F} = \begin{pmatrix} 1153 & 0 & 1778 \\ 991 & 0 & 1510 \\ -70 & 1 & -109 \end{pmatrix}.$$

Then

$$|\mathbf{E}| = |\mathbf{F}|.$$

By the theorem, using the only nonzero element $f_{32} = 1$ of column 2 of **F**, we have

$$|\mathbf{F}| = (-1)^{3+2}(1)|\mathbf{G}| = -|\mathbf{G}|$$

in which **G** is the 2×2 matrix obtained by deleting row 3 and column 2 of **F**:

$$\mathbf{G} = \begin{pmatrix} 1153 & 1778 \\ 991 & 1510 \end{pmatrix}.$$

At the 2×2 state, we evaluate the determinant directly:

$$|\mathbf{G}| = (1153)(1510) - (1778)(991) = -20{,}968.$$

Working back, we now have

$$|\mathbf{A}| = |\mathbf{B}| = |\mathbf{C}| = |\mathbf{D}| = -|\mathbf{E}| = -|\mathbf{F}| = |\mathbf{G}| = -20{,}968. \quad \blacksquare$$

The method is actually quicker to apply than might appear from this example, because we included comments as we proceeded with the calculations. Here is a second example of this method.

EXAMPLE 7.2

Evaluate $|\mathbf{A}|$, where

$$\mathbf{A} = \begin{pmatrix} -2 & 6 & 5 & -3 \\ 4 & 4 & -8 & 2 \\ 6 & 3 & 3 & -6 \\ 8 & 9 & -11 & 4 \end{pmatrix}.$$

Again, one can choose different rows and columns with which to work, but here is one way to proceed. First multiply row 1 by $-\frac{1}{2}$ to form $\mathbf{B}$:

$$\mathbf{B} = \begin{pmatrix} 1 & -3 & -\frac{5}{2} & \frac{3}{2} \\ 4 & 4 & -8 & 2 \\ 6 & 3 & 3 & -6 \\ 8 & 9 & -11 & 4 \end{pmatrix}.$$

This is a Type I operation, and

$$|\mathbf{B}| = -\tfrac{1}{2}|\mathbf{A}|.$$

Form $\mathbf{C}$ from $\mathbf{B}$ by adding -4(row 1) to row 2, -6(row 1) to row 3, and -8(row 1) to row 4:

$$\mathbf{C} = \begin{pmatrix} 1 & -3 & -\frac{5}{2} & \frac{3}{2} \\ 0 & 16 & 2 & -4 \\ 0 & 21 & 18 & -15 \\ 0 & 33 & 9 & -8 \end{pmatrix}.$$

Then

$$|\mathbf{B}| = |\mathbf{C}|$$

because a Type III operation does not change the determinant. Further,

$$|\mathbf{C}| = |\mathbf{D}|,$$

where $\mathbf{D}$ is the 3×3 matrix obtained by deleting row 1 and column 1 of $\mathbf{C}$:

$$\mathbf{D} = \begin{pmatrix} 16 & 2 & -4 \\ 21 & 18 & -15 \\ 33 & 9 & -8 \end{pmatrix}.$$

Multiply row 1 of $\mathbf{D}$ by $\frac{1}{2}$ to get

$$\mathbf{E} = \begin{pmatrix} 8 & 1 & -2 \\ 21 & 18 & -15 \\ 33 & 9 & -8 \end{pmatrix}.$$

Then

$$|\mathbf{E}| = \tfrac{1}{2}|\mathbf{D}|.$$

Next add -8(column 2) to column 1 of $\mathbf{E}$, and 2(column 2) to column 3, to get

$$\mathbf{F} = \begin{pmatrix} 0 & 1 & 0 \\ -123 & 18 & 21 \\ -39 & 9 & 10 \end{pmatrix}.$$

Then

$$|\mathbf{E}| = |\mathbf{F}| = (-1)^{1+2}(1)\begin{vmatrix} -123 & 21 \\ -39 & 10 \end{vmatrix}$$

$$= (-1)(-411) = 411.$$

Working back through the intermediary steps, we finally have

$$|\mathbf{A}| = -2|\mathbf{B}| = -2|\mathbf{C}| = -2|\mathbf{D}|$$

$$= -2[2|\mathbf{E}|] = -4|\mathbf{F}| = -4(411) = -1644. \quad \blacksquare$$

SECTION 7.4   PROBLEMS

In each of Problems 1 through 10, use the strategy of this section to evaluate the determinant of the matrix.

1. $\begin{pmatrix} -2 & 4 & 1 \\ 1 & 6 & 3 \\ 7 & 0 & 4 \end{pmatrix}$

2. $\begin{pmatrix} 2 & -3 & 7 \\ 14 & 1 & 1 \\ -13 & -1 & 5 \end{pmatrix}$

3. $\begin{pmatrix} -4 & 5 & 6 \\ -2 & 3 & 5 \\ 2 & -2 & 6 \end{pmatrix}$

4. $\begin{pmatrix} 2 & -5 & 8 \\ 4 & 3 & 8 \\ 13 & 0 & -4 \end{pmatrix}$

5. $\begin{pmatrix} 17 & -2 & 5 \\ 1 & 12 & 0 \\ 14 & 7 & -7 \end{pmatrix}$

6. $\begin{pmatrix} -3 & 3 & 9 & 6 \\ 1 & -2 & 15 & 6 \\ 7 & 1 & 1 & 5 \\ 2 & 1 & -1 & 3 \end{pmatrix}$

7. $\begin{pmatrix} 0 & 1 & 1 & -4 \\ 6 & -3 & 2 & 2 \\ 1 & -5 & 1 & -2 \\ 4 & 8 & 2 & 2 \end{pmatrix}$

8. $\begin{pmatrix} 2 & 7 & -1 & 0 \\ 3 & 1 & 1 & 8 \\ -2 & 0 & 3 & 1 \\ 4 & 8 & -1 & 0 \end{pmatrix}$

9. $\begin{pmatrix} 10 & 1 & -6 & 2 \\ 0 & 3 & 3 & 9 \\ 0 & 1 & 1 & 7 \\ -2 & 6 & 8 & 8 \end{pmatrix}$

10. $\begin{pmatrix} -7 & 16 & 2 & 4 \\ 1 & 0 & 0 & 5 \\ 0 & 3 & -4 & 4 \\ 6 & 1 & 1 & -5 \end{pmatrix}$

7.5   Cofactor Expansions

Theorem 7.5 suggests the following. If we select any row i of a square matrix $\mathbf{A}$, we can write

$$
\begin{vmatrix}
a_{11} & a_{12} & \cdots & \cdots & a_{1n} \\
\vdots & \vdots & \vdots & \vdots & \vdots \\
a_{i1} & a_{i2} & \cdots & \cdots & a_{in} \\
\vdots & \vdots & \vdots & \vdots & \vdots \\
a_{n1} & a_{n2} & \cdots & \cdots & a_{nn}
\end{vmatrix}
$$

$$
= \begin{vmatrix}
a_{11} & a_{12} & \cdots & \cdots & a_{1n} \\
\vdots & \vdots & \vdots & \vdots & \vdots \\
a_{i1} & 0 & \cdots & \cdots & 0 \\
\vdots & \vdots & \vdots & \vdots & \vdots \\
a_{n1} & a_{n2} & \cdots & \cdots & a_{nn}
\end{vmatrix}
+ \begin{vmatrix}
a_{11} & a_{12} & \cdots & \cdots & a_{1n} \\
\vdots & \vdots & \vdots & \vdots & \vdots \\
0 & a_{i2} & \cdots & \cdots & 0 \\
\vdots & \vdots & \vdots & \vdots & \vdots \\
a_{n1} & a_{n2} & \cdots & \cdots & a_{nn}
\end{vmatrix}
$$

$$
+ \cdots + \begin{vmatrix}
a_{11} & a_{12} & \cdots & \cdots & a_{1n} \\
\vdots & \vdots & \vdots & \vdots & \vdots \\
0 & 0 & \cdots & \cdots & a_{in} \\
\vdots & \vdots & \vdots & \vdots & \vdots \\
a_{n1} & a_{n2} & \cdots & \cdots & a_{nn}
\end{vmatrix}. \tag{7.2}
$$

Each of the determinants on the right of equation (7.2) has a row in which every element but possibly one is zero, so Theorem 7.7 applies to each of these determinants. The first determinant on the right is $(-1)^{i+1}a_{i1}$ times the determinant of the matrix obtained by deleting row i and column 1 from $\mathbf{A}$. The second determinant on the right is $(-1)^{i+2}a_{i2}$ times the determinant of the matrix obtained by deleting row i and column 2 from $\mathbf{A}$. And so on, until the last matrix on the right is $(-1)^{i+n}a_{in}$ times the determinant of the matrix obtained by deleting row i and column n from $\mathbf{A}$. We can put all of this more succinctly by introducing the following standard terminology.

DEFINITION 7.2

Minor

If $\mathbf{A}$ is an $n \times n$ matrix, the minor of a_{ij} is denoted M_{ij} and is the determinant of the $(n-1) \times (n-1)$ matrix obtained by deleting row i and column j of $\mathbf{A}$.

Cofactor
The number $(-1)^{i+j}M_{ij}$ is called the cofactor of a_{ij}.

We can now state the following formula for a determinant.

THEOREM 7.8   *Cofactor Expansion by a Row*

If $\mathbf{A}$ is $n \times n$, then for any integer i with $1 \le i \le n$,

$$
|\mathbf{A}| = \sum_{j=1}^{n}(-1)^{i+j}a_{ij}M_{ij}. \quad \blacksquare \tag{7.3}
$$

This is just equation (7.2) in the notation of cofactors. The sum (7.3) is called the *cofactor expansion* of |**A**| by row *i* because it is the sum, across this row, of each matrix element times its cofactor. This yields |**A**| no matter which row is used. Of course, if some $a_{ik} = 0$, then we need not calculate that term in equation (7.3), so it is to our advantage to expand by a row having as many zero elements as possible. The strategy of the preceding subsection was to create such a row using row and column operations, resulting in what was a cofactor expansion, by a row having only one (possibly) nonzero element.

EXAMPLE 7.3

Let

$$\mathbf{A} = \begin{pmatrix} -6 & 3 & 7 \\ 12 & -5 & -9 \\ 2 & 4 & -6 \end{pmatrix}.$$

If we expand by row 1, we get

$$|\mathbf{A}| = \sum_{j=1}^{3}(-1)^{1+j}M_{1j}$$

$$= (-1)^{1+1}(-6)\begin{vmatrix} -5 & -9 \\ 4 & -6 \end{vmatrix} + (-1)^{1+2}(3)\begin{vmatrix} 12 & -9 \\ 2 & -6 \end{vmatrix} + (-1)^{1+3}(7)\begin{vmatrix} 12 & -5 \\ 2 & 4 \end{vmatrix}$$

$$= (-6)(30 + 36) - 3(-72 + 18) + 7(48 + 10) = 172.$$

Just for illustration, expand by row 3:

$$|\mathbf{A}| = \sum_{j=1}^{3}(-1)^{3+j}a_{3j}M_{3j}$$

$$= (-1)^{3+1}2\begin{vmatrix} 3 & 7 \\ -5 & -9 \end{vmatrix} + (-1)^{3+2}(4)\begin{vmatrix} -6 & 7 \\ 12 & -9 \end{vmatrix} + (-1)^{3+3}(-6)\begin{vmatrix} -6 & 3 \\ 12 & -5 \end{vmatrix}$$

$$= 2(-27 + 35) - 4(54 - 84) - 6(30 - 36) = 172. \blacksquare$$

Because, for purposes of evaluating determinants, row and column operations can both be used, we can also develop a cofactor expansion of |**A**| by column *j*. In this expansion, we move down a column of a matrix and sum each term of the column times its cofactor.

THEOREM 7.9 *Cofactor Expansion by a Column*

Let **A** be an $n \times n$ matrix. Then for any *j* with $1 \le j \le n$,

$$|\mathbf{A}| = \sum_{i=1}^{n}(-1)^{i+j}a_{ij}M_{ij}. \blacksquare \tag{7.4}$$

This differs from the expansion (7.3) in that the latter expands across a row, while the sum (7.4) expands down a column. All of these expansions, by any row or column of **A**, yield |**A**|.

EXAMPLE 7.4

Consider again

$$\mathbf{A} = \begin{pmatrix} -6 & 3 & 7 \\ 12 & -5 & -9 \\ 2 & 4 & -6 \end{pmatrix}.$$

Expanding by column 1 gives us

$$|\mathbf{A}| = \sum_{i=1}^{3} (-1)^{i+1} a_{i1} M_{i1}$$

$$= (-1)^{1+1}(-6) \begin{vmatrix} -5 & -9 \\ 4 & -6 \end{vmatrix} + (-1)^{2+1}(12) \begin{vmatrix} 3 & 7 \\ 4 & -6 \end{vmatrix} + (-1)^{3+1}(2) \begin{vmatrix} 3 & 7 \\ -5 & -9 \end{vmatrix}$$

$$= (-6)(30 + 36) - 12(-18 - 28) + 2(-27 + 35) = 172.$$

If we expand by column 2 we get

$$|\mathbf{A}| = \sum_{i=1}^{3} (-1)^{i+2} a_{i2} M_{i2}$$

$$= (-1)^{1+2}(3) \begin{vmatrix} 12 & -9 \\ 2 & -6 \end{vmatrix} + (-1)^{2+2}(-5) \begin{vmatrix} -6 & 7 \\ 2 & -6 \end{vmatrix} + (-1)^{3+2}(4) \begin{vmatrix} -6 & 7 \\ 12 & -9 \end{vmatrix}$$

$$= (-3)(-72 + 18) - 5(36 - 14) - 4(54 - 84) = 172. \quad \blacksquare$$

SECTION 7.5 PROBLEMS

In Problems 1 through 14, use cofactor expansions, combined with elementary row and column operations when this is useful, to evaluate the determinant of the matrix.

1. $\begin{pmatrix} -4 & 2 & -8 \\ 1 & 1 & 0 \\ 1 & -3 & 0 \end{pmatrix}$

2. $\begin{pmatrix} 1 & 1 & 6 \\ 2 & -2 & 1 \\ 3 & -1 & 4 \end{pmatrix}$

3. $\begin{pmatrix} 7 & -3 & 1 \\ 1 & -2 & 4 \\ -3 & 1 & 0 \end{pmatrix}$

4. $\begin{pmatrix} 5 & -4 & 3 \\ -1 & 1 & 6 \\ -2 & -2 & 4 \end{pmatrix}$

5. $\begin{pmatrix} -5 & 0 & 1 & 6 \\ 2 & -1 & 3 & 7 \\ 4 & 4 & -5 & -8 \\ 1 & -1 & 6 & 2 \end{pmatrix}$

6. $\begin{pmatrix} 4 & 3 & -5 & 6 \\ 1 & -5 & 15 & 2 \\ 0 & -5 & 1 & 7 \\ 8 & 9 & 0 & 15 \end{pmatrix}$

7. $\begin{pmatrix} -3 & 1 & 14 \\ 0 & 1 & 16 \\ 2 & -3 & 4 \end{pmatrix}$

8. $\begin{pmatrix} 14 & 13 & -2 & 5 \\ 7 & 1 & 1 & 7 \\ 0 & 2 & 12 & 3 \\ 1 & -6 & 5 & 2 \end{pmatrix}$

9. $\begin{pmatrix} -5 & 4 & 1 & 7 \\ -9 & 3 & 2 & -5 \\ -2 & 0 & -1 & 1 \\ 1 & 14 & 0 & 3 \end{pmatrix}$

10. $\begin{pmatrix} -8 & 5 & 1 & 7 & 2 \\ 0 & 1 & 3 & 5 & -6 \\ 2 & 2 & 1 & 5 & 3 \\ 0 & 4 & 3 & 7 & 2 \\ 1 & 1 & -7 & -6 & 5 \end{pmatrix}$

11. $\begin{pmatrix} 2 & 4 & -6 & 3 & 6 \\ -6 & 16 & 15 & 4 & -6 \\ 0 & 14 & 12 & 9 & 5 \\ -4 & 0 & 22 & 6 & -8 \\ 4 & -4 & 15 & 8 & 10 \end{pmatrix}$

12. $\begin{pmatrix} 5 & 15 & 3 & 1 & 7 & 2 \\ 0 & 0 & 1 & 4 & -5 & 2 \\ 1 & 7 & -1 & 3 & 1 & 9 \\ 0 & 0 & 1 & -3 & -1 & 4 \\ 1 & 1 & 7 & -4 & 1 & 6 \\ 1 & 0 & 0 & 3 & -9 & -4 \end{pmatrix}$

13. $\begin{pmatrix} 22 & -1 & 3 & 0 & 0 \\ 1 & 4 & -5 & 0 & 2 \\ -1 & 1 & 6 & 0 & -5 \\ 4 & 7 & 9 & 1 & -7 \\ 6 & 6 & -3 & 4 & 1 \end{pmatrix}$

14. $\begin{pmatrix} -5 & 6 & 1 & -1 & 0 & 2 \\ 1 & 3 & -1 & -1 & 3 & 1 \\ 4 & 2 & -2 & 1 & 1 & 0 \\ 0 & 0 & 3 & 1 & -2 & 4 \\ 1 & 0 & 0 & -2 & 1 & 7 \\ 0 & 0 & 1 & 1 & -1 & 7 \end{pmatrix}$

15. Show that

$$\begin{vmatrix} 1 & \alpha & \alpha^2 \\ 1 & \beta & \beta^2 \\ 1 & \gamma & \gamma^2 \end{vmatrix} = (\alpha - \beta)(\gamma - \alpha)(\beta - \gamma).$$

This is *Vandermonde's determinant*. This and the next two problems are best done with a little thought in using facts about determinants, rather than a brute-force approach.

16. Show that

$$\begin{vmatrix} \alpha & \beta & \gamma & \delta \\ \beta & \gamma & \delta & \alpha \\ \gamma & \delta & \alpha & \beta \\ \delta & \alpha & \beta & \gamma \end{vmatrix}$$

$$= (\alpha + \beta + \gamma + \delta)(\beta - \alpha + \delta - \gamma) \begin{vmatrix} 0 & 1 & -1 & 1 \\ 1 & \gamma & \delta & \alpha \\ 1 & \delta & \alpha & \beta \\ 1 & \alpha & \beta & \gamma \end{vmatrix}.$$

17. Show that

$$\begin{vmatrix} 1 & 1 & 1 & 1 \\ \alpha & \beta & \gamma & \delta \\ \alpha^2 & \beta^2 & \gamma^2 & \delta^2 \\ \alpha^3 & \beta^3 & \gamma^3 & \delta^3 \end{vmatrix}$$

$$= (\beta - \alpha)(\beta - \gamma)(\beta - \delta)(\gamma - \alpha)(\delta - \alpha)(\delta - \gamma).$$

18. An $n \times n$ matrix, $\mathbf{A}$ is called *block diagonal* if it has the appearance

$$\begin{pmatrix} \mathbf{D}_1 & \mathbf{O} & \cdots & \mathbf{O} \\ \mathbf{O} & \mathbf{D}_2 & \cdots & \mathbf{O} \\ \vdots & \vdots & \vdots & \vdots \\ \mathbf{O} & \mathbf{O} & \cdots & \mathbf{D}_r \end{pmatrix},$$

in which each $\mathbf{D}_j$ is a $k_j \times k_j$ matrix and $k_1 + k_2 + \cdots + k_r = n$. Each $\mathbf{O}$ is a zero matrix of appropriate size. Thus all elements of $\mathbf{A}$ not in some $\mathbf{D}_j$ are zero. Prove that

$$|\mathbf{A}| = |\mathbf{D}_1||\mathbf{D}_2| \cdots |\mathbf{D}_r|.$$

19. In each of the following, use the result of Problem 18 to evaluate the determinant.

(a) $\begin{vmatrix} 22 & -4 & 3 & 0 & 0 \\ 1 & 0 & 6 & 0 & 0 \\ 2 & 2 & 0 & 0 & 0 \\ 0 & 0 & 0 & 1 & -1 \\ 0 & 0 & 0 & 0 & 4 \end{vmatrix}$

(b) $\begin{vmatrix} 8 & 4 & 0 & 0 & 0 & 0 \\ -6 & 2 & 0 & 0 & 0 & 0 \\ 0 & 0 & 3 & 1 & 0 & 0 \\ 0 & 0 & 1 & 2 & 0 & 0 \\ 0 & 0 & 0 & 0 & -4 & 8 \\ 0 & 0 & 0 & 0 & 12 & 14 \end{vmatrix}$

(c) $\begin{vmatrix} 16 & 14 & 0 & 2 & 0 & 0 \\ 1 & 1 & 3 & 2 & 0 & 0 \\ 0 & 0 & 1 & 4 & 0 & 0 \\ 1 & 3 & 6 & 2 & 0 & 0 \\ 0 & 0 & 0 & 0 & -5 & 3 \\ 0 & 0 & 0 & 0 & 4 & -9 \end{vmatrix}$

$$
\text{(d)} \quad \begin{vmatrix}
5 & 3 & 0 & 0 & 0 & 0 & 0 \\
2 & -1 & 0 & 0 & 0 & 0 & 0 \\
0 & 0 & 3 & 9 & 1 & 0 & 0 \\
0 & 0 & 2 & 2 & -4 & 0 & 0 \\
0 & 0 & 1 & -8 & 6 & 0 & 0 \\
0 & 0 & 0 & 0 & 0 & -2 & 3 \\
0 & 0 & 0 & 0 & 0 & 0 & 4
\end{vmatrix}
$$

20. We will prove shortly that a square matrix is nonsingular exactly when its determinant is nonzero. Assuming this for the moment, prove that if **A** is nonsingular, then $|\mathbf{A}| = 1/|\mathbf{A}^{-1}|$.

21. Let **A** be a square matrix such that $\mathbf{A}^{-1} = \mathbf{A}^t$. Prove that $|\mathbf{A}| = \pm 1$.

22. Prove that three points (x_1, y_1), (x_2, y_2), and (x_3, y_3) are collinear (on the same straight line) if and only if

$$
\begin{vmatrix}
1 & x_1 & y_1 \\
1 & x_2 & y_2 \\
1 & x_3 & y_3
\end{vmatrix} = 0.
$$

Hint: This determinant is zero exactly when one row or column is a linear combination of the other two.

7.6   Determinants of Triangular Matrices

The main diagonal of a square matrix **A** consists of the elements $a_{11}, a_{22}, \ldots, a_{nn}$. We call **A** *upper triangular* if all the elements below the main diagonal are zero. That is, $a_{ij} = 0$ if $i > j$. Such a matrix has the appearance

$$
\mathbf{A} = \begin{pmatrix}
a_{11} & a_{12} & a_{13} & \cdots & a_{1,n-1} & a_{1n} \\
0 & a_{22} & a_{23} & \cdots & a_{2,n-1} & a_{2n} \\
0 & 0 & a_{33} & \cdots & a_{3,n-1} & a_{3n} \\
\vdots & \vdots & \vdots & \vdots & \vdots & \vdots \\
0 & 0 & 0 & \cdots & a_{n-1,n-1} & a_{n-1,n} \\
0 & 0 & 0 & 0 & 0 & a_{nn}
\end{pmatrix}.
$$

If we expand $|\mathbf{A}|$ by cofactors down the first column, we have

$$
|\mathbf{A}| = a_{11} \begin{vmatrix}
a_{22} & a_{23} & \cdots & a_{2,n-1} & a_{2n} \\
0 & a_{33} & \cdots & a_{3,n-1} & a_{3n} \\
\vdots & \vdots & \vdots & \vdots & \vdots \\
0 & 0 & \cdots & a_{n-1,n-1} & a_{n-1,n} \\
0 & 0 & \cdots & 0 & a_{nn}
\end{vmatrix},
$$

and the determinant on the right is again upper triangular, so expand by its first column to get

$$
|\mathbf{A}| = a_{11}a_{22} \begin{vmatrix}
a_{33} & a_{34} & \cdots & a_{3n} \\
0 & a_{44} & \cdots & a_{4n} \\
\vdots & \vdots & \cdots & \vdots \\
0 & 0 & \cdots & a_{nn}
\end{vmatrix}
$$

with another upper triangular determinant on the right. Continuing in this way, we obtain

$$
|\mathbf{A}| = a_{11}a_{22} \cdots a_{nn}.
$$

The determinant of an upper triangular matrix is the product of its main diagonal elements.

The same conclusion holds for lower triangular matrices (all elements above the main diagonal are zero). Now we can expand the determinant along the top row, each time obtaining just one minor that is, again, lower triangular.

EXAMPLE 7.5

$$
\begin{vmatrix}
15 & -7 & 4 & 7 & 3 \\
0 & 12 & -6 & 3 & 9 \\
0 & 0 & \sqrt{2} & 15 & -4 \\
0 & 0 & 0 & \pi & 22 \\
0 & 0 & 0 & 0 & e
\end{vmatrix} = (15)(12)\sqrt{2}\pi e = 180\sqrt{2}\pi e. \quad \blacksquare
$$

SECTION 7.6   PROBLEMS

Evaluate the following determinants.

1.
$$
\begin{vmatrix}
-4 & 0 & 0 & 0 & 0 & 0 \\
12 & 7 & 0 & 0 & 0 & 0 \\
3 & -4 & 2 & 0 & 0 & 0 \\
0 & 1 & 1 & -2 & 0 & 0 \\
1 & -4 & 16 & 1 & 5 & 0 \\
10 & -4 & 16 & 1 & 17 & 4
\end{vmatrix}
$$

2.
$$
\begin{vmatrix}
6 & 1 & -1 & 2 & 2 & 1 \\
0 & -4 & 2 & 2 & -3 & 1 \\
0 & 0 & -5 & 10 & 1 & -7 \\
0 & 0 & 0 & 14 & 0 & 0 \\
0 & 0 & 0 & 0 & 13 & -4 \\
0 & 0 & 0 & 0 & 0 & 3
\end{vmatrix}
$$

3.
$$
\begin{vmatrix}
3 & 0 & 0 & 0 & 0 \\
2 & -6 & 0 & 0 & 0 \\
17 & 14 & 2 & 0 & 0 \\
22 & -2 & 15 & 8 & 0 \\
43 & 12 & 1 & -1 & 5
\end{vmatrix}
$$

7.7   A Determinant Formula for a Matrix Inverse

Determinants can be used to tell whether a matrix is singular or nonsingular. In the latter case, there is a way of writing the inverse of a matrix by using determinants.

First, here is a simple test for nonsingularity. We will use the fact that we reduce a matrix by using elementary row operations, whose effects on determinants are known (Type I operations change the sign, Type II operations multiply the determinant by a nonzero constant, and Type III operations do not change the determinant at all). This means that, for any square matrix $\mathbf{A}$, $|\mathbf{A}| = \alpha |\mathbf{A}_R|$ for some nonzero constant α.

THEOREM 7.10

Let $\mathbf{A}$ be an $n \times n$ matrix. Then $\mathbf{A}$ is nonsingular if and only if $|\mathbf{A}| \neq 0$.

Proof Suppose first that $|\mathbf{A}| \neq 0$. Since $|\mathbf{A}| = \alpha |\mathbf{A}_R|$ for some nonzero constant α, $\mathbf{A}_R$ can have no zero row, so $\mathbf{A}_R = \mathbf{I}_n$. Then $rank(\mathbf{A}) = n$, so $\mathbf{A}$ is nonsingular by Theorems 6.26 and 6.15.

Conversely, suppose $\mathbf{A}$ is nonsingular. Then $\mathbf{A}_R = \mathbf{I}_n$. Then $|\mathbf{A}| = \alpha |\mathbf{A}_R| \neq 0$. $\quad \blacksquare$

Using this result, we can give a short proof of Theorem 6.25(5). Suppose $\mathbf{A}$ and $\mathbf{B}$ are $n \times n$ matrices and $\mathbf{AB}$ is singular. Then

$$|\mathbf{AB}| = |\mathbf{A}||\mathbf{B}| = 0,$$

so $|\mathbf{A}| = 0$ or $|\mathbf{B}| = 0$, hence either $\mathbf{A}$ or $\mathbf{B}$ (or possibly both) must be singular.

We will now write a formula for the inverse of a square matrix, in terms of cofactors of the matrix.

THEOREM 7.11

Let $\mathbf{A}$ be an $n \times n$ nonsingular matrix. Define an $n \times n$ matrix $\mathbf{B}$ by putting

$$b_{ij} = \frac{1}{|\mathbf{A}|}(-1)^{i+j} M_{ji}.$$

Then, $\mathbf{B} = \mathbf{A}^{-1}$. ■

That is, the i, j element of $\mathbf{A}^{-1}$ is the cofactor of a_{ji} (not a_{ij}), divided by the determinant of $\mathbf{A}$.

Proof   By the way $\mathbf{B}$ is defined, the i, j element of $\mathbf{AB}$ is

$$(\mathbf{AB})_{ij} = \sum_{k=1}^{n} a_{ik} b_{kj} = \frac{1}{|\mathbf{A}|}\sum_{k=1}^{n}(-1)^{j+k} a_{ik} M_{jk}.$$

Now examine the sum on the right. If $i = j$, we get

$$(\mathbf{AB})_{ii} = \frac{1}{|\mathbf{A}|}\sum_{k=1}^{n}(-1)^{i+k} a_{ik} M_{ik}$$

and the summation is exactly the cofactor expansion of $|\mathbf{A}|$ by row i. Therefore,

$$(\mathbf{AB})_{ii} = \frac{|\mathbf{A}|}{|\mathbf{A}|} = 1.$$

If $i \neq j$, then the summation in the expression for $(\mathbf{AB})_{ij}$ is the cofactor expansion, by row j, of the determinant of the matrix formed from $\mathbf{A}$ by replacing row j by row i. But this matrix then has two identical rows, hence has determinant zero. Then $(\mathbf{AB})_{ij} = 0$ if $i \neq j$, and we conclude that $\mathbf{AB} = \mathbf{I}_n$. A similar argument shows that $\mathbf{BA} = \mathbf{I}_n$, hence $\mathbf{B} = \mathbf{A}^{-1}$. ■

This method of computing a matrix inverse is not as efficient in general as the reduction method discussed previously. Nevertheless, it works well for small matrices, and in some discussions it is useful to have a formula for the elements of a matrix inverse.

EXAMPLE 7.6

Let

$$\mathbf{A} = \begin{pmatrix} -2 & 4 & 1 \\ 6 & 3 & -3 \\ 2 & 9 & -5 \end{pmatrix}.$$

Then

$$\begin{vmatrix} -2 & 4 & 1 \\ 6 & 3 & -3 \\ 2 & 9 & -5 \end{vmatrix} = 120,$$

so **A** is nonsingular. Compute the nine elements of the inverse matrix **B**:

$$b_{11} = \frac{1}{120} M_{11} = \frac{1}{120} \begin{vmatrix} 3 & -3 \\ 9 & -5 \end{vmatrix} = \frac{12}{120} = \frac{1}{10},$$

$$b_{12} = \frac{1}{120}(-1) M_{21} = -\frac{1}{120} \begin{vmatrix} 4 & 1 \\ 9 & -5 \end{vmatrix} = \frac{29}{120},$$

$$b_{13} = \frac{1}{120} M_{31} = \frac{1}{120} \begin{vmatrix} 4 & 1 \\ 3 & -3 \end{vmatrix} = -\frac{1}{8},$$

$$b_{21} = -\frac{1}{120} M_{12} = -\frac{1}{120} \begin{vmatrix} 6 & -3 \\ 2 & -5 \end{vmatrix} = \frac{1}{5},$$

$$b_{22} = \frac{1}{120} \begin{vmatrix} -2 & 1 \\ 2 & -5 \end{vmatrix} = \frac{1}{15},$$

$$b_{23} = -\frac{1}{120} M_{32} = -\frac{1}{120} \begin{vmatrix} -2 & 1 \\ 6 & -3 \end{vmatrix} = 0,$$

$$b_{31} = \frac{1}{120} M_{13} = \frac{1}{120} \begin{vmatrix} 6 & 3 \\ 2 & 9 \end{vmatrix} = \frac{2}{5},$$

$$b_{32} = -\frac{1}{120} M_{23} = -\frac{1}{120} \begin{vmatrix} -2 & 4 \\ 2 & 9 \end{vmatrix} = \frac{13}{60},$$

$$b_{33} = \frac{1}{120} \begin{vmatrix} -2 & 4 \\ 6 & 3 \end{vmatrix} = -\frac{1}{4}.$$

Then

$$\mathbf{B} = \mathbf{A}^{-1} = \begin{pmatrix} \frac{1}{10} & \frac{29}{120} & -\frac{1}{8} \\ \frac{1}{5} & \frac{1}{15} & 0 \\ \frac{2}{5} & \frac{13}{60} & -\frac{1}{4} \end{pmatrix}. \blacksquare$$

SECTION 7.7   PROBLEMS

In each of Problems 1 through 10, use Theorem 7.10 to determine whether the matrix is nonsingular. If it is, use Theorem 7.11 to find its inverse.

1. $\begin{pmatrix} 2 & -1 \\ 1 & 6 \end{pmatrix}$

2. $\begin{pmatrix} 3 & 0 \\ 1 & 4 \end{pmatrix}$

3. $\begin{pmatrix} -1 & 1 \\ 1 & 4 \end{pmatrix}$

4. $\begin{pmatrix} 2 & 5 \\ -7 & -3 \end{pmatrix}$

5. $\begin{pmatrix} 6 & -1 & 3 \\ 0 & 1 & -4 \\ 2 & 2 & -3 \end{pmatrix}$

6. $\begin{pmatrix} -14 & 1 & -3 \\ 2 & -1 & 3 \\ 1 & 1 & 7 \end{pmatrix}$

7. $\begin{pmatrix} 0 & -4 & 3 \\ 2 & -1 & 6 \\ 1 & -1 & 7 \end{pmatrix}$

8. $\begin{pmatrix} 11 & 0 & -5 \\ 0 & 1 & 0 \\ 4 & -7 & 9 \end{pmatrix}$

9. $\begin{pmatrix} 3 & 1 & -2 & 1 \\ 4 & 6 & -3 & 9 \\ -2 & 1 & 7 & 4 \\ 13 & 0 & 1 & 5 \end{pmatrix}$

10. $\begin{pmatrix} 7 & -3 & -4 & 1 \\ 8 & 2 & 0 & 0 \\ 1 & 5 & -1 & 7 \\ 3 & -2 & -5 & 9 \end{pmatrix}$

7.8   Cramer's Rule

Cramer's rule is a determinant formula for solving a system of equations $\mathbf{AX} = \mathbf{B}$ when $\mathbf{A}$ is $n \times n$ and nonsingular. In this case, the system has the unique solution $\mathbf{X} = \mathbf{A}^{-1}\mathbf{B}$. We can, therefore, find $\mathbf{X}$ by computing $\mathbf{A}^{-1}$ and then $\mathbf{A}^{-1}\mathbf{B}$. Here is another way to find $\mathbf{X}$.

THEOREM 7.12   Cramer's Rule

Let $\mathbf{A}$ be a nonsingular $n \times n$ matrix of numbers. Then the unique solution of $\mathbf{AX} = \mathbf{B}$ is

$$\begin{pmatrix} x_1 \\ x_2 \\ \vdots \\ x_n \end{pmatrix},$$

where

$$x_k = \frac{1}{|\mathbf{A}|}|\mathbf{A}(k; \mathbf{B})|$$

and $\mathbf{A}(k; \mathbf{B})$ is the matrix obtained from $\mathbf{A}$ by replacing column k of $\mathbf{A}$ by $\mathbf{B}$. ■

Here is a heuristic argument to suggest why this works. Let

$$\mathbf{B} = \begin{pmatrix} b_1 \\ b_2 \\ \vdots \\ b_n \end{pmatrix}.$$

Multiply column k of $\mathbf{A}$ by x_k. The determinant of the resulting matrix is $x_k|\mathbf{A}|$, so

$$x_k|\mathbf{A}| = \begin{vmatrix} a_{11} & a_{12} & \cdots & a_{1k}x_k & \cdots & a_{1n} \\ a_{21} & a_{22} & \cdots & a_{2k}x_k & \cdots & a_{2n} \\ \vdots & \vdots & \vdots & \vdots & \vdots & \vdots \\ a_{n1} & a_{n2} & \cdots & a_{nk}x_k & \cdots & a_{nn} \end{vmatrix}.$$

For each $j \neq k$, add x_j times column j to column k. This Type III operation does not change the value of the determinant, and we get

$$x_k|\mathbf{A}| = \begin{vmatrix} a_{11} & a_{12} & \cdots & a_{11}x_1 + a_{12}x_2 + \cdots + a_{1n}x_n & \cdots & a_{1n} \\ a_{21} & a_{22} & \cdots & a_{21}x_1 + a_{22}x_2 + \cdots + a_{2n}a_n & \cdots & a_{2n} \\ \vdots & \vdots & \vdots & \vdots & & \vdots \\ a_{n1} & a_{n2} & \cdots & a_{n1}x_1 + a_{n2}x_2 + \cdots + a_{nn}x_n & \cdots & a_{nn} \end{vmatrix}$$

$$= \begin{vmatrix} a_{11} & a_{12} & \cdots & b_1 & \cdots & a_{1n} \\ a_{21} & a_{22} & \cdots & b_2 & \cdots & a_{2n} \\ \vdots & \vdots & \vdots & \vdots & \vdots & \vdots \\ a_{n1} & a_{n2} & \cdots & b_n & \cdots & a_{nn} \end{vmatrix} = |\mathbf{A}(k; \mathbf{B})| \,.$$

Solving for x_k yields the conclusion of Cramer's rule.

EXAMPLE 7.7

Solve the system

$$x_1 - 3x_2 - 4x_3 = 1$$
$$-x_1 + x_2 - 3x_3 = 14$$
$$x_2 - 3x_3 = 5.$$

The matrix of coefficients is

$$\mathbf{A} = \begin{pmatrix} 1 & -3 & -4 \\ -1 & 1 & -3 \\ 0 & 1 & -3 \end{pmatrix}.$$

We find that $|\mathbf{A}| = 13$. By Cramer's rule,

$$x_1 = \tfrac{1}{13} \begin{vmatrix} 1 & -3 & -4 \\ 14 & 1 & -3 \\ 5 & 1 & -3 \end{vmatrix} = -\tfrac{117}{13} = -9,$$

$$x_2 = \tfrac{1}{13} \begin{vmatrix} 1 & 1 & -4 \\ -1 & 14 & -3 \\ 0 & 5 & -3 \end{vmatrix} = -\tfrac{10}{13},$$

$$x_3 = \tfrac{1}{13} \begin{vmatrix} 1 & -3 & 1 \\ -1 & 1 & 14 \\ 0 & 1 & 5 \end{vmatrix} = -\tfrac{25}{13}. \; \blacksquare$$

Cramer's rule is not as efficient as the Gauss–Jordan reduction. Gauss–Jordan also applies to homogeneous systems, and to systems with different numbers of equations than unknowns. However, Cramer's rule does provide a formula for the solution, and this is useful in some contexts.

SECTION 7.8 PROBLEMS

In each of Problems 1 through 10, either find the solution by Cramer's rule or show that the rule does not apply.

1. $15x_1 - 4x_2 = 5$
$8x_1 + x_2 = -4$

2. $x_1 + 4x_2 = 3$
$x_1 + x_2 = 0$

3. $8x_1 - 4x_2 + 3x_3 = 0$
$x_1 + 5x_2 - x_3 = -5$
$-2x_1 + 6x_2 + x_3 = -4$

4. $5x_1 - 6x_2 + x_3 = 4$
$-x_1 + 3x_2 - 4x_3 = 5$
$2x_1 + 3x_2 + x_3 = -8$

5. $x_1 + x_2 - 3x_3 = 0$
$x_2 - 4x_3 = 0$
$x_1 - x_2 - x_3 = 5$

6. $6x_1 + 4x_2 - x_3 + 3x_4 - x_5 = 7$
$x_1 - 4x_2 + x_5 = -5$
$x_1 - 3x_2 + x_3 - 4x_5 = 0$
$-2x_1 + x_3 - 2x_5 = 4$
$x_3 - x_4 - x_5 = 8$

7. $2x_1 - 4x_2 + x_3 - x_4 = 6$
$x_2 - 3x_3 = 10$
$x_1 - 4x_3 = 0$
$x_2 - x_3 + 2x_4 = 4$

8. $2x_1 - 3x_2 + x_4 = 2$
$x_2 - x_3 + x_4 = 2$
$x_3 - 2x_4 = 5$
$x_1 - 3x_2 + 4x_3 = 0$

9. $14x_1 - 3x_3 = 5$
$2x_1 - 4x_3 + x_4 = 2$
$x_1 - x_2 + x_3 - 3x_4 = 1$
$x_3 - 4x_4 = -5$

10. $x_2 - 4x_4 = 18$
$x_1 - x_2 + 3x_3 = -1$
$x_1 + x_2 - 3x_3 + x_4 = 5$
$x_2 + 3x_4 = 0$

7.9 The Matrix Tree Theorem

In 1847, G. R. Kirchhoff published a classic paper in which he derived many of the electrical circuit laws that bear his name. One of these is the *matrix tree theorem*, which we will discuss now.

Figure 7.1 shows a typical electrical circuit. The underlying geometry of the circuit is shown in Figure 7.2. This diagram of points and connecting lines is called a *graph*, and was seen in the context of the movement of atoms in crystals in Section 6.1.5. A *labeled graph* has symbols attached to the points.

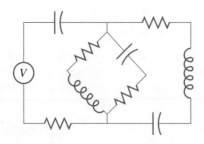

FIGURE 7.1

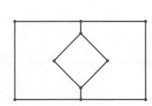

FIGURE 7.2

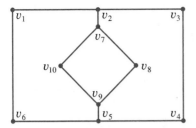 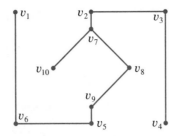 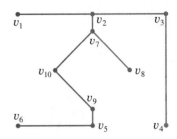

FIGURE 7.3 *A labeled graph and two of its spanning trees.*

Some of Kirchhoff's results depend on geometric properties of the circuit's underlying graph. One such property is the arrangement of the closed loops. Another is the number of spanning trees in the labeled graph. A *spanning tree* is a collection of lines in the graph forming no closed loops, but containing a path between any two points of the graph. Figure 7.3 shows a labeled graph and two spanning trees in this graph.

Kirchhoff derived a relationship between determinants and the number of labeled spanning trees in a graph.

THEOREM 7.13 **Matrix Tree Theorem**

Let G be a graph with vertices labeled $v_1, \ldots, v_n$. Form an $n \times n$ matrix $\mathbf{T} = [t_{ij}]$ as follows. If $i = j$, then t_{ii} is the number of lines to v_i in the graph. If $i \neq j$, then $t_{ij} = 0$ if there is no line between v_i and v_j in G, and $t_{ij} = -1$ if there is such a line. Then, all cofactors of $\mathbf{T}$ are equal, and their common value is the number of spanning trees in G. ∎

EXAMPLE 7.8

For the labeled graph of Figure 7.4, $\mathbf{T}$ is the 7×7 matrix

$$
T = \begin{pmatrix}
3 & -1 & 0 & 0 & 0 & -1 & -1 \\
-1 & 3 & -1 & -1 & 0 & 0 & 0 \\
0 & -1 & 3 & -1 & 0 & -1 & 0 \\
0 & -1 & -1 & 4 & -1 & 0 & -1 \\
0 & 0 & 0 & -1 & 3 & -1 & -1 \\
-1 & 0 & -1 & 0 & -1 & 4 & -1 \\
-1 & 0 & 0 & -1 & -1 & -1 & 4
\end{pmatrix}.
$$

Evaluate any cofactor of $\mathbf{T}$. For example, covering up row 1 and column 1, we have

$$
(-1)^{1+1} M_{11} = \begin{vmatrix}
3 & -1 & -1 & 0 & 0 & 0 \\
-1 & 3 & -1 & 0 & -1 & 0 \\
-1 & -1 & 4 & -1 & 0 & -1 \\
0 & 0 & -1 & 3 & -1 & -1 \\
0 & -1 & 0 & -1 & 4 & -1 \\
0 & 0 & -1 & -1 & -1 & 4
\end{vmatrix}
$$

$$
= 386.
$$

Evaluation of any cofactor of $\mathbf{T}$ yields the same result. ∎

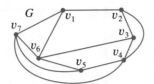

FIGURE 7.4   *Graph G*

Even with this small graph, it would clearly be impractical to enumerate the spanning trees by attempting to list them all.

SECTION 7.9   PROBLEMS

1. Find the number of spanning trees in the graph of Figure 7.5.

FIGURE 7.5

2. Find the number of spanning trees in the graph of Figure 7.6.

FIGURE 7.6

3. Find the number of spanning trees in the graph of Figure 7.7.

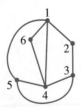

FIGURE 7.7

4. Find the number of spanning trees in the graph of Figure 7.8.

FIGURE 7.8

5. Find the number of spanning trees in the graph of Figure 7.9.

FIGURE 7.9

6. A complete graph on n points consists of n points with a line between each pair of points. This graph is often denoted K_n. With the points labeled $1, 2, \ldots, n$, show that the number of spanning trees in K_n is n^{n-2} for $n = 3, 4, \ldots,$.

CHAPTER 8

Eigenvalues, Diagonalization, and Special Matrices

Suppose $\mathbf{A}$ is an $n \times n$ matrix of real numbers. If we write an n-vector $\mathbf{E}$ as a column

$$\mathbf{E} = \begin{pmatrix} \alpha_1 \\ \alpha_2 \\ \vdots \\ \alpha_n \end{pmatrix},$$

then $\mathbf{AE}$ is an $n \times 1$ matrix, which we may also think of as an n-vector. We may therefore consider $\mathbf{A}$ as an operator that moves vectors about in R^n. Because $\mathbf{A}(a\mathbf{E}_1 + b\mathbf{E}_2) = a\mathbf{AE}_1 + b\mathbf{AE}_2$, $\mathbf{A}$ is called a *linear operator*.

Vectors have directions associated with them. Depending on $\mathbf{A}$, the direction of $\mathbf{AE}$ will generally be different from that of $\mathbf{E}$. It may happen, however, that for some vector $\mathbf{E}$, $\mathbf{AE}$ and $\mathbf{E}$ are parallel. In this event, there is a number λ such that $\mathbf{AE} = \lambda\mathbf{E}$. Then λ is called an *eigenvalue* of $\mathbf{A}$, with $\mathbf{E}$ an associated *eigenvector*.

The idea of an operator moving a vector to a parallel position is simple and geometrically appealing. It also has powerful ramifications in a variety of contexts. Eigenvalues contain important information about the solutions of systems of differential equations, and in models of physical phenomena they may have physical significance as well (such as the modes of vibration of a mechanical system, or the energy states of an atom).

8.1 Eigenvalues and Eigenvectors

Let $\mathbf{A}$ be an $n \times n$ matrix of real or complex numbers.

DEFINITION 8.1 *Eigenvalues and Eigenvectors*
_____

A real or complex number λ is an eigenvalue of $\mathbf{A}$ if there is a nonzero $n \times 1$ matrix (vector) $\mathbf{E}$ such that

$$\mathbf{AE} = \lambda\mathbf{E}.$$

Any nonzero vector $\mathbf{E}$ satisfying this relationship is called an eigenvector associated with the *eigenvalue* λ.

Eigenvalues are also known as *characteristic values* of a matrix, and eigenvectors can be called *characteristic vectors*.

We will typically write eigenvectors as column matrices and think of them as vectors in R^n. If an eigenvector has complex components, we may think of it as a vector in C^n, which consists of n-tuples of complex numbers. Since an eigenvector must be a nonzero vector, at least one component is nonzero.

If α is a nonzero scalar and $\mathbf{AE} = \lambda\mathbf{E}$, then

$$\mathbf{A}(\alpha\mathbf{E}) = \alpha(\mathbf{AE}) = \alpha(\lambda\mathbf{E}) = \lambda(\alpha\mathbf{E}).$$

This means that nonzero scalar multiples of eigenvectors are again eigenvectors.

EXAMPLE 8.1

Since

$$\begin{pmatrix} 1 & 0 \\ 0 & 0 \end{pmatrix}\begin{pmatrix} 0 \\ 4 \end{pmatrix} = \begin{pmatrix} 0 \\ 0 \end{pmatrix} = 0\begin{pmatrix} 0 \\ 4 \end{pmatrix},$$

0 is an eigenvalue of this matrix, with $\begin{pmatrix} 0 \\ 4 \end{pmatrix}$ an associated eigenvector. Although the zero vector cannot be an eigenvector, the number zero can be an eigenvalue of a matrix. For any scalar $\alpha \neq 0$, $\begin{pmatrix} 0 \\ 4\alpha \end{pmatrix}$ is also an eigenvector associated with the eigenvalue 0. ∎

EXAMPLE 8.2

Let

$$\mathbf{A} = \begin{pmatrix} 1 & -1 & 0 \\ 0 & 1 & 1 \\ 0 & 0 & -1 \end{pmatrix}.$$

Then 1 is an eigenvalue with associated eigenvector $\begin{pmatrix} 6 \\ 0 \\ 0 \end{pmatrix}$, because

$$\mathbf{A}\begin{pmatrix} 6 \\ 0 \\ 0 \end{pmatrix} = \begin{pmatrix} 6 \\ 0 \\ 0 \end{pmatrix} = 1\begin{pmatrix} 6 \\ 0 \\ 0 \end{pmatrix}.$$

Because any nonzero multiple of an eigenvector is an eigenvector, then $\begin{pmatrix} \alpha \\ 0 \\ 0 \end{pmatrix}$ is also an eigenvector associated with eigenvalue 1, for any nonzero number α.

Another eigenvalue of **A** is -1, with associated eigenvector $\begin{pmatrix} 1 \\ 2 \\ -4 \end{pmatrix}$, because

$$
\begin{pmatrix} 1 & -1 & 0 \\ 0 & 1 & 1 \\ 0 & 0 & -1 \end{pmatrix} \begin{pmatrix} 1 \\ 2 \\ -4 \end{pmatrix} = \begin{pmatrix} -1 \\ -2 \\ 4 \end{pmatrix} = -1 \begin{pmatrix} 1 \\ 2 \\ -4 \end{pmatrix}.
$$

Again, any vector $\begin{pmatrix} \alpha \\ 2\alpha \\ -4\alpha \end{pmatrix}$, with $\alpha \neq 0$, is an eigenvector associated with -1. ∎

We would like a way of finding all of the eigenvalues of a matrix **A**. The machinery to do this is at our disposal, and we reason as follows. For λ to be an eigenvalue of **A**, there must be an associated eigenvector **E**, and $\mathbf{AE} = \lambda \mathbf{E}$. Then $\lambda \mathbf{E} - \mathbf{AE} = \mathbf{O}$, or

$$\lambda \mathbf{I}_n \mathbf{E} - \mathbf{AE} = \mathbf{O}.$$

The identity matrix was inserted so we could write the last equation as

$$(\lambda \mathbf{I}_n - \mathbf{A})\mathbf{E} = \mathbf{O}.$$

This makes **E** a nontrivial solution of the $n \times n$ system of linear equations

$$(\lambda \mathbf{I}_n - \mathbf{A})\mathbf{X} = \mathbf{O}.$$

But this system can have a nontrivial solution if and only if the coefficient matrix has determinant zero, that is, $|\lambda \mathbf{I}_n - \mathbf{A}| = 0$. Thus, λ is an eigenvalue of **A** exactly when $|\lambda \mathbf{I}_n - \mathbf{A}| = 0$. This is the equation

$$
\begin{vmatrix} \lambda - a_{11} & -a_{12} & \cdots & -a_{1n} \\ -a_{21} & \lambda - a_{22} & \cdots & -a_{2n} \\ \vdots & \vdots & \vdots & \vdots \\ -a_{n1} & -a_{n2} & \cdots & \lambda - a_{nn} \end{vmatrix} = 0.
$$

When the determinant on the left is expanded, it is a polynomial of degree n in λ, called the *characteristic polynomial of* **A**. The roots of this polynomial are the eigenvalues of **A**. Corresponding to any root λ, any nontrivial solution **E** of $(\lambda \mathbf{I}_n - \mathbf{A})\mathbf{X} = \mathbf{O}$ is an eigenvector associated with λ. We will summarize these conclusions.

THEOREM 8.1

Let **A** be an $n \times n$ matrix of real or complex numbers. Then,

1. λ is an eigenvalue of **A** if and only if $|\lambda \mathbf{I}_n - \mathbf{A}| = 0$.
2. If λ is an eigenvalue of **A**, then any nontrivial solution of $(\lambda \mathbf{I}_n - \mathbf{A})\mathbf{X} = \mathbf{O}$ is an associated eigenvector. ∎

DEFINITION 8.2　*Characteristic Polynomial*

The polynomial $|\lambda I_n - A|$ is the characteristic polynomial of A, and is denoted $p_A(\lambda)$.

If A is $n \times n$, then $p_A(\lambda)$ is an nth degree polynomial with real or complex coefficients determined by the elements of A. This polynomial therefore has n roots, though some may be repeated. An $n \times n$ matrix A always has n eigenvalues $\lambda_1, \ldots, \lambda_n$, in which each eigenvalue is listed according to its multiplicity as a root of the characteristic polynomial. For example, if

$$p_A(\lambda) = (\lambda - 1)(\lambda - 3)^2(\lambda - i)^4$$

we list seven eigenvalues: $1, 3, 3, i, i, i, i$. The eigenvalue 3 has multiplicity 2 and i has multiplicity 4.

EXAMPLE 8.3

Let

$$A = \begin{pmatrix} 1 & -1 & 0 \\ 0 & 1 & 1 \\ 0 & 0 & -1 \end{pmatrix}$$

as in Example 8.2. The characteristic polynomial is

$$p_A(\lambda) = \begin{vmatrix} \lambda - 1 & 1 & 0 \\ 0 & \lambda - 1 & -1 \\ 0 & 0 & \lambda + 1 \end{vmatrix}$$

$$= (\lambda - 1)^2(\lambda + 1).$$

The eigenvalues of A are $1, 1, -1$.

To find eigenvectors associated with eigenvalue 1, solve

$$(1I_3 - A)X = \begin{pmatrix} 0 & 1 & 0 \\ 0 & 0 & -1 \\ 0 & 0 & 2 \end{pmatrix} X = O.$$

This has general solution

$$\begin{pmatrix} \alpha \\ 0 \\ 0 \end{pmatrix}$$

and these are the eigenvectors associated with eigenvalue 1, with $\alpha \neq 0$.

For eigenvectors associated with -1, solve

$$(-1I_3 - A)X = \begin{pmatrix} -2 & 1 & 0 \\ 0 & -2 & -1 \\ 0 & 0 & 0 \end{pmatrix} X = O.$$

The general solution is

$$\begin{pmatrix} \beta \\ 2\beta \\ -4\beta \end{pmatrix}$$

and these are the eigenvectors associated with eigenvalue -1, as long as $\beta \neq 0$. ∎

EXAMPLE 8.4

Let

$$\mathbf{A} = \begin{pmatrix} 1 & -2 \\ 2 & 0 \end{pmatrix}.$$

The characteristic polynomial is

$$p_\mathbf{A}(\lambda) = \left| \lambda \begin{pmatrix} 1 & 0 \\ 0 & 1 \end{pmatrix} - \begin{pmatrix} 1 & -2 \\ 2 & 0 \end{pmatrix} \right|$$

$$= \begin{vmatrix} \lambda - 1 & 2 \\ -2 & \lambda \end{vmatrix} = \lambda(\lambda - 1) + 4 = \lambda^2 - \lambda + 4.$$

This has roots $(1 + \sqrt{15}i)/2$ and $(1 - \sqrt{15}i)/2$, and these are the eigenvalues of $\mathbf{A}$. Even though $\mathbf{A}$ has real elements, the eigenvalues may be complex.

To find eigenvectors associated with $(1 + \sqrt{15}i)/2$, solve the system $(\lambda \mathbf{I}_2 - \mathbf{A})\mathbf{X} = \mathbf{O}$, which for this λ is

$$\left[\frac{1 + \sqrt{15}i}{2} \begin{pmatrix} 1 & 0 \\ 0 & 1 \end{pmatrix} - \begin{pmatrix} 1 & -2 \\ 2 & 0 \end{pmatrix} \right] \mathbf{X} = \mathbf{O}.$$

This is the system

$$\begin{pmatrix} \frac{1+\sqrt{15}i}{2} - 1 & 2 \\ -2 & \frac{1+\sqrt{15}i}{2} \end{pmatrix} \begin{pmatrix} x_1 \\ x_2 \end{pmatrix} = \begin{pmatrix} 0 \\ 0 \end{pmatrix}$$

or

$$\frac{-1 + \sqrt{15}i}{2} x_1 + 2x_2 = 0$$

$$-2x_1 + \frac{1 + \sqrt{15}i}{2} x_2 = 0.$$

We find the general solution of this system to be

$$\alpha \begin{pmatrix} 1 \\ \frac{1-\sqrt{15}i}{4} \end{pmatrix},$$

and this is an eigenvector associated with the eigenvalue $1 + \sqrt{15}i/2$ for any nonzero scalar α.
Corresponding to the eigenvalue $1 - \sqrt{15}i/2$, solve the system

$$\begin{pmatrix} \frac{1-\sqrt{15}i}{2} - 1 & 2 \\ -2 & \frac{1-\sqrt{15}i}{2} \end{pmatrix} \mathbf{X} = \mathbf{O},$$

obtaining the general solution

$$\beta \begin{pmatrix} 1 \\ \frac{1+\sqrt{15}i}{4} \end{pmatrix}.$$

This is an eigenvector corresponding to the eigenvalue $1 - \sqrt{15}i/2$ for any $\beta \neq 0$. ∎

Finding the eigenvalues of a matrix is equivalent to finding the roots of an nth degree polynomial, and if $n \geq 3$ this may be difficult. There are efficient computer routines that are usually based on the idea of putting the matrix through a sequence of transformations, the effect of which on the eigenvalues is known. This strategy was used previously to evaluate determinants. There are also approximation techniques, but these are sensitive to error. A number that is very close to an eigenvalue may not behave like an eigenvalue.

We will conclude this section with a theorem due to Gerschgorin. If real eigenvalues are plotted on the real line, and complex eigenvalues as points in the plane, Gerschgorin's theorem enables us to delineate regions of the plane containing the eigenvalues.

8.1.1   Gerschgorin's Theorem

THEOREM 8.2   *Gerschgorin*

Let $\mathbf{A}$ be an $n \times n$ matrix of real or complex numbers. For $k = 1, \ldots, n$, let

$$r_k = \sum_{j=1, j \neq k}^{n} |a_{kj}|.$$

Let C_k be the circle of radius r_k centered at (α_k, β_k), where $a_{kk} = \alpha_k + i\beta_k$. Then each eigenvalue of $\mathbf{A}$, when plotted as a point in the complex plane, lies on or within one of the circles $C_1, \ldots, C_n$. ∎

The circles C_k are called *Gerschgorin circles*. For the radius of C_k, read across row k and add the magnitudes of the row elements, omitting the diagonal element a_{kk}. The center of C_k is a_{kk}, plotted as a point in the complex plane. If the Gerschgorin circles are drawn and the disks they bound are shaded, then we have a picture of a region containing all of the eigenvalues of $\mathbf{A}$.

EXAMPLE 8.5

Let

$$\mathbf{A} = \begin{pmatrix} 12i & 1 & 9 & -4 \\ 1 & -6 & 2+i & -1 \\ 4 & 1 & -1 & 4i \\ 1-3i & -9 & 1 & 4-7i \end{pmatrix}.$$

$\mathbf{A}$ has characteristic polynomial

$$p_{\mathbf{A}}(\lambda) = \lambda^4 + (3 - 5i)\lambda^3 + (18 - 4i)\lambda^2 + (290 + 90i)\lambda + 1374 - 1120i.$$

It is not clear what the roots of this polynomial are. Form the Gerschgorin circles. Their radii are

$$r_1 = 1 + 9 + 4 = 14,$$
$$r_2 = 1 + \sqrt{5} + 1 = 2 + \sqrt{5},$$
$$r_3 = 4 + 1 + 4 = 9$$

and

$$r_4 = \sqrt{10} + 9 + 1 = 10 + \sqrt{10}.$$

C_1 has radius 14 and center $(0, 12)$, C_2 has radius $2 + \sqrt{5}$ and center $(-6, 0)$, C_3 has radius 9 and center $(-1, 0)$, and C_4 has radius $10 + \sqrt{10}$ and center $(4, -7)$. Figure 8.1 shows the Gerschgorin circles containing the eigenvalues of $\mathbf{A}$. ■

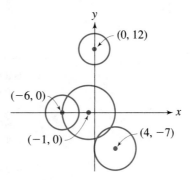

FIGURE 8.1 *Gerschgorin circles.*

Gerschgorin's theorem is not intended as an approximation scheme, since the Gerschgorin circles may have large radii. For some problems, however, just knowing some information about possible locations of eigenvalues can be important. For example, in studies of the stability of fluid flow, it is important to know whether there are eigenvalues in the right half-plane.

SECTION 8.1 PROBLEMS

In each of Problems 1 through 20, (a) find the eigenvalues of the matrix; (b) corresponding to each eigenvalue, find an eigenvector; and (c) sketch the Gerschgorin circles and (approximately) locate the eigenvalues as points in the plane.

1. $\begin{pmatrix} 1 & 3 \\ 2 & 1 \end{pmatrix}$

2. $\begin{pmatrix} -2 & 0 \\ 1 & 4 \end{pmatrix}$

3. $\begin{pmatrix} -5 & 0 \\ 1 & 2 \end{pmatrix}$

4. $\begin{pmatrix} 6 & -2 \\ -3 & 4 \end{pmatrix}$

5. $\begin{pmatrix} 1 & -6 \\ 2 & 2 \end{pmatrix}$

6. $\begin{pmatrix} 0 & 1 \\ 0 & 0 \end{pmatrix}$

7. $\begin{pmatrix} 2 & 0 & 0 \\ 1 & 0 & 2 \\ 0 & 0 & 3 \end{pmatrix}$

8. $\begin{pmatrix} -2 & 1 & 0 \\ 1 & 3 & 0 \\ 0 & 0 & -1 \end{pmatrix}$

9. $\begin{pmatrix} -3 & 1 & 1 \\ 0 & 0 & 0 \\ 0 & 1 & 0 \end{pmatrix}$

10. $\begin{pmatrix} 0 & 0 & -1 \\ 0 & 0 & 1 \\ 2 & 0 & 0 \end{pmatrix}$

11. $\begin{pmatrix} -14 & 1 & 0 \\ 0 & 2 & 0 \\ 1 & 0 & 2 \end{pmatrix}$

12. $\begin{pmatrix} 3 & 0 & 0 \\ 1 & -2 & -8 \\ 0 & -5 & 1 \end{pmatrix}$

13. $\begin{pmatrix} 1 & -2 & 0 \\ 0 & 0 & 0 \\ -5 & 0 & 7 \end{pmatrix}$

14. $\begin{pmatrix} -2 & 1 & 0 & 0 \\ 1 & 0 & 0 & 1 \\ 0 & 0 & 0 & 0 \\ 0 & 0 & 0 & 0 \end{pmatrix}$

15. $\begin{pmatrix} -4 & 1 & 0 & 1 \\ 0 & 1 & 0 & 0 \\ 0 & 0 & 2 & 0 \\ 1 & 0 & 0 & 3 \end{pmatrix}$

16. $\begin{pmatrix} 5 & 1 & 0 & 9 \\ 0 & 1 & 0 & 9 \\ 0 & 0 & 0 & 9 \\ 0 & 0 & 0 & 0 \end{pmatrix}$

17. $\begin{pmatrix} -6 & 0 & 0 & 0 \\ 4 & 1 & 0 & 2 \\ 0 & 1 & 1 & 2 \\ 1 & 0 & 0 & -3 \end{pmatrix}$

18. $\begin{pmatrix} 0 & 0 & 1 & 2 \\ 1 & 0 & 1 & 3 \\ 0 & 0 & 1 & 4 \\ 0 & 0 & 0 & 1 \end{pmatrix}$

19. $\begin{pmatrix} 5 & 0 & 1 & 1 & 2 \\ 0 & 1 & 0 & 0 & 0 \\ 0 & 0 & 0 & -2 & 0 \\ 0 & 0 & 0 & 4 & 0 \\ 0 & 0 & 0 & 0 & 5 \end{pmatrix}$

20. $\begin{pmatrix} 0 & -2 & 0 & -2 & 1 \\ -1 & 0 & 0 & 0 & 0 \\ 0 & 0 & -2 & 0 & 0 \\ 0 & 0 & 0 & 0 & 0 \\ 0 & 0 & 0 & 0 & 0 \end{pmatrix}$

21. Show that the eigenvalues of $\begin{pmatrix} \alpha & \beta \\ \beta & \gamma \end{pmatrix}$, in which $\alpha, \beta,$ and γ are real numbers, are real.

22. Show that the eigenvalues of $\begin{pmatrix} \alpha & \beta & \gamma \\ \beta & \delta & \epsilon \\ \gamma & \epsilon & \zeta \end{pmatrix}$ are real if all of the matrix elements are real.

23. Let λ be an eigenvalue of $\mathbf{A}$ with eigenvector $\mathbf{E}$. Show that, for any positive integer k, λ^k is an eigenvalue of $\mathbf{A}^k$, with eigenvector $\mathbf{E}$.

24. Let λ be an eigenvalue of $\mathbf{A}$ with eigenvector $\mathbf{E}$, and μ an eigenvalue of $\mathbf{A}$ with eigenvector $\mathbf{L}$. Suppose $\lambda \neq \mu$. Show that $\mathbf{E}$ and $\mathbf{L}$ are linearly independent as vectors in R^n.

25. Is it possible for a matrix having at least one complex, nonreal element to have only real eigenvalues? If not, give a proof; if it is possible, give an example.

26. Find the general form of all 2×2 matrices with real elements having eigenvalues 4 and -2. *Hint:* Let the matrix be $\begin{pmatrix} \alpha & \beta \\ \gamma & \delta \end{pmatrix}$. Write the characteristic polynomial and use the fact that the roots are 4 and -2 to determine relationships between the matrix elements.

27. Let $\mathbf{A}$ be an $n \times n$ matrix. Prove that the constant term of $p_{\mathbf{A}}(x)$ is $(-1)^n |\mathbf{A}|$. Use this to show that any singular matrix must have zero as one of its eigenvalues.

28. Prove Gerschgorin's theorem. *Hint:* Let λ be an eigenvalue of $\mathbf{A}$ and let $\mathbf{E} = \begin{pmatrix} e_1 \\ e_2 \\ \vdots \\ e_n \end{pmatrix}$ be an eigenvector associated with λ. Suppose the element of $\mathbf{E}$ having largest absolute value is e_K. (If two or more elements have this value, pick one.) Use the fact that $\mathbf{AE} = \lambda \mathbf{E}$ to show that

$$(\lambda - a_{KK})e_K = \sum_{j=1, j \neq K}^{n} a_{Kj} e_j.$$

Now reason that

$$|(\lambda - a_{KK})e_K| \leq \sum_{j=1, j \neq K}^{n} |a_{Kj}||e_j|$$

$$\leq |e_K| \sum_{j=1, j \neq K}^{n} |a_{Kj}|,$$

and hence show that $|\lambda - a_{KK}| \leq r_K$.

8.2 Diagonalization of Matrices

We have referred to the elements a_{ii} of a square matrix as its *main diagonal elements*. All other elements are called *off-diagonal elements*.

DEFINITION 8.3 **Diagonal Matrix**

A square matrix having all off-diagonal elements equal to zero is called a diagonal matrix.

We often write a diagonal matrix having main diagonal elements $d_1, \ldots, d_n$ as

$$\begin{pmatrix} d_1 & & & O \\ & d_2 & & \\ & & \ddots & \\ O & & & d_n \end{pmatrix},$$

with O in the upper right and lower left corners to indicate that all off-diagonal elements are zero. Here are some properties of diagonal matrices that make them pleasant to work with.

THEOREM 8.3

Let

$$\mathbf{D} = \begin{pmatrix} d_1 & & & O \\ & d_2 & & \\ & & \ddots & \\ O & & & d_n \end{pmatrix} \quad \text{and} \quad \mathbf{W} = \begin{pmatrix} w_1 & & & O \\ & w_2 & & \\ & & \ddots & \\ O & & & w_n \end{pmatrix}.$$

Then,

1.

$$\mathbf{DW} = \mathbf{WD} = \begin{pmatrix} d_1 w_1 & & & O \\ & d_2 w_2 & & \\ & & \ddots & \\ O & & & d_n w_n \end{pmatrix}.$$

2. $|\mathbf{D}| = d_1 d_2 \cdots d_n$.

3. $\mathbf{D}$ is nonsingular if and only if each main diagonal element is nonzero.

4. If each $d_j \neq 0$, then

$$\mathbf{D}^{-1} = \begin{pmatrix} 1/d_1 & & & O \\ & 1/d_2 & & \\ & & \ddots & \\ O & & & 1/d_n \end{pmatrix}.$$

5. The eigenvalues of $\mathbf{D}$ are its main diagonal elements.

6. An eigenvector associated with d_j is

$$\begin{pmatrix} 0 \\ \vdots \\ 0 \\ 1 \\ 0 \\ \vdots \\ 0 \end{pmatrix},$$

with 1 in row j and all other elements zero. ∎

We leave a proof of these conclusions to the student. Notice that (2) follows from the fact that a diagonal matrix is upper (and lower) triangular.

"Most" square matrices are not diagonal matrices. However, some matrices are related to diagonal matrices in a way that enables us to utilize the nice features of diagonal matrices.

DEFINITION 8.4   *Diagonalizable Matrix*

An $n \times n$ matrix $\mathbf{A}$ is diagonalizable if there exists an $n \times n$ nonsingular matrix $\mathbf{P}$ such that $\mathbf{P}^{-1}\mathbf{A}\mathbf{P}$ is a diagonal matrix.

When such $\mathbf{P}$ exists, we say that $\mathbf{P}$ diagonalizes $\mathbf{A}$.

The following theorem not only tells us when a matrix is diagonalizable, but also how to find a matrix $\mathbf{P}$ that diagonalizes it.

THEOREM 8.4   *Diagonalizability*

Let $\mathbf{A}$ be an $n \times n$ matrix. Then $\mathbf{A}$ is diagonalizable if it has n linearly independent eigenvectors. Further, if $\mathbf{P}$ is the $n \times n$ matrix having these eigenvectors as columns, then $\mathbf{P}^{-1}\mathbf{A}\mathbf{P}$ is the diagonal matrix having the corresponding eigenvalues down its main diagonal. ∎

Here is what this means. Suppose $\lambda_1, \ldots, \lambda_n$ are the eigenvalues of $\mathbf{A}$ (some possibly repeated), and $\mathbf{V}_1, \ldots, \mathbf{V}_n$ are corresponding eigenvectors. If these eigenvectors are linearly independent, we can form a nonsingular matrix $\mathbf{P}$ using $\mathbf{V}_j$ as column j. It is the linear independence of the eigenvectors that makes $\mathbf{P}$ nonsingular. We claim that $\mathbf{P}^{-1}\mathbf{A}\mathbf{P}$ is the diagonal matrix having the eigenvalues of $\mathbf{A}$ down its main diagonal, in the order corresponding to the order the eigenvectors were listed as columns of $\mathbf{P}$.

EXAMPLE 8.6

Let

$$\mathbf{A} = \begin{pmatrix} -1 & 4 \\ 0 & 3 \end{pmatrix}.$$

$\mathbf{A}$ has eigenvalues $-1, 3$ and corresponding eigenvectors $\begin{pmatrix} 1 \\ 0 \end{pmatrix}$ and $\begin{pmatrix} 1 \\ 1 \end{pmatrix}$, respectively. Form

$$\mathbf{P} = \begin{pmatrix} 1 & 1 \\ 0 & 1 \end{pmatrix}.$$

Because the eigenvectors are linearly independent, this matrix is nonsingular (note that $|\mathbf{P}| \neq 0$). We find that

$$\mathbf{P}^{-1} = \begin{pmatrix} 1 & -1 \\ 0 & 1 \end{pmatrix}.$$

Now compute

$$\mathbf{P}^{-1}\mathbf{AP} = \begin{pmatrix} 1 & -1 \\ 0 & 1 \end{pmatrix} \begin{pmatrix} -1 & 4 \\ 0 & 3 \end{pmatrix} \begin{pmatrix} 1 & 1 \\ 0 & 1 \end{pmatrix}$$
$$= \begin{pmatrix} -1 & 0 \\ 0 & 3 \end{pmatrix},$$

which has the eigenvalues down the main diagonal, corresponding to the order in which the eigenvectors were written as columns of $\mathbf{P}$.

If we use the other order in writing the eigenvectors as columns and define

$$\mathbf{Q} = \begin{pmatrix} 1 & 1 \\ 1 & 0 \end{pmatrix},$$

then we get

$$\mathbf{Q}^{-1}\mathbf{AQ} = \begin{pmatrix} 3 & 0 \\ 0 & -1 \end{pmatrix}. \blacksquare$$

Any linearly independent eigenvectors can be used in this diagonalization process. For example, if we use $\begin{pmatrix} 6 \\ 0 \end{pmatrix}$ and $\begin{pmatrix} -4 \\ -4 \end{pmatrix}$, which are simply nonzero scalar multiples of the previously used eigenvectors, then we can define

$$\mathbf{S} = \begin{pmatrix} 6 & -4 \\ 0 & -4 \end{pmatrix}$$

and now

$$\mathbf{S}^{-1}\mathbf{AS} = \begin{pmatrix} -1 & 0 \\ 0 & 3 \end{pmatrix}.$$

EXAMPLE 8.7

Here is an example with more complicated arithmetic, but the idea remains the same. Let

$$\mathbf{A} = \begin{pmatrix} -1 & 1 & 3 \\ 2 & 1 & 4 \\ 1 & 0 & -2 \end{pmatrix}.$$

The eigenvalues are $-1, -\frac{1}{2} + \frac{1}{2}\sqrt{29}, -\frac{1}{2} - \frac{1}{2}\sqrt{29}$, with corresponding eigenvectors, respectively,

$$\begin{pmatrix} 1 \\ -3 \\ 1 \end{pmatrix}, \begin{pmatrix} 3 + \sqrt{29} \\ 10 + 2\sqrt{29} \\ 2 \end{pmatrix}, \begin{pmatrix} 3 - \sqrt{29} \\ 10 - 2\sqrt{29} \\ 2 \end{pmatrix}.$$

These are linearly independent. Form the matrix

$$\mathbf{P} = \begin{pmatrix} 1 & 3 + \sqrt{29} & 3 - \sqrt{29} \\ -3 & 10 + 2\sqrt{29} & 10 - 2\sqrt{29} \\ 1 & 2 & 2 \end{pmatrix}.$$

We find that

$$\mathbf{P}^{-1} = \frac{\sqrt{29}}{812} \begin{pmatrix} \frac{232}{\sqrt{29}} & -\frac{116}{\sqrt{29}} & \frac{232}{\sqrt{29}} \\ -16 - 2\sqrt{29} & -1 + \sqrt{29} & -19 + 5\sqrt{29} \\ -16 - 2\sqrt{29} & 1 + \sqrt{29} & 19 + 5\sqrt{29} \end{pmatrix}.$$

Then

$$\mathbf{P}^{-1}\mathbf{A}\mathbf{P} = \begin{pmatrix} -1 & 0 & 0 \\ 0 & \frac{-1+\sqrt{29}}{2} & 0 \\ 0 & 0 & \frac{-1-\sqrt{29}}{2} \end{pmatrix}. \ \blacksquare$$

In this example, although we found $\mathbf{P}^{-1}$ explicitly, we did not actually need it to diagonalize $\mathbf{A}$. Theorem 8.4 assures us that $\mathbf{P}^{-1}\mathbf{A}\mathbf{P}$ is a diagonal matrix with the eigenvalues of $\mathbf{A}$ down its main diagonal. All we really needed was to know that $\mathbf{A}$ had three linearly independent eigenvectors. This is a useful fact to keep in mind, particularly if $\mathbf{P}$ and $\mathbf{P}^{-1}$ are cumbersome to compute.

EXAMPLE 8.8

Let

$$\mathbf{A} = \begin{pmatrix} -1 & -4 \\ 3 & -2 \end{pmatrix}.$$

The eigenvalues are $(-3 + \sqrt{47}i)/2$ and $(-3 - \sqrt{47}i)/2$. Corresponding eigenvectors are, respectively,

$$\begin{pmatrix} 8 \\ 1 - \sqrt{47}i \end{pmatrix}, \begin{pmatrix} 8 \\ 1 + \sqrt{47}i \end{pmatrix}.$$

Since these eigenvalues are linearly independent, there is a nonsingular 2×2 matrix $\mathbf{P}$ that diagonalizes $\mathbf{A}$:

$$\mathbf{P}^{-1}\mathbf{A}\mathbf{P} = \begin{pmatrix} \frac{-3+\sqrt{47}i}{2} & 0 \\ 0 & \frac{-3-\sqrt{47}i}{2} \end{pmatrix}.$$

Of course, if we need $\mathbf{P}$ for some other calculation, as will occur later, we can write it down:

$$\mathbf{P} = \begin{pmatrix} 8 & 8 \\ 1 - \sqrt{47}i & 1 + \sqrt{47}i \end{pmatrix}.$$

And, if we wish, we can compute

$$\mathbf{P}^{-1} = \frac{\sqrt{47}i}{752} \begin{pmatrix} -1 - i\sqrt{47} & 8 \\ 1 - i\sqrt{47} & -8 \end{pmatrix}.$$

However, even without explicitly writing $\mathbf{P}^{-1}$, we know what $\mathbf{P}^{-1}\mathbf{A}\mathbf{P}$ is. $\blacksquare$

EXAMPLE 8.9

It is not necessary that $\mathbf{A}$ have n distinct eigenvalues in order to have n linearly independent eigenvectors. For example, let

$$\mathbf{A} = \begin{pmatrix} 5 & -4 & 4 \\ 12 & -11 & 12 \\ 4 & -4 & 5 \end{pmatrix}.$$

The eigenvalues are 1, 1, -3, with 1 having multiplicity 2. Associated with -3 we find an eigenvector

$$\begin{pmatrix} 1 \\ 3 \\ 1 \end{pmatrix}.$$

To find eigenvectors associated with 1 we must solve the system

$$(\mathbf{I}_3 - \mathbf{A})\mathbf{X} = \begin{pmatrix} -4 & 4 & -4 \\ -12 & 12 & -12 \\ -4 & 4 & -4 \end{pmatrix} \begin{pmatrix} x_1 \\ x_2 \\ x_3 \end{pmatrix} = \begin{pmatrix} 0 \\ 0 \\ 0 \end{pmatrix}.$$

This system has general solution

$$\alpha \begin{pmatrix} 1 \\ 0 \\ -1 \end{pmatrix} + \beta \begin{pmatrix} 0 \\ 1 \\ 1 \end{pmatrix}.$$

We can therefore find two linearly independent eigenvectors associated with eigenvalue 1, for example,

$$\begin{pmatrix} 1 \\ 0 \\ -1 \end{pmatrix} \quad \text{and} \quad \begin{pmatrix} 0 \\ 1 \\ 1 \end{pmatrix}.$$

We can now form the nonsingular matrix

$$\mathbf{P} = \begin{pmatrix} 1 & 1 & 0 \\ 3 & 0 & 1 \\ 1 & -1 & 1 \end{pmatrix}$$

that diagonalizes $\mathbf{A}$:

$$\mathbf{P}^{-1}\mathbf{AP} = \begin{pmatrix} -3 & 0 & 0 \\ 0 & 1 & 0 \\ 0 & 0 & 1 \end{pmatrix}. \blacksquare$$

Here is a proof of Theorem 8.4, explaining why a matrix $\mathbf{P}$ formed from linearly independent eigenvectors must diagonalize $\mathbf{A}$. The proof makes use of an observation we have made before. When multiplying two $n \times n$ matrices $\mathbf{A}$ and $\mathbf{B}$,

$$\text{column } j \text{ of } \mathbf{AB} = \mathbf{A}(\text{column } j \text{ of } \mathbf{B}).$$

Proof Let the eigenvalues of $\mathbf{A}$ be $\lambda_1, \ldots, \lambda_n$ and corresponding eigenvectors be $\mathbf{V}_1, \ldots, \mathbf{V}_n$. These form the columns of $\mathbf{P}$.

Since these eigenvectors are assumed to be linearly independent, the dimension of the column space of $\mathbf{P}$ is n. Therefore, $rank(\mathbf{P}) = n$ and $\mathbf{P}$ is nonsingular by Theorems 6.15 and 6.26. Now compute $\mathbf{P}^{-1}\mathbf{AP}$ as follows. First,

$$\text{column } j \text{ of } \mathbf{AP} = \mathbf{A}(\text{column } j \text{ of } \mathbf{P}) = \mathbf{AV}_j = \lambda_j \mathbf{V}_j.$$

Thus the columns of $\mathbf{AP}$ are $\lambda_1 \mathbf{V}_1, \ldots, \lambda_n \mathbf{V}_n$ and $\mathbf{AP}$ has the form

$$\mathbf{AP} = \begin{pmatrix} | & | & \cdots & | \\ \lambda_1 \mathbf{V}_1 & \lambda_2 \mathbf{V}_2 & \cdots & \lambda_n \mathbf{V}_n \\ | & | & \cdots & | \end{pmatrix}.$$

Then

$$\text{column } j \text{ of } \mathbf{P}^{-1}\mathbf{AP} = \mathbf{P}^{-1}(\text{column } j \text{ of } \mathbf{AP})$$
$$= \mathbf{P}^{-1}[\lambda_j \mathbf{V}_j] = \lambda_j \mathbf{P}^{-1}\mathbf{V}_j.$$

But $\mathbf{V}_j$ is column j of $\mathbf{P}$, so

$$\mathbf{P}^{-1}\mathbf{V}_j = \text{column } j \text{ of } \mathbf{P}^{-1}\mathbf{P} = \begin{pmatrix} 0 \\ 0 \\ \vdots \\ 1 \\ \vdots \\ 0 \end{pmatrix}$$

in which the column matrix on the right has all zero elements except 1 in row j. Combining the last two equations, we have

$$\text{column } j \text{ of } \mathbf{P}^{-1}\mathbf{AP} = \lambda_j \mathbf{P}^{-1}\mathbf{V}_j = \lambda_j \begin{pmatrix} 0 \\ 0 \\ \vdots \\ 1 \\ \vdots \\ 0 \end{pmatrix} = \begin{pmatrix} 0 \\ 0 \\ \vdots \\ \lambda_j \\ \vdots \\ 0 \end{pmatrix}.$$

We now know the columns of $\mathbf{P}^{-1}\mathbf{AP}$, and putting them together gives us

$$\mathbf{P}^{-1}\mathbf{AP} = \begin{pmatrix} \lambda_1 & 0 & 0 & \cdots & 0 \\ 0 & \lambda_2 & 0 & \cdots & 0 \\ 0 & 0 & \lambda_3 & \cdots & 0 \\ \vdots & \vdots & \vdots & \vdots & \vdots \\ 0 & 0 & 0 & \cdots & \lambda_n \end{pmatrix}. \blacksquare$$

We can strengthen the conclusions of Theorem 8.4. So far, if $\mathbf{A}$ has n linearly independent eigenvectors, then we can diagonalize $\mathbf{A}$. We will now show that this is the only time $\mathbf{A}$ can be diagonalized. Further, if, for any $\mathbf{Q}$, $\mathbf{Q}^{-1}\mathbf{AQ}$ is a diagonal matrix, then $\mathbf{Q}$ must have linearly independent eigenvectors of $\mathbf{A}$ as its columns.

THEOREM 8.5

Let $\mathbf{A}$ be an $n \times n$ diagonalizable matrix. Then $\mathbf{A}$ has n linearly independent eigenvectors. Further, if $\mathbf{Q}^{-1}\mathbf{A}\mathbf{Q}$ is a diagonal matrix, then the diagonal elements of $\mathbf{Q}^{-1}\mathbf{A}\mathbf{Q}$ are the eigenvalues of $\mathbf{A}$ and the columns of $\mathbf{Q}$ are corresponding eigenvectors.

Proof   Suppose that

$$\mathbf{Q}^{-1}\mathbf{A}\mathbf{Q} = \begin{pmatrix} d_1 & & & O \\ & d_2 & & \\ & & \ddots & \\ O & & & d_n \end{pmatrix} = \mathbf{D}.$$

Denote column j of $\mathbf{Q}$ as $\mathbf{V}_j$. Then $\mathbf{V}_1, \ldots, \mathbf{V}_n$ are linearly independent, because $\mathbf{Q}$ is nonsingular. We will show that d_j is an eigenvalue of $\mathbf{A}$, with corresponding eigenvector $\mathbf{V}_j$.

Write $\mathbf{A}\mathbf{Q} = \mathbf{Q}\mathbf{D}$ and compute both sides of this product separately. First, since the columns of $\mathbf{Q}$ are $\mathbf{V}_1, \ldots, \mathbf{V}_n$,

$$\mathbf{Q}\mathbf{D} = \begin{pmatrix} | & | & \cdots & | \\ \mathbf{V}_1 & \mathbf{V}_2 & \cdots & \mathbf{V}_n \\ | & | & \cdots & | \end{pmatrix} \mathbf{D}$$

$$= \begin{pmatrix} | & | & \cdots & | \\ d_1\mathbf{V}_1 & d_2\mathbf{V}_2 & \cdots & d_n\mathbf{V}_n \\ | & | & \cdots & | \end{pmatrix},$$

a matrix having $d_j\mathbf{V}_j$ as column j. Now compute

$$\mathbf{A}\mathbf{Q} = \mathbf{A} \begin{pmatrix} | & | & \cdots & | \\ \mathbf{V}_1 & \mathbf{V}_2 & \cdots & \mathbf{V}_n \\ | & | & \cdots & | \end{pmatrix}$$

$$= \begin{pmatrix} | & | & \cdots & | \\ A\mathbf{V}_1 & A\mathbf{V}_2 & \cdots & A\mathbf{V}_n \\ | & | & \cdots & | \end{pmatrix},$$

a matrix having $\mathbf{A}\mathbf{V}_j$ as column j. Since $\mathbf{A}\mathbf{Q} = \mathbf{Q}\mathbf{D}$, then column j of $\mathbf{A}\mathbf{Q}$ equals column j of $\mathbf{Q}\mathbf{D}$, so

$$\mathbf{A}\mathbf{V}_j = d_j\mathbf{V}_j$$

which proves that d_j is an eigenvalue of $\mathbf{A}$ with associated eigenvector $\mathbf{V}_j$. ∎

As a consequence of this theorem, we see that not every matrix is diagonalizable.

EXAMPLE 8.10

Let

$$\mathbf{B} = \begin{pmatrix} 1 & -1 \\ 0 & 1 \end{pmatrix}.$$

B has eigenvalues 1, 1, and every eigenvector has the form $\alpha \begin{pmatrix} 1 \\ 0 \end{pmatrix}$. There are not two linearly independent eigenvectors, so **B** is not diagonalizable.

We could also proceed here by contradiction. If **B** were diagonalizable, then for some **P**,

$$\mathbf{P}^{-1}\mathbf{A}\mathbf{P} = \begin{pmatrix} 1 & 0 \\ 0 & 1 \end{pmatrix}.$$

From Theorem 8.5, the columns of **P** must be eigenvectors, so **P** must have the form

$$\mathbf{P} = \begin{pmatrix} \alpha & \beta \\ 0 & 0 \end{pmatrix}.$$

But this matrix is singular (it has zero determinant and its columns are multiples of each other, hence linearly dependent). Thus no matrix can diagonalize **B**. ∎

The key to diagonalization of an $n \times n$ matrix **A** is therefore the existence of n linearly independent eigenvectors. We saw (Example 8.9) that this does not require that the eigenvalues be distinct. However, if **A** does have n distinct eigenvalues, we claim that it must have n linearly independent eigenvectors, hence must be diagonalizable.

THEOREM 8.6

Let the $n \times n$ matrix **A** have n distinct eigenvalues. Then corresponding eigenvectors are linearly independent.

Proof   We will show by induction that any k distinct eigenvalues have associated with them k linearly independent eigenvectors. For $k = 1$, an eigenvector associated with a single eigenvalue is linearly independent, being a nonzero vector. Now suppose that any $k - 1$ distinct eigenvalues have associated with them $k - 1$ linearly independent eigenvectors. Suppose we have distinct eigenvalues $\lambda_1, \ldots, \lambda_k$. Let $\mathbf{V}_1, \ldots, \mathbf{V}_k$ be associated eigenvectors. We want to show that $\mathbf{V}_1, \ldots, \mathbf{V}_k$ are linearly independent.

If these eigenvectors were linearly dependent, there would be numbers $c_1, \ldots, c_k$ not all zero such that

$$c_1\mathbf{V}_1 + \cdots + c_k\mathbf{V}_k = \mathbf{O}.$$

By relabeling if necessary, we may suppose for convenience that $c_1 \neq 0$. Now

$$(\lambda_1\mathbf{I}_n - \mathbf{A})(c_1\mathbf{V}_1 + \cdots + c_k\mathbf{V}_k) = \mathbf{O}$$

$$= c_1(\lambda_1\mathbf{I}_n - \mathbf{A})\mathbf{V}_1 + c_2(\lambda_1\mathbf{I}_n - \mathbf{A})\mathbf{V}_2 + \cdots + c_k(\lambda_1\mathbf{I}_n - \mathbf{A})\mathbf{V}_k$$

$$= c_1(\lambda_1\mathbf{V}_1 - \mathbf{A}\mathbf{V}_1) + c_2(\lambda_1\mathbf{V}_2 - \mathbf{A}\mathbf{V}_2) + \cdots + c_k(\lambda_1\mathbf{V}_k - \mathbf{A}\mathbf{V}_k)$$

$$= c_1(\lambda_1\mathbf{V}_1 - \lambda_1\mathbf{V}_1) + c_2(\lambda_1\mathbf{V}_2 - \lambda_2\mathbf{V}_2) + \cdots + c_k(\lambda_1\mathbf{V}_k - \lambda_k\mathbf{V}_k)$$

$$= c_2(\lambda_1 - \lambda_2)\mathbf{V}_2 + \cdots + c_k(\lambda_1 - \lambda_k)\mathbf{V}_k.$$

But $\mathbf{V}_2, \ldots, \mathbf{V}_k$ are linearly independent by the inductive hypothesis, so each of these coefficients must be zero. Since $\lambda_1 - \lambda_j \neq 0$ for $j = 2, \ldots, k$ by the assumption that the eigenvalues are distinct, then

$$c_2 = \cdots = c_k = 0.$$

But then $c_1\mathbf{V}_1 = \mathbf{O}$. Since $\mathbf{V}_1$ is an eigenvector and cannot be **O**, then $c_1 = 0$ also, a contradiction. Therefore, $\mathbf{V}_1, \ldots, \mathbf{V}_k$ are linearly independent. ∎

COROLLARY 8.1

If an $n \times n$ matrix $\mathbf{A}$ has n distinct eigenvalues, then $\mathbf{A}$ is diagonalizable. ∎

EXAMPLE 8.11

Let

$$
\mathbf{A} = \begin{pmatrix} -2 & 0 & 0 & 5 \\ 1 & 3 & 0 & 0 \\ 0 & 4 & 4 & 0 \\ 2 & 0 & 0 & -3 \end{pmatrix}.
$$

The eigenvalues of $\mathbf{A}$ are $3, 4, -\frac{5}{2} + \frac{1}{2}\sqrt{41}, -\frac{5}{2} - \frac{1}{2}\sqrt{41}$. Because these are distinct, $\mathbf{A}$ is diagonalizable. For some $\mathbf{P}$,

$$
\mathbf{P}^{-1}\mathbf{AP} = \begin{pmatrix} 3 & 0 & 0 & 0 \\ 0 & 4 & 0 & 0 \\ 0 & 0 & -\frac{5}{2} + \frac{1}{2}\sqrt{41} & 0 \\ 0 & 0 & 0 & -\frac{5}{2} - \frac{1}{2}\sqrt{41} \end{pmatrix}.
$$

We do not need to actually produce $\mathbf{P}$ explicitly to conclude this. ∎

SECTION 8.2 PROBLEMS

In each of Problems 1 through 12, produce a matrix that diagonalizes the given matrix or show that this matrix is not diagonalizable.

1. $\begin{pmatrix} 0 & -1 \\ 4 & 3 \end{pmatrix}$

2. $\begin{pmatrix} 5 & 3 \\ 1 & 3 \end{pmatrix}$

3. $\begin{pmatrix} 1 & 0 \\ -4 & 1 \end{pmatrix}$

4. $\begin{pmatrix} -5 & 3 \\ 0 & 9 \end{pmatrix}$

5. $\begin{pmatrix} 5 & 0 & 0 \\ 1 & 0 & 3 \\ 0 & 0 & -2 \end{pmatrix}$

6. $\begin{pmatrix} 0 & 0 & 0 \\ 1 & 0 & 2 \\ 0 & 1 & 3 \end{pmatrix}$

7. $\begin{pmatrix} -2 & 0 & 1 \\ 1 & 1 & 0 \\ 0 & 0 & -2 \end{pmatrix}$

8. $\begin{pmatrix} 2 & 0 & 0 \\ 0 & 2 & 1 \\ 0 & -1 & 2 \end{pmatrix}$

9. $\begin{pmatrix} 1 & 0 & 0 & 0 \\ 0 & 4 & 1 & 0 \\ 0 & 0 & -3 & 1 \\ 0 & 0 & 1 & -2 \end{pmatrix}$

10. $\begin{pmatrix} -2 & 0 & 0 & 0 \\ -4 & -2 & 0 & 0 \\ 0 & 0 & -2 & 0 \\ 0 & 0 & 0 & -2 \end{pmatrix}$

11. $\begin{pmatrix} 8 & -7 & 1 & 0 \\ 0 & 1 & 0 & 0 \\ 0 & 0 & 0 & 0 \\ 1 & 0 & 0 & 0 \end{pmatrix}$

12.
$$\begin{pmatrix} -7 & 0 & 1 & 0 \\ 0 & 1 & 1 & 0 \\ -4 & 0 & 2 & 0 \\ 0 & 0 & 0 & 0 \end{pmatrix}$$

13. Suppose $\mathbf{A}^2$ is diagonalizable. Prove that $\mathbf{A}$ is diagonalizable.

14. Let $\mathbf{A}$ have eigenvalues $\lambda_1, \ldots, \lambda_n$ and suppose $\mathbf{P}$ diagonalizes $\mathbf{A}$. Prove that, for any positive integer k,

$$\mathbf{A}^k = \mathbf{P} \begin{pmatrix} \lambda_1^k & & & O \\ & \lambda_2^k & & \\ & & \ddots & \\ O & & & \lambda_n^k \end{pmatrix} \mathbf{P}^{-1}.$$

In each of Problems 15 through 18, compute the indicated power of the matrix, using the idea of Problem 14.

15. $\mathbf{A} = \begin{pmatrix} -1 & 0 \\ 1 & -5 \end{pmatrix}$; $\mathbf{A}^{18}$

16. $\mathbf{A} = \begin{pmatrix} -3 & -3 \\ -2 & 4 \end{pmatrix}$; $\mathbf{A}^{16}$

17. $\mathbf{A} = \begin{pmatrix} 0 & -2 \\ 1 & 0 \end{pmatrix}$; $\mathbf{A}^{43}$

18. $\mathbf{A} = \begin{pmatrix} -2 & 3 \\ 3 & -4 \end{pmatrix}$; $\mathbf{A}^{31}$

19. Let $\mathbf{A}$ be a 2×2 real matrix. Prove that there is a nonsingular matrix $\mathbf{P}$ such that $\mathbf{P}^{-1}\mathbf{A}\mathbf{P}$ has one of the following forms:

$$\begin{pmatrix} \alpha & 0 \\ 0 & \beta \end{pmatrix} \quad \text{with } \alpha \neq \beta,$$

$$\text{or} \quad \begin{pmatrix} \alpha & 0 \\ 0 & \alpha \end{pmatrix},$$

$$\text{or} \quad \begin{pmatrix} \alpha & -1 \\ 0 & \alpha \end{pmatrix}.$$

20. Suppose $\mathbf{L}^{-1}\mathbf{A}\mathbf{L} = \mathbf{B}$ for some matrix $\mathbf{L}$. Prove that $\mathbf{A}$ and $\mathbf{B}$ must both be diagonalizable or that neither is diagonalizable. If both are diagonalizable, and $\mathbf{P}$ diagonalizes $\mathbf{A}$ and $\mathbf{Q}$ diagonalizes $\mathbf{B}$, how are $\mathbf{P}^{-1}\mathbf{A}\mathbf{P}$ and $\mathbf{Q}^{-1}\mathbf{B}\mathbf{Q}$ related?

8.3   Orthogonal and Symmetric Matrices

Recall that the transpose of a matrix is obtained by interchanging the rows with the columns. For example, if

$$\mathbf{A} = \begin{pmatrix} -6 & 3 \\ 1 & -7 \end{pmatrix}$$

then

$$\mathbf{A}^t = \begin{pmatrix} -6 & 1 \\ 3 & -7 \end{pmatrix}.$$

Usually $\mathbf{A}^t$ is simply another square matrix. However, in the special circumstance that the transpose of a matrix is its inverse, we call $\mathbf{A}$ an *orthogonal matrix*.

DEFINITION 8.5   Orthogonal Matrix

A real square matrix $\mathbf{A}$ is orthogonal if and only if $\mathbf{A}\mathbf{A}^t = \mathbf{A}^t\mathbf{A} = \mathbf{I}_n$.

An orthogonal matrix is therefore nonsingular, and we find its inverse simply by taking its transpose.

EXAMPLE 8.12

Let

$$
A = \begin{pmatrix} 0 & \frac{1}{\sqrt{5}} & \frac{2}{\sqrt{5}} \\ 1 & 0 & 0 \\ 0 & \frac{2}{\sqrt{5}} & -\frac{1}{\sqrt{5}} \end{pmatrix}.
$$

Then

$$
AA^t = \begin{pmatrix} 0 & \frac{1}{\sqrt{5}} & \frac{2}{\sqrt{5}} \\ 1 & 0 & 0 \\ 0 & \frac{2}{\sqrt{5}} & -\frac{1}{\sqrt{5}} \end{pmatrix} \begin{pmatrix} 0 & 1 & 0 \\ \frac{1}{\sqrt{5}} & 0 & \frac{2}{\sqrt{5}} \\ \frac{2}{\sqrt{5}} & 0 & -\frac{1}{\sqrt{5}} \end{pmatrix} = I_3
$$

and a similar calculation gives $A^t A = I_3$. Therefore, this matrix is orthogonal, and

$$
A^{-1} = A^t = \begin{pmatrix} 0 & 1 & 0 \\ \frac{1}{\sqrt{5}} & 0 & \frac{2}{\sqrt{5}} \\ \frac{2}{\sqrt{5}} & 0 & -\frac{1}{\sqrt{5}} \end{pmatrix}. \ \blacksquare
$$

Because the transpose of the transpose of a matrix is the original matrix, a matrix is orthogonal exactly when its transpose is orthogonal.

THEOREM 8.7

A is an orthogonal matrix if and only if A^t is an orthogonal matrix. $\blacksquare$

Orthogonal matrices have several interesting properties. We will show first that the determinant of an orthogonal matrix must be 1 or -1.

THEOREM 8.8

If A is an orthogonal matrix, then $|A| = \pm 1$.

Proof   Since $AA^t = I_n$, then $|AA^t| = 1 = |A||A^t| = |A|^2$. $\blacksquare$

The next property of orthogonal matrices is actually the rationale for the name orthogonal. A set of vectors in R^n is said to be *orthogonal* if any two distinct vectors in the set are orthogonal (that is, their dot product is zero). The set is *orthonormal* if, in addition, each vector has length 1. We claim that the rows of an orthogonal matrix form an orthonormal set of vectors, as do the columns.

This can be seen in the matrix of the last example. The row vectors are

$$
\begin{pmatrix} 0 & \frac{1}{\sqrt{5}} & \frac{2}{\sqrt{5}} \end{pmatrix}, \begin{pmatrix} 1 & 0 & 0 \end{pmatrix}, \begin{pmatrix} 0 & \frac{2}{\sqrt{5}} & -\frac{1}{\sqrt{5}} \end{pmatrix}.
$$

These each have length 1, and each is orthogonal to each of the other two. Similarly, the columns of that matrix are

$$
\begin{pmatrix} 0 \\ 1 \\ 0 \end{pmatrix}, \begin{pmatrix} \frac{1}{\sqrt{5}} \\ 0 \\ \frac{2}{\sqrt{5}} \end{pmatrix}, \begin{pmatrix} \frac{2}{\sqrt{5}} \\ 0 \\ -\frac{1}{\sqrt{5}} \end{pmatrix}.
$$

Each is orthogonal to the other two, and each has length 1.

Not only do the row (column) vectors of an orthogonal matrix form an orthonormal set of vectors in R^n but this property completely characterizes orthogonal matrices.

THEOREM 8.9

Let A be a real $n \times n$ matrix. Then,

1. A is orthogonal if and only if the row vectors form an orthonormal set of vectors in R^n.
2. A is orthogonal if and only if the column vectors form an orthonormal set of vectors in R^n.

Proof   Recall that the i, j element of AB is the dot product of row i of A with column j of B. Further, the columns of A^t are the rows of A. Therefore,

$$i, j \text{ element of } AA^t = (\text{row } i \text{ of } A) \cdot (\text{column } j \text{ of } A^t)$$

$$= (\text{row } i \text{ of } A) \cdot (\text{row } j \text{ of } A).$$

Now suppose that A is an orthogonal matrix. Then $AA^t = I_n$, so the i, j element of AA^t is zero if $i \neq j$. Therefore, the dot product of two distinct rows of A is zero, and the rows form an orthogonal set of vectors. Further, the dot product of row i with itself is the i, i element of AA^t, and this is 1, so the rows form an orthonormal set of vectors.

Conversely, suppose the rows of A form an orthonormal set of vectors. Then the dot product row i with row j is zero if $i \neq j$, so the i, j element of AA^t is zero if $i \neq j$. Further, the i, i element of AA^t is the dot product of row i with itself, and this is 1. Therefore, $AA^t = I_n$. Similarly, $A^t A$ is I_n, so A is an orthogonal matrix. This proves (1). A proof of (2) is similar. ■

We now have a great deal of information about orthogonal matrices. We will use this to completely determine all 2×2 orthogonal matrices. Let

$$Q = \begin{pmatrix} a & b \\ c & d \end{pmatrix}.$$

What do we have to say about a, b, c, and d to make this an orthogonal matrix? First, the two row vectors must be orthogonal (zero dot product) and must have length 1, so

$$ac + bd = 0 \tag{8.1}$$

$$a^2 + b^2 = 1 \tag{8.2}$$

$$c^2 + d^2 = 1. \tag{8.3}$$

The two column vectors must also be orthogonal, so in addition,

$$ab + cd = 0. \tag{8.4}$$

Finally, $|Q| = \pm 1$, so

$$ad - bc = \pm 1.$$

This leads to two cases.

Case 1 $ad - bc = 1$
Multiply equation (8.1) by d to get

$$acd + bd^2 = 0.$$

Substitute $ad = 1 + bc$ into this equation to get

$$c(1 + bc) + bd^2 = 0$$

or

$$c + b(c^2 + d^2) = 0.$$

But $c^2 + d^2 = 1$ from equation (8.3), so $c + b = 0$, hence

$$c = -b.$$

Put this into equation (8.4) to get

$$ab - bd = 0.$$

Then $b = 0$ or $a = d$, leading to two subcases.

Case 1(a) $b = 0$
Then $c = -b = 0$ also, so

$$\mathbf{Q} = \begin{pmatrix} a & 0 \\ 0 & d \end{pmatrix}.$$

But each row vector has length 1, so $a^2 = d^2 = 1$. Further, $|\mathbf{Q}| = ad = 1$ in the present case, so $a = d = 1$ or $a = d = -1$. In these cases,

$$\mathbf{Q} = \mathbf{I}_2 \quad \text{or} \quad \mathbf{Q} = -\mathbf{I}_2.$$

Case 1(b) $b \neq 0$
Then $a = d$, so

$$\mathbf{Q} = \begin{pmatrix} a & b \\ -b & a \end{pmatrix}.$$

Since $a^2 + b^2 = 1$, there is some θ in $[0, 2\pi)$ such that $a = \cos(\theta)$ and $b = \sin(\theta)$. Then,

$$\mathbf{Q} = \begin{pmatrix} \cos(\theta) & \sin(\theta) \\ -\sin(\theta) & \cos(\theta) \end{pmatrix}.$$

This includes the two results of case 1(a) by choosing $\theta = 0$ or $\theta = \pi$.

Case 2 $ad - bc = -1$
By an analysis similar to that just done, we find now that, for some θ,

$$\mathbf{Q} = \begin{pmatrix} \cos(\theta) & \sin(\theta) \\ \sin(\theta) & -\cos(\theta) \end{pmatrix}.$$

These two cases give all the 2×2 orthogonal matrices. For example, with $\theta = \pi/4$ we get the orthogonal matrices

$$\begin{pmatrix} \frac{1}{\sqrt{2}} & \frac{1}{\sqrt{2}} \\ -\frac{1}{\sqrt{2}} & \frac{1}{\sqrt{2}} \end{pmatrix} \quad \text{and} \quad \begin{pmatrix} \frac{1}{\sqrt{2}} & \frac{1}{\sqrt{2}} \\ \frac{1}{\sqrt{2}} & -\frac{1}{\sqrt{2}} \end{pmatrix}$$

and with $\theta = \pi/6$ we get

$$\begin{pmatrix} \frac{\sqrt{3}}{2} & \frac{1}{2} \\ -\frac{1}{2} & \frac{\sqrt{3}}{2} \end{pmatrix} \quad \text{and} \quad \begin{pmatrix} \frac{\sqrt{3}}{2} & \frac{1}{2} \\ \frac{1}{2} & -\frac{\sqrt{3}}{2} \end{pmatrix}.$$

We can recognize the orthogonal matrices

$$\begin{pmatrix} \cos(\theta) & \sin(\theta) \\ -\sin(\theta) & \cos(\theta) \end{pmatrix}$$

as rotations in the plane. If the positive x, y system is rotated counterclockwise θ radians to form a new x', y' system, the coordinates in the two systems are related by

$$\begin{pmatrix} x' \\ y' \end{pmatrix} = \begin{pmatrix} \cos(\theta) & \sin(\theta) \\ -\sin(\theta) & \cos(\theta) \end{pmatrix} \begin{pmatrix} x \\ y \end{pmatrix}.$$

We will now consider another kind of matrix that is related to the class of orthogonal matrices.

DEFINITION 8.6 Symmetric Matrix

A square matrix is symmetric if $\mathbf{A} = \mathbf{A}^t$.

This means that each $a_{ij} = a_{ji}$, or that the matrix elements are the same if reflected across the main diagonal. For example,

$$\begin{pmatrix} -7 & -2 & 1 & 14 \\ -2 & 2 & -9 & 47 \\ 1 & -9 & 6 & \pi \\ 14 & 47 & \pi & 22 \end{pmatrix}$$

is symmetric.

A symmetric matrix need not have real numbers as elements. However, when it does, it has the remarkable property of having only real eigenvalues.

THEOREM 8.10

The eigenvalues of a real, symmetric matrix are real numbers. ■

Before showing why this is true, we will review some facts about complex numbers. A complex number $z = a + ib$ has magnitude $|z| = \sqrt{a^2 + b^2}$. The conjugate of z is defined to be $\bar{z} = a - ib$. When z is represented as the point (a, b) in the plane, $\bar{z}$ is the point $(a, -b)$, which is the reflection of (a, b) across the x axis. A number is real exactly when it equals its own conjugate. Further,

$$z\bar{z} = a^2 + b^2 = |z|^2$$

and

$$\overline{(\bar{z})} = z.$$

We take the conjugate $\overline{\mathbf{A}}$ of a matrix $\mathbf{A}$ by taking the conjugate of each of its elements. The product of a conjugate is the conjugate of a product:

$$\overline{(\mathbf{AB})} = (\overline{\mathbf{A}})(\overline{\mathbf{B}}).$$

Further, the operation of taking the conjugate commutes with the operation of taking a transpose:

$$\overline{\mathbf{C}}^t = \overline{(\mathbf{C}^t)}.$$

For example, if

$$\mathbf{C} = \begin{pmatrix} i & 1-2i \\ 3 & 0 \\ -2+i & 4 \end{pmatrix}$$

then

$$\overline{\mathbf{C}} = \begin{pmatrix} -i & 1+2i \\ 3 & 0 \\ -2-i & 4 \end{pmatrix}$$

and

$$\overline{\mathbf{C}}^t = \begin{pmatrix} -i & 3 & -2-i \\ 1+2i & 0 & 4 \end{pmatrix} = \overline{(\mathbf{C}^t)}.$$

We will now prove that the eigenvalues of a real symmetric matrix must be real.

Proof   Let $\mathbf{A}$ be an $n \times n$ matrix of real numbers. Let λ be an eigenvalue, and let

$$\mathbf{E} = \begin{pmatrix} e_1 \\ e_2 \\ \vdots \\ e_n \end{pmatrix}$$

be an associated eigenvector. Then $\mathbf{AE} = \lambda\mathbf{E}$. Multiply this equation on the left by the $1 \times n$ matrix

$$\overline{\mathbf{E}}^t = \begin{pmatrix} \overline{e_1} & \overline{e_2} & \cdots & \overline{e_n} \end{pmatrix}$$

to get

$$
\begin{aligned}
\overline{\mathbf{E}}^t \mathbf{AE} &= \overline{\mathbf{E}}^t \lambda \mathbf{E} = \lambda \overline{\mathbf{E}}^t \mathbf{E} \\
&= \lambda \begin{pmatrix} \overline{e_1} & \overline{e_2} & \cdots & \overline{e_n} \end{pmatrix} \begin{pmatrix} e_1 \\ e_2 \\ \vdots \\ e_n \end{pmatrix} \\
&= \lambda[\overline{e_1}e_1 + \overline{e_2}e_2 + \cdots + \overline{e_n}e_n] \\
&= \lambda(|e_1|^2 + |e_2|^2 + \cdots + |e_n|^2),
\end{aligned}
$$

(8.5)

which is λ times a real number. Here we are using the standard convention that a 1×1 matrix is identified with its single element.

Now compute

$$\overline{\overline{\mathbf{E}}^t \mathbf{AE}} = (\overline{\mathbf{E}})^t \overline{\mathbf{AE}} = \mathbf{E}^t \mathbf{A}\overline{\mathbf{E}},$$

(8.6)

in which we have used the fact that $\mathbf{A}$ has real elements to write $\overline{\mathbf{A}} = \mathbf{A}$.

Now $\mathbf{E}^t\mathbf{A}\overline{\mathbf{E}}$ is a 1×1 matrix, and so is the same as its transpose. Recalling that the transpose of a product is the product of the transposes in the reverse order, take the transpose of the last equation (8.6) to get

$$\mathbf{E}^t\mathbf{A}\overline{\mathbf{E}} = (\mathbf{E}^t\mathbf{A}\overline{\mathbf{E}})^t = \overline{\mathbf{E}}^t\mathbf{A}(\mathbf{E}^t)^t = \overline{\mathbf{E}}^t\mathbf{A}\mathbf{E}. \tag{8.7}$$

From equations (8.6) and (8.7) we have

$$\overline{\mathbf{E}^t\mathbf{A}\mathbf{E}} = \overline{\mathbf{E}}^t\mathbf{A}\mathbf{E}.$$

Therefore the 1×1 matrix $\overline{\mathbf{E}}^t\mathbf{A}\mathbf{E}$, being equal to its conjugate, is a real number. Now return to equation (8.5). We have just shown that the left side of this equation is real. Therefore the right side must be real. But $(|e_1|^2 + |e_2|^2 + \cdots + |e_n|^2)$ is certainly real. Therefore λ is real, and the theorem is proved. ∎

One ramification of this theorem is that a real, symmetric matrix also has real eigenvectors. We claim that, more than this, eigenvectors from distinct eigenvalues are orthogonal.

THEOREM 8.11

Let $\mathbf{A}$ be a real symmetric matrix. Then eigenvectors associated with distinct eigenvalues are orthogonal.

Proof   Let λ and μ be distinct eigenvalues with, respectively, eigenvectors

$$\mathbf{E} = \begin{pmatrix} e_1 \\ e_2 \\ \vdots \\ e_n \end{pmatrix} \quad \text{and} \quad \mathbf{G} = \begin{pmatrix} g_1 \\ g_2 \\ \vdots \\ g_n \end{pmatrix}.$$

Identifying, as usual, a real number with the 1×1 matrix having this number as its only element, the dot product of these two n-vectors can be written as a matrix product

$$e_1 g_1 + \cdots + e_n g_n = \mathbf{E}^t\mathbf{G}.$$

Since $\mathbf{A}\mathbf{E} = \lambda\mathbf{E}$ and $\mathbf{A}\mathbf{G} = \mu\mathbf{G}$, we have

$$\lambda\mathbf{E}^t\mathbf{G} = (\lambda\mathbf{E})^t\mathbf{G} = (\mathbf{A}\mathbf{E})^t\mathbf{G} = (\mathbf{E}^t\mathbf{A}^t)\mathbf{G}$$

$$= (\mathbf{E}^t\mathbf{A})\mathbf{G} = \mathbf{E}^t(\mathbf{A}\mathbf{G}) = \mathbf{E}^t(\mu\mathbf{G}) = \mu\mathbf{E}^t\mathbf{G}.$$

Then

$$(\lambda - \mu)\mathbf{E}^t\mathbf{G} = \mathbf{0}.$$

But $\lambda \neq \mu$, so $\mathbf{E}^t\mathbf{G} = 0$ and the dot product of these two eigenvectors is zero. These eigenvectors are therefore orthogonal. ∎

EXAMPLE 8.13

Let

$$\mathbf{A} = \begin{pmatrix} 3 & 0 & -2 \\ 0 & 2 & 0 \\ -2 & 0 & 0 \end{pmatrix},$$

a 3×3 real symmetric matrix. The eigenvalues are 2, -1, 4, with associated eigenvectors

$$\begin{pmatrix} 0 \\ 1 \\ 0 \end{pmatrix}, \begin{pmatrix} 1 \\ 0 \\ 2 \end{pmatrix}, \begin{pmatrix} 2 \\ 0 \\ -1 \end{pmatrix}.$$

These form an orthogonal set of vectors. ∎

In this example, the eigenvectors, while orthogonal to each other, are not all of length 1. However, a scalar multiple of an eigenvector is an eigenvector, so we can also write the following eigenvectors of **A**:

$$\begin{pmatrix} 0 \\ 1 \\ 0 \end{pmatrix}, \begin{pmatrix} \frac{1}{\sqrt{5}} \\ 0 \\ \frac{2}{\sqrt{5}} \end{pmatrix}, \begin{pmatrix} \frac{2}{\sqrt{5}} \\ 0 \\ -\frac{1}{\sqrt{5}} \end{pmatrix}.$$

These are still mutually orthogonal (multiplying by a positive scalar does not change orientation), but are now orthonormal. They can therefore be used as columns of an orthogonal matrix

$$\mathbf{Q} = \begin{pmatrix} 0 & \frac{1}{\sqrt{5}} & \frac{2}{\sqrt{5}} \\ 1 & 0 & 0 \\ 0 & \frac{2}{\sqrt{5}} & -\frac{1}{\sqrt{5}} \end{pmatrix}.$$

These column vectors, being orthogonal to each other, are linearly independent by Theorem 5.14. But whenever we form a matrix from linearly independent eigenvectors of **A**, this matrix diagonalizes **A**. Further, since **Q** is an orthogonal matrix, $\mathbf{Q}^{-1} = \mathbf{Q}^t$. Therefore, as we can easily verify in this example,

$$\mathbf{Q}^{-1}\mathbf{A}\mathbf{Q} = \begin{pmatrix} 2 & 0 & 0 \\ 0 & -1 & 0 \\ 0 & 0 & 4 \end{pmatrix}.$$

The idea we have just illustrated forms the basis for the following result.

THEOREM 8.12

Let **A** be a real, symmetric matrix. Then there is a real, orthogonal matrix that diagonalizes **A**. ∎

EXAMPLE 8.14

Let

$$\mathbf{A} = \begin{pmatrix} 2 & 1 & 0 \\ 1 & -2 & 4 \\ 0 & 4 & 2 \end{pmatrix}.$$

The eigenvalues are $\sqrt{21}$, $-\sqrt{21}$, and 2, with associated eigenvectors, respectively

$$\begin{pmatrix} 1 \\ \sqrt{21} - 2 \\ 4 \end{pmatrix}, \begin{pmatrix} 1 \\ -\sqrt{21} - 2 \\ 4 \end{pmatrix}, \begin{pmatrix} -4 \\ 0 \\ 1 \end{pmatrix}.$$

These eigenvectors are mutually orthogonal, but not orthonormal. Divide each eigenvector by its length to get the three new eigenvectors:

$$\frac{1}{\alpha}\begin{pmatrix} 1 \\ \sqrt{21} - 2 \\ 4 \end{pmatrix}, \quad \frac{1}{\alpha}\begin{pmatrix} 1 \\ -\sqrt{21} - 2 \\ 4 \end{pmatrix}, \quad \frac{1}{\beta}\begin{pmatrix} -4 \\ 0 \\ 1 \end{pmatrix}$$

where

$$\alpha = \sqrt{42 - 4\sqrt{21}} \quad \text{and} \quad \beta = \sqrt{17}.$$

The orthogonal matrix $\mathbf{Q}$ having these normalized eigenvectors as columns diagonalizes $\mathbf{A}$. ∎

SECTION 8.3   PROBLEMS

In each of Problems 1 through 12, find the eigenvalues of the matrix and, for each eigenvalue, a corresponding eigenvector. Check that eigenvectors associated with distinct eigenvalues are orthogonal. Find an orthogonal matrix that diagonalizes the matrix.

1. $\begin{pmatrix} 4 & -2 \\ -2 & 1 \end{pmatrix}$

2. $\begin{pmatrix} -3 & 5 \\ 5 & 4 \end{pmatrix}$

3. $\begin{pmatrix} 6 & 1 \\ 1 & 4 \end{pmatrix}$

4. $\begin{pmatrix} -13 & 1 \\ 1 & 4 \end{pmatrix}$

5. $\begin{pmatrix} 0 & 1 & 0 \\ 1 & -2 & 0 \\ 0 & 0 & 3 \end{pmatrix}$

6. $\begin{pmatrix} 0 & 1 & 1 \\ 1 & 2 & 0 \\ 1 & 0 & 2 \end{pmatrix}$

7. $\begin{pmatrix} 5 & 0 & 2 \\ 0 & 0 & 0 \\ 2 & 0 & 0 \end{pmatrix}$

8. $\begin{pmatrix} 2 & -4 & 0 \\ -4 & 0 & 0 \\ 0 & 0 & 0 \end{pmatrix}$

9. $\begin{pmatrix} 0 & 0 & 0 \\ 0 & 1 & -2 \\ 0 & -2 & 0 \end{pmatrix}$

10. $\begin{pmatrix} 1 & 3 & 0 \\ 3 & 0 & 1 \\ 0 & 1 & 1 \end{pmatrix}$

11. $\begin{pmatrix} 0 & 0 & 0 & 0 \\ 0 & 1 & -2 & 0 \\ 0 & -2 & 1 & 0 \\ 0 & 0 & 0 & 0 \end{pmatrix}$

12. $\begin{pmatrix} 5 & 0 & 0 & 0 \\ 0 & 0 & -1 & 0 \\ 0 & -1 & 0 & 0 \\ 0 & 0 & 0 & 0 \end{pmatrix}$

13. A real symmetric matrix is *positive definite* if all of its eigenvalues are positive. Prove that a real symmetric matrix $\mathbf{A}$ is positive definite if and only if there is a nonsingular matrix $\mathbf{P}$ such that $\mathbf{P}^t\mathbf{P} = \mathbf{A}$.

14. Let $\mathbf{A}$ be a real symmetric matrix. Prove that $\mathbf{A}$ is positive definite if and only if $\mathbf{E}^t\mathbf{AE} > 0$ for every nonzero $n \times 1$ matrix $\mathbf{E}$. ($\mathbf{E}^t\mathbf{AE}$ is a 1×1 matrix, and when we say that this matrix is positive, we mean that its single real entry is positive.)

15. Use the result of Problem 13 to show that every real, nonsingular matrix $\mathbf{A}$ can be written in the form $\mathbf{A} = \mathbf{QS}$, where $\mathbf{S}$ is real, symmetric, and positive definite and $\mathbf{Q}$ is orthogonal. *Hint:* First note that $\mathbf{A}^t\mathbf{A}$ is positive definite and symmetric. For some orthogonal matrix $\mathbf{W}$, $\mathbf{W}^t(\mathbf{A}^t\mathbf{A})\mathbf{W} = \mathbf{D}$, a diagonal matrix whose main diagonal elements are the eigenvalues of $\mathbf{A}^t\mathbf{A}$. Let $\mathbf{M}$ be the diagonal matrix whose main diagonal elements are the square roots of these eigenvalues. Choose $\mathbf{S} = \mathbf{WMW}^t$ and $\mathbf{Q} = \mathbf{AS}^{-1}$.

8.4   Quadratic Forms

A (complex) *quadratic form* is an expression

$$\sum_{j=1}^{n}\sum_{k=1}^{n} a_{jk}\overline{z}_j z_k, \tag{8.8}$$

in which each a_{jk} and z_j is a complex number.

For $n = 2$, this quadratic form is

$$a_{11}\overline{z}_1 z_1 + a_{12}\overline{z}_1 z_2 + a_{21}z_1\overline{z}_2 + a_{22}z_2\overline{z}_2.$$

The terms involving $z_j z_k$ with $j \neq k$ are the *mixed product terms*.

The quadratic form is real if each a_{jk} and z_j is real. In this case, we usually write z_j as x_j. Since $\overline{x}_j = x_j$ when x_j is real, the form (8.8) in this case is

$$\sum_{j=1}^{n}\sum_{k=1}^{n} a_{jk}x_j x_k.$$

For $n = 2$, this is

$$a_{11}x_1^2 + (a_{12} + a_{21})x_1 x_2 + a_{22}x_2^2.$$

The terms involving x_1^2 and x_2^2 are the squared terms in this real quadratic form, and x_{12} is the mixed product term.

It is often convenient to write a quadratic form (8.8) in matrix form. If $\mathbf{A} = [a_{ij}]$ and

$$\mathbf{Z} = \begin{pmatrix} z_1 \\ z_2 \\ \vdots \\ z_n \end{pmatrix}$$

then

$$\overline{\mathbf{Z}}^t \mathbf{A} \mathbf{Z} = \begin{pmatrix} \overline{z}_1 & \overline{z}_2 & \cdots & \overline{z}_n \end{pmatrix} \begin{pmatrix} a_{11} & a_{12} & \cdots & a_{1n} \\ a_{21} & a_{22} & \cdots & a_{2n} \\ \vdots & \vdots & \vdots & \vdots \\ a_{n1} & a_{n2} & \cdots & a_{nn} \end{pmatrix} \begin{pmatrix} z_1 \\ z_2 \\ \vdots \\ z_n \end{pmatrix}$$

$$= \begin{pmatrix} a_{11}\overline{z}_1 + \cdots + a_{n1}\overline{z}_n & \cdots & a_{1n}\overline{z}_1 + \cdots + a_{nn}\overline{z}_n \end{pmatrix} \begin{pmatrix} z_1 \\ z_2 \\ \vdots \\ z_n \end{pmatrix}$$

$$= a_{11}\overline{z_1}z_1 + \cdots + a_{n1}\overline{z_n}z_1 + \cdots + a_{1n}\overline{z_1}z_n + \cdots + a_{nn}\overline{z_n}z_n$$

$$= \sum_{j=1}^{n}\sum_{k=1}^{n} a_{jk}\overline{z_j}z_k.$$

Similarly, any real quadratic form can be written in matrix form as $\mathbf{X}^t\mathbf{A}\mathbf{X}$.

Given a quadratic form, we may choose different matrices $\mathbf{A}$ such that the form is $\mathbf{Z}^t\mathbf{A}\mathbf{Z}$.

EXAMPLE 8.15

Let

$$\mathbf{A} = \begin{pmatrix} 1 & 4 \\ 3 & 2 \end{pmatrix}.$$

Then

$$\begin{pmatrix} x_1 & x_2 \end{pmatrix} \begin{pmatrix} 1 & 4 \\ 3 & 2 \end{pmatrix} \begin{pmatrix} x_1 \\ x_2 \end{pmatrix} = \begin{pmatrix} x_1 + 3x_2 & 4x_1 + 2x_2 \end{pmatrix} \begin{pmatrix} x_1 \\ x_2 \end{pmatrix}$$

$$= x_1^2 + 3x_1x_2 + 4x_1x_2 + 2x_2^2$$

$$= x_1^2 + 7x_1x_2 + 2x_2^2.$$

But we can also write this quadratic form as

$$x_1^2 + \tfrac{7}{2}x_1x_2 + \tfrac{7}{2}x_2x_1 + 2x_2^2 = \begin{pmatrix} x_1 & x_2 \end{pmatrix} \begin{pmatrix} 1 & \tfrac{7}{2} \\ \tfrac{7}{2} & 2 \end{pmatrix} \begin{pmatrix} x_1 \\ x_2 \end{pmatrix}.$$

The advantage of the latter formulation is that the quadratic form is $\mathbf{X}^t\mathbf{A}\mathbf{X}$ with $\mathbf{A}$ a symmetric matrix. ∎

There is an expression involving a quadratic form that gives the eigenvalues of a matrix in terms of an associated eigenvector. We will have use for this shortly.

LEMMA 8.1

Let $\mathbf{A}$ be an $n \times n$ matrix of real or complex numbers. Let λ be an eigenvalue with eigenvector $\mathbf{Z}$. Then

$$\lambda = \frac{\overline{\mathbf{Z}}^t\mathbf{A}\mathbf{Z}}{\overline{\mathbf{Z}}^t\mathbf{Z}}.$$

Proof   Since $\mathbf{A}\mathbf{Z} = \lambda\mathbf{Z}$, then $\overline{\mathbf{Z}}^t\mathbf{A}\mathbf{Z} = \lambda\overline{\mathbf{Z}}^t\mathbf{Z}$. ∎

Using a calculation done in equation (8.5), we can write

$$\lambda = \frac{1}{\sum_{j=1}^{n}|z_j|^2} \sum_{j=1}^{n}\sum_{k=1}^{n} a_{jk}\overline{z_j}z_k.$$

Quadratic forms arise in a variety of contexts. In mechanics, the kinetic energy of a particle is a real quadratic form, and in analytic geometry, a conic is the locus of points in the plane for which a quadratic form in the coordinates is equal to some constant. For example,

$$x_1^2 + \tfrac{1}{4}x_2^2 = 9$$

is the equation of an ellipse in the x_1, x_2 plane.

In some problems involving quadratic forms, calculations are simplified if we transform from the $x_1, x_2, \ldots, x_n$ coordinate system to a $y_1, y_2, \ldots, y_n$ system in which there are no mixed product terms. That is, we want to choose $y_1, \ldots, y_n$ so that

$$\sum_{j=1}^{n} \sum_{k=1}^{n} a_{ij} x_j x_k = \sum_{j=1}^{n} \beta_j y_j^2. \tag{8.9}$$

The $y_1, \ldots, y_n$ coordinates are called *principal axes* for the quadratic form.

This kind of transformation is commonly done in analytic geometry, where a rotation of axes is used to eliminate mixed product terms in the equation of a conic. For example, the change of variables

$$x_1 = \tfrac{1}{\sqrt{2}} y_1 + \tfrac{1}{\sqrt{2}} y_2$$

$$x_2 = \tfrac{1}{\sqrt{2}} y_1 - \tfrac{1}{\sqrt{2}} y_2$$

transforms the quadratic form

$$x_1^2 - 2x_1 x_2 + x_2^2$$

to

$$2y_2^2,$$

with no mixed product term. Using this transformed form, we could analyze the graph of

$$x_1^2 - 2x_1 x_2 + x_2^2 = 4$$

in the x_1, x_2 system in terms of the graph of

$$y_2^2 = 2$$

in the y_1, y_2 system. In the y_1, y_2 plane, it is clear that the graph consists of two horizontal straight lines $y_2 = \pm\sqrt{2}$.

We will now show that a transformation that eliminates the mixed product terms of a real quadratic form always exists.

THEOREM 8.13 *Principal Axis Theorem*

Let $\mathbf{A}$ be a real symmetric matrix with eigenvalues $\lambda_1, \ldots, \lambda_n$. Let $\mathbf{Q}$ be an orthogonal matrix that diagonalizes $\mathbf{A}$. Then the change of variables $\mathbf{X} = \mathbf{QY}$ transforms $\sum_{j=1}^{n} \sum_{k=1}^{n} a_{jk} x_j x_k$ to

$$\sum_{j=1}^{n} \lambda_j y_j^2.$$

Proof   The proof is a straightforward calculation:

$$\sum_{j=1}^{n}\sum_{k=1}^{n} a_{ij}x_j x_k = \mathbf{X}^t\mathbf{AX}$$

$$= (\mathbf{QY})^t\mathbf{A}(\mathbf{QY}) = (\mathbf{Y}^t\mathbf{Q}^t)\mathbf{A}(\mathbf{QY})$$

$$= \mathbf{Y}^t(\mathbf{Q}^t\mathbf{AQ})\mathbf{Y}$$

$$= \begin{pmatrix} y_1 & \cdots & y_n \end{pmatrix} \begin{pmatrix} \lambda_1 & & & O \\ & \lambda_2 & & \\ & & \ddots & \\ O & & & \lambda_n \end{pmatrix} \begin{pmatrix} y_1 \\ \vdots \\ y_n \end{pmatrix}$$

$$= \lambda_1 y_1^2 + \cdots + \lambda_n y_n^2. \quad \blacksquare$$

The expression $\lambda_1 y_1^2 + \cdots + \lambda_n y_n^2$ is called the *standard form* of the quadratic form $\mathbf{X}^t\mathbf{AX}$.

EXAMPLE 8.16

Consider again $x_1^2 - 2x_1 x_2 + x_2^2$. This is $\mathbf{X}^t\mathbf{AX}$ with

$$\mathbf{A} = \begin{pmatrix} 1 & -1 \\ -1 & 1 \end{pmatrix}.$$

The eigenvalues of $\mathbf{A}$ are 0 and 2, with corresponding eigenvectors

$$\begin{pmatrix} 1 \\ 1 \end{pmatrix} \quad \text{and} \quad \begin{pmatrix} 1 \\ -1 \end{pmatrix}.$$

Dividing each eigenvector by its length, we obtain the eigenvectors

$$\begin{pmatrix} \frac{1}{\sqrt{2}} \\ \frac{1}{\sqrt{2}} \end{pmatrix} \quad \text{and} \quad \begin{pmatrix} \frac{1}{\sqrt{2}} \\ -\frac{1}{\sqrt{2}} \end{pmatrix}.$$

These form the columns of an orthogonal matrix $\mathbf{Q}$ that diagonalizes $\mathbf{A}$:

$$\mathbf{Q} = \begin{pmatrix} \frac{1}{\sqrt{2}} & \frac{1}{\sqrt{2}} \\ \frac{1}{\sqrt{2}} & -\frac{1}{\sqrt{2}} \end{pmatrix}.$$

The transformation defined by $\mathbf{X} = \mathbf{QY}$ is

$$\begin{pmatrix} x_1 \\ x_2 \end{pmatrix} = \begin{pmatrix} \frac{1}{\sqrt{2}} & \frac{1}{\sqrt{2}} \\ \frac{1}{\sqrt{2}} & -\frac{1}{\sqrt{2}} \end{pmatrix} \begin{pmatrix} y_1 \\ y_2 \end{pmatrix},$$

which gives exactly the transformation used above to reduce the quadratic form $x_1^2 - 2x_1 x_2 + x_2^2$ to the standard form $2y_2^2$. $\blacksquare$

EXAMPLE 8.17

Analyze the conic $4x_1^2 - 3x_1 x_2 + 2x_2^2 = 8$. First write the quadratic form as $\mathbf{X}^t\mathbf{AX} = 8$, where

$$\mathbf{A} = \begin{pmatrix} 4 & -\frac{3}{2} \\ -\frac{3}{2} & 2 \end{pmatrix}.$$

FIGURE 8.2

The eigenvalues of $\mathbf{A}$ are $(6 \pm \sqrt{13})/2$. By the principal axis theorem there is an orthogonal matrix $\mathbf{Q}$ that transforms the equation of the conic to standard form:

$$\frac{6 + \sqrt{13}}{2} y_1^2 + \frac{6 - \sqrt{13}}{2} y_2^2 = 8.$$

This is an ellipse in the y_1, y_2 plane. Figure 8.2 shows a graph of this ellipse. ■

SECTION 8.4 PROBLEMS

In each of Problems 1 through 6, find a matrix $\mathbf{A}$ such that the quadratic form is $\mathbf{X}^t\mathbf{AX}$.

1. $x_1^2 + 2x_1x_2 + 6x_2^2$

2. $3x_1^2 + 3x_2^2 - 4x_1x_2 - 3x_1x_3 + 2x_2x_3 + x_3^2$

3. $x_1^2 - 4x_1x_2 + x_2^2$

4. $2x_1^2 - x_2^2 + 2x_1x_2$

5. $-x_1^2 + x_4^2 - 2x_1x_4 + 3x_2x_4 - x_1x_3 + 4x_2x_3$

6. $x_1^2 - x_2^2 - x_1x_3 + 4x_2x_3$

In Problems 7 through 13, find the standard form of the quadratic form.

7. $-5x_1^2 + 4x_1x_2 + 3x_2^2$

8. $4x_1^2 - 12x_1x_2 + x_2^2$

9. $-3x_1^2 + 4x_1x_2 + 7x_2^2$

10. $4x_1^2 - 4x_1x_2 + x_2^2$

11. $-6x_1x_2 + 4x_2^2$

12. $5x_1^2 + 4x_1x_2 + 2x_2^2$

13. $-2x_1x_2 + 2x_3^2$

In each of Problems 14 through 18, use the principal axis theorem to analyze the conic.

14. $x_1^2 - 2x_1x_2 + 4x_2^2 = 6$

15. $3x_1^2 + 5x_1x_2 - 3x_2^2 = 5$

16. $-2x_1^2 + 3x_2^2 + x_1x_2 = 5$

17. $4x_1^2 - 4x_2^2 + 6x_1x_2 = 8$

18. $6x_1^2 + 2x_1x_2 + 5x_2^2 = 14$

In each of Problems 19 through 22, write the quadratic form defined by the matrix.

19. $\begin{pmatrix} -2 & 1 \\ 1 & 6 \end{pmatrix}$

20. $\begin{pmatrix} 14 & -3 & 0 \\ -3 & 2 & 1 \\ 0 & 1 & 7 \end{pmatrix}$

21. $\begin{pmatrix} 6 & 1 & -7 \\ 1 & 2 & 0 \\ -7 & 0 & 1 \end{pmatrix}$

22. $\begin{pmatrix} 7 & 1 & -2 \\ 1 & 0 & -1 \\ -2 & -1 & 3 \end{pmatrix}$

23. Is there a real, 3×3 matrix that cannot be the coefficient matrix of a real quadratic form?

24. Let $\mathbf{A}$ be a real symmetric matrix. Prove that there exists a matrix $\mathbf{P}$ such that $\mathbf{X} = \mathbf{PY}$ transforms $\mathbf{X}^t\mathbf{AX}$ into the form

$$y_1^2 + \cdots + y_p^2 - y_{p+1}^2 - \cdots - y_r^2.$$

Here $r = rank(\mathbf{A})$ and p is the number of positive eigenvalues of $\mathbf{A}$. This form of $\mathbf{X}^t\mathbf{AX}$ is called its *canonical form.* (It is not required that $\mathbf{P}$ be an orthogonal matrix.)

25. Prove that the quadratic forms $\mathbf{X}^t\mathbf{AX}$ and $\mathbf{X}^t\mathbf{BX}$ have the same canonical form if and only if there is a matrix $\mathbf{P}$ such that $\mathbf{B} = \mathbf{P}^{-1}\mathbf{AP}$.

8.5 Unitary, Hermitian, and Skew-Hermitian Matrices

If $\mathbf{U}$ is a nonsingular complex matrix, then $\mathbf{U}^{-1}$ exists and is generally also a complex matrix. We claim that the operations of taking the complex conjugate and of taking a matrix inverse can be performed in either order.

LEMMA 8.2

$$\overline{\mathbf{U}}^{-1} = \overline{\mathbf{U}^{-1}}.$$

Proof We know that the conjugate of a product is the product of the conjugates, so

$$\mathbf{I}_n = \overline{\mathbf{I}_n} = \overline{\mathbf{U}\mathbf{U}^{-1}} = \overline{\mathbf{U}}\,\overline{\mathbf{U}^{-1}}.$$

This implies that $\overline{\mathbf{U}^{-1}}$ is the inverse of $\overline{\mathbf{U}}$. ∎

Now define a matrix to be *unitary* if the inverse of its conjugate (or conjugate of its inverse) is equal to its transpose.

DEFINITION 8.8 Unitary Matrix

An $n \times n$ complex matrix $\mathbf{U}$ is unitary if and only if $\overline{\mathbf{U}}^{-1} = \mathbf{U}^t$.

This condition is equivalent to saying that

$$\overline{\mathbf{U}}\mathbf{U}^t = \mathbf{I}_n.$$

EXAMPLE 8.18

Let

$$\mathbf{U} = \begin{pmatrix} i/\sqrt{2} & 1/\sqrt{2} \\ -i/\sqrt{2} & 1/\sqrt{2} \end{pmatrix}.$$

Then $\mathbf{U}$ is unitary because

$$\overline{\mathbf{U}}\mathbf{U}^t = \begin{pmatrix} -i/\sqrt{2} & 1/\sqrt{2} \\ i/\sqrt{2} & 1/\sqrt{2} \end{pmatrix} \begin{pmatrix} i/\sqrt{2} & -i/\sqrt{2} \\ 1/\sqrt{2} & 1/\sqrt{2} \end{pmatrix}$$

$$= \begin{pmatrix} 1 & 0 \\ 0 & 1 \end{pmatrix}. \quad ∎$$

If $\mathbf{U}$ is a real matrix, then the unitary condition $\overline{\mathbf{U}}\mathbf{U}^t = \mathbf{I}_n$ becomes $\mathbf{U}\mathbf{U}^t = \mathbf{I}_n$, which makes $\mathbf{U}$ an orthogonal matrix. Unitary matrices are the complex analogues of orthogonal matrices. Since the rows (or columns) of an orthogonal matrix form an orthonormal set of vectors, we will develop the complex analogue of the concept of orthonormality.

Recall that for two vectors $(x_1, \ldots, x_n)$ and $(y_1, \ldots, y_n)$ in R^n, we can define the column matrices

$$\mathbf{X} = \begin{pmatrix} x_1 \\ x_2 \\ \vdots \\ x_n \end{pmatrix} \quad \text{and} \quad \mathbf{Y} = \begin{pmatrix} y_1 \\ y_2 \\ \vdots \\ y_n \end{pmatrix}$$

and obtain the dot product $\mathbf{X} \cdot \mathbf{Y}$ as $\mathbf{X}^t \mathbf{Y}$. In particular, this gives the square of the length of $\mathbf{X}$ as

$$\mathbf{X}^t \mathbf{X} = x_1^2 + x_2^2 + \cdots + x_n^2.$$

To generalize this to the complex case, suppose we have complex n-vectors $(z_1, z_2, \ldots, z_n)$ and $(w_1, w_2, \ldots, w_n)$. Form the column matrices

$$\mathbf{Z} = \begin{pmatrix} z_1 \\ z_2 \\ \vdots \\ z_n \end{pmatrix} \quad \text{and} \quad \mathbf{W} = \begin{pmatrix} w_1 \\ w_2 \\ \vdots \\ w_n \end{pmatrix}.$$

It is tempting to define the dot product of these complex vectors as $\mathbf{Z}^t \mathbf{W}$. The problem with this is that then we get

$$\mathbf{Z}^t \mathbf{Z} = z_1^2 + z_2^2 + \cdots + z_n^2$$

and this will in general be complex. We want to interpret the dot product of a vector with itself as the square of its length, and this should be a nonnegative real number. We get around this by defining the dot product of complex $\mathbf{Z}$ and $\mathbf{W}$ to be

$$\mathbf{Z} \cdot \mathbf{W} = \overline{\mathbf{Z}}^t \mathbf{W} = \overline{z_1} w_1 + \overline{z_2} w_2 + \cdots + \overline{z_n} w_n.$$

In this way the dot product of $\mathbf{Z}$ with itself is

$$\overline{\mathbf{Z}}^t \mathbf{Z} = \overline{z_1} z_1 + \overline{z_2} z_2 + \cdots + \overline{z_n} z_n = |z_1|^2 + |z_2|^2 + \cdots + |z_n|^2,$$

a nonnegative real number. With this as background, we will define the complex analogue of an orthonormal set of vectors.

DEFINITION 8.9 *Unitary System of Vectors*

Complex n-vectors $\mathbf{F}_1, \ldots, \mathbf{F}_r$ form a unitary system if $\mathbf{F}_j \cdot \mathbf{F}_k = 0$ for $j \neq k$, and each $\mathbf{F}_j \cdot \mathbf{F}_j = 1$.

If each $\mathbf{F}_j$ has all real components, then this corresponds exactly to an orthonormal set of vectors in R^n. We can now state the analogue of Theorem 8.9 for unitary matrices.

THEOREM 8.14

Let $\mathbf{U}$ be an $n \times n$ complex matrix. Then $\mathbf{U}$ is unitary if and only if its row vectors form a unitary system. ∎

The proof is like that of Theorem 8.9 and is left to the student. It is not difficult to show that $\mathbf{U}$ is also unitary if and only if its column vectors form a unitary system.

EXAMPLE 8.19

Consider again

$$\mathbf{U} = \begin{pmatrix} i/\sqrt{2} & 1/\sqrt{2} \\ -i/\sqrt{2} & 1/\sqrt{2} \end{pmatrix}.$$

The row vectors, written as 2×1 matrices, are

$$\mathbf{F}_1 = \begin{pmatrix} i/\sqrt{2} \\ 1/\sqrt{2} \end{pmatrix} \quad \text{and} \quad \mathbf{F}_2 = \begin{pmatrix} -i/\sqrt{2} \\ 1/\sqrt{2} \end{pmatrix}.$$

Then

$$\mathbf{F}_1 \cdot \mathbf{F}_2 = \begin{pmatrix} -i/\sqrt{2} & 1/\sqrt{2} \end{pmatrix} \begin{pmatrix} -i/\sqrt{2} \\ 1/\sqrt{2} \end{pmatrix} = 0,$$

$$\mathbf{F}_1 \cdot \mathbf{F}_1 = \begin{pmatrix} -i/\sqrt{2} & 1/\sqrt{2} \end{pmatrix} \begin{pmatrix} i/\sqrt{2} \\ 1/\sqrt{2} \end{pmatrix} = 1,$$

and

$$\mathbf{F}_2 \cdot \mathbf{F}_2 = \begin{pmatrix} i/\sqrt{2} & 1/\sqrt{2} \end{pmatrix} \begin{pmatrix} -i/\sqrt{2} \\ 1/\sqrt{2} \end{pmatrix} = 1. \quad \blacksquare$$

We will show that the eigenvalues of a unitary matrix must lie on the unit circle in the complex plane.

THEOREM 8.15

Let λ be an eigenvalue of the unitary matrix $\mathbf{U}$. Then $|\lambda| = 1$.

Proof   Let $\mathbf{E}$ be an eigenvector associated with λ. Then $\mathbf{UE} = \lambda\mathbf{E}$, so $\overline{\mathbf{UE}} = \overline{\lambda}\overline{\mathbf{E}}$. Then

$$(\overline{\mathbf{UE}})^t = \overline{\lambda}\overline{\mathbf{E}}^t,$$

so

$$\overline{\mathbf{E}}^t\overline{\mathbf{U}}^t = \overline{\lambda}\overline{\mathbf{E}}^t.$$

But $\mathbf{U}$ is unitary, so $\overline{\mathbf{U}}^t = \mathbf{U}^{-1}$, and

$$\overline{\mathbf{E}}^t\mathbf{U}^{-1} = \overline{\lambda}\overline{\mathbf{E}}^t.$$

Multiply both sides of this equation on the right by $\mathbf{UE}$ to get

$$\overline{\mathbf{E}}^t\mathbf{E} = \overline{\lambda}\overline{\mathbf{E}}^t\mathbf{UE} = \overline{\lambda}\overline{\mathbf{E}}^t\lambda\mathbf{E} = \overline{\lambda}\lambda\overline{\mathbf{E}}^t\mathbf{E}.$$

Now $\overline{\mathbf{E}}^t\mathbf{E}$ is the dot product of the eigenvector with itself, and so is a positive number. Dividing the last equation by $\overline{\mathbf{E}}^t\mathbf{E}$ gives $\overline{\lambda}\lambda = 1$. But then $|\lambda|^2 = 1$, so $|\lambda| = 1$. $\blacksquare$

We have defined a matrix to be unitary if its transpose is the conjugate of its inverse. A matrix is *hermitian* if its transpose is equal to its conjugate. If the transpose equals the negative of its conjugate, the matrix is called *skew-hermitian*.

DEFINITION 8.10

---

1. *Hermitian Matrix*
   An $n \times n$ complex matrix $\mathbf{H}$ is hermitian if and only if $\overline{\mathbf{H}} = \mathbf{H}^t$.

2. *Skew-Hermitian Matrix*
   An $n \times n$ complex matrix $\mathbf{S}$ is skew-hermitian if and only if $\overline{\mathbf{S}} = -\mathbf{S}^t$.

In the case that $\mathbf{H}$ has real elements, hermitian is the same as symmetric, because in this case $\overline{\mathbf{H}} = \mathbf{H}$.

EXAMPLE 8.20

Let

$$
\mathbf{H} = \begin{pmatrix} 15 & 8i & 6 - 2i \\ -8i & 0 & -4 + i \\ 6 + 2i & -4 - i & -3 \end{pmatrix}.
$$

Then

$$
\overline{\mathbf{H}} = \begin{pmatrix} 15 & -8i & 6 + 2i \\ 8i & 0 & -4 - i \\ 6 - 2i & -4 + i & -3 \end{pmatrix} = \mathbf{H}^t,
$$

so $\mathbf{H}$ is hermitian.

If

$$
\mathbf{S} = \begin{pmatrix} 0 & 8i & 2i \\ 8i & 0 & 4i \\ 2i & 4i & 0 \end{pmatrix}
$$

then $\mathbf{S}$ is skew-hermitian because

$$
\overline{\mathbf{S}} = \begin{pmatrix} 0 & -8i & -2i \\ -8i & 0 & -4i \\ -2i & -4i & 0 \end{pmatrix} = -\mathbf{S}^t. \ \blacksquare
$$

The following theorem says something about quadratic forms with hermitian or skew-hermitian matrices.

THEOREM 8.16

Let

$$
\mathbf{Z} = \begin{pmatrix} z_1 \\ z_2 \\ \vdots \\ z_n \end{pmatrix}
$$

be a complex matrix.

Then,

    1. If $\mathbf{H}$ is hermitian, then $\overline{\mathbf{Z}}^t\mathbf{HZ}$ is real.

    2. If $\mathbf{S}$ is skew-hermitian, then $\overline{\mathbf{Z}}^t\mathbf{SZ}$ is zero or pure imaginary.

Proof For (1), suppose $\mathbf{H}$ is hermitian. Then $\overline{\mathbf{H}}^t = \mathbf{H}$, so

$$\overline{\overline{\mathbf{Z}}^t\mathbf{HZ}} = \overline{\mathbf{Z}}^t\,\overline{\mathbf{HZ}} = \mathbf{Z}^t\overline{\mathbf{HZ}}.$$

But $\overline{\mathbf{Z}}^t\mathbf{HZ}$ is a 1×1 matrix and so equals its own transpose. Continuing from the last equation, we have

$$\mathbf{Z}^t\overline{\mathbf{HZ}} = (\mathbf{Z}^t\overline{\mathbf{HZ}})^t = \overline{\mathbf{Z}}^t\,\overline{\mathbf{H}}^t\,(\mathbf{Z}^t)^t = \overline{\mathbf{Z}}^t\mathbf{HZ}.$$

Therefore,

$$\overline{\overline{\mathbf{Z}}^t\mathbf{HZ}} = \overline{\mathbf{Z}}^t\mathbf{HZ}.$$

Since $\overline{\mathbf{Z}}^t\mathbf{HZ}$ equals its own conjugate, then $\overline{\mathbf{Z}}^t\mathbf{HZ}$ is real.

    To prove (2), suppose $\mathbf{S}$ is skew-hermitian. Then $\overline{\mathbf{S}}^t = -\mathbf{S}$. By an argument like that done in the proof of (1), we get

$$\overline{\overline{\mathbf{Z}}^t\mathbf{SZ}} = -\overline{\mathbf{Z}}^t\mathbf{SZ}.$$

Now write $\overline{\mathbf{Z}}^t\mathbf{SZ} = \alpha + i\beta$. The last equation becomes

$$\alpha - i\beta = -\alpha - i\beta.$$

Then $\alpha = -\alpha$, so $\alpha = 0$ and $\overline{\mathbf{Z}}^t\mathbf{SZ}$ is pure imaginary. ∎

    Using these results on quadratic forms, we can say something about the eigenvalues of hermitian and skew-hermitian matrices.

THEOREM 8.17

    1. The eigenvalues of a hermitian matrix are real.

    2. The eigenvalues of a skew-hermitian matrix are zero or pure imaginary.

Proof For (1), let λ be an eigenvalue of the hermitian matrix $\mathbf{H}$, with associated eigenvector $\mathbf{E}$. By Lemma 8.1,

$$\lambda = \frac{\overline{\mathbf{E}}^t\mathbf{HE}}{\overline{\mathbf{E}}^t\mathbf{E}}.$$

But by (1) of the preceding theorem, the numerator of this quotient is real. The denominator is the square of the length of $\mathbf{E}$ and so is also real. Therefore, λ is real.

    For (2), let λ be an eigenvalue of the skew-hermitian matrix $\mathbf{S}$, with associated eigenvector $\mathbf{E}$. Again by Lemma 8.1,

$$\lambda = \frac{\overline{\mathbf{E}}^t\mathbf{SE}}{\overline{\mathbf{E}}^t\mathbf{E}}.$$

By (2) of the preceding theorem, the numerator of this quotient is either zero or pure imaginary. Since the denominator is a positive real number, then λ is either zero or pure imaginary. ∎

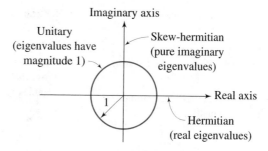

FIGURE 8.3 *Eigenvalue locations.*

Figure 8.3 shows a graphical representation of the conclusions of Theorems 8.15 and 8.17. When plotted as points in the complex plane, eigenvalues of a unitary matrix lie on the unit circle about the origin, eigenvalues of a hermitian matrix lie on the horizontal (real) axis, and eigenvalues of a skew-hermitian matrix lie on the vertical (imaginary) axis.

SECTION 8.5 **PROBLEMS**

In each of Problems 1 through 9, determine whether the matrix is unitary, hermitian, skew-hermitian, or none of these. Find the eigenvalues of each matrix and an associated eigenvector for each eigenvalue. Determine which matrices are diagonalizable. If a matrix is diagonalizable, produce a matrix that diagonalizes it.

1. $\begin{pmatrix} 0 & 2i \\ 2i & 4 \end{pmatrix}$

2. $\begin{pmatrix} 3 & 4i \\ 4i & -5 \end{pmatrix}$

3. $\begin{pmatrix} 0 & 1 & 0 \\ -1 & 0 & 1-i \\ 0 & -1-i & 0 \end{pmatrix}$

4. $\begin{pmatrix} 1/\sqrt{2} & i/\sqrt{2} & 0 \\ -1/\sqrt{2} & i/\sqrt{2} & 0 \\ 0 & 0 & 1 \end{pmatrix}$

5. $\begin{pmatrix} 3 & 2 & 0 \\ 2 & 0 & i \\ 0 & -i & 0 \end{pmatrix}$

6. $\begin{pmatrix} -1 & 0 & 3-i \\ 0 & 1 & 0 \\ 3+i & 0 & 0 \end{pmatrix}$

7. $\begin{pmatrix} i & 1 & 0 \\ -1 & 0 & 2i \\ 0 & 2i & 0 \end{pmatrix}$

8. $\begin{pmatrix} 3i & 0 & 0 \\ -1 & 0 & i \\ 0 & -i & 0 \end{pmatrix}$

9. $\begin{pmatrix} 8 & -1 & i \\ -1 & 0 & 0 \\ -i & 0 & 0 \end{pmatrix}$

10. Let **A** be unitary, hermitian, or skew-hermitian. Prove that $\overline{\mathbf{A}\mathbf{A}^t} = \overline{\mathbf{A}}\mathbf{A}$.

11. Prove that the main diagonal elements of a skew-hermitian matrix must be zero or pure imaginary.

12. Prove that the main diagonal elements of a hermitian matrix must be real.

13. Prove that the product of two unitary matrices is unitary.

14. Let $\mathbf{A}$ be hermitian and unitary. Prove that $\mathbf{A} = \mathbf{A}^{-1}$. Give a 3×3 example of such a matrix, different from $\mathbf{I}_3$.

15. Prove that any square complex matrix is the sum of a hermitian and skew-hermitian matrix.

16. Let $\mathbf{H}$ be hermitian and let α be any real number. Prove that $\alpha\mathbf{H}$ is hermitian. Does this conclusion remain true if α is complex and not real?

17. Let $\mathbf{S}$ be skew-hermitian and let α be any real number. Prove that $\alpha\mathbf{S}$ is skew-hermitian. Does this hold if α is complex and not real?

$$\frac{d}{dt}e^{At} = \frac{d}{dt}\left[I_n + At + \frac{1}{2!}A^2t^2 + \frac{1}{3!}A^3t^3 + \cdots\right] = A + A^2t + \frac{1}{2!}A^3t^2 + \cdots$$

PART 3

Systems of Differential Equations and Qualitative Methods

CHAPTER 9
Systems of Linear Differential Equations

CHAPTER 10
Qualitative Methods and Systems of Nonlinear Differential Equations

We will now use matrices to study systems of differential equations. These arise in modeling mechanical and electrical systems having more than one component, and in many other contexts as well. We will mention two.

Ecologists often use systems of differential equations to analyze the behavior of populations, which might be various kinds of trees in a forest, fish in an ocean, strains of virus in an infection, or people in a country or geographic region. Observations of the populations lead to an assessment of factors influencing population growth or decline, resulting in equations for the rates of change (derivatives) of the populations. Specific examples can be found in predator/prey and competing species population models developed in Chapter 10. The object is to solve the system to obtain functions for the populations, enabling us to better understand the dynamics of the population interactions and changes, and to predict how various factors will influence populations in the future. In many instances public policies in managing resources or treating illnesses or epidemics have been influenced by such studies.

In a different arena, economists have begun to model the behavior of stock markets by systems of differential equations. The ramifications of such analyses may become clear over the next few years.

We will separate our study of systems into two chapters. Chapter 9 is devoted to systems of linear differential equations. For these, powerful matrix methods can be brought to bear to write solutions. For systems of nonlinear differential equations, for which we usually cannot write explicit solutions, we must develop a different set of tools designed to determine qualitative properties of solutions. This is done in Chapter 10.

THEORY OF SYSTEMS OF LINEAR FIRST ORDEF
FERENTIAL EQUATIONS SOLUTION OF $\mathbf{X}' = \mathbf{A}\mathbf{X}$
A IS CONSTANT SOLUTION OF $\mathbf{X}' = \mathbf{A}\mathbf{X} + \mathbf{G}$ T
ORY OF SYSTEMS OF LINEAR FIRST ORDER DII

CHAPTER 9

Systems of Linear Differential Equations

Before beginning to study linear systems, recall from Section 2.6.4 that a linear differential equation of order n always gives rise to a system of n first-order linear differential equations, in such a way that the solution of the system gives the solution of the original nth order equation. Systems can be treated using matrix techniques, which are now at our disposal. For this reason we did not spend time on differential equations of order higher than 2 in Part 1. We will assume familiarity with vectors in R^n, matrix algebra, determinants, and eigenvalues and eigenvectors. These can be reviewed as needed from Part 2.

We begin by laying the foundations for the use of matrices to solve linear systems of differential equations.

9.1 Theory of Systems of Linear First-Order Differential Equations

In this chapter we will consider systems of n first-order linear differential equations in n unknown functions:

$$x_1'(t) = a_{11}(t)x_1(t) + a_{12}(t)x_2(t) + \cdots + a_{1n}(t)x_n(t) + g_1(t)$$
$$x_2'(t) = a_{21}(t)x_1(t) + a_{22}(t)x_2(t) + \cdots + a_{2n}(t)x_n(t) + g_2(t)$$
$$\vdots$$
$$x_n'(t) = a_{n1}(t)x_1(t) + a_{n2}(t)x_2(t) + \cdots + a_{nn}(t)x_n(t) + g_n(t)$$

Let

$$\mathbf{A}(t) = \begin{pmatrix} a_{11}(t) & a_{12}(t) & \cdots & a_{1n}(t) \\ a_{21}(t) & a_{22}(t) & \cdots & a_{2n}(t) \\ \vdots & \vdots & \cdots & \vdots \\ a_{n1}(t) & a_{n2}(t) & \cdots & a_{nn}(t) \end{pmatrix},$$

$$\mathbf{X}(t) = \begin{pmatrix} x_1(t) \\ x_2(t) \\ \vdots \\ x_n(t) \end{pmatrix}, \quad \text{and} \quad \mathbf{G}(t) = \begin{pmatrix} g_1(t) \\ g_2(t) \\ \vdots \\ g_n(t) \end{pmatrix}.$$

Differentiate a matrix by differentiating each element, so

$$\mathbf{X}'(t) = \begin{pmatrix} x_1'(t) \\ x_2'(t) \\ \vdots \\ x_n'(t) \end{pmatrix}.$$

Matrix differentiation follows the "normal" rules we learn in calculus. For example,

$$(\mathbf{X}(t)\mathbf{Y}(t))' = \mathbf{X}'(t)\mathbf{Y}(t) + \mathbf{X}(t)\mathbf{Y}'(t),$$

in which the order of the factors must be maintained.

Now the system of differential equations is

$$\mathbf{X}'(t) = \mathbf{A}(t)\mathbf{X}(t) + \mathbf{G}(t) \tag{9.1}$$

or

$$\mathbf{X}' = \mathbf{A}\mathbf{X} + \mathbf{G}.$$

This system is *nonhomogeneous* if $\mathbf{G}(t) \neq \mathbf{O}$ for at least some t, in which $\mathbf{O}$ denotes the $n \times 1$ zero matrix

$$\begin{pmatrix} 0 \\ 0 \\ \vdots \\ 0 \end{pmatrix}.$$

If $\mathbf{G}(t) = \mathbf{O}$ for all the relevant values of t, then the system is *homogeneous*, and we write just

$$\mathbf{X}' = \mathbf{A}\mathbf{X}.$$

A *solution* of $\mathbf{X}' = \mathbf{A}\mathbf{X} + \mathbf{G}$ is any $n \times 1$ matrix of functions that satisfies this matrix equation.

EXAMPLE 9.1

The 2×2 system

$$x_1' = 3x_1 + 3x_2 + 8$$
$$x_2' = x_1 + 5x_2 + 4e^{3t}$$

can be written

$$\begin{pmatrix} x_1 \\ x_2 \end{pmatrix}' = \begin{pmatrix} 3 & 3 \\ 1 & 5 \end{pmatrix} \begin{pmatrix} x_1 \\ x_2 \end{pmatrix} + \begin{pmatrix} 8 \\ 4e^{3t} \end{pmatrix}.$$

One solution is

$$\mathbf{X}(t) = \begin{pmatrix} 3e^{2t} + e^{6t} - 4e^{3t} - \frac{10}{3} \\ -e^{2t} + e^{6t} + \frac{2}{3} \end{pmatrix},$$

On February 20, 1962, John Glenn became the first American to orbit the Earth. His flight lasted nearly five hours and included three complete circuits of the globe. This and subsequent Mercury orbitings paved the way for the space shuttle program, which now includes shuttles launched from the NASA Kennedy Space Center to carry out experiments under zero gravity, as well as delivery of personnel and equipment to the developing international space station. Ultimate goals of space shuttle missions include studying how humans function in a zero-gravity environment over extended periods of time, scientific observations of phenomena in space and on our own planet, and the commercial development of space. Computation of orbits and forces involved in shuttle missions involves the solution of large systems of differential equations.

as can be verified by substitution into the system. In terms of individual components, this solution is

$$x_1(t) = 3e^{2t} + e^{6t} - 4e^{3t} - \tfrac{10}{3}$$
$$x_2(t) = -e^{2t} + e^{6t} + \tfrac{2}{3}. \quad \blacksquare$$

Initial conditions for system (9.1) have the form

$$\mathbf{X}(t_0) = \begin{pmatrix} x_1(t_0) \\ x_2(t_0) \\ \vdots \\ x_n(t_0) \end{pmatrix} = \mathbf{X}^0,$$

in which $\mathbf{X}^0$ is a given $n \times 1$ matrix of constants. The initial value problem we will consider for systems is the problem:

$$\mathbf{X}' = \mathbf{AX} + \mathbf{G}; \qquad \mathbf{X}(t_0) = \mathbf{X}^0. \tag{9.2}$$

This is analogous to the initial value problem

$$x' = ax + g; \qquad x(t_0) = x_0$$

for single first-order equations. Theorem 1.3 gave criteria for existence and uniqueness of solutions of this initial value problem. The analogous result for the initial value problem (9.2) is given by the following.

THEOREM 9.1   *Existence and Uniqueness*

Let I be an open interval containing t_0. Suppose each $a_{ij}(t)$ and $g_j(t)$ are continuous on I. Let $\mathbf{X}^0$ be a given $n \times 1$ matrix of real numbers. Then the initial value problem

$$\mathbf{X}' = \mathbf{AX} + \mathbf{G}; \qquad \mathbf{X}(t_0) = \mathbf{X}^0$$

has a unique solution defined for all t in I. ∎

EXAMPLE 9.2

Consider the initial value problem

$$x_1' = x_1 + tx_2 + \cos(t)$$
$$x_2' = t^3 x_1 - e^t x_2 + 1 - t$$
$$x_1(0) = 2, \quad x_2(0) = -5.$$

This is the system

$$\mathbf{X}' = \begin{pmatrix} 1 & t \\ t^3 & -e^t \end{pmatrix} \mathbf{X} + \begin{pmatrix} \cos(t) \\ 1 - t \end{pmatrix},$$

with

$$\mathbf{X}^0 = \begin{pmatrix} 2 \\ -5 \end{pmatrix}.$$

This initial value problem has a unique solution defined for all real t, because each $a_{ij}(t)$ and $g_j(t)$ are continuous for all real t. ∎

We will now determine what we must look for to find all solutions of $\mathbf{X}' = \mathbf{AX} + \mathbf{G}$. This will involve a program that closely parallels that for the single first-order equation $x' = ax + g$, beginning with the homogeneous case.

9.1.1   Theory of the Homogeneous System $\mathbf{X}' = \mathbf{AX}$

We begin with the homogeneous system $\mathbf{X}' = \mathbf{AX}$. Because solutions of $\mathbf{X}' = \mathbf{AX}$ are $n \times 1$ matrices of real functions, these solutions have an algebraic structure, and we can form linear combinations (finite sums of scalar multiples of solutions). In the homogeneous case, any linear combination of solutions is again a solution.

THEOREM 9.2

Let $\boldsymbol{\Phi}_1, \ldots, \boldsymbol{\Phi}_k$ be solutions of $\mathbf{X}' = \mathbf{A}\mathbf{X}$, all defined on some open interval I. Let $c_1, \ldots, c_k$ be any real numbers. Then the linear combination $c_1\boldsymbol{\Phi}_1 + \cdots + c_k\boldsymbol{\Phi}_k$ is also a solution of $\mathbf{X}' = \mathbf{A}\mathbf{X}$, defined on I.

Proof Compute

$$(c_1\boldsymbol{\Phi}_1 + \cdots + c_k\boldsymbol{\Phi}_k)' = c_1\boldsymbol{\Phi}_1' + \cdots + c_k\boldsymbol{\Phi}_k' = c_1\mathbf{A}\boldsymbol{\Phi}_1 + \cdots + c_k\mathbf{A}\boldsymbol{\Phi}_k = \mathbf{A}(c_1\boldsymbol{\Phi}_1 + \cdots + c_k\boldsymbol{\Phi}_k). \quad \blacksquare$$

Because of this, the set of all solutions of $\mathbf{X}' = \mathbf{A}\mathbf{X}$ has the structure of a vector space, called the *solution space* of this system. It is not necessary to have a background in vector spaces to follow the discussion of solutions of $\mathbf{X}' = \mathbf{A}\mathbf{X}$ that we are about to develop. However, for those who do have this background, we will make occasional reference to show how ideas fit into this algebraic framework.

In a linear combination $c_1\boldsymbol{\Phi}_1 + \cdots + c_k\boldsymbol{\Phi}_k$ of solutions, any $\boldsymbol{\Phi}_j$ that is already a linear combination of the other solutions is unnecessary. For example, suppose $\boldsymbol{\Phi}_1 = a_2\boldsymbol{\Phi}_2 + \cdots + a_k\boldsymbol{\Phi}_k$. Then

$$c_1\boldsymbol{\Phi}_1 + \cdots + c_k\boldsymbol{\Phi}_k = c_1(a_2\boldsymbol{\Phi}_2 + \cdots + a_k\boldsymbol{\Phi}_k) + c_2\boldsymbol{\Phi}_2 + \cdots + c_k\boldsymbol{\Phi}_k$$

$$= (c_1a_2 + c_2)\boldsymbol{\Phi}_2 + \cdots + (c_1a_k + c_k)\boldsymbol{\Phi}_k.$$

In this case, any linear combination of $\boldsymbol{\Phi}_1, \boldsymbol{\Phi}_2, \ldots, \boldsymbol{\Phi}_k$ is actually a linear combination of just $\boldsymbol{\Phi}_2, \ldots, \boldsymbol{\Phi}_k$, and $\boldsymbol{\Phi}_1$ is not needed. $\boldsymbol{\Phi}_1$ is redundant in the sense that if we have $\boldsymbol{\Phi}_2, \ldots, \boldsymbol{\Phi}_k$, then we have $\boldsymbol{\Phi}_1$ also. We describe this situation by saying that the functions $\boldsymbol{\Phi}_1, \boldsymbol{\Phi}_2, \ldots, \boldsymbol{\Phi}_k$ are linearly dependent. If no one of the functions is a linear combination of the others, then these functions are called *linearly independent*.

DEFINITION 9.1

Linear Dependence
Solutions $\boldsymbol{\Phi}_1, \boldsymbol{\Phi}_2, \ldots, \boldsymbol{\Phi}_k$ of $\mathbf{X}' = \mathbf{A}\mathbf{X}$, defined on an interval I, are linearly dependent on I if one solution is a linear combination of the others on this interval.

Linear Independence
Solutions $\boldsymbol{\Phi}_1, \boldsymbol{\Phi}_2, \ldots, \boldsymbol{\Phi}_k$ of $\mathbf{X}' = \mathbf{A}\mathbf{X}$, defined on an interval I, are linearly independent on I if no solution in this list is a linear combination of the others on this interval.

Thus a set of solutions is linearly independent if it is not linearly dependent.

Linear dependence of functions is a stronger condition than linear dependence of vectors. For vectors in R^n, $\mathbf{V}_1$ is a linear combination of $\mathbf{V}_2$ and $\mathbf{V}_3$ if $\mathbf{V}_1 = a\mathbf{V}_2 + b\mathbf{V}_3$ for some real numbers a and b. In this case, $\mathbf{V}_1, \mathbf{V}_2, \mathbf{V}_3$ are linearly dependent. But for solutions $\boldsymbol{\Phi}_1, \boldsymbol{\Phi}_2, \boldsymbol{\Phi}_3$ of $\mathbf{X}' = \mathbf{A}\mathbf{X}$, $\boldsymbol{\Phi}_1$ is a linear combination of $\boldsymbol{\Phi}_2$ and $\boldsymbol{\Phi}_3$ if there are numbers a and b such that $\boldsymbol{\Phi}_1(t) = a\boldsymbol{\Phi}_2(t) + b\boldsymbol{\Phi}_2(t)$ for all t in the relevant interval, perhaps the entire real line. It is not enough to have this condition hold for just some values of t.

EXAMPLE 9.3

Consider the system

$$\mathbf{X}' = \begin{pmatrix} 1 & -4 \\ 1 & 5 \end{pmatrix} \mathbf{X}.$$

It is routine to check that

$$\mathbf{\Phi}_1(t) = \begin{pmatrix} -2e^{3t} \\ e^{3t} \end{pmatrix} \quad \text{and} \quad \mathbf{\Phi}_2(t) = \begin{pmatrix} (1-2t)e^{3t} \\ te^{3t} \end{pmatrix}$$

are solutions, defined for all real values of t. These solutions are linearly independent on the entire real line, since neither is a constant multiple of the other (for all real t).

The function

$$\mathbf{\Phi}_3(t) = \begin{pmatrix} (11-6t)e^{3t} \\ (-4+3t)e^{3t} \end{pmatrix}$$

is also a solution. However, $\mathbf{\Phi}_1, \mathbf{\Phi}_2, \mathbf{\Phi}_3$ are linearly dependent, because, for all real t,

$$\mathbf{\Phi}_3(t) = -4\mathbf{\Phi}_1(t) + 3\mathbf{\Phi}_2(t).$$

This means that $\mathbf{\Phi}_3$ is a linear combination of $\mathbf{\Phi}_1$ and $\mathbf{\Phi}_2$, and the list of solutions $\mathbf{\Phi}_1, \mathbf{\Phi}_2, \mathbf{\Phi}_3$, although longer, carries no more information about the solution of $\mathbf{X}' = \mathbf{AX}$ than the list of solutions $\mathbf{\Phi}_1, \mathbf{\Phi}_2$. ■

If $\mathbf{\Phi}$ is a solution of $\mathbf{X}' = \mathbf{AX}$, then $\mathbf{\Phi}$ is an $n \times 1$ column matrix of functions:

$$\mathbf{\Phi}(t) = \begin{pmatrix} f_1(t) \\ f_2(t) \\ \vdots \\ f_n(t) \end{pmatrix}.$$

For any choice of t, say $t = t_0$, this is an $n \times 1$ matrix of real numbers that can be thought of as a vector in R^n. This point of view, and some facts about determinants, provides us with a test for linear independence of solutions of $\mathbf{X}' = \mathbf{AX}$.

The following theorem reduces the question of linear independence of n solutions of $\mathbf{X}' = \mathbf{AX}$, to a question of whether an $n \times n$ determinant of real numbers is nonzero.

THEOREM 9.3   *Test for Linear Independence of Solutions*

Suppose that

$$\mathbf{\Phi}_1(t) = \begin{pmatrix} \varphi_{11}(t) \\ \varphi_{21}(t) \\ \vdots \\ \varphi_{n1}(t) \end{pmatrix}, \quad \mathbf{\Phi}_2(t) = \begin{pmatrix} \varphi_{12}(t) \\ \varphi_{22}(t) \\ \vdots \\ \varphi_{n2}(t) \end{pmatrix}, \dots, \quad \mathbf{\Phi}_n(t) = \begin{pmatrix} \varphi_{1n}(t) \\ \varphi_{2n}(t) \\ \vdots \\ \varphi_{nn}(t) \end{pmatrix}$$

are solutions of $\mathbf{X}' = \mathbf{AX}$ on an open interval I. Let t_0 be any number in I. Then

1. $\mathbf{\Phi}_1, \mathbf{\Phi}_2, \dots, \mathbf{\Phi}_n$ are linearly independent on I if and only if $\mathbf{\Phi}_1(t_0), \dots, \mathbf{\Phi}_n(t_0)$ are linearly independent, when considered as vectors in R^n.

2. $\mathbf{\Phi}_1, \mathbf{\Phi}_2, \dots, \mathbf{\Phi}_n$ are linearly independent on I if and only if

$$\begin{vmatrix} \varphi_{11}(t_0) & \varphi_{12}(t_0) & \cdots & \varphi_{1n}(t_0) \\ \varphi_{21}(t_0) & \varphi_{22}(t_0) & \cdots & \varphi_{2n}(t_0) \\ \vdots & \vdots & \cdots & \vdots \\ \varphi_{n1}(t_0) & \varphi_{n2}(t_0) & \cdots & \varphi_{nn}(t_0) \end{vmatrix} \neq 0. \ ■$$

Conclusion (2) is an effective test for linear independence of n solutions of $\mathbf{X}' = \mathbf{AX}$ on an open interval. Evaluate each solution at some point of the interval. Each $\mathbf{\Phi}_j(t_0)$ is an $n \times 1$ (constant) column matrix. Evaluate the determinant of the $n \times n$ matrix having these columns. If this determinant is nonzero, then the solutions are linearly independent; if it is zero, they are linearly dependent.

Another way of looking at (2) of this theorem is that it reduces a question of linear independence of n solutions of $\mathbf{X}' = \mathbf{AX}$ to a question of linear independence of n vectors in R^n. This is because the determinant in (2) is nonzero exactly when its row (or column) vectors are linearly independent.

EXAMPLE 9.4

From the preceding example,

$$\mathbf{\Phi}_1(t) = \begin{pmatrix} -2e^{3t} \\ e^{3t} \end{pmatrix} \quad \text{and} \quad \mathbf{\Phi}_2(t) = \begin{pmatrix} (1 - 2t)e^{3t} \\ te^{3t} \end{pmatrix}$$

are solutions of

$$\mathbf{X}' = \begin{pmatrix} 1 & -4 \\ 1 & 5 \end{pmatrix} \mathbf{X}$$

on the entire real line, which is an open interval. Evaluate these solutions at some convenient point, say $t = 0$:

$$\mathbf{\Phi}_1(0) = \begin{pmatrix} -2 \\ 1 \end{pmatrix} \quad \text{and} \quad \mathbf{\Phi}_2(0) = \begin{pmatrix} 1 \\ 0 \end{pmatrix}.$$

Use these as columns of a 2×2 matrix and evaluate its determinant:

$$\begin{vmatrix} -2 & 1 \\ 1 & 0 \end{vmatrix} = -1 \neq 0.$$

Therefore, $\mathbf{\Phi}_1$ and $\mathbf{\Phi}_2$ are linearly independent solutions. ■

A proof of Theorem 9.3 makes use of the uniqueness of solutions of the initial value problem (Theorem 9.1).

Proof For (1), let t_0 be any point in I. Suppose first that $\mathbf{\Phi}_1, \ldots, \mathbf{\Phi}_n$ are linearly dependent on I. Then one of the solutions is a linear combination of the others. By reordering if necessary, say $\mathbf{\Phi}_1$ is a linear combination of $\mathbf{\Phi}_2, \ldots, \mathbf{\Phi}_n$. Then there are numbers $c_2, \ldots, c_n$ such that

$$\mathbf{\Phi}_1(t) = c_2 \mathbf{\Phi}_2(t) + \cdots + c_n \mathbf{\Phi}_n(t)$$

for all t in I. In particular,

$$\mathbf{\Phi}_1(t_0) = c_2 \mathbf{\Phi}_2(t_0) + \cdots + c_n \mathbf{\Phi}_n(t_0).$$

This implies that the vectors $\mathbf{\Phi}_1(t_0), \ldots, \mathbf{\Phi}_n(t_0)$ are linearly dependent vectors in R^n.

Conversely, suppose that $\mathbf{\Phi}_1(t_0), \ldots, \mathbf{\Phi}_n(t_0)$ are linearly dependent in R^n. Then one of these vectors is a linear combination of the others. Again, as a convenience, suppose $\mathbf{\Phi}_1(t_0)$ is a linear combination of $\mathbf{\Phi}_2(t_0), \ldots, \mathbf{\Phi}_n(t_0)$. Then there are numbers $c_2, \ldots, c_n$ such that

$$\mathbf{\Phi}_1(t_0) = c_2 \mathbf{\Phi}_2(t_0) + \cdots + c_n \mathbf{\Phi}_n(t_0).$$

Define

$$\mathbf{\Phi}(t) = \mathbf{\Phi}_1(t) - c_2 \mathbf{\Phi}_2(t) - \cdots - c_n \mathbf{\Phi}_n(t)$$

for all t in I. Then $\boldsymbol{\Phi}$ is a linear combination of solutions of $\mathbf{X}' = \mathbf{AX}$, hence is a solution. Further,

$$\boldsymbol{\Phi}(t_0) = \begin{pmatrix} 0 \\ 0 \\ \vdots \\ 0 \end{pmatrix}.$$

Therefore, on I, $\boldsymbol{\Phi}$ is a solution of the initial value problem

$$\mathbf{X}' = \mathbf{AX}; \qquad \mathbf{X}(t_0) = \mathbf{O}.$$

But the zero function

$$\boldsymbol{\Psi}(t) = \begin{pmatrix} 0 \\ 0 \\ \vdots \\ 0 \end{pmatrix}$$

is also a solution of this initial value problem. Since this initial value problem has a unique solution, then for all t in I,

$$\boldsymbol{\Phi}(t) = \boldsymbol{\Psi}(t) = \begin{pmatrix} 0 \\ 0 \\ \vdots \\ 0 \end{pmatrix}.$$

Therefore,

$$\boldsymbol{\Phi}(t) = \boldsymbol{\Phi}_1(t) - c_2\boldsymbol{\Phi}_2(t) - \cdots - c_n\boldsymbol{\Phi}_n(t) = \begin{pmatrix} 0 \\ 0 \\ \vdots \\ 0 \end{pmatrix}$$

for all t in I, which means that

$$\boldsymbol{\Phi}_1(t) = c_2\boldsymbol{\Phi}_2(t) + \cdots + c_n\boldsymbol{\Phi}_n(t)$$

for all t in I. Therefore, $\boldsymbol{\Phi}_1$ is a linear combination of $\boldsymbol{\Phi}_2, \ldots, \boldsymbol{\Phi}_n$, hence $\boldsymbol{\Phi}_1, \boldsymbol{\Phi}_2, \ldots, \boldsymbol{\Phi}_n$ are linearly dependent on I.

Conclusion (2) follows from (1) and the fact that n vectors in R^n are linearly independent if and only if the determinant of the $n \times n$ matrix having these vectors as columns is nonzero. ■

Thus far we know how to test n solutions of $\mathbf{X}' = \mathbf{AX}$ for linear independence, if $\mathbf{A}$ is $n \times n$. We will now show that n linearly independent solutions are enough to determine all solutions of $\mathbf{X}' = \mathbf{AX}$ on an open interval I. We saw a result like this previously when it was found that two linear independent solutions of $y'' + p(x)y' + q(x)y = 0$ determine all solutions of this equation.

THEOREM 9.4

Let $\mathbf{A} = [a_{ij}(t)]$ be an $n \times n$ matrix of functions that are continuous on an open interval I. Then

1. The system $\mathbf{X}' = \mathbf{AX}$ has n linearly independent solutions defined on I.

2. Given any n linearly independent solutions $\boldsymbol{\Phi}_1, \ldots, \boldsymbol{\Phi}_n$ defined on I, every solution on I is a linear combination of $\boldsymbol{\Phi}_1, \ldots, \boldsymbol{\Phi}_n$. ■

By (2), every solution of $\mathbf{X}' = \mathbf{AX}$, defined on I, must be of the form $c_1\mathbf{\Phi}_1 + c_2\mathbf{\Phi}_2 + \cdots + c_n\mathbf{\Phi}_n$. For this reason, this linear combination, with $\mathbf{\Phi}_1, \ldots, \mathbf{\Phi}_n$ any n linearly independent solutions, is called the *general solution* of $\mathbf{X}' = \mathbf{AX}$ on I.

Proof To prove that there are n linearly independent solutions, define the $n \times 1$ constant matrices

$$\mathbf{E}^{(1)} = \begin{pmatrix} 1 \\ 0 \\ 0 \\ \vdots \\ 0 \end{pmatrix}, \quad \mathbf{E}^{(2)} = \begin{pmatrix} 0 \\ 1 \\ 0 \\ \vdots \\ 0 \end{pmatrix}, \ldots, \quad \mathbf{E}^{(n)} = \begin{pmatrix} 0 \\ 0 \\ 0 \\ \vdots \\ 1 \end{pmatrix}.$$

Pick any t_0 in I. We know from Theorem 9.1 that the initial value problem

$$\mathbf{X}' = \mathbf{AX}; \qquad \mathbf{X}(t_0) = \mathbf{E}^{(j)}$$

has a unique solution $\mathbf{\Phi}_j$ defined on I, for $j = 1, 2, \ldots, n$. These solutions are linearly independent by Theorem 9.3, because the way the initial conditions were chosen, the $n \times n$ matrix whose columns are these solutions evaluated at t_0 is $\mathbf{I}_n$, with determinant 1. This proves (1).

To prove (2), suppose now that $\mathbf{\Psi}_1, \ldots, \mathbf{\Psi}_n$ are any n linearly independent solutions of $\mathbf{X}' = \mathbf{AX}$, defined on I. Let $\mathbf{\Lambda}$ be any solution. We want to prove that $\mathbf{\Lambda}$ is a linear combination of $\mathbf{\Psi}_1, \ldots, \mathbf{\Psi}_n$.

Pick any t_0 in I. We will first show that there are numbers $c_1, \ldots, c_n$ such that

$$\mathbf{\Lambda}(t_0) = c_1\mathbf{\Psi}_1(t_0) + \cdots + c_n\mathbf{\Psi}_n(t_0).$$

Now $\mathbf{\Lambda}(t_0)$, and each $\mathbf{\Psi}_j(t_0)$, is an $n \times 1$ column matrix of constants. Form the $n \times n$ matrix $\mathbf{S}$, using $\mathbf{\Psi}_1(t_0), \ldots, \mathbf{\Psi}_n(t_0)$ as its columns, and consider the system of n linear algebraic equations in n unknowns

$$\mathbf{S} \begin{pmatrix} c_1 \\ c_2 \\ \vdots \\ c_n \end{pmatrix} = \mathbf{\Lambda}(t_0). \tag{9.3}$$

The columns of $\mathbf{S}$ are linearly independent vectors in R^n, because $\mathbf{\Psi}_1, \ldots, \mathbf{\Psi}_n$ are linearly independent. Therefore, $\mathbf{S}$ is nonsingular, and the system (9.3) has a unique solution. This solution gives constants $c_1, \ldots, c_n$ such that

$$\mathbf{\Lambda}(t_0) = c_1\mathbf{\Psi}_1(t_0) + \cdots + c_n\mathbf{\Psi}_n(t_0).$$

We now claim that

$$\mathbf{\Lambda}(t) = c_1\mathbf{\Psi}_1(t) + \cdots + c_n\mathbf{\Psi}_n(t)$$

for all t in I. But observe that $\mathbf{\Lambda}$ and $c_1\mathbf{\Psi}_1 + \cdots + c_n\mathbf{\Psi}_n$ are both solutions of the initial value problem

$$\mathbf{X}' = \mathbf{AX}; \qquad \mathbf{X}(t_0) = \mathbf{\Lambda}(t_0).$$

Since this problem has a unique solution, then $\mathbf{\Lambda}(t) = c_1\mathbf{\Psi}_1(t) + \cdots + c_n\mathbf{\Psi}_n(t)$ for all t in I, and the proof is complete. ∎

In the language of linear algebra, the solution space of $\mathbf{X}' = \mathbf{AX}$ has dimension n, the order of the coefficient matrix $\mathbf{A}$. Any n linearly independent solutions form a basis for this vector space.

EXAMPLE 9.5

Previously we saw that

$$\Phi_1(t) = \begin{pmatrix} -2e^{3t} \\ e^{3t} \end{pmatrix} \quad \text{and} \quad \Phi_2(t) = \begin{pmatrix} (1-2t)e^{3t} \\ te^{3t} \end{pmatrix}$$

are linearly independent solutions of

$$X' = \begin{pmatrix} 1 & -4 \\ 1 & 5 \end{pmatrix} X.$$

Because $\mathbf{A}$ is 2×2 and we have two linearly independent solutions, the general solution of this system is

$$\Phi(t) = c_1 \begin{pmatrix} -2e^{3t} \\ e^{3t} \end{pmatrix} + c_2 \begin{pmatrix} (1-2t)e^{3t} \\ te^{3t} \end{pmatrix}.$$

The expression on the right contains every solution of this system. In terms of components,

$$x_1(t) = -2c_1e^{3t} + c_2(1-2t)e^{3t},$$

$$x_2(t) = c_1e^{3t} + c_2te^{3t}. \quad \blacksquare$$

We will now make a useful observation. In the last example, form a 2×2 matrix $\boldsymbol{\Omega}(t)$ having $\Phi_1(t)$ and $\Phi_2(t)$ as columns:

$$\boldsymbol{\Omega}(t) = \begin{pmatrix} -2e^{3t} & (1-2t)e^{3t} \\ e^{3t} & te^{3t} \end{pmatrix}.$$

Now observe that if $\mathbf{C} = \begin{pmatrix} c_1 \\ c_2 \end{pmatrix}$, then

$$\boldsymbol{\Omega}(t)\mathbf{C} = \begin{pmatrix} -2e^{3t} & (1-2t)e^{3t} \\ e^{3t} & te^{3t} \end{pmatrix} \begin{pmatrix} c_1 \\ c_2 \end{pmatrix}$$

$$= \begin{pmatrix} c_1[-2e^{3t}] + c_2[(1-2t)e^{3t}] \\ c_1[e^{3t}] + c_2[te^{3t}] \end{pmatrix}$$

$$= c_1 \begin{pmatrix} -2e^{3t} \\ e^{3t} \end{pmatrix} + c_2 \begin{pmatrix} (1-2t)e^{3t} \\ te^{3t} \end{pmatrix}$$

$$= c_1 \Phi_1(t) + c_2 \Phi_2(t).$$

The point is that we can write the general solution $c_1\Phi_1 + c_2\Phi_2$ compactly as $\boldsymbol{\Omega}(t)\mathbf{C}$, with $\boldsymbol{\Omega}(t)$ a square matrix having the independent solutions as columns and $\mathbf{C}$ a column matrix of arbitrary constants. We call a matrix $\boldsymbol{\Omega}$ formed in this way a *fundamental matrix* for the system $\mathbf{X}' = \mathbf{A}\mathbf{X}$. In terms of this fundamental matrix, the general solution is $\mathbf{X}(t) = \boldsymbol{\Omega}(t)\mathbf{C}$.

We can see that $\boldsymbol{\Omega}(t)\mathbf{C}$ also satisfies the matrix differential equation $\mathbf{X}' = \mathbf{A}\mathbf{X}$. Recall that we differentiate a matrix by differentiating each element of the matrix. Then, because $\mathbf{C}$ is a constant matrix,

$$(\boldsymbol{\Omega}(t)\mathbf{C})' = \boldsymbol{\Omega}(t)'\mathbf{C} = \begin{pmatrix} -6e^{3t} & e^{3t} - 6te^{3t} \\ 3e^{3t} & e^{3t} + 3te^{3t} \end{pmatrix} \mathbf{C}.$$

Now compute

$$\mathbf{A}(\mathbf{\Omega}(t)\mathbf{C}) = (\mathbf{A}\mathbf{\Omega}(t))\mathbf{C}$$

$$= \begin{pmatrix} 1 & -4 \\ 1 & 5 \end{pmatrix} \begin{pmatrix} -2e^{3t} & (1-2t)e^{3t} \\ e^{3t} & te^{3t} \end{pmatrix} \mathbf{C}$$

$$= \begin{pmatrix} -2e^{3t} - 4e^{3t} & (1-2t)e^{3t} - 4te^{3t} \\ -2e^{3t} + 5e^{3t} & (1-2t)e^{3t} + 5te^{3t} \end{pmatrix} \mathbf{C}$$

$$= \begin{pmatrix} -6e^{3t} & (1-6t)e^{3t} \\ 3e^{3t} & e^{3t} + 3te^{3t} \end{pmatrix} \mathbf{C}.$$

Therefore,

$$(\mathbf{\Omega}(t)\mathbf{C})' = \mathbf{A}(\mathbf{\Omega}(t)\mathbf{C}),$$

as occurs if $\mathbf{\Omega}(t)\mathbf{C}$ is a solution of $\mathbf{X}' = \mathbf{A}\mathbf{X}$.

DEFINITION 9.2

$\mathbf{\Omega}$ is a fundamental matrix for the $n \times n$ system $\mathbf{X}' = \mathbf{A}\mathbf{X}$ if the columns of $\mathbf{\Omega}$ are linearly independent solutions of this system.

Writing the general solution of $\mathbf{X}' = \mathbf{A}\mathbf{X}$ as $\mathbf{X}(t) = \mathbf{\Omega}\mathbf{C}$ is particularly convenient for solving initial value problems.

EXAMPLE 9.6

Solve the initial value problem

$$\mathbf{X}' = \begin{pmatrix} 1 & -4 \\ 1 & 5 \end{pmatrix}; \qquad \mathbf{X}(0) = \begin{pmatrix} -2 \\ 3 \end{pmatrix}.$$

We know from Example 9.5 that the general solution if $\mathbf{X}(t) = \mathbf{\Omega}(t)\mathbf{C}$, where

$$\mathbf{\Omega}(t) = \begin{pmatrix} -2e^{3t} & (1-2t)e^{3t} \\ e^{3t} & te^{3t} \end{pmatrix}.$$

We need to choose $\mathbf{C}$ so that

$$\mathbf{X}(0) = \mathbf{\Omega}(0)\mathbf{C} = \begin{pmatrix} -2 \\ 3 \end{pmatrix}.$$

Putting $t = 0$ into $\mathbf{\Omega}$, we must solve the algebraic system

$$\begin{pmatrix} -2 & 1 \\ 1 & 0 \end{pmatrix} \mathbf{C} = \begin{pmatrix} -2 \\ 3 \end{pmatrix}.$$

The solution is

$$\mathbf{C} = \begin{pmatrix} -2 & 1 \\ 1 & 0 \end{pmatrix}^{-1} \begin{pmatrix} -2 \\ 3 \end{pmatrix}$$

$$= \begin{pmatrix} 0 & 1 \\ 1 & 2 \end{pmatrix} \begin{pmatrix} -2 \\ 3 \end{pmatrix} = \begin{pmatrix} 3 \\ 4 \end{pmatrix}.$$

The unique solution of the initial value problem is therefore

$$\mathbf{\Phi}(t) = \mathbf{\Omega}(t) \begin{pmatrix} 3 \\ 4 \end{pmatrix}$$

$$= \begin{pmatrix} -2e^{-3t} - 8te^{3t} \\ 3e^{3t} + 4te^{3t} \end{pmatrix}. \quad \blacksquare$$

In this example, $\mathbf{\Omega}(0)^{-1}$ could be found by linear algebra methods (Sections 6.9.1 and 7.8) or by using a software package.

9.1.2 General Solution of the Nonhomogeneous System $\mathbf{X}' = \mathbf{A}\mathbf{X} + \mathbf{G}$

Solutions of the nonhomogeneous system $\mathbf{X}' = \mathbf{A}\mathbf{X} + \mathbf{G}$ do not have the algebraic structure of a vector space, because linear combinations of solutions are not solutions. However, we will show that the general solution of this system (an expression containing all possible solutions) is the sum of the general solution of the homogeneous system $\mathbf{X}' = \mathbf{A}\mathbf{X}$ and any particular solution of the nonhomogeneous system. This is completely analogous to Theorem 2.5 for the second-order equation $y'' + p(x)y' + q(x)y = f(x)$.

THEOREM 9.5

Let $\mathbf{\Omega}$ be a fundamental matrix for $\mathbf{X}' = \mathbf{A}\mathbf{X}$ and let $\mathbf{\Psi}_p$ be any solution of $\mathbf{X}' = \mathbf{A}\mathbf{X} + \mathbf{G}$. Then the general solution of $\mathbf{X}' = \mathbf{A}\mathbf{X} + \mathbf{G}$ is $\mathbf{X} = \mathbf{\Omega}\mathbf{C} + \mathbf{\Psi}_p$, in which $\mathbf{C}$ is an $n \times 1$ matrix of arbitrary constants.

Proof   First, $\mathbf{\Omega}\mathbf{C} + \mathbf{\Psi}_p$ is a solution of the nonhomogeneous system, because

$$(\mathbf{\Omega}\mathbf{C} + \mathbf{\Psi}_p)' = (\mathbf{\Omega}\mathbf{C})' + \mathbf{\Psi}_p'$$

$$= \mathbf{A}(\mathbf{\Omega}\mathbf{C}) + \mathbf{A}\mathbf{\Psi}_p + \mathbf{G} = \mathbf{A}(\mathbf{\Omega}\mathbf{C} + \mathbf{\Psi}_p) + \mathbf{G}.$$

Now let $\mathbf{\Phi}$ be any solution of $\mathbf{X}' = \mathbf{A}\mathbf{X} + \mathbf{G}$. We claim that $\mathbf{\Phi} - \mathbf{\Psi}_p$ is a solution of $\mathbf{X}' = \mathbf{A}\mathbf{X}$. To see this, calculate

$$(\mathbf{\Phi} - \mathbf{\Psi}_p)' = \mathbf{\Phi}' - \mathbf{\Psi}_p'$$

$$= \mathbf{A}\mathbf{\Phi} + \mathbf{G} - (\mathbf{A}\mathbf{\Psi}_p + \mathbf{G}) = \mathbf{A}(\mathbf{\Phi} - \mathbf{\Psi}_p).$$

Since $\mathbf{\Omega}\mathbf{C}$ is the general solution of $\mathbf{X}' = \mathbf{A}\mathbf{X}$, there is a constant $n \times 1$ matrix $\mathbf{K}$ such that $\mathbf{\Phi} - \mathbf{\Psi}_p = \mathbf{\Omega}\mathbf{K}$. Then, $\mathbf{\Phi} = \mathbf{\Omega}\mathbf{K} + \mathbf{\Psi}_p$, completing the proof. $\blacksquare$

We now know what to look for in solving a system of n linear, first-order differential equations in n unknown functions. For the homogeneous system, $\mathbf{X}' = \mathbf{A}\mathbf{X}$, we look for n linearly independent solutions to form a fundamental matrix $\mathbf{\Omega}(t)$. For the nonhomogeneous system $\mathbf{X}' = \mathbf{A}\mathbf{X} + \mathbf{G}$, we first find the general solution $\mathbf{\Omega}\mathbf{C}$ of $\mathbf{X}' = \mathbf{A}\mathbf{X}$ and any particular solution $\mathbf{\Psi}_p$ of $\mathbf{X}' = \mathbf{A}\mathbf{X} + \mathbf{G}$. The general solution of $\mathbf{X}' = \mathbf{A}\mathbf{X} + \mathbf{G}$ is then $\mathbf{\Omega}\mathbf{C} + \mathbf{\Psi}_p$.

This is an overall strategy. Now we need ways of implementing it and actually producing fundamental matrices and particular solutions for given systems.

SECTION 9.1 PROBLEMS

In each of Problems 1 through 5, (a) verify that the given functions satisfy the system; (b) form a fundamental matrix $\Omega(t)$ for the system; (c) write the general solution in the form $\Omega(t)\mathbf{C}$, carry out this product, and verify that the rows of $\Omega(t)\mathbf{C}$ are the components of the given solution; and (d) find the unique solution satisfying the initial conditions.

1. $x_1' = 5x_1 + 3x_2, x_2' = x_1 + 3x_2$

   $x_1(t) = -c_1 e^{2t} + 3c_2 e^{6t}, x_2(t) = c_1 e^{2t} + c_2 e^{6t}$

   $x_1(0) = 0, x_2(0) = 4$

2. $x_1' = 2x_1 + x_2, x_2' = -3x_1 + 6x_2$

   $x_1(t) = c_1 e^{4t} \cos(t) + c_2 e^{4t} \sin(t),$

   $x_2(t) = 2c_1 e^{4t}[\cos(t) - \sin(t)]$

   $\qquad + 2c_2 e^{4t}[\cos(t) + \sin(t)]$

   $x_1(0) = -2, x_2(0) = 1$

3. $x_1'(t) = 3x_1 + 8x_2, x_2'(t) = x_1 - x_2$

   $x_1(t) = 4c_1 e^{(1+2\sqrt{3})t} + 4c_2 e^{(1-2\sqrt{3})t},$

   $x_2(t) = (-1 + \sqrt{3})c_1 e^{(1+2\sqrt{3})t}$

   $\qquad + (-1 - \sqrt{3})c_2 e^{(1-2\sqrt{3})t}$

   $x_1(0) = 2, x_2(0) = 2$

4. $x_1' = x_1 - x_2, x_2' = 4x_1 + 2x_2$

   $$x_1(t) = 2e^{3t/2}\left[c_1 \cos\left(\frac{\sqrt{15}t}{2}\right) + c_2 \sin\left(\frac{\sqrt{15}t}{2}\right)\right],$$

   $$x_2(t) = c_1 e^{3t/2}\left[-\cos\left(\frac{\sqrt{15}t}{2}\right) + \sqrt{15}\sin\left(\frac{\sqrt{15}t}{2}\right)\right]$$

   $$\qquad - c_2 e^{3t/2}\left[\sin\left(\frac{\sqrt{15}t}{2}\right) + \sqrt{15}\cos\left(\frac{\sqrt{15}t}{2}\right)\right],$$

   $x_1(0) = -2, x_2(0) = 7$

5. $x_1' = 5x_1 - 4x_2 + 4x_3, x_2' = 12x_1 - 11x_2 + 12x_3,$

   $x_3' = 4x_1 - 4x_2 + 5x_3,$

   $x_1(t) = c_1 e^t + c_3 e^{-3t}, x_2(t) = c_2 e^{2t} + 3c_3 e^{-3t},$

   $x_3(t) = (c_2 - c_1)e^t + c_3 e^{-3t},$

   $x_1(0) = 1, x_2(0) = -3, x_3(0) = 5$

9.2 Solution of X′ = AX When A Is Constant

Consider the system $\mathbf{X}' = \mathbf{AX}$, with $\mathbf{A}$ an $n \times n$ matrix of real numbers. In the case $y' = ay$, with a constant, we get exponential solutions $y = ce^{ax}$. This suggests we try a similar solution for the system.

Try $\mathbf{X} = \boldsymbol{\xi}e^{\lambda t}$, with $\boldsymbol{\xi}$ an $n \times 1$ matrix of constants to be determined and λ a number to be determined. Substitute this proposed solution into the differential equation to get

$$\boldsymbol{\xi}\lambda e^{\lambda t} = \mathbf{A}(\boldsymbol{\xi}e^{\lambda t}).$$

This requires that

$$\mathbf{A}\boldsymbol{\xi} = \lambda\boldsymbol{\xi}.$$

We should therefore choose λ as an eigenvalue of $\mathbf{A}$, and $\boldsymbol{\xi}$ as an associated eigenvector.

We will summarize this discussion.

THEOREM 9.6

Let $\mathbf{A}$ be an $n \times n$ matrix of real numbers. Then $\boldsymbol{\xi}e^{\lambda t}$ is a nontrivial solution of $\mathbf{X}' = \mathbf{AX}$ if and only if λ is an eigenvalue of $\mathbf{A}$, with associated eigenvector $\boldsymbol{\xi}$. ∎

We need n linearly independent solutions to form a fundamental matrix. We will have these if we can find n linearly independent eigenvectors, whether or not some eigenvalues may be repeated.

THEOREM 9.7

Let $\mathbf{A}$ be an $n \times n$ matrix of real numbers. Suppose $\mathbf{A}$ has eigenvalues $\lambda_1, \ldots, \lambda_n$, and suppose there are associated eigenvectors $\boldsymbol{\xi}_1, \ldots, \boldsymbol{\xi}_n$ that are linearly independent. Then $\boldsymbol{\xi}_1 e^{\lambda_1 t}, \ldots, \boldsymbol{\xi}_n e^{\lambda_n t}$ are linearly independent solutions of $\mathbf{X}' = \mathbf{A}\mathbf{X}$, on the entire real line.

Proof   We know that each $\boldsymbol{\xi}_j e^{\lambda_j t}$ is a nontrivial solution. The question is whether these solutions are linearly independent. Form the $n \times n$ matrix having these solutions, evaluated at $t = 0$, as its columns. This matrix has n linearly independent columns $\boldsymbol{\xi}_1, \ldots, \boldsymbol{\xi}_n$ and therefore has a nonzero determinant. By Theorem 9.3(2), $\boldsymbol{\xi}_1 e^{\lambda_1 t}, \ldots, \boldsymbol{\xi}_n e^{\lambda_n t}$ are linearly independent on the real line. ∎

EXAMPLE 9.7

Consider the system

$$\mathbf{X}' = \begin{pmatrix} 4 & 2 \\ 3 & 3 \end{pmatrix} \mathbf{X}.$$

$\mathbf{A}$ has eigenvalues 1 and 6, with corresponding eigenvectors $\begin{pmatrix} 1 \\ -\frac{3}{2} \end{pmatrix}$ and $\begin{pmatrix} 1 \\ 1 \end{pmatrix}$. These eigenvectors are linearly independent (originating from distinct eigenvalues), so we have two linearly independent solutions,

$$\begin{pmatrix} 1 \\ -\frac{3}{2} \end{pmatrix} e^t \quad \text{and} \quad \begin{pmatrix} 1 \\ 1 \end{pmatrix} e^{6t}.$$

We can write the general solution as

$$\mathbf{X}(t) = c_1 \begin{pmatrix} 1 \\ -\frac{3}{2} \end{pmatrix} e^t + c_2 \begin{pmatrix} 1 \\ 1 \end{pmatrix} e^{6t}.$$

Equivalently, we can write the fundamental matrix

$$\boldsymbol{\Omega}(t) = \begin{pmatrix} e^t & e^{6t} \\ -\frac{3}{2} e^t & e^{6t} \end{pmatrix}.$$

In terms of $\boldsymbol{\Omega}$, the general solution is $\mathbf{X}(t) = \boldsymbol{\Omega}(t)\mathbf{C}$.

In terms of components,

$$x_1(t) = c_1 e^t + c_2 e^{6t}$$

$$x_2(t) = -\tfrac{3}{2} c_1 e^t + c_2 e^{6t}. \quad \blacksquare$$

EXAMPLE 9.8

Solve the system

$$\mathbf{X}' = \begin{pmatrix} 5 & -4 & 4 \\ 12 & -11 & 12 \\ 4 & -4 & 5 \end{pmatrix} \mathbf{X}.$$

The eigenvalues of $\mathbf{A}$ are $-3, 1, 1$. Even though one eigenvalue is repeated, $\mathbf{A}$ has three linearly independent eigenvectors. They are

$$\begin{pmatrix} 1 \\ 3 \\ 1 \end{pmatrix} \quad \text{associated with eigenvalue } -3$$

and

$$\begin{pmatrix} 1 \\ 1 \\ 0 \end{pmatrix} \quad \text{and} \quad \begin{pmatrix} -1 \\ 0 \\ 1 \end{pmatrix} \quad \text{associated with 1.}$$

This gives us three linearly independent solutions

$$\begin{pmatrix} 1 \\ 3 \\ 1 \end{pmatrix} e^{-3t}, \quad \begin{pmatrix} 1 \\ 1 \\ 0 \end{pmatrix} e^{t}, \quad \begin{pmatrix} -1 \\ 0 \\ 1 \end{pmatrix} e^{t}.$$

A fundamental matrix is

$$\mathbf{\Omega}(t) = \begin{pmatrix} e^{-3t} & e^{t} & -e^{t} \\ 3e^{-3t} & e^{t} & 0 \\ e^{-3t} & 0 & e^{t} \end{pmatrix}.$$

The general solution is $\mathbf{X}(t) = \mathbf{\Omega}(t)\mathbf{C}$. ∎

EXAMPLE 9.9 A Mixing Problem

Two tanks are connected by a series of pipes, as shown in Figure 9.1. Tank 1 initially contains 20 liters of water in which 150 grams of chlorine are dissolved. Tank 2 initially contains 50 grams of chlorine dissolved in 10 liters of water.

Beginning at time $t = 0$, pure water is pumped into tank 1 at a rate of 3 liters per minute, while chlorine/water solutions are interchanged between the tanks and also flow out of both tanks at the rates shown. The problem is to determine the amount of chlorine in each tank at any time $t > 0$.

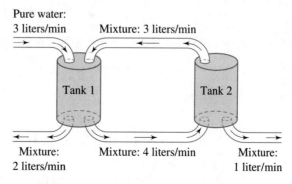

FIGURE 9.1

At the given rates of input and discharge of solutions, the amount of solution in each tank will remain constant. Therefore, the ratio of chlorine to chlorine/water solution in each tank should, in the long run, approach that of the input, which is pure water. We will use this observation as a check of the analysis we are about to do.

Let $x_j(t)$ be the number of grams of chlorine in tank j at time t. Reading from Figure 9.1,

rate of change of $x_1(t) = x_1'(t) =$ rate in minus rate out

$$= 3\left(\frac{\text{liter}}{\text{min}}\right) \cdot 0\left(\frac{\text{gram}}{\text{liter}}\right) + 3\left(\frac{\text{liter}}{\text{min}}\right) \cdot \frac{x_2}{10}\left(\frac{\text{gram}}{\text{liter}}\right)$$

$$- 2\left(\frac{\text{liter}}{\text{min}}\right) \cdot \frac{x_1}{20}\left(\frac{\text{gram}}{\text{liter}}\right) - 4\left(\frac{\text{liter}}{\text{min}}\right) \cdot \frac{x_1}{20}\left(\frac{\text{gram}}{\text{liter}}\right)$$

$$= -\frac{6}{20}x_1 + \frac{3}{10}x_2.$$

Similarly, with the dimensions excluded,

$$x_2'(t) = 4\frac{x_1}{20} - 3\frac{x_2}{10} - \frac{x_2}{10} = \frac{4}{20}x_1 - \frac{4}{10}x_2.$$

The system is $\mathbf{X}' = \mathbf{A}\mathbf{X}$, with

$$\mathbf{A} = \begin{pmatrix} -\frac{3}{10} & \frac{3}{10} \\ \frac{1}{5} & -\frac{2}{5} \end{pmatrix}.$$

The initial conditions are

$$x_1(0) = 150, \quad x_2(0) = 50,$$

or

$$\mathbf{X}(0) = \begin{pmatrix} 150 \\ 50 \end{pmatrix}.$$

The eigenvalues of $\mathbf{A}$ are $-\frac{1}{10}$ and $-\frac{3}{5}$, and corresponding eigenvectors are, respectively,

$$\begin{pmatrix} \frac{3}{2} \\ 1 \end{pmatrix} \quad \text{and} \quad \begin{pmatrix} -1 \\ 1 \end{pmatrix}.$$

These are linearly independent, and we can write the fundamental matrix

$$\mathbf{\Omega}(t) = \begin{pmatrix} \frac{3}{2}e^{-t/10} & -e^{-3t/5} \\ e^{-t/10} & e^{-3t/5} \end{pmatrix}.$$

The general solution is $\mathbf{X}(t) = \mathbf{\Omega}(t)\mathbf{C}$. To solve the initial value problem, we must find $\mathbf{C}$ so that

$$\mathbf{X}(0) = \begin{pmatrix} 150 \\ 50 \end{pmatrix} = \mathbf{\Omega}(0)\mathbf{C} = \begin{pmatrix} \frac{3}{2} & -1 \\ 1 & 1 \end{pmatrix}\mathbf{C}.$$

Then

$$\mathbf{C} = \begin{pmatrix} \frac{3}{2} & -1 \\ 1 & 1 \end{pmatrix}^{-1} \begin{pmatrix} 150 \\ 50 \end{pmatrix}$$

$$= \begin{pmatrix} \frac{2}{5} & \frac{2}{5} \\ -\frac{2}{5} & \frac{3}{5} \end{pmatrix} \begin{pmatrix} 150 \\ 50 \end{pmatrix} = \begin{pmatrix} 80 \\ -30 \end{pmatrix}.$$

The solution of the initial value problem is

$$\mathbf{X}(t) = \begin{pmatrix} \frac{3}{2}e^{-t/10} & -e^{-3t/5} \\ e^{-t/10} & e^{-3t/5} \end{pmatrix} \begin{pmatrix} 80 \\ -30 \end{pmatrix}$$

$$= \begin{pmatrix} 120e^{-t/10} + 30e^{-3t/5} \\ 80e^{-t/10} - 30e^{-3t/5} \end{pmatrix}.$$

Notice that $x_1(t) \to 0$ and $x_2(t) \to 0$ as $t \to \infty$, as we expected. ∎

9.2.1 Solution of X′ = AX When A Has Complex Eigenvalues

Consider a system $\mathbf{X}' = \mathbf{AX}$. If $\mathbf{A}$ is a real matrix, the characteristic polynomial of $\mathbf{A}$ has real coefficients. It may, however, have some complex roots. Suppose $\lambda = \alpha + i\beta$ is a complex eigenvalue, with eigenvector $\boldsymbol{\xi}$. Then $\mathbf{A}\boldsymbol{\xi} = \lambda\boldsymbol{\xi}$, so

$$\overline{\mathbf{A}\boldsymbol{\xi}} = \overline{\lambda}\,\overline{\boldsymbol{\xi}}.$$

But $\overline{\mathbf{A}} = \mathbf{A}$ if $\mathbf{A}$ has real elements, so

$$\mathbf{A}\overline{\boldsymbol{\xi}} = \overline{\lambda}\,\overline{\boldsymbol{\xi}}.$$

This means that $\overline{\lambda} = \alpha - i\beta$ is also an eigenvalue, with eigenvector $\overline{\boldsymbol{\xi}}$. This means that $\boldsymbol{\xi}e^{\lambda t}$ and $\overline{\boldsymbol{\xi}}e^{\overline{\lambda}t}$ can be used as two of the n linearly independent solutions needed to form a fundamental matrix.

This resulting fundamental matrix will contain some complex entries. There is nothing wrong with this. However, sometimes it is convenient to have a fundamental matrix with only real entries. We will show how to replace these two columns involving complex numbers with two other linearly independent solutions involving only real quantities. This can be done for any pair of columns arising from a pair of complex conjugate eigenvalues.

THEOREM 9.8

Let $\mathbf{A}$ be an $n \times n$ real matrix. Let $\alpha + i\beta$ be a complex eigenvalue with corresponding eigenvector $\mathbf{U} + i\mathbf{V}$, in which $\mathbf{U}$ and $\mathbf{V}$ are real $n \times 1$ matrices. Then

$$e^{\alpha t}[\mathbf{U}\cos(\beta t) - \mathbf{V}\sin(\beta t)]$$

and

$$e^{\alpha t}[\mathbf{U}\sin(\beta t) + \mathbf{V}\cos(\beta t)]$$

are real, linearly independent solutions of $\mathbf{X}' = \mathbf{AX}$. ∎

EXAMPLE 9.10

Solve the system $\mathbf{X}' = \mathbf{AX}$, with

$$\mathbf{A} = \begin{pmatrix} 2 & 0 & 1 \\ 0 & -2 & -2 \\ 0 & 2 & 0 \end{pmatrix}.$$

The eigenvalues are $2, -1 + \sqrt{3}i, -1 - \sqrt{3}i$. Corresponding eigenvectors are, respectively,

$$\begin{pmatrix} 1 \\ 0 \\ 0 \end{pmatrix}, \begin{pmatrix} 1 \\ -2\sqrt{3}i \\ -3 + \sqrt{3}i \end{pmatrix}, \begin{pmatrix} 1 \\ 2\sqrt{3}i \\ -3 - \sqrt{3}i \end{pmatrix}.$$

One solution is

$$\begin{pmatrix} 1 \\ 0 \\ 0 \end{pmatrix} e^{2t}$$

and two other solutions are

$$\begin{pmatrix} 1 \\ -2\sqrt{3}i \\ -3 + \sqrt{3}i \end{pmatrix} e^{(-1+\sqrt{3}i)t} \quad \text{and} \quad \begin{pmatrix} 1 \\ 2\sqrt{3}i \\ -3 - \sqrt{3}i \end{pmatrix} e^{(-1-\sqrt{3}i)t}.$$

These three solutions are linearly independent and can be used as columns of a fundamental matrix

$$\mathbf{\Omega}_1(t) = \begin{pmatrix} e^{2t} & e^{(-1+\sqrt{3}i)t} & e^{(-1-\sqrt{3}i)t} \\ 0 & -2\sqrt{3}ie^{(-1+\sqrt{3}i)t} & 2\sqrt{3}ie^{(-1-\sqrt{3}i)t} \\ 0 & (-3+\sqrt{3}i)e^{(-1+\sqrt{3}i)t} & (-3-\sqrt{3}i)e^{(-1-\sqrt{3}i)t} \end{pmatrix}.$$

However, we can also produce a real fundamental matrix as follows. First write

$$\begin{pmatrix} 1 \\ -2\sqrt{3}i \\ -3 + \sqrt{3}i \end{pmatrix} = \begin{pmatrix} 1 \\ 0 \\ -3 \end{pmatrix} + i \begin{pmatrix} 0 \\ -2\sqrt{3} \\ \sqrt{3} \end{pmatrix} = \mathbf{U} + i\mathbf{V}$$

with

$$\mathbf{U} = \begin{pmatrix} 1 \\ 0 \\ -3 \end{pmatrix} \quad \text{and} \quad \mathbf{V} = \begin{pmatrix} 0 \\ -2\sqrt{3} \\ \sqrt{3} \end{pmatrix}.$$

Then

$$\begin{pmatrix} 1 \\ -2\sqrt{3}i \\ -3 + \sqrt{3}i \end{pmatrix} e^{(-1+\sqrt{3}i)t} = (\mathbf{U} + i\mathbf{V})[e^{-t}\cos(\sqrt{3}t) + ie^{-t}\sin(\sqrt{3}t)]$$

$$= \mathbf{U}e^{-t}\cos(\sqrt{3}t) - \mathbf{V}e^{-t}\sin(\sqrt{3}t) \tag{9.4}$$

$$+ i[\mathbf{V}e^{-t}\cos(\sqrt{3}t) + \mathbf{U}e^{-t}\sin(\sqrt{3}t)],$$

and

$$\begin{pmatrix} 1 \\ 2\sqrt{3}i \\ -3 - \sqrt{3}i \end{pmatrix} e^{(-1-\sqrt{3}t)t} = (\mathbf{U} - i\mathbf{V})[e^{-t}\cos(\sqrt{3}t) - ie^{-t}\sin(\sqrt{3}t)]$$

$$= \mathbf{U}e^{-t}\cos(\sqrt{3}t) - \mathbf{V}e^{-t}\sin(\sqrt{3}t) \tag{9.5}$$

$$- i[\mathbf{V}e^{-t}\cos(\sqrt{3}t) + \mathbf{U}e^{-t}\sin(\sqrt{3}t)].$$

The functions (9.4) and (9.5) are solutions, so any linear combination of these is also a solution. Taking their sum and dividing by 2 yields the solution

$$\mathbf{\Phi}_1(t) = \mathbf{U}e^{-t}\cos(\sqrt{3}t) - \mathbf{V}e^{-t}\sin(\sqrt{3}t).$$

And taking their difference and dividing by $2i$ yields the solution

$$\mathbf{\Phi}_2(t) = \mathbf{V}e^{-t}\cos(\sqrt{3}t) + \mathbf{U}e^{-t}\sin(\sqrt{3}t).$$

Using these, together with the solution found from the eigenvalue 2, we can form the fundamental matrix

$$\mathbf{\Omega}_2(t) = \begin{pmatrix} e^{2t} & e^{-t}\cos(\sqrt{3}t) & e^{-t}\sin(\sqrt{3}t) \\ 0 & 2\sqrt{3}e^{-t}\sin(\sqrt{3}t) & -2\sqrt{3}e^{-t}\cos(\sqrt{3}t) \\ 0 & e^{-t}[-3\cos(\sqrt{3}t) - \sqrt{3}\sin(\sqrt{3}t)] & e^{-t}[\sqrt{3}\cos(\sqrt{3}t) - 3\sin(\sqrt{3}t)] \end{pmatrix}.$$

Either fundamental matrix can be used to write the general solution, $\mathbf{X}(t) = \mathbf{\Omega}_1(t)\mathbf{C}$ or $\mathbf{X}(t) = \mathbf{\Omega}_2(t)\mathbf{K}$. However, the latter involves only real numbers and real-valued functions. ∎

A proof of the theorem follows the reasoning of the example, and is left to the student.

9.2.2 Solution of X′ = AX When A Does Not Have *n* Linearly Independent Eigenvectors

We know how to produce a fundamental matrix for $\mathbf{X}' = \mathbf{AX}$ when $\mathbf{A}$ has n linearly independent eigenvectors. This certainly occurs if $\mathbf{A}$ has n distinct eigenvalues and may even occur when $\mathbf{A}$ has repeated eigenvalues. However, we may encounter a matrix $\mathbf{A}$ having repeated eigenvalues for which there are not n linearly independent eigenvectors. In this case, we cannot yet write a fundamental matrix. This section is devoted to a procedure to follow in this case to find a fundamental matrix.

We will begin with two examples and then make some general remarks.

EXAMPLE 9.11

We will solve the system $\mathbf{X}' = \mathbf{AX}$, with

$$\mathbf{A} = \begin{pmatrix} 1 & 3 \\ -3 & 7 \end{pmatrix}.$$

A has one eigenvalue 4 of multiplicity 2. Eigenvectors all have the form $\alpha \begin{pmatrix} 1 \\ 1 \end{pmatrix}$, with $\alpha \neq 0$. A does not have two linearly independent eigenvectors.

We can immediately write one solution

$$\mathbf{\Phi}_1(t) = \begin{pmatrix} 1 \\ 1 \end{pmatrix} e^{4t}.$$

We need another solution. Write $\mathbf{E}_1 = \begin{pmatrix} 1 \\ 1 \end{pmatrix}$ and attempt a second solution

$$\mathbf{\Phi}_2(t) = \mathbf{E}_1 t e^{4t} + \mathbf{E}_2 e^{4t},$$

in which $\mathbf{E}_2$ is a 2×1 constant matrix to be determined. For this to be a solution, we need to have $\mathbf{\Phi}_2'(t) = \mathbf{A}\mathbf{\Phi}_2(t)$:

$$\mathbf{E}_1[e^{4t} + 4te^{4t}] + 4\mathbf{E}_2 e^{4t} = \mathbf{A}\mathbf{E}_1 t e^{4t} + \mathbf{A}\mathbf{E}_2 e^{4t}.$$

Divide this equation by e^{4t} to get

$$\mathbf{E}_1 + 4\mathbf{E}_1 t + 4\mathbf{E}_2 = \mathbf{A}\mathbf{E}_1 t + \mathbf{A}\mathbf{E}_2.$$

But $\mathbf{A}\mathbf{E}_1 = 4\mathbf{E}_1$, so the terms having t as a factor cancel and we are left with

$$\mathbf{A}\mathbf{E}_2 - 4\mathbf{E}_2 = \mathbf{E}_1.$$

Write this equation as

$$(\mathbf{A} - 4\mathbf{I}_2)\mathbf{E}_2 = \mathbf{E}_1.$$

If $\mathbf{E}_2 = \begin{pmatrix} a \\ b \end{pmatrix}$, this is the linear system of two equations in two unknowns:

$$(\mathbf{A} - 4\mathbf{I}_2)\begin{pmatrix} a \\ b \end{pmatrix} = \begin{pmatrix} 1 \\ 1 \end{pmatrix},$$

or

$$\begin{pmatrix} -3 & 3 \\ -3 & 3 \end{pmatrix}\begin{pmatrix} a \\ b \end{pmatrix} = \begin{pmatrix} 1 \\ 1 \end{pmatrix}.$$

This system has general solution $\mathbf{E}_2 = \begin{pmatrix} s \\ \frac{1+3s}{3} \end{pmatrix}$, in which s can be any nonzero number. Let $s = 1$ to get $\mathbf{E}_2 = \begin{pmatrix} 1 \\ \frac{4}{3} \end{pmatrix}$ and hence the second solution

$$\boldsymbol{\Phi}_2(t) = \mathbf{E}_1 t e^{4t} + \mathbf{E}_2 e^{4t} = \begin{pmatrix} 1 \\ 1 \end{pmatrix} t e^{4t} + \begin{pmatrix} 1 \\ \frac{4}{3} \end{pmatrix} e^{4t}$$

$$= \begin{pmatrix} 1 + t \\ \frac{4}{3} + t \end{pmatrix} e^{4t}.$$

If we use $\boldsymbol{\Phi}_1(0)$ and $\boldsymbol{\Phi}_2(0)$ as columns to form the matrix

$$\begin{pmatrix} 1 & 1 \\ 1 & \frac{4}{3} \end{pmatrix},$$

then this matrix has determinant $\frac{1}{3}$, hence $\boldsymbol{\Phi}_1$ and $\boldsymbol{\Phi}_2$ are linearly independent by Theorem 9.3(2). Therefore, $\boldsymbol{\Phi}_1(t)$ and $\boldsymbol{\Phi}_2(t)$ can be used as columns of a fundamental matrix

$$\boldsymbol{\Omega}(t) = \begin{pmatrix} e^{4t} & (1+t)e^{4t} \\ e^{4t} & \left(\frac{4}{3} + t\right) e^{4t} \end{pmatrix}.$$

The general solution of $\mathbf{X}' = \mathbf{A}\mathbf{X}$ is $\mathbf{X}(t) = \boldsymbol{\Omega}(t)\mathbf{C}$. ∎

The procedure followed in this example is similar in spirit to solving the differential equation $y'' - 5y' + 6y = e^{3x}$ by undetermined coefficients. We are tempted to try $y_p(x) = ae^{3x}$, but this will not work because e^{3x} is a solution of $y'' - 5y' + 6y = 0$. We therefore try $y_p(x) = axe^{3x}$, multiplying the first attempt ae^{3x} by x. The analogous step for the system was to try the second solution $\boldsymbol{\Phi}_2(t) = \mathbf{E}_1 t e^{4t} + \mathbf{E}_2 e^{4t}$.

We will continue to explore the case of repeated eigenvalues with another example.

EXAMPLE 9.12

Consider the system $\mathbf{X}' = \mathbf{AX}$, in which

$$\mathbf{A} = \begin{pmatrix} -2 & -1 & -5 \\ 25 & -7 & 0 \\ 0 & 1 & 3 \end{pmatrix}.$$

$\mathbf{A}$ has eigenvalue -2 with multiplicity 3, and corresponding eigenvectors are all nonzero scalar

multiples of $\begin{pmatrix} -1 \\ -5 \\ 1 \end{pmatrix}$. Denoting $\begin{pmatrix} -1 \\ -5 \\ 1 \end{pmatrix} = \mathbf{E}_1$, we have one solution

$$\boldsymbol{\Phi}_1(t) = \begin{pmatrix} -1 \\ -5 \\ 1 \end{pmatrix} e^{-2t} = \mathbf{E}_1 e^{-2t}.$$

We need three linearly independent solutions. We will try a second solution of the form

$$\boldsymbol{\Phi}_2(t) = \mathbf{E}_1 t e^{-2t} + \mathbf{E}_2 e^{-2t},$$

in which $\mathbf{E}_2$ is a 3×1 matrix to be determined. Substitute this proposed solution into $\mathbf{X}' = \mathbf{AX}$ to get

$$\mathbf{E}_1[e^{-2t} - 2te^{-2t}] + \mathbf{E}_2[-2e^{-2t}] = \mathbf{AE}_1 t e^{-2t} + \mathbf{AE}_2 e^{-2t}.$$

Upon dividing by the common factor of e^{-2t}, and recalling that $\mathbf{AE}_1 = -2\mathbf{E}_1$, this equation becomes

$$\mathbf{E}_1 - 2t\mathbf{E}_1 - 2\mathbf{E}_2 = -2t\mathbf{E}_1 + \mathbf{AE}_2,$$

or

$$\mathbf{AE}_2 + 2\mathbf{E}_2 = \mathbf{E}_1.$$

We can write this equation as

$$(\mathbf{A} + 2\mathbf{I}_3)\mathbf{E}_2 = \mathbf{E}_1,$$

or

$$\begin{pmatrix} 0 & -1 & -5 \\ 25 & -5 & 0 \\ 0 & 1 & 5 \end{pmatrix} \mathbf{E}_2 = \begin{pmatrix} -1 \\ -5 \\ 1 \end{pmatrix}.$$

With $\mathbf{E}_2 = \begin{pmatrix} \alpha \\ \beta \\ \gamma \end{pmatrix}$, this is the nonhomogeneous system

$$\begin{pmatrix} 0 & -1 & -5 \\ 25 & -5 & 0 \\ 0 & 1 & 5 \end{pmatrix} \begin{pmatrix} \alpha \\ \beta \\ \gamma \end{pmatrix} = \begin{pmatrix} -1 \\ -5 \\ 1 \end{pmatrix},$$

with general solution $\begin{pmatrix} -s \\ 1 - 5s \\ s \end{pmatrix}$, in which s can be any number. For a specific solution, choose $s = 1$ and let

$$\mathbf{E}_2 = \begin{pmatrix} -1 \\ -4 \\ 1 \end{pmatrix}.$$

This gives us the second solution

$$\mathbf{\Phi}_2(t) = \mathbf{E}_1 t e^{-2t} + \mathbf{E}_2 e^{-2t}$$

$$= \begin{pmatrix} -1 \\ -5 \\ 1 \end{pmatrix} t e^{-2t} + \begin{pmatrix} -1 \\ -4 \\ 1 \end{pmatrix} e^{-2t}$$

$$= \begin{pmatrix} -1 - t \\ -4 - 5t \\ 1 + t \end{pmatrix} e^{-2t}.$$

We need one more solution. Try for a solution of the form

$$\mathbf{\Phi}_3(t) = \tfrac{1}{2} \mathbf{E}_1 t^2 e^{-2t} + \mathbf{E}_2 t e^{-2t} + \mathbf{E}_3 e^{-2t}.$$

We want to solve for $\mathbf{E}_3$. Substitute this proposed solution into $\mathbf{X}' = \mathbf{A}\mathbf{X}$ to get

$$\mathbf{E}_1[t e^{-2t} - t^2 e^{-2t}] + \mathbf{E}_2[e^{-2t} - 2t e^{-2t}] + \mathbf{E}_3[-2e^{-2t}] = \tfrac{1}{2}\mathbf{A}\mathbf{E}_1 t^2 e^{-2t} + \mathbf{A}\mathbf{E}_2 t e^{-2t} + \mathbf{A}\mathbf{E}_3 e^{-2t}.$$

Divide this equation by e^{-2t} and use the fact that $\mathbf{A}\mathbf{E}_1 = -2\mathbf{E}_1$ and $\mathbf{A}\mathbf{E}_2 = \begin{pmatrix} 1 \\ 3 \\ -1 \end{pmatrix}$ to get

$$\mathbf{E}_1 t - \mathbf{E}_1 t^2 + \mathbf{E}_2 - 2\mathbf{E}_2 t - 2\mathbf{E}_3 = -\mathbf{E}_1 t^2 + \begin{pmatrix} 1 \\ 3 \\ -1 \end{pmatrix} t + \mathbf{A}\mathbf{E}_3. \tag{9.6}$$

Now

$$\mathbf{E}_1 t - 2\mathbf{E}_2 t = \left[\begin{pmatrix} -1 \\ -5 \\ 1 \end{pmatrix} - 2 \begin{pmatrix} -1 \\ -4 \\ 1 \end{pmatrix} \right] t = \begin{pmatrix} 1 \\ 3 \\ -1 \end{pmatrix} t,$$

so equation (9.6) has three terms cancel, and it reduces to

$$\mathbf{E}_2 - 2\mathbf{E}_3 = \mathbf{A}\mathbf{E}_3.$$

Write this equation as

$$(\mathbf{A} + 2\mathbf{I}_3)\mathbf{E}_3 = \mathbf{E}_2$$

or

$$\begin{pmatrix} 0 & -1 & -5 \\ 25 & -5 & 0 \\ 0 & 1 & 5 \end{pmatrix} \mathbf{E}_3 = \begin{pmatrix} -1 \\ -4 \\ 1 \end{pmatrix},$$

with general solution

$$\begin{pmatrix} \frac{1-25s}{25} \\ 1 - 5s \\ s \end{pmatrix}$$

in which s can be any number. Choosing $s = 1$, we can let

$$\mathbf{E}_3 = \begin{pmatrix} -\frac{24}{25} \\ -4 \\ 1 \end{pmatrix}.$$

A third solution is

$$\boldsymbol{\Phi}_3(t) = \frac{1}{2} \begin{pmatrix} -1 \\ -5 \\ 1 \end{pmatrix} t^2 e^{-2t} + \begin{pmatrix} -1 \\ -4 \\ 1 \end{pmatrix} t e^{-2t} + \begin{pmatrix} -\frac{24}{25} \\ -4 \\ 1 \end{pmatrix} e^{-2t}$$

$$= \begin{pmatrix} -\frac{24}{25} - t - \frac{1}{2}t^2 \\ -4 - 4t - \frac{5}{2}t^2 \\ 1 + t + \frac{1}{2}t^2 \end{pmatrix} e^{-2t}.$$

To show that $\boldsymbol{\Phi}_1$, $\boldsymbol{\Phi}_2$, and $\boldsymbol{\Phi}_3$ are linearly independent, Theorem 9.3(2) is convenient. Form the 3×3 matrix having these solutions, evaluated at $t = 0$, as columns:

$$\begin{pmatrix} -1 & -1 & -\frac{24}{25} \\ -5 & -4 & -4 \\ 1 & 1 & 1 \end{pmatrix}.$$

The determinant of this matrix is $-\frac{1}{25}$, so this matrix is nonsingular and the solutions are linearly independent. We can use these solutions as columns of a fundamental matrix

$$\boldsymbol{\Omega}(t) = \begin{pmatrix} -e^{-2t} & (-1 - t)e^{-2t} & \left(-\frac{24}{25} - t - \frac{1}{2}t^2\right) e^{-2t} \\ -5e^{-2t} & (-4 - 5t)e^{-2t} & \left(-4 - 4t - \frac{5}{2}t^2\right) e^{-2t} \\ e^{-2t} & (1 + t)e^{-2t} & \left(1 + t + \frac{1}{2}t^2\right) e^{-2t} \end{pmatrix}.$$

The general solution of $\mathbf{X}' = \mathbf{AX}$ is $\mathbf{X}(t) = \boldsymbol{\Omega}(t)\mathbf{C}$. ∎

These examples suggest a procedure that we will now outline in general. Begin with a system $\mathbf{X}' = \mathbf{AX}$, with $\mathbf{A}$ an $n \times n$ matrix of real numbers. We want the general solution, so we need n linearly independent solutions.

Case 1 **A** has n linearly independent eigenvectors.
Use these eigenvectors to write n linearly independent solutions and use these as columns of a fundamental matrix. (This case may occur even if **A** does not have n distinct eigenvalues).

Case 2 **A** does not have n linearly independent eigenvectors.
Let the eigenvalues of **A** be $\lambda_1, \ldots, \lambda_n$. At least one eigenvalue must be repeated, because if **A** has n distinct eigenvectors, the corresponding eigenvectors must be linearly independent, putting us back in case 1. Suppose $\lambda_1, \ldots, \lambda_r$ are the distinct eigenvalues, while $\lambda_{r+1}, \ldots, \lambda_n$ repeat some

of these first r eigenvalues. If $\mathbf{V}_j$ is an eigenvector corresponding to λ_j for $j = 1, \ldots, r$, we can immediately write r linearly independent solutions

$$\mathbf{\Psi}_1(t) = \mathbf{V}_1 e^{\lambda_1 t}, \ldots, \mathbf{\Psi}_r(t) = \mathbf{V}_r e^{\lambda_r t}.$$

Now work with the repeated eigenvalues. Suppose μ is a repeated eigenvalue, say $\mu = \lambda_1$ with multiplicity k. We already have one solution corresponding to μ, namely $\mathbf{\Psi}_1$. To be consistent in notation with the examples just done, denote $\mathbf{V}_1 = \mathbf{E}_1$ and $\mathbf{\Psi}_1 = \mathbf{\Phi}_1$. Then

$$\mathbf{\Phi}_1(t) = \mathbf{V}_1 e^{\lambda_1 t} = \mathbf{E}_1 e^{\mu t}$$

is one solution corresponding to μ. For a second solution corresponding to μ, let

$$\mathbf{\Phi}_2(t) = \mathbf{E}_1 t e^{\mu t} + \mathbf{E}_2 e^{\mu t}.$$

Substitute this proposed solution into $\mathbf{X}' = \mathbf{AX}$ and solve for $\mathbf{E}_2$. If $k = 2$, this yields a second solution corresponding to μ and we move on to another multiple eigenvalue. If $k \geq 3$, we do not yet have all the solutions corresponding to μ, so we attempt

$$\mathbf{\Phi}_3(t) = \frac{1}{2} \mathbf{E}_1 t^2 e^{\mu t} + \mathbf{E}_2 t e^{\mu t} + \mathbf{E}_3 e^{\mu t}.$$

Substitute $\mathbf{\Phi}_3(t)$ into the differential equation and solve for $\mathbf{E}_3$ to get a third solution corresponding to μ. If $\mu \geq 4$, continue with

$$\mathbf{\Phi}_4(t) = \frac{1}{3!} \mathbf{E}_1 t^3 e^{\mu t} + \frac{1}{2!} \mathbf{E}_2 t^2 e^{\mu t} + \mathbf{E}_3 t e^{\mu t} + \mathbf{E}_4 e^{\mu t},$$

substitute into the differential equation, and solve for $\mathbf{E}_4$, and so on. Eventually, we reach

$$\mathbf{\Phi}_k(t) = \frac{1}{(k-1)!} \mathbf{E}_1 t^{k-1} e^{\mu t} + \frac{1}{(k-2)!} \mathbf{E}_2 t^{k-2} e^{\mu t} + \cdots + \mathbf{E}_{k-1} t e^{\mu t} + \mathbf{E}_k e^{\mu t};$$

substitute into the differential equation and solve for $\mathbf{E}_k$.

This procedure gives, for an eigenvalue μ of multiplicity k, k linearly independent solutions of $\mathbf{X}' = \mathbf{AX}$. Repeat the procedure for each eigenvalue until n linearly independent solutions have been found.

9.2.3   Solution of $\mathbf{X}' = \mathbf{AX}$ by Diagonalizing $\mathbf{A}$

We now take a different tack and attempt to exploit diagonalization.

Consider the system

$$\mathbf{X}' = \begin{pmatrix} -2 & 0 & 0 \\ 0 & 4 & 0 \\ 0 & 0 & -6 \end{pmatrix} \mathbf{X}.$$

The constant coefficient matrix $\mathbf{A}$ is a diagonal matrix, and this system really consists of three independent differential equations, each involving just one of the variables:

$$x_1' = -2x_1,$$
$$x_2' = 4x_2,$$
$$x_3' = -6x_3,$$

Such a system is said to be *uncoupled*. Each equation is easily solved independent of the others, obtaining

$$x_1 = c_1 e^{-2t}, \quad x_2 = c_2 e^{4t}, \quad x_3 = c_3 e^{-6t}.$$

The system is uncoupled because the coefficient matrix $\mathbf{A}$ is diagonal. Because of this, we can immediately write the eigenvalues $-2, 4, -6$ of $\mathbf{A}$ and find the corresponding eigenvectors

$$\begin{pmatrix} 1 \\ 0 \\ 0 \end{pmatrix}, \begin{pmatrix} 0 \\ 1 \\ 0 \end{pmatrix}, \begin{pmatrix} 0 \\ 0 \\ 1 \end{pmatrix}$$

Therefore, $\mathbf{X}' = \mathbf{AX}$ has fundamental matrix

$$\boldsymbol{\Omega}(t) = \begin{pmatrix} e^{-2t} & 0 & 0 \\ 0 & e^{4t} & 0 \\ 0 & 0 & e^{-6t} \end{pmatrix}$$

and the general solution is $\mathbf{X}(t) = \boldsymbol{\Omega}(t)\mathbf{C}$. However we wish to approach this system, the point is that it is easy to solve because $\mathbf{A}$ is a diagonal matrix.

Now in general $\mathbf{A}$ need not be diagonal. However, $\mathbf{A}$ may be diagonalizable (Section 8.2). This will occur exactly when $\mathbf{A}$ has n linearly independent eigenvectors. In this event, we can form a matrix $\mathbf{P}$, whose columns are eigenvectors of $\mathbf{A}$, such that

$$\mathbf{P}^{-1}\mathbf{AP} = \mathbf{D} = \begin{pmatrix} \lambda_1 & 0 & \cdots & 0 \\ 0 & \lambda_2 & \cdots & 0 \\ \vdots & \vdots & \cdots & \vdots \\ 0 & 0 & \cdots & \lambda_n \end{pmatrix}.$$

$\mathbf{D}$ is the diagonal matrix having the eigenvalues $\lambda_1, \ldots, \lambda_n$ down its main diagonal. This will hold even if some of the eigenvalues have multiplicities greater than 1, provided that $\mathbf{A}$ has n linearly independent eigenvectors.

Now make the change of variables $\mathbf{X} = \mathbf{PZ}$ in the differential equation $\mathbf{X}' = \mathbf{AX}$. First compute

$$\mathbf{X}' = (\mathbf{PZ})' = \mathbf{PZ}' = \mathbf{AX} = \mathbf{APZ},$$

so

$$\mathbf{Z}' = \mathbf{P}^{-1}\mathbf{APZ} = \mathbf{DZ}.$$

The uncoupled system $\mathbf{Z}' = \mathbf{DZ}$ can be solved by inspection. A fundamental matrix for $\mathbf{Z}' = \mathbf{DZ}$ is

$$\boldsymbol{\Omega}_{\mathbf{D}}(t) = \begin{pmatrix} e^{\lambda_1 t} & 0 & \cdots & \cdots & 0 \\ 0 & e^{\lambda_2 t} & \cdots & \cdots & 0 \\ \vdots & \vdots & \vdots & \cdots & \vdots \\ 0 & 0 & \cdots & e^{\lambda_{n-1} t} & 0 \\ 0 & 0 & \cdots & 0 & e^{\lambda_n t} \end{pmatrix}$$

and the general solution of $\mathbf{Z}' = \mathbf{DZ}$ is

$$\mathbf{Z}(t) = \boldsymbol{\Omega}_{\mathbf{D}}(t)\mathbf{C}.$$

Then

$$\mathbf{X}(t) = \mathbf{PZ}(t) = \mathbf{P}\boldsymbol{\Omega}_{\mathbf{D}}(t)\mathbf{C}$$

is the general solution of the original system $\mathbf{X}' = \mathbf{AX}$. That is, $\boldsymbol{\Omega}(t) = \mathbf{P}\boldsymbol{\Omega}_{\mathbf{D}}(t)$ is a fundamental matrix for $\mathbf{X}' = \mathbf{AX}$.

In this process we need $\mathbf{P}$, whose columns are eigenvectors of $\mathbf{A}$, but we never actually need to calculate $\mathbf{P}^{-1}$.

EXAMPLE 9.13

Solve

$$
\mathbf{X}' = \begin{pmatrix} 3 & 3 \\ 1 & 5 \end{pmatrix} \mathbf{X}.
$$

The eigenvalues and associated eigenvectors of $\mathbf{A}$ are

$$
2, \begin{pmatrix} -3 \\ 1 \end{pmatrix} \quad \text{and} \quad 6, \begin{pmatrix} 1 \\ 1 \end{pmatrix}.
$$

Because $\mathbf{A}$ has distinct eigenvalues, $\mathbf{A}$ is diagonalizable. Make the change of variables $\mathbf{X} = \mathbf{PZ}$, where

$$
\mathbf{P} = \begin{pmatrix} -3 & 1 \\ 1 & 1 \end{pmatrix}.
$$

This transforms $\mathbf{X}' = \mathbf{AX}$ to $\mathbf{Z}' = \mathbf{DZ}$, where

$$
\mathbf{D} = \begin{pmatrix} 2 & 0 \\ 0 & 6 \end{pmatrix}.
$$

This uncoupled system has fundamental matrix

$$
\mathbf{\Omega_D}(t) = \begin{pmatrix} e^{2t} & 0 \\ 0 & e^{6t} \end{pmatrix}.
$$

Then $\mathbf{X}' = \mathbf{AX}$ has fundamental matrix

$$
\mathbf{\Omega}(t) = \mathbf{P}\mathbf{\Omega_D}(t) = \begin{pmatrix} -3 & 1 \\ 1 & 1 \end{pmatrix} \begin{pmatrix} e^{2t} & 0 \\ 0 & e^{6t} \end{pmatrix}
$$

$$
= \begin{pmatrix} -3e^{2t} & e^{6t} \\ e^{2t} & e^{6t} \end{pmatrix}.
$$

The general solution of $\mathbf{X}' = \mathbf{AX}$ is $\mathbf{X}(t) = \mathbf{\Omega}(t)\mathbf{C}$. ∎

9.2.4   Exponential Matrix Solutions of $\mathbf{X}' = \mathbf{AX}$

A first-order differential equation $y' = ay$ has general solution $y(x) = ce^{ax}$. At the risk of stretching the analogy too far, we might conjecture whether there might be a solution $e^{\mathbf{A}t}\mathbf{C}$ for a matrix differential equation $\mathbf{X}' = \mathbf{AX}$. We will now show how to define the exponential matrix $e^{\mathbf{A}t}$ to make sense out of this conjecture. For this section, let $\mathbf{A}$ be an $n \times n$ matrix of real numbers.

The Taylor expansion of the real exponential function,

$$
e^t = 1 + t + \tfrac{1}{2!}t^2 + \tfrac{1}{3!}t^3 + \cdots,
$$

suggests the following.

DEFINITION 9.3   *Exponential Matrix*

The exponential matrix $e^{\mathbf{A}t}$ is the $n \times n$ matrix defined by

$$
e^{\mathbf{A}t} = \mathbf{I}_n + \mathbf{A}t + \tfrac{1}{2!}\mathbf{A}^2 t^2 + \tfrac{1}{3!}\mathbf{A}^3 t^3 + \cdots.
$$

It can be shown that this series converges for all real t, in the sense that the infinite series of elements in the i, j place converges.

Care must be taken in computing with exponential matrices, because matrix multiplication is not commutative. The analogue of the relationship $e^{at}e^{bt} = e^{(a+b)t}$ is given by the following.

THEOREM 9.9

Let **B** be an $n \times n$ real matrix. Suppose $\mathbf{AB} = \mathbf{BA}$. Then

$$e^{(\mathbf{A}+\mathbf{B})t} = e^{\mathbf{A}t}e^{\mathbf{B}t}. \quad \blacksquare$$

A proof is outlined in the exercises.

Because **A** is a constant matrix,

$$\frac{d}{dt}e^{\mathbf{A}t} = \frac{d}{dt}\left[\mathbf{I}_n + \mathbf{A}t + \tfrac{1}{2!}\mathbf{A}^2t^2 + \tfrac{1}{3!}\mathbf{A}^3t^3 + \tfrac{1}{4!}\mathbf{A}^4t^4 + \cdots\right]$$

$$= \mathbf{A} + \mathbf{A}^2t + \tfrac{1}{2!}\mathbf{A}^3t^2 + \tfrac{1}{3!}\mathbf{A}^4t^3 + \cdots$$

$$= \mathbf{A}\left[\mathbf{I}_n + \mathbf{A}t + \tfrac{1}{2!}\mathbf{A}^2t^2 + \tfrac{1}{3!}\mathbf{A}^3t^3 + \tfrac{1}{4!}\mathbf{A}^4t^4 + \cdots\right]$$

$$= \mathbf{A}e^{\mathbf{A}t}.$$

The derivative of $e^{\mathbf{A}t}$, obtained by differentiating each element, is the product $\mathbf{A}e^{\mathbf{A}t}$ of two $n \times n$ matrices and has the same form as the derivative of the scalar exponential function e^{at}. One ramification of this derivative formula is that, for any $n \times 1$ constant matrix **K**, $e^{\mathbf{A}t}\mathbf{K}$ is a solution of $\mathbf{X}' = \mathbf{A}\mathbf{X}$.

LEMMA 9.1

For any real $n \times 1$ constant matrix **K**, $e^{\mathbf{A}t}\mathbf{K}$ is a solution of $\mathbf{X}' = \mathbf{A}\mathbf{X}$.

Proof Compute

$$\mathbf{\Phi}'(t) = (e^{\mathbf{A}t}\mathbf{K})' = \mathbf{A}e^{\mathbf{A}t}\mathbf{K} = \mathbf{A}\mathbf{\Phi}(t). \quad \blacksquare$$

Even more, $e^{\mathbf{A}t}$ is a fundamental matrix for $\mathbf{X}' = \mathbf{A}\mathbf{X}$.

THEOREM 9.10

$e^{\mathbf{A}t}$ is a fundamental matrix for $\mathbf{X}' = \mathbf{A}\mathbf{X}$.

Proof Let $\mathbf{E}_j$ be the $n \times 1$ matrix with 1 in the j, 1 place and all other entries zero:

$$\mathbf{E}_j = \begin{pmatrix} 0 \\ 0 \\ \vdots \\ 1 \\ \vdots \\ 0 \\ 0 \end{pmatrix}.$$

Then $e^{At}\mathbf{E}_j$ is the jth column of e^{At}. This column is a solution of $\mathbf{X}' = \mathbf{AX}$ by the lemma. Further, the columns of e^{At} are linearly independent by Theorem 9.3(2), because if we put $t = 0$, we get $e^{At} = \mathbf{I}_n$, which has a nonzero determinant. Thus e^{At} is a fundamental matrix for $\mathbf{X}' = \mathbf{AX}$. ∎

In theory, then, we can find the general solution $e^{At}\mathbf{C}$ of $\mathbf{X}' = \mathbf{AX}$ if we can compute e^{At}. This, however, can be a daunting task. As an example, for an apparently simple matrix such as

$$\mathbf{A} = \begin{pmatrix} 1 & 2 \\ -2 & 4 \end{pmatrix},$$

we find using a software package that $e^{At} =$

$$\begin{pmatrix} e^{5t/2}\cos(\sqrt{7}t/2) + \sqrt{7}e^{5t/2}\sin(\sqrt{7}t/2) & \frac{4}{\sqrt{7}}e^{5t/2}\sin(\sqrt{7}t/2) \\ -\frac{4}{\sqrt{7}}e^{5t/2}\sin(\sqrt{7}t/2) & e^{5t/2}\cos(\sqrt{7}t/2) - \sqrt{7}e^{5t/2}\sin(\sqrt{7}t/2) \end{pmatrix}.$$

This is a fundamental matrix for $\mathbf{X}' = \mathbf{AX}$. It would be at least as easy, for this $\mathbf{A}$, to find the eigenvalues of $\mathbf{A}$, which are $\frac{5}{2} \pm \frac{1}{2}i\sqrt{7}$, then find corresponding eigenvectors, $\begin{pmatrix} 4 \\ 3 \pm i\sqrt{7} \end{pmatrix}$, and use these to obtain a fundamental matrix.

We will now pursue an interesting line of thought. We claim that even though e^{At} may be tedious or even impractical to compute for a given $\mathbf{A}$, it is often possible to compute the product $e^{At}\mathbf{K}$, for carefully chosen $\mathbf{K}$, as a finite sum, and hence generate solutions to $\mathbf{X}' = \mathbf{AX}$. To do this we need the following.

LEMMA 9.2

Let $\mathbf{A}$ be an $n \times n$ real matrix and $\mathbf{K}$ an $n \times 1$ real matrix. Let μ be any number. Then

1. $e^{\mu \mathbf{I}_n t}\mathbf{K} = e^{\mu t}\mathbf{K}$.
2. $e^{At}\mathbf{K} = e^{\mu t}e^{(\mathbf{A}-\mu \mathbf{I}_n)t}\mathbf{K}$.

Proof For (1), since $(\mathbf{I}_n)^m = \mathbf{I}_n$ for any positive integer m, we have

$$e^{\mu \mathbf{I}_n t}\mathbf{K} = \left[\mathbf{I}_n + \mu \mathbf{I}_n t + \tfrac{1}{2!}(\mu \mathbf{I}_n)^2 t^2 + \tfrac{1}{3!}(\mu \mathbf{I}_n)^3 t^3 + \cdots\right]\mathbf{K}$$
$$= \left[1 + \mu t + \tfrac{1}{2!}\mu^2 t^2 + \tfrac{1}{3!}\mu^3 t^3 + \cdots\right]\mathbf{I}_n\mathbf{K} = e^{\mu t}\mathbf{K}.$$

For (2), first observe that $\mu \mathbf{I}_n$ and $\mathbf{A} - \mu \mathbf{I}_n$ commute, since

$$\mu \mathbf{I}_n(\mathbf{A} - \mu \mathbf{I}_n) = \mu(\mathbf{I}_n\mathbf{A} - \mu(\mathbf{I}_n)^2)$$
$$= \mu(\mathbf{A} - \mu \mathbf{I}_n) = (\mathbf{A} - \mu \mathbf{I}_n)(\mu \mathbf{I}_n).$$

Then, using Theorem 9.9,

$$e^{At}\mathbf{K} = e^{At + \mu \mathbf{I}_n t - \mu \mathbf{I}_n t}\mathbf{K} = e^{(\mathbf{A}-\mu \mathbf{I}_n)t}e^{\mu \mathbf{I}_n t}\mathbf{K}$$
$$= e^{(\mathbf{A}-\mu \mathbf{I}_n)t}e^{\mu t}\mathbf{K} = e^{\mu t}e^{(\mathbf{A}-\mu \mathbf{I}_n)t}\mathbf{K}. \quad ∎$$

Now suppose we want to solve $\mathbf{X}' = \mathbf{AX}$. Let $\lambda_1, \ldots, \lambda_r$ be the distinct eigenvalues of $\mathbf{A}$ and let λ_j have multiplicity m_j. Then

$$m_1 + \cdots + m_r = n.$$

For each λ_j, find as many linearly independent eigenvectors as possible. For λ_j, this can be any number from 1 to m_j inclusive. If this yields n linearly independent eigenvectors, then we can write the general solution as a sum of eigenvectors ξ_j times exponential functions $e^{\lambda_j t}$ and we do not need $e^{\mathbf{A}t}$.

Thus, suppose some λ_j has multiplicity $m_j \geq 2$, but there are fewer than m_j linearly independent eigenvectors. Find an $n \times 1$ constant matrix $\mathbf{K}_1$ that is linearly independent from the eigenvectors found for λ_j and such that

$$(\mathbf{A} - \lambda_j \mathbf{I}_n)\mathbf{K}_1 \neq \mathbf{O} \quad \text{but} \quad (\mathbf{A} - \lambda_j \mathbf{I}_n)^2 \mathbf{K}_1 = \mathbf{O}.$$

Then $e^{\mathbf{A}t}\mathbf{K}_1$ is a solution of $\mathbf{X}' = \mathbf{A}\mathbf{X}$. Further, because of the way $\mathbf{K}_1$ was chosen,

$$e^{\mathbf{A}t}\mathbf{K}_1 = e^{\lambda_j t}e^{(\mathbf{A} - \lambda_j \mathbf{I}_n)t}\mathbf{K}_1 = e^{\lambda_j t}[\mathbf{K}_1 + (\mathbf{A} - \lambda_j \mathbf{I}_n)\mathbf{K}_1 t],$$

with all other terms of the series for $e^{(\mathbf{A} - \lambda_j \mathbf{I}_n)t}\mathbf{K}_1$ vanishing because $(\mathbf{A} - \lambda_j \mathbf{I}_n)^2 \mathbf{K}_1 = \mathbf{O}$ forces $(\mathbf{A} - \lambda_j \mathbf{I}_n)^m \mathbf{K}_1 = \mathbf{O}$ for $m \geq 2$. We can therefore compute $e^{\lambda_j t}\mathbf{K}_1$ as a sum of just two terms.

If we now have m_j solutions corresponding to λ_j, then leave this eigenvalue and move on to any others that do not yet have as many linearly independent solutions as their multiplicity. If we do not yet have m_j solutions corresponding to λ_j, then find a constant $n \times 1$ matrix $\mathbf{K}_2$ such that

$$(\mathbf{A} - \lambda_j \mathbf{I}_n)\mathbf{K}_2 \neq \mathbf{O} \quad \text{and} \quad (\mathbf{A} - \lambda_j \mathbf{I}_n)^2 \mathbf{K}_2 \neq \mathbf{O}$$

but

$$(\mathbf{A} - \lambda_j \mathbf{I}_n)^3 \mathbf{K}_2 = \mathbf{O}.$$

Then $e^{\mathbf{A}t}\mathbf{K}_2$ is a solution of $\mathbf{X}' = \mathbf{A}\mathbf{X}$, and we can compute this solution as a sum of just three terms:

$$e^{\mathbf{A}t}\mathbf{K}_2 = e^{\lambda_j t}e^{(\mathbf{A} - \lambda_j \mathbf{I}_n)t}\mathbf{K}_2 = e^{\lambda_j t}\left[\mathbf{K}_2 + (\mathbf{A} - \lambda_j \mathbf{I}_n)\mathbf{K}_2 t + \frac{1}{2!}(\mathbf{A} - \lambda_j \mathbf{I}_n)^2 t^2\right].$$

The other terms in the infinite series for $e^{\mathbf{A}t}\mathbf{K}_2$ vanish because

$$(\mathbf{A} - \lambda_j \mathbf{I}_n)^3 \mathbf{K}_2 = (\mathbf{A} - \lambda_j \mathbf{I}_n)^4 \mathbf{K}_2 = \cdots = \mathbf{O}.$$

If this gives us m_j solutions associated with λ_j, move on to another eigenvalue for which we do not yet have as many solutions as the multiplicity of the eigenvalue. If not, produce an $n \times 1$ constant matrix $\mathbf{K}_3$ such that

$$(\mathbf{A} - \lambda_j \mathbf{I}_n)\mathbf{K}_3 \neq \mathbf{O}, (\mathbf{A} - \lambda_j \mathbf{I}_n)^2 \mathbf{K}_3 \neq \mathbf{O}, \quad \text{and} \quad (\mathbf{A} - \lambda_j \mathbf{I}_n)^3 \mathbf{K}_3 \neq \mathbf{O}$$

but

$$(\mathbf{A} - \lambda_j \mathbf{I}_n)^4 \mathbf{K}_3 = \mathbf{0}.$$

Then $e^{\mathbf{A}t}\mathbf{K}_3$ can be computed as a sum of four terms.

Keep repeating this process. For λ_j it must terminate after at most $m_j - 1$ steps, because we began with at least one eigenvector associated with λ_j and then produced more solutions to obtain a total of m_j linearly independent solutions associated with λ_j. Once these are obtained, we move on to another eigenvalue for which we have fewer solutions than its multiplicity and repeat this process for that eigenvalue, and so on. Eventually we generate a total of n linearly independent solutions, thus obtaining the general solution of $\mathbf{X}' = \mathbf{A}\mathbf{X}$.

EXAMPLE 9.14

Consider $\mathbf{X}' = \mathbf{A}\mathbf{X}$, where

$$\mathbf{A} = \begin{pmatrix} 2 & 1 & 0 & 3 \\ 0 & 2 & 1 & 1 \\ 0 & 0 & 2 & 4 \\ 0 & 0 & 0 & 4 \end{pmatrix}.$$

The eigenvalues are 4, 2, 2, 2. Associated with 4 we find the eigenvector $\begin{pmatrix} 9 \\ 6 \\ 8 \\ 4 \end{pmatrix}$, so one solution

of $\mathbf{X}' = \mathbf{A}\mathbf{X}$ is

$$\boldsymbol{\Phi}_1(t) = \begin{pmatrix} 9 \\ 6 \\ 8 \\ 4 \end{pmatrix} e^{4t}.$$

Associated with 2 we find that every eigenvector has the form

$$\begin{pmatrix} \alpha \\ 0 \\ 0 \\ 0 \end{pmatrix}.$$

A second solution is

$$\boldsymbol{\Phi}_2(t) = \begin{pmatrix} 1 \\ 0 \\ 0 \\ 0 \end{pmatrix} e^{2t}.$$

Now find a 4×1 constant matrix $\mathbf{K}_1$ such that $(\mathbf{A} - 2\mathbf{I}_4)\mathbf{K}_1 \neq \mathbf{O}$, but $(\mathbf{A} - 2\mathbf{I}_4)^2\mathbf{K}_1 = \mathbf{O}$. First compute

$$(\mathbf{A} - 2\mathbf{I}_4)^2 = \begin{pmatrix} 0 & 1 & 0 & 3 \\ 0 & 0 & 1 & 1 \\ 0 & 0 & 0 & 4 \\ 0 & 0 & 0 & 2 \end{pmatrix}^2$$

$$= \begin{pmatrix} 0 & 0 & 1 & 7 \\ 0 & 0 & 0 & 6 \\ 0 & 0 & 0 & 8 \\ 0 & 0 & 0 & 4 \end{pmatrix}.$$

Solve $(\mathbf{A} - 2\mathbf{I}_4)^2\mathbf{K}_1 = \mathbf{O}$ to find solutions of the form

$$\begin{pmatrix} \alpha \\ \beta \\ 0 \\ 0 \end{pmatrix}.$$

We will choose the solution

$$\mathbf{K}_1 = \begin{pmatrix} 0 \\ 1 \\ 0 \\ 0 \end{pmatrix}$$

to avoid duplicating the eigenvector already found associated with 2. Then

$$(\mathbf{A} - 2\mathbf{I}_4)\mathbf{K}_1 = \begin{pmatrix} 0 & 1 & 0 & 3 \\ 0 & 0 & 1 & 1 \\ 0 & 0 & 0 & 4 \\ 0 & 0 & 0 & 2 \end{pmatrix} \begin{pmatrix} 0 \\ 1 \\ 0 \\ 0 \end{pmatrix} = \begin{pmatrix} 1 \\ 0 \\ 0 \\ 0 \end{pmatrix} \neq \mathbf{0},$$

as required. Thus form the third solution:

$$\boldsymbol{\Phi}_3(t) = e^{\mathbf{A}t}\mathbf{K}_1 = e^{2t}\left[\mathbf{K}_1 + (\mathbf{A} - 2\mathbf{I}_4)\mathbf{K}_1 t\right]$$

$$= e^{2t}\left[\begin{pmatrix} 0 \\ 1 \\ 0 \\ 0 \end{pmatrix} + \begin{pmatrix} 0 & 1 & 0 & 3 \\ 0 & 0 & 1 & 1 \\ 0 & 0 & 0 & 4 \\ 0 & 0 & 0 & 2 \end{pmatrix}\begin{pmatrix} 0 \\ 1 \\ 0 \\ 0 \end{pmatrix} t\right]$$

$$= e^{2t}\left[\begin{pmatrix} 0 \\ 1 \\ 0 \\ 0 \end{pmatrix} + \begin{pmatrix} 1 \\ 0 \\ 0 \\ 0 \end{pmatrix} t\right] = \begin{pmatrix} t \\ 1 \\ 0 \\ 0 \end{pmatrix} e^{2t}.$$

The three solutions found up to this point are linearly independent. Now we need a fourth solution. It must come from the eigenvalue 2, because 4 has multiplicity 1 and we have one solution corresponding to this eigenvalue. Look for $\mathbf{K}_2$ such that

$$(\mathbf{A} - 2\mathbf{I}_4)\mathbf{K}_2 \neq \mathbf{0} \quad \text{and} \quad (\mathbf{A} - 2\mathbf{I}_4)^2\mathbf{K}_2 \neq \mathbf{0}$$

but

$$(\mathbf{A} - 2\mathbf{I}_4)^3\mathbf{K}_2 = \mathbf{0}.$$

First compute

$$(\mathbf{A} - 2\mathbf{I}_4)^3 = \begin{pmatrix} 0 & 0 & 0 & 18 \\ 0 & 0 & 0 & 12 \\ 0 & 0 & 0 & 16 \\ 0 & 0 & 0 & 8 \end{pmatrix}.$$

Solutions of $(\mathbf{A} - 2\mathbf{I}_4)^3\mathbf{K}_2 = \mathbf{0}$ are of the form

$$\begin{pmatrix} \alpha \\ \beta \\ \gamma \\ 0 \end{pmatrix}.$$

We will choose

$$\mathbf{K}_2 = \begin{pmatrix} 1 \\ 1 \\ 1 \\ 0 \end{pmatrix}$$

to avoid duplicating previous choices. Of course, other choices are possible. It is routine to verify that $(\mathbf{A} - 2\mathbf{I}_4)\mathbf{K}_2 \neq \mathbf{O}$ and $(\mathbf{A} - 2\mathbf{I}_4)^2\mathbf{K}_2 \neq \mathbf{O}$. Thus form the fourth solution:

$$\boldsymbol{\Phi}_4(t) = e^{\mathbf{A}t}\mathbf{K}_2 = e^{2t}\left[\mathbf{K}_2 + (\mathbf{A} - 2\mathbf{I}_n)\mathbf{K}_2 t + \frac{1}{2!}(\mathbf{A} - 2\mathbf{I}_n)^2 t^2\right]$$

$$= e^{2t}\left[\begin{pmatrix}1\\1\\1\\0\end{pmatrix} + \begin{pmatrix}0&1&0&3\\0&0&1&1\\0&0&0&4\\0&0&0&2\end{pmatrix}\begin{pmatrix}1\\1\\1\\0\end{pmatrix}t + \frac{1}{2}\begin{pmatrix}0&0&1&7\\0&0&0&6\\0&0&0&8\\0&0&0&4\end{pmatrix}\begin{pmatrix}1\\1\\1\\0\end{pmatrix}t^2\right]$$

$$= \begin{pmatrix}1 + t + \frac{1}{2}t^2\\1 + t\\1\\0\end{pmatrix}e^{2t}.$$

We now have four linearly independent solutions, hence the general solution. We can also write the fundamental matrix

$$\boldsymbol{\Omega}(t) = \begin{pmatrix}9e^{4t} & e^{2t} & te^{2t} & \left(1 + t + \frac{1}{2}t^2\right)e^{2t}\\6e^{4t} & 0 & e^{2t} & (1 + t)e^{2t}\\8e^{4t} & 0 & 0 & e^{2t}\\4e^{4t} & 0 & 0 & 0\end{pmatrix}. \quad \blacksquare$$

SECTION 9.2   PROBLEMS

In each of Problems 1 through 7, find a fundamental matrix for the system and use it to write the general solution. These coefficient matrices for these systems have real, distinct eigenvalues.

1. $x_1' = 3x_1, x_2' = 5x_1 - 4x_2$

2. $x_1' = 4x_1 + 2x_2, x_2' = 3x_1 + 3x_2$

3. $x_1' = x_1 + x_2, x_2' = x_1 + x_2$

4. $x_1' = 2x_1 + x_2 - 2x_3, x_2' = 3x_1 - 2x_2,$

   $x_3' = 3x_1 - x_2 - 3x_3$

5. $x_1' = x_1 + 2x_2 + x_3, x_2' = 6x_1 - x_2, x_3' = -x_1 - 2x_2 - x_3$

6. $x_1' = 6x_1 + 2x_2, x_2' = 4x_1 + 4x_2, x_3' = 2x_3 + 2x_4,$

   $x_4' = x_3 + 3x_4$

7. $x_1' = x_1 - x_2 + 4x_3, x_2' = 3x_1 + 2x_2 - x_3,$

   $x_3' = 2x_1 + x_2 - x_3$

In each of Problems 8 through 13, find a fundamental matrix for the system and use it to solve the initial value problem. The matrices of these systems have real, distinct eigenvalues.

8. $x_1' = 3x_1 - 4x_2, x_2' = 2x_1 - 3x_2; x_1(0) = 7, x_2(0) = 5$

9. $x_1' = x_1 - 2x_2, x_2' = -6x_1; x_1(0) = 1, x_2(0) = -19$

10. $x_1' = 2x_1 - 10x_2, x_2' = -x_1 - x_2;$

    $x_1(0) = -3, x_2(0) = 6$

11. $x_1' = 3x_1 - x_2 + x_3, x_2' = x_1 + x_2 - x_3,$

    $x_3' = x_1 - x_2 + x_3;$

    $x_1(0) = 1, x_2(0) = 5, x_3(0) = 1$

12. $x_1' = 2x_1 + x_2 - 2x_3, x_2' = 3x_1 - 2x_2,$

    $x_3' = 3x_1 + x_2 - 3x_3;$

    $x_1(0) = 1, x_2(0) = 7, x_3(0) = 3$

13. $x_1' = 2x_1 + 3x_2 + 3x_3, x_2' = -x_2 - 3x_3, x_3' = 2x_3;$

    $x_1(0) = 9, x_2(0) = -1, x_3(0) = 3$

14. Show that the change of variables $z = \ln(t)$ for $t > 0$ transforms the system

    $$tx_1' = ax_1 + bx_2, \quad tx_2' = cx_1 + dx_2$$

    into a linear system $\mathbf{X}' = \mathbf{A}\mathbf{X}$, assuming that $a, b, c,$ and d are real constants.

15. Use the idea of Problem 14 to solve the system

    $$tx_1' = 6x_1 + 2x_2, \quad tx_2' = 4x_1 + 4x_2.$$

16. Solve the system

$$tx_1' = -x_1 - 3x_2, \quad tx_2' = x_1 - 5x_2.$$

In each of Problems 17 through 23, find a real-valued fundamental matrix for the system $\mathbf{X}' = \mathbf{AX}$, with $\mathbf{A}$ the given matrix.

17. $\begin{pmatrix} 2 & -4 \\ 1 & 2 \end{pmatrix}$

18. $\begin{pmatrix} 0 & 5 \\ -1 & -2 \end{pmatrix}$

19. $\begin{pmatrix} 3 & -5 \\ 1 & -1 \end{pmatrix}$

20. $\begin{pmatrix} 1 & -1 & 1 \\ 1 & -1 & 0 \\ 1 & 0 & -1 \end{pmatrix}$

21. $\begin{pmatrix} -2 & 1 & 0 \\ -5 & 0 & 0 \\ 0 & 3 & -2 \end{pmatrix}$

22. $\begin{pmatrix} 3 & 0 & 1 \\ 9 & -1 & 2 \\ -9 & 4 & -1 \end{pmatrix}$

23. $\begin{pmatrix} 3 & -2 & 0 & 0 \\ 5 & -3 & 0 & 0 \\ 0 & 0 & 3 & -2 \\ 0 & 0 & 5 & -3 \end{pmatrix}$

In each of Problems 24 through 28, find a real-valued fundamental matrix for the system $\mathbf{X}' = \mathbf{AX}$, with $\mathbf{A}$ the given matrix. Use this to solve the initial value problem, with $\mathbf{X}(0)$ the given $n \times 1$ matrix.

24. $\begin{pmatrix} 3 & 2 \\ -5 & 1 \end{pmatrix}; \begin{pmatrix} 2 \\ 8 \end{pmatrix}$

25. $\begin{pmatrix} 3 & -2 \\ 5 & -3 \end{pmatrix}; \begin{pmatrix} 2 \\ 10 \end{pmatrix}$

26. $\begin{pmatrix} 2 & -5 \\ 1 & -2 \end{pmatrix}; \begin{pmatrix} 5 \\ 0 \end{pmatrix}$

27. $\begin{pmatrix} 3 & -3 & 1 \\ 2 & -1 & 0 \\ 1 & -1 & 1 \end{pmatrix}; \begin{pmatrix} 7 \\ 4 \\ 3 \end{pmatrix}$

28. $\begin{pmatrix} 2 & -5 & 0 \\ 2 & -4 & 0 \\ 4 & -5 & -2 \end{pmatrix}; \begin{pmatrix} 5 \\ 5 \\ 9 \end{pmatrix}$

29. Solve the initial value problem

$$tx_1' = 5x_1 - 4x_2, \quad tx_2' = 2x_1 + x_2;$$
$$x_1(1) = 6, \quad x_2(1) = 5.$$

(Note Problem 14).

30. Can a matrix with at least one complex, nonreal element have only real eigenvalues? If not, give a proof. If it can, give an example.

In each of Problems 31 through 38, find a fundamental matrix for the system $\mathbf{X}' = \mathbf{AX}$, using the method of Section 9.2.2, with $\mathbf{A}$ the given matrix.

31. $\begin{pmatrix} 3 & 2 \\ 0 & 3 \end{pmatrix}$

32. $\begin{pmatrix} 2 & 0 \\ 5 & 2 \end{pmatrix}$

33. $\begin{pmatrix} 2 & -4 \\ 1 & 6 \end{pmatrix}$

34. $\begin{pmatrix} 5 & -3 \\ 3 & -1 \end{pmatrix}$

35. $\begin{pmatrix} 2 & 5 & 6 \\ 0 & 8 & 9 \\ 0 & -1 & 2 \end{pmatrix}$

36. $\begin{pmatrix} 1 & 5 & 0 \\ 0 & 1 & 0 \\ 4 & 8 & 1 \end{pmatrix}$

37. $\begin{pmatrix} 1 & 5 & -2 & 6 \\ 0 & 3 & 0 & 4 \\ 0 & 3 & 0 & 4 \\ 0 & 0 & 0 & 1 \end{pmatrix}$

38. $\begin{pmatrix} 0 & 1 & 0 & 0 \\ 0 & 0 & 1 & 0 \\ 0 & 0 & 0 & 1 \\ -1 & 0 & -2 & 0 \end{pmatrix}$

In each of Problems 39 through 44, find the general solution of the system $\mathbf{X}' = \mathbf{AX}$, with $\mathbf{A}$ the given matrix, and use this general solution to solve the initial value problem, for the given $n \times 1$ matrix $\mathbf{X}(0)$. Use the method of Section 9.2.2 for these problems.

39. $\begin{pmatrix} 7 & -1 \\ 1 & 5 \end{pmatrix}; \begin{pmatrix} 5 \\ 3 \end{pmatrix}$

40. $\begin{pmatrix} 2 & 0 \\ 5 & 2 \end{pmatrix}; \begin{pmatrix} 4 \\ 3 \end{pmatrix}$

41. $\begin{pmatrix} -4 & 1 & 1 \\ 0 & 2 & -5 \\ 0 & 0 & -4 \end{pmatrix}; \begin{pmatrix} 0 \\ 4 \\ 12 \end{pmatrix}$

42. $\begin{pmatrix} -5 & 2 & 1 \\ 0 & -5 & 3 \\ 0 & 0 & -5 \end{pmatrix}; \begin{pmatrix} 2 \\ -3 \\ 4 \end{pmatrix}$

43. $\begin{pmatrix} 1 & -2 & 0 & 0 \\ 1 & -1 & 0 & 0 \\ 0 & 0 & 5 & -3 \\ 0 & 0 & 3 & -1 \end{pmatrix}; \begin{pmatrix} 2 \\ -2 \\ 1 \\ 4 \end{pmatrix}$

44. $\begin{pmatrix} 1 & 4 & 0 & 0 \\ 0 & 1 & 0 & 0 \\ 0 & 0 & 1 & 0 \\ 1 & -3 & 2 & 0 \end{pmatrix}; \begin{pmatrix} 7 \\ 1 \\ -4 \\ -6 \end{pmatrix}$

45. Let $\mathbf{A} = [a_{ij}]$ be an $n \times n$ diagonal matrix of numbers. Let $\mathbf{\Omega}(t) = [\omega_{ij}(t)]$ be the $n \times n$ diagonal matrix defined by $\omega_{ij}(t) = 0$ for all t if $i \neq j$, and $\omega_{jj}(t) = e^{a_{jj}t}$. Prove that $\mathbf{\Omega}$ is a fundamental matrix for $\mathbf{X}' = \mathbf{AX}$.

In each of Problems 46 through 54, find the general solution of the system by diagonalizing the coefficient matrix.

46. $x_1' = -2x_1 + x_2, x_2' = -4x_1 + 3x_2$

47. $x_1' = 3x_1 + 3x_2, x_1' = x_1 + 5x_2$

48. $x_1' = x_1 + x_2, x_2' = x_1 + x_2$

49. $x_1' = 6x_1 + 5x_2, x_2' = x_1 + 2x_2$

50. $x_1' = 3x_1 - 2x_2, x_2' = 9x_1 - 3x_2$

51. $x_1' = 2x_1 + x_2 - 2x_3, x_2' = 3x_1 - 2x_2,$
    $x_3' = 3x_1 + x_2 - 3x_3$

52. $x_1' = 3x_1 - x_2 + x_3, x_2' = x_1 + x_2 - x_3, x_3' = x_1 - x_2 + x_3$

53. $x_1' = x_1 - x_2 - x_3, x_2' = x_1 - x_2, x_3' = x_1 - x_3$

54. $x_1' = x_1 + x_2, x_2' = x_1 + x_2, x_3' = 4x_3 + 2x_4,$
    $x_4' = 3x_3 + 3x_4$

In each of Problems 55 through 62, solve the system $\mathbf{X}' = \mathbf{AX}$, with $\mathbf{A}$ the matrix of the indicated problem, by finding $e^{\mathbf{A}t}$.

55. $\mathbf{A}$ as in Problem 31.
56. $\mathbf{A}$ as in Problem 32.
57. $\mathbf{A}$ as in Problem 33.
58. $\mathbf{A}$ as in Problem 34.
59. $\mathbf{A}$ as in Problem 35.
60. $\mathbf{A}$ as in Problem 36.
61. $\mathbf{A}$ as in Problem 37.
62. $\mathbf{A}$ as in Problem 38.

In each of Problems 63 through 68, solve the initial value problem of the referred problem, using the exponential matrix.

63. Problem 39.
64. Problem 40.
65. Problem 41.
66. Problem 42.
67. Problem 43.
68. Problem 44.

69. Suppose $\mathbf{A}$ and $\mathbf{B}$ are $n \times n$ matrices such that $\mathbf{AB} = \mathbf{BA}$. Prove that $(\mathbf{A} + \mathbf{B})^2 = \mathbf{A}^2 + 2\mathbf{AB} + \mathbf{B}^2$. Show by an example that this equation can fail to hold if $\mathbf{AB} \neq \mathbf{BA}$.

70. Suppose $\mathbf{A}$ and $\mathbf{B}$ are $n \times n$ matrices such that $\mathbf{AB} = \mathbf{BA}$. Prove that for any positive integer k,
$$(\mathbf{A} + \mathbf{B})^k = \sum_{j=0}^{k} \frac{k!}{j!(k-j)!} \mathbf{A}^j \mathbf{B}^{k-j}.$$

71. Suppose $\mathbf{A}$ and $\mathbf{B}$ are $n \times n$ matrices such that $\mathbf{AB} = \mathbf{BA}$. Prove that $e^{(\mathbf{A}+\mathbf{B})t} = e^{\mathbf{A}t} e^{\mathbf{B}t}$.

9.3   Solution of $\mathbf{X}' = \mathbf{AX} + \mathbf{G}$

We now turn to the nonhomogeneous system $\mathbf{X}'(t) = \mathbf{A}(t)\mathbf{X}(t) + \mathbf{G}(t)$, assuming that the elements of the $n \times n$ matrix $\mathbf{A}(t)$, and the $n \times 1$ matrix $\mathbf{G}(t)$, are continuous on some interval I, which may be the entire real line.

Recall that the general solution of $\mathbf{X}' = \mathbf{AX} + \mathbf{G}$ has the form $\mathbf{X}(t) = \mathbf{\Omega}(t)\mathbf{C} + \mathbf{\Psi}_p(t)$, where $\mathbf{\Omega}(t)$ is an $n \times n$ fundamental matrix for the homogeneous system $\mathbf{X}' = \mathbf{AX}$, $\mathbf{C}$ is an $n \times 1$ matrix of arbitrary constants, and $\mathbf{\Psi}_p$ is a particular solution of $\mathbf{X}' = \mathbf{AX} + \mathbf{G}$. At least when $\mathbf{A}$ is a real, constant matrix, we have a strategy for finding $\mathbf{\Omega}$. We will concentrate in this section on strategies for finding a particular solution $\mathbf{\Psi}_p$.

9.3.1   Variation of Parameters

Recall the variation of parameters method for second-order differential equations. If $y_1(x)$ and $y_2(x)$ form a fundamental set of solutions for

$$y''(x) + p(x)y'(x) + q(x)y(x) = 0,$$

then the general solution of this homogeneous equation is

$$y_h(x) = c_1 y_1(x) + c_2 y_2(x).$$

To find a particular solution $y_p(x)$ of the nonhomogeneous equation

$$y''(x) + p(x)y'(x) + q(x)y(x) = f(x),$$

replace the constants in y_h by functions and attempt to choose $u(x)$ and $v(x)$ so that

$$y_p(x) = u(x)y_1(x) + v(x)y_2(x)$$

is a solution.

The variation of parameters method for the matrix equation $\mathbf{X}' = \mathbf{AX} + \mathbf{G}$ follows the same idea. Suppose we can find a fundamental matrix for the homogeneous system $\mathbf{X}' = \mathbf{AX}$. The general solution of this homogeneous system is then $\mathbf{X}_h(t) = \mathbf{\Omega}(t)\mathbf{C}$, in which $\mathbf{C}$ is an $n \times 1$ matrix of arbitrary constants. Look for a particular solution of $\mathbf{X}' = \mathbf{AX} + \mathbf{G}$ of the form

$$\mathbf{\Psi}_p(t) = \mathbf{\Omega}(t)\mathbf{U}(t),$$

in which $\mathbf{U}(t)$ is an $n \times 1$ matrix of functions of t, which is to be determined.

Substitute this proposed solution into the differential equation to get

$$(\mathbf{\Omega U})' = \mathbf{A}(\mathbf{\Omega U}) + \mathbf{G},$$

or

$$\mathbf{\Omega}'\mathbf{U} + \mathbf{\Omega U}' = (\mathbf{A\Omega})\mathbf{U} + \mathbf{G}.$$

Now $\mathbf{\Omega}$ is a fundamental matrix for $\mathbf{X}' = \mathbf{AX}$, so $\mathbf{\Omega}' = \mathbf{A\Omega}$. Therefore, $\mathbf{\Omega}'\mathbf{U} = (\mathbf{A\Omega})\mathbf{U}$ and the last equation reduces to

$$\mathbf{\Omega U}' = \mathbf{G}.$$

Since $\mathbf{\Omega}$ is a fundamental matrix, the columns of $\mathbf{\Omega}$ are linearly independent. This means that $\mathbf{\Omega}$ is nonsingular, so the last equation can be solved for $\mathbf{U}'$ to get

$$\mathbf{U}' = \mathbf{\Omega}^{-1}\mathbf{G}.$$

As in the case of second-order differential equations, we now have the derivative of the function we want. Then

$$\mathbf{U}(t) = \int \mathbf{\Omega}^{-1}(t)\mathbf{G}(t)\,dt,$$

in which we integrate a matrix by integrating each element of the matrix.

Once we find a suitable $\mathbf{U}(t)$, we have a particular solution $\mathbf{\Psi}_p(t) = \mathbf{\Omega}(t)\mathbf{U}(t)$ of $\mathbf{X}' = \mathbf{AX} + \mathbf{G}$. The general solution of this nonhomogeneous equation is then

$$\mathbf{X}(t) = \mathbf{\Omega}(t)\mathbf{C} + \mathbf{\Omega}(t)\mathbf{U}(t),$$

in which $\mathbf{C}$ is an $n \times 1$ matrix of constants.

EXAMPLE 9.15

Solve the system

$$\mathbf{X}' = \begin{pmatrix} 1 & -10 \\ -1 & 4 \end{pmatrix} \mathbf{X} + \begin{pmatrix} e^t \\ \sin(t) \end{pmatrix}.$$

First we need a fundamental matrix for $\mathbf{X}' = \mathbf{AX}$. The eigenvalues of $\mathbf{A}$ are -1 and 6, with associated eigenvectors, respectively, $\begin{pmatrix} 5 \\ 1 \end{pmatrix}$ and $\begin{pmatrix} -2 \\ 1 \end{pmatrix}$. Therefore, a fundamental matrix for $\mathbf{X}' = \mathbf{AX}$ is

$$\mathbf{\Omega}(t) = \begin{pmatrix} 5e^{-t} & -2e^{6t} \\ e^{-t} & e^{6t} \end{pmatrix}.$$

We find (details provided at the end of the example) that

$$\mathbf{\Omega}^{-1}(t) = \frac{1}{7} \begin{pmatrix} e^t & 2e^t \\ -e^{-6t} & 5e^{-6t} \end{pmatrix}.$$

Compute

$$\mathbf{U}'(t) = \mathbf{\Omega}^{-1}(t)\mathbf{G}(t) = \frac{1}{7} \begin{pmatrix} e^t & 2e^t \\ -e^{-6t} & 5e^{-6t} \end{pmatrix} \begin{pmatrix} e^t \\ \sin(t) \end{pmatrix}$$

$$= \frac{1}{7} \begin{pmatrix} e^{2t} + 2e^t \sin(t) \\ -e^{-5t} + 5e^{-6t} \sin(t) \end{pmatrix}.$$

Then

$$\mathbf{U}(t) = \int \mathbf{\Omega}^{-1}(t)\mathbf{G}(t)\, dt = \frac{1}{7} \begin{pmatrix} \int e^{2t}\, dt + 2\int e^t \sin(t)\, dt \\ -\int e^{-5t}\, dt + 5\int e^{-6t} \sin(t)\, dt \end{pmatrix}$$

$$= \begin{pmatrix} \frac{1}{14}e^{2t} + \frac{1}{7}e^t[\sin(t) - \cos(t)] \\ \frac{1}{35}e^{-5t} + \frac{5}{259}e^{-6t}[-6\sin(t) - \cos(t)] \end{pmatrix}.$$

The general solution of $\mathbf{X}' = \mathbf{AX} + \mathbf{G}$ is

$$\mathbf{X}(t) = \mathbf{\Omega}(t)\mathbf{C} + \mathbf{\Omega}(t)\mathbf{U}(t) = \begin{pmatrix} 5e^{-t} & -2e^{6t} \\ e^{-t} & e^{6t} \end{pmatrix} \mathbf{C}$$

$$+ \begin{pmatrix} 5e^{-t} & -2e^{6t} \\ e^{-t} & e^{6t} \end{pmatrix} \begin{pmatrix} \frac{1}{14}e^{2t} + \frac{1}{7}e^t[\sin(t) - \cos(t)] \\ \frac{1}{35}e^{-5t} + \frac{5}{259}e^{-6t}[-6\sin(t) - \cos(t)] \end{pmatrix}$$

$$= \begin{pmatrix} 5e^{-t} & -2e^{6t} \\ e^{-t} & e^{6t} \end{pmatrix} \mathbf{C} + \begin{pmatrix} \frac{3}{10}e^t + \frac{35}{37}\sin(t) - \frac{25}{37}\cos(t) \\ \frac{1}{10}e^t + \frac{1}{37}\sin(t) - \frac{6}{37}\cos(t) \end{pmatrix}.$$

If we want to write the solution in terms of the component functions, let $\mathbf{C} = \begin{pmatrix} c_1 \\ c_2 \end{pmatrix}$ to obtain

$$x_1(t) = 5c_1 e^{-t} - 2c_2 e^{6t} + \frac{3}{10}e^t + \frac{35}{37}\sin(t) - \frac{25}{37}\cos(t),$$

$$x_2(t) = c_1 e^{-t} + c_2 e^{6t} + \frac{1}{10}e^t + \frac{1}{37}\sin(t) - \frac{6}{37}\cos(t). \quad \blacksquare$$

Although the coefficient matrix **A** in this example was constant, this is not a requirement of the variation of parameters method.

In the example we needed $\boldsymbol{\Omega}^{-1}(t)$. Standard software packages will produce this inverse. We could also proceed as follows, reducing $\boldsymbol{\Omega}(t)$ and recording the row operations beginning with the identity matrix, as discussed in Section 6.9.1:

$$\begin{pmatrix} 1 & 0 & | & 5e^{-t} & -2e^{6t} \\ 0 & 1 & | & e^{-t} & e^{6t} \end{pmatrix}$$

add $-\frac{1}{5}$(row 1) to row 2:

$$\begin{pmatrix} 1 & 0 & | & 5e^{-t} & -2e^{6t} \\ -\frac{1}{5} & 1 & | & 0 & \frac{7}{5}e^{6t} \end{pmatrix}$$

multiply row 1 by $\frac{1}{5}e^{t}$:

$$\begin{pmatrix} \frac{1}{5}e^{t} & 0 & | & 1 & -\frac{2}{5}e^{7t} \\ -\frac{1}{5} & 1 & | & 0 & \frac{7}{5}e^{6t} \end{pmatrix}$$

multiply row 2 by $\frac{5}{7}e^{-6t}$:

$$\begin{pmatrix} \frac{1}{5}e^{t} & 0 & | & 1 & -\frac{2}{5}e^{7t} \\ -\frac{1}{7}e^{-6t} & \frac{5}{7}e^{-6t} & | & 0 & 1 \end{pmatrix}$$

add $\frac{2}{5}e^{7t}$(row 2) to row 1:

$$\begin{pmatrix} \frac{1}{7}e^{t} & \frac{2}{7}e^{t} & | & 1 & 0 \\ -\frac{1}{7}e^{-6t} & \frac{5}{7}e^{-6t} & | & 0 & 1 \end{pmatrix}.$$

Since the last two columns form $\mathbf{I}_2$, the first two columns are $\boldsymbol{\Omega}^{-1}(t)$.

EXAMPLE 9.16

Suppose we want to find the currents in the circuit of Figure 9.2. Assume that all currents and charges are zero until the switch is closed at time zero.

By Kirchhoff's laws, using the two interior loops,

$$10i_1 + 4(i_1' - i_2') = 4$$

$$4(i_1' - i_2') = 100q_2.$$

Using the external loop,

$$10i_1 + 100q_2 = 4.$$

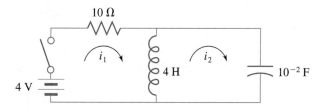

10 Ω

i_1

i_2

4 H

10^{-2} F

4 V

FIGURE 9.2

Any two of these equations are enough to find the currents. To avoid a mix of terms involving charge and derivatives of the current, use the first equation and the derivative of the third to get

$$i_1' = -10i_2$$

$$2i_1' - 2i_2' = -5i_1 + 2.$$

Write this system as

$$\begin{pmatrix} 1 & 0 \\ 2 & -2 \end{pmatrix} \begin{pmatrix} i_1 \\ i_2 \end{pmatrix}' = \begin{pmatrix} 0 & -10 \\ -5 & 0 \end{pmatrix} \begin{pmatrix} i_1 \\ i_2 \end{pmatrix} + \begin{pmatrix} 0 \\ 2 \end{pmatrix}. \tag{9.7}$$

Compute

$$\begin{pmatrix} 1 & 0 \\ 2 & -2 \end{pmatrix}^{-1} = \begin{pmatrix} 1 & 0 \\ 1 & -\frac{1}{2} \end{pmatrix}$$

and multiply equation (9.7) on the left by this inverse to get

$$\begin{pmatrix} i_1 \\ i_2 \end{pmatrix}' = \begin{pmatrix} 1 & 0 \\ 1 & -\frac{1}{2} \end{pmatrix} \begin{pmatrix} 0 & -10 \\ -5 & 0 \end{pmatrix} \begin{pmatrix} i_1 \\ i_2 \end{pmatrix} + \begin{pmatrix} 1 & 0 \\ 1 & -\frac{1}{2} \end{pmatrix} \begin{pmatrix} 0 \\ 2 \end{pmatrix},$$

or

$$\begin{pmatrix} i_1 \\ i_2 \end{pmatrix}' = \begin{pmatrix} 0 & -10 \\ \frac{5}{2} & -10 \end{pmatrix} \begin{pmatrix} i_1 \\ i_2 \end{pmatrix} + \begin{pmatrix} 0 \\ -1 \end{pmatrix}.$$

If we write $\mathbf{i} = \begin{pmatrix} i_1 \\ i_2 \end{pmatrix}$, then this is the system

$$\mathbf{i}' = \begin{pmatrix} 0 & -10 \\ \frac{5}{2} & -10 \end{pmatrix} \mathbf{i} + \begin{pmatrix} 0 \\ -1 \end{pmatrix}.$$

This equation is in the standard form $\mathbf{X}' = \mathbf{AX} + \mathbf{G}$ that we have considered. First solve the homogeneous system $\mathbf{i}' = \mathbf{Ai}$. Now $\begin{pmatrix} 0 & -10 \\ \frac{5}{2} & -10 \end{pmatrix}$ has a single eigenvalue -5 with multiplicity 2. Corresponding to this eigenvalue is an eigenvector $\begin{pmatrix} 2 \\ 1 \end{pmatrix}$. Thus one solution of $\mathbf{i}' = \mathbf{Ai}$ is

$$\mathbf{\Phi}_1(t) = \begin{pmatrix} 2 \\ 1 \end{pmatrix} e^{-5t}.$$

For a second solution, try

$$\mathbf{\Phi}_2(t) = \begin{pmatrix} 2 \\ 1 \end{pmatrix} t e^{-5t} + \mathbf{E}_2 e^{-5t}.$$

Substitute into $\mathbf{i}' = \mathbf{Ai}$ to get

$$\begin{pmatrix} 2 \\ 1 \end{pmatrix} e^{-5t} - 5 \begin{pmatrix} 2 \\ 1 \end{pmatrix} t e^{-5t} - 5\mathbf{E}_2 e^{-5t} = \begin{pmatrix} 0 & -10 \\ \frac{5}{2} & -10 \end{pmatrix} \begin{pmatrix} 2 \\ 1 \end{pmatrix} t e^{-5t} + \begin{pmatrix} 0 & -10 \\ \frac{5}{2} & -10 \end{pmatrix} \mathbf{E}_2 e^{-5t}.$$

The terms involving $t e^{-5t}$ cancel. Divide the remaining terms of this equation by e^{-5t} to get

$$\begin{pmatrix} 2 \\ 1 \end{pmatrix} - 5\mathbf{E}_2 = \begin{pmatrix} 0 & -10 \\ \frac{5}{2} & -10 \end{pmatrix} \mathbf{E}_2.$$

Then

$$
\binom{2}{1} = \begin{pmatrix} 0 & -10 \\ \frac{5}{2} & -10 \end{pmatrix} \mathbf{E}_2 + 5 \begin{pmatrix} 1 & 0 \\ 0 & 1 \end{pmatrix} \mathbf{E}_2
$$

$$
= \begin{pmatrix} 5 & -10 \\ \frac{5}{2} & -5 \end{pmatrix} \mathbf{E}_2.
$$

This has infinitely many solutions for $\mathbf{E}_2$, all of the form $\mathbf{E}_2 = \begin{pmatrix} \alpha \\ \frac{1}{10}(5\alpha - 2) \end{pmatrix}$, with α any number.

Since we need only one choice for $\mathbf{E}_2$, take $\alpha = 1$ and $\mathbf{E}_2 = \begin{pmatrix} 1 \\ \frac{3}{10} \end{pmatrix}$. A second solution of $\mathbf{i}' = \mathbf{A}\mathbf{i}$
is

$$
\mathbf{\Phi}_2(t) = \binom{2}{1} t e^{-5t} + \begin{pmatrix} 1 \\ \frac{3}{10} \end{pmatrix} e^{-5t}
$$

$$
= \begin{pmatrix} (1 + 2t)e^{-5t} \\ \left(\frac{3}{10} + t\right) e^{-5t} \end{pmatrix}.
$$

Using $\mathbf{\Phi}_1(t)$ and $\mathbf{\Phi}_2(t)$ as columns, we have the fundamental matrix

$$
\mathbf{\Omega}(t) = \begin{pmatrix} 2e^{-5t} & (1 + 2t)e^{-5t} \\ e^{-5t} & \left(\frac{3}{10} + t\right) e^{-5t} \end{pmatrix}
$$

for the homogeneous system $\mathbf{i}' = \mathbf{A}\mathbf{i}$. Now use variation of parameters to find a particular solution $\mathbf{\Psi}_p(t)$ of $\mathbf{i}' = \mathbf{A}\mathbf{i} + \mathbf{G}$. First, find

$$
\mathbf{\Omega}^{-1}(t) = \begin{pmatrix} -\frac{1}{4}(3 + 10t)e^{5t} & \frac{5}{2}(1 + 2t)e^{5t} \\ \frac{5}{2}e^{5t} & -5e^{5t} \end{pmatrix}.
$$

Next, calculate

$$
\mathbf{U}(t) = \int \mathbf{\Omega}^{-1}(t)\mathbf{G}(t)\, dt
$$

$$
= \int \begin{pmatrix} -\frac{1}{4}(3 + 10t)e^{5t} & \frac{5}{2}(1 + 2t)e^{5t} \\ \frac{5}{2}e^{5t} & -5e^{5t} \end{pmatrix} \binom{0}{-1} dt
$$

$$
= \int \begin{pmatrix} -\frac{5}{2}(1 + 2t)e^{5t} \\ 5e^{5t} \end{pmatrix} dt = \begin{pmatrix} -\frac{5}{2}\int(1 + 2t)e^{5t}\, dt \\ \int 5e^{5t}\, dt \end{pmatrix}
$$

$$
= \begin{pmatrix} -\frac{3}{10}e^{5t} - e^{5t}t \\ e^{5t} \end{pmatrix}.
$$

A particular solution is

$$
\mathbf{\Psi}_p(t) = \mathbf{\Omega}(t)\mathbf{U}(t) = \begin{pmatrix} 2e^{-5t} & (1 + 2t)e^{-5t} \\ e^{-5t} & \left(\frac{3}{10} + t\right) e^{-5t} \end{pmatrix} \begin{pmatrix} -\frac{3}{10}e^{5t} - e^{5t}t \\ e^{5t} \end{pmatrix}
$$

$$
= \begin{pmatrix} \left(-\frac{3}{5} - 2t\right) + (1 + 2t) \\ \left(-\frac{3}{10} - t\right) + \left(\frac{3}{10} + t\right) \end{pmatrix} = \begin{pmatrix} \frac{2}{5} \\ 0 \end{pmatrix}.
$$

The general solution of $\mathbf{i}' = \mathbf{Ai} + \mathbf{G}$ is

$$\mathbf{i}(t) = \begin{pmatrix} 2e^{-5t} & (1+2t)e^{-5t} \\ e^{-5t} & \left(\frac{3}{10}+t\right)e^{-5t} \end{pmatrix} \mathbf{C} + \begin{pmatrix} \frac{2}{5} \\ 0 \end{pmatrix}, \tag{9.8}$$

in which $\mathbf{C}$ is a 2×1 matrix of arbitrary constants.

Now consider initial conditions. The current $i_1 - i_2$ through the inductor is zero prior to the switch being closed, and so in the limit is zero just after. Then $i_1(0+) = i_2(0+)$. Since the charge on the capacitor is zero prior to time zero, then from the external loop we have

$$10i_1(0+) + 100q_2(0+) = 10i_1(0+) = 4,$$

so

$$i_1(0+) = i_2(0+) = \tfrac{2}{5}.$$

The initial condition is

$$\mathbf{i}(0) = \begin{pmatrix} \frac{2}{5} \\ \frac{2}{5} \end{pmatrix}.$$

Use this to solve for $\mathbf{C}$ in equation (9.8):

$$\mathbf{i}(0) = \begin{pmatrix} \frac{2}{5} \\ \frac{2}{5} \end{pmatrix} = \begin{pmatrix} 2 & 1 \\ 1 & \frac{3}{10} \end{pmatrix} \mathbf{C} + \begin{pmatrix} \frac{2}{5} \\ 0 \end{pmatrix}.$$

Then

$$\begin{pmatrix} 2 & 1 \\ 1 & \frac{3}{10} \end{pmatrix} \mathbf{C} = \begin{pmatrix} 0 \\ \frac{2}{5} \end{pmatrix}$$

and

$$\mathbf{C} = \begin{pmatrix} 2 & 1 \\ 1 & \frac{3}{10} \end{pmatrix}^{-1} \begin{pmatrix} 0 \\ \frac{2}{5} \end{pmatrix} = \begin{pmatrix} 1 \\ -2 \end{pmatrix}.$$

The solution of the initial value problem for the current is

$$\mathbf{i}(t) = \begin{pmatrix} 2e^{-5t} & (1+2t)e^{-5t} \\ e^{-5t} & \left(\frac{3}{10}+t\right)e^{-5t} \end{pmatrix} \begin{pmatrix} 1 \\ -2 \end{pmatrix} + \begin{pmatrix} \frac{2}{5} \\ 0 \end{pmatrix}$$

$$= \begin{pmatrix} -4te^{-5t} + \frac{2}{5} \\ \left(\frac{2}{5} - 2t\right)e^{-5t} \end{pmatrix}. \blacksquare$$

Variation of Parameters and the Laplace Transform

There is a connection between the variation of parameters method and the Laplace transform. Suppose we want a particular solution Ψ_p of $\mathbf{X}' = \mathbf{AX} + \mathbf{G}$, in which $\mathbf{A}$ is an $n \times n$ real matrix. The variation of parameters method is to find a particular solution $\Psi_p(t) = \Omega(t)\mathbf{U}(t)$, where $\Omega(t)$ is a fundamental matrix for $\mathbf{X}' = \mathbf{AX}$. Explicitly,

$$\mathbf{U}(t) = \int \Omega^{-1}(t)\mathbf{G}(t)\,dt.$$

We can choose a particular $\mathbf{U}(t)$ by carrying out this integration from 0 to t:

$$\mathbf{U}(t) = \int_0^t \mathbf{\Omega}^{-1}(s)\mathbf{G}(s)\,ds.$$

Then

$$\mathbf{\Psi}_p(t) = \mathbf{\Omega}(t)\int_0^t \mathbf{\Omega}^{-1}(s)\mathbf{G}(s)\,ds = \int_0^t \mathbf{\Omega}(t)\mathbf{\Omega}^{-1}(s)\mathbf{G}(s)\,ds.$$

In this equation, $\mathbf{\Omega}$ can be any fundamental matrix for $\mathbf{X}' = \mathbf{AX}$. In particular, suppose we choose $\mathbf{\Omega}(t) = e^{\mathbf{A}t}$. This is sometimes called the *transition matrix* for $\mathbf{X}' = \mathbf{AX}$, since it is a fundamental matrix such that $\mathbf{\Omega}(0) = \mathbf{I}_n$. Now $\mathbf{\Omega}^{-1}(s) = e^{-\mathbf{A}s}$, so

$$\mathbf{\Omega}(t)\mathbf{\Omega}^{-1}(s) = e^{\mathbf{A}t}e^{-\mathbf{A}s} = e^{\mathbf{A}(t-s)} = \mathbf{\Omega}(t - s)$$

and

$$\mathbf{\Psi}_p(t) = \int_0^t \mathbf{\Omega}(t - s)\mathbf{G}(s)\,ds.$$

This equation has the same form as the Laplace transform convolution of $\mathbf{\Omega}$ and $\mathbf{G}$, except that in the current setting these are matrix functions. Now define the Laplace transform of a matrix to be the matrix obtained by taking the Laplace transform of each of its elements. This extended Laplace transform has many of the same computational properties as the Laplace transform for scalar functions. In particular, we can define the convolution integral

$$\mathbf{\Omega}(t) * \mathbf{G}(t) = \int_0^t \mathbf{\Omega}(t - s)\mathbf{G}(s)\,ds.$$

In terms of this convolution,

$$\mathbf{\Psi}_p(t) = \mathbf{\Omega}(t) * \mathbf{G}(t).$$

This is a general formula for a particular solution of $\mathbf{X}' = \mathbf{AX} + \mathbf{G}$ when $\mathbf{\Omega}(t) = e^{\mathbf{A}t}$.

EXAMPLE 9.17

Consider the system

$$\mathbf{X}' = \begin{pmatrix} 1 & -4 \\ 1 & 5 \end{pmatrix}\mathbf{X} + \begin{pmatrix} e^{2t} \\ t \end{pmatrix}.$$

We find that

$$e^{\mathbf{A}t} = \begin{pmatrix} (1 - 2t)e^{3t} & -4te^{3t} \\ te^{3t} & (1 + 2t)e^{3t} \end{pmatrix} = \mathbf{\Omega}(t).$$

A particular solution of $\mathbf{X}' = \mathbf{AX} + \mathbf{G}$ is given by

$$\mathbf{\Psi}_p(t) = \int_0^t \mathbf{\Omega}(t - s)\mathbf{G}(s)\,ds$$

$$= \int_0^t \begin{pmatrix} (1 - 2(t - s))e^{3(t-s)} & -4(t - s)e^{3(t-s)} \\ (t - s)e^{3(t-s)} & (1 + 2(t - s))e^{3(t-s)} \end{pmatrix}\begin{pmatrix} e^{2s} \\ s \end{pmatrix}\,dt$$

$$= \int_0^t \begin{pmatrix} (1 - 2t + 2s)e^{3t}e^{-s} - 4(t - s)e^{3t}se^{-3s} \\ (t - s)e^{3t}e^{-s} + (1 + 2t - 2s)e^{3t}se^{-3s} \end{pmatrix}\,ds$$

$$= \begin{pmatrix} \int_0^t [(1 - 2t + 2s)e^{3t}e^{-s} - 4(t - s)e^{3t}se^{-3s}]\,ds \\ \int_0^t [(t - s)e^{3t}e^{-s} + (1 + 2t - 2s)e^{3t}se^{-3s}]\,ds \end{pmatrix}$$

$$= \begin{pmatrix} -3e^{2t} + \frac{89}{27}e^{3t} - \frac{22}{9}te^{3t} - \frac{4}{9}t - \frac{8}{27} \\ e^{2t} + \frac{11}{9}te^{3t} - \frac{28}{27}e^{3t} - \frac{1}{9}t + \frac{1}{27} \end{pmatrix}.$$

The general solution of $X' = AX + G$ is

$$X(t) = \begin{pmatrix} (1 - 2t)e^{3t} & -4te^{3t} \\ te^{3t} & (1 + 2t)e^{3t} \end{pmatrix} C + \begin{pmatrix} -3e^{2t} + \frac{89}{27}e^{3t} - \frac{22}{9}te^{3t} - \frac{4}{9}t - \frac{8}{27} \\ e^{2t} + \frac{11}{9}te^{3t} - \frac{28}{27}e^{3t} - \frac{1}{9}t + \frac{1}{27} \end{pmatrix}. \quad \blacksquare$$

9.3.2 Solution of $X' = AX + G$ by Diagonalizing A

Consider the case that A is a constant, diagonalizable matrix. Then A has n linearly independent eigenvectors. These form columns of a nonsingular matrix P such that

$$P^{-1}AP = D = \begin{pmatrix} \lambda_1 & 0 & \cdots & 0 \\ 0 & \lambda_2 & \cdots & 0 \\ \vdots & \vdots & \cdots & \vdots \\ 0 & 0 & \cdots & \lambda_n \end{pmatrix},$$

with eigenvalues down the main diagonal in the order corresponding to the eigenvector columns of P. As we did in the homogeneous case $X' = AX$, make the change of variables $X = PZ$. Then the system $X' = AX + G$ becomes

$$X' = PZ' = A(PZ) + G,$$

or

$$PZ' = (AP)Z + G.$$

Multiply this equation on the left by P^{-1} to get

$$Z' = (P^{-1}AP)Z + P^{-1}G = DZ + P^{-1}G.$$

This is an uncoupled system of the form

$$z'_1 = \lambda_1 z_1 + f_1(t)$$
$$z'_2 = \lambda_2 z_2 + f_2(t)$$
$$\vdots$$
$$z'_n = \lambda_n z_n + f_n(t),$$

where

$$P^{-1}G(t) = \begin{pmatrix} f_1(t) \\ f_2(t) \\ \vdots \\ f_n(t) \end{pmatrix}.$$

Solve these n first-order differential equations independently, form $Z(t) = \begin{pmatrix} z_1(t) \\ z_2(t) \\ \vdots \\ z_n(t) \end{pmatrix}$, and then

the solution of $X' = AX + G$ is $X(t) = PZ(t)$.

Unlike diagonalization in solving the homogeneous system $X' = AX$, in this nonhomogeneous case we must explicitly calculate P^{-1} in order to determine $P^{-1}G(t)$, the nonhomogeneous term in the transformed system.

EXAMPLE 9.18

Consider

$$X' = \begin{pmatrix} 3 & 3 \\ 1 & 5 \end{pmatrix} X + \begin{pmatrix} 8 \\ 4e^{3t} \end{pmatrix}.$$

The eigenvalues of A are 2 and 6, with eigenvectors, respectively, $\begin{pmatrix} -3 \\ 1 \end{pmatrix}$ and $\begin{pmatrix} 1 \\ 1 \end{pmatrix}$. Let

$$P = \begin{pmatrix} -3 & 1 \\ 1 & 1 \end{pmatrix}.$$

Then

$$P^{-1}AP = \begin{pmatrix} 2 & 0 \\ 0 & 6 \end{pmatrix}.$$

Compute

$$P^{-1} = \begin{pmatrix} -\frac{1}{4} & \frac{1}{4} \\ \frac{1}{4} & \frac{3}{4} \end{pmatrix}.$$

The transformation $X = PZ$ transforms the original system into

$$Z' = \begin{pmatrix} 2 & 0 \\ 0 & 6 \end{pmatrix} Z + P^{-1} \begin{pmatrix} 8 \\ 4e^{3t} \end{pmatrix}$$

$$= \begin{pmatrix} 2 & 0 \\ 0 & 6 \end{pmatrix} Z + \begin{pmatrix} -\frac{1}{4} & \frac{1}{4} \\ \frac{1}{4} & \frac{3}{4} \end{pmatrix} \begin{pmatrix} 8 \\ 4e^{3t} \end{pmatrix},$$

or

$$Z' = \begin{pmatrix} 2 & 0 \\ 0 & 6 \end{pmatrix} Z + \begin{pmatrix} -2 + e^{3t} \\ 2 + 3e^{3t} \end{pmatrix}.$$

This is the uncoupled system

$$z_1' = 2z_1 - 2 + e^{3t}$$
$$z_2' = 6z_2 + 2 + 3e^{3t}.$$

Solve these linear first-order differential equations independently:

$$z_1(t) = c_1 e^{2t} + e^{3t} + 1,$$
$$z_2(t) = c_2 e^{6t} - e^{3t} - \tfrac{1}{3}.$$

Then

$$Z(t) = \begin{pmatrix} c_1 e^{2t} + e^{3t} + 1 \\ c_2 e^{6t} - e^{3t} - \frac{1}{3} \end{pmatrix}.$$

Then

$$\mathbf{X}(t) = \mathbf{PZ}(t) = \begin{pmatrix} -3 & 1 \\ 1 & 1 \end{pmatrix} \begin{pmatrix} c_1 e^{2t} + e^{3t} + 1 \\ c_2 e^{6t} - e^{3t} - \frac{1}{3} \end{pmatrix}$$

$$= \begin{pmatrix} -3c_1 e^{2t} + c_2 e^{6t} - 4e^{3t} - \frac{10}{3} \\ c_1 e^{2t} + c_2 e^{6t} + \frac{2}{3} \end{pmatrix} \tag{9.9}$$

$$= \begin{pmatrix} -3e^{2t} & e^{6t} \\ e^{2t} & e^{6t} \end{pmatrix} \mathbf{C} + \begin{pmatrix} -4e^{3t} - \frac{10}{3} \\ \frac{2}{3} \end{pmatrix}.$$

This is the general solution of $\mathbf{X}' = \mathbf{AX} + \mathbf{G}$. It is the general solution of $\mathbf{X}' = \mathbf{AX}$, plus a particular solution of $\mathbf{X}' = \mathbf{AX} + \mathbf{G}$. Indeed,

$$\begin{pmatrix} -3e^{2t} & e^{6t} \\ e^{2t} & e^{6t} \end{pmatrix}$$

is a fundamental matrix for the homogeneous system $\mathbf{X}' = \mathbf{AX}$. ∎

To illustrate the solution of an initial value problem, suppose we want the solution of

$$\mathbf{X}' = \begin{pmatrix} 3 & 3 \\ 1 & 5 \end{pmatrix} \mathbf{X} + \begin{pmatrix} 8 \\ 4e^{3t} \end{pmatrix}; \qquad \mathbf{X}(0) = \begin{pmatrix} 2 \\ -7 \end{pmatrix}.$$

Since we have the general solution (9.9) of this system, all we need to do is determine $\mathbf{C}$ so that this initial condition is satisfied. We need

$$\mathbf{X}(0) = \begin{pmatrix} -3 & 1 \\ 1 & 1 \end{pmatrix} \mathbf{C} + \begin{pmatrix} -4 - \frac{10}{3} \\ \frac{2}{3} \end{pmatrix}$$

$$= \begin{pmatrix} -3 & 1 \\ 1 & 1 \end{pmatrix} \mathbf{C} + \begin{pmatrix} -\frac{22}{3} \\ \frac{2}{3} \end{pmatrix} = \begin{pmatrix} 2 \\ -7 \end{pmatrix}.$$

This is the equation

$$\mathbf{PC} = \begin{pmatrix} 2 + \frac{22}{3} \\ -7 - \frac{2}{3} \end{pmatrix} = \begin{pmatrix} \frac{28}{3} \\ -\frac{23}{3} \end{pmatrix}.$$

We already have $\mathbf{P}^{-1}$, so

$$\mathbf{C} = \mathbf{P}^{-1} \begin{pmatrix} \frac{28}{3} \\ -\frac{23}{3} \end{pmatrix} = \begin{pmatrix} -\frac{1}{4} & \frac{1}{4} \\ \frac{1}{4} & \frac{3}{4} \end{pmatrix} \begin{pmatrix} \frac{28}{3} \\ -\frac{23}{3} \end{pmatrix}$$

$$= \begin{pmatrix} -\frac{17}{4} \\ -\frac{41}{12} \end{pmatrix}.$$

The initial value problem has the unique solution

$$\mathbf{X} = \begin{pmatrix} -3e^{2t} & e^{6t} \\ e^{2t} & e^{6t} \end{pmatrix} \begin{pmatrix} -\frac{17}{4} \\ -\frac{41}{12} \end{pmatrix} + \begin{pmatrix} -4e^{3t} - \frac{10}{3} \\ \frac{2}{3} \end{pmatrix}$$

$$= \begin{pmatrix} \frac{51}{4} e^{2t} - \frac{41}{12} e^{6t} - 4e^{3t} - \frac{10}{3} \\ -\frac{17}{4} e^{2t} - \frac{41}{12} e^{6t} + \frac{2}{3} \end{pmatrix}.$$

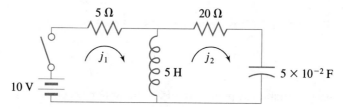

FIGURE 9.3

Application to the Analysis of a Circuit

We will determine the current in each loop of the circuit of Figure 9.3, assuming that all the currents and charges are zero until the switch is closed at time zero.

Apply Kirchhoff's laws to the three loops of the circuit to get

$$5j_1 + 5(j_1' - j_2') = 10$$

$$5(j_1' - j_2') = 10j_2 + \frac{q_2}{5(10^{-2})}$$

$$5j_1 + 20j_2 + \frac{q_2}{5(10^{-2})} = 10.$$

Here we are using j for the currents.

Any two of these equations have enough information to solve the problem. We will use the first and third, because to eliminate q_2 in the second equation we would have to differentiate (since $q_2' = j_2$), and this would introduce second derivatives of j_1 and j_2. Write the first and third equations as

$$j_1' - j_2' = -j_1 + 2$$

$$j_1' + 4j_2' = -4j_2.$$

Write this set of equations as

$$\begin{pmatrix} 1 & -1 \\ 1 & 4 \end{pmatrix} \begin{pmatrix} j_1 \\ j_2 \end{pmatrix}' = \begin{pmatrix} -1 & 0 \\ 0 & -4 \end{pmatrix} \begin{pmatrix} j_1 \\ j_2 \end{pmatrix} + \begin{pmatrix} 2 \\ 0 \end{pmatrix}.$$

This is a nonhomogeneous system of the form $\mathbf{BJ}' = \mathbf{DJ} + \mathbf{K}$. To put this into the standard form we have been considering, calculate

$$\begin{pmatrix} 1 & -1 \\ 1 & 4 \end{pmatrix}^{-1} = \begin{pmatrix} \frac{4}{5} & \frac{1}{5} \\ -\frac{1}{5} & \frac{1}{5} \end{pmatrix}$$

Then

$$\begin{pmatrix} j_1 \\ j_2 \end{pmatrix}' = \begin{pmatrix} \frac{4}{5} & \frac{1}{5} \\ -\frac{1}{5} & \frac{1}{5} \end{pmatrix} \begin{pmatrix} -1 & 0 \\ 0 & -4 \end{pmatrix} \begin{pmatrix} j_1 \\ j_2 \end{pmatrix} + \begin{pmatrix} \frac{4}{5} & \frac{1}{5} \\ -\frac{1}{5} & \frac{1}{5} \end{pmatrix} \begin{pmatrix} 2 \\ 0 \end{pmatrix}$$

or

$$\begin{pmatrix} j_1 \\ j_2 \end{pmatrix}' = \begin{pmatrix} -\frac{4}{5} & -\frac{4}{5} \\ \frac{1}{5} & -\frac{4}{5} \end{pmatrix} \begin{pmatrix} j_1 \\ j_2 \end{pmatrix} + \begin{pmatrix} \frac{8}{5} \\ -\frac{2}{5} \end{pmatrix}.$$

This is in the form $\mathbf{J}' = \mathbf{AJ} + \mathbf{G}$. The eigenvalues of $\mathbf{A}$ are $(-4 + 2i)/5$ and $(-4 - 2i)/5$, with corresponding eigenvectors, respectively, $\begin{pmatrix} 2 \\ -i \end{pmatrix}$ and $\begin{pmatrix} 2 \\ i \end{pmatrix}$. We can diagonalize $\mathbf{A}$ by the matrix

$$\mathbf{P} = \begin{pmatrix} 2 & 2 \\ -i & i \end{pmatrix}.$$

Change variables by putting $\mathbf{J} = \mathbf{PZ}$. The system becomes

$$\mathbf{PZ}' = \mathbf{A}(\mathbf{PZ}) + \mathbf{G}$$

or

$$\mathbf{Z}' = (\mathbf{P}^{-1}\mathbf{AP})\mathbf{Z} + \mathbf{P}^{-1}\mathbf{G}.$$

We know that

$$\mathbf{P}^{-1}\mathbf{AP} = \begin{pmatrix} \frac{-4+2i}{5} & 0 \\ 0 & \frac{-4-2i}{5} \end{pmatrix}.$$

Because of the $\mathbf{P}^{-1}\mathbf{G}$ term, we also need

$$\mathbf{P}^{-1} = \begin{pmatrix} \frac{1}{4} & \frac{1}{2}i \\ \frac{1}{4} & -\frac{1}{2}i \end{pmatrix}.$$

Then

$$\mathbf{Z}' = \begin{pmatrix} \frac{-4+2i}{5} & 0 \\ 0 & \frac{-4-2i}{5} \end{pmatrix} \mathbf{Z} + \begin{pmatrix} \frac{1}{4} & \frac{1}{2}i \\ \frac{1}{4} & -\frac{1}{2}i \end{pmatrix} \begin{pmatrix} \frac{8}{5} \\ -\frac{2}{5} \end{pmatrix}$$

or

$$\mathbf{Z}' = \begin{pmatrix} \frac{-4+2i}{5} & 0 \\ 0 & \frac{-4-2i}{5} \end{pmatrix} \mathbf{Z} + \begin{pmatrix} \frac{2-i}{5} \\ \frac{2+i}{5} \end{pmatrix}.$$

This matrix equation represents the uncoupled system

$$z_1'(t) + \frac{4-2i}{5}z_1 = \frac{2-i}{5},$$

$$z_2'(t) + \frac{4+2i}{5}z_2 = \frac{2+i}{5}.$$

These are first-order linear differential equations with complex coefficients and are easily solved to get

$$z_1(t) = c_1 e^{-(4-2i)t/5} + \frac{1}{2},$$

$$z_2(t) = c_2 e^{-(4+2i)t/5} + \frac{1}{2}.$$

In matrix form,

$$\mathbf{Z}(t) = \begin{pmatrix} c_1 e^{-(4-2i)t/5} + \frac{1}{2} \\ c_2 e^{-(4+2i)t/5} + \frac{1}{2} \end{pmatrix}.$$

Then

$$\mathbf{J}(t) = \mathbf{PZ}(t) = \begin{pmatrix} 2 & 2 \\ -i & i \end{pmatrix} \begin{pmatrix} c_1 e^{-(4-2i)t/5} + \frac{1}{2} \\ c_2 e^{-(4+2i)t/5} + \frac{1}{2} \end{pmatrix}$$

$$= \begin{pmatrix} 2c_1 e^{-(4-2i)t/5} + 2c_2 e^{-(4+2i)t/5} + 2 \\ -ic_1 e^{-(4-2i)t/5} + ic_2 e^{-(4+2i)t/5} \end{pmatrix}.$$

This is the general solution for $\mathbf{J}(t)$.

We need to specify the initial conditions at $t = 0+$, just after the switch is thrown. Now $j_1(0-) = j_2(0-) = q_2(0-) = 0$, because these quantities were zero before the switch was thrown. Therefore, $j_1(0+) - j_2(0+) = 0$, because the current $j_1 - j_2$ through the inductor is assumed to be continuous. By Kirchhoff's law,

$$5j_1(0+) + 20j_2(0+) + \frac{q_2(0+)}{5(10^{-2})} = 10$$

so

$$25j_1(0+) + 20q_2(0+) = 10.$$

Assuming that the charge on the capacitor is continuous, $q_2(0+) = q_2(0-) = 0$, so $25j_1(0+) = 10$ and we conclude that

$$j_1(0+) = \frac{10}{25} = \frac{2}{5}.$$

Therefore, $j_2(0+) = \frac{2}{5}$ also and the initial condition is

$$\mathbf{J}(0) = \begin{pmatrix} \frac{2}{5} \\ \frac{2}{5} \end{pmatrix}.$$

Putting $t = 0$ into the general solution, we have

$$\mathbf{J}(0) = \begin{pmatrix} \frac{2}{5} \\ \frac{2}{5} \end{pmatrix} = \begin{pmatrix} 2c_1 + 2c_2 + 2 \\ -ic_1 + ic_2 \end{pmatrix}.$$

Then

$$2c_1 + 2c_2 = -\frac{8}{5},$$

$$-ic_1 + ic_2 = \frac{2}{5}.$$

Solve these for the constants to obtain

$$c_1 = \frac{-2+i}{5}, \quad c_2 = \frac{-2-i}{5}.$$

The solution for the currents is therefore given by

$$\mathbf{J}(t) = \begin{pmatrix} \frac{2}{5}(-2+i)e^{-(4-2i)t/5} + \frac{2}{5}(-2-i)e^{-(4+2i)t/5} + 2 \\ -\frac{i}{5}(-2+i)e^{-(4-2i)t/5} + \frac{i}{5}(-2-i)e^{-(4+2i)t/5} \end{pmatrix}.$$

Some manipulation, using Euler's formula, gives the solution in the form

$$j_1(t) = 2 - \frac{4}{5}e^{-4t/5}[2\cos(2t/5) + \sin(2t/5)],$$

$$j_2(t) = \frac{2}{5}e^{-4t/5}[\cos(2t/5) - 2\sin(2t/5)].$$

SECTION 9.3   PROBLEMS

In each of Problems 1 through 5, use variation of parameters to find the general solution of $\mathbf{X}' = \mathbf{AX} + \mathbf{G}$, with $\mathbf{A}$ and $\mathbf{G}$ as given.

1. $\begin{pmatrix} 5 & 2 \\ -2 & 1 \end{pmatrix}, \begin{pmatrix} -3e^t \\ e^{3t} \end{pmatrix}$

2. $\begin{pmatrix} 2 & -4 \\ 1 & -2 \end{pmatrix}, \begin{pmatrix} 1 \\ 3t \end{pmatrix}$

3. $\begin{pmatrix} 7 & -1 \\ 1 & 5 \end{pmatrix}, \begin{pmatrix} 2e^{6t} \\ 6te^{6t} \end{pmatrix}$

4. $\begin{pmatrix} 2 & 0 & 0 \\ 0 & 6 & -4 \\ 0 & 4 & -2 \end{pmatrix}; \begin{pmatrix} e^{2t}\cos(3t) \\ -2 \\ -2 \end{pmatrix}$

5. $\begin{pmatrix} 1 & 0 & 0 & 0 \\ 4 & 3 & 0 & 0 \\ 0 & 0 & 3 & 0 \\ -1 & 2 & 9 & 1 \end{pmatrix}, \begin{pmatrix} 0 \\ -2e^{t} \\ 0 \\ e^{t} \end{pmatrix}$

In each of Problems 6 through 9, use variation of parameters to solve the initial value problem $X' = AX + G$; $X(0) = X^0$, with A, G, and X^0 as given (in that order).

6. $\begin{pmatrix} 2 & 0 \\ 5 & 2 \end{pmatrix}, \begin{pmatrix} 2 \\ 10t \end{pmatrix}; \begin{pmatrix} 0 \\ 3 \end{pmatrix}$

7. $\begin{pmatrix} 5 & -4 \\ 4 & -3 \end{pmatrix}, \begin{pmatrix} 2e^{t} \\ 2e^{t} \end{pmatrix}; \begin{pmatrix} -1 \\ 3 \end{pmatrix}$

8. $\begin{pmatrix} 2 & -3 & 1 \\ 0 & 2 & 4 \\ 0 & 0 & 1 \end{pmatrix}, \begin{pmatrix} 10e^{2t} \\ 6e^{2t} \\ -e^{2t} \end{pmatrix}; \begin{pmatrix} 5 \\ 11 \\ -2 \end{pmatrix}$

9. $\begin{pmatrix} 1 & -3 & 0 \\ 3 & -5 & 0 \\ 4 & 7 & -2 \end{pmatrix}, \begin{pmatrix} te^{-2t} \\ te^{-2t} \\ t^{2}e^{-2t} \end{pmatrix}; \begin{pmatrix} 6 \\ 2 \\ 3 \end{pmatrix}$

10. Recall that a transition matrix for $X' = AX$ is a fundamental matrix $\Phi(t)$ such that $\Phi(0) = I_n$.

(a) Prove that for a transition matrix $\Omega(t)$, $\Omega^{-1}(t) = \Omega(-t)$ and $\Omega(t + s) = \Omega(t)\Omega(s)$ for real s and t.

(b) Suppose $\Omega(t)$ is any fundamental matrix for $X' = AX$. Prove that $\Phi(t) = \Omega(t)\Omega^{-1}(0)$ is a transition matrix. That is, $\Phi(t)$ is a fundamental matrix and $\Phi(0) = I_n$.

In each of Problems 11, 12, and 13, verify that the given matrix is a fundamental matrix for the system and use it to find a transition matrix.

11. $x_1' = 4x_1 + 2x_2, x_2' = 3x_1 + 3x_2$;

$\begin{pmatrix} 2e^{t} & e^{6t} \\ -3e^{t} & e^{6t} \end{pmatrix}$

12. $x_1' = -10x_2, x_2' = \frac{5}{2}x_1 - 10x_2$;

$\begin{pmatrix} 2e^{-5t} & (1 + 5t)e^{-5t} \\ e^{-5t} & \frac{5}{2}te^{-5t} \end{pmatrix}$

13. $x_1' = 5x_1 - 4x_2 + 4x_3, x_2' = 12x_1 - 11x_2 + 12x_3,$

$x_3' = 4x_1 - 4x_2 + 5x_3; \begin{pmatrix} e^{-3t} & e^{t} & 0 \\ 3e^{-3t} & 0 & e^{t} \\ e^{-3t} & -e^{t} & e^{t} \end{pmatrix}$

In each of Problems 14 through 22, find the general solution of the system by diagonalization. The general solutions of the associated homogeneous systems $X' = AX$ were requested in Problems 46 through 54 of Section 9.2.

14. $x_1' = -2x_1 + x_2, x_2' = -4x_1 + 3x_2 + 10\cos(t)$

15. $x_1' = 3x_1 + 3x_2 + 8, x_1' = x_1 + 5x_2 + 4e^{3t}$

16. $x_1' = x_1 + x_2 + 6e^{3t}, x_2' = x_1 + x_2 + 4$

17. $x_1' = 6x_1 + 5x_2 - 4\cos(3t), x_2' = x_1 + 2x_2 + 8$

18. $x_1' = 3x_1 - 2x_2 + 3e^{2t}, x_2' = 9x_1 - 3x_2 + e^{2t}$

19. $x_1' = 2x_1 + x_2 - 2x_3 - 2, x_2' = 3x_1 - 2x_2 + 5e^{2t},$

$x_3' = 3x_1 + x_2 - 3x_3 + 9t$

20. $x_1' = 3x_1 - x_2 + x_3 + 12e^{4t}, x_2' = x_1 + x_2 - x_3$

$+ 4\cos(2t), x_3' = x_1 - x_2 + x_3 + 4\cos(2t)$

21. $x_1' = x_1 - x_2 - x_3 + 4e^{t}, x_2' = x_1 - x_2 + 2e^{-3t},$

$x_3' = x_1 - x_3 - 2e^{-3t}$

22. $x_1' = x_1 + x_2 - e^{2t}, x_2' = x_1 + x_2 - e^{2t},$

$x_3' = 4x_3 + 2x_4 + 10e^{6t}, x_4' = 3x_3 + 3x_4 + 15e^{6t}$

In each of Problems 23 through 29, solve the initial value problem by diagonalization.

23. $x_1' = x_1 + x_2 + 6e^{2t}, x_2' = x_1 + x_2 + 2e^{2t}$;

$x_1(0) = 6, x_2(0) = 0$

24. $x_1' = x_1 - 2x_2 + 2t, x_2' = -x_1 + 2x_2 + 5$;

$x_1(0) = 13, x_2(0) = 12$

25. $x_1' = 2x_1 - 5x_2 + 5\sin(t), x_2' = x_1 - 2x_2$;

$x_1(0) = 10, x_2(0) = 5$

26. $x_1' = 5x_1 - 4x_2 + 4x_3 - 3e^{-3t}, x_2' = 12x_1 - 11x_2$

$+ 12x_3 + t, x_3' = 4x_1 - 4x_2 + 5x_3$;

$x_1(0) = 1, x_2(0) = -1, x_3(0) = 2$

27. $x_1' = 3x_1 - x_2 - x_3, x_2' = x_1 + x_2 - x_3 + t,$

$x_3' = x_1 - x_2 + x_3 + 2e^{t}$;

$x_1(0) = 1, x_2(0) = 2, x_3(0) = -2$

28. $x_1' = 3x_1 - 4x_2 + 2, x_2' = 2x_1 - 3x_2 + 4t,$

$x_3' = x_2 - 2x_3 + 14$;

$x_1(0) = -5, x_2(0) = -1, x_3(0) = 2$

29. $x_1' = -2x_2 + t/2, x_2' = x_1 + 2x_2 - t/2$;

$x_1(0) = x_2(0) = 0$

CHAPTER 10

Qualitative Methods and Systems of Nonlinear Differential Equations

10.1 Nonlinear Systems and Existence of Solutions

The preceding chapter was devoted to matrix methods for solving systems of differential equations. Matrices are suited to linear problems. In algebra, the equations we solve by matrix methods are linear, and in differential equations the systems we solve by matrices are also linear.

However, many interesting problems in mathematics, the sciences, engineering, economics, business, and other areas involve systems of nonlinear differential equations, or nonlinear systems. We will consider such systems having the special form:

$$
\begin{aligned}
x_1'(t) &= F_1(t, x_1, x_2, \ldots, x_n), \\
x_2'(t) &= F_2(t, x_1, x_2, \ldots, x_n), \\
&\vdots \\
x_n'(t) &= F_n(t, x_1, x_2, \ldots, x_n).
\end{aligned}
\tag{10.1}
$$

This assumes that each equation of the system can be written with a first derivative isolated on one side and a function of t and the unknown functions $x_1(t), \ldots, x_n(t)$ on the other.

Initial conditions for this system have the form

$$
x_1(t_0) = x_1^0, x_2(t_0) = x_2^0, \ldots, x_n(t_0) = x_n^0,
\tag{10.2}
$$

in which t_0 is a given number and $x_1^0, \ldots, x_n^0$ are given numbers. An initial value problem consists of finding a solution of the system (10.1) satisfying the initial conditions (10.2).

We will state an existence/uniqueness result for this initial value problem. In the statement, an open rectangular parallelepiped in $(n + 1)$-dimensional $t, x_1, \ldots, x_n$-space consists of all points $(t, x_1, \ldots, x_n)$ in R^{n+1} whose coordinates satisfy inequalities

$$
\alpha < t < \beta, \alpha_1 < x_1 < \beta_1, \ldots, \alpha_n < t < \beta_n.
$$

If $n = 1$, this is an open rectangle in the t, x plane, and in three space these points form an open three-dimensional box in 3-space. "Open" means that only points in in the interior of parallelepiped, and no points on the bounding faces, are included.

THEOREM 10.1 *Existence/Uniqueness for Nonlinear Systems*

Let $F_1, \ldots, F_n$ and their first partial derivatives be continuous at all points of an open rectangular parallelepiped K in R^{n+1}. Let $(t_0, x_1^0, \ldots, x_n^0)$ be a point of K. Then there exists a positive number h such that the initial value problem consisting of the system (10.1) and the initial conditions (10.2) has a unique solution

$$x_1 = \varphi_1(t), x_2 = \varphi_2(t), \ldots, x_n = \varphi_n(t)$$

defined for $t_0 - h < t < t_0 + h$. ∎

Many systems we encounter are nonlinear and cannot be solved in terms of elementary functions. This is why we need an existence theorem, and why we will shortly develop qualitative methods to determine properties of solutions without having them explicitly in hand.

As we develop ideas and methods for analyzing nonlinear systems, it will be helpful to have some examples to fall back on and against which to measure new ideas. Here are two examples that are important and that come with some physical intuition about how solutions should behave.

EXAMPLE 10.1 **The Simple Damped Pendulum**

We will derive a system of differential equations describing the motion of a simple pendulum, as shown in Figure 10.1. Although we have some intuition about how a pendulum bob should move, nevertheless the system of differential equations describing this motion is nonlinear and cannot be solved in closed form.

Suppose the pendulum bob has mass m and is at the end of a rod of length L. The rod is assumed to be so light that its weight does not figure into the motion of the bob. It serves only to constrain the bob to remain at fixed distance L from the point of suspension. The position of the bob at any time is described by its displacement angle $\theta(t)$ from the vertical. At some time we call $t = 0$, the bob is displaced by an angle θ_0 and released from rest.

To describe the motion of the bob, we must analyze the forces acting on it. Gravity acts downward with a force of magnitude mg. The damping force (air resistance, friction of the bar at its pivot point) is assumed to have magnitude $c\theta'(t)$ for some positive constant c. By Newton's laws of motion, the rate of change of angular momentum about any point, with respect to time, equals the moment of the resultant force about that point. The angular momentum is $mL^2\theta'(t)$. From the diagram, the horizontal distance between the bob and the vertical center position at time t is $L\sin(\theta(t))$. Then,

$$mL^2\theta''(t) = -cL\theta'(t) - mgL\sin(\theta(t)).$$

The negative signs on the right take into account the fact that if the bob is displaced to the right, these forces tend to make the bob rotate clockwise, which is the negative orientation.

It is customary to write this differential equation as

$$\theta'' + \gamma\theta' + \omega^2 \sin(\theta) = 0, \tag{10.3}$$

with $\gamma = c/mL$ and $\omega^2 = g/L$.

Convert this second-order equation to a system as follows. Let

$$x = \theta, \quad y = \theta'.$$

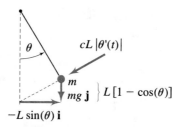

FIGURE 10.1 *Simple, damped pendulum.*

Then the pendulum equation (10.3) becomes

$$x' = y,$$

$$y' + \gamma y + \omega^2 \sin(x) = 0,$$

or

$$x' = y$$

$$y' = -\omega^2 \sin(x) - \gamma y.$$

This is a nonlinear system because of the $\sin(x)$ term. We cannot write a solution of this system in closed form. However, we will soon have methods to analyze the behavior of solutions and hence the motion itself.

In matrix form, the pendulum system is

$$\mathbf{X}' = \begin{pmatrix} 0 & 1 \\ 0 & -\gamma \end{pmatrix} \mathbf{X} + \begin{pmatrix} 0 \\ -\omega^2 \sin(x) \end{pmatrix},$$

in which $\mathbf{X} = \begin{pmatrix} x \\ y \end{pmatrix}$. ∎

EXAMPLE 10.2 Nonlinear Spring

Consider an object of mass m attached to a spring. If the object is displaced and released, its motion is governed by Hooke's law, which states that the force exerted on the mass by the spring is $F(r) = -kr$, with k a positive constant and r the distance displaced from the equilibrium position (position at which the object is at rest). Figure 2.4 shows a typical such mass/spring system, with r used here for the displacement instead of y used in Chapter 2.

This is a linear model, since F is a linear function (a constant times r to the first power). The spring model becomes nonlinear if $F(r)$ is nonlinear. Simple nonlinear models are achieved by adding terms to $-kr$. What kind of terms should we add? Intuition tells us that the spring should not care whether we displace an object left or right before releasing it. Since displacements in opposite directions carry opposite signs, this means that we want $F(-r) = -F(r)$, so F should be an odd function. This suggests adding multiples of odd powers of r. The simplest such model is

$$F(r) = -kr + \alpha r^3.$$

If we also allow a damping force that in magnitude is proportional to the velocity, then by Newton's law this spring motion is governed by the second-order differential equation

$$mr'' = -kr + \alpha r^3 - cr'.$$

To convert this to a system, let $x = r$ and $y = r'$. The system is

$$x' = y,$$

$$y' = -\frac{k}{m}x + \frac{\alpha}{m}x^3 - \frac{c}{m}y.$$

In matrix form, this system is

$$\mathbf{X}' = \begin{pmatrix} 0 & 1 \\ -k/m & -c/m \end{pmatrix} \mathbf{X} + \begin{pmatrix} 0 \\ \alpha x^3/m \end{pmatrix}. \quad \blacksquare$$

SECTION 10.1 PROBLEMS

1. Apply the existence/uniqueness theorem to the system for the simple, damped pendulum, with initial conditions $x(0) = a$, $y(0) = b$. What are the physical interpretations of the initial conditions? Are there any restrictions on the numbers a and b in applying the theorem to assert the existence of a unique solution in some interval $(-h, h)$?

2. Apply the existence/uniqueness theorem to the system for the nonlinear spring system, with initial conditions

$x(0) = a$, $y(0) = b$. What are the physical interpretations of the initial conditions? Are there any restrictions on the numbers a and b in applying the theorem to assert the existence of a unique solution in some interval $(-h, h)$?

3. Suppose the driving force for the nonlinear spring has additional terms, say $F(r) = -kr + \alpha r^3 + \beta r^5$. Does this problem still have a unique solution in some interval $(-h, h)$?

10.2   The Phase Plane, Phase Portraits, and Direction Fields

Throughout this chapter, we will consider systems of two first-order differential equations in two unknowns. In this case, it is convenient to denote the variables as x and y rather than x_1 and x_2. Thus, consider the system

$$x'(t) = f(x(t), y(t)),$$
$$y'(t) = g(x(t), y(t)),$$
(10.4)

in which f and g are continuous, with continuous first partial derivatives, in some part of the plane. We often write this system as

$$\mathbf{X}' = \mathbf{F}(x(t), y(t)),$$

where

$$\mathbf{X} = \begin{pmatrix} x \\ y \end{pmatrix} \quad \text{and} \quad \mathbf{F}(x, y) = \begin{pmatrix} f(x, y) \\ g(x, y) \end{pmatrix}.$$

The system (10.4) is a special case of the system (10.1). We assume in (10.4) that neither f nor g has an explicit dependence on t. Rather, f and g depend only on x and y, and t appears only through dependencies of these two variables on t. We refer to such a system as *autonomous*.

Working in the plane will allow us the considerable advantage of geometric intuition. If $x = \varphi(t)$, $y = \psi(t)$ is a solution of (10.4), the point $(\varphi(t), \psi(t))$ traces out a curve in the plane as t varies. Such a curve is called a *trajectory*, or *orbit*, of the system. A copy of the plane containing drawings of trajectories is called a *phase portrait* for the system (10.4). In this context, the x, y plane is called the *phase plane*.

We may consider trajectories as oriented, with $(\varphi(t), \psi(t))$ moving along the trajectory in a certain direction as t increases. If we think of t as time, then $(\varphi(t), \psi(t))$ traces out the path of motion of a particle, moving under the influence of the system (10.4), as time increases. In the case of orbits that are closed curves, we take counterclockwise orientation as the positive orientation, unless specific exception is made.

In some phase portraits, short arrows are also drawn. The arrow at any point is along the tangent to the trajectory through that point and in the direction of motion along this trajectory. This type of drawing combines the phase portrait with a direction field and gives an overall sense of the flow of the trajectories, as well as graphs of some specific trajectories.

One way to construct trajectories is to write

$$\frac{dy}{dx} = \frac{dy/dt}{dx/dt} = \frac{g(x, y)}{f(x, y)}.$$

Because the system is autonomous, this is a differential equation in x and y and we can attempt to solve it and graph solutions. If the system is nonautonomous, then f/g may depend explicitly on t and we cannot use this strategy to generate trajectories.

EXAMPLE 10.3

Consider the autonomous system

$$x' = y = f(x, y)$$
$$y' = x^2 y^2 = g(x, y).$$

Then

$$\frac{dy}{dx} = \frac{x^2 y^2}{y} = x^2 y,$$

a separable differential equation we write as

$$\frac{1}{y}\, dy = x^2\, dx.$$

Integrate to get

$$\ln|y| = \tfrac{1}{3}x^3 + C,$$

or

$$y = Ae^{x^3/3}.$$

Graphs of these curves for various values of A form trajectories of this system, some of which are shown in Figure 10.2. ∎

EXAMPLE 10.4

For the autonomous system

$$x' = -2y - x\sin(xy),$$
$$y' = 2x + y\sin(xy) \tag{10.5}$$

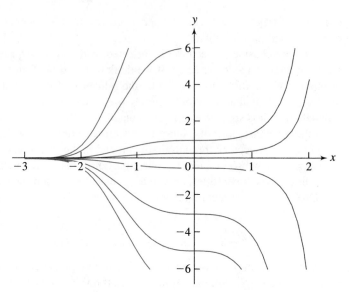

FIGURE 10.2   *Some trajectories of the system* $\begin{cases} x' = y \\ y' = x^2 y^2 \end{cases}$.

we have

$$\frac{dy}{dx} = -\frac{2x + y\sin(xy)}{2y + x\sin(xy)}.$$

This is not separable, but we can write

$$(2x + y\sin(xy))\,dx + (2y + x\sin(xy))\,dy = 0,$$

which is exact. We find the potential function $H(x, y) = x^2 + y^2 - \cos(xy)$, and the general solution of this differential equation is defined implicitly by

$$H(x, y) = x^2 + y^2 - \cos(xy) = C,$$

in which C is an arbitrary constant. Figure 10.3 shows a phase portrait for this system (10.5), consisting of graphs of these curves for various choices of C. ▨

Usually, we will not be so fortunate as to be able to solve $dy/dx = g(x, y)/f(x, y)$ in closed form. In such a case, we may still be able to use a software package to generate a phase portrait. Figure 10.4 is a phase portrait of the system

$$x' = x\cos(y)$$
$$y' = x^2 - y^3 + \sin(x - y)$$

generated in this way. Figure 10.5 (p. 432) is a phase portrait for a damped pendulum with $\omega^2 = 10$ and $\gamma = 0.3$, and Figure 10.6 (p. 432) is a phase portrait for a nonlinear spring system with $\alpha = 0.2, k/m = 4$, and $c/m = 2$. We will consider phase portraits for the damped pendulum and nonlinear spring system in more detail when we treat almost linear systems.

If $x = \varphi(t)$, $y = \psi(t)$ is a solution of (10.4) and c is a constant, we call the pair $\varphi(t + c)$, $\psi(t + c)$ a *translation* of φ and ψ. We will use the following fact.

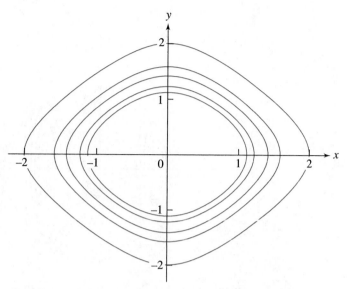

FIGURE 10.3 *Phase portrait for* $\begin{cases} x' = -2y - x\sin(xy) \\ y' = 2x + y\sin(xy) \end{cases}$.

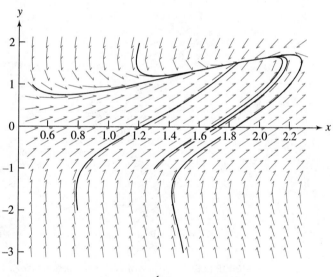

FIGURE 10.4 *Phase portrait for* $\begin{cases} x' = x\cos(y) \\ y' = x^2 - y^3 + \sin(x-y) \end{cases}$.

LEMMA 10.1

A translation of a solution of the system (10.4) is also a solution of this system.

Proof   Suppose $x = \varphi(t)$, $y = \psi(t)$ is a solution. This means that

$$x'(t) = \varphi'(t) = f(\varphi(t), \psi(t)) \quad \text{and} \quad y'(t) = \psi'(t) = g(\varphi(t), \psi(t)).$$

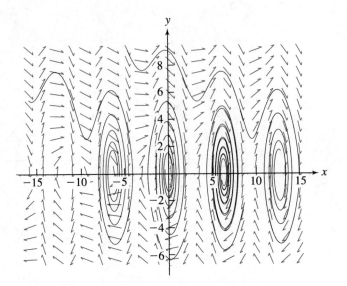

FIGURE 10.5 *Phase portrait for a damped pendulum.*

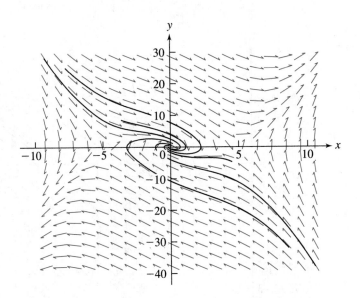

FIGURE 10.6 *Phase portrait for a nonlinear spring.*

Let

$$\widetilde{x}(t) = \varphi(t + c) \quad \text{and} \quad \widetilde{y}(t) = \psi(t + c)$$

for some constant c. By the chain rule,

$$\frac{d\widetilde{x}}{dt} = \frac{d\varphi(t + c)}{d(t + c)} \frac{d(t + c)}{dt} = \varphi'(t + c)$$

$$= f(\varphi(t + c), x(t + c)) = f(\widetilde{x}(t), \widetilde{y}(t))$$

and, similarly,

$$\frac{d\tilde{y}}{dt} = \psi'(t+c) = g(\varphi(t+c), \psi(t+c)) = g(\tilde{x}(t), \tilde{y}(t)).$$

Therefore, $x = \tilde{x}(t)$, $y = \tilde{y}(t)$ is also a solution. ∎

We may think of a translation as a reparametrization of the trajectory, which of course does not alter the fact that it is a trajectory. If we think of the point $(\varphi(t), \psi(t))$ as moving along the orbit, a translation simply means rescheduling the point to change the times at which it passes through given points of the orbit.

We will need the following facts about trajectories.

THEOREM 10.2

Let f and g be continuous, with continuous first partial derivatives, in the (x, y) plane. Then,

1. If (a, b) is a point in the plane, there is a trajectory through (a, b).

2. Two trajectories passing through the same point must be translations of each other.

Proof Conclusion (1) follows immediately from Theorem 10.1, since the initial value problem

$$x' = f(x, y), \ y' = g(x, y); \qquad x(0) = a, \ y(0) = b$$

has a solution, and the graph of this solution is a trajectory through (a, b).

For (2), suppose $x = \varphi_1(t)$, $y = \psi_1(t)$, and $x = \varphi_2(t)$, $y = \psi_2(t)$ are trajectories of the system (10.4). Suppose both trajectories pass through (a, b). Then for some t_0,

$$\varphi_1(t_0) = a \quad \text{and} \quad \psi_1(t_0) = b$$

and for some t_1,

$$\varphi_2(t_1) = a \quad \text{and} \quad \psi_2(t_1) = b.$$

Let $c = t_0 - t_1$ and define $\tilde{x}(t) = \varphi_1(t + c)$ and $\tilde{y}(t) = \psi_1(t + c)$. Then $x = \tilde{x}(t)$, $y = \tilde{y}(t)$ is a trajectory, by Lemma 10.1. Further,

$$\tilde{x}(t_1) = \varphi_1(t_0) = a \quad \text{and} \quad \tilde{y}(t_1) = \psi(t_0) = b.$$

Therefore, $x = \tilde{x}(t)$, $y = \tilde{y}(t)$ is the unique solution of the initial value problem

$$x' = f(x, y), \ y' = g(x, y); \qquad x(t_1) = a, \ y(t_1) = b.$$

But $x = \varphi_2(t)$, $y = \psi_2(t)$ is also the solution of this problem. Therefore, for all t,

$$\varphi_2(t) = \tilde{x}(t) = \varphi_1(t + c) \quad \text{and} \quad \psi_2(t) = \tilde{y}(t) = \psi_1(t + c).$$

This proves that the two trajectories $x = \varphi_1(t)$, $y = \psi_1(t)$ and $x = \varphi_2(t)$, $y = \psi_2(t)$ are translations of each other. ∎

If we think of translations of trajectories as the same trajectory (just a change in the parameter), then conclusion (2) states that distinct trajectories cannot cross each other. This would violate the uniqueness of the solution of the system that passes through the point of intersection.

Conclusion (2) of Theorem 10.2 does not hold for systems that are not autonomous.

EXAMPLE 10.5

Consider the system

$$x'(t) = \frac{1}{t}x = f(t, x, y)$$

$$y'(t) = -\frac{1}{t}y + x = g(t, x, y).$$

This is nonautonomous, since f and g have explicit t-dependencies. We can solve this system. The first equation is separable. Write

$$\frac{1}{x}\,dx = \frac{1}{t}\,dt$$

to obtain $x(t) = ct$. Substitute this into the second equation to get

$$y' + \frac{1}{t}y = ct,$$

a linear first-order differential eqaution. This equation can be written

$$ty' + y = ct^2,$$

or

$$(ty)' = ct^2.$$

Integrate to get

$$ty = \frac{c}{3}t^3 + d.$$

Hence

$$y(t) = \frac{c}{3}t^2 + d\frac{1}{t}.$$

Now observe that conclusion (2) of Theorem 10.2 fails for this system. For example, for any number t_0, the trajectory

$$x(t) = \frac{1}{t_0}t, \quad y(t) = \frac{1}{3t_0}t^2 - \frac{t_0^2}{3}\frac{1}{t}.$$

passes through $(1, 0)$ at time t_0. Because t_0 is arbitrary, this gives many trajectories passing through $(1, 0)$ at different times, and these trajectories are not translations of each other. ∎

We now have some of the vocabulary and tools needed to analyze 2×2 nonlinear autonomous systems of differential equations. First, however, we will reexamine linear systems, which we know how to solve explicitly. This will serve two purposes. It will give us some experience with phase portraits, as well as insight into significant features that solutions of a system might have. In addition, we will see shortly that some nonlinear systems can be thought of as perturbations of linear systems (that is, as linear systems with "small" nonlinear terms added). In such a case, knowledge of solutions of the linear system yields important information about solutions of the nonlinear system.

SECTION 10.2 **PROBLEMS**

In each of Problems 1 through 6, find the general solution of the system and draw a phase portrait containing at least six trajectories of the system.

1. $x' = 4x + y$, $y' = -17x - 4y$

2. $x' = 2x$, $y' = 8x + 2y$

3. $x' = 4x - 7y$, $y' = 2x - 5y$

4. $x' = 3x - 2y$, $y' = 10x - 5y$

5. $x' = 5x - 2y$, $y' = 4y$

6. $x' = -4x - 6y$, $y' = 2x - 11y$

In each of Problems 7 through 18, use the method of Examples 10.3 and 10.4 to draw some integral curves (at least six) for the system. In Problems 13 through 18, the trajectories are conic sections. Identify them as ellipses, hyperbolas, or parabolas.

7. $x' = 9y$, $y' = -4x$

8. $x' = 2xy$, $y' = y^2 - x^2$

9. $x' = y + 2$, $y' = x - 1$

10. $x' = \csc(x)$, $y' = y$

11. $x' = x$, $y' = x + y$

12. $x' = x^2$, $y' = y$

13. $x' = 36x - 52y$, $y' = 73x - 36y$

14. $x' = x - 2y$, $y' = 2x - y$

15. $x' = 3\sqrt{3}x - 2y$, $y' = 8x - 3\sqrt{3}y$

16. $x' = 12x - 2y$, $y' = 9x - 12y$

17. $x' = 3\sqrt{3}x - 7y$, $y' = 13x - 3\sqrt{3}y$

18. $x' = 12x - 9y$, $y' = 16x - 12y$

19. How would phase portraits for the following systems compare with each other?

   (a) $x' = F(x, y)$, $y' = G(x, y)$

   (b) $x' = -F(x, y)$, $y' = -G(x, y)$

10.3 Phase Portraits of Linear Systems

In preparation for studying the nonlinear autonomous system (10.4), we will thoroughly analyze the linear system

$$\mathbf{X}' = \mathbf{AX}, \tag{10.6}$$

in which $\mathbf{A}$ is a 2×2 real matrix and $\mathbf{X} = \begin{pmatrix} x \\ y \end{pmatrix}$. We assume that $\mathbf{A}$ is nonsingular, so the equation $\mathbf{AX} = \mathbf{O}$ has only the trivial solution.

For the linear system $\mathbf{X}' = \mathbf{AX}$, we actually have the solutions in hand. We will examine these solutions to prepare for the analysis of nonlinear systems, for which we are unlikely to have explicit solutions.

The origin $(0, 0)$ stands apart from other points in the plane in the following respect. The trajectory through the origin is the solution of

$$\mathbf{X}' = \mathbf{AX}; \qquad \mathbf{X}(0) = \mathbf{O} = \begin{pmatrix} 0 \\ 0 \end{pmatrix},$$

and this is the constant trajectory

$$x(t) = 0, \, y(t) = 0 \quad \text{for all } t.$$

The graph of this trajectory is the single point $(0, 0)$. For this reason, the origin is called an *equilibrium point* of the system, and the constant solution $\mathbf{X} = \mathbf{O}$ is called an *equilibrium solution*. The origin is also called a *critical point* of $\mathbf{X}' = \mathbf{AX}$. By Theorem 10.1, no other trajectory can

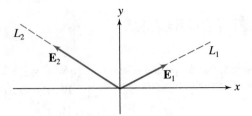

FIGURE 10.7 *Eigenvectors* $\mathbf{E}_1$ *and* $\mathbf{E}_2$ *of*
$\mathbf{X}' = \mathbf{A}\mathbf{X}$, *for distinct negative eigenvalues of* $\mathbf{A}$.

pass through this point. As we proceed, observe how the behavior of trajectories of $\mathbf{X}' = \mathbf{A}\mathbf{X}$ near this critical point is the key to understanding the behavior of trajectories throughout the entire plane. The critical point, then, will be the focal point in drawing a phase portrait of the system and analyzing the behavior of solutions.

We will draw the phase portrait for $\mathbf{X}' = \mathbf{A}\mathbf{X}$ in all cases that can occur. Because the general solution of (10.6) is completely determined by the eigenvalues of $\mathbf{A}$, we will use these eigenvalues to distinguish cases.

Case 1   Real, distinct eigenvalues λ and μ of the same sign.
Let associated eigenvectors be, respectively, $\mathbf{E}_1$ and $\mathbf{E}_2$. Since λ and μ are distinct, $\mathbf{E}_1$ and $\mathbf{E}_2$ are linearly independent. The general solution is

$$\mathbf{X}(t) = \begin{pmatrix} x(t) \\ y(t) \end{pmatrix} = c_1\mathbf{E}_1 e^{\lambda t} + c_2\mathbf{E}_2 e^{\mu t}.$$

Since $\mathbf{E}_1$ and $\mathbf{E}_2$ are vectors in the plane, we can represent them as arrows from the origin, as in Figure 10.7. Draw half-lines L_1 and L_2 from the origin along these eigenvectors, respectively, as shown. These half-line lines are parts of trajectories, and so do not pass through the origin, which is itself a trajectory.

Now consider subcases.

Case 1(a)   The eigenvalues are negative, say $\lambda < \mu < 0$.
Since $e^{\lambda t} \to 0$ and $e^{\mu t} \to 0$ as $t \to \infty$, then $X(t) \to (0, 0)$ and every trajectory approaches the origin as $t \to \infty$. However, this can happen in three ways, depending on an initial point P_0: (x_0, y_0) we choose for a trajectory to pass through at time $t = 0$. Here are the three possibilities.

If P_0 is on L_1, then $c_2 = 0$ and

$$\mathbf{X}(t) = c_1\mathbf{E}_1 e^{\lambda t},$$

which for any t is a scalar multiple of $\mathbf{E}_1$. The trajectory through P_0 is the half-line from the origin along L_1 through P_0, and the arrows toward the origin indicate that points on this trajectory approach the origin along L_1 as t increases. This is the trajectory T_1 of Figure 10.8.

If P_0 is on L_2, then $c_1 = 0$ and now

$$\mathbf{X}(t) = c_2\mathbf{E}_2 e^{\mu t}.$$

This trajectory is a half-line from the origin along L_2 through P_0. Again, the arrows indicate that points on this trajectory also approach the origin along L_2 as $t \to \infty$. This is the trajectory T_2 of Figure 10.8.

If P_0 is on neither L_1 or L_2, then the trajectory is a curve through P_0 having the parametric form

$$\mathbf{X}(t) = c_1\mathbf{E}_1 e^{\lambda t} + c_2\mathbf{E}_2 e^{\mu t}.$$

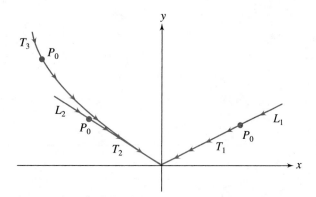

FIGURE 10.8 *Trajectories along* $\mathbf{E}_1$, *along* $\mathbf{E}_2$, *or asymptotic to* $\mathbf{E}_2$ *in the case* $\lambda < \mu < 0$.

Write this as

$$\mathbf{X}(t) = e^{\mu t}[c_1\mathbf{E}_1 e^{(\lambda-\mu)t} + c_2\mathbf{E}_2].$$

Because $\lambda - \mu < 0$, $e^{(\lambda-\mu)t} \to 0$ as $t \to \infty$ and the term $c_1\mathbf{E}_1 e^{(\lambda-\mu)t}$ exerts increasingly less influence on $\mathbf{X}(t)$. In this case, $\mathbf{X}(t)$ still approaches the origin, but also approaches the line L_2, as $t \to \infty$. A typical such trajectory is shown as the curve T_3 of Figure 10.8.

A phase portrait of $\mathbf{X}' = \mathbf{AX}$ in this case therefore has some trajectories approaching the origin along the lines through the eigenvectors of $\mathbf{A}$ and all others approaching the origin along curves that approach one of these lines asymptotically. In this case, the origin is called a *nodal sink* of the system $\mathbf{X}' = \mathbf{AX}$. We can think of particles flowing along the trajectories and toward the origin.

The following example and phase portrait are typical of nodal sinks.

EXAMPLE 10.6

Consider the system $\mathbf{X}' = \mathbf{AX}$, in which

$$\mathbf{A} = \begin{pmatrix} -6 & -2 \\ 5 & 1 \end{pmatrix}.$$

$\mathbf{A}$ has eigenvalues and corresponding eigenvectors

$$-1, \begin{pmatrix} 2 \\ -5 \end{pmatrix} \quad \text{and} \quad -4, \begin{pmatrix} -1 \\ 1 \end{pmatrix}.$$

In the notation of the discussion, $\lambda = -4$ and $\mu = -1$. The general solution is

$$\mathbf{X}(t) = c_1 \begin{pmatrix} -1 \\ 1 \end{pmatrix} e^{-4t} + c_2 \begin{pmatrix} 2 \\ -5 \end{pmatrix} e^{-t}.$$

L_1 is the line through $(-1, 1)$, and L_2 the line through $(2, -5)$. Figure 10.9 shows a phase portrait for this system, with the origin a nodal sink. ∎

Case 1(b) The eigenvalues are positive, say $0 < \mu < \lambda$.
The discussion of Case 1(a) can be replicated with one change. Now $e^{\lambda t}$ and $e^{\mu t}$ approach ∞ instead of zero as t increases. The phase portrait is like that of the previous case, except all the

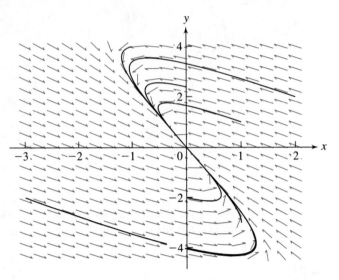

FIGURE 10.9 *Phase portrait showing a nodal sink of* $\begin{cases} x' = -6x - 2y \\ y' = 5x + y \end{cases}$.

arrows are reversed and trajectories flow away from the origin instead of into the origin as time increases. As we might expect, now the origin is called a *nodal source*. Particles are flowing away from the origin.

Here is a typical example of a nodal source.

EXAMPLE 10.7

Consider the system

$$\mathbf{X}' = \begin{pmatrix} 3 & 3 \\ 1 & 5 \end{pmatrix}.$$

This has eigenvalues and eigenvectors

$$2, \begin{pmatrix} -3 \\ 1 \end{pmatrix} \quad \text{and} \quad 6, \begin{pmatrix} 1 \\ 1 \end{pmatrix}.$$

Now $\lambda = 6$ and $\mu = 2$, and the general solution is

$$\mathbf{X}(t) = c_1 \begin{pmatrix} -3 \\ 1 \end{pmatrix} e^{2t} + c_2 \begin{pmatrix} 1 \\ 1 \end{pmatrix} e^{6t}.$$

Figure 10.10 shows a phase portrait for this system, exhibiting the behavior expected for a nodal source at the origin. ■

Case 2 Real, distinct eigenvalues of opposite sign.
Suppose the eigenvalues are λ and μ with $\mu < 0 < \lambda$. The general solution still has the appearance

$$\mathbf{X}(t) = c_1 \mathbf{E}_1 e^{\lambda t} + c_2 \mathbf{E}_2 e^{\mu t},$$

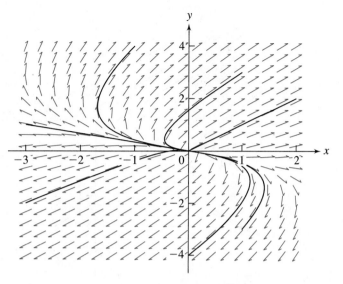

FIGURE 10.10 *Nodal source of* $\begin{cases} x' = 3x + 3y \\ y' = x + 5y \end{cases}$.

and we start to draw a phase portrait by again drawing half-lines L_1 and L_2 from the origin along the eigenvectors.

If P_0 is on L_1, then $c_2 = 0$ and $\mathbf{X}(t)$ moves on this half-line away from the origin as t increases, because $\lambda > 0$ and $e^{\lambda t} \to \infty$ as $t \to \infty$. But if P_0 is on L_2, then $c_1 = 0$ and $\mathbf{X}(t)$ moves along this half-line toward the origin, because $\mu < 0$ and $e^{\mu t} \to 0$ as $t \to \infty$.

The arrows along the half-lines along the eigenvectors therefore have opposite directions, toward the origin along L_2 and away from the origin along L_1. This is in contrast to Case 1, in which solutions starting out on the half-lines through the eigenvectors either both approached the origin or both moved away from the origin as time increased.

If P_0 is on neither L_1 nor L_2, then the trajectory through P_0 does not come arbitrarily close to the origin for any times, but rather approaches the direction determined by the eigenvector $\mathbf{E}_1$ as $t \to \infty$ (in which case $e^{\mu t} \to 0$) or the direction determined by $\mathbf{E}_2$ as $t \to -\infty$ (in which case $e^{\lambda t} \to 0$). The phase portrait therefore has typical trajectories as shown in Figure 10.11. The lines along the eigenvectors determine four trajectories that separate the plane into four regions. A trajectory starting in one of these regions must remain in it because distinct trajectories cannot cross each other, and such a trajectory is asymptotic to both of the lines bounding its region.

The origin in this case is called a *saddle point*.

EXAMPLE 10.8

Consider $\mathbf{X}' = \mathbf{A}\mathbf{X}$ with

$$\mathbf{A} = \begin{pmatrix} -1 & 3 \\ 2 & -2 \end{pmatrix}, \begin{pmatrix} -1 & 3 \\ 2 & -2 \end{pmatrix}.$$

Eigenvalues and eigenvectors of $\mathbf{A}$ are

$$-4, \begin{pmatrix} -1 \\ 1 \end{pmatrix} \quad \text{and} \quad 1, \begin{pmatrix} 3 \\ 2 \end{pmatrix}$$

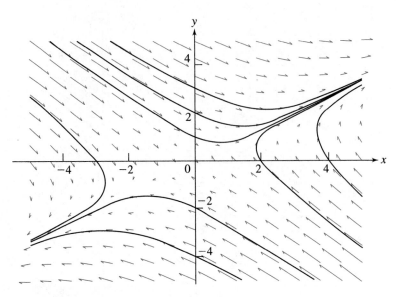

FIGURE 10.11 *Typical phase portrait for a saddle point at the origin.*

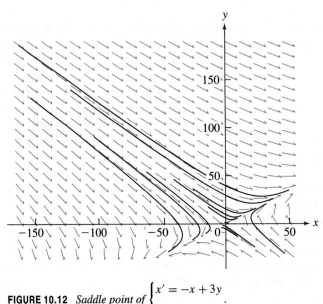

FIGURE 10.12 *Saddle point of* $\begin{cases} x' = -x + 3y \\ y' = 2x - 2y \end{cases}$.

The general solution is

$$\mathbf{X}(t) = c_1 \begin{pmatrix} -1 \\ 1 \end{pmatrix} e^{-4t} + c_2 \begin{pmatrix} 3 \\ 2 \end{pmatrix} e^{t},$$

and a phase portrait is given in Figure 10.12. In this case of a saddle point at the origin, trajectories do not enter or leave the origin, but asymptotically approach the lines determined by the eigenvectors. ∎

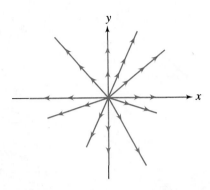

FIGURE 10.13 *Typical proper node with positive eigenvalue of* **A**.

Case 3 Equal eigenvalues.
Suppose **A** has the real eigenvalue λ of multiplicity 2. There are two possibilities.

Case 3(a) **A** has two linearly independent eigenvectors $\mathbf{E}_1$ and $\mathbf{E}_2$.
Now the general solution of $\mathbf{X}' = \mathbf{AX}$ is

$$\mathbf{X}(t) = (c_1\mathbf{E}_1 + c_2\mathbf{E}_2)e^{\lambda t}.$$

If

$$\mathbf{E}_1 = \begin{pmatrix} a \\ b \end{pmatrix} \quad \text{and} \quad \mathbf{E}_2 = \begin{pmatrix} h \\ k \end{pmatrix}$$

then, in terms of components,

$$x(t) = (c_1 a + c_2 h)e^{\lambda t}, \quad y(t) = (c_1 b + c_2 k)e^{\lambda t}.$$

Now

$$\frac{y(t)}{x(t)} = \text{constant}.$$

This means that all trajectories in this case are half-lines from the origin. If $\lambda > 0$, arrows along these trajectories are away from the origin, as in Figure 10.13. If $\lambda < 0$, they move toward the origin, reversing the arrows in Figure 10.13.

The origin in Case 3(a) is called a *proper node*.

Case 3(b) **A** does not have two linearly independent eigenvectors.
In this case, there is an eigenvector **E** and the general solution has the form

$$\mathbf{X}(t) = c_1(\mathbf{E}t + \mathbf{W})e^{\lambda t} + c_2\mathbf{E}e^{\lambda t}$$

$$= [(c_1\mathbf{W} + c_2\mathbf{E}) + c_1\mathbf{E}t]e^{\lambda t}.$$

To visualize the trajectories, begin with arrows from the origin representing the vectors **W** and **E**. Now, for selected constants c_1 and c_2, draw the vector $c_1\mathbf{W} + c_2\mathbf{E}$, which may have various orientations relative to **W** and **E**, depending on the signs and magnitudes of c_1 and c_2. Some possibilities are displayed in Figure 10.14. For given c_1 and c_2, the vector

$$c_1\mathbf{W} + c_2\mathbf{E} + c_1\mathbf{E}t$$

drawn as an arrow from the origin, sweeps out a straight line L as t varies over all real values. For a given t, $\mathbf{X}(t)$ is the vector $c_1\mathbf{W} + c_2\mathbf{E} + c_1\mathbf{E}t$ from the origin to a point on L, with length

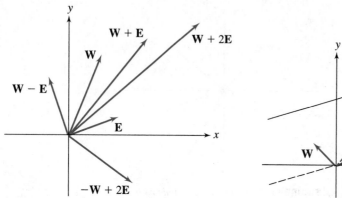

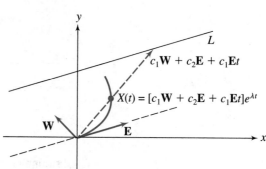

FIGURE 10.14 *Vectors* $c_1\mathbf{W} + c_2\mathbf{E}$ *in the case of an improper node.*

FIGURE 10.15 *Typical trajectory near an improper node.*

adjusted by a factor $e^{\lambda t}$. If λ is negative, then this length goes to zero as $t \to \infty$ and the vector $\mathbf{X}(t)$ sweeps out a curve as shown in Figure 10.15, approaching the origin tangent to $\mathbf{E}$. If $\lambda > 0$, we have the same curve (now $e^{\lambda t} \to 0$ as $t \to -\infty$), except that the arrow indicating direction of the flow on the trajectory is reversed.

The origin in this case is called an *improper node* of the system $\mathbf{X}' = \mathbf{AX}$. The following example has a phase portrait that is typical of improper nodes.

EXAMPLE 10.9

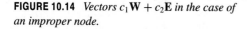

Let

$$\mathbf{A} = \begin{pmatrix} -10 & 6 \\ -6 & 2 \end{pmatrix}.$$

Then $\mathbf{A}$ has eigenvalue -4, and every eigenvector is a real constant multiple of $\mathbf{E} = \begin{pmatrix} 1 \\ 1 \end{pmatrix}$. A routine calculation gives

$$\mathbf{W} = \begin{pmatrix} 1 \\ \frac{7}{6} \end{pmatrix},$$

and the general solution is

$$\mathbf{X}(t) = c_1 \begin{pmatrix} t+1 \\ t+\frac{7}{6} \end{pmatrix} e^{-4t} + c_2 \begin{pmatrix} 1 \\ 1 \end{pmatrix} e^{-4t}.$$

Figure 10.16 is a phase portrait for this system. We can see that the trajectories approach the origin tangent to $\mathbf{E}$ in this case of an improper node at the origin, with negative eigenvalue for $\mathbf{A}$. ∎

Case 4 Complex eigenvalues with nonzero real part.

We know that the complex eigenvalues must be complex conjugates, say $\lambda = \alpha + i\beta$ and $\mu = \alpha - i\beta$. The complex eigenvectors are also conjugates. Write these, respectively, as $\mathbf{U} + i\mathbf{V}$ and $\mathbf{U} - i\mathbf{V}$. Then the general solution of $\mathbf{X}' = \mathbf{AX}$ is

$$\mathbf{X}(t) = c_1 e^{\alpha t}[\mathbf{U}\cos(\beta t) - \mathbf{V}\sin(\beta t)] + c_2 e^{\alpha t}[\mathbf{U}\sin(\beta t) + \mathbf{V}\cos(\beta t)].$$

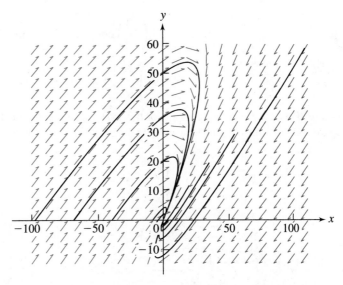

FIGURE 10.16 *Phase portrait for the improper node of* $\begin{cases} x' = -10x + 6y \\ y' = -6x + 2y \end{cases}$.

Suppose first that $\alpha < 0$. The trigonometric terms in this solution cause $\mathbf{X}(t)$ to rotate about the origin as t increases, while the factor $e^{\alpha t}$ causes $\mathbf{X}(t)$ to move closer to the origin (or, equivalently, the length of the vector $\mathbf{X}(t)$ to decrease to zero) as $t \to \infty$. This suggests a trajectory that spirals inward toward the origin as t increases.

Since t varies over the entire real line, taking on both negative and positive values, the trajectories when $\alpha > 0$ have the same spiral appearance, but now the arrows are reversed and $\mathbf{X}(t)$ moves outward, away from the origin, as $t \to \infty$.

The origin in this case is called a *spiral point*. When $\alpha < 0$, the origin is a *spiral sink* because the flow defined by the trajectories is spiraling into the origin. When $\alpha > 0$, the origin is a *spiral source* because now the origin appears to be spewing material outward in a spiral pattern.

The phase portrait in the following example is typical of a spiral sink.

EXAMPLE 10.10

Let

$$\mathbf{A} = \begin{pmatrix} -1 & -2 \\ 4 & 3 \end{pmatrix},$$

with eigenvalues $1 + 2i$ and $1 - 2i$ and eigenvectors, respectively, $\begin{pmatrix} -1+i \\ 2 \end{pmatrix}$ and $\begin{pmatrix} -1-i \\ 2 \end{pmatrix}$. Let

$$\mathbf{U} = \begin{pmatrix} -1 \\ 2 \end{pmatrix} \quad \text{and} \quad \mathbf{V} = \begin{pmatrix} 1 \\ 0 \end{pmatrix}$$

so that the eigenvectors are $\mathbf{U} + i\mathbf{V}$ and $\mathbf{U} - i\mathbf{V}$. The general solution of $\mathbf{X}' = \mathbf{A}\mathbf{X}$ is

$$\mathbf{X}(t) = c_1 e^t \left[\begin{pmatrix} -1 \\ 2 \end{pmatrix} \cos(2t) - \begin{pmatrix} 1 \\ 0 \end{pmatrix} \sin(2t) \right]$$

$$+ c_2 e^t \left[\begin{pmatrix} -1 \\ 2 \end{pmatrix} \sin(2t) + \begin{pmatrix} 1 \\ 0 \end{pmatrix} \cos(2t) \right].$$

Figure 10.17 gives a phase portrait for this system, showing trajectories spiraling away from the spiral source at the origin because the real part of the eigenvalues is positive. ▮

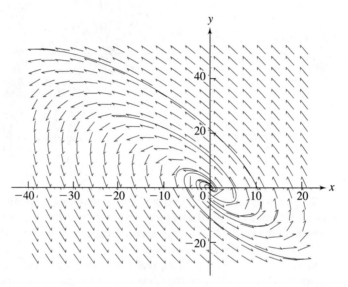

FIGURE 10.17   *Spiral source of the system* $\begin{cases} x' = -x - 2y \\ y' = 4x + 3y \end{cases}$.

Case 5   Pure imaginary eigenvalues.
Now trajectories have the form

$$\mathbf{X}(t) = c_1 [\mathbf{U} \cos(\beta t) - \mathbf{V} \sin(\beta t)] + c_2 [\mathbf{U} \sin(\beta t) + \mathbf{V} \cos(\beta t)].$$

Because of the trigonometric terms, this trajectory moves about the origin. Unlike the preceding case, however, there is no exponential factor to decrease or increase distance from the origin as t increases. This trajectory is a closed curve about the origin, representing a periodic solution of the system. The origin in this case is called a *center* of $\mathbf{X}' = \mathbf{A}\mathbf{X}$. In general, any closed trajectory of $\mathbf{X}' = \mathbf{A}\mathbf{X}$ represents a periodic solution of this system.

EXAMPLE 10.11

Let

$$\mathbf{A} = \begin{pmatrix} 3 & 18 \\ -1 & -3 \end{pmatrix}.$$

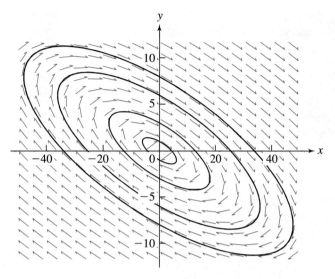

FIGURE 10.18 *Center of the system* $\begin{cases} x' = 3x + 18y \\ y' = -x - 3y \end{cases}$.

A has eigenvalues $3i$ and $-3i$, with respective eigenvectors $\begin{pmatrix} -3 - 3i \\ 1 \end{pmatrix}$ and $\begin{pmatrix} -3 + 3i \\ 1 \end{pmatrix}$. A phase portrait is given in Figure 10.18, showing closed trajectories about the center (origin). If we wish, we can write the general solution

$$\mathbf{X}(t) = c_1 \left[\begin{pmatrix} -3 \\ 1 \end{pmatrix} \cos(3t) + \begin{pmatrix} 3 \\ 0 \end{pmatrix} \sin(3t) \right]$$

$$+ c_2 \left[\begin{pmatrix} -3 \\ 1 \end{pmatrix} \sin(3t) + \begin{pmatrix} -3 \\ 0 \end{pmatrix} \cos(3t) \right]. \quad \blacksquare$$

We now have a complete description of the behavior of trajectories for the 2×2 constant coefficient system $\mathbf{X}' = \mathbf{AX}$. The general appearance of the phase portrait is completely determined by the eigenvalues of $\mathbf{A}$, and the critical point $(0, 0)$ is the primary point of interest, with the following correspondences:

Real, distinct eigenvalues of the same sign—$(0, 0)$ is a nodal source (Figure 10.10, p. 439) or sink (Figure 10.9, p. 438).

Real, distinct eigenvalues of opposite sign—$(0, 0)$ is a saddle point (Figure 10.12, p. 440).

Equal eigenvalues, two linearly independent eigenvectors—$(0, 0)$ is a proper node (Figure 10.13, p. 441).

Equal eigenvalues, all eigenvectors a multiple of a single eigenvector—$(0, 0)$ is an improper node (Figure 10.16, p. 443)

Complex eigenvalues with nonzero real part—$(0, 0)$ is a spiral point (Figure 10.17).

Pure imaginary eigenvalues—$(0, 0)$ is a center (Figure 10.18).

When we speak of a *classification* of the origin of a linear system, we mean a determination of the origin as a nodal source or sink, saddle point, proper or improper node, spiral point, or center.

SECTION 10.3 PROBLEMS

In each of Problems 1 through 16, use the eigenvalues of the matrix of the system to classify the origin of the system. Draw a phase portrait for the system. It is assumed here that software is available to do this, and it is not necessary to solve the system to generate the phase portrait.

1. $x' = 3x - 5y, y' = 5x - 7y$

2. $x' = x + 4y, y' = 3x$

3. $x' = x - 5y, y' = x - y$

4. $x' = 9x - 7y, y' = 6x - 4y$

5. $x' = 7x - 17y, y' = 2x + y$

6. $x' = 2x - 7y, y' = 5x - 10y$

7. $x' = 4x - y, y' = x + 2y$

8. $x' = 3x - 5y, y' = 8x - 3y$

9. $x' = -2x - y, y' = 3x - 2y$

10. $x' = -6x - 7y, y' = 7x - 20y$

11. $x' = 2x + y, y' = x - 2y$

12. $x' = 3x - y, y' = 5x + 3y$

13. $x' = x - 3y, y' = 3x - 7y$

14. $x' = 10x - y, y' = x + 12y$

15. $x' = 6x - y, y' = 13x - 2y$

16. $x' = 3x - 2y, y' = 11x - 3y$

10.4   Critical Points and Stability

A complete knowledge of the possible phase portraits of linear 2×2 systems is good preparation for the analysis of nonlinear systems. In this section, we will introduce the concept of critical point for a nonlinear system, define stability of critical points, and prepare for the qualitiative analysis of nonlinear systems, in which we attempt to draw conclusions about how solutions will behave, without having explicit solutions in hand.

We will consider the 2×2 autonomous system

$$x'(t) = f(x(t), y(t)),$$

$$y'(t) = g(x(t), y(t)),$$

or, more compactly,

$$x' = f(x, y),$$

$$y' = g(x, y).$$

This is the system (10.4) discussed in Section 10.2. We will assume that f and g are continuous with continuous first partial derivatives in some region D of the x, y plane. In specific cases D may be the entire plane.

This system can be written in matrix form as

$$\mathbf{X}' = \mathbf{F}(\mathbf{X}),$$

where

$$\mathbf{X}(t) = \begin{pmatrix} x(t) \\ y(t) \end{pmatrix} \quad \text{and} \quad \mathbf{F}(\mathbf{X}) = \begin{pmatrix} f(x, y) \\ g(x, y) \end{pmatrix}.$$

Taking the lead from the linear system $\mathbf{X}' = \mathbf{AX}$, we make the following definition.

In August of 1999, the Petronas Towers was officially opened. Designed by the American firm of Cesar Pelli and Associates, in collaboration with Kuala Lumpur City Center architects, the graceful towers have an elegant slenderness (height to width) ratio of 9:4. This was made possible by modern materials and building techniques, featuring high-strength concrete that is twice as effective as steel in sway reduction. The towers are supported by 75-foot by 75-foot concrete cores and an outer ring of super columns. The 88 floors stand 452 meters above street level, and include 65,000 square meters of stainless steel cladding and 77,000 square meters of vision glass. Computations of stability of structures involve the analysis of critical points of systems of nonlinear differential equations.

DEFINITION 10.1 *Critical Point*

A point (x_0, y_0) in D is a critical point (or equilibrium point) of $\mathbf{X}' = \mathbf{F}(\mathbf{X})$ if

$$f(x_0, y_0) = g(x_0, y_0) = 0.$$

We see immediately one significant difference between the linear and nonlinear cases. The linear system $\mathbf{X}' = \mathbf{A}\mathbf{X}$, with $\mathbf{A}$ nonsingular, has exactly one critical point, the origin. A nonlinear system $\mathbf{X}' = \mathbf{F}(\mathbf{X})$ can have any number of critical points. We will, however, only consider systems in which critical points are *isolated*. This means that about any critical point, there is a circle that contains no other critical point of the system.

EXAMPLE 10.12

Consider the damped pendulum (Example 10.1), whose motion is governed by the system

$$x' = y,$$
$$y' = -\omega^2 \sin(x) - \gamma y.$$

Here

$$f(x, y) = y$$

and

$$g(x, y) = -\omega^2 \sin(x) - \gamma y.$$

The critical points are solutions of

$$y = 0$$

and

$$-\omega^2 \sin(x) - \gamma y = 0.$$

These equations are satisfied by all points $(n\pi, 0)$, in which $n = 0, \pm1, \pm2, \ldots$. These critical points are isolated. About any point $(n\pi, 0)$, we can draw a circle (for example, of radius $\frac{1}{4}$) that does not contain any other critical point.

For this problem, the critical points split naturally into two classes. Recall that $x = \theta$ is the angle of displacement of the pendulum from the vertical downward position, with the bob at the bottom, and $y = d\theta/dt$. When n is even, then $x = \theta = 2k\pi$ for k any integer. Each critical point $(2k\pi, 0)$ corresponds to the bob pointing straight down, with zero velocity (because $y = x' = \theta' = 0$). When n is odd, then $x = \theta = (2k + 1)\pi$ for k any integer. The critical point $((2k + 1)\pi, 0)$ corresponds to the bob in the vertical upright position, with zero velocity.

Without any mathematical analysis, there is an obvious and striking difference between these two kinds of critical points. At, for example, $x = 0$, the bob hangs straight down from the point of suspension. If we displace it slightly from this position and then release it, the bob will go through some oscillations of decreasing amplitude, after which it will return to its downward position and remain there. This critical point, and all critical points $(2n\pi, 0)$, are what we will call *stable*. Solutions of the pendulum equation for initial values near this critical point remain close to the constant equilibrium solution for all later times.

By contrast, consider the critical point $(\pi, 0)$. This has the bob initially balanced vertically upward. If the bob is displaced, no matter how slightly, it will swing downward and oscillate back and forth some number of times but never return to this vertical position. Solutions near this constant equilibrium solution (bob vertically up) do not remain near this position, but move away from it. This critical point, and any critical point $((2k + 1)\pi, 0)$, is *unstable*. ■

EXAMPLE 10.13

Consider the damped nonlinear spring of Example 10.2. The system of differential equations governing the motion is

$$x' = y,$$

$$y' = -\frac{k}{m}x + \frac{\alpha}{m}x^3 - \frac{c}{m}y.$$

The critical points are $(0, 0)$, $(\sqrt{k/\alpha}, 0)$, and $(-\sqrt{k/\alpha}, 0)$. Recall that x measures the position of the spring, from the equilibrium (rest) position, and $y = dx/dt$ is the velocity of the spring.

We will do a mathematical analysis of this system shortly, but for now look at these critical points from the point of view of our experience with how springs behave. If we displace the spring very slightly from the equilibrium solution $(0, 0)$ and then release it, we expect it to undergo some motion back and forth and then come to rest, approaching the equilibrium point. In this sense $(0, 0)$ is a stable critical point. However, if we displace the spring slightly from a position very nearly at distance $\sqrt{k/\alpha}$ to the right or left of the equilibrium position and then release it, the spring may or may not return to this position, depending on the relative sizes of the damping constant c and the coefficients in the nonlinear spring force function, particularly α. In this sense these equilibrium points may be stable or may not be. In the next section we will develop the tools for a more definitive analysis of these critical points. ∎

Taking a cue from these examples, we will define a concept of stability of critical points. Recall that

$$\|\mathbf{V}\| = \sqrt{v_1^2 + v_2^2}$$

is the length (or norm) of a vector $\mathbf{V} = (v_1, v_2)$ in the plane. If $\mathbf{W} = (w_1, w_2)$ is also a vector in the plane, then $\|\mathbf{V} - \mathbf{W}\|$ is the length of the vector from $\mathbf{W}$ to $\mathbf{V}$. If $\mathbf{W} = (w_1, w_2)$, then

$$\|\mathbf{V} - \mathbf{W}\| = ((v_1 - w_1)^2 + (v_2 - w_2)^2)^{1/2}$$

is also the distance between the points (v_1, v_2) and (w_1, w_2).

Finally, if $\mathbf{X}_0$ is a given vector, then the locus of points (vectors) $\mathbf{X}$ such that

$$\|\mathbf{X} - \mathbf{X}_0\| < r$$

for any positive r, is the set of points $\mathbf{X}$ on the circle of radius r about $\mathbf{X}_0$. These are exactly the points at fixed distance r from $\mathbf{X}_0$.

DEFINITION 10.2 *Stability of a Critical Point*

Let $\mathbf{X}_0 = (x_0, y_0)$ be a critical point of $\mathbf{X}' = \mathbf{F}(\mathbf{X})$. Then $\mathbf{X}_0$ is stable if and only if, given any positive number ϵ there exists a positive number δ_ϵ such that if $\mathbf{X} = \boldsymbol{\Phi}(t)$ is a solution of $\mathbf{X}' = \mathbf{F}(\mathbf{X})$ and

$$\|\boldsymbol{\Phi}(0) - \mathbf{X}_0\| < \delta_\epsilon,$$

then $\boldsymbol{\Phi}(t)$ exists for all $t \geq 0$, and

$$\|\boldsymbol{\Phi}(t) - \mathbf{X}_0\| < \epsilon \quad \text{for all } t \geq 0.$$

We say that $\mathbf{X}_0$ is unstable if this point is not stable.

Keep in mind that the constant solution $\mathbf{X}(t) = \mathbf{X}_0$ is the unique solution through this critical point. That is, the trajectory through a critical point is just this point itself. A critical point $\mathbf{X}_0$ is stable if solutions that are initially (at $t = 0$) close (within δ_ϵ) to $\mathbf{X}_0$ remain close (within ϵ) for all later times. In terms of trajectories, this means that a trajectory that starts out sufficiently close to $\mathbf{X}_0$ at time zero must remain close to this equilibrium solution at all later times. Figure 10.19 illustrates this idea.

This does not imply that solutions that start near $\mathbf{X}_0$ approach this point as a limit as $t \to \infty$. They may simply remain within a small disk about $\mathbf{X}_0$, without approaching $\mathbf{X}_0$ in a limiting sense. If, however, solutions initially near $\mathbf{X}_0$ also approach $\mathbf{X}_0$ as a limit, then we call $\mathbf{X}_0$ an *asymptotically stable* critical point.

DEFINITION 10.3 *Asymptotically Stable Critical Point*

$\mathbf{X}_0$ is an asymptotically stable critical point of $\mathbf{X}' = \mathbf{F}(\mathbf{X})$ if and only if $\mathbf{X}_0$ is a stable critical point and there exists a positive number δ such that if a solution $\mathbf{X} = \boldsymbol{\Phi}(t)$ satisfies $\|\boldsymbol{\Phi}(0) - \mathbf{X}_0\| < \delta$, then $\lim_{t \to \infty} \boldsymbol{\Phi}(t) = \mathbf{X}_0$.

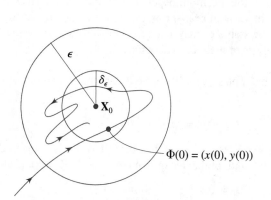

FIGURE 10.19 *Stable critical point of* $\mathbf{X}' = \mathbf{F}(\mathbf{X})$.

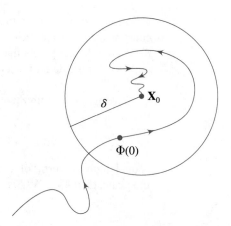

FIGURE 10.20 *Asymptotically stable critical point of* $\mathbf{X}' = \mathbf{F}(\mathbf{X})$.

This concept is illustrated in Figure 10.20. Stability does not imply asymptotic stability. It is less obvious that asymptotic stability does not imply stability. A solution might start "close enough" to the critical point and actually approach the critical point in the limit as $t \to \infty$, but for some arbitrarily large positive times move arbitrarily far from $\mathbf{X}_0$ (before bending back to approach it in the limit).

In the case of the damped pendulum, critical points $(2n\pi, 0)$ are asymptotically stable. If the bob is displaced slightly from the vertical downward position and then released, it will eventually approach this vertical downward position in the limit as $t \to \infty$.

To get some experience with stability and asymptotic stability, and also to prepare for nonlinear systems that are in some sense "nearly" linear, we will review the critical point $(0, 0)$ for the linear system $\mathbf{X}' = \mathbf{A}\mathbf{X}$ in the context of stability.

Nodal Source or Sink

This occurs when the eigenvalues of $\mathbf{A}$ are real and distinct, but of the same sign—a nodal sink when they are negative, and a nodal source when they are positive. From the phase portrait in Figure 10.9 (p. 438), $(0, 0)$ is stable and asymptotically stable when the eigenvalues are negative (nodal sink), because then all trajectories tend toward the origin as time increases. However, $(0, 0)$ is unstable when the eigenvalues are positive (nodal source), because in this case all trajectories move away from the origin with increasing time (Figure 10.10; see p. 439).

Saddle Point

The origin is a saddle point when $\mathbf{A}$ has real eigenvalues of opposite sign. A saddle point is unstable. This is apparent in Figure 10.12 (p. 440), in which we can see that trajectories do not remain near the origin as time increases, nor do they approach the origin as a limit.

Proper Node

The origin is a proper node when the eigenvalues of $\mathbf{A}$ are equal and $\mathbf{A}$ has two linearly independent eigenvectors. Figure 10.13 (p. 441) shows a typical proper node. When the arrows are toward the origin (negative eigenvalues), this node is stable and asymptotically stable. When the trajectories are oriented away from the origin, this node is not stable.

Improper Node

The origin is an improper node when the eigenvalues of $\mathbf{A}$ are equal and $\mathbf{A}$ does not have two linearly independent eigenvectors. Now the origin is a stable and asymptotically stable critical

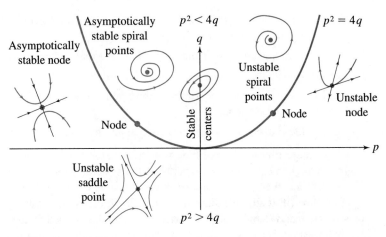

FIGURE 10.21 *Classification of* $(0, 0)$ *for* $\mathbf{X}' = \mathbf{A}\mathbf{X}$.

point if the eigenvalue is negative, and unstable if the eigenvalue is positive. Figure 10.16 shows trajectories near a stable improper node (negative eigenvalue). If the eigenvalue is positive, the trajectories have orientation away from the origin, and then this node is unstable.

Spiral Point
The origin is a spiral point when the eigenvalues are complex conjugates with nonzero real part. When this real part is positive, the origin is a spiral source (trajectories spiral away from the origin, as in Figure 10.17), and in this case the origin is unstable. When this real part is negative, the origin is a stable and asymptotically stable spiral sink (trajectories spiraling into the origin). The phase portrait of such a sink has the same appearance as a spiral source, with arrows on the trajectories reversed.

Center
The origin is a center when the eigenvalues of $\mathbf{A}$ are pure imaginary. A center is stable, but not asymptotically stable (Figure 10.18).

There is a succinct graphical way of summarizing the classifications and stability type of the critical point $(0, 0)$ for the linear system $\mathbf{X}' = \mathbf{A}\mathbf{X}$. Let

$$\mathbf{A} = \begin{pmatrix} a & b \\ c & d \end{pmatrix}.$$

The eigenvalues equation of $\mathbf{A}$ are solutions of

$$\lambda^2 - (a + d)\lambda + ad - bc = 0.$$

Let $p = -(a + d)$ and $q = ad - bc$ to write this equation as

$$\lambda^2 + p\lambda + q = 0.$$

The eigenvalues of $\mathbf{A}$ are

$$\frac{-p \pm \sqrt{p^2 - 4q}}{2}.$$

These are real or complex depending on whether $p^2 - 4q \geq 0$ or $p^2 - 4q < 0$. In the p, q plane of Figure 10.21, the boundary between these two cases is the parabola $p^2 = 4q$. Now the p, q plane gives a summary of conclusions as follows:

Above this parabola ($p^2 < 4q$) but not in the q axis, the eigenvalues are complex conjugates with nonzero real parts (spiral point).

On the parabola ($p^2 = 4q$), the eigenvalues are real and equal (proper or improper node).

On the q axis, the eigenvalues are pure imaginary (center).

Between the p axis and the parabola, the eigenvalues are real and distinct, with the same sign (nodal source or sink).

Below the p axis, the eigenvalues are real and have opposite sign (saddle point).

It is interesting to observe how sensitive the classification and stability type of a critical point are to changes in the coefficients of the system. Suppose we begin with a linear system $\mathbf{X}' = \mathbf{AX}$ and then perturb one or more elements of $\mathbf{A}$ by "small" amounts to form a new system. How (if at all) will this change the classification and stability of the critical point? The classification and stability of $(0, 0)$ are completely determined by the eigenvalues, so the issue is really how small changes in the matrix elements affect the eigenvalues. The eigenvalues of $\mathbf{A}$ are $(-p \pm \sqrt{p^2 - 4q})/2$, which is a continuous function of p and q. Thus small changes in p and q (caused by small changes in a, b, c, and d) result in small changes in the eigenvalues. There are two cases in which arbitrarily small changes in $\mathbf{A}$ will change the nature of the critical point.

1. If the origin is a center (pure imaginary eigenvalues), then $p = -a - d = 0$. Arbitrarily small changes in a and d can change this, resulting in a new matrix whose eigenvalues have positive or negative real parts. For the new, perturbed system, $(0, 0)$ is no longer a center. This means that centers are sensitive to arbitrarily small changes in $\mathbf{A}$.

2. The other sensitive case is that both eigenvalues are the same, which occurs when $p^2 - 4q = 0$. Again, arbitrarily small changes in $\mathbf{A}$ can result in this quantity becoming positive or negative, changing the classification of the critical point. However, the stability or instability of $(0, 0)$ is determined by the sign of p, and sufficiently small changes in $\mathbf{A}$ will leave this sign unchanged. Thus in this case the classification of the kind of critical point the system has is more sensitive to change than its stability or instability.

These considerations should be kept in mind when we state Theorem 10.3 in the next section.

With this background on linear systems and the various characteristics of its critical point, we are ready to analyze systems that are in some sense approximated by linear systems.

SECTION 10.4 PROBLEMS

1–16. For $j = 1, \ldots, 16$, classify the critical point of the system of Problem j, Section 10.3, as to being stable and asymptotically stable, stable and not asymptotically stable, or unstable.

17. Consider the system $\mathbf{X}' = \mathbf{AX}$, where

$$\mathbf{A} = \begin{pmatrix} 1 & -3 \\ 2 & -1 + \epsilon \end{pmatrix}, \text{ with } \epsilon > 0.$$

(a) Show that when $\epsilon = 0$, the critical point is a center, stable but not asymptotically stable. Generate a phase portrait for this system.

(b) Show that when $\epsilon \neq 0$, the critical point is not a center, no matter how small ϵ is chosen. Generate a phase portrait for this system with $\epsilon = \frac{1}{10}$.

This problem illustrates the sensitivity of trajectories of the system to small changes in the coefficients, in the case of pure imaginary eigenvalues.

18. Consider the system $\mathbf{X}' = \mathbf{AX}$, where

$$\mathbf{A} = \begin{pmatrix} 2 + \epsilon & 5 \\ -5 & -8 \end{pmatrix} \text{ and } \epsilon > 0.$$

(a) Show that when $\epsilon = 0$, $\mathbf{A}$ has equal eigenvalues and does not have two linearly independent eigenvectors. Classify the type of critical point at the origin and its stability characteristics. Generate a phase portrait for this system.

(b) Show that if ϵ is not zero (but can be arbitrarily small in magnitude), then $\mathbf{A}$ has real and distinct eigenvalues. Classify the type of critical point at the origin in this case, as well as its stability characteristics. Generate a phase portrait for the case $\epsilon = \frac{1}{10}$.

This problem illustrates the sensitivity of trajectories to small changes in the coefficients, in the case of equal eigenvalues.

10.5 Almost Linear Systems

Suppose $\mathbf{X}' = \mathbf{F}(\mathbf{X})$ is a nonlinear system. We want to define a sense in which this system may be thought of as "almost linear."

Suppose the system has the special form

$$\mathbf{X}' = \mathbf{AX} + \mathbf{G}(\mathbf{X}). \qquad (10.7)$$

This is a linear system $\mathbf{X}' = \mathbf{AX}$, with another term, $\mathbf{G}(\mathbf{X}) = \begin{pmatrix} p(x, y) \\ q(x, y) \end{pmatrix}$ added. Any nonlinearity of the system (10.7) is in $\mathbf{G}(\mathbf{X})$. We refer to the system $\mathbf{X}' = \mathbf{AX}$ as the *linear part* of the system (10.7).

Assume that

$$p(0, 0) = q(0, 0) = 0,$$

so the system (10.7) has a critical point at the origin. The idea we want to pursue is that if the nonlinear term is "small enough," then the behavior of solutions of the linear system $\mathbf{X}' = \mathbf{AX}$ near the origin may give us information about the behavior of solutions of the original, nonlinear system near this critical point. The question is: How small is "small enough"?

We will assume in this discussion that $\mathbf{A}$ is a nonsingular, 2×2 matrix of real numbers and that p and q are continuous at least within some disk about the origin. In the following definition, we refer to partial derivatives of $\mathbf{G}$, by which we mean

$$\mathbf{G}_x = \frac{\partial \mathbf{G}}{\partial x} = \begin{pmatrix} p_x \\ q_x \end{pmatrix} \quad \text{and} \quad \mathbf{G}_y = \frac{\partial \mathbf{G}}{\partial y} = \begin{pmatrix} p_y \\ q_y \end{pmatrix}.$$

DEFINITION 10.4 *Almost Linear*

The system (10.7) is almost linear in a neighborhood of $(0, 0)$ if $\mathbf{G}$ and its first partial derivatives are continuous within some circle about the origin, and

$$\lim_{\mathbf{X} \to \mathbf{0}} \frac{\|\mathbf{G}(\mathbf{X})\|}{\|\mathbf{X}\|} = 0. \qquad (10.8)$$

This condition (10.8) means that as $\mathbf{X}$ is chosen closer to the origin, $\mathbf{G}(\mathbf{X})$ must become small in magnitude faster than $\mathbf{X}$ does. This gives a precise measure of "how small" the nonlinear term must be near the origin for the system (10.7) to qualify as almost linear.

If we write

$$\mathbf{X} = \begin{pmatrix} x \\ y \end{pmatrix}, \mathbf{A} = \begin{pmatrix} a & b \\ c & d \end{pmatrix} \quad \text{and} \quad \mathbf{G}(\mathbf{X}) = \mathbf{G}(x, y) = \begin{pmatrix} p(x, y) \\ q(x, y) \end{pmatrix},$$

then the system (10.7) is

$$x' = ax + by + p(x, y)$$
$$y' = cx + dy + q(x, y).$$

Condition (10.8) now becomes

$$\lim_{(x,y)\to(0,0)} \frac{p(x,y)}{\sqrt{x^2+y^2}} = \lim_{(x,y)\to(0,0)} \frac{q(x,y)}{\sqrt{x^2+y^2}} = 0.$$

These limits, in terms of the components of $\mathbf{G}(\mathbf{X})$, are sometimes easier to deal with than the limit of $\|\mathbf{G}(\mathbf{X})\|/\|\mathbf{X}\|$ as $\mathbf{X}$ approaches the origin, although the two formulations are equivalent.

EXAMPLE 10.14

The system

$$\mathbf{X}' = \begin{pmatrix} 4 & -2 \\ 1 & 6 \end{pmatrix}\mathbf{X} + \begin{pmatrix} -4xy \\ -8x^2 y \end{pmatrix}$$

is almost linear. To verify this, compute

$$\lim_{(x,y)\to(0,0)} \frac{-4xy}{\sqrt{x^2+y^2}} \quad \text{and} \quad \lim_{(x,y)\to(0,0)} \frac{-8x^2 y}{\sqrt{x^2+y^2}}.$$

There are various ways of showing that these limits are zero, but here is a device worth remembering. Express (x, y) in polar coordinates by putting $x = r\cos(\theta)$ and $y = r\sin(\theta)$. Then

$$\frac{-4xy}{\sqrt{x^2+y^2}} = -\frac{4r^2\cos(\theta)\sin(\theta)}{r} = -4r\cos(\theta)\sin(\theta) \to 0$$

as $r \to 0$, which must occur if $(x, y) \to (0, 0)$. Similarly,

$$\frac{-8x^2 y}{\sqrt{x^2+y^2}} = -8\frac{r^3\cos^2(\theta)\sin(\theta)}{r} = -8r^2\cos^2(\theta)\sin(\theta) \to 0$$

as $r \to 0$. ■

Figure 10.22(a) shows a phase portrait of this system. For comparison, a phase portrait of the linear part of $\mathbf{X}' = \mathbf{A}\mathbf{X}$ is given in Figure 10.22(b). Notice a qualitative similarity between the phase portraits near the origin. This is the rationale for the definition of almost linear systems. We will now display a correspondence between the type of critical point, and its stability properties, for the almost linear system $\mathbf{X}' = \mathbf{A}\mathbf{X} + \mathbf{G}$ and its linear part $\mathbf{X}' = \mathbf{A}\mathbf{X}$. The behavior is not always the same. Nevertheless, in some cases that we will identify, properties of the critical point for the linear system carry over to either the same properties for the almost linear system or, if not the same, at least to important information about the nonlinear system.

THEOREM 10.3

Let λ and μ be the eigenvalues of $\mathbf{A}$. Assume that $\mathbf{X}' = \mathbf{A}\mathbf{X} + \mathbf{G}$ is almost linear. Then the following conclusions hold for the system $\mathbf{X}' = \mathbf{A}\mathbf{X} + \mathbf{G}$.

1. If λ and μ are unequal and negative, then the origin is an asymptotically stable nodal sink of $\mathbf{X}' = \mathbf{A}\mathbf{X} + \mathbf{G}$. If these eigenvalues are unequal and positive, then the origin is an unstable nodal source of $\mathbf{X}' = \mathbf{A}\mathbf{X} + \mathbf{G}$.

2. If λ and μ are of opposite sign, then the origin is an unstable saddle point of $\mathbf{X}' = \mathbf{A}\mathbf{X} + \mathbf{G}$.

3. If λ and μ are complex with a negative real part, then the origin is an asymptotically stable spiral point of $\mathbf{X}' = \mathbf{A}\mathbf{X} + \mathbf{G}$. If these eigenvalues have a positive real part, then the origin is an unstable spiral point.

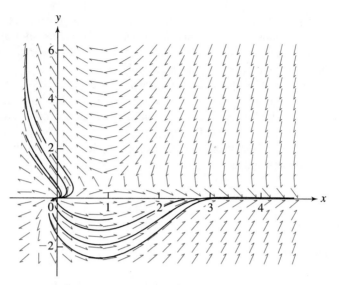

FIGURE 10.22(a) *Phase portrait for* $\begin{cases} x' = 4x - 2y - 4xy \\ y' = x + 6y - 8x^2y \end{cases}$.

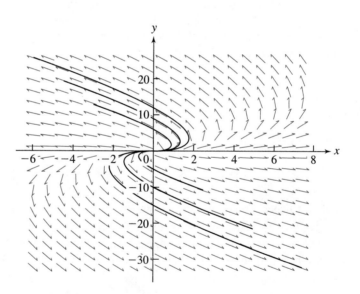

FIGURE 10.22(b) *Phase portrait for the linear part of the system of Figure 10.22(a).*

4. If λ and μ are equal and negative, then the linear system has an asymptotically stable proper or improper node, while the almost linear system has an asymptotically stable node or spiral point. If λ and μ are equal and positive, then the linear system has an unstable proper or improper node, while the almost linear system has an unstable node or spiral point.

5. If λ and μ are pure imaginary (conjugates of each other), then the origin is a center of $\mathbf{X}' = \mathbf{A}\mathbf{X}$, but may be a center or spiral point of the almost linear system $\mathbf{X}' = \mathbf{A}\mathbf{X} + \mathbf{G}$. Further, in the case of a spiral point of the almost linear system, the critical point may be unstable or asymptotically stable. ∎

The only case in which the linear system fails to provide definitive information of some kind about the almost linear system is that when the eigenvalues of **A** are pure imaginary. In this event, the linear system has a stable center, while the almost linear system can have a stable center or a spiral point that may be stable or unstable.

In light of this theorem, when we ask for an analysis of a critical point of an almost linear system, we mean a determination of whether the point is an asymptotically stable nodal sink, an unstable nodal source, an unstable saddle point, an asymptotically stable spiral point or unstable spiral point, or, from (5) of the theorem, either a center or spiral point.

A proof of this theorem requires some delicate analysis that we will avoid. The rest of this section is devoted to examples and phase portraits.

EXAMPLE 10.15

The system

$$\mathbf{X}' = \begin{pmatrix} -1 & -1 \\ -1 & -3 \end{pmatrix} \mathbf{X} + \begin{pmatrix} x^2 y^2 \\ x^3 - y^2 \end{pmatrix}$$

is almost linear and has only one critical point, $(0, 0)$. The eigenvalues of **A** are $-2 + \sqrt{2}$ and $-2 - \sqrt{2}$, which are distinct and negative. The origin is an asymptotically stable nodal sink of the linear system $\mathbf{X}' = \mathbf{AX}$, and hence is also a stable and asymptotically stable nodal sink of the almost linear system. Figure 10.23(a) and (b) shows a phase portrait of the almost linear system and its linear part, respectively. ■

EXAMPLE 10.16

The system

$$\mathbf{X}' = \begin{pmatrix} 3 & -4 \\ 6 & 2 \end{pmatrix} \mathbf{X} + \begin{pmatrix} x^2 \cos(y) \\ y^3 \end{pmatrix}$$

is almost linear. The only critical point is $(0, 0)$. The eigenvalues of **A** are $\frac{5}{2} + \frac{1}{2} i \sqrt{95}$ and $\frac{5}{2} - \frac{1}{2} i \sqrt{95}$. The linear part has an unstable spiral point at the origin. The origin is therefore an unstable spiral point of the almost linear system. Phase portraits for the given nonlinear system and its linear part are shown in Figure 10.24(a) and (b), respectively (see p. 458). ■

EXAMPLE 10.17

The system

$$\mathbf{X}' = \begin{pmatrix} -1 & 2 \\ 2 & 3 \end{pmatrix} \mathbf{X} + \begin{pmatrix} x \sin(y) \\ 8 \sin(x) \end{pmatrix}$$

is almost linear, and the origin is a critical point. The eigenvalues of **A** are $1 + 2\sqrt{2}$ and $1 - 2\sqrt{2}$, which are real and of opposite sign. The origin is an unstable saddle point of the linear part, hence also of the given system. Phase portraits of both systems are shown in Figure 10.25(a) (nonlinear system) and (b) (linear part) (see p. 459). ■

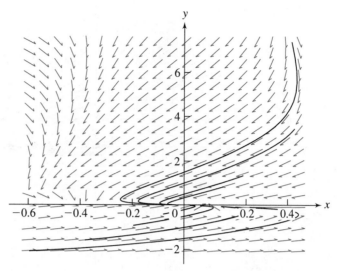

FIGURE 10.23(a) *Phase portrait for* $\begin{cases} x' = -x - y + x^2 y^2 \\ y' = -x - 3y + x^3 - y^2 \end{cases}$.

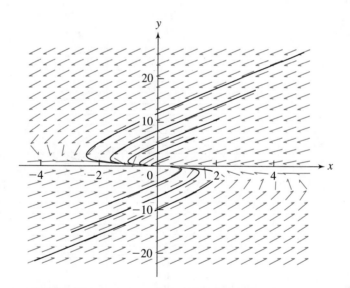

FIGURE 10.23(b) *Phase portrait for the linear part of the system of Figure 10.23(a).*

EXAMPLE 10.18

The system

$$\mathbf{X}' = \begin{pmatrix} 4 & 11 \\ -2 & -4 \end{pmatrix} \mathbf{X} + \begin{pmatrix} x \sin(y) \\ \sin(y) \end{pmatrix}$$

is almost linear, and $(0, 0)$ is a critical point. The eigenvalues of $\mathbf{A}$ are $\sqrt{6}i$ and $-\sqrt{6}i$. The origin is a stable, but not asymptotically stable, center for the linear part. The theorem does not allow us

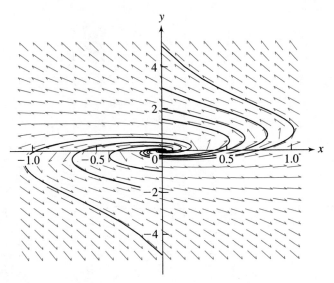

FIGURE 10.24(a)   *Phase portrait for* $\begin{cases} x' = 3x - 4y + x^2\cos(y) \\ y' = 6x + 2y + y^3 \end{cases}$.

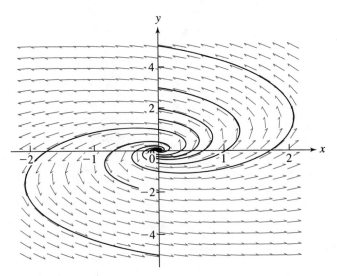

FIGURE 10.24(b)   *Phase portrait for the linear part of the system of Figure 10.24(a).*

to draw a definitive conclusion about the almost linear system, which might have a center or spiral point at the origin. Figure 10.26(a) and (b) shows phase portraits for the almost linear system and its linear part, respectively (see p. 460). ▮

EXAMPLE 10.19

Consider the system

$$\mathbf{X}' = \begin{pmatrix} \alpha & -1 \\ 1 & -\alpha \end{pmatrix} \mathbf{X} + \begin{pmatrix} hx(x^2 + y^2) \\ ky(x^2 + y^2) \end{pmatrix},$$

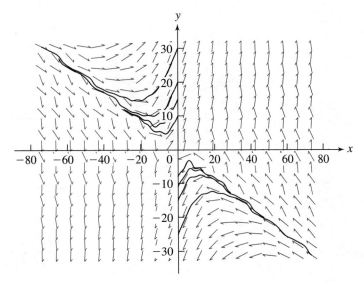

FIGURE 10.25(a) *Phase portrait for* $\begin{cases} x' = -x + 2y + x\sin(y) \\ y' = 2x + 3y + 8\sin(x) \end{cases}$.

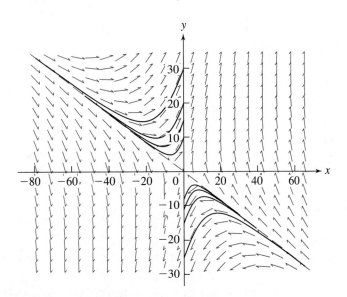

FIGURE 10.25(b) *Phase portrait for the linear part of the system of Figure 10.25(a).*

in which α, h, and k are constants. The eigenvalues of the matrix of the linear part are $\sqrt{\alpha^2 - 1}$ and $-\sqrt{\alpha^2 - 1}$. Consider cases.

If $0 < |\alpha| < 1$, then these eigenvalues are pure imaginary. The origin is a center of the linear part but may be a center or spiral point of the almost linear system.

If $|\alpha| > 1$, then the eigenvalues are real and of opposite sign, so the origin is an unstable saddle point of both the linear part and the original almost linear system.

If $\alpha = \pm 1$, then **A** is singular and the system is not almost linear.

Figure 10.27(a) shows a phase portrait for this system with $h = 0.4$, $k = 0.7$, and $\alpha = \frac{1}{3}$. Figure 10.27(b) has $\alpha = 2$. ∎

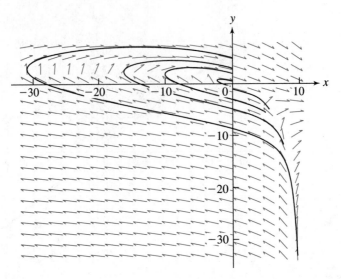

FIGURE 10.26(a) *Phase portrait for* $\begin{cases} x' = 4x + 11y + x\sin(y) \\ y' = -2x - 4y + \sin(y) \end{cases}$.

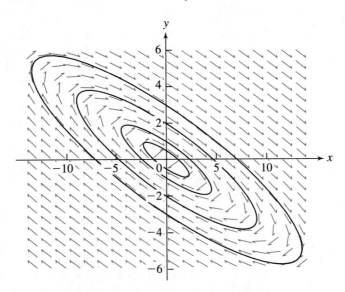

FIGURE 10.26(b) *Phase portrait for the linear part of the system of Figure 10.26(a).*

The next example demonstrates the sensitivity of Case (5) of Theorem 10.3.

EXAMPLE 10.20

Let ϵ be a real number and consider the system

$$\mathbf{X}' = \begin{pmatrix} y + \epsilon x(x^2 + y^2) \\ -x + \epsilon y(x^2 + y^2) \end{pmatrix}.$$

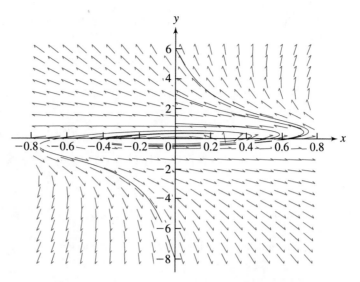

FIGURE 10.27(a) *Phase portrait for* $\begin{cases} x' = \frac{1}{3}x - y + 0.4x(x^2 + y^2) \\ y' = x - \frac{1}{3}y + 0.7y(x^2 + y^2) \end{cases}$.

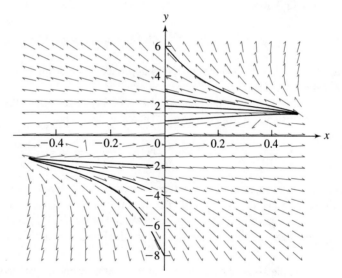

FIGURE 10.27(b) *Phase portrait for* $\begin{cases} x' = 2x - y + 0.4x(x^2 + y^2) \\ y' = x - 2y + 0.7y(x^2 + y^2) \end{cases}$.

We can write this in the form $\mathbf{X}' = \mathbf{A}\mathbf{X} + \mathbf{G}$ as

$$\mathbf{X}' = \begin{pmatrix} 0 & 1 \\ -1 & 0 \end{pmatrix} \mathbf{X} + \begin{pmatrix} \epsilon x(x^2 + y^2) \\ \epsilon y(x^2 + y^2) \end{pmatrix}.$$

The origin is a critical point of this almost linear system. The eigenvalues of $\mathbf{A}$ are i and $-i$, so the linear part of this system has a center at the origin. This is the case in which Theorem 10.3 does not give a definitive conclusion for the nonlinear system.

To analyze the nature of this critical point for the nonlinear system, use polar coordinates r and θ. Since

$$r^2 = x^2 + y^2$$

then

$$rr' = xx' + yy'$$
$$= x[y + \epsilon x(x^2 + y^2)] + y[-x + \epsilon y(x^2 + y^2)]$$
$$= \epsilon(x^2 + y^2)(x^2 + y^2) = \epsilon r^4.$$

Then

$$\frac{dr}{dt} = \epsilon r^3.$$

This is a separable equation for r, which we solve to get

$$r(t) = \frac{1}{\sqrt{k - 2\epsilon t}},$$

in which k is constant determined by initial conditions (a point the trajectory is to pass through). Now consider cases.

If $\epsilon < 0$, then

$$r(t) = \frac{1}{\sqrt{k + 2|\epsilon|t}} \to 0$$

as $t \to \infty$. In this case the trajectory approaches the origin in the limit as $t \to \infty$, and $(0, 0)$ is asymptotically stable.

However, watch what happens if $\epsilon > 0$. Say $r(0) = \rho$, so the trajectory starts at a point at a positive distance ρ from the origin. Then $k = 1/\rho^2$ and

$$r(t) = \frac{1}{\sqrt{(1/\rho^2) - 2\epsilon t}}.$$

In this case, as t increases from 0 and approaches $1/(2\epsilon\rho^2)$, $r(t) \to \infty$. This means that at finite times, the trajectory is arbitrarily far away from $(0, 0)$, hence $(0, 0)$ is unstable when ϵ is positive.

A phase portrait for $\epsilon = -0.2$ is given in Figure 10.28(a), and for $\epsilon = 0.2$ in Figure 10.28(b). Figure 10.28(c) gives a phase portrait for the linear part of this system (see p. 464). ■

Example 10.20 shows how sensitive an almost linear system can be when the eigenvalues of the linear part are pure imaginary. In this example, ϵ can be chosen arbitrarily small. Still, when ϵ is negative, the origin is asymptotically stable, and when ϵ is positive, regardless of magnitude, the origin becomes unstable.

Thus far the discussion has been restricted to nonlinear systems in the special form $\mathbf{X}' = \mathbf{AX} + \mathbf{G}$, with the origin as a critical point. However, in general a nonlinear system comes in the form $\mathbf{X}' = \mathbf{F}(\mathbf{X})$, and there may be critical points other than the origin. We will now show how to translate a critical point (x_0, y_0) to the origin so that $\mathbf{X}' = \mathbf{F}(\mathbf{X})$ translates to a system $\mathbf{X}' = \mathbf{AX} + \mathbf{G}$. This makes the linear part of the translated system transparent. Further, since Theorem 10.3 is set up to deal with critical points at the origin, we can apply it to $\mathbf{X}' = \mathbf{AX} + \mathbf{G}$ whenever this system is almost linear.

Thus suppose (x_0, y_0) is a critical point of $\mathbf{X}' = \mathbf{F}(\mathbf{X})$, where $\mathbf{F} = \begin{pmatrix} f \\ g \end{pmatrix}$. Assume that f and g are continuous with continuous first and second partial derivatives at least within some circle

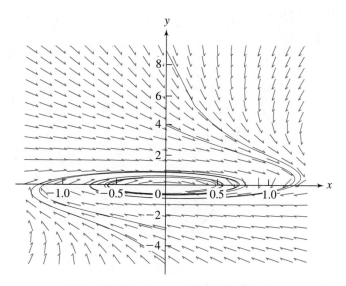

FIGURE 10.28(a) *Phase portrait for* $\begin{cases} x' = y + \epsilon x(x^2 + y^2) \\ y' = -x + \epsilon y(x^2 + y^2) \end{cases}$,
with $\epsilon = -0.2$.

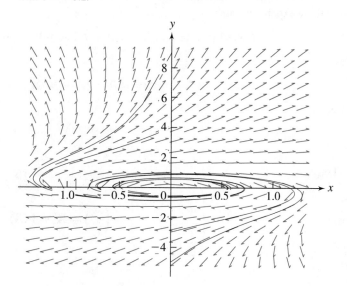

FIGURE 10.28(b) $\epsilon = 0.2$.

about (x_0, y_0). By Taylor's theorem for functions of two variables, we can write for (x, y) within some circle about (x_0, y_0)

$$f(x, y) = f(x_0, y_0) + f_x(x_0, y_0)(x - x_0) + f_y(x_0, y_0)(y - y_0) + \alpha(x, y)$$

and

$$g(x, y) = g(x_0, y_0) + g_x(x_0, y_0)(x - x_0) + g_y(x_0, y_0)(y - y_0) + \beta(x, y),$$

where

$$\lim_{(x,y)\to(x_0,y_0)} \frac{\alpha(x, y)}{\sqrt{(x - x_0)^2 + (y - y_0)^2}} = \lim_{(x,y)\to(x_0,y_0)} \frac{\beta(x, y)}{\sqrt{(x - x_0)^2 + (y - y_0)^2}} = 0. \quad (10.9)$$

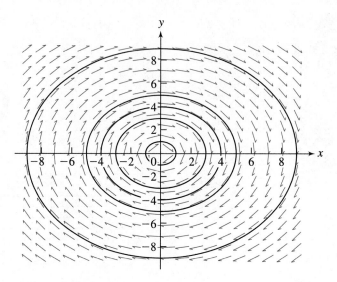

FIGURE 10.28(c) *The linear part ($\epsilon = 0$).*

Now (x_0, y_0) is assumed to be a critical point of $\mathbf{X}' = \mathbf{F}(\mathbf{X})$, so $f(x_0, y_0) = g(x_0, y_0) = 0$, and these expansions are

$$f(x, y) = f_x(x_0, y_0)(x - x_0) + f_y(x_0, y_0)(y - y_0) + \alpha(x, y)$$

and

$$g(x, y) = g_x(x_0, y_0)(x - x_0) + g_y(x_0, y_0)(y - y_0) + \beta(x, y).$$

Let

$$\widetilde{\mathbf{X}} = \begin{pmatrix} x - x_0 \\ y - y_0 \end{pmatrix}.$$

Then

$$\frac{d}{dt}\widetilde{\mathbf{X}} = \begin{pmatrix} \dfrac{d}{dt}(x - x_0) \\ \dfrac{d}{dt}(y - y_0) \end{pmatrix} = \begin{pmatrix} x' \\ y' \end{pmatrix} = \mathbf{X}' = \mathbf{F}(\mathbf{X})$$

$$= \begin{pmatrix} f(x, y) \\ g(x, y) \end{pmatrix}$$

$$= \begin{pmatrix} f_x(x_0, y_0) & f_y(x_0, y_0) \\ g_x(x_0, y_0) & g_y(x_0, y_0) \end{pmatrix} \begin{pmatrix} x - x_0 \\ y - y_0 \end{pmatrix} + \begin{pmatrix} \alpha(x, y) \\ \beta(x, y) \end{pmatrix}$$

$$= \mathbf{A}_{(x_0, y_0)}\widetilde{\mathbf{X}} + \mathbf{G}.$$

Because of the condition (10.9), this system is almost linear. Omitting the tilde notation for simplicity, this puts the translated system into the form $\mathbf{X}' = \mathbf{A}_{(x_0, y_0)}\mathbf{X} + \mathbf{G}$, with the critical point (x_0, y_0) of $\mathbf{X}' = \mathbf{F}(\mathbf{X})$ translated to the origin as the critical point of the almost linear system $\mathbf{X}' = \mathbf{A}_{(x_0, y_0)}\mathbf{X} + \mathbf{G}$. Now we can apply the preceding discussion and Theorem 10.3 to the translated system $\mathbf{X}' = \mathbf{A}_{(x_0, y_0)}\mathbf{X} + \mathbf{G}$ at the origin and hence draw conclusions about the behavior of solutions of $\mathbf{X}' = \mathbf{F}(\mathbf{X})$ near (x_0, y_0). We use the notation $\mathbf{A}_{(x_0, y_0)}$ for the matrix of the linear part of the translated system for two reasons. First, it reminds us that this is the translated system (since we dropped the $\widetilde{\mathbf{X}}$ notation). Second, when we are analyzing several critical points of the same system, this notation reminds us which critical point is under consideration and clearly distinguishes the linear part associated with one critical point from that associated with another.

In carrying out this strategy, it is important to realize that we do not have to explicitly compute $\alpha(x, y)$ or $\beta(x, y)$, which in some cases would be quite tedious or not even practical. The point is that we know that the translated system $\mathbf{X}' = \mathbf{A}_{(x_0, y_0)}\mathbf{X} + \mathbf{G}$ is almost linear if $\mathbf{F}$ has continuous first and second partial derivatives, a condition that is usually easy to verify.

EXAMPLE 10.21

Consider the system

$$\mathbf{X}' = \mathbf{F}(\mathbf{X}) = \begin{pmatrix} \sin(\pi x) - x^2 + y^2 \\ \cos\left((x + y + 1)\dfrac{\pi}{2}\right) \end{pmatrix}.$$

Here, $f(x, y) = \sin(\pi x) - x^2 + y^2$ and $g(x, y) = \cos((x + y + 1)\pi/2)$. This is an almost linear system because f and g are continuous, with continuous first and second partial derivatives throughout the plane.

For the critical points, solve

$$\sin(\pi x) - x^2 + y^2 = 0,$$

$$\cos\left((x + y + 1)\frac{\pi}{2}\right) = 0.$$

Certainly $x = y = n$ is a solution for every integer n. Every point (n, n) in the plane is a critical point. There may be other critical points as well, but other solutions of $f(x, y) = g(x, y) = 0$ are not obvious. We will need the partial derivatives

$$f_x = \pi \cos(\pi x) - 2x, \quad f_y = 2y$$

$$g_x = -\frac{\pi}{2} \sin\left((x + y + 1)\frac{\pi}{2}\right), \quad g_y = -\frac{\pi}{2} \sin\left((x + y + 1)\frac{\pi}{2}\right).$$

Now consider a typical critical point (n, n). We can translate this point to the origin and write the translated system as $\mathbf{X}' = \mathbf{A}_{(n,n)}\mathbf{X} + \mathbf{G}$ with

$$\mathbf{A}_{(n,n)} = \begin{pmatrix} f_x(n, n) & f_y(n, n) \\ g_x(n, n) & g_y(n, n) \end{pmatrix}$$

$$= \begin{pmatrix} \pi \cos(\pi n) - 2n & 2n \\ -\dfrac{\pi}{2} \sin\left((2n + 1)\dfrac{\pi}{2}\right) & -\dfrac{\pi}{2} \sin\left((2n + 1)\dfrac{\pi}{2}\right) \end{pmatrix}$$

$$= \begin{pmatrix} \pi(-1)^n - 2n & 2n \\ (-1)^{n+1}\dfrac{\pi}{2} & (-1)^{n+1}\dfrac{\pi}{2} \end{pmatrix}$$

and

$$\mathbf{G}(\mathbf{X}) = \begin{pmatrix} \alpha(x, y) \\ \beta(x, y) \end{pmatrix}.$$

We need not actually compute $\alpha(x, y)$ or $\beta(x, y)$. Because the system is almost linear, the qualitative behavior of trajectories of the nonlinear system near (n, n) is (with exceptions noted in Theorem 10.3) determined by the behavior of trajectories of the linear system $\mathbf{X}' = \mathbf{A}_{(n,n)}\mathbf{X}$. We are therefore led to consider the eigenvalues of $\mathbf{A}_{(n,n)}$, which are

$$\tfrac{1}{4}\pi(-1)^n - n \pm \tfrac{1}{4}\sqrt{9\pi^2 - 40n\pi(-1)^n + 16n^2}.$$

We will consider several values for n. For $n = 0$, the eigenvalues are π and $-\pi/2$, so the origin is an unstable saddle point of the linear system and also of the nonlinear system.

For $n = 1$, the eigenvalues of $A_{(1,1)}$ are

$$-\tfrac{1}{4}\pi - 1 \pm \tfrac{1}{4}\sqrt{9\pi^2 + 40\pi + 16},$$

which are approximately 2.0101 and -5.5809. Therefore, $(1, 1)$ is also an unstable saddle point.

For $n = 2$, the eigenvalues are

$$\tfrac{1}{4}\pi - 2 \pm \tfrac{1}{4}\sqrt{9\pi^2 - 80\pi + 64},$$

which are approximately $-1.2146 + 2.4812i$ and $-1.2146 - 2.4812i$. These are complex conjugates with negative real part, so $(2, 2)$ is an asymptotically stable spiral point.

For $n = 3$, the eigenvalues are

$$-\tfrac{1}{4}\pi - 3 \pm \tfrac{1}{4}\sqrt{9\pi^2 + 120\pi + 144},$$

which are approximately -9.959 and 2.3882. Thus, $(3, 3)$ is an unstable saddle point.

For $n = 4$, the eigenvalues are

$$\tfrac{1}{4}\pi - 4 \pm \tfrac{1}{4}\sqrt{9\pi^2 - 160\pi + 256},$$

approximately $-3.2146 + 3.1407i$ and $-3.2146 - 3.1407i$. We conclude that $(4, 4)$ is an asymptotically stable spiral point.

For $n = 5$, the eigenvalues are

$$-\tfrac{1}{4}\pi - 5 \pm \tfrac{1}{4}\sqrt{9\pi^2 + 200\pi + 400},$$

approximately 2.5705 and -14.141, so $(5, 5)$ is an unstable saddle point.

The pattern suggested by the cases $n = 2$ and $n = 4$ is continued with $n = 6$. Now the eigenvalues are

$$\tfrac{1}{4}\pi - 6 \pm \tfrac{1}{4}\sqrt{9\pi^2 - 240\pi + 576},$$

approximately $-5.2146 \pm 2.3606i$, so $(6, 6)$ is also an asymptotically stable spiral point. With $n = 7$ we get eigenvalues

$$-\tfrac{1}{4}\pi - 7 \pm \tfrac{1}{4}\sqrt{9\pi^2 + 280\pi + 784},$$

approximately 2.6802 and -18.251, so $(7, 7)$ is an unstable saddle point. A new pattern seems to be forming with the next case. If $n = 8$ the eigenvalues are

$$\tfrac{1}{4}\pi - 8 \pm \tfrac{1}{4}\sqrt{9\pi^2 - 320\pi + 16(64)},$$

approximately -9.8069 and -4.6223, so $(8, 8)$ is a stable node.

Figures 10.29, 10.30, and 10.31 show phase portraits of this system, focusing on trajectories near selected critical points. The student should experiment with phase portraits near some of the other critical points, for example, those with negative coordinates. ∎

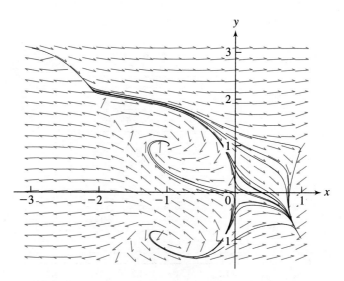

FIGURE 10.29 *Trajectories of the system of Example 10.21 near the origin.*

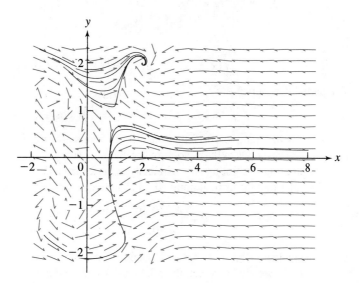

FIGURE 10.30 *Trajectories of the system of Example 10.21 near* $(1, 1)$.

EXAMPLE 10.22 Damped Pendulum

The system for the damped pendulum is

$$x' = y$$
$$y' = -\omega^2 \sin(x) - \gamma y.$$

In matrix form, this system is

$$\mathbf{X}' = \mathbf{F}(\mathbf{X}) = \begin{pmatrix} y \\ -\omega^2 \sin(x) - \gamma y \end{pmatrix}.$$

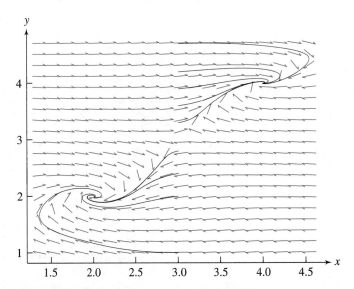

FIGURE 10.31 *Trajectories of the system of Example 10.21 near* $(2, 2)$
and $(4, 4)$.

Here,

$$f(x, y) = y \quad \text{and} \quad g(x, y) = -\omega^2 \sin(x) - \gamma y.$$

The partial derivatives are

$$f_x = 0, \quad f_y = 1, \quad g_x = -\omega^2 \cos(x), \quad g_y = -\gamma.$$

We saw in Example 10.12 that the critical points are $(n\pi, 0)$ with n any integer. When n is even, this corresponds to the pendulum bob hanging straight down, and when n is odd, to the bob initially pointing straight up. We will analyze these critical points.

Consider first the critical point $(0, 0)$. The linear part of the system has matrix

$$\mathbf{A}_{(0,0)} = \begin{pmatrix} f_x(0, 0) & f_y(0, 0) \\ g_x(0, 0) & g_y(0, 0) \end{pmatrix} = \begin{pmatrix} 0 & 1 \\ -\omega^2 & -\gamma \end{pmatrix},$$

with eigenvalues $-\frac{1}{2}\gamma + \frac{1}{2}\sqrt{\gamma^2 - 4\omega^2}$ and $-\frac{1}{2}\gamma - \frac{1}{2}\sqrt{\gamma^2 - 4\omega^2}$. Recall that $\gamma = c/mL$ and $\omega^2 = g/L$. As we might expect, the relative sizes of the damping force, the mass of the bob, and the length of the pendulum will determine the nature of the motion.

The following cases occur.

1. If $\gamma^2 - 4\omega^2 > 0$, then the eigenvalues are real, unequal, and negative, so the origin is an asymptotically stable nodal sink. This happens when $c > 2m\sqrt{gL}$. This gives a measure of how large the damping force must be, compared to the mass of the bob and length of the pendulum, to have trajectories curving toward the equilibrium solution $(0, 0)$. In this case, after release following a small displacement from the vertical downward position, the bob moves toward this position with decreasing velocity, oscillating back and forth through this position and eventually coming to rest in the limit as $t \to \infty$. Figure 10.32 shows a phase portrait for the pendulum with $\gamma^2 = 0.8$ and $\omega = 0.44$.

2. If $\gamma^2 - 4\omega^2 = 0$, then the eigenvalues are equal and negative, corresponding to an asymptotically stable proper or improper node of the linear system. This is the case in which Theorem 10.3 does not give a definitive conclusion, and the origin could be an asymptotically stable node or spiral point of the nonlinear pendulum. This case occurs when $c = 2m\sqrt{gL}$,

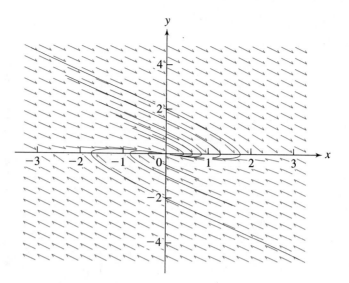

FIGURE 10.32 *Phase portrait for the damped pendulum with*
$\gamma^2 = 0.8$ *and* $\omega = 0.44 (\gamma^2 - 4\omega^2 > 0)$.

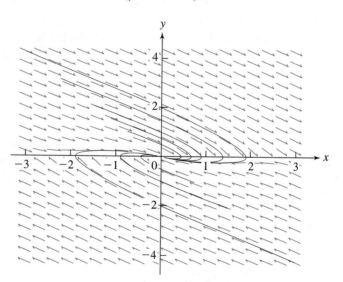

FIGURE 10.33 *Damped pendulum with* $\gamma^2 = 0.8$ *and*
$\omega^2 = 0.2 (\gamma^2 - 4\omega^2 = 0)$.

a delicate balance between the damping force, mass, and pendulum length. In the case of an asymptotically stable node, the bob, when released, moves with decreasing velocity toward the vertical equilibrium position, approaching it as $t \to \infty$ but not oscillating through it. Figure 10.33 gives a phase portrait for this case, in which $\gamma^2 = 0.8$ and $\omega^2 = 0.2$.

3. If $\gamma^2 - 4\omega^2 < 0$, then the eigenvalues are complex conjugates with negative real part. Hence the origin is an asymptotically stable spiral point of both the linear part and the nonlinear pendulum system. This happens when $c < 2m\sqrt{gL}$. Figure 10.34 displays this case, with $\gamma^2 = 0.6$ and $\omega^2 = 0.3$.

It is routine to check that each critical point $(2n\pi, 0)$, in which the first coordinate is an even integer multiple of π, has the same characteristics as the origin.

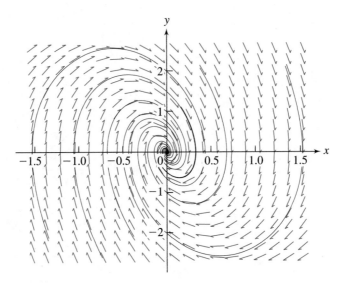

FIGURE 10.34 *Damped pendulum with $\gamma^2 = 0.6$ and $\omega^2 = 0.3(\gamma^2 - 4\omega^2 < 0)$.*

Now consider critical points $((2n + 1)\pi, 0)$, with the first coordinate an odd integer multiple of π. To be specific, consider $(\pi, 0)$. Now the linear part of the system (with $(\pi, 0)$ translated to the origin) is

$$\mathbf{A}_{(\pi,0)} = \begin{pmatrix} f_x(\pi, 0) & f_y(\pi, 0) \\ g_x(\pi, 0) & g_y(\pi, 0) \end{pmatrix} = \begin{pmatrix} 0 & 1 \\ -\omega^2 \cos(\pi) & -\gamma \end{pmatrix} = \begin{pmatrix} 0 & 1 \\ \omega^2 & -\gamma \end{pmatrix}.$$

The eigenvalues are $-\frac{1}{2}\gamma + \frac{1}{2}\sqrt{\gamma^2 + 4\omega^2}$ and $-\frac{1}{2}\gamma - \frac{1}{2}\sqrt{\gamma^2 + 4\omega^2}$. These are real and of opposite sign, so $(\pi, 0)$ is an unstable saddle point. The other critical points $((2n + 1)\pi, 0)$ exhibit the same behavior. This is what we would expect of a pendulum in which the bob is initially in the vertical upward position, since arbitrarily small displacements will result in the bob moving away from this position and never returning to it. The analysis is the same for each critical point $((2n + 1)\pi, 0)$. ■

EXAMPLE 10.23 Nonlinear Damped Spring

The nonlinear damped spring equation is

$$x' = y,$$

$$y' = -\frac{k}{m}x + \frac{\alpha}{m}x^3 - \frac{c}{m}y.$$

This is

$$\mathbf{X}' = \mathbf{F}(\mathbf{X}) = \begin{pmatrix} y \\ -(k/m)x + (\alpha/m)x^3 - (c/m)y \end{pmatrix}.$$

Here,

$$f(x, y) = y \quad \text{and} \quad g(x, y) = -\frac{k}{m}x + \frac{\alpha}{m}x^3 - \frac{c}{m}y.$$

There are three critical points, $(0, 0)$, $(\sqrt{k/\alpha}, 0)$, and $(-\sqrt{k/\alpha}, 0)$. The partial derivatives are

$$f_x = 0, \quad f_y = 1, \quad g_x = -\frac{k}{m} + 3\frac{\alpha}{m}x^2, \quad g_y = -\frac{c}{m}.$$

First consider the behavior of trajectories near the origin. The linear part of the system has matrix

$$\mathbf{A}_{(0,0)} = \begin{pmatrix} 0 & 1 \\ -k/m & -c/m \end{pmatrix},$$

with eigenvalues $1/2m(-c + \sqrt{c^2 - 4mk})$ and $1/2m(-c - \sqrt{c^2 - 4mk})$. This yields three cases, depending, as we might expect, on the relative magnitudes of the mass, damping constant, and spring constant.

1. If $c^2 - mk > 0$, then $\mathbf{A}_{(0,0)}$ has real, distinct, negative eigenvalues, so the origin is an asymptotically stable nodal sink. Small disturbances from the equilibrium position result in a motion that dies out with time, with the mass approaching the equilibrium position.

2. If $c^2 - mk = 0$, then $\mathbf{A}_{(0,0)}$ has equal real, negative eigenvalues, so the origin is an asymptotically stable proper or improper node of the linear system. Hence the origin is an asymptotically stable node or spiral point of the nonlinear system.

3. If $c^2 - mk < 0$, then $\mathbf{A}_{(0,0)}$ has complex conjugate eigenvalues with negative real part, and the origin is an asymptotically stable spiral point.

Figure 10.35 shows a phase portrait for case (3), with $c = 2$, $k = 5$, $\alpha = 1$, and $m = 3$.

Next, consider the critical point $(\sqrt{k/\alpha}, 0)$. Now the linear part of the system obtained by translating this point to the origin is

$$\mathbf{A}_{(\sqrt{k/\alpha},0)} = \begin{pmatrix} 0 & 1 \\ k/m & -c/m \end{pmatrix},$$

with eigenvalues $(1/2m)(-c + \sqrt{c^2 + 4mk})$ and $(1/2m)(-c - \sqrt{c^2 + 4mk})$. The first eigenvalue is positive and the second negative, so $(\sqrt{k/\alpha}, 0)$ is an unstable saddle point. A similar analysis holds for the critical point $(-\sqrt{k/\alpha}, 0)$. ■

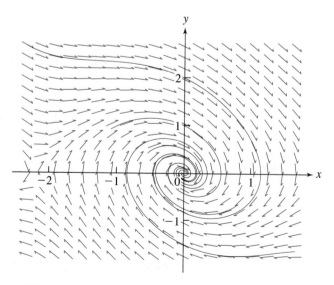

FIGURE 10.35 *Nonlinear spring system with $c = 2$, $k = 5$, $\alpha = 1$, and $m = 3$ ($c^2 - mk < 0$).*

EXAMPLE 10.24 Motion of a Magnet

Suppose a magnet of mass m is suspended from the lower end of a spring having spring modulus k. The magnetic weight comes to rest in an equilibrium position L units above a small piece of iron. The weight is then pulled down and released from rest. Assume that the attractive force between the magnet and the iron has magnitude K/r^2 when they are separated by r units. We want to determine the smallest value of K that will prevent oscillations of the magnet.

Before setting the weight in motion, we will establish the equilibrium position. If the magnetic weight is in this position, then the sum of the external forces acting on it is zero. The downward forces are the weight mg of the magnet and an attractive force of magnitude K/L^2. Balancing the sum of these downward forces must be an upward force, which, by Hooke's Law, equals the product of k and the amount s the spring is stretched beyond its natural length. Thus,

$$ks = mg + \frac{K}{L^2}.$$

Let z be the directed distance from the weight to the equilibrium position, with downward chosen as the positive direction. By Newton's second law of motion,

$$mz'' = mg + \frac{K}{(L-z)^2} - k(s+z).$$

Since $-ks = -mg - K/L^2$, then

$$mz'' = -kz + \frac{K}{(L-z)^2} - \frac{K}{L^2}.$$

This gives us

$$z'' = -\frac{k}{m}z + \frac{K}{m}\left[\frac{1}{(L-z)^2} - \frac{1}{L^2}\right].$$

Convert this second-order differential equation to a system of first-order equations by putting $x = z$ and $y = z'$. The system is

$$x' = y,$$

$$y' = -\frac{k}{m}x + \frac{K}{m}\left[\frac{1}{(L-x)^2} - \frac{1}{L^2}\right].$$

The analysis of this system is made easier if we use the Taylor expansion

$$\frac{1}{(L-x)^2} - \frac{1}{L^2} = \frac{1}{L^2}\left[2\left(\frac{1}{L}x\right) + 3\left(\frac{1}{L}x\right)^2 + 4\left(\frac{1}{L}x\right)^3 + \cdots\right].$$

Now the system is

$$x' = y,$$

$$y' = -\frac{k}{m}x + \frac{K}{mL^2}\left[2\left(\frac{1}{L}x\right) + 3\left(\frac{1}{L}x\right)^2 + 4\left(\frac{1}{L}x\right)^3 + \cdots\right].$$

This almost linear system has one critical point, $(0, 0)$. If we write the system as

$$\mathbf{X}' = \begin{pmatrix} 0 & 1 \\ (2K/mL^3) - k/m & 0 \end{pmatrix}\mathbf{X} + \frac{K}{mL^2}\begin{pmatrix} 3((1/L)x)^2 + 4((1/L)x)^3 + \cdots \\ 0 \end{pmatrix}$$

then it is clear that

$$\mathbf{A}_{(0,0)} = \begin{pmatrix} 0 & 1 \\ (2K/mL^3) - k/m & 0 \end{pmatrix}.$$

The eigenvalues of $\mathbf{A}_{(0,0)}$ are

$$\frac{1}{\sqrt{m}L^{3/2}}\sqrt{2K - kL^3}, \qquad -\frac{1}{\sqrt{m}L^{3/2}}\sqrt{2K - kL^3}.$$

Oscillations occur only if the eigenvalues are complex, which can occur only if

$$2K < kL^3.$$

To prevent oscillations, then, we must make K large enough that

$$2K \geq kL^3,$$

or

$$K \geq \tfrac{1}{2}kL^3.$$

Thus $K = kL^3/2$ is the smallest value of K that will prevent oscillations. ∎

SECTION 10.5   PROBLEMS

In each of Problems 1 through 10, (a) show that the system is almost linear, (b) determine the critical points, (c) use Theorem 10.3 to analyze the nature of the critical point, or state why no conclusion can be drawn, and (d) generate a phase portrait for the system.

1. $x' = x - y + x^2,\ y' = x + 2y$

2. $x' = x + 3y - x^2\sin(y),\ y' = 2x + y - xy^2$

3. $x' = -2x + 2y,\ y' = x + 4y + y^2$

4. $x' = -2x - 3y - y^2,\ y' = x + 4y$

5. $x' = 3x + 12y,\ y' = -x - 3y + x^3$

6. $x' = 2x - 4y + 3xy,\ y' = x + y + x^2$

7. $x' = -3x - 4y + x^2 - y^2,\ y' = x + y$

8. $x' = -3x - 4y,\ y' = -x + y - x^2y$

9. $x' = -2x - y + y^2,\ y' = -4x + y$

10. $x' = 2x - y - x^3\sin(x),\ y' = -2x + y + xy^2$

11. Theorem 10.3 is inconclusive in the case that the critical point of an almost linear system is a center of the associated linear system. Verify that no conclusion is possible in this case in general by considering the following two systems:

$$x' = y - x\sqrt{x^2 + y^2}, \quad y' = -x - y\sqrt{x^2 + y^2}$$

and

$$x' = y + x\sqrt{x^2 + y^2}, \quad y' = -x + y\sqrt{x^2 + y^2}.$$

(a) Show that the origin is a center for the associated linear system of both systems.

(b) Show that each system is almost linear.

(c) Introduce polar coordinates, with $x = r\cos(\theta)$ and $y = r\sin(\theta)$, and use the chain rule to obtain

$$x' = \frac{dx}{dr}\frac{dr}{dt} = \cos(\theta)r'(t)$$

and

$$y' = \frac{dy}{dr}\frac{dr}{dt} = \sin(\theta)r'(t).$$

Use these to evaluate $xx' + yy'$ in terms of r and r', where $r' = dr/dt$. Thus convert each system to a system in terms of $r(t)$ and $\theta(t)$.

(d) Use the polar coordinate version of the first system to obtain a separable differential equation for $r(t)$. Conclude from this that $r'(t) < 0$ for all t. Solve for $r(t)$ and show that $r(t) \to 0$ as $t \to \infty$. Thus conclude that for the first system the origin is asymptotically stable.

(e) Follow the procedure of (d), using the second system. However, now find that $r'(t) > 0$ for all t. Solve for $r(t)$ with the initial condition $r(t_0) = r_0$. Show that $r(t) \to \infty$ as $t \to t_0 + 1/r$ from the left. Conclude that the origin is unstable for the second system.

12. An iron weight of mass m is suspended from the lower end of a spring with modulus k and allowed to come to rest in an equilibrium position. A magnetic ring of radius r is then placed in a horizontal position with the weight at its center. The attractive force between the weight and the ring is K/z^2 when they are separated by a distance z. The weight also has a dashpot attached to it that gives the system a damping constant c. The weight is pulled down y_0 units and released from rest.

(a) Write a differential equation that describes the motion of the weight.

(b) Convert this differential equation into a system of first-order differential equations. (See Section 2.6.4.)

(c) Classify the critical point(s) for various values of K.

(d) Draw some phase portraits for this system (inserting different values for the constants).

13. Suppose the weight in Problem 12 is replaced with a magnet of the same mass as the magnetic ring, but of opposite polarity, so that the force between the ring and the weight is now a repellent one.

(a) Write a differential equation for the motion of the weight.

(b) Convert this differential equation into a system of first-order differential equations.

(c) Classify the critical point(s) for various values of K.

(d) Draw some phase portraits for this system (inserting values for the constants).

10.6 Predator/Prey Population Models

Suppose a certain region has two primary populations. The prey is a source of food for the predator. Examples might be wolf and rabbit populations, or hawks and mice, or people and deer. There may be other populations in this region as well, but attention is focused on the two forming the direct predator/prey relationship.

Suppose, at time t, the prey population is $x(t)$ and the predator population is $y(t)$. We will use a system of differential equations to model the interaction of these populations and how they both increase or decrease based on this interaction. We begin with a relatively simple system and then move to a more sophisticated (and more accurate) model.

10.6.1 A Simple Predator/Prey Model

First, assume that the prey population tends to increase at a rate proportional to its numbers $x(t)$, but that it also decreases at a rate proportional to $x(t)y(t)$, which accounts for interactions between the two species in which the predator kills the prey. We therefore let

$$x'(t) = ax(t) - bx(t)y(t)$$

in which a and b are positive constants. Notice that if the predator population were eliminated, then $y(t) = 0$ and the prey would increase according to $x' = ax$, an exponential growth.

Next, assume that the predator population increases at a rate proportional to the product $x(t)y(t)$, since the predator requires encounters with the prey for food, and that it also decreases at a rate proportional to itself, $y(t)$, because in the absence of prey the predator has no food source and its population decreases. Thus let

$$y'(t) = cx(t)y(t) - ky(t),$$

with c and k positive constants. If $x(t) = 0$, the predator population satisfies $y' = -ky$ and decreases exponentially with time.

The system is

$$\mathbf{X}' = \begin{pmatrix} x \\ y \end{pmatrix}' = \begin{pmatrix} ax - bxy \\ cxy - ky \end{pmatrix}. \tag{10.10}$$

Since $x(t)$ and $y(t)$ are populations, they must be nonnegative. Two extreme cases are that one or the other population is zero. If $x(0) = \alpha > 0$ and $y(0) = 0$ (no predator), then the solution of this system is

$$x(t) = \alpha e^{at}, \quad y(t) = 0.$$

The trajectory in the plane corresponding to this solution is the positive part of the horizontal axis, oriented away from the origin because this population increases with time.

If $x(0) = 0$ and $y(0) = \beta > 0$, then the solution is

$$x(t) = 0, \quad y(t) = \beta e^{-kt},$$

with the predator population dying out in the absence of food. This trajectory is the positive part of the vertical axis in the phase plane, oriented toward the origin.

All other trajectories, in which both populations have positive initial values, lie in the first quadrant $x > 0$, $y > 0$.

The predator/prey system (10.10) has two critical points, $(0, 0)$ and $(k/c, a/b)$. An analysis of a typical predator/prey situation will reveal the main features of this model.

EXAMPLE 10.25

Consider the predator/prey model

$$\mathbf{X}' = \begin{pmatrix} 0.2x - 0.02xy \\ 0.02xy - 1.2y \end{pmatrix}.$$

This system has two critical points, $(0, 0)$ and $(60, 10)$. For this system, $f(x, y) = 0.2x - 0.02xy$ and $g(x, y) = 0.02xy - 1.2y$, so

$$f_x = 0.2 - 0.02y, \quad f_y = -0.02x,$$

$$g_x = 0.02y, \quad g_y = 0.02x - 1.2.$$

First consider $(0, 0)$. The linear part of this equation at the origin has matrix

$$\mathbf{A}_{(0,0)} = \begin{pmatrix} 0.2 & 0 \\ 0 & -1.2 \end{pmatrix}$$

with eigenvalues 0.2 and -1.2. The origin is an unstable saddle point of the predator/prey system.

For the critical point $(60, 10)$, the linear part of the system has matrix

$$\mathbf{A}_{(60,10)} = \begin{pmatrix} 0 & -1.2 \\ 0.2 & 0 \end{pmatrix}$$

with pure imaginary eigenvalues $\frac{1}{5}\sqrt{6}i$ and $-\frac{1}{5}\sqrt{6}i$. This is a case in which Theorem 10.3 does not offer a definite conclusion. We can say that this critical point is either a center or a spiral point, and the latter could be stable or unstable. ■

We can obtain the trajectories of the simple predator/prey system (10.10) by computing $dy/dx = g(x, y)/f(x, y)$. In Example 10.25,

$$\frac{dy}{dx} = \frac{g(x, y)}{f(x, y)} = \frac{0.02xy - 1.2y}{0.2x - 0.02xy}.$$

or

$$\frac{dy}{dx} = \frac{y(x - 60)}{x(10 - y)}.$$

This is a separable differentiable equation, which we write as

$$\frac{10 - y}{y} \, dy = \frac{x - 60}{x} \, dx.$$

Integrate to get

$$10 \ln(y) - y = x - 60 \ln(x) + K$$

for x and y positive. After some rearrangement, this equation can be written

$$y^{10} x^{60} e^{-x-y} = \alpha,$$

in which α is constant. Various choices of initial conditions $x(0)$ and $y(0)$ determine α, and then this equation can be used to graph the corresponding trajectory.

A phase portrait for this system in Example 10.25 is shown in Figure 10.36. The trajectories are closed curves representing periodic solutions enclosing the critical point $(60, 10)$. In interpreting these trajectories to yield information about the predator and prey populations, it is useful to draw a vertical and horizontal line through this critical point. These (dashed) lines, which are not themselves trajectories, separate the first quadrant into four regions, labeled I through IV in Figure 10.36. Follow one of the trajectories in the diagram. In region I, both populations are increasing, and the trajectory is moving upward to the right. Then the trajectory turns, moving upward to the left. The predator population has grown sufficiently that, in region II, the prey population is being decimated. After a time, so much of the prey population is lost that there is not enough food for the predator, and the trajectory begins a downward turn (region III) as both populations decline. Finally, in region IV, the predator population has declined enough that the prey begins making a recovery. This behavior then repeats in cycles, as the populations increase and decrease out of phase with each other but are dependent on the success or failure of the other population.

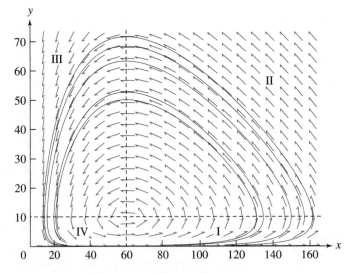

FIGURE 10.36 *Phase portrait for the predator/prey system of Example 10.25.*

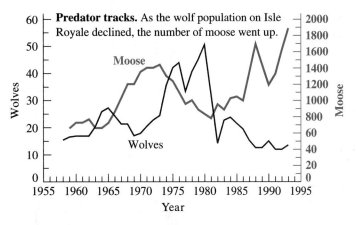

FIGURE 10.37 *(Reprinted from* Science, *vol. 261, August 1993.)*

Exactly this kind of behavior has been observed in predator/prey settings for which good records have been kept. One is the lynx/snowshoe rabbit population in Canada. The Hudson Bay Company has kept records of pelts traded at its stations since about the middle of the nineteenth century. Assume that the populations of lynx and hare are proportional to the number of respective pelts obtained. This seems reasonable. When the lynx are plentiful, trappers are more successful, and when lynx are in decline, fewer pelts are taken. The records for 1845 through 1935 show a clear periodic variation in lynx and hare populations, having about a ten-year cycle. As predicted by the model, the two populations are out of phase, with the lynx reaching a peak as the hares decline, followed by an increase in the hare population when the lynx decline for want of food.

Even today we occasionally find a close-to-ideal predator/prey environment. Michigan's Isle Royale is an untamed island having a length of about 45 miles and was at one time overrun with moose, which had no local natural enemy. The winter of 1949 was particularly harsh, and wolves from Canada crossed a frozen stretch of Lake Superior to the island, searching for food. When the ice melted in the spring, this wolf population was trapped on the island. This resulted in a classic predator/prey confrontation, with moose originally flourishing on the island but suddenly faced with a predator.

These populations on Isle Royale were studied by Purdue biologist Durward Allen, yielding a wealth of information about wolves in particular and more generally about the population fluctuations predicted by the predator/prey model. The experiment turned out to be flawed by two circumstances. First, the original wolf population had too narrow a genetic base because of the small number that had crossed the ice to the island. But more significant was a canine virus that decimated the wolf population. Despite this unexpected twist, the graphs in Figure 10.37 show relative changes in the wolf and moose populations, with a clear out-of-phase increase/decrease relationship in the two populations over the period 1957 through 1993.

10.6.2   An Extended Predator/Prey Model

One significant feature of the model (10.10) is that the encounters $x(t)y(t)$ have a negative influence on the rate of increase of the prey but stimulate the rate of increase of the predator. Encounters are good for the predator and bad for the prey. However, the system (10.10) does not include any terms to account for factors internal to each population that may influence its ability to grow in numbers. For example, the food source for the prey may be abundant or scarce, and whichever it is makes a difference, independent of the prey's problem with the predator. The model does not take such factors into account.

There are various ways to model internal influences. One is to assume that the total effect of such influences (including illness, natural disasters, perhaps hunting by humans, and so on) are proportional to the square of the population. This yields the following predator/prey model:

$$\mathbf{X}' = \begin{pmatrix} a_1 x - a_2 x^2 - a_3 xy \\ b_1 xy - b_2 y - b_3 y^2 \end{pmatrix}, \tag{10.11}$$

in which the coefficients are positive constants. The following example suggests some possibilities from this model.

EXAMPLE 10.26

Consider

$$\mathbf{X}' = \begin{pmatrix} x - \frac{1}{4}x^2 - \frac{1}{4}xy \\ xy - y - \frac{1}{2}y^2 \end{pmatrix}.$$

The critical points are $(0, 0)$, $(0, -2)$, $(4, 0)$, and $(2, 2)$. The critical point $(0, -2)$ is irrelevant for this model, since populations must be nonnegative. We will make use of

$$f_x = 1 - \tfrac{1}{2}x - \tfrac{1}{4}y, \quad f_y = -\tfrac{1}{4}x, \quad g_x = y, \quad g_y = x - 1 - y.$$

Now consider each critical point in the first quadrant or its boundary.

$(0, 0)$—Since we can write the system as

$$\mathbf{X}' = \begin{pmatrix} 1 & 0 \\ 0 & -1 \end{pmatrix} \mathbf{X} + \begin{pmatrix} -\frac{1}{4}x^2 - \frac{1}{4}xy \\ xy - \frac{1}{2}y^2 \end{pmatrix},$$

the linear part has matrix

$$\mathbf{A}_{(0,0)} = \begin{pmatrix} 1 & 0 \\ 0 & -1 \end{pmatrix},$$

with eigenvalues 1 and -1. The origin is an unstable saddle point. If a population pair (x_0, y_0) begins sufficiently close to the origin, one population will increase without bound and the other die out with time.

$(4, 0)$—The linear part of the system with $(4, 0)$ translated to the origin is

$$\mathbf{A}_{(4,0)} = \begin{pmatrix} -1 & -1 \\ 0 & 3 \end{pmatrix},$$

with eigenvalues -1 and 3. Therefore, $(4, 0)$ is also an unstable saddle point. Again, if begun near this point, one population flourishes, the other dies.

$(2, 2)$—Now the linearized part has matrix

$$\mathbf{A}_{(2,2)} = \begin{pmatrix} -\frac{1}{2} & -\frac{1}{2} \\ 2 & -1 \end{pmatrix},$$

with eigenvalues $-\frac{3}{4} + \frac{1}{4}\sqrt{15}i$ and $-\frac{3}{4} - \frac{1}{4}\sqrt{15}i$. Therefore, $(2, 2)$ is an asymptotically stable spiral point. Populations starting sufficiently near this point tend toward this point as a limit.

Figure 10.38 shows a phase portrait for this system, showing the spiral point at $(2, 2)$. ∎

Very sophisticated extensions of predator/prey systems of differential equations are currently used in a wide variety of settings, including ecological considerations and studies of AIDS. In the latter setting, the various strains of AIDS virus are considered the predator against the body's T-cell population. Empirical data from medical studies are used to determine the constants in the differential equations, and the objective is to develop a sufficiently accurate model to be of use in

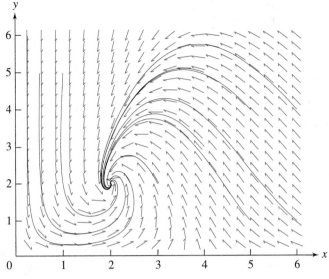

FIGURE 10.38 *Predator/prey model* $\begin{cases} x' = x - \frac{1}{4}x^2 - \frac{1}{4}xy \\ y' = xy - y - \frac{1}{2}y^2 \end{cases}$.

developing strategies for treatment. One important question is whether to introduce large doses of medication early in the treatment (possibly stimulating virus mutation) or to wait until some later point when the patient has reached a certain stage of illness. A thorough treatment of this topic occurs in *Virus Dynamics: Mathematical Principles of Immunology and Virology*, by Martin A. Nowak and Robert M. May (Oxford University Press, 2000).

SECTION 10.6   PROBLEMS

In each of Problems 1 through 5, (a) analyze the nature of the critical points of the system, (b) draw a phase portrait, (c) interpret the model in terms of prospects for survival of each species and how the populations increase and decrease with respect to each other's success or failure.

1. $x' = 6x - 3xy, \; y' = xy - 12y$

2. $x' = 3x - 8xy, \; y' = 2xy - 5y$

3. $x' = 10x - 2xy, \; y' = 12xy - 7y$

4. $x' = 8x - 3xy, \; y' = 15xy - 8y$

5. $x' = 6x - 6xy, \; y' = 14xy - 4y$

6. Starting with the simple predator/prey model (10.10), let $X = x - k/c$ and $Y = y - a/b$ to obtain a new nonlinear system with the critical point $(k/c, a/b)$ of the original system translated to the origin of the new system. Show that the linear part of the translated system has elliptical trajectories. *Hint:* Divide one equation of the linear part by the

other and obtain a separable differential equation that can be integrated.

7. Continuing from Problem 6, show that the linear part of the translated system has complex eigenvalues. Explain why this yields no conclusion about trajectories of the nonlinear translated system.

8. Still continuing from Problem 6, express the trajectories of the linear part of the translated system in terms of sines and cosines. Show that every trajectory of the linear part of the original system has the same period even though the length of the closed path will vary with different choices of constants in the general solution.

9. Find the average value of the predator population in the system (10.10). *Hint:* Let T be the time of a cycle. The average value of the predator population is $(1/T) \int_0^T y(t) \, dt$. Write $(1/x)x'(t) = a - by$ and integrate with respect to t to get $\int_0^T (1/x) \times$

$(dx/dt)\,dt = \int_0^T (a - by)\,dt$. Now consider $x(T) - x(0)$ and evaluate the second integral to solve for the average value.

10. Compute the average prey population in the model (10.10).

In Problems 11 through 16, (a) analyze each critical point of the extended predator/prey model, (b) draw a phase portrait for the system, and (c) discuss the ramifications of the model for the survival of each species.

11. $x' = x(1 - x - 0.5y),\ y' = y(1 - 0.5y - 0.15x)$

12. $x' = x(1 - x - 0.2y),\ y' = y(1 - 0.4y - 0.25x)$

13. $x' = x(2 - x - 0.2y),\ y' = y(2 - 0.5y - 0.4x)$

14. $x' = x(1 - 0.5x - y),\ y' = y(2 - 0.5y - 0.4x)$

15. $x' = x(4 - 0.4x - 0.5y),\ y' = y(2 - 0.1y - 0.7x)$

16. $x' = x(4 - x - y),\ y' = y(2 - 0.6y - 0.3x)$

10.7   Competing Species Models

In a competing species setting, neither species preys on the other, but both compete for some commonly needed resource, such as food. We will consider two systems of differential equations modeling this competition.

10.7.1   A Simple Competing Species Model

We can derive a simple model of competing species behavior as follows. It seems reasonable that an increase in either population will result in a reduction of food available for each species and should therefore cause a decline in the growth rate of both populations. This suggests a model of the form

$$x'(t) = ax(t) - bx(t)y(t),$$

$$y'(t) = ky(t) - cx(t)y(t).$$

The coefficients are positive constants.

Unlike the predator/prey model, now a term proportional to the product of the populations is subtracted in each equation, because competition hurts both species. If there were no competition for x, say $y(t) = 0$, then $x' = ax$ and the x-population would grow exponentially. Similarly, if there were no x-population, then $y' = ky$ and this population would grow exponentially.

In matrix form, the system is

$$\mathbf{X}' = \begin{pmatrix} ax - bxy \\ ky - cxy \end{pmatrix}.$$

The critical points are $(0, 0)$ and $(k/c, a/b)$. The trajectories are confined to the first quadrant, since the coordinates represent nonnegative populations.

This competing species model has two critical points, $(0, 0)$ and $(k/c, a/b)$, which in form appear the same as the critical points of the predator/prey model. However, the behavior of trajectories near $(k/c, a/b)$ is quite different for competing species. Here is an example that shows typical features of this behavior.

EXAMPLE 10.27

Consider the competing species model

$$\mathbf{X}' = \begin{pmatrix} 2x - 0.3xy \\ 4y - 0.7xy \end{pmatrix}.$$

The critical points are $(0, 0)$ and $(\frac{40}{7}, \frac{20}{3})$. With $f(x, y) = 2x - 0.3xy$ and $g(x, y) = 4y - 0.7xy$, we have

$$f_x = 2 - 0.3y, \quad f_y = -0.3x, \quad g_x = -0.7y, \quad g_y = 4 - 0.7x.$$

For the critical point at the origin, consider the matrix

$$\mathbf{A}_{(0,0)} = \begin{pmatrix} 2 & 0 \\ 0 & 4 \end{pmatrix},$$

with positive eigenvalues 2 and 4. The origin is an unstable nodal source for this system.

For the critical point $(\frac{40}{7}, \frac{20}{7})$, the matrix of the linear part of the translated system is

$$\mathbf{A}_{(40/7, 20/3)} = \begin{pmatrix} 0 & -\frac{12}{7} \\ -\frac{14}{3} & 0 \end{pmatrix},$$

with eigenvalues $2\sqrt{2}$ and $-2\sqrt{2}$. These are real and opposite in sign, so this critical point is an unstable saddle point.

Figure 10.39(a) shows a part of the phase portrait for this system that features the behavior of trajectories starting near the origin. Figure 10.39(b) shows a part of the phase portrait having trajectories starting near the critical point $(\frac{40}{7}, \frac{20}{3})$, suggesting their behavior around this saddle point. The phase portraits enable us to predict the fate of each species. For example, from Figure 10.39(b), initial populations of, say, $x(0) = 15$, $y(0) = 5$ would result in $x(t)$ prospering, while $y(t)$ becomes extinct in the limit as $t \to \infty$. ■

As with the predator/prey system, trajectories for a competing species system can also be obtained by solving $dy/dx = g(x, y)/f(x, y)$. If we do this for Example 10.27, we get

$$\frac{dy}{dx} = \frac{4y - 0.7xy}{2x - 0.3xy} = \frac{y(40 - 7x)}{x(20 - 3y)}.$$

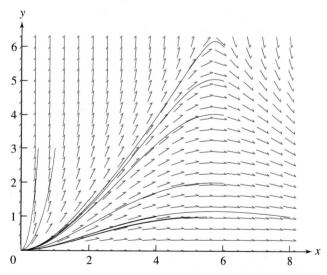

FIGURE 10.39(a) *Trajectories near the origin for* $\begin{cases} x' = 2x - 0.3xy \\ y' = 4y - 0.7xy \end{cases}$.

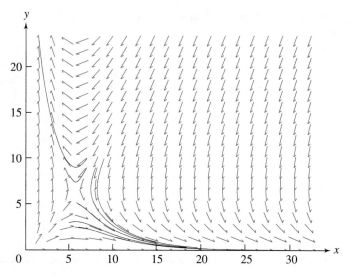

FIGURE 10.39(b) Trajectories near $(\frac{40}{7}, \frac{20}{3})$.

This is separable, and we write

$$\frac{20 - 3y}{y} \, dy = \frac{40 - 7x}{x} \, dx.$$

Integrate to get

$$20 \ln(y) - 3y = 40 \ln(x) - 7x + K.$$

After some rearrangement of terms, we get

$$\frac{y^{20}}{x^{40}} e^{7x - 3y} = \alpha,$$

in which α is constant. Selection of $x(0)$ and $y(0)$ determines α, and the graph of this equation gives the trajectory through $(x(0), y(0))$.

In general, the asymptotes of trajectories in this simple competing species model divide the first quadrant into four regions, I through IV in Figure 10.40. If the initial population $(x(0), y(0))$ is in I or IV, then the x-population will increase in time while the y-population dies out in the limit as $t \to \infty$. If the initial population is in region II or III, then the y-population survives and the x-population dies out. The coefficients $a, b, k,$ and c play a key role in determining these regions, so just having a large initial population is not enough to win.

10.7.2 An Extended Competing Species Model

One observation that stands out about the preceding competing species model is that it is not given to compromise. One species survives, and the other dies (in the limit as $t \to \infty$), depending on initial conditions. Of course, we know that not all competing species have this outcome. We therefore seek a more sophisticated model.

One way to do this is to add a term that accounts for factors within each population that tend to limit its growth, independent of the interaction with the other species, which is accounted for by the xy term. If we assume that such self-limiting factors are proportional to the population

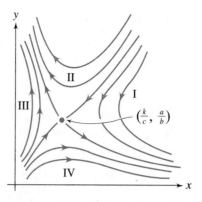

FIGURE 10.40 *Competing species phase portrait.*

squared, then we have a system

$$\mathbf{X}' = \begin{pmatrix} Gx - Dx^2 - Cxy \\ gy - dy^2 - cxy \end{pmatrix}$$

$$= \begin{pmatrix} x(G - Dx - Cy) \\ y(g - dy - cx) \end{pmatrix}$$

(10.12)

in which all constants are positive. This is the system

$$x' = Gx - Dx^2 - Cxy,$$

$$y' = gy - dy^2 - cxy.$$

For the x-population, in the absence of competition, the growth rate increases proportional to the population itself and decreases proportional to the square of the population. Thus, even without competition, there is no exponential and unlimited growth, and similarly for the y-population. We think of D and d as internal growth-limiting factors, while C and c are the competition factors in the model. It will be interesting to derive some condition involving interplay of these constants, which determines scenarios for long-term coexistence of both species as opposed to the eventual extinction of one or both populations.

We will look at two examples before undertaking a general analysis.

EXAMPLE 10.28

Consider the competing species model

$$\mathbf{X}' = \begin{pmatrix} \frac{3}{5}x - \frac{1}{6}x^2 - \frac{1}{4}xy \\ y - \frac{1}{4}y^2 - \frac{1}{2}xy \end{pmatrix}.$$

Here $f(x, y) = \frac{3}{5}x - \frac{1}{6}x^2 - \frac{1}{4}xy$ and $g(x, y) = y - \frac{1}{4}y^2 - \frac{1}{2}xy$. For later use,

$$f_x = \frac{3}{5} - \frac{1}{3}x - \frac{1}{4}y, \quad f_y = -\frac{1}{4}x$$

$$g_x = -\frac{1}{2}y, \quad g_y = 1 - \frac{1}{2}y - \frac{1}{2}x.$$

Now solve for the critical points. Write

$$f(x, y) = x \left(\tfrac{3}{5} - \tfrac{1}{6}x - \tfrac{1}{4}y \right) = 0$$

and

$$g(x, y) = y \left(1 - \tfrac{1}{4}y - \tfrac{1}{2}x \right) = 0.$$

Clearly, $x = y = 0$ is one solution. If $x = 0$ but $y \neq 0$, then from the second equation, $y = 4$. If $y = 0$ but $x \neq 0$, then from the first equation $x = \tfrac{18}{5}$. And if $x \neq 0$ and $y \neq 0$, then we get $x = \tfrac{6}{5}, y = \tfrac{8}{5}$. The critical points are

$$(0, 0), (0, 4), \left(\tfrac{18}{5}, 0 \right) \quad \text{and} \quad \left(\tfrac{6}{5}, \tfrac{8}{5} \right).$$

Immediately we see a difference between this model and the preceding competing species model—there are more critical points. We will characterize them in turn, then examine the phase portrait to determine their significance for the two populations.

$(0, 0)$—The origin itself corresponds to the equilibrium solution in which both populations are zero at time zero. The linear part of the translated system has matrix

$$\mathbf{A}_{(0,0)} = \begin{pmatrix} f_x(0, 0) & f_y(0, 0) \\ g_x(0, 0) & g_y(0, 0) \end{pmatrix} = \begin{pmatrix} \tfrac{3}{5} & 0 \\ 0 & 1 \end{pmatrix}$$

with eigenvalues 1 and $\tfrac{3}{5}$. These are real, distinct, and positive, so the origin is an unstable nodal source.

$(0, 4)$—The linear part of the system obtained by translating this point to the origin is

$$\mathbf{A}_{(0,4)} = \begin{pmatrix} f_x(0, 4) & f_y(0, 4) \\ g_x(0, 4) & g_y(0, 4) \end{pmatrix} = \begin{pmatrix} -\tfrac{2}{5} & 0 \\ -2 & -1 \end{pmatrix},$$

with eigenvalues $-\tfrac{2}{5}$ and -1. Therefore, $(0, 4)$ is an asymptotically stable nodal sink. Solutions beginning near this equilibrium solution tend toward it as time approaches infinity.

$\left(\tfrac{18}{5}, 0 \right)$—The linear part has matrix

$$\mathbf{A}_{(18/5,0)} = \begin{pmatrix} -\tfrac{3}{5} & -\tfrac{9}{10} \\ 0 & -\tfrac{4}{5} \end{pmatrix},$$

with eigenvalues $-\tfrac{3}{5}$ and $-\tfrac{4}{5}$. This critical point is also an asymptotically stable nodal sink. Solutions beginning near this equilibrium solution also tend toward it as $t \to \infty$.

$\left(\tfrac{6}{5}, \tfrac{8}{5} \right)$—The linear part has matrix

$$\mathbf{A}_{(6/5,8/5)} = \begin{pmatrix} -\tfrac{1}{5} & -\tfrac{3}{10} \\ -\tfrac{4}{5} & -\tfrac{2}{5} \end{pmatrix},$$

with eigenvalues $-\tfrac{4}{5}$ and $\tfrac{1}{5}$. This critical point is an unstable saddle point.

Figure 10.41 shows a phase portrait for this system. Trajectories appear to flow outward away from the origin and, depending on where they start at time zero, to flow toward the critical point $(0, 4)$ (so y survives and x becomes extinct) or the critical point $\left(\tfrac{18}{5}, 0 \right)$ (in which case x survives and y does not). These trajectories, as well as the directed lineal elements, also reveal the behavior near the saddle point $\left(\tfrac{6}{5}, \tfrac{8}{5} \right)$. Because this point is unstable, it is possible to find initial points arbitrarily close to $\left(\tfrac{6}{5}, \tfrac{8}{5} \right)$ in which the x-species survives and the y-species dies out, or points from which the reverse occurs.

What does this tell us about survival prospects for both species? In this model, one species must lose. Depending on the initial populations, trajectories tend toward $(0, 4)$, in which the x-population dies out, or toward $\left(\tfrac{18}{5}, 0 \right)$, with the y-population becoming extinct. ∎

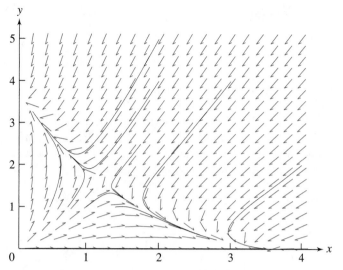

FIGURE 10.41 *Competing species model* $\begin{cases} x' = \frac{3}{5}x - \frac{1}{6}x^2 - \frac{1}{4}xy \\ y' = y - \frac{1}{4}y^2 - \frac{1}{2}xy \end{cases}$.

EXAMPLE 10.29

Consider the competing species model

$$\mathbf{X}' = \begin{pmatrix} x\left(3 - x - \frac{1}{4}y\right) \\ y\left(2 - \frac{1}{2}y - \frac{1}{6}x\right) \end{pmatrix}.$$

The critical points are $(0, 0)$, $(0, 4)$, $(3, 0)$ and $(\frac{24}{11}, \frac{36}{11})$. For reference,

$$f_x = 3 - 2x - \tfrac{1}{4}y, \quad f_y = -\tfrac{1}{4}x, \quad g_x = -\tfrac{1}{6}y, \quad g_y = 2 - y - \tfrac{1}{6}x.$$

Now examine each critical point.

(0, 0)—Compute

$$\mathbf{A}_{(0,0)} = \begin{pmatrix} 3 & 0 \\ 0 & 2 \end{pmatrix},$$

with eigenvalues 2 and 3. The origin is an unstable nodal source.

(0, 4)—Now

$$\mathbf{A}_{(0,4)} = \begin{pmatrix} 2 & 0 \\ -\frac{2}{3} & -2 \end{pmatrix},$$

with eigenvalues 2 and -2. Therefore, $(0, 4)$ is an unstable saddle point.

(3, 0)—Now

$$\mathbf{A}_{(3,0)} = \begin{pmatrix} -3 & -\frac{3}{4} \\ 0 & \frac{3}{2} \end{pmatrix},$$

with eigenvalues -3 and $\frac{3}{2}$. This implies that $(3, 0)$ is also an unstable saddle point.

$(\frac{24}{11}, \frac{36}{11})$—We find that

$$\mathbf{A}_{(24/11,36/11)} = \begin{pmatrix} -\frac{24}{11} & -\frac{6}{11} \\ -\frac{6}{11} & -\frac{18}{11} \end{pmatrix},$$

with eigenvalues $-\frac{21}{11} + \frac{3}{11}\sqrt{5}$ and $-\frac{21}{11} - \frac{3}{11}\sqrt{5}$. These are both negative, so $(\frac{24}{11}, \frac{36}{11})$ is an asymptotically stable nodal sink.

Figure 10.42 shows a phase portrait for this system. In this model, trajectories approach the asymptotically stable critical point $(\frac{24}{11}, \frac{36}{11})$, which is approximately $(2.18, 3.27)$. This means that as time increases, the species are tending toward a stable population, with x approaching 2 and y approaching 3 (the populations must be integers). ■

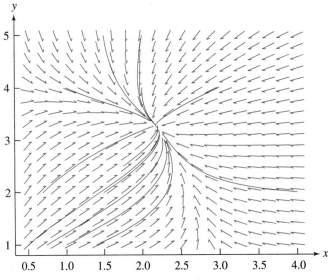

FIGURE 10.42 *Competing species model* $\begin{cases} x' = x(3 - x - \frac{1}{4}y) \\ y' = y(2 - \frac{1}{2}y - \frac{1}{6}x) \end{cases}$.

We will now do a general analysis of the competing species model (10.12) with the purpose of determining the influence of the constants G, D, C, g, d, and c on the prospects for survival, and even growth, of each species. Thus consider

$$\mathbf{X}' = \begin{pmatrix} x(G - Dx - Cy) \\ y(g - dy - cx) \end{pmatrix}. \tag{10.13}$$

It will be helpful to consider the critical points from a geometric point of view. The critical points are simultaneous solutions of

$$x(G - Dx - Cy) = 0,$$

$$y(g - cx - dy) = 0.$$

The solutions are intersections of pairs of lines, namely the x axis, the y axis, and the lines

$$Dx + Cy = G \tag{10.14}$$

and

$$cx + dy = g. \tag{10.15}$$

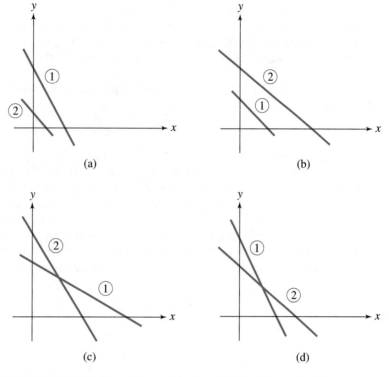

FIGURE 10.43 *Relative positions of the lines* (1) $Dx + Cy = G$ *and* (2) $cx + dy = g$.

There are four relative positions of the last two lines, and these are shown in Figure 10.43(a) through (d). These positions will play an important role shortly in interpreting mathematical conclusions. The critical points are

$$(0, 0), (G/D, 0), (0, g/d) \quad \text{and} \quad \left(\frac{Gd - Cg}{dD - cC}, \frac{Dg - cG}{dD - cC} \right).$$

We will need the partial derivatives

$$f_x = G - 2Dx - Cy, \quad f_y = -Cx,$$

$$g_x = -cy, \quad g_y = g - 2\,dy - cx.$$

Now examine each critical point in turn.

$(0, 0)$—Write the system as

$$\mathbf{X}' = \begin{pmatrix} G & 0 \\ 0 & g \end{pmatrix} \mathbf{X} + \begin{pmatrix} -Dx^2 - Cxy \\ -dy^2 - cxy \end{pmatrix}.$$

The matrix of the linear part has eigenvalues G and g, which are positive. If these are unequal, then the origin is an unstable spiral point. If they are equal, then the origin may be either an unstable node or spiral point.

$(G/D, 0)$—The translated system's linear part has matrix

$$\mathbf{A}_{(G/D, 0)} = \begin{pmatrix} -G & -CG/D \\ 0 & g - cG/D \end{pmatrix},$$

with eigenvalues $-G$ and $g - cG/D$. Certainly, $-G < 0$. If $g/c > G/D$, then the second eigenvalue is positive and the origin is an unstable saddle point. If $g/c < G/D$, then both eigenvalues are negative. If they are distinct, then $(G/D, 0)$ is an asymptotically stable node. If they are equal, then the almost linear system has an asymptotically stable node or spiral point at $(G/D, 0)$.

$(0, g/d)$—Now the linear part of the translated system has matrix

$$\mathbf{A}_{(0, g/d)} = \begin{pmatrix} G - Cg/d & 0 \\ -cg/d & -g \end{pmatrix},$$

with eigenvalues $-g$ and $G - Cg/d$. Now $-g < 0$. If $G/C > g/d$, then the second eigenvalue is positive and $(0, g/d)$ is an unstable saddle point. If $G/C < g/d$, then both eigenvalues are negative. If they are distinct, then $(0, g/d)$ is an asymptotically stable node. If they are equal, then the almost linear system has an asymptotically stable node or spiral point at $(0, g/d)$.

$(Gd - Cg)/(dD - cC), (Dg - cG)/(dD - cD)$—Denote this point $(\tilde{x}, \tilde{y})$. This is the point of intersection of the lines (10.14) and (10.15). To be meaningful in the context of a competing species model, this point must lie in the first quadrant, which occurs in Figure 10.43(c) and (d). In these cases, we can see from the intersections of the lines with the axes that

$$\frac{G}{D} > \frac{g}{c} \quad \text{and} \quad \frac{g}{d} > \frac{G}{C}$$

as in Figure 10.43(c), or

$$\frac{g}{c} > \frac{G}{D} \quad \text{and} \quad \frac{G}{C} > \frac{g}{d}$$

as in Figure 10.43(d).

Using the partial derivatives of f and g, we obtain

$$\mathbf{A}_{(\tilde{x}, \tilde{y})} = \begin{pmatrix} G - 2D\tilde{x} - C\tilde{y} & -C\tilde{x} \\ -c\tilde{y} & g - 2d\tilde{y} - c\tilde{x} \end{pmatrix}.$$

But recall that

$$G = D\tilde{x} + C\tilde{y} \quad \text{and} \quad g = c\tilde{x} + d\tilde{y}$$

so

$$\mathbf{A}_{(\tilde{x}, \tilde{y})} = \begin{pmatrix} -D\tilde{x} & -C\tilde{x} \\ -c\tilde{y} & -d\tilde{y} \end{pmatrix},$$

with eigenvalues

$$\tfrac{1}{2}(-(D\tilde{x} + d\tilde{y}) \pm \sqrt{(D\tilde{x} + d\tilde{y})^2 - 4(Dd - Cc)\tilde{x}\tilde{y}}),$$

which we write as

$$\tfrac{1}{2}(-(D\tilde{x} + d\tilde{y}) \pm \sqrt{(D\tilde{x} - d\tilde{y})^2 + 4Cc\tilde{x}\tilde{y}}).$$

From the latter formulation, it is clear that these eigenvalues are real. Two cases occur.

If $Dd - Cc < 0$, then one eigenvalue is positive and the other negative. In this case, $(\tilde{x}, \tilde{y})$ is an unstable saddle point. If the populations initially start near this critical point, then one population will die out in the limit as t increases, while the other survives. This case corresponds to Figure 10.43(c). The condition $Dd < Cc$ can be interpreted to mean that the product of the internal limiting factors is less than the product of the competition factors. The competition factors tending to decrease the populations are dominant in the model, and only one population survives.

If $Dd - Cc > 0$, then both eigenvalues are negative and $(\tilde{x}, \tilde{y})$ is an asymptotically stable node. This case corresponds to Figure 10.43(d). Now populations beginning near this critical point approach it in the limit, and the competing species manage to survive together. This is the case of coexistence. Now $Cc < Dd$ and the competition factors are less important than internal limiting factors (the x^2 and y^2 terms). With competition playing less of a role in the populations, mutual survival becomes possible.

SECTION 10.7 PROBLEMS

Problems 1 through 6 deal with the simple competing species model. In each problem, (a) determine the critical points and classify their type and stability properties, (b) draw a phase portrait of the system, and (c) interpret the survival prospects for each species as dictated by the model.

1. $x' = x - 4xy$, $y' = 3y - 6xy$

2. $x' = 3x - xy$, $y' = 4y - 10xy$

3. $x' = 3x - 2xy$, $y' = 6y - 2xy$

4. $x' = 8x - 3xy$, $y' = 2y - 7xy$

5. $x' = 3x - 7xy$, $y' = y - 4xy$

6. $x' = 4x - 9xy$, $y' = 5y - 2xy$

Problems 7 through 12 deal with the extended competing species model. In each problem, (a) determine the critical points and classify their type and stability properties, (b) draw a phase portrait for the system, and (c) interpret the survival prospects for each species as dictated by the model.

7. $x' = 7x - 5x^2 - 2xy$, $y' = 3y - 4y^2 - xy$

8. $x' = 4x - 7x^2 - xy$, $y' = y - 2y^2 - 3xy$

9. $x' = 2x - 3x^2 - xy$, $y' = 3y - y^2 - 2xy$

10. $x' = x - 9x^2 - 3xy$, $y' = y - y^2 - xy$

11. $x' = 3x - x^2 - 2xy$, $y' = y - 2y^2 - xy$

12. $x' = x - 6x^2 - 3xy$, $y' = 5y - 2y^2 - 8xy$

10.8   Lyapunov's Stability Criteria

There is a subtle criterion for stability developed by the Russian engineer and mathematician Alexander M. Lyapunov (1857–1918). Suppose $\mathbf{X}' = \mathbf{F}(\mathbf{X})$ is a 2×2 autonomous system of first-order differential equations (not necessarily almost linear) and that $(0, 0)$ is an isolated critical point. Lyapunov's insight was this. Suppose there is a function, commonly denoted V, such that closed curves $V(x, y) = c$ enclose the origin. Further, if the constants are chosen smaller, say $0 < k < c$, then the curve $V(x, y) = k$ lies within the region enclosed by the curve $V(x, y) = c$ (Figure 10.44(a)). So far this has nothing to do with the system of differential equations. However, suppose it also happens that if a trajectory intersects the curve $V(x, y) = c$ at some time, which we can take to be time zero, then it cannot escape from the region bounded by this curve, but must for all later times remain within this region (Figure 10.44(b)). This would force trajectories starting out near the origin (meaning within $V(x, y) = c$) to forever lie at least this close to the origin. But this would imply, by choosing c successively smaller, that the origin is a stable critical point!

If, in addition, trajectories starting at a point on $V(x, y) = c$ point into the region bounded by this curve, then we can further conclude that the trajectories are approaching the origin, hence that the origin is asymptotically stable.

This is the intuition behind an approach to determining whether a critical point is stable or asymptotically stable. We will now develop the vocabulary that will allow us to give substance to this approach. First, we will distinguish certain functions that have been found to serve the role of V in this discussion.

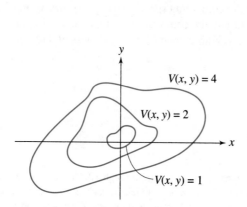

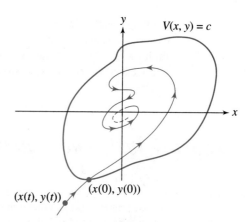

FIGURE 10.44(a) *Closed curves contracting about the origin.*

FIGURE 10.44(b) *Trajectories entering shrinking regions about the origin.*

If $r > 0$, let N_r consist of all (x, y) within distance r from the origin. Thus, (x, y) is in N_r exactly when

$$x^2 + y^2 < r^2.$$

This set is called the r-neighborhood of the origin or, if we need no explicit reference to r, just a neighborhood of the origin.

DEFINITION 10.5 *Positive Definite, Semidefinite*

Let $V(x, y)$ be defined for all (x, y) in some neighborhood N_r of the origin. Suppose V is continuous with continuous first partial derivatives. Then,

1. V is positive definite on N_r if $V(0, 0) = 0$ and $V(x, y) > 0$ for all other points of N_r.
2. V is positive semidefinite on N_r if $V(0, 0) = 0$ and $V(x, y) \geq 0$ for all points of N_r.
3. V is negative definite on N_r if $V(0, 0) = 0$ and $V(x, y) < 0$ for all other points of N_r.
4. V is negative semidefinite on N_r if $V(0, 0) = 0$ and $V(x, y) \leq 0$ for all points of N_r.

For example, $V(x, y) = x^2 + 3xy + 9y^2$ is positive definite on N_r for any positive r, and $-3x^2 + 4xy - 5y^2$ is negative definite on any N_r. The function $(x - y)^4$ is positive semidefinite, being nonnegative but vanishing on the line $y = x$.

The following lemma is useful in producing examples of positive definite and negative definite functions.

LEMMA 10.2

Let $V(x, y) = ax^2 + bxy + cy^2$. Then V is positive definite (on any N_r) if and only if

$$a > 0 \quad \text{and} \quad 4ac - b^2 > 0.$$

V is negative definite (on any N_r) if and only if

$$a < 0 \quad \text{and} \quad 4ac - b^2 > 0.$$

Proof Certainly V is continuous with continuous partial derivatives in the entire x, y plane. Further, $V(0, 0) = 0$, and this is the only point at which $V(x, y)$ vanishes.

Now recall the second derivative test for extrema of a function of two variables. First,

$$V_x(0, 0) = V_y(0, 0) = 0,$$

so the origin is a candidate for a maximum or minimum of V. For a maximum or minimum, we need

$$V_{xx}(0, 0)V_{yy}(0, 0) - V_{xy}(0, 0)^2 > 0.$$

But this condition is the same as

$$(2a)(2c) - b^2 > 0,$$

or

$$4ac - b^2 > 0.$$

Satisfaction of this inequality requires that a and c have the same sign.

When $a > 0$, then $V_{xx}(0, 0) > 0$ and the origin is a point where V has a minimum. In this event, $V(x, y) > V(0, 0) = 0$ for all (x, y) other than $(0, 0)$. Now V is positive definite.

When $a < 0$, then $V_{xx}(0, 0) < 0$ and the origin is a point where V has a maximum. Now $V(x, y) < 0$ for all (x, y) other than $(0, 0)$, and V is negative definite. ∎

If $x = x(t)$, $y = y(t)$ defines a trajectory of $\mathbf{X}' = \mathbf{F}(\mathbf{X})$ and V is a differentiable function of two variables, then $V(x(t), y(t))$ is a differentiable function of t along this trajectory. We will denote the derivative of $V(x(t), y(t))$ with respect to t as $\dot{V}(x, y)$, or just $\dot{V}$. By the chain rule,

$$\dot{V}(x, y) = V_x(x(t), y(t))x'(t) + V_y(x(t), y(t))y'(t),$$

or, more succinctly,

$$\dot{V} = V_x x' + V_y y'.$$

This is called the *derivative* of V along the trajectory, or the *orbital derivative* of V.

The following two theorems show how these ideas about positive and negative definite functions relate to Lyapunov's approach to stable and asymptotically stable critical points. The criteria given in the first theorem constitute *Lyapunov's direct method* for determining the stability or asymptotic stability of a critical point.

THEOREM 10.4 *Lyapunov's Direct Method for Stability*

Let $(0, 0)$ be an isolated critical point of the autonomous 2×2 system $\mathbf{X}' = \mathbf{F}(\mathbf{X})$.

1. If a positive definite function V can be found for some neighborhood N_r of the origin, such that $\dot{V}$ is negative semidefinite on N_r, then the origin is stable.

2. If a positive definite function V can be found for some neighborhood N_r of the origin, such that $\dot{V}$ is negative definite on N_r, then the origin is asymptotically stable. ∎

On the other side of the issue, Lyapunov's second theorem gives a test to determine that a critical point is unstable.

THEOREM 10.5 *Lyapunov's Direct Method for Instability*

Let $(0, 0)$ be an isolated critical point of the autonomous 2×2 system $\mathbf{X}' = \mathbf{F}(\mathbf{X})$. Let V be continuous with continuous first partial derivatives in some neighborhood of the origin, and let $V(0, 0) = 0$.

1. Suppose $R > 0$ and that in every neighborhood N_r, with $0 < r \leq R$, there is a point at which $V(x, y)$ is positive. Suppose $\dot{V}$ is positive definite in N_R. Then $(0, 0)$ is unstable.

2. Suppose $R > 0$ and that in every neighborhood N_r, with $0 < r \leq R$, there is a point at which $V(x, y)$ is negative. Suppose $\dot{V}$ is negative definite in N_R. Then $(0, 0)$ is unstable. ■

Any function V playing the role cited in these theorems is called a *Lyapunov function*. Theorems 10.4 and 10.5 give no suggestion at all as to how a Lyapunov function might be produced, and in attempting to apply them this is the difficult part. Lemma 10.2 is sometimes useful in providing candidates, but, as might be expected, if the differential equation is complicated, the task of finding a Lyapunov function might be insurmountable. In spite of this potential difficulty, Lyapunov's theorems are useful because they do not require solving the system, nor do they require that the system be almost linear.

Adding to the mystique of the theorem is the nonobvious connection between V, $\dot{V}$, and stability characteristics of the critical point. We will give a plausibility argument intended to clarify this connection.

Consider Figure 10.45, which shows a typical curve $V(x, y) = c$ about the origin. Call this curve Γ. Here is how $\dot{V}$ enters the picture. At any point $P : (a, b)$ on this curve, the vector

$$\mathbf{N} = V_x \mathbf{i} + V_y \mathbf{j}$$

is normal (perpendicular) to Γ, by which we mean that it is normal to the tangent to Γ at this point. In addition, consider a trajectory $x = \varphi(t)$, $y = \psi(t)$ passing through P at time $t = 0$, also shown in Figure 10.45. Thus, $\varphi(0) = a$ and $\psi(0) = b$. The vector

$$\mathbf{T} = \varphi'(0)\mathbf{i} + \psi'(0)\mathbf{j}$$

is tangent to this trajectory (not to Γ) at (a, b). Now,

$$\dot{V}(a, b) = V_x(a, b)\varphi'(0) + V_y(a, b)\psi'(0) = \mathbf{N} \cdot \mathbf{T},$$

the dot product of the normal to Γ and the tangent to the trajectory at (a, b). Since the dot product of two vectors is equal to the product of their lengths and the cosine of the angle between them, we obtain

$$\dot{V}(a, b) = \|\mathbf{N}\| \|\mathbf{T}\| \cos(\theta),$$

with θ the angle between $\mathbf{T}$ and $\mathbf{N}$.

Now look at conclusions (1) and (2) of the first Lyapunov theorem. If $\dot{V}$ is negative semidefinite, then $\dot{V}(a, b) \leq 0$, so $\cos(\theta) \leq 0$. Then $\pi/2 \leq \theta \leq 3\pi/2$. This means that the trajectory at this point is moving at this point either into the region enclosed by Γ, or perhaps in the same direction as the tangent to Γ. The effect of this is that the trajectory cannot move away from the region enclosed by Γ. The trajectory cannot escape from this region, and so the origin is stable. If $\dot{V}$ is negative definite, then $\cos(\theta) < 0$, so $\pi/2 < \theta < 3\pi/2$ and now the trajectory actually moves into the region enclosed by Γ, and cannot simply trace out a path around the origin. In this case the origin is asymptotically stable.

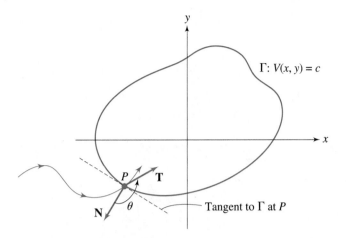

FIGURE 10.45 *Rationale for Lyapunov's direct method.*

We leave it for the student to make a similar geometric argument in support of the second Lyapunov theorem.

EXAMPLE 10.30

Consider the nonlinear system

$$\mathbf{X}' = \begin{pmatrix} -x^3 \\ -4x^2 y \end{pmatrix}.$$

The origin is an isolated critical point. We will attempt to construct a Lyapunov function of the form $V(x, y) = ax^2 + bxy + cy^2$ that will tell us whether the origin is stable or unstable. We may not succeed in this, but it is a good first attempt because at least we know conditions on the coefficients of V to make this function positive definite or negative definite. The key lies in $\dot{V}$, so compute

$$\dot{V} = V_x x' + V_y y'$$
$$= (2ax + by)(-x^3) + (bx + 2cy)(-4x^2 y)$$
$$= -2ax^4 - bx^3 y - 4bx^3 y - 8cx^2 y^2.$$

Now observe that the $-2ax^4$ and $-8cx^2 y^2$ terms will be nonpositive if a and c are positive. The $x^3 y$ term will vary in sign, but we can make $bx^3 y$ vanish by choosing $b = 0$. We can choose a and c as any positive numbers, say $a = c = 1$. Then

$$V(x, y) = x^2 + y^2$$

is positive definite in any neighborhood of the origin, and

$$\dot{V} = -2x^4 - 8x^2 y^2$$

is negative semidefinite in any neighborhood of the origin. By Lyapunov's direct method (Theorem 10.4), the origin is stable. A phase portrait for this system is shown in Figure 10.46. ∎

We can draw no conclusion about asymptotic stability of the origin in the last example. If we had been able to find a Lyapunov function V so that $\dot{V}$ was negative definite (instead of negative

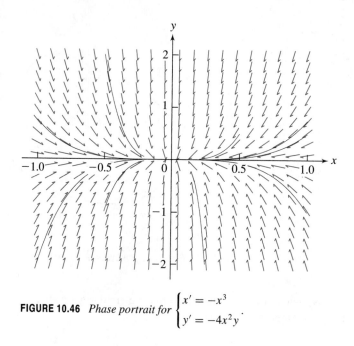

FIGURE 10.46   *Phase portrait for* $\begin{cases} x' = -x^3 \\ y' = -4x^2y \end{cases}$.

semidefinite), then we could have concluded that the origin was asymptotically stable. But we cannot be sure, from the work done, whether there is no such function, or whether we simply did not find one.

EXAMPLE 10.31

It is instructive to consider Lyapunov's theorems as they relate to a simple physical problem, the undamped pendulum (put $c = 0$). The system is

$$x' = y,$$

$$y = -\frac{g}{L}\sin(x).$$

Recall that $x = \theta$, the displacement angle of the bob from the downward vertical rest position.

We have already characterized the critical points of this problem (with damping). However, this example makes an important point. In problems that are drawn from a physical setting, the total energy of the system can often serve as a Lyapunov function. This is a useful observation, since the search for a Lyapunov function constitutes the primary issue in attempting to apply Lyapunov's theorems.

Thus, compute the total energy V of the pendulum. The kinetic energy is

$$\tfrac{1}{2}mL^2(\theta'(t))^2,$$

which in the variables of the system is

$$\tfrac{1}{2}mL^2y^2.$$

The potential energy is the work done in lifting the bob above the lowest position. From Figure 10.1, this is

$$mgL(1 - \cos(\theta)),$$

or

$$mgL(1 - \cos(x)).$$

The total energy is therefore given by

$$V(x, y) = mgL(1 - \cos(x)) + \tfrac{1}{2}mL^2y^2.$$

Clearly, $V(0, 0) = 0$. The rest position of the pendulum (pendulum arm vertical with bob at the low point) has zero energy. Next, compute

$$\dot{V}(x(t), y(t)) = V_x x'(t) + V_y y'(t) = mgL \sin(x)x'(t) + mL^2 yy'(t).$$

Along any trajectory of the system, $x' = y$ and $y' = -(g/L)\sin(x)$, so the orbital derivative is

$$\dot{V}(x(t), y(t)) = mgL \sin(x)y + mL^2 y \left(-\frac{g}{L} \sin(x)\right) = 0.$$

This corresponds to the fact that in a conservative physical setting, the total energy is a constant of the motion.

Now V is positive definite. This is expected because the energy should be a minimum in the rest position, where $V(0, 0) = 0$. Further, $\dot{V}$ is negative semidefinite. Therefore, the origin is stable. ∎

As expected, Lyapunov's theorem did not tell us anything new about the pendulum, which we had already analyzed by other means. However, this example does provide some insight into a line of thought that could have motivated Lyapunov. In any conservative physical system, the total energy must be a constant of the motion, and we expect that any position with the system at rest should be stable if the potential energy is a minimum, and unstable if it is not. This suggests looking at the total energy as a candidate for a Lyapunov function V. In particular, for many mechanical systems the kinetic energy is a quadratic form. One then checks the orbital derivative $\dot{V}$ to see if it is negative definite or semidefinite in a neighborhood of the point of interest.

EXAMPLE 10.32

Consider the system

$$\mathbf{X}' = \begin{pmatrix} x^3 - xy^2 \\ y^3 + 6x^2y \end{pmatrix}.$$

The origin is an isolated critical point. We do not know whether it is stable, asymptotically stable, or unstable, so we will begin by trying to construct a Lyapunov function that fits either of the Lyapunov theorems. Attempt a Lyapunov function

$$V(x, y) = ax^2 + bxy + cx^2.$$

We know how to choose the coefficients to make this positive definite. The question is what happens with the orbital derivative. Compute

$$\begin{aligned} \dot{V} &= V_x x' + V_y y' \\ &= (2ax + by)(x^3 - xy^2) + (bx + 2cy)(y^3 + 6x^2y) \\ &= 2ax^4 + 2cy^4 + (12c - 2a)x^2y^2 + 7bx^3y. \end{aligned}$$

This looks promising because the first three terms can be made strictly positive for $(x, y) \neq (0, 0)$. Thus choose $b = 0$ and $a = c = 1$ to get

$$\dot{V}(x, y) = 2x^4 + 2y^4 + 10x^2y^2 > 0$$

for $(x, y) \neq (0, 0)$. With this choice of the coefficients,

$$V(x, y) = x^2 + y^2$$

is positive definite on any neighborhood of the origin, and $\dot{V}$ is also positive definite. By Lyapunov's second theorem (Theorem 10.5), the origin is unstable. Figure 10.47 shows a phase portrait for this system. ■

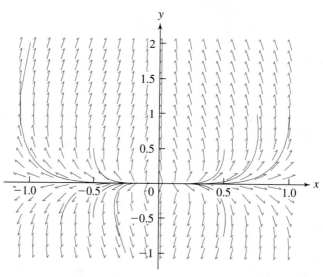

FIGURE 10.47 *Phase portrait for* $\begin{cases} x' = x^3 - xy^2 \\ y' = y^3 + 6x^2y \end{cases}$.

EXAMPLE 10.33

A nonlinear oscillator with linear damping can be modeled by the differential equation

$$z'' + \alpha z' + z + \beta z^2 + \gamma z^3 = 0,$$

in which β and γ are positive and $\alpha \neq 0$. To convert this to a system, let $z = x$ and $z' = y$ to obtain

$$x' = y,$$
$$y' = -\alpha y - x - \beta x^2 - \gamma x^3.$$

This is the system

$$\mathbf{X}' = \begin{pmatrix} y \\ -\alpha y - x - \beta x^2 - \gamma x^3 \end{pmatrix}.$$

We can construct a Lyapunov function by a clever observation. Let

$$V(x, y) = \tfrac{1}{2}y^2 + \tfrac{1}{2}x^2 + \tfrac{1}{3}\beta x^3 + \tfrac{1}{4}\gamma x^4.$$

Then

$$\dot{V} = (x + \beta x^2 + \gamma x^3)y + (y)(-\alpha y - x - \beta x^2 - \gamma x^3)$$
$$= -\alpha y^2.$$

Since $\alpha > 0$, $\dot{V}$ is certainly negative semidefinite in any neighborhood of the origin. It may not be obvious whether V is positive definite in any neighborhood of the origin. Certainly the term $y^2/2$ in V is nonnegative. The other terms are

$$\tfrac{1}{2}x^2 + \tfrac{1}{3}\beta x^3 + \tfrac{1}{4}\gamma x^4,$$

which we can write as

$$x^2 \left(\tfrac{1}{2} + \tfrac{1}{4}\gamma x^2 + \tfrac{1}{3}\beta x \right).$$

Since $g(x) = \tfrac{1}{2} + \tfrac{1}{4}\gamma x^2 + \tfrac{1}{3}\beta x$ is continuous for all x, and $g(0) = 1/2$, there is an interval $(-h, h)$ about the origin such that

$$g(x) > 0 \quad \text{for} \quad -h < x < h.$$

Then, in N_h, $V(x, y) \geq 0$, and $V(x, y) > 0$ if $(x, y) \neq (0, 0)$. Therefore, V is positive definite in this neighborhood.

We now have V positive definite and $\dot{V}$ negative semidefinite in N_h, hence the origin is stable. Figure 10.48 shows a phase portrait for the case $\alpha = 1$, $\beta = \tfrac{1}{4}$, and $\gamma = \tfrac{1}{6}$. It is instructive to try different values of β and γ to get some idea of the effect of the nonlinear terms on the trajectories. ∎

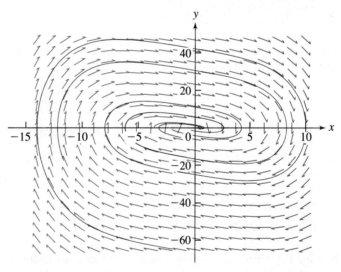

FIGURE 10.48 *Nonlinear oscillator with $\alpha = 1$, $\beta = \tfrac{1}{4}$, $\gamma = \tfrac{1}{6}$.*

EXAMPLE 10.34

Consider the system

$$x' = f(t)y + g(t)x(x^2 + y^2),$$
$$y' = -f(t)x + g(t)y(x^2 + y^2).$$

Assume that the origin is an isolated critical point.

If we attempt a Lyapunov function $V(x, y) = ax^2 + bxy + cy^2$, then

$$\dot{V} = (2ax + by)[f(t)y + g(t)x(x^2 + y^2)] + (bx + 2cy)[-f(t)x + g(t)y(x^2 + y^2)]$$

$$= (2a - 2c)f(t)xy + 2ax^2g(t)(x^2 + y^2) + 2cy^2g(t)(x^2 + y^2)$$

$$+ bf(t)y^2 + 2bg(t)xy(x^2 + y^2) - bf(t)x^2.$$

We can eliminate three terms in the orbital derivative by trying $b = 0$. The term $(2a - 2c)f(t)xy$ vanishes if we choose $a = c$. To have V positive definite we need a and c positive, so let $a = c = 1$. Then

$$V(x, y) = x^2 + y^2,$$

which is positive definite in any neighborhood of the origin, and

$$\dot{V} = 2(x^2 + y^2)^2 g(t).$$

This is negative definite if $g(t) < 0$ for all $t \geq 0$. In this case, the origin is asymptotically stable. If $g(t) \leq 0$ for $t \geq 0$, then $\dot{V}$ is negative semidefinite and then the origin is stable. If $g(t) > 0$ for $t \geq 0$, then the origin is unstable. ∎

SECTION 10.8 PROBLEMS

In each of Problems 1 through 8, use Lyapunov's theorem to determine whether the origin is stable, asymptotically stable, or unstable.

1. $x' = -2xy^2$, $y' = -x^2y$
2. $x' = -x\cos^2(y)$, $y' = (6 - x)y^2$
3. $x' = -2x$, $y' = -3y^3$
4. $x' = -x^2y^2$, $y' = x^2$
5. $x' = xy^2$, $y' = y^3$
6. $x' = x^5(1 + y^2)$, $y' = x^2y + y^3$
7. $x' = x^3(1 + y)$, $y' = y^3(4 + x^2)$
8. $x' = x^3\cot^2(y)$, $y' = y^3(2 + x^4)$

10.9 Limit Cycles and Periodic Solutions

Nonlinear systems of differential equations can give rise to curves called *limit cycles*, which have particularly interesting properties. To see how a limit cycle occurs naturally in a physical setting, draw a circle C of radius R on pavement, with R exceeding the length L between the points where the front and rear wheel of a bicycle touch the ground. Now grab the handlebars and push the bicycle so that its front wheel moves around C. What path does the rear wheel follow?

If you tie a marker to the rear wheel so that it traces out the rear wheel's path as the front wheel moves along C, you find that this path does not approach a particular point. Instead, as the front wheel continues its path around C, the rear wheel asymptotically approaches a circle K concentric with C and having radius $\sqrt{R^2 - L^2}$. If the rear wheel begins outside C, it will spiral inward toward K, while if it begins inside C, it will work its way outward toward K. If the rear wheel begins on K, it will remain on K.

This inner circle K has two properties in common with a stable critical point. Trajectories beginning near K move toward it, and if a trajectory begins on K, it remains there. However, K is not a point, but is instead a closed curve. K is a *limit cycle* of this motion.

DEFINITION 10.6   Limit Cycle

A limit cycle of a 2×2 system $\mathbf{X}' = \mathbf{F}(\mathbf{X})$ is a closed trajectory K having the property that there are trajectories $x = \varphi(t)$, $y = \psi(t)$ of the system such that $(\varphi(t), \psi(t))$ spirals toward K in the limit as $t \to \infty$.

We have already pointed out the analogy between a limit cycle and a critical point. This analogy can be pushed further by defining a concept of stability and asymptotic stability for limit cycles that is modeled after stability and asymptotic stability for critical points.

DEFINITION 10.7

Let K be a limit cycle of $\mathbf{X}' = \mathbf{F}(\mathbf{X})$. Then,

1. K is stable if trajectories starting within a certain distance of K must remain within a fixed distance of K.

2. K is asymptotically stable if every trajectory that starts sufficiently close to K spirals toward K as $t \to \infty$.

3. K is semistable if every trajectory starting on one side of K spirals toward K as $t \to \infty$, while there are trajectories starting on the other side of K that spiral away from K as $t \to \infty$.

4. K is unstable if there are trajectories starting on both sides of K that spiral away from K as $t \to \infty$.

Keep in mind that a closed trajectory of $\mathbf{X}' = \mathbf{F}(\mathbf{X})$ represents a periodic solution. We have seen periodic solutions previously with centers, which are critical points about which trajectories form closed curves. Certain predator/prey models exhibited periodic solutions. A limit cycle is therefore a periodic solution, toward which other solutions approach spirally. Often we are interested in whether a system $\mathbf{X}' = \mathbf{F}(\mathbf{X})$ has a periodic solution, and we will shortly develop some tests to tell whether a system has such a solution, or sometimes to tell that it does not.

EXAMPLE 10.35   Limit Cycle

Consider the almost linear system

$$\mathbf{X}' = \begin{pmatrix} 1 & 1 \\ -1 & 1 \end{pmatrix} \mathbf{X} + \begin{pmatrix} -x\sqrt{x^2 + y^2} \\ -y\sqrt{x^2 + y^2} \end{pmatrix}. \tag{10.16}$$

$(0, 0)$ is the only critical point of this system. The eigenvalues of

$$\begin{pmatrix} 1 & 1 \\ -1 & 1 \end{pmatrix}$$

are $1 \pm i$, so the origin is an unstable spiral point of the system $\mathbf{X}' = \mathbf{AX} + \mathbf{G}$ and also of the linear system $\mathbf{X}' = \mathbf{AX}$. Figure 10.49 shows a phase portrait of the linear system.

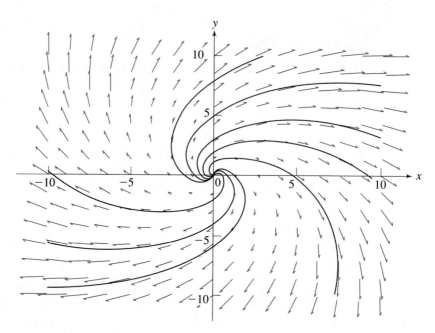

FIGURE 10.49 *Unstable spiral point for $x' = x + y$, $y' = -x + y$.*

Up to this point, whenever we have seen trajectories spiraling outward, they have grown without bound. We will now see that this does not happen with this current system. This will be transparent if we convert the system to polar coordinates. Since

$$r^2 = x^2 + y^2,$$

then

$$rr' = xx' + yy'$$
$$= x[x + y - x\sqrt{x^2 + y^2}] + y[-x + y - y\sqrt{x^2 + y^2}]$$
$$= x^2 + y^2 - (x^2 + y^2)\sqrt{x^2 + y^2} = r^2 - r^3.$$

Then

$$r' = r - r^2 = r(1 - r).$$

This tells us that if $0 < r < 1$, then $r' > 0$, so the distance between a trajectory and the origin is increasing. Trajectories inside the unit circle are moving outward. But if $r > 1$, then $r' < 0$, so the distance between a trajectory and the origin is decreasing. Trajectories outside the unit circle are moving inward.

This does not yet tell us in detail how the trajectories are moving outward or inward. To determine this, we need to bring the polar angle θ into consideration. Differentiate $x = r\cos(\theta)$ and $y = r\sin(\theta)$ with respect to t:

$$x' = r'\cos(\theta) - r\sin(\theta)\theta',$$

$$y' = r'\sin(\theta) + r\cos(\theta)\theta'.$$

Now observe that

$$x'y - xy' = r\sin(\theta)[r'\cos(\theta) - r\sin(\theta)\theta'] - r\cos(\theta)[r'\sin(\theta) + r\cos(\theta)\theta']$$
$$= -r^2[\cos^2(\theta) + \sin^2(\theta)]\theta' = -r^2\theta'.$$

But from the system $\mathbf{X}' = \mathbf{AX} + \mathbf{G}$, we have

$$x'y - xy' = y[x + y - x\sqrt{x^2 + y^2}] - x[-x + y - y\sqrt{x^2 + y^2}]$$
$$= x^2 + y^2 = r^2.$$

Therefore,

$$r^2 = -r^2\theta',$$

from which we conclude that

$$\theta' = -1.$$

We now have an uncoupled system of differential equations for r and θ:

$$r' = r(1 - r), \quad \theta' = -1.$$

This equation for r is separable. For the trajectory through (r_0, θ_0), solve these equations subject to the initial conditions

$$r(0) = r_0, \quad \theta(0) = \theta_0.$$

We get

$$r = \frac{1}{1 - ((r_0 - 1)/r_0)e^{-t}}, \quad \theta = \theta_0 - t.$$

These explicit solutions enable us to conclude the following.

If $r_0 = 1$, then (r_0, θ_0) is on the unit circle. But then $r(t) = 1$ for all t, hence a trajectory that starts on the unit circle remains there for all times. Further, $\theta'(t) = -1$, so the point $(r(t), \theta(t))$ moves clockwise around this circle as t increases.

If $0 < r_0 < 1$, then (r_0, θ_0) is within the disk bounded by the unit circle. Now

$$r(t) = \frac{1}{1 + ((1 - r_0)/r_0)e^{-t}} < 1$$

for all $t > 0$. Therefore, a trajectory starting at a point inside the unit disk remains there forever. Further, $r(t) \to 1$ as $t \to \infty$, so this trajectory approaches the unit circle from within.

Finally, if $r_0 > 1$, then (r_0, θ_0) is outside the unit circle. But now $r(t) > 1$ for all t, so a trajectory starting outside the unit circle remains outside for all time. However, it is still true that $r(t) \to 1$ as $t \to \infty$, so this trajectory approaches the unit circle from without.

In sum, trajectories tend in the limit to wrap around the unit circle, either from within or from without, depending on where they start. The unit circle is a an asymptotically stable limit cycle of $\mathbf{X}' = \mathbf{F}(\mathbf{X})$. A phase portrait is shown in Figure 10.50. ■

Here is an example of a system having infinitely many limit cycles.

EXAMPLE 10.36

The system

$$\mathbf{X}' = \begin{pmatrix} 0 & 1 \\ -1 & 0 \end{pmatrix} \mathbf{X} + \begin{pmatrix} x\sin(\sqrt{x^2 + y^2}) \\ y\sin(\sqrt{x^2 + y^2}) \end{pmatrix}$$

has particularly interesting limit cycles, as can be seen in Figure 10.51. These occur as concentric circles about the origin. Trajectories originating within the innermost circle spiral toward this

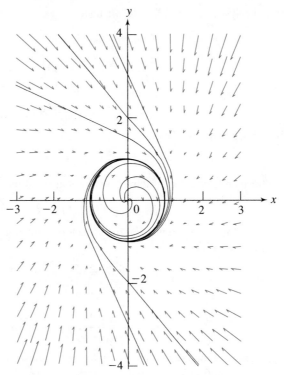

FIGURE 10.50 *Limit cycle of* $\begin{cases} x' = x + y - x\sqrt{x^2 + y^2} \\ y' = -x + y - y\sqrt{x^2 + y^2} \end{cases}$.

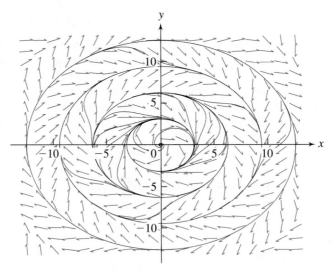

FIGURE 10.51 *Asymptotically stable limit cycles of* $\begin{cases} x' = y + x\sin(\sqrt{x^2 + y^2}) \\ y' = -x + y\sin(\sqrt{x^2 + y^2}) \end{cases}$.

circle, as do trajectories beginning between this first circle and the second one. Trajectories beginning between the second and third circles spiral toward the third circle, as do trajectories originating between the third and fourth circles. This pattern continues throughout the plane. ■

We will now develop some facts about closed trajectories (periodic solutions) and limit cycles. For the remainder of this section,

$$\mathbf{X}' = \mathbf{F}(\mathbf{X}) = \begin{pmatrix} f(x, y) \\ g(x, y) \end{pmatrix}$$

is a 2×2 autonomous system.

The first result states that under commonly encountered conditions, a closed trajectory of $\mathbf{X}' = \mathbf{F}(\mathbf{X})$ must always enclose a critical point.

THEOREM 10.6   *Enclosure of Critical Points*

Let f and g be continuous with continuous first partial derivatives in a region of the plane containing a closed trajectory K of $\mathbf{X}' = \mathbf{F}(\mathbf{X})$. Then K must enclose at least one critical point of $\mathbf{X}' = \mathbf{F}(\mathbf{X})$. ∎

This kind of result can sometimes be used to tell that certain regions of the plane cannot contain closed trajectories of a system of differential equations. For example, suppose the origin is the only critical point of $\mathbf{X}' = \mathbf{F}(\mathbf{X})$. Then we automatically know that there can be no closed trajectory in, for example, one of the quadrants, because such a closed trajectory could not enclose the origin, contradicting the fact that it must enclose a critical point.

Bendixson's theorem, which follows, gives conditions under which $\mathbf{X}' = \mathbf{F}(\mathbf{X})$ has no closed trajectory in a part of the plane. A region of the plane is called *simply connected* if it contains all the points enclosed by any closed curve in the region. For example, the region bounded by the unit circle is simply connected. But the shaded region shown in Figure 10.52 between the curves C and K is not simply connected, because C encloses points not in the region. A simply connected region can have no "holes" in it, because then a closed curve wrapping around a hole encloses points not in the region.

THEOREM 10.7   *Bendixson*

Let f and g be continuous with continuous first partial derivatives in a simply connected region R of the plane. Suppose $f_x + g_y$ has the same sign throughout points of R, either positive or negative. Then $\mathbf{X}' = \mathbf{F}(\mathbf{X})$ has no closed trajectory in R.

Proof   Suppose R contains a closed trajectory C representing the periodic solution $x = \varphi(t)$, $y = \psi(t)$. Suppose $(\varphi(t), \psi(t))$ traverses this curve exactly once as t varies from a to b, and let

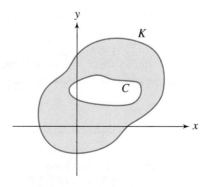

FIGURE 10.52 *Non-simply connected region.*

D be the region enclosed by C. Evaluate the line integral:

$$\oint_C -g(x, y)\, dx + f(x, y)\, dy = \int_a^b [-g(\varphi(t), \psi(t))\varphi'(t) + f(\varphi(t), \psi(t))\psi'(t)]\, dt.$$

So far we have not used the fact that $x = \varphi(t)$, $y = \psi(t)$ is a solution of $\mathbf{X}' = \mathbf{F}(\mathbf{X})$. Using this, we have

$$\varphi'(t) = x' = f(x, y) = f(\varphi(t), \psi(t))$$

and

$$\psi'(t) = y' = g(x, y) = g(\varphi(t), \psi(t)).$$

Then

$$-g(\varphi(t), \psi(t))\varphi'(t) + f(\varphi(t), \psi(t))\psi'(t) = -g(\varphi(t), \psi(t))f(\varphi(t), \psi(t))$$
$$+ f(\varphi(t), \psi(t))g(\varphi(t), \psi(t)) = 0.$$

Therefore,

$$\oint_C -g(x, y)\, dx + f(x, y)\, dy = 0.$$

But by Green's theorem,

$$\oint_C -g(x, y)\, dx + f(x, y)\, dy = \iint_D (f_x + g_y)\, dx\, dy,$$

and this integral cannot be zero because the integrand is continuous and of the same sign throughout D. This contradiction implies that no such closed trajectory C can exist within the region R. ∎

EXAMPLE 10.37

Consider the system

$$\mathbf{X}' = \begin{pmatrix} 3x + 4y + x^3 \\ 5x - 2y + y^3 \end{pmatrix}.$$

Here f and g are continuous, with continuous first partial derivatives, throughout the plane. Further,

$$f_x + g_y = 3 + 3x^2 - 2 + 3y^2 > 0$$

for all (x, y). This system has no closed trajectory, hence no periodic solution. ∎

The last two theorems have been negative, in the sense of providing criteria for $\mathbf{X}' = \mathbf{F}(\mathbf{X})$ to have no periodic solution in some part of the plane. The next theorem, a major result credited dually to Henri Poincaré and Ivar Bendixson, gives a condition under which $\mathbf{X}' = \mathbf{F}(\mathbf{X})$ has a periodic solution.

THEOREM 10.8 *Poincaré–Bendixson*

Let f and g be continuous with continuous first partial derivatives in a region R of the plane that contains no critical point of $\mathbf{X}' = \mathbf{F}(\mathbf{X})$. Let C be a trajectory of $\mathbf{X}' = \mathbf{F}(\mathbf{X})$ that is in R for $t \geq t_0$. Then C must be a periodic solution (closed trajectory), or else C spirals toward a closed trajectory as $t \to \infty$. ∎

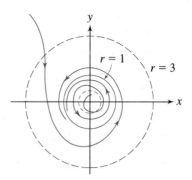

FIGURE 10.53 *Region between the circles $r = \frac{1}{2}$ and $r = 3$, with trajectories approaching the limit cycle $r = 1$.*

In either case, as long as a trajectory enters R at some time, and R contains no critical point, then $\mathbf{X}' = \mathbf{F}(\mathbf{X})$ has a periodic solution, namely this trajectory itself or, if not, a closed trajectory approached spirally by this trajectory.

On the face of it, this result may appear to contradict the conclusion of Theorem 10.6, since any periodic trajectory should enclose a critical point. However, this critical point need not be in the region R of the theorem. To illustrate, consider again the system (10.16). Let R be the region between the concentric circles $r = \frac{1}{2}$ and $r = 3$ shown in Figure 10.53. The only critical point of $\mathbf{X}' = \mathbf{F}(\mathbf{X})$ is the origin, which is not in R. If we choose any trajectory beginning at a point inside the unit circle, then, as we have seen, this trajectory approaches the unit circle, hence eventually enters R. The Poincaré–Bendixson theorem would allow us to assert just from this that R contains a periodic solution of the system.

We will conclude this section with Lienard's theorem, which gives conditions sufficient for a system to have a limit cycle.

THEOREM 10.9   Lienard

Let p and q be continuous and have continuous derivatives on the entire real line. Suppose

1. $q(x) = -q(-x)$ for all x.
2. $q(x) > 0$ for all x.
3. $p(x) = p(-x)$ for all x.

Suppose also the equation $F(x) = 0$ has exactly one positive root, where

$$F(x) = \int_0^x p(\xi)\, d\xi.$$

If this root is denoted γ, suppose $F(x) < 0$ for $0 < x < \gamma$, and $F(x)$ is positive and nondecreasing for $x > \gamma$. Then, the system

$$X' = \begin{pmatrix} y \\ -p(x)y - q(x) \end{pmatrix}$$

has a unique limit cycle enclosing the origin. Further, this limit cycle is asymptotically stable. ■

Under the conditions of the theorem, the system has exactly one periodic solution, and every other trajectory spirals toward this closed curve as $t \to \infty$.

As an illustration, we will use Lienard's theorem to analyze the van der Pol equation.

EXAMPLE 10.38 van der Pol Equation

The second-order differential equation

$$z'' + \alpha(z^2 - 1)z' + z = 0,$$

in which α is a positive constant, is called *van der Pol's equation*. It was derived by the Dutch engineer Balthazar van der Pol in the 1920s in his studies of vacuum tubes. It was of great interest to know whether this equation has periodic solutions, a question to which the answer is not obvious.

First write van der Pol's equation as a system. Let $x = z$ and $y = z'$ to get

$$x' = y,$$

$$y' = -\alpha(x^2 - 1)y - x.$$

This system has exactly one critical point, the origin. Further, this system matches the one in Lienard's theorem if we let

$$p(x) = \alpha(x^2 - 1) \quad \text{and} \quad q(x) = x.$$

Now $q(-x) = -q(x)$, $q(x) > 0$ for $x > 0$, and $p(-x) = p(x)$, as required. Next, let

$$F(x) = \int_0^x p(\xi)\, d\xi = \alpha x \left(\frac{1}{3}x^2 - 1\right).$$

F has exactly one positive zero, $\gamma = \sqrt{3}$. For $0 < x < \sqrt{3}$, $F(x) < 0$. Further, for $x > \sqrt{3}$, $F(x)$ is positive and increasing (hence, nondecreasing). By Lienard's theorem, the van der Pol equation has a unique limit cycle (hence, periodic solution) enclosing the origin. This limit cycle is asymptotically stable, so trajectories beginning at points not on this closed trajectory spiral toward this limit cycle. Figures 10.54 through 10.57 show phase portraits for van der Pol's equation, for various choices of α. ■

SECTION 10.9 PROBLEMS

In each of Problems 1 through 4, use Bendixson's theorem to show that the system has no closed trajectory. Generate a phase portrait for the system.

1. $x' = -2x - y + x^3,\ y' = 10x + 5y + x^2 y - 2y\sin(x)$

2. $x' = -x - 3y + e^{2x},\ y' = x + 2y + \cos(y)$

3. $x' = 3x - 7y + \sinh(x),\ y' = -4y + 5e^{3y}$

4. $x' = y,\ y' = -x + y(9 - x^2 - y^2)$, for (x, y) in the elliptical region bounded by the graph of $x^2 + 9y^2 = 9$

5. Recall that with the transformation from rectangular to polar coordinates, we obtain

$$x\frac{dx}{dt} + y\frac{dy}{dt} = r\frac{dr}{dt}$$

and

$$y\frac{dx}{dt} - x\frac{dy}{dt} = -r^2\frac{d\theta}{dt}.$$

Use these equations to show that the system

$$x' = y + \frac{x}{\sqrt{x^2 + y^2}} f(\sqrt{x^2 + y^2}),$$

$$y' = -x + \frac{y}{\sqrt{x^2 + y^2}} f(\sqrt{x^2 + y^2})$$

has closed trajectories associated with zeros of the function f (a given continuous function of one variable). If the closed trajectory is a limit cycle, what is its direction of orientation?

(Problems continue on p. 508)

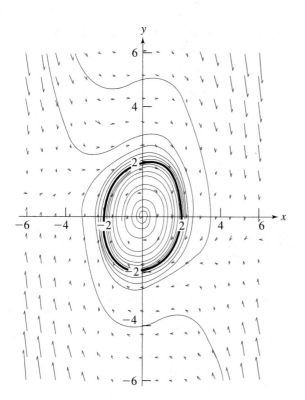

FIGURE 10.54 *Phase portrait for van der Pol's equation for $\alpha = 0.2$.*

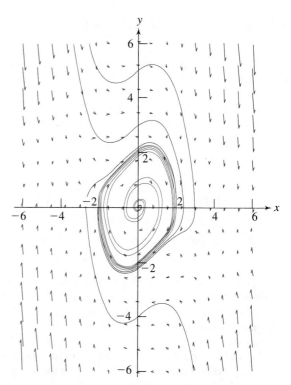

FIGURE 10.55 *Phase portrait for van der Pol's equation for $\alpha = 0.5$.*

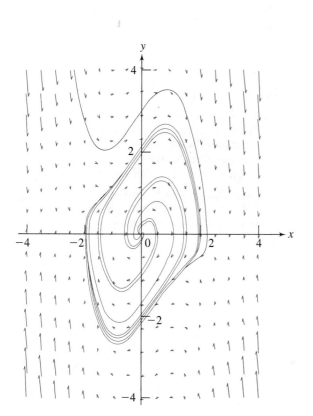

FIGURE 10.56 *Phase portrait for van der Pol's equation for $\alpha = 1$.*

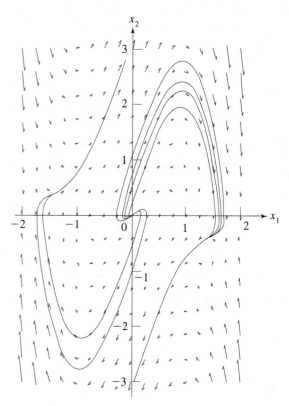

FIGURE 10.57 *Phase portrait for van der Pol's equation for $\alpha = 3$.*

In each of Problems 6 through 9, use a conversion to polar coordinates (see Problem 5) to find all of the closed trajectories of the system and determine which of these are limit cycles. Also classify the stability of each limit cycle. Generate a phase portrait for each system and attempt to identify the limit cycles.

6. $x' = 4y + x \sin(\sqrt{x^2 + y^2})$,

    $y' = -x + y \sin(\sqrt{x^2 + y^2})$

7. $x' = y(1 - x^2 - y^2)$, $y' = -x(1 - x^2 - y^2)$

8. $x' = x(1 - x^2 - y^2)$, $y' = y(1 - x^2 - y^2)$

9. $x' = y + x(1 - x^2 - y^2)(4 - x^2 - y^2)(9 - x^2 - y^2)$,

    $y' = -x + y(1 - x^2 - y^2)(4 - x^2 - y^2)(9 - x^2 - y^2)$

In each of Problems 10 through 13, use the Poincaré–Bendixson theorem to establish the existence of a closed trajectory of the system. In each problem, find an annular region R (region between two concentric circles) about the origin such that solutions within R remain within R. To do this, check the sign of $xx' + yy'$ on circles bounding the annulus. Generate a phase portrait of the system and attempt to identify closed trajectories.

10. $x' = x - y - x\sqrt{x^2 + y^2}$, $y' = x + y - y\sqrt{x^2 + y^2}$

11. $x' = 4x - 4y - x(x^2 + 9y^2)$,

    $y' = 4x + 4y - y(x^2 + 9y^2)$

12. $x' = y$, $x' = -x + y - y(x^2 + 2y^2)$

13. $x' = 4x - 2y - x(4x^2 + y^2)$,

    $y' = 2x + 4y - y(4x^2 + y^2)$

In each of Problems 14 through 22, determine whether the system has a closed trajectory. Generate a phase portrait for the system and attempt to find closed trajectories.

14. $x' = 3x + 4xy + xy^2$, $y' = -2y^2 + x^4y$

15. $x' = -y + x + x(x^2 + y^2)$, $y' = x + y + y(x^2 + y^2)$

16. $x' = -y^2$, $y' = 3x + 2x^3$

17. $x' = y$, $y' = x^2 + e^{\sin(x)}$

18. $x' = y$, $y' = -x + y - x^2y$

19. $x' = x - 5y + y^3$, $y' = x - y + y^3 + 7y^5$

20. $x' = y$, $y' = -x + ye^{-y}$

21. $x' = y$, $y' = -x^3$

22. $x' = 9x - 5y + x(x^2 + 9y^2)$,

    $y' = 5x + 9y - y(x^2 + 9y^2)$

23. Suppose that a solution of the autonomous system $x' = F(x, y)$, $y' = G(x, y)$ has a closed trajectory $C : x = \varphi(t)$, $y = \psi(t)$. Let $\varphi(t_0) = a$ and $\psi(t_0) = b$. Prove that there exists a positive number T such that $\varphi(t_0 + T) = a$ and $\psi(t_0 + T) = b$. (We have asserted that closed trajectories are periodic solutions but have not actually given a proof).

A differential equation $x(t)$ has a periodic solution if there is a solution $x = \mu(t)$ and a positive number T such that $\mu(t + T) = \mu(t)$ for all t. In each of Problems 24 through 30, prove that the differential equation has a periodic solution by converting it to a system and using theorems from this section. Generate a phase portrait for this system and attempt to identify a closed trajectory, which represents a periodic solution.

24. $x'' + (5x^4 - 12x^2)x' + 4x^3 = 0$

25. $x'' + (x^2 - 1)x' + 2\sin(x) = 0$

26. $x'' + (5x^4 + 9x^2 - 4x)x' + \sinh(x) = 0$

27. $x'' + x^3 = 0$

28. $x'' + 4x = 0$

29. $x'' + \dfrac{x}{1 + x^2} = 0$

30. $x'' + 2(x^2 - 1)x' + 5x^3 = 0$

31. Use Bendixson's theorem to show that the van der Pol equation does not have a closed trajectory whose graph is completely contained in any of the following regions: (a) the infinite strip $-1 < x < 1$, (b) the half-plane $x \geq 1$, or (c) the half-plane $x \leq -1$.

32. Prove that the unique closed trajectory of van der Pol's equation meets at least one of the lines $x = 1$ or $x = -1$ in two distinct points.

PART 4

Vector Analysis

CHAPTER 11
Vector Differential Calculus

CHAPTER 12
Vector Integral Calculus

The next two chapters combine vector algebra and geometry with the derivative and integral processes of calculus.

Much of science and engineering deals with the analysis of forces—the force of water on a dam, air turbulence on a wing, tension on bridge supports, wind and weight stresses on buildings, and so on. These forces do not occur in a static state, but vary with position, time and usually a variety of conditions. This leads to the use of vectors that are functions of one or more variables.

Our treatment of vectors is in two parts—vector differential calculus (Chapter 11), and vector integral calculus (Chapter 12).

Vector differential calculus extends out ability to analyze motion problems from the real line to curves and surfaces in 3-space. Tools such as the directional derivative, divergence and curl of a vector, and gradient play significant roles in many applications.

Vector integral calculus generalizes integration to curves and surfaces in 3-space. This will pay many dividends, including the computation of quantities such as mass, center of mass, work, and flux of a vector field, as well as physical interpretations of vector operations. The main results are the integral theorems of Green, Gauss, and Stokes, which have broad applications in such areas as potential theory and the derivation and solution of partial differential equations modeling physical processes.

CHAPTER 11

Vector Differential Calculus

11.1 Vector Functions of One Variable

In vector analysis we deal with functions involving vectors. We will begin with one such class of functions.

DEFINITION 11.1 *Vector Function of One Variable*

A vector function of one variable is a vector, each component of which is a function of the same single variable.

Such a function typically has the appearance

$$\mathbf{F}(t) = x(t)\mathbf{i} + y(t)\mathbf{j} + z(t)\mathbf{k},$$

in which $x(t)$, $y(t)$, and $z(t)$ are the component functions of $\mathbf{F}$. For each t such that the components are defined, $\mathbf{F}(t)$ is a vector. For example, if

$$\mathbf{F}(t) = \cos(t)\mathbf{i} + 2t^2\mathbf{j} + 3t\mathbf{k}, \tag{11.1}$$

then $\mathbf{F}(0) = \mathbf{i}$, $\mathbf{F}(\pi) = -\mathbf{i} + 2\pi^2\mathbf{j} + 3\pi\mathbf{k}$, and $\mathbf{F}(-3) = \cos(-3)\mathbf{i} + 18\mathbf{j} - 9\mathbf{k}$.

A vector function is *continuous at* t_0 if each component function is continuous at t_0. A vector function is *continuous* if each component function is continuous (for those values of t for which they are all defined). For example,

$$\mathbf{G}(t) = \frac{1}{t-1}\mathbf{i} + \ln(t)\mathbf{k}$$

is continuous for all $t > 0$ with $t \neq 1$.

The derivative of a vector function is the vector function formed by differentiating each component. With the function $\mathbf{F}$ of (11.1),

$$\mathbf{F}'(t) = -\sin(t)\mathbf{i} + 4t\mathbf{j} + 3\mathbf{k}.$$

A vector function is differentiable if it has a derivative for all t for which it is defined. The vector function $\mathbf{G}$ just defined is differentiable for t positive and different from 1, and

$$\mathbf{G}'(t) = \frac{-1}{(t-1)^2}\mathbf{i} + \frac{1}{t}\mathbf{k}.$$

We may think of $\mathbf{F}(t)$ as an arrow extending from the origin to $(x(t), y(t), z(t))$. Since $\mathbf{F}(t)$ generally varies with t, we must think of this arrow as having adjustable length and pivoting at the origin to swing about as $(x(t), y(t), z(t))$ moves. In this way, the arrow sweeps out a curve in 3-space as t varies. This curve has parametric equations $x = x(t)$, $y = y(t)$, $z = z(t)$. $\mathbf{F}(t)$ is called a *position vector* for this curve. Figure 11.1 shows a typical such curve.

The derivative of the position vector is the tangent vector to this curve. To see why this is true, observe from Figure 11.2 and the parallelogram law that the vector

$$\mathbf{F}(t_0 + \Delta t) - \mathbf{F}(t_0)$$

is represented by the arrow from $(x(t_0), y(t_0), z(t_0))$ to $(x(t_0 + \Delta t), y(t_0 + \Delta t), z(t_0 + \Delta t))$. Since Δt is a nonzero scalar, the vector

$$\frac{1}{\Delta t}[\mathbf{F}(t_0 + \Delta t) - \mathbf{F}(t_0)] \tag{11.2}$$

is along the line between these points. In terms of components, this vector is

$$\frac{x(t_0 + \Delta t) - x(t_0)}{\Delta t}\mathbf{i} + \frac{y(t_0 + \Delta t) - y(t_0)}{\Delta t}\mathbf{j} + \frac{z(t_0 + \Delta t) - z(t_0)}{\Delta t}\mathbf{k}.$$

In the limit as $\Delta t \to 0$, this vector approaches $x'(t_0)\mathbf{i} + y'(t_0)\mathbf{j} + z'(t_0)\mathbf{k}$, which is $\mathbf{F}'(t_0)$. In this limit, the vector (11.2) moves into a position tangent to the curve at $(x(t_0), y(t_0), z(t_0))$, as suggested by Figure 11.3. This leads us to interpret $\mathbf{F}'(t_0)$ as the tangent vector to the curve at $(x(t_0), y(t_0), z(t_0))$. This assumes that $\mathbf{F}'(t_0)$ is not the zero vector, which has no direction.

We usually represent the tangent vector $\mathbf{F}'(t_0)$ as an arrow from the point $(x(t_0), y(t_0), z(t_0))$ on the curve having $\mathbf{F}(t)$ as position vector.

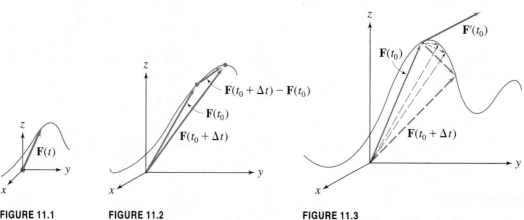

FIGURE 11.1
Position vector for a curve.

FIGURE 11.2

FIGURE 11.3
$$\mathbf{F}'(t_0) = \lim_{\Delta t \to 0} \frac{\mathbf{F}(t_0 + \Delta t) - \mathbf{F}(t_0)}{\Delta t}.$$

EXAMPLE 11.1

Let

$$\mathbf{H}(t) = t^2\mathbf{i} + \sin(t)\mathbf{j} - t^2\mathbf{k}.$$

$\mathbf{H}(t)$ is the position vector for the curve given parametrically by $x(t) = t^2$, $y(t) = \sin(t)$, $z(t) = -t^2$, part of whose graph is given in Figure 11.4. The tangent vector is

$$\mathbf{H}'(t) = 2t\mathbf{i} + \cos(t)\mathbf{j} - 2t\mathbf{k}.$$

The tangent vector at the origin is

$$\mathbf{H}'(0) = \mathbf{j}.$$

The tangent vector at $(1, \sin(1), -1)$ is

$$\mathbf{H}'(1) = 2\mathbf{i} + \cos(1)\mathbf{j} - 2\mathbf{k}. \ \blacksquare$$

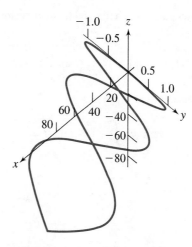

FIGURE 11.4 *Part of the graph of*
$x = t^2,\ y = \sin(t),\ z = -t^2.$

From calculus, we know that the length of a curve given parametrically by $x = x(t)$, $y = y(t)$, $z = z(t)$ for $a \leq t \leq b$ is

$$\text{length} = \int_a^b \sqrt{(x'(t))^2 + (y'(t))^2 + (z'(t))^2}\ dt$$

in which it is assumed that x', y', and z' are continuous on $[a, b]$. Now

$$\|\mathbf{F}'(t)\| = \sqrt{(x'(t))^2 + (y'(t))^2 + (z'(t))^2}$$

is the length of the tangent vector. Thus, in terms of the position vector $\mathbf{F}(t) = x(t)\mathbf{i} + y(t)\mathbf{j} + z(t)\mathbf{k}$,

$$\text{length} = \int_a^b \|\mathbf{F}'(t)\|\ dt.$$

The length of a curve having a tangent at each point is the integral of the length of the tangent vector over the curve.

EXAMPLE 11.2

Consider the curve given by the parametric equations

$$x = \cos(t), \quad y = \sin(t), \quad z = \frac{t}{3}$$

for $-4\pi \leq t \leq 4\pi$. The position vector for this curve is

$$\mathbf{F}(t) = \cos(t)\mathbf{i} + \sin(t)\mathbf{j} + \frac{1}{3}t\mathbf{k}.$$

The graph of the curve is part of a helix wrapping around the cylinder $x^2 + y^2 = 1$, centered about the z axis. The tangent vector at any point is

$$\mathbf{F}'(t) = -\sin(t)\mathbf{i} + \cos(t)\mathbf{j} + \frac{1}{3}\mathbf{k}.$$

Figure 11.5 shows part of the helix and tangent vectors at various points. The length of the tangent vector is

$$\|\mathbf{F}'(t)\| = \sqrt{\sin^2(t) + \cos^2(t) + \frac{1}{9}} = \frac{1}{3}\sqrt{10}.$$

The length of this curve is

$$\text{length} = \int_{-4\pi}^{4\pi} \|\mathbf{F}'(t)\| \, dt = \int_{-4\pi}^{4\pi} \frac{1}{3}\sqrt{10} \, dt = \frac{8\pi}{3}\sqrt{10}. \; \blacksquare$$

FIGURE 11.5 *Part of a circular helix and some of its tangent vectors.*

Sometimes it is convenient to write the position vector of a curve in such a way that the tangent vector at each point has length 1. Such a tangent is called a *unit tangent*. We will show how this can be done (at least in theory) if the coordinate functions of the curve have continuous

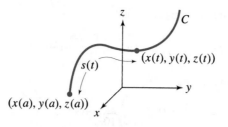

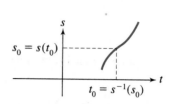

FIGURE 11.6 *Length function along a curve.*

FIGURE 11.7 *A length function has an inverse.*

derivatives. Let

$$\mathbf{F}(t) = x(t)\mathbf{i} + y(t)\mathbf{j} + z(t)\mathbf{k}$$

for $a \le t \le b$, and suppose x', y', and z' are continuous. Define the real-valued function

$$s(t) = \int_a^t \|\mathbf{F}'(\xi)\| \, d\xi.$$

As suggested by Figure 11.6, $s(t)$ is the length of the part of the curve from its initial point $(x(a), y(a), z(a))$ to $(x(t), y(t), z(t))$. As t moves from a to b, $s(t)$ increases from $s(a) = 0$ to $s(b) = L$, which is the total length of the curve. By the fundamental theorem of calculus, $s(t)$ is differentiable wherever $\mathbf{F}(t)$ is continuous, and

$$\frac{ds}{dt} = \|\mathbf{F}'(t)\| = \sqrt{x'(t)^2 + y'(t)^2 + z'(t)^2}.$$

Because s is strictly increasing as a function of t, we can, at least in theory, solve for t in terms of s, giving the inverse function $t(s)$ (see Figure 11.7). Now define

$$\mathbf{G}(s) = \mathbf{F}(t(s)) = x(t(s))\mathbf{i} + y(t(s))\mathbf{j} + z(t(s))\mathbf{k}$$

for $0 \le s \le L$. Then $\mathbf{G}$ is a position function for the same curve as $\mathbf{F}$. As t varies from a to b, $\mathbf{F}(t)$ sweeps out the same curve as $\mathbf{G}(s)$, as s varies from 0 to L. However, $\mathbf{G}$ has the advantage that the tangent vector $\mathbf{G}'$ always has length 1. To see this, use the chain rule to compute

$$\mathbf{G}'(s) = \frac{d}{ds}\mathbf{F}(t(s)) = \frac{d}{dt}\mathbf{F}(t)\frac{dt}{ds}$$

$$= \frac{1}{ds/dt}\mathbf{F}'(t) = \frac{1}{\|\mathbf{F}'(t)\|}\mathbf{F}'(t),$$

and this vector has length 1.

EXAMPLE 11.3

Consider again the helix having position function

$$\mathbf{F}(t) = \cos(t)\mathbf{i} + \sin(t)\mathbf{j} + \frac{1}{3}t\mathbf{k}$$

for $-4\pi \le t \le 4\pi$. We have already calculated

$$\|\mathbf{F}'(t)\| = \frac{1}{3}\sqrt{10}.$$

Therefore, the length function along this curve is

$$s(t) = \int_{-4\pi}^{t} \frac{1}{3}\sqrt{10}\, d\xi = \frac{1}{3}\sqrt{10}(t + 4\pi).$$

Solve for the inverse function to write

$$t = t(s) = \frac{3}{\sqrt{10}}s - 4\pi.$$

Substitute this into the position vector to define

$$\mathbf{G}(s) = \mathbf{F}(t(s)) = \mathbf{F}\left(\frac{3}{\sqrt{10}}s - 4\pi\right)$$

$$= \cos\left(\frac{3}{\sqrt{10}}s - 4\pi\right)\mathbf{i} + \sin\left(\frac{3}{\sqrt{10}}s - 4\pi\right)\mathbf{j} + \frac{1}{3}\left(\frac{3}{\sqrt{10}}s - 4\pi\right)\mathbf{k}$$

$$= \cos\left(\frac{3}{\sqrt{10}}s\right)\mathbf{i} + \sin\left(\frac{3}{\sqrt{10}}s\right)\mathbf{j} + \left(\frac{1}{\sqrt{10}}s - \frac{4}{3}\pi\right)\mathbf{k}.$$

Now compute

$$\mathbf{G}'(s) = -\frac{3}{\sqrt{10}}\sin\left(\frac{3}{\sqrt{10}}s\right)\mathbf{i} + \frac{3}{\sqrt{10}}\cos\left(\frac{3}{\sqrt{10}}s\right)\mathbf{j} + \frac{1}{\sqrt{10}}\mathbf{k}.$$

This is a tangent vector to the helix, and it has length 1. ∎

Assuming that the derivatives exist, then

1. $[\mathbf{F}(t) + \mathbf{G}(t)]' = \mathbf{F}'(t) + \mathbf{G}'(t)$.
2. $[f(t)\mathbf{F}(t)]' = f'(t)\mathbf{F}(t) + f(t)\mathbf{F}'(t)$ if f is a differentiable real-valued function.
3. $[\mathbf{F}(t) \cdot \mathbf{G}(t)]' = \mathbf{F}'(t) \cdot \mathbf{G}(t) + \mathbf{F}(t) \cdot \mathbf{G}'(t)$.
4. $[\mathbf{F}(t) \times \mathbf{G}(t)]' = \mathbf{F}'(t) \times \mathbf{G}(t) + \mathbf{F}(t) \times \mathbf{G}'(t)$.
5. $[\mathbf{F}(f(t))]' = f'(t)\mathbf{F}'(f(t))$.

Items (2), (3) and (4) are all "product rules." Rule (2) is for the derivative of a product of a scalar function with a vector function; (3) is for the derivative of a dot product; and (4) is for the derivative of a cross product. In each case, the rule has the same form as the familiar calculus formula for the derivative of a product of two functions. However, in (4), order is important, since $\mathbf{F} \times \mathbf{G} = -\mathbf{G} \times \mathbf{F}$. Rule (5) is a vector version of the chain rule.

In the next section we will use vector functions to develop the concepts of velocity and acceleration, which we will apply to the geometry of curves in 3-space.

SECTION 11.1 PROBLEMS

In each of Problems 1 through 8, compute the requested derivative (a) by carrying out the vector operation and differentiating the resulting vector or scalar, and (b) by using one of the differentiation rules (1) through (5) stated at the end of this section.

1. $\mathbf{F}(t) = \mathbf{i} + 3t^2\mathbf{j} + 2t\mathbf{k}$, $f(t) = 4\cos(3t)$;
   $(d/dt)[f(t)\mathbf{F}(t)]$

2. $\mathbf{F}(t) = t\mathbf{i} - 3t^2\mathbf{k}$, $\mathbf{G}(t) = \mathbf{i} + \cos(t)\mathbf{k}$;
   $(d/dt)[\mathbf{F}(t) \cdot \mathbf{G}(t)]$

3. $\mathbf{F}(t) = t\mathbf{i} + \mathbf{j} + 4\mathbf{k}$, $\mathbf{G}(t) = \mathbf{i} - \cos(t)\mathbf{j} + t\mathbf{k}$;
   $(d/dt)[\mathbf{F}(t) \times \mathbf{G}(t)]$

4. $\mathbf{F}(t) = \sinh(t)\mathbf{j} - t\mathbf{k}$, $\mathbf{G}(t) = t\mathbf{i} + t^2\mathbf{j} - t^2\mathbf{k}$;
   $(d/dt)[\mathbf{F}(t) \times \mathbf{G}(t)]$

5. $\mathbf{F}(t) = t\mathbf{i} - \cosh(t)\mathbf{j} + e^t\mathbf{k}$, $f(t) = 1 - 2t^3$; $(d/dt)[f(t)\mathbf{F}(t)]$

6. $\mathbf{F}(t) = t\mathbf{i} - t\mathbf{j} + t^2\mathbf{k}$, $\mathbf{G}(t) = \sin(t)\mathbf{i} - 4\mathbf{j} - t^3\mathbf{k}$; $(d/dt)[\mathbf{F}(t) \cdot \mathbf{G}(t)]$

7. $\mathbf{F}(t) = -9\mathbf{i} + t^2\mathbf{j} + t^2\mathbf{k}$, $\mathbf{G}(t) = e^t\mathbf{i}$; $(d/dt)[\mathbf{F} \times \mathbf{G}(t)]$

8. $\mathbf{F}(t) = -4\cos(t)\mathbf{k}$, $\mathbf{G}(t) = -t^2\mathbf{i} + 4\sin(t)\mathbf{k}$; $(d/dt)[\mathbf{F}(t) \cdot \mathbf{G}(t)]$

9. Let $\mathbf{F}(t) = -t^2\mathbf{i} + \cos(t)\mathbf{k}$ and $f(t) = 2t^2$.

(a) Calculate $\mathbf{F}(f(t))$ and use this to compute $(d/dt)[\mathbf{F}(f(t))]$.

(b) Use the chain rule to compute $(d/dt)[\mathbf{F}(f(t))]$ and compare the result with that of (a).

10. Let $\mathbf{F}(t) = -(2t + \cos(t))\mathbf{i} + 4\mathbf{j} - t^4\mathbf{k}$ and $f(t) = 1/t$.

(a) Calculate $\mathbf{F}(f(t))$ and use this to compute $(d/dt)[\mathbf{F}(f(t))]$.

(b) Calculate $(d/dt)[\mathbf{F}(f(t))]$ by the chain rule and compare the result with that of (a).

11. Prove that $(\mathbf{F} + \mathbf{G})' = \mathbf{F}' + \mathbf{G}'$.

12. Prove that $(\mathbf{F} \times \mathbf{G})' = \mathbf{F}' \times \mathbf{G} + \mathbf{F} \times \mathbf{G}'$.

13. Prove that $(\mathbf{F} \cdot \mathbf{G})' = \mathbf{F}' \cdot \mathbf{G} + \mathbf{F} \cdot \mathbf{G}'$.

In each of Problems 14 through 17, (a) write the position vector and tangent vector for the curve whose parametric equations are given, (b) find a length function $s(t)$ for the curve, (c) write the position vector as a function of s, and

(d) verify that the resulting position vector has a derivative of length 1.

14. $x = t$, $y = \cosh(t)$, $z = 1$; $(0 \le t \le \pi)$

15. $x = \sin(t)$, $y = \cos(t)$, $z = 45t$; $(0 \le t \le 2\pi)$

16. $x = y = z = t^3$; $(-1 \le t \le 1)$

17. $x = 2t^2$, $y = 3t^2$, $z = 4t^2$; $(1 \le t \le 3)$

18. Let $\mathbf{F}(t) = x(t)\mathbf{i} + y(t)\mathbf{j} + z(t)\mathbf{k}$. Suppose x, y, and z are differentiable functions of t. Think of $\mathbf{F}(t)$ as the position function of a particle moving along a curve in 3-space. Suppose $\mathbf{F} \times \mathbf{F}' = \mathbf{O}$. Prove that the particle always moves in the same direction.

19. Let $\mathbf{F}(t) = x(t)\mathbf{i} + y(t)\mathbf{j} + z(t)\mathbf{k}$ with x, y, and z twice differentiable. Think of $\mathbf{F}(t)$ as the position vector of a particle.

(a) Prove that $\|\mathbf{F}(t) \times [\mathbf{F}(t + \Delta t) - \mathbf{F}(t)]\|$ is equal to twice the area of the sector bounded by the curve, $\mathbf{F}(t)$ and $\mathbf{F}(t + \Delta t) - \mathbf{F}(t)$.

(b) Suppose $\mathbf{F} \times \mathbf{F}'' = \mathbf{O}$. Prove that the particle moves so that equal areas are swept out by the position vector in equal times. Compare this result with Kepler's second law.

20. Assuming that $\mathbf{F}$, $\mathbf{G}$, and $\mathbf{H}$ are differentiable vector functions, derive a formula for $(d/dt)[\mathbf{F}, \mathbf{G}, \mathbf{H}]$, in which $[\mathbf{F}, \mathbf{G}, \mathbf{H}]$ is the scalar triple product of these vectors. (See Problem 39 of Section 5.3 for the scalar triple product).

11.2   Velocity, Acceleration, Curvature, and Torsion

Imagine a particle moving along a path having position vector $\mathbf{F}(t) = x(t)\mathbf{i} + y(t)\mathbf{j} + z(t)\mathbf{k}$ as t varies from a to b. We want to relate $\mathbf{F}$ to the dynamics of the particle.

For calculations we will be doing, we assume that x, y, and z are twice differentiable. We will also make use of the distance function along the curve,

$$s(t) = \int_a^t \|\mathbf{F}'(\xi)\| \, d\xi.$$

DEFINITION 11.2   *Velocity, Speed*

1. The velocity $\mathbf{v}(t)$ of the particle at time t is defined to be

$$\mathbf{v}(t) = \mathbf{F}'(t).$$

2. The speed $v(t)$ of the particle at time t is the magnitude of the velocity.

Velocity is therefore a vector, having magnitude and direction. If $\mathbf{v}(t)$ is not the zero vector, then the velocity is tangent to the curve of motion of the particle. Thus, at any instant the particle may be thought of as moving in the direction of the tangent to the path of motion.

The speed at time t is a real-valued function, given by

$$v(t) = \|\mathbf{v}(t)\| = \|\mathbf{F}'(t)\| = \frac{ds}{dt}.$$

This is consistent with the familiar idea of speed as the rate of change of distance (along the path of motion) with respect to time.

DEFINITION 11.3 *Acceleration*

The acceleration $\mathbf{a}(t)$ of the particle is the rate of change of the velocity with respect to time:

$$\mathbf{a}(t) = \mathbf{v}'(t).$$

Alternatively,

$$\mathbf{a}(t) = \mathbf{F}''(t).$$

As with velocity, acceleration is a vector.

EXAMPLE 11.4

Let

$$\mathbf{F}(t) = \sin(t)\mathbf{i} + 2e^{-t}\mathbf{j} + t^2\mathbf{k}.$$

The path of the particle is the curve whose parametric equations are

$$x = \sin(t), \quad y = 2e^{-t}, \quad z = t^2.$$

Part of the graph of this curve is shown in Figure 11.8. The velocity and acceleration are, respectively,

$$\mathbf{v}(t) = \cos(t)\mathbf{i} - 2e^{-t}\mathbf{j} + 2t\mathbf{k}$$

and

$$\mathbf{a}(t) = -\sin(t)\mathbf{i} + 2e^{-t}\mathbf{j} + 2\mathbf{k}.$$

The speed of the particle is

$$v(t) = \sqrt{\cos^2(t) + 4e^{-2t} + 4t^2}. \quad \blacksquare$$

If $\mathbf{F}'(t)$ is not the zero vector, then this vector is tangent to the curve at $(x(t), y(t), z(t))$. We may obtain a unit tangent by dividing this vector by its length:

$$\mathbf{T}(t) = \frac{1}{\|\mathbf{F}'(t)\|}\mathbf{F}'(t) = \frac{1}{ds/dt}\mathbf{F}'(t).$$

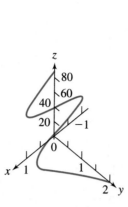

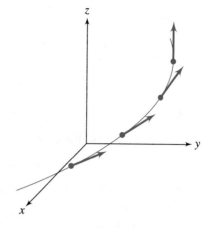

FIGURE 11.8 *Part of the graph of $x = \sin(t)$, $y = 2e^{-t}$, $z = t^2$.*

FIGURE 11.9 *Increasing curvature corresponds to an increasing rate of change of the tangent vector along the curve.*

Equivalently,

$$\mathbf{T}(t) = \frac{1}{\|\mathbf{v}(t)\|}\mathbf{v}(t) = \frac{1}{v(t)}\mathbf{v}(t),$$

the velocity divided by the speed. If arc length s along the path of motion is used as the parameter in the position function, then we have seen that $\|\mathbf{F}'(s)\| = 1$ automatically, so in this case the speed is identically 1 and this unit tangent is just the velocity vector. We will use this unit tangent to define a function that quantifies the "amount of bending" of a curve at a point.

DEFINITION 11.4 Curvature

The curvature κ of a curve is the magnitude of the rate of change of the unit tangent with respect to arc length along the curve:

$$\kappa(s) = \left\| \frac{d\mathbf{T}}{ds} \right\|.$$

The definition is motivated by the intuition (Figure 11.9) that the more a curve bends at a point, the faster the tangent vector is changing there.

If the unit tangent vector has been written using s as parameter, then computing $\mathbf{T}'(s)$ is straightforward. More often, however, the unit tangent is parametrized by some other variable, and then the derivative defining the curvature must be computed by the chain rule:

$$\kappa(t) = \left\| \frac{d\mathbf{T}}{dt}\frac{dt}{ds} \right\|.$$

This gives the curvature as a function of the parameter used in the position vector. Since

$$\frac{dt}{ds} = \frac{1}{ds/dt} = \frac{1}{\|\mathbf{F}'(t)\|},$$

we often write

$$\kappa(t) = \frac{1}{\|\mathbf{F}'(t)\|} \|\mathbf{T}'(t)\| . \qquad (11.3)$$

EXAMPLE 11.5

Consider a straight line having parametric equations

$$x = a + bt, \quad y = c + dt, \quad z = e + ht,$$

in which a, b, c, d, e, and h are constants. The position vector of this line is

$$\mathbf{F}(t) = (a + bt)\mathbf{i} + (c + dt)\mathbf{j} + (e + ht)\mathbf{k}.$$

We will compute the curvature using equation (11.3). First,

$$\mathbf{F}'(t) = b\mathbf{i} + d\mathbf{j} + h\mathbf{k},$$

so

$$\|\mathbf{F}'(t)\| = \sqrt{b^2 + d^2 + h^2}.$$

The unit tangent vector is

$$\mathbf{T}(t) = \frac{1}{\|\mathbf{F}'(t)\|} \mathbf{F}(t)$$

$$= \frac{1}{\sqrt{b^2 + d^2 + h^2}} (b\mathbf{i} + d\mathbf{j} + h\mathbf{k}).$$

This is a constant vector, so $\mathbf{T}'(t) = \mathbf{O}$ and the curvature is

$$\kappa(t) = \frac{1}{\|\mathbf{F}'(t)\|} \|\mathbf{T}'(t)\| = 0.$$

This is consistent with our intuition that a straight line should have curvature zero. ∎

EXAMPLE 11.6

Let C be the circle of radius 4 about the origin in the plane $y = 3$. Using polar coordinates, this curve has parametric equations

$$x = 4\cos(\theta), \quad y = 3, \quad z = 4\sin(\theta)$$

for $0 \le t \le 2\pi$. The circle has position vector

$$\mathbf{F}(\theta) = 4\cos(\theta)\mathbf{i} + 3\mathbf{j} + 4\sin(\theta)\mathbf{k}.$$

Then

$$\mathbf{F}'(\theta) = -4\sin(\theta)\mathbf{i} + 4\cos(\theta)\mathbf{k},$$

so

$$\|\mathbf{F}'(\theta)\| = 4.$$

The unit tangent is

$$\mathbf{T}(\theta) = \frac{1}{4}[-4\sin(\theta)\mathbf{i} + 4\cos(\theta)\mathbf{k}] = -\sin(\theta)\mathbf{i} + \cos(\theta)\mathbf{k}.$$

Then

$$\mathbf{T}'(\theta) = -\cos(\theta)\mathbf{i} - \sin(\theta)\mathbf{k}.$$

The curvature is

$$\kappa(\theta) = \frac{1}{\|\mathbf{F}'(\theta)\|}\,\|\mathbf{T}'(\theta)\| = \frac{1}{4}.$$

The curvature of this circle is constant, again consistent with intuition. ■

One can show that a circle of radius r has curvature $1/r$. Not only does a circle have constant curvature but this curvature also decreases the larger the radius is chosen, as we should expect.

EXAMPLE 11.7

Let C have parametric representation

$$x = \cos(t) + t\sin(t), \quad y = \sin(t) - t\cos(t), \quad z = t^2$$

for $t > 0$. Figure 11.10 shows part of the graph of C. We will compute the curvature.

We can write the position vector

$$\mathbf{F}(t) = [\cos(t) + t\sin(t)]\,\mathbf{i} + [\sin(t) - t\cos(t)]\,\mathbf{j} + t^2\mathbf{k}.$$

A tangent vector is given by

$$\mathbf{F}'(t) = t\cos(t)\mathbf{i} + t\sin(t)\mathbf{j} + 2t\mathbf{k}$$

and

$$\|\mathbf{F}'(t)\| = \sqrt{5}\,t.$$

Next, the unit tangent vector is

$$\mathbf{T}(t) = \frac{1}{\|\mathbf{F}'(t)\|}\mathbf{F}'(t) = \frac{1}{\sqrt{5}}\,[\cos(t)\mathbf{i} + \sin(t)\mathbf{j} + 2\mathbf{k}]\,.$$

Compute

$$\mathbf{T}'(t) = \frac{1}{\sqrt{5}}\,[-\sin(t)\mathbf{i} + \cos(t)\mathbf{j}]\,.$$

We can now use equation (11.3) to compute

$$\kappa(t) = \frac{1}{\|\mathbf{F}'(t)\|}\,\|\mathbf{T}'(t)\|$$

$$= \frac{1}{\sqrt{5}t}\sqrt{\tfrac{1}{5}\left[\sin^2(t) + \cos^2(t)\right]} = \frac{1}{5t}$$

for $t > 0$. ■

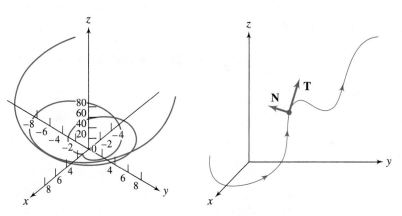

FIGURE 11.10 *Part of the graph of* $x = \cos(t) + t \sin(t)$, $y = \sin(t) - t \cos(t)$, $z = t^2$.

FIGURE 11.11 *Tangent and normal vector to a curve at a point.*

We will now introduce another vector of interest in studying motion along a curve.

DEFINITION 11.5 *Unit Normal Vector*

Using arc length s as parameter on the curve, the unit normal vector $\mathbf{N}(s)$ is defined by

$$\mathbf{N}(s) = \frac{1}{\kappa(s)} \mathbf{T}'(s),$$

provided that $\kappa(s) \neq 0$.

The name given to this vector is motivated by two observations. First, $\mathbf{N}(s)$ is a unit vector. Since $\kappa(s) = \left\| \mathbf{T}'(s) \right\|$, then

$$\left\| \mathbf{N}(s) \right\| = \frac{1}{\left\| \mathbf{T}'(s) \right\|} \left\| \mathbf{T}'(s) \right\| = 1.$$

Second, $\mathbf{N}(s)$ is orthogonal to the unit tangent vector. To see this, begin with the fact that $\mathbf{T}(s)$ is a unit tangent, hence $\left\| \mathbf{T}(s) \right\| = 1$. Then

$$\left\| \mathbf{T}(s) \right\|^2 = \mathbf{T}(s) \cdot \mathbf{T}(s) = 1.$$

Differentiate this equation to get

$$\mathbf{T}'(s) \cdot \mathbf{T}(s) + \mathbf{T}(s) \cdot \mathbf{T}'(s) = 2\mathbf{T}(s) \cdot \mathbf{T}'(s) = 0,$$

hence

$$\mathbf{T}(s) \cdot \mathbf{T}'(s) = 0,$$

which means that $\mathbf{T}(s)$ is orthogonal to $\mathbf{T}'(s)$. But $\mathbf{N}(s)$ is a positive scalar multiple of $\mathbf{T}'(s)$ and so is in the same direction as $\mathbf{T}'(s)$. We conclude that $\mathbf{N}(s)$ is orthogonal to $\mathbf{T}(s)$.

At any point of a curve with differentiable coordinate functions (not all vanishing for the same parameter value), we may now place a tangent vector to the curve, and a normal vector that is perpendicular to the tangent vector (Figure 11.11).

EXAMPLE 11.8

Consider again the curve with position function

$$\mathbf{F}(t) = [\cos(t) + t\sin(t)]\,\mathbf{i} + [\sin(t) - t\cos(t)]\,\mathbf{j} + t^2\mathbf{k}$$

for $t > 0$. In Example 11.7 we computed the unit tangent and the curvature as functions of t. We will write the position vector as a function of arc length and compute the unit tangent $\mathbf{T}(s)$ and the unit normal $\mathbf{N}(s)$.

First, using $\|\mathbf{F}'(t)\| = \sqrt{5}\,t$ from Example 11.7,

$$s(t) = \int_0^t \|\mathbf{F}'(\xi)\|\,d\xi = \int_0^t \sqrt{5}\,\xi\,d\xi = \frac{\sqrt{5}}{2}t^2.$$

Solve for t as a function of s:

$$t = \frac{\sqrt{2}}{5^{1/4}}s = \alpha\sqrt{s},$$

in which $\alpha = \sqrt{2}/5^{1/4}$. In terms of s, the position vector is

$$\mathbf{G}(s) = \mathbf{F}(t(s))$$
$$= \left[\cos(\alpha\sqrt{s}) + \alpha\sqrt{s}\sin(\alpha\sqrt{s})\right]\mathbf{i} + \left[\sin(\alpha\sqrt{s}) - \alpha\sqrt{s}\cos(\alpha\sqrt{s})\right]\mathbf{j} + \alpha^2 s\mathbf{k}.$$

The unit tangent is

$$\mathbf{T}(s) = \mathbf{G}'(s)$$
$$= \frac{1}{2}\alpha^2\cos(\alpha\sqrt{s})\mathbf{i} + \frac{1}{2}\alpha^2\sin(\alpha\sqrt{s})\mathbf{j} + \alpha^2\mathbf{k}.$$

This vector does indeed have length 1:

$$\|\mathbf{T}(s)\|^2 = \frac{1}{4}\alpha^4 + \alpha^4 = \frac{5}{4}\left(\frac{\sqrt{2}}{5^{1/4}}\right)^4 = 1.$$

Now

$$\mathbf{T}'(s) = -\frac{\alpha^3}{4\sqrt{s}}\sin(\alpha\sqrt{s})\mathbf{i} + \frac{\alpha^3}{4\sqrt{s}}\cos(\alpha\sqrt{s})\mathbf{j}$$

so the curvature is

$$\kappa(s) = \|\mathbf{T}'(s)\| = \left(\frac{\alpha^6}{16s}\right)^{1/2} = \frac{1}{4\sqrt{s}}\left(\frac{2^{1/2}}{5^{1/4}}\right)^3 = \frac{1}{5^{3/4}}\frac{1}{\sqrt{2}}\frac{1}{\sqrt{s}}$$

for $s > 0$. Since $s = \sqrt{5}t^2/2$, then in terms of t we have $\kappa = 1/5t$, consistent with Example 11.7.

Now compute the unit normal vector

$$\mathbf{N}(s) = \frac{1}{\kappa(s)}\mathbf{T}'(s) = \frac{4\sqrt{s}}{\alpha^3}\left[-\frac{\alpha^3}{4\sqrt{s}}\sin(\alpha\sqrt{s})\mathbf{i} + \frac{\alpha^3}{4\sqrt{s}}\cos(\alpha\sqrt{s})\mathbf{j}\right]$$
$$= -\sin(\alpha\sqrt{s})\mathbf{i} + \cos(\alpha\sqrt{s})\mathbf{j}.$$

This is a unit vector orthogonal to $\mathbf{T}(s)$. ∎

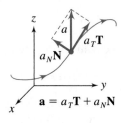

FIGURE 11.12

11.2.1   Tangential and Normal Components of Acceleration

At any point on the trajectory of a particle, the tangent and normal vectors are orthogonal. We will now show how to write the acceleration at a point as a linear combination of the tangent and normal vectors there:

$$\mathbf{a} = a_T \mathbf{T} + a_N \mathbf{N}.$$

This is illustrated in Figure 11.12.

THEOREM 11.1

$$\mathbf{a} = \frac{dv}{dt}\mathbf{T} + v^2 \kappa \mathbf{N}. \ \blacksquare$$

Thus, $a_T = dv/dt$ and $a_N = v^2\kappa$. The tangential component of the acceleration is the derivative of the speed, while the normal component is the curvature at the point, times the square of the speed there.

Proof   First observe that

$$\mathbf{T}(t) = \frac{1}{\|\mathbf{F}'(t)\|}\,\mathbf{F}'(t) = \frac{1}{v}\mathbf{v}.$$

Therefore,

$$\mathbf{v} = v\mathbf{T}.$$

Then

$$\mathbf{a} = \frac{d}{dt}\,\mathbf{v} = \frac{dv}{dt}\mathbf{T} + v\mathbf{T}'(t)$$

$$= \frac{dv}{dt}\mathbf{T} + v\frac{ds}{dt}\frac{d\mathbf{T}}{ds}$$

$$= \frac{dv}{dt}\mathbf{T} + v^2\mathbf{T}'(s)$$

$$= \frac{dv}{dt}\mathbf{T} + v^2\kappa\mathbf{N}. \ \blacksquare$$

Here is one use of this decomposition of the acceleration. Since $\mathbf{T}$ and $\mathbf{N}$ are orthogonal unit vectors, then

$$\|\mathbf{a}\|^2 = \mathbf{a} \cdot \mathbf{a} = (a_T \mathbf{T} + a_N \mathbf{N}) \cdot (a_T \mathbf{T} + a_N \mathbf{N})$$
$$= a_T^2 \mathbf{T} \cdot \mathbf{T} + 2a_T a_N \mathbf{T} \cdot \mathbf{N} + a_N^2 \mathbf{N} \cdot \mathbf{N}$$
$$= a_T^2 + a_N^2.$$

From this, whenever two of $\|\mathbf{a}\|$, a_T, and a_N are known, we can compute the third quantity.

EXAMPLE 11.9

Return again to the curve C having position function

$$\mathbf{F}(t) = [\cos(t) + t \sin(t)]\mathbf{i} + [\sin(t) - t \cos(t)]\mathbf{j} + t^2\mathbf{k}$$

for $t > 0$. We will compute the tangential and normal components of the acceleration. First,

$$\mathbf{v}(t) = \mathbf{F}'(t) = t \cos(t)\mathbf{i} + t \sin(t)\mathbf{j} + 2t\mathbf{k},$$

so the speed is

$$v(t) = \|\mathbf{F}'(t)\| = \sqrt{5}t.$$

The tangential component of the acceleration is therefore

$$a_T = \frac{dv}{dt} = \sqrt{5},$$

a constant for this curve. The acceleration vector is

$$\mathbf{a} = \mathbf{v}' = [\cos(t) - t \sin(t)]\mathbf{i} + [\sin(t) + t \cos(t)]\mathbf{j} + 2\mathbf{k},$$

and a routine calculation gives

$$\|\mathbf{a}\| = \sqrt{5 + t^2}.$$

Then,

$$a_N^2 = \|\mathbf{a}\|^2 - a_T^2 = 5 + t^2 - 5 = t^2.$$

Since $t > 0$, the normal component of acceleration is

$$a_N = t.$$

The acceleration may therefore be written as

$$\mathbf{a} = \sqrt{5}\mathbf{T} + t\mathbf{N}.$$

If we know the normal component a_N and the speed v, it is easy to compute the curvature, since

$$a_N = t = \kappa v^2 = 5t^2 \kappa$$

implies that

$$\kappa = \frac{1}{5t},$$

as we computed in Example 11.7 directly from the tangent vector.

Now the unit tangent and normal vectors are easy to compute in terms of t. First,

$$\mathbf{T}(t) = \frac{1}{v}\mathbf{v} = \frac{1}{\sqrt{5}}[\cos(t)\mathbf{i} + \sin(t)\mathbf{j} + 2\mathbf{k}].$$

This is usually easy to compute, since $\mathbf{v} = \mathbf{F}'(t)$ is a straightforward calculation. But in addition, we now have the unit normal vector (as a function of t)

$$\mathbf{N}(t) = \frac{1}{\kappa} \mathbf{T}'(s) = \frac{1}{\kappa} \frac{dt}{ds} \frac{d\mathbf{T}}{dt} = \frac{1}{\kappa v} \mathbf{T}'(t)$$

$$= \frac{5t}{\sqrt{5}t} \frac{1}{\sqrt{5}} [-\sin(t)\mathbf{i} + \cos(t)\mathbf{j}] = -\sin(t)\mathbf{i} + \cos(t)\mathbf{j}.$$

This calculation does not require the explicit computation of $s(t)$ and its inverse function. ■

11.2.2   Curvature as a Function of t

Equation (11.3) gives the curvature in terms of the parameter t used for the position function:

$$\kappa(t) = \frac{1}{\|\mathbf{F}'(t)\|} \|\mathbf{T}'(t)\|.$$

This is a handy formula because it does not require introduction of the distance function $s(t)$. We will now derive another expression for the curvature that is sometimes useful in calculating κ directly from the position function.

THEOREM 11.2

Let $\mathbf{F}$ be the position function of a curve, and suppose that the components of $\mathbf{F}$ are twice differentiable functions. Then

$$\kappa = \frac{\|\mathbf{F}' \times \mathbf{F}''\|}{\|\mathbf{F}'\|^{3/2}}. \quad ■$$

This states that the curvature is the magnitude of the cross product of the first and second derivatives of $\mathbf{F}$, divided by the cube of the length of $\mathbf{F}'$.

Proof   First write

$$\mathbf{a} = a_T \mathbf{T} + \kappa \left(\frac{ds}{dt}\right)^2 \mathbf{N}.$$

Take the cross product of this equation with the unit tangent vector:

$$\mathbf{T} \times \mathbf{a} = a_T \mathbf{T} \times \mathbf{T} + \kappa \left(\frac{ds}{dt}\right)^2 (\mathbf{T} \times \mathbf{N}) = \kappa \left(\frac{ds}{dt}\right)^2 (\mathbf{T} \times \mathbf{N}),$$

since the cross product of any vector with itself is the zero vector. Then

$$\|\mathbf{T} \times \mathbf{a}\| = \kappa \left(\frac{ds}{dt}\right)^2 \|\mathbf{T} \times \mathbf{N}\|$$

$$= \kappa \left(\frac{ds}{dt}\right)^2 \|\mathbf{T}\| \|\mathbf{N}\| \sin(\theta),$$

where θ is the angle between $\mathbf{T}$ and $\mathbf{N}$. But these are orthogonal unit vectors, so $\theta = \pi/2$ and $\|\mathbf{T}\| = \|\mathbf{N}\| = 1$. Therefore,

$$\|\mathbf{T} \times \mathbf{a}\| = \kappa \left(\frac{ds}{dt}\right)^2.$$

Then

$$\kappa = \frac{\|\mathbf{T} \times \mathbf{a}\|}{(ds/dt)^2}.$$

But $\mathbf{T} = \mathbf{F}'/\|\mathbf{F}'\|$, $\mathbf{a} = \mathbf{T}' = \mathbf{F}''$, and $ds/dt = \|\mathbf{F}'\|$, so

$$\kappa = \frac{1}{\|\mathbf{F}'\|^2} \left\| \frac{1}{\|\mathbf{F}'\|} \mathbf{F}' \times \mathbf{F}'' \right\| = \frac{\|\mathbf{F}' \times \mathbf{F}''\|}{\|\mathbf{F}'\|^3}. \quad \blacksquare$$

EXAMPLE 11.10

Let C have position function

$$\mathbf{F}(t) = t^2\mathbf{i} - t^3\mathbf{j} + t\mathbf{k}.$$

We want the curvature of C. Compute

$$\mathbf{F}'(t) = 2t\mathbf{i} - 3t^2\mathbf{j} + \mathbf{k},$$

$$\mathbf{F}''(t) = 2\mathbf{i} - 6t\mathbf{j},$$

and

$$\mathbf{F}' \times \mathbf{F}'' = \begin{vmatrix} \mathbf{i} & \mathbf{j} & \mathbf{k} \\ 2t & -3t^2 & 1 \\ 2 & -6t & 0 \end{vmatrix}$$

$$= 6t\mathbf{i} + 2\mathbf{j} - 6t^2\mathbf{k}.$$

The curvature is

$$\kappa(t) = \frac{\|6t\mathbf{i} + 2\mathbf{j} - 6t^2\mathbf{k}\|}{\|2t\mathbf{i} - 3t^2\mathbf{j} + \mathbf{k}\|^3}$$

$$= \frac{\sqrt{36t^2 + 4 + 36t^4}}{\left(4t^2 + 9t^4 + 1\right)^{3/2}}. \quad \blacksquare$$

11.2.3 The Frenet Formulas

If $\mathbf{T}$ and $\mathbf{N}$ are the unit tangent and normal vectors, the binormal vector $\mathbf{B} = \mathbf{T} \times \mathbf{N}$ is also a unit vector and is orthogonal to $\mathbf{T}$ and $\mathbf{N}$. At any point on the curve where these three vectors are defined and nonzero, the triple $\mathbf{T}, \mathbf{N}, \mathbf{B}$ forms a right-handed rectangular coordinate system (as in Figure 11.13). We can in effect put an x, y, z coordinate system at any point P on C, with the positive x axis along $\mathbf{T}$, the positive y axis along $\mathbf{N}$, and the positive z axis along $\mathbf{B}$. Of course, this system twists and changes orientation in space as P moves along C (Figure 11.14).

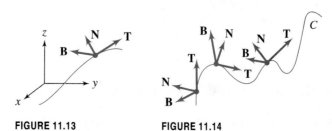

FIGURE 11.13   FIGURE 11.14

Since $\mathbf{N} = (1/\kappa)\mathbf{T}'(s)$, then

$$\frac{d\mathbf{T}}{ds} = \kappa\mathbf{N}.$$

Further, it can be shown that there is a scalar-valued function τ such that

$$\frac{d\mathbf{N}}{ds} = -\kappa\mathbf{T} + \tau\mathbf{B},$$

and

$$\frac{d\mathbf{B}}{ds} = -\tau\mathbf{N}.$$

These three equations are called the *Frenet formulas*. The scalar quantity $\tau(s)$ is the *torsion* of C at $(x(s), y(s), z(s))$. If we look along C at the coordinate system formed at each point by $\mathbf{T}$, $\mathbf{N}$, and $\mathbf{B}$, the torsion measures how this system twists about the curve as the point moves along the curve.

SECTION 11.2 PROBLEMS

In each of Problems 1 through 10, a position vector is given. Determine the velocity, speed, acceleration, tangential and normal components of the acceleration, the curvature, the unit tangent, unit normal and binormal vectors.

1. $\mathbf{F} = 3t\mathbf{i} - 2\mathbf{j} + t^2\mathbf{k}$

2. $\mathbf{F} = t\sin(t)\mathbf{i} + t\cos(t)\mathbf{j} + \mathbf{k}$

3. $\mathbf{F} = 2t\mathbf{i} - 2t\mathbf{j} + t\mathbf{k}$

4. $\mathbf{F} = e^t\sin(t)\mathbf{i} - \mathbf{j} + e^t\cos(t)\mathbf{k}$

5. $\mathbf{F} = 3e^{-t}(\mathbf{i} + \mathbf{j} - 2\mathbf{k})$

6. $\mathbf{F} = \alpha\cos(t)\mathbf{i} + \beta t\mathbf{j} + \alpha\sin(t)\mathbf{k}$

7. $\mathbf{F} = 2\sinh(t)\mathbf{j} - 2\cosh(t)\mathbf{k}$

8. $\mathbf{F} = \ln(t)(\mathbf{i} - \mathbf{j} + 2\mathbf{k})$

9. $\mathbf{F} = \alpha t^2\mathbf{i} + \beta t^2\mathbf{j} + \gamma t^2\mathbf{k}$

10. $\mathbf{F} = 3t\cos(t)\mathbf{j} - 3t\sin(t)\mathbf{k}$

11. Suppose we are given the position vector of a curve and find that the unit tangent vector is a constant vector. Prove that the curve is a straight line.

12. It is easy to verify that the curvature of any straight line is zero. Suppose C is a curve with twice-differentiable coordinate functions and curvature zero. Does it follow that C is a straight line?

13. Show that $\tau(s) = -\mathbf{N}(s) \cdot \mathbf{B}'(s)$.

14. Show that $\tau(s) = [\mathbf{T}(s), \mathbf{N}(s), \mathbf{N}'(s)]$, a scalar triple product.

15. Show that $\tau(s) = [\mathbf{F}'(s), \mathbf{F}''(s), \mathbf{F}'''(s)]/\kappa^2$.

16. Let $\mathbf{F}(t) = \alpha[\cos(\omega t)\mathbf{i} + \sin(\omega t)\mathbf{j}]$ be the position vector of a particle moving in a circle in the x, y plane. Here α and ω are positive constants.

    (a) Show that the angular speed of the particle (speed divided by the radius of the circle) is ω.

    (b) Show that the acceleration $\mathbf{a}$ is directed toward the origin and has constant magnitude $\alpha\omega^2$. (This acceleration vector is the *centripetal acceleration*. The centripetal force is $m\mathbf{a}$ and the *centrifugal force* is $-m\mathbf{a}$, where m is the mass of the particle.)

11.3   Vector Fields and Streamlines

We now turn to the analysis of vector functions of more than one variable.

DEFINITION 11.6   *Vector Field*

A vector field in 3-space is a 3-vector whose components are functions of three variables. A vector field in the plane is a 2-vector whose components are functions of two variables.

A vector field in 3-space has the appearance

$$\mathbf{G}(x, y, z) = f(x, y, z)\mathbf{i} + g(x, y, z)\mathbf{j} + h(x, y, z)\mathbf{k},$$

and, in the plane,

$$\mathbf{K}(x, y) = f(x, y)\mathbf{i} + g(x, y)\mathbf{j}.$$

The term *vector field* is geometrically motivated. At each point P for which a vector field $\mathbf{G}$ is defined, we can represent the vector $\mathbf{G}(P)$ as an arrow from P. It is often useful in working with a vector field $\mathbf{G}$ to draw arrows $\mathbf{G}(P)$ at points through the region where $\mathbf{G}$ is defined. This drawing is also referred to as a vector field (think of arrows growing at points). The variations in length and orientation of these arrows gives some sense of the flow of the vector field, and its variations in strength, just as a direction field helps us visualize trajectories of a system $\mathbf{X}' = \mathbf{F}(\mathbf{X})$ of differential equations. Figures 11.15 and 11.16 show the vector fields $\mathbf{G}(x, y) = xy\mathbf{i} + (x - y)\mathbf{j}$ and $\mathbf{H}(x, y) = y\cos(x)\mathbf{i} + (x^2 - y^2)\mathbf{j}$, respectively, in the plane. Figures 11.17, 11.18 and 11.19 show the vector fields $\mathbf{F}(x, y, z) = \cos(x + y)\mathbf{i} - x\mathbf{j} + (x - z)\mathbf{k}$, $\mathbf{Q}(x, y, z) = -y\mathbf{i} + z\mathbf{j} + (x + y + z)\mathbf{k}$ and $\mathbf{M}(x, y, z) = \cos(x)\mathbf{i} + e^{-x}\sin(y)\mathbf{j} + (z - y)\mathbf{k}$, respectively, in 3-space.

A vector field is continuous if each of its component functions is continuous. A partial derivative of a vector field is the vector field obtained by taking the partial derivative of each component function. For example, if

$$\mathbf{F}(x, y, z) = \cos(x + y)\mathbf{i} - x\mathbf{j} + (x - z)\mathbf{k},$$

then

$$\frac{\partial \mathbf{F}}{\partial x} = \mathbf{F}_x = -\sin(x + y)\mathbf{i} - \mathbf{j} + \mathbf{k}, \quad \frac{\partial \mathbf{F}}{\partial y} = \mathbf{F}_y = -\sin(x + y)\mathbf{i}, \quad \text{and} \quad \frac{\partial \mathbf{F}}{\partial z} = \mathbf{F}_z = -\mathbf{k}.$$

If

$$\mathbf{G}(x, y) = xy\mathbf{i} + (x - y)\mathbf{j},$$

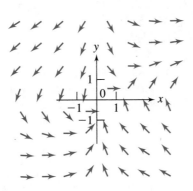

FIGURE 11.15 *Representation of the vector field* $\mathbf{G}(x, y) = xy\mathbf{i} + (x - y)\mathbf{j}$.

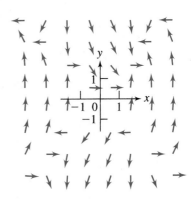

FIGURE 11.16 *Representation of the vector field* $\mathbf{H}(x, y) = y\cos(x)\mathbf{i} + (x^2 - y^2)\mathbf{j}$.

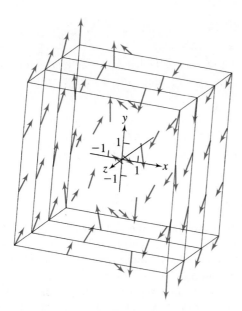

FIGURE 11.17 *Representation of the vector field* $\mathbf{F}(x, y, z) = \cos(x + y)\mathbf{i} - x\mathbf{j} + (x - z)\mathbf{k}$ *as arrows in planes* $z = constant$.

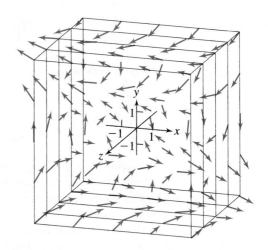

FIGURE 11.18 *Representation of the vector field* $\mathbf{Q}(x, y, z) = -y\mathbf{i} + z\mathbf{j} + (x + y + z)\mathbf{k}$ *in planes* $z = constant$.

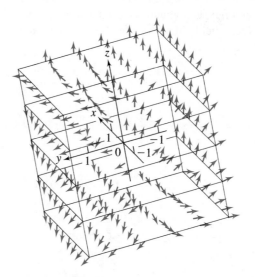

FIGURE 11.19 *The vector field* $\mathbf{M}(x, y, z) = \cos(x)\mathbf{i} + e^{-x} \sin(y)\mathbf{j} + (z - y)\mathbf{k}$ *in planes* $z = constant$.

then

$$\frac{\partial \mathbf{G}}{\partial x} = \mathbf{G}_x = y\mathbf{i} + \mathbf{j} \quad \text{and} \quad \frac{\partial \mathbf{G}}{\partial y} = \mathbf{G}_y = x\mathbf{i} - \mathbf{j}.$$

Streamlines of a vector field $\mathbf{F}$ are curves with the property that at each point (x, y, z), the vector $\mathbf{F}(x, y, z)$ is tangent to the curve through this point.

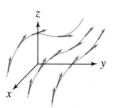

FIGURE 11.20
*Streamlines of a
vector field.*

DEFINITION 11.7 *Streamlines*

Let $\mathbf{F}$ be a vector field, defined for all (x, y, z) in some region Ω of 3-space. Let $\mathfrak{F}$ be a set of curves with the property that through each point P of Ω, there passes exactly one curve from $\mathfrak{F}$. The curves in $\mathfrak{F}$ are streamlines of $\mathbf{F}$ if at each (x, y, z) in Ω, the vector $\mathbf{F}(x, y, z)$ is tangent to the curve in $\mathfrak{F}$ passing through (x, y, z).

Streamlines are also called *flow lines* or *lines of force*, depending on context. If $\mathbf{F}$ is the velocity field for a fluid, the streamlines are often called *flow lines* (paths of particles in the fluid). If $\mathbf{F}$ is a magnetic field, the streamlines are called *lines of force*. If you put iron filings on a piece of cardboard and then hold a magnet underneath, the filings will be aligned by the magnet along the lines of force of the field.

Given a vector field $\mathbf{F}$, we would like to find the streamlines. This is the problem of constructing a curve through each point of a region, given the tangent to the curve at each point. Figure 11.20 shows typical streamlines of a vector field, together with some of the tangent vectors. We want to determine the curves from the tangents.

To solve this problem, suppose C is a streamline of $\mathbf{F} = f\mathbf{i} + g\mathbf{j} + h\mathbf{j}$. Let C have parametric equations

$$x = x(\xi), \quad y = y(\xi), \quad z = z(\xi).$$

The position vector for this curve is

$$\mathbf{R}(\xi) = x(\xi)\mathbf{i} + y(\xi)\mathbf{j} + z(\xi)\mathbf{k}.$$

Now

$$\mathbf{R}'(\xi) = x'(\xi)\mathbf{i} + y'(\xi)\mathbf{j} + z'(\xi)\mathbf{k}$$

is tangent to C at $(x(\xi), y(\xi), z(\xi))$. But for C to be a streamline of $\mathbf{F}$, $\mathbf{F}(x(\xi), y(\xi), z(\xi))$ is also tangent to C at this point, hence must be parallel to $\mathbf{R}'(\xi)$. These vectors must therefore be scalar multiples of each other. For some scalar t (which may depend on ξ),

$$\mathbf{R}'(\xi) = t\mathbf{F}(x(\xi), y(\xi), z(\xi)).$$

But then

$$\frac{dx}{d\xi}\mathbf{i} + \frac{dy}{d\xi}\mathbf{j} + \frac{dz}{d\xi}\mathbf{k} = tf(x(\xi), y(\xi), z(\xi))\mathbf{i} + tg(x(\xi), y(\xi), z(\xi))\mathbf{j} + th(x(\xi), y(\xi), z(\xi))\mathbf{k}.$$

This implies that

$$\frac{dx}{d\xi} = tf, \quad \frac{dy}{d\xi} = tg, \quad \frac{dz}{d\xi} = th. \tag{11.4}$$

Since f, g, and h are given functions, these equations constitute a system of differential equations for the coordinate functions of the streamlines. If f, g, and h are nonzero, then t can be eliminated to write the system in differential form as

$$\frac{dx}{f} = \frac{dy}{g} = \frac{dz}{h}. \tag{11.5}$$

EXAMPLE 11.11

We will find the streamlines of the vector field $\mathbf{F} = x^2\mathbf{i} + 2y\mathbf{j} - \mathbf{k}$.

The system (11.4) is

$$\frac{dx}{d\xi} = tx^2, \quad \frac{dy}{d\xi} = 2ty, \quad \frac{dz}{d\xi} = -t.$$

If x and y are not zero, this can be written in the form of equations (11.5):

$$\frac{dx}{x^2} = \frac{dy}{2y} = \frac{dz}{-1}.$$

These equations can be solved in pairs. To begin, integrate

$$\frac{dx}{x^2} = -dz$$

to get

$$-\frac{1}{x} = -z + c.$$

in which c is an arbitrary constant.

Next, integrate

$$\frac{dy}{2y} = -dz$$

to get

$$\frac{1}{2}\ln|y| = -z + k.$$

It is convenient to express two of the variables in terms of the third. If we solve for x and y in terms of z, we get

$$x = \frac{1}{z - c}, \quad y = ae^{-2z},$$

in which $a = e^{2k}$ is an arbitrary (positive) constant. This gives us parametric equations of the streamlines, with z as parameter:

$$x = \frac{1}{z - c}, \quad y = ae^{-2z}, \quad z = z.$$

To find the streamline through a particular point, we must choose c and a appropriately. For example, suppose we want the streamline through $(-1, 6, 2)$. Then $z = 2$ and we need to choose c and a so that

$$-1 = \frac{1}{2 - c} \quad \text{and} \quad 6 = ae^{-4}.$$

Then $c = 3$ and $a = 6e^4$, so the streamline through $(-1, 6, 2)$ has parametric equations

$$x = \frac{1}{z - 3}, \quad y = 6e^{4-2z}, \quad z = z.$$

A graph of this streamline is shown in Figure 11.21. ∎

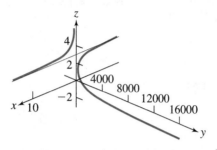

FIGURE 11.21 *Part of the graph of* $x = 1/(z - 3)$, $y = 6e^{4-2z}$, $z = z$.

EXAMPLE 11.12

Suppose we want the streamlines of $\mathbf{F}(x, y, z) = -y\mathbf{j} + z\mathbf{k}$. Here the **i**-component is zero, so we must begin with equations (11.4), not (11.5). We have

$$\frac{dx}{d\xi} = 0, \quad \frac{dy}{d\xi} = -ty, \quad \frac{dz}{d\xi} = tz.$$

The first equation implies that $x = $ constant. This simply means that all the streamlines are in planes parallel to the y, z plane. The other two equations yield

$$\frac{dy}{-y} = \frac{dz}{z}$$

and an integration gives

$$-\ln(y) + c = \ln(z).$$

Then

$$\ln(zy) = c,$$

implying that

$$zy = k,$$

in which k is constant. The streamlines are given by the equations

$$x = c, \quad z = \frac{k}{y},$$

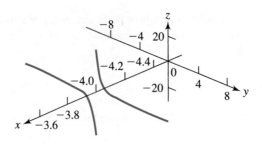

FIGURE 11.22 *Part of the graph of $x = -4$,*
$z = 7/y$.

in which c and k are arbitrary constants and y is the parameter. For example, to find the streamline through $(-4, 1, 7)$, choose $c = -4$ and k so that $z = k/y$ passes through $y = 1$, $z = 7$. We need

$$7 = \frac{k}{1} = k,$$

and the streamline has equations

$$x = -4, \quad z = \frac{7}{y}.$$

The streamline is a hyperbola in the plane $x = -4$ (Figure 11.22).

SECTION 11.3 PROBLEMS

In each of Problems 1 through 5, compute the two first partial derivatives of the vector field and make a diagram in which each indicated vector is drawn as an arrow from the point at which the vector is evaluated.

1. $\mathbf{G}(x, y) = 3x\mathbf{i} - 4xy\mathbf{j}$; $\mathbf{G}(0, 1)$, $\mathbf{G}(1, 3)$, $\mathbf{G}(1, 4)$, $\mathbf{G}(-1, -2)$, $\mathbf{G}(-3, 2)$

2. $\mathbf{G}(x, y) = e^x\mathbf{i} - 2x^2y\mathbf{j}$; $\mathbf{G}(0, 0)$, $\mathbf{G}(1, 0)$, $\mathbf{G}(0, 1)$, $\mathbf{G}(2, -3)$, $\mathbf{G}(-1, -3)$

3. $\mathbf{G}(x, y) = 2xy\mathbf{i} + \cos(x)\mathbf{j}$; $\mathbf{G}(\pi/2, 0)$, $\mathbf{G}(0, 0)$, $\mathbf{G}(-1, 1)$, $\mathbf{G}(\pi, -3)$, $\mathbf{G}(-\pi/4, -2)$

4. $\mathbf{G}(x, y) = \sin(2xy)\mathbf{i} + (x^2 + y)\mathbf{j}$; $\mathbf{G}(-\pi/2, 0)$, $\mathbf{G}(0, 2)$, $\mathbf{G}(\pi/4, 4)$, $\mathbf{G}(1, 1)$, $\mathbf{G}(-2, 1)$

5. $\mathbf{G}(x, y) = 3x^2\mathbf{i} + (x - 2y)\mathbf{j}$; $\mathbf{G}(1, -1)$, $\mathbf{G}(0, 2)$, $\mathbf{G}(-3, 2)$, $\mathbf{G}(-2, -2)$, $\mathbf{G}(2, 5)$

In each of Problems 6 through 10, compute the three first partial derivatives of the vector field.

6. $\mathbf{F} = e^{xy}\mathbf{i} - 2x^2y\mathbf{j} + \cosh(z + y)\mathbf{k}$

7. $\mathbf{F} = 4z^2\cos(x)\mathbf{i} - x^3yz\mathbf{j} + x^3y\mathbf{k}$

8. $\mathbf{F} = 3xy^3\mathbf{i} + \ln(x + y + z)\mathbf{j} + \cosh(xyz)\mathbf{k}$

9. $\mathbf{F} = -z^4\sin(xy)\mathbf{i} + 3xy^4z\mathbf{j} + \cosh(z - x)\mathbf{k}$

10. $\mathbf{F} = (14x - 2y)\mathbf{i} + (x^2 - y^2 - z^2)\mathbf{j} + 5xy\mathbf{k}$

In each of Problems 11 through 20, find the streamlines of the vector field, then find the particular streamline through the given point.

11. $\mathbf{F} = \mathbf{i} - y^2\mathbf{j} + z\mathbf{k}$; $(2, 1, 1)$

12. $\mathbf{F} = \mathbf{i} - 2\mathbf{j} + \mathbf{k}$; $(0, 1, 1)$

13. $\mathbf{F} = (1/x)\mathbf{i} + e^x\mathbf{j} - \mathbf{k}$; $(2, 0, 4)$

14. $\mathbf{F} = \cos(y)\mathbf{i} + \sin(x)\mathbf{j}$; $(\pi/2, 0, -4)$

15. $\mathbf{F} = 2e^z\mathbf{j} - \cos(y)\mathbf{k}$; $(3, \pi/4, 0)$

16. $\mathbf{F} = 3x^2\mathbf{i} - y\mathbf{j} + z^3\mathbf{k}$; $(2, 1, 6)$

17. $\mathbf{F} = e^z\mathbf{i} - x^2\mathbf{k}$; $(4, 2, 0)$

18. $\mathbf{F} = x^2\mathbf{i} + y^2\mathbf{j} - z^2\mathbf{k}$; $(1, 1, 1)$

19. $\mathbf{F} = \sec(x)\mathbf{i} - \cot(y)\mathbf{j} + \mathbf{k}$; $(\pi/4, 0, 1)$

20. $\mathbf{F} = -3\mathbf{i} + 2e^z\mathbf{j} - \cos(y)\mathbf{k}$; $(2, \pi/4, 0)$

21. Construct a vector field whose streamlines are straight lines.

22. Construct a vector field in the x, y plane whose stream-lines are circles about the origin.

23. Is it possible for a vector field in three variables to have streamlines lying only in the x, y plane?

11.4 The Gradient Field and Directional Derivatives

Let $\varphi(x, y, z)$ be a real-valued function of three variables. In the context of vectors, such a function is called a *scalar field*. We will define an important vector field manufactured from φ.

DEFINITION 11.8 *Gradient*

The gradient of a scalar field φ is the vector field $\nabla \varphi$ given by

$$\nabla \varphi = \frac{\partial \varphi}{\partial x} \mathbf{i} + \frac{\partial \varphi}{\partial y} \mathbf{j} + \frac{\partial \varphi}{\partial z} \mathbf{k},$$

wherever these partial derivatives are defined.

The symbol $\nabla \varphi$ is read "del phi", and ∇ is called the *del operator*. It operates on a scalar field to produce a vector field. For example, if $\varphi(x, y, z) = x^2 y \cos(yz)$, then

$$\nabla \varphi = 2xy \cos(yz)\mathbf{i} + [x^2 \cos(yz) - x^2 yz \sin(yz)]\mathbf{j} - x^2 y^2 \sin(yz)\mathbf{k}.$$

The gradient field evaluated at a point P is denoted $\nabla \varphi(P)$. For the gradient just computed,

$$\nabla \varphi(1, -1, 3) = -2 \cos(3)\mathbf{i} + [\cos(3) - 3 \sin(3)]\mathbf{j} + \sin(3)\mathbf{k}.$$

If φ is a function of just x and y, then $\nabla \varphi$ is a vector in the x, y plane. For example, if $\varphi(x, y) = (x - y)\cos(y)$, then

$$\nabla \varphi(x, y) = \cos(y)\mathbf{i} + [-\cos(y) - (x - y) \sin(y)]\mathbf{j}.$$

At $(2, \pi)$ this gradient is

$$\nabla \varphi(2, \pi) = -\mathbf{i} + \mathbf{j}.$$

The gradient has the obvious properties

$$\nabla (\varphi + \psi) = \nabla \varphi + \nabla \psi$$

and, if c is a number, then

$$\nabla (c\varphi) = c\nabla \varphi.$$

We will now define the directional derivative and relate this to the gradient. Suppose $\varphi(x, y, z)$ is a scalar field. Let $\mathbf{u} = a\mathbf{i} + b\mathbf{j} + c\mathbf{k}$ be a unit vector (length 1). Let $P_0 = (x_0, y_0, z_0)$. Represent $\mathbf{u}$ as an arrow from P_0, as in Figure 11.23. We want to define a quantity that measures the rate of change of $\varphi(x, y, z)$ as (x, y, z) varies from P_0, in the direction of $\mathbf{u}$. To do this, notice that if $t > 0$, then the point

$$P : (x_0 + at, y_0 + bt, z_0 + ct)$$

is on the line through P_0 in the direction of $\mathbf{u}$. Further, the distance from P_0 to P along this direction is exactly t, because the vector from P_0 to P is

$$(x_0 + at - x_0)\mathbf{i} + (y_0 + bt - y_0)\mathbf{j} + (z_0 + ct - z_0)\mathbf{k},$$

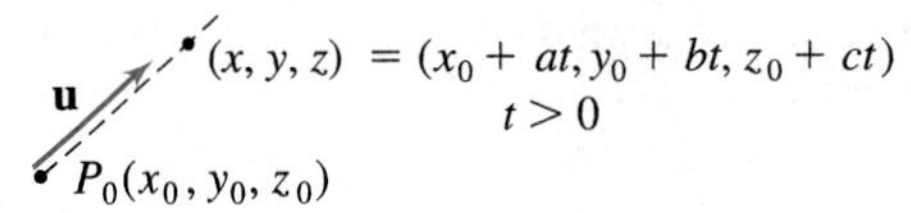

FIGURE 11.23

and this is just $t\mathbf{u}$. The derivative

$$\frac{d}{dt}\,\varphi(x + at, y + bt, z + ct)$$

is the rate of change of $\varphi(x + at, y + bt, z + ct)$ with respect to this distance t, and

$$\frac{d}{dt}\,\varphi(x + at, y + bt, z + ct)\bigg]_{t=0}$$

is this rate of change evaluated at P_0. This derivative gives the rate of change of $\varphi(x, y, z)$ at P_0 in the direction of $\mathbf{u}$. We will summarize this discussion in the following definition.

DEFINITION 11.9   *Directional Derivative*

The directional derivative of a scalar field φ at P_0 in the direction of the unit vector $\mathbf{u}$ is denoted $D_{\mathbf{u}}\varphi(P_0)$ and is given by

$$D_{\mathbf{u}}\varphi(P_0) = \frac{d}{dt}\varphi(x + at, y + bt, z + ct)\bigg]_{t=0}.$$

We usually compute a directional derivative using the following.

THEOREM 11.3

If φ is a differentiable function of two or three variables, and $\mathbf{u}$ is a constant unit vector, then

$$D_{\mathbf{u}}\varphi(P_0) = \nabla\varphi(P_0) \cdot \mathbf{u}.$$

Proof   Let $\mathbf{u} = a\mathbf{i} + b\mathbf{j} + c\mathbf{k}$. By the chain rule,

$$\frac{d}{dt}\,\varphi(x + at, y + bt, z + ct) = \frac{\partial\varphi}{\partial x}a + \frac{\partial\varphi}{\partial y}b + \frac{\partial\varphi}{\partial z}c.$$

Since $(x_0 + at, y_0 + bt, z_0 + ct) = (x_0, y_0, z_0)$ when $t = 0$, then

$$D_{\mathbf{u}}\varphi(P_0) = \frac{\partial\varphi}{\partial x}(P_0)a + \frac{\partial\varphi}{\partial y}(P_0)b + \frac{\partial\varphi}{\partial z}(P_0)k$$

$$= \nabla\varphi(P_0) \cdot \mathbf{u}. \quad\blacksquare$$

EXAMPLE 11.13

Let $\varphi(x, y, z) = x^2y - xe^z$, $P_0 = (2, -1, \pi)$, and $u = (1/\sqrt{6})(\mathbf{i} - 2\mathbf{j} + \mathbf{k})$. Then the rate of change of $\varphi(x, y, z)$ at P_0 in the direction of $\mathbf{u}$ is

$$D_{\mathbf{u}}\varphi(2, -1, \pi) = \nabla\varphi(2, -1, \pi) \cdot \mathbf{u}$$

$$= \varphi_x(2, -1, \pi)\frac{1}{\sqrt{6}} + \varphi_y(2, -1, \pi)\left(-\frac{2}{\sqrt{6}}\right) + \varphi_z(2, -1, \pi)\frac{1}{\sqrt{6}}$$

$$= \frac{1}{\sqrt{6}}\left([2xy - e^z]_{(2,-1,\pi)} - 2[x^2]_{(-2,1,\pi)} + [-xe^z]_{(2,-1,\pi)}\right)$$

$$= \frac{1}{\sqrt{6}}(-4 - e^\pi - 8 - 2e^\pi) = \frac{-3}{\sqrt{6}}(4 + e^\pi). \ \blacksquare$$

In working with directional derivatives, care must be taken that the direction is given by a unit vector. If a vector $\mathbf{w}$ of length other than 1 is used to specify the direction, then use the unit vector $\mathbf{w}/\|\mathbf{w}\|$ in computing the directional derivative. Of course, $\mathbf{w}$ and $\mathbf{w}/\|\mathbf{w}\|$ have the same direction. A unit vector is used with directional derivatives so that the vector specifies only direction, without contributing a factor of magnitude.

Suppose now that $\varphi(x, y, z)$ is defined at least for all points within some sphere about P_0. Imagine standing at P_0 and looking in various directions. We may see $\varphi(x, y, z)$ increasing in some, decreasing in others, and perhaps remaining constant in some directions. In what direction does $\varphi(x, y, z)$ have its greatest, or least, rate of increase from P_0? We will now show that the gradient vector $\nabla\varphi(P_0)$ points in the direction of maximum rate of increase at P_0, and $-\nabla\varphi(P_0)$ in the direction of minimum rate of increase.

THEOREM 11.4

Let φ and its first partial derivatives be continuous in some sphere about P_0, and suppose that $\nabla\varphi(P_0) \neq \mathbf{O}$. Then

1. At P_0, $\varphi(x, y, z)$ has its maximum rate of change in the direction of $\nabla\varphi(P_0)$. This maximum rate of change is $\|\nabla\varphi(P_0)\|$.

2. At P_0, $\varphi(x, y, z)$ has its minimum rate of change in the direction of $-\nabla\varphi(P_0)$. This minimum rate of change is $-\|\nabla\varphi(P_0)\|$.

Proof Let $\mathbf{u}$ be any unit vector. Then

$$D_{\mathbf{u}}\varphi(P_0) = \nabla\varphi(P_0) \cdot \mathbf{u}$$

$$= \|\nabla\varphi(P_0)\| \, \|\mathbf{u}\| \cos(\theta) = \|\nabla\varphi(P_0)\| \cos(\theta),$$

because $\mathbf{u}$ has length 1. θ is the angle between $\mathbf{u}$ and $\nabla\varphi(P_0)$. The direction $\mathbf{u}$ in which φ has its greatest rate of increase from P_0 is the direction in which this directional derivative is a maximum. Clearly, the maximum occurs when $\cos(\theta) = 1$, hence when $\theta = 0$. But this occurs when $\mathbf{u}$ is along $\nabla\varphi(P_0)$. Therefore, this gradient is the direction of maximum rate of change of $\varphi(x, y, z)$ at P_0. This maximum rate of change is $\|\nabla\varphi(P_0)\|$.

For (2), observe that the directional derivative is a minimum when $\cos(\theta) = -1$, hence when $\theta = \pi$. This occurs when $\mathbf{u}$ is opposite $\nabla\varphi(P_0)$, and this minimum rate of change is $-\|\nabla\varphi(P_0)\|$. $\ \blacksquare$

EXAMPLE 11.14

Let $\varphi(x, y, z) = 2xz + e^y z^2$. We will find the maximum and minimum rates of change of $\varphi(x, y, z)$ from $(2, 1, 1)$. First,

$$\nabla\varphi(x, y, z) = 2z\mathbf{i} + e^y z^2\mathbf{j} + (2x + 2ze^y)\mathbf{k},$$

so

$$\nabla\varphi(P_0) = 2\mathbf{i} + e\mathbf{j} + (4 + 2e)\mathbf{k}.$$

The maximum rate of increase of $\varphi(x, y, z)$ at $(2, 1, 1)$ is in the direction of this gradient, and this maximum rate of change is

$$\sqrt{4 + e^2 + (4 + 2e)^2}.$$

The minimum rate of increase is in the direction of $-2\mathbf{i} - e\mathbf{j} - (4 + 2e)\mathbf{k}$ and is $-\sqrt{4 + e^2 + (4 + 2e)^2}$. ■

11.4.1 Level Surfaces, Tangent Planes, and Normal Lines

Depending on the function φ and the constant k, the locus of points $\varphi(x, y, z) = k$ may form a surface in 3-space. Any such surface is called a *level surface* of φ. For example, if $\varphi(x, y, z) = x^2 + y^2 + z^2$ and $k > 0$, then the level surface $\varphi(x, y, z) = k$ is a sphere of radius $\sqrt{k}$. If $k = 0$ this locus is just a single point, the origin. If $k < 0$ this locus is empty. There are no points whose coordinates satisfy this equation. The level surface $\varphi(x, y, z) = 0$ of $\varphi(x, y, z) = z - \sin(xy)$ is shown from three perspectives in Figures 11.24 (a), (b), and (c).

Now consider a point $P_0 : (x_0, y_0, z_0)$ on a level surface $\varphi(x, y, z) = k$. Assume that there are smooth (having continuous tangents) curves on the surface passing through P_0, such as C_1 and C_2 in Figure 11.25 (p. 540). Each such curve has a tangent vector at P_0. These tangent vectors determine a plane Π at P_0, called the *tangent plane* to the surface at P_0. A vector normal (perpendicular) to Π at P_0, in the sense of being normal to each of these tangent vectors, is called a *normal vector*, or just *normal*, to the surface at P_0. We would like to be able to determine the tangent plane and normal to a surface at a point. Recall that we can find the equation of a plane through a given point if we are given a normal vector to the plane. Thus the normal vector is the key to finding the tangent plane.

THEOREM 11.5 *Gradient as a Normal Vector*

Let φ and its first partial derivatives be continuous. Then $\nabla\varphi(P)$ is normal to the level surface $\varphi(x, y, z) = k$ at any point P on this surface at which this gradient vector is nonzero. ■

We will outline an argument suggesting why this is true. Consider a point $P_0 : (x_0, y_0, z_0)$ on the level surface $\varphi(x, y, z) = k$. Suppose a smooth curve C on this surface passes through P_0, as in Figure 11.26(a) (see p. 540). Suppose C has parametric equations

$$x = x(t), \quad y = y(t), \quad z = z(t).$$

Since P_0 is on this curve, for some t_0,

$$x(t_0) = x_0 \quad, y(t_0) = y_0, \quad z(t_0) = z_0.$$

Further, since the curve lies on the level surface, then

$$\varphi(x(t), y(t), z(t)) = k$$

for all t. Then

$$\frac{d}{dt}\varphi(x(t), y(t), z(t)) = 0$$

$$= \varphi_x x'(t) + \varphi_y y'(t) + \varphi_z z'(t)$$

$$= \nabla\varphi \cdot [x'(t)\mathbf{i} + y'(t)\mathbf{j} + z'(t)\mathbf{k}].$$

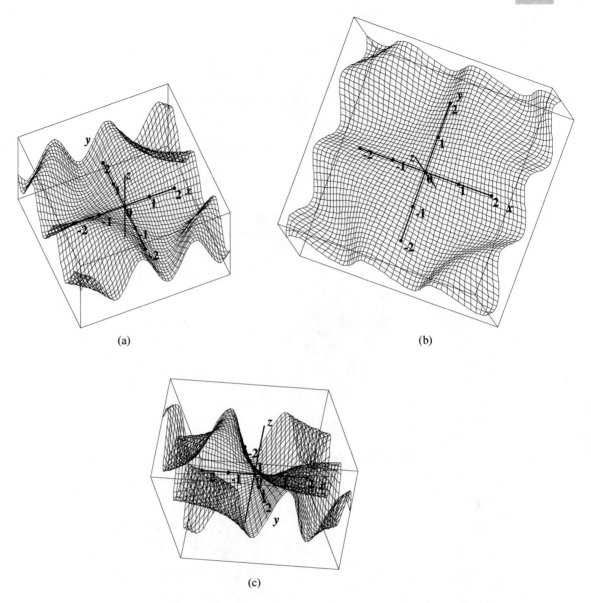

(a)

(b)

(c)

FIGURE 11.24 *Different perspectives of graphs of $z = \sin(xy)$.*

Now $x'(t)\mathbf{i} + y'(t)\mathbf{j} + z'(t)\mathbf{k} = \mathbf{T}(t)$ is a tangent vector to C. In particular, letting $t = t_0$, then $\mathbf{T}(t_0)$ is a tangent vector to C at P_0, and we have

$$\nabla\varphi(P_0) \cdot \mathbf{T}(t_0) = 0.$$

This means that $\nabla\varphi(P_0)$ is normal to the tangent to C at P_0. But this is true for any smooth curve on the surface and passing through P_0. Therefore, $\nabla\varphi(P_0)$ is normal to the surface at P_0 (Figure 11.26(b)).

Once we have this normal vector, finding the equation of the tangent plane is straightforward. If (x, y, z) is any other point on the tangent plane (Figure 11.27), then the vector

$$(x - x_0)\mathbf{i} + (y - y_0)\mathbf{j} + (z - z_0)\mathbf{k}$$

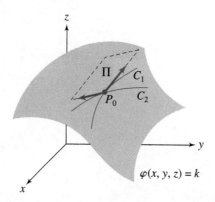

FIGURE 11.25 *Tangents to curves on the surface through P_0 determine the tangent plane Π.*

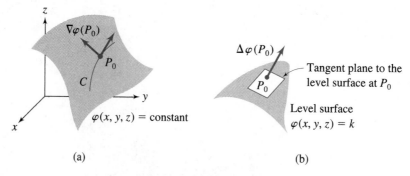

(a)   (b)

FIGURE 11.26 $\nabla \varphi(P_0)$ *is normal to the level surface at P_0.*

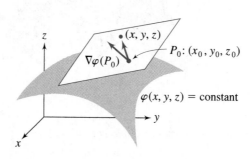

FIGURE 11.27
$$\nabla(P_0) \cdot [(x - x_0)\mathbf{i} + (y - y_0)\mathbf{j} + (z - z_0)\mathbf{k}] = 0.$$

is in this plane, hence is orthogonal to the normal vector. Then

$$\nabla \varphi(P_0) \cdot [(x - x_0)\mathbf{i} + (y - y_0)\mathbf{j} + (z - z_0)\mathbf{k}] = 0.$$

Then

$$\frac{\partial \varphi}{\partial x}(P_0)(x - x_0) + \frac{\partial \varphi}{\partial y}(P_0)(y - y_0) + \frac{\partial \varphi}{\partial z}(P_0)(z - z_0) = 0. \tag{11.6}$$

This equation is satisfied by every point on the tangent plane. Conversely, if (x, y, z) satisfies this equation, then $(x - x_0)\mathbf{i} + (y - y_0)\mathbf{j} + (z - z_0)\mathbf{k}$ is normal to the normal vector, hence lies in the tangent plane, implying that (x, y, z) is a point in this plane. We call equation (11.6) the *equation of the tangent plane* to $\varphi(x, y, z) = k$ at P_0.

EXAMPLE 11.15

Consider the level surface $\varphi(x, y, z) = z - \sqrt{x^2 + y^2} = 0$. This surface is the cone shown in Figure 11.28. We will find the normal vector and tangent plane to this surface at $(1, 1, \sqrt{2})$.

First compute the gradient vector:

$$\nabla\varphi = -\frac{x}{\sqrt{x^2 + y^2}}\mathbf{i} - \frac{y}{\sqrt{x^2 + y^2}}\mathbf{j} + \mathbf{k},$$

provided that x and y are not both zero. Figure 11.29 shows $\nabla\varphi$ at a point on the cone determined by the position vector

$$\mathbf{R}(x, y, z) = x\mathbf{i} + y\mathbf{j} + \sqrt{x^2 + y^2}\mathbf{k}.$$

Then

$$\nabla\varphi(1, 1, \sqrt{2}) = -\frac{1}{\sqrt{2}}\mathbf{i} - \frac{1}{\sqrt{2}}\mathbf{j} + \mathbf{k}.$$

This is the normal vector to the cone at $(1, 1, \sqrt{2})$. The tangent plane at this point has equation

$$-\frac{1}{\sqrt{2}}(x - 1) - \frac{1}{\sqrt{2}}(y - 1) + z - \sqrt{2} = 0,$$

or

$$x + y - \sqrt{2}z = 0.$$

The cone has no tangent plane or normal vector at the origin, where the surface has a "sharp point." This is analogous to a graph in the plane having no tangent vector where it has a sharp point (for example, $y = |x|$ at the origin). ■

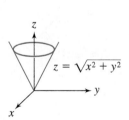

FIGURE 11.28

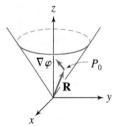

FIGURE 11.29 *Cone* $z = \sqrt{x^2 + y^2}$ *and normal at P_0.*

EXAMPLE 11.16

Consider the surface $z = \sin(xy)$. If we let $\varphi(x, y, z) = \sin(xy) - z$, then this surface is the level surface $\varphi(x, y, z) = 0$. The gradient vector is

$$\nabla\varphi = y\cos(xy)\mathbf{i} + x\cos(xy)\mathbf{j} - \mathbf{k}.$$

This vector field is shown in Figure 11.30, with the gradient vectors drawn as arrows from selected points on the surface.

The tangent plane at any point (x_0, y_0, z_0) on this surface has equation

$$y_0 \cos(x_0 y_0)(x - x_0) + x_0 \cos(x_0 y_0)(y - y_0) - (z - z_0) = 0.$$

For example, the tangent plane at $(2, 1, \sin(2))$ has equation

$$\cos(2)(x - 2) + 2\cos(2)(y - 1) - z + \sin(2) = 0,$$

or

$$\cos(2)x + 2\cos(2)y - z = 4\cos(2) - \sin(2).$$

A patch of this tangent plane is shown in Figure 11.31. Similarly, the tangent plane at $(-1, -2, \sin(2))$ has equation

$$2\cos(2)x + \cos(2)y + z = -4\cos(2) + \sin(2).$$

Part of this tangent plane is shown in Figure 11.32. ∎

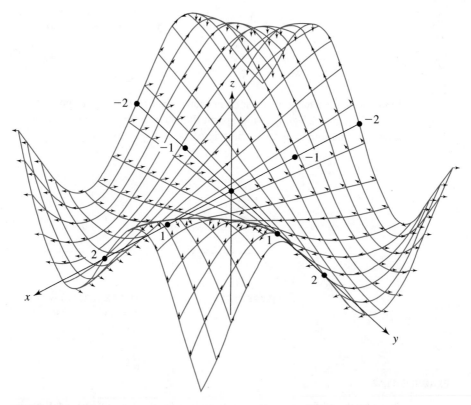

FIGURE 11.30 *Gradient field* $\nabla \varphi = y\cos(xy)\mathbf{i} + x\cos(xy)\mathbf{j} - \mathbf{k}$ *represented as a vector field on the surface* $z = \sin(xy)$.

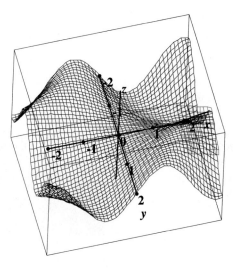

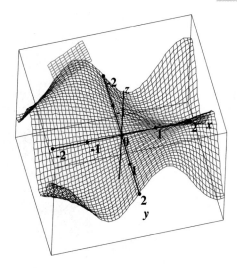

FIGURE 11.31 *Part of the tangent plane to* $z = \sin(xy)$ *at* $(2, 1, \sin(2))$.

FIGURE 11.32 *Part of the tangent plane to* $z = \sin(xy)$ *at* $(-1, -2, \sin(2))$.

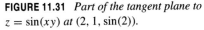

EXAMPLE 11.17

Consider the level surface $\varphi(x, y, z) = \sin(x^2 + y^2) - z = 0$. Graphs of this surface are shown from different perspectives in Figures 11.33(a), (b), and (c). The gradient is

$$\nabla\varphi = 2\cos(x^2 + y^2)[x\mathbf{i} + y\mathbf{j}] - \mathbf{k}.$$

and this vector field is shown in Figure 11.34. The tangent plane at $(2, 1, \sin(5))$ has equation

$$4\cos(5)(x - 2) + 2\cos(5)(y - 1) - (z - \sin(5)) = 0.$$

Part of this tangent plane is shown in Figure 11.35 (p. 545). ∎

A straight line through P_0 and parallel to $\nabla\varphi(P_0)$ is called the *normal line* to the level surface $\varphi(x, y, z) = k$ at P_0, assuming that this gradient vector is not zero. This idea is illustrated in Figure 11.36 (p. 545).

To write the equation of the normal line, let (x, y, z) be any point on it. Then the vector

$$(x - x_0)\mathbf{i} + (y - y_0)\mathbf{j} + (z - z_0)\mathbf{k}$$

is along this line, hence is parallel to $\nabla\varphi(P_0)$. This means that for some scalar t, either of these vectors is t times the other, say

$$(x - x_0)\mathbf{i} + (y - y_0)\mathbf{j} + (z - z_0)\mathbf{k} = t\nabla\varphi(P_0).$$

The components on the left must equal the respective components on the right:

$$x - x_0 = \frac{\partial\varphi}{\partial x}(P_0)t, \quad y - y_0 = \frac{\partial\varphi}{\partial y}(P_0)t, \quad z - z_0 = \frac{\partial\varphi}{\partial z}(P_0)t.$$

These are parametric equations of the normal line. As t varies over the real line, these equations give coordinates of points (x, y, z) on the normal line.

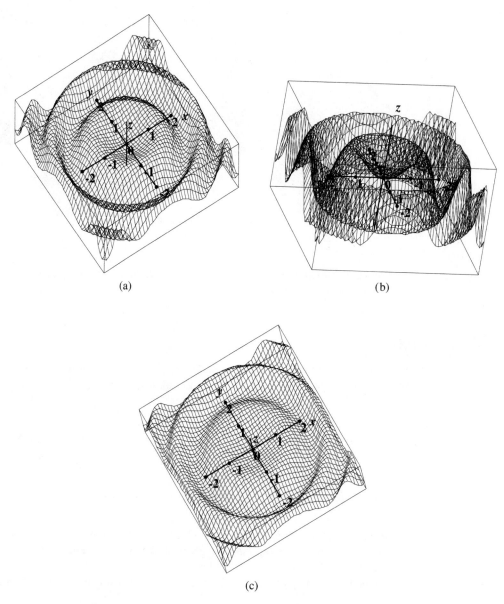

(a)

(b)

(c)

FIGURE 11.33 *Different perspectives of the surface $z = \sin(x^2 + y^2)$.*

EXAMPLE 11.18

Consider again the cone $\varphi(x, y, z) = z - \sqrt{x^2 + y^2} = 0$. In Example 11.15 we computed the gradient vector at $(1, 1, \sqrt{2})$, obtaining

$$-\frac{1}{\sqrt{2}}\mathbf{i} - \frac{1}{\sqrt{2}}\mathbf{j} + \mathbf{k}.$$

The normal line through this point has parametric equations

$$x - 1 = -\frac{1}{\sqrt{2}}t, \quad y - 1 = -\frac{1}{\sqrt{2}}t, \quad z - \sqrt{2} = t.$$

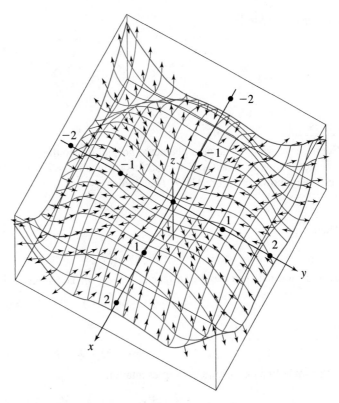

FIGURE 11.34 *Gradient field of $\varphi = \sin(x^2 + y^2) - z$ as normal vectors at points of the surface $z = \sin(x^2 + y^2)$.*

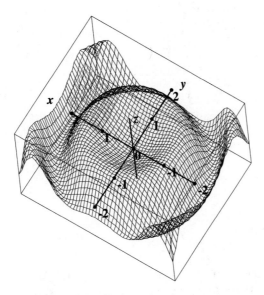

FIGURE 11.35 *Part of the tangent plane to $z = \sin(x^2 + y^2)$ at $(2, 1, \sin(5))$.*

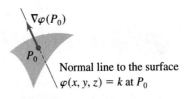

FIGURE 11.36

We can also write

$$x = 1 - \frac{1}{\sqrt{2}}t, \quad y = 1 - \frac{1}{\sqrt{2}}t, \quad z = \sqrt{2} + t. \ \blacksquare$$

EXAMPLE 11.19

Suppose we want the tangent plane and normal line to the surface $z = x^2 + y^2$ at $(2, -2, 8)$.

Let $\varphi(x, y, z) = x^2 + y^2 - z$, so this surface is the level surface $\varphi(x, y, z) = 0$. Part of the surface is shown in Figure 11.37. The gradient is

$$\nabla \varphi = 2x\mathbf{i} + 2y\mathbf{j} - \mathbf{k}.$$

Then

$$\nabla \varphi(2, -2, 8) = 4\mathbf{i} - 4\mathbf{j} - \mathbf{k}.$$

This is a normal vector to the surface at $(2, -2, 8)$. The tangent plane has equation

$$4(x - 2) - 4(y + 2) - (z - 8) = 0,$$

or

$$4x - 4y - z = 8.$$

The normal line has parametric representation

$$x = 2 + 4t, \quad y = -2 - 4t, \quad z = 8 - t.$$

The gradient also tells us other information about this surface. At $(2, -2, 8)$, $\varphi(x, y, z)$ has its greatest rate of increase in the direction of $4\mathbf{i} - 4\mathbf{j} - \mathbf{k}$, and this rate of increase is the magnitude of this vector, $\sqrt{33}$. The least rate of increase at $(2, -2, 8)$ is in the direction of $-4\mathbf{i} + 4\mathbf{j} + \mathbf{k}$ and is $-\sqrt{33}$. $\blacksquare$

FIGURE 11.37 *Part of the surface $z = x^2 + y^2$.*

SECTION 11.4   PROBLEMS

In each of Problems 1 through 10, compute the gradient of the function and evaluate this gradient at the given point. Determine at this point the maximum and minimum rate of change of the function.

1. $\varphi(x, y, z) = xyz$; $(1, 1, 1)$

2. $\varphi(x, y, z) = x^2 y - \sin(xz)$; $(1, -1, \pi/4)$

3. $\varphi(x, y, z) = 2xy + xe^z$; $(-2, 1, 6)$

4. $\varphi(x, y, z) = \cos(xyz); \ (-1, 1, \pi/2)$

5. $\varphi(x, y, z) = \cosh(2xy) - \sinh(z); \ (0, 1, 1)$

6. $\varphi(x, y, z) = \sqrt{x^2 + y^2 + z^2}; \ (2, 2, 2)$

7. $\varphi(x, y, z) = \ln(x + y + z); \ (3, 1, -2)$

8. $\varphi(x, y, z) = -1/\|x\mathbf{i} + y\mathbf{j} + z\mathbf{k}\| \ ; \ (-2, 1, 1)$

9. $\varphi(x, y, z) = e^x \cos(y) \cos(z); \ (0, \pi/4, \pi/4)$

10. $\varphi(x, y, z) = 2x^3 y + ze^y; \ (1, 1, 2)$

In each of Problems 11 through 16, compute the directional derivative of the function in the direction of the given vector.

11. $\varphi(x, y, z) = 8xy^2 - xz; \ (1/\sqrt{3})(\mathbf{i} + \mathbf{j} + \mathbf{k})$

12. $\varphi(x, y, z) = \cos(x - y) + e^z; \ \mathbf{i} - \mathbf{j} + 2\mathbf{k}$

13. $\varphi(x, y, z) = x^2 yz^3; \ 2\mathbf{j} + \mathbf{k}$

14. $\varphi(x, y, z) = yz + xz + xy; \ \mathbf{i} - 4\mathbf{k}$

15. $\varphi(x, y, z) = \sin(x - y + 2z); \ -\mathbf{i} + \mathbf{j} + \mathbf{k}$

16. $\varphi(x, y, z) = 1 - x^2 - y^2 - xyz; \ \mathbf{i} + 3\mathbf{k}$

In each of Problems 17 through 26, find the equations of the tangent plane and normal line to the surface at the point.

17. $x^2 + y^2 + z^2 = 4; \ (1, 1, \sqrt{2})$

18. $z = x^2 + y; \ (-1, 1, 2)$

19. $z^2 = x^2 - y^2; \ (1, 1, 0)$

20. $\sinh(x + y + z) = 0; \ (0, 0, 0)$

21. $2x - 4y^2 + z^3 = 0; \ (-4, 0, 2)$

22. $x^2 - y^2 + z^2 = 0; \ (1, 1, 0)$

23. $2x - \cos(xyz) = 3; \ (1, \pi, 1)$

24. $3x^4 + 3y^4 + 6z^4 = 12; \ (1, 1, 1)$

25. $x^2 - 2y^2 + z^4 = 0; \ (1, 1, 1)$

26. $\cos(x) - \sin(y) + z = 1; \ (0, \pi, 0)$

In each of Problems 27 through 30, find the angle between the two surfaces at the given point of intersection. (Compute this angle as the angle between the normals to the surfaces at this point.)

27. $z = 3x^2 + 2y^2, \ -2x + 7y^2 - z = 0; \ (1, 1, 5)$

28. $x^2 + y^2 + z^2 = 4, \ z^2 + x^2 = 2; \ (1, \sqrt{2}, 1)$

29. $z = \sqrt{x^2 + y^2}, \ x^2 + y^2 = 8; \ (2, 2, \sqrt{8})$

30. $x^2 + y^2 + 2z^2 = 10, \ x + y + z = 5; \ (2, 2, 1)$

31. Suppose $\nabla\varphi = \mathbf{i} + \mathbf{k}$. What can be said about level surfaces of φ? Prove that the streamlines of $\nabla\varphi$ are orthogonal to the level surfaces of φ.

11.5 Divergence and Curl

The gradient operator ∇ produces a vector field from a scalar field. We will now discuss two other vector operations. One produces a scalar field from a vector field, and the other a vector field from a vector field.

DEFINITION 11.10 *Divergence*

The divergence of a vector field $\mathbf{F}(x, y, z) = f(x, y, z)\mathbf{i} + g(x, y, z)\mathbf{j} + h(x, y, z)\mathbf{k}$ is the scalar field

$$\text{div } \mathbf{F} = \frac{\partial f}{\partial x} + \frac{\partial g}{\partial y} + \frac{\partial h}{\partial z}.$$

For example, if $\mathbf{F} = 2xy\mathbf{i} + (xyz^2 - \sin(yz))\mathbf{j} + ze^{x+y}\mathbf{k}$, then

$$\text{div } \mathbf{F} = 2y + xz^2 - z\cos(yz) + e^{x+y}.$$

We read div $\mathbf{F}$ as the divergence of $\mathbf{F}$, or just "div $\mathbf{F}$."

DEFINITION 11.11   Curl

The curl of a vector field $\mathbf{F}(x, y, z) = f(x, y, z)\mathbf{i} + g(x, y, z)\mathbf{j} + h(x, y, z)\mathbf{k}$ is the vector field

$$\text{curl } \mathbf{F} = \left(\frac{\partial h}{\partial y} - \frac{\partial g}{\partial z}\right)\mathbf{i} + \left(\frac{\partial f}{\partial z} - \frac{\partial h}{\partial x}\right)\mathbf{j} + \left(\frac{\partial g}{\partial x} - \frac{\partial f}{\partial y}\right)\mathbf{k}.$$

This vector is read "curl of $\mathbf{F}$," or just "curl $\mathbf{F}$." For example, if $\mathbf{F} = y\mathbf{i} + 2xz\mathbf{j} + ze^x\mathbf{k}$, then

$$\text{curl } \mathbf{F} = -2x\mathbf{i} - ze^x\mathbf{j} + (2z - 1)\mathbf{k}.$$

Divergence, curl, and gradient can all be thought of in terms of the vector operations of multiplication of a vector by a scalar, dot product, and cross product, using the *del operator* ∇. This is defined by

$$\nabla = \frac{\partial}{\partial x}\mathbf{i} + \frac{\partial}{\partial y}\mathbf{j} + \frac{\partial}{\partial z}\mathbf{k}.$$

The symbol ∇, which is read "del," is treated like a vector in carrying out calculations, and the "product" of $\partial/\partial x$, $\partial/\partial y$, and $\partial/\partial z$ with a function $\varphi(x, y, z)$ is interpreted to mean, respectively, $\partial\varphi/\partial x$, $\partial\varphi/\partial y$, and $\partial\varphi/\partial z$. In this way, the gradient of φ is the product of the vector ∇ with the scalar function φ:

$$\left(\frac{\partial}{\partial x}\mathbf{i} + \frac{\partial}{\partial y}\mathbf{j} + \frac{\partial}{\partial z}\mathbf{k}\right)\varphi = \frac{\partial\varphi}{\partial x}\mathbf{i} + \frac{\partial\varphi}{\partial y}\mathbf{j} + \frac{\partial\varphi}{\partial z}\mathbf{k}$$

$$= \nabla\varphi = \text{gradient of } \varphi.$$

The divergence of a vector is the dot product of del with the vector:

$$\nabla \cdot \mathbf{F} = \left(\frac{\partial}{\partial x}\mathbf{i} + \frac{\partial}{\partial y}\mathbf{j} + \frac{\partial}{\partial z}\mathbf{k}\right) \cdot (f\mathbf{i} + g\mathbf{j} + h\mathbf{k})$$

$$= \frac{\partial f}{\partial x} + \frac{\partial g}{\partial y} + \frac{\partial h}{\partial z} = \text{divergence of } \mathbf{F}.$$

And the curl of a vector is the cross product of del with the vector:

$$\nabla \times \mathbf{F} = \begin{vmatrix} \mathbf{i} & \mathbf{j} & \mathbf{k} \\ \partial/\partial x & \partial/\partial y & \partial/\partial z \\ f & g & h \end{vmatrix}$$

$$= \left(\frac{\partial h}{\partial y} - \frac{\partial g}{\partial z}\right)\mathbf{i} + \left(\frac{\partial f}{\partial z} - \frac{\partial h}{\partial x}\right)\mathbf{j} + \left(\frac{\partial g}{\partial x} - \frac{\partial f}{\partial y}\right)\mathbf{k}$$

$$= \text{curl } \mathbf{F}.$$

Informally,

$$\text{del times} = \text{gradient,}$$

$$\text{del dot} = \text{divergence,}$$

and

$$\text{del cross} = \text{curl.}$$

This provides a way of thinking of gradient, divergence, and curl in terms of familiar vector operations involving del and will prove to be an efficient tool in carrying out computations.

There are two fundamental relationships between gradient, divergence, and curl. The first states that the curl of a gradient is the zero vector.

THEOREM 11.6   *Curl of a Gradient*

Let φ be continuous with continuous first and second partial derivatives. Then

$$\nabla \times (\nabla \varphi) = \mathbf{O}. \ \blacksquare$$

This conclusion can also be written

$$\operatorname{curl}(\nabla \varphi) = \mathbf{O}.$$

The zero on the right is the zero vector, since the curl of a scalar field is a vector.

Proof   By direct computation,

$$\nabla \times (\nabla \varphi) = \nabla \times \left(\frac{\partial \varphi}{\partial x}\mathbf{i} + \frac{\partial \varphi}{\partial y}\mathbf{j} + \frac{\partial \varphi}{\partial z}\mathbf{k} \right)$$

$$= \begin{vmatrix} \mathbf{i} & \mathbf{j} & \mathbf{k} \\ \partial/\partial x & \partial/\partial y & \partial/\partial z \\ \partial\varphi/\partial x & \partial\varphi/\partial y & \partial\varphi/\partial z \end{vmatrix}$$

$$= \left(\frac{\partial^2 \varphi}{\partial y\, \partial z} - \frac{\partial^2 \varphi}{\partial z\, \partial y} \right)\mathbf{i} + \left(\frac{\partial^2 \varphi}{\partial z\, \partial x} - \frac{\partial^2 \varphi}{\partial x\, \partial z} \right)\mathbf{j} + \left(\frac{\partial^2 \varphi}{\partial x\, \partial y} - \frac{\partial^2 \varphi}{\partial y\, \partial x} \right)\mathbf{k} = \mathbf{O}$$

because the paired mixed partial derivatives in each set of parentheses are equal. $\blacksquare$

The second relationship states that the divergence of a curl is the number zero.

THEOREM 11.7

Let $\mathbf{F}$ be a continuous vector field whose components have continuous first and second partial derivatives. Then

$$\nabla \cdot (\nabla \times \mathbf{F}) = 0. \ \blacksquare$$

We may also write

$$\operatorname{div}(\operatorname{curl} \mathbf{F}) = 0.$$

Proof   As with the preceding theorem, proceed by direct computation:

$$\nabla \cdot (\nabla \times \mathbf{F})$$

$$= \frac{\partial}{\partial x}\left(\frac{\partial h}{\partial y} - \frac{\partial g}{\partial z} \right) + \frac{\partial}{\partial y}\left(\frac{\partial f}{\partial z} - \frac{\partial h}{\partial x} \right) + \frac{\partial}{\partial z}\left(\frac{\partial g}{\partial x} - \frac{\partial f}{\partial y} \right)$$

$$= \frac{\partial^2 h}{\partial x\, \partial y} - \frac{\partial^2 g}{\partial x\, \partial z} + \frac{\partial^2 f}{\partial y\, \partial z} - \frac{\partial^2 h}{\partial y\, \partial x} + \frac{\partial^2 g}{\partial z\, \partial x} - \frac{\partial^2 f}{\partial z\, \partial y} = 0$$

because equal mixed partials appear in pairs with opposite signs. $\blacksquare$

Divergence and curl have physical interpretations, two of which we will now develop.

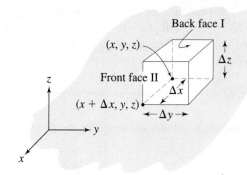

FIGURE 11.38

11.5.1 A Physical Interpretation of Divergence

Suppose $\mathbf{F}(x, y, z, t)$ is the velocity of a fluid at point (x, y, z) and time t. Time plays no role in computing divergence, but is included here because normally a velocity vector does depend on time.

Imagine a small rectangular box within the fluid, as in Figure 11.38. We would like some measure of the rate per unit volume at which fluid flows out of this box across its faces, at any given time.

First look at the front face II and the back face I in the diagram. The normal vector pointing out of the box from face II is $\mathbf{i}$. The flux of the flow out of the box across face II is the normal component of the velocity (dot product of $\mathbf{F}$ with $\mathbf{i}$), multiplied by the area of this face:

$$\text{flux outward across face II} = \mathbf{F}(x + \Delta x, y, z, t) \cdot \mathbf{i} \Delta y \Delta z.$$
$$= f(x + \Delta x, y, z, t) \Delta y \Delta z.$$

On face I, the unit outer normal is $-\mathbf{i}$, so

$$\text{flux outward across face I} = \mathbf{F}(x, y, z, t) \cdot (-\mathbf{i}) \Delta y \Delta z$$
$$= -f(x, y, z, t) \Delta y \Delta z.$$

The total outward flux across faces I and II is therefore

$$[f(x + \Delta x, y, z, t) - f(x, y, z, t)] \Delta y \Delta z.$$

A similar calculation can be done for the other two pairs of sides. The total flux of fluid out of the box across its faces is

$$[f(x + \Delta x, y, z, t) - f(x, y, z, t)] \Delta y \Delta z + [g(x, y + \Delta y, z, t) - g(x, y, z, t)] \Delta x \Delta z$$
$$+ [h(x, y, z + \Delta z, t) - h(x, y, z, t)] \Delta x \Delta y.$$

The flux per unit volume is obtained by dividing this flux by the volume $\Delta x \Delta y \Delta z$ of the box:

$$\text{flux per unit volume out of the box} = \frac{f(x + \Delta x, y, z, t) - f(x, y, z, t)}{\Delta x}$$
$$+ \frac{g(x, y + \Delta y, z, t) - g(x, y, z, t)}{\Delta y}$$
$$+ \frac{h(x, y, z + \Delta z, t) - h(x, y, z, t)}{\Delta z}.$$

Now take the limit as $\Delta x \to 0$, $\Delta y \to 0$, and $\Delta z \to 0$. The box shrinks to the point (x, y, z) and the flux per unit volume approaches

$$\frac{\partial f}{\partial x} + \frac{\partial g}{\partial y} + \frac{\partial h}{\partial z},$$

which is the divergence of $\mathbf{F}(x, y, z, t)$ at time t. We may therefore intrepret the divergence of $\mathbf{F}$ as a measure of the outward flow or expansion of the fluid from this point.

11.5.2   A Physical Interpretation of Curl

Suppose an object rotates with uniform angular speed ω about a line L as in Figure 11.39. The angular velocity vector $\boldsymbol{\Omega}$ has magnitude ω and is directed along L as a right-handed screw would progress if given the same sense of rotation as the object.

Put L through the origin as a convenience, and let $\mathbf{R} = x\mathbf{i} + y\mathbf{j} + z\mathbf{k}$ for any point (x, y, z) on the rotating object. Let $\mathbf{T}(x, y, z)$ be the tangential linear velocity. Then

$$\|\mathbf{T}\| = \omega \|\mathbf{R}\| \sin(\theta) = \|\boldsymbol{\Omega} \times \mathbf{R}\|,$$

where θ is the angle between $\mathbf{R}$ and $\boldsymbol{\Omega}$. Since $\mathbf{T}$ and $\boldsymbol{\Omega} \times \mathbf{R}$ have the same direction and magnitude, we conclude that $\mathbf{T} = \boldsymbol{\Omega} \times \mathbf{R}$. Now write $\boldsymbol{\Omega} = a\mathbf{i} + b\mathbf{j} + c\mathbf{k}$ to obtain

$$\mathbf{T} = \boldsymbol{\Omega} \times \mathbf{R} = (bz - cy)\mathbf{i} + (cx - az)\mathbf{j} + (ay - bx)\mathbf{k}.$$

Then

$$\nabla \times \mathbf{T} = \begin{vmatrix} \mathbf{i} & \mathbf{j} & \mathbf{k} \\ \partial/\partial x & \partial/\partial y & \partial/\partial z \\ bz - cy & cx - az & ay - bx \end{vmatrix}$$

$$= 2a\mathbf{i} + 2b\mathbf{j} + 2c\mathbf{k} = 2\boldsymbol{\Omega}.$$

Therefore

$$\boldsymbol{\Omega} = \tfrac{1}{2}\nabla \times \mathbf{T}.$$

The angular velocity of a uniformly rotating body is a constant times the curl of the linear velocity. Because of this interpretation, *curl* was once written *rot* (for rotation), particularly in British treatments of mechanics. This is also the motivation for the term *irrotational* for a vector field whose curl is zero.

Other interpretations of divergence and curl follow from vector integral theorems we will see in the next chapter.

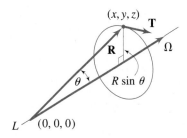

FIGURE 11.39 *Angular velocity as the curl of the linear velocity.*

SECTION 11.5 PROBLEMS

In each of Problems 1 through 6, compute $\nabla \cdot \mathbf{F}$ and $\nabla \times \mathbf{F}$ and verify explicitly that $\nabla \cdot (\nabla \times \mathbf{F}) = 0$.

1. $\mathbf{F} = x\mathbf{i} + y\mathbf{j} + 2z\mathbf{k}$

2. $\mathbf{F} = \sinh(xyz)\mathbf{j}$

3. $\mathbf{F} = 2xy\mathbf{i} + xe^y\mathbf{j} + 2z\mathbf{k}$

4. $\mathbf{F} = \sinh(x)\mathbf{i} + \cosh(xyz)\mathbf{j} - (x+y+z)\mathbf{k}$

5. $\mathbf{F} = x^2\mathbf{i} + y^2\mathbf{j} + z^2\mathbf{k}$

6. $\mathbf{F} = \sinh(x-z)\mathbf{i} + 2y\mathbf{j} + (z-y^2)\mathbf{k}$

In each of Problems 7 through 12, compute $\nabla\varphi$ and verify explicitly that $\nabla \times (\nabla\varphi) = \mathbf{O}$.

7. $\varphi(x, y, z) = x - y + 2z^2$

8. $\varphi(x, y, z) = 18xyz + e^x$

9. $\varphi(x, y, z) = -2x^3yz^2$

10. $\varphi(x, y, z) = \sin(xz)$

11. $\varphi(x, y, z) = x\cos(x+y+z)$

12. $\varphi(x, y, z) = e^{x+y+z}$

13. Let $\varphi(x, y, z)$ be a scalar field and $\mathbf{F}$ a vector field. Derive expressions for $\nabla \cdot (\varphi\mathbf{F})$ and $\nabla \times (\varphi\mathbf{F})$ in terms of operations applied to $\varphi(x, y, z)$ and $\mathbf{F}$.

14. Let $\mathbf{F} = f\mathbf{i} + g\mathbf{j} + h\mathbf{k}$ be a vector field. Define

$$\mathbf{F} \cdot \nabla = \left(f\frac{\partial}{\partial x}\right)\mathbf{i} + \left(g\frac{\partial}{\partial y}\right)\mathbf{j} + \left(h\frac{\partial}{\partial z}\right)\mathbf{k}.$$

Let $\mathbf{G}$ be a vector field. Show that

$$\nabla(\mathbf{F} \cdot \mathbf{G}) = (\mathbf{F} \cdot \nabla)\mathbf{G} + (\mathbf{G} \cdot \nabla)\mathbf{F} + \mathbf{F} \times (\nabla \times \mathbf{G})$$
$$+ \mathbf{G} \times (\nabla \times \mathbf{F}).$$

15. Let $\mathbf{F}$ and $\mathbf{G}$ be vector fields. Prove that

$$\nabla \cdot (\mathbf{F} \times \mathbf{G}) = \mathbf{G} \cdot (\nabla \times \mathbf{F}) - \mathbf{F} \cdot (\nabla \times \mathbf{G}).$$

16. Let $\varphi(x, y, z)$ and $\psi(x, y, z)$ be scalar fields. Prove that $\nabla \cdot (\nabla\varphi \times \nabla\psi) = 0$.

17. Let $\mathbf{R} = x\mathbf{i} + y\mathbf{j} + z\mathbf{k}$ and let $r = \|\mathbf{R}\|$.

(a) Prove that $\nabla r^n = nr^{n-2}\mathbf{R}$ for $n = 1, 2, \ldots$.

(b) Let φ be a real-valued function of one variable. Prove that $\nabla \times (\varphi(r)\mathbf{R}) = \mathbf{O}$.

18. Let $\mathbf{F}$ be a vector field. Prove that

$$\nabla \times (\nabla \times \mathbf{F}) = \nabla(\nabla \cdot \mathbf{F}) - \nabla^2\mathbf{F},$$

where $\nabla^2\mathbf{F} = (\partial^2\mathbf{F}/\partial x^2) + (\partial^2\mathbf{F}/\partial y^2) + (\partial^2\mathbf{F}/\partial z^2)$.

19. Let $\mathbf{A}$ be a constant vector and $\mathbf{R} = x\mathbf{i} + y\mathbf{j} + z\mathbf{k}$.

(a) Prove that $\nabla(\mathbf{A} \cdot \mathbf{R}) = \mathbf{A}$.

(b) Prove that $\nabla \cdot (\mathbf{R} - \mathbf{A}) = 3$.

(c) Prove that $\nabla \times (\mathbf{R} - \mathbf{A}) = \mathbf{O}$.

20. Let $\mathbf{F}$ and $\mathbf{G}$ be vector fields. Prove that

$$\nabla \times (\mathbf{F} \times \mathbf{G}) = (\mathbf{G} \cdot \nabla)\mathbf{F} - (\mathbf{F} \cdot \nabla)\mathbf{G}$$
$$+ (\nabla \cdot \mathbf{G})\mathbf{F} - (\nabla \cdot \mathbf{F})\mathbf{G}.$$

CHAPTER 12

Vector Integral Calculus

This chapter is devoted to integrals of vector fields over curves and surfaces, and relationships between such integrals. These have important uses in solving partial differential equations and in constructing models used in the sciences and engineering.

12.1 Line Integrals

We begin with the integral of a vector field over a curve. This requires some background on curves.

Suppose a curve C in 3-space is given by parametric equations

$$x = x(t), y = y(t), z = z(t) \quad \text{for } a \leq t \leq b.$$

We call $x(t)$, $y(t)$, and $z(t)$ the *coordinate functions* of C.

We will think of C not only as a geometric locus of points $(x(t), y(t), z(t))$ but also as having an orientation or direction, given by the direction this point moves along C as t increases from a to b. Denote this orientation by putting arrows on the graph of the curve (Figure 12.1). Trajectories of a system of differential equations are also oriented curves, with the particle moving along the geometric locus in a direction dictated by the flow of the system. We call $(x(a), y(a), z(a))$ the *initial point* of C, and $(x(b), y(b), z(b))$ the *terminal point*. A curve is *closed* if the initial and terminal points are the same.

EXAMPLE 12.1

Let C be given by

$$x = 2\cos(t), y = 2\sin(t), z = 4; \qquad 0 \leq t \leq 2\pi.$$

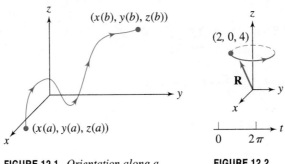

FIGURE 12.1 *Orientation along a curve.*

FIGURE 12.2

A graph of C is shown in Figure 12.2. This graph is the circle of radius 2 about the origin in the plane $z = 4$. The arrow on the curve indicates its orientation (the direction of motion of $(2\cos(t), 2\sin(t), 4)$ around the graph as t varies from 0 to 2π). The initial point is $(2, 0, 4)$, obtained at $t = 0$, and the terminal point is also $(2, 0, 4)$, obtained at $t = 2\pi$. This curve is closed.

Contrast C with the curve K given by

$$x = 2\cos(t), \; y = 2\sin(t), \; z = 4; \qquad 0 \le t \le 3\pi.$$

K has the same graph and counterclockwise orientation as C but is a different curve. K is not closed. The initial point of K is $(2, 0, 4)$, but its terminal point is $(-2, 0, 4)$. A particle moving about K goes around the circle once and then makes another half-circle. It would presumably take more energy to move a particle about K than about C. K is also longer than C. ∎

A curve is

continuous if its coordinate functions are continuous on the parameter interval,

differentiable if its coordinate functions are differentiable, and

smooth if its coordinate functions have continuous derivatives, which are not all zero for the same value of t.

Because $x'(t)\mathbf{i} + y'(t)\mathbf{j} + z'(t)\mathbf{k}$ is tangent to the curve if this is not the zero vector, a smooth curve is one that has a continuous tangent vector.

The *graph* of a curve consists of all points $(x(t), y(t), z(t))$ as t varies over the parameter interval. As Example 12.1 shows, there is more to a curve than just its graph. We are dealing with curves both as geometric objects and as oriented objects, having a sense of direction along the graph.

We will now define the line integral of functions over a smooth curve.

DEFINITION 12.1 *Line Integral*

Suppose a smooth curve C has coordinate functions $x = x(t)$, $y = y(t)$, $z = z(t)$ for $a \le t \le b$. Let $f(x, y, z)$, $g(x, y, z)$, and $h(x, y, z)$ be continuous at least on the graph of C. Then, the line integral

$$\int_C f(x, y, z)\, dx + g(x, y, z)\, dy + h(x, y, z)\, dz$$

is defined by

$$\int_C f(x, y, z)\, dx + g(x, y, z)\, dy + h(x, y, z)\, dz$$

(12.1)

$$= \int_a^b \left[f(x(t), y(t), z(t)) \frac{dx}{dt} + g(x(t), y(t), z(t)) \frac{dy}{dt} + h(x(t), y(t), z(t)) \frac{dz}{dt} \right] dt.$$

We can write this line integral more compactly as

$$\int_C f\, dx + g\, dy + h\, dz.$$

To evaluate $\int_C f\, dx + g\, dy + h\, dz$, substitute the coordinate functions $x = x(t)$, $y = y(t)$, and $z = z(t)$ into $f(x, y, z)$, $g(x, y, z)$, and $h(x, y, z)$, obtaining functions of t. Further, substitute

$$dx = \frac{dx}{dt}\, dt, dy = \frac{dy}{dt}\, dt, \quad \text{and} \quad dz = \frac{dz}{dt}\, dt.$$

This results in the Riemann integral on the right side of equation (12.1) of a function of t over the range of values of this parameter.

EXAMPLE 12.2

Evaluate the line integral $\int_C x\, dx - yz\, dy + e^z\, dz$ if C is given by

$$x = t^3, y = -t, z = t^2; \qquad 1 \le t \le 2.$$

Here,

$$f(x, y, z) = x, g(x, y, z) = -yz, \quad \text{and} \quad h(x, y, z) = e^z$$

and, on C,

$$dx = 3t^2\, dt, dy = -dt, \quad \text{and} \quad dz = 2t\, dt.$$

Then

$$\int_C x\, dx - yz\, dy + e^z\, dz$$

$$= \int_1^2 \left[t^3(3t^2) - (-t)(t^2)(-1) + e^{t^2}(2t) \right] dt$$

$$= \int_1^2 \left[3t^5 - t^3 + 2te^{t^2} \right] dt = \frac{111}{4} + e^4 - e. \ \blacksquare$$

EXAMPLE 12.3

Evaluate $\int_C xyz\, dx - \cos(yz)\, dy + xz\, dz$ over the straight-line segment from $(1, 1, 1)$ to $(-2, 1, 3)$.

Here we are left to find the coordinate functions of the curve. Parametric equations of the line through these points are

$$x(t) = 1 - 3t, \quad y(t) = 1, \quad z(t) = 1 + 2t.$$

We must let t vary from 0 to 1 for the initial point to be $(1, 1, 1)$ and the terminal point to be $(-2, 1, 3)$.

Now

$$\int_C xyz \, dx - \cos(yz) \, dy + xz \, dz$$

$$= \int_0^1 [(1 - 3t)(1)(1 + 2t)(-3) - \cos(1 + 2t)(0) + (1 - 3t)(1 + 2t)(2)] \, dt$$

$$= \int_0^1 \left(-1 + t + 6t^2\right) dt = \frac{3}{2}. \quad \blacksquare$$

If C is a smooth curve in the x, y plane (zero z-component), and $f(x, y)$ and $g(x, y)$ are continuous on C, then we can write a line integral

$$\int_C f(x, y) \, dx + g(x, y) \, dy,$$

which we refer to as a *line integral in the plane*. We evaluate this according to Definition 12.1, except now there is no z-component.

EXAMPLE 12.4

Evaluate $\int_C xy \, dx - y \sin(x) \, dy$ if C is given by $x(t) = t^2$ and $y(t) = t$ for $-1 \leq t \leq 4$.
Proceed:

$$\int_C xy \, dx - y \sin(x) \, dy$$

$$= \int_{-1}^4 [t^2 t (2t) - t \sin(t^2)(1)] \, dt$$

$$= 410 + \tfrac{1}{2} \cos(16) - \tfrac{1}{2} \cos(1). \quad \blacksquare$$

In these examples we have included all the terms to follow equation (12.1) very literally. However, with some experience there are obvious shortcuts one can take. In Example 12.3, for example, y is constant on the curve, so $dy = 0$ and the term $g(x(t), y(t), z(t)) \, dy/dt$ could have been simply omitted.

Thus far we can integrate only over a smooth curve. This requirement can be relaxed as follows. A curve C is *piecewise smooth* if $x'(t)$, $y'(t)$, and $z'(t)$ are continuous, and not all zero for the same value of t, at all but possibly finitely many values of t. Since $x'(t)\mathbf{i} + y'(t)\mathbf{j} + z'(t)\mathbf{k}$ is the tangent vector to C if this is not the zero vector, this condition means that a piecewise smooth curve has a continuous tangent at all but finitely many points. Such a curve typically has the appearance of Figure 12.3, with smooth pieces $C_1, \ldots, C_n$ connected at points where the curve may have no tangent. We will refer to a piecewise smooth curve as a *path*.

In Figure 12.3 the terminal point of C_j is the initial point of C_{j+1} for $j = 1, \ldots, n - 1$. The segments $C_1, \ldots, C_n$ are in order as one moves from the initial to the terminal point of C. This

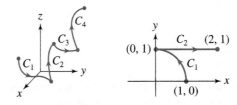

FIGURE 12.3 *Typical piecewise smooth curve (path).*   **FIGURE 12.4**

is indicated by the arrows showing orientation along the smooth pieces of C. If f, g, and h are continuous over each C_j, then we define

$$\int_C f \, dx + g \, dy + h \, dz$$

$$= \int_{C_1} f \, dx + g \, dy + h \, dz + \cdots + \int_{C_n} f \, dx + g \, dy + h \, dz.$$

This allows us to take line integrals over paths, rather than restricting the integral to smooth curves.

EXAMPLE 12.5

Let C be the curve consisting of the quarter circle $x^2 + y^2 = 1$ in the x, y-plane, from $(1, 0)$ to $(0, 1)$, followed by the horizontal line segment from $(0, 1)$ to $(2, 1)$. Compute $\int_C x^2 y \, dx + y^2 \, dy$.

C is piecewise smooth and consists of two smooth pieces (Figure 12.4). Parametrize these as follows. The quarter circle part is

$$C_1 : x = \cos(t), \ y = \sin(t); \qquad 0 \le t \le \pi/2.$$

The straight segment is

$$C_2 : x = p, \ y = 1; \qquad 0 \le p \le 2.$$

We have used different names for the parameters on the curves to help distinguish them.

Now evaluate the line integral along each of the two curves making up the path. For the line integral over C_1, compute

$$\int_{C_1} x^2 y \, dx + y^2 \, dy = \int_0^{\pi/2} [\cos^2(t) \sin(t)(-\sin(t)) + \sin^2(t) \cos(t)] \, dt$$

$$= \int_0^{\pi/2} \left(-\cos^2(t) \sin^2(t) + \sin^2(t) \cos(t) \right) dt = -\frac{1}{16}\pi + \frac{1}{3}.$$

Next evaluate the line integral over C_2. On C_2, $x = p$ and $y = 1$, so $dy = 0$ and

$$\int_{C_2} x^2 y \, dx + y^2 \, dy = \int_0^2 p^2 \, dp = \frac{8}{3}.$$

Then

$$\int_C x^2 y \, dx + y^2 \, dy = -\frac{1}{16}\pi + \frac{1}{3} + \frac{8}{3} = -\frac{\pi}{16} + 3. \quad \blacksquare$$

It is sometimes useful to think of a line integral in terms of vector operations, particularly in the next section when we deal with potential functions. Consider $\int_C f \, dx + g \, dy + h \, dz$. Form a vector field

$$\mathbf{F}(x, y, z) = f(x, y, z)\mathbf{i} + g(x, y, z)\mathbf{j} + h(x, y, z)\mathbf{k}.$$

If C has coordinate functions $x = x(t)$, $y = y(t)$, $z = z(t)$, we can form the position vector, for C:

$$\mathbf{R}(t) = x(t)\mathbf{i} + y(t)\mathbf{j} + z(t)\mathbf{k}.$$

At any time t, the vector $\mathbf{R}(t)$ can be represented by an arrow from the origin to the point $(x(t), y(t), z(t))$ on C. As t varies, this vector pivots about the origin and adjusts its length to sweep out the curve. If C is smooth, then the tangent vector $\mathbf{R}'(t)$ is continuous. Now $d\mathbf{R} = dx\,\mathbf{i} + dy\,\mathbf{j} + dz\,\mathbf{k}$, so

$$\mathbf{F} \cdot d\mathbf{R} = f(x, y, z)\, dx + g(x, y, z)\, dy + h(x, y, z)\, dy$$

and

$$\int_C f(x, y, z)\, dx + g(x, y, z)\, dy + h(x, y, z)\, dz = \int_C \mathbf{F} \cdot d\mathbf{R}.$$

This is just another way of writing a line integral in terms of vector operations.

Line integrals arise in many contexts. For example, consider a force $\mathbf{F}$ causing a particle to move along a smooth curve C having position function $\mathbf{R}(t)$, where t varies from a to b. At any point $(x(t), y(t), z(t))$ of C, the particle may be thought of as moving in the direction of the tangent to its trajectory, and this tangent is $\mathbf{R}'(t)$. Now $\mathbf{F}(x(t), y(t), z(t)) \cdot \mathbf{R}'(t)$ is the dot product of a force with a direction and so has the dimensions of work. By integrating this function from a to b, we "sum" the work being done by $\mathbf{F}$ over the entire path of motion. This suggests that $\int_C \mathbf{F} \cdot d\mathbf{R}$, or $\int_C f \, dx + g \, dy + h \, dz$, can be interpreted as the work done by $\mathbf{F}$ in moving the particle over the path.

EXAMPLE 12.6

Calculate the work done by $\mathbf{F} = \mathbf{i} - y\mathbf{j} + xyz\mathbf{k}$ in moving a particle from $(0, 0, 0)$ to $(1, -1, 1)$ along the curve $x = t$, $y = -t^2$, $z = t$ for $0 \le t \le 1$.

The work is

$$\text{work} = \int_C \mathbf{F} \cdot d\mathbf{R} = \int_C dx - y \, dy + xyz \, dz$$

$$= \int_0^1 \left(1 + t^2(-2t) - t^4\right) dt$$

$$= \int_0^1 \left(1 - 2t^3 - t^4\right) dt = \frac{3}{10}.$$

The correct units (such as foot-pounds) would have to be provided from context. $\blacksquare$

Line integrals have some of the usual properties we associate with integrals.

THEOREM 12.1

Let C be a path having position vector $\mathbf{R}$. Let $\mathbf{F}$ and $\mathbf{G}$ be vector fields that are continuous at points of C. Then

1.

$$\int_C (\mathbf{F} + \mathbf{G}) \cdot d\mathbf{R} = \int_C \mathbf{F} \cdot d\mathbf{R} + \int_C \mathbf{G} \cdot d\mathbf{R}.$$

2. For any number α,

$$\int_C \alpha \mathbf{F} \cdot d\mathbf{R} = \alpha \int_C \mathbf{F} \cdot d\mathbf{R}. \quad \blacksquare$$

This theorem illustrates the efficiency of the vector notation for line integrals. We could also write conclusion (1) as

$$\int_C (f + f^*) \, dx + (g + g^*) \, dy + (h + h^*) \, dz$$

$$= \int_C f \, dx + g \, dy + h \, dz + \int_C f^* \, dx + g^* \, dy + h^* \, dz.$$

For Riemann integrals, reversing the limits of integration changes the sign of the integral: $\int_a^b f(x) \, dx = - \int_b^a f(x) \, dx$. The analogous property for line integrals involves reversing orientation on C. Given C with an orientation from an initial point P to a terminal point Q, let $-C$ denote the curve obtained from C by reversing the orientation to go from Q to P (Figure 12.5). Here is a more careful definition of this orientation reversal.

DEFINITION 12.2

Let C be a smooth curve with coordinate functions $x = x(t)$, $y = y(t)$, $z = z(t)$, for $a \le t \le b$. Then $-C$ denotes the curve having coordinate functions

$$\tilde{x}(t) = x(a + b - t), \quad \tilde{y}(t) = y(a + b - t), \quad \tilde{z}(t) = z(a + b - t)$$

for $a \le t \le b$.

The initial point of $-C$ is

$$(\tilde{x}(a), \tilde{y}(a), \tilde{z}(a)) = (x(b), y(b), z(b)),$$

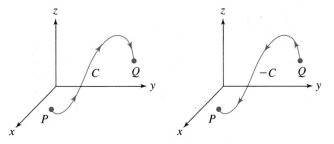

FIGURE 12.5 *Reversing orientation of a curve.*

the terminal point of C. And the terminal point of $-C$ is

$$(\tilde{x}(b), \tilde{y}(b), \tilde{z}(b)) = (x(a), y(a), z(a)),$$

the initial point of C. By the chain rule, $-C$ is piecewise smooth if C is piecewise smooth.

We will now show that the line integral of a vector field over $-C$ is the negative of the line integral of the vector field over C.

THEOREM 12.2

Let C be a smooth curve with coordinate functions $x = x(t)$, $y = y(t)$, $z = z(t)$. Let f, g, and h be continuous on C. Then

$$\int_{-C} f\,dx + g\,dy + h\,dz = -\int_C f\,dx + g\,dy + h\,dz.$$

Proof First,

$$\int_C f\,dx + g\,dy + h\,dz$$

$$= \int_a^b \left[f(x(t), y(t), z(t)) \frac{dx}{dt} + g(x(t), y(t), z(t)) \frac{dy}{dt} + h(x(t), y(t), z(t)) \frac{dz}{dt} \right] dt.$$

Similarly,

$$\int_{-C} f\,dx + g\,dy + h\,dz$$

$$= \int_a^b \left[f(\tilde{x}(t), \tilde{y}(t), \tilde{z}(t)) \frac{d\tilde{x}}{dt} + g(\tilde{x}(t), \tilde{y}(t), \tilde{z}(t)) \frac{d\tilde{y}}{dt} + h(\tilde{x}(t), \tilde{y}(t), \tilde{z}(t)) \frac{d\tilde{z}}{dt} \right] dt.$$

Change variables in the last integral by putting $s = a + b - t$. When $t = a$, $s = b$, and when $t = b$, $s = a$. Further,

$$\frac{d\tilde{x}}{dt} = \frac{d}{dt} x(a + b - t) = \frac{d}{dt} x(s) = \frac{dx}{ds} \frac{ds}{dt} = -\frac{dx}{ds}$$

and, similarly,

$$\frac{d\tilde{y}}{dt} = -\frac{dy}{ds}, \quad \frac{d\tilde{z}}{dt} = -\frac{dz}{ds}.$$

Finally,

$$dt = -ds.$$

Then

$$\int_{-C} f\,dx + g\,dy + h\,dz$$

$$= \int_b^a \left[-f(x(s), y(s), z(s)) \frac{dx}{ds} - g(x(s), y(s), z(s)) \frac{dy}{ds} - h(x(s), y(s), z(s)) \frac{dz}{ds} \right] (-1)\,ds$$

$$= -\int_a^b \left[f(x(s), y(s), z(s)) \frac{dx}{ds} + g(x(s), y(s), z(s)) \frac{dy}{ds} + h(x(s), y(s), z(s)) \frac{dz}{ds} \right] ds$$

$$= -\int_C f\,dx + g\,dy + h\,dz. \quad \blacksquare$$

In view of this theorem, the easiest way to evaluate $\int_{-C} f\,dx + g\,dy + h\,dz$ is usually to take the negative of $\int_C f\,dx + g\,dy + h\,dz$. We need not actually write the coordinate functions of $-C$, as was done in the proof.

EXAMPLE 12.7

A force $\mathbf{F}(x, y, z) = x^2\mathbf{i} - zy\mathbf{j} + x\cos(z)\mathbf{k}$ moves a particle along the path C given by $x = t^2$, $y = t$, $z = \pi t$ for $0 \le t \le 3$. The initial point is $P : (0, 0, 0)$ and the terminal point of C is $Q : (9, 3, 3\pi)$. Suppose we want the work done in moving the particle along this path from Q to P.

Since we want to go from the terminal to the initial point of C, the work done is $\int_{-C} \mathbf{F} \cdot d\mathbf{R}$. However, we do not need to formally define $-C$ in terms of new coordinate functions. We can simply calculate $\int_C \mathbf{F} \cdot d\mathbf{R}$, and take the negative of this. Calculate

$$\int_C \mathbf{F} \cdot d\mathbf{R} = \int_C f\,dx + g\,dy + h\,dz$$

$$= \int_0^3 \left[t^4(2t) - \pi t(t)(1) + t^2 \cos(\pi t)(\pi) \right] dt$$

$$= 243 - 9\pi - \frac{6}{\pi}.$$

The work done in moving the particle along the path from Q to P is

$$\frac{6}{\pi} + 9\pi - 243. \quad \blacksquare$$

12.1.1   Line Integral with Respect to Arc Length

Line integrals with respect to arc length occur in some uses of line integrals. Here is the definition of this kind of line integral.

DEFINITION 12.3   *Line Integral with Respect to Arc Length*

Let C be a smooth curve with coordinate functions $x = x(t)$, $y = y(t)$, $z = z(t)$ for $a \le t \le b$. Let φ be a real-valued function that is continuous on the graph of C. Then the integral of φ over C with respect to arc length is

$$\int_C \varphi(x, y, z)\,ds = \int_a^b \varphi(x(t), y(t), z(t))\sqrt{x'(t)^2 + y'(t)^2 + z'(t)^2}\,dt.$$

The rationale behind this definition is that the length function along C is

$$s(t) = \int_a^t \sqrt{x'(\xi)^2 + y'(\xi)^2 + z'(\xi)^2}\ d\xi.$$

Then

$$ds = \sqrt{x'(t)^2 + y'(t)^2 + z'(t)^2}\ dt,$$

suggesting the integral in the definition.

EXAMPLE 12.8

Evaluate $\int_C xy\,ds$ over the curve given by

$$x = 4\cos(t),\; y = 4\sin(t),\; z = -3 \quad \text{for } 0 \le t \le \pi/2.$$

Compute

$$\int_C xy\,ds = \int_0^{\pi/2} 4\cos(t)[4\sin(t)]\sqrt{16\sin^2(t) + 16\cos^2(t)}\; dt$$

$$= \int_0^{\pi/2} 64\cos(t)\sin(t)\,dt = 32. \quad \blacksquare$$

Line integrals with respect to arc length occur in calculations of mass, density, and various other quantities for one-dimensional objects. Suppose, for example, we want the mass of a thin wire bent into the shape of a piecewise smooth curve C having coordinate functions

$$x = x(t),\; y = y(t),\; z = z(t) \quad \text{for } a \le t \le b.$$

The wire is one-dimensional in the sense that (ideally) it has length but not area or volume.

We will derive an expression for the mass of the wire as follows. Let $\delta(x, y, z)$ be the density of the wire at any point. Partition $[a, b]$ into subintervals by inserting points

$$a = t_0 < t_1 < \cdots < t_{n-1} < t_n = b.$$

Choose these points Δt units apart, so $t_j - t_{j-1} = \Delta t$. These partition points of $[a, b]$ determine points

$$P_j : (x(t_j), y(t_j), z(t_j))$$

along C, as shown in Figure 12.6. Assuming that the density function is continuous, we can choose Δt sufficiently small that on the piece of wire between P_{j-1} and P_j, the values of the density function are approximated to whatever accuracy we wish by $\delta(P_j)$. The length of the segment of wire between P_{j-1} and P_j is $\Delta s = s(P_j) - s(P_{j-1})$, which is approximated by

$$ds = \sqrt{x'(t_j)^2 + y'(t_j)^2 + z'(t_j)^2}\,\Delta t.$$

The density of this piece of wire between P_{j-1} and P_j is therefore approximately the "nearly" constant value of the density on this piece, times the length of this piece of wire, this product being

$$\delta(x(t_j), y(t_j), z(t_j))\sqrt{x'(t_j)^2 + y'(t_j)^2 + z'(t_j)^2}\,\Delta t.$$

The mass of the entire length of wire is approximately the sum of the masses of these pieces:

$$\text{mass} \approx \sum_{j=1}^{n} \delta(x(t_j), y(t_j), z(t_j))\sqrt{x'(t_j)^2 + (y'(t_j)^2 + z'(t_j)^2}\,\Delta t,$$

in which $\approx$ means "approximately equal." Recognize this as the Riemann sum for a definite integral to obtain, in the limit as $\Delta t \to 0$,

$$\text{mass} = \int_a^b \delta(x(t), y(t), z(t))\sqrt{x'(t)^2 + y'(t)^2 + z'(t)^2}\; dt$$

$$= \int_C \delta(x, y, z)\,ds.$$

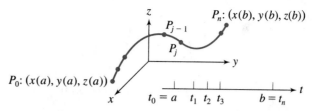

FIGURE 12.6

A similar argument leads to coordinates $(\bar{x}, \bar{y}, \bar{z})$ of the center of mass of the wire:

$$\bar{x} = \frac{1}{m} \int_C x\delta(x, y, z)\, ds, \quad \bar{y} = \frac{1}{m} \int_C y\delta(x, y, z)\, ds, \quad \bar{z} = \frac{1}{m} \int_C z\delta(x, y, z)\, ds.$$

in which m is the mass of the wire.

EXAMPLE 12.9

A wire is bent into the shape of the quarter circle C given by

$$x = 2\cos(t), \ y = 2\sin(t), \ z = 3 \quad \text{for } 0 \le t \le \pi/2.$$

The density function is $\delta(x, y, z) = xy^2$ grams/centimeter. We want the mass and center of mass of the wire.

The mass is

$$m = \int_C xy^2\, ds$$

$$= \int_0^{\pi/2} 2\cos(t)[2\sin(t)]^2\sqrt{4\sin^2(t) + 4\cos^2(t)}\ dt$$

$$= \int_0^{\pi/2} 16\cos(t)\sin^2(t)\, dt = \frac{16}{3} \text{ grams.}$$

Now compute the coordinates of the center of mass. First,

$$\bar{x} = \frac{1}{m} \int_C x\delta(x, y, z)\, ds$$

$$= \frac{3}{16} \int_0^{\pi/2} [2\cos(t)]^2[2\sin(t)]^2\sqrt{4\sin^2(t) + 4\cos^2(t)}\ dt$$

$$= 6 \int_0^{\pi/2} \cos^2(t)\sin^2(t)\, dt = \frac{3\pi}{8}.$$

Next,

$$\bar{y} = \frac{1}{m} \int_C y\delta(x, y, z)\, ds$$

$$= \frac{3}{16} \int_0^{\pi/2} [2\cos(t)][2\sin(t)]^3\sqrt{4\sin^2(t) + 4\cos^2(t)}\ dt$$

$$= 6 \int_0^{\pi/2} \cos(t)\sin^3(t)\, dt = \frac{3}{2}.$$

Finally,

$$\bar{z} = \frac{1}{m} \int_C z\delta(x, y, z) \, ds$$

$$= \frac{3}{16} \int_0^{\pi/2} 3[2\cos(t)][2\sin(t)]^2 \sqrt{4\sin^2(t) + 4\cos^2(t)} \, dt$$

$$= 9 \int_0^{\pi/2} \sin^2(t)\cos(t) \, dt = 3.$$

The last result could have been anticipated, since the z-component on the curve is constant. The center of mass is

$$(3\pi/8, 3/2, 3) . \quad \blacksquare$$

SECTION 12.1   PROBLEMS

In each of Problems 1 through 25, evaluate the line integral

1. $\int_C x \, dx - dy + z \, dz$, with C given by $x(t) = t$, $y(t) = t$, $z(t) = t^3$ for $0 \le t \le 1$

2. $\int_C -4x \, dx + y^2 \, dy - yz \, dz$, with C given by $x(t) = -t^2$, $y(t) = 0$, $z(t) = -3t$ for $0 \le t \le 1$

3. $\int_C (x + y) \, ds$, where C is given by $x = y = t$, $z = t^2$ for $0 \le t \le 2$

4. $\int_C x^2z \, dz$, where C is the line segment from $(0, 1, 1)$ to $(1, 2, -1)$

5. $\int_C \mathbf{F} \cdot d\mathbf{R}$, where $\mathbf{F} = \cos(x)\mathbf{i} - y\mathbf{j} + xz\mathbf{k}$ and $\mathbf{R} = t\mathbf{i} - t^2\mathbf{j} + \mathbf{k}$ for $0 \le t \le 3$

6. $\int_C 4xy \, ds$, with C given by $x = y = t$, $z = 2t$ for $1 \le t \le 2$

7. $\int_C \mathbf{F} \cdot d\mathbf{R}$, with $\mathbf{F} = x\mathbf{i} + y\mathbf{j} - z\mathbf{k}$ and C the circle $x^2 + y^2 = 4$, $z = 0$, going around once counterclockwise

8. $\int_C yz \, ds$, with C the parabola $z = y^2$, $x = 1$ for $0 \le y \le 2$

9. $\int_C -xyz \, dz$, with C the curve $x = 1$, $y = \sqrt{z}$ for $4 \le z \le 9$

10. $\int_C xz \, dy$, with C the curve $x = y = t$, $z = -4t^2$ for $1 \le t \le 3$

11. $\int_C 8z^2 \, ds$, with C the curve $x = y = 2t^2$, $z = 1$ for $1 \le t \le 2$

12. $\int_C \mathbf{F} \cdot d\mathbf{R}$, with $\mathbf{F} = \mathbf{i} - x\mathbf{j} + \mathbf{k}$ and $\mathbf{R} = \cos(t)\mathbf{i} - \sin(t)\mathbf{j} + t\mathbf{k}$ for $0 \le t \le \pi$

13. $\int_C 8x^2 \, dy$, with C given by $x = e^t$, $y = -t^2$, $z = t$ for $1 \le t \le 2$

14. $\int_C x \, dy - y \, dx$, C the curve $x = y = 2t$, $z = e^{-t}$ for $0 \le t \le 3$

15. $\int_C \sin(x) \, ds$, with C given by $x = t$, $y = 2t$, $z = 3t$ for $1 \le t \le 3$

16. $\int_C \mathbf{F} \cdot d\mathbf{R}$, with $\mathbf{F} = x\mathbf{i} + y\mathbf{j} - xyz\mathbf{k}$ and C the curve $x = y = t$, $z = -3t^2$ for $-1 \le t \le 3$

17. $\int_C 3y^3 \, ds$, with C given by $x = t$, $y = -2t$, $z = 5t$ for $1 \le t \le 4$

18. $\int_C \mathbf{F} \cdot d\mathbf{R}$, with $\mathbf{F} = \cos(xy)\mathbf{k}$ and C given by $x = 1$, $y = t$, $z = 2t - 1$ for $0 \le t \le \pi$

19. $\int_C (x + y + z^2) \, dx$, with C given by $x = 2y = z$ for $4 \le x \le 8$

20. $\int_C (x^2 - yz) \, dy$, with C given by $x = t$, $y = z = \sqrt{t}$ for $1 \le t \le 4$

21. $\int_C \mathbf{F} \cdot d\mathbf{R}$, with $\mathbf{F} = -y\mathbf{i} + xy\mathbf{j} + x^2\mathbf{k}$ and C given by $x = \sqrt{t}$, $y = 2t$, $z = t$ for $1 \le t \le 4$

22. $\int_C (x - y + 2z) \, ds$, with C given by $x = 3\cos(t)$, $y = 2$, $z = 3\sin(t)$ for $0 \le t \le \pi$

23. $\int_C \sin(z) \, dy$, with C given by $x = 1 - t$, $y = -t$, $z = 1 - t$ for $2 \le t \le 5$

24. $\int_C \mathbf{F} \cdot d\mathbf{R}$, with $\mathbf{F} = \sin(x)\mathbf{i} + 2z\mathbf{j} - \mathbf{k}$ and C given by $x = t$, $y = t^2$, $z = t^3$ for $0 \le t \le 2$

25. $\int_C 3y^3 \, ds$, with C given by $x = z = t^2$, $y = 1$ for $0 \le t \le 3$

26. Find the mass and center of mass of a thin, straight wire from the origin to $(3, 3, 3)$ if $\delta(x, y, z) = x + y + z$ grams per centimeter.

27. Find the work done by $\mathbf{F} = x^2\mathbf{i} - 2yz\mathbf{j} + z\mathbf{k}$ in moving an object along the line segment from $(1, 1, 1)$ to $(4, 4, 4)$.

28. Find the mass and center of mass of a wire bent into a rectangle with vertices $(1, 1, 3)$, $(1, 4, 3)$, $(6, 1, 3)$, and $(6, 4, 3)$, if the density is 3 grams per centimeter at each point on the sides between $(1, 4, 3)$ and $(1, 1, 3)$, and between $(1, 1, 3)$ and $(6, 1, 3)$, and 5 grams per centimeter at each point on the other two sides.

29. Suppose $\mathbf{F} = \nabla\varphi$.

(a) Let C be a path from P_0 to P_1 and suppose that φ and its first partial derivatives are continuous at points

of C. Show that

$$\int_C \mathbf{F} \cdot d\mathbf{R} = \varphi(P_1) - \varphi(P_0).$$

(b) Show that $\int_C \mathbf{F} \cdot d\mathbf{R} = 0$ if C is a closed curve.

30. Show that any Riemann integral $\int_a^b f(x)\,dx$ is equal to a line integral $\int_C \mathbf{F} \cdot d\mathbf{R}$ for appropriate choices of $\mathbf{F}$ and C. In this sense the line integral generalizes the Riemann integral.

12.2   Green's Theorem

Green's theorem was developed independently by the self-taught British amateur natural philosopher George Green and the Ukrainian mathematician Michel Ostrogradsky. They were studying potential theory (electric potentials, potential functions), and they obtained an important relationship between double integrals and line integrals in the plane.

Let C be a piecewise smooth curve in the plane, having coordinate functions $x = x(t)$, $y = y(t)$ for $a \le t \le b$. We will be interested in this section in C being a closed curve, so the initial and terminal points coincide.

C is *positively oriented* if $(x(t), y(t))$ moves around C counterclockwise as t varies from a to b. If $(x(t), y(t))$ moves clockwise, then we say that C is *negatively oriented*. For example, let $x(t) = \cos(t)$ and $y(t) = \sin(t)$ for $0 \le t \le 2\pi$. Then $(x(t), y(t))$ moves counterclockwise once around the unit circle as t varies from 0 to 2π, so C is positively oriented. If, however, K has coordinate functions $x(t) = -\cos(t)$ and $y(t) = \sin(t)$ for $0 \le t \le 2\pi$, then K is negatively oriented, because now $(x(t), y(t))$ moves in a clockwise sense. However, C and K have the same graph. A closed curve in the plane is positively oriented if, as you walk around it, the region it encloses is over your left shoulder.

A curve is *simple* if the same point cannot be on the graph for different values of the parameter. This means that $x(t_1) = x(t_2)$ and $y(t_1) = y(t_2)$ can occur only if $t_1 = t_2$. If we envision the graph of a curve as a train track, this means that the train does not return to the same location at a later time. Figure 12.7 shows the graph of a curve that is not simple.

This would prevent a closed curve from being simple, but we make an exception of the initial and terminal points. If these are the only points obtained for different values of the parameter, then a closed curve is also called simple. For example, the equations $x = \cos(t)$ and $y = \sin(t)$ for $0 \le t \le 2\pi$ describe a simple closed curve. However, consider M given by $x = \cos(t)$, $y = \sin(t)$ for $0 \le t \le 4\pi$. This is a closed curve, beginning and ending at $(1, 0)$, but $(x(t), y(t))$ traverses the unit circle twice counterclockwise as t varies from 0 to 4π. M is a closed curve but it is not simple.

It is a subtle theorem of topology, the *Jordan curve theorem*, that a simple closed curve C in the plane separates the plane into two regions having C as common boundary. One region contains points arbitrarily far from the origin and is called the *exterior* of C. The other region is called the *interior* of C. These regions are displayed for a typical closed curve in Figure 12.8. The interior of C has finite area, while the exterior does not.

Finally, when a line integral is taken around a closed curve, we often use the symbol $\oint_C$ in place of $\int_C$. This notation is optional and is simply a reminder that C is closed. It does not alter in any way the meaning of the integral.

We are now ready to state the first fundamental theorem of vector integral calculus. Recall that a path is a piecewise smooth curve (having a continuous tangent at all but finitely many points).

FIGURE 12.7   *Graph of a curve that is not simple.*

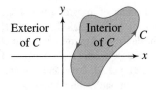

FIGURE 12.8   *The Jordan curve theorem.*

THEOREM 12.3   *Green*

Let C be a simple, closed, positively oriented path in the plane. Let D consist of all points on C and in its interior. Let f, g, $\partial g/\partial x$, and $\partial f/\partial y$ be continuous on D. Then

$$\oint_C f(x, y)\,dx + g(x, y)\,dy = \iint_D \left(\frac{\partial g}{\partial x} - \frac{\partial f}{\partial y} \right) dA. \quad \blacksquare$$

The significance of Green's theorem is that it relates an object that deals with a curve, which is one-dimensional, to an object related to a planar region, which is two-dimensional. This will have important implications when we discuss independence of path of line integrals in the next section, and later when we develop partial differential equations and complex analysis. Green's theorem will also lead shortly to Stokes's theorem and Gauss's theorem, which are its generalizations to 3-space.

We will prove the theorem under restricted conditions at the end of this section. For now, here are two computational examples.

EXAMPLE 12.10

Sometimes we use Green's theorem as a computational aid to convert one kind of integral into another, possibly simpler, one. As an illustration, suppose we want to compute the work done by the force

$$\mathbf{F}(x, y) = (y - x^2 e^x)\mathbf{i} + (\cos(2y^2) - x)\mathbf{j}$$

in moving a particle about the rectangular path C of Figure 12.9.

If you try to evaluate $\oint_C \mathbf{F} \cdot d\mathbf{R}$ as a sum of line integrals over the straight-line sides of this rectangle, you will find that the integrations cannot be done in elementary form. However, apply Green's theorem, with D the region bounded by the rectangle. We obtain

$$\text{work} = \oint_C \mathbf{F} \cdot d\mathbf{R} = \iint_D \left(\frac{\partial}{\partial x}(\cos(2y^2) - x) - \frac{\partial}{\partial y}(y - x^2 e^x) \right) dA$$

$$= \iint_D -2\,dA = (-2)(\text{area of } D) = -4. \quad \blacksquare$$

EXAMPLE 12.11

Another typical use of Green's theorem is in deriving very general results. To illustrate, suppose we want to evaluate

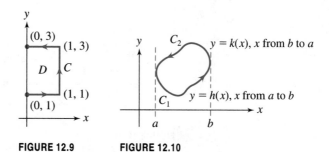

FIGURE 12.9   **FIGURE 12.10**

$$\oint_C 2x\cos(2y)\,dx - 2x^2\sin(2y)\,dy$$

for every positively oriented, simple, closed path C in the plane.

This may appear to be a daunting task, since there are infinitely many different such paths. However, observe the form of $f(x, y)$ and $g(x, y)$ in the line integral. In particular,

$$\frac{\partial}{\partial x}\left(-2x^2\sin(2y)\right) - \frac{\partial}{\partial y}\left(2x\cos(2y)\right) = -4x\sin(2y) + 4x\sin(2y) = 0$$

for all x and y. Therefore, Green's theorem gives us

$$\oint_C 2x\cos(2y)\,dx - 2x^2\sin(2y)\,dy = \iint_D 0\,dA = 0.$$

In the next section we will see how the vanishing of this line integral for any closed curve allows an important conclusion about line integrals $\int_K 2x\cos(2y)\,dx - 2x^2\sin(2y)\,dy$ when K is not closed. ■

We will conclude this section with a proof of Green's theorem under special conditions on the region bounded by C. Assume that D can be described in two ways.

First, D consists of all points (x, y) with

$$a \leq x \leq b$$

and, for each x,

$$h(x) \leq y \leq k(x).$$

Graphs of the curves $y = h(x)$ and $y = k(x)$ form, respectively, the lower and upper parts of the boundary of D (see Figure 12.10).

Second, D consists of all points (x, y) with

$$c \leq y \leq d$$

and, for each y,

$$F(y) \leq x \leq G(y).$$

In this description, the graphs of $x = F(y)$ and $x = G(y)$ form, respectively, the left and right parts of the boundary of D (see Figure 12.11).

Using these descriptions of D and the boundary of D, we can demonstrate Green's theorem by evaluating the integrals involved. First, let C_1 be the lower part of C (graph of $y = h(x)$) and

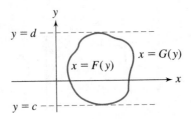

FIGURE 12.11

C_2 the upper part (graph of $y = k(x)$). Then

$$\oint_C f(x, y)\, dx = \int_{C_1} f(x, y)\, dx + \int_{C_2} f(x, y)\, dx$$

$$= \int_a^b f(x, h(x))\, dx + \int_b^a f(x, k(x))\, dx$$

$$= \int_a^b -[f(x, k(x)) - f(x, h(x))]\, dx.$$

The upper and lower limits of integration in the second line maintain a counterclockwise orientation on C.

Now compute

$$\iint_D \frac{\partial f}{\partial y}\, dA = \int_a^b \int_{h(x)}^{k(x)} \frac{\partial f}{\partial y}\, dy\, dx$$

$$= \int_a^b [f(x, y)]_{h(x)}^{k(x)}\, dx$$

$$= \int_a^b [f(x, k(x)) - f(x, h(x))]\, dx.$$

Therefore,

$$\oint_C f(x, y)\, dx = - \iint_D \frac{\partial f}{\partial y}\, dA.$$

Using the other description of D, a similar computation shows that

$$\oint_C g(x, y)\, dy = \iint_D \frac{\partial g}{\partial x}\, dA.$$

Upon adding the last two equations we obtain the conclusion of Green's theorem.

SECTION 12.2   PROBLEMS

1. A particle moves once counterclockwise about the triangle with vertices $(0, 0)$, $(4, 0)$, and $(1, 6)$, under the influence of the force $\mathbf{F} = xy\mathbf{i} + x\mathbf{j}$. Calculate the work done by this force.

2. A particle moves once counterclockwise about the circle of radius 6 about the origin, under the influence of

the force $\mathbf{F} = (e^x - y + x \cosh(x))\mathbf{i} + (y^{3/2} + x)\mathbf{j}$. Calculate the work done.

3. A particle moves once counterclockwise about the rectangle with vertices $(1, 1)$, $(1, 7)$, $(3, 1)$, and $(3, 7)$, under the influence of the force $\mathbf{F} = (-\cosh(4x^4) + xy)\mathbf{i} + (e^{-y} + x)\mathbf{j}$. Calculate the work done.

In each of Problems 4 through 14, use Green's theorem to evaluate $\oint_C \mathbf{F} \cdot d\mathbf{R}$. All curves are oriented counterclockwise.

4. $\mathbf{F} = 2y\mathbf{i} - x\mathbf{j}$, C is the circle of radius 4 about $(1, 3)$

5. $\mathbf{F} = x^2\mathbf{i} - 2xy\mathbf{j}$, C is the triangle with vertices $(1, 1), (4, 1), (2, 6)$

6. $\mathbf{F} = (x + y)\mathbf{i} + (x - y)\mathbf{j}$, C is the ellipse $x^2 + 4y^2 = 1$

7. $\mathbf{F} = 8xy^2\mathbf{j}$, C is the circle of radius 4 about the origin

8. $\mathbf{F} = (x^2 - y)\mathbf{i} + (\cos(2y) - e^{3y} + 4x)\mathbf{j}$, with C any square with sides of length 5

9. $\mathbf{F} = e^x \cos(y)\mathbf{i} - e^x \sin(y)\mathbf{j}$, C is any simple, closed, piecewise smooth curve in the plane

10. $\mathbf{F} = x^2y\mathbf{i} - xy^2\mathbf{j}$, C the boundary of the region $x^2 + y^2 \leq 4, x \geq 0, y \geq 0$

11. $\mathbf{F} = xy\mathbf{i} + (xy^2 - e^{\cos(y)})\mathbf{j}$, C the triangle with vertices $(0, 0), (3, 0), (0, 5)$

12. $\mathbf{F} = y\cos(x)\mathbf{i} - y^3\mathbf{j}$, C the square with vertices $(-1, 0), (0, 0), (0, 1), (-1, 1)$

13. $\mathbf{F} = (e^{\sin(x)} - y)\mathbf{i} + (\sinh(y^3) - 4x)\mathbf{j}$, C the circle of radius 2 about $(-8, 0)$

14. $\mathbf{F} = (x^2+y^2)\mathbf{i} + (x^2-y^2)\mathbf{j}$, C the ellipse $4x^2 + y^2 = 16$

15. Let C be a positively oriented, simple closed path with interior D.

(a) Show that the area of D equals $\oint_C -y\,dx$.

(b) Show that the area of D equals $\oint_C x\,dy$.

(c) Show that the area of D equals $\frac{1}{2}\oint -y\,dx + x\,dy$.

16. Let C be a positively oriented, simple, closed path with interior D. Show that the centroid of D is $(\tilde{x}, \tilde{y})$, where

$$\tilde{x} = \frac{1}{2A}\oint_C x^2\,dy \quad \text{and} \quad \tilde{y} = \frac{1}{2A}\oint_C y^2\,dx,$$

with A the area of D.

17. Let $u(x, y)$ be continuous with continuous first and second partial derivatives on a simple, closed path C and throughout the interior D of C. Show that

$$\oint_C -\frac{\partial u}{\partial y}\,dx + \frac{\partial u}{\partial x}\,dy = \iint_D \left[\frac{\partial^2 u}{\partial x^2} + \frac{\partial^2 u}{\partial y^2}\right]dA.$$

18. Let C be a simple closed path made up of the graph of $r = f(\theta)$ for $\alpha \leq \theta \leq \beta$ (polar coordinates) and parts of the lines $\theta = \alpha$ and $\theta = \beta$. Let $f'(\theta)$ be continuous on $[\alpha, \beta]$. Show that the area of D is equal to $\frac{1}{2}\oint_C [f(\theta)]^2\,d\theta$. *Hint:* Calculate the area using the result of Problem 15(c), using r as parameter on the line segment parts of C, and θ on the portion given by the graph of $r = f(\theta)$.

12.2.1 An Extension of Green's Theorem

There is an extension of Green's theorem to include the case that there are finitely many points enclosed by C at which $f, g, \partial f/\partial y$, and/or $\partial g/\partial x$ are not continuous, or perhaps are not even defined. The idea is to excise the "bad points," as we will now describe.

Suppose C is a simple, closed, positively oriented path in the plane enclosing a region D. Suppose $f, g, \partial f/\partial y$, and $\partial g/\partial x$ are continuous on C and throughout D except at points $P_1, \ldots, P_n$. Green's theorem does not apply to this region. But with a little imagination, we can still draw an interesting conclusion.

Enclose each P_j with a circle K_j of sufficiently small radius that none of these circles intersects either C or each other (Figure 12.12). Next, cut a channel in D from C to K_1, then from K_1 to K_2, and so on until finally a channel is cut from K_{n-1} to K_n. A typical case is shown in Figure 12.13. Form the closed path C^* consisting of C (with a small segment cut out where the channel to K_1 was made), each of the K_j's (with small cuts removed where the channels entered and exited), and the segments forming the connections between C and the successive K_j's. Figure 12.14 shows C^*, which encloses the region D^*.

By the way C^* was formed, the points $P_1, \ldots, P_n$ are *external* to C^* (Figure 12.15). Further, $f, g, \partial f/\partial y$, and $\partial g/\partial x$ are continuous on C^* and throughout D^*. We can therefore apply Green's theorem to C^* and D^* to conclude that

$$\oint_{C^*} f(x, y)\,dx + g(x, y)\,dy = \iint_{D^*}\left(\frac{\partial g}{\partial x} - \frac{\partial f}{\partial y}\right)dA. \tag{12.2}$$

Now imagine that the channels that were cut become narrower, merging to form segments between C and successive K_j's. Then C^* approaches the curve $\hat{C}$ of Figure 12.16, and D^* approaches

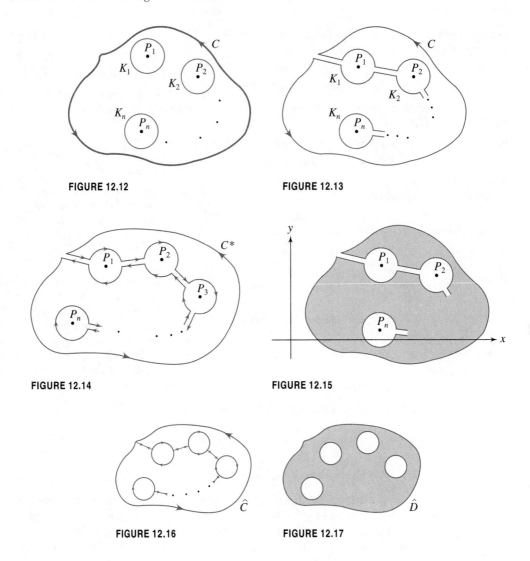

FIGURE 12.12

FIGURE 12.13

FIGURE 12.14

FIGURE 12.15

FIGURE 12.16

FIGURE 12.17

the region $\hat{D}$ shown in Figure 12.17. $\hat{D}$ consists of D with the disks bounded by $K_1, \ldots, K_n$ cut out. In this limit process, equation (12.2) approaches

$$\oint_C f(x, y)\, dx + g(x, y)\, dy + \sum_{j=1}^{n} \oint_{K_j} f(x, y)\, dx + g(x, y)\, dy = \iint_{\hat{D}} \left(\frac{\partial g}{\partial x} - \frac{\partial f}{\partial y} \right) dA.$$

On the left side of this equation, line integrals over the internal segments connecting C and the K_j's cancel because the integration is carried out twice over each segment, once in each direction. Further, the orientation on C is counterclockwise, but the orientation on each K_j is clockwise because of the way the boundaries were traversed (Figure 12.14). If we reverse the orientations on these circles, the line integrals over them change sign and we can write

$$\oint_C f(x, y)\, dx + g(x, y)\, dy = \sum_{j=1}^{n} \oint_{K_j} f(x, y)\, dx + g(x, y)\, dy + \iint_{\hat{D}} \left(\frac{\partial g}{\partial x} - \frac{\partial f}{\partial y} \right) dA,$$

in which all the integrals are in the positive, counterclockwise sense about the curves C and $K_1, \ldots, K_n$. This is the generalization of Green's theorem that we sought.

EXAMPLE 12.12

Suppose we are interested in

$$\oint_C \frac{-y}{x^2 + y^2} \, dx + \frac{x}{x^2 + y^2} \, dy,$$

with C any simple, closed, positively oriented path in the plane, but not passing through the origin.
With

$$f(x, y) = \frac{-y}{x^2 + y^2} \quad \text{and} \quad g(x, y) = \frac{x}{x^2 + y^2},$$

we have

$$\frac{\partial g}{\partial x} = \frac{y^2 - x^2}{(x^2 + y^2)^2} = \frac{\partial f}{\partial y}.$$

$f, g, \partial f/\partial y$, and $\partial g/\partial x$ are continuous at every point of the plane except the origin. This leads us to consider two cases.

Case 1 C does not enclose the origin.
Now Green's theorem applies and

$$\oint_C \frac{-y}{x^2 + y^2} \, dx + \frac{x}{x^2 + y^2} \, dy = \iint_D \left(\frac{\partial g}{\partial x} - \frac{\partial f}{\partial y} \right) dA = 0.$$

Case 2 C encloses the origin.
Draw a circle K centered at the origin, with radius sufficiently small that K does not intersect C (Figure 12.18). By the extension of Green's theorem,

$$\oint_C f(x, y) \, dx + g(x, y) \, dy$$

$$= \oint_K f(x, y) \, dx + g(x, y) \, dy + \iint_D \left(\frac{\partial g}{\partial x} - \frac{\partial f}{\partial y} \right) dA$$

$$= \oint_K f(x, y) \, dx + g(x, y) \, dy,$$

where $\hat{D}$ is the region between K and C. Both of these line integrals are in the counterclockwise sense about the respective curves.

The last line integral can be evaluated explicitly because we know K. Parametrize K by

$$x = r \cos(\theta), \, y = r \sin(\theta) \quad \text{for } 0 \leq \theta \leq 2\pi.$$

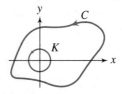

FIGURE 12.18

Then

$$\oint_K f(x, y)\, dx + g(x, y)\, dy$$

$$= \int_0^{2\pi} \left(\frac{-r\sin(\theta)}{r^2}[-r\sin(\theta)] + \frac{r\cos(\theta)}{r^2}[r\cos(\theta)] \right) d\theta$$

$$= \int_0^{2\pi} d\theta = 2\pi.$$

We conclude that

$$\oint_C f(x, y)\, dx + g(x, y)\, dy = \begin{cases} 0 & \text{if } C \text{ does not enclose the origin} \\ 2\pi & \text{if } C \text{ encloses the origin} \end{cases} . \quad \blacksquare$$

SECTION 12.2   ADDITIONAL PROBLEMS

In each of Problems 1 through 5, evaluate $\oint_C \mathbf{F} \cdot d\mathbf{R}$ over any simple, closed path in the x, y plane that does not pass through the origin.

1. $\mathbf{F} = \dfrac{x}{x^2 + y^2}\mathbf{i} + \dfrac{y}{x^2 + y^2}\mathbf{j}$

2. $\mathbf{F} = \left(\dfrac{1}{x^2 + y^2} \right)^{3/2}(x\mathbf{i} + y\mathbf{j})$

3. $\mathbf{F} = \left(\dfrac{-y}{x^2 + y^2} + x^2 \right)\mathbf{i} + \left(\dfrac{x}{x^2 + y^2} - 2y \right)\mathbf{j}$

4. $\mathbf{F} = \left(\dfrac{-y}{x^2 + y^2} + 3x \right)\mathbf{i} + \left(\dfrac{x}{x^2 + y^2} - y \right)\mathbf{j}$

5. $\mathbf{F} = \left(\dfrac{x}{\sqrt{x^2 + y^2}} + 2x \right)\mathbf{i} + \left(\dfrac{y}{\sqrt{x^2 + y^2}} - 3y^2 \right)\mathbf{j}$

12.3   Independence of Path and Potential Theory in the Plane

In physics, a conservative force field is one that is derivable from a potential. We will use the same terminology.

DEFINITION 12.4   *Conservative Vector Field*

Let D be a set of points in the plane. A vector field $\mathbf{F}(x, y)$ is conservative on D if for some real-valued $\varphi(x, y)$, $\mathbf{F} = \nabla\varphi$ for all (x, y) in D. In this event, φ is a potential function for $\mathbf{F}$ on D.

If φ is a potential function for $\mathbf{F}$, then so is $\varphi + c$ for any constant c, because $\nabla(\varphi + c) = \nabla\varphi$. For this reason, we often speak of *a* potential function for $\mathbf{F}$, rather than *the* potential function.

Recall that if $\mathbf{F}(x, y) = f(x, y)\mathbf{i} + g(x, y)\mathbf{j}$ and $\mathbf{R}(t) = x(t)\mathbf{i} + y(t)\mathbf{j}$ is a position function for C, then $\int_C \mathbf{F} \cdot d\mathbf{R}$ is another way of writing $\int_C f(x, y)\, dx + g(x, y)\, dy$. We will make frequent use of the notation $\int_C \mathbf{F} \cdot d\mathbf{R}$ throughout this section because we want to examine the effect on this integral when $\mathbf{F}$ has a potential function, and for this we will use vector notation.

First, the line integral of a conservative vector field can be evaluated directly in terms of a potential function. For suppose C is smooth, with coordinate functions $x = x(t)$, $y = y(t)$ for

$a \leq t \leq b$. If $\mathbf{F} = \nabla\varphi$, then

$$\mathbf{F}(x, y) = \frac{\partial\varphi}{\partial x}\mathbf{i} + \frac{\partial\varphi}{\partial y}\mathbf{j}$$

and

$$\int_C \mathbf{F} \cdot d\mathbf{R} = \int_C \frac{\partial\varphi}{\partial x}\, dx + \frac{\partial\varphi}{\partial y}\, dy$$

$$= \int_a^b \left(\frac{\partial\varphi}{\partial x}\frac{dx}{dt} + \frac{\partial\varphi}{\partial y}\frac{dy}{dt} \right) dt$$

$$= \int_a^b \frac{d}{dt}\, \varphi(x(t), y(t))\, dt$$

$$= \varphi(x(b), y(b)) - \varphi(x(a), y(a)).$$

Denoting $P_1 = (x(b), y(b))$ and $P_0 = (x(a), y(a))$, this result states that

$$\int_C \mathbf{F} \cdot d\mathbf{R} = \varphi(P_1) - \varphi(P_0)$$

$$= \varphi(\text{terminal point of } C) - \varphi(\text{initial point of } C).$$

(12.3)

The line integral of a conservative vector field over a path is the difference in values of a potential function at endpoints of the path. This is familiar from physics. If a particle moves along a path under the influence of a conservative force field, then the work done is equal to the difference in the potential energy at the ends of the path.

One ramification of equation (12.3) is that the actual path itself does not influence the outcome, only the endpoints of the path. If we chose a different path K between the same endpoints, we would obtain the same result for $\int_K \mathbf{F} \cdot d\mathbf{R}$. This suggests the concept of independence of path of a line integral.

DEFINITION 12.5 *Independence of Path*

$\int_C \mathbf{F} \cdot d\mathbf{R}$ is independent of path on a set D of points in the plane if for any points P_0 and P_1 in D, the line integral has the same value over any paths in D having initial point P_0 and terminal point P_1.

The discussion preceding the definition may now be summarized.

THEOREM 12.4

Let φ and its first partial derivatives be continuous for all (x, y) in a set D of points in the plane. Let $\mathbf{F} = \nabla\varphi$. Then $\int_C \mathbf{F} \cdot d\mathbf{R}$ is independent of path in D. Further, if C is a simple, closed path in D, then $\oint_C \mathbf{F} \cdot d\mathbf{R} = 0$. ∎

The independence of path follows from equation (12.3), which states that when $\mathbf{F} = \nabla\varphi$, the value of $\int_C \mathbf{F} \cdot d\mathbf{R}$ depends only on the values of $\varphi(x, y)$ at the end points of the path, and not on where the path goes in between. For the last conclusion of the theorem, if C is a closed path in D, then the initial and terminal points coincide, hence the difference between the values of φ at the terminal and initial points is zero.

EXAMPLE 12.13

Let $\mathbf{F}(x, y) = 2x \cos(2y)\mathbf{i} - 2x^2 \sin(2y)\mathbf{j}$. It is routine to check that $\varphi(x, y) = x^2 \cos(2y)$ is a potential function for $\mathbf{F}$. Since φ is continuous with continuous partial derivatives over the entire plane, we can let D consist of all points in the plane in the definition of independence of path. For example, if C is any path in the plane from $(0, 0)$ to $(1, \pi/8)$, then

$$\int_C \mathbf{F} \cdot d\mathbf{R} = \varphi\left(1, \frac{\pi}{8}\right) - \varphi(0, 0) = \frac{\sqrt{2}}{2}.$$

Further, if K is any simple closed path in D, then $\oint_K \mathbf{F} \cdot d\mathbf{R} = 0$. ∎

It is clearly to our advantage to know whether a vector field is conservative and, if it is, to be able to produce a potential function. Let

$$\mathbf{F}(x, y) = f(x, y)\mathbf{i} + g(x, y)\mathbf{j}.$$

$\mathbf{F}$ is conservative exactly when, for some φ,

$$\mathbf{F} = \nabla\varphi = \frac{\partial\varphi}{\partial x}\mathbf{i} + \frac{\partial\varphi}{\partial y}\mathbf{j},$$

and this requires that

$$\frac{\partial\varphi}{\partial x} = f(x, y) \quad \text{and} \quad \frac{\partial\varphi}{\partial y} = g(x, y).$$

To attempt to find such a φ, begin with either of these equations and integrate with respect to the variable of the derivative, keeping the other variable fixed. The constant of integration is then actually a function of the other (fixed) variable. Finally, use the second equation to attempt to find this function.

EXAMPLE 12.14

Consider the vector field

$$\mathbf{F}(x, y) = 2x \cos(2y)\mathbf{i} - [2x^2 \sin(2y) + 4y^2]\mathbf{j}.$$

We want a real-valued function φ such that

$$\frac{\partial\varphi}{\partial x} = 2x \cos(2y) \quad \text{and} \quad \frac{\partial\varphi}{\partial y} = -2x^2 \sin(2y) - 4y^2.$$

Choose one of these equations. If we pick the first, then integrate with respect to x, holding y fixed:

$$\varphi(x, y) = \int 2x \cos(2y)\, dx = x^2 \cos(2y) + g(y).$$

The "constant" of integration is allowed to involve y because we are reversing a partial derivative, and for any function of y,

$$\frac{\partial}{\partial x}[x^2 \cos(2y) + g(y)] = 2x \cos(2y),$$

as we require. We now have $\varphi(x, y)$ to within some function $g(y)$. From the second equation we need

$$\frac{\partial \varphi}{\partial y} = -2x^2 \sin(2y) - 4y^2 = \frac{\partial}{\partial y}[x^2 \cos(2y) + g(y)].$$

Then

$$-2x^2 \sin(2y) - 4y^2 = -2x^2 \sin(2y) + g'(y),$$

so

$$g'(y) = -4y^2.$$

Choose $g(y) = -4y^3/3$ to obtain the potential function

$$\varphi(x, y) = x^2 \cos(2y) - \tfrac{4}{3}y^3.$$

It is easy to check that $\mathbf{F} = \nabla \varphi$. ∎

EXAMPLE 12.15

Let

$$\mathbf{F}(x, y) = \left(ye^{xy} + xy^2e^{xy} + 2x\right)\mathbf{i} + (xe^{xy} + x^2ye^{xy} - 2y)\mathbf{j}.$$

If there is a potential function φ for this vector field, then it must satisfy the two equations:

$$\frac{\partial \varphi}{\partial x} = ye^{xy} + xy^2e^{xy} + 2x, \quad \frac{\partial \varphi}{\partial y} = xe^{xy} + x^2ye^{xy} - 2y.$$

Pick one of these equations, say the second, and integrate with respect to y:

$$\varphi(x, y) = \int (xe^{xy} + x^2ye^{xy} - 2y)\, dy = xye^{xy} - y^2 + h(x),$$

in which now $h(x)$ is the "constant" of the integration with respect to y. Now use the other equation to write

$$\frac{\partial \varphi}{\partial x} = ye^{xy} + xy^2e^{xy} + 2x$$

$$= \frac{\partial}{\partial x}[xye^{xy} - y^2 + h(x)] = ye^{xy} + xy^2e^{xy} + h'(x).$$

We must choose h so that $h'(x) = 2x$, so we can let $h(x) = x^2$. Then

$$\varphi(x, y) = xye^{xy} - y^2 + x^2.$$

Then $\mathbf{F} = \nabla \varphi$. ∎

Is every vector field in the plane conservative? As the following example shows, the answer is no.

EXAMPLE 12.16

Let $\mathbf{F}(x, y) = (2xy^2 + y)\mathbf{i} + (2x^2y + e^x y)\mathbf{j}$. If this vector field were conservative, there would be a potential φ such that

$$\frac{\partial \varphi}{\partial x} = 2xy^2 + y, \quad \frac{\partial \varphi}{\partial y} = 2x^2y + e^x y.$$

Integrate the first equation with respect to x to get

$$\varphi(x, y) = \int (2xy^2 + y)\, dx = x^2 y^2 + xy + f(y).$$

From the second equation,

$$\frac{\partial \varphi}{\partial y} = 2x^2 y + e^x y = \frac{\partial}{\partial y}(x^2 y^2 + xy + f(y)) = 2x^2 y + x + f'(y).$$

But this would imply that

$$f'(y) = e^x y - x,$$

and we cannot find a function of y alone satisfying this equation. Therefore, $\mathbf{F}$ has no potential. ■

Because not every vector field is conservative, we need some test to determine whether or not a given vector field is conservative. The following theorem provides such a test.

THEOREM 12.5   *Test for a Convervative Field*

Let f and g be continuous in a region D bounded by a rectangle having its sides parallel to the axes. Then $\mathbf{F}(x, y) = f(x, y)\mathbf{i} + g(x, y)\mathbf{j}$ is conservative on D if and only if, for all (x, y) in D,

$$\frac{\partial g}{\partial x} = \frac{\partial f}{\partial y}. \quad ■ \tag{12.4}$$

Sometimes the conditions of the theorem hold throughout the plane, and in this event the vector field is conservative for all (x, y) when equation (12.4) is satisfied.

EXAMPLE 12.17

Consider the vector field $\mathbf{F}(x, y) = \left(y e^{xy} + x y^2 e^{xy} + 2x \right)\mathbf{i} + (x e^{xy} + x^2 y e^{xy} - 2y)\mathbf{j}$ of Example 12.15. Here

$$f(x, y) = y e^{xy} + x y^2 e^{xy} + 2x \quad \text{and} \quad g(x, y) = x e^{xy} + x^2 y e^{xy} - 2y.$$

Compute

$$\frac{\partial f}{\partial y} = e^{xy} + 3xy e^{xy} + x^2 y^2 e^{xy}$$

and

$$\frac{\partial g}{\partial x} = e^{xy} + 3xy e^{xy} + x^2 y^2 e^{xy}$$

so equation (12.4) is satisfied throughout the plane and $\mathbf{F}$ is conservative. We found a potential function in Example 12.15. ■

EXAMPLE 12.18

Consider again $\mathbf{F}(x, y) = (2xy^2 + y)\mathbf{i} + (2x^2 y + e^x y)\mathbf{j}$ from Example 12.16. Compute

$$\frac{\partial f}{\partial y} = 4xy + 1 \quad \text{and} \quad \frac{\partial g}{\partial x} = 4xy + e^x y$$

and these are unequal on any rectangular region of the plane. This vector field is not conservative. We showed in Example 12.16 that no potential function can exist for this field. ∎

12.3.1 A More Critical Look at Theorem 12.5

The condition (12.4) derived in Theorem 12.5 can be written

$$\frac{\partial g}{\partial x} - \frac{\partial f}{\partial y} = 0.$$

But the combination

$$\frac{\partial g}{\partial x} - \frac{\partial f}{\partial y}$$

also occurs in Green's theorem. This must be more than coincidence. In this section we will explore connections between independence of path, Green's theorem, condition (12.4), and existence of a potential function.

The following example is instructive. Let D consist of all points in the plane except the origin. Thus, D is the plane with the origin punched out. Let

$$\mathbf{F}(x, y) = \frac{-y}{x^2 + y^2}\mathbf{i} + \frac{x}{x^2 + y^2}\mathbf{j} = f(x, y)\mathbf{i} + g(x, y)\mathbf{j}.$$

Then $\mathbf{F}$ is defined on D, and f and g are continuous with continuous partial derivatives on D. Further, we saw in Example 12.12 that

$$\frac{\partial f}{\partial y} - \frac{\partial g}{\partial x} = 0 \quad \text{for } (x, y) \text{ in } D.$$

Now evaluate $\int_C f(x, y)\, dx + g(x, y)\, dy$ over two paths from $(1, 0)$ to $(-1, 0)$. First, let C be the top half of the unit circle, given by

$$x = \cos(\theta), y = \sin(\theta) \quad \text{for } 0 \leq \theta \leq \pi.$$

Then

$$\int_C f(x, y)\, dx + g(x, y)\, dy$$

$$= \int_0^\pi [(-\sin(\theta))(-\sin(\theta)) + \cos(\theta)\cos(\theta)]\, d\theta = \int_0^\pi d\theta = \pi.$$

Next, let K be the path from $(1, 0)$ to $(-1, 0)$ along the bottom half of the unit circle, given by

$$x = \cos(\theta), y = -\sin(\theta) \quad \text{for } 0 \leq \theta \leq \pi.$$

Then

$$\int_K f(x, y)\, dx + g(x, y)\, dy$$

$$= \int_0^\pi [\sin(\theta)(-\sin(\theta)) + \cos(\theta)(-\cos(\theta))]\, d\theta = -\int_0^\pi d\theta = -\pi.$$

This means that $\int_C f(x, y)\, dx + g(x, y)\, dy$ is not independent of path in D. The path chosen between two given points makes a difference. This also means that $\mathbf{F}$ is not conservative over D. There is no potential function for $\mathbf{F}$ (by Theorem 12.4, if there were a potential function, then the line integral would have to be independent of path).

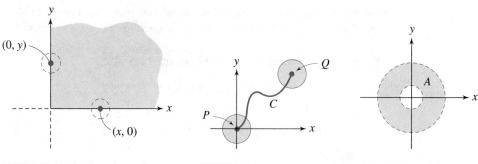

FIGURE 12.19 *Right quarter plane* $x \geq 0$, $y \geq 0$.

FIGURE 12.20

FIGURE 12.21 *The region between two concentric circles.*

This example suggests that there is something about the conditions specified on the set D in Theorem 12.5 that makes a difference. The rectangular set in the theorem, where condition (12.4) is necessary and sufficient for existence of a potential, must have some property or properties that the set in this example lacks. We will explore this line of thought.

Let D be a set of points in the plane. We call D a *domain* if it satisfies two conditions:

1. If P_0 is any point of D, there is a circle about P_0 such that every point enclosed by this circle is also in D.

2. Between any two points of D, there is a path lying entirely in D.

For example, the right quarter plane S consisting of points (x, y) with $x \geq 0$ and $y \geq 0$ enjoys property (2), but not (1). There is no circle that can be drawn about a point $(x, 0)$ with $x \geq 0$ that contains only points with nonnegative coordinates (Figure 12.19). Similarly, any circle drawn about a point $(0, y)$ in S must contain points outside of S.

The shaded set M of points in Figure 12.20 does not satisfy condition (2). Any path C connecting the indicated points P and Q must at some time go outside of M.

Figure 12.21 shows the set A of points between the circles of radius 1 and 3 about the origin. Thus, (x, y) is in A exactly when

$$1 < x^2 + y^2 < 9.$$

This set satisfies conditions (1) and (2) and so is a domain. The boundary circles are drawn as dashed curves to emphasize that points on these curves are not in A.

The conditions defining a domain are enough for the first theorem.

THEOREM 12.6

Let $\mathbf{F}$ be a vector field that is continuous on a domain D. Then $\int_C \mathbf{F} \cdot d\mathbf{R}$ is independent of path on D if and only if $\mathbf{F}$ is conservative.

Proof We know that if $\mathbf{F}$ is conservative, then $\int_C \mathbf{F} \cdot d\mathbf{R}$ is independent of path on D, since the value of this line integral is the difference in the values of a potential function at the terminal and initial points of C. It is the converse that uses the condition that D is a domain.

Conversely, suppose $\int_C \mathbf{F} \cdot d\mathbf{R}$ is independent of path on D. We will produce a potential function. Choose any point $P_0 : (x_0, y_0)$ in D. If $P : (x, y)$ is any point of D, define

$$\varphi(x, y) = \int_C \mathbf{F} \cdot d\mathbf{R},$$

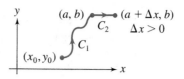

FIGURE 12.22

in which C is any path in D from P_0 to P. There is such a path because D is a domain. Further, because this line integral is independent of path, $\varphi(x, y)$ depends only on (x, y) and P_0 and not on the curve chosen between them. Thus φ is a function. Because $\mathbf{F}$ is continuous on D, φ is also continuous on D.

Now let $\mathbf{F}(x, y) = f(x, y)\mathbf{i} + g(x, y)\mathbf{j}$ and select any point (a, b) in D. We will show that

$$\frac{\partial \varphi}{\partial x}(a, b) = f(a, b) \quad \text{and} \quad \frac{\partial \varphi}{\partial y}(a, b) = g(a, b).$$

For the first of these equations, recall that

$$\frac{\partial \varphi}{\partial x}(a, b) = \lim_{\Delta \to 0} \frac{\varphi(a + \Delta x, b) - \varphi(a, b)}{\Delta x}.$$

Because D is a domain, there is a circle about (a, b) enclosing only points of D. Let r be the radius of such a circle and restrict Δx so that $0 < \Delta x < r$. Let C_1 be any path in D from P_0 to (a, b) and C_2 the horizontal line segment from (a, b) to $(a + \Delta x, b)$, as shown in Figure 12.22. Let C be the path from P_0 to $(a + \Delta x, b)$ consisting of C_1 and then C_2. Now

$$\varphi(a + \Delta x, b) - \varphi(a, b) = \int_C \mathbf{F} \cdot d\mathbf{R} - \int_{C_1} \mathbf{F} \cdot d\mathbf{R} = \int_{C_2} \mathbf{F} \cdot d R.$$

Parametrize C_2 by $x = a + t\Delta x$, $y = b$ for $0 \le t \le 1$. Then

$$\varphi(a + \Delta x, b) - \varphi(a, b)$$

$$= \int_{C_2} \mathbf{F} \cdot d\mathbf{R} = \int_{C_2} f(x, y)\, dx + g(x, y)\, dy$$

$$= \int_0^1 f(a + t\Delta x, b)(\Delta x)\, dt.$$

Then

$$\frac{\varphi(a + \Delta x, b) - \varphi(a, b)}{\Delta x} = \int_0^1 f(a + t\Delta x, b)\, dt.$$

By the mean value theorem for integrals, there is a number ϵ between 0 and 1, inclusive, such that

$$\int_0^1 f(a + t\Delta x, b)\, dt = f(a + \epsilon \Delta x, b).$$

Therefore,

$$\frac{\varphi(a + \Delta x, b) - \varphi(a, b)}{\Delta x} = f(a + \epsilon \Delta x, b).$$

As $\Delta x \to 0$, $f(a + \epsilon \Delta x, b) \to f(a, b)$ by continuity of f, proving that

$$\lim_{\Delta x \to 0+} \frac{\varphi(a + \Delta x, b) - \varphi(a, b)}{\Delta x} = f(a, b).$$

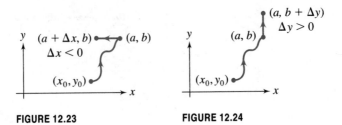

FIGURE 12.23   **FIGURE 12.24**

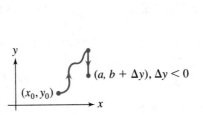

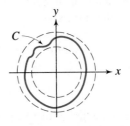

FIGURE 12.25   **FIGURE 12.26** *The set of points between two concentric circles is not simply connected.*

By a similar argument, using the path of Figure 12.23, we can show that

$$\lim_{\Delta x \to 0-} \frac{\varphi(a + \Delta x, b) - \varphi(a, b)}{\Delta x} = f(a, b).$$

Therefore,

$$\frac{\partial \varphi}{\partial x}(a, b) = f(a, b).$$

To prove that $(\partial \varphi / \partial y)(a, b) = g(a, b)$, the reasoning is similar except now use the paths of Figures 12.24 and 12.25.

This completes the proof of the theorem. ■

We have seen that the condition (12.4) is necessary and sufficient for **F** to be conservative within a rectangular region (which is a domain). Although this result is strong enough for many purposes, it is possible to extend it to regions that are not rectangular in shape if another condition is added to the region.

A domain D is called *simply connected* if every simple, closed path in D encloses only points of D. A simply connected domain is one that has no "holes" in it, because a simple, closed path about the hole will enclose points not in the domain. If D is the plane with the origin removed, then D is not simply connected, because the unit circle about the origin encloses a point not in D. We have seen in Example 12.12 that condition (12.4) may be satisfied in this domain by a vector field having no potential function on D. Similarly, the region between two concentric circles is not simply connected, because a closed curve in this region may wrap around the inner circle, hence enclose points not in the region (Figure 12.26).

We will now show that simple connectivity is just what is needed to ensure that condition (12.4) is equivalent to existence of a potential function. The key is that simple connectivity allows the use of Green's theorem.

Let $\mathbf{F}(x, y) = f(x, y)\mathbf{i} + g(x, y)\mathbf{j}$ be a vector field and D a simply connected domain. Suppose $f, g, \partial f/\partial y$, and $\partial g/\partial x$ are continuous on D. Then $\mathbf{F}$ is conservative on D if and only if

$$\frac{\partial f}{\partial y} = \frac{\partial g}{\partial x}$$

for all (x, y) in D.

Proof Suppose first that $\mathbf{F}$ is conservative, with potential φ. Then

$$f(x, y) = \frac{\partial \varphi}{\partial x} \quad \text{and} \quad g(x, y) = \frac{\partial \varphi}{\partial y}.$$

Then

$$\frac{\partial f}{\partial y} = \frac{\partial^2 \varphi}{\partial x\, \partial y} = \frac{\partial^2 \varphi}{\partial y\, \partial x} = \frac{\partial g}{\partial y}$$

for (x, y) in D.

For the converse, suppose that condition (12.4) holds throughout D. We will prove that $\int_C \mathbf{F} \cdot d\mathbf{R}$ is independent of path in D. By the previous theorem, this will imply that $\mathbf{F}$ is conservative. To this end, let P_0 and P_1 be any points of D, and let C and K be paths in D from P_0 to P_1. Suppose first that these paths have only their endpoints in common (Figure 12.27(a)). We can then form a positively oriented, simply closed path J from P_0 to P_0 by moving from P_0 to P_1 along C, then back to P_0 along $-K$ (Figure 12.27(b)). Let J enclose a region D^*. Since D is simply connected, every point in D^* is in D, over which $f, g, \partial f/\partial y$, and $\partial g/\partial x$ are continuous. Apply Green's theorem to write

$$\oint_J \mathbf{F} \cdot d\mathbf{R} = \iint_{D^*} \left(\frac{\partial g}{\partial x} - \frac{\partial f}{\partial y} \right) dA = 0.$$

But then

$$\oint_J \mathbf{F} \cdot d\mathbf{R} = \int_C \mathbf{F} \cdot d\mathbf{R} + \int_{-K} \mathbf{F} \cdot d\mathbf{R}$$

$$= \int_C \mathbf{F} \cdot d\mathbf{R} - \int_K \mathbf{F} \cdot d\mathbf{R} = 0,$$

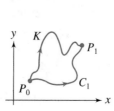

FIGURE 12.27(a)
Paths C and K from
P_0 to P_1.

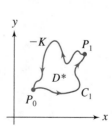

FIGURE 12.27(b) *A*
closed path formed
from C and $-K$.

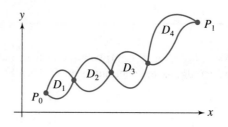

FIGURE 12.28

so

$$\int_C \mathbf{F} \cdot d\mathbf{R} = \int_K \mathbf{F} \cdot d\mathbf{R}.$$

If C and K intersect each other between P_0 and P_1, as in Figure 12.28, then this conclusion can still be drawn by considering closed paths between successive points of intersection. We will not pursue this technical argument.

Once we have independence of path of $\int_C \mathbf{F} \cdot d\mathbf{R}$ on D, then $\mathbf{F}$ is conservative, and the theorem is proved. ∎

In sum:

$$\text{conservative vector field} \Longrightarrow \text{independence of path of } \int_C \mathbf{F} \cdot d\mathbf{R} \text{ on a set } D,$$

$$\text{independence of path on a } \textit{domain} \Longleftrightarrow \text{conservative vector field}$$

and

$$\text{conservative on a } \textit{simply connected domain} \Longleftrightarrow \frac{\partial g}{\partial x} = \frac{\partial f}{\partial y}.$$

SECTION 12.3   PROBLEMS

In each of Problems 1 through 10, determine whether $\mathbf{F}$ is conservative in the given region D. If it is, find a potential function. If D is not defined, it is understood to be the entire plane.

1. $\mathbf{F} = y^3\mathbf{i} + (3xy^2 - 4)\mathbf{j}$

2. $\mathbf{F} = (6y + ye^{xy})\mathbf{i} + (6x + xe^{xy})\mathbf{j}$

3. $\mathbf{F} = 16x\mathbf{i} + (2 - y^2)\mathbf{j}$

4. $\mathbf{F} = 2xy\cos(x^2)\mathbf{i} + \sin(x^2)\mathbf{j}$

5. $\mathbf{F} = \left(\dfrac{2x}{x^2 + y^2}\right)\mathbf{i} + \left(\dfrac{2y}{x^2 + y^2}\right)\mathbf{j}$, D the plane with the origin removed

6. $\mathbf{F} = \sinh(x + y)(\mathbf{i} + \mathbf{j})$

7. $\mathbf{F} = 2\cos(2x)e^y\mathbf{i} + [e^y\sin(2x) - y]\mathbf{j}$

8. $\mathbf{F} = (3x^2y - \sin(x) + 1)\mathbf{i} + (x^3 - e^y)\mathbf{j}$

9. $\mathbf{F} = (y^2 + 3)\mathbf{i} + (2xy + 3x)\mathbf{j}$

10. $\mathbf{F} = (3x^2 - 2y)\mathbf{i} + (12y - 2x)\mathbf{j}$

In each of Problems 11 through 20, evaluate $\oint_C \mathbf{F} \cdot d\mathbf{R}$ for C any path from the first given point to the second.

11. $\mathbf{F} = 3x^2(y^2 - 4y)\mathbf{i} + (2x^3y - 4x^3)\mathbf{j}$; $(-1, 1)$, $(2, 3)$

12. $\mathbf{F} = e^x\cos(y)\mathbf{i} - e^x\sin(y)\mathbf{j}$; $(0, 0)$, $(2, \pi/4)$

13. $\mathbf{F} = 2xy\mathbf{i} + (x^2 - 1/y)\mathbf{j}$; $(1, 3)$, $(2, 2)$ (the path cannot cross the x axis)

14. $\mathbf{F} = \mathbf{i} + (6y + \sin(y))\mathbf{j}$; $(0, 0)$, $(1, 3)$

15. $\mathbf{F} = (3x^2y^2 - 6y^3)\mathbf{i} + (2x^3y - 18xy^2)\mathbf{j}$; $(0, 0)$, $(1, 1)$

16. $\mathbf{F} = \dfrac{y}{x}\mathbf{i} + \ln(x)\mathbf{j}$; $(1, 1)$, $(2, 2)$ (the path must lie in the right half-plane $x > 0$)

17. $\mathbf{F} = (-8e^y + e^x)\mathbf{i} - 8xe^y\mathbf{j}$; $(-1, -1)$, $(3, 1)$

18. $\mathbf{F} = \left(4xy + \dfrac{3}{x^2}\right)\mathbf{i} + 2x^2 j$; $(1, 2)$, $(3, 3)$ (the path must lie in the half-plane $x > 0$)

19. $\mathbf{F} = (-4\cosh(xy) - 4xy\sinh(xy))\mathbf{i} - 4x^2\sinh(xy)\mathbf{j}$; $(1, 0)$, $(2, 1)$

20. $\mathbf{F} = (3y^2 + 3\sin(y))\mathbf{i} + (6xy + 3x\cos(y))\mathbf{j}$; $(0, 0)$, $(-3, \pi)$

21. Prove the law of conservation of energy: The sum of the kinetic and potential energies of an object acted on by a conservative force field is a constant. *Hint:* The kinetic energy is $(m/2)\|\mathbf{R}'(t)\|^2$, where m is the mass and $\mathbf{R}(t)$ the position vector of the particle. The potential energy is $-\varphi(x, y)$, where φ is a potential function for the force. Show that the derivative of the sum of the kinetic and potential energies is zero along any path of motion.

12.4 Surfaces in 3-Space and Surface Integrals

Analogous to the integral of a function over a curve, we would like to develop an integral of a function over a surface. This will require some background on surfaces.

A curve is often given by specifying coordinate functions, each of which is a function of a single variable or parameter. For this reason, we think of a curve as a one-dimensional object, although the graph may be in 2-space or 3-space.

A *surface* may be defined by giving coordinate functions that depend on two independent variables, say

$$x = x(u, v), \quad y = y(u, v), \quad z = z(u, v),$$

with (u, v) varying over some set in the u, v plane. The locus of such points may form a two-dimensional object in the plane or in R^3.

EXAMPLE 12.19

Suppose a surface is given by the coordinate functions

$$x = au\cos(v), \quad y = bu\sin(v), \quad z = u,$$

with u and v any real numbers and a and b nonzero constants. In this case, it is easy to write z in terms of x and y, a tactic that is sometimes useful in visualizing the surface. Notice that

$$\left(\frac{x}{au}\right)^2 + \left(\frac{y}{bu}\right)^2 = \cos^2(v) + \sin^2(v) = 1,$$

so

$$\frac{x^2}{a^2} + \frac{y^2}{b^2} = u^2 = z^2.$$

In the plane $y = 0$ (the x, z plane), $z = \pm x/a$, which are straight lines of slope $\pm 1/a$ through the origin. In the plane $x = 0$ (the y, z plane), $z = \pm y/b$, and these are straight lines of slope $\pm 1/b$ through the origin. The surface intersects a plane $z = c = $ constant $\neq 0$ in an ellipse:

$$\frac{x^2}{a^2} + \frac{y^2}{b^2} = c^2, \quad z = c.$$

This surface is called an *elliptical cone* because it has elliptical cross sections parallel to the x, y plane. ■

EXAMPLE 12.20

Consider the surface having coordinate functions

$$x = u\cos(v), \quad y = u\sin(v), \quad z = \tfrac{1}{2}u^2\sin(2v),$$

in which u and v can be any real numbers. Now

$$z = \tfrac{1}{2}u^2\sin(2v) = u^2\sin(v)\cos(v)$$
$$= [u\cos(v)][u\sin(v)] = xy.$$

This surface intersects any plane $z = c = $ constant $\neq 0$ in the hyperbola $xy = c, z = c$. However, the surface intersects a plane $y = \pm x$ in a parabola $z = \pm x^2$. For this reason this surface is called a *hyperbolic paraboloid*. ∎

Sometimes x and y are used as parameters, and the surface is defined by giving z as a function of x and y, say $z = S(x, y)$. Now the graph of the surface is the locus of points $(x, y, S(x, y))$, as (x, y) varies over some set of points in the x, y plane.

EXAMPLE 12.21

Consider $z = \sqrt{4 - x^2 - y^2}$ for $x^2 + y^2 \leq 4$. By squaring both sides of this equation, we can write

$$x^2 + y^2 + z^2 = 4.$$

This appears to be the equation of a sphere of radius 2 about the origin. However, in the original formulation with z given by the radical, we have $z \geq 0$, so in fact we have not the sphere, but the hemisphere (upper half of the sphere) of radius 2 about the origin. ∎

EXAMPLE 12.22

The equation $z = \sqrt{x^2 + y^2}$ for $x^2 + y^2 \leq 8$ determines a cone having circular cross sections parallel to the x, y plane. The "top" of the cone is the circle $x^2 + y^2 = 8$ in the plane $z = \sqrt{8}$. ∎

EXAMPLE 12.23

The equation $z = x^2 + y^2$ defines a parabolic bowl, extending to infinity in the positive z-direction because there is no restriction on x or y. ∎

These surfaces are easy to visualize and sketch by hand. If the defining function is more complicated, then we usually depend on a software package to sketch all or part of the surface. Examples are given in Figures 12.29 through 12.33.

Just as we can write a position vector to a curve, we can write a position vector

$$\mathbf{R}(u, v) = x(u, v)\mathbf{i} + y(u, v)\mathbf{j} + z(u, v)\mathbf{k}$$

for a surface. For any u and v in the parameter domain, $\mathbf{R}(u, v)$ can be thought of as an arrow from the origin to the point $(x(u, v), y(u, v), z(u, v))$ on the surface.

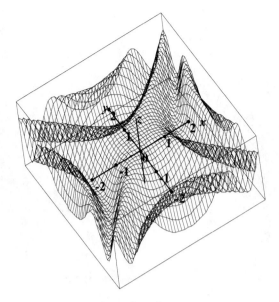

FIGURE 12.29 $z = \cos(x^2 - y^2)$.

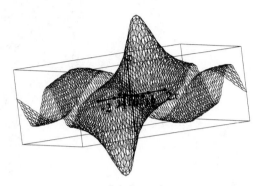

FIGURE 12.30 $z = \dfrac{6\sin(x - y)}{\sqrt{1 + x^2 + y^2}}$.

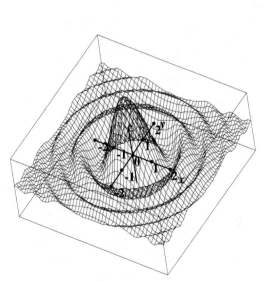

FIGURE 12.31 $z = \dfrac{4\cos(x^2 + y^2)}{1 + x^2 + y^2}$.

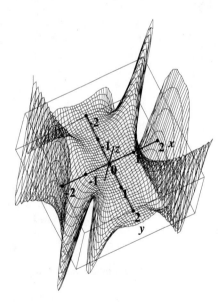

FIGURE 12.32 $z = x^2\cos(x^2 - y^2)$.

A surface is *simple* if $\mathbf{R}(u_1, v_1) = \mathbf{R}(u_2, v_2)$ can occur only if $u_1 = u_2$ and $v_1 = v_2$. A simple surface is one that does not fold over and return to the same point for different values of the parameter pairs.

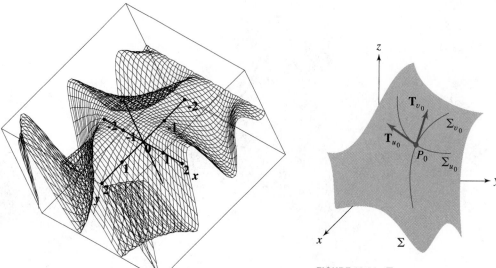

FIGURE 12.33 $z = \cos(xy)\log(4 + y)$.

FIGURE 12.34 *Tangents to curves on $\sum$ at P_0 determine the tangent plane to the surface there.*

12.4.1 Normal Vector to a Surface

Let $\sum$ be a surface with coordinate functions $x = x(u, v)$, $y = y(u, v)$, $z = z(u, v)$. Assume that these functions are continuous with continuous first partial derivatives. Let $P_0 : (x(u_0, v_0), y(u_0, v_0), z(u_0, v_0))$ be a point on $\sum$.

If we fix $v = v_0$, we can define the curve $\sum_{v_0}$ on the surface, having coordinate functions

$$x = x(u, v_0), \quad y = y(u, v_0), \quad z = z(u, v_0).$$

(See Figure 12.34.) This is a curve because its coordinate functions are functions of the single variable u. The tangent vector to $\sum_{v_0}$ at P_0 is

$$\mathbf{T}_{v_0} = \frac{\partial x}{\partial u}(u_0, v_0)\mathbf{i} + \frac{\partial y}{\partial u}(u_0, v_0)\mathbf{j} + \frac{\partial z}{\partial u}(u_0, v_0)\mathbf{k}.$$

Similarly, if we fix $u = u_0$ and use v as parameter, we obtain the curve $\sum_{u_0}$ on the surface (also shown in Figure 12.34). This curve has coordinate functions

$$x = x(u_0, v), \quad y = y(u_0, v), \quad z = z(u_0, v).$$

The tangent vector to $\sum_{u_0}$ at P_0 is

$$\mathbf{T}_{u_0} = \frac{\partial x}{\partial v}(u_0, v_0)\mathbf{i} + \frac{\partial y}{\partial v}(u_0, v_0)\mathbf{j} + \frac{\partial z}{\partial v}(u_0, v_0)\mathbf{k}.$$

Assuming that neither of these tangent vectors is the zero vector, they both lie in the tangent plane to $\sum$ at P_0. Their cross product is therefore normal (orthogonal) to this tangent plane and is the vector we define to be the *normal to $\sum$ at P_0*:

$$\mathbf{N}(P_0) = \left[\frac{\partial x}{\partial u}(u_0, v_0)\mathbf{i} + \frac{\partial y}{\partial u}(u_0, v_0)\mathbf{j} + \frac{\partial z}{\partial u}(u_0, v_0)\mathbf{k}\right]$$

$$\times \left[\frac{\partial x}{\partial v}(u_0, v_0)\mathbf{i} + \frac{\partial y}{\partial v}(u_0, v_0)\mathbf{j} + \frac{\partial z}{\partial v}(u_0, v_0)\mathbf{k}\right]$$

$$
= \begin{vmatrix} \mathbf{i} & \mathbf{j} & \mathbf{k} \\[6pt] \dfrac{\partial x}{\partial u}(u_0, v_0) & \dfrac{\partial y}{\partial u}(u_0, v_0) & \dfrac{\partial z}{\partial u}(u_0, v_0) \\[10pt] \dfrac{\partial x}{\partial v}(u_0, v_0) & \dfrac{\partial y}{\partial v}(u_0, v_0) & \dfrac{\partial z}{\partial v}(u_0, v_0) \end{vmatrix}
$$

$$
= \left(\frac{\partial y}{\partial u}\frac{\partial z}{\partial v} - \frac{\partial z}{\partial u}\frac{\partial y}{\partial v} \right)\mathbf{i} + \left(\frac{\partial z}{\partial u}\frac{\partial x}{\partial v} - \frac{\partial x}{\partial u}\frac{\partial z}{\partial v} \right)\mathbf{j} + \left(\frac{\partial x}{\partial u}\frac{\partial y}{\partial v} - \frac{\partial y}{\partial u}\frac{\partial x}{\partial v} \right)\mathbf{k}, \tag{12.5}
$$

in which all the partial derivatives are evaluated at (u_0, v_0). An expression that is easier to remember is obtained by introducing Jacobian notation. Define the Jacobian determinant (named for the German mathematician Karl Jacobi) of functions f and g to be the 2×2 determinant

$$
\frac{\partial(f, g)}{\partial(u, v)} = \begin{vmatrix} \dfrac{\partial f}{\partial u} & \dfrac{\partial f}{\partial v} \\[10pt] \dfrac{\partial g}{\partial u} & \dfrac{\partial g}{\partial v} \end{vmatrix} = \frac{\partial f}{\partial u}\frac{\partial g}{\partial v} - \frac{\partial g}{\partial u}\frac{\partial f}{\partial v}.
$$

In this notation, the normal vector to $\sum$ at P_0 is

$$
\mathbf{N}(P_0) = \frac{\partial(y, z)}{\partial(u, v)}\mathbf{i} + \frac{\partial(z, x)}{\partial(u, v)}\mathbf{j} + \frac{\partial(x, y)}{\partial(u, v)}\mathbf{k},
$$

with all the partial derivatives evaluated at (u_0, v_0). This notation helps in remembering the normal vector because of the cyclic pattern in the Jacobian symbols. Write

$$
x, y, z
$$

in this order. For the first component of $\mathbf{N}(P_0)$, omit the first letter, x, to obtain $\partial(y, z)/\partial(u, v)$. For the second component, omit y, but maintain the same cyclic direction, moving left to right through x, y, z. This means we start with z, the next letter after y, then back to x, obtaining $\partial(z, x)/\partial(u, v)$. For the third component, omit z, leaving x, y, and the Jacobian $\partial(x, y)/\partial(u, v)$.

Of course, any nonzero real multiple of $\mathbf{N}(P_0)$ is also a normal to $\sum$ at P_0.

EXAMPLE 12.24

Consider again the elliptical cone

$$
x = au\cos(v), \quad y = bu\sin(v), \quad z = u.
$$

Suppose we want the normal vector at $P_0 : (a\sqrt{3}/4, b/4, 1/2)$, obtained when $u = u_0 = 1/2$, $v = v_0 = \pi/6$. Compute the Jacobians:

$$
\frac{\partial(y, z)}{\partial(u, v)}\bigg]_{(1/2,\pi/6)} = \left[\frac{\partial y}{\partial u}\frac{\partial z}{\partial v} - \frac{\partial z}{\partial u}\frac{\partial y}{\partial v} \right]_{(1/2,\pi/6)}
$$

$$
= [b\sin(v)(0) - bu\cos(v)]_{(1/2,\pi/6)} = -\sqrt{3}b/4,
$$

$$
\frac{\partial(z, x)}{\partial(u, v)}\bigg]_{(1/2,\pi/6)} = \left[\frac{\partial z}{\partial u}\frac{\partial x}{\partial v} - \frac{\partial x}{\partial u}\frac{\partial z}{\partial v} \right]_{(1/2,\pi/6)}
$$

$$
= [-au\sin(v) - a\cos(v)(0)]_{(1/2,\pi/6)} = -a/4
$$

and

$$\frac{\partial(x, y)}{\partial(u, v)}\Bigg]_{(1/2,\pi/6)} = \left[\frac{\partial x}{\partial u}\frac{\partial y}{\partial v} - \frac{\partial y}{\partial u}\frac{\partial x}{\partial v}\right]_{(1/2,\pi/6)}$$

$$= [a\cos(v)bu\cos(v) - b\sin(v)(-au\sin(v))]_{(1/2,\pi/6)}$$

$$= ab/2.$$

The normal vector at P_0 is

$$\mathbf{N}(P_0) = -\sqrt{3}\frac{b}{4}\mathbf{i} - \frac{a}{4}\mathbf{j} + \frac{ab}{2}\mathbf{k}. \blacksquare$$

Consider the special case that the surface is given explicitly as $z = S(x, y)$. We may think of $u = x$ and $v = y$ as the parameters for $\sum$ and write the coordinate functions as

$$x = x, \quad y = y, \quad z = S(x, y).$$

Since $\partial x/\partial x = 1 = \partial y/\partial y$ and $\partial x/\partial y = \partial y/\partial x = 0$, we have

$$\frac{\partial(y, z)}{\partial(u, v)} = \frac{\partial(y, z)}{\partial(x, y)} = \begin{vmatrix} 0 & 1 \\ \dfrac{\partial S}{\partial x} & \dfrac{\partial S}{\partial y} \end{vmatrix} = -\frac{\partial S}{\partial x},$$

$$\frac{\partial(z, x)}{\partial(x, y)} = \begin{vmatrix} \dfrac{\partial S}{\partial x} & \dfrac{\partial S}{\partial y} \\ 1 & 0 \end{vmatrix} = -\frac{\partial S}{\partial y},$$

and

$$\frac{\partial(x, y)}{\partial(x, y)} = \begin{vmatrix} 1 & 0 \\ 0 & 1 \end{vmatrix} = 1.$$

The normal at a point $P_0 : (x_0, y_0, S(x_0, y_0))$ in this case is

$$\mathbf{N}(P_0) = -\frac{\partial S}{\partial x}(x_0, y_0)\mathbf{i} - \frac{\partial S}{\partial y}(x_0, y_0)\mathbf{j} + \mathbf{k}$$

$$= -\frac{\partial z}{\partial x}(x_0, y_0)\mathbf{i} - \frac{\partial z}{\partial y}(x_0, y_0)\mathbf{j} + \mathbf{k}. \tag{12.6}$$

We can also denote this vector as $\mathbf{N}(x_0, y_0)$.

EXAMPLE 12.25

Consider the cone given by $z = S(x, y) = \sqrt{x^2 + y^2}$. Then

$$\frac{\partial S}{\partial x} = \frac{x}{\sqrt{x^2 + y^2}} \quad \text{and} \quad \frac{\partial S}{\partial y} = \frac{y}{\sqrt{x^2 + y^2}},$$

except for $x = y = 0$. Take, for example, the point $(3, 1, \sqrt{10})$. The normal vector at this point is

$$N(3, 1, \sqrt{10}) = -\frac{3}{\sqrt{10}}i - \frac{1}{\sqrt{10}}j + k.$$

This normal vector is shown in Figure 12.35, and it points into the cone. In some contexts we want to know whether a normal vector is an inner normal (such as this one) or an outer normal (pointing out of the region bounded by the surface). If we wanted an outer normal at this point, we could take

$$-N(P_0) = \frac{3}{\sqrt{10}}i + \frac{1}{\sqrt{10}}j - k.$$

This cone does not have a normal vector at the origin, which is a "sharp point" of the surface. There is no tangent plane at the origin. ■

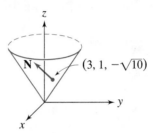

FIGURE 12.35 *Normal to the cone $z = \sqrt{x^2 + y^2}$ at $(3, 1, \sqrt{10})$.*

The normal vector (12.6) could also have been derived using the gradient vector. If $\sum$ is given by $z = S(x, y)$, then $\sum$ is a level surface of the function

$$\varphi(x, y, z) = z - S(x, y).$$

The gradient of this function is a normal vector, so compute

$$\nabla\varphi = \frac{\partial\varphi}{\partial x}i + \frac{\partial\varphi}{\partial y}j + \frac{\partial\varphi}{\partial z}k$$

$$= -\frac{\partial S}{\partial x}i - \frac{\partial S}{\partial y}j + k = N(P).$$

12.4.2 The Tangent Plane to a Surface

If a surface $\sum$ has a normal vector N at a point P_0, then we can use N to determine the equation of the tangent plane to $\sum$ at P_0. Let (x, y, z) be any point on the tangent plane. Then the vector $(x - x_0)i + (y - y_0)j + (z - z_0)k$ is in the tangent plane, hence is orthogonal to N. Then

$$N \cdot [(x - x_0)i + (y - y_0)j + (z - z_0)k] = 0.$$

More explicitly,

$$\left[\frac{\partial(y, z)}{\partial(u, v)}\right]_{(u_0, v_0)} (x - x_0) + \left[\frac{\partial(z, x)}{\partial(u, v)}\right]_{(u_0, v_0)} (y - y_0) + \left[\frac{\partial(x, y)}{\partial(u, v)}\right]_{(u_0, v_0)} (z - z_0) = 0.$$

This is the equation of the tangent plane to $\sum$ at P_0.

EXAMPLE 12.26

Consider again the elliptical cone given by

$$x = au\cos(v), \quad y = bu\sin(v), \quad z = u.$$

We found in Example 12.24 that the normal vector at $P_0 : (a\sqrt{3}/4, b/4, 1/2)$ is $\mathbf{N} = \sqrt{3}(b/4)\mathbf{i} - (a/4)\mathbf{j} + (ab/2)\mathbf{k}$. The tangent plane to $\sum$ at this point has equation

$$-\sqrt{3}\frac{b}{4}\left(x - \frac{a\sqrt{3}}{4}\right) - \frac{a}{4}\left(y - \frac{b}{4}\right) + \frac{ab}{2}\left(z - \frac{1}{2}\right) = 0. \quad \blacksquare$$

In the special case that $\sum$ is given by $z = S(x, y)$, then the normal vector at P_0 is $\mathbf{N} = -(\partial S/\partial x)(x_0, y_0)\mathbf{i} - (\partial S/\partial y)(x_0, y_0)\mathbf{j} + \mathbf{k}$, so the equation of the tangent plane becomes

$$-\frac{\partial S}{\partial x}(x_0, y_0)(x - x_0) - \frac{\partial S}{\partial y}(x_0, y_0)(y - y_0) + (z - z_0) = 0.$$

This equation is usually written

$$z - z_0 = \frac{\partial S}{\partial x}(x_0, y_0)(x - x_0) + \frac{\partial S}{\partial y}(x_0, y_0)(y - y_0).$$

12.4.3 Smooth and Piecewise Smooth Surfaces

Recall that a curve is smooth if it has a continuous tangent. Similarly, a surface is *smooth* if it has a continuous normal vector. A surface is *piecewise smooth* if it consists of a finite number of smooth surfaces. For example, a sphere is smooth, and the surface of a cube is piecewise smooth. A cube consists of six square pieces, which are smooth, but does not have a normal vector along any of its edges.

In calculus, it is shown that the area of a smooth surface $\sum$ given by $z = S(x, y)$ is the integral

$$\text{area of } \sum = \iint_D \sqrt{1 + \left(\frac{\partial S}{\partial x}\right)^2 + \left(\frac{\partial S}{\partial y}\right)^2}\, dA \tag{12.7}$$

where D is the set of points in the x, y plane for which S is defined. This may also be written

$$\text{area of } \sum = \iint_D \sqrt{1 + \left(\frac{\partial z}{\partial x}\right)^2 + \left(\frac{\partial z}{\partial y}\right)^2}\, dx\, dy.$$

Equation (12.7) is the integral of the length of the normal vector (12.6):

$$\text{area of } \sum = \iint_D \|\mathbf{N}(x, y)\|\, dx\, dy.$$

This is analogous to the formula for the length of a curve as the integral of the length of the tangent vector.

More generally, if $\sum$ is given by coordinate functions $x = x(u, v)$, $y = y(u, v)$, and $z = z(u, v)$, with (u, v) varying over some set D in the u, v plane, then

$$\text{area of } \sum = \iint_D \|\mathbf{N}(u, v)\|\, du\, dv, \tag{12.8}$$

the integral of the length of the normal vector, which is given by equation (12.5).

EXAMPLE 12.27

We will illustrate these formulas for surface area for a simple case in which we know the area from elementary geometry. Let $\sum$ be the upper hemisphere of radius 3 about the origin.

We can write $\sum$ as the graph of $z = S(x, y) = \sqrt{9 - x^2 - y^2}$, with $x^2 + y^2 \le 9$. D consists of all points on or inside the circle of radius 3 about the origin in the x, y plane.

We can use equation (12.7). Compute

$$\frac{\partial z}{\partial x} = -\frac{x}{\sqrt{9 - x^2 - y^2}} = -\frac{x}{z}$$

and, by symmetry,

$$\frac{\partial z}{\partial y} = -\frac{y}{z}.$$

Then

$$\text{area of } \sum = \iint_D \sqrt{1 + \left(\frac{x}{z}\right)^2 + \left(\frac{y}{z}\right)^2} \, dx \, dy$$

$$= \iint_D \sqrt{\frac{z^2 + x^2 + y^2}{z^2}} \, dx \, dy = \iint_D \frac{3}{\sqrt{9 - x^2 - y^2}} \, dx \, dy.$$

This is an improper double integral which we can evaluate easily by converting it to polar coordinates. Let $x = r\cos(\theta)$, $y = r\sin(\theta)$. Since D is the disk of radius 3 about the origin, $0 \le r \le 3$ and $0 \le \theta \le 2\pi$. Then

$$\iint_D \frac{3}{\sqrt{9 - x^2 - y^2}} \, dx \, dy = \int_0^{2\pi} \int_0^3 \frac{3}{\sqrt{9 - r^2}} r \, dr \, d\theta$$

$$= 6\pi \int_0^3 \frac{r}{\sqrt{9 - r^2}} \, dr = 6\pi \left[-(9 - r^2)^{1/2}\right]_0^3$$

$$= 6\pi \left[9^{1/2}\right] = 18\pi.$$

This is the area of a hemisphere of radius 3. ∎

We are now prepared to define the integral of a function over a surface.

12.4.4   Surface Integrals

The notion of the integral of a function over a surface is modeled after the line integral, with respect to arc length, of a function over a curve. Recall that if a smooth curve C is given by $x = x(t)$, $y = y(t)$, $z = z(t)$ for $a \le t \le b$, then the arc length along C is

$$s(t) = \int_a^t \sqrt{x'(\xi)^2 + y'(\xi)^2 + z'(\xi)^2} \, d\xi.$$

Then

$$ds = \sqrt{x'(t)^2 + y'(t)^2 + z'(t)^2} \, dt$$

and the line integral of a function f along C, with respect to arc length, is

$$\int_C f(x, y, z)\, ds = \int_a^b f(x(t), y(t), z(t))\sqrt{x'(t)^2 + y'(t)^2 + z'(t)^2}\; dt.$$

We want to lift these ideas up one dimension to integrate over a surface instead of a curve. Now we have coordinate functions that are functions of two independent variables, say u and v, with (u, v) varying over some given set D in the u, v plane. This means that $\int_a^b$ will be replaced by $\iint_D$. The differential element of arc length, ds, which is used in the line integral, will be replaced by the differential element $d\sigma$ of surface area on the surface. By equation (12.8), $d\sigma = \|\mathbf{N}(u, v)\|\, du\, dv$, in which $\mathbf{N}(u, v)$ is the normal vector at the point $(x(u, v), y(u, v), z(u, v))$ on $\sum$.

DEFINITION 12.6   *Surface Integral*

Let $\sum$ be a smooth surface having coordinate functions $x = x(u, v)$, $y = y(u, v)$, $z = z(u, v)$ for (u, v) in some set D of the u, v plane. Let f be continuous on $\sum$. Then the surface integral of f over $\sum$ is denoted $\iint_\sum f(x, y, z)\, d\sigma$, and is defined by

$$\iint_\sum f(x, y, z)\, d\sigma = \iint_D f(x(u, v), y(u, v), z(u, v))\, \|\mathbf{N}(u, v)\|\, du\, dv.$$

If $\sum$ is a piecewise smooth surface having smooth components $\sum_1, \ldots, \sum_n$, with each component either disjoint from the others or intersecting another component in a set of zero area (for example, along a curve), then

$$\iint_\sum f(x, y, z)\, d\sigma = \iint_{\sum_1} f(x, y, z)\, d\sigma + \cdots + \iint_{\sum_n} f(x, y, z)\, d\sigma.$$

For example, we would integrate over the surface of a cube by summing the surface integrals over the six faces. Two such faces either do not intersect or intersect each other along a line segment having zero area. The intersection condition is to prevent the selection of surface components that overlap each other in significant ways. This is analogous to a piecewise smooth curve C formed as the join of smooth curves $C_1, \ldots, C_n$. When we do this, we assume that two of these component curves either do not intersect, or intersect just at an endpoint, not along an arc of both curves.

If $\sum$ is described by $z = S(x, y)$, then

$$\iint_\sum f(x, y, z)\, d\sigma = \iint_D f(x, y, S(x, y))\sqrt{1 + \left(\frac{\partial S}{\partial x}\right)^2 + \left(\frac{\partial S}{\partial y}\right)^2}\; dx\, dy.$$

We will look at some examples of evaluation of surface integrals, then consider uses of surface integrals.

EXAMPLE 12.28

Evaluate $\iint_\sum z\, d\sigma$ if $\sum$ is the part of the plane $x + y + z = 4$ lying above the rectangle $0 \le x \le 2$, $0 \le y \le 1$.

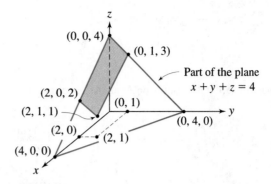

FIGURE 12.36

The surface is shown in Figure 12.36. D consists of all (x, y) with $0 \leq x \leq 2$ and $0 \leq y \leq 1$. With $z = S(x, y) = 4 - x - y$, we have

$$\iint_{\Sigma} z \, d\sigma = \iint_{D} z \sqrt{1 + (-1)^2 + (-1)^2} \, dx \, dy$$

$$= \sqrt{3} \int_{0}^{2} \int_{0}^{1} (4 - x - y) \, dy \, dx.$$

First compute

$$\int_{0}^{1} (4 - x - y) \, dy = (4 - x)y - \tfrac{1}{2}y^2 \big]_{0}^{1}$$

$$= 4 - x - \tfrac{1}{2} = \tfrac{7}{2} - x.$$

Then

$$\iint_{\Sigma} z \, d\sigma = \sqrt{3} \int_{0}^{2} \left(\tfrac{7}{2} - x \right) dx = 5\sqrt{3}. \quad \blacksquare$$

EXAMPLE 12.29

Evaluate $\iint_{\Sigma} (xy/z) \, d\sigma$, with Σ the part of the paraboloid $z = x^2 + y^2$ lying in the first octant for $4 \leq x^2 + y^2 \leq 9$.

Here D is the set in the first quadrant of the x, y plane, consisting of points on or between the circles of radii 2 and 3 about the origin. Part of the paraboloid is shown in Figure 12.37, with Σ the shaded part.

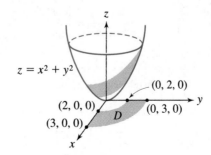

FIGURE 12.37

On Σ, $z = x^2 + y^2 = S(x, y)$, so

$$\iint_\Sigma \frac{xy}{z} \, d\sigma = \iint_D \frac{xy}{x^2 + y^2} \sqrt{1 + (2x)^2 + (2y)^2} \, dy \, dx.$$

Use polar coordinates to describe D as the set of points (r, θ) with $2 \le r \le 3$ and $0 \le \theta \le \pi/2$. Then

$$\iint_\Sigma \frac{xy}{z} \, d\sigma = \int_0^{\pi/2} \int_2^3 \frac{r^2 \cos(\theta) \sin(\theta)}{r^2} \sqrt{1 + 4r^2} r \, dr \, d\theta$$

$$= \int_0^{\pi/2} \cos(\theta) \sin(\theta) \, d\theta \int_2^3 r\sqrt{1 + 4r^2} \, dr$$

$$= \left[\tfrac{1}{2} \sin^2(\theta) \right]_0^{\pi/2} \left[\tfrac{1}{12} (1 + 4r^2)^{3/2} \right]_2^3$$

$$= \tfrac{1}{24} \left[(37)^{3/2} - (17)^{3/2} \right]. \quad \blacksquare$$

EXAMPLE 12.30

Recall the hyperbolic paraboloid of Example 12.20 given by

$$x = u \cos(v), \quad y = u \sin(v), \quad z = \tfrac{1}{2} u^2 \sin(2v).$$

We will compute the surface integral $\iint_\Sigma xyz \, d\sigma$ over the part of this surface corresponding to $1 \le u \le 2, 0 \le v \le \pi$.

First we need the normal vector. The components of $\mathbf{N}(u, v)$ are the Jacobians:

$$\frac{\partial(y, z)}{\partial(u, v)} = \begin{vmatrix} \sin(v) & u\cos(v) \\ u\sin(2v) & u^2\cos(2v) \end{vmatrix}$$

$$= u^2 \left[\sin(v)\cos(2v) - \cos(v)\sin(2v) \right],$$

$$\frac{\partial(z, x)}{\partial(u, v)} = \begin{vmatrix} u\sin(2v) & u^2\cos(2v) \\ \cos(v) & -u\sin(v) \end{vmatrix}$$

$$= -u^2 \left[\sin(v)\sin(2v) + \cos(v)\cos(2v) \right],$$

and

$$\frac{\partial(x, y)}{\partial(u, v)} = \begin{vmatrix} \cos(v) & -u\sin(v) \\ \sin(v) & u\cos(v) \end{vmatrix} = u.$$

Then

$$\|\mathbf{N}(u, v)\|^2 = u^4 \left[\sin(v)\cos(2v) - \cos(v)\sin(2v) \right]^2$$

$$+ u^4 \left[\sin(v)\sin(2v) + \cos(v)\cos(2v) \right]^2 + u^2$$

$$= u^2(1 + u^2),$$

so

$$\|\mathbf{N}(u, v)\| = u\sqrt{1 + u^2}.$$

The surface integral is

$$\iint_{\Sigma} xyz \, d\sigma = \iint_D [u\cos(v)][u\sin(v)]\left[\tfrac{1}{2}u^2 \sin(2v)\right] u\sqrt{1+u^2} \, dA$$

$$= \int_0^{\pi} \cos(v)\sin(v)\sin(2v) \, dv \int_1^2 \tfrac{1}{2}u^5 \sqrt{1+u^2} \, du$$

$$= \frac{\pi}{4}\left(\tfrac{100}{21}\sqrt{5} - \tfrac{11}{105}\sqrt{2}\right). \ \blacksquare$$

As we expect of an integral,

$$\iint_{\Sigma}(f(x,y,z)+g(x,y,z)) \, d\sigma = \iint_{\Sigma}(f(x,y,z) \, d\sigma + \iint_{\Sigma}g(x,y,z)) \, d\sigma$$

and, for any real number α,

$$\iint_{\Sigma}\alpha f(x,y,z) \, d\sigma = \alpha \iint_{\Sigma} f(x,y,z) \, d\sigma.$$

The next section is devoted to some applications of surface integrals.

SECTION 12.4   PROBLEMS

In each of Problems 1 through 10, evaluate $\iint_{\Sigma} f(x,y,z) \, d\sigma$.

1. $f(x,y,z) = x$, Σ is the part of the plane $x + 4y + z = 10$ in the first octant

2. $f(x,y,z) = y^2$, Σ is the part of the plane $z = x$ for $0 \le x \le 2, 0 \le y \le 4$

3. $f(x,y,z) = 1$, Σ is the part of the paraboloid $z = x^2 + y^2$ lying between the planes $z = 2$ and $z = 7$

4. $f(x,y,z) = x + y$, Σ is the part of the plane $4x + 8y + 10z = 25$ lying above the triangle in the x, y plane having vertices $(0,0)$, $(1,0)$, and $(1,1)$

5. $f(x,y,z) = z$, Σ is the part of the cone $z = \sqrt{x^2 + y^2}$ lying in the first octant and between the planes $z = 2$ and $z = 4$

6. $f(x,y,z) = xyz$, Σ is the part of the plane $z = x + y$ with (x,y) lying in the square with vertices $(0,0)$, $(1,0)$, $(0,1)$, and $(1,1)$

7. $f(x,y,z) = y$, Σ is the part of the cylinder $z = x^2$ for $0 \le x \le 2, 0 \le y \le 3$

8. $f(x,y,z) = x^2$, Σ is the part of the paraboloid $z = 4 - x^2 - y^2$ lying above the x, y plane

9. $f(x,y,z) = z$, Σ is the part of the plane $z = x - y$ for $0 \le x \le 1, 0 \le y \le 5$

10. $f(x,y,z) = xyz$, Σ is the part of the cylinder $z = 1 + y^2$ for $0 \le x \le 1, 0 \le y \le 1$

If a surface is given by $x = S(y,z)$ for (y,z) in some region K of the y, z plane, then

$$\iint_{\Sigma} f(x,y,z) \, d\sigma$$

$$= \iint_K f(S(y,z),y,z)\sqrt{1+\left(\frac{\partial x}{\partial y}\right)^2 + \left(\frac{\partial x}{\partial z}\right)^2} \, dA.$$

A similar formulation holds if Σ is given by $y = S(x,z)$ for (x,z) in a region in the x, z plane. Use these formulations to evaluate the following surface integrals:

11. $\iint_{\Sigma} z \, d\sigma$, Σ the cone $x = \sqrt{y^2 + z^2}$ for $y^2 + z^2 \le 9$

12. $\iint_{\Sigma} x \, d\sigma$, Σ the parabolic bowl $y = x^2 + z^2$ for $x^2 + z^2 \le 4$

13. $\iint_{\Sigma} y \, d\sigma$, Σ the rectangle $0 \le x \le 3, 0 \le z \le 4$ in the plane $y = 5$

14. $\iint_{\Sigma} x \, d\sigma$, Σ the part of the cylinder $z = y^2$ for $0 \le y \le 2, 0 \le z \le 4$

15. $\iint_{\Sigma} z^2 \, d\sigma$, Σ the part of the plane $x = y + z$ for $0 \le y \le 1, 0 \le z \le 4$

16. $\iint_{\Sigma} xyz \, d\sigma$, Σ the part of the plane $x = 4z + y$ for $0 \le y \le 2, 0 \le z \le 6$

17. $\iint_{\Sigma} xz \, d\sigma$, Σ the part of the surface $y = x^2 + z^2$ in the first octant for $1 \le x^2 + z^2 \le 4$

12.5 Applications of Surface Integrals

12.5.1 Surface Area

If $\sum$ is a piecewise smooth surface, then

$$\iint_{\Sigma} 1 \, d\sigma = \iint_{D} \|\mathbf{N}(u, v)\| \, du \, dv = \text{area of } \sum.$$

(This assumes a bounded surface having finite area.) Clearly, we do not need the notion of a surface integral to compute the integral on the right and obtain the area of a surface. However, we mention this result because it is in the same spirit as other familiar mensuration formulas:

$$\int_{C} ds = \text{length of } C,$$

$$\iint_{D} dA = \text{area of } D,$$

and, if M is a solid region in 3-space enclosing a volume, then

$$\iiint_{M} dV = \text{volume of } M.$$

12.5.2 Mass and Center of Mass of a Shell

Imagine a thin shell of negligible thickness taking the shape of a smooth surface $\sum$. Let $\delta(x, y, z)$ be the density of the material of the shell at (x, y, z). Assume that δ is continuous. We want to compute the mass of the shell.

Suppose $\sum$ has coordinate functions $x = x(u, v)$, $y = y(u, v)$, $z = z(u, v)$ for (u, v) in D. Form a grid over D in the u, v plane by drawing lines (dashed lines in Figure 12.38) parallel to the axes and retain only those rectangles $R_1, \ldots, R_N$ intersecting D. Let the vertical lines be Δu units apart, and the horizontal lines, Δv units apart. For (u, v) varying over R_j, we obtain a patch or surface element $\sum_j$ on the surface (Figure 12.39). That is, $\sum_j(u, v) = \sum(u, v)$ for (u, v) in R_j. Let (u_j, v_j) be a point in R_j and approximate the density of the surface element $\sum_j$ by the constant

$$\delta_j = \delta(x(u_j, v_j), y(u_j, v_j), z(u_j, v_j)).$$

Because δ is continuous and $\sum_j$ has finite area, we can choose Δu and Δv sufficiently small that δ_j approximates $\delta(x, y, z)$ as closely as we like over $\sum_j$.

Approximate the mass of $\sum_j$ as δ_j times the area of $\sum_j$. Now this area is

$$\text{area of } \sum_j = \iint_{R_j} \|\mathbf{N}(u, v)\| \, du \, dv$$

$$\approx \|\mathbf{N}(u_j, v_j)\| \, \Delta u \Delta v,$$

so

$$\text{mass of } \sum_j \approx \delta_j \|\mathbf{N}(u_j, v_j)\| \, \Delta u \Delta v.$$

In 1927, Congress approved construction of a dam to control the Colorado River for the purpose of fostering agriculture in the American southwest and as a source of hydroelectric power, spurring the growth of Las Vegas and southern California. Construction began in 1931. One major problem of construction was the cooling of concrete as it was poured. Engineers estimated that the amount of concrete required would take 100 years to cool. The solution was to pour it in rows and columns of blocks, through which cooled water was pumped in pipes. Hoover Dam was completed in 1935 and is 727 feet high, 1244 feet long, 660 feet thick at its base, and 45 feet thick at the top. It weighs about 5.5 million tons and contains 3,250,000 cubic yards of concrete. On one side of the dam, Lake Mead is over 500 feet deep. Computation of forces and stresses on parts of the dam surface involve a combination of material science, fluid flow, and vector analysis.

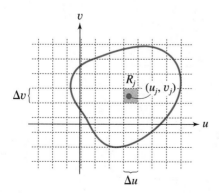

FIGURE 12.38 *Grid over D in the u, v plane.*

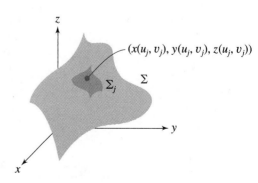

FIGURE 12.39 *Surface element $\sum_j$ on $\sum$.*

The mass of $\sum_j$ is approximately the sum of the masses of the surface elements:

$$\text{mass of } \sum \approx \sum_{j=1}^{N} \delta(x(u_j, v_j), y(u_j, v_j), z(u_j, v_j)) \, \|\mathbf{N}(u_j, v_j)\| \, \Delta u \, \Delta v.$$

This is a Riemann sum for $\iint_D \delta(x(u, v), y(u, v), z(u, v)) \, \|\mathbf{N}(u, v)\| \, du \, dv$. Hence in the limit as $\Delta u \to 0$ and $\Delta v \to 0$, we obtain

$$\text{mass of } \sum = \iint_{\Sigma} \delta(x, y, z) \, d\sigma.$$

The mass of the shell is the surface integral of the density function. This is analogous to the mass of a wire being the line integral of the density function over the wire.

The center of mass of the shell is $(\overline{x}, \overline{y}, \overline{z})$, where

$$\overline{x} = \frac{1}{m} \iint_{\Sigma} x \delta(x, y, z) \, d\sigma, \overline{y} = \frac{1}{m} \iint_{\Sigma} y \delta(x, y, z) \, d\sigma,$$

and

$$\overline{z} = \frac{1}{m} \iint_{\Sigma} z \delta(x, y, z) \, d\sigma,$$

in which m is the mass of the shell.

EXAMPLE 12.31

We will find the mass and center of mass of the cone $z = \sqrt{x^2 + y^2}$, where $x^2 + y^2 \leq 4$ and the density function is $\delta(x, y, z) = x^2 + y^2$.

First calculate the mass m. We will need

$$\frac{\partial z}{\partial x} = \frac{x}{z} \quad \text{and} \quad \frac{\partial z}{\partial y} = \frac{y}{z}.$$

Then

$$m = \iint_{\Sigma} (x^2 + y^2) \, d\sigma$$

$$= \iint_{D} (x^2 + y^2) \sqrt{1 + \frac{x^2}{z^2} + \frac{y^2}{z^2}} \, dy \, dx$$

$$= \int_0^{2\pi} \int_0^2 r^2 \sqrt{2} r \, dr \, d\theta$$

$$= 2\sqrt{2}\pi \frac{1}{4} r^4 \Big]_0^2 = 8\sqrt{2}\pi.$$

By symmetry of the surface and of the density function, we expect the center of mass to lie on the z axis, so $\overline{x} = \overline{y} = 0$. Finally,

$$\overline{z} = \frac{1}{8\sqrt{2}\pi} \iint_{\Sigma} z(x^2 + y^2)d\sigma$$

$$= \frac{1}{8\sqrt{2}\pi} \iint_{D} \sqrt{x^2 + y^2}(x^2 + y^2)\sqrt{1 + \frac{x^2}{z^2} + \frac{y^2}{z^2}} \, dy \, dx$$

$$= \frac{1}{8\pi} \int_{0}^{2\pi} \int_{0}^{2} r(r^2)r \, dr \, d\theta$$

$$= \frac{1}{8\pi} (2\pi) \left[\frac{1}{5}r^5\right]_{0}^{2} = \frac{8}{5}.$$

The center of mass is $(0, 0, \frac{8}{5})$. ∎

12.5.3 Flux of a Vector Field Across a Surface

Suppose a fluid moves in some region of 3-space (for example, through a pipeline), with velocity $\mathbf{V}(x, y, z, t)$. Consider an imaginary surface Σ within the fluid, with continuous *unit* normal vector $\mathbf{N}(u, v, t)$. The flux of $\mathbf{V}$ across Σ is the net volume of fluid, per unit time, flowing across Σ in the direction of $\mathbf{N}$. We would like to calculate this flux.

In a time interval Δt the volume of fluid flowing across a small piece Σ_j of Σ equals the volume of the cylinder with base Σ_j and altitude $V_N \Delta t$, where V_N is the component of $\mathbf{V}$ in the direction of $\mathbf{N}$, evaluated at some point of Σ_j. This volume (Figure 12.40) is $(V_N \Delta t)A_j$, where A_j is the area of Σ_j. Because $\|\mathbf{N}\| = 1$, $V_N = \mathbf{V} \cdot \mathbf{N}$. The volume of fluid flowing across Σ_j per unit time is

$$\frac{V_N(\Delta t)A_j}{\Delta t} = V_N A_j = \mathbf{V} \cdot \mathbf{N}A_j.$$

Sum these quantities over all the pieces of the surface and take the limit as the pieces are chosen smaller, as we did for the mass of the shell. We get

$$\text{flux of } \mathbf{V} \text{ across } \Sigma \text{ in the direction of } \mathbf{N} = \iint_{\Sigma} \mathbf{V} \cdot \mathbf{N} \, d\sigma.$$

The flux of a vector across a surface is therefore computed as the surface integral of the normal component of the vector to the surface.

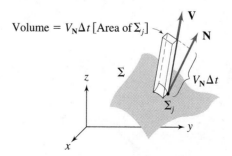

Volume $= V_{\mathbf{N}}\Delta t$ [Area of Σ_j]

FIGURE 12.40

EXAMPLE 12.32

Find the flux of $\mathbf{F} = x\mathbf{i} + y\mathbf{j} + z\mathbf{k}$ across the part of the sphere $x^2 + y^2 + z^2 = 4$ lying between the planes $z = 1$ and $z = 2$.

The surface $\sum$ is shown in Figure 12.41, along with the normal vector (computed next) at a point. We may think of $\sum$ as defined by $z = S(x, y)$, where S is defined by the equation of the sphere and (x, y) varies over a set D in the x, y plane. To determine D, observe that the plane $z = 2$ hits $\sum$ only at its "north pole" $(0, 0, 2)$. The plane $z = 1$ intersects the sphere in the circle $x^2 + y^2 = 3$, $z = 1$. This circle projects onto the x, y plane to the circle of radius $\sqrt{3}$ about the origin. Thus D consists of points (x, y) satisfying $x^2 + y^2 \leq 3$ (shaded in Figure 12.42).

To compute the partial derivatives $\partial z / \partial x$ and $\partial z / \partial y$, we can implicitly differentiate the equation of the sphere to get

$$2x + 2z\frac{\partial z}{\partial x} = 0,$$

so

$$\frac{\partial z}{\partial x} = -\frac{x}{z}$$

and, similarly,

$$\frac{\partial z}{\partial y} = -\frac{y}{z}.$$

A normal vector to the sphere is therefore

$$-\frac{x}{z}\mathbf{i} - \frac{y}{z}\mathbf{j} - \mathbf{k}.$$

Since we need a unit normal in computing flux, we must divide this vector by its length, which is

$$\sqrt{\frac{x^2}{z^2} + \frac{y^2}{z^2} + 1} = \sqrt{\frac{x^2 + y^2 + z^2}{z^2}} = \frac{2}{z}.$$

A unit normal is therefore

$$\mathbf{N} = \frac{z}{2}\left(-\frac{x}{z}\mathbf{i} - \frac{y}{z}\mathbf{j} - \mathbf{k}\right) = -\frac{1}{2}(x\mathbf{i} + y\mathbf{j} + z\mathbf{k}).$$

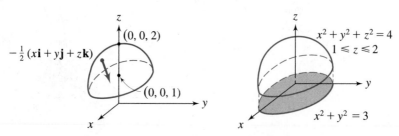

FIGURE 12.41   FIGURE 12.42

This points into the sphere. If we want the flux across $\sum$ from the outside of the sphere toward the inside, this is the normal to use. If we want the flux across $\sum$ from within the sphere, use $-\mathbf{N}$ instead. Now,

$$\mathbf{F} \cdot (-\mathbf{N}) = \frac{1}{2} \left(x^2 + y^2 + z^2 \right).$$

There

$$\begin{aligned}
\text{flux} &= \iint_{\Sigma} \frac{1}{2}(x^2 + y^2 + z^2) \, d\sigma \\
&= \frac{1}{2} \iint_D (x^2 + y^2 + z^2)\sqrt{1 + \frac{x^2}{z^2} + \frac{y^2}{z^2}} \, dA \\
&= \frac{1}{2} \iint_D (x^2 + y^2 + z^2)\sqrt{\frac{x^2 + y^2 + z^2}{z^2}} \, dA \\
&= \frac{1}{2} \iint_D \left(x^2 + y^2 + z^2 \right)^{3/2} \frac{1}{\sqrt{4 - x^2 - y^2}} \, dA \\
&= 4 \iint_D \frac{1}{\sqrt{4 - x^2 - y^2}} \, dA
\end{aligned}$$

because $x^2 + y^2 + z^2 = 4$ on $\sum$. Converting to polar coordinates, we have

$$\begin{aligned}
\text{flux} &= 4 \int_0^{2\pi} \int_0^{\sqrt{3}} \frac{1}{\sqrt{4 - r^2}} r \, dr \, d\theta \\
&= 8\pi \left[-(4 - r^2)^{1/2} \right]_0^{\sqrt{3}} = 8\pi. \quad \blacksquare
\end{aligned}$$

We will see other applications of surface integrals when we discuss the integral theorems of Gauss and Stokes, and again when we study partial differential equations.

SECTION 12.5   PROBLEMS

1. Find the mass and center of mass of the triangular shell having vertices $(1, 0, 0)$, $(0, 3, 0)$, and $(0, 0, 2)$ if $\delta(x, y, z) = xz + 1$.

2. Find the center of mass of the portion of the homogeneous sphere $x^2 + y^2 + z^2 = 9$ lying above the plane $z = 1$. (Homogeneous means that the density function is constant).

3. Find the center of mass of the homogeneous cone $z = \sqrt{x^2 + y^2}$ for $x^2 + z^2 \le 9$.

4. Find the center of mass of the part of the paraboloid $z = 16 - x^2 - y^2$ lying in the first octant and be-

tween the cylinders $x^2 + y^2 = 1$ and $x^2 + y^2 = 9$ if $\delta(x, y, z) = xy/\sqrt{1 + 4x^2 + 4y^2}$.

5. Find the mass and center of mass of the paraboloid $z = 6 - x^2 - y^2 \ge 0$ if $\delta(x, y, z) = \sqrt{1 + 4x^2 + 4y^2}$.

6. Find the center of mass of the part of the homogeneous sphere $x^2 + y^2 + z^2 = 1$ lying in the first octant.

7. Find the flux of $\mathbf{F} = x\mathbf{i} + y\mathbf{j} - z\mathbf{k}$ across the part of the plane $x + 2y + z = 8$ lying in the first octant.

8. Find the flux of $\mathbf{F} = xz\mathbf{i} - y\mathbf{k}$ across the part of the sphere $x^2 + y^2 + z^2 = 4$ lying above the plane $z = 1$.

12.6   Preparation for the Integral Theorems of Gauss and Stokes

The fundamental results of vector integral calculus are the theorems of Gauss and Stokes. In this section we will begin with Green's theorem and explore how natural generalizations lead to these results.

With appropriate conditions on the curve and the functions, the conclusion of Green's theorem is

$$\oint_C f(x, y)\, dx + g(x, y)\, dy = \iint_D \left(\frac{\partial g}{\partial x} - \frac{\partial f}{\partial y} \right) dA$$

in which D is the plane region enclosed by the simple, closed, smooth curve C. Define the vector field

$$\mathbf{F}(x, y) = g(x, y)\mathbf{i} - f(x, y)\mathbf{j}.$$

Then

$$\nabla \cdot \mathbf{F} = \frac{\partial g}{\partial x} - \frac{\partial f}{\partial y}.$$

Now parametrize C by arc length, so the coordinate functions are $x = x(s)$, $y = y(s)$ for $0 \le s \le L$. The unit tangent vector to C is $\mathbf{T}(s) = x'(s)\mathbf{i} + y'(s)\mathbf{j}$, and the unit normal vector is $\mathbf{N}(s) = y'(s)\mathbf{i} - x'(s)\mathbf{j}$. These are shown in Figure 12.43. This normal points away from the interior D of C, and so is an *outer normal*. Now

$$\mathbf{F} \cdot \mathbf{N} = g(x, y)\frac{dy}{ds} + f(x, y)\frac{dx}{ds},$$

so

$$\oint_C f(x, y)\, dx + g(x, y)\, dy = \oint_C \left[f(x, y)\frac{dx}{ds} + g(x, y)\frac{dy}{ds} \right] ds$$

$$= \oint_C \mathbf{F} \cdot \mathbf{N}\, ds.$$

We may therefore write the conclusion of Green's theorem in vector form as

$$\oint_C \mathbf{F} \cdot \mathbf{N}\, ds = \iint_D \nabla \cdot \mathbf{F}\, dA. \tag{12.9}$$

This is a conservation of energy equation. Recall from Section 11.4.1 that the divergence of a vector field at a point is a measure of the flow of the field from that point. Equation (12.9) states that the flux of the vector field outward from D across C (because $\mathbf{N}$ is an outer normal) exactly balances the flow of the field from each point in D.

The reason for writing Green's theorem in this form is that it suggests a generalization to three dimensions. Replace the closed curve C in the plane with a closed surface Σ in 3-space (closed meaning bounding a volume). Replace the line integral over C with a surface integral over Σ and allow the vector field $\mathbf{F}$ to be a function of three variables. We conjecture that equation (12.9) generalizes to

$$\iint_\Sigma \mathbf{F} \cdot \mathbf{N}\, d\sigma = \iiint_M \nabla \cdot \mathbf{F}\, dV,$$

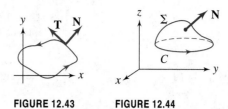

FIGURE 12.43   **FIGURE 12.44**

in which $\mathbf{N}$ is a unit normal to $\sum$ pointing away from the solid region M bounded by $\sum$. We will see that under suitable conditions on $\sum$ and $\mathbf{F}$, this is the conclusion of Gauss's divergence theorem.

Now begin again with Green's theorem. We will pursue a different generalization to three dimensions. This time let

$$\mathbf{F}(x, y, z) = f(x, y)\mathbf{i} + g(x, y)\mathbf{j} + 0\mathbf{k}.$$

The reason for adding the third component is to be able to take the curl:

$$\nabla \times \mathbf{F} = \begin{vmatrix} \mathbf{i} & \mathbf{j} & \mathbf{k} \\ \dfrac{\partial}{\partial x} & \dfrac{\partial}{\partial y} & \dfrac{\partial}{\partial z} \\ f & g & 0 \end{vmatrix} = \left(\frac{\partial g}{\partial x} - \frac{\partial f}{\partial y} \right)\mathbf{k}.$$

Then

$$(\nabla \times \mathbf{F}) \cdot \mathbf{k} = \frac{\partial g}{\partial x} - \frac{\partial f}{\partial y}.$$

Further, with unit tangent $\mathbf{T}(s) = x'(s)\mathbf{i} + y'(s)\mathbf{j}$ to C, we can write

$$\mathbf{F} \cdot \mathbf{T}\, ds = [f(x, y)\mathbf{i} + g(x, y)\mathbf{j}] \cdot \left(\frac{dx}{ds}\mathbf{i} + \frac{dy}{ds}\mathbf{j} \right) ds$$

$$= f(x, y)\, dx + g(x, y)\, dy$$

so the conclusion of Green's theorem can also be written

$$\oint_C \mathbf{F} \cdot \mathbf{T}\, ds = \iint_D (\nabla \times \mathbf{F}) \cdot \mathbf{k}\, dA. \tag{12.10}$$

Now think of D as a flat surface in the x, y plane, with unit normal vector $\mathbf{k}$, and bounded by the closed curve C. To generalize this, allow C to be a curve in 3-space bounding a surface $\sum$ having unit outer normal vector $\mathbf{N}$, as shown in Figure 12.44. Now equation (12.10) suggests that

$$\oint_C \mathbf{F} \cdot \mathbf{T}\, ds = \iint_\Sigma (\nabla \times \mathbf{F}) \cdot \mathbf{N}\, d\sigma.$$

We will see this equation shortly as the conclusion of Stokes's theorem.

SECTION 12.6   PROBLEMS

1. Let C be a simple closed path in the x, y plane, with interior D. Let $\varphi(x, y)$ and $\psi(x, y)$ be continuous with continuous first and second partial derivatives on C and throughout D.

Let

$$\nabla^2 \psi = \frac{\partial^2 \psi}{\partial x^2} + \frac{\partial^2 \psi}{\partial y^2}.$$

Prove that

$$
\iint_D \varphi \nabla^2 \psi \, dA
$$

$$
= \oint_C -\varphi \frac{\partial \psi}{\partial y} \, dx + \varphi \frac{\partial \psi}{\partial x} \, dy - \iint_D \nabla \varphi \cdot \nabla \psi \, dA.
$$

2. Under the conditions of Problem 1, show that

$$
\iint_D \left(\varphi \nabla^2 \psi - \psi \nabla^2 \varphi \right) dA
$$

$$
= \oint_C \left[\psi \frac{\partial \varphi}{\partial y} - \varphi \frac{\partial \psi}{\partial y} \right] dx + \left[\varphi \frac{\partial \psi}{\partial x} - \psi \frac{\partial \varphi}{\partial x} \right] dy.
$$

3. Let C be a simple closed path in the x, y plane, with interior D. Let φ be continuous with continuous first and second partial derivatives on C and at all points of D. Let $\mathbf{N}(x, y)$ be the unit outer normal to C (outer meaning pointing away from D if drawn as an arrow at (x, y) on C). Prove that

$$
\oint_C \varphi_\mathbf{N}(x, y) \, ds = \iint_D \nabla^2 \varphi(x, y) \, dA.
$$

(Here $\varphi_\mathbf{N}(x, y)$ is the directional derivative of φ in the direction of $\mathbf{N}$.)

12.7 The Divergence Theorem of Gauss

We have seen that under certain conditions, the conclusion of Green's theorem is

$$
\oint_C \mathbf{F} \cdot \mathbf{N} \, ds = \iint_D \nabla \cdot \mathbf{F} \, dA.
$$

Now make the following generalizations from the plane to 3-space:

> set D in the plane $\rightarrow$ a 3-dimensional solid M
> a closed curve C bounding D $\rightarrow$ a surface $\sum$ enclosing M
> a unit outer normal $\mathbf{N}$ to C $\rightarrow$ a unit outer normal $\mathbf{N}$ to $\sum$
> a vector field $\mathbf{F}$ in the plane $\rightarrow$ a vector field $\mathbf{F}$ in 3-space
> a line integral $\oint_C \mathbf{F} \cdot \mathbf{N} \, ds$ $\rightarrow$ a surface integral $\iint_\sum \mathbf{F} \cdot \mathbf{N} \, d\sigma$
> a double integral $\iint_D \nabla \cdot \mathbf{F} \, dA$ $\rightarrow$ a triple integral $\iiint_M \nabla \cdot \mathbf{F} \, dV$.

With these correspondences and some terminology, Green's theorem suggests a theorem named for the great nineteenth-century German mathematician and scientist Carl Friedrich Gauss.

A surface $\sum$ is *closed* if it encloses a volume. For example, a sphere is closed, as is a cube, while a hemisphere is not. A surface consisting of the top part of the sphere $x^2 + y^2 + z^2 = a^2$, together with the disk $x^2 + y^2 \leq a^2$ in the x, y plane, is closed. If filled with water (through some opening that is then sealed off), it will hold the water.

A normal vector $\mathbf{N}$ to $\sum$ is an *outer normal* if when represented as an arrow from a point of the surface, it points away from the region enclosed by the surface (Figure 12.45). If $\mathbf{N}$ is also a unit vector, then it is a *unit outer normal*.

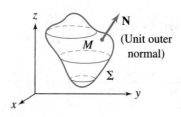

FIGURE 12.45

THEOREM 12.8 *Gauss's Divergence Theorem*

Let $\sum$ be a piecewise smooth closed surface. Let M be the set of points on and enclosed by $\sum$. Let $\sum$ have unit outer normal vector $\mathbf{N}$. Let $\mathbf{F}$ be a vector field whose components are continuous with continuous first and second partial derivatives on $\sum$ and throughout M. Then

$$\iint_{\sum} \mathbf{F} \cdot \mathbf{N} \, d\sigma = \iiint_{M} \nabla \cdot \mathbf{F} \, dV. \blacksquare \qquad (12.11)$$

$\nabla \cdot \mathbf{F}$ is the divergence of the vector field, hence the name "divergence theorem." In the spirit of Green's theorem, Gauss's theorem relates vector operations over objects of different dimensions. A surface is a two-dimensional object (it has area but no volume), while a solid region in 3-space is three-dimensional.

Gauss's theorem has several kinds of applications. One is to replace one of the integrals in equation (12.11) with the other, in the event that this simplifies an integral evaluation. A second is to suggest interpretations of vector operations. A third is to serve as a tool in deriving physical laws. Finally, we will use the theorem in developing relationships to be used in solving partial differential equations.

Before looking at uses of the theorem, here are two purely computational examples to provide some feeling for equation (12.11).

EXAMPLE 12.33

Let $\sum$ be the piecewise smooth closed surface consisting of the surface $\sum_1$ of the cone $z = \sqrt{x^2 + y^2}$ for $x^2 + y^2 \leq 1$, together with the flat cap $\sum_2$ consisting of the disk $x^2 + y^2 \leq 1$ in the plane $z = 1$. This surface is shown in Figure 12.46. Let $\mathbf{F}(x, y, z) = x\mathbf{i} + y\mathbf{j} + z\mathbf{k}$. We will calculate both sides of equation (12.11).

The unit outer normal to $\sum_1$ is

$$\mathbf{N}_1 = \frac{1}{\sqrt{2}} \left(\frac{x}{z}\mathbf{i} + \frac{y}{z}\mathbf{j} - \mathbf{k} \right).$$

Then

$$\mathbf{F} \cdot \mathbf{N}_1 = (x\mathbf{i} + y\mathbf{j} + z\mathbf{k}) \cdot \frac{1}{\sqrt{2}} \left(\frac{x}{z}\mathbf{i} + \frac{y}{z}\mathbf{j} - \mathbf{k} \right)$$

$$= \frac{1}{\sqrt{2}} \left(\frac{x^2}{z} + \frac{y^2}{z} - z \right) = 0$$

because on $\sum_1$, $z^2 = x^2 + y^2$. (One can also see geometrically that $\mathbf{F}$ is orthogonal to $\mathbf{N}_1$.) Then

$$\iint_{\sum_1} \mathbf{F} \cdot \mathbf{N}_1 \, d\sigma = 0.$$

The unit outer normal to $\sum_2$ is $\mathbf{N}_2 = \mathbf{k}$, so

$$\mathbf{F} \cdot \mathbf{N}_2 = z.$$

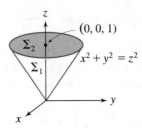

FIGURE 12.46

Since $z = 1$ on $\sum_2$, then

$$\iint_{\Sigma_2} \mathbf{F} \cdot \mathbf{N}_2 \, d\sigma = \iint_{\Sigma_2} z \, d\sigma = \iint_{\Sigma_2} d\sigma$$

$$= \text{area of } \sum_2 = \pi.$$

Therefore,

$$\iint_{\Sigma} \mathbf{F} \cdot \mathbf{N} \, d\sigma = \iint_{\Sigma_1} \mathbf{F} \cdot \mathbf{N}_1 \, d\sigma + \iint_{\Sigma_2} \mathbf{F} \cdot \mathbf{N}_2 \, d\sigma = \pi.$$

Now compute the triple integral. The divergence of $\mathbf{F}$ is

$$\nabla \cdot \mathbf{F} = \frac{\partial}{\partial x} x + \frac{\partial}{\partial y} y + \frac{\partial}{\partial z} z = 3,$$

so

$$\iiint_M \nabla \cdot \mathbf{F} \, dV = \iiint_M 3 \, dV$$

$$= 3 \, [\text{volume of the cone of height 1, radius 1}]$$

$$= 3 \tfrac{1}{3} \pi = \pi. \quad \blacksquare$$

EXAMPLE 12.34

Let $\sum$ be the piecewise smooth surface of the cube having vertices

$$(0, 0, 0), \ (1, 0, 0), \ (0, 1, 0), \ (0, 0, 1),$$

$$(1, 1, 0), \ (0, 1, 1), \ (1, 0, 1), \ (1, 1, 1).$$

Let $\mathbf{F}(x, y, z) = x^2 \mathbf{i} + y^2 \mathbf{j} + z^2 \mathbf{k}$. We would like to compute the flux of this vector field across the faces of the cube.

The flux is $\iint_{\Sigma} \mathbf{F} \cdot \mathbf{N} \, d\sigma$. This integral can certainly be calculated directly, but it requires performing the integration over each of the six smooth faces of $\sum$. It is easier to use the triple integral from Gauss's theorem. Compute the divergence

$$\nabla \cdot \mathbf{F} = 2x + 2y + 2z$$

and then

$$\text{flux} = \iint_{\Sigma} \mathbf{F} \cdot \mathbf{N} \, d\sigma$$

$$= \iiint_M \nabla \cdot \mathbf{F} \, dV = 2 \iiint_M (x + y + z) \, dV$$

$$= \int_0^1 \int_0^1 \int_0^1 (2x + 2y + 2z)\, dz\, dy\, dx$$

$$= \int_0^1 \int_0^1 (2x + 2y + 1)\, dy\, dx$$

$$= \int_0^1 (2x + 2)\, dx = 3. \quad \blacksquare$$

Now we will move to more substantial uses of the theorem.

12.7.1 Archimedes' Principle

Archimedes' principle states that the buoyant force a fluid exerts on a solid object immersed in it is equal to the weight of the fluid displaced. A bar of soap floats or sinks in a full bathtub for the same reason a battleship floats or sinks in the ocean. The issue rests with the weight of the fluid displaced by the object. We will derive this principle.

Consider a solid object M bounded by a piecewise smooth surface Σ. Let ρ be the constant density of the fluid. Draw a coordinate system as in Figure 12.47, with M below the surface of the fluid. Using the fact that pressure is the product of depth and density, the pressure $p(x, y, z)$ at a point on Σ is given by $p(x, y, z) = -\rho z$. The negative sign is used because z is negative in the downward direction and we want pressure to be positive.

Now consider a piece Σ_j of the surface, also shown in Figure 12.47. The force of the pressure on this surface element has magnitude approximately $-\rho z$ times the area A_j of Σ_j. If $\mathbf{N}$ is the unit outer normal to Σ_j, then the force caused by the pressure on Σ_j is approximately $\rho z \mathbf{N} A_j$. The vertical component of this force is the magnitude of the buoyant force acting upward on Σ_j. This vertical component is $\rho z \mathbf{N} \cdot \mathbf{k} A_j$. Sum these vertical components over the entire surface to obtain approximately the net buoyant force on the object, then take the limit as the surface elements are chosen smaller (areas tending to zero). We obtain in this limit that

$$\text{net buoyant force on } \Sigma = \iint_\Sigma \rho z \mathbf{N} \cdot \mathbf{k}\, d\sigma.$$

Write this integral as $\iint_\Sigma \rho z \mathbf{k} \cdot \mathbf{N}\, d\sigma$ and apply Gauss's theorem to convert the surface integral into a triple integral:

$$\text{net buoyant force on } \Sigma = \iiint_M \nabla \cdot (\rho z \mathbf{k})\, dV.$$

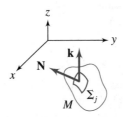

FIGURE 12.47
Pressure force on
$\Sigma_j \approx \rho z \mathbf{N} \cdot A_j.$
Vertical component
$= \rho z \mathbf{N} \cdot \mathbf{k} A_j.$

But $\nabla \cdot (\rho z \mathbf{k}) = \rho$, so

$$\text{net buoyant force on } \sum = \iiint_M \rho \, dV = \rho \, [\text{volume of } M].$$

But this is exactly the weight of the fluid displaced, establishing Archimedes' principle.

12.7.2 The Heat Equation

We will derive a partial differential equation that models heat conduction. Suppose some medium (for example, a metal bar, the air in a room, or water in a pool) has density $\rho(x, y, z)$, specific heat $\mu(x, y, z)$, and coefficient of thermal conductivity $K(x, y, z)$. Let $u(x, y, z, t)$ be the temperature of the medium at time t and point (x, y, z). We want to derive an equation for u.

We will employ a device used frequently in deriving mathematical models. Consider an imaginary smooth closed surface $\sum$ within the medium, bounding a solid region M. The amount of heat energy leaving M across $\sum$ in a time interval Δt is

$$\left(\iint_{\sum} (K \nabla u) \cdot \mathbf{N} \, d\sigma \right) \Delta t.$$

This is the flux of the vector (K times the gradient of u) across this surface, multiplied by the length of the time interval.

But, the change in temperature at (x, y, z) in M in this time interval is approximately $(\partial u / \partial t) \Delta t$, so the resulting heat loss in M is

$$\left(\iiint_M \mu \rho \frac{\partial u}{\partial t} \, dV \right) \Delta t.$$

Assuming that there are no heat sources or losses within M (for example, chemical reactions or radioactivity), the change in heat energy in M over this time interval must equal the heat exchange across $\sum$. Then

$$\left(\iint_{\sum} (K \nabla u) \cdot \mathbf{N} \, d\sigma \right) \Delta t = \left(\iiint_M \mu \rho \frac{\partial u}{\partial t} \, dV \right) \Delta t.$$

Therefore,

$$\iint_{\sum} (K \nabla u) \cdot \mathbf{N} \, d\sigma = \iiint_M \mu \rho \frac{\partial u}{\partial t} \, dV.$$

Apply Gauss's theorem to the surface integral to obtain

$$\iiint_M \nabla \cdot (K \nabla u) \, dV = \iiint_M \mu \rho \frac{\partial u}{\partial t} \, dV.$$

The role of Gauss's theorem here is to convert the surface integral to a triple integral, thus obtaining an equation with the same kind of integral on both sides. This allows us to combine terms and write the last equation as

$$\iiint_M \left(\mu \rho \frac{\partial u}{\partial t} - \nabla \cdot (K \nabla u) \right) dV = 0.$$

Now keep in mind a crucial point—$\sum$ is *any* smooth closed surface within the medium. Assume that the integrand in the last equation is continuous. If this integrand were nonzero at any point P_0 of the medium, then it would be positive or negative at P_0, say positive. By continuity of this integrand, there would be a sphere S, centered at P_0, of small enough radius that the integrand would be strictly positive on and within S. But then we would have

$$\iiint_M \left(\mu \rho \frac{\partial u}{\partial t} - \nabla \cdot (K \nabla u) \right) dV > 0,$$

in which M is the solid ball bounded by S. By choosing $\sum = S$, this is a contradiction. We conclude that

$$\mu\rho\frac{\partial u}{\partial t} - \nabla \cdot (K\nabla u) = 0$$

at all points in the medium, for all times. This gives us the partial differential equation

$$\mu\rho\frac{\partial u}{\partial t} = \nabla \cdot (K\nabla u)$$

for the temperature function at any point and time. This equation is called the *heat equation*.

We can expand

$$\begin{aligned}
\nabla \cdot (K\nabla u) &= \nabla \cdot \left(K\frac{\partial u}{\partial x}\mathbf{i} + K\frac{\partial u}{\partial y}\mathbf{j} + K\frac{\partial u}{\partial z}\mathbf{k} \right) \\
&= \frac{\partial}{\partial x}\left(K\frac{\partial u}{\partial x} \right) + \frac{\partial}{\partial y}\left(K\frac{\partial u}{\partial y} \right) + \frac{\partial}{\partial z}\left(K\frac{\partial u}{\partial z} \right) \\
&= \frac{\partial K}{\partial x}\frac{\partial u}{\partial x} + \frac{\partial K}{\partial y}\frac{\partial u}{\partial y} + \frac{\partial K}{\partial z}\frac{\partial u}{\partial z} + K\left(\frac{\partial^2 u}{\partial x^2} + \frac{\partial^2 u}{\partial y^2} + \frac{\partial^2 u}{\partial z^2} \right) \\
&= \nabla K \cdot \nabla u + K\nabla^2 u,
\end{aligned}$$

in which

$$\nabla^2 u = \frac{\partial^2 u}{\partial x^2} + \frac{\partial^2 u}{\partial y^2} + \frac{\partial^2 u}{\partial z^2}$$

is called the Laplacian of u. (∇^2 is read "del squared.") Now the heat equation can be written

$$\mu\rho\frac{\partial u}{\partial t} = \nabla K \cdot \nabla u + K\nabla^2 u.$$

If K is constant, then its gradient is the zero vector and this equation simplifies to

$$\frac{\partial u}{\partial t} = \frac{K}{\mu\rho}\nabla^2 u.$$

In the case of one space dimension (for example, if $u(x, t)$ is the temperature distribution in a thin bar lying along a segment of the x axis), this is

$$\frac{\partial u}{\partial t} = k\frac{\partial^2 u}{\partial x^2},$$

with $k = K/\mu\rho$.

The steady-state case occurs when u does not change with time. In this case, $\partial u/\partial t = 0$, and the heat equation becomes

$$\nabla^2 u = 0,$$

a partial differential equation called *Laplace's equation*. In Chapters 17 and 18 we will write solutions of the heat equation and Laplace's equation under various conditions.

12.7.3 The Divergence Theorem as a Conservation of Mass Principle

We will derive a model providing a physical interpretation of the divergence of a vector.

Let $\mathbf{F}(x, y, z, t)$ be the velocity of a fluid moving in a region of 3-space at point (x, y, z) and time t. Let P_0 be a point in this region. Place an imaginary sphere $\sum_r$ of radius r about P_0, as in

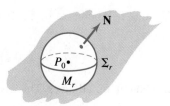

FIGURE 12.48

Figure 12.48. $\sum_r$ bounds a solid ball M_r. Let **N** be the unit outer normal to $\sum_r$. We know that $\iint_{\sum_r} \mathbf{F} \cdot \mathbf{N}\, d\sigma$ is the flux of **F** out of M_r across $\sum_r$.

If r is sufficiently small, then for a given time $\nabla \cdot \mathbf{F}(x, y, z, t)$ is approximated by $\nabla \cdot \mathbf{F}(P_0, t)$ at all points (x, y, z) of M_r, to within any desired tolerance. Therefore,

$$\iiint_{M_r} \nabla \cdot \mathbf{F}(x, y, z, t)\, dV \approx \iiint_{M_r} \nabla \cdot \mathbf{F}(P_0, t)\, dV$$

$$= [\nabla \cdot \mathbf{F}(P_0, t)]\,[\text{volume of } M_r] = \frac{4}{3}\pi r^3 \nabla \cdot \mathbf{F}(P_0, t).$$

Then

$$\nabla \cdot \mathbf{F}(P_0, t) \approx \frac{3}{4\pi r^3} \iiint_{M_r} \nabla \cdot \mathbf{F}(x, y, z, t)\, dV$$

$$= \frac{3}{4\pi r^3} \iint_{\sum} \mathbf{F} \cdot \mathbf{N}\, d\sigma,$$

by Gauss's theorem. Let $r \to 0$. Then $\sum_r$ contracts to its center P_0 and this approximation becomes an equality:

$$\nabla \cdot \mathbf{F}(P_0, t) = \lim_{r \to 0} \frac{3}{4\pi r^3} \iint_{\sum_r} \mathbf{F} \cdot \mathbf{N}\, d\sigma.$$

On the right is the limit, as $r \to 0$, of the flux of **F** across the sphere of radius r, divided by the volume of this sphere. This is the amount per unit volume of fluid flowing out of M_r across $\sum_r$. Since the sphere contracts to P_0 in this limit, we interpret the right side, hence also the divergence of **F** at P_0, as a measure of fluid flow away from P_0. This provides a physical sense of the divergence of a vector field.

In view of this interpretation, the equation

$$\iint_{\sum} \mathbf{F} \cdot \mathbf{N}\, d\sigma = \iiint_M \nabla \cdot \mathbf{F}\, dV$$

states that the flux of **F** out of M across its bounding surface exactly balances the divergence of fluid away from the points of M. This is a conservation of mass statement, in the absence of fluid produced or destroyed within M, and provides a model for the divergence theorem.

12.7.4 Green's Identities

Gauss's theorem is the basis for several integral identities that are used in vector analysis and in the study of partial differential equations.

Let $f(x, y, z)$ and $g(x, y, z)$ be continuous with continuous first and second partial derivatives in some region of 3-space. Let $\sum$ be a piecewise smooth closed surface in this region, bounding a set of points M. Green's first identity is

$$\iint_{\Sigma} f \nabla g \cdot \mathbf{N} \, d\sigma = \iiint_{M} (f \nabla^2 g + \nabla f \cdot \nabla g) \, dV.$$

To prove this, use part of the argument used in the derivation of the heat equation, with f in place of K and g in place of u, to show that

$$\nabla \cdot (f \nabla g) = f \nabla^2 g + \nabla f \cdot \nabla g.$$

Then, by Gauss's theorem,

$$\iint_{\Sigma} f \nabla g \cdot \mathbf{N} \, d\sigma = \iiint_{M} \nabla \cdot (f \nabla g) \, dV$$

$$= \iiint_{M} (f \nabla^2 g + \nabla f \cdot \nabla g) \, dV.$$

If f and g are interchanged in Green's first identity, and the two integrals subtracted, we get

$$\iint_{\Sigma} f \nabla g \cdot \mathbf{N} \, d\sigma - \iint_{\Sigma} g \nabla f \cdot \mathbf{N} \, d\sigma$$

$$= \iiint_{M} (f \nabla^2 g + \nabla f \cdot \nabla g) \, dV - \iiint_{M} (g \nabla^2 f + \nabla g \cdot \nabla f) \, dV$$

$$= \iiint_{M} (f \nabla^2 g - g \nabla^2 f) \, dV,$$

since $\nabla f \cdot \nabla g = \nabla g \cdot \nabla f$. This yields Green's second identity:

$$\iint_{\Sigma} (f \nabla g - g \nabla f) \cdot \mathbf{N} \, d\sigma = \iiint_{M} (f \nabla^2 g - g \nabla^2 f) \, dV.$$

If $f(x, y, z) = 1$ in Green's second identity, we obtain

$$\iint_{\Sigma} \nabla g \cdot \mathbf{N} \, d\sigma = \iiint_{M} \nabla^2 g \, dV.$$

If g satisfies Laplace's equation $\nabla^2 g = 0$, then this equation yields

$$\iint_{\Sigma} \nabla g \cdot \mathbf{N} \, d\sigma = 0.$$

A function satisfying Laplace's equation in a region is said to be *harmonic* in that region. The last equation states that the flux of a harmonic function across a surface is zero. Relationships such as this are used in studying existence and properties of solutions of Laplace's equation under various boundary conditions.

SECTION 12.7 PROBLEMS

In each of Problems 1 through 10, evaluate $\iint_{\Sigma} \mathbf{F} \cdot \mathbf{N} \, d\sigma$ or $\iiint_{M} div(\mathbf{F}) \, dV$, whichever is most convenient.

1. $\mathbf{F} = x\mathbf{i} + y\mathbf{j} - z\mathbf{k}$, $\sum$ the sphere of radius 4 about $(1, 1, 1)$

2. $\mathbf{F} = 4x\mathbf{i} - 6y\mathbf{j} + \mathbf{k}$, $\sum$ the surface of the solid cylinder $x^2 + y^2 \leq 4, 0 \leq z \leq 2$ (the surface includes the end caps of the cylinder)

3. $\mathbf{F} = 2yz\mathbf{i} - 4xz\mathbf{j} + xy\mathbf{k}$, $\sum$ the sphere of radius 5 about $(-1, 3, 1)$

4. $\mathbf{F} = x^3\mathbf{i} + y^3\mathbf{j} + z^3\mathbf{k}$, $\sum$ the sphere of radius 1 about the origin

5. $\mathbf{F} = 4x\mathbf{i} - z\mathbf{j} + x\mathbf{k}$, $\sum$ the hemisphere $x^2 + y^2 + z^2 = 1$, $z \geq 0$, including the base consisting of points (x, y) with $x^2 + y^2 \leq 1$

6. $\mathbf{F} = (x - y)\mathbf{i} + (y - 4xz)\mathbf{j} + xz\mathbf{k}$, $\sum$ the surface of the rectangular box bounded by the coordinate planes $x = 0$, $y = 0$, and $z = 0$ and by the planes $x = 4$, $y = 2$, and $z = 3$

7. $\mathbf{F} = x^2\mathbf{i} + y^2\mathbf{j} + z^2\mathbf{k}$, $\sum$ the cone $z = \sqrt{x^2 + y^2}$ for $x^2 + y^2 \leq 2$, together with the top cap consisting of points $(x, y, \sqrt{2})$ with $x^2 + y^2 \leq 2$

8. $\mathbf{F} = x^2\mathbf{i} - e^z\mathbf{j} + z\mathbf{k}$, $\sum$ the surface bounding the cylinder $x^2 + y^2 \leq 4$, $0 \leq z \leq 2$ (including the top and bottom caps of the cylinder)

9. $\mathbf{F} = 3xy\mathbf{i} + z^2\mathbf{k}$, $\sum$ the sphere of radius 1 about the origin

10. $\mathbf{F} = x^2\mathbf{i} + y^2\mathbf{j} + z^2\mathbf{k}$, $\sum$ the rectangular box bounded by the coordinate planes $x = 0$, $y = 0$, and $z = 0$ and the planes $x = 6$, $y = 2$, and $z = 7$

11. Let $\sum$ be a smooth closed surface and $\mathbf{F}$ a vector field with components that are continuous with continuous first and second partial derivatives through $\sum$ and its interior. Evaluate $\iint_{\sum} (\nabla \times \mathbf{F}) \cdot \mathbf{N} \, d\sigma$.

12. Let $\varphi(x, y, z)$ and $\psi(x, y, z)$ be continuous with continuous first and second partial derivatives on a smooth closed surface $\sum$ and its interior M. Suppose $\nabla \varphi = \mathbf{O}$ in M. Prove that $\iiint_M \varphi \nabla^2 \psi \, dV = 0$.

13. Show that under the conditions of Problem 12, if $\nabla \varphi = \nabla \psi = \mathbf{O}$, then $\iiint_M (\varphi \nabla^2 \psi - \psi \nabla^2 \varphi) \, dV = 0$.

14. Let $\sum$ be a smooth closed surface bounding an interior M. Show that

$$\text{volume of } M = \frac{1}{3} \iint_{\sum} \mathbf{R} \cdot \mathbf{N} \, d\sigma,$$

where $\mathbf{R} = x\mathbf{i} + y\mathbf{j} + z\mathbf{k}$ is a position vector for $\sum$.

15. Let $\sum$ be a smooth closed surface and $\mathbf{K}$ a constant vector. Prove that $\iint_{\sum} \mathbf{K} \cdot \mathbf{N} \, d\sigma = 0$.

16. Find the flux of the vector field $\mathbf{F} = xy^2\mathbf{i} + yz^2\mathbf{j} + zx^2\mathbf{k}$ across the surface $\sum$ bounding the cylinder $2 \leq x^2 + y^2 \leq 4$, $0 \leq z \leq 7$.

17. Prove a special case of Gauss's theorem as follows. Assume that $\sum$ consists of an upper surface $\sum_2$, a lower surface $\sum_1$, and possibly a lateral surface $\sum_3$ joining them (Figure 12.49). If $\sum_1$ and $\sum_2$ meet as in Figure 12.50, omit $\sum_3$. Suppose $\sum_2$ is the graph of $z = K(x, y)$ and $\sum_1$ the graph of $z = H(x, y)$ for (x, y) in D. Let $\mathbf{F} = f\mathbf{i} + g\mathbf{j} + h\mathbf{k}$. Show by consid-

ering the surface integral over each surface that

$$\iint_{\sum} f(x, y, z)\mathbf{i} \cdot \mathbf{N} \, d\sigma = \iiint_M \frac{\partial f}{\partial x} \, dV,$$

$$\iint_{\sum} g(x, y, z)\mathbf{j} \cdot \mathbf{N} \, d\sigma = \iiint_M \frac{\partial g}{\partial y} \, dV,$$

and

$$\iint_{\sum} h(x, y, z)\mathbf{k} \cdot \mathbf{N} \, d\sigma = \iiint_M \frac{\partial h}{\partial z} \, dV.$$

Add these equations to establish the theorem in this case.

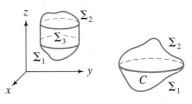

FIGURE 12.49   **FIGURE 12.50**

18. Let $\sum$ be a piecewise smooth closed surface enclosing a region M. Let f be continuous at all points of $\sum$. A *Dirichlet problem* for M consists of finding a function u such that $\nabla^2 u = 0$ for (x, y, z) in M and $u(x, y, z) = f(x, y, z)$ on $\sum$. Prove that a Dirichlet problem can have at most one solution that is continuous and has continuous first and second partial derivatives throughout M. *Hint:* Let u_1 and u_2 be solutions and let $w = u_1 - u_2$. Show that $\nabla^2 w = 0$ throughout M and $w(x, y, z) = 0$ for (x, y, z) on $\sum$. Use Gauss's theorem to show that $w(x, y, z) = 0$ for all (x, y, z) in M.

19. Consider the general heat equation

$$\frac{\partial u}{\partial t} = k\nabla^2 u + \varphi(x, y, z, t).$$

We seek a function u satisfying this partial differential equation and also the condition that $u(x, y, z, t) = f(x, y, z, t)$, a given function, for (x, y, z) on $\sum$. Show that this problem can have at most one solution that is continuous with continuous first and second partial derivatives. *Hint:* Let u_1 and u_2 be solutions and let $w = u_1 - u_2$. Show that $\partial w / \partial t = k\nabla^2 w$ and that $w(x, y, z, t) = 0$ on $\sum$ for $t > 0$, and $w(x, y, z, 0) = 0$ for (x, y, z) in M. Let

$$I(t) = \frac{1}{2} \iiint_M w^2(x, y, z, t) \, dV.$$

Show that $I'(t) = -k \iiint_M \|\nabla w\|^2 \, dV$. Conclude that $I'(t) \leq 0$ for $t > 0$. Next apply the mean value theorem to $I(t)$ on $[0, t]$ to show that $I(t) \leq 0$ for $t > 0$. Thus show that $I(t) = 0$ for $t > 0$ and show from this that $w(x, y, z, t) = 0$ for (x, y, z) in M and $t > 0$.

20. Let $\sum$ be a piecewise smooth closed surface enclosing a region M. Consider the following problem of determining a function u such that

$$\frac{\partial u}{\partial t} = k\nabla^2 u + \varphi(x, y, z, t)$$

for (x, y, z) in M and $t > 0$,

$$\frac{\partial u}{\partial \eta} + hu = f(x, y, z, t)$$

for (x, y, z) on $\sum$ and $t > 0$, and

$$u(x, y, z, 0) = g(x, y, z)$$

for (x, y, z) in M. Here f and g are given continuous functions and h is a positive constant, and $\partial u / \partial \eta = \nabla u \cdot \mathbf{N}$, with $\mathbf{N}$ the normal to $\sum$. (This is the normal derivative of u on $\sum$.) Show that this problem can have at most one solution that is continuous with continuous first and second partial derivatives in M.

21. Suppose f and g satisfy Laplace's equation in a region M bounded by a smooth closed surface $\sum$. Suppose $\partial f / \partial \eta = \partial g / \partial \eta$ on $\sum$. Prove that for some constant k, $f(x, y, z) = g(x, y, z) + k$ for all (x, y, z) in M.

12.8 The Integral Theorem of Stokes

We have seen that the conclusion of Green's theorem can be written

$$\oint_C \mathbf{F} \cdot \mathbf{T} \, ds = \iint_D (\nabla \times \mathbf{F}) \cdot \mathbf{k} \, dA,$$

in which $\mathbf{T}$ is the unit tangent to C, a simple, positively oriented, closed curve enclosing a region D.

Think of D as a flat surface with unit normal vector $\mathbf{k}$ and bounded by C. To generalize to three dimensions, allow C to be a closed curve in 3-space, bounding a smooth surface $\sum$ as in Figure 12.51. Here $\sum$ need not be a closed surface. Let $\mathbf{N}$ be a unit normal to $\sum$.

This raises a subtle point. At any point of $\sum$ there are two normal vectors, as shown in Figure 12.52. Which should we choose? In addition to this decision, we must choose a direction on C. In the plane we chose counterclockwise as positive orientation, but this has no meaning in three dimensions.

First we will give a rule for choosing a particular unit normal to $\sum$ at each point. If $\sum$ has coordinate functions $x = x(u, v)$, $y = y(u, v)$, $z = z(u, v)$, then the normal vector

$$\frac{\partial(y, z)}{\partial(u, v)}\mathbf{i} + \frac{\partial(z, x)}{\partial(u, v)}\mathbf{j} + \frac{\partial(x, y)}{\partial(u, v)}\mathbf{k},$$

divided by its length, yields a unit normal to $\sum$. The negative of this unit normal is also a unit normal at the same point. Choose either this vector or its negative to use as the normal to $\sum$, and call it $\mathbf{n}$. Whichever is chosen as $\mathbf{n}$, use it at all points of $\sum$. That is, do not use $\mathbf{n}$ at one point and $-\mathbf{n}$ at another.

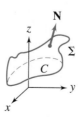

FIGURE 12.51

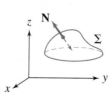

FIGURE 12.52
Normals to a surface at a point.

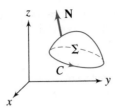

FIGURE 12.53

Now use the choice of **n** to determine an orientation or direction on C to be called the positive orientation. Referring to Figure 12.53, at any point on C, if you stand along **n** (that is, your head is at the tip of **n**), then the positive direction on C is the one in which you walk to have $\sum$ over your left shoulder. The arrow shows the orientation on C obtained in this way. Although this is not a rigorous definition, it is sufficient for our purpose without becoming enmeshed in topological details. With this direction, we say that C has been *oriented coherently* with **n**. If we had chosen the normal in the opposite direction, then we would have reached the opposite orientation on C. The choice of the normal determines the orientation on the curve. There is no intrinsic positive or negative orientation of the curve—simply an orientation coherent with the choice of normal.

With this understanding, we can state Stokes's theorem.

THEOREM 12.9   *Stokes*

Let $\sum$ be a piecewise smooth surface bounded by a piecewise smooth curve C. Suppose a unit normal **n** has been chosen on $\sum$ and that C is oriented coherently with **n**. Let $\mathbf{F}(x, y, z)$ be a vector field whose component functions are continuous with continuous first and second partial derivatives on $\sum$. Then,

$$\oint_C \mathbf{F} \cdot d\mathbf{R} = \iint_{\sum} (\nabla \times \mathbf{F}) \cdot \mathbf{n} \, d\sigma. \quad \blacksquare$$

We will write this conclusion in terms of coordinates and component functions. Let the component functions of **F** be, respectively, f, g, and h. Then

$$\oint_C \mathbf{F} \cdot d\mathbf{R} = \oint_C f(x, y, z) \, dx + g(x, y, z) \, dy + h(x, y, z) \, dz.$$

Next,

$$\nabla \times \mathbf{F} = \begin{vmatrix} \mathbf{i} & \mathbf{j} & \mathbf{k} \\ \dfrac{\partial}{\partial x} & \dfrac{\partial}{\partial y} & \dfrac{\partial}{\partial z} \\ f & g & h \end{vmatrix} = \left(\frac{\partial h}{\partial y} - \frac{\partial g}{\partial z} \right) \mathbf{i} + \left(\frac{\partial f}{\partial z} - \frac{\partial h}{\partial x} \right) \mathbf{j} + \left(\frac{\partial g}{\partial x} - \frac{\partial f}{\partial y} \right) \mathbf{k}.$$

The normal to the surface is given by equation (12.5) as

$$\mathbf{N} = \frac{\partial(y, z)}{\partial(u, v)} \mathbf{i} + \frac{\partial(z, x)}{\partial(u, v)} \mathbf{j} + \frac{\partial(x, y)}{\partial(u, v)} \mathbf{k}.$$

Use this to define the unit normal

$$\mathbf{n}(u, v) = \frac{\mathbf{N}(u, v)}{\|\mathbf{N}(u, v)\|}.$$

Then

$$(\nabla \times \mathbf{F}) \cdot \mathbf{n} = (\nabla \cdot \mathbf{F}) \cdot \frac{\mathbf{N}(u, v)}{\|\mathbf{N}(u, v)\|}$$

$$= \frac{1}{\|\mathbf{N}(u, v)\|} \left[\left(\frac{\partial h}{\partial y} - \frac{\partial g}{\partial z} \right) \frac{\partial(y, z)}{\partial(u, v)} + \left(\frac{\partial f}{\partial z} - \frac{\partial h}{\partial x} \right) \frac{\partial(z, x)}{\partial(u, v)} \right.$$

$$\left. + \left(\frac{\partial g}{\partial x} - \frac{\partial f}{\partial y} \right) \frac{\partial(x, y)}{\partial(u, v)} \right].$$

Then

$$\iint_{\Sigma} (\nabla \times \mathbf{F}) \cdot \mathbf{n} \, d\sigma$$

$$= \iint_{D} \frac{1}{\|\mathbf{N}(u, v)\|} \left[\left(\frac{\partial h}{\partial y} - \frac{\partial g}{\partial z} \right) \frac{\partial(y, z)}{\partial(u, v)} + \left(\frac{\partial f}{\partial z} - \frac{\partial h}{\partial x} \right) \frac{\partial(z, x)}{\partial(u, v)} \right.$$

$$\left. + \left(\frac{\partial g}{\partial x} - \frac{\partial f}{\partial y} \right) \frac{\partial(x, y)}{\partial(u, v)} \right] \|\mathbf{N}(u, v)\| \, du \, dv$$

$$= \iint_{D} \left[\left(\frac{\partial h}{\partial y} - \frac{\partial g}{\partial z} \right) \frac{\partial(y, z)}{\partial(u, v)} + \left(\frac{\partial f}{\partial z} - \frac{\partial h}{\partial x} \right) \frac{\partial(z, x)}{\partial(u, v)} \right.$$

$$\left. + \left(\frac{\partial g}{\partial x} - \frac{\partial f}{\partial y} \right) \frac{\partial(x, y)}{\partial(u, v)} \right] du \, dv.$$

in which the coordinate functions $x(u, v)$, $y(u, v)$, and $z(u, v)$ from Σ are substituted into the integral and D is a set of points (u, v) over which these coordinate functions are defined. Keep in mind that the function to be integrated in Stokes's theorem is $(\nabla \times \mathbf{F}) \cdot \mathbf{n}$, in which $\mathbf{n} = \mathbf{N}/\|\mathbf{N}\|$ is a unit normal. However, in converting the surface integral to a double integral over D, using the definition of surface integral, $(\nabla \times \mathbf{F}) \cdot \mathbf{n}$ must be multiplied by $\|\mathbf{N}(u, v)\|$, with $\mathbf{N}(u, v)$ determined by equation (12.5).

EXAMPLE 12.35

Let $\mathbf{F}(x, y, z) = -y\mathbf{i} + x\mathbf{j} - xyz\mathbf{k}$ and let Σ consist of the part of the cone $z = \sqrt{x^2 + y^2}$ for $x^2 + y^2 \le 9$. We will compute both sides of the conclusion of Stokes's theorem to illustrate the various terms and integrals involved.

The cone is shown in Figure 12.54. Its boundary curve C is the circle around the top of the cone, the curve $x^2 + y^2 = 9$ in the plane $z = 3$. In this example, the surface is described by $z = S(x, y)$. Here x and y are the parameters, varying over the disk D given by $x^2 + y^2 \le 9$. We can compute a normal vector

$$\mathbf{N} = -\frac{\partial z}{\partial x}\mathbf{i} - \frac{\partial z}{\partial y}\mathbf{j} + \mathbf{k}$$

$$= -\frac{x}{z}\mathbf{i} - \frac{y}{z}\mathbf{j} + \mathbf{k}.$$

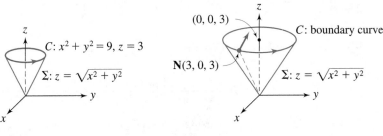

FIGURE 12.54   **FIGURE 12.55**

For $(\nabla \times \mathbf{F}) \cdot \mathbf{n}$ in Stokes's theorem, $\mathbf{n}$ is a unit normal, so compute the norm of $\mathbf{N}$:

$$\|\mathbf{N}\| = \left\| -\frac{x}{z}\mathbf{i} - \frac{y}{z}\mathbf{j} + \mathbf{k} \right\| = \sqrt{\frac{x^2}{z^2} + \frac{y^2}{z^2} + 1}$$

$$= \sqrt{\frac{x^2 + y^2 + (x^2 + y^2)}{x^2 + y^2}} = \sqrt{2}.$$

Use the unit normal

$$\mathbf{n} = \frac{\mathbf{N}}{\|\mathbf{N}\|} = \frac{1}{\sqrt{2}z}\left(-x\mathbf{i} - y\mathbf{j} + z\mathbf{k} \right).$$

This normal is defined at all points of the cone except the origin, where there is no normal. $\mathbf{n}$ is an inner normal, pointing from any point on the cone into the region bounded by the cone.

If we stand along $\mathbf{n}$ at points of C and imagine walking along C in the direction of the arrow in Figure 12.55, then the surface is over our left shoulder. Therefore this arrow orients C coherently with $\mathbf{n}$. If we had used $-\mathbf{n}$ as normal vector, we would orient C in the other direction. We can parametrize C by

$$x = 3\cos(t), y = 3\sin(t), z = 3 \quad \text{for } 0 \le t \le 2\pi.$$

The point $(3\cos(t), 3\sin(t), 3)$ traverses C in the positive direction (as determined by $\mathbf{n}$) as t increases from 0 to 2π.

This completes the preliminary work and we can evaluate the integrals. For the line integral,

$$\oint_C \mathbf{F} \cdot d\mathbf{R} = \oint_C -y\,dx + x\,dy - xyz\,dz$$

$$= \int_0^{2\pi} [-3\sin(t)(-3\sin(t)) + 3\cos(t)3\cos(t)\,dt$$

$$= \int_0^{2\pi} 9\,dt = 18\pi.$$

For the surface integral, first compute the curl of $\mathbf{F}$:

$$\nabla \times \mathbf{F} = \begin{vmatrix} \mathbf{i} & \mathbf{j} & \mathbf{k} \\ \dfrac{\partial}{\partial x} & \dfrac{\partial}{\partial y} & \dfrac{\partial}{\partial z} \\ -y & x & -xyz \end{vmatrix}$$

$$= -xz\mathbf{i} + yz\mathbf{j} + 2\mathbf{k}.$$

Then

$$(\nabla \times \mathbf{F}) \cdot \mathbf{n} = \frac{1}{\sqrt{2}z}(x^2 z - y^2 z + 2z)$$

$$= \frac{1}{\sqrt{2}}(x^2 - y^2 + 2).$$

Then

$$\iint_{\Sigma} (\nabla \times \mathbf{F}) \cdot \mathbf{n}\, d\sigma = \iint_D [(\nabla \times \mathbf{F}) \cdot \mathbf{n}] \, \|\mathbf{N}\| \, dx\, dy$$

$$= \iint_D \frac{1}{\sqrt{2}}(x^2 - y^2 + 2)\sqrt{2}\, dx\, dy$$

$$= \iint_D (x^2 - y^2 + 2)\, dx\, dy.$$

Use polar coordinates on D to write this integral as

$$\int_0^{2\pi} \int_0^3 \left(r^2 \cos^2(\theta) - r^2 \sin^2(\theta) + 2\right) r\, dr\, d\theta$$

$$= \int_0^{2\pi} \int_0^3 r^3 \cos(2\theta)\, dr\, d\theta + \int_0^{2\pi} \int_0^3 2r\, dr\, d\theta$$

$$= \left[\frac{1}{2}\sin(2\theta)\right]_0^{2\pi} \left[\frac{1}{4}r^4\right]_0^3 + 2\pi \left[r^2\right]_0^3 = 18\pi. \quad \blacksquare$$

The following are two applications of Stokes's theorem.

12.8.1 An Interpretation of Curl

We will use Stokes's theorem to argue for a physical interpretation of the curl operation. Think of $\mathbf{F}(x, y, z)$ as the velocity of a fluid and let P_0 be any point in the fluid. Consider a disk $\sum_r$ of radius r about P_0, with unit normal vector $\mathbf{n}$ and boundary circle C_r coherently oriented, as in Figure 12.56. For the disk the normal vector is constant. By Stokes's theorem,

$$\oint_{C_r} \mathbf{F} \cdot d\mathbf{R} = \iint_{\Sigma_r} (\nabla \times \mathbf{F}) \cdot \mathbf{n}\, d\sigma.$$

Since $\mathbf{R}'(t)$ is a tangent vector to C_r, then $\mathbf{F} \cdot \mathbf{R}'$ is the tangential component of the velocity about C_r and $\oint_{C_r} \mathbf{F} \cdot d\mathbf{R}$ measures the circulation of the fluid about C_r.

By choosing r sufficiently small, $\nabla \times \mathbf{F}(x, y, z)$ is approximated by $\nabla \cdot \mathbf{F}(P_0)$ as closely as we like on $\sum_r$. Further, since $\mathbf{n}$ is constant,

$$\text{circulation of } \mathbf{F} \text{ about } C_r \approx \iint_{\Sigma_r} (\nabla \times \mathbf{F})(P_0) \cdot \mathbf{n}\, d\sigma$$

$$= (\nabla \times \mathbf{F})(P_0) \cdot \mathbf{n} \left(\text{area of the disk } \textstyle\sum_r\right)$$

$$= \pi r^2 (\nabla \times \mathbf{F})(P_0) \cdot \mathbf{n}.$$

Therefore,

$$\nabla \times \mathbf{F}(P_0) \cdot \mathbf{n} \approx \frac{1}{\pi r^2} \left(\text{circulation of } \mathbf{F} \text{ about } C_r\right).$$

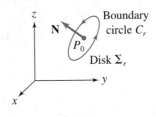

FIGURE 12.56

As $r \to 0$, the disk contracts to its center P_0 and we obtain

$$\nabla \times \mathbf{F}(P_0) \cdot \mathbf{n} = \lim_{r \to 0} \frac{1}{\pi r^2} \text{ (circulation of } \mathbf{F} \text{ about } C_r).$$

Since $\mathbf{n}$ is normal to the plane of C_r, this equation can be read

$$\nabla \times \mathbf{F}(P_0) \cdot \mathbf{n} = \text{circulation of } \mathbf{F} \text{ per unit area in the plane normal to } \mathbf{n}.$$

Thus the curl of $\mathbf{F}$ is a measure of rotation of the fluid at a point. This is the reason why a fluid is called irrotational if the curl of the velocity vector is zero. For example, any conservative vector field is irrotational, because if $\mathbf{F} = \nabla \varphi$, then $\nabla \times \mathbf{F} = \nabla \times (\nabla \varphi) = \mathbf{O}$.

12.8.2 Potential Theory in 3-Space

As in the plane, a vector field $\mathbf{F}(x, y, z)$ in 3-space is conservative if for some potential function φ, $\mathbf{F} = \nabla \varphi$. By exactly the same reasoning as applied in the two-dimensional case, we get

$$\int_C \mathbf{F} \cdot d\mathbf{R} = \varphi(P_1) - \varphi(P_0),$$

if P_0 is the initial point of C and P_1 the terminal point. Therefore, the line integral of a conservative vector field in 3-space is independent of path.

If a potential function exists, we attempt to find it by integration, just as in the plane.

EXAMPLE 12.36

Let

$$\mathbf{F}(x, y, z) = (yze^{xyz} - 4x)\mathbf{i} + (xze^{xyz} + z + \cos(y))\mathbf{j} + (xye^{xyz} + y)\mathbf{k}.$$

For this to be the gradient of a scalar field φ, we must have

$$\frac{\partial \varphi}{\partial x} = yze^{xyz} - 4x, \qquad \frac{\partial \varphi}{\partial y} = xze^{xyz} + z + \cos(y),$$

and

$$\frac{\partial \varphi}{\partial z} = xye^{xyz} + y.$$

Begin with one of these equations, say the last, and integrate with respect to z to get

$$\varphi(x, y, z) = e^{xyz} + yz + k(x, y),$$

where the "constant" of integration may involve the other two variables. Now we need

$$\frac{\partial \varphi}{\partial x} = yze^{xyz} - 4x$$

$$= \frac{\partial}{\partial x}[e^{xyz} + yz + k(x, y)] = yze^{xyz} + \frac{\partial k}{\partial x}.$$

This will be satisfied if

$$\frac{\partial k}{\partial x} = -4x,$$

so $k(x, y)$ must have the form

$$k(x, y) = -2x^2 + c(y).$$

Thus far $\varphi(x, y, z)$ must have the appearance

$$\varphi(x, y, z) = e^{xyz} + yz - 2x^2 + c(y).$$

Finally, we must satisfy

$$\frac{\partial \varphi}{\partial y} = xze^{xyz} + z + \cos(y)$$

$$= \frac{\partial}{\partial y}\left[e^{xyz} + yz - 2x^2 + c(y)\right]$$

$$= xze^{xyz} + z + c'(y).$$

Then

$$c'(y) = \cos(y)$$

and we may choose

$$c(y) = \sin(y).$$

A potential function is given by

$$\varphi(x, y, z) = e^{xyz} + yz - 2x^2 + \sin(y).$$

Of course, for any number a, $\varphi(x, y, z) + a$ is also a potential function for **F**. ∎

As in the plane, in 3-space there are vector fields that are not conservative. We would like to develop a test to determine when **F** has a potential function. The discussion follows that in Section 12.3.1 for the plane. As we saw in the plane, the test requires conditions not only on **F**, but on the set D of points on which we want to find a potential function. We define a set D of points in 3-space to be a *domain* if

1. about every point of D there exists a sphere containing only points of D, and

2. between any two points of D there is a path lying entirely in D.

This definition is analogous to the definition made for sets in the plane. For example, the set of points bounded by two disjoint spheres is not a domain because it fails to satisfy (2), while the set of points on or inside the solid unit sphere about the origin fails to be a domain because of condition (1). The set of points (x, y, z) with $x \geq 0$, $y \geq 0$, and $z \geq 0$ is not a domain because it fails condition (1). For example, there is no sphere about the origin containing only points with nonnegative coordinates. The set of points (x, y, z) with $x > 0$, $y > 0$, and $z > 0$ is a domain.

On a domain, existence of a potential function is equivalent to independence of path.

THEOREM 12.10

Let D be a domain in 3-space, and let $\mathbf{F}$ be continuous on D. Then $\int_C \mathbf{F} \cdot d\mathbf{R}$ is independent of path on D if and only if $\mathbf{F}$ is conservative. ■

We already know that existence of a potential function implies independence of path. For the converse, suppose $\int_C \mathbf{F} \cdot d\mathbf{R}$ is independent of path in D. Choose any P_0 in D. If P is any point of D, there is a path C in D from P_0 to P, and we can define

$$\varphi(P) = \int_C \mathbf{F} \cdot d\mathbf{R}.$$

Because this line integral depends only on the endpoints of any path in D, this defines a function for all (x, y, z) in D. Now the argument used in the proof of Theorem 12.6 can be essentially duplicated to show that $\mathbf{F} = \nabla\varphi$.

With one more condition on the domain D, we can derive a simple test for a vector field to be conservative. A set D of points in 3-space is *simply connected* if every simple, closed path in D is the boundary of a piecewise smooth surface lying in D. This condition enables us to use Stokes's theorem to derive the condition we want.

THEOREM 12.11

Let D be a simply connected domain in 3-space. Let $\mathbf{F}$ and $\nabla \times \mathbf{F}$ be continuous on D. Then $\mathbf{F}$ is conservative if and only if $\nabla \times \mathbf{F} = \mathbf{O}$ in D. ■

Thus, in simply connected domains, the conservative vector fields are the ones having curl zero—that is, the irrotational vector fields.

In one direction, the proof is simple. If $\mathbf{F} = \nabla\varphi$, then $\nabla \times \mathbf{F} = \nabla \times (\nabla\varphi) = \mathbf{O}$, without the requirement of simple connectivity.

In the other direction, suppose $\nabla \times \mathbf{F} = \mathbf{O}$. To prove that $\mathbf{F}$ is conservative, it is enough to prove that $\int_C \mathbf{F} \cdot d\mathbf{R}$ is independent of path in D. Let C and K be two paths from P_0 to P_1 in D. Form a closed path L in D consisting of C and $-K$, as in Figure 12.57. Since D is simply connected, there is a piecewise smooth surface $\sum$ in D having boundary C. Then

$$\oint_L \mathbf{F} \cdot d\mathbf{R} = \int_C \mathbf{F} \cdot d\mathbf{R} - \int_K \mathbf{F} \cdot d\mathbf{R}$$

$$= \iint_{\sum} (\nabla \times \mathbf{F}) \cdot \mathbf{n} \, d\sigma = 0,$$

so

$$\int_C \mathbf{F} \cdot d\mathbf{R} = \int_K \mathbf{F} \cdot d\mathbf{R}$$

and the line integral is independent of path, hence conservative.

If $\mathbf{G}(x, y) = f(x, y)\mathbf{i} + g(x, y)\mathbf{j}$, then we can think of $\mathbf{G}$ as a vector field in 3-space by writing

$$\mathbf{G}(x, y) = f(x, y)\mathbf{i} + g(x, y)\mathbf{j} + 0\mathbf{k}.$$

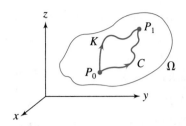

FIGURE 12.57

Then

$$\nabla \times \mathbf{G} = \begin{vmatrix} \mathbf{i} & \mathbf{j} & \mathbf{k} \\ \dfrac{\partial}{\partial x} & \dfrac{\partial}{\partial y} & \dfrac{\partial}{\partial z} \\ f(x,y) & g(x,y) & 0 \end{vmatrix} = \left(\dfrac{\partial g}{\partial x} - \dfrac{\partial f}{\partial y} \right) \mathbf{k},$$

so the condition $\nabla \times \mathbf{G} = \mathbf{O}$ in this two-dimensional case is exactly the condition of Theorem 12.5.

SECTION 12.8   PROBLEMS

In each of Problems 1 through 5, use Stokes's theorem to evaluate $\oint_C \mathbf{F} \cdot d\mathbf{R}$ or $\iint_\Sigma (\nabla \times \mathbf{F}) \cdot \mathbf{N}\,d\sigma$, whichever appears easier.

1. $\mathbf{F} = yx^2\mathbf{i} - xy^2\mathbf{j} + z^2\mathbf{k}$, Σ the hemisphere $x^2 + y^2 + z^2 = 4$, $z \geq 0$

2. $\mathbf{F} = xy\mathbf{i} + yz\mathbf{j} + xz\mathbf{k}$, Σ the paraboloid $z = x^2 + y^2$ for $x^2 + y^2 \leq 9$

3. $\mathbf{F} = z\mathbf{i} + x\mathbf{j} + y\mathbf{k}$, Σ the cone $z = \sqrt{x^2 + y^2}$ for $0 \leq z \leq 4$

4. $\mathbf{F} = z^2\mathbf{i} + x^2\mathbf{j} + y^2\mathbf{k}$, Σ the part of the paraboloid $z = 6 - x^2 - y^2$ above the x, y plane

5. $\mathbf{F} = xy\mathbf{i} + yz\mathbf{j} + xy\mathbf{k}$, Σ the part of the plane $2x + 4y + z = 8$ in the first octant

6. Calculate the circulation of $\mathbf{F} = (x - y)\mathbf{i} + x^2y\mathbf{j} + xza\mathbf{k}$ counterclockwise about the circle $x^2 + y^2 = 1$. Here a is a positive constant. *Hint:* Use Stokes's theorem, with Σ any smooth surface having the circle as boundary.

7. Use Stokes's theorem to evaluate $\oint_C \mathbf{F} \cdot \mathbf{T}\,ds$, where C is the boundary of the part of the plane $x + 4y + z = 12$ lying in the first octant and $\mathbf{F} = (x - z)\mathbf{i} + (y - x)\mathbf{j} + (z - y)\mathbf{k}$.

8. Fill in the details of this proof of a special case of Stokes's theorem. Assume that Σ is the graph of

$z = S(x, y)$. Let $\mathbf{F} = f\mathbf{i} + g\mathbf{j} + h\mathbf{k}$. First show that

$$\iint_\Sigma (\nabla \times \mathbf{F}) \cdot \mathbf{N}\,d\sigma = \iint_D \left[\left(\dfrac{\partial h}{\partial y} - \dfrac{\partial g}{\partial z} \right) \left(-\dfrac{\partial z}{\partial x} \right) \right. $$
$$\left. + \left(\dfrac{\partial f}{\partial z} - \dfrac{\partial h}{\partial x} \right) \left(-\dfrac{\partial z}{\partial y} \right) + \left(\dfrac{\partial g}{\partial x} - \dfrac{\partial f}{\partial y} \right) \right] dA.$$

Now parametrize the boundary C^* of D in the x, y plane by $x = x(t)$, $y = y(t)$ for $a \leq t \leq b$, oriented counterclockwise. Show that the boundary C of Σ can be parametrized by $x = x(t)$, $y = y(t)$, $z = S(x(t), y(t))$ for $a \leq t \leq b$. Show that

$$\oint_C \mathbf{F} \cdot d\mathbf{R} = \oint_{C^*} \left[f + h\dfrac{\partial z}{\partial x} \right] dx + \left[g + h\dfrac{\partial z}{\partial y} \right] dy.$$

Finally, apply Green's theorem to the last line integral and use the first step in this argument to conclude the proof.

In each of Problems 9 through 16, let Ω be all of 3-space (so Ω is a simply connected domain). Test to see if $\mathbf{F}$ is conservative. If it is, find a potential function.

9. $\mathbf{F} = \cosh(x + y)(\mathbf{i} + \mathbf{j} - \mathbf{k})$

10. $\mathbf{F} = 2x\mathbf{i} - 2y\mathbf{j} + 2z\mathbf{k}$

11. $\mathbf{F} = \mathbf{i} - 2\mathbf{j} + \mathbf{k}$

12. $\mathbf{F} = yz\cos(x)\mathbf{i} + (z\sin(x) + 1)\mathbf{j} + y\sin(x)\mathbf{k}$

13. $\mathbf{F} = (x^2 - 2)\mathbf{i} + xyz\mathbf{j} - yz^2\mathbf{k}$

14. $\mathbf{F} = e^{xyz}(1 + xyz)\mathbf{i} + x^2z\mathbf{j} + x^2y\mathbf{k}$

15. $\mathbf{F} = (\cos(x) + y\sin(x))\mathbf{i} + x\sin(xy)\mathbf{j} + \mathbf{k}$

16. $\mathbf{F} = (2x^2 + 3y^2z)\mathbf{i} + 6xyz\mathbf{j} + 3xy^2\mathbf{k}$

In each of Problems 17 through 22, evaluate the line integral of the vector field on any path from the first point to the second by finding a potential function.

17. $\mathbf{F} = \mathbf{i} - 9y^2z\mathbf{j} - 3y^2\mathbf{k}$; $(1, 1, 1)$, $(0, 3, 5)$

18. $\mathbf{F} = (y\cos(xz) - xyz\sin(xz))\mathbf{i} + x\cos(xz)\mathbf{j} - x^2\sin(xz)\mathbf{k}$; $(1, 0, \pi)$, $(1, 1, 7)$

19. $\mathbf{F} = 6x^2e^{yz}\mathbf{i} + 2x^3ze^{yz}\mathbf{j} + 2x^3ye^{yz}\mathbf{k}$; $(0, 0, 0)$, $(1, 2, -1)$

20. $\mathbf{F} = -8y^2\mathbf{i} - (16xy + 4z)\mathbf{j} - 4y\mathbf{k}$; $(-2, 1, 1)$, $(1, 3, 2)$

21. $\mathbf{F} = -\mathbf{i} + 2z^2\mathbf{j} + 4yz\mathbf{k}$; $(0, 0, -4)$, $(1, 1, 6)$

22. $\mathbf{F} = (y - 4xz)\mathbf{i} + x\mathbf{j} + (3z^2 - 2x^2)\mathbf{k}$; $(1, 1, 1)$, $(3, 1, 4)$

PART 5

Fourier Analysis, Orthogonal Expansions, and Wavelets

CHAPTER 13
Fourier Series

CHAPTER 14
The Fourier Integral and Fourier Transforms

CHAPTER 15
Special Functions, Orthogonal Expansions and Wavelets

In 1807 the French mathematician Joseph Fourier (1768–1830) submitted a paper to the Academy of Sciences in Paris. In it he presented a mathematical treatment of problems involving heat conduction. Although the paper was rejected for lack of rigor, it contained ideas that were so rich and widely applicable that they would occupy mathematicians in research to the present day.

One surprising implication of Fourier's work was that many functions can be expanded in infinite series or integrals involving sines and cosines. This revolutionary idea sparked a heated debate among leading mathematicians of the day and led to important advances in mathematics (Cantor's work on cardinals and ordinals, orders of infinity, measure theory, real and complex analysis, differential equations), science and engineering (data compression, signal analysis), and applications undreamed of in Fourier's day (CAT scans, PET scans, nuclear magnetic resonance).

Today the term *Fourier analysis* refers to many extensions of Fourier's original insights, including various kinds of Fourier series and integrals, real and complex Fourier transforms, discrete and finite transforms, and, because of their wide applicability, a variety of computer programs for efficiently computing Fourier coefficients and transforms. The ideas behind Fourier series have also found important generalizations in a broad theory of eigenfunction expansions, in which functions are expanded in series of special functions (Bessel functions, orthogonal polynomials, and other functions generated by differential equations). More recently, wavelet expansions have been developed to provide additional tools in areas such as filtering and signal analysis.

This part is devoted to some of these ideas and their applications.

CHAPTER 13

Fourier Series

13.1   Why Fourier Series?

A Fourier series is a representation of a function as a series of constants times sine and/or cosine functions of different frequencies. In order to see why such a series might be interesting, we will look at a problem of the type that led Fourier to consider them.

Consider a thin bar of length π, constant density, and uniform cross section. Let $u(x, t)$ be the temperature at time t in the cross section of the bar at x, for $0 \leq x \leq \pi$. In Section 12.8.2 we derived a partial differential equation for u:

$$\frac{\partial u}{\partial t} = k\frac{\partial^2 u}{\partial x^2} \quad \text{for } 0 < x < \pi, t > 0, \tag{13.1}$$

in which k is a constant depending on the material of the bar. Suppose the left and right ends of the bar are kept at zero temperature,

$$u(0, t) = u(\pi, t) = 0 \quad \text{for } t > 0, \tag{13.2}$$

and that the temperature throughout the bar at time $t = 0$ is specified

$$u(x, 0) = f(x) = x(\pi - x). \tag{13.3}$$

Intuitively, the heat equation, together with the initial temperature distribution throughout the bar, and the information that the ends are kept at zero degrees for all time, are enough to determine the temperature distribution $u(x, t)$ throughout the bar at any time.

By a process that now bears his name, and which we will develop when we study partial differential equations, Fourier found that functions satisfying the heat equation (13.1) and the conditions at the ends of the bar, equations (13.2), have the form

$$u_n(x, t) = b_n \sin(nx)e^{-kn^2 t}, \tag{13.4}$$

in which n can be any positive integer and b_n can be any real number. We must use these functions to find a function that also satisfies the condition (13.3).

625

Periodic phenomena have long fascinated mankind; our ancient ancestors were aware of the recurrence of phases of the moon and certain planets, the tides of lakes and oceans, and cycles in the weather. Isaac Newton's calculus and law of gravitation enabled him to explain the periodicities of the tides, but it was left to Joseph Fourier and his successors to develop Fourier analysis, which has had profound applications in the study of natural phenomena and the analysis of signals and data.

A single choice of positive integer n_0 and constant b_{n_0} will not do. If we let $u(x, t) = b_{n_0} \sin(n_0 x)e^{-kn_0^2 t}$, then we would need

$$u(x, 0) = x(\pi - x) = b_{n_0} \sin(n_0 x) \quad \text{for } 0 \leq x \leq \pi,$$

an impossibility. A polynomial cannot equal a constant multiple of a sine function over $[0, \pi]$ (or over any nontrivial interval).

The next thing to try is a finite sum of the functions (13.4), say

$$u(x, t) = \sum_{n=1}^{N} b_n \sin(nx)e^{-kn^2 t}. \tag{13.5}$$

Such a function will still satisfy the heat equation and the conditions (13.2). To satisfy the condition (13.3), we must choose N and b_n's so that

$$u(x, 0) = x(\pi - x) = \sum_{n=1}^{N} b_n \sin(nx) \quad \text{for } 0 \leq x \leq \pi.$$

But this is also impossible. A finite sum of constant multiples of sine functions cannot equal a polynomial over $[0, \pi]$.

At this point Fourier had a brilliant insight. Since no finite sum of functions (13.4) can be a solution, attempt an infinite series:

$$u(x, t) = \sum_{n=1}^{\infty} b_n \sin(nx) e^{-kn^2 t}. \tag{13.6}$$

This function will satisfy the heat equation, as well as the conditions $u(x, 0) = u(\pi, 0) = 0$. To satisfy condition (13.3) we must choose the b_n's so that

$$u(x, 0) = x(\pi - x) = \sum_{n=1}^{\infty} b_n \sin(nx) \quad \text{for } 0 \le x \le \pi. \tag{13.7}$$

This is quite different from attempting to represent the polynomial $x(\pi - x)$ by the finite trigonometric sum (13.5). Fourier claimed that equation (13.7) is valid for $0 \le x \le \pi$ if the coefficients are chosen as

$$b_n = \frac{2}{\pi} \int_0^{\pi} x(\pi - x) \sin(nx) \, dx = \frac{4}{\pi} \frac{1 - (-1)^n}{n^3}.$$

By inserting these coefficients into the proposed solution (13.6), Fourier thus claimed that the solution of this heat conduction problem, with the given initial temperature, is

$$u(x, t) = \frac{4}{\pi} \sum_{n=1}^{\infty} \frac{1 - (-1)^n}{n^3} \sin(nx) e^{-kn^2 t}.$$

The claim that

$$\sum_{n=1}^{\infty} \frac{4}{\pi} \frac{1 - (-1)^n}{n^3} \sin(nx) = x(\pi - x) \quad \text{for } 0 \le x \le \pi$$

was too much for mathematicians of Fourier's time to accept. The mathematics of this time was not adequate to proving this kind of assertion. This was the lack of rigor that led the Academy to reject publication of Fourier's paper. But the implications were not lost on Fourier's colleagues. There is nothing unique about $x(\pi - x)$ as an initial temperature distribution, and many different functions could be used. What Fourier was actually claiming was that for a broad class of functions f, coefficients b_n could be chosen so that $f(x) = \sum_{n=1}^{\infty} b_n \sin(nx)$ on $[0, \pi]$.

Eventually this and even more general claims for these series proposed by Fourier were proved. We will now begin an analysis of Fourier's ideas and some of their ramifications.

SECTION 13.1 PROBLEMS

1. On the same set of axes, generate a graph of $x(\pi - x)$ and $\sum_{n=1}^{2} (4/\pi)([1 - (-1)^n]/n^3) \sin(nx)$ for $0 \le x \le \pi$. Repeat this for the partial sums $\sum_{n=1}^{3} (4/\pi)([1 - (-1)^n]/n^3) \sin(nx)$ and $\sum_{n=1}^{4} (4/\pi)([1 - (-1)^n]/n^3) \sin(nx)$. This will give a sense of the correctness of Fourier's intuition in asserting that $x(\pi - x)$ can be accurately represented by $\sum_{n=1}^{\infty} (4/\pi)([1 - (-1)^n]/n^3) \sin(nx)$ on this interval.

2. Prove that a polynomial cannot be a constant multiple of $\sin(nx)$ over $[0, \pi]$ for any positive integer n. *Hint:* One way is to proceed by induction on the degree of the polynomial.

3. Prove that a polynomial cannot be equal to a sum of the form $\sum_{j=0}^{n} c_j \sin(jx)$ for $0 \le x \le \pi$, where the c_j's are real numbers.

13.2   The Fourier Series of a Function

Let $f(x)$ be defined for $-L \leq x \leq L$. For the time being, we assume only that $\int_{-L}^{L} f(x)\,dx$ exists. We want to explore the possibility of choosing numbers $a_0, a_1, \ldots, b_1, b_2, \ldots$ such that

$$f(x) = \frac{1}{2}a_0 + \sum_{n=1}^{\infty} a_n \cos\left(\frac{n\pi x}{L}\right) + b_n \sin\left(\frac{n\pi x}{L}\right) \qquad (13.8)$$

for $-L \leq x \leq L$. We will see that this is sometimes asking too much, but that under certain conditions on f it can be done. However, to get started, we will assume the best of all worlds and suppose for the moment that equation (13.8) holds. What does this tell us about how to choose the coefficients? There is a clever device used to answer this question, which was known to Fourier and others of his time. We will need the following elementary lemma.

LEMMA 13.1

Let n and m be nonnegative integers. Then

1.
$$\int_{-L}^{L} \cos\left(\frac{n\pi x}{L}\right) \sin\left(\frac{m\pi x}{L}\right) dx = 0.$$

2. If $n \neq m$, then
$$\int_{-L}^{L} \cos\left(\frac{n\pi x}{L}\right) \cos\left(\frac{m\pi x}{L}\right) dx = \int_{-L}^{L} \sin\left(\frac{n\pi x}{L}\right) \sin\left(\frac{m\pi x}{L}\right) dx = 0.$$

3. If $n \neq 0$, then
$$\int_{-L}^{L} \cos^2\left(\frac{n\pi x}{L}\right) dx = \int_{-L}^{L} \sin^2\left(\frac{n\pi x}{L}\right) dx = L. \quad \blacksquare$$

The lemma is proved by straightforward integration.

Now, to find a_0, integrate the series (13.8) term-by-term (supposing for now that we can do this):

$$\int_{-L}^{L} f(x)\,dx = \frac{1}{2}a_0 \int_{-L}^{L} dx$$

$$+ \sum_{n=1}^{\infty} a_n \int_{-L}^{L} \cos\left(\frac{n\pi x}{L}\right) dx + b_n \int_{-L}^{L} \sin\left(\frac{n\pi x}{L}\right) dx.$$

All of the integrals on the right are zero, except possibly the first, and this equation reduces to

$$\int_{-L}^{L} f(x)\,dx = La_0.$$

Therefore,

$$a_0 = \frac{1}{L} \int_{-L}^{L} f(x)\,dx.$$

Next, we will determine a_k for any positive integer k. Multiply equation (13.8) by $\cos(k\pi x/L)$ and integrate each term of the resulting series to get

$$\int_{-L}^{L} f(x) \cos\left(\frac{k\pi x}{L}\right) dx = \frac{1}{2} a_0 \int_{-L}^{L} \cos\left(\frac{k\pi x}{L}\right) dx$$

$$+ \sum_{n=1}^{\infty} a_n \int_{-L}^{L} \cos\left(\frac{n\pi x}{L}\right) \cos\left(\frac{k\pi x}{L}\right) dx + b_n \int_{-L}^{L} \sin\left(\frac{n\pi x}{L}\right) \cos\left(\frac{k\pi x}{L}\right) dx.$$

By the lemma, all of the integrals on the right are zero except for $\int_{-L}^{L} \cos(k\pi x/L) \cos(k\pi x/L)\, dx$, which occurs when $n = k$, and this integral equals L. The right side of this equation therefore collapses to just one term, and the equation becomes

$$\int_{-L}^{L} f(x) \cos\left(\frac{k\pi x}{L}\right) dx = a_k L,$$

whereupon

$$a_k = \frac{1}{L} \int_{-L}^{L} f(x) \cos\left(\frac{k\pi x}{L}\right) dx.$$

To determine b_k, return to equation (13.8). This time multiply the equation by $\sin(k\pi x/L)$ and integrate each term to get

$$\int_{-L}^{L} f(x) \sin\left(\frac{k\pi x}{L}\right) dx = \frac{1}{2} a_0 \int_{-L}^{L} \sin\left(\frac{k\pi x}{L}\right) dx$$

$$+ \sum_{n=1}^{\infty} a_n \int_{-L}^{L} \cos\left(\frac{n\pi x}{L}\right) \sin\left(\frac{k\pi x}{L}\right) dx + b_n \int_{-L}^{L} \sin\left(\frac{n\pi x}{L}\right) \sin\left(\frac{k\pi x}{L}\right) dx.$$

Again, by the lemma, all terms on the right are zero except for $\int_{-L}^{L} \sin(n\pi x/L) \sin(k\pi x/L)\, dx$ when $n = k$, and this equation reduces to

$$\int_{-L}^{L} f(x) \sin\left(\frac{k\pi x}{L}\right) dx = b_k L.$$

Therefore,

$$b_k = \frac{1}{L} \int_{-L}^{L} f(x) \sin\left(\frac{k\pi x}{L}\right) dx.$$

We have now "solved" for the coefficients in the trigonometric series expansion (13.8). Of course, this analysis is flawed by the interchange of series and integrals, which is not always justified. However, the argument does tell us how the constants should be chosen, at least under certain conditions, and suggests the following definition.

DEFINITION 13.1 *Fourier Coefficients and Series*

Let f be a Riemann integrable function on $[-L, L]$.

1. The numbers

$$a_n = \frac{1}{L} \int_{-L}^{L} f(x) \cos\left(\frac{n\pi x}{L}\right) dx, \quad \text{for } n = 0, 1, 2, \ldots$$

and

$$b_n = \frac{1}{L} \int_{-L}^{L} f(x) \sin\left(\frac{n\pi x}{L}\right) dx \quad \text{for } n = 1, 2, 3, \ldots$$

are the Fourier coefficients of f on $[-L, L]$.

2. The series

$$\frac{1}{2}a_0 + \sum_{n=1}^{\infty} a_n \cos\left(\frac{n\pi x}{L}\right) + b_n \sin\left(\frac{n\pi x}{L}\right)$$

is the Fourier series of f on $[-L, L]$ when the constants are chosen to be the Fourier coefficients of f on $[-L, L]$.

EXAMPLE 13.1

Let $f(x) = x$ for $-\pi \leq x \leq \pi$. We will write the Fourier series of f on $[-\pi, \pi]$. The coefficients are

$$a_0 = \frac{1}{\pi} \int_{-\pi}^{\pi} x \, dx = 0,$$

$$a_n = \frac{1}{\pi} \int_{-\pi}^{\pi} x \cos(nx) \, dx$$

$$= \left[\frac{1}{n^2\pi} \cos(nx) + \frac{x}{n\pi} \sin(nx)\right]_{-\pi}^{\pi} = 0,$$

and

$$b_n = \frac{1}{\pi} \int_{-\pi}^{\pi} x \sin(nx) \, dx$$

$$= \left[\frac{1}{n^2\pi} \sin(nx) - \frac{x}{n\pi} \cos(nx)\right]_{-\pi}^{\pi}$$

$$= -\frac{2}{n} \cos(n\pi) = \frac{2}{n}(-1)^{n+1},$$

since $\cos(n\pi) = (-1)^n$ if n is an integer. The Fourier series of x on $[-\pi, \pi]$ is

$$\sum_{n=1}^{\infty} \frac{2}{n}(-1)^{n+1} \sin(nx) = 2\sin(x) - \sin(2x) + \frac{2}{3}\sin(3x) - \frac{1}{2}\sin(4x) + \frac{2}{5}\sin(5x) - \cdots.$$

In this example, the constant term and cosine coefficients are all zero, and the Fourier series contains only sine terms. ∎

EXAMPLE 13.2

Let

$$f(x) = \begin{cases} 0 & \text{for } -3 \leq x \leq 0 \\ x & \text{for } 0 \leq x \leq 3 \end{cases}.$$

Here $L = 3$ and the Fourier coefficients are

$$a_0 = \frac{1}{3} \int_{-3}^{3} f(x)\, dx = \frac{1}{3} \int_{0}^{3} x\, dx = \frac{3}{2},$$

$$a_n = \frac{1}{3} \int_{-3}^{3} f(x) \cos\left(\frac{n\pi x}{3}\right) dx$$

$$= \frac{1}{3} \int_{0}^{3} x \cos\left(\frac{n\pi x}{3}\right) dx$$

$$= \frac{3}{n^2\pi^2} \cos\left(\frac{n\pi x}{3}\right) + \frac{x}{n\pi} \sin\left(\frac{n\pi x}{3}\right) \Big]_{0}^{3}$$

$$= \frac{3}{n^2\pi^2}[(-1)^n - 1],$$

and

$$b_n = \frac{1}{3} \int_{-3}^{3} f(x) \sin\left(\frac{n\pi x}{3}\right) dx = \frac{1}{3} \int_{0}^{3} x \sin\left(\frac{n\pi x}{3}\right) dx$$

$$= \frac{3}{n^2\pi^2} \sin\left(\frac{n\pi x}{3}\right) - \frac{x}{n\pi} \cos\left(\frac{n\pi x}{3}\right) \Big]_{0}^{3}$$

$$= \frac{3}{n\pi}(-1)^{n+1}.$$

The Fourier series of f on $[-3, 3]$ is

$$\frac{3}{4} + \sum_{n=1}^{\infty} \left(\frac{3}{n^2\pi^2}[(-1)^n - 1] \cos\left(\frac{n\pi x}{3}\right) + \frac{3}{n\pi}(-1)^{n+1} \sin\left(\frac{n\pi x}{3}\right) \right). \quad \blacksquare$$

EXAMPLE 13.3

Let $f(x) = e^{-4x}$ for $-2 \le x \le 2$. The Fourier coefficients are

$$a_0 = \frac{1}{2} \int_{-2}^{2} e^{-4x}\, dx = \frac{1}{8}\left(e^8 - e^{-8}\right),$$

$$a_n = \frac{1}{2} \int_{-2}^{2} e^{-4x} \cos\left(\frac{n\pi x}{2}\right) dx$$

$$= \left(e^8 - e^{-8}\right) \frac{8(-1)^n}{64 + \pi^2 n^2},$$

and

$$b_n = \frac{1}{2} \int_{-2}^{2} e^{-4x} \sin\left(\frac{n\pi x}{2}\right) dx$$

$$= \left(e^8 - e^{-8}\right) \frac{n\pi(-1)^n}{64 + \pi^2 n^2}.$$

The Fourier series of e^{-4x} on $[-2, 2]$ is

$$\frac{1}{16}(e^8 - e^{-8}) + (e^8 - e^{-8}) \sum_{n=1}^{\infty} \left(\frac{8(-1)^n}{64 + \pi^2 n^2} \cos \left(\frac{n\pi x}{2} \right) + \frac{n\pi (-1)^n}{64 + \pi^2 n^2} \sin \left(\frac{n\pi x}{2} \right) \right). \quad \blacksquare$$

Even when $f(x)$ is fairly simple, $\int_{-L}^{L} f(x) \cos(n\pi x/L)\, dx$ and $\int_{-L}^{L} f(x) \sin(n\pi x/L)\, dx$ can involve considerable labor if done by hand. Use of a software package to evaluate definite integrals is highly recommended.

In these examples, we wrote the Fourier series of f but did not claim that it equaled $f(x)$. For most x it is not obvious what the sum of the Fourier series is. However, in some cases it is obvious that the series does not equal $f(x)$. Consider again $f(x) = x$ on $[-\pi, \pi]$ in Example 13.1. At $x = \pi$ and at $x = -\pi$, every term of the Fourier series is zero, even though $f(\pi) = \pi$ and $f(-\pi) = -\pi$. Even for very simple functions, then, there may be points where the Fourier series does not converge to $f(x)$. Shortly we will determine the sum of the Fourier series of a function. Until this is done, we do not know the relationship between the Fourier series and the function itself.

13.2.1 Even and Odd Functions

Sometimes we can save some work in computing Fourier coefficients by observing special properties of $f(x)$.

DEFINITION 13.2

Even Function
f is an even function on $[-L, L]$ if $f(-x) = f(x)$ for $-L \leq x \leq L$.

Odd Function
f is an odd function on $[-L, L]$ if $f(-x) = -f(x)$ for $-L \leq x \leq L$.

For example, x^2, x^4, $\cos(n\pi x/L)$, and $e^{-|x|}$ are even functions on any interval $[-L, L]$. Graphs of $y = x^2$ and $y = \cos(5\pi x/3)$ are given in Figure 13.1. The graph of such a function for $-L \leq x \leq 0$ is the reflection across the y axis of the graph for $0 \leq x \leq L$ (Figure 13.2).

The functions x, x^3, x^5, and $\sin(n\pi x/L)$ are odd functions on any interval $[-L, L]$. Graphs of $y = x$, $y = x^3$, and $y = \sin(5\pi x/2)$ are shown in Figure 13.3. The graph of an odd function for $-L \leq x \leq 0$ is the reflection across the vertical axis, and then across the horizontal axis, of the graph for $0 \leq x \leq L$ (Figure 13.4). If f is odd, then $f(0) = 0$, since

$$f(-0) = f(0) = -f(0).$$

Of course, "most" functions are neither even nor odd. For example, $f(x) = e^x$ is not even or odd on any interval $[-L, L]$.

Even and odd functions behave like even and odd integers under multiplication:

$$\text{even} \cdot \text{even} = \text{even},$$

$$\text{odd} \cdot \text{odd} = \text{even},$$

and

$$\text{even} \cdot \text{odd} = \text{odd}.$$

For example, $x^2 \cos(n\pi x/L)$ is an even function (product of two even functions); $x^2 \sin(n\pi x/L)$ is odd (product of an even function with an odd function); and $x^3 \sin(n\pi x/L)$ is even (product of two odd functions).

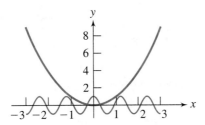

FIGURE 13.1 *Graphs of even functions* $y = x^2$ *and* $y = \cos(5\pi x/3)$.

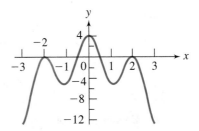

FIGURE 13.2 *Graph of a typical even function, symmetric about the y axis.*

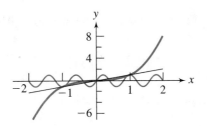

FIGURE 13.3 *Graphis of odd functions* $y = x$, $y = x^3$, *and* $y = \sin(5\pi x/2)$.

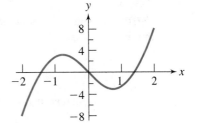

FIGURE 13.4 *Graph of a typical odd function, symmetric through the origin.*

Now recall from calculus that

$$\int_{-L}^{L} f(x)\, dx = 0 \quad \text{if } f \text{ is odd on } [-L, L]$$

and

$$\int_{-L}^{L} f(x)\, dx = 2\int_{0}^{L} f(x)\, dx \quad \text{if } f \text{ is even on } [-L, L].$$

These integrals are suggested by Figures 13.2 and 13.4. In Figure 13.4, f is odd on $[-L, L]$, and the "area" bounded by the graph and the horizontal axis for $-L \leq x \leq 0$ is exactly the negative of that bounded by the graph and the horizontal axis for $0 \leq x \leq L$. This makes $\int_{-L}^{L} f(x)\, dx = 0$. In Figure 13.2, where f is even, the area to the left of the vertical axis, for $-L \leq x \leq 0$, equals that to the right, for $0 \leq x \leq L$, so $\int_{-L}^{L} f(x)\, dx = 2\int_{0}^{L} f(x)\, dx$.

One ramification of these ideas for Fourier coefficients is that if f is an even or odd function, then some of the Fourier coefficients can be seen immediately to be zero, and we need not carry out the integrations explicitly. We saw this in Example 13.1 with $f(x) = x$, which is an odd function on $[-\pi, \pi]$. There we found that the cosine coefficients were all zero, since $x \cos(nx)$ is an odd function.

EXAMPLE 13.4

We will find the Fourier series of $f(x) = x^4$ on $[-1, 1]$. Since f is an even function, $x^4 \sin(n\pi x)$ is odd and we know immediately that all the sine coefficients b_n are zero. For the other coefficients, compute

$$a_0 = \int_{-1}^{1} x^4\, dx = 2\int_{0}^{1} x^4\, dx = \frac{2}{5},$$

and

$$a_n = \int_{-1}^{1} x^4 \cos(n\pi x) \, dx$$

$$= 2 \int_{0}^{1} x^4 \cos(n\pi x) \, dx = 8 \frac{n^2 \pi^2 - 6}{\pi^4 n^4} (-1)^n.$$

The Fourier series of x^4 on $[-1, 1]$ is

$$\frac{1}{5} + \sum_{n=1}^{\infty} 8 \frac{n^2 \pi^2 - 6}{\pi^4 n^4} (-1)^n \cos(n\pi x). \quad \blacksquare$$

To again make the point about convergence, notice that $f(0) = 0$ in this example, but the Fourier series at $x = 0$ is

$$\frac{1}{5} + \sum_{n=1}^{\infty} 8 \frac{n^2 \pi^2 - 6}{\pi^4 n^4} (-1)^n.$$

It is not clear whether this series sums to the function value 0.

EXAMPLE 13.5

Let $f(x) = x^3$ for $-4 \le x \le 4$. Because f is odd on $[-4, 4]$, its Fourier cosine coefficients are all zero. Its Fourier sine coefficients are

$$b_n = \frac{1}{4} \int_{-4}^{4} x^3 \sin\left(\frac{n\pi x}{4}\right) dx$$

$$= \frac{1}{2} \int_{0}^{4} x^3 \sin\left(\frac{n\pi x}{4}\right) dx = (-1)^{n+1} 128 \frac{n^2 \pi^2 - 6}{n^3 \pi^3}.$$

The Fourier series of x^3 on $[-4, 4]$ is

$$\sum_{n=1}^{\infty} (-1)^{n+1} 128 \frac{n^2 \pi^2 - 6}{n^3 \pi^3} \sin\left(\frac{n\pi x}{4}\right). \quad \blacksquare$$

We will make use of this discussion later, so here is a summary of its conclusions:
If f is even on $[-L, L]$, then its Fourier series on this interval is

$$\frac{1}{2} a_0 + \sum_{n=1}^{\infty} a_n \cos\left(\frac{n\pi x}{L}\right), \tag{13.9}$$

in which

$$a_n = \frac{2}{L} \int_{0}^{L} f(x) \cos\left(\frac{n\pi x}{L}\right) dx \quad \text{for } n = 0, 1, 2, \ldots. \tag{13.10}$$

If f is odd on $[-L, L]$, then its Fourier series on this interval is

$$\sum_{n=1}^{\infty} b_n \sin\left(\frac{n\pi x}{L}\right), \tag{13.11}$$

where

$$b_n = \frac{2}{L} \int_{0}^{L} f(x) \sin\left(\frac{n\pi x}{L}\right) dx \quad \text{for } n = 1, 2, \ldots. \tag{13.12}$$

SECTION 13.2 PROBLEMS

In each of Problems 1 through 12, find the Fourier series of the function on the interval.

1. $f(x) = 4, -3 \leq x \leq 3$

2. $f(x) = -x, -1 \leq x \leq 1$

3. $f(x) = \cosh(\pi x), -1 \leq x \leq 1$

4. $f(x) = 1 - |x|, -2 \leq x \leq 2$

5. $f(x) = \begin{cases} -4 & \text{for } -\pi \leq x \leq 0 \\ 4 & \text{for } 0 < x \leq \pi \end{cases}$

6. $f(x) = \sin(2x), -\pi \leq x \leq \pi$

7. $f(x) = x^2 - x + 3, -2 \leq x \leq 2$

8. $f(x) = \begin{cases} -x & \text{for } -5 \leq x < 0 \\ 1 + x^2 & \text{for } 0 \leq x \leq 5 \end{cases}$

9. $f(x) = \begin{cases} 1 & \text{for } -\pi \leq x < 0 \\ 2 & \text{for } 0 \leq x \leq \pi \end{cases}$

10. $f(x) = \cos(x/2) - \sin(x), -\pi \leq x \leq \pi$

11. $f(x) = \cos(x), -3 \leq x \leq 3$

12. $f(x) = \begin{cases} 1 - x & \text{for } -1 \leq x \leq 0 \\ 0 & \text{for } 0 < x \leq 1 \end{cases}$

13. Suppose f and g are integrable on $[-L, L]$ and that $f(x) = g(x)$ except for $x = x_0$, a given point in the interval. How are the Fourier series of f and g related? What does this suggest about the relationship between a function and its Fourier series on an interval?

14. Prove that $\int_{-L}^{L} f(x)\,dx = 0$ if f is odd on $[-L, L]$.

15. Prove that $\int_{-L}^{L} f(x)\,dx = 2\int_{0}^{L} f(x)\,dx$ if f is even on $[-L, L]$.

13.3 Convergence of Fourier Series

It is one thing to be able to write the Fourier coefficients of a function f on an interval $[-L, L]$. This requires only the existence of $\int_{-L}^{L} f(x) \cos(n\pi x/L)\,dx$ and $\int_{-L}^{L} f(x) \sin(n\pi x/L)\,dx$. It is another issue entirely to determine whether the resulting Fourier series converges to $f(x)$—or even that it converges at all! The subtleties of this question were dramatized in 1873 when the French mathematician Paul Du Bois-Reymond gave an example of a function that is continuous on $(-\pi, \pi)$ but whose Fourier series fails to converge at any point of this interval.

However, the obvious utility of Fourier series in solving partial differential equations led in the nineteenth century to an intensive effort to determine their convergence properties. About 1829, Peter Gustav Lejeune-Dirichlet gave conditions on the function f that were sufficient for convergence of the Fourier series of f. Further, Dirichlet's theorem actually gave the sum of the Fourier series at each point, whether or not this sum is $f(x)$.

This section is devoted to conditions on a function that enable us to determine the sum of its Fourier series on an interval. These conditions center about the concept of piecewise continuity.

DEFINITION 13.3 *Piecewise Continuous Function*

Let $f(x)$ be defined on $[a, b]$, except possibly at finitely many points. Then f is piecewise continuous on $[a, b]$ if

1. f is continuous on $[a, b]$ except perhaps at finitely many points.

2. Both $\lim_{x \to a+} f(x)$ and $\lim_{x \to b-} f(x)$ exist and are finite.

3. If x_0 is in (a, b) and f is not continuous at x_0, then $\lim_{x \to x_0+} f(x)$ and $\lim_{x \to x_0-} f(x)$ exist and are finite.

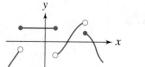

FIGURE 13.5 *A piecewise continuous function.*

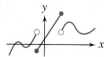

FIGURE 13.6 *Graph of a typical piecewise continuous function.*

Figures 13.5 and 13.6 shows graphs of typical piecewise continuous functions. At points of discontinuity (which we assume are finite in number), the function must have finite one-sided limits. This means that the graph experiences at worst a finite gap at a discontinuity. Points where these occur are called *jump discontinuities* of the function.

As an example of a simple function that is not piecewise continuous, let

$$f(x) = \begin{cases} 0 & \text{for } x = 0 \\ 1/x & \text{for } 0 < x \leq 1 \end{cases}.$$

Then f is continuous on $(0, 1]$ and discontinuous at 0. However, $\lim_{x \to 0+} f(x) = \infty$, so the discontinuity is not a finite jump discontinuity, and f is not piecewise continuous on $[0, 1]$.

EXAMPLE 13.6

Let

$$f(x) = \begin{cases} 5 & \text{for } x = -\pi \\ x & \text{for } -\pi < x < 1 \\ 1 - x^2 & \text{for } 1 \leq x < 2 \\ 4 & \text{for } 2 \leq x \leq \pi \end{cases}.$$

A graph of f is shown in Figure 13.7. This function is discontinuous at $-\pi$, and

$$\lim_{x \to -\pi+} f(x) = -\pi.$$

f is also discontinuous at 1, interior to $[-\pi, \pi]$, and

$$\lim_{x \to 1-} f(x) = 1 \quad \text{and} \quad \lim_{x \to 1+} f(x) = 0.$$

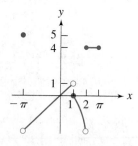

FIGURE 13.7 *Graph of the function of Example 13.6.*

Finally, f is discontinuous at 2, and

$$\lim_{x \to 2-} f(x) = -3 \quad \text{and} \quad \lim_{x \to 2+} f(x) = 4.$$

At each point of discontinuity interior to the interval, the function has finite one-sided limits from both sides. At the point of discontinuity at the endpoint $-\pi$, the function has a finite limit from within the interval. In this example, the other endpoint is not an issue, as f is continuous (from the left) there. Therefore, f is piecewise continuous on $[-\pi, \pi]$. ∎

We will use the following notation for left and right limits of a function at a point:

$$f(x_0+) = \lim_{x \to x_0+} f(x) \quad \text{and} \quad f(x_0-) = \lim_{x \to x_0-} f(x).$$

In Example 13.6,

$$f(1-) = 1 \quad \text{and} \quad f(1+) = 0$$

and

$$f(2-) = -3 \quad \text{and} \quad f(2+) = 4.$$

At the endpoints of an interval, we can still use this notation, except at the left endpoint we consider only the right limit (from inside the interval), and at the right endpoint we use only the left limit (again, so that the limit is taken from within the interval). Again referring to Example 13.6,

$$f(-\pi+) = -\pi \quad \text{and} \quad f(\pi-) = 4.$$

DEFINITION 13.4   *Piecewise Smooth Function*

f is piecewise smooth on $[a, b]$ if f and f' are piecewise continuous on $[a, b]$.

A piecewise smooth function is therefore one that is continuous except possibly for finitely many jump discontinuities and has a continuous derivative at all but finitely many points, where the derivative may not exist but must have finite one-sided limits.

EXAMPLE 13.7

Let

$$f(x) = \begin{cases} 1 & \text{for } -4 \le x < 1 \\ -2x & \text{for } 1 \le x < 2 \\ 9e^{-x} & \text{for } 2 \le x \le 3 \end{cases}.$$

Figure 13.8 shows a graph of f. The function is continuous except for finite jump discontinuities at 1 and 2. Therefore, f is piecewise continuous on $[-4, 3]$. The derivative of f is

$$f'(x) = \begin{cases} 0 & \text{for } -4 < x < 1 \\ -2 & \text{for } 1 < x < 2 \\ -9e^{-x} & \text{for } 2 < x < 3 \end{cases}.$$

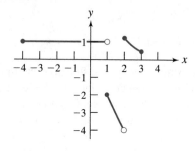

FIGURE 13.8   *Graph of*

$$f(x) = \begin{cases} 1 & \text{for } -4 \le x < 1 \\ -2x & \text{for } 1 \le x < 2 \\ 9e^{-x} & \text{for } 2 \le x \le 3 \end{cases}.$$

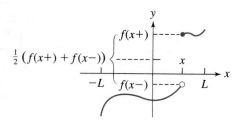

FIGURE 13.9   *Convergence of a Fourier series at a jump discontinuity.*

The derivative is continuous on $[-4, 3]$ except at the points of discontinuity 1 and 2 of f, where $f'(x)$ does not exist. However, at these points $f'(x)$ has finite one-sided limits. Thus f' is piecewise continuous on $[-4, 3]$, so f is piecewise smooth. ■

As suggested by Figure 13.8, a piecewise smooth function is one that has a continuous tangent at all but finitely many points.

We will now state our first convergence theorem.

THEOREM 13.1   *Convergence of Fourier Series*

Let f be piecewise smooth on $[-L, L]$. Then for $-L < x < L$, the Fourier series of f on $[-L, L]$ converges to

$$\tfrac{1}{2}(f(x+) + f(x-)). \quad ■$$

This means that at each point between $-L$ and L, the function converges to the average of its left and right limits. If f is continuous at x, then these left and right limits both equal $f(x)$, so the Fourier series converges to the function value at x. If f has a jump discontinuity at x, then the Fourier series may not converge to $f(x)$, but will converge to the point midway between the ends of the gap in the graph at x (Figure 13.9).

EXAMPLE 13.8

Let

$$f(x) = \begin{cases} 5\sin(x) & \text{for } -2\pi \le x < -\pi/2 \\ 4 & \text{for } x = -\pi/2 \\ x^2 & \text{for } -\pi/2 < x < 2 \\ 8\cos(x) & \text{for } 2 \le x < \pi \\ 4x & \text{for } \pi \le x \le 2\pi \end{cases}.$$

A graph of f is given in Figure 13.10. Since f is piecewise smooth on $[-2\pi, 2\pi]$, we can determine the sum of its Fourier series on this interval. In applying the theorem, we do not actually have to compute this Fourier series. We could do this, but it is not necessary to determine the sum of the Fourier series.

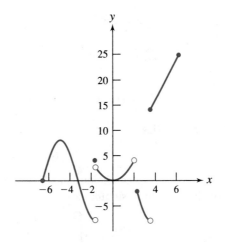

FIGURE 13.10 *Graph of*

$$f(x) = \begin{cases} 5\sin(x) & \text{for } -2\pi \le x < -\pi/2 \\ 4 & \text{for } x = -\pi/2 \\ x^2 & \text{for } -\pi/2 < x < 2 \\ 8\cos(x) & \text{for } 2 \le x < \pi \\ 4x & \text{for } \pi \le x \le 2\pi \end{cases}.$$

For $-2\pi < x < -\pi/2$, f is continuous and the Fourier series converges to $f(x) = 5\sin(x)$.

At $x = -\pi/2$, f has a jump discontinuity and the Fourier series will converge to the average of the left and right limits of $f(x)$ at $-\pi/2$. Compute

$$f(-\pi/2-) = \lim_{x \to -\pi/2-} f(x) = \lim_{x \to -\pi/2-} 5\sin(x) = 5\sin(-\pi/2) = -5$$

and

$$f(-\pi/2+) = \lim_{x \to -\pi/2+} f(x) = \lim_{x \to -\pi/2+} x^2 = \frac{\pi^2}{4}.$$

Therefore, at $x = -\pi/2$, the Fourier series of f converges to

$$\frac{1}{2}\left(\frac{\pi^2}{4} - 5\right).$$

On $(-\pi/2, 2)$ the function is continuous, so the Fourier series converges to x^2 for $-\pi/2 < x < 2$.

At $x = 2$ the function the function has another jump discontinuity. Compute

$$f(2-) = \lim_{x \to 2-} x^2 = 4$$

and

$$f(2+) = \lim_{x \to 2+} 8\cos(x) = 8\cos(2).$$

At $x = 2$ the Fourier series converges to

$$\tfrac{1}{2}(4 + 8\cos(2)).$$

On $(2, \pi)$, f is continuous. At each x with $2 < x < \pi$, the Fourier series converges to $f(x) = 8\cos(x)$.

At $x = \pi$, f has a jump discontinuity. Compute

$$f(\pi-) = \lim_{x \to \pi-} 8\cos(x) = 8\cos(\pi) = -8$$

and

$$f(\pi+) = \lim_{x \to \pi+} 4x = 4\pi.$$

At $x = \pi$ the Fourier series of f converges to

$$\tfrac{1}{2}(4\pi - 8).$$

Finally, on $(\pi, 2\pi)$, f is continuous and the Fourier series converges to $f(x) = 4x$. These conclusions can be summarized:

$$\text{Fourier series converges to}\begin{cases} 5\sin(x) & \text{for } -2\pi < x < -\dfrac{\pi}{2} \\[2mm] \dfrac{1}{2}\left(\dfrac{\pi^2}{4} - 5\right) & \text{for } x = -\dfrac{\pi}{2} \\[2mm] x^2 & \text{for } -\dfrac{\pi}{2} < x < 2 \\[2mm] \dfrac{1}{2}(4 + 8\cos(2)) & \text{for } x = 2 \\[2mm] 8\cos(x) & \text{for } 2 < x < \pi \\[2mm] \dfrac{1}{2}(4\pi - 8) & \text{for } x = \pi \\[2mm] 4x & \text{for } \pi < x < 2\pi \end{cases}.$$

Figure 13.11 shows a graph of this sum of the Fourier series, differing from the function itself on $(-2\pi, 2\pi)$ at the jump discontinuities, where the series converges to the average of the left and right limits. ■

If f is piecewise smooth on $[-L, L]$ and actually continuous on $[-L, L]$, then the Fourier series converges to $f(x)$ for $-L < x < L$.

EXAMPLE 13.9

Let

$$f(x) = \begin{cases} x & \text{for } -2 \le x \le 1 \\ 2 - x^2 & \text{for } 1 < x \le 2 \end{cases}.$$

Then f is continuous on $[-2, 2]$ (Figure 13.12). f is differentiable except at $x = 1$, where $f'(x)$ has finite left and right limits, so f is piecewise smooth. For $-2 < x < 2$, the Fourier series of f converges to $f(x)$. In this example, the Fourier series is an exact representation of the function on $(-2, 2)$. ■

13.3.1   Convergence at the Endpoints

Theorem 13.1 does not address convergence of a Fourier series at the endpoints of the interval. There is a subtlety here that we will now discuss.

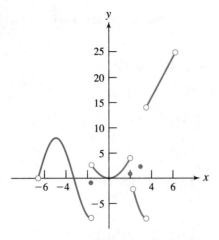

FIGURE 13.11 *Graph of the Fourier series of the function of Figure 13.10.*

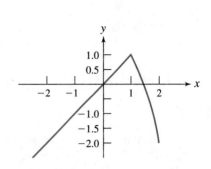

FIGURE 13.12 *Graph of*

$$f(x) = \begin{cases} x & \text{for } -2 \le x \le 1 \\ 2 - x^2 & \text{for } 1 < x \le 2 \end{cases}.$$

The problem is that while the function f of interest may be defined only on $[-L, L]$, its Fourier series

$$\frac{1}{2}a_0 + \sum_{n=1}^{\infty} a_n \cos\left(\frac{n\pi x}{L}\right) + b_n \sin\left(\frac{n\pi x}{L}\right) \tag{13.13}$$

is defined for all real x for which the series converges. Further, the Fourier series is periodic, of period $2L$. The value of the series is unchanged if x is replaced with $x + 2L$. How do we reconcile representing a function that is defined only on an interval by a function that is periodic and may be defined over the entire real line?

The reconciliation lies in a periodic extension of f over the real line. Take the graph of $f(x)$ on $[-L, L)$ and replicate it over successive intervals of length $2L$. This defines a new function, f_p, that agrees with $f(x)$ for $-L \le x < L$ and has period $2L$. This process is illustrated in Figure 13.13 for the function $f(x) = x^2$ for $-2 \le x < 2$. This graph is simply repeated for $2 \le x < 6$, $6 \le x < 10, \ldots, -6 \le x < -2, -10 \le x < -6$, and so on. The reason for using the half-open interval $[-L, L)$ in this extension is that if f_p is to have period $2L$, then

$$f_p(x + 2L) = f_p(x)$$

for all x. But this requires that $f(-L) = f(-L + 2L) = f(L)$, so once $f_p(-L)$ is defined, $f_p(L)$ must equal this value.

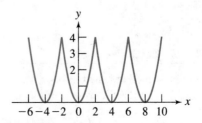

FIGURE 13.13 *Part of the periodic extension, of period 4, of $f(x) = x^2$ for $-2 \le x < 2$.*

If we make this extension, then the convergence theorem applies to $f_p(x)$ at all x. In particular, at $-L$, the series converges to

$$\frac{1}{2}\left(f_p(-L-) + f_p(-L+)\right),$$

which is

$$\frac{1}{2}(f(L-) + f(-L+)).$$

Similarly, at L, the Fourier series converges to

$$\frac{1}{2}\left(f_p(L-) + f_p(L+)\right),$$

which is

$$\frac{1}{2}(f(L-) + f(-L+)).$$

The Fourier series converges to the same value at both L and at $-L$. This can also be seen directly from the series (13.13). If $x = L$, all the sine terms are $\sin(n\pi)$, which vanish, and the cosine terms are $\cos(n\pi)$, so the series at $x = L$ is

$$\frac{1}{2}a_0 + \sum_{n=1}^{\infty} a_n \cos(n\pi).$$

At $x = -L$, again all the sine terms vanish, and because $\cos(-n\pi) = \cos(n\pi)$, the series at $x = -L$ is also

$$\frac{1}{2}a_0 + \sum_{n=1}^{\infty} a_n \cos(n\pi).$$

13.3.2   A Second Convergence Theorem

A second convergence theorem can be framed in terms of one-sided derivatives.

DEFINITION 13.5   *Right Derivative*

Suppose $f(x)$ is defined at least for $c < x < c + r$ for some positive number r. Suppose $f(c+)$ is finite. Then the right derivative of f at c is

$$f_{\mathcal{R}}'(c) = \lim_{h \to 0+} \frac{f(c+h) - f(c+)}{h},$$

if this limit exists and is finite.

DEFINITION 13.6   *Left Derivative*

Suppose $f(x)$ is defined at least for $c - r < x < c$ for some positive number r. Suppose $f(c-)$ is finite. Then the left derivative of f at c is

$$f_{\mathcal{L}}'(c) = \lim_{h \to 0-} \frac{f(c+h) - f(c-)}{h},$$

if this limit exists and is finite.

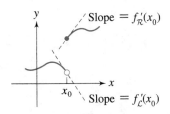

FIGURE 13.14 *One-sided derivatives as slopes from the right or left.*

If $f'(c)$ exists, then f is continuous at c, so $f(c-) = f(c+) = f(c)$, and in this case the left and right derivative are both equal to $f'(c)$. However, Figure 13.14 shows the significance of the left and right derivatives when f has a jump discontinuity at c. The left derivative is the slope of the graph at $x = c$ if we cover up the part of the graph to the right of c and keep only the left side. The right derivative is the slope at c if we cover up the left part and just keep the right part.

EXAMPLE 13.10

Let

$$f(x) = \begin{cases} 1+x & \text{for } -\pi < x < 1 \\ x^2 & \text{for } 1 \le x < \pi \end{cases}.$$

Then f is continuous on $(-\pi, \pi)$ except at 1, where there is a jump discontinuity (Figure 13.15). Further, f is differentiable except at this point of discontinuity. Indeed,

$$f'(x) = \begin{cases} 1 & \text{for } -\pi < x < 1 \\ 2x & \text{for } 1 < x < \pi \end{cases}.$$

From the graph and the slopes of the "left and right pieces" at $x = 1$, we would expect the left derivative at $x = 1$ to be 1 and the right derivative to be 2. Check this from the definition.

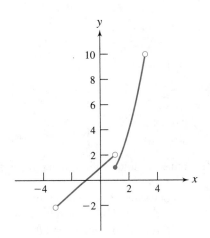

FIGURE 13.15 *Graph of*

$$f(x) = \begin{cases} 1+x & \text{for } -\pi < x < 1 \\ x^2 & \text{for } 1 \le x < \pi \end{cases}.$$

First,

$$f'_{\mathcal{L}}(0) = \lim_{h \to 0-} \frac{f(1+h) - f(1-)}{h}$$

$$= \lim_{h \to 0-} \frac{1 + (1+h) - 2}{h} = \lim_{h \to 0-} \frac{h}{h} = 1,$$

and

$$f'_{\mathcal{R}}(c) = \lim_{h \to 0+} \frac{f(1+h) - f(1+)}{h}$$

$$= \lim_{h \to 0+} \frac{(1+h)^2 - 1}{h} = \lim_{h \to 0+} (2+h) = 2. \ \blacksquare$$

Using the one-sided derivatives, we can state the following convergence theorem.

THEOREM 13.2

Let f be piecewise continuous on $[-L, L]$. Then,

1. If $-L < x < L$ and f has a left and right derivative at x, then the Fourier series of f on $[-L, L]$ converges at x to

$$\frac{1}{2}(f(x+) + f(x-)).$$

2. If $f'_{\mathcal{R}}(-L)$ and $f'_{\mathcal{L}}(L)$ exist, then at both L and $-L$, the Fourier series of f on $[-L, L]$ converges to

$$\frac{1}{2}(f(-L+) + f(L-)). \ \blacksquare$$

As with the first convergence theorem, we need not compute the Fourier series to determine its sum.

EXAMPLE 13.11

Let

$$f(x) = \begin{cases} e^{-x} & \text{for } -2 \leq x < 1 \\ -2x^2 & \text{for } 1 \leq x < 2 \\ 4 & \text{for } x = 2 \end{cases}.$$

We want to determine the sum of the Fourier series of f on $[-2, 2]$. A graph of f is shown in Figure 13.16.

f is piecewise continuous, being continuous except for jump discontinuities at 1 and 2.

For $-2 < x < 1$, f is continuous, and the Fourier series converges to $f(x) = e^{-x}$.

For $1 < x < 2$, f is also continuous and the Fourier series converges to $f(x) = -2x^2$.

At the jump discontinuity $x = 1$, the left and right derivatives exist ($-e^{-1}$ and -4, respectively). We can determine these from the limits in the definitions, but these derivatives are also apparent from looking at the graph of f to the right and left of 1. Therefore, the Fourier series converges at $x = 1$ to

$$\frac{1}{2}(f(1-) + f(1+)),$$

which is

$$\frac{1}{2}\left(e^{-1} - 2\right).$$

This takes care of each point in $(-2, 2)$. Now consider the endpoints. The left derivative of f at 2 is -8 and the right derivative at -2 is $-e^2$. Therefore, at both 2 and at -2, the Fourier series converges to

$$\frac{1}{2}(f(2-) + f(-2+)) = \frac{1}{2}\left(-8 + e^2\right).$$

Figure 13.17 shows a graph of the Fourier series on $[-2, 2]$, and can be compared with the graph of f. The two graphs agree except at the endpoints and at the jump discontinuity. The fact that $f(2) = 4$ does not affect convergence of the Fourier series of $f(x)$ at $x = 2$. ■

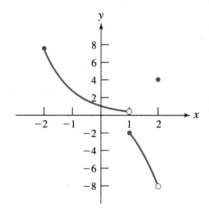

FIGURE 13.16 *Graph of*
$$f(x) = \begin{cases} e^{-x} & \text{for } -2 \le x < 1 \\ -2x^2 & \text{for } 1 \le x < 2 \\ 4 & \text{for } x = 2 \end{cases}.$$

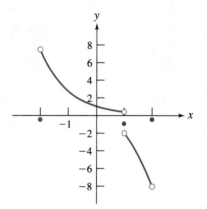

FIGURE 13.17 *Graph of the Fourier series of the function of Figure 13.16.*

A note of caution is warranted in applying the second convergence theorem. The left and right derivatives of a function at a point are relevant only to verify that the hypotheses of the theorem are satisfied at a jump discontinuity of the function. However, these derivatives play no role in the value to which the Fourier series converges at a point. That value involves the left and right limits of the function itself.

13.3.3 Partial Sums of Fourier Series

Fourier's claims for his series were counterintuitive in the sense that functions such as polynomials and exponentials do not seem to be likely candidates to be represented by series of sines and cosines. It is instructive to watch graphs of partial sums of some Fourier series converge to the graph of the function.

EXAMPLE 13.12

Let $f(x) = x$ for $-\pi \le x \le \pi$. We saw in Example 13.1 that the Fourier series is

$$\sum_{n=1}^{\infty} \frac{2}{n}(-1)^{n+1} \sin(nx).$$

We can apply either convergence theorem to show that this series converges to

$$\begin{cases} x & \text{for } -\pi < x < \pi \\ 0 & \text{for } x = \pi \quad \text{and} \quad \text{for } x = -\pi \end{cases}.$$

Figures 13.18(a), (b), and (c) show, respectively, the fourth, tenth, and twentieth partial sums of this series, suggesting how they approach nearer to $f(x) = x$ on $(-\pi, \pi)$ as more terms are included. ■

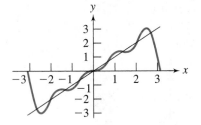

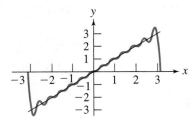

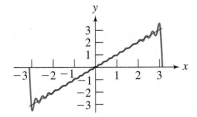

FIGURE 13.18(a) *Fourth partial sum* $S_4(x) = \sum_{n=1}^{4} \frac{2(-1)^{n+1}}{n} \sin(nx)$ *of the Fourier series of* $f(x) = x$ *on* $-\pi \le x \le \pi$.

FIGURE 13.18(b) *Tenth partial sum of the Fourier series of* $f(x) = x$ *on* $[-\pi, \pi]$.

FIGURE 13.18(c) *Twentieth partial sum of the Fourier series of* $f(x) = x$ *on* $[-\pi, \pi]$.

EXAMPLE 13.13

Let $f(x) = e^x$ for $-1 \le x \le 1$. The Fourier series of f on $[-1, 1]$ is

$$\frac{1}{2}\left(e - e^{-1}\right) + \left(e - e^{-1}\right) \sum_{n=1}^{\infty} \left(\frac{(-1)^n}{1 + n^2\pi^2} \cos(n\pi x) - n\pi \frac{(-1)^n}{1 + n^2\pi^2} \sin(n\pi x) \right).$$

This series converges to

$$\begin{cases} e^x & \text{for } -1 < x < 1 \\ \frac{1}{2}(e + e^{-1}) & \text{for } x = 1 \quad \text{and for } x = -1 \end{cases}.$$

Figures 13.19(a) and (b) show the tenth and thirtieth partial sums of this series, compared with the graph of f. ■

EXAMPLE 13.14

Let $f(x) = x^2$ for $-2 \le x \le 2$. The Fourier series of f on this interval is

$$\frac{4}{3} + \sum_{n=1}^{\infty} 16\frac{(-1)^n}{n^2\pi^2} \cos \frac{n\pi x}{2}.$$

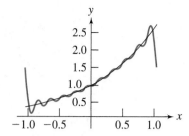

FIGURE 13.19(a) *Tenth partial sum of the Fourier series of $f(x) = e^x$ on $[-1, 1]$.*

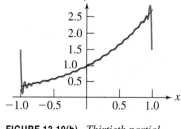

FIGURE 13.19(b) *Thirtieth partial sum of the Fourier series of $f(x) = e^x$ on $[-1, 1]$.*

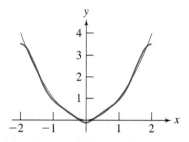

FIGURE 13.20(a) *Third partial sum of the Fourier series of $f(x) = x^2$ on $[-2, 2]$.*

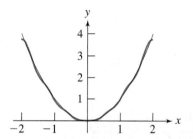

FIGURE 13.20(b) *Sixth partial sum of the Fourier series of $f(x) = x^2$ on $[-2, 2]$.*

This series converges to x^2 for $-2 \le x \le 2$. In this case, $f(2-) = f(-2+) = 4$, so

$$\frac{1}{2}(f(2-) + f(-2+)) = \frac{1}{2}(4 + 4) = 4 = f(2) = f(-2),$$

so the Fourier series converges to $f(x)$ on the entire interval $[-2, 2]$. Figures 13.20(a) and (b) show the third and sixth partial sums of this series compared with the graph of f. ▪

EXAMPLE 13.15

Let $f(x) = x^3$ for $-1 \le x \le 1$. The Fourier series is

$$\sum_{n=1}^{\infty} \left[2 \frac{(n^2\pi^2 - 6)(-1)^{n+1}}{n^3\pi^3} \right] \sin(n\pi x).$$

This converges to x^3 for $-1 < x < 1$ and to 0 at $x = \pm 1$. The fourth and fifteenth partial sums of this series are compared to a graph of the function in Figures 13.21(a) and (b). ▪

EXAMPLE 13.16

Let $f(x) = \sin(x)$ for $-1 \le x \le 1$. The Fourier series of f on $[-1, 1]$ is

$$\sum_{n=1}^{\infty} 2 \frac{n\pi \sin(1)(-1)^{n+1}}{n^2\pi^2 - 1} \sin(n\pi x).$$

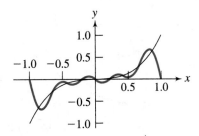

FIGURE 13.21(a) *Fourth partial sum of the Fourier series of $f(x) = x^3$ on $[-1, 1]$.*

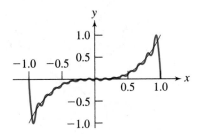

FIGURE 13.21(b) *Fifteenth partial sum of the Fourier series of $f(x) = x^3$ on $[-1, 1]$.*

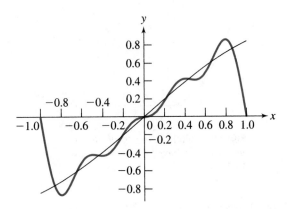

FIGURE 13.22(a) *Fourth partial sum of the Fourier series of $f(x) = \sin(x)$ for $-1 \le x \le 1$.*

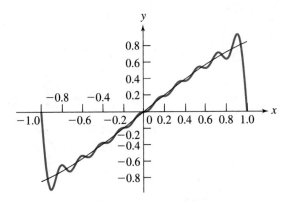

FIGURE 13.22(b) *Tenth partial sum of the function $f(x) = \sin(x)$ for $-1 \le x \le 1$.*

This series converges to

$$
\begin{cases}
\sin(x) & \text{for } -1 < x < 1 \\
0 & \text{for } x = 1 \quad \text{and} \quad \text{for } x = -1
\end{cases}.
$$

Figures 13.22(a) and (b) show partial sums of this series compared with the graph of f. ■

13.3.4 The Gibbs Phenomenon

In 1881 the Michelson-Morley experiment revolutionized physics and helped pave the way for Einstein's theory of general relativity. In a brilliant experiment using their adaptation of the interferometer, Michelson and Morley showed by careful measurements that the postulated "ether," which physicists at that time believed permeated all of space, had no effect on the velocity of light as seen from different directions.

Some years later Michelson was testing a mechanical device he had invented for computing Fourier coefficients and for constructing a function from its Fourier coefficients. In one test he used 80 Fourier coefficients for the function $f(x) = x$ for $-\pi \le x \le \pi$. The machine responded with a graph having unexpected jumps near the endpoints π and $-\pi$. At first Michelson assumed that there was some problem with his machine. Eventually, however, it was found that this behavior is characteristic of Fourier series at jump discontinuities of the function. This became known as

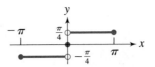

FIGURE 13.23 *Function illustrating the Gibbs phenomenon.*

the Gibbs phenomenon, after the Yale mathematician Josiah Willard Gibbs, who was the first to satisfactorily define and explain it. The phenomenon had also been noticed some 60 years before by the English mathematician Wilbraham, who was, however, unable to analyze it.

To illustrate the phenomenon, consider the function defined by

$$
f(x) = \begin{cases} -\pi/4 & \text{for } -\pi \leq x < 0 \\ 0 & \text{for } x = 0 \\ \pi/4 & \text{for } 0 < x \leq \pi \end{cases}.
$$

Figure 13.23 shows a graph of this function, whose Fourier series is

$$
\sum_{n=1}^{\infty} \frac{1}{2n - 1} \sin((2n - 1)x).
$$

By either convergence theorem, this series converges to $f(x)$ for $-\pi < x < \pi$. There is a jump discontinuity at 0, but

$$
\frac{1}{2}(f(0+) + f(0-)) = \frac{1}{2}\left(-\frac{\pi}{4} + \frac{\pi}{4}\right) = 0 = f(0).
$$

The Nth partial sum of this Fourier series is

$$
S_N(x) = \sum_{n=1}^{N} \frac{1}{2n - 1} \sin((2n - 1)x),
$$

and Figure 13.24 shows graphs of $S_5(x)$, $S_{14}(x)$, and $S_{22}(x)$. Each of these partial sums shows a peak near zero. Intuitively, since the partial sums approach $f(x)$ as $N \to \infty$, we might expect these peaks to flatten out and become smaller as N is chosen larger. But they don't. Instead, the peaks maintain roughly the same height, but move closer to the y axis as N increases. The partial sums do indeed have the function as a limit, but not in quite the way mathematicians expected.

As another example, consider

$$
f(x) = \begin{cases} 0 & \text{for } -2 \leq x < 0 \\ 2 - x & \text{for } 0 \leq x \leq 2 \end{cases}.
$$

This function has a jump discontinuity at 0, and Fourier series

$$
\frac{1}{2} + \sum_{n=1}^{\infty} \left(\frac{2}{n^2 \pi^2}(1 - (-1)^n)) \cos\left(\frac{n\pi x}{2}\right) + \frac{2}{n\pi} \sin\left(\frac{n\pi x}{2}\right) \right).
$$

Figure 13.25 shows the fourth, tenth, and twenty-fifth partial sum of this series. Again, the Gibbs phenomenon shows up at the jump discontinuity. Gibbs showed that this behavior occurs in the Fourier series of a function at every point where it has a jump discontinuity.

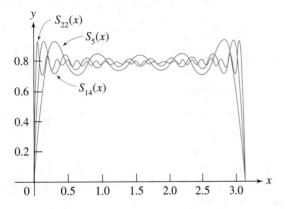

FIGURE 13.24   *Partial sums (for $0 \leq x \leq \pi/4$) showing the Gibbs phenomenon for the function of Figure 13.23.*

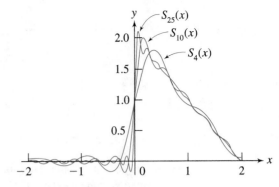

FIGURE 13.25   *Fourth, tenth, and twenty-fifth partial sums of the Fourier series of*

$$f(x) = \begin{cases} 0 & \text{for } -2 \leq x < 0 \\ 2 - x & \text{for } 0 \leq x \leq 2 \end{cases}.$$

SECTION 13.3   PROBLEMS

In each of Problems 1 through 15, use a convergence theorem to determine the sum of the Fourier series of the function on the interval. Whichever theorem is used, verify that the hypotheses are satisfied, assuming familiar facts from calculus about continuous and differentiable functions. It is not necessary to write the series itself to do this.

Next, find the Fourier series of the function and graph f and, for $N = 5, 10, 15, 25$, graph the Nth partial sum of the series, together with the function on the interval. Point out any places where the Gibbs phenomenon is apparent in these graphs.

1. $f(x) = \begin{cases} 2x & \text{for } -3 \leq x < -2 \\ 0 & \text{for } -2 \leq x < 1 \\ x^2 & \text{for } 1 \leq x \leq 3 \end{cases}$

2. $f(x) = x^2$ for $-2 \leq x \leq 2$

3. $f(x) = x^2 e^{-x}$ for $-3 \leq x \leq 3$

4. $f(x) = \begin{cases} 2x - 2 & \text{for } -\pi \leq x \leq 1 \\ 3 & \text{for } 1 < x \leq \pi \end{cases}$

5. $f(x) = \begin{cases} x^2 & \text{for } -\pi \leq x \leq 0 \\ 2 & \text{for } 0 < x \leq \pi \end{cases}$

6. $f(x) = \begin{cases} \cos(x) & \text{for } -2 \leq x < 0 \\ \sin(x) & \text{for } 0 \leq x \leq 2 \end{cases}$

7. $f(x) = \begin{cases} -1 & \text{for } -4 \leq x < 0 \\ 1 & \text{for } 0 \leq x \leq 4 \end{cases}$

8. $f(x) = \begin{cases} 0 & \text{for } -1 \leq x < \dfrac{1}{2} \\ 1 & \text{for } \dfrac{1}{2} \leq x \leq \dfrac{3}{4} \\ 2 & \text{for } \dfrac{3}{4} < x \leq 1 \end{cases}$

9. $f(x) = e^{-|x|}$ for $-\pi \leq x \leq \pi$

10. $f(x) = \begin{cases} -2 & \text{for } -4 \leq x \leq -2 \\ 1 + x^2 & \text{for } -2 < x \leq 2 \\ 0 & \text{for } 2 < x \leq 4 \end{cases}$

11. $f(x) = \begin{cases} -3x & \text{for } -\pi \leq x \leq 0 \\ 5 & \text{for } 0 < x \leq \pi \end{cases}$

12. $f(x) = \cos(4x)$ for $-\pi \leq x \leq \pi$

13. $f(x) = |x|$ for $-5 \leq x \leq 5$

14. $f(x) = \begin{cases} \cos(\pi x) & \text{for } -1 \leq x < 0 \\ -2 & \text{for } 0 \leq x \leq 1 \end{cases}$

15. $f(x) = \begin{cases} -x^2 & \text{for } -3 \leq x < 1 \\ 1 + x & \text{for } 1 < x \leq 3 \end{cases}$

16. Let $f(x) = x^2/2$ for $-\pi \leq x \leq \pi$. Find the Fourier series of $f(x)$ and evaluate it at an appropriately chosen value of x to sum the series $\sum_{n=1}^{\infty} 1/n^2$.

17. Use the Fourier series of Problem 16 to sum the series $\sum_{n=1}^{\infty} (-1)^n/n^2$.

13.4 Fourier Cosine and Sine Series

If $f(x)$ is defined on $[-L, L]$, we may be able to write its Fourier series. The coefficients of this series are completely determined by the function and the interval.

We will now show that if $f(x)$ is defined on the half-interval $[0, L]$, then we have a choice and can write a series containing just cosines or just sines in attempting to represent $f(x)$ on this half-interval.

13.4.1 The Fourier Cosine Series of a Function

Let f be integrable on $[0, L]$. We want to expand $f(x)$ in a series of cosine functions.

We already have the means to do this. Figure 13.26 shows a graph of a typical f. Fold this graph across the y axis to obtain a function f_e defined for $-L \leq x \leq L$:

$$f_e(x) = \begin{cases} f(x) & \text{for } 0 \leq x \leq L \\ f(-x) & \text{for } -L \leq x < 0 \end{cases}.$$

f_e is an even function,

$$f_e(-x) = f(x),$$

and agrees with f on $[0, L]$,

$$f_e(x) = f(x) \quad \text{for } 0 \leq x \leq L.$$

We call f_e the *even extension* of f to $[-L, L]$.

EXAMPLE 13.17

Let $f(x) = e^x$ for $0 \leq x \leq 2$. Then

$$f_e(x) = \begin{cases} e^x & \text{for } 0 \leq x \leq 2 \\ e^{-x} & \text{for } -2 \leq x < 0 \end{cases}.$$

Here we put $f_e(x) = f(-x) = e^{-x}$ for $-2 \leq x < 0$. A graph of f_e is given in Figure 13.27. ∎

Because f_e is an even function on $[-L, L]$, its Fourier series on $[-L, L]$ is

$$\frac{1}{2}a_0 + \sum_{n=1}^{\infty} a_n \cos\left(\frac{n\pi x}{L}\right), \tag{13.14}$$

in which

$$a_n = \frac{2}{L}\int_0^L f_e(x) \cos\left(\frac{n\pi x}{L}\right) dx = \frac{2}{L}\int_0^L f(x) \cos\left(\frac{n\pi x}{L}\right) dx, \tag{13.15}$$

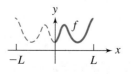

FIGURE 13.26 *Even extension of f to $[-L, L]$.*

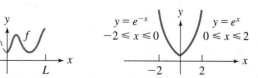

FIGURE 13.27

since $f_e(x) = f(x)$ for $0 \leq x \leq L$. We call the series (13.14) the *Fourier cosine series of f on* $[0, L]$. The coefficients (13.15) are the *Fourier cosine coefficients of f on* $[0, L]$.

The even extension f_e was introduced only to be able to make use of earlier work to derive a series containing just cosines. When we actually write a Fourier cosine series, we just use equation (13.15) to calculate the coefficients, without defining f_e.

The other point to having f_e in the background, however, is that we can use the Fourier convergence theorems to write a convergence theorem for cosine series.

THEOREM 13.3   *Convergence of Fourier Cosine Series*

Let f be piecewise continuous on $[0, L]$. Then,

1. If $0 < x < L$, and f has left and right derivatives at x, then the Fourier cosine series for $f(x)$ on $[0, L]$ converges at x to

$$\frac{1}{2}(f(x-) + f(x+)).$$

2. If f has a right derivative at 0, then the Fourier cosine series for $f(x)$ on $[0, L]$ converges at 0 to $f(0+)$.

3. If f has a left derivative at L, then the Fourier cosine series for $f(x)$ on $[0, L]$ converges at L to $f(L-)$. ∎

Conclusions (2) and (3) follow from Theorem 13.2, applied to f_e. Consider first $x = 0$. The Fourier series of f_e converges at 0 to

$$\frac{1}{2}(f_e(0-) + f_e(0+)).$$

But

$$f_e(0+) = f(0+)$$

and

$$f_e(0-) = f(0+),$$

so at 0 the series converges to

$$\frac{1}{2}(f(0+) + f(0+)) = f(0+).$$

A similar argument proves conclusion (3).

EXAMPLE 13.18

Let $f(x) = e^{2x}$ for $0 \leq x \leq 1$. We will write the Fourier cosine series of f. Compute

$$a_0 = 2 \int_0^1 e^{2x}\, dx = e^2 - 1$$

and

$$a_n = 2 \int_0^1 e^{2x} \cos(n\pi x)\, dx$$

$$= 4\frac{e^2(-1)^n - 1}{4 + n^2\pi^2}.$$

The cosine expansion of f is

$$\frac{1}{2}(e^2 - 1) + \sum_{n=1}^{\infty} 4\frac{e^2(-1)^n - 1}{4 + n^2\pi^2} \cos(n\pi x).$$

This series converges to

$$\begin{cases} e^{2x} & \text{for } 0 < x < 1 \\ 1 & \text{for } x = 0 \\ e^2 & \text{for } x = 1 \end{cases}.$$

Thus this cosine series converges to e^{2x} for $0 \le x \le 1$. Figures 13.28(a) and (b) show a graph of f compared with the fifth and tenth partial sums of this cosine expansion, respectively. ■

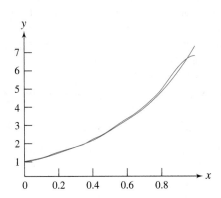

FIGURE 13.28(a) *Fifth partial sum of the cosine expansion of e^{2x} on $[0, 1]$.*

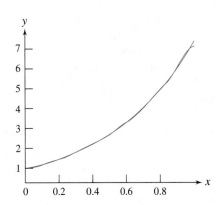

FIGURE 13.28(b) *Tenth partial sum of the cosine expansion of e^{2x} on $[0, 1]$.*

13.4.2 The Fourier Sine Series of a Function

By duplicating the strategy just used for writing a cosine series, except now extending f to an odd function f_o over $[-L, L]$, we can write a Fourier sine series for $f(x)$ on $[0, L]$. In particular, if $f(x)$ is defined on $[0, L]$, let

$$f_o(x) = \begin{cases} f(x) & \text{for } 0 \le x \le L \\ -f(-x) & \text{for } -L \le x < 0 \end{cases}.$$

Then f_o is an odd function, and $f_o(x) = f(x)$ for $0 \le x \le L$. This is the *odd extension* of f to $[-L, L]$. For example, if $f(x) = e^{2x}$ for $0 \le x \le 1$, let

$$f_o(x) = \begin{cases} e^{2x} & \text{for } 0 \le x \le 1 \\ -e^{-2x} & \text{for } -1 \le x < 0 \end{cases}$$

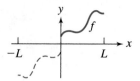

FIGURE 13.29 *Odd extension of f to* $[-L, L]$.

This amounts to folding the graph of f over the vertical axis, then over the horizontal axis (Figure 13.29).

Now write the Fourier series for $f_o(x)$ on $[-L, L]$. By equations (13.11) and (13.12), the Fourier series of f_o is

$$\sum_{n=1}^{\infty} b_n \sin\left(\frac{n\pi x}{L}\right) \tag{13.16}$$

with coefficients

$$b_n = \frac{2}{L} \int_0^L f_o(x) \sin\left(\frac{n\pi x}{L}\right) dx = \frac{2}{L} \int_0^L f(x) \sin\left(\frac{n\pi x}{L}\right) dx. \tag{13.17}$$

We call the series (13.16) the *Fourier sine series* of f on $[0, L]$. The coefficients given by equation (13.17) are the Fourier sine coefficients of f on $[0, L]$. As with cosine series, we do not need to explicitly make the extension to f_o to write the Fourier sine series for f on $[0, L]$.

Again, as with the cosine expansion, we can write a convergence theorem for sine series using the convergence theorem for Fourier series.

THEOREM 13.4   *Convergence of Fourier Sine Series*

Let f be piecewise continuous on $[0, L]$. Then,

1. If $0 < x < L$, and f has left and right derivatives at x, then the Fourier sine series for $f(x)$ on $[0, L]$ converges at x to

$$\frac{1}{2}(f(x-) + f(x+)).$$

2. At 0 and at L, the Fourier sine series for $f(x)$ on $[0, L]$ converges to 0. ∎

Conclusion (2) is immediate because each term of the sine series (13.16) is zero for $x = 0$ and for $x = L$.

EXAMPLE 13.19

Let $f(x) = e^{2x}$ for $0 \le x \le 1$. We will write the Fourier sine series of f on $[0, 1]$. The coefficients are

$$b_n = 2 \int_0^1 e^{2x} \sin(n\pi x)\, dx$$

$$= 2\frac{n\pi(1 - (-1)^n e^2)}{4 + n^2\pi^2}.$$

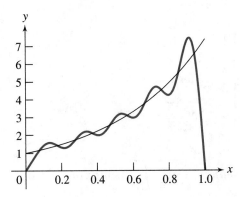

FIGURE 13.30(a) *Tenth partial sum of the sine expansion of e^{2x} on $[0, 1]$.*

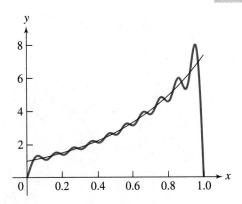

FIGURE 13.30(b) *Fortieth partial sum of the sine expansion of e^{2x} on $[0, 1]$.*

The sine series is

$$\sum_{n=1}^{\infty} 2\frac{n\pi(1 - (-1)^n e^2)}{4 + n^2\pi^2} \sin(n\pi x).$$

This series converges to e^{2x} for $0 < x < 1$, and to zero for $x = 0$ and for $x = 1$. Figures 13.30(a) and (b) show graphs of the tenth and fortieth partial sums of this series. ∎

EXAMPLE 13.20

Let

$$f(x) = \begin{cases} 1 & \text{for } 0 \leq x \leq \pi/2 \\ 2 & \text{for } \pi/2 < x \leq \pi \end{cases}.$$

We will write both the cosine and sine expansions of $f(x)$ on $[0, \pi]$.

For the cosine expansion, compute

$$a_0 = \frac{2}{\pi} \int_0^{\pi/2} dx + \frac{2}{\pi} \int_{\pi/2}^{\pi} 2\, dx = 3$$

and

$$a_n = \frac{2}{\pi} \int_0^{\pi/2} \cos(nx)\, dx + \frac{2}{\pi} \int_{\pi/2}^{\pi} 2\cos(nx)\, dx$$

$$= -\frac{2}{\pi} \frac{\sin\left(\frac{n\pi}{2}\right)}{n}.$$

The cosine series is

$$\frac{3}{2} - \sum_{n=1}^{\infty} \frac{2}{\pi} \frac{\sin\left(\frac{n\pi}{2}\right)}{n} \cos(nx).$$

This series converges to

$$
\begin{cases}
1 & \text{for } 0 \le x < \dfrac{\pi}{2} \\[2mm]
\dfrac{3}{2} & \text{for } x = \dfrac{\pi}{2} \\[2mm]
2 & \text{for } \dfrac{\pi}{2} < x \le \pi
\end{cases}
\;.
$$

For the sine series, compute

$$
b_n = \frac{2}{\pi} \int_0^{\pi/2} \sin(nx)\, dx + \frac{2}{\pi} \int_{\pi/2}^{\pi} 2 \sin(nx)\, dx
$$

$$
= -\frac{2}{\pi} \frac{\cos\left(\dfrac{n\pi}{2}\right) - 1}{n} + \frac{4}{\pi} \frac{-\cos(n\pi) + \cos\left(\dfrac{n\pi}{2}\right)}{n}
$$

$$
= \frac{2}{n\pi}\left(\cos\left(\frac{n\pi}{2}\right) + 1 - 2(-1)^n\right).
$$

The sine series is

$$
\sum_{n=1}^{\infty} \frac{2}{n\pi}\left(\cos\left(\frac{n\pi}{2}\right) + 1 - 2(-1)^n\right) \sin(nx).
$$

This series converges to

$$
\begin{cases}
0 & \text{for } x = 0 \text{ and } x = \pi \\[2mm]
1 & \text{for } 0 < x < \dfrac{\pi}{2} \\[2mm]
\dfrac{3}{2} & \text{for } x = \dfrac{\pi}{2} \\[2mm]
2 & \text{for } \dfrac{\pi}{2} < x < \pi
\end{cases}
\;.
$$

Figure 13.31(a) shows the fortieth partial sum of this sine expansion, and Figure 13.31(b) the thirty-fifth partial sum of the cosine expansion. ∎

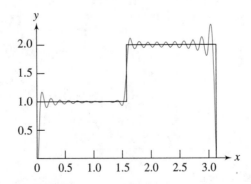

FIGURE 13.31(a) *Fortieth partial sum of the sine expansion of*

$$
f(x) = \begin{cases}
1 & \text{for } 0 \le x \le \pi/2 \\
2 & \text{for } \pi/2 < x \le \pi
\end{cases}
\;.
$$

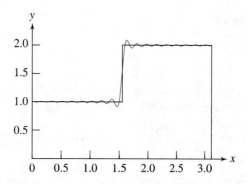

FIGURE 13.31(b) *Thirty-fifth partial sum of the cosine expansion of*

$$f(x) = \begin{cases} 1 & \text{for } 0 \le x \le \pi/2 \\ 2 & \text{for } \pi/2 < x \le \pi \end{cases}.$$

SECTION 13.4 PROBLEMS

In each of Problems 1 through 12, write the Fourier cosine series and the Fourier sine series of the function on the interval. Determine the sum of each series.

1. $f(x) = 4, 0 \le x \le 3$

2. $f(x) = \begin{cases} 1 & \text{for } 0 \le x \le 1 \\ -1 & \text{for } 1 < x \le 2 \end{cases}$

3. $f(x) = \begin{cases} 0 & \text{for } 0 \le x < \pi \\ \cos(x) & \text{for } \pi \le x \le 2\pi \end{cases}$

4. $f(x) = 2x$ for $0 \le x \le 1$

5. $f(x) = x^2$ for $0 \le x \le 2$

6. $f(x) = e^{-x}$ for $0 \le x \le 1$

7. $f(x) = \begin{cases} x & \text{for } 0 \le x \le 2 \\ 2 - x & \text{for } 2 < x \le 3 \end{cases}$

8. $f(x) = \begin{cases} 1 & \text{for } 0 \le x < 1 \\ 0 & \text{for } 1 \le x \le 3 \\ -1 & \text{for } 3 < x \le 5 \end{cases}$

9. $f(x) = \begin{cases} x^2 & \text{for } 0 \le x < 1 \\ 1 & \text{for } 1 \le x \le 4 \end{cases}$

10. $f(x) = 1 - x^3$ for $0 \le x \le 2$

11. $f(x) = \begin{cases} 4x & \text{for } 0 \le x \le 2 \\ -3 & \text{for } 2 < x < 4 \\ 1 & \text{for } 4 \le x \le 7 \end{cases}$

12. $f(x) = \sin(3x)$ for $0 \le x \le \pi$

13. Let $f(x)$ be defined on $[-L, L]$. Prove that f can be written as the sum of an even and an odd function on this interval.

14. Find all functions defined on $[-L, L]$ that are both even and odd.

15. Find the sum of the series $\sum_{n=1}^{\infty}(-1)^n/(4n^2-1)$. *Hint:* Expand $\sin(x)$ in a cosine series on $[0, \pi]$ and choose an appropriate value of x.

13.5 Integration and Differentiation of Fourier Series

In this section we will take a closer look at Fourier coefficients and consider term-by-term differentiation and integration of Fourier series.

Differentiation of Fourier series term-by-term generally leads to absurd results, even for extremely well behaved functions. Consider, for example, $f(x) = x$ for $-\pi \le x \le \pi$. The

Fourier series is

$$\sum_{n=1}^{\infty} \frac{2}{n}(-1)^{n+1} \sin(nx),$$

which converges to x for $-\pi < x < \pi$. Of course, $f'(x) = 1$ for $-\pi < x < \pi$, so f is piecewise smooth. However, if we differentiate the Fourier series term-by-term, we get

$$\sum_{n=1}^{\infty} 2(-1)^{n+1} \cos(nx),$$

which does not even converge on $(-\pi, \pi)$. The term-by-term derivative of this Fourier series is unrelated to the derivative of $f(x)$.

Integration of Fourier series has better prospects.

THEOREM 13.5   *Integration of Fourier Series*
_____

Let f be piecewise continuous on $[-L, L]$, with Fourier series

$$\frac{1}{2}a_0 + \sum_{n=1}^{\infty} a_n \cos\left(\frac{n\pi x}{L}\right) + b_n \sin\left(\frac{n\pi x}{L}\right).$$

Then, for any x with $-L \le x \le L$,

$$\int_{-L}^{x} f(t)\,dt = \frac{1}{2}a_0(x + L) + \frac{L}{\pi}\sum_{n=1}^{\infty} \frac{1}{n}\left[a_n \sin\left(\frac{n\pi x}{L}\right) - b_n\left(\cos\left(\frac{n\pi x}{L}\right) - (-1)^n\right)\right]. \quad \blacksquare$$

The expression on the right in this equation is exactly what we get by integrating the Fourier series term-by-term, from $-L$ to x. This means that for any piecewise continuous function, we can integrate f from $-L$ to x by integrating its Fourier series term-by-term. This holds even if the Fourier series does not converge to $f(x)$ at this particular x (for example, f might have a jump discontinuity at x).

Proof   Define

$$F(x) = \int_{-L}^{x} f(t)\,dt - \frac{1}{2}a_0 x$$

for $-L \le x \le L$. Then F is continuous on $[-L, L]$ and $F(L) = F(-L) = La_0/2$. Further, $F'(x) = f(x) - \frac{1}{2}a_0$ at every point of $[-L, L]$ where f is continuous. Hence F' is piecewise continuous on $[-L, L]$. Therefore, the Fourier series of $F(x)$ converges to $F(x)$ on $[-L, L]$:

$$F(x) = \frac{1}{2}A_0 + \sum_{n=1}^{\infty} A_n \cos\left(\frac{n\pi x}{L}\right) + B_n \sin\left(\frac{n\pi x}{L}\right), \tag{13.18}$$

in which we will use uppercase letters for Fourier coefficients of F and lowercase letters for those of f. Now compute the A_n's and B_n's for $n = 1, 2, \ldots$ by integrating by parts. First,

$$A_n = \frac{1}{L}\int_{-L}^{L} F(t)\cos\left(\frac{n\pi t}{L}\right)dt$$

$$= \frac{1}{L}\left[F(t)\frac{L}{n\pi}\sin\left(\frac{n\pi t}{L}\right)\right]_{-L}^{L} - \frac{1}{L}\int_{-L}^{L}\frac{L}{n\pi}\sin\left(\frac{n\pi t}{L}\right)F'(t)\,dt$$

$$= -\frac{1}{n\pi} \int_{-L}^{L} \left(f(t) - \frac{1}{2}a_0 \right) \sin\left(\frac{n\pi t}{L}\right) dt$$

$$= -\frac{1}{n\pi} \int_{-L}^{L} f(t) \sin\left(\frac{n\pi t}{L}\right) dt + \frac{1}{2n\pi}a_0 \int_{-L}^{L} \sin\left(\frac{n\pi t}{L}\right) dt$$

$$= -\frac{L}{n\pi}b_n,$$

in which b_n is the sine coefficient in the Fourier series of f on $[-L, L]$. Similarly,

$$B_n = \frac{1}{L} \int_{-L}^{L} F(t) \sin\left(\frac{n\pi t}{L}\right) dt$$

$$= \frac{1}{L} \left[F(t) \left(-\frac{L}{n\pi} \cos\left(\frac{n\pi t}{L}\right) \right) \right]_{-L}^{L} - \frac{1}{L} \int_{-L}^{L} F'(t) \left(-\frac{L}{n\pi} \right) \cos\left(\frac{n\pi t}{L}\right) dt$$

$$= \frac{1}{n\pi} \int_{-L}^{L} \left(f(t) - \frac{1}{2}a_0 \right) \cos\left(\frac{n\pi t}{L}\right) dt$$

$$= \frac{1}{n\pi} \int_{-L}^{L} f(t) \cos\left(\frac{n\pi t}{L}\right) dt - \frac{1}{2n\pi}a_0 \int_{-L}^{L} \cos\left(\frac{n\pi t}{L}\right) dt$$

$$= \frac{L}{n\pi}a_n.$$

Therefore, the Fourier series of F is

$$F(x) = \frac{1}{2}A_0 + \frac{L}{n\pi} \sum_{n=1}^{\infty} \left(-b_n \cos\left(\frac{n\pi x}{L}\right) + a_n \sin\left(\frac{n\pi x}{L}\right) \right)$$

for $-L \le x \le L$. Now we must determine A_0. But

$$F(L) = \frac{L}{2}a_0 = \frac{1}{2}A_0 - \frac{L}{n\pi} \sum_{n=1}^{\infty} b_n \cos(n\pi)$$

$$= \frac{1}{2}A_0 - \frac{L}{n\pi} \sum_{n=1}^{\infty} b_n (-1)^n.$$

This gives us

$$A_0 = La_0 + \frac{2L}{n\pi} \sum_{n=1}^{\infty} b_n (-1)^n.$$

Upon substituting these expressions for A_0, A_n, and B_n into the series (13.18), we obtain the conclusion of the theorem.

EXAMPLE 13.21

Let $f(x) = x$ for $-\pi \le x \le \pi$. This function is continuous on $[-\pi, \pi]$, and its Fourier series is

$$\sum_{n=1}^{\infty} \frac{2}{n}(-1)^{n+1} \sin(nx).$$

We have seen that we get nonsense if we differentiate this series term-by-term. However, we can integrate it term-by-term to obtain, for any x in $[-\pi, \pi]$,

$$\int_{-\pi}^{x} t \, dt = \frac{1}{2}(x^2 - \pi^2)$$

$$= \sum_{n=1}^{\infty} \frac{2}{n}(-1)^{n+1} \int_{-\pi}^{x} \sin(nt) \, dt$$

$$= \sum_{n=1}^{\infty} \frac{2}{n}(-1)^{n+1} \left[-\frac{1}{n}\cos(nx) + \frac{1}{n}\cos(n\pi) \right]$$

$$= \sum_{n=1}^{\infty} \frac{2}{n^2}(-1)^{n} \left[\cos(nx) - (-1)^{n} \right]. \ \blacksquare$$

With stronger conditions on f, we can derive a result on term-by-term differentiation of Fourier series.

THEOREM 13.6 *Differentiation of Fourier Series*
_____

Let f be continuous on $[-L, L]$ and suppose $f(L) = f(-L)$. Let f' be piecewise continuous on $[-L, L]$. Then $f(x)$ equals its Fourier series for $-L \leq x \leq L$,

$$f(x) = \frac{1}{2}a_0 + \sum_{n=1}^{\infty} a_n \cos\left(\frac{n\pi x}{L}\right) + b_n \sin\left(\frac{n\pi x}{L}\right),$$

and, at each point in $(-L, L)$ where $f''(x)$ exists,

$$f'(x) = \sum_{n=1}^{\infty} \frac{n\pi}{L}\left(-na_n \sin\left(\frac{\pi x}{L}\right) + b_n \cos\left(\frac{n\pi x}{L}\right)\right). \ \blacksquare$$

We leave a proof of this to the student. The idea is to write the Fourier series of $f'(x)$, noting that this Fourier series converges to $f'(x)$ wherever $f''(x)$ exists. Use integration by parts, as in the proof of Theorem 13.5, to relate the Fourier coefficients of $f'(x)$ to those of $f(x)$.

EXAMPLE 13.22
_____

Let $f(x) = x^2$ for $-2 \leq x \leq 2$. The hypotheses of Theorem 13.6 are satisfied. The Fourier series of f on $[-2, 2]$ is

$$f(x) = \frac{4}{3} + \frac{16}{\pi^2} \sum_{n=1}^{\infty} \frac{(-1)^n}{n^2} \cos\left(\frac{n\pi x}{2}\right),$$

with equality between $f(x)$ and its Fourier series. Because $f'(x) = 2x$ is continuous, and $f''(x) = 2$ exists throughout the interval, then for $-2 < x < 2$,

$$f'(x) = 2x = \frac{8}{\pi} \sum_{n=1}^{\infty} \frac{(-1)^{n+1}}{n} \sin\left(\frac{n\pi x}{2}\right).$$

For example, putting $x = 1$, we get

$$\frac{8}{\pi} \sum_{n=1}^{\infty} \frac{(-1)^{n+1}}{n} \sin\left(\frac{n\pi}{2}\right) = 2,$$

or

$$\sum_{n=1}^{\infty} \frac{(-1)^{n+1}}{n} \sin\left(\frac{n\pi}{2}\right) = \frac{\pi}{4}.$$

Manipulations on Fourier series can sometimes be used to sum series such as this. ▣

We now have conditions under which we can differentiate or integrate a Fourier series term-by-term. We will next consider conditions sufficient for a Fourier series to converge uniformly. First we will derive a set of important inequalities for Fourier coefficients, called *Bessel's inequalities*.

THEOREM 13.7 *Bessel's Inequalities*

Let f be integrable on $[0, L]$. Then

1. The coefficients in the Fourier sine expansion of f on $[0, L]$ satisfy

$$\sum_{n=1}^{\infty} b_n^2 \leq \frac{2}{L} \int_{-L}^{L} f(x)^2 \, dx.$$

2. The coefficients in the Fourier cosine expansion of f on $[0, L]$ satisfy

$$\frac{1}{2} a_0^2 + \sum_{n=1}^{\infty} a_n^2 \leq \frac{2}{L} \int_0^L f(x)^2 \, dx.$$

3. If f is integrable on $[-L, L]$, then the Fourier coefficients of f on $[-L, L]$ satisfy

$$\frac{1}{2} a_0^2 + \sum_{n=1}^{\infty} (a_n^2 + b_n^2) \leq \frac{1}{L} \int_{-L}^{L} f(x)^2 \, dx. \quad ▪$$

In particular, the sum of the squares of the (sine, cosine, or Fourier series) coefficients of f converges. We will prove (1), which is notationally simpler than the other two inequalities but contains the idea of the argument.

Proof   Since $\int_0^L f(x) \, dx$ exists, we can compute the Fourier sine coefficients and write the sine series

$$\sum_{n=1}^{\infty} b_n \sin\left(\frac{n\pi x}{L}\right),$$

where

$$b_n = \frac{2}{L} \int_0^L f(x) \sin\left(\frac{n\pi x}{L}\right) dx.$$

The Nth partial sum of this series is

$$S_N(x) = \sum_{n=1}^{N} b_n \sin\left(\frac{n\pi x}{L}\right).$$

Now consider

$$0 \le \int_0^L (f(x) - S_N(x))^2 \, dx$$

$$= \int_0^L f(x)^2 \, dx - 2 \int_0^L f(x) S_N(x) \, dx + \int_0^L S_N(x)^2 \, dx$$

$$= \int_0^L f(x)^2 \, dx - 2 \int_0^L f(x) \left(\sum_{n=1}^N b_n \sin\left(\frac{n\pi x}{L}\right) \right) dx$$

$$+ \int_0^L \left(\sum_{n=1}^N b_n \sin\left(\frac{n\pi x}{L}\right) \right) \left(\sum_{m=1}^N b_m \sin\left(\frac{m\pi x}{L}\right) \right) dx$$

$$= \int_0^L f(x)^2 \, dx - 2 \sum_{n=1}^N b_n \int_0^L f(x) \sin\left(\frac{n\pi x}{L}\right) dx$$

$$+ \sum_{n=1}^N \sum_{m=1}^N b_n b_m \int_0^L \sin\left(\frac{n\pi x}{L}\right) \sin\left(\frac{m\pi x}{L}\right) dx$$

$$= \int_0^L f(x)^2 \, dx - \sum_{n=1}^N b_n (L b_n) + \sum_{n=1}^N b_n b_n \frac{L}{2},$$

in which we have used the fact that

$$\int_0^L \sin\left(\frac{n\pi x}{L}\right) \sin\left(\frac{m\pi x}{L}\right) dx = \begin{cases} 0 & \text{if } n \ne m \\ L/2 & \text{if } n = m \end{cases}.$$

We therefore have

$$0 \le \int_0^L f(x)^2 \, dx - L \sum_{n=1}^N b_n^2 + \frac{L}{2} \sum_{n=1}^N b_n^2,$$

or

$$\sum_{n=1}^N b_n^2 \le \frac{2}{L} \int_0^L f(x)^2 \, dx.$$

Since the right side is independent of N, we can let $N \to \infty$ to get

$$\sum_{n=1}^\infty b_n^2 \le \frac{2}{L} \int_0^L f(x)^2 \, dx,$$

proving conclusion (1). Conclusions (2) and (3) have similar proofs.

EXAMPLE 13.23

We will use Bessel's inequality to derive an upper bound for an infinite series. Let $f(x) = x^2$ for $-\pi \le x \le \pi$. The Fourier series of f converges to $f(x)$ for all x in $[-\pi, \pi]$:

$$x^2 = \frac{1}{3}\pi^2 + \sum_{n=1}^\infty 4\frac{(-1)^n}{n^2} \cos(nx).$$

Here $a_0 = 2\pi^2/3$, $a_n = 4(-1)^n/n^2$ and $b_n = 0$ (x^2 is an even function). By Bessel's inequality (3) of Theorem 13.7,

$$\frac{1}{2}\left(\frac{2\pi^2}{3}\right)^2 + \sum_{n=1}^{\infty}\left(\frac{4(-1)^n}{n^2}\right)^2 \leq \frac{1}{\pi}\int_{-\pi}^{\pi} x^4\, dx = \frac{2}{5}\pi^4.$$

Then

$$16\sum_{n=1}^{\infty}\frac{1}{n^4} \leq \left(\frac{2}{5}-\frac{2}{9}\right)\pi^4 = \frac{8\pi^4}{45},$$

so

$$\sum_{n=1}^{\infty}\frac{1}{n^4} \leq \frac{\pi^4}{90},$$

which is approximately 1.0823232. ∎

Using Bessel's inequality for coefficients in a Fourier expansion on $[-L, L]$, we can prove a result on uniform convergence of Fourier series.

THEOREM 13.8 *Uniform and Absolute Convergence of Fourier Series*

Let f be continuous on $[-L, L]$ and let f' be piecewise continuous. Suppose $f(-L) = f(L)$. Then, the Fourier series of f on $[-L, L]$ converges absolutely and uniformly to $f(x)$ on $[-L, L]$.

Proof Denote the Fourier coefficients of f by lowercase letters and those of f' by uppercase. Then

$$A_0 = \frac{1}{L}\int_{-L}^{L} f'(\xi)\, d\xi = \frac{1}{L}(f(L) - f(-L)) = 0.$$

For positive integer n, we find by integration by parts, as in the proof of Theorem 13.5, that

$$A_n = \frac{n\pi}{L}b_n \quad \text{and} \quad B_n = -\frac{n\pi}{L}a_n.$$

Now

$$0 \leq \left(|A_n| - \frac{1}{n}\right)^2 = A_n^2 - \frac{2}{n}|A_n| + \frac{1}{n^2}$$

and, similarly,

$$0 \leq B_n^2 - \frac{2}{n}|B_n| + \frac{1}{n^2}.$$

Then

$$\frac{1}{n}|A_n| + \frac{1}{n}|B_n| \leq \frac{1}{2}\left(A_n^2 + B_n^2\right) + \frac{1}{n^2}.$$

Therefore,

$$\frac{\pi}{L}|a_n| + \frac{\pi}{L}|b_n| \leq \frac{1}{2}\left(A_n^2 + B_n^2\right) + \frac{1}{n^2};$$

hence

$$|a_n| + |b_n| \leq \frac{L}{2\pi}\left(A_n^2 + B_n^2\right) + \frac{L}{\pi}\frac{1}{n^2}.$$

Now $\sum_{n=1}^{\infty}(1/n^2)$ converges, and $\sum_{n=1}^{\infty}\left(A_n^2 + B_n^2\right)$ converges by applying Bessel's inequality to the Fourier coefficients of f'. Therefore, by comparison, $\sum_{n=1}^{\infty}\left(|a_n| + |b_n|\right)$ converges also.

But, for $-L \leq x \leq L$,

$$\left|a_n \cos\left(\frac{n\pi x}{L}\right) + b_n \sin\left(\frac{n\pi x}{L}\right)\right| \leq |a_n| + |b_n|.$$

By a theorem of Weierstrass, this implies that the Fourier series of f converges uniformly on $[-L, L]$. Further, the convergence is absolute, since the series of absolute values of terms of the series converges. Finally, by the Fourier convergence theorem, the Fourier series of f converges to $f(x)$ on $[-L, L]$. This completes the proof. ∎

EXAMPLE 13.24

Let $f(x) = e^{-|x|}$ for $-1 \leq x \leq 1$. Then

$$f(x) = \begin{cases} e^x & \text{for } -1 \leq x < 0 \\ e^{-x} & \text{for } 0 \leq x \leq 1 \end{cases}.$$

f is continuous on $[-1, 1]$, and

$$f'(x) = \begin{cases} -e^{-x} & \text{for } 0 < x \leq 1 \\ e^x & \text{for } -1 \leq x < 0 \end{cases}.$$

f has no derivative at $x = 0$, which is a cusp of the graph (Figure 13.32). Thus f' is piecewise continuous on $[-1, 1]$. Finally, $f(1) = f(-1) = e^{-1}$. Therefore, the Fourier series of f converges uniformly and absolutely to $f(x)$ on $[-1, 1]$:

$$f(x) = 1 - e^{-1} + 2\sum_{n=1}^{\infty} \frac{1 - e^{-1}(-1)^n}{1 + \pi^2 n^2} \cos(n\pi x)$$

for $-1 \leq x \leq 1$.

We can integrate this series term-by-term. For example,

$$\int_{-1}^{x} f(t)\, dt = \int_{-1}^{x} (1 - e^{-1})\, dt + 2\sum_{n=1}^{\infty} \frac{1 - e^{-1}(-1)^n}{1 + \pi^2 n^2} \int_{-1}^{x} \cos(n\pi t)\, dt$$

$$= (1 - e^{-1})(x + 1) + 2\sum_{n=1}^{\infty} \frac{1 - e^{-1}(-1)^n}{1 + \pi^2 n^2} \frac{1}{n\pi} \sin(n\pi x).$$

This is a correct equation, but it is not a Fourier series (the right side includes the polynomial term $1 + x$). We may sometimes integrate a Fourier series term-by-term, and the result may be a convergent series, but not necessarily a Fourier series.

We can also differentiate the Fourier series for $f(x)$ term-by-term at any point in $(-1, 1)$ at which $f''(x)$ exists. Thus we can differentiate term-by-term for $-1 < x < 0$ and for $0 < x < 1$. For such x,

$$f'(x) = -2\sum_{n=1}^{\infty} \frac{1 - e^{-1}(-1)^n}{1 + \pi^2 n^2} n\pi \sin(n\pi x). ∎$$

We will conclude this section with Parseval's theorem. Recall that Bessel's inequality for Fourier coefficients on $[-L, L]$ requires only that we be able to compute these coefficients. If,

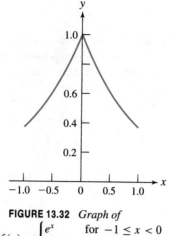

FIGURE 13.32 *Graph of*

$$f(x) = \begin{cases} e^x & \text{for } -1 \le x < 0 \\ e^{-x} & \text{for } 0 \le x \le 1 \end{cases}.$$

however, we place continuity conditions on the function, as in Theorem 13.8, then we turn Bessel's inequality into an equality.

THEOREM 13.9 *Parseval*

Let f be continuous on $[-L, L]$ and let f' be piecewise continuous. Suppose $f(-L) = f(L)$. Then the Fourier coefficients of f on $[-L, L]$ satisfy

$$\frac{1}{2}a_0^2 + \sum_{n=1}^{\infty}(a_n^2 + b_n^2) = \frac{1}{L}\int_{-L}^{L} f(x)^2\, dx.$$

Proof The Fourier series of f on $[-L, L]$ converges to $f(x)$ at each point of this interval:

$$f(x) = \frac{1}{2}a_0 + \sum_{n=1}^{\infty} a_n \cos\left(\frac{n\pi x}{L}\right) + b_n \sin\left(\frac{n\pi x}{L}\right).$$

Then

$$f(x)^2 = \frac{1}{2}a_0 f(x) + \sum_{n=1}^{\infty} a_n f(x) \cos\left(\frac{n\pi x}{L}\right) + b_n f(x) \sin\left(\frac{n\pi x}{L}\right).$$

We can integrate this Fourier series term-by-term, and multiplication of the series by the continuous function $f(x)$ does not change this. Therefore,

$$\int_{-L}^{L} f(x)^2\, dx = \frac{1}{2}a_0 \int_{-L}^{L} f(x)\, dx$$
$$+ \sum_{n=1}^{\infty} a_n \int_{-L}^{L} f(x) \cos\left(\frac{n\pi x}{L}\right) dx + b_n \int_{-L}^{L} f(x) \sin\left(\frac{n\pi x}{L}\right) dx.$$

Recalling the integral formulas for the Fourier coefficients, this equation can be written

$$\int_{-L}^{L} f(x)^2\, dx = \frac{1}{2}a_0 L a_0 + \sum_{n=1}^{\infty}(a_n L a_n + b_n L b_n),$$

and this is equivalent to the conclusion of the theorem. ∎

EXAMPLE 13.25

Parseval's theorem has various applications in deriving other properties of Fourier series. We will encounter it later when we discuss completeness of sets of eigenfunctions. However, one immediate use is in summing certain infinite series. To illustrate, the Fourier coefficients of $\cos(x/2)$ on $[-\pi, \pi]$ are

$$a_0 = \frac{1}{\pi} \int_{-\pi}^{\pi} \cos\left(\frac{x}{2}\right) dx = \frac{4}{\pi}$$

and

$$a_n = \frac{1}{\pi} \int_{-\pi}^{\pi} \cos\left(\frac{x}{2}\right) \cos(nx)\, dx = -\frac{4}{\pi} \frac{(-1)^n}{4n^2 - 1}.$$

By Parseval's theorem,

$$\frac{1}{2}\left(\frac{4}{\pi}\right)^2 + \sum_{n=1}^{\infty} \left(\frac{4}{\pi} \frac{(-1)^n}{4n^2 - 1}\right)^2 = \frac{1}{\pi} \int_{-\pi}^{\pi} \cos^2\left(\frac{x}{2}\right) dx = 1.$$

Then,

$$\sum_{n=1}^{\infty} \frac{1}{(4n^2 - 1)^2} = \frac{\pi^2 - 8}{16}. \blacksquare$$

SECTION 13.5   PROBLEMS

1. Prove Theorem 13.6. An argument can be formulated along the lines discussed following the statement of the theorem.

2. Let $f(x) = |x|$ for $-1 \le x \le 1$.

   (a) Write the Fourier series for $f(x)$ on $[-1, 1]$.

   (b) Show that this series can be differentiated term-by-term to yield the Fourier expansion of $f'(x)$ on $[-1, 1]$.

   (c) Determine $f'(x)$ and write its Fourier series on $[-1, 1]$. Compare this series with that obtained in (b).

3. Let $f(x) = \begin{cases} 0 & \text{for } -\pi \le x \le 0 \\ x & \text{for } 0 < x \le \pi. \end{cases}$

   (a) Write the Fourier series of $f(x)$ on $[-\pi, \pi]$ and show that this series converges to $f(x)$ on $(-\pi, \pi)$.

   (b) Show that this series can be integrated term-by-term.

   (c) Use the results of (a) and (b) to obtain a trigonometric series expansion for $\int_{-\pi}^{x} f(t)\, dt$ on $[-\pi, \pi]$.

4. Let $f(x) = x^2$ for $-3 \le x \le 3$.

   (a) Write the Fourier series for $f(x)$ on $[-3, 3]$.

   (b) Show that this series can be differentiated term-by-term and use this fact to obtain the Fourier expansion of $2x$ on $[-3, 3]$.

   (c) Write the Fourier series of $2x$ on $[-3, 3]$ by computation of the Fourier coefficients and compare the result with that of (b).

5. Let $f(x) = x\sin(x)$ for $-\pi \le x \le \pi$.

   (a) Write the Fourier series for $f(x)$ on $[-\pi, \pi]$.

   (b) Show that this series can be differentiated term-by-term and use this fact to obtain the Fourier expansion of $\sin(x) + x\cos(x)$ on $[-\pi, \pi]$.

   (c) Write the Fourier series of $\sin(x) + x\cos(x)$ on $[-\pi, \pi]$ by computation of the Fourier coefficients and compare the result with that of (b).

6. Carry out the proof of Bessel's inequality for Fourier cosine series on $[0, L]$.

7. Provide the details of a proof of Bessel's inequality for Fourier series on $[-L, L]$.

8. Let f and f' be piecewise continuous on $[-L, L]$. Use Bessel's inequality to show that

$$\lim_{n \to \infty} \int_{-L}^{L} f(x) \cos\left(\frac{n\pi x}{L}\right) dx$$

$$= \lim_{n \to \infty} \int_{-L}^{L} f(x) \sin\left(\frac{n\pi x}{L}\right) dx = 0.$$

This conclusion is known as *Riemann's lemma*.

13.6 The Phase Angle Form of a Fourier Series

A function is *periodic* with period p if $f(x + p) = f(x)$ for all real x. If a function has a period, it has many periods. For example, $\cos(x)$ has periods $2\pi, 4\pi, 6\pi, -2\pi, -4\pi$, and, in fact, $2n\pi$ for any integer n. The smallest positive period of a function is called its *fundamental period*. The fundamental period of $\sin(x)$ and $\cos(x)$ is 2π.

If f has period p, then for any x, and any integer n,

$$f(x + np) = f(x).$$

For example,

$$\cos\left(\frac{\pi}{6}\right) = \cos\left(\frac{\pi}{6} + 2\pi\right) = \cos\left(\frac{\pi}{6} + 4\pi\right) = \cos\left(\frac{\pi}{6} + 6\pi\right) = \cdots$$

$$= \cos\left(\frac{\pi}{6} - 2\pi\right) = \cos\left(\frac{\pi}{6} - 4\pi\right) = \cdots.$$

The graph of periodic $f(x)$ repeats itself over every interval of length p (Figure 13.33). This means that we need only specify $f(x)$ on an interval of length p, say on $[-p/2, p/2)$, to determine $f(x)$ for all x. This specification of function values can be made on any interval $[\alpha, \alpha + p)$ of length p. Since $f(\alpha + p) = f(\alpha)$, the function must have the same value at the endpoints of this interval. This is why we specify values on the half-open interval $[\alpha, \alpha + p)$, since $f(\alpha + p)$ is determined once $f(\alpha)$ is defined.

EXAMPLE 13.26

Let $g(x) = 2x$ for $-1 \leq x < 1$, and suppose g has period 2. Then the graph of g on $[-1, 1)$ is repeated to cover the entire real line, as in Figure 13.34. Knowing the period, and the function values on $[-1, 1)$, are enough to determine the function for all x.

As a specific example, suppose we want to know $g(\frac{7}{2})$. Because g has period 2, $g(x + 2n) = g(x)$ for any x and any integer n. Then

$$g\left(\frac{7}{2}\right) = g\left(-\frac{1}{2} + 4\right) = g\left(-\frac{1}{2}\right) = 2\left(-\frac{1}{2}\right) = -1.$$

Similarly,

$$g(48.3) = g(0.3 + 2(24)) = g(0.3) = 0.6. \ \blacksquare$$

If f has period p and is integrable, then we can calculate its Fourier coefficients on $[-p/2, p/2]$ and write the Fourier series

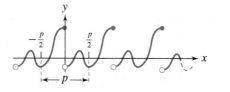

FIGURE 13.33 *Graph of a periodic function of fundamental period p.*

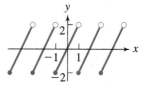

FIGURE 13.34

$$\frac{1}{2}a_0 + \sum_{n=1}^{\infty}\left(a_n \cos\left(\frac{2n\pi x}{p}\right) + b_n \sin\left(\frac{2n\pi x}{p}\right)\right).$$

Here $L = p/2$, so $n\pi x/L = 2n\pi x/p$ in the previous discussion of Fourier series on $[-L, L]$. The Fourier coefficients are

$$a_n = \frac{2}{p}\int_{-p/2}^{p/2} f(x)\cos\left(\frac{2n\pi x}{p}\right)dx \quad \text{for } n = 0, 1, 2, \ldots$$

and

$$b_n = \frac{2}{p}\int_{-p/2}^{p/2} f(x)\sin\left(\frac{2n\pi x}{p}\right)dx \quad \text{for } n = 1, 2, \ldots.$$

Actually, because of the periodicity, we could choose any convenient number α and write

$$a_n = \frac{2}{p}\int_{\alpha}^{\alpha+p} f(x)\cos\left(\frac{2n\pi x}{p}\right)dx \quad \text{for } n = 0, 1, 2, \ldots \qquad (13.19)$$

and

$$b_n = \frac{2}{p}\int_{\alpha}^{\alpha+p} f(x)\sin\left(\frac{2n\pi x}{p}\right)dx \quad \text{for } n = 1, 2, \ldots. \qquad (13.20)$$

Once we compute the coefficients, we can use a convergence theorem to determine where this series represents $f(x)$.

EXAMPLE 13.27

The function f shown in Figure 13.35 has fundamental period 6, and

$$f(x) = \begin{cases} 0 & \text{for } -3 \le x < 0 \\ 1 & \text{for } 0 \le x < 3 \end{cases}.$$

This function is called a *square wave*. Its Fourier series on $[-3, 3]$ is

$$\frac{1}{2} + \sum_{n=1}^{\infty}\frac{1}{n\pi}(1 - (-1)^n)\sin\left(\frac{n\pi x}{3}\right).$$

This series converges to 0 for $-3 < x < 0$, to 1 for $0 < x < 3$, and to $\frac{1}{2}$ at $x = 0$ and $x = \pm 3$. Because of the periodicity, this series also converges to $f(x)$ on $(-6, -3)$ and on $(3, 6)$, on $(-6, -9)$ and on $(6, 9)$, and so on. ■

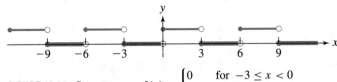

FIGURE 13.35 *Square wave:* $f(x) = \begin{cases} 0 & \text{for } -3 \le x < 0 \\ 1 & \text{for } 0 \le x < 3 \end{cases}$,

and f has period.

Sometimes we write

$$\omega_0 = \frac{2\pi}{p}.$$

Now the Fourier series of f on $[-p/2, p/2]$ is

$$\frac{1}{2}a_0 + \sum_{n=1}^{\infty} (a_n \cos(n\omega_0 x) + b_n \sin(\omega_0 x)), \tag{13.21}$$

where

$$a_n = \frac{2}{p} \int_{-p/2}^{p/2} f(x) \cos(n\omega_0 x)\, dx \quad \text{for } n = 0, 1, 2, \dots$$

and

$$b_n = \frac{2}{p} \int_{-p/2}^{p/2} f(x) \sin(n\omega_0 x)\, dx \quad \text{for } n = 1, 2, \dots.$$

It is sometimes useful to write the Fourier series (13.21) in a different way. We will look for numbers c_n and δ_n so that

$$a_n \cos(n\omega_0 x) + b_n \sin(n\omega_0 x) = c_n \cos(n\omega_0 x + \delta_n).$$

To solve for these constants, write the last equation as

$$a_n \cos(n\omega_0 x) + b_n \sin(n\omega_0 x) = c_n \cos(n\omega_0 x)\cos(\delta_n) - c_n \sin(n\omega_0 x)\sin(\delta_n).$$

One way to satisfy this equation is to have

$$c_n \cos(\delta_n) = a_n$$

and

$$c_n \sin(\delta_n) = -b_n.$$

Solve these for c_n and δ_n. First square both equations and add to obtain

$$c_n^2 = a_n^2 + b_n^2,$$

so

$$c_n = \sqrt{a_n^2 + b_n^2}. \tag{13.22}$$

Next, write

$$\frac{c_n \sin(\delta_n)}{c_n \cos(\delta_n)} = \tan(\delta_n) = -\frac{b_n}{a_n},$$

so

$$\delta_n = \tan^{-1}\left(-\frac{b_n}{a_n}\right),$$

assuming that $a_n \neq 0$. The numbers c_n and δ_n allow us to write the phase angle form of the Fourier series (13.21).

DEFINITION 13.7   *Phase Angle Form*

Let f have fundamental period p. Then the phase angle form of the Fourier series (13.21) of f is

$$\frac{1}{2}a_0 + \sum_{n=1}^{\infty} c_n \cos(n\omega_0 x + \delta_n),$$

in which $\omega_0 = 2\pi/p$, $c_n = \sqrt{a_n^2 + b_n^2}$, and $\delta_n = \tan^{-1}(-b_n/a_n)$ for $n = 1, 2, \ldots$.

The phase angle form of the Fourier series is also called its *harmonic form*. This expression displays the composition of a periodic function (satisfying certain continuity conditions) as a superposition of cosine waves. The term $\cos(n\omega_0 x + \delta_n)$ is the nth *harmonic of* f, c_n is the nth *harmonic amplitude*, and δ_n is the nth *phase angle of* f.

EXAMPLE 13.28

Suppose f has fundamental period $p = 3$, and

$$f(x) = x^2 \quad \text{for } 0 \le x < 3.$$

Since f has fundamental period 3, defining $f(x)$ on any interval $[a, b)$ of length 3 determines $f(x)$ for all x. For example,

$$f(-1) = f(-1+3) = f(2) = 4,$$
$$f(5) = f(2+3) = f(2) = 2^2 = 4,$$

(or observe that $f(5) = f(-1+6) = f(-1+2\cdot 3) = f(-1) = 4$), and

$$f(7) = f(1+6) = f(1) = 1.$$

A graph of f is shown in Figure 13.36.

Care must be taken if we want to write an algebraic expression for $f(x)$ on a different interval. For example, on the symmetric interval $[-\frac{3}{2}, \frac{3}{2})$ about the origin,

$$f(x) = \begin{cases} x^2 & \text{for } 0 \le x < \frac{3}{2} \\ (x+3)^2 & \text{for } -\frac{3}{2} \le x < 0 \end{cases}.$$

To find the Fourier coefficients of f, it is convenient to use equations (13.19) and (13.20) with $\alpha = 0$, since f is given explicitly on $[0, 3)$. Compute

$$a_0 = \frac{2}{3} \int_0^3 x^2 \, dx = 6,$$

$$a_n = \frac{2}{3} \int_0^3 x^2 \cos\left(\frac{2n\pi x}{3}\right) dx = \frac{9}{n^2\pi^2},$$

and

$$b_n = \frac{2}{3} \int_0^3 x^2 \sin\left(\frac{2n\pi x}{3}\right) dx = -\frac{9}{n\pi}.$$

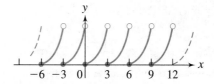

FIGURE 13.36 *Graph of* $f(x) = x^2$ *for* $0 \le x < 3$, *with* $f(x + 3) = f(x)$ *for all x.*

The Fourier series of f is

$$3 + \sum_{n=1}^{\infty} \frac{9}{n\pi} \left(\frac{1}{n\pi} \cos\left(\frac{2n\pi x}{3}\right) - \sin\left(\frac{2n\pi x}{3}\right) \right). \tag{13.23}$$

We can think of this as the Fourier series of f on the symmetric interval $[-\frac{3}{2}, \frac{3}{2}]$ about the origin. By the Fourier convergence theorem, this series converges to

$$\begin{cases} \frac{1}{2}\left(\frac{9}{4} + \frac{9}{4}\right) = \frac{9}{4} & \text{for } x = \pm\frac{3}{2} \\ \frac{9}{2} & \text{for } x = 0 \\ (x+3)^2 & \text{for } -\frac{3}{2} < x < 0 \\ x^2 & \text{for } 0 < x < \frac{3}{2} \end{cases}$$

For the phase angle, or harmonic form, of this Fourier series, compute

$$c_n = \sqrt{a_n^2 + b_n^2} = \frac{9}{n^2\pi^2}\sqrt{1 + n^2\pi^2} \quad \text{for } n = 1, 2, \dots$$

and

$$\delta_n = \tan^{-1}\left(-\frac{-9/n\pi}{9/n^2\pi^2}\right) = \tan^{-1}(n\pi).$$

Since $\omega_0 = 2\pi/3$, the phase angle form of the series (13.23) is

$$3 + \sum_{n=1}^{\infty} \frac{9}{n^2\pi^2}\sqrt{1 + n^2\pi^2} \cos\left(\frac{2n\pi x}{3} + \tan^{-1}(n\pi)\right). \ \blacksquare$$

The *amplitude spectrum* of a periodic function f is a plot of values of $n\omega_0$ on the horizontal axis versus $c_n/2$ on the horizontal axis, for $n = 1, 2, \dots$. Thus the amplitude spectrum consists of points $(n\omega_0, c_n/2)$ for $n = 1, 2, \dots$. It is also common to include the point $(0, |a_0|/2)$ on the vertical axis. Figure 13.37 shows the amplitude spectrum for the function of Example 13.28,

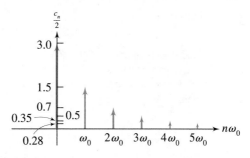

FIGURE 13.37 *Amplitude spectrum of the function of Figure 13.36.*

consisting of points $(0, 3)$ and, for $n = 1, 2, \ldots$,

$$\left(\frac{2n\pi}{3}, \frac{9}{2n^2\pi^2} \sqrt{1 + n^2\pi^2} \right).$$

This graph allows us to envision the magnitude of the harmonics of which the periodic function is composed and clarifies which harmonics dominate in the function. This is useful in signal analysis, in which the function is the signal.

SECTION 13.6 PROBLEMS

1. Let f and g have period p. Show that $\alpha f + \beta g$ has period p for any constants α and β.

2. Let f have period p and let α and β be positive constants. Show that $g(t) = f(\alpha t)$ has period p/α and that $h(t) = f(t/\beta)$ has period βp.

3. Let $f(x)$ be differentiable and have period p. Show that $f'(x)$ has period p.

4. Suppose f has period p. Show that, for any real number α,

$$\int_{\alpha}^{\alpha+p} f(x)\, dx = \int_{0}^{p} f(x)\, dx = \int_{-p/2}^{p/2} f(x)\, dx.$$

In each of Problems 5 through 9, find the phase angle form of the Fourier series of the function. Plot some points of the amplitude spectrum of the function.

5. $f(x) = x$ for $0 \leq x < 2$ and $f(x + 2) = f(x)$ for all x.

6. $f(x) = \begin{cases} 1 & \text{for } 0 \leq x < 1 \\ 0 & \text{for } 1 \leq x < 2 \\ f(x + 2) & \text{for all } x. \end{cases}$

7. $f(x) = 3x^2$ for $0 \leq x < 4$ and $f(x + 4) = f(x)$ for all x.

8. $f(x) = \begin{cases} 1 + x & \text{for } 0 \leq x < 3 \\ 2 & \text{for } 3 \leq x < 4 \\ f(x + 4) & \text{for all } x. \end{cases}$

9. $f(x) = \cos(\pi x)$ for $0 \leq x < 1$ and $f(x) = f(x + 1)$ for all x.

In each of Problems 10 through 14, find the phase angle form of the Fourier series of the function, part of whose graph is given in the indicated diagram. Plot some points of the amplitude spectrum of the function.

10.

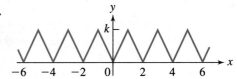

FIGURE 13.38

11.

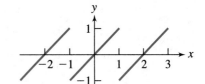

FIGURE 13.39

12.

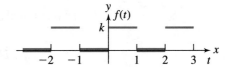

FIGURE 13.40

13.

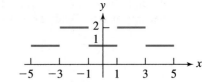

FIGURE 13.41

14.

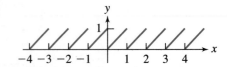

FIGURE 13.42

15. The RMS (root mean square) value of a function f over an interval $[a, b]$ is defined to be

$$\mathrm{RMS}(f) = \sqrt{\frac{\int_a^b [f(x)]^2\, dx}{b - a}}.$$

If f is periodic, this quantity is evaluated over an interval of length equal to the fundamental period of f. Determine $\mathrm{RMS}(E \sin(\omega x))$, with E and ω positive constants.

16. The *power content* of a periodic signal (function) f of period p is

$$\frac{1}{p} \int_{-p/2}^{p/2} [f(x)]^2\, dx.$$

This is motivated by the fact that the power dissipation in a resistor is $i^2 R$, where i is the root mean square value of the current. The root mean square value of a signal is that value of direct current that would dissipate the same power in a resistor. Determine the power content of the function given in Figure 13.39.

17. Determine the Fourier series representation of the steady-state current in the circuit of Figure 13.43 if

$$E(t) = \begin{cases} 100t\,(\pi^2 - t^2) & \text{for } -\pi \le t < \pi \\ E(t + 2\pi) & \text{for all } t \end{cases}.$$

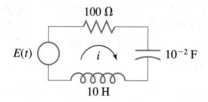

FIGURE 13.43

18. Determine the Fourier series representation of the steady-state current in the circuit shown in Figure 13.44 if $E(t) = |10 \sin(800\pi t)|$. *Hint:* First show that

$$E(t) = \frac{20}{\pi} \left[1 - 2 \sum_{n=1}^{\infty} \frac{\cos(1600 n \pi t)}{4n^2 - 1} \right].$$

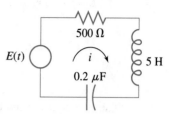

FIGURE 13.44

19. Use the first two nonzero terms in the Fourier series of the current, together with Parseval's identity, to approximate the power dissipation in the resistor of Problem 17 due to the steady-state current.

20. Approximate the power dissipation in the resistor in Problem 18 due to the steady-state current.

13.7 Complex Fourier Series and the Frequency Spectrum

It is often convenient to work in terms of complex numbers, even when the quantities of interest are real. For example, electrical engineers often use equations having complex quantities to compute currents, realizing that at the end the current is the real part of a certain complex expression.

We will cast Fourier series in this setting. Later, complex Fourier series and their coefficients will provide a natural starting point for the development of discrete Fourier transforms.

13.7.1 Review of Complex Numbers

Given a complex number $a + bi$, its *conjugate* is $\overline{a + bi} = a - bi$. If we identify $a + bi$ with the point (a, b) in the plane, then $a - bi$ is $(a, -b)$, the reflection of (a, b) across the horizontal (real) axis (Figure 13.45).

The conjugate of a product is the product of the conjugates:

$$\overline{zw} = \overline{z}\,\overline{w}$$

for any complex numbers z and w.

FIGURE 13.45
*Complex conjugate
as a reflection
across the
horizontal axis.*

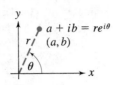

FIGURE 13.46 *Polar
form of a complex
number.*

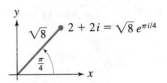

FIGURE 13.47 *Polar form of*
$2 + 2i$.

The *magnitude*, or modulus, of $a + bi$ is $|a + bi| = \sqrt{a^2 + b^2}$, the distance from the origin to (a, b). It is useful to observe that

$$(a + bi)(\overline{a + bi}) = a^2 + b^2 = |a + bi|^2.$$

If we denote the complex number as z, this equation is

$$z\bar{z} = |z|^2.$$

Introduce polar coordinates $x = r\cos(\theta)$, $y = r\sin(\theta)$ to write

$$z = x + iy = r[\cos(\theta) + i\sin(\theta)] = re^{i\theta},$$

by Euler's formula. Then $r = |z|$ and θ is called an *argument* of z. It is the angle between the positive x axis and the point (x, y), or $x + iy$, in the plane (Figure 13.46). The argument is determined to within integer multiples of 2π. For example, $|2 + 2i| = \sqrt{8}$ and the arguments of $2 + 2i$ are the angles $\pi/4 + 2n\pi$, with n any integer (Figure 13.47). Thus we can write

$$2 + 2i = \sqrt{8}e^{i\pi/4}.$$

This is the *polar form* of $2 + 2i$. We can actually write $2 + 2i = \sqrt{8}e^{i(\pi/4 + 2n\pi)}$, but this doesn't contribute anything new to the polar form of $2 + 2i$, since

$$e^{i(\pi/4 + 2n\pi)} = e^{i/4}e^{2n\pi i}$$

and

$$e^{2n\pi i} = \cos(2n\pi) + i\sin(2n\pi) = 1.$$

If we use Euler's formula twice, we can write

$$e^{ix} = \cos(x) + i\sin(x)$$

and

$$e^{-ix} = \cos(x) - i\sin(x).$$

Solve these equations for $\sin(x)$ and $\cos(x)$ to write

$$\cos(x) = \frac{1}{2}\left(e^{ix} + e^{-ix}\right) \quad \text{and} \quad \sin(x) = \frac{1}{2i}\left(e^{ix} - e^{-ix}\right). \tag{13.24}$$

Finally, we will use the fact that if x is a real number, then $\overline{e^{ix}} = e^{-ix}$. This is true because

$$\overline{e^{ix}} = \overline{\cos(x) + i\sin(x)} = \cos(x) - i\sin(x) = e^{-ix}.$$

13.7.2 Complex Fourier Series

We will use these ideas to formulate the Fourier series of a function in complex terms. Let f be a real-valued, periodic function with fundamental period p. Assume that f is integrable on $[-p/2, p/2]$. As we did with the phase angle form of a Fourier series, write the Fourier series of $f(x)$ on this interval as

$$\frac{1}{2}a_0 + \sum_{n=1}^{\infty} [a_n \cos(n\omega_0 x) + b_n \sin(n\omega_0 x)],$$

with $\omega_0 = 2\pi/p$. Use equations (13.24) to write this series as

$$\frac{1}{2}a_0 + \sum_{n=1}^{\infty} \left[a_n \frac{1}{2} \left(e^{in\omega_0 x} + e^{-in\omega_0 x} \right) + b_n \frac{1}{2i} \left(e^{in\omega_0 x} - e^{-in\omega_0 x} \right) \right]$$

$$= \frac{1}{2}a_0 + \sum_{n=1}^{\infty} \left[\frac{1}{2}(a_n - ib_n)e^{in\omega_0 x} + \frac{1}{2}(a_n + ib_n)e^{-in\omega_0 x} \right].$$

(13.25)

In the series (13.25), let

$$d_0 = \frac{1}{2}a_0$$

and, for each positive integer n,

$$d_n = \frac{1}{2}(a_n - ib_n).$$

Then the series (13.25) becomes

$$d_0 + \sum_{n=1}^{\infty} [d_n e^{in\omega_0 x} + \overline{d_n} e^{-in\omega_0 x}] = d_0 + \sum_{n=1}^{\infty} d_n e^{in\omega_0 x} + \sum_{n=1}^{\infty} \overline{d_n} e^{-in\omega_0 x}.$$

(13.26)

Now consider the coefficients. First,

$$d_0 = \frac{1}{2}a_0 = \frac{1}{p} \int_{-p/2}^{p/2} f(t)\, dt.$$

And, for $n = 1, 2, \ldots$,

$$d_n = \frac{1}{2}(a_n - ib_n)$$

$$= \frac{1}{2}\frac{2}{p} \int_{-p/2}^{p/2} f(t) \cos(n\omega_0 t)\, dt - \frac{i}{2}\frac{2}{p} \int_{-p/2}^{p/2} f(t) \sin(n\omega_0 t)\, dt$$

$$= \frac{1}{p} \int_{-p/2}^{p/2} f(t)[\cos(n\omega_0 t) - i \sin(n\omega_0 t)]\, dt$$

$$= \frac{1}{p} \int_{-p/2}^{p/2} f(t)e^{-in\omega_0 t}\, dt.$$

Then

$$\overline{d_n} = \frac{1}{p} \int_{-p/2}^{p/2} f(t)\overline{e^{-in\omega_0 t}}\, dt = \frac{1}{p} \int_{-p/2}^{p/2} f(t)e^{in\omega_0 t}\, dt = d_{-n}.$$

Put these results into the series (13.26) to get

$$d_0 + \sum_{n=1}^{\infty} d_n e^{in\omega_0 x} + \sum_{n=1}^{\infty} \overline{d_n} e^{-in\omega_0 x}$$

$$= d_0 + \sum_{n=1}^{\infty} d_n e^{in\omega_0 x} + \sum_{n=1}^{\infty} d_{-n} e^{-in\omega_0 x}$$

$$= d_0 + \sum_{n=-\infty, n \neq 0}^{\infty} d_n e^{in\omega_0 x} = \sum_{n=-\infty}^{\infty} d_n e^{in\omega_0 x}.$$

We have reached this expression by rearranging terms in the Fourier series of a periodic function f. This suggests the following definition.

DEFINITION 13.8 Complex Fourier Series

Let f have fundamental period p. Let $\omega_0 = 2\pi/p$. Then the complex Fourier series of f is

$$\sum_{n=-\infty}^{\infty} d_n e^{in\omega_0 x},$$

where

$$d_n = \frac{1}{p} \int_{-p/2}^{p/2} f(t) e^{-in\omega_0 t} \, dt$$

for $n = 0, \pm 1, \pm 2, \ldots$. The numbers d_n are the complex Fourier coefficients of f.

In the formula for d_n, the integration can actually be carried out over any interval of length p, because of the periodicity of f. Thus, for any real number α,

$$d_n = \frac{1}{p} \int_{\alpha}^{\alpha+p} f(t) e^{-in\omega_0 t} \, dt.$$

Since the complex Fourier series is just another way of writing the Fourier series, the convergence theorems (13.1) and (13.2) apply without any adjustments.

THEOREM 13.10

Let f be periodic with fundamental period p. Let f be piecewise smooth on $[-p/2, p/2]$. Then at each x the complex Fourier series converges to $\frac{1}{2}(f(x+) + f(x-))$. ∎

The *amplitude spectrum* of the complex Fourier series of a periodic function is a graph of the points $(n\omega_0, |d_n|)$, in which $|d_n|$ is the magnitude of the complex coefficient d_n. Sometimes this amplitude spectrum is referred to as a *frequency spectrum*.

EXAMPLE 13.29

Let $f(x) = 3x/4$ for $0 \le x < 8$, and $f(x + 8) = f(x)$ for all x. Then f is periodic with fundamental period 8. A graph of f is given in Figure 13.48. We will find the complex Fourier series and the amplitude spectrum of f.

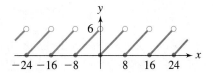

FIGURE 13.48 *Graph of $f(x) = 3x/4$ for $0 \leq x < 8$, if $f(x + 8) = f(x)$ for all real x.*

Here $p = 8$ and $\omega_0 = \pi/4$. Recall in the formulas for the coefficients that the integration can be performed over any interval of length 8. Here it is convenient to use $[0, 8]$ rather than $[-4, 4]$, simply by the way $f(x)$ was defined. Then

$$d_0 = \frac{1}{8} \int_0^8 \frac{3}{4}t \, dt = 3.$$

If we used the interval $[-4, 4]$, then we would calculate

$$d_0 = \frac{1}{8} \int_{-4}^0 \frac{3}{4}(t + 8) \, dt + \frac{1}{8} \int_0^4 \frac{3}{4}t \, dt = 3.$$

Next,

$$d_n = \frac{1}{8} \int_0^8 \frac{3}{4}te^{-n\pi it/4} \, dt = \frac{3i}{n\pi}.$$

The complex Fourier series is

$$3 + \frac{3i}{\pi} \sum_{n=-\infty, n \neq 0}^{\infty} \frac{1}{n} e^{n\pi ix/4}.$$

This series converges to $f(x)$ for $0 < x < 8$, $8 < x < 16$, $16 < x < 24, \ldots, -8 < x < 0$, $-16 < x < -8, \ldots$.

To plot the amplitude spectrum, compute

$$|d_0| = 3, \quad |d_n| = \frac{3}{n\pi}.$$

Since $n\omega_0 = n\pi/4$, the amplitude spectrum is a plot of points

$$\left(\frac{n\pi}{4}, \frac{3}{\pi|n|} \right)$$

for $n = \pm 1, \pm 2, \ldots$, together with the point $(0, 3)$. This plot is given in Figure 13.49. ■

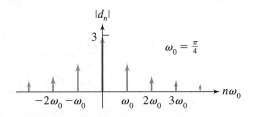

FIGURE 13.49 *Amplitude spectrum of the function of Figure 13.48.*

EXAMPLE 13.30

We will compute the complex Fourier series of the full-wave rectification of $E \sin(\lambda t)$, in which E and λ are positive constants. Note that t is the variable here, not x.

This means that we want the complex Fourier series of $|E \sin(\lambda t)|$, whose graph is shown in Figure 13.50. This function has fundamental period π/λ (even though $E \sin(\lambda t)$ has period $2\pi/\lambda$). In this example, $\omega_0 = 2\pi/(\pi/\lambda) = 2\lambda$. The complex Fourier coefficients are

$$d_n = \frac{\lambda}{\pi} \int_0^{\pi/\lambda} |E \sin(\lambda t)| \, e^{-2n\lambda it} \, dt$$

$$= \frac{E\lambda}{\pi} \int_0^{\pi/\lambda} \sin(\lambda t) e^{-2n\lambda it} \, dt.$$

When $n = 0$, we get

$$d_0 = \frac{E\lambda}{\pi} \int_0^{\pi/\lambda} \sin(\lambda t) \, dt = \frac{2E}{\pi}.$$

When $n \neq 0$, the integration is simplified by putting the sine term in exponential form:

$$d_n = \frac{E\lambda}{\pi} \int_0^{\pi/\lambda} \frac{1}{2i} \left(e^{\lambda it} - e^{-\lambda it} \right) e^{-2n\lambda it} \, dt$$

$$= \frac{E\lambda}{2i\pi} \int_0^{\pi/\lambda} e^{(1-2n)\lambda it} \, dt - \frac{E\lambda}{2i\pi} \int_0^{\pi/\lambda} e^{-(1+2n)\lambda it} \, dt$$

$$= \frac{E\lambda}{2i\pi} \left[\frac{1}{(1-2n)\lambda i} e^{(1-2n)\lambda it} \right]_0^{\pi/\lambda} + \frac{E\lambda}{2i\pi} \left[\frac{1}{(1+2n)\lambda i} e^{-(1+2n)\lambda it} \right]_0^{\pi/\lambda}$$

$$= -\frac{E}{2\pi} \left[\frac{e^{(1-2n)\pi i}}{1-2n} - \frac{1}{1-2n} + \frac{e^{-(1+2n)\pi i}}{1+2n} - \frac{1}{1+2n} \right].$$

Now

$$e^{(1-2n)\pi i} = \cos((1-2n)\pi) + i \sin((1-2n)\pi)$$

$$= (-1)^{1-2n} = -1$$

and

$$e^{-(1+2n)\pi i} = \cos((1+2n)\pi) - i \sin((1+2n)\pi)$$

$$= (-1)^{1+2n} = -1.$$

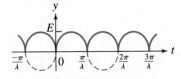

FIGURE 13.50   *Graph of* $|E \sin(\lambda t)|$.

Therefore,

$$d_n = -\frac{E}{2\pi}\left[\frac{-1}{1-2n} - \frac{1}{1-2n} + \frac{-1}{1+2n} + \frac{-1}{1+2n}\right]$$

$$= -\frac{2E}{\pi}\frac{1}{4n^2-1}.$$

When $n = 0$, this yields the correct value for d_0 as well. The complex Fourier series of $|E\sin(\lambda t)|$ is

$$-2\frac{E}{\pi}\sum_{n=-\infty}^{\infty}\frac{1}{4n^2-1}e^{2n\lambda i t}.$$

The amplitude spectrum is a plot of the points

$$\left(2n\lambda, \left|\frac{2E}{(4n^2-1)\pi}\right|\right).$$

Part of this plot is shown in Figure 13.51. ∎

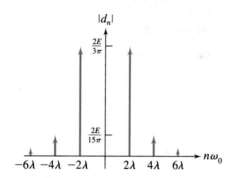

FIGURE 13.51 *Amplitude spectrum of* $|E\sin(\lambda t)|$.

SECTION 13.7 PROBLEMS

In each of Problems 1 through 9, write the complex Fourier series of f, determine what this series converges to, and plot some points of the frequency spectrum. Keep in mind that in specifying a function of period p, it is sufficient to define $f(p)$ on any interval of length p.

1. f has period 3 and $f(x) = 2x$ for $0 \le x < 3$

2. f has period 2 and $f(x) = x^2$ for $0 \le x < 2$

3. f has period 4 and $f(x) = \begin{cases} 0 & \text{for } 0 \le x < 1 \\ 1 & \text{for } 1 \le x < 4 \end{cases}$

4. f has period 6 and $f(x) = 1 - x$ for $0 \le x < 6$

5. f has period 4 and $f(x) = \begin{cases} -1 & \text{for } 0 \le x < 2 \\ 2 & \text{for } 2 \le x < 4 \end{cases}$

6. f has period 5 and $f(x) = e^{-x}$ for $0 \le x < 5$

7. f has period 2 and $f(x) = \begin{cases} x & \text{for } 0 \le x < 1 \\ 2-x & \text{for } 1 \le x < 2 \end{cases}$

8. f has period 1 and $f(x) = \cos(x)$ for $0 \le x < 1$

9. f has period 3 and $f(x) = \begin{cases} x & \text{for } 0 \le x < 2 \\ 0 & \text{for } 2 \le x < 3 \end{cases}$

10. Let f be the periodic function part of whose graph is shown in Figure 13.52. Find the complex Fourier series of f and plot some points of its amplitude spectrum.

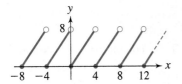

FIGURE 13.52

11. Plot some points of the amplitude spectrum of the function defined by

$$f(x) = 4 + \sum_{n=-\infty, n\neq 0}^{\infty} \frac{26}{n(12-5i)} e^{2nix}.$$

The next two problems involve the *phase spectrum* of f, which is a plot of points $(\varphi_n, n\omega_0)$ for $n = 0, 1, 2, \ldots$. Here $\varphi_n = \tan^{-1}(-b_n/a_n)$ is the nth phase angle of f.

12. Plot the phase spectrum of the periodic function of Problem 5.

13. The graphs of Figures 13.53 and 13.54 define two periodic functions f and g, respectively. Calculate the complex Fourier series of each function. Determine a relationship between the amplitude spectra of these functions and also between their phase spectra.

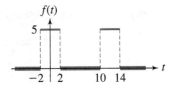

FIGURE 13.53

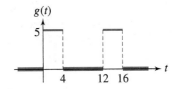

FIGURE 13.54

CHAPTER 14

The Fourier Integral and Fourier Transforms

14.1   The Fourier Integral

If $f(x)$ is defined on an interval $[-L, L]$, we may be able to represent it, at least at "most" points on this interval, by a Fourier series. If f is periodic, then we may be able to represent it by a Fourier series on intervals along the entire real line.

Now suppose $f(x)$ is defined for all x but is not periodic. Then we cannot represent $f(x)$ by a Fourier series over the entire line. However, we may still be able to write a representation in terms of sines and cosines, using an integral instead of a summation. To see how this might be done, suppose f is absolutely integrable, which means that $\int_{-\infty}^{\infty} |f(x)|\, dx$ converges and that f is piecewise smooth on every interval $[-L, L]$. Write the Fourier series of f on an arbitrary interval $[-L, L]$, with the integral formulas for the coefficients included:

$$\frac{1}{2L} \int_{-L}^{L} f(\xi)\, d\xi + \sum_{n=1}^{\infty} \left[\left(\frac{1}{L} \int_{-L}^{L} f(\xi) \cos\left(\frac{n\pi\xi}{L}\right) d\xi \right) \cos\left(\frac{n\pi x}{L}\right) \right.$$

$$\left. + \left(\frac{1}{L} \int_{-L}^{L} f(\xi) \sin\left(\frac{n\pi\xi}{L}\right) d\xi \right) \sin\left(\frac{n\pi x}{L}\right) \right].$$

We want to let $L \to \infty$ to obtain a representation of $f(x)$ over the whole line. To see what limit, if any, this Fourier series approaches, let

$$\omega_n = \frac{n\pi}{L}$$

and

$$\omega_n - \omega_{n-1} = \frac{\pi}{L} = \Delta\omega.$$

681

Then the Fourier series on $[-L, L]$ can be written

$$\frac{1}{2\pi} \left(\int_{-L}^{L} f(\xi) \, d\xi \right) \Delta\omega + \frac{1}{\pi} \sum_{n=1}^{\infty} \left[\left(\int_{-L}^{L} f(\xi) \cos(\omega_n \xi) \, d\xi \right) \cos(\omega_n x) \right.$$

$$\left. + \left(\int_{-L}^{L} f(\xi) \sin(\omega_n \xi) \, d\xi \right) \sin(\omega_n x) \right] \Delta\omega. \tag{14.1}$$

Now let $L \to \infty$, causing $\Delta\omega \to 0$. In the last expression,

$$\frac{1}{2\pi} \left(\int_{-L}^{L} f(\xi) \, d\xi \right) \Delta\omega \to 0$$

because by assumption $\int_{-L}^{L} f(\xi) \, d\xi$ converges. The other terms in the expression (14.1) resemble a Riemann sum for a definite integral, and we assert that, in the limit as $L \to \infty$ and $\Delta\omega \to 0$, this expression approaches the limit

$$\frac{1}{\pi} \int_{0}^{\infty} \left[\left(\int_{-\infty}^{\infty} f(\xi) \cos(\omega\xi) \, d\xi \right) \cos(\omega x) \right.$$

$$\left. + \left(\int_{-\infty}^{\infty} f(\xi) \sin(\omega\xi) \, d\xi \right) \sin(\omega x) \right] d\omega.$$

This is the *Fourier integral* of f on the real line. Under the assumptions made about f, this integral converges to

$$\frac{1}{2} \left(f(x-) + f(x+) \right)$$

at each x. In particular, if f is continuous at x, then this integral converges to $f(x)$.

Often this Fourier integral is written

$$\int_{0}^{\infty} [A_\omega \cos(\omega x) + B_\omega \sin(\omega x)] \, d\omega, \tag{14.2}$$

in which the *Fourier integral coefficients of* f are

$$A_\omega = \frac{1}{\pi} \int_{-\infty}^{\infty} f(\xi) \cos(\omega\xi) \, d\xi$$

and

$$B_\omega = \frac{1}{\pi} \int_{-\infty}^{\infty} f(\xi) \sin(\omega\xi) \, d\xi.$$

This Fourier integral representation of $f(x)$ is entirely analogous to a Fourier series on an interval, with $\int_{0}^{\infty} \cdots d\omega$ replacing $\sum_{n=1}^{\infty}$ and having integral formulas for the coefficients. These coefficients are functions of ω, which is the integration variable in the Fourier integral (14.2).

EXAMPLE 14.1

Let

$$f(x) = \begin{cases} 1 & \text{for } -1 \leq x \leq 1 \\ 0 & \text{for } |x| > 1 \end{cases}.$$

Figure 14.1 is a graph of f. Certainly f is piecewise smooth, and $\int_{-\infty}^{\infty} |f(x)| \, dx$ converges. The Fourier coefficients of f are

$$A_\omega = \frac{1}{\pi} \int_{-1}^{1} \cos(\omega\xi) \, d\xi = \frac{2\sin(\omega)}{\pi\omega}$$

and

$$B_\omega = \frac{1}{\pi} \int_{-1}^{1} f(\xi)\sin(\omega\xi) \, d\xi = 0.$$

The Fourier integral of f is

$$\int_0^\infty \frac{2\sin(\omega)}{\pi\omega} \cos(\omega x) \, d\omega.$$

Because f is piecewise smooth, this converges to $\frac{1}{2}(f(x+) + f(x-))$ for all x. More explicitly,

$$\int_0^\infty \frac{2\sin(\omega)}{\pi\omega} \cos(\omega x) \, d\omega = \begin{cases} 1 & \text{for } -1 < x < 1 \\ \dfrac{1}{2} & \text{for } x = \pm 1 \\ 0 & \text{for } |x| > 1 \end{cases} . \quad \blacksquare$$

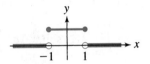

FIGURE 14.1
$$f(x) = \begin{cases} 1 & \text{for } -1 \le x \le 1 \\ 0 & \text{for } |x| > 1 \end{cases} .$$

EXAMPLE 14.2

Let $f(x) = xe^{-|x|}$. Then

$$\int_{-\infty}^{\infty} \left| xe^{-|x|} \right| \, dx = 2.$$

Further, f is continuous and differentiable for all x. A graph of f is given in Figure 14.2.

The Fourier coefficients of f are

$$A_\omega = \frac{1}{\pi} \int_{-\infty}^{\infty} \xi e^{-|\xi|} \cos(\omega\xi) \, d\xi = 0$$

(the integrand is an odd function), and

$$B_\omega = \frac{1}{\pi} \int_{-\infty}^{\infty} \xi e^{-|\xi|} \sin(\omega\xi) \, d\xi = \frac{4}{\pi} \frac{\omega}{\left(1+\omega^2\right)^2}.$$

Then, for all x,

$$xe^{-|x|} = \int_0^\infty \frac{4}{\pi} \frac{\omega}{\left(1+\omega^2\right)^2} \sin(\omega x) \, d\omega. \quad \blacksquare$$

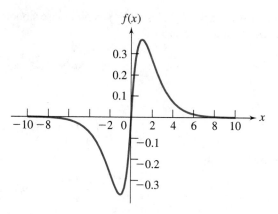

FIGURE 14.2 $f(x) = xe^{-|x|}$.

There is another expression for the Fourier integral of a function that we will sometimes find convenient. Write

$$\int_0^\infty [A_\omega \cos(\omega x) + B_\omega \sin(\omega x)]\, d\omega$$

$$= \int_0^\infty \left[\left(\frac{1}{\pi} \int_{-\infty}^\infty f(\xi) \cos(\omega \xi)\, d\xi \right) \cos(\omega x) \right.$$

$$\left. + \left(\frac{1}{\pi} \int_{-\infty}^\infty f(\xi) \sin(\omega \xi)\, d\xi \right) \sin(\omega x) \right] d\omega \qquad (14.3)$$

$$= \frac{1}{\pi} \int_0^\infty \int_{-\infty}^\infty f(\xi)[\cos(\omega \xi) \cos(\omega x) + \sin(\omega \xi) \sin(\omega x)]\, d\xi\, d\omega$$

$$= \frac{1}{\pi} \int_0^\infty \int_{-\infty}^\infty f(\xi) \cos(\omega(\xi - x))\, d\xi\, d\omega.$$

Of course, this integral has the same convergence properties as the integral expression (14.2), since it is just a rearrangement of that integral.

SECTION 14.1   PROBLEMS

In each of Problems 1 through 10, expand the function in a Fourier integral and determine what this integral converges to.

1. $f(x) = \begin{cases} x & \text{for } -\pi \leq x \leq \pi \\ 0 & \text{for } |x| > \pi \end{cases}$

2. $f(x) = \begin{cases} k & \text{for } -10 \leq x \leq 10 \\ 0 & \text{for } |x| > 10 \end{cases}$

3. $f(x) = \begin{cases} -1 & \text{for } -\pi \leq x \leq 0 \\ 1 & \text{for } 0 < x \leq \pi \\ 0 & \text{for } |x| > \pi \end{cases}$

4. $f(x) = \begin{cases} \sin(x) & \text{for } -4 \leq x \leq 0 \\ \cos(x) & \text{for } 0 < x \leq 4 \\ 0 & \text{for } |x| > 4 \end{cases}$

5. $f(x) = \begin{cases} x^2 & \text{for } -100 \leq x \leq 100 \\ 0 & \text{for } |x| > 100 \end{cases}$

6. $f(x) = \begin{cases} |x| & \text{for } -\pi \leq x \leq 2\pi \\ 0 & \text{for } x < -\pi \text{ and for } x > 2\pi \end{cases}$

7. $f(x) = \begin{cases} \sin(x) & \text{for } -3\pi \leq x \leq \pi \\ 0 & \text{for } x < -3\pi \text{ and for } x > \pi \end{cases}$

8. $f(x) = \begin{cases} \dfrac{1}{2} & \text{for } -5 \le x < 1 \\ 1 & \text{for } 1 \le x \le 5 \\ 0 & \text{for } |x| > 5 \end{cases}$

9. $f(x) = e^{-|x|}$

10. $f(x) = xe^{-|4x|}$

11. Show that the Fourier integral of f can be written

$$\lim_{\omega \to \infty} \frac{1}{\pi} \int_{-\infty}^{\infty} f(t) \frac{\sin(\omega(t-x))}{t-x} \, dt.$$

14.2 Fourier Cosine and Sine Integrals

If f is piecewise smooth on the half-line $[0, \infty)$ and $\int_0^\infty |f(\xi)| \, d\xi$ converges, then we can write a Fourier cosine or sine integral for f that is completely analogous to sine and cosine expansions of a function on an interval $[0, L]$.

To write a cosine integral, extend f to an even function f_e defined on the whole line by setting

$$f_e(x) = \begin{cases} f(x) & \text{for } x \ge 0 \\ f(-x) & \text{for } x < 0 \end{cases}.$$

This reflects the graph for $x \ge 0$ back across the vertical axis. Since f_e is an even function, its Fourier integral has only cosine terms. Since $f_e(x) = f(x)$ for $x \ge 0$, this cosine integral can be defined to be the Fourier cosine integral of f on $[0, \infty)$.

The coefficient of f_e in its Fourier integral expansion is

$$\frac{1}{\pi} \int_{-\infty}^{\infty} f_e(\xi) \cos(\omega\xi) \, d\xi$$

and this is

$$\frac{2}{\pi} \int_0^\infty f(\xi) \cos(\omega\xi) \, d\xi.$$

This suggests the following definition.

DEFINITION 14.1 *Fourier Cosine Integral*

Let f be defined on $[0, \infty)$ and let $\int_0^\infty |f(\xi)| \, d\xi$ converge. The Fourier cosine integral of f is

$$\int_0^\infty A_\omega \cos(\omega x) \, d\omega,$$

in which

$$A_\omega = \frac{2}{\pi} \int_0^\infty f(\xi) \cos(\omega\xi) \, d\xi.$$

By applying the convergence theorem to the integral expansion of f_e, we find that if f is piecewise continuous on each interval $[0, L]$, then its cosine integral expansion converges to $\frac{1}{2}(f(x+) + f(x-))$ for each $x > 0$ and to $f(0)$ for $x = 0$. In particular, at any positive x at which f is continuous, the cosine integral converges to $f(x)$.

By extending f to an odd function f_o, similar to what we did with series, we obtain a Fourier integral for f_o containing only sine terms. Since $f_o(x) = f(x)$ for $x \geq 0$, this gives a sine integral for f on $[0, \infty)$.

DEFINITION 14.2   Fourier Sine Integral

Let f be defined on $[0, \infty)$ and let $\int_0^\infty |f(\xi)| \, d\xi$ converge. The Fourier sine integral of f is

$$\int_0^\infty A_\omega \sin(\omega x) \, d\omega,$$

in which

$$A_\omega = \frac{2}{\pi} \int_0^\infty f(\xi) \sin(\omega \xi) \, d\xi.$$

If f is piecewise smooth on every interval $[0, L]$, then this integral converges to $\frac{1}{2}(f(x+) + f(x-))$ on $(0, \infty)$. As with Fourier sine series on a bounded interval, this Fourier sine integral converges to 0 at $x = 0$.

EXAMPLE 14.3   Laplace's Integrals

Let $f(x) = e^{-kx}$ for $x \geq 0$, with k a positive constant. Then f is continuously differentiable on any interval $[0, L]$, and

$$\int_0^\infty e^{-kx} \, dx = \frac{1}{k}.$$

For the Fourier cosine integral, compute the coefficients

$$A_\omega = \frac{2}{\pi} \int_0^\infty e^{-k\xi} \cos(\omega \xi) \, d\xi = \frac{2}{\pi} \frac{k}{k^2 + \omega^2}.$$

The Fourier cosine integral representation of f converges to e^{-kx} for $x \geq 0$:

$$e^{-kx} = \frac{2k}{\pi} \int_0^\infty \frac{1}{k^2 + \omega^2} \cos(\omega x) \, d\omega.$$

For the sine integral, compute

$$B_\omega = \frac{2}{\pi} \int_0^\infty e^{-k\xi} \sin(k\xi) \, d\xi = \frac{2}{\pi} \frac{\omega}{k^2 + \omega^2}.$$

The sine integral converges to e^{-kx} for $x > 0$ and to 0 for $x = 0$:

$$e^{-kx} = \frac{2}{\pi} \int_0^\infty \frac{\omega}{k^2 + \omega^2} \sin(\omega x) \, d\omega \quad \text{for } x > 0.$$

These integral representations are called *Laplace's integrals* because A_ω is $2/\pi$ times the Laplace transform of $\cos(\omega x)$, while B_ω is $2/\pi$ times the Laplace transform of $\sin(\omega x)$. ■

In each of Problems 1 through 10, find the Fourier sine integral and Fourier cosine integral representations of the function. Determine to what each integral converges.

1. $f(x) = \begin{cases} x^2 & \text{for } 0 \le x \le 10 \\ 0 & \text{for } x > 10 \end{cases}$

2. $f(x) = \begin{cases} \sin(x) & \text{for } 0 \le x \le 2\pi \\ 0 & \text{for } x > 2\pi \end{cases}$

3. $f(x) = \begin{cases} 1 & \text{for } 0 \le x \le 1 \\ 2 & \text{for } 1 < x \le 4 \\ 0 & \text{for } x > 4 \end{cases}$

4. $f(x) = \begin{cases} \cosh(x) & \text{for } 0 \le x \le 5 \\ 0 & \text{for } x > 5 \end{cases}$

5. $f(x) = \begin{cases} 2x + 1 & \text{for } 0 \le x \le \pi \\ 2 & \text{for } \pi < x \le 3\pi \\ 0 & \text{for } x > 3\pi \end{cases}$

6. $f(x) = \begin{cases} x & \text{for } 0 \le x \le 1 \\ x + 1 & \text{for } 1 < x \le 2 \\ 0 & \text{for } x > 2 \end{cases}$

7. $f(x) = e^{-x} \cos(x)$ for $x \ge 0$

8. $f(x) = xe^{-3x}$ for $x \ge 0$

9. $f(x) = \begin{cases} k & \text{for } 0 \le x \le c \\ 0 & \text{for } x > c \end{cases}$

in which k is constant and c is a positive constant

10. $f(x) = e^{-2x} \cos(x)$ for $x \ge 0$

11. Use the Laplace integrals to compute the Fourier cosine integral of $f(x) = 1/(1+x^2)$ and the Fourier sine integral of $g(x) = x/(1 + x^2)$.

12. Suppose $\lim_{x \to \infty} f(x) = \lim_{x \to \infty} f'(x) = 0$ and $f'(0) = 0$ and $f''(x)$ exists for $x \ge 0$. Show that the integrand of the Fourier cosine integral representation of $f''(x)$ is $-\omega^2$ times the integrand of the Fourier cosine integral representation of $f(x)$.

In Problems 13 and 14, assume that

$$f(x) = \frac{1}{\pi} \int_0^\infty A(\omega) \cos(\omega x) \, d\omega,$$

where $A(\omega) = \int_{-\infty}^\infty f(t) \cos(\omega t) \, dt$.

13. Show that

$$x^2 f(x) = \frac{1}{\pi} \int_0^\infty A^*(\omega) \cos(\omega x) \, d\omega,$$

where $A^*(\omega) = -A''(\omega)$.

14. Show that

$$xf(x) = \frac{1}{\pi} \int_0^\infty B^*(\omega) \sin(\omega x) \, d\omega,$$

where $B^*(\omega) = -A'(\omega)$.

14.3 The Complex Fourier Integral and the Fourier Transform

It is sometimes convenient to have a complex form of the Fourier integral. This complex setting will prove a natural platform from which to develop the Fourier transform.

Suppose f is piecewise smooth on each interval $[-L, L]$, and that $\int_{-\infty}^\infty |f(x)| \, dx$ converges. Then, at any x,

$$\frac{1}{2}(f(x+) + f(x-)) = \frac{1}{\pi} \int_0^\infty \int_{-\infty}^\infty f(\xi) \cos(\omega(\xi - x)) \, d\xi \, d\omega,$$

by the expression (14.3). Insert the complex exponential form of the cosine function into this expression to write

$$\frac{1}{2}(f(x+) + f(x-)) = \frac{1}{\pi} \int_0^\infty \int_{-\infty}^\infty f(\xi) \frac{1}{2} \left(e^{i\omega(\xi-x)} + e^{-i\omega(\xi-x)} \right) d\xi \, d\omega$$

$$= \frac{1}{2\pi} \int_0^\infty \int_{-\infty}^\infty f(\xi) e^{i\omega(\xi-x)} d\xi \, d\omega + \frac{1}{2\pi} \int_0^\infty \int_{-\infty}^\infty f(\xi) e^{-i\omega(\xi-x)} d\xi \, d\omega.$$

In the first integral on the last line, put $\omega = -w$ to get

$$\frac{1}{2}(f(x+) + f(x-))$$

$$= \frac{1}{2\pi} \int_{-\infty}^0 \int_{-\infty}^\infty f(\xi) e^{-iw(\xi-x)} d\xi \, dw + \frac{1}{2\pi} \int_0^\infty \int_{-\infty}^\infty f(\xi) e^{-i\omega(\xi-x)} d\xi \, d\omega.$$

Now write the variable of integration in the next to last integral as ω again and combine these two integrals to write

$$\frac{1}{2}(f(x+) + f(x-)) = \frac{1}{2\pi} \int_{-\infty}^\infty \int_{-\infty}^\infty f(\xi) e^{-i\omega(\xi-x)} d\xi \, d\omega. \tag{14.4}$$

This is the *complex Fourier integral representation of f* on the real line. If we let $C_\omega = \int_{-\infty}^\infty f(t) e^{-i\omega t} dt$, then this integral is

$$\frac{1}{2}(f(x+) + f(x-)) = \frac{1}{2\pi} \int_{-\infty}^\infty C_\omega e^{i\omega x} d\omega.$$

We call C_ω the *complex Fourier integral coefficient of f*.

EXAMPLE 14.4

Let $f(x) = e^{-a|x|}$ for all real x, with a a positive constant. We will compute the complex Fourier integral representation of f. First, we have

$$f(x) = \begin{cases} e^{-ax} & \text{for } x \geq 0 \\ e^{ax} & \text{for } x < 0 \end{cases}.$$

Further,

$$\int_{-\infty}^\infty f(x) \, dx = \int_{-\infty}^0 e^{ax} \, dx + \int_0^\infty e^{-ax} \, dx = \frac{2}{a}.$$

Now compute

$$C_\omega = \int_{-\infty}^\infty e^{-a|t|} e^{-i\omega t} \, dt$$

$$= \int_{-\infty}^0 e^{at} e^{-i\omega t} \, dt + \int_0^\infty e^{-at} e^{-i\omega t} \, dt$$

$$= \int_{-\infty}^0 e^{(a-i\omega)t} \, dt + \int_0^\infty e^{-(a+i\omega)t} \, dt$$

$$= \left[\frac{1}{a-i\omega} e^{(a-i\omega)t} \right]_{-\infty}^0 + \left[\frac{-1}{a+i\omega} e^{-(a+i\omega)t} \right]_0^\infty$$

$$= \left(\frac{1}{a+i\omega} + \frac{1}{a-i\omega} \right) = \frac{2a}{a^2 + \omega^2}.$$

The complex Fourier integral representation of f is

$$e^{-a|x|} = \frac{a}{\pi} \int_{-\infty}^{\infty} \frac{1}{a^2 + \omega^2} e^{i\omega x} \, d\omega. \quad \blacksquare$$

The expression on the right side of equation (14.4) leads naturally into the Fourier transform. To emphasize a certain term, write equation (14.4) as

$$\frac{1}{2}(f(x+) + f(x-)) = \frac{1}{2\pi} \int_{-\infty}^{\infty} \left(\int_{-\infty}^{\infty} f(\xi) e^{-i\omega\xi} \, d\xi \right) e^{i\omega x} \, d\omega. \quad (14.5)$$

The term in parentheses is what we will call the *Fourier transform of f*.

DEFINITION 14.3 *Fourier Transform*

Suppose $\int_{-\infty}^{\infty} |f(x)| \, dx$ converges. Then the Fourier transform of f is defined to be the function

$$\mathfrak{F}[f](\omega) = \int_{-\infty}^{\infty} f(t) e^{-i\omega t} \, dt.$$

Thus the Fourier transform of f is the coefficient C_ω in the complex Fourier integral representation of f.

$\mathfrak{F}$ turns a function f into a new function called $\mathfrak{F}[f]$. Because the transform is used in signal analysis, we will often use t (for time) as the variable with f, and ω as the variable of the transformed function $\mathfrak{F}[f]$. The value of the function $\mathfrak{F}[f]$ at ω is $\mathfrak{F}[f](\omega)$, and this number is computed for a given ω by evaluating the integral $\int_{-\infty}^{\infty} f(t) e^{-i\omega t} \, dt$. If we want to keep the variable t before our attention, we sometimes write $\mathfrak{F}[f]$ as $\mathfrak{F}[f(t)]$.

Engineers refer to the variable ω in the transformed function as the *frequency* of the signal f. Later we will discuss how the Fourier transform, and a truncated version called the *windowed Fourier transform*, are used to determine information about the frequency content of a signal.

Because the symbol $\mathfrak{F}[f(t)]$ may be clumsy to use in calculations, we sometimes write the Fourier transform of f as $\hat{f}$. In this notation,

$$\mathfrak{F}[f](\omega) = \hat{f}(\omega).$$

EXAMPLE 14.5

Let a be a positive constant. Then

$$\mathfrak{F}\left[e^{-a|t|}\right](\omega) = \frac{2a}{a^2 + \omega^2}.$$

This follows immediately from Example 14.4, where we calculated the Fourier integral coefficient C_ω of $e^{-a|t|}$. This coefficient is the Fourier transform of f. $\blacksquare$

EXAMPLE 14.6

Let

$$f(t) = \begin{cases} 0 & \text{for } t < 0 \\ e^{-at} & \text{for } t > 0 \end{cases},$$

with a a positive constant. In terms of the Heaviside function, $f(t) = H(t)e^{-at}$. Recall from Section 3.3.2 that the Heaviside function is defined by

$$H(t) = \begin{cases} 1 & \text{for } t \geq 0 \\ 0 & \text{for } t < 0 \end{cases}.$$

A graph of H is given in Figure 3.10, and f is shown in Figure 14.3. f is absolutely integrable, since $\int_{-\infty}^{\infty} f(t)\, dt = 1/a$. The Fourier transform of f is

$$\hat{f}(\omega) = \int_{-\infty}^{\infty} f(t)e^{-i\omega t}\, dt = \int_{0}^{\infty} e^{-at}e^{-i\omega t}\, dt$$

$$= \int_{0}^{\infty} e^{-(a+i\omega)t}\, dt = \left[\frac{-1}{a+i\omega} e^{-(a+i\omega)t} \right]_{0}^{\infty} = \frac{1}{a+i\omega}.$$

We could also write

$$\mathfrak{F}[f](\omega) = \frac{1}{a+i\omega},$$

or

$$\mathfrak{F}[f(t)](\omega) = \frac{1}{a+i\omega}. \quad \blacksquare$$

EXAMPLE 14.7

Let a and k be positive numbers, and let

$$f(t) = \begin{cases} k & \text{for } -a \leq t < a \\ 0 & \text{for } t < -a \quad \text{and for } t \geq a \end{cases}.$$

This pulse function can be written in terms of the Heaviside function as

$$f(t) = k[H(t+a) - H(t-a)],$$

and is graphed in Figure 14.4. The Fourier transform of f is

$$\hat{f}(\omega) = \int_{-\infty}^{\infty} f(t)e^{-i\omega t}\, dt$$

$$= \int_{-a}^{a} ke^{-i\omega t}\, dt = \left[\frac{-k}{i\omega} e^{-i\omega t} \right]_{-a}^{a}$$

$$= -\frac{k}{i\omega} \left[e^{-i\omega a} - e^{i\omega a} \right] = \frac{2k}{\omega} \sin(a\omega).$$

Again, we can also write

$$\mathfrak{F}[f](\omega) = \frac{2k}{\omega} \sin(a\omega),$$

or

$$\mathfrak{F}[f(t)](\omega) = \frac{2k}{\omega} \sin(a\omega). \quad \blacksquare$$

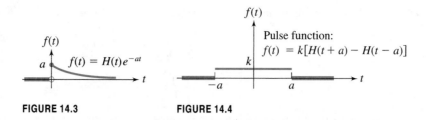

FIGURE 14.3   **FIGURE 14.4**

In view of equation (14.5), the Fourier integral representation of f is

$$\frac{1}{2\pi} \int_{-\infty}^{\infty} \hat{f}(\omega) e^{i\omega t}\, d\omega.$$

If f is continuous, and f' piecewise continuous on every interval $[-L, L]$, then the Fourier integral of f represents f:

$$f(t) = \frac{1}{2\pi} \int_{-\infty}^{\infty} \hat{f}(\omega) e^{i\omega t}\, d\omega. \tag{14.6}$$

We can therefore use equation (14.6) as an inverse Fourier transform, retrieving f from $\hat{f}$. This is important because, in applications, we use the Fourier transform to change a problem involving f from one form to another, presumably easier one, which is solved for $\hat{f}(\omega)$. We must then have some way of getting back to the $f(t)$ that we want, and equation (14.6) is the vehicle that is often used. We write $\mathfrak{F}^{-1}[\hat{f}] = f$ if $\mathfrak{F}[f] = \hat{f}$. As we expect of any integral transform, $\mathfrak{F}$ is linear:

$$\mathfrak{F}[\alpha f + \beta g] = \alpha \mathfrak{F}[f] + \beta \mathfrak{F}[g].$$

The integral defining the transform, and the integral (14.6) giving its inverse, are said to constitute a *transform pair* for the Fourier transform. Under certain conditions on f,

$$\hat{f}(\omega) = \int_{-\infty}^{\infty} f(t) e^{-i\omega t}\, dt \quad \text{and} \quad f(t) = \frac{1}{2\pi} \int_{-\infty}^{\infty} \hat{f}(\omega) e^{i\omega t}\, dt.$$

EXAMPLE 14.8

Let

$$f(t) = \begin{cases} 1 - |t| & \text{for } -1 \leq t \leq 1 \\ 0 & \text{for } t > 1 \quad \text{and for } t < -1 \end{cases}.$$

Then f is continuous and absolutely integrable, and f' is piecewise continuous. Compute

$$\hat{f}(\omega) = \int_{-\infty}^{\infty} f(t) e^{-i\omega t}\, dt$$

$$= \int_{-1}^{1} (1 - |t|) e^{-i\omega t}\, dt = \frac{2(1 - \cos(\omega))}{\omega^2}.$$

This is the Fourier coefficient C_ω in the complex Fourier expansion of $f(t)$.

If we want to go the other way, then by equation (14.6),

$$f(t) = \frac{1}{2\pi} \int_{-\infty}^{\infty} \hat{f}(\omega) e^{i\omega t}\, d\omega$$

$$= \frac{1}{\pi} \int_{-\infty}^{\infty} \frac{(1 - \cos(\omega))}{\omega^2} e^{i\omega t}\, d\omega.$$

We can verify this by explicitly carrying out this integration. A software package yields

$$\frac{1}{\pi} \int_{-\infty}^{\infty} \frac{(1 - \cos(\omega))}{\omega^2} e^{i\omega t}\, d\omega$$

$$= \pi t\, \text{signum}\,(t+1) + \pi\, \text{signum}\,(t+1) + \pi t\, \text{signum}\,(t-1)$$

$$- \pi\, \text{signum}\,(t-1) - 2\, \text{signum}\,(t),$$

in which

$$\text{signum}(\omega) = \begin{cases} 1 & \text{for } \omega > 0 \\ 0 & \text{for } \omega = 0 \\ -1 & \text{for } \omega < 0 \end{cases}.$$

This expression is equal to $1 - |t|$ for $-1 \le t \le 1$, and 0 for $t > 1$ and for $t < -1$, verifying the integral for the inverse. ■

In the context of the Fourier transform, the *amplitude spectrum* is often taken to be a graph of $|\hat{f}(\omega)|$. This is in the same spirit as the use of this term in connection with Fourier series.

EXAMPLE 14.9

In Example 14.6, $f(t) = H(t)e^{-at}$ and $\hat{f}(\omega) = 1/(a + i\omega)$, so

$$|\hat{f}(\omega)| = \frac{1}{\sqrt{a^2 + \omega^2}}.$$

Figure 14.5 shows a graph of $|\hat{f}(\omega)|$. This graph is the amplitude spectrum of f. ■

EXAMPLE 14.10

The amplitude spectrum of the function f of Example 14.7 is a graph of

$$|\hat{f}(\omega)| = 2k \left| \frac{\sin(a\omega)}{\omega} \right|,$$

shown in Figure 14.6. ■

We will now develop some of the important properties and computational rules for the Fourier transform. With each rule, there is also an inverse transform version that we will also state. Throughout, we assume that $\int_{-\infty}^{\infty} |f(t)|\, dt$ converges and, for the inverse version, that f is continuous and f' piecewise continuous on each $[-L, L]$.

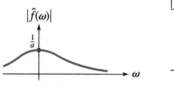

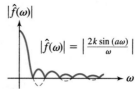

FIGURE 14.5 *Graph of* $|\hat{f}(\omega)| = \dfrac{1}{\sqrt{a^2+\omega^2}}$, *with* $f(t) = H(t)e^{-\alpha t}$.

FIGURE 14.6

THEOREM 14.1 *Time Shifting*

If t_0 is a real number, then

$$\mathfrak{F}[f(t - t_0)](\omega) = e^{-i\omega t_0}\hat{f}(t).$$

That is, if we shift time back t_0 units and replace $f(t)$ by $f(t - t_0)$, then the Fourier transform of this shifted function is the Fourier transform of f, multiplied by the exponential factor $e^{-i\omega t_0}$.

Proof

$$\mathfrak{F}[f(t - t_0)](\omega) = \int_{-\infty}^{\infty} f(t - t_0)e^{-i\omega t}\, dt$$

$$= e^{-i\omega t_0}\int_{-\infty}^{\infty} f(t - t_0)e^{-i\omega(t - t_0)}\, dt.$$

Let $u = t - t_0$ to write

$$\mathfrak{F}[f(t - t_0)](\omega) = e^{-i\omega t_0}\int_{-\infty}^{\infty} f(u)e^{-i\omega u}\, du = e^{-i\omega t_0}\hat{f}(\omega). \blacksquare$$

EXAMPLE 14.11

Suppose we want the Fourier transform of the pulse of amplitude 6 which turns on at time 3 and off at time 7. This is the function

$$g(t) = \begin{cases} 0 & \text{for } t < 3 \quad \text{and for } t \geq 7 \\ 6 & \text{for } 3 \leq t < 7 \end{cases},$$

shown in Figure 14.7. We can certainly compute $\hat{g}(\omega)$ by integration. But we can also observe that the midpoint of the pulse (that is, of the nonzero part) occurs when $t = 5$. Shift the graph 5 units to the left to center the pulse at zero (Figure 14.8). Calling this shifted pulse f, then $f(t) = g(t + 5)$. Shifting f five units to the right again just gets us back to g:

$$g(t) = f(t - 5).$$

The point to this is that we already know the Fourier transform of f from Example 14.7:

$$\mathfrak{F}[f(t)](\omega) = 12\frac{\sin(2\omega)}{\omega}.$$

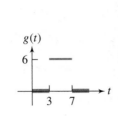

FIGURE 14.7

$$g(x) = \begin{cases} 6 & \text{for } 3 \le t < 7 \\ 0 & \text{for } t < 3 \\ & \text{and for } t \ge 7 \end{cases}.$$

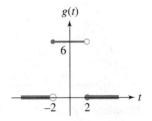

FIGURE 14.8 *The function of Figure 14.7 shifted five units to the left.*

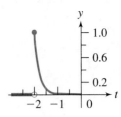

FIGURE 14.9 *Graph of* $H(t+2)e^{-5(t+2)}$.

By the time-shifting theorem,

$$\mathfrak{F}[g(t)](\omega) = \mathfrak{F}[f(t-5)](\omega) = 12e^{-5i\omega}\frac{\sin(2\omega)}{\omega}. \; \blacksquare$$

The inverse version of the time-shifting theorem is

$$\mathfrak{F}^{-1}[e^{-i\omega t_0}f(\omega)](t) = f(t-t_0). \tag{14.7}$$

EXAMPLE 14.12

Suppose we want

$$\mathfrak{F}^{-1}\left[\frac{e^{2i\omega}}{5+i\omega}\right].$$

The presence of the exponential factor suggests the inverse version of the time-shifting theorem. Put $t_0 = -2$ in equation (14.7) to write

$$\mathfrak{F}^{-1}\left[\frac{e^{2i\omega}}{5+i\omega}\right] = f(t-(-2)) = f(t+2),$$

where

$$f(t) = \mathfrak{F}^{-1}\left[\frac{1}{5+i\omega}\right] = H(t)e^{-5t}.$$

Therefore,

$$\mathfrak{F}^{-1}\left[\frac{e^{2i\omega}}{5+i\omega}\right] = f(t+2) = H(t+2)e^{-5(t+2)}.$$

A graph of this function is shown in Figure 14.9. $\blacksquare$

The next result is reminiscent of the first shifting theorem for the Laplace transform (Theorem 3.7).

THEOREM 14.2   *Frequency Shifting*

If ω_0 is any real number, then

$$\mathfrak{F}[e^{i\omega_0 t}f(t)] = \hat{f}(\omega - \omega_0).$$

Proof

$$\mathfrak{F}[e^{i\omega_0 t} f(t)](\omega) = \int_{-\infty}^{\infty} e^{i\omega_0 t} f(t) e^{-i\omega t} \, dt$$

$$= \int_{-\infty}^{\infty} f(t) e^{-i(\omega - \omega_0)t} \, dt = \hat{f}(\omega - \omega_0). \ \blacksquare$$

The inverse version of the frequency-shifting theorem is

$$\mathfrak{F}^{-1}[\hat{f}(\omega - \omega_0)(t)] = e^{i\omega_0 t} f(t).$$

--- **THEOREM 14.3** *Scaling* ---

If a is a nonzero real number, then

$$\mathfrak{F}[f(at)](\omega)] = \frac{1}{|a|} \hat{f}\left(\frac{\omega}{a}\right). \ \blacksquare$$

This can be proved by a straightforward calculation proceeding from the definition. The inverse transform version of this result is

$$\mathfrak{F}^{-1}\left[\hat{f}\left(\frac{\omega}{a}\right)\right](t) = |a| f(at).$$

This conclusion is called a *scaling theorem* because we want the transform not of $f(t)$, but of $f(at)$, in which a can be thought of as a scaling factor. The theorem says that we can compute the transform of the scaled function by replacing ω by ω/a in the transform of the original function, and dividing by the magnitude of the scaling factor.

EXAMPLE 14.13

We know from Example 14.8 that, if

$$f(t) = \begin{cases} 1 - |t| & \text{for } -1 \leq t \leq 1 \\ 0 & \text{for } t > 1 \quad \text{and for } t < -1 \end{cases},$$

then

$$\hat{f}(\omega) = 2 \frac{1 - \cos(\omega)}{\omega^2}.$$

Let

$$g(t) = f(7t) = \begin{cases} 1 - |7t| & \text{for } -\frac{1}{7} \leq t \leq \frac{1}{7} \\ 0 & \text{for } t > \frac{1}{7} \quad \text{and for } t < -\frac{1}{7} \end{cases}.$$

Then

$$\hat{g}(\omega) = \mathfrak{F}[f(7t)](\omega) = \frac{1}{7} \hat{f}\left(\frac{\omega}{7}\right)$$

$$= \frac{2}{7} \frac{1 - \cos(\omega/7)}{(\omega/7)^2} = 14 \frac{1 - \cos(\omega/7)}{\omega^2}. \ \blacksquare$$

THEOREM 14.4 *Time Reversal*

$$\mathfrak{F}[f(-t)](\omega) = \hat{f}(-\omega). \ \blacksquare$$

This result is called time reversal because we replace t by $-t$ in $f(t)$ to get $f(-t)$. The transform of this new function is obtained by simply replacing ω by $-\omega$ in the transform of $f(t)$. This conclusion follows immediately from the scaling theorem by putting $a = -1$. The inverse version of time reversal is

$$\mathfrak{F}^{-1}[\hat{f}(-\omega)](t) = f(-t).$$

THEOREM 14.5 *Symmetry*

$$\mathfrak{F}[\hat{f}(t)](\omega) = 2\pi f(-\omega). \ \blacksquare$$

To understand this conclusion, begin with $f(t)$ and take its Fourier transform $\hat{f}(\omega)$. Replace ω by t and take the transform of the function $\hat{f}(t)$. The symmetry property of the Fourier transform states that the transform of $\hat{f}(t)$ is just the original function $f(t)$ with t replaced by $-\omega$, and then this new function multiplied by 2π.

EXAMPLE 14.14

Let

$$f(t) = \begin{cases} 4 - t^2 & \text{for } -2 \leq t \leq 2 \\ 0 & \text{for } t > 2 \quad \text{and for } t < -2 \end{cases}.$$

Figure 14.10 shows a graph of f. The Fourier transform of f is

$$\hat{f}(\omega) = \int_{-\infty}^{\infty} f(t)e^{-i\omega t}\, d\omega = \int_{-2}^{2}(4 - t^2)e^{-i\omega t}\, dt$$

$$= 4\frac{\sin(2\omega) - 2\omega\cos(2\omega)}{\omega^3}.$$

In this example, $f(-t) = f(t)$, so exchanging $-\omega$ for ω should not make any difference in $\hat{f}(\omega)$, and we can see that this is indeed the case. $\blacksquare$

THEOREM 14.6 *Modulation*

If ω_0 is a real number, then

$$\mathfrak{F}[f(t)\cos(\omega_0 t)](\omega) = \frac{1}{2}\left[\hat{f}(\omega + \omega_0) + \hat{f}(\omega - \omega_0)\right]$$

and

$$\mathfrak{F}[f(t)\sin(\omega_0 t)](\omega) = \frac{1}{2}i\left[\hat{f}(\omega + \omega_0) - \hat{f}(\omega - \omega_0)\right].$$

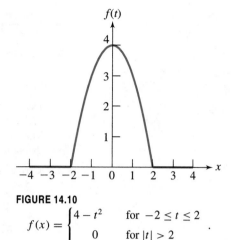

FIGURE 14.10

$$f(x) = \begin{cases} 4 - t^2 & \text{for } -2 \le t \le 2 \\ 0 & \text{for } |t| > 2 \end{cases}.$$

Proof Put $\cos(\omega t) = \frac{1}{2}\left(e^{i\omega_0 t} + e^{-i\omega_0 t}\right)$ and use the linearity of $\mathfrak{F}$ and the frequency-shifting theorem to get

$$\mathfrak{F}[f(t)\cos(\omega_0 t)](\omega) = \mathfrak{F}\left[\frac{1}{2}e^{i\omega_0 t}f(t) + \frac{1}{2}e^{-i\omega_0 t}f(t)\right](\omega)$$
$$= \frac{1}{2}\mathfrak{F}[e^{i\omega_0 t}f(t)](\omega) + \frac{1}{2}\mathfrak{F}[e^{-i\omega_0 t}f(t)](\omega)$$
$$= \frac{1}{2}\hat{f}(\omega - \omega_0) + \frac{1}{2}\hat{f}(\omega + \omega_0).$$

The second conclusion is proved similarly, using $\sin(\omega_0 t) = (1/2i)(e^{i\omega_0 t} - e^{-i\omega_0 t})$. ∎

SECTION 14.3 PROBLEMS

In each of Problems 1 through 8, find the complex Fourier integral of the function and determine what this integral converges to.

1. $f(x) = xe^{-|x|}$

2. $f(x) = \begin{cases} 1 - x & \text{for } -1 \le x \le 1 \\ 0 & \text{for } |x| > 1 \end{cases}$

3. $f(x) = \begin{cases} \sin(\pi x) & \text{for } -5 \le x \le 5 \\ 0 & \text{for } |x| > 5 \end{cases}$

4. $f(x) = \begin{cases} |x| & \text{for } -2 \le x \le 2 \\ 0 & \text{for } |x| > 2 \end{cases}$

5. $f(x) = \begin{cases} x & \text{for } -1 \le x \le 1 \\ e^{-|x|} & \text{for } |x| > 1 \end{cases}$

6. $f(x) = \begin{cases} 1 & \text{for } 0 \le x \le k \\ -1 & \text{for } -k \le x < 0 \\ 0 & \text{for } |x| > k, \end{cases}$

in which k is a positive constant

7. $f(x) = \begin{cases} \cos(x) & \text{for } 0 \le x \le \dfrac{\pi}{2} \\ \sin(x) & \text{for } -\dfrac{\pi}{2} \le x < 0 \\ 0 & \text{for } |x| > \dfrac{\pi}{2} \end{cases}$

8. $f(x) = x^2 e^{-3|x|}$

In each of Problems 9 through 24, find the Fourier transform of the function and graph the amplitude spectrum. Wherever k appears, it is a positive constant. For some problems one or more theorems from this section can be used in conjunction with the following transforms, which can be assumed:

$$\mathfrak{F}[e^{-a|t|}](\omega) = \frac{2a}{a^2 + \omega^2}, \quad \mathfrak{F}[e^{-at^2}](\omega) = \sqrt{\frac{\pi}{a}}e^{-\omega^2/4a},$$

and

$$\mathfrak{F}\left[\frac{1}{a^2 + t^2}\right](\omega) = \frac{\pi}{a}e^{-a|\omega|}.$$

9. $f(t) = \begin{cases} 1 & \text{for } 0 \le t \le 1 \\ -1 & \text{for } -1 \le t < 0 \\ 0 & \text{for } |t| > 1 \end{cases}$

10. $f(t) = \begin{cases} \sin(t) & \text{for } -k \le t \le k \\ 0 & \text{for } |t| > k, \end{cases}$

11. $f(t) = 5[H(t-3) - H(t-11)]$

12. $f(t) = 5e^{-3(t-5)^2}$

13. $f(t) = H(t-k)e^{-t/4}$

14. $f(t) = H(t-k)t^2$

15. $f(t) = 1/(1+t^2)$

16. $f(t) = 3H(t-2)e^{-3t}$

17. $f(t) = 3e^{-4|t+2|}$

18. $f(t) = H(t-3)e^{-2t}$

19. $f(t) = 3e^{-4|t|}\cos(2t)$

20. $f(t) = 8e^{-2t^2}\sin(3t)$

21. $f(t) = \sin(t)/(4+t^2)$

22. $f(t) = 1/(t^2 + 6t + 13)$

23. $f(t) = H(t-1)e^{-5(t+2)}$

24. $f(t) = 4H(t-2)e^{-3t}\cos(t-2)$

In each of Problems 25 through 30, find the inverse Fourier transform of the function.

25. $9e^{-(\omega+4)^2/32}$

26. $\dfrac{e^{(20-4\omega)i}}{3 - (5 - \omega)i}$

27. $\dfrac{e^{(2\omega-6)i}}{5 - (3 - \omega)i}$

28. $\dfrac{10\sin(3\omega)}{\omega + \pi}$

29. $\dfrac{1 + i\omega}{6 - \omega^2 + 5i\omega}$

Hint: Factor the denominator and use partial fractions.

30. $\dfrac{10(4 + i\omega)}{9 - \omega^2 + 8i\omega}$

31. In this section we obtained the result

$$\mathfrak{F}[e^{-a|t|}](\omega) = \frac{2a}{a^2 + \omega^2}.$$

Derive this result again as follows. Set $e^{-a|t|} = H(t)e^{-at} + H(-t)e^{at}$ and note that $f(t) = H(t)e^{-at}$. Then $f(-t) = H(-t)e^{at}$. Now use the time reversal theorem.

32. Derive the result

$$\mathfrak{F}[e^{-at^2}](\omega) = \sqrt{\frac{\pi}{a}}e^{-\omega^2/4a}.$$

Hint: With $\mathfrak{F}[f] = \hat{f}$, compute

$$\hat{f}'(\omega) = -i \int_{-\infty}^{\infty} te^{-at^2}e^{-i\omega t}\, dt.$$

Integrate by parts using $u = e^{-i\omega t}$ and $dv = te^{-at^2}\, dt$ to show that $\hat{f}'(\omega) = -(\omega/2a)\hat{f}(\omega)$. Solve for $\hat{f}(\omega)$, using the fact that $\int_{-\infty}^{\infty} e^{-x^2}\, dx = \sqrt{\pi}$ to evaluate the constant of integration.

33. Prove that $\hat{f}(-\omega) = \overline{\hat{f}(\omega)}$, in which the bar denotes complex conjugation.

For Problems 34 through 37, suppose $\hat{f}(\omega) = R(\omega) + iX(\omega)$.

34. Prove that $R(\omega) = \int_{-\infty}^{\infty} f(t)\cos(\omega t)\, dt$ and $X(\omega) = -\int_{-\infty}^{\infty} f(t)\sin(\omega t)\, dt$. Relate these results to the Fourier integral coefficients of $f(t)$.

35. Show that $R(\omega)$ is an even function and $X(\omega)$ is an odd function.

36. Let $f_e(y) = \frac{1}{2}(f(t) + f(-t))$ and $f_o(t) = \frac{1}{2}(f(t) - f(-t))$. Show that f_e is an even function and f_o an odd function.

37. Show that $\mathfrak{F}[f_e(t)](\omega) = R(\omega)$ and $\mathfrak{F}[f_o(t)](\omega) = iX(\omega)$.

14.4   Additional Properties and Applications of the Fourier Transform

14.4.1   The Fourier Transform of a Derivative

In using the Fourier transform to solve differential equations, we need an expression relating the transform of f' to that of f. The following theorem provides such a relationship for derivatives of any order and is called the *operational rule* for the Fourier transform. A similar issue arises for any integral transform when it is to be used in connection with differential equations (as in Theorems 3.5 and 3.6 for the Laplace transform).

Recall that the kth derivative of f is denoted $f^{(k)}$. As a convenience we may let $k = 0$ in this symbol, with the understanding that $f^{(0)} = f$.

THEOREM 14.7   *Differentiation in the Time Variable*

Let n be a positive integer. Suppose $f^{(n-1)}$ is continuous and $f^{(n)}$ is piecewise continuous on each interval $[-L, L]$. Suppose $\int_{-\infty}^{\infty} |f^{(n-1)}(t)| \, dt$ converges. Suppose

$$\lim_{t \to \infty} f^{(k)}(t) = \lim_{t \to -\infty} f^{(k)}(t) = 0$$

for $k = 0, 1, \ldots, n - 1$. Then

$$\mathfrak{F}[f^{(n)}(t)](\omega) = (i\omega)^n \hat{f}(\omega).$$

Proof   Begin with the first derivative. Integrating by parts, we have

$$\mathfrak{F}[f'](\omega) = \int_{-\infty}^{\infty} f'(t) e^{-i\omega t} \, dt$$

$$= \left[f(t) e^{-i\omega t} \right]_{-\infty}^{\infty} - \int_{-\infty}^{\infty} f(t)(-i\omega) e^{-i\omega t} \, dt.$$

Now $e^{-i\omega t} = \cos(\omega t) - i \sin(\omega t)$ has magnitude 1, and by assumption,

$$\lim_{t \to \infty} f(t) = \lim_{t \to -\infty} f(t) = 0.$$

Therefore,

$$\mathfrak{F}[f^{(n)}(t)](\omega) = i\omega \int_{-\infty}^{\infty} f(t) e^{-i\omega t} \, dt = i\omega \hat{f}(\omega).$$

The conclusion for higher derivatives follows by induction on n and the fact that

$$f'(t) = \frac{d}{dt} f^{(n-1)}(t). \quad \blacksquare$$

The assumption that f is continuous in the operational rule can be relaxed to allow for a finite number of jump discontinuities, if we allow for these in the conclusion by adding appropriate terms. We will state this result for the transform of f'.

THEOREM 14.8

Suppose f is continuous on the real line, except for jump discontinuities at $t_1, \ldots, t_M$. Let f' be piecewise continuous on every $[-L, L]$. Assume that $\int_{-\infty}^{\infty} |f(t)| \, dt$ converges, and that

$$\lim_{t \to \infty} f(t) = \lim_{t \to -\infty} f(t) = 0.$$

Then

$$\mathfrak{F}[f'](\omega) = i\omega \hat{f}(\omega) - \sum_{j=1}^{M} [f(t_j+) - f(t_j-)] e^{-it_j \omega}. \quad \blacksquare$$

Each term $f(t_j+) - f(t_j-)$ is the difference between the one-sided limits of $f(t)$ at the jump discontinuity t_j. This shows up in Figure 14.11 as the size of the jump between ends of the graph at this point.

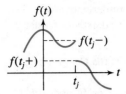

FIGURE 14.11 *The function f has a jump discontinuity at t_j.*

Proof Suppose first that f has a single jump discontinuity, at t_1. In the event of more jump discontinuities, the argument proceeds along the same lines but includes more of the type of calculation we are about to do.

Integrate by parts:

$$\mathcal{F}[f'](\omega) = \int_{-\infty}^{\infty} f'(t)e^{-i\omega t}\, dt$$

$$= \int_{-\infty}^{t_1} f'(t)e^{-i\omega t}\, dt + \int_{t_1}^{\infty} f'(t)e^{-i\omega t}\, dt$$

$$= \left[f(t)e^{-i\omega t}\right]_{-\infty}^{t_1} - \int_{-\infty}^{t_1} f(t)(-i\omega)e^{-i\omega t}\, dt$$

$$+ \left[f(t)e^{-i\omega t}\right]_{t_1}^{\infty} - (-i\omega)\int_{t_1}^{\infty} f(t)e^{-i\omega t}\, dt$$

$$= f(t_1-)e^{-it_1\omega} - f(t_1+)e^{-it_1\omega} + i\omega \int_{-\infty}^{\infty} f(t)e^{-i\omega t}\, dt$$

$$= i\omega\hat{f}(\omega) - [f(t_1+) - f(t_1-)]e^{-it_1\omega}. \ \blacksquare$$

Here is an example of the use of the operational rule in solving a differential equation.

EXAMPLE 14.15

Solve

$$y' - 4y = H(t)e^{-4t},$$

in which H is the Heaviside function. Thus the differential equation is

$$y' - 4y = \begin{cases} e^{-4t} & \text{for } t \geq 0 \\ 0 & \text{for } t < 0 \end{cases}.$$

Apply the Fourier transform to the differential equation to get

$$\mathcal{F}[y'](\omega) - 4\hat{y}(\omega) = \mathcal{F}[H(t)e^{-4t}](\omega).$$

Use Theorem 14.7 and Example 14.6 to write this equation as

$$i\omega\hat{y}(\omega) - 4\hat{y}(\omega) = \frac{1}{4 + i\omega}.$$

Solve for $\hat{y}(\omega)$ to obtain

$$\hat{y}(\omega) = \frac{-1}{16 + \omega^2}.$$

The solution is

$$y(t) = \mathfrak{F}^{-1}\left[\frac{-1}{16 + \omega^2}\right](t) = -\frac{1}{8}e^{-4|t|},$$

which is graphed in Figure 14.12.

The inverse transform just obtained can be derived in several ways. We can use a table of Fourier transforms or a software package that contains this transform. We can also see from Example 14.5 that

$$\mathfrak{F}\left[e^{-a|t|}\right](\omega) = \frac{2a}{a^2 + \omega^2}$$

and choose $a = 4$. ■

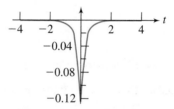

FIGURE 14.12 $y(t) = -\frac{1}{8}e^{-4|t|}$.

There is no arbitrary constant in this solution because the Fourier transform has returned the only solution that is continuous and bounded for all real t. Boundedness is assumed when we use the transform because of the required convergence of $\int_{-\infty}^{\infty} |y(t)|\, dt$.

14.4.2 Frequency Differentiation

The variable ω used for the Fourier transform is the frequency of $f(t)$, since it occurs in the complex exponential $e^{i\omega t}$, which is $\cos(\omega t) + i \sin(\omega t)$. In this context, differentiation of $\hat{f}(\omega)$ with respect to ω is called *frequency differentiation*. We will now relate derivatives of $\hat{f}(\omega)$ and $f(t)$.

THEOREM 14.9 *Frequency Differentiation*

Let n be a positive integer. Let f be piecewise continuous on $[-L, L]$ for every positive number L, and assume that $\int_{-\infty}^{\infty} |t^n f(t)|\, dt$ converges. Then

$$\mathfrak{F}[t^n f(t)](\omega) = i^n \frac{d^n}{d\omega^n} \hat{f}(\omega). \quad ■$$

In particular, under the conditions of the theorem,

$$\mathfrak{F}[tf(t)](\omega) = i\frac{d}{d\omega} \hat{f}(\omega) \quad \text{and} \quad \mathfrak{F}[t^2 f(t)](\omega) = -\frac{d^2}{d\omega^2} \hat{f}(\omega).$$

Proof   We will prove the theorem for $n = 1$. The argument for larger n is similar. Apply Leibniz's rule for differentiation under the integral to write

$$\frac{d}{d\omega} \hat{f}(\omega) = \frac{d}{d\omega} \int_{-\infty}^{\infty} f(t)e^{-i\omega t} \, dt = \int_{-\infty}^{\infty} \frac{\partial}{\partial \omega} \left[f(t)e^{-i\omega t} \right] dt$$

$$= \int_{-\infty}^{\infty} f(t)(-it)e^{-i\omega t} \, dt = -i \int_{-\infty}^{\infty} [tf(t)]e^{-i\omega t} \, dt$$

$$= -i\mathfrak{F}[tf(t)](\omega). \quad \blacksquare$$

EXAMPLE 14.16

Suppose we want to compute $\mathfrak{F}[t^2 e^{-5|t|}]$. Recall from Example 14.5 that

$$\mathfrak{F}[e^{-5|t|}](\omega) = \frac{10}{25 + \omega^2}.$$

By the frequency differentiation theorem,

$$\mathfrak{F}[t^2 e^{-5|t|}](\omega) = i^2 \frac{d^2}{d\omega^2} \left[\frac{10}{25 + \omega^2} \right] = 20 \frac{25 - 3\omega^2}{(25 + \omega^2)^3}. \quad \blacksquare$$

14.4.3   The Fourier Transform of an Integral

The following enables us to take the transform of a function defined by an integral.

THEOREM 14.10

Let f be piecewise continuous on every interval $[-L, L]$. Suppose $\int_{-\infty}^{\infty} |f(t)| \, dt$ converges. Suppose $\hat{f}(0) = 0$. Then

$$\mathfrak{F}\left[\int_{-\infty}^{t} f(\tau) \, d\tau \right] (\omega) = \frac{1}{i\omega} \hat{f}(\omega).$$

Proof   Let $g(t) = \int_{-\infty}^{t} f(\tau) \, d\tau$. Then $g'(t) = f(t)$ for any t at which f is continuous, and $g(t) \to 0$ as $t \to -\infty$. Further,

$$\lim_{t \to \infty} g(t) = \int_{-\infty}^{\infty} f(\tau) \, d\tau = \hat{f}(0) = 0.$$

We can therefore apply Theorem 14.7 to g to obtain

$$\hat{f}(\omega) = \mathfrak{F}[f(t)](\omega) = \mathfrak{F}[g'(t)](\omega)$$

$$= i\omega \mathfrak{F}[g(t)](\omega) = i\omega \mathfrak{F}\left[\int_{-\infty}^{t} f(\tau) \, d\tau \right] (\omega).$$

This is equivalent to the conclusion to be proved. $\blacksquare$

14.4.4   Convolution

There are many transforms defined by integrals, and it is common for such a transformation to have a convolution operation. We saw a convolution for the Laplace transform in Chapter 3. We will now discuss convolution for the Fourier transform.

DEFINITION 14.4 Convolution

Let f and g be functions defined on the real line. Then f has a convolution with g if

1. $\int_a^b f(t)\,dt$ and $\int_a^b g(t)\,dt$ exist for every interval $[a, b]$.
2. For every real number t,

$$\int_{-\infty}^{\infty} |f(t - \tau)g(\tau)|\,d\tau$$

converges. In this event, we define the convolution $f * g$ of f with g to be the function given by

$$(f * g)(t) = \int_{-\infty}^{\infty} f(t - \tau)g(\tau)\,d\tau.$$

In this definition, we wrote $(f * g)(t)$ for emphasis. However, the convolution is a function denoted $f * g$, so we can write just $f * g(t)$ to indicate $f * g$ evaluated at t.

THEOREM 14.11

Suppose f has a convolution with g. Then,

1. *Commutativity of Convolution* g has a convolution with f, and $f * g = g * f$.
2. *Linearity* If f and g both have convolutions with h, and α and β are real numbers, then $\alpha f + \beta g$ also has a convolution with h, and

$$(\alpha f + \beta g) * h = \alpha(f * h) + \beta(g * h).$$

Proof For (1), let $z = t - \tau$ to write

$$f * g(t) = \int_{-\infty}^{\infty} f(t - \tau)g(\tau)\,d\tau$$

$$= \int_{\infty}^{-\infty} f(z)g(t - z)(-1)\,dz = \int_{-\infty}^{\infty} g(t - z)f(z)\,dz = g * f(t).$$

Conclusion (2) follows from elementary properties of integrals, given that the integrals involved converge. ∎

We are now ready for the main results on convolution.

THEOREM 14.12

Suppose f and g are bounded and continuous on the real line and that $\int_{-\infty}^{\infty} |f(t)|\,dt$ and $\int_{-\infty}^{\infty} |g(t)|\,dt$ both converge. Then,

1.

$$\int_{-\infty}^{\infty} f * g(t)\,dt = \int_{-\infty}^{\infty} f(t)\,dt \int_{-\infty}^{\infty} g(t)\,dt.$$

2. *Time Convolution*

$$\widehat{f * g}(\omega) = \hat{f}(\omega)\hat{g}(\omega).$$

3. *Frequency Convolution*

$$\widehat{f(t)g(t)}(\omega) = \frac{1}{2\pi}(\hat{f} * \hat{g})(\omega). \quad \blacksquare$$

The first conclusion is that the integral, over the real line, of the convolution of f with g, is equal to the product of the integrals of f and of g over the line.

Time convolution states that the Fourier transform of a convolution is the product of the transforms of the functions. This formula can be stated

$$\mathfrak{F}[f * g](\omega) = \hat{f}(\omega)\hat{g}(\omega).$$

That is, the Fourier transform of the convolution of f with g, is equal to the product of the transform of f with the transform of g. This has the important inverse version

$$\mathfrak{F}^{-1}\left[\hat{f}(\omega)\hat{g}(\omega)\right](t) = f * g(t).$$

The inverse Fourier transform of the product of two transformed functions is equal to the convolution of these functions. This is sometimes of use in evaluating an inverse Fourier transform. If we want $\mathfrak{F}^{-1}[h(\omega)]$ and are able to factor $h(\omega)$ into $\hat{f}(\omega)\hat{g}(\omega)$, a product of the transforms of two known functions, then the inverse transform of h is the convolution of these known functions.

Frequency convolution can be stated

$$\mathfrak{F}[f(t)g(t)](\omega) = \frac{1}{2\pi}(\hat{f} * \hat{g})(\omega).$$

The Fourier transform of a product of two functions is equal to $(\frac{1}{2\pi})$ times the convolution of the transforms of these functions.

The inverse version of frequency convolution is

$$\mathfrak{F}^{-1}[\hat{f}(\omega) * \hat{g}(\omega)](t) = 2\pi f(t)g(t).$$

Proof   For (1), write

$$\int_{-\infty}^{\infty} f * g(t)\, dt = \int_{-\infty}^{\infty} \left(\int_{-\infty}^{\infty} f(t - \tau)g(\tau)\, d\tau \right) dt$$

$$= \int_{-\infty}^{\infty} \left(\int_{-\infty}^{\infty} f(t - \tau)g(\tau)\, dt \right) d\tau = \int_{-\infty}^{\infty} \left(\int_{-\infty}^{\infty} f(t - \tau)\, dt \right) g(\tau)\, d\tau,$$

assuming the validity of this interchange of the order of integration. Now,

$$\int_{-\infty}^{\infty} f(t - \tau)\, dt = \int_{-\infty}^{\infty} f(t)\, dt$$

for any real number τ. Therefore,

$$\int_{-\infty}^{\infty} f * g(t)\, dt = \int_{-\infty}^{\infty} \left(\int_{-\infty}^{\infty} f(t)\, dt \right) g(\tau)\, d\tau$$

$$= \int_{-\infty}^{\infty} f(t)\, dt \int_{-\infty}^{\infty} g(\tau)\, d\tau = \int_{-\infty}^{\infty} f(t)\, dt \int_{-\infty}^{\infty} g(t)\, dt.$$

For (2), begin by letting $F(t) = e^{-i\omega t} f(t)$ and $G(t) = e^{-i\omega t} g(t)$ for real t and ω. Then,

$$\widehat{f * g}(\omega) = \int_{-\infty}^{\infty} f * g(t) e^{-i\omega t} \, dt$$

$$= \int_{-\infty}^{\infty} \left(\int_{-\infty}^{\infty} f(t - \tau) g(\tau) \, d\tau \right) e^{-i\omega t} \, dt$$

$$= \int_{-\infty}^{\infty} \left(\int_{-\infty}^{\infty} e^{-i\omega t} f(t - \tau) g(\tau) \, d\tau \right) dt$$

$$= \int_{-\infty}^{\infty} \left(\int_{-\infty}^{\infty} e^{-i\omega(t-\tau)} f(t - \tau) e^{-i\omega \tau} g(\tau) \, d\tau \right) dt.$$

Now recognize that the integral in large parentheses in the last line is the convolution of F with G. Then, by (1) of this theorem applied to F and G,

$$\widehat{f * g}(\omega) = \int_{-\infty}^{\infty} F * G(t) \, dt = \int_{-\infty}^{\infty} F(t) \, dt \int_{-\infty}^{\infty} G(t) \, dt$$

$$= \int_{-\infty}^{\infty} f(t) e^{-i\omega t} \, dt \int_{-\infty}^{\infty} g(t) e^{-i\omega t} \, dt = \hat{f}(\omega) \hat{g}(\omega).$$

We leave conclusion (3) to the student. ∎

EXAMPLE 14.17

Suppose we want to compute

$$\mathfrak{F}^{-1} \left[\frac{1}{(4 + \omega^2)(9 + \omega^2)} \right].$$

Recognize the problem as one of computing the inverse transform of a product of functions whose individual transforms we know:

$$\frac{1}{4 + \omega^2} = \mathfrak{F}\left(\frac{1}{4} e^{-2|t|} \right) = \hat{f}(\omega) \quad \text{with } f(t) = \frac{1}{4} e^{-2|t|},$$

and

$$\frac{1}{9 + \omega^2} = \mathfrak{F}\left(\frac{1}{6} e^{-3|t|} \right) = \hat{g}(\omega) \quad \text{with } g(t) = \frac{1}{6} e^{-3|t|}.$$

The inverse version of conclusion (2) tells us that

$$\mathfrak{F}^{-1} \left[\frac{1}{(4 + \omega^2)(9 + \omega^2)} \right](t) = \mathfrak{F}^{-1}[\hat{f}(\omega)\hat{g}(\omega)](t) = f * g(t)$$

$$= \frac{1}{4} e^{-2|t|} * \frac{1}{6} e^{-3|t|} = \frac{1}{24} \int_{-\infty}^{\infty} e^{-2|t-\tau|} e^{-3|\tau|} \, d\tau.$$

We must be careful in evaluating this integral because of the absolute values in the exponents.

First, if $t > 0$, then

$$24[f * g(t)] = \int_{-\infty}^{0} e^{-2|t-\tau|} e^{-3|\tau|} \, d\tau + \int_{0}^{t} e^{-2|t-\tau|} e^{-3|\tau|} \, d\tau + \int_{t}^{\infty} e^{-2|t-\tau|} e^{-3|\tau|} \, d\tau$$

$$= \int_{-\infty}^{0} e^{-2(t-\tau)} e^{3\tau} \, d\tau + \int_{0}^{t} e^{-2(t-\tau)} e^{-3\tau} \, d\tau + \int_{t}^{\infty} e^{-2(\tau-t)} e^{-3\tau} \, d\tau$$

$$= \frac{6}{5} e^{-2t} - \frac{4}{5} e^{-3t}.$$

If $t < 0$, then

$$24[f * g(t)] = \int_{-\infty}^{t} e^{-2|t-\tau|} e^{-3|\tau|} \, d\tau + \int_{t}^{0} e^{-2|t-\tau|} e^{-3|\tau|} \, d\tau + \int_{0}^{\infty} e^{-2|t-\tau|} e^{-3|\tau|} \, d\tau$$

$$= \int_{-\infty}^{t} e^{-2(t-\tau)} e^{3\tau} \, d\tau + \int_{t}^{0} e^{2(t-\tau)} e^{3\tau} \, d\tau + \int_{0}^{\infty} e^{2(t-\tau)} e^{-3\tau} \, d\tau$$

$$= -\frac{4}{5} e^{3t} + \frac{6}{5} e^{2t}.$$

Finally, compute

$$24[f * g](0) = \int_{-\infty}^{\infty} e^{-2|\tau|} e^{-3|\tau|} \, d\tau = \frac{2}{5}.$$

Therefore,

$$\mathfrak{F}^{-1}\left[\frac{1}{(4+\omega^2)(9+\omega^2)}\right](t) = \frac{1}{24}\left(\frac{6}{5} e^{-2|t|} - \frac{4}{5} e^{-3|t|}\right)$$

$$= \frac{1}{20} e^{-2|t|} - \frac{1}{30} e^{-3|t|}. \ \blacksquare$$

14.4.5   Filtering and the Dirac Delta Function

A Dirac delta function is a pulse of infinite magnitude having infinitely short duration. One way to describe such an object mathematically is to form a short pulse

$$\frac{1}{2a}[H(t+a) - H(t-a)],$$

as shown in Figure 14.13, and take the limit as the width of the pulse approaches zero:

$$\delta(t) = \lim_{a \to 0} \frac{1}{2a}[H(t+a) - H(t-a)].$$

This is not a function in the standard sense, but is an object called a *distribution*. Distributions are generalizations of the function concept. For this reason, many theorems do not apply to $\delta(t)$. However, there are some formal manipulations that yield useful results. First, if we take the Fourier transform of the pulse, we get

$$\mathfrak{F}[H(t+a) - H(t-a)] = \int_{-a}^{a} e^{-i\omega t} \, dt = -\frac{1}{i\omega} e^{-i\omega t}\Big]_{-a}^{a}$$

$$= \frac{1}{i\omega}\left(e^{ia\omega} - e^{-ia\omega}\right) = 2\frac{\sin(a\omega)}{\omega}.$$

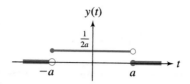

FIGURE 14.13
$y = \frac{1}{2a}[H(t+a) - H(t-a)]$.

By interchanging the limit and the operation of taking the transform, we have

$$\mathcal{F}[\delta(t)](\omega) = \mathcal{F}\left[\lim_{a \to 0} \frac{1}{2a}[H(t+a) - H(t-a)]\right](\omega)$$

$$= \lim_{a \to 0} \frac{1}{2a} \mathcal{F}[H(t+a) - H(t-a)](\omega)$$

$$= \lim_{a \to 0} \frac{\sin(a\omega)}{a\omega} = 1.$$

This leads us to consider the Fourier transform of the delta function to be the function that is identically 1.

Further, putting $\delta(t)$ formally through the convolution, we have

$$\mathcal{F}[\delta * f] = \mathcal{F}[\delta]\mathcal{F}[f] = \mathcal{F}[f]$$

and

$$\mathcal{F}[f * \delta] = \mathcal{F}[f]\mathcal{F}[\delta] = \mathcal{F}[f],$$

suggesting that

$$\delta * f = f * \delta = f.$$

The delta function behaves like the identity under convolution.

The following filtering property enables us to recover a function value by "summing" its values when hit with a shifted delta function.

THEOREM 14.13 *Filtering*

If f has a Fourier transform and is continuous at t_0, then

$$\int_{-\infty}^{\infty} f(t)\,\delta(t - t_0)\,dt = f(t_0). \quad \blacksquare$$

This result can be modified to allow for a jump discontinuity of f at t_0. In this event we get

$$\int_{-\infty}^{\infty} f(t)\delta(t - t_0)\,dt = \frac{1}{2}(f(t_0+) + f(t_0-)).$$

14.4.6 The Windowed Fourier Transform

Suppose f is a signal. This means that f is a function that is defined over the real line and has finite energy $\int_{-\infty}^{\infty} |f(t)|^2\,dt$.

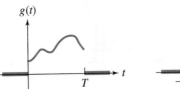

FIGURE 14.14 *Typical window function with compact support* $[0, T]$.

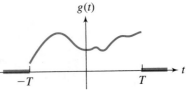

FIGURE 14.15 *Typical window function with compact support* $[-T, T]$.

In analyzing $f(t)$, we sometimes want to localize its frequency content with respect to the time variable. We have mentioned that $\hat{f}(\omega)$ carries information about the frequencies of the signal. However, $\hat{f}(\omega)$ does not particularize information to specific time intervals, since

$$\hat{f}(\omega) = \int_{-\infty}^{\infty} f(t)e^{-i\omega t}\, dt$$

and this integration is over all time. Hence the picture we obtain does not contain information about specific times, but instead enables us only to compute the total amplitude spectrum $\left|\hat{f}(\omega)\right|$. If we think of $f(t)$ as a piece of music being played over time, we would have to wait until the entire piece was done before even computing this amplitude spectrum. However, we can obtain a picture of the frequency content of $f(t)$ within given time intervals by windowing the function before taking its Fourier transform.

To do this, we first need a *window function g*, which is a function taking on nonzero values only on some closed interval, often $[0, T]$ or $[-T, T]$. Figures 14.14 and 14.15 show typical graphs of such functions, one on $[0, T]$ and the other on $[-T, T]$. The interval is called the *support* of g, and in this case that we are dealing with closed intervals, we say that g has *compact support*. The function g has zero values outside of this support interval. We window a function f with g by forming the product $g(t)f(t)$, which vanishes outside of $[-T, T]$.

EXAMPLE 14.18

Consider the window function

$$g(t) = \begin{cases} 1 & \text{for } -4 \leq t \leq 4 \\ 0 & \text{for } |t| > 4 \end{cases},$$

having compact support $[-4, 4]$. This function is graphed in Figure 14.16(a), with the vertical segments at $t = \pm 4$ included to emphasize this interval. Let $f(t) = t \sin(t)$, shown in Figure 14.16(b). To window f with g, form the product $g(t)f(t)$, shown in Figure 14.16(c). This windowed function vanished outside the support of g. For this choice of g, windowing has the effect of turning the signal $f(t)$ on at time -4 and turning it off at $t = 4$. ∎

The *windowed Fourier transform* (with respect to the choice of g) is

$$\mathfrak{F}_{win}[f](\omega) = \widehat{f_{win}}(\omega) = \int_{-\infty}^{\infty} f(t)g(t)e^{-i\omega t}\, dt$$

$$= \int_{-T}^{T} f(t)g(t)e^{-i\omega t}\, dt.$$

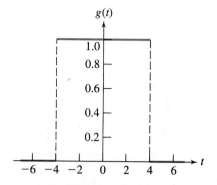

FIGURE 14.16(a) *Window function*

$$g(t) = \begin{cases} 1 & \text{for } |t| \leq 4 \\ 0 & \text{for } |t| > 4 \end{cases}.$$

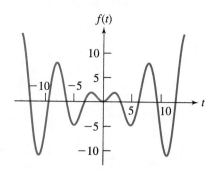

FIGURE 14.16(b) $f(t) = t \sin(t)$.

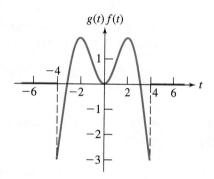

FIGURE 14.16(c) f *windowed with* g.

EXAMPLE 14.19

Let $f(t) = 6e^{-|t|}$. Then,

$$\hat{f}(\omega) = \int_{-\infty}^{\infty} 6e^{-|t|} e^{-i\omega t} \, dt = \frac{12}{1 + \omega^2}.$$

Use the window function

$$g(t) = \begin{cases} 1 & \text{for } -2 \leq t \leq 2 \\ 0 & \text{for } |t| > 2 \end{cases}.$$

Figure 14.17 shows a graph of the windowed function $g(t)f(t)$. The windowed Fourier transform of f is

$$\widehat{f_{win}}(\omega) = \int_{-\infty}^{\infty} 6e^{-|t|} g(t) e^{-i\omega t} \, dt$$

$$= \int_{-2}^{2} 6e^{-|t|} e^{-i\omega t} \, dt$$

$$= 12 \frac{-2e^{-2} \cos^2(\omega) + e^{-2} + e^{-2}\omega \sin(2\omega) + 1}{1 + \omega^2}.$$

This gives the frequency content of the signal f in the time interval $-2 \leq t \leq 2$. ∎

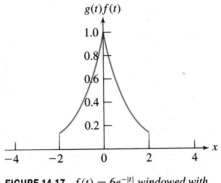

FIGURE 14.17 $f(t) = 6e^{-|t|}$ *windowed with*

$$g(t) = \begin{cases} 1 & \text{for } |t| \le 2 \\ 0 & \text{for } |t| > 2 \end{cases}.$$

Often we use a shifted window function. Suppose the support of g is $[-T, T]$. If $t_0 > 0$, then the graph of $g(t - t_0)$ is the graph of $g(t)$ shifted t_0 units to the right. Now

$$f(t)g(t - t_0) = \begin{cases} f(t)g(t - t_0) & \text{for } t_0 - T \le t \le t_0 + T \\ 0 & \text{for } t < t_0 - T \quad \text{and for } t > t_0 + T \end{cases}.$$

Figures 14.18(a) through (d) illustrate this process. In this case, we take the Fourier transform of the shifted windowed signal to be

$$\widehat{f_{win,t_0}}(\omega) = \mathfrak{F}[f(t)g(t - t_0)](\omega)$$

$$= \int_{t_0-T}^{t_0+T} f(t)g(t - t_0)e^{-i\omega t} \, dt.$$

This gives the frequency content of the signal in the time interval $[t_0 - T, t_0 + T]$.

Engineers sometimes refer to the windowing process as *time-frequency localization*. If g is the window function, the *center* of g is defined to be the point

$$t_C = \frac{\int_{-\infty}^{\infty} t \, |g(t)|^2 \, dt}{\int_{-\infty}^{\infty} |g(t)|^2 \, dt}.$$

The number

$$t_R = \left(\frac{\int_{-\infty}^{\infty} (t - t_C)^2 \, |g(t)|^2 \, dt}{\int_{-\infty}^{\infty} |g(t)|^2 \, dt} \right)^{1/2}$$

is the *radius* of the window function. The width of the window function is $2t_R$ and is referred to as the *RMS duration of the window*. It is assumed in this terminology that the integrals involved all converge.

When we deal with the Fourier transform of the window function, then similar terminology applies:

$$\text{center of } \hat{g} = \omega_C = \frac{\int_{-\infty}^{\infty} \omega \, |\hat{g}(\omega)|^2 \, d\omega}{\int_{-\infty}^{\infty} |\hat{g}(\omega)|^2 \, d\omega}$$

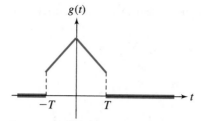

FIGURE 14.18(a) *A window function g on* $[-T, T]$.

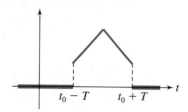

FIGURE 14.18(b) *Shifted window function* $g(t - t_0)$.

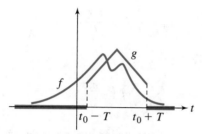

FIGURE 14.18(c) *Typical signal* $f(t)$.

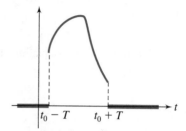

FIGURE 14.18(d) $g(t - t_0)f(t)$.

and

$$\text{radius of } \hat{g} = \omega_R = \left(\frac{\int_{-\infty}^{\infty} (\omega - \omega_C)^2 \, |\hat{g}(\omega)|^2 \, d\omega}{\int_{-\infty}^{\infty} |\hat{g}(\omega)|^2 \, d\omega} \right)^{1/2}.$$

The width of $\hat{g}$ is $2\omega_R$, a number referred to as the *RMS bandwidth* of the window function.

14.4.7 The Shannon Sampling Theorem

We will derive the Shannon sampling theorem, which states that a band-limited signal can be reconstructed from certain sampled values.

A signal f is *band-limited* if its Fourier transform $\hat{f}$ has compact support (has nonzero values only on a closed interval of finite length). This means that, for some L,

$$\hat{f}(\omega) = 0 \quad \text{if } |\omega| > L.$$

Usually we choose L to be the smallest number for which this condition holds. In this event, L is the *bandwidth* of the signal. The total frequency content of such a signal f lies in the band $[-L, L]$.

We will now show that we can reconstruct a band-limited signal from samples taken at appropriately chosen times. Begin with the integral for the inverse Fourier transform, assuming that we can recover $f(t)$ for all real t from its transform:

$$f(t) = \frac{1}{2\pi} \int_{-\infty}^{\infty} \hat{f}(\omega)e^{i\omega t} \, d\omega.$$

Because f is band-limited,

$$f(t) = \frac{1}{2\pi} \int_{-L}^{L} \hat{f}(\omega)e^{i\omega t} \, d\omega. \tag{14.8}$$

Put this aside for the moment and write the complex Fourier series for $\hat{f}(\omega)$ on $[-L, L]$:

$$\hat{f}(\omega) = \sum_{n=-\infty}^{\infty} c_n e^{n\pi i \omega / L}, \tag{14.9}$$

where

$$c_n = \frac{1}{2L} \int_{-L}^{L} \hat{f}(\omega) e^{-n\pi i \omega / L} \, d\omega.$$

Comparing c_n with $f(t)$ in equation (14.8), we conclude that

$$c_n = \frac{\pi}{L} f\left(-\frac{n\pi}{L}\right).$$

Substitute this into equation (14.9) to get

$$\hat{f}(\omega) = \sum_{n=-\infty}^{\infty} \frac{\pi}{L} f\left(-\frac{n\pi}{L}\right) e^{n\pi i \omega / L}.$$

Since n takes on all integer values in this summation, we can replace n with $-n$ to write

$$\hat{f}(\omega) = \frac{\pi}{L} \sum_{n=-\infty}^{\infty} f\left(\frac{n\pi}{L}\right) e^{-n\pi i \omega / L}.$$

Now substitute this series for $\hat{f}(\omega)$ into equation (14.8) to get

$$f(t) = \frac{1}{2\pi} \frac{\pi}{L} \int_{-L}^{L} \sum_{n=-\infty}^{\infty} f\left(\frac{n\pi}{L}\right) e^{-n\pi i \omega / L} e^{i\omega t} \, d\omega.$$

Interchange the sum and the integral to get

$$f(t) = \frac{1}{2L} \sum_{n=-\infty}^{\infty} f\left(\frac{n\pi}{L}\right) \int_{-L}^{L} e^{i\omega(t - n\pi / L)} \, d\omega$$

$$= \frac{1}{2L} \sum_{n=-\infty}^{\infty} f\left(\frac{n\pi}{L}\right) \frac{1}{i(t - n\pi/L)} \left[e^{i\omega(t - n\pi/L)} \right]_{-L}^{L}$$

$$= \frac{1}{2L} \sum_{n=-\infty}^{\infty} f\left(\frac{n\pi}{L}\right) \frac{1}{i(t - n\pi/L)} \left(e^{i(Lt - n\pi)} - e^{-i(Lt - n\pi)} \right) \tag{14.10}$$

$$= \sum_{n=-\infty}^{\infty} f\left(\frac{n\pi}{L}\right) \frac{1}{Lt - n\pi} \frac{1}{2i} \left(e^{i(Lt - n\pi)} - e^{-i(Lt - n\pi)} \right)$$

$$= \sum_{n=-\infty}^{\infty} f\left(\frac{n\pi}{L}\right) \frac{\sin(Lt - n\pi)}{Lt - n\pi}.$$

This means that $f(t)$ is known for all times t if just the function values $f(n\pi/L)$ are determined for all integer values of n. An engineer would sample the signal $f(t)$ at times $0, \pm\pi/L, \pm2\pi/L, \ldots$. Once the values of $f(t)$ are known for these times, then equation (14.10) reconstructs the entire signal. This is actually the way engineers convert digital signals to analog signals, with application to technology such as that involved in making compact discs.

Equation (14.10) is known as the *Shannon sampling theorem*. We will encounter it again when we discuss wavelets. In the case $L = \pi$, the sampling theorem has the simple form

$$f(t) = \sum_{n=-\infty}^{\infty} f(n) \frac{\sin(\pi(t - n))}{\pi(t - n)} \tag{14.11}$$

14.4.8   Lowpass and Bandpass Filters

Consider a signal f, not necessarily band-limited. However, we assume that the signal has finite energy, so

$$\int_{-\infty}^{\infty} |f(t)|^2 \, dt$$

is finite. Such functions are called *square integrable*, and we will also encounter them later with wavelet expansions.

The spectrum of f is given by its Fourier transform

$$\hat{f}(\omega) = \int_{-\infty}^{\infty} f(t) e^{-i\omega t} \, dt.$$

If f is not band-limited, we can replace f with a band-limited signal f_{ω_0} with bandwidth not exceeding a positive number ω_0 by applying a lowpass filter that cuts off $\hat{f}(\omega)$ at frequencies outside the range $[-\omega_0, \omega_0]$. That is, let

$$\widehat{f_{\omega_0}}(\omega) = \begin{cases} \hat{f}(\omega) & \text{for } -\omega_0 \le \omega \le \omega_0 \\ 0 & \text{for } |\omega| > \omega_0 \end{cases}.$$

This defines the transform of the function f_{ω_0}, from which we recover f_{ω_0} by the inverse Fourier transform:

$$f_{\omega_0}(t) = \frac{1}{2\pi} \int_{-\infty}^{\infty} \widehat{f_{\omega_0}}(\omega) e^{i\omega t} \, d\omega = \frac{1}{2\pi} \int_{-\omega_0}^{\omega_0} \widehat{f_{\omega_0}}(\omega) e^{i\omega t} \, d\omega.$$

The process of applying the lowpass filter is carried out mathematically by multiplying by an appropriate function (essentially windowing). Define the characteristic function χ_I of an interval I by

$$\chi_I(t) = \begin{cases} 1 & \text{if } t \text{ is in } I \\ 0 & \text{if } t \text{ is a real number that is not in } I \end{cases}.$$

Now observe that

$$\widehat{f_{\omega_0}}(\omega) = \chi_{[-\omega_0, \omega_0]}(\omega) \hat{f}(\omega), \tag{14.12}$$

or, more succinctly,

$$\widehat{f_{\omega_0}} = \chi_{[-\omega_0, \omega_0]} \hat{f}.$$

In this context, $\chi_{[-\omega_0, \omega_0]}$ is called the *transfer function*. Its graph is shown in Figure 14.19. The inverse Fourier transform of the transfer function is

$$\mathfrak{F}^{-1}[\chi_{[-\omega_0, \omega_0]}](t) = \frac{1}{2\pi} \int_{-\omega_0}^{\omega_0} e^{i\omega t} \, d\omega = \frac{\sin(\omega_0 t)}{\pi t},$$

whose graph is given in Figure 14.20. In the case that $\omega_0 = \pi$, this is the function, evaluated at $t - n$ instead of t, which occurs in the Shannon sampling formula (14.11) that reconstructs $f(t)$

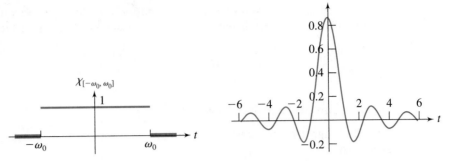

FIGURE 14.19 *Graph of* $\chi_{[-\omega_0,\omega_0]}$.

FIGURE 14.20 *Graph of* $\sin(\omega_0 t)/\pi t$ *for* $\omega_0 = 2.7$.

from sampled values $f(n)$ on the integers. For this reason $\sin(\omega_0 t)/\pi t$ is called the *Shannon sampling function*.

Now recall Theorem 14.12(2) and (3) of Section 14.4.4. Analog filtering in the time variable t is done by convolution. If $\varphi(t)$ is the filter function, then the effect of filtering a function f by φ is a new function g defined by

$$g(t) = (\varphi * f)(t) = \int_{-\infty}^{\infty} \varphi(\xi) f(t - \xi) \, d\xi.$$

Taking the Fourier transform of this equation, we have

$$\hat{g}(\omega) = \hat{\varphi}(\omega) \hat{f}(\omega).$$

We therefore filter in the frequency variable by taking a product of the Fourier transform of the filter function with the transform of the function being filtered.

We can now formulate equation (14.12) as

$$f_{\omega_0}(t) = \left(\frac{\sin(\omega_0 t)}{\pi t} * f(t) \right).$$

This gives the lowpass filtering of f as the convolution of the Shannon sampling function with f.

In lowpass filtering, we produce from the signal f a new signal f_{ω_0} that is band-limited. That is, we filter out the frequencies of the signal outside of $[-\omega_0, \omega_0]$. In a similar kind of filtering, called *bandpass filtering*, we want to filter out the effects of the signal outside of given bandwidths. A band-limited signal f can be decomposed into a sum of signals, each of which carries the information content of f within a certain given frequency band. To see how to do this, let f be a band-limited signal of bandwidth Ω. Consider a finite increasing sequence of frequencies,

$$0 < \omega_1 < \omega_2 < \cdots < \omega_N = \Omega.$$

For $j = 1, \ldots, N$, define a bandwidth filter function β_j by means of its transfer function:

$$\hat{\beta}_j = \chi_{[-\omega_j, -\omega_{j-1}]} + \chi_{[\omega_{j-1}, \omega_j]}.$$

This transfer function, which is a sum of characteristic functions of frequency intervals, is graphed in Figure 14.21. The bandwidth filter function $\beta_j(t)$, which filters the frequency content of $f(t)$ outside of the frequency range $[\omega_{j-1}, \omega_j]$, is obtained by taking the inverse Fourier transform of

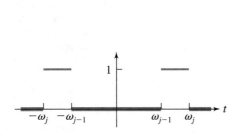

FIGURE 14.21 $\chi_{[-\omega_j,-\omega_{j-1}]} + \chi_{[\omega_{j-1},\omega_j]}$.

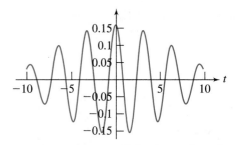

FIGURE 14.22 $\beta_j(t) = \dfrac{\sin(\omega_j t) - \sin(\omega_{j-1}t)}{\pi t}$

with $\omega_j = 2.2$ and $\omega_{j-1} = 1.7$.

$\hat{\beta}_j(\omega)$. We get

$$\beta_j(t) = \frac{\sin(\omega_j t) - \sin(\omega_{j-1}t)}{\pi t},$$

whose graph is shown in Figure 14.22.

Now define functions

$$f_0(t) = \left(\frac{\sin(\omega_0 t)}{\pi t} * f\right)(t)$$

and, for $j = 1, 2, \ldots, N$,

$$f_j(t) = (\beta_j * f)(t)$$

Then, for $j = 1, 2, \ldots, N$, each $f_j(t)$ carries the content of the signal $f(t)$ in the frequency range $\omega_{j-1} \leq \omega \leq \omega_j$, while $f_0(t)$ carries the content in $[0, \omega_0]$, which is the low-frequency range of $f(t)$. Further,

$$f(t) = f_0(t) + f_1(t) + f_2(t) + \cdots + f_N(t), \tag{14.13}$$

giving a decomposition of the signal into components carrying the information of the signal for specific frequency intervals.

SECTION 14.4 PROBLEMS

In each of Problems 1 through 8, determine the Fourier transform of the function.

1. $\dfrac{t}{9 + t^2}$

2. $3te^{-9t^2}$

3. $26H(t)te^{-2t}$

4. $H(t - 3)(t - 3)e^{-4t}$

5. $\dfrac{d}{dt}[H(t)e^{-3t}]$

6. $t[H(t + 1) - H(t - 1)]$

7. $\dfrac{5e^{3it}}{t^2 - 4t + 13}$

8. $H(t - 3)e^{-2t}$

In each of Problems 9, 10, and 11, use convolution to find the inverse Fourier transform of the function.

9. $\dfrac{1}{(1 + i\omega)^2}$

10. $\dfrac{1}{(1 + i\omega)(2 + i\omega)}$

11. $\dfrac{\sin(3\omega)}{\omega(2 + i\omega)}$

In each of Problems 12, 13, and 14, find the inverse Fourier transform of the function.

12. $\dfrac{6e^{4i\omega} \sin(2\omega)}{9 + \omega^2}$

13. $e^{-3|\omega+4|} \cos(2\omega + 8)$

14. $e^{-\omega^2/9} \sin(8\omega)$

15. Prove that
$$\int_{-\infty}^{\infty} f(t)g(t)\, dt = \frac{1}{2\pi} \int_{-\infty}^{\infty} \hat{f}(w)\hat{g}(-w)\, dw.$$

Hint: First use the frequency convolution formula to obtain
$$\mathfrak{F}[f(t)g(t)](\omega) = \frac{1}{2\pi} \int_{-\infty}^{\infty} \hat{f}(w)\hat{g}(\omega - w)\, dw.$$

Conclude that
$$\int_{-\infty}^{\infty} f(t)g(t)e^{-i\omega t}\, dt = \frac{1}{2\pi} \int_{-\infty}^{\infty} \hat{f}(w)\hat{g}(\omega - w)\, dw.$$

Finally, let $\omega = 0$.

16. Let f and g be real-valued functions. Show that
$$\int_{-\infty}^{\infty} f(t)g(t)\, dt = \frac{1}{2\pi} \int_{-\infty}^{\infty} \hat{f}(\omega)\overline{\hat{g}(\omega)}\, d\omega,$$
assuming that the integral on the left converges. *Hint:* Recall that $\overline{\hat{g}(\omega)} = \hat{g}(-\omega)$ from Problem 33 of Section 14.3.

17. Prove the following form of Parseval's theorem:
$$\int_{-\infty}^{\infty} |f(t)|^2\, dt = \frac{1}{2\pi} \int_{-\infty}^{\infty} |\hat{f}(\omega)|^2\, d\omega.$$

18. The *power content* of a signal $f(t)$ is defined to be $\int_{-\infty}^{\infty} |f(t)|^2\, dt$, assuming that this integral converges. Determine the power content of $H(t)e^{-2t}$. (See also Problem 16, Section 13.6)

19. Determine the power content of $(1/t)\sin(3t)$. *Hint:* Use the result of Problem 17.

20. Use the Fourier transform to solve
$$y'' + 6y' + 5y = \delta(t - 3).$$

21. Prove that $|\hat{f}(\omega)| \le \int_{-\infty}^{\infty} |f(t)|\, dt$.

22. Prove the frequency convolution theorem.

23. Evaluate $\int_{-\infty}^{\infty} 8\delta(t - 3)H(t - 3)e^{-5t}\, dt$.

24. Suppose, instead of the definition given, we had defined the Fourier transform of f to be $\int_{-\infty}^{\infty} f(t)e^{-2\pi i\omega t}\, dt$, as is sometimes done. Show that now the operational formula for the transform of a derivative would be
$$\mathfrak{F}[f'(t)](\omega) = 2\pi i\omega \hat{f}(\omega).$$

What would be the corresponding formula for the Fourier transform of $f^{(n)}(t)$?

In each of Problems 25 through 30, compute the windowed Fourier transform of the given function f, using the windowing function g. Also compute the center and RMS bandwidth of the window function.

25. $f(t) = t^2,\ g(t) = \begin{cases} 1 & \text{for } -5 \le t \le 5 \\ 0 & \text{for } |t| > 5 \end{cases}$

26. $f(t) = \cos(at)$,
$$g(t) = \begin{cases} 1 & \text{for } -4\pi \le t \le 4\pi \\ 0 & \text{for } |t| > 4\pi \end{cases}$$

27. $f(t) = e^{-t}$,
$$g(t) = \begin{cases} 1 & \text{for } 0 \le t \le 4 \\ 0 & \text{for } t < 0 \ \text{ and for } t > 4 \end{cases}$$

28. $f(t) = e^t \sin(\pi t),\ g(t) = \begin{cases} 1 & \text{for } -1 \le t \le 1 \\ 0 & \text{for } |t| > 1 \end{cases}$

29. $f(t) = (t + 2)^2,\ g(t) = \begin{cases} 1 & \text{for } -2 \le t \le 2 \\ 0 & \text{for } |t| > 2 \end{cases}$

30. $f(t) = H(t - \pi)$,
$$g(t) = \begin{cases} 1 & \text{for } 3\pi \le t \le 5\pi \\ 0 & \text{for } t < 3\pi \ \text{ and for } t > 5\pi \end{cases}$$

31. This problem illustrates the Shannon sampling theorem. Let
$$f(t) = \begin{cases} \dfrac{1}{2\pi}\dfrac{-8t\cos(2t) + 4\sin(2t)}{t^3} & \text{for } t \ne 0 \\[3mm] \dfrac{16}{3\pi} & \text{for } t = 0 \end{cases}.$$

Assume that f is band-limited with bandwidth 2. Compute $f(n)$ for $n = 0, \pm 1, \ldots, \pm 5$ and use the Shannon sampling theorem, with $L = 2$ and with the sum taken from $n = -5$ to $n = 5$, to obtain a function $g(t)$. Graph g and f on the same set of axes.

32. Let
$$f(t) = \begin{cases} \dfrac{1 - \cos(t)}{\pi t^2} & \text{for } t \ne 0 \\[3mm] \dfrac{1}{2\pi} & \text{for } t = 0 \end{cases}.$$

Assume that f is band-limited with bandwidth 1. Compute $f(n)$ for $n = 0, \pm 1, \ldots, \pm 10$ and use the Shannon sampling theorem, with $L = 2$ and with the sum taken from $n = -10$ to $n = 10$, to obtain a function $g(t)$. Graph g and f on the same set of axes.

14.5   The Fourier Cosine and Sine Transforms

We saw in Section 14.3 how the Fourier integral representation of a function suggested its Fourier transform. We will now show how the Fourier cosine and sine integrals of a function suggest cosine and sine transforms.

Suppose $f(t)$ is piecewise smooth on each interval $[0, L]$ and $\int_0^\infty |f(t)|\, dt$ converges. Then for each t at which f is continuous,

$$f(t) = \int_0^\infty a_\omega \cos(\omega t)\, d\omega,$$

where

$$a_\omega = \frac{2}{\pi} \int_0^\infty f(t) \cos(\omega t)\, dt.$$

Based on these two equations, we make the following.

DEFINITION 14.5   *Fourier Cosine Transform*

The Fourier cosine transform of f is defined by

$$\mathfrak{F}_c[f](\omega) = \int_0^\infty f(t) \cos(\omega t)\, dt. \tag{14.14}$$

Often we will denote $\mathfrak{F}_c[f](\omega) = \hat{f}_C(\omega)$.

Notice that

$$\hat{f}_C(\omega) = \frac{\pi}{2} a_\omega$$

and that

$$f(t) = \frac{2}{\pi} \int_0^\infty \hat{f}_C(\omega) \cos(\omega t)\, d\omega. \tag{14.15}$$

The integrals in expressions (14.14) and (14.15) form the transform pair for the Fourier cosine transform. The latter enables us, under certain conditions, to recover $f(t)$ from $\hat{f}_C(\omega)$.

EXAMPLE 14.20

Let K be a positive number and let

$$f(t) = \begin{cases} 1 & \text{for } 0 \le t \le K \\ 0 & \text{for } t > K \end{cases}.$$

The Fourier transform of f is

$$\hat{f}_C(\omega) = \int_0^\infty f(t) \cos(\omega t)\, dt$$

$$= \int_0^K \cos(\omega t)\, dt = \frac{\sin(K\omega)}{\omega}. \quad \blacksquare$$

The Fourier sine transform is defined in the same spirit.

DEFINITION 14.6   *Fourier Sine Transform*

The Fourier sine transform of f is defined by

$$\mathfrak{F}_S[f](\omega) = \int_0^\infty f(t) \sin(\omega t) \, dt.$$

We also denote this as $\hat{f}_S(\omega)$.

If f is continuous at $t > 0$, then the Fourier integral sine representation is

$$f(t) = \int_0^\infty b_\omega \sin(\omega t) \, d\omega,$$

where

$$b_\omega = \frac{2}{\pi} \int_0^\infty f(t) \sin(\omega t) \, dt.$$

Since

$$\hat{f}_S(\omega) = \frac{\pi}{2} b_\omega$$

then

$$f(t) = \frac{2}{\pi} \int_0^\infty \hat{f}_S(\omega) \sin(\omega t) \, d\omega,$$

and this is the means by which we retrieve $f(t)$ from $\hat{f}_S(\omega)$.

EXAMPLE 14.21

With f the function of Example 14.20,

$$\hat{f}_S(\omega) = \int_0^\infty f(t) \sin(\omega t) \, dt$$

$$= \int_0^K \sin(\omega t) \, dt = \frac{1}{\omega}[1 - \cos(K\omega)]. \ \blacksquare$$

Both of these transforms are linear:

$$\mathfrak{F}_C[\alpha f + \beta g] = \alpha \mathfrak{F}_C[f] + \beta \mathfrak{F}_C[g]$$

and

$$\mathfrak{F}_S[\alpha f + \beta g] = \alpha \mathfrak{F}_S[f] + \beta \mathfrak{F}_S[g],$$

whenever all of these transforms are defined.

When these transforms are used to solve differential equations, the following operational rules play a key role.

—— **THEOREM 14.14** *Operational Rules* ——————————————————————————————————

Let f and f' be continuous on every $[0, L]$, and let $\int_0^\infty |f(t)|\, dt$ converge. Suppose $f(t) \to 0$ and $f'(t) \to 0$ as $t \to \infty$. Suppose f'' is piecewise continuous on every interval $[0, L]$. Then,

1. $$\mathfrak{F}_C[f''(t)](\omega) = -\omega^2 \hat{f}_C(\omega) - f'(0)$$

and

2. $$\mathfrak{F}_S[f''(t)](\omega) = -\omega^2 \hat{f}_S(\omega) + \omega f(0). \quad \blacksquare$$

The theorem is proved by integrating by parts twice for each rule, and we leave the details to the student.

The operational formula dictates which transform is used to solve a given problem. If we seek a function $f(t)$, for $0 \le t < \infty$, and $f(0)$ is specified, then we might consider a Fourier sine transform. If, however, information is given about $f'(0)$, then the cosine transform might be appropriate. When we solve partial differential equations, we will encounter examples where this strategy is invoked.

—— **SECTION 14.5 PROBLEMS** ————————————————————————————

In each of Problems 1 through 6, determine the Fourier cosine transform and the Fourier sine transform of the function.

1. $f(t) = e^{-t}$

2. $f(t) = te^{-at}$, with a any positive number

3. $f(t) = \begin{cases} \cos(t) & \text{for } 0 \le t \le K \\ 0 & \text{for } t > K \end{cases}$,

with K any positive number

4. $f(t) = \begin{cases} 1 & \text{for } 0 \le t < K \\ -1 & \text{for } K \le t < 2K \\ 0 & \text{for } t \ge 2K \end{cases}$

5. $f(t) = e^{-t} \cos(t)$

6. $f(t) = \begin{cases} \sinh(t) & \text{for } K \le t < 2K \\ 0 & \text{for } 0 \le t < K \quad \text{and for } t \ge 2K \end{cases}$

7. Show that under appropriate conditions on f and its derivatives,

$$\mathfrak{F}_S[f^{(4)}(t)](\omega) = \omega^4 \hat{f}_S(\omega) - \omega^3 f(0) + \omega f''(0).$$

Hint: Consider conditions that allow application of the operational formula to $(f''(t))''$.

8. Show that under appropriate conditions on f and its derivatives,

$$\mathfrak{F}_C[f^{(4)}(t)](\omega) = \omega^4 \hat{f}_C(\omega) + \omega^2 f'(0) - f^{(3)}(0).$$

14.6   The Finite Fourier Cosine and Sine Transforms

The Fourier transform, cosine transform, and sine transform are all motivated by the respective integral representations of a function. If we employ essentially the same line of reasoning, but using Fourier cosine and sine series instead of integrals, we obtain what are called *finite transforms*.

Suppose f is piecewise smooth on $[0, \pi]$.

DEFINITION 14.7 *Finite Fourier Cosine Transform*

The finite Fourier cosine transform of f is defined by

$$\mathfrak{C}[f](n) = \tilde{f}_C(n) = \int_0^\pi f(x) \cos(nx)\, dx$$

for $n = 0, 1, 2, \ldots$.

If f is continuous at x in $[0, \pi]$, then $f(x)$ has the Fourier cosine series representation

$$f(x) = \frac{1}{2}a_0 + \sum_{n=1}^\infty a_n \cos(nx),$$

where

$$a_n = \frac{2}{\pi} \int_0^\pi f(x) \cos(nx)\, dx = \frac{2}{\pi} \tilde{f}_C(n).$$

Then

$$f(x) = \frac{1}{\pi} \tilde{f}_C(0) + \frac{2}{\pi} \sum_{n=1}^\infty \tilde{f}_C(n) \cos(nx),$$

an inversion-type of expression from which we can recover $f(x)$ from the finite Fourier cosine transform of f.

By the same token, we can define a finite sine transform.

DEFINITION 14.8 *Finite Fourier Sine Transform*

The finite Fourier sine transform of f is defined by

$$\mathfrak{S}[f](n) = \tilde{f}_S(n) = \int_0^\pi f(x) \sin(nx)\, dx$$

for $n = 1, 2, \ldots$.

For $0 < x < \pi$, if f is continuous at x, then the sine series representation is

$$f(x) = \frac{2}{\pi} \sum_{n=1}^\infty \tilde{f}_S(n) \sin(nx),$$

an inversion formula for the finite sine transform.

EXAMPLE 14.22

Let $f(x) = x^2$ for $0 \le x \le \pi$. For the finite cosine transform, compute

$$\tilde{f}_C(0) = \int_0^\pi x^2\, dx = \frac{1}{3}\pi^3$$

and, for $n = 1, 2, \ldots,$

$$\widetilde{f_C}(n) = \int_0^\pi x^2 \cos(nx)\, dx = 2\pi \frac{(-1)^n}{n^2}.$$

For the finite sine transform, compute

$$\widetilde{f_S}(n) = \int_0^\pi x^2 \sin(nx)\, dx = \frac{(-1)^n[2 - n^2\pi^2] - 2}{n^3}. \ \blacksquare$$

Here are the fundamental operational rules for these transforms.

THEOREM 14.15 *Operational Rules*

Let f and f' be continuous on $[0, \pi]$ and let f'' be piecewise continuous. Then,

1. $$\mathfrak{C}[f''](n) = -n^2 \widetilde{f_C}(n) - f'(0) + (-1)^n f'(\pi),$$

for $n = 1, 2, \ldots,$ and

2. $$\mathfrak{S}[f''](n) = -n^2 \widetilde{f_S}(n) + n f(0) - n(-1)^n f(\pi)$$

for $n = 1, 2, \ldots . \ \blacksquare$

We will see applications of these finite transforms when we discuss partial differential equations. The Fourier transforms covered at this point can be found in Tables 14.1 through 14.5.

TABLE 14.1 **Fourier Transforms**

$f(x)$	$\mathfrak{F}[f](\omega) = \hat{f}(\omega)$
$\dfrac{1}{x}$	$\begin{cases} i & \text{if } \omega > 0 \\ 0 & \text{if } \omega = 0 \\ -i & \text{if } \omega < 0 \end{cases}$
$e^{-a\|x\|} \quad (a > 0)$	$\dfrac{2a}{a^2 + \omega^2}$
$xe^{-a\|x\|} \quad (a > 0)$	$-\dfrac{4ai\omega}{(a^2 + \omega^2)^2}$
$\|x\|e^{-a\|x\|} \quad (a > 0)$	$\dfrac{2(a^2 - \omega^2)}{(a^2 + \omega^2)^2}$
$e^{-a^2 x^2} \quad (a > 0)$	$\dfrac{\sqrt{\pi}}{a} e^{-\omega^2/4a^2}$
$\dfrac{1}{a^2 + x^2} \quad (a > 0)$	$\dfrac{\pi}{a} e^{-a\|\omega\|}$
$\dfrac{x}{a^2 + x^2} \quad (a > 0)$	$-\dfrac{i}{2}\dfrac{\pi}{a}\omega e^{-a\|\omega\|}$
$H(x + a) - H(x - a)$	$\dfrac{2\sin(a\omega)}{\omega}$
$\delta(x)$	1

TABLE 14.2 *Fourier Cosine Transforms*

$f(x)$	$\mathfrak{F}_C[f](\omega) = \hat{f}_C(\omega)$
$x^{r-1} \quad (0 < r < 1)$	$\Gamma(r) \cos\left(\dfrac{\pi r}{2}\right) \omega^{-r}$
$e^{-ax} \quad (a > 0)$	$\dfrac{a}{a^2 + \omega^2}$
$xe^{-ax} \quad (a > 0)$	$\dfrac{a^2 - \omega^2}{(a^2 + \omega^2)^2}$
$e^{-a^2 x^2} \quad (a > 0)$	$\dfrac{\sqrt{\pi}}{2} a^{-1} \omega e^{-\omega^2/4a^2}$
$\dfrac{1}{a^2 + x^2} \quad (a > 0)$	$\dfrac{\pi}{2} \dfrac{1}{a} e^{-a\omega}$
$\dfrac{1}{(a^2 + x^2)^2} \quad (a > 0)$	$\dfrac{\pi}{4} a^{-3} e^{-a\omega}(1 + a\omega)$
$\cos\left(\dfrac{x^2}{2}\right)$	$\dfrac{\sqrt{\pi}}{2}\left[\cos\left(\dfrac{\omega^2}{2}\right) + \sin\left(\dfrac{\omega^2}{2}\right)\right]$
$\sin\left(\dfrac{x^2}{2}\right)$	$\dfrac{\sqrt{\pi}}{2}\left[\cos\left(\dfrac{\omega^2}{2}\right) - \sin\left(\dfrac{\omega^2}{2}\right)\right]$
$\dfrac{1}{2}(1 + x)e^{-x}$	$\dfrac{1}{(1 + \omega^2)^2}$
$\sqrt{\dfrac{2}{\pi x}}$	$\dfrac{1}{\sqrt{\omega}}$
$e^{-x/\sqrt{2}} \sin\left(\dfrac{\pi}{4} + \dfrac{x}{\sqrt{2}}\right)$	$\dfrac{1}{1 + \omega^2}$
$e^{-x/\sqrt{2}} \cos\left(\dfrac{\pi}{4} + \dfrac{x}{\sqrt{2}}\right)$	$\dfrac{\omega^2}{1 + \omega^4}$
$\dfrac{2}{x} e^{-x} \sin(x)$	$\tan^{-1}\left(\dfrac{2}{\omega^2}\right)$
$H(x) - H(x - a)$	$\dfrac{1}{\omega} \sin(a\omega)$

TABLE 14.3   *Fourier Sine Transforms*

$f(x)$	$\mathfrak{F}_S[f](\omega) = \hat{f}_S(\omega)$
$\dfrac{1}{x}$	$\begin{cases} \dfrac{\pi}{2} & \text{if } \omega > 0 \\[2mm] -\dfrac{\pi}{2} & \text{if } \omega < 0 \end{cases}$
$x^{r-1} \quad (0 < r < 1)$	$\Gamma(r) \sin\left(\dfrac{\pi r}{2}\right) \omega^{-r}$
$\dfrac{1}{\sqrt{x}}$	$\sqrt{\dfrac{\pi}{2}}\, \omega^{-1/2}$
$e^{-ax} \quad (a > 0)$	$\dfrac{\omega}{a^2 + \omega^2}$
$xe^{-ax} \quad (a > 0)$	$\dfrac{2a\omega}{(a^2 + \omega^2)^2}$
$xe^{-a^2 x^2} \quad (a > 0)$	$\dfrac{\sqrt{\pi}}{4} a^{-3} \omega e^{-\omega^2/4a^2}$
$x^{-1} e^{-ax} \quad (a > 0)$	$\tan^{-1}\left(\dfrac{\omega}{a}\right)$
$\dfrac{x}{a^2 + x^2} \quad (a > 0)$	$\dfrac{\pi}{2} e^{-a\omega}$
$\dfrac{x}{(a^2 + x^2)^2} \quad (a > 0)$	$\dfrac{\pi}{4a} \omega e^{-a\omega}$
$\dfrac{1}{x(a^2 + x^2)} \quad (a > 0)$	$\dfrac{\pi}{2} a^{-2}(1 - e^{-a\omega})$
$e^{-x/\sqrt{2}} \sin\left(\dfrac{x}{\sqrt{2}}\right)$	$\dfrac{\omega}{1 + \omega^4}$
$\dfrac{2}{\pi} \dfrac{x}{a^2 + x^2}$	$e^{-a\omega}$
$\dfrac{2}{\pi} \tan^{-1}\left(\dfrac{a}{x}\right)$	$\dfrac{1}{\omega}(1 - e^{-a\omega}) \quad (a > 0)$
$\operatorname{erfc}\left(\dfrac{x}{2\sqrt{a}}\right)$	$\dfrac{1}{\omega}(1 - e^{-a\omega^2})$
$\dfrac{4}{\pi} \dfrac{x}{4 + x^4}$	$e^{-\omega} \sin(\omega)$
$\sqrt{\dfrac{2}{\pi x}}$	$\dfrac{1}{\sqrt{\omega}}$

TABLE 14.4 *Finite Fourier Cosine Transforms*

$f(x)$	$\tilde{f}_C(n)$
1	$\begin{cases} 0 & \text{if } n = 1, 2, 3, \ldots \\ \pi & \text{if } n = 0 \end{cases}$
$\begin{cases} 1 & \text{if } 0 \le x \le k \\ -1 & \text{if } k < x < \pi \end{cases}$	$\begin{cases} \dfrac{2}{\pi} \sin(nk) & \text{if } n = 1, 2, 3, \ldots \\ 2k - \pi & \text{if } n = 0 \end{cases}$
x	$\begin{cases} \dfrac{-[1 - (-1)^n]}{n^2} & \text{if } n = 1, 2, 3, \ldots \\ \dfrac{\pi^2}{2} & \text{if } n = 0 \end{cases}$
x^2	$\begin{cases} \dfrac{2\pi(-1)^n}{n} & \text{if } n = 1, 2, 3, \ldots \\ \dfrac{\pi^3}{3} & \text{if } n = 0 \end{cases}$
x^3	$\begin{cases} 3\pi^2 \dfrac{(-1)^n}{n^2} + 6 \dfrac{[1 - (-1)^n]}{n^4} & \text{if } n = 1, 2 \\ \dfrac{\pi^2}{4} & \text{if } n = 0 \end{cases}$
e^{ax}	$\left(\dfrac{(-1)^n e^{a\pi} - 1}{n^2 + a^2} \right) a$
$\sin(ax), \quad a \ne \text{integer}$	$\left(\dfrac{(-1)^n \cos(a\pi) - 1}{n^2 - a^2} \right) a$
$\sin(kx), \quad k = 1, 2, \ldots$	$\begin{cases} \dfrac{(-1)^{n+k} - 1}{n^2 - k^2} & \text{if } n \ne k \\ 0 & \text{if } n = k \end{cases}$
$\cos(kx), \quad k = 1, 2, 3, \ldots$	$\begin{cases} 0 & \text{if } n \ne k \\ \dfrac{\pi}{2} & \text{if } n = k \end{cases}$
$f(\pi - x)$	$(-1)^n f_C(n)$
$\dfrac{1}{\pi} \dfrac{1 - k^2}{1 + k^2 - 2k\cos(x)}$	$k^n \quad (|k| < 1)$
$-\dfrac{1}{\pi} \ln[1 + k^2 - 2k\cos(x)]$	$\dfrac{1}{n} k^n \quad (|k| < 1)$
$\dfrac{\cosh[k(\pi - x)]}{k \sinh(k\pi)}$	$\dfrac{1}{k^2 + n^2} \quad (k \ne 0)$
$\dfrac{1}{\pi} \dfrac{\sinh(y)}{\cosh(y) - \cos(x)}$	$e^{-ny} \quad (y > 0)$
$\dfrac{1}{\pi} [y - \ln(2\cosh(y) - 2\cos(x))]$	$\dfrac{1}{n} e^{-ny} \quad (y > 0)$

TABLE 14.5	*Finite Fourier Sine Transforms*

$f(x)$	$\tilde{f}_S(n)$
$\dfrac{\pi - x}{\pi}$	$\dfrac{1}{n}$
x	$\pi \dfrac{(-1)^{n+1}}{n}$
1	$\dfrac{1 - (-1)^n}{n}$
$\dfrac{x(\pi^2 - x^2)}{6}$	$\dfrac{(-1)^{n+1}}{n^3}$
$\dfrac{x(\pi - x)}{2}$	$\dfrac{1 - (-1)^n}{n^3}$
x^2	$\dfrac{\pi^2 (-1)^{n+1}}{n} - \dfrac{2[1 - (-1)^n]}{n^3}$
x^3	$\pi(-1)^n \left(\dfrac{6}{n^3} - \dfrac{\pi^2}{n} \right)$
e^{ax}	$\dfrac{n}{n^2 + a^2}[1 - (-1)^n e^{a\pi}]$
$\sin(kx), \quad k = 1, 2, 3, \ldots$	$\begin{cases} 0 & \text{if } n \neq k \\ \dfrac{\pi}{2} & \text{if } n = k \end{cases}$
$\cos(ax), \quad a \neq \text{integer}$	$\dfrac{n}{n^2 - a^2}[1 - (-1)^n \cos(a\pi)]$
$\cos(kx), \quad k = 1, 2, 3, \ldots$	$\dfrac{n(1 - (-1)^{n+k})}{n^2 - k^2}$ if $n \neq k$; 0 if $n = k$
$\dfrac{2}{\pi} \tan^{-1}\left(\dfrac{2a \sin(x)}{1 - a^2} \right) \quad (|a| < 1)$	$\dfrac{1 - (-1)^n}{n} a^n$
$f(\pi - x)$	$(-1)^n f_S(n)$
$\dfrac{x}{n}(\pi - x)(2\pi - x)$	$\dfrac{6}{n^3}$
$\cosh(kx)$	$\dfrac{n}{n^2 + k^2}[1 - (-1)^n \cosh(k\pi)]$
$\dfrac{\sin[k(\pi - x)]}{\sin(k\pi)}$	$\dfrac{n}{n^2 - k^2} \quad (k \neq \text{integer})$
$\dfrac{\sinh[k(\pi - x)]}{\sinh(k\pi)}$	$\dfrac{n}{n^2 + k^2} \quad (k \neq 0)$
$\dfrac{2}{\pi} \dfrac{k \sin(x)}{1 + k^2 - 2k \cos(x)}$	$k^n \quad (|k| < 1)$
$\dfrac{2}{\pi} \tan^{-1}\left[\dfrac{k \sin(x)}{1 - k \cos(x)} \right]$	$\dfrac{1}{n} k^n \quad (|k| < 1, n \neq 0)$
$\dfrac{1}{\pi} \dfrac{\sin(x)}{\cosh(y) - \cos(x)}$	$e^{-ny} \quad (y > 0)$
$\dfrac{1}{\pi} \tan^{-1}\left[\dfrac{\sin(x)}{e^y - \cos(x)} \right]$	$\dfrac{1}{n} e^{-ny} \quad (y > 0)$

SECTION 14.6   PROBLEMS

In each of Problems 1 through 7, find the finite Fourier sine transform of the function.

1. K (any constant)

2. x

3. x^2

4. x^5

5. $\sin(ax)$

6. $\cos(ax)$

7. e^{-x}

In each of Problems 8 through 14, find the finite Fourier cosine transform of the function.

8. $f(x) = \begin{cases} 1 & \text{for } 0 \le x < \frac{1}{2} \\ -1 & \text{for } \frac{1}{2} \le x \le \pi \end{cases}$

9. x

10. x^2

11. x^3

12. $\cosh(ax)$

13. $\sin(ax)$

14. e^{-x}

15. Suppose f is continuous on $[0, \pi]$ and f' is piecewise continuous. Prove that

$$(\widetilde{f'})_S(n) = -n\widetilde{f}_C(n)$$

for $n = 1, 2, \ldots$.

16. Let f be continuous and f' piecewise continuous on $[0, \pi]$. Prove that

$$(\widetilde{f'})_C(n) = n\widetilde{f}_S(n) - f(0) + (-1)^n f(\pi)$$

for $n = 0, 1, 2, \ldots$.

In the following, f is piecewise continuous on $[0, \pi]$ and n and m are positive integers. Let $g(x) = f(x)\cos(mx)$ and $h(x) = f(x)\sin(mx)$.

17. Prove that

$$\widetilde{f}_S(n + m) = \widetilde{g}_S(n) + \widetilde{h}_C(n).$$

18. Prove that

$$\widetilde{f}_C(n + m) = \widetilde{g}_C(n) - \widetilde{h}_S(n).$$

19. Prove that, if $m < n$, then

$$\widetilde{g}_C(n) = \frac{1}{2}\left[\widetilde{f}_C(n - m) + \widetilde{f}_C(n + m)\right].$$

20. Prove that, if $m < n$, then

$$\widetilde{h}_C(n) = \frac{1}{2}\left[\widetilde{f}_S(n + m) - \widetilde{f}_S(n - m)\right].$$

14.7   The Discrete Fourier Transform

If f has period p, its complex Fourier series is

$$\sum_{k=-\infty}^{\infty} d_k e^{ik\omega_0 t}.$$

Here $\omega_0 = 2\pi/p$ and the complex Fourier coefficients are given by

$$d_k = \frac{1}{p} \int_\alpha^{\alpha+p} f(t)e^{-ik\omega_0 t}\, dt,$$

in which, because of the periodicity of f, α can be any number. If we substitute the value of ω_0, the complex Fourier series of f is

$$\sum_{k=-\infty}^{\infty} d_k e^{2\pi ikt/p}.$$

Under certain conditions on f, this series converges to $\frac{1}{2}(f(t+) + f(t-))$ at any number t.

We will choose $\alpha = 0$ in the formula for the coefficients, so

$$d_k = \frac{1}{p} \int_0^p f(t) e^{-2\pi i k t/p} \, dt \quad \text{for } k = 0, \pm 1, \pm 2, \dots.$$

To motivate the definition of the discrete Fourier transform, suppose we want to approximate d_k. One way is to subdivide $[0, p]$ into N subintervals of equal length p/N and choose $t_j = jp/N$ in each $[jp/N, (j+1)p/N]$ for $j = 0, 1, \dots, N-1$. Now approximate d_k by a Riemann sum:

$$
\begin{aligned}
d_k &\approx \frac{1}{p} \sum_{j=0}^{N-1} f(t_j) e^{-2\pi i k t_j/p} \frac{p}{N} \\
&= \frac{1}{N} \sum_{j=0}^{N-1} f(t_j) e^{-2\pi i k j/N}.
\end{aligned}
\tag{14.16}
$$

The N-point Fourier transform is a rule that acts on a given sequence of N complex numbers and produces an infinite sequence of complex numbers, one for each integer k (although with periodic repetitions, as we will see later). We will define the transform in such a way that, except for the $1/N$ factor, the approximating sum (14.16) is exactly the N-point discrete Fourier transform of the numbers $f(t_0), f(t_1), \dots, f(t_{N-1})$.

DEFINITION 14.9   *N-Point Discrete Fourier Transform*

Let N be a positive integer. Let $u = \{u_j\}_{j=0}^{N-1}$ be a sequence of N complex numbers. Then the N-point discrete Fourier transform of u is the sequence $\mathbb{D}[u]$ defined by

$$\mathbb{D}[u](k) = \sum_{j=0}^{N-1} u_j e^{-2\pi i j k/N}$$

for $k = 0, \pm 1, \pm 2, \dots.$

To simplify the notation, we will use a convention used with the Laplace transform and denote the N-point discrete Fourier transform of a sequence u by U (lowercase for the given sequence of N numbers, uppercase of the same letter for its N-point discrete Fourier transform). In this notation, if $u = \{u_j\}_{j=0}^{N-1}$, then

$$U_k = \sum_{j=0}^{N-1} u_j e^{-2\pi i j k/N}.$$

We will also abbreviate the phrase "discrete Fourier transform" to DFT.

EXAMPLE 14.23

Consider the constant sequence $u = \{c\}_{j=0}^{N-1}$, in which c is a complex number. The N-point DFT is given by

$$U_k = \sum_{j=0}^{N-1} c e^{-2\pi i j k/N} = c \sum_{j=0}^{N-1} \left(e^{-2\pi i k/N} \right)^j.$$

Now recall that the sum of a finite geometric series is

$$\sum_{j=0}^{N-1} r^j = \frac{1 - r^N}{1 - r}.$$ (14.17)

Applying this to U_k, we have

$$U_k = c\frac{1 - \left(e^{-2\pi ik/N}\right)^N}{1 - e^{-2\pi ik/N}}$$

$$= c\frac{1 - e^{-2\pi ik}}{1 - e^{-2\pi ik/N}} = 0 \quad \text{for } k = 0, \pm 1, \pm 2, \dots$$

because, for any integer k,

$$e^{-2\pi ik} = \cos(2\pi k) - i\sin(2\pi k) = 1.$$

For any positive integer N, the N-point DFT of a constant sequence of N numbers is an infinite sequence of zeros. In more relaxed terms, the N-point DFT of a constant sequence is zero. ∎

EXAMPLE 14.24

Let a be a complex number and N a positive integer. To avoid trivialities, suppose a is not an integer multiple of π. We will find the N-point DFT of the sequence $u = \{\sin(ja)\}_{j=0}^{N-1}$. Denoting this transform by the uppercase letter, we have

$$U_k = \sum_{j=0}^{N-1} \sin(ja)e^{-2\pi ijk/N}.$$

Use the fact that

$$\sin(ja) = \frac{1}{2i}\left(e^{ija} - e^{-ija}\right)$$

to write

$$U_k = \frac{1}{2i}\sum_{j=0}^{N-1} e^{ija}e^{-2\pi ijk/N} - \frac{1}{2i}\sum_{j=0}^{N-1} e^{-ija}e^{-2\pi ijk/N}$$

$$= \frac{1}{2i}\sum_{j=0}^{N-1}\left(e^{ia-2\pi ik/N}\right)^j - \frac{1}{2i}\sum_{j=0}^{N-1}\left(e^{-ia-2\pi ik/N}\right)^j.$$

Upon using equation (14.17) on each sum, we have

$$U_k = \frac{1}{2i}\frac{1 - \left(e^{ia-2\pi ik/N}\right)^N}{1 - e^{ia-2\pi ik/N}} - \frac{1}{2i}\frac{1 - \left(e^{-ia-2\pi ik/N}\right)^N}{1 - e^{-ia-2\pi ik/N}}$$

$$= \frac{1}{2i}\frac{1 - e^{iaN}e^{-2\pi ik}}{1 - e^{ia-2\pi ik/N}} - \frac{1}{2i}\frac{1 - e^{-iaN}e^{-2\pi ik}}{1 - e^{-ia-2\pi ik/N}}$$ (14.18)

$$= \frac{1}{2i}\frac{1 - e^{iaN}}{1 - e^{ia-2\pi ik/N}} - \frac{1}{2i}\frac{1 - e^{-iaN}}{1 - e^{-ia-2\pi ik/N}},$$

since $e^{-2\pi ik} = 1$.

To make the example more explicit, suppose $N = 5$ and $a = \sqrt{2}$. Then the given sequence u is

$$u_0 = 0, \quad u_1 = \sin(\sqrt{2}), \quad u_2 = \sin(2\sqrt{2}), \quad u_3 = \sin(3\sqrt{2}), \quad u_4 = \sin(4\sqrt{2}).$$

The 5-point DFT U has kth term

$$U_k = \frac{1}{2i} \frac{1 - e^{5i\sqrt{2}}}{1 - e^{i\sqrt{2} - 2\pi ik/5}} - \frac{1}{2i} \frac{1 - e^{-5i\sqrt{2}}}{1 - e^{-i\sqrt{2} - 2\pi ik/5}}.$$

For example,

$$U_0 = \frac{1}{2i} \frac{1 - e^{5i\sqrt{2}}}{1 - e^{i\sqrt{2}}} - \frac{1}{2i} \frac{1 - e^{-5i\sqrt{2}}}{1 - e^{-i\sqrt{2}}}$$

$$= \frac{\sin(4\sqrt{2}) + \sin(\sqrt{2}) - \sin(5\sqrt{2})}{2 - 2\cos(\sqrt{2})},$$

$$U_1 = \frac{1}{2i} \frac{1 - e^{5i\sqrt{2}}}{1 - e^{i\sqrt{2} - 2\pi i/5}} - \frac{1}{2i} \frac{1 - e^{-5i\sqrt{2}}}{1 - e^{-i\sqrt{2} - 2\pi i/5}}$$

and

$$U_2 = \frac{1}{2i} \frac{1 - e^{5i\sqrt{2}}}{1 - e^{i\sqrt{2} - 4\pi i/5}} - \frac{1}{2i} \frac{1 - e^{-5i\sqrt{2}}}{1 - e^{-i\sqrt{2} - 4\pi i/5}}. \ \blacksquare$$

We will develop some properties of this transform.

14.7.1 Linearity and Periodicity

If $u = \{u_j\}_{j=0}^{N-1}$ and $v = \{v_j\}_{j=0}^{N-1}$ are sequences of complex numbers and a and b are complex numbers, then

$$au + bv = \{au_j + bv_j\}_{j=0}^{N=1}.$$

Linearity of the N-point DFT is the property:

$$\mathbb{D}[au + bv](k) = aU_k + bV_k.$$

This follows immediately from the definition of the transform, since

$$\mathbb{D}[au + bv](k) = \sum_{j=0}^{N-1}(au_j + bv_j)e^{-2\pi ijk/N}$$

$$= a\sum_{j=0}^{N-1}u_j e^{-2\pi ijk/N} + b\sum_{j=0}^{N-1}v_j e^{-2\pi ijk/N} = aU_k + bV_k.$$

Next we will show that the N-point DFT is periodic of period N. This means that if the given sequence is $u = \{u_j\}_{j=0}^{N-1}$, then for any integer k,

$$U_{k+N} = U_k.$$

This can be seen in the DFT calculated in Example 14.24. In equation (14.18), replace k by $k + N$. In this example, this change shows up only in the term $e^{ia - 2\pi ik/N}$ in the denominator. But, if k is replaced by $k + N$ in this exponential, no change results, since

$$e^{ia - 2\pi i(k+N)/N} = e^{ia}e^{-2\pi ik}e^{-2\pi i} = e^{ia}e^{-2\pi ik}.$$

The argument in general proceeds as follows:

$$U_{k+N} = \sum_{j=0}^{N-1} u_j e^{-2\pi i j(k+N)/N}$$

$$= \sum_{j=0}^{N-1} u_j e^{-2\pi i j k/N} e^{-2\pi i j k} = \sum_{j=0}^{N-1} u_j e^{-2\pi i j k/N} = U_k,$$

since $e^{-2\pi i j k} = 1$.

14.7.2 The Inverse N-Point DFT

Suppose we have an N-point DFT

$$U_k = \sum_{j=0}^{N-1} u_j e^{-2\pi i j k/N}$$

of a sequence $\{u_j\}_{j=0}^{N-1}$ of N numbers. We claim that

$$u_j = \frac{1}{N} \sum_{k=0}^{N-1} U_k e^{2\pi i j k/N} \quad \text{for } j = 0, 1, \ldots, N-1. \tag{14.19}$$

Because this expression retrieves the original N-point sequence from its discrete transform, equation (14.19) is called the *inverse N-point discrete Fourier transform*.

To verify equation (14.19), it is convenient to put $W = e^{-2\pi i/N}$. Then

$$W^N = 1 \quad \text{and} \quad W^{-1} = e^{2\pi i/N}.$$

Now write

$$\frac{1}{N} \sum_{k=0}^{N-1} U_k e^{2\pi i j k/N} = \frac{1}{N} \sum_{k=0}^{N-1} U_k W^{-jk}$$

$$= \frac{1}{N} \sum_{k=0}^{N-1} \left(\sum_{r=0}^{N-1} u_r e^{-2\pi i r k/N} \right) W^{-jk} = \frac{1}{N} \sum_{k=0}^{N-1} \sum_{r=0}^{N-1} u_r W^{rk} W^{-jk} \tag{14.20}$$

$$= \frac{1}{N} \sum_{r=0}^{N-1} u_r \sum_{k=0}^{N-1} W^{rk} W^{-jk}.$$

In the last summation, observe that

$$W^{rk} W^{-jk} = e^{-2\pi i r k/N} e^{2\pi i j k/N} = e^{-2\pi i (r-j)k/N} = W^{(r-j)k}.$$

For given j, if $r \neq j$, then by equation (14.17) for the finite sum of a geometric series,

$$\sum_{k=0}^{N-1} W^{rk} W^{-jk} = \sum_{k=0}^{N-1} W^{(r-j)k} = \sum_{k=0}^{N-1} (W^{r-j})^k = \frac{1 - (W^{r-j})^N}{1 - W^{r-j}} = 0$$

because $(W^{(r-j)})^N = e^{-2\pi i(r-j)} = 1$ and $W^{r-j} = e^{-2\pi i(r-j)/N} \neq 1$. But if $r = j$, then

$$\sum_{k=0}^{N-1} W^{rk} W^{-jk} = \sum_{k=0}^{N-1} W^{jk} W^{-jk} = \sum_{k=0}^{N-1} 1 = N.$$

Therefore, in the last double sum in equation (14.20), we need retain only the term when $r = j$ in the summation with respect to r, yielding

$$\frac{1}{N} \sum_{r=0}^{N-1} u_r \sum_{k=0}^{N-1} W^{rk} W^{-jk} = \frac{1}{N} u_j \sum_{k=0}^{N-1} W^{jk} W^{-jk} = \frac{1}{N} u_j N = u_j,$$

and verifying equation (14.19).

14.7.3 DFT Approximation of Fourier Coefficients

We began this section by defining the N-point DFT so that Riemann sums approximating the Fourier coefficients of a periodic function were $1/N$ times the N-point DFT of the sequence of function values at partition points of the interval. We will now pursue more closely the idea of approximating Fourier coefficients by a discrete Fourier transform, with the idea of sampling partial sums of Fourier series. This approximation also allows the application of DFT software to the approximation of Fourier coefficients.

We will consider a specific example, $f(t) = \sin(t)$ for $0 \le t < 4$, with the understanding that f is extended over the entire real line with period 4. A graph of part of f is shown in Figure 14.23. With $p = 4$, the Fourier coefficients are

$$d_k = \frac{1}{4} \int_0^4 \sin(\xi) e^{-2\pi i k \xi / 4} \, d\xi = \frac{1}{4} \int_0^4 \sin(\xi) e^{-\pi i k \xi / 2} \, d\xi$$

$$= \frac{\cos(4) - 1}{\pi^2 k^2 - 4} + \frac{1}{2} i \frac{\pi k \sin(4)}{\pi^2 k^2 - 4}. \tag{14.21}$$

for $k = 0, \pm 1, \pm 2, \dots$.

Now let N be a positive integer and subdivide $[0, 4]$ into N subintervals of equal length $4/N$. These subintervals are $[4j/N, 4(j+1)/N]$ for $j = 0, 1, \dots, N - 1$. Form N numbers by evaluating $f(t)$ at the left endpoint of each of these subintervals. These points are $4j/N$, so we obtain the N-point sequence

$$u = \left\{ \sin\left(\frac{4j}{N}\right) \right\}_{j=0}^{N-1}.$$

Form the N-point DFT of this sequence:

$$U_k = \sum_{j=0}^{N-1} \sin\left(\frac{4j}{N}\right) e^{-2\pi i j k / 4} = \sum_{j=0}^{N-1} \sin\left(\frac{4j}{N}\right) e^{-\pi i j k / 2}.$$

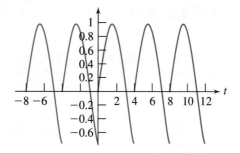

FIGURE 14.23 $f(t) = \sin(t)$ *for* $0 \le t < 4$, *extended periodically over the real line.*

Then

$$\frac{1}{N} U_k = \frac{1}{N} \sum_{j=0}^{N-1} \sin\left(\frac{4j}{N}\right) e^{-\pi i j k / 2}$$

is a Riemann sum for the integral defining d_k. We ask: To what extent does $(1/N)U_k$ approximate d_k? In this example, we have an explicit expression (14.21) for d_k. We will explicitly evaluate $(1/N)U_k$, using $a = 4/N$ in the DFT of $\{\sin(ja)\}_{j=0}^{N-1}$ determined in Example 14.24. This gives us

$$\frac{1}{N} U_k = \frac{1}{N} \left[\frac{1}{2i} \frac{1 - e^{4i}}{1 - e^{4i/N - 2k\pi i/N}} - \frac{1}{2i} \frac{1 - e^{-4i}}{1 - e^{-4i/N - 2k\pi i/N}} \right].$$

Now approximate the exponential terms in the denominator by using the approximation

$$e^x \approx 1 + x$$

for $|x| \ll 1$. Then

$$\frac{1}{N} U_k \approx \frac{1}{N} \left[\frac{1}{2i} \frac{1 - e^{4i}}{1 - [1 + (4i/N - 2k\pi i/N)]} - \frac{1}{2i} \frac{1 - e^{-4i}}{1 - [1 + (-4i/N - 2k\pi i/N)]} \right]$$

$$= -\frac{1}{4} \left[\frac{1 - e^{4i}}{-2 + k\pi} - \frac{1 - e^{-4i}}{2 + k\pi} \right]$$

$$= -\frac{1}{4} \frac{1}{\pi^2 k^2 - 4} \left[4 - \pi k(e^{4i} - e^{-4i}) - 2(e^{4i} + e^{-4i}) \right]$$

$$= -\frac{1}{4} \frac{1}{\pi^2 k^2 - 4} \left[4 - 2\pi k i \sin(4) - 4\cos(4) \right]$$

$$= \frac{\cos(4) - 1}{\pi^2 k^2 - 4} + \frac{1}{2} i \frac{\pi k \sin(4)}{\pi^2 k^2 - 4}.$$

The approximation $e^x \approx 1 + x$ has therefore led to an approximate expression for $(1/N)U_k$ that is exactly equal to d_k. This approximation cannot be valid for all k, however. First, the approximation used for e^x assumes that $|x| \ll 1$, and second, the N-point DFT is periodic of period N, so $U_{k+N} = U_k$, while there is no such periodicity in the d_k's.

In general, it would be very difficult to derive an estimate on relative sizes of $|k|$ and N that would result in $(1/N)U_k$ approximating d_k to within a given tolerance and that would hold for a reasonable class of functions. However, for many science and engineering applications, the empirical rule $|k| \leq N/8$ has proved effective.

SECTION 14.7   PROBLEMS

In each of Problems 1 through 6, compute $\mathbb{D}[u](k)$ for $k = 0, \pm 1, \ldots, \pm 4$ for the given sequence u.

1. $\{\cos(j)\}_{j=0}^5$

2. $\{e^{ij}\}_{j=0}^5$

3. $\{1/(j + 1)\}_{j=0}^5$

4. $\{1/(j + 1)^2\}_{j=0}^5$

5. $\{j^2\}_{j=0}^5$

6. $\{\cos(j) - \sin(j)\}_{j=0}^5$

In each of Problems 7 through 12, a sequence $\{U_k\}_{k=0}^N$ is given. Determine the N-point inverse discrete Fourier transform of this sequence.

7. $U_k = (1+i)^k$, $N = 6$

8. $U_k = i^{-k}$, $N = 5$

9. $U_k = e^{-ik}$, $N = 7$

10. $U_k = k^2$, $N = 5$

11. $U_k = \cos(k)$, $N = 5$

12. $U_k = \ln(k+1)$, $N = 6$

13. $f(t) = \cos(t)$ for $0 \le t \le 2$, f has period 2

14. $f(t) = e^{-t}$ for $0 \le t < 3$, f has period 3

15. $f(t) = t^2$ for $0 \le t < 1$, f has period 1

16. $f(t) = te^{2t}$ for $0 \le t < 4$, f has period 4

17. $f(t) = t \cos(t)$ for $0 \le t < 3\pi$, f has period 3π

18. $f(t) = e^{-t} \cos(2t)$ for $0 < t \le 2$, f has period 2

In each of Problems 13 through 18, compute the first seven complex Fourier coefficients d_0, $d_{\pm 1}$, $d_{\pm 2}$, and $d_{\pm 3}$ of f (see Section 13.7). Then use the DFT to approximate these coefficients, with $N = 128$.

14.8   Sampled Fourier Series

In the preceding subsection we discussed approximation of Fourier coefficients of a periodic function f. This was done by approximating terms of an N-point discrete Fourier transform formed by sampling $f(t)$ at N points of $[0, p]$. We will now discuss the use of an inverse DFT to approximate sampled partial sums of the Fourier series of a periodic function (that is, partial sums evaluated at chosen points).

Consider the partial sum

$$S_M(t) = \sum_{k=-M}^{M} d_k e^{2\pi ikt/p}.$$

Subdivide $[0, p]$ into N subintervals and choose sample points $t_j = jp/N$ for $n = 0, 1, \ldots, N-1$. Form the N-point sequence $u = \{f(jp/N)\}_{j=0}^{N-1}$ and approximate

$$d_k \approx \frac{1}{N} U_k,$$

where

$$U_k = \sum_{j=0}^{N-1} f\left(\frac{jp}{N}\right) e^{-2\pi ijk/N}$$

In order to have $|k| \le N/8$, as mentioned at the end of the preceding subsection, we will require that $M \le N/8$. Thus,

$$S_M(t) \approx \sum_{k=-M}^{M} \frac{1}{N} U_k e^{2\pi ikt/p}.$$

In particular, if we sample this partial sum at the partition points jp/N, then

$$S_M\left(\frac{jp}{N}\right) \approx \frac{1}{N} \sum_{k=-M}^{M} U_k e^{2\pi ijk/N}.$$

We will show that the sum on the right is actually an N-point inverse DFT for a particular N-point sequence, which we will now determine. We will exploit the periodicity of the N-point DFT—that is, $U_{k+N} = U_k$ for all integers k. Write

$$
\begin{aligned}
S_M\left(\frac{jp}{N}\right) &\approx \frac{1}{N}\sum_{k=-M}^{-1} U_k e^{2\pi ijk/N} + \frac{1}{N}\sum_{k=0}^{M} U_k e^{2\pi ijk/N} \\
&= \frac{1}{N}\sum_{k=1}^{M} U_{-k} e^{-2\pi ijk/N} + \frac{1}{N}\sum_{k=0}^{M} U_k e^{2\pi ijk/N} \\
&= \frac{1}{N}\sum_{k=1}^{M} U_{-k+N} e^{2\pi ij(-k+N)/N} + \frac{1}{N}\sum_{k=0}^{M} U_k e^{2\pi ijk/N} \\
&= \frac{1}{N}\sum_{k=N-M}^{N-1} U_k e^{2\pi ijk/N} + \frac{1}{N}\sum_{k=0}^{M} U_k e^{2\pi ijk/N}.
\end{aligned}
\tag{14.22}
$$

In these summations, we use the $2M + 1$ numbers

$$
U_{N-M}, \ldots, U_{N-1}, U_0, \ldots, U_M.
$$

Since $M < N/8$, we must fill in other values to obtain an N-point sequence. One way to do this is to fill in the other places with zeros. Thus define

$$
V_k = \begin{cases} U_k & \text{for } k = 0, 1, \ldots, M \\ 0 & \text{for } k = M+1, \ldots, N-M-1 \\ U_k & \text{for } k = N-M, \ldots, N-1 \end{cases}.
$$

Then the Mth partial sum of the Fourier series of f, sampled at jp/N, is approximated by

$$
S_M\left(\frac{jp}{N}\right) \approx \frac{1}{N}\sum_{k=0}^{N-1} V_k e^{2\pi ijk/N}.
$$

EXAMPLE 14.25

Let $f(t) = t$ for $0 \le t < 2$, and extend f over the entire real line with period 2. Part of the graph of f is shown in Figure 14.24.

The Fourier coefficients of f are

$$
d_k = \frac{1}{2}\int_0^2 te^{-2\pi ikt/2}\, dt = \begin{cases} \dfrac{i}{\pi k} & \text{for } k \ne 0 \\ 1 & \text{for } k = 0 \end{cases}.
$$

and the complex Fourier series is

$$
1 + \sum_{k=-\infty, k\ne 0}^{\infty} \frac{i}{\pi k} e^{\pi ikt}.
$$

This converges to t on $0 < t < 2$ and on periodic extensions of this interval. The Mth partial sum is

$$
S_M(t) = 1 + \sum_{k=-M, k\ne 0}^{M} \frac{i}{\pi k} e^{\pi ikt}.
$$

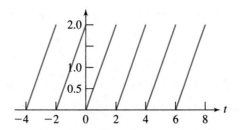

FIGURE 14.24 $f(t) = t$ *for* $0 \leq t < 2$,
periodically extended over the real line.

To be specific, choose $N = 2^7 = 128$ and $M = 10$, so $M \leq N/8$. Sample the partial sum at points $jp/N = j/64$ for $j = 0, 1, \ldots, 127$. Then

$$u = \left\{ f\left(\frac{jp}{N}\right) \right\}_{j=0}^{N-1} = \left\{ \frac{j}{64} \right\}_{j=0}^{127}.$$

The 128-point DFT of u has kth term

$$U_k = \sum_{j=0}^{127} \frac{j}{64} e^{-\pi i jk/64}.$$

Define

$$V_k = \begin{cases} U_k & \text{for } k = 0, 1, \ldots, 10 \\ 0 & \text{for } k = 11, \ldots, 117 \\ U_k & \text{for } k = 118, \ldots, 127 \end{cases}.$$

Then

$$S_{10}\left(\frac{jp}{N}\right) = S_{10}\left(\frac{j}{64}\right) = 1 + \sum_{k=-10, k\neq 0}^{10} \frac{i}{\pi k} e^{\pi i jk/64}$$

$$\approx \frac{1}{128} \sum_{k=0}^{127} V_k e^{\pi i jk/64}. \tag{14.23}$$

In understanding this discussion of approximation of sampled partial sums of a Fourier series, it is worthwhile to see the numbers actually play out in an example. We will do the computation $S_{10}(\frac{1}{2})$, and then of the approximation (14.23) with $j = 32$.

First,

$$S_{10}\left(\frac{1}{2}\right) = 1 + \sum_{k=-10, k\neq 0}^{10} \frac{i}{\pi k} e^{\pi i k/2} = .45847.$$

Now we must compute the V_k's. For these, we need the numbers

$$U_0 = \sum_{j=0}^{127} \frac{j}{64} = 127, \quad U_1 = \sum_{j=0}^{127} \frac{j}{64} e^{-\pi i j/64} = -1.0 + 40.735i,$$

$$U_2 = \sum_{j=0}^{127} \frac{j}{64} e^{-\pi i j/32} = -1.0 + 20.355i, \quad U_3 = -1.0 + 13.557i,$$

$$U_4 = -1.0 + 10.153i, \quad U_5 = -1.0 + 8.1078i, \quad U_6 = -1.0 + 6.7415i,$$

$$U_7 = -1.0 + 5.7631i, \quad U_8 = -1.0 + 5.0273i, \quad U_9 = -1.0 + 4.4532i,$$

$$U_{10} = -1.0 + 3.9922i, \quad U_{118} = -1.0 - 3.9922i, \quad U_{119} = -1.0 - 4.4532i,$$

$$U_{120} = -1.0 - 5.0273i, \quad U_{121} = -1.0 - 5.7631i, \quad U_{122} = -1.0 - 6.7415i,$$

$$U_{123} = -1.0 - 8.1078i, \quad U_{124} = -1.0 - 10.153i, \quad U_{125} = -1.0 - 13.557i,$$

$$U_{126} = -1.0 - 20.355i, \quad U_{127} = -1.0 - 40.735i.$$

Now compute

$$\sum_{k=0}^{127} V_k e^{\pi i k/2} = 127 + (-1.0 + 40.735i)e^{\pi i/2} + (-1.0 + 20.355i)\, e^{\pi i}$$

$$+ (-1.0 + 13.557i)\, e^{3\pi i/2} + (-1.0 + 10.153i)\, e^{2\pi i} + (-1.0 + 8.1078i)\, e^{5\pi i/2}$$

$$+ (-1.0 + 6.7415i)\, e^{3\pi i} + (-1.0 + 5.7631i)\, e^{7\pi i/2} + (-1.0 + 5.0273i)\, e^{4\pi i}$$

$$+ (-1.0 + 4.4532i)\, e^{9\pi i/2} + (-1.0 + 3.9922i)\, e^{5\pi i}$$

$$+ (-1.0 - 3.9922i)\, e^{118\pi i/2} + (-1.0 - 4.4532i)\, e^{119\pi i/2}$$

$$+ (-1.0 - 5.0273i)\, e^{120\pi i/2} + (-1.0 - 5.7631i)\, e^{121\pi i/2}$$

$$+ (-1.0 - 6.7415i)\, e^{122\pi i/2} + (-1.0 - 8.1078i)\, e^{123\pi i/2}$$

$$+ (-1.0 - 10.153i)\, e^{124\pi i/2} + (-1.0 - 13.557i)\, e^{125\pi i/2}$$

$$+ (-1.0 - 20.355i)\, e^{126\pi i/2} + (-1.0 - 40.735i)\, e^{127\pi i/2}$$

$$= 61.04832.$$

Then

$$\frac{1}{128} \sum_{k=0}^{127} V_k e^{\pi i j k/64} = .476\,94.$$

This gives the 128-point DFT approximation 0.47694 to the sampled partial sum $S_{10}(\frac{1}{2})$, which we computed to be 0.45847. The difference is 0.0185. The actual sum of the complex Fourier series at $t = \frac{1}{2}$ is $f(\frac{1}{2}) = 0.50000$.

In practice, we would obtain greater accuracy by using much larger N (allowing larger M) and a software routine to do the computations. ■

14.8.1   Approximation of a Fourier Transform by an *N*-Point DFT

We will show how the discrete Fourier transform can be used to approximate the Fourier transform of a function, under certain conditions. Suppose, to begin, that $\hat{f}(\omega)$ can be approximated to within some acceptable tolerance by an integral over a finite interval:

$$\hat{f}(\omega) = \int_{-\infty}^{\infty} f(\xi)e^{-i\omega\xi}\, d\xi \approx \int_{0}^{2\pi L} f(\xi)e^{-i\omega\xi}\, d\xi.$$

Here we have written the length of the interval as $2\pi L$ for a reason that will reveal itself shortly. Subdivide $[0, 2\pi L]$ into N subintervals of length $2\pi L/N$ and choose partition points $\xi_j = 2\pi j L/N$ for $j = 0, 1, \ldots, N$. We can then approximate the integral on the right by a Riemann sum, obtaining

$$\hat{f}(\omega) \approx \sum_{j=0}^{N-1} f\left(\frac{2\pi j L}{N}\right) e^{-2\pi i j L \omega / N} \left(\frac{2\pi L}{N}\right)$$

$$= \frac{2\pi L}{N} \sum_{j=0}^{N-1} f\left(\frac{2\pi j L}{N}\right) e^{-2\pi i j L \omega / N}.$$

The sum on the right is nearly in the form of a DFT. If we put $\omega = k/L$, with k any integer, then we have

$$\hat{f}(k/L) \approx \frac{2\pi L}{N} \sum_{j=0}^{N-1} f\left(\frac{2\pi j L}{N}\right) e^{-2\pi i j k / N}. \tag{14.24}$$

This gives $\hat{f}(k/L)$, the Fourier transform of f sampled at points k/L, approximated by $2\pi L/N$ times the N-point DFT of the sequence

$$\left\{ f\left(\frac{2\pi j L}{N}\right) \right\}_{j=0}^{N-1}.$$

As noted previously, the DFT is periodic of period N, while $\hat{f}(k/L)$ is not, so we again make the restriction that $|k| \leq N/8$.

EXAMPLE 14.26

We will test the approximation (14.24) for a simple case. Let

$$f(t) = \begin{cases} e^{-t} & \text{for } t \geq 0 \\ 0 & \text{for } t < 0 \end{cases}.$$

Then f has Fourier transform

$$\hat{f}(\omega) = \int_{-\infty}^{\infty} f(\xi) e^{-i\omega\xi} \, d\xi$$

$$= \int_{0}^{\infty} e^{-\xi} e^{-i\omega\xi} \, d\xi = \frac{1 - i\omega}{1 + \omega^2}.$$

Choose $L = 1$, $N = 2^7 = 128$, and $k = 3$ (keep in mind that we want $|k| \leq N/8$). Now $k/L = 3$ and

$$\hat{f}(k/L) = \hat{f}(3) \approx \frac{2\pi}{128} \sum_{j=0}^{127} e^{-\pi j/64} e^{-6\pi i j/128}$$

$$= \frac{\pi}{64} \sum_{j=0}^{127} e^{-\pi j/64} e^{-3\pi i j/64} = 0.12451 - 0.29884i.$$

For comparison,

$$\hat{f}(3) = \frac{1 - 3i}{10} = 0.1 - 0.3i.$$

Suppose we try a larger N, say $N = 2^9 = 512$. Now

$$\hat{f}(3) \approx \frac{2\pi}{512} \sum_{j=0}^{511} e^{-2\pi j/512} e^{-6\pi ij/512}$$

$$= \frac{\pi}{256} \sum_{j=0}^{511} e^{-\pi j/256} e^{-3\pi ij/256} = 0.10595 - 0.2994i,$$

a better approximation than obtained with $N = 128$. ∎

EXAMPLE 14.27

We will continue from the preceding example. There the emphasis was on detailing the idea of approximating a value of $\hat{f}(\omega)$. Now we will use the same function but carry out the approximation at enough points to sketch approximate graphs of Re[$\hat{f}(\omega)$], Im[$\hat{f}(\omega)$], and $|\hat{f}(\omega)|$. Using $L = 4$ and $N = 2^8 = 256$, we obtain the approximation

$$\hat{f}\left(\frac{k}{4}\right) \approx \frac{\pi}{32} \sum_{j=0}^{255} e^{-\pi j/32} e^{-\pi ijk/128}.$$

We should have $|k| \leq N/8 = 32$, although we will only compute approximate values of $\hat{f}(k/4)$ for $k = 1, \ldots, 13$. Because in this example we can compute $\hat{f}(\omega)$ exactly, these values are included in the table to allow comparison.

	DFT approx. of $\hat{f}(\omega)$	$\hat{f}(\omega)$
$(k = 1)\ \hat{f}(\frac{1}{4})$	$.99107 - .23509i$	$.94118 - .23529i$
$(k = 2)\ \hat{f}(\frac{1}{2})$	$.84989 - .3996i$	$.8 - .4i$
$(k = 3)\ \hat{f}(\frac{3}{4})$	$.68989 - .4794i$	$.64 - .48i$
$(k = 4)\ \hat{f}(1)$	$.54989 - .4992i$	$.5 - .5i$
$(k = 5)\ \hat{f}(\frac{5}{4})$	$.44013 - .4868i$	$.39024 - .4878i$
$(k = 6)\ f(\frac{3}{2})$	$.35758 - .46033i$	$.3077 - .4615i$
$(k = 7)\ \hat{f}(\frac{7}{4})$	$.29605 - .42936i$	$.24615 - .43077i$
$(k = 8)\ \hat{f}(2)$	$.24989 - .39839i$	$.2 - .4i$
$(k = 9)\ \hat{f}(\frac{9}{4})$	$.21484 - .36933i$	$.16495 - .37113i$
$(k = 10)\ \hat{f}(\frac{5}{2})$	$.18782 - .34282i$	$.13793 - .34483i$
$(k = 11)\ \hat{f}(\frac{11}{4})$	$.16668 - .31896i$	$.11679 - .32117i$
$(k = 12)\ \hat{f}(3)$	$.14989 - .29759i$	$.1 - .3i$
$(k = 13)\ \hat{f}(\frac{13}{4})$	$.13638 - .27847i$	$.086486 - .28108i$

The real part of $\hat{f}(\omega)$ is consistently approximated in this scheme with an error of about .05, while the imaginary part is approximated in many cases with an error of about .002. Improved accuracy can be achieved by choosing N larger.

In Figures 14.25, 14.26, and 14.27, the approximate values of Re[$\hat{f}(\omega)$], Im[$\hat{f}(\omega)$], and $|\hat{f}(\omega)|$, respectively, are compared with the values obtained from the exact expression for $\hat{f}(\omega)$. The solid dots represent approximate values, and the shaded squares are actual values. In Figure 14.26 the approximation is sufficiently close that the points are nearly indistinguishable (within the resolution of the diagram). ∎

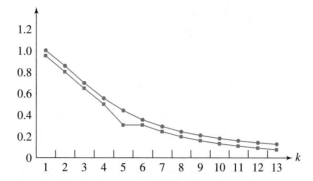

FIGURE 14.25 *Comparison of the DFT approximation of*
$\text{Re}[\hat{f}(\omega)]$ *with actual values for*

$$f(t) = \begin{cases} e^{-t} & \text{for } t \geq 0 \\ 0 & \text{for } t < 0 \end{cases}.$$

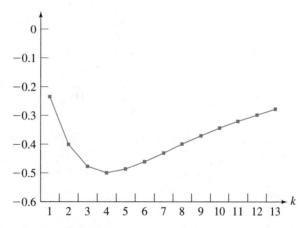

FIGURE 14.26 *Comparison of the DFT approximation of*
$\text{Im}[\hat{f}(\omega)]$ *with actual values for*

$$f(t) = \begin{cases} e^{-t} & \text{for } t \geq 0 \\ 0 & \text{for } t < 0 \end{cases}.$$

Thus far the discussion has centered on functions f for which $\hat{f}(\omega)$ can be approximated by an integral $\int_0^{2\pi L} f(\xi)e^{-i\omega\xi} \, d\xi$. We can extend this idea to the case that $\hat{f}(\omega)$ is approximated by an integral $\int_{-\pi L}^{\pi L} f(\xi)e^{-i\omega\xi} \, d\xi$ over a symmetric interval of length $2\pi L$:

$$\hat{f}(\omega) \approx \int_{-\pi L}^{\pi L} f(\xi)e^{-i\omega\xi} \, d\xi.$$

Then,

$$\hat{f}\left(\frac{k}{L}\right) \approx \int_{-\pi L}^{\pi L} f(\xi)e^{-ik\xi/L} \, d\xi$$

$$= \int_{-\pi L}^{0} f(\xi)e^{-ik\xi/L} \, d\xi + \int_{0}^{\pi L} f(\xi)e^{-ik\xi/L} \, d\xi.$$

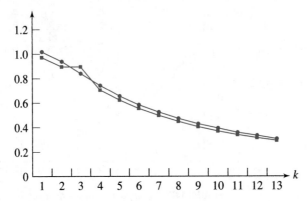

FIGURE 14.27   *Comparison of the DFT approximation of*

$$|\hat{f}(\omega)| \text{ with actual values for } f(t) = \begin{cases} e^{-t} & \text{for } t \geq 0 \\ 0 & \text{for } t < 0 \end{cases}.$$

Upon letting $\zeta = \xi + 2\pi L$ in the first integral of the last line, we have

$$\hat{f}\left(\frac{k}{L}\right) \approx \int_{\pi L}^{2\pi L} f(\zeta - 2\pi L)e^{-ik(\zeta - 2\pi L)/L}\, d\zeta + \int_{0}^{\pi L} f(\xi)e^{-ik\xi/L}\, d\xi$$

$$= \int_{\pi L}^{2\pi L} f(\zeta - 2\pi L)e^{-ik\zeta/L}\, d\zeta + \int_{0}^{\pi L} f(\xi)e^{-ik\xi/L}\, d\xi,$$

since $e^{-2\pi ik} = 1$ if k is an integer. Write ξ for ζ as the variable of integration, obtaining

$$\hat{f}\left(\frac{k}{L}\right) = \int_{\pi L}^{2\pi L} f(\xi - 2\pi L)e^{-ik\xi/L}\, d\xi + \int_{0}^{\pi L} f(\xi)e^{-ik\xi/L}\, d\xi.$$

Now define

$$g(t) = \begin{cases} f(t) & \text{for } 0 \leq t < \pi L \\ \dfrac{1}{2}\left(f(\pi L) + f(-\pi L)\right) & \text{for } t = \pi L \\ f(t - 2\pi L) & \text{for } \pi L < t \leq 2\pi L \end{cases}. \tag{14.25}$$

Then

$$\hat{f}\left(\frac{k}{L}\right) \approx \int_{0}^{2\pi L} g(\xi)e^{-ik\xi/L}\, d\xi$$

$$= \int_{0}^{L} g(2\pi t)e^{-2\pi ikt/L}(2\pi)\, dt \qquad (\text{let } \xi = 2\pi t)$$

$$= 2\pi \int_{0}^{L} g(2\pi t)e^{-2\pi ikt/L}.$$

Finally, approximate the last integral by a Riemann sum, subdividing $[0, L]$ into N subintervals and choosing $t_j = jL/N$ for $j = 0, 1, \ldots, N - 1$. Then

$$\hat{f}\left(\frac{k}{L}\right) \approx \frac{2\pi L}{N} \sum_{j=0}^{N-1} g\left(\frac{2\pi jL}{N}\right) e^{-2\pi ijk/N}.$$

As before, we assume in using this approximation that $|k| \leq N/8$. This approximates $\hat{f}(k/L)$ by a constant multiple of the N-point DFT of the sequence

$$\left\{ g\left(\frac{2\pi j L}{N}\right) \right\}_{j=0}^{N-1}$$

in which points of the sequence are obtained from the function g manufactured from f according to equation (14.25).

14.8.2 Filtering

A periodic signal $f(t)$, of period $2L$, is often filtered for the purpose of canceling out, or diminishing, certain unwanted effects, or perhaps for emphasizing certain effects one wants to study. Suppose $f(t)$ has complex Fourier series

$$\sum_{n=-\infty}^{\infty} d_n e^{n\pi i t/L},$$

where

$$d_n = \frac{1}{2L} \int_{-L}^{L} f(t) e^{-n\pi i t/L}\, dt.$$

Consider the Nth partial sum

$$S_N(t) = \sum_{j=-N}^{N} d_j e^{\pi i j t/L}.$$

A *filtered partial sum* of the Fourier series of f is a sum of the form

$$\sum_{j=-N}^{N} Z\left(\frac{j}{N}\right) d_j e^{\pi i j t/L}, \tag{14.26}$$

in which the filter function Z is a continuous, even function on $[-1, 1]$. In particular applications the object is to choose Z to serve some specific purpose. By way of introduction, we will illustrate filtering for a filter that actually forms a basic approach to the entire issue of convergence of Fourier series.

In the nineteenth century, there was an intense effort to understand the subtleties of convergence of Fourier series. An example of Du Bois-Reymond showed that it is possible for the Fourier series of a continuous function to diverge at every point. In the course of delving into the convergence question, it was observed that in many cases the sequence of averages of partial sums of a Fourier series is better behaved than the sequence of partial sums itself. This led to a consideration of averages of partial sums:

$$\sigma_N(t) = \frac{1}{N} \sum_{k=0}^{N-1} S_k(t) = \frac{1}{N}\left[S_0(t) + S_1(t) + \cdots + S_{N-1}(t)\right].$$

The quantity $\sigma_N(t)$ is called the Nth *Cesàro sum* of f, after the Italian mathematician who studied their properties. It was found that if the partial sums of the Fourier series approach a particular limit at t, then $\sigma_N(t)$ must approach the same limit as $N \to \infty$, but not conversely. It is possible for the Cesàro sums to have a limit for some t, but for the Fourier series to diverge there. It was the 19-year-old prodigy Fejér who proved that if f is periodic of period 2π, and $\int_0^{2\pi} f(t)\, dt$ exists, then $\sigma_N(t) \to f(t)$ wherever f is continuous. This is a stronger result than holds for the partial sums of the Fourier series.

With this as background, write

$$\sigma_N(t) = \frac{1}{N} \sum_{k=0}^{N-1} \left(\sum_{j=-k}^{k} d_j e^{\pi i j t / L} \right).$$

We leave it as an exercise for the student to show that the terms in this double sum can be rearranged to write

$$\sigma_N(t) = \sum_{n=-N}^{N} \left(1 - \left| \frac{n}{N} \right| \right) d_n e^{\pi i n t / L}.$$

This is of the form of equation (14.26) with the *Cesàro filter function*

$$Z(t) = 1 - |t| \quad \text{for } -1 \le t \le 1.$$

The sequence

$$\left\{ Z\left(\frac{n}{N} \right) \right\}_{n=-N}^{N} = \left\{ 1 - \left| \frac{n}{N} \right| \right\}_{n=-N}^{N}$$

is called the *sequence of filter factors for the Cesàro filter*.

One effect of the Cesàro filter is to damp out the Gibbs phenomenon, which is seen in the convergence of the Fourier series of a function at a point of discontinuity. As an example that displays the Gibbs phenomenon very clearly, consider

$$f(t) = \begin{cases} -1 & \text{for } -\pi \le t < 0 \\ 1 & \text{for } 0 \le t < \pi \end{cases},$$

with periodic extension to the real line. Figure 14.28 shows a graph of this periodic extension. Its complex Fourier coefficients are

$$d_0 = \frac{1}{2\pi} \int_{-\pi}^{0} -1 \, dt + \frac{1}{2\pi} \int_{0}^{\pi} dt = 0$$

and

$$d_n = \frac{1}{2\pi} \int_{-\pi}^{0} -e^{-nit} \, dt + \frac{1}{2\pi} \int_{0}^{\pi} e^{-nit} \, dt = \frac{i}{\pi} \frac{-1 + (-1)^n}{n}$$

The Nth partial sum of this series is

$$S_N(t) = \sum_{n=-N, n \neq 0}^{N} \frac{i}{\pi} \frac{-1 + (-1)^n}{n} e^{nit}.$$

If N is odd, then

$$S_N(t) = \frac{4}{\pi} \left(\sin(t) + \frac{1}{3} \sin(3t) + \frac{1}{5} \sin(5t) + \cdots + \frac{1}{N} \sin(Nt) \right).$$

The Nth Cesàro sum (with $L = \pi$) is

$$\sigma_N(t) = \sum_{n=-N}^{N} \left(1 - \left| \frac{n}{N} \right| \right) \frac{i}{\pi} \frac{-1 + (-1)^n}{n} e^{int}.$$

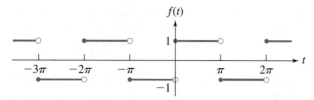

FIGURE 14.28 $f(t) = \begin{cases} -1 & \text{for } -\pi \leq t < 0 \\ 1 & \text{for } 0 \leq t < \pi \end{cases}$, *and*

$f(t + 2\pi) = f(t)$ *for all real* t.

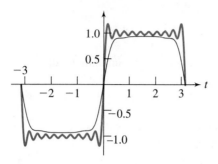

FIGURE 14.29 $S_{10}(t)$ and $\sigma_{10}(t)$ for the function of Figure 14.28.

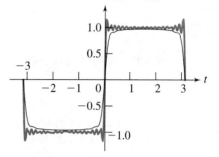

FIGURE 14.30 $S_{20}(t)$ and $\sigma_{20}(t)$ for the function of Figure 14.28.

This can be written

$$\sigma_N(t) = \sum_{n=1}^{N} \left(1 - \frac{n}{N}\right) \left(\frac{-2}{\pi}\right) \left(\frac{(-1) + (-1)^n}{n}\right) \sin(nt).$$

Figure 14.29 shows graphs of $S_{10}(t)$ and $\sigma_{10}(t)$, and Figure 14.30 shows graphs of $S_{20}(t)$ and $\sigma_{20}(t)$. In the partial sums $S_N(t)$, the Gibbs phenomenon is readily apparent near $t = 0$, where f has a jump discontinuity. Even though $S_N(t) \to f(t)$ for $0 < t < \pi$ and for $-\pi < t < 0$, the graphs of $S_N(t)$ have relatively high peaks near zero, which remain at nearly constant height even as N increases (although these peaks move toward the vertical axis as N increases). However, this phenomenon is not seen in the graphs of $\sigma_N(t)$, which accelerates and "smooths out" the convergence of the Fourier series.

The Cesàro filter also damps the effects of the higher frequency terms in the Fourier series, because the Cesàro filter factor $1 - |n/N|$ tends to zero as n increases toward N. This effect is also seen in the graphs of the Cesàro sums.

There are many filters that are used in signal analysis. Two of the more commonly encountered ones are the Hamming and Gauss filters. The *Hamming filter* is named for Richard Hamming, who was for many years a senior scientist and researcher at Bell Labs. It is given by

$$Z(t) = .54 + .46 \cos(\pi t).$$

The filtered Nth partial sum of the complex Fourier series of f, using the Hamming filter, is

$$\sum_{n=-N}^{N} \left(.54 + .46 \cos\left(\frac{\pi n}{N}\right)\right) d_n e^{\pi n i t / L}.$$

Another filter frequently used to filter out background noise in a signal is the *Gauss filter*, named for the nineteenth-century mathematician and scientist Carl Friedrich Gauss. It is given by

$$Z(t) = e^{-a\pi^2 t^2},$$

in which a is a positive constant. The Gauss-filtered partial sum of the complex Fourier series of f is

$$\sum_{n=-N}^{N} e^{-a\pi^2 n^2 / N^2} d_n e^{\pi n i t / L}.$$

Filtering is also applied to Fourier transforms. The filtered Fourier transform of f, using the filter function $Z(t)$, is

$$\int_{-\infty}^{\infty} Z(\xi) f(\xi) e^{-i\omega\xi} \, d\xi.$$

If this integral is approximated by an integral over a finite interval,

$$\int_{-\infty}^{\infty} Z(\xi) f(\xi) e^{-i\omega\xi} \, d\xi \approx \int_{-L}^{L} Z(\xi) f(\xi) e^{-i\omega\xi} \, d\xi,$$

then it is standard practice to approximate the integral on the right using a DFT. The Cesàro, Hamming, and Gauss filters for this integral are, respectively,

$$Z(t) = 1 - \left| \frac{t}{L} \right| \quad \text{(Cesàro)}$$

$$Z(t) = .54 + .46 \cos\left(\frac{\pi t}{L} \right) \quad \text{(Hamming)}$$

and

$$Z(t) = e^{-a(\pi t / L)^2} \quad \text{(Gauss)}.$$

SECTION 14.8 PROBLEMS

In each of Problems 1 through 8, a function is given, having period p. Compute the complex Fourier series of the function and then the 10th partial sum of this series at the indicated point t_0. Then, using $N = 128$, compute a DFT approximation to this partial sum at the point. Approximate the difference between $f_{10}(t_0)$ and the DFT approximation.

1. $f(t) = 1 + t$ for $0 \le t < 2$, $p = 2$, $t_0 = \frac{1}{8}$
2. $f(t) = t^2$ for $0 \le t < 1$, $p = 1$, $t_0 = \frac{1}{2}$
3. $f(t) = \cos(t)$ for $0 \le t < 2$, $p = 2$, $t_0 = \frac{1}{8}$
4. $f(t) = e^{-t}$ for $0 \le t < 4$, $p = 4$, $t_0 = \frac{1}{4}$
5. $f(t) = t^3$ for $0 \le t < 1$, $p = 1$, $t_0 = \frac{1}{4}$
6. $f(t) = t \sin(t)$ for $0 < t \le 4$, $p = 4$, $t_0 = \frac{1}{8}$
7. $f(t) = t$ for $0 \le t < 2$, $p = 2$, $t_0 = \frac{1}{2}$
8. $f(t) = te^t$ for $0 \le t < 2$, $p = 2$, $t_0 = \frac{1}{8}$

In each of Problems 9 through 14, make a DFT approximation to the Fourier transform of f at the given point, using $N = 512$ and the given value of L.

9. $f(t) = \begin{cases} e^{-4t} & \text{for } t \ge 0 \\ 0 & \text{for } t < 0 \end{cases}$,

   $L = 3$; $\hat{f}(4)$

10. $f(t) = \begin{cases} \cos(2t) & \text{for } t \ge 0 \\ 0 & \text{for } t < 0 \end{cases}$,

    $L = 6$, $\hat{f}(2)$

11. $f(t) = \begin{cases} te^{-2t} & \text{for } t \ge 0 \\ 0 & \text{for } t < 0 \end{cases}$,

    $L = 3$, $\hat{f}(12)$

12. $f(t) = \begin{cases} t^2 \cos(t) & \text{for } t \geq 0 \\ 0 & \text{for } t < 0 \end{cases}$,

$L = 4, \hat{f}(4)$

13. $f(t) = \begin{cases} \sin\left(\dfrac{t}{2}\right) & \text{for } t \geq 0 \\ 0 & \text{for } t < 0 \end{cases}$,

$L = 1, \hat{f}(8)$

14. $f(t) = \begin{cases} \cos^2(t) & \text{for } t \geq 0 \\ 0 & \text{for } t < 0 \end{cases}$,

$L = 2, \hat{f}\left(\tfrac{1}{2}\right)$

In each of Problems 15 through 20, use the DFT to approximate graphs of $\text{Re}[\hat{f}(\omega)]$, $\text{Im}[\hat{f}(\omega)]$, and $|\hat{f}(\omega)|$ for $0 \leq \omega \leq 3$, using $N = 256$. For these functions, $\hat{f}(\omega)$ can be computed exactly. Graph each of the approximations of $\text{Re}[\hat{f}(\omega)]$, $\text{Im}[\hat{f}(\omega)]$, and $|\hat{f}(\omega)|$ on the same set of axes with, respectively, the actual function itself.

15. $f(t) = t[H(t-1) - H(t-2)]$

16. $f(t) = 2e^{-4|t|}$

17. $f(t) = H(t) - H(t-1)$

18. $f(t) = e^t[H(t) - H(t-2)]$

19. $f(t) = \cos(\pi t)[H(t) - H(t-1)]$

20. $f(t) = t[H(t) - H(t-1)]$

In each of Problems 21 through 25, graph the function, the fifth partial sum of its Fourier series on the interval, and the fifth Cesàro sum, using the same set of axes. Repeat this process for the tenth and twenty-fifth partial sums. Notice in particular the graphs at points of discontinuity of the function, where the Gibbs phenomenon shows up in the partial sums of the Fourier series, but is filtered from the Cesàro sums.

21. $f(t) = \begin{cases} 1 & \text{for } 0 \leq t < 2 \\ -1 & \text{for } -2 \leq t < 0 \end{cases}$

22. $f(t) = \begin{cases} t^2 & \text{for } -2 \leq t < 1 \\ 2+t & \text{for } 1 \leq t < 2 \end{cases}$

23. $f(t) = \begin{cases} -1 & \text{for } -1 \leq t < -\frac{1}{2} \\ 0 & \text{for } -\frac{1}{2} \leq t < \frac{1}{2} \\ 1 & \text{for } \frac{1}{2} \leq t < 1 \end{cases}$

24. $f(t) = \begin{cases} e^{-t} & \text{for } -3 \leq t < 1 \\ \cos(t) & \text{for } 1 \leq t < 3 \end{cases}$

25. $f(t) = \begin{cases} 2+t & \text{for } -1 \leq t < 0 \\ 7 & \text{for } 0 < t < 1 \end{cases}$

26. Let $f(t) = \begin{cases} 1 & \text{for } 0 \leq t < 2 \\ -1 & \text{for } -2 \leq t < 0 \end{cases}$.

Plot the fifth partial sum of the Fourier series for $f(t)$ on $[-2, 2]$, together with the fifth Cesàro sum, the fifth Hamming filtered partial sum, and the fifth Gauss filtered partial sum on the same set of axes. Repeat this for the tenth sums and the twenty-fifth sums.

27. Let $f(t) = \begin{cases} t & \text{for } -2 \leq t < 0 \\ 2+t & \text{for } 0 \leq t < 2 \end{cases}$.

Plot the fifth partial sum of the Fourier series for $f(t)$ on $[-\pi, \pi]$, together with the fifth Cesàro sum, the fifth Hamming filtered partial sum, and the fifth Gauss filtered partial sum on the same set of axes. Repeat this for the tenth sums and the twenty-fifth sums.

28. Let $f(t) = \begin{cases} \cos(t) & \text{for } -\pi \leq t < 0 \\ 1 & \text{for } 0 \leq t < \pi \end{cases}$.

Plot the fifth partial sum of the Fourier series for $f(t)$ on $[-2, 2]$, together with the fifth Cesàro sum, the fifth Hamming filtered partial sum, and the fifth Gauss filtered partial sum on the same set of axes. Repeat this for the tenth sums and the twenty-fifth sums.

29. Let $f(t) = \begin{cases} -2t & \text{for } -3 \leq t < 1 \\ -1 & \text{for } 1 \leq t < 3 \end{cases}$.

Plot the fifth partial sum of the Fourier series for $f(t)$ on $[-3, 3]$, together with the fifth Cesàro sum, the fifth Hamming filtered partial sum, and the fifth Gauss filtered partial sum on the same set of axes. Repeat this for the tenth sums and the twenty-fifth sums.

14.9 The Fast Fourier Transform

The discrete Fourier transform is a powerful tool for approximating Fourier coefficients, partial sums of Fourier series, and Fourier transforms. However, such a tool is only useful if there are efficient computing techniques for carrying out the large numbers of calculations involved in typical applications. This is where the fast Fourier transform, or FFT, comes in. The FFT is not a transform at all, but rather an efficient procedure for computing discrete Fourier transforms. Its impact in engineering and science over the past 35 years has been profound, because it makes the DFT a practical tool.

The FFT first appeared formally in 1965 in a five-page paper, "An Algorithm for the Machine Calculation of Complex Fourier Coefficients," by James W. Cooley of IBM and John W. Tukey of Princeton University. The catalyst behind preparation and publication of the paper was Richard Garwin, a physicist who had consulted for federal agencies on questions involving weapons and defense policies. Garwin became aware that Tukey had developed an algorithm for computing Fourier transforms, a tool that Garwin needed for his own work. When Garwin took Tukey's ideas to the computer center at IBM Research in Yorktown Heights for the purpose of having them programmed, James Cooley was assigned to assist him. Because of the importance of an efficient method for computing Fourier transforms, word of Cooley's program quickly spread, and it became so much in demand that the Cooley–Tukey paper resulted.

After the paper's publication it was found that some of the concepts underlying the method, or similar to it, had already appeared in other contexts. Tukey himself has related that Phillip Rudnick of the Scripps Oceanographic Institute had reported programming a special case of the algorithm, using ideas from a paper by G. D. Danielson and Cornelius Lanczos. Lanczos, a Hungarian-born physicist/mathematician whose career spanned many areas, had developed the essential ideas around 1938 and in the years following, when he was working on problems in numerical methods and Fourier analysis. Much earlier, Gauss had essentially discovered discrete Fourier analysis in calculating the orbit of Pallas, but of course there were no computers in the Napoleonic era.

Since the FFT is known to many users as a computer routine that they simply key into their analyses, much as we use software packages to plot surfaces or compute integrals, we will concentrate here on the underlying ideas behind the FFT, how it achieves its efficiency, and a numerical measure of what efficiency means in this context.

The problem addressed by the FFT is that of computing the N-point DFT $\sum_{j=0}^{N-1} u_j e^{-2\pi ijk/N}$ of a given sequence $\{u_j\}_{j=0}^{N-1}$. However, we will focus not on this particular sum, but on a general sum having its essential features, with the primary goal of keeping track of the number of arithmetic operations involved. This is what determines the efficiency of an algorithm. Thus, consider a sum of the form

$$A(n) = \sum_{k=0}^{N-1} a(k)e^{-2\pi ikn/N} \tag{14.27}$$

for $n = 0, 1, \ldots, N - 1$. In applying the FFT, N is chosen as a positive integer power of 2.

The idea behind the FFT becomes clear if we examine in detail a simple case, namely $N = 2^2 = 4$. In this case, the quantities to be computed in equation (14.27) are

$$A(0) = a(0) + a(1) + a(2) + a(3)$$
$$A(1) = a(0) + a(1)e^{-2\pi i/4} + a(2)e^{-4\pi i/4} + a(3)e^{-6\pi i/4}$$
$$A(2) = a(0) + a(1)e^{-4\pi i/4} + a(2)e^{-8\pi i/4} + a(3)e^{-12\pi i/4}$$
$$A(3) = a(0) + a(1)e^{-6\pi i/4} + a(2)e^{-12\pi i/4} + a(3)e^{-18\pi i/4}.$$

Let

$$W = e^{-2\pi i/4}.$$

In this simple case, $W = -i$, but we maintain the present notation to make it easier to imagine what happens with larger N. The value $N = 4$ appears in the denominator of the fraction in the exponent.

Notice that

$$W^k = W^{k \bmod(4)}. \tag{14.28}$$

In this symbol, $k \bmod(4)$ is the remainder upon dividing k by 4. For example, $5 \bmod(4) = 1$, since 5 divided by 4 leaves a remainder of 1. And $6 \bmod(4) = 2$, $7 \bmod(4) = 3$, $8 \bmod(4) = 0$, and so on. In general, $k \bmod(4) = 0$ exactly when k is an integer multiple of 4, leaving zero remainder when divided by 4.

We therefore have

$$W^0 = W^4 = W^8 = W^{12} = \cdots = 1,$$

$$W = W^5 = W^9 = W^{13} = W^{17} = \cdots = -i,$$

$$W^2 = W^6 = W^{10} = W^{14} = \cdots = -1,$$

and

$$W^3 = W^7 = W^{11} = W^{15} = \cdots = i.$$

Of course, for larger N these relations would be more complicated but would still follow the pattern similar to that dictated by equation (14.28).

Now write the equations for the $A(n)$'s as

$$A(0) = a(0) + a(1) + a(2) + a(3)$$

$$A(1) = a(0) + a(1)W + a(2)W^2 + a(3)W^3$$

$$A(2) = a(0) + a(1)W^2 + a(2)W^0 + a(3)W^2$$

$$A(3) = a(0) + a(1)W^3 + a(2)W^2 + a(3)W.$$

In the third equation, $W^0 = 1$ but we kept the symbol W^0 to maintain the pattern of the coefficients. Write this system of equations in matrix form:

$$
\begin{pmatrix} A(0) \\ A(2) \\ A(1) \\ A(3) \end{pmatrix}
=
\begin{pmatrix}
1 & 1 & 1 & 1 \\
1 & W & W^2 & W^3 \\
1 & W^2 & W^0 & W^2 \\
1 & W^3 & W^2 & W
\end{pmatrix}
\begin{pmatrix} a(0) \\ a(1) \\ a(2) \\ a(3) \end{pmatrix}.
$$

A straightforward count shows that N^2 (in this case, 16) complex multiplications and $N(N-1)$, or 12, complex additions are needed to compute this matrix product, for a total of $N^2 + N(N-1)$, or 28 operations. It is this number of operations we seek to reduce.

The crucial step is to recognize that the last matrix equation can be written in factored form as

$$
\begin{pmatrix} A(0) \\ A(2) \\ A(1) \\ A(3) \end{pmatrix}
=
\begin{pmatrix}
1 & W^0 & 0 & 0 \\
1 & W^2 & 0 & 0 \\
0 & 0 & 1 & W \\
0 & 0 & 1 & W^3
\end{pmatrix}
\begin{pmatrix}
1 & 0 & W^0 & 0 \\
0 & 1 & 0 & W^0 \\
1 & 0 & W^2 & 0 \\
0 & 1 & 0 & W^2
\end{pmatrix}
\begin{pmatrix} a(0) \\ a(1) \\ a(2) \\ a(3) \end{pmatrix}. \tag{14.29}
$$

This factorization is not obvious but, once given, is easily verified. A price to be paid for the factorization is that $A(2)$ and $A(1)$ have become switched. This is tolerable, as long as we know that it is happening and can keep track of such interchanges as they occur.

Now count the number of arithmetic operations needed to compute the matrix products in equation (14.29). First, compute the product of the last two matrices, letting

$$\begin{pmatrix} b(0) \\ b(1) \\ b(2) \\ b(3) \end{pmatrix} = \begin{pmatrix} 1 & 0 & W^0 & 0 \\ 0 & 1 & 0 & W^0 \\ 1 & 0 & W^2 & 0 \\ 0 & 1 & 0 & W^2 \end{pmatrix} \begin{pmatrix} a(0) \\ a(1) \\ a(2) \\ a(3) \end{pmatrix}.$$

Count the operations needed as follows. First,

$$b(0) = a(0) + a(2)W^0$$

requires one addition and one multiplication. Next,

$$b(1) = a(1) + a(3)W^0$$

requires one addition and one multiplication. Next,

$$b(2) = a(0) + a(2)W^2 = a(0) - a(2)W^0.$$

Since $a(2)W^0$ was already computed for $b(0)$, computing $b(2)$ only costs one addition. Finally,

$$b(3) = a(1) + a(3)W^2 = a(1) - a(3)W^0.$$

Since $a(3)W^0$ was computed for $b(1)$, determining $b(3)$ only costs us one more addition.

This part of the product (14.29) therefore costs four additions and two multiplications. For the rest of the product, we have

$$\begin{pmatrix} A(0) \\ A(2) \\ A(1) \\ A(3) \end{pmatrix} = \begin{pmatrix} 1 & W^0 & 0 & 0 \\ 1 & W^2 & 0 & 0 \\ 0 & 0 & 1 & W \\ 0 & 0 & 1 & W^3 \end{pmatrix} \begin{pmatrix} b(0) \\ b(1) \\ b(2) \\ b(3) \end{pmatrix}.$$

First,

$$A(0) = b(0) + b(1)W^0$$

costs one addition and one multiplication. Next,

$$A(2) = b(0) + b(1)W^2 = b(0) - b(1)W^0,$$

costs one more addition (since $b(1)W^0$ was computed for $A(0)$). Continuing,

$$A(1) = b(2) + b(3)W$$

requires one addition and one multiplication. Finally,

$$A(3) = b(2) + b(3)W^3 = b(2) - b(3)W$$

requires one addition. This second phase of computing the product (14.29) has needed four additions and two multiplications.

In total, computation of the factored product (14.29) requires eight complex additions and four complex multiplications, for a total of twelve arithmetic operations. This is a substantial savings over the twenty-eight required prior to the factorization.

This is the basic idea of the FFT. In general, the $N \times N$ matrix of coefficients in the equations to be computed for the coefficients is factored into a product of N simpler matrices, resulting in a reduction in the number of arithmetic computations needed to complete the matrix product. As we saw even in this small case $N = 4$, one price to pay for improved efficiency is that the coefficients $A(0), A(1), \ldots, A(N - 1)$ become scrambled. However, this happens in a predictable way that enables the programmer to unscramble them at the end.

The FFT has become a standard part of the software of many instruments. For example, the FT–NMR, which stands for Fourier Transform–Nuclear Magnetic Resonance, uses the FFT as part of its data analysis system.

We will develop a quantitative measure of the computational efficiency of the FFT.

14.9.1 Computational Efficiency of the FFT

We will prove that for a sampling of a waveform at N points, the FFT reduces the number of complex computations from a number proportional to N^2 to a number proportional to $N \log_2(N)$. The proof involves several concepts that we will develop first.

Nth Roots of Unity

If N is a positive integer, then an Nth root of unity is a complex number whose Nth power equals 1. Clearly 1 is an Nth root of unity for any N. We expect there to be others as well, since, if z is an Nth root of unity, then $z^N - 1 = 0$, and an Nth degree polynomial has N roots.

We can find the Nth roots of unity by using Euler's formula

$$e^{i\theta} = \cos(\theta) + i \sin(\theta).$$

From this,

$$e^{2k\pi i} = 1$$

for any integer k. Then

$$\left(e^{2k\pi i/N}\right)^N = 1,$$

hence $e^{2\pi i k/N}$ is an Nth root of unity for $k = 0, \pm 1, \pm 2, \ldots$. In particular, if we choose $k = 0, 1, 2, \ldots, N - 1$, we get the N distinct Nth roots of unity:

$$1, e^{2\pi i/N}, e^{4\pi i/N}, \ldots, e^{2\pi i(N-1)/N}. \tag{14.30}$$

Use of any other integer value of k simply reproduces one of the numbers in this list. For example, if we use $k = N$, we get

$$e^{2\pi Ni/N} = e^{2\pi i} = 1,$$

and if we use $k = N + 1$ we get

$$e^{2\pi (N+1)i/N} = e^{2\pi i + 2\pi i/N} = e^{2\pi i} e^{2\pi i/N} = e^{2\pi i/N},$$

the second number in the list (14.30).

EXAMPLE 14.28

We will find the sixth roots of unity. With $N = 6$, these numbers are

$$1, e^{2\pi i/6} = e^{\pi i/3}, e^{4\pi i/6} = e^{2\pi i/3}, e^{6\pi i/6} = e^{\pi i}, e^{8\pi i/6} = e^{3\pi i/2}, \quad \text{and} \quad e^{10\pi i/6} = e^{5\pi i/3}.$$

More explicitly, these roots of unity are 1 and

$$e^{\pi i/3} = \cos\left(\frac{\pi}{3}\right) + i\sin\left(\frac{\pi}{3}\right) = \frac{1}{2} + \frac{\sqrt{3}}{2}i,$$

$$e^{2\pi i/3} = \cos\left(\frac{2\pi}{3}\right) + i\sin\left(\frac{2\pi}{3}\right) = -\frac{1}{2} + \frac{\sqrt{3}}{2}i,$$

$$e^{\pi i} = \cos(\pi) + i\sin(\pi) = -1,$$

$$e^{3\pi i/2} = \cos\left(\frac{3\pi}{2}\right) + i\sin\left(\frac{3\pi}{2}\right) = -\frac{1}{2} - \frac{\sqrt{3}}{2}i,$$

$$e^{5\pi i/3} = \cos\left(\frac{5\pi}{3}\right) + i\sin\left(\frac{5\pi}{3}\right) = \frac{1}{2} - \frac{\sqrt{3}}{2}i.$$

These numbers form the vertices of a regular polygon of six sides, with vertices on the unit circle. ■

Ω and Functions on Ω

If N is a positive integer, denote

$$u = e^{2\pi i/N}.$$

In the discussion of the preceding section, $W = u^{-1}$. In terms of u, the Nth roots of unity are

$$1, u, u^2, \ldots, u^{N-1}.$$

Define Ω to be the set of these N numbers. In the notation commonly used for finite sets,

$$\Omega = \{1, u, \ldots, u^{N-1}\}.$$

This set has the following two properties.

1. A product of two numbers in Ω is in Ω.
   To see why this is true, consider u^j and u^k in Ω, where $0 \le j, k \le N - 1$. Now

$$u^j u^k = e^{2\pi ij/N} e^{2\pi ik/N} = e^{2\pi i(j+k)/N} = u^{j+k}.$$

   If $j + k \le N - 1$, then u^{j+k} is in Ω. If $j + k \ge N$, write $j + k = N + r$, where $0 \le r \le N - 2$. Then

$$u^{j+k} = u^{N+r} = e^{2\pi i(N+r)/N} = e^{2\pi i} e^{2\pi ir/N} = u^r$$

   is in Ω.

2. For any number in Ω, there is a number in Ω such that the product of these two numbers is 1.

   This statement holds if the number chosen is 1, since $1 \cdot 1 = 1$. If we choose one of the other numbers in Ω, say u^j, then $1 \le j \le N - 1$, so u^{N-j} is also in Ω. Further,

$$u^j u^{N-j} = u^N = 1.$$

In the terminology of algebra, Ω forms a group under multiplication.

Recall that the complex conjugate of $a + ib$ is $a - ib$, denoted $\overline{a + ib}$, and that

$$(a + ib)\overline{(a + ib)} = a^2 + b^2 = |a + ib|.$$

The conjugate of u^j is u^{-j}, because

$$\overline{u^j} = \overline{e^{2\pi ij/N}} = \overline{\cos(2\pi j/N) + i\,\sin(2\pi j/N)}$$

$$= \cos\left(\frac{2\pi j}{N}\right) - i\,\sin\left(\frac{2\pi j}{N}\right) = e^{-2\pi ij/N} = u^{-j}.$$

We now want to consider complex-valued functions on Ω. If g is such a function, then $g(u^j)$ is a complex number for each $j = 0, 1, 2, \ldots, N-1$. Letting $\mathfrak{C}$ denote the set of complex numbers, we indicate that g is such a function by writing $g : \Omega \to \mathfrak{C}$. These functions are different from those normally encountered in analysis. They are defined only for a finite collection of numbers, and there is no concept of continuity or differentiability attached to them.

EXAMPLE 14.29

Let $N = 3$, so Ω consists of the third roots of unity. With $u = e^{2\pi i/3}$, these roots are

$$u^0 = 1, \quad u = e^{2\pi i/3} = \cos\left(\frac{2\pi}{3}\right) + i\,\sin\left(\frac{2\pi}{3}\right) = -\frac{1}{2} + \frac{\sqrt{3}}{2}i$$

and

$$u^2 = e^{4\pi/3} = \cos\left(\frac{4\pi}{3}\right) + i\,\sin\left(\frac{4\pi}{3}\right) = -\frac{1}{2} - \frac{\sqrt{3}}{2}i.$$

Then

$$\Omega = \{1, u, u^2\}.$$

There are infinitely many functions we can define from Ω to $\mathfrak{C}$. One example is $g : \Omega \to \mathfrak{C}$ defined by setting

$$g(1) = 1 + i, \quad g(u) = i, \quad g(u^2) = -4.$$

Another example is the function $h : \Omega \to \mathfrak{C}$ defined by

$$h(1) = 2 - 7i, \quad h(u) = 21, \quad h(u^2) = -3i. \ \blacksquare$$

We can define a dot product of two functions from Ω to $\mathfrak{C}$ by

$$g \cdot h = \frac{1}{N} \sum_{j=0}^{N-1} g(u^j)\overline{h(u^j)}. \tag{14.31}$$

To illustrate, for the two functions of the preceding example,

$$g \cdot h = \frac{1}{3}\left[g(1)\overline{h(1)} + g(u)\overline{h(u)} + g(u^2)\overline{h(u^2)} \right]$$

$$= \frac{1}{3}[(1+i)(2+7i) + i(21) + (-4)(3i)] = \frac{1}{3}(-5 + 18i).$$

Although this operation often results in a complex, not a real number, it behaves in ways similar to the dot product of vectors.

THEOREM 14.16

Let $g : \Omega \to \mathfrak{C}$ and $h : \Omega \to \mathfrak{C}$. Then,

1. $g \cdot h = \overline{h \cdot g}$.
2. If α is any (real or complex) number, then $(\alpha g) \cdot h = \alpha(g \cdot h)$.

3. If $k : \Omega \to \mathfrak{C}$, then $(g + h) \cdot k = g \cdot k + h \cdot k$.

4. $g \cdot g \geq 0$, and $g \cdot g = 0$ if and only if each $g(u^j) = 0$.

Proof   For (1),

$$\overline{h \cdot g} = \overline{\frac{1}{N} \sum_{j=0}^{N-1} h(u^j) \overline{g(u^j)}}$$

$$= \frac{1}{N} \sum_{j=0}^{N-1} g(u^j) \overline{h(u^j)} = g \cdot h.$$

For (4),

$$g \cdot g = \frac{1}{N} \sum_{j=0}^{N-1} g(u^j) \overline{g(u^j)} = \frac{1}{N} \sum_{j=0}^{N-1} \left| g(u^j) \right|^2 \geq 0.$$

Further, this sum of nonnegative numbers can be zero only if each term is zero, hence only if each $g(u^j) = 0$. ∎

Now, if k is an integer and $0 \leq k \leq N - 1$, define a function $e_k : \Omega \to \mathfrak{C}$ by

$$e_k(u^j) = u^{kj} \quad \text{for } j = 0, 1, \ldots, N - 1.$$

If $N = 5$, then $u = e^{2\pi i/5}$. Now

$$\Omega = \{1, u, u^2, u^3, u^4\},$$

while higher powers of u repeat these values. For example,

$$u^5 = e^{10\pi i/5} = e^{2\pi i} = 1,$$

$$u^6 = e^{12\pi i/5} = e^{(2+2/5)\pi i} = e^{2\pi i} e^{2\pi i/5} = e^{2\pi i/5} = u,$$

and so on.

With $k = 3$, we have

$$e_3(u^0) = u^{3(0)} = u^0 = 1,$$

$$e_3(u) = e_3(u^1) = u^{3(1)} = u^3,$$

$$e_3(u^2) = u^{3(2)} = u^6 = u,$$

$$e_3(u^3) = u^9 = u^5 u^4 = u^4,$$

and

$$e_3(u^4) = u^{12} = u^2.$$

If we choose $k = 4$, we have the function e_4, with

$$e_4(u^0) = u^0 = 1,$$

$$e_4(u) = u^4,$$

$$e_4(u^2) = u^8 = u^3,$$

$$e_4(u^3) = u^{12} = u^2,$$

and

$$e_4(u^4) = u^{16} = u.$$

The functions e_k have the special property that not only is each $e_k(u^j)$ a complex number but each $e_k(u^j)$ is actually in Ω. Indeed, as can be seen in these two examples, e_k simply rearranges the Nth roots of unity comprising Ω and gives them back in a possibly different order.

We will show that the functions $e_0, e_1, \ldots, e_{N-1}$ behave like mutually orthogonal unit vectors under the dot product given by equation (14.31). That is, the dot product of distinct e_k and e_j is zero, and the dot product of any e_k with itself is 1.

THEOREM 14.17

    1. $e_k \cdot e_j = 0$ if $k \neq j$.
    2. $e_k \cdot e_k = 1$.

Proof   Begin by writing

$$e_k \cdot e_j = \frac{1}{N} \sum_{m=0}^{N-1} e_k(u^m)\overline{e_j(u^m)} = \frac{1}{N} \sum_{m=0}^{N-1} u^{km}\overline{u^{jm}}$$

$$= \frac{1}{N} \sum_{m=0}^{N-1} u^{km}u^{-jm} = \frac{1}{N} \sum_{m=0}^{N-1} u^{(k-j)m}.$$

If $k = j$, then each $u^{(k-j)m} = 1$ and then

$$e_k \cdot e_k = \frac{1}{N} \sum_{m=0}^{N-1} 1 = \frac{N}{N} = 1,$$

proving (2). For (1), suppose $k \neq j$. Then $r = u^{k-j} \neq 1$, so we can sum the finite geometric series:

$$\sum_{m=0}^{N-1} u^{(k-j)m} = \sum_{m=0}^{N-1} r^m = \frac{r^N - 1}{r - 1} = \frac{(u^{k-j})^N - 1}{u^{k-j} - 1}.$$

But

$$u^{(k-j)N} = u^{kN}u^{-jN} = (u^N)^k(u^N)^{-j} = 1,$$

because $u^N = 1$ by choice of u. Therefore, $r^N - 1 = 0$, hence $e_k \cdot e_j = 0$. ∎

In view of conclusion (1), we say that the functions $e_0, e_1, \ldots, e_{N-1}$ form an orthogonal set of functions on Ω into $\mathfrak{C}$. We will now show that any function $g : \Omega \to \mathfrak{C}$ can be written as a sum of constants times these functions.

THEOREM 14.18

Let $g : \Omega \to C$. Then $g = \sum_{j=0}^{N-1} (g \cdot e_j)e_j$.

Proof   Note that both g and $\sum_{j=0}^{N-1} (g \cdot e_j)e_j$ are functions from Ω to $\mathfrak{C}$. We will show that $\sum_{j=0}^{N-1} (g \cdot e_j)e_j$ has the same effect on any u^k in Ω that g does. Thus consider

$$\left(\sum_{j=0}^{N-1}(g \cdot e_j)e_j\right)u^k = \sum_{j=0}^{N-1}(g \cdot e_j)e_j(u^k) = \sum_{j=0}^{N-1}(g \cdot e_j)u^{kj}$$

$$= \sum_{j=0}^{N-1}\left(\frac{1}{N}\sum_{m=0}^{N-1}g(u^m)\overline{e_j(u^m)}\right)u^{kj}$$

$$= \frac{1}{N}\sum_{j=0}^{N-1}\sum_{m=0}^{N-1}g(u^m)\overline{u^{mj}}u^{kj}$$

$$= \frac{1}{N}\sum_{j=0}^{N-1}\sum_{m=0}^{N-1}g(u^m)u^{-mj}u^{kj}$$

$$= \frac{1}{N}\sum_{j=0}^{N-1}\sum_{m=0}^{N-1}g(u^m)u^{(k-m)j}$$

$$= \frac{1}{N}\sum_{m=0}^{N-1}\left(\sum_{j=0}^{N-1}u^{(k-m)j}\right)g(u^m),$$

in which, in the last line, we interchanged the order of the summations. This is valid because both summations are finite. Now, for any m from 0 to $N-1$ inclusive,

$$\sum_{j=0}^{N-1}u^{(k-m)j} = \sum_{j=0}^{N-1}1 = N \quad \text{if } k = m.$$

But if $k \neq m$, then we have the geometric series

$$\sum_{j=0}^{N-1}u^{(k-m)j} = \frac{(u^{k-m})^N - 1}{u^{k-m} - 1} = 0$$

because

$$(u^{k-m})^N = (u^N)^k(u^N)^{-m} = 1^k 1^{-m} = 1.$$

Thus, in $(1/N)\sum_{m=0}^{N-1}\left(\sum_{j=0}^{N-1}u^{(k-m)j}\right)g(u^m)$, the inner series collapses to just one term, N, and this occurs when the index of summation m of the outer summation equals k. Therefore,

$$\frac{1}{N}\sum_{m=0}^{N-1}\left(\sum_{j=0}^{N-1}u^{(k-m)j}\right)g(u^m) = \frac{1}{N}Ng(u^k) = g(u^k).$$

This proves that $\sum_{j=0}^{N-1}(g \cdot e_j)e_j$ and g are the same function. ∎

The expansion of g as a sum of constants times the functions e_j is similar to a Fourier series. In fact, we can derive the coefficients by an argument that exactly parallels that used to motivate the selection of Fourier coefficients in Section 13.2. Suppose we want to write

$$g = \sum_{j=0}^{N-1}a_je_j.$$

Let k be any integer from 0 to $N - 1$ inclusive and take the dot product of this equation with e_k. In view of Theorem 14.17, we get

$$g \cdot e_k = \sum_{j=0}^{N-1} a_j(e_j \cdot e_k) = a_k(e_k \cdot e_k) = a_k,$$

so the coefficient of e_k in this representation must be $g \cdot e_k$. The orthogonality condition $e_j \cdot e_k = 0$ for $j \neq k$, which enables us to find the coefficients in the present case, is exactly analogous to the conditions

$$\int_{-\pi}^{\pi} \sin(jx) \sin(kx)\, dx = \int_{-\pi}^{\pi} \cos(jx) \cos(kx)\, dx = 0 \quad \text{for } k \neq j$$

and

$$\int_{-\pi}^{\pi} \sin(jx) \cos(kx)\, dx = 0,$$

which enabled us to solve for the Fourier coefficients of a function in Section 13.2.

Now define, for any $g : \Omega \to \mathfrak{C}$,

$$\hat{g}(e_j) = \frac{1}{N} \sum_{k=0}^{N-1} g(u^k) u^{-kj}$$

for $j = 0, 1, \ldots, N - 1$. Then $\hat{g}$ is a function, but it operates on functions from $\Omega \to \mathfrak{C}$. The relationship between g and $\hat{g}$ is revealed by the following.

THEOREM 14.19

Let $g : \Omega \to \mathfrak{C}$. Then $g = \sum_{j=0}^{N-1} \hat{g}(e_j) e_j$.

Proof We know that $g = \sum_{j=0}^{N-1} (g \cdot e_j) e_j$. But

$$g \cdot e_j = \frac{1}{N} \sum_{k=0}^{N-1} g(u^k) \overline{e_j(u^k)}$$

$$= \frac{1}{N} \sum_{k=0}^{N-1} g(u^k) u^{-kj} = \hat{g}(e_j). \quad \blacksquare$$

The formulas

$$g = \sum_{j=0}^{N-1} \hat{g}(e_j) e_j$$

and

$$\hat{g}(e_j) = \frac{1}{N} \sum_{k=0}^{N-1} g(u^k) u^{-kj}$$

are analogous to the transform pair for the Fourier transform. They are also completely analogous to the definition of the discrete Fourier transform, and its inverse, equation (14.19).

Because of the finite series expansion of g given by this theorem, we will refer to the numbers $\hat{g}(e_j)$ as the *Fourier coefficients of* g, for $g : \Omega \to \mathfrak{C}$.

Efficient Computation of $\hat{g}(e_j)$

We are now ready for the main issue of this section, efficient computation of the Fourier coefficients $\hat{g}(e_j)$ for any function $g : \Omega \to \mathfrak{C}$.

We have defined Ω to consist of the N numbers $1, u = e^{2\pi i/N}, u^2, \ldots, u^{N-1}$. This means that there is actually a different Ω for each positive integer N. This has caused no difficulty up to this point because the discussion has only dealt with one such Ω. Now, however, we will make an inductive argument that will involve Nth roots of unity for different values of N. Therefore, let

$$u = e^{2\pi i/N} \quad \text{and} \quad \Omega = \{1, u, u^2, \ldots, u^{N-1}\}$$

as before, but also let

$$v = e^{\pi i/N} \quad \text{and} \quad \Psi = \{1, v, v^2, \ldots, v^{2N-1}\}.$$

Since $v = e^{2\pi i/2N}$, v is a $(2N)$th root of unity, and Ψ is the set of $(2N)$th roots of unity. Of course, general statements made previously for Ω hold for Ψ as well, since $2N$ is a positive integer if N is.

EXAMPLE 14.30

Let $N = 3$. Then $u = e^{2\pi i/3}$ and

$$\Omega = \left\{ 1, -\frac{1}{2} + \frac{\sqrt{3}}{2}i, -\frac{1}{2} - \frac{\sqrt{3}}{2}i \right\},$$

as we found in Example 14.29.

Now $v = e^{\pi i/3}$, so

$$\Psi = \{1, v, v^2, v^3, v^4, v^5\} = \left\{ 1, -1, \frac{1}{2} \pm \frac{\sqrt{3}}{2}i, -\frac{1}{2} \pm \frac{\sqrt{3}}{2}i \right\}. \quad \blacksquare$$

We have defined functions $e_m : \Omega \to \mathfrak{C}$ by specifying that

$$e_m(u^j) = u^{mj} \quad \text{for } j = 0, 1, 2, \ldots, N - 1.$$

Similarly, we can define functions $E_m : \Psi \to \mathfrak{C}$ by

$$E_m(v^j) = v^{mj} \quad \text{for } j = 0, 1, 2, \ldots, 2N - 1.$$

Finally, if $h : \Psi \to \mathfrak{C}$, we can manufacture two functions from $\Omega \to \mathfrak{C}$ as follows:

$$h_1(u^j) = h(v^{2j}) \quad \text{for } j = 0, 1, \ldots, N - 1$$

and

$$h_2(u^j) = h(v^{2j+1}) \quad \text{for } j = 0, 1, \ldots, N - 1.$$

The point to these constructions is that we can compute $\hat{h}(E_k)$ in terms of the Fourier coefficients of h_1 and h_2 as follows. For $k = 0, 1, \ldots, 2N - 1$,

$$\hat{h}(E_k) = \frac{1}{2N} \sum_{j=0}^{2N-1} h(v^j) E_k(v^j)$$

$$= \frac{1}{2} \left(\frac{1}{N} \sum_{j=0}^{N-1} h(v^{2j}) E_k(v^{2j}) + \frac{1}{N} \sum_{j=0}^{N-1} h(v^{2j+1}) E_k(v^{2j+1}) \right)$$

$$\text{(14.32)}$$

$$= \frac{1}{2} \left(\frac{1}{N} \sum_{j=0}^{N-1} h_1(u^j) e_k(u^j) + \frac{1}{N} \sum_{j=0}^{N-1} h_2(u^j) e_k(u^j) v^k \right)$$

$$= \frac{1}{2} \left(\widehat{h_1}(e_k) + v^k \widehat{h_2}(e_k) \right).$$

We are now ready to prove a theorem on the number of operations needed for certain computations.

THEOREM 14.20

Let N be a positive integer and assume that $u = e^{2\pi i/N}$ has been computed. Suppose there is a method for computing the Fourier coefficients $\hat{g}(e_j)$ of any $g : \Omega \to \mathfrak{C}$ in no more than M arithmetic operations (addition and multiplication). Then the Fourier coefficients $\hat{h}(E_j)$ of any $h : \Psi \to \mathfrak{C}$ can be computed in no more than $2M + 8N$ operations.

Proof First compute $v, v^2, \ldots, v^{2N-1}$. This requires no more than $2N$ operations. Now let $h : \Psi \to \mathfrak{C}$. Then $h_1 : \Omega \to \mathfrak{C}$ and $h_2 : \Omega \to \mathfrak{C}$. By assumption, we can compute $\widehat{h_1}(e_j)$ for $j = 0, 1, \ldots, N-1$ in no more than M operations, and the same for $\widehat{h_2}(e_j)$. By equation (14.32), we can then compute $\hat{h}(E_k)$ in at most three operations (one addition, two multiplications) for each $k = 0, 1, \ldots, 2N - 1$. Counting all operations needed from the beginning of this computation, we can therefore compute all of the Fourier coefficients $\hat{h}(E_k)$ for h, for $k = 0, 1, \ldots, 2N - 1$, in no more than

$$2N + 2M + 3(2N)$$

operations, or $2M + 8N$ operations.

THEOREM 14.21

Let N be a positive integer power of 2. Then the Fourier coefficients of any $g : \Omega \to C$ can be computed using no more than $4N \log_2(N)$ arithmetic operations.

Proof If $N = 2^n$, let $O(n)$ be the minimal number of operations in which we can compute the Fourier coefficients of any $g : \Omega \to \mathfrak{C}$, assuming that $u = e^{2\pi i/N}$ has been computed. We will show by induction on n that

$$O(n) \le n 2^{n+2}. \tag{14.33}$$

First let $n = 1$, so $N = 2$. Now $u = e^{2\pi i/2} = e^{\pi i} = -1$ and $\Omega = \{1, u\} = \{1, -1\}$. In this simple case, Ω consists of the square roots of unity. If $g : \Omega \to \mathfrak{C}$, then

$$\hat{g}(e_0) = \frac{1}{2} \left(g(u^0) \overline{e_0(u^0)} + g(u^1) \overline{e_0(u^1)} \right)$$

$$= \frac{1}{2} (g(1) + g(-1))$$

and

$$\hat{g}(e_1) = \tfrac{1}{2}\left(g(u^0)\overline{e_1(u^0)} + g(u^1)\overline{e_1(u^1)}\right)$$

$$= \tfrac{1}{2}\left(g(1) - g(-1)\right).$$

Computation of $\hat{g}(e_0)$ and $\hat{g}(e_1)$ therefore requires two additions and two multiplications, hence certainly

$$O(1) = 4 \le 1 \cdot 2^{1+2} = 8.$$

Hence inequality (14.33) holds for $n = 1$.

Now suppose for some positive integer n, $O(n) \le n2^{n+2}$. We want to show that $O(n+1) \le (n+1)2^{n+3}$. But, by the preceding theorem,

$$O(n+1) \le 2O(n) + 8N = 2O(n) + 8 \cdot 2^n \le 2n2^{n+2} + 2^{n+3} = (n+1)2^{n+3}.$$

This proves that $O(n) \le n2^{n+2}$ for every positive integer n. But, $N = 2^n$, so

$$n2^{n+2} = 4N\log_2(N),$$

and the proof is complete. ∎

To see what this might mean in practical terms, consider a typical computation. Suppose we want to compute a DFT approximation of a Fourier transform sampled at chosen points, using the approximation (14.24) in Section 14.8. If we simply compute all of the sums and products involved, we must do $N - 1$ additions and $N + 1$ multiplications, each duplicated N times to get the approximations at N points. This is a total of

$$N(N - 1) + N(N + 1) = 2N^2$$

operations. Suppose, to be specific, $N = 2^{20} = 1,048,576$. Then $2N^2 = 2.1990(10^{12})$. If the computer we are using performs 1 million operations per second, this calculation will require about $2,199,023$ seconds, or about 25.45 days of computer time. Since a given project may require computation of the Fourier transform of many functions, this is intolerable in terms of both time and money.

By contrast, Theorem 14.21 tells us that it is possible to carry out the computation using no more than $n2^{n+3}$ operations. With $n = 20$, this is $16,777,216$ operations. At 1 million operations per second, the computer can do this calculation in a little under 16.8 seconds, a very substantial improvement over the 25.45 days estimated using the brute-force approach.

We will conclude our discussion of the FFT with some cases in which it is used to carry out an analysis.

14.9.2   Use of the FFT in Analyzing Power Spectral Densities of Signals

The FFT is routinely used to display graphs of the power spectral densities of signals. For example, consider the relatively simple signal

$$f(t) = \sin(2\pi(50)t) + 2\sin(2\pi(120)t) + \sin(2\pi(175)t) + \sin(2\pi(210)t).$$

$f(t)$ is written in this way to make the frequencies of the components readily identifiable. By writing $\sin(100\pi t)$ as $\sin(2\pi(50)t)$, we immediately see that this function has frequency 50. Figure 14.31 shows a plot of the power spectral density versus frequency in Hz.

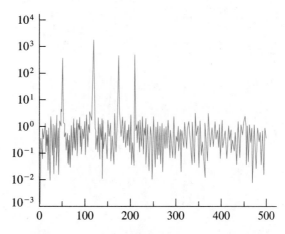

FIGURE 14.31 *FTT display of the power spectral density graph of* $y = \sin(100\pi t) + 2\sin(240\pi t) + \sin(350\pi t) + \sin(420\pi t)$.

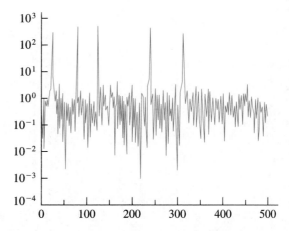

FIGURE 14.32 *FFT display of the power spectral density graph of* $y = \cos(50\pi t) + \cos(160\pi t) + \cos(250\pi t) + \cos(480\pi t) + \cos(630\pi t)$.

Where is the FFT in this? It is in the software that produced the plot. For this example, the graph was drawn using MATLAB and an FFT with $N = 2^{10} = 1024$. Using the same program and choice of N, Figure 14.32 shows the power spectral density graph of

$$g(t) = \cos(2\pi(25)t) + \cos(2\pi(80)t) + \cos(2\pi(125)t) + \cos(2\pi(240)t) + \cos(2\pi(315)t).$$

In both graphs the peaks occur at the primary frequencies of the function.

14.9.3 Filtering Noise from a Signal

The FFT is used sometimes to filter noise from a signal. We discussed filtering previously, but the FFT is the tool for actually carrying it out. To illustrate, consider the signal

$$f(t) = \sin(2\pi(25)t) + \sin(2\pi(80)t) + \sin(2\pi(125)t) + \sin(2\pi(240)t) + \sin(2\pi(315)t).$$

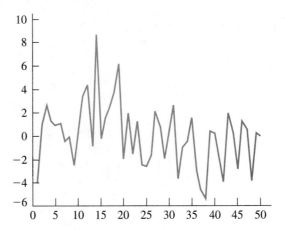

FIGURE 14.33 *A portion of the signal* $y = \sin(50\pi t) + \sin(160\pi t) + \sin(250\pi t) + \sin(480\pi t) + \sin(630\pi t)$ *corrupted with zero-mean random noise.*

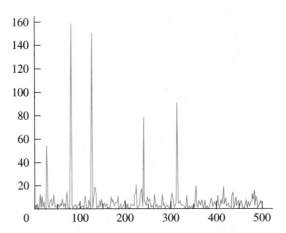

FIGURE 14.34 *FFT calculation of the power spectral density of the signal of Figure 14.33.*

This is a simple signal. However, the signal shown in Figure 14.33 corresponds more closely to reality and was obtained from the graph of $f(t)$ by introducing zero-mean random noise. If we did not know the original signal $f(t)$, it would be difficult to identify from Figure 14.33 the main frequency components of $f(t)$ because of the effect of the noise. However, the Fourier transform sorts out the frequencies. The power spectral density of the noisy signal of Figure 14.33 is shown in Figure 14.34, where the five main frequencies can be identified easily. This particular plot does not reliably give the amplitudes, but the frequencies stand out very well. Figure 14.34 was done using the FFT via MATLAB, with $N = 2^9 = 512$.

14.9.4 Analysis of the Tides in Morro Bay

We will use the DFT and FFT to analyze a set of tidal data, seeking correlations between high and low tides and the relative positions of the sun, earth, and moon.

The forces that cause the tides were of great interest to Isaac Newton as he struggled to understand the world around him, and he devoted considerable space in the *Principia* to this

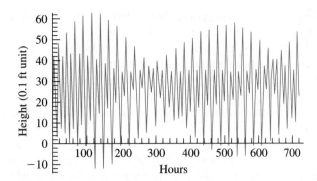

FIGURE 14.35 *Tide profile in Morro Bay from hourly data collected May 1993.*

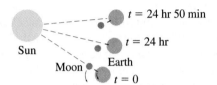

FIGURE 14.36

topic. At one point, Newton required new tables of lunar positions from then Royal Astronomer Flamsteed, who, because of a busy schedule coupled with a personal feud with Newton, was not forthcoming with the data. Newton responded by exerting both professional and political pressure on Flamsteed, through his connections at court, finally forcing Flamsteed to publish the data at his own expense. Years later, Flamsteed came into possession of the remaining copies of this book and is reported to have given vent to his anger with Newton by burning every copy.

It was a triumph of Newton's theory of gravitation, applied to the system consisting of the earth, moon, and sun, that enabled Newton to account for two of the primary tides that occur each day. He was also able to explain why the tides have a twice-monthly maximum and minimum and why the extremes are greatest when the moon is farthest from the earth's equatorial plane. The elliptical orbit of the moon about the earth also accounts for the monthly variation in tide heights resulting from the change in the distance between the earth and moon throughout the month.

Morro Bay is near San Luis Obispo in California. Extensive data have been collected as the Pacific Ocean rolls in and out of the bay and tides wash up on the shore. Figure 14.35 shows a curve drawn through data points giving hourly tide heights for May 1993. We will analyze this data to determine the primary forces causing these tidal variations. As a curiosity, comparison with Figure 2.12 even suggests the presence of beats in periodic tide oscillation! Before carrying out this analysis, we need some background information.

The length of a solar day is 24 hours. This is the time it takes the earth to spin once relative to the sun. The lunar day is 50 minutes longer than this. It takes the earth about 24.8 hours to spin once relative to the moon because the moon is traveling in the direction of the earth's rotation (Figure 14.36).

The sun exerts its primary tidal forces at a point on the earth twice each day, and the moon, twice each 24-hour-and-50-minute period. It is fairly clear why the tide should have a local maximum at a particular location when either the sun or moon is nearly above that point. It is not as obvious, however, that the tide will also rise at a point when either of these bodies is on the opposite side of the earth, as is observed. Newton was able to show that as the earth/moon system travels about its center of mass (which is always interior to the earth), the moon actually exerts an outward force on the opposite side of the earth. The same is true of the earth/sun system. Hence both the sun and moon cause two daily tides.

Tidal forces are proportional to the product of the masses of the bodies involved and inversely proportional to the cube of the distance between them. This enables us to determine the relative tidal forces of the moon and sun on the earth and its waters. Since the sun has a mass of approximately $27(10^6)$ that of the moon and is about 390 times as far from the earth as the earth is from the moon, the sun's influence on the earth's tides is only about 0.46 times that of the moon's influence.

The semidiurnal (twice daily) tides caused by the sun and moon do not just vary between the same highs and lows each day. Other forces change the amplitudes of these highs and lows. These forces are periodic and are responsible for the beats that seem to be present in Figure 14.35. Authorities on tides claim that there are actually about 390 measurably significant partial tides. Depending on the application of the data, usually only seven to twelve of these are used in computing tables of high and low tides. We will focus for the rest of this discussion on three major contributing forces.

First, as the moon orbits the earth, the distance between the two changes from about 222,000 miles at perigee to 253,000 miles at apogee. With the inverse cube law of tidal forces this difference is significant. The time from perigee to apogee is about 27.55 days.

Next, since the moon gains on the sun by about 50 minutes each day, if the three bodies are in conjunction at some time, then they will be in quadrature about seven days later. The twice-daily tides will have large amplitudes when everything is aligned and the smallest variations when the earth/moon/sun angle is 90 degrees. The change from these greatest to smallest tide variations and back again is periodic, with a period of 14.76 days, half the time it takes the moon to circle the earth.

The last tidal force we will consider is that resulting from the moon's orbit being tilted about 5 degrees from the plane containing the earth's orbit about the sun. The result of this deviation can be seen by observing the moon's location in the sky over a one-month period. As the moon traverses the earth in its orbit, it will be above the Northern Hemisphere for a while, helping create high tides in that region. Then it will move in a southerly direction, and while it is in the Southern Hemisphere there is little variation in the tides in the north. It takes 13.66 days for the moon to move from the most northerly point to that farthest south.

The principal periods resulting from these forces are the solar semidiurnal period of 12 hours; the lunar semidiurnal period of 12 hours, 50 minutes, 14 seconds; a lunar–solar diurnal period of 23 hours, 56 minutes, 4 seconds; and a lunar diurnal period of 25 hours, 49 minutes, 10 seconds.

Now consider the actual data used to generate the graph in Figure 14.35 and look for this information. Apply the FFT to calculate the DFT of this set of 720 data points, take absolute values, and plot the resulting points. This results in the amplitude spectrum of Figure 14.37. The units along the horizontal (frequency) axis are cycles per 720 hours.

Begin from the right side of the amplitude spectrum in Figure 14.37 and move left. The first place we see a high point is at about 60, which indicates a term in the data at a frequency of 60/720, or 1/12 cycles per hour. Equivalently, this point denotes the presence of a force that is felt about every 12 hours. This is the solar semidiurnal force.

The next high point in the amplitude spectrum occurs almost immediately to the left of the first, at 58. The height of this data point indicates that this is the largest contribution to the tides. It occurs every 720/58, or 12.4 hours. This is the lunar semidiurnal tide. There is also some other small amplitude activity near this point, about which we will comment shortly.

Continuing to move left in Figure 14.37, there is a large contribution at about 30, indicating a force with a frequency of 30/720, or 1/24, hence a period of about 24 hours. This is the lunar–solar diurnal period.

The only other term of influence that stands out occurs at 28, indicating a frequency of 28/720. This translates into a period of 25.7 hours and indicates the lunar diurnal period.

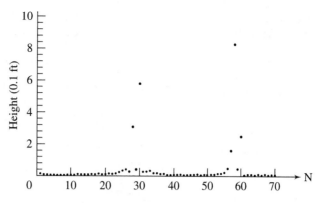

FIGURE 14.37 *Morro Bay tide spectrum.*

Thus, all of the dominant periods are accounted for and no other significant information occurs in the amplitude spectrum, except for the small scattering noted previously in the region around 58. Since the lunar day is not an exact multiple of 1 hour and the data samples were taken hourly, some of the data associated with the moon's tidal forces have leaked onto adjacent points. This also skews the amplitudes, hindering our ability to accurately determine the sun/moon ratio of forces. The same rationale could account for some of the data near 28.

No other discernible information shows up in the amplitude spectrum because all of the remaining forces have periods longer than one month, and this is longer than the time over which the data was taken.

It is interesting to speculate on what Newton would have thought of this graphical verification of his theory. Given his personality, it is possible that he would not have been impressed, having worked it all out to his own satisfaction with his calculus.

SECTION 14.9 PROBLEMS

1. Prove Theorem 14.16 (2).

2. Prove Theorem 14.16 (3).

3. Prove the Cauchy–Schwarz inequality for the dot product defined for functions $g : \Omega \to \mathfrak{C}$. This states that

$$|g \cdot h|^2 \le (g \cdot g)(h \cdot h).$$

Hint: The proof for vectors in R^n applies almost verbatim.

4. If $g : \Omega \to \mathfrak{C}$, define $\|g\| = (g \cdot g)^{1/2}$. Prove that for any number α, $\|\alpha g\| = |\alpha| \|g\|$.

5. Continuing from Problem 4, show that for $g : \Omega \to \mathfrak{C}$ and $h : \Omega \to \mathfrak{C}$, $\|g + h\| \le \|g\| + \|h\|$. This is a triangle inequality. The proof of the triangle inequality for vectors in R^n can be adapted.

6. If $g : \Omega \to C$ and $h : \Omega \to C$, define the convolution of g and h by

$$(g * h)(u^k) = \frac{1}{N} \sum_{j=0}^{N-1} g(u^{k-j}) h(u^j).$$

(a) Prove that $g * h = h * g$.

(b) Prove that $(g * h) * k = g * (h * k)$.

(c) Prove that $g * (h + k) = (g * h) + (g * k)$.

(d) Prove that for any complex number α, $\alpha(g * h) = (\alpha g) * h = g * (\alpha h)$.

7. Let $u = e^{2\pi i/N}$ with N a positive integer. Let m be a positive integer that is not an integer multiple of N. Prove that $\sum_{j=0}^{N-1} u^{mj} = 0$.

8. In the notation of the preceding problem, evaluate $\sum_{j=0}^{N-1} (-1)^{j+1} u^{mj}$.

In each of Problems 9 through 12, use a software package with the FFT to produce a graph of the power spectrum of the function. Use $N = 2^{10}$.

9. $y(t) = 4\sin(80\pi t) - \sin(20\pi t)$

10. $y(t) = 2\cos(40\pi t) + \sin(90\pi t)$

11. $y(t) = 3\cos(90\pi t) - \sin(30\pi t)$

12. $y(t) = \cos(220\pi t) + \cos(70\pi t)$

In each of Problems 13 through 16, corrupt the signal with zero-mean random noise and use the FFT to plot the power density spectrum to identify the frequency components of the original signal.

13. $y(t) = \cos(30\pi t) + \cos(70\pi t) + \cos(140\pi t)$

14. $y(t) = \sin(60\pi t) + 4\sin(130\pi t) + \sin(2405\pi t)$

15. $y(t) = \cos(20\pi t) + \sin(140\pi t) + \cos(240\pi t)$

16. $y(t) = \sin(30\pi t) + 3\sin(40\pi t) + \sin(130\pi t) + \sin(196\pi t) + \sin(220\pi t)$

CHAPTER 15

Special Functions, Orthogonal Expansions, and Wavelets

A function is designated as *special* when it has some distinctive characterics that make it worthwhile determining and recording its properties and behavior. Perhaps the most familiar examples of special functions are $\sin(kx)$ and $\cos(kx)$, which are solutions of an important differential equation, $y'' + k^2 y = 0$, and occur in many other contexts as well.

For us, the primary motivation for studying certain special functions is that they arise in solving ordinary and partial differential equations that model many physical phenomena. Like Fourier series, they constitute necessary items in the toolkit of anyone who wishes to understand and work with such models.

We will begin with Legendre polynomials and Bessel functions. These are important in their own right but also form a model of how to approach special functions and the kinds of properties we should look for. Following these, we will develop parts of Sturm–Liouville theory, which will provide a template for studying certain aspects of special functions in general, for example, eigenfunction expansions, of which Fourier series are a special case. The chapter concludes with a brief introduction to wavelets, in the setting of orthogonal expansions.

15.1 Legendre Polynomials

There are many different approaches to Legendre polynomials. We will begin with Legendre's differential equation

$$(1 - x^2)y'' - 2xy' + \lambda y = 0 \tag{15.1}$$

in which $-1 \leq x \leq 1$ and λ is a real number. This equation has the equivalent form

$$[(1 - x^2)y']' + \lambda y = 0,$$

which we will encounter in Chapter 16 in solving for the steady-state temperature distribution over a solid sphere.

We seek values of λ for which Legendre's equation has nontrivial solutions. Writing Legendre's equation as

$$y'' - \frac{2x}{1 - x^2}y' + \frac{\lambda}{1 - x^2}y = 0,$$

we conclude that 0 is an ordinary point. There are therefore power series solutions

$$y(x) = \sum_{n=0}^{\infty} a_n x^n.$$

Substitute this series into the differential equation to get

$$\sum_{n=2}^{\infty} a_n n(n-1)a_n x^{n-2} - \sum_{n=2}^{\infty} n(n-1)a_n x^n - \sum_{n=1}^{\infty} 2na_n x^n + \sum_{n=0}^{\infty} \lambda a_n x^n = 0.$$

Shift indices in the first summation to write the last equation as

$$\sum_{n=0}^{\infty} (n+2)(n+1)a_{n+2}x^n - \sum_{n=2}^{\infty} n(n-1)a_n x^n - \sum_{n=1}^{\infty} 2na_n x^n + \sum_{n=0}^{\infty} \lambda a_n x^n = 0.$$

Now combine terms for $n \geq 2$ under one summation, writing the $n = 0$ and $n = 1$ terms separately:

$$2a_2 + 6a_3 x - 2a_1 x + \lambda a_0 + \lambda a_1 x + \sum_{n=2}^{\infty} [(n+2)(n+1)a_{n+2} - (n^2 + n - \lambda)]x^n = 0.$$

The coefficient of each power of x must be zero, hence

$$2a_2 + \lambda a_0 = 0, \tag{15.2}$$

$$6a_3 - 2a_1 + \lambda a_1 = 0, \tag{15.3}$$

and, for $n = 2, 3, \ldots,$

$$(n+1)(n+2)a_{n+2} - [n(n+1) - \lambda]a_n = 0$$

from which we get the recurrence relation

$$a_{n+2} = \frac{n(n+1) - \lambda}{(n+1)(n+2)} \quad \text{for } n = 2, 3, \ldots. \tag{15.4}$$

From equation (15.2) we have

$$a_2 = -\frac{\lambda}{2}a_0.$$

From equation (15.4),

$$a_4 = \frac{6-\lambda}{3\cdot4}a_2 = -\frac{\lambda}{2}\frac{6-\lambda}{3\cdot4}a_0 = \frac{-\lambda(6-\lambda)}{4!}a_0,$$

$$a_6 = \frac{20-\lambda}{5\cdot6}a_4 = \frac{-\lambda(6-\lambda)(20-\lambda)}{6!}a_0,$$

and so on. Every even-indexed coefficient a_{2n} is a multiple, involving n and λ, of a_0. Here we have used the factorial notation, in which $n!$ is the product of the integers from 1 through n, if n is a positive integer. For example, $6! = 720$. By convention, $0! = 1$.

From equation (15.3),

$$a_3 = \frac{2-\lambda}{6}a_1 = \frac{2-\lambda}{3!}a_1.$$

Then, from the recurrence relation (15.4),

$$a_5 = \frac{12-\lambda}{4\cdot 5}a_3 = \frac{(2-\lambda)(12-\lambda)}{5!}a_1,$$

$$a_7 = \frac{30-\lambda}{6\cdot 7}a_5 = \frac{(2-\lambda)(12-\lambda)(30-\lambda)}{7!}a_1,$$

and so on. Every odd-indexed coefficient a_{2n+1} is a multiple, also involving n and λ, of a_1.

In this way we can write the solution

$$y(x) = \sum_{n=0}^{\infty} a_n x^n = a_0\left(1 - \frac{\lambda}{2}x^2 - \frac{\lambda(6-\lambda)}{4!}x^4 - \frac{\lambda(6-\lambda)(20-\lambda)}{6!}x^6 + \cdots\right)$$

$$+ a_1\left(x + \frac{2-\lambda}{3!}x^3 + \frac{(2-\lambda)(12-\lambda)}{5!}x^5 + \frac{(2-\lambda)(12-\lambda)(30-\lambda)}{7!}x^7 + \cdots\right).$$

The two series in large parentheses are linearly independent, one containing only even powers of x, the other only odd powers. Put

$$y_e(x) = 1 - \frac{\lambda}{2}x^2 - \frac{\lambda(6-\lambda)}{4!}x^4 - \frac{\lambda(6-\lambda)(20-\lambda)}{6!}x^6 + \cdots$$

and

$$y_o(x) = x + \frac{2-\lambda}{3!}x^3 + \frac{(2-\lambda)(12-\lambda)}{5!}x^5 + \frac{(2-\lambda)(12-\lambda)(30-\lambda)}{7!}x^7 + \cdots.$$

The general solution of Legendre's differential equation is

$$y(x) = a_0 y_e(x) + a_1 y_o(x),$$

in which a_0 and a_1 are arbitrary constants. Some particular solutions are:

with $\lambda = 0$ and $a_1 = 0$,

$$y(x) = a_0.$$

with $\lambda = 2$ and $a_0 = 0$,

$$y(x) = a_1 x.$$

with $\lambda = 6$ and $a_1 = 0$,

$$y(x) = a_0(1 - 3x^2).$$

with $\lambda = 12$ and $a_0 = 0$,

$$y(x) = a_1\left(x - \tfrac{5}{3}x^3\right).$$

with $\lambda = 20$ and $a_1 = 0$,

$$y(x) = a_0\left(1 - 10x^2 + \tfrac{35}{3}x^4\right),$$

and so on.

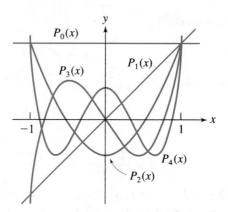

FIGURE 15.1 *The first five Legendre polynomials.*

The values of λ for which solutions are polynomials (finite series) are $\lambda = n(n+1)$ for $n = 0, 1, 2, 3, \ldots$. This should not be surprising, since the recurrence relation (15.4) contains $n(n+1) - \lambda$ in its numerator. If for some nonnegative integer N we choose $\lambda = N(N+1)$, then $a_{N+2} = 0$; hence also $a_{N+4} = a_{N+6} = \cdots = 0$, and one of $y_e(x)$ or $y_o(x)$ will contain only finitely many nonzero terms, hence is a polynomial.

These polynomial solutions of Legendre's differential equation have many applications, for example, in astronomy, analysis of heat conduction, and approximation of solutions of equations $f(x) = 0$. To standardize and tabulate these polynomial solutions, a_0 or a_1 is chosen for each $\lambda = n(n+1)$ so that the polynomial solution has the value 1 at $x = 1$. The resulting polynomials are called *Legendre polynomials* and are usually denoted $P_n(x)$. The first six Legendre polynomials are

$$P_0(x) = 1, \quad P_1(x) = x, \quad P_2(x) = \tfrac{1}{2}(3x^2 - 1), \quad P_3(x) = \tfrac{1}{2}(5x^3 - 3x),$$

$$P_4(x) = \tfrac{1}{8}(35x^4 - 30x^2 + 3), \quad P_5(x) = \tfrac{1}{8}(63x^5 - 70x^3 + 15x).$$

Graphs of some of these polynomials are given in Figure 15.1. $P_n(x)$ is of degree n, and contains only even powers of x if n is even, and only odd powers if n is odd. Although these polynomials are defined for all real x, the relevant interval for Legendre's differential equation is $-1 < x < 1$.

It will also be useful to keep in mind that if $q(x)$ is any polynomial solution of Legendre's equation with $\lambda = n(n+1)$, then $q(x)$ must be a constant multiple of $P_n(x)$.

15.1.1   A Generating Function for the Legendre Polynomials

Many properties of Legendre polynomials can be derived by using a *generating function*, a concept we will now develop. Let

$$L(x, t) = \frac{1}{\sqrt{1 - 2xt + t^2}}.$$

We claim that if $L(x, t)$ is expanded in a power series in powers of t, then the coefficient of t^n is exactly the nth Legendre polynomial.

THEOREM 15.1   *Generating Function for the Legendre Polynomials*

$$L(x, t) = \sum_{n=0}^{\infty} P_n(x)t^n. \quad \blacksquare$$

We will give an argument suggesting why this is true. Write the Maclaurin series for $(1 - w)^{-1/2}$:

$$\frac{1}{\sqrt{1-w}} = 1 + \tfrac{1}{2}w + \tfrac{3}{8}w^2 + \tfrac{15}{48}w^3 + \tfrac{105}{384}w^4 + \tfrac{945}{3840}w^5 + \cdots$$

for $-1 < w < 1$. Put $w = 2xt - t^2$ to obtain

$$\frac{1}{\sqrt{1-2xt+t^2}} = 1 + \tfrac{1}{2}(2xt - t^2) + \tfrac{3}{8}(2xt - t^2)^2 + \tfrac{15}{48}(2xt - t^2)^3$$
$$+ \tfrac{105}{384}(2xt - t^2)^4 + \tfrac{945}{3840}(2xt - t^2)^5 + \cdots.$$

Now expand each of these powers of $2xt - t^2$ and collect the coefficient of each power of t in the resulting expression:

$$\frac{1}{\sqrt{1-2xt+t^2}} = 1 + xt - \tfrac{1}{2}t^2 + \tfrac{3}{2}x^2t^2 - \tfrac{3}{2}xt^3 + \tfrac{3}{8}t^4 + \tfrac{5}{2}x^3t^3 - \tfrac{15}{4}x^2t^4 + \tfrac{15}{8}xt^5 - \tfrac{5}{16}t^6$$
$$+ \tfrac{35}{8}x^4t^4 - \tfrac{35}{4}x^3t^5 + \tfrac{105}{16}x^2t^6 - \tfrac{35}{16}xt^7 + \tfrac{35}{128}t^8 + \tfrac{63}{8}x^5t^5 - \tfrac{315}{16}x^4t^6$$
$$+ \tfrac{315}{16}x^3t^7 - \tfrac{315}{32}x^2t^8 + \tfrac{315}{128}xt^9 - \tfrac{63}{256}t^{10} + \cdots$$
$$= 1 + xt + \left(-\tfrac{1}{2} + \tfrac{3}{2}x^2\right)t^2 + \left(-\tfrac{3}{2}x + \tfrac{5}{2}x^3\right)t^3 + \left(\tfrac{3}{8} - \tfrac{15}{4}x^2 + \tfrac{35}{8}x^4\right)t^4$$
$$+ \left(\tfrac{15}{8}x - \tfrac{35}{4}x^3 + \tfrac{63}{8}x^5\right)t^5 + \cdots$$
$$= P_0(x) + P_1(x)t + P_2(x)t^2 + P_3(x)t^3 + P_4(x)t^4 + P_5(x)t^5 + \cdots.$$

The generating function provides an efficient way of deriving many properties of Legendre polynomials. We will begin by using it to show that

$$P_n(1) = 1 \quad \text{and} \quad P_n(-1) = (-1)^n$$

for $n = 0, 1, 2, \ldots$. First, setting $x = 1$ we have

$$L(1, t) = \frac{1}{\sqrt{1-2t+t^2}} = \frac{1}{\sqrt{(1-t)^2}} = \frac{1}{1-t} = \sum_{n=0}^{\infty} P_n(1)t^n.$$

But, for $-1 < t < 1$,

$$\frac{1}{1-t} = \sum_{n=0}^{\infty} t^n.$$

Since $1/(1 - t)$ has only one Maclaurin expansion, the coefficients in these two series must be the same, hence each $P_n(1) = 1$.

Similarly,

$$L(-1, t) = \frac{1}{\sqrt{1+2t+t^2}} = \frac{1}{\sqrt{(1+t)^2}} = \frac{1}{1+t} = \sum_{n=0}^{\infty} P_n(-1)t^n.$$

But, for $-1 < t < 1$,

$$\frac{1}{1+t} = \sum_{n=0}^{\infty} (-1)^n t^n,$$

so $P_n(-1) = (-1)^n$.

15.1.2 A Recurrence Relation for the Legendre Polynomials

We will use the generating function to derive a recurrence relation for Legendre polynomials.

THEOREM 15.2   *Recurrence Relation for Legendre Polynomials*

For any positive integer n,

$$(n+1)P_{n+1}(x) - (2n+1)xP_n(x) + nP_{n-1}(x) = 0. \tag{15.5}$$

Proof   Begin by differentiating the generating function with respect to t:

$$\frac{\partial L(x,t)}{\partial t} = -\frac{1}{2}(1 - 2xt + t^2)^{-3/2}(-2x + 2t) = \frac{x-t}{(1-2xt+t^2)^{3/2}}.$$

Now notice that

$$(1 - 2xt + t^2)\frac{\partial L(x,t)}{\partial t} - (x-t)L(x,t) = 0.$$

Substitute $L(x,t) = \sum_{n=0}^{\infty} P_n(x)t^n$ into the last equation to obtain

$$(1 - 2xt + t^2)\sum_{n=1}^{\infty} nP_n(x)t^{n-1} - (x-t)\sum_{n=0}^{\infty} P_n(x)t^n = 0.$$

Carry out the indicated multiplications to write

$$\sum_{n=1}^{\infty} nP_n(x)t^{n-1} - \sum_{n=1}^{\infty} 2nxP_n(x)t^n + \sum_{n=1}^{\infty} nP_n(x)t^{n+1} - \sum_{n=0}^{\infty} xP_n(x)t^n + \sum_{n=0}^{\infty} P_n(x)t^{n+1} = 0.$$

Rearrange these series to have like powers of t in each summation:

$$\sum_{n=0}^{\infty}(n+1)P_{n+1}(x)t^n - \sum_{n=1}^{\infty} 2nxP_n(x)t^n + \sum_{n=2}^{\infty}(n-1)P_{n-1}(x)t^n$$

$$-\sum_{n=0}^{\infty} xP_n(x)t^n + \sum_{n=1}^{\infty} P_{n-1}(x)t^n = 0.$$

Combine summations from $n=2$ on, writing the terms for $n=0$ and $n=1$ separately:

$$P_1(x) + 2P_2(x)t - 2xP_1(x)t - xP_0(x) - xP_1(x)t + P_0(x)t$$

$$+ \sum_{n=2}^{\infty}[(n+1)P_{n+1}(x) - 2nxP_n(x) + (n-1)P_{n-1}(x) - xP_n(x) + P_{n-1}(x)]\,t^n = 0.$$

For this power series in t to be zero for all t in some interval about 0, the coefficient of t^n must be zero for $n = 0, 1, 2, \ldots$. Then

$$P_1(x) - xP_0(x) = 0,$$

$$2P_2(x) - 2xP_1(x) - xP_1(x) + P_0(x) = 0,$$

and, for $n = 2, 3, \ldots$,

$$(n+1)P_{n+1}(x) - 2nxP_n(x) + (n-1)P_{n-1}(x) - xP_n(x) + P_{n-1}(x) = 0.$$

These give us

$$P_1(x) = xP_0(x),$$

$$P_2(x) = \tfrac{1}{2}(3xP_1(x) - P_0(x))$$

and, for $n = 2, 3, \ldots$,

$$(n + 1)P_{n+1}(x) - (2n + 1)x P_n(x) + n P_{n-1}(x) = 0.$$

Since this equation is also valid for $n = 1$, this establishes the recurrence relation for all positive integers. ∎

Later we will need to know the coefficient of x^n in $P_n(x)$. We will use the recurrence relation to derive a formula for this number.

THEOREM 15.3

For $n = 1, 2, \ldots$, let A_n be the coefficient of x^n in $P_n(x)$. Then

$$A_n = \frac{1 \cdot 3 \cdot \cdots \cdot (2n - 1)}{n!}. \quad ∎$$

For example,

$$A_1 = 1, \quad A_2 = \frac{1 \cdot 3}{2!} = \frac{3}{2}, \quad \text{and} \quad A_3 = \frac{1 \cdot 3 \cdot 5}{3!} = \frac{5}{2},$$

as can be verified from the explicit expressions derived previously for $P_1(x)$, $P_2(x)$, and $P_3(x)$.

Proof   In the recurrence relation (15.5), the highest power of x that occurs is x^{n+1}, and this term appears in $P_{n+1}(x)$ and in $x P_n(x)$. Thus the coefficient of x^{n+1} in the recurrence relation is

$$(n + 1)A_{n+1} - (2n + 1)A_n.$$

This must equal zero (because the other side of this recurrence equation is zero). Therefore,

$$A_{n+1} = \frac{2n + 1}{n + 1} A_n,$$

and this holds for $n = 0, 1, 2, \ldots$. Now we can work back:

$$A_{n+1} = \frac{2n + 1}{n + 1} A_n = \frac{2n + 1}{n + 1} \frac{2(n - 1) + 1}{(n - 1) + 1} A_{n-1}$$

$$= \frac{2n + 1}{n + 1} \frac{2n - 1}{n} A_{n-1}$$

$$= \frac{2n + 1}{n + 1} \frac{2n - 1}{n} \frac{2(n - 2) + 1}{(n - 2) + 1} A_{n-2}$$

$$= \frac{2n + 1}{n + 1} \frac{2n - 1}{n} \frac{2n - 3}{n - 1} A_{n-2} = \cdots = \frac{2n + 1}{n + 1} \frac{2n - 1}{n} \frac{2n - 3}{n - 1} \cdots \frac{3}{2} A_0.$$

But $A_0 = 1$ because $P_0(x) = 1$, so

$$A_{n+1} = \frac{1 \cdot 3 \cdot 5 \cdot \cdots \cdot (2n - 1)(2n + 1)}{(n + 1)!} \tag{15.6}$$

for $n = 0, 1, 2, \ldots$. The conclusion of the theorem simply states this conclusion in terms of A_n instead of A_{n+1}. ∎

15.1.3   Orthogonality of the Legendre Polynomials

We will prove the following.

THEOREM 15.4   *Orthogonality of the Legendre Polynomials on* [−1, 1]

If n and m are nonnegative integers, then

$$\int_{-1}^{1} P_n(x) P_m(x)\, dx = 0 \quad \text{if } n \neq m. \ \blacksquare \tag{15.7}$$

This integral relationship is called *orthogonality* of the Legendre polynomials on $[-1, 1]$. We have seen this kind of behavior before, with the functions

$$1, \cos(x), \cos(2x), \ldots, \sin(x), \sin(2x), \ldots$$

on the interval $[-\pi, \pi]$. The integral, from $-\pi$ to π, of the product of any two of these (distinct) functions is zero. Because of this property, we were able to find the Fourier coefficients of a function (recall the argument given in Section 13.2). We will pursue a similar idea for Legendre polynomials after establishing equation (15.7).

Proof   Begin with the fact that $P_n(x)$ is a solution of Legendre's equation (15.1) for $\lambda = n(n+1)$. In particular, if n and m are distinct nonnegative integers, then

$$[(1 - x^2) P_n'(x)]' + n(n + 1) P_n(x) = 0$$

and

$$[(1 - x^2) P_m'(x)]' + m(m + 1) P_m(x) = 0.$$

Multiply the first equation by $P_m(x)$ and the second by $P_n(x)$ and subtract the resulting equations to get

$$[(1 - x^2) P_n'(x)]' P_m(x) - [(1 - x^2) P_m'(x)]' P_n(x) + [n(n + 1) - m(m + 1)] P_n(x) P_m(x) = 0.$$

Integrate this equation:

$$\int_{-1}^{1} [(1 - x^2) P_n'(x)]' P_m(x)\, dx - \int_{-1}^{1} [(1 - x^2) P_m'(x)]' P_n(x)\, dx$$

$$= [m(m + 1) - n(n + 1)] \int_{-1}^{1} P_n(x) P_m(x)\, dx.$$

Since $n \neq m$, equation (15.7) will be proved if we can show that the left side of the last equation is zero. But, by integrating the left side by parts, we have

$$\int_{-1}^{1} [(1 - x^2) P_n'(x)]' P_m(x)\, dx - \int_{-1}^{1} [(1 - x^2) P_m'(x)]' P_n(x)\, dx$$

$$= \left[(1 - x^2) P_n'(x) P_m(x) \right]_{-1}^{1} - \int_{-1}^{1} (1 - x^2) P_n'(x) P_m'(x)\, dx$$

$$- \left[(1 - x^2) P_m'(x) P_n(x) \right]_{-1}^{1} + \int_{-1}^{1} (1 - x^2) P_n'(x) P_m'(x)\, dx = 0,$$

and the orthogonality of the Legendre polynomials on $[-1, 1]$ is proved. $\blacksquare$

15.1.4 Fourier–Legendre Series

Suppose $f(x)$ is defined for $-1 \leq x \leq 1$. We want to explore the possibility of expanding $f(x)$ in a series of Legendre polynomials:

$$f(x) = \sum_{n=0}^{\infty} c_n P_n(x). \tag{15.8}$$

We were in a similar situation in Section 13.2, except there we wanted to expand a function defined on $[-\pi, \pi]$ in a series of sines and cosines. We will follow the same reasoning that led to success then. Pick a nonnegative integer m and multiply the proposed expansion by $P_m(x)$, and then integrate the resulting equation, interchanging the series and the integral:

$$\int_{-1}^{1} f(x) P_m(x)\, dx = \sum_{n=0}^{\infty} c_n \int_{-1}^{1} P_n(x) P_m(x)\, dx.$$

Because of equation (15.7), all terms in the summation on the right are zero except when $n = m$. The preceding equation reduces to

$$\int_{-1}^{1} f(x) P_m(x)\, dx = c_m \int_{-1}^{1} P_m^2(x)\, dx.$$

Then

$$c_m = \frac{\int_{-1}^{1} f(x) P_m(x)\, dx}{\int_{-1}^{1} P_m^2(x)\, dx}. \tag{15.9}$$

Taking the lead from Fourier series, we call the expansion $\sum_{n=0}^{\infty} c_n P_n(x)$ *the Fourier–Legendre series*, or expansion, of $f(x)$, when the coefficients are chosen according to equation (15.9). We call these c_n's the *Fourier–Legendre coefficients* of f.

As with Fourier series, we must address the question of convergence of the Fourier–Legendre series of a function. This is done in the following theorem, which is similar in form to the Fourier convergence theorem. As we will see later, this is not a coincidence.

THEOREM 15.5

Let f be piecewise smooth on $[-1, 1]$. Then, for $-1 < x < 1$,

$$\sum_{n=0}^{\infty} c_n P_n(x) = \tfrac{1}{2}\left(f(x+) + f(x-)\right),$$

if the c_n's are the Fourier–Legendre coefficients of f. ∎

This means that under the conditions on f, the Fourier–Legendre expansion of $f(x)$ converges to the average of the left and right limits of $f(x)$ at x, for $-1 < x < 1$. This is midway between the gap at the ends of the graph at x if $f(x)$ has a jump discontinuity there (Figure 15.2). We saw this behavior previously with convergence of Fourier (trigonometric) series. If f is continuous at x, then $f(x+) = f(x-) = f(x)$ and the Fourier–Legendre series converges to $f(x)$.

As a special case of general Fourier–Legendre expansions, any polynomial $q(x)$ is a linear combination of Legendre polynomials. In the case of a polynomial, this linear combination can be obtained by just solving for x^n in terms of $P_n(x)$ and writing each power of x in $q(x)$ in terms of Legendre polynomials.

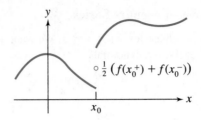

FIGURE 15.2 *Convergence of a Fourier–Legendre expansion at a jump discontinuity of the function.*

For example, let

$$q(x) = -4 + 2x + 9x^2.$$

We can begin with

$$x = P_1(x)$$

and then solve for x^2 in $P_2(x)$:

$$P_2(x) = \tfrac{3}{2}x^2 - \tfrac{1}{2},$$

so

$$x^2 = \tfrac{2}{3}P_2(x) + \tfrac{1}{3} = \tfrac{2}{3}P_2(x) + \tfrac{1}{3}P_0(x).$$

Then

$$-4 + 2x + 9x^2 = -4P_0(x) + 2P_1(x) + 9\left(\tfrac{2}{3}P_2(x) + \tfrac{1}{3}P_0(x)\right)$$
$$= -P_0(x) + 2P_1(x) + 6P_2(x).$$

We can now prove the perhaps surprising result that every Legendre polynomial is orthogonal to every polynomial of lower degree.

THEOREM 15.6

Let $q(x)$ be a polynomial of degree m, and let $n > m$. Then

$$\int_{-1}^{1} q(x)P_n(x)\,dx = 0.$$

Proof Write

$$q(x) = c_0 P_0(x) + c_1 P_1(x) + \cdots + c_m P_m(x).$$

Then

$$\int_{-1}^{1} q(x)P_n(x)\,dx = \sum_{k=0}^{m} c_k \int_{-1}^{1} P_k(x)P_n(x)\,dx = 0,$$

since for $0 \le k \le m < n$, $\int_{-1}^{1} P_k(x)P_n(x)\,dx = 0$. ∎

This result will be useful shortly in obtaining information about the zeros of the Legendre polynomials.

15.1.5 Computation of Fourier–Legendre Coefficients

The equation (15.9) for the Fourier–Legendre coefficients of f has $\int_{-1}^{1} P_n^2(x)\,dx$ in the denominator. We will derive a simple expression for this integral.

THEOREM 15.7

If n is a nonnegative integer, then

$$\int_{-1}^{1} P_n^2(x)\,dx = \frac{2}{2n+1}.$$

Proof   As before, denote the coefficient of x^n in $P_n(x)$ as A_n. We will also denote

$$p_n = \int_{-1}^{1} P_n^2(x)\,dx.$$

The highest power term in $P_n(x)$ is $A_n x^n$, while the highest power term in $P_{n-1}(x)$ is $A_{n-1}x^{n-1}$. This means that all terms involving x^n cancel in the polynomial

$$q(x) = P_n(x) - \frac{A_n}{A_{n-1}}x P_{n-1},$$

and so $q(x)$ has degree at most $n-1$. Write

$$P_n(x) = q(x) + \frac{A_n}{A_{n-1}}x P_{n-1}(x).$$

Then

$$p_n = \int_{-1}^{1} P_n(x)P_n(x)\,dx = \int_{-1}^{1} P_n(x)\left(q(x) + \frac{A_n}{A_{n-1}}x P_{n-1}(x) \right)dx$$

$$= \frac{A_n}{A_{n-1}} \int_{-1}^{1} x P_n(x)P_{n-1}(x)\,dx,$$

because $\int_{-1}^{1} q(x)P_n(x)\,dx = 0$. Now invoke the recurrence relation (15.5) to write

$$x P_n(x) = \frac{n+1}{2n+1} P_{n+1}(x) + \frac{n}{2n+1} P_{n-1}(x).$$

Then

$$x P_n(x) P_{n-1}(x) = \frac{n+1}{2n+1} P_{n+1}(x) P_{n-1}(x) + \frac{n}{2n+1} P_{n-1}^2(x),$$

so

$$p_n = \frac{A_n}{A_{n-1}} \int_{-1}^{1} x P_n(x) P_{n-1}(x)\,dx$$

$$= \frac{A_n}{A_{n-1}} \left[\frac{n+1}{2n+1} \int_{-1}^{1} P_{n+1}(x) P_{n-1}(x)\,dx + \frac{n}{2n+1} \int_{-1}^{1} P_{n-1}^2(x)\,dx \right].$$

Since $\int_{-1}^{1} P_{n+1}(x) P_{n-1}(x)\,dx = 0$, we are left with

$$p_n = \frac{A_n}{A_{n-1}} \frac{n}{2n+1} \int_{-1}^{1} P_{n-1}^2\,dx = \frac{A_n}{A_{n-1}} \frac{n}{2n+1} p_{n-1}.$$

Using the previously obtained value for A_n, we have

$$p_n = \frac{1 \cdot 3 \cdot 5 \cdots (2n-3) \cdot (2n-1)}{n!} \frac{(n-1)!}{1 \cdot 3 \cdot 5 \cdots (2n-3)} \frac{n}{2n+1} p_{n-1} = \frac{2n-1}{2n+1} p_{n-1}.$$

Now work forward:

$$p_1 = \frac{1}{3} p_0 = \frac{1}{3} \int_{-1}^{1} P_0(x)^2 \, dx = \frac{1}{3} \int_{-1}^{1} dx = \frac{2}{3},$$

$$p_2 = \frac{3}{5} p_1 = \frac{3}{5} \frac{2}{3} = \frac{2}{5}, \quad p_3 = \frac{5}{7} p_2 = \frac{2}{7}, \quad p_4 = \frac{7}{9} p_3 = \frac{2}{9},$$

and so on. By induction, we find that

$$p_n = \frac{2}{2n+1},$$

proving the theorem. ∎

This means that the Fourier–Legendre coefficient of f is

$$c_n = \frac{\int_{-1}^{1} f(x) P_n(x) \, dx}{\int_{-1}^{1} P_n^2(x) \, dx} = \frac{2n+1}{2} \int_{-1}^{1} f(x) P_n(x) \, dx.$$

EXAMPLE 15.1

Let $f(x) = \cos(\pi x/2)$ for $-1 \le x \le 1$. Then f and f' are continuous on $[-1, 1]$, so the Fourier–Legendre expansion of f converges to $\cos(\pi x/2)$ for $-1 < x < 1$. The coefficients are

$$c_n = \frac{2n+1}{2} \int_{-1}^{1} \cos\left(\frac{\pi x}{2}\right) P_n(x) \, dx.$$

Because $\cos(x/2)$ is an even function, $\cos(\pi x/2) P_n(x)$ is an odd function for n odd. This means that $c_n = 0$ if n is odd. We need only compute even-indexed coefficients. Some of these are

$$c_0 = \frac{1}{2} \int_{-1}^{1} \cos\left(\frac{\pi x}{2}\right) dx = \frac{2}{\pi},$$

$$c_2 = \frac{5}{2} \int_{-1}^{1} \cos\left(\frac{\pi x}{2}\right) \frac{1}{2}(3x^2 - 1) \, dx = 10 \frac{\pi^2 - 12}{\pi^3},$$

$$c_4 = \frac{9}{2} \int_{-1}^{1} \cos\left(\frac{\pi x}{2}\right) \frac{1}{8}\left(35x^4 - 30x^2 + 3\right) dx = 18 \frac{\pi^4 + 1680 - 180\pi^2}{\pi^5}.$$

Then, for $-1 < x < 1$,

$$\cos\left(\frac{\pi x}{2}\right) = \frac{2}{\pi} + 10 \frac{\pi^2 - 12}{\pi^3} P_2(x) + 18 \frac{\pi^4 + 1680 - 180\pi^2}{\pi^5} P_4(x) + \cdots$$

$$= \frac{2}{\pi} + 5 \frac{\pi^2 - 12}{\pi^3}(3x^2 - 1) + \frac{9}{4} \frac{\pi^4 + 1680 - 180\pi^2}{\pi^5}\left(35x^4 - 30x^2 + 3\right) + \cdots.$$

Although in this example, $f(x)$ is simple enough to compute some Fourier–Legendre coefficients exactly, in a typical application we would use a software package to compute coefficients. The terms we have computed yield the approximation

$$\cos(\pi x/2) \approx 0.63662 - 0.34355(3x^2 - 1) + .0064724\left(35x^4 - 30x^2 + 3\right) + \cdots$$

$$= .99959 - 1.2248x^2 + .22653x^4 + \cdots.$$

Figure 15.3 shows a graph of $\cos(\pi x/2)$ and the first three nonzero terms of its Fourier–Legendre expansion. This series agrees very well with $\cos(\pi x/2)$ for $-1 < x < 1$, but the two diverge from each other outside this interval. This emphasizes the fact that the Fourier–Legendre expansion is only for $-1 < x < 1$. ∎

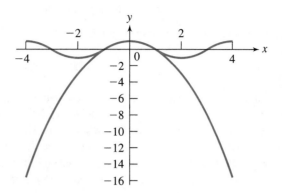

FIGURE 15.3 *Comparison of $\cos(\pi x/2)$ with a partial sum of a series expansion in Legendre polynomials.*

15.1.6   Zeros of the Legendre Polynomials

$P_0(x) = 1$ and has no zeros, while $P_1(x) = x$ has exactly one zero, namely $x = 0$. $P_2(x) = \frac{1}{2}(3x^2 - 1)$ has two real zeros, $\pm 1/\sqrt{3}$. $P_3(x)$ has three real zeros, namely 0 and $\pm\sqrt{3/5}$. After $n = 3$, finding zeros of Legendre polynomials quickly becomes complicated. For example, $P_4(x)$ has four real zeros, and they are

$$\pm\frac{1}{35}\sqrt{\left(525 + 70\sqrt{30}\right)} \quad \text{and} \quad \pm\frac{1}{35}\sqrt{\left(525 - 70\sqrt{30}\right)}.$$

These are approximately ± 0.8611 and ± 0.3400.

Each $P_n(x)$ just tested has n real roots, all lying in the interval $(-1, 1)$. We will show that this is true for all the Legendre polynomials. This includes $P_0(x)$, which of course has no roots. The proof of this assertion is based on the orthogonality of the Legendre polynomials.

THEOREM 15.8   *Zeros of $P_n(x)$*

Let n be a positive integer. Then $P_n(x)$ has n real, distinct roots, all lying in $(-1, 1)$.

Proof   We first show that if $P_n(x)$ has a real root x_0 in $(-1, 1)$, then this root must be simple (that is, not repeated). For suppose x_0 is a repeated root. Then $P_n(x_0) = P'_n(x_0) = 0$. Then $P_n(x)$ is a solution of the initial value problem

$$((1 - x^2)y')' + n(n + 1)y = 0; \qquad y(x_0) = y'(x_0) = 0.$$

But this problem has a unique solution, and the trivial function $y(x) = 0$ is a solution. This implies that $P_n(x)$ is the zero function on an interval containing x_0, and this is false. Hence $P_n(x)$ cannot have a repeated root in $(-1, 1)$.

Now suppose n is a positive integer. Then $P_n(x)$ and $P_0(x)$ are orthogonal on $[-1, 1]$, so

$$\int_{-1}^{1} P_n(x) P_0(x)\, dx = \int_{-1}^{1} P_n(x)\, dx = 0.$$

Therefore, $P_n(x)$ cannot be strictly positive or strictly negative on $(-1, 1)$, hence must change sign in this interval. Since $P_n(x)$ is continuous, there must exist some x_1 in $(-1, 1)$ with $P_n(x_1) = 0$. So far, this gives us one real zero in this interval.

Let $x_1, \ldots, x_m$ be all the zeros of $P_n(x)$ in $(-1, 1)$, with $-1 < x_1 < \cdots < x_m < 1$. Then $1 \le m \le n$. Suppose $m < n$. Then the polynomial

$$q(x) = (x - x_1)(x - x_2) \cdots (x - x_m)$$

has degree less than n, and so is orthogonal to $P_n(x)$:

$$\int_{-1}^{1} q(x) P_n(x)\, dx = 0.$$

But $q(x)$ and $P_n(x)$ change sign at exactly the same points in $(-1, 1)$, namely at $x_1, \ldots, x_m$. Therefore, $q(x)$ and $P_n(x)$ are either of the same sign on each interval $(-1, x_1)$, (x_1, x_2), ..., $(x_m, 1)$ or of opposite sign on each of these intervals. This means that $q(x) P_n(x)$ is either strictly positive or strictly negative on $(-1, 1)$ except at the finitely many points $x_1, \ldots, x_m$ where this product vanishes. But then $\int_{-1}^{1} q(x) P_n(x)\, dx$ must be either positive or negative, a contradiction. We conclude that $m = n$, hence $P_n(x)$ has n simple zeros in $(-1, 1)$. ∎

Referring back to the graphs of $P_0(x)$ through $P_4(x)$ in Figure 15.1, we can see that each of these Legendre polynomials crosses the x axis exactly n times between -1 and 1.

15.1.7   Derivative and Integral Formulas for $P_n(x)$

We will derive two additional formulas for $P_n(x)$ that are sometimes used to further analyze Legendre polynomials. The first gives the nth Legendre polynomial in terms of the nth derivative of $(x^2 - 1)^n$.

THEOREM 15.9   *Rodrigues's Formula*

For $n = 0, 1, 2, \ldots,$

$$P_n(x) = \frac{1}{2^n n!} \frac{d^n}{dx^n}\left((x^2 - 1)^n\right). \ ∎$$

In this statement, it is understood that the zero-order derivative of a function is the function itself. Thus, when $n = 0$, the proposed formula gives

$$\frac{1}{2^0 0!} \frac{d^0}{dx^0}\left((x^2 - 1)^0\right) = (x^2 - 1)^0 = 1 = P_0(x).$$

For $n = 1$ it gives

$$\frac{1}{2(1!)} \frac{d}{dx}(x^2 - 1) = \frac{1}{2}(2x) = x = P_1(x),$$

and for $n = 2$, it gives

$$\frac{1}{2^2(2!)} \frac{d^2}{dx^2}\left((x^2 - 1)^2\right) = \frac{1}{8}(12x^2 - 4) = \frac{3}{2}x^2 - \frac{1}{2} = P_2(x).$$

Proof   Let $w = (x^2 - 1)^n$. Then

$$w' = n(x^2 - 1)^{n-1}(2x).$$

Then

$$(x^2 - 1)w' - 2nxw = 0.$$

If this equation is differentiated $k + 1$ times, it is a routine exercise to verify that we obtain

$$(x^2 - 1) \frac{d^{k+2}w}{dx^{k+2}} - (2n - 2k - 2)x \frac{d^{k+1}w}{dx^{k+1}}$$

$$- [2n + (2n - 2) + \cdots + (2n - 2(k - 1)) + 2n - 2k)] \frac{d^k w}{dx^k} = 0.$$

Putting $k = n$, we have

$$(x^2 - 1) \frac{d^{n+2}w}{dx^{n+2}} + 2x \frac{d^{n+1}w}{dx^{n+1}}$$

$$- [2n + (2n - 2) + \cdots + (2n - 2(n - 1)) + (2n - 2n)] \frac{d^n w}{dx^n} = 0.$$

The quantity in square brackets in this equation is

$$2n + (2n - 2) + \cdots + 2,$$

which is the same as

$$2(1 + 2 + \cdots + n).$$

But this quantity is equal to $n(n + 1)$. (Recall that $\sum_{j=1}^{n} j = \frac{1}{2}n(n + 1)$.) Therefore,

$$(x^2 - 1) \frac{d^{n+2}w}{dx^{n+2}} + 2x \frac{d^{n+1}w}{dx^{n+1}} - n(n + 1) \frac{d^n w}{dx^n} = 0.$$

Upon multiplying this equation by -1, we have

$$(1 - x^2) \frac{d^{n+2}w}{dx^{n+2}} - 2x \frac{d^{n+1}w}{dx^{n+1}} + n(n + 1) \frac{d^n w}{dx^n} = 0.$$

But this means that $d^n w/dx^n$ is a solution of Legendre's equation with $\lambda = n(n + 1)$. Further, repeated differentiation of the polynomial $(x^2 - 1)^n$ yields a polynomial. Therefore, the polynomial solution $d^n w/dx^n$ must be a constant multiple of $P_n(x)$:

$$\frac{d^n w}{dx^n} = c P_n(x). \tag{15.10}$$

Now, the highest-order term in $(x^2 - 1)^n$ is x^{2n}, and the nth derivative of x^{2n} is

$$2n(2n - 1) \cdots (n + 1)x^n.$$

Therefore, the coefficient of the highest power of x in $d^n w/dx^n$ is $2n(2n - 1) \cdots (n + 1)$. The highest-order term in $c P_n(x)$ is $c A_n$, where A_n is the coefficient of x^n in $P_n(x)$. We know A_n, so equation (15.10) gives us

$$2n(2n - 1) \cdots (n + 1) = c A_n = c \frac{1 \cdot 3 \cdot 5 \cdots (2n - 1)}{n!}.$$

Then

$$c = \frac{n!(n + 1) \cdots (2n - 1)(2n)}{1 \cdot 3 \cdot 5 \cdots \cdot (2n - 1)} = \frac{(2n)!}{1 \cdot 3 \cdot 5 \cdots \cdot (2n - 1)}$$

$$= 2 \cdot 4 \cdot 6 \cdots \cdot 2n = 2^n n!.$$

But now equation (15.10) becomes

$$\frac{d^n}{dx^n} (x^2 - 1)^n = 2^n n! P_n(x),$$

which is equivalent to Rodrigues's formula. ∎

Next we will derive an integral formula for $P_n(x)$.

For $n = 0, 1, 2, \ldots$,

$$P_n(x) = \frac{1}{\pi} \int_0^\pi \left(x + \sqrt{x^2 - 1} \cos(\theta) \right)^n d\theta. \quad \blacksquare$$

For example, with $n = 0$ we get

$$\frac{1}{\pi} \int_0^\pi d\theta = 1 = P_0(x).$$

With $n = 1$ we get

$$\frac{1}{\pi} \int_0^\pi \left(x + \sqrt{x^2 - 1} \cos(\theta) \right) d\theta = x = P_1(x),$$

and with $n = 2$ we get

$$\frac{1}{\pi} \int_0^\pi \left(x + \sqrt{x^2 - 1} \cos(\theta) \right)^2 d\theta$$

$$= \frac{1}{\pi} \int_0^\pi \left(x^2 + 2x\sqrt{x^2 - 1} \cos(\theta) + (x^2 - 1) \cos^2(\theta) \right) d\theta$$

$$= \frac{3}{2} x^2 - \frac{1}{2} = P_2(x).$$

Proof Let

$$Q_n(x) = \frac{1}{\pi} \int_0^\pi \left(x + \sqrt{x^2 - 1} \cos(\theta) \right)^n d\theta.$$

The strategy behind the proof is to show that Q_n satisfies the same recurrence relation as the Legendre polynomials. Since $Q_0 = P_0$ and $Q_1 = P_1$, this will imply that $Q_n = P_n$ for all non-negative integers n. Proceed:

$$(n + 1)Q_{n+1}(x) - (2n + 1)xQ_n(x) + nQ_{n-1}(x)$$

$$= \frac{n + 1}{\pi} \int_0^\pi \left(x + \sqrt{x^2 - 1} \cos(\theta) \right)^{n+1} d\theta$$

$$- \frac{2n + 1}{\pi} \int_0^\pi x \left(x + \sqrt{x^2 - 1} \cos(\theta) \right)^n d\theta$$

$$+ \frac{n}{\pi} \int_0^\pi \left(x + \sqrt{x^2 - 1} \cos(\theta) \right)^{n-1} d\theta.$$

After a straightforward but lengthy computation, we find that

$$(n + 1)Q_{n+1}(x) - (2n + 1)xQ_n(x) + nQ_{n-1}(x)$$

$$= \frac{n}{\pi} \int_0^\pi \left(x + \sqrt{x^2 - 1} \cos(\theta) \right)^{n-1} (1 - x^2) \sin^2(\theta) \, d\theta$$

$$+ \frac{1}{\pi} \int_0^\pi \left(x + \sqrt{x^2 - 1} \cos(\theta) \right)^n \sqrt{x^2 - 1} \cos(\theta) \, d\theta.$$

Integrate the second integral by parts, with $u = \left(x + \sqrt{x^2 - 1} \cos(\theta)\right)^n$ and $dv = \sqrt{x^2 - 1}$ $\cos(\theta) \, d\theta$ to get

$$(n + 1)Q_{n+1}(x) - (2n + 1)x Q_n(x) + n Q_{n-1}(x)$$

$$= \frac{n}{\pi} \int_0^\pi \left(x + \sqrt{x^2 - 1} \cos(\theta)\right)^{n-1} (1 - x^2) \sin^2(\theta) \, d\theta$$

$$+ \left[\frac{1}{\pi} \left(x + \sqrt{x^2 - 1} \cos(\theta)\right)^n \sqrt{x^2 - 1} \sin(\theta)\right]_0^\pi$$

$$- \frac{1}{\pi} \int_0^\pi \sqrt{x^2 - 1} \sin(\theta) n \left(x + \sqrt{x^2 - 1} \cos(\theta)\right)^{n-1} \sqrt{x^2 - 1}(-\sin(\theta)) \, d\theta$$

$$= 0,$$

completing the proof. ∎

SECTION 15.1 PROBLEMS

1. For $n = 0, 1, 3, 4, 5$, verify by substitution that $P_n(x)$ is a solution of Legendre's equation corresponding to $\lambda = n(n + 1)$.

2. Use the recurrence relation (Theorem 15.2) and the list of $P_0(x), \ldots, P_5(x)$ given previously to determine $P_6(x)$ through $P_{10}(x)$. Graph these functions and observe the location of their zeros in $[-1, 1]$.

3. Use Rodrigues's formula to obtain $P_1(x)$ through $P_5(x)$.

4. Use Theorem 15.10 to obtain $P_3(x)$, $P_4(x)$, and $P_5(x)$.

5. It can be shown that

$$P_n(x) = \sum_{k=0}^{[n/2]} (-1)^k \frac{(2n - 2k)!}{2^n k!(n - k)!(n - 2k)!} x^{n-2k}.$$

Use this formula to generate $P_0(x)$ through $P_5(x)$. The symbol $[n/2]$ denotes the largest integer not exceeding n.

6. Show that

$$P_n(x) = \sum_{k=0}^{n} \frac{n!}{k!(n - k)!} \frac{d^k}{dx^k} [(x + 1)^n] \frac{d^{n-k}}{dx^{n-k}} [(x - 1)^n].$$

Hint: Put $x^2 - 1 = (x - 1)(x + 1)$ in Rodrigues's formula.

7. Let n be a nonnegative integer. Use reduction of order (Section 2.2) and the fact that $P_n(x)$ is one solution of Legendre's equation with $\lambda = n(n + 1)$ to obtain a second, linearly independent solution:

$$Q_n(x) = P_n(x) \int \frac{1}{[P_n(x)]^2 (1 - x^2)} \, dx.$$

8. Use the result of Problem 7 to show that

$$Q_0(x) = -\frac{1}{2} \ln\left(\frac{1 + x}{1 - x}\right),$$

$$Q_1(x) = 1 - \frac{x}{2} \ln\left(\frac{1 + x}{1 - x}\right),$$

and

$$Q_2(x) = \frac{1}{4}(3x^2 - 1) \ln\left(\frac{1 + x}{1 - x}\right) - \frac{3}{2} x$$

for $-1 < x < 1$.

9. The gravitational potential at a point $P : (x, y, z)$ due to a unit mass at (x_0, y_0, z_0) is

$$\varphi(x, y, z) = \frac{1}{\sqrt{(x - x_0)^2 + (y - y_0)^2 + (z - z_0)^2}}.$$

For some purposes (such as in astronomy) it is convenient to expand $\varphi(x, y, z)$ in powers of r or $1/r$, where $r = \sqrt{x^2 + y^2 + z^2}$. To do this, introduce the angle shown in Figure 15.4. Let $d = \sqrt{x_0^2 + y_0^2 + z_0^2}$ and $R = \sqrt{(x - x_0)^2 + (y - y_0)^2 + (z - z_0)^2}$.

(a) Use the law of cosines to write

$$\varphi(x, y, z) = \frac{1}{d\sqrt{1 - 2(r/d) \cos(\theta) + (r/d)^2}}.$$

(b) From our discussion of the generating function for Legendre polynomials, recall that, if $1/\sqrt{1 - 2at + t^2}$ is expanded in a series about 0, convergent for $|t| < 1$, then the coefficient of t^n is $P_n(a)$.

(c) If $r < d$, let $a = \cos(\theta)$ and $t = r/d$ to obtain

$$\varphi(r) = \sum_{n=0}^{\infty} \frac{1}{d^{n+1}} P_n(\cos(\theta)) r^n.$$

(d) If $r > d$, show that

$$\varphi(r) = \frac{1}{r} \sum_{n=0}^{\infty} d^n P_n(\cos(\theta)) r^{-n}.$$

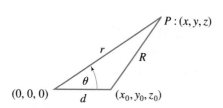

FIGURE 15.4

10. Show that $\sum_{n=0}^{\infty} 1/2^{n+1} P_n\left(\frac{1}{2}\right) = 1/\sqrt{3}$.

11. Prove that, for any positive integer n,

$$n P_n(x) - x P_n'(x) + P_{n-1}'(x) = 0.$$

Hint: Use the generating function. Show that

$$t \frac{\partial L}{\partial t} - (x - t) \frac{\partial L}{\partial x} = 0$$

and substitute $L(x, t) = \sum_{n=0}^{\infty} P_n(x) t^n$.

12. Show that for any positive integer n,

$$(1 - x^2) P_n'(x) = n P_{n-1}(x) - nx P_n(x).$$

13. Prove that for any nonnegative integer n,

$$\int_{-1}^{1} x P_n(x) P_{n-1}(x)\, dx = \frac{2n}{4n^2 - 1}.$$

14. Prove that for any nonnegative integer n,

$$\int_{-1}^{1} x P_n'(x) P_n(x)\, dx = \frac{2n}{2n + 1}.$$

15. Let n be a nonnegative integer. Prove that

$$P_{2n+1}(0) = 0 \quad \text{and} \quad P_{2n}(0) = (-1)^n \frac{(2n)!}{2^{2n}(n!)^2}.$$

16. Expand each of the following in a series of Legendre polynomials:

(a) $1 + 2x - x^2$

(b) $2x + x^2 - 5x^3$

(c) $2 - x^2 + 4x^4$

In each of Problems 17 through 21, find the first five coefficients in the Fourier–Legendre expansion of the function. Graph the function and the sum of the first five terms of this expansion on the same set of axes, for $-3 \le x \le 3$.

Note that, for some functions just five terms provide a good approximation to the function on $[-1, 1]$, while for other functions (such as Problem 21) more terms are needed.

17. $f(x) = \sin(\pi x/2)$

18. $f(x) = e^{-x}$

19. $f(x) = \sin^2(x)$

20. $f(x) = \cos(x) - \sin(x)$

21. $f(x) = \begin{cases} -1 & \text{for } -1 \le x \le 0 \\ 1 & \text{for } 0 < x \le 1 \end{cases}$

22. $f(x) = (x + 1)\cos(x)$

23. Consider N charged beads strung on a wire from -1 to 1 along the x axis. The beads are free to move along the wire on this interval. Each end of the wire has a planar charge of $+1$, and each bead has a charge of $+2$. Show that the equilibrium positions of the beads coincide with the zeros of $P_N(x)$. *Hint:* Particles carrying planar charge q_1 and q_2 and located r units apart exert a force of magnitude $K q_1 q_2/r$ on each other. Now carry out the following steps:

(a) If $N = 1$, there is only one bead and it will come to rest midway between the charges at -1 and $+1$, hence at $x = 0$. This is the zero of $P_1(x)$. Thus the assertion is true for $N = 1$.

(b) Assume that there are two beads and show that they come to rest at $\pm 1/\sqrt{3}$, which are the zeros of $P_2(x)$.

(c) Now assume that there are N beads at $x_1, \ldots, x_N$, with $-1 < x_j < x_{j+1} < 1$. Show that, for $k = 1, 2, \ldots, N$,

$$\sum_{i=1, i \neq k}^{N} \frac{4K}{x_k - x_i} + \frac{2K}{x_k + 1} + \frac{2K}{x_k - 1} = 0.$$

Hence conclude that

$$\frac{2x_k}{1 - x_k^2} = \sum_{i=1, i \neq k}^{N} \frac{2}{x_k - x_i}.$$

(d) Write

$$P_N''(x) - \frac{2x}{1 - x^2} P_N'(x) + \frac{N(N + 1)}{1 - x^2} P_N(x) = 0.$$

Solve this expression for $P_N''(x_k)$ and obtain an expression for $P_N''(x_k)/P_N'(x_k)$. Compare this expression with the result of part (c).

(e) Since $x_1, \ldots, x_N$ are the zeros of $P_N(x)$, then $P_N(x) = A(x - x_1)(x - x_2) \cdots (x - x_N)$ for some constant A, which is the coefficient of x^N. This product is often written

$$P_N(x) = A \prod_{j=1}^{N} (x - x_j),$$

in which $\prod$ denotes a product just as $\sum$ denotes a summation. Show that

$$P_N'(x) = A \sum_{i=1}^{N} \prod_{m=1, m \neq i}^{N} (x - x_m)$$

and

$$P_N''(x) = A \sum_{j=1}^{N} \sum_{i=1, i \neq j}^{N} \prod_{m=1, m \neq i, m \neq j}^{N} (x - x_m).$$

(f) Proceeding from part (e), show that

$$P_N'(x_k) = A \prod_{m=1, m \neq k}^{N} (x_k - x_m) \neq 0$$

and

$$P_N''(x_k) = 2A \sum_{i=1, i \neq k}^{N} \prod_{m=1, m \neq i, m \neq k}^{N} (x_k - x_m).$$

Use these expressions to again evaluate $P_N''(x_k)/P_N'(x_k)$.

(g) Recall that $P_N(x)$ is just one of two linearly independent solutions of Legendre's equation. Use this fact and the results of parts (c), (d), (e), and (f) to complete the proof.

15.2 Bessel Functions

We will now develop the second kind of special function we will use to introduce the general topic of special functions.

Recall from Chapter 4 that the second-order differential equation

$$x^2 y'' + xy' + (x^2 - v^2)y = 0$$

is called Bessel's equation of order v. Thus the term *order* is used in two senses here—the differential equation is of second order, but traditionally we say that the equation has order v to refer to the parameter v occurring in the coefficient of y.

In Example 4.12 of Section 4.3, we used the method of Frobenius to find a series solution

$$y(x) = c_0 \sum_{n=0}^{\infty} \frac{(-1)^n}{2^{2n} n! (1 + v)(2 + v) \cdots (n + v)} x^{2n + v},$$

in which c_0 is a nonzero constant and $v \geq 0$. This solution is valid in some interval $(0, R)$, depending on v.

It will be useful to write this solution in terms of the gamma function, which we will now develop.

15.2.1 The Gamma Function

For $x > 0$, the *gamma function* Γ is defined by

$$\Gamma(x) = \int_0^{\infty} t^{x-1} e^{-t} \, dt.$$

This integral converges for all $x > 0$. The gamma function has a fascinating history and many interesting properties. For us, the most useful is the following.

THEOREM 15.11   *Factorial Property of the Gamma Function*

If $x > 0$, then

$$\Gamma(x + 1) = x\Gamma(x).$$

Proof   If $0 < a < b$, then we can integrate by parts, with $u = t^x$ and $dv = e^{-t} dt$, to get

$$\int_a^b t^x e^{-t} \, dt = \left[t^x (-e^{-t}) \right]_a^b - \int_a^b x t^{x-1} (-1) e^{-t} \, dt$$

$$= -b^x e^{-b} + a^x e^{-a} + x \int_a^b t^{x-1} e^{-t} \, dt.$$

Take the limit of this equation as $x \to 0+$ and $b \to \infty$ to get

$$\int_0^\infty t^x e^{-t} \, dt = \Gamma(x + 1) = x \int_0^\infty t^{x-1} e^{-t} \, dt = x\Gamma(x). \quad\blacksquare$$

The reason why this is called the *factorial property* can be seen by letting $x = n$, a positive integer. By repeated application of the theorem, we get

$$\Gamma(n + 1) = n\Gamma(n) = n\Gamma((n - 1) + 1) = n(n - 1)\Gamma(n - 1)$$

$$= n(n - 1)\Gamma((n - 2) + 1) = n(n - 1)(n - 2)\Gamma(n - 2)$$

$$= \cdots = n(n - 1)(n - 2)\cdots(2)(1)\Gamma(1) = n!\,\Gamma(1).$$

But

$$\Gamma(1) = \int_0^\infty e^{-t} \, dt = 1,$$

so

$$\Gamma(n + 1) = n!$$

for any positive integer n. This is the reason for the term "factorial property" of the gamma function.

It is possible to extend $\Gamma(x)$ to negative (but noninteger) values of x by using the factorial property. For $x > 0$, write

$$\Gamma(x) = \frac{1}{x}\Gamma(x + 1). \tag{15.11}$$

If $-1 < x < 0$, then $x + 1 > 0$ so $\Gamma(x + 1)$ is defined and we use the right side of equation (15.11) to define $\Gamma(x)$.

Once we have extended $\Gamma(x)$ to $-1 < x < 0$, we can let $-2 < x < -1$. Then $-1 < x+1 < 0$ so $\Gamma(x + 1)$ has been defined and we can again use equation (15.11) to define $\Gamma(x)$. In this way we can walk to the left along the real line, defining $\Gamma(x)$ on $(-n - 1, -n)$ as soon as it has been defined on the interval $(-n, -n + 1)$ immediately to the right.

For example,

$$\Gamma\left(-\frac{1}{2}\right) = \frac{1}{-\frac{1}{2}}\Gamma\left(-\frac{1}{2} + 1\right) = -2\Gamma\left(\frac{1}{2}\right),$$

and

$$\Gamma\left(-\frac{3}{2}\right) = \frac{1}{-\frac{3}{2}}\Gamma\left(-\frac{3}{2} + 1\right) = -\frac{2}{3}\Gamma\left(-\frac{1}{2}\right) = \frac{4}{3}\Gamma\left(\frac{1}{2}\right).$$

Figure 15.5(a) shows a graph of $y = \Gamma(x)$ for $0 < x < 5$. Graphs for $-1 < x < 0$, $-2 < x < -1$, and $-3 < x < -2$, respectively, are given in Figures 15.5(b), (c), and (d).

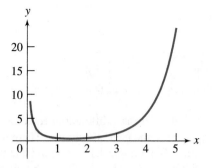

FIGURE 15.5(a) $\Gamma(x)$ *for* $0 < x < 5$.

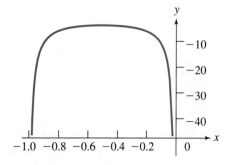

FIGURE 15.5(b) $\Gamma(x)$ *for* $-1 < x < 0$.

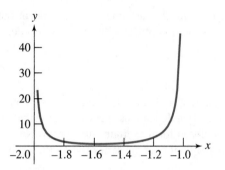

FIGURE 15.5(c) $\Gamma(x)$ *for* $-2 < x < -1$.

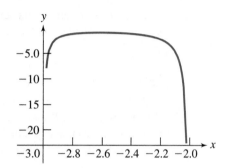

FIGURE 15.5(d) $\Gamma(x)$ *for* $-3 < x < -2$.

15.2.2 Bessel Functions of the First Kind and Solutions of Bessel's Equation

Now return to the Frobenius solution $y(x)$ of Bessel's equation, given above. Part of the denominator in this solution is

$$(1 + v)(2 + v) \cdots (n + v),$$

in which we assume that $v \geq 0$. Now use the factorial property of the gamma function to write

$$\Gamma(n + v + 1) = (n + v)\Gamma(n + v) = (n + v)(n + v - 1)\Gamma(n + v - 1)$$

$$= \cdots = (n + v)(n + v - 1) \cdots (n + v - (n - 1))\Gamma(n + v - (n - 1))$$

$$= (1 + v)(2 + v) \cdots (n - 1 + v)(n + v)\Gamma(v + 1).$$

Therefore,

$$(1 + v)(2 + v) \cdots (n + v) = \frac{\Gamma(n + v + 1)}{\Gamma(v + 1)}$$

and we can write the solution as

$$y(x) = c_0 \sum_{n=0}^{\infty} \frac{(-1)^n \Gamma(v + 1)}{2^{2n} n! \Gamma(n + v + 1)} x^{2n+v}.$$

It is customary to choose

$$c_0 = \frac{1}{2^v \Gamma(v + 1)}.$$

To obtain the solution we will denote as $J_\nu(x)$:

$$J_\nu(x) = \sum_{n=0}^{\infty} \frac{(-1)^n}{2^{2n+\nu} n! \Gamma(n + \nu + 1)} x^{2n+\nu}.$$

J_ν is called a *Bessel function of the first kind of order* ν. The series defining $J_\nu(x)$ converges for all x.

Because Bessel's equation is of second order (as a differential equation), we need a second solution, linearly independent from J_ν, to write the general solution. Theorem 4.4 in Section 4.4 tells us how to proceed to a second solution. In Example 4.12 we found that the indicial equation of Bessel's equation is $r^2 - \nu^2 = 0$, with roots $\pm\nu$. The key lies in the difference, 2ν, between these roots. Omitting the details of the analysis, here are the conclusions.

1. If 2ν is not an integer, then J_ν and $J_{-\nu}$ are linearly independent (neither is a constant multiple of the other), and the general solution of Bessel's equation of order ν is

$$y(x) = a J_\nu(x) + b J_{-\nu}(x),$$

with a and b arbitrary constants.

2. If 2ν is an odd positive integer, say $2\nu = 2n + 1$, then $\nu = n + \frac{1}{2}$ for some positive integer n. In this case, J_ν and $J_{-\nu}$ are still linearly independent. It can be shown that in this case $J_{n+1/2}(x)$ and $J_{-n-1/2}(x)$ can be expressed in closed form as finite sums of terms involving square roots, sines, and cosines. For example, by manipulating the series for $J_\nu(x)$, we find that

$$J_{1/2}(x) = \sqrt{\frac{2}{\pi x}} \sin(x), \quad J_{-1/2}(x) = \sqrt{\frac{2}{\pi x}} \cos(x), \quad J_{3/2}(x) = \sqrt{\frac{2}{\pi x}} \left[\frac{\sin(x)}{x} - \cos(x) \right],$$

and

$$J_{-3/2}(x) = \sqrt{\frac{2}{\pi x}} \left[-\sin(x) - \frac{\cos(x)}{x} \right].$$

In this case, the general solution of Bessel's equation of order ν is

$$y(x) = a J_{n+1/2}(x) + b J_{-n-1/2}(x),$$

with a and b arbitrary constants.

3. 2ν is an integer but is not of the form $n + \frac{1}{2}$ for any positive integer n. In this case, $J_\nu(x)$ and $J_{-\nu}(x)$ are solutions of Bessel's equation, but they are linearly dependent. Indeed, one can check from the series that in this case,

$$J_{-\nu}(x) = (-1)^\nu J_\nu(x).$$

In this case, we must construct a second solution of Bessel's equation, linearly independent from $J_\nu(x)$. This leads us to Bessel functions of the second kind.

15.2.3 Bessel Functions of the Second Kind

In Section 4.4 we derived a second solution for Bessel's equation for the case $\nu = 0$. It was

$$y_2(x) = J_0(x) \ln(x) + \sum_{n=1}^{\infty} \frac{(-1)^{n+1}}{2^{2n} (n!)^2} \emptyset(n) x^{2n},$$

in which

$$\emptyset(n) = 1 + \frac{1}{2} + \cdots + \frac{1}{n}.$$

Instead of using this solution as it is written, it is customary to use a linear combination of $y_2(x)$ and $J_0(x)$, which will of course also be a solution. This combination is denoted $Y_0(x)$ and is defined for $x > 0$ by

$$Y_0(x) = \frac{2}{\pi} [y_2(x) + (\gamma - \ln(2)) J_0(x)],$$

where γ is Euler's constant, defined by

$$\gamma = \lim_{n \to \infty} (\emptyset(n) - \ln(n)) = .577215664901533\ldots.$$

J_0 and Y_0 are linearly independent because of the $\ln(x)$ term in $Y_0(x)$, and the general solution of Bessel's equation of order zero is therefore

$$y(x) = a J_0(x) + b Y_0(x),$$

with a and b arbitrary constants. Y_0 is called a *Bessel function of the second kind of order zero*. With the choice made for the constants in defining Y_0, this function is also called *Neumann's function of order zero*.

If ν is a positive integer, say $\nu = n$, a derivation similar to that of $Y_0(x)$, but with more computational details, yields the second solution

$$Y_n(x) = \frac{2}{\pi} \left[J_n(x) \left[\ln \left(\frac{x}{2} \right) + \gamma \right] + \sum_{k=1}^{\infty} \frac{(-1)^{k+1} [\emptyset(k) + \emptyset(k+1)]}{2^{2k+n+1} k! (k+n)!} x^{2k+n} \right]$$

$$- \frac{2}{\pi} \sum_{k=0}^{n-1} \frac{(n-k-1)!}{2^{2k-n+1} k!} x^{2k-n}.$$

This agrees with $Y_0(x)$ if $n = 0$, with the understanding that in this case the last summation does not appear.

The general solution of Bessel's equation of positive integer order n is therefore

$$y(x) = a J_n(x) + b Y_n(x).$$

Thus far we only have $Y_\nu(x)$ for ν a nonnegative integer. We did not need this Bessel function of the second kind for the general solution of Bessel's equation in other cases. However, it is possible to extend this definition of $Y_\nu(x)$ to include all real values of ν by letting

$$Y_\nu(x) = \frac{1}{\sin(\nu\pi)} [J_\nu(x) \cos(\nu\pi) - J_{-\nu}(x)].$$

For any nonnegative integer n, one can show that

$$Y_n(x) = \lim_{\nu \to n} Y_\nu(x).$$

Y_ν is *Neumann's Bessel function of order ν*. This function is linearly independent from $J_\nu(x)$ for $x > 0$, and it enables us to write the general solution of Bessel's equation of order ν in all cases as

$$y(x) = a J_\nu(x) + b Y_\nu(x).$$

Graphs of some Bessel functions of both kinds are shown in Figures 15.6 and 15.7.

It is interesting to notice that solutions of Bessel's equation illustrate all of the cases of the Frobenius theorem (Theorem 4.4). Case 1 occurs if 2ν is not an integer, case 2 if $\nu = 0$, case 3 with no logarithm term if $\nu = n + \frac{1}{2}$ for some nonnegative integer n, and case 3 with a logarithm term if ν is a positive integer.

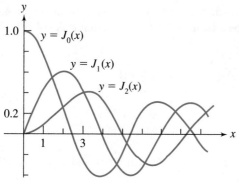

FIGURE 15.6 *Bessel functions of the first kind.*

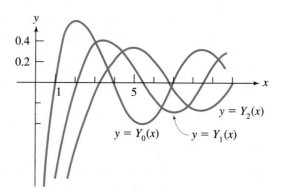

FIGURE 15.7 *Bessel functions of the second kind.*

In applications and models of physical systems, Bessel's equation often occurs in disguised form, requiring a change of variables to write the solution in terms of Bessel functions.

EXAMPLE 15.2

Consider the differential equation

$$9x^2 y'' - 27xy' + (9x^2 + 35)y = 0.$$

Let $y = x^2 u$ and compute

$$y' = 2xu + x^2 u', \quad y'' = 2u + 4xu' + x^2 u''.$$

Substitute these into the differential equation to get

$$18x^2 u + 36x^3 u' + 9x^4 u'' - 54x^2 u - 27x^3 u' + 9x^4 u + 35x^2 u = 0.$$

Collect terms to write

$$9x^4 u'' + 9x^3 u' + (9x^4 - x^2)u = 0.$$

Divide by $9x^2$ to get

$$x^2 u'' + xu' + \left(x^2 - \tfrac{1}{9}\right) u = 0,$$

which is Bessel's equation of order $\nu = \tfrac{1}{3}$. Since 2ν is not an integer, the general solution for u is

$$u(x) = a J_{1/3}(x) + b J_{-1/3}(x).$$

Therefore, the original differential equation has general solution

$$y(x) = ax^2 J_{1/3}(x) + bx^2 J_{-1/3}(x)$$

for $x > 0$. ∎

If a, b, and c are constants and n is any nonnegative integer, then it is routine to show that $x^a J_\nu(bx^c)$ and $x^a Y_\nu(bx^c)$ are solutions of the general differential equation

$$y'' - \left(\frac{2a-1}{x}\right) y' + \left(b^2 c^2 x^{2c-2} + \frac{a^2 - \nu^2 c^2}{x^2}\right) y = 0. \tag{15.12}$$

EXAMPLE 15.3

Consider the differential equation

$$y'' - \left(\frac{2\sqrt{3}-1}{x}\right) y' + \left(784x^6 - \frac{61}{x^2}\right) y = 0.$$

To fit this into the template of equation (15.12), we must clearly choose $a = \sqrt{3}$. Because of the x^6 term, try putting $2c - 2 = 6$, hence $c = 4$. Now we must choose b and ν so that

$$784 = b^2 c^2 = 16b^2,$$

so $b = 7$, and

$$a^2 - \nu^2 c^2 = 3 - 16\nu^2 = -61.$$

This equation is satisfied by $\nu = 2$. The general solution of the differential equation is therefore

$$y(x) = c_1 x^{\sqrt{3}} J_2\left(7x^4\right) + c_2 x^{\sqrt{3}} Y_2(7x^4),$$

for $x > 0$. Here c_1 and c_2 are arbitrary constants. ∎

15.2.4 Modified Bessel Functions

Sometimes a model of a physical phenomenon will require a modified Bessel function for its solution. We will show how these are obtained. Begin with the general solution

$$y(x) = c_1 J_0(kx) + c_2 Y_0(kx)$$

of the zero-order Bessel function

$$y'' + \frac{1}{x} y' + k^2 y = 0.$$

Let $k = i$. Then

$$y(x) = c_1 J_0(ix) + c_2 Y_0(ix)$$

is the general solution of

$$y'' + \frac{1}{x} y' - y = 0$$

for $x > 0$. This is a *modified Bessel equation of order zero*, and $J_0(ix)$ is a *modified Bessel function of the first kind of order zero*. Usually this is denoted

$$I_0(x) = J_0(ix) = 1 + \frac{1}{2^2} x^2 + \frac{1}{2^2 4^2} x^4 + \frac{1}{2^2 4^2 6^2} x^6 + \cdots.$$

Normally $Y_0(ix)$ is not used, but instead the second solution is chosen to be

$$K_0(x) = [\ln(2) - \gamma] I_0(x) - I_0(x) \ln(x) + \frac{1}{4} x^2 + \cdots$$

for $x > 0$. Here γ is Euler's constant. K_0 is a *modified Bessel function of the second kind of order zero*. Figure 15.8 shows graphs of $I_0(x)$ and $K_0(x)$.

The general solution of

$$y'' + \frac{1}{x} y' - y = 0$$

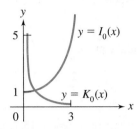

FIGURE 15.8 *Modified Bessel functions.*

is therefore

$$y(x) = c_1 I_0(x) + c_2 K_0(x)$$

for $x > 0$. The general solution of

$$y'' + \frac{1}{x}y' - b^2 y = 0 \tag{15.13}$$

is

$$y(x) = c_1 I_0(bx) + c_2 K_0(bx)$$

for $x > 0$.

By a routine calculation using the series expansion, we find that

$$\int x I_0(\alpha x)\, dx = \frac{x}{\alpha} I_0'(\alpha x) + C \tag{15.14}$$

for any nonzero constant α.

Often we are interested in the behavior of a function when the variable assumes increasing large values. This is called *asymptotic behavior*, and we will treat it later in some detail for Bessel functions in general. However, with just a few lines of work we can get some idea of how $I_0(x)$ behaves for large x. Begin with

$$y'' + \frac{1}{x}y' - y = 0,$$

of which $c I_0(x)$ is a solution for any constant c. Under the change of variables $y = ux^{-1/2}$, this equation transforms to

$$u'' = \left(1 - \frac{1}{4x^2}\right) u,$$

with solution $u(x) = c\sqrt{x} I_0(x)$ for $x > 0$ and c any constant. Transform further by putting $u = ve^x$, obtaining

$$v'' + 2v' + \frac{1}{4x^2} v = 0,$$

with solution $v(x) = c\sqrt{x} e^{-x} I_0(x)$. Since we are interested in the behavior of solutions for large x, attempt a series solution of this differential equation for v of the form

$$v(x) = 1 + c_1 \frac{1}{x} + c_2 \frac{1}{x^2} + c_3 \frac{1}{x^2} + \cdots.$$

Substitute into the differential equation and arrange terms to obtain

$$\left(-2c_1 + \frac{1}{4}\right)\frac{1}{x^2} + \left(2c_1 - 4c_2 + \frac{1}{4}c_1\right)\frac{1}{x^3}$$

$$+ \left(6c_2 - 6c_3 + \frac{1}{4}c_2\right)\frac{1}{x^4} + \left(12c_3 - 8c_4 + \frac{1}{4}c_3\right)\frac{1}{x^5} + \cdots = 0.$$

Each coefficient must vanish, hence

$$-2c_1 + \frac{1}{4} = 0,$$

$$2c_1 - 4c_2 + \frac{1}{4}c_1 = 0,$$

$$6c_2 - 6c_3 + \frac{1}{4}c_2 = 0,$$

$$12c_3 - 8c_4 + \frac{1}{4}c_3 = 0,$$

and so on. Then

$$c_1 = \frac{1}{8},$$

$$c_2 = \frac{9}{16}c_1 = \frac{9}{16}\frac{1}{8} = \frac{3^2}{2 \cdot 8^2},$$

$$c_3 = \frac{25}{24}c_2 = \frac{25}{24}\frac{3^2}{2 \cdot 8^2} = \frac{3^2 5^2}{3! 8^3},$$

$$c_4 = \frac{49}{32}c_3 = \frac{49}{32}\frac{3^2 5^2}{3! 8^3} = \frac{3^2 5^2 7^2}{4! 8^4},$$

and the pattern is clear:

$$v(x) = 1 + \frac{1}{8}\frac{1}{x} + \frac{3^2}{2 \cdot 8^2}\frac{1}{x^2} + \frac{3^2 5^2}{3! 8^3}\frac{1}{x^3} + \frac{3^2 5^2 7^2}{4! 8^4}\frac{1}{x^4} + \cdots.$$

Then, for some constant c,

$$I_0(x) = c\frac{e^x}{\sqrt{x}}\left(1 + \frac{1}{8}\frac{1}{x} + \frac{3^2}{2 \cdot 8^2}\frac{1}{x^2} + \frac{3^2 5^2}{3! 8^3}\frac{1}{x^3} + \frac{3^2 5^2 7^2}{4! 8^4}\frac{1}{x^4} + \cdots\right).$$

The series on the right actually diverges, but the sum of the first N terms approximates $I_0(x)$ as closely as we want, for x sufficiently large. This is called an *asymptotic expansion* of $I_0(x)$. By an analysis we will not carry out, it can be shown that $c = 1/\sqrt{2\pi}$.

These results about modified Bessel functions will be applied shortly to a description of the skin effect in the flow of an alternating current through a wire of circular cross section.

15.2.5 Some Applications of Bessel Functions

We will use Bessel functions in the next chapter to solve certain partial differential equations. However, Bessel functions arise in many different contexts. We will discuss several such settings here.

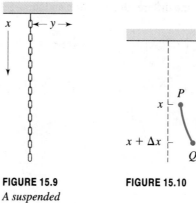

FIGURE 15.9
A suspended chain.

FIGURE 15.10

Displacement of a Suspended Chain

Consider a uniform flexible chain suspended as shown in Figure 15.9 and allowed to come to rest. We are interested in the oscillations caused by a small displacement in a vertical plane.

Assume that each particle of the chain moves in a horizontal straight line. Let m be the mass per unit length of the chain, L its length, and $y(x, t)$ the horizontal displacement at time t of the particle whose distance from the point of suspension is x. Place the origin O of a coordinate system at the point where the chain is fixed and consider the motion of an element PQ of chain between x and $x + \Delta x$ (Figure 15.10). If the tension pulling at the ends of this element has magnitude T at one end and $T + \Delta T$ at the other, then Newton's second law of motion, applied to the horizontal components of the tension at the ends, gives us

$$m(\Delta x) \frac{\partial^2 y}{\partial t^2} = \frac{\partial}{\partial x}\left(T \frac{\partial y}{\partial x}\right) \Delta x.$$

Then

$$m \frac{\partial^2 y}{\partial t^2} = \frac{\partial}{\partial x}\left(T \frac{\partial y}{\partial x}\right).$$

Now $T = mg(L - x)$, the weight of the chain below x. Therefore the last equation becomes

$$\frac{\partial^2 y}{\partial t^2} = -g \frac{\partial y}{\partial x} + g(L - x) \frac{\partial^2 y}{\partial x^2}.$$

Change variables by putting

$$z = L - x \quad \text{and} \quad u(z, t) = y(L - z, t).$$

Then

$$\frac{\partial^2 y}{\partial t^2} = \frac{\partial^2 u}{\partial t^2}$$

and

$$\frac{\partial u}{\partial z} = \frac{\partial y}{\partial x} \frac{\partial x}{\partial z} = -\frac{\partial y}{\partial x}.$$

The partial differential equation for y becomes

$$\frac{\partial^2 u}{\partial t^2} = g \frac{\partial u}{\partial z} + gz \frac{\partial^2 u}{\partial z^2}.$$

There is a method for reducing the solution of this partial differential equation to a problem involving an ordinary differential equation. Since we expect the oscillations to be periodic in the time, attempt a solution of the form

$$u(z, t) = f(z) \cos(\omega t - \delta).$$

Substitute this into the partial differential equation to get

$$-\omega^2 f(z) \cos(\omega t - \delta) = g f'(z) \cos(\omega t - \delta) + g z f''(z) \cos(\omega t - \delta).$$

Divide this equation by $gz \cos(\omega t - \delta)$ to get

$$f''(z) + \frac{1}{z} f'(z) + \frac{\omega^2}{gz} f(z) = 0.$$

This differential equation has the same form as equation (15.12), with $2a - 1 = -1, b^2 c^2 = \omega^2/g$, $2c - 2 = -1$, and $a^2 - \nu^2 c^2 = 0$. Thus choose

$$a = \nu = 0, \quad c = \frac{1}{2}, \quad b = \frac{2\omega}{\sqrt{g}}.$$

The general solution for $f(z)$ is

$$f(z) = c_1 J_0 \left(2\omega \sqrt{\frac{z}{g}} \right) + c_2 Y_0 \left(2\omega \sqrt{\frac{z}{g}} \right).$$

Now $Y_0(2\omega\sqrt{z/g}) \to -\infty$ as $z \to 0+$, because of its logarithm term. But we expect oscillations in a chain to be bounded in amplitude. Thus choose $c_2 = 0$ to obtain

$$f(z) = c_1 J_0 \left(2\omega \sqrt{\frac{z}{g}} \right).$$

Then

$$u(z, t) = c_1 J_0 \left(2\omega \sqrt{\frac{z}{g}} \right) \cos(\omega t - \delta),$$

so

$$y(x, t) = c_1 J_0 \left(2\omega \sqrt{\frac{L - x}{g}} \right) \cos(\omega t - \delta).$$

The chain is fixed at its upper end (the origin), so $y(0, t) = 0$ for all t. Assuming that $c_1 \neq 0$ (certainly reasonable on physical grounds), this requires that

$$J_0 \left(2\omega \sqrt{\frac{L}{g}} \right) = 0.$$

We will soon show that the equation $J_0(\alpha) = 0$ has infinitely many positive roots. If these are denoted $\alpha_1, \alpha_2, \ldots$, in ascending order, then we can solve for ω, since $2\omega\sqrt{L/g}$ must be one of these roots. Thus we have infinitely many possibilities for ω:

$$\omega_j = \frac{1}{2} \alpha_j \sqrt{\frac{g}{L}}.$$

These are the fundamental frequencies of oscillation of the chain. From a table of zeros of Bessel functions, we find that the first five positive roots of $J_0(x)$ are approximately 2.405, 5.520,

8.654, 11.792, and 14.931. These are α_1 through α_5, respectively. Thus the first five fundamental frequencies of the chain are

$$\omega_1 = 1.203\sqrt{\frac{g}{L}}, \quad \omega_2 = 2.76\sqrt{\frac{g}{L}}, \quad \omega_3 = 4.327\sqrt{\frac{g}{L}}, \quad \omega_4 = 5.896\sqrt{\frac{g}{L}}, \quad \text{and} \quad \omega_5 = 7.466\sqrt{\frac{g}{L}}.$$

These represent approximate frequencies of the normal modes of vibration. The period associated with ω_j is $2\pi/\omega_j$.

The Critical Length of a Vertical Rod

Consider a thin elastic rod of uniform density and circular cross section, clamped in a vertical position as in Figure 15.11. If the rod is long enough, and the upper end is given a displacement and held in that position until the rod is at rest, the rod will remain bent or displaced when released. Such a length is referred to as an *unstable length*. At some shorter lengths, however, the rod will return to the vertical position when released, after some small oscillations. These lengths are referred to as *stable lengths* for the rod. We would like to determine the *critical length L_C*, the transition point from stable to unstable.

Suppose the rod has length L and weight w per unit length. Let a be the radius of its circular cross section and E the Young's modulus for the material of the rod. (This is the ratio of stress to the corresponding strain for an elongation or linear compression.) The moment of inertia about a diameter is $I = \pi a^4/4$. Assume that the rod is in equilibrium and is then displaced slightly from the vertical, as in Figure 15.12. The x axis is vertical along the original position of the rod, with downward as positive and the origin at the upper end of the rod at equilibrium. Let $P(x, y)$ and $Q(\xi, \eta)$ be points on the displaced rod, as shown. The moment about P of the weight of an element $w\Delta x$ at Q is $w\Delta x[y(\xi) - y(x)]$. By integrating this expression we obtain the moment about P of the weight of the rod above P. Assume from the theory of elasticity that this moment about P is $EIy''(x)$. Since the part of the rod above P is in equilibrium, then

$$EIy''(x) = \int_0^x w[y(\xi) - y(x)]\,d\xi.$$

Differentiate this equation with respect to x:

$$EIy^{(3)}(x) = w[y(x) - y(x)] - \int_0^x wy'(x)\,d\xi = -wxy'(x).$$

Then

$$y^{(3)}(x) + \frac{w}{EI}xy'(x) = 0.$$

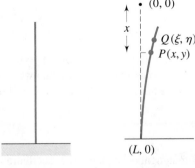

FIGURE 15.11   FIGURE 15.12

Let $u = y'$ to obtain the second-order differential equation

$$u'' + \frac{w}{EI}xu = 0.$$

Compare this equation with equation (15.12). We need

$$2a - 1 = 0, \quad a^2 - v^2c = 0, \quad 2c - 2 = 1, \quad b^2c^2 = \frac{w}{EI}.$$

This leads us to choose

$$a = \frac{1}{2}, \quad c = \frac{3}{2}, \quad v = \frac{1}{3}, \quad b = \frac{2}{3}\sqrt{\frac{w}{EI}}.$$

The general solution for $u(x)$ is

$$u(x) = y'(x) = c_1\sqrt{x}J_{1/3}\left(\frac{2}{3}\sqrt{\frac{w}{EI}}x^{3/2}\right) + c_2\sqrt{x}J_{-1/3}\left(\frac{2}{3}\sqrt{\frac{w}{EI}}x^{3/2}\right).$$

Since there is no bending moment at the top of the rod,

$$y''(0) = 0.$$

We leave it for the student to show that this condition requires $c_1 = 0$. Then

$$y'(x) = c_2\sqrt{x}J_{-1/3}\left(\frac{2}{3}\sqrt{\frac{w}{EI}}x^{3/2}\right).$$

Since the lower end of the rod is clamped vertically, $y'(L) = 0$, so

$$c_2\sqrt{L}J_{-1/3}\left(\frac{2}{3}\sqrt{\frac{w}{EI}}L^{3/2}\right) = 0.$$

Since c_2 must be nonzero to avoid a trivial solution, we need

$$J_{-1/3}\left(\frac{2}{3}\sqrt{\frac{w}{EI}}L^{3/2}\right) = 0.$$

The critical length L_C is the smallest positive number that can be substituted for L in this equation. From a table of Bessel functions, we find that the smallest positive number α such that $J_{-1/3}(\alpha) = 0$ is approximately 1.8663. Therefore,

$$\frac{2}{3}\sqrt{\frac{w}{EI}}L_C^{3/2} \approx 1.8663,$$

so

$$L_C \approx 1.9863\left(\frac{EI}{w}\right)^{1/3}.$$

Alternating Current in a Wire

We will analyze alternating current in a wire of circular cross section, culminating in a mathematical description of the skin effect (at high frequencies, "most" of the current flows through a thin layer at the surface of the wire).

Begin with general principles named for Ampère and Faraday. Ampère's law states that the line integral of the magnetic force around a closed curve (circuit) is equal to 4π times the integral of the electric current through the circuit. Faraday's law states that the line integral of the electric force around a closed circuit equals the negative of the time derivative of the magnetic induction through the circuit.

We want to use these laws to determine the current density at radius r in a wire of circular cross section and radius a. Let ρ be the specific resistance of the wire, μ its permeability, and $x(r, t)$ and $H(r, t)$ the current density and magnetic intensity, respectively, at radius r and time t.

To begin, apply Ampère's law to a circle of radius r having its axis along the axis of the wire. We get

$$2\pi r H = 4\pi \int_0^r x(2\pi \xi)\, d\xi,$$

or

$$r H = 4\pi \int_0^r x \xi\, d\xi. \tag{15.15}$$

Then

$$\frac{\partial}{\partial r}(r H) = 4\pi x r,$$

so

$$\frac{1}{r}\frac{\partial}{\partial r}(r H) = 4\pi x(r, t). \tag{15.16}$$

Now apply Faraday's law to the rectangular circuit of Figure 15.13, having one side of length L along the axis of the cylinder. We get

$$\rho L x(0, t) - \rho L x(r, t) = -\frac{\partial}{\partial t}\int_0^r \mu L H(\xi, t)\, d\xi.$$

Differentiate this equation with respect to r to get

$$\rho \frac{\partial x}{\partial r} = \mu \frac{\partial H}{\partial t}. \tag{15.17}$$

We want to use equations (15.16) and (15.17) to eliminate H. First multiply equation (15.17) by r to get

$$\rho r \frac{\partial x}{\partial r} = \mu r \frac{\partial H}{\partial t}.$$

Differentiate with respect to r:

$$\rho \frac{\partial}{\partial r}\left(r \frac{\partial x}{\partial r}\right) = \mu \frac{\partial}{\partial r}\left(r \frac{\partial H}{\partial t}\right) = \mu \frac{\partial}{\partial t}\left(\frac{\partial}{\partial r}(r H)\right) = \mu \frac{\partial}{\partial t}(4\pi x r) = 4\pi \mu r \frac{\partial x}{\partial t},$$

in which we substituted from equation (15.16) at the next to last step. Then

$$\rho \frac{\partial}{\partial r}\left(r \frac{\partial x}{\partial r}\right) = 4\pi \mu r \frac{\partial x}{\partial t}. \tag{15.18}$$

The idea is to solve this partial differential equation for $x(r, t)$, then obtain $H(r, t)$ from equation (15.15). To do this, assume that the current through the wire is an alternating current given by $C \cos(\omega t)$, with C constant. Thus the period of the current is $2\pi/\omega$. It is convenient to write

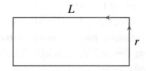

FIGURE 15.13

$z(r, t) = x(r, t) + iy(r, t)$, so $x(r, t) = \text{Re}(z(r, t))$, and to think of the current as the real part of the complex exponential $Ce^{i\omega t}$. The differential equation (15.18), with z in place of x, is

$$\rho \frac{\partial}{\partial r}\left(r \frac{\partial z}{\partial r}\right) = 4\pi \mu r \frac{\partial z}{\partial t}. \tag{15.19}$$

To solve this equation, we will attempt a solution of the form

$$z(r, t) = f(r)e^{i\omega t}.$$

Substitute this proposed solution into equation (15.19) to get

$$\rho \frac{\partial}{\partial r}\left(r f'(r)e^{i\omega t}\right) = 4\pi \mu r f(r)i\omega e^{i\omega t}.$$

Divide by $e^{i\omega t}$ and carry out the differentiations to get

$$f''(r) + \frac{1}{r} f'(r) - b^2 f(r) = 0,$$

where

$$b^2 = \frac{4\pi \mu \omega}{\rho} i.$$

Comparing this equation with equation (15.13), we can write the general solution for $f(r)$ in terms of modified Bessel functions:

$$f(r) = c_1 I_0(br) + c_2 K_0(br),$$

where

$$b = \sqrt{\frac{4\pi \mu \omega}{\rho}} \frac{1 + i}{\sqrt{2}}.$$

Because of the logarithm term in $K_0(r)$, which has infinite limit as $r \to 0$ (center of the wire), choose $c_2 = 0$. Thus $f(r)$ has the form

$$f(r) = c_1 I_0(br)$$

and

$$z(r, t) = c_1 I_0(br)e^{i\omega t}.$$

To determine the constant, use the fact that (the real part of) $Ce^{i\omega t}$ is the total current, hence, using equation (15.14),

$$C = 2\pi c_1 \int_0^a r I_0(br)\, dr = \frac{2\pi a c_1}{b} I_0'(ba).$$

Then

$$c_1 = \frac{bC}{2\pi a} \frac{1}{I_0'(ba)}$$

and

$$z(r, t) = \frac{bC}{2\pi a} \frac{1}{I_0'(ba)} I_0(br)e^{i\omega t}.$$

Then $x(r, t) = \text{Re}(z(r, t))$, and we leave it for the student to show that

$$H(r, t) = \text{Re}\left(\frac{2C}{a I_0'(ba)} I_0(br)e^{i\omega t}\right).$$

We can use the solution for $z(r, t)$ to model the skin effect. The entire current flowing through a cylinder of radius r within the wire (and having the same central axis as the wire) is the real part of

$$\frac{b}{2\pi a I_0'(ba)} C e^{i\omega t} \int_0^r I_0(br) 2\pi r \, dr,$$

and some computation shows that this is the real part of

$$\frac{r I_0'(br)}{a I_0'(ba)} C e^{i\omega t}.$$

Therefore,

$$\frac{\text{current in the cylinder of radius } r}{\text{total current in the wire}} = \frac{r}{a} \frac{I_0'(br)}{I_0'(ba)}.$$

When the frequency ω is large, then the magnitude of b is large, and we can use the asymptotic expansion of $I_0(x)$ given in Section 15.2.4 to write

$$\frac{r}{a} \frac{I_0'(br)}{I_0'(ba)} \approx \frac{r}{a} \frac{e^{br}}{\sqrt{br}} \frac{\sqrt{ba}}{e^{ba}} = \sqrt{\frac{r}{a}} e^{-b(a-r)}.$$

For any r, with $0 < r < a$, we can make $\sqrt{r/a} e^{-b(a-r)}$ as small as we like by taking the frequency ω sufficiently large. This means that for large frequencies, most of the current is flowing near the outer surface of the wire. This is the skin effect.

Poe's Pendulum

In his classic short story "The Pit and the Pendulum," Edgar Allen Poe describes the torment of a prisoner strapped on his back on the floor of a stone dungeon while a pendulum swings back and forth, its razor-sharp blade descending toward him from the ceiling. Here is an excerpt from Poe's narrative, as it appears in *The Portable Poe*, edited by Philip Van Doren Stern, The Viking Portable Library, Penguin Books, 1977.

Looking upward, I surveyed the ceiling of my own prison. It was some thirty or forty feet overhead, and constructed much as the side walls. In one of its panels a very singular figure riveted my whole attention. It was the painted figure of Time as he is commonly represented, save that, in lieu of a scythe, he held what, at a casual glance, I supposed to be the pictured image of a huge pendulum such as we see on antique clocks. There was something wrong, however, in the appearance of this machine which caused me to regard it more attentively. While I gazed directly upward at it (for its position was immediately over my own) I fancied that I saw it in motion. In an instant afterward the fancy was confirmed. Its sweep was brief, and of course slow. I watched it for some minutes, somewhat in fear, but more in wonder. Wearied at length with observing its dull movement, I turned my eyes upon the other objects in the cell.

It might have been half an hour, perhaps even an hour (for I could take but imperfect note of time), before I again cast my eyes upward. What I saw then confounded and amazed me. The sweep of the pendulum had increased in extent by nearly a yard. As a natural consequence, its velocity was also much greater. But what mainly disturbed me was the idea that it had perceptibly descended. I now observed—with what horror it is needless to say—that its nether extremity was formed of a crescent of glittering steel, about a foot in length from horn to horn; the horns upward, and the under edge evidently as keen as that of a razor. Like a razor also, it seemed massy and heavy, tapering from the edge into a solid and broad structure above. It was appended to a weighty rod of brass, and the whole hissed as it swung through the air.

What boots it to tell of the long, long hours of horror more than mortal, during which I counted the rushing vibrations of the steel! Inch by inch—line by line—with a descent only appreciable at intervals that seemed ages—down and still down it came.

The vibration of the pendulum was at right angles to my length. I saw that the crescent was designed to cross the region of my heart. It would fray the serge of my robe—it would return and repeat its operations—again—and again. Notwithstanding its terrifically wide sweep (some thirty feet or more) and the hissing vigor of its descent, sufficient to sunder these very walls of iron ...

Down—steadily down it crept. Down—certainly, relentlessly down!

At first the motion described by the prisoner may seem plausible enough, and certainly the narrative is focused more on the prisoner's predicament than on technical details of the motion. However, the question arises as to whether a descending pendulum will actually behave in the way described. A little experimentation, using a weight on a stretched string, suggests that it will not!

We will consider a study of this question by Roger L. Borrelli, Courtney S. Coleman, and Dana D. Hobson of Harvey Mudd College (*Mathematics Magazine*, vol. 58, no. 2, March 1985). The authors model the pendulum under assumptions aimed at reproducing the motion described by Poe and raise a question that is apparently still open.

To construct a model, consider a pendulum as shown in Figure 15.14 with support wire having length $L(t)$ and making an angle $\theta(t)$ with the vertical. Derive a differential equation for θ as follows. Let $\mathbf{R}(t)$ be the position vector of the blade relative to the pivot. To use Newton's second law of motion, we need the component of the acceleration of the blade in the direction perpendicular to the support wire. Define unit vectors $\mathbf{u}$ and $\mathbf{v}$ as shown. Then

$$\mathbf{v} = \cos(\theta)\mathbf{i} + \sin(\theta)\mathbf{j}$$

and

$$\mathbf{u} = -\sin(\theta)\mathbf{i} + \cos(\theta)\mathbf{j}.$$

Further, $\mathbf{R}(t) = L(t)\mathbf{v}$, and

$$\mathbf{v}'(t) = -\sin(\theta)\theta'(t)\mathbf{i} + \cos(\theta)\theta'(t)\mathbf{j} = \theta'(t)\mathbf{u}$$

and

$$\mathbf{u}'(t) = -\cos(\theta)\theta'(t)\mathbf{i} - \sin(\theta)\theta'(t)\mathbf{j} = -\theta'(t)\mathbf{v}.$$

Newton's law for the present setting is

$$m\alpha = \text{sum of the forces perpendicular to the wire} = -mg\sin(\theta),$$

in which α is the component of the acceleration in the direction of $\mathbf{u}$. To determine α, write

$$\mathbf{R}''(t) = \alpha\mathbf{u} + \beta\mathbf{v}$$

and solve for α. Since $\mathbf{R} = L\mathbf{v}$, compute

$$\mathbf{R}' = L'\mathbf{v} + L\mathbf{v}' = L'\mathbf{v} + L\theta'\mathbf{u},$$

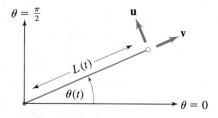

FIGURE 15.14 *Coordinate system for Poe's pendulum.*

and

$$\mathbf{R}'' = L''\mathbf{v} + L'\mathbf{v}' + L'\theta'\mathbf{u} + L\theta''\mathbf{u} + L\theta'\mathbf{u}'$$

$$= L''\mathbf{v} + L'\theta'\mathbf{u} + L'\theta'\mathbf{u} + L\theta''\mathbf{u} - L\left(\theta'\right)^2\mathbf{v}$$

$$= (2L'\theta' + L\theta'')\mathbf{u} + (L'' - L(\theta')^2)\mathbf{v}.$$

Therefore, $\alpha = 2L'\theta' + L\theta''$ and we now have from Newton's second law of motion,

$$m(2L'\theta' + L\theta'') + mg\sin(\theta) = 0.$$

Write this differential equation as

$$L\theta'' + 2L'\theta + g\sin(\theta) = 0.$$

For small θ, approximate $\sin(\theta) \approx \theta$, so the differential equation is approximately

$$L\theta'' + 2L'\theta' + g\theta = 0.$$

It is not clear from the prisoner's description, but we will assume that the pendulum descends at a steady rate. Then $L'(t) = b$ and $L(t) = bt + a$. Now the differential equation becomes

$$(a + bt)\theta'' + 2b\theta' + g\theta = 0.$$

Put

$$x = \frac{2}{b}\sqrt{(a + bt)g} \quad \text{and} \quad y = \sqrt{(a + bt)}\theta.$$

After some computation we get

$$x^2 y''(x) + xy'(x) + (x^2 - 1)y(x) = 0,$$

which is Bessel's equation of order 1. The general solution is $y(x) = c_1 J_1(x) + c_2 Y_1(x)$. Then

$$\theta(t) = \frac{1}{\sqrt{a + bt}} y(x)$$

$$= \frac{1}{\sqrt{a + bt}}\left[c_1 J_1\left(\frac{2}{b}\sqrt{(a + bt)g}\right) + c_2 Y_1\left(\frac{2}{b}\sqrt{(a + bt)g}\right)\right].$$

Some conclusions can be drawn from this general solution. Both $J_1(z)$ and $Y_1(z)$ decay to zero as $z \to \infty$. Therefore, the angle $\theta(t) \to 0$ as t increases. By contrast, the sweep $s(t) = L(t)\theta(t)$ behaves like $(a + bt)^{1/4}$ as t increases, and so increases without bound. Finally, since $L(t) = a + bt$, the curvilinear velocity $[L(t)\theta(t)]'$ decreases to zero with increasing time.

These conclusions do not agree well with the prisoner's troubled narrative. Certainly this model does include back and forth oscillations of the pendulum. However, the fact that the sweep grows at a rate proportional to $(a + bt)^{1/4}$ is at odds with the prisoner's observation of a "... terrifically wide sweep (some thirty feet or more)." It can be shown that if the pendulum length was initially one foot, then by the time the pendulum has dropped 40 feet, the sweep is no more than 10 feet. Finally, the fact that the arc length $L(t)\theta(t)$ decreases with time contradicts the prisoner's claim that the sweep increases in velocity as the pendulum is lowered.

What conclusion can be drawn from this analysis? We assume that Poe wrote his story based on his own conception of how a lowering pendulum would move, perhaps influenced by an interest in dramatic effect, without checking whether this coincided with reality. Certainly a model having length increasing according to the linear rule $L(t) = a + bt$ leads to a motion quite different from that envisioned by Poe. Various other models also lead to a motion differing from that of the story. For example, setting $L(t) = bt^2$ results in an Euler equation

$$bt^2\theta'' + 4bt\theta' + g\theta = 0.$$

The resulting motion is not Poe's pendulum either.

It is interesting to wonder whether it is possible to find a continuously differentiable length function $L(t)$ leading to a motion of the pendulum in agreement with that of the prisoner's description. Borrelli, Coleman, and Hobson conjectured that there is no such function.

15.2.6   A Generating Function for $J_n(x)$

We now return to a development of general properties of Bessel functions. For the Legendre polynomials, we produced a generating function $L(x, t)$ with the property that

$$L(x, t) = \sum_{n=0}^{\infty} P_n(x)t^n.$$

In the same spirit, we will now produce a generating function for the integer order Bessel functions of the first kind.

THEOREM 15.12   *Generating Function for Bessel Functions*

$$e^{x(t-1/t)/2} = \sum_{n=-\infty}^{\infty} J_n(x)t^n. \quad \blacksquare \tag{15.20}$$

To understand why equation (15.20) is true, begin with the familiar Maclaurin expansion of the exponential function to write

$$e^{x(t-1/t)/2} = e^{xt/2}e^{-x/2t}$$

$$= \left(\sum_{m=0}^{\infty} \frac{1}{m!}\left(\frac{xt}{2}\right)^m\right)\left(\sum_{k=0}^{\infty} \frac{1}{k!}(-1)^k\left(\frac{x}{2t}\right)^k\right)$$

$$\left(1 + \frac{xt}{2} + \frac{1}{2!}\frac{x^2t^2}{2^2} + \frac{1}{3!}\frac{x^3t^3}{2^3} + \cdots\right)\left(1 - \frac{x}{2t} + \frac{1}{2!}\frac{x^2}{2^2t^2} - \frac{1}{3!}\frac{x^3}{2^3t^3} + \cdots\right).$$

To illustrate the idea, look for the coefficient of t^4 in this product. We obtain t^4 when $x^4t^4/2^44!$ on the left is multiplied by 1 on the right, and when $x^5t^5/2^55!$ is multiplied by $-x/2t$ on the right, and when $x^6t^6/2^66!$ is multiplied by $x^2/2^22!t^2$ on the right, and so on. In this way we find that the coefficient of t^4 in this product is

$$\frac{1}{2^44!}x^4 - \frac{1}{2^65!}x^5 + \frac{1}{2^82!6!}x^6 - \frac{1}{2^{10}3!7!}x^7 + \cdots = \sum_{n=0}^{\infty} \frac{(-1)^n}{2^{2n+4}n!(n+4)!}x^{2n+4}.$$

Now compare this series with

$$J_4(x) = \sum_{n=0}^{\infty} \frac{(-1)^n}{2^{2n+4}n!\Gamma(n+4+1)}x^{2n+4} = \sum_{n=0}^{\infty} \frac{(-1)^n}{2^{2n+4}n!(n+4)!}x^{2n+4}.$$

Similar reasoning establishes that the coefficient of t^n in equation (15.20) is $J_n(x)$ for any non-negative integer n. For negative integers, we can use the fact that

$$J_{-n}(x) = (-1)^n J_n(x).$$

15.2.7 An Integral Formula for $J_n(x)$

Using the generating function, we can derive an integral formula for $J_n(x)$ when n is a nonnegative integer.

THEOREM 15.13 *Bessel's Integral*

If n is a nonnegative integer, then

$$J_n(x) = \frac{1}{\pi} \int_0^\pi \cos(n\theta - x\sin(\theta))\, d\theta.$$

Proof Begin with the fact that

$$e^{xt/2} e^{-x/2t} = \sum_{n=-\infty}^{\infty} J_n(x) t^n.$$

Since $J_{-n}(x) = (-1)^n J_n(x)$,

$$
\begin{aligned}
e^{xt/2} e^{-x/2t} = e^{x(t-1/t)/2} &= \sum_{n=-\infty}^{-1} J_n(x) t^n + J_0(x) + \sum_{n=1}^{\infty} J_n(x) t^n \\
&= \sum_{n=1}^{\infty} (-1)^n J_n(x) t^{-n} + J_0(x) + \sum_{n=1}^{\infty} J_n(x) t^n \\
&= J_0(x) + \sum_{n=1}^{\infty} J_n(x) \left(t^n + (-1)^n \frac{1}{t^n} \right) \\
&= J_0(x) + \sum_{n=1}^{\infty} J_{2n}(x) \left(t^{2n} + \frac{1}{t^{2n}} \right) + \sum_{n=1}^{\infty} J_{2n-1}(x) \left(t^{2n-1} - \frac{1}{t^{2n-1}} \right).
\end{aligned}
$$

(15.21)

Now let

$$t = e^{i\theta} = \cos(\theta) + i\sin(\theta).$$

Then

$$t^{2n} + \frac{1}{t^{2n}} = e^{2in\theta} + e^{-2in\theta} = 2\cos(2n\theta)$$

and

$$t^{2n-1} - \frac{1}{t^{2n-1}} = e^{i(2n-1)\theta} - e^{-i(2n-1)n\theta} = 2i\sin((2n-1)\theta).$$

Therefore, equation (15.21) becomes

$$
\begin{aligned}
e^{x(t-1/t)/2} &= e^{ix\sin(\theta)} \\
&= \cos(x\sin(\theta)) + i\sin(x\sin(\theta)) \\
&= J_0(x) + 2\sum_{n=1}^{\infty} J_{2n}(x)\cos(2n\theta) + 2i\sum_{n=1}^{\infty} J_{2n-1}(x)\sin((2n-1)\theta).
\end{aligned}
$$

The real part of the left side of this equation must equal the real part of the right side, and similarly for the imaginary parts:

$$\cos(x \sin(\theta)) = J_0(x) + 2 \sum_{n=1}^{\infty} J_{2n}(x) \cos(2n\theta) \tag{15.22}$$

and

$$\sin(x \sin(\theta)) = 2 \sum_{n=1}^{\infty} J_{2n-1}(x) \sin((2n-1)\theta). \tag{15.23}$$

Now recognize that the series on the right in equations (15.22) and (15.23) are Fourier series. Focusing on equation (15.22) for the moment, its Fourier series is therefore

$$\cos(x \sin(\theta)) = \frac{1}{2} a_0 + \sum_{k=1}^{\infty} a_k \cos(k\theta) + b_k \sin(k\theta)$$

$$= J_0(x) + 2 \sum_{n=1}^{\infty} J_{2n}(x) \cos(2n\theta).$$

Since we know the coefficients in a Fourier expansion, we conclude that

$$a_k = \frac{1}{\pi} \int_{-\pi}^{\pi} \cos(x \sin(\theta)) \cos(k\theta) \, d\theta = \begin{cases} 0 & \text{if } k \text{ is odd} \\ 2 J_k(x) & \text{if } k \text{ is even} \end{cases} \tag{15.24}$$

and

$$b_k = \frac{1}{\pi} \int_{-\pi}^{\pi} \cos(x \sin(\theta)) \sin(k\theta) \, d\theta = 0 \quad \text{for } k = 1, 2, 3, \ldots. \tag{15.25}$$

Similarly, from equation (15.23),

$$\sin(x \sin(\theta)) = \frac{1}{2} A_0 + \sum_{k=1}^{\infty} A_k \cos(k\theta) + B_k \sin(k\theta)$$

$$= 2 \sum_{n=1}^{\infty} J_{2n-1}(x) \sin((2n-1)\theta),$$

so these Fourier coefficients are

$$A_k = \frac{1}{\pi} \int_{-\pi}^{\pi} \sin(x \sin(\theta)) \cos(k\theta) \, d\theta = 0 \quad \text{for } k = 0, 1, 2, \ldots \tag{15.26}$$

and

$$B_k = \frac{1}{\pi} \int_{-\pi}^{\pi} \sin(x \sin(\theta)) \sin(k\theta) \, d\theta = \begin{cases} 0 & \text{if } k \text{ is even} \\ 2 J_k(x) & \text{if } k \text{ is odd} \end{cases}. \tag{15.27}$$

Upon adding equations (15.24) and (15.27), we have

$$\frac{1}{\pi} \int_{-\pi}^{\pi} \cos(x \sin(\theta)) \cos(k\theta) \, d\theta + \frac{1}{\pi} \int_{-\pi}^{\pi} \sin(x \sin(\theta)) \sin(k\theta) \, d\theta$$

$$= \frac{1}{\pi} \int_{-\pi}^{\pi} \cos(k\theta - x \sin(\theta)) \, d\theta = \begin{cases} 2 J_k(x) & \text{if } k \text{ is even} \\ 2 J_k(x) & \text{if } k \text{ is odd} \end{cases}.$$

Thus

$$J_k(x) = \frac{1}{2\pi} \int_{-\pi}^{\pi} \cos(k\theta - x\sin(\theta))\, d\theta \quad \text{for } k = 0, 1, 2, 3, \dots.$$

To complete the proof, we have only to observe that $\cos(k\theta - x\sin(\theta))$ is an even function, hence $\int_{-\pi}^{\pi} = 2\int_0^{\pi}$, so

$$J_k(x) = \frac{1}{\pi} \int_0^{\pi} \cos(k\theta - x\sin(\theta))\, d\theta \quad \text{for } k = 0, 1, 2, 3, \dots. \quad \blacksquare$$

15.2.8 A Recurrence Relation for $J_\nu(x)$

We will derive three recurrence-type relationships involving Bessel functions of the first kind. These provide information about the function or its derivative in terms of functions of the same type, but lower index. We begin with two relationships involving derivatives.

THEOREM 15.14

If ν is a real number, then

$$\frac{d}{dx}(x^\nu J_\nu(x)) = x^\nu J_{\nu-1}(x). \tag{15.28}$$

Proof Begin with the case that ν is not a negative integer. By direct computation,

$$\frac{d}{dx}(x^\nu J_\nu(x)) = \frac{d}{dx}\left[x^\nu \sum_{n=0}^{\infty} \frac{(-1)^n}{2^{2n+\nu} n!\, \Gamma(n+\nu+1)} x^{2n+\nu} \right]$$

$$= \frac{d}{dx}\left[\sum_{n=0}^{\infty} \frac{(-1)^n}{2^{2n+\nu} n!\, \Gamma(n+\nu+1)} x^{2n+2\nu} \right]$$

$$= \sum_{n=0}^{\infty} \frac{(-1)^n 2(n+\nu)}{2^{2n+\nu} n!\, (n+\nu)\Gamma(n+\nu)} x^{2n+2\nu-1}$$

$$= x^\nu \sum_{n=0}^{\infty} \frac{(-1)^n}{2^{2n+\nu-1} n!\, \Gamma(n+\nu)} x^{2n+\nu-1} = x^\nu J_{\nu-1}(x).$$

Now extend this result to the case that ν is a negative integer, say $\nu = -m$ with m a positive integer, by using the fact that

$$J_{-m}(x) = (-1)^m J_m(x).$$

We leave this detail to the student. $\blacksquare$

THEOREM 15.15

If ν is a real number, then

$$\frac{d}{dx}(x^{-\nu} J_\nu(x)) = -x^{-\nu} J_{\nu+1}(x). \quad \blacksquare \tag{15.29}$$

Verification of this relationship is similar to that of equation (15.28).

Using these two recurrence formulas involving derivatives, we can derive the following relationship between Bessel functions of the first kind of different orders.

THEOREM 15.16

Let ν be a real number. Then for $x > 0$,

$$\frac{2\nu}{x} J_\nu(x) = J_{\nu+1}(x) + J_{\nu-1}(x). \tag{15.30}$$

Proof   Carry out the differentiations in equations (15.28) and (15.29) to write

$$x^\nu J_\nu'(x) + \nu x^{\nu-1} J_\nu(x) = x^\nu J_{\nu-1}(x)$$

and

$$x^{-\nu} J_\nu'(x) - \nu x^{-\nu-1} J_\nu(x) = -x^\nu J_{\nu+1}(x).$$

Multiply the first equation by $x^{-\nu}$ and the second by x^ν to obtain

$$J_\nu'(x) + \frac{\nu}{x} J_\nu(x) = J_{\nu-1}(x)$$

and

$$J_\nu'(x) - \frac{\nu}{x} J_\nu(x) = -J_{\nu+1}(x).$$

Upon subtracting the second of these equations from the first, we obtain the conclusion of the theorem. ■

EXAMPLE 15.4

Previously we stated that

$$J_{1/2}(x) = \sqrt{\frac{2}{\pi x}} \sin(x), \quad J_{-1/2}(x) = \sqrt{\frac{2}{\pi x}} \cos(x),$$

results obtained by direct reference to the infinite series for these Bessel functions. Putting $\nu = \frac{1}{2}$ into equation (15.30), we get

$$\frac{1}{x} J_{1/2}(x) = J_{3/2}(x) + J_{-1/2}(x).$$

Then

$$J_{3/2}(x) = \frac{1}{x} J_{1/2}(x) - J_{-1/2}(x)$$

$$= \frac{1}{x} \sqrt{\frac{2}{\pi x}} \sin(x) - \sqrt{\frac{2}{\pi x}} \cos(x)$$

$$= \sqrt{\frac{2}{\pi x}} \left(\frac{1}{x} \sin(x) - \cos(x) \right).$$

Then, upon putting $\nu = \frac{3}{2}$ into equation (15.30), we get

$$\frac{3}{x} J_{3/2}(x) = J_{5/2}(x) + J_{1/2}(x).$$

Then

$$J_{5/2}(x) = -J_{1/2}(x) + \frac{3}{x}J_{3/2}(x)$$

$$= -\sqrt{\frac{2}{\pi x}}\sin(x) + \frac{3}{x}\sqrt{\frac{2}{\pi x}}\left(\frac{1}{x}\sin(x) - \cos(x)\right)$$

$$= \sqrt{\frac{2}{\pi x}}\left[-\sin(x) + \frac{3}{x^2}\sin(x) - \frac{3}{x}\cos(x)\right].$$

This process can be continued indefinitely. The point is that this is a better way to generate Bessel functions $J_{n+1/2}(x)$ than by referring to the infinite series each time. ∎

15.2.9   Zeros of $J_\nu(x)$

We have seen in some of the applications that we sometimes need to know where $J_\nu(x) = 0$. Such points are the zeros of $J_\nu(x)$. We will show that $J_\nu(x)$ has infinitely many simple positive zeros and also obtain estimates for their locations.

As a starting point, recall from equation (15.12) that $y = J_\nu(kx)$ is a solution of

$$x^2 y'' + y' + \left(k^2 x^2 - \nu^2\right)y = 0.$$

Let $k > 1$. Now put $u(x) = \sqrt{kx}\,J_\nu(kx)$. Substitute this into Bessel's equation to get

$$u''(x) + \left(k^2 - \frac{\nu^2 - \frac{1}{4}}{x^2}\right)u(x) = 0. \tag{15.31}$$

Our intuition is that, as x increases, the term $\left(\nu^2 - \frac{1}{4}\right)/x^2$ exerts less influence on this equation for u, which begins to look more like $u'' + k^2 u = 0$, with sine and cosine solutions. This suggests that for large x, $J_\nu(kx)$ is approximated by a trigonometric function, divided by $\sqrt{kx}$. Since such a function has infinitely many positive zeros, so must $J_\nu(kx)$.

In order to exploit this intuition, consider the equation

$$v''(x) + v(x) = 0. \tag{15.32}$$

This has solution $v(x) = \sin(x - \alpha)$, with α any positive number. Multiply equation (15.31) by v and equation (15.32) by u and subtract to get

$$uv'' - vu'' = \left(k^2 - \frac{\nu^2 - \frac{1}{4}}{x^2}\right)uv - uv.$$

Write this equation as

$$(uv' - vu')' = \left(k^2 - 1 - \frac{\nu^2 - \frac{1}{4}}{x^2}\right)uv.$$

Now compute

$$\int_\alpha^{\alpha+\pi}(uv' - vu')'\,dx$$

$$= u(\alpha + \pi)v'(\alpha + \pi) - u(\alpha)v'(\alpha) - v(\alpha + \pi)u'(\alpha + \pi) + v(\alpha)u'(\alpha)$$

$$= -u(\alpha + \pi) - u(\alpha)$$

$$= \int_\alpha^{\alpha+\pi}\left(k^2 - 1 - \frac{\nu^2 - \frac{1}{4}}{x^2}\right)u(x)v(x)\,dx.$$

Apply the mean value theorem for integrals to the last integral. There is some number τ between α and $\alpha + \pi$ such that

$$-u(\alpha + \pi) - u(\alpha) = u(\tau) \int_{\alpha}^{\alpha+\pi} \left(k^2 - 1 - \frac{v^2 - \frac{1}{4}}{x^2} \right) \sin(x - \alpha)\, dx.$$

Now $\sin(x - \alpha) > 0$ for $\alpha < x < \alpha + \pi$. Further, we can choose α large enough (depending on v and k) that

$$k^2 - 1 - \frac{v^2 - \frac{1}{4}}{x^2} > 0$$

for $\alpha \le x \le \alpha + \pi$. Therefore, the integral on the right in the last equation is positive. Then $u(\alpha + \pi)$, $u(\alpha)$, and $u(\tau)$ cannot all be of the same sign. Since u is continuous, u must have a zero somewhere between α and $\alpha + \pi$. Since $u(x) = \sqrt{kx}\, J_v(kx)$, this proves that $J_v(kx)$ has at least one zero between α and $\alpha + \pi$.

In general, if α is any sufficiently large number and $k > 1$, then $J_v(x)$ has a zero between α and $\alpha + k\pi$.

We can now state a general result on positive zeros of Bessel functions of the first kind.

THEOREM 15.17 Zeros of $J_v(x)$

Let $k > 1$ and let v be a real number. Then, for α sufficiently large, there is a zero of $J_v(x)$ between $\alpha + kn\pi$ and $\alpha + k(n + 1)\pi$ for $n = 0, 1, 2, \ldots$. Further, each zero is simple.

Proof The argument given prior to the theorem shows that for any number sufficiently large (depending on v and the selected $k > 1$), there is a zero of $J_v(x)$ in the interval from that number to that number plus $k\pi$. Thus there is a zero between α and $\alpha + k\pi$, and then between $(\alpha + k\pi)$ and $(\alpha + (k + 1)\pi)$, and so on.

Further, each zero is simple. For if a zero β has multiplicity greater than 1, then $J_v(\beta) = J_v'(\beta) = 0$. But then $J_v(x)$ is a solution of the initial value problem

$$x^2 y'' + y' + \left(k^2 x^2 - v^2 \right) y = 0; \qquad y(\beta) = y'(\beta) = 0.$$

Since the solution of this problem is unique, and the zero function is a solution, this would imply that $J_v(x) = 0$ for $x > 0$, a contradiction. Thus each zero is simple. ∎

The theorem implies that we can order the positive zeros of $J_v(x)$ in an increasing sequence

$$j_1 < j_2 < j_3 < \cdots,$$

so that $\lim_{n \to \infty} j_n = \infty$.

It can be shown that for $v > -1$, $J_v(x)$ has no complex zeros.

We will show that J_v has no positive zero in common with J_{v+1} or J_{v-1}. However, we claim that both J_{v-1} and J_{v+1} have at least one zero between any pair of positive zeros of J_v. This is the interlacing lemma stated as Theorem 15.18, and it means that the graphs of these three functions weave about each other, as can be seen in Figure 15.15 for $J_7(x)$, $J_8(x)$, and $J_9(x)$. First we need the following.

LEMMA 15.1

Let v be a real number. Then, except possibly for $x = 0$, J_v has no zero in common with either J_{v-1} or with J_{v+1}.

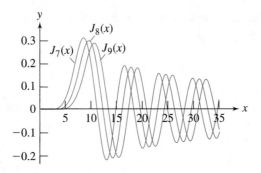

FIGURE 15.15 *Interlacing of $J_7(x)$, $-J_8(x)$, and $-J_9(x)$.*

Proof   Recall from the proof of Theorem 15.16 that

$$J_\nu'(x) + \frac{\nu}{x} J_\nu(x) = J_{\nu-1}(x).$$

If $\beta \neq 0$ and $J_\nu(\beta) = J_{\nu-1}(\beta) = 0$, then $J_\nu'(\beta) = 0$ also. But then β would be a zero of multiplicity of at least 2 for J_ν, a contradiction.

A similar use of the relation

$$J_\nu'(x) - \frac{\nu}{x} J_\nu(x) = J_{\nu+1}(x)$$

shows that J_ν also cannot share a nonzero zero with $J_{\nu+1}$. ∎

THEOREM 15.18   Interlacing Lemma

Let ν be any real number. Let a and b be distinct positive zeros of J_ν. Then $J_{\nu-1}$ and $J_{\nu+1}$ each have at least one zero between a and b.

Proof   Let $f(x) = x^\nu J_\nu(x)$. Then $f(a) = f(b) = 0$. Because f is differentiable at all points between a and b, by Rolle's theorem, there is some c between a and b at which $f'(c) = 0$. But

$$f'(x) = \frac{d}{dx}\left(x^\nu J_\nu(x)\right) = x^\nu J_{\nu-1}(x),$$

so $f'(c) = 0$ implies that $J_{\nu-1}(c) = 0$.

Similar reasoning, applied to $g(x) = x^{-\nu} J_\nu(x)$, and using the relation

$$\frac{d}{dx}\left(x^{-\nu} J_\nu(x)\right) = -x^{-\nu} J_{\nu+1}(x),$$

shows that $J_{\nu+1}$ has a zero between a and b. ∎

The following table gives the first five positive zeros of $J_\nu(x)$ for $\nu = 0, 1, 2, 3, 4$. The numbers here are rounded at the third decimal place. The interlacing property of successive indexed Bessel functions can be seen by looking down the columns. For example, the second positive zero of $J_2(x)$ falls between the second positive zeros of $J_1(x)$ and $J_3(x)$.

	j_1	j_2	j_3	j_4	j_5
$J_0(x)$	2.405	5.520	8.654	11.792	14.931
$J_1(x)$	3.832	7.016	10.173	13.323	16.470
$J_2(x)$	5.135	8.417	11.620	14.796	17.960
$J_3(x)$	6.379	9.760	13.017	16.224	19.410
$J_4(x)$	7.586	11.064	14.373	17.616	20.827

15.2.10   Fourier–Bessel Expansions

Taking a cue from the Legendre polynomials, we might suspect that Bessel functions are orthogonal on some interval. They are not.

However, let ν be any positive number. We know that J_ν has infinitely many positive zeros, which we can arrange in an ascending sequence

$$j_1 < j_2 < j_3 < \cdots .$$

For each such j_n, we can consider the function $\sqrt{x}\, J_\nu(j_n x)$ for $0 \leq x \leq 1$ (so $j_n x$ varies from 0 to j_n). We claim that these functions are orthogonal on [0, 1], in the sense that the integral of the product of any two of these functions over [0, 1], is zero.

THEOREM 15.19   *Orthogonality*

Let $\nu \geq 0$. Then the functions $\sqrt{x}\, J_\nu(j_n x)$, for $n = 1, 2, 3, \ldots$, are orthogonal on [0, 1] in the sense that

$$\int_0^1 x J_\nu(j_n x) J_\nu(j_m x)\, dx = 0 \quad \text{if } n \neq m. \ \blacksquare$$

This is in the same spirit as the orthogonality of the Legendre polynomials on $[-1, 1]$ and orthogonality of the functions

$$1, \cos(x), \cos(2x), \ldots, \sin(x), \sin(2x), \ldots$$

on $[-\pi, \pi]$.

Proof   Again invoking equation (15.12), $u(x) = J_\nu(j_n x)$ satisfies

$$x^2 u'' + x u' + (j_n^2 x^2 - \nu^2) u = 0.$$

And $v(x) = J_\nu(j_m x)$ satisfies

$$x^2 v'' + x v' + (j_m^2 x^2 - \nu^2) v = 0.$$

Multiply the first equation by v and the second by u and subtract the resulting equations to obtain

$$x^2 u'' v + x u' v + (j_n^2 x^2 - \nu^2) uv - x^2 v'' u - x v' u - (j_m^2 x^2 - \nu^2) uv = 0.$$

This equation can be written

$$x^2 (u'' v - uv'') + x(u' v - uv') = (j_m^2 - j_n^2) x^2 uv.$$

Divide by x:

$$x(u'' v - uv'') + (u' v - uv') = (j_m^2 - j_n^2) x uv.$$

Write this equation as

$$[x(u' v - uv')]' = (j_m^2 - j_n^2) x uv.$$

Then

$$\int_0^1 [x(u'v - uv')]' dx = [x(u'v - uv')]_0^1$$

$$= J_\nu'(j_n) J_\nu(j_m) - J_\nu(j_n) J_\nu'(j_m) = 0$$

$$= (j_m^2 - j_n^2) \int_0^1 x J_\nu(j_n x) J_\nu(j_m x)\, dx.$$

Since $j_n \neq j_m$, this proves the orthogonality of these functions on $[0, 1]$. ∎

As usual, whenever we have an orthogonality relationship, we are led to attempt Fourier-type expansions. Let f be defined on $[0, 1]$. How should we choose the coefficients to have an expansion

$$f(x) = \sum_{n=1}^\infty a_n J_\nu(j_n x)?$$

Using a now familiar strategy, multiply this equation by $x J_\nu(j_k x)$ and integrate to get

$$\int_0^1 x f(x) J_\nu(j_k x)\, dx = \sum_{n=1}^\infty a_n \int_0^1 x J_\nu(j_n x) J_\nu(j_k x)\, dx = a_k \int_0^1 x J_\nu^2(j_k x)\, dx.$$

The infinite series of integrals has collapsed to a single term because of the orthogonality. Then

$$a_k = \frac{\int_0^1 x f(x) J_\nu(j_k x)\, dx}{\int_0^1 x J_\nu^2(j_k x)\, dx}.$$

We call these numbers the *Fourier–Bessel coefficients* of f. When these numbers are used in the series, we call $\sum_{n=1}^\infty a_n J_\nu(j_n x)$ the *Fourier–Bessel expansion*, or *Fourier–Bessel series*, of f in terms of the functions $\sqrt{x} J_\nu(j_n x)$.

Sometimes a different point of view is adopted. It is common to say that the functions $J_\nu(j_n x)$ are *orthogonal on* $[0, 1]$ *with respect to the weight function* $\rho(x) = x$. This simply means that the integral of the product of any two of these functions, and also multiplied by $\rho(x)$, is zero over the interval $[0, 1]$:

$$\int_0^1 \rho(x) J_\nu(j_n x) J_\nu(j_m x)\, dx = 0 \quad \text{if } n \neq m.$$

This is the same integral we had before for orthogonality, but the integral is placed in the context of the weight function $\rho(x)$, a viewpoint we will see shortly with Sturm–Liouville theory. Putting $\rho(x) = x$ in this integral has the same effect as putting a factor $\sqrt{x}$ with each $J_\nu(j_n x)$.

As with Fourier and Fourier–Legendre expansions, the fact that we can compute the coefficients and write the series does not mean that it is related to the function in any particular way. The following convergence theorem deals with this issue.

THEOREM 15.20   *Convergence of Fourier–Bessel Series*

Let f be piecewise smooth on $[0, 1]$. Then, for $0 < x < 1$,

$$\sum_{n=1}^\infty a_n J_\nu(j_n x) = \frac{1}{2}(f(x+) + f(x-)),$$

where a_n is the nth Fourier–Bessel coefficient of f. ∎

We will give an example of a Fourier–Bessel expansion after we learn more about the coefficients.

15.2.11 Fourier–Bessel Coefficients

The integral $\int_0^1 x J_\nu^2(j_k x)\, dx$ occurs in the denominator of the expression for the Fourier–Bessel coefficients of any function, so it is useful to have an evaluation of this integral.

THEOREM 15.21

If $\nu \geq 0$, then

$$\int_0^1 x J_\nu^2(j_k x)\, dx = \frac{1}{2} J_{\nu+1}^2(j_k). \quad \blacksquare$$

Notice the importance here of the fact that J_ν and $J_{\nu+1}$ cannot have a positive zero in common. Knowing that $J_\nu(j_k) = 0$ implies that $J_{\nu+1}(j_k) \neq 0$.

Proof   From the preceding discussion,

$$x^2 u'' + x u' + (j_k^2 x^2 - \nu^2) u = 0,$$

where $u(x) = J_\nu(j_k x)$. Multiply this equation by $2u'(x)$ to get

$$2x^2 u' u'' + 2x(u')^2 + 2(j_k^2 x^2 - \nu^2) u u' = 0.$$

We can write this equation as

$$[x^2(u')^2 + (j_k^2 x^2 - \nu^2) u^2]' - 2 j_k^2 x u^2 = 0.$$

Now integrate, keeping in mind that $u(1) = 0$:

$$0 = \int_0^1 [x^2(u')^2 + (j_k^2 x^2 - \nu^2) u^2]'\, dx - 2 j_k^2 \int_0^1 x u^2\, dx$$

$$= [x^2(u')^2 + (j_k^2 x^2 - \nu^2) u^2]_0^1 - 2 j_k^2 \int_0^1 x u^2\, dx$$

$$= (u'(1))^2 - 2 j_k^2 \int_0^1 x u^2\, dx$$

$$= j_k^2 [J_\nu'(j_k)]^2 - 2 j_k^2 \int_0^1 x [J_\nu(j_k x)]^2\, dx.$$

Then

$$\int_0^1 x J_\nu^2(j_k x)\, dx = \frac{1}{2}[J_\nu'(j_k)]^2.$$

Now in general

$$J_\nu'(x) - \frac{\nu}{x} J_\nu(x) = -J_{\nu+1}(x).$$

Then

$$J_\nu'(j_k) - \frac{\nu}{j_k} J_\nu(j_k) = -J_{\nu+1}(j_k)$$

so

$$J_\nu'(j_k) = -J_{\nu+1}(j_k).$$

Therefore,

$$\int_0^1 x J_\nu^2(j_k x)\, dx = \frac{1}{2}[J_{\nu+1}(j_k)]^2. \quad \blacksquare$$

In view of this conclusion, the Fourier–Bessel coefficient of f is

$$a_n = \frac{2}{[J_{\nu+1}(j_k)]^2} \int_0^1 x f(x) J_\nu(j_n x)\, dx.$$

Fourier–Bessel series will occur later when we solve the heat equation for certain kinds of regions. We will then be faced with the task of expanding the initial temperature function in a Fourier–Bessel series. We will also see a Fourier–Bessel expansion when we solve for the normal modes of vibration in a circular membrane.

Generally, Fourier–Bessel coefficients are difficult to compute because Bessel functions are difficult to evaluate at particular points, and even their zeros must be approximated. However, with modern computing power we can often make approximations to whatever degree of accuracy is needed.

EXAMPLE 15.5

Let $f(x) = x(1 - x)$ for $0 \le x \le 1$. Since f is continuous with a continuous derivative, its Fourier–Bessel series will converge to $f(x)$ on $(0, 1)$:

$$x(1 - x) = \sum_{n=1}^\infty a_n J_1(j_n x) \quad \text{for } 0 < x < 1,$$

where

$$a_n = \frac{2}{[J_2(j_n)]^2} \int_0^1 x^2(1 - x) J_1(j_n x)\, dx.$$

We will compute a_1 through a_4, using eight decimal places in the first four zeros of $J_1(x)$:

$$j_1 = 3.83170597, \quad j_2 = 7.01558667, \quad j_3 = 10.17346814, \quad j_4 = 13.32369194.$$

The following calculations were done using MAPLE to carry out the integrations and evaluate the Bessel functions. Although we will use equal signs, the computations are of course approximations:

$$a_1 = \frac{2}{[J_2(3.83170597)]^2} \int_0^1 x^2(1 - x) J_1(3.83170597x)\, dx$$

$$= 12.32930609 \int_0^1 x^2(1 - x) J_1(3.83170597x)\, dx = 0.45221702,$$

$$a_2 = \frac{2}{[J_2(7.01558667)]^2} \int_0^1 x^2(1 - x) J_1(7.01558667x)\, dx$$

$$= 22.20508362 \int_0^1 x^2(1 - x) J_1(7.01558667x)\, dx = -0.03151859,$$

$$a_3 = \frac{2}{[J_2(10.17346814)]^2} \int_0^1 x^2(1-x)J_1(10.17346814x)\, dx$$

$$= 32.07568554 \int_0^1 x^2(1-x)J_1(10.17346814x)\, dx = 0.03201789,$$

and

$$a_4 = \frac{2}{[J_2(13.32369194)]^2} \int_0^1 x^2(1-x)J_1(13.32369194x)\, dx$$

$$= 41.94557796 \int_0^1 x^2(1-x)J_1(13.32369194x)\, dx = -0.00768864.$$

Using just these first four coefficients, approximate

$$x(1-x) \approx 0.45221702 J_1(3.83170597x) - 0.03151859 J_1(7.01558667x)$$

$$+ 0.03201789 J_1(10.17346814x) - 0.00768864 J_1(13.32369194x)$$

for $0 < x < 1$. Figure 15.16 shows a graph of $x(1-x)$ and a graph of this four-term sum of Bessel functions on $[0, 1]$. The graph is drawn on $[-1, \frac{3}{2}]$ to emphasize that outside of $[0, 1]$, there is no claim that $x(1-x)$ is approximated by the Fourier–Bessel series, and indeed the graphs diverge away from each other outside of $[0, 1]$. Accuracy on $[0, 1]$ can be improved by computing more terms in the series. ■

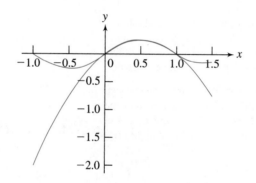

FIGURE 15.16 *Approximation of $x(1-x)$ on $[0, 1]$ by a Fourier–Bessel series.*

SECTION 15.2   PROBLEMS

1. Show that $x^a J_\nu(bx^c)$ is a solution of

$$y'' - \left(\frac{2a-1}{x}\right)y' + \left(b^2c^2x^{2c-2} + \frac{a^2 - \nu^2c^2}{x^2}\right)y = 0.$$

In each of Problems 2 through 11, write the general solution of the differential equation in terms of functions $x^a J_\nu(bx^c)$ and $x^a J_{-\nu}(bx^c)$.

2. $y'' + \frac{1}{3x}y' + \left(1 + \frac{7}{144x^2}\right)y = 0$

3. $y'' + \frac{1}{x}y' + \left(4x^2 - \frac{4}{9x^2}\right)y = 0$

4. $y'' - \frac{5}{x}y' + \left(64x^6 + \frac{5}{x^2}\right)y = 0$

5. $y'' + \dfrac{3}{x}y' + \left(16x^2 - \dfrac{5}{4x^2}\right)y = 0$

6. $y'' - \dfrac{3}{x}y' + 9x^4 y = 0$

7. $y'' - \dfrac{7}{x}y' + \left(36x^4 + \dfrac{175}{16x^2}\right)y = 0$

8. $y'' + \dfrac{1}{x}y' - \dfrac{1}{16x^2}y = 0$

9. $y'' + \dfrac{5}{x}y' + \left(81x^4 + \dfrac{7}{4x^2}\right)y = 0$

10. $y'' - \dfrac{5}{x}y' + \left(32x^2 + \dfrac{8}{x^2}\right)y = 0$

11. $y'' + \dfrac{1}{x}y' + \left(36x^4 - \dfrac{81}{25x^2}\right)y = 0$

12. Use the change of variables $by = (1/u)(du/dx)$ to transform the differential equation

$$\frac{dy}{dx} + by^2 = cx^m$$

into the differential equation

$$\frac{d^2u}{dx^2} - bcx^m u = 0.$$

Use the result of Problem 1 to find the general solution of this differential equation in terms of Bessel functions, and use this solution to solve the original differential equation. Assume that b is a positive constant.

In each of Problems 13 through 20, use the given change of variables to transform the differential equation into one whose general solution can be written in terms of Bessel functions. Use this to write the general solution of the original differential equation.

13. $4x^2 y'' + 4xy' + (x - 9)y = 0$; $z = \sqrt{x}$

14. $4x^2 y'' + 4xy' + (9x^3 - 36)y = 0$; $z = x^{3/2}$

15. $9x^2 y'' + 9xy' + (4x^{2/3} - 16)y = 0$; $z = 2x^{1/3}$

16. $9x^2 y'' - 27xy' + (9x^2 + 35)y = 0$; $u = y/x^2$

17. $36x^2 y'' - 12xy' + (36x^2 + 7)y = 0$; $u = x^{-2/3}y$

18. $4x^2 y'' + 8xy' + (4x^2 - 35)y = 0$; $u = y\sqrt{x}$

19. $4x^2 y'' + 20xy' + (9x + 7)y = 0$; $z = 3\sqrt{x}$, then $u = z^4 W(z)$, where $W(z) = y(z^2/9)$

20. $9x^{4/3} y'' + 33x^{1/3} y' + y = 0$; $z = x^{1/3}$, then $u = z^4 W(z)$, where $W(z) = y(z^3)$

21. Show that $y(x) = \sqrt{x}J_{1/3}(2kx^{3/2}/3)$ is a solution of $y'' + k^2 xy = 0$.

In each of Problems 22 through 31, write the general solution of the differential equation in terms of functions $x^a J_\nu(bx^c)$ and $x^a Y_\nu(bx^c)$.

22. $y'' - \dfrac{3}{x}y' + \left(4 - \dfrac{5}{x^2}\right)y = 0$

23. $y'' - \dfrac{1}{x}y' + \left(1 - \dfrac{3}{x^2}\right)y = 0$

24. $y'' - \dfrac{5}{x}y' + \left(1 - \dfrac{7}{x^2}\right)y = 0$

25. $y'' - \dfrac{3}{x}y' + \left(\dfrac{1}{4x} + \dfrac{3}{x^2}\right)y = 0$

26. $y'' - \dfrac{1}{x}y' + \left(16x^2 - \dfrac{15}{x^2}\right)y = 0$

27. $y'' - \dfrac{7}{x}y' + \left(1 + \dfrac{15}{x^2}\right)y = 0$

28. $y'' - \dfrac{3}{x}y' + \left(1 + \dfrac{4}{x^2}\right)y = 0$

29. $y'' + \dfrac{3}{x}y' + \dfrac{1}{16x}y = 0$

30. $y'' - \dfrac{7}{x}y' + \left(36x^4 - \dfrac{20}{x^2}\right)y = 0$

31. $y'' - \dfrac{3}{x}y' + \left(4x^2 - \dfrac{60}{x^2}\right)y = 0$

32. Show that for any real number ν,

$$x^2 J_\nu''(x) = (\nu^2 - \nu - x^2)J_\nu(x) + xJ_{\nu+1}(x).$$

33. Show that for any real number ν,

$$\left[x^{-\nu}J_\nu(x)\right]' = -x^{-\nu}J_{\nu+1}(x)$$

and

$$[x^\nu J_\nu(x)]' = x^\nu J_{\nu-1}(x).$$

34. Show that for any real number ν,

$$xJ_\nu'(x) = -\nu J_\nu(x) + xJ_{\nu-1}(x)$$

and

$$xJ_\nu'(x) = \nu J_\nu(x) - xJ_{\nu+1}(x).$$

Hint: Use the result of Problem 33.

35. Prove that for any positive integer n,

$$\int x^{n+1} J_n(x)\, dx = x^{n+1} J_{n+1}(x) + C.$$

Hint: Use the result of Problem 33.

36. Show that for any nonnegative integer n,

$$\int x[J_n(x)]^2\, dx$$

$$= \frac{1}{2}x^2[(J_n(x))^2 - J_{n-1}(x)J_{n+1}(x)] + C.$$

37. Show that

$$J_{5/2}(x) = \sqrt{\frac{2}{\pi x}}\left[\left(\frac{3}{x^2} - 1\right)\sin(x) - \frac{3}{x}\cos(x)\right].$$

38. Show that

$$J_{-5/2}(x) = \sqrt{\frac{2}{\pi x}}\left[\left(\frac{3}{x^2} - 1\right)\cos(x) + \frac{3}{x}\sin(x)\right].$$

39. Let α be a positive zero of $J_0(x)$. Show that $\int_0^1 J_1(\alpha x)\,dx = 1/\alpha$.

40. Show that $W[J_\nu(x), J_{-\nu}(x)] = C/x$ for some constant C (depending on ν). Here W denotes the Wronskian. *Hint:* Write Bessel's equation of order ν as $y'' + (1/x)y' + ((x^2 - \nu^2)/x^2)y = 0$ and use Abel's identity, which states that if φ and ψ are solutions of $y'' + p(x)y' + q(x)y = 0$ on an open interval I, then the Wronskian of these two functions is $C\exp(-\int p(x)\,dx)$ for some constant C and all x in I.

41. Let ν be positive, but not an integer. Write the series representations for $J_\nu(x)$, $J_\nu'(x)$, $J_{-\nu}(x)$, and $J_{-\nu}'(x)$ to obtain a series representation for $xW[J_\nu(x), J_{-\nu}(x)]$. Evaluate this series at $x = 0$ to show that the constant C of Problem 40 is $2/\Gamma(-\nu)\Gamma(\nu + 1)$.

42. Let $u(x) = J_0(\alpha x)$ and $v(x) = J_0(\beta x)$.

(a) Show that $xu'' + u' + \alpha^2 xu = 0$. Derive a similar differential equation for v.

(b) Multiply the differential equation for u by v and the differential equation for v by u and subtract to show that

$$[x(u'v - v'u)]' = (\beta^2 - \alpha^2)xuv.$$

(c) Show from part (b) that

$$(\beta^2 - \alpha^2)\int x J_0(\alpha x)J_0(\beta x)\,dx$$

$$= x\left[\alpha J_0'(\alpha x)J_0(\beta x) - \beta J_0'(\beta x)J_0(\alpha x)\right].$$

This is one of a set of formulas called *Lommel's integrals.*

43. Show that $[xI_0'(x)]' = xI_0(x)$.

44. Use the result of Problem 43 to show that for any positive α,

$$\int xI_0(\alpha x)\,dx = \frac{x}{\alpha}I_0'(\alpha x) + C.$$

45. Show that for any positive numbers α and β,

$$(\beta^2 - \alpha^2)\int xI_0(\alpha x)I_0(\beta x)\,dx$$

$$= x\left[\alpha I_0'(\alpha x)I_0(\beta x) - \beta I_0'(\beta x)I_0(\alpha x)\right].$$

Hint: Follow the argument of Problem 42.

46. Show that $Y_0'(x) = -Y_1(x)$.

47. Show that for any ν,

$$J_\nu(x)Y_\nu'(x) - J_\nu'(x)Y_\nu(x) = \frac{2}{\pi x}$$

for $x > 0$. Thus evaluate the Wronskian of $J_\nu(x)$ and $Y_\nu(x)$.

48. In each of (a) through (d), find (approximately) the first five terms in the Fourier–Bessel expansion $\sum_{n=1}^{\infty} a_n J_1(j_n x)$ of $f(x)$, which is defined for $0 \le x \le 1$. Compare the graph of this function with the graph of the sum of the first five terms in the series.

(a) $f(x) = x$

(b) $f(x) = e^{-x}$

(c) $f(x) = xe^{-x}$

(d) $f(x) = x^2 e^{-x}$

49. Carry out the program of Problem 48, except now use an expansion $\sum_{n=1}^{\infty} a_n J_2(j_n x)$.

15.3   Sturm–Liouville Theory and Eigenfunction Expansions

15.3.1   The Sturm–Liouville Problem

We have now seen essentially the same scenario played out three times:

$$\text{differential equation} \Longrightarrow \text{solutions that are orthogonal on } [a, b]$$

$$\Longrightarrow \text{expansions of arbitrary functions in series of these solutions}$$

$$\Longrightarrow \text{convergence theorem for the expansion.}$$

First we had Fourier (trigonometric) series, then Legendre polynomials and Fourier–Legendre series, and then Bessel functions and Fourier–Bessel expansions.

It stretches the imagination to think that the similarities in the convergence theorems for these expansions are mere coincidence. We will now develop a general theory into which these convergence theorems fit naturally. This will also expand our arsenal of tools in preparation for solving partial differential equations.

Consider the differential equation

$$y'' + R(x)y' + (Q(x) + \lambda P(x))y = 0. \tag{15.33}$$

Given an interval (a, b) on which the coefficients are continuous, we seek values of λ for which this equation has nontrivial solutions. As we will see, in some cases there will be boundary conditions solutions must satisfy (conditions specified at a and b), and sometimes not.

First put the differential equation into a convenient standard form. Multiply equation (15.33) by

$$r(x) = e^{\int R(x)\, dx}$$

to get

$$y''e^{\int R(x)\, dx} + R(x)y'e^{\int R(x)\, dx} + (Q(x) + \lambda P(x))e^{\int R(x)\, dx} y = 0.$$

Since $r(x) \neq 0$, this equation has the same solutions as equation (15.33). Now recognize that the last equation can be written

$$(ry')' + (q + \lambda p)y = 0. \tag{15.34}$$

Equation (15.34) is called the *Sturm–Liouville differential equation*, or the *Sturm–Liouville form of equation* (15.33). We will assume that p, q, and r and r' are continuous on $[a, b]$, or at least on (a, b), and $p(x) > 0$ and $r(x) > 0$ on (a, b).

EXAMPLE 15.6

Legendre's differential equation is

$$(1 - x^2)y'' - 2xy' + \lambda y = 0.$$

We can immediately write this in Sturm–Liouville form as

$$((1 - x^2)y')' + \lambda y = 0,$$

for $-1 \leq x \leq 1$. Corresponding to the values $\lambda = n(n + 1)$, with $n = 0, 1, 2, \ldots$, the Legendre polynomials are solutions. As we saw in Section 15.1, there are also nonpolynomial solutions corresponding to other choices for λ. However, these nonpolynomial solutions are not bounded on $[-1, 1]$. ∎

EXAMPLE 15.7

Equation (15.12), with $a = 0, c = 1$, and $b = \sqrt{\lambda}$, can be written

$$(xy')' + \left(\lambda x - \frac{v^2}{x}\right)y = 0.$$

This is the Sturm–Liouville form of Bessel's equation. For $\lambda > 0$, this equation has solutions in terms of the Bessel functions of order v of the first and second kinds, $J_v(\sqrt{\lambda}x)$ and $Y_v(\sqrt{\lambda}x)$. ∎

We will now distinguish three kinds of Sturm–Liouville problems.

The Regular Sturm–Liouville Problem

We want numbers λ for which there are nontrivial solutions of

$$(ry')' + (q + \lambda p)y = 0$$

on an interval $[a, b]$. These solutions must satisfy *regular boundary conditions*, which have the form

$$A_1 y(a) + A_2 y'(a) = 0, \quad B_1 y(b) + B_2 y'(b) = 0.$$

A_1 and A_2 are given constants, not both zero, and similarly for B_1 and B_2.

The Periodic Sturm–Liouville Problem

Now suppose $r(a) = r(b)$. We seek numbers λ and corresponding nontrivial solutions of the Sturm–Liouville equation on $[a, b]$ satisfying the *periodic boundary conditions*

$$y(a) = y(b), \quad y'(a) = y'(b).$$

The Singular Sturm–Liouville Problem

We look for numbers λ and corresponding nontrivial solutions of the Sturm–Liouville equation on (a, b), subject to one of the following three kinds of boundary conditions:

Type 1. $r(a) = 0$ and there is no boundary condition at a, while at b the boundary condition is

$$B_1 y(b) + B_2 y'(b) = 0,$$

where B_1 and B_2 are not both zero.

Type 2. $r(b) = 0$ and there is no boundary condition at b, while at a the condition is

$$A_1 y(a) + A_2 y'(a) = 0,$$

with A_1 and A_2 not both zero.

Type 3. $r(a) = r(b) = 0$, and no boundary condition is specified at a or b. In this case we want solutions that are bounded functions on $[a, b]$.

Each of these problems is a *boundary value problem*, specifying certain conditions at the endpoints of an interval, as contrasted with an initial value problem, which specifies information about the function and its derivative at a point (in the second-order case). Boundary value problems usually do not have unique solutions. Indeed, it is exactly this lack of uniqueness that can be exploited to solve many important problems.

In each of these problems, a number λ for which the Sturm–Liouville differential equation has a nontrivial solution is called an *eigenvalue* of the problem. A corresponding nontrivial solution is called an *eigenfunction associated with this eigenvalue*. The zero function cannot be an eigenfunction. However, any nonzero constant multiple of an eigenfunction associated with a particular eigenvalue is also an eigenfunction for this eigenvalue. In mathematical models of problems in physics and engineering, eigenvalues usually have some physical significance. For example, in studying wave motion the eigenvalues are fundamental frequencies of vibration of the system.

We will consider examples of these kinds of problems. The first two will be important in analyzing problems involving heat conduction and wave propagation.

EXAMPLE 15.8 A Regular Problem

Consider the regular problem

$$y'' + \lambda y = 0; \quad y(0) = y(L) = 0$$

on an interval $[0, L]$. We will find the eigenvalues and eigenfunctions by considering cases on λ. Since we will show later that a Sturm–Liouville problem cannot have a complex eigenvalue, there are three cases.

Case 1 $\lambda = 0$
Then $y(x) = cx + d$ for some constants c and d. Now $y(0) = d = 0$, and $y(L) = cL = 0$ requires that $c = 0$. This means that $y(x) = cx + d$ must be the trivial solution. In the absence of a nontrivial solution, $\lambda = 0$ is not an eigenvalue of this problem.

Case 2 λ is negative, say $\lambda = -k^2$ for $k > 0$.
Now $y'' - k^2 y = 0$ has general solution

$$y(x) = c_1 e^{kx} + c_2 e^{-kx}.$$

Since

$$y(0) = c_1 + c_2 = 0,$$

then $c_2 = -c_1$, so $y = 2c_1 \sinh(kx)$. But then

$$y(L) = 2c_1 \sinh(kL) = 0.$$

Since $kL > 0$, $\sinh(kL) > 0$, so $c_1 = 0$. This case also leads to the trivial solution, so this Sturm–Liouville problem has no negative eigenvalue.

Case 3 λ is positive, say $\lambda = k^2$.
The general solution of $y'' + k^2 y = 0$ is

$$y(x) = c_1 \cos(kx) + c_2 \sin(kx).$$

Now

$$y(0) = c_1 = 0,$$

so $y(x) = c_2 \sin(kx)$. Finally, we need

$$y(L) = c_2 \sin(kL) = 0.$$

To avoid the trivial solution, we need $c_2 \neq 0$. Then we must choose k so that $\sin(kL) = 0$, which means that kL must be a positive integer multiple of π, say $kL = n\pi$. Then

$$\lambda_n = \frac{n^2 \pi^2}{L^2} \quad \text{for } n = 1, 2, 3, \ldots.$$

Each of these numbers is an eigenvalue of this Sturm–Liouville problem. Corresponding to each n, the eigenfunctions are

$$y_n(x) = c \sin\left(\frac{n\pi x}{L}\right),$$

in which c can be any nonzero real number. ■

EXAMPLE 15.9   Another Regular Problem

Consider the regular Sturm–Liouville problem

$$y'' + \lambda y = 0; \qquad y'(0) = y'(L) = 0.$$

Again, consider cases on λ.

Case 1 $\lambda = 0$
Now $y(x) = cx + d$, and $y'(0) = y'(L) = c = 0$. This leaves the constant function $y(x) = d$, which is an eigenfunction for the eigenvalue $\lambda = 0$ if we choose $d \neq 0$.

Case 2 $\lambda < 0$
Write $\lambda = -k^2$, so $y'' - k^2 y = 0$. Then

$$y(x) = c_1 e^{kx} + c_2 e^{-kx}.$$

Now

$$y'(0) = kc_1 - kc_2 = 0$$

implies that $c_1 = c_2$, so $y(x) = 2c_1 \cosh(kx)$. But then

$$y'(L) = 2kc_1 \sinh(kL) = 0$$

forces $c_1 = 0$. Since there is no nontrivial solution in this case, this problem has no negative eigenvalue.

Case 3 $\lambda > 0$, say $\lambda = k^2$.
Now $y'' + k^2 y = 0$, with general solution

$$y(x) = c_1 \cos(kx) + c_2 \sin(kx).$$

Now

$$y'(0) = kc_2 = 0$$

so $c_2 = 0$. Then $y(x) = c_1 \cos(kx)$. But then

$$y'(L) = -kc_1 \sin(kL) = 0.$$

To avoid a trivial solution we need $c_1 \neq 0$, so choose $kL = n\pi$ for n a positive integer. This yields the eigenvalues

$$\lambda_n = \frac{n^2 \pi^2}{L^2}$$

and corresponding eigenfunctions

$$y_n(x) = c \cos\left(\frac{n\pi x}{L}\right),$$

in which c can be any nonzero real number and n any positive integer.

Cases 1 and 3 can be combined by allowing $n = 0$ in Case 3, including the eigenvalue $\lambda = 0$ and the corresponding constant eigenfunction $y(x) = c$. ∎

EXAMPLE 15.10   A Periodic Sturm–Liouville Problem

Consider the problem

$$y'' + \lambda y = 0; \qquad y(-L) = y(L), \, y'(-L) = y'(L)$$

on an interval $[-L, L]$. Comparing this differential equation to equation (15.34), we have $r(x) = 1$, so $r(-L) = r(L)$, as required for a periodic Sturm–Liouville problem. Consider cases on λ.

Case 1 $\lambda = 0$
Then $y = cx + d$. Now

$$y(-L) = -cL + d = y(L) = cL + d$$

implies that $c = 0$. The constant function $y = d$ satisfies both boundary conditions. Thus $\lambda = 0$ is an eigenvalue with nonzero constant eigenfunctions.

Case 2 $\lambda < 0$, say $\lambda = -k^2$.
Now

$$y(x) = c_1 e^{kx} + c_2 e^{-kx}.$$

Since $y(-L) = y(L)$, then

$$c_1 e^{-kL} + c_2 e^{kL} = c_1 e^{kL} + c_2 e^{-kL}. \tag{15.35}$$

And $y'(-L) = y'(L)$ gives us (after dividing out the common factor k)

$$c_1 e^{-kL} - c_2 e^{kL} = c_1 e^{kL} - c_2 e^{-kL}. \tag{15.36}$$

Rewrite equation (15.35) as

$$c_1(e^{-kL} - e^{kL}) = c_2(e^{-kL} - e^{kL}).$$

This implies that $c_1 = c_2$. Then equation (15.36) becomes

$$c_1(e^{-kL} - e^{kL}) = c_1(e^{kL} - e^{-kL})$$

But this implies that $c_1 = -c_1$, hence $c_1 = 0$. The solution is therefore trivial, hence this problem has no negative eigenvalue.

Case 3 λ is positive, say $\lambda = k^2$.
Now

$$y(x) = c_1 \cos(kx) + c_2 \sin(kx).$$

Now

$$y(-L) = c_1 \cos(kL) - c_2 \sin(kL) = y(L) = c_1 \cos(kL) + c_2 \sin(kL).$$

But this implies that

$$c_2 \sin(kL) = 0.$$

Next,

$$y'(-L) = kc_1 \sin(kL) + kc_2 \cos(kL)$$
$$= y'(L) = -kc_1 \sin(kL) + kc_2 \cos(kL).$$

Then

$$kc_1 \sin(kL) = 0.$$

If $\sin(kL) \neq 0$, then $c_1 = c_2 = 0$, leaving the trivial solution. Thus suppose $\sin(kL) = 0$. This requires that $kL = n\pi$ for some positive integer n. Therefore, the numbers

$$\lambda_n = \frac{n^2 \pi^2}{L^2}$$

are eigenvalues for $n = 1, 2, \ldots$, with corresponding eigenfunctions

$$y_n(x) = c_1 \cos\left(\frac{n\pi x}{L}\right) + c_2 \sin\left(\frac{n\pi x}{L}\right),$$

with c_1 and c_2 not both zero.

As in Example 15.9, we can combine Cases 1 and 3 by allowing $n = 0$, so the eigenvalue $\lambda = 0$ has corresponding nonzero constant eigenfunctions. ∎

EXAMPLE 15.11   Bessel Functions as Eigenfunctions of a Singular Problem

Consider Bessel's equation of order ν,

$$(xy')' + \left(\lambda x - \frac{\nu^2}{x}\right) y = 0,$$

on the interval $(0, R)$. Here ν is any given nonnegative real number, and $R > 0$. In the context of the Sturm–Liouville differential equation, $r(x) = x$, and $r(0) = 0$, so there is no boundary condition at 0. Let the boundary condition at R be

$$y(R) = 0.$$

We know that if $\lambda > 0$, then the general solution of Bessel's equation is

$$y(x) = c_1 J_\nu(\sqrt{\lambda}x) + c_2 Y_\nu(\sqrt{\lambda}x).$$

To have a solution that is bounded as $x \to 0+$, we must choose $c_2 = 0$. This leaves solutions of the form $y = c_1 J_\nu(\sqrt{\lambda}x)$. To satisfy the boundary condition at $x = R$, we must have

$$y(R) = c_1 J_\nu(\sqrt{\lambda}R) = 0.$$

We need $c_1 \neq 0$ to avoid the trivial solution, so we must choose λ so that $J_\nu(\sqrt{\lambda}R) = 0$. If $j_1, j_2, \ldots$ are the positive zeros of $J_\nu(x)$, then $\sqrt{\lambda}R$ can be chosen as any j_n. This yields an infinite sequence of eigenvalues

$$\lambda_n = \frac{j_n^2}{R^2},$$

with corresponding eigenfunctions

$$c J_\nu\left(\frac{j_n x}{R}\right),$$

with c constant but nonzero.

This is an example of a type 1 singular Sturm–Liouville problem. ∎

EXAMPLE 15.12   Legendre Polynomials as Eigenfunctions of a Singular Problem

Consider Legendre's differential equation

$$((1 - x^2)y')' + \lambda y = 0.$$

In the setting of Sturm–Liouville theory, $r(x) = 1 - x^2$. On the interval $[-1, 1]$, we have $r(-1) = r(1) = 0$, so there are no boundary conditions and this is a singular Sturm–Liouville problem of type 3. We want bounded solutions on this interval, so choose $\lambda = n(n+1)$, with $n = 0, 1, 2, \ldots$. These are the eigenvalues of this problem. Corresponding eigenfunctions are nonzero constant multiples of the Legendre polynomials $P_n(x)$. ∎

Finally, here is an example with more complicated boundary conditions.

EXAMPLE 15.13

Consider the regular problem

$$y'' + \lambda y = 0; \qquad y(0) = 0, 3y(1) + y'(1) = 0.$$

This problem is defined on $[0, 1]$. To find the eigenvalues and eigenfunctions, consider cases on λ.

Case 1 $\lambda = 0$
Now $y(x) = cx + d$, and $y(0) = d = 0$. Then $y = cx$. But from the second boundary condition,

$$3y(1) + y'(1) = 3c + c = 0$$

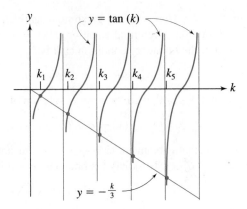

FIGURE 15.17

forces $c = 0$, so this case has only the trivial solution. This means that 0 is not an eigenvalue of this problem.

Case 2 $\lambda < 0$
Write $\lambda = -k^2$ with $k > 0$, so $y'' - k^2 y = 0$, with general solution

$$y(x) = c_1 e^{kx} + c_2 e^{-kx}.$$

Now $y(0) = 0 = c_1 + c_2$, so $c_2 = -c_1$ and $y(x) = 2c_1 \sinh(kx)$. Next,

$$3y(1) + y'(1) = 0 = 6c_1 \sinh(k) + 2c_1 k \cosh(k).$$

But for $k > 0$, $\sinh(k)$ and $k \cosh(k)$ are positive, so this equation forces $c_1 = 0$ and again we obtain only the trivial solution. This problem has no negative eigenvalue.

Case 3 $\lambda > 0$, say $\lambda = k^2$.
Now $y'' + k^2 y = 0$, with general solution

$$y(x) = c_1 \cos(kx) + c_2 \sin(kx).$$

Then $y(0) = c_1 = 0$, so $y(x) = c_2 \sin(kx)$. The second boundary condition gives us

$$0 = 3c_2 \sin(k) + kc_2 \cos(k).$$

We need $c_2 \neq 0$ to avoid the trivial solution, so look for k so that

$$3 \sin(k) + k \cos(k) = 0.$$

This means that

$$\tan(k) = -\frac{k}{3}.$$

This equation cannot be solved algebraically. However, Figure 15.17 shows graphs of $y = \tan(k)$ and $y = -k/3$ on the same set of axes. These graphs intersect infinitely often in the half-plane $k > 0$. Let the k coordinates of these points of intersection be $k_1, k_2, \ldots$. The numbers $\lambda_n = k_n^2$ are the eigenvalues of this problem, with corresponding eigenfunctions $c \sin(k_n x)$ for $c \neq 0$. ∎

15.3.2 The Sturm–Liouville Theorem

With these examples as background, here is the fundamental theorem of Sturm–Liouville theory.

THEOREM 15.22

1. Each regular and each periodic Sturm–Liouville problem has an infinite number of distinct real eigenvalues. If these are labeled $\lambda_1, \lambda_2, \ldots$ so that $\lambda_n < \lambda_{n+1}$, then $\lim_{n \to \infty} \lambda_n = \infty$.

2. If λ_n and λ_m are distinct eigenvalues of any of the three kinds of Sturm–Liouville problems defined on an interval (a, b), and φ_n and φ_m are corresponding eigenfunctions, then

$$\int_a^b p(x)\varphi_n(x)\varphi_m(x)\, dx = 0.$$

3. All eigenvalues of a Sturm–Liouville problem are real numbers.

4. For a regular Sturm–Liouville problem, any two eigenfunctions corresponding to a single eigenvalue are constant multiples of each other. ∎

Conclusion (1) assures us of the existence of eigenvalues, at least for regular and periodic problems. A singular problem may also have an infinite sequence of eigenvalues, as we saw in Example 15.11 with Bessel functions. Conclusion (1) also asserts that the eigenvalues "spread out," so that if arranged in increasing order, they increase without bound. For example, numbers $1 - 1/n$ could not be eigenvalues of a Sturm–Liouville problem, since these numbers approach 1 as $n \to \infty$.

In (2), denote $f \cdot g = \int_a^b p(x) f(x) g(x)\, dx$. This dot product for functions has many of the properties we have seen for the dot product of vectors. In particular, for functions f, g, and h that are integrable on $[a, b]$,

$$f \cdot g = g \cdot f,$$

$$f \cdot (g + h) = f \cdot g + f \cdot h,$$

$$(\alpha f) \cdot g = \alpha(f \cdot g)$$

for any real number α, and

$$f \cdot f \geq 0.$$

The last property relies on the assumption made for the Sturm–Liouville equation that $p(x) > 0$ on (a, b). If f is also continuous on $[a, b]$, then $f \cdot f = 0$ only if f is the zero function, since in this case $\int_a^b p(x) f(x)^2\, dx = 0$ can be true only if $f(x) = 0$ for $a \leq x \leq b$.

This analogy between vectors and functions is useful in visualizing certain processes and concepts, and now is an appropriate time to formalize the terminology.

DEFINITION 15.1

Let p be continuous on $[a, b]$ and $p(x) > 0$ for $a < x < b$.

1. If f and g are integrable on $[a, b]$, then the dot product of f with g, with respect to the weight function p, is given by

$$f \cdot g = \int_a^b p(x) f(x) g(x)\, dx.$$

2. f and g are orthogonal on $[a, b]$, with respect to the weight function p, if $f \cdot g = 0$.

The definition of orthogonality is motivated by the fact that two vectors $\mathbf{F}$ and $\mathbf{G}$ in 3-space are orthogonal exactly when $\mathbf{F} \cdot \mathbf{G} = 0$.

Conclusion (2) may now be stated: Eigenfunctions associated with distinct eigenvalues are orthogonal on $[a, b]$, with weight function $p(x)$. The weight function p is the coefficient of λ in the Sturm–Liouville equation.

As we have seen explicitly for Fourier (trigonometric) series, Fourier–Legendre series, and Fourier–Bessel series, this orthogonality of eigenfunctions is the key to expansions of functions in series of eigenfunctions of a Sturm–Liouville problem. This will become a significant issue when we solve certain partial differential equations modeling wave and radiation phenomena.

Conclusion (3) states that a Sturm–Liouville problem can have no complex eigenvalue. This is consistent with the fact that eigenvalues for certain problems have physical significance, such as measuring modes of vibration of a system.

Finally, conclusion (4) applies only to regular Sturm–Liouville problems. For example, the periodic Sturm–Liouville problem of Example 15.10 has eigenfunctions $\cos(n\pi x/L)$ and $\sin(n\pi x/L)$ associated with the single eigenvalue $n^2\pi^2/L^2$, and these functions are certainly not constant multiples of each other.

We will prove parts of the Sturm–Liouville theorem.

Proof   A proof of (1) requires some delicate analysis that we will not pursue. For (2), we will essentially reproduce arguments made previously for Legendre polynomials and Bessel functions. Begin with the fact that

$$(r\varphi_n')' + (q + \lambda_n p)\varphi_n = 0$$

and

$$(r\varphi_m')' + (q + \lambda_m p)\varphi_m = 0.$$

Multiply the first equation by φ_m and the second by φ_n and subtract to get

$$(r\varphi_n')'\varphi_m - (r\varphi_m')'\varphi_n = (\lambda_m - \lambda_n)p\varphi_n\varphi_m.$$

Then

$$\int_a^b \left[(r(x)\varphi_n'(x))'\varphi_m(x) - (r(x)\varphi_m'(x))'\varphi_n(x) \right] dx = (\lambda_m - \lambda_n) \int_a^b p(x)\varphi_n(x)\varphi_m(x)\, dx.$$

Since $\lambda_n \neq \lambda_m$, conclusion (2) will be proved if we can show that the left side of the last equation is zero. Integrate by parts:

$$\int_a^b (r(x)\varphi_n'(x))'\varphi_m(x)\, dx - \int_a^b (r(x)\varphi_m'(x))'\varphi_n'(x)\, dx$$

$$= \left[\varphi_m(x)r(x)\varphi_n'(x) \right]_a^b - \int_a^b r(x)\varphi_n'(x)\varphi_m'(x)\, dx$$

$$\quad - \left[\varphi_n(x)r(x)\varphi_m'(x) \right]_a^b + \int_a^b r(x)\varphi_n'(x)\varphi_m'(x)\, dx \tag{15.37}$$

$$= r(b)\varphi_m(b)\varphi_n'(b) - r(a)\varphi_m(a)\varphi_n'(a)$$

$$\quad - r(b)\varphi_n(b)\varphi_m'(b) + r(a)\varphi_n(a)\varphi_m'(a)$$

$$= r(b)\left[\varphi_m(b)\varphi_n'(b) - \varphi_n(b)\varphi_m'(b) \right] - r(a)\left[\varphi_m(a)\varphi_n'(a) - \varphi_n(a)\varphi_m'(a) \right].$$

To prove that this quantity is zero, use the boundary conditions that are in effect. Suppose first that we have a regular problem, with boundary conditions

$$A_1 y(a) + A_2 y'(a) = 0, \quad B_1 y(b) + B_2 y'(b) = 0.$$

Applying the boundary condition at a to φ_n and φ_m, we have

$$A_1\varphi_n(a) + A_2\varphi_n'(a) = 0$$

and

$$A_1\varphi_m(a) + A_2\varphi_m'(a) = 0.$$

Since A_1 and A_2 are assumed to be not both zero in the regular problem, then the system of algebraic equations

$$\varphi_n(a)X + \varphi_n'(a)Y = 0,$$

$$\varphi_m(a)X + \varphi_m'(a)Y = 0$$

has a nontrivial solution (namely, $X = A_1$, $Y = A_2$). This requires that the determinant of the coefficients vanish:

$$\begin{vmatrix} \varphi_n(a) & \varphi_n'(a) \\ \varphi_m(a) & \varphi_m'(a) \end{vmatrix} = \varphi_n(a)\varphi_m'(a) - \varphi_m(a)\varphi_n'(a) = 0.$$

Using the boundary condition at b, we obtain

$$\varphi_n(b)\varphi_m'(b) - \varphi_m(b)\varphi_n'(b) = 0.$$

Therefore, the right side of equation (15.37) is zero, proving the orthogonality relationship in the case of a regular Sturm–Liouville problem. The conclusion is proved similarly for the other kinds of Sturm–Liouville problems, by applying the relevant boundary conditions in equation (15.37).

To prove conclusion (3), suppose that a Sturm–Liouville problem has a complex eigenvalue $\lambda = \alpha + i\beta$. Let $\varphi(x) = u(x) + iv(x)$ be a corresponding eigenfunction. Now

$$(r\varphi')' + (q + \lambda p)\varphi = 0.$$

Take the complex conjugate of this equation, noting that $\varphi'(x) = u'(x) + iv'(x)$ and

$$\overline{\varphi'(x)} = u'(x) - iv'(x) = \left(\overline{\varphi(x)}\right)'.$$

Since $r(x)$, $p(x)$, and $q(x)$ are real-valued, these quantities are their own conjugates, and we get

$$(r\overline{\varphi}')' + (q + \overline{\lambda}p)\overline{\varphi} = 0.$$

This means that $\overline{\lambda}$ is also an eigenvalue, with eigenfunction $\overline{\varphi}$. Now, if $\beta \neq 0$, then λ and $\overline{\lambda}$ are distinct eigenvalues, hence

$$\int_a^b p(x)\varphi(x)\overline{\varphi(x)}\,dx = 0.$$

But then

$$\int_a^b p(x)[u(x)^2 + v(x)^2]\,dx = 0.$$

But for a Sturm–Liouville problem, it is assumed that $p(x) > 0$ for $a < x < b$. Therefore, $u(x)^2 + v(x)^2 = 0$, so

$$u(x) = v(x) = 0$$

on $[a, b]$ and $\varphi(x)$ is the trivial solution. This contradicts φ being an eigenfunction. We conclude that $\beta = 0$, so λ is real.

Finally, to prove (4), suppose λ is an eigenvalue of a regular Sturm–Liouville problem and φ and ψ are both eigenfunctions associated with λ. Use the boundary condition at a, and reason as in part of the proof of (2) to show that

$$\varphi(a)\psi'(a) - \psi(a)\varphi'(a) = 0.$$

But then the Wronskian of φ and ψ vanishes at a, so φ and ψ are linearly dependent and one is a constant multiple of the other. ∎

We now have the machinery needed for general eigenfunction expansions.

15.3.3   Eigenfunction Expansions

In solving partial differential equations, we will often encounter the need to expand a function in a series of solutions of an associated ordinary differential equation—a Sturm–Liouville problem. Fourier series, Fourier–Legendre series, and Fourier–Bessel series are examples of such expansions. The function to be expanded will have some special significance in the problem. It might, for example, be an initial temperature function or the initial displacement or velocity of a wave.

To create a unified setting in which such series expansions can be understood, consider an analogy with vectors in 3-space. Given a vector $\mathbf{F}$, we can always find real numbers a, b, and c so that

$$\mathbf{F} = a\mathbf{i} + b\mathbf{j} + c\mathbf{k}.$$

Although the constants are easy to find, we will pursue a formal process in order to identify a pattern. First,

$$\mathbf{F} \cdot \mathbf{i} = a\mathbf{i} \cdot \mathbf{i} + b\mathbf{j} \cdot \mathbf{i} + c\mathbf{k} \cdot \mathbf{i} = a,$$

because

$$\mathbf{i} \cdot \mathbf{i} = 1 \quad \text{and} \quad \mathbf{j} \cdot \mathbf{i} = \mathbf{k} \cdot \mathbf{i} = 0.$$

Similarly,

$$b = \mathbf{F} \cdot \mathbf{j} \quad \text{and} \quad c = \mathbf{F} \cdot \mathbf{k}.$$

The orthogonality of $\mathbf{i}, \mathbf{j}$, and $\mathbf{k}$ provide a convenient mechanism for determining the coefficients in the expansion by means of the dot product.

More generally, suppose $\mathbf{U}, \mathbf{V}$, and $\mathbf{W}$ are any three nonzero vectors in 3-space that are mutually orthogonal, so

$$\mathbf{U} \cdot \mathbf{V} = \mathbf{U} \cdot \mathbf{W} = \mathbf{V} \cdot \mathbf{W} = 0.$$

These vectors need not be unit vectors and do not have to be aligned along the axes. However, because of their orthogonality, we can also easily write $\mathbf{F}$ in terms of these three vectors. Indeed, if

$$\mathbf{F} = \alpha\mathbf{U} + \beta\mathbf{V} + \gamma\mathbf{W}$$

then

$$\mathbf{F} \cdot \mathbf{U} = \alpha\mathbf{U} \cdot \mathbf{U} + \beta\mathbf{V} \cdot \mathbf{U} + \gamma\mathbf{W} \cdot \mathbf{U} = \alpha\mathbf{U} \cdot \mathbf{U},$$

so

$$\alpha = \frac{\mathbf{F} \cdot \mathbf{U}}{\mathbf{U} \cdot \mathbf{U}}.$$

Similarly,

$$\beta = \frac{\mathbf{F} \cdot \mathbf{V}}{\mathbf{V} \cdot \mathbf{V}} \quad \text{and} \quad \gamma = \frac{\mathbf{F} \cdot \mathbf{W}}{\mathbf{W} \cdot \mathbf{W}}. \tag{15.38}$$

Again, we have a simple dot product formula for the coefficients.

The idea of expressing a vector as a sum of constants times mutually orthogonal vectors, with formulas for the coefficients, extends to writing functions in series of eigenfunctions of Sturm–Liouville problems, with a formula similar to equation (15.38) for the coefficients. We have seen three such instances already, which we will briefly review in the context of the Sturm–Liouville theorem.

Fourier Series
The Sturm–Liouville problem is

$$y'' + \lambda y = 0; \qquad y(-L) = y(L) = 0$$

(a periodic problem) with eigenvalues $n^2\pi^2/L^2$ for $n = 0, 1, 2, \ldots$ and eigenfunctions

$$1, \cos(\pi x/L), \cos(2\pi x/L), \ldots, \sin(\pi x/L), \sin(2\pi x/L), \ldots.$$

Here $p(x) = 1$ and the dot product to be used is

$$f \cdot g = \int_{-L}^{L} f(x)g(x)\, dx.$$

If f is piecewise smooth on $[-L, L]$, then for $-L < x < L$,

$$\frac{1}{2}(f(x+) + f(x-)) = \frac{1}{2}a_0 + \sum_{n=1}^{\infty} a_n \cos\left(\frac{n\pi x}{L}\right) + b_n \sin\left(\frac{n\pi x}{L}\right),$$

where

$$a_n = \frac{\int_{-L}^{L} f(x)\cos(n\pi x/L)\, dx}{\int_{-L}^{L} \cos^2(n\pi x/L)\, dx} = \frac{f(x) \cdot \cos(n\pi x/L)}{\cos(n\pi x/L) \cdot \cos(n\pi x/L)} \quad \text{for } n = 0, 1, 2, \ldots$$

and

$$b_n = \frac{\int_{-L}^{L} f(x)\sin(n\pi x/L)\, dx}{\int_{-L}^{L} \sin^2(n\pi x/L)\, dx} = \frac{f(x) \cdot \sin(n\pi x/L)}{\sin(n\pi x/L) \cdot \sin(n\pi x/L)} \quad \text{for } m = 1, 2, \ldots.$$

Fourier–Legendre Series
The Sturm–Liouville problem is

$$((1 - x^2)y')' + \lambda y = 0,$$

with no boundary conditions on $[-1, 1]$ because $r(x) = 1 - x^2$ vanishes at these endpoints. However, we seek bounded solutions. Eigenvalues are $n(n+1)$ with corresponding eigenfunctions the Legendre polynomials $P_0(x), P_1(x), \ldots$. Since $p(x) = 1$, use the dot product

$$f \cdot g = \int_{-1}^{1} f(x)g(x)\, dx.$$

If f is piecewise smooth on $[-1, 1]$, then for $-1 < x < 1$,

$$\frac{1}{2}(f(x+) + f(x-)) = \sum_{n=0}^{\infty} c_n P_n(x)$$

where

$$
c_n = \frac{\int_{-1}^{1} f(x) P_n(x)\, dx}{\int_{-1}^{1} P_n^2(x)\, dx} = \frac{f \cdot P_n}{P_n \cdot P_n}.
$$

Fourier–Bessel Series

Consider the Sturm–Liouville problem

$$
(xy')' + \left(\lambda x - \frac{v^2}{x} \right) y = 0
$$

with boundary condition $y(1) = 0$ on $(0, 1)$. Eigenvalues are $\lambda = j_n^2$ for $n = 1, 2, \ldots$, where $j_1, j_2, \ldots$ are the positive zeros of $J_v(x)$, and eigenfunctions are $J_v(j_n x)$. In this Sturm–Liouville problem, $p(x) = x$ and the dot product is

$$
f \cdot g = \int_0^1 x f(x) g(x)\, dx.
$$

If f is piecewise smooth on $[0, 1]$, then for $0 < x < 1$ we can write the series

$$
\frac{1}{2}(f(x+) + f(x-)) = \sum_{n=1}^{\infty} c_n J_v(j_n x),
$$

where

$$
c_n = \frac{\int_0^1 x f(x) J_v(j_n x)\, dx}{\int_0^1 x J_v^2(j_n x)\, dx} = \frac{f(x) \cdot J_v(j_n x)}{J_v(j_n x) \cdot J_v(j_n x)},
$$

again fitting the template we have seen in the other kinds of expansions.

These expansions are all special cases of a general theory of expansions in series of eigenfunctions of Sturm–Liouville problems.

THEOREM 15.23   *Convergence of Eigenfunction Expansions*

Let $\lambda_1, \lambda_2, \ldots$ be the eigenvalues of a Sturm–Liouville differential equation

$$
(ry')' + (q + \lambda p)y = 0
$$

on $[a, b]$, with one of the sets of boundary conditions specified previously. Let $\varphi_1, \varphi_2, \ldots$ be corresponding eigenfunctions and define the dot product

$$
f \cdot g = \int_a^b p(x) f(x) g(x)\, dx.
$$

Let f be piecewise smooth on $[a, b]$. Then, for $a < x < b$,

$$
\frac{1}{2}(f(x+) + f(x-)) = \sum_{n=1}^{\infty} c_n \varphi_n(x),
$$

where

$$
c_n = \frac{f \cdot \varphi_n}{\varphi_n \cdot \varphi_n}. \quad \blacksquare
$$

We call the numbers

$$
\frac{f \cdot \varphi_n}{\varphi_n \cdot \varphi_n} \tag{15.39}
$$

the *Fourier coefficients of f with respect to the eigenfunctions of this Sturm–Liouville problem.* With this choice of coefficients, $\sum_{n=1}^{\infty} c_n \varphi_n(x)$ is the *eigenfunction expansion of f* with respect to these eigenfunctions.

If the differential equation generating the eigenvalues and eigenfunctions has a special name (such as Legendre's equation, or Bessel's equation), then the eigenfunction expansion is usually called the *Fourier–... series*, for example, Fourier–Legendre series and Fourier–Bessel series.

EXAMPLE 15.14

Consider the Sturm–Liouville problem

$$y'' + \lambda y = 0; \qquad y'(0) = y'(\pi/2) = 0.$$

We find in a routine way that the eigenvalues of this problem are $\lambda = 4n^2$ for $n = 0, 1, 2, \ldots$. Corresponding to $\lambda = 0$, we can choose $\varphi_0(x) = 1$ as an eigenfunction. Corresponding to $\lambda = 4n^2$, $\varphi_n(x) = \cos(2nx)$ is an eigenfunction. This gives us the set of eigenfunctions

$$\varphi_0(x) = 1, \quad \varphi_1(x) = \cos(2x), \quad \varphi_2(x) = \cos(4x), \ldots.$$

Because the coefficient of λ in the differential equation is $p(x) = 1$, and the interval is $[0, \pi/2]$, the dot product for this problem is

$$f \cdot g = \int_0^{\pi/2} f(x)g(x)\,dx.$$

We will write the eigenfunction expansion of $f(x) = x^2(1 - x)$ for $0 \le x \le \pi/2$. Since f and f' are continuous, this expansion will converge to $x^2(1 - x)$ for $0 < x < \pi/2$. The coefficients in this expansion are

$$c_0 = \frac{f \cdot 1}{1 \cdot 1} = \frac{\int_0^{\pi/2} x^2(1 - x)\,dx}{\int_0^{\pi/2} dx} = \frac{-(1/64)\pi^4 + (1/24)\pi^3}{\pi/2} = \pi^2\left(\frac{1}{12} - \frac{\pi}{32}\right)$$

and, for $n = 1, 2, \ldots$,

$$c_n = \frac{f \cdot \cos(2nx)}{\cos(2nx) \cdot \cos(2nx)}$$

$$= \frac{\int_0^{\pi/2} x^2(1 - x)\cos(2nx)\,dx}{\int_0^{\pi/2} \cos^2(2nx)\,dx}$$

$$= -\frac{1}{4\pi n^4}\left(-4\pi n^2(-1)^n + 3\pi^2 n^2(-1)^n - 6(-1)^n + 6\right).$$

Therefore, for $0 < x < \pi/2$,

$$x^2(1 - x) = \pi^2\left(\frac{1}{12} - \frac{\pi}{32}\right)$$

$$-\frac{1}{4\pi}\sum_{n=1}^{\infty}\frac{1}{n^4}\left(-4\pi n^2(-1)^n + 3\pi^2 n^2(-1)^n - 6(-1)^n + 6\right)\cos(2nx).$$

Figure 15.18(a) shows the fifth partial sum of this series, compared with f, and Figure 15.18(b) shows the fifteenth partial sum of this expansion. Clearly this eigenfunction expansion is converging quite rapidly to $x^2(1 - x)$ on this interval. ∎

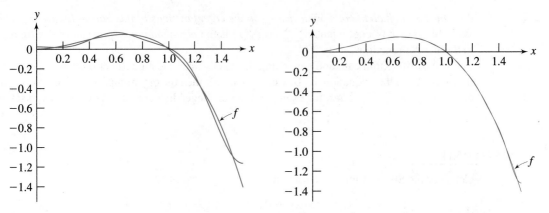

FIGURE 15.18(a) *Fifth partial sum in Example 15.14.*　　**FIGURE 15.18(b)** *Fifteenth partial sum.*

15.3.4 Approximation in the Mean and Bessel's Inequality

In this and the next two sections we will discuss some additional properties of Fourier coefficients, as well as some subtleties in the convergence of Fourier series. For this discussion, let $\varphi_1, \varphi_2, \ldots$ be normalized eigenfunctions of a Sturm–Liouville problem on $[a, b]$. *Normalized* means that each eigenfunction φ_n has been multiplied by a positive constant so that $\varphi_n \cdot \varphi_n = 1$. This can always be done because a nonzero constant multiple of an eigenfunction is again an eigenfunction. We now have

$$\varphi_n \cdot \varphi_m = \int_a^b p(x)\varphi_n(x)\varphi_m(x)\, dx = \begin{cases} 1 & \text{if } n = m \\ 0 & \text{if } n \neq m \end{cases}.$$

For these normalized eigenfunctions, the nth Fourier coefficient is

$$c_n = \frac{f \cdot \varphi_n}{\varphi_n \cdot \varphi_n} = f \cdot \varphi_n. \tag{15.40}$$

We will now define one measure of how well a linear combination $\sum_{n=1}^{N} k_n \varphi_n$ approximates a given function f.

DEFINITION 15.2　*Best Mean Approximation*

Let N be a positive integer and let f be a function that is integrable on $[a, b]$. A linear combination

$$\sum_{n=1}^{N} k_n \varphi_n(x)$$

of $\varphi_1, \varphi_2, \ldots, \varphi_N$ is the best approximation in the mean to f *on* $[a, b]$ if the coefficients $k_1, \ldots, k_N$ minimize the quantity

$$I_N(f) = \int_a^b p(x) \left(f(x) - \sum_{n=1}^{N} k_n \varphi_n(x) \right)^2 dx.$$

$I_N(f)$ is the dot product of $f(x) - \sum_{n=1}^{N} k_n \varphi_n(x)$ with itself (with weight function p). For vectors in R^3, the dot product of a vector $\mathbf{V} = a\mathbf{i} + b\mathbf{j} + c\mathbf{k}$ with itself is the square of its length:

$$\mathbf{V} \cdot \mathbf{V} = a^2 + b^2 + c^2 = (\text{length of } \mathbf{V})^2.$$

This suggests that we define a length for functions by

$$g \cdot g = \int_a^b p(x)g(x)^2 \, dx = (\text{length of } g)^2.$$

Now $I_N(f)$ has the geometric interpretation of being the (square of the) length of $f(x) - \sum_{n=1}^N k_n \varphi_n(x)$. The smaller this length is, the better the linear combination $\sum_{n=1}^N k_n \varphi_n(x)$ approximates $f(x)$ on $[a, b]$. This approximation is an average over the entire interval, as opposed to looking at the approximation at a particular point, hence the term "approximation in the mean." We want to choose the k_n's to make $\sum_{n=1}^N k_n \varphi_n(x)$ the best possible mean approximation to f on $[a, b]$, which means we want to make the length of $f(x) - \sum_{n=1}^N k_n \varphi_n(x)$ as small as possible.

To determine how to choose the k_n's, write

$$0 \leq I_N(f) = \int_a^b p(x) \left(f(x)^2 - 2 \sum_{n=1}^N f(x)\varphi_n(x) + \left(\sum_{n=1}^N k_n \varphi_n(x) \right)^2 \right) dx$$

$$= \int_a^b p(x)f(x)^2 \, dx - 2 \sum_{n=1}^N k_n \int_a^b p(x)f(x)\varphi_n(x) \, dx$$

$$+ \sum_{n=1}^N \sum_{m=1}^N k_n k_m \int_a^b p(x)\varphi_n(x)\varphi_m(x) \, dx$$

$$= f \cdot f - 2 \sum_{n=1}^N k_n f \cdot \varphi_n + \sum_{n=1}^N \sum_{m=1}^N k_n k_m \varphi_n \cdot \varphi_m$$

$$= f \cdot f - 2 \sum_{n=1}^N k_n f \cdot \varphi_n + \sum_{n=1}^N k_n^2 \varphi_n \cdot \varphi_n$$

$$= f \cdot f - 2 \sum_{n=1}^N k_n f \cdot \varphi_n + \sum_{n=1}^N k_n^2,$$

since $\varphi_n \cdot \varphi_n = 1$ for this normalized set of eigenfunctions. Now let $c_n = f \cdot \varphi_n$, the nth Fourier coefficient of f for this set of normalized eigenfunctions. Complete the square by writing the last inequality as

$$0 \leq f \cdot f - 2 \sum_{n=1}^N k_n c_n + \sum_{n=1}^N k_n^2 - \sum_{n=1}^n c_n^2 + \sum_{n=1}^n c_n^2$$

$$= f \cdot f + \sum_{n=1}^N (c_n - k_n)^2 - \sum_{n=1}^N c_n^2. \tag{15.41}$$

In this formulation, it is obvious that the right side achieves its minimum when each $k_n = c_n$. We have proved the following.

THEOREM 15.24

Let f be integrable on $[a, b]$ and N a positive integer. Then, the linear combination $\sum_{n=1}^N k_n \varphi_n$ that is the best approximation in the mean to f on $[a, b]$ is obtained by putting

$$k_n = f \cdot \varphi_n$$

for $n = 1, 2, \ldots$. ∎

Thus, for any given N, the Nth partial sum $\sum_{n=1}^{N}(f \cdot \varphi_n)\varphi_n$ of the Fourier series $\sum_{n=1}^{\infty}(f \cdot \varphi_n)\varphi_n$ of f is the best approximation in the mean to f by a linear combination of $\varphi_1, \varphi_2, \ldots, \varphi_N$.

The argument leading to the theorem has another important consequence. Put $k_n = c_n = f \cdot \varphi_n$ in inequality (15.41) to obtain

$$0 \leq f \cdot f - \sum_{n=1}^{N}(f \cdot \varphi_n)^2,$$

or

$$\sum_{n=1}^{N}(f \cdot \varphi_n)^2 \leq f \cdot f.$$

Since N can be any positive integer, the series of squares of the Fourier coefficients of f converges, and the sum of this series cannot exceed the dot product of f with itself. This is Bessel's inequality, and was proved in Section 13.5 (Theorem 13.7) for Fourier trigonometric series.

THEOREM 15.25   *Bessel's Inequality*

Let f be integrable on $[a, b]$. Then the series of squares of the Fourier coefficients of f with respect to the normalized eigenfunctions $\varphi_1, \varphi_2, \ldots$ converges. Further,

$$\sum_{n=1}^{\infty}(f \cdot \varphi_n)^2 \leq f \cdot f. \blacksquare$$

Under some circumstances, the inequality can be replaced by an equality. This leads us to consider the concept of convergence in the mean.

15.3.5   Convergence in the Mean and Parseval's Theorem

Continuing from the preceding subsection, $\varphi_1, \varphi_2, \ldots$ are assumed to be the normalized eigenfunctions of a Sturm–Liouville problem on $[a, b]$. If f is continuous on $[a, b]$ with a piecewise continuous derivative, then for $a < x < b$,

$$f(x) = \sum_{n=1}^{\infty}(f \cdot \varphi_n)\varphi_n(x).$$

This convergence is called *pointwise convergence*, because it deals with convergence of the Fourier series individually at each x in (a, b). Under some conditions, this series may also converge uniformly.

In addition to these two kinds of convergence, convergence in the mean is often used in the context of eigenfunction expansions.

DEFINITION 15.3   *Convergence in the Mean*

Let f be integrable on $[a, b]$. The Fourier series $\sum_{n=1}^{\infty}(f \cdot \varphi_n)\varphi_n$ of f, with respect to the normalized eigenfunctions $\varphi_1, \varphi_2, \ldots$, is said to converge to f in the mean on $[a, b]$ if

$$\lim_{N \to \infty} \int_a^b p(x) \left(f(x) - \sum_{n=1}^{N}(f \cdot \varphi_n)\varphi_n \right)^2 dx = 0.$$

Convergence in the mean of a Fourier series of f, to f, occurs when the length of $f(x) - \sum_{n=1}^{N}(f \cdot \varphi_n)\varphi_n(x)$ approaches zero as N approaches infinity. This will certainly happen if the Fourier series converges to f, because then $f(x) = \sum_{n=1}^{\infty}(f \cdot \varphi_n)\varphi_n(x)$, and we know that this holds if f is continuous with a piecewise continuous derivative.

For the remainder of this section, let $C'[a, b]$ be the set of functions that are continuous on $[a, b]$, with piecewise continuous derivatives on (a, b).

THEOREM 15.26

1. If $f(x) = \sum_{n=1}^{\infty}(f \cdot \varphi_n)\varphi_n(x)$ for $a < x < b$, then $\sum_{n=1}^{\infty}(f \cdot \varphi_n)\varphi_n$ also converges in the mean to f on $[a, b]$.

2. If f is in $C'[a, b]$, then $\sum_{n=1}^{\infty}(f \cdot \varphi_n)\varphi_n$ converges in the mean to f on $[a, b]$. ∎

The converse of (1) is false. It is possible for the length of $f(x) - \sum_{n=1}^{N}(f \cdot \varphi_n)\varphi_n(x)$ to have limit zero as $N \to \infty$ but for the Fourier series not to converge to $f(x)$ on the interval. This is because the integral in the definition of mean convergence is an averaging process and does not focus on the behavior of the Fourier series at any particular point.

We will show that convergence in the mean for functions in $C'[a, b]$ is equivalent to being able to turn Bessel's inequality into an equality for all functions in this class.

THEOREM 15.27

$\sum_{n=1}^{\infty}(f \cdot \varphi_n)\varphi_n$ converges in the mean to f for every f in $C'[a, b]$ if and only if

$$\sum_{n=1}^{\infty}(f \cdot \varphi_n)^2 = f \cdot f$$

for every f in $C'[a, b]$.

Proof From the calculation done in proving Theorem 15.24, with $k_n = f \cdot \varphi_n$,

$$0 \le I_N(f) = \int_a^b p(x)\left(f(x) - \sum_{n=1}^{N}(f \cdot \varphi_n)\varphi_n\right)^2 dx = f \cdot f - \sum_{n=1}^{N}(f \cdot \varphi_n)^2.$$

Therefore,

$$\lim_{N \to \infty} \int_a^b p(x)\left(f(x) - \sum_{n=1}^{N}(f \cdot \varphi_n)\varphi_n\right)^2 dx = 0$$

if and only if

$$f \cdot f - \sum_{n=1}^{\infty}(f \cdot \varphi_n)^2 = 0. \quad ∎$$

Replacing the inequality with an equality in Bessel's inequality yields the Parseval relationship. We can now state a condition under which this holds.

COROLLARY 15.1 *Parseval's Theorem*

If f is in $C'[a, b]$, then

$$\sum_{n=1}^{\infty}(f \cdot \varphi_n)^2 = f \cdot f. \quad ∎$$

This follows immediately from the last two theorems. We know by Theorem 15.26(2) that if f is in $C'[a, b]$, then the Fourier series of f converges to f in the mean. Then, by Theorem 15.27, $\sum_{n=1}^{\infty} (f \cdot \varphi_n)^2 = f \cdot f$. With more effort, the Parseval equation can be proved under much weaker conditions on f.

15.3.6   Completeness of the Eigenfunctions

Completeness is a concept that is perhaps most easily understood in terms of vectors. In 3-space, the vector $\mathbf{k}$ cannot be written as a linear combination $\alpha \mathbf{i} + \beta \mathbf{j}$, even though $\mathbf{i}$ and $\mathbf{j}$ are orthogonal. The reason for this is that there is another direction in 3-space that is orthogonal to the plane of $\mathbf{i}$ and $\mathbf{j}$, and $\mathbf{i}$ and $\mathbf{j}$ carry no information about the component a vector may have in this third direction. The vectors $\mathbf{i}$ and $\mathbf{j}$ are incomplete in R^3. By contrast, there is no nonzero vector that is orthogonal to each of $\mathbf{i}$, $\mathbf{j}$, and $\mathbf{k}$, so we say that these vectors are complete in R^3. Any 3-vector can be written as a linear combination of $\mathbf{i}$, $\mathbf{j}$, and $\mathbf{k}$.

Now consider the normalized eigenfunctions φ_1, φ_2, ... Think of each φ_j as defining a different direction, or axis, in the space of functions under consideration, which we take to be $C'[a, b]$. We say that these eigenfunctions are *complete* in $C'[a, b]$ if the only function in $C'[a, b]$ that is orthogonal to every eigenfunction is the zero function. If, however, there is a nontrivial function f in $C'[a, b]$ that is orthogonal to every eigenfunction, then we say that the eigenfunctions are *incomplete*. In this case there is another axis, or direction, in $C'[a, b]$ that is not determined by all of the eigenfunctions. A function having a component in this other direction could not possibly be represented in a series of the incomplete eigenfunctions.

We claim that the eigenfunctions are complete in the space of continuous functions with piecewise continuous derivatives on (a, b).

THEOREM 15.28

The normalized eigenfunctions φ_1, φ_2, ... are complete in $C'[a, b]$.

Proof   Suppose the eigenfunctions are not complete. Then there is some nontrivial function f in $C'[a, b]$ that is orthogonal to each φ_n. But because f is orthogonal to each φ_n, each $(f, \varphi_n) = 0$, so

$$f(x) = \sum_{n=1}^{\infty} (f \cdot \varphi_n) \varphi_n(x) = 0 \quad \text{for } a < x < b.$$

This contradiction proves the theorem. ∎

EXAMPLE 15.15

The normalized eigenfunctions of the Sturm–Liouville problem

$$y'' + \lambda y = 0; \qquad y'(0) = y'(\pi/2) = 0$$

are

$$\sqrt{\frac{2}{\pi}}, \quad \frac{2}{\sqrt{\pi}} \cos(2x), \quad \frac{2}{\sqrt{\pi}} \cos(4x), \quad \frac{2}{\sqrt{\pi}} \cos(6x), \dots.$$

The constants were chosen to normalize the eigenfunctions, since

$$\varphi_n \cdot \varphi_n = \int_0^{\pi/2} \varphi_n^2 \, dx = \int_0^{\pi/2} \frac{4}{\pi} \cos^2(2nx) \, dx = 1.$$

This set E of eigenfunctions is complete in $C'[0, \pi/2]$. This means that except for $f(x) \equiv 0$, there is no f in $C'[a, b]$ that is orthogonal to each eigenfunction.

Observe the effect if one eigenfunction is removed. For example, the set E_1 of eigenfunctions

$$\sqrt{\frac{2}{\pi}}, \quad \frac{2}{\sqrt{\pi}}\cos(4x), \quad \frac{2}{\sqrt{\pi}}\cos(6x), \ldots,$$

is formed by removing $f(x) = (2/\sqrt{\pi})\cos(2x)$ from E. Now $\cos(2x)$ has no expansion in terms of E_1, even though $\cos(2x)$ is continuous with a continuous derivative on $(0, \pi/2)$. Indeed, if

$$\cos(2x) = \sqrt{\frac{2}{\pi}}c_0 + \sum_{n=2}^{\infty} c_n \frac{2}{\sqrt{\pi}}\cos(2nx),$$

then

$$c_0 = \sqrt{\frac{2}{\pi}} \cdot k\cos(2x) = 0$$

and, for $n = 2, 3, \ldots,$

$$c_n = k\cos(2x) \cdot \frac{2}{\sqrt{\pi}}\cos(2nx) = 0,$$

implying that $\cos(2x) = 0$ for $0 < x < \pi/2$. This is an absurdity. The deleted set of eigenfunctions E_1, with one function removed from E, is not complete in $C'[0, \pi/2]$. ■

SECTION 15.3 PROBLEMS

In each of Problems 1 through 16, classify the Sturm–Liouville problem as regular, periodic, or singular; state the relevant interval; find the eigenvalues; and, corresponding to each eigenvalue, find an eigenfunction. In some cases eigenvalues may be implicitly defined by an equation.

1. $y'' + \lambda y = 0$; $y(0) = 0$, $y'(L) = 0$

2. $y'' + \lambda y = 0$; $y'(0) = 0$, $y'(L) = 0$

3. $y'' + \lambda y = 0$; $y'(0) = y(4) = 0$

4. $y'' + \lambda y = 0$; $y(0) = y(\pi)$, $y'(0) = y'(\pi)$

5. $y'' + \lambda y = 0$; $y(-3\pi) = y(3\pi)$, $y'(-3\pi) = y'(3\pi)$

6. $y'' + \lambda y = 0$; $y(0) = 0$, $y(\pi) + 2y'(\pi) = 0$

7. $y'' + \lambda y = 0$; $y(0) - 2y'(0) = 0$, $y'(1) = 0$

8. $y'' + 2y' + (1 + \lambda)y = 0$; $y(0) = y(1) = 0$

9. $y'' - 12y' + 4(7 + \lambda)y = 0$; $y(0) = y(5) = 0$

10. $(xy')' + \lambda x^{-1}y = 0$; $y(1) = y(e) = 0$

11. $(xy')' + \lambda x^{-1}y = 0$; $y(1) = y'(e) = 0$

12. $(x^{-3}y')' + (\lambda + 4)x^{-5}y = 0$; $y(1) = y(e^2) = 0$

13. $(e^{2x}y')' + \lambda e^{2x}y = 0$; $y(0) = y(\pi) = 0$

14. $(e^{-6x}y')' + (1 + \lambda)e^{-6x}y = 0$; $y(0) = y(8) = 0$

15. $(x^3 y')' + \lambda xy = 0$; $y(1) = y(e^3) = 0$

16. $(x^{-1}y')' + (4 + \lambda)x^{-3}y = 0$; $y(1) = y(e^4) = 0$

17. Suppose λ is an eigenvalue of a Sturm–Liouville problem on $[a, b]$, with eigenfunction φ. Derive the *Rayleigh quotient* for λ:

$$\lambda =$$

$$-\frac{[r(x)\varphi(x)\varphi'(x)]_a^b + \int_a^b [q(x)(\varphi(x))^2 - r(x)(\varphi'(x))^2]\,dx}{\int_a^b p(x)(\varphi(x))^2\,dx}.$$

Hint: Begin by putting $y = \varphi(x)$ into the Sturm–Liouville equation. Multiply this equation by $\varphi(x)$ and integrate from a to b. Solve the resulting equation for λ and integrate the numerator by parts.

18. It can be shown that the smallest eigenvalue of a Sturm–Liouville problem is the greatest lower bound of the numbers

$$R(\sigma) =$$

$$-\frac{[r(x)\sigma(x)\sigma'(x)]_a^b + \int_a^b [q(x)(\sigma(x))^2 - r(x)(\sigma'(x))^2]\,dx}{\int_a^b p(x)(\sigma(x))^2\,dx},$$

taken over all differentiable functions $\sigma(x)$ satisfying the boundary conditions of the problem (but not necessarily the differential equation). Given any such function, the number $R(\sigma)$ is at least as large as the first (smallest) eigenvalue of the Sturm–Liouville problem under consideration. This fact is the basis of some numerical techniques for approximating the first eigenvalue of a problem. The functions $\sigma(x)$ are called *test functions*.

(a) For Problem 1, select four test functions and evaluate $R(\sigma)$ for each. The numbers thus generated should all be at least as large as the first eigenvalue of the problem.

(b) Carry out (a), except now use Problem 3.

(c) Carry out (a), but now use Problem 5.

19. Use the Rayleigh quotient to get an upper bound for the first eigenvalue of the Sturm–Liouville problem

$$y'' + \lambda x y = 0; \qquad y(1) = 0, \, y'(2) = 0.$$

Use test functions $\sigma_n(x) = (x-1)^n - nx + n$.

20. Use the Rayleigh quotient to get an upper bound for the first eigenvalue of the Sturm–Liouville problem

$$(xy')' + \lambda y = 0; \qquad y(1) = 0, \, y(\pi) = 0.$$

In each of Problems 21 through 28, find the eigenfunction expansion of the given function in the eigenfunctions of the Sturm–Liouville problem. In each case, determine what the eigenfunction expansion converges to on the interval and graph the function and the sum of first N terms of the eigenfunction expansion on the same set of axes for the given interval and the given N. (In Problem 21, do the graph for $L = 1$).

21. $f(x) = 1 - x$ for $0 \le x \le L$
$y'' + \lambda y = 0; \, y(0) = y(L) = 0; \, N = 40$

22. $f(x) = |x|$ for $0 \le x \le \pi$
$y'' + \lambda y = 0; \, y(0) = y'(\pi) = 0; \, N = 30$

23. $f(x) = \begin{cases} -1 & \text{for } 0 \le x \le 2 \\ 1 & \text{for } 2 < x \le 4 \end{cases}$
$y'' + \lambda y = 0; \, y'(0) = y(4) = 0; \, N = 40$

24. $f(x) = \sin(2x)$ for $0 \le x \le \pi$
$y'' + \lambda y = 0; \, y'(0) = y'(\pi) = 0; \, N = 30$

25. $f(x) = x^2$ for $-3\pi \le x \le 3\pi$
$y'' + \lambda y = 0; \, y(-3\pi) = y(3\pi), \, y'(-3\pi) = y'(3\pi);$
$N = 10$

26. $f(x) = \begin{cases} 0 & \text{for } 0 \le x \le \frac{1}{2} \\ 1 & \text{for } \frac{1}{2} < x \le 1 \end{cases}$
$y'' + 2y' + (1+\lambda)y = 0; \, y(0) = y(1) = 0; \, N = 30$

27. $f(x) = e^{-x}$ for $0 \le x \le \pi$
$y'' + \lambda y = 0; \, y'(0) = 0, \, y(\pi) = 0; \, N = 15$

28. $f(x) = 1 + 2x$ for $0 \le x \le 5$
$y'' - 12y' + 4(7+\lambda)y = 0; \, y(0) = y(5) = 0; \, N = 30$

29. Write Bessel's inequality for the function $f(x) = x(4-x)$ for the eigenfunctions of the Sturm–Liouville problem of Problem 3.

30. Write Bessel's inequality for the function $f(x) = e^{-x}$ for the eigenfunctions of the Sturm–Liouville problem of Problem 6.

31. Use Parseval's theorem to derive a formula for an infinite series, using $f(x) = x$ and the eigenfunctions of the Sturm–Liouville problem of $y'' + \lambda y = 0$, $y(0) = y(\pi) = 0$.

32. Use Parseval's theorem to derive a formula for an infinite series, using $f(x) = e^{-2x}$ and the eigenfunctions of the Sturm–Liouville problem of Problem 1.

15.4   Orthogonal Polynomials

Various sets of orthogonal polynomials arise in connection with important differential equations. Legendre polynomials constitute one example, and there are others. In this section we will develop, in outline form, a program for other special polynomials similar to that followed for Legendre polynomials. We will consider the differential equation and certain polynomial solutions, orthogonality of solutions on a certain interval, a generating function, recurrence relation, and other important relationships. Details of verification are left to the student.

15.4.1   Chebyshev Polynomials

Chebyshev is a Russian name with a variety of transliterations, including Chebychev, Chebycheff, Tchebycheff, Tchebyshev, and others. The equation

$$(1 - x^2)y'' - xy' + \lambda y = 0$$

is known as *Chebyshev's differential equation.* To put this into Sturm–Liouville form, first write the differential equation as

$$y'' - \frac{x}{1 - x^2} y' + \frac{\lambda}{1 - x^2} y = 0$$

for $-1 < x < 1$. Compute

$$e^{\int -[x/(1-x^2)]\,dx} = \exp\left(-\frac{1}{2}\ln(1 - x^2)\right) = \frac{1}{\sqrt{1 - x^2}}.$$

Multiply the differential equation by this function to obtain

$$\frac{1}{\sqrt{1 - x^2}} y'' - \frac{x}{(1 - x^2)^{3/2}} y' + \frac{\lambda}{(1 - x^2)^{3/2}} y = 0,$$

or

$$\left(\frac{1}{\sqrt{1 - x^2}} y'\right)' + \frac{\lambda}{(1 - x^2)^{3/2}} y = 0.$$

This is the Sturm–Liouville form of Chebyshev's equation and is also referred to as *Chebyshev's equation on* $(-1, 1)$. The two differential equations have the same solutions.

One can now derive the following facts.

1. For $\lambda = n^2$ ($n = 0, 1, 2, \ldots$) and $-1 < x < 1$, there is a polynomial $T_n(x)$ of degree n, with arbitrary constant chosen so that $T_n(1) = 1$. These are the Chebyshev polynomials. $T_n(x)$ is an eigenfunction of Chebyshev's equation corresponding to the eigenvalue n^2.

2. $T_0(x) = 1$, $T_1(x) = x$, $T_2(x) = 2x^2 - 1$, $T_3(x) = 4x^3 - 3x$, $T_4(x) = 8x^4 - 8x^2 + 1$, $T_5(x) = 16x^5 - 20x^3 + 5x$.

3. Generating function—Let

$$T(x, t) = \frac{1 - xt}{1 - 2xt + t^2}.$$

Then

$$T_n(x, t) = \sum_{n=0}^{\infty} T_n(x) t^n$$

for $-1 < t < 1$.

4. Recurrence relation—For $n = 1, 2, 3, \ldots$,

$$T_{n+1}(x) - 2x T_n(x) + T_{n-1}(x) = 0.$$

5. For $n = 0, 1, 2, \ldots$, and $-1 \le x \le 1$,

$$T_n(x) = \cos(n \arccos(x)).$$

6. If n is a positive integer, then T_n has exactly n zeros in $(-1, 1)$, and these are the numbers

$$\cos\left(\frac{2k - 1}{n} \frac{\pi}{2}\right) \quad \text{for } k = 1, 2, \ldots, n.$$

7. A Rodrigues-type formula—For $n = 1, 2, 3, \ldots$,

$$T_n(x) = \frac{(-1)^n}{1(3)(5) \cdots (2n - 1)} \sqrt{1 - x^2} \frac{d^n}{dx^n}\left((1 - x^2)^{(2n-1)/2}\right).$$

8. The appropriate dot product for dealing with the Chebyshev polynomials is

$$f \cdot g = \int_{-1}^{1} \frac{1}{\sqrt{1 - x^2}} f(x)g(x)\,dx.$$

If n and m are distinct nonnegative integers, then

$$T_n \cdot T_m = \int_{-1}^{1} \frac{1}{\sqrt{1 - x^2}} T_n(x) T_m(x)\,dx = 0.$$

Thus the Chebyshev polynomials are orthogonal on $[-1, 1]$ with weight function $1/\sqrt{1 - x^2}$.

9.

$$T_n \cdot T_n = \begin{cases} \pi & \text{for } n = 0 \\ \pi/2 & \text{for } n = 1, 2, 3, \ldots \end{cases}.$$

10. Fourier–Chebyshev series—If f is piecewise smooth on $[-1, 1]$, then for $-1 < x < 1$,

$$\frac{1}{2}(f(x+) + f(x-)) = \sum_{n=0}^{\infty} c_n T_n(x),$$

where the Fourier–Chebyshev coefficients are given by

$$c_n = \frac{f \cdot T_n}{T_n \cdot T_n}.$$

Here

$$c_0 = \frac{1}{\pi} \int_{-1}^{1} \frac{1}{\sqrt{1 - x^2}} f(x)\,dx$$

and

$$c_n = \frac{2}{\pi} \int_{-1}^{1} \frac{1}{\sqrt{1 - x^2}} f(x) T_n(x)\,dx.$$

15.4.2 Laguerre Polynomials

Laguerre's differential equation is

$$xy'' + (1 - x)y' + \lambda y = 0.$$

This can be put into the Sturm–Liouville form

$$\left(xe^{-x}y'\right)' + \lambda e^{-x}y = 0.$$

This is also called *Laguerre's equation*.

1. For $0 < x < \infty$ and for $\lambda = n = 0, 1, 2, \ldots$, there is a polynomial solution $L_n(x)$ of degree n, with arbitrary constant chosen so that the coefficient of x^n in $L_n(x)$ is 1.

2. Let

$$\mathcal{L}(x, t) = \frac{1}{1 - t} e^{-xt/(1-t)}.$$

This is a generating function for the Laguerre polynomials, in the sense that for $-1 < t < 1$,

$$\mathcal{L}(x, t) = \sum_{n=0}^{\infty} L_n(x) t^n.$$

3. Recurrence relation—For $n = 1, 2, 3, \ldots$,

$$(n+1)L_{n+1}(x) - (2n+1-x)L_n(x) + nL_{n-1}(t) = 0.$$

4. A Rodrigues-type formula:

$$L_n(x) = \frac{1}{n!}e^x \frac{d^n}{dx^n}(x^n e^{-x}).$$

5. Since $p(x) = e^{-x}$ in the Sturm–Liouville form of Laguerre's differential equation, the appropriate dot product is

$$f \cdot g = \int_0^\infty e^{-x} f(x)g(x)\,dx.$$

If n and m are distinct nonnegative integers, then

$$L_n \cdot L_m = 0.$$

6. For any nonnegative integer n,

$$L_n \cdot L_n = 1.$$

7. Fourier–Laguerre series—If $\int_0^\infty |f(x)|\,dx$ converges, and f is piecewise smooth on every interval $[0, L]$, then for $x > 0$,

$$\frac{1}{2}(f(x+) + f(x-)) = \sum_{n=0}^\infty c_n L_n(x),$$

where

$$c_n = f \cdot L_n.$$

15.4.3   Hermite Polynomials

Hermite's differential equation is

$$y'' - 2xy' + \lambda y = 0.$$

This has Sturm–Liouville form

$$(e^{-x^2}y')' + \lambda e^{-x^2} y = 0$$

for $-\infty < x < \infty$.

1. For $\lambda = 2n$, with $n = 0, 1, 2, \ldots$, there is a polynomial solution $H_n(x)$ of degree n.

2. Generating function—Let

$$H(x, t) = e^{2xt - t^2}.$$

Then

$$H(x, t) = \sum_{n=0}^\infty H_n(x)t^n$$

for $-1 < t < 1$.

3. Recurrence relation—For $n = 1, 2, 3, \ldots$,

$$(n+1)H_{n+1}(x) - 2x H_n(x) + 2H_{n-1}(x) = 0.$$

4. Rodrigues-type formula—For $n = 0, 1, 2, \ldots$,

$$H_n(x) = \frac{(-1)^n}{n!} e^{x^2} \frac{d^n}{dx^n} \left(e^{-x^2} \right).$$

5. The appropriate dot product for Hermite polynomials is

$$f \cdot g = \int_{-\infty}^{\infty} e^{-x^2} f(x) g(x) \, dx.$$

If n and m are distinct nonnegative integers, then

$$H_n \cdot H_m = 0.$$

6. If n is a nonnegative integer, then

$$H_n \cdot H_n = \frac{2^n \sqrt{\pi}}{n!}.$$

7. If $\int_{-\infty}^{\infty} |f(x)| \, dx$ converges, and f is piecewise smooth on every interval $[-L, L]$, then for any real x,

$$\frac{1}{2}(f(x+) + f(x-)) = \sum_{n=0}^{\infty} c_n H_n(x),$$

where

$$c_n = \frac{n!}{2^n \sqrt{\pi}} f \cdot H_n.$$

SECTION 15.4   PROBLEMS

1. Use the generating function to derive the first five Chebyshev polynomials.

2. Use the Rodrigues-type formula to derive these first five Chebyshev polynomials again.

3. Derive the recurrence relation for the Chebyshev polynomials. *Hint:* Use the generating function.

4. Find the first five coefficients in the Fourier–Chebyshev expansion of $f(x) = x$ on $[-1, 1]$. Graph $f(x)$ and the sum of the first five terms of this expansion on the same set of axes.

5. Use the generating function to derive the first five Laguerre polynomials.

6. Use the Rodrigues-type formula to derive these first five Laguerre polynomials again.

7. Derive the recurrence relation for Laguerre polynomials.

8. Find the first five coefficients in the Fourier–Laguerre expansion of $f(x) = e^{-4x}$ on $[0, \infty)$. Graph $f(x)$ and the sum of the first five terms of this expansion on the same set of axes for $0 \le x \le 5$.

9. Use the generating function to derive the first five Hermite polynomials.

10. Use the Rodrigues-type formula to derive these first five Hermite polynomials again.

11. Derive the recurrence relation for Hermite polynomials.

12. Find the first five coefficients in the Fourier–Hermite expansion of $f(x) = e^{-4|x|}$ on $(-\infty, \infty)$. Graph $f(x)$ and the sum of the first five terms of this expansion on the same set of axes for $-5 \le x \le 5$.

15.5 Wavelets

15.5.1 The Idea Behind Wavelets

Recent years have seen an explosion in both the mathematical development of wavelets and their applications, which include signal analysis, data compression, filtering, and electromagnetics. Our purpose here is to introduce enough of the ideas behind wavelets to enable the student to pursue more thorough treatments.

Think of a function defined on the real line as a signal. If the signal contains one fundamental frequency ω_0, then f is a periodic function with period $2\pi/\omega_0$ and the Fourier series of $f(t)$ is one tool for analyzing the signal's frequency content. The amplitude spectrum of f consists of a plot of points $(n\omega_0, c_n/2)$ in which

$$c_n = \sqrt{a_n^2 + b_n^2},$$

with a_n and b_n the Fourier coefficients of f. Under certain conditions on f, this enables us to represent the signal as a trigonometric series displaying the natural frequencies:

$$f(t) = \frac{1}{2}a_0 + \sum_{n=1}^{\infty} a_n \cos(n\omega_0 t) + b_n \sin(n\omega_0 t).$$

Often we model the signal by taking a partial sum of the Fourier series:

$$f(t) \approx \frac{1}{2}a_0 + \sum_{n=1}^{N} a_n \cos(n\omega_0 t) + b_n \sin(n\omega_0 t).$$

Although this process has proved useful in many instances, the Fourier trigonometric representation is not always the best device for analyzing signals. First, we may be interested in a signal that is not periodic. More generally, we may have a signal that is defined over the entire real line with no periodicity, and we require only that its energy be finite. This means that $\int_{-\infty}^{\infty}(f(t))^2\,dt$ is finite or, if $f(t)$ is complex valued, that $\int_{-\infty}^{\infty}|f(t)|^2\,dt$ is finite. This integral is the energy content of the signal, and functions having finite energy are said to be *square integrable*. In general, Fourier expansions are not the best tool for the analysis of such functions.

There are other disadvantages to Fourier trigonometric series. For a given f, we may have to choose N very large to model $f(t)$ by a partial sum of a Fourier series. Finally, if we are interested on focusing on the behavior of $f(t)$ in some finite time interval, or near some particular time, we cannot isolate those terms in the Fourier expansion that describe this behavior, but instead have to take the entire Fourier series, or its entire partial sum, if we are modeling the signal.

To illustrate, consider the signal shown in Figure 15.19. Explicitly,

$$
f(t) = \begin{cases}
1 & \text{for } 0 \leq t < \dfrac{1}{4} \\[2mm]
-\dfrac{1}{5} & \text{for } \dfrac{1}{4} \leq t < \dfrac{3}{8} \\[2mm]
\dfrac{11}{5} & \text{for } \dfrac{3}{8} \leq t < \dfrac{1}{2} \\[2mm]
1 & \text{for } \dfrac{1}{2} \leq t < \dfrac{3}{4} \\[2mm]
-3 & \text{for } \dfrac{3}{4} \leq t < 1 \\[2mm]
-\dfrac{4}{5} & \text{for } 1 \leq t < \dfrac{5}{4} \\[2mm]
\dfrac{14}{5} & \text{for } \dfrac{5}{4} \leq t < \dfrac{11}{8} \\[2mm]
\dfrac{4}{5} & \text{for } \dfrac{11}{8} \leq t < \dfrac{3}{2} \\[2mm]
0 & \text{for } t \geq \dfrac{3}{2} \quad \text{and for } t < 0
\end{cases}
$$

The Fourier series of f on $[-\frac{3}{2}, \frac{3}{2}]$ is

$$
\frac{1}{12} + \sum_{n=1}^{\infty} a_n \cos\left(\frac{2n\pi x}{3}\right) + b_n \sin\left(\frac{2n\pi x}{3}\right),
$$

where

$$
a_n = -\frac{1}{5n\pi}\left[-6\sin\left(\frac{n\pi}{6}\right) + 12\sin\left(\frac{n\pi}{4}\right) - 6\sin\left(\frac{n\pi}{3}\right) - 20\sin\left(\frac{n\pi}{2}\right)\right.
$$
$$
\left. + 11\sin\left(\frac{2n\pi}{3}\right) + 18\sin\left(\frac{5n\pi}{6}\right) - 10\sin\left(\frac{11n\pi}{12}\right)\right]
$$

and

$$
b_n = \frac{1}{5n\pi}\left[-6\cos\left(\frac{n\pi}{6}\right) + 5 + 12\cos\left(\frac{n\pi}{4}\right) - 6\cos\left(\frac{n\pi}{3}\right) - 20\cos\left(\frac{n\pi}{2}\right)\right.
$$
$$
\left. + 11\cos\left(\frac{2n\pi}{3}\right) + 18\cos\left(\frac{5n\pi}{6}\right) - 10\cos\left(\frac{11n\pi}{12}\right) - 4\cos(n\pi)\right].
$$

This series converges very slowly to the function. Indeed, Figure 15.20(a) shows the 80th partial sum of this series, and Figure 15.20(b) the 100th partial sum. Even with this number of terms, this partial sum does not model the signal very well. In addition, if we were interested in focusing on just part of the signal, there is no way of distinguishing certain terms of the Fourier series as carrying the most information about this part of the signal. Put another way, Fourier series do not localize information.

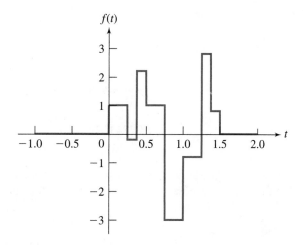

FIGURE 15.19 *The signal $f(t)$.*

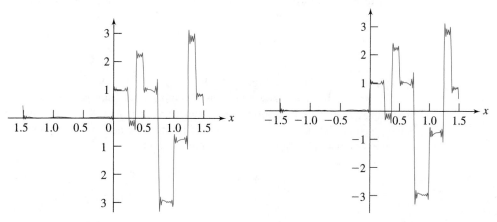

FIGURE 15.20(a) *Eightieth partial sum of the Fourier series of the signal.*

FIGURE 15.20(b) *One-hundredth partial sum of the Fourier series of the signal.*

These considerations suggest that we seek other sets of complete orthogonal functions in which square integrable functions might be expanded and which overcome some of the difficulties just cited for Fourier trigonometric series. This is a primary motivation for wavelets. We will begin our discussion of wavelets by developing one important wavelet in detail, then use this construction to suggest some of the ideas behind wavelets in general.

15.5.2 The Haar Wavelets

We will construct an example that is important both historically and for present-day applications. The Haar wavelets were the first to be found (about 1910) and serve as a model of one approach to the development of other wavelets.

Let $L^2(R)$ denote the set of all real-valued functions that are defined on the entire real line and are square integrable. $L^2(R)$ has the structure of a vector space, since linear combinations $\alpha_1 f_1 + \alpha_2 f_2 + \cdots + \alpha_n f_n$ of square integrable functions are square integrable. The dot product we will use for functions in $L^2(R)$ is

$$f \cdot g = \int_{-\infty}^{\infty} f(t)g(t)\, dt.$$

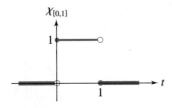

FIGURE 15.21 $\chi_{[0,1]}$

Now consider the characteristic function of an interval I (or of any set of numbers on the real line). This function is denoted χ_I and has the value 1 for t in I and zero for t not in I. That is,

$$
\chi_I(t) = \begin{cases} 1 & \text{if } t \text{ is in } I \\ 0 & \text{if } t \text{ is not in } I \end{cases}.
$$

In particular, we will use the characteristic function of the half-open unit interval:

$$
\chi_{[0,1)}(t) = \begin{cases} 1 & \text{for } 0 \le t < 1 \\ 0 & \text{if } t < 0 \quad \text{or if } t \ge 1 \end{cases}.
$$

A graph of $\chi_{[0,1)}$ is shown in Figure 15.21.

We want to introduce new functions by both scaling and translation, with the objective of producing a complete orthonormal set of functions in $L^2(R)$. Recall that the graph of $f(t-k)$ is the graph of $f(t)$ translated k units to the right if k is positive, and $|k|$ units to the left if k is negative. For example, Figure 15.22(a) shows a graph of

$$
f(t) = \begin{cases} t \sin(t) & \text{for } 0 \le t \le 15 \\ 0 & \text{for } t < 0 \quad \text{and for } t > 15 \end{cases}.
$$

Figure 15.22(b) is a graph of $f(t+5)$ (graph of $f(t)$ shifted five units to the left), and Figure 15.22(c) is a graph of $f(t-5)$ (the graph of $f(t)$ shifted five units to the right). In addition, $f(kt)$ is a scaling of the graph of f. $f(kt)$ compresses (if $k > 1$) or stretches (if $0 < k < 1$) the graph of $f(t)$ for $a \le t \le b$ onto the interval $[a/k, b/k]$. For example, Figure 15.23(a) shows a graph of

$$
f(t) = \begin{cases} t \sin(\pi t) & \text{for } -2 \le t \le 3 \\ 0 & \text{for } t < -2 \quad \text{and for } t > 0 \end{cases}.
$$

Figure 15.23(b) shows a graph of $f(3t)$, compressing the graph of Figure 15.23(a) to the right and left, and Figure 15.23(b) shows a graph of $f(t/3)$, stretching out the graph of Figure 15.23(a).

Let $\varphi(t) = \chi_{[0,1)}(t)$ and define

$$
\psi(t) = \varphi(2t) - \varphi(2t-1) = \begin{cases} 1 & \text{for } 0 \le t < \dfrac{1}{2} \\ -1 & \text{for } \dfrac{1}{2} \le t < 1 \\ 0 & \text{for } t < 0 \quad \text{and for } t \ge 1 \end{cases}.
$$

A graph of ψ is shown in Figure 15.24.

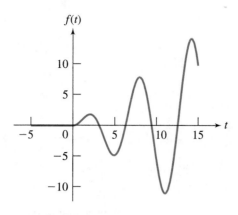

FIGURE 15.22(a)

$$f(t) = \begin{cases} t\sin(t) & \text{for } 0 \leq t \leq 15 \\ 0 & \text{for } t < 0 \quad \text{and for } t > 15 \end{cases}$$

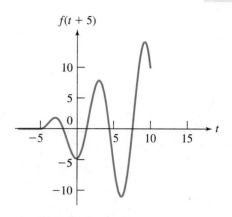

FIGURE 15.22(b) $f(t+5)$.

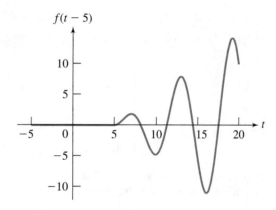

FIGURE 15.22(c) $f(t-5)$.

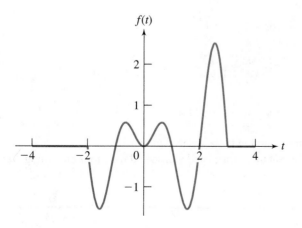

FIGURE 15.23(a)

$$f(t) = \begin{cases} t\sin(\pi t) & \text{for } -2 \leq t \leq 3 \\ 0 & \text{for } t < -2 \quad \text{and for } t > 0 \end{cases}$$

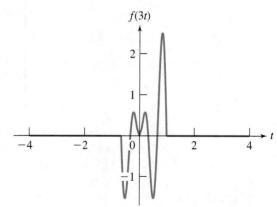

FIGURE 15.23(b) $f(3t)$.

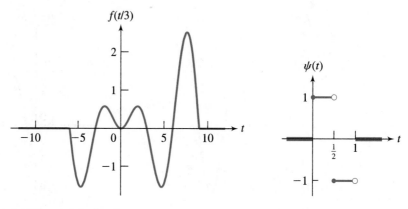

FIGURE 15.23(c) $f(t/3)$.

FIGURE 15.24
$\psi(t) = \varphi(2t) - \varphi(2t-1)$,
with $\varphi(t) = \chi_{[0,1]}$.

Next, consider translations $\psi(t - n)$, in which n is any integer. This is the function

$$\psi(t - n) = \varphi(2(t - n)) - \varphi(2(t - n) - 1)$$

$$= \varphi(2t - 2n) - \varphi(2t - 2n - 1)$$

$$= \begin{cases} 1 & \text{for } n \leq t < n + \dfrac{1}{2} \\[2mm] -1 & \text{for } n + \dfrac{1}{2} \leq t < n + 1 \\[2mm] 0 & \text{for } t < n \quad \text{and for } t \geq n + 1 \end{cases}$$

A graph of $\psi(t - n)$ is shown in Figure 15.25.

Now combine a translation with a scaling. Consider the function

$$\psi(2t - m) = \varphi(2(2t - m)) - \varphi(2(2t - m) - 1)$$

$$= \varphi(4t - 2m) - \varphi(4t - 2m - 1)$$

$$= \begin{cases} 1 & \text{for } \dfrac{m}{2} \leq t < \left(\dfrac{m}{2}\right) + \dfrac{1}{4} \\[2mm] -1 & \text{for } \left(\dfrac{m}{2}\right) + \dfrac{1}{4} \leq t < \dfrac{(m + 1)}{2} \\[2mm] 0 & \text{for } t < \dfrac{m}{2} \quad \text{and for } t \geq \dfrac{(m + 1)}{2} \end{cases}$$

in which m is any integer. A graph of this function is shown in Figure 15.26.

Before proceeding, we will observe that these translated and scaled functions are orthogonal in $L^2(R)$.

LEMMA 15.2

1. For distinct integers n and m,

$$\psi(t - n) \cdot \psi(t - m) = 0$$

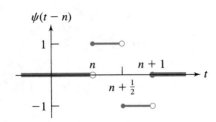

FIGURE 15.25
$\psi(t-n) = \varphi(2(t-n)) - \varphi(2(t-n)-1)$.

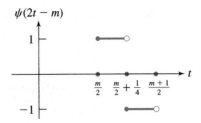

FIGURE 15.26 $\psi(2t-m) =$
$\varphi(2(2t-m)) - \varphi(2(2t-m)-1)$.

and

$$\psi(2t-n) \cdot \psi(2t-m) = 0.$$

2. For any integers n and m,

$$\psi(t-n) \cdot \psi(2t-m) = 0.$$

Proof If $n \neq m$, then the intervals $[n, n+1)$ on which $\psi(t-n)$ takes on its nonzero values, and $[m, m+1)$ on which $\psi(t-m)$ assumes its nonzero values, are disjoint. Then $\psi(t-n)\psi(t-m) = 0$ for all t and

$$\psi(t-n) \cdot \psi(t-m) = \int_{-\infty}^{\infty} \psi(t-n)\psi(t-m)\, dt = 0.$$

Similarly, for $n \neq m$, the intervals $[n/2, (n+1)/2)$ and $[m/2, (m+1)/2)$ on which $\psi(2t-n)$ and $\psi(2t-m)$, respectively, take on their nonzero values, are disjoint, so $\psi(2t-n)\cdot\psi(2t-m) = 0$.

For (2), let n and m be any integers. If the intervals on which $\psi(t-n)$ and $\psi(2t-m)$ have nonzero values are disjoint, then these functions are orthogonal. There are two cases in which these intervals are not disjoint.

Case 1 $n = m/2$
In this case

$$\psi(t-n)\psi(2t-m) = \begin{cases} 1 & \text{for } n \leq t < n + \dfrac{1}{4} \\[2mm] -1 & \text{for } n + \dfrac{1}{4} \leq t < n + \dfrac{1}{2} \\[2mm] 0 & \text{for } t < n \quad \text{and for } t \geq n + \dfrac{1}{2} \end{cases}.$$

Then

$$\psi(t-n) \cdot \psi(2t-m) = \int_n^{n+1/4} dt - \int_{n+1/4}^{n+1/2} dt = 0.$$

Case 2 $n + 1/2 = m/2$

Now

$$
\psi(t-n)\psi(2t-m) = \begin{cases} -1 & \text{for } n + \dfrac{1}{2} \le t < n + \dfrac{3}{4} \\[2ex] 1 & \text{for } n + \dfrac{3}{4} \le t < n + 1 \\[2ex] 0 & \text{for } t < n + \dfrac{1}{2} \text{ and for } t \ge n + \dfrac{3}{4} \end{cases}
$$

so

$$
\psi(t-n) \cdot \psi(2t-m) = - \int_{n+1/2}^{n+3/4} dt + \int_{n+3/4}^{n+1} dt = 0. \quad \blacksquare
$$

However, while the functions $\psi(t-n)$ and $\psi(2t-m)$ are orthogonal in $L^2(R)$, they do not form a complete set, as n and m vary over the integers. We leave it for the student to produce nontrivial (that is, nonzero at least on some interval) square integrable functions that are orthogonal to all of these translated and scaled functions

The idea now is to extend this set of functions by using scaling factors 2^m for integer m to obtain functions that take on nonzero constant values on intervals that can be made shorter (positive m) or longer (negative m). Let

$$
\sigma_{m,n}(t) = \psi(2^m t - n)
$$

for each integer m and each integer n. Then

$$
\sigma_{m,n}(t) = \varphi(2^{m+1}t - 2n) - \varphi(2^{m+1}t - 2n - 1)
$$

$$
= \begin{cases} 1 & \text{for } \dfrac{n}{2^m} \le t < \dfrac{n}{2^m} + \dfrac{1}{2^{m+1}} \\[2ex] -1 & \text{for } \dfrac{n}{2^m} + \dfrac{1}{2^{m+1}} \le t < \dfrac{n}{2^m} + \dfrac{1}{2^m} \\[2ex] 0 & \text{for } t < \dfrac{n}{2^m} \text{ and for } t \ge \dfrac{n}{2^m} + \dfrac{1}{2^m} \end{cases}.
$$

Figure 15.27 shows graphs of $\sigma_{0,n}(t)$, $\sigma_{1,n}(t)$, $\sigma_{2,n}(t)$, and $\sigma_{3,n}(t)$ on the same set of axes, for comparison. Note that n determines how far out the t axis the graph occurs, while m controls the size of the interval over which the function is nonzero (shorter for m increasing and positive, longer for $|m|$ increasing but m negative). In the drawing n is a positive integer, but n can also be chosen negative, in which case the graphs are to the left of the vertical axis.

We claim that these functions form an orthogonal set in $L^2(R)$.

THEOREM 15.29

If n, m, n', and m' are integers, and $(m, n) \ne (m', n')$, then

$$
\sigma_{m,n} \cdot \sigma_{m',n'} = 0. \quad \blacksquare
$$

A proof of this is left to the student.

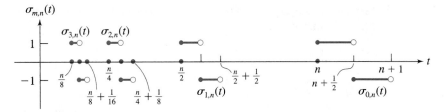

FIGURE 15.27 $\sigma_{m,n}(t)$ *for* $m = 0, 1, 2, 3$.

One last detail before we get to the main point. The $\sigma_{m,n}$'s are orthogonal, but they are not orthonormal. This is easily fixed. Divide each of these functions by its length, as defined by the dot product in $L^2(R)$. Compute

$$\text{(length of } \sigma_{m,n})^2 = \sigma_{m,n} \cdot \sigma_{m,n} = \int_{n/2^m}^{n/2^m + 1/2^m} \sigma_{m,n}^2(t)\, dt = \int_{n/2^m}^{n/2^m + 1/2^m} dt = \frac{1}{2^m}.$$

This suggests that we define the functions

$$\psi_{m,n}(t) = 2^{m/2}\sigma_{m,n}(t) = 2^{m/2}\left[\varphi(2^{m+1}t - 2n) - \varphi(2^{m+1}t - 2n - 1)\right]$$

$$= \begin{cases} 2^{m/2} & \text{for } \dfrac{n}{2^m} \le t < \dfrac{n}{2^m} + \dfrac{1}{2^{m+1}} \\[2mm] -2^{m/2} & \text{for } \dfrac{n}{2^m} + \dfrac{1}{2^{m+1}} \le t < \dfrac{n}{2^m} + \dfrac{1}{2^m} \\[2mm] 0 & \text{for } t < \dfrac{n}{2^m} \text{ and for } t \ge \dfrac{n}{2^m} + \dfrac{1}{2^m} \end{cases}.$$

The functions $\psi_{m,n}$ form an orthonormal set in $L^2(R)$. These functions are the *Haar wavelets*. In the construction, φ is called the *scaling function*, and $\psi(t) = \varphi(2t) - \varphi(2t - 1)$ is the *mother wavelet*. Graphs of these wavelets are similar to the graphs of Figure 15.27, but the segment at height 1 in Figure 15.27 is now at height $2^{m/2}$, and the segment at height -1 in Figure 15.27 is now at height $-2^{m/2}$.

The Haar wavelets are complete in $L^2(R)$. The idea behind this can be envisioned as follows. If f is square integrable, then $f(t)$ can be approximated as accurately as we like by a function g having compact support ($g(t) = 0$ outside some closed interval) and having constant values on half-open intervals of the form $[n/2^m, (n + 1)/2^m)$, with n and m integers. Such intervals are of length $1/2^m$, which can be made longer or shorter by choice of the integer m. In turn, g can be approximated as closely as we like by a sum of constants times Haar wavelets, which are defined on such intervals, with the error in the approximation tending to zero as the number of terms in the sum is taken larger.

15.5.3 A Wavelet Expansion

Suppose f is a square integrable function. We can attempt an expansion of f in a series of the Haar wavelets, which form a complete orthonormal set in $L^2(R)$. Such an expansion has the appearance

$$f(t) = \sum_{m=-\infty}^{\infty} \sum_{n=-\infty}^{\infty} c_{mn} \psi_{m,n}(t).$$

The equality in this expression is taken to mean that the series on the right converges in the mean to $f(t)$. This means that

$$\lim_{M \to \infty} \int_{-\infty}^{\infty} \left(f(t) - \sum_{m=-\infty}^{M} \sum_{n=-\infty}^{\infty} c_{mn} \psi_{m,n}(t) \right)^2 dt = 0.$$

The coefficients c_{mn} can be found in the usual way by using the orthonormality of the Haar wavelets:

$$f \cdot \psi_{m_0,n_0} = \sum_{m=-\infty}^{\infty} \sum_{n=-\infty}^{\infty} c_{mn} \psi_{m,n} \cdot \psi_{m_0,n_0} = c_{m_0 n_0}.$$

We will complete the example begun in Section 15.5.1, in which f is the signal whose graph is shown in Figure 15.19. As we saw in Figures 15.20(a) and (b), we would have to use a very large number of terms to model this signal with the partial sum of its Fourier expansion on $[-\frac{3}{2}, \frac{3}{2}]$. However, if we calculate the coefficients in the Haar expansion, we find that

$$f(t) = \psi_{0,0}(t) + \sqrt{2}\psi_{1,1}(t) - 0.6\psi_{2,1}(t) - 0.4\sqrt{2}\psi_{1,2}(t) + \psi_{2,5}(t).$$

For some purposes we want Fourier trigonometric expansions, but for this signal the Haar wavelets provide a very efficient expansion.

15.5.4   Multiresolution Analysis with Haar Wavelets

The term *multiresolution analysis* refers to a sequence of closed subspaces of $L^2(R)$ that are related to the scaling used in defining a set of wavelets. We will discuss what this means in the context of the Haar wavelets.

Because $L^2(R)$ has the structure of a vector space, the following three conditions hold:

1. Linear combinations $\sum_{j=1}^{N} c_j f_j$ of functions in $L^2(R)$ are also in $L^2(R)$.
2. The zero function, $\theta(t) = 0$ for all t, is in $L^2(R)$ and serves as the zero vector of $L^2(R)$. For any function f in $L^2(R)$, $f + \theta = f$.
3. If f is in $L^2(R)$, $-f$, defined by $(-f)(t) = -f(t)$, is also in $L^2(R)$.

A set S of square integrable functions is said to be a subspace of $L^2(R)$ if S has at least one function in it, and whenever f and g are in S, then $f - g$ is in S. For example, the set of all constant multiples of $\chi_{[0,1]}$ forms a subspace of $L^2(R)$.

A subspace S is closed if convergent sequences of functions in S have their limit functions in S. For example, the subspace of all continuous square integrable functions is not closed, because a limit (in the sense of mean convergence) of continuous functions need not be continuous.

If a subspace S is not closed, we can form the "smallest" subspace of $L^2(R)$ containing all the functions in S, together with all the limits of convergent sequences of functions in S. This subspace, which may be all of $L^2(R)$, is called the *closure* of S, and is denoted $\overline{S}$. $\overline{S}$ is closed, because by its formation it has all the limits of convergent sequences of functions that are in this space.

We will now show how the Haar wavelets generate a sequence of closed subspaces of $L^2(R)$, which can be indexed by the integers so that each is contained in the next one in the list. The spaces are generated by different scalings of the scaling function φ and may be thought of as associated with different degrees of resolution of the signal.

To begin defining these spaces, let S_0 consist of all linear combinations of the translated scaling function. These translated scaling functions have the form $\varphi(t - n)$ for integer n, and a typical function in S_0 has the form

$$\sum_{j=1}^{N} c_j \varphi(t - n_j),$$

where N is a positive integer, the c_j's are real numbers, and each n_j is an integer. Now let V_0 be the closure of S_0:

$$V_0 = \overline{S_0}.$$

Next, let S_m be the space of all linear combinations of the functions $\varphi(2^m t - n)$, where n varies over the integers and m is a fixed integer in defining S_m. Let

$$V_m = \overline{S_m}.$$

From the scaling property of the scaling function,

$$\varphi(t) = \varphi(2t) + \varphi(2t - 1),$$

we find that $f(t)$ is in V_m exactly when $f(2t)$ is in V_{m+1}, and each V_m is contained within V_{m+1} (written $V_m \subset V_{m+1}$). Thus the closed subspaces V_m, with integer m, form an ascending chain:

$$\cdots \subset V_{-2} \subset V_{-1} \subset V_0 \subset V_1 \subset V_2 \subset \cdots$$

This chain has two additional properties of importance. First, there is no nontrivial function contained in every V_m. We say that the intersection of all the closed subspaces V_m consists of just the zero function. And, finally, the ascending chain ends in $L^2(R)$. This means that every function in $L^2(R)$ has a series expansion in terms of the Haar functions, a fact to which we have already alluded.

The spaces V_m are said to form a *multiresolution analysis* of $L^2(R)$. This multiresolution analysis is generated by the scaling function φ.

15.5.5 General Construction of Wavelets and Multiresolution Analysis

The Haar wavelets have been known for nearly a century, together with the chain of subspaces that form a multiresolution analysis of $L^2(R)$. However, it remained unknown for some time whether this construction could be duplicated, resulting in multiresolution analyses starting from different scaling functions. To this end, we will use the hindsight of the Haar construction to make a definition of a scaling function and the associated multiresolution analysis.

DEFINITION 15.4 *Scaling Function and Associated Multiresolution Analysis*

Let φ be in $L^2(R)$. Then φ is a scaling function with multiresolution analysis $\{V_m\}$ if

$$\cdots \subset V_{-2} \subset V_{-1} \subset V_0 \subset V_1 \subset V_2 \subset \cdots$$

is an ascending chain of closed subspaces of $L^2(R)$ satisfying the conditions:

1. The translated functions $\varphi(t - n)$, for integer n, are orthonormal, and every function in V_0 is a linear combination of functions of this form.

2. There is no nontrivial function that belongs to every V_m. (That is, the V_m's have trivial intersection.)

3. $f(t)$ is in V_m exactly when $f(2t)$ is in V_{m+1}.

4. Every function in $L^2(R)$ can be expanded in a series of functions from the V_m's.

V_0 is a subspace of V_1, which contains functions orthogonal to every function in V_0. The subspace of V_1 containing all of these functions is called the *orthogonal complement* of V_0 in V_1. To draw an analogy from vectors in R^3, the constant multiples of $\mathbf{k}$ form a subspace of R^3 that is the orthogonal complement of the plane defined by $\mathbf{i}$ and $\mathbf{j}$. Every vector in this orthogonal complement is orthogonal to each linear combination $a\mathbf{i} + b\mathbf{j}$.

Now use the scaling function to produce a mother wavelet ψ, having the property that every function in this orthogonal complement of V_0 in V_1 is a linear combination of translates $\psi(t-n)$. If there is such a mother wavelet, then we can form the family of wavelets

$$\psi_{mn} = 2^{m/2}\psi(2^m t - n)$$

for integers m and n.

15.5.6   Shannon Wavelets

The Haar wavelets form a prototype for wavelets and multiresolution analysis, partly because they were first and partly because they are relatively easy to work with and visualize. The reason it took many years before other examples of scaling function/wavelet/multiresolution analysis were found is that this involves some fairly heavy analysis. However, there are other relatively simple examples. One consists of the Shannon wavelets. For these, begin with the Fourier transform of a potential scaling function. Let

$$\hat{\varphi}(\omega) = \chi_{[-\pi,\pi)}.$$

Taking the inverse Fourier transform, we obtain

$$\varphi(t) = \frac{\sin(\pi t)}{\pi t}.$$

This function occurs in the Shannon reconstruction theorem, which was proved in Section 14.4.7 for functions of bandwidth $\leq L$. In the case that $L = \pi$, the theorem states that a signal f whose Fourier transform $\hat{f}(\omega)$ vanishes outside the interval $[-\pi, \pi]$ (that is, f has bandwidth $\leq \pi$) can be reconstructed by sampling its values on the integers. Specifically,

$$f(t) = \sum_{n=-\infty}^{\infty} f(n)\frac{\sin(\pi(t-n))}{\pi(t-n)} = \sum_{n=-\infty}^{\infty} f(n)\varphi(t-n).$$

The space V_0 in this context consists of functions in $L^2(R)$ of bandwidth not exceeding π.

By scaling (let $g(t) = f(2t)$) we can consider the space V_1 of functions of bandwidth not exceeding 2π, and so on, forming a multiresolution analysis. Thus φ is a scaling function. We now need a mother wavelet ψ that is orthogonal to each $\varphi(t-n)$ for integer n. By an argument we will not carry out (but whose conclusions can be verified in straightforward fashion), we obtain a suitable ψ from φ in this case by setting

$$\psi(t) = \varphi\left(t - \tfrac{1}{2}\right) - 2\varphi(2t - 1) = \frac{\sin(2\pi t) - \cos(\pi t)}{\pi\left(t - \tfrac{1}{2}\right)}.$$

The frequency content of this function is obtained from its Fourier transform,

$$\hat{\psi}(\omega) = -e^{-i\omega/2}\chi_A(\omega),$$

where A consists of all ω in $[-2\pi, -\pi)$, together with all ω in $(\pi, 2\pi]$. That is, on each of these intervals, $\hat{\psi}(\omega) = -e^{-i\omega/2}$, and for ω outside of these intervals, $\hat{\psi}(\omega) = 0$. Figure 15.28 shows a graph of the mother wavelet ψ, and Figure 15.29 a graph of its amplitude spectrum. This gives the frequency content of ψ.

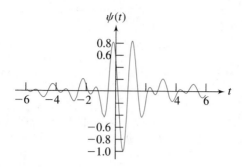

FIGURE 15.28 *Shannon mother wavelet*
$$\psi(t) = \frac{\sin(2\pi t) - \cos(\pi t)}{\pi \left(t - \frac{1}{2}\right)}.$$

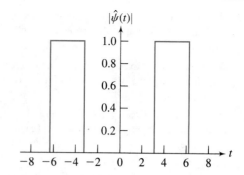

FIGURE 15.29 *Amplitude spectrum of the Shannon mother wavelet.*

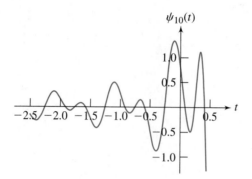

FIGURE 15.30(a) *Shannon wavelet $\psi_{10}(t)$.*

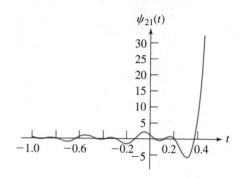

FIGURE 15.30(b) *Shannon wavelet $\psi_{21}(t)$.*

The Shannon wavelets are the functions

$$\psi_{mn}(t) = 2^{m/2}\psi(2^m t - n)$$

$$= \frac{2^{m/2}}{\pi \left(t - \frac{1}{2}\right)} \left(\sin(2\pi(2^m t - n)) - \cos(\pi(2^m t - n))\right).$$

We leave it for the student to explore properties of these wavelets. Graphs of $\psi_{10}(t)$ and $\psi_{21}(t)$ are given in Figures 15.30(a) and (b).

There are many other families of wavelets, including Meyer wavelets, Daubechies wavelets, and Stömberg wavelets. These require a good deal more preliminary work for their definitions. Different wavelets are constructed for specific purposes, and they have applications in such areas as signal analysis, data compression, and solution of integral equations. For an application to the problem of using color patterns in the iris of the eye as a means of identification, see the article *Iris Recognition*, by John Daugman, appearing in *American Scientist*, July–August, 2001, pages 326–333.

SECTION 15.5 PROBLEMS

1. Show that $\sigma_{m,n}(t) \cdot \sigma_{m',n'}(t) = 0$ if $(m, n) \neq (m', n')$.

2. On the same set of axes, graph $\sigma_{1,1}(t)$ and $\sigma_{1,2}(t)$. Explain from the graph why these two functions are orthogonal.

3. On the same set of axes, graph $\sigma_{1,3}(t)$ and $\sigma_{-2,1}(t)$. Explain from the graph why these two functions are orthogonal.

4. On the same set of axes, graph $\sigma_{2,1}(t)$ and $\sigma_{1,1}(t)$. Explain from the graph why these two functions are orthogonal.

5. On the same set of axes, graph $\sigma_{-2,-1}(t)$ and $\sigma_{-3,-1}(t)$. Explain from the graph why these two functions are orthogonal.

6. Graph $\psi(t - 3)$.

7. Graph $\psi(2t - 3)$.

8. Graph $\psi(2t + 6)$.

9. Let $f(t) = 4\sigma_{-3,-2}(t) + 6\sigma_{-1,1}(t)$. Write the Fourier series of $f(t)$ on $[-16, 16]$. Graph the fiftieth partial

sum of this series on the same set of axes with a graph of $f(t)$.

10. Let $f(t) = -3\sigma_{2,-2}(t) + 4\sigma_{2,0}(t) + 7\sigma_{1,-1}(t)$. Write the Fourier series of $f(t)$ on $[-2, 2]$. Graph the fiftieth partial sum of this series on the same set of axes with a graph of $f(t)$.

11. Let $f(t) = 3\sigma_{-4,-1}(t) + 8\sigma_{-2,1}(t)$. Write the Fourier series of $f(t)$ on $[-16, 16]$. Graph the fiftieth partial sum of this series on the same set of axes with a graph of $f(t)$.

12. Let $f(t) = \sigma_{-2,-2}(t) + 4\sigma_{1,3}(t) + 2\sigma_{1,-2}(t)$. Write the Fourier series of $f(t)$ on $[-8, 8]$. Graph the fiftieth partial sum of this series on the same set of axes with a graph of $f(t)$.

$$\mu\rho \frac{\partial u}{\partial t} = K\left(\frac{\partial^2 u}{\partial x^2} + \frac{\partial^2 u}{\partial y^2} + \frac{\partial^2 u}{\partial z^2}\right) + \nabla K \cdot \nabla u, \qquad \mu\rho \frac{\partial u}{\partial t} = K\left(\frac{\partial^2 u}{\partial x^2} + \frac{\partial^2 u}{\partial y^2} + \frac{\partial^2 u}{\partial z^2}\right) +$$

PART 6

Partial Differential Equations

CHAPTER 16
The Wave Equation

CHAPTER 17
The Heat Equation

CHAPTER 18
The Potential Equation

CHAPTER 19
Canonical Forms, Existence and Uniqueness of Solutions, and Well-Posed Problems

A differential equation in which partial derivatives occur is called a *partial differential equation*. Mathematical models of physical phenomena involving more than one independent variable often include partial differential equations. They also arise in such diverse areas as epidemiology (for example, multivariable predator/prey models of AIDS), traffic flow studies, and the analysis of economies.

We will be primarily concerned in this part with three broadly defined kinds of phenomena: wave motion, radiation or conduction of energy, and potential theory. Models of these phenomena involve partial differential equations called, respectively, the wave equation, the heat equation, and the potential equation, or Laplace's equation. We will consider each of these in turn, deriving solutions under a variety of boundary and initial conditions describing different settings.

The solution of partial differential equations requires a broad array of mathematical tools, including Fourier series, integrals and transforms, special functions, and eigenfunction expansions. These were covered in Part 5 and can be referred to as needed.

CHAPTER 16

The Wave Equation

16.1   The Wave Equation and Initial and Boundary Conditions

Vibrations in a membrane or drumhead, or oscillations induced in a guitar or violin string, are governed by a partial differential equation called the *wave equation*. We will derive this equation in a simple setting.

Consider an elastic string stretched between two pegs, as on a guitar. We want to describe the motion of the string if it is given a small displacement and released to vibrate in a plane.

Place the string along the x axis from 0 to L and assume that it vibrates in the x, y plane. We want a function $y(x, t)$ such that at any time $t > 0$, the graph of the function $y = y(x, t)$ of x is the shape of the string at that time. Thus, $y(x, t)$ allows us to take a snapshot of the string at any time, showing it as a curve in the plane. For this reason $y(x, t)$ is called the *position function* for the string. Figure 16.1 shows a typical configuration.

To begin with a simple case, neglect damping forces such as air resistance and the weight of the string and assume that the tension $\mathbf{T}(x, t)$ in the string always acts tangentially to the string and that individual particles of the string move only vertically. Also assume that the mass ρ per unit length is constant.

Now consider a typical segment of string between x and $x + \Delta x$ and apply Newton's second law of motion to write

net force on this segment due to the tension

= acceleration of the center of mass of the segment times its mass.

This is a vector equation. For Δx small, the vertical component of this equation (Figure 16.2) gives us approximately

$$T(x + \Delta x, t) \sin(\theta + \Delta\theta) - T(x, t) \sin(\theta) = \rho \Delta x \frac{\partial^2 y}{\partial t^2}(\overline{x}, t),$$

where $\overline{x}$ is the center of mass of the segment and $T(x, t) = \|\mathbf{T}(x, t)\| = $ magnitude of $\mathbf{T}$. Then

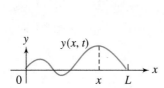

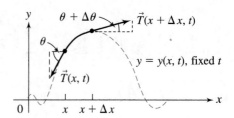

FIGURE 16.1 *String profile at time t.*

FIGURE 16.2

$$\frac{T(x + \Delta x, t)\sin(\theta + \Delta\theta) - T(x, t)\sin(\theta)}{\Delta x} = \rho\frac{\partial^2 y}{\partial t^2}(\overline{x}, t).$$

Now $v(x, t) = T(x, t)\sin(\theta)$ is the vertical component of the tension, so the last equation becomes

$$\frac{v(x + \Delta x, t) - v(x, t)}{\Delta x} = \rho\frac{\partial^2 y}{\partial t^2}(\overline{x}, t).$$

In the limit as $\Delta x \to 0$, we also have $\overline{x} \to x$ and the last equation becomes

$$\frac{\partial v}{\partial x} = \rho\frac{\partial^2 y}{\partial t^2}. \tag{16.1}$$

The horizontal component of the tension is $h(x, t) = T(x, t)\cos(\theta)$, so

$$v(x, t) = h(x, t)\tan(\theta) = h(x, t)\frac{\partial y}{\partial x}.$$

Substitute this into equation (16.1) to get

$$\frac{\partial}{\partial x}\left(h\frac{\partial y}{\partial x}\right) = \rho\frac{\partial^2 y}{\partial t^2}(x, t). \tag{16.2}$$

To compute the left side of this equation, use the fact that the horizontal component of the tension of the segment is zero, so

$$h(x + \Delta x, t) - h(x, t) = 0.$$

Thus h is independent of x and equation (16.2) can be written

$$h\frac{\partial^2 y}{\partial x^2} = \rho\frac{\partial^2 y}{\partial t^2}.$$

Letting $c^2 = h/\rho$, this equation is often written

$$\frac{\partial^2 y}{\partial t^2} = c^2\frac{\partial^2 y}{\partial x^2}.$$

This is the *one-dimensional (1-space dimension) wave equation*. If we use subscript notation for partial derivatives, in which

$$y_x = \frac{\partial y}{\partial x} \quad \text{and} \quad y_t = \frac{\partial y}{\partial t}$$

then the wave equation is

$$y_{tt} = c^2 y_{xx}.$$

This spectacular photo, taken by Ensign John Gay from the U.S.S. Constellation, *shows a shock wave cloud forming over the tail of a U.S. Navy F/A-18 Hornet as it breaks the sound barrier. Current theory is that sound density waves generated by the plane accumulate in a cone at the plane's tail, and a drop in air pressure causes moist air to condense into water droplets there. Shock waves are not yet fully understood, and their mathematical modeling uses advanced techniques from the theory of partial differential equations.*

In order to model the string's motion, we need more than just the wave equation. We must also incorporate information about constraints on the ends of the string and about the initial velocity and position of the string, which will obviously influence the motion.

If the ends of the string are fixed, then

$$y(0, t) = y(L, t) = 0 \quad \text{for } t \geq 0.$$

These are the *boundary conditions*.

The *initial conditions* specify the initial (at time zero) position

$$y(x, 0) = f(x) \quad \text{for } 0 \leq x \leq L$$

and the initial velocity

$$\frac{\partial y}{\partial t}(x, 0) = g(x) \quad \text{for } 0 < x < L,$$

in which f and g are given functions satisfying certain compatibility conditions. For example, if the string is fixed at its ends, then the initial position function must reflect this by satisfying

$$f(0) = f(L) = 0.$$

If the initial velocity is zero (the string is released from rest), then $g(x) = 0$.

The wave equation, together with the boundary and initial conditions, constitute a *boundary value problem* for the position function $y(x, t)$ of the string. These provide enough information to uniquely determine the solution $y(x, t)$.

If there is an external force of magnitude F units of force per unit length acting on the string in the vertical direction, then this derivation can be modified to obtain

$$\frac{\partial^2 y}{\partial t^2} = c^2 \frac{\partial^2 y}{\partial x^2} + \frac{1}{\rho} F.$$

Again, the boundary value problem consists of this wave equation and the boundary and initial conditions.

In 2-space dimensions the wave equation is

$$\frac{\partial^2 z}{\partial t^2} = c^2 \left(\frac{\partial^2 z}{\partial x^2} + \frac{\partial^2 z}{\partial y^2} \right). \tag{16.3}$$

This equation governs vertical displacements $z(x, y, t)$ of a membrane covering a specified region of the plane (for example, vibrations of a drum surface).

Again, boundary and initial conditions must be given to determine a unique solution. Typically, the frame is fixed on a boundary (the rim of the drum surface), so we would have no displacement of points on the boundary:

$$z(x, y, t) = 0 \quad \text{for } (x, y) \text{ on the boundary of the region and } t > 0.$$

Further, the initial displacement and initial velocity must be given. These initial conditions have the form

$$z(x, y, 0) = f(x, y), \quad \frac{\partial z}{\partial t}(x, y, 0) = g(x, y)$$

with f and g given.

We will have occasion to use the two-dimensional wave equation (16.3) expressed in polar coordinates, so we will derive this equation. Let

$$x = r \cos(\theta), \quad y = r \sin(\theta).$$

Then

$$r = \sqrt{x^2 + y^2} \quad \text{and} \quad \theta = \tan^{-1}(y/x).$$

Let

$$z(x, y) = z(r \cos(\theta), r \sin(\theta)) = u(r, \theta).$$

Compute

$$\frac{\partial z}{\partial x} = \frac{\partial u}{\partial r} \frac{\partial r}{\partial x} + \frac{\partial u}{\partial \theta} \frac{\partial \theta}{\partial x}$$

$$= \frac{x}{\sqrt{x^2 + y^2}} \frac{\partial u}{\partial r} - \frac{y}{x^2 + y^2} \frac{\partial u}{\partial \theta}$$

$$= \frac{x}{r} \frac{\partial u}{\partial r} - \frac{y}{r^2} \frac{\partial u}{\partial \theta}.$$

Then

$$\frac{\partial^2 z}{\partial x^2} = \frac{\partial u}{\partial r}\frac{\partial}{\partial x}\left(\frac{x}{r}\right) - \frac{\partial u}{\partial \theta}\frac{\partial}{\partial x}\left(\frac{y}{r^2}\right) + \frac{x}{r}\frac{\partial}{\partial x}\left(\frac{\partial u}{\partial r}\right) - \frac{y}{r^2}\frac{\partial}{\partial x}\left(\frac{\partial u}{\partial \theta}\right)$$

$$= \frac{y^2}{r^3}\frac{\partial u}{\partial r} + \frac{2xy}{r^4}\frac{\partial u}{\partial \theta} + \frac{x^2}{r^2}\frac{\partial^2 u}{\partial r^2} - \frac{2xy}{r^3}\frac{\partial^2 u}{\partial r\partial\theta} + \frac{y^2}{r^4}\frac{\partial^2 u}{\partial\theta^2}.$$

By a similar calculation, we get

$$\frac{\partial z}{\partial y} = \frac{y}{r}\frac{\partial u}{\partial r} + \frac{x}{r^2}\frac{\partial u}{\partial \theta}$$

and

$$\frac{\partial^2 z}{\partial y^2} = \frac{x^2}{r^3}\frac{\partial u}{\partial r} - \frac{2xy}{r^4}\frac{\partial u}{\partial \theta} + \frac{y^2}{r^2}\frac{\partial^2 u}{\partial r^2} + \frac{2xy}{r^3}\frac{\partial^2 u}{\partial r\partial\theta} + \frac{x^2}{r^4}\frac{\partial^2 u}{\partial\theta^2}.$$

Then

$$\frac{\partial^2 z}{\partial x^2} + \frac{\partial^2 z}{\partial y^2} = \frac{\partial^2 u}{\partial r^2} + \frac{1}{r}\frac{\partial u}{\partial r} + \frac{1}{r^2}\frac{\partial^2 u}{\partial\theta^2}.$$

Therefore, in polar coordinates, the two-dimensional wave equation (16.3) is

$$\frac{\partial^2 u}{\partial t^2} = c^2\left(\frac{\partial^2 u}{\partial r^2} + \frac{1}{r}\frac{\partial u}{\partial r} + \frac{1}{r^2}\frac{\partial^2 u}{\partial\theta^2}\right), \tag{16.4}$$

in which $u(r, \theta, t)$ is the vertical displacement of the membrane from the x, y plane at point (r, θ) and time t.

For the rest of this chapter we will solve boundary value problems involving wave motion in a variety of settings, making use of several techniques.

SECTION 16.1 **PROBLEMS**

1. Let $y(x, t) = \sin(n\pi x/L)\cos(n\pi ct/L)$. Show that y satisfies the one-dimensional wave equation for any positive integer n.

2. Show that

$$z(x, y, t) = \sin(nx)\cos(my)\cos(\sqrt{n^2 + m^2}ct)$$

satisfies the two-dimensional wave equation for any positive integers n and m.

3. Let f be any twice-differentiable function of one variable. Show that

$$y(x, t) = \tfrac{1}{2}[f(x + ct) + f(x - ct)]$$

satisfies the one-dimensional wave equation.

4. Show that $y(x, t) = \sin(x)\cos(ct) + (1/c)\cos(x)\sin(ct)$ satisfies the one-dimensional wave equation, together with the boundary conditions

$$y(0, t) = y(2\pi, t) = \frac{1}{c}\sin(ct) \quad \text{for } t > 0$$

and the initial conditions

$$y(x, 0) = \sin(x), \frac{\partial y}{\partial t}(x, 0) = \cos(x) \quad \text{for } 0 < x < 2\pi.$$

5. Formulate a boundary value problem (partial differential equation, boundary and initial conditions) for vibrations of a rectangular membrane occupying a region $0 \le x \le a, 0 \le y \le b$ if the initial position is the graph of $z = f(x, y)$ and the initial velocity (at time zero) is $g(x, y)$. The membrane is fastened to a stiff frame along the rectangular boundary of the region.

6. Formulate a boundary value problem for the motion of an elastic string of length L, fastened at both ends and released from rest with an initial position given by $f(x)$. The string vibrates in the x, y plane. Its motion is opposed by air resistance, which has a force at each point of magnitude proportional to the square of the velocity at that point.

7. Derive the wave equation in 3-space dimensions in spherical coordinates. Recall that the spherical coordinates (ρ, θ, φ) of a point and its rectangular coordinates (x, y, z) are related by $x = \rho \cos(\theta) \sin(\varphi)$, $y = \rho \sin(\theta) \sin(\varphi)$, and $z = \rho \cos(\varphi)$. ρ is the distance from the origin to (x, y, z), θ is the angle between the positive x axis and the projection into the x, y plane of the line from the origin to the point, and φ is the angle of declination from the z axis to the line from the origin to the point.

16.2 Fourier Series Solutions of the Wave Equation

We will begin with problems involving wave motion on a bounded interval. First we will consider the problem when there is an initial displacement but no initial velocity (string released from rest). Following this we will allow an initial velocity but no initial displacement (string given an initial blow, but from its horizontal stretched position). Then we will show how to combine these to allow for both an initial velocity and initial displacement.

16.2.1 Vibrating String with Zero Initial Velocity

Consider an elastic string of length L, fastened at its ends on the x axis at $x = 0$ and $x = L$. The string is displaced, then released from rest to vibrate in the x, y plane. We want to find the displacement function $y(x, t)$, whose graph is a curve in the x, y plane showing the shape of the string at time t. If we took a snapshot of the string at time t, we would see this curve.

The boundary value problem for the displacement function is

$$\frac{\partial^2 y}{\partial t^2} = c^2 \frac{\partial^2 y}{\partial x^2} \quad \text{for } 0 < x < L, t > 0,$$

$$y(0, t) = y(L, t) = 0 \quad \text{for } t \geq 0,$$

$$y(x, 0) = f(x) \quad \text{for } 0 \leq x \leq L,$$

$$\frac{\partial y}{\partial t}(x, 0) = 0 \quad \text{for } 0 \leq x \leq L.$$

The graph of $f(x)$ is the position of the string before release.

The *Fourier method*, or *separation of variables*, consists of attempting a solution of the form $y(x, t) = X(x)T(t)$. Substitute this into the wave equation to obtain

$$XT'' = c^2 X''T,$$

where $T' = dT/dt$ and $X' = dX/dx$. Then

$$\frac{X''}{X} = \frac{T''}{c^2 T}.$$

The left side of this equation depends only on x, and the right only on t. Because x and t are independent, we can choose any t_0 we like and fix the right side of this equation at the constant value $T''(t_0)/c^2 T(t_0)$, while varying x on the left side. Therefore, X''/X must be constant for all x in $(0, L)$. But then $T''/c^2 T$ must equal the same constant for all $t > 0$. Denote this constant $-\lambda$. (The negative sign is customary and convenient, but we would arrive at the same final solution if we used just λ). λ is called the *separation constant*, and we now have

$$\frac{X''}{X} = \frac{T''}{c^2 T} = -\lambda.$$

Then

$$X'' + \lambda X = 0 \quad \text{and} \quad T'' + \lambda c^2 T = 0.$$

The wave equation has separated into two ordinary differential equations.

Now consider the boundary conditions. First,

$$y(0, t) = X(0)T(t) = 0$$

for $t \geq 0$. If $T(t) = 0$ for all $t \geq 0$, then $y(x, t) = 0$ for $0 \leq x \leq L$ and $t \geq 0$. This is indeed the solution if $f(x) = 0$, since in the absence of initial velocity or a driving force, and with zero displacement, the string remains stationary for all time. However, if $T(t) \neq 0$ for any time, then this boundary condition can be satisfied only if

$$X(0) = 0.$$

Similarly,

$$y(L, t) = X(L)T(t) = 0$$

for $t \geq 0$ requires that

$$X(L) = 0.$$

We now have a boundary value problem for X:

$$X'' + \lambda X = 0; \qquad X(0) = X(L) = 0.$$

The values of λ for which this problem has nontrivial solutions are the *eigenvalues* of this problem, and the corresponding nontrivial solutions for X are the *eigenfunctions*. We solved this regular Sturm–Liouville problem in Example 15.8, obtaining the eigenvalues

$$\lambda_n = \frac{n^2 \pi^2}{L^2}.$$

The eigenfunctions are nonzero constant multiples of

$$X_n(x) = \sin\left(\frac{n\pi x}{L}\right)$$

for $n = 1, 2, \ldots$. At this point we therefore have infinitely many possibilities for the separation constant and for $X(x)$.

Now turn to $T(t)$. Since the string is released from rest,

$$\frac{\partial y}{\partial t}(x, 0) = X(x)T'(0) = 0.$$

This requires that $T'(0) = 0$. The problem to be solved for T is therefore

$$T'' + \lambda c^2 T = 0; \qquad T'(0) = 0.$$

However, we now know that λ can take on only values of the form $n^2 \pi^2 / L^2$, so this problem is really

$$T'' + \frac{n^2 \pi^2 c^2}{L^2} T = 0; \qquad T'(0) = 0.$$

The differential equation for T has general solution

$$T(t) = a \cos\left(\frac{n\pi ct}{L}\right) + b \sin\left(\frac{n\pi ct}{L}\right)$$

Now

$$T'(0) = \frac{n\pi c}{L}b = 0,$$

so $b = 0$. We therefore have solutions for $T(t)$ of the form

$$T_n(t) = c_n \cos\left(\frac{n\pi ct}{L}\right)$$

for each positive integer n, with the constants c_n as yet undetermined.

We now have, for $n = 1, 2, \ldots$, functions

$$y_n(x, t) = c_n \sin\left(\frac{n\pi x}{L}\right) \cos\left(\frac{n\pi ct}{L}\right). \tag{16.5}$$

Each of these functions satisfies the wave equation, both boundary conditions, and the initial condition $y_t(x, 0) = 0$. We need to satisfy the condition $y(x, 0) = f(x)$.

It may be possible to choose some n so that $y_n(x, t)$ is the solution for some choice of c_n. For example, suppose the initial displacement is

$$f(x) = 14 \sin\left(\frac{3\pi x}{L}\right).$$

Now choose $n = 3$ and $c_3 = 14$ to obtain the solution

$$y(x, t) = 14 \sin\left(\frac{3\pi x}{L}\right) \cos\left(\frac{3\pi ct}{L}\right).$$

This function satisfies the wave equation, the conditions $y(0) = y(L) = 0$, the initial condition $y(x, 0) = 14 \sin(3\pi x/L)$, and the zero initial velocity condition

$$\frac{\partial y}{\partial t}(x, 0) = 0.$$

However, depending on the initial displacement function, we may not be able to get by simply by picking a particular n and c_n in equation (16.5). For example, if we initially pick the string up in the middle and have initial displacement function

$$f(x) = \begin{cases} x & \text{for } 0 \le x \le \dfrac{L}{2} \\[2ex] L - x & \text{for } \dfrac{L}{2} < x \le L \end{cases}, \tag{16.6}$$

(as in Figure 16.3), then we can never satisfy $y(x, 0) = f(x)$ with one of the y_n's. Even if we try a finite linear combination

$$y(x, t) = \sum_{n=1}^{N} y_n(x, t)$$

we cannot choose $c_1, \ldots, c_N$ to satisfy $y(x, 0) = f(x)$ for this function, since $f(x)$ cannot be written as a finite sum of sine functions.

We are therefore led to attempt an infinite superposition

$$y(x, t) = \sum_{n=1}^{\infty} c_n \sin\left(\frac{n\pi x}{L}\right) \cos\left(\frac{n\pi ct}{L}\right).$$

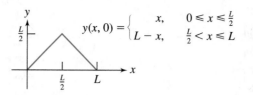

$$y(x, 0) = \begin{cases} x, & 0 \leqslant x \leqslant \frac{L}{2} \\ L - x, & \frac{L}{2} < x \leqslant L \end{cases}$$

FIGURE 16.3

We must choose the c_n's to satisfy

$$y(x, 0) = \sum_{n=1}^{\infty} c_n \sin\left(\frac{n\pi x}{L}\right) = f(x).$$

We can do this! This series is the Fourier sine expansion of $f(x)$ on $[0, L]$. Thus choose the Fourier sine coefficients

$$c_n = \frac{2}{L} \int_0^L f(\xi) \sin\left(\frac{n\pi \xi}{L}\right) d\xi.$$

With this choice, we obtain the solution

$$y(x, t) = \frac{2}{L} \sum_{n=1}^{\infty} \left(\int_0^L f(\xi) \sin\left(\frac{n\pi \xi}{L}\right) d\xi \right) \sin\left(\frac{n\pi x}{L}\right) \cos\left(\frac{n\pi ct}{L}\right). \qquad (16.7)$$

This strategy will work for any initial displacement function f that is continuous with a piecewise continuous derivative on $[0, L]$ and satisfies $f(0) = f(L) = 0$. These conditions ensure that the Fourier sine series of $f(x)$ on $[0, L]$ converges to $f(x)$ for $0 \leq x \leq L$.

In specific instances, where $f(x)$ is given, we can of course explicitly compute the coefficients in this solution. For the initial position function (16.6), compute the coefficients:

$$\frac{2}{L} \int_0^L f(\xi) \sin\left(\frac{n\pi \xi}{L}\right) d\xi$$

$$= \frac{2}{L} \int_0^{L/2} \xi \sin\left(\frac{n\pi \xi}{L}\right) d\xi + \frac{2}{L} \int_{L/2}^L (L - \xi) \sin\left(\frac{n\pi \xi}{L}\right) d\xi$$

$$= 4L \frac{\sin(n\pi/2)}{n^2 \pi^2}.$$

The solution for this initial displacement function, and zero initial velocity, is

$$y(x, t) = \frac{4L}{\pi^2} \sum_{n=1}^{\infty} \frac{\sin(n\pi/2)}{n^2} \sin\left(\frac{n\pi x}{L}\right) \cos\left(\frac{n\pi ct}{L}\right).$$

Since $\sin(n\pi/2) = 0$ if n is even, we can sum over just the odd integers. Further, if $n = 2k - 1$, then

$$\sin(n\pi/2) = \sin((2k - 1)\pi/2) = (-1)^{k+1}.$$

Therefore,

$$y(x, t) = \frac{4L}{\pi^2} \sum_{k=1}^{\infty} \frac{(-1)^{k+1}}{(2k - 1)^2} \sin\left(\frac{(2k - 1)\pi x}{L}\right) \cos\left(\frac{(2k - 1)\pi ct}{L}\right). \qquad (16.8)$$

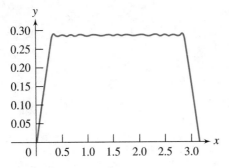

FIGURE 16.4(a) *Approximate string profile at t = 5 seconds.*

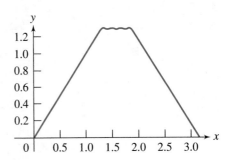

FIGURE 16.4(b) *Profile at t = 6.*

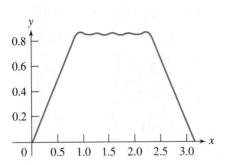

FIGURE 16.4(c) *Profile at t = 7.*

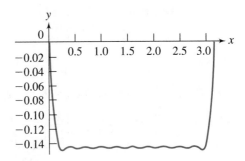

FIGURE 16.4(d) *Profile at t = 8.*

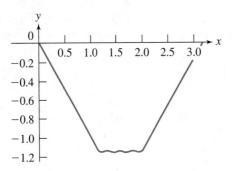

FIGURE 16.4(e) *Profile at t = 9.*

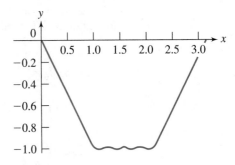

FIGURE 16.4(f) *Profile at t = 10.*

Figures 16.4(a) through (m) shows graphs of partial sums of this series for different values of t, using $L = \pi$ and $c = 1$. These graphs give approximate snapshots of the string at one-second intervals at times $t = 5$ through $t = 17$.

The solution we have derived by separation of variables can be put into the context of Sturm–Liouville theory (Section 15.3). The problem for X, namely

$$X'' + \lambda X = 0; \qquad X(0) = X(L) = 0,$$

is a regular Sturm–Liouville problem, and we found its eigenvalues and corresponding eigenfunctions. The final step in the solution was to expand the initial position function in a series of the eigenfunctions. For this problem this series is the Fourier sine expansion of $f(x)$ on $[0, L]$.

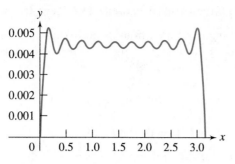

FIGURE 16.4(g) *Profile at t* = 11.

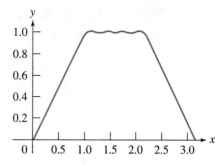

FIGURE 16.4(h) *Profile at t* = 12.

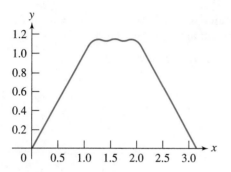

FIGURE 16.4(i) *Profile at t* = 13.

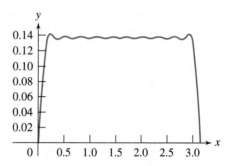

FIGURE 16.4(j) *Profile at t* = 14.

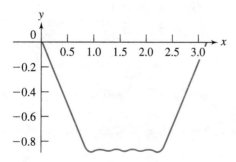

FIGURE 16.4(k) *Profile at t* = 15.

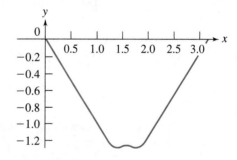

FIGURE 16.4(l) *Profile at t* = 16.

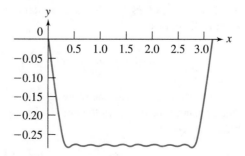

FIGURE 16.4(m) *Profile at t* = 17.

16.2.2   Vibrating String with Given Initial Velocity and Zero Initial Displacement

Now consider the case that the string is released from its horizontal position (zero initial displacement) but with an initial velocity given at x by $g(x)$. The boundary value problem for the displacement function is

$$\frac{\partial^2 y}{\partial t^2} = c^2 \frac{\partial^2 y}{\partial x^2} \quad \text{for } 0 < x < L, t > 0,$$

$$y(0, t) = y(L, t) = 0 \quad \text{for } t \geq 0,$$

$$y(x, 0) = 0 \quad \text{for } 0 \leq x \leq L,$$

$$\frac{\partial y}{\partial t}(x, 0) = g(x) \quad \text{for } 0 \leq x \leq L.$$

We begin as before with separation of variables. Put $y(x, t) = X(x)T(t)$. Since the partial differential equation and boundary conditions are the same as before, we again obtain

$$X'' + \lambda X = 0; \qquad X(0) = X(L) = 0,$$

with eigenvalues

$$\lambda_n = \frac{n^2 \pi^2}{L^2}$$

and eigenfunctions constant multiples of

$$X_n(x) = \sin\left(\frac{n\pi x}{L}\right).$$

Now, however, the problem for T is different and we have

$$y(x, 0) = 0 = X(x)T(0),$$

so $T(0) = 0$. The problem for T is

$$T'' + \frac{n^2 \pi^2 c^2}{L^2} T = 0; \qquad T(0) = 0.$$

(In the case of zero initial velocity we had $T'(0) = 0$.) The general solution of the differential equation for T is

$$T(t) = a \cos\left(\frac{n\pi ct}{L}\right) + b \sin\left(\frac{n\pi ct}{L}\right).$$

Since $T(0) = b = 0$, solutions for $T(t)$ are constant multiples of $\sin(n\pi ct/L)$. Thus, for $n = 1, 2, \ldots$, we have functions

$$y_n(x, t) = c_n \sin\left(\frac{n\pi x}{L}\right) \sin\left(\frac{n\pi ct}{L}\right).$$

Each of these functions satisfies the wave equation, the boundary conditions, and the zero initial displacement condition. To satisfy the initial velocity condition $y_t(x, 0) = g(x)$, we generally must attempt a superposition

$$y(x, t) = \sum_{n=1}^{\infty} c_n \sin\left(\frac{n\pi x}{L}\right) \sin\left(\frac{n\pi ct}{L}\right).$$

Assuming that we can differentiate this series term-by-term, then

$$\frac{\partial y}{\partial t}(x, 0) = \sum_{n=1}^{\infty} c_n \frac{n\pi c}{L} \sin\left(\frac{n\pi x}{L}\right) = g(x).$$

This is the Fourier sine expansion of $g(x)$ on $[0, L]$. Choose the *entire coefficient* of $\sin(n\pi x/L)$ to be the Fourier sine coefficient of $g(x)$ on $[0, L]$:

$$c_n \frac{n\pi c}{L} = \frac{2}{L} \int_0^L g(\xi) \sin\left(\frac{n\pi \xi}{L}\right) d\xi,$$

or

$$c_n = \frac{2}{n\pi c} \int_0^L g(\xi) \sin\left(\frac{n\pi \xi}{L}\right) d\xi.$$

The solution is

$$y(x, t) = \frac{2}{\pi c} \sum_{n=1}^{\infty} \frac{1}{n} \left(\int_0^L g(\xi) \sin\left(\frac{n\pi \xi}{L}\right) d\xi \right) \sin\left(\frac{n\pi x}{L}\right) \sin\left(\frac{n\pi ct}{L}\right). \tag{16.9}$$

For example, suppose the string is released from its horizontal position with an initial velocity given by $g(x) = x(1 + \cos(\pi x/L))$. Compute

$$c_n = \frac{2}{n\pi c} \int_0^L \xi \left(1 + \cos\left(\frac{\pi \xi}{L}\right)\right) \sin\left(\frac{n\pi \xi}{L}\right) d\xi = \begin{cases} \dfrac{2(-1)^n L^2}{n^2 \pi^2 c} \dfrac{1}{n^2 - 1} & \text{for } n = 2, 3, \ldots \\[3mm] \dfrac{3L^2}{2\pi^2 c} & \text{for } n = 1 \end{cases}.$$

The solution for this initial velocity function is

$$y(x, t) = \frac{3L^2}{2\pi^2 c} \sin\left(\frac{\pi x}{L}\right) \sin\left(\frac{\pi ct}{L}\right) + \sum_{n=2}^{\infty} \frac{2(-1)^n L^2}{n^2 \pi^2 c} \frac{1}{n^2 - 1} \sin\left(\frac{n\pi x}{L}\right) \sin\left(\frac{n\pi ct}{L}\right)$$

$$\tag{16.10}$$

If we let $c = 1$ and $L = \pi$, we obtain

$$y(x, t) = \frac{3}{2} \sin(x) \sin(t) + \sum_{n=2}^{\infty} \frac{2(-1)^n}{n^2} \frac{1}{n^2 - 1} \sin(nx) \sin(nt).$$

Figure 16.5 shows graphs of this solution (positions of the string) at times $t = 0.4, 1.2, 1.7, 2.6, 3.5,$ and 4.3.

16.2.3 Vibrating String with Initial Displacement and Velocity

Consider the motion of the string with both initial displacement $f(x)$ and initial displacement $g(x)$.

Formulate two separate problems, the first with initial displacement $f(x)$ and zero initial velocity, and the second with zero initial displacement and initial velocity $g(x)$. We know how to solve both of these. Let $y_1(x, t)$ be the solution of the first problem, and $y_2(x, t)$ the solution of the second. Now let

$$y(x, t) = y_1(x, t) + y_2(x, t).$$

Then y satisfies the wave equation and the boundary conditions. Further,

$$y(x, 0) = y_1(x, 0) + y_2(x, 0) = f(x) + 0 = f(x)$$

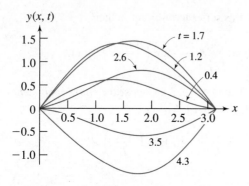

FIGURE 16.5 *String profiles at times $t = 0.4$, 1.2, 1.7, 2.6, 3.5, and 4.3.*

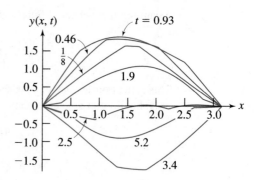

FIGURE 16.6 *Snapshot of the string at times $t = \frac{1}{8}$, 0.46, 0.93, 1.9, 2.5, 3.4, and 5.2.*

and

$$\frac{\partial y}{\partial t}(x, 0) = \frac{\partial y_1}{\partial t}(x, 0) + \frac{\partial y_2}{\partial t}(x, 0) = 0 + g(x) = g(x).$$

Thus $y(x, t)$ is the solution in this case of nonzero initial displacement and velocity functions.

For example, let the initial displacement function be

$$f(x) = \begin{cases} x & \text{for } 0 \le x \le \dfrac{L}{2} \\ L - x & \text{for } \dfrac{L}{2} < x \le L \end{cases},$$

and the initial velocity

$$g(x) = x\left(1 + \cos\left(\frac{\pi x}{L}\right)\right).$$

The solution for the displacement function is the sum of the solution for just displacement $f(x)$, with zero initial velocity, and the solution with zero initial displacement and initial velocity $g(x)$. From equations (16.8) and (16.10),

$$y(x, t) = \frac{4L}{\pi^2} \sum_{k=1}^{\infty} \frac{(-1)^{k+1}}{(2k-1)^2} \sin\left(\frac{(2k-1)\pi x}{L}\right) \cos\left(\frac{(2k-1)\pi ct}{L}\right)$$

$$+ \frac{3L^2}{2\pi^2 c} \sin\left(\frac{\pi x}{L}\right) \sin\left(\frac{\pi ct}{L}\right) + \sum_{n=2}^{\infty} \frac{2(-1)^n L^2}{n^2 \pi^2 c} \frac{1}{n^2 - 1} \sin\left(\frac{n\pi x}{L}\right) \sin\left(\frac{n\pi ct}{L}\right).$$

Figure 16.6 shows graphs of this solution for $t = \frac{1}{8}$, 0.46, 0.93, 1.9, 2.5, 3.4 and 5.2, with $L = \pi$ and $c = 1$. The student can experiment with this motion at fixed times for different values of c to gauge the influence of this constant on the motion.

16.2.4   Verification of Solutions

In the solutions we have obtained thus far we have had to use an infinite series

$$y(x, t) = \sum_{n=1}^{\infty} y_n(x, t)$$

and determine the coefficients in the y_n's by using a Fourier expansion. The question now is whether this infinite sum is indeed a solution of the boundary value problem.

To be specific, consider the problem with initial position function $f(x)$ and zero initial velocity. We derived the proposed solution

$$y(x, t) = \sum_{n=1}^{\infty} c_n \sin\left(\frac{n\pi x}{L}\right) \cos\left(\frac{n\pi ct}{L}\right), \tag{16.11}$$

in which

$$c_n = \frac{2}{L} \int_0^L f(\xi) \sin\left(\frac{n\pi \xi}{L}\right) d\xi.$$

Certainly, $y(0, t) = y(L, t) = 0$, because every term in the series for $y(x, t)$ vanishes at $x = 0$ and at $x = L$. Further, under reasonable conditions on f, the Fourier sine series of $f(x)$ converges to $f(x)$ on $[0, L]$, so $y(x, 0) = f(x)$.

It is not obvious, however, that $y(x, t)$ satisfies the wave equation, even though each term in the series certainly does. The reason for this uncertainty is that we cannot justify term-by-term differentiation of the proposed series solution.

We will now demonstrate a remarkable fact, which has other ramifications as well. We will show that the series in equation (16.11) can be summed in closed form. To do this, first write

$$\sin\left(\frac{n\pi x}{L}\right) \cos\left(\frac{n\pi ct}{L}\right) = \frac{1}{2}\left[\sin\left(\frac{n\pi(x + ct)}{L}\right) + \sin\left(\frac{n\pi(x - ct)}{L}\right)\right].$$

Then equation (16.11) becomes

$$y(x, t) = \frac{1}{2}\left\{\sum_{n=1}^{\infty} c_n \sin\left(\frac{n\pi(x + ct)}{L}\right) + \sum_{n=1}^{\infty} c_n \sin\left(\frac{n\pi(x - ct)}{L}\right)\right\}. \tag{16.12}$$

If the Fourier sine series for $f(x)$ converges to $f(x)$ on $[0, L]$, as might normally be expected of a function that can be a displacement function for a string, then

$$f(x) = \sum_{n=1}^{\infty} c_n \sin\left(\frac{n\pi x}{L}\right)$$

for $0 \leq x \leq L$, and equation (16.12) becomes

$$y(x, t) = \frac{1}{2}[f(x + ct) + f(x - ct)].$$

If f is twice differentiable, we can use the chain rule to verify directly that $y(x, t)$ given by this expression satisfies the wave equation, wherever $f(x + ct)$ and $f(x - ct)$ are defined.

This raises a difficulty, however, since $f(x)$ is defined only for $0 \leq x \leq L$. But t can be any nonnegative number, so the numbers $x + ct$ and $x - ct$ can vary over the entire real line. How then can we evaluate $f(x + ct)$ and $f(x - ct)$?

This difficulty can be overcome in two steps. First, extend f to an odd function f_o defined on $[-L, L]$ by setting

$$f_o(x) = \begin{cases} f(x) & \text{for } 0 \leq x \leq L \\ -f(-x) & \text{for } -L < x < 0 \end{cases}.$$

Notice that $f_o(0) = f_o(L) = f_o(-L) = 0$ because the ends of the string are fixed.

Now extend f_o to a periodic function F of period $2L$ by replicating the graph of f_o on successive intervals $[L, 3L]$, $[3L, 5L]$, ..., $[-3L, -L]$, $[-5L, -3L]$, ... Figure 16.7(a) displays the odd extension of f defined on $[0, L]$ to f_o defined on $[-L, L]$, and Figure 16.7(b) shows the periodic extension of f_o to the real line.

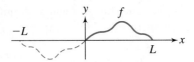

FIGURE 16.7(a) *Odd extension of f to [−L, L].*

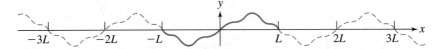

FIGURE 16.7(b) *Periodic extension F of f_o to the real line.*

We now have

$$y(x,t) = \frac{1}{2}[F(x+ct) + F(x-ct)] \tag{16.13}$$

for $0 \le x \le L$ and $t > 0$. Assuming that f is twice differentiable, and that the joins at the ends of intervals where f has been extended to produce F are sufficiently smooth, then F is also twice differentiable, and the chain rule can be used to directly verify that $y(x,t)$ satisfies the wave equation. This is an elegant expression for the solution in terms of the initial displacement function and the number c, which depends on the material from which the string is made. It is reasonable that the motion should be determined by these quantities.

In practice, there will often be finitely many points in $[0, L]$ at which f is not differentiable. For example, $f(x)$ as given by equation (16.6) is not differentiable at $L/2$. In such a case, $y(x,t)$ given by equation (16.13) is the solution in a restricted sense, as there are isolated points at which it does not satisfy all the conditions of the boundary value problem.

Equation (16.13) has an appealing physical interpretation. If we think of $F(x)$ as a wave, then $F(x+ct)$ is this wave translated ct units to the left, and $F(x-ct)$ is the wave translated ct units to the right. The motion of the string (in this case with zero initial velocity) is a sum of two waves, one moving to the right with velocity c, the other to the left with velocity c, and both waves are determined by the initial displacement function. We will say more about this when we discuss d'Alembert's solution for the motion of an infinitely long string.

16.2.5 Transformation of Boundary Value Problems Involving the Wave Equation

There are boundary value problems involving the wave equation for which separation of variables does not lead to the solution. This can occur because of the form of the wave equation (for example, there may be an external forcing term), or because of the form of the boundary conditions. Here is an example of such a problem and a strategy for overcoming the difficulty.

Consider the boundary value problem

$$\frac{\partial^2 y}{\partial t^2} = \frac{\partial^2 y}{\partial x^2} + Ax \quad \text{for } 0 < x < L, t > 0,$$

$$y(0,t) = y(L,t) = 0 \quad \text{for } t \ge 0,$$

$$y(x,0) = 0, \frac{\partial y}{\partial t}(x,0) = 1 \quad \text{for } 0 < x < L.$$

A is a positive constant. The term Ax in the wave equation represents an external force which at x has magnitude Ax. We have let $c = 1$ in this problem.

If we put $y(x, t) = X(x)T(t)$ into the partial differential equation, we get

$$XT'' = X''T + Ax,$$

and there is no way to separate the t dependency on one side of the equation and the x dependent terms on the other.

We will transform this problem into one for which separation of variables works. Let

$$y(x, t) = Y(x, t) + \psi(x).$$

The idea is to choose ψ to reduce the given problem to one we have already solved. Substitute $y(x, t)$ into the partial differential equation to get

$$\frac{\partial^2 Y}{\partial t^2} = \frac{\partial^2 Y}{\partial x^2} + \psi''(x) + Ax.$$

This will be simplified if we choose ψ so that

$$\psi''(x) + Ax = 0.$$

There are many such choices. By integrating twice, we get

$$\psi(x) = -A\frac{x^3}{6} + Cx + D,$$

with C and D constants we can still choose any way we like. Now look at the boundary conditions. First,

$$y(0, t) = Y(0, t) + \psi(0) = 0.$$

This will be just $y(0, t) = Y(0, t)$ if we choose

$$\psi(0) = D = 0.$$

Next,

$$y(L, t) = Y(L, t) + \psi(L) = Y(L, t) - A\frac{L^3}{6} + CL = 0.$$

This will reduce to $y(L, t) = Y(L, t)$ if we choose C so that

$$\psi(L) = -A\frac{L^3}{6} + CL = 0$$

or

$$C = \frac{1}{6}AL^2.$$

This means that

$$\psi(x) = -\frac{1}{6}Ax^3 + \frac{1}{6}AL^2x = \frac{1}{6}Ax\left(L^2 - x^2\right).$$

With this choice of ψ,

$$Y(0, t) = Y(L, t) = 0.$$

Now relate the initial conditions for y to initial conditions for Y. First,

$$Y(x, 0) = y(x, 0) - \psi(x) = -\psi(x) = \frac{1}{6}Ax(x^2 - L^2)$$

and

$$\frac{\partial Y}{\partial t}(x, 0) = \frac{\partial y}{\partial t}(x, 0) = 1.$$

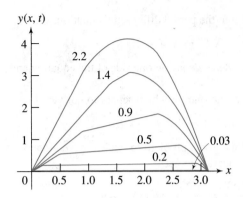

FIGURE 16.8(a) *Position of the string at times*
t = 0.03, 0.2, 0.5, 0.9, 1.4, and 2.2.

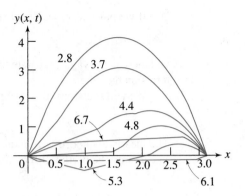

FIGURE 16.8(b) *Position at times t = 2.8, 3.7,*
4.4, 4.8, 5.3, 6.1, and 6.7.

We now have a boundary value problem for $Y(x, t)$:

$$\frac{\partial^2 Y}{\partial t^2} = \frac{\partial^2 Y}{\partial x^2} \quad \text{for } 0 < x < L, t > 0,$$

$$Y(0, t) = 0, Y(L, t) = 0 \quad \text{for } t > 0,$$

$$Y(x, 0) = \frac{1}{6}Ax(x^2 - L^2), \frac{\partial Y}{\partial t}(x, 0) = 1 \quad \text{for } 0 < x < L.$$

Using equations (16.7) and (16.9), we immediately write the solution

$$Y(x, t) = \frac{2}{L}\sum_{n=1}^{\infty}\left(\int_0^L \frac{1}{6}A\xi(\xi^2 - L^2)\sin\left(\frac{n\pi\xi}{L}\right)d\xi\right)\sin\left(\frac{n\pi x}{L}\right)\cos\left(\frac{n\pi t}{L}\right)$$

$$+ \frac{2}{\pi}\sum_{n=1}^{\infty}\frac{1}{n}\left(\int_0^L \sin\left(\frac{n\pi\xi}{L}\right)d\xi\right)\sin\left(\frac{n\pi x}{L}\right)\sin\left(\frac{n\pi t}{L}\right)$$

$$= \frac{2AL^3}{\pi^3}\sum_{n=1}^{\infty}\frac{(-1)^n}{n^3}\sin\left(\frac{n\pi x}{L}\right)\cos\left(\frac{n\pi t}{L}\right)$$

$$+ \frac{2L}{\pi^2}\sum_{n=1}^{\infty}\frac{1 - (-1)^n}{n^2}\sin\left(\frac{n\pi x}{L}\right)\sin\left(\frac{n\pi t}{L}\right).$$

The solution of the original problem is

$$y(x, t) = Y(x, t) + \frac{1}{6}Ax\left(L^2 - x^2\right).$$

Figure 16.8(a) shows snapshots of the string at times $t = 0.03, 0.2, 0.5, 0.9, 1.4,$ and 2.2.
Figure 16.8(b) has times $2.8, 3.7, 4.4, 4.8, 5.3, 6.1,$ and 6.7. These use $L = \pi$ and $c = 1$.

16.2.6 Effects of Initial Conditions and Constants on the Motion

Using separation of variables, we have obtained series solutions of problems involving the vibrating string on a bounded interval. It is interesting to examine the effects that constants occuring in the problem have on the solution. We begin with an example investigating the effect of the constant c in the motion of the string.

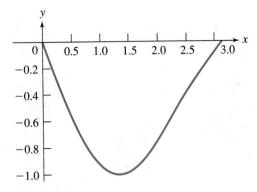

FIGURE 16.9(a) $t = 5.3$ *and* $c = 1.05$.

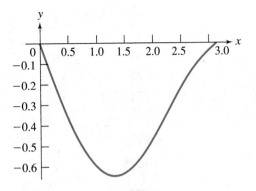

FIGURE 16.9(b) $t = 5.3$ *and* $c = 1.1$.

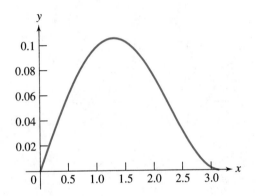

FIGURE 16.9(c) $t = 5.3$ *and* $c = 1.2$.

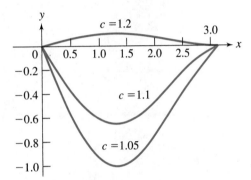

FIGURE 16.9(d) *String profile at time* $t = 5.3$ *with c having values* 1.05, 1.1, *and* 1.2.

EXAMPLE 16.1

Consider again the problem of the wave equation with zero initial displacement and initial velocity given by

$$g(x) = x \left(1 + \cos \left(\frac{\pi x}{L} \right) \right).$$

The solution obtained previously, with $L = \pi$, is

$$y(x, t) = \frac{3}{2c} \sin(x) \sin(ct) + \sum_{n=2}^{\infty} \frac{2(-1)^n}{n^2 c} \frac{1}{n^2 - 1} \sin(nx) \sin(nct)$$

Figure 16.5 showed graphs of the string's position at different times, with $c = 1$. Now we want to focus on how c influences the motion. Figure 16.9(a) shows the string profile at time $t = 5.3$ seconds, with $c = 1.05$. Figures 16.9(b) and (c) show the profile at the same time, but with $c = 1.1$ and $c = 1.2$, respectively. These graphs are placed on the same set of axes for comparison in Figure 16.9(d). The student is invited to select other times and graph the solution for different values of c. ∎

Next, consider a problem in which the initial data of the problem depend on a parameter.

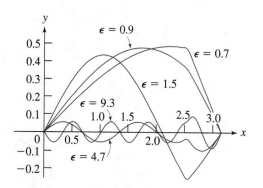

FIGURE 16.10(a) *String profiles at t = 0.5 for ε equal to 0.7, 0.9, 1.5, 4.7, and 9.3.*

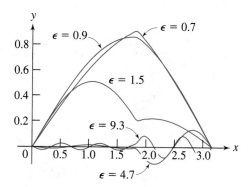

FIGURE 16.10(b) *String profiles at t = 1.1.*

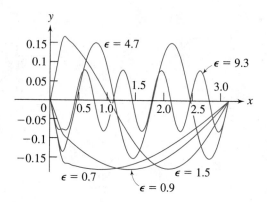

FIGURE 16.10(c) *String profiles at t = 2.8.*

EXAMPLE 16.2

Consider the problem

$$\frac{\partial^2 y}{\partial t^2} = 1.44 \frac{\partial^2 y}{\partial x^2} \quad \text{for } 0 < x < \pi, t > 0,$$

$$y(0, t) = y(\pi, t) = 0 \quad \text{for } t \geq 0,$$

$$y(x, 0) = 0, \frac{\partial y}{\partial t}(x, 0) = \sin(\epsilon x) \quad \text{for } 0 < x < \pi,$$

in which ϵ is a positive number that is not an integer. It is routine to write the solution

$$y(x, t) = \frac{5}{3\pi} \sum_{n=1}^{\infty} \frac{\sin(\pi \epsilon)(-1)^{n+1}}{n^2 - \epsilon^2} \sin(nx) \sin(1.2nt).$$

Now compare graphs of this solution at various times, with different choices of ϵ. Figure 16.10(a) shows the string profile at $t = 0.5$ for ϵ equal to 0.7, 0.9, 1.5, 4.7, and 9.3. Figure 16.10(b) shows the graphs for these values of ϵ at $t = 1.1$, and Figure 16.10(c) shows the graphs at $t = 2.8$. We can also follow the motion of the string at different times for the same value of ϵ. Figure 16.11(a)

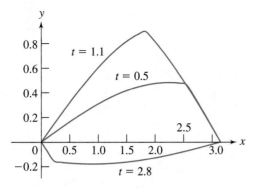

FIGURE 16.11(a) *Graphs of the string with* $\epsilon = 0.7$ *for times* $t = 0.5$, 1.1, *and* 2.8.

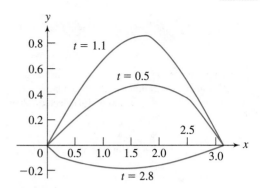

FIGURE 16.11(b) $\epsilon = 0.9$.

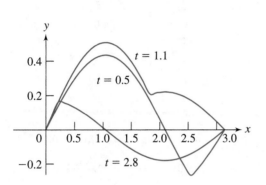

FIGURE 16.11(c) $\epsilon = 1.5$.

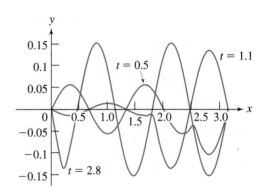

FIGURE 16.11(d) $\epsilon = 4.7$.

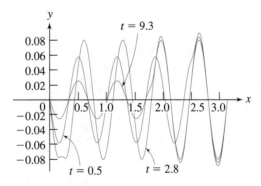

FIGURE 16.11(e) $\epsilon = 9.3$.

shows the string profiles for $\epsilon = 0.7$ at times $t = 0.5$, 1.1, and 2.8. Figures 16.11(b), (c), (d), and (e) each show the string profile for a given ϵ and for these three times. ∎

In some of the following exercises we will ask the student to employ a graphics package to exhibit string profiles at different times and under different conditions.

SECTION 16.2 PROBLEMS

In each of Problems 1 through 10, solve the boundary value problem using separation of variables. Graph some of the partial sums of the series solution, for selected values of the time.

1.
$$\frac{\partial^2 y}{\partial t^2} = c^2 \frac{\partial^2 y}{\partial x^2} \quad \text{for } 0 < x < 2, t > 0,$$

$$y(0, t) = y(2, t) = 0 \quad \text{for } t \geq 0,$$

$$y(x, 0) = 0, \frac{\partial y}{\partial t}(x, 0) = g(x) \quad \text{for } 0 \leq x \leq 2,$$

where $g(x) = \begin{cases} 2x & \text{for } 0 \leq x \leq 1 \\ 0 & \text{for } 1 < x < 2 \end{cases}$

2.
$$\frac{\partial^2 y}{\partial t^2} = 9 \frac{\partial^2 y}{\partial x^2} \quad \text{for } 0 < x < 4, t > 0,$$

$$y(0, t) = y(4, t) = 0 \quad \text{for } t \geq 0,$$

$$y(x, 0) = 2\sin(\pi x), \frac{\partial y}{\partial t}(x, 0) = 0 \quad \text{for } 0 \leq x \leq 4$$

3.
$$\frac{\partial^2 y}{\partial t^2} = 4 \frac{\partial^2 y}{\partial x^2} \quad \text{for } 0 < x < 3, t > 0,$$

$$y(0, t) = y(3, t) = 0 \quad \text{for } t \geq 0,$$

$$y(x, 0) = 0, \frac{\partial y}{\partial t}(x, 0) = x(3 - x) \quad \text{for } 0 \leq x \leq 3$$

4.
$$\frac{\partial^2 y}{\partial t^2} = 9 \frac{\partial^2 y}{\partial x^2} \quad \text{for } 0 < x < \pi, t > 0,$$

$$y(0, t) = y(\pi, t) = 0 \quad \text{for } t \geq 0,$$

$$y(x, 0) = \sin(x), \frac{\partial y}{\partial t}(x, 0) = 1 \quad \text{for } 0 \leq x \leq \pi$$

5.
$$\frac{\partial^2 y}{\partial t^2} = 8 \frac{\partial^2 y}{\partial x^2} \quad \text{for } 0 < x < 2\pi, t > 0,$$

$$y(0, t) = y(2\pi, t) = 0 \quad \text{for } t \geq 0,$$

$$y(x, 0) = f(x), \frac{\partial y}{\partial t}(x, 0) = 0 \quad \text{for } 0 \leq x \leq 2\pi,$$

where

$$f(x) = \begin{cases} 3x & \text{for } 0 \leq x \leq \pi \\ 6\pi - 3x & \text{for } \pi < x \leq 2\pi \end{cases}$$

6.
$$\frac{\partial^2 y}{\partial t^2} = 4 \frac{\partial^2 y}{\partial x^2} \quad \text{for } 0 < x < 5, t > 0,$$

$$y(0, t) = y(5, t) = 0 \quad \text{for } t \geq 0,$$

$$y(x, 0) = 0, \frac{\partial y}{\partial t}(x, 0) = g(x) \quad \text{for } 0 \leq x \leq 4,$$

where

$$g(x) = \begin{cases} 0 & \text{for } 0 \leq x < 4 \\ 5 - x & \text{for } 4 \leq x \leq 5 \end{cases}$$

7.
$$\frac{\partial^2 y}{\partial t^2} = 9 \frac{\partial^2 y}{\partial x^2} \quad \text{for } 0 < x < 2, t > 0,$$

$$y(0, t) = y(2, t) = 0 \quad \text{for } t \geq 0,$$

$$y(x, 0) = x(x - 2), \frac{\partial y}{\partial t}(x, 0) = g(x) \quad \text{for } 0 \leq x \leq 2,$$

where

$$g(x) = \begin{cases} 0 & \text{for } 0 \leq x < \frac{1}{2} \text{ and for } 1 < x \leq 2 \\ 3 & \text{for } \frac{1}{2} \leq x \leq 1 \end{cases}$$

8.
$$\frac{\partial^2 y}{\partial t^2} = 16 \frac{\partial^2 y}{\partial x^2} \quad \text{for } 0 < x < 5, t > 0,$$

$$y(0, t) = y(5, t) = 0 \quad \text{for } t \geq 0,$$

$$y(x, 0) = -3x^2(x - 5), \frac{\partial y}{\partial t}(x, 0) = 0 \quad \text{for } 0 \leq x \leq 5$$

9.
$$\frac{\partial^2 y}{\partial t^2} = 8 \frac{\partial^2 y}{\partial x^2} \quad \text{for } 0 < x < 4, t > 0,$$

$$y(0, t) = y(4, t) = 0 \quad \text{for } t \geq 0,$$

$$y(x, 0) = x^2(x - 4), \frac{\partial y}{\partial t}(x, 0) = 1 \quad \text{for } 0 \leq x \leq 4$$

10.
$$\frac{\partial^2 y}{\partial t^2} = 25 \frac{\partial^2 y}{\partial x^2} \quad \text{for } 0 < x < \pi, t > 0,$$

$$y(0, t) = y(\pi, t) = 0 \quad \text{for } t \geq 0,$$

$$y(x, 0) = \sin(2x), \frac{\partial y}{\partial t}(x, 0) = \pi - x \quad \text{for } 0 \leq x \leq \pi$$

11. Solve the boundary value problem

$$\frac{\partial^2 y}{\partial t^2} = 3 \frac{\partial^2 y}{\partial x^2} + 2x \quad \text{for } 0 < x < 2, t > 0,$$

$$y(0, t) = y(2, t) = 0 \quad \text{for } t \geq 0,$$

$$y(x, 0) = 0, \frac{\partial y}{\partial t}(x, 0) = 0 \quad \text{for } 0 \leq x \leq 2.$$

Graph some partial sums of the series solution. *Hint:* Upon putting $y(x, t) = X(x)T(t)$, we find that the variables do not separate. Put $Y(x, t) = y(x, t) + h(x)$

and choose h to obtain a boundary value problem that can be solved by Fourier series.

12. Solve

$$\frac{\partial^2 y}{\partial t^2} = 9\frac{\partial^2 y}{\partial x^2} + x^2 \quad \text{for } 0 < x < 4, t > 0,$$

$$y(0, t) = y(4, t) = 0 \quad \text{for } t \geq 0,$$

$$y(x, 0) = 0, \frac{\partial y}{\partial t}(x, 0) = 0 \quad \text{for } 0 \leq x \leq 4.$$

Graph some partial sums of the solution for values of t.

13. Solve

$$\frac{\partial^2 y}{\partial t^2} = \frac{\partial^2 y}{\partial x^2} - \cos(x) \quad \text{for } 0 < x < 2\pi, t > 0,$$

$$y(0, t) = y(2\pi, t) = 0 \quad \text{for } t \geq 0,$$

$$y(x, 0) = 0, \frac{\partial y}{\partial t}(x, 0) = 0 \quad \text{for } 0 \leq x \leq 2\pi.$$

Graph some partial sums of the solution for selected values of the time.

14. Solve the vibrating string problem when there is damping proportional to the velocity. With zero initial velocity and initial displacement function $f(x)$, the boundary value problem is

$$\frac{\partial^2 y}{\partial t^2} = c^2\frac{\partial^2 y}{\partial x^2} - \alpha\frac{\partial y}{\partial t} \quad \text{for } 0 < x < L, t > 0,$$

$$y(0, t) = y(L, t) = 0 \quad \text{for } t \geq 0,$$

$$y(x, 0) = f(x), \frac{\partial y}{\partial t}(x, 0) = 0 \quad \text{for } 0 \leq x \leq L.$$

Assume that $0 < \alpha < 2\pi c/L$.

15. Solve the following problem, which models longitudinal displacements in a cylindrical bar of length L, if, at time zero, the bar is stretched by an amount AL and released:

$$\frac{\partial^2 y}{\partial t^2} = c^2\frac{\partial^2 y}{\partial x^2} \quad \text{for } 0 < x < L, t > 0,$$

$$\frac{\partial y}{\partial x}(0, t) = \frac{\partial y}{\partial x}(L, t) = 0 \quad \text{for } t \geq 0,$$

$$y(x, 0) = (A + 1)x, \frac{\partial y}{\partial t}(x, 0) = 0 \quad \text{for } 0 \leq x \leq L.$$

In this model, $c^2 = E/\delta$, where E is the modulus of elasticity of the bar and δ is the mass per unit volume, assumed constant.

16. Transverse vibrations in a homogeneous rod of length π are modeled by the partial differential equation

$$a^2\frac{\partial^4 u}{\partial x^4} + \frac{\partial^2 u}{\partial t^2} = 0 \quad \text{for } 0 < x < \pi, t > 0.$$

Here $u(x, t)$ is the displacement at time t of the cross section through x perpendicular to the x axis, and $a^2 = EI/\rho A$, where E is Young's modulus, I is the moment of inertia of a cross section perpendicular to the x axis, ρ is the constant density, and A the cross-sectional area, assumed constant.

(a) Let $u(x, t) = X(x)T(t)$ to separate the variables.

(b) Solve for values of the separation constant and for X and T in the case of free ends:

$$\frac{\partial^2 u}{\partial x^2}(0, t) = \frac{\partial^2 u}{\partial x^2}(\pi, t) = \frac{\partial^3 u}{\partial x^3}(0, t) = \frac{\partial^3 u}{\partial x^3}(\pi, t) = 0$$

for $t > 0$.

(c) Solve for values of the separation constant and for X and T in the case of supported ends:

$$u(0, t) = u(\pi, t) = \frac{\partial^2 u}{\partial x^2}(0, t) = \frac{\partial^2 u}{\partial x^2}(\pi, t) = 0$$

for $t > 0$.

17. Solve the *telegraph equation*

$$\frac{\partial^2 u}{\partial t^2} + A\frac{\partial u}{\partial t} + Bu = c^2\frac{\partial^2 u}{\partial x^2} \quad \text{for } 0 < x < L, t > 0.$$

Here A and B are positive constants. The boundary conditions are

$$u(0, t) = u(L, t) = 0 \quad \text{for } t \geq 0.$$

The initial conditions are

$$u(x, 0) = f(x), \frac{\partial u}{\partial t}(x, 0) = 0 \quad \text{for } 0 \leq x \leq L.$$

Assume that $A^2 L^2 < 4(BL^2 + c^2\pi^2)$.

18. The current $i(x, t)$ and voltage $v(x, t)$ at time t and distance x from one end of a transmission line satisfy the system

$$-\frac{\partial v}{\partial x} = Ri + L\frac{\partial i}{\partial t},$$

$$-\frac{\partial i}{\partial x} = Sv + K\frac{\partial v}{\partial t},$$

where R is the resistance, L the inductance, S the leakage conductance, and K the capacitance to ground, all per unit length and all assumed constant. By differentiating appropriately and eliminating i, show that v satisfies the telegraph equation (Problem 17). Similarly, show that i also satisfies the telegraph equation.

19. Let $\theta(x, t)$ be the angular displacement at time t of the cross section at x of a homogeneous cylindrical shell about the x axis. Then

$$\frac{\partial^2 \theta}{\partial t^2} = c^2\frac{\partial^2 \theta}{\partial x^2} \quad \text{for } 0 < x < L, t > 0.$$

Solve this equation subject to

$$\theta(x, 0) = f(x), \frac{\partial \theta}{\partial t}(x, 0) = 0 \quad \text{for } 0 \le x \le L$$

if the ends of the shell are fixed elastically:

$$\frac{\partial \theta}{\partial x}(0, t) - \alpha \theta(0, t) = 0, \frac{\partial \theta}{\partial x} + \alpha \theta(L, t) = 0 \quad \text{for } t \ge 0,$$

with α a positive constant.

20. Consider the boundary value problem

$$\frac{\partial^2 y}{\partial t^2} = 9 \frac{\partial^2 y}{\partial x^2} + 5x^3 \quad \text{for } 0 < x < 4, t > 0,$$

$$y(0, t) = y(4, t) = 0 \quad \text{for } t \ge 0$$

$$y(x, 0) = \cos(\pi x), \frac{\partial y}{\partial t}(x, 0) = 0 \quad \text{for } 0 \le x \le 4.$$

(a) Write a series solution.

(b) Find a series solution when the term $5x^3$ is removed from the wave equation.

(c) In order to gauge the effect of the forcing term on the motion, graph the 40th partial sum of the solution for (a) and (b) on the same set of axes at time $t = 0.4$ seconds. Repeat this procedure successively for times $t = 0.8, 1.4, 2, 2.5, 3,$ and 4 seconds.

21. Consider the boundary value problem

$$\frac{\partial^2 y}{\partial t^2} = 9 \frac{\partial^2 y}{\partial x^2} + \cos(\pi x) \quad \text{for } 0 < x < 4, t > 0,$$

$$y(0, t) = y(4, t) = 0 \quad \text{for } t \ge 0$$

$$y(x, 0) = x(4 - x), \frac{\partial y}{\partial t}(x, 0) = 0 \quad \text{for } 0 \le x \le 4.$$

(a) Write a series solution.

(b) Find a series solution when the term $\cos(\pi x)$ is removed from the wave equation.

(c) In order to gauge the effect of the forcing term on the motion, graph the 40th partial sum of the solution for (a) and (b) on the same set of axes at time $t = 0.6$ seconds. Repeat this procedure for $t = 0.5, 4.9, 6.4,$ 9.8, and other times.

22. Consider the boundary value problem

$$\frac{\partial^2 y}{\partial t^2} = 9 \frac{\partial^2 y}{\partial x^2} - e^{-x} \quad \text{for } 0 < x < 4, t > 0,$$

$$y(0, t) = y(4, t) = 0 \quad \text{for } t \ge 0$$

$$y(x, 0) = \sin(\pi x), \frac{\partial y}{\partial t}(x, 0) = 0 \quad \text{for } 0 \le x \le 4.$$

(a) Write a series solution.

(b) Find a series solution when the term e^{-x} is removed from the wave equation.

(c) In order to gauge the effect of the forcing term on the motion, graph the 40th partial sum of the solution for (a) and (b) on the same set of axes at time $t = 0.6$ seconds. Repeat this procedure successively for times $t = 1, 1.4, 2, 3, 5,$ and 7 seconds.

23. Consider the boundary value problem

$$\frac{\partial^2 y}{\partial t^2} = 9 \frac{\partial^2 y}{\partial x^2} + x^2 \quad \text{for } 0 < x < 1, t > 0,$$

$$y(0, t) = y(1, t) = 0 \quad \text{for } t \ge 0$$

$$y(x, 0) = 0, \frac{\partial y}{\partial t}(x, 0) = 1 \quad \text{for } 0 \le x \le 1.$$

(a) Write a series solution.

(b) Find a series solution when the term $4x$ is removed from the wave equation.

(c) In order to gauge the effect of the forcing term on the motion, graph the 40th partial sum of the solution for (a) and (b) on the same set of axes at time $t = 0.6$ seconds. Repeat this procedure successively for times $t = 0.5, 1.4, 2, 2.5, 4,$ and 7 seconds.

24. Consider the boundary value problem

$$\frac{\partial^2 y}{\partial t^2} = 9 \frac{\partial^2 y}{\partial x^2} + xe^{-x} \quad \text{for } 0 < x < 4, t > 0,$$

$$y(0, t) = 0, y(4, t) = 0 \quad \text{for } t \ge 0$$

$$y(x, 0) = x(4 - x), \frac{\partial y}{\partial t}(x, 0) = x \quad \text{for } 0 \le x \le 4.$$

(a) Write a series solution.

(b) Find a series solution when the condition xe^{-x} is removed from the wave equation.

(c) Graph the 40th partial sum of the solution for (a) and (b) on the same set of axes at time $t = 0.6$ second. Repeat this procedure successively for times $t = 0.3,$ 1.2, 2.5, 3, 5, and 7 seconds.

25. Consider the boundary value problem

$$\frac{\partial^2 y}{\partial t^2} = 4 \frac{\partial^2 y}{\partial x^2} + 2 \quad \text{for } 0 < x < 9, t > 0,$$

$$y(0, t) = 0, y(9, t) = 0 \quad \text{for } t \ge 0$$

$$y(x, 0) = 0, \frac{\partial y}{\partial t}(x, 0) = x \quad \text{for } 0 \le x \le 9.$$

(a) Write a series solution.

(b) Find a series solution when the term 2 is removed from the wave equation.

(c) Graph the 40th partial sum of the solution for (a) and (b) on the same set of axes at time $t = 0.6$ seconds. Repeat this procedure for times $t = 0.5, 1.2, 2.6, 3.5,$ 5, and 8 seconds.

16.3 Wave Motion Along Infinite and Semi-infinite Strings

16.3.1 Wave Motion Along an Infinite String

If long distances are involved (such as with sound waves in the ocean used to monitor temperature changes), wave motion is sometimes modeled by an infinite string, in which case there are no boundary conditions. As with the finite string, we will consider separately the cases of zero initial velocity and zero initial displacement.

Zero Initial Velocity
Consider the initial value problem

$$\frac{\partial^2 y}{\partial t^2} = c^2 \frac{\partial^2 y}{\partial x^2} \quad \text{for } -\infty < x < \infty, t > 0$$

$$y(x, 0) = f(x), \frac{\partial y}{\partial t}(x, 0) = 0 \quad \text{for } \infty < x < \infty.$$

There are no boundary conditions, but we will impose the condition that the solution be a bounded function.

To separate the variables, let $y(x, t) = X(x)T(x)$ and obtain, as before,

$$X'' + \lambda X = 0, \quad T'' + \lambda c^2 T = 0.$$

Consider cases on λ.

Case 1 $\quad \lambda = 0$
Now $X(x) = ax + b$. This is a bounded solution if $a = 0$. Thus $\lambda = 0$ is an eigenvalue, with nonzero constant eigenfunctions.

Case 2 $\quad \lambda < 0$
Write $\lambda = -\omega^2$ with $\omega > 0$. Then $X'' - \omega^2 X = 0$, with general solution

$$X(x) = ae^{\omega x} + be^{-\omega x}.$$

But $e^{\omega x}$ is unbounded on $(0, \infty)$, so we must choose $a = 0$. And $e^{-\omega x}$ is unbounded on $(-\infty, 0)$, so we must choose $b = 0$, leaving only the zero solution. This problem has no negative eigenvalue.

Case 3 $\quad \lambda > 0$, say $\lambda = \omega^2$ with $\omega > 0$.
Now $X'' + \omega^2 X = 0$, with general solution

$$X_\omega(x) = a_\omega \cos(\omega x) + b_\omega \sin(\omega x).$$

These functions are bounded for all $\omega > 0$. Thus every positive number $\lambda = \omega^2$ is an eigenvalue, with corresponding eigenfunction $a \cos(\omega x) + b \sin(\omega x)$ for a and b not both zero.

We can include Case 1 in Case 3, since $a \cos(\omega x) + b \sin(\omega x) = $ constant if $\omega = 0$.

Now consider the equation for T, which we can now write as $T'' + c^2 \omega^2 T = 0$ for $\omega \geq 0$. This has general solution

$$T(t) = a \cos(\omega c t) + b \sin(\omega c t).$$

Now

$$\frac{\partial y}{\partial t}(x, 0) = X(t)T'(0) = X(t)\omega c b = 0,$$

so $b = 0$. Thus solutions for T are constant multiples of

$$T_\omega(t) = \cos(\omega ct).$$

For any $\omega \geq 0$, we now have a function

$$y_\omega(x, t) = X_\omega(x)T_\omega(t) = [a_\omega \cos(\omega x) + b_\omega \sin(\omega x)] \cos(\omega ct)$$

which satisfies the wave equation and the condition

$$\frac{\partial y}{\partial t}(x, 0) = 0.$$

We need to satisfy the condition $y(x, 0) = f(x)$. For the similar problem on $[0, L]$, we had a function $y_n(x, t)$ for each positive integer n, and we attempted a superposition $\sum_{n=1}^{\infty} y_n(x, t)$. Now the eigenvalues fill out the entire nonnegative real line, so replace $\sum_{n=1}^{\infty}$ with $\int_0^{\infty} \cdots d\omega$ in forming the superposition:

$$y(x, t) = \int_0^{\infty} y_\omega(x, t)\, d\omega = \int_0^{\infty} [a_\omega \cos(\omega x) + b_\omega \sin(\omega x)] \cos(\omega ct)\, d\omega. \qquad (16.14)$$

The initial displacement condition requires that

$$y(x, 0) = \int_0^{\infty} [a_\omega \cos(\omega x) + b_\omega \sin(\omega x)]\, d\omega = f(x).$$

The integral on the left is the Fourier integral representation of $f(x)$ for $-\infty < x < \infty$. Thus choose the constants as the Fourier integral coefficients:

$$a_\omega = \frac{1}{\pi} \int_{-\infty}^{\infty} f(\xi) \cos(\omega\xi)\, d\xi$$

and

$$b_\omega = \frac{1}{\pi} \int_{-\infty}^{\infty} f(\xi) \sin(\omega\xi)\, d\xi.$$

With this choice of the coefficients, and certain conditions on f (see the convergence theorem for Fourier integrals in Section 14.1), equation (16.14) is the solution of the problem.

EXAMPLE 16.3

Consider the problem

$$\frac{\partial^2 y}{\partial t^2} = c^2 \frac{\partial^2 y}{\partial x^2} \quad \text{for } -\infty < x < \infty, t > 0$$

$$y(x, 0) = e^{-|x|}, \frac{\partial y}{\partial t}(x, 0) = 0 \quad \text{for } \infty < x < \infty.$$

A graph of the initial position of the string is given in Figure 16.12.

To use equation (16.14), compute the Fourier integral coefficients:

$$a_\omega = \frac{1}{\pi} \int_{-\infty}^{\infty} e^{-|\xi|} \cos(\omega\xi)\, d\xi = \frac{2}{\pi (1 + \omega^2)}$$

and

$$b_\omega = \frac{1}{\pi} \int_{-\infty}^{\infty} e^{-|\xi|} \sin(\omega\xi)\, d\xi = 0.$$

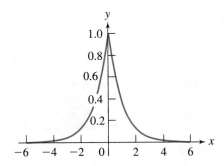

FIGURE 16.12 *Graph of $y = e^{-|x|}$.*

(For b_ω we need not actually carry out the integration because the integrand is an odd function). The solution is

$$y(x, t) = \frac{2}{\pi} \int_0^\infty \frac{1}{1 + \omega^2} \cos(\omega x) \cos(\omega c t) \, d\omega.$$ ▪

The solution (16.14) may be written in more compact form as follows. Insert the integral formulas for the coefficients:

$$y(x, t) = \int_0^\infty [a_\omega \cos(\omega x) + b_\omega \sin(\omega x)] \cos(\omega c t) \, d\omega$$

$$= \frac{1}{\pi} \int_0^\infty \left[\left(\int_{-\infty}^\infty f(\xi) \cos(\omega \xi) \, d\xi \right) \cos(\omega x) \right.$$

$$\left. + \left(\int_{-\infty}^\infty f(\xi) \sin(\omega \xi) \, d\xi \right) \sin(\omega x) \right] \cos(\omega c t) \, d\omega \qquad (16.15)$$

$$= \frac{1}{\pi} \int_{-\infty}^\infty \int_0^\infty [\cos(\omega \xi) \cos(\omega x) + \sin(\omega \xi) \sin(\omega x)] f(\xi) \cos(\omega c t) \, d\omega \, d\xi$$

$$= \frac{1}{\pi} \int_{-\infty}^\infty \int_0^\infty \cos(\omega(\xi - x)) f(\xi) \cos(\omega c t) \, d\omega \, d\xi.$$

Zero Initial Displacement

Suppose now the string is released from the horizontal position (zero initial displacement), with initial velocity $g(x)$. The initial value problem for the displacement function is

$$\frac{\partial^2 y}{\partial t^2} = c^2 \frac{\partial^2 y}{\partial x^2} \quad \text{for } -\infty < x < \infty, t > 0$$

$$y(x, 0) = 0, \frac{\partial y}{\partial t}(x, 0) = g(x) \quad \text{for } \infty < x < \infty.$$

Let $y(x, t) = X(x)T(t)$ and proceed exactly as in the case of initial displacement $f(x)$ and zero initial velocity, obtaining eigenvalues $\lambda = \omega^2$ for $\omega \geq 0$ and eigenfunctions

$$X_\omega(x) = a_\omega \cos(\omega x) + b_\omega \sin(\omega x).$$

Turning to T, we obtain, again as before,

$$T(t) = a \cos(\omega c t) + b \sin(\omega c t).$$

However, this problem differs from the preceding one in the initial condition on $T(t)$. Now we have

$$y(x, 0) = X(x)T(0) = 0,$$

so $T(0) = 0$ and hence $a = 0$. Thus for each $\omega \geq 0$, $T(t)$ is a constant multiple of $\sin(\omega ct)$. This gives us functions

$$y_\omega(x, t) = [a_\omega \cos(\omega x) + b_\omega \sin(\omega x)] \sin(\omega ct).$$

Now use the superposition

$$y(x, t) = \int_0^\infty [a_\omega \cos(\omega x) + b_\omega \sin(\omega x)] \sin(\omega ct) \, d\omega \qquad (16.16)$$

in order to satisfy the initial condition. Compute

$$\frac{\partial y}{\partial t} = \int_0^\infty [a_\omega \cos(\omega x) + b_\omega \sin(\omega x)] \omega c \cos(\omega ct) \, d\omega.$$

We need

$$\frac{\partial y}{\partial t}(x, 0) = \int_0^\infty [\omega c a_\omega \cos(\omega x) + \omega c b_\omega \sin(\omega x)] \, d\omega = g(x).$$

This is a Fourier integral expansion of the initial velocity function. With conditions on g (such as are given in the convergence theorem for Fourier integrals), choose

$$\omega c a_\omega = \frac{1}{\pi} \int_{-\infty}^\infty g(\xi) \cos(\omega \xi) \, d\xi$$

and

$$\omega c b_\omega = \frac{1}{\pi} \int_{-\infty}^\infty g(\xi) \sin(\omega \xi) \, d\xi.$$

Then

$$a_\omega = \frac{1}{\pi c \omega} \int_{-\infty}^\infty g(\xi) \cos(\omega \xi) \, d\xi \quad \text{and} \quad b_\omega = \frac{1}{\pi c \omega} \int_{-\infty}^\infty g(\xi) \sin(\omega \xi) \, d\xi.$$

With these coefficients, equation (16.16) is the solution of the problem.

EXAMPLE 16.4

Suppose the initial displacement is zero and the initial velocity is given by

$$g(x) = \begin{cases} e^x & \text{for } 0 \leq x \leq 1 \\ 0 & \text{for } x < 0 \quad \text{and for } x > 1 \end{cases}.$$

A graph of this function is shown in Figure 16.13. To use equation (16.16) to write the displacement function, compute the coefficients:

$$a_\omega = \frac{1}{\pi c \omega} \int_{-\infty}^\infty g(\xi) \cos(\omega \xi) \, d\xi = \frac{1}{\pi c \omega} \int_0^1 e^\xi \cos(\omega \xi) \, d\xi$$

$$= \frac{1}{\pi c \omega} \frac{e \cos(\omega) + e \omega \sin(\omega) - 1}{1 + \omega^2}$$

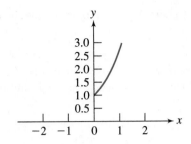

FIGURE 16.13

$$g(x) = \begin{cases} e^x & \text{for } 0 \le x \le 1 \\ 0 & \text{for } x < 0 \quad \text{and for } x > 1 \end{cases}.$$

and

$$b_\omega = \frac{1}{\pi c \omega} \int_{-\infty}^{\infty} g(\xi) \sin(\omega \xi) \, d\xi = \frac{1}{\pi c \omega} \int_0^1 e^\xi \sin(\omega \xi) \, d\xi$$

$$= -\frac{1}{\pi c \omega} \frac{e\omega \cos(\omega) - e \sin(\omega) - \omega}{1 + \omega^2}$$

The solution is

$$y(x, t) = \int_0^\infty \left(\frac{1}{\pi c \omega} \frac{e \cos(\omega) + e\omega \sin(\omega) - 1}{1 + \omega^2} \right) \cos(\omega x) \sin(\omega c t) \, d\omega$$

$$- \int_0^\infty \left(\frac{1}{\pi c \omega} \frac{e\omega \cos(\omega) - e \sin(\omega) - \omega}{1 + \omega^2} \right) \sin(\omega x) \sin(\omega c t) \, d\omega. \; \blacksquare$$

As in the case of wave motion on $[0, L]$, the solution of a problem with nonzero initial velocity and displacement can be obtained as the sum of the solutions of two problems, in one of which there is no initial displacement, and in the other, zero initial velocity.

16.3.2 Wave Motion Along a Semi-infinite String

We will now consider the problem of wave motion along a string fastened at the origin and stretched along the nonnegative x axis. Unlike the case of the string along the entire line, there is now one boundary condition, at $x = 0$. The problem is

$$\frac{\partial^2 y}{\partial t^2} = c^2 \frac{\partial^2 y}{\partial x^2} \quad \text{for } 0 < x < \infty, t > 0,$$

$$y(0, t) = 0 \quad \text{for } t \ge 0,$$

$$y(x, 0) = f(x), \frac{\partial y}{\partial t}(x, 0) = g(x) \quad \text{for } 0 < x < \infty.$$

Again, we want a bounded solution.

Let $y(x, t) = X(x)T(t)$ and obtain

$$X'' + \lambda X = 0, \quad T'' + \lambda c^2 T = 0.$$

In this problem we have a boundary condition:

$$y(0, t) = X(0)T(t) = 0,$$

implying that $X(0) = 0$. Begin by looking for the eigenvalues λ and corresponding eigenfunctions. Consider cases on λ.

Case 1 $\lambda = 0$
Now $X(x) = ax + b$. Since $X(0) = b = 0$, then $X(x) = ax$. This is unbounded on $[0, \infty)$ unless $a = 0$, so $\lambda = 0$ yields no bounded nontrivial solution for X, and 0 is not an eigenvalue.

Case 2 λ is negative.
Now write $\lambda = -\omega^2$ to obtain $X'' - \omega^2 X = 0$. This has general solution

$$X(x) = ae^{\omega x} + be^{-\omega x}.$$

Now

$$X(0) = a + b = 0$$

implies that $b = -a$, so $X(x) = 2a \sinh(\omega x)$. This is unbounded for $x > 0$ unless $a = 0$, so this problem has no negative eigenvalue.

Case 3 λ is positive.
Now write $\lambda = \omega^2$ and obtain

$$X(x) = a \cos(\omega x) + b \sin(\omega x).$$

Since $X(0) = a = 0$, only the sine terms remain. Thus every positive number is an eigenvalue, with corresponding eigenfunctions nonzero constant multiples of $\sin(\omega x)$.

Now the problem for T is $T'' + c^2 \omega^2 T = 0$, with general solution

$$T(t) = a \cos(\omega c t) + b \sin(\omega c t).$$

At this point we must isolate the problem into one with zero initial displacement or zero initial velocity. Suppose, to be specific, that $g(x) = 0$. Then $T'(0) = 0$, so $b = 0$ and $T(t)$ must be a constant multiple of $\cos(\omega c t)$. We therefore have functions

$$y_\omega(x, t) = c_\omega \sin(\omega x) \cos(\omega c t)$$

for each $\omega > 0$. Define the superposition

$$y(x, t) = \int_0^\infty c_\omega \sin(\omega x) \cos(\omega c t) \, d\omega.$$

Each such function satisfies the wave equation and the boundary condition, as well as $y_t(x, 0) = 0$ for $x > 0$. To satisfy the condition on initial displacement, we must choose the coefficients so that

$$y(x, 0) = \int_0^\infty c_\omega \sin(\omega x) \, d\omega = f(x).$$

This is the Fourier sine integral expansion of $f(x)$ on $[0, \infty)$, so choose

$$c_\omega = \frac{2}{\pi} \int_0^\infty f(\xi) \sin(\omega \xi) \, d\xi.$$

The solution of the problem is

$$y(x, t) = \frac{2}{\pi} \int_0^\infty \left(\int_0^\infty f(\xi) \sin(\omega \xi) \, d\xi \right) \sin(\omega x) \cos(\omega c t) \, d\omega.$$

If the problem has zero initial displacement, and initial velocity $g(x)$, then a similar analysis leads to the solution

$$y(x, t) = \int_0^\infty c_\omega \sin(\omega x) \sin(\omega c t) \, d\omega,$$

where

$$c_\omega = \frac{2}{\pi c \omega} \int_0^\infty g(\xi) \sin(\omega \xi) \, d\xi.$$

EXAMPLE 16.5

Consider wave motion along the half-line governed by the problem:

$$\frac{\partial^2 y}{\partial t^2} = 16 \frac{\partial^2 y}{\partial x^2} \quad \text{for } x > 0, t > 0,$$

$$y(0, t) = 0 \quad \text{for } t \geq 0,$$

$$\frac{\partial y}{\partial t}(x, 0) = 0, \, y(x, 0) = \begin{cases} \sin(\pi x) & \text{for } 0 \leq x \leq 4 \\ 0 & \text{for } x > 4 \end{cases}.$$

Here, $c = 4$. To write the solution, we need only compute the coefficients

$$c_\omega = \frac{2}{\pi} \int_0^\infty f(\xi) \sin(\omega \xi) \, d\xi$$

$$= \frac{2}{\pi} \int_0^4 \sin(\pi \xi) \sin(\omega \xi) \, d\xi = 8 \sin(\omega) \cos(\omega) \frac{2 \cos^2(\omega) - 1}{\omega^2 - \pi^2}.$$

The solution is

$$y(x, t) = \int_0^\infty 8 \sin(\omega) \cos(\omega) \frac{2 \cos^2(\omega) - 1}{\omega^2 - \pi^2} \sin(\omega x) \cos(4\omega t) \, d\omega. \quad \blacksquare$$

16.3.3   Fourier Transform Solution of Problems on Unbounded Domains

It is useful to have a variety of tools and methods available to solve boundary value problems. To this end, we will revisit problems of wave motion on the line and half-line and approach the solution through the use of a Fourier transform.

First, here is a brief description of what is involved in using a transform.

1. The range of values for the variable in which the transform will be performed is one determining factor in choosing a transform. Another is how the information given in the boundary value problem fits into the operational formula for the transform. For example, the operational formula for the Fourier sine transform is

$$\mathfrak{F}_S[f''(x)](\omega) = -\omega^2 \hat{f}_S(\omega) + \omega f(0),$$

so we must be given information about $f(0)$ in the problem to make use of this transform.

2. If the transform is performed with respect to a variable α of the boundary value problem, we obtain a differential equation involving the other variable(s). This differential equation must be solved subject to other information given in the problem. This solution gives the transform of the solution of the original boundary value problem.

3. Once we have the transform of the solution of the boundary value problem, we must invert it to obtain the solution of the boundary value problem itself.

Finally, the Fourier transform of a real-valued function is often complex-valued. If the solution is real-valued, then the real part of the expression obtained using the Fourier transform is the solution. However, because expressions such as $e^{-i\omega x}$ are often easier to manipulate than $\cos(\omega x)$ and $\sin(\omega x)$, we often retain the entire complex expression as the "solution," extracting the real part when we need numerical values, graphs, or other information.

For reference, we will summarize (without conditions on the functions) some facts about the Fourier transform and the Fourier sine and cosine transforms.

Fourier Transform

$$\mathfrak{F}[f](\omega) = \hat{f}(\omega) = \int_{-\infty}^{\infty} f(x) e^{-i\omega x}\, dx$$

$$f(x) = \frac{1}{2\pi} \int_{-\infty}^{\infty} \hat{f}(\omega) e^{i\omega x}\, d\omega$$

$$\mathfrak{F}[f''](\omega) = -\omega^2 \hat{f}(\omega)$$

Fourier Cosine Transform

$$\mathfrak{F}_C[f](\omega) = \hat{f}_C(\omega) = \int_0^{\infty} f(x) \cos(\omega x)\, dx$$

$$f(x) = \frac{2}{\pi} \int_0^{\infty} \hat{f}_C(\omega) \cos(\omega x)\, d\omega$$

$$\mathfrak{F}_C[f''](\omega) = -\omega^2 \hat{f}_C(\omega) - f'(0)$$

Fourier Sine Transform

$$\mathfrak{F}_S[f](\omega) = \hat{f}_S(\omega) = \int_0^{\infty} f(x) \sin(\omega x)\, dx$$

$$f(x) = \frac{2}{\pi} \int_0^{\infty} \hat{f}_S(\omega) \sin(\omega x)\, d\omega$$

$$\mathfrak{F}_S[f''](\omega) = -\omega^2 \hat{f}_S(\omega) + \omega f(0)$$

Fourier Transform Solution of the Wave Equation on the Line
Consider again the problem

$$\frac{\partial^2 y}{\partial t^2} = c^2 \frac{\partial^2 y}{\partial x^2} \quad \text{for } -\infty < x < \infty, t > 0$$

$$y(x, 0) = f(x), \ \frac{\partial y}{\partial t}(x, 0) = 0 \quad \text{for } -\infty < x < \infty.$$

Because x varies over the entire line, we can try the Fourier transform in the x variable. To do this, transform $y(x, t)$ as a function of x, leaving t as a parameter. First apply $\mathfrak{F}$ to the wave equation:

$$\mathfrak{F}\left[\frac{\partial^2 y}{\partial t^2}\right](\omega) = c^2 \mathfrak{F}\left[\frac{\partial^2 y}{\partial x^2}\right](\omega).$$

Because we are transforming in x, leaving t alone, we have

$$\mathfrak{F}\left[\frac{\partial^2 y}{\partial t^2}\right](\omega) = \int_{-\infty}^{\infty} \frac{\partial^2 y}{\partial t^2}(x, t) e^{-i\omega x}\, dx = \frac{\partial^2}{\partial t^2} \int_{-\infty}^{\infty} y(x, t) e^{-i\omega x}\, dx = \frac{\partial^2}{\partial t^2} \hat{y}(\omega, t),$$

where $\hat{y}(\omega, t)$ is the Fourier transform, with respect to x, of $y(x, t)$. The partial derivative with respect to t passes through the integral with respect to x because x and t are independent.

For the Fourier transform, in x, of $\partial^2 y / \partial x^2$, use the operational formula:

$$\mathfrak{F}\left[\frac{\partial^2 y}{\partial x^2}\right](\omega) = -\omega^2 \hat{y}(\omega, t).$$

The transformed wave equation is therefore

$$\frac{\partial^2}{\partial t^2} \hat{y}(\omega, t) = -c^2 \omega^2 \hat{y}(\omega, t),$$

or

$$\frac{\partial^2}{\partial t^2} \hat{y}(\omega, t) + c^2 \omega^2 \hat{y}(\omega, t) = 0.$$

Think of this as an ordinary differential equation for $\hat{y}(\omega, t)$ in t, with ω carried along as a parameter. The general solution has the form

$$\hat{y}(\omega, t) = a_\omega \cos(\omega c t) + b_\omega \sin(\omega c t).$$

We obtain the coefficients by transforming the initial data. First,

$$\hat{y}(\omega, 0) = a_\omega = \mathfrak{F}[y(x, 0)](\omega) = \mathfrak{F}[f](\omega) = \hat{f}(\omega),$$

the transform of the initial position function. Next

$$\omega c b_\omega = \frac{\partial \hat{y}}{\partial t}(\omega, 0) = \mathfrak{F}\left[\frac{\partial y}{\partial t}(x, 0)\right](\omega) = \mathfrak{F}[0](\omega) = 0$$

because the initial velocity is zero. Therefore, $b_\omega = 0$ and

$$\hat{y}(\omega, t) = \hat{f}(\omega) \cos(\omega c t).$$

We now know the transform of the solution $y(x, t)$. Invert this to find $y(x, t)$:

$$y(x, t) = \frac{1}{2\pi} \int_{-\infty}^{\infty} \hat{f}(\omega) \cos(\omega c t) e^{i\omega x} \, d\omega. \tag{16.17}$$

This is an integral formula for the solution, since $\hat{f}(\omega)$ is presumably known to us because we were given f. Since $e^{i\omega x}$ is complex-valued, we must actually take the real part of this integral to obtain $y(x, t)$. However, the integral is often left in the form of equation (16.17) with the understanding that $y(x, t)$ is the real part.

We will show that the solutions of this problem obtained by Fourier transform and Fourier integral are the same. Write the solution just obtained by transform as

$$\begin{aligned}
y_{tr}(x, t) &= \frac{1}{2\pi} \int_{-\infty}^{\infty} \hat{f}(\omega) \cos(\omega c t) e^{i\omega x} \, d\omega \\
&= \frac{1}{2\pi} \int_{-\infty}^{\infty} \left(\int_{-\infty}^{\infty} f(\xi) e^{-i\omega\xi} \, d\xi \right) \cos(\omega c t) e^{i\omega x} \, d\omega \\
&= \frac{1}{2\pi} \int_{-\infty}^{\infty} \int_{-\infty}^{\infty} e^{-i\omega(\xi - x)} \cos(\omega c t) f(\xi) \, d\omega \, d\xi \\
&= \frac{1}{2\pi} \int_{-\infty}^{\infty} \int_{-\infty}^{\infty} [\cos(\omega(\xi - x)) - i \sin(\omega(\xi - x))] \cos(\omega c t) f(\xi) \, d\omega \, d\xi.
\end{aligned}$$

Since the displacement function is real-valued, we must take the real part of this integral, obtaining

$$y(x, t) = \frac{1}{2\pi} \int_{-\infty}^{\infty} \int_{-\infty}^{\infty} \cos(\omega(\xi - x)) \cos(\omega ct) f(\xi) \, d\omega \, d\xi.$$

Finally, this integrand is an even function of ω, so

$$\frac{1}{2\pi} \int_{-\infty}^{\infty} \cdots d\omega = 2\frac{1}{2\pi} \int_{0}^{\infty} \cdots d\omega = \frac{1}{\pi} \int_{0}^{\infty} \cdots d\omega,$$

yielding

$$y(x, t) = \frac{1}{\pi} \int_{-\infty}^{\infty} \int_{0}^{\infty} \cos(\omega(\xi - x)) \cos(\omega ct) f(\xi) \, d\omega \, d\xi$$

This agrees with the solution (16.15) obtained by Fourier integral.

EXAMPLE 16.6

Solve for the displacement function on the real line if the initial velocity is zero and the initial displacement function is given by

$$f(x) = \begin{cases} \cos(x) & \text{for } -\frac{\pi}{2} \le x \le \frac{\pi}{2} \\ \\ 0 & \text{for } |x| > \frac{\pi}{2} \end{cases}.$$

To use the solution (16.17) we must compute

$$\hat{f}(\omega) = \int_{-\infty}^{\infty} f(\xi) e^{-i\omega\xi} \, d\xi = \int_{-\pi/2}^{\pi/2} \cos(\xi) e^{-i\omega\xi} \, d\xi$$

$$= \begin{cases} 2\dfrac{\cos(\pi\omega/2)}{1 - \omega^2} & \text{for } \omega \ne 1 \\ \\ \dfrac{\pi}{2} & \text{for } \omega = 1 \end{cases}.$$

$\hat{f}(\omega)$ is continuous, since

$$\lim_{\omega \to 1} \frac{2\cos(\pi\omega/2)}{1 - \omega^2} = \frac{\pi}{2}.$$

The solution can be written

$$y(x, t) = \frac{1}{\pi} \int_{-\infty}^{\infty} \frac{\cos(\pi\omega/2)}{1 - \omega^2} \cos(\omega ct) e^{i\omega x} \, d\omega,$$

with the understanding that $y(x, t)$ is the real part of the integral on the right. If we explicitly take this real part, then

$$y(x, t) = \frac{1}{\pi} \int_{-\infty}^{\infty} \frac{\cos(\pi\omega/2)}{1 - \omega^2} \cos(\omega x) \cos(\omega ct) \, d\omega. \ \blacksquare$$

EXAMPLE 16.7

In some instances, a clever use of the Fourier transform can yield a closed-form solution. Consider the problem

$$\frac{\partial^2 y}{\partial t^2} = 9 \frac{\partial^2 y}{\partial x^2} \quad \text{for } -\infty < x < \infty, t \geq 0,$$

$$y(x, 0) = 4e^{-5|x|} \quad \text{for } -\infty < x < \infty,$$

$$\frac{\partial y}{\partial t}(x, 0) = 0.$$

Take the transform of the differential equation, obtaining as in the discussion

$$\frac{\partial^2 \hat{y}}{\partial t^2}(\omega, t) = -9\omega^2 \hat{y}(\omega, t),$$

with general solution

$$\hat{y}(\omega, t) = a_\omega \cos(3\omega t) + b_\omega \sin(3\omega t).$$

Now use the initial conditions. Using the initial position function we have

$$\hat{y}(\omega, 0) = a_\omega = \mathfrak{F}[y(x, 0)](\omega) = \mathfrak{F}\left[4e^{-5|x|}\right](\omega) = \frac{40}{25 + \omega^2}.$$

Next, using the initial velocity, write

$$\frac{\partial \hat{y}}{\partial t}(\omega, 0) = 3\omega b_\omega = \mathfrak{F}\left[\frac{\partial y}{\partial t}(x, 0)\right](\omega) = 0,$$

so $b_\omega = 0$. Then

$$\hat{y}(\omega, t) = \frac{40}{25 + \omega^2} \cos(3\omega t).$$

We can now write the solution in integral form as

$$y(x, t) = \mathfrak{F}^{-1}[\hat{y}(\omega, t)](x) = \frac{1}{2\pi} \int_{-\infty}^{\infty} \frac{40}{25 + \omega^2} \cos(3\omega t) e^{i\omega x} \, d\omega.$$

However, in this case we can explicitly invert $\hat{y}(\omega, t)$, using some facts about the Fourier transform. Begin by using the convolution theorem to write

$$y(x, t) = \mathfrak{F}^{-1}\left[\frac{40}{25 + \omega^2} \cos(3\omega t)\right]$$

$$= \mathfrak{F}^{-1}\left[\frac{40}{25 + \omega^2}\right] * \mathfrak{F}^{-1}[\cos(3\omega t)] \tag{16.18}$$

$$= 4e^{-5|x|} * \mathfrak{F}^{-1}[\cos(3\omega t)].$$

We need to compute the inverse Fourier transform of $\cos(3\omega t)$. Here ω is the variable of the transformed function, with t carried along as a parameter. The variable of the inverse transform will be x. Combine the fact that $F[\delta(t)](\omega) = 1$ from Section 14.4.5, with the modulation theorem (Theorem 14.6 in Section 14.3) to get

$$\mathfrak{F}[\cos(\omega_0 x)] = \pi[\delta(\omega + \omega_0) + \delta(\omega - \omega_0)],$$

in which δ is the Dirac delta function. By the symmetry theorem (Theorem 14.5 of Section 14.3),

$$\mathfrak{F}[\pi[\delta(x + \omega_0) + \delta(x - \omega_0)]] = 2\pi \cos(\omega_0 \omega).$$

Therefore,

$$\mathfrak{F}^{-1}[\cos(\omega_0 \omega)](x) = \tfrac{1}{2}[\delta(x + \omega_0) + \delta(x - \omega_0)].$$

Now put $\omega_0 = 3t$ to get

$$\mathfrak{F}^{-1}[\cos(3\omega t)](x) = \tfrac{1}{2}[\delta(x + 3t) + \delta(x - 3t)].$$

Therefore equation (16.18) gives

$$y(x, t) = 4e^{-5|x|} * \tfrac{1}{2}[\delta(x + 3t) + \delta(x - 3t)]$$

$$= 2 \left(e^{-5|x|} * \delta(x + 3t) + e^{-5|x|} * \delta(x - 3t) \right)$$

$$= 2 \int_{-\infty}^{\infty} e^{-5|x-\xi|} \delta(\xi + 3t)\, d\xi + 2 \int_{-\infty}^{\infty} e^{-5|x-\xi|} \delta(\xi - 3t)\, d\xi$$

$$= 2e^{-5|x+3t|} + 2e^{-5|x-3t|},$$

in which the last line was obtained by using the filtering property of the Delta function (Theorem 14.13 of Section 14.4.5). This is the closed form of the solution and is easily verified directly. ■

Transform Solution of the Wave Equation on a Half-Line

We will use a transform to solve a wave problem on a half-line, with the left end fixed at $x = 0$. This time we will take the case of zero initial displacement, but a nonzero initial velocity:

$$\frac{\partial^2 y}{\partial t^2} = c^2 \frac{\partial^2 y}{\partial x^2} \quad \text{for } 0 < x < \infty, t > 0,$$

$$y(0, t) = 0 \quad \text{for } t \geq 0,$$

$$y(x, 0) = 0, \frac{\partial y}{\partial t}(x, 0) = g(x) \quad \text{for } 0 < x < \infty.$$

Now the Fourier transform is inappropriate because both x and t range only over the nonnegative real numbers. We can try the Fourier sine or cosine transform in x. The operational formula for the sine transform requires the value of the solution at $x = 0$, while the formula for the cosine transform uses the value of the derivative at the origin. Since we are given the condition $y(0, t) = 0$ (fixed left end of the string), we are led to try the sine transform. Let $\hat{y}_S(\omega, t)$ be the sine transform of $y(x, t)$ in the x variable. Take the sine transform of the wave equation. The partial derivatives with respect to t pass through the transform, and we use the operational formula for the transform of the second derivative with respect to x:

$$\frac{\partial^2 \hat{y}_S}{\partial t^2} = c^2 \mathfrak{F}_S \left[\frac{\partial^2 y}{\partial x^2} \right] = -c^2 \omega^2 \hat{y}_S(\omega, t) + \omega c^2 y(0, t) = -c^2 \omega^2 \hat{y}_S(\omega, t).$$

Then

$$\hat{y}_S(\omega, t) = a_\omega \cos(\omega c t) + b_\omega \sin(\omega c t).$$

Now

$$a_\omega = \hat{y}_S(\omega, 0) = \mathfrak{F}_S[y(x, 0)](\omega) = \mathfrak{F}_S[0](\omega) = 0,$$

and

$$\frac{\partial \hat{y}_S}{\partial t}(\omega, 0) = \omega c b_\omega = \hat{g}_S(\omega),$$

so

$$b_\omega = \frac{1}{\omega c} \hat{g}_S(\omega).$$

Therefore,

$$\hat{y}(\omega, t) = \frac{1}{\omega c} \hat{g}_S(\omega) \sin(\omega c t).$$

This is the sine transform of the solution. The solution is obtained by inverting:

$$y(x, t) = \mathfrak{F}_S^{-1} \left[\frac{1}{\omega c} \hat{g}_S(\omega) \sin(\omega c t) \right](x) = \frac{2}{\pi} \int_0^\infty \frac{1}{\omega c} \hat{g}_S(\omega) \sin(\omega x) \sin(\omega c t) \, d\omega.$$

EXAMPLE 16.8

Consider the following problem on a half-line

$$\frac{\partial^2 y}{\partial t^2} = 25 \frac{\partial^2 y}{\partial x^2} \quad \text{for } x > 0, t > 0,$$

$$y(0, t) = 0 \quad \text{for } t \geq 0,$$

$$y(x, 0) = 0, \frac{\partial y}{\partial t}(x, 0) = g(x) \quad \text{for } 0 < x < \infty,$$

where

$$g(x) = \begin{cases} 9 - x^2 & \text{for } 0 \leq x \leq 3 \\ 0 & \text{for } x > 3 \end{cases}.$$

If we use the Fourier sine transform, then the solution is

$$y(x, t) = \frac{2}{\pi} \int_0^\infty \frac{1}{5\omega} \hat{g}_S(\omega) \sin(\omega x) \sin(5\omega t) \, d\omega.$$

All that is left to do is compute

$$\hat{g}_S(\omega) = \int_0^\infty g(\xi) \sin(\omega x) \, dx$$

$$= \int_0^3 (9 - x^2) \sin(\omega x) \, dx$$

$$= \frac{-8 \cos^3(\omega) + 6 \cos(\omega) - 24\omega \sin(\omega) \cos^2(\omega) + 6\omega \sin(\omega) + 9\omega^2 + 2}{\omega^3},$$

yielding an integral expression for the solution. ∎

SECTION 16.3 PROBLEMS

In each of Problems 1 through 10, consider the wave equation $\partial^2 y / \partial t^2 = c^2 (\partial^2 y / \partial x^2)$ on the line for the given value of c and the given initial conditions $y(x, 0) = f(x)$ and $(\partial y / \partial t)(x, 0) = g(x)$. Solve the problem using the Fourier integral and then again using the Fourier transform.

1. $c = 12$, $f(x) = e^{-5|x|}$, $g(x) = 0$

2. $c = 8$, $f(x) = \begin{cases} 8 - x & \text{for } 0 \leq x \leq 8 \\ 0 & \text{for } x < 0 \quad \text{and for } x > 0 \end{cases}$

   $g(x) = 0$

3. $c = 4$, $f(x) = 0$,

$$g(x) = \begin{cases} \sin(x) & \text{for } -\pi \le x \le \pi \\ 0 & \text{for } |x| > \pi \end{cases}$$

4. $c = 1$, $f(x) = \begin{cases} 2 - |x| & \text{for } -2 \le x \le 2 \\ 0 & \text{for } |x| > 2 \end{cases}$

$g(x) = 0$

5. $c = 3$, $f(x) = 0$, $g(x) = \begin{cases} e^{-2x} & \text{for } x \ge 1 \\ 0 & \text{for } x < 1 \end{cases}$

6. $c = 2$, $f(x) = 0$,

$$g(x) = \begin{cases} 1 & \text{for } 0 \le x \le 2 \\ -1 & \text{for } -2 \le x < 0 \\ 0 & \text{for } x > 2 \quad \text{and for } x < -2 \end{cases}$$

7. $c = 4$, $f(x) = \begin{cases} 16 - x^4 & \text{for } -2 \le x \le 2 \\ 0 & \text{for } |x| > 2 \end{cases}$

$g(x) = 0$

8. $c = 6$, $f(x) = 0$,

$$g(x) = \begin{cases} \cos(\pi x) & \text{for } -\dfrac{3}{2} \le x \le \dfrac{3}{2} \\ 0 & \text{for } |x| > \dfrac{3}{2} \end{cases}$$

9. $c = 2$,

$$f(x) = \begin{cases} \sin(x) & \text{for } -2\pi \le x \le \pi \\ 0 & \text{for } x < -2\pi \quad \text{and for } x > \pi \end{cases}$$

$g(x) = e^{-4|x|}$

10. $c = 3$, $f(x) = xe^{-|x|}$,

$$g(x) = \begin{cases} 1 & \text{for } 0 \le x \le 1 \\ 0 & \text{for } x < 0 \quad \text{and for } x > 0 \end{cases}$$

In each of Problems 11 through 17, consider the wave equation $\partial^2 y/\partial t^2 = c^2(\partial^2 y/\partial x^2)$ on the half-line, with $y(0, t) = 0$ for $t > 0$, and for the given value of c and the given boundary initial conditions $y(x, 0) = f(x)$ and $(\partial y/\partial t)(x, 0) = g(x)$ for $x \ge 0$. Solve the problem using separation of variables (the Fourier sine integral) and then again using the Fourier sine transform.

11. $c = 3$, $f(x) = \begin{cases} x(1 - x) & \text{for } 0 \le x \le 1 \\ 0 & \text{for } x > 1 \end{cases}$

$g(x) = 0$

12. $c = 3$, $f(x) = 0$, $g(x) = \begin{cases} 0 & \text{for } 0 \le x < 4 \\ 2 & \text{for } 4 \le x \le 11 \\ 0 & \text{for } x > 11 \end{cases}$

13. $c = 2$, $f(x) = 0$,

$$g(x) = \begin{cases} \cos(x) & \text{for } \dfrac{\pi}{2} \le x \le \dfrac{5\pi}{2} \\ 0 & \text{for } 0 \le x < \dfrac{\pi}{2} \quad \text{and for } x > \dfrac{5\pi}{2} \end{cases}$$

14. $c = 6$, $f(x) = -2e^{-x}$, $g(x) = 0$

15. $c = 14$, $f(x) = 0$,

$$g(x) = \begin{cases} x^2(3 - x) & \text{for } 0 \le x \le 3 \\ 0 & \text{for } x > 3 \end{cases}$$

16. $c = 7$, $f(x) = \begin{cases} x \sin(x) & \text{for } 0 \le x \le 3\pi \\ 0 & \text{for } x > 3\pi \end{cases}$

$g(x) = 5e^{-3x}$

17. $c = 4$, $f(x) = \begin{cases} (x - 5)^2 & \text{for } 0 \le x \le 5 \\ 0 & \text{for } x > 5 \end{cases}$

$$g(x) = \begin{cases} 4x^2 & \text{for } 0 \le x \le 2 \\ 0 & \text{for } x > 2 \end{cases}$$

Sometimes the Laplace transform is effective in solving boundary value problems involving the wave equation. Use the Laplace transform to solve the following.

18.

$$\frac{\partial^2 y}{\partial t^2} = c^2 \frac{\partial^2 y}{\partial x^2} \quad \text{for } x > 0, t > 0$$

$$y(0, t) = \begin{cases} \sin(2\pi t) & \text{for } 0 \le t \le 1 \\ 0 & \text{for } t > 0 \end{cases}$$

$$y(x, 0) = \frac{\partial y}{\partial t}(x, 0) = 0 \quad \text{for } x > 0$$

19.

$$\frac{\partial^2 y}{\partial t^2} = c^2 \frac{\partial^2 y}{\partial x^2} \quad \text{for } x > 0, t > 0$$

$$y(0, t) = t \quad \text{for } t > 0$$

$$y(x, 0) = 0, \frac{\partial y}{\partial t}(x, 0) = A \quad \text{for } x > 0$$

20.

$$\frac{\partial^2 y}{\partial t^2} = 4 \frac{\partial^2 y}{\partial x^2} \quad \text{for } x > 0, t > 0$$

$$y(0, t) = \sin(t) \quad \text{for } t > 0$$

$$y(x, 0) = 0, \frac{\partial y}{\partial t}(x, 0) = e^{-x} \quad \text{for } x > 0$$

16.4 Characteristics and d'Alembert's Solution

This section will involve repeated chain rule differentiations, which are efficiently written using subscript notation for partial derivatives. For example, $\partial u/\partial t = u_t$, $\partial u/\partial x = u_x$, $\partial^2 u/\partial t^2 = u_{tt}$, and so on. Our objective is to examine a different perspective on the problem

$$u_{tt} = c^2 u_{xx} \quad \text{for } -\infty < x < \infty, t > 0,$$

$$u(x, 0) = f(x), u_t(x, 0) = g(x) \quad \text{for } -\infty < x < \infty.$$

Here we are using $u(x, t)$ as the position function because we will be changing variables from the (x, y) plane to a (ξ, η) plane, and we do not want to confuse the solution function with coordinates of points.

This boundary value problem, which we have solved using the Fourier integral and again using the Fourier transform, is referred to as the *Cauchy problem for the wave equation.* We will write a solution that dates to the eighteenth century. The lines

$$x - ct = k_1, \quad x + ct = k_2,$$

with k_1 and k_2 any real constants, are called *characteristics* of the wave equation. These form two families of lines, one consisting of parallel lines with slope $1/c$, the other of parallel lines with slope $-1/c$. Figure 16.14 shows some of these characteristics. We will see that these lines are closely related to the wave motion. However, our first use of them will be to write an explicit solution of the wave equation in terms of the initial data.

Define a change of coordinates

$$\xi = x - ct, \quad \eta = x + ct.$$

This transformation is invertible, since

$$x = \frac{1}{2}(\xi + \eta), \quad t = \frac{1}{2c}(-\xi + \eta).$$

Define

$$U(\xi, \eta) = u(x(\xi, \eta), y(\xi, \eta)).$$

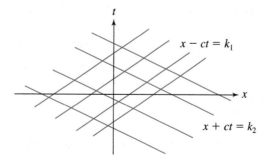

FIGURE 16.14 *Characteristics of the wave equation.*

Now compute derivatives:

$$u_x = U_\xi \xi_x + U_\eta \eta_x = U_\xi + U_\eta,$$

$$u_{xx} = U_{\xi\xi}\xi_x + U_{\xi\eta}\eta_x + U_{\eta\xi}\xi_x + U_{\eta\eta}\eta_x$$

$$= U_{\xi\xi} + 2U_{\xi\eta} + U_{\eta\eta}$$

$$u_t = U_\xi(-c) + U_\eta(c)$$

and

$$u_{tt} = -c[U_{\xi\xi}(-c) + U_{\xi\eta}(c)] + c[U_{\eta\xi}(-c) + U_{\eta\eta}(c)] = c^2 U_{\xi\xi} - 2c^2 U_{\xi\eta} + c^2 U_{\eta\eta}.$$

Then

$$u_{tt} - c^2 u_{xx} = -4c^2 U_{\xi\eta}.$$

In the new coordinates, the wave equation is

$$U_{\xi\eta} = 0.$$

This is called the *canonical form of the wave equation*, and it is an easy equation to solve. First write it as

$$(U_\eta)_\xi = 0.$$

This means that U_η is independent of ξ, say

$$U_\eta = h(\eta).$$

Integrate to get

$$U(\xi, \eta) = \int h(\eta)\,d\eta + F(\xi),$$

in which $F(\xi)$ is the "constant" of integration of the partial derivative with respect to η. Now $\int h(\eta)\,d\eta$ is just another function of η, which we will write as $G(\eta)$. Thus

$$U(\xi, \eta) = F(\xi) + G(\eta),$$

where F and G must be twice continuously differentiable functions of one variable, but are otherwise arbitrary. We have shown that the solution of $u_{tt} = c^2 u_{xx}$ has the form

$$u(x, t) = F(x - ct) + G(x + ct). \tag{16.19}$$

Equation (16.19) is called *d'Alembert's solution* of the wave equation, after the French mathematician Jean le Rond d'Alembert (1717–1783). Every solution of $u_{tt} = c^2 u_{xx}$ must have this form.

Now we will show how to choose F and G to satisfy the initial conditions. First,

$$u(x, 0) = F(x) + G(x) = f(x) \tag{16.20}$$

and

$$u_t(x, 0) = -cF'(x) + cG'(x) = g(x). \tag{16.21}$$

Integrate equation (16.21) and rearrange terms to obtain

$$-F(x) + G(x) = \frac{1}{c}\int_0^x g(\xi)\,d\xi - F(0) + G(0).$$

Add this equation to equation (16.20) to get

$$2G(x) = f(x) + \frac{1}{c} \int_0^x g(\xi) \, d\xi - F(0) + G(0).$$

Therefore,

$$G(x) = \frac{1}{2} f(x) + \frac{1}{2c} \int_0^x g(\xi) \, d\xi - \frac{1}{2} F(0) + \frac{1}{2} G(0). \tag{16.22}$$

But then, from equation (16.20),

$$F(x) = f(x) - G(x) = \frac{1}{2} f(x) - \frac{1}{2c} \int_0^x g(\xi) \, d\xi + \frac{1}{2} F(0) - \frac{1}{2} G(0). \tag{16.23}$$

Finally, use equations (16.22) and (16.23) to write the solution as

$$u(x, t) = F(x - ct) + G(x + ct)$$

$$= \frac{1}{2} f(x - ct) - \frac{1}{2c} \int_0^{x-ct} g(\xi) \, d\xi + \frac{1}{2} F(0) - \frac{1}{2} G(0)$$

$$+ \frac{1}{2} f(x + ct) + \frac{1}{2c} \int_0^{x+ct} g(\xi) \, d\xi - \frac{1}{2} F(0) + \frac{1}{2} G(0),$$

or, after cancellations,

$$u(x, t) = \frac{1}{2} \left(f(x - ct) + f(x + ct) \right) + \frac{1}{2c} \int_{x-ct}^{x+ct} g(\xi) \, d\xi. \tag{16.24}$$

Equation (16.24) is *d'Alembert's formula* for the solution of the Cauchy problem for the wave equation on the entire line. It is an explicit formula for the solution of the Cauchy problem, in terms of the given initial position and velocity functions.

EXAMPLE 16.9

We will solve the boundary value problem

$$u_{tt} = 4u_{xx} \quad \text{for } -\infty < x < \infty, t > 0,$$

$$u(x, 0) = e^{-|x|}, u_t(x, 0) = \cos(4x) \quad \text{for } -\infty < x < \infty.$$

By d'Alembert's formula, we immediately have

$$u(x, t) = \frac{1}{2} \left(e^{-|x-2t|} + e^{-|x+2t|} \right) + \frac{1}{4} \int_{x-2t}^{x+2t} \cos(4\xi) \, d\xi$$

$$= \frac{1}{2} \left(e^{-|x-2t|} + e^{-|x+2t|} \right) + \frac{1}{16} [\sin(4(x + 2t)) - \sin(4(x - 2t))]$$

$$= \frac{1}{2} \left(e^{-|x-2t|} + e^{-|x+2t|} \right) + \frac{1}{8} \cos(4x) \sin(8t). \quad \blacksquare$$

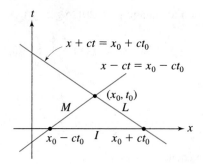

FIGURE 16.15 *Characteristic triangle.*

16.4.1 A Nonhomogeneous Wave Equation

Using the characteristics, we will write an expression for the solution of the nonhomogeneous problem:

$$u_{tt} = c^2 u_{xx} + F(x, t) \quad \text{for } -\infty < x < \infty, t > 0,$$

$$u(x, 0) = f(x), u_t(x, 0) = g(x) \quad \text{for } -\infty < x < \infty.$$

This problem is called *nonhomogeneous* because of the term $F(x, t)$, which we assume to be continuous for all real x and $t \geq 0$. $F(x, t)$ can be thought of as an external driving or damping force acting on the string.

Suppose we want the solution at (x_0, t_0). Recall that the characteristics of the wave equation are straight lines in the x, t plane. There are exactly two characteristics through this point, and these are the lines

$$x - ct = x_0 - ct_0 \quad \text{and} \quad x + ct = x_0 + ct_0.$$

Segments of these characteristics, together with the interval $[x_0 - ct_0, x_0 + ct_0]$, form a *characteristic triangle* Δ, shown in Figure 16.15. Label the sides of Δ as L, M, and I. Since Δ is a region in the x, t plane, we can compute the double integral of $-F(x, t)$ over Δ:

$$-\iint_\Delta F(x, t) \, dA = \iint_\Delta (c^2 u_{xx} - u_{tt}) \, dA = \iint_\Delta \left(\frac{\partial}{\partial x}(c^2 u_x) - \frac{\partial}{\partial t}(u_t) \right) dA.$$

Apply Green's theorem to the last integral, with x and t as the independent variables instead of x and y. This converts the double integral to a line integral around the boundary C of Δ. This piecewise smooth curve, which consists of three line segments, is oriented counterclockwise. We obtain by Green's theorem,

$$-\iint_\Delta F(x, t) \, dA = \oint_C u_t \, dx + c^2 u_x \, dt.$$

Now evaluate the line integral on the right by evaluating it on each segment of C in turn.

On I, $t = 0$, so $dt = 0$, and x varies from $x_0 - ct_0$ to $x_0 + ct_0$, so

$$\int_I u_t \, dx + c^2 u_x \, dt = \int_{x_0 - ct_0}^{x_0 + ct_0} u_t(x, 0) \, dx = \int_{x_0 - ct_0}^{x_0 + ct_0} g(\xi) \, d\xi.$$

On L, $x + ct = x_0 + ct_0$, so $dx = -c\,dt$ and

$$\int_L u_t\,dx + c^2 u_x\,dt = \int_L u_t(-c)\,dt + c^2 u_x\left(-\frac{1}{c}\right)dx = -c\int_L du$$

$$= -c\left[u(x_0, t_0) - u(x_0 + ct_0, 0)\right].$$

Finally, on M, $x - ct = x_0 - ct_0$, so $dx = c\,dt$ and

$$\int_M u_t\,dx + c^2 u_x\,dt = \int_L u_t(c)\,dt + c^2 u_x\left(\frac{1}{c}\right)dx = c\int_M du$$

$$= c\left[u(x_0 - ct_0, 0) - u(x_0, t_0)\right].$$

M has initial point (x_0, t_0) and terminal point $(x_0 - ct_0, 0)$ because of the counterclockwise orientation on the boundary of Δ.

Upon summing these line integrals, we obtain

$$-\iint_\Delta F(x, t)\,dA = \int_{x_0 - ct_0}^{x_0 + ct_0} g(\xi)\,d\xi$$

$$- c\left[u(x_0, t_0) - u(x_0 + ct_0, 0)\right] + c\left[u(x_0 - ct_0, 0) - u(x_0, t_0)\right].$$

Then

$$-\iint_\Delta F(x, t)\,dA = \int_{x_0 - ct_0}^{x_0 + ct_0} g(\xi)\,d\xi - 2cu(x_0, t_0) + cu(x_0 + ct_0, 0) + cu(x_0 - ct_0, 0)$$

$$= \int_{x_0 - ct_0}^{x_0 + ct_0} g(\xi)\,d\xi - 2cu(x_0, t_0) + c\left[f(x_0 + ct_0) + f(x_0 - ct_0)\right].$$

Solve this equation for $u(x_0, t_0)$ to obtain

$$u(x_0, t_0) = \frac{1}{2}\left[f(x_0 - ct_0) + f(x_0 + ct_0)\right] + \frac{1}{2c}\int_{x_0 - ct_0}^{x_0 + ct_0} g(\xi)\,d\xi + \frac{1}{2c}\iint_\Delta F(x, t)\,dA.$$

We have used the subscript 0 on (x_0, t_0) to focus attention on the point at which we are evaluating the solution. However, this can be any point with x_0 real and $t_0 > 0$. Thus the solution at an arbitrary point (x, t) is

$$u(x, t) = \frac{1}{2}\left[f(x - ct) + f(x + ct)\right] + \frac{1}{2c}\int_{x-ct}^{x+ct} g(\xi)\,d\xi + \frac{1}{2c}\iint_\Delta F(\xi, \eta)\,d\xi\,d\eta.$$

The solution at (x, t) of the problem with the forcing term $F(x, t)$ is therefore d'Alembert's solution for the homogeneous problem (no forcing term), plus $(1/2c)$ times the double integral of the forcing term over the characteristic triangle having (x, t) as a vertex.

EXAMPLE 16.10

Consider the problem

$$u_{tt} = 25u_{xx} + x^2 t^2 \quad \text{for } -\infty < x < \infty, t > 0,$$

$$u(x, 0) = x\cos(x), u_t(x, 0) = e^{-x} \quad \text{for } -\infty < x < \infty.$$

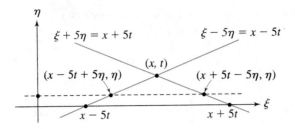

FIGURE 16.16

The solution at any point x and time t has the form

$$u(x, t) = \frac{1}{2}[(x - 5t)\cos(x - 5t) + (x + 5t)\cos(x + 5t)] + \frac{1}{10}\int_{x-5t}^{x+5t} e^{-\xi}\, d\xi$$

$$+ \frac{1}{10}\iint_\Delta \xi^2\eta^2\, d\xi\, d\eta.$$

All we have to do is evaluate the integrals. First,

$$\frac{1}{10}\int_{x-5t}^{x+5t} e^{-\xi}\, d\xi = -\frac{1}{10}e^{-x-5t} + \frac{1}{10}e^{-x+5t}.$$

For the double integral of the forcing term, proceed from Figure 16.16:

$$\frac{1}{10}\iint_\Delta \xi^2\eta^2\, d\xi\, d\eta = \frac{1}{10}\int_0^t \int_{x-5t+5\eta}^{x+5t-5\eta} \xi^2\eta^2\, d\xi\, d\eta$$

$$= \frac{1}{12}t^4 x^2 + \frac{5}{36}t^6.$$

The solution is

$$u(x, t) = \frac{1}{2}[(x - 5t)\cos(x - 5t) + (x + 5t)\cos(x + 5t)]$$

$$- \frac{1}{10}e^{-x-5t} + \frac{1}{10}e^{-x+5t} + \frac{1}{12}t^4 x^2 + \frac{5}{36}t^6. \ \blacksquare$$

In the last example, $u(x, t)$ gives the position function of the string at any given time t. The graph of $u(x, t)$ in the x, t plane is not a snapshot of the string at any time. Rather, a picture of the string at time t is the graph of points $(x, u(x, t))$, with t fixed at the time of interest. Figure 16.17(a) shows a segment of the string at time $t = 0.3$, for both the forced and unforced motion. Figure 16.17(b) shows a segment of the string for $t = 0.6$, again for both unforced and forced motion.

This method of characteristics can also be used to solve boundary value problems involving the wave equation on a bounded interval $[0, L]$. However, this is a good deal more involved than the solution on the entire line, so we will leave this to a more advanced treatment of partial differential equations.

16.4.2 Forward and Backward Waves

Continuing with the boundary value problem for the wave equation on the entire real line, we can write d'Alembert's formula (16.24) for the solution as

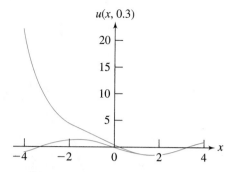

FIGURE 16.17(a) *Profile of the forced and unforced string at t = 0.3.*

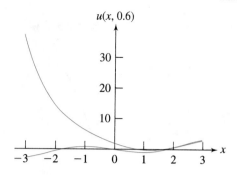

FIGURE 16.17(b) $t = 0.6$.

$$u(x, t) = \frac{1}{2}\left(f(x - ct) - \frac{1}{c}\int_0^{x-ct} g(\xi)\,d\xi\right)$$
$$+ \frac{1}{2}\left(f(x + ct) + \frac{1}{c}\int_0^{x+ct} g(\xi)\,d\xi\right)$$
$$= \varphi(x - ct) + \beta(x + ct),$$

where

$$\varphi(x) = \frac{1}{2}f(x) - \frac{1}{2c}\int_0^x g(\xi)\,d\xi$$

and

$$\beta(x) = \frac{1}{2}f(x) + \frac{1}{2c}\int_0^x g(\xi)\,d\xi.$$

We call $\varphi(x - ct)$ a forward (or right) wave, and $\beta(x + ct)$ a backward (or left) wave. The graph of $\varphi(x - ct)$ is the graph of $\varphi(x)$ translated ct units to the right. We may therefore think of $\varphi(x - ct)$ as the graph of $\varphi(x)$ moving to the right with velocity c. The graph of $\beta(x + ct)$ is the graph of $\beta(x)$ translated ct units to the left. Thus $\beta(x + ct)$ is the graph of $\beta(x)$ moving to the left with velocity c. The string profile at time t, given by the graph of $y = u(x, t)$ as a function of x, is the sum of these forward and backward waves at time t.

As an example of this process, consider the boundary value problem in which $c = 1$,

$$f(x) = \begin{cases} 4 - x^2 & \text{for } -2 \leq x \leq 2 \\ 0 & \text{for } |x| > 2 \end{cases}$$

and $g(x) = 0$. This initial position function is shown in Figure 16.18(a). The solution is a sum of a forward and a backward wave:

$$u(x, t) = \varphi(x + ct) + \beta(x - ct) = \tfrac{1}{2}f(x + t) + \tfrac{1}{2}f(x - t).$$

At any time t, the motion consists of the initial position function translated t units to the right, superimposed on the initial position function translated t units to the left. We see the motion as the initial position function (Figure 16.18(a)) moving simultaneously right and left. Because $f(x)$ vanishes outside of $[-2, 2]$, these forward and backward waves actually separate and become disjoint, one continuing to move to the right, and the other to the left on the real line. This process is shown in Figures 16.18(b) through (h).

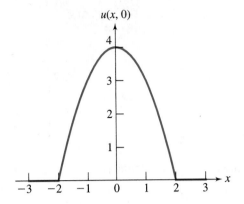

FIGURE 16.18(a)

$$f(x) = \begin{cases} 4 - x^2 & \text{for } -2 \le x \le 2 \\ 0 & \text{for } |x| > 2 \end{cases}.$$

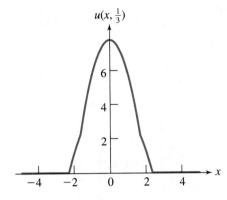

FIGURE 16.18(b) *Superposition of forward and backward waves at $t = \frac{1}{3}$.*

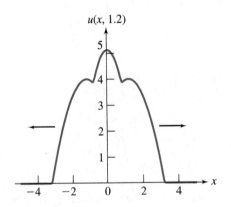

FIGURE 16.18(c) $t = 1.2$.

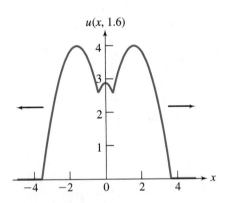

FIGURE 16.18(d) $t = 1.6$.

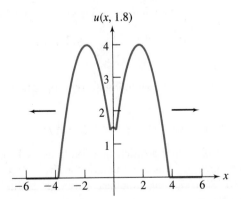

FIGURE 16.18(e) $t = 1.8$.

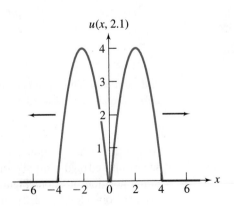

FIGURE 16.18(f) $t = 2.1$.

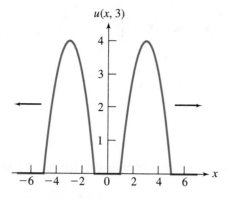

FIGURE 16.18(g) $t = 3$.

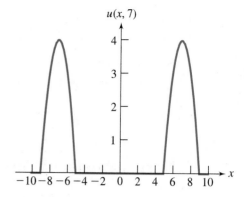

FIGURE 16.18(h) $t = 7$.

SECTION 16.4   PROBLEMS

In each of Problems 1 through 10, determine the characteristics of the wave equation for the problem

$$u_{tt} = c^2 u_{xx} \quad \text{for } -\infty < x < \infty, t > 0,$$

$$u(x, 0) = f(x), u_t(x, 0) = g(x) \quad \text{for } -\infty < x < \infty$$

for the given value of c and write the d'Alembert solution.

1. $c = 1$, $f(x) = x^2$, $g(x) = -x$
2. $c = 4$, $f(x) = x^2 - 2x$, $g(x) = \cos(x)$
3. $c = 7$, $f(x) = \cos(\pi x)$, $g(x) = 1 - x^2$
4. $c = 5$, $f(x) = \sin(2x)$, $g(x) = x^3$
5. $c = 14$, $f(x) = e^x$, $g(x) = x$
6. $c = 12$, $f(x) = -5x + x^2$, $g(x) = 3$
7. $c = 4$, $f(x) = xe^x$, $g(x) = \cosh(x)$
8. $c = 8$, $f(x) = x^2 - 2x$, $g(x) = 1 + x^2$
9. $c = 15$, $f(x) = |x|$, $g(x) = e^{2x}$
10. $c = 2$, $f(x) = x^3 - 1$, $g(x) = \cos(\pi x)$

In each of Problems 11 through 20, solve the problem

$$u_{tt} = c^2 u_{xx} + F(x, t) \quad \text{for } -\infty < x < \infty, t > 0,$$

$$u(x, 0) = f(x), u_t(x, 0) = g(x) \quad \text{for } -\infty < x < \infty$$

for the given c, $f(x)$, and $g(x)$.

11. $c = 4$, $f(x) = x$, $g(x) = e^{-x}$, $F(x, t) = x + t$
12. $c = 2$, $f(x) = \sin(x)$, $g(x) = 2x$, $F(x, t) = 2xt$
13. $c = 8$, $f(x) = x^2 - x$, $g(x) = \cos(2x)$, $F(x, t) = xt^2$
14. $c = 4$, $f(x) = x^2$, $g(x) = xe^{-x}$, $F(x, t) = x \sin(t)$
15. $c = 3$, $f(x) = \cosh(x)$, $g(x) = 1$, $F(x, t) = 3xt^3$

16. $c = 7$, $f(x) = 1 + x$, $g(x) = \sin(x)$,
    $F(x, t) = x - \cos(t)$
17. $c = 2$, $f(x) = \cos(2x)$, $g(x) = 1 - \cos(x)$,
    $F(x, t) = t^2$
18. $c = 1$, $f(x) = x^3$, $g(x) = \sin^2(x)$,
    $F(x, t) = e^{-x} \cos(t)$
19. $c = 2$, $f(x) = x \sin(x)$, $g(x) = e^{-x}$, $F(x, t) = xt$
20. $c = 6$, $f(x) = 1 - x^2$, $g(x) = x \sin(x)$,
    $F(x, t) = t \sin(x)$

In each of Problems 21 through 26, write the solution of the problem

$$u_{tt} = u_{xx} \quad \text{for } -\infty < x < \infty, t > 0,$$

$$u(x, 0) = f(x), u_t(x, 0) = 0 \quad \text{for } -\infty < x < \infty.$$

as a sum of a forward and backward wave. Graph the initial position function and then graph the solution at selected times, showing the solution as a superposition of forward and backward waves moving in opposite directions along the real line.

21. $f(x) = \begin{cases} \sin(2x) & \text{for } -\pi \leq x \leq \pi \\ 0 & \text{for } |x| > \pi \end{cases}$

22. $f(x) = \begin{cases} 1 - |x| & \text{for } -1 \leq x \leq 1 \\ 0 & \text{for } |x| > 1 \end{cases}$

23. $f(x) = \begin{cases} \cos(x) & \text{for } -\dfrac{\pi}{2} \leq x \leq \dfrac{\pi}{2} \\ 0 & \text{for } |x| > \dfrac{\pi}{2} \end{cases}$

24. $f(x) = \begin{cases} 1 - x^2 & \text{for } |x| \le 1 \\ 0 & \text{for } |x| > 1 \end{cases}$

26. $f(x) = \begin{cases} x^3 - x^2 - 4x + 4 & \text{for } -2 \le x \le 2 \\ 0 & \text{for } |x| > 2 \end{cases}$

25. $f(x) = \begin{cases} x^2 - x - 2 & \text{for } -1 \le x \le 2 \\ 0 & \text{for } x < -1 \quad \text{and for } x > 2 \end{cases}$

16.5   Normal Modes of Vibration of a Circular Elastic Membrane

We will analyze the motion of a membrane (such as a drumhead) fastened onto a circular frame and set in motion with given initial position and velocity. Let the rest position of the membrane be in the x, y plane with the origin at the center, and let the membrane have radius R. Using polar coordinates, the particle of membrane at (r, θ) is assumed to vibrate vertical to the x, y plane, and its displacement from the rest position at time t is $z(x, y, t)$.

Equation (16.4) gives the wave equation for this displacement function:

$$\frac{\partial^2 z}{\partial t^2} = c^2 \left(\frac{\partial^2 z}{\partial r^2} + \frac{1}{r} \frac{\partial z}{\partial r} + \frac{1}{r^2} \frac{\partial^2 z}{\partial \theta^2} \right).$$

We will assume for the moment that the motion of the membrane is symmetric about the origin, in which case z depends only on r and t. Now the wave equation is

$$\frac{\partial^2 z}{\partial t^2} = c^2 \left(\frac{\partial^2 z}{\partial r^2} + \frac{1}{r} \frac{\partial z}{\partial r} \right).$$

Let the initial displacement be given by $z(r, 0) = f(r)$, and let the initial velocity be

$$\frac{\partial z}{\partial t}(r, 0) = g(r).$$

Attempt a solution

$$z(r, t) = F(r)T(\theta).$$

We obtain, after a routine calculation,

$$T'' + \lambda T = 0 \quad \text{and} \quad F'' + \frac{1}{r} F' + \frac{\lambda}{c^2} F = 0.$$

If $\lambda > 0$, say $\lambda = \omega^2$, the equation for F is a zero-order Bessel equation, with general solution

$$F(r) = a J_0 \left(\frac{\omega}{c} r \right) + b Y_0 \left(\frac{\omega}{c} r \right).$$

Since $Y_0(\omega r/c) \to -\infty$ as $r \to 0$ (the center of the membrane), choose $b = 0$. Now the equation for T is

$$T'' + \omega^2 T = 0,$$

with general solution

$$T(t) = d \cos(\omega t) + k \sin(\omega t).$$

We have, for each $\omega > 0$, a function

$$z_\omega(r, t) = a_\omega J_0 \left(\frac{\omega}{c} r \right) \cos(\omega t) + b_\omega J_0 \left(\frac{\omega}{c} r \right) \sin(\omega t).$$

Since the membrane is fixed on a circular frame,

$$z_\omega(R, t) = a_\omega J_0\left(\frac{\omega}{c}R\right)\cos(\omega t) + b_\omega J_0\left(\frac{\omega}{c}R\right)\sin(\omega t) = 0$$

for $t > 0$. This condition is satisfied if $J_0(\omega R/c) = 0$. Let $j_1, j_2, \ldots$ be the positive zeros of J_0, with

$$j_1 < j_2 < \cdots,$$

and choose

$$\frac{\omega R}{c} = j_n$$

or

$$\omega_n = \frac{j_n c}{R}$$

for $n = 1, 2, \ldots$. This yields the eigenvalues of this problem:

$$\lambda_n = \omega_n^2 = \frac{j_n^2 c^2}{R^2}.$$

We now have

$$z_n(r, t) = a_n J_0\left(\frac{j_n r}{R}\right)\cos\left(\frac{j_n ct}{R}\right) + b_n J_0\left(\frac{j_n r}{R}\right)\sin\left(\frac{j_n ct}{R}\right).$$

All of these functions satisfy the boundary condition $z(R, t) = 0$. To satisfy the initial conditions, attempt a superposition

$$z(r, t) = \sum_{n=1}^{\infty}\left[a_n J_0\left(\frac{j_n r}{R}\right)\cos\left(\frac{j_n ct}{R}\right) + b_n J_0\left(\frac{j_n r}{R}\right)\sin\left(\frac{j_n ct}{R}\right)\right]. \qquad (16.25)$$

Now

$$z(r, 0) = f(r) = \sum_{n=1}^{\infty} a_n J_0\left(\frac{j_n r}{R}\right),$$

a Fourier–Bessel expansion of $f(r)$. Let $s = r/R$ to convert this series to

$$f(Rs) = \sum_{n=1}^{\infty} a_n J_0(j_n s),$$

in which s varies from 0 to 1. We know from Section 15.3.3 that the coefficients in this expansion are given by

$$a_n = \frac{2}{[J_1(j_n)]^2}\int_0^1 sf(Rs)J_0(j_n s)\,ds$$

for $n = 1, 2, \ldots$.

Next we must solve for the b_n's. Compute

$$\frac{\partial z}{\partial t}(r, 0) = g(r) = \sum_{n=1}^{\infty} b_n \frac{j_n c}{R}J_0\left(\frac{j_n r}{R}\right).$$

This is a Fourier–Bessel expansion of $g(r)$. Again referring to Section 15.3.3, we must choose

$$b_n \frac{j_n c}{R} = \frac{2}{[J_1(j_n)]^2}\int_0^1 sg(Rs)J_0(j_n s)\,ds,$$

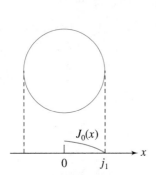

FIGURE 16.19(a) *First normal mode.*

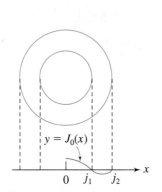

FIGURE 16.19(b) *Second normal mode.*

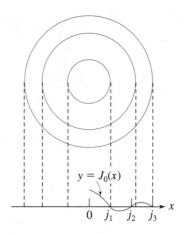

FIGURE 16.19(c) *Third normal mode.*

or

$$b_n = \frac{2R}{c j_n [J_1(j_n)]^2} \int_0^1 s g(Rs) J_0(j_n s) \, ds$$

for $n = 1, 2, \ldots$. With these coefficients, equation (16.25) is the solution for the position function of the membrane.

The numbers $\omega_n = j_n c / R$ are the frequencies of normal modes of vibration, which have periods $2\pi / \omega_n = 2\pi R / j_n c$. The normal modes of vibration are the functions $z_n(r, t)$. Often these functions are written in phase angle form as

$$z_n(r, t) = A_n J_0 \left(\frac{j_n r}{R} \right) \cos(\omega_n t + \delta_n)$$

in which A_n and δ_n are constants.

The first normal mode is

$$z_1(r, t) = A_1 J_0 \left(\frac{j_1 r}{R} \right) \cos(\omega_1 t + \delta_1).$$

As r varies from 0 to R, $j_1 r / R$ varies from 0 to j_1. At any time t, a radial section through the membrane takes the shape of the graph of $J_0(x)$ for $0 \le x \le j_1$ (Figure 16.19(a)).

The second normal mode is

$$z_2(r, t) = A_2 J_0 \left(\frac{j_2 r}{R} \right) \cos(\omega_2 t + \delta_2).$$

Now as r varies from 0 to R, $j_2 r / R$ varies from 0 to j_2, passing through j_1 along the way. Since $J_0(j_2 r / R) = 0$ when $j_2 r / R = j_1$, this mode has a nodal circle (fixed in the motion) at radius

$$r = \frac{j_1 R}{j_2}.$$

A section through the membrane takes the shape of the graph of $J_0(x)$ for $0 \le x \le j_2$ (Figure 16.19(b)).

Similarly, the third normal mode is

$$z_3(r, t) = A_3 J_0 \left(\frac{j_3 r}{R} \right) \cos(\omega_3 t + \delta_3),$$

and this mode had two nodes, one at $r = j_1 R / j_3$ and the second at $r = j_2 R / j_3$. Now a radial section has the shape of the graph of $J_0(x)$ for $0 \leq x \leq j_3$ (Figure 16.19(c)).

In general, the nth normal mode has $n - 1$ nodes (fixed circles in the motion of the membrane), occurring at $j_1 R / j_n, \ldots, j_{n-1} R / j_n$.

In the next section we will revisit this problem, this time retaining the θ dependence of the displacement function. This will lead us to a solution involving a double Fourier sine series.

SECTION 16.5 PROBLEMS

1. Let $c = R = 1$, $f(r) = 1 - r$ and $g(r) = 0$. Using material from Section 15.2 (Bessel functions), approximate the coefficients a_1 through a_5 in the solution given by equation (16.25) and graph the fifth partial sum of the solution for a selection of times.

2. Repeat Problem 1, except now use $f(r) = 1 - r^2$ and $g(r) = 0$.

3. Repeat Problem 1, but now use $f(r) = \sin(\pi r)$ and $g(r) = 0$.

16.6 Vibrations of a Circular Elastic Membrane, Revisited

We will continue from the last section with vibrations of an elastic membrane fixed on a circular frame. Now, however, retain the θ dependence of the displacement function and consider the entire wave equation

$$\frac{\partial^2 z}{\partial t^2} = c^2 \left(\frac{\partial^2 z}{\partial r^2} + \frac{1}{r} \frac{\partial z}{\partial r} + \frac{1}{r^2} \frac{\partial^2 z}{\partial \theta^2} \right)$$

for $0 \leq r < R, -\pi \leq \theta \leq \pi, t > 0$. We will use the initial conditions

$$z(r, \theta, 0) = f(r, \theta), \quad \frac{\partial z}{\partial t}(r, \theta, 0) = 0,$$

so the membrane is released from rest with the given initial displacement.

In cylindrical coordinates, θ can be replaced by $\theta + 2n\pi$ for any integer n, so we will also impose the *periodicity conditions*

$$z(r, -\pi, t) = z(r, \pi, t) \quad \text{and} \quad \frac{\partial z}{\partial \theta}(r, -\pi, t) = \frac{\partial z}{\partial \theta}(r, \pi, t)$$

for $0 \leq r < R$ and $t > 0$.

Put $z(r, \theta, t) = F(r)\Theta(\theta)T(r)$ in the wave equation to get

$$\frac{T''}{c^2 T} = \frac{F'' + (1/r)F'}{F} + \frac{1}{r^2} \frac{\Theta''}{\Theta} = -\lambda$$

for some constant λ since the left side depends only on t, and the right side only on r and θ. Then

$$T'' + \lambda c^2 T = 0$$

and

$$\frac{r^2 F'' + r F'}{F} + \lambda r^2 = -\frac{\Theta''}{\Theta}.$$

Because the left side depends only on r and the right side only on θ, and these are independent, for some constant μ,

$$\frac{r^2 F'' + r F'}{F} + \lambda r^2 = -\frac{\Theta''}{\Theta} = \mu.$$

Then

$$\Theta'' + \mu\Theta = 0$$

and

$$r^2 F'' + r F' + (\lambda r^2 - \mu) F = 0.$$

In solving these differential equations for $T(t)$, $F(r)$, and $\Theta(\theta)$, we have the following boundary conditions. First, by periodicity,

$$\Theta(-\pi) = \Theta(\pi) \quad \text{and} \quad \Theta'(-\pi) = \Theta'(\pi).$$

Next, because the membrane is fixed on the circular frame,

$$F(R) = 0.$$

Finally, because the initial velocity of the membrane is zero,

$$T'(0) = 0.$$

The problem for $\Theta(\theta)$ is a periodic Sturm–Liouville problem, which was solved in Section 15.3.1 (Example 15.10). The eigenvalues are

$$\mu_n = n^2 \quad \text{for } n = 0, 1, 2, \ldots,$$

and eigenfunctions are

$$\Theta_n(\theta) = a_n \cos(n\theta) + b_n \sin(n\theta).$$

With $\mu = n^2$, the problem for F is

$$r^2 F''(r) + r F'(r) + (\lambda r^2 - n^2) F(r) = 0; \qquad F(R) = 0.$$

We have seen (Section 14.2.2) that this differential equation has general solution

$$F(r) = a J_n(\sqrt{\lambda} r) + b Y_n(\sqrt{\lambda} r),$$

in terms of Bessel functions of order n of the first and second kinds. Because $Y_n(\sqrt{\lambda} r)$ is unbounded as $r \to 0+$, choose $b = 0$ to have a bounded solution. This leaves $F(r) = a J_n(\sqrt{\lambda} r)$. To find admissable values of λ, we need

$$F(R) = a J_n(\sqrt{\lambda} R) = 0.$$

We want to satisfy this with a nonzero to avoid a trivial solution. Thus $\sqrt{\lambda} R$ must be one of the positive zeros of J_n. Let these positive zeros be

$$j_{n1} < j_{n2} < \cdots,$$

doubly indexed because this derivation depends on the choice of $\mu = n^2$. Then

$$\lambda_{nk} = \frac{j_{nk}^2}{R^2},$$

with j_{nk} the kth positive zero of $J_n(x)$. The λ_{nk}'s are the eigenvalues. Corresponding eigenfunctions are nonzero multiples of

$$J_n\left(\frac{j_{nk}}{R} r\right) \quad \text{for } n = 0, 1, 2, \ldots \quad \text{and} \quad k = 1, 2, \ldots.$$

With these values of λ, the problem for T is

$$T'' + c^2 \left(\frac{j_{nk}}{R}\right)^2 T = 0; \qquad T'(0) = 0$$

with solutions constant multiples of

$$T_{nk}(t) = \cos\left(\frac{j_{nk}}{R}ct\right).$$

We can now form functions

$$z_{nk}(r, \theta, t) = [a_{nk}\cos(n\theta) + b_{nk}\sin(n\theta)]J_n\left(\frac{j_{nk}}{R}r\right)\cos\left(\frac{j_{nk}}{R}ct\right)$$

for $n = 0, 1, 2, \ldots$ and $k = 1, 2, \ldots$. Each of these functions satisfies the wave equation and the boundary conditions, together with the condition of zero initial velocity. To satisfy the condition that the initial position is given by f, write a superposition

$$z(r, \theta, t) = \sum_{n=0}^{\infty}\sum_{k=1}^{\infty}[a_{nk}\cos(n\theta) + b_{nk}\sin(n\theta)]J_n\left(\frac{j_{nk}}{R}r\right)\cos\left(\frac{j_{nk}}{R}ct\right). \tag{16.26}$$

Now we need

$$z(r, \theta, 0) = f(r, \theta) = \sum_{n=0}^{\infty}\sum_{k=1}^{\infty}[a_{nk}\cos(n\theta) + b_{nk}\sin(n\theta)]J_n\left(\frac{j_{nk}}{R}r\right).$$

To see how to choose these coefficients, first write this equation in the form

$$f(r, \theta) = \sum_{k=1}^{\infty}a_{0k}J_0\left(\frac{j_{0k}}{R}r\right) + \sum_{n=1}^{\infty}\left(\left[\sum_{k=1}^{\infty}a_{nk}J_n\left(\frac{j_{nk}}{R}r\right)\right]\cos(n\theta)\right.$$

$$\left. + \left[\sum_{k=1}^{\infty}b_{nk}J_n\left(\frac{j_{nk}}{R}r\right)\right]\sin(n\theta)\right).$$

For a given r, think of $f(r, \theta)$ as a function of θ. The last equation is the Fourier series expansion, on $[-\pi, \pi]$, of this function of θ. Since we know the coefficients in the Fourier expansion of a function of θ, we can immediately write

$$\sum_{k=1}^{\infty}a_{0k}J_0\left(\frac{j_{0k}}{R}r\right) = \frac{1}{2\pi}\int_{-\pi}^{\pi}f(r, \theta)\,d\theta = \frac{1}{2}\alpha_0(r),$$

and, for $n = 1, 2, \ldots$,

$$\sum_{k=1}^{\infty}a_{nk}J_n\left(\frac{j_{nk}}{R}r\right) = \frac{1}{\pi}\int_{-\pi}^{\pi}f(r, \theta)\cos(n\theta)\,d\theta = \alpha_n(r)$$

and

$$\sum_{k=1}^{\infty}b_{nk}J_n\left(\frac{j_{nk}}{R}r\right) = \frac{1}{\pi}\int_{-\pi}^{\pi}f(r, \theta)\sin(n\theta)\,d\theta = \beta_n(r)$$

Now recognize that for each $n = 0, 1, 2, \ldots$, the last three equations are expansions of functions of r in series of Bessel functions, with sets of coefficients, respectively, a_{0k}, a_{nk}, and b_{nk}. From Section 15.3.3, we know the coefficients in these expansions:

$$a_{0k} = \frac{1}{[J_1(j_{0k})]^2}\int_0^1 \xi\alpha_0(R\xi)J_0(j_{0k}\xi)\,d\xi \quad \text{for } k = 1, 2, \ldots,$$

and, for $n = 1, 2, \ldots,$

$$a_{nk} = \frac{2}{[J_{n+1}(j_{nk})]^2} \int_0^1 \xi \alpha_n(R\xi) J_n(j_{nk}\xi)\, d\xi \quad \text{for } k = 1, 2, \ldots,$$

and

$$b_{nk} = \frac{2}{[J_{n+1}(j_{nk})]^2} \int_0^1 \xi \beta_n(R\xi) J_n(j_{nk}\xi)\, d\xi \quad \text{for } k = 1, 2, \ldots.$$

The idea in calculating the coefficients is to first perform the integrations with respect to θ to obtain the functions $\alpha_0(r)$, $\alpha_n(r)$, and $\beta_n(r)$, written as Fourier–Bessel series. We then obtain the coefficients in these series, which are the a_{nk}'s and the b_{nk}'s, by evaluating the integrals for the coefficients in this type of eigenfunction expansion. In practice, these integrals must be approximated because the zeros of the Bessel functions of order n can only be approximated.

SECTION 16.6 PROBLEMS

1. Calculate each $\alpha_n(r)$ and $\beta_n(r)$ for the case $c = R = 2$, $f(r, \theta) = (4 - r^2)\sin^2(\theta)$, and $g(r, \theta) = \theta$. Use a table of zeros of Bessel functions to approximate the first five terms of $z(r, \theta, t)$ that are independent of θ, and the first four terms having a θ dependence.

2. Use the solution given in the section to prove the plausible fact that the center of the membrane remains undeflected for all time if the initial displacement is an odd function of θ (that is, $f(r, -\theta) = -f(r, \theta)$). *Hint:* The only integer-order Bessel function that is different from zero at $r = 0$ is J_0.

16.7 Vibrations of a Rectangular Membrane

Consider an elastic membrane stretched across a rectangular frame to which it is fixed. Suppose the frame and the rectangle it encloses occupy the region of the x, y plane defined by $0 \le x \le L$, $0 \le y \le K$. The membrane is given an initial displacement and released with a given initial velocity. We want to determine the vertical displacement function $z(x, y, t)$. At any time t, the graph of $z = z(x, y, t)$ for $0 < x < L, 0 < y < K$ is a snapshot of the membrane's position at that time. If we had a film of this function evolving over time, we would have a motion picture of the membrane.

The boundary value problem for z is

$$\frac{\partial^2 z}{\partial t^2} = c^2 \left(\frac{\partial^2 z}{\partial x^2} + \frac{\partial^2 z}{\partial y^2} \right) \quad \text{for } 0 < x < L, 0 < y < K, t > 0,$$

$$z(x, 0, t) = z(x, K, t) = 0 \quad \text{for } 0 < x < L, t > 0,$$

$$z(0, y, t) = y(L, y, t) = 0 \quad \text{for } 0 < y < K, t > 0,$$

$$z(x, y, 0) = f(x, y) \quad \text{for } 0 < x < L, 0 < y < K,$$

$$\frac{\partial z}{\partial t}(x, y, 0) = g(x, y) \quad \text{for } 0 < x < L, 0 < y < K.$$

We will solve this problem for the case of zero initial velocity, $g(x, y) = 0$.

Attempt a separation of variables, $z(x, y, t) = X(x)Y(y)T(t)$. We get

$$XYT'' = c^2[X''YT + XY''T],$$

or

$$\frac{T''}{c^2 T} - \frac{Y''}{Y} = \frac{X''}{X}.$$

We are unable to isolate three variables on different sides of an equation. However, we can argue that the left side is a function of just y and t, and the right side just of x, and these three variables are independent. Therefore, for some constant λ,

$$\frac{T''}{c^2 T} - \frac{Y''}{Y} = \frac{X''}{X} = -\lambda.$$

Now we have

$$X'' + \lambda X = 0 \quad \text{and} \quad \frac{T''}{c^2 T} + \lambda = \frac{Y''}{Y}.$$

In the last equation, the left side depends only on t and the right side only on y, so for some constant μ,

$$\frac{T''}{c^2 T} + \lambda = \frac{Y''}{Y} = -\mu.$$

Then

$$Y'' + \mu Y = 0 \quad \text{and} \quad T'' + c^2(\lambda + \mu)T = 0.$$

The variables have been separated, at the cost of introducing two separation constants. Now use the boundary conditions:

$$z(0, y, t) = X(0)Y(y)T(t) = 0$$

implies that $X(0) = 0$. Similarly,

$$X(L) = 0, Y(0) = 0 \quad \text{and} \quad Y(K) = 0.$$

The two problems for X and Y are

$$X'' + \lambda X = 0; \qquad X(0) = X(L) = 0$$

and

$$Y'' + \mu Y = 0; \qquad Y(0) = Y(K) = 0.$$

These have solutions:

$$\lambda_n = \frac{n^2 \pi^2}{L^2}, \quad X_n(x) = \sin\left(\frac{n \pi x}{L}\right)$$

and

$$\mu_m = \frac{m^2 \pi^2}{K^2}, \quad Y_m(x) = \sin\left(\frac{m \pi y}{K}\right),$$

with n and m varying independently over the positive integers. The problem for T now becomes

$$T'' + c^2\left(\frac{n^2 \pi^2}{L^2} + \frac{m^2 \pi^2}{K^2}\right)T = 0$$

Further, because of the assumption of zero initial velocity,

$$\frac{\partial z}{\partial t}(x, y, 0) = X(x)Y(y)T'(0) = 0,$$

so $T'(0) = 0$. Then T must be a constant multiple of

$$\cos\left(\sqrt{\frac{n^2}{L^2} + \frac{m^2}{K^2}}\,\pi ct\right).$$

For each positive integer n and m, we now have a function

$$z_{nm}(x, y, t) = a_{nm} \sin\left(\frac{n\pi x}{L}\right) \sin\left(\frac{m\pi y}{K}\right) \cos\left(\sqrt{\frac{n^2}{L^2} + \frac{m^2}{K^2}}\,\pi ct\right)$$

that satisfies all of the conditions of the problem, except possibly the initial condition $z(x, y, 0) = f(x, y)$. For this, use a superposition

$$z(x, y, t) = \sum_{n=1}^{\infty} \sum_{m=1}^{\infty} a_{nm} \sin\left(\frac{n\pi x}{L}\right) \sin\left(\frac{m\pi y}{K}\right) \cos\left(\sqrt{\frac{n^2}{L^2} + \frac{m^2}{K^2}}\,\pi ct\right).$$

We must choose the constants to satisfy

$$z(x, y, 0) = f(x, y) = \sum_{n=1}^{\infty} \sum_{m=1}^{\infty} a_{nm} \sin\left(\frac{n\pi x}{L}\right) \sin\left(\frac{m\pi y}{K}\right).$$

We can do this by exploiting a trick we used when introducing Fourier series. Pick a positive integer m_0 and multiply both sides of this equation by $\sin(m_0 \pi y / K)$ to get

$$f(x, y) \sin\left(\frac{m_0 \pi y}{K}\right) = \sum_{n=1}^{\infty} \sum_{m=1}^{\infty} a_{nm} \sin\left(\frac{n\pi x}{L}\right) \sin\left(\frac{m\pi y}{K}\right) \sin\left(\frac{m_0 \pi y}{K}\right).$$

Now integrate from 0 to K in the y variable, leaving terms in x alone. We get

$$\int_0^K f(x, y) \sin\left(\frac{m_0 \pi y}{K}\right) dy = \sum_{n=1}^{\infty} \sum_{m=1}^{\infty} a_{nm} \sin\left(\frac{n\pi x}{L}\right) \int_0^K \sin\left(\frac{m\pi y}{K}\right) \sin\left(\frac{m_0 \pi y}{K}\right) dy.$$

Now

$$\int_0^K \sin\left(\frac{m\pi y}{K}\right) \sin\left(\frac{m_0 \pi y}{K}\right) dy = \begin{cases} \dfrac{K}{2} & \text{if } m = m_0 \\ 0 & \text{if } m \neq m_0 \end{cases}$$

and the last equation becomes

$$\int_0^K f(x, y) \sin\left(\frac{m_0 \pi y}{K}\right) dy = \sum_{n=1}^{\infty} \frac{K}{2} a_{nm_0} \sin\left(\frac{n\pi x}{L}\right).$$

The left side of this equation is a function of x. Pick any positive integer n_0 and multiply this equation by $\sin(n_0 \pi x / L)$:

$$\int_0^K f(x, y) \sin\left(\frac{n_0 \pi x}{L}\right) \sin\left(\frac{m_0 \pi y}{K}\right) dy = \sum_{n=1}^{\infty} \frac{K}{2} a_{nm_0} \sin\left(\frac{n\pi x}{L}\right) \sin\left(\frac{n_0 \pi x}{L}\right).$$

Integrate, this time in the x variable:

$$\int_0^L \int_0^K f(x, y) \sin\left(\frac{n_0 \pi x}{L}\right) \sin\left(\frac{m_0 \pi y}{K}\right) dy\, dx$$

$$= \sum_{n=1}^{\infty} \frac{K}{2} a_{nm_0} \int_0^L \sin\left(\frac{n\pi x}{L}\right) \sin\left(\frac{n_0 \pi x}{L}\right) dx.$$

All the terms on the right are zero except when $n = n_0$, and this term is $L/2$. The last equation becomes

$$\int_0^L \int_0^K f(x, y) \sin\left(\frac{n_0 \pi x}{L}\right) \sin\left(\frac{m_0 \pi y}{K}\right) dy\, dx = \frac{K}{2} \frac{L}{2} a_{n_0 m_0}.$$

Dropping the zero subscripts, which were just for ease in keeping track of which integers were fixed, we now have

$$a_{nm} = \frac{4}{LK} \int_0^L \int_0^K f(x, y) \sin\left(\frac{n \pi x}{L}\right) \sin\left(\frac{m \pi y}{K}\right) dy\, dx.$$

With this choice of the coefficients, we have the solution for the displacement function.

EXAMPLE 16.11

Suppose the initial displacement is given by

$$z(x, y, 0) = x(L - x)y(K - y),$$

and the initial velocity is zero. The coefficients in the double Fourier expansion are

$$a_{nm} = \frac{4}{LK} \int_0^L \int_0^K x(L - x)y(K - y) \sin\left(\frac{n \pi x}{L}\right) \sin\left(\frac{m \pi y}{K}\right) dy\, dx$$

$$= \frac{4}{LK} \left(\int_0^L x(L - x) \sin\left(\frac{n \pi x}{L}\right) dx\right) \left(\int_0^K y(K - y) \sin\left(\frac{m \pi y}{K}\right) dy\right)$$

$$= \frac{16L^2 K^2}{(nm\pi^2)^3} [(-1)^n - 1][(-1)^m - 1].$$

The solution for the displacement function in this case is

$$z(x, y, t) = \sum_{n=1}^{\infty} \sum_{m=1}^{\infty} \left[\frac{16L^2 K^2}{(nm\pi^2)^3} [(-1)^n - 1][(-1)^m - 1] \sin\left(\frac{n \pi x}{L}\right) \sin\left(\frac{m \pi y}{K}\right) \cos\left(\sqrt{\frac{n^2}{L^2} + \frac{m^2}{K^2}} \pi c t\right)\right]. \quad \blacksquare$$

SECTION 16.7   PROBLEMS

1. Solve

$$\frac{\partial^2 z}{\partial t^2} = \frac{\partial^2 z}{\partial x^2} + \frac{\partial^2 z}{\partial y^2} \quad \text{for } 0 < x < 2\pi,$$
$$0 < y < 2\pi, t > 0,$$

$z(x, 0, t) = z(x, 2\pi, t) = 0 \quad$ for $0 < x < 2\pi, t > 0,$

$z(0, y, t) = z(2\pi, y, t) = 0 \quad$ for $0 < y < 2\pi, t > 0,$

$z(x, y, 0) = x^2 \sin(y) \quad$ for $0 < x < 2\pi, 0 < y < 2\pi,$

$\dfrac{\partial z}{\partial t}(x, y, 0) = 0 \quad$ for $0 < x < 2\pi, 0 < y < 2\pi.$

2. Solve

$$\frac{\partial^2 z}{\partial t^2} = 9\left(\frac{\partial^2 z}{\partial x^2} + \frac{\partial^2 z}{\partial y^2}\right) \quad \text{for } 0 < x < \pi,$$
$$0 < y < \pi, t > 0,$$

$z(x, 0, t) = z(x, \pi, t) = 0 \quad$ for $0 < x < \pi, t > 0,$

$z(0, y, t) = z(\pi, y, t) = 0 \quad$ for $0 < y < \pi, t > 0,$

$z(x, y, 0) = \sin(x)\cos(y) \quad$ for $0 < x < \pi, 0 < y < \pi,$

$\dfrac{\partial z}{\partial t}(x, y, 0) = xy \quad$ for $0 < x < \pi, 0 < y < \pi.$

3. Solve

$$\frac{\partial^2 z}{\partial t^2} = 4\left(\frac{\partial^2 z}{\partial x^2} + \frac{\partial^2 z}{\partial y^2}\right) \quad \text{for } 0 < x < 2\pi,$$

$$0 < y < 2\pi, t > 0,$$

$$z(x, 0, t) = z(x, 2\pi, t) = 0 \quad \text{for } 0 < x < 2\pi, t > 0,$$

$$z(0, y, t) = z(2\pi, y, t) = 0 \quad \text{for } 0 < y < 2\pi, t > 0,$$

$$z(x, y, 0) = 0 \quad \text{for } 0 < x < 2\pi, 0 < y < 2\pi,$$

$$\frac{\partial z}{\partial t}(x, y, 0) = 1 \quad \text{for } 0 < x < 2\pi, 0 < y < 2\pi.$$

4. Solve

$$\frac{\partial^2 u}{\partial t^2} = c^2\left(\frac{\partial^2 u}{\partial x^2} + \frac{\partial^2 u}{\partial y^2} + \frac{\partial^2 u}{\partial z^2}\right) \quad \text{for } 0 < x < L,$$

$$0 < y < K,$$

$$0 < z < M, t > 0,$$

$$u(0, y, z, t) = u(L, y, z, t) = 0,$$

$$u(x, 0, z, t) = u(x, K, z, t) = 0,$$

$$u(x, y, 0, t) = u(x, y, M, t) = 0,$$

$$u(x, y, z, 0) = 0, \quad \frac{\partial u}{\partial t}(x, y, z, 0) = g(x, y, z).$$

This problem models sound waves in a rectangular room, assuming a given initial velocity and zero initial disturbance.

5. Find an expression for the motion of sound waves in a rectangular room, assuming a zero initial velocity and an initial displacement $f(x, y, z)$.

6. Solve for the displacement function for a rectangular membrane attached to a fixed frame in the x, y plane if the initial position is $f(x, y)$ and the initial velocity is $g(x, y)$.

CHAPTER 17

The Heat Equation

Heat and radiation phenomena are often modeled by a partial differential equation called the *heat equation*. We derived a three-dimensional version of the heat equation, using Gauss's divergence theorem. We will now examine the heat equation more closely and solve it under a variety of conditions, following a program that parallels the one just carried out for the wave equation.

17.1 The Heat Equation and Initial and Boundary Conditions

Let $u(x, y, z, t)$ be the temperature at time t and location (x, y, z) in a region of space. In Section 12.8.2, we showed that u satisfies the partial differential equation

$$\mu\rho \frac{\partial u}{\partial t} = K \left(\frac{\partial^2 u}{\partial x^2} + \frac{\partial^2 u}{\partial y^2} + \frac{\partial^2 u}{\partial z^2} \right) + \nabla K \cdot \nabla u,$$

in which $K(x, y, z)$ is the thermal conductivity of the medium, $\mu(x, y, z)$ is the specific heat and $\rho(x, y, z)$ is the density. The term $\nabla K \cdot \nabla u$ is the dot product of the gradients of K and u. This is the heat equation in three space variables and the time.

If the thermal conductivity of the medium is constant, then ∇K is the zero vector and the term $\nabla K \cdot \nabla u = 0$. Now the three-dimensional heat equation is

$$\mu\rho \frac{\partial u}{\partial t} = K \left(\frac{\partial^2 u}{\partial x^2} + \frac{\partial^2 u}{\partial y^2} + \frac{\partial^2 u}{\partial z^2} \right).$$

The 1-dimensional heat equation is

$$\frac{\partial u}{\partial t} = \frac{K}{\mu\rho} \frac{\partial^2 u}{\partial x^2}.$$

915

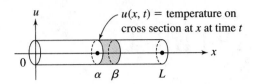

FIGURE 17.1

This equation often applies, for example, to heat conduction in a thin bar or wire whose length is much larger than its other dimensions. To get some feeling for what is involved in the 1-dimensional heat equation, we will give a separate derivation of it from basic principles.

Consider a straight, thin bar of constant density ρ and constant cross-sectional area A placed along the x axis from 0 to L. Assume that the sides of the bar are insulated and do not allow heat loss and that the temperature on the cross section of the bar perpendicular to the x axis at x is a function $u(x, t)$ of x and t only. Let the specific heat μ and the thermal conductivity K be constant.

Consider a typical segment of the bar between $x = \alpha$ and $x = \beta$, as in Figure 17.1. By the definition of specific heat, the rate at which heat energy accumulates in this segment is

$$\int_{\alpha}^{\beta} \mu \rho A \frac{\partial u}{\partial t}\, dx.$$

By Newton's law of cooling, heat energy flows within this segment from the warmer to the cooler end at a rate equal to K times the negative of the temperature gradient (difference in temperature at the ends of the segment). Therefore, the net rate at which heat energy enters this segment of the bar at time t is

$$K A \frac{\partial u}{\partial x}(\beta, t) - K A \frac{\partial u}{\partial x}(\alpha, t).$$

Assume that no energy is produced within the segment. Such production could occur, for example, if there is radiation or a heat source such as a chemical reaction. These would also change the mass of the segment with time. In the absence of these effects, the rate at which heat energy accumulates within the segment must balance the rate at which it enters the segment. Therefore,

$$\int_{\alpha}^{\beta} \mu \rho A \frac{\partial u}{\partial t} dx = K A \left(\frac{\partial u}{\partial x}(\beta, t) - \frac{\partial u}{\partial x}(\alpha, t) \right) = K A \int_{\alpha}^{\beta} \frac{\partial^2 u}{\partial x^2}\, dx,$$

so

$$\int_{\alpha}^{\beta} \left(\mu \rho \frac{\partial u}{\partial t} - K \frac{\partial^2 u}{\partial x^2} \right) dx = 0.$$

This equation must be true for every α and β with $0 \le \alpha < \beta \le L$. If the term in parentheses in this integral were nonzero at any x_0 and t_0, then by continuity we could choose an interval (α, β) about x_0 throughout which this term would be strictly positive or strictly negative. But then this integral of a positive or negative function over (α, β) would be, respectively, positive or negative, a contradiction. We conclude that

$$\mu \rho \frac{\partial u}{\partial t} - K \frac{\partial^2 u}{\partial x^2} = 0$$

for $0 < x < L$ and for $t > 0$. This is the 1-dimensional heat equation. Often this partial differential equation is written

$$\frac{\partial u}{\partial t} = k \frac{\partial^2 u}{\partial x^2},$$

where $k = K/\mu\rho$ is a positive constant depending on the material of the bar. The number k is called the *diffusivity* of the bar.

This equation certainly does not determine the temperature function $u(x, t)$ uniquely. For example, if $u(x, t)$ is one solution, so is $u(x, t) + c$ for any real number c. For uniqueness of the solution, which we expect in models of physical phenomena, we need *boundary conditions*, specifying information at the ends of the bar at all times, and *initial conditions*, giving the temperature throughout the bar at some time usually designated as time zero. The heat equation, together with certain initial and boundary conditions, uniquely determines the temperature distribution throughout the bar at all later times.

For example, we might have the boundary value problem

$$\frac{\partial u}{\partial t} = k\frac{\partial^2 u}{\partial x^2} \quad \text{for } 0 < x < L, t > 0,$$

$$u(0, t) = T_1, u(L, t) = T_2 \quad \text{for } t \geq 0,$$

$$u(x, 0) = f(x) \quad \text{for } 0 \leq x \leq L.$$

This problem models the temperature distribution in a bar of length L whose left end is kept at constant temperature T_1 and right end at constant temperature T_2 and whose initial temperature in the cross section at x is $f(x)$. The conditions at the ends of the bar are the boundary conditions, and the temperature at time zero is the initial condition.

As a second example, consider the boundary value problem

$$\frac{\partial u}{\partial t} = k\frac{\partial^2 u}{\partial x^2} \quad \text{for } 0 < x < L, t > 0,$$

$$\frac{\partial u}{\partial x}(0, t) = \frac{\partial u}{\partial x}(L, t) = 0 \quad \text{for } t \geq 0,$$

$$u(x, 0) = f(x) \quad \text{for } 0 \leq x \leq L.$$

This problem models the temperature distribution in a bar having no heat loss across its ends. The boundary conditions given in this problem are called *insulation conditions*.

Still other kinds of boundary conditions can be specified. For example, we might have a combination of fixed temperature and insulation conditions. If the left end is kept at constant temperature T and the right end is insulated, then

$$u(0, t) = T \quad \text{and} \quad \frac{\partial u}{\partial x}(L, t) = 0.$$

Or we might have free radiation (convection), in which the bar loses heat by radiation from its ends into the surrounding medium, which is assumed to be maintained at constant temperature T. Now the model consists of the heat equation, the initial temperature function, and the boundary conditions

$$\frac{\partial u}{\partial x}(0, t) = A[u(0, t) - T], \quad \frac{\partial u}{\partial x}(L, t) = -A[u(L, t) - T]$$

for $t \geq 0$. Here A is a positive constant. Notice that if the bar is kept hotter than the surrounding medium, then the heat flow, as measured by $\partial u/\partial x$, must be positive at the left end and negative at the right end.

Boundary conditions

$$u(0, t) = T_1, \quad \frac{\partial u}{\partial x}(L, t) = -A[u(L, t) - T_2]$$

are used if the left end is kept at the constant temperature T_1 while the right end radiates heat energy into a medium of constant temperature T_2.

In 2-space dimensions, with constant thermal conductivity, the heat equation is

$$\frac{\partial u}{\partial t} = k\left(\frac{\partial^2 u}{\partial x^2} + \frac{\partial^2 u}{\partial y^2}\right),$$

while in 3-space dimensions it is

$$\frac{\partial u}{\partial t} = k\left(\frac{\partial^2 u}{\partial x^2} + \frac{\partial^2 u}{\partial y^2} + \frac{\partial^2 u}{\partial z^2}\right).$$

SECTION 17.1   PROBLEMS

1. Formulate a boundary value problem modeling heat conduction in a thin bar of length L if the left end is kept at temperature zero and the right end is insulated. The initial temperature in the cross section at x is $f(x)$.

2. Formulate a boundary value problem modeling heat conduction in a thin bar of length L if the left end is kept at temperature $\alpha(t)$ and the right end at tempera-

ture $\beta(t)$. The initial temperature in the cross section at x is $f(x)$.

3. Formulate a boundary value for the temperature function in a thin bar of length L if the left end is kept insulated and the right end at temperature $\beta(t)$. The initial temperature in the cross section at x is $f(x)$.

17.2   Fourier Series Solutions of the Heat Equation

In this section we will solve several boundary value problems modeling heat conduction on a bounded interval. For this setting we will use separation of variables and Fourier series.

17.2.1   Ends of the Bar Kept at Temperature Zero

Suppose we want the temperature distribution $u(x, t)$ in a thin, homogeneous (constant density) bar of length L, given that the initial temperature in the bar at time zero in the cross section at x perpendicular to the x axis is $f(x)$. The ends of the bar are maintained at temperature zero for all time.

The boundary value problem modeling this temperature distribution is

$$\frac{\partial u}{\partial t} = k\frac{\partial^2 u}{\partial x^2} \quad \text{for } 0 < x < L, t > 0,$$

$$u(0, t) = u(L, t) = 0 \quad \text{for } t \geq 0,$$

$$u(x, 0) = f(x) \quad \text{for } 0 \leq x \leq L.$$

We will use separation of variables. Substitute $u(x, t) = X(x)T(t)$ into the heat equation to get

$$XT' = kX''T$$

or

$$\frac{T'}{kT} = \frac{X''}{X}.$$

The left side depends only on time, and the right side only on position, and these variables are independent. Therefore for some constant λ,

$$\frac{T'}{kT} = \frac{X''}{X} = -\lambda.$$

Now

$$u(0, t) = X(0)T(t) = 0.$$

If $T(t) = 0$ for all t, then the temperature function has the constant value zero, which occurs if the initial temperature $f(x) = 0$ for $0 \le x \le L$. Otherwise, $T(t)$ cannot be identically zero, so we must have $X(0) = 0$. Similarly, $u(L, t) = X(L)T(t) = 0$ implies that $X(L) = 0$. The problem for X is therefore

$$X'' + \lambda X = 0; \qquad X(0) = X(L) = 0.$$

We seek values of λ (the eigenvalues) for which this problem for X has nontrivial solutions (the eigenfunctions).

This problem for X is exactly the same one encountered for the space-dependent function in separating variables in the wave equation. There we found that the eigenvalues are

$$\lambda_n = \frac{n^2\pi^2}{L^2}$$

for $n = 1, 2, \ldots$, and corresponding eigenfunctions are nonzero constant multiples of

$$X_n(x) = \sin\left(\frac{n\pi x}{L}\right).$$

The problem for T becomes

$$T' + \frac{n^2\pi^2 k}{L^2} T = 0,$$

which has general solution

$$T_n(t) = c_n e^{-n^2\pi^2 kt/L^2}.$$

For $n = 1, 2, \ldots$, we now have functions

$$u_n(x, t) = c_n \sin\left(\frac{n\pi x}{L}\right) e^{-n^2\pi^2 kt/L^2}$$

which satisfy the heat equation on $[0, L]$ and the boundary conditions $u(0, t) = u(L, t) = 0$. There remains to find a solution satisfying the initial condition. We can choose n and c_n so that

$$u_n(x, 0) = c_n \sin\left(\frac{n\pi x}{L}\right) = f(x)$$

only if the given initial temperature function is a multiple of this sine function. This need not be the case. In general, we must attempt to construct a solution using the superposition

$$u(x, t) = \sum_{n=1}^{\infty} c_n \sin\left(\frac{n\pi x}{L}\right) e^{-n^2\pi^2 kt/L^2}.$$

Now we need

$$u(x, 0) = \sum_{n=1}^{\infty} c_n \sin\left(\frac{n\pi x}{L}\right) = f(x),$$

which we recognize as the Fourier sine expansion of $f(x)$ on $[0, L]$. Thus choose

$$c_n = \frac{2}{L} \int_0^L f(\xi) \sin\left(\frac{n\pi\xi}{L}\right) d\xi.$$

With this choice of the coefficients, we have the solution for the temperature distribution function:

$$u(x, t) = \frac{2}{L} \sum_{n=1}^{\infty} \left(\int_0^L f(\xi) \sin\left(\frac{n\pi\xi}{L}\right) d\xi\right) \sin\left(\frac{n\pi x}{L}\right) e^{-n^2\pi^2 kt/L^2}. \qquad (17.1)$$

EXAMPLE 17.1

Suppose the initial temperature function is constant A for $0 < x < L$, while the temperature at the ends is maintained at zero. To write the solution for the temperature distribution function, we need to compute

$$c_n = \frac{2}{L} \int_0^L A \sin\left(\frac{n\pi\xi}{L}\right) d\xi = \frac{2A}{n\pi}[1 - \cos(n\pi)] = \frac{2A}{n\pi}[1 - (-1)^n].$$

The solution (17.1) is

$$u(x, t) = \frac{2A}{\pi} \sum_{n=1}^{\infty} \frac{1 - (-1)^n}{n} \sin\left(\frac{n\pi x}{L}\right) e^{-n^2\pi^2 kt/L^2}.$$

Since $1 - (-1)^n$ is zero if n is even, and equals 2 if n is odd, we need only sum over the odd integers and can write

$$u(x, t) = \frac{4A}{\pi} \sum_{n=1}^{\infty} \frac{1}{2n - 1} \sin\left(\frac{(2n - 1)\pi x}{L}\right) e^{-(2n-1)^2\pi^2 kt/L^2}. \quad \blacksquare$$

Verification of the Solution

The function given by equation (17.1) clearly satisfies the boundary and initial conditions of the problem. Each term vanishes at $x = 0$ and at $x = L$, and the coefficients were chosen so that $u(x, 0) = f(x)$. If we could differentiate this series term-by-term, it would also be easy to show that $u(x, t)$ satisfies the heat equation, since each term does.

When we were faced with this problem with the wave equation, we used a trigonometric identity to sum the series. Here, because of the rapidly decaying exponential function in $u(x, t)$, we can easily prove that the series converges uniformly. Choose any $t_0 > 0$. Then, for $t \geq t_0$,

$$\left|\frac{1}{2n - 1} \sin\left(\frac{(2n - 1)\pi x}{L}\right) e^{-(2n-1)^2\pi^2 kt/L^2}\right| \leq \frac{1}{2n - 1} e^{-(2n-1)^2\pi^2 kt_0/L^2}.$$

Because the series

$$\sum_{n=1}^{\infty} \frac{1}{2n - 1} e^{-(2n-1)^2\pi^2 kt_0/L^2}$$

converges, the series for $u(x, t)$ converges uniformly for $0 \leq x \leq L$ and $t \geq t_0$, by a theorem of Weierstrass often referred to as the *M-test*.

By a similar argument, we can show that the series obtained by differentiating $u(x, t)$ term-by-term, once with respect to t or twice with respect to x, also converge uniformly. We can therefore differentiate this series term-by-term, once with respect to t and twice with respect to

x. Since each term in the series satisfies the heat equation, so does $u(x, t)$, verifying the solution (17.1).

We will now consider the problem of heat conduction in a bar with insulated ends.

17.2.2 Temperature in a Bar with Insulated Ends

Consider heat conduction in a bar with insulated ends, hence no energy loss across the ends. If the initial temperature is $f(x)$, the temperature function is modeled by the boundary value problem

$$\frac{\partial u}{\partial t} = k \frac{\partial^2 u}{\partial x^2} \quad \text{for } 0 < x < L, t > 0,$$

$$\frac{\partial u}{\partial x}(0, t) = \frac{\partial u}{\partial x}(L, t) = 0 \quad \text{for } t > 0,$$

$$u(x, 0) = f(x) \quad \text{for } 0 \leq x \leq L.$$

Attempt a separation of variables by putting $u(x, t) = X(x)T(t)$. We obtain, as in the preceding subsection,

$$X'' + \lambda X = 0, \quad T' + \lambda k T = 0.$$

Now

$$\frac{\partial u}{\partial x}(0, t) = X'(0)T(t) = 0$$

implies (except in the trivial case of zero temperature) that $X'(0) = 0$. Similarly,

$$\frac{\partial u}{\partial x}(L, t) = X'(L)T(t) = 0$$

implies that $X'(L) = 0$. The problem for $X(x)$ is therefore

$$X'' + \lambda X = 0; \quad X'(0) = X'(L) = 0.$$

We have encountered this problem before (for example, in Section 15.3.1). The eigenvalues are

$$\lambda_n = \frac{n^2 \pi^2}{L^2}$$

for $n = 0, 1, 2, \ldots$, with eigenfunctions nonzero constant multiples of

$$X_n(x) = \cos\left(\frac{n\pi x}{L}\right).$$

The equation for T is now

$$T' + \frac{n^2 \pi^2 k}{L^2} T = 0.$$

When $n = 0$, we get

$$T_0(t) = \text{constant}.$$

For $n = 1, 2, \ldots,$

$$T_n(t) = c_n e^{-n^2 \pi^2 k t / L^2}.$$

We now have functions

$$u_n(x, t) = c_n \cos\left(\frac{n\pi x}{L}\right) e^{-n^2 \pi^2 k t / L^2}$$

for $n = 0, 1, 2, \ldots$, each of which satisfies the heat equation and the insulation boundary conditions. To satisfy the initial condition, we must generally use a superposition

$$u(x, t) = \frac{1}{2}c_0 + \sum_{n=1}^{\infty} c_n \cos\left(\frac{n\pi x}{L}\right) e^{-n^2\pi^2 kt/L^2}.$$

Here we wrote the constant term ($n = 0$) as $c_0/2$ in anticipation of a Fourier cosine expansion. Indeed, we need

$$u(x, 0) = f(x) = \frac{1}{2}c_0 + \sum_{n=1}^{\infty} c_n \cos\left(\frac{n\pi x}{L}\right), \tag{17.2}$$

the Fourier cosine expansion of $f(x)$ on $[0, L]$. (This is also the expansion of the initial temperature function in the eigenfunctions of this problem.) We therefore choose

$$c_n = \frac{2}{L}\int_0^L f(\xi)\cos\left(\frac{n\pi\xi}{L}\right) d\xi.$$

With this choice of coefficients, equation (17.2) gives the solution of this boundary value problem.

EXAMPLE 17.2

Suppose the left half of the bar is initially at temperature A and the right half is kept at temperature zero. Thus

$$f(x) = \begin{cases} A & \text{for } 0 \le x \le \dfrac{L}{2} \\ 0 & \text{for } \dfrac{L}{2} < x \le L \end{cases}.$$

Then

$$c_0 = \frac{2}{L}\int_0^{L/2} A\, d\xi = A$$

and, for $n = 1, 2, \ldots$,

$$c_n = \frac{2}{L}\int_0^{L/2} A\cos\left(\frac{n\pi\xi}{L}\right) d\xi = \frac{2A}{n\pi}\sin\left(\frac{n\pi}{2}\right).$$

The solution for this temperature function is

$$u(x, t) = \frac{1}{2}A + \frac{2A}{\pi}\sum_{n=1}^{\infty} \frac{1}{n}\sin\left(\frac{n\pi}{2}\right)\cos\left(\frac{n\pi x}{L}\right) e^{-n^2\pi^2 kt/L^2}.$$

Now $\sin(n\pi/2)$ is zero if n is even. Further, if $n = 2j - 1$ is odd, then $\sin(n\pi/2) = (-1)^{j+1}$. The solution may therefore be written

$$u(x, t) = \frac{1}{2}A + \frac{2A}{\pi}\sum_{n=1}^{\infty} \frac{(-1)^{n+1}}{2n-1}\cos\left(\frac{(2n-1)\pi x}{L}\right) e^{-(2n-1)^2\pi^2 kt/L^2}. \blacksquare$$

17.2.3   Temperature Distribution in a Bar with Radiating End

Consider a thin, homogeneous bar of length L, with the left end maintained at zero temperature, while the right end radiates energy into the surrounding medium, which is kept at temperature zero.

If the initial temperature in the bar's cross section at x is $f(x)$, then the temperature distribution is modeled by the boundary value problem

$$\frac{\partial u}{\partial t} = k\frac{\partial^2 u}{\partial x^2} \quad \text{for } 0 < x < L, t > 0,$$

$$u(0, t) = 0, \frac{\partial u}{\partial x}(L, t) = -Au(L, t) \quad \text{for } t > 0,$$

$$u(x, 0) = f(x) \quad \text{for } 0 \le x \le L.$$

The boundary condition at L assumes that heat energy radiates from this end at a rate proportional to the temperature at that end of the bar. A is a positive constant called the *transfer coefficient*.

Let $u(x, t) = X(x)T(t)$ and obtain

$$X'' + \lambda X = 0, \quad T' + \lambda kT = 0.$$

Since $u(0, t) = X(0)T(t) = 0$, then

$$X(0) = 0.$$

The condition at the right end of the bar implies that

$$X'(L)T(t) = -AX(L)T(t),$$

hence

$$X'(L) + AX(L) = 0.$$

The problem for X is therefore

$$X'' + \lambda X = 0, \quad X(0) = 0, \quad X'(L) + AX(L) = 0.$$

This is a regular Sturm–Liouville problem, which we solved in Example 15.13 for the case $A = 3$ and $L = 1$, with $y(x)$ in place of $X(x)$. We will find the eigenvalues and eigenfunctions in this more general setting by following that analysis. Consider cases on λ.

Case 1 $\lambda = 0$

Then $X(x) = cx + d$. Since $X(0) = d = 0$, then $X(x) = cx$. But then

$$X'(L) = c = -AX(L) = -AcL$$

implies that $c(1 + AL) = 0$. But $1 + AL > 0$, so $c = 0$ and this case has only the trivial solution. Hence 0 is not an eigenvalue of this problem.

Case 2 $\lambda < 0$

Write $\lambda = -\alpha^2$ with $\alpha > 0$. Then $X'' - \alpha^2 X = 0$, with general solution

$$X(x) = ce^{\alpha x} + de^{-\alpha x}.$$

Now

$$X(0) = c + d = 0$$

so $d = -c$. Then $X(x) = 2c \sinh(\alpha x)$. Next,

$$X'(L) = 2\alpha c \cosh(\alpha L) = -AX(L) = -2Ac \sinh(\alpha L).$$

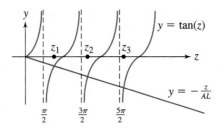

FIGURE 17.2 *Eigenvalues of the problem for a bar with a radiating end.*

Now $\alpha L > 0$, so $2\alpha c \cosh(\alpha L) > 0$ and $-2Ac \sinh(\alpha L) < 0$, so this equation is impossible unless $c = 0$. This case therefore yields only the trivial solution for X, so this problem has no negative eigenvalue.

Case 3 $\lambda > 0$

Now write $\lambda = \alpha^2$ with $\alpha > 0$. Now $X'' + \alpha^2 X = 0$, so

$$X(x) = c\cos(\alpha x) + d\sin(\alpha x).$$

Then

$$X(0) = c = 0,$$

so $X(x) = d\sin(\alpha x)$. Next,

$$X'(L) = d\alpha\cos(\alpha L) = -AX(L) = -Ad\sin(\alpha L).$$

Then $d = 0$ or

$$\tan(\alpha L) = -\frac{\alpha}{A}.$$

We can therefore have a nontrivial solution for X if α is chosen to satisfy this equation. Let $z = \alpha L$ to write this condition as

$$\tan(z) = -\frac{1}{AL}z.$$

Figure 17.2 shows graphs of $y = \tan(z)$ and $y = -z/AL$ in the z, y plane (with z as the horizontal axis). These graphs have infinitely many points of intersection to the right of the vertical axis. Let the z coordinates of these points of intersection be $z_1, z_2, \ldots$, written in increasing order. Since $\alpha = z/L$, then

$$\lambda_n = \alpha_n^2 = \frac{z_n^2}{L^2}$$

are the eigenvalues of this problem, for $n = 1, 2, \ldots$. Eigenfunctions are nonzero constant multiples of $\sin(\alpha_n x)$ or $\sin(z_n x/L)$.

The eigenvalues here are obtained as solutions of a transcendental equation that we cannot solve exactly. Nevertheless, from Figure 17.2 it is clear that there is an infinite number of positive eigenvalues, and these can be approximated as closely as we like by numerical techniques.

Now the equation for T is

$$T' + \frac{z_n^2 k}{L^2}T = 0$$

with general solution

$$T_n(t) = c_n e^{-z_n^2 kt/L^2}.$$

For each positive integer n, let

$$u_n(x, t) = X_n(x) T_n(t) = c_n \sin\left(\frac{z_n x}{L}\right) e^{-z_n^2 kt/L^2}.$$

Each of these functions satisfies the heat equation and the boundary conditions. To satisfy the initial condition, we must generally employ a superposition

$$u(x, t) = \sum_{n=1}^{\infty} c_n \sin\left(\frac{z_n x}{L}\right) e^{-z_n^2 kt/L^2}$$

and choose the c_ns so that

$$u(x, 0) = \sum_{n=1}^{\infty} c_n \sin\left(\frac{z_n x}{L}\right) = f(x).$$

This is not a Fourier sine series. It is, however, an expansion of the initial temperature function in eigenfunctions of the Sturm–Liouville problem for X. From Section 15.3.3, choose

$$c_n = \frac{\int_0^L f(\xi) \sin(z_n \xi/L) \, d\xi}{\int_0^L \sin^2(z_n \xi/L) \, d\xi}.$$

The solution is

$$u(x, t) = \sum_{n=1}^{\infty} \left(\frac{\int_0^L f(\xi) \sin(z_n \xi/L) \, d\xi}{\int_0^L \sin^2(z_n \xi/L) \, d\xi} \right) \sin\left(\frac{z_n x}{L}\right) e^{-z_n^2 kt/L^2}.$$

If we want to compute numerical values of the temperature at different points and times, we must make approximations. As an example, suppose $A = L = 1$ and $f(x) = 1$ for $0 < x < 1$. Use Newton's method to solve $\tan(z) = -z$ approximately to obtain

$$z_1 \approx 2.0288, \quad z_2 \approx 4.9132, \quad z_3 \approx 7.9787, \quad z_4 \approx 11.0855.$$

Using these values, perform numerical integrations to obtain

$$c_1 \approx 1.9207, \quad c_2 \approx 2.6593, \quad c_3 \approx 4.1457, \quad c_4 \approx 5.6329.$$

Using just the first four terms, we have the approximation

$$u(x, t) \approx 1.9027 \sin(2.0288x) e^{-4.1160kt} + 2.6593 \sin(4.9132x) e^{-24.1395kt}$$

$$+ 4.1457 \sin(7.9787x) e^{-63.6597kt} + 5.6329 \sin(11.0855x) e^{-122.8883kt}.$$

Depending on the magnitude of k, these exponentials may be decaying so fast that these first few terms would suffice for some applications.

17.2.4 Transformations of Boundary Value Problems Involving the Heat Equation

Depending on the partial differential equation and the boundary conditions, it may be impossible to separate the variables in a boundary value problem involving the heat equation. Here are two examples of strategies that work for some problems.

Heat Conduction in a Bar with Ends at Different Temperatures

Consider a thin, homogeneous bar extending from $x = 0$ to $x = L$. The left end is maintained at constant temperature T_1, and the right end at constant temperature T_2. The initial temperature throughout the bar in the cross section at x is $f(x)$.

The boundary value problem modeling this setting is

$$\frac{\partial u}{\partial t} = k \frac{\partial^2 u}{\partial x^2} \quad \text{for } 0 < x < L, t > 0,$$

$$u(0, t) = T_1, u(L, t) = T_2 \quad \text{for } t > 0,$$

$$u(x, 0) = f(x) \quad \text{for } 0 \leq x \leq L.$$

We assume that T_1 and T_2 are not both zero.

Attempt a separation of variables by putting $u(x, t) = X(x)T(t)$ into the heat equation to obtain

$$X'' + \lambda X = 0, \quad T' + \lambda kT = 0.$$

The variables have been separated. However, we must satisfy

$$u(0, t) = X(0)T(t) = T_1.$$

If $T_1 = 0$, this equation is satisfied by making $X(0) = 0$. If, however, $T_1 \neq 0$, then $T(t) = T_1/X(0) = $ constant. Similarly, $u(L, t) = X(L)T(t) = T_2$, so $T(t) = T_2/X(L) = $ constant. These conditions are impossible to satisfy except in trivial cases (such as $f(x) = 0$ and $T_1 = T_2 = 0$).

We will perturb the temperature distribution function with the idea of obtaining a more tractable problem for the perturbed function. Set

$$u(x, t) = U(x, t) + \psi(x).$$

Substitute this into the heat equation to get

$$\frac{\partial U}{\partial t} = k \left(\frac{\partial^2 U}{\partial x^2} + \psi''(x) \right).$$

This is the standard heat equation if we choose ψ so that

$$\psi''(x) = 0.$$

This means that ψ must have the form

$$\psi(x) = cx + d.$$

Now

$$u(0, t) = T_1 = U(0, t) + \psi(0)$$

becomes the more friendly condition $U(0, t) = 0$ if $\psi(0) = T_1$. Thus choose

$$d = T_1.$$

So far, $\psi(x) = cx + T_1$. Next,

$$u(L, t) = T_2 = U(L, t) + \psi(L)$$

becomes $U(L, t) = 0$ if $\psi(L) = cL + T_1 = T_2$, so choose

$$c = \frac{1}{L}(T_2 - T_1).$$

Thus let

$$\psi(x) = \frac{1}{L}(T_2 - T_1)x + T_1.$$

Finally,

$$u(x, 0) = f(x) = U(x, 0) + \psi(x)$$

becomes the following initial condition for U:

$$U(x, 0) = f(x) - \psi(x).$$

We now have a boundary value problem for U:

$$\frac{\partial U}{\partial t} = k\frac{\partial^2 U}{\partial x^2},$$

$$U(0, t) = U(L, t) = 0,$$

$$U(x, 0) = f(x) - \frac{1}{L}(T_2 - T_1)x - T_1.$$

We know the solution of this problem (equation (17.1)) and can immediately write

$$U(x, t) = \frac{2}{L}\sum_{n=1}^{\infty}\left(\int_0^L \left[f(\xi) - \frac{1}{L}(T_2 - T_1)\xi - T_1\right]\sin\left(\frac{n\pi\xi}{L}\right)d\xi\right)\sin\left(\frac{n\pi x}{L}\right)e^{-n^2\pi^2 kt/L^2}.$$

Once we obtain $U(x, t)$, the solution of the original problem is

$$u(x, t) = U(x, t) + \frac{1}{L}(T_2 - T_1)x + T_1.$$

Physically we may regard this solution as a decomposition of the temperature distribution into a transient part and a steady-state part. The transient part is $U(x, t)$, which decays to zero as t increases. The other term, $\psi(x)$, equals $\lim_{t\to\infty} u(x, t)$ and is the steady-state part. This part is independent of the time, representing the limiting value that the temperature approaches in the long-term.

Such decompositions are seen in many physical systems. For example, in a typical electrical circuit the current can be written as a transient part, which decays to zero as time increases, and a steady-state part, which is the limit of the current function as $t \to \infty$.

EXAMPLE 17.3

Suppose, in the preceding discussion, $T_1 = 1$, $T_2 = 2$, and $f(x) = \frac{3}{2}$ for $0 < x < L$. Compute

$$\int_0^L \left(f(\xi) - \frac{1}{L}(T_2 - T_1)\xi - T_1\right)\sin\left(\frac{n\pi\xi}{L}\right)d\xi = \int_0^L \left(\frac{1}{2} - \frac{1}{L}\xi\right)\sin\left(\frac{n\pi\xi}{L}\right)d\xi$$

$$= \frac{1}{2}L\frac{1 + (-1)^n}{n\pi}.$$

The solution in this case is

$$u(x, t) = \sum_{n=1}^{\infty}\left(\frac{1 + (-1)^n}{n\pi}\right)\sin\left(\frac{n\pi x}{L}\right)e^{-n^2\pi^2 kt/L^2} + \frac{1}{L}x + 1. \quad\blacksquare$$

Diffusion of Charges in a Transistor

Suppose a transistor extends over an interval $0 \leq x \leq L$ and that $h(x, t)$ is the concentration of positive charge carriers at time t and position x. It can be shown that $h(x, t)$ is closely approximated by the solution of the boundary value problem

$$\frac{\partial h}{\partial t} = k \left(\frac{\partial^2 h}{\partial x^2} - \frac{\eta}{L} \frac{\partial h}{\partial x} \right) \quad \text{for } 0 < x < L, t > 0,$$

$$h(0, t) = h(L, t) = 0 \quad \text{for } t > 0,$$

$$h(x, 0) = \frac{KL}{k\eta} \left[1 - e^{-\eta(1 - x/L)} \right] \quad \text{for } 0 \leq x \leq L,$$

in which η is a positive constant.

The preceding problem, in which the ends of the bar were kept at nonzero temperatures, was nonstandard because of the nonhomogeneous boundary conditions. This problem is nonstandard because of the first derivative term in x occurring in the partial differential equation. One way to handle this term is to make the transformation

$$h(x, t) = e^{\alpha x + \beta t} u(x, t).$$

The idea is to choose α and β to transform the problem for h into a simpler problem for u.

Upon substituting this expression for $h(x, t)$ into the partial differential equation, and dividing out the common factor of $e^{\alpha x + \beta t}$, we obtain

$$u_t = k u_{xx} + \left(2\alpha k - \frac{k\eta}{L} \right) u_x + \left(\alpha^2 k - \frac{\alpha k \eta}{L} - \beta \right) u.$$

This equation is simplified if the coefficients of u_x and u vanish. This will happen if

$$2\alpha k - \frac{k\eta}{L} = 0$$

and

$$\alpha^2 k - \frac{\alpha k \eta}{L} - \beta = 0.$$

Choose

$$\alpha = \frac{\eta}{2L} \quad \text{and} \quad \beta = -\frac{k\eta^2}{4L^2}.$$

Next,

$$h(0, t) = e^{\beta t} u(0, t) = 0$$

implies that $u(0, t) = 0$. Similarly, $u(L, t) = 0$.

Finally,

$$h(x, 0) = \frac{KL}{k\eta} \left[1 - e^{-\eta(1 - x/L)} \right] = e^{\alpha x} u(x, 0) = e^{\eta x/2L} u(x, 0),$$

so

$$u(x, 0) = \frac{KL}{k\eta} e^{-\eta x/2L} \left[1 - e^{-\eta(1 - x/L)} \right] = \frac{2KL}{k\eta} e^{-\eta/2} \sinh \left(\frac{\eta}{2} \left(1 - \frac{x}{L} \right) \right).$$

The boundary value problem for u is

$$\frac{\partial u}{\partial t} = k\frac{\partial^2 u}{\partial x^2} \quad \text{for } 0 < x < L, t > 0,$$

$$u(0, t) = u(L, t) = 0 \quad \text{for } t > 0,$$

$$u(x, 0) = \frac{2KL}{k\eta}e^{-\eta/2}\sinh\left(\frac{\eta}{2}\left(1 - \frac{x}{L}\right)\right) \quad \text{for } 0 \le x \le L.$$

We have solved this kind of problem before. The solution is

$$u(x, t) = \sum_{n=1}^{\infty} c_n \sin\left(\frac{n\pi x}{L}\right)e^{-n^2\pi^2 kt/L^2},$$

where

$$c_n = \frac{2}{L}\frac{2KL}{k\eta}e^{-\eta/2}\int_0^L \sinh\left(\frac{\eta}{2}\left(1 - \frac{\xi}{L}\right)\right)\sin\left(\frac{n\pi\xi}{L}\right)d\xi$$

$$= \frac{4K}{k\eta}e^{-\eta/2}\left[2Ln\pi\left(1 - e^{-\eta}\right)\frac{e^{\eta/2}}{4n^2\pi^2 + \eta^2}\right]$$

$$= \frac{8KL}{k\eta}(1 - e^{-\eta})\frac{n\pi}{4n^2\pi^2 + \eta^2}.$$

The original problem for $h(x, t)$ therefore has solution

$$h(x, t) = e^{\alpha x + \beta t}u(x, t)$$

$$= \frac{8KL}{k\eta}(1 - e^{-\eta})e^{\eta x/2L}e^{-\eta^2 kt/4L^2}\sum_{n=1}^{\infty}\frac{n\pi}{4n^2\pi^2 + \eta^2}\sin\left(\frac{n\pi x}{L}\right)e^{-n^2\pi^2 kt/L^2}.$$

The collapse of positive charge carriers at x produces a current $I(t)$, called the *emitter discharge current* and given by

$$I(t) = c\frac{k}{K}\frac{\partial h}{\partial x}(0, t),$$

in which c is constant. A routine calculation gives

$$I(t) = \frac{2c}{\eta}(1 - e^{-\eta})\sum_{n=1}^{\infty}\frac{n^2\pi^2}{n^2\pi^2 + \eta^2/4}\exp\left[-\frac{k}{L^2}\left(n^2\pi^2 + \frac{\eta^2}{4}\right)t\right].$$

The *reclaimable charge* is defined to be

$$Q = \int_0^{\infty} I(t)\,dt,$$

an improper integral at both limits of integration. If we integrate the series for $I(t)$ term-by-term from r to R, we get

$$\int_r^R I(t)\, dt$$

$$= \frac{2c}{\eta}(1 - e^{-\eta}) \sum_{n=1}^{\infty} \frac{n^2\pi^2}{n^2\pi^2 + \eta^2/4} \left(-\frac{L^2}{k(n^2\pi^2 + \eta^2/4)}\right) \exp\left[-\frac{k}{L^2}\left(n^2\pi^2 + \frac{\eta^2}{4}\right)R\right]$$

$$- \frac{2c}{\eta}(1 - e^{-\eta}) \sum_{n=1}^{\infty} \frac{n^2\pi^2}{n^2\pi^2 + \eta^2/4} \left(-\frac{L^2}{k(n^2\pi^2 + \eta^2/4)}\right) \exp\left[-\frac{k}{L^2}\left(n^2\pi^2 + \frac{\eta^2}{4}\right)r\right].$$

In the limit as $r \to 0+$ and $R \to \infty$, we get

$$Q = \frac{2cL^2}{k\eta}(1 - e^{-\eta}) \sum_{n=1}^{\infty} \left(\frac{n\pi}{n^2\pi^2 + \eta^2/4}\right)^2.$$

Perhaps surprisingly, this infinite series can be summed. Go back to the Fourier sine expansion made previously for $u(x, 0)$:

$$u(x, 0) = \frac{KL}{k\eta} e^{-\eta x/2L}\left[1 - e^{-\eta(1 - x/L)}\right] = \frac{2KL}{k\eta}(1 - e^{-\eta}) \sum_{n=1}^{\infty} \frac{n\pi}{n^2\pi^2 + \eta^2/4} \sin\left(\frac{n\pi x}{L}\right).$$

Then

$$\sum_{n=1}^{\infty} \frac{n\pi}{n^2\pi^2 + \eta^2/4} \sin\left(\frac{n\pi x}{L}\right) = \frac{k}{2KL}\frac{\eta}{1 - e^{-\eta}} u(x, 0)$$

$$= \frac{1}{2(1 - e^{-\eta})} e^{-\eta x/2L}\left(1 - e^{-\eta(1 - x/L)}\right) = f(x).$$

The series on the left is the Fourier sine expansion on $[0, L]$ of the function denoted $f(x)$. Now recognize that

$$\sum_{n=1}^{\infty} \left(\frac{n\pi}{n^2\pi^2 + \eta^2/4}\right)^2$$

is the sum of the squares of the coefficients of this sine expansion of $f(x)$. By Parseval's theorem for Fourier sine series, this sum of the squares of the coefficients equals $(2/L)$ times the integral of the square of the function:

$$\sum_{n=1}^{\infty} \left(\frac{n\pi}{n^2\pi^2 + \eta^2/4}\right)^2 = \frac{2}{L} \int_0^L (f(x))^2\, dx$$

$$= \frac{2}{L} \int_0^L \frac{1}{4}\frac{1}{(1 - e^{-\eta})^2} e^{-\eta x/L}\left(1 - e^{-\eta(1 - x/L)}\right)^2 dx$$

$$= \frac{1}{2L}\frac{1}{(1 - e^{-\eta})^2}\left[L\frac{1 - 2e^{-\eta}\eta - e^{-2\eta}}{\eta}\right].$$

Therefore,

$$Q = \frac{2cL^2}{k\eta}(1 - e^{-\eta})\frac{1}{2L}\frac{1}{(1 - e^{-\eta})^2}\left[L\frac{1 - 2e^{-\eta}\eta - e^{-2\eta}}{\eta}\right].$$

After some routine manipulation, we obtain

$$Q = \frac{cL^2}{k}\left(\frac{1-e^{-\eta}}{\eta}\right)\left[\frac{(\sinh(\eta)/\eta)-1}{\cosh(\eta)-1}\right],$$

a closed-form expression for the reclaimable charge on the transistor.

17.2.5   A Nonhomogeneous Heat Equation

In this section we will consider a nonhomogeneous heat conduction problem on a finite interval:

$$\frac{\partial u}{\partial t} = k\frac{\partial^2 u}{\partial x^2} + F(x,t) \quad \text{for } 0 < x < L, t > 0,$$

$$u(0,t) = u(L,t) = 0 \quad \text{for } t \geq 0,$$

$$u(x,0) = f(x) \quad \text{for } 0 \leq x \leq L.$$

The term $F(x,t)$ could, for example, account for a heat source within the medium. It is easy to check that separation of variables does not work for this heat equation. To develop another approach, go back to the simple case that $F(x,t) = 0$. In this event we found a solution

$$u(x,t) = \sum_{n=1}^{\infty} b_n \sin\left(\frac{n\pi x}{L}\right) e^{-n^2\pi^2 kt/L^2},$$

in which b_n is the nth coefficient in the Fourier sine expansion of $f(x)$ on $[0,L]$. Taking a cue from this, we will attempt a solution of the current problem of the form

$$u(x,t) = \sum_{n=1}^{\infty} T_n(t) \sin\left(\frac{n\pi x}{L}\right). \tag{17.3}$$

The problem is to determine each $T_n(t)$. The strategy for doing this is to derive a differential equation for $T_n(t)$.

If t is fixed, then the left side of equation (17.3) is just a function of x, and the right side is its Fourier sine expansion on $[0,L]$. We know the coefficients in this expansion, so

$$T_n(t) = \frac{2}{L}\int_0^L u(\xi,t)\sin\left(\frac{n\pi\xi}{L}\right)d\xi. \tag{17.4}$$

Now assume that for any choice of $t \geq 0$, $F(x,t)$, thought of as a function of x, can also be expanded in a Fourier sine series on $[0,L]$:

$$F(x,t) = \sum_{n=1}^{\infty} B_n(t)\sin\left(\frac{n\pi x}{L}\right), \tag{17.5}$$

where

$$B_n(t) = \frac{2}{L}\int_0^L F(\xi,t)\sin\left(\frac{n\pi\xi}{L}\right)d\xi. \tag{17.6}$$

This coefficient may of course depend on t.

Differentiate equation (17.4) to get

$$T_n'(t) = \frac{2}{L}\int_0^L \frac{\partial u}{\partial t}(\xi,t)\sin\left(\frac{n\pi\xi}{L}\right)d\xi. \tag{17.7}$$

Substitute for $\partial u/\partial t$ from the heat equation to get

$$T_n'(t) = \frac{2k}{L} \int_0^L \frac{\partial^2 u}{\partial x^2}(\xi, t) \sin\left(\frac{n\pi\xi}{L}\right) d\xi + \frac{2}{L} \int_0^L F(\xi, t) \sin\left(\frac{n\pi\xi}{L}\right) d\xi.$$

In view of equation (17.5), this equation becomes

$$T_n'(t) = \frac{2k}{L} \int_0^L \frac{\partial^2 u}{\partial x^2}(\xi, t) \sin\left(\frac{n\pi\xi}{L}\right) d\xi + B_n(t). \tag{17.8}$$

Now apply integration by parts twice to the integral on the right side of equation (17.8), at the end making use of the boundary conditions and of equation (17.4):

$$\int_0^L \frac{\partial^2 u}{\partial x^2}(\xi, t) \sin\left(\frac{n\pi\xi}{L}\right) d\xi = \left[\frac{\partial u}{\partial x}(\xi, t) \sin\left(\frac{n\pi\xi}{L}\right)\right]_0^L - \int_0^L \frac{n\pi}{L} \frac{\partial u}{\partial x}(\xi, t) \cos\left(\frac{n\pi\xi}{L}\right) d\xi$$

$$= -\frac{n\pi}{L} \int_0^L \frac{\partial u}{\partial x}(\xi, t) \cos\left(\frac{n\pi\xi}{L}\right) d\xi$$

$$= -\frac{n\pi}{L} \left[u(\xi, t) \cos\left(\frac{n\pi\xi}{L}\right)\right]_0^L$$

$$+ \frac{n\pi}{L} \int_0^L -\frac{n\pi}{L} u(\xi, t) \sin\left(\frac{n\pi\xi}{L}\right) d\xi$$

$$= -\frac{n^2\pi^2}{L^2} \int_0^L u(\xi, t) \sin\left(\frac{n\pi\xi}{L}\right) d\xi$$

$$= -\frac{n^2\pi^2}{L^2} \frac{L}{2} T_n(t) = -\frac{n^2\pi^2}{2L} T_n(t).$$

Substitute this into equation (17.8) to get

$$T_n'(t) = -\frac{n^2\pi^2 k}{L^2} T_n(t) + B_n(t).$$

For $n = 1, 2, \ldots$, we now have a first-order ordinary differential equation for $T_n(t)$:

$$T_n'(t) + \frac{n^2\pi^2 k}{L^2} T_n(t) = B_n(t).$$

Next, use equation (17.4) to get the condition

$$T_n(0) = \frac{2}{L} \int_0^L u(\xi, 0) \sin\left(\frac{n\pi\xi}{L}\right) d\xi = \frac{2}{L} \int_0^L f(\xi) \sin\left(\frac{n\pi\xi}{L}\right) d\xi = b_n,$$

the nth coefficient in the Fourier sine expansion of $f(x)$ on $[0, L]$. Solve the differential equation for $T_n(t)$ subject to this condition to get

$$T_n(t) = \int_0^t e^{-n^2\pi^2 k(t-\tau)/L^2} B_n(\tau) \, d\tau + b_n e^{-n^2\pi^2 kt/L^2}.$$

Finally, substitute this into equation (17.3) to obtain the solution

$$u(x, t) = \sum_{n=1}^{\infty} \left(\int_0^t e^{-n^2\pi^2 k(t-\tau)/L^2} B_n(\tau) \, d\tau\right) \sin\left(\frac{n\pi x}{L}\right)$$

$$+ \frac{2}{L} \sum_{n=1}^{\infty} \left(\int_0^L f(\xi) \sin\left(\frac{n\pi\xi}{L}\right) d\xi\right) \sin\left(\frac{n\pi x}{L}\right) e^{-n^2\pi^2 kt/L^2}. \tag{17.9}$$

Notice that the last term is the solution of the problem if the term $F(x, t)$ is missing, while the first term is the effect of the source term on the solution.

EXAMPLE 17.4

Solve the problem

$$\frac{\partial u}{\partial t} = 4\frac{\partial^2 u}{\partial x^2} + xt \quad \text{for } 0 < x < \pi, t > 0,$$

$$u(0, t) = u(\pi, t) = 0 \quad \text{for } t \geq 0,$$

$$u(x, 0) = f(x) = \begin{cases} 20 & \text{for } 0 \leq x \leq \frac{\pi}{4} \\ 0 & \text{for } \frac{\pi}{4} < x \leq \pi \end{cases}.$$

Since we have a formula for the solution, we need only carry out the required integrations. First compute

$$B_n(t) = \frac{2}{\pi} \int_0^\pi \xi t \sin(n\xi) \, d\xi = 2\frac{(-1)^{n+1}}{n} t.$$

Now we can evaluate

$$\int_0^t e^{-4n^2(t-\tau)} B_n(\tau) \, d\tau = \int_0^t 2\frac{(-1)^{n+1}}{n} \tau e^{-4n^2(t-\tau)} \, d\tau$$

$$= \frac{1}{8}(-1)^{n+1} \frac{-1 + 4n^2 t + e^{-4n^2 t}}{n^5}.$$

Finally, we need

$$b_n = \frac{2}{\pi} \int_0^\pi f(\xi) \sin(n\xi) \, d\xi = \frac{40}{\pi} \int_0^{\pi/4} \sin(n\xi) \, d\xi = \frac{40}{\pi} \frac{1 - \cos(n\pi/4)}{n}.$$

We can now write the solution

$$u(x, t) = \sum_{n=1}^\infty \left(\frac{1}{8}(-1)^{n+1} \frac{-1 + 4n^2 t + e^{-4n^2 t}}{n^5}\right) \sin(nx)$$

$$+ \sum_{n=1}^\infty \frac{40}{\pi} \frac{1 - \cos(n\pi/4)}{n} \sin(nx)e^{-4n^2 t}.$$

The second term on the right is the solution of the problem with the term xt deleted in the heat equation. Denote this "nonsource" solution as

$$u_0(x, t) = \sum_{n=1}^\infty \frac{40}{\pi} \frac{1 - \cos(n\pi/4)}{n} \sin(nx)e^{-4n^2 t}.$$

The solution with the source term is

$$u(x, t) = u_0(x, t) + \sum_{n=1}^\infty \left(\frac{1}{8}(-1)^{n+1} \frac{-1 + 4n^2 t + e^{-4n^2 t}}{n^5}\right) \sin(nx).$$

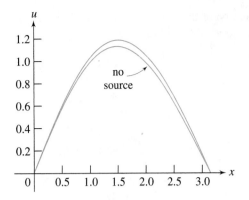

FIGURE 17.3(a) *Comparison of solutions with and without a source term for* $t = 0.3$.

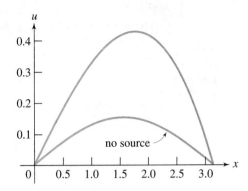

FIGURE 17.3(b) $t = 0.8$.

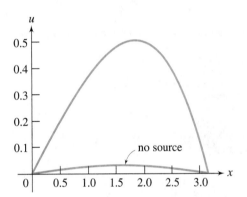

FIGURE 17.3(c) $t = 1.2$.

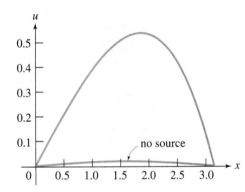

FIGURE 17.3(d) $t = 1.32$.

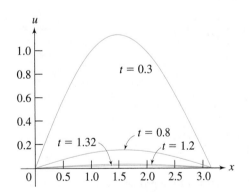

FIGURE 17.4 $u_0(x, t)$ *at times* $t = 0.3, 0.8, 1.2,$ *and* 1.32.

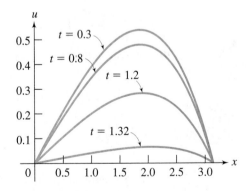

FIGURE 17.5 $u(x, t)$ *at times* $t = 0.3, 0.8, 1.2,$ *and* 1.32.

To gauge the effect on the solution of the term xt in the heat equation, Figures 17.3(a) through (d) show graphs of $u(x, t)$ and $u_0(x, t)$ at times $t = 0.3, 0.8, 1.2,$ and 1.32. Both solutions decay to zero quite rapidly as time increases. This is shown in Figure 17.4, which shows the evolution of $u_0(x, t)$ over these times, and Figure 17.5, which follows $u(x, t)$. The effect of the xt term is to retard this decay. Other terms $F(x, t)$ would of course have different effects. ∎

17.2.6 Effects of Boundary Conditions and Constants on Heat Conduction

We have solved several problems involving heat conduction in a thin, homogeneous bar of finite length. As we did with wave motion on an interval, computing power enables us to examine the effects of various constants or terms appearing in these problems on the behavior of the solutions.

EXAMPLE 17.5

Consider a thin bar of length π, whose initial temperature is given by $f(x) = x^2 \cos(x/2)$. The ends of the bar are assumed to be maintained at zero temperature. The temperature function satisfies

$$\frac{\partial u}{\partial t} = k \frac{\partial^2 u}{\partial x^2} \quad \text{for } 0 < x < \pi, t > 0,$$

$$u(0, t) = u(\pi, t) = 0 \quad \text{for } t > 0,$$

$$u(x, 0) = x^2 \cos(x/2) \quad \text{for } 0 \le x \le \pi.$$

The solution is

$$u(x, t) = \frac{2}{\pi} \sum_{n=1}^{\infty} \left(\int_0^{\pi} \xi^2 \cos\left(\frac{\xi}{2}\right) \sin(n\xi) \, d\xi \right) \sin(nx) e^{-n^2 kt}$$

$$= \frac{4}{\pi} \sum_{n=1}^{\infty} \left(\frac{16\pi n (-1)^n - 64\pi n^3 (-1)^n - 48n - 64n^3}{64n^6 - 48n^4 + 12n^2 - 1} \right) \sin(nx) e^{-n^2 kt}.$$

We can examine the effects of the diffusivity constant k on this solution by drawing graphs of $y = u(x, t)$ for various times, with different choices of this constant. Figure 17.6(a) shows the temperature distributions at time $t = 0.2$, for k taking the values 0.3, 0.6, 1.1, and 2.7. Figure 17.6(b) shows the temperature distributions at time $t = 1.2$ for these values of k. ∎

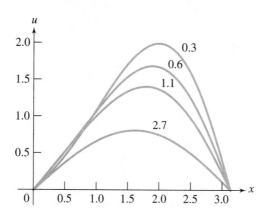

FIGURE 17.6(a) *Solution at time $t = 0.2$ with $k = 0.3, 0.6, 1.1,$ and 2.7.*

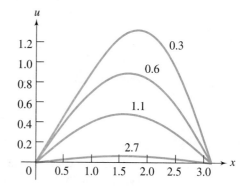

FIGURE 17.6(b) *Solution at time $t = 1.2$ with $k = 0.3, 0.6, 1.1,$ and 2.7.*

EXAMPLE 17.6

Consider an insulated bar of length π, with initial temperature function $f(x) = x^2$. The temperature distribution is

$$u(x, t) = \frac{1}{3}\pi^2 + \sum_{n=1}^{\infty}\left(4\frac{(-1)^n}{n^2}\right)\cos(nx)e^{-n^2 kt}.$$

Figure 17.7(a) shows the temperature distribution for $k = 0.4$, at times $t = 0.3, 0.7, 1.2$, and 1.7. Figures 17.7(b), (c), and (d) show the same times, but, respectively, $k = 0.9, 1.5$, and 2.4. ∎

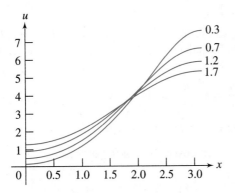

FIGURE 17.7(a) *Temperature distribution with $k = 0.4$ at times $t = 0.3, 0.7, 1.2$, and 1.7.*

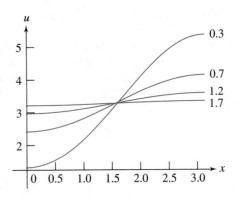

FIGURE 17.7(b) $k = 0.9$.

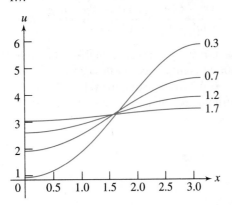

FIGURE 17.7(c) $k = 1.5$.

FIGURE 17.7(d) $k = 2.4$.

EXAMPLE 17.7

What difference does it make in the temperature distribution whether the ends are insulated or kept at temperature zero? Consider an initial temperature function $f(x) = x^2(\pi - x)$, with a bar of length π. Let the diffusivity be $k = \frac{1}{4}$. The solution if the ends are kept at temperature zero is

$$u_1(x, t) = \sum_{n=1}^{\infty}\left(\frac{8(-1)^{n+1} - 4}{n^3}\right)\sin(nx)e^{-n^2 t/4}$$

The solution if the ends are insulated is

$$u_2(x, t) = \frac{1}{12}\pi^3 + \sum_{n=1}^{\infty}\left(\frac{n^2\pi^2(-1)^{n+1} + 6(-1)^n - 6}{n^4}\right)\cos(nx)e^{-n^2 t/4}.$$

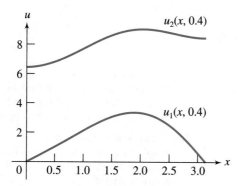

FIGURE 17.8(a) *Comparison of the solution with insulated ends, with the solution having ends kept at zero temperature, at time $t = 0.4$.*

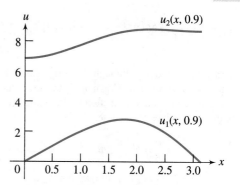

FIGURE 17.8(b) $t = 0.9$.

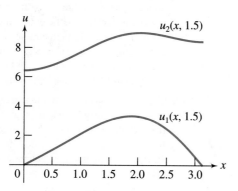

FIGURE 17.8(c) $t = 1.5$.

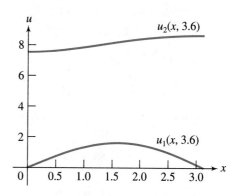

FIGURE 17.8(d) $t = 3.6$.

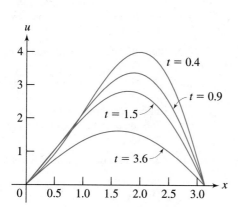

FIGURE 17.9(a) *Evolution with time of the solution with ends kept at zero temperature.*

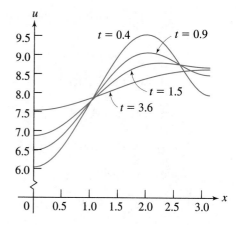

FIGURE 17.9(b) *Evolution of the solution with insulated ends.*

Figures 17.8(a) through (d) compare these two solutions for different values of the time. Figure 17.9(a) shows the evolution of the solution with zero end temperatures at different times, and Figure 17.9(b) shows this evolution for the solution with insulated ends. ▪

SECTION 17.2 PROBLEMS

In each of Problems 1 through 7, write a solution of the boundary value problem. Graph the 20th partial sum of the temperature distribution function on the same set of axes for different values of the time.

1.
$$\frac{\partial u}{\partial t} = k\frac{\partial^2 u}{\partial x^2} \quad \text{for } 0 < x < L, t > 0$$

$$u(0, t) = u(L, t) = 0 \quad \text{for } t \geq 0$$

$$u(x, 0) = x(L - x) \quad \text{for } 0 \leq x \leq L$$

2.
$$\frac{\partial u}{\partial t} = 4\frac{\partial^2 u}{\partial x^2} \quad \text{for } 0 < x < L, t > 0$$

$$u(0, t) = u(L, t) = 0 \quad \text{for } t \geq 0$$

$$u(x, 0) = x^2(L - x) \quad \text{for } 0 \leq x \leq L$$

3.
$$\frac{\partial u}{\partial t} = 3\frac{\partial^2 u}{\partial x^2} \quad \text{for } 0 < x < L, t > 0$$

$$u(0, t) = u(L, t) = 0 \quad \text{for } t \geq 0$$

$$u(x, 0) = L\left[1 - \cos\left(\frac{2\pi x}{L}\right)\right] \quad \text{for } 0 \leq x \leq L$$

4.
$$\frac{\partial u}{\partial t} = \frac{\partial^2 u}{\partial x^2} \quad \text{for } 0 < x < \pi, t > 0$$

$$\frac{\partial u}{\partial x}(0, t) = \frac{\partial u}{\partial x}(\pi, t) = 0 \quad \text{for } t \geq 0$$

$$u(x, 0) = \sin(x) \quad \text{for } 0 \leq x \leq \pi$$

5.
$$\frac{\partial u}{\partial t} = 4\frac{\partial^2 u}{\partial x^2} \quad \text{for } 0 < x < 2\pi, t > 0$$

$$\frac{\partial u}{\partial x}(0, t) = \frac{\partial u}{\partial x}(2\pi, t) = 0 \quad \text{for } t \geq 0$$

$$u(x, 0) = x(2\pi - x) \quad \text{for } 0 \leq x \leq 2\pi$$

6.
$$\frac{\partial u}{\partial t} = 4\frac{\partial^2 u}{\partial x^2} \quad \text{for } 0 < x < 3, t > 0$$

$$\frac{\partial u}{\partial x}(0, t) = \frac{\partial u}{\partial x}(3, t) = 0 \quad \text{for } t \geq 0$$

$$u(x, 0) = x^2 \quad \text{for } 0 \leq x \leq 3$$

7.
$$\frac{\partial u}{\partial t} = 2\frac{\partial^2 u}{\partial x^2} \quad \text{for } 0 < x < 6, t > 0$$

$$\frac{\partial u}{\partial x}(0, t) = \frac{\partial u}{\partial x}(6, t) = 0 \quad \text{for } t \geq 0$$

$$u(x, 0) = e^{-x} \quad \text{for } 0 \leq x \leq 6$$

8. A thin, homogeneous bar of length L has insulated ends and initial temperature B, a positive constant. Find the temperature distribution in the bar.

9. A thin, homogeneous bar of length L has initial temperature equal to a constant B, and the right end ($x = L$) is insulated, while the left end is kept at temperature zero. Find the temperature distribution in the bar.

10. A thin, homogeneous bar of thermal diffusivity 9 and length 2 cm and insulated sides has its left end maintained at temperature zero, while its right end is perfectly insulated. The bar has an initial temperature $f(x) = x^2$ for $0 \leq x \leq 2$. Determine the temperature distribution in the bar. What is $\lim_{t \to \infty} u(x, t)$?

11. A thin, homogeneous bar of thermal diffusivity 4 and length 6 cm with insulated sides has its left end maintained at temperature zero. Its right end is radiating (with transfer coefficient $\frac{1}{2}$) into the surrounding medium, which has temperature zero. The bar has an initial temperature $f(x) = x(6 - x)$. Approximate the temperature distribution $u(x, t)$ by finding the fourth partial sum of the series representation for $u(x, t)$.

12. Consider a long, thin, homogeneous bar of length L with sides poorly insulated. Heat radiates freely from the bar along its length. Assuming a positive transfer coefficient A and a constant temperature T in the surrounding medium, the equation for the temperature function is

$$\frac{\partial u}{\partial t} = k\frac{\partial^2 u}{\partial x^2} - A(u - T).$$

Assume insulated ends and an initial temperature $f(x)$. Solve for $u(x, t)$. *Hint:* Transform this boundary value problem into a familiar one by setting $v(x, t) = e^{At}u(x, t)$.

13. Show that the partial differential equation

$$\frac{\partial u}{\partial t} = k\left(\frac{\partial^2 u}{\partial x^2} + A\frac{\partial u}{\partial x} + Bu\right)$$

can be transformed into a standard heat equation by choosing α and β appropriately and letting $u(x, t) = e^{\alpha x + \beta t}v(x, t)$.

14. Use the idea of Problem 13 to solve

$$\frac{\partial u}{\partial t} = \left(\frac{\partial^2 u}{\partial x^2} + 4\frac{\partial u}{\partial x} + 2u\right) \quad \text{for } 0 < x < \pi, t > 0$$

$$u(0, t) = u(\pi, t) = 0 \quad \text{for } t \geq 0$$

$$u(x, 0) = x(\pi - x) \quad \text{for } 0 \leq x \leq \pi.$$

15. Use the idea of Problem 13 to solve

$$\frac{\partial u}{\partial t} = \left(\frac{\partial^2 u}{\partial x^2} + 6\frac{\partial u}{\partial x}\right) \quad \text{for } 0 < x < 4, t > 0$$

$$u(0, t) = u(4, t) = 0 \quad \text{for } t \geq 0$$

$$u(x, 0) = 1.$$

Graph the 20th partial sum of the solution for a selection of times.

16. Use the idea of Problem 13 to solve

$$\frac{\partial u}{\partial t} = \left(\frac{\partial^2 u}{\partial x^2} - 6\frac{\partial u}{\partial x}\right) \quad \text{for } 0 < x < \pi, t > 0$$

$$u(0, t) = u(\pi, t) = 0 \quad \text{for } t \geq 0$$

$$u(x, 0) = x^2(\pi - x) \quad \text{for } 0 \leq x \leq \pi.$$

Graph the 20th partial sum of the solution for selected times.

17. Solve

$$\frac{\partial u}{\partial t} = 16\frac{\partial^2 u}{\partial x^2} \quad \text{for } 0 < x < 1, t > 0$$

$$u(0, t) = 2, u(1, t) = 5 \quad \text{for } t \geq 0$$

$$u(x, 0) = x^2.$$

Graph the 20th partial sum of the solution for selected times.

18. Solve

$$\frac{\partial u}{\partial t} = k\frac{\partial^2 u}{\partial x^2} \quad \text{for } 0 < x < L, t > 0$$

$$u(0, t) = T, u(L, t) = 0 \quad \text{for } t \geq 0$$

$$u(x, 0) = x(L - x) \quad \text{for } 0 \leq x \leq L.$$

19. Solve

$$\frac{\partial u}{\partial t} = 9\frac{\partial^2 u}{\partial x^2} \quad \text{for } 0 < x < 5, t > 0$$

$$u(0, t) = 0, u(5, t) = 3 \quad \text{for } t \geq 0$$

$$u(x, 0) = 0 \quad \text{for } 0 \leq x \leq 5.$$

Graph the 20th partial of the solution for selected times.

20. Solve

$$\frac{\partial u}{\partial t} = k\frac{\partial^2 u}{\partial x^2} \quad \text{for } 0 < x < 9, t > 0$$

$$u(0, t) = T_1, u(9, t) = T_2 \quad \text{for } t \geq 0$$

$$u(x, 0) = x^2 \quad \text{for } 0 < x < 9.$$

Let $k = 1$ and choose a value of the time. On the same set of axes, graph the 20th partial sum of the solution for $T_2 = 1$ and T_1 taking on the values 2, 5, 9, and 14. Do this for different times to gauge the effect of the

constant temperature at the left end on the behavior of the solution.

21. Solve

$$\frac{\partial u}{\partial t} = 4\frac{\partial^2 u}{\partial x^2} - Au \quad \text{for } 0 < x < 9, t > 0$$

$$u(0, t) = u(9, t) = 0 \quad \text{for } t \geq 0$$

$$u(x, 0) = 0 \quad \text{for } 3x \leq x \leq 9.$$

Here A is a positive constant.

Choose $A = \frac{1}{4}$ and graph the 20th partial sum of the solution for a selection of times, using the same set of axes. Repeat this for the values $A = \frac{1}{2}, A = 1$, and $A = 3$. This gives some sense of the effect of the $-Au$ term in the heat equation on the behavior of the temperature distribution.

22. Solve

$$\frac{\partial u}{\partial t} = 9\frac{\partial^2 u}{\partial x^2} \quad \text{for } 0 < x < L, t > 0$$

$$u(0, t) = T, u(L, t) = 0 \quad \text{for } t \geq 0$$

$$u(x, 0) = 0 \quad \text{for } 0 \leq x \leq 2\pi.$$

23. A thin bar of thermal diffusivity 9 and length 2 m and insulated sides has its left end maintained at a temperature of $100°F$ and its right end maintained at $200°$. The bar has an initial temperature $u(x, 0) = x$ for $0 < x < 2$. Determine the temperature distribution in the bar. What is $\lim_{t \to \infty} u(x, t)$? This limit is called the *steady-state temperature distribution*. Graph the solution for selected times.

24. Determine the temperature distribution in a long, thin rod of length L that has insulated ends but radiates heat along its length into the surrounding medium with transfer coefficient A. This medium has temperature T, and the initial temperature is $f(x)$. *Hint:* Note Problem 12.

In each of Problems 25 through 29, solve the problem

$$\frac{\partial u}{\partial t} = k\frac{\partial^2 u}{\partial x^2} + F(x, t) \quad \text{for } 0 < x < L, t > 0,$$

$$u(0, t) = u(L, t) = 0 \quad \text{for } t \geq 0,$$

$$u(x, 0) = f(x) \quad \text{for } 0 \leq x \leq L$$

for the given F, k, L, and f. In each, choose a value of the time and, on the same set of axes, graph the 20th partial sum of the solution of the given problem, together with the 20th partial sum of the solution of the problem with the source term $F(x, t)$ removed. Repeat this for other times. This yields some sense of the significance of $F(x, t)$ on the behavior of the temperature distribution.

25. $k = 4, F(x, t) = t, f(x) = x(\pi - x), L = \pi$

26. $k = 1$, $F(x, t) = x \sin(t)$, $f(x) = 1$, $L = 4$

27. $k = 1$, $F(x, t) = t \cos(x)$, $f(x) = x^2(5 - x)$, $L = 5$

28. $k = 4$, $F(x, t) = \begin{cases} K & \text{for } 0 \le x \le 1 \\ 0 & \text{for } 1 < x \le 2 \end{cases}$

$f(x) = \sin(\pi x/2)$, $L = 2$

29. $k = 16$, $F(x, t) = xt$, $f(x) = K$, $L = 3$

30. Adapt the method of Section 17.2.5 to obtain a series solution of the problem

$$\frac{\partial u}{\partial t} = k\frac{\partial^2 u}{\partial x^2} + F(x, t) \quad \text{for } 0 < x < L, t > 0,$$

$$\frac{\partial u}{\partial x}(0, t) = \frac{\partial u}{\partial x}(L, t) = 0 \quad \text{for } t \ge 0,$$

$$u(x, 0) = f(x) \quad \text{for } 0 \le x \le L.$$

Hint: Begin with

$$u(x, t) = \frac{1}{2}T_0(t) + \sum_{n=1}^{\infty} T_n(t)\cos\left(\frac{n\pi x}{L}\right),$$

where

$$T_n(t) = \frac{2}{L}\int_0^L u(\xi, t)\cos\left(\frac{n\pi\xi}{L}\right)d\xi.$$

Write

$$F(x, t) = \frac{1}{2}A_0(t) + \sum_{n=1}^{\infty} A_n(t)\cos\left(\frac{n\pi x}{L}\right)$$

with

$$A_n(t) = \frac{2}{L}\int_0^L F(\xi, t)\cos\left(\frac{n\pi\xi}{L}\right)d\xi.$$

Now follow the reasoning used in the section.

31. Solve the boundary value problem of Problem 30 for the case $k = 16$, $F(x, t) = xt$, $f(x) = 1$, and $L = 5$.

32. Solve

$$\frac{\partial u}{\partial t} = k\frac{\partial^2 u}{\partial x^2} \quad \text{for } 0 < x < L, t > 0,$$

$$u(0, t) = \alpha(t), \frac{\partial u}{\partial x}(L, t) = \beta(t) \quad \text{for } t > 0,$$

$$u(x, 0) = f(x) \quad \text{for } 0 \le x \le L.$$

Hint: Attempt a solution

$$u(x, t) = \sum_{n=1}^{\infty} T_n(t)\sin\left(\frac{(2n - 1)\pi x}{2L}\right).$$

Let $\lambda_n = ((2n - 1)\pi/L)^2$. Show that $T_n'(t) + k\lambda_n T_n(t) = b_n(t)$, where

$$b_n(t) = \frac{2k}{L}\left[\sqrt{\lambda_n}\alpha(t) + (-1)^{n+1}\beta(t)\right].$$

Solve for $T_n(t)$ subject to the initial condition $T_n(0) = 0$.

33. Solve the boundary value problem of Problem 32 with $k = 1$, $\alpha(t) = 1$, $\beta(t) = t$, and $L = 1$.

17.3 Heat Conduction in Infinite Media

We will now consider problems involving the heat equation with the space variable extending over the entire real line or half-line.

17.3.1 Heat Conduction in an Infinite Bar

For a setting in which the length of the medium is very much greater than the other dimensions, it is sometimes suitable to model heat conduction by imagining the space variable free to vary over the entire real line. Consider the problem

$$\frac{\partial u}{\partial t} = k\frac{\partial^2 u}{\partial x^2} \quad \text{for } -\infty < x < \infty, t > 0,$$

$$u(x, 0) = f(x) \quad \text{for } -\infty < x < \infty.$$

There are no boundary conditions, so we impose the physically realistic condition that solutions should be bounded.

Separate the variables by putting $u(x, t) = X(x)T(t)$ to obtain

$$X'' + \lambda X = 0, \quad T' + \lambda kT = 0.$$

The problem for X is the same as that encountered with the wave equation on the line, and the same analysis yields eigenvalues $\lambda = \omega^2$ for $\omega \geq 0$ and eigenfunctions of the form $a_\omega \cos(\omega) + b_\omega \sin(\omega x)$.

The problem for T is $T' + \omega^2 k T = 0$, with general solution $de^{-\omega^2 kt}$. This is bounded for $t \geq 0$.

We now have, for $\omega \geq 0$, functions

$$u_\omega(x, t) = [a_\omega \cos(\omega) + b_\omega \sin(\omega x)] e^{-\omega^2 kt}$$

that satisfy the heat equation and are bounded on the real line. To satisfy the initial condition, attempt a superposition of these functions over all $\omega \geq 0$, which takes the form of an integral:

$$u(x, t) = \int_0^\infty [a_\omega \cos(\omega x) + b_\omega \sin(\omega x)] e^{-\omega^2 kt} \, d\omega. \tag{17.10}$$

We need

$$u(x, 0) = \int_0^\infty [a_\omega \cos(\omega) + b_\omega \sin(\omega x)] \, d\omega = f(x).$$

This is the Fourier integral of $f(x)$ on the real line, leading us to choose the coefficients

$$a_\omega = \frac{1}{\pi} \int_{-\infty}^\infty f(\xi) \cos(\omega \xi) \, d\xi$$

and

$$b_\omega = \frac{1}{\pi} \int_{-\infty}^\infty f(\xi) \sin(\omega \xi) \, d\xi.$$

EXAMPLE 17.8

Suppose the initial temperature function is $f(x) = e^{-|x|}$. Compute the coefficients

$$a_\omega = \frac{1}{\pi} \int_{-\infty}^\infty e^{-|\xi|} \cos(\omega \xi) \, d\xi = \frac{2}{\pi} \frac{1}{1 + \omega^2}$$

and

$$b_\omega = \frac{1}{\pi} \int_{-\infty}^\infty e^{-|\xi|} \sin(\omega \xi) \, d\xi = 0.$$

The solution for this initial temperature distribution is

$$u(x, t) = \frac{2}{\pi} \int_0^\infty \frac{1}{1 + \omega^2} \cos(\omega x) e^{-\omega^2 kt} \, d\omega. \ \blacksquare$$

The integral (17.10) for the solution is sometimes written in more compact form, reminiscent of the calculation in Section 16.3.1 for Fourier integral solutions of the wave equation on the entire line. Substitute the integrals for the coefficients into the integral for the solution to write

$$u(x, t) = \int_0^\infty \left[\frac{1}{\pi} \int_{-\infty}^\infty f(\xi) \cos(\omega \xi) \, d\xi \cos(\omega x) \right.$$

$$\left. + \frac{1}{\pi} \int_{-\infty}^\infty f(\xi) \sin(\omega \xi) \, d\xi \sin(\omega x) \right] e^{-\omega^2 kt} \, d\omega$$

$$= \frac{1}{\pi} \int_0^\infty \int_{-\infty}^\infty [\cos(\omega\xi)\cos(\omega x) + \sin(\omega\xi)\sin(\omega x)] f(\xi)\, d\xi\, e^{-\omega^2 kt}\, d\omega$$

$$= \frac{1}{\pi} \int_0^\infty \int_{-\infty}^\infty \cos(\omega(\xi - x)) f(\xi) e^{-\omega^2 kt}\, d\xi\, d\omega.$$

A Single Integral Expression for the Solution on the Real Line

Consider again the problem

$$\frac{\partial u}{\partial t} = k\frac{\partial^2 u}{\partial x^2} \quad \text{for } -\infty < x < \infty, t > 0,$$

$$u(x, 0) = f(x) \quad \text{for } -\infty < x < \infty.$$

We have solved this problem to obtain the double integral expression

$$u(x, t) = \frac{1}{\pi} \int_0^\infty \int_{-\infty}^\infty \cos(\omega(\xi - x)) f(\xi) e^{-\omega^2 kt}\, d\xi\, d\omega.$$

Since the integrand is an even function in ω, then $\int_0^\infty \cdots d\omega = \frac{1}{2}\int_{-\infty}^\infty \cdots d\omega$ and this solution can also be written

$$u(x, t) = \frac{1}{2\pi} \int_{-\infty}^\infty \int_{-\infty}^\infty \cos(\omega(\xi - x)) f(\xi) e^{-\omega^2 kt}\, d\xi\, d\omega.$$

We will show how this solution can be put in terms of a single integral. We need the following.

LEMMA 17.1

For real α and β, with $\beta \neq 0$,

$$\int_{-\infty}^\infty e^{-\zeta^2} \cos\left(\frac{\alpha\zeta}{\beta}\right) d\zeta = \sqrt{\pi}\, e^{-\alpha^2/4\beta^2}.$$

Proof Let

$$F(x) = \int_0^\infty e^{-\zeta^2} \cos(x\zeta)\, d\zeta.$$

One can show that this integral converges for all x, as does the integral obtained by interchanging d/dx and $\int_0^\infty \cdots d\zeta$. We can therefore compute

$$F'(x) = \int_0^\infty -e^{-\zeta^2} \zeta \sin(x\zeta)\, d\zeta.$$

Integrate by parts to get

$$F'(x) = -\frac{x}{2} F(x).$$

Then

$$\frac{F'(x)}{F(x)} = -\frac{x}{2}$$

and an integration yields

$$\ln|F(x)| = -\frac{1}{4}x^2 + c.$$

Then

$$F(x) = Ae^{-x^2/4}.$$

To evaluate the constant A, use the fact that

$$F(0) = A = \int_0^\infty e^{-\zeta^2} d\zeta = \frac{\sqrt{\pi}}{2},$$

a result found in many integral tables. Therefore,

$$\int_0^\infty e^{-\zeta^2} \cos(x\zeta) d\zeta = \frac{\sqrt{\pi}}{2} e^{-x^2/4}.$$

Finally, let $x = \alpha/\beta$ and use the fact that the integrand is even with respect to ζ to obtain

$$\int_{-\infty}^\infty e^{-\zeta^2} \cos\left(\frac{\alpha\zeta}{\beta}\right) d\zeta = 2 \int_0^\infty e^{-\zeta^2} \cos\left(\frac{\alpha\zeta}{\beta}\right) d\zeta = \sqrt{\pi} e^{-\alpha^2/4\beta^2}. \blacksquare$$

Now let

$$\zeta = \sqrt{kt}\,\omega, \alpha = x - \xi, \quad \text{and} \quad \beta = \sqrt{kt}.$$

Then

$$\frac{\alpha\zeta}{\beta} = \omega(x - \xi)$$

and

$$\int_{-\infty}^\infty e^{-\zeta^2} \cos\left(\frac{\alpha\zeta}{\beta}\right) d\zeta = \int_{-\infty}^\infty e^{-\omega^2 kt} \cos(\omega(x - \xi)) \sqrt{kt}\, d\omega = \sqrt{\pi} e^{-(x-\xi)^2/4kt}.$$

Then

$$\int_{-\infty}^\infty e^{-\omega^2 kt} \cos(\omega(x - \xi))\, d\omega = \frac{\sqrt{\pi}}{\sqrt{kt}} e^{-(x-\xi)^2/4kt}.$$

The solution of the heat conduction on the real line is therefore

$$u(x, t) = \frac{1}{2\pi} \int_{-\infty}^\infty \int_{-\infty}^\infty f(\xi) \cos(\omega(\xi - x)) e^{-\omega^2 kt}\, d\xi\, d\omega$$

$$= \frac{1}{2\pi} \int_{-\infty}^\infty \frac{\sqrt{\pi}}{\sqrt{kt}} e^{-(x-\xi)^2/4kt} f(\xi)\, d\xi.$$

After some manipulation, this solution is

$$u(x, t) = \frac{1}{2\sqrt{\pi kt}} \int_{-\infty}^\infty e^{-(x-\xi)^2/4kt} f(\xi)\, d\xi.$$

This is simpler than the previously stated solution in the sense of containing only one integral.

17.3.2 Heat Conduction in a Semi-infinite Bar

If we consider heat conduction in a bar extending from 0 to infinity, then there is a boundary condition at the left end. If the temperature is maintained at zero at this end, then the problem is

$$\frac{\partial u}{\partial t} = k\frac{\partial^2 u}{\partial x^2} \quad \text{for } 0 < x < \infty, t > 0,$$

$$u(0, t) = 0 \quad \text{for } t \geq 0,$$

$$u(x, 0) = f(x) \quad \text{for } 0 < x < \infty.$$

Letting $u(x, t) = X(x)T(t)$, the problems for X and T are

$$X'' + \lambda X = 0, \quad T' + \lambda k T = 0.$$

If we proceed as we did for the real line, we obtain $\lambda = \omega^2$ for $\omega \geq 0$ and functions

$$X_\omega(x) = a_\omega \cos(\omega x) + b_\omega \sin(\omega x).$$

Now, however, we also have the condition

$$u(0, t) = X(0)T(t) = 0,$$

implying that

$$X(0) = 0.$$

Thus we must choose each $a_\omega = 0$, leaving $X_\omega(x) = b_\omega \sin(\omega x)$. Solutions for T have the form of constants times $e^{-\omega^2 kt}$, so for each $\omega > 0$ we have functions

$$u_\omega(x, t) = b_\omega \sin(\omega x)e^{-\omega^2 kt}.$$

Each of these functions satisfies the heat equation and the boundary condition $u(0, t) = 0$. To satisfy the initial condition, write a superposition

$$u(x, t) = \int_0^\infty b_\omega \sin(\omega x)e^{-\omega^2 kt} \, d\omega. \tag{17.11}$$

Now the initial condition requires that

$$u(x, 0) = \int_0^\infty b_\omega \sin(\omega x) \, d\omega,$$

so choose the b_ω's as the coefficients in the Fourier sine integral of $f(x)$ on $[0, \infty)$:

$$b_\omega = \frac{2}{\pi} \int_0^\infty f(\xi) \sin(\omega \xi) \, d\xi.$$

With this choice of coefficients, the function given by equation (17.11) is the solution of the problem.

EXAMPLE 17.9

Suppose the initial temperature function is given by

$$f(x) = \begin{cases} \pi - x & \text{for } 0 \leq x \leq \pi \\ 0 & \text{for } x > \pi \end{cases}.$$

The coefficients in solution (17.11) are

$$b_\omega = \frac{2}{\pi} \int_0^\pi (\pi - \xi) \sin(\omega \xi) \, d\xi = \frac{2}{\pi} \frac{\pi \omega - \sin(\pi \omega)}{\omega^2}.$$

The solution for this initial temperature function is

$$u(x, t) = \frac{2}{\pi} \int_0^\infty \left(\frac{\pi \omega - \sin(\pi \omega)}{\omega^2} \right) \sin(\omega x)e^{-\omega^2 kt} \, d\omega. \quad \blacksquare$$

17.3.3   Integral Transform Methods for the Heat Equation in an Infinite Medium

As we did with the wave equation on an unbounded domain, we will illustrate the use of Fourier transforms in problems involving the heat equation.

Heat Conduction on the Line

Consider again the problem

$$\frac{\partial u}{\partial t} = k\frac{\partial^2 u}{\partial x^2} \quad \text{for } -\infty < x < \infty, t > 0,$$

$$u(x, 0) = f(x) \quad \text{for } -\infty < x < \infty,$$

which we have solved by separation of variables. Since x varies over the real line, we can attempt to use the Fourier transform in the x variable. Take the transform of the heat equation to get

$$\mathfrak{F}\left[\frac{\partial u}{\partial t}\right] = k\mathfrak{F}\left[\frac{\partial^2 u}{\partial x^2}\right].$$

Because x and t are independent, the transform passes through the partial derivative with respect to t:

$$\mathfrak{F}\left[\frac{\partial u}{\partial t}\right](\omega) = \int_{-\infty}^{\infty} \frac{\partial u(\xi, t)}{\partial t} e^{-i\omega\xi}\, d\xi = \frac{\partial}{\partial t}\int_{-\infty}^{\infty} u(\xi, t) e^{-i\omega\xi}\, d\xi = \frac{\partial}{\partial t}\hat{u}(\omega, t).$$

For the transform, in the x variable, of the second partial derivative of u with respect to x, use the operational formula:

$$\mathfrak{F}\left[\frac{\partial^2 u}{\partial x^2}\right](\omega) = -\omega^2\hat{u}(\omega, t).$$

The transform of the heat equation is therefore

$$\frac{\partial}{\partial t}\hat{u}(\omega, t) + k\omega^2\hat{u}(\omega, t) = 0,$$

with general solution

$$\hat{u}(\omega, t) = a_\omega e^{-\omega^2 kt}.$$

To determine the coefficient a_ω, take the transform of the initial condition to get

$$\hat{u}(\omega, 0) = \hat{f}(\omega) = a_\omega.$$

Therefore,

$$\hat{u}(\omega, t) = \hat{f}(\omega)e^{-\omega^2 kt}.$$

This is the Fourier transform of the solution of the problem. To retrieve the solution, apply the inverse Fourier transform:

$$u(x, t) = \mathfrak{F}^{-1}\left[\hat{f}(\omega)e^{-\omega^2 kt}\right](x) = \frac{1}{2\pi}\int_{-\infty}^{\infty} \hat{f}(\omega)e^{-\omega^2 kt}e^{i\omega x}\, d\omega.$$

Of course, the real part of this expression is $u(x, t)$. To see that this solution agrees with that obtained by separation of variables, insert the integral for $\hat{f}(\omega)$ to obtain

$$\frac{1}{2\pi} \int_{-\infty}^{\infty} \hat{f}(\omega) e^{-\omega^2 kt} e^{i\omega x} \, d\omega = \frac{1}{2\pi} \int_{-\infty}^{\infty} \left(\int_{-\infty}^{\infty} f(\xi) e^{-i\omega\xi} \, d\xi \right) e^{i\omega x} e^{-\omega^2 kt} \, d\omega$$

$$= \frac{1}{2\pi} \int_{-\infty}^{\infty} \int_{-\infty}^{\infty} f(\xi) e^{-i\omega(\xi - x)} e^{-\omega^2 kt} \, d\xi \, d\omega$$

$$= \frac{1}{2\pi} \int_{-\infty}^{\infty} \int_{-\infty}^{\infty} f(\xi) \cos(\omega(\xi - x)) e^{-\omega^2 kt} \, d\xi \, d\omega$$

$$- \frac{i}{2\pi} \int_{-\infty}^{\infty} \int_{-\infty}^{\infty} f(\xi) \sin(\omega(\xi - x)) e^{-\omega^2 kt} \, d\xi \, d\omega.$$

Taking the real part of this expression, we have

$$u(x, t) = \frac{1}{2\pi} \int_{-\infty}^{\infty} \int_{-\infty}^{\infty} f(\xi) \cos(\omega(\xi - x)) e^{-\omega^2 kt} \, d\xi \, d\omega,$$

the solution obtained by separation of variables.

Heat Conduction on the Half-Line

Consider again the problem

$$\frac{\partial u}{\partial t} = k \frac{\partial^2 u}{\partial x^2} \quad \text{for } 0 < x < \infty, t > 0,$$

$$u(0, t) = 0 \quad \text{for } t \geq 0,$$

$$u(x, 0) = f(x) \quad \text{for } 0 < x < \infty,$$

which we have solved by separation of variables. To illustrate the transform technique, we will solve this problem again using the Fourier sine transform. Take the sine transform of the heat equation with respect to x, using the operational formula for the transform of the $\partial^2 u/\partial x^2$ term, to get

$$\frac{\partial}{\partial t} \hat{u}_S(\omega, t) = -\omega^2 k \hat{u}_S(\omega, t) + \omega k u(0, t).$$

Since $u(0, t) = 0$, this is

$$\frac{\partial}{\partial t} \hat{u}_S(\omega, t) = -\omega^2 k \hat{u}_S(\omega, t),$$

with general solution

$$\hat{u}_S(\omega, t) = b_\omega e^{-\omega^2 kt}.$$

Now $u(x, 0) = f(x)$, so

$$\hat{u}_S(\omega, 0) = \hat{f}_S(\omega) = b_\omega$$

and therefore

$$\hat{u}_S(\omega, t) = \hat{f}_S(\omega) e^{-\omega^2 kt}.$$

This is the sine transform of the solution. For the solution, apply the inverse Fourier sine transform to obtain

$$u(x, t) = \frac{2}{\pi} \int_0^\infty \hat{f}_S(\omega) e^{-\omega^2 kt} \sin(\omega x)\, d\omega.$$

We leave it for the student to insert the integral expression for $\hat{f}_S(\omega)$ and show that this solution agrees with that obtained by separation of variables.

Laplace Transform Solution of a Boundary Value Problem

We have illustrated the use of the Fourier transform and Fourier sine transform in solving heat conduction problems. Here is an example in which the Laplace transform is the natural transform to use.

Consider the problem on a half-line:

$$\frac{\partial u}{\partial t} = k \frac{\partial^2 u}{\partial x^2} \quad \text{for } x > 0, t > 0,$$

$$u(x, 0) = A \quad \text{for } x > 0,$$

$$u(0, t) = \begin{cases} B & \text{for } 0 \le t \le t_0 \\ 0 & \text{for } t > t_0 \end{cases},$$

in which A, B, and t_0 are positive constants. This specifies a problem with a nonzero constant initial temperature and a discontinuous temperature distribution at the left end of the bar.

We can write the boundary condition more neatly in terms of the Heaviside function H defined in Section 3.3.2:

$$u(0, t) = B[1 - H(t - t_0)].$$

Because of the discontinuity in $u(0, t)$, we think of trying a Laplace transform in t. Denote

$$\mathcal{L}[u(x, t)](s) = U(x, s),$$

with s the variable of the transformed function and x carried along as a parameter. Take the Laplace transform of the heat equation:

$$\mathcal{L}\left[\frac{\partial u}{\partial t}\right] = k\mathcal{L}\left[\frac{\partial^2 u}{\partial x^2}\right].$$

For the transform of $\partial u/\partial t$, the derivative of the transformed variable, use the operational formula for the Laplace transform:

$$\mathcal{L}\left[\frac{\partial u}{\partial t}\right](s) = sU(x, s) - u(x, 0) = sU(x, s) - A.$$

The transform passes through $\partial^2 u/\partial x^2$ because x and t are independent:

$$\mathcal{L}\left[\frac{\partial^2 u}{\partial x^2}\right](s) = \int_0^\infty e^{-st} \frac{\partial^2 u}{\partial x^2}(x, t)\, dt = \frac{\partial^2}{\partial x^2} \int_0^\infty e^{-st} u(x, t)\, dt = \frac{\partial^2 U(x, s)}{\partial x^2}.$$

Transforming the heat equation therefore yields

$$sU(x, s) - A = k \frac{\partial^2 U(x, s)}{\partial x^2}.$$

Write this equation as

$$\frac{\partial^2 U(x, s)}{\partial x^2} - \frac{s}{k} U(x, s) = -\frac{A}{k},$$

a differential equation in x, for each $s > 0$. The general solution of this equation is

$$U(x, s) = a_s e^{\sqrt{s/k}x} + b_s e^{-\sqrt{s/k}x} + \frac{A}{s}.$$

The notation reflects the fact that the coefficients will in general depend on s. Now, to have a bounded solution we need $a_s = 0$, since $e^{\sqrt{s/k}x} \to \infty$ as $x \to \infty$. Therefore,

$$U(x, s) = b_s e^{-\sqrt{s/k}x} + \frac{A}{s}. \tag{17.12}$$

To obtain b_s, take the Laplace transform of $u(0, t) = B[1 - H(t - t_0)]$ to get

$$U(0, s) = B\mathcal{L}[1](s) - B\mathcal{L}[H(t - t_0)](s) = B\frac{1}{s} - B\frac{1}{s}e^{-t_0 s}.$$

Then

$$U(0, s) = B\frac{1}{s} - B\frac{1}{s}e^{-t_0 s} = b_s + \frac{A}{s},$$

so

$$b_s = \frac{B - A}{s} - \frac{B}{s}e^{-t_0 s}.$$

Put this into equation (17.12) to get

$$U(x, s) = \left[\frac{B - A}{s} - \frac{B}{s}e^{-t_0 s}\right]e^{-\sqrt{s/k}x} + \frac{A}{s}.$$

The solution is now obtained by using the inverse Laplace transform:

$$u(x, t) = \mathcal{L}^{-1}[U(x, s)].$$

This inverse can be calculated using standard tables and makes use of the error function and complementary error function, which see frequent use in statistics. These functions are defined by

$$\text{erf}(x) = \frac{2}{\sqrt{\pi}} \int_0^x e^{-\xi^2} d\xi$$

and

$$\text{erfc}(x) = \frac{2}{\sqrt{\pi}} \int_x^\infty e^{-\xi^2} d\xi = 1 - \text{erf}(x).$$

We obtain

$$u(x, t) = \left(A\,\text{erf}\left(\frac{x}{2\sqrt{kt}}\right) + B\,\text{erfc}\left(\frac{x}{2\sqrt{kt}}\right)\right)(1 - H(t - t_0))$$

$$+ \left(A\,\text{erf}\left(\frac{x}{2\sqrt{kt}}\right) + B\,\text{erfc}\left(\frac{x}{2\sqrt{kt}}\right) - B\,\text{erfc}\left(\frac{x}{2\sqrt{k(t - t_0)}}\right)\right)H(t - t_0).$$

SECTION 17.3 PROBLEMS

In each of Problems 1 through 6, consider the problem

$$\frac{\partial u}{\partial t} = k\frac{\partial^2 u}{\partial x^2} \quad \text{for } -\infty < x < \infty, t > 0$$

$$u(x, 0) = f(x) \quad \text{for } -\infty < x < \infty.$$

Obtain a solution first by separation of variables (Fourier integral) and then again by Fourier transform.

1. $f(x) = e^{-4|x|}$

2. $f(x) = \begin{cases} \sin(x) & \text{for } |x| \leq \pi \\ 0 & \text{for } |x| > \pi \end{cases}$

3. $f(x) = \begin{cases} x & \text{for } 0 \leq x \leq 4 \\ 0 & \text{for } x < 0 \quad \text{and for } x > 4 \end{cases}$

4. $f(x) = \begin{cases} e^{-x} & \text{for } -1 \leq x \leq 1 \\ 0 & \text{for } |x| > 1 \end{cases}$

5. $f(x) = \begin{cases} 1 & \text{for } 0 \leq x \leq 1 \\ -1 & \text{for } -1 \leq x < 0 \\ 0 & \text{for } |x| > 1 \end{cases}$

6. $f(x) = \begin{cases} -2 & \text{for } -1 \leq x \leq 0 \\ x & \text{for } 0 < x \leq 1 \\ 0 & \text{for } |x| > 1 \end{cases}$

In each of Problems 7 through 12, solve the problem

$$\frac{\partial u}{\partial t} = k\frac{\partial^2 u}{\partial x^2} \quad \text{for } 0 < x < \infty, t > 0,$$

$$u(0, t) = 0 \quad \text{for } t \geq 0,$$

$$u(x, 0) = f(x) \quad \text{for } 0 < x < \infty.$$

7. $f(x) = e^{-\alpha x}$, with α any positive constant.

8. $f(x) = xe^{-\alpha x}$, with $\alpha > 0$.

9. $f(x) = \begin{cases} 1 & \text{for } 0 \leq x \leq h \\ 0 & \text{for } x > h \end{cases}$

   with h any positive number.

10. $f(x) = \begin{cases} x & \text{for } 0 \leq x \leq 2 \\ 0 & \text{for } x > 2 \end{cases}$

11. $f(x) = \begin{cases} 1 & \text{for } 0 \leq x \leq h \\ -1 & \text{for } h < x \leq 2h \\ 0 & \text{for } x > 2h \end{cases}$

    with h any positive number.

12. $f(x) = e^{-x}\sin(x)$ for $x \geq 0$, and $k = 4$

In each of Problems 13 and 14, use a Fourier transform on the half-line to obtain a solution.

13.
$$\frac{\partial u}{\partial t} = \frac{\partial^2 u}{\partial x^2} - tu \quad \text{for } x > 0, t > 0$$

$$u(x, 0) = xe^{-x} \quad \text{for } x > 0$$

$$u(0, t) = 0 \quad \text{for } t \geq 0$$

14.
$$\frac{\partial u}{\partial t} = \frac{\partial^2 u}{\partial x^2} - u \quad \text{for } x > 0, t > 0$$

$$u(x, 0) = 0 \quad \text{for } x > 0$$

$$\frac{\partial u}{\partial x}(0, t) = f(t) \quad \text{for } t \geq 0$$

In each of Problems 15 and 16, use the Laplace transform to obtain a solution.

15.
$$\frac{\partial u}{\partial t} = k\frac{\partial^2 u}{\partial x^2} \quad \text{for } x > 0, t > 0,$$

$$u(0, t) = t^2 \quad \text{for } t \geq 0,$$

$$u(x, 0) = 0 \quad \text{for } x > 0$$

16.
$$\frac{\partial u}{\partial t} = k\frac{\partial^2 u}{\partial x^2} \quad \text{for } x > 0, t > 0,$$

$$u(0, t) = 0 \quad \text{for } t \geq 0,$$

$$u(x, 0) = e^{-x} \quad \text{for } x > 0$$

17.4   Heat Conduction in an Infinite Cylinder

We will consider the problem of determining the temperature distribution function in a solid, infinitely long, homogeneous cylinder of radius R. Let the axis of the cylinder be along the z axis in x, y, z space (Figure 17.10). If $u(x, y, z, t)$ is the temperature function, then u satisfies the

3-dimensional heat equation

$$\frac{\partial u}{\partial t} = k\left(\frac{\partial^2 u}{\partial x^2} + \frac{\partial^2 u}{\partial y^2} + \frac{\partial^2 u}{\partial z^2}\right).$$

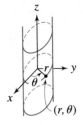

It is convenient to use cylindrical coordinates, which consist of polar coordinates in the plane together with the usual z coordinate, as in the diagram. With $x = r\cos(\theta)$ and $y = r\sin(\theta)$, let

$$u(x, y, z, t) = U(r, \theta, z, t).$$

FIGURE 17.10

We saw in Section 16.1 that

$$\frac{\partial^2 u}{\partial x^2} + \frac{\partial^2 u}{\partial y^2} = \frac{\partial^2 U}{\partial r^2} + \frac{1}{r}\frac{\partial U}{\partial r} + \frac{1}{r^2}\frac{\partial^2 U}{\partial \theta^2}.$$

Thus, in cylindrical coordinates, with $U(r, \theta, z, t)$ the temperature in the cylinder at point (r, θ, z) and time t, U satisfies:

$$\frac{\partial U}{\partial t} = k\left(\frac{\partial^2 U}{\partial r^2} + \frac{1}{r}\frac{\partial U}{\partial r} + \frac{1}{r^2}\frac{\partial^2 U}{\partial \theta^2} + \frac{\partial^2 U}{\partial z^2}\right).$$

This is a formidable equation to engage at this point, so we will assume that the temperature at any point in the cylinder depends only on the time t and the horizontal distance r from the z axis. This symmetry assumption means that $\partial U/\partial\theta = \partial U/\partial z = 0$, and the heat equation becomes

$$\frac{\partial U}{\partial t} = k\left(\frac{\partial^2 U}{\partial r^2} + \frac{1}{r}\frac{\partial U}{\partial r}\right) \quad \text{for } 0 \le r < R, t > 0.$$

In this case we will write $U(r, t)$ instead of $U(r, \theta, z, t)$.

The boundary condition is

$$U(R, t) = 0 \quad \text{for } t > 0.$$

This means that the outer surface of the cylinder is kept at zero temperature.

The initial condition is

$$U(r, 0) = f(r) \quad \text{for } 0 \le r < R.$$

Separate the variables in the heat equation by putting $U(r, t) = F(r)T(t)$. We obtain

$$F(r)T'(t) = k\left(F''(r)T(t) + \frac{1}{r}F'(r)T(t)\right).$$

Because r and t are independent variables, this yields

$$\frac{T'}{kT} = \frac{F'' + (1/r)F'}{F} = -\lambda,$$

in which λ is the separation constant. Then

$$T' + \lambda kT = 0 \quad \text{and} \quad F'' + \frac{1}{r}F' + \lambda F = 0.$$

Further, $U(R, t) = F(R)T(t) = 0$ for $t > 0$, so we have the boundary condition

$$F(R) = 0.$$

The problem for F is a singular Sturm–Liouville problem (see Section 15.3.1) on $[0, R]$, with only one boundary condition. We impose the condition that the solution must be bounded. Consider cases on λ.

Case 1 $\lambda = 0$
Now

$$F'' + \frac{1}{r}F' = 0.$$

To solve this, put $w = F'(r)$ to get

$$w'(r) + \frac{1}{r}w(r) = 0,$$

or

$$rw' + w = (rw)' = 0.$$

This has general solution

$$rw(r) = c,$$

so

$$w(r) = \frac{c}{r} = F'(r).$$

Then

$$F(r) = c\ln(r) + d.$$

Now $\ln(r) \to -\infty$ as $r \to 0+$ (center of the cylinder), so choose $c = 0$ to have a bounded solution. This means that $F(r) = $ constant for $\lambda = 0$. The equation for T in this case is $T' = 0$, with $T = $ constant also. In this event, $U(r, t) = $ constant. Since $U(R, t) = 0$, this constant must be zero. In fact, $U(r, t) = 0$ is the solution in the case that $f(r) = 0$. If the temperature on the surface is maintained at zero, and the temperature throughout the cylinder is initially zero, then the temperature distribution remains zero at all later times, in the absence of heat sources.

Case 2 $\lambda < 0$
Write $\lambda = -\omega^2$ with $\omega > 0$. Now $T' - k\omega^2 T = 0$ has general solution

$$T(t) = ce^{\omega^2 kt},$$

which is unbounded unless $c = 0$, leading again to $U(r, t) = 0$. This case leads only to the trivial solution.

Case 3 $\lambda > 0$, say $\lambda = \omega^2$.
Now $T' + k\omega^2 T = 0$ has solutions that are constant multiples of $e^{-\omega^2 kt}$, and these are bounded for $t > 0$. The equation for F is

$$F''(r) + \frac{1}{r}F'(r) + \omega^2 F(r) = 0,$$

or

$$r^2 F''(r) + rF'(r) + \omega^2 r^2 F(r) = 0.$$

In this form we recognize Bessel's equation of order zero, with general solution

$$F(r) = cJ_0(\omega r) + dY_0(\omega r).$$

J_0 is Bessel's function of the first kind of order zero, and Y_0 is Bessel's function of the second kind of order zero (see Section 15.2.3). Since $Y_0(\omega r) \to -\infty$ as $r \to 0+$, we must have $d = 0$. However, $J_0(\omega r)$ is bounded on $[0, R]$, so $F(r)$ is a constant multiple of $J_0(\omega R)$.

The condition $F(R) = 0$ now requires that this constant be zero (in which case we get the trivial solution) or that ω be chosen so that

$$J_0(\omega R) = 0.$$

This can be done. Recall that $J_0(x)$ has infinitely many positive zeros, which we arrange as

$$0 < j_1 < j_2 < \cdots.$$

We can therefore have $J_0(\omega R) = 0$ if ωR is any one of these numbers. Thus choose

$$\omega_n = \frac{j_n}{R}.$$

The numbers

$$\lambda_n = \omega_n^2 = \frac{j_n^2}{R^2}$$

are the eigenvalues of this problem, and the eigenfunctions are nonzero constant multiples of $J_0(j_n r / R)$.

We now have, for each positive integer n, a function

$$U_n(r, t) = a_n J_0\left(\frac{j_n r}{R}\right) e^{-j_n^2 k t / R^2}.$$

To satisfy the initial condition $U(r, 0) = f(r)$ we must generally use a superposition

$$U(r, t) = \sum_{n=1}^{\infty} a_n J_0\left(\frac{j_n r}{R}\right) e^{-j_n^2 k t / R^2}.$$

We now must choose the coefficients so that

$$U(r, 0) = \sum_{n=1}^{\infty} a_n J_0\left(\frac{j_n r}{R}\right) = f(r).$$

This is an eigenfunction expansion of $f(r)$ in terms of the eigenfunctions of the singular Sturm–Liouville problem for $F(r)$. We know from Section 15.3.3 how to find the coefficients. Let $\xi = r / R$. Then

$$f(R\xi) = \sum_{n=1}^{\infty} a_n J_0(j_n \xi),$$

and

$$a_n = \frac{2}{[J_1(j_n)]^2} \int_0^1 \xi f(R\xi) J_0(j_n \xi)\, d\xi.$$

The solution of the problem is

$$U(r, t) = \sum_{n=1}^{\infty} \left(\frac{2}{[J_1(j_n)]^2} \int_0^1 \xi f(R\xi) J_0(j_n \xi)\, d\xi\right) J_0\left(\frac{j_n r}{R}\right) e^{-j_n^2 k t / R^2}.$$

SECTION 17.4 PROBLEMS

1. Suppose the cylinder has radius $R = 1$ and, in polar coordinates, initial temperature $U(r, 0) = f(r) = r$

for $0 \leq r < 1$. Assume that $U(1, t) = 0$ for $t > 0$. Approximate the integral in the solution and write the first five terms in the series solution for $U(r, t)$, with $k = 1$. (The first five zeros of $J_0(x)$ are given in Section 15.2.) Graph this sum of five terms for different values of t.

2. Suppose the cylinder has radius $R = 3$ and, in polar coordinates, initial temperature $U(r, 0) = f(r) = e^r$ for $0 \leq r < 3$. Assume that $U(3, t) = 0$ for $t > 0$. Approximate the integral in the solution and write the first five terms in the series solution for $U(r, t)$, with $k = 16$. Graph this sum of five terms for different values of t.

3. Suppose the cylinder has radius $R = 3$ and, in polar coordinates, initial temperature $U(r, 0) = f(r) = 9 - r^2$ for $0 \leq r < 3$. Assume that $U(3, t) = 0$ for $t > 0$. Approximate the integral in the solution and write the

first five terms in the series solution for $U(r, t)$, with $k = \frac{1}{2}$. Graph this sum of five terms for different values of t.

4. Determine the temperature distribution in a homogeneous circular cylinder of radius R with insulated top and bottom caps under the assumption that the temperature is independent of both the radial angle and height. Assume that heat is radiating from the lateral surface into the surrounding medium, which has temperature zero, with transfer coefficient A. The initial temperature is $U(r, 0) = f(r)$. *Hint:* It will be necessary to know that an equation of the form $k J_0'(x) + A J_0(x) = 0$ has infinitely many positive solutions. This can be proved, but assume it here. Solutions of this equation yield the eigenvalues for this problem.

17.5   Heat Conduction in a Rectangular Plate

Consider the temperature distribution $u(x, y, t)$ in a flat, square, homogeneous plate covering the region $0 \leq x \leq 1, 0 \leq y \leq 1$ in the plane. The edges are kept at temperature zero and the interior temperature at time zero at (x, y) is given by

$$f(x, y) = x(1 - x^2)y(1 - y).$$

The problem for u is

$$\frac{\partial u}{\partial t} = k\left(\frac{\partial^2 u}{\partial x^2} + \frac{\partial^2 u}{\partial y^2}\right) \quad \text{for } 0 \leq x \leq 1, 0 \leq y \leq 1, t > 0,$$

$$u(x, 0, t) = u(x, 1, t) = 0 \quad \text{for } 0 < x < 1, t > 0$$

$$u(0, y, t) = u(1, y, t) = 0 \quad \text{for } 0 < y < 1, t > 0,$$

$$u(x, y, 0) = x(1 - x^2)y(1 - y).$$

Let $u(x, y, t) = X(x)Y(y)T(t)$ and obtain

$$X'' + \lambda X = 0, \quad Y'' + \mu Y = 0, \quad T' + (\lambda + \mu)kT = 0,$$

where λ and μ are the separation constants. The boundary conditions imply in the usual way that

$$X(0) = X(1) = 0, \quad Y(0) = Y(1) = 0.$$

The eigenvalues and eigenfunctions are

$$\lambda_n = n^2\pi^2, \quad X_n(x) = \sin(n\pi x),$$

for $n = 1, 2, \ldots$ and

$$\mu_m = m^2\pi^2, \quad Y_m(y) = \sin(m\pi y)$$

for $m = 1, 2, \ldots$. The problem for T is now

$$T' + (n^2 + m^2)\pi^2 kT = 0,$$

with general solution

$$T_{nm}(t) = c_{nm}e^{-(n^2+m^2)\pi^2 kt}.$$

For each positive integer n and each positive integer m, we now have functions

$$u_{nm}(x, y, t) = c_{nm} \sin(n\pi x) \sin(m\pi y)e^{-(n^2+m^2)\pi^2 kt}$$

which satisfy the heat equation and the boundary conditions. To satisfy the initial condition, let

$$u(x, y, t) = \sum_{n=1}^{\infty}\sum_{m=1}^{\infty} c_{nm} \sin(n\pi x) \sin(m\pi y)e^{-(n^2+m^2)\pi^2 kt}.$$

We must choose the coefficients so that

$$u(x, y, 0) = x(1 - x^2)y(1 - y) = \sum_{n=1}^{\infty}\sum_{m=1}^{\infty} c_{nm} \sin(n\pi x) \sin(m\pi y).$$

We find (as in Section 16.7) that

$$c_{nm} = 4\int_0^1\int_0^1 x(1 - x^2)y(1 - y) \sin(n\pi x) \sin(m\pi y)\, dx\, dy$$

$$= 4\left(\int_0^1 x(1 - x^2) \sin(n\pi x)\, dx\right)\left(\int_0^1 y(1 - y) \sin(m\pi y)\, dy\right)$$

$$= 48\left(\frac{(-1)^n}{n^3\pi^3}\right)\left(\frac{(-1)^m - 1}{m^3\pi^3}\right).$$

The solution is

$$u(x, y, z)$$

$$= \frac{48}{\pi^6}\sum_{n=1}^{\infty}\sum_{m=1}^{\infty}\left(\frac{-1)^n}{n^3}\right)\left(\frac{(-1)^n - 1}{m^3}\right) \sin(n\pi x) \sin(m\pi y)e^{-(n^2+m^2)\pi^2 kt}.$$

SECTION 17.5   PROBLEMS

1. Taking a cue from the problem just solved, write a double series solution for the following more general problem:

$$\frac{\partial u}{\partial t} = k\left(\frac{\partial^2 u}{\partial x^2} + \frac{\partial^2 u}{\partial y^2}\right) \quad \text{for } 0 < x < L,$$
$$0 < y < K, t > 0,$$

$$u(x, 0, t) = u(x, K, t) = 0 \quad \text{for } 0 \leq x \leq L, t > 0$$

$$u(0, y, t) = u(L, y, t) = 0 \quad \text{for } 0 \leq y \leq K, t > 0,$$

$$u(x, y, 0) = f(x, y).$$

2. Write the solution for Problem 1 in the case that $k = 4$, $L = 2$, $K = 3$, and $f(x, y) = x^2(L - x) \sin(y)(K - y)$.

3. Write the solution for Problem 1 in the case that $k = 1$, $L = \pi$, $K = \pi$, and $f(x, y) = \sin(x)y\cos(y/2)$.

CHAPTER 18

The Potential Equation

18.1 Harmonic Functions and the Dirichlet Problem

The partial differential equation

$$\frac{\partial^2 u}{\partial x^2} + \frac{\partial^2 u}{\partial y^2} = 0$$

is called *Laplace's equation* in two dimensions. In three dimensions this equation is

$$\frac{\partial^2 u}{\partial x^2} + \frac{\partial^2 u}{\partial y^2} + \frac{\partial^2 u}{\partial z^2} = 0.$$

The Laplacian ∇^2 (read "del squared") is defined in two dimensions by

$$\nabla^2 u = \frac{\partial^2 u}{\partial x^2} + \frac{\partial^2 u}{\partial y^2}$$

and in three dimensions by

$$\nabla^2 u = \frac{\partial^2 u}{\partial x^2} + \frac{\partial^2 u}{\partial y^2} + \frac{\partial^2 u}{\partial z^2}.$$

In this notation, Laplace's equation is $\nabla^2 u = 0$.

A function satisfying Laplace's equation in a certain region is said to be *harmonic* on that region. For example,

$$x^2 - y^2$$

and

$$2xy$$

are both harmonic over the entire plane.

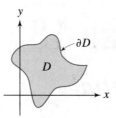

FIGURE 18.1 *Typical boundary ∂D of a region D.*

Laplace's equation is encountered in problems involving potentials, such as potentials for force fields in mechanics or electromagnetic or gravitational fields. Laplace's equation is also known as the *steady-state heat equation*. The heat equation in 2- or 3-space dimensions is

$$\frac{\partial u}{\partial t} = k\nabla^2 u.$$

In the steady-state case (the limit as $t \to \infty$), the solution becomes independent of t, so $\partial u/\partial t = 0$ and the heat equation becomes Laplace's equation.

In problems involving Laplace's equation there are no initial conditions. However, we often encounter the problem of solving

$$\nabla^2 u(x, y) = 0$$

for (x, y) in some region D of the plane, subject to the condition that

$$u(x, y) = f(x, y)$$

for (x, y) on the boundary of D. This boundary is denoted ∂D. Here f is a function having given values on ∂D, which is often a curve or made up of several curves (Figure 18.1). The problem of determining a harmonic function having given boundary values is called a *Dirichlet problem*, and f is called the *boundary data* of the problem. There are versions of this problem in higher dimensions, but we will be concerned primarily with dimension 2.

The difficulty of a Dirichlet problem is usually dependent on how complicated the region D is. In general, we have a better chance of solving a Dirichlet problem for a region that possesses some type of symmetry, such as a disk or rectangle. We will begin by solving the Dirichlet problem for some familiar regions in the plane.

SECTION 18.1 PROBLEMS

1. Let f and g be harmonic on a set D of points in the plane. Show that $f + g$ is harmonic, as well as αf for any real number α.

2. Show that the following functions are harmonic on the entire plane:

   (a) $x^3 - 3xy^2$

   (b) $3x^2 y - y^3$

   (c) $x^4 - 6x^2 y^2 + y^4$

   (d) $4x^3 y - 4xy^3$

   (e) $\sin(x)\cosh(y)$

   (f) $\cos(x)\sinh(y)$

   (g) $e^{-x}\cos(y)$

3. Show that $\ln(x^2 + y^2)$ is harmonic on the plane with the origin removed.

4. Show that $r^n \cos(n\theta)$ and $r^n \sin(n\theta)$, in polar coordinates, are harmonic on the plane, for any positive integer n. *Hint:* Look up Laplace's equation in polar coordinates.

5. Show that for any positive integer n, $r^{-n} \cos(n\theta)$, and $r^{-n} \sin(n\theta)$ are harmonic on the plane with the origin removed.

18.2 Dirichlet Problem for a Rectangle

Let R be a solid rectangle, consisting of points (x, y) with $0 \le x \le L, 0 \le y \le K$. We want to find a function that is harmonic at points interior to R and takes on prescribed values on the four sides of R, which form the boundary ∂R of R.

This kind of problem can be solved by separation of variables if the boundary data is nonzero on only one side of the rectangle. We will illustrate this kind of problem and then outline a strategy to follow if the boundary data is nonzero on more than one side.

EXAMPLE 18.1

Consider the Dirichlet problem

$$\nabla^2 u(x, y) = 0 \quad \text{for } 0 < x < L, 0 < y < K,$$

$$u(x, 0) = 0 \quad \text{for } 0 \le x \le L,$$

$$u(0, y) = u(L, y) = 0 \quad \text{for } 0 \le y \le K,$$

$$u(x, K) = (L - x)\sin(x) \quad \text{for } 0 \le x \le L.$$

Figure 18.2 shows the region and the boundary data.

Let $u(x, y) = X(x)Y(y)$ and substitute into Laplace's equation to obtain

$$\frac{X''}{X} = -\frac{Y''}{Y} = -\lambda.$$

Then

$$X'' + \lambda X = 0 \quad \text{and} \quad Y'' - \lambda Y = 0.$$

From the boundary conditions,

$$u(x, 0) = X(x)Y(0) = 0$$

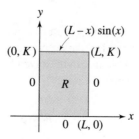

FIGURE 18.2 *Boundary data given on boundary sides of the rectangle.*

so $Y(0) = 0$. Similarly,

$$X(0) = X(L) = 0.$$

The problem for $X(x)$ is a familiar one, with eigenvalues $\lambda_n = n^2\pi^2/L^2$ and eigenfunctions that are nonzero constant multiples of $\sin(n\pi x/L)$.

The problem for Y is now

$$Y'' - \frac{n^2\pi^2}{L^2}Y = 0; \qquad Y(0) = 0.$$

Solutions of this problem are constant multiples of $\sinh(n\pi y/L)$.

For each positive integer $n = 1, 2,\ldots$, we now have functions

$$u_n(x, y) = b_n \sin\left(\frac{n\pi x}{L}\right) \sinh\left(\frac{n\pi y}{L}\right),$$

which are harmonic on the rectangle and satisfy the zero boundary conditions on the left, bottom, and right sides of the rectangle. To satisfy the boundary condition on the side $y = K$, we must use a superposition

$$u(x, y) = \sum_{n=1}^{\infty} b_n \sin\left(\frac{n\pi x}{L}\right) \sinh\left(\frac{n\pi y}{L}\right).$$

Choose the coefficients so that

$$u(x, K) = \sum_{n=1}^{\infty} b_n \sin\left(\frac{n\pi x}{L}\right) \sinh\left(\frac{n\pi K}{L}\right) = (L - x)\sin(x).$$

This is a Fourier sine expansion of $(L-x)\sin(x)$ on $[0, L]$, so we must choose the entire coefficient to be the sine coefficient:

$$b_n \sinh\left(\frac{n\pi K}{L}\right) = \frac{2}{L}\int_0^L (L - \xi)\sin(\xi)\sin\left(\frac{n\pi\xi}{L}\right)d\xi$$

$$= 4L^2\frac{n\pi[1 - (-1)^n\cos(L)]}{L^4 - 2L^2n^2\pi^2 + n^4\pi^4}.$$

Then

$$b_n = \frac{4L^2}{\sinh(n\pi K/L)}\frac{n\pi[1 - (-1)^n\cos(L)]}{(L^2 - n^2\pi^2)^2}.$$

The solution is

$$u(x, y) = \sum_{n=1}^{\infty} \frac{4L^2}{\sinh(n\pi K/L)}\frac{n\pi[1 - (-1)^n\cos(L)]}{(L^2 - n^2\pi^2)^2}\sin\left(\frac{n\pi x}{L}\right)\sinh\left(\frac{n\pi y}{L}\right). \quad\blacksquare$$

If nonzero boundary data is prescribed on all four sides of R, define four Dirichlet problems, in each of which the boundary data is nonzero on only one side. This process is outlined in Figure 18.3. Each of these problems can be solved by separation of variables. If $u_j(x, y)$ is the solution of the jth problem, then

$$u(x, y) = \sum_{j=1}^{4} u_j(x, y)$$

is the solution of the original problem. This sum will satisfy the original boundary data because each $u_j(x, y)$ satisfies the nonzero data on one side and is zero on the other three.

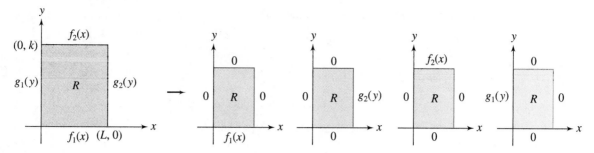

FIGURE 18.3 $u(x, y) = \Sigma_{j=1}^{4} u_j(x, y)$.

SECTION 18.2 PROBLEMS

In each of Problems 1 through 5, solve the Dirichlet problem for the rectangle, with the given boundary conditions.

1. $u(0, y) = u(1, y) = 0$ for $0 \le y \le \pi$, $u(x, \pi) = 0$, and $u(x, 0) = \sin(\pi x)$ for $0 \le x \le 1$

2. $u(0, y) = y(2 - y)$, $u(3, y) = 0$ for $0 \le y \le 2$, and $u(x, 0) = u(x, 2) = 0$ for $0 \le x \le 3$

3. $u(0, y) = u(1, y) = 0$ for $0 \le y \le 4$, and $u(x, 0) = 0$, $u(x, 4) = x \cos(\pi x/2)$ for $0 \le x \le 1$

4. $u(0, y) = \sin(y)$, $u(\pi, y) = 0$ for $0 \le y \le \pi$, and $u(x, 0) = x(\pi - x)$, $u(x, \pi) = 0$ for $0 \le x \le \pi$

5. $u(0, y) = 0$, $u(2, y) = \sin(y)$ for $0 \le y \le \pi$, and $u(x, 0) = 0$, $u(x, \pi) = x \sin(\pi x)$ for $0 \le x \le 2$

6. Apply separation of variables to solve the following mixed boundary value problem ("mixed" means that some boundary conditions are given on the function and others on its partial derivatives):

$$\nabla^2 u(x, y) = 0 \quad \text{for } 0 < x < a, 0 < y < b$$

$$u(x, 0) = \frac{\partial u}{\partial y}(x, b) = 0 \quad \text{for } 0 \le x \le a$$

$$u(0, y) = 0, u(a, y) = g(y) \quad \text{for } 0 \le y \le b.$$

7. Apply separation of variables to solve the following mixed boundary value problem:

$$\nabla^2 u(x, y) = 0 \quad \text{for } 0 < x < a, 0 < y < b$$

$$u(x, 0) = 0, u(x, b) = f(x) \quad \text{for } 0 \le x \le a$$

$$u(0, y) = \frac{\partial u}{\partial x}(a, y) = 0 \quad \text{for } 0 \le y \le b.$$

8. Solve for the steady-state temperature distribution in a thin, flat plate covering the rectangle $0 \le x \le a$, $0 \le y \le b$ if the temperature on the vertical sides and bottom side are kept at zero and the temperature along the top side is $f(x) = x(x - a)^2$.

9. Solve for the steady-state temperature distribution in a thin, flat plate covering the rectangle $0 \le x \le 4$, $0 \le y \le 1$ if the temperature on the horizontal sides is zero, while on the left side it is $f(y) = \sin(\pi y)$ and on the right side it is $f(y) = y(1 - y)$.

18.3   Dirichlet Problem for a Disk

We will solve the Dirichlet problem for a disk of radius R centered about the origin. In polar coordinates, the problem is

$$\nabla^2 u(r, \theta) = 0 \quad \text{for } 0 \le r < R, -\pi \le \theta \le \pi,$$

$$u(R, \theta) = f(\theta) \quad \text{for } -\pi \le \theta \le \pi.$$

Laplace's equation in polar coordinates is

$$\frac{\partial^2 u}{\partial r^2} + \frac{1}{r}\frac{\partial u}{\partial r} + \frac{1}{r^2}\frac{\partial^2 u}{\partial \theta^2} = 0.$$

It is easy to check that the functions

$$1, r^n \cos(n\theta), \quad \text{and} \quad r^n \sin(n\theta)$$

are all harmonic on the entire plane. Thus attempt a solution

$$u(r, \theta) = \frac{1}{2}a_0 + \sum_{n=1}^{\infty} a_n r^n \cos(n\theta) + b_n r^n \sin(n\theta).$$

To satisfy the boundary condition, we need to choose the coefficients so that

$$u(R, \theta) = f(\theta) = \frac{1}{2}a_0 + \sum_{n=1}^{\infty} a_n R^n \cos(n\theta) + b_n R^n \sin(n\theta).$$

But this is just the Fourier expansion of $f(\theta)$ on $[-\pi, \pi]$, leading us to choose

$$a_0 = \frac{1}{\pi}\int_{-\pi}^{\pi} f(\xi)\, d\xi,$$

$$a_n R^n = \frac{1}{\pi}\int_{-\pi}^{\pi} f(\xi)\cos(n\xi)\, d\xi,$$

and

$$b_n R^n = \frac{1}{\pi}\int_{-\pi}^{\pi} f(\xi)\sin(n\xi)\, d\xi.$$

Then

$$a_n = \frac{1}{\pi R^n}\int_{-\pi}^{\pi} f(\xi)\cos(n\xi)\, d\xi$$

and

$$b_n = \frac{1}{\pi R^n}\int_{-\pi}^{\pi} f(\xi)\cos(n\xi)\, d\xi.$$

The solution is

$$u(r, \theta) = \frac{1}{2\pi}\int_{-\pi}^{\pi} f(\xi)\, d\xi$$

$$+ \frac{1}{\pi}\sum_{n=1}^{\infty}\left(\frac{r}{R}\right)^n \left(\int_{-\pi}^{\pi} f(\xi)\cos(n\xi)\, d\xi \cos(n\theta) + \int_{-\pi}^{\pi} f(\xi)\sin(n\xi)\, d\xi \sin(n\theta)\right).$$

This can also be written

$$u(r, \theta) = \frac{1}{2\pi}\int_{-\pi}^{\pi} f(\xi)\, d\xi + \frac{1}{\pi}\sum_{n=1}^{\infty}\left(\frac{r}{R}\right)^n \int_{-\pi}^{\pi} f(\xi)\cos(n(\xi - \theta))\, d\xi,$$

or

$$u(r, \theta) = \frac{1}{2\pi}\int_{-\pi}^{\pi}\left[1 + 2\sum_{n=1}^{\infty}\left(\frac{r}{R}\right)^n \cos(n(\xi - \theta))\right] f(\xi)\, d\xi.$$

EXAMPLE 18.2

Solve the Dirichlet problem

$$\nabla^2 u(r, \theta) = 0 \quad \text{for } 0 \le r < 4, -\pi \le \theta \le \pi,$$

$$u(4, \theta) = f(\theta) = \theta^2 \quad \text{for } -\pi \le \theta \le \pi.$$

The solution is

$$u(r, \theta) = \frac{1}{2\pi} \int_{-\pi}^{\pi} \xi^2 \, d\xi + \frac{1}{\pi} \sum_{n=1}^{\infty} \left(\frac{r}{4}\right)^n \int_{-\pi}^{\pi} \xi^2 \cos(n(\xi - \theta)) \, d\xi$$

$$= \frac{1}{3}\pi^2 + \sum_{n=1}^{\infty} \frac{4(-1)^n}{n^2} \left(\frac{r}{4}\right)^n \cos(n\theta). \quad \blacksquare$$

EXAMPLE 18.3

Solve the Dirichlet problem

$$\nabla^2 u(x, y) = 0 \quad \text{for } x^2 + y^2 < 9,$$

$$u(x, y) = x^2 y^2 \quad \text{for } x^2 + y^2 = 9.$$

Convert the problem to polar coordinates, using $x = r\cos(\theta)$ and $y = r\sin(\theta)$. Let

$$u(x, y) = u(r\cos(\theta), r\sin(\theta)) = U(r, \theta).$$

The condition on the boundary, where $r = 3$, becomes

$$U(3, \theta) = 9\cos^2(\theta) 9\sin^2(\theta) = 81\cos^2(\theta)\sin^2(\theta) = f(\theta).$$

The solution is

$$U(r, \theta) = \frac{1}{2\pi} \int_{-\pi}^{\pi} 81 \sin^2(\xi) \cos^2(\xi) \, d\xi$$

$$+ \frac{1}{\pi} \sum_{n=1}^{\infty} \left(\frac{r}{3}\right)^n \left[\int_{-\pi}^{\pi} 81 \cos^2(\xi) \sin^2(\xi) \cos(n\xi) \, d\xi \, \cos(n\theta) \right.$$

$$\left. + \int_{-\pi}^{\pi} 81 \cos^2(\xi) \sin^2(\xi) \sin(n\xi) \, d\xi \, \sin(n\theta) \right].$$

Now

$$\int_{-\pi}^{\pi} 81 \sin^2(\xi) \cos^2(\xi) \, d\xi = \frac{81}{4}\pi,$$

$$\int_{-\pi}^{\pi} 81 \cos^2(\xi) \sin^2(\xi) \cos(n\xi) \, d\xi = \begin{cases} 0 & \text{if } n \ne 4 \\ -\frac{81\pi}{8} & \text{for } n = 4 \end{cases},$$

and

$$\int_{-\pi}^{\pi} 81 \cos^2(\xi) \sin^2(\xi) \sin(n\xi) \, d\xi = 0.$$

Therefore,

$$U(r, \theta) = \frac{1}{2\pi} \frac{81\pi}{4} - \frac{1}{\pi} \frac{81\pi}{8} \left(\frac{r}{3}\right)^4 \cos(4\theta) = \frac{81}{8} - \frac{1}{8} r^4 \cos(4\theta).$$

To convert this solution back to rectangular coordinates, use the fact that

$$\cos(4\theta) = 8\cos^4 \theta - 8\cos^2 \theta + 1.$$

Then

$$U(r, \theta) = \frac{81}{8} - \frac{1}{8} \left(8r^4 \cos^4(\theta) - 8r^2 r^2 \cos^2(\theta) + r^4\right)$$

$$= \frac{81}{8} - \frac{1}{8} \left(8x^4 - 8(x^2 + y^2)x^2 + (x^2 + y^2)^2\right) = u(x, y). \quad \blacksquare$$

SECTION 18.3 PROBLEMS

In each of Problems 1 through 8, find a series solution for the Dirichlet problem for a disk of the given radius, with the given boundary data (in polar coordinates).

1. $R = 3$, $f(\theta) = 1$

2. $R = 3$, $f(\theta) = 8\cos(4\theta)$

3. $R = 2$, $f(\theta) = \theta^2 - \theta$

4. $R = 5$, $f(\theta) = \theta \cos(\theta)$

5. $R = 4$, $f(\theta) = e^{-\theta}$

6. $R = 1$, $f(\theta) = \sin^2(\theta)$

7. $R = 8$, $f(\theta) = 1 - \theta^3$

8. $R = 4$, $f(\theta) = \theta e^{2\theta}$

By converting to polar coordinates, write a solution for each of the following Dirichlet problems.

9. $\nabla^2 u(x, y) = 0$ for $x^2 + y^2 < 16$
   $u(x, y) = x^2$ for $x^2 + y^2 = 16$

10. $\nabla^2 u(x, y) = 0$ for $x^2 + y^2 < 9$
    $u(x, y) = x - y$ for $x^2 + y^2 = 9$

11. $\nabla^2 u(x, y) = 0$ for $x^2 + y^2 < 4$
    $u(x, y) = x^2 - y^2$ for $x^2 + y^2 = 4$

12. $\nabla^2 u(x, y) = 0$ for $x^2 + y^2 < 25$
    $u(x, y) = xy$ for $x^2 + y^2 = 25$

18.4 Poisson's Integral Formula for the Disk

We have a series formula for the solution of the Dirichlet problem for a disk. In this section we will derive an integral formula for this solution. The problem for a disk of radius 1, in polar coordinates, is

$$\nabla^2 u(r, \theta) = 0 \quad \text{for } 0 \le r < 1, -\pi \le \theta \le \pi,$$

$$u(1, \theta) = f(\theta) \quad \text{for } -\pi \le \theta < \pi.$$

The series solution from the preceding section, with $R = 1$, is

$$u(r, \theta) = \frac{1}{2\pi} \int_{-\pi}^{\pi} \left[1 + 2\sum_{n=1}^{\infty} r^n \cos(n(\xi - \theta))\right] f(\xi) \, d\xi. \tag{18.1}$$

The quantity

$$\frac{1}{2\pi}\left[1 + 2\sum_{n=1}^{\infty} r^n \cos(n\zeta)\right]$$

is called the *Poisson kernel* and is denoted $P(r, \zeta)$. In terms of the Poisson kernel, the solution is

$$u(r, \theta) = \int_{-\pi}^{\pi} P(r, \xi - \theta) f(\xi)\, d\xi.$$

We will now obtain a closed form for the Poisson kernel, yielding an integral formula for the solution. Let $z = re^{i\zeta}$. By Euler's formula,

$$z^n = r^n e^{in\zeta} = r^n \cos(n\zeta) + ir^n \sin(n\zeta),$$

so $r^n \cos(n\zeta)$, which appears in the Poisson kernel, is the real part of z^n, denoted $\text{Re}(z^n)$. Then

$$1 + 2\sum_{n=1}^{\infty} r^n \cos(n\zeta) = \text{Re}\left(1 + 2\sum_{n=1}^{\infty} z^n\right).$$

But $|z| = r < 1$ in the unit disk, so the geometric series $\sum_{n=1}^{\infty} z^n$ converges. Further,

$$\sum_{n=1}^{\infty} z^n = \frac{z}{1-z}.$$

Then

$$1 + 2\sum_{n=1}^{\infty} r^n \cos(n\zeta) = \text{Re}\left(1 + 2\sum_{n=1}^{\infty} z^n\right)$$

$$= \text{Re}\left(1 + 2\frac{z}{1-z}\right) = \text{Re}\left(\frac{1+z}{1-z}\right) = \text{Re}\left(\frac{1+re^{i\zeta}}{1-re^{i\zeta}}\right).$$

The rest is just computation to help us extract this real part:

$$\frac{1+re^{i\zeta}}{1-re^{i\zeta}} = \left(\frac{1+re^{i\zeta}}{1-re^{i\zeta}}\right)\left(\frac{1-re^{-i\zeta}}{1-re^{-i\zeta}}\right)$$

$$= \frac{1-r^2 + r\left(e^{i\zeta} - e^{-i\zeta}\right)}{1+r^2 - r\left(e^{i\zeta} + e^{-i\zeta)}\right)} = \frac{1-r^2 + 2ir\sin(\zeta)}{1+r^2 - 2r\cos(\zeta)}.$$

Therefore,

$$1 + 2\sum_{n=1}^{\infty} r^n \cos(n\zeta) = \text{Re}\left(1 + 2\sum_{n=1}^{\infty} z^n\right) = \frac{1-r^2}{1+r^2 - 2r\cos(\zeta)},$$

hence the solution given by equation (18.1) is

$$u(r, \theta) = \frac{1}{2\pi} \int_{-\pi}^{\pi} \frac{1-r^2}{1+r^2 - 2r\cos(\xi - \theta)} f(\xi)\, d\xi.$$

This is *Poisson's integral formula* for the solution of the Dirichlet problem for the unit disk. For a disk of radius R, a simple change of variables yields the solution

$$u(r, \theta) = \frac{1}{2\pi} \int_{-\pi}^{\pi} \frac{R^2 - r^2}{R^2 + r^2 - 2Rr\cos(\xi - \theta)} f(\xi)\, d\xi. \tag{18.2}$$

This integral, for the disk of radius R, is also known as *Poisson's formula*.

EXAMPLE 18.4

Revisit the problem

$$\nabla^2 u(r, \theta) = 0 \quad \text{for } 0 \le r < 4, -\pi \le \theta \le \pi,$$

$$u(4, \theta) = f(\theta) = \theta^2 \quad \text{for } -\pi \le \theta \le \pi,$$

which was solved by Fourier series in the preceding section. The Poisson integral formula for the solution is

$$u(r, \theta) = \frac{1}{2\pi} \int_{-\pi}^{\pi} \frac{16 - r^2}{16 + r^2 - 8r \cos(\xi - \theta)} \xi^2 \, d\xi$$

$$= \frac{16 - r^2}{2\pi} \int_{-\pi}^{\pi} \frac{\xi^2}{16 + r^2 - 8r \cos(\xi - \theta)} \, d\xi.$$

This integral cannot be evaluated in closed form, but is often more suitable for numerical approximations than the infinite series solution. ∎

SECTION 18.4   PROBLEMS

In each of Problems 1 through 4, write an integral formula for the solution of the Dirichlet problem for a disk of radius R about the origin with the given data function on the boundary. Use the integral solution to approximate the value of $u(r, \theta)$ at the given points.

1. $R = 1$, $f(\theta) = \theta$; $(1/2, \pi)$, $(3/4, \pi/3)$, $(0.2, \pi/4)$

2. $R = 4$, $f(\theta) = \sin(4\theta)$; $(1, \pi/6)$, $(3, 7\pi/2)$, $(1, \pi/4)$, $(2.5, \pi/12)$

3. $R = 15$, $f(\theta) = \theta^3 - \theta$; $(4, \pi)$, $(12, 3\pi/2)$, $(8, \pi/4)$, $(7, 0)$

4. $R = 6$, $f(\theta) = e^{-\theta}$; $(5.5, 3\pi/5)$, $(4, 2\pi/7)$, $(1, \pi)$, $(4, \pi/4)$

5. Dirichlet's integral formula can sometimes be used to evaluate quite general integrals. As an example, let n be a positive integer and let $u(r, \theta) = r^n \sin(n\theta)$ for $0 \le r < R$, $0 \le \theta \le 2\pi$. We know that u is harmonic on the entire plane. We may therefore think of u as the solution of the Dirichlet problem on the disk $r \le R$ satisfying $u(R, \theta) = f(\theta) = R^n \sin(n\theta)$. Use the Poisson integral formula of equation (18.2) (knowing in this case the solution) to write

$$r^n \sin(n\theta)$$

$$= \frac{1}{2\pi} \int_0^{2\pi} \frac{R^2 - r^2}{r^2 - 2rR \cos(\xi - \theta) + R^2} R^n \sin(n\xi) \, d\xi.$$

Now evaluate $u(R/2, \pi/2)$ to derive the integral formula

$$\int_0^{2\pi} \frac{\sin(n\xi)}{5 - 4 \sin(\xi)} \, d\xi = \frac{\pi}{3(2^{n-1})} \sin\left(\frac{n\pi}{2}\right).$$

6. In Problem 5, evaluate $u(R/2, \pi)$. What integral is obtained?

7. Use the strategy outlined in Problem 5, but now use $u(r, \theta) = r^n \cos(n\theta)$. Obtain integrals by evaluating $u(R/2, \pi/2)$ and $u(R/2, \pi)$.

8. What integral formula is obtained by setting $u(r, \theta) = 1$ in Poisson's integral formula?

18.5   Dirichlet Problems in Unbounded Regions

We will consider the Dirichlet problem for some regions that are unbounded in the sense of containing points arbitrarily far from the origin. For such problems, the Fourier integral, Fourier transform, or Fourier sine or cosine transform may be a good means to a solution.

18.5.1   Dirichlet Problem for the Upper Half-Plane

Consider the problem

$$\nabla^2 u(x, y) = 0 \quad \text{for } -\infty < x < \infty, \, y > 0,$$

$$u(x, 0) = f(x) \quad \text{for } -\infty < x < \infty.$$

We want a function that is harmonic on the upper half-plane and takes on given values along the x axis. Let $u(x, y) = X(x)Y(y)$ and separate the variables in Laplace's equation to obtain

$$X'' + \lambda X = 0, \quad Y'' - \lambda Y = 0.$$

We want a bounded solution. Consider cases on λ.

Case 1 $\lambda = 0$
Now $X(x) = ax + b$, and we obtain a bounded solution by choosing $a = 0$. Thus 0 is an eigenvalue of this problem, with constant eigenfunctions.

Case 2 $\lambda = -\omega^2 < 0$
Now $X(x) = ae^{\omega x} + be^{-\omega x}$. But $e^{\omega x} \to \infty$ as $x \to \infty$, so we must choose $a = 0$. And $e^{-\omega x} \to \infty$ as $x \to -\infty$, so we must let $b = 0$ also, leaving the trivial solution. This problem has no negative eigenvalue.

Case 3 $\lambda = \omega^2 > 0$
Now $X(x) = a \cos(\omega x) + b \sin(\omega x)$, a bounded function for any constants a and b.

The equation for Y now becomes $Y'' - \omega^2 Y = 0$, with general solution $Y(y) = ae^{\omega y} + be^{-\omega y}$. Since $y > 0$ and $\omega > 0$, $e^{\omega y} \to \infty$ as $y \to \infty$, so we need $a = 0$. However, $e^{-\omega y}$ is bounded for $y > 0$, so $Y(y) = be^{-\omega y}$.

For each $\omega \geq 0$, we now have a function

$$u_\omega(x, y) = [a_\omega \cos(\omega x) + b_\omega \sin(\omega x)]e^{-\omega y}$$

that satisfies Laplace's equation. Attempt a solution of the problem with the superposition

$$u(x, y) = \int_0^\infty [a_\omega \cos(\omega x) + b_\omega \sin(\omega x)]e^{-\omega y} \, d\omega.$$

To satisfy the boundary condition, choose the coefficients so that

$$u(x, 0) = f(x) = \int_0^\infty [a_\omega \cos(\omega x) + b_\omega \sin(\omega x)] \, d\omega.$$

This is the Fourier integral expansion of $f(x)$, so

$$a_\omega = \frac{1}{\pi} \int_{-\infty}^\infty f(\xi) \cos(\omega \xi) \, d\xi$$

and

$$b_\omega = \frac{1}{\pi} \int_{-\infty}^\infty f(\xi) \sin(\omega \xi) \, d\xi.$$

With these coefficients, we have the solution, which can be written in a compact form, involving only one integral, as follows. Write

$$u(x, y) = \frac{1}{\pi} \int_0^\infty \left[\left(\int_{-\infty}^\infty f(\xi) \cos(\omega\xi)\, d\xi \right) \cos(\omega x) \right.$$

$$\left. + \left(\int_{-\infty}^\infty f(\xi) \sin(\omega\xi)\, d\xi \right) \sin(\omega x) \right] e^{-\omega y}\, d\omega$$

$$= \frac{1}{\pi} \int_0^\infty \int_{-\infty}^\infty [\cos(\omega\xi)\cos(\omega x) + \sin(\omega\xi)\sin(\omega x)]\, f(\xi) e^{-\omega y}\, d\xi\, d\omega$$

$$= \frac{1}{\pi} \int_{-\infty}^\infty \left[\int_0^\infty \cos(\omega(\xi - x)) e^{-\omega y}\, d\omega \right] f(\xi)\, d\xi.$$

The inner integral can be evaluated explicitly:

$$\int_0^\infty \cos(\omega(\xi - x)) e^{-\omega y}\, d\omega = \left[\frac{e^{-\omega y}}{y^2 + (\xi - x)^2} [-y \cos(\omega(\xi - x)) + (\xi - x)\sin(\omega(\xi - x))] \right]_0^\infty$$

$$= \frac{y}{y^2 + (\xi - x)^2}.$$

Therefore the solution of the Dirichlet problem for the upper half-plane is

$$u(x, y) = \frac{y}{\pi} \int_{-\infty}^\infty \frac{f(\xi)}{y^2 + (\xi - x)^2}\, d\xi. \tag{18.3}$$

To illustrate the technique, we will solve this problem again using the Fourier transform.

Solution Using the Fourier Transform

Apply the Fourier transform in the x variable to Laplace's equation. Now $\partial/\partial y$ passes through the transform, and we can use the operational rule to take the transform of the derivative with respect to x. We get

$$\mathfrak{F}\left(\frac{\partial^2 u}{\partial y^2} \right) + \mathfrak{F}\left(\frac{\partial^2 u}{\partial x^2} \right) = \frac{\partial^2 \hat{u}}{\partial y^2}(\omega, y) - \omega^2 \hat{u}(\omega, y) = 0.$$

The general solution of this differential equation in the y variable is

$$\hat{u}(\omega, y) = a_\omega e^{\omega y} + b_\omega e^{-\omega y}.$$

Keep in mind that here ω varies over the real line (unlike in the solution by Fourier integral, where ω designated a variable of integration over the half-line). Because $e^{\omega y} \to \infty$ as $y \to \infty$, we must have $a_\omega = 0$ for positive ω. But $e^{-\omega y} \to \infty$ as $y \to \infty$ if $\omega < 0$, so $b_\omega = 0$ for negative ω. Thus,

$$\hat{u}(\omega, y) = \begin{cases} b_\omega e^{-\omega y} & \text{if } \omega \geq 0 \\ a_\omega e^{\omega y} & \text{if } \omega < 0 \end{cases}.$$

We can consolidate this notation by writing

$$\hat{u}(\omega, y) = c_\omega e^{-|\omega| y}.$$

To solve for c_ω, use the fact that $u(x, 0) = f(x)$ to get

$$\hat{u}(\omega, 0) = \hat{f}(\omega) = c_\omega.$$

The Fourier transform of the solution is

$$\hat{u}(\omega, y) = \hat{f}(\omega) e^{-|\omega| y}.$$

To obtain $u(x, y)$, apply the inverse Fourier transform to this function:

$$
\begin{aligned}
u(x, y) &= \mathfrak{F}^{-1}\left[\hat{f}(\omega)e^{-|\omega|y}\right](x) \\
&= \frac{1}{2\pi}\int_{-\infty}^{\infty}\hat{f}(\omega)e^{-|\omega|y}e^{i\omega x}\,d\omega \\
&= \frac{1}{2\pi}\int_{-\infty}^{\infty}\left(\int_{-\infty}^{\infty}f(\xi)e^{-i\omega\xi}\,d\xi\right)e^{-|\omega|y}e^{i\omega x}\,d\omega \\
&= \frac{1}{2\pi}\int_{-\infty}^{\infty}\left(\int_{-\infty}^{\infty}e^{-|\omega|y}e^{-i\omega(\xi-x)}\,d\omega\right)f(\xi)\,d\xi.
\end{aligned}
$$

Now

$$
e^{-i\omega(\xi-x)} = \cos(\omega(\xi-x)) - i\sin(\omega(\xi-x))
$$

and a routine integration gives

$$
\int_{-\infty}^{\infty}e^{-|\omega|y}e^{-i\omega(\xi-x)}\,d\omega = \frac{2y}{y^2 + (\xi-x)^2}.
$$

The solution by Fourier transform is

$$
u(x, y) = \frac{y}{\pi}\int_{-\infty}^{\infty}\frac{f(\xi)}{y^2 + (\xi-x)^2}\,d\xi,
$$

in agreement with the solution obtained by using separation of variables.

18.5.2 Dirichlet Problem for the Right Quarter-Plane

Sometimes we can use the solution of one problem to produce a solution of another problem. We will illustrate this with the Dirichlet problem for the right quarter-plane:

$$
\nabla^2 u(x, y) = 0 \quad \text{for } x > 0,\ y > 0,
$$

$$
u(x, 0) = f(x) \quad \text{for } x \geq 0,
$$

$$
u(0, y) = 0 \quad \text{for } y \geq 0.
$$

The boundary of the right quarter-plane consists of the nonnegative x axis together with the nonnegative y axis, and information about the function sought must be given on both segments. In this case we are prescribing zero values on the vertical part and given values $f(x)$ on the horizontal part of the boundary.

We could solve this problem by separation of variables. However, if we fold the upper half-plane across the vertical axis, we obtain the right quarter-plane, suggesting that we explore the possibility of using the solution for the upper half-plane to obtain the solution for the right quarter-plane. To do this, let

$$
g(x) = \begin{cases} f(x) & \text{for } x \geq 0 \\ \text{anything} & \text{for } x < 0 \end{cases}.
$$

By "anything," we mean that for the moment we do not care what values are given $g(x)$ for $x < 0$, but reserve the right to assign these values later.

The Dirichlet problem

$$
\nabla^2 u(x, y) = 0 \quad \text{for } -\infty < x < \infty,\ y > 0,
$$

$$
u(x, 0) = g(x) \quad \text{for } -\infty < x < \infty
$$

for the upper half-plane has the solution

$$u_{hp}(x, y) = \frac{y}{\pi} \int_{-\infty}^{\infty} \frac{g(\xi)}{y^2 + (\xi - x)^2} \, d\xi.$$

Write this as

$$u_{hp}(x, y) = \frac{y}{\pi} \left[\int_{-\infty}^{0} \frac{g(\xi)}{y^2 + (\xi - x)^2} \, d\xi + \int_{0}^{\infty} \frac{g(\xi)}{y^2 + (\xi - x)^2} \, d\xi \right].$$

Change variables in the leftmost integral by letting $w = -\xi$. This integral becomes

$$\int_{-\infty}^{0} \frac{g(\xi)}{y^2 + (\xi - x)^2} \, d\xi = \int_{\infty}^{0} \frac{g(-w)}{y^2 + (w + x)^2} (-1) \, dw.$$

Now replace the integration variable by ξ again to write

$$u_{hp}(x, y) = \frac{y}{\pi} \left[\int_{\infty}^{0} \frac{g(-\xi)}{y^2 + (\xi + x)^2} (-1) \, d\xi + \int_{0}^{\infty} \frac{g(\xi)}{y^2 + (\xi - x)^2} \, d\xi \right]$$

$$= \frac{y}{\pi} \int_{0}^{\infty} \left(\frac{g(-\xi)}{y^2 + (\xi + x)^2} + \frac{f(\xi)}{y^2 + (\xi - x)^2} \right) d\xi,$$

where in the last integral we have used the fact that $g(\xi) = f(\xi)$ if $\xi \geq 0$. Now fill in the "anything" in the definition of g. Observe that the last integral will vanish on the positive y axis, at points $(0, y)$, if $f(\xi) + g(-\xi) = 0$ for $\xi \geq 0$. This will occur if $g(-\xi) = -f(\xi)$. That is, make g the odd extension of f to the entire real line, obtaining

$$u_{hp}(x, y) = \frac{y}{\pi} \int_{0}^{\infty} \left(\frac{1}{y^2 + (\xi - x)^2} - \frac{1}{y^2 + (\xi + x)^2} \right) f(\xi) \, d\xi.$$

This is the solution of this particular Dirichlet problem for the upper half-plane. But this function is also harmonic on the right quarter-plane, vanishes when $x = 0$, and equals $f(x)$ if $x \geq 0$ and $y = 0$. Therefore, $u_{hp}(x, y)$ is also the solution of this Dirichlet problem for the right quarter-plane.

EXAMPLE 18.5

Consider the problem

$$\nabla^2 u = 0 \quad \text{for } x > 0, y > 0,$$
$$u(0, y) = 0 \quad \text{for } y > 0,$$
$$u(x, 0) = xe^{-x} \quad \text{for } x > 0.$$

The solution is

$$u(x, y) = \frac{y}{\pi} \int_{0}^{\infty} \left(\frac{1}{y^2 + (\xi - x)^2} - \frac{1}{y^2 + (\xi + x)^2} \right) \xi e^{-\xi} \, d\xi. \ \blacksquare$$

EXAMPLE 18.6

We will solve the problem

$$\nabla^2 u = 0 \quad \text{for } x > 0, y > 0,$$
$$u(0, y) = 0 \quad \text{for } y > 0,$$
$$u(x, 0) = 1 \quad \text{for } x > 0.$$

The solution is

$$u(x, y) = \frac{y}{\pi} \int_0^\infty \frac{1}{y^2 + (\xi - x)^2} \, d\xi - \frac{y}{\pi} \int_0^\infty \frac{1}{y^2 + (\xi + x)^2} \, d\xi.$$

These integrals can be evaluated in closed form. For the first,

$$\frac{y}{\pi} \int_0^\infty \frac{1}{y^2 + (\xi - x)^2} \, d\xi = \frac{y}{\pi} \int_0^\infty \frac{1}{y^2 \left[1 + ((\xi - x)/y)^2 \right]} \, d\xi$$

$$= \frac{y}{\pi} \frac{1}{y} \left(\frac{\pi}{2} - \arctan\left(-\frac{x}{y} \right) \right) = \frac{1}{2} + \frac{1}{\pi} \arctan\left(\frac{x}{y} \right).$$

By a similar calculation,

$$\frac{y}{\pi} \int_0^\infty \frac{1}{y^2 + (\xi + x)^2} \, d\xi = \frac{1}{2} - \frac{1}{\pi} \arctan\left(\frac{x}{y} \right).$$

Then

$$u(x, y) = \frac{2}{\pi} \arctan\left(\frac{x}{y} \right).$$

This function is harmonic on the right quarter-plane and $u(0, y) = 0$ for $y > 0$. Further, if $x > 0$,

$$\lim_{y \to 0+} \frac{2}{\pi} \arctan\left(\frac{x}{y} \right) = \frac{2}{\pi} \frac{\pi}{2} = 1,$$

as required. ■

18.5.3 An Electrostatic Potential Problem

Consider the problem

$$\nabla^2 u(x, y) = -h \quad \text{for } 0 < x < \pi, y > 0,$$

$$u(0, y) = 0, u(\pi, y) = 1 \quad \text{for } y > 0,$$

$$u(x, 0) = 0 \quad \text{for } 0 < x < \pi.$$

This is a Dirichlet problem if $h = 0$, but we will assume that h is a positive constant. This problem models the electrostatic potential in the strip consisting of all (x, y) with $0 < x < \pi$ and $y > 0$, assuming a uniform distribution of charge having density $h/4\pi$ throughout this region. The partial differential equation $\nabla^2 u = -h$ is called *Poisson's equation*. The boundary of the strip consists of the half-lines $x = 0$ and $x = \pi$ with $y \geq 0$ and the segment on the x axis with $0 \leq x \leq \pi$. The strip and its boundary are shown in Figure 18.4.

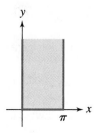

FIGURE 18.4
Strip $0 \leq x \leq \pi$,
$y \geq 0$.

Consider the possibilities for approaching this problem. Since $y > 0$, we might consider a Fourier sine or cosine transform in y. The difficulty here is that in transforming Poisson's equation, we would have to take the transform of $-h$, and a constant does not have a sine or cosine transform. For example, if we try to compute the Fourier sine transform, we must evaluate

$$\int_0^\infty -h \sin(\omega x)\, dx,$$

and this integral diverges.

Since x varies from 0 to π, we might try a finite Fourier sine or cosine transform in x. If we try the finite Fourier cosine transform, then the operational formula requires that we have information about the derivative of the function at the origin, and we have no such information. However, the finite sine transform's operational formula requires information about the function at the ends of the interval, and this is given in the boundary conditions for $y > 0$. We will therefore attempt a solution using this transform. Denote the finite Fourier sine transform in the x variable as

$$\mathcal{S}[u(x, y)](n) = \tilde{u}_S(n, y).$$

Now apply the transform with respect to x to Poisson's equation:

$$\mathcal{S}\left[\frac{\partial^2 u}{\partial x^2}\right] + \mathcal{S}\left[\frac{\partial^2 u}{\partial y^2}\right] = \mathcal{S}[-h].$$

By the operational formula,

$$\mathcal{S}\left[\frac{\partial^2 u}{\partial x^2}\right](n) = -n^2 \tilde{u}_S(n, y) - n(-1)^n u(\pi, y) + nu(0, y).$$

Because x and y are independent,

$$\mathcal{S}\left[\frac{\partial^2 u}{\partial y^2}\right] = \int_0^\pi \frac{\partial^2 u}{\partial y^2}(x, y) \sin(nx)\, dx = \frac{\partial^2}{\partial y^2} \int_0^\pi u(x, y) \sin(nx)\, dx = \frac{\partial^2}{\partial y^2} \tilde{u}_S(n, y).$$

Finally,

$$\mathcal{S}[-h] = \int_0^\pi -h \sin(nx)\, dx = -\frac{h}{n}[1 - (-1)^n].$$

Therefore Poisson's equation transforms to

$$-n^2 \tilde{u}_S(n, y) - n(-1)^n u(\pi, y) + nu(0, y) + \frac{\partial^2}{\partial y^2} \tilde{u}_S(n, y) = -\frac{h}{n}[1 - (-1)^n].$$

Now $u(\pi, y) = 1$ and $u(0, y) = 0$, so this equation can be written as

$$\frac{\partial^2}{\partial y^2} \tilde{u}_S(n, y) - n^2 \tilde{u}_S(n, y) = n(-1)^n - \frac{h}{n}[1 - (-1)^n].$$

For $n = 1, 2, \ldots$, this equation has general solution

$$\tilde{u}_S(n, y) = a_n e^{ny} + b_n e^{-ny} + \frac{(-1)^{n+1}}{n} + \frac{h}{n^3}[1 - (-1)^n].$$

For this function to remain bounded for $y > 0$, choose $a_n = 0$ for $n = 1, 2, \ldots$. Then

$$\tilde{u}_S(n, y) = b_n e^{-ny} + \frac{(-1)^{n+1}}{n} + \frac{h}{n^3}[1 - (-1)^n].$$

To solve for b_n, take the transform of the condition $u(x, 0) = 0$ and let $y = 0$ in the preceding equation to get

$$0 = \tilde{u}_S(n, 0) = b_n + \frac{(-1)^{n+1}}{n} + \frac{h}{n^3}[1 - (-1)^n].$$

Then

$$b_n = \frac{(-1)^n}{n} - \frac{h}{n^3}[1 - (-1)^n].$$

We therefore have

$$\tilde{u}_S(n, y) = \left[\frac{(-1)^n}{n} - \frac{h}{n^3}[1 - (-1)^n]\right]e^{-ny} + \frac{(-1)^{n+1}}{n} + \frac{h}{n^3}[1 - (-1)^n]$$

$$= \left[\frac{(-1)^n}{n} - \frac{h}{n^3}[1 - (-1)^n]\right](e^{-ny} - 1).$$

By the inversion formula, these are the coefficients in the Fourier sine series (in x) of the solution, so the solution is

$$u(x, y) = \frac{2}{\pi}\sum_{n=1}^{\infty}\left[\frac{(-1)^n}{n} - \frac{h}{n^3}[1 - (-1)^n]\right](e^{-ny} - 1)\sin(nx).$$

SECTION 18.5 PROBLEMS

1. Write an integral solution for the Dirichlet problem for the upper half-plane if the boundary data is

$$f(x) = \begin{cases} -1 & \text{for } -4 \le x < 0 \\ 1 & \text{for } 0 \le x \le 4 \\ 0 & \text{for } |x| > 4 \end{cases}.$$

2. Write an integral solution for the Dirichlet problem for the upper half-plane if the boundary data is $f(x) = e^{-|x|}$.

3. Write an integral solution for the Dirichlet problem for the right quarter-plane if

$$u(x, 0) = e^{-x}\cos(x) \quad \text{for } x > 0$$

and

$$u(0, y) = 0 \quad \text{for } y > 0.$$

4. Write an integral solution for the Dirichlet problem for the right quarter-plane if

$$u(x, 0) = 0 \quad \text{for } x > 0$$

and

$$u(0, y) = g(y) \quad \text{for } y > 0.$$

Derive a solution first using separation of variables, and then by using an appropriate Fourier transform.

5. Write an integral solution for the Dirichlet problem for the right quarter-plane if

$$u(x, 0) = f(x) \quad \text{for } x > 0$$

and

$$u(0, y) = g(y) \quad \text{for } y > 0.$$

6. Write an integral solution for the Dirichlet problem for the lower half-plane $y < 0$.

7. Obtain an integral solution for the Dirichlet problem for the quarter-plane $x < 0, y < 0$, if the boundary data is

$$u(x, 0) = f(x) \quad \text{for } x < 0$$

and

$$u(0, y) = 0 \quad \text{for } y < 0.$$

8. Solve the electrostatic potential problem for the strip $0 < x < \pi, y > 0$ with the boundary data

$$u(0, y) = 4, u(\pi, y) = 0 \quad \text{for } y > 0,$$

$$u(x, 0) = 0 \quad \text{for } 0 < x < \pi.$$

9. Solve the electrostatic potential problem for the strip $0 < x < \pi$, $y > 0$ with the boundary data

$$u(0, y) = 0, u(\pi, y) = 0 \quad \text{for } y > 0,$$

$$u(x, 0) = B \sin(x) \quad \text{for } 0 < x < \pi.$$

Here B is a positive constant.

10. Solve the Dirichlet problem for the strip $-\infty < x < \infty, 0 < y < 1$ if

$$u(x, 0) = 0 \quad \text{for } x < 0$$

and

$$u(x, 0) = e^{-\alpha x} \quad \text{for } x > 0,$$

with α a positive number.

11. Solve the Dirichlet problem for the strip $0 < x < \pi$, $y > 0$ if

$$u(0, y) = 0 \quad \text{and} \quad u(\pi, y) = 2 \quad \text{for } y > 0,$$

and

$$u(x, 0) = -4 \quad \text{for } 0 < x < \pi.$$

12. Solve the following problem, in which data on the boundary is a mixture of values of the function and values of a partial derivative of the function:

$$\nabla^2 u = 0 \quad \text{for } 0 < x < \pi, 0 < y < 2,$$

$$u(0, y) = 0 \quad \text{and} \quad u(\pi, y) = 4 \quad \text{for } 0 < y < 2,$$

and

$$\frac{\partial u}{\partial y}(x, 0) = u(x, 2) = 0 \quad \text{for } 0 < x < \pi.$$

13. Solve the Dirichlet problem for the strip $x > 0$, $0 < y < 1$ if

$$u(0, y) = y^2(1 - y) \quad \text{for } 0 < y < 1$$

and

$$u(x, 0) = u(x, 1) = 0 \quad \text{for } x > 0.$$

14. Solve

$$\nabla^2 u = 0 \quad \text{for } -\infty < x < \infty, 0 < y < 1,$$

$$\frac{\partial u}{\partial y}(x, 0) = 0$$

and

$$u(x, 1) = e^{-x^2} \quad \text{for } -\infty < x < \infty.$$

15. Find the steady-state temperature distribution in a thin, homogeneous, flat plate extending over the right quarter-plane $x \geq 0, y \geq 0$ if the temperature at y on the vertical side is e^{-y} and the temperature on the horizontal side is zero.

16. Solve for the steady-state temperature distribution in a homogeneous, infinite, flat plate covering the half-plane $x \geq 0$ if the temperature on the boundary $x = 0$ is $f(y)$, where

$$f(y) = \begin{cases} 1 & \text{for } |y| \leq 1 \\ 0 & \text{for } |y| > 1 \end{cases}.$$

17. Solve for the steady-state temperature distribution in an infinite, homogeneous, flat plate covering the half-plane $y \geq 0$ if the temperature on the boundary $y = 0$ is zero for $x < 4$, constant A for $4 \leq x \leq 8$, and zero for $x > 8$.

18. Write a general expression for the steady-state temperature distribution in an infinite, homogeneous, flat plate covering the strip $0 \leq y \leq 1, x \geq 0$ if the temperatures on the left boundary and on the bottom side are zero and the temperature on the top part of the boundary is $f(x)$.

19. Find the steady-state temperature distribution for a flat, homogeneous plate covering the quarter plane $x \geq 0$, $y \geq 0$ if the temperature on the horizontal side is xe^{-x}, and on the vertical side, e^{-y}.

20. Find the steady-state temperature distribution for a homogeneous, flat plate extending over the strip $x \geq 0$, $0 \leq y \leq 2$ if the left and bottom sides are insulated and the top side has temperature function $f(x)$.

21. Find the steady-state temperature distribution in a thin, homogeneous, flat plate covering the quarter-plane $x \geq 0, y \geq 0$ if the temperature on the vertical side is e^{-y} and the bottom side is insulated.

22. Find the steady-state temperature distribution for a homogeneous, flat plate extending over the strip $x \geq 0$, $0 \leq y \leq 1$ if the temperature on the horizontal sides is zero and the temperature on the vertical side is 2.

18.6   A Dirichlet Problem for a Cube

We will illustrate a Dirichlet problem in 3-space. Consider:

$$\nabla^2 u(x, y, z) = 0 \quad \text{for } 0 < x < A, 0 < y < B, 0 < z < C,$$

$$u(x, y, 0) = u(x, y, C) = 0,$$

$$u(0, y, z) = u(A, y, z) = 0,$$

$$u(x, 0, z) = 0, \quad u(x, B, z) = f(x, z).$$

We want a function that is harmonic on the cube (which may have edges of unequal length), and zero on five sides, but with prescribed values $f(x, z)$ on the sixth side.

Let $u(x, y, z) = X(x)Y(y)Z(z)$ to obtain

$$\frac{X''}{X} = -\frac{Y''}{Y} - \frac{Z''}{Z} = -\lambda,$$

and then, after a second separation,

$$\frac{Z''}{Z} = \lambda - \frac{Y''}{Y} = -\mu.$$

Then

$$X'' + \lambda X = 0, \; Z'' + \mu Z = 0, \quad \text{and} \quad Y'' - (\lambda + \mu)Y = 0.$$

From the boundary conditions,

$$X(0) = X(A) = 0,$$

$$Z(0) = Z(C) = 0,$$

and

$$Y(0) = 0.$$

The problems for X and Z are familiar ones, and we obtain eigenvalues and eigenfunctions:

$$\lambda_n = \frac{n^2 \pi^2}{A^2}, \quad X_n(x) = \sin\left(\frac{n\pi x}{A}\right),$$

and

$$\mu_m = \frac{m^2 \pi^2}{C^2}; \quad Z_m(z) = \sin\left(\frac{m\pi z}{C}\right),$$

with n and m independently varying over the positive integers.

The differential equation for $Y(y)$ becomes

$$Y'' - \left(\frac{n^2 \pi^2}{A^2} + \frac{m^2 \pi^2}{C^2}\right) Y = 0; \quad Y(0) = 0.$$

This has solutions that are constant multiples of $\sinh(\beta_{nm} y)$, where

$$\beta_{nm} = \sqrt{\frac{n^2 \pi^2}{A^2} + \frac{m^2 \pi^2}{C^2}}.$$

For each positive integer n and m, we now have a function

$$u_{nm}(x, y, z) = c_{nm} \sin\left(\frac{n\pi x}{A}\right) \sin\left(\frac{m\pi z}{C}\right) \sinh(\beta_{nm} y),$$

which satisfies Laplace's equation and the zero boundary conditions given on five of the faces of the cube. To satisfy the condition on the sixth face, we generally must use a superposition

$$u(x, y, z) = \sum_{n=1}^{\infty} \sum_{m=1}^{\infty} c_{nm} \sin\left(\frac{n\pi x}{A}\right) \sin\left(\frac{m\pi z}{C}\right) \sinh(\beta_{nm} y).$$

Now we must choose the coefficients so that

$$u(x, B, z) = f(x, z) = \sum_{n=1}^{\infty} \sum_{m=1}^{\infty} c_{nm} \sin\left(\frac{n\pi x}{A}\right) \sin\left(\frac{m\pi z}{C}\right) \sinh(\beta_{nm} B).$$

We have encountered this kind of double Fourier sine expansion previously, in treating vibrations of a fixed-frame, rectangular, elastic membrane. From that experience, we can write

$$c_{nm} = \frac{4}{AC \sinh(\beta_{nm} B)} \int_0^A \int_0^C f(\xi, \zeta) \sin\left(\frac{n\pi \xi}{A}\right) \sin\left(\frac{m\pi \zeta}{C}\right) d\zeta \, d\xi.$$

As usual, if nonzero data is prescribed on more than one face, then we split the Dirichlet problem into a sum of problems, on each of which there is nonzero data on only one face.

SECTION 18.6   PROBLEMS

1. Solve

$$\nabla^2 u(x, y, z) = 0 \quad \text{for } 0 < x < 1, 0 < y < 1,$$
$$0 < z < 1$$
$$u(0, y, z) = u(1, y, z) = u(x, 0, z) = u(x, 1, z)$$
$$= u(x, y, 0) = 0$$
$$u(x, y, 1) = xy.$$

2. Solve

$$\nabla^2 u(x, y, z) = 0 \quad \text{for } 0 < x < 2\pi, 0 < y < 2\pi,$$
$$0 < z < 1$$
$$u(0, y, z) = u(x, 0, z) = u(x, 2\pi, z) = u(x, y, 0)$$
$$= u(x, y, 1) = 0$$
$$u(2\pi, y, z) = 2.$$

3. Solve

$$\nabla^2 u(x, y, z) = 0 \quad \text{for } 0 < x < 1, 0 < y < 2\pi,$$
$$0 < z < \pi,$$
$$u(0, y, z) = u(1, y, z) = u(x, 0, z) = u(x, y, 0) = 0$$
$$u(x, 2\pi, z) = 2, u(x, y, \pi) = 1.$$

4. Solve

$$\nabla^2 u(x, y, z) = 0 \quad \text{for } 0 < x < 1, 0 < y < 2,$$
$$0 < z < \pi,$$
$$u(x, y, 0) = x^2(1 - x)y(2 - y), u(x, y, \pi) = 0,$$
$$u(0, y, z) = 0, u(1, y, z) = \sin(\pi y) \sin(z),$$
$$u(x, 0, z) = u(x, 2, z) = 0.$$

18.7   The Steady-State Heat Equation for a Solid Sphere

Consider a solid sphere of radius R, centered at the origin. We want to solve for the steady-state temperature distribution, given the temperature at all times on the surface.

In the steady-state case, $\partial u/\partial t = 0$ and the heat equation is Laplace's equation $\nabla^2 u = 0$. We will use spherical coordinates (ρ, θ, φ), in which ρ is the distance from the origin to (x, y, z), θ is the polar angle between the positive x axis and the projection onto the x, y plane of the line from the origin to (x, y, z), and φ is the angle of declination from the positive z axis to this line (Figure 18.5). We will also assume symmetry about the z axis, so u is a function of ρ and φ only. Then $\partial u/\partial \theta = 0$, and Laplace's equation becomes

$$\nabla^2 u(\rho, \varphi) = \frac{\partial^2 u}{\partial \rho^2} + \frac{2}{\rho} \frac{\partial u}{\partial \rho} + \frac{1}{\rho^2} \frac{\partial^2 u}{\partial \varphi^2} + \frac{\cot(\varphi)}{\rho^2} \frac{\partial u}{\partial \varphi} = 0.$$

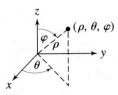

FIGURE 18.5 *Spherical coordinates.*

The temperature on the surface is

$$u(R, \varphi) = f(\varphi).$$

To separate variables in the differential equation, let $u(\rho, \varphi) = X(\rho)\Phi(\varphi)$ to obtain

$$X''\Phi + \frac{2}{\rho}X'\Phi + \frac{1}{\rho^2}X\Phi'' + \frac{\cot(\varphi)}{\rho^2}X\Phi' = 0.$$

Then

$$\frac{\Phi''}{\Phi} + \cot(\varphi)\frac{\Phi'}{\Phi} = -\rho^2\frac{X''}{X} - 2\rho\frac{X'}{X} = -\lambda.$$

Then

$$\rho^2 X'' + 2\rho X' - \lambda X = 0 \quad \text{and} \quad \Phi'' + \cot(\varphi)\Phi' + \lambda\Phi = 0.$$

The differential equation for Φ can be written

$$\frac{1}{\sin(\varphi)}[\Phi'\sin(\varphi)]' + \lambda\Phi = 0. \tag{18.4}$$

Change variables by putting

$$x = \cos(\varphi).$$

Then $\varphi = \arccos(x)$. Let

$$G(x) = \Phi(\arccos(x)).$$

Since $0 \leq \varphi \leq \pi$, then $-1 \leq x \leq 1$. Compute

$$\Phi'(\varphi)\sin(\varphi) = \sin(\varphi)\frac{d\Phi}{dx}\frac{dx}{d\varphi}$$

$$= \sin(\varphi)G'(x)[-\sin(\varphi)]$$

$$= -\sin^2(\varphi)G'(x) = -[1 - \cos^2(\varphi)]G'(x)$$

$$= -(1 - x^2)G'(x).$$

Then

$$\frac{d}{d\varphi}[\Phi'(\varphi)\sin(\varphi)] = -\frac{d}{d\varphi}[(1 - x^2)G'(x)]$$

$$= -\frac{d}{dx}[(1 - x^2)G'(x)]\frac{dx}{d\varphi}$$

$$= -\frac{d}{dx}[(1 - x^2)G'(x)](-\sin(\varphi)).$$

Then

$$\frac{1}{\sin(\varphi)} \frac{d}{d\varphi} [\Phi'(\varphi)\sin(\varphi)] = \frac{d}{dx}[(1-x^2)G'(x)],$$

and equation (18.4) transforms to

$$[(1-x^2)G'(x)]' + \lambda G(x) = 0.$$

This is Legendre's differential equation (Section 15.1). For bounded solutions, choose $\lambda = n(n+1)$ for $n = 0, 1, 2, \ldots$. These are the eigenvalues of this problem. The eigenfunctions are nonzero constant multiples of the Legendre polynomials $P_n(x)$.

For $n = 0, 1, 2, \ldots$, we now have a solution of the differential equation for Φ:

$$\Phi_n(\varphi) = G(\cos(\varphi)) = P_n(\cos(\varphi)).$$

Now that we know the admissible values for λ, the differential equation for X becomes

$$\rho^2 X'' + 2\rho X' - n(n+1)X = 0.$$

This is a second-order Euler differential equation, with general solution

$$X(\rho) = a\rho^n + b\rho^{-n-1}.$$

We must choose $b = 0$ to have a bounded solution at the center of the sphere, because $\rho^{-n-1} \to \infty$ as $\rho \to 0+$. Thus

$$X_n(\rho) = a_n \rho^n.$$

For each nonnegative integer n, we now have a function

$$u_n(\rho, \varphi) = a_n \rho^n P_n(\cos(\varphi))$$

that satisfies Laplace's equation. To satisfy the boundary condition, write a superposition of these functions:

$$u(\rho, \varphi) = \sum_{n=0}^{\infty} a_n \rho^n P_n(\cos(\varphi)).$$

We must choose the coefficients to satisfy

$$u(R, \varphi) = \sum_{n=0}^{\infty} a_n R^n P_n(\cos(\varphi)) = f(\varphi).$$

To put this into the setting of Fourier–Legendre expansions, let $\varphi = \arccos(x)$ to write

$$\sum_{n=0}^{\infty} a_n R^n P_n(x) = f(\arccos(x)).$$

This is a Fourier–Legendre series for the known function $f(\arccos(x))$. From Section 15.1.5, the coefficients are

$$a_n R^n = \frac{2n+1}{2} \int_{-1}^{1} f(\arccos(x)) P_n(x)\, dx,$$

or

$$a_n = \frac{2n+1}{2R^n} \int_{-1}^{1} f(\arccos(x)) P_n(x)\, dx.$$

The steady-state temperature distribution is

$$u(\rho, \varphi) = \sum_{n=0}^{\infty} \frac{2n+1}{2} \left(\int_{-1}^{1} f(\arccos(x)) P_n(x) \, dx \right) \left(\frac{\rho}{R} \right)^n P_n(\cos(\varphi)).$$

EXAMPLE 18.7

Consider this solution in a specific case, with $f(\varphi) = \varphi$. Now

$$u(\rho, \varphi) = \sum_{n=0}^{\infty} \frac{2n+1}{2} \left(\int_{-1}^{1} \arccos(x) P_n(x) \, dx \right) \left(\frac{\rho}{R} \right)^n P_n(\cos(\varphi)).$$

We will approximate some of these coefficients by approximating the integrals. From Section 15.1, the first six Legendre polynomials are

$$P_0(x) = 1, \quad P_1(x) = x, \quad P_2(x) = \frac{1}{2}(3x^2 - 1)$$

$$P_3(x) = \frac{1}{2}(5x^3 - 3x), \quad P_4(x) = \frac{1}{8}(35x^4 - 30x^2 + 3),$$

$$P_5(x) = \frac{1}{8}(63x^5 - 70x^3 + 15x).$$

Approximate:

$$\int_{-1}^{1} \arccos(x) \, dx \approx \pi, \int_{-1}^{1} x \arccos(x) \, dx \approx -.785\,4,$$

$$\int_{-1}^{1} \frac{1}{2}(3x^2 - 1) \arccos(x) \, dx = 0,$$

$$\int_{-1}^{1} \frac{1}{2}(5x^3 - 3x) \arccos(x) \, dx \approx -4.9087 \times 10^{-2}$$

$$\int_{-1}^{1} \frac{1}{8}(35x^4 - 30x^2 + 3) \arccos(x) \, dx = 0,$$

and

$$\int_{-1}^{1} \frac{1}{8}(63x^5 - 70x^3 + 15x) \arccos(x) \, dx \approx -1.2272 \times 10^{-2}.$$

Taking the first six terms of the series as an approximation to the solution, we obtain

$$u(\rho, \varphi) \approx \frac{1}{2}\pi - \frac{3}{2}(.7854)\frac{\rho}{R}\cos(\varphi) - \frac{7}{2}(.049087)\frac{1}{2}\left(\frac{\rho}{R}\right)^3 \left(5\cos^3(\varphi) - 3\cos(\varphi)\right)$$

$$- \frac{11}{2}(.012272)\left(\frac{\rho}{R}\right)^5 \frac{1}{8}\left(63\cos^5(\varphi) - 70\cos^3(\varphi) + 15\cos(\varphi)\right).$$

Some of these terms can be combined, but we have written them all out to indicate how they arise. ∎

We will return to the Dirichlet problem again when we treat complex analysis. There we will be in a position to exploit conformal mappings. The idea will be to map the region of interest in a certain way to the unit disk. Since we can solve the Dirichlet problem for the disk (that is, we know a formula for the solution), this maps the original problem to a problem we can solve. We then attempt to invert the map to transform the solution for the disk back into the solution for the original region.

We will conclude this chapter with a brief discussion of the Neumann problem.

SECTION 18.7 PROBLEMS

1. Write a solution for the steady-state temperature distribution in the sphere if the initial data is given by $f(\varphi) = A\varphi^2$, in which A is a positive constant. Write the first six terms of the series solution.

2. Carry out the program of Problem 1 for the initial data function $f(\varphi) = \sin(\varphi)$ for $0 \le \varphi \le \pi$.

3. Carry out the program of Problem 1 for the initial data function $f(\varphi) = \varphi^3$.

4. Carry out the program of Problem 1 for the initial data function $f(\varphi) = 2 - \varphi^2$.

5. Solve for the steady-state temperature distribution in a hollowed sphere, given in spherical coordinates by $R_1 \le \rho \le R_2$. The inner surface $\rho = R_1$ is kept at constant temperature T_1, while the outer surface $\rho = R_2$ is kept at temperature zero. Assume that the temperature distribution is a function of ρ and φ only.

6. Approximate the solution of Problem 5 by writing the first six terms of the series solution, carrying out any required integrations by a numerical method.

7. Solve for the steady-state temperature distribution in a solid, closed hemisphere, which in spherical coordinates is given by $0 \le \rho \le R$, $0 \le \theta \le 2\pi$, $0 \le \varphi \le \pi/2$. The base disk is kept at temperature zero and the hemispherical surface is kept at constant temperature A. Assume that the distribution is independent of θ.

8. Redo Problem 7, except now the base is insulated instead of being kept at temperature zero.

9. Redo Problem 7 for the case that the temperature on the hemispherical surface is $u(R, \varphi) = f(\varphi)$, not necessarily constant.

18.8 The Neumann Problem

A Neumann problem in the plane consists of finding a function that is harmonic on a given region D and whose normal derivative on the boundary of the region is given. This problem has the form

$$\nabla^2 u(x, y) = 0 \quad \text{for } (x, y) \text{ in } D,$$

$$\frac{\partial u}{\partial n}(x, y) = g(x, y) \quad \text{for } (x, y) \text{ in } \partial D,$$

where, as usual, ∂D denotes the boundary of D. This boundary is often a piecewise smooth curve in the plane (but not necessarily a closed curve). The normal derivative is defined by

$$\frac{\partial u}{\partial n} = \nabla u \cdot \mathbf{n},$$

the dot product of the gradient of u with the unit outer normal to the curve (Figure 18.6). If this normal is $\mathbf{n} = n_1 \mathbf{i} + n_2 \mathbf{j}$, then $\partial u / \partial n$ is

$$\frac{\partial u}{\partial n} = n_1 \frac{\partial u}{\partial x} + n_2 \frac{\partial u}{\partial y}.$$

We will use the following.

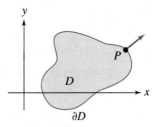

FIGURE 18.6 *Outer normal* **n** *at a point on* ∂D.

LEMMA 18.1 *Green's First Identity*

Let D be a bounded region in the plane whose boundary ∂D is a closed, piecewise smooth curve. Let k and h be continuous with continuous first and second partial derivatives on D and its boundary. Then

$$\oint_{\partial D} k\frac{\partial h}{\partial n}\,ds = \iint_D (k\nabla^2 h + \nabla k \cdot \nabla h)\,dA. \quad \blacksquare$$

In this line integral, ds denotes integration with respect to arc length along the curve bounding D. A 3-dimensional version was proved in Section 12.8.4 using Gauss's divergence theorem. Here is a proof for this version in the plane.

Proof By Green's theorem write

$$\oint_{\partial D} k\frac{\partial h}{\partial n}\,ds = \oint_{\partial D} (k\nabla h) \cdot \mathbf{n}\,ds = \iint_D div\,(k\nabla h)\,dA.$$

Now,

$$div(k\nabla h) = div\left(k\frac{\partial h}{\partial x}\mathbf{i} + k\frac{\partial h}{\partial y}\mathbf{j}\right)$$

$$= \frac{\partial}{\partial x}\left(k\frac{\partial h}{\partial x}\right) + \frac{\partial}{\partial y}\left(k\frac{\partial h}{\partial y}\right)$$

$$= k\left(\frac{\partial^2 h}{\partial x^2} + \frac{\partial^2 h}{\partial y^2}\right) + \frac{\partial k}{\partial x}\frac{\partial h}{\partial x} + \frac{\partial k}{\partial y}\frac{\partial h}{\partial y}$$

$$= k\nabla^2 h + \nabla k \cdot \nabla h. \quad \blacksquare$$

Use this result as follows. If $k = 1$ and $h = u$, a harmonic function on D, then the double integral is zero because its integrand vanishes, and the line integral is just the line integral of the normal derivative of u over the boundary of the region. But on ∂D, $\partial u/\partial n = g$, a given function. We conclude that

$$\oint_{\partial D} \frac{\partial u}{\partial n}\,ds = \oint_{\partial D} g\,ds = 0.$$

This means that a necessary condition for a Neumann problem to have a solution is for the integral of the given normal derivative about the boundary of the region to be zero. This conclusion can be extended to the case that ∂D is not a closed curve. For example, the boundary of the upper half-plane is the horizontal axis, which is not a closed curve.

EXAMPLE 18.8

Solve the Neumann problem for a square:

$$\nabla^2 u(x, y) = 0 \quad \text{for } 0 \le x \le 1, 0 \le y \le 1,$$

subject to

$$\frac{\partial u}{\partial n} = 0$$

on the left side and top and bottom sides, while

$$\frac{\partial u}{\partial n}(1, y) = y^2 \quad \text{for } 0 \le y \le 1.$$

Since

$$\oint_{\partial D} \frac{\partial u}{\partial n} \, ds = \int_0^1 y^2 \, dy = \frac{1}{3} \ne 0,$$

this problem has no solution. ∎

Existence can also be a question for a Dirichlet problem. However, for a Dirichlet problem, if the function given on the boundary is well behaved (for example, continuous) and the region is "simple" (such as a disk, rectangle, half-plane, and the like), then the Dirichlet problem has a solution. For Neumann problems, even for simple regions and apparently well-behaved data given for the normal derivative, there may be no solution if the integral of the data function about the boundary is not zero.

We will now solve two Neumann problems to illustrate what is involved.

18.8.1   A Neumann Problem for a Rectangle

Consider the problem

$$\nabla^2 u(x, y) = 0 \quad \text{for } 0 < x < a, 0 < y < b,$$

$$\frac{\partial u}{\partial y}(x, 0) = \frac{\partial u}{\partial y}(x, b) = 0 \quad \text{for } 0 \le x \le a,$$

$$\frac{\partial u}{\partial x}(0, y) = 0 \quad \text{for } 0 \le y \le b,$$

$$\frac{\partial u}{\partial x}(a, y) = g(y) \quad \text{for } 0 \le y \le b.$$

For the rectangle, the normal derivative is $\partial u / \partial x$ on the vertical sides, and $\partial u / \partial y$ on the horizontal sides. As a necessary (but not sufficient) condition for existence of a solution, we require that

$$\int_0^b g(y) \, dy = 0.$$

This example will clarify why there can be no solution without this condition.

Let $u(x, y) = X(x)Y(y)$ and obtain

$$X'' + \lambda X = 0, \quad Y'' - \lambda Y = 0.$$

Now

$$\frac{\partial u}{\partial y}(x, 0) = X(x)Y'(0) = 0$$

implies that $Y'(0) = 0$. Similarly,

$$\frac{\partial u}{\partial y}(x, b) = X(x)Y'(b) = 0$$

implies that $Y'(b) = 0$. The problem for Y is

$$Y'' - \lambda Y = 0; \qquad Y'(0) = Y'(b) = 0.$$

This familiar Sturm–Liouville has eigenvalues and eigenfunctions

$$\lambda_n = -\frac{n^2\pi^2}{b^2} \quad \text{and} \quad Y_n(y) = \cos\left(\frac{n\pi y}{b}\right)$$

for $n = 0, 1, 2, \ldots$.

Now the problem for X is

$$X'' - \frac{n^2\pi^2}{b^2}X = 0.$$

Further,

$$\frac{\partial u}{\partial x}(0, y) = X'(0)Y(y) = 0$$

implies that $X'(0) = 0$.

For $n = 0$, the differential equation for X is just $X'' = 0$, so $X(x) = cx + d$. Then $X'(0) = c = 0$, so $X(x) = $ constant in this case.

If n is a positive integer, then the differential equation for X has general solution

$$X(x) = ce^{n\pi x/b} + de^{-n\pi x/b}.$$

Now

$$X'(0) = \frac{n\pi}{b}c - \frac{n\pi}{b}d = 0$$

implies that $c = d$. This gives us

$$X_n(x) = \cosh\left(\frac{n\pi x}{b}\right).$$

We now have functions

$$u_0(x, y) = \text{constant}$$

and, for each positive integer n,

$$u_n(x, y) = c_n \cosh\left(\frac{n\pi x}{b}\right)\cos\left(\frac{n\pi y}{b}\right).$$

To satisfy the last boundary condition (on the right side of the rectangle), use a superposition

$$u(x, y) = c_0 + \sum_{n=1}^{\infty} c_n \cosh\left(\frac{n\pi x}{b}\right)\cos\left(\frac{n\pi y}{b}\right).$$

We need

$$\frac{\partial u}{\partial x}(a, y) = g(y) = \sum_{n=1}^{\infty} \frac{n\pi}{b}c_n \sinh\left(\frac{n\pi a}{b}\right)\cos\left(\frac{n\pi y}{b}\right),$$

a Fourier cosine expansion of $g(y)$ on $[0, b]$. Notice that the constant term in this expansion of $g(y)$ is zero. This constant term is

$$\frac{1}{b} \int_0^b g(y)\, dy,$$

which we have assumed to be zero. If this integral were not zero, then the cosine expansion of $g(y)$ would have a nonzero constant term, contradicting the fact that that it does not. In this event this Neumann problem would have no solution.

For the other coefficients in this cosine series, we have

$$\frac{n\pi}{b} c_n \sinh\left(\frac{n\pi a}{b}\right) = \frac{2}{b} \int_0^b g(\xi) \cos\left(\frac{n\pi \xi}{b}\right) d\xi,$$

so

$$c_n = \frac{2}{n\pi \sinh(n\pi a/b)} \int_0^b g(\xi) \cos\left(\frac{n\pi \xi}{b}\right) d\xi.$$

With this choice of coefficients, the solution of this Neumann problem is

$$u(x, y) = c_0 + \sum_{n=1}^\infty c_n \cosh\left(\frac{n\pi x}{b}\right) \cos\left(\frac{n\pi y}{b}\right).$$

The number c_0 is undetermined and remains arbitrary. This is because Neumann problems do not have unique solutions. If u is any solution of a Neumann problem, so is $u + c$ for any constant c, because the boundary condition is on the normal derivative and c vanishes in this differentiation.

18.8.2   A Neumann Problem for a Disk

We will solve the Neumann problem for a disk of radius R centered about the origin. In polar coordinates, the problem is

$$\nabla^2 u(r, \theta) = 0 \quad \text{for } 0 \le r < R, -\pi \le \theta \le \pi,$$

$$\frac{\partial u}{\partial r}(R, \theta) = f(\theta) \quad \text{for } -\pi \le \theta \le \pi.$$

The normal derivative here is $\partial/\partial r$, since the line from the origin to a point on this circle is in the direction of the outer normal vector to the circle at that point.

A necessary condition for existence of a solution is that

$$\int_{-\pi}^{\pi} f(\theta)\, d\theta = 0,$$

and we assume that f satisfies this condition.

As we did with the Dirichlet problem for a disk, attempt a solution

$$u(r, \theta) = \frac{1}{2} a_0 + \sum_{n=1}^\infty a_n r^n \cos(n\theta) + b_n r^n \sin(n\theta).$$

We need

$$\frac{\partial u}{\partial r}(R, \theta) = f(\theta) = \sum_{n=1}^\infty n a_n R^{n-1} \cos(n\theta) + n b_n R^{n-1} \sin(n\theta).$$

This is a Fourier expansion of $f(\theta)$ on $[-\pi, \pi]$. The constant term in this expansion is

$$\frac{1}{\pi} \int_{-\pi}^{\pi} f(\theta)\, d\theta,$$

and this must be zero because this Fourier series for $(\partial u/\partial r)(R, \theta)$ has a zero constant term. The assumption that this integral is zero is therefore consistent with this boundary condition.

For the other coefficients, we need

$$na_n R^{n-1} a_n = \frac{1}{\pi} \int_{-\pi}^{\pi} f(\xi) \cos(n\xi) \, d\xi$$

and

$$na_n R^{n-1} b_n = \frac{1}{\pi} \int_{-\pi}^{\pi} f(\xi) \sin(n\xi) \, d\xi.$$

Thus choose

$$a_n = \frac{1}{n\pi R^{n-1}} \int_{-\pi}^{\pi} f(\xi) \cos(n\xi) \, d\xi$$

and

$$b_n = \frac{1}{n\pi R^{n-1}} \int_{-\pi}^{\pi} f(\xi) \sin(n\xi) \, d\xi.$$

Upon inserting these coefficients, the solution is

$$u(r, \theta) = \frac{1}{2} a_0 + \frac{R}{\pi} \sum_{n=1}^{\infty} \frac{1}{n} \left(\frac{r}{R} \right)^n \int_{-\pi}^{\pi} [\cos(n\xi) \cos(n\theta) + \sin(n\xi) \sin(n\theta)] f(\xi) \, d\xi.$$

We can also write this solution as

$$u(r, \theta) = \frac{1}{2} a_0 + \frac{R}{\pi} \sum_{n=1}^{\infty} \frac{1}{n} \left(\frac{r}{R} \right)^n \int_{-\pi}^{\pi} \cos(n(\xi - \theta)) f(\xi) \, d\xi.$$

The term $a_0/2$ is an arbitrary constant. The factor of $\frac{1}{2}$ in this arbitrary constant is just customary.

EXAMPLE 18.9

Solve the Neumann problem

$$\nabla^2 u(x, y) = 0 \quad \text{for } x^2 + y^2 < 1,$$

$$\frac{\partial u}{\partial n}(x, y) = xy^2 \quad \text{for } x^2 + y^2 = 1.$$

Switch to polar coordinates, letting $u(r \cos(\theta), r \sin(\theta)) = U(r, \theta)$. Then

$$\nabla^2 U(r, \theta) = 0 \quad \text{for } 0 \le r < 1, -\pi \le \theta \le \pi,$$

$$\frac{\partial U}{\partial r}(1, \theta) = \cos(\theta) \sin^2(\theta).$$

First, compute

$$\int_{-\pi}^{\pi} \cos(\theta) \sin^2(\theta) \, d\theta = 0,$$

so it is worthwhile to try to solve this problem. Write the solution

$$U(r, \theta) = \frac{1}{2} a_0 + \frac{1}{\pi} \sum_{n=1}^{\infty} \frac{1}{n} (r)^n \int_{-\pi}^{\pi} \cos(n(\xi - \theta)) \cos(\xi) \sin^2(\xi) \, d\xi.$$

Evaluate

$$\int_{-\pi}^{\pi} \cos(n(\xi - \theta)) \cos(\xi) \sin^2(\xi) \, d\xi = \begin{cases} 0 & \text{for } n = 2, 4, 5, 6, \ldots \\[2mm] \dfrac{\pi \cos(\theta)}{4} & \text{if } n = 1 \\[2mm] -\pi \cos^3(\theta) + \dfrac{3\pi \cos(\theta)}{4} & \text{if } n = 3 \end{cases}.$$

The solution is therefore

$$U(r, \theta) = \tfrac{1}{2}a_0 + \tfrac{1}{4}r \cos(\theta) + \tfrac{1}{3}r^3 \left(-\cos^3(\theta) + \tfrac{3}{4}\cos(\theta) \right)$$

$$= \tfrac{1}{2}a_0 + \tfrac{1}{4}r \cos(\theta) - \tfrac{1}{3}r^3 \cos^3(\theta) + \tfrac{1}{4}r^3 \cos(\theta).$$

To obtain the solution in rectangular coordinates, use $x = r \cos(\theta)$ and $r^2 = x^2 + y^2$ to write

$$u(x, y) = \tfrac{1}{2}a_0 + \tfrac{1}{4}x - \tfrac{1}{3}x^3 + \tfrac{1}{4}x(x^2 + y^2). \quad \blacksquare$$

Again, the solution has an arbitrary constant, which is written with a factor of $\tfrac{1}{2}$ simply as a matter of custom.

18.8.3 A Neumann Problem for the Upper Half-Plane

As an illustration of a Neumann problem for an unbounded domain, consider:

$$\nabla^2 u(x, y) = 0 \quad \text{for } -\infty < x < \infty, y > 0,$$

$$\frac{\partial u}{\partial y}(x, 0) = f(x) \quad \text{for } -\infty < x < \infty.$$

The boundary of the region is the real axis, and $\partial/\partial y$ is the derivative normal to this line.

We require that $\int_{-\infty}^{\infty} f(x) \, dx = 0$ as a necessary condition for a solution to exist.

There is an elegant device for reducing this problem to one we have already solved. Let $v = \partial u/\partial y$. Then

$$\nabla^2 v = \frac{\partial^2}{\partial x^2}\left(\frac{\partial u}{\partial y}\right) + \frac{\partial^2}{\partial y^2}\left(\frac{\partial u}{\partial y}\right) = \frac{\partial}{\partial y}\left(\frac{\partial^2 u}{\partial x^2} + \frac{\partial^2 u}{\partial y^2}\right) = 0,$$

so v is harmonic wherever u is. Further

$$v(x, 0) = \frac{\partial u}{\partial y}(x, 0) = f(x) \quad \text{for } -\infty < x < \infty.$$

Therefore v is the solution of a Dirichlet problem for the upper half-plane. But we know the solution of this problem:

$$v(x, y) = \frac{y}{\pi} \int_{-\infty}^{\infty} \frac{f(\xi)}{y^2 + (\xi - x)^2} \, d\xi.$$

Now recover u from v by integrating:

$$u(x, y) = \int \frac{\partial u}{\partial y} \, dy = \int \frac{y}{\pi} \int_{-\infty}^{\infty} \frac{f(\xi)}{y^2 + (\xi - x)^2} \, d\xi \, dy$$

$$= \frac{1}{\pi} \int_{-\infty}^{\infty} \left(\int \frac{y}{y^2 + (\xi - x)^2} \, dy \right) f(\xi) \, d\xi$$

$$= \frac{1}{2\pi} \int_{-\infty}^{\infty} \ln(y^2 + (\xi - x)^2) f(\xi) \, d\xi + c,$$

in which c is an arbitrary constant. This gives the solution of the Neumann problem for the upper half-plane.

SECTION 18.8 PROBLEMS

1. Solve

$$\nabla^2 u(x, y) = 0 \quad \text{for } 0 < x < 1, 0 < y < 1,$$

$$\frac{\partial u}{\partial y}(x, 0) = 4\cos(\pi x),$$

$$\frac{\partial u}{\partial y}(x, 1) = 0 \quad \text{for } 0 \leq x \leq 1,$$

$$\frac{\partial u}{\partial x}(0, y) = \frac{\partial u}{\partial x}(1, y) = 0 \quad \text{for } 0 \leq y \leq 1.$$

2. Solve

$$\nabla^2 u(x, y) = 0 \quad \text{for } 0 < x < 1, 0 < y < \pi,$$

$$\frac{\partial u}{\partial y}(x, 0) = \frac{\partial u}{\partial y}(x, \pi) = 0 \quad \text{for } 0 \leq x \leq 1,$$

$$\frac{\partial u}{\partial x}(0, y) = y - \frac{\pi}{2} \quad \text{for } 0 \leq y \leq \pi,$$

$$\frac{\partial u}{\partial x}(\pi, y) = \cos(y) \quad \text{for } 0 \leq y \leq \pi.$$

3. Solve

$$\nabla^2 u(x, y) = 0 \quad \text{for } 0 < x < \pi, 0 < y < \pi,$$

$$\frac{\partial u}{\partial y}(x, 0) = \cos(3x),$$

$$\frac{\partial u}{\partial y}(x, \pi) = 6x - 3\pi \quad \text{for } 0 \leq x \leq \pi,$$

$$\frac{\partial u}{\partial x}(0, y) = \frac{\partial u}{\partial x}(\pi, y) = 0 \quad \text{for } 0 \leq y \leq \pi.$$

4. Use separation of variables to solve the mixed boundary value problem

$$\nabla^2 u(x, y) = 0 \quad \text{for } 0 < x < \pi, 0 < y < \pi,$$

$$u(x, 0) = f(x), u(x, \pi) = 0 \quad \text{for } 0 \leq x \leq \pi,$$

$$\frac{\partial u}{\partial x}(0, y) = \frac{\partial u}{\partial x}(\pi, y) = 0 \quad \text{for } 0 \leq y \leq \pi.$$

Does this problem have a unique solution?

5. Attempt a separation of variables to solve

$$\nabla^2 u(x, y) = 0 \quad \text{for } 0 < x < 1, 0 < y < 1,$$

$$u(x, 0) = u(x, 1) = 0 \quad \text{for } 0 \leq x \leq 1,$$

$$\frac{\partial u}{\partial x}(0, y) = 3y^2 - 2y,$$

$$\frac{\partial u}{\partial x}(1, y) = 0 \quad \text{for } 0 \leq y \leq 1.$$

6. Write a series solution for

$$\nabla^2 u(r, \theta) = 0 \quad \text{for } 0 \leq r < R, -\pi \leq \theta \leq \pi,$$

$$\frac{\partial u}{\partial r}(R, \theta) = \sin(3\theta) \quad \text{for } -\pi \leq \theta \leq \pi.$$

7. Write a solution for

$$\nabla^2 u(r, \theta) = 0 \quad \text{for } 0 \leq r < R, -\pi \leq \theta \leq \pi,$$

$$\frac{\partial u}{\partial r}(R, \theta) = \cos(2\theta) \quad \text{for } -\pi \leq \theta \leq \pi.$$

8. Solve

$$\nabla^2 u(r, \theta) = 0 \quad \text{for } 0 \leq r < R, -\pi \leq \theta \leq \pi,$$

$$\frac{\partial u}{\partial r}(R, \theta) = \theta \quad \text{for } -\pi \leq \theta \leq \pi.$$

9. Solve

$$\nabla^2 u(x, y) = 0 \quad \text{for } x^2 + y^2 < 9,$$

$$\frac{\partial u}{\partial n}(x, y) = 4xy \quad \text{for } x^2 + y^2 = 9.$$

10. Solve

$$\nabla^2 u(x, y) = 0 \quad \text{for } x^2 + y^2 < 1,$$

$$\frac{\partial u}{\partial n}(x, y) = x \quad \text{for } x^2 + y^2 = 1.$$

11. Solve

$$\nabla^2 u(x, y) = 0 \quad \text{for } x^2 + y^2 < 1,$$

$$\frac{\partial u}{\partial n}(x, y) = x^2 y^2 \quad \text{for } x^2 + y^2 = 1.$$

12. Solve the following Neumann problem for the upper half-plane:

$$\nabla^2 u(x, y) = 0 \quad \text{for } -\infty < x < \infty, \, y > 0,$$

$$\frac{\partial u}{\partial y}(x, 0) = xe^{-|x|} \quad \text{for } -\infty < x < \infty.$$

13. Solve the following Neumann problem for the upper half-plane:

$$\nabla^2 u(x, y) = 0 \quad \text{for } -\infty < x < \infty, \, y > 0,$$

$$\frac{\partial u}{\partial y}(x, 0) = e^{-|x|} \sin(x) \quad \text{for } -\infty < x < \infty.$$

14. Solve the following Neumann problem for the lower half-plane:

$$\nabla^2 u(x, y) = 0 \quad \text{for } -\infty < x < \infty, \, y < 0,$$

$$\frac{\partial u}{\partial y}(x, 0) = f(x) \quad \text{for } -\infty < x < \infty.$$

15. Solve the following Neumann problem for the right quarter-plane:

$$\nabla^2 u(x, y) = 0 \quad \text{for } x > 0, \, y > 0,$$

$$\frac{\partial u}{\partial x}(0, y) = 0 \quad \text{for } y \geq 0,$$

$$\frac{\partial u}{\partial y}(x, 0) = f(x) \quad \text{for } 0 \leq x < \infty.$$

16. Solve the following mixed problem:

$$\nabla^2 u(x, y) = 0 \quad \text{for } x > 0, \, y > 0,$$

$$u(0, y) = 0 \quad \text{for } y \geq 0,$$

$$\frac{\partial u}{\partial y}(x, 0) = f(x) \quad \text{for } 0 \leq x < \infty.$$

CHAPTER 19

Canonical Forms, Existence and Uniqueness of Solutions, and Well-Posed Problems

Thus far we have concentrated on techniques for solving important partial differential equations in settings that occur frequently. This chapter touches upon some additional topics and theoretical aspects of interest in studying partial differential equations.

19.1 Canonical Forms

Consider the general constant coefficient second-order differential equation

$$Au_{xx} + 2Bu_{xy} + Cu_{yy} + Du_x + Eu_y + Fu = G(x, y) \tag{19.1}$$

in which $A, \ldots, F$ are constants and $A \neq 0$ (so there is at least one second partial derivative with respect to one of the variables).

We will show that equation (19.1) can be transformed by a change of variables into one of three standard forms, depending on the sign of the number $B^2 - AC$. This standard form is called its *canonical form* and is sometimes easier to work with than the original equation. The strategy is to try to obtain the solution of the transformed equation, then apply the inverse transformation to solve the original partial differential equation.

Consider three cases.

Case 1 $B^2 - AC > 0$

In this case the partial differential equation (19.1) is called *hyperbolic*. The wave equation,

$$c^2 u_{xx} - u_{tt} = 0,$$

with t for time in place of y, is hyperbolic, since $A = c^2$, $B = 0$, and $C = -1$, so $B^2 - AC = c^2 > 0$.

987

In general, in this hyperbolic case, there are associated with equation (19.1) two families of straight lines called *characteristics*. The first family consists of the lines

$$y = \left(\frac{B + \sqrt{B^2 - AC}}{A}\right) x + k_1$$

and the second family consists of lines

$$y = \left(\frac{B - \sqrt{B^2 - AC}}{A}\right) x + k_2,$$

with k_1 and k_2 arbitrary constants. In the case of the wave equation, we exploited the characteristics

$$x - ct = k_1, \quad x + ct = k_2$$

to derive the d'Alembert solution for the Cauchy problem on the entire real line. Here is another example of characteristics, with x and y as the variables instead of x and t.

EXAMPLE 19.1

Consider the second-order equation

$$9u_{xx} + 10u_{xy} + u_{yy} + 2u_x - u_y + xy = 0.$$

Here $A = 9$, $B = 5$, and $C = 1$, so $B^2 - AC = 25 - 9 = 16$. The characteristics of this equation are the lines

$$y = x + k_1$$

and

$$y = \tfrac{1}{9}x + k_2.$$

These may also be written as the families of lines

$$x - y = c_1, \quad x - 9y = c_2,$$

with c_1 and c_2 arbitrary constants. Some of these characteristics are shown in Figure 19.1. ∎

The families of characteristics can be used to define a change of variables that transforms equation (19.1) to a standard (canonical) form. Rather than carry out an abstract argument, we will do the calculation for the equation of the last example.

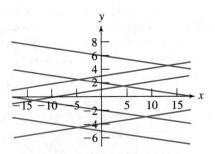

FIGURE 19.1 *Some characteristics of*
$9u_{xx} + 10u_{xy} + u_{yy} + 2u_x - u_y + xy = 0.$

Use the equations of the characteristics to define new variables

$$\xi = x - y, \quad \eta = x - 9y.$$

The inverse of this transformation is

$$x = \tfrac{1}{8}(9\xi - \eta), \quad y = \tfrac{1}{8}(\xi - \eta).$$

Now define

$$U(\xi, \eta) = u(x(\xi, \eta), y(\xi, \eta)).$$

Compute the partial derivatives:

$$u_x = U_\xi + U_\eta,$$
$$u_{xx} = U_{\xi\xi} + 2U_{\xi\eta} + U_{\eta\eta},$$
$$u_y = -U_\xi - 9U_\eta,$$
$$u_{yy} = U_{\xi\xi} + 18U_{\xi\eta} + 81U_{\eta\eta},$$

and

$$u_{xy} = -U_{\xi\xi} - 10U_{\xi\eta} - 9U_{\eta\eta}.$$

Substitute these into $9u_{xx} + 10u_{xy} + u_{yy} + 2u_x - u_y + xy = 0$ to obtain

$$9\left[U_{\xi\xi} + 2U_{\xi\eta} + U_{\eta\eta}\right] + 10\left[-U_{\xi\xi} - 10U_{\xi\eta} - 9U_{\eta\eta}\right]$$
$$+ \left[U_{\xi\xi} + 18U_{\xi\eta} + 81U_{\eta\eta}\right] + 2\left[U_\xi + U_\eta\right] - \left[-U_\xi - 9U_\eta\right]$$
$$+ \left(\tfrac{1}{8}(9\xi - \eta)\right)\left(\tfrac{1}{8}(\xi - \eta)\right) = 0.$$

Some terms cancel and we are left with

$$-64U_{\xi\eta} + 3U_\xi + 11U_\eta + \tfrac{1}{64}(9\xi - \eta)(\xi - \eta) = 0.$$

Finally, divide by $-\tfrac{1}{64}$ to get

$$U_{\xi\eta} - \tfrac{3}{64}U_\xi - \tfrac{11}{64}U_\eta - \tfrac{1}{4096}(9\xi - \eta)(\xi - \eta) = 0.$$

This is the canonical form of the original differential equation.

If we proceed in this way, using the characteristics to define a linear change of variables, we can always transform equation (19.1) in this hyperbolic case to the canonical form

$$U_{\xi\eta} + \Phi(\xi, \eta, U, U_\xi, U_\eta) = 0.$$

This canonical form contains just one second partial derivative, the mixed partial $U_{\xi\eta}$, together with terms involving ξ, η, and possibly first partial derivatives.

EXAMPLE 19.2

The partial differential equation

$$2u_{xx} + 6u_{xy} - 5u_{yy} = 1$$

is hyperbolic. Here $A = 2$, $B = 3$, and $C = -5$, so $B^2 - AC = 19$. The characteristics are the families of lines

$$y = \frac{3 + \sqrt{19}}{2}x + k_1$$

and

$$y = \frac{3 - \sqrt{19}}{2}x + k_2.$$

Define the change of variables

$$\xi = \frac{3 + \sqrt{19}}{2}x - y,$$

$$\eta = \frac{3 - \sqrt{19}}{2}x - y.$$

Let $U(\xi, \eta) = u(x(\xi, \eta), y(\xi, \eta))$ and compute the partial derivatives. For convenience, let $\alpha = (3 + \sqrt{19})/2$ and $\beta = (3 - \sqrt{19})/2$. We get

$$u_x = \alpha U_\xi + \beta U_\eta,$$

$$u_{xx} = \alpha \left(\alpha U_{\xi\xi} + \beta U_{\xi\eta} \right) + \beta \left(\alpha U_{\eta\xi} + \beta U_{\eta\eta} \right)$$

$$= \alpha^2 U_{\xi\xi} + 2\alpha\beta U_{\xi\eta} + \beta^2 U_{\eta\eta},$$

$$u_{xy} = \alpha \left(U_{\xi\xi}(-1) + U_{\xi\eta}(-1) \right) + \beta \left(U_{\eta\xi}(-1) + U_{\eta\eta}(-1) \right)$$

$$= -\alpha U_{\xi\xi} - (\alpha + \beta)U_{\xi\eta} - \beta U_{\eta\eta},$$

$$u_y = -U_\xi - U_\eta,$$

and

$$u_{yy} = -\left(-U_{\xi\xi} - U_{\xi\eta} \right) - \left(-U_{\eta\xi} - U_{\eta\eta} \right)$$

$$= U_{\xi\xi} + 2U_{\xi\eta} + U_{\eta\eta}.$$

Substitute into the partial differential to obtain the transformed equation:

$$2 \left(\alpha^2 U_{\xi\xi} + 2\alpha\beta U_{\xi\eta} + \beta^2 U_{\eta\eta} \right) + 6 \left(-\alpha U_{\xi\xi} - (\alpha + \beta)U_{\xi\eta} - \beta U_{\eta\eta} \right)$$

$$- 5 \left(U_{\xi\xi} + 2U_{\xi\eta} + U_{\eta\eta} \right) = 1.$$

This reduces to the canonical form

$$U_{\xi\eta} = -\tfrac{1}{38}.$$

To solve this equation, write it as

$$(U_\xi)_\eta = -\tfrac{1}{38}$$

and integrate with respect to η to get

$$U_\xi = -\tfrac{1}{38}\eta + F(\xi).$$

Integrate this equation with respect to ξ to get

$$U(\xi, \eta) = -\tfrac{1}{38}\xi\eta + \int F(\xi)\, d\xi + G(\eta).$$

Since $\int F(\xi)\, d\xi$ is just another function of ξ, we can write

$$U(\xi, \eta) = -\tfrac{1}{38}\xi\eta + H(\xi) + G(\eta),$$

where G and H are any twice continuously differentiable functions of a single variable defined on the entire line.

To obtain the general solution of the original differential equation, substitute the equations of the transformation to get

$$u(x, y) = -\frac{1}{38}(\alpha x - y)(\beta x - y) + H(\alpha x - y) + G(\beta x - y). \quad \blacksquare$$

Case 2 $B^2 - AC = 0$

When $B^2 - AC = 0$, equation (19.1) is *parabolic*. The heat equation

$$u_t = k u_{xx} + P(x, t)$$

(with x and t instead of x and y) is parabolic. We may write this equation as

$$k u_{xx} - u_t = P(x, t),$$

so $A = k$ and $B = C = 0$.

We will show that whenever equation (19.1) is parabolic, there is a change of variables transforming it to the canonical form

$$U_{\eta\eta} + \Phi(\xi, \eta, U, U_\xi, U_\eta) = 0.$$

This canonical form has just one second derivative term, the second partial derivative of the transformed function with respect to one of the new variables, η.

As we saw with the (hyperbolic) wave equation, putting the partial differential equation into canonical form can be a significant step toward obtaining solutions. Here is an example.

EXAMPLE 19.3

We will solve

$$u_{xx} + 8u_{xy} + 16u_{yy} = 0.$$

This is parabolic, since $B = 4$, $A = 1$, and $C = 16$. The equation

$$\frac{dy}{dx} = \frac{B}{A} = 4$$

is called the *characteristic* equation for this differential equation. The characteristics are the curves formed from solutions of the characteristic equation, in this case

$$y - 4x = k$$

with k any constant. While hyperbolic equations have two families of characteristics, parabolic equations have only one. Use this family to define the new variable

$$\xi = y - 4x.$$

There is great latitude in defining the second equation of the transformation to new variables. The guiding principle is that the Jacobian of the transformation should not vanish in the region of the plane of interest. For example, we could try the simple

$$\eta = x.$$

The Jacobian of this transformation is

$$J = \begin{vmatrix} \xi_x & \xi_y \\ \eta_x & \eta_y \end{vmatrix} = \begin{vmatrix} -4 & 1 \\ 1 & 0 \end{vmatrix} = -1 \neq 0.$$

We can use this transformation, which has inverse

$$x = \eta, \quad y = \xi + 4\eta.$$

Let $U(\xi, \eta) = u(x(\xi, \eta), y(\xi, \eta))$ and compute partial derivatives:

$$u_x = U_\xi(-4) + U_\eta,$$

$$u_{xx} = -4\left[U_{\xi\xi}(-4) + U_{\xi\eta}\right] + U_{\eta\xi}(-4) + U_{\eta\eta},$$

$$u_{xy} = -4\left[U_{\xi\xi}\right] + U_{\eta\xi},$$

$$u_y = U_\xi, u_{yy} = U_{\xi\xi}.$$

Substitute into the differential equation to get

$$16U_{\xi\xi} - 8U_{\xi\eta} + U_{\eta\eta} + 8\left[-4U_{\xi\xi} + U_{\xi\eta}\right] + 16U_{\xi\xi} = 0.$$

This reduces to

$$U_{\eta\eta} = 0,$$

which is the canonical form of this partial differential equation. Write this as

$$(U_\eta)_\eta = 0.$$

Then U_η must be independent of η, say

$$U_\eta = F(\xi).$$

Then

$$U(\xi, \eta) = \int F(\xi) \, d\eta = \eta F(\xi) + G(\xi),$$

in which F and G can be any twice differentiable functions of a single variable. Therefore,

$$u(x, y) = xF(y - 4x) + G(y - 4x)$$

is the general solution of the original partial differential equation. ∎

Case 3 $B^2 - AC < 0$

The third case is $B^2 - AC < 0$. Now equation (19.1) is *elliptic*. Laplace's equation $u_{xx} + u_{yy} = 0$ is elliptic, with $A = C = 1$ and $B = 0$.

The elliptic equation does not have characteristics, so we do not have the device used in the parabolic and hyperbolic cases to define a transformation to a canonical form. However, we will develop a strategy to transform any elliptic partial differential equation to the canonical form

$$U_{\xi\xi} + U_{\eta\eta} + \Phi(\xi, \eta, U, U_\xi, U_\eta) = 0.$$

Laplace's equation is already in this form.

We will devise an appropriate transformation in two stages. First, transform equation (19.1) to a new partial differential equation in which there is no mixed second derivative term $U_{\xi\eta}$. To do this, suppose

$$\xi = \xi(x, y), \quad \eta = \eta(x, y)$$

is a change of variables and let $U(\xi, \eta) = u(x, y)$, as usual. By chain rule differentiations, the coefficient of $U_{\xi\eta}$ in the transformed equation is

$$\xi_x(A\eta_x + B\eta_y) + \xi_y(B\eta_x + C\eta_y)$$

and this will be zero if we choose ξ and η so that

$$-\frac{\xi_x}{\xi_y} = \frac{B\eta_x + C\eta_y}{A\eta_x + B\eta_y}.$$

But, if $\xi(x, y) = k$, then

$$\xi_x \, dx + \xi_y \, dy = 0,$$

so

$$-\frac{\xi_x}{\xi_y} = \frac{dy}{dx} = \frac{B\eta_x + C\eta_y}{A\eta_x + B\eta_y}. \tag{19.2}$$

Choose any "reasonable" $\eta(x, y)$, for example $\eta = x$, then obtain the general solution $\xi(x, y) = k$ of this differential equation. This transforms equation (19.1) to a new partial differential equation with no mixed second derivative term.

EXAMPLE 19.4

Consider

$$2u_{xx} + 4u_{xy} + 5u_{yy} - u = 0.$$

Here $A = 2$, $B = 2$, and $C = 5$, so $B^2 - AC < 0$. Begin by trying a simple function for η, say $\eta(x, y) = x$. Now equation (19.2) is

$$\frac{dy}{dx} = \frac{2}{2} = 1$$

with general solution $y - x = k$. Thus let $\xi = -x + y$. The change of variables is therefore given by

$$\xi = -x + y, \quad \eta = x.$$

This transformation has Jacobian

$$J = \begin{vmatrix} -1 & 1 \\ 1 & 0 \end{vmatrix} = -1 \neq 0,$$

so the transformation is one-to-one and should achieve our objective. Compute

$$u_x = U_\xi(-1) + U_\eta,$$

$$u_{xx} = -(U_{\xi\xi}(-1) + U_{\xi\eta}) + U_{\eta\xi}(-1) + U_{\eta\eta},$$

$$u_{xy} = -U_{\xi\xi} + U_{\eta\xi},$$

$$u_y = U_\xi, u_{yy} = U_{\xi\xi}.$$

Substitute these into the partial differential equation to obtain

$$2\left(U_{\xi\xi} - 2U_{\xi\eta} + U_{\eta\eta}\right) + 4\left(-U_{\xi\xi} + U_{\xi\eta}\right) + 5U_{\xi\xi} - U = 0.$$

This is

$$3U_{\xi\xi} + 2U_{\eta\eta} - U = 0. \quad \blacksquare$$

Thus far we have a transformation of an elliptic equation to one having no mixed partial derivative term. We now want to transform from this stage to an elliptic partial differential equation in which the coefficients of $U_{\xi\xi}$ and $U_{\eta\eta}$ are both 1.

To this end, suppose the first stage has been completed and the transformed equation has no mixed second derivative term. To keep the notation simple, begin again with an equation

$$Au_{xx} + Cu_{yy} + Du_x + Eu_y + F = G(x, y),$$

having no u_{xy} term. We want to transform from this to an equation in which the coefficient of $U_{\xi\xi}$ and $U_{\eta\eta}$ are the same and there is no $U_{\xi\eta}$ term. If $\xi = \xi(x, y)$, $\eta = \eta(x, y)$ is any one-to-one transformation, then setting the coefficients of $U_{\xi\xi}$ and $U_{\eta\eta}$ equal gives us the condition

$$A\xi_x^2 + C\xi_y^2 = A\eta_x^2 + C\eta_y^2 \tag{19.3}$$

and requiring the absence of a $U_{\xi\eta}$ term gives us

$$A\xi_x\eta_x + C\xi_y\eta_y = 0. \tag{19.4}$$

One way to try to satisfy equation (19.3) is to look for functions ξ and η satisfying

$$\sqrt{A}\xi_x = \sqrt{C}\eta_y \quad \text{and} \quad \sqrt{C}\xi_y = -\sqrt{A}\eta_x. \tag{19.5}$$

If we can find η, then we can determine ξ as a line integral

$$\xi(x, y) = \int_{(x_0, y_0)}^{(x, y)} \sqrt{\frac{C}{A}}\eta_y \, dx - \sqrt{\frac{A}{C}}\eta_x \, dy, \tag{19.6}$$

in which (x_0, y_0) is a conveniently chosen point. This function will satisfy equations (19.5) when this integral is independent of path, which occurs when

$$\left(-\sqrt{\frac{A}{C}}\eta_x\right)_x = \left(\sqrt{\frac{C}{A}}\eta_y\right)_y. \tag{19.7}$$

In practice, we begin with equation (19.7) to find η, then use this in equation (19.6) to find ξ. This leads to the canonical form

$$U_{\xi\xi} + U_{\eta\eta} + \Phi(\xi, \eta, U, U_\xi, U_\eta)$$

for this elliptic equation.

EXAMPLE 19.5

Continue from the preceding example. Beginning with the elliptic equation

$$2u_{xx} + 4u_{xy} + 5u_{yy} - u = 0,$$

we obtained the transformed equation

$$3U_{\xi\xi} + 2U_{\eta\eta} - U = 0.$$

This is not yet in canonical form, because the coefficients of the second derivative terms are unequal. For simplicity of notation, begin all over by rewriting the last equation as

$$3u_{xx} + 2u_{yy} - u = 0.$$

This keeps us from having to define a new set of variables. For this equation, $A = 3$, $B = 0$ (as a result of the first transformation), and $C = 2$. This transformed equation is still elliptic. Now write equation (19.7):

$$\left(-\sqrt{\frac{3}{2}}\eta_x\right)_x = \left(\sqrt{\frac{2}{3}}\eta_y\right)_y.$$

Many functions satisfy this equation. One simple choice is $\eta(x, y) = x + y$. Now equation (19.6) is

$$\xi = \int_{(0,0)}^{(x,y)} \sqrt{\frac{2}{3}}\, dx - \sqrt{\frac{3}{2}}\, dy = \sqrt{\frac{2}{3}}x - \sqrt{\frac{3}{2}}y.$$

Thus set

$$\xi = \sqrt{\frac{2}{3}}x - \sqrt{\frac{3}{2}}y \quad \text{and} \quad \eta = x + y.$$

We leave it for the reader to check that this transforms $3u_{xx} + 2u_{yy} - u = 0$ to the canonical form

$$U_{\xi\xi} + U_{\eta\eta} - \tfrac{1}{4}U = 0. \quad \blacksquare$$

Although we have restricted our attention here to the constant coefficient case, we can classify any second-order partial differential equation according to the preceding scheme. However, in the case that the coefficients are functions of x and y, the type of the equation may vary from one region to another. For example, Tricomi's equation is

$$yu_{xx} + u_{yy} = 0.$$

Here $A = y$, $B = 0$, and $C = 1$. The equation is hyperbolic if $y < 0$, parabolic if $y = 0$, and elliptic if $y > 0$.

Classification is important in understanding the kinds of initial and boundary data that must be provided to have a unique solution to a problem. For the wave equation (hyperbolic) on a finite interval, we found it physically plausible to specify two boundary conditions (at the ends of the string) and two initial conditions (initial position and velocity), because these clearly influence the resulting motion. This type of information is typical of hyperbolic problems. For the heat equation (parabolic), we needed a boundary condition at each end, but only one initial condition, the initial temperature. Physically, this is enough to determine the temperature function and is typical of parabolic problems. For Laplace's equation (elliptic), there are no initial conditions, but values the solution must take on the boundary are specified. This is typical of problems involving potentials.

Classification is also important in determining numerical techniques for approximating solutions. Different techniques apply to elliptic, parabolic, and hyperbolic equations.

SECTION 19.1 PROBLEMS

For each of the following, classify the partial differential equation and determine its characteristics (if it has any) and its canonical form.

1. $2u_{xx} + 10u_{xy} + 8u_{yy} + xu_x - yu_y = 0$

2. $4u_{xx} + 2u_{xy} + u_{yy} - u_x + xyu_y + u = 0$

3. $3u_{xx} + 2u_{xy} - u_{yy} + yu_x - u_y = 0$

4. $2u_{xx} - 4u_{xy} + 2u_{yy} - y^2u_x + u_y - xu = 0$

5. $4u_{xx} - 16u_{xy} + 7u_{yy} + (x + y)u_y = 0$

6. $3u_{xx} + 6u_{xy} + 4u_{yy} - u_x - xu_y + xyu = 0$

7. $u_{xx} - 4u_{xy} + 4u_{yy} + u_x + u_y = 0$

8. $6u_{xx} + 5u_{yy} - x^2u_x + yu_y - xu = 0$

9. $2u_{xx} - 10u_{xy} + 8u_{yy} + u_x - u_y = 0$

10. $2u_{xx} - 2u_{xy} - 3u_{yy} + y^2u_x - u = 0$

19.2 Existence and Uniqueness of Solutions

When we solve a problem explicitly, existence is not a question. In general settings, however, existence of a solution of a partial differential equation, satisfying certain conditions, may not be obvious (or even true). We will give an example of an existence result.

A *Cauchy problem* in three variables (x, y, t) consists of solving the partial differential equation

$$u_t(x, y, t) = F(x, y, t, u_x(x, y, t), u_y(x, y, t))$$

for $u(x, y, t)$ in some region of 3-space, subject to the condition

$$u(x, y, 0) = f(x, y).$$

The following theorem is named for Augustin-Louis Cauchy and Sonia Kovalevski.

THEOREM 19.1 *Cauchy–Kovalevski*

Let F be analytic in a six-dimensional sphere about the point

$$(0, 0, 0, f(0, 0), f_x(0, 0), f_y(0, 0))$$

in R^6. Suppose f is analytic in some disk about $(0, 0)$ in the x, y plane. Then the Cauchy problem has a unique solution $u(x, y, t)$ defined in some sphere about the origin in x, y, t-space. ∎

In this statement, a function is analytic about a point if it has a power series representation about the point. The Cauchy–Kovalevski theorem was the first general result on existence of solutions of a broad class of problems involving partial differential equations. Since then, a large body of research has been devoted to existence results and conditions on F and f that are sufficient to guarantee that a solution exists.

Usually existence is more difficult to establish than uniqueness, which has to do with how many solutions a partial differential equation can have that satisfy certain boundary and/or initial conditions. We will illustrate a uniqueness argument for the following heat conduction problem. Consider the heat equation

$$\frac{\partial u}{\partial t} = k\nabla^2 u + \varphi(x, y, z, t),$$

for (x, y, z) in some region D of 3-space bounded by a piecewise smooth surface $\sum$ and for $t > 0$. We will show that there can be only one continuous solution, with continuous first partial derivatives with respect to time and continuous second partial derivatives with respect to the space variables, satisfying

$$u(x, y, z, 0) = f(x, y, z) \quad \text{for } (x, y, z) \text{ in } D \text{ or in } \sum$$

and

$$u(x, y, z, t) = g(x, y, z, t) \quad \text{for } (x, y, z) \text{ in } \sum \text{ and } t \geq 0.$$

This proof will not imply existence of a such a solution—only that if there is a solution satisfying these conditions, then there can be only one.

Suppose that u_1 and u_2 are solutions. Let $w(x, y, z, t) = u_1(x, y, z, t) - u_2(x, y, z, t)$. It is routine to verify that

$$\frac{\partial w}{\partial t} = k\nabla^2 w \quad \text{for } (x, y, z) \text{ in } D \text{ and } t > 0,$$

$$w(x, y, z, 0) = 0 \quad \text{for } (x, y, z) \text{ in } D \text{ or in } \sum,$$

$$w(x, y, z, t) = 0 \quad \text{for } (x, y, z) \text{ in } \sum \text{ and } t \geq 0.$$

Let $\overline{D}$ consist of all points in D or on its boundary surface $\sum$. Define

$$I(t) = \frac{1}{2} \iiint_{\overline{D}} w^2(x, y, z, t) \, dV \quad \text{for } t \geq 0.$$

Then $I(t)$ is continuous for $t \geq 0$ and differentiable for $t > 0$. Further, $I(0) = 0$ because $w(x, y, z, 0) = 0$. Compute

$$I'(t) = \iiint_{\overline{D}} w \frac{\partial w}{\partial t} \, dV = \iiint_{\overline{D}} kw\nabla^2 w \, dV.$$

Apply Gauss's divergence theorem to the last integral to obtain

$$I'(t) = k \iint_{\sum} w\nabla w \cdot \mathbf{N} \, d\sigma - k \iiint_{\overline{D}} \left[\left(\frac{\partial w}{\partial x}\right)^2 + \left(\frac{\partial w}{\partial y}\right)^2 + \left(\frac{\partial w}{\partial z}\right)^2 \right] dV.$$

Since $w = 0$ on $\sum$, the surface integral in the last equation is zero, and

$$I'(t) = -k \iiint_{\overline{D}} \left[\left(\frac{\partial w}{\partial x}\right)^2 + \left(\frac{\partial w}{\partial y}\right)^2 + \left(\frac{\partial w}{\partial z}\right)^2 \right] dV \leq 0.$$

Now apply the mean value theorem to I on the interval $[0, t]$ for any $t > 0$. For some τ between 0 and t,

$$I'(\tau) = \frac{1}{t}[I(t) - I(0)] = \frac{1}{t}I(t).$$

But we have seen that $I'(t) \leq 0$ for all $t > 0$, hence $I'(\tau) \leq 0$ and, since $t > 0$, we also must have $I(t) \leq 0$. But $I(t)$ is the integral of a nonnegative function, hence $I(t) \geq 0$. We conclude that $I(t) = 0$ for $t \geq 0$.

Since $w(x, y, z, t)^2$ is continuous, the only way the integral defining $I(t)$ can be zero for all t is that $w(x, y, z, t) = 0$ for all (x, y, z) in $\overline{D}$ and $t > 0$. We conclude that $u_1 = u_2$, and the uniqueness of the solution is established.

SECTION 19.2   PROBLEMS

1. Consider the boundary value problem

$$\frac{\partial^2 y}{\partial t^2} = c^2 \frac{\partial^2 y}{\partial x^2} \quad \text{for } 0 < x < L, t > 0$$

$$y(0, t) = \alpha(t), y(L, t) = \beta(t) \quad \text{for } t > 0$$

$$y(x, 0) = f(x), \frac{\partial y}{\partial t}(x, 0) = g(x) \quad \text{for } 0 < x < L.$$

Assume that α and β are differentiable on $[0, \infty)$ and that f is differentiable on $[0, L]$. Prove that there can be at most one continuous solution having continuous first and second partial derivatives. *Hint:* Let y_1 and y_2 be solutions and let $w = y_1 - y_2$. Derive a boundary value problem of which w is a solution. Let

$$I(t) = \frac{1}{2} \int_0^L \left[\left(\frac{\partial w}{\partial x} \right)^2 + \frac{1}{c^2} \left(\frac{\partial w}{\partial t} \right)^2 \right] dx.$$

Prove that $I'(t) = 0$ for $t > 0$ and conclude that $I(t) =$ constant. By letting $t = 0$, prove that this constant is zero.

2. In the boundary value problem of Problem 1, replace the boundary conditions at $x = 0$ and $x = L$ with the conditions

$$\frac{\partial y}{\partial t}(0, t) = \alpha(t) \quad \text{and} \quad \frac{\partial y}{\partial t}(L, t) = \beta(t).$$

Prove that there can be at most one continuous solution having continuous first and second partial derivatives.

3. Consider the Dirichlet problem

$$\nabla^2 u(x, y, z) = 0$$

for all (x, y, z) in a region M bounded by a piecewise smooth surface $\sum$ and suppose the boundary condition

is

$$u(x, y, z) = f(x, y, z) \quad \text{for } (x, y, z) \text{ on } \sum.$$

Prove that there can be at most one solution that is continuous and has continuous first and second partial derivatives throughout M and on $\sum$. *Hint:* Suppose there are solutions u_1 and u_2 and let $w = u_1 - u_2$. Determine a Dirichlet problem satisfied by w. Use Gauss's theorem to show that $w(x, y, z) = 0$ for all (x, y, z) in M and on $\sum$.

4. Show that the conclusion of Problem 3 remains valid if the boundary conditions are replaced by

$$a\frac{\partial u}{\partial \eta}(x, y, z) + bu(x, y, z) = f(x, y, z)$$

$$\text{for } (x, y, z) \text{ on } \sum.$$

Here a and b are positive constants, f is a given function that is continuous on $\sum$, and $\partial u/\partial \eta$ is the normal derivative of u on $\sum$.

19.3 Well-Posed Problems

A problem involving a partial differential equation, and boundary and/or initial conditions, is said to be *well posed* if there exists a unique solution that depends continuously on the data of the problem. This means that "small" changes in initial and boundary information result in "small" changes in the solution.

As a specific example, consider the following problem for the wave equation on a line:

$$\frac{\partial^2 u}{\partial t^2} = c^2 \frac{\partial^2 u}{\partial x^2} \quad \text{for } -\infty < x < \infty, t > 0,$$

$$u(x, 0) = f(x), \frac{\partial u}{\partial t}(x, 0) = g(x) \quad \text{for } -\infty < x < \infty.$$

Suppose f is twice continuously differentiable and g is continuous. We know that the solution of this problem is given by d'Alembert's formula:

$$u(x, t) = \frac{1}{2}(f(x - ct) + f(x + ct)) + \frac{1}{2c} \int_{x-ct}^{x+ct} g(\xi) \, d\xi.$$

This proves both existence and uniqueness of the solution.

To show that the solution depends continuously on the data, suppose u_j is the solution with initial position f_j and initial velocity g_j, for $j = 1, 2$. Let $\epsilon > 0$ and $T > 0$. We will show that there is a positive number δ such that if

$$|f_1(x) - f_2(x)| < \delta \quad \text{and} \quad |g_1(x) - g_2(x)| < \delta$$

for all x, then

$$|u_1(x, t) - u_2(x, t)| < \epsilon$$

for all x and for $0 \le t \le T$. This means that the solutions can be made arbitrarily close by choosing the initial positions and initial velocities, respectively, close enough.

To prove this result, use d'Alembert's formula:

$$
|u_1(x, t) - u_2(x, t)| = \left| \frac{1}{2}(f_1(x - ct) + f_1(x + ct)) + \frac{1}{2c} \int_{x-ct}^{x+ct} g_1(\xi)\, d\xi \right.
$$

$$
\left. - \frac{1}{2}(f_2(x - ct) + f_2(x + ct)) - \frac{1}{2c} \int_{x-ct}^{x+ct} g_2(\xi)\, d\xi \right|
$$

$$
\le \left| \frac{1}{2}(f_1(x - ct) - f_2(x - ct)) \right| + \left| \frac{1}{2}(f_1(x + ct) - f_2(x + ct)) \right|
$$

$$
+ \frac{1}{2c} \int_{x-ct}^{x+ct} |g_1(\xi) - g_2(\xi)|\, d\xi
$$

$$
\le \frac{1}{2}\delta + \frac{1}{2}\delta + \frac{1}{2c}(x + ct - (x - ct))\delta
$$

$$
= \delta + \frac{1}{2c}2ct\delta \le \delta + T\delta = (1 + T)\delta < \epsilon
$$

if we choose δ as any positive number satisfying

$$
0 < \delta < \frac{\epsilon}{1 + T}.
$$

Continuous dependence on data is something we might intuitively expect. However, an example due to Jacques Hadamard shows that this intuition is incorrect. We will pursue this idea in the exercises.

SECTION 19.3 PROBLEMS

1. In the spirit of the discussion, devise a definition of continuous dependence on data for solutions of the boundary value problem

$$
\frac{\partial^2 y}{\partial t^2} = c^2 \frac{\partial^2 y}{\partial x^2} \quad \text{for } 0 < x < L, t > 0
$$

$$
y(0, t) = y(L, t) = 0 \quad \text{for } t > 0
$$

$$
y(x, 0) = f(x), \frac{\partial y}{\partial t}(x, 0) = g(x) \quad \text{for } 0 < x < L.
$$

2. Devise a definition of continuous dependence on data for the problem

$$
\frac{\partial y}{\partial t} = k \frac{\partial^2 y}{\partial x^2} \quad \text{for } 0 < x < L, t > 0
$$

$$
u(0, t) = u(L, t) = 0 \quad \text{for } t > 0
$$

$$
u(x, 0) = f(x) \quad \text{for } 0 < x < L.
$$

3. (*Hadamard's Example*) Assume that all boundary value problems in this example have unique solutions. First, consider the problem

$$
\nabla^2 u(x, y) = 0 \quad \text{for all real } x, y
$$

$$
u(x, 0) = 0, \frac{\partial u}{\partial y}(x, 0) = \frac{1}{n}\sin(nx) \quad \text{for all real } x.
$$

Here n is any positive integer.

(a) Show that $u(x, y) = (1/n^2)\sin(nx)\sinh(ny)$ is a solution of this problem.

(b) Let $A(x, y)$ be a solution of

$$
\nabla^2 u(x, y) = 0 \quad \text{for all real } x, y
$$

$$
u(x, 0) = f(x), \frac{\partial u}{\partial y}(x, 0) = g(x) \quad \text{for all real } x.
$$

Let $B(x, y)$ be a solution of

$$\nabla^2 u(x, y) = 0 \quad \text{for all real } x, y$$

$$u(x, 0) = f(x), \frac{\partial u}{\partial y}(x, 0) = g(x) + \frac{1}{n} \sin(nx)$$

for all real x.

Prove that

$$B(x, y) - A(x, y) = \frac{1}{n^2} \sin(nx) \sinh(ny).$$

(c) Conclude that the first boundary value problem stated in part (b) does not depend continuously on the data.

$$\lim_{h \to 0} \frac{f(z+h) - f(z)}{h} = \lim_{h \to 0} \frac{(z+h) - f(z)}{h} = \lim_{h \to 0} \frac{f(z+h) - f(z)}{h} = \lim_{h \to 0} \frac{\overline{z+h} - \overline{z}}{h}$$

PART 7

Complex Analysis

CHAPTER 20
*Geometry and Arithmetic
of Complex Numbers*

CHAPTER 21
Complex Functions

CHAPTER 22
Complex Integration

CHAPTER 23
Series Representation of Functions

CHAPTER 24
*Singularities and the Residue
Theorem*

CHAPTER 25
Conformal Mappings

This part is devoted to the calculus of functions that act on complex numbers to produce complex numbers.

Often our first encounter with complex numbers is in finding roots of polynomials. For example, the equation $x^2 + 1 = 0$ has no real solutions, but is easily solved if we use complex quantities. Once complex numbers were understood (this took until the nineteenth century), it was not long before complex functions and an associated calculus of derivatives and integrals were developed.

Although complex analysis is interesting in its own right, there are also important and sometimes surprising applications to problems we might initially think of as belonging to the real domain. This includes the evaluation of real integrals and summation of series by complex integration techniques, as well as the use of complex quantities such as the Fourier transform and discrete Fourier transform to solve problems in wave motion, heat propagation, and signal analysis, whose solutions are real-valued. The use of conformal mappings between domains in the complex plane is also a significant technique for solving certain partial differential equations.

We will begin with the algebra and geometry of the complex plane.

CHAPTER 20

Geometry and Arithmetic of Complex Numbers

20.1 Complex Numbers

A *complex number* is a symbol of the form $x + iy$, or $x + yi$, in which x and y are real numbers and $i^2 = -1$. Arithmetic of complex numbers is defined by

$$\text{equality: } a + ib = c + id \text{ exactly when } a = c \text{ and } b = d,$$

$$\text{addition: } (a + ib) + (c + id) = (a + c) + i(b + d),$$

and

$$\text{multiplication: } (a + ib)(c + id) = ac - bd + i(ad + bc).$$

In multiplying complex numbers, we proceed exactly as we would with first-order polynomials $a + bx$ and $c + dx$, but with i in place of x:

$$(a + bi)(c + di) = ac + adi + bci + bdi^2$$
$$= ac - bd + (ad + bc)i$$

because $i^2 = -1$. For example,

$$(6 - 4i)(8 + 13i) = (6)(8) - (-4)(13) + i[(6)(13) + (-4)(8)] = 100 + 46i.$$

The real number a is called the real part of $a + bi$ and is denoted $\text{Re}(a + bi)$. The real number b is the imaginary part, denoted $\text{Im}(a + bi)$. For example,

$$\text{Re}(-23 + 7i) = -23 \quad \text{and} \quad \text{Im}(-23 + 7i) = 7.$$

Both the real and imaginary part of any complex number are real numbers.

We may think of the complex number system as an extension of the real number system in the sense that every real number a is a complex number $a + 0i$. This extension of the reals to the complex numbers has profound consequences, both for algebra and analysis. For example,

1003

the polynomial equation $x^2 + 1 = 0$ has no real solution, but it has two complex solutions, i and $-i$. More generally, the fundamental theorem of algebra states that every polynomial of positive degree n, having complex coefficients (some or all of which may be real), has exactly n roots in the complex numbers, counting repeated roots. This means that we need never extend beyond the complex numbers to find roots of polynomials having complex coefficients, as we have to extend beyond the reals to find the roots of a simple polynomial such as $x^2 + 1$.

Complex addition obeys many of the same rules of arithmetic as real numbers. Specifically, for any complex numbers z, w, and u,

$z + w = w + z$ (commutativity of addition)

$zw = wz$ (commutativity of multiplication)

$z + (w + u) = (z + w) + u$ (associativity of addition)

$z(wu) = (zw)u$ (associativity of multiplication)

$z(w + u) = zw + zu$ (distributivity)

$z + 0 = 0 + z$

$z \cdot 1 = 1 \cdot z.$

20.1.1 The Complex Plane

Complex numbers admit two natural geometric intepretations.

First, we may identify the complex number $a + bi$ with the point (a, b) in the plane, as in Figure 20.1. In this interpretation, each real number a, or $a + 0i$, is identified with the point $(a, 0)$ on the horizontal axis, which is therefore called the *real axis*. A number $0 + bi$, or just bi, is called a *pure imaginary number* and is associated with the point $(0, b)$ on the vertical axis. This axis is called the *imaginary axis*. Because of this correspondence between complex numbers and points in the plane, we often refer to the x, y plane as the *complex plane*.

When complex numbers were first noticed (in solving polynomial equations), mathematicians were suspicious of them, and even the great eighteenth-century Swiss mathematician Leonhard Euler, who used them in calculations with unparalleled proficiency, did not recognize them as "legitimate" numbers. It was the nineteenth-century German mathematician Carl Friedrich Gauss who fully appreciated their geometric significance and used his standing in the scientific community to promote their legitimacy to other mathematicians and natural philosophers.

The second geometric interpretation of complex numbers is in terms of vectors. The complex number $z = a + bi$, or the point (a, b), may be thought of as a vector $a\mathbf{i} + b\mathbf{j}$ in the plane, which may in turn be represented as an arrow from the origin to (a, b), as in Figure 20.2. The first component of this vector is $\text{Re}(z)$ and the second component is $\text{Im}(z)$. In this interpretation, the definition of addition of complex numbers is equivalent to the parallelogram law for vector addition, since we add two vectors by adding their respective components (Figure 20.3).

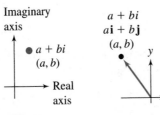

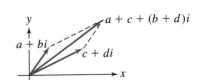

FIGURE 20.1
The complex plane.

FIGURE 20.2
Complex numbers as vectors in a plane.

FIGURE 20.3 *Parallelogram law for addition of complex numbers.*

20.1.2   Magnitude and Conjugate

DEFINITION 20.1   *Magnitude*

The magnitude of $a + bi$ is denoted $|a + bi|$ and is defined by

$$|a + bi| = \sqrt{a^2 + b^2}.$$

Of course, the magnitude of zero is zero. As Figure 20.4 suggests, if $z = a + ib$ is a nonzero complex number, then $|z|$ is the distance from the origin to the point (a, b). Alternatively, $|z|$ is the length of the vector $a\mathbf{i} + b\mathbf{j}$ representing z. For example,

$$|2 - 5i| = \sqrt{4 + 25} = \sqrt{29}.$$

The magnitude of a complex number is also called its *modulus*.

DEFINITION 20.2   *Conjugate*

The complex conjugate (or just conjugate) of $a + bi$ is the number denoted $\overline{a + bi}$ and defined by

$$\overline{a + bi} = a - bi.$$

We get the conjugate of z by changing the sign of the imaginary part of z. For example,

$$\overline{3 - 8i} = 3 + 8i, \overline{i} = -i, \quad \text{and} \quad \overline{-25} = -25.$$

This operation does not change the real part of z. We have

$$\text{Re}(\overline{a + ib}) = a = \text{Re}(a + ib)$$

and

$$\text{Im}(\overline{a + ib}) = -b = -\text{Im}(a + ib).$$

The operation of taking the conjugate can be interpreted as a reflection across the real axis, because the point $(a, -b)$ associated with $a - ib$ is the reflection across the horizontal axis of the point (a, b) associated with $a + ib$ (Figure 20.5).

Here are some computational rules for magnitude and conjugate.

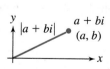

FIGURE 20.4
Magnitude of a complex number.

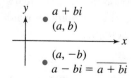

FIGURE 20.5 *Conjugate of a complex number.*

— **THEOREM 20.1** ————————————————————————————————————

Let z and w be complex numbers. Then

1. $\overline{\overline{z}} = z$.
2. $\overline{z + w} = \overline{z} + \overline{w}$.
3. $\overline{zw} = (\overline{z})(\overline{w})$.
4. $|z| = |\overline{z}|$.
5. $|zw| = |z|\,|w|$.
6. $\operatorname{Re}(z) = \frac{1}{2}(z + \overline{z})$ and $\operatorname{Im}(z) = \frac{1}{2i}(z - \overline{z})$.
7. $|z| \geq 0$, and $|z| = 0$ if and only if $z = 0$.
8. $z\overline{z} = |z|^2$.

Proof   Conclusion (1) states that taking the conjugate of a conjugate returns the original number. This is geometrically evident, since a reflection of (x, y) to $(x, -y)$, followed by a reflection of $(x, -y)$ to (x, y), returns to the original point. For an analytical argument, write

$$\overline{\overline{a + ib}} = \overline{a - ib} = a + ib.$$

For conclusion (5), let $z = a + ib$ and $w = c + id$. Then

$$
\begin{aligned}
|zw| &= |(ac - bd) + i(ad + bc)| \\
&= \sqrt{(ac - bd)^2 + (ad + bc)^2} \\
&= \sqrt{a^2c^2 + b^2d^2 - 2acbd + a^2d^2 + b^2c^2 + 2adbc} \\
&= \sqrt{a^2c^2 + a^2d^2 + b^2c^2 + b^2d^2} \\
&= \sqrt{a^2 + b^2}\sqrt{c^2 + d^2} = |z|\,|w|.
\end{aligned}
$$

A much neater proof of (5) will be available when we know about the polar form of a complex number.

The other parts of the theorem are left to the student. ∎

20.1.3   Complex Division

Suppose we want to form the quotient z/w, where $w \neq 0$. This quotient is the complex number u such that $wu = z$. This is not very helpful in actually finding u, however. Here is a computationally effective way of performing complex division. Let $z = a + ib$ and $w = c + id$ and write

$$\frac{a + ib}{c + id} = \frac{a + ib}{c + id}\frac{c - id}{c - id} = \frac{ac + bd + i(bc - ad)}{c^2 + d^2}.$$

By multiplying and dividing the original fraction by the conjugate of the denominator, we obtain an expression in which the real and imaginary parts of the quotient are apparent. The reason for this is that the denominator is $w\overline{w}$, which is the real number $|w|^2$.

For example,

$$\frac{2 - 7i}{8 + 3i} = \frac{2 - 7i}{8 + 3i}\frac{8 - 3i}{8 - 3i} = \frac{-5 - 62i}{64 + 9} = -\frac{5}{73} - \frac{62}{73}i,$$

so the real part of this quotient is $-5/73$ and the imaginary part is $-62/73$.

20.1.4   Inequalities

There are several inequalities we will have occasion to use.

THEOREM 20.2

Let z and w be complex numbers. Then

1. $|\text{Re}(z)| \leq |z|$ and $|\text{Im}(z)| \leq |z|$.
2. $|z + w| \leq |z| + |w|$.
3. $||z| - |w|| \leq |z - w|$.

Proof   If $z = a + ib$, then

$$|\text{Re}(z)| = |a| \leq \sqrt{a^2 + b^2} = |z|$$

and

$$|\text{Im}(z)| = |b| \leq \sqrt{a^2 + b^2} = |z|.$$

Conclusion (2), which is called the *triangle inequality*, was proved for vectors. Here is a separate proof in the context of complex numbers:

$$0 \leq |z + w|^2 = (z + w)(\overline{z + w}) = (z + w)(\overline{z} + \overline{w}) = z\overline{z} + z\overline{w} + w\overline{z} + w\overline{w}$$

$$= |z|^2 + z\overline{w} + \overline{z\overline{w}} + |w|^2 = |z|^2 + 2\,\text{Re}(z\overline{w}) + |w|^2 \leq |z|^2 + 2\,|z\overline{w}| + |w|^2$$

$$= |z|^2 + 2\,|z|\,|\overline{w}| + |w|^2 = |z|^2 + 2\,|z|\,|w| + |w|^2 = (|z| + |w|)^2.$$

In summary,

$$0 \leq |z + w|^2 \leq (|z| + |w|)^2.$$

Upon taking the square root of these nonnegative quantities, we obtain the triangle inequality.

For (3), use the triangle inequality to write

$$|z| = |(z + w) - w| \leq |z + w| + |w|,$$

hence

$$|z| - |w| \leq |z + w|.$$

By interchanging z and w,

$$|w| - |z| \leq |z + w|.$$

Therefore,

$$-|z + w| \leq |z| - |w| \leq |z + w|,$$

so

$$||z| - |w|| \leq |z + w|. \blacksquare$$

20.1.5   Argument and Polar Form of a Complex Number

Let $z = a + ib$ be a nonzero complex number. Then (a, b) is a point other than the origin in the plane. Let this point have polar coordinates (r, θ). As is standard with polar coordinates, the polar angle θ of (a, b) is not uniquely determined. If we walk out the real axis r units to the right from the origin and rotate this segment θ_0 radians until the segment ends at (a, b), as in Figure 20.6,

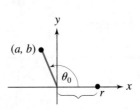

FIGURE 20.6 (a, b) *has polar coordinates* $(r, \theta_0 + 2n\pi)$, *n any integer.*

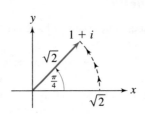

FIGURE 20.7 *Polar coordinates of* $1 + i$ *are* $(\sqrt{2}, \pi/4 + 2n\pi)$, *n any integer.*

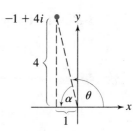

FIGURE 20.8 $\pi - \tan^{-1}(4)$ *is an argument of* $-1 + 4i$.

then the polar angle θ for (a, b) is any number $\theta_0 + 2n\pi$, in which n is any integer. A positive choice for n corresponds to rotating the segment from 0 to r an initial θ_0 radians to reach (a, b) and then continuing counterclockwise an additional n full circles, which again lands us on (a, b). A negative choice for n corresponds to rotating the segment from 0 to r an initial θ_0 radians and then going around an additional n circles clockwise, again ending at (a, b). Thus, by convention, we think of counterclockwise rotations as having positive orientation in the plane and clockwise rotations as having negative orientation.

To illustrate, consider $z = 1 + i$. The point $(1, 1)$ has polar coordinates $(\sqrt{2}, \pi/4)$, since $1 + i$ is $\sqrt{2}$ units from the origin, and the segment from the origin to $1 + i$ makes an angle $\pi/4$ radians with the positive real axis (Figure 20.7). All of the polar coordinates of $(1, 1)$ have the form

$$(\sqrt{2}, \pi/4 + 2n\pi),$$

in which n can be any integer.

If nonzero z has polar coordinates (r, θ), then $r = |z|$. The angle θ (which we will always express in radians) is called an *argument* of z. Any nonzero number has infinitely many arguments, and they differ from each other by integer multiples of 2π. The arguments of $1 + i$ are $\pi/4 + 2n\pi$, for n any integer.

Now recall Euler's formula

$$e^{i\theta} = \cos(\theta) + i \sin(\theta).$$

If θ is any argument of $z = a + ib$, then (a, b) has polar coordinates (r, θ), so $a = r \cos(\theta)$ and $b = r \sin(\theta)$. Combining this fact with Euler's formula, we have

$$z = a + ib = r \cos(\theta) + ir \sin(\theta) = re^{i\theta}.$$

This exponential form for z is called the *polar form* of z. Any argument of z can be used in this polar form, because any two arguments θ_0 and θ_1 differ by some integer multiple of 2π. If, say $\theta_1 = \theta_0 + 2k\pi$, then

$$re^{i\theta_1} = r[\cos(\theta_0 + 2k\pi) + i \sin(\theta_0 + 2k\pi)]$$

$$= r[\cos(\theta_0) + i \sin(\theta_0)] = re^{i\theta_0}.$$

EXAMPLE 20.1

We will find the polar form of $-1 + 4i$. First, $r = |-1 + 4i| = \sqrt{17}$. Now consider Figure 20.8. θ is an argument of $-1 + 4i$, and α will be handy in determining θ. From the diagram, $\tan(\alpha) = 4$,

so

$$\theta = \pi - \alpha = \pi - \tan^{-1}(4).$$

We can therefore write the polar form

$$-1 + 4i = \sqrt{17}e^{i(\pi - \tan^{-1}(4))}. \quad \blacksquare$$

EXAMPLE 20.2

We will find the polar form of $3 + 3i$. As indicated in Figure 20.9, $\pi/4$ is an argument of $3 + 3i$. The polar form is

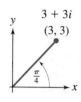

$$3 + 3i = \sqrt{18}e^{i\pi/4}. \quad \blacksquare$$

For any real θ,

$$\left|e^{i\theta}\right| = \cos^2(\theta) + \sin^2(\theta) = 1.$$

FIGURE 20.9
$\pi/4$ *is an argument of* $3 + 3i$.

This means that in writing the polar form $z = re^{i\theta}$, the magnitude of z is wholly contained in the factor r, while $e^{i\theta}$, which has magnitude 1, contributes all the information about the direction of (nonzero) z from the origin.

Because of the properties of the exponential function, some computations with complex numbers are simplified if polar forms are used. To illustrate, suppose we want to prove that $|zw| = |z||w|$, something we did before by algebraic manipulation. Put $z = re^{i\theta}$ and $w = \rho e^{i\xi}$ to immediately get

$$|zw| = \left|r\rho e^{i\theta} e^{i\xi}\right| = r\rho \left|e^{i(\theta + \xi)}\right| = r\rho = |z||w|.$$

The fact that $e^{i(\theta+\xi)} = e^{i\theta}e^{i\xi}$ also means that the argument of a product is the sum of the arguments of the factors, to within an integer multiple of 2π. Put more carefully, if θ_0 is any argument of z, and θ_1 is any argument of w, and Θ is any argument of zw, then for some integer n,

$$\Theta = \theta_0 + \theta_1 + 2n\pi.$$

Multiplying two complex numbers has the effect of adding their arguments, to within integer multiples of 2π.

EXAMPLE 20.3

Let $z = i$ and $w = 2 - 2i$. One argument of z is $\theta_0 = \pi/2$, and one argument of w is $\theta_1 = 7\pi/4$ (Figure 20.10). Now

$$zw = i(2 - 2i) = 2 + 2i,$$

and one argument of $2 + 2i$ is $\Theta = \pi/4$. With this choice of arguments,

$$\theta_0 + \theta_1 = \frac{\pi}{2} + \frac{7\pi}{4} = \frac{9\pi}{4} = \Theta + 2\pi.$$

If we had chosen $\theta_0 = \pi/2$ and $\theta_1 = -\pi/4$, then we would get

$$\theta_0 + \theta_1 = \frac{\pi}{2} - \frac{\pi}{4} = \frac{\pi}{4} = \Theta. \quad \blacksquare$$

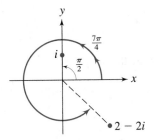

FIGURE 20.10 $\pi/2$ *is an argument of* i, *and* $7\pi/4$ *an argument of* $2 - 2i$.

20.1.6 Ordering

Given any two distinct real numbers a and b, exactly one of $a < b$ or $b < a$ must be true. The real numbers are said to be ordered. We will show that there is no ordering of complex numbers.

To understand why this is true, we must investigate the idea behind the ordering of the reals. The ordering of the real numbers is actually a partitioning of the nonzero real numbers into two mutually exclusive sets, N and P, with the following properties:

1. If x is a nonzero real number, then x is in P or $-x$ is in P, but not both.

2. If x and y are in P, then $x + y$ and xy are in P.

Think of P as the set of positive numbers and N as the set of negative numbers. The existence of such a partition of the nonzero reals satisfying conditions (1) and (2) is the reason why we can order the reals. An ordering is established by defining $x < y$ if and only if $y - x$ is in P. For example, $2 < 5$ because $5 - 2 = 3$ is positive.

Does there exist a partition of the nonzero complex numbers into two sets, P and N, having properties (1) and (2)? If so, we can order the complex numbers.

Suppose such a partition exists. Then either i is in P or $-i$ is in P, but not both. If i is in P, then $i^2 = -1$ is in P by (2), so $(-1)(i) = -i$ is in P. But this violates condition (1). If $-i$ is in P, then $(-i)(-i) = i^2 = -1$ is in P, so $(-1)(-i) = i$ is in P, again violating (1).

This shows that no such partition exists, and the complex numbers cannot be ordered. Whenever we write $z < w$, we are assuming that z and w are real numbers.

20.1.7 Binomial Expansion of $(z + w)^n$

If x and y are real numbers and n is a positive integer, then

$$(x + y)^n = \sum_{j=0}^{n} \binom{n}{j} x^{n-j} y^j$$

$$= x^n + nx^{n-1}y + \frac{n(n-1)}{2}x^{n-2}y^2 + \frac{n(n-1)(n-2)}{6}x^{n-3}y^3$$

$$+ \cdots + \frac{n(n-1)}{2}x^2 y^{n-2} + nxy^{n-1} + y^n.$$

This is the binomial expansion of $(x + y)^n$. The numbers $\binom{n}{j}$ are the binomial coefficients, defined by

$$\binom{n}{j} = \frac{n!}{j!(n-j)!}.$$

In combinatorics, $\binom{n}{j}$ is the number of ways of choosing j objects from n objects, if the order of selection is disregarded in the counting.

For example,

$$(x+y)^4 = \sum_{j=0}^{4} \binom{4}{j} x^{4-j} y^j$$

$$= \binom{4}{0} x^4 + \binom{4}{1} x^3 y + \binom{4}{2} x^2 y^2 + \binom{4}{3} xy^3 + \binom{4}{4} y^4$$

$$= x^4 + 4x^3 y + 6x^2 y^2 + 4xy^3 + y^4.$$

This is exactly what we would get if we simply multiplied out $(x + y)^4$, but the binomial formula gives a systematic way of writing this product, tells us what the coefficient of $x^{n-j} y^j$ is, and is particularly useful if n is large.

This expansion remains valid for complex numbers:

$$(z+w)^n = \sum_{j=0}^{n} \binom{n}{j} z^{n-j} w^j.$$

The formula can be proved by mathematical induction.

SECTION 20.1 PROBLEMS

In each of Problems 1 through 10, carry out the indicated calculation.

1. $(3 - 4i)(6 + 2i)$

2. $i(6 - 2i) + |1 + i|$

3. $\dfrac{2+i}{4-7i}$

4. $\dfrac{(2+i) - (3-4i)}{(5-i)(3+i)}$

5. $(17 - 6i)(\overline{-4 - 12i})$

6. $\left| \dfrac{3i}{-4 + 8i} \right|$

7. $i^3 - 4i^2 + 2$

8. $(3 + i)^3$

9. $\left(\dfrac{-6 + 2i}{1 - 8i} \right)^2$

10. $(-3 - 8i)(2i)(4 - i)$

11. Prove that, for any positive integer n,
$$i^{4n} = 1, i^{4n+1} = i, i^{4n+2} = -1, \quad \text{and} \quad i^{4n+3} = -i.$$

12. Let $z = a + ib$. Determine $\text{Re}(z^2)$ and $\text{Im}(z^2)$.

13. Let $z = a + ib$. Determine $\text{Re}(z^2 - iz + 1)$ and $\text{Im}(z^2 - iz + 1)$.

14. Prove that $z^2 = \bar{z}^2$ if and only if z is either real or pure imaginary.

15. Let z, w, and u be complex numbers. Prove that when represented as points in the plane, these numbers form the vertices of an equilateral triangle if and only if
$$z^2 + w^2 + u^2 = zw + zu + wu.$$

16. Prove that $\text{Re}(iz) = -\text{Im}(z)$ and $\text{Im}(iz) = \text{Re}(z)$.

In each of Problems 17 through 22, determine $\arg(z)$. The answer should include all arguments of the number.

17. $3i$

18. $-2 + 2i$

19. $-3 + 2i$

20. $8 + i$

21. -4

22. $3 - 4i$

In each of Problems 23 through 28, write the complex number in polar form.

23. $-2 + 2i$

24. $-7i$

25. $5 - 2i$

26. $-4 - i$

27. $8 + i$

28. $-12 + 3i$

29. Let z and w be complex numbers such that $\bar{z}w \neq 1$, but such that either z or w has magnitude 1. Prove that

$$\left| \frac{z - w}{1 - \bar{z}w} \right| = 1.$$

Hint: In problems involving magnitude, it is often useful to recall Theorem 20.1(8). To apply this result, square both sides of the proposed equality.

30. Prove that for any complex numbers z and w,

$$|z + w|^2 + |z - w|^2 = 2 \left(|z|^2 + |w|^2 \right).$$

Hint: Keep Theorem 20.1(8) in mind.

31. Prove that for any complex numbers z and w and any positive integer n,

$$(z + w)^n = \sum_{j=0}^{n} \binom{n}{j} z^{n-j} w^j.$$

32. Use the result of Problem 31 to calculate $(2 + i)^8$.

33. Use the result of Problem 31 to calculate $(1 - i)^6$.

34. Prove *de Moivre's theorem*: If n is any integer, then

$$[\cos(\theta) + i \sin(\theta)]^n = \cos(n\theta) + i \sin(n\theta).$$

Hint: Use Euler's formula.

35. Express $\cos(3\theta)$ as a sum of products of powers of $\sin(\theta)$ and $\cos(\theta)$. *Hint:* Compute $[\cos(\theta) + i \sin(\theta)]^3$ first as a product and then by de Moivre's theorem, and equate the real part of the left side with the real part of the right side.

36. Prove *Lagrange's trigonometric identity*: If $\sin(\theta/2) \neq 0$ and n is a positive integer, then

$$\sum_{j=0}^{n} \cos(j\theta) = \frac{1}{2} + \frac{\sin((n + 1/2)\theta)}{2 \sin(\theta/2)}.$$

Hint: Begin with the algebraic identity

$$\sum_{j=0}^{n} z^j = \frac{1 - z^{n+1}}{1 - z} \quad \text{if } z \neq 1.$$

This may be assumed. Let $z = \cos(\theta) + i \sin(\theta)$ and use de Moivre's theorem, separating the real and imaginary parts.

20.2 Loci and Sets of Points in the Complex Plane

Sometimes complex notation is very efficient in specifying loci of points in the plane. In this section we will illustrate this, and also discuss the complex representation of certain sets that occur frequently in discussions of complex integrals and derivatives.

20.2.1 Distance

If $z = a + ib$ is any complex number, $|z| = \sqrt{a^2 + b^2}$ is the distance from the origin to z (point (a, b)) in the complex plane. If $w = c + id$ is also a complex number, then

$$|z - w| = |(a - c) + i(b - d)|$$
$$= \sqrt{(a - c)^2 + (b - d)^2}$$

is the distance between z and w in the complex plane (Figure 20.11). This is the standard formula from geometry for the distance between points (a, b) and (c, d).

20.2.2 Circles and Disks

If a is a complex number and r is a positive (hence real) number, the equation

$$|z - a| = r$$

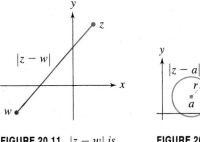

FIGURE 20.11 $|z - w|$ *is the distance between z and w.*

FIGURE 20.12 *The circle of radius r about a.*

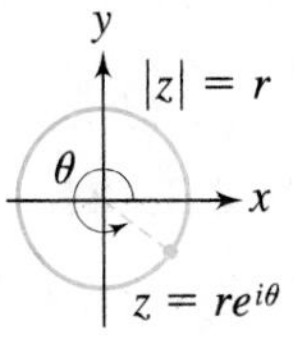

FIGURE 20.13 *The circle of radius r about the origin.*

is satisfied by exactly those z whose distance from a is r. The locus of points satisfying this condition is the circle of radius r about a (Figure 20.12). This is the way we specify circles in the complex plane, and we often refer to "the circle $|z - a| = r$."

If $a = 0$, then any point on the circle $|z| = r$ has polar form

$$z = re^{i\theta},$$

where θ is the angle from the positive real axis to the line from the origin through z (Figure 20.13). As θ varies from 0 to 2π, the point $z = re^{i\theta}$ moves once counterclockwise around this circle, starting at $z = r$ on the positive real axis when $\theta = 0$, reaching ri when $\theta = \pi/2$, $-r$ when $\theta = \pi$, $-ri$ when $\theta = 3\pi/2$, and returning to r when $\theta = 2\pi$.

If $a \neq 0$, then the center of the circle $|z - a| = r$ is a instead of the origin. Now a point on the circle has the form

$$z = a + re^{i\theta},$$

which is simply a polar coordinate system translated to have a as its origin (Figure 20.14). As θ varies from 0 to 2π, this point moves once counterclockwise about this circle. For example, the equation $|z - 3 + 7i| = 4$ defines the circle of radius 4 about the point $(3, -7)$ in the plane. The complex number $3 - 7i$ is the center of the circle. A typical point on the circle has the form $z = 3 - 7i + 4e^{i\theta}$ (Figure 20.15).

An inequality $|z - a| < r$ specifies all points within the disk of radius r about a. Such a set is called an *open disk*. "Open" here means that points on the circumference of the circle bounding this disk are not in this set. A point on this circle would satisfy $|z - a| = r$, not $|z - a| < r$. We often indicate in a drawing that a disk is open by drawing a dashed boundary circle (Figure 20.16). For example, $|z - i| < 8$ specifies the points within the open disk of radius 8 about i.

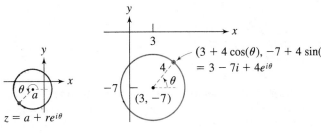

FIGURE 20.14

FIGURE 20.15 *The circle $|z - 3 + 7i| = 4$.*

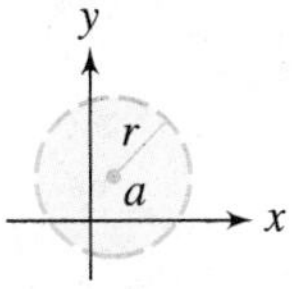

FIGURE 20.16 $|z - a| < r$, *the open disk of radius r about a.*

A *closed disk* of radius r and center a consists of all points on or within the circle of radius r about a. This set is specified by the weak inequality $|z - a| \leq r$. In a drawing of such a set, we often draw a solid circle as a boundary to indicate that these points are included in the closed disk (Figure 20.17).

20.2.3 The Equation $|z - a| = |z - b|$

Let w_1 and w_2 be distinct complex numbers. An equation

$$|z - w_1| = |z - w_2|$$

can be verbalized as "the distance between z and w_1 must equal the distance between z and w_2." As Figure 20.18 suggests, this requires that z be on the perpendicular bisector of the line segment connecting w_1 and w_2. The equation $|z - w_1| = |z - w_2|$ may therefore be considered the equation of this line.

EXAMPLE 20.4

The equation

$$|z + 6i| = |z - 1 + 3i|$$

is satisfied by all points on the perpendicular bisector of the segment between $-6i$ and $1 - 3i$. This is the segment connecting $(0, -6)$ and $(1, -3)$, as shown in Figure 20.19.

We can obtain the "standard" equation of this line as follows. First write

$$|z + 6i|^2 = |z - 1 + 3i|^2,$$

or

$$(z + 6i)(\bar{z} - 6i) = (z - 1 + 3i)(\bar{z} - 1 - 3i).$$

This eliminates the absolute value signs. Carry out the multiplications to obtain

$$z\bar{z} + 6i(\bar{z} - z) + 36 = z\bar{z} - z - 3iz - \bar{z} + 1 + 3i + 3i\bar{z} - 3i + 9.$$

Let $z = x + iy$. Then $\bar{z} - z = (x - iy) - (x + iy) = -2iy$ and $-\bar{z} - z = -2x$, so the last equation becomes

$$6i(-2iy) + 36 = -2x + 3i(-2iy) + 10,$$

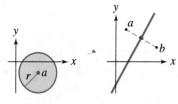

FIGURE 20.17
$|z - a| \leq r$,
the closed disk
of radius r
about a.

FIGURE 20.18
$|z-a| = |z-b|$
is satisfied by
all z on the
perpendicular
bisector of the
segment $\overline{ab}$.

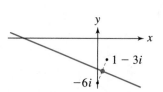

FIGURE 20.19 *The locus of the*
equation $|z+6i| = |z-1+3i|$.

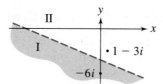

FIGURE 20.20 *Region I consists of points satisfying* $|z + 6i| < |z - 1 + 3i|$.

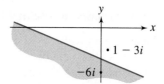

FIGURE 20.21 *The region given by* $|z + 6i| \leq |z - 1 + 3i|$.

FIGURE 20.22 *Locus of points z with* $|z|^2 + 3 \operatorname{Re}(z^2) = 4$.

or

$$12y = -2x + 6y - 26.$$

This is the line

$$y = -\tfrac{1}{3}(x + 13). \quad \blacksquare$$

Now consider the inequality

$$|z + 6i| < |z - 1 + 3i|.$$

We already know that the equation $|z + 6i| = |z - 1 + 3i|$ describes a line separating the plane into two sets, having this line as boundary (Figure 20.19). The given inequality holds for points in one of these sets, on one side or the other of this line. Clearly z is closer to $-6i$ than to $1 - 3i$ if z is below the boundary line. Thus the inequality specifies all z below this line, the shaded region in Figure 20.20. The boundary line itself is drawn dashed because points on this line do not belong to this region.

The weak inequality $|z + 6i| \leq |z - 1 + 3i|$ specifies all points in the shaded region of Figure 20.21, together with all points on the boundary line itself.

20.2.4 Other Loci

When a geometric argument is not apparent, we try to determine a locus by substituting $z = x + iy$ into the given equation or inequality.

EXAMPLE 20.5

Consider the equation

$$|z|^2 + 3 \operatorname{Re}(z^2) = 4.$$

If $z = x + iy$, this equation becomes

$$x^2 + y^2 + 3(x^2 - y^2) = 4,$$

or

$$4x^2 - 2y^2 = 4.$$

The graph of this equation is the hyperbola of Figure 20.22. A complex number satisfies the given equation if and only if its representation as a point in the plane lies on this hyperbola. $\blacksquare$

20.2.5 Interior Points, Boundary Points, and Open and Closed Sets

In the development of the calculus of complex functions, certain kinds of sets and points will be important. For this section let S be a set of complex numbers. A number is an interior point of S if it is in a sense completely surrounded by points of S.

DEFINITION 20.3 Interior Point

A complex number z_0 is an interior point of S if there is an open disk about z_0 containing only points of S.

This means that, for some positive r, all points satisfying $|z - z_0| < r$ are in S. Clearly this forces z_0 to be in S as well.

DEFINITION 20.4 Open Set

S is open if every point of S is an interior point.

EXAMPLE 20.6

Let K be the open disk $|z - a| < r$ (Figure 20.23). Every point of K is an interior point because about any point in K we can draw a disk of small enough radius to contain only points in K. Thus K is an open set, justifying the terminology "open disk" used previously for a disk that does not include any points on its bounding circle. ∎

EXAMPLE 20.7

Let L consist of all points satisfying $|z - a| \leq r$. Now L contains points that are not interior points, specifically those on the circle $|z - a| = r$. Any open disk drawn about such a point will contain points outside of the disk L (Figure 20.24). This set is not an open set. ∎

EXAMPLE 20.8

Let V consist of all $z = x + iy$ with $x > 0$. This is the right half-plane, not including the imaginary axis that forms the boundary between the left and right half-planes. As suggested by Figure 20.25, every point of V is an interior point, because about any point $z_0 = x_0 + iy_0$ with $x_0 > 0$, we can draw a small enough disk that all points it encloses will also have positive real parts. Since every point of V is an interior point, V is an open set. ∎

EXAMPLE 20.9

Let M consist of all $z = x + iy$ with $x \geq 0$. Every point in M with positive real part is an interior point, just as in the preceding example. But not every point of M is interior. A point $z = iy$ on the imaginary axis is in M but cannot be enclosed in a disk that contains only points in M, having

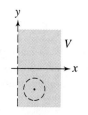
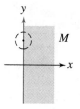

FIGURE 20.23
An open disk is an open set (all points are interior points).

FIGURE 20.24
Points on $|z - a| = r$ are not interior points of $|z - a| \leq r$.

FIGURE 20.25
Half-plane $\mathrm{Re}(z) > 0$ (an open set).

FIGURE 20.26
Half-plane $\mathrm{Re}(z) \geq 0$ (not an open set).

nonnegative real part (Figure 20.26). Since M contains points that are not interior points, M is not open. ■

EXAMPLE 20.10

Let W consist of all points on the real axis. Then no point of W is an interior point. Any disk, no matter how small the radius, drawn about a point on the real axis will contain points not on this axis, hence not in W. No point of W is an interior point of W. ■

Returning again to the general discussion, boundary points of a set S are complex numbers that are in some sense on the "edge" of S.

DEFINITION 20.5   *Boundary Point*

A point z_0 is a boundary point of S if every open disk about z_0 contains at least one point in S and at least one point not in S.

A boundary point itself may or may not be in S. Because the definitions of interior point and boundary point are exclusive, no point can be an interior point and a boundary point of the same set. The set of all boundary points of S is called the *boundary* of S, and is denoted ∂S.

EXAMPLE 20.11

The sets K and L of Examples 20.6 and 20.7 have the same boundary, namely, the points on the circle $|z - a| = r$. K does not contain any of its boundary points, while L contains them all. ■

EXAMPLE 20.12

The set V of Example 20.8 has all the points on the imaginary axis as its boundary points. This set does not contain any of its boundary points. By contrast, M of Example 20.9 has the same boundary points as V, namely, all points on the imaginary axis, but M contains all of these boundary points. ■

EXAMPLE 20.13

For the real line (W in Example 20.10), every point of W is a boundary point. If we draw any open disk about a real number x, this disk contains a point of W, namely, x, and many points not in W. There are no other boundary points of W. ■

EXAMPLE 20.14

Let E consist of all complex numbers $z = x + iy$ with $y > 0$, together with the point $-23i$ (Figure 20.27). Then $-23i$ is a boundary point of E, because every disk about $-23i$ certainly contains points not in E, but also contains a point of E, namely $-23i$ itself. Every real number (horizontal axis) is also a boundary point of E. ■

A careful reading of the definition shows that every point of a set is either an interior point or a boundary point.

THEOREM 20.3

Let S be a set of complex numbers and let z be in S. Then z is either a boundary point of S or an interior point of S.

Proof Suppose z is in S, but is not an interior point. If D is any open disk about z, then D cannot contain only points of S, and so must contain at least one point not in S. But D also contains a point of S, namely, z itself. Thus z must be a boundary point of S. ■

We emphasize, however, that a set may have boundary points that are not in the set, as occurs in some of the preceding examples.

We also observe that an open set cannot contain any of its boundary points.

THEOREM 20.4

Let S be a set of complex numbers. If S is open, then S can contain no boundary point.

Proof Suppose z is in S and S is open. Then some open disk D about z contains only points of S. But then this disk does not contain any points not in S, so z cannot be a boundary point of S. ■

DEFINITION 20.6 *Closed Set*

A set of complex numbers is closed if it contains all of its boundary points.

For example, the closed disk $|z - z_0| \leq r$ is a closed set. The boundary points are all the points on the circle $|z - z_0| = r$, and these are all in the closed disk. The set M of Example 20.9 is closed. Its boundary points are all the points on the imaginary axis, and these all belong to the set. The set W of Example 20.10 is closed, since every point in the set is a boundary point, and the set has no other boundary points.

The terms *closed* and *open* are not mutually exclusive, and one is not the opposite of the other. A set may be both closed and open, or closed and not open, or open and not closed, or neither open nor closed. For example, the set $\mathfrak{C}$ of all complex numbers is open (every point is interior) and closed (there are no boundary points, so $\mathfrak{C}$ contains them all). A closed disk is closed

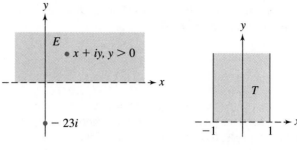

FIGURE 20.27 $-23i$ *is a boundary point of* E.

FIGURE 20.28 *The strip consisting of all* $z = x + iy$ *with* $-1 \leq x \leq 1$, $y > 0$.

but not open, and an open disk is open and not closed. The following example gives a set that is neither open nor closed.

EXAMPLE 20.15

Let T consist of all points $z = x + iy$ with $-1 \leq x \leq 1$ and $y > 0$. This is the infinite strip shown in Figure 20.28. The boundary points are all points $-1 + iy$ with $y \geq 0$, all points $1 + iy$ with $y \geq 0$, and all points x with $-1 \leq x \leq 1$. Some of these points are in T, for example, the boundary points $-1 + iy$ with $y > 0$. This means that T cannot be open. But some of these boundary points are not in T, for example, the points x with $-1 \leq x \leq 1$. Thus T is not closed. ∎

20.2.6 Limit Points

A number z_0 is a *limit point* of S if there are points of S arbitrarily close to z_0 but different from z_0.

> **DEFINITION 20.7** *Limit Point*
>
> A complex number z_0 is a limit point of a set S if every open disk about z_0 contains at least one point of S different from z_0.

Limit point differs from boundary point in requiring that every open disk about the point contain something from S *other than the point itself*. In Example 20.14, $-23i$ is a boundary point of W, but not a limit point of W, because there are open disks about $-23i$ that contain no other point of W.

EXAMPLE 20.16

For the set V of Example 20.8, every point on the vertical axis is a limit point. Given any such point $z_0 = iy_0$, every disk about z_0 contains points of V other than iy_0. Thus z_0 is both a boundary point and a limit point. This example shows that a limit point of a set need not belong to the set.

This set has many other limit points. For example, every number $x + iy$ with $x > 0$ is a limit point that also belongs to V. ∎

EXAMPLE 20.17

Let Q consist of the numbers i/n for $n = 1, 2, \ldots$. Every open disk about 0, no matter how small the radius, contains points i/n in Q if we choose n large enough. Therefore 0 is a limit point of Q. In this example, 0 is also a boundary point of Q (its only boundary point). ∎

EXAMPLE 20.18

Let N consist of all in, with n an integer. Then N has no limit points. An open disk of radius $\frac{1}{2}$ about in can have only one point in common with N, namely, in itself. ∎

As these examples show, a limit point of a set may or may not be in the set. We claim that closed sets are exactly those that contain all of their limit points, with the understanding that a set contains all of its limit points in a vacuous sense if it has no limit points.

THEOREM 20.5

Let S be a set of complex numbers. Then S is closed if and only if S contains all of its limit points.

Proof Suppose first that S is closed and let w be a limit point of S. We will show that w is in S. Suppose w is not in S. We know that any disk $|z - w| < r$ must contain a point z_r of S other than w. But then this disk contains a point in S (namely z_r) and a point not in S (namely w itself). Therefore w is a boundary point of S. But S is closed, and so contains all of its boundary points, in particular w. This contradiction shows that w must be in S, hence S contains all of its limit points.

Conversely, suppose if w is a limit point of S, then w is in S. We want to show that S is closed. To do this, we will show that S contains its boundary points. Let b be a boundary point of S. Suppose b is not in S. If $|z - b| < r$ is an open disk about b, then this disk contains a point of S, because b is a boundary point. But this point is not b, because we have assumed that b is not in S. Then every open disk about b contains a point of S other than b, hence b is a limit point of S. But we have supposed that every limit point of S is in S, so b is in S. This contradiction proves that S contains all of its boundary points, hence S is closed. ∎

Here are some additional examples of limit points.

EXAMPLE 20.19

Let X consist of all numbers $2 - i/n$, with $n = 1, 2, \ldots$. Then 2 is a limit point (and boundary point) of X. There are no other limit points of X. ∎

EXAMPLE 20.20

Let Q consist of all complex numbers $a + ib$ with a and b rational numbers. Then every complex number is both a limit point and a boundary point of Q. Some limit points of Q are in Q (if a and b are rational), and some are not (if a or b is irrational). ∎

EXAMPLE 20.21

Let P consist of all complex numbers $x + iy$ with $-1 \le y < 1$. Then each point of P is a limit point, and the points $x + i$ are also limit points of P that do not belong to P. ■

EXAMPLE 20.22

Let D be the open disk $|z - z_0| < r$. Every point in D is a limit point. However, the points on the boundary circle $|z - z_0| = r$, which do not belong to D, are also limit points, as well as boundary points, of D. ■

20.2.7 Complex Sequences

The notion of a complex sequence is a straightforward adaptation from the concept of real sequence.

DEFINITION 20.8 *Sequence*

A complex sequence $\{z_n\}$ is an assignment of a complex number z_n to each positive integer n.

The number z_n is the nth term of the sequence. For example, $\{i^n\}$ has nth term i^n.

We often indicate a sequence by listing the first few terms, including enough terms so that the pattern becomes clear and one can predict what z_n is for any n. For example, we might write $\{i^n\}$ as

$$i, i^2, i^3, \ldots, i^n, \ldots.$$

Convergence of complex sequences is also modeled after convergence of real sequences.

DEFINITION 20.9 *Convergence*

The complex sequence $\{z_n\}$ converges to the number L if, given any positive number ϵ, there is a positive number N such that

$$|z_n - L| < \epsilon \quad \text{if } n \ge N.$$

This means that we can make each term z_n as close as we like to L by choosing n at least as large as some number N. Put another way, given any open disk D about z_0, we can find some term of the sequence so that all terms beyond this one in the list (that is, for large enough index) lie in D (Figure 20.29). This is the same idea as that behind convergence of real sequences, except on the real line open intervals replace open disks. When $\{z_n\}$ converges to L, we write $z_n \to L$, or $\lim_{n\to\infty} z_n = L$.

If a sequence does not converge to any number, then we say that the sequence *diverges*.

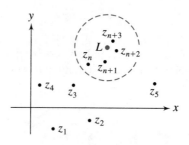

FIGURE 20.29 *Convergence of $\{z_n\}$ to L.*

EXAMPLE 20.23

The sequence $\{i^n\}$ diverges. This is the sequence

$$i, -1, -i, 1, i, -1, -i, 1, \ldots$$

and there is no point in the sequence beyond which all the terms approach one specific number arbitrarily closely. For example, if we take the disk $|z - i| < \frac{1}{2}$, then the first term of the sequence, and every fourth term after this, is in this disk, but no other term is in this disk. ■

EXAMPLE 20.24

The sequence $\{1 + i/n\}$ converges to 1. This follows from the definition since, if $\epsilon > 0$, then

$$\left| \left(1 + \frac{i}{n} \right) - 1 \right| = \left| \frac{i}{n} \right| = \frac{1}{n} < \epsilon$$

if n is chosen larger than $1/\epsilon$. Given $\epsilon > 0$, we can choose $N = 1/\epsilon$ in the definition of convergence. ■

Convergence of a complex sequence can always be reduced to an issue of convergence of two real sequences.

THEOREM 20.6

Let $z_n = x_n + iy_n$ and $L = a + ib$. Then $z_n \to L$ if and only if $x_n \to a$ and $y_n \to b$. ■

For example, let $z_n = (1 + 1/n)^n + ((n + 2)/n)i$. We know that

$$\lim_{n \to \infty} \left(1 + \frac{1}{n} \right)^n = e$$

and

$$\lim_{n \to \infty} \frac{n + 2}{n} = 1.$$

Then

$$\lim_{n \to \infty} z_n = e + i.$$

Proof Suppose first that $z_n \to a + bi$. Let $\epsilon > 0$. For some N, $|z_n - L| < \epsilon$ if $n \geq N$. Then, by Theorem 20.2(1), for $n \geq N$,

$$|x_n - a| = |\text{Re}(z_n - L)| \leq |z_n - L| < \epsilon,$$

so $x_n \to a$. Similarly, if $n \geq N$,

$$|y_n - b| = |\text{Im}(z_n - L)| \leq |z_n - L| < \epsilon,$$

so $y_n \to b$.

Conversely, suppose $x_n \to a$ and $y_n \to b$. Let $\epsilon > 0$. For some N_1,

$$|x_n - a| < \frac{\epsilon}{2} \quad \text{if } n \geq N_1.$$

For some N_2,

$$|y_n - b| < \frac{\epsilon}{2} \quad \text{if } n \geq N_2.$$

Then, for $n \geq N_1 + N_2$,

$$|z_n - L| = |(x_n - a) + i(y_n - b)| \leq |x_n - a| + |y_n - b| < \frac{\epsilon}{2} + \frac{\epsilon}{2} = \epsilon,$$

proving that $z_n \to L$. ∎

The notion of convergence of a complex sequence is intimately tied to the concept of a limit point of a set.

THEOREM 20.7

Let K be a set of complex numbers and let w be a complex number. Then w is a limit point of K if and only if there is a sequence $\{k_n\}$ of points in K, with each $k_n \neq w$, that converges to w. ∎

This is the rationale for the term *limit point*. A number w can be a limit point of a set only if w is the limit of a sequence of points in the set, all different from w. This holds whether or not w itself is in the set.

For example, consider the open unit disk $|z| < 1$. We know that i is a limit point of this disk, because any open disk about i contains points of the unit disk different from i. But we can also find a sequence of points in the unit disk converging to i, for example, $z_n = (1 - 1/n)i$.

Proof Suppose first that w is a limit point of K. Then for each positive integer n, the open disk of radius $1/n$ about w must contain a point of K different from w. Choose such a point and call it k_n. Then each $k_n \neq w$, k_n is in K, and $|k_n - w| < 1/n$. Since $1/n \to 0$ as $n \to \infty$, then $\{k_n\}$ converges to w.

Conversely, suppose there is a sequence of points k_n in K, all different from w and converging to w. Let D be any open disk about w, say of radius ϵ. Because $k_n \to w$, D must contain all k_n for n larger than some number N. But then D contains points of K different from w, and hence w is a limit point of K. ∎

20.2.8 Subsequences

A subsequence of a sequence is formed by picking out certain terms to form a new sequence.

DEFINITION 20.10 *Subsequence*

A sequence $\{w_j\}$ is a subsequence of $\{z_n\}$ if there are positive integers

$$n_1 < n_2 < \cdots$$

such that

$$w_j = z_{n_j}.$$

The subsequence is therefore formed from $\{z_n\}$ by listing the terms of this sequence,

$$z_1, z_2, z_3, \ldots$$

and then choosing, in order from left to right, some of the z_j's to form a new sequence. A subsequence is a sequence in its own right but consists of selected terms of an initially given sequence.

EXAMPLE 20.25

Let $z_n = i^n$. We can define many subsequences of $\{z_n\}$, but here is one. Let

$$w_j = z_{4j}$$

for $j = 1, 2, \ldots$. Then each $w_j = z_{4j} = i^{4j} = 1$, and every term of this subsequence equals 1. Here $n_j = 4j$ in the definition. ∎

If a sequence converges, then every subsequence of it converges to the same limit. To see this, suppose $z_n \to L$. Let D be an open disk about L. Then "eventually" (that is, for large enough n), every z_n is in D. If $\{w_j\}$ is a subsequence, then each $w_j = z_{n_j}$, so eventually all of these terms will also be in D and the subsequence also converges to L.

However, a subsequence of a divergent sequence may diverge, or it may converge, as Example 20.25 shows. The sequence $\{i^n\}$ diverges, but we were able to choose a subsequence having every term equal to 1, and this subsequence converges to 1.

It is also possible for a divergent sequence to have no convergent subsequence. For example, let $z_n = ni$. This sequence diverges, and we cannot choose a subsequence that converges. Now matter what subsequence is chosen, its terms will simply increase without bound the further out we go in the subsequence.

In this example, the sequence $\{ni\}$ is unbounded, hence divergent, and any subsequence is also unbounded and divergent. We get a more interesting result with bounded sequences.

DEFINITION 20.11   *Bounded Sequence*

$\{z_n\}$ is a bounded sequence if for some number M, $|z_n| \le M$ for $n = 1, 2, \ldots$.

Alternatively, a sequence is bounded if there is some disk that contains all of its terms.

We claim that every bounded sequence, convergent or not, has a convergent subsequence.

THEOREM 20.8

Let $\{z_n\}$ be a bounded sequence. Then $\{z_n\}$ has a convergent subsequence. ∎

This result has important consequences, for example, in treating the Cauchy integral theorem. The proof for real sequences is sketched in Problem 42. Using this, we can prove the theorem for complex sequences.

Proof   Let $z_n = x_n + iy_n$ form a bounded sequence. Then $\{x_n\}$ is a bounded real sequence, and so has a subsequence $\{x_{n_j}\}$ that converges to some real number a. But then $\{y_{n_j}\}$ is also a bounded real sequence, and so has a convergent subsequence $\{y_{n_{jk}}\}$ that converges to some real number b. Using these indices, form the subsequence $\{x_{n_{jk}}\}$ of $\{x_{n_j}\}$. This subsequence also converges to a. Then $\{x_{n_{jk}} + iy_{n_{jk}}\}$ is a subsequence of $\{z_n\}$ that converges to $a + ib$. ∎

20.2.9   Compactness and the Bolzano–Weierstrass Theorem

DEFINITION 20.12   *Bounded Set*

A set K of complex numbers is bounded if, for some number M, $|z| \leq M$ for all z in K.

A bounded set is therefore one whose points cannot be arbitrarily far from the origin. Certainly any finite set is bounded, as is any open or closed disk. The set of points in, for integer n, is not bounded.

The concepts of closed set and bounded set are independent. However, when combined, they characterize sets that have properties that are important in the analysis of complex functions. Such sets are called *compact*.

DEFINITION 20.13   *Compact Set*

A set K of complex numbers is compact if it is closed and bounded.

Any closed disk is compact, while an open disk is not (it is not closed). The set of points in for integer n is not compact because it is not bounded (even though it is closed). Any finite set is compact.

We will now show that any infinite compact set must contain at least one limit point. This is a remarkable result, since closed sets need not contain (or even have) any limit points, and bounded sets need not have limit points.

THEOREM 20.9   *Bolzano–Weierstrass* ——————————————————————————

Let K be an infinite compact set of complex numbers. Then K contains a limit point.

Proof   Since K is closed, any limit point of K must be in K. We will therefore concentrate on showing that there is a limit point of K.

Choose any number z_1 in K. Because K is infinite, we can choose a second number z_2 in K, distinct from z_1. Next choose some z_3 in K distinct from z_1 and z_2 and continue this process. In this way generate an infinite sequence $\{z_n\}$ of distinct points in K. Because K is a bounded set, this sequence is bounded. Therefore $\{z_n\}$ contains a subsequence $\{z_{n_j}\}$ that converges to some number L. Because each term of this sequence is distinct from all the others, we can choose the subsequence so that no z_{n_j} equals L. By Theorem 20.7, L is a limit point of K. ∎

We are now ready to begin the calculus of complex functions.

SECTION 20.2   PROBLEMS

In each of Problems 1 through 17, determine the set of all z satisfying the given equation or inequality. In some cases it may be convenient to specify the set by a clearly labeled diagram.

1. $|z - 8 + 4i| = 9$
2. $|z| = |z - i|$
3. $|z|^2 + \text{Im}(z) = 16$

4. $|z - i| + |z| = 9$

5. $|z| + \text{Re}(z) = 0$

6. $z + \bar{z}^2 = 4$

7. $\text{Im}(z - i) = \text{Re}(z + 1)$

8. $|z| = \text{Im}(z - i)$

9. $|z + 1 + 6i| = |z - 3 + i|$

10. $|z - 4i| \leq |z + 1|$

11. $|z + 2 + i| > |z - 1|$

12. $|z + 3| \geq |z - 2i|$

13. $-1 \leq \text{Re}(z) < 2$

14. $|z + i| > |z|$ and $|z| < 1$. *Hint:* Take the set of points common to the sets determined by each inequality.

15. $|z - 2i| \leq |z + 1 + i|$ and $|z| > 4$

16. $1 < |z| \leq 4$

17. $0 < \text{Im}(z + 3i) < 4$

In each of Problems 18 through 27, a set of points (complex numbers) is given. Determine whether the set is open, closed, open and closed, or neither open nor closed. Determine all limit points of the set, all boundary points, and the boundary of the set. Also determine whether the set is compact.

18. S is the set of all z with $|z| > 2$.

19. K is the set of all z satisfying $|z - 1| \leq |z + 4i|$.

20. T is the set of all z with $4 \leq |z + i| \leq 8$.

21. M consists of all z with $\text{Im}(z) < 7$.

22. R is the set of all complex numbers $1/m + (1/n)i$, in which m and n may be any positive integers.

23. U is the set of all z such that $1 < \text{Re}(z) \leq 3$.

24. V is the set of all z such that $2 < \text{Re}(z) \leq 3$ and $-1 < \text{Im}(z) < 1$.

25. W consists of all z such that $\text{Re}(z) > (\text{Im}(z))^2$.

26. S is the set of all complex numbers $x + (1/n)i$ with x any real number and n any positive integer.

27. K is the set of all numbers $x + iy$ with x and y any rational numbers.

28. Prove that for any set S of complex numbers, the set consisting of all points of S, together with all limit points of S, is a closed set.

29. Let S be a set of complex numbers.

(a) Prove that S is open if and only if the set of complex numbers not in S is closed.

(b) Prove that S is closed if and only if the set of complex numbers not in S is open.

30. Suppose S is a finite set of complex numbers, say consisting of numbers $z_1, z_2, \ldots, z_n$.

(a) Show that S has no limit point.

(b) Show that every z_j is a boundary point of S.

(c) Show that S is closed.

In each of Problems 31 through 37, find the limit of the sequence, or state that the sequence diverges.

31. $\left\{1 + \dfrac{2in}{n + 1}\right\}$

32. $\{i^{2n}\}$

33. $\left\{\dfrac{1 + 2n^2}{n^2} - \dfrac{n - 1}{n}i\right\}$

34. $\{e^{n\pi i/3}\}$

35. $\{-(i^{4n})\}$

36. $\{\sin(n)i\}$

37. $\left\{\dfrac{1 + 3n^2 i}{2n^2 - n}\right\}$

38. Prove that every convergent sequence is bounded.

39. Suppose $\{z_n\}$ is a convergent sequence of complex numbers. Prove that $\{|z_n|\}$ converges. Show by an example that convergence of $\{|z_n|\}$ does not imply convergence of $\{z_n\}$.

40. Consider the sequence $\{e^{n\pi i/3}\}$ of Problem 34. Find two different convergent subsequences of this sequence.

41. Find two convergent subsequences of the sequence $\{i^{2n}\}$ of Problem 32.

42. Fill in the details of the following proof that a bounded sequence of real numbers has a convergent subsequence. Let $\{x_n\}$ be a bounded sequence of real numbers. Let S_1 consist of all the numbers x_n in the sequence, for $n = 1, 2, \ldots$. For each positive number $m \geq 2$, let S_m be the set of all the numbers in the sequence, except for $x_1, \ldots, x_{m-1}$. For example, S_3 consists of the numbers $x_3, x_4, \ldots$.

(a) Show that for some positive number M, $-M < x_n < M$ for $n = 1, 2, \ldots$.

(b) Suppose first that each S_j has a number x_{n_j} in it that is at least as large as all the other numbers in S_j. Show that $\{x_{n_j}\}$ is a nonincreasing sequence of real numbers that is bounded below, hence converges.

(c) If some S_N has no largest number, then for each x_k in S_N, there is some x_{k+j} in S_N with $x_{k+j} > x_k$. Continuing this process starting with x_{k+j}, generate a nondecreasing subsequence of $\{x_n\}$ that is bounded above, hence converges.

CHAPTER 21

Complex Functions

21.1 Limits, Continuity, and Derivatives

A *complex function* is a function that is defined for complex numbers in some set S and takes on complex values. If $\mathfrak{C}$ denotes the set of complex numbers, and f is such a function, then we write $f : S \to \mathfrak{C}$. This simply means that $f(z)$ is a complex number for each z in S. The set S is called the *domain* of f. For example, let S consist of all z with $|z| < 1$ and define $f(z) = z^2$ for z in S. Then $f : S \to \mathfrak{C}$ and f is a complex function.

Often we define a function by some explicit expression in z, for example,

$$f(z) = \frac{z + i}{z^2 + 4}.$$

In the absence of specifying a domain S, we agree to allow all z for which the expression for $f(z)$ is defined. This function is therefore defined for all complex z except $2i$ and $-2i$.

21.1.1 Limits

The notion of limit for a complex function is modeled after that for real-valued functions, with disks replacing intervals.

DEFINITION 21.1   *Limit*

Let $f : S \to \mathfrak{C}$ be a complex function and let z_0 be a limit point of S. Let L be a complex number. Then

$$\lim_{z \to z_0} f(z) = L$$

---

if and only if, given $\epsilon > 0$, there exists a positive number δ such that

$$|f(z) - L| < \epsilon$$

for all z in S such that

$$0 < |z - z_0| < \delta.$$

When $\lim_{z \to z_0} f(z) = L$, we call L the limit of $f(z)$ as z approaches z_0.

Thus, $\lim_{z \to z_0} f(z) = L$ if function values $f(z)$ can be made to lie arbitrarily close (within ϵ) to L by choosing z in S (so $f(z)$ is defined) and close enough (within δ) to z_0, but not actually equal to z_0. The condition $0 < |z - z_0| < \delta$ excludes z_0 itself from consideration. We are only interested in the behavior of $f(z)$ at other points close to z_0.

Put another way, given an open disk D_ϵ of radius ϵ about L, we must be able to find an open disk D_δ of radius δ about z_0 so that every point in D_δ, except z_0 itself, that is also in S is sent by the function into D_ϵ. This is illustrated in Figure 21.1.

While it is not required in this definition that $f(z_0)$ be defined, we do require that there be points arbitrarily close to z_0 at which $f(z)$ is defined. This is assured by making z_0 a limit point of S and is the reason for making this a requirement in the definition. It makes no sense to speak of a limit of $f(z)$, as z approaches z_0, if $f(z)$ is not defined as z nears z_0.

Even if $f(z_0)$ is defined, there is no requirement in this limit that $f(z_0) = L$.

EXAMPLE 21.1

Let

$$f(z) = \begin{cases} z^2 & \text{for } z \neq i \\ 0 & \text{for } z = i \end{cases}.$$

Then $\lim_{z \to i} f(z) = -1$, but the limit does not equal $f(0)$. Indeed, even if $f(0)$ were not defined, we would still have $\lim_{z \to i} f(z) = -1$. ∎

Often we write

$$f(z) \to L \quad \text{as } z \to z_0$$

when $\lim_{z \to z_0} f(z) = L$.

Many limit theorems from real calculus hold for complex functions as well. Suppose $\lim_{z \to z_0} f(z) = L$ and $\lim_{z \to z_0} g(z) = K$. Then

$$\lim_{z \to z_0} [f(z) + g(z)] = L + K,$$

$$\lim_{z \to z_0} [f(z) - g(z)] = L - K,$$

$$\lim_{z \to z_0} cf(z) = cL \quad \text{for any number } c,$$

$$\lim_{z \to z_0} [f(z)g(z)] = LK,$$

and, if $K \neq 0$,

$$\lim_{z \to z_0} \frac{f(z)}{g(z)} = \frac{L}{K}.$$

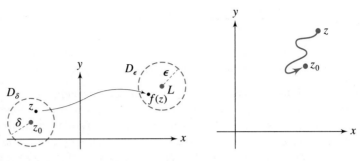

FIGURE 21.1 $\lim_{z \to a} f(z) = L$.

FIGURE 21.2 *z approaches* z_0 *along any path in defining* $\lim_{z \to z_0} f(z)$.

One significant difference between limits of complex functions and limits of real functions is the way the variable can approach the point. For a real function g, $\lim_{x \to a} g(x)$ involves the behavior of $g(x)$ as x approaches a from either side. On the line there are only two ways x can approach a. But $\lim_{z \to z_0} f(z) = L$ involves the behavior of $f(z)$ as z approaches z_0 in the complex plane (or in a specified set S of allowable values) and this may involve z approaching z_0 from any direction (Figure 21.2). The numbers $f(z)$ must approach L along every path of approach of z to z_0 in S. If along a single path of approach of z to z_0 $f(z)$ does not approach L, then $f(z)$ does not have limit L there. This makes $\lim_{z \to z_0} f(z) = L$ in the complex plane a stronger statement than its real counterpart, requiring more of $f(z)$ for z near z_0 than is required of real functions. We will exploit this fact later to derive facts about complex functions.

21.1.2 Continuity

DEFINITION 21.2

A complex function $f : S \to \mathfrak{C}$ is *continuous* at a number z_0 in S if and only if

$$\lim_{z \to z_0} f(z) = f(z_0).$$

We say that f is continuous on a set K if f is continuous at each point of K. In particular, if f is continuous at all z for which $f(z)$ is defined, then f is a continuous function.

Many familiar functions are continuous. Any polynomial is continuous for all z, and any rational function (quotient of polynomials) is continuous wherever its denominator is nonzero. When we have complex versions of the trigonometric and exponential functions, we will see that they are also continuous.

If f is continuous at z_0, so is $|f|$. We should expect this. If as z is chosen closer to z_0, $f(z)$ becomes closer to $f(z_0)$, then it is reasonable that $|f(z)|$ approaches $|f(z_0)|$. More rigorously,

$$0 \leq ||f(z)| - |f(z_0)|| \leq |f(z) - f(z_0)| \to 0$$

if $\lim_{z \to z_0} f(z) = f(z_0)$.

If $\{z_n\}$ is a sequence of complex numbers and each $f(z_n)$ is defined, then $\{f(z_n)\}$ is also a complex sequence. For example, if $f(z) = 2z^2$ and $z_n = 1/n$, then $f(z_n) = 2/n^2$. We claim that $\{f(z_n)\}$ converges if $\{z_n\}$ does, when f is continuous. Another way of saying this is that continuity preserves convergence of sequences.

THEOREM 21.1

Let $f : S \to \mathbb{C}$ be continuous, and let $\{z_n\}$ be a sequence of complex numbers in S. If $\{z_n\}$ converges to a number w in S, then $\{f(z_n)\}$ converges to $f(w)$. ∎

Here is the idea behind the theorem. Since f is continuous at w, then $\lim_{z \to w} f(z) = f(w)$. This means that $f(z)$ must approach $f(w)$ along any path of approach of z to w in S. But, if $z_n \to w$, we can think of the z_n's as determining a path of approach of the variable z to w. Then $f(z)$ must approach $f(w)$ along this path, and hence $f(z_n) \to f(w)$.

A converse of Theorem 21.1 can also be proved. If $f(z_n) \to f(w)$ for *every* sequence $\{z_n\}$ of points of S converging to w, then f is continuous at w.

We will now develop a significant property of continuous functions. First, define a complex function (continuous or not) to be bounded if the numbers $f(z)$ do not become arbitrarily large in magnitude.

DEFINITION 21.3 *Bounded Function*

Let $f : S \to \mathbb{C}$. Then f is a bounded function if there is a positive number M such that

$$|f(z)| \leq M$$

for all z in S.

Alternatively, f is bounded if there is a disk about the origin containing all the numbers $f(z)$ for z in S.

A continuous function need not be bounded (look at $f(z) = 1/z$ for $z \neq 0$). We claim, however, that a continuous function defined on a compact set is bounded. This is analogous to the result that a real function that is continuous on a closed interval is bounded. On the real line, closed intervals are compact sets.

THEOREM 21.2

Let $f : S \to \mathbb{C}$. Suppose S is compact and f is continuous on S. Then f is bounded.

Proof Suppose f is not bounded. Then, if n is a positive integer, the disk of radius n about the origin cannot contain all $f(z)$ for z in S. This means that there is some z_n in S such that $|f(z_n)| > n$.

Now $\{z_n\}$ is a sequence of points in the bounded set S, hence has a convergent subsequence $\{z_{n_j}\}$. Let this subsequence converge to w. Then w is a limit point of S, and S is closed, so w is in S also.

Because f is continuous, $f(z_{n_j}) \to f(w)$. Then, for some N, we can make each $f(z_{n_j})$ lie in the open disk of radius 1 about w by choosing $n_j \geq N$. But this contradicts the fact that each $\left| f(z_{n_j}) \right| > n_j$. Therefore, f must be a bounded function. ∎

We can improve on this theorem as follows. In addition to being bounded, we claim that, under the conditions of the preceding theorem, $|f(z)|$, which is real-valued, actually assumes a maximum and a minimum on S.

THEOREM 21.3

Let $f : S \to \mathbb{C}$ be continuous, and suppose S is compact. Then there are numbers z_1 and z_2 in S such that, for all z in S,

$$|f(z_1)| \leq |f(z)| \leq |f(z_2)|. \ \blacksquare$$

The proof of this theorem depends on the concepts of least upper bound and greatest lower bound of a set of real numbers. The reader unfamiliar with these concepts can omit the proof but should understand the conclusion of the theorem for use later.

Proof   By Theorem 21.2, for some M,

$$0 \leq |f(z)| \leq M$$

for all z in S. Let F be the set of all numbers $|f(z)|$ for z in S. F is a set of real numbers, and every number in F is real and nonnegative and less than or equal to M. Therefore, F has a least upper bound U and a greatest lower bound L. For all z in S,

$$L \leq |f(z)| \leq U.$$

We will show that there are numbers z_1 and z_2 in S such that $L = f(z_1)$ and $U = f(z_2)$.

Let n be any positive integer. There is some β_n in S such that

$$U - \frac{1}{n} < |f(\beta_n)| \leq U.$$

The reason for this is that if there were no such number in S, then we would have $|f(z)| \leq U - 1/n$ for all z in S. But then $U - 1/n$ would be a smaller upper bound for F than its least upper bound U, a contradiction.

Now $U - 1/n \to U$ as $n \to \infty$, hence

$$|f(\beta_n)| \to U.$$

Because S is a bounded set, $\{\beta_n\}$ is a bounded sequence and therefore has a subsequence β_{n_j} converging to some number z_2. Because f is continuous, $f(\beta_{n_j}) \to f(z_2)$. Therefore, $\left| f(\beta_{n_j}) \right| \to |f(z_2)|$, so $U = |f(z_2)|$.

Similarly, by definition of the greatest lower bound, argue that for some α_n in S,

$$L \leq |f(\alpha_n)| < L + \frac{1}{n}.$$

Then $|f(\alpha_n)| \to L$. But $\{\alpha_n\}$ has a convergent subsequence $\{\alpha_{n_j}\}$, which must converge to some number z_1 in S. Then $\left| f(\alpha_{n_j}) \right| \to |f(z_1)|$, so $L = |f(z_1)|$.

Finally,

$$|f(z_1)| \leq |f(z)| \leq |f(z_2)|$$

for z in S, completing the proof. $\blacksquare$

21.1.3 The Derivative of a Complex Function

DEFINITION 21.4 **Derivative**

Let $f : S \to \mathbb{C}$ and suppose S is an open set. Let z_0 be in S. Then f is differentiable at z_0 if, for some complex number L,

$$\lim_{h \to 0} \frac{f(z_0 + h) - f(z_0)}{h} = L.$$

In this event we call L the derivative of f at z_0 and denote it $f'(z_0)$.

If f is differentiable at each point of a set, then we say that f is differentiable on this set.

The reason for having S open in this definition is to be sure that there is some open disk about z_0 throughout which $f(z)$ is defined. When the complex number h is small enough in magnitude, then $z_0 + h$ is in this disk and $f(z_0 + h)$ is defined. This allows h to approach zero from any direction in the limit defining the derivative. This will have important ramifications shortly in the Cauchy–Riemann equations.

EXAMPLE 21.2

Let $f(z) = z^2$ for all complex z. Then

$$f'(z) = \lim_{h \to 0} \frac{(z + h)^2 - z^2}{h} = \lim_{h \to 0} (2z + h) = 2z$$

for all z. ∎

For familiar functions such as polynomials, the usual rules for taking derivatives apply. For example, if n is a positive integer and $f(z) = z^n$, then $f'(z) = nz^{n-1}$. When we develop the complex sine function $f(z) = \sin(z)$, we will see that $f'(z) = \cos(z)$. Other familiar derivative formulas are:

$$(f + g)'(z) = f'(z) + g'(z),$$

$$(f - g)'(z) = f'(z) - g'(z),$$

$$(cf)'(z) = cf'(z),$$

$$(fg)'(z) = f(z)g'(z) + f'(z)g(z),$$

and

$$\left(\frac{f}{g}\right)'(z) = \frac{g(z)f'(z) - f(z)g'(z)}{[g(z)]^2} \quad \text{if } g(z) \neq 0.$$

These conclusions assume that the derivatives involved exist. There is also a complex version of the chain rule. Recall that the composition of two functions is defined by

$$(f \circ g)(z) = f(g(z)).$$

The chain rule for differentiating a composition is

$$(f \circ g)'(z) = f'(g(z))g'(z),$$

assuming that g is differentiable at z and f is differentiable at $g(z)$.

Often $f'(z)$ is denoted using the Leibniz notation

$$\frac{df}{dz}.$$

In this notation, the chain rule is

$$\frac{d}{dz}(f(g(z))) = \frac{df}{dw}\frac{dw}{dz},$$

where $w = g(z)$.

Not all functions are differentiable.

EXAMPLE 21.3

Let $f(z) = \bar{z}$. We will show that f is not differentiable at any point. To see why this is true, compute

$$\frac{f(z+h) - f(z)}{h} = \frac{\overline{z+h} - \bar{z}}{h} = \frac{\bar{h}}{h}.$$

We want the limit of this quotient as $h \to 0$. But this limit is in the complex plane, and the complex number h must be allowed to approach zero along any path. If h approaches zero along the real axis, then h is real, $\bar{h} = h$ and $\bar{h}/h = 1 \to 1$. But if h approaches zero along the imaginary axis, then $h = ik$ for k real, and

$$\frac{\bar{h}}{h} = \frac{-ik}{ik} = -1 \to -1$$

as $k \to 0$. The quotient $\bar{h}/h$ approaches different numbers as h approaches zero along different paths. This means that

$$\lim_{h \to 0} \frac{\bar{h}}{h}$$

does not exist, so f has no derivative at any point. ∎

As with real functions, a complex function is continuous wherever it is differentiable.

THEOREM 21.4

Let f be differentiable at z_0. Then f is continuous at z_0.

Proof We know that

$$\lim_{h \to 0}\left(\frac{f(z_0 + h) - f(z_0)}{h} - f'(z_0)\right) = 0.$$

Let

$$\epsilon(h) = \frac{f(z_0 + h) - f(z_0)}{h} - f'(z_0).$$

Then $\lim_{h \to 0} \epsilon(h) = 0$. Further,

$$f(z_0 + h) - f(z_0) = hf'(z_0) + h\epsilon(h).$$

Since the right side has limit zero as $h \to 0$, then

$$\lim_{h \to 0}[f(z_0 + h) - f(z_0)] = 0.$$

This is the same as

$$\lim_{h \to 0} f(z_0 + h) = f(z_0),$$

and this in turn implies that $\lim_{z \to z_0} f(z) = f(z_0)$. Therefore, f is continuous at z_0. ∎

21.1.4   The Cauchy–Riemann Equations

We will derive a set of partial differential equations that must be satisfied by the real and imaginary parts of a differentiable complex function. These equations also play a role in potential theory and in treatments of the Dirichlet problem.

Let f be a complex function. If $z = x + iy$, we can always write

$$f(z) = f(x + iy) = u(x, y) + iv(x, y),$$

in which u and v are real-valued functions of the two real variables x and y. Then

$$u(x, y) = \text{Re}[f(z)] \quad \text{and} \quad v(x, y) = \text{Im}[f(z)].$$

EXAMPLE 21.4

Let $f(z) = z^2$. Then

$$f(x + iy) = (x + iy)^2 = x^2 - y^2 + 2ixy,$$

so

$$u(x, y) = x^2 - y^2 \quad \text{and} \quad v(x, y) = 2xy. \ \blacksquare$$

EXAMPLE 21.5

Let $f(z) = 1/z$ for $z \neq 0$. Then

$$f(x + iy) = \frac{1}{x + iy} = \frac{1}{x + iy}\frac{x - iy}{x - iy} = \frac{x}{x^2 + y^2} - i\frac{y}{x^2 + y^2}.$$

For this function

$$u(x, y) = \frac{x}{x^2 + y^2} \quad \text{and} \quad v(x, y) = -\frac{y}{x^2 + y^2}. \ \blacksquare$$

We will now derive a relationship between partial derivatives of u and v at any point where f is differentiable.

THEOREM 21.5   *Cauchy–Riemann Equations*

Let $f : S \to \mathfrak{C}$, with S an open set. Write $f = u + iv$. Suppose $z = x + iy$ is a point of S and $f'(z)$ exists. Then, at (x, y),

$$\frac{\partial u}{\partial x} = \frac{\partial v}{\partial y} \quad \text{and} \quad \frac{\partial v}{\partial x} = -\frac{\partial u}{\partial y}.$$

Proof   Begin with

$$f'(z) = \lim_{h \to 0} \frac{f(z + h) - f(z)}{h}.$$

We know that this limit exists, hence must have the same value, $f'(z)$, however h approaches zero. Consider two paths of approach of h to the origin.

First, let $h \to 0$ along the real axis (Figure 21.3). Now h is real, and $z + h = x + h + iy$. Then

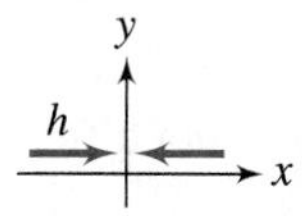

FIGURE 21.3

$$f'(z) = \lim_{h \to 0} \frac{u(x + h, y) + iv(x + h, y) - u(x, y) - iv(x, y)}{h}$$

$$= \lim_{h \to 0} \left(\frac{u(x + h, y) - u(x, y)}{h} + i\frac{v(x + h, y) - v(x, y)}{h} \right)$$

$$= \frac{\partial u}{\partial x} + i\frac{\partial v}{\partial x}.$$

Next, take the limit along the imaginary axis (Figure 21.4). Put $h = ik$ with k real, so $h \to 0$ as $k \to 0$. Now $z = x + i(y + k)$ and

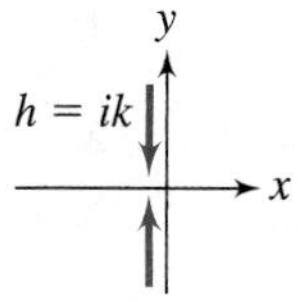

FIGURE 21.4

$$f'(z) = \lim_{k \to 0} \frac{u(x, y + k) + iv(x, y + k) - u(x, y) - iv(x, y)}{ik}$$

$$= \lim_{k \to 0} \left(\frac{1}{i}\frac{u(x, y + k) - u(x, y)}{k} + \frac{v(x, y + k) - v(x, y)}{k} \right)$$

$$= -i\frac{\partial u}{\partial y} + \frac{\partial v}{\partial y},$$

in which we have used the fact that $1/i = -i$.

We now have two expressions for $f'(z)$, so they must be equal:

$$\frac{\partial u}{\partial x} + i\frac{\partial v}{\partial x} = -i\frac{\partial u}{\partial y} + \frac{\partial v}{\partial y}.$$

Setting the real part of the left side equal to the real part of the right, and then the imaginary part of the left side to the imaginary part of the right, yields the Cauchy–Riemann equations. ∎

One extra dividend of this proof is that we have also derived formulas for $f'(z)$ in terms of the real and imaginary parts of $f(z)$. For example, if $f(z) = z^3$, then

$$f(z) = f(x + iy) = (x + iy)^3 = x^3 - 3xy^2 + i(3x^2y - y^3).$$

Then

$$u(x, y) = x^3 - 3xy^2 \quad \text{and} \quad v(x, y) = 3x^2y - y^3,$$

so

$$f'(z) = \frac{\partial u}{\partial x} + i\frac{\partial v}{\partial x} = \left(3x^2 - 3y^2\right) + i\left(6xy\right).$$

This automatically displays the real and imaginary parts of $f'(z)$. Of course, for this simple function it is just as easy to write directly

$$f'(z) = 3z^2 = 3(x + iy)^2 = 3(x^2 - y^2) + 6xyi.$$

The Cauchy–Riemann equations constitute a necessary condition for f to be differentiable at a point. If they are not satisfied, then $f'(z)$ does not exist at this point.

EXAMPLE 21.6

Let $f(z) = \bar{z}$. Then $f(z) = x - iy$ and $u(x, y) = x, v(x, y) = -y$. Now

$$\frac{\partial u}{\partial x} = 1 \neq \frac{\partial v}{\partial y},$$

so the Cauchy–Riemann equations do not hold for f, at any point, and therefore f is not differentiable at any point. ∎

EXAMPLE 21.7

Let $f(z) = z\operatorname{Re}(z)$. Then

$$f(x + iy) = (x + iy)x = x^2 + ixy,$$

so $u(x, y) = x^2$ and $v(x, y) = xy$. Now

$$\frac{\partial u}{\partial x} = 2x, \quad \frac{\partial v}{\partial y} = x$$

and

$$\frac{\partial u}{\partial y} = 0, \quad \frac{\partial v}{\partial x} = y.$$

The Cauchy–Riemann equations do not hold at any point except $z = 0$. This means that f is not differentiable at z if $z \neq 0$, but may have a derivative at 0. In fact, this function is differentiable at 0, since

$$f'(0) = \lim_{h \to 0} \frac{f(h) - f(0)}{h} = \lim_{h \to 0} \operatorname{Re}(h) = 0. \ \blacksquare$$

While the Cauchy–Riemann equations are necessary for differentiability, they are not sufficient. If the Cauchy–Riemann equations hold at a point z, then f may or may not be differentiable at z. In the preceding example, the Cauchy–Riemann equations held at the origin and $f'(0)$ existed. Here is an example in which the Cauchy–Riemann equations are satisfied at the origin, but f has no derivative there.

EXAMPLE 21.8

Let

$$f(z) = \begin{cases} z^5 / |z|^4 & \text{for } z \neq 0 \\ 0 & \text{for } z = 0 \end{cases}.$$

We will show that the Cauchy–Riemann equations are satisfied at $z = 0$ but that f is not differentiable at 0. First do some algebra to obtain

$$u(x, y) = \frac{5xy^4 - 10x^3y^2 + x^5}{(x^2 + y^2)^2} \quad \text{if } (x, y) \neq (0, 0),$$

$$v(x, y) = \frac{y^5 - 10x^2y^3 + 5x^4y}{(x^2 + y^2)^2} \quad \text{if } (x, y) \neq (0, 0),$$

and

$$u(0, 0) = v(0, 0) = 0.$$

Compute the partial derivatives at the origin:

$$\frac{\partial u}{\partial x}(0, 0) = \lim_{h \to 0} \frac{u(h, 0) - u(0, 0)}{h} = \lim_{h \to 0} \frac{h^5}{hh^4} = 1;$$

$$\frac{\partial u}{\partial y}(0, 0) = \lim_{h \to 0} \frac{u(0, h) - u(0, 0)}{h} = \lim_{h \to 0} 0 = 0;$$

$$\frac{\partial v}{\partial x}(0, 0) = \lim_{h \to 0} \frac{v(h, 0) - v(0, 0)}{h} = \lim_{h \to 0} 0 = 0;$$

and

$$\frac{\partial v}{\partial y}(0, 0) = \lim_{h \to 0} \frac{v(0, h) - v(0, 0)}{h} = \lim_{h \to 0} \frac{h^5}{hh^4} = 1.$$

Therefore the Cauchy–Riemann equations are satisfied at the origin. However, f is not differentiable at 0. Consider

$$\frac{f(0 + h) - f(0)}{h} = \frac{h^5}{h|h|^4} = \frac{h^5}{h(h\bar{h})^2} = \frac{h^2}{(\bar{h})^2} = \left(\frac{h}{\bar{h}}\right)^2.$$

We claim that $(h/\bar{h})^2$ has no limit as $h \to 0$. This is easily seen by converting to polar form. If $h = re^{i\theta}$, then $\bar{h} = re^{-i\theta}$ and

$$\left(\frac{h}{\bar{h}}\right)^2 = \frac{r^2 e^{2i\theta}}{r^2 e^{-2i\theta}} = e^{4i\theta}.$$

On the line making an angle θ with the positive real axis (Figure 21.5), the difference quotient

$$\frac{f(0 + h) - f(0)}{h}$$

has the constant value $e^{4i\theta}$, and so approaches this number as $h \to 0$. The difference quotient therefore approaches different values along different paths, and so has no limit as $h \to 0$. ∎

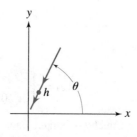

FIGURE 21.5

This example means that some condition(s) must be added to the Cauchy-Riemann equations to guarantee existence of a derivative at a point. The following theorem gives sufficient conditions for differentiability.

THEOREM 21.6

Let $f : S \to C$ be a complex function, with S an open set. Let $f = u + iv$. Suppose u, v, and their first partial derivatives are continuous on S. Suppose also that u and v satisfy the Cauchy–Riemann equations on S. Then f is differentiable at each point of S. ■

In real calculus, a function whose derivative is zero throughout an interval must be constant on that interval. Here is the complex analogue of this result, together with another result we will need later.

THEOREM 21.7

Let f be differentiable on an open disk D. Let $f = u + iv$ and suppose u and v satisfy the Cauchy–Riemann equations and are continuous with continuous first partial derivatives in D. Then,

1. If $f'(z) = 0$ for all z in D, then $f(z)$ is constant on D.
2. If $|f(z)|$ is constant in D, so is $f(z)$.

Proof To prove (1), recall from the proof of Theorem 21.5 that for z in D,

$$f'(z) = 0 = \frac{\partial u}{\partial x} + i\frac{\partial v}{\partial x}.$$

But then $\partial u/\partial x$ and $\partial v/\partial x$ are zero throughout D. By the Cauchy–Riemann equations, $\partial u/\partial y$ and $\partial v/\partial y$ are also zero at each point of D. Then $u(x, y)$ and $v(x, y)$ are constant on D, hence so is $f(z)$.

For (2), suppose $|f(z)| = k$ for all z in D. Then

$$|f(z)|^2 = u(x, y)^2 + v(x, y)^2 = k^2 \tag{21.1}$$

for (x, y) in D. If $k = 0$, then $|f(z)| = 0$ for all z in D, hence $f(z) = 0$ on D. If $k \neq 0$, differentiate equation (21.1) with respect to x to get

$$u\frac{\partial u}{\partial x} + v\frac{\partial v}{\partial x} = 0 \tag{21.2}$$

Differentiate equation (21.1) with respect to y to get

$$u\frac{\partial u}{\partial y} + v\frac{\partial v}{\partial y} = 0. \tag{21.3}$$

Using the Cauchy–Riemann equations, equations (21.2) and (21.3) can be written

$$u\frac{\partial u}{\partial x} - v\frac{\partial u}{\partial y} = 0 \tag{21.4}$$

and

$$u\frac{\partial u}{\partial y} + v\frac{\partial u}{\partial x} = 0. \tag{21.5}$$

Multiply equation (21.4) by u and equation (21.5) by v and add the resulting equations to get

$$(u^2 + v^2)\frac{\partial u}{\partial x} = k^2\frac{\partial u}{\partial x} = 0.$$

Therefore,

$$\frac{\partial u}{\partial x} = 0$$

for all (x, y) in D. By the Cauchy–Riemann equations,

$$\frac{\partial v}{\partial y} = 0$$

throughout D. Now a similar manipulation shows that

$$\frac{\partial u}{\partial y} = \frac{\partial v}{\partial x} = 0$$

on D. Therefore, $u(x, y)$ and $v(x, y)$ are constant on D, so $f(z)$ is constant also. ∎

SECTION 21.1 PROBLEMS

In each of Problems 1 through 16, find u and v so that $f(z) = u(x, y) + iv(x, y)$, determine all points (if any) at which the Cauchy–Riemann equations are satisfied, and determine all points at which the function is differentiable. Familiar facts about continuity of real-valued functions of two variables may be assumed.

1. $f(z) = z - i$
2. $f(z) = z^2 - iz$
3. $f(z) = |z|$
4. $f(z) = \dfrac{2z + 1}{z}$
5. $f(z) = i|z|^2$
6. $f(z) = z + \text{Im}(z)$
7. $f(z) = \dfrac{z}{\text{Re}(z)}$
8. $f(z) = z^3 - 8z + 2$
9. $f(z) = \bar{z}^2$
10. $f(z) = iz + |z|$
11. $f(z) = -4z + \dfrac{1}{z}$
12. $f(z) = \dfrac{z - i}{z + i}$
13. $f(z) = \text{Re}(z) - \text{Im}(z)$
14. $f(z) = 2iz^3$
15. $f(z) = \text{Im}\left(\dfrac{2z}{z + 1}\right)$
16. $f(z) = \dfrac{z}{|z|}$

21.2 Power Series

We now know some facts about continuity and differentiability. However, the only complex functions we have at this point are polynomials and rational functions. A complex polynomial is a function

$$p(z) = a_0 + a_1 z + a_2 z^2 + \cdots + a_n z^n,$$

in which the a_j's are complex numbers, and a rational function is a quotient of polynomials,

$$R(z) = \frac{a_0 + a_1 z + \cdots + a_n z^n}{b_0 + b_1 z + \cdots + b_m z^m}.$$

Polynomials are differentiable for all z, and a rational function is differentiable for all z at which the denominator does not vanish.

The vehicle for expanding our catalog of functions, obtaining exponential and trigonometric functions, logarithms, power functions, and others, is the power series. We will precede a development of complex power series with some facts about series of constants.

21.2.1 Series of Complex Numbers

We will assume standard results about series of real numbers. Consider a complex series $\sum_{n=1}^{\infty} c_n$, with each c_n a complex number. The Nth *partial sum* of this series is the finite sum $\sum_{n=1}^{N} c_n$. The sequence $\left\{ \sum_{n=1}^{N} c_n \right\}$ is the *sequence of partial sums* of this series, and the series converges if and only if this sequence of partial sums converges.

If $c_n = a_n + i b_n$, then

$$\sum_{n=1}^{N} c_n = \sum_{n=1}^{N} a_n + i \sum_{n=1}^{N} b_n,$$

so $\left\{ \sum_{n=1}^{N} c_n \right\}$ converges if and only if the real partial sums $\sum_{n=1}^{N} a_n$ and $\sum_{n=1}^{N} b_n$ converge as $N \to \infty$. Further, if $\sum_{n=1}^{\infty} a_n = A$ and $\sum_{n=1}^{\infty} b_n = B$, then

$$\sum_{n=1}^{\infty} c_n = A + i B.$$

We can therefore study any series of complex constants by considering two series of real constants, for which tests are available (ratio test, root test, integral test, comparison test, and others).

As with real series, if $\sum_{n=1}^{\infty} c_n$ converges, then necessarily $\lim_{n \to \infty} c_n = 0$.

In some instances we can not only show that a series converges but we can find its sum. The geometric series is an important illustration of this that we will use often.

EXAMPLE 21.9

Consider the series $\sum_{n=1}^{\infty} z^n$, with z a given complex number. A series that adds successive powers of a single number is called a *geometric series*. We can sum this series as follows. Let

$$S_N = \sum_{n=1}^{N} z^n = z + z^2 + z^3 + \cdots + z^{N-1} + z^N.$$

Then

$$z S_N = z^2 + z^3 + \cdots + z^N + z^{N+1},$$

If we subtract these finite sums, most terms cancel and we are left with

$$S_N - zS_N = (1-z)S_N = z - z^{N+1}.$$

Then, for $z \neq 1$,

$$S_N = \frac{z}{1-z} - \frac{1}{1-z}z^{N+1}.$$

If $|z| < 1$, then $|z|^{N+1} \to 0$ as $N \to \infty$, hence $z^{N+1} \to 0$ also and in this case the geometric series converges:

$$\sum_{n=1}^{\infty} z^n = \lim_{N\to\infty} S_N = \frac{z}{1-z}.$$

If $|z| \geq 1$, the geometric series diverges. ■

Sometimes we have a geometric series with a first term equal to 1. This is the series

$$\sum_{n=0}^{\infty} z^n = 1 + \sum_{n=1}^{\infty} z^n = 1 + \frac{z}{1-z} = \frac{1}{1-z} \quad \text{if } |z| < 1.$$

The series $\sum_{n=1}^{\infty} c_n$ is said to *converge absolutely* if the real series $\sum_{n=1}^{\infty} |c_n|$ converges.

THEOREM 21.8

If $\sum_{n=1}^{\infty} c_n$ converges absolutely, then this series converges.

Proof Let $c_n = a_n + ib_n$. Suppose $\sum_{n=1}^{\infty} |c_n|$ converges. Since $0 \leq |a_n| \leq |c_n|$, then by comparison, $\sum_{n=1}^{\infty} a_n$ converges. Similarly, $0 \leq |b_n| \leq |c_n|$, so $\sum_{n=1}^{\infty} b_n$ converges. Then $\sum_{n=1}^{\infty} a_n + ib_n = \sum_{n=1}^{\infty} c_n$ converges. ■

EXAMPLE 21.10

Consider the series

$$\sum_{n=1}^{\infty} (-1)^n \frac{2-i}{(1+i)^n}.$$

Compute

$$\left| (-1)^n \frac{2-i}{(1+i)^n} \right| = \frac{\sqrt{5}}{(\sqrt{2})^n}.$$

Now the real series $\sum_{n=1}^{\infty} \sqrt{5}/(\sqrt{2})^n$ converges. This is $\sqrt{5}$ times the real geometric series $\sum_{n=1}^{\infty} 1/(\sqrt{2})^n$, which converges because $1/\sqrt{2} < 1$. Therefore the given complex series converges absolutely, hence converges. ■

The point to Theorem 21.8 is that $\sum_{n=1}^{\infty} |c_n|$ is a real series, and we have methods to test real series for convergence or divergence. We can therefore (in the case of absolute convergence) test a complex series for convergence by testing a real series. This approach is not all-inclusive, however, because a series may converge, but not converge absolutely. Such a series is said to *converge conditionally*. For example, the series

$$\sum_{n=1}^{\infty} \frac{(-1)^n}{n}$$

is well known to converge, but the series of absolute values of its terms is the divergent harmonic series $\sum_{n=1}^{\infty}(1/n)$.

With this background on complex series, we can take up power series.

21.2.2   Power Series

DEFINITION 21.5

A *power series* is a series of the form

$$\sum_{n=0}^{\infty} c_n(z - z_0)^n,$$

in which z_0 and each c_n is a given complex number.

The summation in a power series begins at $n = 0$ to allow for a constant term:

$$\sum_{n=0}^{\infty} c_n(z - z_0)^n = c_0 + c_1(z - z_0) + c_2(z - z_0)^2 + \cdots.$$

The number z_0 is the *center* of the series, and the numbers c_n are its *coefficients*.

Given a power series, we want to know for what values of z, if any, it converges. Certainly any power series converges at its center $z = z_0$, because then the series is just c_0. The following theorem provides the key to determining if there are other values of z for which it converges. It says that if we find a point $z_1 \neq z_0$ where the power series converges, then the series must converge absolutely at least for all points closer to z_0 than z_1. (This gives (absolute) convergence at least at points interior to the disk of Figure 21.6.)

THEOREM 21.9

Suppose $\sum_{n=0}^{\infty} c_n(z - z_0)^n$ converges for some z_1 different from z_0. Then the power series converges absolutely for all z satisfying

$$|z - z_0| < |z_1 - z_0|.$$

Proof   Suppose $\sum_{n=0}^{\infty} c_n(z_1 - z_0)^n$ converges. Then $\lim_{n \to \infty} c_n(z_1 - z_0)^n = 0$. Then, for some N,

$$\left| c_n(z_1 - z_0)^n \right| < 1 \quad \text{if } n \geq N.$$

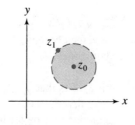

FIGURE 21.6
*Convergence at $z_1 \neq z_0$
implies convergence on
$|z - z_0| < r = |z_1 - z_0|$.*

Then, for $n \geq N$,

$$\left| c_n(z-z_0)^n \right| = |c_n| \left| \frac{(z-z_0)^n}{(z_1-z_0)^n} \right| \left| (z_1-z_0)^n \right| \leq \left| \frac{(z-z_0)^n}{(z_1-z_0)^n} \right| = \left| \frac{z-z_0}{z_1-z_0} \right|^n.$$

But if $|z-z_0| < |z_1-z_0|$, then

$$\left| \frac{z-z_0}{z_1-z_0} \right| < 1$$

and the geometric series

$$\sum_{n=1}^{\infty} \left| \frac{z-z_0}{z_1-z_0} \right|^n$$

converges. By comparison (since these are series of real numbers),

$$\sum_{n=N}^{\infty} \left| c_n(z-z_0)^n \right|$$

converges. But then

$$\sum_{n=0}^{\infty} \left| c_n(z-z_0)^n \right|$$

converges, so $\sum_{n=0}^{\infty} c_n(z-z_0)^n$ converges absolutely, as we wanted to prove. ∎

Apply this conclusion as follows. Imagine standing at z_0 in the complex plane. Looking out in all directions, we may see no other points at which the power series converges. In this event, the series converges only for $z = z_0$. This is an uninteresting power series.

A second possibility is that we see only points at which the power series converges. Now the power series converges for all z.

The third possibility is that we see some points at which the series converges and some at which it diverges. Let R be the distance from z_0 to the nearest point, say ζ, at which the power series diverges. The distance R is critical in the following sense.

If z is further from z_0 than ζ, then the power series must diverge at z. For if it converged, then it would converge at all points closer to z_0 than z, and hence would converge at ζ by Theorem 21.9.

If z is closer to z_0 than ζ, then the power series must converge at z, since ζ is the point closest to z_0 at which the series diverges.

This means that in this third case, the power series

$$\text{converges for all } z \text{ with } |z-z_0| < R,$$

and

$$\text{diverges for all } z \text{ with } |z-z_0| > R.$$

The number R is called the *radius of convergence* of the power series, and the open disk $|z-z_0| < R$ is called the *open disk of convergence*. The series converges inside this disk and diverges outside the closed disk $|z-z_0| \leq R$. At points on the boundary of this disk, $|z-z_0| = R$, the series might converge or diverge.

If the power series converges only for $z = z_0$, we let the radius of convergence be $R = 0$. In this case we do not speak of an open disk of convergence.

If the power series converges for all z, let $R = \infty$. Now the open disk of convergence is the entire complex plane. In this case it is convenient to denote the disk of convergence as $|z-z_0| < \infty$.

Sometimes the radius of convergence can be calculated for a power series by using the ratio test.

EXAMPLE 21.11

Consider the power series

$$\sum_{n=0}^{\infty}(-1)^n \frac{2^n}{n+1}(z-1+2i)^{2n}.$$

The center is $z_0 = 1 - 2i$. We want the radius of convergence of this series.
Consider the magnitude of the ratio of successive terms of this series:

$$\left|\frac{(-1)^{n+1}\frac{2^{n+1}}{n+2}(z-1+2i)^{2n+2}}{(-1)^n\frac{2^n}{n+1}(z-1+2i)^{2n}}\right| = \frac{2(n+1)}{n+2}|z-1+2i|^2$$

$$\to 2|z-1+2i|^2 \quad \text{as } n \to \infty.$$

From the ratio test for real series, the power series will converge absolutely if this limit is less than 1 and diverge if this limit is greater than 1. Thus, the power series converges absolutely if

$$2|z-1+2i|^2 < 1,$$

or

$$|z-1+2i| < \frac{1}{\sqrt{2}}.$$

And the series diverges if

$$2|z-1+2i|^2 > 1,$$

or

$$|z-1+2i| > \frac{1}{\sqrt{2}}.$$

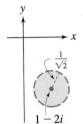

The radius of convergence is $1/\sqrt{2}$ and the open disk of convergence is $|z-1+2i| < 1/\sqrt{2}$ (Figure 21.7). ∎

FIGURE 21.7

EXAMPLE 21.12

Consider

$$\sum_{n=1}^{\infty}\left(\frac{2}{3i}\right)^n (z-i)^n.$$

Take the magnitude of the ratio of successive terms:

$$\left|\frac{(2/3i)^{n+1}(z-i)^{n+1}}{(2/3i)^n(z-i)^n}\right| = \frac{2}{3}|z-i| \to \frac{2}{3}|z-i|$$

FIGURE 21.8

as $n \to \infty$. The power series converges absolutely if $|z-i| < \frac{3}{2}$ and diverges if $|z-i| > \frac{3}{2}$. The radius of convergence is $\frac{3}{2}$ and the open disk of convergence is the open disk of radius $\frac{3}{2}$ centered at i (Figure 21.8). ∎

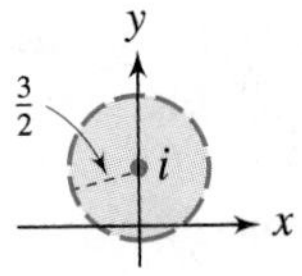

Suppose now that a power series has a positive or infinite radius of convergence. For each z in the open disk of convergence, let

$$f(z) = \sum_{n=1}^{\infty} c_n(z - z_0)^n.$$

This defines a function f over this disk. We want to explore the properties of this function, in particular whether it is differentiable. Answering this question requires the following technical lemma.

LEMMA 21.1

The power series $\sum_{n=0}^{\infty} c_n(z - z_0)^n$ and $\sum_{n=1}^{\infty} nc_n(z - z_0)^{n-1}$ have the same radius of convergence. ∎

A proof of this lemma, which is outlined in the exercises, can be omitted, but the conclusion is important. It states that term-by-term differentiation of a power series does not change the radius of convergence. This means that within the open disk of convergence, a power series defines a differentiable function whose derivative can be obtained by term-by-term differentiation.

THEOREM 21.10

Let $\sum_{n=0}^{\infty} c_n(z - z_0)^n$ have positive or infinite radius of convergence. For each z in the open disk of convergence, let

$$f(z) = \sum_{n=0}^{\infty} c_n(z - z_0)^n.$$

Then f is differentiable on this open disk, and

$$f'(z) = \sum_{n=1}^{\infty} nc_n(z - z_0)^{n-1}. \quad \blacksquare$$

Using this theorem, we know how to differentiate a function defined by a power series. But there is more to Theorem 21.10 than this. The series $\sum_{n=1}^{\infty} nc_n(z - z_0)^{n-1}$ is a power series in its own right, having the same radius of convergence as the series $\sum_{n=0}^{\infty} c_n(z - z_0)^n$. We can therefore apply the theorem to this differentiated series and obtain

$$f''(z) = \sum_{n=2}^{\infty} n(n - 1)c_n(z - z_0)^{n-2}$$

within the open disk of convergence. Further, we can continue to differentiate as many times as we like within this disk. If $f^{(k)}(z)$ denotes the kth derivative, then

$$f^{(3)}(z) = \sum_{n=3}^{\infty} n(n - 1)(n - 2)c_n z^{n-3},$$

and in general

$$f^{(k)}(z) = \sum_{n=k}^{\infty} n(n - 1)(n - 2) \cdots (n - k + 1)c_n(z - z_0)^{n-k}.$$

If the kth derivative is evaluated at z_0, then all terms of the series for $f^{(k)}(z_0)$ having positive powers of $z - z_0$ vanish, leaving just the constant first term in this differentiated series. In this way, we get

$$f(z_0) = c_0,$$
$$f'(z_0) = c_1,$$
$$f''(z_0) = 2c_2,$$
$$f^{(3)}(z_0) = 3(2)c_3$$

and, in general,

$$f^{(k)}(z_0) = k(k - 1)(k - 2) \cdots (1)c_k.$$

We can solve these equations for the coefficients in terms of the function and its derivatives at z_0:

$$c_k = \frac{1}{k!} f^{(k)}(z_0) \quad \text{for } k = 0, 1, 2, \ldots, \tag{21.6}$$

where $k!$ is the product of the integers from 1 through k, $0! = 1$ by convention, and the zeroeth derivative $f^{(0)}(z)$ is just $f(z)$. This notation enables us to write one formula for the coefficients without considering the case $k = 0$ separately. The numbers given by equation (21.6) are the *Taylor coefficients* of f at z_0, and the power series

$$\sum_{n=0}^{\infty} \frac{1}{n!} f^{(n)}(z_0)(z - z_0)^n$$

is called the *Taylor series* for f at (or about) z_0.

We have shown that if a function f is defined in a disk by a power series centered at z_0, then the coefficients in this power series must be the Taylor coefficients, and the power series must be the Taylor series of f about z_0.

We are now in a position to define some of the elementary complex functions, including exponential and trigonometric functions and power functions.

SECTION 21.2   PROBLEMS

In each of Problems 1 through 10, determine the radius of convergence and open disk of convergence of the power series.

1. $\sum_{n=0}^{\infty} \frac{n+1}{2^n}(z + 3i)^n$

2. $\sum_{n=0}^{\infty}(-1)^n \frac{1}{(2n+1)^2}(z - i)^{2n}$

3. $\sum_{n=0}^{\infty} \frac{n^n}{(n+1)^n}(z - 1 + 3i)^n$

4. $\sum_{n=0}^{\infty} \left(\frac{2i}{5+i}\right)^n (z + 3 - 4i)^n$

5. $\sum_{n=0}^{\infty} \frac{i^n}{2^{n+1}}(z + 8i)^n$

6. $\sum_{n=0}^{\infty} \frac{(1-i)^n}{n+2}(z - 3)^n$

7. $\sum_{n=0}^{\infty} \frac{n^2}{2n+1}(z + 6 + 2i)^n$

8. $\sum_{n=0}^{\infty} \frac{n^3}{4^n}(z + 2i)^{3n}$

9. $\sum_{n=0}^{\infty} \left(\frac{e^{in}}{2n+1}\right)(z + 4)^n$

10. $\sum \left(\frac{1-i}{1+i}\right)^n (z + 2i)^n$

11. Is it possible for $\sum c_n(z - 2i)^n$ to converge at 0 and diverge at i?

12. Is it possible for $\sum_{n=0}^{\infty} c_n(z - 4 + 2i)^n$ to converge at i and diverge at $1 + i$?

13. Suppose $\sum_{n=0}^{\infty} c_n z^n$ has radius of convergence R and $\sum_{n=0}^{\infty} c_n^* z^n$ has radius of convergence R^*. What can be

said about the radius of convergence of $\sum_{n=0}^{\infty} (c_n + c_n^*) z^n$?

14. Consider $\sum_{n=0}^{\infty} c_n (z - z_0)^n = \sum_{n=0}^{\infty} d_n (z - z_0)^n$ for all z in some disk $|z - z_0| < R$, where R is some positive number. Prove that $c_n = d_n$ for $n = 0, 1, 2, \ldots$.

15. Consider $\sum_{n=0}^{\infty} c_n z^n$, where $c_n = 2$ if n is even and $c_n = 1$ if n is odd. Show that the radius of convergence of this power series is 1, but that this number cannot be computed using the ratio test. (This simply means that it is not always possible to use this test to determine the radius of convergence of a power series.)

21.3 The Exponential and Trigonometric Functions

We want to define the complex exponential function e^z so that it agrees with the real exponential function when z is real. For all real x,

$$e^x = \sum_{n=0}^{\infty} \frac{1}{n!} x^n.$$

Replace x with z in this series to obtain the power series

$$\sum_{n=0}^{\infty} \frac{1}{n!} z^n.$$

Compute

$$\lim_{n \to \infty} \left| \frac{z^{n+1}/(n+1)!}{z^n/n!} \right| = \lim_{n \to \infty} \frac{1}{n+1} |z| = 0.$$

Because this limit is less than 1 for all z, this power series converges for all z, and we can make the following definition.

DEFINITION 21.6 *Exponential Function*

For complex z, define the complex exponential function e^z by

$$e^z = \sum_{n=0}^{\infty} \frac{1}{n!} z^n.$$

THEOREM 21.11

For every complex number, and every positive integer k, the kth derivative of e^z is

$$f^{(k)}(z) = e^z.$$

Proof Compute

$$f'(z) = \sum_{n=1}^{\infty} \frac{1}{n!} n z^{n-1} = \sum_{n=1}^{\infty} \frac{1}{(n-1)!} z^{n-1} = \sum_{n=0}^{\infty} \frac{1}{n!} z^n = e^z.$$

Therefore $f'(z) = e^z$. Continued differentiation now gives $f^{(k)}(z) = e^z$ for any positive integer k. ∎

We will list properties of the complex exponential function, many of which are familiar from the real case. Conclusion (8) gives the real and imaginary parts of e^z, enabling us to write

$e^z = u(x, y) + iv(x, y)$. Conclusion (9) is perhaps the main surprise we find when we extend the real exponential function to the complex plane. The complex exponential function is periodic! This period does not manifest itself in the real case because it is pure imaginary.

THEOREM 21.12

---

1. $e^0 = 1$.
2. If g is differentiable at z, then so is $e^{g(z)}$, and

$$\frac{d}{dz} e^{g(z)} = g'(z) e^{g(z)}.$$

3. $e^{z+w} = e^z e^w$ for all complex z and w.
4. $e^z \neq 0$ for all z.
5. $e^{-z} = 1/e^z$.
6. If z is real, then e^z is real and $e^z > 0$.
7. (Euler's Formula) If y is real, then

$$e^{iy} = \cos(y) + i \sin(y).$$

8. If $z = x + iy$, then

$$e^z = e^x \cos(y) + i e^x \sin(y).$$

9. e^z is periodic with period $2n\pi i$ for any integer n.

Proof   (1) is obvious and (2) follows from the chain rule for differentiation.

To prove (3), fix any complex number u and define $f(z) = e^z e^{u-z}$, for all complex z. Then

$$f'(z) = e^z e^{u-z} - e^z e^{u-z} = 0$$

for all z. By Theorem 21.7, on any open disk $D : |z| < R$, $f(z)$ is constant. For some number K, $f(z) = K$ for $|z| < R$. But then $f(0) = K = e^0 e^u = e^u$, so for all z in D,

$$e^z e^{u-z} = e^u.$$

Now let $u = z + w$ to get

$$e^z e^w = e^{z+w}.$$

Since R can be as large as we want, this holds for all complex z and w.

To prove (4), suppose $e^\alpha = 0$. Then

$$1 = e^0 = e^{\alpha - \alpha} = e^\alpha e^{-\alpha} = 0,$$

a contradiction.

For (5), argue as in (4) that

$$1 = e^0 = e^{z-z} = e^z e^{-z},$$

so

$$e^{-z} = 1/e^z.$$

To prove (7), write

$$e^{iy} = \sum_{n=0}^{\infty} \frac{1}{n!}(iy)^n = \sum_{n=0}^{\infty} \frac{1}{(2n)!}(iy)^{2n} + \sum_{n=0}^{\infty} \frac{1}{(2n+1)!}(iy)^{2n+1}$$

$$= \sum_{n=0}^{\infty} \frac{1}{(2n)!}i^{2n}y^{2n} + \sum_{n=0}^{\infty} \frac{1}{(2n+1)!}i^{2n+1}y^{2n+1}.$$

Now

$$i^{2n} = (i^2)^n = (-1)^n$$

and

$$i^{2n+1} = i(i^{2n}) = i(-1)^n,$$

so

$$e^{iy} = \sum_{n=0}^{\infty} \frac{(-1)^n}{(2n)!}y^{2n} + i\sum_{n=0}^{\infty} \frac{(-1)^n}{(2n+1)!}y^{2n+1} = \cos(y) + i\sin(y),$$

in which we have used the (real) Maclaurin expansions of $\cos(y)$ and $\sin(y)$ for real y.

For (8), use (7) to write

$$e^z = e^{x+iy} = e^x e^{iy} = e^x(\cos(y) + i\sin(y)).$$

Finally, for conclusion (9), for any integer n,

$$e^{z+2n\pi i} = e^{x+i(y+2n\pi)} = e^x(\cos(y+2n\pi) + i\sin(y+2n\pi))$$

$$= e^x\cos(y) + ie^x\sin(y) = e^z.$$

Thus for any nonzero integer n, $2n\pi i$ is a period of e^z. ∎

Conclusion (8) actually gives the polar form of e^z in terms of x and y. It implies that the magnitude of e^z is e^x and that an argument of e^z is y. We may state these conclusions:

$$|e^z| = e^{\text{Re}(z)} = e^x$$

and

$$\arg(e^z) = \text{Im}(z) + 2n\pi = y + 2n\pi.$$

It is also easy to verify that

$$\overline{e^z} = e^{\bar{z}}.$$

To see this, write

$$\overline{e^z} = \overline{e^x(\cos(y) + i\sin(y))} = e^x(\cos(y) - i\sin(y)) = e^{x-iy} = e^{\bar{z}}.$$

For example,

$$\overline{e^{2+6i}} = e^{\overline{2+6i}} = e^{2-6i} = e^2(\cos(6) - i\sin(6)).$$

Conclusion (9) can be improved. Not only is $2n\pi i$ a period of e^z but these numbers are the only periods. This is part (4) of the next theorem.

THEOREM 21.13

1. $e^z = 1$ if and only if $z = 2n\pi i$ for some integer n.
2. $e^z = -1$ if and only if $z = (2n+1)\pi i$ for some integer n.

3. $e^z = e^w$ if and only if $z - w = 2n\pi i$ for some integer n.

4. If p is a period of e^z, then $p = 2n\pi i$ for some integer n.

Contrast conclusion (2) of this theorem with conclusion (6) of the preceding theorem. If x is real, then e^x is a positive real number. However, the complex exponential function can assume negative values. Conclusion (2) of this theorem gives all values of z such that e^z assumes the value -1.

Proof For (1), suppose first that $e^z = 1$. Then

$$e^z = 1 = e^x \cos(y) + ie^x \sin(y).$$

Then

$$e^x \cos(y) = 1 \quad \text{and} \quad e^x \sin(y) = 0.$$

Now x is real, so $e^x > 0$ and the second equation requires that $\sin(y) = 0$. Since this is the real sine function, we know all of its zeros and can conclude that $y = k\pi$ for integer k. Now we must have

$$e^x \cos(y) = e^x \cos(k\pi) = 1.$$

But $\cos(k\pi) = (-1)^k$ for integer k, so

$$e^x(-1)^k = 1.$$

For this to be satisfied, we first need $(-1)^k$ to be positive, hence k must be an even integer, say $k = 2n$. This leaves us with

$$e^x = 1,$$

so $x = 0$. Therefore $z = x + iy = 2n\pi i$.

Conversely, suppose $z = 2n\pi i$ for some integer n. Then

$$e^z = \cos(2n\pi) + i \sin(2n\pi) = 1.$$

Conclusion (2) can be proved by an argument that closely parallels that just done for (1). For (3), if $z - w = 2n\pi i$, then

$$e^{z-w} = \frac{e^z}{e^w} = e^{2n\pi i} = 1,$$

so

$$e^z = e^w.$$

Conversely, suppose $e^z = e^w$. Then $e^{z-w} = 1$, so by (1), $z - w = 2n\pi i$ for some integer n. Finally, for (4), suppose p is a period of e^z. Then

$$e^{z+p} = e^z$$

for all z. But then

$$e^z e^p = e^z$$

so $e^p = 1$ and, by (1), $p = 2n\pi i$ for some integer n. ∎

Using the properties we have derived for e^z, we can sometimes solve equations involving this function.

EXAMPLE 21.13

Find all z such that

$$e^z = 1 + 2i.$$

To do this, let $z = x + iy$, so

$$e^x \cos(y) + i e^x \sin(y) = 1 + 2i.$$

Then

$$e^x \cos(y) = 1$$

and

$$e^x \sin(y) = 2.$$

Add the squares of these equations to get

$$e^{2x}(\cos^2(y) + \sin^2(y)) = e^{2x} = 5.$$

Then

$$x = \tfrac{1}{2}\ln(5),$$

in which $\ln(5)$ is the real natural logarithm of 5. Next, divide

$$\frac{e^x \sin(y)}{e^x \cos(y)} = \tan(y) = 2,$$

so $y = \tan^{-1}(2)$. One solution of the given equation is $z = \tfrac{1}{2}\ln(5) + i\tan^{-1}(2)$, or approximately $0.8047 + 1.1071i$. ∎

We are now ready to extend the trigonometric functions from the real line to the complex plane. We want to define $\sin(z)$ and $\cos(z)$ for all complex z so that these functions agree with the real sine and cosine functions when z is real. Following the method used to extend the exponential function from the real line to the complex plane, we begin with power series.

DEFINITION 21.7

For all complex z, let

$$\sin(z) = \sum_{n=0}^{\infty} \frac{(-1)^n}{(2n+1)!} z^{2n+1} \quad \text{and} \quad \cos(z) = \sum_{n=0}^{\infty} \frac{(-1)^n}{(2n)!} z^{2n}.$$

The definition presupposes that these series converge for all complex z, a fact that is easy to show.

From the power series, it is immediate that

$$\cos(-z) = \cos(z) \quad \text{and} \quad \sin(-z) = -\sin(z).$$

By differentiating the series term-by-term, we find that for all z,

$$\frac{d}{dz}\sin(z) = \cos(z) \quad \text{and} \quad \frac{d}{dz}\cos(z) = -\sin(z).$$

Euler's formula states that, for real y, $e^{iy} = \cos(y) + i\sin(y)$. We will now extend this to the entire complex plane.

THEOREM 21.14

For every complex number z,

$$e^{iz} = \cos(z) + i\sin(z). \quad \blacksquare$$

The proof follows that of Theorem 21.12(7), with z in place of y.

We can express $\sin(z)$ and $\cos(z)$ in terms of the exponential function as follows. First, from Theorem 21.14,

$$e^{iz} = \cos(z) + i\sin(z)$$

and

$$e^{-iz} = \cos(z) - i\sin(z).$$

Solve these equations for $\sin(z)$ and $\cos(z)$ to obtain

$$\cos(z) = \frac{1}{2}(e^{iz} + e^{-iz}) \quad \text{and} \quad \sin(z) = \frac{1}{2i}(e^{iz} - e^{-iz}).$$

Formulas such as these reveal one of the benefits of extending these familiar functions to the complex plane. On the real line, there is no apparent connection between e^x, $\sin(x)$, and $\cos(x)$. These formulations are also convenient for carrying out many manipulations involving $\sin(z)$ and $\cos(z)$. For example, to derive the identity

$$\sin(2z) = 2\cos(z)\sin(z),$$

we have immediately that

$$2\sin(z)\cos(z) = 2\frac{1}{2}(e^{iz} + e^{-iz})\frac{1}{2i}(e^{iz} - e^{-iz})$$

$$= \frac{1}{2i}(e^{2iz} - e^{-2iz} + 1 - 1) = \frac{1}{2i}(e^{2iz} - e^{-2iz}) = \sin(2z).$$

Identities involving real trigonometric functions remain true in the complex case, and we will often use them without proof. For example,

$$\sin(z + w) = \sin(z)\cos(w) + \cos(z)\sin(w).$$

Not all properties of the real sine and cosine are passed along to their complex extensions. Recall that $|\cos(x)| \leq 1$ and $|\sin(x)| \leq 1$ for real x. Contrast this with the following.

THEOREM 21.15

$\cos(z)$ and $\sin(z)$ are unbounded in the complex plane. $\quad \blacksquare$

The proof consists of simply showing that both functions can be made arbitrarily large in magnitude by certain choices of z. Let $z = iy$ with y real. Then

$$\sin(z) = \sin(iy) = \frac{1}{2i}(e^{-y} - e^{y})$$

so

$$|\sin(z)| = \frac{1}{2}|e^y - e^{-y}|,$$

and the right side can be made as large as we like by choosing y sufficiently large in magnitude. That is, as z moves away from the origin in either direction along the vertical axis, $|\sin(z)|$ increases in magnitude without bound. It is easy to check that $|\cos(z)|$ exhibits the same behavior.

It is often useful to know the real and imaginary parts of these functions.

THEOREM 21.16

Let $z = x + iy$. Then

$$\cos(z) = \cos(x)\cosh(y) - i\sin(x)\sinh(y)$$

and

$$\sin(z) = \sin(x)\cosh(y) + i\cos(x)\sinh(y). \quad \blacksquare$$

These expressions are routine to derive starting from the exponential expressions for $\sin(z)$ and $\cos(z)$.

We will now show that the complex sine and cosine functions have exactly the same periods and zeros as their real counterparts.

THEOREM 21.17
---

1. $\sin(z) = 0$ if and only if $z = n\pi$ for some integer n.
2. $\cos(z) = 0$ if and only if $z = (2n + 1)\pi/2$ for some integer n.
3. $\sin(z)$ and $\cos(z)$ are periodic with periods $2n\pi$, for n any nonzero integer. Further, these are the only periods of these functions. $\quad \blacksquare$

Conclusion (3) means that

$$\cos(z + 2n\pi) = \cos(z) \quad \text{and} \quad \sin(z + 2n\pi) = \sin(z)$$

for all complex z and, conversely, if

$$\cos(z + p) = \cos(z) \quad \text{for all } z,$$

then $p = 2n\pi$, and if

$$\sin(z + q) = \sin(z) \quad \text{for all } z,$$

then $q = 2n\pi$. This guarantees that the sine and cosine functions do not pick up additional periods when extended to the complex plane, as occurs with the complex exponential function.

Proof For (1), if n is an integer, then

$$\sin(n\pi) = \frac{1}{2i}(e^{n\pi i} - e^{-n\pi i}) = \frac{1}{2i}(1 - 1) = 0.$$

Thus every $z = n\pi$, with n an integer, is a zero of $\sin(z)$. To show that these are the only zeros, suppose $\sin(z) = 0$. Let $z = x + iy$. Then

$$\sin(x)\cosh(y) + i\cos(x)\sinh(y) = 0.$$

Then

$$\sin(x)\cosh(y) = 0 \quad \text{and} \quad \cos(x)\sinh(y) = 0.$$

Since $\cosh(y) > 0$ for all real y, then $\sin(x) = 0$, and for this real sine function, this means that $x = n\pi$ for some integer n. Then

$$\cos(x)\sinh(y) = \cos(n\pi)\sinh(y) = 0.$$

But $\cos(n\pi) = (-1)^n \neq 0$, so $\sinh(y) = 0$ and this forces $y = 0$. Thus $z = n\pi$.

(2) can be proved by an argument similar to that used for (1).

For (3), if n is an integer, then

$$\sin(z + 2n\pi) = \frac{1}{2i}(e^{i(z+2n\pi)} - e^{-i(z+2n\pi)})$$

$$= \frac{1}{2i}(e^{iz}e^{2n\pi i} - e^{-iz}e^{-2n\pi i}) = \frac{1}{2i}(e^{iz} - e^{-iz}) = \sin(z),$$

so each even integer multiple of π is a period of $\sin(z)$. To show that there are no other periods, suppose p is a period of $\sin(z)$. Then

$$\sin(z + p) = \sin(z)$$

for all complex z. In particular, this must hold for $z = 0$, so $\sin(p) = 0$ and then by (1), $p = n\pi$ for integer n. But we can also put $z = i$ to have

$$\sin(i + n\pi) = \sin(i).$$

Then

$$e^{i(i+n\pi)} - e^{-i(i+n\pi)} = e^{-1} - e.$$

Therefore,

$$e^{-1}\cos(n\pi) - e\cos(n\pi) = e^{-1} - e.$$

If n is even, then $\cos(n\pi) = 1$ and this equation is true. If n is odd, then $\cos(n\pi) = -1$ and this equation becomes

$$-e^{-1} + e = e^{-1} - e,$$

an impossibility. Therefore, n is even, and the only periods of $\sin(z)$ are even integer multiples of π.

A similar argument establishes the same result for periods of $\cos(z)$. ■

Here is an example in which facts about $\cos(z)$ are used to solve an equation.

EXAMPLE 21.14

Solve $\cos(z) = i$.

Let $z = x + iy$, so

$$\cos(x)\cosh(y) - i\sin(x)\sinh(y) = i.$$

Then

$$\cos(x)\cosh(y) = 0 \quad \text{and} \quad \sin(x)\sinh(y) = -1.$$

Since $\cosh(y) > 0$ for all real y, the first equation implies that $\cos(x) = 0$, so

$$x = \frac{2n+1}{2}\pi,$$

in which (so far) n can be any integer. From the second equation,

$$\sin\left(\frac{2n+1}{2}\pi\right)\sinh(y) = -1.$$

Now $\sin((2n+1)\pi/2) = (-1)^n$, so

$$\sinh(y) = (-1)^{n+1},$$

with n any integer. Thus $y = \sinh^{-1}((-1)^{n+1})$. The solutions of $\cos(z) = i$ are therefore the complex numbers

$$\frac{2n+1}{2}\pi + i\sinh^{-1}(-1) \quad \text{for } n \text{ an even integer,}$$

and

$$\frac{2n+1}{2}\pi + i\sinh^{-1}(1) \quad \text{for } n \text{ an odd integer.}$$

A standard formula for the inverse hyperbolic sine function gives

$$\sinh^{-1}(\beta) = \ln\left(\beta + \sqrt{\beta^2 + 1}\right)$$

for β real. Therefore the solutions can be written

$$\frac{2n+1}{2}\pi + i\ln(-1+\sqrt{2}) \quad \text{for } n \text{ an even integer,}$$

and

$$\frac{2n+1}{2}\pi + i\ln(1+\sqrt{2}) \quad \text{for } n \text{ an even integer.} \quad \blacksquare$$

The other trigonometric functions are defined by

$$\sec(z) = \frac{1}{\cos(z)}, \quad \csc(z) = \frac{1}{\sin(z)}, \quad \tan(z) = \frac{\sin(z)}{\cos(z)}, \quad \cot(z) = \frac{\cos(z)}{\sin(z)},$$

in each case for all z for which the denominator does not vanish. Properties of these functions can be derived from properties of $\sin(z)$ and $\cos(z)$.

SECTION 21.3 PROBLEMS

In each of Problems 1 through 10, write the function value in the form $a + bi$.

1. e^i

2. $\sin(1 - 4i)$

3. $\cos(3 + 2i)$

4. $\tan(3i)$

5. e^{5+2i}

6. $\cot\left(1 - \dfrac{\pi i}{4}\right)$

7. $\sin^2(1+i)$

8. $\cos(2-i) - \sin(2-i)$

9. $e^{\pi i/2}$

10. $\sin(e^i)$

11. Find $u(x, y)$ and $v(x, y)$ such that $e^{z^2} = u(x, y) + iv(x, y)$. Show that u and v satisfy the Cauchy–Riemann equations for all complex z.

12. Find $u(x, y)$ and $v(x, y)$ such that $e^{1/z} = u(x, y) + iv(x, y)$. Show that u and v satisfy the Cauchy–Riemann equations for all z except zero.

13. Find $u(x, y)$ and $v(x, y)$ such that $\tan(z) = u(x, y) + iv(x, y)$. Determine where these functions are defined and show that they satisfy the Cauchy–Riemann equations for these points (x, y).

14. Find $u(x, y)$ and $v(x, y)$ such that $\sec(z) = u(x, y) + iv(x, y)$. Determine where these functions are defined and show that they satisfy the Cauchy–Riemann equations for all such points.

15. Prove that $\sin^2(z) + \cos^2(z) = 1$ for all complex z.

16. Let z and w be complex numbers.

    (a) Prove that $\sin(z + w) = \sin(z)\cos(w) + \cos(z)\sin(w)$.

    (b) Prove that $\cos(z + w) = \cos(z)\cos(w) - \sin(z)\sin(w)$.

17. Define the hyperbolic sine and hyperbolic cosine complex functions by

$$\sinh(z) = \tfrac{1}{2}(e^z - e^{-z}), \quad \cosh(z) = \tfrac{1}{2}(e^z + e^{-z}).$$

Show that both of these functions can be written in terms of trigonometric functions.

18. For real x, $\cosh(x)$ is always positive. Is it possible in the complex plane for $\cosh(z)$ to be zero?

19. Determine all complex numbers such that $\sinh(z) = 0$.

20. Find all solutions of $\cos(z) = \frac{3}{2}$.

21. Find all solutions of $e^z = 2i$.

22. Find all solutions of $\sin(z) = i$.

23. Find all solutions of $e^z = -2$.

24. Let $z = x + iy$. Show that $|\sinh(y)| \le |\sin(z)| \le \cosh(y)$.

25. The product of power series $\sum_{n=0}^{\infty} c_n(z - z_0)^n$ and $\sum_{n=0}^{\infty} d_n(z - z_0)^n$ is defined to be the power series $\sum_{n=0}^{\infty} p_n(z - z_0)^n$, where

$$p_n = \sum_{j=0}^{n} c_j d_{n-j}.$$

It can be shown that if the first series converges to $f(z)$ and the second to $g(z)$ in some disk $|z - z_0| < R$, then this product series converges to $f(z)g(z)$ in this disk. Using this product series, multiply the series for e^z and e^w and show that this yields the series for e^{z+w}. *Hint:* In looking at the sum forming p_n, keep in mind the binomial expansion of $(a + b)^n$.

26. Using the product series defined in Problem 25, prove the identity $2\sin(z)\cos(z) = \sin(2z)$ by multiplying the series for $\sin(z)$ and $\cos(z)$ and showing that this yields one-half the series for $\sin(2z)$.

21.4 The Complex Logarithm

In real calculus, the natural logarithm is the inverse of the exponential function for $x > 0$,

$$y = \ln(x) \quad \text{if and only if } x = e^y.$$

In this way, the real natural logarithm can be thought of as the solution of the equation $x = e^y$ for y in terms of x.

We can attempt this approach in seeking a definition of the complex logarithm. Given $z \ne 0$, we ask whether there are complex numbers w such that

$$e^w = z.$$

To answer this question, put z in polar form as $z = re^{i\theta}$. Let $w = u + iv$. Then

$$z = re^{i\theta} = e^w = e^u e^{iv}. \tag{21.7}$$

Since θ and v are real, $\left|e^{i\theta}\right| = \left|e^{iv}\right| = 1$ and equation (21.7) gives us $r = |z| = e^u$. Hence

$$u = \ln(r),$$

the real natural logarithm of the positive number r.

But now equation (21.7) implies that $e^{i\theta} = e^{iv}$, so by Theorem 2.13(3),

$$iv = i\theta + 2n\pi i$$

and therefore

$$v = \theta + 2n\pi,$$

in which n can be any integer.

In summary, given nonzero complex number $z = re^{i\theta}$, there are infinitely many complex numbers w such that $e^w = z$, and these numbers are

$$w = \ln(r) + i\theta + 2n\pi i,$$

with n any integer. Since θ is any argument of z, and all arguments of z are contained in the expression $\theta + 2n\pi$ for n integer, then in terms of z,

$$w = \ln(|z|) + i\arg(z),$$

with the understanding that there are infinitely many different values for $\arg(z)$. Each of these numbers is called a *complex logarithm* of z.

Each nonzero complex number therefore has infinitely many logarithms. To emphasize this, we often write

$$\log(z) = \{\ln(|z| + i\arg(z)\}.$$

This is read, "the logarithm of z is the set of all numbers $\ln(|z|) + i\theta$, where θ varies over all arguments of z."

EXAMPLE 21.15

Let $z = 1 + i$. Then $z = \sqrt{2}e^{i(\pi/4 + 2n\pi)}$. Then

$$\log(z) = \left\{\ln(\sqrt{2}) + i\frac{\pi}{4} + 2n\pi i\right\}.$$

Some of the logarithms of $1 + i$ are

$$\ln(\sqrt{2}) + \frac{\pi}{4}i, \quad \ln(\sqrt{2}) + \frac{9\pi}{4}i, \quad \ln(\sqrt{2}) - \frac{7\pi}{4}i, \dots \blacksquare$$

EXAMPLE 21.16

Let $z = -3$. An argument of z is π, and in polar form $z = 3e^{i(\pi + 2n\pi)} = 3e^{(2n+1)\pi i}$. Then

$$\log(z) = \{\ln(3) + (2n + 1)\pi i\}.$$

Some values of $\log(-3)$ are $\ln(3) + \pi i, \ln(3) + 3\pi i, \ln(3) + 5\pi i, \dots, \ln(3) - \pi i, \ln(3) - 3\pi i$, and so on. $\blacksquare$

The complex logarithm is not a function because with each nonzero z it associates infinitely many different complex numbers. Nevertheless, $\log(z)$ exhibits some of the properties we are accustomed to with real logarithm functions, if properly understood.

THEOREM 21.18

Let $z \neq 0$. If w is any value of $\log(z)$, then $e^w = z$. ∎

This is the complex function equivalent of the fact that in real calculus, $e^{\ln(x)} = x$. This is the condition we used to reason to a definition of $\log(z)$.

THEOREM 21.19

Let z and w be nonzero complex numbers. Then each value of $\log(zw)$ is a sum of values of $\log(z)$ and $\log(w)$.

Proof   Let $z = re^{i\theta}$ and $w = \rho e^{i\varphi}$. Then $zw = r\rho e^{i(\theta+\varphi)}$. If α is any value of $\log(zw)$, then for some integer N,

$$\alpha = \ln(\rho r) + i(\theta + \varphi + 2N\pi) = [\ln(r) + i\theta] + [\ln(\rho) + i(\varphi + 2N\pi)].$$

But $\ln(r) + i\theta$ is one value of $\log(z)$, and $\ln(\rho) + i(\varphi + 2N\pi)$ is one value of $\log(w)$, proving the theorem. ∎

Here is an example of the use of the logarithm to solve an equation involving the exponential function.

EXAMPLE 21.17

Solve for all z such that $e^z = 1 + 2i$.

In Example 21.14 we found one solution by separating the real and imaginary parts of e^z. Using the logarithm, we obtain all solutions as follows:

$$e^z = 1 + 2i$$

means that

$$z = \log(1 + 2i) = \ln(|1 + 2i|) + i\arg(1 + 2i) = \tfrac{1}{2}\ln(5) + i(\arctan(2) + 2n\pi),$$

in which n is any positive integer. ∎

Sometimes it is convenient to agree on a particular logarithm to use for nonzero complex numbers. This can be done by choosing an argument. For example, we could define, for $z \neq 0$,

$$\mathrm{Log}(z) = \ln(|z|) + i\theta,$$

where $0 \leq \theta < 2\pi$. This assigns to the symbol $\mathrm{Log}(z)$ that particular value of $\log(z)$ corresponding to the argument of z lying in $[0, 2\pi)$. For example,

$$\mathrm{Log}(1 + i) = \ln(\sqrt{2}) + i\frac{\pi}{4}$$

and

$$\mathrm{Log}(-3) = \ln(3) + i\pi.$$

If this is done, then care must be taken in doing computations. For example, in general $\mathrm{Log}(zw) \neq \mathrm{Log}(z) + \mathrm{Log}(w)$.

In each of Problems 1 through 6, determine all values of $\log(z)$ and also the value of $\text{Log}(z)$ defined in the discussion.

1. $-4i$

2. $2 - 2i$

3. -5

4. $1 + 5i$

5. $-9 + 2i$

6. 5

7. Let z and w be nonzero complex numbers. Show that each value of $\log(z/w)$ is equal to a value of $\log(z)$ minus a value of $\log(w)$.

8. Give an example to show that in general $\text{Log}(zw) \neq \text{Log}(z) + \text{Log}(w)$ for all nonzero complex z and w.

21.5   Powers

We want to assign a meaning to the symbol z^w when w and z are complex numbers and $z \neq 0$. We will build this idea in steps. Throughout this section z is a nonzero complex number.

21.5.1   Integer Powers

Integer powers present no problem. Define $z^0 = 1$. If n is a positive integer, then $z^n = z \cdot z \cdots z$, a product of n factors of z. For example,

$$(1 + i)^4 = (1 + i)(1 + i)(1 + i)(1 + i) = -4.$$

If n is a negative integer, then $z^n = 1/z^{|n|}$. For example,

$$(1 + i)^{-4} = \frac{1}{(1 + i)^4} = -\frac{1}{4}.$$

21.5.2 $z^{1/n}$ for Positive Integer n

Let n be a positive integer. A number u such that $u^n = z$ is called an nth *root* of z, and is denoted $z^{1/n}$. Like the logarithm and argument, this is a symbol that denotes more than one number. In fact, we will see that every nonzero complex number has exactly n distinct nth roots.

To determine these nth roots of z, let $z = re^{i\theta}$, with $r = |z|$ and θ any argument of z. Then

$$z = re^{i(\theta + 2k\pi)},$$

in which k can be any integer. Then

$$z^{1/n} = r^{1/n}e^{i(\theta + 2k\pi)/n}. \tag{21.8}$$

Here $r^{1/n}$ is the unique real nth root of the positive number r. As k varies over the integers, the expression on the right side of equation (21.8) produces complex numbers whose nth powers equal z. Let us see how many such numbers it produces.

For $k = 0, 1, \ldots, n - 1$, we get n distinct nth roots of z. They are

$$r^{1/n}e^{i\theta/n}, r^{1/n}e^{i(\theta + 2\pi)/n}, r^{1/n}e^{i(\theta + 4\pi)/n}, \ldots, r^{1/n}e^{i(\theta + 2(n-1)\pi)/n}. \tag{21.9}$$

We claim that other choices of k simply reproduce one of these nth roots. For example, if $k = n$, then equation (21.8) yields

$$r^{1/n}e^{i(\theta + 2n\pi)/n} = r^{1/n}e^{i\theta/n}e^{2\pi i} = r^{1/n}e^{i\theta/n},$$

the first number in the list (21.9). If $k = n + 1$, we get

$$r^{1/n}e^{i(\theta+2(n+1)\pi)/n} = r^{1/n}e^{i(\theta+2\pi)/n}e^{2\pi i} = r^{1/n}e^{i(\theta+2\pi)/n},$$

the second number in the list (21.9), and so on.

To sum up, for any positive integer n, the number of nth roots of any nonzero complex number z, is n. These nth roots are

$$r^{1/n}e^{i(\theta+2k\pi)/n} \quad \text{for } k = 0, 1, \ldots, n - 1,$$

or

$$r^{1/n}\left(\cos\left(\frac{\theta + 2k\pi}{n}\right) + i\sin\left(\frac{\theta + 2k\pi}{n}\right)\right) \quad \text{for } k = 0, 1, \ldots, n - 1.$$

EXAMPLE 21.18

We will find the cube roots of 8. First, $|8| = 8$, and an argument of 8 is $\theta = 0$. In polar form,

$$8 = 8e^{i(0+2k\pi)} = 8e^{2k\pi i}.$$

We get all cube roots of 8 as

$$8^{1/3}e^{2k\pi i/3} \quad \text{for } k = 0, 1, 2.$$

These numbers are

$$2, 2e^{2\pi i/3} \quad \text{and} \quad 2e^{4\pi i/3}.$$

We can also use Euler's formula to write these roots as

$$2, 2\left(\cos\left(\frac{2\pi}{3}\right) + i\sin\left(\frac{2\pi}{3}\right)\right), \quad \text{and} \quad 2\left(\cos\left(\frac{4\pi}{3}\right) + i\sin\left(\frac{4\pi}{3}\right)\right),$$

or, more explicitly,

$$2, -1 + \sqrt{3}i, -1 - \sqrt{3}i. \quad \blacksquare$$

The conclusion that the number of nth roots of nonzero z is n is consistent with facts we know from algebra. The cube roots of 8 are solutions of the polynomial equation $z^3 - 8 = 0$, and we know that this equation has three roots. Further, since the coefficients of this equation are real, the complex conjugate of each root should be a root, and we see that this is the case: $\overline{-1 - \sqrt{3}i} = -1 + \sqrt{3}i$.

EXAMPLE 21.19

Find the fourth roots of $1 + i$.

Since one argument of $1 + i$ is $\pi/4$ and $|1 + i| = \sqrt{2}$, we have the polar form

$$1 + i = \sqrt{2}e^{i(\pi/4+2k\pi)}.$$

The fourth roots are

$$(\sqrt{2})^{1/4}e^{i(\pi/4+2k\pi)/4} \quad \text{for } k = 0, 1, 2, 3.$$

These numbers are

$$2^{1/8} e^{\pi i/16}, \quad 2^{1/8} e^{i(\pi/4 + 2\pi)/4}, \quad 2^{1/8} e^{i(\pi/4 + 4\pi)/4}, \quad 2^{1/8} e^{i(\pi/4 + 6\pi)/4},$$

or

$$2^{1/8} \left(\cos \left(\frac{\pi}{16} \right) + i \sin \left(\frac{\pi}{16} \right) \right),$$

$$2^{1/8} \left(\cos \left(\frac{9\pi}{16} \right) + i \sin \left(\frac{9\pi}{16} \right) \right),$$

$$2^{1/8} \left(\cos \left(\frac{17\pi}{16} \right) + i \sin \left(\frac{17\pi}{16} \right) \right),$$

$$2^{1/8} \left(\cos \left(\frac{25\pi}{16} \right) + i \sin \left(\frac{25\pi}{16} \right) \right). \quad \blacksquare$$

EXAMPLE 21.20

The nth roots of 1 are called the nth roots of unity. These numbers have many uses, for example, in connection with the fast Fourier transform. Since 1 has magnitude 1, and an argument of 1 is zero, the nth roots of unity are

$$e^{2k\pi i/n} \quad \text{for } k = 0, 1, \ldots, n-1.$$

If we put $\omega = e^{2\pi i/n}$, then these nth roots of unity are $1, \omega, \omega^2, \ldots, \omega^{n-1}$.

For example, the fifth roots of unity are

$$1, e^{2\pi i/5}, e^{4\pi i/5}, e^{6\pi i/5}, \quad \text{and} \quad e^{8\pi i/5}.$$

These are

$$1, \cos \left(\frac{2\pi}{5} \right) + i \sin \left(\frac{2\pi}{5} \right), \cos \left(\frac{4\pi}{5} \right) + i \sin \left(\frac{4\pi}{5} \right),$$

$$\cos \left(\frac{6\pi}{5} \right) + i \sin \left(\frac{6\pi}{5} \right), \cos \left(\frac{8\pi}{5} \right) + i \sin \left(\frac{8\pi}{5} \right). \quad \blacksquare$$

If plotted as points in the plane, the nth roots of unity form vertices of a regular polygon with vertices on the unit circle $|z| = 1$ and having one vertex at $(1, 0)$. Figure 21.9 shows the fifth roots of unity displayed in this way.

FIGURE 21.9

If n is a negative integer, then

$$z^{1/n} = \frac{1}{z^{1/|n|}},$$

in the sense that the n numbers represented by the symbol on the left are calculated by taking the n numbers produced on the right. These are just the reciprocals of the nth roots of z.

21.5.3 Rational Powers

A rational number is a quotient of integers, say $r = m/n$. We may assume that n is positive and that m and n have no common factors. Write

$$z^r = z^{m/n} = (z^m)^{1/n},$$

the nth roots of z^m.

It is routine to check that we get the same numbers if we first take the nth roots of z, then raise each to power m. This is because

$$\left(z^m\right)^{1/n} = \left(r^m e^{im(\theta+2k\pi)}\right)^{1/n} = r^{m/n} e^{im(\theta+2k\pi)/n} = \left(r^{1/n} e^{i(\theta+2k\pi)/n}\right)^m = \left(z^{1/n}\right)^m.$$

EXAMPLE 21.21

We will find all values of $(2 - 2i)^{3/5}$.

First, $(2-2i)^3 = -16 - 16i$. Thus we want the fifth roots of $-16 - 16i$. Now $|-16 - 16i| = \sqrt{512}$, and $5\pi/4$ is an argument of $-16 - 16i$. Then

$$-16 - 16i = (512)^{1/2} e^{i(5\pi/4 + 2k\pi)}$$

and

$$(-16 - 16i)^{1/5} = (512)^{1/10} e^{i(5\pi/4 + 2k\pi)/5}.$$

Letting $k = 0, 1, 2, 3, 4$, we obtain the numbers

$$(512)^{1/10} e^{5\pi i/4}, \quad (512)^{1/10} e^{13\pi i/20}, \quad (512)^{1/10} e^{21\pi i/20}, \quad (512)^{1/10} e^{29\pi i/20}, \quad (512)^{1/10} e^{37\pi i/20}.$$

These are all values of $(2 - 2i)^{3/5}$. ∎

21.5.4 Powers z^w

Suppose $z \neq 0$ and let w be any complex number. We want to define the symbol z^w.

In the case of real powers, a^b is defined to be $b\ln(a)$. For example, $2^\pi = e^{\pi \ln(2)}$, and this is defined because $\ln(2)$ is defined. We will take the same approach to z^w, except now we must allow for the fact that $\log(z)$ denotes an infinite set of complex numbers. We therefore define z^w to be the set of all numbers $e^{w \log(z)}$.

If $w = m/n$, a rational number with common factors divided out, then $e^{w \log(z)}$ has n distinct values. If w is not a rational number, then z^w is an infinite set of complex numbers.

EXAMPLE 21.22

We will compute all values of 2^i.

We need all the values of $e^{i \log(2)}$. First,

$$\log(2) = \{\ln(2) + i \arg(2)\} = \{\ln(2) + 2n\pi i\},$$

in which n varies over all the integers. The values of 2^i are

$$e^{i\log(2)} = e^{i[\ln(2)+2n\pi i]} = e^{-2n\pi}e^{i\ln(2)} = e^{-2n\pi}\left(\cos(\ln(2)) + i\sin(\ln(2))\right),$$

in which n is any integer. To emphasize that 2^i has infinitely many complex values, we can write

$$2^i = \{e^{-2n\pi}\left(\cos(\ln(2)) + i\sin(\ln(2))\right)\},$$

as we did with argument and the complex logarithm. ■

EXAMPLE 21.23

We will find all values of $(1 - i)^{1+i}$.

These numbers are obtained as $e^{(1+i)\log(1-i)}$. First, $|1 - i| = \sqrt{2}$ and $-\pi/4$ is an argument of $1 - i$ (we reach the point $(1, -1)$ by rotating $\pi/4$ radians clockwise from the positive real axis). Therefore, in polar form,

$$1 - i = \sqrt{2}e^{i(-\pi/4+2n\pi)}.$$

Thus all values of $\log(1 - i)$ are given by

$$\ln(\sqrt{2}) + i\left(-\frac{\pi}{4} + 2n\pi\right).$$

Every value of $(1 - i)^{1+i}$ is contained in the expression

$$e^{(1+i)[\ln(\sqrt{2})+i(-\pi/4+2n\pi)]} = e^{\ln(\sqrt{2})+\pi/4-2n\pi}e^{i(\ln(\sqrt{2})-\pi/4+2n\pi)}$$

$$= \sqrt{2}e^{\pi/4-2n\pi}\left(\cos(\ln(\sqrt{2} - \pi/4 + 2n\pi)) + i\sin(\ln(\sqrt{2} - \pi/4 + 2n\pi))\right)$$

$$= \sqrt{2}e^{\pi/4-2n\pi}\left(\cos(\ln(\sqrt{2} - \pi/4)) + i\sin(\ln(\sqrt{2} - \pi/4))\right).$$

As n varies over all integer values, this expression gives all values of $(1 - i)^{1+i}$. ■

SECTION 21.5   PROBLEMS

In each of Problems 1 through 20, determine all values of z^w.

1. i^{1+i}

2. $(1 + i)^{2i}$

3. i^i

4. $(1 + i)^{2-i}$

5. $(-1 + i)^{-3i}$

6. $(1 - i)^{1/3}$

7. $i^{1/4}$

8. $16^{1/4}$

9. $(-4)^{2-i}$

10. 6^{-2-3i}

11. $(-16)^{1/4}$

12. $\left(\dfrac{1+i}{1-i}\right)^{1/3}$

13. $1^{1/6}$

14. $(7i)^{3i}$

15. i^{2-4i}

16. 2^{3-i}

17. $(3 + 3i)^{4/5}$

18. $(1 - i)^{-3/7}$

19. $(8 - 2i)^{1+3i}$

20. $(1 + 4i)^{1/6}$

21. Let n be a positive integer, and let $u_1, \ldots, u_n$ be the nth roots of unity. Prove that $\sum_{j=1}^{n} u_j = 0$. *Hint:* Write each nth root of unity as a power of $e^{2\pi i/n}$.

22. Let n be a positive integer, and $\omega = e^{2\pi i/n}$. Evaluate $\sum_{j=0}^{n-1}(-1)^j \omega^j$.

23. Show that the numbers

$$\frac{-b \pm (b^2 - 4ac)^{1/2}}{2a}$$

are the solutions of the equation $az^2 + bz + c = 0$, with a, b, and c complex and $a \neq 0$.

24. Prove that if $p(z)$ is a polynomial with real coefficients, and the complex number z is a root, then $\bar{z}$ is also a root.

25. Find all solutions of $(1 - i)z^2 + 2z - 1 = 0$. Are the roots of this polynomial conjugates of each other? How does this result reconcile with the conclusion of Problem 24?

26. Find the solutions of $iz^2 - (1 + i)z + 2 = 0$.

CHAPTER 22

Complex Integration

We now know some important complex functions, as well as some facts about derivatives of complex functions. Next we want to develop an integral for complex functions.

Real functions are defined over sets of real numbers and are usually integrated over intervals. Complex functions are defined over sets of points in the plane and are integrated over curves. Before defining this integral, we will review some facts about curves. For reference, real line integrals are discussed in Chapter 12.

22.1 Curves in the Plane

A *curve* in the complex plane is a function $\Gamma : [a, b] \to \mathfrak{C}$, defined on a real interval $[a, b]$ and having complex values. For each number t in $[a, b]$, $\Gamma(t)$ is a complex number, or point in the plane. The locus of such points is the *graph* of the curve. However, the curve is more than just a locus of points in the plane. Γ has a natural orientation, which is the direction the point $\Gamma(t)$ moves along the graph as t increases from a to b. In this sense, it is natural to refer to $\Gamma(a)$ as the *initial point* of the curve, and $\Gamma(b)$ as the *terminal point*. r is closed if $r(a) = r(b)$.

If $\Gamma(t) = x(t) + iy(t)$, then the graph of Γ is the locus of points $(x(t), y(t))$ for $a \leq t \leq b$. The initial point of Γ is $(x(a), y(a))$ and the terminal point is $(x(b), y(b))$, and $(x(t), y(t))$ moves from the initial to the terminal point as t varies from a to b. The functions $x(t)$ and $y(t)$ are the *coordinate functions* of Γ.

EXAMPLE 22.1

Let $\Gamma(t) = 2t + t^2 i$ for $0 \leq t \leq 2$. Then

$$\Gamma(t) = x(t) + iy(t),$$

where $x(t) = 2t$ and $y(t) = t^2$. The graph of this curve is the part of the parabola $y = (x/2)^2$, shown in Figure 22.1. As t varies from 0 to 2, the point $\Gamma(t) = (2t, t^2)$ moves along this graph from the initial point $\Gamma(0) = (0, 0)$ to the terminal point $\Gamma(2) = (4, 4)$. The arrows on the graph indicate this orientation. ■

EXAMPLE 22.2

Let $\Psi(t) = e^{it}$ for $0 \le t \le 3\pi$. Then $\Psi(t) = \cos(t) + i\sin(t) = x(t) + iy(t)$, so

$$x(t) = \cos(t), \quad y(t) = \sin(t).$$

Since $x^2 + y^2 = 1$, every point on this curve is on the unit circle about the origin. However, the initial point of Ψ is $\Psi(0) = 1$ and the terminal point is $\Psi(3\pi) = e^{3\pi i} = -1$. This curve is not closed. If this were a racetrack, the race begins at point 1 in Figure 22.2 and ends at -1. A circular racetrack does not mean that the starting and ending points of the race are the same. This is not apparent from the graph itself. Ψ is oriented counterclockwise, as the arrow indicates. ■

EXAMPLE 22.3

Let $\Theta(t) = e^{it}$ for $0 \le t \le 4\pi$. This curve is closed, since $\Theta(0) = 1 = \Theta(4\pi)$. However, the point $(x(t), y(t))$ moves around the unit circle $x^2 + y^2 = 1$ twice as t varies from 0 to 4π. This is also not apparent from just the graph itself (Figure 22.3). ■

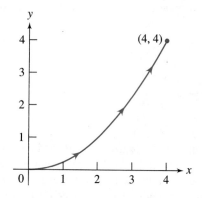

FIGURE 22.1 $x = 2t$, $y = t^2$ for $0 \le t \le 2$.

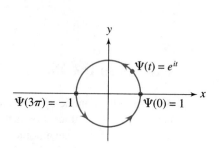

FIGURE 22.2 $\Psi(t) = e^{it}$ for $0 \le t \le 3\pi$.

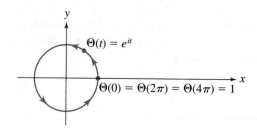

FIGURE 22.3 $\Theta(t) = e^{it}$ for $0 \le t \le 4\pi$.

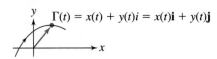

FIGURE 22.4 *Position vector of a curve.*

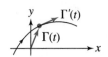

FIGURE 22.5
Tangent vector to a curve.

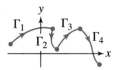

FIGURE 22.6 *The join $\Gamma_1 \oplus \Gamma_2 \oplus \Gamma_3 \oplus \Gamma_4$.*

A curve Γ is *simple* if $\Gamma(t_1) \neq \Gamma(t_2)$ whenever $t_1 \neq t_2$. This means that the same point is never repeated at different times. An exception is made for closed curves, which require that $\Gamma(a) = \Gamma(b)$. If this is the only point at which $\Gamma(t_1) = \Gamma(t_2)$ with $t_1 \neq t_2$, then we call Γ a *simple closed curve*. The curve Θ of Example 22.3 is closed, but not simple. If we define $\Lambda(t) = e^{it}$ for $0 \leq t \leq 2\pi$, then $(x(t), y(t))$ goes around the circle exactly once as t varies from 0 to 2π, and Λ is a simple closed curve.

A curve $\Gamma : [a, b] \to \mathfrak{C}$ is *continuous* if each of its coordinate functions is continuous on $[a, b]$. If $x(t)$ and $y(t)$ are differentiable on $[a, b]$, we call Γ a *differentiable curve*. If $x'(t)$ and $y'(t)$ are continuous, and do not both vanish for the same value of t, we call Γ a *smooth curve*. All the curves in the preceding examples are smooth.

In vector terms, we can write $\Gamma(t) = x(t)\mathbf{i} + y(t)\mathbf{j}$ (Figure 22.4). If Γ is differentiable, and $x'(t)$ and $y'(t)$ are not both zero, then $\Gamma'(t) = x'(t)\mathbf{i} + y'(t)\mathbf{j}$ is the tangent vector to the curve at the point $(x(t), y(t))$ (Figure 22.5). If Γ is smooth, then $x'(t)$ and $y'(t)$ are continuous, so this tangent vector is continuous. A smooth curve is therefore one having a continuous tangent. To illustrate, in Example 22.3, $\Theta(t) = \cos(t) + i \sin(t)$, so $\Theta'(t) = -\sin(t) + i \cos(t)$. We can leave this as it is or write the tangent vector $\Theta'(t) = -\sin(t)\mathbf{i} + \cos(t)\mathbf{j}$, exploiting the natural correspondence between complex numbers and vectors in the plane.

Sometimes we form a curve Γ by joining several curves $\Gamma_1, \ldots, \Gamma_n$ in succession, with the understanding that the terminal point of Γ_{j-1} must be the same as the initial point of Γ_j for $j = 2, \ldots, n$ (Figure 22.6). Such a curve is called the *join* of $\Gamma_1, \ldots, \Gamma_n$ and is denoted

$$\Gamma = \Gamma_1 \oplus \Gamma_2 \oplus \cdots \oplus \Gamma_n.$$

The curves Γ_j are the *components* of this join. If each component of a join is smooth, then the join is *piecewise smooth*. It has a continuous tangent at each point, except perhaps at the seams where Γ_{j-1} is joined to Γ_j. If the seams join in a smooth fashion, a join can even have a tangent at each of these points and itself be smooth.

EXAMPLE 22.4

Let $\Gamma_1(t) = e^{it}$ for $0 \leq t \leq \pi$ and let $\Gamma_2(t) = -1 + ti$ for $0 \leq t \leq 3$. Then $\Gamma_1(\pi) = -1 = \Gamma_2(0)$, so the terminal point of Γ_1 is the initial point of Γ_2. Figure 22.7 shows a graph of $\Gamma_1 \oplus \Gamma_2$. This curve is piecewise smooth, being a join of two smooth curves. The join has a tangent at each point except -1, where the connection is made to form the join. ■

We will define a kind of equivalence of curves. Suppose

$$\Gamma : [a, b] \to \mathfrak{C} \quad \text{and} \quad \Phi : [A, B] \to \mathfrak{C}$$

are two smooth curves. We call these curves *equivalent* if one can be obtained from the other by a change of variables defined by a differentiable, increasing function. This means that there is a function φ taking points of $[A, B]$ to $[a, b]$ such that

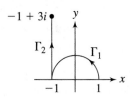

FIGURE 22.7 *The join of*
$\Gamma_1(t) = e^{it}$ *for* $0 \leq t \leq \pi$,
with $\Gamma_2(t) = -1 + it$
for $0 \leq t \leq 3$.

1. $\varphi'(p) > 0$ for $A < p < B$,
2. $\varphi(A) = a$ and $\varphi(B) = b$, and
3. $\Phi(p) = \Gamma(\varphi(p))$ for $A \leq p \leq B$.

If we think of $t = \varphi(p)$, then $\Gamma(t) = \Phi(p)$. The curves have the same initial and terminal points and the same graph and orientation, but $\Gamma(t)$ moves along the graph as t varies from a to b, while $\Phi(p)$ moves along the same graph in the same direction as p varies from A to B. Informally, two curves are equivalent if one is just a reparametrization of the other.

EXAMPLE 22.5

Let

$$\Gamma(t) = t^2 - 2ti \quad \text{for } 0 \leq t \leq 1,$$

and

$$\Phi(p) = \sin^2(p) - 2\sin(p)i \quad \text{for } 0 \leq p \leq \pi/2.$$

Both of these curves have the same graph (Figure 22.8), extending from initial point 0 to terminal point $1 - 2i$. Let

$$t = \varphi(p) = \sin(p) \quad \text{for } 0 \leq p \leq \pi/2.$$

Then φ is a differentiable, increasing function that takes $[0, \pi/2]$ onto $[0, 1]$. Further, for $0 \leq p \leq \pi/2$,

$$\Gamma(\sin(p)) = \sin^2(p) - 2\sin(p)i = \Phi(p).$$

These curves are therefore equivalent. ∎

Informally, we will often describe a curve geometrically and speak of a curve and its graph interchangeably. When this is done, it is important to keep track of the orientation along the curve and whether or not the curve is closed.

For example, suppose Γ is the straight line from $1 + i$ to $3 + 3i$ (Figure 22.9). This gives the graph and its orientation, and from this we can find Γ. Since the graph is the segment of the straight line from $(1, 1)$ to $(3, 3)$, the coordinate functions are

$$x = t, y = t \quad \text{for } 1 \leq t \leq 3.$$

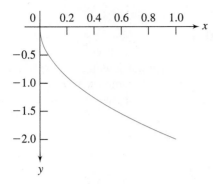

FIGURE 22.8 *Graph of* $\Gamma(t) = t^2 - 2it$ *for* $0 \leq t \leq 1$.

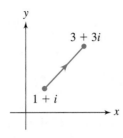

FIGURE 22.9 *Directed line from* $1 + i$ *to* $3 + 3i$.

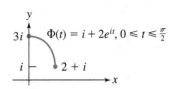

FIGURE 22.10

Then

$$\Gamma(t) = x(t) + y(t)i = (1 + i)t \quad \text{for } 1 \leq t \leq 3$$

is one representation of the curve that has been described. There are, of course, other, equivalent representations.

As another example, suppose Φ is the quarter-circle of radius 2 about i, from $2 + i$ to $3i$ (Figure 22.10). Again, we have given the graph and its orientation. Using polar coordinates centered at $i = (0, 1)$, we can write the coordinate functions

$$x(t) = 2\cos(t), y(t) = 1 + 2\sin(t) \quad \text{for } 0 \leq t \leq \pi/2.$$

As a function, this curve can be written

$$\Phi(t) = 2\cos(t) + 2i\sin(t) + i = i + 2e^{it} \quad \text{for } 0 \leq t \leq \pi/2.$$

Other, equivalent representations can also be used.

Finally, we will often make statements such as "f is continuous on Γ," by which we mean that f is a complex function that is continuous at all points on the graph of Γ. And when we refer to "z on Γ," we mean a complex number z lying on the graph of Γ.

Curves are the objects over which we integrate complex functions. We will now define this integral.

SECTION 22.1   PROBLEMS

In each of Problems 1 through 10, graph the curve, determine its initial and terminal points, whether it is closed or not closed, whether or not it is simple, and the tangent to the curve at each point where the tangent exists. This tangent may be expressed as a vector or as a complex function.

1. $\Gamma(t) = 4 - 2i + 2e^{it}$ for $0 \leq t \leq \pi$

2. $\Gamma(t) = ie^{2it}$ for $0 \leq t \leq 2\pi$

3. $\Gamma(t) = t + t^2 i$ for $1 \leq t \leq 3$

4. $\Gamma(t) = 3\cos(t) + 5\sin(t)i$ for $0 \leq t \leq 2\pi$

5. $\Theta(t) = 3\cos(t) + 5\sin(t)i$ for $0 \leq t \leq 4\pi$

6. $\Lambda(t) = 4\sin(t) - 2\cos(t)i$ for $-\pi \leq t \leq \pi/2$

7. $\Psi(t) = t - t^2 i$ for $-2 \leq t \leq 4$

8. $\Phi(t) = (2t + 1) - \frac{1}{2}t^2 i$ for $-3 \leq t \leq -1$

9. $\Gamma(t) = \cos(t) - 2\sin(2t)i$ for $0 \leq t \leq 2$

10. $\Delta(t) = t^2 - t^4 i$ for $-1 \leq t \leq 1$

22.2 The Integral of a Complex Function

We will define the integral of a complex function in two stages, beginning with the special case that f is a complex function defined on an interval $[a, b]$ of real numbers. An example of such a function is $f(x) = x^2 + \sin(x)i$ for $0 \le x \le \pi$. It is natural to integrate such a function as

$$\int_0^\pi f(x)\,dx = \int_0^\pi x^2\,dx + i \int_0^\pi \sin(x)\,dx = \frac{1}{3}\pi^3 + 2i.$$

This is the model we follow in general for such functions.

DEFINITION 22.1

Let $f : [a, b] \to \mathfrak{C}$ be a complex function. Let $f(x) = u(x) + iv(x)$ for $a \le x \le b$. Then

$$\int_a^b f(x)\,dx = \int_a^b u(x)\,dx + i \int_a^b v(x)\,dx.$$

Both of the integrals on the right are Riemann integrals of real-valued functions over $[a, b]$.

EXAMPLE 22.6

Let $f(x) = x - ix^2$ for $1 \le x \le 2$. Then

$$\int_1^2 f(x)\,dx = \int_1^2 x\,dx - i \int_1^2 x^2\,dx = \frac{3}{2} - \frac{7}{3}i. \quad \blacksquare$$

EXAMPLE 22.7

Let $f(x) = \cos(2x) + i\sin(2x)$ for $0 \le x \le \pi/4$. Then

$$\int_0^{\pi/4} f(x)\,dx = \int_0^{\pi/4} \cos(2x)\,dx + i \int_0^{\pi/4} \sin(2x)\,dx = \frac{1}{2} + \frac{1}{2}i. \quad \blacksquare$$

In the last example, it is tempting to let $f(x) = e^{2ix}$ and adapt the fundamental theorem of calculus to complex functions to obtain

$$\int_0^{\pi/4} f(x)\,dx = \int_0^{\pi/4} e^{2ix}\,dx = \left[\frac{1}{2i}e^{2ix}\right]_0^{\pi/4} = \frac{1}{2i}\left(e^{\pi i/2} - 1\right)$$

$$= \frac{1}{2i}\left(\cos\left(\frac{\pi}{2}\right) + i\sin\left(\frac{\pi}{2}\right) - 1\right) = \frac{1}{2i}(-1 + i) = \frac{1}{2}(1 + i).$$

We will justify this calculation shortly.

We can now define the integral of a complex function over a curve in the plane.

DEFINITION 22.2

Let f be a complex function. Let $\Gamma : [a, b] \rightarrow \mathfrak{C}$ be a smooth curve in the plane. Assume that f is continuous at all points on Γ. Then the integral of f over Γ is defined to be

$$\int_\Gamma f(z)\, dz = \int_a^b f(\Gamma(t))\Gamma'(t)\, dt.$$

Since $z = \Gamma(t)$ on the curve, this integral is often written

$$\int_\Gamma f(z)\, dz = \int_a^b f(z(t))z'(t)\, dt.$$

This formulation has the advantage of suggesting the way $\int_\Gamma f(z)\, dz$ is evaluated—replace z with $z(t)$ on the curve, let $dz = z'(t)\, dt$, and integrate over the interval $a \leq t \leq b$.

EXAMPLE 22.8

Evaluate $\int_\Gamma \bar{z}\, dz$ if $\Gamma(t) = e^{it}$ for $0 \leq t \leq \pi$.

The graph of Γ is the upper half of the unit circle, oriented counterclockwise from 1 to -1 (Figure 22.11). On Γ, $z(t) = e^{it}$ and $z'(t) = ie^{it}$. Further, $f(z(t)) = \overline{z(t)} = e^{-it}$ because t is real. Then

$$\int_\Gamma f(z)\, dz = \int_0^\pi e^{-it} ie^{it}\, dt = i \int_0^\pi dt = \pi i. \quad \blacksquare$$

EXAMPLE 22.9

Evaluate $\int_\Phi z^2\, dz$ if $\Phi(t) = t + it$ for $0 \leq t \leq 1$.

The graph of Φ is the straight line segment from the origin to $(1, 1)$, as shown in Figure 22.12. On the curve, $z(t) = (1 + i)t$. Since $f(z) = z^2$,

$$f(z(t)) = (z(t))^2 = (1 + i)^2 t^2 = 2it^2$$

and

$$z'(t) = 1 + i.$$

Then

$$\int_\Phi z^2\, dz = \int_0^1 2it^2(1 + i)\, dt = (-2 + 2i) \int_0^1 t^2\, dt = \frac{2}{3}(-1 + i). \quad \blacksquare$$

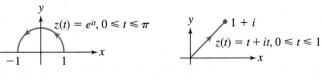

FIGURE 22.11   **FIGURE 22.12**

EXAMPLE 22.10

Evaluate $\int_\Gamma z \operatorname{Re}(z)\, dz$ if $\Gamma(t) = t - it^2$ for $0 \le t \le 2$.

Here $f(z) = z \operatorname{Re}(z)$, and on this curve, $z(t) = t - it^2$, so

$$f(z(t)) = z(t) \operatorname{Re}(z(t)) = (t - it^2)t = t^2 - it^3.$$

Further,

$$z'(t) = 1 - 2it,$$

so

$$\int_\Gamma z \operatorname{Re}(z)\, dz = \int_0^2 (t^2 - it^3)(1 - 2it)\, dt = \int_0^2 (t^2 - 3it^3 - 2t^4)\, dt$$

$$= \int_0^2 (t^2 - 2t^4)\, dt - 3i \int_0^2 t^3\, dt = -\frac{152}{15} - 12i. \ \blacksquare$$

EXAMPLE 22.11

Sometimes we integrate over a curve specified by a description, not a function. In this case we have to find the coordinate functions of the curve ourselves. For example, suppose we want to integrate $f(z) = z$ over the straight line segment L from 3 to $2 + i$ (Figure 22.13). This is part of the line $y = 3 - x$, and we can parametrize L as $z(t) = (5 - t) + (t - 2)i$ for $2 \le t \le 3$. Notice that $z(2) = 3$ and $z(3) = 2 + i$, so the orientation is correct. Now $z'(t) = -1 + i$ and

$$\int_L z\, dz = \int_2^3 [(5 - t) + (t - 2)i](-1 + i)\, dt = \int_2^3 [-3 + 7i - 2ti]\, dt = -3 + 2i. \ \blacksquare$$

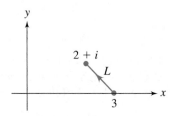

FIGURE 22.13

We will show that the integrals of a function over equivalent curves are equal. This is important because we can parametrize a given curve infinitely many different ways, and this should not change the value of the integral of a given function over the curve.

THEOREM 22.1

Let Γ and Φ be equivalent curves and let f be continuous on their graph. Then

$$\int_\Gamma f(z)\, dz = \int_\Phi f(z)\, dz.$$

Proof   Suppose $\Gamma : [a, b] \to \mathbb{C}$ and $\Phi : [A, B] \to \mathbb{C}$. Because these curves are equivalent, there is a continuous function φ with positive derivative on $[A, B]$ such that $\varphi(A) = a$ and $\varphi(B) = b$

and $\Phi(p) = \Gamma(\varphi(p))$ for $A \leq t \leq B$. By the chain rule,

$$\Phi'(p) = \Gamma'(\varphi(p))\varphi'(p).$$

Then

$$\int_\Phi f(z)\,dz = \int_A^B f(\Phi(p))\Phi'(p)\,dp = \int_A^B f(\Gamma(\varphi(p)))\Gamma'(\varphi(p))\varphi'(p)\,dp.$$

Let $t = \varphi(p)$. Then t varies from a to b as p varies from A to B. Continuing from the last equation, we have

$$\int_\Phi f(z)\,dz = \int_a^b f(\Gamma(t))\Gamma'(t)\,dt = \int_\Gamma f(z)\,dz. \quad \blacksquare$$

Thus far we can integrate only over smooth curves. We can extend the definition to an integral over piecewise smooth curves by adding the integrals over the components of the join.

DEFINITION 22.3

Let $\Gamma = \Gamma_1 \oplus \Gamma_2 \oplus \cdots \oplus \Gamma_n$ be a join of smooth curves. Let f be continuous on each Γ_j. Then

$$\int_\Gamma f(z)\,dz = \sum_{j=1}^n \int_{\Gamma_j} f(z)\,dz.$$

EXAMPLE 22.12

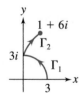

FIGURE 22.14

Let $\Gamma_1(t) = 3e^{it}$ for $0 \leq t \leq \pi/2$, and let $\Gamma_2(t) = t^2 + 3i(t+1)$ for $0 \leq t \leq 1$. Γ_1 is the quarter circle of radius 3 about the origin, extending counterclockwise from 3 to $3i$, and Γ_2 is the part of the parabola $x = (y-3)^2/9$ from $3i$ to $1+6i$. Figure 22.14 shows a graph of $\Gamma = \Gamma_1 \oplus \Gamma_2$. We will evaluate $\int_\Gamma \text{Im}(z)\,dz$.

On Γ_1, write $z(t) = 3e^{it} = 3\cos(t) + 3i\sin(t)$. Then

$$\int_{\Gamma_1} \text{Im}(z)\,dz = \int_0^{\pi/2} \text{Im}(z(t))z'(t)\,dt = \int_0^{\pi/2} 3\sin(t)[-3\sin(t) + 3i\cos(t)]\,dt$$

$$= -9\int_0^{\pi/2} \sin^2(t)\,dt + 9i\int_0^{\pi/2} \sin(t)\cos(t)\,dt = -\frac{9}{4}\pi + \frac{9}{2}i.$$

On Γ_2, $z(t) = t^2 + 3i(t+1)$ and $z'(t) = 2t + 3i$, so

$$\int_{\Gamma_2} \text{Im}(z)\,dz = \int_0^1 \text{Im}[t^2 + 3i(t+1)][2t + 3i]\,dt$$

$$= \int_0^1 3(t+1)(2t+3i)\,dt = \int_0^1 (6t^2 + 6t + 9it + 9i)\,dt$$

$$= \int_0^1 (6t^2 + 6t)\,dt + 9i\int_0^1 (t+1)\,dt = 5 + \frac{27}{2}i.$$

Then

$$\int_\Gamma f(z)\, dz = -\frac{9}{4}\pi + \frac{9}{2}i + 5 + \frac{27}{2}i = 5 - \frac{9}{2}\pi + 18i. \quad\blacksquare$$

22.2.1 The Complex Integral in Terms of Real Integrals

It is possible to think of the integral of a complex function over a curve as a sum of line integrals of real-valued functions of two real variables over the curve. Let $f(z) = u(x, y) + iv(x, y)$ and, on the curve Γ, suppose $z(t) = x(t) + iy(t)$ for $a \le t \le b$. Now

$$f(z(t)) = u(x(t), y(t)) + iv(x(t), y(t))$$

and

$$z'(t) = x'(t) + iy'(t)$$

so

$$f(z(t))z'(t) = [u(x(t), y(t)) + iv(x(t), y(t))][x'(t) + iy'(t)]$$
$$= u(x(t), y(t))x'(t) - v(x(t), y(t))y'(t)$$
$$+ i[v(x(t), y(t))x'(t) + u(x(t), y(t))y'(t)].$$

Then

$$\int_\Gamma f(z)\, dz = \int_a^b \left[u(x(t), y(t))x'(t) - v(x(t), y(t))y'(t)\right] dt$$
$$+ i \int_a^b [v(x(t), y(t))x'(t) + u(x(t), y(t))y'(t)]\, dt.$$

In the notation of real line integrals,

$$\int_\Gamma f(z)\, dz = \int_\Gamma u\, dx - v\, dy + i \int_\Gamma v\, dx + u\, dy. \tag{22.1}$$

This formulation allows a perspective that is sometimes useful in developing properties of complex integrals.

EXAMPLE 22.13

Evaluate $\int_\Gamma iz^2\, dz$ if $\Gamma(t) = 4\cos(t) + i\sin(t)$ for $0 \le t \le \pi/2$. Figure 22.15 shows a graph of Γ, which is part of the ellipse

$$\frac{x^2}{16} + y^2 = 1.$$

To evaluate $\int_\Gamma iz^2\, dz$ in terms of real line integrals, first compute

$$f(z) = iz^2 = -2xy + i(x^2 - y^2) = u + iv,$$

where

$$u(x, y) = -2xy \quad \text{and} \quad v(x, y) = x^2 - y^2.$$

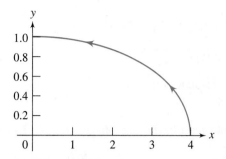

FIGURE 22.15 $x = 4\cos(t)$, $y = \sin(t)$ for $0 \le t \le \pi/2$.

On the curve, $x(t) = 4\cos(t)$ and $y(t) = \sin(t)$. Now equation (22.1) gives us

$$\int_{\Gamma} iz^2 \, dz = \int_0^{\pi/2} (-8\cos(t)\sin(t))(-4\sin(t)) \, dt - \int_0^{\pi/2} \left(16\cos^2(t) - \sin^2(t)\right)\cos(t) \, dt$$

$$+ i\left[\int_0^{\pi/2} \left(16\cos^2(t) - \sin^2(t)\right)(-4\sin(t)) \, dt + \int_0^{\pi/2} (-8\cos(t)\sin(t))\cos(t) \, dt\right]$$

$$= \frac{1}{3} - \frac{64}{3}i. \quad \blacksquare$$

We will have an easier way of evaluating simple line integrals such as $\int_{\Gamma} iz^2 \, dz$, when we have more properties of complex integrals.

22.2.2 Properties of Complex Integrals

We will develop some properties of $\int_{\Gamma} f(z) \, dz$.

THEOREM 22.2 *Linearity*

Let Γ be a piecewise smooth curve and let f and g be continuous on Γ. Let α and β be complex numbers. Then

$$\int_{\Gamma} (\alpha f(z) + \beta g(z)) \, dz = \alpha \int_{\Gamma} f(z) \, dz + \beta \int_{\Gamma} g(z) \, dz. \quad \blacksquare$$

This conclusion is certainly something we expect of anything called an integral. The result extends to arbitrary finite sums:

$$\int_{\Gamma} \sum_{j=1}^{n} \alpha_i f_j(z) \, dz = \sum_{j=1}^{n} \alpha_j \int_{\Gamma} f_j(z) \, dz.$$

Orientation plays a significant role in the complex integral, because it is an intrinsic part of the curve over which the integral is taken. Suppose $\Gamma : [a, b] \to \mathbb{C}$ is a smooth curve, as typically shown in Figure 22.16. The arrow indicates orientation. We can reverse this orientation by defining the new curve

$$\Gamma_r(t) = \Gamma(a + b - t) \quad \text{for } a \le t \le b.$$

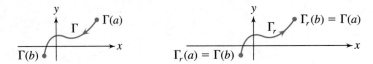

FIGURE 22.16 *Reversing orientation on a curve.*

Γ_r is a smooth curve having the same graph as Γ. However,

$$\Gamma_r(a) = \Gamma(b) \quad \text{and} \quad \Gamma_r(b) = \Gamma(a).$$

Γ_r starts where Γ ends, and Γ_r ends where Γ begins. The orientation has been reversed. We claim that reversing orientation changes the sign of the integral.

THEOREM 22.3   *Reversal of Orientation*

Let $\Gamma : [a, b] \to C$ be a smooth curve. Let f be continuous on Γ. Then

$$\int_{\Gamma} f(z)\, dz = -\int_{\Gamma_r} f(z)\, dz.$$

Proof   Let $u = a + b - t$. By the chain rule,

$$\frac{d}{dt}\Gamma_r(t) = \frac{d}{dt}\Gamma(a + b - t) = \Gamma'(u)u'(t) = -\Gamma'(u) = -\Gamma'(a + b - t).$$

Then

$$\int_{\Gamma_r} f(z)\, dz = \int_a^b f(\Gamma_r(t))\Gamma_r'(t)\, dt = -\int_a^b f(\Gamma(a + b - t))\Gamma'(a + b - t)\, dt.$$

Now change variables by putting $s = a + b - t$. Then

$$\int_{\Gamma_r} f(z)\, dz = -\int_b^a f(\Gamma(s))\Gamma'(s)(-1)\, ds = -\int_a^b f(\Gamma(s))\Gamma'(s)\, ds = -\int_{\Gamma} f(z)\, dz. \ \blacksquare$$

We need not actually define Γ_r to reverse orientation in specific integrals—just integrate from b to a instead of from a to b. This reverses the roles of the initial and terminal points and hence the orientation. Or we can integrate from a to b and take the negative of the result.

We will next state a complex version of the fundamental theorem of calculus. It states that if f has a continuous antiderivative F, then the value of $\int_{\Gamma} f(z)\, dz$ is the value of F at the terminal point of Γ minus the value of F at the initial point.

THEOREM 22.4

Let f be continuous on an open set G and suppose $F'(z) = f(z)$ for z in G. Let $\Gamma : [a, b] \to G$ be a smooth curve in G. Then

$$\int_a^b f(z)\, dz = F(\Gamma(b)) - F(\Gamma(a)).$$

Proof With $\Gamma(t) = z(t) = x(t) + iy(t)$ and $F(z) = U(x, y) + iV(x, y)$,

$$\int_\Gamma f(z)\,dz = \int_a^b f(z(t))z'(t)\,dt = \int_a^b F'(z(t))z'(t)\,dt = \int_a^b \frac{d}{dt}F(z(t))\,dt.$$

$$= \int_a^b \frac{d}{dt}U(x(t), y(t))\,dt + i\int_a^b \frac{d}{dt}V(x(t), y(t))\,dt.$$

Now we can apply the fundamental theorem of calculus to the two real integrals on the right to obtain

$$\int_\Gamma f(z)\,dz = U(x(b), y(b)) + iV(x(b), y(b)) - [U(x(a), y(a)) + iV(x(a), y(a))]$$

$$= F(x(b), y(b)) - F(x(a), y(a)) = F(\Gamma(b)) - F(\Gamma(a)). \quad\blacksquare$$

EXAMPLE 22.14

We will compute $\int_\Gamma (z^2 + iz)\,dz$ if $\Gamma(t) = t^5 - t\cos(t)i$ for $0 \le t \le 1$.

This is an elementary but tedious calculation if done by computing $\int_0^1 f(z(t))z'(t)\,dt$. However, if we let G be the entire complex plane, then G is open, and $F(z) = z^3/3 + iz^2/2$ satisfies $F'(z) = f(z)$. The initial point of Γ is $\Gamma(0) = 0$ and the terminal point is $\Gamma(1) = 1 - \cos(1)i$. Therefore,

$$\int_\Gamma (z^2 + iz)\,dz = F(\Gamma(1)) - F(\Gamma(0)) = F(1 - \cos(1)i) - F(0)$$

$$= \frac{1}{3}(1 - \cos(1)i)^3 + \frac{i}{2}(1 - \cos(1)i)^2$$

$$= (1 - \cos(1)i)^2\left(\frac{1}{3}(1 - \cos(1)i) + \frac{1}{2}i\right). \quad\blacksquare$$

EXAMPLE 22.15

Evaluate $\int_\Gamma \cos(2z)\,dz$ if $\Gamma(t) = i - 5e^{3it}$ for $0 \le t \le \pi/2$.

Let $F(z) = \frac{1}{2}\sin(2z)$. Then $F'(z) = \cos(2z)$ for all z. We can take G in Theorem 22.4 to be the entire complex plane. Now $\Gamma(0) = i - 5$ is the initial point of Γ, and $\Gamma(\pi/2) = i - 5e^{3\pi i/2} = 6i$ is the terminal point. Therefore,

$$\int_\Gamma \cos(2z)\,dz = F(6i) - F(-5 + i) = \frac{1}{2}[\sin(12i) - \sin(-10 + 2i)].$$

The result can be left in this form, or we can write the real and imaginary part explicitly:

$$\int_\Gamma \cos(2z)\,dz = \frac{1}{4i}\left[e^{-12} - e^{12}\right] - \frac{1}{4i}\left[e^{-10i-2} - e^{10i+2}\right]$$

$$= \frac{i}{2}\sinh(12) + \frac{i}{4}[e^{-2}(\cos(10) - i\sin(10)) - e^2(\cos(10) + i\sin(10))]$$

$$= \frac{i}{2}\sinh(12) + \frac{1}{2}\cosh(2)\sin(10) - \frac{i}{2}\sinh(2)\cos(10). \quad\blacksquare$$

One ramification of Theorem 22.4 is that under the given conditions, the value of $\int_\Gamma f(z)\,dz$ depends only on the initial and terminal points of the curve. If Φ is also a smooth curve in G having the same initial point as Γ and the same terminal point as Γ, then

$$\int_\Gamma f(z)\,dz = \int_\Phi f(z)\,dz.$$

This is called *independence of path*, about which we will say more later.

Another consequence is that if Γ is a closed curve in G, then the initial and terminal points coincide and

$$\int_\Gamma f(z)\,dz = 0.$$

We will consider this circumstance in more detail when we consider Cauchy's theorem.

The next result is used in bounding the magnitude of an integral, as we sometimes need to do in making estimates in equations or inequalities.

THEOREM 22.5

Let $\Gamma : [a, b] \to C$ be a smooth curve and let f be continuous on Γ. Then

$$\left| \int_\Gamma f(z)\,dz \right| \le \int_a^b |f(z(t))| \, |z'(t)| \, dt.$$

If, in addition, there is a positive number M such that $|f(z)| \le M$ for all z on Γ, then

$$\left| \int_\Gamma f(z)\,dz \right| \le ML,$$

where L is the length of Γ.

Proof   Write the complex number $\int_\Gamma f(z)\,dz$ in polar form:

$$\int_\Gamma f(z)\,dz = re^{i\theta}.$$

Then

$$r = e^{-i\theta} \int_\Gamma f(z)\,dz = e^{-i\theta} \int_a^b f(z(t))z'(t)\,dt.$$

Since r is real,

$$r = \text{Re}(r) = \text{Re}\left[e^{-i\theta} \int_a^b f(z(t))z'(t)\,dt \right] = \int_a^b \text{Re}\left[e^{-i\theta} f(z(t))z'(t) \right] dt.$$

Now for any complex number w, $\text{Re}(w) \le |w|$. Therefore,

$$\text{Re}\left[e^{-i\theta} f(z(t))z'(t) \right] \le \left| e^{-i\theta} f(z(t))z'(t) \right| = \left| f(z(t))z'(t) \right|,$$

since $\left| e^{-i\theta} \right| = 1$ for θ real. Then

$$\left| \int_\Gamma f(z)\,dz \right| = r = \left| \int_a^b f(z(t))z'(t)\,dt \right| \le \int_a^b \left| f(z(t))z'(t) \right| dt = \int_a^b |f(z(t))| \, |z'(t)| \, dt,$$

as we wanted to show.

If now $|f(z)| \leq M$ on Γ, then

$$\left| \int_{\Gamma} f(z)\,dz \right| \leq \int_{a}^{b} |f(z(t))|\, |z'(t)|\,dt \leq M \int_{a}^{b} |z'(t)|\,dt.$$

If $\Gamma(t) = x(t) + iy(t)$, then

$$|z'(t)| = |x'(t) + iy'(t)| = \sqrt{(x'(t))^2 + (y'(t))^2},$$

so

$$\left| \int_{\Gamma} f(z)\,dz \right| \leq M \int_{a}^{b} \sqrt{(x'(t))^2 + (y'(t))^2}\,dt = ML.$$

EXAMPLE 22.16

We will obtain a bound on $\left| \int_{\Gamma} e^{\mathrm{Re}(z)}\,dz \right|$, where Γ is the circle of radius 2 about the origin, traversed once counterclockwise.

On Γ we can write $z(t) = 2\cos(t) + 2i\sin(t)$ for $0 \leq t \leq 2\pi$. Now

$$\left| e^{\mathrm{Re}(z(t))} \right| = e^{2\cos(t)} \leq e^2$$

for $0 \leq t \leq 2\pi$. Since the length of Γ is 4π, then

$$\left| \int_{\Gamma} e^{\mathrm{Re}(z)}\,dz \right| \leq 4\pi e^2. \ \blacksquare$$

This number bounds the magnitude of the integral. It is not claimed to be an approximation to the value of the integral to any degree of accuracy.

22.2.3 Integrals of Series of Functions

We often want to interchange an integral and a series. We would like conditions under which

$$\int_{\Gamma} \left(\sum_{n=1}^{\infty} f_n(z)\,dz \right) \overset{?}{=} \sum_{n=1}^{\infty} \int_{\Gamma} f_n(z)\,dz.$$

We will show that if we can bound each $f_n(z)$, for z on the curve, by a positive constant M_n so that $\sum_{n=1}^{\infty} M_n$ converges, then we can interchange the summation and the integral and integrate the series term-by-term.

THEOREM 22.6 *Term-by-Term Integration*

Let Γ be a smooth curve and let f_n be continuous on Γ for $n = 1, 2, \ldots$. Suppose for each positive integer n there is a positive number M_n such that $\sum_{n=1}^{\infty} M_n$ converges and, for all z on Γ,

$$|f_n(z)| \leq M_n.$$

Then $\sum_{n=1}^{\infty} f_n(z)$ converges absolutely for all z on Γ. Further, if we denote $\sum_{n=1}^{\infty} f_n(z) = g(z)$, then

$$\int_{\Gamma} g(z)\,dz = \sum_{n=1}^{\infty} \int_{\Gamma} f_n(z)\,dz.$$

Proof   For each z on Γ, the real series $\sum_{n=1}^{\infty} |f_n(z)|$ converges by comparison with the convergent series $\sum_{n=1}^{\infty} M_n$. Now let L be the length of Γ and consider the partial sum

$$F_N(z) = \sum_{n=1}^{N} f_n(z).$$

Each F_N is continuous on Γ and

$$\left| \int_\Gamma g(z)\, dz - \sum_{n=1}^{N} \int_\Gamma f_n(z)\, dz \right| = \left| \int_\Gamma [g(z)\, dz - F_N(z)]\, dz \right|$$

$$\leq L \left(\max_{z \text{ on } \Gamma} |g(z) - F_N(z)| \right).$$

Now for all z on Γ,

$$|g(z) - F_n(z)| = \left| \sum_{n=N+1}^{\infty} f_n(z) \right| \leq \left| \sum_{n=N+1}^{\infty} M_n \right|.$$

If ϵ is any positive number, we can choose N large enough that $\sum_{n=N+1}^{\infty} M_n < \epsilon/L$, because $\sum_{n=1}^{\infty} M_n$ converges. But then

$$\max_{z \text{ on } \Gamma} |g(z) - F_N(z)| < \frac{\epsilon}{L},$$

so

$$\left| \int_\Gamma g(z)\, dz - \sum_{n=1}^{N} \int_\Gamma f_n(z)\, dz \right| < L\frac{\epsilon}{L} = \epsilon$$

for N sufficiently large. This proves that

$$\lim_{N \to \infty} \sum_{n=1}^{N} \int_\Gamma f_n(z)\, dz = \int_\Gamma g(z)\, dz,$$

as we wanted to show. ∎

Of course, the theorem applies to a power series within its circle of convergence, with $f_n(z) = c_n(z - z_0)^n$.

SECTION 22.2   PROBLEMS

In each of Problems 1 through 20, evaluate $\int_\Gamma f(z)\, dz$. All closed curves are oriented counterclockwise unless specific exception is made.

1. $f(z) = 1$; $\Gamma(t) = t^2 - it$ for $1 \leq t \leq 3$

2. $f(z) = z^2 - iz$; Γ is the quarter circle about the origin from 2 to $2i$

3. $f(z) = \text{Re}(z)$; Γ is the line segment from 1 to $2 + i$

4. $f(z) = 1/z$; Γ is the part of the half-circle of radius 4 about the origin, from $4i$ to $-4i$

5. $f(z) = z - 1$; Γ is any piecewise smooth curve from $2i$ to $1 - 4i$

6. $f(z) = iz^2$; Γ is the line segment from $1 + 2i$ to $3 + i$

7. $f(z) = \sin(2z)$; Γ is the line segment from $-i$ to $-4i$

8. $f(z) = 1 + z^2$; Γ is the part of the circle of radius 3 about the origin from $-3i$ to $3i$

9. $f(z) = -i\cos(z)$; Γ is any piecewise smooth curve from 0 to $2 + i$

10. $f(z) = |z|^2$; Γ is the line segment from -4 to i

11. $f(z) = (z - i)^3$; $\Gamma(t) = t - it^2$ for $0 \leq t \leq 2$

12. $f(z) = e^{iz}$; Γ is any piecewise smooth curve from -2 to $-4 - i$

13. $f(z) = i\overline{z}$; Γ is the line segment from 0 to $-4 + 3i$

14. $f(z) = \text{Im}(z)$; Γ is the circle of radius 1 about the origin

15. $f(z) = |z|^2$; Γ is the line segment from $-i$ to 1

16. $f(z) = z^2 - z + 8$; Γ is any piecewise smooth curve from 1 to $2 + 2i$

17. $f(z) = z^2 \cos(z^3)$; Γ is any piecewise smooth curve from 0 to i

18. $f(z) = 1/\overline{z}$; Γ is the line segment from 1 to 6

19. $f(z) = ze^{-z^2}$; Γ is the circle of radius 3 about -4

20. $f(z) = (1 - i)z^2 + 2iz - 4$; Γ is the line segment from 1 to $2i$

21. Find a bound for $\left| \int_\Gamma \cos(z^2)\, dz \right|$, if Γ is the circle of radius 4 about the origin.

22. Find a bound for $\left| \int_\Gamma (1/(1 + z))\, dz \right|$, if Γ is the line segment from $2 + i$ to $4 + 2i$.

23. Let $f(z) = \sum_{n=0}^\infty ((-1)^n/(3n)!)(z - 2 + i)^n$. Find an infinite series for $\int_\Gamma f(z)\, dz$, if Γ is the line segment from 0 to $1 + i$.

24. Let $f(z) = \sum_{n=0}^\infty (1/5i)^n (z - 2i)^n$. Determine $\int_\Gamma f(z)\, dz$ if Γ is the circle of radius 2 about the origin.

22.3 Cauchy's Theorem

Cauchy's theorem (or the Cauchy integral theorem) is considered the fundamental theorem of complex integration and is named for the early nineteenth-century French mathematician and engineer Augustin-Louis Cauchy. He had the idea of the theorem, as well as many of its consequences, but was able to prove it only under what were later found to be unnecessarily restrictive conditions. Edouard Goursat proved the theorem as it is now usually stated, and for this reason it is sometimes called the Cauchy–Goursat theorem.

The statement of the theorem implicitly makes use of the Jordan curve theorem, discussed previously in Section 12.2 in connection with Green's theorem. It states that a continuous, simple, closed curve Γ in the plane separates the plane into two open sets. One of these sets is unbounded and is called the *exterior* of Γ, and the other set is bounded and is called the *interior* of Γ. The (graph of the) curve itself does not belong to either of these sets, but forms the boundary for both. Figure 22.17 illustrates the theorem. Although this conclusion may seem obvious for closed curves we might routinely sketch, it is difficult to prove because of the generality of its statement.

Some terminology will make the statement of Cauchy's theorem more efficient.

DEFINITION 22.4 *Path*

A path is a simple, piecewise smooth curve. A path in a set S is a path whose graph lies in S.

Thus, a path is a join of smooth curves with no self-intersections.

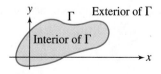

FIGURE 22.17 *Jordan curve theorem.*

DEFINITION 22.5 **Connected Set**

A set S of complex numbers is connected if, given any two points z and w in S, there is a path in S having z and w as endpoints.

S is connected if it is possible to get from any point of S to any other point by moving along some path lying completely in S. An open disk is connected, as is a closed disk, while the set consisting of the two open disks $|z| < 1$ and $|z - 10i| < 1$ is not (Figure 22.18), because we cannot get from 0 to $10i$ without going outside the set.

DEFINITION 22.6 **Domain**

An open, connected set of complex numbers is called a domain.

D is a domain if

1. about any z in D, there is some open disk containing only points of D, and
2. we can get from any point in D to any other point in D by a path in D.

For example, any open disk is a domain, as is the upper half-plane consisting of all z with $\text{Im}(z) > 0$. A closed disk is not a domain (it is connected but not open), and a set consisting of two disjoint open disks is not a domain (it is open but not connected).

DEFINITION 22.7 **Simply Connected**

A set S of complex numbers is simply connected if every closed path in S encloses only points of S.

Every open disk is simply connected (Figure 22.19). If we draw a closed path in an open disk, this closed path will enclose only points in the open disk. The annulus of Figure 22.20, consisting of points between two concentric circles, is not simply connected, even though it is connected. We can draw a closed path contained in the annulus, but enclosing the inner boundary circle of the annulus. This curve encloses points not in the annulus, namely those enclosed by the inner boundary circle.

We are now ready to state one version of Cauchy's theorem.

THEOREM 22.7 *Cauchy's Theorem*

Let f be differentiable on a simply connected domain G. Let Γ be a closed path in G. Then

$$\int_{\Gamma} f(z)\, dz = 0. \quad \blacksquare$$

Often integrals around closed paths are denoted $\oint$. In this notation, the conclusion of the theorem reads $\oint_{\Gamma} f(z)\, dz = 0$. The oval on the integral sign is just a reminder that the path is closed and does not alter the way we operate with the integral or the way it is evaluated.

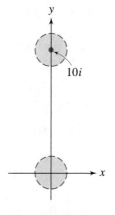

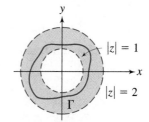

FIGURE 22.18
Disjoint open disks form a set that is not connected.

FIGURE 22.19
An open disk is simply connected.

FIGURE 22.20 *The set of points between two concentric circles is not simply connected.*

Informally, Cauchy's theorem states that $\oint_{\Gamma} f(z)\, dz = 0$ if f is differentiable on the curve and on all points enclosed by the curve. We will discuss a proof after looking at two examples. As a convention, closed curves are understood to be oriented positively (counterclockwise), unless specific exception is made.

EXAMPLE 22.17

Evaluate $\oint_{\Gamma} e^{z^2}\, dz$, where Γ is any closed path in the plane.

Figure 22.21 shows a typical Γ. Here $f(z) = e^{z^2}$ is differentiable for all z, and the entire plane is a simply connected domain. Therefore,

$$\oint_{\Gamma} e^{z^2}\, dz = 0. \quad \blacksquare$$

EXAMPLE 22.18

Evaluate

$$\oint_{\Gamma} \frac{2z+1}{z^2 + 3iz}\, dz,$$

where Γ is the circle $|z + 3i| = 2$ of radius 2 and center $-3i$ (Figure 22.22).

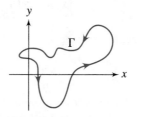

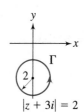

FIGURE 22.21 *A simple closed path Γ.*

FIGURE 22.22

We can parametrize $\Gamma(t) = -3i + 2e^{it}$ for $0 \le t \le 2\pi$. $\Gamma(t)$ traverses the circle once counterclockwise as t varies from 0 to 2π.

First observe that $f(z)$ is differentiable except at points where the denominator vanishes, 0 and $-3i$. Use a partial fractions decomposition to write

$$f(z) = \frac{1}{3i}\frac{1}{z} + \left(\frac{6+i}{3}\right)\frac{1}{z+3i}.$$

Since $1/z$ is differentiable on and within the simply connected domain enclosed by Γ, by Cauchy's theorem,

$$\oint_\Gamma \frac{1}{3i}\frac{1}{z}\,dz = 0.$$

However, $1/(z+3i)$ is not differentiable in the simply connected domain enclosed by Γ, so Cauchy's theorem does not apply to an integral of this function. We will evaluate this integral directly by writing $z(t) = -3i + 2e^{it}$:

$$\oint_\Gamma \left(\frac{6+i}{3}\right)\frac{1}{z+3i}\,dz = \frac{6+i}{3}\int_0^{2\pi}\frac{1}{z(t)+3i}z'(t)\,dt$$

$$= \frac{6+i}{3}\int_0^{2\pi}\frac{1}{2e^{it}}2ie^{it}\,dt = \frac{6+i}{3}\int_0^{2\pi}i\,dt = \frac{6+i}{3}(2\pi i).$$

Therefore,

$$\oint_\Gamma \frac{2z+1}{z^2+3iz}\,dz = \frac{6+i}{3}(2\pi i) = \left(-\frac{2}{3}+4i\right)\pi.\ \blacksquare$$

We will see more impressive ramifications of Cauchy's theorem shortly. The rest of this section is devoted to a proof of the theorem. This proof can be omitted without compromising applications of the theorem in the sections that follow.

22.3.1 Proof of Cauchy's Theorem for a Special Case

If we add an additional hypothesis, Cauchy's theorem is easy to prove. Let $f(z) = u(x, y) + iv(x, y)$ and assume that u and v and their first partial derivatives are continuous on G. Now we obtain Cauchy's theorem immediately by applying Green's theorem and the Cauchy–Riemann equations to equation (22.1). If D consists of all points on and enclosed by Γ, then

$$\oint_\Gamma f(z)\,dz = \oint_\Gamma u\,dx - v\,dy + i\oint_\Gamma v\,dx + u\,dy$$

$$= \iint_D \left(\frac{\partial(-v)}{\partial x} - \frac{\partial u}{\partial y}\right)dA + i\iint_D \left(\frac{\partial u}{\partial x} - \frac{\partial v}{\partial y}\right)dA = 0,$$

because, by the Cauchy–Riemann equations,

$$\frac{\partial u}{\partial x} = \frac{\partial v}{\partial y} \quad\text{and}\quad \frac{\partial u}{\partial y} = -\frac{\partial v}{\partial x}.$$

This argument is good enough for many settings in which Cauchy's theorem is used. However, it is not an optimal argument because it makes the additional assumption about continuity of the partial derivatives of u and v. A rigorous proof of the theorem as stated involves topological subtleties we do not wish to engage here. However, to get some feeling for what is involved, we will present a proof for the case that the curve is a rectangle. Surprisingly, the theorem for this special case is actually enough for many applications.

22.3.2 Proof of Cauchy's Theorem for a Rectangle

For this case, we will need two lemmas.

LEMMA 22.1

If Γ is a closed path, then

$$\oint_\Gamma dz = \oint_\Gamma z\, dz = 0.$$

Proof The first integral follows from the fact that $f(z) = 1$ is differentiable for all z, with antiderivative $F(z) = z$. The second follows similarly, since $f(z) = z$ is differentiable for all z, with antiderivative $F(z) = z^2/2$. ∎

LEMMA 22.2

Let $R^{(n)}$ be the set of points on and interior to a rectangle in the plane, for $n = 1, 2, \ldots$. Suppose $R^{(n+1)}$ is contained within $R^{(n)}$. Suppose also that if d_n is the diameter of $R^{(n)}$, then $d_n \to 0$ as $n \to \infty$. Then there is exactly one point that is common to every $R^{(n)}$. ∎

This lemma is a special case of a theorem of Georg Cantor. Figures 22.23 and 22.24 show typical rectangles as described in the lemma. This lemma does not by itself involve complex functions, but we will use it shortly to prove Cauchy's theorem for a rectangle. The sets $R^{(n)}$ are all closed, and each is contained in the preceding set in the sequence. We say that these sets form a *nested chain of closed sets*. In the case that the diameters tend to zero, the lemma asserts that the sets collapse to a single point that is in every $R^{(n)}$.

Proof Choose any z_1 in $R^{(1)}$. Next, choose any z_2 in $R^{(2)}$ but different from z_1. Choose any z_3 in $R^{(3)}$ different from z_1 and z_2, and so on. This generates a sequence $\{z_n\}$ of distinct points, with each z_j in $R^{(j)}$.

Now $\{z_n\}$ is a bounded sequence, because every z_n is in $R^{(1)}$, a bounded set. Therefore, $\{z_n\}$ contains a bounded subsequence $\{z_{n_j}\}$ converging to some number ζ. We will prove that ζ is the unique point belonging to every $R^{(n)}$.

Consider any rectangle $R^{(N)}$. We will first show that ζ is in this rectangle. Let D be any open disk about ζ. Then for some K, z_{n_j} is in D for $n_j > K$, because $z_{n_j} \to \zeta$. There are infinitely many $n_j > N$, and for these n_j, z_{n_j} is in $R^{(n_j)}$. But $R^{(N)}$ contains $R^{(n_j)}$ for $n_j > N$, so each such z_{n_j} is also in $R^{(N)}$. Therefore, D contains points of $R^{(N)}$ other than ζ for any open disk D about

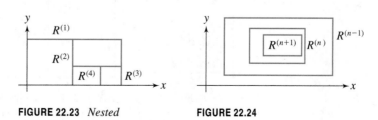

FIGURE 22.23 *Nested rectangles.*

FIGURE 22.24

ζ, making ζ a limit point of $R^{(N)}$. But $R^{(N)}$ is closed, hence contains all of its limit points, and in particular, ζ.

This proves that ζ belongs to all of the rectangles. We claim that no other point can belong to all of the rectangles. Suppose ξ is in every $R^{(n)}$. If $\xi \neq \zeta$, then $r = |\zeta - \xi| > 0$. But, for some M, the diameter of $R^{(M)}$ is less than r, because the diameters tend to zero. But then both ζ and ξ cannot be in $R^{(M)}$, because the distance between these points exceeds the diameter of $R^{(M)}$.

This contradiction shows that only one point can be common to all of the rectangles and completes the proof of the lemma. ■

Now consider Cauchy's theorem for the case that Γ is a rectangle having sides parallel to the axes, as in Figure 22.25. Form four rectangles by inserting line segments connecting the midpoints of opposite sides of Γ. Label the smaller rectangles $\Gamma_1, \ldots, \Gamma_4$. Each Γ_j is oriented counterclockwise, and

$$\oint_\Gamma f(z)\,dz = \sum_{j=1}^{4} \oint_{\Gamma_j} f(z)\,dz.$$

Each of the four interior segments is part of two interior rectangles, and in this sum of integrals, each interior segment is integrated over in both directions. Therefore, in the sum on the right the total contribution of the integrals over the newly inserted interior lines is zero. Since

$$\left| \oint_\Gamma f(z)\,dz \right| \leq \sum_{j=1}^{4} \left| \oint_{\Gamma_j} f(z)\,dz \right|,$$

for at least one of the subrectangles, call it $\Gamma^{(1)}$,

$$\frac{1}{4} \left| \oint_\Gamma f(z)\,dz \right| \leq \left| \oint_{\Gamma^{(1)}} f(z)\,dz \right|.$$

Now begin again and subdivide $\Gamma^{(1)}$ by connecting midpoints of its opposite sides, as in Figure 22.26. This produces four subrectangles within $\Gamma^{(1)}$ and, for at least one of these, call it $\Gamma^{(2)}$,

$$\frac{1}{4} \left| \oint_{\Gamma^{(1)}} f(z)\,dz \right| \leq \left| \oint_{\Gamma^{(2)}} f(z)\,dz \right|.$$

The accumulated effect so far is that

$$\frac{1}{4^2} \left| \oint_\Gamma f(z)\,dz \right| \leq \left| \oint_{\Gamma^{(2)}} f(z)\,dz \right|.$$

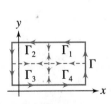

FIGURE 22.25

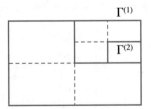

FIGURE 22.26 *Subdividing the rectangle into nested rectangles* $\Gamma^{(1)}, \Gamma^{(2)}, \ldots$.

Continuing in this way, we produce an infinite sequence of rectangles $\Gamma^{(1)}, \Gamma^{(2)}, \ldots$. Let $R^{(n)}$ be the solid rectangular region consisting of all points on $\Gamma^{(n)}$ and in the domain bounded by this curve. Then the sets $R^{(n)}$ form a nested chain of closed sets. It is routine to check that $R^{(n)}$ has diameter $d/2^n$ and perimeter $p/2^n$, where d is the diameter of Γ and p its perimeter. Thus the diameters of $R^{(n)}$ tend to zero as $n \to \infty$. Further,

$$\frac{1}{4^n} \left| \oint_\Gamma f(z)\, dz \right| \le \left| \oint_{\Gamma^{(n)}} f(z)\, dz \right|. \tag{22.2}$$

By Lemma 22.2, there is exactly one point ζ common to every $R^{(n)}$. By assumption, f is differentiable at ζ. Since $f(\zeta)$ and $f'(\zeta)$ are numbers, then by Lemma 22.1, for any positive integer j,

$$\oint_{\Gamma^{(j)}} f(z)\, dz = \oint_{\Gamma^{(j)}} [f(z) - f(\zeta) - (z - \zeta) f'(\zeta)]\, dz. \tag{22.3}$$

Now,

$$f'(\zeta) = \lim_{z \to \zeta} \frac{f(z) - f(\zeta)}{z - \zeta}.$$

Let ϵ be any positive number. Then, for some δ,

$$\left| \frac{f(z) - f(\zeta)}{z - \zeta} - f'(\zeta) \right| < \epsilon \quad \text{if } 0 < |z - \zeta| < \delta.$$

Then

$$\left| f(z) - f(\zeta) - f'(\zeta)(z - \zeta) \right| < \epsilon |z - \zeta| \quad \text{if } 0 < |z - \zeta| < \delta. \tag{22.4}$$

Choose N large enough that $R^{(N)}$ is contained within the open disk of radius δ about ζ. Then

$$0 < |z - \zeta| < \delta$$

for all z on $\Gamma^{(n)}$. From equation (22.3) and inequality (22.4),

$$\left| \oint_{\Gamma^{(n)}} f(z)\, dz \right| \le \oint_{\Gamma^{(N)}} \epsilon |z - \zeta|\, dz < \epsilon \frac{d}{2^N} \frac{p}{2^N},$$

in which $p/2^N$ is the length of $\Gamma^{(N)}$. This inequality also uses the fact that $|z - \zeta| < d/2^N$. Finally, invoke inequality (22.2) to get

$$\frac{1}{4^N} \left| \oint_\Gamma f(z)\, dz \right| \le \left| \oint_{\Gamma^{(N)}} f(z)\, dz \right| < \frac{\epsilon\, dp}{4^N}.$$

Therefore,

$$\left| \oint_\Gamma f(z)\, dz \right| < \epsilon\, dp.$$

This implies that the nonnegative number $\left| \oint_\Gamma f(z)\, dz \right|$ is less than any positive number, since ϵ can be made as small as we like. We conclude that

$$\oint_\Gamma f(z)\, dz = 0,$$

as we wanted to prove.

In the next section we will develop some important consequences of Cauchy's theorem.

SECTION 22.3 PROBLEMS

In each of Problems 1 through 15, evaluate the integral of the function over the given closed path. All paths are positively oriented (counterclockwise). In some cases Cauchy's theorem applies, and in some it does not.

1. $f(z) = \sin(3z)$; Γ is the circle $|z| = 4$

2. $f(z) = \dfrac{2z}{z-i}$; Γ is the circle $|z - i| = 3$

3. $f(z) = \dfrac{1}{(z-2i)^3}$; Γ is given by $|z - 2i| = 2$

4. $f(z) = z^2 \sin(z)$; Γ is the square having vertices 0, 1, $1 + i$, and i

5. $f(z) = \bar{z}$; Γ is the unit circle about the origin

6. $f(z) = 1/\bar{z}$; Γ is the circle of radius 5 about the origin

7. $f(z) = ze^z$; Γ is the circle $|z - 3i| = 8$

8. $f(z) = z^2 - 4z + i$; Γ is the rectangle with vertices 1, 8, $8 + 4i$, and $1 + 4i$

9. $f(z) = |z|^2$; Γ is the circle of radius 7 about the origin

10. $f(z) = \sin(1/z)$; Γ is the circle $|z - 1 + 2i| = 1$

11. $f(z) = \text{Re}(z)$; Γ is given by $|z| = 2$

12. $f(z) = z^2 + \text{Im}(z)$; Γ is the square with vertices 0, $-2i, 2 - 2i$, and 2

13. $f(z) = z - |z|$; Γ is the circle $|z| = 4$

14. $f(z) = \dfrac{z^2 - 2z + 4}{z^8 - 2}$; Γ is the circle $|z - 12| = 2$

15. $f(z) = \dfrac{z^2 - 8z}{\cos(z)}$; Γ is the rectangle with vertices $-\dfrac{\pi i}{4}$, $\dfrac{\pi(1-i)}{4}, \dfrac{\pi(1+i)}{4}$, and $\dfrac{\pi i}{4}$

16. Use partial fractions and Cauchy's theorem to evaluate $\displaystyle\oint_\Gamma \frac{1}{1+z^2}\, dz$, with Γ the circle $|z - i| = \dfrac{1}{4}$.

17. Use partial fractions and Cauchy's theorem to evaluate $\displaystyle\oint_\Gamma \frac{1}{z^2 + 8z}\, dz$, with Γ the unit circle about the origin.

22.4 Consequences of Cauchy's Theorem

This section lays out some of the main results of complex integration, with profound implications for understanding the behavior and properties of complex functions as well as for applications of the integral. As usual, all integrals over closed curves are taken with a counterclockwise orientation unless otherwise noted.

22.4.1 Independence of Path

In Section 22.2.1 we mentioned independence of path, according to which, under certain conditions on f, the value of $\int_\Gamma f(z)\, dz$ depends only on the endpoints of the curve, and not on the particular curve chosen between those endpoints.

Independence of path can also be viewed from the perspective of Cauchy's theorem. Suppose f is differentiable on a simply connected domain G, and z_0 and z_1 are points of G. Let Γ_1 and Γ_2 be piecewise smooth curves in G having initial point z_0 and terminal point z_1 (Figure 22.27). If we reverse the orientation on Γ_2, we obtain the new curve, $-\Gamma_2$, going from z_1 to z_0. Further, the join of Γ_1 and $-\Gamma_2$ forms a closed curve Γ, having initial and terminal point z_0 (Figure 22.28). By Cauchy's theorem and Theorem 22.3,

$$\oint_\Gamma f(z)\, dz = 0 = \oint_{\Gamma_1 \oplus (-\Gamma_2)} f(z)\, dz = \oint_{\Gamma_1} f(z)\, dz - \oint_{\Gamma_2} f(z)\, dz,$$

implying that

$$\oint_{\Gamma_1} f(z)\, dz = \oint_{\Gamma_2} f(z)\, dz.$$

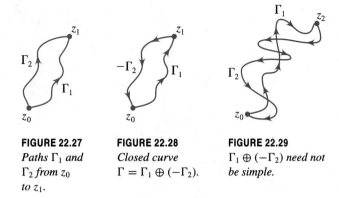

FIGURE 22.27
Paths Γ_1 and Γ_2 from z_0 to z_1.

FIGURE 22.28
Closed curve $\Gamma = \Gamma_1 \oplus (-\Gamma_2)$.

FIGURE 22.29
$\Gamma_1 \oplus (-\Gamma_2)$ need not be simple.

This means that the integral does not depend on the particular curve (in G) between z_0 and z_1 and is therefore independent of path.

This argument is not entirely rigorous, because $\Gamma_1 \oplus (-\Gamma_2)$ need not be a simple curve (Figure 22.29). In fact, Γ_1 and Γ_2 may cross each other any number of times as they progress from z_0 to z_1. Nevertheless, we wanted to point out the connection between Cauchy's theorem and the concept of independence of path of an integral that was discussed previously.

If $\int_{\Gamma} f(z)\, dz$ is independent of path in G, and Γ is any path from z_0 to z_1, we sometimes write

$$\oint_{\Gamma} f(z)\, dz = \int_{z_0}^{z_1} f(z)\, dz.$$

The symbol on the right has the value of the line integral on the left, with Γ any path from z_0 to z_1 in G.

22.4.2   The Deformation Theorem

The deformation theorem enables us, under certain conditions, to replace one closed path of integration with another, perhaps more convenient one.

THEOREM 22.8   *Deformation Theorem*
_____

Let Γ and γ be closed paths in the plane, with γ in the interior of Γ. Let f be differentiable on an open set containing both paths and all points between them. Then,

$$\int_{\Gamma} f(z)\, dz = \int_{\gamma} f(z)\, dz. \quad \blacksquare$$

Figure 22.30 shows the setting of the theorem. We may think of deforming one curve, say γ, to the other (Figure 22.29). Imagine γ as made of rubber, and continuously deform it into the shape of Γ. In doing this, it is necessary that the intermediate stages of the deformation from γ to Γ only cross over points at which f is differentiable, hence the hypothesis about f being differentiable on some open set containing both paths and all points between them.

The theorem states that the integral of f has the same value over both paths when one can be deformed into the other, moving only over points at which the function is differentiable. This means that we can replace Γ with another path γ that may be more convenient to use in evaluating the integral. Consider the following example.

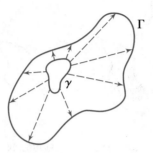

FIGURE 22.30 *Deforming γ continuously into Γ.*

EXAMPLE 22.19

Evaluate

$$\oint_{\Gamma} \frac{1}{z-a}\, dz$$

over any closed path enclosing the given complex number a.

Figure 22.31 shows a typical such path. We cannot parametrize Γ because we do not know it specifically—it is simply any path enclosing a. Let γ be a circle of radius r about a, with r small enough that γ is enclosed by Γ (Figure 22.32). Now $f(z) = 1/(z-a)$ is differentiable at all points except a, hence on both curves and the region between them. By the deformation theorem,

$$\oint_{\Gamma} \frac{1}{z-a}\, dz = \oint_{\gamma} \frac{1}{z-a}\, dz.$$

But γ is easily parametrized: $\gamma(t) = a + re^{it}$ for $0 \le t \le 2\pi$. Then

$$\oint_{\gamma} \frac{1}{z-a}\, dz = \int_0^{2\pi} \frac{1}{re^{it}} ire^{it}\, dt = \int_0^{2\pi} i\, dt = 2\pi i.$$

Therefore,

$$\oint_{\Gamma} \frac{1}{z-a}\, dz = 2\pi i.$$

The point is that by means of the deformation theorem, we can evaluate this integral over any path enclosing a. Of course, if Γ does not enclose a, and a is not on Γ, then $1/(z-a)$ is differentiable on Γ and the set it encloses, so $\oint_{\Gamma}[1/(z-a)]\, dz = 0$ by Cauchy's theorem. ∎

The proof of the theorem employs a technique we will find useful in several settings.

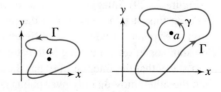

FIGURE 22.31   **FIGURE 22.32**

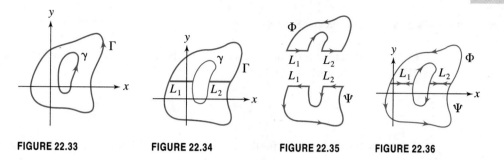

FIGURE 22.33   FIGURE 22.34   FIGURE 22.35   FIGURE 22.36

Proof   Figure 22.33 shows graphs of typical paths Γ and γ. Insert lines L_1 and L_2 between Γ and γ (Figure 22.34) and use these to form two closed paths Φ and Ψ (shown separated for emphasis in Figure 22.35). One path, Φ, consists of parts of Γ and γ, together with L_1 and L_2, with orientation on each piece as shown in order to have positive orientation on Φ. The other path, Ψ, consists of the rest of Γ and γ, again with L_1 and L_2, with the orientation chosen on each piece so that Ψ has positive orientation. Figure 22.36 shows the paths more realistically, sharing the inserted segments L_1 and L_2. In Figure 22.36, Γ is oriented counterclockwise, but γ is clockwise, due to their orientations as parts of Φ and Ψ.

Because f is differentiable on both Φ and Ψ and the sets they enclose, Cauchy's theorem yields

$$\oint_{\Phi} f(z)\,dz = \oint_{\Psi} f(z)\,dz = 0.$$

Then

$$\oint_{\Phi} f(z)\,dz + \oint_{\Psi} f(z)\,dz = 0. \tag{22.5}$$

In this sum of integrals, each of L_1 and L_2 is integrated over in one direction as part of Φ and the opposite direction as part of Ψ. The contributions from these segments therefore cancel in the sum (22.5). Next observe that in adding these integrals, we obtain the integral over all of Γ, oriented counterclockwise, together with the integral over all of γ, oriented clockwise. In view of Theorem 22.3, equation (22.5) becomes

$$\oint_{\Phi} f(z)\,dz - \oint_{\gamma} f(z)\,dz = 0,$$

or

$$\oint_{\Phi} f(z)\,dz = \oint_{\gamma} f(z)\,dz,$$

in which the orientation on both Γ and γ in these integrals is positive (counterclockwise). This proves the theorem. ∎

22.4.3   Cauchy's Integral Formula

We will now state a remarkable result that gives an integral formula for the values of a differentiable function.

THEOREM 22.9 *Cauchy Integral Formula*

Let f be differentiable on an open set G. Let Γ be a closed path in G enclosing only points of G. Then, for any z_0 enclosed by Γ,

$$f(z_0) = \frac{1}{2\pi i} \oint_\Gamma \frac{f(z)}{z - z_0}\, dz. \quad \blacksquare$$

We will see many uses for this theorem, but one is immediate. Write the formula as

$$\oint_\Gamma \frac{f(z)}{z - z_0}\, dz = 2\pi i f(z_0).$$

This gives, under the conditions of the theorem, an evaluation of the integral on the left as a constant multiple of the function value on the right.

EXAMPLE 22.20

Evaluate

$$\oint_\Gamma \frac{e^{z^2}}{z - i}\, dz$$

for any closed path that does not pass through i.

Let $f(z) = e^{z^2}$. Then f is differentiable for all z. There are two cases.

Case 1 Γ does not enclose i. Now $\oint_\Gamma (e^{z^2}/(z - i))\, dz = 0$ by Cauchy's theorem, because $e^{z^2}/(z - 2)$ is differentiable on and within Γ.

Case 2 Γ encloses i. By Cauchy's integral formula, with $z_0 = i$,

$$\oint_\Gamma \frac{e^{z^2}}{z - i}\, dz = 2\pi i f(i) = 2\pi i e^{-1}. \quad \blacksquare$$

EXAMPLE 22.21

Evaluate

$$\oint_\Gamma \frac{e^{2z} \sin(z^2)}{z - 2}\, dz$$

over any path not passing through 2.

Let $f(z) = e^{2z} \sin(z^2)$. Then f is differentiable for all z. This leads to two cases.

Case 1 If Γ does not enclose 2, then $f(z)/(z - 2)$ is differentiable on the curve and at all points it encloses. Now the integral is zero by Cauchy's theorem.

Case 2 If Γ encloses 2, then by the integral formula,

$$\oint_\Gamma \frac{e^{2z} \sin(z^2)}{z - 2}\, dz = 2\pi i f(2) = 2\pi i e^4 \sin(4). \quad \blacksquare$$

Observe the distinction between the roles of $f(z)$ in Cauchy's theorem and in Cauchy's integral representation. Cauchy's theorem is concerned with $\oint_\Gamma f(z)\, dz$. The integral representation is concerned with integrals of the form $\oint_\Gamma [f(z)/(z - z_0)]\, dz$, with $f(z)$ given, but multiplied by a factor $1/(z - z_0)$, which is not defined at z_0. If Γ does not enclose z_0, then $f(z)/(z - z_0) = g(z)$ may be differentiable at z_0 and we can attempt to apply Cauchy's theorem to $\oint_\Gamma g(z)\, dz$. If z_0 is enclosed by Γ, then under the appropriate conditions, the integral formula gives $\oint_\Gamma g(z)\, dz$ in terms of $f(z_0)$.

Here is a proof of the integral representation.

Proof First use the deformation theorem to replace Γ with a circle γ of radius r about z_0, as in Figure 22.37. Then

$$\oint_\Gamma \frac{f(z)}{z - z_0}\, dz = \oint_\gamma \frac{f(z)}{z - z_0}\, dz = \oint_\gamma \frac{f(z) - f(z_0) + f(z_0)}{z - z_0}\, dz$$

$$= f(z_0) \oint_\gamma \frac{1}{z - z_0}\, dz + \oint_\gamma \frac{f(z) - f(z_0)}{z - z_0}\, dz,$$

in which $f(z_0)$ could be brought outside the first integral because $f(z_0)$ is constant. By Example 22.19,

$$\oint_\gamma \frac{1}{z - z_0}\, dz = 2\pi i$$

because γ encloses z_0. Therefore,

$$\oint_\Gamma \frac{f(z)}{z - z_0}\, dz = 2\pi i f(z_0) + \oint_\gamma \frac{f(z) - f(z_0)}{z - z_0}\, dz.$$

The integral representation is proved if we can show that the last integral is zero. Write $\gamma(t) = z_0 + re^{it}$ for $0 \le t \le 2\pi$. Then

$$\left| \oint_\gamma \frac{f(z) - f(z_0)}{z - z_0}\, dz \right| = \left| \int_0^{2\pi} \frac{f(z_0 + re^{it}) - f(z_0)}{re^{it}} ire^{it}\, dt \right|$$

$$= \left| \int_0^{2\pi} [f(z_0 + re^{it}) - f(z_0)]\, dt \right|$$

$$\le \int_0^{2\pi} \left| f(z_0 + re^{it}) - f(z_0) \right| dt$$

$$\le 2\pi \left(\max_{0 \le t \le 2\pi} \left| f(z_0 + re^{it}) - f(z_0) \right| \right).$$

By continuity of $f(z)$ at z_0, $f(z_0 + re^{it}) \to f(z_0)$ as $r \to 0$, so the term on the right in this inequality has limit zero as $r \to 0$. Therefore we can make

$$\left| \oint_\gamma \frac{f(z) - f(z_0)}{z - z_0}\, dz \right|$$

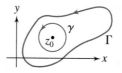

FIGURE 22.37

arbitrarily small by choosing r sufficiently small. But this integral is independent of r by the deformation theorem. Hence

$$\left| \oint_\gamma \frac{f(z) - f(z_0)}{z - z_0} \, dz \right| = 0,$$

so

$$\oint_\gamma \frac{f(z) - f(z_0)}{z - z_0} \, dz = 0$$

and the theorem is proved. ■

The integral representation gives some idea of how strong the condition of differentiability is for complex functions. The integral

$$\oint_\Gamma \frac{f(z)}{z - z_0} \, dz$$

equals $2\pi i f(z_0)$ and so determines $f(z_0)$ at each z_0 enclosed by Γ. But the value of this integral depends only on the values of $f(z)$ on Γ. Thus, for a differentiable function, knowing function values on Γ determines the values of the function at all points enclosed by Γ. There is no analogous result for differentiable real functions. Knowing the values of a differentiable real function $\varphi(x)$ at the endpoints of an interval in general gives no information about values of this function throughout the interval.

22.4.4   Cauchy's Integral Formula for Higher Derivatives

We will now show that a complex function that is differentiable on an open set must have derivatives of all orders on this set. There is no result like this for real functions. A real function that is differentiable need not have a second derivative. And if it has a second derivative, it need not have a third, and so on.

Not only does a differentiable complex function have derivatives of all orders, we will show that the nth derivative of the function at a point is also given by an integral formula, very much in the spirit of Cauchy's integral formula.

THEOREM 22.10   *Cauchy's Integral Formula for Higher Derivatives*

Let f be differentiable on an open set G. Then f has derivatives of all orders at each point of G. Further, if Γ is a closed path in G enclosing only points of G, and z_0 is any point enclosed by Γ, then

$$f^{(n)}(z_0) = \frac{n!}{2\pi i} \oint_\Gamma \frac{f(z)}{(z - z_0)^{n+1}} \, dz. \quad ■$$

The integral on the right is exactly what we would obtain by differentiating Cauchy's integral formula for $f(z_0)$, n times with respect to z_0, under the integral sign.

As with the integral formula, this conclusion is often used to evaluate integrals.

EXAMPLE 22.22

Evaluate

$$\oint_\Gamma \frac{e^{z^3}}{(z - i)^3} \, dz$$

with Γ any path not passing through i.

If Γ does not enclose i then this integral is zero by Cauchy's theorem, since the only point at which $e^{z^3}/(z-i)^3$ fails to be differentiable is i. Thus suppose Γ encloses i. Because the factor $z-i$ occurs to the third power in the denominator, use $n=2$ in the theorem, with $f(z)=e^{z^3}$, to get

$$\oint_{\Gamma} \frac{e^{z^3}}{(z-i)^3}\, dz = \frac{2\pi i}{2!} f^{(2)}(i) = \pi i f''(i).$$

Now

$$f'(z) = 3z^2 e^{z^3} \quad \text{and} \quad f''(z) = 6z e^{z^3} + 9z^4 e^{z^3},$$

so

$$\oint_{\Gamma} \frac{e^{z^3}}{(z-i)^3}\, dz = \pi i \left[6i e^{-i} + 9 e^{-i}\right] = (-6 + 9i)\pi e^{-i}. \quad \blacksquare$$

The theorem can be proved by induction on n, but we will not carry out the details.

22.4.5 Bounds on Derivatives

Cauchy's integral formula for higher derivatives can be used to obtain bounds on derivatives of all orders.

THEOREM 22.11

Let f be differentiable on an open set G. Let z_0 be a point of G and let the open disk of radius r about z_0 be in G. Suppose

$$|f(z)| \le M$$

for z on the circle of radius r about z_0. Then, for any positive integer n,

$$\left| f^{(n)}(z_0) \right| \le \frac{Mn!}{r^n}.$$

Proof Let $\gamma(t) = z_0 + r e^{it}$ for $0 \le t \le 2\pi$. Then $\left| f(z_0 + r e^{it}) \right| \le M$ for $0 \le t \le 2\pi$. By Theorems 22.10 and 22.5,

$$\left| f^{(n)}(z_0) \right| = \frac{n!}{2\pi} \left| \oint_{\gamma} \frac{f(z)}{(z-z_0)^{n+1}}\, dz \right| = \frac{n!}{2\pi} \left| \int_0^{2\pi} \frac{f(z_0 + r e^{it})}{r^{n+1} e^{i(n+1)t}} i r e^{it}\, dt \right|$$

$$\le \frac{n!}{2\pi} \int_0^{2\pi} \frac{\left| f(z_0 + r e^{it}) \right|}{r^n}\, dt \le \frac{n!}{2\pi} (2\pi) M \frac{1}{r^n} = \frac{Mn!}{r^n}. \quad \blacksquare$$

As an application of this theorem, we will prove Liouville's theorem on bounded, differentiable functions.

THEOREM 22.12 *Liouville*

Let f be a bounded function that is differentiable for all z. Then f is a constant function. $\blacksquare$

Previously we noted that $\sin(z)$ is not a bounded function in the complex plane as it is on the real line. This is consistent with Liouville's theorem. Since $\sin(z)$ is differentiable for all z and is clearly not a constant function, it cannot be bounded.

Here is a proof of Liouville's theorem.

Proof   Suppose $|f(z)| \leq M$ for all complex z. Choose any number z_0 and any $r > 0$. By Theorem 22.11, with $n = 1$,

$$\left|f'(z_0)\right| \leq \frac{M}{r}.$$

Since f is differentiable in the entire complex plane, r may be chosen as large as we like, so $\left|f'(z_0)\right|$ must be less than any positive number. We conclude that $\left|f'(z_0)\right| = 0$, hence $f'(z_0) = 0$. Since z_0 is any number, then $f(z) =$ constant. ∎

Liouville's theorem can be used to give a simple proof of the fundamental theorem of algebra. This theorem states that any nonconstant complex polynomial $p(z) = a_0 + a_1 z + \cdots + a_n z^n$ has a complex root. That is, for some number ζ, $p(\zeta) = 0$. From this it can be further shown that if $a_n \neq 0$, then $p(z)$ must have exactly n roots, counting each root k times in the list if its multiplicity is k. For example, $p(z) = z^2 - 6z + 9$ has exactly two roots, 3 and 3 (a root of multiplicity 2).

This fundamental theorem assumes only elementary terminology for its statement and is usually included, in some form, in the high school mathematics curriculum. The leading nineteenth-century mathematician Carl Friedrich Gauss considered the theorem so important that he devised many proofs (some say nearly 20) over his lifetime. But even today rigorous proofs of the theorem require mathematical terms and devices far beyond those needed to state it.

To prove the theorem using Liouville's theorem, suppose $p(z)$ is a nonconstant complex polynomial and that $p(z) \neq 0$ for all z. Then $1/p(z)$ is differentiable for all z.

Let $p(z) = a_0 + a_1 z + \cdots + a_n z^n$ with $n \geq 1$ and $a_n \neq 0$. We will show that $1 \neq |p(z)|$ is bounded for all z. Since

$$a_n z^n = p(z) - a_0 - a_1 z - \cdots - a_{n-1} z^{n-1},$$

then

$$|a_n| \, |z|^n \leq |p(z)| + |a_0| + |a_1| \, |z| + \cdots + |a_{n-1}| \, |z|^{n-1}.$$

Then, for $|z| \geq 1$,

$$|p(z)| \geq |a_n| \, |z|^n - \left(|a_0| + |a_1| \, |z| + \cdots + |a_{n-1}| \, |z|^{n-1}\right)$$

$$= |z|^{n-1} \left(|a_n| \, |z| - \frac{|a_0|}{|z|^{n-1}} - \frac{|a_1|}{|z|^{n-2}} - \cdots - \frac{|a_{n-1}|}{|z|^{n-n}}\right)$$

$$\geq |z|^{n-1} \left(|a_n| \, |z| - |a_0| - |a_1| - \cdots - |a_{n-1}|\right).$$

But then

$$\frac{1}{|p(z)|} \leq \frac{1}{|z|^{n-1} \left(|a_n| \, |z| - |a_0| - |a_1| - \cdots - |a_{n-1}|\right)} \to 0$$

as $|z| \to \infty$. Therefore, $\lim_{n \to \infty} 1/|p(z)| = 0$. This implies that for some positive number R,

$$\frac{1}{|p(z)|} < 1 \quad \text{if } |z| > R.$$

But the closed disk $|z| \leq R$ is compact, and $1/|p(z)|$ is continuous, so $1/|p(z)|$ is bounded on this disk by Theorem 21.1. Therefore, for some M,

$$\frac{1}{|p(z)|} \leq M \quad \text{for } |z| \leq R.$$

Now we have $1/|p(z)|$ bounded both inside and outside $|z| \leq R$. Putting these bounds together,

$$\frac{1}{|p(z)|} \leq M + 1 \quad \text{for all } z,$$

both in $|z| \leq R$ and in $|z| \geq R$. This makes $1/p(z)$ a bounded function that is differentiable for all z. By Liouville's theorem, $1/p(z)$ must be constant, a contradiction. Therefore there must be some complex ζ such that $p(\zeta) = 0$, proving the fundamental theorem of algebra.

Complex analysis provides several proofs of this theorem. Later we will see one using Rouché's theorem and then another using a technique for evaluating real integrals of rational functions involving sines and cosines.

22.4.6 An Extended Deformation Theorem

The deformation theorem allows us to deform one closed path of integration, Γ, into another, γ, without changing the value of the line integral of a differentiable function f. A crucial condition for this process is that no stage of the deformation should pass over a point at which f fails to be differentiable. This means that f needs to be differentiable on both curves and the region between them. We will now extend this result to the case that Γ encloses any finite number of disjoint closed paths. As usual, unless explicitly stated otherwise, all closed paths are assumed to be oriented positively (counterclockwise).

THEOREM 22.13 *Extended Deformation Theorem*

Let Γ be a closed path. Let $\gamma_1, \ldots, \gamma_n$ be closed paths enclosed by Γ. Assume that no two of Γ, $\gamma_1, \ldots, \gamma_n$ intersect, and no point interior to any γ_j is interior to any other γ_k. Let f be differentiable on an open set containing Γ, each γ_j, and all points that are both interior to Γ and exterior to each γ_j. Then,

$$\oint_\Gamma f(z)\, dz = \sum_{j=1}^n \oint_{\gamma_j} f(z)\, dz. \quad \blacksquare$$

This is the deformation theorem in the case $n = 1$. Figure 22.38 shows a typical scenario covered by this theorem. With the curves as shown (and the differentiability assumption on f), the integral of f about Γ is the sum of the integrals of f about each of the closed curves $\gamma_1, \ldots, \gamma_n$.

We will outline a proof after an illustration of a typical use of the theorem.

EXAMPLE 22.23

Consider

$$\oint_\Gamma \frac{z}{(z+2)(z-4i)}\, dz,$$

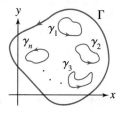

FIGURE 22.38

where Γ is a closed path enclosing -2 and $4i$. We will evaluate this integral using the extended deformation theorem. Place a circle γ_1 about -2 and a circle γ_2 about $4i$ of sufficiently small radii that neither circle intersects the other or Γ and each is enclosed by Γ (Figure 22.39). Then

$$\oint_\Gamma \frac{z}{(z+2)(z-4i)}\,dz = \oint_{\gamma_1} \frac{z}{(z+2)(z-4i)}\,dz + \oint_{\gamma_2} \frac{z}{(z+2)(z-4i)}\,dz.$$

Use a partial fractions decomposition to write

$$\frac{z}{(z+2)(z-4i)} = \frac{\frac{1}{5}-\frac{2}{5}i}{z+2} + \frac{\frac{4}{5}+\frac{2}{5}i}{z-4i}.$$

Then

$$\oint_\Gamma \frac{z}{(z+2)(z-4i)}\,dz = \left(\frac{1}{5}-\frac{2}{5}i\right)\oint_{\gamma_1}\frac{1}{z+2}\,dz + \left(\frac{4}{5}+\frac{2}{5}i\right)\oint_{\gamma_1}\frac{1}{z-4i}\,dz$$

$$+ \left(\frac{1}{5}-\frac{2}{5}i\right)\oint_{\gamma_2}\frac{1}{z+2}\,dz + \left(\frac{4}{5}+\frac{2}{5}i\right)\oint_{\gamma_2}\frac{1}{z-4i}\,dz.$$

On the right, the second and third integrals are zero by Cauchy's theorem (γ_1 does not enclose $4i$ and γ_2 does not enclose -2). The first and fourth integrals equal $2\pi i$ by Example 22.19. Therefore,

$$\oint_\Gamma \frac{z}{(z+2)(z-4i)}\,dz = 2\pi i\left[\left(\frac{1}{5}-\frac{2}{5}i\right)+\left(\frac{4}{5}+\frac{2}{5}i\right)\right] = 2\pi i. \ \blacksquare$$

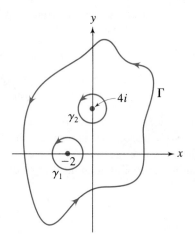

FIGURE 22.39

A proof of the theorem can be modeled after the proof of the deformation theorem.

Proof As suggested in Figure 22.40, draw line segments L_1 from Γ to γ_1, L_2 from γ_1 to $\gamma_2,\ldots,L_n$ from γ_{n-1} to γ_n, and, finally, L_{n+1} from γ_n to Γ. Form the closed paths Φ and Δ shown separately in Figures 22.41 and 22.42, and as they actually appear in Figure 22.43. Then

$$\oint_\Phi f(z)\,dz + \oint_\Delta f(z)\,dz = 0,$$

both integrals being zero by Cauchy's theorem. (By the hypotheses of the theorem, f is differentiable on and inside both Φ and Δ.)

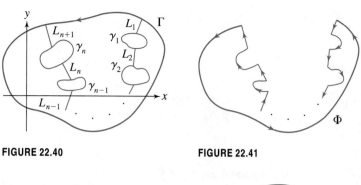

FIGURE 22.40

FIGURE 22.41

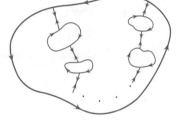

FIGURE 22.42

FIGURE 22.43

In this sum of integrals over Φ and Δ, each line segment L_j is integrated over in both directions, hence the contributions of the integrals over these segments is zero. Further, in this sum we retrieve the integral of $f(z)$ over all of Γ and all of each γ_j, with the orientation counterclockwise on Γ and clockwise on each γ_j (note the orientations in Figure 22.43). Reversing the orientations on the γ_j's, so that all paths are oriented counterclockwise, the last sum becomes

$$\oint_\Gamma f(z)\,dz - \sum_{j=1}^n \oint_{\gamma_j} f(z)\,dz = 0,$$

yielding the conclusion of the theorem. ■

SECTION 22.4 PROBLEMS

In each of Problems 1 through 20, evaluate $\int f(z)\,dz$ for the given function and path. These problems may involve Cauchy's theorem, Cauchy's integral formulas, and/or the deformation theorems.

1. $f(z) = \dfrac{z^4}{z - 2i}$; Γ is any closed path enclosing $2i$

2. $f(z) = \dfrac{\sin(z^2)}{z - 5}$; Γ is any closed path enclosing 5

3. $f(z) = \dfrac{z^2 - 5z + i}{z - 1 + 2i}$; Γ is the circle $|z| = 3$

4. $f(z) = \dfrac{2z^3}{(z - 2)^2}$; Γ is the rectangle having vertices $4 \pm i$ and $-4 \pm i$

5. $f(z) = \dfrac{ie^z}{(z - 2 + i)^2}$; Γ is the circle $|z - 1| = 4$

6. $f(z) = \dfrac{\cos(z - i)}{(z + 2i)^3}$; Γ is any closed path enclosing $-2i$

7. $f(z) = \dfrac{z \sin(3z)}{(z + 4)^3}$; Γ is the circle $|z - 2i| = 9$

8. $f(z) = 2i\bar{z}\,|z|$; Γ is the line segment from 1 to $-i$

9. $f(z) = -\dfrac{(2 + i)\sin(z^4)}{(z + 4)^2}$; Γ is any closed path enclosing -4

10. $f(z) = (z - i)^2$; Γ is the semicircle of radius 1 about 0 from i to $-i$

11. $f(z) = \text{Re}(z+4)$; Γ is the line segment from $3+i$ to $2-5i$

12. $f(z) = \dfrac{3z^2 \cosh(z)}{(z+2i)^2}$; Γ is the circle of radius 8 about 1

13. $f(z) = z + i\,|z|$; Γ is the line segment from 0 to $-2i$

14. $f(z) = 1 - z^2$; Γ is the square having vertices 0, 1, $1+i$, and i

15. $f(z) = \text{Im}(iz)$; Γ is the unit circle about the origin

16. $f(z) = 3iz^2 - 1$; Γ is the line segment from 1 to $2+5i$

17. $f(z) = \dfrac{\cosh(\sinh(z))}{z+3}$; Γ is any closed path enclosing -3

18. $f(z) = \dfrac{z^2 - \cos(3z)}{(z+4i)^4}$; Γ is the triangle having vertices $0, -1 - 8i$, and $1 - 8i$

19. $f(z) = \dfrac{\cos(z) - \sin(z)}{(z+i)^4}$; Γ is the circle of radius 7 about the origin

20. $f(z) = \dfrac{1}{e^z(z+3i)}$; Γ is the circle of radius 5 about $-2i$

21. Evaluate

$$\int_0^{2\pi} e^{\cos(\theta)} \cos(\sin(\theta))\, d\theta.$$

Hint: Consider $\oint_\Gamma (e^z/z)\, dz$, with Γ the unit circle about the origin. Evaluate this integral once using Cauchy's integral formula, then again directly by using the coordinate functions for Γ.

22. Prove Morera's theorem. Let f be continuous on a simply connected domain S. Suppose $\oint_\Gamma f(z)\, dz = 0$ for every closed path in S. Then f is differentiable on S. (This is nearly a converse of Cauchy's theorem.) *Hint:* Choose any a in S. Show that if Γ and Θ are any paths in S from a to z in S, then $\int_\Gamma f(w)\, dw = \int_\Theta f(w)\, dw$. Let $F(z) = \int_\Gamma f(w)\, dw$. This defines a complex function on S. Show that $F'(z) = f(z)$.

23. Let f be differentiable on an open set G containing a closed path Γ and all points enclosed by Γ. Prove that

$$\oint_\Gamma \frac{f'(z)}{z-a}\, dz = \oint_\Gamma \frac{f(z)}{(z-a)^2}\, dz$$

for any a enclosed by Γ.

24. Use the extended form of the deformation theorem to evaluate $\oint_\Gamma ((z-4i)/(z^3 + 4z))\, dz$, where Γ is a closed path enclosing the origin, $2i$, and $-2i$.

CHAPTER 23

Series Representations of Functions

We will now develop two kinds of representations of an $f(z)$ in series of powers of $z - z_0$. The first series will contain only nonnegative integer powers, hence is a power series, and applies when f is differentiable at z_0. The second will contain negative integer powers of $z - z_0$ as well and will be used when f is not differentiable at z_0.

23.1 Power Series Representations

We know that a power series that converges in an open disk, or perhaps the entire plane, defines a function that is infinitely differentiable within this disk or the plane. We will now go the other way and show that a function that is differentiable on an open disk is represented by a power series expansion about the center of this disk. This will have important applications, including information about zeros of functions and the maximum value that can be taken by the modulus $|f(z)|$ of a differentiable function.

THEOREM 23.1 *Taylor Series*

Let f be differentiable on an open disk D about z_0. Then, for each z in D,

$$f(z) = \sum_{n=0}^{\infty} \frac{f^{(n)}(z_0)}{n!}(z - z_0)^n. \quad \blacksquare$$

The series on the right is the *Taylor series* of $f(z)$ about z_0, and the number $f^{(n)}(z_0)/n!$ is the nth *Taylor coefficient* of $f(z)$ at z_0. The theorem asserts that the Taylor series of $f(z)$ converges to $f(z)$, hence represents $f(z)$, within this disk.

1101

Proof   Let z be any point of D and R the radius of D. Choose a number r with $|z - z_0| < r < R$ and let γ be the circle of radius r about z_0 (Figure 23.1). By the Cauchy integral formula, using w for the variable of integration on γ,

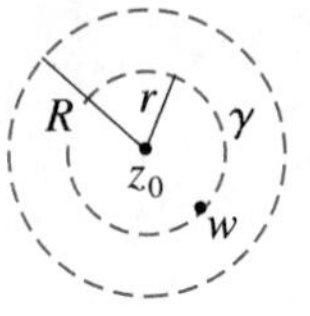

$$f(z) = \frac{1}{2\pi i} \oint_\gamma \frac{f(w)}{w - z} \, dw.$$

Now

FIGURE 23.1

$$\frac{1}{w - z} = \frac{1}{w - z_0 - (z - z_0)} = \frac{1}{w - z_0} \frac{1}{1 - (z - z_0)/(w - z_0)}.$$

Since w is on γ, and z is enclosed by γ, then

$$\left| \frac{z - z_0}{w - z_0} \right| < 1,$$

so we can use a convergent geometric series to write

$$\frac{1}{w - z} = \frac{1}{w - z_0} \sum_{n=0}^{\infty} \left(\frac{z - z_0}{w - z_0} \right)^n = \sum_{n=0}^{\infty} \frac{1}{(w - z_0)^{n+1}} (z - z_0)^n.$$

Then

$$\frac{f(w)}{w - z} = \sum_{n=0}^{\infty} \frac{f(w)}{(w - z_0)^{n+1}} (z - z_0)^n. \tag{23.1}$$

Since f is continuous on γ, for some M, $|f(w)| \le M$ for w on γ. Further, $|w - z_0| = r$, so

$$\left| \frac{f(w)}{(w - z_0)^{n+1}} (z - z_0)^n \right| \le M \frac{1}{r} \left(\frac{|z - z_0|}{r} \right)^n.$$

Designate

$$M \frac{1}{r} \left(\frac{|z - z_0|}{r} \right)^n = M_n.$$

Then $\sum_{n=0}^{\infty} M_n$ converges (this series is a constant times a convergent geometric series). By Theorem 22.6, the series in equation (23.1) can be integrated term-by-term to yield

$$f(z) = \frac{1}{2\pi i} \oint_\gamma \frac{f(w)}{w - z} \, dw = \frac{1}{2\pi i} \oint_\gamma \left(\sum_{n=0}^{\infty} \frac{f(w)}{(w - z_0)^{n+1}} (z - z_0)^n \right) dw$$

$$= \sum_{n=0}^{\infty} \left(\frac{1}{2\pi i} \oint_\gamma \frac{f(w)}{(w - z_0)^{n+1}} \, dw \right) (z - z_0)^n = \sum_{n=0}^{\infty} \frac{f^{(n)}(z_0)}{n!} (z - z_0)^n,$$

in which we used Cauchy's integral formula for the nth derivative to write the coefficient in the last series. This proves the theorem. ∎

A complex function is said to be *analytic* at z_0 if it has a power series expansion in some open disk about z_0. We have just proved that a function that is differentiable on an open disk about a point is analytic at that point.

We only compute the coefficients of a Taylor series by derivative or integral formulas when other means fail. When possible, we use known series and operations such as differentiation and integration to derive a series representation. This strategy makes use of the uniqueness of power series representations.

Suppose, in some disk $|z - z_0| < r$,

$$\sum_{n=0}^{\infty} c_n (z - z_0)^n = \sum_{n=0}^{\infty} d_n (z - z_0)^n.$$

Then, for $n = 0, 1, 2, \ldots, c_n = d_n$.

Proof   If we let $f(z)$ be the function defined in this disk by both power series, then

$$c_n = \frac{f^{(n)}(z_0)}{n!} = d_n. \quad \blacksquare$$

This means that no matter what method is used to find a power series for $f(z)$ about z_0, the end result is the Taylor series.

EXAMPLE 23.1

Find the Taylor expansion of e^z about i.

We know that for all z,

$$e^z = \sum_{n=0}^{\infty} \frac{1}{n!} z^n.$$

For an expansion about i, the power series must be in terms of powers of $z - i$. Thus write

$$e^z = e^{z-i+i} = e^i e^{z-i} = \sum_{n=0}^{\infty} e^i \frac{1}{n!} (z - i)^n.$$

This series converges for all z. $\blacksquare$

In this example, it would have been just as easy to compute the Taylor coefficients directly:

$$c_n = \frac{f^{(n)}(i)}{n!} = \frac{e^i}{n!}.$$

EXAMPLE 23.2

Write the Maclaurin series for $\cos(z^3)$.

A Maclaurin expansion is a Taylor series about zero. For all z,

$$\cos(z) = \sum_{n=0}^{\infty} \frac{(-1)^n}{(2n)!} z^{2n}.$$

All we need to do is replace z with z^3:

$$\cos(z^3) = \sum_{n=0}^{\infty} \frac{(-1)^n}{(2n)!} (z^3)^{2n} = \sum_{n=0}^{\infty} \frac{(-1)^n}{(2n)!} z^{6n}.$$

Since this is an expansion about the origin, it is the expansion we seek. $\blacksquare$

EXAMPLE 23.3

Expand

$$\frac{2i}{4 + iz}$$

in a Taylor series about $-3i$.

We want a series in powers of $z + 3i$. Do some algebraic manipulation and then use a geometric series. In order to get $z + 3i$ into the picture, write

$$\frac{2i}{4 + iz} = \frac{2i}{4 + i(z + 3i) + 3} = \frac{2i}{7 + i(z + 3i)} = \frac{2i}{7} \frac{1}{1 + (i/7)(z + 3i)}.$$

If $|t| < 1$, then

$$\frac{1}{1 + t} = \frac{1}{1 - (-t)} = \sum_{n=0}^{\infty} (-t)^n = \sum_{n=0}^{\infty} (-1)^n t^n.$$

With $t = (z + 3i)i/7$, we have

$$\frac{1}{1 + (i/7)(z + 3i)} = \sum_{n=0}^{\infty} (-1)^n \left(\frac{i}{7}(z + 3i) \right)^n.$$

Therefore,

$$\frac{2i}{4 + iz} = \frac{2i}{7} \sum_{n=0}^{\infty} (-1)^n \left(\frac{i}{7} \right)^n (z + 3i)^n = \sum_{n=0}^{\infty} \frac{2(-1)^n i^{n+1}}{7^{n+1}} (z + 3i)^n.$$

Because this is a series expansion of the function about $-3i$, it is the Taylor series about $-3i$. This series converges for

$$\left| \frac{i}{7}(z + 3i) \right| < 1,$$

or

$$|z + 3i| < 7.$$

Thus z must be in the open disk of radius 7 about $-3i$. The radius of convergence of this series is 7. ∎

From Section 21.2, we can differentiate a Taylor series term-by-term within its open disk of convergence. This is sometimes useful in deriving the Taylor expansion of a function.

EXAMPLE 23.4

Find the Taylor expansion of $f(z) = 1/(1 - z)^3$ about the origin.

We could do this by algebraic manipulation, but it is easier to begin with the familiar geometric series.

$$g(z) = \frac{1}{1 - z} = \sum_{n=0}^{\infty} z^n \quad \text{for } |z| < 1.$$

Then

$$g'(z) = \frac{1}{(1 - z)^2} = \sum_{n=1}^{\infty} n z^{n-1}$$

and

$$g''(z) = \frac{2}{(1-z)^3} = \sum_{n=2}^{\infty} n(n-1)z^{n-2} = \sum_{n=0}^{\infty}(n+1)(n+2)z^n$$

for $|z| < 1$. Then

$$f(z) = \sum_{n=0}^{\infty} \frac{1}{2}(n+1)(n+2)z^n \quad \text{for } |z| < 1. \quad \blacksquare$$

When $f(z)$ is expanded in a power series about z_0, the radius of convergence of the series will be the distance from z_0 to the nearest point at which $f(z)$ is not differentiable. Think of a disk expanding uniformly from z_0, free to continue its expansion until it hits a point at which $f(z)$ is not differentiable.

For example, suppose $f(z) = 2i/(4+iz)$ and we want a Taylor expansion about $-3i$. The only point at which $f(z)$ is not defined is $4i$, so the radius of convergence of this series will be the distance between $-3i$ and $4i$, or 7. We obtained this result previously from the Taylor expansion $\sum_{n=0}^{\infty}(2(-1)^n i^{n+1}/7^{n+1})(z+3i)^n$ of $f(z)$.

EXAMPLE 23.5

We will find the radius of convergence of the Taylor series of $\csc(z)$ about $3 - 4i$.

Since $\csc(z) = 1/\sin(z)$, this function is differentiable except at $z = n\pi$, with n any integer. As Figure 23.2 illustrates, π is the nearest point to $3 - 4i$ at which $\csc(z)$ is not differentiable. The distance between π and $3 - 4i$ is

$$\sqrt{(\pi - 3)^2 + 16},$$

and this is the radius of convergence of the expansion of $\csc(z)$ about $3 - 4i$. $\blacksquare$

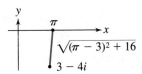

FIGURE 23.2

Existence of a power series expansion implies the existence of an antiderivative.

THEOREM 23.3

Let f be differentiable in an open disk D about z_0. Then there is a differentiable function F such that $F'(z) = f(z)$ for all z in D.

Proof We know that f has a power series expansion in D:

$$f(z) = \sum_{n=0}^{\infty} c_n (z - z_0)^n.$$

Let

$$F(z) = \sum_{n=0}^{\infty} \frac{1}{n+1} c_n (z - z_0)^{n+1}$$

for z in D. It is routine to check that this power series has a radius of convergence at least as large as the radius of D and that

$$F'(z) = f(z)$$

for z in D. ∎

23.1.1 Isolated Zeros and the Identity Theorem

We will use the Taylor series representation of a differentiable function to derive important information about its zeros.

DEFINITION 23.1

A number ζ is a zero of f if $f(\zeta) = 0$.
 A zero ζ of f is isolated if there is an open disk about ζ containing no other zero of f.

For example, the zeros of $\sin(z)$ are $n\pi$, with n any integer. These zeros are all isolated. By contrast, let

$$f(z) = \begin{cases} \sin(1/z) & \text{if } z \neq 0 \\ 0 & \text{if } z = 0 \end{cases}.$$

The zeros of f are 0 and the numbers $1/n\pi$, with n any nonzero integer. However, 0 is not an isolated zero, since any open disk about zero contains other zeros, $1/n\pi$, for n sufficiently large.

We will show that the behavior of the zeros in this example disqualifies f from being differentiable at 0.

THEOREM 23.4

Let f be differentiable on an open set G and let ζ be a zero of f in G. Then either ζ is an isolated zero of f or there is an open disk about ζ on which $f(z)$ is identically zero. ∎

This means that a differentiable function that is not identically zero on some disk can have only isolated zeros.

Proof Write the power series expansion of $f(z)$ about ζ,

$$f(z) = \sum_{n=0}^{\infty} c_n (z - \zeta)^n,$$

in some open disk D centered at ζ. Now consider two cases.

Case 1 If each $c_n = 0$, then $f(z) = 0$ throughout D.

Case 2 Suppose, instead, that some coefficients in the power series are not zero. Let m be the smallest integer such that $c_m \neq 0$. Then $c_0 = c_1 = \cdots = c_{m-1} = 0$ and $c_m \neq 0$. For z in D,

$$f(z) = \sum_{n=m}^{\infty} c_n (z - \zeta)^n = \sum_{n=0}^{\infty} c_{n+m} (z - \zeta)^{n+m} = (z - \zeta)^m \sum_{n=0}^{\infty} c_{n+m} (z - \zeta)^n.$$

Now $\sum_{n=0}^{\infty} c_{n+m}(z - \zeta)^n$ is a power series and so defines a differentiable function $g(z)$ for z in D. Further,

$$f(z) = (z - \zeta)^m g(z).$$

But $g(\zeta) = c_m \neq 0$, so there is some open disk K about ζ in which $g(z) \neq 0$. Therefore, for z in K, $f(z) \neq 0$ also, hence ζ is an isolated zero of f. ∎

This proof contained additional information about zeros that will be useful later. The number m in the proof is called the *order* of the zero ζ of $f(z)$. It is the smallest integer m such that the coefficient c_m in the expansion of $f(z)$ about ζ is nonzero. Now recall that $c_n = f^{(n)}(\zeta)/n!$. Thus $c_0 = c_1 = \cdots = c_{m-1} = 0$ implies that

$$f(\zeta) = f'(\zeta) = \cdots = f^{(m-1)}(\zeta) = 0,$$

while $c_m \neq 0$ implies that

$$f^{(m)}(\zeta) \neq 0.$$

In summary, an isolated zero ζ of f has order m if the function and its first $m - 1$ derivatives vanish at ζ, but the mth derivative at ζ is nonzero. Put another way, the order of the zero ζ is the lowest-order derivative of f that does not vanish at ζ.

We also derived the important fact that if ζ is an isolated zero of order m of f, then we can write

$$f(z) = (z - \zeta)^m g(z),$$

where g is also differentiable in some disk about ζ, and $g(\zeta) \neq 0$.

EXAMPLE 23.6

Consider $f(z) = z^2 \cos(z)$. 0 is an isolated zero of this differentiable function. Compute

$$f'(z) = 2z \cos(z) - z^2 \sin(z)$$

and

$$f''(z) = 2\cos(z) - 4z \sin(z) + z^2 \cos(z).$$

Observe that $f(0) = f'(0) = 0$ while $f''(0) \neq 0$. Thus 0 is a zero of order 2 of f. In this case, we already have

$$f(z) = (z - 0)^2 g(z)$$

with $g(0) \neq 0$, since we can choose $g(z) = \cos(z)$. ∎

EXAMPLE 23.7

We have to exercise some care in identifying the order of a zero. Consider $f(z) = z^2 \sin(z)$. 0 is an isolated zero of f. Compute

$$f'(z) = 2z \sin(z) + z^2 \cos(z),$$

$$f''(z) = 2\sin(z) + 4z \cos(z) - z^2 \sin(z),$$

$$f^{(3)}(z) = 2\cos(z) + 4\cos(z) - 4z \sin(z) - 2z \sin(z) - z^2 \cos(z)$$

$$= 6\cos(z) - 6z \sin(z) - z^2 \cos(z).$$

Then

$$f(0) = f'(0) = f''(0) = 0$$

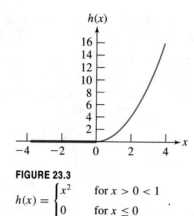

FIGURE 23.3

$$h(x) = \begin{cases} x^2 & \text{for } x > 0 < 1 \\ 0 & \text{for } x \leq 0 \end{cases}.$$

while

$$f^{(3)}(0) \neq 0.$$

This means that 0 is a zero of order 3. We can write

$$f(z) = z^2 \sum_{n=0}^{\infty} \frac{(-1)^n}{(2n+1)!} z^{2n+1} = z^3 \sum_{n=0}^{\infty} \frac{(-1)^n}{(2n+1)!} z^{2n} = z^3 g(z),$$

where

$$g(z) = \sum_{n=0}^{\infty} \frac{(-1)^n}{(2n+1)!} z^{2n}$$

is differentiable (in this example, for all z) and $g(0) = 1 \neq 0$. ∎

As a result of Theorem 23.3, we can show that if a differentiable complex function vanishes on a convergent sequence of points in a domain (open, connected set), then the function is *identically zero throughout the entire domain*. This is a very strong result, for which there is no analogue for real functions. For example, consider

$$h(x) = \begin{cases} 0 & \text{for } x \leq 0 \\ x^2 & \text{for } x > 0 \end{cases},$$

whose graph is shown in Figure 23.3. This function is differentiable for all real x and is identically zero on half the line, but is not identically zero over the entire line. Another difference between differentiability for real and complex functions is evident in this example. While h is differentiable, it does not have a power series expansion about 0. By contrast, a complex function that is differentiable on an open set has a power series expansion about each point of the set.

THEOREM 23.5

Let f be differentiable on a domain G. Suppose $\{z_n\}$ is a sequence of distinct zeros of f in G, converging to a point of G. Then $f(z) = 0$ for all z in G.

Proof   Suppose $z_n \to \zeta$ in G. Since f is continuous, $f(z_n) \to f(\zeta)$. But each $f(z_n) = 0$, so $f(\zeta) = 0$ also and ζ must also be a zero of f in G. This means that ζ is not an isolated zero, so by Theorem 23.4, there must be an open disk D about ζ in which $f(z)$ is identically zero (Figure 23.4).

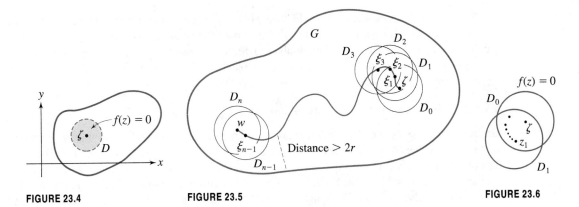

FIGURE 23.4   FIGURE 23.5   FIGURE 23.6

We want to prove that this forces $f(z) = 0$ for all z in G. To do this, let w be any point of G. We will show that $f(w) = 0$.

Since G is connected, there is a path Γ in G from ζ to w. Choose a number r such that every point of Γ is at a distance at least $2r$ from the boundary of G, and also choose r to be less than the radius of D. Now walk along Γ from ζ to w, along the way selecting points at distance less than r from each other. This yields points $\zeta = \xi_0, \xi_2, \ldots, \xi_n = w$ on Γ, as in Figure 23.5. Form an open disk D_j of radius r about each ξ_j. (By choice of r, none of these disks reaches the boundary of G.) Each ξ_{j-1} is in D_{j-1}, D_j, and D_{j+1} for $j = 1, \ldots, n - 1$. Further, $\xi_0 = \zeta$ is in D_0 and D_1, and w is in D_{n-1} and D_n.

Since ξ_1 is in D_0 and D_1, there is a sequence of points in both D_0 and D_1 converging to ξ_1 (Figure 23.6). But $f(z)$ is identically zero on D_0, so $f(z)$ vanishes on this sequence. Since this sequence is also in D_1, $f(z) = 0$ for all z in D_1.

Now ξ_2 is in D_1 and D_2. Choose a sequence of points common to both of these disks and converging to ξ_2. Since $f(z)$ is identically zero on D_1, then $f(z) = 0$ at each point of this sequence. But since this sequence is also in D_2, then $f(z)$ is identically zero on D_2.

Continuing in this way, walk along Γ from ζ to w. We find that $f(z)$ is identically zero on each of the disks along the way. Finally, $f(z)$ is zero on D_n. But w is in D_n, so $f(w) = 0$, and therefore $f(z) = 0$ for all z in G. ∎

This theorem leads immediately to the conclusion that two differentiable functions that agree on a convergent sequence in a domain must be the same function. This is called the *identity theorem*.

COROLLARY 23.1 *Identity Theorem*

Let f and g be differentiable on a domain G. Suppose $f(z)$ and $g(z)$ agree on a convergent sequence of distinct points of G. Then $f(z) = g(z)$ for all z in G.

Proof Apply Theorem 23.5 to the differentiable function $f - g$. ∎

To get some idea of the power of this result, consider the problem of defining the complex sine function $\sin(z)$ so that it agrees with the real sine function when z is real. How many ways can this be done?

Put another way, is it possible to invent two distinct differentiable complex functions, f and g, defined for all z, such that, when x is real,

$$f(x) = g(x) = \sin(x)?$$

If this could be done, then we would have $f(z) = g(z)$ on a convergent sequence of complex numbers (chosen along the real line) in a domain (the entire plane), so necessarily $f = g$. There can be only one extension of a differentiable function from the real to the complex domain.

This is why, when we extend a real function (such as the exponential or trigonometric functions) to the complex plane, we can be assured that this extension is unique.

23.1.2   The Maximum Modulus Theorem

Suppose $f : S \to \mathbb{C}$, and S is a compact set. We know from Theorem 21.3 that $|f(z)|$ assumes a maximum value on S. This means that for at least one ζ in S, $|f(z)| \le |f(\zeta)|$ for all z in S. But this does not give any information about where in S the point ζ might be. We will now prove that any such ζ must lie on the boundary of S if f is a differentiable function. This is called the *maximum modulus theorem*. The name of the theorem derives from the fact that the real-valued function $|f(z)|$ is called the *modulus* of $f(z)$, and we are concerned with the maximum that the modulus of $f(z)$ has as z varies over a set S.

We will first show that a differentiable function that is not constant on an open disk cannot have its maximum modulus at the center of the disk.

LEMMA 23.1

Let f be differentiable and not constant on an open disk D centered at z_0. Then, for some z in this disk,

$$|f(z)| > |f(z_0)|.$$

Proof   Suppose, instead, that $|f(z)| \le |f(z_0)|$ for all z in D. We will derive a contradiction.

Let $\gamma(t) = z_0 + re^{it}$ for $0 \le t \le 2\pi$. Suppose r is small enough that this circle is contained in D. By Cauchy's integral theorem,

$$f(z_0) = \frac{1}{2\pi i} \oint_\gamma \frac{f(z)}{z - z_0}\, dz = \frac{1}{2\pi} \int_0^{2\pi} f(z_0 + re^{it})\, dt.$$

Then

$$|f(z_0)| \le \frac{1}{2\pi} \int_0^{2\pi} |f(z_0 + re^{it})|\, dt.$$

But $z_0 + re^{it}$ is in D for $0 \le t \le 2\pi$, so $\left|f(z_0 + re^{it})\right| \le |f(z_0)|$. Then

$$\frac{1}{2\pi} \int_0^{2\pi} |f(z_0 + re^{it})|\, dt \le \frac{1}{2\pi} \int_0^{2\pi} |f(z_0)|\, dt = |f(z_0)|.$$

The last two inequalities imply that

$$\frac{1}{2\pi} \int_0^{2\pi} |f(z_0 + re^{it})|\, dt = |f(z_0)|.$$

But then

$$\frac{1}{2\pi} \int_0^{2\pi} \left(|f(z_0)| - \left|f(z_0 + re^{it})\right|\right) dt = 0.$$

This integrand is continuous and nonnegative for $0 \le t \le 2\pi$. If it were positive for any t, then there would be some subinterval of $[0, 2\pi]$ on which the integrand would be positive and then this integral would be positive, a contradiction. Therefore the integrand must be identically zero:

$$\left|f(z_0 + re^{it})\right| = |f(z_0)| \quad \text{for } 0 \le t \le 2\pi.$$

This says that $|f(z)|$ has the constant value $|f(z_0)|$ on every circle about z_0 and contained in D. But every point in D is on some circle about z_0. Therefore, $|f(z)| = |f(z_0)| = $ constant for all z in D. Then by Theorem 21.7, $f(z) = $ constant on D. This contradiction proves the lemma. ∎

We can now derive the maximum modulus theorem.

THEOREM 23.6 *Maximum Modulus Theorem*

Let S be a compact, connected set of complex numbers. Let f be continuous on S and differentiable at each interior point of S. Then $|f(z)|$ achieves its maximum value at a boundary point of S. Further, if f is not a constant function, then $|f(z)|$ does not achieve its maximum at an interior point of S.

Proof Because S is compact and f is continuous, we know from Theorem 21.3 that $|f(z)|$ achieves a maximum value at some (perhaps many) points of S. Let ζ be such a point. If ζ is an interior point, then there is an open disk D about ζ that contains only points of S. But then $|f(z)|$ achieves its maximum on this disk at its center. Now there are two cases.

Case 1 $f(z)$ is constant on this disk. By the identity theorem, $f(z)$ is constant on S. In this event $|f(z)|$ is constant on S.

Case 2 $f(z)$ is not constant on this disk. Then $|f(z)| \leq |f(\zeta)|$ for z in this disk, contradicting Lemma 23.1. In this case $|f(z)|$ cannot achieve a maximum in the interior of S and so must achieve its maximum at a boundary point. ∎

EXAMPLE 23.8

Let $f(z) = \sin(z)$. We will determine the maximum value of $|f(z)|$ on the square $0 \leq x \leq \pi$, $0 \leq y \leq \pi$.

First, it is convenient to work with $|f(z)|^2$, since this will have its maximum at the same value of z that $|f(z)|$ does. Now

$$f(z) = \sin(z) = \sin(x)\cosh(y) + i\,\cos(x)\sinh(y),$$

so

$$|f(z)|^2 = \sin^2(x)\cosh^2(y) + \cos^2(x)\sinh^2(y).$$

By the maximum modulus theorem, $|f(z)|^2$ must achieve its maximum value (for this square) on one of the sides of the square. Look at each side in turn.

On the bottom side, $y = 0$ and $0 \leq x \leq \pi$, so $|f(z)|^2 = \sin^2(x)$ achieves a maximum value of 1.

On the right side, $x = \pi$ and $0 \leq y \leq \pi$, so $|f(z)|^2 = \sinh^2(y)$ achieves a maximum value of $\sinh^2(\pi)$. This is because $\cos^2(\pi) = 1$ and $\sinh(y)$ is a strictly increasing function on $[0, \pi]$.

On the top side of the square, $y = \pi$ and $0 \leq x \leq \pi$. Now $|f(z)|^2 = \sin^2(x)\cosh^2(\pi) + \cos^2(x)\sinh^2(\pi)$. We need to know where this achieves its maximum value for $0 \leq x \leq \pi$. This is a problem in single-variable calculus. Let

$$g(x) = \sin^2(x)\cosh^2(\pi) + \cos^2(x)\sinh^2(\pi).$$

Then

$$g'(x) = 2\sin(x)\cos(x)\cosh^2(\pi) - 2\cos(x)\sin(x)\sinh^2(\pi)$$
$$= \sin(2x)[\cosh^2(\pi) - \sinh^2(\pi)] = \sin(2x).$$

This derivative is zero on $(0, \pi)$ at $x = \pi/2$, so this is the critical points of g. Further,

$$g\left(\frac{\pi}{2}\right) = \cosh^2(\pi).$$

At the ends of the interval, we have

$$g(0) = g(\pi) = \sinh^2(\pi) < \cosh^2(\pi).$$

Therefore, on the top side of the square, $|f(z)|^2$ achieves the maximum value of $\cosh^2(\pi)$.

Finally, on the left side of the square, $x = 0$ and $0 \le y \le \pi$, so $|f(z)|^2 = \sinh^2(y)$, with maximum $\sinh^2(\pi)$ on $0 \le y \le \pi$.

We conclude that on this square, $|f(z)|^2$ has its maximum value of $\cosh^2(\pi)$, which is the maximum value of $|f(z)|^2$ on the boundary of the square. Therefore, $|f(z)|$ has a maximum value of $\cosh(\pi)$ on this square. ∎

SECTION 23.1 PROBLEMS

In each of Problems 1 through 18, find the Taylor series of the function about the point. Also determine the radius of convergence and open disk of convergence of the series.

1. $\cos(2z); z = 0$

2. $e^{-z}; z = -3i$

3. $\dfrac{1}{1-z}; 4i$

4. $\sin(z^2); 0$

5. $\dfrac{1}{(1-z)^2}; 0$

6. $\dfrac{1}{2+z}; 1 - 8i$

7. $z^2 - 3z + i; 2 - i$

8. $1 + \dfrac{1}{2+z^2}; i$

9. $(z-9)^2; 1 + i$

10. $e^z - i \sin(z); 0$

11. $\sin(z+i); -i$

12. $\dfrac{3}{z - 4i}; -5$

13. $\dfrac{1}{z - 2 - 4i}; -2i$

14. $e^{3-z}; i$

15. $\cos(z^2) - \sin(z); 0$

16. $\cos(z); i$

17. $e^z \cos(z); 0$

18. $\cos^2(z); 0$

19. Suppose f is differentiable in an open disk about zero and satisfies $f''(z) = 2f(z) + 1$. Suppose $f(0) = 1$ and $f'(0) = i$. Find the Maclaurin expansion of $f(z)$.

20. Find the first three terms of the Maclaurin expansion of $\sin^2(z)$ in three ways as follows:

(a) First, compute the Taylor coefficients at 0.

(b) Find the first three terms of the product of the Maclaurin series for $\sin(z)$ with itself.

(c) Write $\sin^2(z)$ in terms of the exponential function and use the Maclaurin expansion of this function.

21. Derive the complex Fourier series expansion of $f(z)$ as follows. Begin with a Maclaurin expansion $f(z) = \sum_{n=0}^{\infty} c_n z^n$ for $|z| < R$. For z in this disk, write $z = re^{i\theta}$ and obtain $f(z) = \sum_{n=0}^{\infty} c_n r^n e^{in\theta}$, where

$$c_n = \frac{1}{2\pi R^n} \int_0^{2\pi} f(Re^{i\theta}) e^{-in\theta} \, d\theta.$$

22. Continuing Problem 21, show that

$$\sum_{n=0}^{\infty} |c_n|^2 r^{2n} = \frac{1}{2\pi} \int_0^{2\pi} |f(re^{i\theta})|^2 \, d\theta.$$

This is a form of *Parseval's theorem*.

23. Show that

$$\sum_{n=0}^{\infty} \frac{1}{(n!)^2} z^{2n} = \frac{1}{2\pi} \int_0^{2\pi} e^{2z\cos(\theta)} \, d\theta.$$

Hint: First show that

$$\left(\frac{z^n}{n!}\right)^2 = \frac{1}{2\pi i} \oint_\Gamma \frac{z^n}{n! w^{n+1}} e^{zw} \, dw$$

for $n = 0, 1, 2, \ldots$ and Γ the unit circle about the origin.

24. Let $g(x) = e^x$ for x real. Prove that there exists exactly one complex function $E(z)$ that is analytic for all z and such that $E(x) = g(x)$ for x real.

25. Let f be a complex function that is analytic for all complex z. Suppose $f(1/n) = \sin(1/n)$ for every positive integer n. Show that $f(z) = \sin(z)$ for all z.

26. Find the maximum value of $|\cos(z)|$ on the square $0 \le x \le \pi, 0 \le y \le \pi$.

27. Find the maximum value of $|e^z|$ on the rectangle $0 \le x \le 1, 0 \le y \le \pi$.

28. Find the maximum value of $|\sin(z)|$ on the rectangle $0 \le x \le 2\pi, 0 \le y \le 1$.

23.2 The Laurent Expansion

If f is differentiable in some disk about z_0, then $f(z)$ has a Taylor series representation about z_0. If a function is not differentiable at z_0, it may have a different kind of series expansion about z_0, a Laurent expansion. This will have profound implications in analyzing properties of functions and in such applications as summing series and evaluating real and complex integrals.

First we need some terminology. The open set of points between two concentric circles is called an *annulus*. Typically an annulus is described by inequalities

$$ r < |z - z_0| < R, $$

in which r is the radius of the inner circle and R the radius of the outer circle (Figure 23.7). We allow $r = 0$ in this inequality, in which case the annulus $0 < |z - z_0| < R$ is a *punctured disk* (open disk with the center removed).

We also allow $R = \infty$. The annulus $r < |z - z_0| < \infty$ consists of all points outside the inner circle of radius r. An annulus $0 < |z - z_0| < \infty$ consists of all complex z except z_0.

We can now state the main result on Laurent series.

THEOREM 23.7

Let $0 \le r < R \le \infty$. Suppose f is differentiable in the annulus $r < |z - z_0| < R$. Then, for each z in this annulus,

$$ f(z) = \sum_{n=-\infty}^{\infty} c_n(z - z_0)^n, $$

where, for each integer n,

$$ c_n = \frac{1}{2\pi i} \oint_\Gamma \frac{f(z)}{(z - z_0)^{n+1}} \, dz, $$

and Γ is any closed path about z_0 lying entirely in the annulus. ∎

A typical such Γ is shown in Figure 23.8. The series in the theorem, which may include both positive and negative powers of $z - z_0$, is the *Laurent expansion*, or *Laurent series*, for $f(z)$ about z_0 in the given annulus. This expansion has the appearance

$$ \cdots + \frac{c_{-2}}{(z - z_0)^2} + \frac{c_{-1}}{z - z_0} + c_0 + c_1(z - z_0) + c_2(z - z_0)^2 + \cdots . $$

The function need not be differentiable, or even defined, at z_0 or at other points within the inner circle of the annulus. The numbers c_n are the *Laurent coefficients* of f about z_0.

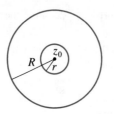

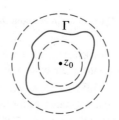

FIGURE 23.7
*Circles $|z - z_0| = r$
and $|z - z_0| = R$
bounding the open
annulus
$r < |z - z_0| < R$.*

FIGURE 23.8 *Closed
path Γ enclosing z_0
and lying in the
annulus
$r < |z - z_0| < R$.*

A Laurent series is a decomposition of $f(z)$ into a sum

$$f(z) = \sum_{n=-\infty}^{-1} c_n(z - z_0)^n + \sum_{n=0}^{\infty} c_n(z - z_0)^n = \sum_{n=1}^{\infty} \frac{c_{-n}}{(z - z_0)^n} + \sum_{n=0}^{\infty} c_n(z - z_0)^n = h(z) + g(z).$$

The part containing only nonnegative powers of $z - z_0$ defines a function $g(z)$ that is differentiable on $|z - z_0| < R$ (because this part is a Taylor expansion). The part containing only negative powers of $z - z_0$ defines a function $h(z)$ that is not defined at z_0. This part determines the behavior of $f(z)$ about a point z_0 where f is not differentiable.

We will now prove the theorem.

Proof Let z be in the annulus and choose r_1 and r_2 so that $r < r_1 < r_2 < R$ and so that the circle $\gamma_1 : |z - z_0| = r_1$ does not enclose z, while the $\gamma_2 : |z - z_0| = r_2$ does enclose z (Figure 23.9). Insert line segments L_1 and L_2 between γ_1 and γ_2 as shown in Figure 23.10, forming two closed paths Γ_1 and Γ_2 in the annulus. These paths are shown separately in Figure 23.11.

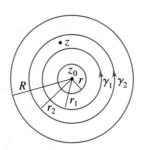

FIGURE 23.9

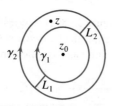

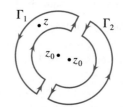

FIGURE 23.10

FIGURE 23.11

f is differentiable on and in the interior of Γ_1, and this curve encloses z. By Cauchy's integral formula,

$$f(z) = \frac{1}{2\pi i} \oint_{\Gamma_1} \frac{f(w)}{w-z}\, dw.$$

Further, since Γ_2 does not enclose z, $f(w)/(w-z)$ is differentiable on and in the interior of Γ_2. By Cauchy's theorem,

$$\oint_{\Gamma_2} \frac{f(w)}{w-z}\, dw = 0.$$

Add these two equations (after multiplying the second by $1/2\pi i$) to get

$$f(z) = \frac{1}{2\pi i} \oint_{\Gamma_1} \frac{f(w)}{w-z}\, dw + \frac{1}{2\pi i} \oint_{\Gamma_2} \frac{f(w)}{w-z}\, dw.$$

As usual, the orientation on both paths is counterclockwise.

From Figure 23.11, in this sum of integrals, each of the inserted line segments L_1 and L_2 is integrated over in both directions. The contribution of the integrals over these segments to the last equation is therefore zero, and we are left with the integrals over γ_1 and γ_2, which make up the rest of Γ_1 and Γ_2. However, as parts of Γ_1 and Γ_2, the orientation on γ_2 is counterclockwise, but that on γ_1 is clockwise. Therefore, reversing this orientation so that all integrations are counterclockwise, the last equation becomes

$$f(z) = \frac{1}{2\pi i} \oint_{\gamma_2} \frac{f(w)}{w-z}\, dw - \frac{1}{2\pi i} \oint_{\gamma_1} \frac{f(w)}{w-z}\, dw. \tag{23.2}$$

For the first integral in equation (23.2), write

$$\frac{1}{w-z} = \frac{1}{w-z_0-(z-z_0)} = \frac{1}{w-z_0}\frac{1}{1-(z-z_0)/(w-z_0)}$$

$$= \frac{1}{w-z_0}\sum_{n=0}^{\infty}\left(\frac{z-z_0}{w-z_0}\right)^n = \sum_{n=0}^{\infty}\frac{1}{(w-z_0)^{n+1}}(z-z_0)^n.$$

This geometric series is valid because for w on γ_2, $|(z-z_0)/(w-z_0)| < 1$.

For the second integral in equation (23.2), write

$$\frac{1}{w-z} = \frac{1}{w-z_0-(z-z_0)} = \frac{-1}{z-z_0}\frac{1}{1-(w-z_0)/(z-z_0)}$$

$$= -\frac{1}{z-z_0}\sum_{n=0}^{\infty}\left(\frac{w-z_0}{z-z_0}\right)^n = -\sum_{n=0}^{\infty}(w-z_0)^n\frac{1}{(z-z_0)^{n+1}}.$$

This geometric expansion is valid because, for w on γ_1, $|(w-z_0)/(z-z_0)| < 1$.

Now substitute these series into the appropriate integrals in equation (23.2) and use Theorem 22.6 to get

$$f(z) = \frac{1}{2\pi i} \oint_{\gamma_2} \left(\sum_{n=0}^{\infty} \frac{f(w)}{(w - z_0)^{n+1}} \, dw \right) (z - z_0)^n$$

$$+ \frac{1}{2\pi i} \oint_{\gamma_1} \left(\sum_{n=0}^{\infty} f(w)(w - z_0)^n \, dw \right) \frac{1}{(z - z_0)^{n+1}}$$

$$= \sum_{n=0}^{\infty} \left(\frac{1}{2\pi i} \oint_{\gamma_2} \frac{f(w)}{(w - z_0)^{n+1}} \, dw \right) (z - z_0)^n$$

$$+ \sum_{n=0}^{\infty} \left(\frac{1}{2\pi i} \oint_{\gamma_1} f(w)(w - z_0)^n \, dw \right) \frac{1}{(z - z_0)^{n+1}}.$$

Put $n = -m - 1$ in the last summation to get

$$f(z) = \sum_{n=0}^{\infty} \left(\frac{1}{2\pi i} \oint_{\gamma_2} \frac{f(w)}{(w - z_0)^{n+1}} \, dw \right) (z - z_0)^n$$

$$+ \sum_{m=-1}^{-\infty} \left(\frac{1}{2\pi i} \oint_{\gamma_1} \frac{f(w)}{(w - z_0)^{m+1}} \right) (z - z_0)^{-m}.$$

Now use the deformation theorem to replace both γ_1 and γ_2 by any closed path Γ enclosing z_0 and lying entirely in the annulus. This gives

$$f(z) = \sum_{n=-\infty}^{\infty} c_n(z - z_0)^n,$$

with

$$c_n = \frac{1}{2\pi i} \oint_{\Gamma} \frac{f(z)}{(z - z_0)^{n+1}} \, dz,$$

completing the proof. ∎

As with Taylor series, we rarely compute the coefficients in a Laurent expansion using this integral formula (quite the contrary, we will use one of these coefficients to evaluate integrals). Instead, we use known series and algebraic or analytic manipulations. This requires that we be assured that the Laurent expansion of a function in an annulus about a point is unique and does not change with the method of derivation.

THEOREM 23.8

Let f be differentiable in an annulus $r < |z - z_0| < R$. Suppose, for z in this annulus,

$$f(z) = \sum_{n=-\infty}^{\infty} b_n(z - z_0)^n.$$

Then the b_n's are the Laurent coefficients c_n of f, and this series is the Laurent expansion of $f(z)$ in this annulus.

Proof Choose γ as a circle about z_0 in the annulus. Let k be any integer. Using Theorem 22.6, we get

$$2\pi i c_k = \oint_\gamma \frac{f(w)}{(w-z_0)^{k+1}}\, dw = \oint_\gamma \left(\sum_{n=-\infty}^{\infty} b_n(z-z_0)^n \right) \frac{1}{(w-z_0)^{k+1}}\, dw$$

$$= \sum_{n=-\infty}^{\infty} b_n \oint_\gamma \frac{1}{(w-z_0)^{k-n+1}}\, dw. \tag{23.3}$$

Now, on γ, $w = z_0 + re^{it}$ for $0 \le t \le 2\pi$, with r the radius of γ. Then

$$\oint_\gamma \frac{1}{(w-z_0)^{k-n+1}}\, dw = \int_0^{2\pi} \frac{1}{r^{k-n+1}(e^{it})^{k-n+1}} ire^{it}\, dt$$

$$= \frac{i}{r^{k-n}} \int_0^{2\pi} e^{i(n-k)t}\, dt = \begin{cases} 0 & \text{if } k \neq n \\ 2\pi i & \text{if } k = n \end{cases}.$$

Thus in equation (23.3), all terms in the last series vanish except the term with $n = k$, and the equation reduces to

$$2\pi i c_k = 2\pi i b_k.$$

Hence for each integer k, $b_k = c_k$. ∎

Here are some examples of Laurent expansions.

EXAMPLE 23.9

$e^{1/z}$ is differentiable in the annulus $0 < |z| < \infty$, the plane with the origin removed. Since

$$e^z = \sum_{n=0}^{\infty} \frac{1}{n!} z^n,$$

then, in this annulus,

$$e^{1/z} = \sum_{n=0}^{\infty} \frac{1}{n!} \left(\frac{1}{z}\right)^n = 1 + \frac{1}{z} + \frac{1}{2}\frac{1}{z^2} + \frac{1}{6}\frac{1}{z^3} + \frac{1}{24}\frac{1}{z^4} + \cdots.$$

This is the Laurent expansion of $e^{1/z}$ about 0, and it converges for all nonzero z. This expansion contains a constant term and infinitely many negative integer powers of z, but no positive powers. ∎

EXAMPLE 23.10

We will find the Laurent expansion of $\cos(z)/z^5$ about zero. For all z,

$$\cos(z) = \sum_{n=0}^{\infty} \frac{(-1)^n}{(2n)!} z^{2n}.$$

For $z \neq 0$,

$$\frac{\cos(z)}{z^5} = \sum_{n=0}^{\infty} \frac{(-1)^n}{(2n)!} z^{2n-5} = \frac{1}{z^5} - \frac{1}{2}\frac{1}{z^3} + \frac{1}{24}\frac{1}{z} - \frac{1}{720}z + \frac{1}{40,320}z^3 - \cdots.$$

This is the Laurent expansion of $\cos(z)/z^5$ about 0. This expansion has exactly three terms containing negative powers of z, and the remaining terms contain only positive powers. We can think of $\cos(z)/z^5 = h(z) + g(z)$, where

$$g(z) = -\frac{1}{720}z + \frac{1}{40,320}z^3 - \cdots$$

is a differentiable function (it is a power series about the origin), and

$$h(z) = \frac{1}{z^5} - \frac{1}{2}\frac{1}{z^3} + \frac{1}{24}\frac{1}{z}.$$

It is $h(z)$ that determines the behavior of $\cos(z)/z^5$ near the origin. ∎

EXAMPLE 23.11

We will find the Laurent expansion of $1/(1 + z^2)$ about $-i$.

We want a series of powers of $z + i$. The function is differentiable in the annulus $0 < |z + i| < 2$ centered at $-i$. This annulus has radius 2 because this is the distance between $-i$, the center of the proposed expansion, and i, the other point at which $1/(1 + z^2)$ is not defined.

To get powers of $z + i$, first use partial fractions:

$$\frac{1}{1 + z^2} = \frac{1}{(z + i)(z - i)} = \frac{i}{2}\frac{1}{z + i} - \frac{i}{2}\frac{1}{z - i}.$$

The next to last term is already a power of $z + i$, so we will keep it as it is and rearrange the last term on the right:

$$\frac{1}{z - i} = \frac{1}{-2i + (z + i)} = \frac{-1}{2i}\frac{1}{1 - (z + i)/2i}$$

$$= \frac{-1}{2i}\sum_{n=0}^{\infty}\left(\frac{z + i}{2i}\right)^n = -\sum_{n=0}^{\infty}\frac{1}{(2i)^{n+1}}(z + i)^n.$$

This geometric series expansion is valid for $|(z + i)/2i| < 1$, or $|z + i| < 2$, as we obtained earlier by looking at the distance between points where $1/(1 + z^2)$ is not defined. We now have

$$\frac{1}{1 + z^2} = \left(\frac{i}{2}\right)\frac{1}{z + i} - \left(\frac{i}{2}\right)\left(-\sum_{n=0}^{\infty}\frac{1}{(2i)^{n+1}}(z + i)^n\right)$$

$$= \frac{i}{2}\frac{1}{z + i} + \frac{i}{2}\sum_{n=0}^{\infty}\frac{1}{(2i)^{n+1}}(z + i)^n.$$

This is the Laurent expansion of $1/(1 + z^2)$ in the annulus $0 < |z + i| < 2$. It expresses the function as a power series part and a part containing a negative power of $z + i$. ∎

EXAMPLE 23.12

Find the Laurent expansion of

$$\frac{1}{(z+1)(z-3i)}$$

about -1.

Use partial fractions to write

$$\frac{1}{(z+1)(z-3i)} = \frac{-1+3i}{10}\frac{1}{z+1} + \frac{1-3i}{10}\frac{1}{z-3i}.$$

$1/(z+1)$ is already expanded around -1, so concentrate on the last term:

$$\frac{1}{z-3i} = \frac{1}{-1-3i+(z+1)} = \frac{1}{-1-3i}\frac{1}{1-(z+1)/(1+3i)} = \frac{-1}{1+3i}\sum_{n=0}^{\infty}\left(\frac{z+1}{1+3i}\right)^n$$

$$= -\sum_{n=0}^{\infty}\frac{1}{(1+3i)^{n+1}}(z+1)^n.$$

This expansion is valid for $|(z+1)/(1+3i)| < 1$, or $|z+1| < \sqrt{10}$. The Laurent expansion of $1/(z+1)(z-3i)$ about -1 is

$$\frac{1}{(z+1)(z-3i)} = \frac{-1+3i}{10}\frac{1}{z+1} - \frac{1-3i}{10}\sum_{n=0}^{\infty}\frac{1}{(1+3i)^{n+1}}(z+1)^n,$$

and this representation is valid in the annulus $0 < |z+1| < \sqrt{10}$.

Notice that $\sqrt{10}$ is the distance from -1, the center of the Laurent expansion, to the other point, $3i$, at which the function is not differentiable. ∎

In the next chapter we will use the Laurent expansion to develop the powerful residue theorem, which has many applications, including the evalution of real and complex integrals and summation of series.

SECTION 23.2 PROBLEMS

In each of Problems 1 through 13, write the Laurent expansion of the function in an annulus $0 < |z - z_0| < R$ about the point.

1. $\dfrac{2z}{1+z^2}; i$

2. $\dfrac{\sin(z)}{z^2}; 0$

3. $\dfrac{1-\cos(2z)}{z^2}; 0$

4. $z^2\cos\left(\dfrac{i}{z}\right); 0$

5. $\dfrac{z^2}{1-z}; 1$

6. $\dfrac{z^2+1}{2z-1}; \dfrac{1}{2}$

7. $\dfrac{e^{z^2}}{z^2}; 0$

8. $\dfrac{\sin(4z)}{z}; 0$

9. $\dfrac{z+i}{z-i}; i$

10. $\sinh\left(\dfrac{1}{z^3}\right); 0$

11. $e^{1/(z+i)}; -i$

12. $\dfrac{2i}{z-1+i}; 1-i$

13. $\dfrac{1}{z^2-4}; -2$

14. The complex Bessel function $J_n(z)$ of order n can be defined for integer n by

$$e^{z(w-1/w)/2} = \sum_{n=-\infty}^{\infty} J_n(z)w^n.$$

That is, $J_n(z)$ is the coefficient of w^n in the Laurent expansion of $e^{z(w-1/w)/2}$, as a function of w, about 0.

(a) Use the integral formula for the Laurent coefficients to show that

$$J_n(z) = \frac{1}{\pi}\int_0^\pi \cos(n\theta - z\sin(\theta))\, d\theta.$$

(b) Write

$$e^{z(w-1/w)/2} = e^{zw/2}e^{-z/2w}$$

and multiply the series expansions of these functions about 0 to obtain

$$J_n(z) = \sum_{k=0}^{\infty}(-1)^k \frac{1}{k!(n+k)!}\left(\frac{z}{2}\right)^{n+2k}$$

for $n = 0, 1, 2, \ldots$.

CHAPTER 24

Singularities and the Residue Theorem

As a prelude to the residue theorem, we will use the Laurent expansion to classify points at which a function is not differentiable.

24.1 Singularities

DEFINITION 24.1 *Isolated Singularity*

A complex function f has an isolated singularity at z_0 if f is differentiable in an annulus $0 < |z - z_0| < R$, but not at z_0 itself.

For example, $1/z$ has an isolated singularity at 0, and $\sin(z)/(z - \pi)$ has an isolated singularity at π.

We will now identify singularities as being of different types, depending on the terms appearing in the Laurent expansion of the function about the singularity.

DEFINITION 24.2 *Classification of Singularities*

Let f have an isolated singularity at z_0. Let the Laurent expansion of $f(z)$ in an annulus $0 < |z - z_0| < R$ be

$$f(z) = \sum_{n=-\infty}^{\infty} c_n (z - z_0)^n.$$

Then:

1. z_0 is a removable singularity of f if $c_n = 0$ for $n = -1, -2, \ldots$.
2. z_0 is a pole of order m (m a positive integer) if $c_{-m} \neq 0$ and $c_{-m-1} = c_{-m-2} = \cdots = 0$.
3. z_0 is an essential singularity of f if $c_{-n} \neq 0$ for infinitely many positive integers n.

These three types cover all the possibilities for an isolated singularity.

In the case of a removable singularity, the Laurent expansion has no negative powers of $z - z_0$ and is therefore

$$f(z) = \sum_{n=0}^{\infty} c_n (z - z_0)^n,$$

a power series about z_0. In this case we can assign $f(z_0)$ the value c_0 to obtain a function that is differentiable in the open disk $|z - z_0| < r$.

EXAMPLE 24.1

Let

$$f(z) = \frac{1 - \cos(z)}{z}$$

for $0 < |z| < \infty$. Since

$$\cos(z) = 1 - \frac{z^2}{2!} + \frac{z^4}{4!} - \frac{z^6}{6!} + \cdots$$

for all z, then

$$f(z) = \frac{1 - \cos(z)}{z} = \frac{z}{2!} - \frac{z^3}{4!} + \frac{z^5}{6!} + \cdots = \sum_{n=1}^{\infty} \frac{(-1)^n}{(2n)!} z^{2n-1}$$

for $z \neq 0$. The series on the right is actually a power series, having the value 0 at $z = 0$. We can therefore define a new function

$$g(z) = \begin{cases} (1 - \cos(z))/z & \text{for } z \neq 0 \\ 0 & \text{for } z = 0 \end{cases},$$

which agrees with $f(z)$ for $z \neq 0$ but is defined at 0 in such a way as to be differentiable there, because $g(z)$ has a power series expansion about 0. Because it is possible to extend f to a function g that is differentiable at 0, we call 0 a *removable singularity* of f. ∎

Thus, a removable singularity is one that can be "removed" by appropriately assigning the function a value at the point.

EXAMPLE 24.2

$f(z) = \sin(z)/(z - \pi)$ has a removable singularity at π. To see this, first write the Laurent expansion of $f(z)$ in $0 < |z - \pi| < \infty$. An easy way to do this is to begin with

$$\sin(z - \pi) = \sin(z)\cos(\pi) - \cos(z)\sin(\pi) = -\sin(z),$$

so

$$\sin(z) = -\sin(z-\pi) = \sum_{n=0}^{\infty} \frac{(-1)^{n+1}}{(2n+1)!}(z-\pi)^{2n+1}.$$

Then, for $z \neq \pi$,

$$\frac{\sin(z)}{z-\pi} = \sum_{n=0}^{\infty} \frac{(-1)^{n+1}}{(2n+1)!}(z-\pi)^{2n} = -1 + \frac{1}{6}(z-\pi)^2 - \frac{1}{120}(z-\pi)^4 + \cdots.$$

Although $f(\pi)$ is not defined, the series on the right is defined for $z = \pi$, and is equal to -1 there. We therefore extend f to a differentiable function g defined over the entire plane by assigning the new function the value -1 when $z = \pi$:

$$g(z) = \begin{cases} f(z) & \text{for } z \neq \pi \\ -1 & \text{for } z = \pi \end{cases}.$$

This extension "removes" the singularity of f at π, since $f(z) = g(z)$ for $z \neq \pi$ and $g(\pi) = -1$. ■

For f to have a pole at z_0, the Laurent expansion of f about z_0 must have terms with negative powers of $z - z_0$, but only finitely many such terms. If the pole has order m, then this Laurent expansion has the form

$$f(z) = \frac{c_{-m}}{(z-z_0)^m} + \frac{c_{-m+1}}{(z-z_0)^{m-1}} + \cdots + \frac{c_{-1}}{z-z_0} + \sum_{n=0}^{\infty} c_n(z-z_0)^n,$$

with $c_{-m} \neq 0$. This expansion is valid in some annulus $0 < |z - z_0| < R$.

EXAMPLE 24.3

Let $f(z) = 1/(z + i)$. This function is its own Laurent expansion about $-i$ and $c_{-1} = 1$, while all other coefficients are zero. Thus $-i$ is a pole of order 1 of f.

This singularity is not removable. There is no way to assign a value to $f(-i)$ so that the extended function is differentiable at $-i$. ■

EXAMPLE 24.4

$g(z) = 1/(z + i)^3$, then g has a pole of order 3 at $-i$. Here the function is its own Laurent expansion about $-i$, and the coefficient of $1/(z+i)^3$ is nonzero, while all other coefficients are zero. ■

DEFINITION 24.3 *Simple and Double Poles*

A pole of order 1 is called a simple pole. A pole of order 2 is a double pole.

EXAMPLE 24.5

Let

$$f(z) = \frac{\sin(z)}{z^3}.$$

For $z \neq 0$,

$$f(z) = \frac{1}{z^3} \sum_{n=0}^{\infty} \frac{(-1)^n}{(2n+1)!} z^{2n+1} = \sum_{n=0}^{\infty} \frac{(-1)^n}{(2n+1)!} z^{2n-2}$$

$$= \frac{1}{z^2} - \frac{1}{6} + \frac{1}{120} z^2 - \frac{1}{5040} z^4 + \cdots .$$

Therefore f has a double pole at 0. ∎

EXAMPLE 24.6

$e^{1/z}$ is defined for all nonzero z, and, for $z \neq 0$,

$$e^{1/z} = \sum_{n=0}^{\infty} \frac{1}{n!} \frac{1}{z^n}.$$

Since this Laurent expansion has infinitely many negative powers of z, 0 is an essential singularity of $e^{1/z}$. ∎

We will discuss several results that are useful in identifying poles of a function.

THEOREM 24.1 *Condition for a Pole of Order m*

Let f be differentiable in the annulus $0 < |z - z_0| < R$. Then f has a pole of order m at z_0 if and only if

$$\lim_{z \to z_0} (z - z_0)^m f(z)$$

exists finite and is nonzero.

Proof Expand $f(z)$ in a Laurent series in this annulus:

$$f(z) = \sum_{n=-\infty}^{\infty} c_n (z - z_0)^n \quad \text{for } 0 < |z - z_0| < R.$$

Suppose f has a pole of order m at z_0. Then $c_{-m} \neq 0$ and $c_{-m-1} = c_{-m-2} = \cdots = 0$, so the Laurent series is

$$f(z) = \sum_{n=-m}^{\infty} c_n (z - z_0)^n.$$

Then

$$(z - z_0)^m f(z) = \sum_{n=-m}^{\infty} c_n (z - z_0)^{n+m} = \sum_{n=0}^{\infty} c_{n-m} (z - z_0)^n$$

$$= c_{-m} + c_{-m+1}(z - z_0) + c_{-m+2}(z - z_0)^2 + \cdots .$$

Then

$$\lim_{z \to z_0} (z - z_0)^m f(z) = c_{-m} \neq 0.$$

Conversely, suppose $\lim_{z \to z_0} (z - z_0)^m f(z) = L \neq 0$. We want to show that f has a pole of order m at z_0.

Let $\epsilon > 0$. Because of this limit, there is a positive $\delta < R$ such that

$$\left|(z - z_0)^m f(z) - L\right| < \epsilon \quad \text{if } 0 < |z - z_0| < \delta.$$

Then, for such z,

$$\left|(z - z_0)^m f(z)\right| < |L| + \epsilon.$$

In particular, if $|z - z_0| = \delta$, then

$$\left|(z - z_0)^{-n-1} f(z)\right| < (|L| + \epsilon) |z - z_0|^{-n-m-1} = (|L| + \epsilon) \delta^{-n-m-1}.$$

The coefficients in the Laurent expansion of $f(z)$ about z_0 are given by

$$c_n = \frac{1}{2\pi i} \oint_\Gamma \frac{f(z)}{(z - z_0)^{n+1}} \, dz,$$

in which we can choose Γ to be circle of radius δ about z_0. Then

$$|c_n| \leq \frac{1}{2\pi}(2\pi\delta) \max_{z \text{ on } \Gamma} \left|f(z)(z - z_0)^{-n-1}\right| < \delta\,(|L| + \epsilon)\,\delta^{-n-m-1} = (|L| + \epsilon)\,\delta^{-n-m}.$$

Now δ^{-n-m} can be made as small as we like by choosing δ small, if $n < -m$. We conclude that $|c_n| = 0$, hence $c_n = 0$, if $n < -m$. Thus the Laurent expansion of $f(z)$ about z_0 has the form

$$f(z) = \frac{c_{-m}}{(z - z_0)^m} + \frac{c_{-m+1}}{(z - z_0)^{m-1}} + \cdots + \frac{c_{-1}}{z - z_0} + \sum_{n=0}^{\infty} c_n(z - z_0)^n,$$

and therefore f has a pole of order m at z_0, as we wanted to show. ∎

EXAMPLE 24.7

Look again at Example 24.3. Since

$$\lim_{z \to -i} (z + i) f(z) = \lim_{z \to -i} (z + i) \frac{1}{z + i} = 1 \neq 0,$$

f has a simple pole at $-i$.

In Example 24.4,

$$\lim_{z \to -i} (z + i)^3 g(z) = \lim_{z \to -i} (z + i)^3 \frac{1}{(z + i)^3} = 1 \neq 0,$$

so g has a pole of order 3 at $-i$.

In Example 24.5,

$$\lim_{z \to 0} z^2 \frac{\sin(z)}{z^3} = \lim_{z \to 0} \frac{\sin(z)}{z} = 1 \neq 0,$$

so $\sin(z)/z^3$ has a double pole at 0. It is a common error to think that this function has a pole of order 3 at zero because the denominator has a zero of order 3 there. However,

$$\lim_{z \to 0} z^3 \frac{\sin(z)}{z^3} = \lim_{z \to 0} \sin(z) = 0,$$

so by Theorem 24.1, the function cannot have a third-order pole at 0. ∎

If $f(z)$ is a quotient of functions, it is natural to look for poles at places where the denominator vanishes. Our first result along these lines deals with a quotient in which the denominator vanishes at z_0 but the numerator does not. Recall that $g(z)$ has a zero of order k at z_0 if $g(z_0) = \cdots = g^{(k-1)}(z_0) = 0$, but $g^{(k)}(z_0) \neq 0$. The order of the zero is the order of the lowest-order derivative that does not vanish at the point.

THEOREM 24.2

Let $f(z) = h(z)/g(z)$, where h and g are differentiable in some open disk about z_0. Suppose $h(z_0) \neq 0$, but g has a zero of order m at z_0. Then f has a pole of order m at z_0. ■

We leave a proof of this result to the student.

EXAMPLE 24.8

$$f(z) = \frac{1 + 4z^3}{\sin^6(z)}$$

has a pole of order 6 at 0, because the numerator does not vanish at 0, and the denominator has a zero of order 6 at 0. By the same token, f has a pole of order 6 at $n\pi$ for any integer n. ■

Theorem 24.2 does not apply if the numerator also vanishes at z_0. The example $f(z) = \sin(z)/z^3$ is instructive. The numerator has a zero of order 1 at 0 and the denominator a zero of order 3 at 0, and we saw in Example 24.5 that the quotient has a pole of order 2. It would appear that the orders of the zeros of numerator and denominator subtract (or cancel) to give the order of a pole at the point. This is indeed the case.

THEOREM 24.3   *Poles of Quotients*

Let $f(z) = h(z)/g(z)$ and suppose h and g are differentiable in some open disk about z_0. Let h have a zero of order k at z_0 and g a zero of order m at z_0, with $m > k$. Then f has a pole of order $m - k$ at z_0. ■

A proof of this is left to the student. By allowing $k = 0$, this theorem includes the case that the numerator $h(z)$ has no zero at z_0.

EXAMPLE 24.9

Consider

$$f(z) = \frac{(z - 3\pi/2)^4}{\cos^7(z)}.$$

The numerator has a zero of order 4 at $3\pi/2$ and the denominator has a zero of order 7 there, so the quotient f has a pole of order 3 at $3\pi/2$. ■

EXAMPLE 24.10

Let $f(z) = \tan^3(z)/z^9$. The numerator has a zero of order 3 at 0 and the denominator has a zero of order 9 at 0. Therefore, f has a pole of order 6 at 0. ■

There are also some useful results stated in terms of products, rather than quotients. We claim that the order of a pole of a product is the sum of the orders of the poles of the factors at a given point.

THEOREM 24.4 Poles of Products

Let f have a pole of order m at z_0 and let g have a pole of order n at z_0. Then fg has a pole of order $m + n$ at z_0. ∎

EXAMPLE 24.11

Let

$$f(z) = \frac{1}{\cos^4(z)(z - \pi/2)^2}.$$

Here $f(z)$ is a product, which we write for emphasis as

$$f(z) = \left[\frac{1}{\cos^4(z)}\right]\left[\frac{1}{(z - \pi/2)^2}\right].$$

Now $1/\cos^4(z)$ has a pole of order 4 at $\pi/2$ and $1/(z - \pi/2)^2$ has a pole of order 2 there, so f has a pole of order 6 at $\pi/2$. f also has poles of order 4 (not 6) at $z = (2n + 1)\pi/2$ for n any nonzero. ∎

We are now prepared to develop the powerful residue theorem.

SECTION 24.1 PROBLEMS

In each of Problems 1 through 20, determine all singularities of the function and classify each singularity as removable, a pole of a certain order, or an essential singularity.

1. $\dfrac{\cos(z)}{z^2}$

2. $\dfrac{4\sin(z + 2)}{(z + i)^2(z - i)}$

3. $e^{1/z}(z + 2i)$

4. $\dfrac{\sin(z)}{z - \pi}$

5. $\dfrac{\cos(2z)}{(z - 1)^2(1 + z^2)}$

6. $\dfrac{z}{(z + 1)^2}$

7. $\dfrac{z - i}{z^2 + 1}$

8. $\dfrac{\sin(z)}{\sinh(z)}$

9. $\dfrac{z}{z^4 - 1}$

10. $\tan(z)$

11. $\dfrac{1}{\cos(z)}$

12. $e^{1/z(z+1)}$

13. $\dfrac{1}{z^2}e^{iz}$

14. $\dfrac{\sin(z)}{z(z - \pi)(z - i)^2}$

15. $\dfrac{e^{2\pi}}{(z + 1)^4}$

16. $\dfrac{\sinh(z)}{z^4}$

17. $\cosh(z)$

18. $\dfrac{2i - 1}{(z^2 + 2z - 3)^2}$

19. $\dfrac{\sin(z)}{(z+\pi)^2}$

20. $\csc^3(2z)$

21. Let f be differentiable at z_0 and let g have a pole of order m at z_0. Let $f(z_0) \neq 0$. Prove that fg has a pole of order m at z_0.

22. Let h and g be differentiable at z_0, $g(z_0) \neq 0$, and h have a zero of order 2 at z_0. Prove that $g(z)/h(z)$ has a pole of order 2 at z_0.

23. Suppose h and g are differentiable at z_0 and $g(z_0) \neq 0$, while h has a zero of order 3 at z_0. Prove that $g(z)/h(z)$ has a pole of order 3 at z_0.

24.2 The Residue Theorem

To see a connection between Laurent series and the integral of a function, suppose f has a Laurent expansion

$$f(z) = \sum_{n=-\infty}^{\infty} c_n(z-z_0)^n$$

in some annulus $0 < |z - z_0| < R$. Let Γ be a closed path in this annulus and enclosing z_0. According to Theorem 23.6, the Laurent coefficients are given by an integral formula. In particular, the coefficient of $1/(z - z_0)$ is

$$c_{-1} = \frac{1}{2\pi i} \oint_\Gamma f(z) \, dz.$$

Therefore,

$$\oint_\Gamma f(z) \, dz = 2\pi i c_{-1}. \tag{24.1}$$

Knowing this one coefficient in the Laurent expansion yields the value of this integral. This fact gives this coefficient a special importance, so we will give it a name.

DEFINITION 24.4 Residue

Let f have an isolated singularity at z_0 and Laurent expansion $f(z) = \sum_{n=-\infty}^{\infty} c_n(z-z_0)^n$ in some annulus $0 < |z - z_0| < R$. Then the coefficient c_{-1} is called the residue of f at z_0 and is denoted $\mathrm{Re}\,s(f, z_0)$.

We will now extend the idea behind equation (24.1) to include the case that Γ may enclose any finite number of points at which f is not differentiable.

THEOREM 24.5 Residue Theorem

Let Γ be a closed path and let f be differentiable on Γ and all points enclosed by Γ, except for $z_1, \ldots, z_n$, which are all the isolated singularities of f enclosed by Γ. Then

$$\oint_\Gamma f(z) \, dz = 2\pi i \sum_{j=1}^{n} \mathrm{Re}\,s(f, z_j). \quad \blacksquare$$

In words, the value of this integral is $2\pi i$ times the sum of the residues of f at the singularities of f enclosed by Γ.

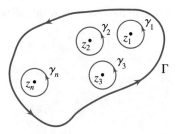

FIGURE 24.1

Proof Enclose each singularity z_j with a closed path γ_j (Figure 24.1) so that each γ_j is in the interior of Γ, encloses exactly one singularity, and does not intersect any other γ_k. By the extended deformation theorem,

$$\oint_\Gamma f(z)\,dz = \sum_{j=1}^n \oint_{\gamma_j} f(z)\,dz = 2\pi i \sum_{j=1}^n \mathrm{Re}\,s(f, z_j). \ \blacksquare$$

The residue theorem is only as effective as our efficiency in evaluating residues of a function at singularities. If we had to actually write the Laurent expansion of f about each singularity to pick off the coefficient of the $1/(z - z_j)$ term, the theorem would be difficult to apply in many instances. What adds to its importance is that, at least for poles, there are efficient ways of calculating residues. We will now develop some of these.

THEOREM 24.6 *Residue at a Simple Pole*

If f has a simple pole at z_0, then

$$\mathrm{Re}\,s(f, z_0) = \lim_{z \to z_0} (z - z_0) f(z).$$

Proof If f has a simple pole at z_0, then its Laurent expansion about z_0 is

$$f(z) = \frac{c_{-1}}{z - z_0} + \sum_{n=0}^\infty c_n (z - z_0)^n$$

in some annulus $0 < |z - z_0| < R$. Then

$$(z - z_0) f(z) = c_{-1} + \sum_{n=0}^\infty c_n (z - z_0)^{n+1},$$

so

$$\lim_{z \to z_0} (z - z_0) f(z) = c_{-1} = \mathrm{Re}\,s(f, z_0). \ \blacksquare$$

EXAMPLE 24.12

$f(z) = \sin(z)/z^2$ has a simple pole at 0, and

$$\mathrm{Re}\,s(f, 0) = \lim_{z \to 0} z \frac{\sin(z)}{z^2} = \lim_{z \to 0} \frac{\sin(z)}{z} = 1.$$

If Γ is any closed path in the plane enclosing the origin, then by the residue theorem,

$$\oint_{\Gamma} \frac{\sin(z)}{z^2} \, dz = 2\pi i \operatorname{Re} s(f, 0) = 2\pi i. \quad \blacksquare$$

EXAMPLE 24.13

Let

$$f(z) = \frac{z - 6i}{(z - 2)^2(z + 4i)}.$$

Then f has a simple pole at $-4i$ and a double pole at 2. Theorem 24.6 will not help us with the residue of f at 2, but at the simple pole,

$$\operatorname{Re} s(f, -4i) = \lim_{z \to -4i} (z + 4i) \frac{z - 6i}{(z - 2)^2(z + 4i)} = \lim_{z \to -4i} \frac{z - 6i}{(z - 2)^2} = \frac{-4i - 6i}{(-4i - 2)^2}$$

$$= -\frac{2}{5} + \frac{3}{10}i. \quad \blacksquare$$

Before looking at residues at poles of order greater than 1, the following version of Theorem 24.6 is sometimes handy.

COROLLARY 24.1

Let $f(z) = h(z)/g(z)$, where h is continuous at z_0 and $h(z_0) \neq 0$. Suppose g is differentiable at z_0 and has a simple zero there. Then f has a simple pole at z_0 and

$$\operatorname{Re} s(f, z_0) = \frac{h(z_0)}{g'(z_0)}.$$

Proof f has a simple pole at z_0 by Theorem 24.2. By Theorem 24.6,

$$\operatorname{Re} s(f, z_0) = \lim_{z \to z_0} (z - z_0) \frac{h(z)}{g(z)} = \lim_{z \to z_0} \frac{h(z)}{((g(z) - g(z_0))/(z - z_0))} = \frac{h(z_0)}{g'(z_0)}. \quad \blacksquare$$

EXAMPLE 24.14

Let

$$f(z) = \frac{4iz - 1}{\sin(z)}.$$

Then f has a simple pole at π, and, by Corollary 24.1,

$$\operatorname{Re} s(f, \pi) = \frac{4i\pi - 1}{\cos(\pi)} = 1 - 4\pi i.$$

In fact, f has a simple pole at $n\pi$ for any integer n, and

$$\text{Re}\, s(f, n\pi) = \frac{4in\pi - 1}{\cos(n\pi)} = (-1)^n(-1 + 4n\pi i). \quad \blacksquare$$

EXAMPLE 24.15

Evaluate

$$\oint_\Gamma \frac{4iz - 1}{\sin(z)}\, dz$$

with Γ the closed path of Figure 24.2.

Γ encloses the poles $0, \pi, 2\pi$, and $-\pi$ but no other singularities of f. By the residue theorem and Example 24.14,

$$\oint_\Gamma \frac{4iz - 1}{\sin(z)}\, dz = 2\pi i [\text{Re}\, s(f, 0) + \text{Re}\, s(f, \pi) + \text{Re}\, s(f, 2\pi) + \text{Re}\, s(f, -\pi)]$$

$$= 2\pi i\, [-1 + (1 - 4\pi i) + (-1 + 8\pi i) + (1 + 4\pi i)] = -16\pi^2. \quad \blacksquare$$

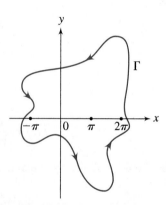

FIGURE 24.2 Γ *encloses only the singularities* $-\pi, 0, \pi, 2\pi$, *and* 3π *of* $\frac{4iz-1}{\sin(z)}$.

Here is a formula for the residue of a function at a pole of order greater than 1.

THEOREM 24.7 *Residue at a Pole of Order m*

Let f have a pole of order m at z_0. Then

$$\text{Re}\, s(f, z_0) = \frac{1}{(m - 1)!} \lim_{z \to z_0} \frac{d^{m-1}}{dz^{m-1}}[(z - z_0)^m f(z)]. \quad \blacksquare$$

If $m = 1$ (simple pole), then $(m - 1)! = 0! = 1$ by definition, and the $(m-1)$-order derivative is defined to be just the function itself. With these conventions, the conclusion of the theorem reduces to the result for residues at simple poles when $m = 1$.

Proof   In some annulus about z_0,

$$f(z) = \frac{c_{-m}}{(z - z_0)^m} + \frac{c_{-m+1}}{(z - z_0)^{m-1}} + \cdots + \frac{c_{-1}}{z - z_0} + \sum_{n=0}^{\infty} c_n (z - z_0)^n.$$

It is c_{-1} we want. Write

$$(z - z_0)^m f(z) = c_{-m} + c_{-m+1}(z - z_0) + \cdots + c_{-1}(z - z_0)^{m-1} + \sum_{n=0}^{\infty} c_n (z - z_0)^{n+m}.$$

The right side of this equation is a power series about z_0 and can be differentiated any number of times within its open disk of convergence. Compute

$$\frac{d^{m-1}}{dz^{m-1}}[(z - z_0)^m f(z)]$$

$$= (m - 1)!c_{-1} + \sum_{n=0}^{\infty}(n + m)(n + m - 1) \cdots (n + 1)(z - z_0)^{n+1}.$$

In the limit as $z \to z_0$, this equation yields

$$\lim_{z \to z_0} \frac{d^{m-1}}{dz^{m-1}}[(z - z_0)^m f(z)] = (m - 1)!c_{-1} = (m - 1)! \, \mathrm{Re}\, s(f, z_0). \quad \blacksquare$$

EXAMPLE 24.16

Let

$$f(z) = \frac{\cos(z)}{(z + i)^3}.$$

Then f has a pole of order 3 at $-i$. By Theorem 24.7,

$$\mathrm{Re}\, s(f, -i) = \frac{1}{2!} \lim_{z \to -i} \frac{d^2}{dz^2}\left((z + i)^3 \frac{\cos(z)}{(z + i)^3}\right)$$

$$= \frac{1}{2} \lim_{z \to -i} \frac{d^2}{dz^2} \cos(z) = -\frac{1}{2}\cos(-i) = -\frac{1}{2}\cos(i). \quad \blacksquare$$

Here are some examples of the residue theorem in evaluating complex integrals.

EXAMPLE 24.17

Let

$$f(z) = \frac{2iz - \cos(z)}{z^3 + z}.$$

We want to evaluate $\oint_{\Gamma} f(z)\, dz$, with Γ a closed path that does not pass through any singularity of f.

The singularities of f are simple poles at 0, i, and $-i$. First compute the residue of f at each of these points. Here it is convenient to use Corollary 24.1:

$$\mathrm{Re}\,s(f,0) = \frac{-\cos(0)}{1} = -1,$$

$$\mathrm{Re}\,s(f,i) = \frac{2i^2 - \cos(i)}{3(i)^2 + 1} = \frac{-2 - \cos(i)}{-2} = 1 + \frac{1}{2}\cos(i),$$

and

$$\mathrm{Re}\,s(f,-i) = \frac{2i(-i) - \cos(-i)}{3(-i)^2 + 1} = -1 + \frac{1}{2}\cos(i).$$

Now consider cases.

1. If Γ does not enclose any of the singularities, then $\oint_\Gamma f(z)\,dz = 0$ by Cauchy's theorem.
2. If Γ encloses 0 but not i or $-i$, then

$$\oint_\Gamma f(z)\,dz = 2\pi i\,\mathrm{Re}\,s(f,0) = -2\pi i.$$

3. If Γ encloses i but not 0 or $-i$, then

$$\oint_\Gamma f(z)\,dz = 2\pi i\left(1 + \frac{1}{2}\cos(i)\right).$$

4. If Γ encloses $-i$ but not 0 or i, then

$$\oint_\Gamma f(z)\,dz = 2\pi i\left(-1 + \frac{1}{2}\cos(i)\right).$$

5. If Γ encloses 0 and i but not $-i$, then

$$\oint_\Gamma f(z)\,dz = 2\pi i\left(-1 + 1 + \frac{1}{2}\cos(i)\right) = \pi i\cos(i).$$

6. If Γ encloses 0 and $-i$ but not i, then

$$\oint_\Gamma f(z)\,dz = 2\pi i\left(-1 - 1 + \frac{1}{2}\cos(i)\right) = 2\pi i\left(-2 + \frac{1}{2}\cos(i)\right).$$

7. If Γ encloses i and $-i$ but not 0, then

$$\oint_\Gamma f(z)\,dz = 2\pi i\left(1 + \frac{1}{2}\cos(i) - 1 + \frac{1}{2}\cos(i)\right) = 2\pi i\cos(i).$$

8. If Γ encloses all three singularities, then

$$\oint_\Gamma f(z)\,dz = 2\pi i\left(-1 + 1 + \frac{1}{2}\cos(i) - 1 + \frac{1}{2}\cos(i)\right) = 2\pi i\left(-1 + \cos(i)\right). \quad\blacksquare$$

EXAMPLE 24.18

Let

$$f(z) = \frac{e^z}{(z + 2i)^3(z + i)}.$$

We will evaluate $\oint_\Gamma f(z)\,dz$ with Γ a closed path enclosing $-i$ and $-2i$.
The pole at $-i$ is simple, and

$$\mathrm{Re}\,s(f,-i) = \lim_{z\to -i}(z+i)\frac{e^z}{(z+2i)^3(z+i)} = \frac{e^{-i}}{(i)^3} = ie^{-i}.$$

f has a pole of order 3 at $-2i$, so

$$\mathrm{Re}\,s(f,-2i) = \frac{1}{2}\lim_{z\to -2i}\frac{d^2}{dz^2}\left(\frac{e^z}{z+i}\right)$$

$$= \frac{1}{2}\lim_{z\to -2i}\left(e^z\left((z+i)^{-1} - 2(z+i)^{-2} + 2(z+i)^{-3}\right)\right) = \frac{1}{2}e^{-2i}(2-i).$$

Then

$$\oint_\Gamma \frac{e^z}{(z+2i)^3(z+i)}\,dz = 2\pi i\left(ie^{-i} + \frac{1}{2}e^{-2i}(2-i)\right). \quad\blacksquare$$

EXAMPLE 24.19

Let

$$f(z) = \frac{\sin(z)}{z^2(z^2+4)}.$$

We want to evaluate $\oint_\Gamma f(z)\,dz$, where Γ is a closed path enclosing 0 and $2i$ but not $-2i$.

By Theorem 24.3, f has a simple pole at 0, not a double pole, because $\sin(z)$ has a simple zero at 0. f also has simple poles at $2i$ and $-2i$. Only the poles at 0 and $2i$ are of interest in using the residue theorem, because Γ does not enclose $-2i$.

Compute

$$\mathrm{Re}\,s(f,0) = \lim_{z\to 0} zf(z) = \lim_{z\to 0}\frac{\sin(z)}{z}\frac{1}{z^2+4} = \frac{1}{4},$$

and

$$\mathrm{Re}\,s(f,2i) = \lim_{z\to 2i}(z-2i)f(z) = \lim_{z\to 2i}\frac{\sin(z)}{z^2(z+2i)} = \frac{\sin(2i)}{(-4)(4i)} = \frac{i}{16}\sin(2i).$$

Then

$$\oint_\Gamma \frac{\sin(z)}{z^2(z^2+4)}\,dz = 2\pi i\left(\frac{1}{4} + \frac{i}{16}\sin(2i)\right). \quad\blacksquare$$

EXAMPLE 24.20

We will evaluate

$$\oint_\Gamma e^{1/z}\,dz$$

for Γ any closed path not passing through the origin.

There are two cases. If Γ does not enclose the origin, then $\oint_\Gamma e^{1/z}\, dz = 0$ by Cauchy's theorem.

If Γ does enclose the origin, then use the residue theorem. We need $\operatorname{Re} s(e^{1/z}, 0)$. As we found in Example 24.6, 0 is an essential singularity of $e^{1/z}$. There is no simple general formula for the residue of a function at an essential singularity. However,

$$e^{1/z} = \sum_{n=0}^{\infty} \frac{1}{n!} \frac{1}{z^n}$$

is the Laurent expansion of $e^{1/z}$ about 0, and the coefficient of $1/z$ is 1. Thus $\operatorname{Re} s(e^{1/z}, 0) = 1$ and

$$\oint_\Gamma e^{1/z}\, dz = 2\pi i. \ \blacksquare$$

Next we will look at a variety of applications of the residue theorem.

SECTION 24.2 PROBLEMS

In each of Problems 1 through 27, use the residue theorem to evaluate the integral over the given path.

1. $\oint_\Gamma \dfrac{1 + z^2}{(z-1)^2(z+2i)}\, dz$; Γ is the circle of radius 7 about $-i$

2. $\oint_\Gamma \dfrac{2z}{(z-i)^2}\, dz$; Γ is the circle of radius 3 about 1

3. $\oint_\Gamma \left(\dfrac{e^z}{z}\right)\, dz$; Γ is the circle of radius 2 about $-3i$

4. $\oint_\Gamma \dfrac{\cos(z)}{4 + z^2}\, dz$; Γ is the square of side length 3 and sides parallel to the axes, centered at $-2i$

5. $\oint_\Gamma \dfrac{z+i}{z^2+6}\, dz$; Γ is the square of side length 8 and sides parallel to the axes, centered at the origin

6. $\oint_\Gamma \dfrac{z-i}{2z+1}\, dz$; Γ is the circle of radius 1 about the origin

7. $\oint_\Gamma \dfrac{z}{\sinh^2(z)}\, dz$; Γ is the circle of radius 1 about $\dfrac{1}{2}$

8. $\oint_\Gamma \dfrac{\cos(z)}{ze^z}\, dz$; Γ is the circle of radius $\dfrac{1}{2}$ about $\dfrac{i}{8}$

9. $\oint_\Gamma \dfrac{iz}{(z^2+9)(z-i)}\, dz$; Γ is the circle of radius 2 about $-3i$

10. $\oint_\Gamma e^{2/z^2}\, dz$; Γ is the square with sides parallel to the axes and of length 3, centered at $-i$

11. $\oint_\Gamma \dfrac{8z - 4i + 1}{z + 4i}\, dz$; Γ is the circle of radius 2 about $-i$

12. $\oint_\Gamma \dfrac{z^2}{z - 1 + 2i}\, dz$; Γ is the square of side length 4 and sides parallel to the axes, centered at $1 - 2i$

13. $\oint_\Gamma \coth(z)\, dz$; Γ is the circle of radius 2 about i

14. $\oint_\Gamma \dfrac{(1-z)^2}{z^3 - 8}\, dz$; Γ is the circle of radius 2 about 2

15. $\oint_\Gamma \dfrac{e^{2z}}{z(z - 4i)}\, dz$; Γ is any closed path enclosing 0 and $4i$

16. $\oint_\Gamma \left(\dfrac{z}{z-1}\right)^2\, dz$; Γ is any closed path enclosing 1

17. $\oint_\Gamma \dfrac{\cosh(3z)}{z(z^2 - 9)}\, dz$; Γ is any closed path enclosing 0 and -3 but not 3

18. $\oint_\Gamma \dfrac{\sin(z)\cos(2z)}{1 + z^2}\, dz$; Γ is the circle of radius 2 about the origin

19. $\oint_\Gamma \dfrac{e^z(z^2 - 4)^2}{(z + i)^2}\, dz$; Γ is the circle $|z - 1 + 2i| = 4$

20. $\oint_\Gamma \dfrac{(z - i)^2}{\sin^2(z)}\, dz$; Γ is the square with vertices $4 \pm 4i$ and $-4 \pm 4i$

21. $\oint_\Gamma \dfrac{ze^{-z}}{z^4 + i}\, dz$; Γ is the circle $|z - 5| = 1$

22. $\oint_\Gamma z^{-4}e^{2z}\, dz$; Γ is the circle $|z - \pi i| = 2\pi$

23. $\oint_\Gamma \dfrac{(z - 2i)^2}{z^2 - 2z + 2}\, dz$; Γ is the circle $|z| = 8$

24. $\oint_\Gamma (1 + z^5)^{-1}\, dz$; Γ is the circle $|z + 1| = 1$

25. $\oint_\Gamma \dfrac{2z - 3}{(z + i)^3}\, dz$; Γ is the circle $|z - 2i| = 4$

26. $\oint_\Gamma \dfrac{(z + 1)\cosh(z)}{(z - i)(z + 2i)}\, dz$; Γ is the circle $|z - 3i| = \dfrac{1}{2}$

27. $\oint_\Gamma \dfrac{z - \cos(4iz)}{(1 + z^2)(z^2 - 1)}\, dz$; Γ is the circle $|z + i| = 1$

28. With h and g as in Problem 22 of Section 24.1, show that

$$\mathrm{Re}\,s\left(\frac{g(z)}{h(z)}, z_0\right) = 2\frac{g'(z_0)}{h''(z_0)} - \frac{2}{3}\frac{g(z_0)h^{(3)}(z_0)}{[h''(z_0)]^2}.$$

29. With h and g as in Problem 23 of Section 24.1, show that

$$\mathrm{Re}\,s\left(\frac{g(z)}{h(z)}, z_0\right) = 3\frac{g''(z_0)}{h'''(z_0)} - \frac{3}{10}\frac{g(z_0)h^{(5)}(z_0)}{(h'''(z_0))^2}$$

$$+ 9\left(\frac{g(z_0)h^{(4)}(z_0)}{24} - \frac{g'(z_0)h'''(z_0)}{6}\right)\frac{h^{(4)}(z_0)}{(h'''(z_0))^3}$$

30. Let g and h be differentiable at z_0. Suppose $g(z_0) \neq 0$ and let h have a zero of order k at z_0. Prove that $g(z)/h(z)$ has a pole of order k at z_0, and

$$\mathrm{Re}\,s\left(\frac{g(z)}{h(z)}, z_0\right) = \left(\frac{k!}{h^{(k)}(z_0)}\right)^k$$

$$\times \begin{vmatrix} H_k & 0 & 0 & \cdots & 0 & G_0 \\ H_{k+1} & H_k & 0 & \cdots & 0 & G_1 \\ H_{k+2} & H_{k+1} & H_k & \cdots & 0 & G_2 \\ \vdots & \vdots & \vdots & \vdots & \vdots & \vdots \\ H_{2k-1} & H_{2k-2} & H_{2k-3} & \cdots & H_{k+1} & G_{k-1} \end{vmatrix},$$

where

$$H_j = \frac{h^{(j)}(z_0)}{j!} \quad \text{and} \quad G_j = \frac{g^{(j)}(z_0)}{j!}.$$

24.3   Some Applications of the Residue Theorem

24.3.1   The Argument Principle

The argument principle is an integral formula for the difference between the number of zeros and the number of poles of a function (counting multiplicities) enclosed by a given closed path Γ.

THEOREM 24.8   *Argument Principle*

Let f be differentiable on a closed path Γ and at all points in the set G of points enclosed by Γ, except at possibly a finite number of poles of f in G. Let Z be the number of zeros of f in G, and P the number of poles of f in G, with each pole and zero counted k times if its multiplicity is k. Then,

$$\oint_\Gamma \frac{f'(z)}{f(z)}\, dz = 2\pi i(Z - P).$$

Proof   Observe first that the only points in G where f'/f might possibly have a singularity are the zeros and poles of f in G.

Now suppose that f has a zero of order k at z_0 in G. We will show that f'/f must have a simple pole at z_0, and that $\mathrm{Re}\,s(f'/f, z_0) = k$. To see this, first note that, because z_0 is a zero of order k,

$$f(z_0) = f'(z_0) = \cdots = f^{(k-1)}(z_0) = 0$$

while $f^{(k)}(z_0) \neq 0$. Then, in some open disk about z_0, the Taylor expansion of $f(z)$ is

$$f(z) = \sum_{n=k}^{\infty} c_n (z - z_0)^n = \sum_{n=0}^{\infty} c_{n+k} (z - z_0)^{n+k}$$

$$= (z - z_0)^k \sum_{n=0}^{\infty} c_{n+k} (z - z_0)^n = (z - z_0)^k g(z),$$

where g is differentiable at z_0 (because it has a Taylor expansion there) and $g(z_0) = c_k \neq 0$. Now, in some annulus $0 < |z - z_0| < R$,

$$\frac{f'(z)}{f(z)} = \frac{k(z - z_0)^{k-1} g(z) + (z - z_0)^k g'(z)}{(z - z_0)^k g(z)} = \frac{k}{z - z_0} + \frac{g'(z)}{g(z)}.$$

Since $g'(z)/g(z)$ is differentiable at z_0, then $f'(z)/f(z)$ has a simple pole at z_0 and $\mathrm{Re}\,s(f'/f, z_0) = k$.

Next, suppose f has a pole of order m at z_1. In some annulus about z_0 $f(z)$ has Laurent expansion

$$f(z) = \sum_{n=-m}^{\infty} d_n (z - z_1)^n,$$

with $d_{-m} \neq 0$. Then

$$(z - z_1)^m f(z) = \sum_{n=-m}^{\infty} d_n (z - z_1)^{n+m} = \sum_{n=0}^{\infty} d_{n-m} (z - z_1)^n = h(z),$$

with h differentiable at z_1 and $h(z_1) = d_{-m} \neq 0$. Then $f(z) = (z - z_1)^{-m} h(z)$, so in some annulus about z_1,

$$\frac{f'(z)}{f(z)} = \frac{-m(z - z_1)^{-m-1} h(z) + (z - z_1)^{-m} h'(z)}{(z - z_1)^{-m} h(z)} = \frac{-m}{z - z_1} + \frac{h'(z)}{h(z)}.$$

Therefore, f'/f has a simple pole at z_1, with $\mathrm{Re}\,s(f'/f, z_1) = -m$.

Therefore, the sum of the residues of $f'(z)/f(z)$ at singularities of this function in G counts the zeros of f in G, according to multiplicity, and the negative of the number of poles of f in G, again according to multiplicity. ■

EXAMPLE 24.21

We will evaluate $\oint_\Gamma \cot(z)\, dz$, with Γ the closed path of Figure 24.3.

Write

$$\cot(z) = \frac{\cos(z)}{\sin(z)} = \frac{f'(z)}{f(z)}$$

where $f(z) = \sin(z)$. Since f has five simple zeros and no poles enclosed by Γ, the argument principle yields

$$\oint_\Gamma \cot(z)\, dz = 2\pi i (5 - 0) = 10\pi i. \quad ■$$

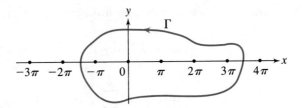

FIGURE 24.3

24.3.2   Rouché's Theorem

Finding the zeros of a function is a fundamental but potentially difficult problem, depending on the function. Rouché's theorem is sometimes useful in determining the number of zeros of a function that are enclosed by a given closed path.

THEOREM 24.9   Rouché

Let f and g be differentiable on a closed path Γ and on the set G of points enclosed by Γ. Suppose

$$|g(z)| < |f(z)| \quad \text{for all } z \text{ on } \Gamma.$$

Then f and $f + g$ have the same number of zeros in G, with the understanding that each zero is counted k times if its multiplicity is k. ■

In practice, we have a function whose zeros we want to know something about, and we attempt to write it as $f + g$, where f and g satisfy the hypotheses of the theorem. If we can choose f so that we know its zeros (or at least their numbers and multiplicities), then we can determine information about the zeros of $f + g$.

EXAMPLE 24.22

We will estimate the number of solutions of the equation

$$3z^6 - 14z^4 + 6z^2 - z + 1 = 0$$

enclosed by the circle $\Gamma : |z| = 2$.

Let $f(z) = -14z^4$ and $g(z) = 3z^6 + 6z^2 - z + 1$. Then the polynomial we are interested in is $f(z) + g(z)$. For z on Γ, $|z| = 2$, so

$$|f(z)| = 14\left|z^4\right| = 14(2^4) = 224$$

and

$$|g(z)| = \left|3z^6 + 6z^2 - z + 1\right| \le 3\,|z|^6 + 6\,|z|^2 + |z| + 1 = 219.$$

On Γ,

$$|g(z)| < |f(z)|.$$

In $|z| < 2$, f has exactly four zeros (the zero 0 of multiplicity 4). Therefore, $f(z) + g(z)$ has exactly four zeros in this open disk. ■

In this example, we let $f + g$ be the function of interest and chose $f(z) = -14z^4$ because (1) its zeros are obvious, and (2) we can easily bound $|f(z)|$ on Γ. This strategy works only if $|g(z)| < |f(z)|$ on Γ, and this was indeed found to be the case. Note that this inequality is required only at points on the curve, generally a much simpler requirement to ask of g than having such an inequality throughout the region bounded by Γ.

Here is a proof of Rouché's theorem.

Proof Define

$$h(t) = \frac{1}{2\pi i} \oint_\Gamma \frac{(f(z) + tg(z))'}{f(z) + tg(z)} \, dz \quad \text{for } 0 \le t \le 1.$$

By the argument principle, $h(1)$ is the number of zeros of $f + g$ in G, and $h(0)$ is the number of zeros of f in G. Further, $h(t)$ is an integer for $0 \le t \le 1$.

The theorem will be proved if we can show that $h(0) = h(1)$. To do this, we will begin by showing that h is continuous. Let t_1 and t_2 be in $[0, 1]$. A routine calculation yields

$$h(t_1) - h(t_2) = \frac{t_1 - t_2}{2\pi i} \oint_\Gamma \frac{g'(z)f(z) - f'(z)g(z)}{[f(z) + t_1 g(z)][f(z) + t_2 g(z)]} \, dz. \tag{24.2}$$

Because Γ is compact, by Theorem 21.3, there are positive numbers M and m such that, for all z on Γ,

$$\left| g'(z)f(z) - f'(z)g(z) \right| \le M, \quad |g(z)| \le M,$$

and

$$m \le |f(z) + t_1 g(z)|.$$

Then, for $|t_1 - t_2| \le m/2M$,

$$|f(z) + t_2 g(z)| = |f(z) + t_1 g(z) - t_1 g(z) + t_2 g(z)|$$

$$\ge |f(z) + t_1 g(z)| - |t_2 - t_1| \, |g(z)| \ge m/2.$$

Then, by equation (24.2) and Theorem 22.5,

$$|h(t_1) - h(t_2)| \le \frac{1}{\pi} |t_1 - t_2| \frac{M}{m} (\text{length of } \Gamma),$$

and this can be made as small as we like by choosing $|t_1 - t_2|$ sufficiently small. This proves that h is continuous on $[0, 1]$.

We can now show that $h(0) = h(1)$. Suppose $h(0) \ne h(1)$. By the intermediate value theorem, if r is any noninteger real number chosen between $h(0)$ and $h(1)$, there must be some t_0 in $(0, 1)$ such that $h(t_0) = r$, and this is impossible because $h(t)$ is integer valued. We conclude that $h(0) = h(1)$, proving the theorem. ∎

24.3.3 Summation of Real Series

A complex series $\sum_{n=1}^\infty a_n + ib_n$ can be treated by analyzing the two real series $\sum_{n=1}^\infty a_n$ and $\sum_{n=1}^\infty b_n$. However, even when a real series can be shown to converge, it is usually difficult or impossible to determine its sum. We will use the residue theorem to derive formulas for the sums of two types of real series.

Let $P(z)$ be a polynomial of degree at least 2, with real coefficients and no integer zeros. Let $z_1, \ldots, z_m$ be the distinct zeros of $P(z)$. These may be real or complex numbers, but not integers. We claim that

$$\sum_{n=-\infty}^\infty \frac{1}{P(n)} = -\pi \sum_{j=1}^m \text{Res} \left(\frac{\cot(\pi z)}{P(z)}, z_j \right) \tag{24.3}$$

and

$$\sum_{n=-\infty}^{\infty} (-1)^n \frac{1}{P(n)} = -\pi \sum_{j=1}^{m} \operatorname{Re} s\left(\frac{\csc(\pi z)}{P(z)}, z_j\right). \tag{24.4}$$

These formulas provide often practical means of summing real series of the form

$$\cdots + \frac{1}{P(-2)} + \frac{1}{P(-1)} + \frac{1}{P(0)} + \frac{1}{P(1)} + \frac{1}{P(2)} + \cdots$$

and

$$\cdots + \frac{1}{P(-2)} - \frac{1}{P(-1)} + \frac{1}{P(0)} - \frac{1}{P(1)} + \frac{1}{P(2)} + \cdots.$$

We will consider two examples and then show how formulas (24.3) and (24.4) arise.

EXAMPLE 24.23

We will sum the series

$$\sum_{n=0}^{\infty} \frac{1}{a^2 + n^2},$$

in which a is any positive constant. This is the series

$$\frac{1}{a^2} + \frac{1}{a^2 + 1} + \frac{1}{a^2 + 4} + \frac{1}{a^2 + 9} + \cdots.$$

Let $P(z) = a^2 + z^2$. Then P has degree 2 and zeros $\pm ai$, which are not integers. Equation (24.3) applies. Let

$$f(z) = \frac{\cot(\pi z)}{a^2 + z^2} = \frac{\cos(\pi z)}{(a^2 + z^2)\sin(\pi z)} = \frac{\cos(\pi z)}{(z + ai)(z - ai)\sin(\pi z)}.$$

f has simple poles at ai and $-ai$. Compute the residues:

$$\operatorname{Re} s(f, ai) = \lim_{z \to ai} (z - ai) f(z) = \lim_{z \to ai} \frac{\cos(\pi z)}{(z + ai)\sin(\pi z)} = \frac{\cos(\pi ai)}{2ai \sin(\pi ai)}$$

and

$$\operatorname{Re} s(f, -ai) = \lim_{z \to -ai} (z + ai) f(z) = \frac{\cos(-\pi ai)}{-2ai \sin(-\pi ai)} = \frac{\cos(\pi ai)}{2ai \sin(\pi ai)}.$$

By equation (24.3),

$$\sum_{n=-\infty}^{\infty} \frac{1}{a^2 + n^2} = -\pi\left(\frac{\cos(\pi ai)}{2ai \sin(\pi ai)} + \frac{\cos(\pi ai)}{2ai \sin(\pi ai)}\right) = \frac{\pi i \cos(\pi ai)}{a \sin(\pi ai)}.$$

Since we expect a real series to converge to a real number, we will simplify this result. First,

$$\cos(\pi ai) = \frac{1}{2}\left(e^{-\pi a} + e^{\pi a}\right) = \cosh(\pi a)$$

and

$$\sin(\pi ai) = \frac{1}{2i}\left(e^{-\pi a} - e^{\pi a}\right) = i \sinh(\pi a).$$

Then

$$\sum_{n=-\infty}^{\infty} \frac{1}{a^2 + n^2} = \frac{\pi i}{a} \frac{\cosh(\pi a)}{i \sinh(\pi a)} = \frac{\pi}{a} \frac{\cosh(\pi a)}{\sinh(\pi a)}.$$

Now write

$$\sum_{n=-\infty}^{\infty} \frac{1}{a^2 + n^2} = \sum_{n=-\infty}^{0} \frac{1}{a^2 + n^2} - \frac{1}{a^2} + \sum_{n=0}^{\infty} \frac{1}{a^2 + n^2}$$

$$= -\frac{1}{a^2} + 2 \sum_{n=0}^{\infty} \frac{1}{a^2 + n^2} = \frac{\pi}{a} \frac{\cosh(\pi a)}{\sinh(\pi a)}.$$

The term $-1/a^2$ in this equation balances the fact that $1/a^2$ appears as the $n = 0$ term in both summations, and it should appear only once in the equation. Thus,

$$\sum_{n=0}^{\infty} \frac{1}{a^2 + n^2} = \frac{1}{2a^2} + \frac{\pi}{2a} \frac{\cosh(\pi a)}{\sinh(\pi a)}.$$

For example, with $a = 1$ we have

$$\sum_{n=0}^{\infty} \frac{1}{1 + n^2} = 1 + \frac{1}{2} + \frac{1}{5} + \frac{1}{10} + \frac{1}{17} + \cdots = \frac{1}{2} + \frac{\pi}{2} \frac{\cosh(\pi)}{\sinh(\pi)},$$

which is approximately 2.0767. ∎

EXAMPLE 24.24

We will determine

$$\sum_{n=0}^{\infty} \frac{(-1)^n}{(2n + 1)^3},$$

which is

$$1 - \frac{1}{27} + \frac{1}{125} - \cdots.$$

Let $P(z) = (2z + 1)^3$. Then P has a third-order zero at $z = -\frac{1}{2}$ and no integer zeros, so we can use equation (24.4). Let

$$f(z) = \frac{\csc(\pi z)}{(2z + 1)^3}.$$

Compute

$$\operatorname{Re} s\left(f, -\frac{1}{2}\right) = \lim_{z \to -1/2} \frac{1}{2} \frac{d^2}{dz^2}\left[\left(z + \frac{1}{2}\right)^3 \frac{\csc(\pi z)}{(2z + 1)^3}\right]$$

$$= \frac{1}{2} \lim_{z \to -1/2} \frac{d^2}{dz^2}\left[\left(z + \frac{1}{2}\right)^3 \frac{\csc(\pi z)}{8\left(z + \frac{1}{2}\right)^3}\right]$$

$$= \frac{1}{16} \lim_{z \to -1/2} \frac{d^2}{dz^2} \csc(\pi z)$$

$$= \frac{1}{16} \lim_{z \to -1/2} \pi^2[\csc(\pi z)\cot^2(\pi z) + \csc^3(\pi z)]$$

$$= \frac{\pi^2}{16}\left[\csc\left(-\frac{\pi}{2}\right)\cot^2\left(-\frac{\pi}{2}\right) + \csc^3\left(-\frac{\pi}{2}\right)\right] = -\frac{\pi^2}{16}.$$

Then

$$\sum_{n=-\infty}^{\infty} (-1)^n \frac{1}{(2n + 1)^3} = (-\pi)\frac{-\pi^2}{16} = \frac{\pi^3}{16}.$$

Now it is routine to verify that

$$\sum_{n=-\infty}^{\infty} (-1)^n \frac{1}{(2n + 1)^3} = 2\sum_{n=0}^{\infty}(-1)^n \frac{1}{(2n + 1)^3},$$

so

$$\sum_{n=0}^{\infty}(-1)^n \frac{1}{(2n + 1)^3} = \frac{\pi^3}{32}. \quad \blacksquare$$

We will give the idea behind equation (24.3). Specifically, we want to clarify the connection between the summation and a sum of residues of the function appearing in equation (24.3).

Consider

$$\oint_\Gamma \frac{\pi \cot(\pi z)}{P(z)} dz,$$

with Γ_N the rectangle of Figure 24.4, having sides parallel to the axes and passing through $\pm(N + \frac{1}{2})$ on the real axis and $\pm Ni$ on the imaginary axis. Choose N large enough that Γ encloses all the zeros of $P(z)$. The singularities of $\pi \cot(\pi z)/P(z)$ enclosed by Γ are simple poles at $0, \pm 1, \ldots, \pm N$ due to the $\cot(\pi z)$ factor and also poles at the zeros of $P(z)$. By the residue theorem,

$$\oint_\Gamma \frac{\pi \cot(\pi z)}{P(z)} dz = 2\pi i \left[\sum_{j=-N}^{N} \operatorname{Re} s\left(\frac{\pi \cot(\pi z)}{P(z)}, j\right) + \sum_p \operatorname{Re} s\left(\frac{\pi \cot(\pi z)}{P(z)}, p\right)\right],$$

in which the second summation is over all poles of $\pi \cot(\pi z)/P(z)$ occurring at zeros of $P(z)$.

By estimating $|\cot(\pi z)/P(z)|$ on each of the straight-line sides of Γ, we can show that the integral on the left side of the last equation has limit 0 as $N \to \infty$. By evaluating the residues

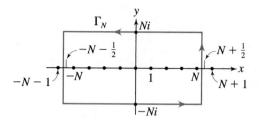

FIGURE 24.4

on the right side of the equation, we obtain equation (24.3). The details of this are left to the student.

24.3.4 An Inversion Formula for the Laplace Transform

If f is a complex function defined at least for all z on $[0, \infty)$, the Laplace transform of f is

$$\mathcal{L}[f](z) = \int_0^\infty e^{-zt} f(t)\, dt,$$

for all z such that this integral is defined and finite. If $\mathcal{L}[f] = F$, then F is the Laplace transform of f, and f is an inverse Laplace transform of F. Sometimes we write $f = \mathcal{L}^{-1}[f]$, although this requires additional conditions for uniqueness because there are in general many functions having a given F as their Laplace transform.

We will give a formula for $\mathcal{L}^{-1}[f]$ in terms of the sum of the residues of $e^{zt}F(z)$ at the poles of f.

THEOREM 24.10   *Inverse Laplace Transform*
_____

Let F be differentiable for all z except for a finite number of points $z_1, \ldots, z_n$, which are all poles of F. Suppose for some real σ, F is differentiable for all z with $\mathrm{Re}(z) > \sigma$. Suppose also that there are numbers M and R such that

$$|zF(z)| \leq M \quad \text{for } |z| > R.$$

For $t \geq 0$, let

$$f(t) = \sum_{j=1}^{n} \mathrm{Re}\, s(e^{zt} F(z), z_j).$$

Then

$$\mathcal{L}[f](z) = F(z) \quad \text{for } \mathrm{Re}(z) > \sigma. \quad \blacksquare$$

The condition that F is differentiable for $\mathrm{Re}(z) > \sigma$ means that $F'(z)$ exists for all z to the right of the vertical line $x = \sigma$. It is also assumed that $zF(z)$ is a bounded function for z outside some sufficiently large circle about the origin. For example, this condition is satisfied by any rational function (quotient of polynomials) in which the degree of the denominator exceeds that of the numerator.

We will sketch a proof after looking at two examples.

EXAMPLE 24.25

Let $a > 0$. We want an inverse Laplace transform of $F(z) = 1/(a^2 + z^2)$.

This can be found in tables of Laplace transforms. To use the theorem, F has simple poles at $\pm ai$. Compute

$$\operatorname{Re} s\left(\frac{e^{zt}}{a^2 + z^2}, ai\right) = \frac{e^{ati}}{2ai}$$

and

$$\operatorname{Re} s\left(\frac{e^{zt}}{a^2 + z^2}, -ai\right) = \frac{e^{-ati}}{-2ai}.$$

An inverse Laplace transform of F is given by

$$f(t) = \frac{1}{2ai}\left(e^{ati} - e^{-ati}\right) = \frac{1}{a}\sin(at)$$

for $t \geq 0$. ∎

EXAMPLE 24.26

We want a function whose Laplace transform is

$$F(z) = \frac{1}{(z^2 - 4)(z - 1)^2}.$$

F has simple poles at ± 2 and a double pole at 1. Compute

$$\operatorname{Re} s\left(\frac{e^{zt}}{(z^2 - 4)(z - 1)^2}, 2\right) = \lim_{z \to 2} \frac{e^{zt}}{(z + 2)(z - 1)^2} = \frac{1}{4}e^{2t},$$

$$\operatorname{Re} s\left(\frac{e^{zt}}{(z^2 - 4)(z - 1)^2}, -2\right) = \lim_{z \to -2} \frac{e^{zt}}{(z - 2)(z - 1)^2} = -\frac{1}{36}e^{-2t},$$

and

$$\operatorname{Re} s\left(\frac{e^{zt}}{(z^2 - 4)(z - 1)^2}, 1\right) = \lim_{z \to 1} \frac{d}{dz}\left(\frac{e^{zt}}{z^2 - 4}\right)$$

$$= \lim_{z \to 1} e^{zt}\frac{tz^2 - 4t - 2z}{\left(z^2 - 4\right)^2} = -\frac{1}{3}te^t - \frac{2}{9}e^t.$$

An inverse Laplace transform of F is given by

$$f(t) = -\frac{1}{3}te^t - \frac{2}{9}e^t + \frac{1}{4}e^{2t} - \frac{1}{36}e^{-2t},$$

for $t > 2$ (since all poles of F occur on or to the left of the line $\operatorname{Re}(z) = 2$). ∎

Here is an outline of a proof of Theorem 24.10. Let $a > \sigma$ and consider $\oint_\Gamma e^{tz}F(z)\,dz$, with Γ the closed path of Figure 24.5. Γ is defined to enclose all the poles of F. Let Γ intersect the axes

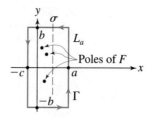

FIGURE 24.5

at $x = -c$, $x = a$, $y = b$, and $y = -b$, with b and c positive numbers. By the residue theorem,

$$\oint_\Gamma e^{tz} F(z)\, dz = 2\pi i \sum_{j=1}^{n} \operatorname{Res}\left(e^{tz} F(z), z_j\right).$$

For $|z|$ sufficiently large, $|zF(z)| \le M$, and one can show by making estimates along the sides of the path that

$$\oint_\Gamma e^{tz} F(z)\, dz \to \int_{L_a} e^{tz} F(z)\, dz$$

as $c \to \infty$ and $b \to \infty$. Here L_a is the vertical line $\operatorname{Re}(z) = a$, and in the limit the rectangle expands over the entire half-plane $\operatorname{Re}(z) \le a$. For $\operatorname{Re}(z) = a$, then,

$$2\pi i f(t) = \oint_\Gamma e^{tz} F(z)\, dz.$$

Now

$$\int_0^r e^{-tz} \left[\oint_\Gamma e^{\xi t} F(\xi)\, d\xi\right] dt = \oint_\Gamma \int_0^r e^{-tz} e^{\xi t} F(\xi)\, dt\, d\xi$$

$$= \oint_\Gamma \int_0^r e^{(\xi - z)t} F(\xi)\, d\xi = \oint_\Gamma \frac{1}{\xi - z} \left[e^{(\xi - z)t}\right]_0^r F(\xi)\, d\xi$$

$$= \oint_\Gamma \frac{1}{\xi - z} \left[e^{(\xi - z)r - 1}\right] F(\xi)\, d\xi.$$

If ξ is on Γ, then $\operatorname{Re}(\xi) < \operatorname{Re}(z)$, so

$$e^{(\xi - z)r} \to 0 \quad \text{as } r \to \infty.$$

Therefore,

$$2\pi i \mathfrak{L}[f](z) = -\oint_\Gamma \frac{F(\xi)}{\xi - z}\, d\xi.$$

As $c \to \infty$ and $b \to \infty$,

$$-\oint_\Gamma \frac{F(\xi)}{\xi - z}\, d\xi \to -\oint_{L_a} \frac{F(\xi)}{\xi - z}\, d\xi.$$

Now reflect Γ about the line $x = a$ to obtain the closed path Γ^* shown in Figure 24.6. By an argument similar to that just sketched, we get

$$\oint_{\Gamma^*} \frac{F(\xi)}{\xi - z}\, d\xi \to -\oint_{L_a} \frac{F(\xi)}{\xi - z}\, d\xi.$$

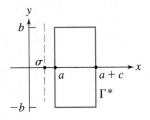

FIGURE 24.6

But, by Cauchy's integral formula,

$$\oint_{\Gamma^*} \frac{F(\xi)}{\xi - z} d\xi = 2\pi i F(z).$$

Therefore,

$$F(z) = \mathfrak{L}[f](z),$$

as we wanted to show.

In these sections we can see a theme developing. A variety of problems (zeros of functions, sums of series, inverse Laplace transforms, others to be discussed) can be approached by integrating an appropriately chosen complex function over an appropriately chosen path. The function and path must be selected so that the integral gives us the quantity we want to calculate, perhaps after some limit process. We can then use the residue theorem to explicitly evaluate the integral. Depending on the problem, choosing the right function and the right path can be a nontrivial task, but at least this method provides an approach.

24.3.5 Evaluation of Real Integrals

We will illustrate the use of the residue theorem in evaluating several general classes of real integrals.

Integrals of $\int_0^{2\pi} K(\cos(\theta), \sin(\theta))\, d\theta$

Let $K(x, y)$ be a quotient of polynomials in x and y, for example,

$$\frac{x^3 y - 2xy^2 + x - 2y}{x^4 + x^3}.$$

If we replace x with $\cos(\theta)$ and y with $\sin(\theta)$, we obtain a quotient involving sums of products of positive integer powers of $\cos(\theta)$ and $\sin(\theta)$. We are interested in evaluating integrals of the form

$$\int_0^{2\pi} K(\cos(\theta), \sin(\theta))\, d\theta.$$

The idea will be to show that this real integral is equal to an integral of a certain complex function over the unit circle. We then use the residue theorem to evaluate this complex integral, obtaining the value of the real integral.

To execute this strategy, let γ be the unit circle, oriented counterclockwise as usual. Parametrize γ by $\gamma(\theta) = e^{i\theta}$ for $0 \le \theta \le 2\pi$. On this curve, $z = e^{i\theta}$ and $\bar{z} = e^{-i\theta} = 1/z$, so

$$\cos(\theta) = \frac{1}{2}\left(e^{i\theta} + e^{-i\theta}\right) = \frac{1}{2}\left(z + \frac{1}{z}\right)$$

and

$$\sin(\theta) = \frac{1}{2i}\left(e^{i\theta} - e^{-i\theta}\right) = \frac{1}{2i}\left(z - \frac{1}{z}\right).$$

Further, on γ,

$$dz = ie^{i\theta}\,d\theta = iz\,d\theta,$$

so

$$d\theta = \frac{1}{iz}\,dz.$$

Now we have

$$\oint_\gamma K\left(\frac{1}{2}\left(z + \frac{1}{z}\right), \frac{1}{2i}\left(z - \frac{1}{z}\right)\right)\frac{1}{iz}\,dz = \int_0^{2\pi} K(\cos(\theta), \sin(\theta))\frac{1}{ie^{i\theta}}ie^{i\theta}\,d\theta$$

$$= \int_0^{2\pi} K(\cos(\theta), \sin(\theta))\,d\theta.$$

This converts the real integral we want to evaluate into the integral of a complex function $f(z)$ over the unit circle, where

$$f(z) = K\left(\frac{1}{2}\left(z + \frac{1}{z}\right), \frac{1}{2i}\left(z - \frac{1}{z}\right)\right)\frac{1}{iz}.$$

Use the residue theorem to evaluate $\oint_\gamma f(z)\,dz$, obtaining

$$\int_0^{2\pi} K(\cos(\theta), \sin(\theta))\,d\theta = 2\pi i \sum_p \mathrm{Re}\,s(f, p). \tag{24.5}$$

The sum on the right is over all of the poles p of $f(z)$ enclosed by the unit circle. Poles occurring outside the unit circle are not included in the calculation. Finally, equation (24.5) assumes that $f(z)$ has no singularities on the unit circle.

The procedure for evaluating $\int_0^{2\pi} K(\cos(\theta), \sin(\theta))\,d\theta$, then, is to compute $f(z)$, determine its poles within the unit circle, evaluate the residues there, and apply equation (24.5). This is a very powerful method that often yields closed-form evaluations of integrals for which standard techniques of integration from real calculus are inadequate.

EXAMPLE 24.27

We will evaluate

$$\int_0^{2\pi} \frac{\sin^2(\theta)}{2 + \cos(\theta)}\,d\theta.$$

The function K in the preceding discussion is

$$K(x, y) = \frac{y^2}{2 + x}.$$

The first step is to replace $x = \cos(\theta)$ with $(z + (1/z))/2$ and $y = \sin(\theta)$ with $(z - (1/z))/2i$, and then multiply by $1/iz$, to produce the complex function

$$f(z) = K\left(\frac{1}{2}\left(z + \frac{1}{z}\right), \frac{1}{2i}\left(z - \frac{1}{z}\right)\right)\frac{1}{iz}$$

$$= \frac{[(1/2i)(z - (1/z))]^2}{2 + \frac{1}{2}(z + (1/z))}\frac{1}{iz} = \frac{i}{2}\frac{z^4 - 2z^2 + 1}{z^2(z^2 + 4z + 1)}.$$

f has a double pole at 0 and simple poles at zeros of $z^2 + 4z + 1$, which are $-2 + \sqrt{3}$ and $-2 - \sqrt{3}$. Of these two simple poles of f, the first is enclosed by γ and the second is not, so discard $-2 - \sqrt{3}$. By equation (24.5),

$$\int_0^{2\pi} \frac{\sin^2(\theta)}{2 + \cos(\theta)}\, d\theta = 2\pi i\left[\mathrm{Re}\, s(f, 0) + \mathrm{Re}\, s(f, -2 + \sqrt{3})\right].$$

Now

$$\mathrm{Re}\, s(f, 0) = \lim_{z \to 0}\frac{d}{dz}\left[z^2 f(z)\right] = \lim_{z \to 0}\frac{d}{dz}\frac{i}{2}\frac{z^4 - 2z^2 + 1}{z^2 + 4z + 1}$$

$$= \frac{i}{2}\lim_{z \to 0}\left(2\frac{z^5 + 6z^4 - 4z^2 - 3z + 2z^3 - 2}{\left(z^2 + 4z + 1\right)^2}\right) = -2i$$

and

$$\mathrm{Re}\, s(f, -2 + \sqrt{3}) = \frac{i}{2}\left[\frac{z^4 - 2z^2 + 1}{2z(z^2 + 4z + 1) + z^2(2z + 4)}\right]_{z=-2+\sqrt{3}}$$

$$= \frac{i}{2}\frac{42 - 24\sqrt{3}}{-12 + 7\sqrt{3}}.$$

Then

$$\int_0^{2\pi} \frac{\sin^2(\theta)}{2 + \cos(\theta)}\, d\theta = 2\pi i\left[-2i + \frac{i}{2}\frac{42 - 24\sqrt{3}}{-12 + 7\sqrt{3}}\right] = \frac{90 - 52\sqrt{3}}{12 - 7\sqrt{3}}\pi,$$

approximately 1.68357. ∎

In applying this method, if a nonreal number results, check the calculations, because a real integral must have a real value.

EXAMPLE 24.28

We will evaluate

$$\int_0^\pi \frac{1}{\alpha + \beta \cos(\theta)}\, d\theta,$$

where $0 < \beta < \alpha$.

Since the method we have developed deals with integrals over $[0, 2\pi]$, we must first decide how to accommodate an integral over $[0, \pi]$. Write

$$\int_0^{2\pi} \frac{1}{\alpha + \beta \cos(\theta)}\, d\theta = \int_0^\pi \frac{1}{\alpha + \beta \cos(\theta)}\, d\theta + \int_\pi^{2\pi} \frac{1}{\alpha + \beta \cos(\theta)}\, d\theta.$$

Let $w = 2\pi - \theta$ in the last integral to obtain

$$\int_\pi^{2\pi} \frac{1}{\alpha + \beta \cos(\theta)} \, d\theta = \int_\pi^0 \frac{1}{\alpha + \beta \cos(2\pi - w)} (-1) \, dw = \int_0^\pi \frac{1}{\alpha + \beta \cos(w)} \, dw.$$

Therefore,

$$\int_0^\pi \frac{1}{\alpha + \beta \cos(\theta)} \, d\theta = \frac{1}{2} \int_0^{2\pi} \frac{1}{\alpha + \beta \cos(\theta)} \, d\theta,$$

and we can concentrate on the integral over $[0, 2\pi]$. First produce the function

$$f(z) = \frac{1}{\alpha + \frac{\beta}{2} \left(z + \frac{1}{z}\right)} \frac{1}{iz} = \frac{-2i}{\beta z^2 + 2\alpha z + \beta}.$$

f has simple poles at

$$z = \frac{-\alpha \pm \sqrt{\alpha^2 - \beta^2}}{\beta}.$$

Since $\alpha > \beta$, these numbers are real. Only one of them,

$$z_1 = \frac{-\alpha + \sqrt{\alpha^2 - \beta^2}}{\beta},$$

is enclosed by γ. The other is outside the unit disk and is irrelevant for our purposes. Then

$$\int_0^\pi \frac{1}{\alpha + \beta \cos(\theta)} \, d\theta = \frac{1}{2} \int_0^{2\pi} \frac{1}{\alpha + \beta \cos(\theta)} \, d\theta = \frac{1}{2} 2\pi i \operatorname{Re} s(f, z_1)$$

$$= \pi i \frac{-2i}{2\beta z_1 + 2\alpha} = \frac{\pi}{\sqrt{\alpha^2 - \beta^2}}. \quad \blacksquare$$

Before continuing with other kinds of real integrals we can evaluate using the residue theorem, we will take a brief excursion and give another, perhaps surprising, proof of the fundamental theorem of algebra. This argument is originally due to N. C. Ankeny, and the version we give appeared in *Lion Hunting and Other Mathematical Pursuits*, by R. P. Boas (The Mathematical Association of America Dolciani Mathematical Expositions, Vol. 15).

Let $p(z)$ be a nonconstant polynomial with complex coefficients. We want to show that for some number z, $p(z) = 0$.

First, we may assume that $p(x)$ is real if x is real. To see why this is true, let

$$p(z) = a_0 + a_1 z + \cdots + a_n z^n,$$

where $a_n \neq 0$. Denote

$$\overline{p}(z) = \overline{a_0} + \overline{a_1} z + \cdots + \overline{a_n} z^n.$$

Then $q(z) = p(z)\overline{p}(z)$ is a nonconstant polynomial. Further, if $z = x$ is real, then $\overline{x} = x$ and

$$q(x) = p(x)\overline{p}(x)$$

$$= (a_0 + a_1 x + \cdots + a_n x^n)(\overline{a_0} + \overline{a_1} x + \cdots + \overline{a_n} x^n)$$

$$= (a_0 + a_1 x + \cdots + a_n x^n)\overline{\left(a_0 + a_1 x + \cdots + a_n x^n\right)}$$

$$= |a_0 + a_1 x + \cdots + a_n x^n|^2$$

is real. We could then use $q(z)$ in our argument in place of $p(z)$, since $q(z)$ is a polynomial with no zero if $p(z)$ has no zero. Thus suppose that $p(z) \neq 0$ for all z, and $p(x)$ is real if x is real.

Because $p(x)$ is continuous and never zero for real x, $p(x)$ must be strictly positive or strictly negative for all real x. But then

$$\int_0^{2\pi} \frac{1}{p(2\cos(\theta))}\, d\theta \neq 0.$$

But, by the method we have just discussed, with γ the unit circle, we conclude that

$$\int_0^{2\pi} \frac{1}{p(2\cos(\theta))}\, d\theta = \oint_\gamma \frac{1}{p(z+1/z)}\frac{1}{iz}\, dz = \frac{1}{i}\oint_\gamma \frac{z^{n-1}}{r(z)}\, dz \neq 0,$$

where

$$r(z) = z^n p\left(z + \frac{1}{z}\right)$$

$$= z^n\left[a_0 + a_1\left(z + \frac{1}{z}\right) + a_2\left(z + \frac{1}{z}\right)^2 + \cdots + a_n\left(z + \frac{1}{z}\right)^n\right]$$

$$= z^n\left[a_0 + a_1\frac{z^2+1}{z} + a_2\frac{(z^2+1)^2}{z^2} + \cdots + a_n\frac{(z^2+1)^n}{z^n}\right].$$

From this it is clear that $r(z)$ is a polynomial. If $r(\zeta) = 0$ for some $\zeta \neq 0$, then we would have $p(\zeta + 1/\zeta) = 0$, so $\zeta + 1/\zeta$ would be a zero of p, a contradiction. Further, $r(0) = a_n \neq 0$ because p has degree n. Therefore, $r(z) \neq 0$ for all z, so $z^{n-1}/r(z)$ is a differentiable function for all z. But then, by Cauchy's theorem,

$$\frac{1}{i}\oint_\gamma \frac{z^{n-1}}{r(z)}\, dz = 0,$$

a contradiction. We conclude that $p(z) = 0$ for some number z, proving the fundamental theorem of algebra.

Evaluation of $\int_{-\infty}^{\infty}[p(x)/q(x)]\, dx$

We will now consider real integrals of the form

$$\int_{-\infty}^{\infty} \frac{p(x)}{q(x)}\, dx,$$

in which p and q are polynomials with real coefficients and no common factors, q has no real zeros, and the degree of q exceeds the degree of p by at least 2. These conditions are sufficient to ensure convergence of this improper integral.

As with the preceding class of integrals, the strategy is to devise a complex integral that is equal to this real integral, then evaluate the complex integral using the residue theorem. To do this, first observe that $q(z)$ has real coefficients, so its zeros occur in complex conjugate pairs. Suppose the zeros of q are $z_1, \overline{z_1}, z_2, \overline{z_2}, \ldots, z_m, \overline{z_m}$, with each z_j in the upper half-plane $\text{Im}(z) > 0$ and each $\overline{z_j}$ in the lower half-plane $\text{Im}(z) < 0$. Let Γ be the curve shown in Figure 24.7, consisting of a semicircle γ of radius R and the segment S from $-R$ to R on the real axis, with R large enough that Γ encloses all the poles $z_1, \ldots, z_m$ of $p(z)/q(z)$ in the upper half-plane. Then

$$\oint_\Gamma \frac{p(z)}{q(z)}\, dz = 2\pi i \sum_{j=1}^m \text{Res}(p/q, z_j) = \int_S \frac{p(z)}{q(z)}\, dz + \int_\gamma \frac{p(z)}{q(z)}\, dz. \tag{24.6}$$

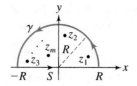

FIGURE 24.7

On S, $z = x$ for $-R \leq x \leq R$, so

$$\int_S \frac{p(z)}{q(z)} \, dz = \int_{-R}^{R} \frac{p(x)}{q(x)} \, dx.$$

Next consider the integral over γ. Since the degree of $q(z)$ exceeds that of $p(z)$ by at least 2,

$$\text{degree of } z^2 p(z) \leq \text{degree of } q(z).$$

This means that for sufficiently large R, $z^2 p(z)/q(z)$ is bounded for $|z| \geq R$. That is, for some number M,

$$\left| \frac{z^2 p(z)}{q(z)} \right| \leq M \quad \text{for } |z| \geq R.$$

Then

$$\left| \frac{p(z)}{q(z)} \right| \leq \frac{M}{|z|^2} \leq \frac{M}{R^2} \quad \text{for } |z| \geq R,$$

so

$$\left| \int_\gamma \frac{p(z)}{q(z)} \, dz \right| \leq \frac{M}{R^2} \, (\text{length of } \gamma)$$

$$= \frac{M}{R^2} (\pi R) = \frac{\pi M}{R} \to 0 \quad \text{as } R \to \infty.$$

Thus, in the limit as $R \to \infty$ in equation (24.6), the first integral on the right has limit $\int_{-\infty}^{\infty} (p(x)/q(x)) \, dx$ and the second integral has limit zero. In the limit as $R \to \infty$, equation (24.6) yields

$$\int_{-\infty}^{\infty} \frac{p(x)}{q(x)} \, dx = 2\pi i \sum_{j=1}^{m} \text{Re} \, s\left(\frac{p(z)}{q(z)}, z_j \right). \tag{24.7}$$

Equation (24.7) provides a general method for evaluating integrals of rational functions over the real line, under the assumptions made earlier. It is not necessary to repeat the derivation of this equation each time it is used—simply determine the zeros of $q(z)$ in the upper half-plane, evaluate the residue of p/q at each such zero (which is a pole of p/q whose order must be determined), and apply equation (24.7).

EXAMPLE 24.29

We will evaluate

$$\int_{-\infty}^{\infty} \frac{1}{x^6 + 64} \, dx.$$

Here $p(z) = 1$ and $q(z) = z^6 + 64$. The degree of q exceeds that of p by 6, and q has no real zeros. The zeros of $z^6 + 64$ are the sixth roots of -64. To find these, put -64 in polar form:

$$-64 = 64e^{i(\pi + 2n\pi)},$$

in which n can be any integer. The six sixth roots of -64 are

$$2e^{i(\pi + 2n\pi)/6} \quad \text{for } n = 0, 1, 2, 3, 4, 5.$$

The three roots in the upper half-plane are

$$z_1 = 2e^{\pi i/6}, z_2 = 2e^{\pi i/2} = 2i, \quad \text{and} \quad z_3 = 2e^{5\pi i/6}.$$

We need the residue of $1/(z^6 + 64)$ at each of these simple poles. Corollary 24.1 is convenient to use here:

$$\text{Res}\left(\frac{1}{z^6 + 64}, 2e^{\pi i/6}\right) = \frac{1}{6(2e^{\pi i/6})^5} = \frac{1}{192}e^{-5\pi i/6},$$

$$\text{Res}\left(\frac{1}{z^6 + 64}, 2i\right) = \frac{1}{6(2i)^5} = -\frac{i}{192},$$

and

$$\text{Res}\left(\frac{1}{z^6 + 64}, 2e^{5\pi i/6}\right) = \frac{1}{6(2e^{5\pi i/6})^5} = \frac{1}{192}e^{-25\pi i/6} = \frac{1}{192}e^{-\pi i/6}.$$

Then

$$\int_{-\infty}^{\infty} \frac{1}{x^6 + 64} dx = \frac{2\pi i}{192}\left[e^{-5\pi i/6} - i + e^{-\pi i/6}\right]$$

$$= \frac{\pi i}{96}\left[\cos\left(\frac{5\pi}{6}\right) - i\sin\left(\frac{5\pi}{6}\right) - i + \cos\left(\frac{\pi}{6}\right) - i\sin\left(\frac{\pi}{6}\right)\right].$$

Now

$$\cos\left(\frac{5\pi}{6}\right) + \cos\left(\frac{\pi}{6}\right) = 0$$

and

$$\sin\left(\frac{5\pi}{6}\right) + \sin\left(\frac{\pi}{6}\right) = 1$$

so

$$\int_{-\infty}^{\infty} \frac{1}{x^6 + 64} dx = \frac{\pi i}{96}(-2i) = \frac{\pi}{48}. \quad \blacksquare$$

EXAMPLE 24.30

Let α and β be distinct positive numbers. We will evaluate

$$\int_{-\infty}^{\infty} \frac{1}{(x^2 + \alpha^2)(x^2 + \beta^2)} dx.$$

The zeros of the denominator in the upper half-plane are αi and βi, and these are simple. By Corollary 24.1,

$$
\operatorname{Res}\left(\frac{1}{(z^2+\alpha^2)(z^2+\beta^2)}, \alpha i\right)
$$

$$
=\left[\frac{1}{2z(z^2+\beta^2)+(z^2+\alpha^2)(2z)}\right]_{z=\alpha i}=\frac{1}{2\alpha i(\beta^2-\alpha^2)}.
$$

By symmetry,

$$
\operatorname{Res}\left(\frac{1}{(z^2+\alpha^2)(z^2+\beta^2)}, \beta i\right)=\frac{1}{2\beta i(\alpha^2-\beta^2)}.
$$

Then

$$
\int_{-\infty}^{\infty}\frac{1}{(x^2+\alpha^2)(x^2+\beta^2)}\,dx=2\pi i\left[\frac{1}{2\alpha i(\beta^2-\alpha^2)}+\frac{1}{2\beta i(\alpha^2-\beta^2)}\right]
$$

$$
=\frac{\pi}{\alpha\beta(\alpha+\beta)}. \quad\blacksquare
$$

Integrals of $\int_{-\infty}^{\infty}[p(x)/q(x)]\cos(cx)\,dx$ and $\int_{-\infty}^{\infty}[p(x)/q(x)]\sin(cx)\,dx$

Suppose p and q are polynomials with real coefficients and no common factors, that the degree of q exceeds the degree of p by at least 2, and that q has no real zeros. We want to evaluate integrals

$$
\int_{-\infty}^{\infty}\frac{p(x)}{q(x)}\cos(cx)\,dx \quad\text{and}\quad \int_{-\infty}^{\infty}\frac{p(x)}{q(x)}\sin(cx)\,dx,
$$

in which c is any positive number.

Again, we proceed by looking for the integral of a suitably chosen complex function over a suitably chosen closed curve. Consider

$$
\oint_\Gamma \frac{p(z)}{q(z)}e^{icz}\,dz,
$$

where Γ is the closed path of the preceding subsection, enclosing all the zeros $z_1,\ldots,z_m$ of q lying in the upper half-plane. Here is why this integral is promising. With Γ consisting of the semicircle γ and the segment S on the real axis, as before, then

$$
\oint_\Gamma \frac{p(z)}{q(z)}e^{icz}\,dz = \int_\gamma \frac{p(z)}{q(z)}e^{icz}\,dz + \int_{-R}^{R}\frac{p(x)}{q(x)}e^{icx}\,dx
$$

$$
=\oint_\gamma \frac{p(z)}{q(z)}e^{icz}\,dz + \int_{-R}^{R}\frac{p(x)}{q(x)}\cos(cx)\,dx + i\int_{-R}^{R}\frac{p(x)}{q(x)}\sin(cx)\,dx
$$

$$
=2\pi i\sum_{j=1}^{m}\operatorname{Res}\left(\frac{p(z)}{q(z)}e^{icz}, z_j\right).
$$

As $R\to\infty$, one can show that $\int_\gamma[p(z)/q(z)]e^{icz}\,dz\to 0$, leaving

$$
\int_{-\infty}^{\infty}\frac{p(x)}{q(x)}\cos(cx)\,dx + i\int_{-\infty}^{\infty}\frac{p(x)}{q(x)}\sin(cx)\,dx=2\pi i\sum_{j=1}^{m}\operatorname{Res}\left(\frac{p(z)}{q(z)}e^{icz}, z_j\right). \quad (24.8)
$$

The real part of the right side of equation (24.8) is $\int_{-\infty}^{\infty}[p(x)/q(x)]\cos(cx)\,dx$, and the imaginary part is $\int_{-\infty}^{\infty}[p(x)/q(x)]\sin(cx)\,dx$.

EXAMPLE 24.31

We will evaluate

$$\int_{-\infty}^{\infty} \frac{\cos(cx)}{(x^2+\alpha^2)(x^2+\beta^2)}\,dx,$$

in which c, α, and β are positive numbers and $\alpha \neq \beta$.

The zeros of the denominator in the upper half-plane are αi and βi, and these are simple poles of

$$f(z) = \frac{e^{icz}}{(z^2+\alpha^2)(z^2+\beta^2)}.$$

Compute

$$\mathrm{Re}\,s(f,\alpha i) = \frac{e^{ic\alpha i}}{2\alpha i(\beta^2-\alpha^2)} = \frac{e^{-c\alpha}}{2\alpha i(\beta^2-\alpha^2)}$$

and

$$\mathrm{Re}\,s(f,\beta i) = \frac{e^{-c\beta}}{2\beta i(\alpha^2-\beta^2)}.$$

Then

$$\int_{-\infty}^{\infty} \frac{\cos(cx)}{(x^2+\alpha^2)(x^2+\beta^2)}\,dx + i\int_{-\infty}^{\infty} \frac{\sin(cx)}{(x^2+\alpha^2)(x^2+\beta^2)}\,dx$$

$$= 2\pi i\left[\frac{e^{-c\alpha}}{2\alpha i(\beta^2-\alpha^2)} + \frac{e^{-c\beta}}{2\beta i(\alpha^2-\beta^2)}\right] = \frac{\pi}{\beta^2-\alpha^2}\left(\frac{e^{-c\alpha}}{\alpha} - \frac{e^{-c\beta}}{\beta}\right).$$

Separating real and imaginary parts, we have

$$\int_{-\infty}^{\infty} \frac{\cos(cx)}{(x^2+\alpha^2)(x^2+\beta^2)}\,dx = \frac{\pi}{\beta^2-\alpha^2}\left(\frac{e^{-c\alpha}}{\alpha} - \frac{e^{-c\beta}}{\beta}\right)$$

and

$$\int_{-\infty}^{\infty} \frac{\sin(cx)}{(x^2+\alpha^2)(x^2+\beta^2)}\,dx = 0.$$

The latter is obvious because the integrand is an odd function. ∎

Integrals Using Indented Contours

Equation (24.7) enables us to evaluate certain improper integrals of quotients of polynomials, assuming that the denominator has no real zeros. We will extend this result to the case that the

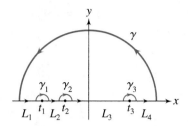

FIGURE 24.8

denominator has simple real zeros. Consider

$$\int_{-\infty}^{\infty} \frac{p(x)}{q(x)} \, dx$$

in which p and q are polynomials with real coefficients and no common factors and the degree of q exceeds that of p by at least 2. Suppose q has complex zeros $z_1, \ldots, z_m$ in the upper half-plane as well as simple real zeros $t_1, \ldots, t_k$.

Let Γ be the path of Figure 24.8, including a semicircle γ of radius R about the origin, small semicircles γ_j of radius ϵ centered at each real zero t_j, and segments L_j along the real line connecting these semicircles as shown. We call such a path an *indented* path because of the small semicircles about the real zeros of $q(x)$. Let ϵ be small enough that the half-disks determined by γ_j and γ_k do not intersect if $j \neq k$, and no z_j is enclosed by any γ_k. Also suppose R is large enough that Γ encloses each z_j. Note that each t_j is outside Γ.

By the residue theorem,

$$\oint_{\Gamma} \frac{p(z)}{q(z)} \, dz = 2\pi i \sum_{j=1} \mathrm{Re}\, s\left(\frac{p(z)}{q(z)}, z_j\right)$$

$$= \int_{\gamma} \frac{p(z)}{q(z)} \, dz + \sum_{j=1}^{k} \int_{\gamma_j} \frac{p(z)}{q(z)} \, dz + \sum_{j=1}^{k+1} \int_{L_j} \frac{p(x)}{q(x)} \, dx. \tag{24.9}$$

We want to investigate what happens in equation (24.9) when $R \to \infty$ and $\epsilon \to 0$.

When $R \to \infty$, we claim that $\int_{\gamma} [p(z)/q(z)] \, dz \to 0$ by an argument like that we have done before. The sum of the residues at zeros of q is unchanged in this limit.

As $\epsilon \to 0$, the semicircles γ_j contract to t_j, and the segments L_j expand to cover the interval $[-R, R]$ and then the entire real line as $R \to \infty$. This means that in equation (24.9),

$$\sum_{j=1}^{k+1} \int_{L_j} \frac{p(z)}{q(z)} \, dz \to \int_{-\infty}^{\infty} \frac{p(x)}{q(x)} \, dx$$

It is not yet clear what happens to each integral $\int_{\gamma_j} [p(z)/q(z)] \, dz$ in this process. We will show that each of these integrals approaches πi times the residue of $p(z)/q(z)$ at the real simple pole t_j.

To see this, write the Laurent expansion of $p(z)/q(z)$ about t_j:

$$\frac{p(z)}{q(z)} = \frac{c_{-1}}{z - t_j} + \sum_{s=0}^{\infty} c_s (z - t_j)^s = \frac{c_{-1}}{z - t_j} + g(z),$$

where g is differentiable at t_j. On γ_j, $z = t_j + \epsilon e^{it}$, where t varies from π to 0 (for counterclockwise orientation on Γ). Then

$$\int_{\gamma_j} \frac{p(z)}{q(z)} \, dz = c_{-1} \int_{\gamma_j} \frac{1}{z - t_j} \, dz + \int_{\gamma_j} g(z) \, dz$$

$$= c_{-1} \int_{\pi}^{0} \frac{1}{\epsilon e^{it}} i\epsilon e^{it} \, dt + \int_{\pi}^{0} g(t_j + \epsilon e^{it}) i\epsilon e^{it} \, dt$$

$$= -\pi i c_{-1} + i\epsilon \int_{\pi}^{0} g(t_j + \epsilon e^{it}) e^{it} \, dt$$

$$= -\pi i \operatorname{Re} s\left(\frac{p(z)}{q(z)}, t_j \right) + i\epsilon \int_{\pi}^{0} g(t_j + \epsilon e^{it}) e^{it} \, dt.$$

Now $i\epsilon \int_{\pi}^{0} g(t_j + \epsilon e^{it}) e^{it} \, dt \to 0$ as $\epsilon \to 0$. Therefore,

$$\int_{\gamma_j} \frac{p(z)}{q(z)} \, dz \to -\pi i \operatorname{Re} s\left(\frac{p(z)}{q(z)}, t_j \right).$$

Therefore, as $R \to \infty$ and $\epsilon \to 0$ in equation (24.9), we get

$$2\pi i \sum_{j=1}^{k} \operatorname{Re} s\left(\frac{p(z)}{q(z)}, z_j \right) = -\pi i \sum_{j=1}^{k} \operatorname{Re} s\left(\frac{p(z)}{q(z)}, t_j \right) + \int_{-\infty}^{\infty} \frac{p(x)}{q(x)} \, dx,$$

hence

$$\int_{-\infty}^{\infty} \frac{p(x)}{q(x)} \, dx = \pi i \sum_{j=1}^{k} \operatorname{Re} s\left(\frac{p(z)}{q(z)}, t_j \right) + 2\pi i \sum_{j=1}^{k} \operatorname{Re} s\left(\frac{p(z)}{q(z)}, z_j \right). \tag{24.10}$$

In a sense, the simple poles of $p(z)/q(z)$ on the real line contribute "half residues," having been enclosed by semicircles instead of circles, while the poles in the upper half-plane contribute "full residues" to this sum.

EXAMPLE 24.32

Evaluate

$$\int_{-\infty}^{\infty} \frac{3x + 2}{x(x - 4)(x^2 + 9)}.$$

Here

$$f(z) = \frac{3z + 2}{z(z - 4)(z^2 + 9)}.$$

The denominator has simple real zeros at 0 and 4 and simple complex zeros $-3i$ and $3i$. Only $3i$ is in the upper half-plane. Compute the residues:

$$\operatorname{Re} s(f, 0) = \lim_{z \to 0} z f(z) = \frac{2}{-36} = -\frac{1}{18},$$

$$\operatorname{Re} s(f, 4) = \lim_{z \to 4} (z - 4) f(z) = \frac{14}{100} = \frac{7}{50}$$

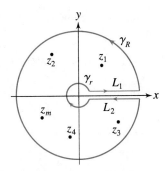

FIGURE 24.9

and

$$\mathrm{Re}\, s(f, 3i) = \lim_{z \to 3i} (z - 3i) f(z) = \frac{9i + 2}{3i(3i - 4)(6i)} = \frac{2 + 9i}{72 - 54i}.$$

Then

$$\int_{-\infty}^{\infty} \frac{3x + 2}{x(x - 4)(x^2 + 9)} = \pi i \left(-\frac{1}{18} + \frac{7}{50}\right) + 2\pi i \left(\frac{2 + 9i}{72 - 54i}\right) = -\frac{14}{75}\pi.$$

Integrals $\int_0^\infty x^a [p(x)/q(x)]\, dx$

Let $0 < a < 1$. We will consider integrals of the form

$$\int_0^\infty x^a \frac{p(x)}{q(x)}\, dx,$$

in which p and q are polynomials with real coefficients and no common factors, q has no positive zeros, and the degree of q exceeds the degree of p by at least 1. We also assume that either $q(0) \neq 0$ or $q(z)$ has a simple zero at the origin.

Let the nonzero zeros of q be $z_1, \ldots, z_m$. These are all the nonzero zeros of q, not just those in the upper half-plane. Since the coefficients of q are real, this list includes complex conjugate pairs.

Choose r small enough and R large enough that the closed path Γ shown in Figure 24.9 encloses $z_1, \ldots, z_m$. Γ consists of γ_R ("most" of the circle of radius R about 0), γ_r ("most" of the circle of radius r about 0), and the line segments L_1 and L_2 connecting γ_r and γ_R. We will eventually let $r \to 0$ and $R \to \infty$, but some work is required first.

We must agree on a meaning for z^a, since this symbol generally denotes a (possibly infinite) set of different numbers. Write $z = \rho e^{i\theta}$ for some θ in $[0, 2\pi)$ and define

$$z^a = \rho^a e^{ia\theta}.$$

As z approaches L_1,

$$f(z) = \frac{z^a p(z)}{q(z)} \to \frac{x^a p(x)}{q(x)},$$

where $r < x < R$. But as z approaches L_2, the lower side of the positive real axis,

$$f(z) \to \frac{x^a e^{2\pi a i} p(x)}{q(x)}.$$

The reason for this is that the argument increases by 2π as z approaches the positive real axis from below, and then

$$z^a = \rho^a e^{i(\theta + 2\pi)a} = \rho^a e^{ia\theta} e^{2\pi ai}.$$

By the residue theorem,

$$\oint_\Gamma \frac{z^a p(z)}{q(z)} \, dz = 2\pi i \sum_{j=1}^m \mathrm{Re}\, s(f, z_j)$$

$$= \int_{\gamma_R} \frac{z^a p(z)}{q(z)} \, dz + \int_{\gamma_r} \frac{z^a p(z)}{q(z)} \, dz + \int_{L_1} \frac{x^a p(x)}{q(x)} \, dx + \int_{L_2} \frac{x^a e^{2\pi ai} p(x)}{q(x)} \, dx.$$

On L_1, x varies from r to R, while on L_2, x varies from R to r to maintain counterclockwise orientation on Γ. The last equation becomes

$$2\pi i \sum_{j=1}^m \mathrm{Re}\, s(f, z_j) = \int_{\gamma_R} \frac{z^a p(z)}{q(z)} \, dz + \int_{\gamma_r} \frac{z^a p(z)}{q(z)} \, dz$$

$$+ \int_r^R \frac{x^a p(x)}{q(x)} \, dx + \int_R^r \frac{x^a e^{2\pi ai} p(x)}{q(x)} \, dx.$$

By making estimates on the two circular arcs, we can show that the first two integrals in the last equation tend to zero as $r \to 0$ and $R \to \infty$. In this limit, the last equation becomes

$$2\pi i \sum_{j=1}^m \mathrm{Re}\, s(f, z_j) = \int_0^\infty \frac{x^a p(x)}{q(x)} \, dx + \int_\infty^0 \frac{x^a e^{2\pi ai} p(x)}{q(x)} \, dx.$$

From this we obtain

$$\int_0^\infty \frac{x^a p(x)}{q(x)} \, dx = \frac{2\pi i}{1 - e^{2\pi ai}} \sum_{j=1}^m \mathrm{Re}\, s(f, z_j). \tag{24.11}$$

EXAMPLE 24.33

We will use equation (24.11) to evaluate

$$\int_0^\infty \frac{x^{1/3}}{x(x^2 + 1)} \, dx.$$

Here $p(z) = 1$, $a = \frac{1}{3}$, and $q(z) = z(1 + z^2)$, with simple zeros at 0, i, and $-i$. Compute

$$\mathrm{Re}\, s\left(\frac{z^{1/3}}{q(z)}, i \right) = \frac{i^{1/3}}{2i^2} = \frac{(e^{\pi i/2})^{1/3}}{-2} = -\frac{1}{2} e^{\pi i/6}$$

and

$$\mathrm{Re}\, s\left(\frac{z^{1/3}}{q(z)}, -i \right) = \frac{(e^{3\pi i/2})^{1/3}}{-2} = -\frac{1}{2} e^{\pi i/2}.$$

Then

$$\int_0^\infty \frac{x^{1/3}}{x(x^2+1)} \, dx = \frac{2\pi i}{1 - e^{2\pi i/3}} \left(-\frac{1}{2}\right) \left(e^{\pi i/6} + e^{\pi i/2}\right)$$

$$= \frac{-\pi i}{1 + \frac{1}{2} - \frac{\sqrt{3}}{2}i} \left(\frac{\sqrt{3}}{2} + \frac{1}{2}i + i\right) = \pi. \quad \blacksquare$$

Many other kinds of real integrals can be evaluated using complex integral techniques. Some of these require considerable ingenuity in seeking the right function to integrate over a path selected to get the result that is wanted.

The Cauchy Principal Value

Since we have been dealing with improper integrals, we will mention the Cauchy principal value.

An integral

$$I = \int_{-\infty}^\infty g(x) \, dx$$

is defined to be

$$\lim_{r \to -\infty} \int_r^0 g(x) \, dx + \lim_{R \to \infty} \int_0^R g(x) \, dx,$$

if both of these integrals converge. These limits are independent of each other.

The *Cauchy principal value* of I is defined to be

$$\text{CPV} \left(\int_{-\infty}^\infty g(x) \, dx \right) = \lim_{R \to \infty} \int_{-R}^R g(x) \, dx.$$

This is a special case of the two independent limits defining I.

In the event that $\int_{-\infty}^\infty g(x) \, dx$ converges, certainly the value of I agrees with the Cauchy principal value of the integral. However, it is possible for an integral to have a finite CPV but to diverge in the broader sense of the definition of I. This occurs with $\int_{-\infty}^\infty x \, dx$, which certainly diverges. However, this integral has Cauchy principal value 0, because for any positive R,

$$\int_{-R}^R x \, dx = 0.$$

In some of the examples we have discussed, we were actually computing Cauchy principal values whenever we took a limit of $\int_{-R}^R g(x) \, dx$ as $R \to \infty$. In these examples the conditions imposed ensured that the improper integral converged in the more general sense as well.

SECTION 24.3   PROBLEMS

1. Evaluate $\oint_\Gamma \frac{z}{1+z^2} \, dz$, with Γ the circle $|z| = 2$, first using the residue theorem and then by the argument principle.

2. Evaluate $\oint_\Gamma \tan(z) \, dz$, with Γ the circle $|z| = \pi$, first by the residue theorem and then by the argument principle.

3. Evaluate $\oint_\Gamma \frac{z+1}{z^2 + 2z + 4} \, dz$, with Γ the circle $|z| = 1$, first by the residue theorem and then by the argument principle.

4. Let $p(z) = (z - z_1)(z - z_2) \cdots (z - z_n)$, with $z_1, \ldots, z_n$ distinct complex numbers. Let Γ be a positively oriented closed path enclosing each of the z_j's. Prove that

$$\oint_\Gamma \frac{p'(z)}{p(z)} \, dz = 2\pi i n,$$

first by the residue theorem and then by the argument principle.

5. Let Γ be a positively oriented closed path in the plane and let f be differentiable on an open set containing Γ and all points interior to Γ. Let a be any complex number and assume that $f(z) \neq a$ for all z on the graph of Γ. Prove that

$$\frac{1}{2\pi i} \oint \frac{f'(z)}{f(z) - a} \, dz$$

is equal to the number of points z in G for which $f(z) = a$.

6. Use Rouché's theorem to determine the number of zeros of $z^4 + 4z + 1$ interior to the unit circle about the origin.

7. Let n be a positive integer. Prove that the equation $e^z = 3z^n$ has n solutions enclosed by the unit circle. *Hint:* Let $f(z) = -3z^n$ and $g(z) = e^z$ in Rouché's theorem.

8. Use Rouché's theorem to prove that a complex polynomial of degree $n \geq 1$ has exactly n zeros (counting multiplicities). *Hint:* If $p(z) = a_0 + a_1 z + \cdots + a_n z^n$, let $f(z) = a_n z^n$ and $g(z) = p(z) - f(z)$. This provides another proof of the fundamental theorem of algebra.

9. Determine the number of zeros of $z^5 - 6i z^4 - 4z + 1$ lying within the circle $|z| = 3$.

In each of Problems 10 through 13, sum the series. Wherever it appears, a is a positive constant.

10. $\displaystyle \sum_{n=-\infty}^{\infty} \frac{1}{a^4 + n^4}$

11. $\displaystyle \sum_{n=-\infty}^{\infty} (-1)^n \frac{1}{a^4 + n^4}$

12. $\displaystyle \sum_{n=0}^{\infty} \frac{1}{(2n^2 + 1)^2}$

13. $\displaystyle \sum_{n=1}^{\infty} (-1)^n \frac{1}{n^2 - 2}$

14. Show that

$$\sum_{n=-\infty}^{\infty} (-1)^n \frac{1}{a^2 + n^2} = \pi^2 \csc(\pi a) \cot(\pi a)$$

if a is not an integer.

15. Fill in the details of the argument suggested to derive equation (24.3).

16. Derive equation (24.4).

In each of Problems 17 through 26, find an inverse Laplace transform of the function, using residues.

17. $\dfrac{z}{z^2 + 9}$

18. $\dfrac{1}{(z + 3)^2}$

19. $\dfrac{1}{(z - 2)^2(z + 4)}$

20. $\dfrac{1}{(z^2 + 9)(z - 2)^2}$

21. $(z + 5)^{-3}$

22. $\dfrac{\sin(z)}{z^2 + 16}$

23. $\dfrac{1}{z^4 + 1}$

24. $\dfrac{1}{e^z(z - 1)}$

25. $\dfrac{z}{(z - 2)^3}$

26. $\dfrac{z + 3}{(z^3 - 1)(z + 2)}$

27. Use Theorem 24.10 to derive the Heaviside formulas for the Laplace transform (Problem 51, Section 3.3).

In each of Problems 28 through 54, evaluate the integral.

28. $\displaystyle \int_0^{2\pi} \frac{1}{6 + \sin(\theta)} \, d\theta$

29. $\displaystyle \int_0^{2\pi} \frac{1}{2 - \cos(\theta)} \, d\theta$

30. $\displaystyle \int_{-\infty}^{\infty} \frac{1}{x^4 + 1} \, dx$

31. $\displaystyle \int_{-\infty}^{\infty} \frac{1}{x^6 + 1} \, dx$

32. $\displaystyle \int_{-\infty}^{\infty} \frac{1}{x^2 - 2x + 6} \, dx$

33. $\displaystyle \int_{-\infty}^{\infty} \frac{x \sin(2x)}{x^4 + 16} \, dx$

34. $\displaystyle \int_0^{2\pi} \frac{2 \sin(\theta)}{2 + \sin^2(\theta)} \, d\theta$

35. $\displaystyle \int_{-\infty}^{\infty} \frac{1}{x(x + 4)(x^2 + 16)} \, dx$

36. $\displaystyle\int_{-\infty}^{\infty} \frac{\sin(x)}{x^2 - 4x + 5}\, dx$

37. $\displaystyle\int_{-\infty}^{\infty} \frac{\cos^2(x)}{(x^2+4)^2}\, dx$

38. $\displaystyle\int_{0}^{2\pi} \frac{\sin(\theta) + \cos(\theta)}{2 - \cos(\theta)}\, d\theta$

39. $\displaystyle\int_{-\infty}^{\infty} \frac{1}{(x-4)(x^5+1)}\, dx$

40. $\displaystyle\int_{0}^{\infty} \frac{x^{3/4}}{x^4+1}\, dx$

41. $\displaystyle\int_{-\infty}^{\infty} \frac{x^2}{(x^2+1)^2}\, dx$

42. $\displaystyle\int_{0}^{2\pi} \frac{1}{\cos^2(\theta) + 2\sin^2(\theta)}\, d\theta$

43. $\displaystyle\int_{-\infty}^{\infty} \frac{\cos(3x)}{x^2+9}\, dx$

44. $\displaystyle\int_{-\infty}^{\infty} \frac{x+2}{(x^2-4)(x^2+4)}\, dx$

45. $\displaystyle\int_{0}^{2\pi} \frac{\sin(\theta)}{4 + \sin(\theta)}\, d\theta$

46. $\displaystyle\int_{0}^{2\pi} \frac{\cos(\theta) - \sin(\theta)}{2 - \cos(\theta)}\, d\theta$

47. $\displaystyle\int_{-\infty}^{\infty} \frac{x^2\cos(x)}{1+x^6}\, dx$

48. $\displaystyle\int_{0}^{2\pi} \frac{1}{1 + \cos^2(\theta)}\, d\theta$

49. $\displaystyle\int_{-\infty}^{\infty} \frac{x^4}{1+x^6}\, dx$

50. $\displaystyle\int_{0}^{2\pi} \frac{\sin^2(\theta)}{3 + \cos^2(\theta)}\, d\theta$

51. $\displaystyle\int_{0}^{2\pi} \frac{1}{\sin^2(\theta) + \cos^2(\theta)}\, d\theta$

52. $\displaystyle\int_{-\infty}^{\infty} \frac{2}{(x+2)(x-10)(x^2+1)}\, dx$

53. $\displaystyle\int_{0}^{\infty} \frac{x^{1/4}}{(x+1)(x^2+4)}\, dx$

54. $\displaystyle\int_{-\infty}^{\infty} \frac{\cos(4x)}{(x^2+1)^2}\, dx$

55. Let α be a positive number. Show that
$$\int_{-\infty}^{\infty} \frac{\cos(\alpha x)}{x^2+1}\, dx = \pi e^{-\alpha}.$$

56. Let α and β be positive numbers. Show that
$$\int_{-\infty}^{\infty} \frac{\cos(\alpha x)}{x^2 + \beta^2}\, dx = \frac{\pi}{2\beta^3}(1 + \alpha\beta)e^{-\alpha\beta}.$$

57. Let α and β be distinct positive numbers. Show that
$$\int_{0}^{2\pi} \frac{1}{\alpha^2\cos^2(\theta) + \beta^2\sin^2(\theta)}\, d\theta = \frac{2\pi}{\alpha\beta}.$$

58. Let α be a positive number. Show that
$$\int_{0}^{\pi/2} \frac{1}{\alpha + \sin^2(\theta)}\, d\theta = \frac{\pi}{2\sqrt{\alpha(1+\alpha)}}.$$

59. Evaluate $\displaystyle\int_{-\infty}^{\infty} \frac{1}{(1+x^2)^n}$ for $n = 1, 2, 3, 4, 5$. From these results, conjecture a value for $\displaystyle\int_{-\infty}^{\infty} \frac{1}{(1+x^2)^n}\, dx$ for positive integer n.

60. Let n be a positive integer. Evaluate
$$\int_{0}^{2\pi} \cos^{2n}(\theta)\, d\theta.$$

Hint: After converting this to a complex integral, the integrand has a pole of order $2n + 1$ at the origin. The residue of the integrand at this pole is the coefficient of z^{2n} in the binomial expansion of $(1 + z^2)^{2n}$.

61. Let β be a positive number. Show that
$$\int_{0}^{\infty} e^{-x^2}\cos(2\beta x)\, dx = \frac{\sqrt{\pi}}{2}e^{-\beta^2}.$$

Hint: Integrate e^{-z^2} about the rectangular path having corners at $\pm R$ and $\pm R + \beta i$. Use Cauchy's theorem to evaluate this integral, set this equal to the sum of the integrals on the sides of the rectangle, and take the limit as $R \to \infty$. Assume the standard result that
$$\int_{0}^{\infty} e^{-x^2}\, dx = \frac{\sqrt{\pi}}{2}.$$

62. Prove *Jordan's lemma:* $\int_{0}^{\pi/2} e^{-R\sin(\theta)}\, d\theta < (\pi/2)R$ for any positive number R. *Hint:* First show that
$$\frac{2\theta}{\pi} < \sin(\theta) \le 1$$
for $0 \le \theta \le \pi/2$. Then use the fact that $e^{-R\sin(\theta)} \le e^{-2R\theta/\pi}$.

63. Derive *Fresnel's integrals:*
$$\int_{0}^{\infty} \cos(x^2)\, dx = \int_{0}^{\infty} \sin(x^2)\, dx = \frac{1}{2}\sqrt{\frac{\pi}{2}}.$$

Hint: Integrate e^{iz^2} over the closed path bounding the sector $0 \leq x \leq R$, $0 \leq \theta \leq \pi/4$, shown in Figure 24.10. Use Cauchy's theorem to evaluate this integral, then evaluate it as the sum of integrals over the boundary segments of the sector. Use Jordan's lemma to show that the integral over the circular arc tends to zero as $R \rightarrow \infty$, and use the integrals over the straight segments to obtain Fresnel's integrals.

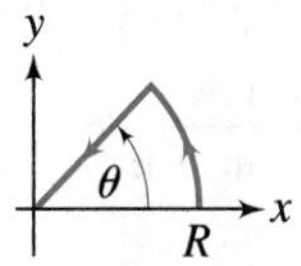

FIGURE 24.10

64. Let α and β be positive numbers. Show that

$$\int_0^\infty \frac{x \sin(\alpha x)}{x^4 + \beta^4} \, dx = \frac{\pi}{2\beta^2} e^{-\alpha\beta/\sqrt{2}} \sin\left(\frac{\alpha\beta}{\sqrt{2}}\right).$$

65. Let $0 < \beta < \alpha$. Show that

$$\int_0^\pi \frac{1}{(\alpha + \beta \cos(\theta))^2} \, d\theta = \frac{\alpha\pi}{(\alpha^2 - \beta^2)^{3/2}}.$$

CHAPTER 25

Conformal Mappings

In the calculus of real functions of a single real variable, we may gain some insight into the function's behavior by sketching its graph. For complex functions we cannot make the same kind of graph, since a complex variable $z = x + iy$ by itself has two variables. However, we can set $w = f(z)$ and make two copies of the complex plane, one for z and the other for image points w. As z traces out a path or varies over a set S in the z plane, we can plot image points $w = f(z)$ in the w plane, yielding a picture of how the function acts on this path or on points in S. The set of all image points $f(z)$ for z in S is denoted $f(S)$. When looked at in this way, a function is called a *mapping* or *transformation*. This idea is diagrammed in Figure 25.1.

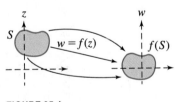

FIGURE 25.1

Thinking of a function as a mapping can be a powerful tool in solving certain kinds of problems, including the analysis of fluid motion and the solution of partial differential equations, particularly Dirichlet problems. We will develop some ideas about mappings, then turn to applications.

25.1 Functions as Mappings

First we need some terminology. Let f be a complex function and D a set of points in the plane on which $f(z)$ is defined. Let D^* also be a set of complex numbers.

DEFINITION 25.1

---

1. f maps D into D^* if $f(z)$ is in D^* for every z in D. In this event, we write $f : D \rightarrow D^*$.

2. f maps D onto D^* if $f(z)$ is in D^* for every z in D and, conversely, if w is in D^*, then there is some z in D such that $w = f(z)$. In this event, we call f an *onto* mapping.

Thus, $f : D \rightarrow D^*$ is onto if every point of D^* is the image under f of some point in D.

EXAMPLE 25.1

Let $f(z) = iz$ for $|z| \leq 1$. Then f acts on points in the closed unit disk $D : |z| \leq 1$. If z is in D, then $|f(z)| = |iz| = |z| \leq 1$, so the image of any point in this disk is also in this disk. Here f maps D into D (so $D^* = D$ in the definition).

This mapping is onto. If w is in D, then $z = w/i$ is also in D, and

$$f(z) = f(w/i) = i(w/i) = w.$$

Every point in the unit disk is the image of some point in this disk under this mapping.

We can envision this mapping geometrically. Since $i = e^{i\pi/2}$, if $z = re^{i\theta}$, then

$$f(z) = iz = re^{i\pi/2}e^{i\theta} = re^{i(\theta+\pi/2)},$$

so f takes z and adds $\pi/2$ to its argument. This rotates the line from the origin to z by $\pi/2$ radians counterclockwise. The action of f on z can be envisioned as in Figure 25.2. Since this function is simply a counterclockwise rotation by $\pi/2$ radians, it is clear why f maps the unit disk onto itself. ∎

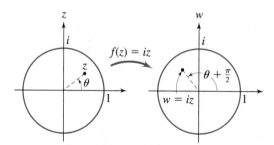

FIGURE 25.2   *The mapping $f(z) = iz$ for $|z| \leq 1$.*

Often we have the function f and the set D of complex numbers to which we want to apply this mapping. We then must analyze $f(z)$ to determine the image of D under the mapping. In effect, we are finding D^* so that f is a mapping of D onto D^*.

EXAMPLE 25.2

Let $f(z) = z^2$ for z in the wedge D shown in Figure 25.3. D consists of all complex numbers on or between the nonnegative real axis and the line $y = x$.

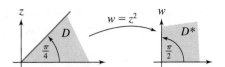

FIGURE 25.3 $w = z^2$ maps D one-to-one onto D^*.

In polar form, $z = re^{i\theta}$ is in D if $0 \le \theta \le \pi/4$. Then $f(z) = z^2 = r^2 e^{2i\theta}$, so f has the effect of squaring the magnitude of z and doubling its argument. If z has argument between 0 and $\pi/4$, then z^2 has argument between 0 and $\pi/2$. This spreads the wedge D out to cover the entire right quarter-plane, consisting of points on or between the nonnegative real and imaginary axes. If we call this right quarter-plane D^*, then f maps D onto D^*. ∎

Some functions map more than one point to the same image. For example, $f(z) = \sin(z)$ maps all integer multiples of π to zero. If each image point is the unique target of exactly one number, then the mapping is called *one-to-one*.

DEFINITION 25.2

A mapping $f : D \rightarrow D^*$ is one-to-one if distinct points of D map to distinct points in D^*.

Thus f is one-to-one (or $1 - 1$) if $z_1 \ne z_2$ implies that $f(z_1) \ne f(z_2)$.

The notions of one-to-one and onto are independent of each other. A mapping may have one of these properties, or both, or neither. The mapping $f(z) = z^2$ of Example 25.2 maps the wedge $0 \le \arg(z) \le \pi/4$ in one-to-one fashion onto the right quarter-plane. However, $f(z) = z^2$ does not map the entire complex plane to the complex plane in a one-to-one manner, since $f(-z) = f(z)$. This function does map the plane onto itself, since, given any complex number w, there is some z such that $f(z) = z^2 = w$.

EXAMPLE 25.3

Let $h(z) = z^2$ for all z. h maps the entire plane onto itself but is not one-to-one.

If $z = x + iy$, then

$$h(z) = x^2 - y^2 + 2ixy = u + iv,$$

where $u = x^2 - y^2$ and $v = 2xy$. We will use this formulation to determine the image under f of a vertical line $x = a$. Any point on this line has the form $z = a + iy$, and maps to

$$h(a + iy) = u + iv = a^2 - y^2 + 2iay.$$

Points on the line $x = a$ map to points (u, v) with $u = a^2 - y^2$ and $v = 2ay$. Write $y = v/2a$ (assuming that $a \ne 0$) to obtain

$$u = a^2 - \frac{v^2}{4a^2}$$

or

$$v^2 = 4a^2(a^2 - u),$$

the equation of a parabola in the u, v plane. h maps vertical lines $x = a \ne 0$ to parabolas.

If $a = 0$, the vertical line $x = a$ is the imaginary axis, consisting of points $z = iy$. Now $h(z) = -y^2$, so h maps the imaginary axis in the x, y plane to the nonpositive part of the real axis in the u, v plane.

Figure 25.4 shows the parabolic image of a line $x = a \neq 0$. The larger a is, the more the parabola opens up to the left (intersects the v axis further from the origin). As a is chosen smaller, approaching 0, these parabolas become "flatter," approaching the nonpositive real axis in the u, v plane.

A horizontal line $y = b$ consists of points $z = x + ib$, which map to

$$h(z) = (x + ib)^2 = x^2 - b^2 + 2ixb = u + iv.$$

Now $u = x^2 - b^2$ and $v = 2xb$, so, for $b \neq 0$,

$$v^2 = 4b^2(b^2 + u).$$

A typical such parabola is shown in Figure 25.5, opening up to the right. These parabolas also open up more the larger b is. If $b = 0$, the line $y = b$ is the real axis in the z plane, and this maps to $h(x) = x^2$, giving the nonnegative real axis in the u, v plane as x takes on all real values. ∎

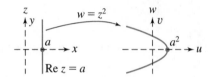

FIGURE 25.4 $w = z^2$ maps vertical lines to parabolas opening left.

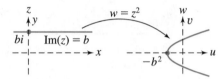

FIGURE 25.5 $w = z^2$ maps horizontal lines to parabolas opening right.

EXAMPLE 25.4

We will look at the exponential function $w = E(z) = e^z$ as a mapping. Write

$$w = u + iv = e^{x+iy} = e^x \cos(y) + ie^x \sin(y),$$

so

$$u = e^x \cos(y) \quad \text{and} \quad v = e^x \sin(y).$$

As a mapping of the entire plane to itself, E is not onto (no number maps to zero), and E is also not one-to-one (all points $z + 2n\pi i$ have the same image, for n any integer).

Consider a vertical line $x = a$ in the x, y plane. The image of this line consists of points $u + iv$ with

$$u = e^a \cos(y), \quad v = e^a \sin(y).$$

Then

$$u^2 + v^2 = e^{2a},$$

so the line $x = a$ maps to the circle of radius e^a about the origin in the u, v plane. Actually, as the point $z = a + iy$ moves along this vertical line, the image point $u + iv$ makes one complete circuit around the circle every time y varies over an interval of length 2π, since $\cos(y + 2n\pi) = \cos(y)$ and $\sin(y + 2n\pi) = \sin(y)$. We may therefore think of a vertical line as infinitely many intervals of length 2π strung together, and the exponential function wraps each segment once around the circle $u^2 + v^2 = e^{2a}$ (Figure 25.6).

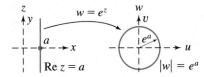

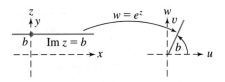

FIGURE 25.6 $w = e^z$ *wraps a vertical line around a circle, covering the circle once for every interval of length* 2π.

FIGURE 25.7 $w = e^z$ *maps horizontal lines to half-rays from the origin.*

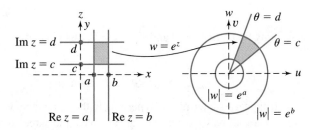

FIGURE 25.8 $w = e^z$ *maps the rectangle shown to a wedge bounded by two half-rays and two circles.*

The image of a point $z = x + ib$ on the horizontal line $y = b$ is a point $u + iv$ with

$$u = e^x \sin(b), \quad y = e^x \cos(b).$$

As x varies over the real line, e^x varies from 0 to ∞ over the positive real axis. The point $(e^x \sin(b), e^x \cos(b))$ moves along a half-line from the origin to infinity, making an angle b radians with the positive real axis (Figure 25.7). In polar coordinates, this half-line is $\theta = b$.

Using these results, we can find the image of any rectangle in the x, y plane, having sides parallel to the axes. Let the rectangle have sides on the lines $x = a$, $x = b$, $y = c$, and $y = d$ (in the x, y plane in Figure 25.8). These lines map, respectively, to the circles

$$u^2 + v^2 = e^{2a}, \quad u^2 + v^2 = e^{2b}$$

and the half-lines

$$\theta = c \quad \text{and} \quad \theta = d.$$

The wedge in the u, v plane in Figure 25.8 is the image of the rectangle under this exponential mapping. ∎

EXAMPLE 25.5

Consider the mapping $f(z) = \cos(z)$. This is a mapping of the complex plane onto itself but is not one-to-one. Write

$$w = \cos(x + iy) = \cos(x)\cosh(y) - i\sin(x)\sinh(y).$$

A vertical line $x = a$ has image

$$w = \cos(a)\cosh(y) - i\sin(a)\sinh(y) = u + iv.$$

so $u = \cos(a)\cosh(y)$ and $v = -\sin(a)\sinh(y)$.

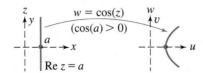
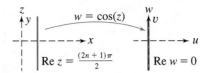

FIGURE 25.9 $w = \cos(z)$ *maps a vertical line, $\operatorname{Re} z = a$, to the right branch of a hyperbola if $\cos(a) > 0$ and $\sin(a) \neq 0$.*

FIGURE 25.10 $w = \cos(z)$ *maps vertical lines $\operatorname{Re} z = (2n + 1)\pi/2$ to the imaginary axis.*

Assuming that $\cos(a) \neq 0$ and $\sin(a) \neq 0$, this gives us

$$\frac{u^2}{\cos^2(a)} - \frac{v^2}{\sin^2(a)} = 1.$$

This would appear to determine a hyperbola in the u, v plane. However, $u = \cos(a)\cosh(y)$, and $\cosh(y) \geq 1$ for all real y, so u must always have the same sign as $\cos(a)$. The image of $x = a$ is therefore one branch of this hyperbola, the right branch if $\cos(a) > 0$ and the left branch if $\cos(a) < 0$ (see Figure 25.9).

If $\cos(a) = 0$, so $a = (2n + 1)\pi/2$ for some integer n, then $\sin(a) = (-1)^n$. Now

$$u = 0 \quad \text{and} \quad v = (-1)^{n+1} \sinh(y).$$

Since $\sinh(y)$ takes on all real values as y varies over the real line, then so does v. The image of the line $x = (2n + 1)\pi/2$ is therefore the imaginary axis in the u, v plane (Figure 25.10).

If $\sin(a) = 0$ then $a = n\pi$ for some integer n. Now $\cos(a) = (-1)^n$ and the image of the line $x = n\pi$ is given by

$$u = (-1)^n \cosh(y), \quad v = 0.$$

Since $\cosh(y) \geq 1$, this is the interval $[1, \infty)$ if n is even, and $(-\infty, -1]$ if n is odd. These cases are shown in Figure 25.11(a) and (b).

This completes all cases for the image of a vertical line $x = a$. Now consider the image of a horizontal line $y = b$. In this case,

$$u = \cos(x)\cosh(b) \quad \text{and} \quad v = -\sin(x)\sinh(b).$$

If $b \neq 0$, then

$$\frac{u^2}{\cosh^2(b)} + \frac{v^2}{\sinh^2(b)} = 1,$$

an ellipse (Figure 25.12). If $b = 0$, so that the horizontal line in the x, y plane is the real axis, then

$$u = \cos(x) \quad \text{and} \quad v = 0.$$

Since $\cos(x)$ varies from -1 to 1 inclusive as x takes on all real values, the image of the real axis is the interval $[-1, 1]$ on the real axis in the w plane (Figure 25.13). ∎

Given a mapping f and a domain D, here is a strategy that is often successful in determining $f(D)$. Suppose D has a boundary made up of curves $\gamma_1, \ldots, \gamma_n$. Find the images of these curves, $f(\gamma_1), \ldots, f(\gamma_n)$. These form curves in the w plane, bounding two sets, labeled I and II in Figure 25.14. $f(D)$ is one of these two sets. To determine which it is, choose a convenient point ζ in D and locate $f(\zeta)$. This point will be in $f(D)$.

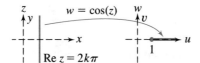

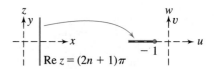

FIGURE 25.11(a) $w = \cos(z)$ *maps vertical lines* $\mathrm{Re}\, z = 2n\pi$ *to the real axis from 1 to* ∞.

FIGURE 25.11(b) $w = \cos(z)$ *maps vertical lines* $\mathrm{Re}\, z = (2n + 1)\pi$ *to the real axis from* $-\infty$ *to* -1.

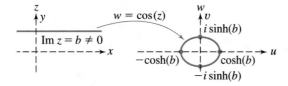

FIGURE 25.12 $w = \cos(z)$ *maps horizontal lines* $\mathrm{Im}\, z = b \neq 0$ *to ellipses.*

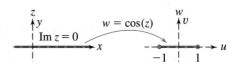

FIGURE 25.13 $w = \cos(z)$ *maps the real axis* $\mathrm{Im}\, z = 0$ *to the real interval* $[-1, 1]$.

FIGURE 25.14

EXAMPLE 25.6

We will determine the image, under the mapping $w = f(z) = \sin(z)$, of the strip S consisting of all z with $-\pi/2 < \mathrm{Re}(z) < \pi/2$ and $\mathrm{Im}(z) > 0$. S is shown in Figure 25.15.

The boundary of S consists of the segment $-\pi/2 \leq x \leq \pi/2$ on the real axis, together with the half-lines $x = -\pi/2$ and $x = \pi/2$ for $y \geq 0$. We will carry out the strategy of looking at images of the lines bounding S. First,

$$w = u + iv = \sin(x)\cosh(y) + i\cos(x)\sinh(y).$$

If $x = -\pi/2$, then

$$w = u + iv = -\cosh(y).$$

Since $0 \leq y < \infty$ on this part of the boundary of S, then $\cosh(y)$ varies from 1 to ∞. The image of the left vertical boundary of S is therefore the interval $(-\infty, -1]$ on the real axis in the u, v plane.

If $x = \pi/2$, a similar analysis shows that the image of the right vertical boundary of S is $[1, \infty)$ on the real axis in the u, v plane.

Finally, if $y = 0$, then

$$w = \sin(x).$$

As x varies from $-\pi/2$ to $\pi/2$, $\sin(x)$ varies from -1 to 1. Thus $[-\pi/2, \pi/2]$ maps to $[-1, 1]$ in the u, v plane.

Figure 25.16 shows these results. The boundary of S maps onto the entire real axis in the u, v plane. This axis is the boundary of two sets in the w plane, the upper half-plane and the lower

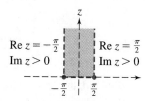

FIGURE 25.15 *Strip bounded by vertical lines $x = -\pi/2$ and $x = \pi/2$ and the x axis.*

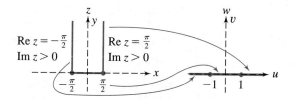

FIGURE 25.16 $w = \sin(z)$ *maps* $x = -\pi/2,\ y \geq 0,\ to$ $u \leq -1;\ -\pi/2 \leq x \leq \pi/2\ to\ -1 \leq u \leq 1;\ and$ $x = \pi/2,\ y \geq 0,\ to\ u \geq 1.$

half-plane. Choose any convenient z in S, say $z = i$. Its image is

$$w = \sin(i) = i \sinh(1),$$

lying in the upper half-plane. Therefore the image of S is the upper half-plane.

Orientation plays a role in these mappings. Imagine walking along the boundary of S in a counterclockwise sense. This means that we start somewhere up the left boundary $x = -\pi/2$, walk down this line to the real axis, then turn left and walk along this axis to $x = \pi/2$, then turn left again and proceed up the right boundary line. Follow the movement of the image point $f(z)$ as z takes this route. As z moves down the line $x = -\pi/2$, $f(z) = \sin(z)$ begins somewhere to the left of -1 on the real axis in the w plane, and moves toward $w = -1$. As w turns the first corner and moves along the real axis in the z plane from $-\pi/2$ to $\pi/2$, $f(z)$ continues through -1 and proceeds along the real axis to $w = +1$. Finally, z turns upward and moves along the line $x = \pi/2$, and $f(z)$ moves through $w = 1$ and out the real axis in the w plane. As z traverses the boundary of the strip in a counterclockwise sense (interior of S to the left), $f(z)$ traverses the boundary of the upper half-plane from left to right in a counterclockwise sense (interior of this half-plane to the left). ∎

In this example, as z moves over the boundary of D in a positive (counterclockwise) sense, $f(z)$ moves over the boundary of $f(D)$ in a positive sense. In the next section we will discuss mappings that preserve angles and sense of rotation.

SECTION 25.1 PROBLEMS

1. In each of (a) through (e), find the image of the given rectangle under the mapping $w = e^z$. Sketch the rectangle in the z plane and its image in the w plane.

   (a) $0 \leq x \leq \pi, 0 \leq y \leq \pi$

   (b) $-1 \leq x \leq 1, -\dfrac{\pi}{2} \leq y \leq \dfrac{\pi}{2}$

   (c) $0 \leq x \leq 1, 0 \leq y \leq \dfrac{\pi}{4}$

   (d) $1 \leq x \leq 2, 0 \leq y \leq \pi$

   (e) $-1 \leq x \leq 2, \dfrac{-\pi}{2} \leq y \leq \dfrac{\pi}{2}$

2. In each of (a) through (e), find the image of the rectangle under the mapping $w = \cos(z)$. Sketch the rectangle and its image in each case.

   (a) $0 \leq x \leq 1, 1 \leq y \leq 2$

   (b) $\dfrac{\pi}{2} \leq x \leq \pi, 1 \leq y \leq 3$

   (c) $0 \leq x \leq \pi, \dfrac{\pi}{2} \leq y \leq \pi$

   (d) $\pi \leq x \leq 2\pi, 1 \leq y \leq 2$

   (e) $0 \leq x \leq \dfrac{\pi}{2}, 0 \leq y \leq 1$

3. In each of (a) through (e), find the image of the rectangle under the mapping $w = 4\sin(z)$. Sketch the rectangle and its image in each case.

   (a) $0 \leq x \leq \dfrac{\pi}{2}, 0 \leq y \leq \dfrac{\pi}{2}$

   (b) $\dfrac{\pi}{4} \leq x \leq \dfrac{\pi}{2}, 0 \leq y \leq \dfrac{\pi}{4}$

(c) $0 \le x \le 1, 0 \le y \le \dfrac{\pi}{6}$

(d) $\dfrac{\pi}{2} \le x \le \dfrac{3\pi}{2}, 0 \le y \le \dfrac{\pi}{2}$

(e) $1 \le x \le 2, 1 \le y \le 2$

4. Determine the images of lines parallel to the axes under the mapping $w = \sinh(z)$.

5. Determine the images of lines parallel to the axes under the mapping $w = \cosh(z)$.

6. Determine the image of the sector $\pi/4 \le \theta \le 5\pi/4$ under the mapping $w = z^2$. Sketch the sector and its image.

7. Determine the image of the sector $\pi/6 \le \theta \le \pi/3$ under the mapping $w = z^3$. Sketch the sector and its image.

8. Show that the mapping

$$w = \frac{1}{2}\left(z + \frac{1}{z}\right)$$

maps the circle $|z| = r$ onto an ellipse with foci 1 and -1 in the w plane. Sketch a typical circle and its image.

9. Show that the mapping of Problem 8 maps a half-line $\theta = $ constant onto a hyperbola with foci ± 1 in the w plane. Sketch a typical half-line and its image.

10. Determine the image of the exterior of the unit circle about the origin under the mapping $w = 1/z$. Sketch the set and its image.

11. Determine the image of the right quarter-plane $\operatorname{Re}(z) \ge 0$, $\operatorname{Im}(z) \ge 0$ under the mapping $w = z^3$. Sketch the set and its image.

12. Consider the mapping $w = e^{\alpha z}$, in which $\alpha = a + ib$ is a constant with nonzero real and imaginary parts.

(a) Show that a horizontal line $\operatorname{Im}(z) = c$ maps to the curve (in polar coordinates) $r = e^d e^{a\theta/b}$, in which $d = -c \, |\alpha| \, /b$.

(b) Show that a vertical line $\operatorname{Re}(z) = c$ maps to the curve $r = e^{-bd/a} e^{-b\theta/a}$.

(c) The image curves in parts (a) and (b) are called *logarithmic spirals*. Sketch their graphs for $a = b = c = 1$.

13. Find the image of the upper half-plane $\operatorname{Im}(z) \ge 0$ under the mapping $w = z^{1/2}$. Sketch the set and its image.

14. Show that the mapping $w = 1/z$ maps every straight line to a circle or straight line, and every circle to a circle or straight line. Give an example of a circle that maps to a line, and a line that maps to a circle.

15. Determine the image of the infinite strip $0 \le \operatorname{Im}(z) \le 2\pi$ under the mapping $w = e^z$.

16. Let D consist of all z in the rectangle having vertices $\pm \alpha i$ and $\pi \pm \alpha i$, with α a positive number.

(a) Determine the image of D under the mapping $w = \cos(z)$. Sketch D and its image.

(b) Determine the image of D under the mapping $w = \sin(z)$. Sketch this image.

(c) Determine the image of D under the mapping $w = 2z^2$. Sketch this image.

17. Determine the image of the sector $\pi/9 \le \theta \le \pi/3$ under the mapping $w = z^{1/3}$. Sketch the sector and its image.

18. Determine the image of the sector $0 \le \theta \le \pi$ under the mapping $w = z^{1/4}$. Sketch the sector and its image.

25.2 Conformal Mappings

Let $f : D \to D^*$ be a mapping.

DEFINITION 25.3 *Angle-Preserving Mapping*

We say that f preserves angles if for any z_0 in D, two smooth curves in D intersect at z_0, and the angle between these curves at z_0 is θ, then the images of these curves intersect at the same angle θ at $f(z_0)$.

This idea is illustrated in Figure 25.17. The images of γ_1 and γ_2 are curves $f(\gamma_1)$ and $f(\gamma_2)$ in D^*. Suppose γ_1 and γ_2 intersect at z_0 and that their tangents have an angle θ between them

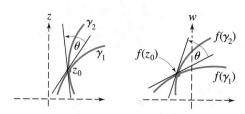

FIGURE 25.17　Angle-preserving mapping.

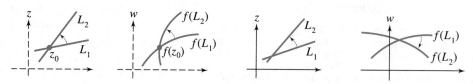

FIGURE 25.18　Orientation-preserving mapping.

FIGURE 25.19　A nonorientation-preserving mapping.

there. We require that the tangents to $f(\gamma_1)$ and $f(\gamma_2)$ intersect at $f(z_0)$ at the same angle. If this condition holds for all smooth curves passing through each point of D, then f is angle preserving on D.

DEFINITION 25.4　*Orientation-Preserving Mapping*

f preserves orientation if a counterclockwise rotation in D is mapped by f to a counterclockwise rotation in D^*.

This idea is illustrated in Figure 25.18. If L_1 and L_2 are lines through any point z_0 in D and the sense of rotation from L_1 to L_2 is counterclockwise, then the sense of rotation from $f(L_1)$ to $f(L_2)$ through $f(z_0)$ in D^* must also be counterclockwise. Of course, $f(L_1)$ and $f(L_2)$ need not be straight lines, but one can still consider a sense of rotation from the tangent to $f(L_1)$ to the tangent to $f(L_2)$ at $f(z_0)$. By contrast, Figure 25.19 illustrates a mapping that does not preserve orientation.

Preservation of angles and orientation are independent concepts. A mapping may preserve one but not the other. If $f : D \to D^*$ preserves both, we say that f is *conformal*.

DEFINITION 25.5　*Conformal Mapping*

$f : D \to D^*$ is a conformal mapping if f is angle preserving and orientation preserving.

The following theorem generates many examples of conformal mappings.

THEOREM 25.1

Let $f : D \to D^*$ be a differentiable function defined on a domain D. Suppose $f'(z) \neq 0$ for all z in D. Then f is conformal. ■

Thus, a differentiable function with a nonvanishing derivative on a domain (connected, open set) maps this set in such a way as to preserve both angles and orientation. We will sketch an argument

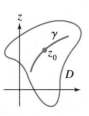
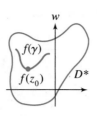

FIGURE 25.20

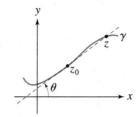
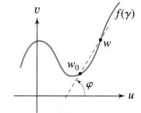

FIGURE 25.21

showing why this is true. Let z_0 be in D and let γ be a smooth curve in D through z_0. Then $f(\gamma)$ is a smooth curve through $f(z_0)$ in D^* (Figure 25.20). If $w = f(z)$ and $w_0 = f(z_0)$, then

$$w - w_0 = \frac{f(z) - f(z_0)}{z - z_0}(z - z_0).$$

Now recall that argument behaves like a logarithm, in the sense that any argument of a product is a sum of arguments of the individual factors. Then

$$\arg(w - w_0) = \arg\left(\frac{f(z) - f(z_0)}{z - z_0}\right) + \arg(z - z_0), \tag{25.1}$$

to within integer multiples of 2π. In Figure 25.21, θ is the angle between the positive real axis and the line through z and z_0 and is an argument of $z - z_0$. The angle φ between the positive real axis and the line through w and w_0 in the w plane is an argument of $w - w_0$. In the limit as $z \to z_0$, equation (25.1) gives us

$$\varphi = \arg[f'(z_0)] + \theta.$$

It is here that we use the assumption that $f'(z_0) \neq 0$, because 0 has no argument.

If γ^* is another smooth curve through z_0, then by the same reasoning,

$$\varphi^* = \arg[f'(z_0)] + \theta^*.$$

Then

$$\varphi - \varphi^* = \theta - \theta^*,$$

(to within integer multiples of 2π). But $\theta - \theta^*$ is the angle between the tangents to γ and γ^* at z_0, and $\varphi - \varphi^*$ is the angle between the tangents to $f(\gamma)$ and $f(\gamma^*)$ at $f(z_0)$. Therefore f preserves angles.

The last "equation" also implies that f preserves orientation, since the sense of rotation from γ to γ^* is the same as the sense of rotation from $f(\gamma)$ to $f(\gamma^*)$. A reversal of the sense of rotation would be implied if we had found that $\varphi - \varphi^* = \theta^* - \theta$. For example, $w = \sin(z)$ is differentiable, with a nonvanishing derivative, on the strip $-\pi/2 < \text{Re}(z) < \pi/2$, and so is a conformal mapping of this strip onto a set in the w plane.

A composition of conformal mappings is conformal. Suppose f maps D conformally onto D^*, and g maps D^* conformally onto D^{**}. Then $g \circ f$ maps D conformally onto D^{**} (Figure 25.22), because angles and orientation are preserved at each stage of the mapping.

We will now consider an important class of conformal mappings.

25.2.1 Linear Fractional Transformations

Often we have domains D and D^* (for example, representing areas of a fluid flow), and we want to produce a conformal mapping of D onto D^*. This can be a formidable task. Linear fractional transformations are relatively simple conformal mappings that will sometimes serve this purpose.

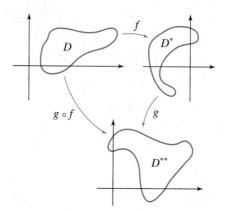

FIGURE 25.22 *A composition of conformal mappings is conformal.*

DEFINITION 25.6 *Linear Fractional Transformation*

A linear fractional transformation is a function

$$T(z) = \frac{az + b}{cz + d}$$

in which a, b, c, and d are given complex numbers and $ad - bc \neq 0$.

Other names for this kind of function are *Möbius transformation* and *bilinear transformation*. The function is defined except at $z = -d/c$, which is a simple pole of T. Further,

$$T'(z) = \frac{ad - bc}{(cz + d)^2},$$

and this is nonzero if $z \neq -d/c$. T is therefore a conformal mapping of the plane with the point $z = -d/c$ removed.

The condition $ad - bc \neq 0$ ensures that T is one-to-one, hence invertible. If we set $w = (az + b)/(cz + d)$, then the inverse mapping is

$$z = \frac{dw - b}{-cw + a},$$

which is also a linear fractional transformation.

We will look at some special kinds of linear fractional transformations.

EXAMPLE 25.7

Let $w = T(z) = z + b$, with b constant. This is called a *translation* because T shifts z horizontally by $\text{Re}(b)$ units and vertically by $\text{Im}(z)$ units. For example, if $T(z) = z + 2 - i$, then T takes z and moves it two units to the right and one unit down (Figure 25.23). We can see this with the following points and their images:

$$0 \to 2 - i, \quad 1 \to 3 - i, \quad i \to 2, \quad 4 + 3i \to 6 + 2i. \quad \blacksquare$$

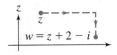

FIGURE 25.23

EXAMPLE 25.8

Let $w = T(z) = az$, with a a nonzero constant. This is called a *rotation/magnification*. To see why, first observe that

$$|w| = |a|\,|z|\,.$$

If $|a| > 1$, this transformation lengthens a complex number, in the sense that w is farther from the origin than z is. If $|a| < 1$, it shortens this distance. Thus the term *magnification*.

Now write the polar forms $z = re^{i\theta}$ and $a = Ae^{i\alpha}$. Then

$$T(z) = Are^{i(\theta+\alpha)},$$

so the transformation adds α to the argument of any nonzero complex number. This rotates the number counterclockwise through an angle α. This is the reason for the term *rotation*.

The total effect of the transformation is therefore a scaling and a rotation. As a specific example, consider

$$w = (2 + 2i)z.$$

This will map

$$i \rightarrow -2 + 2i, 1 \rightarrow 2 + 2i, \quad \text{and} \quad 1 + i \rightarrow 4i,$$

as shown in Figure 25.24. As Figure 25.25 suggests, in general the image of z is obtained by multiplying the magnitude of z by $|2 + 2i| = \sqrt{8}$ and rotating the line from the origin to z counterclockwise through $\pi/4$ radians. ▧

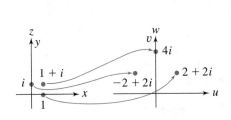

FIGURE 25.24

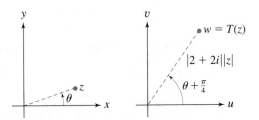

FIGURE 25.25 *The mapping $T(z) = (2 + 2i)z$.*

If $|a| = 1$, $T(z) = az$ is called a *pure rotation*, since in this case there is no magnification effect, just a rotation through an argument of a.

EXAMPLE 25.9

Let $w = T(z) = 1/z$. This mapping is called an *inversion*. For $z \neq 0$,

$$|w| = \frac{1}{|z|}$$

and

$$\arg(w) = \arg(1) - \arg(z) = -\arg(z)$$

(within integer multiples of 2π). This means that we arrive at $T(z)$ by moving $1/|z|$ units from the origin along the line from 0 to z and then reflecting this point across the real axis (Figure 25.26). This maps points inside the unit disk to the exterior of it and points exterior to the interior, while points on the unit circle remain on the unit circle (but get moved around this circle, except for 1 and -1). For example, if $z = (1+i)/\sqrt{2}$, then $1/z = (1-i)/\sqrt{2}$ (Figure 25.27). ■

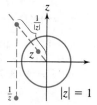

FIGURE 25.26
*Image of z under
an inversion.*

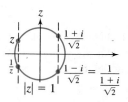

FIGURE 25.27 *Image of
a point on the unit circle
under an inversion.*

We will now show that translations, rotation/magnifications, and inversions are the fundamental linear fractional transformations, in the sense that any such mapping can be achieved as a sequence of transformations of these three kinds. To see how to do this, begin with

$$T(z) = \frac{az + b}{cz + d}.$$

If $c = 0$, then

$$T(z) = \frac{a}{d}z + \frac{b}{d}$$

is a rotation/magnification followed by a translation:

$$z \underset{\text{rot/mag}}{\to} \frac{a}{d}z \underset{\text{trans}}{\to} \frac{a}{d}z + \frac{b}{d}.$$

If $c \neq 0$, then T is the result of the following sequence:

$$z \underset{\text{rot/mag}}{\to} cz \underset{\text{trans}}{\to} cz + d \underset{\text{inv}}{\to} \frac{1}{cz + d}$$

$$\underset{\text{rot/mag}}{\to} \frac{bc - ad}{c}\frac{1}{cz + d} \to \frac{bc - ad}{c}\frac{1}{cz + d} + \frac{a}{c} = \frac{az + b}{cz + d} = T(z).$$
$$\underset{\text{trans}}{}$$

This way of breaking a linear fractional transformation into simpler components has two purposes. First, we can analyze general properties of these transformations by analyzing the simpler component transformations. Perhaps more important, we sometimes use this sequence to build conformal mappings between given domains.

The following is a fundamental property of linear fractional transformations. We use the term *line* to mean straight line.

THEOREM 25.2

A linear fractional transformation maps any circle to a circle or line and any line to a circle or line.

Proof Because of the preceding discussion, we need to verify this only for translations, rotation/magnifications, and inversions.

It is obvious geometrically that a translation maps a circle to a circle and a line to a line. Similarly, a rotation/magnification maps a circle to a circle and a line to a line.

Now we need to determine the effect of an inversion on a circle or line. Begin with the fact that any circle or line in the plane is the graph of an equation

$$A(x^2 + y^2) + Bx + Cy + R = 0,$$

in which A, B, C, and R are real numbers. This graph is a circle if $A \neq 0$ and a line if $A = 0$ and B and C are not both zero. With $z = x + iy$, this equation becomes

$$A\,|z|^2 + \frac{B}{2}(z + \overline{z}) + \frac{C}{2i}(z - \overline{z}) + R = 0.$$

Now let $w = 1/z$. The image in the w plane of this locus is the graph of

$$A\frac{1}{|w|^2} + \frac{B}{2}\left(\frac{1}{w} + \frac{1}{\overline{w}}\right) + \frac{C}{2i}\left(\frac{1}{w} - \frac{1}{\overline{w}}\right) + R = 0.$$

Multiply this equation by $w\overline{w}$ (the same as $|w|^2$) to get

$$R\,|w|^2 + \frac{B}{2}(w + \overline{w}) - \frac{C}{2i}(w - \overline{w}) + A = 0.$$

In the w plane, this is the equation of a circle if $R \neq 0$, and a line if $R = 0$ and B and C are not both zero. ∎

As the proof shows, translations and rotation/magnifications actually map circles to circles and lines to lines, while an inversion may map a circle to a circle or line and a line to a circle or line.

EXAMPLE 25.10

Let $w = T(z) = i(z - 2) + 3$. This is the sequence

$$z \rightarrow z - 2 \rightarrow i(z - 2) \rightarrow i(z - 2) + 3 = w, \tag{25.2}$$

a translation by 2 to the left, followed by a counterclockwise rotation by $\pi/2$ radians (an argument of i is $\pi/2$), and then a translation by 3 to the right. Since this mapping does not involve an inversion, it maps circles to circles and lines to lines.

As a specific example, consider the circle K given by

$$(x - 2)^2 + y^2 = 9,$$

with radius 3 and center $(2, 0)$. Write this equation as

$$x^2 + y^2 - 4x - 5 = 0$$

or

$$|z|^2 - 2(z + \bar{z}) - 5 = 0.$$

Solve $w = i(z - 2) + 3$ to get $z = -i(w - 3) + 2$ and substitute into the last equation to get

$$|i(w - 3) + 2|^2 - 2\left(-i(w - 3) + 2 + \overline{-i(w - 3) + 2}\right) - 5 = 0.$$

After some routine manipulation, this gives us

$$|w|^2 - 3(w + \overline{w}) = 0.$$

With $w = u + iv$, this is

$$(u - 3)^2 + v^2 = 9,$$

a circle of radius 3 and center $(3, 0)$ in the u, v plane. This result could have been predicted geometrically from the sequence of elementary mappings 25.2, shown in stages in Figure 25.28. The sequence first moves the circle 2 units to the left, then (multiplication by i) rotates it by $\pi/2$ radians (which leaves the center and radius the same), and then translates it 3 units right. The result is a circle of radius 3 about $(3, 0)$. ∎

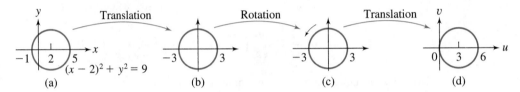

FIGURE 25.28

EXAMPLE 25.11

We will examine the effects of an inversion $w = 1/z$ on the vertical line $\text{Re}(z) = a \neq 0$. On this line, $z = a + iy$ and its image consists of points

$$w = \frac{1}{a + iy} = \frac{a}{a^2 + y^2} - \frac{y}{a^2 + y^2}i = u + iv.$$

It is routine to check that

$$\left(u - \frac{1}{2a}\right)^2 + v^2 = \frac{1}{4a^2},$$

so the image of this vertical line is a circle in the u, v plane having center $(1/2a, 0)$ and radius $1/2a$. ∎

In preparation for constructing mappings between given domains, we will show that we can always produce a linear fractional transformation mapping three given points to three given points.

THEOREM 25.3   *Three-Point Theorem*

Let z_1, z_2, and z_3 be three distinct points in the z plane and w_1, w_2, and w_3 three distinct points in the w plane. Then there is a linear fractional transformation T of the z plane to the w plane such that

$$T(z_1) = w_1, T(z_2) = w_2, \quad \text{and} \quad T(z_3) = w_3.$$

Proof Let $w = T(z)$ be the solution for w in terms of z and the six given points in the equation

$$(w_1 - w)(w_3 - w_2)(z_1 - z_2)(z_3 - z) = (z_1 - z)(z_3 - z_2)(w_1 - w_2)(w_3 - w). \quad (25.3)$$

Substitution of $z = z_j$ into this equation yields $w = w_j$ for $j = 1, 2, 3$. ∎

EXAMPLE 25.12

We will find a linear fractional transformation that maps

$$3 \rightarrow i, 1 - i \rightarrow 4, \quad \text{and} \quad 2 - i \rightarrow 6 + 2i.$$

Put

$$z_1 = 3, \quad z_2 = 1 - i, \quad z_3 = 2 - i$$

and

$$w_1 = i, \quad w_2 = 4, \quad w_3 = 6 + 2i$$

in equation (25.3) to get

$$(i - w)(2 + 2i)(2 + i)(2 - i - z) = (3 - z)(1)(i - 4)(6 + 2i - w).$$

Solve for w to get

$$w = T(z) = \frac{(20 + 4i)z - (68 + 16i)}{(6 + 5i)z - (22 + 7i)}.$$

Then each $T(z_j) = w_j$. ∎

One can show that specification of three points and their images uniquely determines a linear fractional transformation. In the last example, then, T is the only linear fractional transformation mapping the three given points to their given images.

When dealing with mappings, it is sometimes convenient to replace the complex plane with the complex sphere. To visualize how this is done, consider the 3-dimensional coordinate system in Figure 25.29. A sphere of radius 1 is placed with its south pole at the origin and north pole at $(0, 0, 2)$. Think of the x, y plane as the complex plane. For any (x, y) in this plane, the line from $(0, 0, 2)$ to (x, y) intersects the sphere in exactly one point $S(x, y)$. This associates with each point on the sphere except $(0, 0, 2)$, a unique point in the complex plane, and conversely. This mapping is called the *stereographic projection* of the sphere (minus its north pole) onto the plane. This punctured sphere is called the *complex sphere*. The point $(0, 0, 2)$ plays the role of a point at infinity. This is motivated by the fact that as (x, y) is chosen further from the origin in the x, y plane, $S(x, y)$ moves closer to $(0, 0, 2)$ on the sphere. The point $(0, 0, 2)$ is not associated with any complex number, but gives a way of envisioning infinity as a point, something we cannot do in the plane. The *extended complex plane* (consisting of all complex numbers, together with infinity) is in a one-to-one correspondence with this sphere, including its north pole.

To get some feeling for the point at infinity, consider the line $y = x$ in the x, y plane. This consists of complex numbers $x + xi$. If we let $x \rightarrow \infty$, the point $(1 + i)x$ moves out this line indefinitely far away from the origin. The image of this line on the complex sphere is part of a great circle, and the image point $S(x, y)$ on the sphere approaches $(0, 0, 2)$ as $x \rightarrow \infty$. This

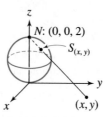

FIGURE 25.29
Stereographic projection identifying the complex sphere with the extended complex plane.

enables us to think of $(1 + i)x$ as approaching a specific location we can point to in this limit process, instead of just going further from the origin.

In defining a linear fractional transformation, it is sometimes convenient to map one of the three given points in the last theorem to infinity. This can be done by deleting the factors in equation (25.3) involving w_3.

THEOREM 25.4

Let z_1, z_2, z_3 be three distinct complex numbers and w_1, w_2 distinct complex numbers. Then there is a linear fractional transformation $w = T(z)$ that maps

$$T(z_1) = w_1, T(z_2) = w_2, \quad \text{and} \quad T(z_3) = \infty.$$

Proof   Such a transformation is obtained by solving for w in the equation

$$(w_1 - w)(z_1 - z_2)(z_3 - z) = (z_1 - z)(w_1 - w_2)(z_3 - z_2). \tag{25.4}$$

EXAMPLE 25.13

We will find a linear fractional transformation mapping

$$i \to 4i, \quad 1 \to 3 - i, \quad 2 + i \to \infty.$$

Solve for w in the equation

$$(4i - w)(i - 1)(2 + i - z) = (i - z)(-3 + 5i)(1 + i)$$

to get

$$w = T(z) = \frac{(5 - i)z - 1 + 3i}{-z + 2 + i}. \ \blacksquare$$

Some other properties of linear fractional transformations are pursued in the exercises. We now turn to the problem of constructing conformal mappings between given domains.

SECTION 25.2 PROBLEMS

In each of Problems 1 through 10, find a linear fractional transformation taking the given points to the indicated images.

1. $1 \to 1, 2 \to -i, 3 \to 1+i$

2. $i \to i, 1 \to -i, 2 \to 0$

3. $1 \to 1+i, 2i \to 3-i, 4 \to \infty$

4. $-5+2i \to 1, 3i \to 0, -1 \to \infty$

5. $6+i \to 2-i, i \to 3i, 4 \to -i$

6. $10 \to 8-i, 3 \to 2i, 4 \to 3-i$

7. $1 \to 6-4i, 1+i \to 2, 3+4i \to \infty$

8. $8-i \to 1-i, i \to 1+i, 3i \to \infty$

9. $2i \to \dfrac{1}{2}, 4i \to \dfrac{1}{3}, i \to 6i$

10. $-1 \to -i, i \to 2-3i, 1-2i \to \infty$

In each of Problems 11 through 18, find the image of the given circle or line under the linear fractional transformation.

11. $w = \dfrac{2z+i}{z-i}$; $\text{Im}(z) = 2$

12. $w = \dfrac{2i}{z}$; $\text{Re}(z) = -4$

13. $w = 2iz - 4$; $\text{Re}(z) = 5$

14. $w = \dfrac{z-i}{iz}$; $\dfrac{1}{2}(z+\bar{z}) + \dfrac{1}{2i}(z-\bar{z}) = 4$

15. $w = \dfrac{z-1+i}{2z+1}$; $|z| = 4$

16. $w = 3z - i$; $|z-4| = 3$

17. $w = \dfrac{2z-5}{z+i}$; $z+\bar{z} - \dfrac{3}{2i}(z-\bar{z}) - 5 = 0$

18. $w = \dfrac{(1+3i)z-2}{z}$; $|z-i| = 1$

19. Determine the image of any line $\text{Im}(z) = a \neq 0$ under the inversion $w = 1/z$.

20. Let $w = f(z)$ be a conformal mapping of a domain D onto a domain D^* in the w plane. Let $w^* = g(w)$ be a conformal mapping of D^* onto a domain D^{**} in the w^* plane. Prove that $g \circ f$ is a conformal mapping of D onto D^{**}.

21. Prove that the mapping $w = \bar{z}$ is not conformal.

22. Prove that the composition of two linear fractional transformations is a linear fractional transformation.

23. Prove that every linear fractional transformation has an inverse and that this inverse is also a linear fractional

transformation. (T^* is an inverse of T if $T \circ T^*$ and $T^* \circ T$ are both the identity mapping, taking each point to itself.)

24. Let
$$w = a\frac{z-b}{z-\bar{b}},$$
where $|a| = 1$ and $\text{Im}(b) > 0$. Prove that this mapping takes the real line onto the unit circle $|w| = 1$. Show that points in the upper half-plane $\text{Im}(z) \geq 0$ map to points in the closed unit disk $|w| \leq 1$.

25. Let
$$w = a\frac{z-b}{\bar{b}z-1},$$
where $|a| = 1$ and $|b| < 1$. Prove that this transformation maps the unit disk $|z| \leq 1$ onto the unit disk $|w| \leq 1$.

26. Show that there is no linear fractional transformation mapping the open disk $|z| < 1$ onto the set of points bounded by the ellipse $u^2/4 + v^2 = 1/16$.

In Problems 27 and 28, the setting is the extended complex plane, which includes the point at infinity.

27. A point z_0 is a *fixed point* of a mapping f if $f(z_0) = z_0$. Suppose f is a linear fractional transformation that is neither a translation nor the identity mapping $f(z) = z$. Prove that f must have either one or two fixed points but cannot have three. Why does this conclusion fail to hold for translations? How many fixed points can a translation have?

28. Let f be a linear fractional transformation with three fixed points. Prove that f is the identity mapping.

In each of Problems 29 through 34, write the linear fractional transformation as the end result of a sequence of mappings, each of which is a translation, rotation/magnification, or inversion.

29. $w = \dfrac{iz-4}{z}$

30. $w = \dfrac{z-4}{2z+i}$

31. $w = i(z+6) - 2 + i$

32. $w = \dfrac{z-1}{z+3+i}$

33. $w = \dfrac{(-2+3i)z}{z+4}$

34. $w = \dfrac{4i}{z+6}$

25.3   Construction of Conformal Mappings Between Domains

A strategy for solving some kinds of problems (for example, Dirichlet problems) is to find the solution for a "simple" domain (for example, the unit disk), then map this domain conformally to the domain of interest. This mapping may carry the solution for the disk to a solution for the latter domain. Of course, this strategy is predicated on two steps: finding a domain for which we can solve the problem, and being able to map this domain to the domain for which we want the solution. We will now discuss the latter problem.

Although in practice it may be an imposing task to find a conformal mapping between given domains, the following result assures us that such a mapping exists, with one exception.

THEOREM 25.5   *Riemann Mapping Theorem*

Let D^* be a domain in the w plane and assume that D^* is not the entire w plane. Then there exists a one-to-one conformal mapping of the unit disk $|z| < 1$ onto D^*. ∎

This powerful result implies the existence of a conformal mapping between given domains. Suppose we want to map D onto D^* (neither of which is the entire plane). Insert a third plane, the ζ plane, between the z plane and the w plane, as in Figure 25.30. By Riemann's theorem, there is a one-to-one conformal mapping g of the unit disk $|\zeta| < 1$ onto D^*. Similarly, there is a one-to-one conformal mapping f of $|\zeta| < 1$ onto D. Then $g \circ f^{-1}$ is a one-to-one conformal mapping of D onto D^*.

In theory, then, two domains, neither of which is the entire plane, can be mapped conformally in one-to-one fashion onto one another. This does not, however, make such mappings easy to find. In attempting to find such mappings, the following observation is useful.

A conformal mapping of a domain D onto a domain D^* will map the boundary of D to the boundary of D^*. We use this as follows. Suppose D is bounded by a path C (not necessarily closed) that separates the z plane into two domains, D and $\mathfrak{D}$. These are called *complementary domains*. Similarly, suppose D^* is bounded by a path C^* that separates the w plane into complementary domains D^* and $\mathfrak{D}^*$ (Figure 25.31). Try to find a conformal mapping f that sends points of C to points of C^*. This may be easier than trying to find a mapping of the entire domain. This mapping will then send D to either D^* or to $\mathfrak{D}^*$. To see which, choose a point z_0 in D and see whether $f(z_0)$ is in D^* or $\mathfrak{D}^*$. If $f(z_0)$ is in D^* (Figure 25.32(a)), then $f : D \rightarrow D^*$ and we have our conformal mapping. If $f(z_0)$ is in $\mathfrak{D}^*$ as in Figure 25.32(b), then $f : D \rightarrow \mathfrak{D}^*$. This is not yet the mapping we want, but sometimes we can take another step and use f to manufacture a conformal mapping from D to D^*.

We will now construct some conformal mappings, beginning with very simple ones and building up to more difficult problems.

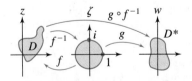

FIGURE 25.30 *Mapping D onto D^* through the unit disk.*

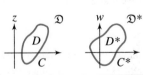

FIGURE 25.31

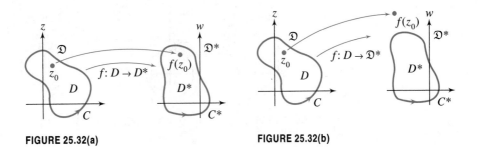

FIGURE 25.32(a)   **FIGURE 25.32(b)**

EXAMPLE 25.14

Suppose we want to map the unit disk $D : |z| < 1$ conformally onto the disk $D^* : |w| < 3$.

Clearly a magnification $w = f(z) = 3z$ will do this, because all we have to do is expand the unit disk to a disk of radius 3 (Figure 25.33). Notice that this mapping carries the boundary of D onto the boundary of D^*. ∎

EXAMPLE 25.15

Map the unit disk $D : |z| < 1$ conformally onto the domain $|w| > 3$.

Here we are mapping D to the complementary domain of the preceding example. We already know that $f(z) = 3z$ maps D conformally onto $|w| < 3$. Combine this map with an inversion, letting

$$g(z) = f\left(\frac{1}{z}\right) = \frac{3}{z}.$$

This maps $|z| < 1$ to $|w| > 3$ (Figure 25.34). Again, the boundary of the unit disk maps to the boundary of $|w| > 3$, which is the circle of radius 3 about the origin in the w plane. ∎

EXAMPLE 25.16

We will map the unit disk $D : |z| < 1$ onto the disk $D^* : |w - i| < 3$, of radius 3 and centered at i in the w plane.

Figure 25.35 suggests one way to construct this map. We want to expand the unit disk's radius by a factor of 3, then translate the resulting disk up one unit. Thus, map in steps:

$$z \rightarrow 3z \rightarrow 3z + i,$$

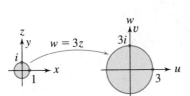

FIGURE 25.33 *Mapping of* $|z| < 1$ *onto* $|w| < 3$.

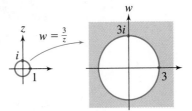

FIGURE 25.34 *Mapping of* $|z| < 1$ *onto* $|w| > 3$.

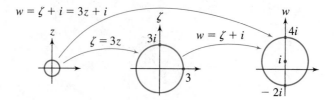

FIGURE 25.35 *Mapping of $|z| < 1$ onto $|w - i| < 3$.*

a magnification followed by a translation. The mapping is

$$w = f(z) = 3z + i.$$

This maps the unit circle $|z| = 1$ to the circle $|w - i| = 3$ because

$$|w - i| = |3z| = 3\,|z| = 3.$$

Further, the origin in the z plane (center of D) maps to i in the w plane, and i is the center of D^*, so $f : D \to D^*$. ∎

EXAMPLE 25.17

Suppose we want to map the right half-plane $D : \text{Re}(z) > 0$ onto the unit disk $D^* : |w| < 1$.

The domains are shown in Figure 25.36. The boundary of D is the imaginary axis $\text{Re}(z) = 0$. We will map this to the boundary of the unit disk $|w| = 1$. To do this, pick three points on $\text{Re}(z) = 0$ and three on $|w| = 1$ and use these to define a linear fractional transformation. There is a subtlety, however. To maintain positive orientation (counterclockwise on closed curves), choose three points in succession down the imaginary axis, so a person walking along these points sees the right half-plane on the left. Map these to points in order counterclockwise around $|w| = 1$.

For example, we can choose

$$z_1 = i, z_2 = 0, \quad \text{and} \quad z_3 = -i$$

on the imaginary axis in the z plane and map these in order to

$$w_1 = 1, \quad w_2 = i, \quad w_3 = -1.$$

From equation (25.3), we have

$$(1 - w)(-1 - i)(i)(-i - z) = (i - z)(-i)(1 - i)(-1 - w).$$

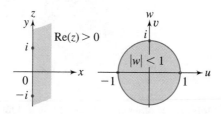

FIGURE 25.36

Solve for w:

$$w = T(z) = -i\left(\frac{z-1}{z+1}\right).$$

This conformal mapping must take the right half-plane to the interior or exterior of the unit disk in the w plane. To see which it is, pick a point in $\text{Re}(z) > 0$, say $z = 1$. Since $T(1) = 0$ is in D^*, T maps the right half-plane to the unit disk $|w| < 1$, as we want. ■

EXAMPLE 25.18

Suppose we want to map the right half-plane to the exterior of the unit disk, the domain $|w| > 1$. We have $T : \text{Re}(z) > 0 \to |w| < 1$ from the preceding example. If we follow this map (taking $\text{Re}(z) > 0$ onto the unit disk) with an inversion (taking the unit disk to the exterior of the unit disk), we will have the map we want. Thus, with $T(z)$ as in the last example, let

$$f(z) = \frac{1}{T(z)} = i\left(\frac{z+1}{z-1}\right).$$

As a check, $1 + i$ is in the right half-plane, and

$$f(1+i) = i\left(\frac{2+i}{i}\right) = 2 + i$$

is exterior to the unit disk in the w plane. ■

EXAMPLE 25.19

We will map the right half-plane $\text{Re}(z) > 0$ conformally onto the disk $|w - i| < 3$.

We can do this as a composition of mappings we have already constructed. Put an intermediate ζ plane between the z and w planes (Figure 25.37). From Example 25.17, map $\text{Re}(z) > 0$ onto the unit disk $|\zeta| < 1$ by

$$\zeta = f(z) = -i\left(\frac{z-1}{z+1}\right).$$

Now use the mapping of Example 25.16 to send the unit disk $|\zeta| < 1$ onto the disk $|w - i| < 3$:

$$w = g(\zeta) = 3\zeta + i.$$

The composition $g \circ f$ is a conformal mapping of $\text{Re}(z) > 0$ onto $|w - i| < 3$:

$$w = (g \circ f)(z) = g(f(z)) = 3f(z) + i = 3\left[-i\left(\frac{z-1}{z+1}\right)\right] + i = \frac{2i(-z+2)}{z+1}. \quad ■$$

EXAMPLE 25.20

We will map the infinite strip $S : -\pi/2 < \text{Im}(z) < \pi/2$ onto the unit disk $|w| < 1$.

Recall from Example 25.4 that the exponential function maps horizontal lines to half-lines from the origin. The boundary of S consists of two horizontal lines, $\text{Im}(z) = -\pi/2$ and $\text{Im}(z) = \pi/2$. On the lower boundary line, $z = x - i\pi/2$, so

$$e^z = e^x e^{-i\pi/2} = -ie^x,$$

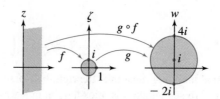

FIGURE 25.37   *Mapping of* $\text{Re}(z) > 0$
onto $|w - i| < 3$.

which varies over the negative imaginary axis as x takes on all real values. On the upper boundary of S, $z = x + i\pi/2$, and

$$e^z = i e^x$$

varies over the positive part of the imaginary axis as x varies over the real line.

The imaginary axis forms the boundary of the right half-plane $\text{Re}(w) > 0$, as well as of the left half-plane $\text{Re}(w) < 0$ in the w plane. The mapping $w = e^z$ must map S to one of these complementary domains. However, this mapping sends 0 to 1, in the right half-plane, so the mapping $w = f(z) = e^z$ maps S to the right half-plane.

We want to map S onto the unit disk. But now we know a mapping of S onto the right half-plane, and we also know a mapping of the right half-plane onto the unit disk. All we have to do is put these together.

In Figure 25.38, put a ζ plane between the z and w planes. Map

$$\zeta = f(z) = e^z$$

taking S onto the right half-plane $\text{Re}(\zeta) > 0$. Next map

$$w = g(\zeta) = -i\left(\frac{\zeta - 1}{\zeta + 1}\right),$$

taking the right half-plane $\text{Re}(\zeta) > 0$ onto the unit disk $|w| < 1$. Therefore, the function

$$w = (g \circ f)(z) = g(f(z)) = g(e^z) = -i\left(\frac{e^z - 1}{e^z + 1}\right)$$

is a conformal mapping of S onto $|w| < 1$. In terms of hyperbolic functions, this mapping can be written

$$w = -i \tanh\left(\frac{z}{2}\right). \quad \blacksquare$$

The conformal mapping of the last example is not a linear fractional transformation. These are convenient to use whenever possible. However, even when we know from the Riemann mapping

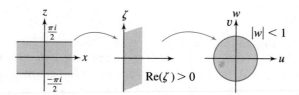

FIGURE 25.38 $|\text{Im}(z)| < \frac{\pi}{2} \rightarrow \text{Re}(\zeta) > 0 \rightarrow |w| < 1$
yields a mapping of $|\text{Im}(z)| < \frac{\pi}{2}$ *onto* $|w| < 1$.

theorem that a conformal mapping exists between two domains, we are not guaranteed that we can always find such a mapping in the form of a linear fractional transformation.

EXAMPLE 25.21

We will map the disk $|z| < 2$ onto the domain $D^* : u + v > 0$ in the u, v plane. These domains are shown in Figure 25.39.

Consider mappings we already have that relate to this problem. First, we can map $|z| < 2$ to $|\zeta| < 1$ by a simple magnification (multiply by $\frac{1}{2}$). But we also know a mapping from the unit disk to the right half-plane. Finally, we can obtain D^* from the right half-plane by a counterclockwise rotation through $\pi/4$ radians, an effect achieved by multiplying by $e^{i\pi/4}$. This suggests the strategy of constructing the mapping we want in the stages shown in Figure 25.40:

$$|z| < 2 \rightarrow |\zeta| < 1 \rightarrow \text{Re}(\xi) > 0 \rightarrow u + v > 0.$$

The first step is achieved by

$$\zeta = \frac{1}{2}z.$$

Next, use the inverse of the mapping from Example 25.17 and name the variables ζ and ξ to get

$$\xi = \frac{1 + i\zeta}{1 - i\zeta}.$$

This maps

$$|\zeta| < 1 \rightarrow \text{Re}(\xi) > 0.$$

Finally, perform the rotation:

$$w = e^{i\pi/4}\xi.$$

In sum,

$$w = e^{i\pi/4}\xi = e^{i\pi/4}\left(\frac{1 + i\zeta}{1 - i\zeta}\right) = e^{i\pi/4}\left(\frac{1 + i(z/2)}{1 - i(z/2)}\right) = \frac{2 + iz}{2 - iz}e^{i\pi/4}.$$

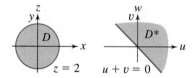

FIGURE 25.39

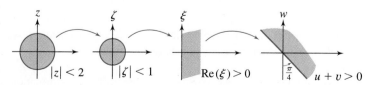

FIGURE 25.40

This maps the disk $|z| < 2$ conformally onto the half-plane $u + v > 0$. For example, 0 is in the disk, and

$$w(0) = e^{i\pi/4} = \frac{\sqrt{2}}{2}(1 + i)$$

is in $u + v > 0$. ■

EXAMPLE 25.22

We want to map the unit disk $|z| < 1$ onto the domain D^* exterior to the ellipse

$$\frac{u^2}{\cosh^2(4)} + \frac{v^2}{\sinh^2(4)} = 1.$$

From Example 25.5, $w = \cos(z)$ maps the line $\text{Im}(z) = 4$ onto this ellipse. If we knew how to map the unit circle to the line $\text{Im}(z) = 4$, we could compose mappings and construct the mapping we want.

Thus put an intermediate ζ plane between the z and w planes (Figure 25.41). We can map $|z| = 1$ to $\text{Im}(z) = 4$ by a linear fractional transformation. Choose three points in counterclockwise order on $|z| = 1$, say

$$z_1 = 1, \quad z_2 = i, \quad z_3 = -1.$$

Pick two points going left to right on $\text{Im}(\zeta) = 4$, say

$$\zeta_1 = -1 + 4i, \quad \zeta_2 = 4i.$$

We will map $z_1 \rightarrow \zeta_1, z_2 \rightarrow \zeta_2$, and, to save a little computation, $z_3 \rightarrow \infty$. From equation (25.4) we have

$$(-1 - 4i - \zeta)(1 - i)(-1 - z) = (1 - z)(-1)(-1 - i).$$

Solve for ζ to get

$$\zeta = -i\left(\frac{z - 1}{z + 1}\right) - 1 + 4i.$$

This maps $|z| = 1$ to $\text{Im}(\zeta) = 4$. The composition

$$w = \cos(\zeta) = \cos\left(-i\left(\frac{z - 1}{z + 1}\right) - 1 + 4i\right) = f(z)$$

therefore maps $|z| = 1$ to the ellipse. This ellipse determines two complementary domains, I and II, in Figure 25.42. Choose a point interior to the unit circle, say $z = 0$. Then

$$w(0) = \cos(-1 + 5i) = \cos(1)\cosh(5) + i\sin(1)\sinh(5).$$

This point is exterior to the ellipse because

$$\frac{\cos^2(1)\cosh^2(5)}{\cosh^2(4)} + \frac{\sin^2(1)\sinh^2(5)}{\sinh^2(4)} > 1$$

(the quantity on the left is approximately 7.3908). Therefore, f is a conformal mapping of the unit disk to the exterior of the ellipse, as we wanted. ■

We will briefly discuss the Schwarz–Christoffel transformation, which may be used when a domain has a polygon as a boundary.

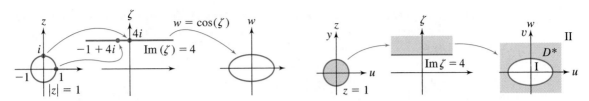

FIGURE 25.41

FIGURE 25.42 $|z| < 1 \to \text{Im}(\zeta) > 4 \to D^*$ *yields a mapping of* $|z| < 1$ *onto* D^*.

25.3.1 Schwarz–Christoffel Transformation

Suppose we want a conformal mapping of the upper half-plane $\mathfrak{H}$ to the interior $\mathfrak{P}$ of a polygon P, which could be a triangle, rectangle, pentagon, or other polygon. A linear fractional transformation will not do this. However, the Schwarz–Christoffel transformation was constructed just for this purpose.

Let P have vertices $w_1, \ldots, w_n$ in the w plane (Figure 25.43). Let the exterior angles of P be $\pi\alpha_1, \ldots, \pi\alpha_n$. We claim that there are constants z_0, a, and b, with $\text{Im}(z_0) > 0$, and real numbers $x_1, \ldots, x_n$ such that the function

$$f(z) = a \int_{z_0}^{z} (\xi - x_1)^{-\alpha_1}(\xi - x_2)^{-\alpha_2} \cdots (\xi - x_n)^{-\alpha_n} \, d\xi + b \tag{25.5}$$

is a conformal mapping of $\mathfrak{H}$ onto $\mathfrak{P}$. This integral is taken over any path in $\mathfrak{H}$ from z_0 to z in $\mathfrak{H}$. The factors $(\xi - x_j)^{-\alpha_j}$ are defined using the complex logarithm obtained by taking the argument lying in $[0, 2\pi)$.

Any function of the form of equation (25.5) is called a *Schwarz–Christoffel transformation*. To see the idea behind this function, suppose each $x_j < x_{j+1}$. If z is in $\mathfrak{H}$, let

$$g(z) = a(z - x_1)^{-\alpha_1}(z - x_2)^{-\alpha_2} \cdots (z - x_n)^{-\alpha_n}.$$

Then $f'(z) = g(z)$ and

$$\arg[f'(z)] = \arg(z) - \alpha_1 \arg(z - x_1) - \cdots - \alpha_n \arg(z - x_n).$$

As we saw in the discussion of Theorem 25.1, $\arg[f'(z)]$ is the number of radians by which the mapping f rotates tangent lines, if $f'(z) \neq 0$.

Now imagine z moving from left to right along the real axis (Figure 25.44), which is the boundary of $\mathfrak{H}$. On $(-\infty, x_1)$, $f(z)$ moves along a straight line (no change in the angle). As z passes over x_1, however, $\arg[f'(z)]$ changes by $\alpha_1\pi$. This angle remains fixed as z moves from x_1 toward x_2. As z passes over x_2, $\arg[f'(z)]$ changes by $\alpha_2\pi$, then remains at this value until z reaches x_3, where $\arg[f'(z)]$ changes by $\alpha_3\pi$, and so on. Thus $\arg[f'(z)]$ remains constant on

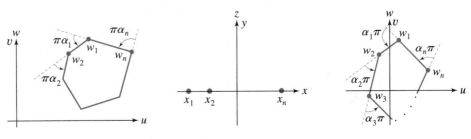

FIGURE 25.43

FIGURE 25.44

intervals (x_{j-1}, x_j) and increases by $\alpha_j \pi$ as z passes over x_j. The net result is that the real axis is mapped to a polygon P^* having exterior angles $\alpha_1 \pi, \ldots, \alpha_n \pi$. These numbers are actually determined by $\alpha_1, \ldots, \alpha_{n-1}$, since

$$\sum_{j=1}^{n} \alpha_j \pi = 2\pi.$$

P^* has the same exterior angles as P but need not be the same as P because of its location and size. We may have to rotate, translate, and/or magnify P^* to obtain P. These effects are achieved by choosing $x_1, \ldots, x_n$ to make P^* similar to P, and then choosing a (rotation/magnification) and b (translation) to obtain P.

If we choose $z_n = \infty$, then $z_1, \ldots, z_{n-1}, \infty$ are mapped to the vertices of P. In this case the Schwarz–Christoffel transformation is

$$f(z) = a \int_{z_0}^{z} (\xi - x_1)^{-\alpha_1} (\xi - x_2)^{-\alpha_2} \cdots (\xi - x_n)^{-\alpha_{n-1}} \, d\xi + b. \tag{25.6}$$

It can be shown that any conformal mapping of $\mathfrak{H}$ onto a polygon must have the form of a Schwarz–Christoffel transformation.

In practice a Schwarz–Christoffel transformation can be difficult or impossible to determine in closed form because of the integration.

EXAMPLE 25.23

We will map the upper half-plane $\mathfrak{H}$ onto a triangle.

Let the triangle have vertices w_1, w_2, w_3. If we map ∞ to w_3, the transformation has the form

$$f(z) = a \int_{z_0}^{z} (\xi - x_1)^{-\alpha_1} (\xi - x_2)^{-\alpha_2} \, d\xi + b.$$

There is not much we can do with this unless we make specific choices. Suppose $x_1 = -1$, $x_2 = 1$, and $x_3 = \infty$. Suppose also the triangle is equilateral, so $\alpha_1 = \alpha_2 = \alpha_3 = \frac{2}{3}$. If we choose $a = z_0 = 1$ and $b = 0$, then

$$f(z) = \int_{1}^{z} (\xi + 1)^{-2/3} (\xi - 1)^{-2/3} \, d\xi.$$

Now $f(1) = 0$, so x_2 maps to $w_2 = 0$. To find $f(-1)$, notice that when $z = -1$ in the integral, $\xi = x$ (real). When $-1 < x < 1$, then $x + 1 > 0$ and $\arg(x + 1) = 0$. But $\arg(x - 1) = \pi$ because $|x - 1| = 1 - x$. Therefore,

$$f(-1) = \int_{1}^{-1} (x + 1)^{-2/3} (x - 1)^{-2/3} \, dx$$

$$= \int_{1}^{-1} (x + 1)^{-2/3} [(-1)(1 - x)]^{-2/3} \, dx$$

$$= \int_{1}^{-1} (x + 1)^{-2/3} [e^{\pi i} (1 - x)]^{-2/3} \, dx$$

$$= \int_{1}^{-1} (x + 1)^{-2/3} (1 - x)^{-2/3} e^{-2\pi i / 3} \, dx$$

$$= -e^{-2\pi i/3} \int_{-1}^{1} (1+x)^{-2/3} (1-x)^{-2/3} \, dx$$

$$= e^{\pi i/3} \int_{-1}^{1} (1-x^2)^{-2/3} \, dx.$$

This integral is a beta function evaluated at $(\frac{1}{2}, \frac{1}{3})$ and can be evaluated in terms of the gamma function as

$$\frac{\Gamma(\frac{1}{2})\Gamma(\frac{1}{3})}{\Gamma(\frac{5}{6})}.$$

Therefore,

$$w_1 = f(-1) = \frac{\Gamma(\frac{1}{2})\Gamma(\frac{1}{3})}{\Gamma(\frac{5}{6})} e^{\pi i/3}.$$

Finally, the image of ∞ is

$$w_3 = \int_{1}^{\infty} (x+1)^{-2/3} (x-1)^{-2/3} \, dx = \int_{1}^{\infty} (x^2-1)^{-2/3} \, dx = \frac{\Gamma(\frac{1}{2})\Gamma(\frac{1}{3})}{\Gamma(\frac{5}{6})}.$$

The image of $\mathfrak{H}$ is the equilateral triangle with side length $\Gamma(\frac{1}{2})\Gamma(\frac{1}{3})/\Gamma(\frac{5}{6})$ and having one vertex at the origin and one side on the positive real axis, as shown in Figure 25.45. ■

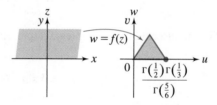

FIGURE 25.45

EXAMPLE 25.24

We will map the upper half-plane $\mathfrak{H}$ onto a rectangle.

Choose $x_1 = 0$, $x_2 = 1$, and x_3 as any real number greater than 1. The Schwarz–Christoffel transformation of equation (25.6) has the form

$$f(z) = a \int_{z_0}^{z} \frac{1}{\sqrt{\xi(\xi-1)(\xi-x_3)}} \, d\xi + b,$$

with a and b chosen to fit the dimensions of the rectangle and its orientation with respect to the axes. The radical appears because the exterior angles of a rectangle are all equal to $\pi/2$, so $\sum_{j=1}^{4} \alpha_j = 4\alpha_k = 2$. This integral is an elliptic integral and cannot be evaluated in closed form. ■

EXAMPLE 25.25

We will map $\mathfrak{H}$ onto the strip $S : \text{Im}(w) > 0$, $-c < \text{Re}(w) < c$ in the w plane. Here c is a positive constant.

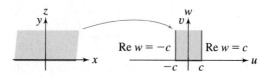

FIGURE 25.46

$\mathfrak{H}$ and the strip S are shown in Figure 25.46. To use the Schwarz–Christoffel transformation, we must think of S as a polygon with vertices $-c$, c, and ∞. Choose $x_1 = -1$ to map to $-c$ and $x_2 = 1$ to map to c. Map ∞ to ∞. The exterior angles of the strip are $\pi/2$ and $\pi/2$, so $\alpha_1 = \alpha_2 = \frac{1}{2}$. The transformation has the form

$$ w = f(z) = a \int_{z_0}^{z} (\xi + 1)^{-1/2} (\xi - 1)^{-1/2} \, d\xi + b. $$

We will choose $z_0 = 0$ and $b = 0$. Write

$$ (\xi - 1)^{-1/2} = [-(1 - \xi)]^{-1/2} = -i(1 - \xi)^{-1/2}. $$

With $-ai = A$, we have

$$ w = f(z) = A \int_{0}^{z} \frac{1}{(1 - \xi^2)^{1/2}} \, d\xi. $$

This integral is reminiscent of the real integral representation of the inverse sine function. Indeed, we can write

$$ w = A \sin^{-1}(z), $$

by which we mean that

$$ z = \sin\left(\frac{w}{A}\right). $$

To choose A so that -1 maps to $-c$ and 1 to c, we need

$$ \sin\left(\frac{c}{A}\right) = 1. $$

Thus choose $c/A = \pi/2$, or

$$ A = \frac{2c}{\pi}. $$

The mapping is

$$ w = \frac{2c}{\pi} \sin^{-1}(z). $$

If we choose $c = \pi/2$, this mapping is just $w = \sin^{-1}(z)$, mapping $\mathfrak{H}$ onto the strip $\text{Im}(w) > 0$, $-\pi/2 < \text{Re}(w) < \pi/2$. This is consistent with the result of Example 25.6. ∎

In each of Problems 1 through 10, find a linear fractional transformation mapping the first domain onto the second.

1. $|z| < 3$ onto $|w - 1 + i| < 6$

2. $|z| < 3$ onto $|w - 1 + i| > 6$

3. $|z + 2i| < 1$ onto $|w - 3| > 2$

4. $\text{Re}(z) > 1$ onto $\text{Im}(w) > -1$

5. $\text{Re}(z) < 0$ onto $|w| < 4$

6. $\text{Im}(z) > -4$ onto $|w - i| > 2$

7. $\text{Re}(z) > 0$ onto $\text{Im}(w) < 3$

8. $\text{Re}(z) < -4$ onto $|w + 1 - 2i| > 3$

9. $|z - 1| > 4$ onto $\text{Im}(w) < 2$

10. $|z - 1 + 3i| > 1$ onto $\text{Re}(w) < -5$

11. Find a conformal mapping of the upper half-plane $\text{Im}(z) > 0$ onto the wedge $0 < \arg(w) < \pi/3$.

12. Let $w = Log(z)$, in which the logarithm is given a unique value for each nonzero z by rectricting the argument of z to lie in $[0, 2\pi)$. Show that this mapping takes $\text{Im}(z) > 0$ onto the strip $0 < \text{Im}(w) < \pi$.

13. Let $w = \frac{1}{2}(z + (1/z))$.

    (a) Show that the interior and exterior of the unit circle are both mapped onto the complex w plane with the interval $[-1, 1]$ of the real axis removed.

    (b) Show that $|z| = r$ maps to an ellipse if $0 < r \le 1$.

14. Let

$$w = f(z) = \left(\frac{z - z_0}{1 - \overline{z_0} z} \right) e^{i\theta},$$

    where z_0 is constant, $|z_0| \le 1$, and θ is a fixed number in $[0, 2\pi]$. Show that f is a conformal mapping of the unit disk onto itself.

15. Show that the Schwarz–Christoffel transformation

$$f(z) = 2i \int_0^z (\xi + 1)^{-1/2}(\xi - 1)^{-1/2}\xi^{-1/2}\, d\xi$$

maps the upper half-plane onto the rectangle with vertices $0, c, c + ic$, and ic, where $c = \Gamma(\frac{1}{2})\Gamma(\frac{1}{4})/\Gamma(\frac{3}{4})$. Here Γ is the gamma function.

16. Define the *cross ratio* of z_1, z_2, z_3, and z_4 to be the image of z_1 under the linear fractional transformation that maps $z_2 \to 1, z_3 \to 0, z_4 \to \infty$. Denote this cross ratio as $[z_1, z_2, z_3, z_4]$. Suppose T is any linear fractional transformation. Show that T preserves the cross ratio. That is,

$$[z_1, z_2, z_3, z_4] = [Tz_1, Tz_2, Tz_3, Tz_4].$$

17. Prove that $[z_1, z_2, z_3, z_4]$ is the image of z_1 under the linear fractional transformation defined by

$$w = 1 - \frac{z_3 - z_4}{z_3 - z_2} \frac{z - z_2}{z - z_4}.$$

18. Prove that $[z_1, z_2, z_3, z_4]$ is real if and only if the z_j's are on the same circle or straight line.

19. Let z and z^* be complex numbers and let C be a circle passing through given points z_1, z_2, and z_3. Define z and z^* to be *symmetric with respect to C* if and only if $[z^*, z_1, z_2, z_3] = \overline{[z, z_1, z_2, z_3]}$. A point is *self-symmetric* with respect to C if z is symmetric with itself with respect to C (i.e., $z = z^*$). Prove that the points that are self-symmetric with respect to a circle are exactly the points on the circle.

20. Prove the *symmetry principle*: Suppose a linear fractional transformation T maps a circle C onto a circle K. Let z and z^* be symmetric with respect to C. Then $T(z)$ and $T(z^*)$ are symmetric with respect to K. *Hint:* Note the conclusion of Problem 16.

25.4   Harmonic Functions and the Dirichlet Problem

Given a set D of points in the plane, let ∂D denote the boundary of D. A Dirichlet problem for D is to find a solution of Laplace's equation

$$\frac{\partial^2 u}{\partial x^2} + \frac{\partial^2 u}{\partial y^2} = 0$$

for (x, y) in D, satisfying the boundary condition

$$u(x, y) = f(x, y) \quad \text{for } (x, y) \text{ in } \partial D.$$

Here f is a given function, usually assumed to be continuous on the boundary of D.

A function satisfying Laplace's equation in a set is said to be *harmonic* on that set. Thus the Dirichlet problem for a set is to find a function that is harmonic on that set and satisfies given data on the boundary of the set.

Chapter 18 is devoted to solutions of Dirichlet problems using methods from real analysis. Our purpose here is to apply complex function methods to Dirichlet problems. The connection between a Dirichlet problem and complex function theory is given by the following.

THEOREM 25.6

Let D be an open set in the plane, and let $f(z) = u(x, y) + iv(x, y)$ be differentiable on D. Then u and v are harmonic on D. ∎

That is, the real and imaginary parts of a differentiable complex function are harmonic.

Proof   By the Cauchy–Riemann equations,

$$\frac{\partial^2 u}{\partial x^2} + \frac{\partial^2 u}{\partial y^2} = \frac{\partial}{\partial x}\left(\frac{\partial u}{\partial x}\right) + \frac{\partial}{\partial y}\left(\frac{\partial u}{\partial y}\right)$$

$$= \frac{\partial}{\partial x}\left(\frac{\partial v}{\partial y}\right) + \frac{\partial}{\partial y}\left(-\frac{\partial v}{\partial x}\right) = \frac{\partial^2 v}{\partial x \partial y} - \frac{\partial^2 v}{\partial y \partial x} = 0.$$

Therefore, u is harmonic on D. The proof that v is harmonic is similar. ∎

Conversely, given a harmonic function u, there is a harmonic function v so that $f(z) = u(x, y) + iv(x, y)$ is differentiable. Such a v is called a *harmonic conjugate* for u.

THEOREM 25.7

Let u be harmonic on a domain D. Then, for some v, $u(x, y) + iv(x, y)$ defines a differentiable complex function for $z = x + iy$ in D.

Proof   Let

$$g(z) = \frac{\partial u}{\partial x} - i\frac{\partial u}{\partial y}$$

for (x, y) in D. Using the Cauchy–Riemann equations and Theorem 21.6, we find that g is differentiable on D. Then, for some function G, $G'(z) = g(z)$ for z in D. Write

$$G(z) = U(x, y) + iV(x, y).$$

Now

$$G'(z) = \frac{\partial U}{\partial x} - i\frac{\partial U}{\partial y} = g(z) = \frac{\partial u}{\partial x} - i\frac{\partial u}{\partial y}.$$

Therefore,

$$\frac{\partial U}{\partial x} = \frac{\partial u}{\partial x} \quad \text{and} \quad \frac{\partial U}{\partial y} = \frac{\partial u}{\partial y}$$

on D. Then, for some constant K,

$$U(x, y) = u(x, y) + K.$$

Let $f(z) = G(z) - K$. Then f is differentiable at all points of D. Further,

$$f(z) = G(z) - K = U(x, y) + iV(x, y) - K = u(x, y) + iv(x, y).$$

We may therefore choose $v(x, y) = V(x, y)$, proving the theorem. ∎

Given a harmonic function u, we will not be interested in actually producing a harmonic conjugate v. However, we will exploit the fact that there exists such a function to produce a differentiable complex function $f = u + iv$, given harmonic u. This enables us to apply complex function methods to Dirichlet problems. As a preliminary, we will derive two important properties of harmonic functions.

THEOREM 25.8   *Mean Value Property*

Let u be harmonic on a domain D. Let (x_0, y_0) be any point of D and let C be a circle of radius r centered at (x_0, y_0), contained in D and enclosing only points of D. Then

$$u(x_0, y_0) = \frac{1}{2\pi} \int_0^{2\pi} u(x_0 + r\cos(\theta), y_0 + r\sin(\theta)) \, d\theta. \quad ∎$$

As θ varies from 0 to 2π, $(x_0 + r\cos(\theta), y_0 + r\sin(\theta))$ moves once about the circle of radius r about (x_0, y_0). The conclusion of the theorem is called the *mean value property* because it states that the value of a harmonic function at the center of any circle in the domain is the average of its values on the circle.

Proof   For some v, $f = u + iv$ is differentiable on D. Let $z_0 = x_0 + iy_0$. By Cauchy's integral formula,

$$u(x_0, y_0) + iv(x_0, y_0) = f(z_0) = \frac{1}{2\pi i} \oint_C \frac{f(z)}{z - z_0} \, dz$$

$$= \frac{1}{2\pi i} \int_0^{2\pi} \frac{f(z_0 + re^{i\theta})}{re^{i\theta}} ire^{i\theta} \, d\theta$$

$$= \frac{1}{2\pi} \int_0^{2\pi} u(x_0 + r\cos(\theta), y_0 + r\sin(\theta)) \, d\theta$$

$$+ \frac{i}{2\pi} \int_0^{2\pi} v(x_0 + r\cos(\theta), y_0 + r\sin(\theta)) \, d\theta.$$

By taking the real and imaginary part of both sides of this equation, we get the conclusion of the theorem. ∎

If D is a bounded domain, then the set $cl(D)$ consisting of D together with all boundary points of D is called the *closure* of D. The set $cl(D)$ closed and bounded, hence is a compact set.

If $u(x, y)$ is continuous on $cl(D)$, then $u(x, y)$ must achieve a maximum value on $cl(D)$. If u is also harmonic on D, we claim that this maximum must occur at a boundary point of D. This is reminiscent of the maximum modulus theorem, from which it follows.

THEOREM 25.9

Let D be a bounded domain. Suppose u is continuous on $cl(D)$ and harmonic on D. Then $u(x, y)$ achieves its maximum value on $cl(D)$ at a boundary point of D.

Proof   First produce v so that $f = u + iv$ is differentiable on D. Define

$$g(z) = e^{f(z)}$$

for all z in $cl(D)$. Then g is differentiable on D. By the maximum modulus theorem, $|g(z)|$ achieves its maximum at a boundary point of D. But

$$|g(z)| = \left|e^{u(x,y)+iv(x,y)}\right| = e^{u(x,y)}.$$

Since $e^{u(x,y)}$ is a strictly increasing real function, $e^{u(x,y)}$ and $u(x, y)$ must achieve their maximum values at the same point. Therefore, $u(x, y)$ must achieve its maximum at a boundary point of D. ∎

For example, $u(x, y) = x^2 - y^2$ is harmonic on the open unit disk $x^2 + y^2 < 1$, and continuous on its closure $x^2 + y^2 \leq 1$. This function must therefore achieve its maximum value for $x^2 + y^2 \leq 1$ at a boundary point of this disk, namely at a point for which $x^2 + y^2 = 1$. We find that this maximum value is 1, occurring at $(1, 0)$ and at $(-1, 0)$.

25.4.1   Solution of Dirichlet Problems by Conformal Mapping

We want to use conformal mappings to solve Dirichlet problems. The strategy is to first solve the Dirichlet problem for a disk. Once we have this, we can attempt to solve a Dirichlet problem for another domain D by constructing a conformal mapping between the unit disk and D and applying this mapping to the solution we have for the disk.

In Section 18.3 we derived Poisson's integral formula for the solution of the Dirichlet problem for a disk, using Fourier methods. We will use complex function methods to obtain a form of this solution that is particularly suited to using with conformal mappings.

We want a function u that is harmonic on the disk $\widetilde{D} : |z| < 1$ and assumes given values $u(x, y) = g(x, y)$ on the boundary circle. Suppose u is harmonic on the slightly larger disk $|z| < 1 + \epsilon$. If v is a harmonic conjugate of u, then $f = u + iv$ is differentiable on this disk. We may, by adding a constant if necessary, choose v so that $v(0, 0) = 0$.

Expand f in a Maclaurin series

$$f(z) = \sum_{n=0}^{\infty} a_n z^n. \tag{25.7}$$

Then

$$u(x, y) = \mathrm{Re}(f(x + iy)) = \frac{1}{2}\left(f(z) + \overline{f(z)}\right)$$

$$= \frac{1}{2}\sum_{n=0}^{\infty}\left(a_n z^n + \overline{a_n}\,\overline{z}^n\right) = a_0 + \frac{1}{2}\sum_{n=1}^{\infty}\left(a_n z^n + \overline{a_n}\,\overline{z}^n\right).$$

Now let ζ be on the unit circle γ. Then $|\zeta|^2 = \zeta\overline{\zeta} = 1$, so $\overline{\zeta} = 1/\zeta$ and the series is

$$u(\zeta) = a_0 + \frac{1}{2}\sum_{n=1}^{\infty}\left(a_n\zeta^n + \overline{a_n}\zeta^{-n}\right).$$

Multiply this equation by $\zeta^m/2\pi i$ and integrate over γ. Within the open disk of convergence, the series and the integral can be interchanged. We get

$$\frac{1}{2\pi i}\oint_{\gamma}u(\zeta)\zeta^m\,d\zeta = \frac{a_0}{2\pi i}\oint_{\gamma}\zeta^m\,d\zeta + \frac{1}{2}\frac{1}{2\pi i}\sum_{n=1}^{\infty}\left(a_n\oint_{\gamma}\zeta^{n+m}\,d\zeta + \overline{a_n}\oint_{\gamma}\zeta^{-n+m}\,d\zeta\right). \quad (25.8)$$

Recall that

$$\oint_{\gamma}\zeta^k\,d\zeta = \begin{cases} 0 & \text{if } k \neq -1 \\ 2\pi i & \text{if } k = -1 \end{cases}.$$

Therefore, if $m = -1$ in equation (25.8), we have

$$\frac{1}{2\pi i}\oint_{\gamma}u(\zeta)\frac{1}{\zeta}\,d\zeta = a_0.$$

If $m = -n - 1$ with $n = 1, 2, 3, \ldots$, we obtain

$$\frac{1}{2\pi i}\oint_{\gamma}u(\zeta)\zeta^{-n-1} = \frac{1}{2}a_n.$$

Substitute these coefficients into equation (25.7) to get

$$f(z) = \sum_{n=0}^{\infty}a_nz^n = \frac{1}{2\pi i}\oint_{\gamma}u(\zeta)\frac{1}{\zeta}\,d\zeta + \sum_{n=1}^{\infty}\left(\frac{1}{\pi i}\oint_{\gamma}u(\zeta)\zeta^{-n-1}\right)d\zeta\,z^n$$

$$= \frac{1}{2\pi i}\oint_{\gamma}\left[1 + 2\sum_{n=1}^{\infty}\left(\frac{z}{\zeta}\right)^n\right]\frac{u(\zeta)}{\zeta}\,d\zeta.$$

Since $|z| < 1$ and $|\zeta| = 1$, then $|z/\zeta| < 1$ and the geometric series in this equation converges:

$$\sum_{n=1}^{\infty}\left(\frac{z}{\zeta}\right)^n = \frac{z/\zeta}{1 - z/\zeta} = \frac{z}{\zeta - z}.$$

Then

$$f(z) = \frac{1}{2\pi i}\oint_{\gamma}u(\zeta)\left[1 + \frac{2z}{\zeta - z}\right]\frac{1}{\zeta}\,d\zeta = \frac{1}{2\pi i}\oint_{\gamma}u(\zeta)\left(\frac{\zeta + z}{\zeta - z}\right)\frac{1}{\zeta}\,d\zeta.$$

If $u(\zeta) = g(\zeta)$, given values for u on the boundary of the unit disk, then, for $|z| < 1$,

$$u(x, y) = \text{Re}[f(z)] = \text{Re}\left(\frac{1}{2\pi i}\oint_{\gamma}g(\zeta)\left(\frac{\zeta + z}{\zeta - z}\right)\frac{1}{\zeta}\,d\zeta\right). \quad (25.9)$$

This is an integral formula for the solution of the Dirichlet problem for the unit disk. We leave it as an exercise for the student to retrieve the Poisson integral formula from this expression by putting $z = re^{i\theta}$ and $\zeta = e^{i\varphi}$.

Equation (25.9) is well suited to solving certain Dirichlet problems by conformal mappings. Suppose we know a differentiable, one-to-one conformal mapping $T : D \to \widetilde{D}$, where $\widetilde{D}$ is the unit disk $|w| < 1$ in the w plane. Assume that T maps C, the boundary of D, onto the unit circle $\widetilde{C}$ bounding $\widetilde{D}$ and that T^{-1} is also a differentiable conformal mapping.

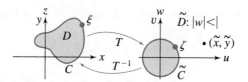

FIGURE 25.47

To help follow the discussion, we will use ζ for an arbitrary point of $\widetilde{C}$, ξ for a point on C, and $(\widetilde{x}, \widetilde{y})$ for a point in the w plane (Figure 25.47).

Now consider a Dirichlet problem for D:

$$\frac{\partial^2 u}{\partial x^2} + \frac{\partial^2 u}{\partial y^2} = 0 \quad \text{for } (x, y) \text{ in } D,$$

$$u(x, y) = g(x, y) \quad \text{for } (x, y) \text{ in } C.$$

If $w = T(z)$, then $z = T^{-1}(w)$ and we define

$$\widetilde{g}(w) = g(T^{-1}(w)) = g(z).$$

In the w plane, we now have a Dirichlet problem for the unit disk:

$$\frac{\partial^2 \widetilde{u}}{\partial x^2} + \frac{\partial^2 \widetilde{u}}{\partial y^2} = 0 \quad \text{for } (\widetilde{x}, \widetilde{y}) \text{ in } \widetilde{D},$$

$$\widetilde{u}(\widetilde{x}, \widetilde{y}) = \widetilde{g}(\widetilde{x}, \widetilde{y}) \quad \text{for } (\widetilde{x}, \widetilde{y}) \text{ on } \widetilde{C}.$$

From equation (25.9) the solution of this problem for the unit disk is the real part of

$$\widetilde{f}(w) = \frac{1}{2\pi i} \oint_{\widetilde{C}} \widetilde{g}(\zeta) \left(\frac{\zeta + w}{\zeta - w} \right) \frac{1}{\zeta} d\zeta.$$

Finally, recalling that T maps C onto $\widetilde{C}$, let $\zeta = T(\xi)$ for ξ on C to obtain

$$u(x, y) = \text{Re}[f(z)] = \text{Re}\left[\frac{1}{2\pi i} \int_C \widetilde{g}(T(\xi)) \left(\frac{T(\xi) + T(z)}{T(\xi) - T(z)} \right) \frac{1}{T(\xi)} T'(\xi) \, d\xi \right].$$

Since $\widetilde{g}(T(\xi)) = g(T^{-1}(T(\xi))) = g(\xi)$, we have the solution

$$u(x, y) = \text{Re}[f(z)] = \text{Re}\left[\frac{1}{2\pi i} \int_C g(\xi) \left(\frac{T(\xi) + T(z)}{T(\xi) - T(z)} \right) \frac{T'(\xi)}{T(\xi)} \, d\xi \right]. \tag{25.10}$$

This solves the Dirichlet problem for the original domain D.

To illustrate this technique, we will solve the Dirichlet problem for the right half-plane:

$$\frac{\partial^2 u}{\partial x^2} + \frac{\partial^2 u}{\partial y^2} = 0 \quad \text{for } x > 0, -\infty < y < \infty,$$

$$u(0, y) = g(y) \quad \text{for } -\infty < y < \infty.$$

We need a conformal mapping from the right half-plane to the unit disk. There are many such mappings. From Example 25.17, we can use

$$w = T(z) = -i \left(\frac{z - 1}{z + 1} \right).$$

Compute

$$T'(z) = \frac{-2i}{(z+1)^2}.$$

From equation (25.10), the solution is the real part of

$$f(z) = \frac{1}{2\pi i} \int_C u(\xi) \left(\frac{-i(\xi-1)/(\xi+1) - i(z-1)/(z+1)}{-i(\xi-1)/(\xi+1) + i(z-1)/(z+1)} \right) \frac{1}{-i(\xi-1)/(\xi+1)} \frac{-2i}{(\xi+1)^2} \, d\xi$$

$$= \frac{1}{\pi i} \int_C u(\xi) \left(\frac{\xi z - 1}{\xi - z} \right) \frac{1}{\xi^2 - 1} \, d\xi.$$

The boundary C of the right half-plane is the imaginary axis and is not a closed curve. Parametrize C as $\xi = (0, t) = it$, with t varying from ∞ to $-\infty$ for positive orientation on C (as we walk down this axis, D is over our left shoulder). We get

$$f(z) = \frac{1}{\pi i} \int_{\infty}^{-\infty} u(0, t) \left(\frac{itz - 1}{it - z} \right) \left(\frac{-1}{1 + t^2} \right) i \, dt$$

$$= \frac{1}{\pi} \int_{-\infty}^{\infty} u(0, t) \left(\frac{itz - 1}{it - z} \right) \left(\frac{1}{1 + t^2} \right) dt.$$

The solution is the real part of this integral. Now t, $u(0, \xi)$, and $1/(1+t^2)$ are real, so concentrate on the term containing i and $z = x + iy$:

$$\frac{itz - 1}{it - z} = \frac{itx - ty - 1}{it - x - iy} = \frac{itx - ty - 1}{it - x - iy} \frac{-it - x + iy}{-it - x + iy}$$

$$= \frac{tx(t - y) - itx^2 + ity(t - y) + txy + i(t - y) + x}{x^2 + (t - y)^2}.$$

The real part of this expression is

$$\frac{x(1 + t^2)}{x^2 + (t - y)^2}.$$

Therefore,

$$u(x, y) = \text{Re}[f(z)] = \frac{1}{\pi} \int_{-\infty}^{\infty} u(0, t) \frac{x(1 + t^2)}{x^2 + (t - y)^2} \frac{1}{1 + t^2} \, dt$$

$$= \frac{1}{\pi} \int_{-\infty}^{\infty} g(t) \frac{x}{x^2 + (t - y)^2} \, dt.$$

This is an integral formula for the solution of the Dirichlet problem for the right half-plane.

SECTION 25.4   PROBLEMS

1. Using complex function methods, write an integral solution for the Dirichlet problem for the upper half-plane $\text{Im}(z) > 0$.

2. Using complex function methods, write an integral solution for the Dirichlet problem for the right quarter-plane $\text{Re}(z) > 0$, $\text{Im}(z) > 0$, if the boundary conditions are $u(x, 0) = f(x)$ and $u(0, y) = 0$.

3. Write an integral solution for the Dirichlet problem for the disk $|z - z_0| < R$.

4. Write a formula for the solution of the Dirichlet problem for the right half-plane if the boundary condition is given by

$$u(0, y) = \begin{cases} 1 & \text{for } -1 \le y \le 1 \\ 0 & \text{for } |y| > 1 \end{cases}.$$

5. Write a formula for the solution of the Dirichlet problem for the unit disk if the boundary condition is given by $u(x, y) = x - y$ for (x, y) on the unit circle.

6. Write a formula for the solution of the Dirichlet problem for the unit disk if the boundary condition is given

by

$$u(e^{i\theta}) = \begin{cases} 1 & \text{for } 0 \le \theta \le \pi/4 \\ 0 & \text{for } \pi/4 < \theta < 2\pi \end{cases}.$$

7. Write a formula for the solution of the Dirichlet problem for the strip $-1 < \text{Im}(z) < 1$, $\text{Re}(z) > 0$ if the boundary condition is given by

$$u(x, 1) = u(x, -1) = 0 \quad \text{for } 0 < x < \infty$$
$$u(0, y) = 1 - |y| \quad \text{for } -1 \le y \le 1.$$

8. Write an integral formula for the solution of the Dirichlet problem for the strip $-1 < \text{Re}(z) < 1$, $\text{Im}(z) > 0$ if the boundary condition is given by

$$u(x, 0) = 1 \quad \text{for } -1 < x < 1$$
$$u(-1, y) = u(1, y) = e^{-y} \quad \text{for } 0 < y < \infty.$$

25.5 Complex Function Models of Plane Fluid Flow

We will discuss how complex functions and integration are used in modeling and analyzing the flow of fluids.

Consider an incompressible fluid, such as water under normal conditions. Assume that we are given a velocity field $\mathbf{V}(x, y)$ in the plane. By assuming that the flow depends only on two variables, we are taking the flow to be the same in all planes parallel to the complex plane. Such a flow is called *plane-parallel*. This velocity vector is also assumed to be independent of time, a circumstance described by saying that the flow is *stationary*.

Write

$$\mathbf{V}(x, y) = u(x, y)\mathbf{i} + v(x, y)\mathbf{j}.$$

Since we can identify vectors and complex numbers, we will by a mild abuse of notation write the velocity vector as a complex function

$$V(z) = V(x + iy) = u(x, y) + iv(x, y).$$

Given $V(z)$, think of the complex plane as divided into two sets. The first is the domain D on which V is defined. The complement of D consists of all complex numbers not in D. Think of the complement as comprising channels confining the fluid to D or as barriers through which the fluid cannot flow. This enables us to model fluid flow through a variety of configurations and around barriers of various shapes.

Suppose γ is a closed path in D. From vector analysis, if we parametrize γ by $x = x(s)$, $y = y(s)$, with s arc length along the path, then the vector $x'(s)\mathbf{i} + y'(s)\mathbf{j}$ is a unit tangent vector to γ, and

$$(u\mathbf{i} + v\mathbf{j}) \cdot \left(\frac{dx}{ds}\mathbf{i} + \frac{dy}{ds}\mathbf{j}\right) ds = u\, dx + v\, dy.$$

This is the dot product of the velocity with the tangent to a path trajectory, leading us to interpret

$$\oint_{\gamma} u\, dx + v\, dy$$

The Panama Canal, completed in 1914, extends 51 miles between Limón Bay on the Atlantic and the Bay of Panama on the Pacific edge of the Isthmus of Panama. Considered one of the great engineering successes of the world, it shortened a ship's transit from New York to San Francisco by nearly 8000 miles. The canal has three sets of locks, chambers that can be filled with water to raise or lower ships from one level of the canal to another. Each lock is 1000 feet long and 100 feet wide, with a depth of 70 feet. Some of the cranes and heavy equipment still in use date back to 1912. The flow of fluids in vessels such as these locks is often analyzed using complex function techniques.

as a measure of the velocity of the fluid along γ. The value of this integral is called the *circulation* of the fluid along γ.

The vector $-y'(s)\mathbf{i} + x'(s)\mathbf{j}$ is a unit normal vector to γ, being perpendicular to the tangent vector (Figure 25.48). Therefore

$$ -\oint_\gamma (u\mathbf{i} + v\mathbf{j}) \cdot \left(-\frac{dy}{dx}\mathbf{i} + \frac{dx}{ds}\mathbf{j} \right) ds = \oint_\gamma -v\,dx + u\,dy $$

is the negative of the integral of the normal component of the velocity along the path. When this integral is not zero, it is called the *flux* of the fluid across the path. This gives a measure of fluid flowing across γ out from the region bounded by γ. When this flux is zero for every closed path in the domain of the fluid, the fluid is called *solenoidal*.

A point $z_0 = (x_0, y_0)$ is a vortex of the fluid if the circulation has a nonzero constant value on every closed path about z_0 in the interior of some punctured disk $0 < |z - z_0| < r$. The constant value of the circulation is the *strength of the vortex*.

If $\oint_\gamma -v\,dx + u\,dy$ has the same positive value k for all closed paths about z_0 in some punctured disk about z_0, then we call z_0 a *source* of strength k; if k is negative, z_0 is a *sink* of strength $|k|$.

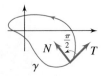

FIGURE 25.48

The connection between the velocity field of a fluid and complex functions is provided by the following.

THEOREM 25.10

Let u and v be continuous with continuous first and second partial derivatives in a simply connected domain D. Suppose $u\mathbf{i} + v\mathbf{j}$ is irrotational and solenoidal in D. Then u and $-v$ satisfy the Cauchy–Riemann equations in D, and $f(z) = u(x, y) - iv(x, y)$ is a differentiable complex function on D.

Conversely, if u and $-v$ satisfy the Cauchy–Riemann equations on D, then $u\mathbf{i} + v\mathbf{j}$ defines an irrotational, solenoidal flow on D.

Proof   Let γ be any closed path in D. If M is the interior of γ, then every point in M is also in D by the assumption that D is simply connected. By Green's theorem,

$$\oint_\gamma u\,dx + v\,dy = \iint_M \left(\frac{\partial v}{\partial x} - \frac{\partial u}{\partial y} \right) dA = 0$$

because the flow is irrotational. But the flow is also solenoidal, so, again by Green's theorem,

$$\oint_\gamma -v\,dx + u\,dy = \iint_M \left(\frac{\partial u}{\partial x} + \frac{\partial v}{\partial y} \right) dA = 0.$$

Because M can be any set of points in D bounded by a closed path, the integrands in both of these double integrals must be zero throughout D, so

$$\frac{\partial u}{\partial x} = \frac{\partial}{\partial y}(-v) \quad \text{and} \quad \frac{\partial u}{\partial y} = -\frac{\partial}{\partial x}(-v).$$

By Theorem 21.6, $f(z) = u(x, y) - iv(x, y)$ is differentiable on D.

The converse follows by a similar argument.   ∎

Theorem 25.10 provides some insight into irrotational, solenoidal flows. If the flow is irrotational, then

$$curl(u\mathbf{i} + v\mathbf{j}) = \left(\frac{\partial v}{\partial x} - \frac{\partial u}{\partial y} \right)\mathbf{k} = \mathbf{0},$$

as shown in the proof of the theorem. This curl is a vector normal to the plane of the flow. From the discussion of Section 11.4.2, the curl of $u\mathbf{i} + v\mathbf{j}$ is twice the angular velocity of the particle of fluid at (x, y). The fact that this curl is zero for an irrotational flow means that the fluid particles may experience translations and distortions in their motion, but no rotation. There is no swirling effect in the fluid.

If the flow is solenoidal, then

$$div(u\mathbf{i} + v\mathbf{j}) = \frac{\partial u}{\partial x} + \frac{\partial v}{\partial y} = 0.$$

A further connection between flows and complex functions is provided by the following.

Let f be a differentiable function defined on a domain D. Then $\overline{f'(z)}$ is an irrotational, solenoidal flow on D.

Conversely, if $\mathbf{V} = u\mathbf{i} + v\mathbf{j}$ is an irrotational, solenoidal vector field on a simply connected domain D, then there is a differentiable complex function f defined on D such that $\overline{f'(z)} = V$. Further, if $f(z) = \varphi(x, y) + i\psi(x, y)$, then

$$\frac{\partial \varphi}{\partial x} = u, \frac{\partial \varphi}{\partial y} = v, \frac{\partial \psi}{\partial x} = -v, \quad \text{and} \quad \frac{\partial \psi}{\partial y} = u. \ \blacksquare$$

We leave a proof of this result to the student. In view of the fact that $\overline{f'(z)}$ is the velocity of the flow, we call f a *complex potential* of the flow.

Theorem 25.11 implies that any differentiable function $f(z) = \varphi(x, y) + i\psi(x, y)$ defined on a simply connected domain determines an irrotational, solenoidal flow

$$\overline{f'(z)} = \frac{\partial \varphi}{\partial x} + i\frac{\partial \psi}{\partial x} = \overline{u(x, y) - iv(x, y)} = u(x, y) + iv(x, y).$$

We call φ the *velocity potential* of the flow, and curves $\varphi(x, y) = k$ are called *equipotential curves*. The function ψ is called the *stream function* of the flow, and curves $\psi(x, y) = c$ are called *streamlines*.

We may think of $w = f(z)$ as a conformal mapping wherever $f'(z) \neq 0$. A point at which $f'(z) = 0$ is called a *stagnation point* of the flow. Thinking of f as a mapping, in the w plane, we have

$$w = f(z) = \varphi(x, y) + i\psi(x, y) = \alpha + i\beta.$$

Equipotential curves $\varphi(x, y) = k$ map under f to vertical lines $\alpha = k$, and streamlines $\psi(x, y) = c$ map to horizontal lines $\beta = c$. Since these sets of vertical and horizontal lines are mutually orthogonal in the w plane, the streamlines and equipotential curves in the z plane also form orthogonal families. Every streamline is orthogonal to each equipotential curve at any point where they intersect. This conclusion fails at a stagnation point, where the mapping may not be conformal.

Along an equipotential curve $\varphi(x, y) = k$,

$$d\varphi = \frac{\partial \varphi}{\partial x} dx + \frac{\partial \varphi}{\partial y} dy = u\, dx + v\, dy = 0.$$

Now $u\mathbf{i} + v\mathbf{j}$ is the velocity of the flow at (x, y), and $x'(s)\mathbf{i} + y'(s)\mathbf{j}$ is a unit tangent to the equipotential curve through (x, y). Since the dot product of these two vectors is zero from the fact that $d\varphi = 0$ along the equipotential curve, we conclude that the velocity is orthogonal to the equipotential curve through (x, y), provided that (x, y) is not a stagnation point.

Similarly, along a streamline $\psi(x, y) = c$,

$$d\psi = \frac{\partial \psi}{\partial x} dx + \frac{\partial \psi}{\partial y} dy = -v\, dx + u\, dy = 0,$$

so the normal to the velocity vector is orthogonal to the streamline. This means that the velocity is tangent to the streamline and justifies the interpretation that the particle of fluid at (x, y) is moving in the direction of the streamline at that point. We therefore interpret streamlines as the trajectories of particles in the fluid. If we sat on a particular particle, we would ride along a streamline. For this reason, graphs of streamlines form a picture of the motions of the particles of fluid.

The remainder of this section is devoted to some illustrations of these ideas.

EXAMPLE 25.26

Let $f(z) = -Ke^{i\theta}z$, in which K is a positive constant and $0 \le \theta \le 2\pi$.
  Write

$$f(z) = -K[\cos(\theta) + i\sin(\theta)][x + iy]$$
$$= -K[x\cos(\theta) - y\sin(\theta)] - iK[y\cos(\theta) + x\sin(\theta)].$$

If $f(z) = \varphi(x, y) + i\psi(x, y)$, then

$$\varphi(x, y) = -K[x\cos(\theta) - y\sin(\theta)]$$

and

$$\psi(x, y) = -K[y\cos(\theta) + x\sin(\theta)].$$

Equipotential curves are graphs of

$$\varphi(x, y) = -K[x\cos(\theta) - y\sin(\theta)] = \text{constant}.$$

Since K is constant, equipotential curves are graphs of

$$x\cos(\theta) - y\sin(\theta) = k,$$

or

$$y = \cot(\theta)x + b,$$

in which $b = k\csc(\theta)$ is constant. These are straight lines with slope $\cot(\theta)$.
  Streamlines are graphs of

$$y = -\tan(\theta)x + d,$$

straight lines with slope $-\tan(\theta)$. These lines make an angle $\pi - \theta$ with the positive real axis, as in Figure 25.49. These are the trajectories of the flow, which may be thought of as moving along these straight lines. The streamlines and equipotential curves are orthogonal, their slopes being negative reciprocals.
  Now compute

$$\overline{f'(z)} = \overline{-Ke^{i\theta}} = -Ke^{-i\theta}.$$

This implies that the velocity is of constant magnitude K.
  In summary, f models a uniform flow with velocity of constant magnitude K and making an angle $\pi - \theta$ with the positive real axis, as in Figure 25.49. ∎

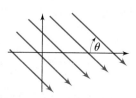

FIGURE 25.49
*Streamlines of the flow
with complex potential
$f(z) = -Ke^{i\theta}z$.*

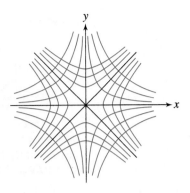

FIGURE 25.50 *Equipotential curves and streamlines of the flow with complex potential $f(z) = z^2$.*

EXAMPLE 25.27

Consider the flow represented by the complex potential $f(z) = z^2$. This function is differentiable for all z, but $f'(0) = 0$, so the origin is a stagnation point. We will see what effect this has on the flow and determine the trajectories.

With $z = x + iy$, $f(z) = x^2 - y^2 + 2ixy$, so

$$\varphi(x, y) = x^2 - y^2 \quad \text{and} \quad \psi(x, y) = 2xy.$$

Equipotential curves are hyperbolas

$$x^2 - y^2 = k$$

if $k \neq 0$. Streamlines are hyperbolas

$$xy = c$$

if $c \neq 0$. Some curves of these families are shown in Figure 25.50.

If $k = 0$ the equipotential curves are graphs of $x^2 - y^2 = 0$, which are two straight lines $y = x$ and $y = -x$ through the origin. If $c = 0$ the streamlines are the axes $x = 0$ and $y = 0$.

The velocity of the flow is $\overline{f'(z)} = 2\overline{z}$. f models a nonuniform flow having velocity of magnitude $2|z|$ at z. We can envision this flow as a fluid moving along the streamlines. In any quadrant, the particles move along the hyperbolas $xy = c$, with the axes acting as barriers of the flow (think of sides of a container holding the fluid). ∎

EXAMPLE 25.28

We will analyze the complex potential

$$f(z) = \frac{iK}{2\pi} \text{Log}(z).$$

Here K is a positive number and $\text{Log}(z)$ denotes that branch of the logarithm defined by $\text{Log}(z) = \frac{1}{2} \ln(x^2 + y^2) + i\theta$, where θ is the argument of z lying in $0 \leq \theta < 2\pi$ for $z \neq 0$. If $z = x + iy$, then

$$f(z) = \frac{iK}{2\pi} \left(\frac{1}{2} \ln(x^2 + y^2) + i\theta \right) = -\frac{K\theta}{2\pi} + \frac{iK}{4\pi} \ln(x^2 + y^2).$$

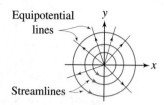

FIGURE 25.51 *Streamlines and equipotential lines of the flow having complex potential* $f(z) = (iK/2\pi)\text{Log}(z)$.

Now

$$\varphi(x, y) = -\frac{K}{2\pi}\theta \quad \text{and} \quad \psi(x, y) = \frac{K}{4\pi}\ln(x^2 + y^2).$$

Equipotential curves are graphs of $\theta = $ constant, and these are half-lines from the origin making an angle θ with the positive real axis. Streamlines are graphs of $\psi(x, y) = $ constant, and these are circles about the origin. Since streamlines are trajectories of the fluid, particles are moving on circles about the origin. Some streamlines and equipotential curves are shown in Figure 25.51.

It is easy to check that $f'(z) = (iK/2\pi)(1/z)$ if $z \neq 0$. On a circle $|z| = r$, the magnitude of the velocity of the fluid is

$$\left|\overline{f'(z)}\right| = \frac{K}{2\pi}\frac{1}{|z|} = \frac{K}{2\pi r}.$$

This velocity increases as $r \to 0$, so we have particles of fluid swirling about the origin, with increasing velocity toward the center (origin) (Figure 25.52). The origin is a vortex of the flow.

To calculate the circulation of the flow about the origin, write

$$\overline{f'(z)} = -\frac{iK}{2\pi}\frac{1}{\overline{z}} = -\frac{iK}{2\pi}\frac{z}{|z|^2} = \frac{K}{2\pi}\frac{y}{x^2 + y^2} - \frac{iK}{2\pi}\frac{x}{x^2 + y^2} = u + iv.$$

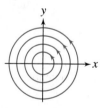

FIGURE 25.52
Streamlines of the flow having complex potential $f(z) = (iK/2\pi)\text{Log}(z)$.

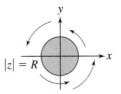

FIGURE 25.53 *Flow around a cylindrical barrier of radius R.*

If γ is the circle of radius r about the origin, then on γ, $x = r\cos(\theta)$ and $y = r\sin(\theta)$, so

$$\oint_{\gamma} u\,dx + v\,dy = \int_0^{2\pi} \left[\frac{K}{2\pi} \frac{r\sin(\theta)}{r^2}(-r\sin(\theta)) - \frac{K}{2\pi} \frac{r\cos(\theta)}{r^2}(r\cos(\theta)) \right] d\theta$$

$$= \frac{K}{2\pi} \int_0^{2\pi} -d\theta = -K.$$

This is the value of the circulation on any circle about the origin.

By a similar calculation, we get

$$\oint_{\gamma} -v\,dx + u\,dy = 0,$$

so the origin is neither a source nor a sink.

In this example we may restrict $|z| > R$ and think of a solid cylinder about the origin as a barrier, with the fluid swirling about this cylinder (Figure 25.53). ■

EXAMPLE 25.29

We can interchange the roles of streamlines and equipotential curves in the preceding example by setting

$$f(z) = K\,\mathrm{Log}(z),$$

with K a positive constant. Now

$$f(z) = \frac{K}{2}\ln(x^2 + y^2) + iK\theta,$$

so

$$\varphi(x, y) = \frac{K}{2}\ln(x^2 + y^2) \quad \text{and} \quad \psi(x, y) = K\theta.$$

The equipotential curves are circles about the origin and the streamlines are half-lines emanating from the origin (Figure 25.54). As they must, these circles and lines form orthogonal families of curves. The velocity of this flow is

$$\overline{f'(z)} = \frac{K}{\overline{z}} = K\frac{x}{x^2 + y^2} + iK\frac{y}{x^2 + y^2} = u + iv.$$

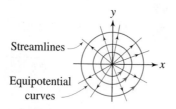

FIGURE 25.54 *Equipotential curves and streamlines for the potential $f(z) = K \operatorname{Log}(z)$.*

Let γ be a circle of radius r about the origin. Now we find that

$$\oint_\gamma u\, dx + v\, dy = 0$$

and

$$\oint_\gamma -v\, dx + u\, dy = 2\pi K.$$

The origin is a source of strength $2\pi K$. We can think of particles of fluid streaming out from the origin, moving along straight lines with decreasing velocity as their distance from the origin increases. ∎

EXAMPLE 25.30

We will model flow around an elliptical barrier. From Example 25.28, the complex potential $f(z) = (iK/2\pi)\operatorname{Log}(z)$ for $|z| > R$ models flow with circulation $-K$ about a cylindrical barrier of radius R about the origin. To model flow about an elliptical barrier, conformally map the circle $|z| = R$ to an ellipse. For this, consider the mapping

$$w = z + \frac{a^2}{z},$$

in which a is a positive constant. This is called a *Joukowski transformation*, and is used in analyzing fluid flow around airplane wings because of the different images of the circle that result by making different choices of a.

Let $z = x + iy$ and $w = X + iY$. We find that the circle $x^2 + y^2 = R^2$ is mapped to the ellipse

$$\frac{X^2}{1 + (a/R)^2} + \frac{Y^2}{1 - (a/R)^2} = R^2,$$

provided that $a \neq R$. This ellipse is shown in Figure 25.55. If $a = R$, the circle maps to $[-2a, 2a]$ on the real axis.

Solve for z in the Joukowski transformation. As a quadratic equation, this yields two solutions, and we choose

$$z = \frac{w + \sqrt{w^2 - 4a^2}}{2}.$$

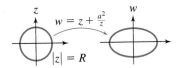

FIGURE 25.55 *Joukowski transformation mapping a circle to an ellipse.*

Compose this mapping with the complex potential function for the circular barrier in Example 25.27. We get

$$F(w) = f(z(w)) = \frac{iK}{2\pi}\pi \, \text{Log}\left(\frac{w + \sqrt{w^2 - 4a^2}}{2}\right).$$

This is the complex potential for flow in the w plane about an elliptical barrier if $R > a$, and about the flat plate $-2a \le X \le 2a$, $Y = 0$ if $R = a$. ∎

We will conclude this section with an application of complex integration to fluid flow. Suppose f is a complex potential for a flow about a barrier whose boundary is a closed path γ. Let the thrust of the fluid outside the barrier be the vector $A\mathbf{i} + B\mathbf{j}$. Then a theorem of Blasius asserts that

$$A - iB = \frac{1}{2}i\rho \oint_\gamma [f'(z)]^2 \, dz,$$

in which ρ is the constant density of the fluid. Further, the moment of the thrust about the origin is given by

$$\text{Re}\left(-\frac{1}{2}\rho \oint_\gamma z[f'(z)]^2 \, dz\right).$$

In practice, these integrals are usually evaluated using the residue theorem.

SECTION 25.5 PROBLEMS

1. Analyze the flow given by the complex potential $f(z) = az$, in which a is a nonzero complex constant. Sketch some equipotential curves and streamlines, and determine the velocity and whether the flow has any sources or sinks.

2. Analyze the flow having complex potential $f(z) = z^3$. Sketch some equipotential curves and streamlines.

3. Sketch some equipotential curves and streamlines for the flow having potential $f(z) = \cos(z)$.

4. Sketch some equipotential curves and streamlines for the flow having potential $f(z) = z + iz^2$.

5. Analyze the flow having potential $f(z) = K\,\text{Log}(z - z_0)$, in which K is a nonzero real constant and z_0 is a given complex number. Show that z_0 is a source for

this flow if $K > 0$ and a sink if $K < 0$. Sketch some equipotential curves and streamlines of this flow.

6. Analyze the flow having potential $f(z) = K\,\text{Log}((z-a)/(z-b))$, where K is a nonzero real number and a and b are distinct complex numbers. Sketch some equipotential curves and streamlines for this flow.

7. Let $f(z) = k(z + (1/z))$, with k a nonzero real constant. Sketch some equipotential curves and streamlines for this flow. Show that f models flow around the upper half of the unit circle.

8. Let

$$f(z) = \frac{m - ik}{2\pi} \, \text{Log}\left(\frac{z-a}{z-b}\right),$$

in which m and k are nonzero real numbers and a and b are distinct complex numbers. Show that this flow has a source or sink of strength m and a vortex of strength k at both a and b. (A point combining properties of a source (or sink) and a vortex is called a *spiral vortex*.) Sketch some equipotential curves and streamlines for this flow.

9. Analyze the flow having potential

$$f(z) = k\left(z + \frac{1}{z}\right) + \frac{ib}{2\pi}\,\text{Log}(z),$$

in which k and b are nonzero real constants. Sketch some equipotential curves and streamlines for this flow.

10. Analyze the flow having potential

$$f(z) = iKa\sqrt{3}\,\text{Log}\left(\frac{2z - ia\sqrt{3}}{2z + ia\sqrt{3}}\right),$$

with K and a positive constants. Show that this potential models an irrotational flow around a cylinder $4x^2 + 4(y - a)^2 = a^2$ with a flat boundary along the y axis. Sketch some equipotential curves and streamlines for this flow.

11. Use Blasius's theorem to show that the force per unit width on the cylinder in Problem 10 has vertical component $2\sqrt{3}\pi\rho a K^2$, with ρ the constant density of the fluid.

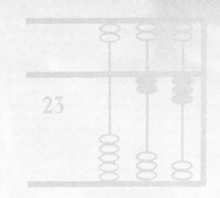

PART 8

Historical Notes

CHAPTER 26
*Development of Areas
of Mathematics*

CHAPTER 27
Biographical Sketches

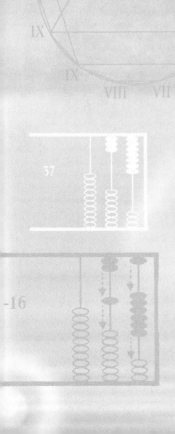

*Development
of Areas
of Mathematics*

26.1   Ordinary Differential Equations

Isaac Newton (1642–1727) and Gottfried Wilhelm Leibniz (1646–1716) are credited with independently discovering the fundamental ideas of calculus, including differentiation, integration, and the fundamental theorem that relates them. Many others made discoveries or posed questions that prepared the way, and some historians believe that some ideas of calculus may be found in ancient Greek and Indian writings, such as treatises by Archimedes dealing with areas and volumes.

For Newton, calculus was a means of studying motion and describing order in physical phenomena, particularly in astronomy and mechanics. His immediate successors, who sought to extend his legacy, lacked an overall theory such as we enjoy today and used derivatives and differentials without distinction. They obtained particular equations for particular problems, which they would attempt to solve by any method they could devise. Solutions might be analytical or geometrical and often combined insights from analysis, geometry, algebra, intuition, and the physical setting itself. In many instances operations that we would immediately dismiss today were included in solutions. For example, some investigators believed that the infinite series $1 - 1 + 1 - 1 + 1 - 1 + \cdots$ converges to $\frac{1}{2}$. They reasoned that the even partial sums equal 0, and the odd partial sums equal 1, and $\frac{1}{2}$ is the average of 0 and 1.

It is interesting to read some of the original writings (usually translated from Latin) of early investigators such as Newton, Leibniz, the Bernoullis, Euler, and others. These provide insight into their motivations and how they thought about problems, as well as some fascinating personal rivalries and disputes. One collection of early papers may be found in David Eugene Smith's *A Source Book in Mathematics* (Mineola, N.Y., Dover Publications, Inc., 1959).

At the beginning of the eighteenth century, several categories of problems dominated scientific investigations, leading to differential equations and methods of solution in special cases. One class dealt with elasticity, which is concerned with properties of bodies that deform under the stress of loading but return to their original configurations when the stress is removed. Analysis of the bending of beams subjected to different kinds of stress was important in the design of buildings

The Valley of the Kings was also known to the ancient Egyptians as The Place of Truth, and is located on the west bank of the Nile, near Thebes. Pharaohs of the Eighteenth, Nineteenth, and Twentieth Dynasties planned their tombs here, seeking the immortality that was theirs as living gods of Egypt. These pyramids, vast in scale and elegant in design, have stood for over 3000 years as an enigmatic tribute not only to a magnificent civilization, but also to the superb engineers and architects who conceived them and oversaw their construction.

and bridges and was one of the sciences discussed by Galileo in his *Dialogues Concerning Two New Sciences*.

Another problem of great interest was the shape a string or cable would take under various conditions (such as suspended from two fixed points or vertically suspended and caused to vibrate). Scientists also wondered what shape a rod would take if it were fixed at its ends and subjected to a load.

Pendulums attracted much attention. Galileo conducted many experiments with pendulums and stated a formula for the period of a simple pendulum. It soon became clear, however, that the period is not independent of the mass of the bob, and a more careful series of investigations followed.

Astronomy was another source of problems leading to differential equations. Newton's *Principia* had provided the inverse square law of gravitational attraction and explained for the first time the elliptical orbits of planets that had been established empirically by Kepler. Differential equations resulted from attempts to study planetary motions analytically. As the mathematics evolved, improved observational techniques detected perturbations in planetary orbits. The orbits were not quite the ellipses predicted by theory, and attempts at explanations required new analytical methods, a line of research that continues today.

In the late seventeenth and early eighteenth centuries, some results were published in journals, usually sponsored by royal societies, but many ideas were also communicated by private correspondence, or even by word of mouth. Some discoveries also led to disputes over priority, and today it is not always clear how to assign credit for methods, theory, or terminology.

James Bernoulli (1654–1705) was a member of a prominent Swiss family that produced several generations of outstanding lawyers, physicians, merchants, mathematicians, and scientists. He was among the first to explicitly use calculus to solve a differential equation. In 1690 he found the curve along which a pendulum takes the same time to complete one period of its motion, regardless of the length of arc through which it swings. The curve is an arc of a cycloid, and in deriving it Bernoulli made the first use of the word *integral* in connection with the solution of a differential equation.

In the same investigation, James Bernoulli posed the problem of finding the shape taken by a flexible cord suspended freely between two points. Galileo had thought that the curve took the form of an arc of a parabola. The solution, known today as a *catenary*, is part of a hyperbolic cosine curve, and was derived in 1691 by Leibniz, James Bernoulli's brother John (1667–1748), and the Dutch scientist Christian Huygens (1629–1695). Huygens, who also did pioneering work in optics, based his solution on geometrical reasoning, while Bernoulli and Leibniz adopted more analytical approaches.

The method of separation of variables is credited to Leibniz, who described it in a letter to Huygens, dated 1691. Leibniz also knew how to solve the homogeneous differential equation $y' = f(y/x)$ by reducing it to a separable differential equation through the transformation $u = y/x$, as we do today. In 1694, Leibniz showed how to obtain the general solution of the linear differential equation $y' + p(x)y = q(x)$. In 1695, James Bernoulli considered the differential equation $y' = P(x)y + Q(x)y^\alpha$, which bears his name. Leibniz showed, in 1696, that this equation can be reduced to a linear equation by the change of variables $z = y^{1-\alpha}$.

In 1694, James Bernoulli posed the problem of determining curves intersecting given curves in given angles. The problem of determining the orthogonal trajectories of a family of curves was considered by Leibniz and John Bernoulli, and in 1715 Newton revealed his general method for finding orthogonal trajectories.

A singular solution of a first-order differential equation cannot be obtained from the general solution by any choice of the constant. Such solutions had been noticed by the British mathematician Brook Taylor (1685–1731), for whom Taylor series are named. A systematic study of singular solutions was undertaken by Joseph-Louis Lagrange (1736–1813), who showed how to obtain a singular solution in certain cases by eliminating the constant from the general solution. Lagrange also showed that a singular solution could be interpreted as the envelope of the family of integral curves.

In 1734, the French mathematician Alexis-Claude Clairaut (1713–1765) studied the differential equation $y' = xy + f(y')$, now known as Clairaut's equation. This was one of the first studies of a differential equation from a broad perspective (that is, with f arbitrary).

In the early eighteenth century, second-order differential equations came to the fore in a variety of problems. In 1733, John Bernoulli's son Daniel (1700–1782) considered the problem of describing the motion of a heavy, oscillating hanging chain. The displacement $y(x)$ of the point on the chain at distance x from one end was found to satisfy the equation $\alpha(xy'(x))' + y(x) = 0$, with α constant. Bernoulli derived an infinite series solution. If the thickness of the chain is not uniform, y satisfies $\alpha(g(x)y'(x))' + yg'(x) = 0$, in which $g(x)$ described the distribution of mass along the chain. Bernoulli also obtained one solution of this equation as a series we would recognize as a Frobenius series.

John Bernoulli, in a letter to Daniel, considered a weightless elastic string loaded with equally spaced objects of equal mass. By passing to the limit as the number of masses is increased, he considered a continuous string and calculated the fundamental frequencies with which the string vibrates when set in motion.

About 1728, the Swiss mathematician Leonhard Euler (1707–1783) began to study second-order differential equations. At the time, Euler was a resident mathematician in the St. Petersburg Academy in Russia. In 1734, Daniel Bernoulli wrote to Euler concerning a fourth-order differential equation describing the bending of a loaded beam. In response to this inquiry, and some work of his own, Euler initiated a systematic study of second- and higher-order differential equations, particularly linear, constant coefficient equations. In this work he introduced the exponential function. Euler also observed the phenomenon of resonance while studying the equation $y'' + ky = \alpha \sin(\omega t)$.

Lagrange attempted to extend Euler's results to linear equations with nonconstant coefficients. Closely related to Lagrange's work was a study of Riccati's equation $y'' = p(x)y^2 + q(x)y + r(x)$, introduced about 1724 by Count Jacopo Francesco Riccati of Venice (1676–1754). Lagrange, who was of French and Italian descent, is credited with developing the method of variation of parameters. He formalized much of Newton's work in his 1788 masterpiece *Mécanique analytique*.

Reduction of order is sometimes credited to Jean le Rond d'Alembert (1717–1783), who also contributed to the development of partial derivatives and partial differential equations. He obtained the solution of the wave equation that is named for him.

Series solutions were used from Newton's time, although, for want of a theory of convergence, without much rigor. It was Euler who systematized their use and even had in his possession an ad hoc form of the method of Frobenius. Georg Frobenius (1849–1917) formalized the method named for him much later. In this way, Euler derived solutions of what we now call *Bessel's equation* $y'' + (1/x)y' + (1 - n^2/x^2)y = 0$. Euler also considered the hypergeometric equation $x(1-x)y'' + [c - (a+b+1)x]y' - aby = 0$, although the name "hypergeometric" is attributed to his friend Johann Pfaff (1765–1825). Solutions of this equation were studied in the early nineteenth century by the German mathematician Carl Friedrich Gauss (1777–1855).

Systems of differential equations arise naturally when Newton's laws of motion, which deal with vectors, are written in component form, and they occur in dealing with mechanical and, later, electrical systems. Much of the early work on systems appears in Lagrange's *Mécanique analytique*. In this book, Lagrange considered the classical three-body problem, which is to describe the motion of three planets, each of which exerts a gravitational force on the other. Even for three planets this problem is difficult, and research continues on it today.

Lagrange also applied the method of variation of parameters to differential equations of order n, although the method is suggested in Newton's *Principia*. Further work on the n-body problem was done by Pierre-Simon de Laplace (1749–1827) in his *Mécanique céleste*.

The nineteenth century saw an expanded use of series solutions. Certain differential equations, which occurred in a variety of applications, came under study. Bessel equations (named for Friedrich Wilhelm Bessel (1784–1846), director of the observatory at Königsberg) fell into this category. Solutions of these special equations became known as *special functions*, and properties and numerical values of these functions were catalogued and tabulated. Today special functions include not only Bessel functions, but also various kinds of orthogonal polynomials (Legendre polynomials, Hermite polynomials, Laguerre polynomials, and many others).

Legendre polynomials, which arise in connection with potential theory in spherical coordinates, appeared in the writings of Laplace and Adrien-Marie Legendre (1752–1833). Legendre was a mathematician at the École Militaire. The first systematic treatise on Legendre polynomials was written by Robert Murphy of Cambridge University in 1833.

Hermite polynomials are named for Charles Hermite (1822–1901), a professor at the École Polytechnique and the Sorbonne in Paris. Laguerre polynomials are named for Edmond Laguerre (1834–1866) of the Collège de France.

Sturm–Liouville theory originated in the work of Charles Sturm (1803–1855), a professor of mechanics at the Sorbonne, and Joseph Liouville (1809–1882), a professor of mathematics at the Collège de France. Problems associated with eigenvalues, and the idea of eigenfunction expansions, were first considered about 1750 and occurred in connection with special functions and studies of heat conduction. Sturm and Liouville were the first to construct a general theory for a class of differential equations.

The Laplace transform can be found in writings of Laplace in the 1780s and Siméon-Denis Poisson (1781–1840) in the 1820s, as well as in Joseph Fourier's classic 1811 paper on heat conduction. The popularity of the Laplace transform as a computational method in differential equations and especially electrical engineering owes much to the efforts of Oliver Heaviside (1850–1925), a telephone and telegraph engineer who spent the later years of his career developing vector and transform methods for use in solving engineering problems.

Stability and the qualitative theory of systems form a relatively new area of activity. Seminal work was done by Alexander Mikhailovich Lyapunov (1857–1918), a Russian mathematician and mechanical engineer. His work began to appear about 1892 and gave birth to an approach to stability through what are known today as *Lyapunov functions*.

Alfred Lienard (1869–1958) was director of the School of Mines in Paris and worked on spiral approaches of trajectories in the phase plane. The Poincaré–Bendixson theorem deals with the existence of closed trajectories in the phase plane and is named for Ivar Bendixson (1861–1935), a professor at the University of Stockholm in Sweden, and Jules Henri Poincaré (1854–1912), a highly original mathematician who did seminal work in differential equations, dynamical systems, mathematical physics, stability, and an emerging field known as topology. Van der Pol's equation was developed by Balthazar van der Pol (1889–1959), a Dutch radio engineer who derived his equation in the late 1920s.

26.2   Matrices and Determinants

Matrices and determinants, particularly in connection with transformations of geometric objects and the solution of systems of linear equations, developed out of the mathematics of the eighteenth and nineteenth centuries.

The early emphasis was actually on the determinant, not the matrix. Many properties of determinants were discovered by talented amateurs and figures of minor importance in the history of mathematics, although such giants as Gauss and Cauchy also made contributions. Cauchy stated the correct definition of the product of two matrices and showed that the determinant of a product of square matrices is the product of their determinants. Cauchy also proved Laplace's formula for the cofactor expansion of a determinant. Gauss's main interest in determinants was connected to the study of quadratic forms.

James Joseph Sylvester (1814–1897) was a leading figure in the development of determinants. Sylvester's work touched upon determinant applications to the solution of polynomial equations as well as geometric aspects of quadratic forms. Out of this, and Lagrange's work on planetary motion, came the determinant formulation of the characteristic equation of a matrix, which was then called the *secular equation*. Sylvester's career included positions at the University of Virginia and at Johns Hopkins University in Baltimore, as well as periods as a lawyer and actuary in his native England. His primary motivation for visiting American universities was that his Jewish ancestry mitigated against his winning tenured teaching posts at British universities.

Carl Gustav Jacob Jacobi (1804–1851) and Eugene Catalan (1814–1894) are credited with discovering the determinant used in transforming multiple integrals. This determinant, now known as the *Jacobian*, is related to the action of the transformation on elements of area, volume, or their higher dimensional analogues. The Wronskian, a determinant involving solutions of differential equations and their derivatives, is named for the Polish mathematician H. Wronski (1778–1853).

Lewis Carroll, author of *Alice's Adventures in Wonderland*, is the subject of an anecdote involving determinants. Lewis Carroll was the pseudonym of the Reverend Charles Lutwidge Dodgson (1832–1898), a fellow of Christ Church College, Oxford, as well as an Anglican clergyman and tutor in mathematics. His popular stories about Alice so delighted the queen that she commanded Dodgson to send her his next work. This was a treatise on determinants. Her reaction is not recorded.

The term *matrix* was apparently coined by Sylvester, about 1850. Arthur Cayley (1821–1895) pointed out the rationale for thinking of matrices as preceding determinants in the logical presentation of linear algebra, even though the historical development was the reverse and many properties of determinants were known before the study of matrices was formalized. Cayley collaborated with Sylvester in the study of matrices, emphasizing the geometry of n-dimensional space, invariants of quadratic forms, and linear transformations. Cayley and Sylvester must have made a curious pair. Sylvester was a lively, well-traveled man of many accomplishments, while Cayley was as calm and even in temperament as Sylvester was excitable. Cayley spent his entire mathematical career at Cambridge, except for the year 1892 when he visited his friend at Johns Hopkins.

The fact that real, symmetric matrices have real eigenvalues was known to Cauchy and was proved later by Charles Hermite. Rudolf Friedrich Alfred Clebsch (1833–1872) showed that the nonzero eigenvalues of a real, skew-hermitian matrix are pure imaginary. Hermite had the idea of orthogonal matrices, but Georg Frobenius (1849–1917) formulated the definition we use today.

The work of Sir William Rowan Hamilton (1805–1865), who developed quaternions in an attempt to generalize complex numbers to three dimensions, is seen in linear algebra through the Cayley–Hamilton theorem, which asserts that a square matrix satisfies its own characteristic equation.

Camille Jordan (1828–1922) was a French mathematician who made fundamental contributions to the growing fields of modern algebra and topology. In linear algebra, he is remembered for the Jordan canonical form, a way of decomposing a matrix that gives information about its eigenvalues and subspaces generated by eigenvectors. In topology he is known for the Jordan curve theorem encountered in connection with Green's theorem in vector analysis and Cauchy's theorem in complex analysis.

26.3   Vector Analysis

The origins of vectors can be traced to problems connected with the arithmetic of complex numbers. As we now know, both complex numbers and vectors in the plane can be identified with points in the plane.

Addition of complex numbers, given by

$$(a + ib) + (c + id) = (a + c) + i(b + d),$$

is consistent with the component-wise addition of 2-vectors:

$$(a\mathbf{i} + b\mathbf{j}) + (c\mathbf{i} + d\mathbf{j}) = (a + c)\mathbf{i} + (b + d)\mathbf{j}.$$

The practice of identifying vectors with complex numbers has its drawbacks, however. Vectors are used to represent forces, which are not always confined to a plane. This led to the problem of representing vectors in three dimensions, which led to attempts to generalize complex numbers to three dimensions. This became the obsession of Sir William Rowan Hamilton (1805–1865), who proposed a solution in the form of *quaternions*. A quaternion is a symbol,

$$a + bi + cj + dk,$$

in which a, b, c, and d are real numbers. Quaternions are added as we might expect, by adding respective components. For example,

$$(2 + 7i - 4j + k) + (3 - 5i - 2j + 6k) = 5 + 2i - 6j + 7k.$$

Multiplication of quaternions was defined by the rules

$$jk = i = -kj, \quad ki = j = -ik, \quad ij = k = -ji$$

and

$$i^2 = j^2 = k^2 = -1.$$

The first set of product rules is reminiscent of vector cross product, and the second reminds us of the imaginary unit i, since $i^2 = -1$. A quaternion therefore attempts to combine a vector part and a scalar part in one quantity, with multiplication that sometimes behaves like cross product and sometimes like a product of imaginary numbers. In this way Hamilton believed he could represent vectors in three dimensions, while retaining an algebraic and geometric behavior like that exhibited by complex numbers. Hamilton believed that quaternions would provide the natural vehicle for the expression of physical laws and that they would supplant vectors. In this he was wrong, and today we do vector analysis instead of quaternion analysis. Quaternions do have interesting properties but are most appreciated in the context of modern algebra.

Soon after Hamilton began to publish his views on quaternions, a debate developed among scientists and mathematicians over his claims for them. James Clerk Maxwell (1831–1879), a professor of physics at Cambridge and the author of the fundamental equations governing electric and magnetic fields, separated the scalar and vector parts of quaternions in his thinking. To him, a was the scalar part (a real number) and $bi + cj + dk$ was the vector part. Maxwell was an important and influential figure, and his emphasis was soon taken up by others. Eventually, attention focused on just $bi + cj + dk$ in representing vectors. These came to be written $a\mathbf{i} + b\mathbf{j} + c\mathbf{k}$, with the standard unit vectors we use today. The cross product rules for $\mathbf{i}, \mathbf{j}$, and $\mathbf{k}$ are retained, while the requirement that $i^2 = j^2 = k^2 = -1$ has been dropped.

Maxwell was the first to use the term *rotation* for what we have called the *curl* of a vector, in connection with describing swirl in fluid motion. He also introduced Laplace's operator

$$\nabla = \frac{\partial}{\partial x}\mathbf{i} + \frac{\partial}{\partial y}\mathbf{j} + \frac{\partial}{\partial z}\mathbf{k}$$

and recorded the identities

$$div(curl\ \mathbf{F}) = 0, \quad curl(grad\ \varphi) = \mathbf{O}.$$

William Kingdom Clifford (1845–1903), of Queen's College in London, apparently originated the term *divergence of a vector*.

The debate on quaternions versus vectors continued until the work of Yale's Josiah Willard Gibbs (1839–1903), whose name is attached to the Gibbs phenomenon in Fourier analysis. His book, *Vector Analysis*, which Gibbs coauthored in 1901 with E. B. Wilson, based largely on Gibbs's lectures, helped to popularize vector notation and algebra. Also influential in establishing the use of vectors was Oliver Heaviside's *Electromagnetic Theory*, the first volume of which contained an introduction to vector algebra. Heaviside (1850–1925) was an electrical engineer who promoted the use of the Laplace transform and vector algebra in the mathematical analysis of engineering problems.

Today we recognize the main results of vector calculus as the integral theorems named for Green, Ostrogradsky, Gauss, and Stokes. George Green (1793–1841) was a self-taught English mathematician who made it his life's work to undertake a mathematical study of electricity and magnetism. In 1828 he published, at his own expense, a booklet entitled *An Essay on the Application of Mathematical Analysis in the Theories of Electricity and Magnetism*. The booklet contained what we have called Green's theorem, as well as vector integral identities, which are often called *Green's lemmas*. Because Green had no scientific reputation, his work went largely unnoticed until Sir William Thomson (Lord Kelvin, (1824–1907)) happened upon his booklet and arranged for its publication in the *Journal für Mathematik*. Much of Green's work was done independently by the Ukrainian mathematician Michel Ostrogradsky (1801–1861), who presented his version of Green's lemmas to the St. Petersburg Academy of Sciences in 1828. The divergence theorem, popularly known as *Gauss's theorem*, was also discovered independently by Ostrogradsky, who published his version in 1831. Stokes's theorem is named for Sir George Gabriel Stokes (1819–1903), professor of mathematics at Cambridge University. The theorem can be traced to a letter dated July 2, 1850, sent to Stokes by Lord Kelvin. Stokes used it as a question in an examination for the Smith Prize at Cambridge in 1854. This was a competition held annually for the leading mathematics students at Cambridge, and the result became known as *Stokes's theorem*.

26.4   Fourier Analysis

Fourier analysis, which is the study of Fourier series, integrals, and transforms, is named for Joseph Fourier (1768–1830), a French mathematician who lived during the Napoleonic era. In addition to his mathematical achievements, Fourier accompanied Napoleon's expedition to Egypt, held a high position (roughly equivalent to a governor) in the French government, and narrowly escaped death by the guillotine at least once during the French Revolution under Robespierre.

Although Fourier is justly honored by having his name attached to this important branch of analysis, many of his contemporaries and immediate predecessors contributed to his achievement. Leonhard Euler (1707–1783) was a Swiss mathematician, sometimes called the greatest analyst who ever lived. Off and on for many years, beginning about 1729, Euler considered the interpolation problem of finding a continuous function f, defined on $[1, n]$ for some positive integer n, and taking on prescribed values at $1, 2, \ldots, n$. This problem arose in the course of some calculations Euler was doing while studying planetary perturbations, and it led him to a trigonometric series. He found that the solution of an equation of the form

$$f(x) = f(x - 1) + F(x)$$

can be written

$$f(x) = \int_0^x F(t)\,dt + 2\sum_{n=1}^{\infty}\left[\int_0^x F(t)\cos(2n\pi t)\,dt\right]\cos(2n\pi x)$$

$$+ 2\sum_{n=1}^{\infty}\left[\int_0^x F(t)\sin(2n\pi t)\,dt\right]\sin(2n\pi x).$$

Soon after this observation, about 1754, Jean le Rond d'Alembert (1717–1783) obtained a trigonometric cosine expansion for the reciprocal of the distance between two planets, in terms of the angle between the vectors from the origin of the coordinate system to the planets. The integral formulas for the Fourier coefficients appear in d'Alembert's work.

Other expressions involving series of trigonometric functions began to appear in a variety of contexts. Using a geometric series, Euler found that

$$\frac{a\cos(x) - a^2}{1 - 2a\cos(x) + a^2} = \sum_{n=1}^{\infty} a^n \cos(nx)$$

and

$$\frac{a \sin(x)}{1 - 2a \cos(x) + a^2} = \sum_{n=1}^{\infty} a^n \sin(nx).$$

These expansions are indeed valid if $|a| < 1$, a restriction not fully appreciated at the time. Euler set $a = 1$ to conclude incorrectly that

$$\sum_{n=1}^{\infty} \cos(nx) = -\frac{1}{2}. \tag{26.1}$$

In fact, the series on the left diverges. If this "equation" is integrated term-by-term from x to π, we get

$$\sum_{n=1}^{\infty} \frac{1}{n} \sin(nx) = \frac{1}{2}(\pi - x),$$

which is the correct Fourier sine expansion of $(\pi - x)/2$ on $(0, \pi)$. Equation (26.1) also appeared in a study of sound waves by Joseph-Louis Lagrange (1736–1813).

Thus, as early as the middle of the eighteenth century, trigonometric series were beginning to appear. Leading mathematicians encountered them and calculated with them, but their significance was not fully understood, and some of the calculations were incorrect. By the 1750s, the integral formulas for the Fourier coefficients were known, but not always trusted, and mathematicians such as Euler preferred deriving trigonometric series by other means, correct or not.

The continuing appearance of trigonometric series in connection with important problems raised questions and led to their further investigation. One question was how a nonperiodic function could be represented by a series of sines and cosines, which are periodic. This point alone touched off an intense and important debate engaging the leading mathematicians of the period, including Euler, Lagrange, d'Alembert, Daniel Bernoulli, and later Laplace. The definition of function came under scrutiny, along with the critical issue of identifying those functions that could be expanded in trigonometric series.

Soon it became clear that the concept of function itself was little understood or agreed upon. The term had usually been taken for granted, but upon closer investigation it was found that several different meanings were used. To some, a function had to be continuous, to others, differentiable. To some, a function had to be given by an analytic expression, such as $x^2 - \cos(x)$. Questions also arose about the meaning of an infinite series, and various interpretations were given for convergence. These were difficult questions and the tools for resolving them had not yet been developed.

In 1768, at the height of this debate, Joseph Fourier was born. As a student he showed a talent for mathematics but turned to it as a profession only when his common heritage (he was the son of tailor) made it impossible for him to obtain a military commission.

The mathematical description of heat conduction was an outstanding problem at the turn of the nineteenth century. In 1807, Fourier submitted a paper on this subject to the prestigious Academy of Sciences of Paris, in competition for a prize that had been offered for the most successful attack on this problem. Such mathematical giants as Laplace, Lagrange, and Legendre evaluated Fourier's work and rejected it for lack of precision. They did, however, encourage Fourier to continue his research and to attempt to supply some of the details he had omitted. In 1811, Fourier submitted a revised version of his paper and was awarded the prize by the committee. However, the academy still declined to publish the paper because many details were still unclear. Finally, in 1822, Fourier published his now classic work *Théorie analytique de la chaleur*, incorporating most of his results from 1811 together with some new ones.

In this paper, Fourier considered the 3-dimensional heat equation

$$\frac{\partial u}{\partial t} = k \left[\frac{\partial^2 u}{\partial x^2} + \frac{\partial^2 u}{\partial y^2} + \frac{\partial^2 u}{\partial z^2} \right],$$

in which $u(x, y, z, t)$ is the temperature of an object at time t and point (x, y, z). By what is now known as separation of variables, he derived trigonometric series representations of solutions. In these derivations he made explicit use of the formulas for the coefficients now known as *Fourier coefficients*.

It is interesting to ask how Fourier's work differed from that of Euler and others who had achieved some results with trigonometric series.

First, Fourier's general methods for solving the heat equation proved to be of fundamental and lasting importance in solving boundary value problems of partial differential equations. Specific investigations that preceded his did not lead to such general results.

Second, while others occasionally used Fourier coefficients, none fully appreciated their importance. Fourier was the first, although not for entirely rigorous reasons, to use them confidently and extensively and to assert the generality with which an "arbitrary" function could be represented on an interval by a Fourier series. In some technical details, Fourier was sometimes wrong, and he glossed over sensitive convergence questions that he had no way to treat. The mathematical machinery needed to address these details was not available in his day. Nevertheless, he took the major steps on which others built, and for this he deserves the credit reflected in the terms *Fourier series*, *Fourier integral*, and *Fourier transform*.

Finally, the importance of a piece of mathematics is often judged by its ramifications as seen from a later perspective. Questions and results arising from the study of Fourier series have had a profound influence on the development of mathematics. Some areas affected by this work include the solution of partial differential equations, the development of set theory and measure theory, harmonic analysis, and the critical examination of the concepts underlying convergence of series, the definition of a function, and continuous and differentiable functions. Georg Cantor (1845–1918), who is now known for his theories of transcendental numbers and infinite sets, but was in his day rejected for his revolutionary ideas about orders of infinity, was initially led to his discoveries by investigations into convergence properties of Fourier series.

Sufficient conditions for the convergence of a Fourier series were first established about 1829 by Peter Gustav Lejeune-Dirichlet (1805–1859). Dirichlet's conditions are essentially those of Theorem 13.1. The fact that the Fourier coefficients have limit zero as $n \rightarrow \infty$ (if f is integrable on $[-L, L]$) was discovered by Georg Friedrich Bernhard Riemann (1826–1866), a student of Gauss who worked with Dirichlet for a time. Riemann's name is attached to many concepts, including the Riemann integral and Riemann sums, Riemann surfaces, and the Riemann zeta function.

In 1873, Paul Du Bois-Reymond (1831–1889) gave an example of a function that is continuous on $(-\pi, \pi)$ but whose Fourier series fails to converge at any point of this interval. This example greatly surprised the mathematical community at a time when functions were still thought of as relatively well-behaved objects specified by analytic expressions. Examples of this kind of behavior led to the recognition of deep properties of the structure of the real number system and also to the development of measure theory and real analysis.

The Fourier integral appeared near the end of Fourier's 1811 paper, when he attempted to extend his results to functions defined over the half-line. A separate treatment appeared in Cauchy's 1816 prize-winning paper on surface waves in a fluid. Augustin-Louis Cauchy (1789–1857) was a leading French mathematician and engineer who established the basis for much of complex function theory. Still a third treatment appeared in Poisson's 1816 paper on water waves.

Siméon-Denis Poisson (1781–1840) was a prominent French mathematical physicist for whom the partial differential equation $\nabla^2 u = f$ is named. This equation arises in studying electrical potentials.

The Fourier transform is found in early writings of Cauchy and Laplace (from about 1782 on). A formula for the cosine transform and its inverse appears in Fourier's 1811 paper. The fast Fourier transform, an algorithm for efficiently computing complex Fourier coefficients, is largely responsible for the widespread success of Fourier transform methods today in treating a broad range of problems in the sciences and engineering. Part of the history of the FFT is traced in Section 14.9.

26.5   Partial Differential Equations

Perhaps surprisingly, we may think of the wave equation as belonging to the eighteenth century and the heat equation to the nineteenth century.

In 1727, the Swiss mathematician John Bernoulli (1667–1748) treated the vibrating string problem by imagining the string as a flexible thread having a finite number of equally spaced weights or beads placed along it. The differential equation he derived in this way was time independent, however, and so was actually an ordinary differential equation. It was the French mathematician Jean le Rond d'Alembert (1717–1783) who introduced the time variable and let the number of weights in Bernoulli's treatment go to infinity, deriving the one-dimensional wave equation. His work, which appeared about 1746, included the solution named for him today.

In the 1750s, the Swiss mathematician Leonhard Euler (1707–1783) considered what we would call Fourier series solutions of the wave equation. In 1781, he used what was essentially a Fourier–Bessel series to solve the vibrating membrane problem for a circular drum.

Laplace's equation, or the potential equation, arose in the study of gravitational attraction. The main figures in its early development and solution in special cases were Pierre-Simon de Laplace (1749–1827) and Adrien-Marie Legendre (1752–1833). Both were professors of mathematics at the École Militaire in France. Legendre also developed what are known today as *Legendre polynomials*.

In 1807, Joseph Fourier (1768–1830) submitted a paper to the Academy of Sciences in Paris, deriving the heat equation and exploiting his separation of variables method. The paper was rejected for publication because of lack of rigor, as was his 1811 revision. Finally, in 1822 Fourier published his classic *Théorie analytique de la chaleur*, laying the foundations not only for applications of separation of variables and Fourier series but for the Fourier integral and transform as well.

Pioneering work on properties of harmonic functions was done by George Green (1793–1841), a self-taught British mathematician. Green's main interest was in the mathematical treatment of electricity and magnetism. Much of his work was discovered independently by Michel Ostrogradsky (1801–1861), a Ukrainian mathematician.

26.6   Complex Function Theory

The acceptance of complex numbers in algebra and analysis came only after much controversy. In 1770, the Swiss mathematician Leonhard Euler wrote, "Because all conceivable numbers are either greater than zero, or less than zero, or equal to zero, then it is clear that the square roots of negative numbers cannot be included among the possible numbers." He went on to refer to complex numbers as impossible, or fancied, numbers.

Complex numbers continued to occur, however, in the solution of many problems, such as finding roots of polynomials or solutions of differential equations. It would only be a matter of time before they were placed on a more solid and acceptable foundation.

The major factors in the acceptance of complex numbers came in the nineteenth century as understanding of their geometric significance evolved. Caspar Wessel (1745–1818), a Norwegian surveyor, wrote a paper in 1797, "On the Analytic Representation of Directions: An Attempt," which expressed the basic ideas of representing complex numbers as points in the plane and of adding them by the parallelogram law. His work went largely unnoticed for many years, but the germ of the idea was in the air.

In 1806, a Swiss bookkeeper, Jean-Robert Argand (1768–1822) wrote a short book on the geometric representation of complex numbers. Today, the complex plane is sometimes referred to as the Argand diagram.

The major credit for the acceptance of complex numbers goes to Carl Friedrich Gauss (1777–1855), who had the prestige due the leading mathematician of his time. As early as 1799, he had incorporated many of his ideas on complex numbers into his proof of the fundamental theorem of algebra. By 1815, it was clear that Gauss thoroughly understood the geometry of complex numbers, and by the 1830s he included them freely in his writings.

Questions concerning complex functions and integrals began to be raised by Gauss and Siméon-Denis Poisson (1781–1840). Poisson originally worked on the recently developed Fourier series, but he was also the first to integrate complex functions along curves in the plane. It was left for Cauchy, however, to formulate and develop many of the properties of complex functions and integrals.

Augustin-Louis Cauchy (1789–1857) was born in Paris and became a professor at the École Polytechnique, the Sorbonne, and the Collège de France, following a brief career as a military engineer. He made important contributions to the mechanics of waves in elastic media and to the theory of light, but his most lasting work was in mathematics, where he authored more than 700 papers, second only to Euler. His personal life was greatly influenced by his royalist sympathies, and he supported the Bourbons during a time of political upheaval in France. In recognition of Cauchy's stature as one of the leading mathematicians in Europe, Napoleon III excused him from the oath of allegiance required of state employees. Cauchy responded by donating his salary from the Sorbonne to the poor of Sceaux, the town in which he resided at the time.

In a sequence of papers from 1814 to the early 1840s, Cauchy formulated his integral theorem and many of its consequences, including the concept of independence of path and the integral representations of a function and its derivatives. He also grasped the idea of poles and residues and the residue theorem and worked with power series and multiple-valued functions. (At this time the only singularities that were understood were poles. The idea of an essential singularity came later.)

Pierre-Alphonse Laurent (1813–1854) developed the Laurent series about 1843. The result had also been known to the German mathematician Karl Weierstrass (1815–1897), who did not publish his ideas in this area. Weierstrass studied the subtleties of power series representations of functions and helped establish the concept of uniform convergence. His main contributions were in placing complex analysis on a rigorous foundation.

Georg Friedrich Bernhard Riemann (1826–1866) may have more important theorems and concepts named for him than any other mathematician. Most of Riemann's work was in areas beyond an introductory treatment, but his name is included with Cauchy's in the fundamental Cauchy–Riemann equations. Riemann was the first to realize that the increment must be allowed to approach zero along any path in the limit of the difference quotient defining the derivative of a complex function.

CHAPTER 27

Biographical Sketches

This section contains biographical sketches of some of the main figures in the development of mathematics, particularly appied mathematics and differential equations. Obviously this list is far from complete.

27.1   Galileo Galilei (1564–1642)

Galileo is honored by many today as the father of modern science. Many conflicting stories have been chronicled about this fascinating man, and much about him remains a mystery. He is recorded as a master experimentalist who was among the first to design careful experiments and collect data from which to draw conclusions. Others claim that he had little patience for such things and drew conclusions based purely upon reason. The latter view is in some measure supported by part of the dialogue in his work *Dialogues on the Two Great Systems of the World*, in which the character Simplicio asks whether Galileo had performed a certain experiment. Galileo replies, "No, and I do not need it, as without any experience I can affirm that it is so, because it cannot be otherwise." Some historians have also claimed that Galileo was selective in pruning his experimental data, retaining those supporting his theories.

Disagreements over details of Galileo's beliefs and practices span the entire period of his life. Because of his persecution by the Catholic church toward the end of his life, he is regarded by some as one of the supreme "martyrs of science," while others believe that the church's sentence was mild and served as only a minor and symbolic rebuke of Galileo's beliefs.

Galileo's main interests lay in physics, astronomy, and mathematics, which he pursued as a professor at Padua, a position he took at the age of twenty-eight. His troubles with the dominant Roman Catholic religion were inevitable from his acceptance of the Copernican theory, which disputed the church's view that the earth lay at the center of the universe. This disagreement with established doctrine took a practical turn when he learned of the development of the telescope in the Netherlands. It did not take long for Galileo to construct a telescope for his own use, and he was astonished to observe that our planet's moon was not the smooth, perfect sphere that the church taught, but was in fact "… full of inequalities, uneven, full of hollows and protuberances, just like the surface of the Earth itself…." Further observations led Galileo to see differences between stars and planets (the former seem nearly stationary when viewed over short periods of time, while planets move noticeably about the heavens). He studied the Milky Way, noting that it appeared to be a conglomeration of "innumerable stars." In 1610, he observed Jupiter and was surprised to see that it appeared to have four moons.

All of these observations were interpreted by the church as an effort to support the Copernican theory, bringing Galileo into direct conflict with the Roman Inquisition. Galileo's astronomical observations, coupled with his theories on the mechanics of moving objects, contested the literal interpretation of the Bible and the accepted Aristotelian view of the universe, which held that the earth lay at the center of the universe, designated specifically for the use of mankind. These were dangerous times for inquisitive minds, as Galileo knew from the experiences of others and would learn firsthand in his later years.

Galileo's work in mechanics immediately preceded Newton's (he died in the year of Newton's birth) and had great influence on Newton's development of his laws of motion. His most celebrated discovery was the physical law governing the motion of an object falling under the influence of the earth's gravity. Aristotle had held, and the world believed, that the speed with which an object falls depends on its weight, with heavier objects falling faster than lighter ones. There is a famous story that Galileo tested this theory by dropping various objects from the leaning tower of Pisa. This may be false. In fact, Galileo did carry out some experiments using an inclined plane to test his belief in the familiar law $s = gt^2/2$ governing the acceleration of objects falling near the earth under the influence of gravity. These expriments confirmed his formula within experimental discrepancies, which he was quite willing to discard because he had derived, and believed, his formula independent of experiment. Somewhat in a manner later followed by Einstein, he liked to work with "thought experiments," in which he attempted to formulate physical laws based on an understanding of the forces involved and theories about how these forces were related.

Writing a formula or equation to describe a physical phenomenon embodied a new and striking idea—that mathematics constituted a way of describing nature. In his 1610 work *The Assayer*, Galileo wrote one of his most quoted passages:

> *Philosophy is written in that great book which ever lies before our eyes—I mean the universe—but we cannot understand it if we do not first learn the language and grasp the symbols in which it is written. The book is written in the mathematical language, and the symbols are triangles, circles and other geometrical figures, without whose help it is impossible to comprehend a single word of it; without which one wanders in vain through a dark labyrinth.*

Here Galileo affirms his belief in mathematics as the vehicle for understanding the workings of "philosophy," by which he meant nature. In this sense, Galileo may have been the first mathematical physicist.

Galileo also carried out experiments on trajectories of propelled objects, such as cannon shells, and on the motion of pendulums. With regard to projectiles, he was aware that the vertical and horizontal components of the motion are independent. He also had a concept of inertia, which is implicit in his calculations that an object in uniform motion remains in motion in a straight line unless acted upon by an external force.

At the age of sixty-nine, Galileo was finally brought before the court of the Inquisition. The price of condemnation could be severe, and Galileo knew that the court had burned Giordano Bruno at the stake. Finally Galileo confessed, signing a statement which asserted, among other things,

> *I am willing to remove from the minds of your Eminences, and of every Catholic Christian, this vehement suspicion [or heresy] rightly entertained towards me, therefore, with a sincere heart and unfeigned faith, I abjure, curse and detest the said errors and heresies, and generally every other*

error and sect contrary to the said Holy Church; and I swear that I will never more in future say, or assert anything, verbal or in writing, which may give rise to a similar suspicion of me. . . .

Despite such a confession, and living under what was essentially house arrest, Galileo completed his last work, *Two New Sciences*, and arranged for its publication by his survivors.

27.2   Isaac Newton (1642–1727)

The year of Galileo's death saw the birth of the man who would establish the "system of the world." Isaac Newton was born on Christmas Day in 1642 in Woolsthorpe, a small farming village in England. On the Continent, where the Gregorian calendar was used, the date of birth was officially January 4, 1642. Civil war in England had just begun, and Stuart King Charles I was on the throne. John Milton was thirty-four years of age, and in the Netherlands, Rembrandt was producing his masterpiece, *The Night Watch*. René Descartes was living in the Netherlands and working on his *Principles of Philosophy*, which later influenced Newton's own physical theories.

Newton's father died before he was born, but his mother married again while he was very young. He had a lonely childhood and spent much of his time reading from a small collection of books belonging to a relative. At the age of twelve he was enrolled in King's School in Grantham, a village about seven miles north of Woolsthorpe. He boarded there with a local pharmacist, who treated his unusual boarder kindly, and was encouraged by the school's headmaster, Henry Stokes. In June 1661, Newton entered Trinity College, Cambridge, as a subsizar, a student working to earn tuition and lodging. His mathematics instructor was Isaac Barrow, who held the professorship in geometry. At least in the beginning, Newton was a student of very ordinary appearance.

In 1664 and 1665, Cambridge closed because of the plague in nearby London, and Newton returned to Woolsthorpe, where he laid the foundations for his theory of light and colors, his laws of motion, and the ideas of differential and integral calculus. It is conjectured that no man has ever had two years of sustained mental effort of this magnitude or produced results of such depth, importance, and variety in such a short period of time. (Einstein's achievements culminating in his famous 1905 papers come to mind.) He returned to Cambridge in 1666 and was elected a fellow of Trinity in 1667 in a vote that was near to a rejection because he had not yet published his ideas or otherwise distinguished himself to the faculty.

Apparently Newton was a poor lecturer and not very successful as a teacher, but he pursued research in many areas, including biblical history and alchemy. More than once the small furnaces he used in his alchemy experiments started fires in his quarters, and some historians believe that during one of these he lost some papers he regarded as highly important.

Newton's unusual personality combined a desire for recognition with extreme reluctance to publish or expose himself to possible criticism. With the editorial assistance of Edmund Halley, he published his great work *Mathematical Principles of Natural Philosophy*, whose original Latin title is usually abbreviated to *Principia*. In it, Newton developed the fundamental concepts of calculus and the physics of motion, the inverse square law of gravitational attraction, theories of planetary perturbations and of the flattening of the earth at the poles, ideas about the influence of the moon on the tides and motions of various shaped objects moving through various media, and many other subjects. The book was widely praised for the variety and originality of its ideas, and it firmly established Newton as a leading scientific figure of the period. The *Principia* is still being studied today for its wealth of ideas, and occasionally one sees new insights into what Newton was thinking as he developed his ideas. The Russian mathematician and Fields medalist V. I. Arnol'd, and the Nobel laureate S. Chandrasekhar have been active in this study.

As his influence grew, Newton became president of the Royal Society and engaged in a wide correspondence on scientific and mathematical matters. Not all of this was of a friendly nature, as Newton was a private and suspicious person. He had a particularly strong dislike for Robert Hooke, dating back to Newton's first presentation of his theory of optics to the Royal Society. The Society employed Hooke to set up its experiments, and Newton felt that Hooke had put his ideas on optics at a disadvantage through lack of attention in his presentation. Hooke also claimed to have inspired Newton to the inverse square law of gravitation and had made something of a public show of a careless error Newton had made in a private communication to Hooke. Newton's hatred of Hooke serves as a good illustration of Newton's darker side. Newton's famous comment "If I have seen farther than others, it is because I have stood upon the shoulders of giants" appears to pay tribute to his predecessors, but in fact was a sarcastic comment about Hooke, who was born with a deformed spine and was small of stature. Later in his life, when Newton controlled the Royal Society, he had Hooke's name expunged from the membership and destroyed every portrait of Hooke that he could find. Today there is no surviving portrait giving us Hooke's likeness.

Another example of Newton's adversarial nature occurred in the Flamsteed affair. Flamsteed was Astronomer Royal, and had some tables of positions of the moon that Newton wanted for a revised edition of the *Principia*. When Flamsteed delayed in producing the tables and then submitted carelessly done work, Newton used his influence at court to force Flamsteed to publish the tables, at Flamsteed's expense. It is reported that much later, Flamsteed gathered all the copies of this book that he could find and in his anger with Newton, burned them on the lawn of the Royal Observatory at Greenwich.

In his later years, Newton moved more into the area of public service. He was elected to Parliament, representing Cambridge, and later moved to London and became warden, then master, of the Mint. In this he showed a talent for administration and brought the Mint's affairs into good order. The potential impact of a poor coinage system, which invited counterfeiting and lack of confidence in the crown, should not be underestimated, and Newton's success in reorganizing and strengthening the Mint was a great service to England. He was a strong supporter of capital punishment for offenses against his domain, such as counterfeiting or shaving precious metals from coins.

Although his scientific research ceased during this period, Newton's mind remained sharp and he retained his mathematical power. In 1696, the Swiss mathematician John Bernoulli joined with Leibniz to challenge the mathematical world with two problems, one of which was the brachistochrone problem. In what shape should one bend a wire so that a bead sliding down it will reach the bottom in the least time? After some of the great minds of Europe had foundered on the problem, Newton heard of it and solved it after dinner and a long day at the Mint. He sent his solution anonymously to the Royal Society. According to one story, Bernoulli, upon seeing the solution, immediately recognized it as the work of Newton—"the Lion is known by his claw."

Perhaps the most unfortunate aspect of Newton's later years was his acrimonious dispute with Leibniz over primacy in developing the calculus. In this, the two men became embroiled in a quarrel which transcended them both and became a matter of national pride, England versus Germany, and later the entire continent. Some believe that the intransigence of English mathematicians in maintaining Newton's cumbersome notation and then distancing themselves from their continental colleagues, particularly in France and Germany, was a serious disservice to English mathematics and science over the next hundred years.

Newton died in his sleep on March 20, 1727, at the age of eight-five, and was buried in Westminster Abbey, a tribute England reserves for its most distinguished subjects.

27.3 Gottfried Wilhelm Leibniz (1646–1716)

Leibniz was born in Leipzig on July 1, 1646. His father was a professor of moral philosophy, and at an early age Leibniz distinguished himself in the classics. At age fifteen he entered the

University of Leipzig and would have received his doctorate at age twenty but was denied by a faculty already jealous of his intellectual achievements. At this time, Newton had returned to Woolsthorpe from Cambridge and was formulating his theories.

From this disappointment at Leipzig, Leibniz proceeded to the University of Altdorf, where he received his degree and an offer of a professorship. He declined this offer and set out on a career as a lawyer and diplomat, pursuing philosophy, logic, and mathematics as his intellectual interests. He traveled widely, meeting many of the leading philosophers and scientists of his day, including Newton during a trip to London to visit the Royal Society.

At one point, Leibniz was employed by the prominent Brunswick family as a lawyer. One of his assignments was to produce records (by any means necessary) to validate the family's claims to certain positions and fortunes. Not one to engage in matters in a small way, Leibniz actually attempted to reunite the Protestant and Catholic churches, a feat which would have provided his clients substantial personal gain.

By about 1676, Leibniz was certainly working on original ideas in what would become the calculus. He published ideas about the fundamental theorem some 11 years after Newton was in possession of them, but before Newton had published anything about them himself. Amidst all of this activity, Leibniz founded the Berlin Academy of Sciences near the turn of the century.

Leibniz died in 1716 while still engaged in rewriting certain aspects of the Brunswick family history. In all of his dealings, he was a curious mixture of accuracy and industrious intelligence, with a tendency to bend the rules of ethical behavior when this served his cause or that of his employer. He seemed to find everyone interesting and had a kind word for nearly everyone.

27.4 The Bernoulli Family

Switzerland in the seventeenth and eighteenth centuries saw the rise to prominence of the remarkable Bernoulli family, which produced some of the outstanding mathematicians, lawyers, merchants, and diplomats of the time. Nicholas Bernoulli (1623–1708) had three sons, James (1654–1705), Nicholas (1662–1716), and John (1667–1748). John in turn had three sons, Nicholas (1695–1726), Daniel (1700–1782), and John (1710–1790). Finally, John produced two mathematicians, John (1746–1807) and James (1759–1789).

The Bernoullis had moved from Antwerp, Belgium, in 1583 to avoid religious persecution, settling in Basel, Switzerland. There, the elder Nicholas and his father and grandfather amassed a fortune as merchants. Nicholas's son James taught himself calculus from Leibniz's writings and made important contributions in probability and in the calculus of variations, which deals with constrained max/min problems. James's brother John began as a doctor and learned mathematics from James. Together they discovered (about 1697) that the solution of the brachistochrone problem is an arc of the cycloid. They were an interesting pair, carrying on a long and at times bitter feud over mathematics. Each often accused the other of stealing solutions or ideas, and sometimes these charges were accurate. Their other brother, Nicholas, took a doctorate in philosophy at age sixteen and at twenty added a degree in law.

John had three sons, Nicholas, Daniel, and John. Daniel became a medical doctor, then a mathematician, working on models of hydrodynamics, the theory of vibrating strings, and the kinetic theory of gases. He was a collaborator with Euler in St. Petersburg and is known today as the father of mathematical physics. Daniel's brother John began in law, then took up mathematics and its applications to physics. His son John took a doctorate in philosophy at age thirteen, then studied law and finally became astronomer royal in Berlin. John's other son James began with the law, then turned to experimental physics and mathematics, which he, like Daniel, pursued at the St. Petersburg Academy. He died by drowning at age thirty.

27.5   Leonhard Euler (1707–1783)

Euler was born in Basel, Switzerland, on April 15, 1707. His father was a Calvinist minister and intended him for the ministry as well. He began to study theology at Basel but took up mathematics with John Bernoulli, the father of Daniel and Nicholas.

From about age fourteen, Euler pursued a mathematical career. Daniel and Nicholas Bernoulli were at the St. Petersburg Academy, and Euler joined them in 1727. When Daniel returned to Switzerland six years later, Euler officially assumed the position Daniel had vacated. While in Russia, Euler married and eventually had thirteen children, of whom five survived childhood.

Euler's powers of memory and concentration were prodigious. At one point, he solved in three days a problem in astronomy expected by experts to take several months. Perhaps it was the strain of this and similar efforts that made him ill, and he lost the sight in his right eye.

While in Russia, Euler extended methods of calculus and applied them to a wide variety of problems. He also assumed tasks assigned him by the Russian government, including writing elementary mathematics textbooks and redesigning the system of weights and measures.

In 1740, Euler and his family moved to the Berlin Academy at the invitation of Friedrich the Great. Eventually he fell out of favor there, partly because of his habit of arguing over philosophical issues, about which he knew little. At the invitation of Catherine the Great, he returned to Russia in 1766. Although he soon lost the sight in his other eye, he was extended every consideration by Catherine, and was able to continue his mathematical work, with his sons copying formulas and text onto a slate as he dictated to them. He often spent his evenings surrounded by his children and grandchildren, carrying out his work in his mind while inventing games for them to play.

In 1776, Euler's wife died, but he soon remarried. Shortly thereafter, an operation was thought to have restored his sight, but an infection set in and he again became totally blind. On September 18, 1783, he suffered a stroke and died while playing with a grandson.

Euler's lifetime mathematical productivity was varied, high in quality, and unmatched in quantity. He wrote introductory texts on algebra, calculus, geometry, and calculus of variations that were standard books for nearly a century. He also did original and significant work in calculus, differential equations, and the calculus of variations and brought a high level of analytical skill to problems in what we would now call applied mathematics. Unlike some of his predecessors, he did not originate new branches of mathematics, but he greatly expanded the areas he touched. The exact extent of his work is not known. In 1909, the Swiss Association for Natural Science began a project of publishing his collected works, which were estimated to encompass 70 to 80 large volumes. But then another collection of manuscripts was discovered in St. Petersburg. Perhaps still more of Euler's papers lie somewhere awaiting rediscovery.

27.6   Carl Friedrich Gauss (1777–1855)

Gauss was born to a poor family in Braunschweig (known in English as Brunswick), Germany, on April 30, 1777. His brilliance was apparent at an early age, and in the local grammar school he astounded his teacher with his calculating ability. Encouraged by this teacher, the young Gauss began to study mathematics.

Gauss had the good fortune to be brought to the attention of the Duke of Brunswick, head of a local family of wealth and position, and the Duke became his patron. He sponsored Gauss at the local Collegium Carolinum in Brunswick, which Gauss entered in 1792. By this time he had mastered the binomial theorem and understood the idea of convergence of an infinite series.

At school, Gauss learned the classics and languages and read and understood Newton's *Principia*. His first discoveries in mathematics were in number theory and geometry, and at age twenty he decided on mathematics as a career. At this point, he began to keep a diary in which he recorded brief, often cryptic remarks, preserving his thoughts on various topics in mathematics. The existence of this diary became generally known only after Gauss's death, and it was loaned by a grandson to the Royal Society of Göttingen. It contains 146 brief inscriptions in Latin and mathematical notation and provides clear proof that Gauss had in his possession many of the great mathematical discoveries that would be made and published by others over the next century. Some of the entries consist of brief remarks whose meaning is still unknown. For example, an entry for October 11, 1796, reads "Vicimus GEGAN."

Gauss studied at the University of Göttingen but earned his doctorate from Helmstädt in 1799, where he worked with Johann Friedrich Pfaff. He did his dissertation on the fundamental theorem of algebra, a topic upon which he placed great importance and to which he returned many times throughout his career.

Gauss became interested in astronomy and calculated the orbit of Ceres. In 1809, he published a book on the motion of the planets. In the meantime, he married and had three children, Joseph, Minna, and Louis. His wife died after Louis's birth, and he soon remarried. His sons eventually emigrated to the United States, and it is said that one became a prosperous merchant in the St. Louis area.

After the death of the Duke of Brunswick, who was mortally wounded in a battle with Napoleon's forces, Gauss was offered a position at the Royal Academy in St. Petersburg but instead became director of the observatory at the University of Göttingen, where he spent the rest of his life. Gauss was a mediocre teacher but gave careful attention to students of talent and collaborated with physicist Wilhelm Weber in studies of electricity and magnetism (the gauss is a unit of magnetic field intensity). He also did important work in geometry, number theory, complex analysis, and differential equations, as well as specialized areas such as elliptic functions. He kept much of this work to himself, even when the results were rediscovered and announced by others. It was his habit to publish only what he regarded as "finished products," and he often left it for others to retrace his steps and even to claim the credit for results he left unannounced. But his diary gives silent evidence of the remarkable breadth of his achievements.

In 1855, Gauss began to suffer from shortness of breath, probably caused by an enlarged heart. He died on February 23, 1855, and is regarded by many as the greatest mathematician of his century.

27.7 Joseph-Louis Lagrange (1736–1813)

Lagrange was born in Turin, Italy, and was of French and Italian extraction. While still a schoolboy, he read Edmund Halley's treatise on the merits of Newton's calculus and became interested in mathematics. By the age of nineteen, he had mastered much of what was known in analysis at that time and became a professor at the Royal Artillery School of Turin.

Lagrange's chief interest lay in applying Newton's law of gravitational attraction to an analysis of planetary motion. His crowning achievement was his *Mécanique analytique*, an attempt at a rigorous treatment of the mathematics of planetary motion. The book first appeared in 1788, with the second edition following in 1811.

In 1766, when Euler left Berlin to return to the St. Petersburg Academy, he recommended to Friedrich the Great that Lagrange be offered his position. This was done, and Lagrange lived in Berlin for 20 years. After Friedrich's death, Lagrange was invited by Louis XVI to return to Paris, where he was given an apartment in the Louvre.

In 1772, Lagrange received a prize for his paper "Essai sur le problème des trois corps," which dealt with the three-body problem. This is the problem of determining the orbits of three spherical bodies, each acting on the other through gravitational attraction. Although a general solution defies a complete analysis, special cases can be treated. In one such case, Lagrange analyzed the orbits if the bodies begin at vertices of an equilateral triangle. In 1906, it was observed that this case actually occurs in our solar system, with the three bodies being the sun, Jupiter, and the asteroid Achilles. Of course, these bodies are also acted upon by numerous other gravitational influences, so it is not strictly speaking a three-body problem in isolation from other bodies.

Lagrange also studied planetary perturbations. If two spherical bodies act on each other through gravitational attraction, they will describe orbits that are conic sections. For example, in theory a planet moves on an elliptical orbit with the sun at one focus. Any deviation from this elliptical orbit is called a *perturbation*. In reality, planets do not describe perfectly elliptical orbits, because they are influenced by the gravitational attraction of many other bodies, particularly those nearby in our own solar system. Lagrange applied himself to the analysis of such perturbed motion. (Although this did not involve Lagrange, it was an analysis of perturbations in the orbit of Neptune that led astronomers to suspect that there was another planet influencing this orbit, leading to the discovery of Uranus.)

Lagrange's work was characterized not only by its importance but also by its elegance and clarity. His masterpiece contained much that is still of value today, including Lagrange's equations of motion, which are studied in classical mechanics.

27.8　Pierre-Simon de Laplace (1749–1827)

Laplace was born to moderately wealthy parents in Beaumont, France. At sixteen, he entered the University of Caen with the intention of studying for the priesthood but soon became interested in mathematics. From Caen he proceeded to Paris with a letter of introduction to d'Alembert, a leading French mathematician. D'Alembert was a busy person and paid little attention to the young, unknown Laplace. Shortly thereafter, Laplace wrote d'Alembert a letter outlining some of his work on mechanics. Upon reading this letter, d'Alembert sent for Laplace and recommended him for a professorship at the École Militaire in Paris. It is interesting to note that, in 1783, Laplace became an examiner at the École, examining, among others, a young student named Napoleon. (Napoleon's path crossed that of more than one great mathematician. Joseph Fourier accompanied Napoleon on his Egyptian campaign.)

During the Revolution, Laplace was made a member of the Commission on Weights and Measures but was later removed for failing to embrace certain Republican ideals. He moved to Melun, near Paris, to work on his *Exposition du Système du Monde*, which appeared in 1796.

After the Revolution, Laplace became a professor at the École Normale, along with Lagrange. Like Newton, he later assumed administrative positions, becoming minister of the interior and then a member, and later chancellor, of the Senate. Napoleon made him a count, but Laplace was a royalist who supported Louis XVIII, and when Louis returned to power Laplace was made a marquis and peer of France.

During the period 1799 to 1825, Laplace published his five-volume *Mécanique céleste*, in which he attempted to obtain analytic solutions of the equations of motion for various bodies in the solar system. He also worked on hydrodynamics, the propagation of sound waves, and the motion of the tides. His primary interest was in using mathematics to describe nature, and in this he placed less importance on the elegance that distinguished Lagrange's work.

27.9   Augustin-Louis Cauchy (1789–1857)

Cauchy was born in Paris on August 21, 1789, shortly after the fall of the Bastille. He was the oldest of six children, having a brother and four sisters. His early childhood was a mixture of deprivation during the terror of the French Revolution combined with an offensive sense of holiness and piety generated by the extreme Catholic devotion of his parents.

Cauchy's parents moved the family to the countryside, in the village of Arcueil, to insulate them from the immediate dangers of the Revolution. Due to poor nutrition, Cauchy was in frail health as a child, only recovering to some extent in his early twenties while he was in the military. During this time in Arcueil, the schools were closed and the parents educated the children.

It happened that Arcueil was near the estates of the mathematician Pierre-Simon de Laplace and also the chemist Claude-Louis Berthellot. Laplace, in particular, noticed the young boy's intellectual curiosity and preoccupation with books in general and mathematics in particular.

In 1800, the family's fortunes took a turn for the better as the elder Cauchy was called to Paris to serve as secretary of the Senate. His son shared his office in the Luxembourg Palace, which was visited often by Joseph-Louis Lagrange, a professor at the École Polytechnique. Lagrange was also quick to recognize Augustin-Louis's mathematical talent. At about age thirteen, Cauchy entered the Central School of the Pantheon and then went on to the École Polytechnique, where he studied mathematics and civil engineering.

Cauchy's first commission as an engineer was as part of a huge project to expand the port of Cherbourg. Napoleon envisioned an invasion of England, and this would require a large fleet, which would have to be built and protected while preparations were completed. Three years of intense work on this project seemed to do wonders for Cauchy's health, and for a time he flourished. Around 1810, he began to spend more of his time on mathematics, reviewing ideas he had been taught and pondering how they could be extended, as well as considering new avenues for investigation. He published some of his ideas, primarily in geometry and in an area involving symmetric functions.

By the time Napoleon's military fortunes were reversed at Leipzig in 1813, Cauchy had left the Cherbourg project and had immersed himself in studies of algebraic structures, particularly what are now known as finite groups (permutation groups). This and previous work on geometry brought him to prominence among French mathematicians of his day.

About 1814, Cauchy began to study integrals of complex functions, initiating a long and successful period of innovation in what would become complex function theory and deriving the fundamental result now known as Cauchy's theorem. There is evidence that Gauss had this result about 1811. However, Gauss preferred to publish only completed theories, and he chose to keep this theorem to himself and go on to other things, while Cauchy published it and justly received credit for its discovery. This was followed by many of the results that today bear Cauchy's name, including the integral representations of a differentiable function and its higher derivatives.

During his development of complex analysis, Cauchy also did significant work in other areas. In 1816, he won the Grand Prize of the French Academy of Sciences for a long treatise on water waves. By the age of thirty, he had received numerous honors and offers of prestigious positions, becoming professor of mathematics at the École Polytechnique, the Sorbonne, and the Collège de France.

Cauchy's personal fortunes took a downturn in 1830 when the Revolution removed Charles from the throne. As an outspoken royalist, Cauchy was expelled from the academy. While his family remained in Paris, he went for a time to Switzerland and then took a professorship in Turin. At one point, he had the misfortune to be given responsibility for the education of Charles's thirteen-year-old heir, the Duke of Bordeaux. Despite many distractions, he was able to continue his work, producing a theory of light in which light is caused by vibrations in solids.

By the age of fifty, Cauchy managed to disentangle himself from the Duke and return to Paris. He was allowed a special dispensation from the oath of allegiance to the French government and resumed his seat in the academy. There followed a period of tremendous creative activity, touching upon many areas of mathematics and physics.

As political leadership changed in France, the specter of the oath of allegiance presented itself again. Finally, by a tacit agreement between Cauchy and the government of Napoleon III, Cauchy was excused from the oath once again and continued in his academic position undisturbed by political turmoil.

Cauchy died in 1857 from respiratory problems from which he was expected to recover. Although he was an unpopular figure among his colleagues, partly due to his intense and un-welcome promotion of his religious convictions, he was recognized as being in the front rank of the great mathematicians. The volume, quality, and diversity of his research, and his impact on mathematics, secured his place in history. As he is reported to have said to the archbishop of Paris shortly before his death, "Men pass away but their deeds abide."

27.10   Joseph Fourier (1768–1830)

Joseph Fourier was born in 1768 in Auxerre, France, the son of a tailor. Orphaned at age ten, he won a scholarship at a Benedictine school nearby. Upon graduation he sought a military career, only to find that officer ranks were closed to him because of his common birth.

While considering entering the Benedictine order, Fourier became caught up in the French Revolution, rising to the position of president of the revolutionary committee of Auxerre. But these were treacherous times, and soon Fourier found himself under arrest and about to be executed. Changing fortunes in Paris and the fall of Robespierre saved his life.

At this point, Fourier, who had taught at the Benedictine school in Auxerre, was nominated for a position in a new college, the École Normale, established in Paris to train teachers. But his past associations with the Revolution in Auxerre soon caught up with him, and he was arrested again.

As Fourier's experience demonstrates, the politics of France at this time were very unstable, and if one could stay alive, it was possible to go from prison to a university teaching post, and back again, with surprising speed. As had happened before, Fourier was released from prison and into a position at the newly formed École Polytechnique. There he succeeded Lagrange in the chair of analysis and mechanics, examined an ambitious young military officer named Napoleon, and might have aspired to a quiet life of scholarship had not Napoleon embarked on his Eastern campaign. Fourier was named as part of a committee of intellectuals assigned to accompany the army to Egypt.

Lord Nelson's destruction of the French fleet offset Napoleon's victories on land, and the general deserted his army, returned to France, and initiated the coup that put him on the throne as emperor. Fourier and others of his company were left in Egypt, where Fourier held various administrative posts. Upon his return to France in 1801, Napoleon appointed him prefect of the Department of the Isère. Departments were districts into which Napoleon had subdivided France for administrative purposes. There were 13 in all, and Fourier's district included the important area around Grenoble.

At this point, Fourier embarked on a successful and varied administrative career. Two of his major projects were reclamation of swamp lands and construction of a road (still in use today) through the Alps. He also drew upon his Egyptian experience to write a history of Egypt, which was included as part of a book written by the intellectuals who had accompanied Napoleon's expedition. This contribution was so well received that Fourier's reputation among Egyptologists is as a historian, not a mathematician. As another indication of Fourier's curious ability to be in interesting places at the right time, he also encouraged a young Champollion, a talented student of languages, to concentrate on Egyptian writing. It was Champollion who later deciphered the Rosetta stone, providing the key to Egyptian hieroglyphics. The Rosetta stone had been discovered by a French soldier during the campaign.

Throughout all of these activities, Fourier had done relatively unimportant work in mathematics. A turning point came in 1804 when he began his study of heat conduction. In 1807, he submitted a paper to the Academy of Sciences in Paris, and it was read by some of the great men of the time (Lagrange and Laplace, among others). It was not well received for two reasons. One was that Fourier did not have available to him many of the concepts and methods known today and could not justify his results. Thus, lack of rigor was one vulnerability. The other was more personal. Laplace had worked on problems involving heat conduction, but from a different point of view, and was not prepared to accept Fourier's approach as either superior or even tractable. Neither Laplace nor Lagrange would believe the ramifications of Fourier's work, that familiar, nonperiodic functions could be expanded in series of sines and cosines.

In 1811, Fourier submitted his revised work to the academy, seeking a prize it had put forward for the best mathematical treatment of heat conduction. This memoir received the prize, but not the full acceptance of the committee, which still refused publication. In the meantime, Fourier's political fortunes experienced several more oscillations, beginning in 1814 when Napoleon was defeated and Louis XVIII gained the throne. Fourier was allowed to remain as prefect, but then Napoleon returned from his exile at Elba and briefly regained control of France. Giving his support to Napoleon, Fourier became prefect of the Rhone, a position with little longevity because Nopoleon was soon defeated at Waterloo. The returning King Louis did not look with favor upon Fourier's change of allegiance to Napoleon, and Fourier was left unemployed and without a pension.

Politically, Fourier rose from the ashes one last time. A connection in Paris had him appointed director of the Statistical Bureau of the Seine, and in 1816 he was elected to the Academy of Sciences. Although the king refused confirmation, Fourier was put forward again the next year and this time was approved for membership.

Now in his fifties, Fourier's ideas in mathematics began to receive wide attention, and in 1822 his monograph on heat conduction, rejected years past by the Academy, was published. He was elected permanent secretary of the mathematical section of the Academy, and then to the Académie Française. He lived the remainder of his years as a respected elder statesman of French mathematics.

One last note is of curious interest. There lived at this time a French sociologist, Charles Fourier, whose public recognition exceeded that of Joseph Fourier. The author Victor Hugo wrote, "There was at the Academy of Sciences a celebrated Fourier whom posterity has forgotten." Today, scientists, mathematicians, and engineers throughout the world know the name Fourier.

27.11   Henri Poincaré (1854–1912)

Jules Henri Poincaré was a professor of mathematics at the University of Paris. He and the German mathematician David Hilbert were considered the leading mathematicians of the latter part of the nineteenth century and early part of the twentieth. He is considered the last "universal mathematician," which means that he embraced all of the areas of mathematics known in his time, and he contributed to a wide variety of fields. In addition to many papers of remarkable depth and originality, his ideas have developed into fields now known as topology and dynamical systems, and he produced the three-volume work *Les méthodes nouvelles de la mécanique céleste.*

Poincaré also did important work on eigenvalue problems in partial differential equations and in the area of combinatorial structures arising in the qualitative study of differential equations. His work in this area was continued by Ivar Bendixson and others.

Answers and Solutions to Selected Odd-Numbered Problems

CHAPTER 1

Section 1.1

1. Yes, since $2\varphi\varphi' = 2(\sqrt{x-1})\left(\dfrac{1}{2\sqrt{x-1}}\right) = 1$ for $x > 1$. **3.** Yes **5.** Yes

7. No; $x^2\varphi\varphi' = \dfrac{x^4 - 16}{4x}$, whereas $-1 - x\varphi^2 = -\dfrac{x^4 - 8x^2 + 4x + 16}{4x}$ **9.** Yes

11. $\dfrac{d}{dx}(y^2 + xy - 2x^2 - 3x - 2y) = (y - 4x - 3) + (2y + x - 2)y' = 0$

17. $y = 3 - e^{-x}$ **19.** $y = 2\sin^2(x) - 2$

21. Direction field for $y' = x + y$;
solution satisfying $y(2) = 2$

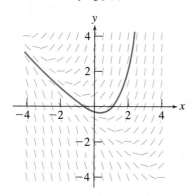

23. Direction field for $y' = xy$;
solution satisfying $y(0) = 2$

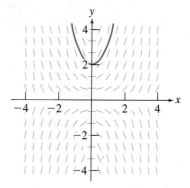

25. Direction field for $y' = \sin(y)$;
solution satisfying $y(1) = \pi/2$

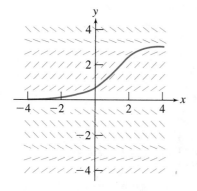

27. Direction field for $y' = e^{-x} + 2xy$;
solution satisfying $y(0) = 1$

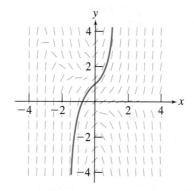

29. Direction field for $y' = y\sin(x) - 3x^2$; solution satisfying $y(0) = 1$

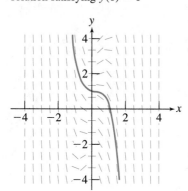

31. Direction field for $y' - y\cos(x) = 1 - x^2$; solution satisfying $y(2) = 2$

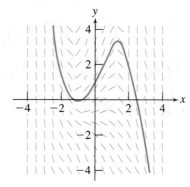

33. *Hint:* The lineal element to the solution curve through (x_0, z) has equation $y = [q(x_0) - p(x_0)z](x - x_0) + z$.

Section 1.2

1. $2x^2 = y^3 + c$ **3.** Not separable   **5.** $y = \dfrac{1}{1 - cx}$; also $y = 0$ **7.** $\sec(y) = Ax$ **9.** Not separable

11. $\frac{1}{2}y^2 - y + \ln(y + 1) = \ln(x) - 2$ **13.** $(\ln(y))^2 = 3x^2 - 3$ **15.** $3y\sin(3y) + \cos(3y) = 9x^2 - 5$

17. $45°F$ **19.** $V = \left[V_0^{1/3} + k\left(\dfrac{4\pi}{3}\right)^{1/3} t \right]^3$; $V_0 =$ initial volume   **21.** 8.57 kg   **23.** $\frac{1}{2}\sqrt{\pi}e^{-6}$

27. $3888\sqrt{2}$ s, or approximately 91 min, 39 s   **29.** (a) 576 s   (b) 489 s (approximate)   **31.** No

33. (a) 1920 s   (b) $1920 - 480\sqrt{2}$ s (about 1241 seconds)

35. $\frac{1}{2}(\sqrt{5} - 1)$ h, or about 37 min before noon (about 11:23 A.M.)

Section 1.3

1. $y = cx^3 + 2x^3 \ln|x|$ **3.** $y = \frac{1}{2}x - \frac{1}{4} + ce^{-2x}$ **5.** $y = 4x^2 + 4x + 2 + ce^{2x}$

7. $y = e^{-x} \int \left(\dfrac{x - 1}{x^2}\right) e^x \, dx + ce^{-x}$ **9.** $y = \dfrac{1}{4}x^3 + x + \dfrac{c}{x}$ **11.** $y = -\dfrac{3}{2x}e^{-x^2} + \dfrac{c}{x}$ **13.** $xe^{2y} - y = c$

15. $y = x^2 - x - 2$ **17.** $y = x + 1 + 4(x + 1)^{-2}$ **19.** $y = \frac{2}{3}e^{4x} - \frac{11}{3}e^x$ **21.** $y = \frac{2}{5}x - \frac{22}{5}x^{-4}$

23. $y = -2x^2 + cx$ **25.** $1000 - \frac{39}{\sqrt{5}}$, or about 983 pounds, at $t = 200$

27. $A_1(t) = 50 - 30e^{-t/20}$, $A_2(t) = 75 + 90e^{-t/20} - 75e^{-t/30}$; $A_2(t)$ has its minimum value of $\frac{5450}{81}$ pounds at $60\ln(\frac{9}{5})$ minutes.

Section 1.4

1. $2xy^2 + e^{xy} + y^2 = c$ **3.** Not exact   **5.** $y^3 + xy + \ln|x| = c$ **7.** $\cosh(x)\sinh(y) = c$ **9.** $x^y = c$

11. $xy + e^x = c$ **13.** $3xy^4 - x = 47$ **15.** $x\sin(2y - x) = \dfrac{\pi}{24}$ **17.** Not exact   **19.** $x^2 + y^3 + \cos(xy) = 9$

21. $\alpha = -3$; $x^2y^3 - 3xy - 3y^2 = c$ **23.** $\dfrac{\partial}{\partial x}(\varphi + c) = \dfrac{\partial \varphi}{\partial x} = M$ and $\dfrac{\partial}{\partial y}(\varphi + c) = \dfrac{\partial \varphi}{\partial y} = N$

Section 1.5

1. $\dfrac{1}{M}\left(\dfrac{\partial N}{\partial x} - \dfrac{\partial M}{\partial y}\right)$ is independent of x.

3. (a) $\dfrac{\partial M}{\partial y} = 1$, $\dfrac{\partial N}{\partial x} = -1$, so this equation is not exact.   (b) $\mu(x) = \dfrac{1}{x^2}$ (c) $\nu(y) = \dfrac{1}{y^2}$

(d) $\eta(x, y) = x^a y^b$ for all a and b satisfying $a + b = -2$

5. e^{3y}; $xe^{3y} - e^y = c$ **7.** x^2y; $x^4y^2 + 2x^3y^3 = c$ **9.** $\dfrac{1}{y + 1}$; $x^2y = c$ or $y = -1$ **11.** e^{2y-x}; $(x^2 - 2xy)e^{2y-x} = c$

13. $e^{-3x}y^{-4}$; $y^3 - 1 = ky^3 e^{3x}$ **15.** $\dfrac{1}{x}$; $y = 4 - \ln|x|$ **17.** x; $x^2(y^3 - 2) = -9$ **19.** $\dfrac{1}{y}$; $y = 4e^{-x^2/3}$

21. e^x; $e^x \sin(x - y) = \frac{1}{2}$ **23.** $\dfrac{\partial}{\partial y}(c\mu M) = c\dfrac{\partial}{\partial y}(\mu M) = c\dfrac{\partial}{\partial x}(\mu N) = \dfrac{\partial}{\partial x}(c\mu N)$ **27.** *Hint:* See Problems 3 and 26.

Section 1.6

1. $y = x + \dfrac{x}{c - \ln|x|}$ **3.** $y = 1/(1 + e^{x^2/2})$ **5.** $y \ln|y| - x = cy$ **7.** $xy - x^2 - y^2 = c$

9. $y = x^{-1}\left(c - \frac{7}{5}x^{-5/4}\right)^{4/7}$ **11.** $y = 2 + \dfrac{2}{cx^2 - 1}$ **13.** $y = \dfrac{2e^x}{ce^{2x} - 1}$ **15.** $h = \dfrac{ce - br}{ae - bd}$, $k = \dfrac{ar - dc}{ae - bd}$

17. $3(x - 2)^2 - 2(x - 2)(y + 3) - (y + 3)^2 = c$ **19.** $(2x + y - 3)^2 = c(y - x + 3)$

21. $(x - y + 3)^2 = 2x + c$ **23.** $3(x - 2y) - 8\ln|x - 2y + 4| = x + c$

25. Let the dog start at $(A, 0)$, and suppose the man walks upward into the upper half-plane. The dog's path is along the graph of $y = -\sqrt{A}\sqrt{A - x} + \dfrac{1}{3\sqrt{A}}(A - x)^{3/2} + \dfrac{2}{3}A$, and the dog catches the man at $(A, 2A/3)$ at time $t = 2A/3v$ units.

27. (a) $r = a - \dfrac{v\theta}{\omega}$, a spiral   (b) $\dfrac{a\omega}{2\pi v}$ revolution   (c) Distance $= \dfrac{1}{2}\left[\dfrac{a\omega}{v^2}\sqrt{v^2 + a^2\omega^2} + \ln\left(\dfrac{a\omega + \sqrt{v^2 + a^2\omega^2}}{v}\right)\right]$

29. With polar coordinates placed so the origin is at the point of the original sub sighting and the destroyer on the polar axis at $(9, 0)$ when the sub is sighted, the destroyer should sail 6 miles toward the point where the sub was first sighted and then follow the spiral $r = f(\theta) = 3e^{\theta/\sqrt{3}}$ to pass directly over the sub. The time at which this passing occurs is not known, hence this strategy will not help sink the sub (unless depth charges have a fairly large destructive radius).

37. $f(x) = \dfrac{f(0)}{(1 + kx)^2}$

Section 1.7

1. Velocity $= \dfrac{8}{9}\sqrt{\dfrac{1358}{3}}$, about 18.91 ft/s   **3.** Velocity $= 8\sqrt{30}$, about 43.82 ft/s

5. Velocity $= 12\sqrt{5}$, about 26.84 ft/s; time $= \frac{\sqrt{3}}{2}\ln(6 + \sqrt{35})$, about 2.15 s

7. Time $= \dfrac{\sqrt{3}}{8}\displaystyle\int_{10}^{40} \dfrac{x}{\sqrt{x^3 - 1000}}\, dx$, about 1.7117 seconds   **9.** Velocity $= 2\sqrt{210}$, about 28.98 ft/s

11. Max height $= 342.25$ ft; object hits the ground at 148 ft/s, 4.75 s after drop

13. Velocity $= \frac{44}{3}(1 - e^{-t/1200})$ ft/s $= 10(1 - e^{-3t})$ mi/h; terminal velocity $= 10$ mi/h; reaches 8.5 mi/h approximately 6.82 mi from starting point

15. $v(t) = 18 - 16e^{-2t/3}$ ft/s, $s(t) = 18t + 24(e^{-2t/3} - 1)$ ft

17. $v(t) = 32 - 32e^{-t}$ for $0 \le t \le 4$, $v(t) = 8(1 + ke^{-8t})/(1 - ke^{-8t})$ ft/s for $t \ge 4$, with $k = e^{32}(3e^4 - 4)/(5e^4 - 4)$; $\lim_{t \to \infty} v(t) = 8$ ft/s; $s(t) = 32(t + e^{-t} - 1)$ for $0 \le t \le 4$, $s(t) = 8t + 2\ln(1 - ke^{-8t}) + 64 + 32e^{-4} - 2\ln(2e^4/(5e^4 - 4))$ for $t \ge 4$

19. 17.5 ft/s   **21.** 64 lb at 990 mi and 26.8 lb at 3690 mi   **23.** $t = 2\sqrt{R/g}$, where $R =$ radius of the circle

25. $M\dfrac{dv}{dt} = -kv^\alpha$ for $0 < \alpha < 1$, $v(t) = \left[V_0^{1-\alpha} - \dfrac{k}{M}(1 - \alpha)t\right]^{1/(1-\alpha)}$, and $v\left(\dfrac{MV_0^{1-\alpha}}{k(1 - \alpha)}\right) = 0$; if $\alpha \ge 1$, then $v(t) > 0$ for all t

27. $V_C = 76$ V when $t = \ln(20)/2$; $i\left(\frac{1}{2}\ln(20)\right) = 16\ \mu\text{A}$ **29.** $i_1(0+) = i_2(0+) = 3/20$ amp; $i_3(0+) = 0$

31. (a) $q(t) = EC + (q_0 - EC)e^{-t/RC}$ (b) EC (d) $-RC\ln\left(\dfrac{0.01EC}{q_0 - EC}\right)$ **33.** $i(t) = \dfrac{A}{23}\left(e^{-2t/25} - e^{-t}\right)$

35. $i(t) = \dfrac{AL}{R^2 + L^2\omega_1^2}\left[\dfrac{R}{L}\sin(\omega_1 t) - \omega_1 \cos(\omega_1 t)\right] + \dfrac{BL}{R^2 + L^2\omega_2^2}\left[\dfrac{R}{L}\cos(\omega_2 t) + \omega_2 \sin(\omega_2 t)\right]$
$\qquad + \left[\dfrac{AL\omega_1}{R^2 + L^2\omega_1^2} - \dfrac{BR}{R^2 + L^2\omega_2^2}\right]e^{-Rt/L}$

37. $i(t) = \dfrac{1}{R} + \dfrac{1}{R - L}e^{-t} - \left(\dfrac{1}{R} + \dfrac{e^{-t}}{RL}\right)e^{-Rt/L}$ if $R \ne L$; $i(t) = \dfrac{1}{L} + \dfrac{1}{L}(t - 1)e^{-t}$ if $R = L$

39. $y = c - \frac{3}{4}\ln|x|$ **41.** $x^2 + 2y^2 - 4y = c$ **43.** $y^2(\ln(y^2) - 1) + 2x^2 = c$ **45.** $\frac{4}{3}y^{3/2} = c - x$

47. $y^2 = \ln|x| - \frac{1}{2}x^2 + c$

Section 1.8

7. (b) $y = 2 - e^{-x}$ (c) $y_n = 1 + x - \frac{1}{2}x^2 + \frac{1}{6}x^3 - \frac{1}{24}x^4 + \cdots + \frac{(-1)^{n+1}}{n!}x^n$

    (d) $y = 2 - \sum\limits_{n=0}^{\infty} \frac{(-1)^n}{n!}x^n = 1 + x - \frac{1}{2}x^2 + \frac{1}{6}x^3 - \frac{1}{24}x^4 + \cdots + \frac{(-1)^{n+1}}{n!}x^n + \cdots$

9. (b) $y = \frac{7}{3} + \frac{2}{3}x^3$ (c) $y_0 = 3,\ y_1 = \frac{7}{3} + \frac{2}{3}x^3 = y_2 = y_3 = \cdots = y_n$

    (d) $y = 3 + 2(x - 1) + 2(x - 1)^2 + \frac{2}{3}(x - 1)^3 = y_1$

Additional Problems

1. $x^4 y - 6xe^y = c$ **3.** $2y^2 - 4xy + 7x^2 = k$ **5.** $(\ln(y/x))^2 = 2\ln(x) + k$ **7.** $(y - x)^2(x + y + 2)^3 = c$

9. $y^2 = 3x^2 + c$ **11.** $y^5 + cy^2 = x$, together with the trivial solution $y = 0$

13. $4x + y\ln|y| = cy$, and the trivial solution   **15.** $\ln|y/x| - 1/y = c$ and the trivial solution

17. $y^2 = 2x - x^2 + k$ **19.** $x + 7y^2 = kx^8$ **21.** $y^4 - x = cy^3$ and the trivial solution

23. $y = -x\cos(x) + 4\sin(x) + \dfrac{12}{x}\cos(x) - \dfrac{24}{x^2}\sin(x) - \dfrac{24}{x^3}\cos(x) + \dfrac{c}{x^3}$

25. $2x + \sin(x) = cy$, and the trivial solution   **27.** $x + y = \tan(x + c)$ **29.** $\dfrac{2}{\sqrt{7}}\tan^{-1}\left(\dfrac{2y - x}{\sqrt{7}x}\right) = \ln|x| + c$

CHAPTER 2

Section 2.2

1. $y = \cosh(2x)$ **3.** $y = \frac{12}{5}e^{-3x} - \frac{7}{5}e^{-8x}$ **5.** $y = 2x^4 - 4x^4\ln|x|$

7. y_1 and y_2 are solutions of $x^2 y'' - 4xy' + 6y = 0$. We must write this equation as $y'' - \dfrac{4}{x}y' + \dfrac{6}{x^2}y = 0$ to have the form to which the theorem applies, and then we must have $x \neq 0$. On an interval not containing 0, the Wronskian of these solutions is nonzero.

9. $y'' - y' - 2y = 0$ has solutions $y_1 = e^{-x}$ and $y_2 = e^{2x}$, but $y_1 y_2 = e^x$ is not a solution. Many other examples also work.

11. *Hint:* Recall that at a relative extremum x_0, $y'(x_0) = 0$.   **13.** *Hint:* Consider $W(x_0)$.

Section 2.3

1. $y = c_1\cos(2x) + c_2\sin(2x)$ **3.** $y = c_1 e^{5x} + c_2 x e^{5x}$ **5.** $y = c_1 x^2 + c_2 x^2 \ln|x|$ **7.** $y = c_1 x^4 + c_2 x^{-2}$

9. $y = c_1\left(\dfrac{\cos(x)}{\sqrt{x}}\right) + c_2\left(\dfrac{\sin(x)}{\sqrt{x}}\right)$ **11.** $y = c_1 e^{-ax} + c_2 x e^{-ax}$

13. If curvature $\kappa = 0$, then $y = c_1 x + c_2$. If $\kappa \neq 0$, then $(x - c_1)^2 + (y - c_2)^2 = 1/\kappa^2$.

15. (a) $y^4 = c_1 x + c_2$ (b) $(y - 1)e^y = c_1 x + c_2$ or $y = c_3$ (c) $y = c_1 e^{c_1 x}/(c_2 - e^{c_1 x})$ or $y = 1/(c_3 - x)$

    (d) $y = \ln|\sec(x + c_1)| + c_2$ (e) $y = \ln|c_1 x + c_2|$

19. (a) $y = c_1 x^{-1} + c_2 x^{-1/2}$ (b) $y = x^{-1}(c_1 + c_2\ln|x|)$ (c) $y = c_1 x^{-1} + c_2 x^{-2}$ (d) $y = e^{-3x}[c_1\int x^{-1}e^{3x}\,dx + c_2]$

    (e) $y = xe^{-x^2/8}[c_1\int x^{-2}e^{x^2/8}\,dx + c_2]$

Section 2.4

1. $y = c_1 e^{-2x} + c_2 e^{3x}$ **3.** $y = e^{-3x}[c_1 + c_2 x]$ **5.** $y = e^{-5x}[c_1\cos(x) + c_2\sin(x)]$

7. $y = e^{-3x/2}[c_1\cos(3\sqrt{7}x/2) + c_2\sin(3\sqrt{7}x/2)]$ **9.** $y = e^{7x}(c_1 + c_2 x)$ **11.** $y = e^{-2x}(c_1\cos(\sqrt{5}x) + c_2\sin(\sqrt{5}x))$

13. $y = e^{x/2}(c_1 e^{\sqrt{5}x/2} + c_2 e^{-\sqrt{5}x/2})$ **15.** $y = e^{9x/2}(c_1 e^{3\sqrt{13}x/2} + c_2 e^{-3\sqrt{13}x/2})$ **17.** $y = 5 - 2e^{-3x}$ **19.** $y(x) = 0$

21. $y = \frac{9}{7}e^{3(x-2)} + \frac{5}{7}e^{-4(x-2)}$ **23.** $y = e^{x-1}(29 - 17x)$

25. $y = e^{(x+2)/2}\left[\cos\left(\frac{\sqrt{15}}{2}(x + 2)\right) + \frac{5}{\sqrt{15}}\sin\left(\frac{\sqrt{15}}{2}(x + 2)\right)\right]$

29. (a) $\psi(x) = e^{ax}[c + (d - ac)x]$ (b) $\psi_\epsilon(x) = \dfrac{1}{2\epsilon}e^{ax}[(d - ac + \epsilon c)e^{\epsilon x} + (ac - d + \epsilon c)e^{-\epsilon x}]$

    (c) $\lim_{\epsilon \to 0}\psi_\epsilon(x) = \psi(x)$ (use l'Hôpital's rule)

31. (a) $8\sqrt{3}\cos\left(6x + \dfrac{4\pi}{3}\right)$ (b) $2\sqrt{3}\cos\left(\pi x + \dfrac{5\pi}{6}\right)$ (c) $8\sqrt{2}\cos\left(2x + \dfrac{\pi}{4}\right)$ (d) $5\cos\left(4x + \dfrac{3\pi}{2}\right)$

Section 2.5

1. $y = c_1 x^2 + c_2 x^{-3}$ **3.** $y = c_1\cos(2\ln(x)) + c_2\sin(2\ln(x))$ **5.** $y = c_1 x^4 + c_2 x^{-4}$ **7.** $y = c_1 x^{-2} + c_2 x^{-3}$

9. $y = x^{-12}(c_1 + c_2\ln(x))$ **11.** $y = x^{3/2}[c_1\cos(\sqrt{39}\ln(x)/2) + c_2\sin(\sqrt{39}\ln(x)/2)]$

13. $y = c_1 x^{-3+\sqrt{11}} + c_2 x^{-3-\sqrt{11}}$ **15.** $y = c_1 x^{(9+\sqrt{113})/2} + c_2 x^{(9-\sqrt{113})/2}$

17. $y = x^{-2}(3\cos(4\ln(-x)) - 2\sin(4\ln(-x)))$ **19.** $y = -3 + 2x^2$ **21.** $y = -x^{-3}\cos(2\ln(-x))$

23. $y = -4x^{-12}(1 + 12\ln(x))$ **25.** $y = \frac{11}{4}x^2 + \frac{17}{4}x^{-2}$ **29.** $y = e^{-e^x/2}[c_1\cos(\sqrt{3}e^x/2) + c_2\sin(\sqrt{3}e^x/2)]$

31. $y = e^{-x^2/4}[c_1\cos(\sqrt{3}x^2/4) + c_2\sin(\sqrt{3}x^2/4)]$ **33.** $x^2 y'' + 5xy' + 13y = 0$

Section 2.6

1. $y = c_1\cos(x) + c_2\sin(x) - \cos(x)\ln|\sec(x) + \tan(x)|$

3. $y = c_1\cos(3x) + c_2\sin(3x) + 4x\sin(3x) + \frac{4}{3}\ln|\cos(3x)|\cos(3x)$ **5.** $y = c_1 e^x + c_2 e^{2x} - e^{2x}\cos(e^{-x})$

7. $y = c_1 e^{2x} + c_2 e^{-x} - x^2 + x - 4$ **9.** $y = e^x[c_1\cos(3x) + c_2\sin(3x)] + 2x^2 + x - 1$ **11.** $y = c_1 e^{2x} + c_2 e^{4x} + e^x$

13. $y = c_1 e^x + c_2 e^{2x} + 3\cos(x) + \sin(x)$ **15.** $y = e^{2x}[c_1\cos(3x) + c_2\sin(3x)] + \frac{1}{3}e^{2x} - \frac{1}{2}e^{3x}$

17. $y = c_1 + c_2 e^{4x} - 3x^3 - 2x^2 - 7x$ **19.** $y = c_1 + c_2 e^{-12x} + \frac{1}{51}\cos(3x) - \frac{4}{51}\sin(3x) - \frac{3}{74}\cos(2x) - \frac{1}{148}\sin(2x)$

21. $y = e^{-x}(c_1 + c_2 x) - \frac{3}{2}x^2 e^{-x} + \frac{4}{3}x^3 e^{-x} + 1$ **23.** $y = c_1\cos(2x) + c_2\sin(2x) + \frac{5}{8}\sinh(2x)$

25. $y = e^{-2x}(c_1 + c_2 x) + \frac{7}{4}(x - 1) - \frac{3}{8}\sin(2x) + \frac{5}{6}x^3 e^{-2x}$ **27.** $y = c_1 e^{2x} + c_2 e^{-x} + \frac{1}{3}xe^{2x}$

29. $y = c_1 e^{2x} + c_2 e^{-3x} - \frac{1}{6}x - \frac{1}{36}$ **31.** $y = c_1 x^2 + c_2 x^4 + x$

33. $y = c_1\cos(2\ln(x)) + c_2\sin(2\ln(x)) - \frac{1}{4}\cos(2\ln(x))\ln(x)$ **35.** $y = c_1 e^{2x} + c_2 xe^{2x} + e^{3x} - \frac{1}{4}$

37. $y = c_1 x + c_2 x\ln(x) + 3x[\ln(x)]^2$ **39.** $y = c_1 x^{-1} + c_2 x^{-1}\ln(x) + x^2 + 2x + 5$

41. $y = c_1 x^4 + c_2 x^{-3} - \frac{1}{3}x$ **43.** $y = c_1 x^2 + c_2 x^3 - \frac{1}{2}x - x^2\ln(x)$ **45.** $y = \frac{7}{4}e^{2x} - \frac{3}{4}e^{-2x} - \frac{7}{4}xe^{2x} - \frac{1}{4}x$

47. $y = \frac{3}{8}e^{-2x} - \frac{19}{120}e^{-6x} + \frac{1}{5}e^{-x} + \frac{7}{12}$ **49.** $y = 2e^{4x} + 2e^{-2x} - 2e^{-x} - e^{2x}$

51. $y = -\frac{17}{4}e^{2x} + \frac{55}{13}e^{3x} + \frac{1}{52}\cos(2x) - \frac{5}{52}\sin(2x)$

53. $y = \frac{1}{11}e^{4(x+1)}\left[(55 - e)\cosh(\sqrt{14}(x + 1)) + \frac{1}{\sqrt{14}}(5e - 198)\sinh(\sqrt{14}(x + 1))\right] + \frac{1}{11}e^{-x}$

55. $y = 4e^{-x} - \sin^2(x) - 2$ **57.** $y = 2x^3 + x^{-2} - 2x^2$ **59.** $y = x - x^2 + 3\cos(\ln(x)) + \sin(\ln(x))$

61. $y_p = \displaystyle\sum_{n=1}^{\infty}\left[-\frac{\sin(nx)}{n(n^2 + 4)}\right]$ **63.** $y = c_1 x^2 + c_2(x - 1) + x^4 - 4x^3$

Section 2.7

1. $y = e^{-2t}[5\cosh(\sqrt{2}t) + \frac{10}{\sqrt{2}}\sinh(\sqrt{2}t)]$;
$y = \frac{5}{\sqrt{2}}e^{-2t}\sinh(\sqrt{2}t)$ (graphed below)

3. $y = \frac{5}{2}e^{-t}[2\cos(2t) + \sin(2t)]$;
$y = \frac{5}{2}e^{-t}\sin(2t)$ (graphed below)

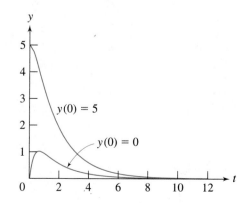

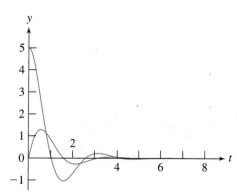

5. $y = \dfrac{A}{\sqrt{2}} e^{-2t} \sinh(\sqrt{2}t)$ (graphed below)

7. $y = Ate^{-2t}$ (graphed below)

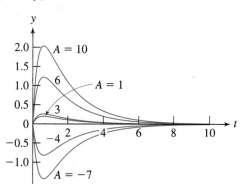

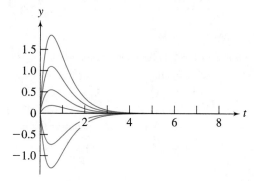

9. $y = \dfrac{A}{2} e^{-t} \sin(2t)$ (graphed below)

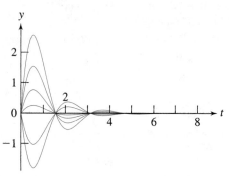

11. $y = \frac{1}{60}(57e^{-8t} - 52e^{-3t})$ down positive

13. At most once; no condition on only $y(0)$ is enough, as one needs to specify $y'(0)$ also.

15. Increasing C decreases the frequency   **21.** *Hint:* Show that $(m/k)^2 = (d/a)^2$ **23.** $\dfrac{A}{2m\omega_0} t \sin(\omega_0 t)$

25. (a) $y = \frac{1}{373}\left[e^{-3t}\left(2266\cosh(\sqrt{7}t) + \frac{6582}{\sqrt{7}}\sinh(\sqrt{7}t)\right) - 28\cos(3t) + 72\sin(3t)\right]$

   (b) $y = \frac{1}{373}\left[e^{-3t}\left(28\cosh(\sqrt{7}t) + \frac{2106}{\sqrt{7}}\sinh(\sqrt{7}t)\right) - 28\cos(3t) + 72\sin(3t)\right]$

   These functions are graphed below.

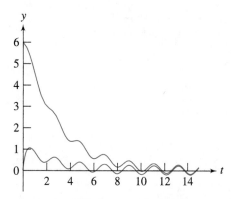

27. (a) $y = \frac{1}{15}\left[e^{-t/2}\left(98\cos(\sqrt{11}t/2) + \frac{74}{\sqrt{11}}\sin(\sqrt{11}t/2)\right) - 8\cos(3t) + 4\sin(3t)\right]$

   (b) $y = \frac{1}{15}\left[e^{-t/2}\left(8\cos(\sqrt{11}t/2) + \frac{164}{\sqrt{11}}\sin(\sqrt{11}t/2)\right) - 8\cos(3t) + 4\sin(3t)\right]$

These functions are graphed below.

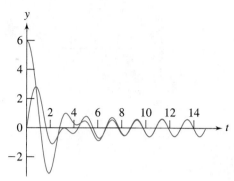

29. $y(t) = y_c(t) + \dfrac{1}{k}t - \dfrac{c}{k^2}$, where y_c is the general solution of $my'' + cy' + ky = 0$

33. *Hint:* Show that $y_{ss}(t) = \dfrac{A}{\sqrt{(k - m\omega^2)^2 + c^2\omega^2}} \cos(\omega t + \varphi)$ and find the value of ω^2 that minimizes the quantity in the radical.

35. amplitude $= 2$ ft; frequency $= \dfrac{\sqrt{2}}{\pi}$ cycles/s

37. $i(t) = .015e^{-0.0625t} - 5.4 \times 10^{-7}e^{-3333.27t} + 0.015\cos(20t) - 0.00043\sin(20t)$

39. $i(t) = 0.001633e^{-t} + 0.00161e^{-0.3177t} + 0.000023e^{-t}\cos(6t) - 0.000183e^{-t}\sin(6t)$

Additional Problems

1. $y = x^5[c_1 + c_2 \ln|x|]$ **3.** $y = [c_1 + c_2 x]e^{7x}$ **5.** $y = c_1 e^{-x} + c_2 e^{2x} + \frac{1}{2}e^{4x} - 3x + \frac{3}{2}$

7. $y = x^4[c_1 + c_2 \ln|x|]$ **9.** $y = \dfrac{1}{c_1 x + c_2}$ **11.** $y = c_1 \cos(x) + c_2 \sin(x) - \cos(x) \ln|\sec(x) + \tan(x)|$

13. $y = c_1 e^x + c_2 e^{-3x} - \frac{7}{5}\cos(2x) + \frac{4}{5}\sin(2x)$ **15.** $y = \frac{1}{25}[e^x(47 - 10x) + 3\cos(3x) - 4\sin(3x)]$

17. $y = c_1 e^{-4x} + c_2 x e^{-4x} + \frac{3}{100}x\cos(2x) + \frac{1}{25}x\sin(2x) - \frac{1}{250}\cos(2x) - \frac{11}{500}\sin(2x)$

19. $y = 3x^{1+\sqrt{3}} + 3x^{1-\sqrt{3}} + x^3 - 2\ln|x| + 2$

CHAPTER 3

Section 3.1

1. $\dfrac{1}{s - 1} - \dfrac{1}{s + 1}$ **3.** $\dfrac{16s}{(s^2 + 4)^2}$ **5.** $\dfrac{1}{s^2} - \dfrac{s}{s^2 + 25}$ **7.** $\dfrac{2}{s^3} + \dfrac{8}{s^2} + \dfrac{16}{s}$ **9.** $\dfrac{6}{s^4} - \dfrac{3}{s^2} + \dfrac{s}{s^2 + 16}$ **11.** $-2e^{-16t}$

13. $2\cos(4t) - \frac{5}{4}\sin(4t)$ **15.** $3e^{7t} + t$ **17.** $e^{4t} - 6te^{4t}$ **25.** $\dfrac{5}{s(1 + e^{-3s})}$

27. From the graph, $f(t) = 0$ if $0 < t \le 5$, and $f(t) = 5$ if $5 < t \le 10$, and $f(t) = 0$ if $10 < t \le 25$. Further, $f(t + 25) = f(t)$, so f is periodic of period $T = 25$. Thus $\mathfrak{L}[f](s) = \dfrac{5e^{-5s}(1 - e^{-5s})}{s(1 - e^{-25s})}$.

29. $\dfrac{E\omega}{s^2 + \omega^2} \dfrac{1}{1 - e^{-\pi s/\omega}}$ **31.** $f(t) = h$ if $0 < t \le a$ and $f(t) = 0$ if $a < t \le 2a$. Further, $f(t + 2a) = f(t)$, so f is periodic of period $2a$, and $L[f](s) = \dfrac{h}{s(1 + e^{-as})}$.

Section 3.2

1. $y = \frac{1}{4} - \frac{13}{4}e^{-4t}$ **3.** $y = -\frac{4}{17}e^{-4t} + \frac{4}{17}\cos(t) + \frac{1}{17}\sin(t)$ **5.** $y = -\frac{1}{4} + \frac{1}{2}t + \frac{17}{4}e^{2t}$

7. $y = \frac{22}{25}e^{2t} - \frac{13}{5}te^{2t} + \frac{3}{25}\cos(t) - \frac{4}{25}\sin(t)$ **9.** $y = \frac{1}{16} + \frac{1}{16}t - \frac{33}{16}\cos(4t) + \frac{15}{64}\sin(4t)$

Section 3.3

1. $\dfrac{6}{(s+2)^4} - \dfrac{3}{(s+2)^2} + \dfrac{2}{s+2}$ **3.** $\dfrac{1}{s}(1 - e^{-7s}) + \dfrac{s}{s^2+1}\cos(7)e^{-7s} - \dfrac{1}{s^2+1}\sin(7)e^{-7s}$

5. $\dfrac{1}{s^2} - \dfrac{11}{s}e^{-3s} - \dfrac{4}{s^2}e^{-3s}$ **7.** $\dfrac{1}{s+1} - \dfrac{2}{(s+1)^3} + \dfrac{1}{(s+1)^2+1}$ **9.** $\dfrac{s}{s^2+1} + \left(\dfrac{2}{s} - \dfrac{s}{s^2+1} - \dfrac{1}{s^2+1}\right)e^{-2\pi s}$

11. $\dfrac{s^2+4s-5}{(s^2+4s+13)^2}$ **13.** $\dfrac{1}{s^2} - \dfrac{2}{s} - \left(\dfrac{1}{s^2} + \dfrac{15}{s}\right)e^{-16s}$ **15.** $\dfrac{24}{(s+5)^5} + \dfrac{4}{(s+5)^3} + \dfrac{1}{(s+5)^2}$ **17.** $e^{2t}\sin(t)$

19. $\cos(3(t-2))H(t-2)$ **21.** $\dfrac{1}{\sqrt{2}}e^{-3t}\sinh(\sqrt{2}t)$ **23.** $e^{-3t}\cosh(2\sqrt{2}t) - \dfrac{1}{2\sqrt{2}}e^{-3t}\sinh(2\sqrt{2}t)$

25. $\dfrac{1}{16}[1 - \cos(4(t-21))]H(t-21)$ **27.** $2e^{2t} + 8te^{2t}$ **29.** $e^{7t}\cosh(4\sqrt{3}t) + \dfrac{7}{4\sqrt{3}}e^{7t}\sinh(4\sqrt{3}t)$

31. $y = \cos(2t) + \dfrac{3}{4}[1 - \cos(2(t-4))]H(t-4)$ **33.** $y = [-\dfrac{1}{4} + \dfrac{1}{12}e^{2(t-6)} + \dfrac{1}{6}e^{-(t-6)}\cos(\sqrt{3}(t-4))]H(t-6)$

35. $y = -\dfrac{1}{4} + \dfrac{2}{5}e^t - \dfrac{3}{20}\cos(2t) - \dfrac{1}{5}\sin(2t) + [-\dfrac{1}{4} + \dfrac{2}{5}e^{t-5} + \dfrac{3}{20}\cos(2(t-5)) - \dfrac{1}{5}\sin(2(t-5))]H(t-5)$

37. $y = -\dfrac{1}{28}[(4 - \sqrt{2})e^{-(1+2\sqrt{2})t} + (4 + \sqrt{2})e^{-(1-2\sqrt{2})t}] - \dfrac{1}{28}[8 - (4 - \sqrt{2})e^{-(1+2\sqrt{2})(t-5)} - (4 + \sqrt{2})e^{-(1-2\sqrt{2})(t-5)}]H(t-5)$

39. $y = \dfrac{1}{4} + \dfrac{3}{4}e^{-2t} + \dfrac{7}{2}te^{-2t} + [-\dfrac{1}{4} + \dfrac{1}{4}e^{-2(t-2)} + \dfrac{1}{2}(t-2)e^{-2(t-2)}]H(t-2)$ **41.** $\varphi(t) = [1 - 2e^{t-3} + e^{2(t-3)}]H(t-3)$

43. $E_{\text{out}} = 5e^{-4t} + 10[(1 - e^{-4(t-5)})H(t-5)]$ **45.** $i(t) = \dfrac{k}{R}(1 - e^{-Rt/L}) - \dfrac{k}{R}(1 - e^{-R(t-5)/L})H(t-5)$

47. $\mathcal{L}[KH(t-a) - KH(t-b)](s) = \dfrac{K}{s}e^{-as} - \dfrac{K}{s}e^{-bs}$

49. $\mathcal{L}\left[h\left(\dfrac{t-a}{b-a}\right)H(t-a) + h\left(\dfrac{c-t}{c-b} - \dfrac{t-a}{b-a}\right)H(t-b) - h\left(\dfrac{c-t}{c-b}\right)H(t-c)\right]$

$= \dfrac{h}{b-a}\dfrac{e^{-as}}{s^2} - \dfrac{h(c-a)}{(c-b)(b-a)}\dfrac{e^{-bs}}{s^2} + \dfrac{h}{c-b}\dfrac{e^{-cs}}{s^2}$

51. (a) $2e^{2t} - e^t$ (b) $\dfrac{10}{7}e^{6t} - \dfrac{3}{7}e^{-t}$ (c) $\dfrac{2}{7}e^{2t} - \dfrac{1}{2}e^{-3t} + \dfrac{3}{14}e^{-5t}$ (d) $\dfrac{37}{66}e^{3t} - \dfrac{13}{30}e^{-3t} + \dfrac{48}{55}e^{-8t}$

Section 3.4

1. $\dfrac{1}{16}[\sinh(2t) - \sin(2t)]$ **3.** $\dfrac{\cos(at) - \cos(bt)}{(b-a)(b+a)}$ if $b^2 \neq a^2$; $\dfrac{t\sin(at)}{2a}$ if $b^2 = a^2$

5. $\dfrac{1}{a^4}[1 - \cos(at)] - \dfrac{1}{2a^3}t\sin(at)$ **7.** $(\dfrac{1}{2} - \dfrac{1}{2}e^{-2(t-4)})H(t-4)$ **9.** $y(t) = e^{3t} * f(t) - e^{2t} * f(t)$

11. $y(t) = \dfrac{1}{4}e^{6t} * f(t) - \dfrac{1}{4}e^{2t} * f(t) + 2e^{6t} - 5e^{2t}$ **13.** $y(t) = \dfrac{1}{3}\sin(3t) * f(t) - \cos(3t) + \dfrac{1}{3}\sin(3t)$

15. $y(t) = \dfrac{4}{3}e^t - \dfrac{1}{4}e^{2t} - \dfrac{1}{12}e^{-2t} - \dfrac{1}{3}e^t * f(t) + \dfrac{1}{4}e^{2t} * f(t) + \dfrac{1}{12}e^{-2t} * f(t)$ **17.** $f(t) = \dfrac{1}{2}e^{-2t} - \dfrac{3}{2}$

19. $f(t) = \cosh(t)$ **21.** $f(t) = 3 + \dfrac{2}{5}\sqrt{15}e^{t/2}\sin(\sqrt{15}t/2)$ **23.** $f(t) = \dfrac{1}{4}e^{-2t} + \dfrac{3}{4}e^{-6t}$

Section 3.5

1. $y = 3[e^{-2(t-2)} - e^{-3(t-2)}]H(t-2) - 4[e^{-2(t-5)} - e^{-3(t-5)}]H(t-5)$ **3.** $y = 6(e^{-2t} - e^{-t} + te^{-t})$

5. $\varphi(t) = (B+9)e^{-2t} - (B+6)e^{-3t}$; $\varphi(0) = 3$, $\varphi'(0) = B$ **7.** $3/\pi$ **9.** 4

11. $E_{\text{out}} = 10e^{-4(t-2)}H(t-2) - 10e^{-4(t-3)}H(t-3)$ **13.** 0 if $t < a$; $f(t-a)$ if $t \geq a$

15. $y(x) = \dfrac{1}{6}F_0x^3 - \dfrac{x}{6L}\left(F_0L^3 + \dfrac{M}{EI}(L-a)^3\right) + \dfrac{M}{6EI}(x-a)^3H(x-a)$ **17.** $y(t) = \sqrt{\dfrac{m}{k}}v_0\sin\left(\sqrt{\dfrac{k}{m}}t\right)$

19. $y'(0) = 4$ ft/s upward; frequency $= 4/\pi$ cycles/s and amplitude of $\dfrac{1}{2}$ foot

Section 3.6

1. $x(t) = -2 + 2e^{t/2} - t$, $y(t) = -1 + e^{t/2} - t$ **3.** $x(t) = \dfrac{4}{9} + \dfrac{1}{3}t - \dfrac{4}{9}e^{3t/4}$, $y(t) = -\dfrac{2}{3} + \dfrac{2}{3}e^{3t/4}$

5. $x(t) = \dfrac{3}{4} - \dfrac{3}{4}e^{2t/3} + \dfrac{1}{2}t^2 + \dfrac{1}{2}t$, $y(t) = -\dfrac{3}{2}e^{2t/3} + t + \dfrac{3}{2}$ **7.** $x(t) = e^{-t}\cos(t) + t - 1$, $y(t) = e^{-t}\sin(t) + t^2 - t$

9. $x(t) = 1 - e^{-t} - 2te^{-t}$, $y(t) = 1 - e^{-t}$

11. $y_1(t) = \dfrac{1}{2}e^t + \dfrac{1}{2}e^{-t} - 1 - t$, $y_2(t) = -\dfrac{1}{4}t^2 - \dfrac{1}{2}t$, $y_3(t) = -\dfrac{1}{6}e^t + \dfrac{1}{6}e^{-t} - \dfrac{1}{3}t$

13. $i_1(t) = \dfrac{1}{5}(1 - \dfrac{1}{2}e^{-t/2}) - \dfrac{2}{85}[e^{-(t-4)/2} - \cos(2(t-4)) + \dfrac{9}{2}\sin(2(t-4))]H(t-4)$, and

$i_2(t) = \dfrac{1}{10}e^{-t/2} + \dfrac{2}{85}[e^{-(t-4)/2} - \cos(2(t-4)) - 4\sin(2(t-4))]H(t-4)$

15. $x_1(t) = \frac{5}{36} - \frac{1}{20}\cos(2t) - \frac{4}{45}\cos(3t) - [\frac{5}{36} - \frac{1}{20}\cos(2(t-2)) - \frac{4}{45}\cos(3(t-2))]H(t-2)$, and
$\qquad x_2(t) = \frac{1}{18} - \frac{1}{10}\cos(2t) + \frac{2}{45}\cos(3t) - [\frac{1}{18} - \frac{1}{10}\cos(2(t-2)) + \frac{2}{45}\cos(3(t-2))]H(t-2)$

17. $m_1 y_1'' = k(y_2 - y_1)$, $m_2 y_2'' = -k(y_2 - y_1)$; $y_1(0) = y_1'(0) = y_2'(0) = 0$; $y_2(0) = d$. Then $(m_1 s^2 + k)Y_1 - kY_2 = 0$ and
$\qquad (m_2 s^2 + k)Y_2 - kY_1 = m_2 ds$. Replace Y_2 with $\dfrac{m_1 s^2 + k}{k} Y_1$ in the second equation to get

$$Y_1(s) = \frac{kd}{m_1 s \left[s^2 + k\left((m_1 + m_2)/m_1 m_2\right)\right]}.$$ The quadratic term in the denominator indicates that the objects will

oscillate with period $2\pi \sqrt{\dfrac{m_1 m_2}{k(m_1 + m_2)}}$.

19. $i_1 = [\frac{1}{10}e^{-(t-1)} + \frac{3}{20}e^{-(t-1)/6}]H(t-1)$, $i_2 = [-\frac{1}{10}e^{-(t-1)} + \frac{1}{10}e^{-(t-1)/6}]H(t-1)$

21. $x_1(t) = 9e^{-t/100} + e^{-3t/50} + 3[e^{-(t-3)/100} - e^{-3(t-3)/50}]H(t-3)$;
$\qquad x_2(t) = 6e^{-t/100} - e^{-3t/50} + [2e^{-(t-3)/100} + 3e^{-3(t-3)/50}]H(t-3)$

Section 3.7

1. $y = -1 + ce^{-2/t}$ **3.** $y = 7t^2$ **5.** $y = ct^2 e^{-t}$ **7.** $y = 4$ **9.** $y = 3t^2/2$
11. The transformed equation is $s^2 Y'' + (4 - B)sY' + (C - B + 2)Y = 0$, which is again an Euler equation.

13. $y(t) = \sum_{n=0}^{\infty} \frac{(-1)^n}{(n!)^2}\left(\frac{t}{2}\right)^{2n}$ **17.** $\frac{1}{2}\ln\left(\frac{s+1}{s-1}\right)$ **19.** $y = t^2 - 8\frac{(t-1)^{-3}}{t}H(t-1)$

CHAPTER 4

Section 4.1

1. $y = -2 - \frac{1}{3}x^3 + \frac{1}{12}x^4 - \frac{1}{60}x^5 - \frac{1}{120}x^6 + \cdots$ **3.** $y = 3 + \frac{5}{2}(x-1)^2 + \frac{5}{6}(x-1)^3 + \frac{5}{24}(x-1)^4 + \frac{1}{6}(x-1)^5 + \cdots$
5. $y = 7 + 3(x-1) - 2(x-1)^2 - (x-1)^3 + (x-1)^4 + \cdots$ **7.** $y = -3 + x + 4x^2 + \frac{7}{6}x^3 + \frac{1}{3}x^4 + \cdots$
9. $y = 1 + x - \frac{1}{4}x^2 + \frac{1}{4}x^3 + \frac{1}{32}x^4 + \cdots$

11. $y = 1 + \pi(x-1) + \dfrac{1}{6}(\pi+1)(x-1)^3 - \dfrac{\pi}{24}(x-1)^4 + \dfrac{3+5\pi}{120}(x-1)^5 + \cdots$

13. $y = 3 + 3x + \frac{7}{4}x^2 + \frac{23}{24}x^3 + \frac{71}{96}x^4 + \cdots$
15. $y = a_0[1 - \frac{1}{2}x^2 + \frac{1}{6}x^4 - \frac{31}{720}x^6 + \frac{379}{40320}x^8] - [\frac{1}{2}x^2 - \frac{1}{8}x^4 + \frac{5}{144}x^6 - \frac{43}{5760}x^8 + \frac{1741}{1209600}x^{10} + \cdots]$
17. $y = x - 1 + A\left[1 - \frac{1}{2}x^2 + \frac{1}{4(2!)}x^4 - \frac{1}{8(3!)}x^6 + \frac{1}{16(4!)}x^8 + \cdots\right]$;

$\qquad y = x - 1 + A\sum_{n=0}^{\infty}(-1)^n \frac{1}{n!2^n}x^{2n} = x - 1 + (A+1)e^{-x^2/2}$

19. $a_{n+2} = -\dfrac{a_{n-1}}{(n+1)(n+2)}$ for $n \geq 1$; $(a_1 = 0)$ $y_1 = 1 - \frac{1}{6}x^3 + \frac{1}{180}x^6 - \frac{1}{12960}x^9 + \frac{1}{1710720}x^{12} - \cdots$; $(a_0 = 0)$
$\qquad y_2 = x - \frac{1}{12}x^4 + \frac{1}{504}x^7 - \frac{1}{45360}x^{10} + \frac{1}{7076160}x^{13} - \cdots$

21. $y = a_0\left[1 + \frac{1}{4(5)}x^5 + \frac{1}{4(5)(9)(10)}x^{10} + \frac{1}{4(5)(9)(10)(14)(15)}x^{15} + \cdots\right] + a_1\left[x + \frac{1}{5(6)}x^6 + \frac{1}{5(6)(10)(11)}x^{11} + \right.$

$\qquad \left. \frac{1}{5(6)(10)(11)(15)(16)}x^{16} + \cdots\right] + \frac{1}{2}x^2 + \frac{1}{2(6)(7)}x^7 + \frac{1}{2(6)(7)(11)(12)}x^{12} + \frac{1}{2(6)(7)(11)(12)(16)(17)}x^{17} + \cdots$

23. $y = a_0[1 + \frac{1}{12}x^4 - \frac{1}{60}x^5 + \frac{1}{360}x^6 - \frac{1}{2520}x^7 + \cdots] + a_1[x - \frac{1}{2}x^2 + \frac{1}{6}x^3 - \frac{1}{24}x^4 + \frac{7}{120}x^5 - \cdots]$
25. $y = a_0[1 - \frac{1}{3}x^4 - \frac{4}{45}x^6 + \frac{1}{210}x^8 + \frac{68}{14175}x^{10} + \cdots] + a_1[x + \frac{1}{3}x^3 - \frac{1}{10}x^5 - \frac{1}{18}x^7 - \frac{17}{3240}x^9 + \cdots]$
27. $y = a_0 + a_1[x - x^2 + \frac{2}{3}x^3 - \frac{1}{4}x^4 + \frac{23}{360}x^6 + \cdots] + [\frac{1}{6}x^3 - \frac{1}{12}x^4 + \frac{1}{30}x^5 + \frac{1}{180}x^6 - \frac{1}{105}x^7 + \cdots]$
29. $y = a_0[1 + \frac{1}{2}x^2 - \frac{1}{2}x^3 + \frac{5}{12}x^4 - \frac{13}{40}x^5 + \cdots] + a_1[x + \frac{1}{6}x^3 - \frac{1}{4}x^4 + \frac{7}{30}x^5 - \frac{7}{40}x^6 + \cdots] + [\frac{1}{6}x^4 + \frac{1}{180}x^6$
$\qquad - \frac{1}{84}x^7 + \frac{17}{1260}x^8 - \frac{109}{10080}x^9 + \cdots]$

31. $y = a[1 - \frac{1}{2(3)}x^3 + \frac{1}{2(3)(5)(6)}x^6 - \frac{1}{2(3)(5)(6)(8)(9)}x^9 + \cdots] + b[x - \frac{1}{3(4)}x^4 + \frac{1}{3(4)(6)(7)}x^7 - \frac{1}{(3)(4)(6)(7)(9)(10)}x^{10} + \cdots]$

33. $y = 2 - 3e^{-x} = -1 + 3x - \dfrac{3}{2}x^2 + \dfrac{3x^3}{3!} - \dfrac{1}{8}x^4 + \cdots$

35. $y = \frac{1}{5} + \frac{1}{5}e^{2x}[32\sin(x) - 6\cos(x)] = -1 + 4x + 11x^2 + \frac{34}{3}x^3 + \frac{27}{4}x^4 + \cdots$

Section 4.2

1. a_0 arbitrary, $a_1 = 1$, $2a_2 - a_0 = -1$, $a_n = \dfrac{1}{n}a_{n-2}$ for $n \geq 3$,

$y = a_0 + (a_0 - 1)\left[\frac{1}{2}x^2 + \frac{1}{2(4)}x^4 + \frac{1}{2(4)(6)}x^6 + \frac{1}{2(4)(6)(8)}x^8 + \cdots\right] + x + \frac{1}{3}x^3 + \frac{1}{3(5)}x^5 + \frac{1}{3(5)(7)}x^7 + \frac{1}{3(5)(7)(9)}x^9 + \cdots$

3. a_0 arbitrary, $a_1 + a_0 = 0$, $2a_2 + a_1 = 1$, $a_{n+1} = \dfrac{1}{n+1}(-a_n + a_{n-2})$ for $n \geq 2$;

$y = a_0\left[1 - x + \frac{1}{2!}x^2 + \frac{1}{3!}x^3 - \frac{7}{4!}x^4 + \cdots\right] + \frac{1}{2!}x^2 - \frac{1}{3!}x^3 + \frac{1}{4!}x^4 + \frac{11}{5!}x^5 - \frac{31}{6!}x^6 + \cdots$

5. a_0, a_1 arbitrary, $a_2 = \frac{1}{2}(3 - a_0)$, $a_{n+2} = \dfrac{(n-1)}{(n+1)(n+2)}a_n$ for $n \geq 1$;

$y = a_0 + a_1 x + (3 - a_0)\left[\frac{1}{2!}x^2 + \frac{1}{4!}x^4 + \frac{3}{6!}x^6 + \frac{3(5)}{8!}x^8 + \frac{3(5)(7)}{10!}x^{10} + \cdots\right]$

7. a_0, a_1 arbitrary, $a_2 + a_0 = 0$, $6a_3 + 2a_1 = 1$, $a_{n+2} = \dfrac{(n-1)a_{n-1} - 2a_n}{(n+1)(n+2)}$ for $n \geq 2$;

$y = a_0 + a_1 x - a_0 x^2 + \left(\frac{1}{6} - \frac{1}{3}a_1\right)x^3 + \left(\frac{1}{6}a_0 + \frac{1}{12}a_1\right)x^4 + \cdots$

9. a_0, a_1 arbitrary, $2a_2 + a_1 + 2a_0 = 1$, $6a_3 + 2a_2 + a_1 = 0$;

$a_{n+2} = \dfrac{(n-2)a_n - (n+1)a_{n+1}}{(n+1)(n+2)}$ for $n \geq 3$,

$y(x) = a_0 + a_1 x + \left(\frac{1}{2} - a_0 - \frac{1}{2}a_1\right)x^2 + \left(\frac{1}{3}a_0 - \frac{1}{6}\right)x^3 - \left(\frac{1}{24} + \frac{1}{12}a_0\right)x^4 + \cdots$

11. a_0 arbitrary, $a_1 = 1$, $a_{2k} = -\frac{1}{2k}a_{2k-2}$ for $k \geq 1$; $a_{2k+1} = \dfrac{-a_{2k-1} + (1/(2k)!)(-1)^k}{2k+1}$ for $k \geq 1$;

$y = a_0\left[1 - \frac{1}{2}x^2 + \frac{1}{2(4)}x^4 - \frac{1}{2(4)(6)}x^6 + \cdots\right] + x - \frac{1}{2!}x^3 + \frac{13}{5!}x^5 - \cdots$

13. a_0, a_1 arbitrary, $2a_2 + a_1 - a_0 = 1$, $6a_3 + 2a_2 - a_1 = 0$, $12a_4 + 3a_3 - a_2 = -1$,

$20a_5 + 4a_4 - a_3 = 0$, $30a_6 + 5a_5 - a_4 = 1$; $a_{n+2} = \dfrac{a_n - (n+1)a_{n+1}}{(n+1)(n+2)}$ for $n \geq 5$;

$y = a_0 + a_1 x + \left(\frac{1}{2} + \frac{1}{2}a_0 - \frac{1}{2}a_1\right)x^2 - \left(\frac{1}{6} + \frac{1}{6}a_0 - \frac{1}{3}a_1\right)x^3 + \left(\frac{1}{12}a_0 - \frac{1}{8}a_1\right)x^4 + \cdots$

15. a_0, a_1 arbitrary, $a_{2k} = -\dfrac{1}{2k(2k-1)}a_{2k-2}$ for $k \geq 1$; $a_{2k+1} = \dfrac{(1/(2k-2)!)(-1)^{k+1} - a_{2k-1}}{(2k)(2k+1)}$ for $k \geq 1$,

$y = a_0 + a_1 x - \frac{1}{2}a_0 x^2 - \frac{1}{6}(a_1 + 1)x^3 + \frac{1}{24}a_0 x^4 + \cdots$

Section 4.3

1. 0, regular, 3, regular   **3.** 0, regular, 2, regular   **5.** 0, irregular, 2, regular

7. $4r^2 - 2r = 0$; $c_n = \dfrac{1}{2(n+r)(2n+2r-1)}c_{n-1}$ for $n \geq 1$; $y_1 = \displaystyle\sum_{n=0}^{\infty}\dfrac{1}{(2n+1)!}x^{n+1/2}$, $y_2 = \displaystyle\sum_{n=0}^{\infty}\dfrac{1}{(2n)!}x^n$

9. $9r^2 - 9r + 2 = 0$; $c_n = \dfrac{-4}{9(n+r)(n+r-1)+2}c_{n-1}$ for $n \geq 1$;

$y_1 = x^{2/3}\left[1 - \frac{4}{3(4)}x + \frac{4^2}{3(4)(6)(7)}x^2 - \frac{4^3}{3(4)(6)(7)(9)(10)}x^3 + \frac{4^4}{3(4)(6)(7)(9)(10)(12)(13)}x^4 - \frac{4^5}{3(4)(6)(7)(9)(10)(12)(13)(15)(16)}x^6 + \cdots\right]$

$= x^{2/3} + \displaystyle\sum_{n=1}^{\infty}\dfrac{2(5)(8)\cdots(3n-1)(-1)^n 4^n}{(3n+1)!}x^{n+2/3}$; $y_2 = x^{1/3} + \displaystyle\sum_{n=1}^{\infty}\dfrac{1(4)(7)\cdots(3n-2)(-1)^n 4^n}{(3n)!}x^{n+1/3}$

11. $2r^2 - r = 0$; $c_n = \dfrac{-2(n+r-1)}{(n+r)(2n+2r-1)}c_{n-1}$ for $n \geq 1$;

$y_1 = x^{1/2} + \displaystyle\sum_{n=1}^{\infty}\dfrac{(-1)^n(2n+3)}{3(n!)}x^{n+1/2}$; $y_2 = 1 + \displaystyle\sum_{n=1}^{\infty}\dfrac{(-1)^n 2^n(n+1)}{1(3)(5)\cdots(2n-1)}x^n$

13. $2r^2 - r - 1 = 0$; $c_1 = -\dfrac{2}{2r+3}c_0$; $c_n = \dfrac{2c_{n-2} - 2(n+r-1)c_{n-1}}{2(n+r)^2 - (n+r) - 1}$ for $n \geq 2$;

$y_1 = x - \frac{2}{1!(5)}x^2 + \frac{18}{2!(5)(7)}x^3 - \frac{164}{3!(5)(7)(9)}x^4 + \frac{2284}{4!(5)(7)(9)(11)}x^5 - \frac{37272}{5!(5)(7)(9)(11)(13)}x^6 + \cdots$;

$y_2 = x^{-1/2} - x^{1/2} + \frac{3}{2}x^{3/2} - \frac{13}{3!(3)}x^{5/2} + \frac{119}{4!(3)(5)}x^{7/2} - \frac{1353}{5!(3)(5)(7)}x^{9/2} + \cdots$

15. $9r^2 - 4 = 0$; $(9r^2 + 18r + 5)c_1 = 0$; $c_n = \dfrac{-9}{9(n+r)^2 - 4}c_{n-2}$ for $n \geq 2$;

$$y_1 = x^{2/3} - \frac{3}{2^2 1!(5)}x^{8/3} + \frac{3^2}{2^4 2!(5)(8)}x^{14/3} - \frac{3^3}{2^6 3!(5)(8)(11)}x^{20/3}$$
$$+ \frac{3^4}{2^8 4!(5)(8)(11)(14)}x^{26/3} - \frac{3^5}{2^{10} 5!(5)(8)(11)(14)(17)}x^{32/3} + \cdots;$$

$$y_2 = x^{-2/3} - \frac{3}{2^2 1!(1)}x^{4/3} + \frac{3^2}{2^4 2!(1)(4)}x^{10/3} - \frac{3^3}{2^6 3!(1)(4)(7)}x^{16/3}$$
$$+ \frac{3^4}{2^8 4!(1)(4)(7)(10)}x^{22/3} - \frac{3^5}{2^{10} 5!(1)(4)(7)(10)(13)}x^{28/3} + \cdots$$

17. $2r^2 - r = 0$; $(2r^2 + 3r + 1)c_1 = 0$; $c_n = \dfrac{-6}{2(n+r)^2 - (n+r)}c_{n-2}$ for $n \geq 2$;

$$y_1 = x^{1/2} - \frac{3}{1!(5)}x^{5/2} + \frac{3^2}{2!(5)(9)}x^{9/2} - \frac{3^3}{3!(5)(9)(13)}x^{13/2}$$
$$+ \frac{3^4}{4!(5)(9)(13)(17)}x^{17/2} - \frac{3^5}{5!(5)(9)(13)(17)(21)}x^{21/2} + \cdots;$$

$$y_2 = 1 - \frac{3}{1!3}x^2 + \frac{3^2}{2!3(7)}x^4 - \frac{3^3}{4!3(7)(11)}x^6 + \frac{3^4}{4!3(7)(11)(15)}x^8 - \frac{3^5}{5!3(7)(11)(15)(19)}x^{10} + \cdots$$

19. $2r^2 - 5r - 3 = 0$; $c_1 = 0$; $c_n = \dfrac{2}{2(n+r)^2 - 5(n+r) - 3}c_{n-2}$ for $n \geq 2$;

$$y_1 = x^3\left(1 + \tfrac{1}{11}x^2 + \tfrac{1}{2!11(15)}x^4 + \tfrac{1}{3!11(15)(19)}x^6 + \tfrac{1}{4!11(15)(19)(23)}x^8 + \cdots\right),$$
$$y_2 = x^{-1/2}\left(1 - \tfrac{1}{3}x^2 - \tfrac{1}{2!3}x^4 - \tfrac{1}{3!3(5)}x^6 - \tfrac{1}{4!3(5)(9)}x^8 - \cdots\right)$$

21. $r^3 + (c-1)r = 0$; $r = 0$ or $r = 1 - c$; $c_n = \dfrac{(n+r)^2 + (a+b-2)(n+r) + 1 - a - b + ab}{(n+r)^2 + (c-1)(n+r)}c_{n-1}$ for $n \geq 1$;

$$y_1 = 1 + \sum_{n=1}^{\infty} \frac{a(a+1)(a+2)\cdots(a+n-1)b(b+1)(b+2)\cdots(b+n-1)}{n!c(c+1)(c+2)\cdots(c+n-1)}x^n,$$
$$y_2 = x^{1-c}\left[1 + \sum_{n=1}^{\infty} \frac{(a-c+1)(a-c+2)\cdots(a-c+n)(b-c+1)\cdots(b-c+n)}{n!(2-c)(3-c)\cdots(n+1-c)}x^n\right]$$

Section 4.4

1. $y_1 = c_0(x-1)$, $y_2 = c_0^*[(x-1)\ln(x) - 3x + \tfrac{1}{4}x^2 + \tfrac{1}{36}x^3 + \tfrac{1}{288}x^4 + \tfrac{1}{2400}x^5 + \cdots]$

3. $y_1 = c_0[x^4 + 2x^5 + 3x^6 + 4x^7 + 5x^8 + \cdots] = c_0\dfrac{x^4}{(x-1)^2}$, $y_2 = c_0^*\dfrac{3-4x}{(x-1)^2}$

5. $y_1 = c_0\left[x^{1/2} - \dfrac{1}{2\cdot 1!3}x^{3/2} + \dfrac{1}{2^2 2!3(5)}x^{5/2} - \dfrac{1}{2^3 3!3(5)(7)}x^{7/2} + \dfrac{1}{2^4 4!3(5)(7)(9)}x^{9/2} + \cdots\right]$,
$$y_2 = c_0^*\left[1 - \dfrac{1}{2}x + \dfrac{1}{2^2 2!3}x^2 - \dfrac{1}{2^3 3!3(5)}x^3 + \dfrac{1}{2^4 4!3(5)(7)}x^4 - \cdots\right]$$

7. $y_1 = c_0\left[x^2 + \tfrac{1}{3!}x^4 + \tfrac{1}{5!}x^6 + \tfrac{1}{7!}x^8 + \tfrac{1}{9!}x^{10} + \cdots\right]$, $y_2 = c_0^*\left[x - x^2 + \tfrac{1}{2!}x^3 - \tfrac{1}{3!}x^4 + \tfrac{1}{4!}x^5 - \cdots\right]$

9. $y_1 = c_0(1-x)$; $y_2 = c_0^*[1 + \tfrac{1}{2}(x-1)\ln((x-2)/x))]$

11. $y_1 = c_0[x^2 - x^3 + \tfrac{1}{3}x^4 - \tfrac{7}{36}x^5 - \tfrac{7}{720}x^6 + \cdots]$, $y_2 = c_0^*[y_1\ln(x) - x + \tfrac{3}{2}x^3 - \tfrac{31}{36}x^4 + \tfrac{65}{432}x^5 + \tfrac{61}{4320}x^6 + \cdots]$

13. $y_1 = c_0 x^{2+2\sqrt{2}}\left[1 + \tfrac{3+2\sqrt{2}}{1+4\sqrt{2}}x + \tfrac{(3+2\sqrt{2})(4+2\sqrt{2})}{(1+4\sqrt{2})(2+4\sqrt{2})}x^2 + \cdots\right]$,
$$y_2 = c_0^* x^{2-2\sqrt{2}}\left[1 - \tfrac{3-2\sqrt{2}}{-5-2\sqrt{2}}x + \tfrac{(3-2\sqrt{2})(4-2\sqrt{2})}{(-5-2\sqrt{2})(-2-4\sqrt{2})}x^2 - \tfrac{(3-2\sqrt{2})(4-2\sqrt{2})(5-2\sqrt{2})}{(-5-2\sqrt{2})(-2-4\sqrt{2})(3-6\sqrt{2})}x^3 + \cdots\right]$$

15. (a) $25r^2 + 15r - 4 = 0$, with roots $r_1 = \tfrac{1}{5}, r_2 = -\tfrac{4}{5}$

(b) $y_1 = \displaystyle\sum_{n=0}^{\infty} c_n x^{n+1/5}$, $c_0 \neq 0$, and $y_2 = ky_1(x)\ln(x) + \displaystyle\sum_{n=0}^{\infty} c_n^* x^{n-4/5}$

17. (a) $48r^2 - 20r - 8 = 0$, with roots $r_1 = \tfrac{2}{3}, r_2 = -\tfrac{1}{4}$

(b) $y_1 = \displaystyle\sum_{n=0}^{\infty} c_n x^{n+2/3}$ with $c_0 \neq 0$, $y_2 = \displaystyle\sum_{n=0}^{\infty} c_n^* x^{n-1/4}$ with $c_0^* \neq 0$

19. (a) $4r^2 - 10r = 0$, with roots $r_1 = \frac{5}{2}, r_2 = 0$ (b) $y_1 = \sum_{n=0}^{\infty} c_n x^{n+5/2}$ with $c_0 \neq 0$, $y_2 = \sum_{n=0}^{\infty} c_n^* x^n$ with $c_0^* \neq 0$

Section 4.5

1. $R = 1; (3, 5]$ **3.** $R = 1; [-2, 0)$ **5.** $R = 1; (-1, 1)$ **7.** $R = \frac{2}{3}; (\frac{11}{6}, \frac{19}{6})$ **9.** $R = 1; (-1, 1)$

11. $R = 3; [-1, 5]$ **13.** $R = 1; [-1, 1)$ **15.** $\sum_{n=1}^{\infty} \frac{1}{2n+2}(-1)^n x^n$ **17.** $\sum_{n=0}^{\infty} \frac{2n+5}{n+1} x^n$

19. $1 + 2x + \sum_{n=2}^{\infty} (2^{n-1} + n + 1)x^n$ **21.** $1 + \frac{1}{2}x + \sum_{n=2}^{\infty} \left(\frac{n!}{n+1} + 2^n \right) x^n$ **23.** $-27 + \sum_{n=1}^{\infty} \left(\frac{1}{2n} - (n+3)^{n+3} \right) x^n$

25. $1 + \frac{1}{2}x + \frac{1}{3}x^2 + \frac{1}{4}x^3 + \sum_{n=4}^{\infty} \frac{2n^2 - 5n - 6}{n+1} x^n$

29. $\frac{1}{1+x} = \frac{1}{3} \frac{1}{1+(x-2)/3} = \frac{1}{3} \sum_{n=0}^{\infty} (-1)^n \left(\frac{x-2}{3} \right)^n = \sum_{n=0}^{\infty} \frac{(-1)^n}{3^{n+1}} (x-2)^n$ for $\left| \frac{x-2}{3} \right| < 1$, or $-1 < x < 5$

CHAPTER 5

Section 5.1

1. $(2 + \sqrt{2})i + 3j; (2 - \sqrt{2})i - 9j + 10k; \sqrt{38}; \sqrt{63}; 4i - 6j + 10k; 3\sqrt{2}i + 18j - 15k$
3. $3i - k; i - 10j + k; \sqrt{29}; 3\sqrt{3}; 4i - 10j; 3i + 15j - 3k$
5. $3i - j + 3k; -i + 3j - k; \sqrt{3}; 2\sqrt{3}; 2i + 2j + 2k; 6i - 6j + 6k$

7. $F + G = 3i - 2j$ $F - G = i$ **9.** $F + G = 2i - 5j$ $F - G = j$

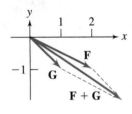   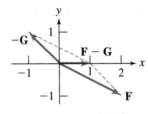   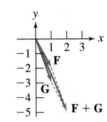   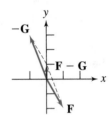

11. $\alpha F = -\frac{1}{2}i - \frac{1}{2}j$ **13.** $\alpha F = 12j$ **15.** $\alpha F = -9i + 6j$

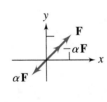   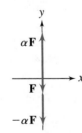   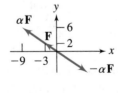

17. $x = 3 + 6t, y = -t, z = 0; -\infty < t < \infty$ **19.** $x = 0, y = 1 + t, z = 3 + 2t; -\infty < t < \infty$
21. $x = 2 + 3t, y = -3 - 9t, z = 6 + 2t; -\infty < t < \infty$ **23.** $x = 3 + t, y = 3 + 9t, z = -5 - 6t; -\infty < t < \infty$
25. $x = 4 + 5t, y = -8 - 8t, z = 1 + t; -\infty < t < \infty$ **27.** $F = 3i + 3\sqrt{3}j$ **29.** $F = \frac{15}{\sqrt{2}}i - \frac{15}{\sqrt{2}}j$

Section 5.2

1. $F \cdot G = 2; \cos(\theta) = 2/\sqrt{14}$; not orthogonal; $|F \cdot G| = 2 < \sqrt{14} = \|F\| \|G\|$
3. $-23; -23/\sqrt{29}\sqrt{41}$; not orthogonal; $23 < \sqrt{29}\sqrt{41}$ **5.** $-18; -9/10$; not orthogonal; $18 < \sqrt{10}\sqrt{40} = 20$
7. $3x - y + 4z = 4$ **9.** $4x - 3y + 2z = 25$ **11.** $7x + 6y - 5z = -26$ **13.** $112/11\sqrt{105}$
15. $113/5\sqrt{590}$ **17.** $F = 0$ **19.** *Hint:* $|F \cdot U| = \|F\| \cos(\theta)$, which is a maximum when $\cos(\theta) = 1$.

21. *Hint:* Take $\mathbf{F} = \left(\dfrac{\mathbf{A} \cdot \mathbf{B}}{\|\mathbf{B}\|^2}\right)\mathbf{B}$ and $\mathbf{G} = \mathbf{A} - \mathbf{F}$.   **23.** All properties would still hold except now $\mathbf{F} \cdot \mathbf{F} \neq \|\mathbf{F}\|^2$.

25. Let $\mathbf{A}$ and $\mathbf{B}$ be the vectors of two adjacent sides of the rhombus. Then $\|\mathbf{A}\| = \|\mathbf{B}\|$. Further, the diagonals are the vectors $\mathbf{A} + \mathbf{B}$ and $\mathbf{A} - \mathbf{B}$. Now compute $(\mathbf{A} + \mathbf{B}) \cdot (\mathbf{A} - \mathbf{B})$.

27. With $\mathbf{F} = f_1\mathbf{i} + f_2\mathbf{j} + f_3\mathbf{k}$ and $\mathbf{G} = g_1\mathbf{i} + g_2\mathbf{j} + g_3\mathbf{k}$, we have
$\mathbf{F} \cdot \mathbf{G} = f_1g_1 + f_2g_2 + f_3g_3 = g_1f_1 + g_2f_2 + g_3f_3 = \mathbf{G} \cdot \mathbf{F}$.

29. With $\mathbf{F} = f_1\mathbf{i} + f_2\mathbf{j} + f_3\mathbf{k}$, $\|\mathbf{F}\|^2 = f_1^2 + f_2^2 + f_3^2$. This sum of nonnegative real numbers is zero if and only if each term is zero, hence $f_1 = f_2 = f_3 = 0$, so $\mathbf{F} = \mathbf{O}$.

Section 5.3

In Problems 1 through 9, the value is given in order for $\mathbf{F} \times \mathbf{G}$, $\cos(\theta)$, $\sin(\theta)$, and the common value of $\|\mathbf{F}\|\,\|\mathbf{G}\|\sin(\theta)$ and $\|\mathbf{F} \times \mathbf{G}\|$.

1. $8\mathbf{i} + 2\mathbf{j} + 12\mathbf{k}$; $-4/\sqrt{69}$; $\sqrt{53}/\sqrt{69}$; $2\sqrt{53}$ **3.** $-8\mathbf{i} - 12\mathbf{j} - 5\mathbf{k}$; $-12/\sqrt{29}\sqrt{13}$; $\sqrt{233}/\sqrt{29}\sqrt{13}$; $\sqrt{233}$

5. $18\mathbf{i} + 50\mathbf{j} - 60\mathbf{k}$; $62/5\sqrt{2}\sqrt{109}$; $\sqrt{1606}/5\sqrt{2}\sqrt{109}$; $2\sqrt{1606}$

7. $-85\mathbf{i} + 18\mathbf{j} + 396\mathbf{k}$; $-70/\sqrt{349}\sqrt{485}$; $\sqrt{164,365}/\sqrt{349}\sqrt{485}$; $\sqrt{164,365}$

9. $12\mathbf{i} + 34\mathbf{j} + 8\mathbf{k}$; $19/3\sqrt{6}\sqrt{13}$; $\sqrt{341}/3\sqrt{6}\sqrt{13}$; $2\sqrt{341}$ **11.** Not collinear; $x - 2y + z = 3$

13. Not collinear; $2x - 11y + z = 0$ **15.** Not collinear; $29x + 37y - 12z = 30$ **17.** $7\sqrt{2}$ **19.** $2\sqrt{209}$ **21.** 92

23. 98   **25.** 22   **27.** $\mathbf{i} - \mathbf{j} + 2\mathbf{k}$ **29.** $7\mathbf{i} + \mathbf{j} - 7\mathbf{k}$

31. $\mathbf{F} \times (\mathbf{G} + \mathbf{H}) = \begin{vmatrix} \mathbf{i} & \mathbf{j} & \mathbf{k} \\ f_1 & f_2 & f_3 \\ g_1 + h_1 & g_2 + h_2 & g_3 + h_3 \end{vmatrix} = \begin{vmatrix} \mathbf{i} & \mathbf{j} & \mathbf{k} \\ f_1 & f_2 & f_3 \\ g_1 & g_2 & g_3 \end{vmatrix} + \begin{vmatrix} \mathbf{i} & \mathbf{j} & \mathbf{k} \\ f_1 & f_2 & f_3 \\ h_1 & h_2 & h_3 \end{vmatrix} = (\mathbf{F} \times \mathbf{G}) + (\mathbf{F} \times \mathbf{H})$

35. *Hint:* $\mathbf{F} \cdot (\mathbf{G} \times \mathbf{H}) = \begin{vmatrix} f_1 & f_2 & f_3 \\ g_1 & g_2 & g_3 \\ h_1 & h_2 & h_3 \end{vmatrix}$. Now use properties of determinants and of the dot product.

39. $[\mathbf{F}, \mathbf{G}, \mathbf{H}] = (a_1\mathbf{i} + b_1\mathbf{j} + c_1\mathbf{k}) \cdot \begin{vmatrix} \mathbf{i} & \mathbf{j} & \mathbf{k} \\ a_2 & b_2 & c_2 \\ a_3 & b_3 & c_3 \end{vmatrix} = a_1(b_2c_3 - c_2b_3) + b_1(c_2a_3 - a_2c_3) + c_1(a_2b_3 - b_2a_3)$

$= \begin{vmatrix} a_1 & b_1 & c_1 \\ a_2 & b_2 & c_2 \\ a_3 & b_3 & c_3 \end{vmatrix}$

41. $[\alpha\mathbf{F} + \beta\mathbf{K}, \mathbf{G}, \mathbf{H}] = (\alpha\mathbf{F} + \beta\mathbf{K}) \cdot (\mathbf{G} \times \mathbf{H}) = \alpha\mathbf{F} \cdot (\mathbf{G} \times \mathbf{H}) + \beta\mathbf{K} \cdot (\mathbf{G} \times \mathbf{H}) = \alpha[\mathbf{F}, \mathbf{G}, \mathbf{H}] + \beta[\mathbf{K}, \mathbf{G}, \mathbf{H}]$

Section 5.4

In Problems 1 through 6, we give in order the sum of the two vectors, their dot product, and the cosine of the angle between them.

1. $5\mathbf{e}_1 + 5\mathbf{e}_2 + 6\mathbf{e}_3 + 5\mathbf{e}_4 + \mathbf{e}_5$; 0; 0   **3.** $17\mathbf{e}_1 - 4\mathbf{e}_2 + 3\mathbf{e}_3 + 6\mathbf{e}_4$; 24; $6/\sqrt{30}\sqrt{17}$

5. $6\mathbf{e}_1 + 7\mathbf{e}_2 - 6\mathbf{e}_3 - 2\mathbf{e}_5 - \mathbf{e}_6 + 11\mathbf{e}_7 + 4\mathbf{e}_8$; -94; $-47/\sqrt{37}\sqrt{303}$ **7.** S is a subspace of R^5.

9. S is not a subspace of R^6.   **11.** S is a subspace of R^4.   **13.** S is a subspace of R^4.   **15.** S is a subspace of R^7.

17. *Hint:* Suppose to be specific that $a \neq 0$. Then $x = -\dfrac{b}{a}y - \dfrac{c}{a}z$, and for $\mathbf{V}$ to be in S, we must have

$\mathbf{V} = \left(-\dfrac{b}{a}y - \dfrac{c}{a}z, y, z\right)$. Now verify conditions of the definition.

19. $\|\mathbf{F} + \mathbf{G}\|^2 + \|\mathbf{F} - \mathbf{G}\|^2 = (\mathbf{F} + \mathbf{G}) \cdot (\mathbf{F} + \mathbf{G}) + (\mathbf{F} - \mathbf{G}) \cdot (\mathbf{F} - \mathbf{G})$
$= \mathbf{F} \cdot \mathbf{F} + 2\mathbf{F} \cdot \mathbf{G} + \mathbf{G} \cdot \mathbf{G} + \mathbf{F} \cdot \mathbf{F} - 2\mathbf{F} \cdot \mathbf{G} + \mathbf{G} \cdot \mathbf{G}$
$= 2\mathbf{F} \cdot \mathbf{F} + 2\mathbf{G} \cdot \mathbf{G} = 2\left(\|\mathbf{F}\|^2 + \|\mathbf{G}\|^2\right)$

21. Using part of the calculation from Problem 19, $\|\mathbf{F} + \mathbf{G}\|^2 = \|\mathbf{F}\|^2 + 2\mathbf{F} \cdot \mathbf{G} + \|\mathbf{G}\|^2$. If also $\|\mathbf{F} + \mathbf{G}\|^2 = \|\mathbf{F}\|^2 + \|\mathbf{G}\|^2$, then $\mathbf{F} \cdot \mathbf{G} = 0$ and the vectors are orthogonal.

23. Again using part of the calculation from Problem 19, $\|\mathbf{F} + \mathbf{G}\|^2 = \|\mathbf{F}\|^2 + 2\mathbf{F} \cdot \mathbf{G} + \|\mathbf{G}\|^2 \leq \|\mathbf{F}\|^2 + 2\,|\mathbf{F} \cdot \mathbf{G}| + \|\mathbf{G}\|^2$ and, by using the Cauchy–Schwarz inequality, this gives $\|\mathbf{F} + \mathbf{G}\|^2 \leq \|\mathbf{F}\|^2 + 2\,\|\mathbf{F}\|\,\|\mathbf{G}\| + \|\mathbf{G}\|^2 = (\|\mathbf{F}\| + \|\mathbf{G}\|)^2$. Since all quantities involved are nonnegative, the conclusion follows by taking square roots.

25. Define $\mathbf{F} = (x_1, x_2, \ldots, x_n)$ and $\mathbf{G} = (y_1, y_2, \ldots, y_n)$. These are vectors in R^n. By the Cauchy–Schwarz inequality

applied to these vectors, $|\mathbf{F} \cdot \mathbf{G}| = \left| \sum_{j=1}^{n} x_j y_j \right| \leq \|\mathbf{F}\| \, \|\mathbf{G}\| = \left(\sum_{j=1}^{n} x_j^2 \right)^{1/2} \left(\sum_{j=1}^{n} y_j^2 \right)^{1/2}$.

Section 5.5

1. Independent   **3.** Independent   **5.** Dependent   **7.** Dependent   **9.** Independent   **13.** Dependent; $[\mathbf{F}, \mathbf{G}, \mathbf{H}] = 0$

15. Independent; $[\mathbf{F}, \mathbf{G}, \mathbf{H}] = -44 \neq 0$ **17.** A basis consists of $(1, 0, 0, -1)$ and $(0, 1, -1, 0)$; the dimension is 2.

19. A basis consists of $(1, 0, -2)$ and $(0, 1, 1)$; the dimension is 2.

21. A basis consists of $(1, 0, 0, 0)$, $(0, 0, 1, 0)$ and $(0, 0, 0, 1)$; the dimension is 3.

23. A basis consists of $(1, 4)$; the dimension is 1.   **25.** The dimension is 1 in both R^2 and R^3.

29. Let $\mathbf{F}_1, \ldots, \mathbf{F}_{k-1}$ be any $k - 1$ of the original vectors, renumbered if necessary, and suppose
$\alpha_1 \mathbf{F}_1 + \cdots + \alpha_{k-1}\mathbf{F}_{k-1} = \mathbf{O}$. Then $\alpha_1 \mathbf{F}_1 + \cdots + \alpha_{k-1}\mathbf{F}_{k-1} + 0\mathbf{F}_k = \mathbf{O}$. But $\mathbf{F}_1, \ldots, \mathbf{F}_k$ are linearly independent, so
$\alpha_1 = \cdots = \alpha_{k-1} = 0$. Therefore, $\mathbf{F}_1, \ldots, \mathbf{F}_{k-1}$ are linearly independent.

31. For any $\mathbf{V}$ in R^3, $\mathbf{F}, \mathbf{G}, \mathbf{H}$, and $\mathbf{V}$ are linearly dependent (see Problem 28). Thus there are numbers, not all zero, such
that $\alpha\mathbf{F} + \beta\mathbf{G} + \gamma\mathbf{H} + \delta\mathbf{V} = \mathbf{O}$. If $\delta = 0$, then this equation would imply that $\mathbf{F}, \mathbf{G}$, and $\mathbf{H}$ are linearly dependent, a
contradiction. Thus $\delta \neq 0$. Write $\mathbf{V} = -\dfrac{\alpha}{\delta}\mathbf{F} - \dfrac{\beta}{\delta}\mathbf{G} - \dfrac{\gamma}{\delta}\mathbf{H} = a\mathbf{F} + b\mathbf{G} + c\mathbf{H}$. To solve for these coefficients, we have

$[\mathbf{V}, \mathbf{G}, \mathbf{H}] = [a\mathbf{F} + b\mathbf{G} + c\mathbf{H}, \mathbf{G}, \mathbf{H}] = a[\mathbf{F}, \mathbf{G}, \mathbf{H}] + b[\mathbf{G}, \mathbf{G}, \mathbf{H}] + c[\mathbf{H}, \mathbf{G}, \mathbf{H}] = a[\mathbf{F}, \mathbf{G}, \mathbf{H}]$, so $a = \dfrac{[\mathbf{V}, \mathbf{G}, \mathbf{H}]}{[\mathbf{F}, \mathbf{G}, \mathbf{H}]}$.

Similarly, $b = \dfrac{[\mathbf{F}, \mathbf{V}, \mathbf{H}]}{[\mathbf{F}, \mathbf{G}, \mathbf{H}]]}$ and $c = \dfrac{[\mathbf{F}, \mathbf{G}, \mathbf{V}]}{[\mathbf{F}, \mathbf{G}, \mathbf{H}]}$.

Section 5.6

1. P_m satisfies the conditions for a vector space. The polynomials $1, x, \ldots, x^m$ form a basis, so the dimension of P_m is
$m + 1$.

3. If f and g are in V, then $(f + g)(b) = f(b) + g(b) = 1 + 1 = 2$, so $f + g$ is not in V. Thus V is not a vector space.

5. The sum of two polynomials of degree n may be of degree less than n, so D_n is not a vector space. Other conditions
also fail. For example, the zero polynomial is not in D_n.

7. A sum of solutions of this equation is not a solution.

9. Conditions (2), (3), (4), and (10) of the definition fail to hold, so this is not a vector space.

11. Y is a vector space.   **13.** Q is not a vector space because condition (6) does not hold if α is an irrational number.

15. P is not a vector space. Conditions (1), (4), and (6) fail to hold.

17. P is not a vector space. Conditions (7), (9), and (10) fail to hold.

CHAPTER 6

Section 6.1

1. $\begin{pmatrix} 14 & -2 & 6 \\ 10 & -5 & -6 \\ -26 & -43 & -8 \end{pmatrix}$ **3.** $\begin{pmatrix} 2 + 2x - x^2 & -12x + (1 - x)(x + e^x + 2\cos(x)) \\ 4 + 2x + 2e^x + 2xe^x & -22 - 2x + e^{2x} + 2e^x \cos(x) \end{pmatrix}$

5. $\begin{pmatrix} -36 & 0 & 68 & 196 & 20 \\ 128 & -40 & -36 & -8 & 72 \end{pmatrix}$ **7.** $\mathbf{AB} = \begin{pmatrix} -10 & -34 & -16 & -30 & -14 \\ 10 & -2 & -11 & -8 & -45 \\ -5 & 1 & 15 & 61 & -63 \end{pmatrix}$; $\mathbf{BA}$ is not defined.

9. $\mathbf{AB} = (115)$; $\mathbf{BA} = \begin{pmatrix} 3 & -18 & -6 & -42 & 66 \\ -2 & 12 & 4 & 28 & -44 \\ -6 & 36 & 12 & 84 & -132 \\ 0 & 0 & 0 & 0 & 0 \\ 4 & -24 & -8 & -56 & 88 \end{pmatrix}$.   **11.** $\mathbf{AB}$ is not defined; $\mathbf{BA} = \begin{pmatrix} 410 & 36 & -56 & 227 \\ 17 & 253 & 40 & -1 \end{pmatrix}$.

13. **AB** is not defined; **BA** $= (-16 \quad -13 \quad -5)$.   **15.** $\mathbf{AB} = \begin{pmatrix} 39 & -84 & 21 \\ -23 & 38 & 3 \end{pmatrix}$; **BA** is not defined.

17. $\mathbf{AB} = \begin{pmatrix} 14 & -26 & 42 \\ 4 & 46 & -28 \\ 6 & 10 & 2 \end{pmatrix}$; $\mathbf{BA} = \begin{pmatrix} 16 & -16 & 32 \\ 3 & 42 & -28 \\ 7 & 6 & 4 \end{pmatrix}$

19. $\mathbf{AB} = \begin{pmatrix} -1 & 5 & -10 & 9 & -22 & 14 \\ -2 & 2 & 28 & 18 & 28 & -4 \end{pmatrix}$; **BA** is not defined.

21. **AB** is 14×14; **BA** is 21×21.   **23.** **AB** is not defined; **BA** is 4×2.   **25.** **AB** is not defined; **BA** is 7×6.

27. Let **A** be an $r \times s$ matrix. Then $\mathbf{A}^t$ is $s \times r$, so $\mathbf{A} = \mathbf{A}^t$ implies that $r = s$.

29. (a) $\begin{pmatrix} 1 & 2 \\ 2 & 3 \end{pmatrix} \begin{pmatrix} -1 & 1 \\ 1 & 2 \end{pmatrix} = \begin{pmatrix} 1 & 5 \\ 1 & 8 \end{pmatrix}$

(b) Let **A** and **B** be symmetric and first suppose **AB** is symmetric. Then $\mathbf{AB} = (\mathbf{AB})^t = \mathbf{B}^t \mathbf{A}^t = \mathbf{BA}$. Conversely, suppose $\mathbf{AB} = \mathbf{BA}$. Then $(\mathbf{AB})^t = (\mathbf{BA})^t = \mathbf{A}^t \mathbf{B}^t = \mathbf{AB}$, hence **AB** is symmetric.

31. $\mathbf{A}^3 = \begin{pmatrix} 2 & 7 & 7 & 4 & 4 \\ 7 & 8 & 9 & 9 & 9 \\ 7 & 9 & 8 & 9 & 9 \\ 4 & 9 & 9 & 6 & 7 \\ 4 & 9 & 9 & 7 & 6 \end{pmatrix}$ and $\mathbf{A}^4 = \begin{pmatrix} 14 & 17 & 17 & 18 & 18 \\ 17 & 34 & 33 & 26 & 26 \\ 17 & 33 & 34 & 26 & 26 \\ 18 & 26 & 26 & 25 & 24 \\ 18 & 26 & 26 & 24 & 25 \end{pmatrix}$.

The number of distinct $v_1 - v_4$ walks of length 3 is $(\mathbf{A}^3)_{14}$, or 4; the number of $v_1 - v_4$ walks of length 4 is $(\mathbf{A}^4)_{14} = 18$. The number of distinct $v_2 - v_3$ walks of length 3 is $(\mathbf{A}^3)_{23} = 9$; the number of distinct $v_2 - v_4$ walks of length 4 is 26.

33. $\mathbf{A}^2 = \begin{pmatrix} 4 & 2 & 3 & 3 & 2 \\ 2 & 3 & 2 & 2 & 3 \\ 3 & 2 & 4 & 3 & 2 \\ 3 & 2 & 3 & 4 & 2 \\ 2 & 3 & 2 & 2 & 3 \end{pmatrix}$, $\mathbf{A}^3 = \begin{pmatrix} 10 & 10 & 11 & 11 & 10 \\ 10 & 6 & 10 & 10 & 6 \\ 11 & 10 & 10 & 11 & 10 \\ 11 & 10 & 11 & 10 & 10 \\ 10 & 6 & 10 & 10 & 6 \end{pmatrix}$, and $\mathbf{A}^4 = \begin{pmatrix} 42 & 32 & 41 & 41 & 32 \\ 32 & 30 & 32 & 32 & 30 \\ 41 & 32 & 42 & 41 & 32 \\ 41 & 32 & 41 & 42 & 32 \\ 32 & 30 & 32 & 32 & 30 \end{pmatrix}$.

The number of distinct $v_4 - v_5$ walks of length 2 is 2, the number of $v_2 - v_3$ walks of length 3 is 10, the number of $v_1 - v_2$ walks of length 4 is 32, and the number of $v_4 - v_5$ walks of length 4 is 32.

Section 6.2

In Problems 1 through 11, the first matrix is $\Omega\mathbf{A}$, the second is Ω.

1. $\begin{pmatrix} -2 & 1 & 4 & 2 \\ 0 & \sqrt{3} & 16\sqrt{3} & 3\sqrt{3} \\ 1 & -2 & 4 & 8 \end{pmatrix}$, $\begin{pmatrix} 1 & 0 & 0 \\ 0 & \sqrt{3} & 0 \\ 0 & 0 & 1 \end{pmatrix}$ **3.** $\begin{pmatrix} 40 & 5 & -15 \\ -2+2\sqrt{13} & 14+9\sqrt{13} & 6+5\sqrt{13} \\ 2 & 9 & 5 \end{pmatrix}$, $\begin{pmatrix} 0 & 5 & 0 \\ 1 & 0 & \sqrt{13} \\ 0 & 0 & 1 \end{pmatrix}$

5. $\begin{pmatrix} 30 & 120 \\ -3+2\sqrt{3} & 15+8\sqrt{3} \end{pmatrix}$, $\begin{pmatrix} 0 & 15 \\ 1 & \sqrt{3} \end{pmatrix}$ **7.** $\begin{pmatrix} -1 & 0 & 3 & 0 \\ -36 & 28 & -20 & 28 \\ -13 & 3 & 44 & 9 \end{pmatrix}$; $\begin{pmatrix} 1 & 0 & 0 \\ 0 & 0 & 4 \\ 14 & 1 & 0 \end{pmatrix}$

9. $\begin{pmatrix} -3 & 7 & 1 & 1 \\ -10 & -5 & 25 & -15 \\ -6 & 17 & 5 & -3 \end{pmatrix}$, $\begin{pmatrix} 1 & 0 & 0 \\ 0 & 0 & -5 \\ 2 & 1 & 0 \end{pmatrix}$ **11.** $\begin{pmatrix} -3 & 15 & 0 \\ 10 & -15 & 5 \\ 1 & -5 & 0 \end{pmatrix}$, $\begin{pmatrix} 0 & 1 & -3 \\ 5 & 0 & 0 \\ 0 & 0 & 1 \end{pmatrix}$

Section 6.3

In Problems 1 through 19, the reduced matrix $\mathbf{A}_R$ is given first, then a matrix Ω such that $\Omega\mathbf{A} = \mathbf{A}_R$.

1. $\begin{pmatrix} 1 & 0 & 5 \\ 0 & 1 & 2 \\ 0 & 0 & 0 \end{pmatrix}$, $\begin{pmatrix} 1 & 1 & 0 \\ 0 & 1 & 0 \\ 0 & 0 & 1 \end{pmatrix}$ **3.** $\begin{pmatrix} 1 & -4 & -1 & 0 \\ 0 & 0 & 0 & 1 \\ 0 & 0 & 0 & 0 \\ 0 & 0 & 0 & 0 \end{pmatrix}$, $\begin{pmatrix} -1 & 0 & 0 & 1 \\ 0 & 0 & 0 & 1 \\ 0 & 0 & 1 & 0 \\ 0 & 1 & 0 & 0 \end{pmatrix}$ **5.** $\begin{pmatrix} 1 & 0 \\ 0 & 1 \\ 0 & 0 \\ 0 & 0 \end{pmatrix}$, $\begin{pmatrix} 0 & 0 & 1 & -3 \\ 0 & 0 & 0 & 1 \\ 1 & 0 & -6 & 17 \\ 0 & 1 & 0 & 0 \end{pmatrix}$

7. $\begin{pmatrix} 1 & 0 & 0 \\ 0 & 1 & 0 \\ 0 & 0 & 1 \end{pmatrix}, \dfrac{1}{270}\begin{pmatrix} -8 & -2 & 38 \\ 37 & 43 & -7 \\ 19 & -29 & 11 \end{pmatrix}$ **9.** $\begin{pmatrix} 1 & 0 & 0 & 0 \\ 0 & 1 & \frac{3}{2} & \frac{1}{2} \end{pmatrix}, \begin{pmatrix} 0 & 1 \\ \frac{1}{2} & \frac{1}{2} \end{pmatrix}$ **11.** $\begin{pmatrix} 1 & 0 & 0 \\ 0 & 1 & 0 \\ 0 & 0 & 1 \end{pmatrix}, \begin{pmatrix} 0 & \frac{1}{2} & -1 \\ 0 & 0 & 1 \\ -\frac{1}{7} & \frac{2}{7} & -\frac{3}{7} \end{pmatrix}$

13. $\begin{pmatrix} 1 & 0 & 0 & -\frac{6}{5} & \frac{2}{5} \\ 0 & 1 & 0 & \frac{21}{10} & \frac{14}{5} \\ 0 & 0 & 1 & -\frac{3}{10} & -\frac{7}{5} \end{pmatrix}, \dfrac{1}{20}\begin{pmatrix} -16 & 12 & -4 \\ 53 & -31 & 17 \\ -9 & 3 & -1 \end{pmatrix}$ **15.** $(1 \quad 0 \quad -\frac{1}{4} \quad -\frac{1}{4} \quad \frac{5}{4}), (-\frac{1}{4})$

17. $\begin{pmatrix} 1 & 0 & 0 \\ 0 & 1 & 0 \\ 0 & 0 & 1 \\ 0 & 0 & 0 \end{pmatrix}, \dfrac{1}{342}\begin{pmatrix} 0 & -38 & 0 & 38 \\ 0 & -59 & 36 & -4 \\ 0 & -3 & 54 & -6 \\ 342 & 243 & -270 & 144 \end{pmatrix}$ **19.** $\begin{pmatrix} 1 & 0 & 0 \\ 0 & 1 & 0 \\ 0 & 0 & 1 \end{pmatrix}, \dfrac{1}{178}\begin{pmatrix} 41 & 37 & 9 \\ 18 & 64 & 30 \\ 3 & -19 & 5 \end{pmatrix}$

Section 6.4

In Problems 1 through 19, first the reduced matrix is given, then its rank, a basis for its row space (as row vectors), the dimension of the row space, a basis for the column space (as column vectors) and the dimension of the column space.

1. $\begin{pmatrix} 1 & 0 & -\frac{3}{5} \\ 0 & 1 & \frac{3}{5} \end{pmatrix}$; 2; $(-4, 1, 3), (2, 2, 0)$; 2; $\begin{pmatrix} -4 \\ 2 \end{pmatrix}, \begin{pmatrix} 1 \\ 2 \end{pmatrix}$; 2   **3.** $\begin{pmatrix} 1 & 0 \\ 0 & 1 \\ 0 & 0 \end{pmatrix}$; 2; $(-3, 1), (2, 2)$; 2; $\begin{pmatrix} -3 \\ 2 \\ 4 \end{pmatrix}, \begin{pmatrix} 1 \\ 2 \\ -3 \end{pmatrix}$; 2

5. $\begin{pmatrix} 1 & 0 & -\frac{1}{4} & \frac{1}{2} \\ 0 & 1 & -\frac{5}{4} & \frac{1}{2} \end{pmatrix}$; 2; $(8, -4, 3, 2), (1, -1, 1, 0)$; 2; $\begin{pmatrix} 8 \\ 1 \end{pmatrix}, \begin{pmatrix} -4 \\ -1 \end{pmatrix}$; 2

7. $\begin{pmatrix} 1 & 0 & 0 \\ 0 & 1 & 0 \\ 0 & 0 & 1 \\ 0 & 0 & 0 \end{pmatrix}$; 3; $(2, 2, 1), (1, -1, 3), (0, 0, 1)$; 3; $\begin{pmatrix} 2 \\ 1 \\ 0 \\ 4 \end{pmatrix}, \begin{pmatrix} 2 \\ -1 \\ 0 \\ 0 \end{pmatrix}, \begin{pmatrix} 1 \\ 3 \\ 1 \\ 7 \end{pmatrix}$; 3

9. $\begin{pmatrix} 1 & 0 & 0 \\ 0 & 1 & 0 \\ 0 & 0 & 1 \end{pmatrix}$; 3; $(0, 4, 3), (6, 1, 0), (2, 2, 2)$; 3; $\begin{pmatrix} 0 \\ 6 \\ 2 \end{pmatrix}, \begin{pmatrix} 4 \\ 1 \\ 2 \end{pmatrix}, \begin{pmatrix} 3 \\ 0 \\ 2 \end{pmatrix}$; 3

11. $\begin{pmatrix} 1 & 0 & 0 \\ 0 & 1 & 0 \\ 0 & 0 & 1 \end{pmatrix}$; 3; $(-3, 2, 2), (1, 0, 5), (0, 0, 2)$; 3; $\begin{pmatrix} -3 \\ 1 \\ 0 \end{pmatrix}, \begin{pmatrix} 2 \\ 0 \\ 0 \end{pmatrix}, \begin{pmatrix} 2 \\ 5 \\ 2 \end{pmatrix}$; 3

13. $\begin{pmatrix} 1 & 0 & -11 \\ 0 & 1 & -3 \\ 0 & 0 & 0 \end{pmatrix}$; 2; $(-2, 5, 7), (0, 1, -3)$; 2; $\begin{pmatrix} -2 \\ 0 \\ -4 \end{pmatrix}, \begin{pmatrix} 5 \\ 1 \\ 11 \end{pmatrix}$; 2

15. $\begin{pmatrix} 1 & 0 & 1 & \frac{1}{7} \\ 0 & 1 & 3 & \frac{3}{2} \\ 0 & 0 & 0 & 0 \\ 0 & 0 & 0 & 0 \end{pmatrix}$; 2; $(7, -2, 1, -2), (0, 2, 6, 3)$; 2; $\begin{pmatrix} 7 \\ 0 \\ 7 \\ 7 \end{pmatrix}, \begin{pmatrix} -2 \\ 2 \\ 2 \\ 0 \end{pmatrix}$; 2

17. $\begin{pmatrix} 1 & 0 & 0 & -1 \\ 0 & 1 & 0 & 6 \\ 0 & 0 & 1 & -1 \end{pmatrix}$; 3; $(4, 1, -3, 5), (2, 0, 0, -2), (13, 2, 0, -1)$; 3; $\begin{pmatrix} 4 \\ 2 \\ 13 \end{pmatrix}, \begin{pmatrix} 1 \\ 0 \\ 2 \end{pmatrix}, \begin{pmatrix} -3 \\ 0 \\ 0 \end{pmatrix}$; 3

19. $\begin{pmatrix} 1 & 0 & -\frac{1}{2} & \frac{1}{2} & -\frac{3}{2} \\ 0 & 1 & -\frac{15}{4} & -\frac{7}{4} & -\frac{17}{4} \\ 0 & 0 & 0 & 0 & 0 \end{pmatrix}$; 2; $(5, -2, 5, 6, 1), (-2, 0, 1, -1, 3)$; 2; $\begin{pmatrix} 5 \\ -2 \\ -1 \end{pmatrix}, \begin{pmatrix} -2 \\ 0 \\ -2 \end{pmatrix}$; 2

Section 6.5

1. $\alpha \begin{pmatrix} -1 \\ 1 \\ 1 \\ 0 \end{pmatrix} + \beta \begin{pmatrix} 1 \\ -1 \\ 0 \\ 1 \end{pmatrix}$ **3.** $\alpha \begin{pmatrix} 0 \\ 0 \\ 0 \end{pmatrix}$ (only the trivial solution) **5.** $\alpha \begin{pmatrix} -\frac{9}{4} \\ -\frac{7}{4} \\ -\frac{5}{8} \\ \frac{13}{8} \\ 1 \end{pmatrix}$ **7.** $\alpha \begin{pmatrix} -\frac{5}{6} \\ -\frac{2}{3} \\ -\frac{8}{3} \\ -\frac{2}{3} \\ 1 \\ 0 \end{pmatrix} + \beta \begin{pmatrix} -\frac{5}{9} \\ -\frac{10}{9} \\ -\frac{13}{9} \\ -\frac{1}{9} \\ 0 \\ 1 \end{pmatrix}$

9. $\alpha \begin{pmatrix} \frac{5}{14} \\ \frac{11}{7} \\ \frac{6}{7} \\ 1 \end{pmatrix}$ **11.** $\alpha \begin{pmatrix} 1 \\ 1 \\ 0 \\ 1 \\ 1 \\ 0 \\ 0 \end{pmatrix} + \beta \begin{pmatrix} -2 \\ -\frac{3}{2} \\ \frac{2}{3} \\ -\frac{4}{3} \\ 0 \\ 1 \\ 0 \end{pmatrix} + \gamma \begin{pmatrix} 0 \\ \frac{1}{2} \\ -3 \\ 0 \\ 0 \\ 0 \\ 1 \end{pmatrix}$ **13.** $\alpha \begin{pmatrix} 20 \\ 5 \\ \frac{1}{2} \\ 1 \\ 0 \end{pmatrix} + \beta \begin{pmatrix} -25 \\ -6 \\ 0 \\ 0 \\ 1 \end{pmatrix}$ **15.** $\alpha \begin{pmatrix} \frac{2}{5} \\ 5 \\ 1 \\ 0 \\ 0 \end{pmatrix} + \beta \begin{pmatrix} \frac{1}{5} \\ 1 \\ 0 \\ 1 \\ 0 \end{pmatrix} + \gamma \begin{pmatrix} -\frac{3}{5} \\ -7 \\ 0 \\ 0 \\ 1 \end{pmatrix}$

17. $\alpha \begin{pmatrix} \frac{23}{9} \\ \frac{13}{9} \\ -8 \\ 1 \end{pmatrix}$ **19.** $\alpha \begin{pmatrix} \frac{21}{13} \\ -1 \\ \frac{23}{13} \\ 1 \\ 0 \\ 0 \end{pmatrix} + \beta \begin{pmatrix} \frac{6}{13} \\ -6 \\ -\frac{12}{13} \\ 0 \\ 1 \\ 0 \end{pmatrix} + \gamma \begin{pmatrix} -\frac{44}{13} \\ 1 \\ -\frac{55}{13} \\ 0 \\ 0 \\ 1 \end{pmatrix}$

Section 6.6

1. 2 **3.** 0 **5.** 1 **7.** 2 **9.** 1 **11.** 3 **13.** 2 **15.** 3 **17.** 1 **19.** 3
21. Yes, provided that $rank(\mathbf{A}) <$ number of unknowns in the system.

Section 6.7

In each of Problems 1 through 19, the unique or general solution is given, whichever applies, or it is noted that the system has no solution.

1. $\begin{pmatrix} 1 \\ \frac{1}{2} \\ 4 \end{pmatrix}$ **3.** $\alpha \begin{pmatrix} 1 \\ 1 \\ \frac{3}{2} \\ 1 \\ 0 \\ 0 \end{pmatrix} + \beta \begin{pmatrix} 0 \\ 0 \\ \frac{1}{2} \\ 0 \\ 1 \\ 0 \end{pmatrix} + \gamma \begin{pmatrix} -\frac{17}{2} \\ -6 \\ -\frac{51}{4} \\ 0 \\ 0 \\ 1 \end{pmatrix} + \begin{pmatrix} \frac{9}{2} \\ 3 \\ \frac{25}{4} \\ 0 \\ 0 \\ 0 \end{pmatrix}$ **5.** $\alpha \begin{pmatrix} 2 \\ 2 \\ 7 \\ \frac{3}{2} \\ 1 \\ 0 \end{pmatrix} + \beta \begin{pmatrix} -2 \\ -1 \\ -\frac{9}{2} \\ -\frac{3}{4} \\ 0 \\ 1 \end{pmatrix} + \begin{pmatrix} -4 \\ -4 \\ -38 \\ -\frac{11}{2} \\ 0 \\ 0 \end{pmatrix}$

7. $\alpha \begin{pmatrix} -\frac{1}{2} \\ -1 \\ 3 \\ 1 \\ 0 \end{pmatrix} + \beta \begin{pmatrix} -\frac{3}{4} \\ 1 \\ -2 \\ 0 \\ 1 \end{pmatrix} + \begin{pmatrix} \frac{9}{8} \\ 2 \\ 0 \\ 0 \\ 0 \end{pmatrix}$ **9.** $\alpha \begin{pmatrix} -1 \\ 1 \\ 0 \\ 0 \\ 0 \\ 0 \end{pmatrix} + \beta \begin{pmatrix} 1 \\ 0 \\ 0 \\ 1 \\ 0 \\ 0 \end{pmatrix} + \gamma \begin{pmatrix} -\frac{3}{14} \\ 0 \\ \frac{3}{14} \\ 0 \\ 1 \\ 0 \end{pmatrix} + \delta \begin{pmatrix} -1 \\ 0 \\ 0 \\ 0 \\ 0 \\ 1 \end{pmatrix} + \epsilon \begin{pmatrix} \frac{1}{14} \\ 0 \\ -\frac{1}{14} \\ 0 \\ 0 \\ 0 \end{pmatrix} + \begin{pmatrix} -\frac{29}{7} \\ 0 \\ \frac{1}{7} \\ 0 \\ 0 \\ 0 \end{pmatrix}$

11. $\alpha \begin{pmatrix} -\frac{19}{15} \\ 3 \\ \frac{67}{15} \\ 1 \end{pmatrix} + \begin{pmatrix} \frac{22}{15} \\ -5 \\ -\frac{121}{15} \\ 0 \end{pmatrix}$ **13.** $\begin{pmatrix} \frac{16}{57} \\ \frac{99}{57} \\ \frac{23}{57} \end{pmatrix}$ **15.** No solution   **17.** $\alpha \begin{pmatrix} -\frac{17}{45} \\ \frac{4}{45} \\ \frac{97}{45} \\ 1 \end{pmatrix} + \begin{pmatrix} \frac{109}{90} \\ \frac{67}{90} \\ -\frac{344}{90} \\ 0 \end{pmatrix}$ **19.** $\alpha \begin{pmatrix} \frac{2}{29} \\ \frac{7}{29} \\ \frac{18}{29} \\ 1 \end{pmatrix} + \begin{pmatrix} \frac{67}{29} \\ -\frac{7}{87} \\ \frac{110}{29} \\ 0 \end{pmatrix}$

21. $\mathbf{AX} = \mathbf{B}$ has a solution if and only if $rank(\mathbf{A}) = rank([\mathbf{A} \vdots \mathbf{B}])$. Since $\mathbf{A}_R$ is a reduced matrix, this occurs only if the last $n - r$ rows of $\mathbf{B}$ are zero rows, hence $b_{r+1} = b_{r+2} = \cdots = b_n = 0$.

Section 6.8

1. $\alpha \begin{pmatrix} 4 \\ -13 \\ -5 \end{pmatrix}$ **3.** $\alpha \begin{pmatrix} -3 \\ -5 \\ 5 \end{pmatrix} + \begin{pmatrix} 1 \\ 4 \\ 0 \end{pmatrix}$ **5.** $\alpha \begin{pmatrix} 5 \\ -4 \\ 3 \\ 2 \end{pmatrix} + \begin{pmatrix} -3 \\ \frac{7}{2} \\ 0 \\ 0 \end{pmatrix}$ **7.** No solution   **9.** $\alpha \begin{pmatrix} -1 \\ 1 \\ 0 \end{pmatrix} + \begin{pmatrix} \frac{65}{23} \\ 0 \\ -\frac{10}{23} \end{pmatrix}$

11. $\begin{pmatrix} \frac{73}{101} \\ \frac{49}{101} \\ \frac{22}{101} \end{pmatrix}$ **13.** $\alpha \begin{pmatrix} -203 \\ 344 \\ 85 \\ 48 \end{pmatrix} + \begin{pmatrix} \frac{5}{8} \\ 0 \\ -\frac{3}{8} \\ 0 \end{pmatrix}$ **15.** $\begin{pmatrix} \frac{7}{5} \\ \frac{13}{10} \\ \frac{1}{10} \end{pmatrix}$

Section 6.9

1. $\frac{1}{5} \begin{pmatrix} -1 & 2 \\ 2 & 1 \end{pmatrix}$ **3.** $\frac{1}{12} \begin{pmatrix} -2 & 2 \\ 1 & 5 \end{pmatrix}$ **5.** $\frac{1}{12} \begin{pmatrix} 3 & -2 \\ -3 & 6 \end{pmatrix}$ **7.** $\frac{1}{31} \begin{pmatrix} -6 & 11 & 2 \\ 3 & 10 & -1 \\ 1 & -7 & 10 \end{pmatrix}$ **9.** $\frac{1}{12} \begin{pmatrix} -6 & 6 & 0 \\ 3 & 9 & -2 \\ -3 & 3 & 2 \end{pmatrix}$

11. $\frac{1}{14} \begin{pmatrix} 48 & 2 & 12 \\ -4 & 1 & -1 \\ -14 & 0 & 0 \end{pmatrix}$ **13.** $\frac{1}{2} \begin{pmatrix} -2 & 85 & -55 & 2 \\ 0 & -9 & 7 & 2 \\ 0 & 6 & -4 & 0 \\ 0 & -1 & 1 & 0 \end{pmatrix}$ **15.** $\mathbf{A}$ is singular because $\mathbf{A}_R$ has a zero row.

17. $\frac{1}{11} \begin{pmatrix} -23 \\ -75 \\ -9 \\ 14 \end{pmatrix}$ **19.** $\frac{1}{7} \begin{pmatrix} 22 \\ 27 \\ 30 \end{pmatrix}$ **21.** $\frac{1}{5} \begin{pmatrix} -21 \\ 14 \\ 0 \end{pmatrix}$

CHAPTER 7

Section 7.1

1. 1, 2, 3 (the identity permutation) is even, as are 2, 3, 1 and 3, 1, 2; the permutations 1, 3, 2 and 2, 1, 3 and 3, 2, 1 are odd.

Section 7.2

1. -3 **3.** -62

Section 7.3

1. $\det(\mathbf{B}) = \sum_p (-1)^{sgn(p)} b_{1p(1)} b_{2p(2)} \cdots b_{np(n)} = \sum_p (-1)^{sgn(p)} (\alpha a_{1p(1)})(\alpha a_{2p(2)}) \cdots (\alpha a_{np(n)})$

$= (\alpha)^n \sum_p (-1)^{sgn(p)} a_{1p(1)} a_{2p(2)} \cdots a_{np(n)} = \alpha^n \det(\mathbf{A})$

3. Using the result of Problem 1, $\det(\mathbf{A}) = (-\mathbf{A}^t) = (-1)^n \det(\mathbf{A}^t) = (-1)^n \det(\mathbf{A})$. If n is odd, then $(-1)^n = -1$, so $\det(\mathbf{A}) = -\det(\mathbf{A})$, hence $\det(\mathbf{A}) = 0$.

Section 7.4

1. -22 **3.** -14 **5.** -2247 **7.** -122 **9.** -72

Section 7.5

1. 32   **3.** 3   **5.** -773 **7.** -152 **9.** 1693   **11.** 21,024   **13.** -5987 **19.** (a) -1224 (c) 9900

21. Using the result of Problem 20, $1 = \det(\mathbf{I}_n) = \det(\mathbf{A}\mathbf{A}^{-1}) = \det(\mathbf{A})\det(\mathbf{A}^{-1}) = \det(\mathbf{A})\det(\mathbf{A}') = [\det(\mathbf{A})]^2$, hence $\det(\mathbf{A}) = \pm 1$.

Section 7.6

1. 2240   **3.** -1440

Section 7.7

1. $\dfrac{1}{13}\begin{pmatrix} 6 & 1 \\ -1 & 2 \end{pmatrix}$ **3.** $\dfrac{1}{5}\begin{pmatrix} -4 & 1 \\ 1 & 1 \end{pmatrix}$ **5.** $\dfrac{1}{32}\begin{pmatrix} 5 & 3 & 1 \\ -8 & -24 & 24 \\ -2 & -14 & 6 \end{pmatrix}$ **7.** $\dfrac{1}{29}\begin{pmatrix} -1 & 25 & -21 \\ -8 & -3 & 6 \\ -1 & -4 & 8 \end{pmatrix}$

9. $\dfrac{1}{378}\begin{pmatrix} 210 & -42 & 42 & 0 \\ 899 & -124 & 223 & -135 \\ 275 & -64 & 109 & -27 \\ -601 & 122 & -131 & 81 \end{pmatrix}$

Section 7.8

1. $x_1 = -11/47, x_2 = -100/47$ **3.** $x_1 = -1/2, x_2 = -19/22, x_3 = 2/11$ **5.** $x_1 = 5/6, x_2 = -10/3, x_3 = -5/6$

7. $x_1 = -86, x_2 = -109/2, x_3 = -43/2, x_4 = 37/2$ **9.** $x_1 = 11/31, x_2 = -409/93, x_3 = -1/93, x_4 = 116/93$

Section 7.9

1. 21   **3.** 61   **5.** 61

CHAPTER 8

Section 8.1

In the following, when an eigenvalue is listed twice in succession, this eigenvalue has multiplicity 2, but does not have two linearly independent eigenvectors (hence only one eigenvector is listed for such an eigenvalue). In Problems 1–19, MAPLE code for the Gerschgorin circles is included.

1. $1 + \sqrt{6}, \begin{pmatrix} \sqrt{6}/2 \\ 1 \end{pmatrix}; 1 - \sqrt{6}, \begin{pmatrix} -\sqrt{6}/2 \\ 1 \end{pmatrix}$

```
plot({[1+(3)*sin(t),3*cos(t), t=-Pi..Pi],[1+(2)*sin(t),2*cos(t), t=-Pi..Pi]},
  scaling=CONSTRAINED); # and plot POINTS at   (1,0)
```

3. $-5, \begin{pmatrix} 7 \\ -1 \end{pmatrix}; 2, \begin{pmatrix} 0 \\ 1 \end{pmatrix}$

```
plot({[2+ sin(t),cos(t), t=-Pi..Pi]},
  scaling=CONSTRAINED); # and plot POINTS at   (-5,0) and (2,0)
```

5. $\dfrac{1}{2}(3 + \sqrt{47}i), \begin{pmatrix} -1 + \sqrt{47}i \\ 4 \end{pmatrix}; \dfrac{1}{2}(3 - \sqrt{47}i), \begin{pmatrix} -1 - \sqrt{47}i \\ 4 \end{pmatrix}$

```
plot({[1+ (6)*sin(t),6*cos(t), t=Pi..Pi],[2+ (2)*sin(t),2*cos(t), t=-Pi..Pi]},
  scaling=CONSTRAINED); # and plot POINTS at   (1,0) and  (2,0)
```

7. $0, \begin{pmatrix} 0 \\ 1 \\ 0 \end{pmatrix}; 2, \begin{pmatrix} 2 \\ 1 \\ 0 \end{pmatrix}; 3, \begin{pmatrix} 0 \\ 2 \\ 3 \end{pmatrix}$

```
plot({[3*sin(t),3*cos(t), t=-Pi..Pi]},
scaling=CONSTRAINED); # and plot POINTS at    (2,0), (0,0) and (3,0)
```

9. $0, 0, \begin{pmatrix} 1 \\ 0 \\ 3 \end{pmatrix}; -3, \begin{pmatrix} 1 \\ 0 \\ 0 \end{pmatrix}$

```
plot({[-3+(2)*sin(2),2*cos(t),t=-Pi..Pi],[sin(t),cos(t),t=-Pi..Pi]},
scaling=CONSTRAINED); # and plot
POINTS at    (-3,0) and   (0,0)
```

11. $2, 2, \begin{pmatrix} 0 \\ 0 \\ 1 \end{pmatrix}; -14, \begin{pmatrix} -16 \\ 0 \\ 1 \end{pmatrix}$

```
plot({[-14+sin(t),cos(t),t=-Pi..Pi],[2+sin(t),cos(t),t=-Pi..Pi]},
scaling=CONSTRAINED); # and plot
POINTS at    (-14,0) and   (2,0)
```

13. $0, \begin{pmatrix} 14 \\ 7 \\ 10 \end{pmatrix}; 1, \begin{pmatrix} 6 \\ 0 \\ 5 \end{pmatrix}; 7, \begin{pmatrix} 0 \\ 0 \\ 1 \end{pmatrix}$

```
plot({[1+(2)*sin(t),2*cos(t),t=-Pi..Pi],[7+(5)*sin(t),5*cos(t),t=-Pi..Pi]},
scaling=CONSTRAINED); # and
plot POINTS at    (1,0), (0,0) and   (7,0)
```

15. $1, \begin{pmatrix} -2 \\ -11 \\ 0 \\ 1 \end{pmatrix}; 2, \begin{pmatrix} 0 \\ 0 \\ 1 \\ 0 \end{pmatrix}; \frac{1}{2}(-1+\sqrt{53}), \begin{pmatrix} -7+\sqrt{53} \\ 0 \\ 0 \\ 2 \end{pmatrix}; \frac{1}{2}(-1-\sqrt{53}), \begin{pmatrix} -7-\sqrt{53} \\ 0 \\ 0 \\ 2 \end{pmatrix}$

```
plot({[-4+(2)*sin(t),2*cos(t),t=-Pi..Pi],[3+sin(t),cos(t),t=-Pi..Pi]},
scaling=CONSTRAINED); # and plot
POINTS at    (-4,0), (1,0), (2,0) and   (3,0)
```

17. $-6, \begin{pmatrix} -3 \\ \frac{10}{7} \\ -\frac{24}{49} \\ 1 \end{pmatrix}; -3, \begin{pmatrix} 0 \\ -\frac{1}{2} \\ -\frac{3}{8} \\ 1 \end{pmatrix}; 1, 1, \begin{pmatrix} 0 \\ 0 \\ 1 \\ 0 \end{pmatrix}$

```
plot({[1+(6)*sin(t),6*cos(t), t=-Pi..Pi],[1+ (3)*sin(t),3*cos(t),
t=-Pi..Pi],[-3+ sin(t),cos(t), t=-Pi..Pi]},
scaling=CONSTRAINED); # and plot POINTS at    (-6,0), (1,0) and (-3,0)
```

19. $0, \begin{pmatrix} 1 \\ 0 \\ -5 \\ 0 \\ 0 \end{pmatrix}; 1, \begin{pmatrix} 0 \\ 1 \\ 0 \\ 0 \\ 0 \end{pmatrix}; 4, \begin{pmatrix} 1 \\ 0 \\ 1 \\ -2 \\ 0 \end{pmatrix}; 5, 5, \begin{pmatrix} 1 \\ 0 \\ 0 \\ 0 \\ 0 \end{pmatrix}$

```
plot({[5+4*sin(t),4*cos(t),t=-Pi..Pi],[2*sin(t),2*cos(t),t=-Pi..Pi]},
scaling=CONSTRAINED); # and plot POINTS at    (5,0), (1,0), (0,0) and   (4,0)
```

21. The characteristic polynomial is $|\lambda \mathbf{I}_2 - \mathbf{A}| = \lambda^2 - (\alpha + \gamma)\lambda + \alpha\gamma - \beta^2$. Now
$(\alpha + \gamma)^2 - 4(\alpha\gamma - \beta^2) = (\alpha - \gamma)^2 + 4\beta^2 \geq 0$, so the eigenvalues are real.

23. Suppose $\mathbf{AE} = \lambda \mathbf{E}$ and $\mathbf{E} \neq \mathbf{O}$. Then $\mathbf{A}^2\mathbf{E} = \mathbf{A}(\mathbf{AE}) = \mathbf{A}(\lambda \mathbf{E}) = \lambda(\mathbf{AE}) = \lambda^2\mathbf{E}$, so λ^2 is an eigenvalue of $\mathbf{A}^2$ with eigenvector $\mathbf{E}$. The same idea holds for $k > 2$.

25. Yes; $\begin{pmatrix} i & \sqrt{2} \\ \sqrt{2} & -i \end{pmatrix}$ has eigenvalues $1, -1$.

27. The constant term of $p_A(\lambda) = |\lambda I_n - A|$ is found by setting $\lambda = 0$, yielding $|-A|$, which equals $(-1)^n |A|$. 0 can be an eigenvalue of A if and only if this constant term is zero, which occurs exactly when $|A| = 0$, and this occurs exactly when A is singular.

Section 8.2

1. $\dfrac{1}{8} \begin{pmatrix} 3+\sqrt{7}i & 3-\sqrt{7}i \\ -8 & -8 \end{pmatrix}$ **3.** Not diagonalizable **5.** $\begin{pmatrix} 0 & 5 & 0 \\ 1 & 1 & -\frac{3}{2} \\ 0 & 0 & 1 \end{pmatrix}$ **7.** Not diagonalizable

9. $\begin{pmatrix} 1 & 0 & 0 & 0 \\ 0 & 1 & \frac{2-3\sqrt{5}}{41} & \frac{2+3\sqrt{5}}{41} \\ 0 & 0 & \frac{-1+\sqrt{5}}{2} & \frac{-1-\sqrt{5}}{2} \\ 0 & 0 & 1 & 1 \end{pmatrix}$ **11.** Not diagonalizable

13. *Hint:* A^2 has n linearly independent eigenvectors $X_1, \ldots, X_n$ with associated eigenvalues $\lambda_1, \ldots, \lambda_n$. Thus $(A^2 - \lambda_j I_n)X_j = (A - \sqrt{\lambda_j} I_n)(A + \sqrt{\lambda_j} I_n)X_j = O$. Now show that either $(A - \sqrt{\lambda_j} I_n)X_j = O$ or $(A + \sqrt{\lambda_j} I_n)X_j = O$.

15. $\begin{pmatrix} 1 & 0 \\ \frac{1}{4}(1 - 5^{18}) & 5^{18} \end{pmatrix}$ **17.** $\begin{pmatrix} 0 & 2^{22} \\ 2^{21} & 0 \end{pmatrix}$

Section 8.3

1. $0, \begin{pmatrix} 1 \\ 2 \end{pmatrix}; 5, \begin{pmatrix} -2 \\ 1 \end{pmatrix}; \dfrac{1}{\sqrt{5}} \begin{pmatrix} 1 & -2 \\ 2 & 1 \end{pmatrix}$ **3.** $5+\sqrt{2}, \begin{pmatrix} 1+\sqrt{2} \\ 1 \end{pmatrix}; 5-\sqrt{2}, \begin{pmatrix} 1-\sqrt{2} \\ 1 \end{pmatrix}; \begin{pmatrix} \frac{1+\sqrt{2}}{\sqrt{4+2\sqrt{2}}} & \frac{1-\sqrt{2}}{\sqrt{4-2\sqrt{2}}} \\ \frac{1}{\sqrt{4+2\sqrt{2}}} & \frac{1}{\sqrt{4-2\sqrt{2}}} \end{pmatrix}$

5. $3, \begin{pmatrix} 0 \\ 0 \\ 1 \end{pmatrix}; -1+\sqrt{2}, \begin{pmatrix} 1+\sqrt{2} \\ 1 \\ 0 \end{pmatrix}; -1-\sqrt{2}, \begin{pmatrix} 1-\sqrt{2} \\ 1 \\ 0 \end{pmatrix}; \begin{pmatrix} 0 & \frac{1+\sqrt{2}}{\sqrt{4+2\sqrt{2}}} & \frac{1-\sqrt{2}}{\sqrt{4-2\sqrt{2}}} \\ 0 & \frac{1}{\sqrt{4+2\sqrt{2}}} & \frac{1}{\sqrt{4-2\sqrt{2}}} \\ 1 & 0 & 0 \end{pmatrix}$

7. $0, \begin{pmatrix} 0 \\ 1 \\ 0 \end{pmatrix}; \dfrac{1}{2}(5+\sqrt{41}), \begin{pmatrix} 5+\sqrt{41} \\ 0 \\ 4 \end{pmatrix}; \dfrac{1}{2}(5-\sqrt{41}), \begin{pmatrix} 5-\sqrt{41} \\ 0 \\ 4 \end{pmatrix}; \begin{pmatrix} 0 & \frac{5+\sqrt{41}}{\sqrt{82+10\sqrt{41}}} & \frac{5-\sqrt{41}}{\sqrt{82-10\sqrt{41}}} \\ 1 & 0 & 0 \\ 0 & \frac{4}{\sqrt{82+10\sqrt{41}}} & \frac{4}{\sqrt{82-10\sqrt{41}}} \end{pmatrix}$

9. $0, \begin{pmatrix} 1 \\ 0 \\ 0 \end{pmatrix}; \dfrac{1}{2}(1+\sqrt{17}), \begin{pmatrix} 0 \\ -1-\sqrt{17} \\ 4 \end{pmatrix}; \dfrac{1}{2}(1-\sqrt{17}), \begin{pmatrix} 0 \\ -1+\sqrt{17} \\ 4 \end{pmatrix}; \begin{pmatrix} 1 & 0 & 0 \\ 0 & -\frac{1+\sqrt{17}}{\sqrt{34+2\sqrt{17}}} & -\frac{1-\sqrt{17}}{\sqrt{34-2\sqrt{17}}} \\ 0 & \frac{4}{\sqrt{34+2\sqrt{17}}} & \frac{4}{\sqrt{34+2\sqrt{17}}} \end{pmatrix}$

11. $0, \begin{pmatrix} 1 \\ 0 \\ 0 \\ 0 \end{pmatrix}; 0, \begin{pmatrix} 0 \\ 0 \\ 0 \\ 1 \end{pmatrix}; -1, \begin{pmatrix} 0 \\ 1 \\ 1 \\ 0 \end{pmatrix}; 3, \begin{pmatrix} 0 \\ -1 \\ 1 \\ 0 \end{pmatrix}; \begin{pmatrix} 1 & 0 & 0 & 0 \\ 0 & 0 & \frac{1}{\sqrt{2}} & -\frac{1}{\sqrt{2}} \\ 0 & 0 & \frac{1}{\sqrt{2}} & \frac{1}{\sqrt{2}} \\ 0 & 1 & 0 & 0 \end{pmatrix}$

Section 8.4

1. $\begin{pmatrix} 1 & 1 \\ 1 & 6 \end{pmatrix}$ **3.** $\begin{pmatrix} 1 & -2 \\ -2 & 1 \end{pmatrix}$ **5.** $\begin{pmatrix} -1 & 0 & -\frac{1}{2} & -1 \\ 0 & 0 & 2 & \frac{3}{2} \\ -\frac{1}{2} & 2 & 0 & 0 \\ -1 & \frac{3}{2} & 0 & 1 \end{pmatrix}$ **7.** $(-1+2\sqrt{5})y_1^2 + (-1-2\sqrt{5})y_2^2$

9. $(2+\sqrt{29})y_1^2 + (2-\sqrt{29})y_2^2$ **11.** $(2+\sqrt{13})y_1^2 + (2-\sqrt{13})y_2^2$ **13.** $-y_1^2 + y_2^2 + 2y_3^2$

15. $\frac{1}{2}\sqrt{61}\,y_1^2 - \frac{1}{2}\sqrt{61}\,y_2^2 = 5$; hyperbola   **17.** $5y_1^2 - 5y_2^2 = 8$; hyperbola   **19.** $-2x_1^2 + 2x_1x_2 + 6x_2^2$

21. $6x_1^2 + 2x_1x_2 - 14x_1x_3 + 2x_2^2 + x_3^2$

Section 8.5

1. None of these; 2, 2, $\begin{pmatrix} i \\ 1 \end{pmatrix}$; not diagonalizable

3. Skew-hermitian; 0, $\begin{pmatrix} 1 \\ 0 \\ \frac{1+i}{2} \end{pmatrix}$; $\sqrt{3}i$, $\begin{pmatrix} 1 \\ \sqrt{3}i \\ -1-i \end{pmatrix}$; $-\sqrt{3}i$, $\begin{pmatrix} 1 \\ -\sqrt{3}i \\ -1-i \end{pmatrix}$; $\begin{pmatrix} 1 & 1 & 1 \\ 0 & \sqrt{3}i & -\sqrt{3}i \\ \frac{1+i}{2} & -1-i & -1-i \end{pmatrix}$

5. Hermitian; eigenvalues satisfy $\lambda^3 - 3\lambda^2 - 5\lambda + 3 = 0$. Eigenvalues and corresponding eigenvectors are,

approximately, 4.051374, $\begin{pmatrix} 1 \\ 0.525687 \\ -0.129755i \end{pmatrix}$; 0.482696, $\begin{pmatrix} 1 \\ -1.258652 \\ 2.607546i \end{pmatrix}$; -1.53407, $\begin{pmatrix} 1 \\ -2.267035 \\ -1.477791i \end{pmatrix}$;

$\begin{pmatrix} 1 & 1 & 1 \\ 0.525687 & -1.258652 & -2.267035 \\ -0.129755i & 2.607546i & -1.477791i \end{pmatrix}$

7. Skew-hermitian; eigenvalues satisfy $\lambda^3 - i\lambda^2 + 5\lambda - 4i = 0$. Eigenvalues and eigenvectors are, approximately,

$-2.164248i$, $\begin{pmatrix} -i \\ -3.164248 \\ 2.924109 \end{pmatrix}$; $0.772866i$, $\begin{pmatrix} i \\ 0.227134 \\ 0.587772 \end{pmatrix}$; $2.391382i$, $\begin{pmatrix} i \\ -1.391382 \\ -1.163664 \end{pmatrix}$;

$\begin{pmatrix} -i & i & i \\ -3.164248 & 0.227134 & -1.391382 \\ 2.924109 & 0.587772 & -1.163664 \end{pmatrix}$

9. Hermitian; 0, $\begin{pmatrix} 0 \\ i \\ 1 \end{pmatrix}$; $4+3\sqrt{2}$, $\begin{pmatrix} 4+3\sqrt{2} \\ -1 \\ -i \end{pmatrix}$; $4-3\sqrt{2}$, $\begin{pmatrix} 4-3\sqrt{2} \\ -1 \\ -i \end{pmatrix}$; $\begin{pmatrix} 0 & 4+3\sqrt{2} & 4-3\sqrt{2} \\ i & -1 & -1 \\ 1 & -i & -i \end{pmatrix}$

15. *Hint:* Define $\mathbf{H} = \frac{1}{2}(\mathbf{A} + \overline{\mathbf{A}}^t)$, $\mathbf{S} = \frac{1}{2}(\mathbf{A} - \overline{\mathbf{A}}^t)$. Then $\mathbf{A} = \mathbf{H} + \mathbf{S}$. Show that $\mathbf{H}^t = \overline{\mathbf{H}}$ and $\mathbf{S}^t = -\overline{\mathbf{S}}$.

CHAPTER 9

Section 9.1

1. $\Omega(t) = \begin{pmatrix} -e^{2t} & 3e^{6t} \\ e^{2t} & e^{6t} \end{pmatrix}$; $x_1(t) = -3e^{2t} + 3e^{6t}$, $x_2(t) = 3e^{2t} + e^{6t}$

3. $\Omega(t) = \begin{pmatrix} 4e^{(1+2\sqrt{3})t} & 4e^{(1-2\sqrt{3})t} \\ (-1+\sqrt{3})e^{(1+2\sqrt{3})t} & (-1-\sqrt{3})e^{(1-2\sqrt{3})t} \end{pmatrix}$;

$x_1(t) = \left(1 + \frac{5}{3}\sqrt{3}\right)e^{(1+2\sqrt{3})t} + \left(1 - \frac{5}{3}\sqrt{3}\right)e^{(1-2\sqrt{3})t}$, $x_2(t) = \left(1 - \frac{1}{6}\sqrt{3}\right)e^{(1+2\sqrt{3})t} + \left(1 + \frac{1}{6}\sqrt{3}\right)e^{(1-2\sqrt{3})t}$

5. $\Omega(t) = \begin{pmatrix} e^t & 0 & e^{-3t} \\ 0 & e^t & 3e^{-3t} \\ -e^t & e^t & e^{-3t} \end{pmatrix}$; $x_1(t) = 10e^t - 9e^{-3t}$, $x_2(t) = 24e^t - 27e^{-3t}$, $x_3(t) = 14e^t - 9e^{-3t}$

Section 9.2

1. $\Omega(t) = \begin{pmatrix} 7e^{3t} & 0 \\ 5e^{3t} & e^{-4t} \end{pmatrix}$ is a fundamental matrix; general solution is $\mathbf{X}(t) = \Omega(t)\mathbf{C} = \begin{pmatrix} 7c_1e^{3t} \\ 5c_1e^{3t} + c_2e^{-4t} \end{pmatrix}$.

3. $\Omega(t) = \begin{pmatrix} 1 & e^{2t} \\ -1 & e^{2t} \end{pmatrix}; \mathbf{X}(t) = \Omega(t)\mathbf{C} = \begin{pmatrix} c_1 + c_2 e^{2t} \\ -c_1 + c_2 e^{2t} \end{pmatrix}$

5. $\Omega(t) = \begin{pmatrix} 1 & 2e^{3t} & -e^{-4t} \\ 6 & 3e^{3t} & 2e^{-4t} \\ -13 & -2e^{3t} & e^{-4t} \end{pmatrix}; \mathbf{X}(t) = \begin{pmatrix} c_1 + 2c_2 e^{3t} - c_3 e^{-4t} \\ 6c_1 + 3c_2 e^{3t} + 2c_3 e^{-4t} \\ -13c_1 - 2c_2 e^{3t} + c_3 e^{-4t} \end{pmatrix}$

7. $\Omega(t) = \begin{pmatrix} -e^t & e^{-2t} & e^{3t} \\ 4e^t & -e^{-2t} & 2e^{3t} \\ e^t & -e^{-2t} & e^{3t} \end{pmatrix}; \mathbf{X}(t) = \begin{pmatrix} -c_1 e^t + c_2 e^{-2t} + c_3 e^{3t} \\ 4c_1 e^t - c_2 e^{-2t} + 2c_3 e^{3t} \\ c_1 e^t - c_2 e^{-2t} + c_3 e^{3t} \end{pmatrix}$

9. $\Omega(t) = \begin{pmatrix} 2e^{4t} & e^{-3t} \\ -3e^{4t} & 2e^{-3t} \end{pmatrix}$ is a fundamental matrix; the solution of the initial value problem is

$\mathbf{X}(t) = \begin{pmatrix} 6e^{4t} - 5e^{-3t} \\ -9e^{4t} - 10e^{-3t} \end{pmatrix}.$

11. $\Omega(t) = \begin{pmatrix} 0 & e^{2t} & 3e^{3t} \\ 1 & e^{2t} & e^{3t} \\ 1 & 0 & e^{3t} \end{pmatrix}; \mathbf{X}(t) = \begin{pmatrix} 4e^{2t} - 3e^{3t} \\ 2 + 4e^{2t} - e^{3t} \\ 2 - e^{3t} \end{pmatrix}$

13. $\Omega(t) = \begin{pmatrix} 3e^{-t} & 0 & e^{4t} \\ -5e^{-t} & -e^{2t} & 0 \\ 0 & e^{2t} & 0 \end{pmatrix}; \mathbf{X}(t) = \begin{pmatrix} -\frac{6}{5}e^{-t} + \frac{51}{5}e^{4t} \\ 2e^{-t} - 3e^{2t} \\ 3e^{2t} \end{pmatrix}$

15. $\begin{pmatrix} c_1 t^8 + c_2 t^2 \\ c_1 t^8 - 2c_2 t^2 \end{pmatrix}$

17. $\Omega(t) = \begin{pmatrix} 2e^{2t} \cos(2t) & 2e^{2t} \sin(2t) \\ e^{2t} \sin(2t) & -e^{2t} \cos(2t) \end{pmatrix}$

19. $\begin{pmatrix} 5e^t \cos(t) & 5e^t \sin(t) \\ e^t[2\cos(t) + \sin(t)] & e^t[2\sin(t) - \cos(t)] \end{pmatrix}$

21. $\begin{pmatrix} 0 & e^{-t}\cos(2t) & e^{-t}\sin(2t) \\ 0 & e^{-t}[\cos(2t) - 2\sin(2t)] & e^{-t}[\sin(2t) + 2\cos(2t)] \\ e^{-2t} & 3e^{-t}\cos(2t) & 3e^{-t}\sin(2t) \end{pmatrix}$

23. $\begin{pmatrix} 2\cos(t) & 2\sin(t) & 0 & 0 \\ 3\cos(t) + \sin(t) & 3\sin(t) - \cos(t) & 0 & 0 \\ 0 & 0 & 2\cos(t) & 2\sin(t) \\ 0 & 0 & 3\cos(t) + \sin(t) & 3\sin(t) - \cos(t) \end{pmatrix}$

25. $\begin{pmatrix} 2\cos(t) - 14\sin(t) \\ 10\cos(t) - 20\sin(t) \end{pmatrix}$

27. $\begin{pmatrix} 2e^t + 5e^t[\cos(t) + \sin(t)] \\ 2e^t + e^t[2\cos(t) + 6\sin(t)] \\ 2e^t + e^t[\cos(t) + 3\sin(t)] \end{pmatrix}$

29. $\begin{pmatrix} 6t^3 \cos(2\ln(t)) - 4t^3 \sin(2\ln(t)) \\ 5t^3 \cos(2\ln(t)) + t^3 \sin(2\ln(t)) \end{pmatrix}$

31. $\Omega(t) = \begin{pmatrix} e^{3t} & 2te^{3t} \\ 0 & e^{3t} \end{pmatrix}$; general solution $\mathbf{X}(t) = \begin{pmatrix} c_1 e^{3t} + 2c_2 te^{3t} \\ c_2 e^{3t} \end{pmatrix}$

33. $\Omega(t) = \begin{pmatrix} 2e^{4t} & (1 - 2t)e^{4t} \\ -e^{4t} & te^{4t} \end{pmatrix}; \mathbf{X}(t) = \begin{pmatrix} 2c_1 e^{4t} + c_2(1 - 2t)e^{4t} \\ -c_1 e^{4t} + c_2 te^{4t} \end{pmatrix}$

35. $\begin{pmatrix} e^{2t} & 3e^{5t} & 27te^{5t} \\ 0 & 3e^{5t} & (3 + 27t)e^{5t} \\ 0 & -e^{5t} & (2 - 9t)e^{5t} \end{pmatrix}; \mathbf{X}(t) = \begin{pmatrix} c_1 e^{2t} + [3c_2 + 27c_3 t]e^{5t} \\ [3c_2 + (3 + 27t)c_3]e^{5t} \\ [-c_2 + (2 - 9t)c_3]e^{5t} \end{pmatrix}$

37. $\begin{pmatrix} 2 & 3e^{3t} & e^t & 0 \\ 0 & 2e^{3t} & 0 & -2e^t \\ 1 & 2e^{3t} & 0 & -2e^t \\ 0 & 0 & 0 & e^t \end{pmatrix}; \mathbf{X}(t) = \begin{pmatrix} 2c_1 + 3c_2 e^{3t} + c_3 e^t \\ 2c_2 e^{3t} - 2c_4 e^t \\ c_1 + 2c_2 e^{3t} - 2c_4 e^t \\ c_4 e^t \end{pmatrix}$

39. $\mathbf{X}(t) = \begin{pmatrix} (5 + 2t)e^{6t} \\ (3 + 2t)e^{6t} \end{pmatrix}$

41. $\mathbf{X}(t) = \begin{pmatrix} -e^{2t} + (1 + 22t)e^{-4t} \\ -6e^{2t} + 10e^{-4t} \\ 12e^{-4t} \end{pmatrix}$

43. $\mathbf{X}(t) = \begin{pmatrix} 2\cos(t) + 6\sin(t) \\ -2\cos(t) + 4\sin(t) \\ (1 - 9t)e^{2t} \\ (4 - 9t)e^{2t} \end{pmatrix}$

47. $\mathbf{X}(t) = \begin{pmatrix} 3c_1 e^{2t} + c_2 e^{6t} \\ -c_1 e^{2t} + c_2 e^{6t} \end{pmatrix}$

49. $\mathbf{X}(t) = \begin{pmatrix} c_1 e^t + 5c_2 e^{7t} \\ -c_1 e^t + c_2 e^{7t} \end{pmatrix}$ **51.** $\mathbf{X}(t) = \begin{pmatrix} c_1 e^t + c_2 e^{-t} + c_3 e^{-3t} \\ c_1 e^t + 3c_2 e^{-t} - 3c_3 e^{-3t} \\ c_1 e^t + 3c_2 e^{-t} + c_3 e^{-3t} \end{pmatrix}$

53. $\mathbf{X}(t) = \begin{pmatrix} (c_2 - c_3)\cos(t) - (c_2 + c_3)\sin(t) \\ -c_1 e^{-t} + c_2 \cos(t) - c_3 \sin(t) \\ c_1 e^{-t} + c_2 \cos(t) - c_3 \sin(t) \end{pmatrix}$ **55.** $e^{\mathbf{A}t} = \begin{pmatrix} e^{3t} & 2te^{3t} \\ 0 & e^{3t} \end{pmatrix}$ **57.** $e^{\mathbf{A}t} = \begin{pmatrix} (1 - 2t)e^{4t} & -4te^{4t} \\ te^{4t} & (1 + 2t)e^{4t} \end{pmatrix}$

59. $e^{\mathbf{A}t} = \begin{pmatrix} e^{2t} & (\frac{2}{3} + 3t)e^{5t} - \frac{2}{3}e^{2t} & (-1 + 9t)e^{5t} + e^{2t} \\ 0 & (1 + 3t)e^{5t} & 9te^{5t} \\ 0 & -te^{5t} & (1 - 3t)e^{5t} \end{pmatrix}$ **61.** $e^{\mathbf{A}t} = \begin{pmatrix} e^t & \frac{1}{2}e^t - 2 + \frac{3}{2}e^{3t} & 2 - 2e^t & 3e^{3t} - 3e^t \\ 0 & e^{3t} & 0 & 2e^{3t} - 2e^t \\ 0 & -1 + e^{3t} & 1 & 2e^{3t} - 2e^t \\ 0 & 0 & 0 & e^t \end{pmatrix}$

69. $(\mathbf{A} + \mathbf{B})^2 = (\mathbf{A} + \mathbf{B})(\mathbf{A} + \mathbf{B}) = \mathbf{A}(\mathbf{A} + \mathbf{B}) + \mathbf{B}(\mathbf{A} + \mathbf{B}) = \mathbf{A}^2 + \mathbf{AB} + \mathbf{BA} + \mathbf{B}^2 = \mathbf{A}^2 + 2\mathbf{AB} + \mathbf{B}^2 = (\mathbf{A} + \mathbf{B})^2$ if $\mathbf{AB} = \mathbf{BA}$. If $\mathbf{AB} \neq \mathbf{BA}$, and this equality may fail, for example, let $\mathbf{A} = \begin{pmatrix} 1 & 2 \\ 3 & -1 \end{pmatrix}$ and $\mathbf{B} = \begin{pmatrix} 1 & -2 \\ -1 & 1 \end{pmatrix}$.

Section 9.3

1. $\mathbf{X}(t) = \begin{pmatrix} [c_1(1 + 2t) + 2c_2 t + t^2]e^{3t} \\ [-2c_1 t + (1 - 2t)c_2 + t - t^2]e^{3t} + \frac{3}{2}e^t \end{pmatrix}$ **3.** $\mathbf{X}(t) = \begin{pmatrix} [c_1 + (1 + t)c_2 + 2t + t^2 - t^3]e^{6t} \\ [c_1 + c_2 t + 4t^2 - t^3]e^{6t} \end{pmatrix}$

5. $\mathbf{X}(t) = \begin{pmatrix} c_2 e^t \\ (1 - 2c_2)e^t + (c_3 - 9c_4)e^{3t} \\ 2c_4 e^{3t} \\ (c_1 - 5c_2 t + 1 + 3t)e^t + c_3 e^{3t} \end{pmatrix}$ **7.** $\mathbf{X}(t) = \begin{pmatrix} (-1 - 14t)e^t \\ (3 - 14t)e^t \end{pmatrix}$

9. $\mathbf{X}(t) = \begin{pmatrix} (6 + 12t + \frac{1}{2}t^2)e^{-2t} \\ (2 + 12t + \frac{1}{2}t^2)e^{-2t} \\ (3 + 38t + 66t^2 + \frac{13}{6}t^3)e^{-2t} \end{pmatrix}$ **11.** $\Omega(t)\Omega^{-1}(0) = \begin{pmatrix} \frac{2}{5}e^t + \frac{3}{5}e^{6t} & -\frac{2}{5}e^t + \frac{2}{5}e^{6t} \\ -\frac{3}{5}e^t + \frac{3}{5}e^{6t} & \frac{3}{5}e^t + \frac{2}{5}e^{6t} \end{pmatrix}$ is a transition matrix.

13. $\begin{pmatrix} -e^{-3t} + 2e^t & e^{-3t} - e^t & -e^{-3t} + e^t \\ -3e^{-3t} + 3e^t & 3e^{-3t} - 2e^t & -3e^{-3t} + 3e^t \\ -e^{-3t} + e^t & e^{-3t} - e^t & -e^{-3t} + 2e^t \end{pmatrix}$ **15.** $\mathbf{X}(t) = \begin{pmatrix} 3c_1 e^{2t} + c_2 e^{6t} - 4e^{3t} - \frac{10}{3} \\ -c_1 e^{2t} + c_2 e^{6t} + \frac{2}{3} \end{pmatrix}$

17. $\mathbf{X}(t) = \begin{pmatrix} c_1 e^t + 5c_2 e^{7t} + \frac{68}{145}\cos(3t) - \frac{54}{145}\sin(3t) + \frac{40}{7} \\ -c_1 e^t + c_2 e^{7t} + \frac{2}{145}\cos(3t) + \frac{24}{145}\sin(3t) - \frac{48}{7} \end{pmatrix}$ **19.** $\mathbf{X}(t) = \begin{pmatrix} c_1 e^t + c_2 e^{-t} + c_3 e^{-3t} + e^{2t} + 12t + 6 \\ c_1 e^t + 3c_2 e^{-t} - 3c_3 e^{-3t} + 2e^{2t} + 18t \\ c_1 e^t + 3c_2 e^{-t} + c_3 e^{-3t} + e^{2t} + 21t - 1 \end{pmatrix}$

21. $\mathbf{X}(t) = \begin{pmatrix} (c_2 - c_3)\cos(t) - (c_2 + c_3)\sin(t) + 4e^t \\ -c_1 e^{-t} + c_2 \cos(t) - c_3 \sin(t) - e^{-3t} + 2e^t \\ c_1 e^{-t} + c_2 \cos(t) - c_3 \sin(t) + 2e^t + e^{-3t} \end{pmatrix}$ **23.** $\mathbf{X}(t) = \begin{pmatrix} 2 + 4(1 + t)e^{2t} \\ -2 + 2(1 + 2t)e^{2t} \end{pmatrix}$

25. $\mathbf{X}(t) = \begin{pmatrix} 10\cos(t) + \frac{5}{2}t\sin(t) - 5t\cos(t) \\ 5\cos(t) + \frac{5}{2}\sin(t) - \frac{5}{2}t\cos(t) \end{pmatrix}$ **27.** $\mathbf{X}(t) = \begin{pmatrix} -\frac{1}{4}e^{2t} + (2 + 2t)e^t - \frac{3}{4} - \frac{1}{2}t \\ e^{2t} + (2 + 2t)e^t - 1 - t \\ -\frac{5}{4}e^{2t} + 2te^t - \frac{3}{4} - \frac{1}{2}t \end{pmatrix}$

29. $\mathbf{X}(t) = \begin{pmatrix} \frac{1}{4}e^t[-\cos(t) + \sin(t)] + \frac{1}{4} \\ \frac{1}{4}t - \frac{1}{4}e^t \sin(t) \end{pmatrix}$

CHAPTER 10

Section 10.1

1. a gives the initial value of θ (hence initial displacement); b gives the initial rate of change of θ with respect to time. There are no restrictions (at least mathematically) on a and b.

3. Yes

Section 10.2

1. $\mathbf{X}(t) = \begin{pmatrix} -c_1 \cos(t) - c_2 \sin(t) \\ (4c_1 - c_2) \cos(t) + (c_1 + 4c_2) \sin(t) \end{pmatrix}$; a phase portrait is shown below.

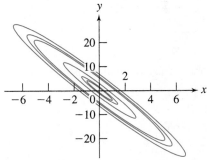

3. $\mathbf{X}(t) = \begin{pmatrix} c_1 e^{-3t} + 7c_2 e^{2t} \\ c_1 e^{-3t} + 2c_2 e^{2t} \end{pmatrix}$; a phase portrait is shown below.

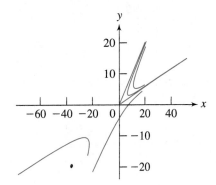

5. $\mathbf{X}(t) = \begin{pmatrix} c_1 e^{5t} + 2c_2 e^{4t} \\ c_2 e^{4t} \end{pmatrix}$; a phase portrait is shown below.

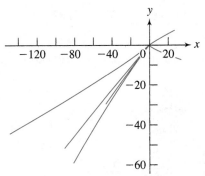

7. $\dfrac{dy}{dx} = -\dfrac{4x}{9y}$; $4x^2 + 9y^2 = c$; integral curves are ellipses with center $(0, 0)$.

9. $\dfrac{dy}{dx} = \dfrac{x - 1}{y + 2}$; $(x - 1)^2 - (y + 2)^2 = c$; integral curves are hyperbolas with center $(1, -2)$.

11. $\dfrac{dy}{dx} = \dfrac{x + y}{x}$; $y = x \ln|cx|$ **13.** $73x^2 - 72xy + 52y^2 = c$; graphs are ellipses.

15. $4x^2 - 3\sqrt{3}xy + y^2 = c$; hyperbolas   **17.** $13x^2 - 6\sqrt{3}xy + 7y^2 = c$; ellipses

19. These systems have the same graphs of trajectories, but with opposite orientations.

Section 10.3

1. eigenvalues $-2, -2$; improper nodal sink (phase portrait given below)

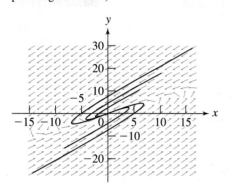

3. $\pm 2i$, center (phase portrait given below)

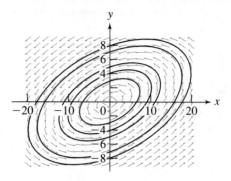

5. $4 \pm 5i$; spiral point (phase portrait given below)

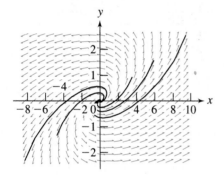

7. 3, 3; improper nodal source (phase portrait given below)

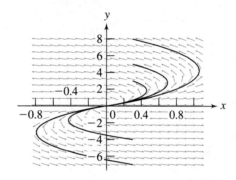

9. $-2 \pm \sqrt{3}i$; spiral point (phase portrait given below)

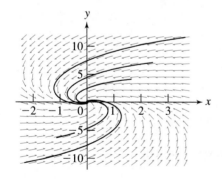

11. $\pm\sqrt{5}$; saddle point (phase portrait given below)

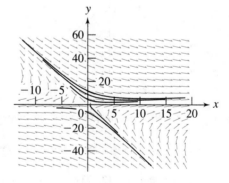

13. $-3 \pm \sqrt{7}$; nodal sink (phase portrait given below)   **15.** $2 \pm \sqrt{3}$; nodal source (phase portrait given below)

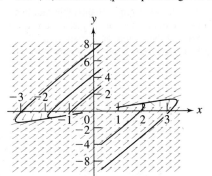
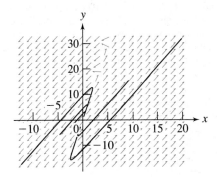

Section 10.4

1. stable and asymptotically stable improper node   **3.** stable but not asymptotically stable center
5. unstable spiral point   **7.** unstable improper node   **9.** stable and asymptotically stable spiral point
11. unstable saddle point   **13.** stable and asymptotically stable nodal sink   **15.** unstable node

Section 10.5

1. $(0, 0)$ is an unstable spiral point; $(-3/2, 3/4)$ is an unstable saddle point (phase portrait given below).

3. $(0, 0)$ is an unstable saddle point; $(-5, -5)$ is an asymptotically stable nodal sink (phase portrait given below).

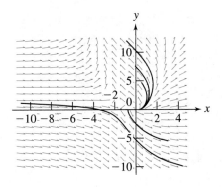
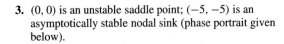
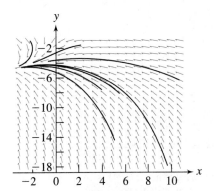

5. $(0, 0)$ is a center of the linear system, and may be a center or spiral point of the nonlinear system; $(1/2, -1/8)$ and $(-1/2, 1/8)$ are unstable saddle points (phase portrait given below).

7. $(0, 0)$ is an asymptotically stable node or spiral point (phase portrait given below).

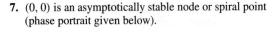
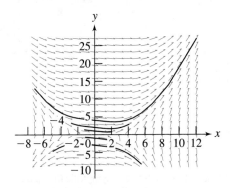
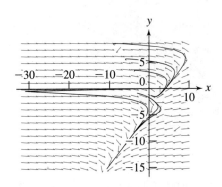

9. $(0, 0)$ is an unstable saddle point; $(3/8, 3/2)$ is an asymptotically stable spiral point.

Section 10.6

1. Critical points are $(0, 0)$, $(12, 2)$. The origin is an unstable saddle point; $(12, 2)$ is a center of the linear system (complex eigenvalues), so is either a center or spiral point of the nonlinear system. In this case $(12, 2)$ is a stable center. A phase portrait is given below.

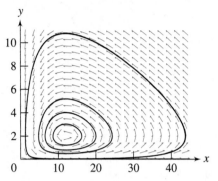

5. $(0, 0)$, $(2/7, 1)$; the origin is an unstable saddle point; $(2/7, 1)$ is a stable center. A phase portrait is given below.

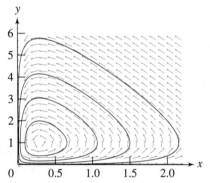

13. $(0, 0)$ is an unstable node; $(0, 4)$ is an unstable saddle point; $(2, 0)$ is an unstable nodal source; $(1.4286, 2.8571)$ is an asymptotically stable nodal sink. A phase portrait is given below.

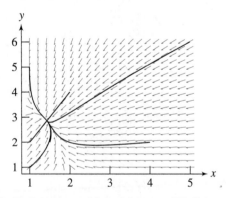

3. $(0, 0)$, $(7/12, 5)$; the origin is an unstable saddle point, $(7/12, 5)$ is a center of the linear system, so is a spiral point or center of the nonlinear system (in this case, a stable center). A phase portrait is given below.

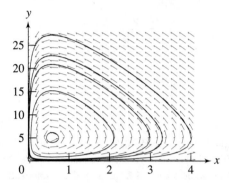

11. The critical points are $(0, 0)$, $(0, 2)$, $(1, 0)$. The origin is an unstable node, $(0, 2)$ is an asymptotically stable nodal sink, $(1, 0)$ is an unstable saddle point. A phase portrait is given below.

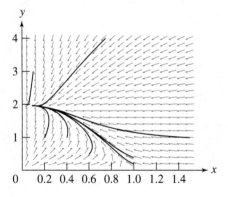

15. $(0, 0)$ is an unstable nodal source; $(10, 0)$ is an asymptotically stable nodal sink; $(0, 20)$ is an asymptotically stable nodal sink; $(1.9355, 6.4516)$ is an unstable saddle point.

Section 10.7

1. $(0, 0)$ is an unstable nodal source; $(1/2, 1/4)$ is an unstable saddle point.

3. $(0, 0)$ is an unstable nodal source; $(3, 3/2)$ is an unstable saddle point.

5. $(0, 0)$ is an unstable nodal source; $(1/4, 3/7)$ is an unstable saddle point.

7. $(0, 0)$ is an unstable nodal source; $(7/5, 0)$ is an unstable saddle point; $(0, 3/4)$ is an unstable saddle point; $(11/9, 4/9)$ is an asymptotically stable nodal sink.

9. $(0, 0)$ is an unstable nodal source; $(2/3, 0)$ is an unstable saddle point; $(0, 3)$ is an asymptotically stable nodal sink.

11. $(0, 0)$ is an unstable nodal source; $(3, 0)$ is an asymptotically stable nodal sink; $(0, 1/2)$ is an unstable saddle point.

Section 10.8

1. stable   **3.** asymptotically stable   **5.** unstable   **7.** unstable

Section 10.9

1. $\dfrac{\partial f}{\partial x} + \dfrac{\partial g}{\partial y} = 3 + 4x^2 - 2\sin(x) \geq 1 + 4x^2 > 0$ **3.** $\dfrac{\partial f}{\partial x} + \dfrac{\partial g}{\partial y} = \cosh(x) - 1 + 15e^{3y} > 0$

7. $\dfrac{dr}{dt} = 0, \dfrac{d\theta}{dt} = r^2 - 1$; trajectories are the origin (a critical point), all points on the circle $r = 1$ (each is a critical point), and the closed circles $r = a \neq 1$. Each closed trajectory $r = a \neq 1$ is a stable limit cycle.

9. $\dfrac{dr}{dt} = r(1 - r^2)(4 - r^2)(9 - r^2), \dfrac{d\theta}{dt} = -1$; trajectories are the origin (a critical point), and the circles $r = 1, r = 2,$ and $r = 3$ are each closed trajectories. The circles $r = 1$ and $r = 3$ are asymptotically stable limit cycles, and $r = 2$ is an unstable limit cycle.

11. $xx' + yy' = (x^2 + y^2)\left[4 - (x^2 + 9y^2)\right]$. The ellipse $x^2 + 9y^2 = 4$ is a closed trajectory, and the annular region $\frac{4}{9} \leq r^2 \leq 4$ has the stated property.

13. The ellipse $4x^2 + y^2 = 4$ is a limit cycle; the region defined by $1 \leq r^2 \leq 4$ has the stated property.

15. No; $\dfrac{dr}{dt} = r(1 + r^2) > 0$ for $r > 0$.   **17.** No; $\dfrac{dy}{dt} > 0$ and the system has no critical points, hence no trajectory.

19. No; $\dfrac{\partial f}{\partial x} + \dfrac{\partial g}{\partial y} > 0$ **21.** $x^4 + 2y^2 = C$ is a closed trajectory for all positive C.

25. Apply Lienard's theorem with $p(x) = x^2 - 1$ and $q(x) = 2x + \sin(x)$.

27. $x^4 + 2y^2 = C$ are closed trajectories for each positive C.

29. $\ln(1 + x^2) + y^2 = C$ are closed trajectories for each positive C.

31. $\dfrac{\partial f}{\partial x} + \dfrac{\partial g}{\partial y} = -\alpha(x^2 - 1)$, having one sign on each of the three given regions.

CHAPTER 11

Section 11.1

1. $\dfrac{d}{dt}[f(t)\mathbf{F}(t)] = -12\sin(3t)\mathbf{i} + 12t[2\cos(3t) - 3t\sin(3t)]\mathbf{j} + 8[\cos(3t) - 3t\sin(3t)]\mathbf{k}$

3. $\dfrac{d}{dt}[\mathbf{F}(t) \times \mathbf{G}(t)] = [1 - 4\sin(t)]\mathbf{i} - 2t\mathbf{j} - [\cos(t) - t\sin(t)]\mathbf{k}$

5. $\dfrac{d}{dt}[f(t)\mathbf{F}(t)] = (1 - 8t^3)\mathbf{i} + [6t^2\cosh(t) - (1 - 2t^3)\sinh(t)]\mathbf{j} + [-6t^2 e^t + e^t(1 - 2t^3)]\mathbf{k}$

7. $\dfrac{d}{dt}[\mathbf{F}(t) \times \mathbf{G}(t)] = te^t(2 + t)[\mathbf{j} - \mathbf{k}]$ **9.** $\dfrac{d}{dt}[\mathbf{F}(f(t))] = -16t^3\mathbf{i} - 4t\sin(2t^2)\mathbf{k}$

15. position: $\mathbf{F}(t) = \sin(t)\mathbf{i} + \cos(t)\mathbf{j} + 45t\mathbf{k}$; tangent: $\mathbf{F}'(t) = \cos(t)\mathbf{i} - \sin(t)\mathbf{j} + 45\mathbf{k}$; length: $s(t) = \sqrt{2026}\,t$; position: $\mathbf{G}(s) = \mathbf{F}(t(s)) = \sin\left(\dfrac{s}{\sqrt{2026}}\right)\mathbf{i} + \cos\left(\dfrac{s}{\sqrt{2026}}\right)\mathbf{j} + \dfrac{45s}{\sqrt{2026}}\mathbf{k}$

17. position: $\mathbf{F}(t) = 2t^2\mathbf{i} + 3t^2\mathbf{j} + 4t^2\mathbf{k} = t^2(2\mathbf{i} + 3\mathbf{j} + 4\mathbf{k})$; tangent: $\mathbf{F}'(t) = 2t(2\mathbf{i} + 3\mathbf{j} + 4\mathbf{k})$; length: $s(t) = \sqrt{29}(t^2 - 1)$; position: $\mathbf{G}(s) = \left(1 + \dfrac{s}{\sqrt{29}}\right)(2\mathbf{i} + 3\mathbf{j} + 4\mathbf{k})$

20. $\dfrac{d}{dt}[\mathbf{F}, \mathbf{G}, \mathbf{H}] = \mathbf{F}'(t) \cdot [\mathbf{G}(t) \times \mathbf{H}(t)] + \mathbf{F}(t) \cdot [\mathbf{G}'(t) \times \mathbf{H}(t)] + \mathbf{F}(t) \cdot [\mathbf{G}(t) \times \mathbf{H}'(t)]$

Section 11.2

1. $\mathbf{v}(t) = 3\mathbf{i} + 2t\mathbf{k}$, $v(t) = \sqrt{9 + 4t^2}$, $\mathbf{a}(t) = 2\mathbf{k}$, $a_T = \dfrac{4t}{\sqrt{9 + 4t^2}}$, $a_N = \dfrac{6}{\sqrt{9 + 4t^2}}$, $\kappa = \dfrac{6}{(9 + 4t^2)^{3/2}}$,

$\mathbf{T} = \dfrac{1}{\sqrt{9 + 4t^2}}[3\mathbf{i} + 2t\mathbf{k}]$, $\mathbf{N} = \dfrac{1}{\sqrt{9 + 4t^2}}[-2t\mathbf{i} + 3\mathbf{k}]$, $\mathbf{B} = -\mathbf{j}$

3. $\mathbf{v} = 2\mathbf{i} - 2\mathbf{j} + \mathbf{k}$, $v = 3$, $\mathbf{a} = \mathbf{O}$, $a_T = a_N = \kappa = 0$, $\mathbf{T} = \frac{1}{3}(2\mathbf{i} - 2\mathbf{j} + \mathbf{k})$,

$\mathbf{N} = \dfrac{1}{\sqrt{2}}(\mathbf{i} + \mathbf{j})$ (or any unit vector perpendicular to $\mathbf{T}$), $\mathbf{B} = \frac{1}{6}\sqrt{2}(-\mathbf{i} + \mathbf{j} + 4\mathbf{k})$

5. $\mathbf{v} = -3e^{-t}(\mathbf{i} + \mathbf{j} - 2\mathbf{k})$, $v = 3\sqrt{6}e^{-t}$, $\mathbf{a} = 3e^{-t}(\mathbf{i} + \mathbf{j} - 2\mathbf{k})$, $a_T = -3\sqrt{6}e^{-t}$,

$a_N = 0$, $\kappa = 0$, $\mathbf{T} = \dfrac{1}{\sqrt{6}}(-\mathbf{i} - \mathbf{j} + 2\mathbf{k})$, $\mathbf{N} = \dfrac{1}{\sqrt{2}}(\mathbf{i} - \mathbf{j})$, or any unit vector perpendicular to $\mathbf{T}$, $\mathbf{B} = \dfrac{1}{\sqrt{3}}(\mathbf{i} + \mathbf{j} + \mathbf{k})$

7. $\mathbf{v} = 2\cosh(t)\mathbf{j} - 2\sinh(t)\mathbf{k}$, $v = 2\sqrt{\cosh(2t)}$, $\mathbf{a} = 2\sinh(t)\mathbf{j} - 2\cosh(t)\mathbf{k}$,

$a_T = \dfrac{2\sinh(2t)}{\sqrt{\cosh(2t)}}$, $a_N = \dfrac{2}{\sqrt{\cosh(2t)}}$, $\kappa = \dfrac{1}{2[\cosh(2t)]^{3/2}}$,

$\mathbf{T} = \dfrac{1}{\sqrt{\cosh(2t)}}[\cosh(t)\mathbf{j} - \sinh(t)\mathbf{k}]$, $\mathbf{N} = \dfrac{1}{\sqrt{\cosh(2t)}}[-\sinh(t)\mathbf{j} - \cosh(t)\mathbf{k}]$, $\mathbf{B} = -\mathbf{i}$

9. $\mathbf{v} = 2t(\alpha\mathbf{i} + \beta\mathbf{j} + \gamma\mathbf{k})$, $v = 2|t|\sqrt{\alpha^2 + \beta^2 + \gamma^2}$, $\mathbf{a} = 2(\alpha\mathbf{i} + \beta\mathbf{j} + \gamma\mathbf{k})$, $a_T = 2sgn(t)\sqrt{\alpha^2 + \beta^2 + \gamma^2}$, where $sgn(t)$

equals 1 if $t \geq 0$ and -1 if $t < 0$, $a_N = \kappa = 0$, $\mathbf{T} = \dfrac{1}{\sqrt{\alpha^2 + \beta^2 + \gamma^2}}(\alpha\mathbf{i} + \beta\mathbf{j} + \gamma\mathbf{k})$, $\mathbf{N}$ is any unit vector

perpendicular to $\mathbf{T}$, and $\mathbf{B} = \mathbf{T} \times \mathbf{N}$.

13. From the third Frenet formula, $\mathbf{B}'(s) = -\tau\mathbf{N}(s)$, so $\mathbf{N} \cdot \mathbf{B}' = -\tau\mathbf{N} \cdot \mathbf{N} = -\tau \|\mathbf{N}\|^2 = -\tau$, since $\mathbf{N}$ is a unit vector.

15. With all the functions expressed in terms of s and derivatives taken with respect to s, we have

$\mathbf{F}' = \mathbf{T}$, $\mathbf{F}'' = \mathbf{T}' = \kappa\mathbf{N}$, $\mathbf{F}''' = \kappa'\mathbf{N} + \kappa\mathbf{N}' = \kappa'\mathbf{N} + \kappa(-\kappa\mathbf{T} + \tau\mathbf{B}) = \kappa'\mathbf{N} - \kappa^2\mathbf{T} + \kappa\tau\mathbf{B}$. Therefore,

$[\mathbf{F}', \mathbf{F}'', \mathbf{F}'''] = \kappa^2\tau$, and so $\tau = \dfrac{1}{\kappa^2}[\mathbf{F}', \mathbf{F}'', \mathbf{F}''']$.

Section 11.3

1. $\dfrac{\partial\mathbf{G}}{\partial x} = 3\mathbf{i} - 4y\mathbf{j}$, $\dfrac{\partial\mathbf{G}}{\partial y} = -4x\mathbf{j}$ **3.** $\dfrac{\partial\mathbf{G}}{\partial x} = 2y\mathbf{i} - \sin(x)\mathbf{j}$, $\dfrac{\partial\mathbf{G}}{\partial y} = 2x\mathbf{i}$

5. $\dfrac{\partial\mathbf{G}}{\partial x} = 6x\mathbf{i} + \mathbf{j}$, $\dfrac{\partial\mathbf{G}}{\partial y} = -2\mathbf{j}$ **7.** $\dfrac{\partial\mathbf{F}}{\partial x} = -4z^2\sin(x)\mathbf{i} - 3x^2yz\mathbf{j} + 3x^2y\mathbf{k}$, $\dfrac{\partial\mathbf{F}}{\partial y} = -x^3z\mathbf{j} + x^3\mathbf{k}$, $\dfrac{\partial\mathbf{F}}{\partial z} = 8z\cos(x)\mathbf{i} - x^3y\mathbf{j}$

9. $\dfrac{\partial\mathbf{F}}{\partial x} = -yz^4\cos(xy)\mathbf{i} + 3y^4z\mathbf{j} - \sinh(z - x)\mathbf{k}$, $\dfrac{\partial\mathbf{F}}{\partial y} = -xz^4\cos(xy)\mathbf{i} + 12xy^3z\mathbf{j}$,

$\dfrac{\partial\mathbf{F}}{\partial z} = -4z^3\sin(xy)\mathbf{i} + 3xy^4\mathbf{j} + \sinh(z - x)\mathbf{k}$

11. $x = x$, $y = \dfrac{1}{x + c}$, $z = e^{x+k}$; $x = x$, $y = \dfrac{1}{x - 1}$, $z = e^{x-2}$

13. $x = x$, $y = e^x(x - 1) + c$, $x^2 = -2z + k$; $x = x$, $y = e^x(x - 1) - e^2$, $z = \frac{1}{2}(12 - x^2)$

15. $x = c$, $y = y$, $2e^z = k - \sin(y)$; $x = 3$, $y = y$, $z = \ln(1 + \frac{1}{4}\sqrt{2} - \frac{1}{2}\sin(y))$

17. $x = x$, $y = c$, $\frac{1}{3}x^3 + k = -e^z$; $x = x$, $y = 2$, $z = \ln(\frac{1}{3}(67 - x^3))$

19. $\sin(x) + c = \ln|\cos(y)|$, $y = y$, $z = \ln|\cos(y)| + k$;

$x = \sin^{-1}(\frac{1}{2}\sqrt{2} + \ln|\cos(y)|)$, $y = y$, $z = 1 + \ln|\cos(y)|$

21. Take any constant $\mathbf{F}$, say $\mathbf{F} = \mathbf{i} + \mathbf{j} + \mathbf{k}$.   **23.** This is impossible.

Section 11.4

1. $yz\mathbf{i} + xz\mathbf{j} + xy\mathbf{k}$; $\mathbf{i} + \mathbf{j} + \mathbf{k}$; $\sqrt{3}$ and $-\sqrt{3}$

3. $(2y + e^z)\mathbf{i} + 2x\mathbf{j} + xe^z\mathbf{k}$; $(2 + e^6)\mathbf{i} - 4\mathbf{j} - 2e^6\mathbf{k}$, $\sqrt{20 + 4e^6 + 5e^{12}}$ and $-\sqrt{20 + 4e^6 + 5e^{12}}$

5. $2y\sinh(2xy)\mathbf{i} + 2x\sinh(2xy)\mathbf{j} - \cosh(z)\mathbf{k}$; $-\cosh(1)\mathbf{k}$; $\cosh(1)$ and $-\cosh(1)$

7. $\dfrac{1}{x+y+z}(\mathbf{i}+\mathbf{j}+\mathbf{k})$; $\frac{1}{2}(\mathbf{i}+\mathbf{j}+\mathbf{k})$; $\frac{1}{2}\sqrt{3}$ and $-\frac{1}{2}\sqrt{3}$

9. $e^x\cos(y)\cos(z)\mathbf{i}-e^x\sin(y)\cos(z)\mathbf{j}-e^x\cos(y)\sin(z)\mathbf{k}$; $\frac{1}{2}(\mathbf{i}-\mathbf{j}-\mathbf{k})$; $\frac{1}{2}\sqrt{3}$ and $-\frac{1}{2}\sqrt{3}$

11. $\dfrac{1}{\sqrt{3}}(8y^2-z+16xy-x)$ **13.** $\dfrac{1}{\sqrt{5}}(2x^2z^3+3x^2yz^2)$

15. 0 **17.** $x+y+\sqrt{2}z=4$; $x=y=1+2t$, $z=\sqrt{2}(1+2t)$ **19.** $x=y$; $x=1+2t$, $y=1-2t$, $z=0$

21. $x+6z=8$; $x=-4+2t$, $y=0$, $z=2+12t$ **23.** $x=1$; $x=1+2t$, $y=\pi$, $z=1$

25. $x-2y+2z=1$; $x=1+2t$, $y=1-4t$, $z=1+4t$ **27.** $\cos^{-1}(45/\sqrt{10653})\approx 1.11966$ rad

29. $\cos^{-1}(1/\sqrt{2})$, or $\pi/4$ rad   **31.** Level surfaces are planes $x+z=k$. The streamlines of $\nabla\varphi$ are parallel to $\nabla\varphi$.

Section 11.5

In 1, 3, and 5, $\nabla\cdot\mathbf{F}$ is given first, then $\nabla\times\mathbf{F}$.

1. 4, $\mathbf{O}$ **3.** $2y+xe^y+2$; $(e^y-2x)\mathbf{k}$ **5.** $2(x+y+z)$; $\mathbf{O}$

In 7, 9, and 11, $\nabla\varphi$ is given.

7. $\mathbf{i}-\mathbf{j}+4z\mathbf{k}$ **9.** $-6x^2yz^2\mathbf{i}-2x^3z^2\mathbf{j}-4x^3yz\mathbf{k}$

11. $[\cos(x+y+z)-x\sin(x+y+z)]\mathbf{i}-x\sin(x+y+z)\mathbf{j}-x\sin(x+y+z)\mathbf{k}$

13. $\nabla\cdot(\varphi\mathbf{F})=\nabla\varphi\cdot\mathbf{F}+\varphi\nabla\cdot\mathbf{F}$; $\nabla\times(\varphi\mathbf{F})=\nabla\varphi\times\mathbf{F}+\varphi\nabla\times\mathbf{F}$

15. *Hint:* Let $\mathbf{F}=f_1\mathbf{i}+f_2\mathbf{j}+f_3\mathbf{k}$ and $\mathbf{G}=g_1\mathbf{i}+g_2\mathbf{j}+g_3\mathbf{k}$ and compute both sides of the proposed identity.

17. Let $r=\|\mathbf{R}\|=(x^2+y^2+z^2)^{1/2}$. Then $r^n=(x^2+y^2+z^2)^{n/2}$ and $\dfrac{\partial}{\partial x}r^n=\dfrac{n}{2}(x^2+y^2+z^2)^{(n/2)-1}(2x)=nxr^{n-2}$,

and similarly for $\dfrac{\partial}{\partial y}r^n$ and $\dfrac{\partial}{\partial z}r^n$. Thus, for (a), we have $\nabla r^n=nr^{n-2}(x\mathbf{i}+y\mathbf{j}+z\mathbf{k})=nr^{n-2}\mathbf{R}$. For (b), the first

component of $\nabla\times(\varphi(r)\mathbf{R})$ is $\dfrac{\partial}{\partial y}[\varphi(r)z]-\dfrac{\partial}{\partial z}[\varphi(r)y]=z\varphi'(r)\dfrac{\partial r}{\partial y}-y\varphi'(r)\dfrac{\partial r}{\partial z}=\varphi'(r)[zy(x^2+y^2+z^2)^{-1/2}$

$-yz(x^2+y^2+z^2)^{-1/2}]=0$. Similarly for the second and third components, hence $\nabla\times(\varphi(r)\mathbf{R})=\mathbf{O}$.

CHAPTER 12

Section 12.1

1. 0 **3.** $\frac{26}{3}\sqrt{2}$ **5.** $\sin(3)-\frac{81}{2}$ **7.** 0 **9.** $-\frac{422}{5}$ **11.** $48\sqrt{2}$ **13.** $-12e^4+4e^2$ **15.** $\sqrt{14}[\cos(1)-\cos(3)]$

17. $-1530\sqrt{30}$ **19.** $\frac{556}{3}$ **21.** $\frac{1573}{30}$ **23.** $\cos(1)-\cos(4)$ **25.** $27\sqrt{2}$ **27.** $-\frac{27}{2}$

Section 12.2

1. -8 **3.** -12 **5.** -40 **7.** 512π **9.** 0 **11.** $\frac{95}{4}$ **13.** -12π

Section 12.2.1

1. 0 **3.** 2π if C encloses the origin, 0 if C does not.   **5.** 0

Section 12.3

1. Conservative; $\varphi(x,y)=xy^3-4y$.   **3.** Conservative; $\varphi(x,y)=8x^2+2y-\frac{1}{3}y^3$.

5. Conservative; $\varphi(x,y)=\ln(x^2+y^2)$.   **7.** Conservative; $\varphi(x,y)=e^y\sin(2x)-\frac{1}{2}y^2$.

9. Not conservative.   **11.** -27 **13.** $5+\ln(3/2)$ **15.** -5 **17.** $e^3-24e-9e^{-1}$ **19.** $-8\cosh(2)+4$

Section 12.4

1. $125\sqrt{2}$ **3.** $\dfrac{\pi}{6}[(29)^{3/2}-27]$ **5.** $\dfrac{28\pi}{3}\sqrt{2}$ **7.** $\frac{9}{8}[\ln(4+\sqrt{17})+4\sqrt{17}]$ **9.** $-10\sqrt{3}$ **11.** 0 **13.** 60

15. $\dfrac{64\sqrt{3}}{3}$ **17.** $\frac{1}{240}[391\sqrt{17}-25\sqrt{5}]$

Section 12.5

1. $\frac{49}{12}$; $(\frac{12}{35}, \frac{33}{35}, \frac{24}{35})$ **3.** $9\pi\sqrt{2}$; $(0, 0, 2)$ **5.** 78π; $(0, 0, \frac{27}{13})$ **7.** 32

Section 12.6

1. *Hint:* Apply Green's theorem to $\oint_C -\varphi\dfrac{\partial\psi}{\partial y}\,dx + \varphi\dfrac{\partial\psi}{\partial x}\,dy$.

3. *Hint:* The unit tangent to C is $x'(s)\mathbf{i} + y'(s)\mathbf{j}$. Show that the unit outer normal is $\mathbf{N} = y'(s)\mathbf{i} - x'(s)\mathbf{j}$. Then
$D_{\mathbf{N}}\varphi(x, y) = \nabla\varphi \cdot \mathbf{N} = \dfrac{\partial\varphi}{\partial x}\dfrac{dy}{ds} - \dfrac{\partial\varphi}{\partial y}\dfrac{dx}{ds}$. But then $D_{\mathbf{N}}\varphi(x, y)\,ds = -\dfrac{\partial\varphi}{\partial y}\,dx + \dfrac{\partial\varphi}{\partial x}\,dy$. Now apply Green's theorem to
$\oint_C -\dfrac{\partial\varphi}{\partial y}\,dx + \dfrac{\partial\varphi}{\partial x}\,dy$.

Section 12.7

1. $\frac{256}{3}\pi$ **3.** 0   **5.** $\frac{8}{3}\pi$ **7.** 2π **9.** 0   **11.** 0 because $\nabla \cdot (\nabla \times \mathbf{F}) = 0$.   **13.** *Hint:* Use Green's second identity.
15. *Hint:* Apply Gauss's theorem and notice that $\nabla \cdot \mathbf{K} = 0$.

Section 12.8

1. -8π **3.** -16π **5.** $-\frac{32}{3}$ **7.** -108 **9.** Not conservative.   **11.** $\varphi(x, y) = x - 2y + z$ **13.** Not conservative.
15. Not conservative.   **17.** -403 **19.** $2e^{-2}$ **21.** 71

CHAPTER 13

Section 13.1

1. Graphs of the second, third, and fourth partial sums are shown below.

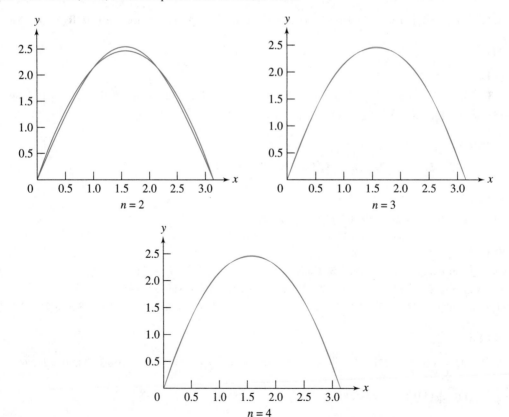

$n = 2$

$n = 3$

$n = 4$

Section 13.2

1. 4   **3.** $\dfrac{1}{\pi}\sinh(\pi) + \dfrac{2}{\pi}\sinh(\pi)\displaystyle\sum_{n=1}^{\infty}\dfrac{(-1)^n}{n^2+1}\cos(n\pi x)$ **5.** $\dfrac{16}{\pi}\displaystyle\sum_{n=1}^{\infty}\dfrac{1}{2n-1}\sin((2n-1)x)$

7. $\dfrac{13}{3} + \displaystyle\sum_{n=1}^{\infty}(-1)^n\left[\dfrac{16}{n^2\pi^2}\cos\left(\dfrac{n\pi x}{2}\right) + \dfrac{4}{n\pi}\sin\left(\dfrac{n\pi x}{2}\right)\right]$ **9.** $\dfrac{3}{2} + \dfrac{2}{\pi}\displaystyle\sum_{n=1}^{\infty}\dfrac{1}{2n-1}\sin((2n-1)x)$

11. $\dfrac{1}{3}\sin(3) + 6\sin(3)\displaystyle\sum_{n=1}^{\infty}\dfrac{(-1)^n}{n^2\pi^2-9}\cos\left(\dfrac{n\pi x}{3}\right)$ **13.** The Fourier coefficients of f and g are the same.

15. *Hint:* If f is odd, then $f(-x) = -f(x)$.

Section 13.3

1. The Fourier series is $\dfrac{11}{18} + \displaystyle\sum_{n=1}^{\infty}\left\{\dfrac{1}{n\pi}\left[4\sin\left(\dfrac{2n\pi}{3}\right) - \sin\left(\dfrac{n\pi}{3}\right)\right] + \dfrac{6}{n^2\pi^2}\left[\cos\left(\dfrac{2n\pi}{3}\right) - \cos\left(\dfrac{n\pi}{3}\right) + 2(-1)^n\right] + \right.$

$\left.\dfrac{18}{n^3\pi^3}\sin\left(\dfrac{n\pi}{3}\right)\right\}\cos\left(\dfrac{n\pi x}{3}\right) + \displaystyle\sum_{n=1}^{\infty}\left\{\dfrac{1}{n\pi}\left[4\cos\left(\dfrac{2n\pi}{3}\right) + \cos\left(\dfrac{n\pi}{3}\right) - 15(-1)^n\right] - \dfrac{6}{n^2\pi^2}\left[\sin\left(\dfrac{2n\pi}{3}\right) + \sin\left(\dfrac{n\pi}{3}\right)\right] \right.$

$\left. - \dfrac{18}{n^3\pi^3}\left[\cos\left(\dfrac{n\pi}{3}\right) - (-1)^n\right]\right\}\sin\left(\dfrac{n\pi x}{3}\right)$; this series converges to $\frac{3}{2}$ if $x = 3$ or if $x = -3$, to $2x$ if $-3 < x < -2$,

to -2 if $x = -2$, to 0 if $-2 < x < 1$, to $\frac{1}{2}$ if $x = 1$, and to x^2 if $1 < x < 3$.

3. Let $\alpha_n = n\pi/3$. The Fourier series is $\dfrac{11}{3}\sinh(3) - 2\cosh(3) + \displaystyle\sum_{n=1}^{\infty}(-1)^n\left\{\sinh(3)\left[\dfrac{1}{1+\alpha_n^2} + \dfrac{4(1-3\alpha_n^2)}{3(1+\alpha_n^2)^3}\right] + \right.$

$\left.\dfrac{4(\alpha_n^2-1)\cosh(3)}{(1+\alpha_n^2)^2}\right\}\cos(\alpha_n x) + \displaystyle\sum_{n=1}^{\infty}(-1)^n\left\{\sinh(3)\left[\dfrac{6\alpha_n}{1+\alpha_n^2} + \dfrac{4\alpha_n(\alpha_n^2-3)}{3(1+\alpha_n^2)^2}\right] - \dfrac{8\alpha_n\cosh(3)}{(1+\alpha_n^2)^2}\right\}\sin(\alpha_n x)$; this

converges to $18\cosh(3)$ if $x = -3$ or $x = 3$, and to $x^2 e^{-x}$ if $-3 < x < 3$.

5. $\dfrac{6+\pi^2}{6} + 2\displaystyle\sum_{n=1}^{\infty}\dfrac{(-1)^n}{n^2}\cos(nx) + \dfrac{1}{\pi}\displaystyle\sum_{n=1}^{\infty}\left[\left(\dfrac{2}{n^3}+\dfrac{2}{n}\right)(1-(-1)^n) + \dfrac{\pi^2}{n}(-1)^n\right]\sin(nx)$; this converges to $\frac{1}{2}(\pi^2+2)$

for $x = \pi$ or $x = -\pi$, to x^2 if $-\pi < x < 0$, to 1 if $x = 0$, and to 2 if $0 < x < \pi$.

7. $\dfrac{4}{\pi}\displaystyle\sum_{n=1}^{\infty}\dfrac{1}{2n-1}\sin\left(\dfrac{(2n-1)\pi x}{4}\right)$; converges to -1 if $-4 < x < 0$, to 0 if $x = -4, 0$ or 4, and to 1 if $0 < x < 4$.

9. $\dfrac{1-e^{-\pi}}{\pi} + \dfrac{2}{\pi}\displaystyle\sum_{n=1}^{\infty}\dfrac{1-(-1)^n e^{-\pi}}{n^2+1}\cos(nx)$; converges to $e^{-|x|}$ for $-\pi \le x \le \pi$.

11. $\dfrac{10+3\pi}{4} - \dfrac{1}{\pi}\displaystyle\sum_{n=1}^{\infty}\left[\dfrac{3(1-(-1)^n)}{n^2}\cos(nx) - \dfrac{5+(3\pi-5)(-1)^n}{n}\sin(nx)\right]$; converges to $\frac{1}{2}(3\pi+5)$ if $x = -\pi$ or

$x = \pi$, to $\frac{5}{2}$ if $x = 0$, to $-3x$ if $-\pi < x < 0$, and to 5 if $0 < x < \pi$.

13. $\dfrac{5}{2} - \dfrac{20}{\pi^2}\displaystyle\sum_{n=1}^{\infty}\dfrac{1}{(2n-1)^2}\cos\left((2n-1)\dfrac{\pi x}{5}\right)$; this converges to $|x|$ for $-5 \le x \le 5$.

15. $-\dfrac{5}{9} + \displaystyle\sum_{n=1}^{\infty}\left[\left(\dfrac{18}{n^3\pi^3}-\dfrac{3}{n\pi}\right)\sin\left(\dfrac{n\pi}{3}\right) - \dfrac{9}{n^2\pi^2}\cos\left(\dfrac{n\pi}{3}\right) - \dfrac{15(-1)^n}{n^2\pi^2}\right]\cos\left(\dfrac{n\pi x}{3}\right) +$

$\displaystyle\sum_{n=1}^{\infty}\left[\left(\dfrac{3}{n\pi}-\dfrac{18}{n^3\pi^3}\right)\cos\left(\dfrac{n\pi}{3}\right) - \dfrac{9}{n^2\pi^2}\sin\left(\dfrac{n\pi}{3}\right) + (-1)^n\left(\dfrac{18}{n^3\pi^3}-\dfrac{13}{n\pi}\right)\right]\sin\left(\dfrac{n\pi x}{3}\right)$; this converges to $-\frac{5}{2}$ if

$x = -3$ or $x = 3$, to $-x^2$ for $-3 < x < 1$, to 0 if $x = 1$, and to $1 + x$ if $1 < x < 3$.

17. $-\pi^2/12$

Section 13.4

1. cosine series: 4 (this function is its own Fourier cosine expansion), converging to 4 for $0 \le x \le 3$; sine series:

$\dfrac{16}{\pi}\displaystyle\sum_{n=1}^{\infty}\dfrac{1}{2n-1}\sin\left(\dfrac{(2n-1)\pi x}{3}\right)$, converging to 0 if $x=0$ or $x=3$, and to 4 for $0<x<3$

3. cosine series: $\dfrac{1}{2}\cos(x)-\dfrac{2}{\pi}\displaystyle\sum_{n=1}^{\infty}\dfrac{(-1)^n(2n-1)}{(2n-3)(2n+1)}\cos\left(\dfrac{(2n-1)x}{2}\right)$, converging to 0 if $0\le x<\pi$ or $x=2\pi$, to $-\dfrac{1}{2}$

if $x=\pi$, and to $\cos(x)$ if $\pi<x<2\pi$; sine series: $\dfrac{2}{\pi}\displaystyle\sum_{n=1}^{\infty}\dfrac{(-1)^n(2n-1)}{(2n-3)(2n+1)}\sin\left(\dfrac{(2n-1)x}{2}\right)-\dfrac{2}{\pi}\displaystyle\sum_{n=2}^{\infty}\dfrac{n}{n-1}\sin(nx)$,

converging to 0 if $0\le x<\pi$ or $x=2\pi$, to $-\dfrac{1}{2}$ if $x=\pi$, and to $\cos(x)$ if $\pi<x<2\pi$

5. cosine series: $\dfrac{4}{3}+\dfrac{16}{\pi^2}\displaystyle\sum_{n=1}^{\infty}\dfrac{(-1)^n}{n^2}\cos\left(\dfrac{n\pi x}{2}\right)$, converging to x^2 for $0\le x\le 2$; sine series: $-\dfrac{8}{\pi}\displaystyle\sum_{n=1}^{\infty}\left[\dfrac{(-1)^n}{n}+\right.$

$\left.\dfrac{2(1-(-1)^n)}{n^3\pi^2}\right]\sin\left(\dfrac{n\pi x}{2}\right)$, converging to x^2 for $0\le x<2$ and to 0 for $x=2$

7. cosine series: $\dfrac{1}{2}+\displaystyle\sum_{n=1}^{\infty}\left[\dfrac{4}{n\pi}\sin\left(\dfrac{2n\pi}{3}\right)+\dfrac{12}{n^2\pi^2}\cos\left(\dfrac{2n\pi}{3}\right)-\dfrac{6}{n^2\pi^2}(1+(-1)^n)\right]\cos\left(\dfrac{n\pi x}{3}\right)$, converging to x if

$0\le x\le 2$, to 1 if $x=2$, and to $2-x$ if $2<x\le 3$; sine series: $\displaystyle\sum_{n=1}^{\infty}\left[\dfrac{12}{n^2\pi^2}\sin\left(\dfrac{2n\pi}{3}\right)-\dfrac{4}{n\pi}\cos\left(\dfrac{2n\pi}{3}\right)+\right.$

$\left.\dfrac{2}{n\pi}(-1)^n\right]\sin\left(\tfrac{n\pi x}{3}\right)$, converging to x if $0\le x<2$, to 1 if $x=2$, to $2-x$ if $2<x<3$, and to 0 if $x=3$

9. cosine series: $\dfrac{5}{6}+\dfrac{16}{\pi^2}\displaystyle\sum_{n=1}^{\infty}\left[\dfrac{1}{n^2}\cos\left(\dfrac{n\pi}{4}\right)-\dfrac{4}{n^3\pi}\sin\left(\dfrac{n\pi}{4}\right)\right]\cos\left(\dfrac{n\pi x}{4}\right)$, converging to x^2 if $0\le x\le 1$, and to 1 if

$1<x\le 4$; sine series: $\displaystyle\sum_{n=1}^{\infty}\left[\dfrac{16}{n^2\pi^2}\sin\left(\dfrac{n\pi}{4}\right)+\dfrac{64}{n^3\pi^3}\left(\cos\left(\dfrac{n\pi}{4}\right)-1\right)-\dfrac{2(-1)^n}{n\pi}\right]\sin\left(\dfrac{n\pi x}{4}\right)$, converging to x^2 if

$0\le x\le 1$, to 1 if $1<x\le 4$, and to 0 if $x=4$

11. cosine series: $\dfrac{5}{7}+\dfrac{2}{\pi}\displaystyle\sum_{n=1}^{\infty}\dfrac{1}{n}\left\{11\sin\left(\dfrac{2n\pi}{7}\right)-4\sin\left(\dfrac{4n\pi}{7}\right)-\dfrac{28}{n\pi}\left[1-\cos\left(\dfrac{2n\pi}{7}\right)\right]\right\}\cos\left(\dfrac{n\pi x}{7}\right)$, converging to

$4x$ if $0\le x\le 2$, to -3 if $2<x<4$, to -1 if $x=4$, and to 1 if $4<x\le 7$; sine series:

$\dfrac{2}{\pi}\displaystyle\sum_{n=1}^{\infty}\dfrac{1}{n}\left[-11\cos\left(\dfrac{2n\pi}{7}\right)+4\cos\left(\dfrac{4n\pi}{7}\right)+\dfrac{28}{n\pi}\sin\left(\dfrac{2n\pi}{7}\right)-(-1)^n\right]\sin\left(\dfrac{n\pi x}{7}\right)$, converging to $4x$ if $0\le x<2$,

to $\dfrac{5}{2}$ if $x=2$, to -3 if $2<x<4$, to -1 if $x=4$, to 1 if $4<x<7$, and to 0 if $x=7$

13. Let $g(x)=\frac{1}{2}(f(x)+f(-x))$ and $h(x)=\frac{1}{2}(f(x)-f(-x))$. Then g is even and h is odd, and $f(x)=g(x)+h(x)$.

15. $\dfrac{1}{2}-\dfrac{\pi}{4}$

Section 13.5

3. (a) The Fourier series of f on $[-\pi,\pi]$ is $\dfrac{1}{4}\pi+\displaystyle\sum_{n=1}^{\infty}\left[\dfrac{(-1)^n-1}{\pi n^2}\cos(nx)+\dfrac{(-1)^{n+1}}{n}\sin(nx)\right]$. This series converges to

0 for $-\pi<x<0$, to x for $0<x<\pi$, and to $\frac{1}{2}(f(0+)+f(0-))$, or 0, at $x=0$.
(b) f is continuous, hence piecewise continuous on $[-\pi,\pi]$. By Theorem 13.5, its Fourier series can be integrated term-by-term to yield the integral of the sum of the Fourier series.

(c) First, $\displaystyle\int_{-\pi}^{x}f(t)\,dt=\begin{cases}0 & \text{if }-\pi\le x\le 0\\ \frac{1}{2}x^2 & \text{if }0<x\le\pi\end{cases}$.

This function is represented by the series obtained by integrating the Fourier series term-by-term from $-\pi$ to x to

obtain: $\dfrac{1}{4}x\pi+\dfrac{1}{4}\pi^2+\displaystyle\sum_{n=1}^{\infty}\left[\dfrac{(-1)^n-1}{\pi n^2}\dfrac{1}{n}\sin(nx)+\dfrac{(-1)^{n+1}}{n}\dfrac{1}{n}(-\cos(nx)+(-1)^n)\right]$.

5. (a) For $-\pi\le x\le\pi$, $x\sin(x)=\pi-\dfrac{1}{2}\pi\cos(x)+2\pi\displaystyle\sum_{n=2}^{\infty}\dfrac{(-1)^{n+1}}{n^2-1}\cos(nx)$.

(b) f is continuous with continuous first and second derivatives on $[-\pi,\pi]$, and $f(-\pi)=f(\pi)$. Theorem 13.6 gives us $x \cos(x) + \sin(x) = \frac{1}{2}\pi \sin(x) + 2\pi \sum_{n=2}^{\infty} \frac{(-1)^n}{n^2 - 1} n \sin(nx)$ for $-\pi < x < \pi$.

(c) The Fourier series of $x \cos(x) + \sin(x)$ on $[-\pi, \pi]$ is $\frac{1}{2}\pi \sin(x) + \sum_{n=2}^{\infty} 2n(-1)^n \pi \frac{1}{n^2 - 1} \sin(nx)$.

Section 13.6

3. *Hint:* Write the definition of $f'(x + p)$ and use the periodicity of f.   **5.** $1 - \frac{2}{\pi} \sum_{n=1}^{\infty} \frac{1}{n} \cos\left(n\pi x - \frac{\pi}{2}\right)$

7. $16 + \frac{48}{\pi^2} \sum_{n=1}^{\infty} \frac{1}{n^2} \sqrt{1 + n^2\pi^2} \cos\left(\frac{n\pi x}{2} + \tan^{-1}(n\pi)\right)$ **9.** $\frac{8}{\pi} \sum_{n=1}^{\infty} \frac{n}{4n^2 - 1} \cos\left(2n\pi x - \frac{\pi}{2}\right)$

11. $\frac{2}{\pi} \sum_{n=1}^{\infty} \frac{1}{n} \cos\left(n\pi x + \frac{\pi}{2}(-1)^n\right)$ **13.** $\frac{3}{2} + \frac{2}{\pi} \sum_{n=1}^{\infty} \frac{1}{2n - 1} \cos\left(\frac{(2n - 1)\pi x}{2} + \frac{\pi}{2}(1 - (-1)^n)\right)$

15. The fundamental period of $E \sin(\omega x)$ is $2\pi/\omega$. Then $\mathrm{RMS}(E \sin(\omega x)) = \sqrt{\frac{1}{(2\pi/\omega)} \int_0^{2\pi/\omega} E^2 \sin^2(\omega x)\, dx} = \frac{E}{\sqrt{2}}$.

17. $i(t) = \sum_{n=1}^{\infty} \frac{120(-1)^{n+1}(10 - n^2)\pi}{n^2[100n^2 + (10 - n^2)^2]} \cos(nt) + \frac{1200(-1)^{n+1}\pi}{n[100n^2 + (10 - n^2)^2]} \sin(nt)$

19. power $= i^2 R \approx 50(120\pi)^2(\frac{1}{181} + \frac{1}{6976}) \approx 40,279$ W

Section 13.7

1. $3 + \frac{3i}{\pi} \sum_{n=-\infty, n\neq 0}^{\infty} \frac{1}{n} e^{2n\pi i x/3}$ **3.** $\frac{3}{4} - \frac{1}{2\pi} \sum_{n=-\infty, n\neq 0}^{\infty} \frac{1}{n}\left[\sin\left(\frac{n\pi}{2}\right) + i\left(\cos\left(\frac{n\pi}{2}\right) - 1\right)\right] e^{n\pi i x/2}$

5. $\frac{1}{2} + \frac{3i}{\pi} \sum_{n=-\infty, n\neq 0}^{\infty} \frac{1}{2n - 1} e^{(2n-1)\pi i x/2}$ **7.** $\frac{1}{2} - \frac{2}{\pi^2} \sum_{n=-\infty, n\neq 0}^{\infty} \frac{1}{(2n - 1)^2} e^{(2n-1)\pi i x}$

9. $\frac{2}{3} + \frac{1}{4\pi^2} \sum_{n=-\infty, n\neq 0}^{\infty} \frac{1}{n^2}\left\{\left[4n\pi \sin\left(\frac{4n\pi}{3}\right) - 3 + 3\cos\left(\frac{4n\pi}{3}\right)\right] + i\left[4n\pi \cos\left(\frac{4n\pi}{3}\right) - 3\sin\left(\frac{4n\pi}{3}\right)\right]\right\} e^{2n\pi i x/3}$

11. The frequency spectrum consists of the point $(0, 4)$ and points $(2n, 2/|n|)$.

13. $f(x) = \frac{5}{3} + \frac{5}{\pi} \sum_{n=-\infty, n\neq 0}^{\infty} \frac{1}{n} \sin\left(\frac{n\pi}{3}\right) e^{n\pi i x/6}$ and $g(x) = \frac{5}{3} + \frac{5i}{\pi} \sum_{n=-\infty, n\neq 0}^{\infty} \frac{1}{n}\left\{\left[\cos\left(\frac{2n\pi}{3}\right) - 1\right] - i\sin\left(\frac{2n\pi}{3}\right)\right\} e^{n\pi i x/6}$. f and g have the same frequency spectra but different phase spectra.

CHAPTER 14

Section 14.1

1. $\int_0^{\infty} \left[\frac{2\sin(\pi\omega)}{\pi\omega^2} - \frac{2\cos(\pi\omega)}{\omega}\right] \sin(\omega x)\, d\omega$, converging to $-\frac{\pi}{2}$ if $x = -\pi$, to x for $-\pi < x < \pi$, to $\frac{\pi}{2}$ if $x = \pi$, and to 0 if $|x| > \pi$.

3. $\int_0^{\infty} \left(\frac{2}{\pi\omega}(1 - \cos(\pi\omega))\right) \sin(\omega x)\, d\omega$, converging to $-\frac{1}{2}$ if $x = -\pi$, to -1 if $-\pi < x < 0$, to $x = 0$ if $x = 0$, to 1 if $0 < x < \pi$, to $\frac{1}{2}$ if $x = \pi$, and to 0 if $|x| > \pi$.

5. $\int_0^{\infty} \frac{1}{\pi\omega^3}[400\omega \cos(100\omega) + (20,000\omega^2 - 4)\sin(100\omega)]\cos(\omega x)\, d\omega$, converging to x^2 if $-100 < x < 100$, to 5000 if $x = \pm 100$, and to 0 if $|x| > 100$.

7. $\int_0^{\infty} \frac{2}{\pi(\omega^2 - 1)}[-\sin(\pi\omega)\sin(2\pi\omega)\cos(\omega x) - \cos(\pi\omega)\sin(2\pi\omega)\sin(\omega x)]\, d\omega$, converging to $\sin(x)$ if $-3\pi \leq x \leq \pi$, and to 0 if $x < -3\pi$ or $x > \pi$.

9. $\int_0^\infty \dfrac{2}{\pi(1+\omega^2)}\cos(\omega x)\,d\omega$, converging to $e^{-|x|}$ for all real x.

Section 14.2

1. sine integral: $\displaystyle\int_0^\infty \dfrac{4}{\pi\omega^3}[10\omega\sin(10\omega)-(50\omega^2-1)\cos(10\omega)-1]\sin(\omega x)\,d\omega$;

cosine integral: $\displaystyle\int_0^\infty \dfrac{4}{\pi\omega^3}[10\omega\cos(10\omega)-(50\omega^2-1)\sin(10\omega)]\cos(\omega x)\,d\omega$; both integrals converge to x^2 for
$0\le x<10$, to 50 if $x=10$, and to 0 for $x>10$.

3. sine integral: $\displaystyle\int_0^\infty \dfrac{2}{\pi\omega}[1+\cos(\omega)-2\cos(4\omega)]\sin(\omega x)\,d\omega$; cosine integral: $\displaystyle\int_0^\infty \dfrac{2}{\pi\omega}[2\sin(4\omega)-\sin(\omega)]\cos(\omega x)\,d\omega$;

both integrals converge to 1 for $0<x<1$, to $\dfrac{3}{2}$ for $x=1$, to 2 for $1<x<4$, to 1 for $x=4$, and to 0 for $x>4$. The
cosine integral converges to 1 at $x=0$, while the sine integral converges to 0 at $x=0$.

5. sine integral: $\displaystyle\int_0^\infty \left\{\dfrac{2}{\pi\omega}[1+(1-2\pi)\cos(\pi\omega)-2\cos(3\pi\omega)]+\dfrac{4}{\pi\omega^2}\sin(\pi\omega)]\right\}\sin(\omega x)\,d\omega$; cosine integral:

$\displaystyle\int_0^\infty \left\{\dfrac{2}{\pi\omega}[(2\pi-1)\sin(\pi\omega)+2\sin(3\pi\omega)]+\dfrac{4}{\pi\omega^2}[\cos(\pi\omega)-1]\right\}\cos(\omega x)\,d\omega$; both integrals converge to $1+2x$ for
$0<x<\pi$, to $\frac{1}{2}(3+2\pi)$ for $x=\pi$, to 2 for $\pi<x<3\pi$, to 1 for $x=3\pi$, and to 0 for $x>3\pi$. The sine integral
converges to 0 for $x=0$, while the cosine integral converges to 1 for $x=0$.

7. sine integral: $\displaystyle\int_0^\infty \dfrac{2}{\pi}\left(\dfrac{\omega^3}{4+\omega^4}\right)\sin(\omega x)\,d\omega$; cosine integral: $\displaystyle\int_0^\infty \dfrac{2}{\pi}\left(\dfrac{2+\omega^2}{4+\omega^4}\right)\cos(\omega x)\,d\omega$; both integrals converge
to $e^{-x}\cos(x)$ for $x>0$. The cosine integral converges to 1 for $x=0$, and the sine integral converges to 0 for $x=0$.

9. sine integral: $\displaystyle\int_0^\infty \dfrac{2k}{\pi\omega}[1-\cos(c\omega)]\sin(\omega x)\,d\omega$; cosine integral: $\displaystyle\int_0^\infty \dfrac{2k}{\pi\omega}\sin(c\omega)\cos(\omega x)\,d\omega$; both integrals
converge to k for $0<x<c$, to $\frac{1}{2}k$ for $x=c$, and to 0 for $x>c$. The cosine integral converges to k for $x=0$, and the
sine integral to 0 for $x=0$.

11. For all x, $\displaystyle\int_0^\infty e^{-\omega}\cos(\omega x)\,d\omega=\dfrac{1}{1+x^2}$ and $\displaystyle\int_0^\infty e^{-\omega}\sin(\omega x)\,d\omega=\dfrac{x}{1+x^2}$.

Section 14.3

1. $\dfrac{i}{\pi}\displaystyle\int_{-\infty}^\infty \left(-2\omega\dfrac{(1-\omega^2)^2}{\left((1-\omega^2)^2+4\omega^2\right)^2}-8\dfrac{\omega^3}{\left((1-\omega^2)^2+4\omega^2\right)^2}\right)e^{i\omega x}\,d\omega$; converging to $xe^{-|x|}$ for all real x.

3. $i\displaystyle\int_{-\infty}^\infty \left(\dfrac{\sin(5\omega)}{\omega^2-\pi^2}\right)e^{i\omega x}\,d\omega$; converging to $\sin(\pi x)$ for $-5<x<5$, and to 0 for $|x|\ge 5$.

5. $\dfrac{1}{\pi}\displaystyle\int_{-\infty}^\infty \left[-e^{-1}\dfrac{\omega}{\omega^2+1}\sin(\omega)+e^{-1}\dfrac{1}{\omega^2+1}\cos(\omega)+\dfrac{i}{\omega^2}[\omega\cos(\omega)-\sin(\omega)]\right]e^{i\omega x}\,d\omega$; converging to x for
$-1<x<1$, to $\frac{1}{2}(1+e^{-1})$ for $x=1$, to $\frac{1}{2}(-1+e^{-1})$ for $x=-1$, and to $e^{-|x|}$ for $|x|>1$.

7. $\dfrac{1}{2\pi}\displaystyle\int_{-\infty}^\infty \left[-\dfrac{\cos(\pi\omega/2)}{\omega^2-1}+i\dfrac{\sin(\pi\omega/2)-\omega}{\omega^2-1}+\dfrac{1-\omega\sin(\pi\omega/2)}{\omega^2-1}+i\dfrac{\omega}{\omega^2-1}\cos(\pi\omega/2)\right]e^{i\omega x}\,d\omega$; converging to
$\cos(x)$ for $0<x<\pi/2$, to $\sin(x)$ for $-\pi/2<x<0$, to 0 for $|x|>\pi/2$, to $\frac{1}{2}$ at $x=0$, to $-\frac{1}{2}$ at $x=-\pi/2$, and to 0
at $x=\pi/2$.

9. $\dfrac{2i}{\omega}[\cos(\omega)-1]$ **11.** $-\dfrac{10}{\omega}e^{-2\omega i}\sin(\omega)$ **13.** $\dfrac{4}{1+4i\omega}e^{-(1+4i\omega)k/4}$ **15.** $\pi e^{-|\omega|}$ **17.** $\dfrac{24}{16+\omega^2}e^{2i\omega}$

19. $\dfrac{12}{16+(\omega-2)^2}+\dfrac{12}{16+(\omega+2)^2}$ **21.** $\dfrac{\pi i}{4}[e^{-2|\omega+1|}-e^{-2|\omega-1|}]$ **23.** $\dfrac{1}{5+i\omega}e^{-15-i\omega}$ **25.** $18\sqrt{\dfrac{2}{\pi}}e^{-8t^2}e^{-4it}$

27. $H(t+2)e^{-10-(5-3i)t}$ **29.** $H(t)[2e^{-3t}-e^{-2t}]$ **33.** $\hat f(-\omega)=\displaystyle\int_{-\infty}^\infty f(t)e^{-i(-\omega)t}\,dt=\overline{\int_{-\infty}^\infty f(t)e^{-i\omega t}\,dt}=\overline{\hat f(\omega)}$

35. $R(-\omega)=\displaystyle\int_{-\infty}^\infty f(t)\cos(-\omega t)\,dt=\int_{-\infty}^\infty f(t)\cos(\omega t)\,dt=R(\omega)$, so R is an even function.

$X(-\omega)=\displaystyle\int_{-\infty}^\infty f(t)\sin(-\omega t)\,dt=-\int_{-\infty}^\infty f(t)\sin(\omega t)\,dt=-X(\omega)$, so X is an odd function.

Section 14.4

1. $\pi i[H(-\omega)e^{3\omega} - H(\omega)e^{-3\omega}]$ **3.** $\dfrac{26}{(2+i\omega)^2}$ **5.** $\dfrac{i\omega}{3+i\omega} - 1$ **7.** $\dfrac{5\pi}{3}e^{-2i(\omega-3)}e^{-3|\omega-3|}$ **9.** $H(t)te^{-t}$

11. $\frac{1}{4}\left[1 - e^{-2(t+3)}\right]H(t+3) - \frac{1}{4}\left[1 - e^{-2(t-3)}\right]H(t-3)$ **13.** $\dfrac{3}{2\pi}e^{-4it}\left[\dfrac{1}{9+(t+2)^2} + \dfrac{1}{9+(t-2)^2}\right]$

17. Let $g(t) = f(t)$ in Problem 16. **19.** 3π

21. $\left|\hat{f}(\omega)\right| = \left|\displaystyle\int_{-\infty}^{\infty} f(t)e^{-i\omega t}\,dt\right| \le \displaystyle\int_{-\infty}^{\infty} |f(t)|\left|e^{-i\omega t}\right|\,dt = \int_{-\infty}^{\infty} |f(t)|\,dt$ because $\left|e^{-i\omega t}\right| = 1$ for ω and t real.

23. $4e^{-15}$ **25.** $\dfrac{1}{\omega^3}[50\omega^2 \sin(5\omega) + 20\omega \cos(5\omega) - 4\sin(5\omega)]$, $t_C = 0$, $t_R = \dfrac{25}{3}$

27. $e^{-4}\sin(4\omega)\dfrac{\omega}{\omega^2+1} - \dfrac{e^{-4}\cos(4\omega) - 1}{\omega^2+1} + i\left([e^{-4}\cos(4\omega) - 1]\dfrac{\omega}{\omega^2+1} + e^{-4}\dfrac{\sin(4\omega)}{\omega^2+1}\right)$, $t_C = 2$, $t_R = \dfrac{4}{3}$

29. $-\dfrac{2}{\omega^3}[-8\omega^2 \sin(2\omega) - 4\omega \cos(2\omega) + 2\sin(2\omega)] + 2i\dfrac{1}{\omega^3}[8\omega^2 \cos(2\omega) - 4\omega \sin(2\omega)]$, $t_C = 0$, $t_R = \dfrac{4}{3}$

Section 14.5

1. $\hat{f}_C(\omega) = \dfrac{1}{1+\omega^2}$, $\hat{f}_S(\omega) = \dfrac{\omega}{1+\omega^2}$

3. $\hat{f}_C(\omega) = \dfrac{1}{2}\left[\dfrac{\sin(K(1-\omega))}{1-\omega} + \dfrac{\sin(K(1+\omega))}{1+\omega}\right]$

$\hat{f}_S(\omega) = \dfrac{\omega}{\omega^2-1} - \dfrac{1}{2}\left[\dfrac{\cos(K(1+\omega))}{1+\omega} - \dfrac{\cos(K(1-\omega))}{1-\omega}\right]$

5. $\hat{f}_C(\omega) = \dfrac{1}{2}\left[\dfrac{1}{1+((1+\omega)^2} + \dfrac{1}{1+(1-\omega)^2}\right]$

$\hat{f}_S(\omega) = \dfrac{1}{2}\left[\dfrac{1+\omega}{1+(1+\omega)^2} - \dfrac{1-\omega}{1+(1-\omega)^2}\right]$

7. Sufficient conditions are: f'' and $f^{(3)}$ continuous on $[0, \infty)$; $f^{(4)}$ piecewise continuous on $[0, L]$ for every positive L; and $f(t) \to 0$, $f'(t) \to 0$, $f''(t) \to 0$; and $f^{(3)}(t) \to 0$ as $t \to \infty$. Also needed are $\displaystyle\int_0^\infty |f(t)|\,dt$ and $\displaystyle\int_0^\infty |f''(t)|\,dt$ convergent.

Section 14.6

1. $\dfrac{K}{n}[1 - (-1)^n]$ for $n = 1, 2, \ldots$ **3.** $-\dfrac{2}{n^3} + (-1)^n\left(\dfrac{2}{n^3} - \dfrac{\pi^2}{n}\right)$ for $n = 1, 2, \ldots$

5. $\dfrac{(-1)^n n \sin(a\pi)}{a^2 - n^2}$ for $n = 1, 2, \ldots$ if a is not an integer; if $a = m$, a positive integer, then $\tilde{f}_S(n) = \begin{cases} 0 & \text{if } n \neq m \\ \pi/2 & \text{if } n = m \end{cases}$.

7. $\dfrac{n}{n^2+1}[1 - (-1)^n e^{-\pi}]$ for $n = 1, 2, \ldots$ **9.** $\tilde{f}_C(n) = \begin{cases} \pi^2/2 & \text{if } n = 0 \\ [(-1)^n - 1]/n^2 & \text{if } n = 1, 2, \ldots \end{cases}$.

11. $\tilde{f}_C(0) = \dfrac{1}{4}\pi^4$, $\tilde{f}_C(n) = \dfrac{6}{n^4} + (-1)^n\left(\dfrac{3\pi^2}{n^2} - \dfrac{6}{n^4}\right)$ if $n = 1, 2, \ldots$.

13. $\dfrac{a}{a^2 - n^2}[1 - (-1)^n]\cos(a\pi)$ for $n = 0, 1, 2, \ldots$, if a is not an integer.

15. Write $\tilde{f'}_S(n) = \displaystyle\int_0^\pi f'(x)\sin(nx)\,dx$ and integrate by parts.

17. $\tilde{f}_S(n+m) = \displaystyle\int_0^\pi f(x)\sin((n+m)x)\,dx = \int_0^\pi [f(x)\cos(mx)]\sin(nx)\,dx + \int_0^\pi [f(x)\sin(mx)]\cos(nx)\,dx = \tilde{g}_S(n) + \tilde{h}_C(n)$

19. $\tilde{g}_C(n) = \displaystyle\int_0^\pi f(x)\cos(mx)\cos(nx)\,dx = \int_0^\pi f(x)\dfrac{1}{2}[\cos((n-m)x) + \cos((n+m)x)]\,dx = \dfrac{1}{2}\left[\tilde{f}_C(n-m) + \tilde{f}_C(n+m)\right]$

Section 14.7

1. $\mathbb{D}[u](0) = \sum_{j=0}^{5} \cos(j) \approx -.23582$, $\mathbb{D}[u](1) = \sum_{j=0}^{5} \cos(j)e^{-\pi ij/3} \approx 2.9369 - .42794i$,

$\mathbb{D}[u](2) = \sum_{j=0}^{5} \cos(j)e^{-2\pi ij/3} \approx .13292 - 1.6579 \times 10^{-2}i$, $\mathbb{D}[u](3) = \sum_{j=0}^{5} \cos(j)e^{-\pi ij} \approx 9.6238 \times 10^{-2}$,

$\mathbb{D}[u](4) = \sum_{j=0}^{5} \cos(j)e^{-4\pi ij/3} \approx .13292 + 1.6579 \times 10^{-2}i$, $\mathbb{D}[u](-1) = \sum_{j=0}^{5} \cos(j)e^{\pi ij/3} \approx 2.9369 + .42794i$,

$\mathbb{D}[u](-2) = \sum_{j=0}^{5} \cos(j)e^{2\pi ij/3} \approx .13292 + 1.6579 \times 10^{-2}i$, $\mathbb{D}[u](-3) = \sum_{j=0}^{5} \cos(j)e^{\pi ij} \approx 9.6238 \times 10^{-2}$,

$\mathbb{D}[u](-4) = \sum_{j=0}^{5} \cos(j)e^{4\pi ij/3} \approx .13292 - 1.6579 \times 10^{-2}i$

3. $\mathbb{D}[u](0) = \sum_{j=0}^{5} \frac{1}{j+1} = 2.45$, $\mathbb{D}[u](1) = \sum_{j=0}^{5} \frac{1}{j+1}e^{-\pi ij/3} \approx .81667 - .40415i$, $\mathbb{D}[u](2) \approx .65 - .17321i$,
$\mathbb{D}[u](3) \approx .61667$, $\mathbb{D}[u](4) \approx .65 + .17321i$, $\mathbb{D}[u](-1) \approx .81667 + .40415i$, $\mathbb{D}[u](-2) \approx .65 + .17321i$,
$\mathbb{D}[u](-3) \approx .61667$, $\mathbb{D}[u](-4) \approx .65 - .17321i$

5. $\mathbb{D}[u](0) = 55$, $\mathbb{D}[u](1) \approx -6.0 + 31.177i$, $\mathbb{D}[u](2) \approx -14.0 + 10.392i$, $\mathbb{D}[u](3) = 15$, $\mathbb{D}[u](4) \approx -14.0 - 10.392i$,
$\mathbb{D}[u](-1) \approx -6.0 - 31.177i$, $\mathbb{D}[u](-2) \approx -14.0 - 10.392i$, $\mathbb{D}[u](-3) \approx -15$, $\mathbb{D}[u](-4) \approx -14.0 + 10.392i$

7. The inverse is $\{u_j\}_{j=0}^{5}$, where $u_0 = \frac{1}{6}\sum_{k=0}^{5}(1+i)^k \approx -1.3333 + .16667i$, $u_1 = \frac{1}{6}\sum_{k=0}^{5}(1+i)^k e^{\pi ik/3} \approx -.42703$
$+ .54904i$, $u_2 \approx -1.6346 \times 10^{-2} + .561i$, $u_3 \approx .33333 + .5i$, $u_4 \approx .84968 + .27233i$, and $u_5 \approx 1.5937 - 2.049i$

9. $u_0 = \frac{1}{7}\sum_{k=0}^{6} e^{-ik} \approx .10348 + 1.4751 \times 10^{-2}i$, $u_1 = \frac{1}{7}\sum_{k=0}^{6} e^{-ik}e^{2\pi ik/7} \approx .93331 - .29609i$,
$u_2 \approx -9.4163 \times 10^{-2} + 8.8785 \times 10^{-2}i$, $u_3 \approx -2.3947 \times 10^{-2} + 6.2482 \times 10^{-2}i$,
$u_4 \approx 4.3074 \times 10^{-3} + 5.1899 \times 10^{-2}i$, $u_5 \approx 2.5788 \times 10^{-2} + 4.3852 \times 10^{-2}i$, $u_6 \approx 5.1222 \times 10^{-2} + 3.4325 \times 10^{-2}i$

11. $u_0 \approx -.1039$, $u_1 \approx .42051 + .29456i$, $u_2 \approx .13143 + 3.1205 \times 10^{-2}i$,
$u_3 \approx .13143 - 3.1205 \times 10^{-2}i$, $u_4 \approx .42051 - .29456i$

13. The Fourier coefficients are $d_k = \frac{1}{2}\int_0^2 \cos(\xi)e^{-\pi ki\xi}\,d\xi = -\frac{1}{2}\frac{\sin(2)}{\pi^2 k^2 - 1} + \frac{k\pi i}{2}\frac{\cos(2)-1}{\pi^2 k^2 - 1}$. For the DFT

approximation, choose $N = 128$, and approximate d_k by $f_k = \frac{1}{128}\sum_{j=0}^{127}\cos(j/64)e^{-\pi ijk/64}$.

Then
$d_0 = \frac{1}{2}\sin(2) \approx 0.45465$, $f_0 \approx 0.46017$
$d_1 \approx -5.1259 \times 10^{-2} - .2508i$ and $f_1 \approx -4.5737 \times 10^{-2} - .25075i$
$d_2 \approx -1.1816 \times 10^{-2} - .11562i$ and $f_2 \approx -6.2931 \times 10^{-3} - .11553i$
$d_3 \approx -5.1767 \times 10^{-3} - 7.5984 \times 10^{-2}i$ and $f_3 \approx 3.4589 \times 10^{-4} - 7.5849 \times 10^{-2}i$
$d_{-1} \approx -5.1259 \times 10^{-2} + .2508i$ and $f_{-1} \approx -4.5737 \times 10^{-2} + .25075i$
$d_{-2} \approx -1.1816 \times 10^{-2} + .11562i$, $f_{-2} \approx -6.2931 \times 10^{-3} + .11553i$
$d_{-3} \approx -5.1767 \times 10^{-3} + 7.5984 \times 10^{-2}i$, $f_{-3} \approx 3.4589 \times 10^{-4} + 7.5849 \times 10^{-2}i$

15. Now, $d_k = \int_0^1 \xi^2 e^{-2\pi ki\xi}\,d\xi = \frac{1}{2\pi^2 k^2} + i\frac{1}{2\pi k}$ if $k \neq 0$, and $d_0 = \frac{1}{3}$. The DFT approximation is

$f_k = \frac{1}{128}\sum_{j=0}^{127}\left(\frac{j}{128}\right)^2 e^{-\pi ijk/64}$.

$d_0 = \frac{1}{3}$, $f_0 \approx .32944$
$d_1 \approx 5.0661 \times 10^{-2} + .15915i$, $f_1 \approx 4.6765 \times 10^{-2} + .15912i$
$d_2 \approx 1.2665 \times 10^{-2} + 7.9577 \times 10^{-2}i$, $f_2 \approx 8.7691 \times 10^{-3} + 7.9514 \times 10^{-2}i$
$d_3 \approx 5.629 \times 10^{-3} + 5.3052 \times 10^{-2}i$. $f_3 \approx 1.7329 \times 10^{-3} + 5.2956 \times 10^{-2}i$
$d_{-1} \approx 5.0661 \times 10^{-2} - .15915i$, $f_{-1} \approx 4.6765 \times 10^{-2} - .15912i$
$d_{-2} \approx 1.2665 \times 10^{-2} - 7.9577 \times 10^{-2}i$, $f_{-2} \approx 8.7691 \times 10^{-3} - 7.9514 \times 10^{-2}i$
$d_{-3} \approx 5.629 \times 10^{-3} - 5.3052 \times 10^{-2}i$, $f_{-3} \approx 1.7329 \times 10^{-3} - 5.2956 \times 10^{-2}i$

17. Now, $d_k = \dfrac{1}{3\pi} \displaystyle\int_0^{3\pi} \xi \cos(\xi)e^{-2\pi k i \xi/3\pi}\, d\xi = \dfrac{3}{\pi}\dfrac{-8k^2 - 18}{(2k-3)^2\,(2k+3)^2} - 3i\dfrac{8k^3 - 18k}{(2k-3)^2\,(2k+3)^2}$ and the DFT

approximation is $f_k = \dfrac{1}{128}\displaystyle\sum_{j=0}^{128}\left(\dfrac{3\pi j}{128}\right)\cos(3\pi j/128)e^{-\pi i j k/64}$.

$d_0 \approx -.21221,\; f_0 \approx -0.17549$
$d_1 \approx -.99313 + 1.2i,\; f_1 \approx -0.95641 + 1.2003i$
$d_2 \approx -.97442 - 1.7143i,\; f_2 \approx -0.93770 - 1.7137i$
$d_3 \approx -.11789 - .66667i,\; f_3 \approx -8.1773 \times 10^{-2} - .66576i$
$d_{-1} \approx -.99313 - 1.2i,\; f_{-1} \approx -0.95641 - 1.2003i$
$d_{-2} \approx -.97442 + 1.7143i,\; f_{-2} \approx -0.93770 + 1.7137i$
$d_{-3} \approx -.11789 + .66667i,\; f_{-3} \approx -8.1173 \times 10^{-2} + .66576i$

Section 14.8

1. The complex Fourier series is $2 + \displaystyle\sum_{k=-\infty, k\neq 0}^{\infty}\dfrac{i}{\pi k}e^{\pi i k t}$, so $S_{10}(\tfrac{1}{8}) = 1.0207 + 1.6653 \times 10^{-16}i$. Approximate

$S_{10}(\tfrac{1}{8}) \approx \tfrac{1}{128}\sum_{k=-0}^{127} V_k e^{\pi i k/8}$, which is $1.0552 + 10^{-14}i$;
$\left|1.0207 + 1.6653 \times 10^{-16}i - (1.0552 - 2.0983 \times 10^{-16}i)\right| = 0.003452$.

3. The complex Fourier series is $\displaystyle\sum_{k=-\infty}^{\infty}\left[-\dfrac{1}{2}\dfrac{\sin(2)}{\pi^2 k^2 - 1} + \dfrac{i}{2}k\pi\dfrac{\cos(2)-1}{\pi^2 k^2 - 1}\right]e^{\pi i k t}$. Then

$S_{10}(\tfrac{1}{8}) = 1.0672 - 3.4694 \times 10^{-18}i$. Approximate $S_{10}(\dfrac{1}{8}) \approx \dfrac{1}{128}\displaystyle\sum_{k=0}^{127} V_k e^{\pi i k/8}$, which works out to

$1.0428 + 3.8025 \times 10^{-15}i;\; \left|1.0672 - 3.4694 \times 10^{-18}i - (1.0428 + 3.8025 \times 10^{-15}i)\right| = 0.002440$.

5. The complex Fourier series is $\dfrac{1}{4} + \displaystyle\sum_{k=-\infty, k\neq 0}^{\infty}\left(\dfrac{3}{4}\dfrac{1}{\pi^2 k^2} + \dfrac{i}{4}\dfrac{2k^2\pi^2 - 3}{\pi^3 k^3}\right)e^{2k\pi i t}$, and

$S_{10}(\tfrac{1}{4}) = -7.2901 \times 10^{-4} + 1.0408 \times 10^{-17}i$. Approximate $S_{10}(\dfrac{1}{4}) \approx \dfrac{1}{128}\displaystyle\sum_{k=0}^{127} V_k e^{\pi k i/2}$, which works out to

$3.4826 \times 10^{-3} + 9.1593 \times 10^{-16}i;\; \left|-7.2901 \times 10^{-4} + 1.0408 \times 10^{-17}i - (3.4826 \times 10^{-3} + 9.1593 \times 10^{-16}i)\right|$
≈ 0.004212.

7. The complex Fourier series is $1 + \displaystyle\sum_{k=-\infty, k\neq 0}^{\infty}\dfrac{i}{\pi k}e^{\pi i k t}$, so $S_{10}(\tfrac{1}{2}) = .46847$. Approximate $S_{10}(\tfrac{1}{2}) \approx \dfrac{1}{128}\displaystyle\sum_{k=0}^{127} V_k e^{\pi i k/2} =$

$0.4693 + 2.3809 \times 10^{-15}i$. The difference is $\left|0.46847 - (0.47693 + 2.3809 \times 10^{-15}i)\right| = 0.008454$.

9. $0.14386 - 0.12455i$ **11.** $-6.5056 \times 10^{-3} - 2.191 \times 10^{-3}i$ **13.** $-1.5699 \times 10^{-2} - 4.8753 \times 10^{-17}i$

In the graphs associated with Problems 15, 17, and 19, the series 1 points are the actual values computed from the transform, and the series two points are the DFT approximations. In all cases the approximations can be improved by choosing N larger.

15. $\hat{f}(\omega) = \dfrac{2\omega\sin(2\omega) + \cos(2\omega) - \omega\sin(\omega) - \cos(\omega)}{\omega^2} + i\dfrac{2\omega\cos(2\omega) - \sin(2\omega) - \omega\cos(\omega) + \sin(\omega)}{\omega^2}$. Note: in the

sum of equation (14.24), $11 \leq j \leq 20$ because $f(t)$ is zero outside the interval $[1, 2)$. Generate the following table using $L = 4$:

k	$\hat{f}(k/4)$	DFT $\hat{f}(k/4)$	$\left|\hat{f}(k/4)\right|$	$\left|\text{DFT } \hat{f}(k/4)\right|$
1	$1.3845 - 0.5673i$	$1.3764 - 0.5714i$	1.4962	1.4903
2	$1.0579 - 1.0421i$	$1.0445 - 1.048i$	1.485	1.4796
3	$0.5761 - 1.3488i$	$0.5558 - 1.3521i$	1.4667	1.4619
4	$0.2068 - 1.4404i$	$-.0573 - 1.4372i$	1.4406	1.4383
5	$-0.5162 - 1.3098i$	$-0.5453 - 1.2959i$	1.4078	1.406
6	$-0.9483 - 0.9865i$	$-0.9749 - 0.9602i$	1.3684	1.3684
7	$-1.2108 - 0.5325i$	$-1.2288 - 0.4945i$	1.3227	1.3247
8	$-1.2708 - 0.0295i$	$-1.2751 + .0170i$	1.2711	1.2752
9	$-1.1323 + 0.4386i$	$-1.1194 + 0.4866i$	1.2143	1.2206
10	$-0.8329 + 0.7966i$	$-0.8026 + 0.8393i$	1.1525	1.1613
11	$-0.4359 + 0.9953i$	$-0.3909 + 1.0258i$	1.0866	1.0978
12	$-0.0166 + 1.0168i$	$.0374 + 1.0299i$	1.0169	1.0306

The plots below compare the (a) real parts, (b) imaginary parts, and (c) absolute values, of $\hat{f}(k/4)$ and the DFT approximations.

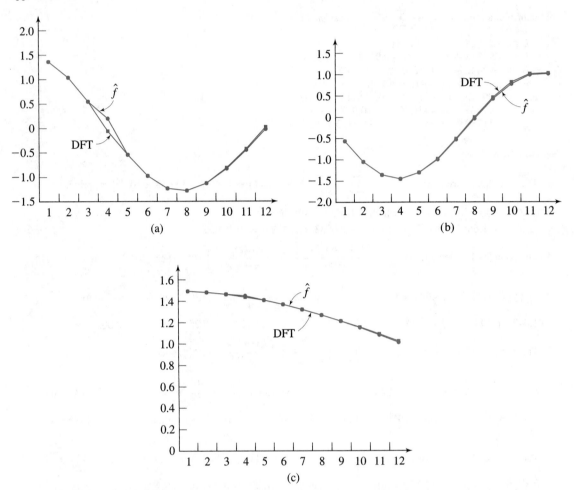

17. $\hat{f}(\omega) = \dfrac{\sin(\omega)}{\omega} + i\,\dfrac{\cos(\omega) - 1}{\omega}$. With $L = 4$ we obtain the following table (with j summing from 0 to 10 in equation (14.24)):

k	$\hat{f}(k/4)$	DFT $\hat{f}(k/4)$	$\left|\hat{f}(k/4)\right|$	$\left|\text{DFT } \hat{f}(k/4)\right|$
1	$.9896 - .1244i$	$1.0686 - .1318i$	.99739	1.0767
2	$.9589 - .2448i$	$1.035 - .25925i$	.98965	1.067
3	$.9089 - .3577i$	$.98047 - .37821i$	.97675	1.0509
4	$.8415 - .4597i$	$.90716 - .48489i$	.95888	1.0286
5	$.7592 - .5477i$	$.81791 - .57604i$	.93614	1.0004
6	$.6650 - .6195i$	$.71616 - .64909i$	.90885	.96654
7	$.5623 - .6733i$	$.60574 - .70222i$	.87722	.92738
8	$.4546 - .7081i$	$.49076 - .73447i$	.84147	.88334
9	$.3458 - .7236i$	$.37537 - .74574i$	.80198	.83488
10	$.2394 - .7205i$	$.26362 - .73678i$	.75923	.78252
11	$.1388 - .6997i$	$.15924 - .70914i$	.71333	.7268
12	$.0470 - .6633i$	$.0655 - .66506i$	.66496	.66828

The plots below compare the (a) real parts, (b) imaginary parts, and (c) absolute values, of $\hat{f}(k/4)$ and the DFT approximations.

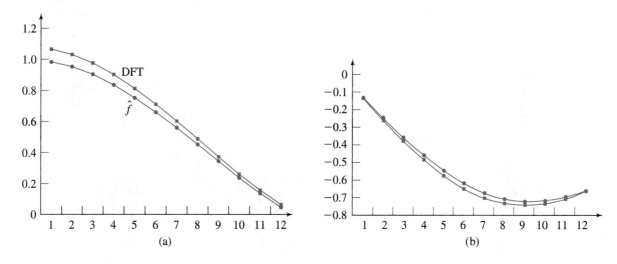

(a)

(b)

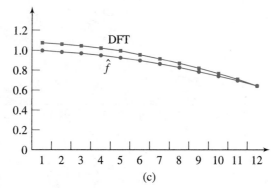

(c)

19. $\hat{f}(\omega) = \omega \dfrac{\sin(\omega)}{\pi^2 - \omega^2} + i\omega \dfrac{\cos(\omega) + 1}{\pi^2 - \omega^2}$. With $L = 4$, obtain

k	$\hat{f}(k/4)$	DFT $\hat{f}(k/4)$	$\left|\hat{f}(k/4)\right|$	$\left|\text{DFT } \hat{f}(k/4)\right|$
1	$.0063 + .0502i$	$.02545 + .0581i$	.0506	.06339
2	$.0249 + .0976i$	$.0469 + .11285i$	.10073	.12220
3	$.0549 + .1395i$	$.0814 + .16126i$	.14997	.18066
4	$.0949 + .1737i$	$.12732 + .20052i$	.19789	.23753
5	$.1428 + .1979i$	$.18222 + .2283i$	.24406	.2921
6	$.1964 + .2108i$	$.24338 + .24285i$	.28808	.34382
7	$.2530 + .2113i$	$.3077 + .24309i$	.32958	.39213
8	$.3098 + .1989i$	$.37193 + .22863i$	.36820	.43658
9	$.3642 + .1740i$	$.43284 + .19983i$	.40363	.47674
10	$.4134 + .1373i$	$.48736 + .15774i$	.4356	.51225
11	$.4549 + .0902i$	$.53272 + .10408i$	.46379	.54279
12	$.4868 + .0345i$	$.56663 + .0411i$	.48806	.56812

The plots below compare the (a) real parts, (b) imaginary parts, and (c) absolute values, of $\hat{f}(k/4)$ and the DFT approximations.

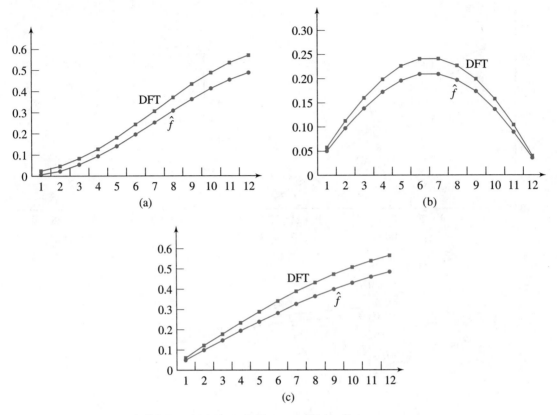

(a)

(b)

(c)

21. The Fourier series of f is $\displaystyle\sum_{n=-\infty, n\neq 0}^{\infty} \frac{i}{n\pi}(-1+(-1)^n)e^{n\pi it/2}$, or $\displaystyle\sum_{n=1}^{\infty} \frac{4}{(2n-1)\pi} \sin\left(\frac{(2n-1)\pi t}{2}\right)$.

Then $S_N(t) = \displaystyle\sum_{n=1}^{N} \frac{2}{n\pi}(1-(-1)^n) \sin\left(\frac{n\pi t}{2}\right)$. The Nth Cesàro sum is

$\sigma_N(t) = \displaystyle\sum_{n=-N}^{N} \left(1-\left|\frac{n}{N}\right|\right)\frac{i}{n\pi}(-1+(-1)^n)e^{n\pi it/2}$, or $\sigma_N(t) = \displaystyle\sum_{n=1}^{N} \left(1-\left|\frac{n}{N}\right|\right)\frac{2}{n\pi}(1-(-1)^n) \sin\left(\frac{n\pi t}{2}\right)$.

Graphs are drawn for $N = 5, 10, 25$.

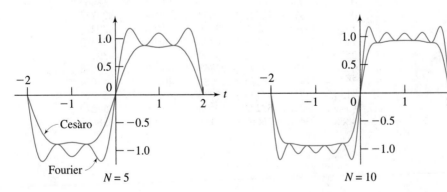

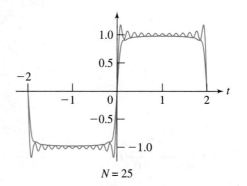

$N = 25$

23. $S_N(t) = \displaystyle\sum_{n=-N, n\neq 0}^{N} \frac{i}{n\pi}\left((-1)^n - \cos(n\pi/2)\right)e^{n\pi it} = \sum_{n=1}^{N} \frac{2}{n\pi}\left(\cos(n\pi/2) - (-1)^n\right)\sin(n\pi t),\ \sigma_N(t) =$

$\displaystyle\sum_{n=-N, n\neq 0}^{N}\left(1 - \left|\frac{n}{N}\right|\right)\frac{i}{n\pi}\left((-1)^n - \cos(n\pi/2)\right)e^{n\pi it} = \sum_{n=1}^{N}\left(1 - \left|\frac{n}{N}\right|\right)\frac{2}{n\pi}\left(\cos(n\pi/2) - (-1)^n\right)\sin(n\pi t)$

Graphs are drawn for $N = 5, 10, 25$.

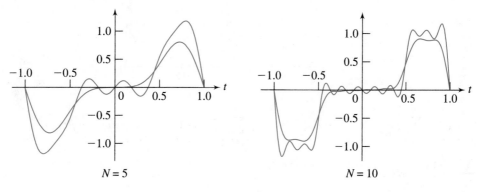

$N = 5$ $N = 10$

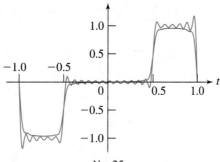

$N = 25$

25. $S_N(t) = \dfrac{17}{4} + \displaystyle\sum_{n=-N, n\neq 0}^{N}\left(\frac{1}{2}\frac{1-(-1)^n}{n^2\pi^2} + \frac{i}{n\pi}\left(1 - \frac{1}{2}(-1)^n + \frac{7}{2}((-1)^n - 1)\right)\right)e^{n\pi it} =$

$\dfrac{17}{4} + \displaystyle\sum_{n=1}^{N}\frac{1-(-1)^n}{n^2\pi^2}\cos(n\pi t) - \frac{2}{n\pi}\left(-\frac{5}{2} + 3(-1)^n\right)\sin(n\pi t),$

$\sigma_N(t) = \dfrac{17}{4} + \displaystyle\sum_{n=-N, n\neq 0}^{N}\left(1 - \left|\frac{n}{N}\right|\right)\left[\frac{1}{2}\frac{1-(-1)^n}{n^2\pi^2} + \frac{i}{n\pi}\left(1 - \frac{1}{2}(-1)^n + \frac{7}{2}[(-1)^n - 1]\right)\right]e^{n\pi it} =$

$\dfrac{17}{4} + \displaystyle\sum_{n=1}^{N}\left(1 - \left|\frac{n}{N}\right|\right)\left[\frac{1-(-1)^n}{n^2\pi^2}\cos(n\pi t) - \frac{2}{n\pi}\left(-\frac{5}{2} + 3(-1)^n\right)\sin(n\pi t)\right]$

Graphs are drawn for $N = 5, 10, 25$ (see next page).

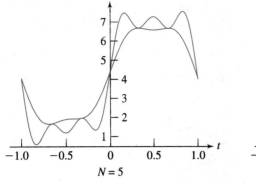

$N = 5$

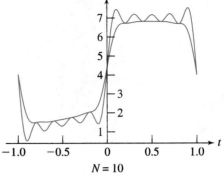

$N = 10$

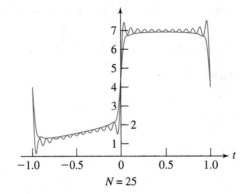

$N = 25$

27. $S_N(t) = 1 + \sum_{n=-N, n\neq 0}^{N} \frac{i}{n\pi} (3(-1)^n - 1) e^{n\pi i t/2} = 1 + \sum_{n=1}^{N} \frac{-2}{n\pi} (3(-1)^n - 1) \sin(n\pi t/2), \sigma_N(t) =$

$1 + \sum_{n=1}^{N} \frac{-2}{n\pi} \left(1 - \left|\frac{n}{N}\right|\right) (3(-1)^n - 1) \sin(n\pi t/2)$

Hamming filtered partial sum: $H_N(t) = 1 + \sum_{n=1}^{N} \frac{-2}{n\pi} (0.54 + 0.46 \cos(\pi n/N)) (3(-1)^n - 1) \sin(n\pi t/2)$

Gauss filtered partial sum: $G_N(t) = 1 + \sum_{n=1}^{N} -\frac{2}{n\pi} e^{-a\pi^2 n^2/N^2} (3(-1)^n - 1) \sin(n\pi t/2)$

Graphs are drawn for $N = 5, 10, 25$, with $a = \frac{1}{2}$ in the Gauss filter.

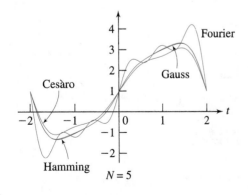

$N = 5$

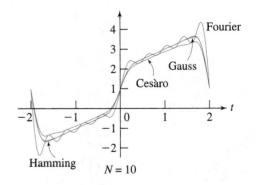

$N = 10$

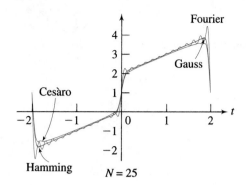

29. $S_N(t) = 1 + \displaystyle\sum_{n=-N,n\neq0}^{n=N} \left[\frac{1}{n^2\pi^2} (-n\pi \sin(n\pi/3) - 3\cos(n\pi/3) + 3(-1)^n) + \frac{1}{2}\frac{\sin(n\pi/3)}{n\pi} \right.$

$\left. + i\left(-\frac{1}{n^2\pi^2}(n\pi\cos(n\pi/3) - 3\sin(n\pi/3) + 3n\pi(-1)^n) + \frac{1}{2}\frac{(-1)^{n+1} + \cos(n\pi/3)}{n\pi}\right)\right] e^{n\pi it/3}$. This can be written

$S_N(t) = 1 + \displaystyle\sum_{n=1}^{N}\left[\frac{2}{n^2\pi^2}(-n\pi\sin(n\pi/3) - 3\cos(n\pi/3) + 3(-1)^n) + \frac{\sin(n\pi/3)}{n\pi}\right]\cos(n\pi t/3)$

$+ \displaystyle\sum_{n=1}^{N}\left[\frac{2}{n^2\pi^2}(n\pi\cos(n\pi/3) - 3\sin(n\pi/3) + 3n\pi(-1)^n) - \frac{(-1)^{n+1} + \cos(n\pi/3)}{n\pi}\right]\sin(n\pi t/3)$. If this is written

$S_N(t) = 1 + \sum_{n=-N,n\neq0}^{N} d_n e^{n\pi it/3}$, then the Cesàro, Hamming, and Gauss partial sums are, respectively,

$\sigma_N(t) = 1 + \displaystyle\sum_{n=-N,n\neq0}^{N} d_n\left(1 - \left|\frac{n}{N}\right|\right)e^{n\pi it/3}$, $H_N(t) = 1 + \displaystyle\sum_{n=-N,n\neq0}^{N} d_n(0.54 + 0.46\cos(n\pi/N))e^{n\pi it/3}$, and

$G_N(t) = 1 + \displaystyle\sum_{n=-N,n\neq0}^{N} d_n e^{-an^2\pi^2/N^2}e^{n\pi it/3}$, with $a = 1/2$. Graphs are shown below for $N = 5, 10, 25$.

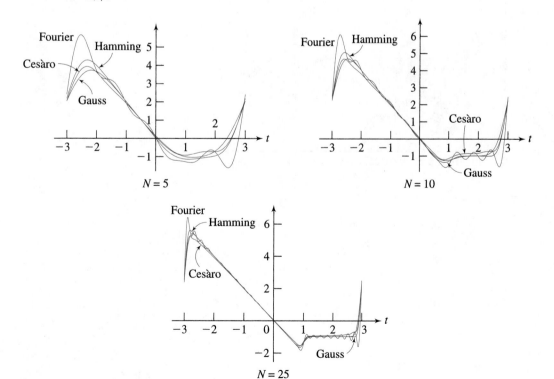

Section 14.9

9. Power spectrum of $y(t) = 4\sin(80\pi t) - \sin(20\pi t)$

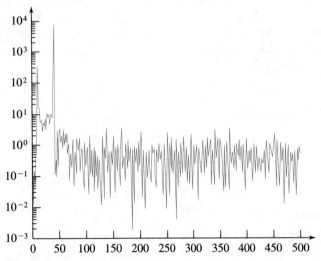

11. Power spectrum of $y(t) = 3\cos(90\pi t) - \sin(30\pi t)$

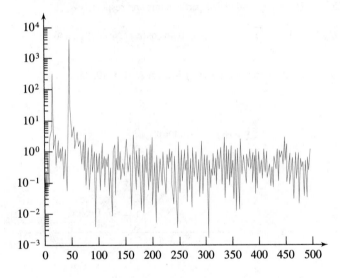

13. Corrupted signal of $y(t) = \cos(30\pi t) + \cos(70\pi t) + \cos(140\pi t)$

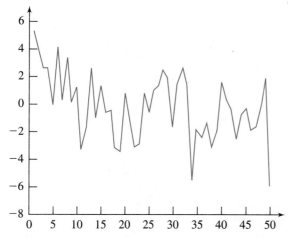

Frequency components of the corrupted signal

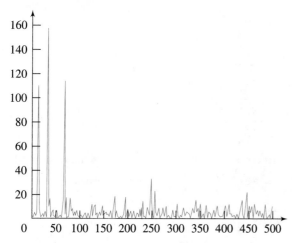

15. Corrupted signal of $y(t) = \cos(20\pi t) + \sin(140\pi t) + \cos(240\pi t)$

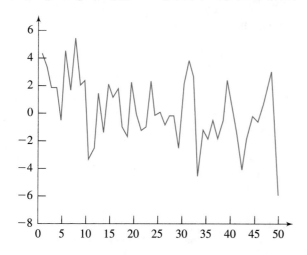

Frequency components of the corrupted signal

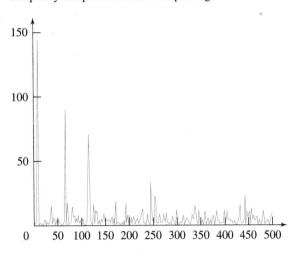

CHAPTER 15

Section 15.1

3. We will illustrate for $P_3(x)$. By Rodrigues's formula,

$$P_3(x) = \frac{1}{2^3 3!} \frac{d^3}{dx^3}((x^2 - 1)^3) = \frac{1}{48}(120x^3 - 72x) = \frac{1}{2}(5x^3 - 3x).$$

5. For example, with $n = 3$ we have $\left[\frac{3}{2}\right] = 1$ and $\displaystyle\sum_{k=0}^{1}(-1)^k \frac{(6 - 2k)!}{8k!(3 - k)!(3 - 2k)!}x^{3-2k} =$

$$\frac{6!}{8(3!)(3!)}x^3 - \frac{4!}{8(2!)(1!)}x = \frac{5}{2}x^3 - \frac{3}{2}x = P_3(x).$$

7. Substitute $Q(x) = P_n(x)u(x)$ in Legendre's equation. After cancellations because $P_n(x)$ is one solution, we obtain

$$u'' + \left(2\frac{P_n'}{P_n} - \frac{2x}{1 - x^2}\right)u' = 0. \text{ Let } v = u' \text{ to write } \frac{v'(x)}{v(x)} = -2\frac{P_n'(x)}{P_n(x)} + \frac{2x}{1 - x^2}. \text{ Integrate to obtain}$$

$$\ln(v(x)) = -2\ln(P_n(x)) - \ln(1 - x^2), \text{ so } v(x) = \frac{1}{P_n(x)^2(1 - x^2)}. \text{ Then } u(x) = \int \frac{1}{P_n(x)^2(1 - x^2)}dx \text{ and}$$

$$Q(x) = P_n(x)u(x).$$

17. For $-1 < x < 1$, $\sin(\pi x/2) = \dfrac{12}{\pi^2}x + 168\dfrac{\pi^2 - 10}{\pi^4}\dfrac{1}{2}(5x^3 - 3x) + 660\dfrac{-112\pi^2 + \pi^4 + 1008}{\pi^6}\dfrac{1}{8}(63x^5 - 70x^3 +$

$15x) + \cdots$. In the graph, the sum of these terms is indistinguishable from the function.

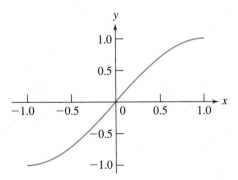

19. For $-1 < x < 1$, $\sin^2(x) = \left[-\frac{1}{2}\cos(1)\sin(1) + \frac{1}{2}\right] + \left[-\frac{5}{8}\cos(1)\sin(1) + \frac{15}{8} - \frac{15}{4}\cos^2(1)\right]\frac{1}{2}(3x^2 - 1) +$
$\left[\frac{531}{32}\cos(1)\sin(1) - \frac{585}{32} + \frac{585}{16}\cos^2(1)\right]\frac{1}{8}(35x^4 - 30x^2 + 3) + \cdots$. In the graph, the sum of these terms is indistinguishable from the function.

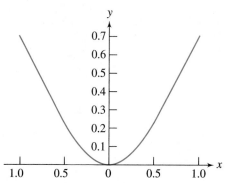

21. For $-1 < x < 0$ and for $0 < x < 1$, $f(x) = \frac{3}{2}x - \frac{7}{8}\frac{1}{2}(5x^3 - 3x) + \frac{11}{16}\frac{1}{8}(63x^5 - 70x^3 + 15x) + \cdots$. The graph below shows the function and the sum of these three terms of the Fourier–Legendre expansion.

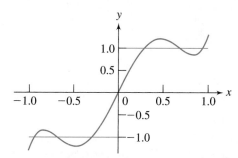

Note that in 17 and 19, just the first three terms of the eigenfunction expansion approximate the function so closely that the two graphs are virtually indistinguishable. In 21, it is clear that we must include more terms of the eigenfunction expansion to reasonably approximate $f(x)$.

23. (b) If the charges are located at x_1 and x_2 with $x_1 < x_2$, then the total force on q_1 is

$$2K\left[\frac{1}{x_1 + 1} - \frac{2}{x_2 - x_1} + \frac{1}{x_1 - 1}\right] = 0 \text{ and the force on } q_2 \text{ is } 2K\left[\frac{1}{x_2 + 1} + \frac{2}{x_2 - x_1} + \frac{1}{x_2 - 1}\right] = 0. \text{ Add these two}$$

equations to get $\dfrac{2x_1}{x_1^2 - 1} + \dfrac{2x_2}{x_2^2 - 1} = 0$. Then $2(x_1 x_2 - 1)(x_1 + x_2) = 0$, hence $x_1 = -x_2$. But then

$$\frac{1}{x_1 + 1} + \frac{1}{x_1} + \frac{1}{x_1 - 1} = 0, \text{ so } 3x_1^2 - 1 = 0 \text{ and } x_1 = \pm 1/\sqrt{3}. \text{ Since } x_2 = -x_1, \text{ the charges are at the zeros of } P_2(x).$$

(d) From Legendre's equation, $P_n''(x) = \dfrac{2x}{1 - x^2}P_n'(x) - \dfrac{n(n+1)}{1 - x^2}P_n(x)$. Since $P_n(x)$ has simple zeros at x_k for

$k = 1, 2, \ldots, N$, then $P_n'(x_k) \neq 0$. Then $\dfrac{P_n''(x_k)}{P_n'(x_k)} = \dfrac{2x_k}{1 - x_k^2}$.

Section 15.2

1. Let $y = x^a J_\nu(bx^c)$ and compute $y' = ax^{a-1} J_\nu(bx^c) + x^a bcx^{c-1} J_\nu'(bx^c)$ and
$y'' = a(a-1)x^{a-2} J_\nu(bx^c) + [2ax^{a-1}bcx^{c-1} + x^a bc(c-1)x^{c-2}]J_\nu'(bx^c) + x^a b^2 c^2 x^{2c-2} J_\nu''(bx^c)$. Substitute these into the differential equation and simplify to get $c^2 x^{a-2}\{(bx^c)^2 J_\nu''(bx^c) + bx^c J_\nu'(bx^c) + [(bx^c)^2 - \nu^2]J_\nu(bx^c)\} = 0$.
3. $y = c_1 J_{1/3}(x^2) + c_2 J_{-1/3}(x^2)$ **5.** $y = c_1 x^{-1} J_{3/4}(2x^2) + c_2 x^{-1} J_{-3/4}(2x^2)$
7. $y = c_1 x^4 J_{3/4}(2x^3) + c_2 x^4 J_{-3/4}(2x^3)$ **9.** $y = c_1 x^{-2} J_{1/2}(3x^3) + c_2 x^{-2} J_{-1/2}(3x^3)$
11. $y = c_1 J_{3/5}(2x^3) + c_2 J_{-3/5}(2x^3)$ **13.** $y_1 = c_1 J_3(\sqrt{x}) + c_2 Y_3(\sqrt{x})$ **15.** $y = c_1 J_4(2x^{1/3}) + c_2 Y_4(2x^{1/3})$
17. $y = c_1 x^{2/3} J_{1/2}(x) + c_2 x^{2/3} J_{-1/2}(x)$ **19.** $y = c_1 x^{-2} J_3(3\sqrt{x}) + c_2 x^{-2} Y_3(3\sqrt{x})$
21. Substitute into the differential equation and use the fact that $J_{1/3}(z)$ satisfies $z^2 J_{1/3}'' + z J_{1/3}' + (z^2 - \frac{1}{9})J_{1/3} = 0$.
23. $y = c_1 x J_2(x) + c_2 x Y_2(x)$ **25.** $y = c_1 x^2 J_2(\sqrt{x}) + c_2 x^2 Y_2(\sqrt{x})$ **27.** $y = c_1 x^4 J_1(x) + c_2 x^4 Y_1(x)$
29. $y = c_1 x^{-1} J_2(\sqrt{x}/2) + c_2 x^{-1} Y_2(\sqrt{x}/2)$ **31.** $y = c_1 x^2 J_4(x^2) + c_2 x^2 Y_4(x^2)$
49. (a) The sum of the first five terms of the Fourier–Bessel expansion is
$\approx 1.67411 J_2(5.135x) - 0.77750 J_2(8.417x) + 0.8281 J_2(11.620x) - 0.6201 J_2(14.796x) + 0.6281 J_2(17.960x)$. From the graphs of x and the first five terms of this expansion, more terms would have to be included to achieve reasonable accuracy.

(c) The sum of the first five terms of the Fourier–Bessel expansion is $0.85529 J_2(5.135x) - 0.21338 J_2(8.417x)$ $+ 0.35122 J_2(11.620x) - 0.20338 J_2(14.796x) + 0.025800 J_2(17.960x)$. As the graphs indicate, more terms are needed to reasonably approximate the function by the partial sum of the Fourier–Bessel expansion.

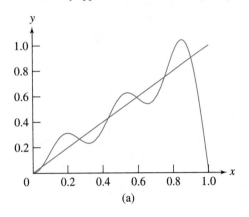

(a)

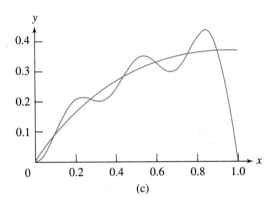

(c)

Section 15.3

1. regular problem on $[0, L]$; eigenvalues $\left(\dfrac{2n - 1}{2L}\pi\right)^2$ for $n = 1, 2, \ldots$; eigenfunctions nonzero constant multiples of $\sin\left(\dfrac{2n - 1}{2L}\pi x\right)$ **3.** regular on $[0, 4]$; $\left(\dfrac{2n - 1}{2}\dfrac{\pi}{4}\right)^2$; $\cos\left(\dfrac{2n - 1}{2}\dfrac{\pi}{4}x\right)$

5. periodic on $[-3\pi, 3\pi]$; 0 is an eigenvalue with eigenfunction 1; for $n = 1, 2, \ldots$, $\frac{1}{9}n^2$ is an eigenvalue with eigenfunction $a_n \cos(nx/3) + b_n \sin(nx/3)$, not both a_n and b_n zero

7. regular on $[0, 1]$; eigenvalues are positive solutions of $\tan(\sqrt{\lambda}) = \frac{1}{2}\sqrt{\lambda}$. There are infinitely many such solutions, of which the first four are approximately 0.43, 10.84, 40.47, and 89.82. Corresponding to an eigenvalue λ_n, eigenfunctions have the form $2\sqrt{\lambda}\cos(\sqrt{\lambda_n}x) + \sin(\sqrt{\lambda}x)$.

9. regular on $[0, 5]$; $2 + \dfrac{n^2\pi^2}{100}$ for $n = 1, 2, \ldots$; $e^{6x}\sin(n\pi x/5)$

11. regular on $[1, e]$; $\left(\dfrac{2n - 1}{2}\pi\right)^2$ for $n = 1, 2, \ldots$; $\sin\left(\left(\dfrac{2n - 1}{2}\pi\right)\ln(x)\right)$ for $n = 1, 2, \ldots$

13. regular on $[0, \pi]$; $1 + n^2$ for $n = 1, 2, \ldots$; $e^{-x}\sin(nx)$ for $n = 1, 2, \ldots$

15. regular on $[1, e^3]$; $1 + \dfrac{n^2\pi^2}{9}$ for $n = 1, 2, \ldots$; $x^{-1}\sin\left(\dfrac{n\pi}{3}\ln(x)\right)$ for $n = 1, 2, \ldots$

19. With $n = 3$ we get the upper bound 1.44672.

21. For $0 < x < 1$, $1 - x = \displaystyle\sum_{n=1}^{\infty} \dfrac{2}{n\pi}\sin(n\pi x)$. The tenth partial sum of the series is compared to the function in the graph.

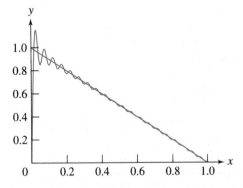

23. The expansion is $\displaystyle\sum_{n=1}^{\infty} \frac{4}{\pi} \frac{\sqrt{2}\cos(n\pi/2) - \sqrt{2}\sin(n\pi/2) - (-1)^n}{2n-1} \cos\left(\frac{2n-1}{8}\pi x\right)$. This converges to -1 for $0 < x < 2$ and 1 for $2 < x < 4$, and to 0 if $x = 2$. The tenth partial sum of the series is compared to the function in the graph.

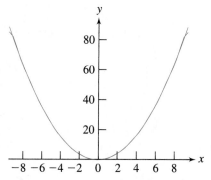

25. For $-3\pi < x < 3\pi$, $x^2 = 3\pi^2 + 36 \displaystyle\sum_{n=1}^{\infty} \frac{(-1)^n}{n^2} \cos(nx/3)$. The tenth partial sum of the series is compared to the function in the graph.

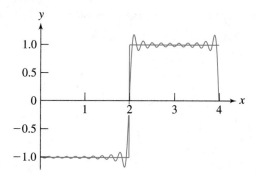

27. For $0 < x < \pi$, $e^{-x} = \displaystyle\sum_{n=1}^{\infty} \frac{4}{\pi}\left(\frac{2 + e^{-\pi}(-1)^n - 2e^{-\pi}n(-1)^n}{4n^2 - 4n + 5}\right) \cos\left(\frac{2n-1}{2}x\right)$. The tenth partial sum of the series is compared to the function in the graph.

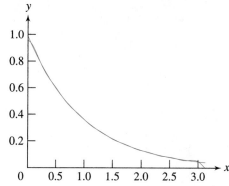

29. Normalized eigenfunctions are $\varphi_n(x) = \dfrac{1}{\sqrt{2}} \cos\left(\dfrac{2n-1}{8}\pi x\right)$. Now, $(f \cdot \varphi_n) =$

$$\int_0^4 x(4-x)\frac{1}{\sqrt{2}}\cos\left(\frac{2n-1}{8}\pi x\right) dx = -128\sqrt{2}\frac{4(-1)^n + (2n-1)\pi}{\pi^3(2n-1)^3}, \text{ so } \sum_{n=1}^{\infty}\left((128\sqrt{2})\frac{4(-1)^n + (2n-1)\pi}{\pi^3(2n-1)^3}\right)^2$$

$$\leq f \cdot f = \int_0^4 x^2(4-x)^2\, dx = \tfrac{512}{15}, \text{ or } \sum_{n=1}^{\infty}\left(\frac{4(-1)^n + (2n-1)\pi}{\pi^3(2n-1)^3}\right)^2 \leq \frac{512}{15(128\sqrt{2})^2} = \frac{1}{960}.$$

31. $\displaystyle\sum_{n=1}^{\infty}(x \cdot \varphi_n)^2 = 2\pi \sum_{n=1}^{\infty}\frac{1}{n^2} = (f \cdot f) = \frac{1}{3}\pi^3$, so $\displaystyle\sum_{n=1}^{\infty}\frac{1}{n^2} = \frac{1}{6}\pi^2$.

Section 15.5

3. The interval on which $\sigma_{1,3}(t)$ is nonzero is disjoint from the interval on which $\sigma_{-2,1}(t)$ is nonzero, so $\sigma_{1,3}(t)\sigma_{-2,1}(t)$ is identically zero, hence $\displaystyle\int_{-\infty}^{\infty}\sigma_{1,3}(t)\sigma_{-2,1}(t)\,dt = 0$. Graphs of $\sigma_{1,3}(t)$ and $\sigma_{-2,1}(t)$ are shown below.

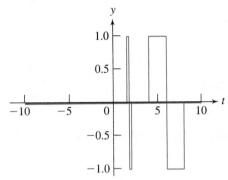

5. Notice that $\sigma_{-2,-1}(t)\sigma_{-3,-1}(t) = \begin{cases} -1 & \text{if } -4 < t < -2 \\ 1 & \text{if } -2 < t < 0 \\ 0 & \text{if } t \le -4 \text{ or } t \ge 0 \end{cases}$. Therefore, $\displaystyle\int_{-\infty}^{\infty}\sigma_{-2,-1}(t)\sigma_{-3,-1}(t)\,dt = 0$. Graphs of $\sigma_{-2,-1}(t)$ and $\sigma_{-3,-1}(t)$ are given below.

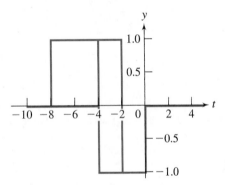

7. A graph of $\psi(2t - 3)$ is shown below.

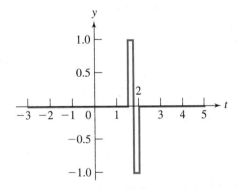

9. The Fourier series of $f(t)$ on $[-16, 16]$ is

$$\sum_{n=1}^{\infty} \left[\frac{4}{n\pi} \left(-2\sin\left(\frac{3n\pi}{4}\right) + \sin\left(\frac{n\pi}{2}\right) \right) - \frac{6}{n\pi} \left(\sin\left(\frac{n\pi}{8}\right) - 2\sin\left(\frac{3n\pi}{16}\right) \right) + \right.$$

$$\left. \sin\left(\frac{n\pi}{4}\right) \right] \cos\left(\frac{n\pi t}{16}\right) + \left[-\frac{4}{n\pi} \left(-\cos(n\pi) + 2\cos\left(\frac{3n\pi}{4}\right) - \cos\left(\frac{n\pi}{2}\right) \right) + \frac{6}{n\pi} \left(\cos\left(\frac{n\pi}{8}\right) - 2\cos\left(\frac{3n\pi}{16}\right) \right) + \right.$$

$$\left. \cos\left(\frac{n\pi}{4}\right) \right] \sin\left(\frac{n\pi t}{16}\right).$$ The graph below compares the function with the fiftieth partial sum of its Fourier series.

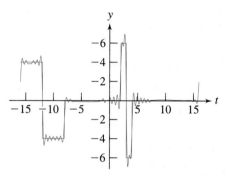

11. The Fourier series of $f(t)$ on $[-16, 16]$ is $\displaystyle\sum_{n=1}^{\infty} \frac{1}{n\pi} \left[-14\sin\left(\frac{n\pi}{2}\right) - 8\sin\left(\frac{n\pi}{4}\right) + 16\sin\left(\frac{3n\pi}{8}\right) \right] \cos\left(\frac{n\pi t}{16}\right) +$

$\displaystyle \frac{1}{n\pi} \left[3(-1)^n + 2\cos\left(\frac{n\pi}{2}\right) + 3 + 8\cos\left(\frac{n\pi}{4}\right) - 16\cos\left(\frac{3n\pi}{8}\right) \right] \sin\left(\frac{n\pi t}{16}\right).$ The graph below compares the function with the fiftieth partial sum of its Fourier series.

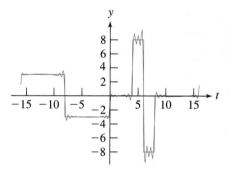

CHAPTER 16

Section 16.1

1. Compute $\dfrac{\partial^2 y}{\partial x^2} = -\dfrac{n^2\pi^2}{L^2} \sin(n\pi x/L) \cos(n\pi ct/L)$ and $\dfrac{\partial^2 y^2}{\partial t} = -\dfrac{n^2\pi^2 c^2}{L^2} \sin(n\pi x/L) \cos(n\pi ct/L).$

3. Compute $\dfrac{\partial^2 y}{\partial x^2} = \dfrac{1}{2} \left[f''(x + ct) + f''(x - ct) \right]$ and $\dfrac{\partial^2 y}{\partial t^2} = \dfrac{1}{2} c^2 [f''(x + ct) + f''(x - ct)].$

5. The problem for the displacement function $z(x, y, t)$ is $\dfrac{\partial^2 z}{\partial t^2} = c^2 \left(\dfrac{\partial^2 z}{\partial x^2} + \dfrac{\partial^2 z}{\partial y^2} \right)$ for $0 < x < a, 0 < y < b,$

$$z(x, y, 0) = f(x, y), \frac{\partial z}{\partial t}(x, y, 0) = 0 \text{ for } 0 < x < a, 0 < y < b,$$

$$z(0, y, t) = z(a, y, t) = z(x, 0, t) = z(x, b, t) = 0.$$

7. Let $v(\rho, \theta, \varphi, t) = u(\rho \cos(\theta) \sin(\varphi), \rho \sin(\theta) \sin(\varphi), \rho \cos(\varphi), t)$. Suppose u satisfies $u_{tt} = c^2(u_{xx} + u_{yy} + u_{zz})$. Here we use subscript notation for partial derivatives. Use the chain rule to compute

$$v_\rho = u_x \cos(\theta) \sin(\varphi) + u_y \sin(\theta) \sin(\varphi) + u_z \cos(\varphi),$$

$$v_\theta = -u_x \rho \sin(\theta) \sin(\varphi) + u_y \rho \cos(\theta) \sin(\varphi),$$

$$v_\varphi = u_x \rho \cos(\theta) \cos(\varphi) + u_y \rho \sin(\theta) \cos(\varphi) - u_z \rho \sin(\varphi),$$

$$\begin{aligned}
v_{\rho\rho} = {} & \cos(\theta) \sin(\varphi)[u_{xx} \cos(\theta) \sin(\varphi) + u_{xy} \sin(\theta) \sin(\varphi) + u_{xz} \cos(\varphi)] \\
& + \sin(\theta) \sin(\varphi)[u_{yx} \cos(\theta) \sin(\varphi) + u_{yy} \sin(\theta) \sin(\varphi) + u_{yz} \cos(\varphi)] \\
& + \cos(\varphi)[u_{zx} \cos(\theta) \sin(\varphi) + u_{zy} \sin(\theta) \sin(\varphi) + u_{zz} \cos(\varphi)],
\end{aligned}$$

$$\begin{aligned}
v_{\theta\theta} = {} & -u_x \rho \cos(\theta) \sin(\varphi) - u_y \rho \sin(\theta) \sin(\varphi) \\
& - \rho \sin(\theta) \sin(\varphi)[u_{xx}(-\rho \sin(\theta) \sin(\varphi)) + u_{xy} \rho \cos(\theta) \sin(\varphi)] \\
& + \rho \cos(\theta) \sin(\varphi)[u_{yx}(-\rho \sin(\theta) \sin(\varphi)) + u_{yy} \rho \cos(\theta) \sin(\varphi)],
\end{aligned}$$

and

$$\begin{aligned}
v_{\varphi\varphi} = {} & -u_x \rho \cos(\theta) \sin(\varphi) - u_y \rho \sin(\theta) \sin(\varphi) - u_z \rho \cos(\varphi) \\
& + \rho \cos(\theta) \cos(\varphi)[u_{xx} \rho \cos(\theta) \cos(\varphi) + u_{xy} \rho \sin(\theta) \cos(\varphi) - u_{xz} \rho \sin(\varphi)] \\
& + \rho \sin(\theta) \cos(\varphi)[u_{yx} \rho \cos(\theta) \cos(\varphi) + u_{yy} \rho \sin(\theta) \cos(\varphi) - u_{yz} \rho \sin(\varphi)] \\
& - \rho \sin(\varphi)[u_{zx} \rho \cos(\theta) \cos(\varphi) + u_{zy} \rho \sin(\theta) \cos(\varphi) - u_{zz} \rho \sin(\varphi)].
\end{aligned}$$

Now it is tedious but routine to check that

$$v_{tt} = c^2 \frac{1}{\rho^2} \left[\rho \frac{\partial^2}{\partial \rho^2}(\rho v) + \frac{1}{\sin(\varphi)} \frac{\partial}{\partial \varphi}\left(\sin(\varphi) \frac{\partial v}{\partial \varphi}\right) + \frac{1}{\sin^2(\varphi)} \frac{\partial^2 v}{\partial \theta^2} \right].$$

Section 16.2

1. $y(x, t) = \displaystyle\sum_{n=1}^\infty \frac{16(-1)^n}{(2n-1)^3 \pi^3 c} \sin\left(\frac{(2n-1)\pi x}{2}\right) \sin\left(\frac{(2n-1)\pi ct}{2}\right) + \sum_{n=1}^\infty \frac{2(-1)^n}{n^2 \pi^2 c} \sin(n\pi x) \sin(n\pi ct)$

3. $y(x, t) = \displaystyle\sum_{n=1}^\infty \frac{108}{(2n-1)^4 \pi^4} \sin\left(\frac{(2n-1)\pi x}{3}\right) \sin\left(\frac{2(2n-1)\pi t}{3}\right)$

5. $y(x, t) = \displaystyle\sum_{n=1}^\infty \frac{24}{(2n-1)^2 \pi}(-1)^{n+1} \sin\left(\frac{(2n-1)x}{2}\right) \cos\left((2n-1)\sqrt{2}t\right)$

7. $y(x, t) = \displaystyle\sum_{n=1}^\infty \frac{-32}{(2n-1)^3 \pi^3} \sin\left(\frac{(2n-1)\pi x}{2}\right) \cos\left(\frac{3(2n-1)\pi t}{2}\right)$

$\qquad + \displaystyle\sum_{n=1}^\infty \frac{4}{n^2 \pi^2} \sin\left(\frac{n\pi x}{2}\right)\left[\cos\left(\frac{n\pi}{4}\right) - \cos\left(\frac{n\pi}{2}\right)\right] \sin\left(\frac{3n\pi t}{2}\right)$

9. $y(x, t) = \displaystyle\sum_{n=1}^\infty \frac{256}{n^3 \pi^3}[2(-1)^n + 1] \sin\left(\frac{n\pi x}{4}\right) \cos\left(\frac{n\pi t}{\sqrt{2}}\right) + \sum_{n=1}^\infty \frac{8}{\sqrt{2}(2n-1)^2 \pi^2} \sin\left(\frac{(2n-1)\pi x}{4}\right) \sin\left(\frac{(2n-1)\pi t}{\sqrt{2}}\right)$

11. Let $Y(x, t) = y(x, t) + h(x)$ and substitute into the problem to choose $h(x) = \frac{1}{9}x^3 - \frac{4}{9}x$. The problem for Y becomes

$$\frac{\partial^2 Y}{\partial t^2} = 3 \frac{\partial^2 Y}{\partial x^2},$$

$$Y(0, t) = Y(2, t) = 0,$$

$$Y(x, 0) = \frac{1}{9}x^3 - \frac{4}{9}x, \quad \frac{\partial Y}{\partial t}(x, 0) = 0.$$

We find that $Y(x, t) = \displaystyle\sum_{n=1}^\infty \frac{32}{3n^3 \pi^3}(-1)^n \sin\left(\frac{n\pi x}{2}\right) \cos\left(\frac{n\pi \sqrt{3}t}{2}\right)$, and then $y(x, t) = Y(x, t) - h(x)$.

13. Let $Y(x, t) = y(x, t) + h(x)$ and choose $h(x) = \cos(x) - 1$. The problem for Y is

$$\frac{\partial^2 Y}{\partial t^2} = \frac{\partial^2 Y}{\partial x^2},$$

$$Y(0, t) = Y(2\pi, t) = 0,$$

$$Y(x, 0) = \cos(x) - 1, \quad \frac{\partial Y}{\partial t}(x, 0) = 0.$$

This problem has solution $Y(x, t) = \sum_{n=1}^{\infty} \frac{16}{\pi} \frac{1}{(2n-1)[(2n-1)^2 - 4]} \sin\left(\frac{(2n-1)x}{2}\right) \cos\left(\frac{(2n-1)t}{2}\right)$, and then
$y(x, t) = Y(x, t) + 1 - \cos(x)$.

15. $y(x, t) = \frac{1}{2}(A+1)L - \sum_{n=1}^{\infty} \frac{4(A+1)L}{(2n-1)^2\pi^2} \cos\left(\frac{(2n-1)\pi x}{L}\right) \cos\left(\frac{(2n-1)\pi ct}{L}\right)$

17. $u(x, t) = e^{-At/2} \sum_{n=1}^{\infty} C_n \sin\left(\frac{n\pi x}{L}\right) \left[\frac{1}{AL} r_n \cos\left(\frac{r_n t}{2L}\right) + \sin\left(\frac{r_n t}{2L}\right)\right]$, where $C_n = \frac{2A}{r_n} \int_0^L f(x) \sin\left(\frac{n\pi x}{L}\right) dx$ and
$r_n = \sqrt{4(BL^2 + n^2\pi^2 c^2) - A^2 L^2}$

19. With $\theta(x, t) = X(x)T(t)$, we get $X'' + \lambda X = 0$; $X'(0) - \alpha X(0) = 0$, $X'(L) + \alpha X(L) = 0$, and $T'' + a^2\lambda T = 0$.
Solve this regular Sturm–Liouville problem for X to get $X(x) = A\cos(\sqrt{\lambda}x) + B\sin(\sqrt{\lambda}x)$ for $\lambda > 0$. This problem
has no negative eigenvalue, and zero is not an eigenvalue. Put $\lambda = k^2$ to get $0 = X'(0) - \alpha X(0) = kB - \alpha A$ and
$0 = X'(L) + \alpha X(L) = -Ak\sin(kL) + Bk\cos(kL) + \alpha A\cos(kL) + \alpha B\sin(kL)$. For a nontrivial solution for X we
need $A \neq 0$, and k must satisfy $\tan(kL) = 2ak/(k^2 - \alpha^2)$. A sketch of graphs of $\tan(kL)$ and $2ak/(k^2 - \alpha^2)$ as
functions of k indicates the existence of an infinite number of positive solutions $k_1 < k_2 < \cdots$ for k. The eigenvalues
are $\lambda_n = k_n^2$, with eigenfunctions $\varphi_n(x) = k_n \cos(k_n x) + \alpha \sin(k_n x)$. These eigenfunctions are orthogonal, meaning that
$\int_0^L \varphi_n(x)\varphi_m(x)\, dx = 0$ if $n \neq m$. Obtain the solution by an eigenfunction expansion:

$\theta(x, t) = \sum_{n=1}^{\infty} A_n \varphi_n(x) \cos(ck_n t)$, where $A_n = \dfrac{\int_0^L f(x)\varphi_n(x)\, dx}{\int_0^L \varphi_n^2(x)\, dx}$. One can obtain an approximate solution by
computing the first few eigenvalues numerically and approximating the first few coefficients in this series.

21. (a) The solution with the forcing term is $y_f(x, t) = \sum_{n=1}^{\infty} \frac{64}{\pi^3} \left(\frac{2}{(2n-1)^3} - \frac{1}{9} \frac{1}{(2n-1)[(2n-1)^2 - 16]}\right) \sin\left(\frac{n\pi x}{4}\right)$
$\cos\left(\frac{3(2n-1)\pi t}{4}\right) + \frac{1}{9\pi^2}[\cos(\pi x) - 1]$.

(b) Without the forcing term, the solution is $y(x, t) = \sum_{n=1}^{\infty} \frac{128}{\pi^3(2n-1)^3} \sin\left(\frac{(2n-1)\pi x}{4}\right) \cos\left(\frac{3(2n-1)\pi t}{4}\right)$.
Both solutions are graphed together for times $t = 0.5, 0.6, 4.9$, and 9.8.

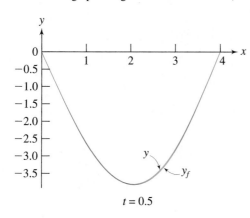

$t = 0.5$

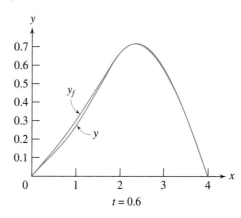

$t = 0.6$

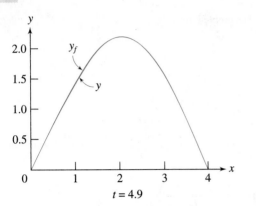

$$t = 4.9$$

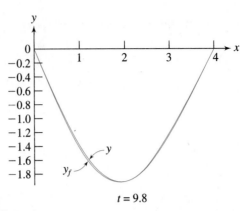

$$t = 9.8$$

23. (a) The solution with the forcing term is $y_f(x, t) = \sum_{n=1}^{\infty} \frac{1}{18} \frac{4n^2\pi^2(-1)^n - 8(-1)^n + 8}{n^5\pi^5} \sin(n\pi x)$

$\cos(3n\pi t) + \frac{2}{3} \sum_{n=1}^{\infty} \left(\frac{1-(-1)^n}{n^2\pi^2} \right) \sin(n\pi x) \sin(3n\pi t) + \frac{1}{108} x(1 - x^3).$

(b) Without the forcing term, the solution is $y(x, t) = \frac{2}{3} \sum_{n=1}^{\infty} \left(\frac{1-(-1)^n}{n^2\pi^2} \right) \sin(n\pi x) \sin(3n\pi t).$

Both solutions are graphed together for times $t = 0.6, 0.5, 1.4$ and 2.5.

25. (a) The solution with the forcing term is $y_f(x, t) = \sum_{n=1}^{\infty} 81 \frac{(-1)^n - 1}{n^3\pi^3} \sin\left(\frac{n\pi x}{9} \right) \cos\left(\frac{2n\pi t}{9} \right) + \frac{1}{4} x(9 - x)$

$+ \sum_{n=1}^{\infty} \frac{81}{\pi^2} \frac{(-1)^{n+1}}{n^2} \sin\left(\frac{n\pi x}{9} \right) \sin\left(\frac{2n\pi t}{9} \right).$

(b) The solution without the forcing term is $y(x, t) = \sum_{n=1}^{\infty} \frac{81}{\pi^2} \frac{(-1)^{n+1}}{n^2} \sin\left(\frac{n\pi x}{9} \right) \sin\left(\frac{2n\pi t}{9} \right).$

Both solutions are graphed together for times $t = 0.6, 0.5, 1.2, 2.6, 3.5, 5,$ and 8.

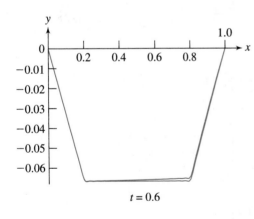

$$t = 0.6$$

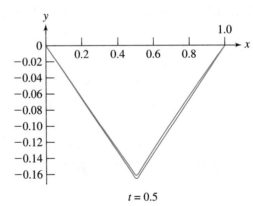

$$t = 0.5$$

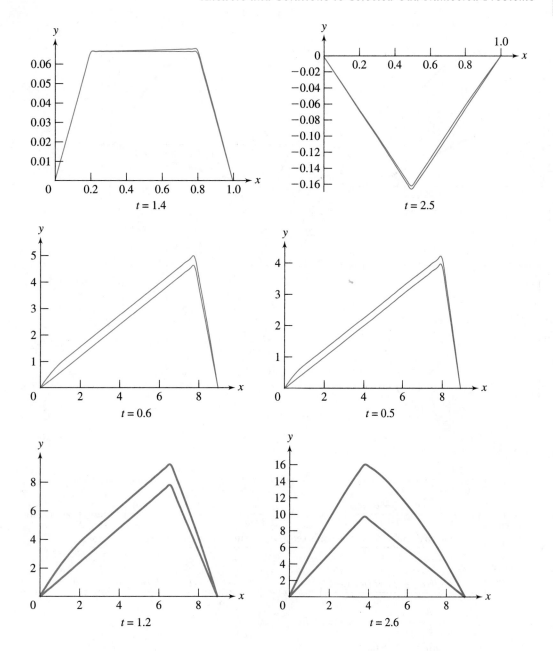

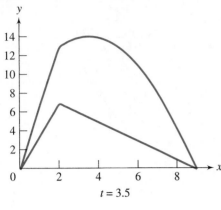

$t = 3.5$

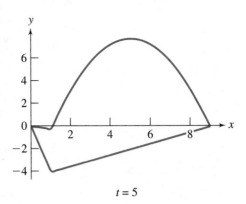

$t = 5$

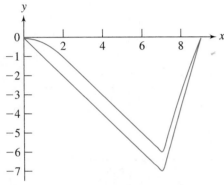

$t = 8$

Section 16.3

1. $y(x, t) = \displaystyle\int_0^\infty \frac{10}{\pi \left(25 + \omega^2\right)} \cos(\omega x) \cos(12\omega t) d\omega$ **3.** $y(x, t) = \displaystyle\int_0^\infty \frac{-1}{2\pi\omega} \frac{\sin(\pi\omega)}{\omega^2 - 1} \sin(\omega x) \sin(4\omega t) d\omega$

5. $y(x, t) = \displaystyle\int_0^\infty \left[\left(\frac{1}{3\pi\omega} e^{-2} \frac{2\cos(\omega) - \omega\sin(\omega)}{4 + \omega^2} \right) \cos(\omega x) + \left(\frac{1}{3\pi\omega} e^{-2} \frac{\omega\cos(\omega) + 2\sin(\omega)}{4 + \omega^2} \right) \sin(\omega x) \right] \sin(3\omega t) d\omega$

7. $y(x, t) = \displaystyle\int_0^\infty \left(-\frac{32}{\pi} \frac{4\omega^3 \cos^2(\omega) - 2\omega^3 - 6\omega^2 \sin(\omega)\cos(\omega) + 3\sin(\omega)\cos(\omega) - 6\omega\cos^2(\omega) + 3\omega}{\omega^5} \right)$
$\cos(\omega x) \cos(4\omega t) \, d\omega$

9. $y(x, t) = \displaystyle\int_0^\infty \left[\left(-\frac{1}{\pi} \frac{\cos(\pi\omega) + 2\cos^2(\pi\omega) - 1}{-1 + \omega^2} \right) \cos(\omega x) + \left(\frac{1}{\pi} \sin(\pi\omega) \frac{-1 + 2\cos(\pi\omega)}{-1 + \omega^2} \right) \sin(\omega x) \right]$
$\cos(2\omega t) \, d\omega + \displaystyle\int_0^\infty \frac{4}{\pi\omega \left(16 + \omega^2\right)} \cos(\omega x) \sin(2\omega t) \, d\omega$

11. $y(x, t) = \displaystyle\int_0^\infty \frac{2}{\pi} \frac{2 - \omega\sin(\omega) - 2\cos(\omega)}{\omega^3} \sin(\omega x) \cos(3\omega t) \, d\omega$

13. $y(x, t) = \displaystyle\int_0^\infty \frac{1}{\pi\omega} \frac{\sin(\pi\omega/2) - \sin(5\pi\omega/2)}{\omega^2 - 1} \sin(\omega x) \sin(2\omega t) \, d\omega$

15. $y(x, t) = \displaystyle\int_0^\infty -\frac{3}{7\pi\omega^5} \left[16\omega\cos^3(\omega) - 12\omega\cos(\omega) + 12\omega^2 \sin(\omega)\cos^2(\omega) - 3\omega^2 \sin(\omega) - 8\sin(\omega)\cos^2(\omega) \right.$
$\left. + 2\sin(\omega) + 2\omega \right] \sin(\omega x) \sin(14\omega t) \, d\omega$

17. $y(x, t) = \displaystyle\int_0^\infty \frac{2}{\pi} \frac{32\cos^5(\omega) - 40\cos^3(\omega) + 10\cos(\omega) - 2 + 25\omega^2}{\omega^3} \sin(\omega x) \cos(4\omega t) \, d\omega$
$+ \displaystyle\int_0^\infty -\frac{8}{\pi\omega^4} \left(2\omega^2 \cos^2(\omega) - \omega^2 - \cos^2(\omega) + 1 - 2\omega\sin(\omega)\cos(\omega) \right) \sin(\omega x) \sin(4\omega t) \, d\omega$

19. $y(x, t) = At + (1 - A)\left(t - \dfrac{x}{c}\right) H\left(t - \dfrac{x}{c}\right)$

Section 16.4

1. Characteristics are lines $x - t = k_1$, $x + t = k_2$;
$$y(x,t) = \tfrac{1}{2}\left[(x-t)^2 + (x+t)^2\right] + \tfrac{1}{2}\int_{x-t}^{x+t} -\xi \, d\xi = x^2 + t^2 - xt$$

3. Characteristics are $x - 7t = k_1$, $x + 7t = k_2$;
$$y(x,t) = \frac{1}{2}\left[\cos(\pi(x-7t)) + \cos(\pi(x+7t))\right] + t - x^2 t - \frac{49}{3}t^3$$

5. Characteristics are $x - 14t = k_1$, $x + 14t = k_2$;
$$y(x,t) = \tfrac{1}{2}\left[e^{x-14t} + e^{x+14t}\right] + xt$$

7. Characteristics are $x - 4t = k_1$, $x + 4t = k_2$;
$$y(x,t) = \tfrac{1}{2}\left[(x-4t)e^{x-4t} + (x+4t)e^{x+4t}\right] + \tfrac{1}{8}\left[\sinh(x+4t) - \sinh(x-4t)\right]$$

9. Characteristics are $x - 15t = k_1$, $x + 15t = k_2$;
$$y(x,t) = \tfrac{1}{2}\left[|x-15t| + |x+15t|\right] + \tfrac{1}{60}\left(e^{2(x+15t)} - e^{2(x-15t)}\right)$$

11. $y(x,t) = x + \tfrac{1}{8}\left(e^{-x+4t} - e^{-x-4t}\right) + \tfrac{1}{2}xt^2 + \tfrac{1}{6}t^3$

13. $y(x,t) = x^2 + 64t^2 - x + \tfrac{1}{32}\left(\sin(2(x+8t)) - \sin(2(x-8t))\right) + \tfrac{1}{12}t^4 x$

15. $y(x,t) = \tfrac{1}{2}\left[\cosh(x-3t) + \cosh(x+3t)\right] + t + \tfrac{1}{4}xt^4$

17. $y(x,t) = \tfrac{1}{2}\left[\cos(2(x-2t)) + \cos(2(x+2t))\right] + t - \tfrac{1}{8}\left[2\sin(x+2t) - 2\sin(x-2t)\right] + \tfrac{1}{12}t^4$

19. $y(x,t) = \tfrac{1}{2}\left[(x-2t)\sin(x-2t) + (x+2t)\sin(x+2t)\right] - \tfrac{1}{4}e^{-x-2t} + \tfrac{1}{4}e^{-x+2t} + \tfrac{1}{6}xt^3$

In each of 21, 23, and 25, the graphs show a progression of the motion as a sum of forward and backward waves.

21.

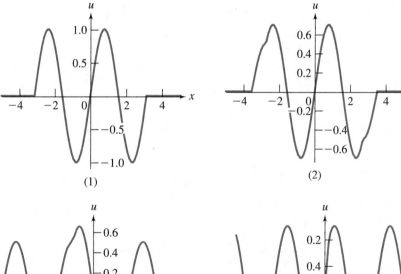

(1)

(2)

(3)

(4)

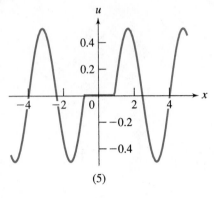

(5)

23.

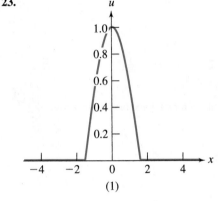

(1)

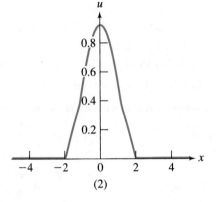

(2)

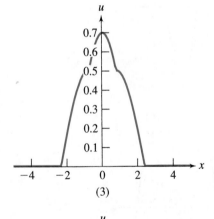

(3)

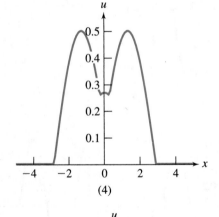

(4)

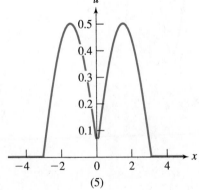

(5)

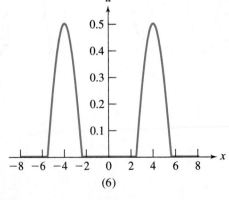

(6)

25.

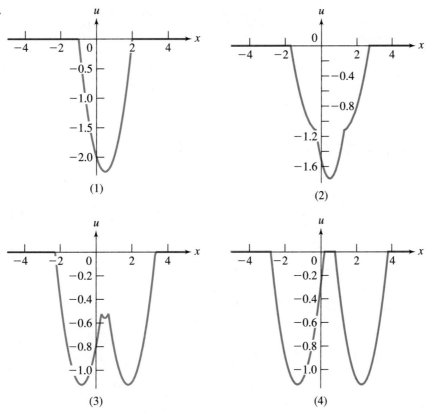

(1) (2)

(3) (4)

Section 16.5

1. We find that (approximately) $a_1 = \dfrac{2\int_0^1 x J_0(2.405x)\,dx}{[J_1(2.405)]^2} = 2\dfrac{0.1057}{0.2695} = 0.78442$, $a_2 = 0.04112$, $a_3 = -8.1366$, $a_4 = -375.2$, $a_5 = -6470.9$. The fifth partial sum of the series gives the approximation $z(r,t) \approx$
$0.78442 J_0(2.405r)\cos(2.405t) + 0.04112 J_0(5.520r)\cos(5.520t) - 8.1366 J_0(8.654r)\cos(8.654t) - 375.2 J_0(11.792r)\cos(11.792t) - 6470.9 J_0(14.931r)\cos(14.931t)$.
The graph below shows $z(r,t)$ at various times.

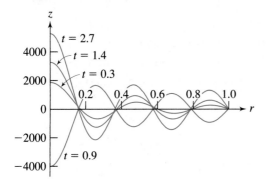

3. Approximately, $z(r, t) \approx 1.2534 J_0(2.405r) \cos(2.405t) - 0.88824 J_0(5.520r) \cos(5.520t) - 24.89 J_0(8.654r) \cos(8.654t) - 1133.6 J_0(11.792r) \cos(11.792t) - 19523 J_0(14.931r) \cos(14.931t)$. The graph below shows $z(r, t)$ at selected times.

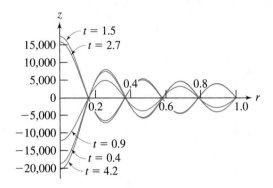

Section 16.6

1. Compute

$$\frac{1}{2}\alpha_0(r) = \frac{1}{2\pi} \int_{-\pi}^{\pi} (4 - r^2) \sin^2(\theta)\, d\theta = \frac{1}{2}\left(4 - r^2\right)$$

$$\alpha_n(r) = \frac{4 - r^2}{\pi} \int_{-\pi}^{\pi} \sin^2(\theta) \cos(n\theta)\, d\theta = \begin{cases} 0 & \text{if } n \neq 2 \\ \dfrac{4 - r^2}{\pi}\left(-\dfrac{1}{2}\pi\right) & \text{if } n = 2 \end{cases}$$

$$\beta_n(r) = \frac{4 - r^2}{\pi} \int_{-\pi}^{\pi} \sin^2(\theta) \sin(n\theta)\, d\theta = 0.$$

Thus, $z(r, \theta, t) = \displaystyle\sum_{k=1}^{\infty} \frac{2}{[J_1(j_{0k})]^2}\left(\int_0^1 \xi(1 - \xi^2) J_0(j_{0k}\xi)\, d\xi\right) J_0\left(\frac{1}{2} j_{0k} r\right) \cos(j_{0k} t)$

$$+ \sum_{k=1}^{\infty} \frac{4}{[J_3(j_{2k})]^2}\left(\int_0^1 \xi(\xi^2 - 1) J_2(j_{2k}\xi)\, d\xi\right) J_2\left(\frac{1}{2} j_{2k} r\right) \cos(2\theta) \cos(j_{2k} t)$$

$$+ \sum_{p=1}^{\infty} \sin(p\theta) \sum_{q=1}^{\infty} \frac{4(-1)^{p+1}}{p j_{pq}[J_{p+1}(j_{pq})]^2} \int_0^1 \xi J_p(j_{pq}\xi)\, d\xi) J_\theta\left(\frac{1}{2} j_{pq} r\right) \sin(j_{pq} t)$$

$$\approx 1.1081 J_0(1.2025r) \cos(2.40483t) - 0.13975 J_0(2.760r) \cos(5.52008t)$$

$$+ 0.4555 J_0(4.3270r) \cos(8.65373t) - 0.02105 J_0(5.8960r) \cos(11.7915t)$$

$$+ 0.01165 J_0(7.4655r) \cos(14.43092t) + \cdots - 2.9777 J_2(2.5675r) \cos(2\theta) \cos(5.1356t)$$

$$- 1.4035 J_2(4.2085r) \cos(2\theta) \cos(8.41724t) - 1.1405 J_2(5.8100r) \cos(2\theta) \cos(11.6198t)$$

$$- 0.83271 J_2(7.398r) \cos(2\theta) \cos(14.7960t) - \cdots.$$

Section 16.7

1. $z(x, y, t) = \dfrac{1}{\pi} \displaystyle\sum_{n=1}^{\infty} \left[\dfrac{8(-1)^{n+1}\pi^2}{n} + \dfrac{16}{n^3}[(-1)^n - 1]\right] \sin\left(\dfrac{nx}{2}\right) \sin(y) \cos\left(\dfrac{1}{2}\sqrt{n^2 + 4t}\right)$

3. $z(x, y, t) = \displaystyle\sum_{n=1}^{\infty} \sum_{m=1}^{\infty} \left[\dfrac{16}{\pi^2(2n - 1)(2m - 1)\sqrt{(2n - 1)^2 + (2m - 1)^2}}\right] \cos\left(\dfrac{(2n - 1)x}{2}\right) \sin\left(\dfrac{(2m - 1)x}{2}\right)$

$$\times \sin\left(\sqrt{(2n - 1)^2 + (2m - 1)^2}\, t\right)$$

5. If the room occupies $0 \le x \le L, 0 \le y \le K, 0 \le z \le M$, then $u(x, y, z, t) = \sum_{n=1}^{\infty} \sum_{m=1}^{\infty} \sum_{p=1}^{\infty} \alpha_{nmp} \sin\left(\frac{n\pi x}{L}\right) \sin\left(\frac{m\pi y}{K}\right)$

$\times \sin\left(\frac{p\pi z}{M}\right) \sin\left(\sqrt{\frac{n^2}{L^2} + \frac{m^2}{K^2} + \frac{p^2}{M^2}}\, \pi c t\right)$, where $\alpha_{nmp} = \dfrac{8}{LKM} \dfrac{1}{\sqrt{\frac{n^2}{L^2} + \frac{m^2}{K^2} + \frac{p^2}{M^2}}\, \pi c} \displaystyle\int_0^L \int_0^K \int_0^M f(x, y, z)$

$\times \sin\left(\frac{n\pi x}{L}\right) \sin\left(\frac{m\pi y}{K}\right) \sin\left(\frac{p\pi z}{M}\right) dz\, dy\, dx.$

CHAPTER 17

Section 17.1

1. $\dfrac{\partial u}{\partial t} = k\dfrac{\partial^2 u}{\partial x^2}$ for $0 < x < L, t > 0$; $u(0, t) = \dfrac{\partial u}{\partial x}(L, t) = 0$ for $t \ge 0$; $u(x, 0) = f(x)$ for $0 \le x \le L$.

3. $\dfrac{\partial u}{\partial t} = k\dfrac{\partial^2 u}{\partial x^2}$ for $0 < x < L, t > 0$; $\dfrac{\partial u}{\partial x}(0, t) = 0$ and $u(L, t) = \beta(t)$ for $t \ge 0$; $u(x, 0) = f(x)$ for $0 \le x \le L$.

Section 17.2

In these solutions, $\exp(A) = e^A$.

1. $u(x, t) = \displaystyle\sum_{n=1}^{\infty} \frac{8L^2}{(2n-1)^3 \pi^3} \sin\left(\frac{(2n-1)\pi x}{L}\right) \exp\left(\frac{-(2n-1)^2 \pi^2 kt}{L^2}\right)$

3. $u(x, t) = \displaystyle\sum_{n=1}^{\infty} \frac{-16L}{(2n-1)\pi[(2n-1)^2 - 4]} \sin\left(\frac{(2n-1)\pi x}{L}\right) \exp\left(\frac{-3(2n-1)^2 \pi^2 t}{L^2}\right)$

5. $u(x, t) = \dfrac{2}{3}\pi^2 - \displaystyle\sum_{n=1}^{\infty} \frac{4}{n^2} \cos(nx) e^{-4n^2 t}$ 　**7.** $u(x, t) = \dfrac{1}{6}\left(1 - e^{-6}\right) + \displaystyle\sum_{n=1}^{\infty} 12\left(\frac{1 - e^{-6}(-1)^n}{36 + n^2\pi^2}\right) \cos\left(\frac{n\pi x}{6}\right) e^{-n^2\pi^2 t/18}$

9. $u(x, t) = \displaystyle\sum_{n=1}^{\infty} \frac{4B}{(2n-1)\pi} \sin\left(\frac{(2n-1)\pi x}{2L}\right) \exp\left(\frac{-(2n-1)^2 \pi^2 kt}{4L^2}\right)$

11. The eigenvalues are $\lambda_n = z_n^2/36$, where the z_ns are the positive solutions of $\tan(z_n) = -z_n/3$. Eigenfunctions are

$\sin(z_n x/6)$, so $u(x, t) = \displaystyle\sum_{n=1}^{\infty} A_n \sin\left(\frac{z_n x}{6}\right) e^{-z_n^2 kt}$, where $A_n = \dfrac{\int_0^6 x(6 - x) \sin(z_n x/6)\, dx}{\int_0^6 \sin^2(z_n x/6)\, dx}$. Using the fourth partial

sum, $u(x, t) \approx \displaystyle\sum_{n=1}^{4} A_n \sin\left(\frac{z_n x}{6}\right) e^{-z_n^2 kt}$.

13. Substitute $e^{\alpha x + \beta t} v(x, t)$ into the partial differential equation and solve for α and β so that $v_t = kv_{xx}$. We get $\alpha = -A/2$ and $\beta = k(B - A^2/4)$.

15. Let $u(x, t) = e^{-3x - 9t} v(x, t)$. Then $v_t = v_{xx}$, $v(0, t) = v(4, t) = 0$, and $v(x, 0) = e^{3x}$. Then $v(x, t) =$

$\displaystyle\sum_{n=1}^{\infty} \left(2n\pi \frac{1 - e^{12}(-1)^n}{144 + n^2\pi^2}\right) \sin\left(\frac{n\pi x}{4}\right) e^{-n^2\pi^2 t/16}$. Graphs of the solution are shown for times $t = 0.003, 0.02, 0.08$, and 1.3.

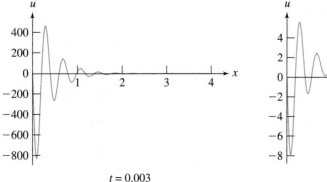

$t = 0.003$

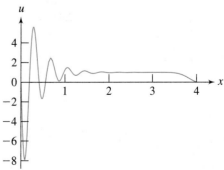

$t = 0.02$

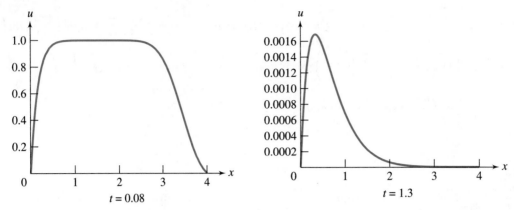

$t = 0.08$ $t = 1.3$

17. Let $u(x, t) = v(x, t) + f(x)$ and choose $f(x) = 3x + 2$ to have $v_t = 16v_{xx}$, $v(0, t) = v(1, t) = 0$, and

$$v(x, 0) = x^2 - f(x). \text{ Then } v(x, t) = \sum_{n=1}^{\infty} 2 \left(\frac{4n^2\pi^2(-1)^n + 2(-1)^n - 2 - 2n^2\pi^2}{n^3\pi^3} \right) \sin(n\pi x)e^{-16n^2\pi^2 t} \text{ and}$$

$u(x, t) = v(x, t) + 3x + 2.$ Graphs of the solution are shown for times $t = 0.005, 0.009,$ and 0.01.

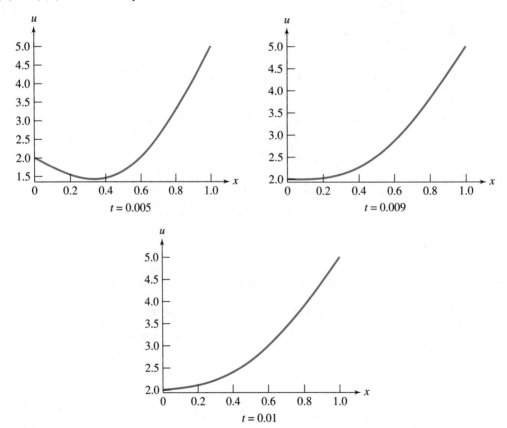

$t = 0.005$ $t = 0.009$

$t = 0.01$

19. Let $v(x, t) = u(x, t) + f(x)$ and choose $f(x) = -3x/5$. Then $v_t = 9v_{xx}$, $v(0, t) = v(5, t) = 0$, and

$v(x, 0) = -3x/5$. Obtain $v(x, t) = \displaystyle\sum_{n=1}^{\infty} \frac{6}{n\pi} (-1)^n \sin\left(\frac{n\pi x}{5}\right) e^{-9n^2\pi^2 t/25}$. The graph below compares the solution at

times $t = 0.005, 0.04, 0.01$, and 1.1.

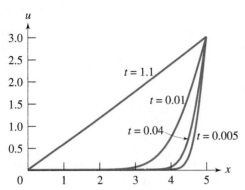

21. Let $u(x, t) = e^{-At} w(x, t)$. Then $w_t = 4w_{xx}$, $w(0, t) = w(9, t) = 0$, and $w(x, 0) = 3x$. Obtain

$w(x, t) = \displaystyle\sum_{n=1}^{\infty} \frac{54(-1)^{n+1}}{n\pi} \sin\left(\frac{n\pi x}{9}\right) e^{-4n^2\pi^2 t/81}$. The graphs compare solutions at times $t = 0.008, 0.04$, and 0.6 for

$A = \frac{1}{4}, 1$, and 3.

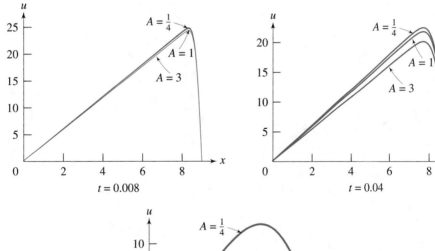

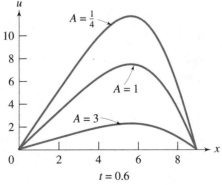

23. Let $v(x, t) = u(x, t) + f(x)$ and choose $f(x) = -50x - 100$ so the equation for v becomes $v_t = 9v_{xx}$, $v(0, t) = v(2, t) = 0$, $v(x, 0) = -49x - 100$. Obtain $v(x, t) = \sum_{n=1}^{\infty} \frac{4}{n\pi}[99(-1)^n - 50]\sin\left(\frac{n\pi x}{2}\right)e^{-9n^2\pi^2 t/4}$. Then $u(x, t) = v(x, t) + 50x + 100$. The steady-state temperature is $\lim_{t\to\infty} u(x, t) = 50x + 100$. The graph shows the solution at times $t = 0.003, 0.009, 0.04$, and 0.1.

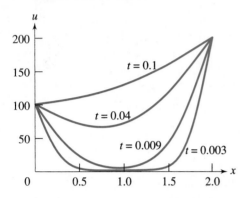

25. $u(x, t) = \sum_{n=1}^{\infty}\left[\frac{1}{8\pi}\frac{1 - (-1)^n}{n^5}\left(-1 + 4n^2 t + e^{-4n^2 t}\right)\right]\sin(nx) + \frac{4}{\pi}\sum_{n=1}^{\infty}\left(\frac{1 - (-1)^n}{n^3}\right)\sin(nx)e^{-4n^2 t}$. The graphs compare solutions with and without the source term, for times $t = 0.8, 0.4$, and 1.3.

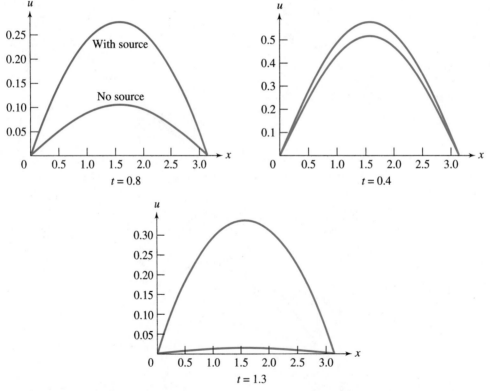

27. $u(x, t) = \sum_{n=1}^{\infty}\frac{50}{n^3\pi^3}\frac{1 - (\cos 5)(-1)^n}{n^2\pi^2 - 25}\left(-25 + n^2\pi^2 t + 25e^{-n^2\pi^2 t/25}\right)\sin\left(\frac{n\pi x}{5}\right)$

$+ \sum_{n=1}^{\infty}\left(-250\frac{4(-1)^n + 2}{n^3\pi^3}\right)\sin\left(\frac{n\pi x}{5}\right)e^{-n^2\pi^2 t/25}$. The graphs compare solutions with and without the source term, for times $t = 0.7, 1.5, 2.6$, and 4.2.

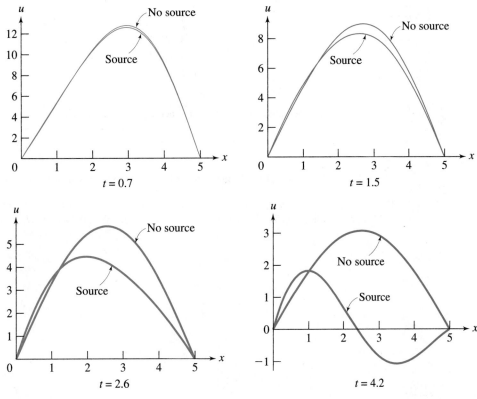

29. $u(x,t) = \displaystyle\sum_{n=1}^{\infty} \frac{27\,(-1)^{n+1}}{128} \left(\frac{16n^2\pi^2 t + 9e^{-16n^2\pi^2 t/9} - 9}{n^5\pi^5} \right) \sin\left(\frac{n\pi x}{3}\right) + 2K \sum_{n=1}^{\infty} \left(\frac{1 - (-1)^n}{n\pi} \right) \sin\left(\frac{n\pi x}{3}\right) e^{-16n^2\pi^2 t/9}.$

The graphs show the solution with and without the source term, at times $t = 0.05$ and 0.2, with $K = \frac{1}{2}$.

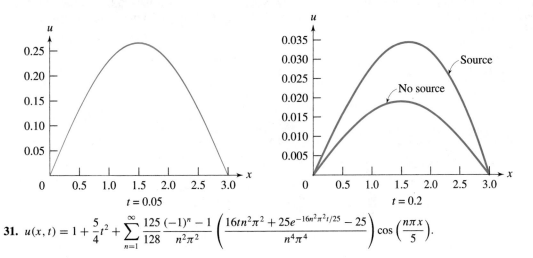

31. $u(x,t) = 1 + \dfrac{5}{4}t^2 + \displaystyle\sum_{n=1}^{\infty} \frac{125}{128} \frac{(-1)^n - 1}{n^2\pi^2} \left(\frac{16tn^2\pi^2 + 25e^{-16n^2\pi^2 t/25} - 25}{n^4\pi^4} \right) \cos\left(\frac{n\pi x}{5}\right).$

Section 17.3

1. $u(x,t) = \dfrac{1}{\pi} \displaystyle\int_0^{\infty} \frac{8}{(16 + \omega^2)} \cos(\omega x) e^{-\omega^2 kt}\, d\omega$

3. $u(x,t) = \displaystyle\int_0^{\infty} \left[\left(\frac{8}{\pi} \cos(\omega) \frac{\cos^3(\omega) - \cos(\omega) + 4\omega \sin(\omega)\cos^2(\omega) - 2\omega \sin(\omega)}{\omega^2} \right) \cos(\omega x) \right.$
$\left. - \left(\frac{4}{\pi} \frac{-2\sin(\omega)\cos^3(\omega) + \sin(\omega)\cos(\omega) + 8\omega\cos^4(\omega) - 8\omega\cos^2(\omega) + \omega}{\omega^2} \right) \sin(\omega x) \right] e^{-\omega^2 kt}\, d\omega$

5. $u(x, t) = \dfrac{2}{\pi} \displaystyle\int_0^\infty \dfrac{\cos(\omega) - 1}{\omega} \sin(\omega x) e^{-\omega^2 kt} \, d\omega$ **7.** $u(x, t) = \dfrac{2}{\pi} \displaystyle\int_0^\infty \dfrac{\omega}{\alpha^2 + \omega^2} \sin(\omega x) e^{-\omega^2 kt} \, d\omega$

9. $u(x, t) = \dfrac{2}{\pi} \displaystyle\int_0^\infty \dfrac{1 - \cos(h\omega)}{\omega} \sin(\omega x) e^{-\omega^2 kt} \, d\omega$ **11.** $u(x, t) = \dfrac{4}{\pi} \displaystyle\int_0^\infty \left(\dfrac{\cos^2(h\omega) - \cos(h\omega)}{\omega} \right) \sin(\omega x) e^{-\omega^2 kt} \, d\omega$

13. $u(x, t) = \dfrac{4}{\pi} \displaystyle\int_0^\infty \dfrac{\omega}{(1 + \omega^2)^2} \sin(\omega x) e^{-\omega^2 t} e^{-t^2/2} \, d\omega$

15. $u(x, t) = \displaystyle\int_0^t 2(t - \tau) \operatorname{erfc}\left(\dfrac{x}{2\sqrt{k\tau}} \right) d\tau$, in which erfc is the complementary error function.

Section 17.4

1. $U(r, t) = \displaystyle\sum_{n=1}^\infty \dfrac{2}{[J_1(j_n)]^2} \left(\int_0^1 \xi^2 J_0(j_n \xi) \, d\xi \right) J_0(j_n r) e^{-j_n^2 t}$; the fifth partial sum, with approximate values inserted, is

$U(r, t) \approx .8170 J_0(2.405r)e^{-5.785t} - 1.1394 J_0(5.520r)e^{-30.47t} + 0.7983 J_0(8.654r)e^{-74.89t} - 0.747 J_0(11.792r)e^{-139.04t} + 0.6315 J_0(14.931r)e^{-222.93t}$. A graph of this function is shown for times $t = 0.003, 0.009, 0.04$, and 0.7.

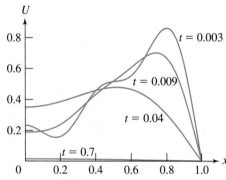

3. $U(r, t) = \displaystyle\sum_{n=1}^\infty \dfrac{2}{[J_1(j_n)]^2} \left(\int_0^1 \xi(9 - 9\xi^2) J_0(j_n \xi) \, d\xi \right) J_0\left(\dfrac{j_n}{3} r \right) e^{-j_n^2 t/18}$; the fifth partial sum, with approximate values

inserted, is $U(r, t) \approx 9.9722 J_0(2.405r/3)e^{-5.78t/18} - 1.258 J_0(5.520r/3)e^{-30.47t/18} + 0.4093 J_0(8.654r/3)e^{-74.89t/18} - 0.1889 J_0(11.792r/3)e^{-139.04t/18} + 0.1048 J_0(14.931r/3)e^{-222.93t/18}$. A graph of this partial sum is shown for times $t = 0.003, 0.009, 0.08$, and 0.4.

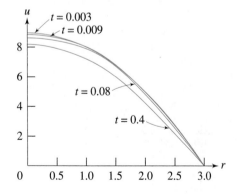

Section 17.5

1. $u(x, y, t) = \displaystyle\sum_{n=1}^\infty \sum_{m=1}^\infty b_{nm} \sin\left(\dfrac{n\pi x}{L} \right) \sin\left(\dfrac{m\pi y}{K} \right) e^{-\beta_{nm} kt}$, where $\beta_{nm} = \left(\dfrac{n^2}{L^2} + \dfrac{m^2}{K^2} \right) \pi^2$ and

$\beta_{nm} = \dfrac{4}{LK} \displaystyle\int_0^K \int_0^L f(x, y) \sin(n\pi x/L) \sin(m\pi y/K) \, dx \, dy$.

3. $u(x, y) = \displaystyle\sum_{m=1}^\infty \dfrac{4}{\pi} \dfrac{8(-1)^{m+1} m}{(2m + 1)^2(2m - 1)^2} \sin(x) \sin(my) e^{-(1+m^2)t}$.

CHAPTER 18

Section 18.1

1. $\nabla^2(f+g) = (f+g)_{xx} + (f+g)_{yy} = (f_{xx} + f_{yy}) + (g_{xx} + g_{yy}) = \nabla^2 f + \nabla^2 g$ and
$\nabla^2(\alpha f) = (\alpha f)_{xx} + (\alpha f)_{yy} = \alpha \left(f_{xx} + f_{yy} \right) = \alpha \nabla^2 f$.

3. Compute $\dfrac{\partial}{\partial x} \ln(x^2 + y^2) = \dfrac{2x}{x^2 + y^2}$ and $\dfrac{\partial}{\partial x} \left(\dfrac{2x}{x^2 + y^2} \right) = 2 \dfrac{y^2 - x^2}{(x^2 + y^2)^2}$.

Similarly, $\dfrac{\partial^2}{\partial y^2} \left(\dfrac{2x}{x^2 + y^2} \right) = 2 \dfrac{x^2 - y^2}{(x^2 + y^2)^2}$. Then $\nabla^2 \ln(x^2 + y^2) = 0$, provided that $x^2 + y^2 \neq 0$.

5. Recall that, in polar coordinates, Laplace's equation is $\dfrac{\partial^2 u}{\partial r^2} + \dfrac{1}{r} \dfrac{\partial u}{\partial r} + \dfrac{1}{r^2} \dfrac{\partial^2 u}{\partial \theta^2} = 0$. It is routine to verify by substitution that the given functions are harmonic.

Section 18.2

1. $u(x, y) = \dfrac{-1}{\sinh(\pi^2)} \sin(\pi x) \sinh(\pi(y - \pi)))$ **3.** $u(x, y) = \displaystyle\sum_{n=1}^{\infty} \dfrac{32}{\pi^2 \sinh(4n\pi)} \dfrac{n(-1)^{n+1}}{(2n-1)^2 (2n+1)^2} \sin(n\pi x) \sinh(n\pi y)$

5. $u(x, y) = \dfrac{1}{\sinh(\pi^2)} \sin(\pi x) \sinh(\pi y) + \displaystyle\sum_{n=1, n\neq 2}^{\infty} 16n \dfrac{(-1)^n - 1}{\pi^2(n-2)^2(n+2)} \dfrac{1}{\sinh(n\pi^2/2)} \sin\left(\dfrac{n\pi x}{2} \right) \sinh \left(\dfrac{n\pi y}{2} \right)$
$\quad + \dfrac{1}{\sinh(2)} \sin(y) \sinh(x)$

7. $u(x, y) = \displaystyle\sum_{n=1}^{\infty} c_n \sin \left(\dfrac{(2n-1)\pi x}{2a} \right) \sinh \left(\dfrac{(2n-1)\pi y}{2a} \right)$, where $c_n =$
$\dfrac{2}{a \sinh[(2n-1)\pi b/2a]} \displaystyle\int_0^a f(x) \sin \left(\dfrac{(2n-1)\pi x}{2a} \right) dx$.

9. $u(x, y) = \dfrac{-1}{\sinh(4\pi)} \sin(\pi y) \sinh(\pi(x-4)) + \displaystyle\sum_{n=1}^{\infty} \dfrac{2}{\sinh(4n\pi)} \left(2 \dfrac{1-(-1)^n}{\pi^3 n^3} \right) \sin(n\pi y) \sinh(n\pi x)$

Section 18.3

1. $u(r, \theta) = 1$ **3.** $u(r, \theta) = \dfrac{1}{3}\pi^2 + \displaystyle\sum_{n=1}^{\infty} \left(\dfrac{r}{2} \right)^n 2(-1)^n \dfrac{1}{n^2} [2\cos(n\theta) + n \sin(n\theta)]$

5. $u(r, \theta) = \dfrac{1}{\pi} \sinh(\pi) + \dfrac{1}{\pi} \displaystyle\sum_{n=1}^{\infty} \left(\dfrac{r}{4} \right)^n \dfrac{e^{-\pi}(-1)^n}{n^2 + 1} [-\cos(n\theta) - n \sin(n\theta) + e^{2\pi} \cos(n\theta) + e^{2\pi} n \sin(n\theta)]$

7. $u(r, \theta) = 1 + \displaystyle\sum_{n=1}^{\infty} \left(\dfrac{r}{8} \right)^n \left(\dfrac{2}{n^3} \right) [n^2 \pi^2 (-1)^n \sin(n\theta) - 6(-1)^n \sin(n\theta)]$

9. In polar coordinates, the problem is $\nabla^2 U(r, \theta) = 0$ for $r < 4$, $U(4, \theta) = 16 \cos^2(\theta)$. This has solution $U(r, \theta) = 8 + r^2(\cos^2 \theta - \frac{1}{2})$. In rectangular coordinates, the solution is $u(x, y) = \frac{1}{2}(x^2 - y^2) + 8$.

11. In polar coordinates, the solution is $U(r, \theta) = r^2 \left(2\cos^2 \theta - 1 \right)$, so $u(x, y) = x^2 - y^2$.

Section 18.4

1. $u \left(\dfrac{1}{2}, \pi \right) = \dfrac{3}{8\pi} \displaystyle\int_0^{2\pi} \dfrac{\xi}{5/4 - \cos(\xi - \pi)} \, d\xi = 9.8696/\pi$; $u \left(\dfrac{3}{4}, \pi/3 \right) \approx 4.813941647/\pi$,
$u(0.2, \pi/4) \approx 8.843875590/\pi$

3. $u(4, \pi) \approx 155.25/\pi$, $u(12, 3\pi/2) \approx 302/\pi$, $u(8, \pi/4) \approx 111.56/\pi$, $u(7, 0) \approx 248.51/\pi$

5. With $u(r, \theta) = r^n \sin(n\theta)$, compute $u(R/2, \pi/2) = \dfrac{R^n}{2^n} \sin(n\pi/2) =$
$\dfrac{1}{2\pi} \displaystyle\int_0^{2\pi} \dfrac{R^2 - R^2/4}{R^2 + R^2/4 - R^2 \cos(\xi - \pi/2)} R^n \sin(n\xi) \, d\xi$. Upon dividing out common powers of R and solving for the
integral, we obtain $\dfrac{1}{2^n} \dfrac{2\pi}{3} \sin(n\pi/2) = \displaystyle\int_0^{2\pi} \dfrac{\sin(n\xi)}{5 - 4\sin(\xi)} \, d\xi$.

7. $\dfrac{\pi}{3(2^{n-1})}\cos\left(\dfrac{n\pi}{2}\right) = \displaystyle\int_0^{2\pi}\dfrac{1}{5-4\sin(\xi)}\cos(n\xi)\,d\xi$ $\dfrac{\pi}{3(2^{n-1})}(-1)^n = \displaystyle\int_0^{2\pi}\dfrac{1}{5+4\cos(\xi)}\cos(n\xi)\,d\xi$

Section 18.5

1. $u(x,y) = \dfrac{1}{\pi}\left[\arctan\left(\dfrac{4-x}{y}\right) - \arctan\left(\dfrac{4+x}{y}\right)\right]$ for $-\infty < x < \infty,\ y > 0$

3. $u(x,y) = \dfrac{y}{\pi}\displaystyle\int_0^\infty\left(\dfrac{1}{y^2+(\xi-x)^2} - \dfrac{1}{y^2+(\xi+x)^2}\right)e^{-\xi}\cos(\xi)\,d\xi$

5. $u(x,y) = \dfrac{2}{\pi}\displaystyle\int_0^\infty\left(\int_0^\infty f(\xi)\sin(\omega\xi)\,d\xi\right)\sin(\omega x)e^{-\omega y}\,d\omega + \dfrac{2}{\pi}\displaystyle\int_0^\infty\left(\int_0^\infty g(\xi)\sin(\omega\xi)\,d\xi\right)\sin(\omega y)e^{-\omega x}\,d\omega$

7. $u(x,y) = \dfrac{y}{\pi}\displaystyle\int_{-\infty}^0\left[\dfrac{1}{y^2+(\xi+x)^2} - \dfrac{1}{y^2+(\xi-x)^2}\right]f(\xi)\,d\xi$

9. $u(x,y) = Be^{-y}\sin(x) + \displaystyle\sum_{n=1}^\infty\dfrac{2}{\pi}\dfrac{h}{n^3}\left(1-(-1)^n\right)\left(1-e^{-ny}\right)\sin(nx)$

11. Using a finite Fourier sine transform in x, we get $u(x,y) = \dfrac{2}{\pi}\displaystyle\sum_{n=1}^\infty\left[\left(-\dfrac{4}{n}+6\dfrac{(-1)^n}{n}\right)e^{-ny} - 2\dfrac{(-1)^n}{n}\right]\sin(nx)$.

13. Using a Fourier sine transform in x, we get $u(x,y) = \dfrac{2}{\pi}\displaystyle\int_0^\infty\left[\dfrac{2}{\omega^3}\left(\dfrac{2+e^{-\omega}}{e^\omega - e^{-\omega}}\right)e^{\omega y} - \dfrac{2}{\omega^3}\left(\dfrac{2+e^\omega}{e^\omega - e^{-\omega}}\right)e^{-\omega y} \right.$

$\left. + \dfrac{1}{\omega}y^2(1-y) + \dfrac{2-6y}{\omega^3}\right]\sin(\omega x)\,d\omega$.

15. $u(x,y) = \dfrac{2}{\pi}\displaystyle\int_0^\infty\left(\dfrac{\omega}{1+\omega^2}\right)\sin(\omega y)e^{-\omega x}\,d\omega$

17. $u(x,y) = \dfrac{y}{\pi}\displaystyle\int_4^8\dfrac{A}{y^2+(\xi-x)^2}\,d\xi = \dfrac{A}{\pi}\left[-\arctan\left(\dfrac{x-8}{y}\right) + \arctan\left(\dfrac{x-4}{y}\right)\right]$

19. $u(x,y) = \dfrac{4}{\pi}\displaystyle\int_0^\infty\dfrac{\omega}{\left(1+\omega^2\right)^2}\sin(\omega x)e^{-\omega y}\,d\omega + \dfrac{2}{\pi}\displaystyle\int_0^\infty\dfrac{\omega}{1+\omega^2}\sin(\omega y)e^{-\omega x}\,d\omega$

21. $u(x,y) = \displaystyle\int_0^\infty\dfrac{2}{\pi\left(1+\omega^2\right)}\cos(\omega y)e^{-\omega x}\,d\omega$

Section 18.6

1. $u(x,y,z) = \displaystyle\sum_{n=1}^\infty\sum_{m=1}^\infty\dfrac{4(-1)^{n+m}}{nm\pi^2\sinh\left(\pi\sqrt{n^2+m^2}\right)}\sin(n\pi x)\sin(m\pi y)\sinh(\pi\sqrt{n^2+m^2}\,z)$

3. $u(x,y,z) = \displaystyle\sum_{n=1}^\infty\sum_{m=1}^\infty\left[\dfrac{16}{\pi^2(2n-1)(2m-1)\sinh(2\pi\sqrt{(2m-1)^2+\pi^2(2n-1)^2})}\right.$

$\left. \times\sin((2n-1)\pi x)\sin((2m-1)z)\sinh(\sqrt{(2m-1)^2+\pi^2(2n-1)^2}\,y)\right]$

$+ \displaystyle\sum_{n=1}^\infty\sum_{m=1}^\infty\left[\dfrac{16}{\pi^2(2n-1)(2m-1)}\dfrac{1}{\sinh\left(\pi\sqrt{\dfrac{(2m-1)^2}{4}+\pi^2(2n-1)^2}\right)}\right.$

$\left. \times\sin((2n-1)\pi x)\sin\left(\dfrac{(2m-1)y}{2}\right)\sinh\left(\sqrt{\dfrac{(2m-1)^2}{4}+\pi^2(2n-1)^2}\,z\right)\right]$

Section 18.7

1. $u(\rho, \varphi) = \sum_{n=0}^{\infty} \frac{(2n+1)A}{2} \left(\int_{-1}^{1} (\arccos(\xi))^2 P_n(\xi)\, d\xi \right) \left(\frac{\rho}{R} \right)^n P_n(\cos(\varphi))$

$\approx 2.9348A - 3.7011A \left(\frac{\rho}{R} \right) P_1(\cos(\varphi)) + 1.1111A \left(\frac{\rho}{R} \right)^2 P_2(\cos(\varphi))$

$- 0.5397A \left(\frac{\rho}{R} \right)^3 P_3(\cos(\varphi)) + 0.3200A \left(\frac{\rho}{R} \right)^4 P_4(\cos(\varphi)) - 0.2120 \left(\frac{\rho}{R} \right)^5 P_5(\cos(\varphi)) + \cdots$

3. $u(\rho, \varphi) \approx 6.0784 - 9.8602 \left(\frac{\rho}{R} \right) P_1(\cos(\varphi)) + 5.2360 \left(\frac{\rho}{R} \right)^2 P_2(\cos(\varphi))$

$- 2.4044 \left(\frac{\rho}{R} \right)^3 P_3(\cos(\varphi)) + 1.5080 \left(\frac{\rho}{R} \right)^4 P_4(\cos(\varphi)) - 0.9783 \left(\frac{\rho}{R} \right)^5 P_5(\cos(\varphi)) + \cdots$

5. $u(\rho, \varphi) = \frac{1}{R_2 - R_1} (T_1 R_1) \left[\frac{1}{\rho} R_2 - 1 \right]$

7. $u(\rho, \varphi) = \sum_{n=1}^{\infty} a_{2n-1} \rho^{2n-1} P_{2n-1}(\cos(\varphi))$, where $a_{2n-1} = \frac{\int_0^1 A P_{2n-1}(x)\, dx}{R^{2n-1} \int_0^1 (P_{2n-1}(x))^2\, dx} = \frac{(4n-1)A}{R^{2n-1}} \int_0^1 P_{2n-1}(x)\, dx$.

9. $u(\rho, \varphi) = \sum_{n=1}^{\infty} a_{2n-1} \rho^{2n-1} P_{2n-1}(\cos(\varphi))$, where $a_{2n-1} = \frac{4n-1}{R^{2n-1}} \int_0^1 f(\arccos(x)) P_{2n-1}(x)\, dx$.

Section 18.8

1. Since $\int_0^1 4\cos(\pi x)\, dx = 0$, a solution may exist. We find $u(x, y) = \frac{4}{-\pi \sinh(\pi)} \cos(\pi x) \cosh(\pi(1-y)) + C$.

3. Since $\int_0^\pi \cos(3x)\, dx = \int_0^\pi (6x - 3\pi)\, dx = 0$, a solution may exist. We find $u(x, y) =$

$\frac{1}{-3 \sinh(3\pi)} \cos(3x) \cosh(3(\pi - y)) + \sum_{n=1}^{\infty} \frac{12}{\pi} \frac{(-1)^n - 1}{n^3 \sinh(n\pi)} \cos(nx) \cosh(ny) + C$.

5. $u(x, y) = \sum_{n=1}^{\infty} 2 \left(\frac{n^2 \pi^2 (-1)^n + 6(1 - (-1)^n)}{n^4 \pi^4 \sinh(\pi)} \right) \sin(n\pi y) \cosh(n\pi(1-x))$

7. $u(r, \theta) = \frac{1}{2} a_0 + \frac{R}{2} \left(\frac{r}{R} \right)^2 (2\cos^2 \theta - 1)$

9. In polar coordinates, the boundary condition is $\frac{\partial U}{\partial r}(3, \theta) = 18 \sin(2\theta)$, with solution $U(r, \theta) = \frac{1}{2} a_0 + 3r^2 \sin(2\theta)$. In rectangular coordinates, the solution is $u(x, y) = 6xy$.

11. In polar coordinates, $\frac{\partial U}{\partial r}(1, \theta) = \frac{1}{4} \sin^2(2\theta)$, and the solution is $U(r, \theta) = \frac{1}{2} a_0 + \frac{1}{4} r^4 \left(-\cos^4 \theta + \cos^2 \theta - \frac{1}{8} \right)$. In rectangular coordinates, $u(x, y) = \frac{1}{2} a_0 + \frac{1}{4} \left(x^2 y^2 - \frac{1}{8} (x^2 + y^2)^2 \right)$.

13. $u(x, y) = \frac{1}{2\pi} \int_{-\infty}^{\infty} \ln(y^2 + (\xi - x)^2) e^{-|\xi|} \sin(\xi)\, d\xi + c$

15. $u(x, y) = \int_0^{\infty} a_\omega \cos(\omega x) e^{-\omega y}\, d\omega + c$, with $a_\omega = -\frac{2}{\pi \omega} \int_0^{\infty} f(\xi) \cos(\omega \xi)\, d\xi$

CHAPTER 19

Section 19.1

1. $B^2 - AC = 25 - 16 > 0$, so the equation is hyperbolic; characteristics are lines $y = 4x + c$, $y = x + k$. The canonical form is $U_{\xi\eta} - \frac{1}{54}(\xi - \eta)(4U_\xi + U_\eta) - \frac{1}{54}(\xi - 4\eta)(U_\xi + U_\eta) = 0$.

3. hyperbolic $(B^2 - AC = 4)$; $x - y = c$, $x + 3y = k$; $U_{\xi\eta} + \frac{1}{64}(\eta - \xi)(U_\xi + U_\eta) + \frac{1}{16} U_\xi - \frac{3}{16} U_\eta = 0$

5. hyperbolic; $7x + 2y = c$, $x + 2y = k$; $U_{\xi\eta} - \frac{1}{864}(\xi + 5\eta)(U_\xi + U_\eta) = 0$

7. parabolic; $y = -2x + c$; $U_{\eta\eta} + 3U_\xi + U_\eta = 0$ **9.** hyperbolic; $4x + y = c$, $x + y = k$; $U_{\xi\eta} - \frac{1}{6} U_\xi = 0$

Section 19.2

1. Compute

$$I'(t) = \frac{1}{2} \int_0^L \left[2 \frac{\partial w}{\partial x} \frac{\partial^2 w}{\partial x \partial t} + \frac{2}{c^2} \frac{\partial w}{\partial t} \frac{\partial^2 w}{\partial t^2} \right] dx$$

$$= \int_0^L \left[\frac{\partial w}{\partial x} \frac{\partial^2 w}{\partial x \partial t} + \frac{1}{c^2} \frac{\partial w}{\partial t} c^2 \frac{\partial^2 w}{\partial x^2} \right] dx = \int_0^L \frac{\partial}{\partial x} \left(\frac{\partial w}{\partial x} \frac{\partial w}{\partial t} \right) dx$$

$$= \frac{\partial w}{\partial x}(L, t) \frac{\partial w}{\partial t}(L, t) - \frac{\partial w}{\partial x}(0, t) \frac{\partial w}{\partial t}(0, t).$$

Now

$$\frac{\partial w}{\partial t}(L, t) = \frac{\partial y_1}{\partial t}(L, t) - \frac{\partial y_2}{\partial t}(L, t) = \beta'(t) - \beta'(t) = 0$$

and

$$\frac{\partial w}{\partial t}(0, t) = \frac{\partial y_1}{\partial t}(0, t) - \frac{\partial y_2}{\partial t}(0, t) = \alpha'(t) - \alpha'(t) = 0,$$

so $I'(t) = 0$ for $t > 0$. Since I is continuous, then $I(t) = $ constant for $t \geq 0$. But

$$I(0) = \frac{1}{2} \int_0^L \left[\left(\frac{\partial w}{\partial x}(x, 0) \right)^2 + \frac{1}{c^2} \left(\frac{\partial w}{\partial t}(x, 0) \right)^2 \right] dx,$$

and

$$\frac{\partial w}{\partial x}(x, 0) = \frac{\partial y_1}{\partial x}(x, 0) - \frac{\partial y_2}{\partial x}(x, 0) = f'(x) - f'(x) = 0$$

and

$$\frac{\partial w}{\partial t}(x, 0) = \frac{\partial y_1}{\partial t}(x, 0) - \frac{\partial y_2}{\partial t}(x, 0) = g(x) - g(x) = 0,$$

so $I(0) = 0$. Thus, $I(t) = 0$ for $t \geq 0$. Therefore $\frac{\partial w}{\partial x} = 0$ and $\frac{\partial w}{\partial t} = 0$ for $0 \leq x \leq L$ and $t \geq 0$. Therefore, $w(x, t) = $ constant. But $w(x, 0) = f(x) - f(x) = 0$, so $w(x, t) = 0$ and $y_1(x, t) = y_2(x, t)$.

3. w satisfies $\nabla^2 w = 0$ in M and $\frac{\partial w}{\partial n} = 0$ on ∂M. By one of Green's identities (which follow from Gauss's divergence theorem),

$$\iint_{\partial M} w \nabla w \cdot \mathbf{N} \, d\sigma = \iiint_M w \nabla^2 w \, dV + \iiint_M \left[\left(\frac{\partial w}{\partial x} \right)^2 + \left(\frac{\partial w}{\partial y} \right)^2 + \left(\frac{\partial w}{\partial z} \right)^2 \right] dV.$$

The integral on the left is zero because $w = 0$ on ∂M. On the right, the first triple integral is zero because $\nabla^2 w = 0$ in M. Therefore, the last integral is zero, forcing $\frac{\partial w}{\partial x} = \frac{\partial w}{\partial y} = \frac{\partial w}{\partial z} = 0$ in M. But then $w(x, y, z) = $ constant on M. Since $w = 0$ on ∂M, then w is identically zero on M.

Section 19.3

1. This problem is well posed if, given $\epsilon > 0$, there exists a positive number δ such that if $|f_1(x) - f_2(x)| < \delta$ and $|g_1(x) - g_2(x)| < \delta$ for $0 \leq x \leq L$, then $|y_1(x, t) - y_2(x, t)| < \epsilon$ for $0 \leq x \leq L, t \geq 0$, where y_1 is the solution of the problem with initial conditions

$$y(x, 0) = f_1(x), \quad \frac{\partial y}{\partial t}(x, 0) = g_1(x) \quad \text{for } 0 < x < L,$$

and y_2 is the solution with

$$y(x, 0) = f_2(x), \quad \frac{\partial y}{\partial t}(x, 0) = g_2(x) \quad \text{for } 0 < x < L.$$

CHAPTER 20

Section 20.1

1. $26 - 18$ **3.** $\frac{1}{65}(1 + 18i)$ **5.** $4 + 228i$ **7.** $6 - i$ **9.** $\frac{1}{4225}(-1632 + 2024i)$

11. $i^{4n} = ((i^2)^2)^n = 1^n = 1$; $i^{4n+1} = ii^{4n} = i$, since $i^{4n} = 1$; $i^{4n+2} = i^{4n}(i^2) = -1$; $i^{4n+3} = i^{4n}i^2i = -i$

13. $a^2 - b^2 + b + 1$; $2ab - a$ **17.** $\dfrac{\pi}{2} + 2n\pi$ **19.** $\pi - \tan^{-1}(\frac{2}{3}) + 2n\pi$ **21.** $\pi + 2n\pi$

23. $2\sqrt{2}[\cos(3\pi/4) + i\sin(3\pi/4)]$ **25.** $\sqrt{29}\left[\cos(-\tan^{-1}(\frac{2}{5})) + i\sin(-\tan^{-1}(\frac{2}{5}))\right]$

27. $\sqrt{65}\left[\cos(\tan^{-1}(\frac{1}{8})) + i\sin(\tan^{-1}(\frac{1}{8}))\right]$

29. *Hint:* If $|z| = 1$, then $z\bar{z} = 1$ and $\left|\dfrac{z - w}{1 - \bar{z}w}\right| = \left|\dfrac{z - w}{z\bar{z} - w\bar{z}}\right| = \dfrac{1}{|\bar{z}|}\left|\dfrac{z - w}{z - w}\right| = 1$. **33.** $8i$

35. $\cos^3(\theta) - 3\cos(\theta)\sin^2(\theta)$

Section 20.2

1. Circle of radius 9 with center $(8, -4)$ **3.** Circle of radius $\frac{1}{2}\sqrt{65}$ with center $(0, -\frac{1}{2})$ **5.** The real axis for $x \leq 0$

7. The line $y = x + 2$ **9.** The line $8x + 10y + 27 = 0$ **11.** The half-plane $3x + y + 2 > 0$

13. The vertical strip consisting of all (x, y) with $-1 \leq x < 2$

15. The points on the half-plane $x + 3y - 1 \geq 0$ that are also outside the circle $x^2 + y^2 = 16$

17. The horizontal strip of the xy plane with $-3 < y < 1$

19. K is the closed half-plane $2x + 8y + 15 \geq 0$; every point of K is a limit point of K, and there are no other limit points; boundary points of K are those points on the line $2x + 8y + 15 = 0$; K is closed but not compact (because K is not bounded).

21. M consists of all points below the line $y = 7$; limit points are points of M and points on the line $y = 7$; boundary points are points $x + 7i$; M is open; M is not compact.

23. U consists of all points $x + iy$ with $1 < x \leq 3$; limit points are points of U and points on the line $x = 1$; boundary points are points on the lines $x = 1$ and $x = 3$; U is neither open nor closed; U is not compact (neither closed nor bounded).

25. W consists of all $x + iy$ with $x > y^2$. These are points (x, y) inside and to the right of the parabola $x = y^2$. This set is open, and not compact. Limit points are all points of W and points on the parabola; boundary points are the points on the parabola.

27. K is neither open nor closed. Every complex number is both a limit point and a boundary point. K is not compact.

29. (a) Let S^C be the set of complex numbers not in S.

First suppose S is open. Let z be a boundary point of S^C. If N is any open disk about z, then N must contain a point of S^C and a point not in S^C. But a point not in S^C is in S, so N contains at least one point of S and a point of S^C. Therefore, z is a boundary point of S. But S is open, and so cannot contain a boundary point. Therefore, z is in S^C. Then S^C contains all of its boundary points, and so is closed.

Conversely, suppose S^C is closed. Suppose S is not open. Then for some z in S, every open disk about z contains a point in S and a point in S^C. Then z is a boundary point of S^C. But S^C is closed, so S^C contains its boundary points. Then z is in S^C. This contradicts z being in S. Therefore S is open.

The proof of (b) is similar.

31. $1 + 2i$ **33.** $2 - i$ **35.** -1 **37.** $\frac{3}{2}i$

39. Suppose $\{z_n\}$ converges to L. If $\epsilon > 0$, there is some N such that $|z_n - L| < \epsilon$ if $n \geq N$. But then $||z_n| - |L|| < \epsilon$ if $n \geq N$ by Theorem 20.2(3). For the example, $\{i^n\}$ diverges, but $\{|i|^n\} = \{1\}$ converges.

41. If n is even, say $n = 2m$, then $i^{2n} = i^{4m} = 1$, so $\{1\}$ is one convergent subsequence; if $n = 2m + 1$, then $i^{2n} = i^{4m}i^2 = i^2 = -1$, so $\{-1\}$ is another convergent subsequence. There are others.

CHAPTER 21

Section 21.1

1. $u = x$, $v = y - 1$; Cauchy–Riemann equations hold at all points; f is differentiable for all complex z.

3. $u = \sqrt{x^2 + y^2}$, $v = 0$; nowhere; nowhere

5. $u = 0$, $v = x^2 + y^2$; the Cauchy–Riemann equations hold at $(0, 0)$; nowhere

7. $u = 1$, y/x; nowhere; nowhere **9.** $u = x^2 - y^2$, $v = -2xy$; $(0, 0)$; nowhere

11. $u = -4x + \dfrac{x}{x^2 + y^2}$, $v = -4y - \dfrac{y}{x^2 + y^2}$; Cauchy–Riemann equations hold for all nonzero z; differentiable for $z \neq 0$

13. $u = x - y$, $v = 0$; nowhere; nowhere   **15.** $u = \dfrac{2y}{(x+1)^2 + y^2}$, $v = 0$; nowhere; nowhere

Section 21.2

1. 2; $|z + 3i| < 2$ **3.** 1; $|z - 1 + 3i| < 1$ **5.** 2; $|z + 8i| < 2$ **7.** 1; $|z + 6 + 2i| < 1$ **9.** 1; $|z + 4| < 1$

11. No; i is closer to $2i$ than 0 is, so if the series converged at 0 it would have to converge at i.

13. $\displaystyle\sum_{n=0}^{\infty} (c_n + c_n^*) z^n$ has radius of convergence at least as large as the minimum of R and R^*.

15. c_{n+1}/c_n is either 2 or $\frac{1}{2}$, depending on whether n is odd or even, so c_{n+1}/c_n has no limit. However, $\displaystyle\lim_{n \to \infty} |c_n|^{1/n} = 1$, so

the radius of convergence is 1 by the nth root test applied to $\displaystyle\sum_{n=0}^{\infty} |c_n z^n|$.

Section 21.3

1. $\cos(1) + i \sin(1)$ **3.** $\cos(3) \cosh(2) - i \sin(3) \sinh(2)$ **5.** $e^5 \cos(2) + i e^5 \sin(2)$

7. $\frac{1}{2}[1 - \cos(2) \cosh(2)] + \frac{1}{2} i \sin(2) \sinh(2)$ **9.** i **11.** $u = e^{x^2 - y^2} \cos(2xy)$, $v = e^{x^2 - y^2} \sin(2xy)$

13. $u = \dfrac{\sin(x) \cos(x)}{\cos^2(x) \cosh^2(y) + \sin^2(x) \sinh^2(y)}$, $v = \dfrac{\cosh(y) \sinh(y)}{\cos^2(x) \cosh^2(y) + \sin^2(x) \sinh^2(y)}$

15. $\sin^2(z) + \cos^2(z) = \left(\dfrac{1}{2i}(e^{iz} - e^{-iz})\right)^2 + \left(\dfrac{1}{2}(e^{iz} + e^{-iz})\right)^2 = 1$ **17.** $\sinh(z) = -i \sin(iz)$, $\cosh(z) = \cos(iz)$

19. $\sinh(z) = 0$ only if $\sin(iz) = 0$, so $iz = n\pi$, or $z = n\pi i$ with n any integer.

21. $z = \ln(2) + i\left(\dfrac{\pi}{2} + 2n\pi\right)$, n any integer   **23.** $z = \ln(2) + (2n + 1)\pi i$, n any integer

25. $e^z e^w = \left(\displaystyle\sum_{n=0}^{\infty} \dfrac{1}{n!} z^n\right)\left(\displaystyle\sum_{n=0}^{\infty} \dfrac{1}{n!} w^n\right) = \displaystyle\sum_{n=0}^{\infty} \left(\displaystyle\sum_{k=0}^{n} \dfrac{1}{k!} z^k \dfrac{1}{(n-k)!} w^{n-k}\right)$. Now use the binomial formula to recognize that

$$\sum_{k=0}^{n} \dfrac{1}{k!} z^k \dfrac{1}{(n-k)!} w^{n-k} = \dfrac{1}{n!} (z + w)^n.$$

Section 21.4

1. $\ln(4) + \dfrac{4n - 1}{2} \pi i$, $\ln(4) + \dfrac{3}{2} \pi i$ **3.** $\ln(5) + (2n + 1)\pi i$, $\ln(5) + \pi i$

5. $\ln(\sqrt{85}) + [(2n + 1)\pi + \tan^{-1}(-\frac{2}{9})]i$, $\ln(\sqrt{85}) + (\pi + \tan^{-1}(-\frac{2}{9}))i$ **7.** *Hint:* In polar form, $\dfrac{z}{w} = \left|\dfrac{z}{w}\right| e^{i(\arg(z) - \arg(w))}$.

Section 21.5

1. $ie^{-(2n\pi + \pi/2)}$ **3.** $e^{-(2n\pi + \pi/2)}$ **5.** $e^{3(2n\pi + 3\pi/4)}\left[\cos\left(\dfrac{3\ln(2)}{2}\right) - i \sin\left(\dfrac{3\ln(2)}{2}\right)\right]$

7. $\cos\left(\dfrac{\pi}{8} + \dfrac{n\pi}{2}\right) + i \sin\left(\dfrac{\pi}{8} + \dfrac{n\pi}{2}\right)$ **9.** $16e^{(2n+1)\pi}[\cos(\ln(4)) - i \sin(\ln(4))]$

11. $2\left[\cos\left(\dfrac{(2n + 1)\pi}{4}\right) + i \sin\left(\dfrac{(2n + 1)\pi}{4}\right)\right]$ **13.** $\cos(n\pi/3) + i \sin(n\pi/3)$

15. $-e^{2(4n+1)\pi}$ **17.** $(324)^{1/5}\left[\cos\left(\dfrac{(2n + 1)\pi}{5}\right) + i \sin\left(\dfrac{(2n + 1)\pi}{5}\right)\right]$

19. $\sqrt{68} e^{3\tan^{-1}(1/4) - 6n\pi}[\cos(3\ln(\sqrt{68})) - \tan^{-1}(\frac{1}{4})) + i \sin(3\ln(\sqrt{68})) - \tan^{-1}(\frac{1}{4}))]$

21. The nth roots of unity are $\omega_k = e^{2k\pi i/n}$ for $k = 0, 1, \ldots, n - 1$. Now use the fact that for $z \neq 1$, $\displaystyle\sum_{k=0}^{n-1} z^k = \dfrac{z^n - 1}{z - 1}$, with

$z = e^{2\pi i/n}$.

23. *Hint:* Complete the square on $z^2 + \dfrac{b}{a} z = -\dfrac{c}{a}$.

25. $\frac{1}{2}\{[-1 \pm 5^{1/4}[\cos(\theta) + \sin(\theta)] + i[-1 \pm 5^{1/4}[\cos(\theta) - \sin(\theta)]]\}$, where $\theta = \frac{1}{2} \tan^{-1}(1/2)$

CHAPTER 22

Section 22.1

The graph of the curves in Problems 1, 3, 5, 7, and 9 are shown below.

1. initial point $6 - 2i$, terminal point $2 - 2i$; simple and not closed; tangent $\Gamma'(t) = 2ie^{it} = -2\sin(t)\mathbf{i} + 2\cos(t)\mathbf{j}$

3. $1 + i, 3 + 9i$; simple and not closed; $\Gamma'(t) = 1 + 2ti = \mathbf{i} + 2t\mathbf{j}$

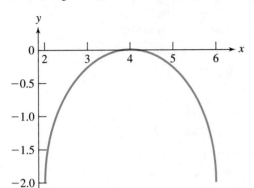

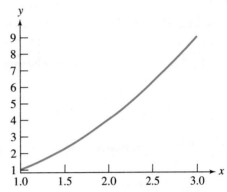

5. $3, 3$; closed but not simple; $\Theta'(t) = -3\sin(t) + 5\cos(t)i = -3\sin(t)\mathbf{i} + 5\cos(t)\mathbf{j}$

7. $-2 - 4i, 4 - 16i$; simple and not closed; $\Psi'(t) = 1 - ti = \mathbf{i} - t\mathbf{j}$

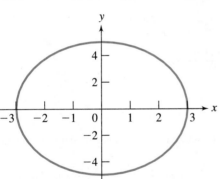

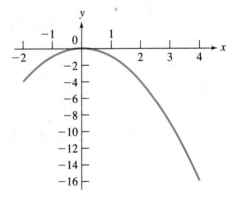

9. $1, \cos(2) - 2\sin(4)i$; simple and not closed; $\Gamma'(t) = -\sin(t) - 4\cos(2t)i = -\sin(t)\mathbf{i} - 4\cos(2t)\mathbf{j}$

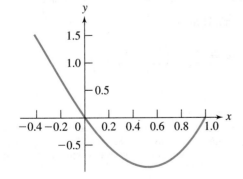

Section 22.2

1. $8 - 2i$ **3.** $\frac{3}{2}(1 + i)$ **5.** $\frac{1}{2}(-13 + 4i)$ **7.** $-\frac{1}{2}[\cosh(8) - \cosh(2)]$

9. $-\frac{1}{2}[e^{-1}(\cos(2) + i\sin(2)) - e(\cos(2) - i\sin(2))]$ **11.** $10 + 210i$ **13.** $\frac{25}{2}i$ **15.** $\frac{2}{3}(1 + i)$ **17.** $-\frac{i}{3}\sinh(1)$

19. 0   **21.** $8\pi e^{16}$ (or any larger number)   **23.** $\displaystyle\sum_{n=0}^{\infty} \frac{(-1)^n}{(3n)!} \frac{1}{n+1}\left[(-1+2i)^{n+1} - (-2+i)^{n+1}\right]$

Section 22.3

1. 0   **3.** 0   **5.** $2\pi i$ **7.** 0   **9.** 0   **11.** $4\pi i$ **13.** 0   **15.** 0   **17.** $\frac{1}{4}\pi i$

Section 22.4

1. $32\pi i$ **3.** $2\pi i(-8+7i)$ **5.** $-2\pi e^2(\cos(1) - i\sin(1))$ **7.** $\pi i(6\cos(12) - 36\sin(12))$

9. $-512\pi(1-2i)\cos(256)$ **11.** $-\frac{13}{2} - 39i$ **13.** 0   **15.** πi **17.** $2\pi i\cosh(\sinh(3))$ **19.** $\dfrac{\pi}{3}(\sinh(1) + i\cosh(1))$

21. 2π **23.** By Cauchy's integral formula, both integrals equal $2\pi i f'(a)$.

CHAPTER 23

Section 23.1

1. $\displaystyle\sum_{n=0}^{\infty} \frac{(-1)^n}{(2n)!} 2^{2n} z^{2n}$; $|z| < \infty$ (that is, the series converges for all complex z)

3. $\displaystyle\sum_{n=0}^{\infty} \frac{1}{(1-4i)^{n+1}}(z - 4i)^n$; $|z - 4i| < \sqrt{17}$ **5.** $\displaystyle\sum_{n=0}^{\infty}(n + 1)z^n$; $|z| < 1$

7. $-3 + (1 - 2i)(z - 2 + i) + (z - 2 + i)^2$; $|z| < \infty$ **9.** $63 - 16i + (-16 + 2i)(z - 1 - i) + (z - 1 - i)^2$; $|z| < \infty$

11. $\displaystyle\sum_{n=0}^{\infty} \frac{(-1)^n}{(2n+1)!}(z + i)^{2n+1}$; $|z| < \infty$ **13.** $\displaystyle\sum_{n=0}^{\infty} \frac{-1}{(2+6i)^{n+1}}(z + 2i)^n$; $|z + 2i| < \sqrt{40}$

15. $\displaystyle\sum_{n=0}^{\infty} \frac{(-1)^n}{(2n)!} z^{4n} - \sum_{n=0}^{\infty} \frac{(-1)^n}{(2n+1)!} z^{2n+1}$; $|z| < \infty$ **17.** $\displaystyle\sum_{n=0}^{\infty} \frac{(1+i)^n + (1-i)^n}{2n!} z^n$; $|z| < \infty$

19. $1 + iz + \displaystyle\sum_{n=1}^{\infty}\left(\frac{2^n + 2^{n-1}}{(2n)!} z^{2n} + i\frac{2^n}{(2n+1)!} z^{2n+1}\right)$; $|z| < \infty$

21. Put $z = re^{i\theta}$ into $f(z) = \displaystyle\sum_{n=0}^{\infty} c_n z^n$ to get $f(re^{i\theta}) = \displaystyle\sum_{n=0}^{\infty} c_n r^n e^{in\theta}$. By Cauchy's integral formula for derivatives, with C

the circle of radius r about the origin, $c_n = \dfrac{1}{n!} f^{(n)}(0) = \dfrac{1}{n!}\dfrac{n!}{2\pi i}\displaystyle\int_C \frac{f(\xi)}{\xi^{n+1}} d\xi = \dfrac{1}{2\pi i}\displaystyle\int_0^{2\pi} \frac{f(re^{i\xi})}{r^{n+1}e^{i(n+1)\xi}} ire^{i\xi} d\xi =$

$\dfrac{1}{2\pi r^n}\displaystyle\int_0^{2\pi} f(re^{i\xi})e^{-in\xi} d\xi$.

23. Fix z and think of w as the variable. Define $f(w) = e^{zw}$. Then $f^{(n)}(w) = z^n e^{zw}$. By Cauchy's integral formula,

$f^{(n)}(0) = z^n = \dfrac{n!}{2\pi i}\displaystyle\int_\Gamma \frac{e^{zw}}{w^{n+1}} dw$, with Γ the unit circle about the origin. Then $\dfrac{z^n}{n!} = \dfrac{1}{2\pi i}\displaystyle\int_\Gamma \frac{e^{zw}}{w^{n+1}} dw$, so $\left(\dfrac{z^n}{n!}\right)^2 =$

$\dfrac{1}{2\pi i}\displaystyle\int_\Gamma \frac{z^n}{n!w^{n+1}} e^{zw} dw$. Then $\displaystyle\sum_{n=0}^{\infty}\left(\frac{z^n}{n!}\right)^2 = \dfrac{1}{2\pi i}\displaystyle\sum_{n=0}^{\infty}\int_\Gamma \frac{z^n}{n!w^{n+1}} e^{zw} dw = \dfrac{1}{2\pi i}\displaystyle\int_\Gamma\left(\sum_{n=0}^{\infty}\frac{1}{n!}\left(\frac{z}{w}\right)n\right)\frac{e^{zw}}{w} dw =$

$\dfrac{1}{2\pi i}\displaystyle\int_\Gamma e^{z(w+1/w)}\frac{1}{w} dw$. Now let $w = e^{i\theta}$ on Γ to derive the result.

25. Let $g(z) = \sin(z)$. Then f and g agree on the convergent sequence $\{1/n\}_{n=1}^{\infty}$, so $f = g$.

27. The maximum must occur at a boundary point of the rectangle. Consider each side. On the left vertical side, $x = 0$ and $|e^z| = |e^{iy}| = 1$. On the right vertical side, $|e^z| = e^1|e^{iy}| = e$ has maximum e. On the lower horizontal side, $|e^z| = e^x$ for $0 \le x \le 1$, with maximum e. On the upper horizontal side, $|e^z| = e^x$ has maximum e. Thus the maximum of $|e^z|$ on this rectangle is e.

Section 23.2

1. $\dfrac{1}{z - i} + \displaystyle\sum_{n=0}^{\infty} \frac{(-1)^n}{(2i)^{n+1}}(z - i)^n$, for $0 < |z - i| < 2$ **3.** $\displaystyle\sum_{n=1}^{\infty} \frac{(-1)^{n+1}4^n}{(2n)!} z^{2n-2}$; $|z| < \infty$

5. $-\dfrac{1}{z-1} - 2 - (z-1); 0 < |z-1| < \infty$ **7.** $\dfrac{1}{z^2} + \displaystyle\sum_{n=0}^{\infty} \dfrac{z^{2n}}{(n+1)!}; 0 < |z| < \infty$ **9.** $1 + \dfrac{2i}{z-i}; 0 < |z-i| < \infty$

11. $\displaystyle\sum_{n=0}^{\infty} \dfrac{1}{n!}(z+i)^{-n}; 0 < |z+i| < \infty$ **13.** $-\dfrac{1}{4}\dfrac{1}{z+2} - \dfrac{1}{4}\displaystyle\sum_{n=0}^{\infty}\dfrac{1}{4^{n+1}}(z+2)^n;$ for $0 < |z-2| < 4$

CHAPTER 24

Section 24.1

1. pole of order 2 at $z = 0$ **3.** essential singularity at $z = 0$ **5.** simple poles at i and $-i$, pole of order 2 at 1

7. simple pole at $-i$, removable singularity at i **9.** simple poles at $1, -1, i,$ and $-i$ **11.** simple poles at $\dfrac{(2n+1)}{2}\pi$

13. double pole at 0 **15.** pole of order 4 at -1 **17.** no singularities **19.** simple pole at $-\pi$

21. f has a Taylor expansion $f(z) = \displaystyle\sum_{n=0}^{\infty} a_n(z-z_0)^n$ in some open disk about z_0, and g has a Laurent expansion of the

form $g(z) = \dfrac{b_{-1}}{z-z_0} + \displaystyle\sum_{n=0}^{\infty} b_n(z-z_0)^n$ in some annulus $0 < |z-z_0| < r$. Then fg has an expansion of the form

$\dfrac{b_{-1}a_0}{z-z_0} + \displaystyle\sum_{n=0}^{\infty} c_n(z-z_0)^n$ in this annulus, and $b_{-1}a_0 \neq 0$ because $b_{-1} \neq 0$ and $a_0 = f(z_0) \neq 0$.

23. Write $h(z) = (z-z_0)^3 q(z)$, where q is analytic at z_0 and $q(z_0) \neq 0$. Then $\dfrac{g(z)}{h(z)} = \dfrac{1}{(z-z_0)^3}\dfrac{g(z)}{q(z)}$ in some annulus

$0 < |z-z_0| < r$. Now g/q is analytic at z_0, and so has Taylor expansion $g(z)/q(z) = \displaystyle\sum_{n=0}^{\infty} c_n(z-z_0)^n$ in some disk

about z_0. Further, $c_0 \neq 0$ because $q(z_0) \neq 0$ and $g(z_0) \neq 0$. Then, in some annulus about z_0,

$\dfrac{g(z)}{h(z)} = \dfrac{c_0}{(z-z_0)^3} + \displaystyle\sum_{n=1}^{\infty} c_n(z-z_0)^{n-2}.$

Section 24.2

1. The residue at 1 is $\frac{1}{25}(16-12i)$, and at $-2i$, $\frac{1}{25}(9+12i)$; the value of the integral is therefore $2\pi i$.

3. 0 **5.** $2\pi i$ **7.** $2\pi i$ **9.** $-\pi i/4$ **11.** 0 **13.** $2\pi i$ **15.** $\dfrac{\pi}{2}(e^{8i}-1)$ **17.** $\dfrac{\pi i}{9}(\cosh(9)-2)$

19. $\pi(-40+50i)e^{-i}$ **21.** 0 **23.** $4\pi(2+i)$ **25.** 0 **27.** $\dfrac{\pi}{2}(-i-\cos(4))$

29. Write $g(z) = \sum_{n=0}^{\infty} a_n(z-z_0)^n$ and $h(z) = \displaystyle\sum_{n=3}^{\infty} b_n(z-z_0)^n$, with $a_0 \neq 0$ and $b_3 \neq 0$. From Problem 23, Section 24.1,

$\dfrac{g(z)}{h(z)} = \displaystyle\sum_{n=-3}^{\infty} d_n(z-z_0)^n$, with $d_{-3} \neq 0$. Write $g(z) = \displaystyle\sum_{n=0}^{\infty} a_n(z-z_0)^n = \left(\displaystyle\sum_{n=3}^{\infty} b_n(z-z_0)^n\right)\left(\displaystyle\sum_{n=-3}^{\infty} d_n(z-z_0)^n\right)$ and

equate the coefficient of $(z-z_0)^n$ on the left with the coefficient of $(z-z_0)^n$ in the product on the right. We get
$a_0 = d_{-3}b_3, a_1 = d_{-3}b_4 + d_{-2}b_3, a_2 = d_{-3}b_5 + d_{-2}b_4 + d_{-1}b_3$. Use these to solve for d_{-1} in terms of coefficients a_0,
$a_1, a_2, b_1, \ldots, b_4$ and use the fact that $a_n = \dfrac{1}{n!}g^{(n)}(z_0), b_n = \dfrac{1}{n!}h^{(n)}(z_0)$.

Section 24.3

1. $2\pi i$ **3.** 0 **5.** Let $g(z) = f(z) - a$ in the Argument Principle.
7. On the unit circle, $|g(z)| = |e^z| = e$ and $|f(z)| = 3$, so by Rouché's theorem, g has the same number of zeros in the
unit disk as does $3z^n$, which has n zeros (counting multiplicities).

9. 4 **11.** $\dfrac{\pi}{a^3}\sqrt{2}\left(\dfrac{\sin(\pi a/\sqrt{2})\cosh(\pi a/\sqrt{2}) + \cos(\pi a/\sqrt{2})\sinh(\pi a/\sqrt{2})}{\cosh^2(\sqrt{2}\pi a) - \cos^2(\sqrt{2}\pi a)}\right)$ **13.** $\frac{1}{4}(1 - \pi\sqrt{2}\csc(\sqrt{2}\pi))$

17. $\cos(3t)$ **19.** $\frac{1}{36}e^{-4t} - \frac{1}{36}e^{2t} + \frac{1}{6}te^{2t}$ **21.** $\frac{1}{2}t^2 e^{-5t}$ **23.** $\dfrac{1}{\sqrt{2}}[\sin(t/\sqrt{2})\cosh(t/\sqrt{2}) - \cos(t/\sqrt{2})\sinh(t/\sqrt{2})]$

25. $\frac{1}{2}t^2e^{2t}\cos^2(2) - e^{2t}\cos(4) - 2te^{2t}\sin(4)$

27. Use the fact that if $F(z) = P(z)/Q(z)$, with no common zeros, and Q has a simple zero at a, then
$$\operatorname{Re} s(e^{tz}F(z), a_j) = \frac{e^{ta_j}P(a_j)}{Q'(a_j)}.$$

29. $2\pi/\sqrt{3}$ **31.** $2\pi/3$ **33.** $\frac{1}{4}\pi e^{-2\sqrt{2}}\sin(2\sqrt{2})$ **35.** $-\pi/128$ **37.** $\frac{\pi}{32}(1 + 5e^{-4})$

39. $-\frac{16\pi}{5}\sin\left(\frac{3\pi}{5}\right)\frac{17\cos(\pi/5) + 16\cos(3\pi/5)}{289 + 168\cos(2\pi/5) + 136\cos(4\pi/5) + 32\cos(6\pi/5)}$ **41.** $\pi/2$ **43.** $\frac{\pi}{3}e^{-9}$

45. $2\pi\left(1 - \frac{4}{\sqrt{15}}\right)$ **47.** $-\frac{\pi}{3}e^{-1} + \frac{2\pi}{3}e^{-1/2}\cos\left(\frac{\sqrt{3}}{2}\right)$ **49.** $2\pi/3$ **51.** 2π

53. $\frac{\pi}{10}2^{1/4}\left[3\sin\left(\frac{3\pi}{8}\right) + \cos\left(\frac{3\pi}{8}\right) - 2^{5/4}\right]$

55. $\operatorname{Re} s(e^{i\alpha z}/(z^2 + 1), i) = -\frac{1}{2}ie^{-\alpha}$, so $\int_{-\infty}^{\infty}\frac{\cos(\alpha x)}{x^2 + 1}\,dx = 2\pi i\left(-\frac{1}{2}ie^{-\alpha}\right) = \pi e^{-\alpha}$.

57. With the trigonometric substitutions, we obtain $-4i\int_{\Gamma}\frac{z}{(\alpha^2 - \beta^2)z^2 + 2(\alpha^2 + \beta^2)z + (\alpha^2 - \beta^2)}\,dz$, with Γ the unit circle. The two poles within the unit disk are $z = \pm\sqrt{\frac{\beta - \alpha}{\beta + \alpha}}$, and the residue at each is $-i/2\alpha\beta$. Therefore, the value of the integral is $2\pi i(-i/\alpha\beta)$, or $2\pi/\alpha\beta$.

59. For $n = 1, 2, 3, 4, 5$, obtain, respectively, the value $\pi, \pi/2, 3\pi/8, 5\pi/16$ and $35\pi/128$ for the integral. In general,
$$\int_{-\infty}^{\infty}\frac{1}{(1 + x^2)^n}\,dx = \frac{\pi(2n - 2)!}{2^{2n-2}[(n - 1)!]^2}.$$

61. By Cauchy's theorem, $\int_{\Gamma}e^{-z^2}\,dz = 0$, where Γ is the rectangular path. Writing the integral over each piece of the boundary (starting on the bottom and going counterclockwise), we have $\int_{-R}^{R}e^{-x^2}\,dx + \int_{0}^{\beta}e^{-(R+it)^2}i\,dt +$
$\int_{R}^{-R}e^{-(x+\beta i)^2}\,dx + \int_{\beta}^{0}e^{-(-R+it)^2}i\,dt = 0$. Let $R \to \infty$. The second integral is $e^{-R^2}\int_{0}^{\beta}e^{t^2}e^{-2iRt}\,dt$, and this goes to zero as $R \to \infty$. Similarly, the fourth integral has limit zero. The first and third integrals give $\int_{-\infty}^{\infty}e^{-x^2}\,dx -$
$\int_{-\infty}^{\infty}e^{-x^2}e^{\beta^2}e^{-2\beta ix}\,dx = 0$. Then $\sqrt{\pi} = e^{\beta^2}\int_{-\infty}^{\infty}e^{-x^2}\cos(2\beta x)\,dx + ie^{-\beta^2}\int_{-\infty}^{\infty}e^{-x^2}\sin(2\beta x)\,dx$. The last integral is zero because the integral is odd. Finally, write $\sqrt{\pi} = 2e^{\beta^2}\int_{0}^{\infty}e^{-x^2}\cos(2\beta x)\,dx$, since this integrand is even. Now solve for the integral.

63. By Cauchy's theorem, $\int_{\Gamma}e^{iz^2}\,dz = 0$. Now integrate over each piece of Γ, going counterclockwise and beginning with the segment $[0, R]$: $\int_{0}^{R}e^{ix^2}\,dx + \int_{0}^{\pi/4}e^{iR^2e^{2i\xi}}Rie^{i\xi}\,d\xi + \int_{R}^{0}e^{i(re^{i\pi/4})^2}e^{i\pi/4}\,dr = 0$. The second integral tends to zero as $R \to \infty$, and in the limit the last equation becomes
$\int_{0}^{\infty}[\cos(x^2) + i\sin(x^2)]\,dx = \frac{\sqrt{2}}{2}(1 + i)\int_{0}^{\infty}e^{-r^2}\,dr = \frac{\sqrt{2}}{2}(1 + i)\frac{1}{2}\sqrt{\pi}$. Equate the real part of each side and the imaginary part of each side to evaluate Fresnel's integrals.

65. $\int_{0}^{2\pi}\frac{1}{(\alpha + \beta\cos(\theta))^2}\,d\theta = \int_{\Gamma}\frac{1}{\left(\alpha + \frac{\beta}{2}\left(z + \frac{1}{z}\right)\right)^2}\frac{1}{iz}\,dz = -4i\int_{\Gamma}\frac{z}{(\beta z^2 + 2\alpha z + \beta)^2}\,dz$. The integrand has double poles at $\frac{1}{\beta}\left(-\alpha \pm \sqrt{\alpha^2 - \beta^2}\right)$, but only $\frac{1}{\beta}\left(-\alpha + \sqrt{\alpha^2 - \beta^2}\right)$ is enclosed by the unit circle Γ. Compute
$\operatorname{Re} s\left(\frac{-4iz}{(\beta z^2 + 2\alpha z + \beta)^2}, \frac{1}{\beta}\left(-\alpha + \sqrt{\alpha^2 - \beta^2}\right)\right) = -\frac{\alpha i}{(\alpha^2 - \beta^2)^{3/2}}$. Then $\int_{0}^{2\pi}\frac{1}{(\alpha + \beta\cos(\theta))^2}\,d\theta = \frac{2\pi\alpha}{(\alpha^2 - \beta^2)^{3/2}}$.

Finally, check that $\int_{0}^{2\pi}\frac{1}{(\alpha + \beta\cos(\theta))^2}\,d\theta = 2\int_{0}^{\pi}\frac{1}{(\alpha + \beta\cos(\theta))^2}\,d\theta$.

CHAPTER 25

Section 25.1

1. The images are given by the following diagrams.

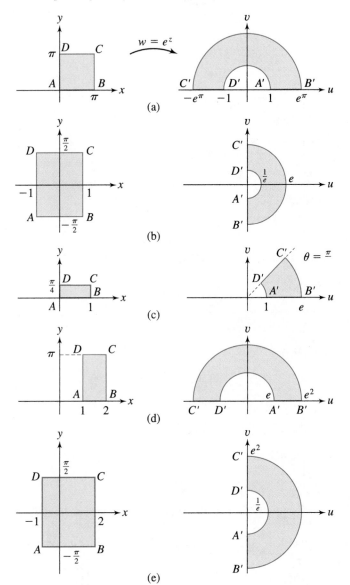

3. The images are given by the following diagrams.

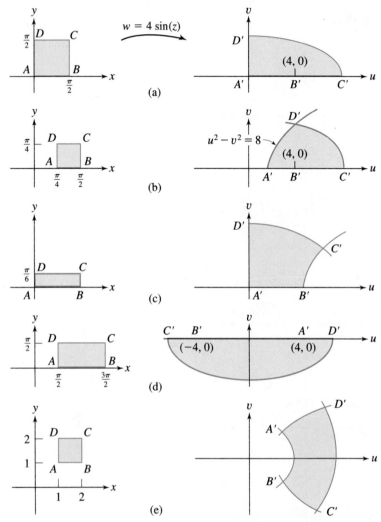

5. Lines $x = a \neq 0$ map onto ellipses $\dfrac{u^2}{\cosh^2(a)} + \dfrac{v^2}{\sinh^2(a)} = 1$. The line $x = 0$ maps onto the segment $-1 \le u \le 1$.

Lines $y = b \neq n\pi/2$ (for integer n) map onto hyperbolas $\dfrac{u^2}{\cos^b(b)} - \dfrac{v^2}{\sin^2(b)} = 1$. Lines $y = n\pi$ for integer n map

onto $|u| \ge 1$. Lines $y = (2n + 1)\pi/2$ map onto $u = 0$.

7. The sector consisting of all w with an argument in $[\pi/2, \pi]$.

9. *Hint:* Put $z = re^{i\theta}$ and obtain $u = \dfrac{1}{2}\left(r + \dfrac{1}{r}\right)\cos(k)$, $v = \dfrac{1}{2}\left(r - \dfrac{1}{r}\right)\sin(k)$.

11. The closed first, second, and third quadrants of the w plane.   **13.** The closed first quadrant of the w plane.

15. The entire w plane with the origin excluded.   **17.** The sector consisting of all w with an argument in $[\pi/27, \pi/9]$.

Section 25.2

1. $w = \dfrac{3 + 8i - (1 + 4i)z}{4 + 7i - (2 + 3i)z}$ **3.** $w = \dfrac{16 - 16i + (-7 + 13i)z}{4 - 8i + (-1 + 2i)z}$ **5.** $w = \dfrac{4 - 75i + (3 + 22i)z}{-21 + 4i + (2 + 3i)z}$

7. $w = \dfrac{8 + 38i - (24 + 10i)z}{-4 + 3i - iz}$ **9.** $w = \dfrac{1/3 + (3 + i/3)z}{5/3 - 12i + (12 + 7i/6)z}$ **11.** $(u - \tfrac{7}{2})^2 + v^2 = \tfrac{9}{4}$

13. $v = 10$ **15.** $(u - \frac{11}{21})^2 + (v + \frac{1}{63})^2 = \frac{208}{3969}$ **17.** $(u - 1)^2 + (v + \frac{19}{4})^2 = \frac{377}{16}$

19. $u^2 + \left(v + \dfrac{1}{2a}\right)^2 = \dfrac{1}{4a^2}$ **21.** $w = \bar{z}$ reverses orientation.

23. If $w = \dfrac{az + b}{cz + d}$ and $ad - bc \neq 0$, then $z = -\dfrac{dw - b}{cw - a}$ is also a linear fractional transformation.

25. For $|z| = |a| = 1$ and $|b| < 1$, $|w| = \dfrac{|a|\,|z - b|}{|z|\,|\bar{b} - \bar{z}|} = 1$, and $|w(0)| = |b| < 1$.

27. *Hint:* Show that $z = \dfrac{az + b}{cz + d}$ has either one or two solutions, depending on whether or not c is zero. A translation has no fixed point.

In each of Problems 29 through 33, there are many solutions, of which one is given here.

29. $z \to \dfrac{1}{z} \to -4\dfrac{1}{z} \to -4\dfrac{1}{z} + i = w$ **31.** $z \to iz \to iz - (2 - 7i) = w$

33. $z \to z + 4 \to \dfrac{1}{z + 4} \to 4(2 - 3i)\dfrac{1}{z + 4} \to 4(2 - 3i)\dfrac{1}{z + 4} - (2 - 3i) = w$

Section 25.3

In Problems 1 through 10, there are many mappings having the stated property. We give one such mapping in each case.

1. $w = 2z + 1 - i$ **3.** $w = \dfrac{3z + 2 + 6i}{z + 2i}$ **5.** $w = \dfrac{-4i(z + 1)}{z - 1}$

7. $w = -3i - iz$ **9.** $w = \dfrac{z - 1 + 12i}{-4 + iz - i}$ **11.** $w = z^{1/3}$

13. (a) For $|z| = 1$, $w = \dfrac{1}{2}\left(z + \dfrac{1}{z}\right) = \dfrac{1}{2}(z + \bar{z}) = \mathrm{Re}(z) = x$, so $-1 \leq w \leq 1$. Conversely, if $-1 \leq w \leq 1$, then

$\sqrt{w^2 - 1}$ is zero or pure imaginary. Solve $w = \dfrac{1}{2}\left(z + \dfrac{1}{z}\right)$ for z to get $z = -w \pm \sqrt{w^2 - 1}$, so

$\overline{-w + \sqrt{w^2 - 1}} = -w - \sqrt{w^2 - 1}$. Then $|z|^2 = z\bar{z} = (-w + \sqrt{w^2 - 1})(-w - \sqrt{w^2 - 1}) = 1$. Therefore, $-1 \leq w \leq 1$ if and only if $|z| = 1$.

(b) For $|z| = r < 1$, $w = \dfrac{1}{2}\left(z + \dfrac{1}{z}\right) = \dfrac{1}{2}\left(x + iy + \dfrac{x - iy}{r^2}\right) = \left(\dfrac{r^2 + 1}{2r}\right)\dfrac{x}{r} + i\left(\dfrac{r^2 - 1}{2r}\right)\dfrac{y}{r} = u + iv$. Thus,

$\dfrac{u^2}{\left(\dfrac{r^2 + 1}{2r}\right)^2} + \dfrac{v^2}{\left(\dfrac{r^2 - 1}{2r}\right)^2} = 1$, which is the equation of an ellipse.

15. *Hint:* Evaluate $f(1)$, $f(-1)$, $f(0)$, and $f(\infty)$ and then use the result that $\displaystyle\int_0^1 t^{m-1}(1 - t)^{n-1}\,dt = \dfrac{\Gamma(m)\Gamma(n)}{\Gamma(m + n)}$ for positive integers m and n.

Section 25.4

1. $u(x, y) = \dfrac{y}{\pi}\displaystyle\int_{-\infty}^{\infty} \dfrac{g(t)}{(t - x)^2 + y^2}\,dt$, where $u(x, 0) = g(x)$.

3. $u(x, y) = \dfrac{1}{2\pi}\displaystyle\int_0^{2\pi} g(x_0 + R\cos(t), y_0 + R\sin(t))$

$\times \left[\dfrac{R^2 - (x - x_0)^2 - (y - y_0)^2}{R^2 + (x - x_0)^2 + (y - y_0)^2 - 2R(x - x_0)\cos(t) - 2R(y - y_0)\sin(t)}\right]dt$

5. $u(r\cos(\theta), r\sin(\theta)) = \dfrac{1}{2\pi}\displaystyle\int_0^{2\pi} \dfrac{[r\cos(t) - r\sin(t)](1 - r^2)}{1 + r^2 - 2r\cos(t - \theta)}\,dt$

7. $u(x, y) = \dfrac{1}{8}\displaystyle\int_{-1}^{1} \dfrac{(1 - |t|)\cos(\pi t/2)}{1 + \sin^2(\pi t/2)}$

$\times \left[\dfrac{4\sinh(\pi x/2)\cos(\pi y/2)[1 + \sin^2(\pi t/2)] + \sinh(\pi x)\sin(\pi y)[1 - \sin(\pi t/2)]}{\sinh^2(\pi x/2) + \sin^2(\pi y/2) - 2\cosh(\pi x/2)\sin(\pi y/2)\sin(\pi t/2) + \sin^2(\pi t/2)}\right]dt$

Section 25.5

1. With $a = Ke^{i\theta}$, equipotential curves are $\varphi(x, y) = K[x\cos(\theta) - y\sin(\theta)] = $ constant. These are lines of the form $y = \cot(\theta)x + b$. Streamlines are $\psi(x, y) = K[y\cos(\theta) + x\sin(\theta)] = $ constant, which are lines $y = -\tan(\theta)x + b$. Velocity $= \overline{f'(z)} = Ke^{-i\theta}$. There are no sources or sinks.

3. $\varphi(x, y) = \cos(x)\cosh(y)$, $\psi(x, y) = -\sin(x)\sinh(y)$. Equipotential curves are graphs of $y = \cosh^{-1}(K/\cos(x))$, streamlines are graphs of $y = \sinh^{-1}(C/\sin(x))$.

5. $\varphi(x, y) = K\ln|z - z_0|$, $\psi(x, y) = K\arg(z - z_0)$. Equipotential curves are circles $|z - z_0| = r$ and streamlines are rays emanating from z_0; $\displaystyle\int_C -v\,dx + u\,dv = 2\pi K$, with C the circle of radius r about z_0.

7. $f(z) = k\left[x + \dfrac{x}{x^2 + y^2} + i\left(y - \dfrac{y}{x^2 + y^2}\right)\right]$. Equipotential curves are graphs of $x + \dfrac{x}{x^2 + y^2} = c$, streamlines are graphs of $y - \dfrac{y}{x^2 + y^2} = d$.

9. Equipotential curves are graphs of $K\left[x + \dfrac{x}{x^2 + y^2}\right] - \dfrac{b}{2\pi}\arg(z) = c$. Streamlines are graphs of $k\left[y - \dfrac{y}{x^2 + y^2}\right] + \dfrac{b}{4\pi}\ln|z| = d$. Stagnation points occur where $f'(z) = k\left(1 - \dfrac{1}{z^2}\right) + \dfrac{ib}{2\pi z} = 0$, or $z = -\dfrac{ib}{4k\pi} \pm \sqrt{1 - \dfrac{b^2}{16\pi^2 k^2}}$.

Index

Numbers in parentheses refer to problems in the text.

absolute convergence of Fourier series, 663
abstract vector space, 235
Academy of Sciences
    in Paris, 1235
    of St. Petersburg, 1230, 1231
acceleration, 518
addition
    of complex numbers, 1003
    of matrices, 243
    of vectors, 204, 222
adjacency matrix, 252
adjacent points in a graph, 251
alchemy, 1227
algebra of vectors, 204
Alice's Adventures in Wonderland, 1217
Allen, Durward, 476
almost linear, 453
alternating current, 795
Ampère's law, 795
amplitude spectrum, 671, 676, 692
analog filtering, 714
analytic, 164, 193, 1102
Angel, Roger, 58(10)
angle
    between two lines, 212
    between two vectors, 211
Ankeny, N. C., 1149
annulus, 1113
antiderivative, 1076
Archimedes' principle, 607
arc length parametrization, 517
Arcueil, 1233
area of a parallelogram, 219
Argand, Jean-Robert, 1224
Argand diagram, 1224
argument, 674, 1008
argument principle, 1136
Arnol'd, V. I., 1227
The Assayer, 1226
associative law for vector addition, 204
asymptotic expansion, 791
asymptotically stable critical point, 449

asymptotic behavior, 790
asymptotic stability, 450
autonomous system, 428
Auxerre, 1234

backward wave, 903
band-limited signal, 711
bandpass filtering, 714
bandwidth, 711
Barrow, Isaac, 1227
Basel, 1229, 1230
basis, 233
beats, 108
Bendixson, Ivar, 1217, 1236
Bendixson's theorem, 503
Berlin Academy of Sciences, 1230
Bernoulli
    Daniel, 1215, 1221, 1229
    family, 1229
    James, 1215, 1229
    Nicholas, 1229
Bernoulli differential equation, 44
Berthellot, Claude-Louis, 1233
Bessel
    equation of order v, 179, 783
    equation of order zero, 161(13)
    function of the first kind, 180, 786
    function of the second kind, 787
    inequalities, 661, 832
Bessel's integral, 802
best mean square approximation, 830
bilinear transformation, 1174
binomial expansion, 1010
binormal, 527
Blasius's theorem, 1209
Boas, R. P., 1149
Bolzano–Weierstrass theorem, 1025
Borrelli, Roger, 799
bound
    for derivatives, 1095
    for an integral, 1078
boundary, 1017
    conditions, 861, 919
    data, 958
    of a plane region, 565

point, 1017
    value problem, 162(19), 817
bounded
    function, 1030
    sequence, 1024
    set, 1025
brachistochrone problem, 1228
Broughton bridge, 106
Bruno, Giordano, 1226
Brunswick family, 1229

Cambridge University, 1227
canonical form
    of an elliptic equation, 994
    of a hyperbolic equation, 991
    of a parabolic equation, 993
    of a subitemratic form, 367(24)
Cantor, Georg, 623, 1085
capacitor, 54
carbon dating, 18
carbon-14, 17
Carroll, Lewis, 1217
Catalan, Eugene, 1217
catenary, 1215
Catherine the Great, 1230
Cauchy
    formula for higher derivatives, 1094
    integral formula, 1092
    integral theorem, 1082
    principal value, 1159
    problem, 897
Cauchy, Augustin-Louis, 1081, 1217,
        1218, 1222, 1224, 1233
Cauchy–Goursat theorem, 1081
Cauchy–Kovalevsky theorem, 998
Cauchy–Riemann equations, 1035
Cauchy–Schwarz inequality, 215
    in n-space, 224
Cauchy's inequality, 683(27)
Cayley–Hamilton theorem, 1218
center
    of mass of a shell, 598
    of a power series, 189, 1042
    of mass of a wire, 563
    of a window function, 710
centripetal acceleration, 570(16)

centripetal force, 570(16)
Ceres, 1100
Cesàro filter, 742
Champollion, 1235
Chandrasekhar, S., 1227
characteristic
    equation, 77
    function of an interval, 844
    polynomial, 340
    triangle, 900
    values, 338
    vectors, 338
characteristics of a partial differential
        equation, 897
    for a hyperbolic equation, 990
    for a parabolic equation, 993
Chebyshev
    equation, 836
    polynomials, 837
Christ Church College, 1217
circulation, 617, 1201
Clairaut, Alexis-Claude, 1215
Clairaut's equation, 64(32)
classification
    of partial differential equations, 989
    of singularities, 445
Clebsch, Rudolf Friedrich Alfred, 1218
Clifford, William Kingdom, 1219
closed
    curve, 553
    disk, 1014
    form, 25
    set, 1018
    surface, 604
closure, 1195
coefficient of thermal conductivity, 608
cofactor, 324
    expansion by rows, 324
    expansion by columns, 325
coherent orientation, 613
Coleman, Courtney S., 799
Collège de France, 1216
Collegium Carolinum, 1230
collinear, 219
column
    space, 272
    vector, 250
commutative law
    for the dot product, 281
    for vector addition, 274
commutativity for convolution
    for the Fourier transform, 703
    for the Laplace transform, 144
compact
    set, 1025
    support, 708
competing species model
    extended, 482
    simple, 480
complementary
    domain, 1182

error function, 950
complete
    graph, 336(6)
    pivoting, 282
completeness of eigenfunctions, 834
complex
    addition, 1003
    Bessel function, 1120(14)
    conjugate, 673, 1005
    cosine function, 1051
    division, 1006
    exponential function, 1047
    Fourier coefficients, 676
    Fourier integral, 688
    Fourier series, 676
    function, 1027
    integral, 1070, 1071, 1073
    logarithm, 1057
    number, 1003
    plane, 1004
    potential, 1203
    powers, 1059
    sequence, 1021
    sine function, 1051
    sphere, 1179
components
    of a piecewise smooth curve, 1067
    of a vector, 202
computational efficiency of the FFT, 757
condition for a pole of order m, 1124
conformal mapping, 1172
connected set, 1082
conservation
    of energy, 583(21), 111(20), 583(21)
    of mass, 610
conservative vector field
    in 3-space, 618
    in the plane, 572
consistent system of equations, 292
constant coefficient differential equation,
        77
continuous
    complex function, 1029
    curve, 554
    vector field, 529
    vector function, 511
convection conditions, 919
convergence
    absolute, 1041
    conditional, 1041
    in the mean, 832
    of a complex sequence, 1021
    of a complex series, 1040
    of an eigenfunction expansion, 828
    of a Fourier–Bessel series, 810
    of a Fourier cosine integral, 685
    of a Fourier cosine series, 652
    of a Fourier integral, 682
    of a Fourier–Legendre series, 773
    of a Fourier series, 638, 644
    of a Fourier sine integral, 686

    of a Fourier sine series, 654
    of a power series, 1042
convolution
    for the Fourier transform, 703
    for the Laplace transform, 142
    theorem for the Fourier transform, 703
    theorem for the Laplace transform, 142
Cooley, James W., 746
coordinate functions, 553, 1065
Copernican, 1226
cosmic rays, 17
Cotes, Roger, 80
Cramer's rule, 332
critical
    length, 794
    point, 435, 447
critically damped, forced motion, 104
critically damped, unforced motion, 101
cross product, 216
cross ratio, 1193(16)
crystal, 251
curl, 548
current law (Kirchhoff), 54
curvature, 519
curve, 1065
    closed, 1065
    continuous, 1067
    differentiable, 1067
    negatively oriented, 565
    piecewise smooth, 1067
    simple, 1067
    simple closed, 1067
    smooth, 1067

d'Alembert, Jean le Rond, 1216, 1220,
        1223, 1232
d'Alembert's formula, 899
d'Alembert's solution, 898
damped
    pendulum, 467
    spring motion, 99
Danielson, G. C., 746
deflection of a beam, 151(14)
deformation theorem, 1089
degree of a point in a graph, 256(34)
del operator, 535
delta function, 147
de Moivre's theorem, 1012(34)
dependent unknown, 296
derivative
    of a complex function, 1032
    of a vector function, 512
Descartes, René, 1227
determinant, 313
diagonalizable, 346
diagonalizes, 346
*Dialogue on the Two Great Systems of the
    World*, 1225
Dialogues Concerning Two New Sciences,
    1214
dielectric, 54

differentiable
  complex function, 1032
  curve, 554
  vector function, 512
differential equation, 1
  of higher order, 96
differentiation in the time variable, 699
diffusivity, 919
dimension, 234
Dirac, P. A. M., 147
Dirac delta function, 147
directional derivative, 536
direction field, 8
Dirichlet, Peter Gustrav Lejeune, 635, 1222
Dirichlet problem, 612(18), 958
  for a cube, 974
  for a disk, 961
  for a rectangle, 959
  for the right quarter-plane, 969
  for the upper half-plane, 967
Dirichlet's conditions, 1093
discrete Fourier transform, 727
distance, 1012
distribution, 147, 706
distributive law for the dot product, 210
divergence
  theorem, 605
  of a vector field, 547
Dodgson, Reverend C. L., 1217
domain
  in the complex plane, 1082
  in the plane, 578
  in 3-space, 619
dot product, 209, 238
  of functions, 823
  of n-vectors, 223
double pole, 1123
Du Bois-Reymond, Paul, 635, 741, 1222
Duke of Brunswick, 1230

École
  Militaire, 1232
  Normale, 1232
  Polytechnique, 1233
eigenfunction
  expansions, 826
  of a Sturm–Liouville problem, 817
eigenvalue
  of a matrix, 338
  of a Sturm–Liouville problem, 817
eigenvector, 338
Electromagnetic Theory, 1219
elementary
  column operations, 318
  row operations, 256
element of a matrix, 317
elliptical cone, 583
elliptic
  equation, 994
  integral, 1191

emitter discharge current, 931
entry of a matrix, 242
envelope, 64(32)
equal vectors, 202
equation of a tangent plane, 541
equilibrium
  point, 435
  position, 98
  solution, 435
equipotential curves, 1203
equivalent curves, 1067
error function, 950
essential singularity, 1122
Euler, Leonhard, 1004, 1215, 1220, 1223, 1230
Euler's
  differential equation, 83
  formula, 80
even
  extension, 651
  function, 632
exact differential equation
  first-order, 30
  second-order, 77(18)
existence
  of analytic solutions of a differential
    equation, 164
  of analytic solutions of an initial value
    problem, 167
  of the Laplace transform, 119
  of solutions of a Cauchy problem, 998
  of solutions of a first-order initial value
    problem, 62
  of solutions of a first-order system, 380,
    426
  of solutions of the second-order linear
    initial value problem, 68
exponential
  complex function, 1047
  matrix, 402
  model of population growth, 22(25)
Exposition du Systeme du Monde, 1232
extended complex plane, 1179
extended deformation theorem, 1097
extension of Green's theorem, 569
exterior of a curve, 565, 1081

factorial property, 784
family
  of curves, 56
  of orthogonal trajectories, 56
Faraday's law, 795
fast Fourier transform, 745
filtered partial sum, 741
filtering, 707, 741
  property of the Dirac delta function, 147
finite
  Fourier cosine transform, 720
  Fourier sine transform, 720
first
  component, 202

shifting theorem, 127
first-order differential equation, 3
fixed point, 1181(27)
Flamsteed, 1228
flow lines, 531
flux, 599, 1201
forced motion, 103
Fourier, Charles, 1235
Fourier, Joseph, 623, 1220, 1223, 1234
Fourier
  –Bessel coefficients, 810
  –Bessel expansion, 810
  –Bessel series, 810
  coefficients, 630
  complex integral coefficients, 688
  cosine coefficients, 652
  cosine integral, 685
  cosine series, 652
  cosine transform, 717
  –Chebyshev coefficients, 837
  –Chebyshev series, 837
  integral, 682
  integral coefficients, 682
  –Laguerre coefficients, 839
  –Laguerre series, 839
  –Legendre coefficients, 773, 775
  –Legendre series, 773
  method, 864
  series on $[-L, L]$, 630
  sine coefficients, 654
  sine integral, 686
  sine series, 654
  sine transform, 718
  transform, 689
forward wave, 903
free
  falling particle, 18
  radiation conditions, 919
Frenet formulas, 528
frequency
  convolution, 704
  differentiation, 701
  shifting, 694
  spectrum, 676
Fresnel's integrals, 1161(63)
Friedrich the Great, 1230
Frobenius series, 176
full wave rectification, 678
fundamental
  matrix, 387
  period, 667
  set of solutions, 71
  theorem of algebra, 1096, 1149

Galileo, 1214, 1225
gamma function, 783
Garwin, Richard, 746
Gauss, Carl Friedrich, 1004, 1216, 1217, 1224, 1230, 1233
Gauss filter, 744
Gauss–Jordan reduction, 282

Gauss's divergence theorem, 605
general solution of a
  first-order differential equation, 4
  homogeneous system of algebraic
    equations, 283
  nonhomogeneous system of algebraic
    equations, 291
  second-order linear homogeneous
    differential equation, 72
  second-order linear nonhomogeneous
    differential equation, 72
  system of differential equations, 385,
    388
generating function for
  Bessel functions, 801
  for Chebyshev polynomials, 837
  for Hermite polynomials, 839
  for Laguerre polynomials, 838
  for Legendre polynomials, 768
geometric series, 1040
Gerschgorin circles, 1040
Gerschgorin's theorem, 342
Gibbs, Josiah Willard, 649, 1219
Gibbs phenomenon, 649, 743
Glenn, John, 379
Göttingen, 1231
Goursat, Edouard, 1081
gradient, 535
graph, 251, 334
  of a curve, 1065
Green, George, 565, 1065, 1219, 1223
Green's
  first identity, 611, 981
  second identity, 611
  theorem, 566
Gregorian calendar, 1227

Hadamard's example, 1001(3)
Haar wavelets, 849
half-life, 17
Halley, Edmund, 1227
Hamilton, Sir William Rowan, 1218
Hamming filter, 743
Hamming, Richard, 743
hare population in Canada, 476
harmonic
  amplitude, 670
  conjugate, 1194
  form, 82(31), 670
  function, 957, 1194
heat equation, 608, 917
  for a bar of finite length, 920, 923, 924
  for an infinite bar, 942
  for an infinite cylinder, 951
  for a rectangular plate, 955
  for a semi-infinite bar, 945
  nonhomogeneous, 933
Heaviside, Oliver, 130, 1217, 1219
Heaviside
  formulas, 142(51)
  function, 130

Helmstädt, 1231
Hermite, Charles, 1216, 1218
Hermite
  equation, 839
  polynomials, 839
higher-order differential equations, 96
Hilbert, David, 1235
Hobson, Dana D., 799
homogeneous
  first-order differential equation, 40
  second-order differential equation, 68
  system of algebraic equations, 278
  system of differential equations, 378
Hooke, Robert, 1228
Hooke's law, 59(20), 98, 427
Hoover Dam, 597
Hudson Bay Company, 476
Huygens, Christian, 1215
hyperbolic
  equation, 989
  paraboloid, 584

identity theorem, 1109
image, 1032
imaginary
  axis, 1004
  part, 1003
  unit, 1219
inclined plane, 52
incompressible fluid, 1200
inconsistent system of equations, 292
indented contour, 1155
independence of path
  in 3-space, 620
  in the plane, 573, 1078, 1088
independent unknown, 296
indicial equation, 177
inductance, 54
inductive time constant, 60
initial
  conditions, 6, 861, 919
  point, 553, 1065
  position, 861
  temperature, 919
  value problem, 6
  velocity, 861
inner normal, 589
input frequency, 105
insulation conditions, 919
integral curve, 5
integral formula for Legendre
  polynomials, 780
integrating factor, 24, 35
interior of a curve, 565, 1081
interior point, 1016
interlacing lemma, 808
interval of convergence, 190
inverse
  discrete Fourier transform, 730
  finite Fourier cosine transform, 720
  finite Fourier sine transform, 720

Fourier cosine transform, 717
Fourier sine transform, 718
Fourier transform, 691
Laplace transform, 120
N-point DFT, 730
inverse version of the
  convolution theorem for the Laplace
    transform, 144
  first shifting theorem, 128
  second shifting theorem, 133
inversion, 1175
irregular singular point, 175
irrotational
  flow, 1202
  vector field, 551
Isle Royale, 476
isolated
  critical point, 447
  singularity, 1121
  zero, 1106

Jacobi, Carl Gustav Jacob, 1217
Jacobian, 587
Johns Hopkins, 1217
join, 1067
Jordan, Camille, 1218
Jordan curve theorem, 565, 1081
Joukowski transformation, 1208
jump discontinuity, 129, 636

Kepler, Johannes, 1214
kinetic energy, 19
Kirchhoff, G. R., 334
Kirchhoff's
  current law, 54
  voltage law, 54

labeled graph, 334
Lagrange, Joseph-Louis, 1215, 1221,
  1231, 1233
Lagrange's identity, 221(37)
Lagrange's trigonometric identity,
  1012(36)
Laguerre, Edmond, 1216
Laguerre polynomials, 838
Laguerre's equation, 838
Lanczos, Cornelius, 746
Laplace, Pierre Simon, 1216, 1221, 1223,
  1232
Laplace
  integrals, 686
  transform, 113
    of a derivative, 123
    of higher derivatives, 123
Laplace's equation, 609, 957
  in polar coordinates, 962
  in spherical coordinates, 976
Laplacian, 609
Laurent, Pierre-Alphonse, 1224
Laurent
  coefficients, 1113

expansion, 1113
series, 1113
law of cosines, 210
leading entry, 263
left
  derivative, 642
  wave, 903
Legendre, Adrien-Marie, 1216, 1221, 1223
Legendre polynomials, 768
Legendre's differential equation, 765
Leibniz, Gottfried Wilhelm, 1213, 1228
length
  function, 517
  of an 3-vector, 202
  of an n-vector, 223
Lerch's theorem, 120
Les methodes nouvelles de la mécanique céleste, 1236
level surface, 538
Lienard, Alfred, 1217
Lienard's theorem, 505
limit
  of a complex function, 1027
  of a complex sequence, 1021
  point, 1019
limit cycle, 499
  asymptotically stable, 499
  semistable, 499
  stable, 499
  unstable, 499
limit point, 1019
lineal element, 8
linear
  combination, 68, 228
  dependence, 68, 69, 229
    of solutions of a system of differential equations, 381
  differential equation, 23, 65, 861
  first-order equation, 23
  fractional transformation, 1174
  independence, 69, 229
    of solutions of a system of differential equations, 381
  operator, 337
  second-order ordinary differential equation, 65
line integral, 554
  in the plane, 556
  with respect to arc length, 561
lines of force, 531
Liouville, Joseph, 1216
Liouville's theorem, 1095
local theorem, 62
logistic
  equation, 21(25)
  model, 21(35)
Lommel's integrals, 815(42)
longitudinal displacements in a bar, 881
lunar
  diurnal period, 762

semidiurnal period, 762
solar diurnal period, 762
Luxembourg Palace, 1233
Lyapunov function, 492
Lyapunov's
  direct method for stability, 491
  direct method for unstability, 492
lynx population in Canada, 476

Maclaurin series, 194
magnitude
  of a complex number, 674, 1005
  of a jump discontinuity, 129
magnification, 1175
main diagonal, 345
main diagonal elements, 345
MAPLE, 6
mapping, 1163
  angle-preserving, 1171
  conformal, 1172
  into, 1164
  one to one, 11655
  onto, 1164
  orientation-preserving, 1172
mass
  of a shell, 598
  of a wire, 562
MATHEMATICA, 6
mathematical modeling, 20
Mathematical Principles of Natural Philosophy, 1227
MATLAB, 6
matrix, 242
  addition, 243,
  adjacency, 252
  augmented, 292
  block diagonal, 327(18)
  of coefficients, 246, 264, 290
  diagonal, 345
  diagonalizable, 346
  element, 242
  elementary, 257
  entry, 242
  equal, 242
  exponential, 402
  hermitian, 371
  identify, 248
  inverse, 304
  lower triangular, 328
  nonsingular, 305
  orthogonal, 354
  positive definite, 362(13)
  product, 243
  product with a scalar, 243
  reduced, 264
  row equivalent, 261
  singular, 305
  skew-hermitian, 371
  skew-symmetric, 319(3)
  square, 242
  symmetric, 358

tree theorem, 335
unitary, 368
upper triangular, 328
zero, 247
maximum modulus theorem, 1111
Maxwell, James Clerk, 1219
mean value property, 1195
Mécanique analytique, 1231
Mécanique céleste, 1232
median, 213
method
  of Frobenius, 176
  of reduction of order, 74
  of undetermined coefficients, 89
  of variation of parameters, 87
Michelson, Albert A., 648
Michelson–Morley experiment, 648
Milton, John, 1227
minor, 324
mixed product terms, 363
mixing problem, 25, 391
Möbius transformation, 1174
modified
  Bessel functions of order zero, 789
  Bessel function of the first kind, 789
  Bessel function of the second kind, 789
modulation, 696
modulus, 674, 1005
Morera's theorem, 1100(22)
Morro Bay, 761
multiple Fourier series, 914
multiplication
  of complex numbers, 1003
  of matrices, 243
  of power series, 1056(25)
multiresolution analysis, 850, 851
mutually orthogonal, 232
Murphy, Robert, 1216

Napoleon, 1233, 1234
NASA Kennedy Space Center, 379
natural
  frequency, 105
  length, 98
  period, 111(19)
nearly homogeneous, 46(15)
negative
  definite function, 490
  semidefinite function, 490
nested chain of closed sets, 1085
Neumann problem, 980
  for a disk, 984
  for a rectangle, 982
  for the upper half-plane, 986
Neumann's
  Bessel function of order ν, 787
  function of order zero, 787
Newton, Sir Isaac, 1213, 1227
Newton's
  law of cooling, 918
  second law of motion, 48

nodal
  sink, 437
  source, 438
node
  improper, 442
  proper, 441
nonautonomous system, 434
nonhomogeneous, 72, 86, 290
nonlinear
  damped spring, 470
  spring, 427
nonzero row, 263
norm, 201
  of an n-vector, 223
normal
  component of acceleration, 524
  derivative, 613(20), 980
  form of a line, 206
  inner, 589
  line, 543
  modes of vibration, 908
  outer, 589
  vector, 218, 538, 586
normalized eigenfunctions, 830
N-point discrete Fourier transform, 727
n-space, 222
nth
  Cesàro sum, 741
  harmonic, 670
  phase angle, 670
  roots of unity, 749, 1061
  Taylor coefficient, 1101
n-vector, 222

odd extension, 653
odd function, 632
off-diagonal element, 345
open disk, 1013
  of convergence, 1043
open
  interval of convergence, 190
  set, 1016
operational rule
  for the Fourier cosine transform, 719
  for the finite Fourier transform, 721
  for the Fourier sine transform, 719
  for the Fourier transform, 698
orbit, 428
orbital derivative, 491
ordering, 1010
order
  of a differential equation, 2
  of a zero, 1107
ordinary point, 174
orientation reversal, 559
orientation of a curve, 1065
oriented coherently, 613
orthogonal
  trajectories, 55
  vectors, 213, 355
  with respect to a weight function, 810,
    823

orthogonality
  of Bessel functions, 809
  of Chebyshev polynomials, 837
  of Legendre polynomials, 772
  with respect to a weight function, 823
orthonormal vectors, 355
Ostrogradsky, Michel, 565, 1219, 1223
outer normal, 602, 604
overdamping
  forced, 104
  unforced, 100

Panama Canal lock, 1201
parabolic equation, 993
paraboloid of revolution, 58
parallelogram law, 204
  for complex addition, 1004
parallel vectors, 203
parametric equations of a line, 206
Parseval's theorem, 665, 833, 1112(22)
partial differential equation, 857
partial sum of a power series, 1040
particular solution, 4, 87
path, 1081
Pearl and Reed, 22(25)
permutation, 311
  even, 312
  odd, 312
pendulum, 416
periodic
  boundary conditions, 817
  extension, 641
  function, 67, 121(20)
  Sturm–Liouville problem, 817
periodicity conditions, 857
perpendicular vectors, 213
perturbation, 1232
Petronas Towers, 447
Pfaff, Johann, 1216, 1231
phase
  angle form (Fourier series), 670
  angle form of a solution, 82(31)
  plane, 429
  portrait, 429
  spectrum, 680(12,13)
Picard iterates, 63
piecewise continuous, 118, 635
piecewise smooth
  curve, 556
  function, 637
  surface, 590
Pisa, 1226
plane parallel flow, 1200
Poe's pendulum, 798
Poincaré–Bendixson theorem, 504, 1235
Poincaré, Jules Henri, 1217
pointwise convergence, 832
Poisson, Simeon-Denis, 1216, 1222, 1224
Poisson's
  equation, 971
  integral formula, 965, 1197
polar form, 674, 1008

pole, 1122
  double, 1123
  simple, 1123
poles
  of products, 1127
  of quotients, 1126
position vector, 512, 584
position function for a string, 859
positive
  definite function, 490
  semidefinite function, 490
positively oriented curve, 565
potential
  energy, 58(8), 111(20)
  function, 30, 572
power
  of a complex number, 1059
  content, 673(16)
  dissipation, 673(16)
  series (complex), 1042
  series (real), 189
  spectral density, 758
predator/prey model
  extended, 477
  simple, 474
principal axis theorem, 365
  Principia, 760, 1214, 1227
principle
  axes, 365
  of conservation of energy, 111(20)
  of superposition, 95
Principles of Philosophy, 1227
product
  of a vector with a scalar, 202
  of power series, 192
properties
  of the cross product, 217
  of the dot product, 210
pulse, 131
punctured disk, 1113
pure
  imaginary, 1004
  rotation, 1175
pursuit problem, 41
Pythagoras's theorem, 227(20)

quadratic form, 363
quadrilateral, 207
quasi period, 111(19)
quaternion, 1218
quotient of complex numbers, 1006

radioactive
  carbon, 17
  decay, 16
radius of a window function, 710
radius of convergence
  of a complex power series, 1043
  of a real power series, 190
random walk, 251
rank, 275

rational
    function, 1040
    powers, 1062
ratio test, 190
Rayleigh quotient, 835(17)
real
    axis, 1004
    part, 1003
reclaimable charge, 931
rectangular parallelepiped, 219
recurrence relation, 170
    for Bessel functions, 804
    for Chebyshev polynomials, 837
    for Hermite polynomials, 839
    for Laguerre polynomials, 839
    for Legendre polynomials, 770
reduced
    form, 263
    matrix, 264
    row echelon form, 263
    row echelon matrix, 264
    system, 279
reducing a matrix, 266
reduction of order, 74
regular
    boundary conditions, 817
    singular point, 175
    Sturm–Liouville problem, 816
Rembrandt, 1227
removable singularity, 1122
residue, 1128
    at a simple pole, 1129
    at a pole of order m, 1129
residue theorem, 1128
resonance, 105
reversal of orientation, 1075
Riccati, Count Jacopo Francesco, 1216
Riccati equation, 45
Riemann, Georg Friedrich Bernhard,
    1222, 1224
Riemann's lemma, 666(8)
Riemann mapping theorem, 1182
right
    derivative, 642
    hand rule, 218
    wave, 903
RMS
    bandwidth, 711
    duration of a window, 710
Robespierre, 1234
Rodrigues's formula, 778
    for Chebyshev polynomials, 837
Roman Inquisition, 1096
root mean square, 673(15)
roots of unity, 749, 1061
Rosetta stone, 1235
rotation, 1175
rotation/magnification, 1175
Rouché's theorem, 1138
row
    equivalence, 261
    space, 271

vector, 271
row operations, 256
Royal
    Artillery School of Turin, 1231
    Observatory, 1228
    Society
        of Berlin, 1100
        of Göttingen, 1231
        (of London), 1227
Rudnick, Phillip, 746

saddle point, 439
scalar, 199
    field, 535
    triple product, 220, 221(39–42)
scaling
    function for Haar wavelets, 849
    theorem, 695
Schwarz–Christoffel transformation, 1189
second
    component, 202
    -order differential equation, 65
    shifting theorem, 132
    solution in the method of Frobenius,
        181
secular equation, 1217
self-symmetric, 1193(19)
semidiurnal tides, 762
separable equation, 11
separation constant, 864
separation of variables, 864
sequence
    of partial sums, 1040
    of Cesàro filter factors, 742
Shannon
    sampling function, 714
    sampling theorem, 713
    wavelets, 852
shifted
    function, 132
    Heaviside function, 130
shifting
    indices, 195
    in the s variable, 127
    in the t variable, 132
shock wave, 861
simple
    closed curve, 565
    curve, 565
    pendulum, 426
    pole, 1123
    surface, 585
simply connected
    in 3-space, 620
    in the plane, 503, 580
singular
    point, 174
    Sturm–Liouville problem, 817
singularity
    essential, 1122
    isolated, 1121
    removable, 1122

sink, 1201
skew-hermitian matrix, 371
skin effect, 798
Smith, David Eugene, 1213
Smith prize, 1220
smooth
    curve, 554
    surface, 590
solar day, 761
solenoidal flow, 1201
solution
    explicit, 4
    implicitly defined, 4
    of a differential equation, 2
    of a first-order differential equation, 3
    of a second-order differential equation,
        65
    of a system of ordinary differential
        equations, 378
    singular, 12
    space, 287, 381
    steady state part, 27
    transient part, 27
Sorbonne, 1233
source, 1201
A Source Book in Mathematics, 1213
spanning
    set, 229
    tree, 335
special function, 765
specific heat, 608
speed, 517
spherical coordinates, 864(7)
spin casting, 58(10)
spiral
    point, 443
    sink, 443
    source, 443
    vortex, 1210(8)
spring constant, 98
square
    integrable function, 713, 841
    terms, 363
    wave, 668
stable
    critical point, 449
    limit cycle, 499
stagnation point, 1203
standard
    form of a subitemratic form, 366
    representation of a vector
        in n-space, 224
        in 3-space, 205
stationary flow, 1200
steady-state
    equation for a solid sphere, 976
    heat equation, 958
    part, 929
    temperature distribution, 941(23)
stereographic projection, 1179
Stokes, Sir George Gabriel, 1220
Stokes's theorem, 614

stream function, 1203
streamlines, 531, 1203
strength of a vortex, 1201
Stuart King Charles I, 1227
Sturm, Charles, 1216
Sturm–Liouville
 differential equation, 816
 form, 816
 problem, 816
 theorem, 823
Sturm separation theorem, 73(18)
subsequence, 1023
subsizar, 1227
subspace
 of n-space, 225
 of the plane, 226
 of 3-space, 227
summation of series, 1139
support, 708
surface, 583
 area, 596
 integral, 592
 piecewise smooth, 590
 simple, 585
 smooth, 590
Sylvester, James Joseph, 1217
symmetric with respect to a circle, 1193
symmetry, 696
symmetry principle, 1193(20)
system of ordinary differential equations,
 377

Tacoma Narrows Bridge, 106
tangent
 plane, 538, 541, 589
 vector, 512
tangential component of acceleration, 524
Taylor, Brook, 1215
Taylor
 coefficients, 193, 1046
 series, 193, 1046, 1101
 in two variables, 463
telegraph equation, 881(17)
term by term
 differentiation of Fourier series, 660
 integration of Fourier series, 658
 integration of series of functions, 1079
terminal
 point, 553, 1065
 velocity, 50
test
 for a conservative field, 576
 for independence of a system of
 differential equations, 381
 for exactness, 31
Théorie analytique de la chaleur, 1221,
 1223
thermal conductivity, 608
third component, 202
Thomson, Sir William (Lord Kelvin),
 1219
thought experiments, 1226

three-body problem, 1232
three-point theorem, 1178
time
 convolution, 703
 frequency localization, 710
 reversal, 696
 shifting, 693
Torricelli's law, 18
torsion, 528
tractrix, 47(28)
trajectory, 428
transfer coefficient, 925
transfer function, 713
transform
 finite Fourier cosine, 720
 finite Fourier sine, 720
 Fourier, 689
 Fourier cosine, 717
 Fourier sine, 718
 Laplace, 113
 pair
 for the Fourier cosine transform, 712
 for the Fourier sine transform, 718
 for the Fourier transform, 691
transformation, 1163
 bilinear, 1174
 linear fractional, 1174
 Möbius, 1174
transient, 149
transient part, 27, 929
transition matrix, 417
translation, 1174
 of a trajectory, 430
transpose, 249
triangle inequality, 1007
Tricomi's equation, 997
Trinity College, 1227
trivial
 solution, 288
 subspace, 225
Tukey, John W., 746
Two New Sciences, 1227

uncoupled system of equations, 400
underdamping
 forced motion, 104
 unforced motion, 101
undetermined coefficients, 89
unforced motion, 99
uniform convergence of Fourier series,
 663
unit
 normal, 522
 outer normal, 604
 step function, 129
 tangent, 514
 vector, 205
unitary system, 369
universal gravitational constant, 49
unstable
 critical point, 449
 length, 794

limit cycle, 499
 stable length, 794
uranium-238, 21(20)

Valley of the Kings, 1214
Vandermonde's determinant, 327(15)
van der Pol, Balthazar, 1217
van der Pol's equation, 506
variation of parameters, 87, 411
vector, 201
 field, 329
 function, 511
 position, 512
 space, 223
 sum, 204
Vector Analysis, 1000
velocity, 517
 potential, 1203
Verhulst, 21(25)
vibration absorber, 156(16)
vibrating string
 infinite, 883
 nonzero initial displacement and
 velocity, 871, 887
 semi-infinite, 887
 zero initial displacement, 870, 885
 zero initial velocity, 864, 883
voltage law (Kirchhoff), 54
volume of a parallelepiped, 220
vortex, 1206

walk, 252
wave equation
 driven by an external force, 900
 for a circular elastic membrane, 906,
 909
 for an infinite string, 883, 885
 for a rectangular elastic membrane, 912
 for a semi-infinite string, 887
 in 1-space dimension, 860
 in polar coordinates, 863
 in 2-space dimensions, 862
 nonhomogeneous, 900
 on a bounded interval, 864, 870
 wavelet expansion, 849
Weber, Wilhelm, 860
Weierstrass, Karl, 1224
weight function, 823
well-posed problem, 1000
Wessel, Caspar, 1224
Westminster Abbey, 1228
Wilbraham, 649
William Lowell Putnam Mathematical
 Competition, 21(24), 48(37)
windowed Fourier transform, 689, 708
window function, 708
Woolsthorpe, 1227
work, 558
Wronskian, 69, 1217
 test, 70
Wronski, H., 1217

Young's modulus, 151(14)

zero
matrix, 247
mean random noise, 760
of a function, 1106

row, 263
solution, 288
vector, 236
zeros
isolated, 1106
of Bessel functions, 807

of Chebyshev polynomials, 837
of Legendre polynomials, 777

Credits